Encyclopedia of Chromatography

Encyclopedia of Chromatography

edited by
Jack Cazes
Florida Atlantic University
Boca Raton, Florida

MARCEL DEKKER, INC. NEW YORK • BASEL

ISBN: 0-8247-0511-4

This book is printed on acid-free paper.

Headquarters
Marcel Dekker, Inc.
270 Madison Avenue, New York, NY 10016
tel: 212-696-9000; fax: 212-685-4540

Eastern Hemisphere Distribution
Marcel Dekker AG
Hutgasse 4, Postfach 812, CH-4001 Basel, Switzerland
tel: 41-61-261-8482; fax: 41-61-261-8896

World Wide Web
http://www.dekker.com

The publisher offers discounts on this book when ordered in bulk quantities. For more information, write to Special Sales/Professional Marketing at the headquarters address above.

Current printing (last digit):
10 9 8 7 6 5 4 3 2 1

PRINTED IN THE UNITED STATES OF AMERICA

Preface

The twentieth century has seen monumental advances in the development of new analytical technologies and methodologies. By far, chromatography takes its place at the top of the list. Chromatographic methods have become the method of choice for the solution of analytical problems in virtually all areas of application, including biotechnology, pharmaceuticals, environmental sciences, polymers, food additives and nutrients, pathology, toxicology, fossil fuels, nuclear chemistry, and many more.

The development of new chromatographic technologies has proceeded by leaps and bounds; a chromatographic scientist of the mid-1900s would be overwhelmed by the wealth of information that has been published dealing with new kinds of chromatography and unique applications of these new technologies.

The *Encyclopedia of Chromatography* fulfills a need for a practical, single-volume source of information on chromatographic techniques and methodologies. It is by no means complete; rather, it is the basis for an ongoing compendium of information that will serve to introduce novices as well as seasoned chromatographers to specific topics for which a leading reference, an introductory understanding, or starting point is needed, and to lead one to further reading on the subject.

The encyclopedia will become an often used source of information on chromatographic techniques. Included in this volume are topics dealing with high-performance liquid chromatography, gas chromatography, thin-layer chromatography, supercritical fluid chromatography, countercurrent chromatography, capillary electrophoresis and its subtechnologies, capillary electrochromatography, field-flow fractionation, size-exclusion chromatography, affinity chromatography, and more. The text and associated references will provide the latest innovations and refresh one's memory about techniques which have now become standard.

The Editor heartily thanks all those who have helped to make the *Encyclopedia of Chromatography* an outstanding reference work by contributing their valuable time and expertise to this complex project.

Jack Cazes

Contents

Contributors

Hassan Y. Aboul-Enein Pharmaceutical Analysis Laboratory, King Faisal Specialist Hospital and Research Centre, Riyadh, Saudi Arabia

Ibrahim A. Al-Duraibi Pharmaceutical Analysis Laboratory, King Faisal Specialist Hospital and Research Centre, Riyadh, Saudi Arabia

Serge Alex Centre d'Etudes des Procédés Chimiques du Québec, Montreal, Canada

Juan G. Alvarez Department of Obstetrics and Gynecology, Beth Israel Deaconess Medical Center, Harvard Medical School, Boston, Massachusetts

Victor P. Andreev Institute for Analytical Instrumentation, Russian Academy of Sciences, St. Petersburg, Russia

Christine M. Aurigemma Pfizer Global Research and Development, La Jolla, California

John Austin Chemistry Department, University of Nebraska, Lincoln, Nebraska

Yoshinobu Baba Department of Medicinal Chemistry, University of Tokushima, Tokushima, Japan

James J. Bao Advanced Medicine, Inc., South San Francisco, California

M. A. Barbirato Laboratorio de Cromatografia, Instituto de Quimica de São Carlos, Universidade de São Paulo, São Carlos/SP, Brazil

Damià Barceló Department of Environmental Chemistry, IIQAB-CSIC, Barcelona, Spain

Csaba Barta Institute of Medical Chemistry, Molecular Biology, and Pathobiochemistry, Semmelweis University, Budapest, Hungary

I. Bataille Institut Galilee, Université Paris Nord, Villetaneuse, France

S. Battu Laboratoire de Chimie Analytique et Bromatologie, Université de Limoges, Limoges, France

Ronald Beckett Department of Chemistry, Monash University, Melbourne, Australia

Philippe J. Berny Unité de Toxicologie, Ecole Nationale Veterinaire de Lyon, Marcy l'Etoile, France

Alain Berthod Laboratoire des Sciences Analytiques, CNRS, Université de Lyon I, Villeurbanne, France

Jacques Bodennec Laboratory of Tumor Glycobiology, University Claude Bernard Lyon I, Oullins, France

Frederic Bonfils CIRAD-CP, Programme Hevea, Montpellier, France

Michael Breslav R. W. Johnson Pharmaceutical Research Institute, Spring House, Pennsylvania

Yefim Brun Waters Corporation, Milford, Massachusetts

Jean-Pierre Busnel Université du Maine, U.M.R. 6120/CNRS, Le Mans, France

Yong Cai Department of Chemistry and Southeast Environmental Research Center, Florida International University, Miami, Florida

Ping Cao Biology Department, Tularik, Inc., South San Francisco, California

Wenjie Cao Huntsman Polymers Corporation, Odessa, Texas

Sylvain Caravieilhes Laboratoire des Sciences Analytiques, CNRS, Université de Lyon I, Villeurbanne, France

Philippe Cardot Laboratoire de Chimie Analytique et Bromatologie, Université de Limoges, Limoges, France

M. Caude Analytical Chemistry Department, ESPCI, Paris, France

Teresa Cecchi Dipartamento di Scienze Chimiche, Universitá degli Studi di Camerino, Camerino, Italy

Jeffrey J. Chalmers Department of Chemical Engineering, The Ohio State University, Columbus, Ohio

Bezhan Chankvetadze Department of Physical Chemistry, Tbilisi State University, Tbilisi, Georgia

Bailin Chen Department of Chemistry, Monash University, Melbourne, Australia

T. Chianea Laboratoire de Chimie Analytique et Bromatologie, Université de Limoges, Limoges, France

Oscar Chiantore Dipartimento di Chimica IFM, Università degli Studi di Torino, Torino, Italy

Josef Chmelík Institute of Analytical Chemistry, Academy of Sciences of the Czech Republic, Brno, Czech Republic

Irena Choma Department of Chemical Physics, Marie Curie Sklodowska University, Lublin, Poland

Gabriela Cimpan Analytical Chemistry Department, "Babes-Bolyai" University, Cluj-Napoca, Romania

Alessandra Cincinelli Departimento di Sanità Pubblica Epidemiologia e Chimica Analitica Ambientale, Università degli Studi di Firenze, Firenze, Italy

William Clarke Department of Chemistry, University of Nebraska, Lincoln, Nebraska

Christa L. Colyer Department of Chemistry, Wake Forest University, Winston-Salem, North Carolina

Danilo Corradini Institute of Chromatography, Rome, Italy

Tibor Cserháti Institute of Chemistry, Chemical Research Center, Hungarian Academy of Sciences, Budapest, Hungary

James Curry International Specialty Products, Wayne, New Jersey

Claude De Bellefon Laboratoire de Génie des Procédés Catalytiques, CNRS, CPE Lyon I, Villeurbanne, France

M. de Moraes Laboratorio de Cromatografia, Instituto de Quimica de São Carlos, Universidade de São Paulo, São Carlos/SP, Brazil

Richard DeMuro Shimadzu Scientific Instruments, Inc., Columbia, Maryland

Yulin Deng Neuropsychiatry Research Unit, University of Saskatchewan, Saskatoon, Saskatchewan, Canada

N. Dimov Analytical Department, NIHFI, Sofia, Bulgaria

Jahangir Emrani Novartis Crop Protection, Inc., Greensboro, North Carolina

William P. Farrell Pfizer Global Research and Development, La Jolla, California

Petr S. Fedotov Vernadsky Institute of Geochemistry and Analytical Chemistry, Russian Academy of Sciences, Moscow, Russia

Sam J. Ferrito Analytical Services Department, Cooper Power Systems, Franksville, Wisconsin

John C. Ford Department of Chemistry, Indiana University of Pennsylvania, Indiana, Pennsylvania

Esther Forgács Institute of Chemistry, Chemical Research Center, Hungarian Academy of Sciences, Budapest, Hungary

George M. Frame II Consultant, Halfmoon, New York

Kenneth G. Furton Department of Chemistry, International Forensic Research Institute (IFRI), Florida International University, Miami, Florida

Kalliopi A. Georga Laboratory of Analytical Chemistry, Chemistry Department, Aristotle University of Thessaloniki, Thessaloniki, Greece

Árpád Gerstner Institute of Medical Chemistry, Molecular Biology, and Pathobiochemistry, Semmelweis University, Budapest, Hungary

Michel Girard Bureau of Biologics and Radiopharmaceuticals, Health Canada, F. G. Banting Research Centre, Ottawa, Ontario, Canada

Ivan Gitsov Faculty of Chemistry, College of Environmental Science and Forestry, State University of New York, Syracuse, New York

Kazimierz Głowniak Department of Pharmacognosy, Medical University, Lublin, Poland

Simion Gocan Department of Analytical Chemistry, "Babes-Bolyai" University, Cluj-Napoca, Romania

Karen M. Gooding Eli Lilly and Company, Indianapolis, Indiana

Mohan Gownder Huntsman Polymers Corporation, Odessa, Texas

Susan V. Greene Ethyl Petroleum Additives Corporation, Richmond, Virginia

Nelu Grinberg Analytical Research Department, Merck Research Laboratories, Rahway, New Jersey

András Guttman Torrey Mesa Research Institute, La Jolla, California

David S. Hage Department of Chemistry, University of Nebraska, Lincoln, Nebraska

J. E. Haky Department of Chemistry and Biochemistry, Florida Atlantic University, Boca Raton, Florida

Susana Maria Halpine Consultant, Playa Del Rey, California

Jamel S. Hamada Southern Regional Research Center, USDA-ARS, New Orleans, Louisiana

Martin Hassellöv Department of Analytical and Marine Chemistry, Goteborg University, Gothenburg, Sweden

Michael P. Henry Advanced Technology Center, Beckman Coulter, Inc., Fullerton, California

Gordon S. Hunter Gilson, Inc., Middleton, Wisconsin

W. Jeffrey Hurst Hershey Foods Technical Center, Hershey, Pennsylvania

Robert J. Hurtubise Department of Chemistry, University of Wyoming, Laramie, Wyoming

Christine Hürzeler Postnova Analytics, Munich, Germany

Radovan Hynek Department of Biochemistry and Microbiology, Institute of Chemical Technology, Prague, Czech Republic

Gunawan Indrayanto Faculty of Pharmacy, Airlangga State University, Surabaya, Indonesia

Haleem J. Issaq NCI-Frederick Cancer Research and Development Center, Frederick, Maryland

Yoichiro Ito National Heart, Lung, and Blood Institute, National Institutes of Health, Bethesda, Maryland

Josef Janča Departement de Chimie, Université de La Rochelle, La Rochelle, France

A. Jardy Analytical Chemistry Department, ESPCI, Paris, France

Alfonso Jiménez Migallon Department of Analytical Chemistry, University of Alicante, Alicante, Spain

Kiyokatsu Jinno School of Materials Science, Toyohashi University of Technology, Toyohashi, Japan

Harald John IPF PharmaCeuticals GmbH, Hannover, Germany

Brian Jones Selerity Technologies, Inc., Salt Lake City, Utah

Huba Kalász Department of Pharmacology and Pharmacotherapy, Semmelweis University, Budapest, Hungary

George Karaiskakis Department of Chemistry, University of Patras, Patras, Greece

Jan Káš Department of Biochemistry and Microbiology, Institute of Chemical Technology, Prague, Czech Republic

Galina Kassalainen Department of Chemistry and Geochemistry, Colorado School of Mines, Golden, Colorado

Sarah Kazmi Department of Chemistry, Northeastern University, Boston, Massachusetts

Ernst Kenndler Institute for Analytical Chemistry, University of Vienna, Vienna, Austria

Eileen Kennedy Novartis Crop Protection, Inc., Greensboro, North Carolina

Yuriko Kiba Department of Medicinal Chemistry, University of Tokushima, Tokushima, Japan

Peter Kilz Polymer Standards Service, Mainz, Germany

Peter T. Kissinger Bioanalytical Systems, Inc., and Purdue University, West Lafayette, Indiana

Eiichi Kitazume Laboratory of Chemistry, Iwate University, Morioka, Japan

Thorsten Klein Postnova Analytics, Munich, Germany

Oliver Klett Institute of Chemistry, Uppsala University, Uppsala, Sweden

Athanasia Koliadima Department of Chemistry, University of Patras, Patras, Greece

Vadim L. Kononenko Institute of Biochemical Physics, Russian Academy of Sciences, Moscow, Russia

Teresa Kowalska Institute of Chemistry, Silesian University, Katowice, Poland

Anna Kozak Department of Biochemistry and Microbiology, Institute of Chemical Technology, Prague, Czech Republic

Ira S. Krull Department of Chemistry, Northeastern University, Boston, Massachusetts

Silvia Lacorte Department of Environmental Chemistry, IIQAB-CSIC, Barcelona, Spain

Vaishali Soneji Lafita Abbott Laboratories, Inc., Abbott Park, Illinois

Fernando M. Lanças Laboratorio de Cromatografia, Instituto de Quimica de São Carlos, Universidade de São Paulo, São Carlos/SP, Brazil

David Y. W. Lee McLean Hospital, Harvard Medical School, Belmont, Massachusetts

Seungho Lee Department of Chemistry, Hannam University, Taejon, Korea

Luciano Lepri Dipartimento di Sanità Pubblica Epidemiologia e Chimica Analitica Ambientale, Università degli Studi di Firenze, Firenze, Italy

James Lesec Laboratoire Physique et Chimie Macromoleculaire, CNRS-ESPCI, Paris, France

Rosario LoBrutto* Chemistry Department, Seton Hall University, South Orange, New Jersey

E. S. M. Lutz Department of DMPK and Bioanalytical Chemistry, AstraZeneca R&D Mölndal, Mölndal, Sweden

Ying Ma National Heart, Lung, and Blood Institute, National Institutes of Health, Bethesda, Maryland

Edward Malawer International Specialty Products, Wayne, New Jersey

M. L. Marín Department of Analytical Chemistry, University of Alicante, Alicante, Spain

Wojciech Markowski Department of Inorganic and Analytical Chemistry, Medical University, Lublin, Poland

T. Maryutina Vernadsky Institute of Geochemistry and Analytical Chemistry, Russian Academy of Sciences, Moscow, Russia

Maria T. Matyska Department of Chemistry, San Jose State University, San Jose, California

Gregorio R. Meira INTEC, Universidad Nacional del Litoral and CONICET, Santa Fe, Argentina

Raniero Mendichi Istituto di Chimica delle Macromolecole (CNR), Milano, Italy

Jean-Michel Menet Process Development Chemistry, Aventis Pharma, Vitry-sur-Seine, France

Ivan Mikšík Institute of Physiology, Academy of Sciences of the Czech Republic, Prague, Czech Republic

Myeong Hee Moon Department of Chemistry, Pusan National University, Pusan, Korea

J. J. S. Moreira Laboratorio de Cromatografia, Instituto de Quimica de São Carlos, Universidade de São Paulo, São Carlos/SP, Brazil

Sadao Mori PAC Research Institute, Mie University, Nagoya, Japan

Mark Moskovitz Scientific Adsorbents, Inc., Atlanta, Georgia

Tomasz Mroczek Department of Pharmacognosy, Medical University, Lublin, Poland

Sanjay Mukherjee Department of Chemistry, University of Nebraska, Lincoln, Nebraska

Muhammad Mulja Faculty of Pharmacy, Airlangga State University, Surabaya, Indonesia

D. Muller Institut Galilee, Université Paris Nord, Villetaneuse, France

S. Muralidharan Chemistry Department, Western Michigan University, Kalamazoo, Michigan

Roy A. Musil Althea Technologies, Inc., San Diego, California

Ron Myers Wyatt Technology Corporation, Santa Barbara, California

Noh-Hong Myoung Institute of Health and Environment, Seoul, Korea

* *Current affiliation*: Merck Research Laboratories, Rahway, New Jersey.

Tim Nadler Applied Biosystems, Framingham, Massachusetts

Monica J. S. Nadler Beth Israel Deaconess Medical Center and Harvard Medical School, Boston, Massachusetts

Tuan Q. Nguyen Department of Materials Science, Swiss Federal Institute of Technology, Lausanne, Switzerland

Boryana Nikolova-Damyanova Institute of Organic Chemistry, Bulgarian Academy of Sciences, Sofia, Bulgaria

Hisao Oka Aichi Prefectural Institute of Public Health, Nagoya, Japan

Koji Otsuka Department of Material Science, Himeji Institute of Technology, Hyogo, Japan

Paul K. Owens Pharmaceutical Research and Development, AstraZeneca R&D Mölndal, Mölndal, Sweden

Anders Palm Cell and Molecular Biology, AstraZeneca, Lund, Sweden

Ioannis N. Papadoyannis Laboratory of Analytical Chemistry, Chemistry Department, Aristotle University of Thessaloniki, Thessaloniki, Greece

Joseph J. Pesek Department of Chemistry, San Jose State University, San Jose, California

Miroslav Petro Symyx Technologies, Santa Clara, California

Terry M. Phillips Ultramicro Analytical Immunochemistry Resource, DBEPS, ORS, OD, NIH, Rockville, Maryland

Jacques Portoukalian Laboratory of Tumor Glycobiology, University Claude Bernard Lyon I, Oullins, France

K. R. Preston Grain Research Laboratory, Canadian Grain Commission, Winnipeg, Canada

Wojciech Prus Technical University of Łódź, Branch in Bielsko-Biała, Bielsko-Biała, Poland

Waraporn Putalun Graduate School of Pharmaceutical Sciences, Kyushu University, Fukuoka, Japan

Alina Pyka Faculty of Pharmacy, Silesian Academy of Medicine, Sosnowiec, Poland

Fred M. Rabel EM Science, Gibbstown, New Jersey

Chitra K. Ratnayake Bioresearch Systems Development Center, Beckman Coulter, Inc., Fullerton, California

Jetse C. Reijenga Department of Chemical Engineering and Chemistry, Eindhoven University of Technology, Eindhoven, The Netherlands

Pierluigi Reschiglian Department of Chemistry "G. Ciamician," University of Bologna, Bologna, Italy

Mark P. Richards Growth Biology Laboratory, USDA-ARS-ANRI, Beltsville, Maryland

M.-C. Rolet-Menet Laboratoire de Chimie Analytique, UFR des Sciences Pharmacochimie et Biologie, Paris, France

Jan K. Różyło Department of Adsorption and Planar Chromatography, Marie Curie-Sklodowska University, Lublin, Poland

Jirí Sajdok Department of Biochemistry and Microbiology, Institute of Chemical Technology, Prague, Czech Republic

Peter Sajonz Merck Research Laboratories, Rahway, New Jersey

Victoria F. Samanidou Laboratory of Analytical Chemistry, Chemistry Department, Aristotle University of Thessaloniki, Thessaloniki, Greece

Mária Sasvári-Székely Institute of Medical Chemistry, Molecular Biology, and Pathobiochemistry, Semmelweis University, Budapest, Hungary

Wes Schafer Merck Research Laboratories, Rahway, New Jersey

Martin E. Schimpf Chemistry Department, Boise State University, Boise, Idaho

Oliver Schmitz Division of Molecular Toxicology, German Cancer Research Center, Heidelberg, Germany

Raymond P. W. Scott Scientific Detectors Ltd., Banbury, Oxfordshire, England

Stephen L. Secreast Pharmaceutical Sciences, Pharmacia Corporation, Kalamazoo, Michigan

H. Seegulum Department of Chemistry and Biochemistry, Florida Atlantic University, Boca Raton, Florida

S. N. Semenov Institute of Biochemical Physics RAS, Moscow, Russia

Larry Senak International Specialty Products, Wayne, New Jersey

Vince Serignese Pharmaceutical Analysis Laboratory, King Faisal Specialist Hospital and Research Centre, Riyadh, Saudi Arabia

Joanne Severs Bayer Pharmaceuticals, Berkeley, California

Joseph Sherma Chemistry Department, Lafayette College, Easton, Pennsylvania

Yoichi Shibusawa School of Pharmacy, Tokyo University of Pharmacy and Life Science, Tokyo, Japan

Zak K. Shihabi Department of Pathology, Wake Forest University, Winston-Salem, North Carolina

Kazufusa Shinomiya College of Pharmacy, Nihon University, Chiba, Japan

Yukihiro Shoyama Graduate School of Pharmaceutical Sciences, Kyushu University, Fukuoka, Japan

Edward Soczewinski Department of Inorganic and Analytical Chemistry, Medical University, Lublin, Poland

Boris Ya. Spivakov Vernadsky Institute of Geochemistry and Analytical Chemistry, Russian Academy of Sciences, Moscow, Russia

Raluca-Ioana Stefan Department of Chemistry, University of Pretoria, Pretoria, South Africa

S. G. Stevenson Grain Research Laboratory, Canadian Grain Commission, Winnipeg, Canada

André M. Striegel Solutia, Inc., Springfield, Massachusetts

Ian A. Sutherland Brunel Institute for Bioengineering, Brunel University, Uxbridge, Middlesex, United Kingdom

Hiroyuki Tanaka Graduate School of Pharmaceutical Sciences, Kyushu University, Fukuoka, Japan

M. C. H. Tavares Laboratorio de Cromatografia, Instituto de Quimica de São Carlos, Universidade de São Paulo, São Carlos/SP, Brazil

D. A. Teifer Department of Chemistry and Biochemistry, Florida Atlantic University, Boca Raton, Florida

Shigeru Terabe Department of Material Science, Himeji Institute of Technology, Hyogo, Japan

Iwao Teraoka Department of Chemistry, Chemical Engineering, and Materials Science, Polytechnic University, Brooklyn, New York

Gerald J. Terfloth Research and Development Division, SmithKline Beecham Pharmaceuticals, King of Prussia, Pennsylvania

Georgios A. Theodoridis Laboratory of Analytical Chemistry, Chemistry Department, Aristotle University of Thessaloniki, Thessaloniki, Greece

Richard Thompson Analytical Research Department, Merck Research Laboratories, Rahway, New Jersey

Niem Tri Department of Chemistry, Monash University, Melbourne, Australia

Anant Vailaya Merck Research Laboratories, Rahway, New Jersey

Jacobus F. van Staden Department of Chemistry, University of Pretoria, Pretoria, South Africa

Jorge R. Vega INTEC, Universidad Nacional del Litoral and CONICET, Santa Fe, Argentina

Manuel C. Ventura Pfizer Global Research and Development, La Jolla, California

J. Vial Analytical Chemistry Department, ESPCI, Paris, France

Nikolay Vladimirov Hercules, Inc., Wilmington, Delaware

Qin-Sun Wang National Key Laboratory of Elemento-Organic Chemistry, Nankai University, Tianjin, People's Republic of China

Tao Wang Merck Research Laboratories, Rahway, New Jersey

Teresa Wawrzynowicz Department of Inorganic and Analytical Chemistry, Medical University, Lublin, Poland

Robert Weinberger CE Technologies, Inc., Chappaqua, New York

Adrian Weisz Office of Cosmetics and Colors, Center for Food Safety and Applied Nutrition, USFDA, Washington, DC

S. Kim Ratanathanawongs Williams Department of Chemistry and Geochemistry, Colorado School of Mines, Golden, Colorado

P. Stephen Williams Department of Biomedical Engineering, The Cleveland Clinic Foundation, Cleveland, Ohio

Chi-san Wu International Specialty Products, Wayne, New Jersey

Philip J. Wyatt Wyatt Technology Corporation, Santa Barbara, California

Yu Yang Department of Chemistry, East Carolina University, Greenville, North Carolina

Maciej Zborowski Department of Biomedical Engineering, The Cleveland Clinic Foundation, Cleveland, Ohio

Igor G. Zenkevich Chemical Research Institute, St. Petersburg State University, St. Petersburg, Russia

Ji-Feng Zhang Biotechnology Process Engineering Center, Massachusetts Institute of Technology, Cambridge, Massachusetts

L. Zhang National Key Laboratory of Elemento-Organic Chemistry, Nankai University, Tianjin, People's Republic of China

Lifeng Zhang Environmental Technology Institute, Innovation Centre (NTU), Singapore

Weihua Zhang Department of Chemistry and Southeast Environmental Research Center, Florida International University, Miami, Florida

Xi-Chun Zhou Department of Chemistry, Cambridge University, Cambridge, England

Anastasia Zotou Laboratory of Analytical Chemistry, Chemistry Department, Aristotle University of Thessaloniki, Thessaloniki, Greece

Absorbance Detection in Capillary Electrophoresis

On-Capillary Detection

Most forms of detection in High-Performance Capillary Electrophoresis (HPCE) employ on-capillary detection. Exceptions are techniques that use a sheath flow such as laser-induced fluorescence [1] and electrospray ionization mass spectrometry [2].

In high-performance liquid chromatography (HPLC), postcolumn detection is generally used. This means that all solutes are traveling at the same velocity when they pass through the detector flow cell. In HPCE with on-capillary detection, the velocity of the solute determines the residence time in the flow cell. This means that slowly migrating solutes spend more time in the optical path and thus accumulate more area counts [3].

Because peak areas are used for quantitative determinations, the areas must be normalized when quantitating without standards. Quantitation without standards is often used when determining impurity profiles in pharmaceuticals, chiral impurities, and certain DNA applications. The correction is made by normalizing (dividing) the raw peak area by the migration time. When a matching standard is used, it is unnecessary to perform this correction. If the migration times are not reproducible, the correction may help, but it is better to correct the situation causing this problem.

Limits of Detection

The limit of detection (LOD) of a system can be defined in two ways: the concentration limit of detection (CLOD) and the mass limit of detection (MLOD). The CLOD of a typical peptide is about 1 μg/mL using absorbance detection at 200 nm. If 10 nL are injected, this translates to an MLOD of 10 pg at three times the baseline noise. The MLOD illustrates the measuring capability of the instrument. The more important parameter is the CLOD, which relates to the sample itself. The CLOD for HPCE is relatively poor, whereas the MLOD is quite good, especially when compared to HPLC. In HPLC, the injection size can be 1000 times greater compared to HPCE.

The CLOD can be calculated using Beer's Law:

$$\text{CLOD} = \frac{A}{ab} = \frac{5 \times 10^{-5}}{(5000)(5 \times 10)^{-3}} = 2 \times 10^{-6} M \qquad (1)$$

where A is the absorbance (AU), a is the molar absorptivity (AU/cm/M), b is the capillary diameter or optical path length (cm), and CLOD is the concentration (M). The noise of a good detector is typically 5×10^{-5} AU. A modest chromophore has a molar absorptivity of 5000. Then in a 50-μm-inner diameter (i.d.) capillary, a CLOD of 2×10^{-6} M is obtained at a signal-to-noise ratio of 1, assuming no other sources of band broadening.

Detector Linear Dynamic Range

The noise level of the best detectors is about 5×10^{-5} AU. Using a 50-μm-i.d. capillary, the maximum signal that can be obtained while yielding reasonable peak shape is 5×10^{-1} AU. This provides a linear dynamic range of about 10^4. This can be improved somewhat through the use of an extended path-length flow cell. In any event, if the background absorbance of the electrolyte is high, the noise of the system will increase regardless of the flow cell utilized.

Classes of Absorbance Detectors

Ultraviolet/visible absorption detection is the most common technique found in HPCE. Several types of absorption detectors are available on commercial instrumentation, including the following:

1. Fixed-wavelength detector using mercury, zinc, or cadmium lamps with wavelength selection by filters
2. Variable-wavelength detector using a deuterium or tungsten lamp with wavelength selection by a monochromator
3. Filter photometer using a deuterium lamp with wavelength selection by filters
4. Scanning ultraviolet (UV) detector
5. Photodiode array detector

Each of these absorption detectors have certain attributes that are useful in HPCE. Multiwavelength detectors such as the photodiode array or scanning UV detector are valuable because spectral as well as electrophoretic information can be displayed. The filter photometer is invaluable for low-UV detection. The use of the 185-nm mercury line becomes practical in HPCE with phosphate buffers because the short optical path length minimizes the background absorption.

Photoacoustic, thermo-optical, or photothermal detectors have been reported in the literature [4]. These detectors measure the nonradiative return of the excited molecule to the ground state. Although these can be quite sensitive, it is unlikely that they will be used in commercial instrumentation.

Optimization of Detector Wavelength

Because of the short optical path length defined by the capillary, the optimal detection wavelength is frequently much lower into the UV compared to HPLC. In HPCE with a variable-wavelength absorption detector, the optimal signal-to-noise (S/N) ratio for peptides is found at 200 nm. To optimize the detector wavelength, it is best to plot the S/N ratio at various wavelengths. The optimal S/N is then easily selected.

Extended Path-Length Capillaries

Increasing the optical path length of the capillary window should increase S/N simply as a result of Beer's Law. This has been achieved using a z cell (LC Packings, San Francisco CA) [5], bubble cell (Agilent Technologies, Wilmington, DE), or a high-sensitivity cell (Agilent Technologies). Both the z cell and bubble cell are integral to the capillary. The high-sensitivity cell comes in three parts: an inlet capillary, an outlet capillary, and the cell body. Careful assembly permits the use of this cell without current leakage. The bubble cell provides approximately a threefold improvement in sensitivity using a 50-μm capillary, whereas the z cell or high-sensitivity cell improves things by an order of magnitude. This holds true only when the background electrolyte (BGE) has low absorbance at the monitoring wavelength.

Indirect Absorbance Detection

To determine ions that do not absorb in the UV, indirect detection is often utilized [6]. In this technique, a UV-absorbing reagent of the same charge (a co-ion) as the solutes is added to the BGE. The reagent elevates the baseline, and when nonabsorbing solute ions are present, they displace the additive. As the separated ions migrate past the detector window, they are measured as negative peaks relative to the high baseline. For anions, additives such as trimellitic acid, phthalic acid, or chromate ions are used at 2–10 m*M* concentrations. For cations, creatinine, imidazole, or copper(II) are often used. Other buffer materials are either not used or added in only small amounts to avoid interfering with the detection process.

It is best to match the mobility of the reagent to the average mobilities of the solutes to minimize electrodispersion, which causes band broadening [7]. When anions are determined, a cationic surfactant is added to the BGE to slow or even reverse the electro-osmotic flow (EOF). When the EOF is reversed, both electrophoresis and electro-osmosis move in the same direction. Anion separations are performed using reversed polarity.

Indirect detection is used to determine simple ions such as chloride, sulfate, sodium, and potassium. The technique is also applicable to aliphatic amines, aliphatic carboxylic acids, and simple sugars [8].

References

1. Y. F. Cheng and N. J. Dovichi, *SPIE*, 910: 111 (1988).
2. E. C. Huang, T. Wachs, J. J. Conboy, and J. D. Henion, *Anal. Chem. 62*: 713 (1990).
3. X. Huang, W. F. Coleman, and R. N. Zare, *J. Chromatogr.* 480: 95 (1989).
4. J. M. Saz and J. C. Diez-Masa, *J. Liq. Chromatogr. 17*: 499 (1994).

5. J. P. Chervet, R. E. J. van Soest, and M. Ursem, *J. Chromatogr. 543*: 439 (1991).
6. P. Jandik, W. R. Jones, A. Weston, and P. R. Brown, *LC–GC 9*: 634 (1991).
7. R. Weinberger, *Am. Lab. 28*: 24 (1996).
8. X. Xu, W. T. Kok, and H. Poppe, *J. Chromatogr. A 716*: 231 (1995).

Robert Weinberger

Acoustic Field-Flow Fractionation for Particle Separation

Field-flow fractionation (FFF) is a suite of elution methods suitable for the separation and sizing of macromolecules and particles [1]. It relies on the combined effects of an applied force interacting with sample components and the parabolic velocity profile of carrier fluid in the channel. For this to be effective, the channel is unpacked and the flow must be under laminar conditions. Field or gradients that are commonly used in generating the applied force are gravity, centrifugation, fluid flow, temperature gradient, and electrical and magnetic fields. Each field or gradient produces a different subtechnique of FFF, which separates samples on the basis of a particular property of the molecules or particles.

The potential for using acoustic radiation forces generated by ultrasonic waves to extend the versatility of FFF seems very promising. Although only very preliminary experiments have been performed so far, the possibility of using such a gentle force would appear to have huge potential in biology, medicine, and environmental studies.

Acoustic radiation or ultrasonic waves are currently being exploited as a noncontact particle micromanipulation technique [2]. The main drive to develop such techniques comes from the desire to manipulate biological cells and blood constituents in biotechnology and fine powders in material engineering.

In a propagating wave, the acoustic force, F_{ac}, acting on a particle is a function of size given by [1]

$$F_{ac} = \pi r^2 E Y_p \tag{1}$$

where r is the particle radius, E is the sound energy density, and Y_p is a complicated function depending on the characteristics of the particle which approaches unity if the wavelength used is much smaller than the particle. Particles in a solution subjected to a propagating sound wave will be pushed in the direction of sound propagation. Therefore, sized-based separations may be possible if this force is applied to generate selective transport of different components in a mixture. In a FFF channel, it is likely that the receiving wall will reflect at least some of the emitted wave. If the channel thickness corresponds exactly to one-half wavelength, then a single standing wave will be created (see Fig. 1). For a single standing wave, it is interesting to note that three pressure (force) nodes are generated, one at each wall and one in the center of the channel.

Yasuda and Kamakura [3] and Mandralis and co-workers [4] have demonstrated that it is possible to generate standing-wave fields between a transducer and a reflecting wall, although of much larger dimensions (1–20 cm) than across a FFF channel. Sound travels at a velocity of 1500 m/s through water, which translates to a wave of frequency of approximately 6 MHz for a 120-μm thick FFF channel.

The force experienced by a particle in a stationary acoustic wave was reported by Yosioka and Kawasima [5] to be

$$F_{ac} = 4\pi r^3 \kappa E_{ac} A \sin(2\kappa x) \tag{2}$$

where r is the particle radius, κ is the wave number, E_{ac} is the time-averaged acoustic energy density, and A is the acoustic contrast factor given by

$$A = \frac{1}{3}\left(\frac{5\rho_p - 2\rho_l}{\rho_l + 2\rho_p} - \frac{\gamma_p}{\gamma_l}\right) \tag{3}$$

where ρ_p and γ_p are the particle density and compressibility, respectively, and ρ_l and γ_l are the liquid density and compressibility, respectively. Thus, in a propagating wave, the force on a particle has a second-order dependence, and in a standing wave, the force is third order. This should give rise to increased selectivity for separations being carried out in a standing wave [6].

Due to the nature of the acoustic fields, the distribution of the particles will depend on the particle size and the compressibility and density of the particle relative to the fluid medium. Closer examination of the acoustic contrast factor shows that is may be negative (usually applicable to

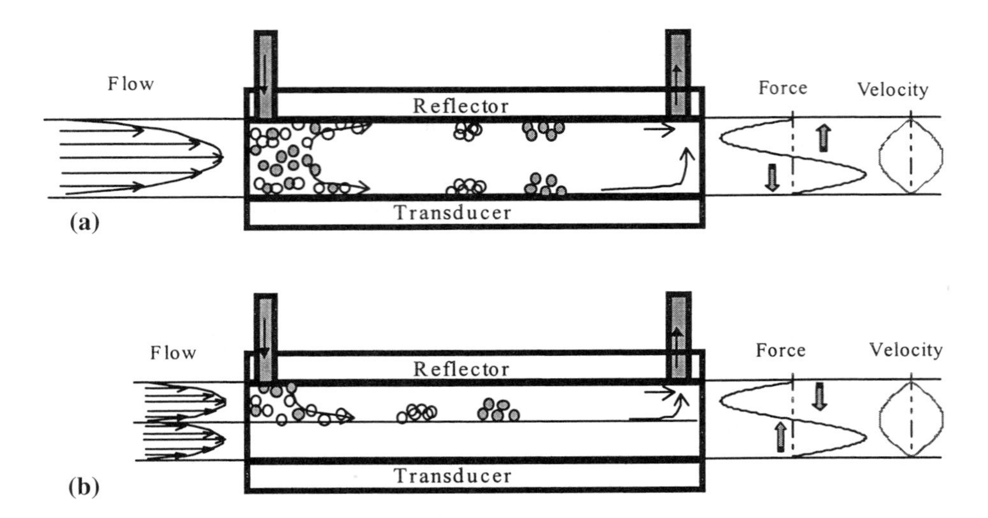

Fig. 1 Acoustic FFF channels suitable for particles with (a) $A < 0$ and (b) $A > 0$, utilizing a divided acoustic FFF channel.

biological cells which are more compressible and less dense relative to the surrounding medium) or positive (as is in many inorganic and polymer colloids). Therefore, acoustic FFF (AcFFF) has tremendous potential in very clean separations of cells from other particles. One important application may be for the separation of bacterial and algal cells in soils and sediments.

If the acoustic contrast factor $A < 0$, then a conventional FFF channel will enable normal and steric mode FFF separations to be carried out (Fig. 1a).

However, if $A > 0$, then the particles will migrate toward the center of the channel. In this case, a divided FFF cell could be used as shown in Fig. 1b. This ensures that particles are driven to an accumulation wall rather than the center of the channel where the velocity profile is quite flat and selectivity would be minimal.

Johnson and Feke [7] effectively demonstrated that latex spheres migrate to the nodes (center of the cell) and Hawkes and co-workers [8] showed that yeast cells migrate to the antinodes (walls of the cell). These authors used a method similar to SPLITT, which is another technique closely related to FFF, also originally developed by Giddings [9]. Semyonov and Maslow [10] demonstrated that acoustic fields in a FFF channel affected the retention time of a sphere of 3.8 μm diameter when subjected to varying acoustic fields. However, the high resolution inherent in FFF has not yet been exploited.

Naturally, with some design modifications to the FFF channel, SPLITT cells could be used for sample concentration or fluid clarification.

References

1. J. C. Giddings, *J. Chem. Phys. 49*: 81 (1968).
2. T. Kozuka, T. Tuziuti, H. Mitome, and T. Fukuda, *Proc. IEEE* 435 (1996).
3. K. Yasuda and T. Kamakura, *Appl. Phys. Lett. 71*: 1771 (1997).
4. Z. Mandralis, W. Bolek, W. Burger, E. Benes, and D. L. Feke, *Ultrasonics 32*: 113 (1994).
5. K. Yosioka and Y. Kawasima, *Acustica 5*: 167 (1955).
6. A. Berthod and D. W. Armstrong, *Anal. Chem. 59*: 2410 (1987).
7. D. A. Johnson and D. L. Feke, *Separ. Technol. 5*: 251 (1995).
8. J. J. Hawkes, D. Barrow, and W. T. Coakley, *Ultrasonics 36*: 925 (1998).
9. J. C. Giddings, *Anal. Chem. 57*: 945 (1985).
10. S. N. Semyonov and K. I. Maslow, *J. Chromatogr. 446*: 151 (1998).

Niem Tri
Ronald Beckett

Adhesion of Colloids on Solid Surfaces by Field-Flow Fractionation

Introduction

The adhesion of colloids on solid surfaces, which is of great significance in filtration, corrosion, detergency, coatings, and so forth, depends on the total potential energy of interaction between the colloidal particles and the solid surfaces. The latter, which is the sum of the attraction potential energy and that of repulsion, depends on particle size, the Hamaker constant, the surface potential, and the Debye–Huckel reciprocal distance, which is immediately related to the ionic strength of carrier solution. With the aid of the field-flow fractionation technique, the adhesion and detachment processes of colloidal materials on and from solid surfaces can be studied. As model samples for the adhesion of colloids on solid surfaces (e.g., Hastelloy-C), hematite (a-Fe_2O_3) and titanium dioxide (TiO_2) submicron spherical particles, as well as hydroxyapatite [$Ca_5(PO_4)_3OH$] submicron irregular particles were used. The experimental conditions favoring the adhesion process were those decreasing the surface potential of the particles through the pH and ionic-strength variation, as well as increasing the effective Hamaker constant between the particles and the solid surfaces through the surface-tension variation. On the other hand, the detachment of the same colloids from the solid surfaces can be favored under the experimental conditions decreasing the potential energy of attraction and increasing the repulsion potential energy.

Methodology

Field-flow fractionation (FFF) technology is applicable to the characterization and separation of particulate species and macromolecules. Separations in FFF take place in an open flow channel over which a field is applied perpendicular to the flow. Among the various FFF subtechniques, depending on the kind of the applied external fields, sedimentation FFF (SdFFF) is the most versatile and accurate, as it is based on simple physical phenomena that can be accurately described mathematically. SdFFF, which uses a centrifugal gravitational force field, is a flow-modified equilibrium sedimentation-separation method. Solute layers that are poorly resolved under static equilibrium sedimentation become well separated as they are eluted by the laminar flow profile in the SdFFF channel.

In normal SdFFF, where the colloidal particles under study do not interact with the channel wall, the potential energy of a spherical particle, $\varphi(x)$, is related to the particle radius, a, to the density difference, $\Delta\rho$, between the particle (ρ_s) and the liquid phase (ρ), and to the sedimentation field strength expressed in acceleration, G:

$$\varphi(x) = \frac{4}{3}\pi a^3 \Delta\rho G x \tag{1}$$

where x is the coordinate position of the center of particle mass.

When the colloidal particles interact with the SdFFF channel wall, the total potential energy, φ_{tot}, of a spherical particle is given by

$$\varphi_{\text{tot}} = \frac{4}{3}\pi a^3 \Delta\rho G x + \frac{A_{132}}{6}\left[\ln\left(\frac{h+2a}{h}\right) - \frac{2a(h+a)}{h(h+a)}\right] + 16\varepsilon a\left(\frac{kT}{e}\right)^2 \tan h\left(\frac{e\psi_1}{4kT}\right)\tan h\left(\frac{e\psi_2}{4kT}\right)e^{-\kappa x} \tag{2}$$

where the second and third terms of Eq. (2) accounts for the contribution of the van der Waals attraction potential and of the double-layer repulsion potential between the particle and the wall, respectively, A_{132} is the effective Hamaker constant for media 1 and 2 interacing across medium 3, h is the separation distance between the sphere and the channel wall, ε is the dielectric constant of the suspending medium, e is the electronic charge, ψ_1 and ψ_2 are the surface potentials of the particles and the solid wall, respectively, k is Boltzmann's constant, T is the absolute temperature, and κ is the Debye–Huckel reciprocal length, which is immediately related to the ionic strength, I, of the medium.

Equation (2) shows that the total potential energy of interaction between a colloidal particle and a solid substrate is a function of the particle radius and surface potential, the ionic strength and dielectric constant of the suspending medium, the value of the effective Hamaker constant, and the temperature. Adhesion of colloidal particles on solid surfaces is increased by a decrease in the particle radius, surface potential, the dielectric constant of the medium and by an increase in the effective Hamaker constant, the ionic strength of the dispersing liquid, or the temperature. For a given particle and a medium with a known dielectric constant, the adhesion and detachment processes depend on the following three parameters:

1. The surface potential of the particles, which can be varied experimentally by various quantities one of which is the suspension pH

2. The ionic strength of the solution, which can be varied by adding to the suspension various amounts of an indifferent electrolyte
3. The Hamaker constant, which can be easily varied by adding to the suspending medium various amounts of a detergent. The later results in a variation of the medium surface tension.

Applications

The critical electrolyte (KNO_3) concentrations found by SdFFF for the adhesion of a-Fe_2O_3(I) (with nominal diameter 0.148 μm), a-Fe_2O_3(II) (with nominal diameter 0.248 μm), and TiO_2 (with nominal diameter 0.298 μm) monodisperse spherical particles on the Hastelloy-C channel wall were $8 \times 10^{-2}M$, $3 \times 10^{-2}M$, and $3 \times 10^{-2}M$, respectively. The values for the same sample (a-Fe_2O_3) depend on the particle size, in accordance with the theoretical predictions, whereas the same values are identical for various samples [a-Fe_2O_3(II) and TiO_2] having different particle diameters. The latter indicates that these values depend also, apart from the size, on the sample's physicochemical properties, as is predicted by Eq. (2). The detachment of the whole number of particles of the above samples from the channel wall was succeeded by decreasing the ionic strength of the carrier solution.

The critical KNO_3 concentration for the detachment process was $3 \times 10^{-2}M$ for the a-Fe_2O_3(I) sample and $1 \times 10^{-3}M$ for the samples of a-Fe_2O_3(II) and TiO_2. Those obtained by SdFFF particle diameters after the detachment of the adherent particles [0.148 μm for a-Fe_2O_3(I), 0.245 μm for a-Fe_2O_3(II), and 0.302 μm for TiO_2] are in excellent agreement with the corresponding nominal particle diameters obtained by transmission electron microscopy. The desorption of all of the adherent particles was verified by the fact that no elution peak was obtained, even when the field strength was reduced to zero. A second indication for the desorption of all of the adherent material was that the sample peaks after adsorption and desorption emerge intact and without degradation.

In a second series of experiments, the adhesion and detachment processes of hydroxyapatite (HAP) polydisperse particles with number average diameter of 0.261 μm on and from the Hastelloy-C channel wall were succeeded by the variation of the suspension pH, whereas the medium's ionic strength was kept constant ($10^{-3}M$ KNO_3). At a suspension pH of 6.8, the whole number of injected HAP particles was adhered at the beginning of the SdFFF channel wall, which was totally released when the pH increased to 9.7, showing that, except for the ionic strength, the pH of the suspending medium is also a principal quantity influencing the interaction energy between colloidal particles and solid surfaces. The number-average diameter of the HAP particles found by SdFFF after the detachment of the adherent particles ($d_N = 0.262$ μm) was also in good agreement with that obtained when the particles were injected into the channel with a carrier solution in which no adhesion occurs ($d_N = 0.261$ μm).

The variation of the potential energy of interaction between colloidal particles and solid surfaces can be also succeeded by the addition of a detergent to the suspending medium, which leads to a decrease in the Hamaker constant and, consequently, in the potential energy of attraction.

In conclusion, field-flow fractionation is a relatively simple technique for the study of adhesion and detachment of submicrometer or supramicrometer colloidal particles on and from solid surfaces.

Future Developments

Looking to the future, it is reasonable to expect more experimental and theoretical work in order to quantitatively investigate the adhesion/detachment phenomena of colloids on and from solid surfaces by measuring the corresponding rate constants with the aid of FFF.

Suggested Further Reading

Athanasopoulou, A. and G. Karaiskakis, *Chromatographia 43*: 369 (1996).
Giddings, J. C., M. N. Myers, K. D. Caldwell, and S. R. Fisher, in *Methods of Biochemical Analysis Vol. 26*, D. Glick (ed.), John Wiley & Sons, New York, 1980, p. 79.
Giddings, J. C., G. Karaiskakis, K. D. Caldwell, and M. N. Myers, *J. Colloid Interf. Sci.* 92(1): 66 (1983).
Hansen, M. E. and J. C. Giddings, *Anal. Chem. 61*: 811 (1989).
Hiemenz, P. C., *Principles of Colloid and Surface Chemistry*, Marcel Dekker, Inc., New York, 1977.
Karaiskakis, G. and J. Cazes (eds.), *J. Liq. Chromatogr. Rel. Technol. 20*(16 & 17) (1997).
Karaiskakis, G., A. Athanasopoulou, and A. Koliadima, *J. Micro. Separ. 9*: 275 (1997).
Koliadima, A. and G. Karaiskakis, *J. Chromatogr. 517*: 345 (1990).
Ruckenstein, E. and D. C. Prieve, *AIChE J. 22*(2): 276 (1976).

George Karaiskakis

Adsorption Chromatography

In essence, the original chromatographic technique was adsorption chromatography. It is frequently referred to as liquid–solid chromatography. Tswett developed the technique around 1900 and demonstrated its use by separating plant pigments. Open-column chromatography is a classical form of this type of chromatography, and the open-bed version is called thin-layer chromatography.

Adsorption chromatography is one of the more popular modern high-performance liquid chromatographic techniques today. However, open-column chromatography and thin-layer chromatography are still widely used [1]. The adsorbents (stationary phases) used are silica, alumina, and carbon. Although some bonded phases have been considered to come under adsorption chromatography, these bonded phases will not be discussed. By far, silica and alumina are more widely used than carbon. The mobile phases employed are less polar than the stationary phases, and they usually consist of a signal or binary solvent system. However, ternary and quaternary solvent combinations have been used.

Adsorption chromatography has been employed to separate a very wide range of samples. Most organic samples are readily handled by this form of chromatography. However, very polar samples and ionic samples usually do not give very good separation results. Nevertheless, some highly polar multifunctional compounds can be separated by adsorption chromatography. Compounds and materials that are not very soluble in water or water–organic solvents are usually more effectively separated by adsorption chromatography compared to reversed-phase liquid chromatography.

When one has an interest in the separation of different types of compound, silica or alumina, with the appropriate mobile phase, can readily accomplish this. Also, isomer separation frequently can easily be accomplished with adsorption chromatography; for example, 5,6-benzoquinoline can be separated from 7,8-benzoquinoline with silica as the stationary phase and 2-propanol:hexane (1:99). This separation is difficult with reversed-phase liquid chromatography [1].

Stationary Phases

Silica is the most widely used stationary phase in adsorption chromatography [2]. However, the extensive work of Snyder [3] involved investigations with both silica and alumina. Much of Snyder's earlier work was with alumina. Even though the surface structures of the two adsorbents have distinct differences, they are sufficiently similar. Thus, many of the fundamental principles developed for alumina are applicable to silica. The general elution order for these two adsorbents is as follows [1]: saturated hydrocarbons (small retention time) $<$ olefins $<$ aromatic hydrocarbons $<$ aromatic hydrocarbons $\approx$ organic halides $<$ sulfides $<$ ethers $<$ nitro-compounds $<$ esters $\approx$ aldehydes $\approx$ ketones $<$ alcohols $\approx$ amines $<$ sulfones $<$ sulfoxides $<$ amides $<$ carboxylic acids (long retention time). There are several reasons why silica is more widely used than alumina. Some of these are that a higher sample loading is permitted, fewer unwanted reactions occur during separation, and a wider range of chromatographic forms of silica are available.

Chromatographic silicas are amorphous and porous and they can be prepared in a wide range of surface areas and average pore diameters. The hydroxyl groups in silica are attached to silicon, and the hydroxyl groups are mainly either free or hydrogen-bonded. To understand some of the details of the chromatographic processes with silica, it is necessary to have a good understanding of the different types of hydroxyl groups in the adsorbent [1,3]. Chromatographic alumina is usually γ-alumina. Three specific adsorption sites are found in alumina: (a) acidic, (b) basic, and (c) electron-acceptor sites. It is difficult to state specifically the exact nature of the adsorption sites. However, it has been postulated that the adsorption sites are exposed aluminum atoms, strained Al—O bonds, or cationic sites [4]. Table 1 gives some of the properties of silica and alumina.

The adsorbent water content is particularly important in adsorption chromatography. Without the deactivation of strong adsorption sites with water, nonreproducible retention times will be obtained, or irreversible adsorption of solutes can occur. Prior to using an adsorbent for open-column chromatography, the adsorbent is dried, a specified amount of water is added to the adsorbent, and then the adsorbent is allowed to stand for 8–16 h to permit the equilibration of water [3,4]. If one is using a high-performance column, it is a good idea to consider adding water to the mobile phase to deactivate the stronger adsorption sites on the adsorbent. Some of the benefits are less variation in retention times, partial compensation for lot-to-lot differences in the adsorbent, and reduced band tailing [1]. However, there can be some problems in adding water to the mobile phase, such as how much water to add to the mobile phase for optimum performance.

Table 1 Some Adsorbents Used in Adsorption Chromatography

Type	Name	Form	Average particle size (μm)	Surface area (m^2/g)
Silica[a]	BioSil A	Bulk	2–10	400
	μPorasil	Column	10	400+
Silica[b]	Hypersil	Bulk	5–7	200
	Zobax Sil	Bulk or column	6	350
Alumina[a]	ICN Al-N	Bulk	3–7, 7–12	200
	MicroPak Al	Bulk or column	5, 10	79
Alumina[b]	Spherisorb AY	—	5, 10, 20	95

[a]Irregular
[b]Spherical
Source: Adapted from Ref. 1.

Snyder and Kirkland [1] have discussed several of these aspects in detail.

Mobile Phases

To vary sample retention, it is necessary to change the mobile-phase composition. Thus, the mobile phase plays a major role in adsorption chromatography. In fact, the mobile phase can give tremendous changes in sample retention characteristics. Solvent strength controls the capacity factor's values of all the sample bands. A solvent strength parameter (ε^0), which has been widely used over the years, can be employed quantitatively for silica and alumina. The solvent strength parameter is defined as the adsorption energy of the solvent on the adsorbent per unit area of solvent [1,3]. Table 2 gives the solvent strength values for selected solvents that have been used in adsorption chromatography. The smaller values of ε^0 indicate weaker solvents, whereas the larger values of ε^0 indicate stronger solvents. The solvents listed in Table 2 are single solvents. Normally, solvents are selected by mixing two solvents with large differences in their ε^0 values, which would permit a continuous change in the solvent strength of the binary solvent mixture. Thus, some specific combination of the two solvents would provide the appropriate solvent strength. In adsorption chromatography, the solvent strength increases with solvent polarity, and the solvent strength is used to obtain the proper capacity factor values, usually in the range of 1–5 or 1–10. It should be realized that the solvent strength does not vary linearly over a wide range of solvent compositions, and several guidelines and equations that allow one to calculate the solvent strength of binary solvents have been developed for acquiring the correct solvent strength in adsorption chromatography [1,3]. However, it frequently happens that the solvent strength is such that all of the solutes are not separated in a sample. Thus, one needs to consider solvent selectivity, which is discussed below.

Table 2 Selected Solvents Used in Adsorption Chromatography

	Solvent strength (ε_0)	
	Silica	Alumina
Solvent		
n-Hexane	0.01	0.01
1-Chlorobutane	0.20	0.26
Chloroform	0.26	0.40
Isopropyl ether	0.34	0.28
Ethyl acetate	0.38	0.58
Tetrahydrofuran	0.44	0.57
Acetonitrile	0.50	0.65

Source: Adapted from Ref. 1.

To change the solvent selectivity, the solvent strength is held constant and the composition of the mobile phase is varied. It should be realized that because the solvent strength is directly related to the polarity of the solvent and polarity is the total of the dispersion, dipole, hydrogen-bonding, and dielectric interactions of the sample and solvent, one would not expect that solvent strength alone could be used to fine-tune a separation. A trial-and-error approach can be employed by using different solvents of equal ε^0. However, there are some guidelines that have been developed that permit improved selectivity. These are the "B-concentration" rule and the "hydrogen-bonding" rule [1]. In general, with the B-concentration rule, the largest change in selectivity is obtained when a very dilute or a very concentrated solution of B (stronger solvent) in a weak solvent (A) is used. The hydrogen-bonding rule states that any change in the mobile phase that results in a change in hydrogen-bonding between sample and mobile-phase molecules usually results in a large change in selectivity. A more comprehensive means for improving selectivity is the solvent-selectivity triangle [1,5]. The solvent-selectivity triangle classifies solvents according to their relative dipole moments, basic properties, and acidic properties. For example, if an initial chromatographic experiment does not separate all the components with a binary mobile phase, then the solvent-selectivity triangle can be used to choose another solvent for the binary system that has properties that are very different than one of the solvents in the original solvent system. A useful publication that discusses the properties of numerous solvents and also considers many chromatographic applications is Ref. 6.

Mechanistic Aspects in Adsorption Chromatography

Models for the interactions of solutes in adsorption chromatography have been discussed extensively in the literature [7–9]. Only the interactions with silica and alumina will be considered here. However, various modifications to the models for the previous two adsorbents have been applied to modern high-performance columns (e.g., amino-silica and cyano-silica). The interactions in adsorption chromatography can be very complex. The model that has emerged which describes many of the interactions is the displacement model developed by Snyder [1,3,4,7,8]. Generally, retention is assumed to occur by a displacement process. For example, an adsorbing solute molecule X displaces n molecules of previously adsorbed mobile-phase molecules M [8]:

$$X_n + nM_a \rightleftharpoons X_a + nM_n$$

The subscripts n and a in the above equation represent a molecule in a nonsorbed and adsorbed phase, respectively. In other words, retention in adsorption chromatography involves a competition between sample and solvent molecules for sites on the adsorbent surface. A variety of interaction energies are involved, and the various energy terms have been described in the literature [7,8]. One fundamental equation that has been derived from the displacement model is

$$\log\left(\frac{k_1}{k_2}\right) = \alpha' A_S(\varepsilon_2 - \varepsilon_1)$$

where k_1 and k_2 are the capacity factors of a solute in two different mobile phases, α' is the surface activity of the adsorbent (relative to a standard adsorbent), A_S is the cross-sectional area of the solute on the adsorbent surface, and ε_1 and ε_2 are the solvent strengths of the two different mobile phases. This equation is valid in situations where the solute and solvent molecules are considered nonlocalizing. This condition is fulfilled with nonpolar or moderately polar solutes and mobile phases. If one is dealing with multisolvent mobile phases, the solvent strength of those solvents can be related to the solvent strengths of the pure solvents in the solvent system. The equations for calculating solvents strengths for multisolvent mobile phases have been discussed in the literature [8].

As the polarities of the solute and solvent molecules increase, the interactions of these molecules become much stronger with the adsorbent, and they adsorb with localization. The net result is that the fundamental equation for adsorption chromatography with relatively nonpolar solutes and solvents has to be modified. Several localization effects have been elucidated, and the modified equations that take these factors into consideration are rather complex [7,8,10]. Nevertheless, the equations provide a very important framework in understanding the complexities of adsorption chromatography and in selecting mobile phases and stationary phases for the separation of solutes.

Applications

There have been thousands of articles published on the application of adsorption chromatography over the decades. Today, adsorption chromatography is used around the world in all areas of chemistry, environmental problem solving, medical research, and so forth. Only a few examples will be discussed in this section. Gogou et al. [11] developed methods for the determination of organic molecular markers in marine aerosols and sediment. They used a one-step flash chromatography compound-class fractionation method to isolate compound-class fractions. Then, they employed gas chromatography/mass spectrometry and/or gas chromatography/flame ionization detection analysis of the fractions. The key adsorption chromatographic step prior to the gas chromatography was the one-step flash chromatography. For example, an organic extract of marine aerosol or sediment was applied on the top of a 30 × 0.7-cm column containing 1.5 g of silica. The following solvent systems were used to elute the different compound classes: (a) 15 mL of *n*-hexane (aliphatics); (b) 15 mL toluene:*n*-hexane (5.6:9.4) (polycyclic aromatic hydrocarbons and nitro-polycyclic aromatic hydrocarbons); (c) 15 mL *n*-hexane:methylene chloride (7.5:7.5) (carbonyl compounds); (d) 20 mL ethyl acetate:*n*-hexane (8:12) (*n*-alkanols and sterols); (e) 20 mL (4%, v/v) pure formic acid in methanol (free fatty acids). This example illustrates very well how adsorption chromatography can be used for compound-class separation.

Hanson and Unger [12] have discussed the application of nonporous silica particles in high-performance liquid chromatography. Nonporous silica packings can be used for the rapid chromatographic analysis of biomolecules because the particles lack pore diffusion and have very effective mass-transfer capabilities. Several of the advantages of nonporous silica are maximum surface accessibility, controlled topography of ligands, better preservation of biological activity caused by shorter residence times on the column, fast column regeneration, less solvent consumption, and less susceptibility to compression during packing. The very low external surface area of the nonporous supports is a disadvantage because it gives considerably lower capacity compared with porous materials. This drawback is counterbalanced partially by the high packing density compared to porous silica. The smooth surface of the nonporous silica offers better biocompatibility relative to porous silica. Well-defined nonporous silicas are now commercially available.

References

1. L. R. Snyder and J. J. Kirkland, *Introduction to Modern Liquid Chromatography*, 2nd ed., John Wiley & Sons, New York, 1979.
2. J. H. Knox (ed.), *High-Performance Liquid Chromatography*, Edinburgh University Press, Edinburgh, 1980.
3. L. R. Snyder, *Principles of Adsorption Chromatography*, Marcel Dekker, Inc., New York, 1968.
4. L. R. Snyder, in *Chromatography: A Laboratory Handbook of Chromatographic and Electrophoretic Methods*, 3rd ed., E. Heftmann (ed.), Van Nostrand Reinhold, New York, 1975, pp. 46–76.
5. L. R. Snyder, J. L. Glajch, and J. J. Kirkland, *Practical HPLC Method Development*, John Wiley & Sons, New York, 1988, pp. 36–39.
6. P. C. Sadek, *The HPLC Solvent Guide*, John Wiley & Sons, New York, 1996.
7. L. R. Snyder and H. Poppe, *J. Chromatogr. 184*: 363 (1980).
8. L. R. Snyder, in *High-Performance Liquid Chromatography, Vol. 3*, C. Horvath (ed.), Academic Press, New York, 1983, pp. 157–223.
9. R. P. W. Scott and P. Kucera, *J. Chromatogr. 171*: 37 (1979).
10. L. R. Snyder and J. L. Glajch, *J. Chromatogr. 248*: 165 (1982).
11. A. I. Gogou, M. Apostolaki, and E. G. Stephanou, *J. Chromatogr. A, 799*: 215 (1998).
12. M. Hanson and K. K. Unger, *LC–GC 15*: 364 (1997).

Robert J. Hurtubise

Adsorption Studies by Field-Flow Fractionation

Adsorption is an important process in many industrial, biological, and environmental systems. One compelling reason to study adsorption phenomena is because an understanding of colloid stability depends on the availability of adequate theories of adsorption from solution and of the structure and behavior of adsorbed layers. Another example is the adsorption of pollutants, such as metals, toxic organic compounds, and nutrients, onto fine particles and their consequent transport and fate, which has great environmental implications. Often, these systems are quite complex and it is often favorable to separate these into specific size for subsequent study.

A new technique able to separate such complex mixtures is field-flow fractionation [1–3]. Field-flow fractionation (FFF) is easily adaptable to a large choice of field forces (such as gravitational, centrifugal, fluid cross-flows, electrical, magnetic and thermal fields or gradients) to effect high-resolution separations. Although the first uses for FFF were for sizing of polymer and colloidal samples, recent advances have demonstrated that well-designed FFF experiments can be used in adsorption studies [4,5].

Although the theory of FFF for the characterisation and fractionation of polymers and colloids has been outlined elsewhere, two important features of FFF need to be emphasized here. The first is the versatility of FFF, which is partly due to the diverse range of operating fields that may be used and the fact that each field is capable of delivering different information about a colloidal sample.

For example, an electrical field separates particles on the basis of both size and charge, whereas a centrifugal field (sedimentation FFF) separates particles on the basis of buoyant mass (i.e. size and density). The second important feature is that this information can usually be measured directly from the retention data using rigorous theory. This is in contrast to most forms of chromatography (size-exclusion chromatography exempted), where the retention time of a given component must be identified by running standards.

In 1991, both Beckett et al. [4] and Li and Caldwell [5] published articles demonstrating novel but powerful uses for sedimentation FFF in probing the characteristics of adsorbed layers or films on colloidal particles. Beckett et al's article demonstrated that it is possible to measure the mass of an adsorbed coating down to a few attograms (10^{-18} g), which translates to a mean coating thickness of human γ-globulin, ovalbumin, RNA, and cortisone ranging from 0.1 to 20 nm. A discussion of the theory and details of the experiment is beyond the scope of this article. However, it is possible to appreciate how such high sensitivities arise by considering the linear approximation of retention time, t_r, of an eluting particle in sedimentation FFF with the field-induced force on the particle, F.

$$t_r = t_0 \frac{Fw}{6kT} \tag{1}$$

where w is the thickness of the channel (typically 100–500 μm), k is the Boltzmann constant, and T is the temperature in Kelvin. F is the force on the individual particle and is the product of the applied field and the buoyant mass of the particle (relative mass of the particle in the surrounding liquid medium).

The highest sensitivity of retention time to changes in the surface coating was found to occur when the density of the core particle was equal to that of the surrounding medium (i.e., the buoyant mass diminishes to zero and no retention is observed for the bare particle). If a thin film of a much denser material is adsorbed onto the particles, then the small increment in mass due to the adsorbed film causes a significant change in the particle's buoyant mass (see Fig. 1a). Consequently, the force felt by the particle is now sufficient to effect retention by an observable amount. Incidentally, analogous behavior is also possible if the coatings are less dense than the carrier liquid. If the diameter of the bare particle is known (from independent experiments) so that the surface area can be estimated, then it is also possible to calculate the thickness of the adsorbed film, provided the density of the film is the same as the bulk density of the material being adsorbed (i.e., no solvation of the adsorbed layer). In some systems, it may be possible to alter the solvent density to match the core particle density by the addition of sucrose or other density modifiers to the FFF carrier solution.

Using the above approach with experimental results from centrifugal FFF, adsorption isotherms were constructed by directly measuring the mass of adsorbate deposited onto the polymer latex particle surface at different solution concentrations. It was found that for human globulin and ovalbumin adsorbates, Langmuir isotherms were obtained. The measured limiting adsorption density was found to agree with values measured using conventional solution uptake techniques.

The model used in the above studies ignores the departure from the bulk density of the adsorbate brought about by the interaction of the two interfaces. Li and Caldwell's article addresses this issue by introducing a three-component model consisting of a core particle, a flexible macromolecular substance with affinity toward the particle, and a solvation shell (see Fig. 1b).

In this model, the buoyant mass is then the sum of the buoyant mass of the three components, assuming that these are independent of the mass of solvent occupied in the solvation shell. Thus, the mass of the adsorbed shell can be calculated if information about the mass and density of the core particle and the density of the macromolecule and solvent are known. Photon correlation spectroscopy, electron microscopy, flow FFF, or other sizing techniques can readily provide some independent information on the physical or hydrodynamic particle size, and pycnometry can be used to measure the densities of the colloidal suspension, polymer solution, and pure liquid.

The above measurements were combined to estimate the mass of the polymer coating, a surface coverage density, and the solvated layer thickness. These results showed good agreement with the adsorption data derived from conventional polymer radiolabeling experiments.

Another approach for utilizing FFF techniques in the study of adsorption processes is to use the following general protocol:

1. Expose the suspension to the adsorbate
2. Run the sample through an FFF separation and collect fractions at designated elution volume intervals corresponding to specific size ranges
3. Analyze the size fractions for the amount of adsorbate

It must be emphasized that only strongly adsorbed material will be retained on the particles as the sample is constantly washed by the carrier solution during the FFF separation. Unless adsorbent is added to the carrier, these

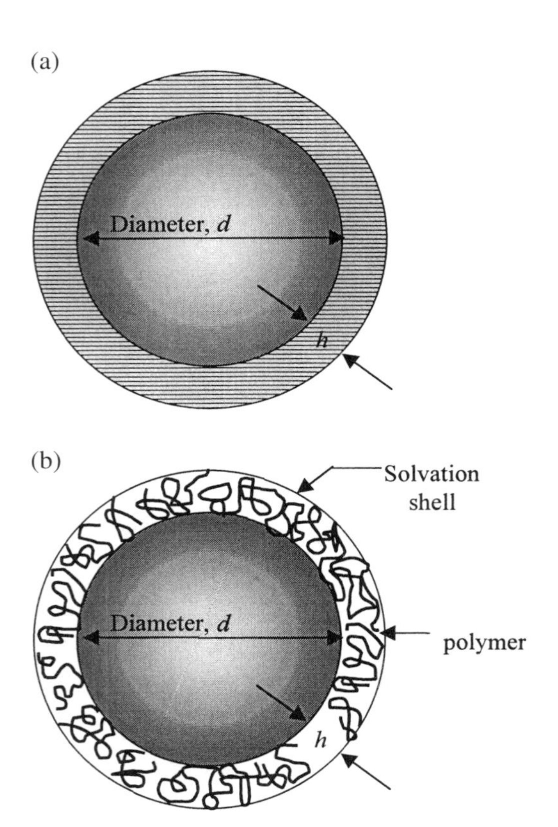

Fig. 1 Schematic representation of the adsorption complex proposed by (a) Beckett et al. [4] showing the core particle with a dense nonhydrated adsorbed film and by (b) Li and Caldwell [5] showing the core particle with an adsorbed polymer and the associated solvation shell.

experiments will not represent the reversible equilibrium adsorption situation.

This approach was first outlined by Beckett et al. [6], where radiolabelled pollutants (^{32}P as orthophosphate, ^{14}C in atrazine, and glyphosate) were adsorbed to two Australian river colloid samples. Sedimentation FFF was used to fractionate the samples and the radioactivity of each fraction was measured. From this, it was possible to generate a surface adsorption density distribution (SADD) across the size range of the sample. The SADD is a plot of the amount of compound adsorbed per unit particle surface area as a function of the particle size. It was shown that the adsorption density was not always constant, indicating perhaps a change in particle mineralogy, surface chemistry, shape, or texture as a function of particle size.

The above method is currently being extended to use other sensitive analytical techniques such as inductively coupled plasma–mass spectrometry (ICP–MS), graphite furnace atomic absorption (GFAAS), and inductively coupled plasma–atomic emission spectrophotometry (ICP–AES). With multielement techniques, it is not only possible to measure the amount adsorbed but changes in the particle composition with size can be monitored [7], which is most useful in interpreting the adsorption results [8]. Hassellov et al. [9] showed that using sedimentation FFF coupled to ICP–MS, it was possible to study both the major elements Al, Si, Fe, and Mn but also the Cs, Cd, Cu, Pb, Zn, and La. It was shown that it was possible to distinguish between the weaker and stronger binding sites as well as between different adsorption and ion-exchange mechanisms.

In electrical FFF, samples are separated on the basis of surface charge and even minute amount of adsorbate will significantly be reflected in electrical FFF data, as demonstrated by Dunkel et al. [10]. However, this technique is severely limited by the generation of polarization products at the channel wall due to the applied voltages.

In conclusion, the versatility and power of FFF are not restricted to its ability to effect high-resolution separations and sizing of particles and macromolecules. FFF can also be used to probe the surface properties of colloidal samples. Such studies have great potential to provide detailed insight into the nature of adsorption phenomena.

References

1. K. D. Caldwell, *Anal. Chem. 60*: 959A (1988).
2. J. C. Giddings, *Science 260*: 1456 (1993).
3. R. Beckett and B. T. Hart, in *Environmental Particles*, J. Buffle and H. P. van Leeuwen (eds.), Lewis Publishers, 1993, Vol. 2, pp. 165–205.
4. R. Beckett, Y. Ho, Y. Jiang, and J. C. Giddings, *Langmuir 7*: 2040 (1991).
5. J.-T. Li and K. D. Caldwell, *Langmuir 7*: 2034 (1991).
6. R. Beckett, D. M. Hotchin, and B. T. Hart, *J. Chromatogr. 517*: 435 (1990).
7. J. F. Ranville, F. Shanks, R. J. F. Morrison, P. Harris, F. Doss, and R. Beckett, *Anal. Chem. Acta 381*: 315 (1999).
8. J. Vanberkel and R. Beckett, *J. Liq. Chromatogr. Related Technol. 20*: 2647 (1997).
9. M. Hassellov, B. Lyven, and R. Beckett, *Environ. Sci. Technol. 33*: 4528 (1999).
10. M. Dunkel, N. Tri, R. Beckett, and K. D. Caldwell, *J. Micro. Separ. 9*: 177 (1997).

Niem Tri
Ronald Beckett

Affinity Chromatography of Cells

Affinity cell separations techniques are based on principles similar to those described in procedures for the isolation of molecules and are used to quickly and efficiently isolate specific cell types from heterogeneous cellular suspensions. The procedure (Fig. 1) involves making a single-cell suspension and passing it through a column packed with a support to which a selective molecule (ligand) has been immobilized. As the cells pass over the immobilized ligand-coated support (Fig. 1a), the ligand interacts with specific molecules on the cell surface, thus capturing the cell of interest (Fig. 1b). This cell is retained by the ligand-coated support while nonreactive cells are washed through the column. Finally, the captured cell is released (Fig. 1c) by disrupting the bond between the ligand and its selected molecule, allowing a homogeneous population of cells to be harvested.

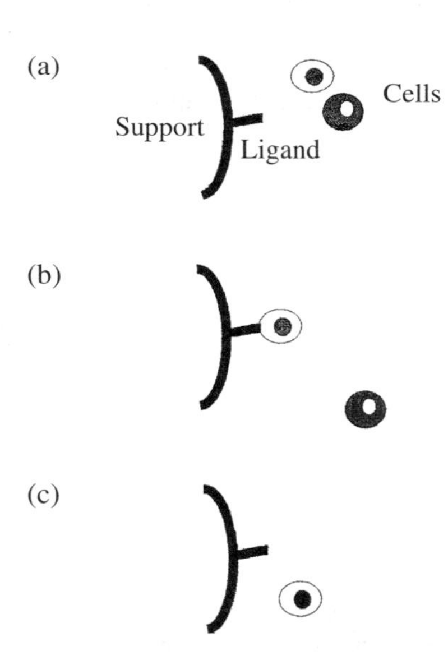

Fig. 1 Affinity isolation of specific cells. (a) The cell suspension containing the cell of interest (clear cytoplasm) and another cell type (dark cytoplasm) are passed over the support bearing a selective ligand immobilized to its surface. (b) The ligand interacts with its target molecule on the cell of interest, thus capturing it. The other cell type is not bound and passes through the column. (c) The bound cell is released by the addition of an elution agent to the running buffer of the column. This agent disrupts the binding between the ligand and the cell, thus releasing the cell. The free cell is now washed through the column and harvested as a homogeneous population.

Although affinity chromatography of cells is essentially performed in a similar manner to other affinity techniques, it is commonly used for both negative and positive selection. Negative selection removes specific cell types from the sample population, whereas positive selection isolates a single cell type from the sample. In the latter situation, the selected cells are recovered by elution from the immobilized ligand, thus yielding an enriched population. However, unlike molecules, cells are often quite delicate and care must be exercised when choosing the chromatographic support and the method of retrieval. The support matrix must exhibit minimal nonspecific cell adhesion but be sufficiently porous to allow cells to pass through without physically trapping them or creating undue sheer forces likely to cause cell injury or death. Usually, the support matrices of choice are loosely packed fibers, large-pore cross-linked dextrans or agarose, and large plastic or glass beads. The elution agent must also be carefully selected. It must be able to either disrupt the binding of the ligand to the cell surface molecule or it must be able to compete with the cell molecule for ligand binding. In many cases, such as lectin affinity chromatography, the elution agent is easy to select – it is usually a higher concentration of the sugar to which the ligand binds. Elution agents for other techniques, such as immunoaffinity, are harder to select. Harsh acid or alkaline conditions, although efficient at breaking antibody–antigen binding, are usually detrimental to cells. Elution in these cases is often achieved using mild acids, cell molecule competition (like the lectins), or mild chaotropic ion elution.

Immunologists have long used the relatively nonspecific affinity of charged nylon wool to fractionate lymphocytes into different subpopulations. Such separations are achieved because certain subpopulations of lymphocytes express an affinity for the charged fibers, whereas others do not. This negative selection process has been used to prepare pure suspensions of T-lymphocytes for many years but has recently been replaced by the more selective immunoaffinity procedures. A good review of the early history of affinity cell separation is provided by Sharma and Mahendroo [1]; however, the review focuses primarily on the application of lectins as the selective ligands for cell affinity chromatography. Tlaskalova-Hogenova et al. [2] demonstrated the usefulness of affinity cell chromatography to isolate T- and B-lymphocytes from human tissues. These authors describe comparative studies on three popular approaches to the isolation of lymphocyte subpopulations, namely nylon wool columns, immunoaffinity cell panning (a batch technique using antibodies immobilized to the bottom of culture dishes), and immunoaffinity using anti-human immunoglobulins attached to Sephron (hydroxyethyl methacrylate) or Sepharose supports. These studies clearly indicate that the selectiveness of immobilized antibodies were superior for isolating defined subpopulations of cells.

Immobilized antibody ligands or immunoaffinity chromatography is now the approach of choice for cell separation procedures. Kondorosi et al. [3] prepared columns packed with a support coated with nonimmune rat immunoglobulin and used these columns to isolate cells expressing surface Fc or immunoglobulin receptors, whereas van Overveld et al. [4] used anti-human IgE-coated Sepharose beads as an immunoaffinity chromatography step when fractionating human mast cells from lung tissue.

Plant lectins are one of the most popular ligands for affinity cell separations. These molecules express selective affinities for certain sugar moieties (Table 1), different lectins being used as selective agents for specific sugars. Whitehurst et al. [5] found that the lectin *Pisum sativum* agglutinin could bind feline B-lymphocytes much more readily than T-lymphocytes and used lectin-coated supports to obtain pure subpopulations of T-lymphocytes by negative selection. Additionally, the retained cells were recovered by elution from the immobilized lectin with a suitable sugar. Lectins are efficient ligands for cell selection, but, in many cases, their interaction with the selected cell surface molecule is highly stable and efficient, requiring mechanical agitation of the packing before recovery of the cells can be achieved. Pereira and Kabat [6] have reported the use of lectins immobilized to Sephadex or Sepharose beads for the isolation of erythrocytes.

Another useful ligand is protein A, which is a protein derived from the wall of certain *Staphylococcus* species of bacteria. This reagent binds selected classes of IgG immunoglobulin via their Fc or tail portion making it an excellent ligand for binding immunoglobulins attached to cell surfaces, making it an ideal general-purpose reagent. Ghetie et al. [7] demonstrated that protein A-coated Sepharose beads were useful for cell separations following initial incubation of the cells with IgG antibodies directed against specific cell surface markers. Surface IgG-bearing cells mouse spleen cells were pretreated with rabbit antibodies to mouse IgG prior to passage over the protein A-coated support. The cells of interest were then isolated by positive selection chromatography.

In addition to bacterial proteins, other binding proteins such as chicken egg white avidin have become popular reagents for affinity chromatography. These supports work on the principle that immobilized avidin binds biotin, which can be chemically attached to a variety of ligands including antibodies. Tassi et al. [8] used a column with an avidin-coated polyacrylamide support to bind and retain cells marked with biotinylated antibodies. Human bone marrow samples were incubated with monoclonal mouse antibodies directed against the surface marker CD34, followed by a second incubation with biotinylated

Table 1 Lectins and Their Reactive Sugar Moieties

Common name	Latin name	Reactive sugar residues
Castor bean RCA_{120}	*Ricinus communis*	β-D-Galactosyl
Fava bean	*Vicia faba*	D-Mannose, D-Glucose
Gorse UEA I	*Ulex europaeus*	α-L-Fucose,
UEA II		*N,N'*-Diacetylchitobiose
Jacalin	*Artocarpus integrifolia*	α-D-Galactosyl, β-(1,3) *n*-Acetyl galactosamine
Concanavalin A	*Canavalia ensiformis*	α-D-Mannosyl, α-D-Glucosyl
Jequirity bean	*Abrus precatorius*	α-D-Galactose
Lentil	*Lens culinaris*	α-D-Mannosyl, α-D-Glucosyl
Mistletoe	*Viscum album*	β-D-Galactosyl
Mung bean	*Vigna radiata*	α-D-Galactosyl
Osage orange	*Maclura pomifera*	α-D-Galactosyl, *N*-Acetyl-D-galactosaminyl
Pea	*Pisum sativum*	α-D-Glucosyl, α-D-Mannosyl
Peanut	*Arachis hypogaea*	β-D-Galactosyl
Pokeweed	*Phytolacca americana*	*N*-acetyl-β-D-Glucosamine oligomers
Snowdrop	*Galanthus nivalis*	Nonreducing terminal end of α-D-mannosyl
Soybean	*Glycine max*	*N*-Acetyl-D-galactosamine
Wheatgerm	*Triticum vulgaris*	*N*-Acetyl-β-D-glucosaminyl, *N*-Acetyl-β-D-glucosamine oligomers

goat anti-mouse immunoglobulins. Binding of the biotin to the avidin support effectively isolated the antibody-coated cells.

A wide variety of immobilized antigens, chemicals, and receptor molecules have been used effectively for affinity cell chromatography. Sepharose beads coated with thyroglobulin have been used to separate thyroid follicular and para-follicular cells, and immobilized insulin on Sepharose beads has been used to isolate adipocytes by affinity chromatography. Dvorak et al. [9] reported the successful retrieval of a 95% pure fraction of chick embryonic neuronal cells using an affinity chromatography approach utilizing α-bungarotoxin immobilized to Sepharose beads.

References

1. S. K. Sharma and P. P. Mahendroo, *J. Chromatogr. 184*: 471 (1980).
2. H. Tlaskalova-Hogenova, V. Vetvicka, M. Pospisil, L. Fornusek, L. Prokesova, J. Coupek, A. Frydrychova, J. Kopecek, H. Fiebig, and J. Brochier, *J. Chromatogr. 376*: 401 (1986).
3. E. Kondorosi, J. Nagy, and G. Denes, *J. Immunol. Methods 16*: 1 (1977).
4. F. J. van Overveld, G. K. Terpstra, P. L. Bruijnzeel, J. A. Raaijmakers, and J. Kreukniet, *Scand. J. Immunol. 27*: 1 (1988).
5. C. E. Whitehurst, N. K. Day, and N. Gengozian, *J. Immunol. Methods 175*: 189 (1994).
6. M. E. Pereira and E.A. Kabat, *J. Cell Biol. 82*: 185 (1979).
7. V. Ghetie, G. Mota, and J. Sjoquist, *J. Immunol. Methods 21*: 133 (1978).
8. C. Tassi, A. Fortuna, A. Bontadini, R. M. Lemoli, M. Gobbi, and P. L. Tazzari, *Haematologica 76*(Suppl. 1): 41 (1991).
9. D. J. Dvorak, E. Gipps, and C. Kidson, *Nature 271*: 564 (1978).

Terry M. Phillips

Affinity Chromatography with Immobilized Antibodies

Introduction

Antibodies are serum proteins that are generated by the immune system which bind specifically to introduced antigens. The high degree of specificity of the antibody–antigen interaction plays a central role in an immune response, directing the removal of antigens in concert with complement lysis (humoral immunity). Importantly, this high degree of specific binding has been exploited as an analytical tool: Antigens can be detected, quantified, and purified from sources in which they are in low abundance with numerous contaminants. Examples include enzyme-linked immunosorbent assays (ELISAs), Ouchterlony assays, and Western blots. Antibodies that are specifically immobilized on high-performance chromatographic media offer a means of both detection and purification that is unparalleled in specificity, versatility, and speed.

We will focus, here, on the use of immobilized antibodies for analytical affinity chromatography, which offers a number of advantages over standard partition chromatography. The first advantage is the specificity imparted by the antibody itself, which allows an antigen to be completely separated from any contaminants. During a chromatographic run with an antibody affinity column, all of the contaminants wash through the column unbound, and the bound antigen is subsequently eluted, resulting in only two peaks generated in the chromatogram (contaminants in the flow-through step and antigen in the elution step). With antibodies which are immobilized on high-speed media such as perfusive media [1,2], typical analytical chromatograms can be generated in less than 5 min and columns can last for hundreds of analyses. In Fig. 1, an example of 5 consecutive analytical affinity chromatography assays are shown, followed by the results of the last 5 assays of a set of 5000. Note that, here, the cycle time for loading, washing out the unbound material, eluting the bound material, and reequilibration of the affinity column is only 0.1 min (6 s). Also note that the calibration curve has changed little between the first analysis and after 5000 analyses, demonstrating both the durability and reproducibility of this analytical technique. Although many soft-gel media are also available for antibody immobilization, these media do not withstand high linear velocity and, therefore, are not suited for high-performance affinity chromatography.

Affinity chromatography using immobilized antibodies offers several advantages over conventional chromatographic assay development. First, assay development can be very rapid because specificity is an inherent property of antibody and solvent mobile-phase selection is limited

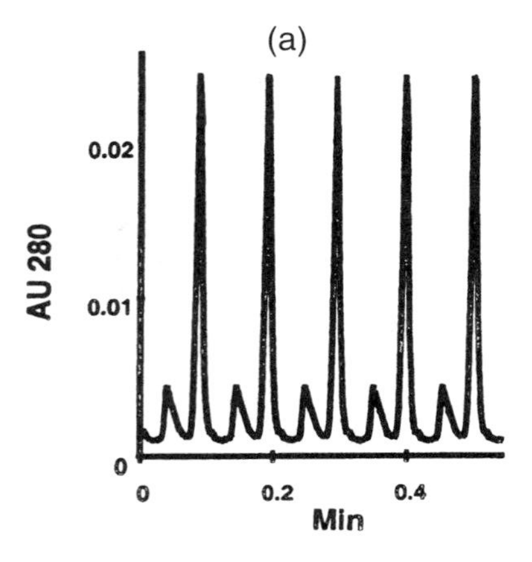

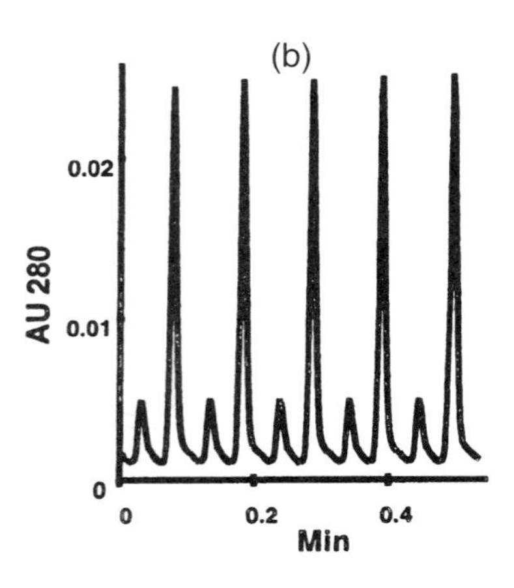

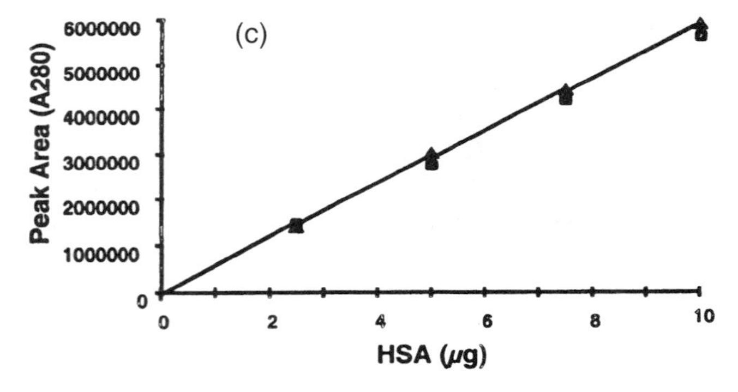

Fig. 1 Examples of affinity chromatography with an epoxy-immobilized polyclonal anti-human serum albumin (HSA) antibody in a 2.1-mm-inner diameter × 30-mm POROS CO column run at 5 mL/min (8000 cm/h) using phosphate-buffered saline for loading and 12 m*M* HCl with 150 m*M* NaCl for elution. The sample was 10 μg HSA at 1 mg/mL. (a) shows the first five analyses of a relatively pure sample of HSA, where the first small peak is the unbound contaminant and the larger peak is the elution of the HSA from the affinity column. (b) shows the results of the last 5 analyses from a set of 5000 and (c) shows the calibration curve before (squares) the 5000 analyses and after (triangles).

to a capture buffer and an elution buffer, which is often the same from one antibody to the next. Therefore, there is less "column scouting" for appropriate conditions. In addition, the assays are fast (see above) and chromatograms yield only two peaks instead of multiple peaks. Furthermore, the two peaks in the affinity chromatogram indicate both antigen concentration (from the eluted peak) and purity (from the ratio of the eluted peak to the total peak area). Thus, affinity chromatography with immobilized antibodies allows both fast assay development and rapid analysis times.

The limitations of immobilized antibody affinity chromatography are few. First, plentiful amounts of antibody, usually milligram quantities, are required to get reasonable ligand density on a useful amount of chromatographic media. Also, it is optimal if the antibody is antigen affinity purified, so that when it is immobilized, no other contaminating proteins with competing specificity dilute the antibody's concentration. Finally, the antibody must be amenable to affinity chromatography such that it is not irreversibly denatured by the immobilization process and can withstand many cycles of antigen capture and elution. Both monoclonal and polyclonal antibodies have been used successfully.

Immobilization Chemistries

Many different chemistries can be used to immobilize antibodies onto chromatographic media and only a few will be discussed. In most cases, the chromatographic media is coated with the active chemistry, which will then react with the antibody. These include amine reactive chemistries such as epoxide-, aldehyde-, and cyanogen bromide (CNBr)-activated media, carboxyl reactive chemistries such as carbodiimides, aldehyde-reactive chemistries such as amino and hydrazide, and thiol-reactive chemistries such as iodoacetyl and reduce thiol media. Although there are several antibody isotypes (IgA, IgE, IgG, IgM), the most common antibody immobilized for affinity chromatography is IgG, which is composed of four polypeptide chains (two heavy and two light) which are disulfide linked to form a Y-shaped structure capable of binding two antigens. For best results, it is also important to antigen affinity purify the antibody prior to immobilization to yield optimum binding capacity and a wider dynamic range for analytical work. Also note that the antibody may be digested with pepsin or papain to separate the constant region from the antigen-binding domains, which may then be immobilized.

Antibodies are very often immobilized through their amino groups either through the N-terminal amines or the epsilon amino groups of lysine. Reactions with epoxide-activated media are performed under alkaline conditions and lead to extremely stable linkages between the chromatographic support and the antibody. Similarly, immobilization using an aldehyde-activated media first proceeds through a Schiff base intermediate which must then be reduced (often by sodium cyanoborohydride) to yield a very stable carbon–nitrogen bond linking the antibody to the media. *N*-Hydroxy-succinimide-activated media also couples via primary amines and leads to a stable linkage in a single-step reaction. The major advantage of these chemistries is that they are extremely stable due to the formation of covalent bonds to the media. Although less stable but easy to use is CNBr-activated media, which also immobilizes antibodies through their primary amines.

Antibodies can also be immobilized through their carboxyl groups by first treating them with a carbodiimide such as EDC (1-ethyl-3-[3-dimethlaminopropyl]-carbodiimide) followed by immobilization on an amine-activated chromatographic resin. It is important to note that EDC does not add a linker chain between the antibody and the media, but simply facilitates the formation of an amide bond between the antibody's carboxyl and the amine on the media. Coupling through sulfhydryls on free cysteines can be accomplished with thiol-activated media by formation of disulfide bonds between the media and the antibody. However, this coupling is not stable to reducing conditions and a more stable iodoacetyl-activated media is often preferred because the resulting carbon–sulfur bond is more stable. Free cysteines can be generated in the antibody by use of mild reducing agents (e.g., 2-mercaptoethylamine), which can selectively reduce disulfide bonds in the hinge region of the antibody. Alternatively, antibodies may also be immobilized through their carbohydrate moieties. One method involves oxidation of the carbohydrate with sodium periodate to generate two aldehydes in the place of vicinyl hydroxyls. These aldehydes may then be coupled either directly to hydrazide-activated media or through amine-activated media with the addition of sodium cyanoborohydride to reduce the Schiff base. The primary advantages of these chemistries is to offer alternative linkages to the antibody beyond primary amines.

In addition, antibodies may also be coupled to other previously immobilized proteins. For example, the antibody may be first captured on protein A or protein G media and then cross-linked to the immobilized protein A or G with reagents such as glutaraldehyde or dimethyl pimelimidate. The advantage here is that the antibody need not be pure prior to coupling because the protein A or protein G will selectively bind only antibody and none of the other serum proteins. The disadvantage is that free protein A or protein G will still be available to cross-react with any free antibody in samples to be analyzed, which will only be problematic with serum-based samples.

Antibody coupling does not need to be covalent to be effective. For example, biotinylated antibodies can be coupled to immobilized streptavidin. The avidin–biotin interaction is extremely strong and will not break under normal antigen elution conditions. The advantage of this immobilization protocol is that many different biotinylation reagents are available in a wide range of chemistries and linker chain lengths. Once biotinylated and free biotin are removed, the antibody is simply injected onto the streptavidin column and it is ready for use. Immobilization can be accomplished through hydrophobic interaction by simply injecting the antibody onto a reversed-phase column and then blocking with an appropriate protein solution such as albumin, gelatin, or milk. This is analogous to techniques used to coat ELISA plates and perform Western blots, and although this noncovalent coupling is not stable to organic solvents and detergents, it can last for hundreds of analyses under the normal aqueous analysis conditions. The advantage of this immobilization is that it can be done very quickly (in several minutes) by simply injecting an antibody first and then a blocking agent.

Operation

A wide range of buffers can be used for loading the sample and eluting the bound antigen; however, for best analytical performance, a buffer system that has low a low ultraviolet (UV) cutoff and rapid reequilibration properties is desirable. One of the better examples is phosphate-buffered saline (PBS) for loading and 12 m*M* HCl with 150 m*M* NaCl. The NaCl is not required in the elution buffer but helps to minimize baseline disturbances due to the refractive index change between the PBS loading buffer and the elution buffer because both will contain about 150 m*M* NaCl. UV detection is well suited for these assays and wavelengths at 214 or 280 nm are commonly used.

For analytical work, large binding capacities are not required, but increased capacity does increase the dynamic range of the analysis. However, the dynamic range can be increased by injecting a smaller volume of sample onto the column at the expense of sensitivity at the low end of the calibration curve. Likewise, sensitivity can be increased by injecting more sample volume.

Application Examples

The most obvious way to use immobilized antibodies for analytical affinity chromatography is to simply use it in a traditional single-column method to determine an antigen's concentration and/or purity. However, there are a number of ways this technique can be advanced to more sophisticated analyses. For example, instead of immobilizing an antibody, the antigen may be immobilized to quantify the antibody as has been done with the Lewis Y antigen [3]. However, the analysis is still a single-column method.

Immobilized antibodies have also been used extensively in multidimensional liquid chromatography (MDLC) analyses. As shown in Fig. 2, an affinity column with immobilized anti-HSA is used to capture all of the human serum albumin in a sample, allowing all of the other components to flow through to waste. Then, the affinity chromatography column is eluted directly into a size-exclusion column where albumin monomers and aggregates are separated and quantified. In this example, neither mode of chromatography would be sufficient by itself. The affinity media does not distinguish between monomer and aggregate, and the size-exclusion column would not be able to discriminate between albumin and the other coeluting proteins in the sample. Other MDLC applications employing immobilized antibodies include an acetylcholine esterase assay utilizing size-exclusion chromatography [4], combinations of immobilized antibodies with reversed-phase analysis [5–7], protein variant determination using immobilized antibodies to select hemoglobin from a biological sample followed by on-column proteolytic digestion, and liquid chromatography–mass spectrometry peptide mapping [8].

There are many more examples of immobilized antibodies used for affinity chromatography which are not mentioned here, but it was the goal of this section to present some of the capabilities of this technique for analytical chromatographic applications.

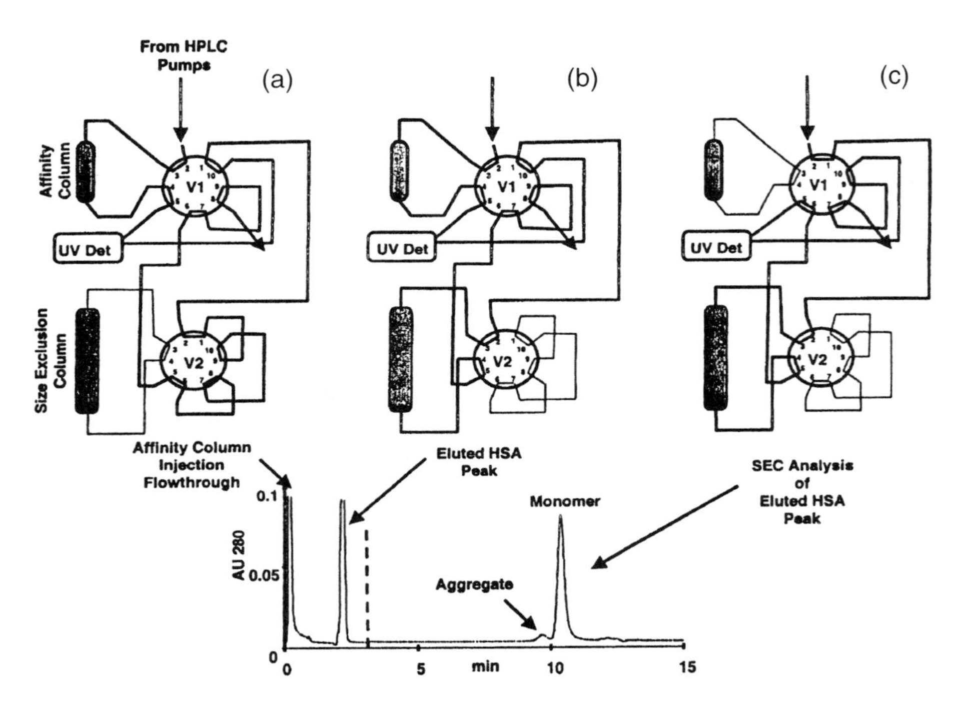

Fig. 2 Example of a multidimensional liquid chromatographic analysis for albumin aggregates using immobilized antibody affinity chromatography with size-exclusion chromatography. (a) Shows the flow path during the loading of the sample to capture the albumin monomer and aggregates while allowing all other proteins to elute to waste. (b) Shows the transfer of the albumin and its aggregates to the size-exclusion column. (c) Shows the flow path used to elute the size-exclusion column to separate the aggregate and monomer. (d) Shows the UV trace from this analysis. Note that in this plumbing configuration, the albumin passes through the detector twice, once as it is transferred from the affinity to the size-exclusion column and again as the albumin elutes from the size-exclusion column. The affinity column is a 2.1-mm-inner diameter (i.d.) × 30 mm POROS XL column to which anti-human serum albumin has been covalently cross-linked, run at 1 mL/min, loaded in PBS, and eluted with 12 m*M* HCl. The size-exclusion column is a 7.5-mm-i.d. × 300-mm Ultrasphere OG run at 1 mL/min with 100 m*M* potassium phosphate with 100 m*M* sodium phosphate, pH 7.0. The sample was 100 μg heat-treated albumin.

References

1. N. B. Afeyan, N. F. Gordon, I. Mazsaroff, L. Varady, S. P. Fulton, Y. B. Yang, and F. E. Regnier, Flow-through particles for the high-performance liquid chromatographic separation of biomolecules: Perfusion chromatography, *J. Chromatogr.* 519(1): 1 (1990).
2. N. B. Afeyan, N. F. Gordon, and F. E. Regnier, "Automated real-time immunoassay of biomolecules," *Nature 358*(6387): 603 (1992).
3. M. A. Schenerman and T. J. Collins, "Determination of a monoclonal antibody binding activity using immunodetection," *Anal. Biochem.* 217(2): 241 (1994).
4. M. Vanderlaan, R. Lotti, G. Siek, D. King, and M. Goldstein, Perfusion immunoassay for acetylcholinesterase: analyte detection based on intrinsic activity, *J. Chromatogr. A* 711(1): 23 (1995).
5. B. Y. Cho, H. Zou, R. Strong, D. H. Fisher, J. Nappier, and I. S. Krull, Immunochromatographic analysis of bovine growth hormone releasing factor involving reversed-phase high-performance liquid chromatography-immunodetection, *J. Chromatogr. A* 743(1): 181 (1996).
6. J. E. Battersby, M. Vanderlaan, and A. J. Jones, Purification and quantitation of tumor necrosis factor receptor immunoadhesin using a combination of immunoaffinity and reversed-phase chromatography, *J. Chromatogr. B* 728(1): 21 (1999).
7. C. K. Holtzapple, S. A. Buckley, and L. H. Stanker, Determination of four fluoroquinolones in milk by on-line immunoaffinity capture coupled with reversed-phase liquid chromatography, *J. AOAC Int.* 82(3): 607 (1999).
8. Y. L. Hsieh, H. Wang, C. Elicone, J. Mark, S. A. Martin, and F. Regnier, Automated analytical system for the examination of protein primary structure, *Anal. Chem.* 68(3): 455 (1996).

Monica J. S. Nadler
Tim Nadler

Affinity Chromatography: An Overview

Introduction

Affinity chromatography is a liquid chromatographic technique that uses a "biologically related" agent as a stationary phase for the purification or analysis of sample components [1–4]. The retention of solutes in this method is generally based on the same types of specific, reversible interactions that are found in biological systems, such as the binding of an enzyme with a substrate or an antibody with an antigen. These interactions are exploited in affinity chromatography by immobilizing (or adsorbing) one of a pair of interacting molecules onto a solid support and using this as a stationary phase. This immobilized molecule is known as the *affinity ligand* and is what gives an affinity column the ability to bind to particular compounds in a sample.

Affinity chromatography is a valuable tool in areas such as biochemistry, pharmaceutical science, clinical chemistry, and environmental testing, where it has been used for both the purification and analysis of compounds in complex sample mixtures [1–5]. The strong and relatively specific binding that characterizes many affinity ligands allows solutes that are quantitated or purified by these ligands to be separated with little or no interferences from other sample components. Often, the solute of interest can be isolated in one or two steps, with purification yields of 100-fold to several thousand-fold being common [2]. Similar selectivity has been observed when using affinity chromatography for compound quantitation in such samples as serum, plasma, urine, food, cell cultures, water, and soil extracts [3–5].

General Formats for Affinity Chromatography

The most common scheme for performing affinity chromatography is by using a step gradient for elution, as shown in Fig. 1. This involves injecting a sample onto the affinity column in the presence of a mobile phase that has the right pH and solvent composition for solute–ligand binding. This solvent, which represents the weak mobile phase of the affinity column, is called the *application buffer*. During the application phase of the separation, compounds which are complementary to the affinity ligand will bind as the sample is carried through the column by the application buffer. However, due to the high selectivity of the solute–ligand interaction, the remainder of the sample components will pass through the column nonretained. After the nonretained components have been completely washed from the column, the retained solutes

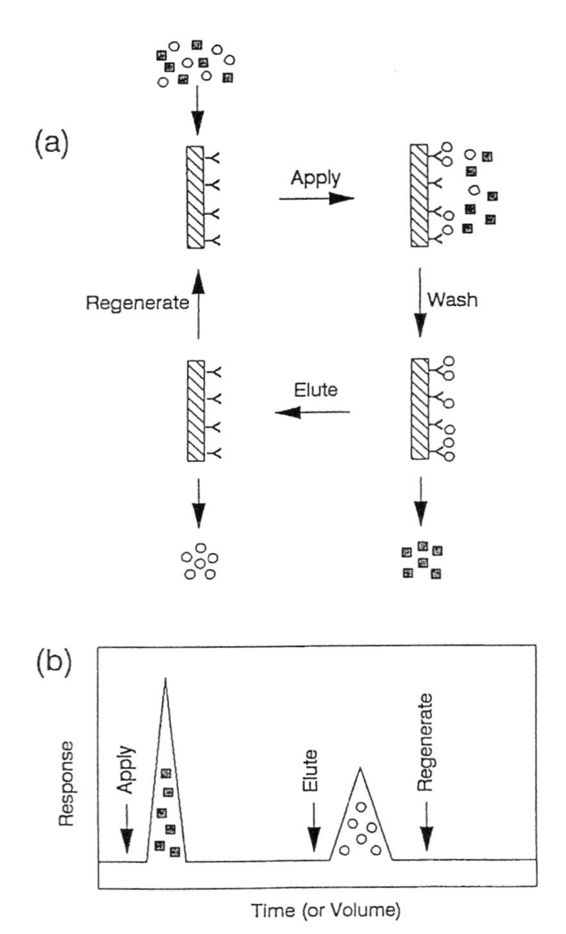

Fig. 1 (a) Typical separation scheme and (b) chromatogram for affinity chromatography. The open circles represent the test analyte and the squares represent other, nonretained sample components. [Reproduced with permission from the Clinical Ligand Assay Society from D. S. Hage, *J. Clin. Ligand Assay 20*: 293 (1998).]

can be eluted by applying a solvent that displaces them from the column or that promotes dissociation of the solute–ligand complex. This solvent represents the strong mobile phase for the column and is known as the *elution buffer*. As the solutes of interest elute from the column, they are either measured or collected for later use. The column is then regenerated by reequilibration with the application buffer prior to injection of the next sample [2–4].

Even though the *step-gradient*, or *"on/off" elution method* illustrated in Fig. 1 is the most common way of performing affinity chromatography, it is sometimes possible to use affinity methods under isocratic conditions. This can be done if a solute's retention is sufficiently weak to allow elution on the minute-to-hour time scale and if the kinetics for its binding and dissociation are fast enough to allow a large number of solute–ligand interactions to occur as the analyte travels through the column. This approach is sometimes called *weak-affinity chromatography* and is best performed if a solute binds to the ligand with an association constant that is less than or equal to about 10^4–$10^6 M^{-1}$ [3,6].

Types of Affinity Ligands

The most important factor in determining the success of any affinity separation is the type of ligand that is used within the column. A number of ligands that are commonly used in affinity chromatography are listed in Table 1. Most of these ligands are of biological origin, but a wide range of natural and synthetic molecules of nonbiological origin can also be used. Regardless of their origin, all of these ligands can be placed into one of two categories: high-specificity ligands or general ligands [2–4].

The term *high-specificity ligand* refers to a compound which binds to only one or a few closely related molecules. This type of ligand is used in affinity systems where the goal is to analyze or purify a specific solute. Examples include antibodies (for binding antigens), substrates or inhibitors (for separating enzymes), and single-stranded nucleic acids (for the retention of a complementary sequence). As this list suggests, most high-specificity ligands

Table 1 Common Ligands Used in Affinity Chromatography

Type of ligand	Examples of retained compounds
High-Specificity Ligands	
Antibodies	Various agents (drugs, hormones, peptides, proteins, viruses, etc.)
Enzyme inhibitors and cofactors	Enzymes
Nucleic acids	Complementary nucleic acid strands and DNA/RNA-binding proteins
General Ligands	
Lectins	Small sugars, polysaccharides, glycoproteins, and glycolipids
Protein A and protein G	Intact antibodies and Fc fragments
Boronates	Catechols and compounds which contain sugar residues, such as polysaccharides and glycoproteins
Synthetic dyes	Dehydrogenases, kinases and other proteins
Metal chelates	Metal-binding amino acids, peptides, or proteins

tend to be of biological origin and often have large association constants for their particular analytes. *General*, or *group-specific*, *ligands* are compounds which bind to a family or class of related molecules. These ligands are used when the goal is to isolate a class of structurally related compounds. General ligands can be of either biological or nonbiological origin and include compounds such as protein A or protein G, lectins, boronates, triazine dyes, and immobilized metal chelates. This class of ligands usually exhibits weaker binding for solutes than is seen with high-specificity ligands; however, some general ligands like protein A and protein G do have association constants that rival those of high-specificity ligands [7].

Support Materials

Another important factor to consider in affinity chromatography is the material used to hold the ligand within the column. Ideally, this support should have low nonspecific binding for sample components, it should be easy to modify for ligand attachment, and it should be stable under the flow-rate, pressure, and solvent conditions that will be employed in the analysis or purification of samples. Depending on what type of support material is being used, affinity chromatography can be characterized as being either a low- or high-performance technique [4].

In *low-performance* (or *column*) *affinity chromatography*, the support is usually a large-diameter, nonrigid material (e.g., a carbohydrate-based gel or one of several synthetic organic-based polymers). The low back pressures that are produced by these supports means that these materials can often be operated under gravity flow or with a peristaltic pump, making them relatively simple and inexpensive to use for affinity purifications or in sample pretreatment. Disadvantages of these materials include their slow mass-transfer properties and their limited stabilities at high flow rates and pressures. These factors tend to limit the direct use of these supports in analytical applications, where both rapid and efficient separations are usually desired [2,5].

High-performance affinity chromatography (*HPAC*) is characterized by a support which consists of small, rigid particles capable of withstanding high flow rates and/or pressures [2–4,8,9]. Examples of affinity supports that are suitable for work under these conditions include modified silica or glass, azalactone beads, and hydroxylated polystyrene media. The stability and efficiency of these supports allows them to be used with standard high-performance liquid chromatography (HPLC) equipment. Although the need for HPLC instrumentation does make HPAC more expensive to perform than low-performance affinity chromatography, the better speed and precision of HPAC makes it the affinity method of choice for many analytical applications.

Immobilization Methods

A third item to consider in affinity chromatography is the way in which the ligand is attached to the solid support, or the *immobilization method*. For a protein or peptide, this generally involves coupling the molecule through free amine, carboxylic acid, or sulfhydryl residues present in its structure. Immobilization of a ligand through other functional sites (e.g., aldehyde groups produced by carbohydrate oxidation) is also possible. All immobilization methods involve at least two steps: (a) an *activation step*, in which the support is converted to a form which can be chemically attached to the ligand, and (b) a *coupling step*, in which the affinity ligand is attached to the activated support. Occasionally, a third step is necessary to remove remaining activated groups.

The method by which an affinity ligand is immobilized is important because it can affect the actual or apparent activity of the final affinity column. If the correct procedure is not used, a decrease in ligand activity can result from multisite attachment, improper orientation, and/or steric hindrance. *Multisite attachment* refers to the coupling of a ligand to the support through more than one functional group, which can lead to distortion of the ligand's active region and a loss of activity. This can be avoided by using a support with a limited number of activated sites or by using a method that couples through groups that occur only a few places in the structure of the ligand. *Improper orientation* can lead to a loss in activity by coupling the ligand to the support through its active region; this can be minimized by coupling through functional groups that are distant from this region. *Steric hindrance* refers to the loss of ligand activity due to the presence of a nearby support or neighboring ligand molecules that interfere with solute binding. This effect can be avoided through the use of a spacer arm or by using supports that contain a relatively low coverage of the ligand.

Application and Elution Conditions

Two other items that must be considered in affinity chromatography are the application and elution buffers. Most application buffers in affinity chromatography are solvents that mimic the pH, ionic strength, and polarity experienced by the solute and ligand in their natural environment. This gives the solute its highest association constant for the ligand and, thus, its highest degree of retention on the column. The application buffer should also be chosen so that it minimizes nonspecific binding due to undesired sample components.

Elution conditions in affinity chromatography are usually chosen so that they promote either the fast or gentle removal of solute from the column. The two approaches used for this are *nonspecific elution* and *biospecific elution*, respectively [2]. Biospecific elution is based on the addition of a competing agent that gently displaces a solute from the column. This is done by either adding an agent that competes with the ligand for solute (i.e., *normal-role elution*) or that competes with the solute for ligand-binding sites (i.e., *reversed-role elution*). Although it is a gentle method, biospecific elution does result in long elution times and broad solute peaks that are difficult to quantitate. Nonspecific elution uses a solvent that directly promotes weak solute–ligand binding. For instance, this is done by changing the pH, ionic strength or polarity of the mobile phase or by adding a denaturing agent or chaotropic substance to the elution buffer. Nonspecific elution tends to be much faster than biospecific elution and results in sharper peaks with lower limits of detection. However, care must be exercised in nonspecific elution to avoid using a buffer which is too harsh and causes solute denaturation or a loss of ligand activity.

Types of Affinity Chromatography

There are many types of affinity chromatography that are in common use. *Bioaffinity chromatography* is probably the broadest category and includes any method that uses a biological molecule as the affinity ligand. *Immunoaffinity chromatography* (IAC) is a special category of bioaffinity chromatography in which the affinity ligand is an antibody or antibody-related agent [3,5,8,10]. This creates a highly specific method that is ideal for use in affinity purification or in analytical methods that involve complex samples. *Immunoextraction* is a subcategory of IAC in which an affinity column is used to isolate compounds from a sample prior to analysis by a second method. IAC can also be used to monitor the elution of analytes from other columns, giving rise to a scheme known as *postcolumn immunodetection*. Another common type of bioaffinity method is that which uses bacterial cell-wall proteins like protein A or protein G for antibody purification. In *lectin affinity chromatography*, immobilized lectins like concanavalin A or wheat germ agglutinin are used for binding to molecules which contain certain sugar residues. Additional types of bioaffinity chromatography are those that make use of ligands which are enzymes, inhibitors, cofactors, nucleic acids, hormones, or cell receptors [1–4].

There are also many types of affinity chromatography that use ligands which are of a nonbiological origin. One example is *dye-ligand affinity chromatography*, which uses an immobilized synthetic dye that mimics the active site of a protein or enzyme. This is a popular tool for enzyme and protein purification. *Immobilized metal-ion affinity chromatography* (IMAC) is an affinity technique in which the ligand is a metal ion which is complexed with an immobilized chelating agent. This is used to separate proteins and peptides that contain amino acids with electron-donor groups. *Boronate affinity chromatography* employs boronic acid or a boronate as the affinity ligand. These ligands are useful in binding to compounds which contain *cis*-diol groups, such as catecholamines and glycoproteins [1–4].

There are a number of other chromatographic methods closely related to traditional affinity chromatography. For instance, affinity chromatography can be adapted as a tool for studying solute–ligand interactions. This application is known as *analytical*, or *quantitative*, *affinity chromatography* and can be used to acquire information regarding the equilibrium and rate constants for biological interactions, as well as the number and type of binding sites that are involved in these interactions [11]. Other methods that are related to affinity chromatography include *hydrophobic interaction chromatography* (HIC) and *thiophilic adsorption*. HIC is based on the interactions of proteins, peptides, and nucleic acids with short nonpolar chains, such as those that were originally used as spacer arms on affinity supports. Thiophilic adsorption, also known as *covalent* or *chemisorption chromatography*, makes use of immobilized thiol groups for solute retention. Applications of this method include the analysis of sulfhydryl-containing peptides or proteins and mercurated polynucleotides. Finally, many types of *chiral liquid chromatography* can be considered affinity methods because they are based on binding agents which are of a biological origin. Examples include columns which use cyclodextrins or immobilized proteins for chiral separations [4].

References

1. J. Turkova, *Affinity Chromatography*, Elsevier, Amsterdam, 1978.
2. R. R. Walters, *Anal. Chem. 57*: 1099A (1985).
3. D. S. Hage, *Handbook of HPLC*, E. Katz, R. Eksteen, and N. Miller (eds.), Marcel Dekker, Inc., New York, 1998, Chap. 13.
4. D. S. Hage, *Clin. Chem. 45*: 593 (1999).
5. D. S. Hage, *J. Chromatogr. B 715*: 3 (1998).
6. D. Zopf and S. Ohlson, *Nature 346*: 87 (1990).

7. R. Lindmark, C. Biriell, and J. Sjoequist, *Scand. J. Immunol. 14*: 409 (1981).
8. M. de Frutos and F. E. Regnier, *Anal. Chem. 65*: 17A (1993).
9. S. Ohlson, L. Hansson, P.-O. Larsson, and K. Mosbach, *FEBS Lett. 93*: 5 (1978).
10. G. J. Calton, *Methods Enzymol. 104*: 381 (1984).
11. I. M. Chaiken (ed.), *Analytical Affinity Chromatography*, CRC Press, Boca Raton, FL, 1987.

David S. Hage
William Clarke

Aggregation of Colloids by Field-Flow Fractionation

Introduction

The separation of the components of complex colloidal materials is one of the most difficult challenges in separation science. Most chromatographic methods fail in the colloidal size range or, if operable, they perform poorly in terms of resolution, recovery, and reproducibility. Therefore, it is desirable to examine alternate means that might solve important colloidal separation and characterization problems encountered in working with biological, industrial, environmental, and geological materials. One of the most important colloidal processes that is generally quite difficult to characterize is the aggregation of single particles to form complexes made up of multiples of the individual particles. Aggregation is a common phenomenon for both natural and industrial colloids. The high degree of stability, which is frequently observed in colloidal systems, is a kinetic phenomenon in that the rate of aggregation of such systems may be practically zero. Thus, in studies of the colloidal state, the kinetics of aggregation are of paramount importance. Although the kinetics of aggregation can be described easily by a bimolecular equation, it is not an easy thing to do experimentally.

One technique for doing this is to count the particles microscopically. In addition to particle size limitation, this is an extraordinarily tedious procedure. Light scattering can be also used for the kinetic study of aggregation, but experimental turbidities must be interpreted in terms of the number and size of the scattering particles.

In the present work, it is shown that the field-flow fractionation (FFF) technique can be used with success to study the aggregation phenomena of colloids.

The techniques of field-flow fractionation appear to be well suited to colloid analysis. The special subtechnique of sedimentation FFF (SdFFF) is particularly effective in dealing with colloidal particles in the diameter range from 0.02 to 1 μm, using the normal or Brownian mode of operation (up to 100 μm using the steric-hyperlayer mode). As a model sample for the observation of aggregate particles by SdFFF, of poly(methyl methacrylate) was used, whereas for the kinetic study of aggregation by SdFFF, the hydroxyapatite sample [$Ca_5(PO_4)_3OH$] consisting of submicron irregularly shaped particles was used. The stability of hydroxyapatite, which is of paramount importance in its applications, is dependent on the total potential energy of interaction between the hydroxyapatite particles. The latter, which is the sum of the attraction potential energy and that of repulsion, depends on particle size, the Hamaker constant, the surface potential, and the Debye–Hückel reciprocal distance, which is immediately related to the ionic strength of carrier solution.

Methodology

Field-flow fractionation is a highly promising tool for the characterization of colloidal materials. It is a dynamic separation technique based on differential elution of the sample constituents by a laminar flow in a flat, ribbonlike channel according to their sensitivity to an external field applied in the perpendicular direction to that of the flow.

The total potential energy of interaction between two colloidal particles, U_{tot}, is given by the sum of the energy of interaction of the double layers, U_R, and the energy of interaction of the particles themselves due to van der Waals forces, U_A. Consequently,

$$U_{tot} = U_R + U_A \tag{1}$$

For identical spherical particles U_R and U_A are defined as follows:

$$U_R = \frac{\varepsilon r \psi_0^2}{2} \ln[1 + \exp(-\kappa H)] \quad (\kappa r \gg 1) \tag{2}$$

$$U_R = \frac{\varepsilon r \psi_0^2}{R} \exp(-\kappa H) \quad (\kappa r \ll 1) \tag{3}$$

$$U_A = -\frac{A_{212}r}{12H} \tag{4}$$

where ψ is the dielectric constant of the dispersing liquid, r is the radius of the particle, ψ_0 is the particle's surface potential, κ is the reciprocal double-layer thickness, R is the distance of the centers of the two particles, A_{212} is the effective Hamaker constant of two particles of type 2 separated by the medium of type 1, and H is the nearest distance between the surfaces of the particles.

Equations (2)–(4) show that the total potential energy of interaction between two colloidal spherical particles depends on the surface potential of the particles, the effective Hamaker constant, and the ionic strength of the suspending medium. It is known that the addition of an indifferent electrolyte can cause a colloid to undergo aggregation. Furthermore, for a particular salt, a fairly sharply defined concentration, called "critical aggregation concentration" (CAC), is needed to induce aggregation.

The following equation gives the rate of diffusion-controlled aggregation, u_r, of spherical particles in a disperse system as a result of collisions in the absence of any energy barrier to aggregate:

$$u_r = -k_r N_0^2 \tag{5}$$

where k_r is the second-order rate constant for diffusion-controlled rapid aggregation and N_0 is the initial number of particles per unit volume.

In the presence of an energy barrier to aggregate, the rate of aggregation, u_s, is

$$u_s = -k_s N_0^2 \tag{6}$$

where k_s is the rate constant of slow aggregation in the presence of an energy barrier.

The stability ratio, w, of a dispersion is defined as the ratio of the rate constants for aggregation in the absence, k_r, and the presence, k_s, of an energy barrier, respectively:

$$w = \frac{k_r}{k_s} \tag{7}$$

The aggregation process is described by the bimolecular kinetic equation

$$\frac{1}{N_i} = \frac{1}{N_0} + k_{\text{app}} t_i \tag{8}$$

where N_i is the total number of particles per unit volume at time t_i and k_{app} is the apparent rate constant for the aggregation process. The measurement of the independent kinetic units per unit volume, N_i, at different times t_i can give the rate constant for the aggregation process.

Considering that d_{N_0} and d_{N_i} are the measured number-average diameters of the particles at times $t = 0$ and t_i, respectively, Eq. (8), for polydisperse samples, gives

$$d_{N_i}^3 = d_{N_0}^3 + d_{N_0}^3 N_0 k_{\text{app}} t_i \tag{9}$$

Equation (9) shows that from the slope of the linear plot of the $d_{N_i}^3$ versus t_i, the apparent rate constant k_{app} can be determined, as the N_0 values can be found from the ratio of the total volume of the injected sample to the volume of the particle, which can be determined from the diameter calculated from the intercept of the above plot.

Applications

The observation of a series of peaks (Fig. 1) while analyzing samples of poly(methyl methacrylate) (PMMA) colloidal latex spheres by SdFFF suggests that part of the la-

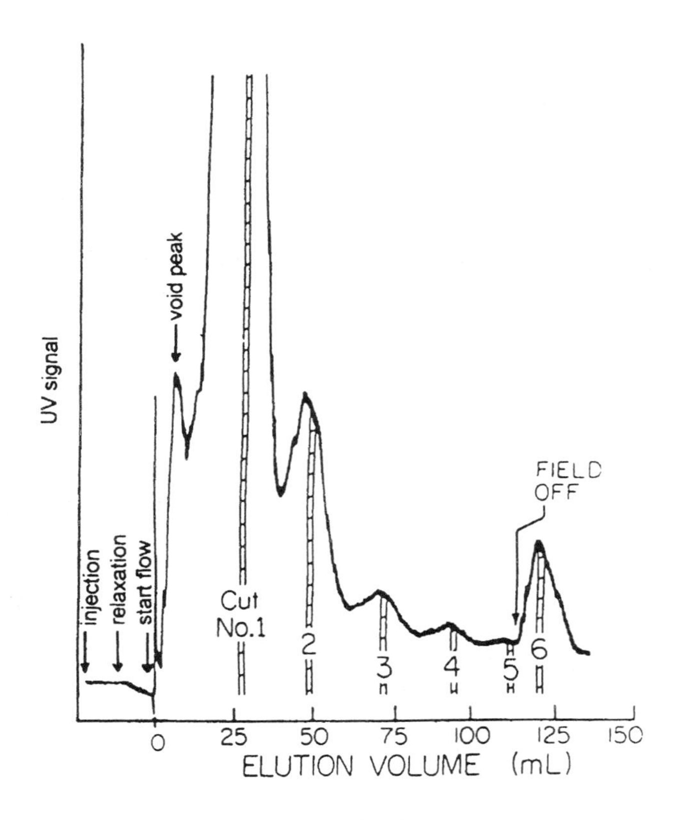

Fig. 1 Sedimentation field-flow fractionation fractogram of 0.207-μm poly(methyl methacrylate) aggregate series from which six cuts were collected and analyzed by electron microscopy. Experimental conditions: field strength of 61.6 g and flow rate of 0.84 mL/min. [Reproduced with permission from H. K. Jones et. al. (1988) *J. Chromatogr. 455*, 1; Copyright Elsevier Science Publishers B. V.]

tex population has aggregated into doublets, triplets, and higher-order particle clusters. The particle diameter of the latex spheres was given as 0.207 μm. The aggregation hypothesis is confirmed by retention calculations and by electron microscopy. For this purpose, narrow fractions or cuts were collected from the first five peaks as shown in Fig. 1. A fraction was also collected for the peak which appeared after the field was turned off. The individual fractions were subjected to electron microscopy and as expected, cut No. 1 yielded singlets, cut No. 2 yielded doublets, cut No. 3 yielded triplets, cut No. 4 yielded quads, cut No. 5 yielded quints, and the cut after the field was turned off yielded clusters from six individual particles.

Sedimentation field-flow fractionation was used also for the kinetic study of hydroxyapatite (HAP) particles' aggregation in the presence of various electrolytes to determine the rate constants for the bimolecular process of aggregation and to investigate the possible aggregation mechanisms describing the experimental data. The HAP sample contained polydisperse, irregular colloidal particles with number-average diameter $d_N = 0.262 \pm 0.046\ \mu$m.

The number-average diameter, d_N, for the HAP particles increases with the electrolyte KNO_3 concentration until the critical aggregation concentration is reached, where the d_N value remains approximately constant. The starting point of the maximum d_N corresponds to the electrolyte concentration called CAC. The last value, which depends on the electrolyte used, was found to be $1.27 \times 10^{-2} M$ for the electrolyte KNO_3.

According to Eq. (9), the plot of $d^3_{N_i}$ versus t_i at various electrolyte concentrations determines the apparent rate constant, k_{app}, of the HAP particles' aggregation. The found k_{app} value for the aggregation of the HAP particles in the presence of $1 \times 10^{-3} M$ KNO_3 is 2.5×10^{-21} cm^3/s. It is possible to make a calculation which shows whether the value of k_{app} is determined by the rate at which two HAP particles can diffuse up to each other (diffusion control) or whether the rate of reaction is limited by other slower processes. The rate constant for the bimolecular collision (k_1) of the HAP particles, can be calculated by the Stokes–Einstein equation:

$$k_1 = \frac{8kT}{3n} \text{ cm}^3/\text{s} \qquad (10)$$

where n is the viscosity of the medium. The calculated value of $k_1 = 1.1 \times 10^{-11}$ cm^3/s is about 10 orders of magnitude greater than the value of k_{app} actually measured. So, the aggregation rates are slower than those expected if the process was simply diffusion controlled when electrostatic repulsion is absent. The latter indicates that the minimal mechanism for the aggregation process of the HAP particles would be

$$\underset{1}{\text{Particle}} + \underset{2}{\text{Particle}} \underset{k_{-1}}{\overset{k_1}{\rightleftarrows}} \underset{\text{complex}}{\text{Intermediate}} \xrightarrow{k_2} \underset{\text{aggregate}}{\text{Stable}} \qquad (11)$$

where k_{-1} is the rate constant for the dissociation of the intermediate aggregate and k_2 is the rate constant for the process representing the rate-determining step in the aggregation reaction. Because k_{app}, describing the overall process, is smaller than the calculated k_1 value, there must be rapid equilibration of the individual particles and their intermediate complexes followed by the slower step of irreversible aggregation. The stability factor, w, of HAP's particles found to be 4.4×10^9 is too high, indicating that the particles are very stable, even in the presence of significant quantity of the electrolyte KNO_3.

As a general conclusion, the FFF method can be used with success to study the aggregation process of colloidal materials.

Future Developments

Looking to the future, it is reasonable to expect continuous efforts to improve the theoretical predictions and more experimental work to investigate the aggregation phenomena of natural and industrial colloids.

Suggested Further Reading

Athanasopoulou, A., G. Karaiskakis, and A. Travlos, *J. Liq. Chromatogr. Related Technol.* 20(16&17): 2525 (1997).

Athanasopoulou, A., D. Gavril, A. Koliadima, and G. Karaiskakis, *J. Chromatogr. A* (in press).

Caldwell, K. D., T. T. Nguyen, J. C. Giddings, and H. M. Mazzone, *J. Virol. Methods 1*: 241 (1980).

Everett, D. H., *Basic Principles of Colloid Science*, Royal Society of Chemistry Paperbacks, London, 1988.

Family, F. and D.P. Landan (eds.), *Kinetic of Aggregation and Gelation*, North-Holland, Amsterdam, 1984.

Jones, H. K., B. N. Barman, and J. C. Giddings, *J. Chromatogr. 455*, 1 (1988).

Koliadima, A., *J. Liq. Chromatogr. Related Technol.* (in press).

Wittgren, B., J. Borgström, L. Piculell, and K. G. Wahlund, *Biopolymers 45*: 85 (1998).

Athanasia Koliadima

Amino Acid Analysis by HPLC

Introduction

Amino acid analysis (AAA) is a classic analytical technique that characterizes proteins and peptides based on the composition of their constituent amino acids. It provides qualitative identification and is essential for the accurate quantification of proteinaceous materials.

Amino acid analysis is widely applied in research, clinical facilities, and industry. It is a fundamental technique in biotechnology, used to determine the concentration of peptide solutions, to confirm protein binding in antibody conjugates, and for end-terminal analysis following enzymatic digestion. Clinical applications include diagnosing metabolic disorders in newborns. In industry, it is used for quality control of products ranging from animal feed to infant formula.

The analysis of a polypeptide typically involves four steps: hydrolysis (or deproteination with physiological samples), separation, derivatization, and detection. Hydrolysis breaks the peptide bonds and releases free amino acids, which are then separated by side group using column chromatography. Derivatization with a chromogenic reagent enhances the separation and spectral properties of the amino acids and is required for sensitive detection. A data processing system compares the resulting chromatogram, based on peak area or peak height, to a calibrated standard (see Fig. 1a). The results, expressed as mole percent and microgram of residue per sample, determine the percent composition of each amino acid as well as the total amount of protein in the sample. Unknown proteins may be identified by comparing their amino acid composition with those in protein databases. Successful identification of unknown proteins may be achieved using Internet search programs.

Other techniques, such as capillary electrophoresis and [matric-assisted laser desorption ionization] mass spectrometry, provide qualitative analyses, often with greater speed and sensitivity. Nevertheless, AAA by high-performance liquid chromatography (HPLC) complements other structural analysis techniques, such as peptide sequencing, and remains indispensable for quantifying proteinaceous materials.

Peptide Hydrolysis

Conventional hydrolysis exposes the polypeptide to 6*M* HCl acid under vacuum at 110°C for 20–24 h [1,2]. Protective agents, such as 0.1% phenol, are added to improve recovery. Gas-phase hydrolysis, in which the acid is delivered as a vapor, gives comparable results to the liquid phase. Additionally, the gas phase minimizes acid contaminants and allows parallel hydrolysis of standards and samples within the same chamber.

Acid hydrolysis yields 16 of the 20 coded amino acids; tryptophan is destroyed, cysteine recovery is unreliable, and asparagine and glutamine are converted to aspartic acid and glutamic acid, respectively. Furthermore, some side groups, such as the hydroxyl in serine, promote the breakdown of the residue, whereas aliphatic amino acids, protected by stearic hindrance, require longer hydrolysis time. This variation in yield can be overcome by hydrolyzing samples for 24, 48, and 72 h and extrapolating the results to zero time point.

The reaction rate doubles with every 10°C increase, so that hydrolysis at 145°C for 4 h gives results comparable to the conventional method. Microwave hydrolysis reduces analysis time to 30–45 min. Alternative hydrolysis agents include sulfonic acid, which often gives better recovery but is nonvolatile, and alkaline hydrolysis, used in the analysis of tryptophan, proteoglycans, and proteolipids.

Careful sample preparation and handling during the hydrolysis step are critical for maintaining accurate and reproducible results. Salts, metal ions, and other buffer components remaining in a sample may accelerate hydrolysis, producing unreliable quantification. The Maillard reaction between amino acids and carbohydrates results in colored condensation products (humin) and decreased yield. Routine method calibration with proteins and amino acid standards, use of an internal standard (1 nmol norleucine is used for sensitive analysis), and control blanks are strongly recommended, along with steps to minimize background contaminants (see Fig. 1b). The practical limit for high-sensitivity hydrolysis is 10–50 ng of sample; below this amount, background contaminants begin to play a larger role.

Derivatizing Reagents for Analysis of Amino Acids by HPLC

The first automated analyzer was developed by Moore, Stein, Spackman, and Hamilton in the 1950s. Hydrolysates were separated on an ion-exchange column, followed by postcolumn reaction with ninhydrin. Although

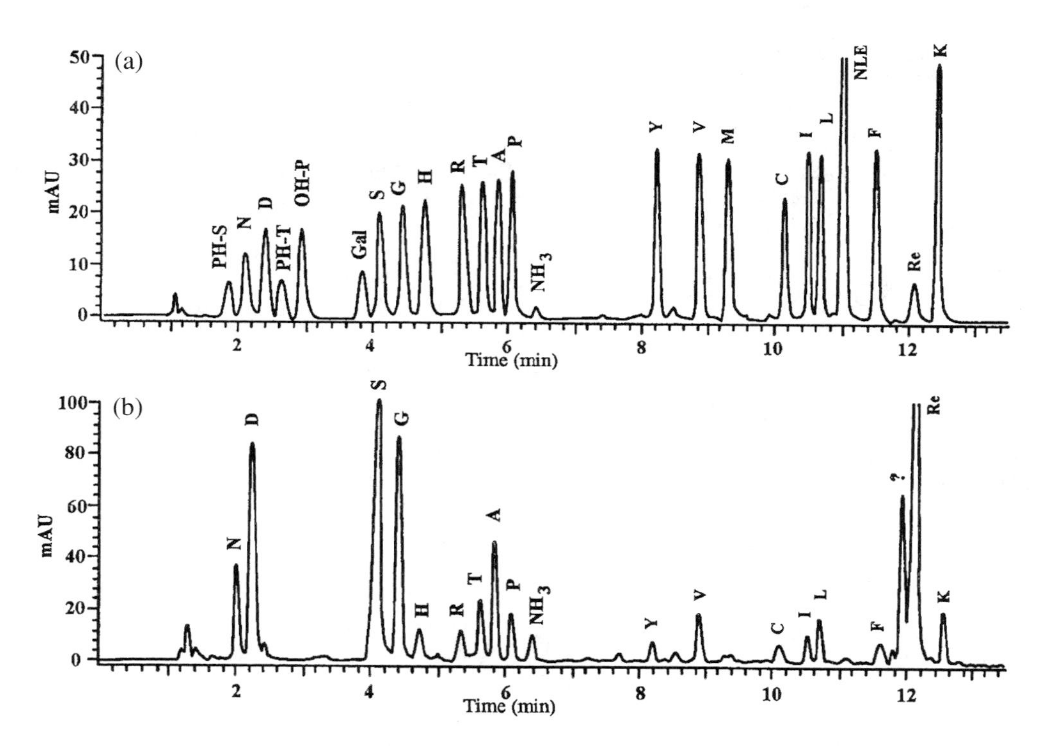

Fig. 1 (a) 200 pmol of PTC-amino acid standard, including phosphoserine (PH-S), aspartate (N), glutamate (D), phosphothreonine (PH-T), hydroxyproline (OH-P), galactosamine (Gal), serine (S), glycine (G), histidine (H), arginine (R), threonine (T), alanine (A), proline (P), tyrosine (Y), valine (V), methionine (M), cysteine (C), isoleucine (I), leucine (L), norleucine (NLE, 1 nmol internal standard), phenylalanine (F), excess reagent (Re), and lysine (K). (b) Analysis of a human fingerprint, taken up from watchglass using a mixture of water and ethanol. (Courtesy of National Gallery of Art and the Andrew W. Mellon Foundation.)

this system remains the standard method, its major drawback is low sensitivity. Several methods have since been developed offering high sensitivity and faster analyses without sacrificing reproducibility [1–4].

Amino acids react with many reagents to form stable derivatives and strong chromophores (see Table 1). Derivatization can precede (precolumn) or follow (in-line postcolumn) the chromatographic separation. Both precolumn and postcolumn systems are currently employed: ninhydrin and PITC analyzers are widely used, whereas AQC, OPA and OPA–FMOC systems provide the highest sensitivity.

Postcolumn systems typically use cation-exchange columns with either sodium citrate (for hydrolysates) or lithium citrate (for physiological samples) as mobile phases. Contaminating salts and detergents are better tolerated because the samples are "cleaned up" before reaction with the reagent. The additional reagent pump, however, may lead to sample dilution, peak broadening, baseline fluctuations, and longer analysis time (30–90 min). Fluorescent reagents are compatible with a wider range of buffers, but the buffers must be amine-free if used with postcolumn methods.

Since the 1980s, precolumn derivatization methods have gained wider acceptance due to simpler preparation, faster analyses, and better resolution. The separation on reversed-phase C-8 or C-18 columns typically requires low-ultraviolet (UV) mobile phases, such as sodium phosphate or sodium acetate buffers, with acetonitrile or methanol as organic solvents. Separation times range from 15 to 50 min.

Improved Recovery of Sensitive Amino Acids

Cysteine and tryptophan require special treatment for quantitative analysis [5]. Cystine/cysteine can be determined using three equally successful methods: oxidation, alkylation, and disulfide exchange. Oxidation to cysteic acid is commonly carried out by pretreatment with performic acid. Alkylation using pretreatment with 4-vinylpyridine or iodoacetate produces piridylethylcys-

Table 1 Amino Acid Derivatization Reagents

Reagent	Chromophore	Detection limit	Separation time	Drawbacks	Advantages
AQC (6-aminoquinolyl-*N*-hydroxysuccinimidyl carbamate)	Fluorescent (ex. 245 nm, em. 395 nm); UV 245 nm	160 fmol	35–50 min precolumn	Quaternary gradient elution required for complex, non-hydrolysate samples	Tolerates salts and detergents, rapid reaction, stable product, good reagent separation, high sensitivity and accuracy
Dabsyl chloride (4-*N*,*N*-dimethylaminoazobenzene-4′-sulfonyl chloride)	Visible 436 nm	Low fmol	18–44 min precolumn	Multiple products, critical concentration	Stable product, good separation, high sensitivity
Dansyl chloride (5,*N*,*N*-dimethylaminonaphthalene-1-sulfonyl chloride)	Fluorescent (ex. 360–385 nm, em. 460–495 nm); UV 254 nm	Low pmol	60–90 min precolumn	Multiple products, critical concentration, difficult separation leads to long separation time	Stable product
Fluorescamine (4-phenyl-spiro[furan-2(3H), 1′-phthalan]-3,3′-dione)	Fluorescent (ex. 390 nm, em. 475 nm)	20–100 pmol	30–90 min postcolumn	Secondary amine pretreatment, critical concentration, may give background interference	Rapid reaction, stable product, good reagent separation
FMOC (9-fluorenylmethylchloroformate)	Fluorescent (ex. 265 nm, em. 320 nm); UV 265 nm	1 pmol	20–45 min precolumn	Multiple products, extraction of excess reagent	Stable product, used with OPA for detection of secondary amine
Ninhydrin (triketohydrindene hydrate)	Primary amine (440 nm), secondary amine (570 nm)	100 pmol	30 min postcolumn	Low sensitivity and resolution	Good reproducibility
OPA (ortho-phthalaldehyde)	Fluorescent (ex. 340 nm, em. 455 nm)	50 fmol	90 min postcolumn, 17–35 precolumn	Secondary amine pretreatment, slow reaction, unstable derivative, background interference	Good reagent separation, high sensitivity and reproducibility with automated system
PITC (phenylisothiocyanate)	UV 254 nm	1 pmol	15–27 min precolumn	Salt interference, requires refrigeration, excess reagent removed under vacuum	Ease of use, flexibility, good separation, reproducibility enhanced with automation

Source: Refs. 1 and 2.

teine (PEC) and carboxymethylcysteine (CMC), respectively. Disulfide exchange is achieved by adding reagents such as dithiodipropionic acid, dithiodiglycolic acid, or dimethylsulfoxide (DMSO) to the HCl acid during hydrolysis. The latter treatment offers both ease of use as well as accurate yields.

The superior approach to tryptophan analysis involves the addition of dodecanethiol to HCl acid, especially when combined with automatic vapor-phase hydrolysis. Alternative hydrolysis agents such as methane sulfonic acid, mercaptoethanesulfonic acid, or thioglycolic acid can produce 90% or greater yields. Acid hydrolysis additives and alkaline hydrolysis using 4.2*M* NaOH are also used with varying results.

Qualitative analysis of glycopeptides and phosphoamino acids is achieved through a separate, partial hydrolysis with 6*N* HCl acid at 110°C for 1 and 1.5 h, respectively. Separation of cysteine, tryptophan, and amino sugars requires minimal chromatographic adjustments; phosphoamino acid separation is straightforward using the reversed phase, but it is cumbersome using ion exchange.

Analysis of Free and Modified Amino Acids

Blood, urine, cerebrospinal, and other physiological fluids contain a great number of posttranslationally modified amino acids (approximately 170 have been studied to date) and in a wider range of concentrations than protein hydrolysates [6]. Additionally, plant sources produce about 500 nonprotein amino acids and, in geological samples, highly unusual amino acids may indicate extraterrestrial origin [7, 8].

Although the free amino acids in these samples do not require hydrolysis, blood plasma and cerebrospinal fluid must be deproteinated before analysis. Proteins may bind irreversibly to ion-exchange columns, resulting in loss of resolution. Furthermore, any peptide hydrolases must be inactivated. For ion-exchange analysates, a protein precipitant is added before centrifugation. Sulfosalicylic acid, a common precipitating agent, is added in solid form to avoid sample dilution. For reversed-phase analysates, ultrafiltration, size-exclusion chromatography, or organic solvent extraction is recommended. Samples with low protein and high amino acid concentrations, such as urine and amniotic fluid, need only to be diluted before analysis.

Precolumn derivatives are more tolerant to lipid-rich samples. Changing the guard-column routinely is recommended to avoid column buildup, especially for reversed-phase systems.

Amino Acid Racemization Analysis

Racemization, the interconversion of amino acid enantiomers, occurs slowly in biological and geological systems. Although the L-form is the most prevalent, D-amino acids are found in fossils and living organisms. The rate increases with extreme pH values, high temperature, and high ionic strength. Rates also vary among amino acids: at 25°C, the racemization half-life of serine is about 400 years, whereas that of isoleucine is 40,000. Enantiomer analysis is used to confirm the bioactivity of synthetic peptides and for geological dating [1,3,8].

Hydrolysis itself accelerates racemization. Shorter acid exposure at higher temperatures, such as 160°C for 1 h, decreases racemization by about 50% compared to conventional hydrolysis. Liquid-phase methanesulfonic acid, conventional, and microwave hydrolysis produce progressively higher rates of racemization. Addition of phenol, however, significantly reduces racemization during microwave hydrolysis [1].

The three general approaches to enantiomer separation entail a chiral stationary phase, a chiral mobile phase, or a chiral reagent. Tandem columns, with reversed and chiral stationary phases, were used to separate 18 D–L pairs of PTC-amino acids in 150 min. OPA–amino acid enantiomers have been separated on both ion-exchange and reversed-phase columns using a sodium acetate buffer with a L-proline–cupric acetate additive. Chiral reagents, such as Marphey's reagent and OPA/IBLC (*N*-isobutiril-L cysteine), were successfully used for racemization analysis within 80 min.

Acknowledgments

The author would like to thank Dr. Steven Birken, Dr. Chun-Hsien Huang, Dr. Stacy C. Marsella, and Dr. Conceicao Minetti for their assistance in proofreading this article.

References

1. M. Fountoulakis and H.-W. Lahm, Hydrolysis and amino acid composition analysis of proteins, *J. Chromatogr. A 826*: 109 (1998).
2. C. Fini, A. Floridi, V. N. Finelli, and B. Wittman-Liebold (eds.), *Laboratory Methodology in Biochemistry, Amino Acid Analysis and Protein Sequencing*. CRC Press, Boca Raton, FL, 1990.
3. G. C. Barrett (ed.), *Chemistry and Biochemistry of Amino Acids*, Chapman and Hall, London, 1985.
4. W. S. Hancock (ed.), *CRC Handbook of HPLC for the Separation of Amino Acids, Peptides, and Proteins, Vol. I*, CRC Press, Boca Raton, FL, 1984.
5. D. J. Strydom, T. T. Andersen, I. Apostol, J. W. Fox, R. J. Paxton, and J. W. Crabb, Cysteine and tryptophan amino acid analysis of ABRF92-AAA, in *Techniques in Protein Chemistry IV*, R. Hogue Angeletti (ed.), Academic Press, San Diego, 1993, pp. 279–288.
6. P. A. Haynes, D. Sheumack, L. G. Greig, J. Kibby, and J. W. Redwood, Applications of automated amino acid analysis using 9-fluorenylmethyl chloroformate, *J. Chromatogr. 588*: 107 (1991).
7. G. Rosenthal, *Plant Nonprotein Amino and Imino Acids: Biological, Biochemical, and Toxicological Properties*, Academic Press, New York, 1982.
8. P. E. Hare, T. C. Hoering, and K. King (eds.), *Biogeochemistry of Amino Acids*, John Wiley & Sons, New York, 1980.

Susana Maria Halpine

Amino Acids and Derivatives: Analysis by TLC

Amino acids are derivatives of the carboxylic acids in which a hydrogen atom in the side chain (usually on the alpha carbon) has been replaced by an amino group and, therefore, they are amphoteric.

In weak acid solution (about pH = 6.0) the carboxyl group of a neutral amino acid (one amino group and one carboxyl group) is dissociated, and the amino group binds a proton to give a dipolar ion (zwitterion). The pH at which the concentration of the dipolar ion is a maximum is called the isoelectric point (pI) of that amino acid. The isoelectric point of an amino acid is calculated from the relationship

$$pI = \frac{1}{2}(pK_1 + pK_2)$$

where pK_1 and pK_2 refer to the dissociation of carboxyl group and protonated amino group, respectively.

Amino acids constitute the basic units of all proteins. The number of α-amino acids obtained from various proteins is about 40, but only 20 are present in all proteins in varying amounts.

Thin-layer chromatography (TLC) is one of the most promising separation methods for such compounds that are not amenable to gas chromatographic analysis.

Preparation of Test Solutions

Amino acids should be as free from impurities as possible because they also exhibit a pronounced capacity for binding metal ions. The analysis of amino acids in natural fluids or extracts requires the removal of interfering compounds prior to chromatographic separation, in order to prevent tailing and deformation of the spots (i.e., in urine samples and hydrolizates of proteins or peptides, high salt concentrations occur). Salts can be conveniently removed by passing the sample through a cation-exchange resin column.

Enrichment of amino acids in urine can be performed by extracting the lyophilized sample (10 mL) with 1mL methanol–1M HCl mixture (4:1 v/v) and applying an aliquot of supernatant liquid to the plate after centrifugation.

Chromatographic Techniques for Amino Acid Separation

Untreated Amino Acids

Standard solutions of amino acids have been prepared in aqueous–alcoholic solvents (e.g., 70% ethanol), with the addition of hydrochloric acid (0.1M) for the dissolution of relatively insoluble amino acids (i.e., tyrosine and cystine). Detection is generally performed with a ninhydrin reagent. After color development with the ninhydrin reagent, the treatment of the layer with a complex-forming cation (e.g., Cu^{II}, Cd^{II}, Ni^{II}) changes the blue color to red and increases colorfastness considerably. More specific coloration of amino acids can be achieved by adding bases such as collidine and cyclohexylamine to the detecting agent solution.

Amino acids have been separated on layers of a wide variety of inorganic and organic adsorbents, ion exchangers, and impregnated plates. The two most commonly used adsorbents are silica gel and cellulose.

Separation on Silica Gel and Cellulose Layers

It is interesting to note that by using neutral eluents such as ethanol or n-propanol–water, the acidic amino acids (Glu, Asp) travel much faster on silica gel than basic amino acids (Lys, Arg, His), which, indeed, show very small R_f values. The difference is likely due to cation exchange between the protonated amino groups of basic amino acids (two or more amino groups and one carboxyl group) and the acidic groups present on silica gel. The strong retention observed for these compounds when eluting with acidic solvents (see Table 1) confirms such a hypothesis. A similar phenomenon is also observed on cellulose plates and might be due to the cellulose carboxyl groups.

Furthermore, it is seen that a hydroxyl group in the molecule does not necessarily reduce the R_f value as the chromatographic behavior of serine with respect to glycine on both layers of silica gel and cellulose shows (see Table 1). Some of the numerous eluents that have been used for the separation of amino acids on silica gel are acetone–water–acetic acid–formic acid (50:15:12:3), ethylacetate–pyridine–acetic acid–water (30:20:6:11), 96% ethanol–water–diethylamine (70:29:1), chloroform–formic acid (20:1), chloroform–methanol (9:1), isopropanol–5% ammonia (7:3), and phenol–0.06M borate buffer pH 9.30 (9:1). On cellulose plates, ethylacetate–pyridine–acetic acid–water (5:5:1:3), n-butanol–acetic acid–water (4:1:1 and 10:3:9), n-butanol–acetone–ammonia–water (20:20:4:1), collidine–n-butanol–acetone–water (2:10:10:5), phenol–methanol–water (7:10:3), ethanol–acetic acid–water (2:1:2), and cyclohexanol–acetone–diethylamine–water (10:5:2:5), have also

Table 1 hR_f Values of the 20 Common Amino Acids in Different Experimental Conditions (Ascending Technique)

Amino acid and symbol	Silica gel G (A)	Micro-crystalline cellulose (B)	Fixion 50-X8 (Na^+) (C)	Silanized silica gel +4% HDBS (D)	p*I*
Glycine (Gly)	18	15	56	83	6.0
Alanine (Ala)	22	29	51	74	6.0
Serine (Ser)	18	16	67	85	5.7
Threonine (Thr)	20	21	67	83	6.5
Leucine (Leu)	44	64	22	26	6.0
Isoleucine (Ile)	43	60	28	31	6.1
Valine (Val)	32	48	43	54	6.0
Methionine (Met)	35	23	28	42	5.8
Cysteine (Cys)	7	3	56	—	5.0
Proline (Pro)	14	34	—	63	6.3
Phenylanine (Phe)	43	55	14	21	5.5
Tyrosine (Tyr)	41	36	12	45	5.7
Tryptophan (Trp)	47	36	2	13	5.9
Aspartic acid (Asp)	17	15	72	86	3.0
Asparagine (Asn)	14	—	—	85	5.4
Glutamic acid (Glu)	24	27	35	83	3.2
Glutamine (Gln)	15	—	—	—	5.7
Arginine (Arg)	6	11	2	28	10.8
Histidine (His)	5	7	11	40	7.6
Lysine (Lys)	3	7	8	47	9.8

Note: Eluents: A = *n*-butanol–acetic acid–water (80 +20 +20 v/v/v); B = 2-butanol–acetic acid–water (3:1:1 v/v/v); C = 84 g citric acid + 16 g NaOH + 5.8 g NaCl + 54 g ethylene glycol + 4 mL conc. HCl (pH = 3.3); D = 0.5*M* HCl + 1*M* CH_3COOH in 30% methanol (pH = 0.7).

been used as eluents. Separation efficiency can be increased by two-dimensional (TD) chromatography. Several solvent systems are suitable for TD separations and the combination of chloroform–methanol–17% ammonium hydroxide (40+40+20 v/v/v) and phenol–water (75 g+25 g) will separate all protein amino acids, except leucine and isoleucine, on silica plates.

Separation on Ion Exchangers and Impregnated Plates

Cellulose ion exchangers (e.g., diethylaminoethyl cellulose) and ion-exchange resins have been widely used as stationary phases for TLC separation of untreated amino acids.

Fixion 50-X8 commercial plates, which contain Dowex 50-X8 type resin, have been tested on both Na^+ and H^+ forms for 30 amino acids and the results obtained for the 20 common protein amino acids are reported in Table 1. The isomer pair leucine and isoleucine is well separated by this method. In addition, the hydroxyl group notably increases the R_f values owing to the hydrophobic properties of the resin and the pairs serine and glycine, and threonine and alanine can be resolved.

Many studies have been recently focused on impregnated plates. The methods used for impregnation depend on whether the plates are home-made or commercially available. In the first case, the impregnation reagent is usually added to a slurry of the adsorbent, whereas ready-to-use plates are dipped in the solution of the reagent.

The resolution of amino acids has been reported by using different metal ions as impregnating agents at various concentrations. On silica gel impregnated with Ni(II) salts, the results indicate a predominant role of partitioning phenomenon when eluting with acidic aqueous and nonaqueous solutions (e.g., *n*-butanol–acetic acid–water and *n*-butanol–acetic acid–chloroform in the 3:1:1 v/v/v ratio). The impregnation of silanized silica gel with 4% dodecylbenzensulfonic acid (HDBS) solution on both home-made and ready-to-use plates is particularly useful in resolving amino acids [1,2]. The parameters affecting the retention of amino acids on these layers are type of adsorbent, concentration and properties of the impregnating agent, percentage and kind of organic modifier, pH, and ionic strength of the eluent.

The data of Table 1 show that complete resolution of basic amino acids (Arg, His, Lys) and of neutral amino acids which differ in their side-chain carbon atom number (e.g., Gly, Ala, Met, Val, Leu, or Ile) are possible on homemade plates of silanized silica gel (C_2) impregnated with a 4% solution of dodecylbenzensulfonic acid in 95% ethanol. More compact spots can be obtained on RP-18 ready-to-use plates dipped in the same solution of the surfactant agent.

Resolution of Amino Acid Derivatives

The identification of N-terminal amino acids in peptides and proteins is of considerable practical importance because it constitutes an essential step in the process of sequential analysis of peptide structures.

Many *N*-amino acid derivatives have been proposed to this purpose and the most commonly studied by TLC are 2,4-dinitrophenyl (DNP)- and 5-dimethylamino-1-naphthalene-sulfonyl (Dansyl,Dns)-amino acids, and 3-phenyl-2-thiohydantoins (PTH-amino acids).

Recently, 4-(dimethylamino)azobenzene-4′-isothiocyanate (DABITC) derivatives of amino acids have been also investigated.

DNP-Amino Acids

The dinitrophenylation of amino acids, peptides, and proteins and their separation by one-dimensional and two-dimensional TLC have been reviewed by Rosmus and Deyl [3]. DNP-amino acids are divided into those which are ether extractable and those which remain in the aqueous phase.

Water-soluble α-DNP-Arg, α-DNP-His, ε-DNP-Lys, bis-DNP-His, O-DNP-Tyr, DNP-cysteic acid ($CySO_3H$), and DNP-cystine $(Cys)_2$ have been identified on silica gel plates in the *n*-propanol–34% ammonia (7:3 v/v) system. Although separation of DNP-Arg and ε-DNP-Lys is incomplete (R_f values of 0.43 and 0.44, respectively), both of them can be detected because of the color difference produced in the ninhydrin reaction.

Ether-soluble DNP-amino acids have been investigated by one-dimensional and TD chromatography. This last technique offers the possibility of almost complete separation of the two groups of derivatives.

The yellow color of DNP-amino acids deepens upon exposure to ammonia vapor and it is sufficiently intense that 0.1 μg can be visualized. The detection limit is lower (about 0.02 μg) under ultraviolet (UV) light (360 nm with dried plates and 254 nm with wet ones), but it increases for TD chromatography (about 0.5 μg). At present, the applications of DNP-amino acids are limited.

PTH-Amino Acids

The formation of PTH-amino acids by the Edman degradation [4] of peptides and proteins or by successive modifications of the method constitutes the most commonly used technique for the study of the structure of biologically active polypeptides today. Identification of PTH-amino acids in mixtures may be successfully achieved by TLC. Quantitative determination is based on UV absorption (detection limit 0.1 μg at 270 nm). An alternative is offered by the chlorine/tolidine test, which is very useful because the minimal amount required for detection is about 0.5 μg.

Using one-dimensional chromatography on alumina, polyamide, and silica gel, difficulties are encountered in resolving Leu/Ile and Glu/Asp pairs as well as other combinations of PTH-amino acids (e.g., Phe/Val/Met/Thr). The most common solvents used on polyamide plates are *n*-heptane–*n*-butanol–acetic acid (40:30:9), toluene–*n*-pentane–acetic acid (60:30:35), ethylene chloride–acetic acid (90:16), and ethylacetate–*n*-butanol–acetic acid (35:10:1), and those on silica gel are *n*-heptane–methylene chloride–propionic acid (45:25:30), xylene–methanol (80:10), chloroform–ethanol (98:2), chloroform–ethanol–methanol (89.25:0.75:10), chloroform–*n*-butylacetate (90:10), diisopropil ether–ethanol (95:5), methylene chloride–ethanol–acetic acid (90:8:2), *n*-hexane–*n*-butanol (29:1); *n*-hexane–*n*-butylacetate (4:1), pyridine–benzene (2.5:20), methanol–carbon tetrachloride (1:20), and acetone–methylene dichloride (0.3:8). The complete resolution of specific mixtures is possible with TD chromatography by the use of certain solvent systems given above.

The characteristic colors of PTH-amino acids, following ninhydrin spray and the colored spots observed under UV light on polyamide plates containing fluorescent additives, are very useful in identifying those amino acids that have nearly identical R_f values.

Thin-layer chromatography of PTH-amino acids has been reviewed by Rosmus and Deyl [3].

Dns-Amino Acids

Dansylation reaction in bicarbonate buffer (pH = 9.5) is widely used for identification of N-terminal amino acids in proteins and it is the most sensitive method for quantitative determination of amino acids because dansyl derivatives are fluorescent under an UV lamp (254 nm).

Much research has been focused on silica gel and polyamide plates, using both one-dimensional and TD chromatography.

No solvent system resolves all the Dns-amino acids by one-dimensional chromatography and, also, TD chromatography requires more than two runs for a complete resolution. The most common used eluents on polyamide layers are benzene–acetic acid (9:1), toluene–acetic acid (9:1), toluene–ethanol–acetic acid (17:1:2), water–formic acid (200:3), water–ethanol–ammonium hydroxide (17:2:1 and 14:15:1), ethylacetate–ethanol–ammonium hydroxide (20:5:1), *n*-heptane–*n*-butanol–acetic acid (3:3:1), chlorobenzene–acetic acid (9:1), and ethylacetate–acetic acid–methanol (20:1:1).

A widely employed chromatographic system is based on polyamide plates eluted with water–formic acid (200: 3 v/v) in the first direction and benzene–acetic acid (9: 1 v/v) in the second direction.

A third run in 1*M* ammonia–ethanol (1:1, v/v) or ethylacetate–acetic acid–methanol (20:1:1, v/v/v) resolves especially basic Dns-amino acids or Glu/Asp and Thr/Ser pairs, respectively.

DABTH-Amino Acids

These derivatives are obtained in basic medium by the reaction of DABITC with the primary amino group of N-terminal amino acids in peptides. The color difference between DABITC (or its degradation products) and DABTH-amino acids facilitates identification on TLC. These derivatives are colored compounds and, because of their stability and sensitivity, are usually used for qualitative and quantitative analyses of amino compounds such as amino acids and amines.

All DABTH-amino acids except the Leu/Ile pair can be separated by TD chromatography on layers of polyamide, with water–acetic acid (2:1, v/v) and toluene–*n*-hexane–acetic acid (2:1:1, v/v/v) as solvents 1 and 2, respectively.

Resolution of the DABTH-Leu/DABTH-Ile pair on polyamide is possible with formic acid–ethanol (10:9, v/v) and on silica plates using chloroform–ethanol (100:3, v/v), as eluent.

Cbo- and BOC-Amino Acids

Carbobenzoxy (Cbo) and *tert*-butyloxycarbonyl (BOC) amino acids are very useful in the synthesis of peptides and, consequently, their separation from each other and from unreacted components used in their preparation is very important.

For this separation, various mixtures of *n*-butanol–acetic acid–5%ammonium hydroxide and of *n*-butanol–acetic acid–pyridine with or without the addition of water were used on silica gel.

The BOC-amino acids give a negative nihydrin test; however, if the plates are heated at 130°C for 25 min and immediately sprayed with a 0.25% solution of ninhydrin in butanol, a positive test is obtained.

Resolution of Enantiomeric Amino Acids and Their Derivatives

Amino acids are optically active, and the separation of the enantiomeric pairs is an important task. The topic is discussed in the article Enantiomeric Separations by TLC.

References

1. L. Lepri, P. G. Desideri, and D. Heimler, Reversed-phase and soap-thin-layer chromatography of amino acids, *J. Chromatogr. 195*: 65 (1980).
2. L. Lepri, P. G. Desideri, and D. Heimler, Thin-layer chromatography of amino acids and dipeptides on RP-2, RP-8 and RP-18 plates impregnated with dodecylbenzensulphonic acid, *J.Chromatogr. 209*: 312 (1981).
3. J. Rosmus and Z. Deyl, The methods for identification of N-terminal amino acids in peptides and proteins. Part B, *J.Chromatogr. 70*: 221 (1972).
4. P. Edman, Method for determination of the amino acid sequence in peptides, *Acta Chem.Scand.* 4: 283 (1950).

Suggested Further Reading

Bhushan, R. and J. Martens, Amino Acids and their Derivatives, in *Handbook of Thin-Layer Chromatography*, J. Sherma and B. Friend (eds.), Marcel Dekker, Inc., New York, 1996, Vol. 71, pp. 389–425.

Luciano Lepri
Alessandra Cincinelli

Amino Acids, Peptides, and Proteins: Analysis by CE

Amino acids, peptides, and proteins are analyzed by a variety of modes of capillary electrophoresis (CE) which employ the same instrumentation, but are different in the mechanism of separation. A fundamental aspect of each mode of CE is the composition of the electrolyte solution. Depending on the specific mode of CE, the electrolyte solution can consist of either a continuous or a discontinuous system. In continuous systems, the composition of the electrolyte solution is constant along the capillary tube, whereas in discontinuous systems, it is varied along the migration path.

Capillary zone electrophoresis (CZE), micellar capillary electrokinetic chromatography (MECC), capillary gel electrophoresis (CGE), and affinity capillary electrophoresis (ACE) are CE modes using continuous electrolyte solution systems. In CZE, the velocity of migration is proportional to the electrophoretic mobilities of the analytes, which depends on their effective charge-to-hydrodynamic radius ratios. CZE appears to be the simplest and, probably, the most commonly employed mode of CE for the separation of amino acids, peptides, and proteins. Nevertheless, the molecular complexity of peptides and proteins and the multifunctional character of amino acids require particular attention in selecting the capillary tube and the composition of the electrolyte solution employed for the separations of these analytes by CZE.

The various functional groups of amino acids, peptides, and proteins can interact with a variety of active sites on the inner surface of fused-silica capillaries, giving rise to peak broadening and asymmetry, irreproducible migration times, low mass recovery, and, in some cases, irreversible adsorption. The detrimental effects of these undesirable interactions are usually more challenging in analyzing proteins than peptides or amino acids, owing to the generally more complex molecular structures of the larger polypeptides. One of the earliest, and still more adopted, strategy to preclude the interactions of peptides and proteins with the wall of bare fused-silica capillaries is the chemical coating of the inner surface of the capillary tube with neutral hydrophilic moieties [1]. The chemical coating has the effect of deactivating the silanol groups by either converting them to inert hydrophilic moieties or by shielding all the active interacting groups on the capillary wall. A variety of alkylsilane, carbohydrate, and neutral polymers can be covalently bonded to the silica capillary wall by silane derivatization [2]. Polyacrylamide (PA), poly(ethylene glycol) (PEG), poly(ethylene oxide) (PEO), and polyvinylpyrrolidone (PVP) can be successfully anchored onto the capillary surface treated with several different silanes, including 3-(methacryloxy)-propyltrimethoxysilane, 3-glycidoxypropyltrimethoxysilane, trimethoxyallylsilane, and chlorodimethyloctylsilane. Alternatively, a polymer can be adsorbed onto the capillary wall and then cross-linked *in situ*. Other procedures are based on simultaneous coupling and cross-linking. Alternative materials to fused silica such as polytetrafluorethylene (Teflon) and poly(methyl methyacrylate) (PMMA) hollow fibers has found limited application.

The deactivation of the silanol groups can also be achieved by the dynamic coating of the inner wall by flushing the capillary tube with a solution containing a coating agent. A number of neutral or charged polymers with the property of being strongly adsorbed at the interface between the capillary wall and the electrolyte solution are employed for the dynamic coating of bare fused-silica capillaries. Modified cellulose and other linear or branched neutral polymers may adsorb at the interface between the capillary wall and the electrolyte solution with the main consequence of increasing the local viscosity in the electric double layer and masking the silanol groups and other active sites on the capillary surface. This results in lowering or suppressing the electro-osmotic flow and in reducing the interactions with the capillary wall.

Polymeric polyamines are also strongly adsorbed in the compact region of the electric double layer as a combination of multisite electrostatic and hydrophobic interactions. The adsorption results in masking the silanol groups and the other adsorption active sites on the capillary wall and in altering the electroosmotic flow, which is lowered and, in most cases, reversed from cathodic to anodic. One of the most widely employed polyamine coating agents is polybrene (or hexadimetrine bromide), a linear hydrophobic polyquaternary amine polymer of the ionene type [3]. Alternative choices are polydimethyldiallylammonium chloride, another linear polyquaternary amine polymer, and polyethylenimine (PEI). Very promising is the efficient dynamic coating obtained with ethylenediamine-derivatized spherical polystyrene nanoparticles of 50–100 nm diameter, which can be successively converted to a more hydrophilic diol coating by *in situ* derivatization of the free amino groups with 2,3-epoxy-1-propanol [4].

In most cases, the electrolyte solution employed in CZE consists of a buffer in aqueous media. Although all buffers can maintain the pH of the electrolyte solution constant and can serve as background electrolytes, they are not equally meritorious in CZE. The chemical nature of the buffer system can be responsible for poor efficiency, asymmetric peaks, and other untoward phenomena arising from the interactions of its components with the sample. In addition, the composition of the electrolyte solution can strongly influence sample solubility and detection, native conformation, molecular aggregation, electrophoretic mobility, and electro-osmotic flow. Consequently, selecting the proper composition of the electrolyte solution is of paramount importance in optimizing the separation of amino acids, peptides, and proteins in CZE. The proper selection of a buffer requires evaluating the physical–chemical properties of all components of the buffer system, including buffering capacity, conductivity, and compatibility with the detection system and with the sample.

Nonbuffering additives are currently incorporated into the electrolyte solution to enhance solubility, break aggregation, modulate selectivity, improve resolution, and allow detection, which is particularly challenging for amino acids and short peptides. In addition, a large number of amino compounds, including monovalent amines, amino sugars, diaminoalkanes, polyamines, and short-chain alkylammonio quaternary salts are successfully employed as additives for the electrolyte solution to aid in minimizing interactions of peptides and proteins with the capillary wall in bare fused-silica capillaries. Other additives effective at preventing the interactions of proteins, peptides, and amino acids with the capillary wall include neutral polymers, zwitterions, and a variety of ionic and nonionic surfactants [5]. Less effective at preventing these untoward interactions are strategies using electrolyte solutions at extreme pH values, whether acidic, to suppress the silanol dissociation, or alkaline, to have both the analytes and the capillary wall negatively charged.

Selectivity in CZE is based on differences in the electrophoretic mobilities of the analytes, which depends on their effective charge-to-hydrodynamic radius ratios. This implies that selectivity is strongly affected by the pH of the electrolyte solution and by any interaction of the analyte with the components of the electrolyte solution which may affect its charge and/or hydrodynamic radius.

Additives can improve selectivity by interacting specifically, or to different extents, with the components of the sample. Most of the additives employed in amino acid, peptide, and protein CZE are amino modifiers, zwitterions, anionic or cationic ion-pairing agents, inclusion complexants (only for amino acids and short peptides), organic solvents, and denaturing agents.

The capability of several compounds to ion-pair with amino acids, peptides, and proteins is the basis for their selection as effective additives for modulating the selectivities of these analytes in CZE [5]. Selective ion-pair formation is expected to enlarge differences in the effective charge-to-hydrodynamic radius ratio of these analytes, leading to enhanced differences in their electrophoretic mobilities, which determine improved selectivity. Several diaminoalkanes, including 1,4-diaminobutane (putrescine), 1,5-diaminopentane (cadaverine), 1,3-diaminopropane, and *N,N,N′,N′*-tetramethyl-1,3-butanediamine (TMBD) can be successfully employed as additives for modulating the selectivity of peptides and proteins (see Figs. 1 and 2). Moreover, several anions, such as phosphate, citrate, and borate, which are components of the buffer solutions employed as the background electrolyte, may also act as ion-pairing agents influencing the electrophoretic mobilities of amino acids, peptides, and proteins

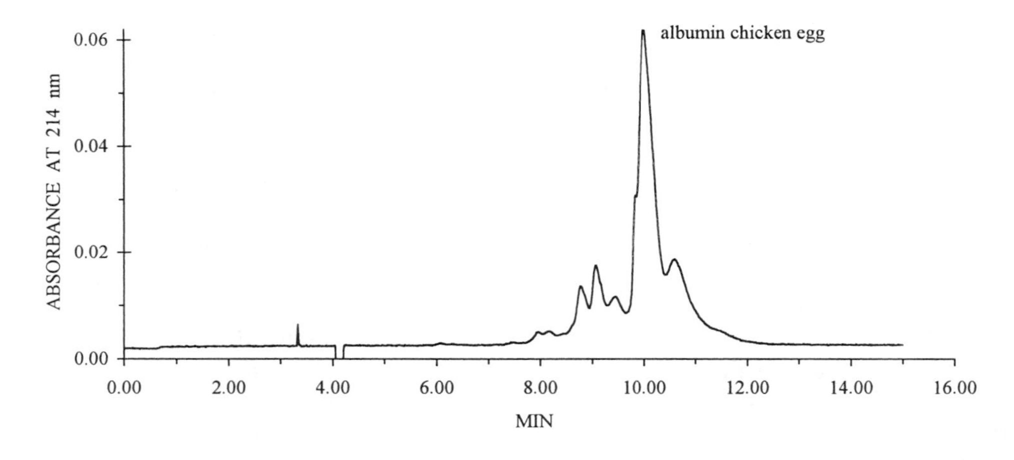

Fig. 1 Detection of microheterogeneity of albumin chicken egg by capillary zone electrophoresis. Capillary, bare fused-silica (50 μm $\times$ 37 cm, 30 cm to the detector); electrolyte solution, 25 m*M* Tris-glycine buffer containing 0.5% (v/v) Tween-20 and 2.0 m*M* putrescine; applied voltage, 20 kV; UV detection at the cathodic end.

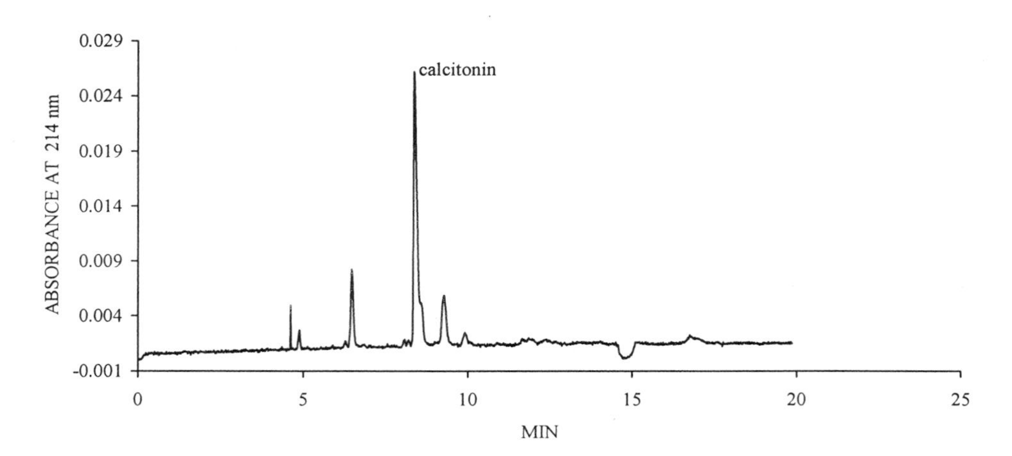

Fig. 2 Separation of impurities from a sample of synthetic human calcitonin for therapeutic use by CZE. Capillary, bare fused-silica (50 μm × 37 cm, 30 cm to the detector); electrolyte solution, 40 m*M N,N,N′,N′*-tetramethyl-1,3-butanediamine (TMBD), titrated to pH 6.5 with phosphoric acid; applied voltage, 15 kV; UV detection at the cathodic end.

and, hence, selectivity and resolution. Other cationic ion-pairing agents include the ionic polymers polydimethyldiallylammonium chloride and polybrene, whereas myoinositol hexakis-(dihydrogen phosphate), commonly known as phytic acid, is an interesting example of a polyanionic ion-pairing agent.

Surfactants have been investigated extensively in CE for the separation of both charged and neutral molecules using a technique based on the partitioning of the analyte molecules between the hydrophobic micelles formed by the surfactant and the electrolyte solution, which is termed micellar electrokinetic capillary chromatography (MECC or MEKC). This technique is widely used for the analysis of a variety of peptides and amino acids [6], but it is less popular for protein analysis [7]. The limited applications of MECC to protein analysis may be attributed to several factors, including the strong interactions between the hydrophobic moieties on the protein molecules and the micelles, the inability of large proteins to penetrate into the micelles, and the binding of the monomeric surfactant to the proteins. The result is that, even though the surfactant concentration in the electrolyte solution exceeds the critical micelle concentration, the protein–surfactant complexes are likely to be not subjected to partitioning in the micelles, as do amino acids, peptides, and other smaller molecules.

However, surfactants incorporated into the electrolyte solution at concentrations below their critical micelle concentration (CMC) may act as hydrophobic selectors to modulate the electrophoretic selectivity of hydrophobic peptides and proteins. The binding of ionic or zwitterionic surfactant molecules to peptides and proteins alters both the hydrodynamic (Stokes) radius and the effective charges of these analytes. This causes a variation in the electrophoretic mobility, which is directly proportional to the effective charge and inversely proportional to the Stokes radius. Variations of the charge-to-hydrodynamic radius ratios are also induced by the binding of nonionic surfactants to peptide or protein molecules. The binding of the surfactant molecules to peptides and proteins may vary with the surfactant species and its concentration, and it is influenced by the experimental conditions such as pH, ionic strength, and temperature of the electrolyte solution. Surfactants may bind to samples, either to the same extent [e.g., protein–sodium dodecyl sulfate (SDS) complexes], or to a different degree, which can enlarge differences in the electrophoretic mobilities of the separands.

In CGE, the separation is based on a size-dependent mechanism similar to that operating in polyacrylamide gel electrophoresis (PAGE), employing as the sieving matrix either entangled polymer solutions or gel-filled capillaries [8]. This CE mode is particularly suitable for analyzing protein complexes with SDS. The separation mechanism is based on the assumption that fully denatured proteins hydrophobically bind a constant amount of SDS (1.4 g of SDS per 1 g of protein), resulting in complexes of approximately constant charge-to-mass ratios and, consequently, identical electrophoretic mobilities. Therefore, in a sieving medium, protein–SDS complexes migrate proportionally to their effective molecular radii and, thus, to the protein molecular weight. Consequently, SDS–CGE can be used to estimate the apparent molecular masses of proteins, using calibration procedures similar to those employed in SDS–PAGE.

Continuous electrolyte solution systems are also employed in ACE [9], where the separation depends on the biospecific interaction between the analyte of interest and a specific selector or ligand. The molecules with bioaffinity for the analyte (the selector or ligand) can be incorporated into the electrolyte solution or can be immobilized, either to an insoluble polymer filled into the capillary or to a portion of the capillary wall. ACE is a useful and sensitive tool for measuring the binding constant of ligands to proteins and characterizing molecular properties of peptides and proteins by analyzing biospecific interactions. Examples of biospecific interactions currently investigated by ACE include molecular recognition between proteins or peptides and low-molecular-mass receptors, antigen–antibody complexes, lectin–sugar interactions, and enzyme–substrate complexes. ACE is also employed for the chiral separation of amino acids using a protein as the chiral selector.

Enantiomeric separations of amino acids and short peptides are performed using either a direct or the indirect approach [10]. The indirect approach employs chiral reagents for diasteromer formation and their subsequent separation by various modes of CE. The direct approach uses a variety of chiral selectors that are incorporated into the electrolyte solution. Chiral selectors are optically pure compounds bearing at least one functional group with a chiral center (usually represented by an asymmetric carbon atom) which allows sterically selective interactions with the two enantiomers. Among others, cyclodextrins (CDs) are the most widely chiral selectors used as additives in chiral CE. These are cyclic polysaccharides built up from D-(+)-glucopyranose units linked by α-(1,4) bonds, whose structure is similar to a truncated cone. Substitution of the hydroxyl groups of the CDs results in new chiral selectors which exhibit improved solubility in aqueous solutions and different chiral selectivity. Other chiral selectors include crown ethers, chiral dicarboxylic acids, macrocyclic antibiotics, chiral calixarenes, ligand-exchange complexes, and natural and semisynthetic linear polysaccharides. Chiral selectors are also commonly employed in combination with ionic and nonionic surfactants for enantiomeric separations of amino acids and peptides by MECC.

In discontinuous systems, the composition of the electrolyte solution is varied along the migration path with the purpose of changing one or more parameters responsible for the electrophoretic mobilities of the analytes. The discontinuous electrolyte solution systems employed in capillary isoelectric focusing (CIEF) have the function of generating a pH gradient inside the capillary tube in order to separate peptides and proteins according to their isoelectric points [11]. Each analyte migrates inside the capillary until it reaches the zone with the local pH value corresponding to its isoelectric point, where it stops moving as a result of the neutralized charge and consequent annihilated electrophoretic mobility. CIEF is successfully employed for the resolution of isoenzymes, to measure the isoelectric point (p*I*) of peptides and proteins, for the analysis of recombinant protein formulation, hemoglobins, human serum, and plasma proteins. Discontinuous electrolyte solution systems are also employed in capillary isotachophoresis (CITP), where the analytes migrate as discrete zones with an identical velocity between a leading and a terminating electrolyte solution having different electrophoretic mobilities. CITP finds large applications as an on-line preconcentration technique prior to CZE, MECC, and CGE. It is also employed for the analysis of serum and plasma proteins and amino acids [12].

The majority of amino acids and short peptides have no, or only negligible, UV absorbance. Detection of these analytes often requires chemical derivatization using reagents bearing UV or fluorescence chromophores. High detection sensitivity, reaching the attomolar (10^{-21}) mass detection limit can be obtained using fluorescence labeling procedures in combination with laser-induced fluorescence detection [13]. A variety of fluorescence and UV labeling reagents are currently employed, including *o*-phthaldehyde (OPA), fluorescein isothiocyanate (FITC), 1-dimethylaminonaphthalene-5-sulfonyl chloride (dansyl chloride), 4-phenylspiro[furan-2(3H)-1′phthalene (fluorescamine), and naphthalene dicarboxaldehyde (NDA). However, derivatization may reduce the charge-to-hydrodynamic radius ratio differences between analytes, making separations difficult to achieve. In addition, precolumn derivatization is not suitable for large peptides and proteins, due to the formation of multilabeled products. These problems can be overcome using postcolumn derivatization procedures.

Another, very attractive alternative is indirect UV detection [14]. This indirect detection procedure makes use of a UV-absorbing compound (or "probe"), having the same charge as the analytes, that is incorporated into the electrolyte solution. Displacement of the probe by the migrating analyte generates a region where the concentration of the UV-absorbing species is less than that in the bulk electrolyte solution, causing a variation in the detector signal. In the indirect UV detection technique, the composition of the electrolyte solution is of critical importance, because it dictates separation performance and detection sensitivity.

Probes currently employed in the indirect UV detection of amino acids include *p*-aminosalicylic acid, benzoic acid, phthalic acid, sodium chromate, 4-(*N*,*N*′-dimethylamino)benzoic acid, 1,2,4,-benzenetricarboxylic acid (trimellitic acid), 1,2,4,5-benzenetetracarboxylic acid (pyromellitic acid), and quinine sulfate. Several of these probes are employed in combination with metal cations and cationic surfactants, which are incorporated into the electrolyte solution as modifiers of the electro-osmotic flow.

Coupling mass spectrometry (MS) to capillary electrophoresis provides detection and identification of amino acids, peptides, and proteins based on the accurate determination of their molecular masses [15]. The most critical part of coupling MS to CE is the interface technique employed to transfer the sample components from the CE capillary column into the vacuum of the MS. Electrospray ionization (ESI) is the dominant interface which allows a direct coupling under atmospheric pressure conditions. Another distinguishing features of this "soft" ionization technique when applied to the analysis of peptides and proteins is the generation of a series of multiple charged, intact ions. These ions are represented in the mass spectrum as a sequence of peaks, the ion of each peak differing by one unit of charge from those of adjacent neighbors in the sequence. The molecular mass is obtained by computation of the measured mass-to-charge ratios for the protonated proteins using a "deconvolution algorithm" that transforms the multiplicity of mass-to-charge ratio signals into a single peak on a real mass scale. Obtaining multiple charged ions is actually advantageous, as it allows the analysis of proteins up to 100–150 kDa using mass spectrometers with an upper mass limit of 1500–4000 amu.

Concentration detection limits in CE–MS with the ESI interface are similar to those with UV detection. Sample sensitivity can be improved by using ion-trapping or time-of-flight (TOF) mass spectrometers. MS analysis can also be performed off-line, after appropriate sample collection, using plasma desorption–mass spectrometry (PD–MS) or matrix-assisted laser desorption–mass spectrometry (MALDI–MS).

References

1. S. Hjerten, *Chromatogr. Rev. 9*: 122 (1967).
2. I. Rodriguez and S. F. Y. Li, *Anal. Chim. Acta 383*: 1 (1999).
3. J. E. Wiktorowicz and J. C. Colburn, *Electrophoresis 11*: 769 (1990).
4. Kleindiest, G., C. G. Huber, D. T. Gjerde, L. Yengoyan, and G. K. Bonn, *Electrophoresis 19*: 262 (1998).
5. D. Corradini, *J. Chromatogr. B 699*: 221 (1997).
6. P. G. Muijselaar, K. Otsuka, and S. Terabe, *J. Chromatogr. A 780*: 41 (1998).
7. M. A. Strege and A. L. Lagu, *J. Chromatogr. A 780*: 285 (1997).
8. A. Guttman, *Electrophoresis 17*: 1333 (1996).
9. H. Kajiwara, *Anal. Chim. Acta 383*: 61 (1999).
10. H. Wan and L. G. Blomberg, *J. Chromatogr. A 875*: 43 (2000).
11. R. Rodriguez-Diaz, T. Wehr, and M. Zhu, *Electrophoresis 18*: 2134 (1997).
12. P. Gebauer and P. Bocek, *Electrophoresis 18*: 2154 (1997).
13. K. Swinney and D. J. Bornhop, *Electrophoresis 21*: 1239 (2000).
14. S. Hjerten, K. Elenbring, F. Kilar, J.-L. Liao, A. J. C. Chen, C. J. Siebert, and M.-D. Zhu, *J. Chromatogr. 403*: 47 (1987).
15. R. D. Smith, J. A. Loo, C. J. Barinaga, C. G. Edmonds, and H. R. Udseth, *J. Chromatogr. 480*: 211 (1989).

Danilo Corradini

Analysis of Alcoholic Beverages by Gas Chromatography

Introduction

Alcoholic beverages have been consumed by a significant range of worldwide population since the beginning of civilization until the present time. Therefore, there should be a great interest on consumption of beverages quality and, consequently, the usage of a suitable analytical technique to verify and control this desirable quality.

Alcoholic beverages are classified, in a general way, in fermented beverages (such as beer, wine, sake, etc.) and distilled ones (vodka, whisky, aguardente, tequila, cognac, liquors, etc.). The main volatile substances present in most alcoholic beverages belongs to the following chemical classes: alcohols (including ethanol, methanol, isobutanol, 3-methyl butan-2-ol, etc.), esters (such as ethyl acetate, methyl acetate, ethyl isobutyrate, isoamyl acetate, etc.), aldehydes (propanal, isobutanal, acetal, furfural, etc.), acids (acetic acid, propionic acid, butyric acid, etc.) and ketones (acetone, diacetyl, etc.).

Some of the substances are of greater concern than the others due to its relative quantities or to its flavored characteristic [1]. As an example, ethanol is the major compound in the group of alcohols being responsible for the formation of various other substances, such as acetaldehyde, resulting from ethanol oxidation, and it is the most abundant of the carbonylic compounds in distilled beverages. For the same reason, acetic acid is the major compound within its group, the carboxylic acids.

Fusel alcohols (e.g., 1-propanol and 3-methyl butan-2-ol) as well as ethanol are also important substances in the alcohols group contributing to the flavor of distilled beverages because their odor is very distinctive and characteristic. There is an enormous variety of substances in small concentrations belonging to the esters group. Even so, ethyl acetate corresponds to more than 50% of the esters within this group.

These compounds are responsible to important characteristics as smell and taste, in which the large fraction of these substances originates from fermentation or during beverages storage [1].

Because gas chromatography (GC) is an analytical technique in which separation and identification of volatile compounds occurs, it might be considered the best technique for this kind of sample [2].

Analysis by Gas Chromatography

Analysis of alcoholic beverages by gas chromatography have as their main objectives to investigate flavoring compounds and contaminants which might be intentional or occasional. Whereas the former ones (the flavors) are analyzed to control their favorable characteristics to the beverage, the adulterants have to be controlled due to their deleterious contribution. Adulteration includes the addition of certain compounds to enhance a desirable flavor. Because these compounds are usually added as a racemic mixture, their presence can be verified using a suitable chiral column for enantiomer separation [3]. On the other hand, occasional contaminants are substances originating in small quantities during beverage production and might be carcinogenic. The main source might be raw materials like grape, sugar cane, and so forth contaminated with pesticides [4] or as a result of the fermentation process itself [5,6].

Comments in each major part of the GC instrumentation as used for beverage analysis are presented.

Sample Introduction

Distilled Beverages

Generally, samples are injected in the chromatographic system without any dilution or pretreatment step, using the split mode (i.e., with sample division), which is suitable for the analysis of the major compounds in beverages. When the objective of analysis is the determination of compounds present in small quantities (μg/L), some extraction and/or concentration step is necessary, followed by the sample injection in the splitless mode (without sample division). This last sample introduction mode is usually combined with extraction techniques such as liquid–liquid extraction (LLE), solid-phase extraction (SPE) [4], and solid-phase microextraction (SPME) [7,8].

Fermented Beverages

Sample introduction is basically the same compared to the distilled one. Nevertheless, in many complex samples (i.e., some kind of wines), it is not advisable to inject them into the chromatograph without a pretreatment step.

Recently, SPME has provided many improvements as the cleanup step of complex samples, particularly for the analysis of volatile compounds by *headspace* techniques [8]. SPME is a solventless extraction and concentration technique which has advantages as a simple and economic technique that reduces health hazards and environmental issues.

Columns

Until 1960, all commercial chromatographic columns were packed in wide-bore tubing and separations had low resolution and low efficiency, taking a long time for a run to be completed. Since then, there have been significant improvements with the introduction of wall-coated open tubular (WCOT) columns, whose inner diameter was smaller than the packed ones are coated with a thin film of the liquid phase [9].

When capillary fused-silica columns arose, a large number of separations of complex samples have obtained success as a result of the higher number of plates (about 30 times over the packed columns in average).

Despite the efficient separations, it has been noticed that some low-boiling compounds of alcoholic beverages coeluted because of the use of a polar stationary phase. This column separates mainly based on the boiling temperatures of chemical substances, but separation becomes very difficult if there are some compounds with similar boiling temperatures. A polar stationary phase like poly(ethylene glycol) (PEG) is a better choice for this sort of problem because these separations are based on compound structures. In all cases, cross-linked or immobilized phases are recommended because they are more thermolabile and also resistant to most solvents. This is particularly important when splitless injection is used in combination with PEG-type phases because otherwise a severe column bleeding might be observed after ~220°C.

Detection Systems

The flame ionization detector (FID) is one of the most commonly used detectors in beverage analysis by GC, as it is suitable to most groups of compounds investigated in alcoholic beverages [9]. This occurs because almost all compounds of interest in such samples are able to burn in the flame, forming ions and producing a potential difference measured by a collector electrode.

In trace analysis of contaminant substances, one can use specific detectors for certain compounds, such as a nitrogen–phosphorus detector (NPD), thus gaining detection ability for nitrogenated and phosphorylated compounds; the electron-capture detector (ECD) shows excellent performance for chlorinated substances and the flame photometric detector (FPD) is the most widely used for sulfur-containing compounds.

Gas chromatography–mass spectrometry (GC–MS) combination has become one of the most important coupling in analytical chemistry used for the confirmation of results obtained by other detectors [9]. This technique is based on the fragmentation of the molecules that arrives into the detector. Ion formation occurs and they are separated by the mass/charge ratio (m/z) generally detected by a electron multiplier. Quantitative analysis can be realized through the single-ion monitoring (SIM) mode, where some characteristic ions are selected and monitored increasing the detection sensibility and selectivity. Another qualitative technique, gas chromatography–sniffing [10] is very much used for flavor analysis, despite discordance among researchers. In this case, the volatile substances from a extract are separated by GC, and as they leave the equipment through a specially designed orifice, a trained analyst is able to smell some of the substances and tentatively identify them.

More than 500 compounds have been found in concentrated flavor extracts in distilled beverages [1]. For this reason, there is an obvious necessity to find and optimize analytical techniques capable of investigating and to keep trading control of the alcoholic beverages. Among them, gas chromatography has been far from any other preferred tool for the analysis of alcoholic beverages.

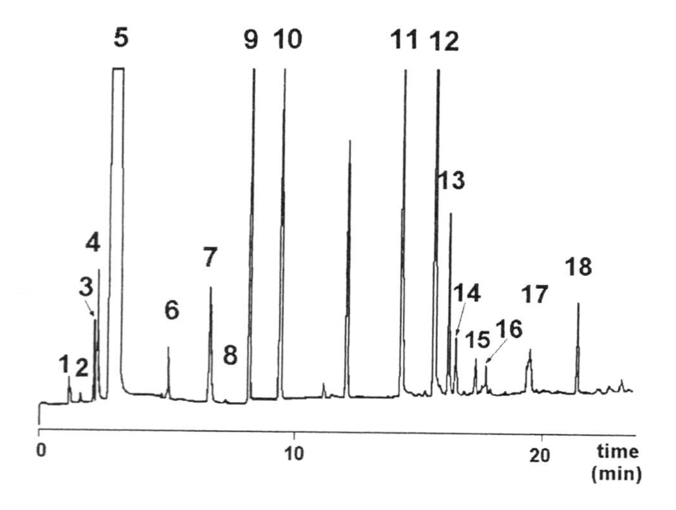

Fig. 1 Gas chromatographic analysis of white wine sample by fused-silica column poly(ethylene glycol) type (15.00 m × 0.53 mm × 1.00 μm). Chromatographic conditions: carrier gas: hydrogen (3.6 mL/min); column temperature: 40°C (4 min) → 8°C/min → 270°C; injector port temperature: 250°C; detector temperature: 300°C. Identity of the selected peaks: (1) acetaldehyde, (2) acetone, (3) ethyl acetate, (4) ethanol, (5) ethanol, (6) propanol, (7) isoamyl acetate, (8) *n*-butanol, (9) heptanone, (10) isoamyl alcohol, (11) acetic acid, (12) propionic acid, (13) isobutyric acid, (14) 2,3 butanediol, (15) 1,2 propanediol, (16) butyric acid, (17) isovaleric acid, (18) valeric acid.

Acknowledgments

Professor Lanças wishes to express his acknowledgments to FAPESP (Fundação de Amparo à Pesquisa do Estado de São Paulo), CNPQ (Conselho Nacional de Desenvolvimento Científico e Tecnológico), and CAPES (Coordenação de Aperfeiçoamento e Pessoal de Nível Superior) for financial support to our laboratory and a fellowship to Marcelo de Moraes.

References

1. L. Nykänen and I. Nykänen. Distilled beverages, in *Volatiles Compounds in Foods and Beverages*, H. Maarse (ed.), Marcel Dekker, Inc., New York, 1991, pp. 547–580.
2. F. M. Lanças and M. S. Galhiane, Fast routine analysis of light components of alcoholic beverage using large bore open tubular fused silica column, *Bol. Soc. Chil. Quim. 38*: 177 (1993).
3. W. A. König, new developments in enantiomer separation by capillary gas chromatography, in *Analysis of Volatiles Methods, Applications*, P. Schreier (ed.), Berlin, 1984, pp. 77–91.
4. A. Kaufmann, fully automated determination of pesticides in wine, *J. AOAC Int. 80*(6): 1302 (1997).
5. J. F. Lawrence, B. D. Page, and B. S. Conacher, the formation and determination of ethyl carbamate in alcoholic beverages, *Adv. Environ. Sci. Technol. 23*: 457 (1990).
6. K. Shiomi, Determination of acetaldehyde, acetal and other volatile congeners in alcoholic beverages, *J. High Resol. Chromatogr. 14*(2): 136 (1991).
7. J. Pawliszyn, *Solid Phase Microextraction — Theory and Practice*, Wiley–VCH, New York, 1997, pp. 1–247.

8. D. de la C. Garcia, M. Reichenächer, K. Danzer, C. Hurlbeck, C. Bartzsch, and K. H. Feller, analysis of wine bouquet components using headspace solid phase micro extraction–capillary gas chromatography, *J. High Resol. Chromatogr.* *21*(7), 373 (1998).
9. R. M. Smith, *Gas and Liquid Chromatography in Analytical Chemistry*, John Wiley & Sons, London, 1988, pp. 1–401.
10. N. Abbott, P. Etievant, D. Langlois, I. Lesschaeve, and S. Issanchou, Evaluation of the representativeness of the odor of beer extracts prior to analysis by GC eluate sniffing, *J. Agric. Food Chem.* *41*(5): 777 (1993).

Fernando M. Lanças
M. de Moraes

Analysis of Mycotoxins by TLC

Thin-layer chromatography (TLC) is a widely used analytical technique for many investigative purposes. Detection of mycotoxins by means of TLC has been in use for many years. Official methods of analysis often rely on these techniques for both identification and quantification of several mycotoxins [1].

Mycotoxins are natural toxins produced by fungi contaminating foods and feeds. They may be extremely toxic [e.g., aflatoxin (*Aspergillus flavus*), which is a potent carcinogen [2]. They may be found in plant-derived products, but also in the tissues of exposed animals, where they can accumulate and, thus, be a serious health hazard for human beings. Consequently, many analytical methods have been developed in order to assess the potential contamination of food derived either from plants or from food-producing animals. A third group of analytical techniques has been developed for diagnostic purposes in animals without any direct public health hazard. In this article, we will briefly review some of the techniques available to address these three goals.

Analysis of Plants and Plant-Derived Products for Mycotoxins

Analysis of plants may yield very interesting results, because they are the major carrier of fungi and of mycotoxins. The presence of a genus of fungus may or may not be associated with the production of mycotoxins, depending on the climatic conditions at the time of harvest, for instance [3]. Therefore, it is extremely interesting to have qualitative as well as quantitative methods for the determination of mycotoxins in these matrices. Numerous techniques have been developed to determine the nature and degree of contamination of plants with mycotoxins; it would not be possible to present all of them in a single article.

Extraction and cleanup procedures usually require solid-phase extraction based on commercially available C_{18} cartridges, for instance, after liquid extraction with common organic solvents (methylene dichloride, chloroform, acetonitrile [3–5]). This step appears to be necessary to remove most of the interfering components such as carotenoids or chlorophyls, which are highly abundant in plants. Table 1 gives a list of some of the mycotoxins which can be analyzed by TLC, together with analytical features (plate material, elution, detection/visualization, limit of detection). Of particular interest are the following products: aflatoxins B1, B2, G1, and G2, zearalenon, ochratoxin, and fumonisins B1 and B2 in maize. Recently, high-performance TLC (HPTLC) techniques have also been applied to forage samples commonly infected with an endophyte considered as symbiotic but responsible for disorders in animals [5]. These mycotoxins include lolitrems and ergot alkaloids.

Thin-layer chromatographic plates used are generally silica gel plates, although C_{18} reversed-phase TLC plates may occasionaly be used [4]. Elution may be monodimensional or bidimensional [3]. The former is more common. Recent automated gradient techniques appear promising for the simultaneous determination of several mycotoxins in a single sample. Solvents used for elution depend on the type of plate used. Most of the time, organic solvents (ethanol chloroform, acetone) are used. One feature of HPTLC is that it uses much less solvent than high-performance liquid chromatography (HPLC) and permits analysis of many samples in a very short time [5].

Detection may be based on several techniques; older systems used postelution derivatization [8] and observation of colored spots. It is more common now to use either

Table 1 Examples of HPTLC Methods for the Determination of Mycotoxins in Biological Samples

Mycotoxin	Plate	Matrix	Elution	Detection	Results	Ref.
Aflatoxins B1, G1, M1	Si60	Muscle, liver, kidney	1-Hexane/tetrahydrofuran/ethanol 2-$CHCl_3$/methanol	Fluorescence	LOD[a]: 0.01 μg/kg (B1, M1) LOD: 0.005 μg/kg (G1)	2
Fumonisin B1, B2	C18	Corn	Methanol/KCl 4% in water	Fluorescamin Fluorescence	LOD: 0.1 (B2), 0.5 (B1) μg/kg, recovery > 80%	4
Lolitrem	Si60	Forage	CH_2Cl_2/acetonitrile	UV (268 nm)	LOD: 0.1 mg/kg, recovery > 90%	5
Patulin	Si60	Fruit	1-Hexane/diethylether, 2-diethylether	UV (273 nm)	LOD: 40–100 μg/kg	7
Zearalenon	Si60	Corn	Toluene/ethylacetate/formic acid	Fluorescence	LOD: 2.6 μg/kg, recovery > 63%	10

[a]LOD = limit of detection.

ultraviolet (UV) [5] or even fluorescence [3, 4] because these techniques allow quantification of mycotoxin residues. More complex systems have been tested (computer imaging [7]) for particular applications, but these techniques are not applied to routine analysis of plants.

Analysis of Animal Tissues

Because animals may accumulate mycotoxins in their tissues, and considering the high toxicity of some products, it is a public health concern to have analytical techniques available for routine detection and quantitation of residues of these compounds in edible tissues of food-producing animals. Analysis of animal tissues may be difficult for several reasons:

1. Mycotoxins are poorly concentrated in many tissues or biological fluids (e.g., muscle or milk) and the analytical method employed should be sensitive enough to detect down to a few micrograms mycotoxins per kilogram sample or even less (aflatoxins in muscle samples, for instance [2]).
2. Tissues and organs may contain interfering substances on chromatograms or densitograms.

For extremely toxic compounds like aflatoxins, acceptable daily intakes (ADI) have been defined and it is, therefore, necessary to check suspected tissues to monitor residues at this level.

Thin-layer chromatography and HPTLC offer many advantages over other conventional methods (gas chromatography, HPLC), such as rapidity, simplicity, and sensitivity. However, it is usually necessary to extract and purify samples before spraying them onto TLC plates. In the case of aflatoxins, for instance, a commonly employed technique is based on liquid extraction by an organic solvent (chloroform), followed by purification on a silica gel column. The column has to be washed to elute interfering substances (with hexane and ether) and aflatoxins are eluted individually or all together with a specific combination of solvents (chloroform and acetone) [2].

If the purpose of the TLC technology is only to screen samples, the results may be merely qualitative. Older methods were usually based on the visual determination of dark spots on a bright fluorescent plate under UV light. Another use of TLC in the determination of mycotoxin residues is as preparative TLC, in which specific spots are visualized, scraped, and resolubilized for further analysis (Association of Official Analytical Chemistry).

More and more, however, quantitative analysis can be performed by means of scanners (UV, fluorescence), and HPTLC does not need to be completed by another analytical technique for the precise determination of residues. Today, most official methods of analysis for the detection of mycotoxins in foods are based on TLC technology.

Diagnostic Purposes

Mycotoxicoses may induce various pathological disorders in animals as well as in human beings. Considering the poor specificity of the signs observed and the very low concentrations of the toxic compounds in most biological tissues or fluids, it is necessary to be able to analyze, promptly and efficiently, biological samples to evaluate the risk of mycotoxicosis. The most common mycotoxins involved include aflatoxins, fumonisins, ochratoxin, zearalenon, and T2 toxin [9]. Analysis is usually based on foods and feeds (cereals, etc.). In swine, for instance, an epidemiological study conducted on feed samples detected ochratoxin (mean 58.3 μg/kg) and zearalenon (mean 30.3 μg/kg) in corn. These concentrations were associated with respiratory disorders and also infertility [10]. The advantage of TLC/HPTLC in such a situation is that it provides rapid and specific analysis of food samples at a low cost and, therefore, a reasonable cost/benefit ratio for breeders. If the diagnosis has to be confirmed on animal/human tissues or fluids, depending on the compound of interest, HPTLC techniques may also be available and convenient (see above).

Conclusion

Thin-layer chromatography and HPTLC offer many possibilites for the determination of mycotoxins in plant or animal samples. Plant samples usually contain higher concentrations of mycotoxins, but analysis of animal tissues may be necessary either to confirm a suspected mycotoxicosis or to detect potential residues for human food. Many official methods are available, based on TLC, and the recent development of HPTLC also offers many possibilities for the detection and quantitation of several mycotoxins in various biological samples.

References

1. R. Stubblefield, J. P. Honstead, and O. L. Shotwell, *J. Assoc. Off. Anal. Chem. 74*(6): 897 (1991).
2. A. Fernandez, R. Belio, J. J. Ramos, M. C. Sanz, and T. Saez, *J. Sci. Food Agric. 74*: 161 (1997).
3. S. I. Phillips, P. W. Wareing, A. Dutta, S. Panigrahi, and V. Medlock, *Mycopathology* 133: 15 (1996).
4. G. E. Rottinghaus, C. E. Coatney, and H. C. Minor, *J. Vet. Diagn. Invest. 4*(3): 326 (1992).
5. P. Berny, P. Jaussaud, A. Durix, C. Ravel, and S. Bony, *J. Chromol. 769*: 343 (1997).
6. Y. Liang, M. E. Baker, B. Todd Yeager, and M. Bonner Denton, *Anal. Chem. 68*: 3885 (1996).
7. P. F. Ross, L. G. Rice, R. D. Plattner, G. D. Osweiler, T. M. Wilson, D. L. Owens, H. A. Nelson, and J. L. Richard, *Mycopathology 114*(3): 129 (1991).
8. G. D. Osweiler, T. L. Carson, W. B. Buck, and G. A. Van Gelder, *Clinical and Diagnostic Veterinary Toxicology*, 3rd ed., Kendall/Hunt, Dubuque, 1988.
9. C. Ewald, A. Rehm, and C. Haupt, *Berlin. Münch. Tierärtzlich. Wochen. 104*(5): 161 (1991).
10. M. Dawlatana, R. D. Coker, M. L. Nagler, G. Bluden, and G. W. O. Oliver, *Chromatographia 47*: 215 (1998).

Philippe J. Berny

Analysis of Plant Toxins by TLC

Introduction

Analysis of plants is a vast and complex field of analytical chemistry. There is a constant need for new compounds or active ingredients for pharmaceutical or other interesting properties. Plant chemistry is such that a wide variety of compounds may be produced within the different organs. Toxins usually represent only a small fraction of the total organic matter of the plant. It is important, however, to be able to analyze and detect those toxins, especially when poisoning cases are suspected. Although vegetal toxins are extremely diverse in nature, a common feature among them is that they are heatunstable. Consequently, standard gas chromatographic (GC) or GC–MS (mass spectrometric) procedures cannot be routinely used to detect them, and the use of liquid chromatography appears of benefit.

Analysis of plant toxins is required in the following circumstances:

1. When plant poisoning is suspected in human beings or in animals: In this situation, the analyst must be able to separate the plant toxin from its plant matrix and/or from an animal tissue of fluid. The toxin is also diluted as compared with the plant product.
2. For research or development purposes: when a family of plants is under investigation and the presence of a toxin is suspected and not expected (therapeutic use).
3. When a family of toxins is wellknown, like the pyrrolizidine alkaloids [1], it is highly interesting to screen suspected plants for their presence when poisoning is suspected in animals and there may be

some residues in food, even in animals which did not present any disorder.

In this article, we will review some examples of these three areas of plant chemistry and see how thin-layer chromatography (TLC) or high-performance thin-layer chromatography (HPTLC) can fulfill the various requirements.

Investigation of Suspected Poisoning Cases

Poisoning by plants can occur with various animal species, including human beings. It obviously occurs most frequently in herbivores like cattle or sheep and examples of analytical investigation are more common in these species. Unfortunately, it is generally necessary to screen rumen content for plant toxins and this matrix is extremely rich in organic constituents, including natural pigments. Careful and adapted cleaning steps are necessary. When alkaline or acidic substances are to be determined, pH-based liquid–liquid separation may be used.

Our first example is based on a very severe acute poisoning case with yew trees (*Taxus* sp.). These trees contain a highly toxic group of toxins known as taxins. Cattle or sheep usually do not eat the leaves or branches of yew trees because of their bitter taste. Unfortunately, bitterness tends to disappear when the leaves are dessicated, but toxicity remains and animals may eat enough to become deadly sick. One published example mentioned poisoning in 43 cattle, with 17 dead before any treatment or diagnosis could be attempted [2]. Diagnosis may rely on the epidemiological evidence of poisoning (cut branches) or on necropsy (branches in the rumen), but these elements may not necessary be conclusive and analytical techniques may represent the only way of confirming a tentative diagnosis. The published method for yew tree analysis in cattle relied on the analysis of rumen contents and identification of taxol. This alcohol is specific of the *Taxus* genus, although it may not be the toxic substance, but it is easily obtained from commercial dealers, whereas taxins have to be extracted and purified.

The development of scanners for HPTLC also offers better potential and we have further developed the method of Panter et al. [2] with ultraviolet (UV) scanning (at 238 nm) and quantification prior to derivatization (confirmatory analysis). Eventually, diagnosis relies on the determination of Taxins (primarily taxin B) in the rumen content. As an example, Fig. 1 provides a densitogram of a rumen content with identification of taxin peaks and the solid-phase UV spectrum of this compound in a case of confirmed yew tree poisoning. The same technique could be applied to determine the most toxic part of the plant, the effect of season on the toxin content, and so forth. For example, it was found that the leaves contained about 0.03% taxol, while the stem and twigs contained around 0.001% and 0.0006%, respectively [2].

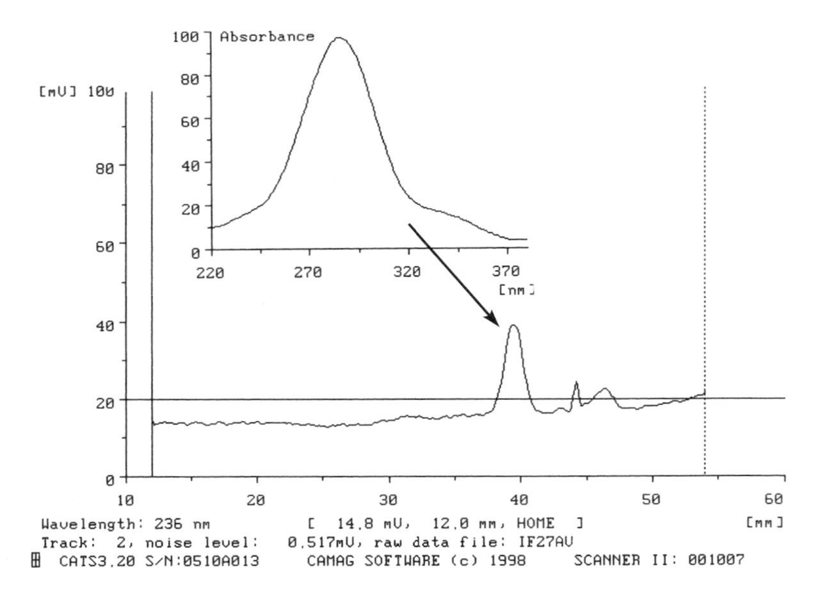

Fig. 1 Rumen content with taxin B (arrow) and its UV spectrum.

Identification of Toxins in Plant Screenings for Research or Development Purposes

Some plants are famous for the presence of toxic substances. One major group of toxicants involved is the pyrrolizidine alkaloids. These substances induce severe poisoning manifested by emaciation, anemia, and skin lesions which may develop over months or years and may eventually result in the death of the poisoned animals. These compounds represent both a toxic and economic threat in some areas of the world, (e.g., in the Himalayans [1]). Screening of plants may be highly desirable to prevent these losses. Winter et al. [1] developed screening methods based either on high-performance liquid chromatography (HPLC) or on TLC. They used silica gel plates with a mixture of ethylacetate, acetone, ethanol, and ammonia (5/3/1/1) and postchromatographic derivatization with *o*-chloranil and Erlich's reagent. This technique was only qualitative but gave results in accordance with reference HPLC techniques. With these two techniques, the authors analyzed over 350 samples (various plants, plant parts, and locations) to determine the presence and amount of alkaloids.

Another example of the use of TLC in research is given by Ma et al. [3], who used TLC as a taxonomic tool to classify plants. Their example was based on the lycopodium alkaloids of plants from the Lycopodiacae family. The TLC technique used was qualitative and also relied on postchromatographic derivatization with Dragendorff's reagent [3]. It should be stated that plant toxins are usually recognized by means of qualitative TLC and visualization based on postchromatographic derivatization procedures. Common solutions rely on common reactions of alkaloids, and the use of Dragendorff's reagent is one of these solutions (combined mixture of tartaric acid, bismuth nitrate, and potassium iodide), and vanillin reagent enables the detection of amines and amino acids.

The Chinese pharmacopoea uses a wide range of plants and herbal medicines and Chinese scientists have been publishing methods and techniques for decades, identifying therapeutic substances or toxic ingredients of traditional remedies. The interested reader should refer preferentially to the *Journal of Chinese Herbal Medicine* to find analytical methods.

Identification of new compounds may start with TLC analysis of plant extracts. For instance, Jakupovic et al. [4] isolated and further identified several diterpenes from *Euphorbia segetalis*. Similarly, chamomile essential oil (*Chamomilla reticulata*) was analyzed with 11 different development systems and the authors discussed both the most efficient (separation power) and the ideal way they are to be used to identify an unknown component in such a complex mixture, using the minimum number of TLC systems [5]. This area of work still has to be investigated, considering the wide variety of the vegetal reign and of potential plant toxins.

Detection of Residues in Food

This part is certainly the least developed, to our knowledge. It is important to remember, however, that an animal may ingest a toxic plant and may also survive. If this is the case, and provided this animal or its production are intended for human consumption, one should be able either to analyze tissues and fluids for toxin residues or to monitor fluids (plasma, milk, urine) to determine whether this animal or its productions can be considered safe for human consumption. There are very limited examples of such occurrences. In our laboratory, we analyzed muscular tissues after a confirmed yew tree poisoning case (*Taxus baccata*). Three animals did not display any significant trouble except for a transient depression, which resolved itself after 12 h. These animals were butchered and muscle samples were analyzed for taxin residues, as they were known to be exposed to it. Our analytical technique (extraction in alkalinized methylene dichloride) followed by TLC development based on a modification of a published technique (9685), showed that the muscle samples contained between 0.012 and 0.015 μg/g taxin (wet weight). The presence of taxin in muscle tissues had never been previously reported in cattle after moderate poisoning. Based on this result, the meat was not considered edible. This example is, simply, to illustrate that residues of plant toxins in food-producing animals should be part of a research or development protocol whenever possible.

Conclusion

Plant toxins represent one of the most important areas of analytical development and the few techniques related herein should only be considered as mere examples of the numerous possibilities of TLC in this field. With the development of densitometry and multiple development systems, the separation power of TLC and its quantitative potential are increasing as well. Considering the usual thermal instability of many plant toxins and their high polarity, HPTLC certainly offers one of the most powerful technologies for the detection, identification, and quantification of plant toxins.

References

1. H. Winter, A. A. Seawright, H. J. Noltie, A. R. Mattocks, R. Jukes, K. Wangdi, and J. B. Gurung, *Vet. Rec. 134*: 135 (1994).
2. K. E. Panter, R. J. Molyneux, R. A. Smart, L. Mitchell, and S. Hansen, *J. Amr. Vet. Med. Assoc. 202*: 1476 (1993).
3. X. Q. Ma, S. H. Jiang, and D. Y. Zhu, *Biochem. Syst. Ecol. 26*: 723 (1998).
4. J. Jakupovic, F. Jeske, T. Morgenstern, J. A. Marco, and W. Berendsohn, *Phytochemistry 47*: 1583 (1998).
5. M. Medic-Saric, G. Stanic, Z. Males, and S. Saric, *J. Chromatogr. A 776*: 355 (1997).

Philippe J. Berny

Antibiotics: Analysis by TLC

Antibiotics are an extremely important class of human and veterinary drugs. Chemically, they constitute a widely diverse group with different functions and modes of operation. They can be derived from living organisms or obtained synthetically. However, all of them exhibit antibacterial properties (i.e., either inhibit the growth of, or kill, bacteria).

Penicillin, the first natural antibiotic produced by genus *Penicillium*, discovered in 1928 by Fleming, as well as sulfonamides, the first chemotherapeutic agents discovered in the 1930s, lead a long list of currently known antibiotics. Besides β-lactams (penicillins and cephalosporines) and sulfonamides, the list includes aminoglycosides, macrolides, tetracyclines, quinolones, peptides, polyether ionophores, rifamycins, linkosamides, coumarins, nitrofurans, nitro heterocytes, chloramphenicol, and others.

In principle, antibiotics should eradicate pathogenic bacteria in the host organism without causing significant damage to it. Nevertheless, most antibiotics are toxic, some of them even highly. The toxicity of antibiotics for humans is not only due to medical treatment but also to absorption of those drugs along with contaminated food. In modern agricultural practice, antibiotics are administered to animals, both for treatment of diseases and for prophylaxis, as well as to promote growth as feed or water additives. When proper withdrawal periods are not observed, unsafe antibiotic residues or their metabolites may be present in edible products (e.g., in milk, eggs, and meat). Some of them, like penicillins, can cause allergic reactions in sensitive individuals. Therefore, monitoring antibiotic residues should be an important task for government authorities.

There are many analytical methods for determining antibiotics in body fluids and food. They can be based on microbiological, immunochemical, and physicochemical principles. The most popular methods belonging to the latter group are chromatographic ones, mainly liquid chromatography, including high-performance liquid chromatography (HPLC) and thin-layer chromatography (TLC).

High-performance liquid chromatography offers high sensitivity and separation efficiencies. However, it requires sophisticated equipment and is expensive. Usually, before HPLC analysis, tedious sample pretreatment is necessary, such as protein precipitation, ultrafiltration, partitioning, metal chelate affinity chromatography (MCAC), solid-phase dispersion (SPD), or solid-phase extraction (SPE). Generally, the sample cleanup procedures used before TLC separation are the same as for HPLC. Nevertheless, they can be strongly limited in the case of screening TLC or when the plates with a concentrating zone are applied.

Thin-layer chromatography is less expensive and less complicated than HPLC, provides high sample throughput, and usually requires limited sample pretreatment. However, the method is generally less sensitive and selective and offers poor resolution. Some of these problems can be solved by high-performance thin-layer chromatography (HPTLC) or forced-flow planar chromatography (FFPC). Lower detection limits can also be achieved using an autosampler for injection, applying special techniques of development and densitometry as a detection method, and/or spraying the plate after development with appropriate reagents.

There is also a possibility of coupling TLC with mass spectrometry (MS). Then, TLC can reach selectivity, sensitivity, and resolution close to those of HPLC.

Thin-layer chromatography stripped of the abovementioned attributes may still serve as a screening method (i.e., one which establishes the presence or absence of antibiotics above a defined level of concentration). Screening TLC methods show sensitivity similar to microbiological assays, which are the most popular screening methods, applied for controlling antibiotic residues in food in many countries. Thin-layer chromatography–bioautography (TLC–B) is one of the TLC screening methods. The developed TLC plates are placed on or immersed in a bacterial growth medium which has been seeded with an appropriate bacteria strain. The locations of zones of growth inhibition provides the information about antibiotic residues.

In relation to extremely diverse nature of antibiotics, a variety of different separation and detection modes is used in analytical practice. Short characteristics and some general rules of separation for the most popular classes of antibiotics are presented next.

Penicillins

The basic structure of penicillins is a thiazolidine ring linked to a β-lactam ring to form 6-aminopenicillanic acid, the so-called "penicillin nucleus." This acid, obtained from *Penicillium chrysogenum* cultures is a precursor for semisynthetic penicillins produced by attaching different side chains to the "nucleus." The most widely used stationary phase for analysis of penicillins is silica gel, but reversed-phase (RP) or cellulose plates have also been employed. It is advantageous to add acetic acid to the mobile phase and/or spotting acetic acid before the sample injection in order to avoid the decomposition of β-lactams on silica gel. RP phases usually contain pH 5–6 buffer and organic solvent(s). The most popular detection is bioautography and ultraviolet (UV) densitometry, often coupled with spraying with appropriate reagents.

Cephalosporines

Cephalosporines are derived from natural cephalosporin C produced by *Cephalosporinum acremonium*. They possess a cephem nucleus (7-amino-cephalosporanic acid) substituted with two side chains. They are commonly divided into three classes differing in their spectra and toxicity. Cephalosporines can be analyzed both by normal and reversed-phase TLC or HPTLC; hence, more efficient separation is obtained on silanized gel than on bare, untreated silica gel. Mobile phases are polar and similar to those used for penicillins. Acetic acid or acetates are very often components of solvents for normal phase (NP) TLC, the ammonium acetate–acetic acid buffer for RP TLC. All cephalosporines can be detected at 254 nm. The detection limit can be diminished by applying reagents such as ninhydrin, iodoplatinate, chloroplatinic acid, or iodine vapor. Alternative to UV detection is bioautography with, for instance, *Neisseria catarrhalis*.

Aminoglycosides

Aminoglycosides consist of two or more amino sugars joined via a glycoside linkage to a hexose nucleus. Streptomycin was isolated in 1943 from *Streptomyces griseus*, then others were discovered in different *Streptomyces* strains. Aminoglycosides are particularly active against aerobic microorganisms and against *Tubercle bacillus*, but because of their potential ototoxicity and nephrotoxicity, they should be carefully administered. Aminoglycosides, due to their extremely polar, hydrophilic character, are analyzed mostly on silica gel. Polar organic solvents (methanol, acetone, chloroform) mixed with 25% aqueous ammonia are the most popular mobile phases. Because the majority of aminoglycosides lack UV absorption, they must be derivatized by spraying or dipping after development with, for instance, fluram, vanillin, or ninhydrin solutions. Bioautography with *Bacillus subtilis, Sarcina lutea*, and *Mycobacterium phlei* is also possible.

Macrolides

Macrolides are bacteriostatic antibiotics composed of a macrocyclic lactone ring and one or more deoxy sugars attached to it. The main representative of the class, erythromycin, was discovered in 1952 as a metabolic product of *Steptomyces erythreus*. Now, erythromycin experiences its renaissance because of its high activity against many new, dangerous bacteria such as *Campylobacter* or *Legionella*. The macrolide antibiotics group is still being expanded due to the search for macrolides of pharmacokinetic properties better than erythromycin. Separation of macrolides is performed on silica gel, kieselguhr, cellulose, and reversed-phase layers. Silica gel and polar mobile phases are very frequently applied, usually with the addition of methanol, ethanol, ammonia, sodium, or ammonium acetate. Because of the absence of chromophore groups, bioautography or postchromatographic derivatization is used, mainly charring by spraying with acid solutions (e.g., anisaldehyde–sulfuric acid–ethanol) and heating.

Tetracyclines

Tetracyclines, consisting of a octahydronaphthacene skeleton, are "broad-spectrum" antibiotics produced by

Streptomyces or obtained semisynthetically. They can be separated both by RP and NP TLC. Cellulose, kiselguhr, or silica gel impregnated with EDTA or Na_2EDTA can be used. Impregnation is necessary due to the very strong interaction of tetracyclines with hydroxyl groups. Also, mobile phases, both for RP or NP TLC, should contain chelating agents such as Na_2EDTA, citric, or oxalic acid. Tetracyclines give fluorescent spots, which can be detected by UV lamp, fixed at 366 nm or by densitometry. Spraying with reagents, for instance with Fast Violet B Salt solution, provides lower detection levels. Tetracyclines can also be detected by fast-atom bombardment–mass spectrometry (FAB–MS) and bioautography.

Quinolones

Nalidixic acid, discovered casually in 1962, was the first member of this class, although of rather minor importance. In the 1980s, synthetic fluoroquinolones were developed and became valid antibiotics with broad spectra and of good tolerance. Quinolones may be analyzed on silica gel plates, preferably impregnated with Na_2EDTA or K_2HPO_4. Multicomponent organic mobile phases are employed, usually with the addition of aqueous solutions of ammonia or acids. Densitometry or fluorescence densitometry are preferred detection methods, sometimes preceded by postchromatographic derivatization.

Peptides

Peptide antibiotics are composed of the peptide chain of amino acids, D and L, covalently linked to other moieties. Most peptides are toxic and are poorly absorbed from the alimentary tract. Peptide antibiotics are difficult to analyze in biological and food samples, as they are similar to matrix components. They can be separated on silica gel, amino silica gel, and silanized silica gel plates. A variety of mobile phases are applied, from a simple one like chloroform–methanol to a multicomponent one like *n*-butanol–butyl acetate–methanol–acetic acid–water. Bioautographic detection can be employed with *Bacillus subtilis* and *Mycobacterium smegmatis* or densitometry as well as fluorescence densitometry after spraying the plate with reagents such as ninhydrin or fluram.

Besides typical antibiotics analysis, focused on the separation of antibiotics belonging to one or different classes, there are many examples of diverse TLC applications such as the following:

1. Purity control of antibiotics
2. Purification of newly discovered antibiotics before further testing
3. Examining stability and breakdown products of antibiotics in solutions and dosage forms
4. Analysis of antibiotic metabolites
5. Studying interactions of antibiotics with cell membrane or human serum albumin
6. Examining reactions of antibiotics with different compounds
7. Separation of antibiotics derivatives, obtained in the process of searching for new antibiotics
8. Quantitation of antibiotics by densitometry without elution with a solvent
9. Thermodynamic study of the retention behavior of antibiotics
10. Determining hydrophobicity parameters of antibiotics by RP TLC
11. Applying some antibiotics as stationary or mobile-phase additives

Suggested Additional Reading

Barker, S. A., and C. C. Walker, *J. Chromatogr. 624*: 195 (1992).
Bobbitt, D. R. and K. W. Ng, *J. Chromatogr. 624*: 153 (1992).
Boison, J. O., *J. Chromatogr. 624*: 171 (1992).
Hoogmartens, J. (ed.), *J. Chromatogr. A 812* (1998) (special issue).
Lambert, H. P. and F. W. O'Grady, *Antibiotics and Chemotherapy*, 6th ed., Longman Group UK Ltd., London, 1992.
Sherma, J., and B. Fried (eds.), *Handbook of Thin-Layer Chromatography*, 2nd ed., Marcel Dekker, Inc., New York, 1996.

Irena Choma

Application of Capillary Electrochromatography to Biopolymers and Pharmaceuticals

Introduction

The Capillary Electrochromatography Technique

Capillary electrochromatography (CEC) has grown considerably over the past few years, due to the developments in column technology and the appearance of several articles demonstrating the high efficiencies possible with this technique [1]. The literature has shown that there can be numerous applications for this technology, which was not possible with the earlier separation methods. Tsuda published an article that discussed the CEC technique in detail [2]. The technique itself is a derivative of high-performance capillary electrophoresis (HPCE) and high-performance liquid chromatography (HPLC), where the separations are performed using fused-silica tubes of 50–100 μm inner diameter (i.d.), that are packed with either a monolithic packing or small (3 μm or smaller) silica-based particles [3–5]. The packing is similar to the conventional HPLC; however, the mobile phase is driven by electro-osmosis, which results from the electric field applied across the capillary rather than by pressurized flow. The mobile phase is made up of aqueous buffers and organic modifiers [e.g., acetronile, (ACN)]. An electroosmotic flow (EOF) of up to 3 mm/s can be generated. It is a plug flow, where the linear velocity is independent of the channel width and there is no column back-pressure [4]. Partitioning or adsorption of the neutral analyte occurs in the same way as in HPLC. The analytes are separated while moving through the column with the EOF. Charged solutes have additional electrophoretic mobility in the applied electric field; therefore, the separation occurs by electrophoresis and partitioning. The selectivity in analysis of the charged analytes is increased by electromigration of the sample molecules. The flat flow profile results in a more efficient radial mass transport compared to the parabolic laminar flow in pressure-driven liquid chromatography (LC), and this results in a significant enhancement in separation performance and shorter analysis times [1,5,6]. The capillaries can be made shorter to offer the same plate count as HPLC, therefore reducing the back-pressure. The packing material is smaller compared to HPLC, so with the high electric fields, the efficiency of this technique is very high (up to about 500,000 plates/m) [1], the peaks are sharp, the resolution is high, and the process is highly selective. An article by Angus et al. [7] demonstrated the separation efficiencies of 200,000–260,000 plates/m that were obtained by CEC and were reproducible from column to column for structurally related, polar neutral compounds of pharmaceutical relevance. The sample capacity in CEC is 10–100 times higher than that of capillary electrophoresis (CE), and this means that more sample volume can be placed on the CEC column to give better sensitivity. The high capacity comes from the high column loadability that results from the stationary phase's retentive mechanism [8]. The absence of additives and predominantly organic mobile phases make CEC better suited for use in mass spectrometry (MS). In fact, nonaqueous CEC is already being practiced by analysts [8]. A recent article by Hansen and Helboe gives a detailed study of the possibility of using CEC to replace gradients or ion-pairing reagents. The group optimized the separation of six nucleotides using a background analyte consisting of 5 mM acetic acid, 3 mM triethylamine (TEA), and 98% acetonitrile and a C_{18} 3-μm column. This was accomplished in half the time taken for a similar separation in HPLC [9].

A variation of gradient CEC is pressurized-flow CEC or PEC (pressurized flow electrochromatography). A pump forms the gradient and then allows part of this pressurized flow to pump the mobile phase through the packed bed. In this way, one can perform isocratic or gradient CEC with part of the mobile-phase driving force being pumped, part electrophoretic and part electroosmotic flows [1,8,10].

Detection of Biomolecules and Pharmaceuticals in CEC

There are many different types of detectors used for pharmaceutical applications in CEC. They vary from indicating just the presence of a sample [fluorescence (FL)], to giving some qualitative information about a sample [photodiode array UV/vis detection (PDA)], to absolute sample determination of the analyte (MS). The methods can be on-column, off-column, and end-column. With on-column, the solutes are detected while still on the capillary, in off-column, the solute is transported from the outlet of the capillary to the detector, and end-column is done with the detector placed right at the end of the capillary. Some modes of detection used in CEC are as follows:

1. UV/vis absorbance detection is widely used in capillary electrophoresis. Absorptivity depends on the chromophore (light-absorbing part) of the solute,

the wavelength of the incident light, and the pH and composition of the run buffer. A photodetector measures light intensities and the detector electronics convert this into absorbance [11].

2. Fluorescence detection. When light energy strikes a molecule, some of that energy may be given off as heat and some as light. Depending on the electronic transitions within a molecule, the light given off may be fluorescent or phosphorescent [12]. Fluorescence occurs when an electron drops from an excited singlet to the ground state, as opposed to phosphorescence, which occurs when an electron's transition is from an excited triplet to the ground state.
3. LIF detection, such as argon ion, helium–cadmium [5], and helium–argon lasers, can be used for this detection method. The criterion for choosing the laser is that the wavelength should be at or near the excitation maxima for the solute to be determined. The higher the power of the laser, the higher the intensity and the peak height and the laser's ability to focus the beam to a small spot.
4. MS detection [4]. This is the only detector that has high sensitivity and selectivity and can be used universally; therefore, the increased interest in interfacing this technology with CEC compared to other detection methods. It can detect all solutes that have a molecular weight within the mass range of the MS. In the selected ion-monitoring mode, it detects only solutes of a given mass, and in the total ion chromatogram mode, it detects all the solutes within a given mass range.

Current applications of CEC use on-column UV or laser-induced fluorescence detection; however, for UV, the path length is quite short, which limits sensitivity, although bubble, Z-shaped, and high-sensitivity cells have helped to improve detection limits. However, UV and fluorescence are only good for samples that fluoresce and absorb light or are amenable to derivatization with fluorescing or absorbing chromophores. These detectors impose difficult cell volumes and sample size limits if high separation efficiencies are to be realized, and they are very expensive. All of these drawbacks are nonexistent for MS techniques, which are expensive but provide more structural information and high sensitivity and appear to have the greatest overall potential [8].

According to an issue of *LC–GC* [8], combining CEC with detection techniques such as MS, MS–MS and inductively coupled plasma (ICP)–MS are easier to accomplish, as the flow rates are at nanoliter per minute levels. Analysts must add makeup solvent after the capillary separation for certain ionization methods, and, because it is added later, users can select solvents that are more compatible with the detection technique. CEC mobile phases have a high organic solvent content that is more amenable to MS. Also, the low CEC flow rates means less maintenance and downtime for MS source cleaning.

In the references to the application of CEC to biopolymers, most of the work discusses CEC–electrospray ionization(ESI)–MS, much less to direct CEC–UV/FL methods. However, much of the work has evolved from the use of commercially available, prepacked capillaries, such as C_{18} or ion exchange or a mixed mode containing both ion exchange and reversed phase. There are very few articles that have actually attempted to develop new phases specifically for biopolymers.

When using MS, the actual CEC conditions never really need to be fully optimized because the MS accomplishes the additional resolution and specific identification, as needed. The specific mobile-phase conditions in CEC–MS may be quite different than for CEC–UV/FL or HPLC, and thus optimization of CEC–MS conditions will be somewhat different than for CEC–UV/FL. This would include, just as for LC–MS, the use of volatile organic solvents and organic buffers, low flow rates, no void volumes, or loss of resolution in the CEC–MS interface and the usual interfacing requirements already developed and optimized for CE–MS [10,13–23].

Few descriptions of quantitation have been reported so far. Most of the literature is qualitative by nature, simply demonstrating suitable, if not fully optimized, experimental conditions that provide evidence of the presence of certain biopolymers and their high resolution from other components in that particular sample. Absolute quantitation and validation needs to be developed and fully optimized for CEC, to become a more valuable and applicable separation mode for biopolymers.

Separation of Proteins

Capillary electrochromatography can accomplish high plate counts, as mentioned earlier; this means a high peak capacity (number of peaks that can be fitted into a typical separation time for a given length of column), therefore highly complex materials can be separated. The implication of better peak capacities is a better resolution of the peaks in a complex analyte. Because of the frequent overlap of peaks due to components in a complex sample, it is difficult to demonstrate peak purity with other methods. It is possible in CEC to quantitatively determine the presence of a particular analyte. CEC techniques have produced the separation of the enantiomers of amino acids [24–27]. This is done with limited use of solvents, buffer additives, salts, organics, chiral species, packing materials, and total time of analysis. Other groups have successfully utilized gradient elution to separate mixtures of dansylated amino acid mixtures on the ODS (octadecylsilane) stationary phase [24]. Also, microprocessor control of pressure flow and voltage, automated sample injection, automated data collection, automated capillary switching, and the ability to interface with a variety of detection instrumentation make CEC an appealing technique for protein separation and peptide mapping.

Proteins and peptides are water-soluble complex molecules that are composed of amino acids linked by peptidic and disulfide bonds. Proteins are really just larger peptides of higher molecular weight, and antibodies are larger proteins of specific conformations, shape, size and immunogenicity, together with antigenic recognition properties [28–31]. The type, number, and sequence of amino acids in the chain determine the chemical characteristics of a peptide. The amino acid sequence determines the electrophoretic properties of the peptide. In addition to the amine terminus of the sequence, the amine and the guanidine residues of lysine and arginine are the main carriers of the positive charges, and the negative charge contribution is associated with the carboxylic acid terminus and the acidic groups of the aspartic and glutamic acids. The isoelectric and isoionic points of the peptide are their important characteristics in electrophoresis. These points are similar in peptides but not identical; the isoelectric point is determined by the given aqueous medium, whereas the isoionic point is related to the interactions with protons. The relation of the electrophoretic mobility of the peptides and their relative molecular weight is described by Offord's equation:

$$\mu_{\text{rel}} = \frac{Z}{[\sqrt[3]{(M^2)}]}$$

where μ_{rel} is the relative mobility, Z is the total net charge, and M is the molar mass in gram per mole. Calculation of the net charge cannot be done easily from the pK values of the acidic and basic groups for large peptides, but additional factors such as conformational differences, primary sequence, chirality, and so forth need to be considered.

The popular methods of analysis of proteins currently are HPLC, HPCE, and MS. However, due to the complexity of proteins, LC approaches show a single, broad, ragged peak, which indicates that the method is unable to resolve the individual species [1]. In CEC, the success of the protein separation requires that the capillary packing material meet certain properties. Depending on the ionic characteristics of the biopolymers, pH-dependent "ideal" packings would be either reversed-phase (RP) or ion-exchange chromatography (IEC) or a combination of both [1,32–46]. There are several references that detail the possible application of size-exclusion chromatographic (SEC) packings in CEC, but these have mainly been applied to synthetic organic polymers and much less to biopolymers [47–50]. Regardless of which packing is actually utilized, it should contain a stationary (bonded, not coated) phase that can successfully interact with the biopolymers, as in RP–HPLC, and prevent any unwanted silanol interactions with the underlying silica or ionic sites. It must also provide additional or programmable EOF, besides that from the uncoated fused-silica capillary walls. Perhaps an ideal packing would combine a cationic-exchange material (cationic-exchange chromatography (CIEC)] together with RP, in order to allow separations based on RP (hydrophobicity) alone, a combination of RP and IEC, or just IEC alone, mobile phase (buffer) dependent. Also, that packing, be it single or mixed mode, should prevent unwanted biopolymer (e.g., peptide amino groups) interaction with the support, such as amine–silanol hydrogen-bonding in RP–HPLC for amine containing analytes (e.g., pharmaceuticals, peptides, and proteins). Additional articles on open tubular CEC (OT–CEC) applications, where a coating is applied on the inner surface of the capillary as in capillary gas chromatography (GC) have appeared [51–54]. There are articles on packed-bed CEC, where there is a real packing in the capillary [55,56]. Also, then there are methods that employ just CEC, without any additional, pressure-driven flow [38,39], as well as pressurized CEC or PEC, with additional pressurized flow [4,51–53]. There is also electro-HPLC that utilizes gradient elution with an applied voltage, but it is mainly conventional HPLC with some voltage applied sporadically or continuously during the HPLC separation [54].

CEC of Biopolymers (Proteins, Peptides, and Antibodies)

The majority of the applications of CEC for biopolymers have dealt with peptides, of varying sizes and complexity, utilizing different modes of CEC (OT–CEC, conventional isocratic CEC, gradient CEC, PEC, and others). The following accounts are listed according to the work of different authors in the field.

Palm and Novotny applied the polymeric gel beds (monoliths) for peptide resolutions in CEC [5]. The peptide separation used the above packing beds with 29% C_{12} as the ligand. Additional CEC conditions are indicated in Fig. 1, which depicts the separation of a series of tyrosine-containing peptides, with detection at 270–280 nm. In this particular study, peptide elution patterns were very sensitive to changes in pH and ACN concentrations. A gradient elution technique, not employed here, would have been more appropriate for such samples of peptides

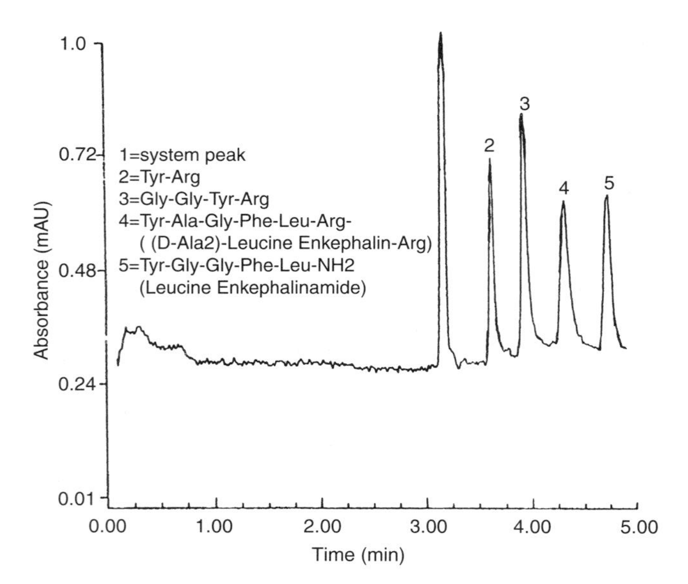

Fig. 1 Isocratic electrochromatography of peptides in a capillary filled with a macroporous polyacrylamide–polyethylene glycol matrix, derivatized with a C_{12} ligand (29%) and containing acrylic acid. Conditions: mobile phase, 47% acetonitrile in a buffer; voltage, 22.5kV (900 V/cm), 7 μm; sample concentration, 4–10 mg/mL; detection, UV absorbance at 270 nm; other conditions are described in Ref. 5. (From Ref. 5; reproduced with permission of the authors and the American Chemical Society.)

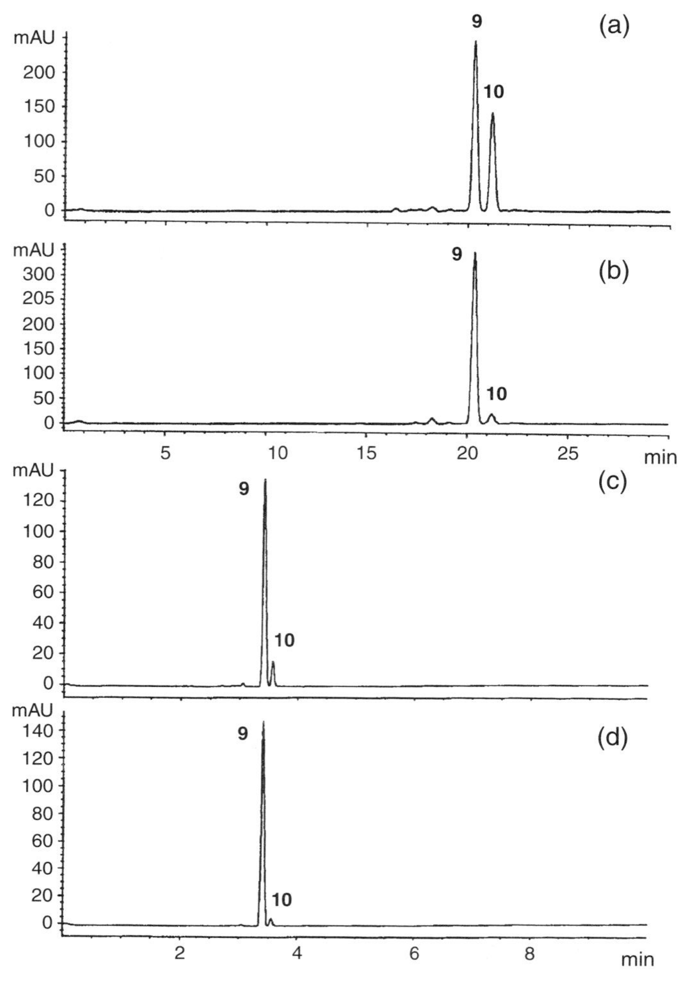

Fig. 2 Separation of synthetic, protected tetrapeptide intermediates: (9) *N*-methyl C- and N-protected tetrapeptide; (10) non-*N*-methyl C- and N-protected tetrapeptide. The structures of these compounds is proprietary information and consequently cannot be disclosed. Detection wavelength of 210 nm with a 10-nm bandwidth and a 1-s rise time. Electrochromatography was performed on a 250 mm × 50 μm i.d. spherisorb ODS-1 packed capillary using an acetonitrile-Tris (50 mmol/L, pH 7.8) buffer 80:20 v/v mobile phase, capillary temperature of 15°C, and an electrokinetic injection of 5 kV/15 s. (a) Synthetic mixture of protected tetrapeptides 9 and 10. Efficiency values of 124,000 and 131,000 plates/m were obtained for analytes 9 and 10, respectively. (b) Chromatogram of synthetically prepared 9, the presence of residual nonmethylated tetrapeptide (10) can be seen. (c) Chromatogram of synthetically prepared 9, spiked with 10% of the nonmethylated tetrapeptide (10). Efficiency values of 83,000 and 101,000 plates/m were obtained for analytes 9 and 10, respectively. (d) Chromatogram of synthetically prepared 9, the presence of residual nonmethylated tetrapeptide (10) can be clearly seen at the 3% level. Additional conditions are indicated in Ref. 38. (From Ref. 38; reproduced with permission of the authors and John Wiley and Sons, Inc.)

having small differences in their constitution. Attempts to elute protein samples were unsuccessful with these particular gel matrices, perhaps due to the high hydrophobicity of the packings [5].

Euerby et al. reported the separation of an N-methylated, C- and N-protected tetrapeptide from its nonmethylated analog, (Fig. 2) [38]. These separations utilized a Spherisorb ODS-1, 3-μm packing material, without pressure-driven flow (true CEC using a commercially available CE instrument), and an ACN buffer. Using nonoptimized, nonpressurized CEC conditions (non-PEC, non-pressure-driven CEC), separation of the two tetrapeptides could be achieved in a run time of 21 min with efficiency values of 124,000 and 131,000 plates/m. In comparison, when a pressurized CE (buffer reservoirs and capillary were pressurized, with pressure-driven flow of buffer) system was used, separation of the components was achieved within 3.5 min. According to Euerby et al. the separation of these two peptides using a pressure-driven HPLC gradient analysis took 30 min and gave comparable peak area results. Although this study illustrated the improved efficiency of both nonpressurized CEC and pressurized CEC over HPLC, no reasonable conclusions can be drawn from this work.

The attempts that have been made to utilize true chemometric optimization of operating conditions in CEC are unclear in most of the studies done utilizing CEC. This has been done for many years in GC and HPLC, as well as in CE, but there are no obvious articles that have appeared which have utilized true chemometric software approaches to optimization in CEC [57–59]. It is not clear that any true method optimization has been performed or what analytical figures of merit were used to define an optimized set of conditions for biopolymer analysis by CEC. It is also unclear as to why a specific stationary phase (packing) was finally selected as the optimal support in these particular CEC applications for biopolymers. In the future, it is hoped that more sophisticated optimization routines, especially computerized chemometrics (expert systems, theoretical software, or simplex/optiplex routines) will be employed from start to finish.

The coupling of ESI and MS with a pressurized CEC system (PEC) has been shown to separate peptides [4]. This particular study of Schmeer et al. utilized a commercial reversed phase silica gel, Gromsil ODS-2, 1.5-μm packing material, already utilized in capillary HPLC for peptide separations. It was never made perfectly clear why this particular packing material was selected or why PEC was selected for MS interfacing over conventional, isocratic CEC conditions. It is possible that the EOF alone with this packing material was insufficient to elute all peptides in a reasonable time frame and, thus, pressurized flow was introduced. No gradient elution PEC was demonstrated in this particular study. A mixture of enkephalin methyl ester and enkephalin amide was separated using the packed capillary column (Fig. 3). The coupling of these two methods showed enhanced sensitivity and detectability. Like the Euerby study, this offered little insight into the capabilities of CEC to separate peptides; however, the study does provide a nice example of a peptide separation based on chromatographic and electrophoretic separation mechanisms, probably occurring simultaneously. This report also described the ability of easily interfacing CEC and PEC with ESI–MS.

The coupling of an MS with CEC or PEC provides several advantages. With the capillary columns of 100 μm inner-diameter (i.d.), flow rates of 1–2 L/min are obtained, which are ideal for electrospray MS [4]. No interface like a liquid sheath flow is required and the sintered silica gel frits allow direct coupling of the packed capillary columns without additional transfer capillaries. The spray is therefore formed directly at the outlet side of the column. Verheij et al. carried out the first coupling of a pseudoelectrochromatography system to a fast-atom bombardment (FAB)–MS in 1991 [6]. However, this required transfer capillaries that caused a loss in efficiency, which was also a problem with other experimentations with this technique.

Lubman's group published several papers on the PEC–MS system [51–54]. Reverse-phase open tubular columns (RP–OTC), which were prepared by a sol–gel process, were coated with an amine that enhanced the EOF in an acidic buffer solution and reduced the nonspecific adsorption between the peptides and the column wall. A six-peptide mixture was separated to baseline within 3 min using this system coupled to an on-line ion-trap storage–time-of-flight mass spectrometer (ITS–TOFMS). A full-range mass detection speed of 8 Hz was used in all these experiments, which was rapid to maintain the high efficiency and ultrafast separation. A high-quality total ion chromatogram could be obtained with only a couple of femtomoles of peptide samples, due to the high-duty cycle of the MS and the column path-length-independent and concentration-sensitive feature of the ESI process. The concentration limit of detection was also improved to about $1 \times 10^{-6}M$ because of the preconcentration capability of the reversed-phase CEC. A tryptic digest of horse myoglobin was successfully separated within 6 min on the gradient CEC system. The use of the MS increased the resolving power of this system by clearly identifying the coeluting components.

Another article by Wu et al. [52] dealt with a PEC coupled to an ion-trap storage/reflectron TOFMS (RTOFMS) for the analysis of peptide mixtures and protein digests. Taking advantage of the EOF, a high separation efficiency has been achieved in PEC due to a relatively flat flow profile and the use of smaller packing materials. With columns only 6 cm long, a tryptic digest of bovine cytochrome-*c* was successfully separated in about 14 min by properly tuning the applied voltage and the supplementary pressure. A relatively complex protein digest (tryptic digest) of chicken albumin gave 20 peaks (resolved) in the total ion current chromatogram in 17 min (Fig. 4). The sample concentrations were also on the order of about $1 \times 10^{-5}M$. The detector increased the resolving power of PEC by unambiguously identifying coeluting components. The CEC was directly interfaced to the MS via an ESI, which provided the molecular-weight information of the protein digest products and structural information via MS–MS (Table 1). The device uses a quadrupole ion trap as a front-end storage device, which converts a continuous electrospray beam for TOF analysis. The storage property of the ion trap provides ion integration for low-intensity signals, whereas the nonscanning property of the TOFMS provides high sensitivity. A description of the MS is provided in an article by Wu et al. [54]. According to an article by Verheij et al., problems that were encountered earlier, like formation of bubbles,

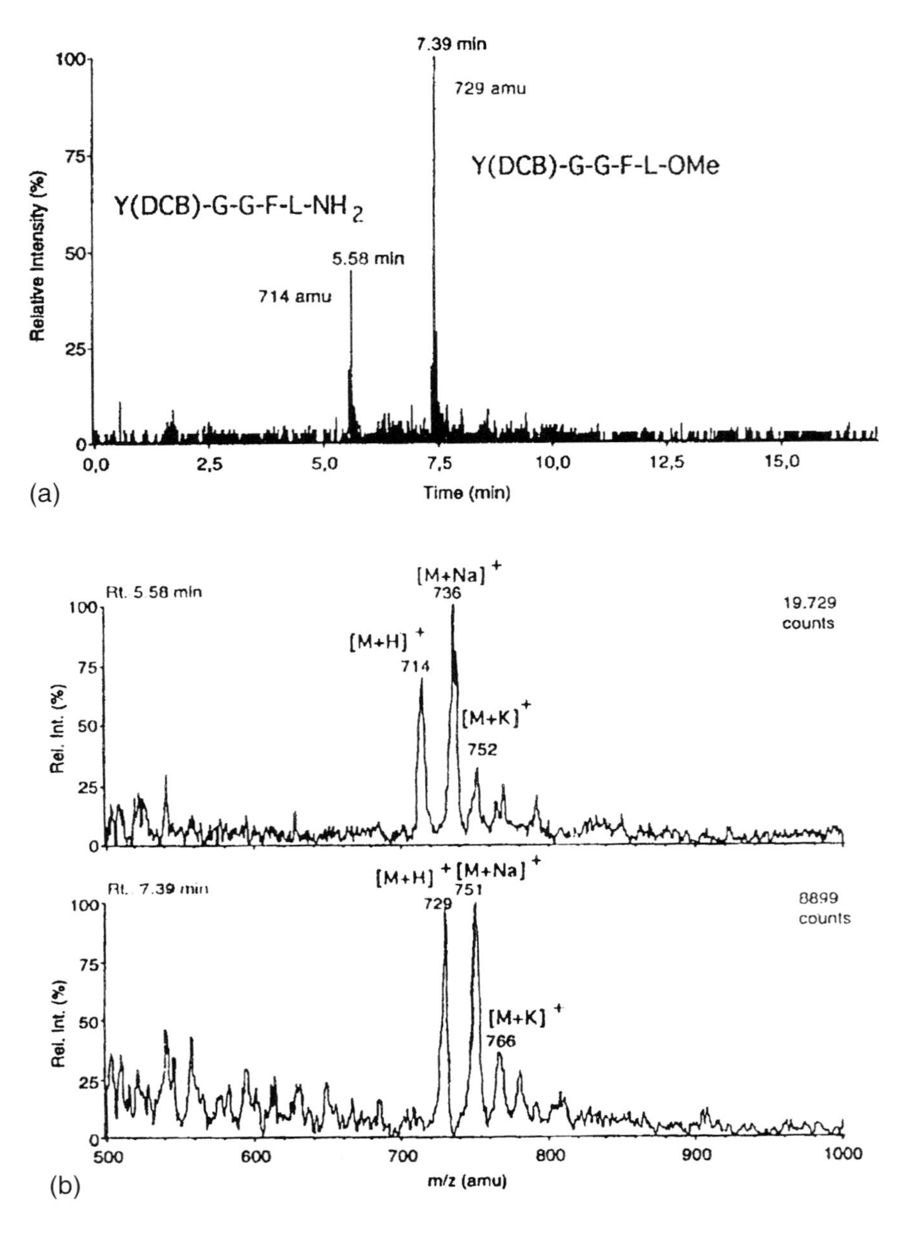

Fig. 3 Interfacing of pressure-driven CEC (PEC) for the separation of two simple peptides, enkephalin methyl ester (5.58 min) and enkephalin amide (7.39 min). (a) Extracted mass chromatogram of m/z 714 and 729 for the on-line peptide separation. Specific operating conditions are indicated in Ref. 4. (b) Mass spectra taken from the chromatographic peaks in (a), illustrating true M_r and the presence of M+H, M+Na, and M+K cations at appropriate m/z (amu) values. (From Ref. 4; reproduced with permission of the authors and the American Chemical Society.)

have been overcome by using liquid junctions to apply the electric field over the column [60].

A recent review article by the Lubman group points out that there are some serious advantages in using an open tubular column (OTC) for CEC as compared with packed-bed CEC [54]. OTCs with inner diameters around 10 μm have been found to have a smaller plate height when compared to packed columns. This is due to the lack of band-broadening effects associated with the presence of packing materials and end-column frits. OTC capillaries

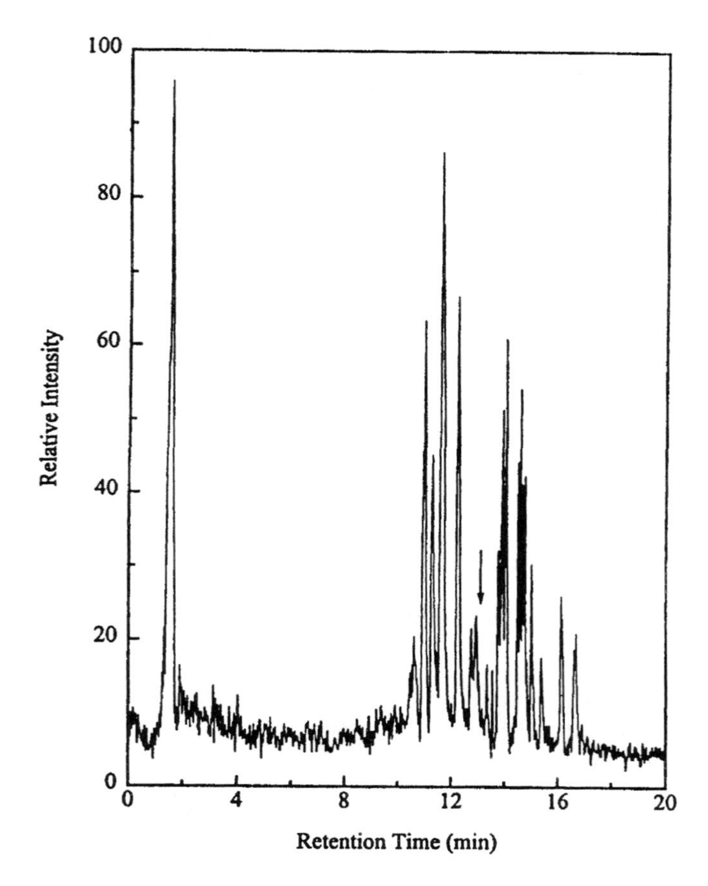

Fig. 4 The total ion chromatography (TIC) of the separation of a tryptic digest of chicken ovalbumin with a sample injection amount of 12 pmol corresponding to the original protein [52]. Column length, 6 cm. Conditions: 20 min, 0–40% acetonitrile gradient: 1000 V applied voltage with a 40-bar supplementary pressure. (From Ref. 52; reproduced with permission of the authors and the American Chemical Society.)

do not require end frits. High concentration sensitivity is another advantage of OTCs, as columns with very small dimensions are used. The small diameters of OTCs allow for the use of a higher voltage in CEC, without significant Joule heating. OTCs can also often provide more rapid separations than packed columns, by eliminating intraparticle diffusion, which is an important elimination for ultrafast separations in packed columns. However, there are some grave difficulties involved in using OTCs, perhaps because of the real difficulties with sample injection and detection. The injection volume of OTCs is in the low nanoliter or even picoliter range. The very small inner diameters of most OTCs make optical detection difficult, but they are very compatible with a concentration-sensitive detection method, such as ESI–IT–TOFMS. With peptide mixtures, however, gradient elution CEC, with or without pressure-driven flow, is almost required over isocratic or step-gradient methods, because small changes in the mobile phase composition results in large changes in peptide retention times.

In a later study, Pesek et al. reported the separation of other proteins using a diol stationary phase [61–64]. The use of a diol stationary phase should result in a surface that is more hydrophilic than a typical alkyl-bonded moiety, like C_{18} or C_8. The overall results showed significant variations in retention times due to differences in solute-bonded phase interactions. Other factors, such as pH, could also influence this interaction, due to its influence on charge and protein conformations. Combining all these factors in the separation of peptides and proteins provides an experimentalist with many decisions to be made in the optimized experimental conditions to be used. Other chemical modifications of etched fused silica need to be studied in order to provide a better understanding of their interactions with proteins and peptides, as well as other classes of biopolymers.

Conclusions

At the present time, although there are several applications of CEC–PEC to biopolymer classes, these are to be considered only preliminary and not necessarily fully optimized in all possible parameters. At times, significant improvements in peak shape, plate counts, resolutions, efficiency, and the time of analysis can be realized. However, final optimizations of these separations have not been realized or possible. Some workers have utilized pressurized flow to solve the problems of obtaining reasonable EOF without silanol–analyte interaction; however, this does not solve the problem. It just forces the analyte to elute and approaches electro-HPLC, rather than true CEC. There are real differences between electro-HPLC, PEC and CEC that need to be recognized. There does not, in general, seem to have been any serious attempt to utilize any chemometric software approaches in CEC–PEC for biopolymer separation optimizations or rationale for doing so. At this time, packings are simply used because they were on the shelf in a laboratory or commercially available and not necessarily because they were really the best for protein–peptide separations in PEC–CEC. There remains a need for research-oriented column choices from commercial vendors to avoid the need to pack capillaries in-house with commercial HPLC supports.

Table 1 Comparison of Calculated and Measured Tryptic Peptides of Chicken Ovalbumin from PEC–MS

No.	Tryptic peptides	Calculated mass[a]	Determined mass[a,b]	Sequence
1	1–16	1709.0	1709.6	GSIGAASMEFCFDVFK
4, 5	47–55	1080.2	1079.7	DSTRTQINK
5	51–55	602.7	602.9	TQINK
6, 7	56–61	781.0	781.4	VVRFDK
7	59–61	408.5	408.4	FDK
10	105–110	779.8	780.1	IYAEER
11	111–122	1465.8	1466.3	YPILPEYLQCVK
12	123–126	579.7	579.7	ELYR
13	127–142	1687.8	1687.5	GGLEPINFQTAADQAR
16	182–186	631.7	631.6	GLWEK
16, 17	182–189	996.1	995.9	GLWEKAFK
17	187–189	364.4	364.5	AFK
18	190–199	1209.3	1209.0	DEDTQAMPFR
20	219–226	821.9	821.7	VASMASEK
21	227–228	277.4	277.6	MK
23	264–276	1581.7	1581.3	LTEWTSSNVMEER
24, 25	277–279	405.5	405.4	KIK
26	280–284	646.8	646.8	VYLPR
26, 27	280–286	924.2	924.4	VYLPRMK
27, 28	285–290	813.0	813.1	MKMEEK
28	287–290	535.6	535.5	MEEK
30	323–339	1773.9	1774.2	ISQAVHAAHAEINEAGR
31	340–359	2009.1	2008.5	EVVGSAEAGVDAASVSEEFR
32	360–369	1190.4	1190.2	ADHPFLFCIK
33	370–381	1345.6	1345.3	HIATNAVLFFGR
33, 34	370–385	1750.1	1749.5	HIATNAVLFFGRCVSP
34	382–385	404.5	404.4	CVSP

[a] Average masses.
[b] Average of all charge states observed.
Source: Ref. 52; reproduced with permission of the authors and the American Chemical Society.

References

1. I. S. Krull, K. Mistry, and R. Stevenson, *Am. Lab. 16A* (August 1998).
2. T. Tsuda, *Anal. Chem. 59*: 521 (1987).
3. J. A. Olivares, N. T. Nguyen, C. R. Yonker, and R. D. Smith, *Anal. Chem. 59*: 1230 (1987).
4. K. Schmeer, B. Behnke, and E. Bayer, *Anal. Chem. 67*: 3656 (1995).
5. A. Palm and M. V. Novotny, *Anal. Chem. 69*: 4499 (1997).
6. E. R. Verheij, U. R. Tjaden, W. A. M. Niessen, and J. van der Greef, *J. Chromatogr. 554*: 339 (1991).
7. P. D. A. Angus, E. Victorino, K. M. Payne, C. W. Demarest, T. Catalano, and J. F. Stobaugh, *Electrophoresis 19*: 2073 (1998).
8. R. E. Majors, *LC–GC* Magazine *16*(2): 96 (1998).
9. S. H. Hansen and T. Helboe, *J. Chromatogr. A 836*: 315 (1999).
10. J. P. Landers (ed.), *CRC Handbook of Capillary Electrophoresis — Principles, Methods, and Applications*, 2nd ed., CRC Press, Boca Raton, FL, 1997.
11. D. R. Baker, *Capillary Electrophoresis*, Techniques in Analytical Chemistry Series, John Wiley & Sons, New York, 1995.
12. D. M. Hercules, *Fluorescence and Phosphorescence Analysis: Principles and Applications*, Interscience, New York, 1966, p. 19.
13. R. L. Cunico, K. M. Gooding, and T. Wehr, *Basic HPLC and CE of Biomolecules*, Bay Bioanalytical Laboratory, Richmond, CA, 1998.
14. Cs. Horvath and J. G. Nikelly (eds.), *Analytical Biotechnology: Capillary Electrophoresis and Chromatography*, ACS

Symposium Series Vol. 434, American Chemical Society, Washington, DC, 1990.
15. S. F. Y. Li, *Capillary Electrophoresis: Principles, Practice and Applications*, Elsevier Science, Amsterdam, 1992.
16. P. D. Grossman and J. C. Colburn (eds.), *Capillary Electrophoresis — Theory and Practice*, Academic Press, San Diego, CA, 1992.
17. P. G. Righetti (ed.), *Capillary Electrophoresis in Analytical Biotechnology*, CRC Series in Analytical Biotechnology, CRC Press, Boca Raton, FL, 1996.
18. P. Camilleri (ed.), *Capillary Electrophoresis: Theory and Practice*, CRC Press, Boca Raton, FL, 1993.
19. R. A. Mosher and W. Thormann, *The Dynamics of Electrophoresis*, VCH, Weinhein, 1992, Chap. 7.
20. K. D. Altria and M. M. Rogan. *Introduction to Quantitative Applications of Capillary Electrophoresis in Pharmaceutical Analysis, A Primer*, Beckman Instruments, Inc., Fullerton, CA, 1995.
21. R. Weinberger and R. Lombardi, *Method Development, Optimization and Troubleshooting for High Performance Capillary Electrophoresis*, Simon and Schuster Custom Publishing, Needham Heights, MA, 1997.
22. B. L. Karger and Wm. Hancock (eds.), *High Resolution Separation and Analysis of Biological Macromolecules*, Methods in Enzymology Series Vol. 270, Part A, Fundamentals, Academic Press, San Diego, CA, 1996.
23. K. D. Altria (ed.), *Capillary Electrophoresis Guidebook, Principles, Operation and Applications*, Methods in Molecular Biology, Humana Press, Totowa, NJ, 1996.
24. I. S. Lurie, R. P. Meyers, and T. S. Conver, *Anal. Chem. 70*: 3255 (1998).
25. P. Sandra, A. Dermaux, V. Ferraz, M. M. Dittman, and G. Rozing., *J. Micro. Separ. 9*: 409 (1997).
26. A. Dermaux, P. Sandra, M. Ksir, and K. F. F. Zarrouck., *J. High Resolut. Chromatogr. 21*: 545 (1998).
27. D. Li, H. H. Knobel, S. Kitagawa, A. Tsuji, H. Watanabe, M. Nakshima, and T. Tsuda, *J. Micro. Separ. 9*: 347 (1997).
28. J. R. Mazzeo and I. S. Krull, *Capillary Electrophoresis — Technology*, N. Guzman (ed.), Marcel Dekker, Inc., New York, 1993, Chap. 29.
29. J. R. Mazzeo, J. Martineau, and I. S. Krull, *CRC Handbook of Capillary Electrophoresis: Principles, Methods, and Applications*, J. P. Landers, (ed.), CRC Press, Boca Raton, FL, 1994, Chap. 18.
30. X. Liu, Z. Sosic, and I. S. Krull, *J. Chromatogr. B 735*: 165 (1996).
31. I. S. Krull, J. Dai, C. Gendreau, and G. Li, *J. Pharm. Biomed. Anal. 16*: 377 (1997).
32. Capillary Electrochromatography, Symposium, San Francisco, CA, organized by the California Separations Science Society, San Francisco, August 1997.
33. Royal Society of Chemistry Analytical Division, Northeast Region, Chromatography and Electrophoresis Group, Symposium on New Developments and Applications in Electrochromatography, University of Bradford, Bradford, U.K. December 3, 1997.
34. T. Tsuda (ed.), *Electric Field Applications in Chromatography, Industrial and Chemical Processes*, VCH, Weinheim, 1995.
35. M. M. Dittmann, K. Weinand, F. Bek, and G. P. Rozing, *LC/GC Mag. 13*(10): 800 (1995).
36. M. M. Dittmann and G. P. Rozing, *J. Chromatogr. A 744*: 63 (1996).
37. G. Ross, M. Dittmann, F. Bek, and G. Rozing, *Am. Lab.* (March 1996), p. 34.
38. M. R. Euerby, D. Gilligan, C. M. Johnson, S. C. P. Roulin, P. Myers, and K. D. Bartle, *J. Micro. Separ. 9*: 373 (1997).
39. M. R. Euerby, C. M. Johnson, K. D. Bartle, P. Myers, and S. C. P. Roulin, *Anal. Commun. 33*: 403 (1996).
40. M. M. Robson, M. G. Cikalo, P. Myers, M. R. Euerby, and K. D. Bartle, *J. Micro. Separ. 9*: 357 (1997).
41. J. H. Miwaya and M. S. Alesandro, *LC/GC Mag. 16*(1): 36 (1998).
42. R. E. Majors, *LC/GC Mag. 16*(1): 12 (1998).
43. I.H. Grant, *Capillary Electrochromatography*, K. D. Altria (ed.), Methods in Molecular Biology Vol. 52, Humana Press, Totowa, NJ, 1996, Chap. 15.
44. M. R. Euerby, C. M. Johnson, and K. D. Bartle, *LC/GC Int.* (January 1998), p. 39.
45. M. G. Cikalo, K. D. Bartle, M. M. Robson, P. Myers, and M. R. Euerby, *Analyst 123*: 87R (1998).
46. W. Wei, G. Luo, and C. Yan, *Am. Lab. 20C* (January 1998).
47. E. C. Peters, K. Lewandowsk, M. Petro, J. M. J. Frechet, and F. Svec, IN HPLC 98, St. Louis, MO, May 1998.
48. E. C. Peters, K. Lewandowski, M. Petro, F. Svec, and J. M. J. Frechet, *Anal. Commun. 35*: 83 (1998).
49. E. C. Peters, M. Petro, F. Svec, and J. M. J. Frechet, *Anal. Chem. 69*: 3646 (1997).
50. E. Venema, J. C. Kraak, H. Poppe, and R. Tijssen, *Chromatographia 48*(5/6): 347 (1998).
51. J. T. Wu, P. Huang, M. X. Li, M. G. Qian, and D. M. Lubman, *Anal. Chem. 69*: 320 (1997).
52. J. T. Wu, P. Huang, M. X. Li, M. G. Qian, and D. M. Lubman, *Anal. Chem. 69*: 2908 (1997).
53. J. T. Wu, P. Huang, M. X. Li, M. G. Qian, and D. M. Lubman, *Anal. Chem. 69*: 2870 (1997).
54. J. T. Wu, P. Huang, M. X. Li, M. G. Qian, and D. M. Lubman, *J. Chomatogr. A 794*: 377 (1998).
55. C. Yang and Z. El Rassi, *Electrophoresis 19*: 2061 (1998).
56. M. Zhang and Z. El Rassi, *Electrophoresis 19*: 2068 (1998).
57. R. J. Bopp, T. J. Wozniak, S. L. Anliker, and J. Palmer, *Pharmaceutical and Biomedical Applications of Liquid Chromatography*, C. M. Riley, W. J. Lough, and I. W. Wainer (eds.), Progress in Pharmaceutical and Biomedical Analysis Vol. 1, Pergamon/Elsevier Science, Amsterdam, 1994, Chap. 10.
58. J. W. Dolan and L. R. Snyder, *Am. Lab. 50* (May 1990).
59. L. R. Snyder, J. J. Kirkland, and J. L. Glajch, *Practical HPLC Method Development*, 2nd ed., John Wiley & Sons, New York, 1997, Chap. 10.
60. E. R. Verheij, U. R. Tjaden, W. A. M. Niessen, and J. van der Greef, *J. Chromatogr. 712*: 201 (1995).
61. J. J. Pesek, M. T. Matyska, J. E. Sandoval, and E. J. Wil-

liamsen, *J. Liq. Chromatogr. Related Technol. 19*(17/18): 2843 (1996).
62. J. J. Pesek, M. T. Matyska, and L. Mauskar, *J. Chromatogr. A 763*: 307 (1997).
63. J. J. Pesek and M. T. Matyska, *J. Chromatogr. A 736*: 255 (1996).
64. J. J. Pesek and M. T. Matyska, *J. Chromatogr. A 736*: 313 (1996).

Ira S. Krull
Sarah Kazmi

Applications of Evaporative Light-Scattering Detection in HPLC

Introduction

High-performance liquid chromatography (HPLC) is mainly carried out using light absorption detectors as ultraviolet (UV) photometers and spectrophotometers (UVD) and, to a lesser extent, refractive index detectors (RID). These detectors constitute the main workhorses in the field [1]. The sensitive detection of compounds having weak absorption bands in the range 200–400 nm, such as sugars and lipids is, however, very difficult with absorption detectors. The use of the more universal RID is also restricted in practice because of its poor detection limit and its high sensitivity to small fluctuations of chromatographic experimental conditions, such as flow rate, solvent composition, and temperature [2]. Moreover, if the separation of complex samples requires the use of gradient elution, the application of RID becomes almost impossible. Although for some solutes the use of either a reaction detector (RD) or a fluorescence detector (FD) is possible, this is not a general solution. In this regard, the analysis of complex mixtures of lipids or sugars by HPLC remains difficult owing to the lack of a suitable detector.

The miniaturization of detector cells is also extremely difficult and the technological problems have not yet been solved because the detection limit should also be decreased or, at least, kept constant [2–6]. Some progress in the design of very small cells for UVD and FD has been reported [3–7], but the miniaturization of RD and RID seems much more difficult in spite of some suggestions [8]. Similarly, the development of open tubular columns is plagued by the lack of a suitable detector with a small contribution to band broadening.

A nonselective detector more sensitive than the RID and easier to use with a small contribution to band broadening is thus desirable in HPLC. The mass spectrometer would be a good solution if it were not so complex [10] and expensive. The electron-capture detector (ECD) [11] and flame-based detectors have been suggested [12]. Both are very sensitive and could be made with very small volumes. Unfortunately, the ECD can be used only with volatile analytes and it is very selective. Both ECD and flame-based detectors are very sensitive to the solvent flow rate, and noisy signals are often produced. The adaptability of these detectors to packed columns is thus difficult. This probably explains why the ECD has been all but abandoned.

The evaporative light-scattering analyzer [13–14], on the other hand, is an alternative solution which seems very attractive for a number of reasons. As most analytes in HPLC have a very low vapor pressure at room temperature and the solvents used as the mobile phase have a significant vapor pressure, some kind of phase separation is conceivable.

Evaporative Light-Scattering Detector

Principle of Operation

The unique detection principle of evaporative light-scattering detectors involves nebulization of the column effluent to form an aerosol, followed by solvent vaporization in the drift tube to produce a cloud of solute droplets (or particles), and then detection of the solute droplets (or particles) in the light-scattering cell.

Detector Components

Nebulizer: The nebulizer is connected directly to the analytical column outlet. In the nebulizer, the column effluent is mixed with a steady stream of nebulizing gas, usually nitrogen, to form an aerosol. The aerosol consists of a uniform dispersion of droplets. Two nebulization properties can be adjusted to regulate the droplet size of the analysis. These properties are gas and mobile-phase flow rates. The lower the mobile-phase flow rate, the less gas and heat are needed to nebulize and evaporate it. Reduction of flow rate by using a 2.1-mm-inner diameter column should be considered when sensitivity is important. The gas flow rate will also regulate the size of the droplets in the aerosol. Larger droplets will scatter more light and increase the sensitivity of the analysis. The lower the gas flow rate, the larger the droplets. It is also important to remember that the larger the droplet, the more difficult it will be to vaporize in the drift tube. An unvaporized mobile phase will increase the baseline noise. There will be an optimum gas flow rate for each method which will produce the highest signal-to-noise ratio.

Drift tube: In the drift tube, volatile components of the aerosol are evaporated. The nonvolatile particles in the mobile phase are not evaporated and continue down the drift tube to the light-scattering cell to be detected. Nonvolatile impurities in the mobile phase or nebulizing gas will produce noise. Using the highest-quality gas, solvents, and volatile buffers, preferably a filter, will greatly reduce the baseline noise. Detector noise will also increase if the mobile phase is not completely evaporated. The sample may also be volatilized if the drift-tube temperature is too high or the sample is too volatile. The optimal temperature in the drift tube should be determined by observing the signal-to-noise ratio with respect to temperature.

Light-scattering cell: The nebulized column effluent enters the light-scattering cell. In the cell, the sample particles scatter the laser light, but the evaporated mobile phase does not. The scattered light is detected by a silicone photodiode located at a 90° from the laser. The photodiode produces a signal which is sent to the analog outputs for collection. A light trap is located 180° from the laser to collect any light not scattered by particles in the aerosol stream.

The signal is related to the solute concentration by the function $A = am^x$, where x is the slope of the response line, m is the mass of the solute injected in the column, and a is the response factor.

Applications

Evaporative light-scattering detection finds wide applicability in the analysis of lipids and sugars. The analysis of lipids and sugars by HPLC has classicaly been hampered due to the lack of absorbing chromophores in these molecules. Accordingly, most analyses are carried out by gas chromatography, requiring derivatization in the case of the sugars or being especially difficult like the separation of the high-molecular-weight triglycerides, or even impossible for the important class of phospholipids, which cannot withstand high temperatures. Specific applications are as follows:

1. Use of evaporative light scattering detector in reversed-phase chromatography of oligomeric surfactants. Y. Mengerink, H. C. De Man, and S. J. Van Der Wal, *J. Chromatogr. 552*: 593 (1991)
2. A rapid method for phospholipid separation by HPLC using a light-scattering detector: W. S. Letter, *J. Liq. Chromatogr. 15*: 253 (1992)
3. Detection of HPLC separation of glycophospholipids: J. V. Amari, P. R. Brown, and J. G. Turcotte, *Am. Lab.* 23 (Feb. 1992)
4. Analysis of fatty acid methyl esters by using supercritical fluid chromatography with mass evaporative light-scattering detection: S. Cooks and R. Smith, *Anal. Proc. 28*, 11 (1991)
5. HPLC analysis of phospholipids by evaporative light-scattering detection: T. L. Mounts, S. L. Abidi, and K. A. Rennick, *J. AOCS 69*: 438 (1992)
6. Determination of cholesterol in milk fat by reversed-phase high-performance liquid chromatography and evaporative light-scattering detection: G. A. Spanos and S. J. Schwartz, *LC–GC 10*(10): 774 (19XX)
7. A qualitative method for triglyceride analysis by HPLC using ELSD: W. S. Letter, *J. Liq. Chromatogr. 16*: 225 (1993)
8. Detect anything your LC separates, P. A. Asmus, *Res. Dev. 2*: 96 (1986).
9. Rapid separation and quantification of lipid classes by HPLC and mass (light scattering) detection: W. H. Christie, *J. Lipid Res. 26*: 507 (1985)

References

1. R. P. W. Scott, *Liquid Chromatography Detectors*, Elsevier, Amsterdam, 1977.
2. H. Colin, A. Krstulovic, and G. Guiochon, *Analysis 11*: 155 (1983).
3. R. P. W. Scott and P. Kucera, *J. Chromatogr. 169*: 155 (1979).
4. H. Knox and M. T. Gilbert, *J. Chromatogr. 186*: 405 (1979).
5. G. Guiochon, *Anal. Chem. 53*: 1318 (1981).
6. G. Guiochon, in *Miniaturization of LC Equipment*, P. Kucera (ed.), Elsevier, Amsterdam, 1983.
7. P. Kucera and H. Umagat, *J. Chromatogr. 255*: 563 (1983).
8. J. W. Jorgensen and E. J. Guthrie, *J. Chromatogr. 255*: 335 (1983).
9. J. W. Jorgensen, S. L. Smiths, and M. Novotny, *J. Chromatogr. 142*: 233 (1977).

10. P. J. Arpino and G. Guiochon, *Anal. Chem. 51*: 682A (1979).
11. F. W. Willmont and R. J. Dolphin, *J. Chromatogr. Sci. 12*: 695 (1974).
12. V. L. McGuffin and M. Novotny, *J. Chromatogr. 218*: 179 (1981).
13. J. H. Charlesworth, *Anal. Chem. 50*: 1414 (1978).
14. R. Macrae and J. Dick, *J. Chromatogr. 210*: 138 (1981).

Juan G. Alvarez

Applied Voltage: Effect on Mobility, Selectivity, and Resolution in Capillary Electrophoresis

Generally, migration times t_m in capillary electrophoresis (CE) are inversely proportional to the applied voltage; in terms of analysis time, the voltage should, therefore, be as large as possible:

$$t_m \cong \frac{1}{V}$$

Under conditions optimized for limited power dissipation, effective mobilities and selectivities (defined as effective mobility ratios) are independent of the applied voltage.

Efficiency is also determined by the applied voltage, but in a much more complicated manner (see Band Broadening in Capillary Electrophoresis). If efficiency is limited by diffusion, a higher voltage also leads to a higher efficiency. Limitations are due to insulation properties and heat dissipation. Voltages larger than 30 kV should always be avoided because of danger of sparking and leaking currents, even more so in cases of significant atmospheric humidity. Excessive heat dissipation leads to an average temperature increase inside the capillary, which can be reduced by forced cooling. What cannot be reduced is the contribution of heat dissipation to band broadening. This can only be reduced by a lower conductivity, a lower current density, or a smaller inner diameter (see Band Broadening in Capillary Electrophoresis). In the case of diffusion-limited efficiency, the efficiency (as given by the plate number) is directly proportional to the applied voltage:

$$N \cong V$$

The ultimate criterion for quality of separation is the resolution R, given by the following relationship:

$$R = \frac{\Delta t_m}{4\sigma}$$

With the definition of plate number, it follows that $R \cong \sqrt{V}$.

Figure 1 shows a computer simulation of the resolution and analysis time of a mixture of anions at 5 10 and 25 kV. In order to better visualize the effect on resolution, a logarithmic x axis was chosen.

Suggested Further Reading

Hjertén, S., *Chromatogr. Rev. 9*: 122 (1967).
Jorgenson, J. W. and K. D. Lucaks, *Science 222*: 266 (1983).
Li, S. F. Y., *Capillary Electrophoresis – Principles, Practice and Applications*, Elsevier, Amsterdam, 1992.
Reijenga, J. C. and E. Kenndler, *J. Chromatogr. A 659*: 403 (1994).
Reijenga, J. C. and E. Kenndler, *J. Chromatogr. A 659*: 417 (1994).

Jetse C. Reijenga

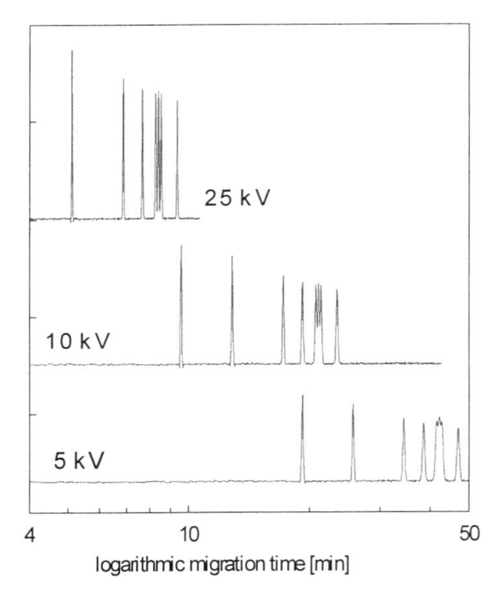

Fig. 1 Analysis of a mixture of weak anions at three different voltages. Suppressed EOF in a 400-mm capillary with negative inlet polarity. *Note*: The time axis is logarithmic.

Aqueous Two-Phase Solvent Systems for Countercurrent Chromatography

Introduction

Aqueous two-phase solvent (ATPS) systems are made of two aqueous liquid phases containing various polymers. Such systems are gentle toward biological materials and they can be used for the partition of biomolecules, membrane vesicles, cellular organites, and whole cells. They are characterized by a high content of water in each phase, by very close densities and refraction indices of the two phases, by a very low interfacial tension, and by high viscosities of the phases. As a result, settling times are particularly long and may last up to 1 h or longer.

The partition of a substance between the two phases depends on many factors. Theoretical studies have been carried out in order to better understand the reasons for the separation in two aqueous phases, thanks to the introduction of various polymers, and the role of various factors on the partition of the substances. However, no global theory is available to predict the observed behaviors. Hopefully, some empirical knowledge has been acquired which will help in the use of these unique solvent systems.

Various devices have been used for the partition of substances in ATPS systems. Countercurrent chromatography (CCC) has again revealed its unique features, because it has enabled the use of such very viscous systems at relatively high flow rates while obtaining a satisfactory efficiency and a good resolution for the separation. Many applications have been described in the literature for the use of ATPS systems with CCC devices.

ATPS Systems

For further information on ATPS systems and the partitioning, we strongly recommend the books by Albertsson [1] and Walter et al. [2], which are reference books in this area.

Polymers Used for ATPS Systems

Many ATPS systems contain a polymer which is sugar based and a second one which is of hydrocarbon ether type. Sugar-based polymers include dextran (Dx), hydroxy propyl dextran (HPDx), Ficoll (Fi) (a polysaccharide), methyl cellulose (MC), or ethylhydroxyethyl cellulose (EHEC). Hydrocarbon ether-type polymers include poly(ethylene glycol) (PEG), poly(propylene glycol) (PPG), or the copolymer of PEG and PPG. Derivatized polymers can also be useful, such as PEG-fatty acids or di-ethylaminoethyl-dextran (Dx-DEAE).

Dextran, or α-1,6-glucose, is available in a mass range from 10,000 to 2,000,000. Dx T500 fractions, also called Dx 48 from Pharmacia (Uppsala, Sweden), are among the best known: their weight-average molecular weight (M_w) varies from 450,000 to 500,000. These white powders contain about 5–10% of water. PEG is a linear synthetic polymer which is available in many molecular weights, the most common being between 300 and 20,000.

Physical Properties of the ATPS Phases

Common characteristics of ATPS phases are their high content of water for both phases, typically 85–99% by weight and very close densities and refraction indices for the two phases. Moreover, both phases are viscous and the interfacial tension is low, from 0.1 to 0.001 dynes/cm. The settling times in the Earth's gravitational field range from 5 min to 1 h.

Practical Use of ATPS

Because ATPS systems are particularly suited for protein separations, many research workers have worked in this area and have tried to model their behavior when varying the composition of these systems. However, there are still no theoretical models to calculate, a priori, the partition coefficient of a protein for a wide range of molecular weights of polymers and concentrations in salts and polymers. However, it remains possible to have qualitative explanations of the role of key factors on the partition of the substances.

Choice of the ATPS System

The general principle for designing an ATPS system is to try various systems, either made from two phases containing polymers or from one phase containing a polymer and the other component a salt. The nature of the substances to be separated shall be taken into account: Fragile solutes may be denaturated by a too high interfacial tension, as encountered in aqueous polymer–salt systems. Some substances may even aggregate in an irreversible way, or be altered in other ways by their contact with some polymers or salts. Moreover, some substances can require the specific use of given salts, or pH, or temperature. When all the previous considerations have been taken into account, the partition coefficients can then be determined in test-tube experiments. Afterward, the composition of the phases can be adjusted.

Systems Suited for the Separation of Molecules

A simple way consists in testing two ATPS systems, dextran 40/PEG-8000 and dextran 500/PEG-8000, which lead to a relatively quick settling and allow reproducible results. For charged macromolecules, the two key parameters are the pH with regard to the isoelectric point of the product and the nature and concentration of the chosen salt. The composition in polymer has a smaller influence, except for some neutral compounds.

Systems Suited for the Separation of Cells and Particles

The main parameter for such separations is the difference of concentrations of each polymer between the two phases. If the concentrations in polymer are quite high, particles tend to adsorb at the interface of the two phases, without any specificity. For instance, all human erythrocytes adsorb at the interface of the dextran 500/PEG-8000 systems with respective concentrations higher than 7.0% and 4.4% (w/w).

The goal is to find a system close to the critical point (in the phase diagram) to achieve the separation, as the partition coefficients all become close to 1. However, this requires an increased attention to the experimental conditions in order to obtain reproducible results. If necessary, a change in the molecular weight of one of the polymers allows one to choose the aqueous phase in which the particle tends to accumulate. For instance, most mammalian cells partition between the interface and the upper phase rich in PEG in dextran 500/PEG-8000 systems, whereas they partition between the interface and the lower phase rich in dextran in dextran 40/PEG-8000 systems.

Adjustment of the Partition Coefficient

We note that the partition coefficient K is defined as the ratio of the concentration of the substance in the upper phase to its concentration in the lower phase. As cells and particles tend to partition between one phase and the interface, only molecules, such as proteins, are the subject of this section.

First, the partition coefficient of the substance should be determined in a standard system, such as dextran 500 (7.0%, w/w)/PEG-8000 (5.0%, w/w) with 5–10 mM of buffer added. Then, the following empirical laws can be used for the adjustment:

1. K is increased by diminishing the molecular weight of the polymer which is predominant in the upper phase (e.g., PEG) or by increasing the molecular weight of the polymer which is predominant in the lower phase (e.g., dextran). Reversing these changes decreases K.
2. K is significantly different than the unit value only if the concentrations of the polymers are high. K tends to the unit value when the concentrations of the polymers are decreased.
3. K can be adjusted by the addition of a salt, as long as the proteins are not close to their isoelectric points. For a negatively charged protein, K is decreased by following the series: phosphate $<$ sulfate $<$ acetate $<$ chloride $<$ thiocyanate $<$ perchlorate and lithium $<$ ammonium $<$ sodium $<$ potassium (for instance, lithium decreases K by a smaller amount than sodium). The influence of the nature of the salt may be amplified by an increase of the pH, which increases the negative net charge of the molecule. Positively charged proteins exhibit the opposite behavior. All these rules apply only for low concentrations of salts. Higher concentrations could, however, be used to favor the partition of the molecules in the upper phase.
4. Charged polymers can also be used; their influence is greater than that of the salts. The most common include charged polymers derived from PEG, such as PEG–TMA (trimethylamino) or PEG–S (sulfonate). Dextran can also be modified.
5. The derivatization by hydrophobic groups can also facilitate the extraction of molecules containing hydrophobic sites. The most common polymer is PEG–P (PEG–palmitate).

K depends on the temperature, but in a complex way, so that its use is difficult for common cases.

Optimization of the Selectivity

Some general rules apply to proteins in PEG/dextran systems and are summarized as follows:

1. The concentration of the polymer is important: Decreasing the concentrations brings the system closer to the critical point (in the phase diagram),

smoothing K values toward the unit value and finally decreasing the selectivity.
2. The nature of the salt is important. The most important effects are encountered for $NaClO_4$, which extracts positively charged molecules in the upper phase, and tetrabutyl ammonium phosphate, which extracts negatively charged proteins in the upper phase.

Applications

These aqueous two-phase solvent systems are more difficult to handle than organic-based solvent systems, so that the number of applications in the literature is quite small as compared to the other systems. However, these applications are really specific, quite often striking in their separation power, and they truly reveal some unique features of countercurrent chromatography.

Former applications of ATPS systems on CCC devices were gathered by Sutherland et al. [3]. For instance, both toroidal and type J [also called a high-speed countercurrent chromatograph (HSCCC)] countercurrent chromatographs were successfully applied for the fractionation of subcellular particles. Using standard rat liver homogenates, plasma membranes, lysosomes, and endoplasmic reticulum were separated by a 3.3% (w/w) dextran T500, 5.4% PEG-6000, 10 m*M* sodium phosphate–phosphoric acid buffer (pH 7.4), 0.26*M* sucrose, 0.05 m*M* Na_2EDTA, and 1 m*M* ethanol. Purification of torpedo electroplax membranes were also carried out, and the separation of various bacterial cells were also described, including the purification of different strains of *Escherichia coli* and the separation of *Salmonella typhirum* cells, using PEG–dextran ATPS systems. Moreover, these CCC devices were also applied to larger cells, such as the separation of various species of red blood cells.

In the same way, the separation of cytochrome-*c* and lysozyme was achieved in 1988 by Ito and Oka [4] using the type J (or HSCCC) device. The chosen ATPS system consisted of 12.5% (w/w) PEG-1000 and 12.5% (w/w) dibasic potassium phosphate in water. The two peaks were resolved in 5 h using a 1-mL/min flow rate, but the retention of the stationary phase was as low as 26%. The limitation of this type of apparatus is definitely the low retention of the stationary phase for ATPS systems.

Several ATPS systems were also used with a centrifugal partition chromatograph (also called Sanki-type from the name of its unique manufacturer). Foucault and Nakanishi [5] tested PEG-1000/ammonium sulfate, PEG-8000/dextran, and other PEG-8000/hydoxypropylated starch on a test separation of crude albumin using a model LLN centrifugal partition chromatograph (CPC) containing six partition cartridges. They demonstrated that the systems could be used with the CPC apparatus, but the efficiency was particularly low (due to very poor mass transfer) and the flow rate was quickly limited by a strong decrease in the retention of the stationary phase (and not by the back-pressure). Afterward, CPC was then not considered as really suited for ATPS systems.

The third type, which is close in principle to the type J high-speed countercurrent chromatograph, was designed in the early eighties and is named "cross-axis coil planet centrifuge." Such a new design has led to successful results with highly viscous polymer phase systems [6]. Indeed, it allows satisfactory retention of the stationary phase of ATPS systems, either in the polymer–salt form or the polymer–polymer form. Such an apparatus eliminates the main drawback of the previous CCC devices, as it allows one to maintain a good retention of the stationary phase with a sufficient flow rate to ensure an acceptable separation or purification time. Using such solvent systems, it has been applied to the separation and purification of various protein samples:

- Mixture of cytochrome-*c*, myoglobin, ovalbumin and hemoglobin [7]
- Histones and serum proteins [8]
- Recombinant uridine phosphorylase from *E. coli* lysate [9]
- Human lipoproteins from serum [10]
- Lactic acid dehydrogenase from a crude bovine heat extract [11]
- Profilin–actin complex from *Acanthamoeba* extract [12]
- Lyzozyme, ovalbumin, and ovotransferrin from chicken egg white [13]
- Acidic fibroblast growth factor from *E. coli* lysate [14]

The cross-axis coil planet centrifuge has, consequently, has been demonstrated to be particularly suited for the use of ATPS systems, leading to satisfactory retention of the stationary phase while keeping a sufficient flow rate of the mobile phase to limit the duration of the experiments.

References

1. P.-A. Albertson, *Partition of Cell Particles and Macromolecules*, 3rd ed., John Wiley & Sons, New York, 1986.
2. H. Walter, D. E. Brooks, and D. Fisher, *Partitioning in Aqueous Two-Phase Systems*, Academic Press, New York, 1985.
3. I. A. Sutherland, D. Heywood-Waddington, and Y. Ito, *J. Chromatogr. 384*: 197 (1987).

4. Y. Ito and H. Oka, *J. Chromatogr. 457*: 393 (1988).
5. A. P. Foucault and K. Nakanishi, *J. Liq. Chromatogr. 13*(12): 2421 (1990).
6. M. Bhatnagar, H. Oka, and Y. Ito, *J. Chromatogr. 463*: 317 (1989).
7. Y. Shibusawa and Y. Ito, *J. Chromatogr. 550*: 695 (1991).
8. Y. Shibusawa and Y. Ito, *J. Liq. Chromatogr. 15*: 2787 (1992).
9. Y. W. Lee, Y. Shibusawa, F. T. Chen, J. Myers, J. M. Schooler, and Y. Ito, *J. Liq. Chromatogr. 15*: 2831 (1992).
10. Y. Shibusawa, Y. Ito, K. Ikewaki, D. J. Rader, and J. Bryan Brewer, Jr. *J. Chromatogr. 596*: 118 (1992).
11. Y. Shibusawa, Y. Eriguchi, and Y. Ito, *J. Chromatogr. B 696*: 25 (1997).
12. Y. Shibusawa and Y. Ito, *Am. Biotechnol. Lab. 15*: 8 (1997).
13. Y. Shibusawa, S. Kihira, and Y. Ito, *J. Chromatogr. B 709*: 301 (1998).
14. J. M. Menet, Thèse de Doctorat de l'Université Paris 6 (1995).

Jean-Michel Menet

Argon Detector

The argon detector was the first of a family of detectors developed by Lovelock [1] in the late 1950s; its function is quite unique. The outer octet of electrons in the noble gases is complete and, as a consequence, collisions between argon atoms and electrons are perfectly elastic. Thus, if a high potential is set up between two electrodes in argon and ionization is initiated by a suitable radioactive source, electrons will be accelerated toward the anode and will not be impeded by energy absorbed from collisions with argon atoms. However, if the potential of the anode is high enough, the electrons will eventually develop sufficient kinetic energy that, on collision with an argon atom, energy can be absorbed and a *metastable* atom can be produced. A metastable atom carries *no* charge but adsorbs its energy from collision with a high-energy electron by the displacement of an electron to an outer orbit. This gives the metastable atom an energy of about 11.6 electron volts. Now 11.6 V is sufficient to ionize most organic molecules. Hence, collision between a metastable argon atom and an organic molecule will result in the outer electron of the metastable atom collapsing back to its original orbit, followed by the expulsion of an electron from the organic molecule. The electrons produced by this process are collected at the anode, generating a large increase in anode current. However, when an ion is produced by collision between a metastable atom and an organic molecule, the electron, simultaneously produced, is immediately accelerated toward the anode. This results in a further increase in metastable atoms and a consequent increase in the ionization of the organic molecules. This cascade effect, unless controlled, results in an exponential increase in ion current with solute concentration.

The relationship between the ionization current and the concentration of vapor was deduced by Lovelock [2,3] to be

$$I = \frac{CA(x + y) + Bx}{CA\{1 - a\exp[b(V - 1)]\} + B}$$

where A, B, a, and b are constants, V is the applied potential, x is the primary electron concentration, and y is the initial concentration of metastable atoms. The rapid increase in current with increasing vapor concentration, as predicted by the equation, is controlled by the use of a high impedance in series with detector power supply. As the current increases, more volts are dropped across the resistance, and less are applied to the detector electrodes.

The Simple or Macro Argon Detector Sensor

A diagram of the macro argon detector sensor is shown in Fig. 1. The cylindrical body is usually made of stainless steel and the insulator made of PTFE or for high-temperature operation, a suitable ceramic. The very first argon detector sensors used a tractor sparking plug as the electrode, the ceramic seal being a very efficient insulator at high temperatures.

Inside the main cavity of the original sensor was a strontium-90 source contained in silver foil. The surface layer on the foil that contained the radioactive material had to be very thin or the β particles would not be able to leave the surface. This tenuous layer protecting the radioactive material is rather vulnerable to mechanical abrasion, which could result in radioactive contamination

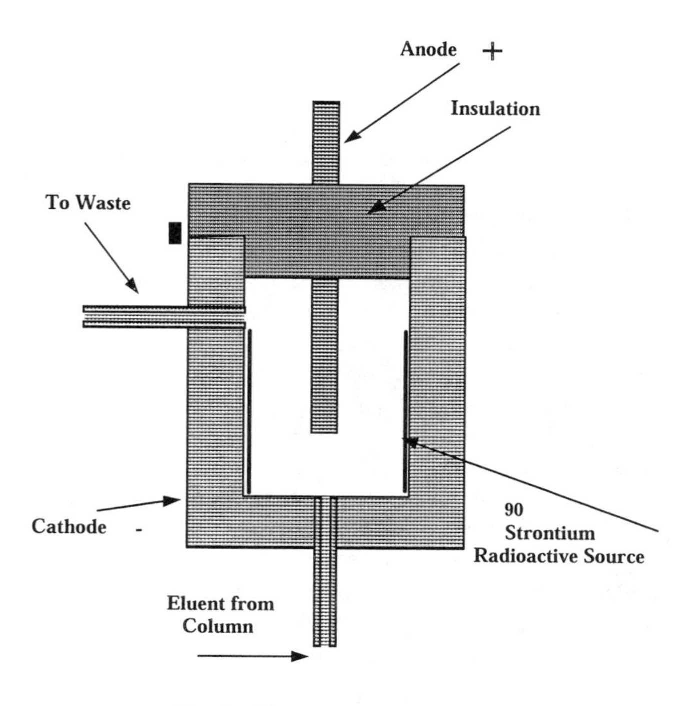

Fig. 1 The macro argon detector.

(strontium-90 has now been replaced by ^{63}Ni). The radioactive strength of the source was about 10 mCu which for strontium-90 can be considered a *hot* source. The source had to be inserted under properly protected conditions. The decay of strontium-90 occurs in two stages, each stage emitting a β particle producing the stable atom of zirconium-90:

$$\underset{\substack{\text{half-life}\\ \text{25 year}}}{^{90}\text{SR}} \;\underset{\substack{0.6\\ \text{MeV}}}{\xrightarrow{\varepsilon}}\; \underset{\substack{\text{half-life}\\ \text{60 h}}}{^{90}\text{Y}} \;\underset{\substack{2.5\\ \text{MeV}}}{\xrightarrow{\varepsilon}}\; \underset{\text{stable}}{^{90}\text{ZR}}$$

The electrons produced by the radioactive source were accelerated under a potential that ranged from 800 to 2000 V, depending on the size of the sensor and the position of the electrodes. The signal is taken across a 1×10^{8}-Ω resistor, and as the standing current from the ionization of the argon is about 2×10^{-8} A, there is a standing voltage of 2 V across it that requires "backing off."

In a typical detector, the primary current is about 10^{11} electrons/s. Taking the charge on the electron as 1.6×10^{-19} C, this gives a current of 1.6×10^{-8} A. According to Lovelock [1], if each of these electrons can generate 10,000 metastables on the way to the electrode, the steady-state concentration of metastables will be about 10^{10} per milliliter (this assumes a life span for the metastables of about 10^{-5} s at NTP). From the kinetic theory of gases, it can be calculated that the probability of collision between a metastable atom and an organic molecule will be about 1.6:1. This would lead to a very high ionization efficiency and Lovelock claims that in the more advanced sensors ionization efficiencies of 10% have been achieved.

The minimum detectable concentration of a well-designed argon detector is about an order of magnitude higher than the FID (i.e., 4×10^{-13} g/mL). Although the argon detector is a very sensitive detector and can achieve ionization efficiencies of greater than 0.5%, the detector was not popular, largely because it was not linear over more than two orders of magnitude of concentration ($0.98 < r > 1.02$) and its response was not predictable. In practice, nearly all organic vapors and most inorganic vapors have ionization potentials of less than 11.6 eV and thus are detected. The short list of substances that are not detected include H_2, N_2, O_2, CO_2, $(CN)_2$, H_2O, and fluorocarbons. The compounds methane, ethane, acetonitrile, and propionitrile have ionization potentials well above 11.6 eV; nevertheless, they do provide a slight response between 1% and 10% of that for other compounds. The poor response to acetonitrile makes this substance a convenient solvent in which to dissolve the sample before injection on the column. It is also interesting to note that the inorganic gases H_2S, NO, NO_2, NH_3, PH_3, BF_3, and many others respond normally in the argon detector. As these are the type of substances that are important in environmental contamination, it is surprising that the argon detector, with its very high sensitivity for these substances, has not been reexamined for use in environmental analysis.

References

1. J. E. Lovelock, *Gas Chromatography 1960*, R. P. W. Scott (ed.), Butterworths Scientific, London 1960, p. 9.
2. J. E. Lovelock, *J. Chromatogr. 1*: 35 (1958).
3. J. E. Lovelock, *Nature* (London) *181*: 1460 (1958).

Suggested Further Reading

R. P. W. Scott, *Chromatographic Detectors*, Marcel Dekker, Inc., New York, 1996.

R. P. W. Scott, *Introduction to Gas Chromatography*, Marcel Dekker, Inc., New York, 1998.

Raymond P. W. Scott

Asymmetric Flow FFF in Biotechnology

Introduction

The research and development in the fields of biochemistry, biotechnology, microbiology, and genetic engineering are fast-growing areas in science and industry. Chromatography, electrophoresis, and ultracentrifugation are the most common separation methods used in these fields. However, even these efficient and widespread analytical methods cannot cover all applications. In this article, asymmetric flow field-flow fractionation (AF4) is introduced as a powerful analytical separation technique for the characterization of biopolymers and bioparticles. Asymmetric flow field-flow fractionation can close the gap between analyzing small and medium-sized molecules/particles [analytical methods: high-performance liquid chromatography (HPLC), GFC, etc.] on the one hand and large particles (analytical methods: sedimentation, centrifugation) on the other hand [1,2], whereas HPLC and GFC are overlapping with asymmetric field-flow fractionation in the lower separation ranges.

First publications about field-flow fractionation (FFF) by Giddings et al. [3] appeared in 1966. From this point, FFF was developed in different directions and, in the following years, various subtechniques of FFF emerged. Well-known FFF subtechniques are sedimentation FFF, thermal FFF, electric FFF, and flow FFF. Each method has its own advantages and gives a different point of view of the examined sample systems. Using sedimentation FFF shows new insights about the size and density of the analytes, thermal FFF gives new information about the chemical composition and the size of the polymers/particles, and electric FFF separates on the basis of different charges. Flow FFF, and especially asymmetric flow FFF (the most powerful version of flow FFF) is the most universal FFF method, because it separates strictly on the basis of the diffusion coefficient (size or molecular weight) 2. and it has the broadest separation range of all the FFF methods. It is usable for a large number of applications in the fields of biotechnology, pharmacology, and genetic engineering.

Separation Principle of Asymmetric Flow Field-Flow Fractionation

All FFF methods work on the same principle and use a special, very flat separation channel without a stationary phase. The separation channel is used instead of the column, which is needed in chromatography. Inside the channel, a parabolic flow is generated, and perpendicular to this parabolic flow, another force is created. In principle, the FFF methods only differ in the nature of this perpendicular force.

The separation channel in asymmetrical flow FFF (AF4) is approximately 30 cm long, 2 cm wide, and between 100 and 500 μm thick. A carrier flow which forms a laminar flow profile streams through the channel. In contrast to the other FFF methods, there is no external force, but the carrier flow is split into two partial flows inside the channel. One partial flow is led to the channel outlet and, afterward, to the detection systems. The other partial flow, called the cross-flow, is pumped out of the channel through the bottom of the channel. In the AF4, the bottom of the separation channel is limited through a special membrane and the top is made of an impermeable plate (glass, stainless steel, etc.). The separation force, therefore, is generated internally, directly inside the channel, and not by an externally applied force.

Under the impact of the cross-flow, the biopolymers/particles are forced in the direction of the membrane. To ensure that the analytes do not pass through the membrane, different pore sizes can be used. In this way, the analytes can be selectively rejected and it is possible to remove low-molecular compounds before the separation. The analytes' diffusion back from this membrane is counteracted by the cross-flow, where, after a time, a dynamic equilibrium is established. The medium equilibrium height for smaller sized analytes is located higher in the channel than for the larger analytes. The smaller sized analytes are traveling in the faster velocity lines of the laminar channel flow and will be eluted first. As a result, fractograms, which show size separation of the fractions, are obtained as an analog to the chromatograms from HPLC or GFC.

Applications of AF4 in Biotechnology

In addition to widespread applications in the field of polymer and material science or environmental research, AF4 can be used in bioanalytics, especially for the characterization of proteins, protein aggregates, polymeric proteins, cells, cell organelles, viruses, liposomes, and various other bioparticles and biopolymers.

Cells and Viruses

The advantage of AF4, in contrast to chromatography, is the capability to separate bioparticles and biopoylmers which usually stick onto chromatography columns. They are more or less filtered out (removed) by the stationary phase. Various applications using AF4 for the separation of shear-force sensitive bioparticles with high molecular weight and size have been reported in the literature. They deal with the efficient and fast separation of viruses [4,5] and bacteria [5]. Reference 4 discusses the investigation of a virus (STNV) with AF4 and the separation of the viral aggregates. In Reference 5, Litzen Wahlund report the separation of a virus (CPMV) together with different other proteins (BSA, Mab). They also present the characterization of bacillus streptococcus faecalis and its aggregates using AF4.

Proteins/Antibodies/DNA

The separation of proteins with AF4 has been demonstrated a number of times. For example, the fractionation of ferritin [7], of HSA and BSA [8], and of monoclonal antibodies [8], including their various aggregates, were published. AF4 is especially suitable for the separation and characterization of large and sensitive proteins and their aggregates because it is fast and gentle and aqueous solvents can be used that achieve maximum bioactivity of the isolated proteins and antibodies. Furthermore, even very large and sticky proteins can be analyzed because of the relatively low surface area and the separation in the absence a stationary phase. Nearly independent of the nature of the bioparticles, AF4 separates by size (diffusion coefficient). Therefore, DNA, RNA, and plasmids can be separated quickly and gently, together with proteins. Reference 6 deals with this issue and presents the AF4 separation of a mixture of cytochrome-*c*, BSA, ferritin, and plasmid DNA.

Artificial Polymeric Proteins

In addition to the characterization of well-known protein substances (serum proteins, aggregates, antibodies, etc.), AF4 is also a very promising separation/characterization technique for a new class of artificially made polymeric proteins from therapeutic/diagnostic applications, such as poly-streptavidin and polymeric hemoglobin [personal communication of the authors]. These proteins usually have very high molecular weights and huge molecular sizes, and they are difficult to analyze by conventional GFC and related techniques. Very often, these proteins are also sticky and show adsorptive effects on the column material. Using AF4 without a stationary phase and without size-exclusion limit, these polymeric proteins can be readily separated and characterized. The application shown in Fig. 1 was done using an AF4 system (HRFFF 10.000 series, Postnova Analytics) and ultraviolet (UV) detection at 210 nm.

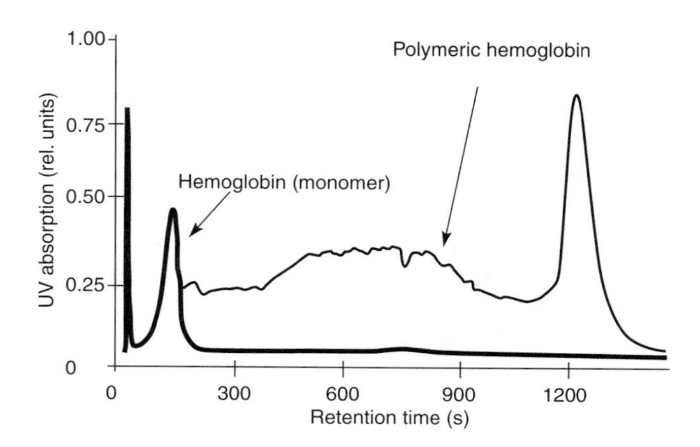

Fig. 1 Pig hemoglobin separated with AF4 and UV detection.

Summary

Asymmetric flow FFF is a new member in the FFF familiy of separation technologies; it is a powerful characterization technique, especially suited for the separation of large and complex biopolymers and bioparticles. AF4 has many of the general benefits of FFF; it adds on several additional characteristics. In particular, these characteristics are as follows:

1. No sample preparation, or only limited sample preparation necessary.
2. Possibility of direct injection of unprepared samples.
3. Large accessible size molecular-weight range, no size-exclusion limit.
4. Very gentle separation conditions in the absence of a stationary phase.
5. Weak or no shear forces inside the flow channel.
6. Rapid analysis times, generally faster than GFC.
7. Fewer sample interactions during separation because of small surface area.
8. On-line sample concentration/large volume injection possible.
9. Gentle and flexible because it uses a wide range of eluents/buffers/detectors.
10. AF4 is a useful analytical tool, and when the limitations of the technology (e.g., sample interactions with membrane or the sample dilution during separation) are carefully observed, samples

can be characterized where other analytical technologies fail or only yield limited information.

References

1. T. Klein, Chemisch–physikalische Charakterisierung von schwermetallhaltigen Hydrokolloiden in natürlichen aquatischen Systemen mit Ultrafiltration und Flow-FFF. Diploma thesis, TU-Munich (1995).
2. T. Klein, Entwicklung und Anwendung einer Asymmetrischen Fluß-Feldflußfraktionierung zur Charakterisierung von Hydrosolen, Ph.D. thesis, TU-Munich (1998).
3. J. C. Giddings, A new separation concept based on a coupling of concentration and flow nonuniformities, *Separ. Sci. 1*: 123 (1966).
4. A. Litzen and K. G. Wahlund. Zone broadening and dilution in rectangular and trapezoidal asymmetrical flow field-flow fractionation channels, *Anal. Chem. 63*: 1001 (1991).
5. A. Litzen and K. G. Wahlund. Effects of temparature, carrier composition and sample load in asymmetrical flow field-flow fractionation, *J. Chromatogr. 548*: 393 (1991).
6. J. J. Kirkland, C. H. Dilks, S. W. Rementer, and W. W. Yau. Asymmetric-channel flow field-flow fractionation with exponential force-field programming. *J. Chromatogr. 593*: 339 (1992).
7. C. Tank and M. Antonietti. Characterization of water-soluble polymers and aqueous colloids with asymmetrical flow field-flow fractionation, *Macromol. Chem. Phys. 197*: 2943 (1996).
8. A. Litzen, J. K. Walter, H. Krischollek, and K. G. Wahlund. Separation and quantitation of monoclonal antibody aggregates by asymmetrical flow field-flow fractionation and comparison to gel permeation chromatography, *Anal. Biochem. 212*: 169 (1993).

Thorsten Klein
Christine Hürzeler

Automation and Robotics in Planar Chromatography

Automation involves the use of systems (instruments) in which an element of nonhuman decision has been interpolated. It is defined as the use of combinations of mechanical and instrumental devices to replace, refine, extend, or supplement human effect and faculties in the performance of a given process, in which at least one major operation is controlled, without human intervention, by a feedback system, A feedback system is defined as an instrumental device combining sensing and commanding elements which can modify the performance of a given act [1].

Three approaches to the automation process can be distinguished, taking into account the criterion of the flexibility of the automation device [2]. The first, denoted as flexible, is characterized by the possibility of adaptation of the instruments to new and varying demands required from the laboratory; examples of these instruments are robots. The second approach, denoted as semiflexible, involves some restrictions for the tasks executed by the instrument; the tasks are controlled by a computer program and its menu. As examples, autosamplers or robots of limited moves can be given. In the third approach, the instruments can execute one or two tasks, without feasibility of new requirements; as examples, supercritical fluid extractors or equipment for dissolution of samples can be given.

Automation processes have several advantages: better reproducibility, increase of the number of samples which can be analyzed, personnel can be utilized for more creative tasks (e.g., planning of experiments and interpretation of results). Harmful conditions in the workplace can be avoided and the equipment of the laboratory can be more effectively utilized.

To illustrate the feasibility of automation in thin-layer chromatography (TLC), the fundamental operations of the chromatographic process are given in Fig. 1 [3].

The first and basic stage of the process, not limited to TLC, but also occurring in other chromatographic techniques, is the preparation of samples. This is the most tedious, time-consuming and error-generating process in the whole analytical cycle which can be fully automated, or the automated stations may be complementary to operations or tasks executed individually. For instance, in a station, a volume of liquid is transferred from the first to a second container, an internal standard is added, and the solution is diluted and mixed. Further actions may be executed manually. Another, more advanced, solution consists in automated execution of the tasks by the station and the sample is transferred from one station to the other by a robot or another transport device.

In a limited version, only the most critical stages are automated by the use of robots with limited, strictly defined movements; examples are automated processes of solid-phase extraction (SPE), heating, and mixing. The

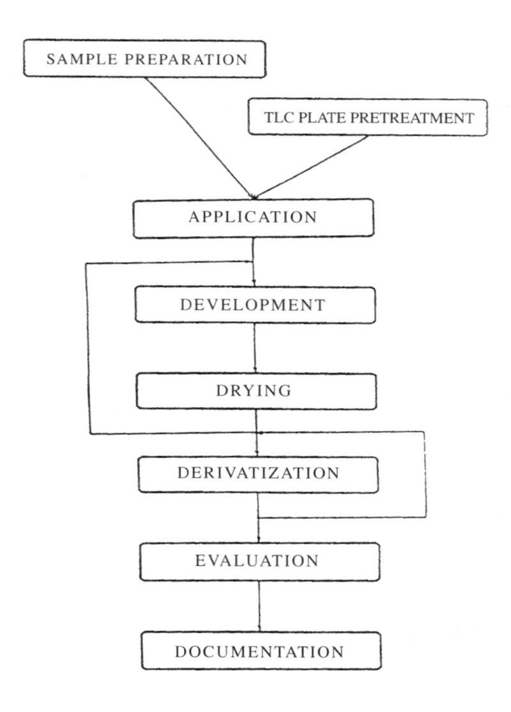

Fig. 1 Survey on operation succession in QTLC.

robots are controlled by a computer and the operator chooses the suitable values of the parameters from given ranges (e.g., autosampler).

Sample Application

The application of the sample onto the thin layer is a critical moment, owing to later localization by the scanner (densitometer) and the beginning of the chromatographic process at the moment of contact of the liquid sample with the chromatographic bed. Therefore, the applicator must warrant the precise localization of the sample and uniform compact cross-section of the starting band. Semiautomatic applicators presently available have the volume range of 20 nL to 10 μL. The sample is delivered from 0.5-, 1.0-, and 10-μL syringes. The piston stroke can be set in a continuous manner. To apply the sample, the piston is stopped and the solution is injected from the end of the capillary; the whole volume of the sample is displaced from the capillary. The position of the end of the capillary is adapted to the layer thickness, the spring-relieved syringe guide warrants that the capillary needle only lightly touches the adsorbent layer, thereby avoiding its damage. The change of position is automatic. The device permits application of spots or streaks at a distance of 5 mm from the edge of the plate; the syringe can be washed twice. A more advanced version is a computer-controlled applicator (e.g., Desaga TLC Applicator AS 30 or Camag Automatic TLC System) composed of an application module, interface, software, and an IBM PC-AT.

The application module dispenses samples from a stainless-steel capillary which is connected to a dosage syringe operated by a stepping motor. Samples can be applied as spots or bands onto TLC plates or sheets up to 20 × 20 cm. Bandwise sample application uses the spray-on technique; for spotwise application, either contact transfer or spraying may be selected. The samples are contained in vials, which may be sealed with regular septa. The vials are arranged in racks with 16 positions; two racks may be inserted per application program. The application pattern can be selected for normal development, for development from both sides, and for circular and anticircular chromatography.

Development of the Chromatogram

The next important stage is the chromatogram development. ADC chambers (Automatic Developing Chamber, Camag), DC-MAT (Byron), TLC-MAT (Desaga) are automatically operating development systems. They increase the reproducibility of the chromatographic results because the development is carried out under controlled conditions. The progress of the solvent front is monitored by a sensor. The development process is terminated as soon as the mobile phase has traveled the programmed distance. Preconditioning, tank or sandwich configuration, solvent migration distance, and the drying conditions are selectable. All relevant parameters are entered via a keypad. The AMD system (Automated Multiple Development, Camag) is a fully automated version of multiple development and stepwise technique with a free choice of mobile-phase gradient. Because the chromatogram is developed repeatedly in the same direction and each individual run is somewhat farther than the last, a focusing of the separated substance zones takes place in the direction of development. The chromatography is reproducible because the mobile phase is removed from the separation chamber after each step and the layer is completely freed from the mobile phase, in vacuum. Then, a fresh mobile phase is introduced for the next run. Provided all parameters, including solvent migration increments, are properly maintained, which is only possible with a fully automatic system, the densitogram of a chromatogram track can be superimposed with a matched-scale diagram of the gradient.

Another device for automated development is the chamber constructed by Tyihak [4], in which the adsorbent layer is placed between two plates and the mobile phase flows under increased applied pressure. It can be operated in the linear or radial mode. Another automated device is the UMRC (Ultra Micro Rotation Chromatograph), where the eluent is delivered to the center of a rotating TLC plate [5]. A simple device was constructed by Delvorde and Postaire [6] in which the liquid is pumped out (by vacuum) which causes the flow of the mobile phase and decreases the vapor pressure.

Derivatization

Derivatization can be carried out both before and after development of the plate. In the latter case, it may be applied before detection or after scanning densitometry. Derivatization may be carried out using the device constructed by Kruzig (Anton Paar KG), where the sprayer moves along a vertical guide while the plate moves horizontally [7]. Another method of derivatization consists in immersion of the plate into a suitable reagent solution. For this purpose, the device available from Camag can be used (Camag Chromatogram Immersion Device III), in which a low velocity motor causes the immersion and removal of the plate from the reagent solution.

Evaluation

Thin-film chromatographic detection, contrary to other chromatographic techniques, requires stopping development, drying of the layer, and scanning with an appropriate detector. There are basically two alternatives for the evaluation of thin-layer chromatograms: elution of the separated substance from the layer, followed by photometric determination (indirect determination), and *in situ* evaluation (scanning) directly on the TLC plate. The *in situ* evaluation of the chromatogram is carried out using a high-resolution chromatogram spectrophotometer (densitometer) to scan each chromatogram track, from start to solvent front in the direction of development, by means of a slit. The measurements are carried out either in the visible-light range for colored or fluorescent substances or in the ultraviolet (UV) range for UV-light-absorbing solutes. The wavelength of maximum absorption is generally selected as measurement wavelength. The scanning process yields absorption or fluorescence scans (peaks) which are also used to assess the quality of chromatographic separation. TLC plates are generally scanned in the reflectance mode (diffuse reflectance), meaning that the monochromatic light is directed by a mirror to the layer surface at 90° and the diffuse reflectance is measured at 45° by means of a detector. The optical pathways used for absorption and fluorescence measurements are identical in commercially available scanners. The only difference is the light source: Visible-light measurements are performed using tungsten lamps, whereas high-pressure mercury lamps are used for fluorescence measurements and deuterium lamps for absorption measurements in the UV range. In the case of fluorometric detection, it is also necessary to place a cutoff filter in front of the detector to prevent the comeasurement of the short-wavelength excitation radiation. All functions of the scanner are controlled from a personal computer that is linked via an RS232 interface. The scanner transmits all measurement data, in digital form, to the computer for processing with the specific software. The final report is based in the following sequence: raw data acquisition–integration–calibration, and calculation. Integration is performed, postrun, from the raw data gathered during scanning (i.e., after all tracks of a chromatogram plate have been measured). Integration results can be influenced by selecting appropriate integration parameters. As all measured raw data remain stored on a disk, reintegration with other parameters is possible at any time. The system automatically defines and corrects the baseline and sets fraction limits. The operator can accept these or can override the automatic process by video integration. All steps can be followed on the screen.

Robots

Laboratory robots are adapted now for linking all of the steps between extraction and obtaining the analysis results. They are able to automate lengthy, routine, multistep analyses. They require electronic communication in real time to know exactly the operating time and possible breakdowns. A robotics system allows space saving and easier integration of equipment in the laboratory. Today, a conventional robot has a movable arm. The purpose of the arm is to extend the capabilities of the human arm. There are five basic parts to every robotic arm: controller, arm, drive, end effector, and sensor. There are also five basic functions: base, shoulder, elbow, pitch, and roll. Most modern robots belong to one of four categories: Cartesian, spherical and cylindrical robots, and revolute arms [8]. In 1989, Prosek et al. [3] developed a planar chromatography robot (Fig. 2). Its arm, supported by a rotating base, executed four degrees of freedom movements. Its work envelope comprised four tanks: the first for cleaning, the second for development, and the last two for derivatization by dipping. Also required was a hot plate, a drying system, and a digital camera for evaluation of the derivatized plate. The system was controlled by an Apple IIe computer.

The planar chromatography automaton was designed by Delvordre and Postaire with the objective of reducing

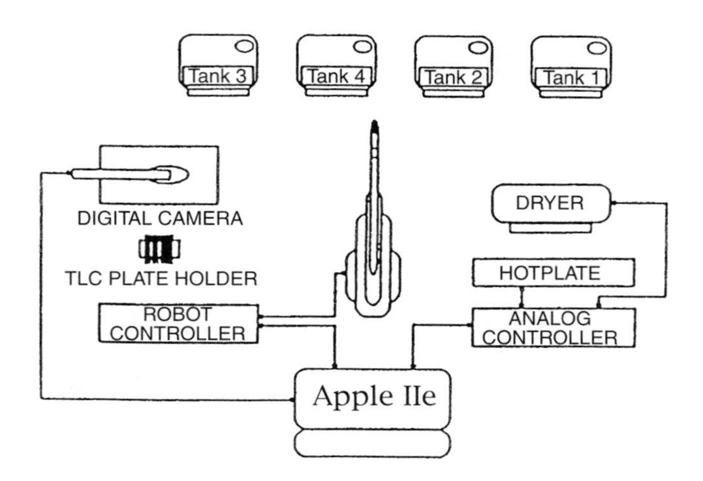

Fig. 2 Prosek robotic TLC apparatus.

the number of human movements required for the handling of precoated plates [8]. This device uses a conveyor-belt-like system to sustain all the chromatographic steps along with their own supply of reagents and tools. The procedure comprises six stages. Using this method, qualitative and quantitative data are obtained 50–150 min after starting the procedure.

Technological progress has enabled automation of planar chromatography and will provide users with a greatly improved technique. Such improvements will now meet requirements of the industrial sector, not only in terms of productivity, effectiveness, reduced cost, GLP, and environmental quality, but also on the technical side (validation, flexibility, evolution). Complete chromatographic automation will bring planar chromatography to the same level as other chromatograpic methods.

References

1. *IUPAC Compendium of Analytical Literature*, Pergamon Press, Oxford, 1978, pp. 22–23.
2. R. E. Majors and B. D. Holden, *LC–GC Int. 6*(9): 530 (1993).
3. M. Prošek, M. Pukl, A. Smidownik, and A. Medja, *J. Planar Chromatogr. 2*(6): 244 (1989).
4. E. Tyihak and E. Mincsovics, *LC-GC Int. 4*: 24 (1991).
5. S. Nyiredy, L. Batz, and O. Sticcer, *J. Planar Chromatogr. 2*: 53 (1989).
6. P. Delvordre, C. Reynault, and E. Postaire, *J. Liq. Chromatogr. 15*: 1673 (1992).
7. F. Kreuzig, *Chromatographia 13*: 238 (1980).
8. E. P. R. Postaire, P. Delvodre, and Ch. Sarbach, in *Handbook of Thin-Layer Chromatography*, 2nd ed., J. Sherma and B. Fried (eds.), Marcel Dekker, Inc., New York, 1997, pp. 373–385.

Wojciech Markowski

Axial Dispersion Correction Methods in GPC–SEC

Effect of Axial Dispersion

In ideal size-exclusion chromatography (SEC), fractionation is exclusively by hydrodynamic volume. Due to axial dispersion, however, a whole distribution of hydrodynamic volumes (and, therefore, of molecular weights) is instantaneously present in the detector cell. Under these conditions, it is assumed that the mass chromatogram $w(V)$ (i.e., the instantaneous mass w versus the elution time or elution volume V) is a broadened version of a true (or corrected) mass chromatogram $w^c(V)$, as follows [1]:

$$w(V) = \int_{-\infty}^{\infty} g(V, V_0) w^c(V_0)\, dV_0 \quad (1)$$

where $g(V, V_0)$ is the (in general, nonuniform and skewed) spreading function and V_0 is a dummy integration variable that also represents the average retention volumes. At each V_0, an (in principle, different) individual $g(V)$ function must be defined. The determination of $g(V, V_0)$ is still a matter of controversy that is outside the scope of the present article.

With narrow standards of known molecular weights, a calibration log $M(V)$ can be obtained; such a calibration is assumed to be unaffected by axial dispersion. If the chromatogram $w(V)$ is combined with the molecular-weight calibration, a broadened molecular weight distribution (MWD) $w(M)$ will be estimated. The corresponding number-average molecular weight $\overline{M}_n$ will result underestimated, and the weight-average molecular weight $\overline{M}_w$ and the polydispersity $\overline{M}_w/\overline{M}_n$ will both be overestimated. If a broad and smooth chromatogram is obtained with a modern high-resolution column set, the axial dispersion effect is expected to be negligible, and no specific corrections will be required. In contrast, axial dispersion correction may be important in the cases of (a) a narrow chromatogram with a width that is close to that of the broadening functions in the same elution volume range and (b) a wide chromatogram, but containing sharp elbows and/or narrow peaks. Equation (1) has been extended to other detectors such as molar mass or specific group sensors. In all cases, the same function $g(V, V_0)$ is applicable; this is because the broadening is assumed independent of the polymer chemical nature. For example, calling $s_w(V)$ the chromatogram obtained from a light-scattering detector, the following may be written [2]:

$$s_w(V) = \int_{-\infty}^{\infty} g(V, V_0) s_w^c(V_0)\, dV_0 \tag{2}$$

where $s_w^c(V)$ is the corrected molar mass chromatogram and $s_w(V)$ is proportional to $[w(V)M_w(V)]$, where $M_w(V)$ is the instantaneous weight-average molecular weight. Even with perfectly accurate sensors, a distorted MWD $w(M_w)$ will be estimated from $s_w(V)$ and $w(V)$. However, in that case, the $\overline{M}_w$ estimate will still be exact, whereas $\overline{M}_n$ and therefore the polydispersity will be both underestimated [3].

Correction Methods for Mass Chromatograms

Consider the numerical inversion of Eq. (1), that is, the calculation of $w^c(V)$ from the knowledge of $w(V)$ and $g(V, V_0)$. To this effect, let us first transform the continuous model of Eq. (1) into the following equivalent discrete model:

$$\mathbf{w} = \mathbf{G}\,\mathbf{w}^c \tag{3a}$$

with

$$\mathbf{G} = \begin{bmatrix} g(1,1) & \cdots & g(1,j) & \cdots & g(1,p) \\ & \ddots & & & \\ \vdots & & g(j,j) & & \vdots \\ & & & \ddots & \\ g(n,1) & & \cdots & & g(n,p) \end{bmatrix}, \quad (n > p) \tag{3b}$$

where $\mathbf{w}$ is a $(n \times 1)$-column vector containing only the nonzero heights of $w(V)$ sampled at regular ΔV intervals, $\mathbf{w}^c$ is a $(p \times 1)$-column vector containing only the nonzero heights of $w^c(V)$, and $\mathbf{G}$ is a rectangular matrix that represents $g(V, V_0)$. Each column of $\mathbf{G}$ contains the heights of the successive (discrete) broadening functions, with the average retention volume heights at $(c + j, j)$. The following is verified: (a) the p elements of $\mathbf{w}^c$ correspond to the central elements of $\mathbf{w}$; (b) $n - p + 1$ is equal to the number of nonzero elements of the discrete version of $g(V)$; and (c) the largest elements of $\mathbf{G}$ are located c rows below the main diagonal, and the elements on the top-right hand corner and bottom-left hand corner are all zeros.

Were it not for the ill-conditioned nature of the problem, $\mathbf{w}^c$ could be simply estimated from $\hat{\mathbf{w}}^c = [\mathbf{G}^T\mathbf{G}]^{-1}\mathbf{G}^T\mathbf{w}$; where the "hat" indicates the estimated value. Unfortunately, $[\mathbf{G}^T\mathbf{G}]^{-1}$ is almost singular; therefore, such a solution is highly oscillatory with negative peaks.

In what follows, several correction techniques originally developed for mass chromatograms are described. Such techniques are (a) the "phenomenological" Methods I–III that consider the system as a "black box" and numerically invert Eq. (1) prior to calculating the MWD, and (b) the more specific Methods IV–V that (avoiding the ill-conditioned nature of the numerical inversion) require an independent molecular-weight calibration and calculate the MWD in a single step. Methods I–III are, in general, suitable for nonuniform and skewed broadening functions.

Method I: The Difference Function [4]

Consider the iterative procedure presented as Method 1 in Ref. 4. The following set of difference equations must be first calculated:

$$\Delta_i \mathbf{w} = \Delta_{i-1}\mathbf{w} - \mathbf{G}\,\Delta_{i-1}\mathbf{w}, \quad \text{with} \quad \Delta_0 \mathbf{w} = \mathbf{w}, \quad (i = 1, 2, \ldots) \tag{4}$$

where i is the iteration step. Then, as $i \to \infty$, it is expected that $\Delta_i \mathbf{w} \to 0$, and the corrected chromatogram is given by $\hat{\mathbf{w}}^c = \sum_{i=1}^{\infty} \Delta_i \mathbf{w}$. In practice, the best solution is obtained in a generally small number of iterations.

Method II: Singular Value Decomposition [5]

The final expression for this least-squares estimation procedure is

$$\hat{\mathbf{w}}^c = \sum_{k=1}^{r} \frac{\mathbf{u}_k^T \mathbf{w}}{\sigma_k} \mathbf{v}_k$$

$$r \leq p, \quad \sigma_1 \geq \sigma_2 \geq \cdots \geq \sigma_r \cdots \geq \sigma_p \geq 0 \tag{5}$$

where $\mathbf{u}_k$ and $\mathbf{v}_k$ are the eigenvectors of $\mathbf{G}\mathbf{G}^T$ and $\mathbf{G}^T\mathbf{G}$, respectively, σ_k are the singular values [5] of $\mathbf{G}$, and p is the full rank of $\mathbf{G}$. In Eq. (5), the number of "effective" terms is limited to r because the lower σ_k's excessively amplify the measurement noise. The lowest admissible σ_r must be larger than the inverse of the lowest signal-to-noise ratio. The lowest signal-to-noise ratio is, in turn, normally found at the chromatogram tails.

Method III: The Kalman Filter [6]

This fast and effective digital algorithm is based on a linear stochastic model that is equivalent to Eq. (1). The theoretical background is beyond the scope of the present article; unfortunately, some knowledge of the basic Kalman filtering theory [5,7] is necessary for an adequate algorithm adjustment. The adjustment requires estimating the variances of the measurement noise and of the expected solution.

In all direct-inversion methods, the best adjustment normally involves a trade-off between an excessively "rich" solution that may contain spurious high-frequency oscillations and negative peaks, and an excessively smoothened solution where the high-frequency information is lost, but it is otherwise acceptable. Consider now some methods that avoid the direct chromatogram inversion.

Method IV: Rotation of the Calibration Curve [8,9]

This method aims at obtaining the corrected MWD in a single step, by simply rotating the linear calibration log M(V) counterclockwise around some selected mid-pivoting point. In this way, the effective molecular-weight range is reduced with respect to that obtained from the original calibration. The technique is numerically robust, but, unfortunately, it is based on the following rather strong assumptions: (a) the axial dispersion must be uniform and Gaussian; (b) the chromatogram itself must be Gaussian or Wesslau [9]; and (c) the original molecular-weight calibration must be linear and of the form $M = D_1 \exp(-D_2 V)$. The rotated calibration $M = D_1' \exp(-D_2' V)$ is obtained from

$$D_1' = D_1 \exp\left\{\frac{D_2\sigma_g^2[D_2(\sigma_w^2 - \sigma_g^2) - 2\overline{V}]}{2\sigma_w^2}\right\} \tag{6a}$$

$$D_2' = D_2 \exp\left(1 - \frac{\sigma_g^2}{\sigma_w^2}\right) \tag{6b}$$

where σ_g^2 and σ_w^2 are the variances of $g(V)$ and $w(V)$, respectively, and the pivoting abscissa $\overline{V}$ is the number-average retention volume of the chromatogram.

Method V: The Approximate "Analytical" Solution [10]

This approach is based on assuming (a) a Gaussian (but nonuniform) axial dispersion with a variance $\sigma_g^2(V)$; (b) a Gaussian (but nonuniform) instantaneous MWD of variance $\sigma_0^2(V)$ and average retention volume $V_0(V)$; and (c) an (in general nonlinear) calibration $M = D_1(V) \exp[-D_2(V)V]$. The corrected chromatogram is obtained from

$$\hat{w}^c(V) = w(V)\left(\frac{\sigma_g^2(V)}{\sigma_0^2(V)}\right) \exp\left(-\frac{(V - V_0(V))^2}{2\sigma_0^2(V)}\right) \tag{7a}$$

with:

$$V_0(V) = V + \frac{1}{D_2(V)} \cdot \ln\left(\frac{w(V + D_2(V)\sigma_g^2(V))}{\sqrt{w(V - D_2(V)\sigma_g^2(V))w(V + D_2(V)\sigma_g^2(V))}}\right) \tag{7b}$$

$$\sigma_0^2(V) = \sigma^2(V) + \frac{1}{D_2^2(V)} \cdot \ln\left(\frac{w(V - D_2(V)\sigma_g^2(V))w(V + D_2(V)\sigma_g^2(V))}{w^2(V)}\right) \tag{7c}$$

Method V assumes that the instantaneous MWDs are Gaussian. However, the instantaneous MWDs are never truly Gaussian, even for Gaussian broadening functions, thus reducing the precision of the approach. (For example, it can be shown that at the chromatogram tails, the instantaneous distributions are always highly skewed with the concavities facing toward the mid-chromatogram section [3].)

An Evaluation Example

Correction methods are best evaluated through synthetic (or numerical) examples. The reason for this is that only in numerical examples, the true solution is a priori known, and therefore the quality of the different estimation algorithms can be adequately compared. (In contrast, in a real SEC measurement, the true corrected chromatogram or MWD are never accurately known.) In what follows, Methods I–V are evaluated on the basis of a synthetic example that has been previously investigated in several occasions. [6,11–13]

The basic data consist of the corrected chromatogram $w^c(V)$ and the (uniform) broadening function $g(V)$ presented in Fig. 1a. The proposed example is particularly demanding because (a) $w^c(V)$ is multipeaked and (b) the variance of $w^c(V)$ is comparable to the variance of $g(V)$. For a uniform broadening, Eq. (1) reduces to a simple convolution integral. By convolution of $w^c(V)$ and $g(V)$, a noise-free "measurement" was first obtained. Then, this function was rounded to the last integer. [13] This procedure is equivalent to adding a zero-mean white noise (with

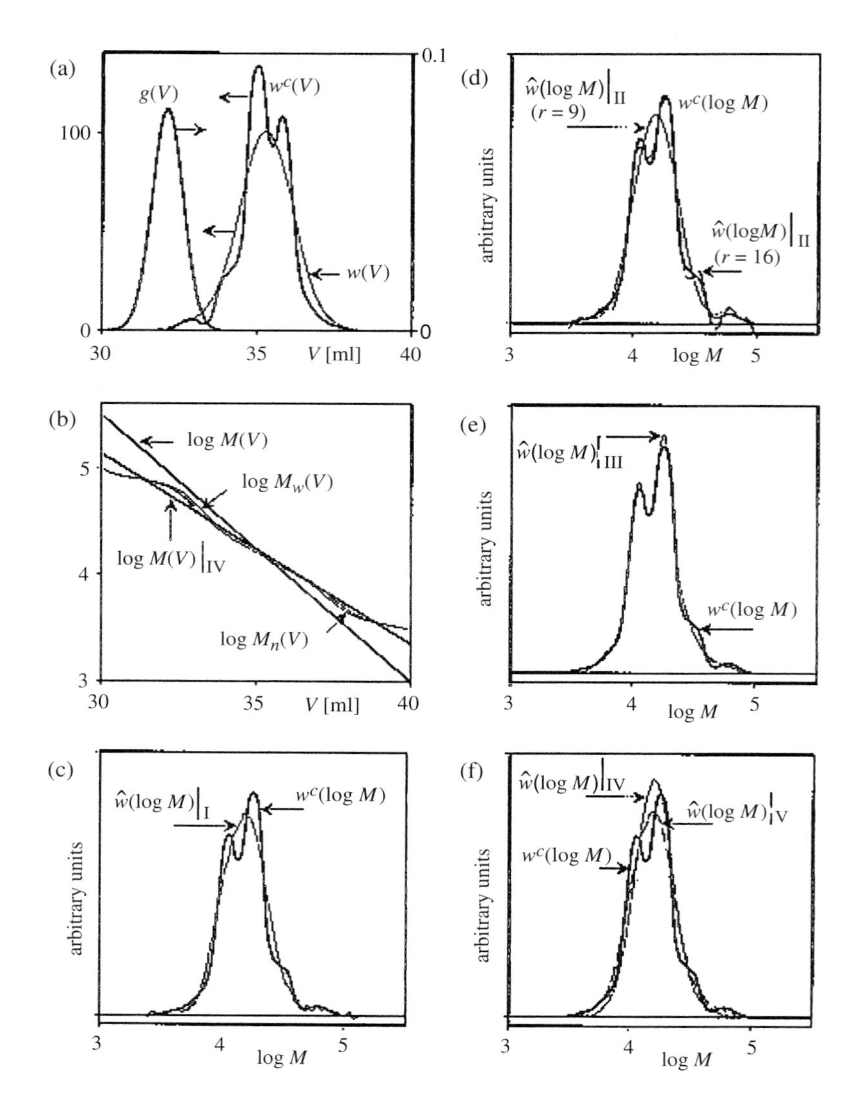

Fig. 1 The numerical example [6,11–13]. (a) "True" mass chromatogram $w^c(V)$, uniform broadening $g(V)$, and resulting "measured" chromatogram $w(V)$. (b) Molecular-weight calibration log $M(V)$, "rotated" calibration according to Method IV [8,9] log $M(V)|_{IV}$, and calibrations assuming ideal molar mass sensors log $M_n(V)$ and log $M_w(V)$. (c–f) Comparison between the true MWD $w^c(\log M)$ and the MWD estimates $\hat{w}^c(\log M)$ obtained from Methods I[4], II[5], III[6], IV[8,9], and V[10].

a rectangular probability distribution in the range ±0.5) to the noise-free measurement. The resulting "chromatogram" is $w(V)$ in Fig. 1a. Note that the multimodality of $w^c(V)$ is lost in $w(V)$.

In the cited works [6,11–13], only the ability of several numerical inversion algorithms for recuperating $w^c(V)$ was evaluated, and the MWD calculation was not considered. In the present article, the purely numerical Methods I–III are compared with Methods IV and V, which require a linear molecular-weight calibration. For this reason, the calibration log $M(V)$ is included in Fig. 1b. From that calibration and $w^c(V)$, the "true" MWD $w^c(\log M)$ of Figs. 1c–1f was obtained. Note that the selection of a uniform and Gaussian broadening is not a limitation for an adequate evaluation of the (more general) Methods I–III.

In Figs. 1c–1f, the MWDs recuperated through Methods I–V are compared with the real distribution. In Table 1, the real and estimated average molecular weights and polydispersities are presented. In Method I, the best results were found when limiting the procedure to only four iterations (Fig. 1c). In Method II, the full rank of **G** was $p = 61$, and the signal-to-noise ratio at the chromato-

Table 1 The Numerical Example[a]: "True" versus Recuperated Average Molecular Weights

	"True" values	Without correct.	Method I ($i = 4$)	Method II ($r = 9$)	($r = 16$)	Method III	Method IV	Method V
$\overline{M}_n$	13,975	13,464	14,029	13,993	14,033	14,084	14,871	14,038
$\overline{M}_w$	17,342	18,041	17,315	17,310	17,313	17,156	17,224	17,335
$\overline{M}_w/\overline{M}_n$	1.241	1.340	1.234	1.237	1.234	1.218	1.158	1.235

[a]From Refs. 6 and 11–13.

gram tails suggested truncating the summation of Eq. (5) at $r = 16$. The obtained solution exhibits a negative peak (Fig. 1d). For comparison, the less "rich" solution with $r = 9$ is also shown in Fig. 1d. In Method III, a (time-invariant) measurement noise variance was estimated from the baseline noise, and a (time-varying) solution variance was estimated by simply squaring the raw chromatogram heights (Fig. 1e). In Method IV, the rotated calibration is shown in Fig. 1b, and the recuperated distribution is given in Fig. 1f. In Method V, it was numerically observed that (for the adopted linear calibration) the solution becomes almost independent of $D_2(V)$ (Fig. 1f).

To verify the validity of Method IV, the real instantaneous MWDs were simulated. From such distributions, the true instantaneous averages $M_n(V)$ and $M_w(V)$ were calculated and represented in Fig. 1b using a logarithmic scale. Clearly, log $M_n(V)$ and log $M_w(V)$ differ from the (linear) rotated calibration. The differences occur because the original chromatogram is non-Gaussian. This illustrates the limitations of applying the technique to a chromatogram of an arbitrary shape.

The following comments can be made. Only Method II [5] (with $r = 16$) and Method III [6] were capable of recuperating the details of the true MWD; all others produce unimodal solutions. In general, the more sophisticated correction techniques such as Methods II and III require a priori information on the measurement noise and/or on the expected solution. This generally yields better estimations, but at the cost of complicating the adjustment procedure. The recuperated average molecular weights presented in Table 1 are, in all cases, quite acceptable.

Correction Methods for Multidetection SEC

A difficulty with multidetection SEC is that downstream chromatograms are shifted (and eventually distorted) with respect to the first-emerging chromatogram. This is a consequence of the interdetector capillaries and the (downstream) detector cell volumes. Such corrections are outside the scope of the present article.

Methods I–III are clearly applicable to chromatograms produced by any SEC sensor. As we have seen, all chromatograms are distorted in a similar fashion through a common broadening function $g(V, V_0)$. For this reason, similar difficulties in their individual numerical inversions are to be expected.

Many problems remain still unsolved in relation to axial dispersion correction in multidetection SEC. For example, when an on-line light-scattering detector is used, the instantaneous weight-average molecular weight is obtained from the ratio between the light scattering and the mass signals. Unfortunately, the measurement errors makes it unfeasible to independently invert each raw chromatogram prior to performing the signals ratio. Furthermore, even the direct signals ratio is only feasible in the mid-chromatogram section. A similar problem is presented when analyzing a linear copolymer by standard dual detection (i.e., employing a UV spectrophotometer and a differential refractometer). In this case, it seems preferable to first calculate the instantaneous mass and copolymer composition directly from the raw chromatograms, and then correct such derived variables for axial dispersion. [14] (The more obvious alternative procedure of first correcting the individual chromatograms for axial dispersion and then calculating the derived variables is not recommendable due to the propagation of errors [14].)

References

1. L. H. Tung, Method of calculating molecular weight distribution function from gel permeation chromatograms. III. Application of the method, *J. Appl. Polym. Sci. 10*: 1271 (1966).
2. M. Netopilik, Correction for axial dispersion in gel permeation chromatography with a detector of molar masses, *Polym. Bull. 7*: 575 (1982).
3. P. I. Prougenes, D. Berek, and G. R. Meira, Size exclusion chromatography of polymers with molar mass detection. Computer simulation study on instrumental broadening biases and proposed correction method, *Polymer 40*: 117 (1998).
4. T. Ishige, S. I. Lee, and A. E. Hamielec, Solution of Tung's

axial dispersion equation by numerical techniques, *J. Appl. Polym. Sci 15*: 1607 (1971).
5. J. M. Mendel, *Lessons in Estimation Theory for Signal Processing, Communications, and Control*, Prentice-Hall, Englewood Cliffs, NJ, 1995.
6. D. Alba and G. R. Meira, Inverse optimal filtering method for the instrumental broadening in SEC, *J. Liq. Chromatogr. 7*: 2833 (1984).
7. A. Felinger, *Data Analysis and Signal Processing in Chromatography*, Data Handling in Science and Technology Vol. 21, Elsevier, Amsterdam, 1998.
8. W. W. Yau, H. J. Stoklosa, and D. D. Bly, Calibration and molecular weight calculations in GPC using a new practical method for dispersion correction-GPCV2, *J. Appl. Polym. Sci. 21*: 1911 (1977).
9. C. Jackson and W. W. Yau, Computer simulation study of size exclusion chromatography with simultaneous viscometry and light scattering measurements, *J. Chromatogr. 645*: 209 (1993).
10. A. E. Hamielec, H. J. Ederer, and K. H. Ebert, Size exclusion chromatography of complex polymers. Generalized analytical corrections for imperfect resolution, *J. Liq. Chromatogr. 4*: 1697 (1981).
11. K. S. Chang and R. Y. M. Huang, A new method for calculating and correcting molecular weight distributions from permeation chromatography, *J. Appl. Polym. Sci. 13*: 1459 (1969).
12. L. M. Gugliotta, D. Alba, and G. R. Meira, Correction for instrumental broadening in SEC through a stochastic matrix approach based on Wiener filtering theory, *ACS Symp. Ser. 352*: 287 (1987).
13. L. M. Gugliotta, J. R. Vega, and G. R. Meira, Instrumental broadening correction in size exclusion chromatography. Comparison of several deconvolution techniques, *J. Liq. Chromatogr. 13*: 1671 (1990).
14. R. O. Bielsa and G. R. Meira, Linear copolymer analysis with dual-detection size exclusion chromatography: Correction for instrumental broadening, *J. Appl. Polym. Sci. 46*: 835 (1992).

Gregorio R. Meira
Jorge R. Vega

B

Band Broadening in Capillary Electrophoresis

Introduction

As in chromatography, band broadening in capillary electrophoresis (CE) is determined by a number of instrumental and sample parameters and has a negative effect on detectability, due to dilution. Also, as in chromatographic techniques, the user can minimize some, but not all, of the parameters contributing to band broadening. In capillary electrophoresis, injection and detection are generally on-column, so that band broadening is limited to on-column effects. As will be shown, several effects are similar in chromatography; others are specific for CE and, in particular, for the potential gradient as a driving force. General equations for CE in open systems are given where the relative contribution of electro-osmosis is given by the electromigration factor f_{em}, given by

$$f_{em} = \frac{\mu_{eff}}{\mu_{eff} + \mu_{EOF}}$$

in which μ_{eff} is the effective mobility and μ_{EOF} is the electro-osmotic flow mobility. This electromigration factor is unity for systems with suppressed EOF.

The band-broadening contributions can be described in the form of a plate-height equation, where one usually assumes, as in chromatography, mutual independence of terms.

Injection

Band broadening due to injection is naturally proportional to the injection volume, relative to the capillary volume, but, in contrast to chromatography, sample stacking or destacking may decrease or respectively increase the injection band broadening thus defined. Without stacking or destacking, the following plate-height term can be used:

$$H_{inj} = \frac{\delta_{inj}^2}{12L_d}$$

in which δ_{inj} is the length in the capillary of the sample plug and L_d is the length of the capillary to the detector. Naturally, the above relationship can be rewritten in terms of sample and capillary volume, which are in the order of 10 nL and 1 μL, respectively.

Diffusion

As in chromatography, the effect of diffusion on band broadening is generally pronounced. It is directly proportional to the diffusion coefficient and the residence time between injection and detection. This effect can, consequently, be reduced by increasing the voltage, or by increasing the electro-osmotic flow, in cases where cations are analyzed at positive inlet polarity, where it further shortens the analysis times. The effect is less at lower temperatures (as the diffusion coefficient decreases approximately 2.5% per degree Celsius of temperature drop), but most significantly decreases with increasing molecular mass of the sample component.

$$H_{diff} = \frac{2Dt_m}{L_d}$$

in which L_d is the capillary length to the detector, D is the diffusion coefficient, and t_m is the migration time. Substituting the diffusion coefficient, using the Nernst–Einstein relation, yields

$$H_{\text{diff}} = \frac{2RTf_{\text{em}}}{z_{\text{eff}}EF}$$

in which R is the gas constant, T is the temperature, z_{eff} is the overall effective charge of the sample ion, E is the electric field strength, and F is the Faraday constant. In this relationship, z_{eff} and E, by definition, have opposite signs for negative values of f_{em} only.

Detection

The detector time constant and detector cell volume are both involved. The slit width along the length of the capillary is proportional to the latter. A value of 200 μm for the slit width in the case of 10^5 plates in a 370-mm capillary has negligible contribution to band broadening:

$$H_{\text{slit}} = \frac{\delta_{\text{det}}^2}{12L_d}$$

where δ_{det} is the detector slit width along the capillary axis. In cases of diode array detection, larger slit widths are usually applied; this reduces the noise level but may affect the peak shape at high plate numbers ($>10^5$).

The contribution of the detector time constant τ is modeled by the following relation:

$$H_\tau = L_d\left(\frac{\tau}{t_m}\right)^2$$

A detector time constant τ of 0.2 s is generally safe.

Thermal Effects

In cases of a relatively high current density, power dissipation in the capillary may result in significant radial temperature profiles. The plate-height contribution is given by

$$H_{\text{ther}} = \frac{f_T^2\kappa^2E^5R_i^6z_{\text{eff}}Ff_{\text{em}}}{1536RT\lambda_s^2}$$

where f_T is the temperature coefficient for conductivity, κ is the specific conductivity of the buffer, E is the electric field strength, R_i is the capillary inner diameter, and λ_s is the thermal conductivity of the solution.

As the effective mobility increases with the temperature at approximately 2.5% per degree, radial mobility differences may accumulate to significant band-broadening effects. The effect increases with increasing current density and capillary inner diameter. In a 75-μm-inner diameter capillary, a power dissipation of 1–2 W/m is generally safe. This value is calculated by multiplying the voltage and the current and dividing by the capillary length. Under these conditions, the radial temperature profile in the capillary is less than 0.5°C and the contribution to peak broadening negligible. In the case of higher conductivity buffers (e.g., a pH 3 phosphate buffer), the power dissipation and temperature profile can be 10 times higher and the effect on peak broadening significant. It should be emphasized that more effective cooling has no effect on thermal band broadening; the only effect is decreased averaged temperature inside the capillary.

Electro-osmotic Effects

Electro-osmosis in open systems is generally considered not to contribute to peak broadening. In hydrodynamically closed systems with nonsuppressed electro-osmosis, or in cases of axially different electro-osmotic regimes, however, a considerable contribution may result. The corresponding plate-height term is

$$H_{\text{EOF}} = \frac{R_i^2\zeta^2\varepsilon^2z_{\text{eff}}EF}{24RT\eta^2\mu_{\text{eff}}^2}$$

where ε is the dielectric constant and η is the local viscosity of the buffer at the plane of shear. This relationship shows that, in closed systems, the ζ-potential should be close to zero and that a viscosity increase near the capillary wall will be advantageous.

Electrophoretic Effects

Peak broadening due to electrophoretic effects are generally proportional to the conductivity (and thus the ionic strength) of the sample solution, relative to that of the buffer. This effect is readily understood when considering that in the case of a high sample concentration, the electric field strength (and, consequently, the linear velocities) in the sample plug are much lower than in the adjacent buffer. Due to this, a dilution (destacking) of the sample occurs. This is illustrated in curve a in Fig. 1—the separation of a concentrated 1 m*M* solution benzenesulfonic, *p*-toluene sulfonic, and benzoic acid, dissolved in a buffer of 1 m*M* propionic acid/Tris to pH 8.

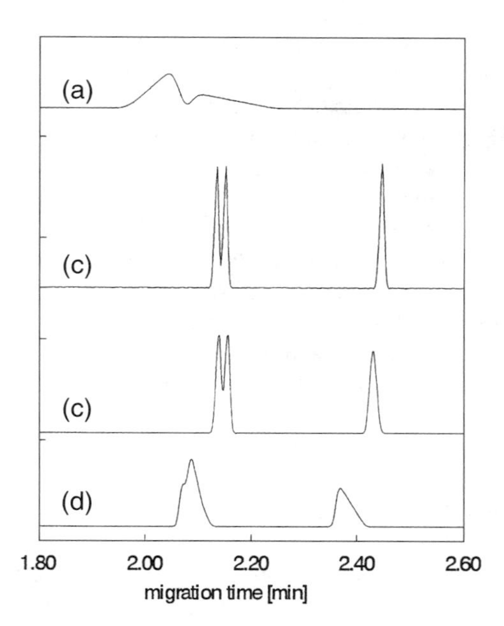

Fig. 1 Electrophoretic bandbroadening effects of benzoates as sample. Destacking trace a (1-m*M* sample in 1-m*M* buffer), stacking trace b (0.01-m*M* sample in 25-m*M* buffer), and trace c (1-m*M* sample in 25-m*M* buffer) and peak triangulation trace d (1-m*M* sample in 25-m*M* chloride buffer).

Alternatively, when injecting a low-conductivity (diluted, 0.01 m*M*) sample in a 25-m*M* buffer of same composition (curve b in Fig. 1—100 times amplified with respect to the others), the local field strength in the sample compartment is higher than in the adjacent buffer, resulting in a rapid focusing of ionic material at the sample–buffer interface (stacking), and resulting in very sharp sample injection plugs and high plate counts. This stacking takes place during the first second after switching on the high voltage. It may thus be advantageous to inject a larger volume of a more diluted sample for better efficiency. Choosing a higher-conductivity buffer also enhances the effect, where one has to consider that this may result in more pronounced band broadening due to other effects. Curve c in Fig. 1 shows that in such a high-conductivity buffer, even the 1-m*M* sample is separated to reasonable extent.

Peak symmetry is another important issue. Generally, capillary zone electrophoresis peaks are non-Gaussian and show nonsymmetry. This peak triangulation increases with increasing concentration overload. It is also proportional to the difference in effective mobility of the sample ion and the co-ion in the buffer. For instance, analyzing the same 1-m*M* sample mixture in a buffer consisting of, for example, 25 m*M* chloride/Tris to pH 8 will give triangular peaks (curve d in Fig. 1) because the effective mobility of benzoic acid is much lower than that of the buffer anion chloride: The buffer co-ion is not properly tuned to the sample component mobilities.

Suggested Further Reading

Giddings, J. C., in *Treatise on Analytical Chemistry*, I. M. Kolthoff and P. J. Elving (eds.), John Wiley & Sons, New York, 1981, Part I, Vol. 5.
Hjertén, S., *Chromatogr. Rev. 9*: 122 (1967).
Jorgenson, J. W., and K. D. Lucaks, *Science 222*: 266 (1983).
Kenndler, E., *J. Capillary Electrophoresis 3*(4): 191 (1996).
Reijenga, J. C. and E. Kenndler, *J. Chromatogr. A 659*: 403 (1994).
Reijenga, J. C. and E. Kenndler, *J. Chromatogr. A 659*: 417 (1994).
Virtanen, R., *Acta Polytech. Scand. 123*: 1 (1974).

Jetse C. Reijenga

Band Broadening in Size-Exclusion Chromatography

In classical chromatography, band broadening (BB), which defines the shape of the chromatogram of a pure solute, is one of the factors limiting the resolution, but individual peaks are generally observable and the discussion of BB extent is direct. In size-exclusion chromatography (SEC), the situation is more complex, as we observe, generally, only the envelope of a large number of individual peaks (Fig. 1). Imperfect resolution and its consequences on results cannot be directly observed. A few years after the pioneer publication on SEC by Moore [1], Tung [2] presented the general mathematical problem of band-broadening correction (BBC). Until 1975, a number of simplified procedures have been proposed in order to compensate for the limited resolution of columns. After 1975, a spectacular increase in column resolution rendered the problem less important, but, recently, there is

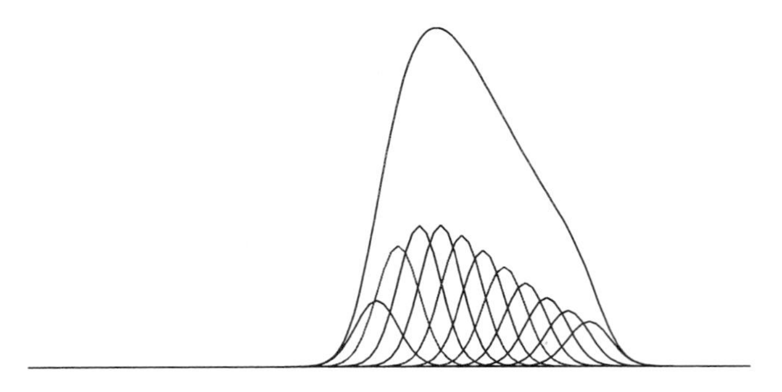

Fig. 1 Example of imperfect resolution: 10 peaks, $R = 0.25$ between neighbors.

a growing interest in BBC as SEC users intend to obtain more and more detailed information on molecular-weight distributions (MWDs) and not only average MW values. For this reason, this discussion is separated into three parts:

- Experimental determination of extent of BB
- Interpretation of BB processes
- Correction methods for BB

Experimental Determination of the Extent of BB

It is useful to choose a solute which is really eluted by a size-exclusion process, without adsorption or any additional interaction phenomena which might modify the shape of the peak. The most trivial method is to analyze the shape of a low-MW pure substance. This is usually used to determine the number of theoretical plates, $N = (V_r/\sigma)^2$, where V_r is the retention volume (volume at peak top) and σ is the standard deviation. σ can be computed from the weighing of each data point of the peak or can be estimated from the width at 10% maximum height ($\sigma = W_{0.1}/4.3$).

For this reason, when using THF as eluent and styrene/DVB gels, methanol or toluene are not good candidates; octadecane is preferred. For aqueous SEC, saccharose is the classical standard.

For polymers, a number of authors have claimed that the peak width increases as the MW increases, but to discuss band broadening for polymers, several precautions are required. First, it is necessary to be sure that the injected solution is sufficiently dilute to prevent any viscous effect. (Practically no viscous effect is observable, even for narrow standards when $[\eta]C < 0.1$; for flexible polymers, this corresponds roughly to a concentration <1 mg/mL for MWs up to 500,000; for a higher MW, it is necessary to reduce the concentration.) Then, the real difficulty is to analyze very narrow standards for which polymolecularity is sufficiently low, so as not to participate in the peak width, or at least which polymolecularity is very precisely known.

Commercial indications on standards are in progress, but suppliers rarely guarantee the exact value for $I_p = M_w/M_n$. A tendency is to guarantee that I_p is lower than a given value, but that is not sufficient for precise BB study. Usual values for medium MW standards are $I_p < 1.03$ and $I_p < 1.05$ for high MW standards; the situation is worse for aqueous SEC with values around 1.1.

Recently, possibilities appeared with the results from thermal gradient interaction chromatography (TGIC) [3]. That method has a much better resolution than SEC and allows a very precise determination of I_p. It is even possible to use it for a preparative scale to obtain extremely narrow standards, but, until now, it is only available for organic-soluble standards.

With all these precautions and when using modern columns, it clearly appears that peaks are not Gaussian, but systematically skewed. New computing facilities allow one to analyze more precisely such peaks by various functions. The best results seems to be obtained by exponentially modified gaussian (EMG) functions [4], which are the convolution of a Gaussian dispersion and an exponential decay. In that case, two parameters define the shape of the peak σ and τ, which allow quantitative mapping of BB characteristics for further correction. Generally, σ and τ are constant or admit a limited increase with MW for samples eluted well after the void volume. However, a dramatic increase of σ and τ occurs near the total exclusion volume [5].

Band-Broadening Interpretation

As previously indicated, this discussion is organized for chromatograms from very narrow polymer standards for which we can consider that the effect of molecular weight distribution is negligible and for which the unique separation process is size exclusion. With these limitations, the contribution to band broadening is conveniently separated into extra column effects, eddy dispersion, static dispersion, and mass transfer. In the most classical chromatographic interpretation, extra-column effects are not discussed and the three other contributions are considered as Gaussian, so there is simply the addition of their variances. The number of theoretical plates is defined as $N = (V_e/\sigma)^2$ and the influence of v, the linear velocity of the eluent, is summarized by the so-called Van Deemter equation:

$$H = \frac{L}{N} = a + \frac{b}{v} + cv$$

This classical interpretation is not sufficient, as experimental results clearly indicate that there is peak skewing; for this reason, it is useful to study each contribution separately.

Static Dispersion

This classical contribution corresponds to the diffusion of the sample along the axis of the column by Brownian motion during the time t_0 spent in the interstitial volume. That spreading effect is Gaussian and its standard deviation σ is related to D, the diffusion coefficient of the solute: $\sigma = (2Dt_0)^{1/2}$. In classical operating conditions in modern liquid chromatography, that contribution is generally a minor one, due to the use of relatively high flow rate. In SEC, the diffusion coefficients of polymers are very low and that contribution becomes negligible.

Extra-Column Effects

Generally, these effects are not discussed in detail, considering it is only necessary to select a chromatographic apparatus with a proper design to render these effects negligible. Recent results indicate that this situation tends to be different for macromolecular solutes. Using a chromatographic apparatus for which the column is replaced by tubings of various lengths, the end of elution is characterized by an exponential decay, the time of which is dependent not only on geometry but also strongly increases for high-MW solutes. The explanation is related to the more or less rapid averaging of radial positions of the solute in a cylindrical tube. In the case of high-molecular-weight solutes, the diffusion coefficient is small. Solute molecules which enter near the center of the tube stay in the high-velocity zone and those which enter near the walls stay in low-velocity zones; this introduces skewing which is much more important than for low-molecular-weight solutes.

Eddy Dispersion

This contribution is related to the variety of channels available for any solute molecule throughout the elution process. These channels are defined by the interstitial volume between the beads of the column package, so they correspond to a variety of shapes and flow velocities. This produces a distribution in elution time which is classically considered as Gaussian and weakly depends on flow rate. As a rule of thumb, the theoretical plate height corresponding to this effect can be considered as being equal to the bead diameter of the packing for well-packed columns.

More detailed results take into account a "wall effect" to explain why elution profiles are skewed, even for non-retained solutes in modern columns. Detailed experimental results were recently presented by Farkas and Guiochon [6] on the radial distribution of flow velocity using local multichannel detection devices. On average, the flow velocity is very homogeneous in the center of the column, but, inevitably, it becomes lower near the walls. Similarly, the peak shape from a local microdetector situated near the wall is clearly distorted and skewed compared with that of a similar detector situated near the center of the column.

Mass Transfer

The simple model of a theoretical plate, which is simply the affirmation of the existence of N successive equilibrium steps, is not satisfactory, as it assumes a Gaussian spreading.

To obtain more realistic information, it is necessary to discuss the rate of exchange between the interstitial volume and pores [7]. Potschka [8] proposed taking into account the competition between diffusion and convection in the special situation of "perfusion chromatography," where very large pores exist inside the beads, which allow some distribution of the solute by convection.

Recently, a model has been proposed for which pores are simply long cylinders and time of residence corresponds to a one-dimension Brownian motion [5]. Exact mathematical expressions are available for describing that process [9] and the distribution of such time of residence is highly skewed. Additionally, for each solute molecule, the number of visited pores obeys a Poisson distribution, and when the average number of visits becomes small, the distribution of elution time becomes wider and more skewed. That explains, precisely, why strong peak distortion is observed for samples eluted near the total exclusion limit.

Band-Broadening Correction

As first stated by Tung, the general starting point is that the experimental chromatogram $H(V)$ (from a concentration detector) is the convolution of $g(V, V_r)$, the spreading function defining the elution of a single species with a peak apex position of V_r, and $w(V_r)\,dV_r$, the weight fraction of species which peak apex, is between V_r and $V_r + dV_r$:

$$H(V) = \int_{V_1}^{V_2} g(V, V_r)w(V_r)\,dV_r$$

In any case, a second step consists of converting $w(V_r)$ into $w(M)$ by defining a calibration curve which correlates M and V_r.

As there is only a finite number of data points and, with some instrumental noise, generally such an inversion problem is ill-defined, the stability of the values of $w(V_r)$ depends on the algorithm which is used. Among the huge number of articles treating such problems, a review by Meira and co-workers [10] gives useful information on the mathematical aspects and a detailed review by Hamielec [11] presents a variety of instrumental situations, from the simplest one (constant Gaussian spreading function and simple concentration detector) to the most complex (general spreading function, multidetection).

To take into account experimental evidence which clearly indicates systematic skewing, this discussion will no longer concern methods limited to Gaussian spreading functions.

Simplest Situation: Constant Spreading Function, Linear Calibration Curve, Single Concentration Detector

This situation corresponds to a useful approximation in many cases, and it is almost strictly exact when the sample has a narrow MWD. From the chromatogram of an ideal isomolecular sample, considering that it is defined by a set of h_i values equidistant on the elution volume axis, classical summations give the uncorrected molecular-weight values:

$$M_{n_{\text{uncorrected}}} = \frac{\Sigma\, h_i}{\Sigma\, (h_i/M_i)} \quad \text{and} \quad M_{w_{\text{uncorrected}}} = \frac{\Sigma\, (h_i M_i)}{\Sigma\, h_i}$$

and the peak apex position gives the real molecular weight: M_{peak}.

Therefore, two correction factors exist for M_n and M_w:

$$K_n = \frac{M_{\text{peak}}}{M_{n_{\text{uncorrected}}}} \quad \text{and} \quad K_w = \frac{M_{\text{peak}}}{M_{w_{\text{uncorrected}}}}$$

For any other isomolecular sample analyzed on the same system, the change is simply a shift along the elution volume axis and each M_i value is multiplied by the same factor. Thus, the correction factors are unchanged. Finally, any broad MW sample analyzed on the same system is the addition of a set of isomolecular species; therefore, when summing, there is factorization of the correction factors and

$$M_{n_{\text{corrected}}} = K_n M_{n_{\text{uncorrected}}}$$
$$M_{w_{\text{corrected}}} = K_w M_{w_{\text{uncorrected}}}$$

This very simple BBC can always be used, at least to give a preliminary indication of the extent of BB. When using extremely narrow standards, as obtained by preparative TGIC, the result is accurate; more easily, it is possible to set the correction factors between two limits: lower one using data from a low-MW pure chemical and a higher limit using data from an imperfect narrow standard.

General Situation: Spreading Function Depends on Elution Volume and Calibration Curve Is Not Linear, Single Concentration Detector

In such cases, as stated by Meira and co-workers [10], the quality of results depends on computational refinements. It is necessary to add specific constraints related to the chromatographic problem: rejection of negative values or unrealistic fluctuations in the weight distribution. The normal way is to invert the large matrix defining the spreading function for any position on the elution volume scale. With modern computational facilities, that becomes easy, but it is still not trivial to obtain stable results, and proper filtering processes are useful.

Good results can be obtained by using a more direct iterative method which can be briefly presented: n equidistant values are chosen on the elution volume scale; let us note these values as V_i (typically, n can be 200). Any sample is arbitrarily defined as the sum of n isomolecular species, whose positions at the peak apex are V_j. For each peak j, n values of heights h_{ij} for each V_i value are computed, normalizing the surface (the peak shape is defined by interpolation from experimental BB data).

The chromatogram corresponding to the sample is defined by a set of n H_i values at positions V_i.

To define the weight distribution of the sample, it is simply necessary to adjust a set of w_i values until the summation converges toward the experimental H_i values. For the first attempt, $w_i = H_i$: This gives, by addition, a chromatogram ($H1_i$) Then, $w_i = H_i/H1_1$; that gives $H2_i$, and so on until convergence

The method is reasonably efficient. Stable convergence is observed except for very large samples for which the problem is too severely ill-conditioned. Applying it to narrow PS standards allows one to find, again, the true polymolecularity index.

Multidetection Problem

The aim of multidetection, especially LS/DRI coupling, is to find a useful calibration curve directly from the sample data and without external information from standards. A crude calibration curve is obtained by plotting, on a semilogarithmic scale, the instantaneous weight average M_{wi} values for each data point at elution volume V_i. At this point, correction for BB is rarely used, as the accuracy of the calibration curve is generally poor and does not justify sophisticated corrections. Additionally, for complex polymers (blends, copolymers, branched polymers, etc.), a variety of molecular weights are eluted at the same position, even in the absence of band broadening [12], so it is still very difficult to propose a general solution for the problem and results are available only for simplified situations. Normally, it would be necessary to find the exact weight distribution w_i, as in the preceding paragraph, simply using the DRI signal; then, it becomes possible to adjust the calibration curve until the calculated M_{wi} values converge toward the experimental set of M_{wi} values.

Conclusion

Band broadening in SEC has several specific aspects, compared with other chromatographic processes. Solutes may have very low diffusion coefficients and that introduces additional tailing due the imperfect averaging of radial positions all along the tubing. Mass transfer can be described from Brownian motion properties, and for samples eluted near total exclusion volume, as the number of visited pores become small, this introduces a significant increase of skewing and a very important loss of resolution.

For correcting band broadening, the main difficulty is in obtaining precise mapping of the spreading function of the system. Normally, this needs very high quality standards and TGIC offers new possibilities in that area. Computational techniques are now sufficiently efficient to solve the general inversion problem associated with band-broadening correction, but it still needs some precautions to obtain stable results without artificial oscillations. Finally, as corrections become very important and unstable near total exclusion volume, it remains very imprudent to interpret data when part of the sample is totally excluded. It is better to first find a well-adapted column set, able to efficiently fractionate the whole sample.

References

1. J. C. Moore, *J. Polym. Sci. A-2* 835 (1964).
2. L. H. Tung, *J. Appl. Polym. Sci. 10*: 375 (1966).
3. W. Lee, H. C. Lee, T. Park, T. Chang, and J. Y. Chang, *Polymer 40*: 7227 (1999).
4. M. S. Jeansonne and J. P. Foley, *J. Chromatogr. Sci. 29*: 258 (1991).
5. J.-P. Busnel, F. Foucault, L. Denis, W. Lee, and T. Chang, *J. Liq. Chromatogr. Related Technol.* (in press).
6. T. Farkas and G. Guiochon, *Anal. Chem 69*: 4592 (1997).
7. D. H. Kim and A. F. Johnson, in *Size Exclusion Chromatography*, T. Provder (ed.), ACS Symposium Series 245, American Chemical Society, Washington, DC, 1984.
8. M. Potschka, *J. Chromatogr 648*: 41 (1993).
9. I. Karatzas and S. Shreve, in *Brownian Motion and Stochastic Calculus*, Springer-Verlag, New York, 1991.
10. L. M. Gugliotta, J. R. Vega, and G. R. Meira, *J. Liq. Chromatogr. 13*: 1671 (1990).
11. A. E. Hamielec, in *Steric Exclusion Liquid Chromatography of Polymers*, J. Janca (ed.), Marcel Dekker, Inc., New York, 1984, pp. 117–160.
12. W. Radke, P. F. W. Simon, and A. H. E. Muller, *Macromolecules 29*: 4926 (1996).

Jean-Pierre Busnel

Binding Constants: Determination by Affinity Chromatography

Introduction

Numerous interactions within cells and the body are characterized by the specific binding that occurs between two or more molecules. Examples include the binding of hormones with hormone receptors, drugs with target enzymes or receptors, antibodies with antigens, and small solutes with transport proteins. The study of these interactions is important in determining the role they play in biological systems. Because of this, there have been numerous methods developed to characterize such reactions. One of these approaches is that of affinity chromatography.

Affinity chromatography is a liquid-chromatographic technique that makes use of an immobilized ligand, usually of biological origin, for the separation and analysis of analytes within a sample. However, it is also possible to use affinity chromatography as a tool for studying the interactions that take place between the ligand and injected solutes. This application is known as *quantitative* or *analytical affinity chromatography*. Some attractive features of this approach include its relative simplicity, good precision and accuracy, and ability to use the same ligand for multiple studies. There are various techniques that are employed for such studies. These include methods for measuring both equilibrium constants and rate constants for biological processes, thus giving data on the thermodynamics and kinetics of these reactions.

Zonal Elution

The method of *zonal elution* is one of the most common techniques used in affinity chromatography to examine biological interactions. An example of this type of experiment is shown in Fig. 1a. In its usual form, zonal elution involves the application of a small amount of analyte (in the absence or presence of a competing agent) to a column that contains an immobilized ligand. The retention of the analyte in this case will depend on how strongly the analyte and competing agent bind to the ligand and on the amount of ligand that is in the column. This makes it possible to measure the equilibrium constants for these binding processes by examining the change in analyte retention as the competing agent's concentration is varied. Zonal elution has been used to examine numerous biological systems, including the binding of drugs with transport proteins, lectins with sugars, enzymes with inhibitors, and hormones with hormone-binding proteins.

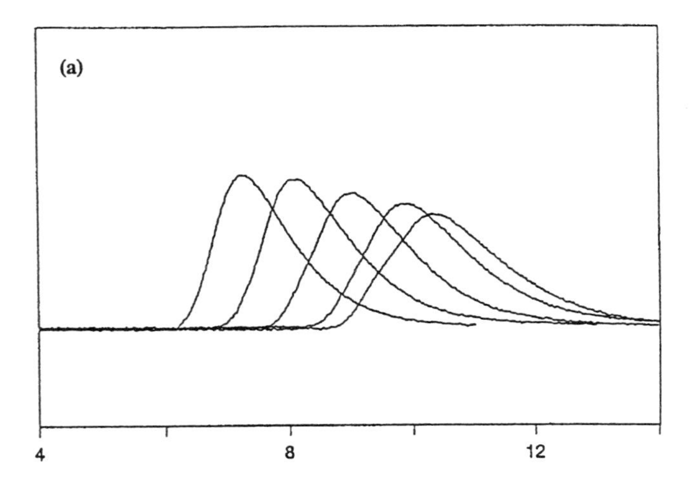

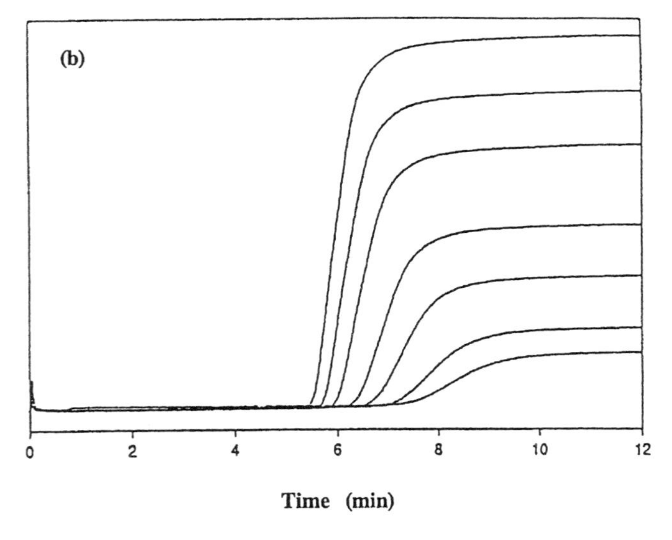

Fig. 1 (a) Zonal elution studies for the injection of R-warfarin onto an immobilized human serum albumin column in the presence (left to right) of 1.90-0 · 10^{-6} M L-reverse triiodothyronine as a competing agent. (b) Frontal analysis studies for the binding of R-warfarin to immobilized human serum albumin at applied analyte concentrations (left to right) of 1.50-0.22 · 10^{-6} M. [Reproduced with permission from Elsevier from B. Loun and D. S. Hage, *J. Chromatogr. B 665*: 303 (1995) and the American Chemical Society from B. Loun and D. S. Hage, *Anal. Chem. 66*: 3814 (1994).]

Equation (1) represents one specific type of zonal elution study, in which the injected analyte and competing agent bind at a single common site on the immobilized ligand:

$$k = \frac{\{K_{a,\mathrm{A}} m_L\}}{\{V_M(1 + K_{a,\mathrm{I}}[\mathrm{I}])\}} \tag{1}$$

Similar equations can be derived for other systems, such as those involving multiple types of binding sites or the presence of both soluble and immobilized forms of the ligand. In Eq. (1), $K_{a,\mathrm{A}}$ and $K_{a,\mathrm{I}}$ are the association equilibrium constants for the binding of the ligand to the analyte (A) and competing agent (I) at their site of competition. The term [I] is the concentration of I that is being applied to the column, m_L is the moles of common ligand sites for A and I, and V_M is the void volume of column. The term k is the retention factor (or capacity factor) that is measured for A, as given by the relationship $k = (t_R/t_M) - 1$, where t_R is the retention time for A and t_M is the column void time. In this case, the values of the association constants $K_{a,\mathrm{A}}$ and/or $K_{a,\mathrm{I}}$ can be obtained by examining how the retention factor for A changes with [I]. If these studies are performed at several temperatures, thermodynamic values can also be obtained for the changes in enthalpy and entropy that occur during these binding processes.

Frontal Analysis

An alternative approach for equilibrium constant measurements is to use the method of *frontal analysis*. In this technique, a solution containing a known concentration of the analyte is continuously applied to an affinity column at a fixed flow rate (see Fig. 1b). As the solute/analyte binds to the immobilized ligand, the ligand becomes saturated with the analyte and the amount of analyte eluting from the column gradually increases. This forms a characteristic breakthrough curve. The volume of the analyte solution required to reach the mean position of this curve is measured. If the association and the dissociation kinetics are fast, the mean position of the breakthrough curve will be related to the concentration of the applied solute, the amount of ligand in the column, and the association equilibrium constants for solute–ligand binding. Frontal analysis experiments have been used to examine such systems as drug–protein binding, antibody–antigen interactions, and enzyme–inhibitor interactions.

A simple example of a frontal analysis system is one where an applied analyte binds to a single type of immobilized ligand site. In this situation, the following equation can be used to relate the true number of active binding sites in the column (m_L) to the apparent moles of analyte ($m_{L,\text{app}}$) required to reach the mean position of the breakthrough curve:

$$\frac{1}{m_{L,\text{app}}} = \frac{1}{K_{a,\text{A}} m_L [\text{A}]} + \frac{1}{m_L} \tag{2}$$

As defined earlier, $K_{a,\text{A}}$ is the association constant for the binding of A to L, and [A] is the molar concentration of analyte applied to the column. Equation (2) predicts that a plot of $1/m_{L,\text{app}}$ versus $1/[\text{A}]$ for a system with single-site binding will give a straight line with a slope of $1/(K_{a,\text{A}} m_L)$ and an intercept of $1/m_L$. The value of $K_{a,\text{A}}$ can be determined by calculating the ratio of the intercept to the slope, and $1/m_L$ is obtained from the inverse of the intercept. Similar relationships can be derived for cases in which there is more than one type of binding site or in which both a competing agent and solute are applied simultaneously to the column.

One disadvantage of frontal analysis is that it requires a relatively large amount of analyte for study. However, frontal analysis does provide information on both the association constant for a solute and its total number of binding sites in a column. This feature makes frontal analysis the method of choice for high accuracy in equilibrium measurements, because the resulting association constants are essentially independent of the number of binding sites in the column.

Band-Broadening Measurements

Another group of methods in analytical affinity chromatography are those that examine the kinetics of biological interactions. *Band-broadening measurements* (also known as the *isocratic method*) represent one such approach. This is really a modification of the zonal elution method in which the widths of the eluting peaks are measured along with their retention times. Systems that have been studied with this method include the binding of lectins with sugars, the interactions of drugs and amino acids with serum albumin, and the kinetics of protein-based chiral stationary phases.

This type of experiment involves injecting a small amount of an analyte onto an affinity column while carefully monitoring the retention time and width of the eluting peak. These injections are performed at several flow rates on both the affinity column and on a column of the same size which contains an identical support but with no immobilized ligand being present. This control column is needed to help correct for any band broadening that occurs due to processes other than the binding and dissociation of the analyte from the immobilized ligand. By comparing plots of the peak widths (or plate heights) for the affinity and control columns, it is possible to determine the value of the dissociation rate constant for the analyte–ligand interaction.

Split-Peak Effect

Another way in which kinetic information can be obtained by affinity chromatography is to use the *split-peak effect*. This effect occurs when the injection of a single solute gives rise to two peaks: the first representing a nonretained fraction and the second representing the retained solute. This effect can be observed even when only a small amount of analyte is injected and is the result of slow adsorption kinetics and/or the slow mass transfer of the analyte within the column. Such an effect can occur in any type of chromatography, but it is most common in affinity columns because of their smaller size, their lower amount of binding sites, and the slower association rates of affinity ligands compared to other types of stationary phases.

One way in split-peak measurements can be performed is by injecting a small amount of analyte onto an affinity column at various flow rates. A plot of the inverse negative log of the measured free fraction is then made versus the flow rate. The slope of this graph is related to the adsorption kinetics and mass-transfer rates within the column. If the system is known to have adsorption-limited retention or if the mass-transfer rates are known, then the association rate constant for analyte binding can be determined. This approach has the advantages of being fast to perform and potentially has greater accuracy and precision than band-broadening measurements. Its disadvantages are that it requires fairly specialized operating conditions that may not be suitable for all analytes. Examples of biological systems that have been examined by the split-peak method include the binding of protein A and protein G to immunoglobulins, and the binding of antibodies with both high- and low-molecular-weight antigens.

Peak-Decay Method

The peak-decay method is a third approach that can be used in affinity chromatography to examine the kinetics of an analyte–ligand interaction. This technique is performed by first equilibrating and saturating a small affinity column with a solution that contains the analyte of interest or an easily detected analog of this analyte. The column is then quickly switched to a mobile phase in which the analyte is not present. The release of the bound analyte is then monitored over time, resulting in a decay curve. This decay is related to the dissociation rate of the analyte and the mass-transfer kinetics within the column. If the mass-transfer rate is known or is fast compared to analyte dissociation, then the decay curve can be used to provide the dissociation rate constant for the analyte from the immobilized ligand. Systems which have been studied with this approach include the dissociation of drugs from transport proteins and the dissociation of sugars from immobilized lectins.

Suggested Further Reading

Chaiken, I. M. (ed.), *Analytical Affinity Chromatography*, CRC Press, Boca Raton, FL, 1987.
Dunn, B. M. and I. M. Chaiken, *Proc. Natl. Acad. Sci. USA 71*: 2372 (1974).
Hage, D. S. and S. A. Tweed, *J. Chromatogr. B 699*: 499 (1997).
Hage, D. S., R. R. Walters, and H. W. Hethcote, *Anal. Chem. 58*: 274 (1986).
Kasai, K.-I. and S.-I. Ishii, *J. Biochem. 78*: 653 (1975).
Loun, B. and D. S. Hage, *Anal. Chem. 68*: 1218 (1996).
Wainer, I. W., *J. Chromatogr. A 666*: 221 (1994).

David S. Hage
Sanjay Mukherjee

Binding Molecules Via —SH Groups

A prerequisite for producing a good affinity support is a firm, stable attachment of the ligand to the surface of the support. There are numerous linkage chemistries available for performing this task, and although the most popular approach is a reaction between the reactive side groups on the support with a primary amine on the ligand, there are a number of supports that can perform similar attachments through free thiol or sulfhydryl groups. Supports containing maleimide reactive side groups are specific for free sulfhydryl groups present in the ligand when the reaction is performed at pH 6.5–7.0. At pH 7.0, the interaction of maleimides with sulfhydryl groups is approximately 1000-fold faster than with amine groups. The stable thioether linkage formed between the maleimide support and the sulfhydryl group on the ligand cannot be easily cleaved under physiological conditions, therefore ensuring a stable affinity matrix. Immobilization of sulfhydryl-containing molecules can also be achieved using either α-haloacetyl or pyridyl sulfide cross-linking agents. The α-haloacetyl cross-linkers [i.e., *N*-succinimidy(4-iodoacetyl) aminobenzoate)] contain a iodacetyl group that is able to react with sulfhydryl groups present in the ligand at physiological pH. During this reaction, the nucleophil substitution of iodine with a thiol takes place, producing a stable thioether linkage. However, a shortcoming of this approach is that the α-haloacetyls interact with other amino acids, especially when a shortage or absence of free sulfhydryl groups exists. Linkage of pyridyl disulfides with aliphatic thiols at pH 4.0–5.0 produces a disulfide bond with the release of pyridine-2-thione as a by-product of the reaction. A disadvantage of this approach is the acidic pH of the reaction, which is essential for optimal linkage. The reaction can be performed at physiological pH, but under these conditions, the reaction is extremely slow.

Ligand immobilization through sulfhydryl groups can be advantageous due to its ability to be site directed. Additionally, depending on the linkage, the ligand support can be cleavable, allowing the same support to be reused. However, many useful affinity ligands do not possess free sulfhydryl groups, and in such cases, free sulfhydryl groups can be engineered into the ligand via a series of commercially available reagents. Traut's reagent (2-iminothiolane) is the most common, although *N*-succinimidyl *S*-acretylthioacetate (SATA) and *N*-succinimidyl-3-(2-pyridyldithio)-propionate (SPDP) can also be used (Fig. 1). Traut's reagent reacts with primary amines present in the ligand introducing exposed sulfhydryl groups for further coupling reactions.

Chrisey et al. [1] described an interesting use of sulfhydryl-mediated immobilization for immobilizing thiol-modified DNA. A hetero-bifunctional cross-linker bearing both thiol and amino reactive groups was used

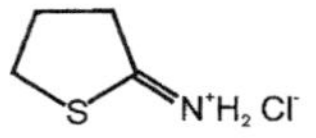

Traut's reagent

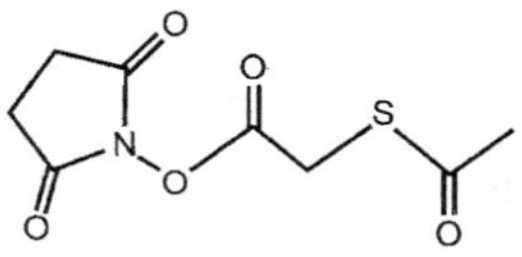

SATA

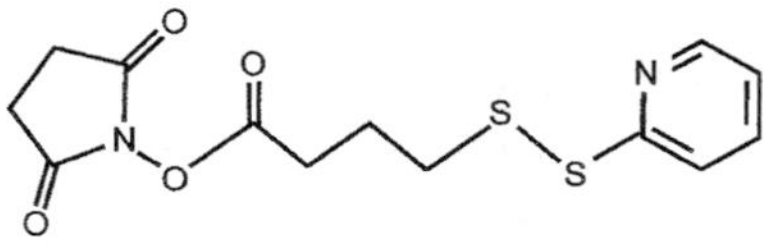

SPDP

Fig. 1 Chemical structures of commercially available reagents for introducing sulfhydryl groups into molecules.

to immobilize thiol-modified DNA oligomers to self-assembled monolayer silane films on fused-silica and oxidized silicon substrates. The advantage of this approach was to use site-directed immobilization to ensure the correct orientation of the DNA molecule.

Cleaving disulfide bonds already present in the ligand can also generate free sulfhydryl groups. The classic example of this approach is the digestion of the IgG antibody molecule to produce two monovalent, reactive FAb fragments, each containing a free sulfhydryl group. In this case, reduction of the disulfide bridge (holding the two FAb arms together) is achieved using Cleland's reagent (DTT: dithiothreitol). The FAb is then attached to free thiol groups present on the support by reforming a disulfide bond [2]. The advantage of this approach is that not only is a covalent linkage formed but also the linkage helps to orient the antigen receptor of the FAb away from the support matrix.

References

1. L. A. Chrisey, G. U. Lee, and C. E. O'Ferrall, *Nucleic Acids Res. 24*: 3031 (1996).
2. T. M. Phillips, *Anal. Chim. Acta 372*: 209 (1998).

Suggested Further Reading

Hermanson, G. T., A. K. Mallia, and P. K. Smith, *Immobilized Affinity Ligand Techniques*, Academic Press, New York, 1992.

Lundblad, R. L., *Techniques in Protein Modification*, CRC Press, Boca Raton, FL, 1995.

Wong, S. S., *Chemistry of Protein Conjugation and Cross-linking*, CRC Press, Boca Raton, FL, 1991.

Terry M. Phillips

Biopharmaceuticals by Capillary Electrophoresis

Introduction

In the relatively short period of time since the introduction of the first commercial instruments in the late 1980s, capillary electrophoresis (CE) has established itself as one of the most versatile analytical techniques. In addition to providing exceptional separation efficiencies, it offers substantial advantages over conventional slab–gel electrophoretic techniques, namely fast separation times, automation, reproducibility, and quantitative capabilities. Furthermore, owing to the different mechanisms by which products are separated in CE, data generated are generally complementary to those obtained by high-performance liquid chromatography (HPLC), thus allowing for more complete product characterization. CE methods have also been successfully validated with respect to well-established analytical criteria (e.g., precision, accuracy, reproducibility, and linearity), making them a source of reliable information. These considerations are of key importance to the pharmaceutical industry in adopting CE as a front-line analytical technique for product characterization to meet the specific requirements associated with the manufacturing and testing of biopharmaceuticals [1].

Historical Perspective in the Development of Biopharmaceuticals

Therapeutic products consisting of biopolymers (e.g., proteins) are generally referred to as biopharmaceuticals. Although, to date, most products on the market are proteins and polypeptides, new therapeutics based on antisense oligonucleotides or DNA fragments are being developed. Traditionally, biopharmaceuticals were obtained from biological sources (e.g., human, animal, plant, or cellular origin) in the form of crude extracts or partially purified components of extracts. Because of the highly complex nature of these mixtures, only minimal physicochemical characterization could be carried out and product evaluation was generally based on a biological response or surrogate bioassays. Although a few traditional products remain in use today, newer production methods based on recombinant DNA or hybridoma technology are now being used for the large-scale production of biopharmaceuticals. These developments have been paralleled by major advances in biomolecular separation techniques and, consequently, have resulted in improvements in product development leading to the preparation of more consistent products with purity levels approaching those of conventional, small-molecule pharmaceuticals. A number of important therapeutic proteins such as human growth hormone (hGH), insulin, interferons (IFN-α, β, and γ), tissue plasminogen activator (tPA), erythropoietin (EPO), and hepatitis B vaccine have been produced in this manner and their approval for human use has been based on a comprehensive chemistry and manufacturing submission with a strong emphasis on high-resolution analytical methodologies including CE.

Application of Capillary Electrophoresis to Biopharmaceuticals

Capillary electrophoresis has widespread applications in the field of biopharmaceuticals, particularly for product characterization. It is a technique particularly well suited for assessing product heterogeneity arising from post-translational modifications, degradation, or genetic variation. There are several CE modes, based on different separation mechanisms, which, alone or in combination, can be used (Table 1). The choice of the most appropriate CE separation mode, or combination thereof, will depend on the nature of the product under study and the type of information required. In the following sections, a brief overview of the use of CE for the characterization of biopharmaceuticals with respect to product identity and purity will be presented.

Product Identity

One of the critical aspects to be considered during the manufacturing of any drug is product identity. Although, in itself, it does not fulfill all of the requirements for a safe and effective drug, product identity testing provides assurance that the product generated is that which is intended and offers a measure of the consistency of the manufacturing process. CE-based methods have been widely applied to confirm product identity of biopharmaceuticals. Approaches usually involve the comparison of

Table 1 Common Capillary Electrophoresis Separation Modes for the Characterization of Biopharmaceuticals

Mode	Separation mechanism	Application
Capillary zone electrophoresis (CZE),	Charge-to-size ratio	Proteins and peptides, peptide mapping, glycoproteins, monoclonal antibodies, carbohydrates and oligosaccharides
Capillary isoelectric focusing (CIEF),	Isoelectric point (p*I*)	Proteins and peptides, glycoproteins, monoclonal antibodies, isoelectric point determination, peptide mapping
Capillary gel electrophoresis (CGE),	Size determination	Protein molecular weight determination, aggregates, oligonucleotides, DNA fragments, polysaccharides
Micellar electrokinetic chromatography (MEKC),	Partition based on hydrophobicity	Peptide mapping, carbohydrates

the property of the substrate to that of a preestablished, well-characterized reference standard with demonstrated efficacy and safety. Aside from performing a simple identity test involving comigration of the substrate with the reference standard, a number of methods have been devised to provide qualitative and quantitative information with respect to specific structural features of the molecule (e.g., primary sequence, molecular weight/size, and carbohydrate profile).

Peptide mapping is one of the most powerful tools for the identification of proteins and it has been successfully adapted to CE [2]. It involves the cleavage of the amino acid chain at specific sites, using proteases or chemicals, to generate a mixture of smaller peptides. The analysis of the resulting peptide digest is generally carried out by capillary zone electrophoresis (CZE), where products are separated based on differences in charge-to-mass ratios. Methods using capillary isoelectric focusing (CIEF) or micellar electrokinetic chromatography (MEKC) have also been reported. The peptide map serves as a fingerprint of the substrate which, when compared to a reference standard, enables the confirmation of the identity and allows the detection and identification of amino acid and peptide modifications. In addition, it may be used to confirm the presence and position of disulfide bridges and glycosylation sites. When linked to mass spectrometry (MS), peptide mapping by CZE can also be used as an effective replacement for protein sequencing. Besides its application to simple proteins, peptide mapping by CZE has been particularly useful for the characterization of monoclonal antibodies (MAbs) [3]. Peptide mapping by CZE is usually faster than by high-performance liquid chromatography (HPLC) and typically provides greater resolution of a larger number of peptides.

Several important therapeutic proteins are glycoproteins (e.g., EPO and tPA) which exist as mixtures of closely related species that differ in their glycosylation patterns (glycoforms). These differences are often the result of both compositional and sequence variations. Moreover the biological activity of glycoproteins is frequently linked to the presence of these carbohydrates and, consequently, the characterization of glycoprotein microheterogeneity represents one of the more challenging tasks in identity testing. Several CE approaches, based mostly on CZE and CIEF, have been devised [4]. For the frequently encountered sialoglycoproteins (i.e., sialic acid-containing glycoproteins), the analysis of the glycoform profile can be performed on intact glycoproteins. Alternatively, an analysis of the oligosaccharide profile may be performed following chemical or enzymatic release from the polypeptide. In both cases, the profile obtained is an indication of the varying number of sialic acid residues on the oligosaccharide chains. Methods have also been developed for the analysis of the monosaccharide composition resulting from hydrolysis. In such a case, the monosaccharides must be derivatized with reagents such as 1-aminopyrene-3,6,8-trisulfonate, which provides both a readily ionizable group and a detectable chromophore. Finally, CE is a valuable tool for the confirmation of the structural integrity of glycoproteins in final drug formulations [5].

Other useful identity tests that may be adequately performed by CE include protein molecular-weight (or DNA size) determination using capillary gel electrophoresis (CGE) [6] and isoelectric point (p*I*) determination using CIEF [7]. Typically, a protein molecular-weight determination is performed under denaturing conditions where sodium dodecyl sulfate (SDS)–protein complexes are formed with net negative charges that are proportional to their masses. These complexes migrate through the gel-filled capillary, acting as a sieving medium, in order of increasing molecular weight. The mobility of the substrate is used to estimate the molecular weight from a preestablished calibration plot of log molecular mass versus mobility prepared from a series of protein standards of known molecular mass. CGE separation of SDS–proteins has the advantage over SDS–polyacrylamide gel electrophoresis (PAGE) of giving higher resolution and requiring shorter analysis time. Similarly, using appropriate standards, CIEF can be used to determine the p*I* of a protein. Product identity techniques such as CGE and CIEF can be of great value to biopharmaceutical manufacturers because they can be incorporated into in-process controls as was recently demonstrated for two recombinant proteins [8].

Several monoclonal antibodies (MAbs) have been prepared for therapeutic purposes. They are among the most complex protein-based molecules, consisting of several light and heavy polypeptide chains, joined by multiple disulfide bridges, and containing a number of glycosylation sites of varying sequences and arrangements. Typically, MAbs are very large molecules with molecular weights around 150,000 Da, a feature that, when combined with their structural complexity, makes high-resolution chromatographic methods for the analysis of the intact molecule of little value. However, CE approaches have been highly successful for the characterization of MAbs [9]. Although all of the major CE separation modes have been applied, CIEF and CGE are particularly useful techniques. For instance, the high resolution achieved in CIEF allows monitoring of the profile of charge isoforms resulting from differential C-terminal processing (at lysine or arginine), a situation that frequently occurs in mammalian cell-derived products. CGE analysis under denaturing conditions has been used to estimate MAbs molecular weight as well as the presence of aggregates. When performed under denaturing and reducing conditions, CGE provides an effective way to monitor the light and heavy chains that make up the typical antibody structure.

Product Purity

Purity determination is an essential component of the assessment of the quality of any drug. However, the purity determination of biopharmaceuticals is not as straightforward as for small-molecule pharmaceuticals because biopharmaceuticals are structurally complex and have a wide range of potential impurities. Approaches usually involve the judicious choice of a combination of methods that will enable the detection and quantitation of impurities from which an overall purity assessment can be made. CE-generated data now play a significant role in such purity assessments.

Proteins are inherently labile molecules, especially when placed under non-physiological conditions, and, consequently, the formation of impurities may occur throughout their manufacturing process. Common protein degradation pathways leading to the formation of several types of impurities have been identified (Table 2) and most of these impurities can be detected by CE [10]. For instance, CE can be used successfully for the separation and detection of low levels of charge variants such as deamidation products of asparagine or glutamine residues as well as clipped forms resulting from proteolytic cleavage of the polypeptide chain. In particular, CZE has proven to be highly effective for simple proteins having no carbohydrate-mediated heterogeneity present. The high efficiency of CZE, in some cases, allows the resolution of multiple-charge variants, such as occur in hGH [11], to be accomplished in a single run. The high resolving power and quantitative properties of CGE can be used to detect non-dissociable aggregates and clipped forms in proteins as well as deletion sequence in antisense oligonucleotides.

Table 2 Typical Impurities in Protein Biopharmaceuticals

Deamidation products
Oxidation products
Disulfide interchange
Proteolytic cleavage products
Aggregates
Amino acid substitutions
N-, C-terminal truncated product

The coupling of CE to high-sensitivity detection devices such as laser-induced fluorescence (LIF) detectors provides substantial enhancement of the detection limits of impurities [12].

Conclusion

The use of capillary electrophoresis has become an integral part of the study of biopharmaceuticals, especially for the monitoring of product identity and purity. It is a powerful technique that, in many instances, has been shown to be superior to the more conventional electrophoretic techniques and complementary to the widely used high-resolution chromatographic techniques. It is particularly well suited to the study of complex mixtures such as glycoproteins and monoclonal antibodies.

References

1. M. Richardson, Biopharmaceutical regulation and product analysis: Origin, reform and the well-characterised product, *J. Biotechnol. Healthcare 3*: 36 (1996).
2. E. C. Rickard and J. K. Towns, The use of capillary electrophoresis for peptide mapping of proteins, in *New Methods in Peptide Mapping for the Characterization of Proteins*, W. S. Hancock (ed.), CRC Press, Boca Raton, FL 1996, pp. 97–118.
3. J. Liu, H. Zhao, K. J. Volk, S. E. Klohr, E. H. Kerns, and M. S. Lee, Analysis of monoclonal antibody and immunoconjugate digests by capillary electrophoresis and capillary liquid chromatography, *J. Chromatogr. A 735*: 357 (1996).
4. K. Kakehi and S. Honda, Analysis of glycoproteins, glycopeptides and glycoprotein-derived oligosaccharides by high performance capillary electrophoresis, *J. Chromatogr. A 220*: 377 (1996).
5. H. P. Bietlot and M. Girard, Analysis of recombinant human erythropoietin in drug formulations by high performance capillary electrophoresis, *J. Chromatogr. A 759*: 177 (1997).
6. B. L. Karger, F. Foret, and J. Berka, Capillary electrophoresis with polymer matrices: DNA and protein separation and analysis, *Methods Enzymol. 271*: 293 (1996).
7. T. Wehr, M. Zhu, and R. Rodriguez-Diaz, Capillary isoelectric focusing, *Methods Enzymol. 270*: 358 (1996).
8. A. Buchacher, P. Schulz, J. Choromanski, H. Schwinn, and D. Josic, High performance capillary electrophoresis for in-process control in the production of antithrombin III and human clotting factor IX, *J. Chromatogr. A 802*: 355 (1998).
9. I. S. Krull, X. Liu, J. Dai, C. Gendreau, and G. Li, HPCE methods for the identification and quantitation of antibodies, their conjugates and complexes, *J. Pharm. Biomed. Anal. 16*: 377 (1997).
10. G. Teshima and S.-L. Wu, Capillary electrophoresis anal-

ysis of recombinant proteins, *Methods Enzymol. 271*: 264 (1996).

11. P. Dupin, F. Galinou, and A. Bayol, Analysis of recombinant human growth hormone and its related impurities by capillary electrophoresis, *J. Chromatogr. A 707*: 396 (1995).
12. T. T. Lee, S. J. Lillard and E. S. Yeung, Screening and characterization of biopharmaceuticals by high performance capillary electrophoresis with laser-induced native fluorescence detection, *Electrophoresis 14*: 429 (1993).

Michel Girard

Bonded Phases in HPLC

The development of chemically bonded stationary phases is one of the major factors that lead to the growth of high-performance liquid chromatography (HPLC) and is responsible for its importance as a separation technique. In its earliest form, gravity flow moved the mobile phase through the column which was generally packed with a solid adsorbent such as silica or alumina. In a few instances, a high-molecular-weight liquid was coated on the solid particle to provide different types of selectivity. Under these circumstances, the column was similar to those used in gas chromatography (GC), where a liquid stationary phase was held in place by physical forces alone. In GC, the requirement for the stationary phase to remain in place for a long time is low volatility. In liquid chromatography, the requirement for durability is insolubility in the mobile phase. However, with the development of reliable high-pressure pumps that could produce stable flow rates for long periods of time, immiscibility with the mobile phase is not sufficient. At the pressures used to force solvents through most packed HPLC columns (from tens to a few hundred atmospheres), the shear forces developed at the interface between the stationary and the mobile phases are high enough to remove even insoluble liquids from the surface of the solid support. The stationary phase then is forced out of the column as an insoluble droplet. Removal of the stationary phase from a chromatography column is usually referred to as "column bleed." Therefore, it was necessary to develop a means of fixing the stationary phase on the solid support through a chemical bond. If the chemical bond between the surface of the solid support and the compound used as the stationary phase is stable under the experimental conditions of the HPLC experiment (temperature and mobile-phase composition), then column bleed will be avoided.

Fortunately, the most common support material used in liquid-chromatography experiments was silica. The chemistry of silica had been investigated for many years so a considerable amount of information was available about possible reactions on its surface. Silica can be considered as a polymer of silicic acid (H_2SiO_3). The terminal groups of the polymer located on the surface of the solid are hydroxide groups. These Si—OH functions are referred to as silanols. Because they come from an acid precursor, they are acidic themselves and generally have a pK_a near 5. This value is variable, depending on other constituents in the silica matrix such as metals. The structure of silica, including its major chemical features, are shown in Fig. 1. The polymeric unit consists of a series of siloxane bonds (—Si—O—Si—) that are slightly hydrophobic in nature. What is generally regarded as the most prominent feature on the surface is the silanol group, as indicated earlier. In a few cases, a single silicon atom will have two hydroxyl groups, which is called a geminal silanol. The silanols exist in two forms. First, they can be independent of other entities around them and are thus referred to as free or isolated silanols. If they are close enough to interact with a neighboring silanol, then these moieties become hydrogen-bonded or associated silanols. All forms

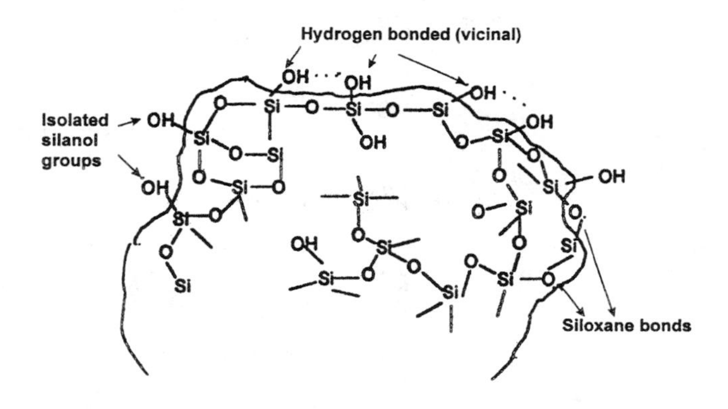

Fig. 1 Structure of silica showing the surface chemical features.

of silanol are polar hydrophilic species. The relative number of free versus hydrogen-bonded silanols also has an influence on the pK_a value of the silica. Finally, because of the polar and hydrogen-bonding characteristics of the silanols, water is strongly adsorbed to the surface. This water is not easily removed, even at prolonged heating above 100°C. It is this complex matrix that must undergo a chemical reaction in order to attach a moiety to the surface as a stationary phase. According to the findings of early investigations on the reactivity of silica, it was determined that the silanol groups were the site of chemical modification on the surface.

The concept of attaching an organic moiety as a stationary phase to a silica surface was first applied in packed-column gas chromatography. The rationale for developing these materials was to prevent column bleed at the high temperatures required for some separations in GC. As long as the chemical bond was stable, the organic moiety would remain fixed to the surface. Some of the reactions utilized for the attachment of organic compounds in the synthesis of bonded stationary phases were originally developed for the modification of ordinary glass surfaces. Therefore, it was known that most of these modified surfaces were reasonably temperature stable and should be applicable to the bonding of organic groups onto the porous silica particles used as supports in chromatography.

The first reaction used for the modification of porous silica in chromatography involves an alcohol as the organic species. This process is referred to as an esterification reaction. This may seem like incorrect nomenclature in order to describe the chemical process taking place between the silianol (Si—OH) and the organic compound (R—OH). However, the OH of the silanol is an acidic species, so the reaction taking place involves an acid and an alcohol, which, in typical organic chemistry nomenclature, is an esterification. The chemical reaction is illustrated in Fig. 2. The product of this reaction can be used as a stationary phase in gas chromatography because the material is thermally stable up to temperatures of approximately 300°C. However, the Si—O—C bond that exists between the surface and the bonded moiety is hydrolytically unstable in the presence of relatively small amounts of water. Therefore, these materials cannot be used for stationary phases in liquid chromatography,

REACTION TYPE	REACTION	SURFACE LINKAGES
ESTERIFICATION	$Si-OH + R-OH \rightarrow Si-OR + H_2O$	Si - O - C
ORGANOSILANE	$Si-OH + X-SiR'_2R \rightarrow Si-O-SiR'_2R + HX$ O / Si - OH / O $+ X_3Si-R \rightarrow$ O / Si - O - Si - R / O (each Si bearing O above and below) $+ 3HX$	Si - O - Si - C
CHLORINATION FOLLOWED BY REACTION OF GRIGNARD REAGENTS OR ORGANOLITHIUM COMPOUNDS	$Si-OH + SOCl_2 \xrightarrow{Toluene} Si-Cl + SO_2 + HCl$ a). $Si-Cl + BrMgR \rightarrow Si-R + MgClBr$ or b). $Si-Cl + Li-R \rightarrow Si-R + LiCl$	Si - C
a). TES SILANIZATION	a). O / Si - OH / O / Si - OH / O / Si - OH / O $\rightarrow$ O / Si - O - Si - H / O / Si - O - Si - H / O / Si - O - Si - H / O	a). Si - H monolayer
b). HYDROSILATION	b). $Si-H + CH_2=CH-R \xrightarrow{Catalyst} Si-CH_2-CH_2-R$	b). Si - C

Fig. 2 Reactions for the modification of silica surfaces.

where water comprises even a small fraction of the mobile phase.

The second reaction shown in Fig. 2 is the most common means used for the modification of silica surfaces. This method is referred to as organosilanization. Within this general reaction scheme, there are two possible approaches, as shown in Fig. 2. The first possibility involves the use of an organosilane reagent (RR′R′SiX) with only a single reactive group (X). The substituents on the silicon atom are as follows: X is a halide, most often Cl, methoxy or ethoxy; R is the organic moiety giving the surface the desired properties (i.e., hydrophobic, hydrophilic, ionic, etc.), and R′ is a small organic group, typically methyl. This reaction leads to a single siloxane bond between the reagent and the surface. Because of the single point of attachment of the reagent, the resulting bonded material is referred to as a monomeric phase. The second approach to organosilanization involves a reagent with the general formula $RSiX_3$. The substituents on the silicon atom in this reagent are defined as above. The basic difference between the approaches (as shown in Fig. 2) is that the reagent with three reactive groups results in bonding to the surface as well as cross-linking among adjacent bonded moieties and is referred to as a polymeric phase. This cross-linking effect provides extra stability to the bonded moiety but is less reproducible than the monomeric method. The one-step organosilanization procedure is relatively easy and the modification of the surface can be done by stirring the reagent continuously with the porous silica support. The reaction mixture is heated for about 1–2 h, then the reagent solution is removed, usually by centrifugation and/or filtration. The bonded phase is then washed with several solvents and dried under vacuum to remove as much of the rinse solutions as possible. Organosilanization accounts for virtually all of the commercially available chemically bonded stationary phases.

Another method that has been reported for the modification of silica supports is based on a chlorination/organometalation two-step reaction sequence. This process is also depicted in Fig. 2. In the first step, the silanols on the porous silica surface are converted to chlorides via a reaction with thionyl chloride. This step must be done under extremely dry conditions because the presence of any water results in the reversal of the reaction with hydroxyl replacing the chloride (Si—Cl), resulting in the regeneration of silanols (Si—OH). If the chlorinated material can be preserved (usually done in a closed vessel purged with a dry gas like nitrogen), then an organic group can be attached to the surface via a Grignard reaction or an organolithium reaction. The main advantage of this process is that it produces a very stable silicon–carbon linkage at the surface. However, the stringent reaction conditions for the first step and the possibility of forming salts that could affect chromatographic properties as byproducts in the second reaction have resulted in relatively little commercial use of this process.

The final method shown in Fig. 2 involves, first, silanization of the silica surface, followed by attachment of the organic group through a hydrosilation reaction. In the first step, the use of triethoxysilane under controlled conditions results in a monolayer of the cross-linked reagent being deposited on the surface. This reaction results in the replacement of hydroxides by hydrides. In the second step, an organic moiety is attached to the surface via the hydride moiety by a hydrosilation reaction using a catalyst such as hexachloroplatinic acid (Speier's catalyst), but other transition metal complexes or a free-radical initiator have been reported as well. This process also results in a silicon–carbon bond at the surface, does not require dry conditions (water is required as a catalyst in the first step), and is applicable to a variety of unsaturated functional groups in the hydrosilation reaction, although terminal olefins are the most common. The silanization/hydrosilation method also has seen limited commercial utilization to date.

In all of the reactions described, the choice of the organic moiety on the reagent (R group) determines the properties of the material as a stationary phase. Therefore, selection of a hydrophobic moiety where R is typically an alkyl group leads to a stationary phase that selectively retains nonpolar analytes. These materials are typically used in reversed-phase chromatography. If the organic moiety contains a polar functional group such as amine, cyano, or diol, then the stationary phase selectively retains polar compounds. These materials are typically used in normal-phase chromatography.

The bonding of the organic group on the surface results in the replacement of silanols whose adsorptive properties are strong, especially for bases, and often nonreproducible. Although it is impossible to replace all silanols, the remaining Si—OH groups are often shielded from solutes by the steric hindrance of the bonded organic moiety. In many cases though, some silanols are accessible to typical solutes. In order to inhibit the interaction between analytes and residual silanols, the bonded phase is subjected to an additional reaction referred to as endcapping. In this case, a small organosilane, often trimethylchlorosilane, penetrates into the spaces between the bonded groups to react with the most accessible silanols. This process generally greatly reduces or eliminates solute interactions with silanols.

After the bonded phase is prepared, it must be packed into a column, usually a stainless-steel tube. In order for the material to form a uniform bed of high density that will not form voids after prolonged use, the packing process must be done under high pressure (> 500 atm). The stationary phase is mixed with a solvent of approximately the same density as silica, so that a slurry is formed. This slurry is then forced into the column at high pressure with another solvent, usually methanol. After packing, most stationary phases require several hours of conditioning, with the mobile phase passing through the column at normal flow rates, before actual chromatographic analysis can begin.

Suggested Further Reading

Iler, R. K., *The Chemistry of Silica*, John Wiley & Sons, New York, 1979.
Marciniec, B., *Comprehensive Handbook on Hydrosilylation*, Pergamon Press, Oxford, 1992.
Nawrocki, J., *Chromatographia 31*: 177 (1991).
Nawrocki, J., *Chromatographia, 31*: 193 (1991).
Pesek, J. J. and M. T. Matyska, *Interf. Sci. 5*: 103 (1997).
Pesek, J. J., M. T. Matyska, J. E. Sandoval, and E. J. Williamsen, *J. Liq. Chromatogr. Related Technol. 19*: 2843 (1996).
Unger, K. K., *Porous Silica*, Elsevier, Amsterdam, 1979.
Vansant, E. F., P. Van Der Voort and K. C. Vrancken, *Characterization and Chemical Modification of Silica*, Elsevier, Amsterdam, 1995.

Joseph J. Pesek
Maria T. Matyska

Buffer Systems for Capillary Electrophoresis

Introduction

The solution contained within the capillary in which the separation occurs is known as the background electrolyte (BGE), carrier electrolyte, or, simply, the buffer. The BGE always contains a buffer because pH control is the most important parameter in electrophoresis. The pH may affect the charge and thus the mobility of an ionizable solute. The electro-osmotic flow (EOF) is also affected by the buffer pH. Table 1 contains a list of buffers that may prove useful in high-performance capillary electrophoresis (HPCE). As will be seen later, only a few of these buffers are necessary for most separations.

Other reagents, known as additives, are often added to the BGE to adjust selectivity (secondary equilibrium), modify the EOF, maintain solubility, and reduce the adherence of the solute or sample matrix components to the capillary wall. Table 2 provides these applications, along with some of the commonly used reagents.

Buffers

The selection of the appropriate buffer is usually straightforward. For acids, start with a borate buffer (pH 9.3), and for bases, a phosphate buffer (pH 2.5). These two buffer systems, along with the appropriate additives will work well for most applications. Both buffers have good buffer capacity and the ultraviolet (UV) absorbance of each is low. If bases are not soluble in phosphate buffer, acetate buffer (pH 4) may be more effective. Higher pHs may be required for basic proteins to avoid solute adherence to the capillary wall. If pH 7 is desired, the phosphate buffer works well at that pH. If necessary, the buffer pH can be fine-tuned using a mobility plot.

Alternative buffer systems include zwitterions and dual-buffering reagents. Zwitterionic buffers such as bicine, tricine, CAPS, MES, and Tris may be useful for protein and peptide separations. An advantage of a zwitterionic buffer is low conductivity when the buffer pH is adjusted to its pI. There is little buffer capacity when the pK_a and pI are separated by more than 2 pH units. When the pI and pK_a are close together, the buffer is known as an isoelectric buffer [1].

Selection of the appropriate counterion is also important. Lithium ion has the lowest mobility of the alkali earth metals. Its use provides for a low-conductivity buffer. Sodium salts are used more frequently due to purity and availability. It makes little sense to ever use a potassium salt. Dual-buffering systems with low-mobility ions and counterions (Tris-phosphate, Tris-borate, aminomethylpropanediol–cacodylic acid) are effective in minimizing buffer conductivity. These buffers are often used in the slab–gel, where low conductivity is particularly important.

The buffer concentration plays an important role in the separation. Typical buffer concentrations range from 20 to 150 mM. At the higher buffer concentrations, the production of heat may require the use of lower field strength

Table 1 Buffers for HPCE

Buffer	pK_a	Buffer	pK_a
Aspartate	1.99	DIPSO	7.5
Phosphate	2.14, 7.10, 13.3	HEPES	7.51
Citrate	3.12, 4.76, 6.40	TAPSO	7.58
β-Alanine	3.55	HEPPSO	7.9
Formate	3.75	EPPS	7.9
Lactate	3.85	POPSO	7.9
Acetate	4.76	DEB	7.91
Creatinine	4.89	Tricine	8.05
MES	6.13	GLYGLY	8.2
ACES	6.75	Bicine	8.25
MOPSO	6.79	TAPS	8.4
BES	7.16	Borate	9.14
MOPS	7.2	CHES	9.55
TES	7.45	CAPS	10.4

or smaller-diameter capillaries (25 μm instead of 50 μm). An Ohm's law plot is used to select the appropriate voltage. The advantages of high-concentration buffers include improved peak shape, fewer wall effects, and increased sample stacking.

Low-concentration buffers (less that 20 mM) provide the fastest separations because solute mobility and EOF is inversely proportional to the square root of the buffer concentration. Because the conductivity of a dilute buffer is low, a high electric field strength can be used as well. Problems with low-concentration buffers are loading capacity, wall effects, and poor stacking. Sawtooth-shaped peaks from a process known as electrodispersion may occur whenever the solute concentration approaches the BGE concentration. It also becomes more likely that proteins will adhere to the capillary wall when the buffer concentration is low. Ionic-strength-mediated sample stacking relies on a high-conductivity BGE and a low-conductivity sample [2]. This important process is less effective at low buffer concentrations. When indirect detec-

Table 2 Buffer additives

Purpose	Reagent	Mechanism
Modify mobility	Borate	Complex with carbohydrates, diols
	Calixarenes	Inclusion complex
	Chelating agents	Complex formation with metals
	Crown ethers	Inclusion complex
	Cyclodextrins	Inclusion complex
	Dendrimers	Inclusion complex
	Macrocyclic antibiotics	Inclusion complex
	Organic solvents	Solvation
	Sulfonic acids	Ion-pair formation
	Surfactants	Micelle interaction
	Transition metals	Complex formation
	Quaternary amines	Ion-pair formation
Modify EOF	Cationic surfactant	Dynamic coating, EOF reversal
	Linear polymers	Dynamic coating
	Organic solvents	Affects viscosity
Reduce wall effects	Cationic surfactant	Dynamic coating, EOF reversal
	Linear polymers	Dynamic coating
Polyamines		Covers silanols
Maintain solubility	Organic solvents	Hydrophobicity
	Urea	"Iceberg effect"

tion is employed, the buffer (indirect detection reagent) concentration must be kept low to optimize sensitivity [3]. Sawtooth peaks are often observed when this technique is used.

It is important to refresh the BGE reservoirs frequently to avoid a process known as buffer depletion [4]. Electrolysis at the respective electrodes produces protons and hydroxide ions. This can cause pH changes in the buffer reservoirs.

High-pH buffers (>pH 11) are used for certain small ion separations and for the separation of carbohydrates using indirect detection. Adsorption of carbon dioxide can cause the buffer pH to decline. It is best to use small containers filled to the top when storing these buffers.

Buffer Additives

Secondary Equilibrium

If two solutes are inseparable based on pH alone, secondary equilibrium can be employed to effect a separation. The following equilibrium expressions can be written [5].

$$A^+ + R \overset{K_A}{\rightleftharpoons} A^+R \tag{1}$$

$$B^+ + R \overset{K_B}{\rightleftharpoons} B^+R \tag{2}$$

If the equilibrium is pushed too far to the left, no separation can occur because A^+ and B^+ are inseparable. When the reagent interacts with the solute, the mobility decreases because the neutral reagent contributes mass without charge. However, if the equilibrium is pushed too far to the right, no separation occurs because A^+R and B^+R are inseparable. Separation only occurs when two conditions are met:

1. K_A does not equal K_B.
2. The equilibrium is not pushed to either extreme.

The next feature to consider is the charge of the reagent and solute. To separate charged solutes, the reagent can be charged or neutral. When the solute is neutral, the reagent must be charged.

Micelles and cyclodextrins are the most common reagents used for this technique. Micellar electrokinetic capillary chromatography (MECC or MEKC) is generally used for the separation of small molecules [6]. Sodium dodecyl sulfate at concentrations from 20 to 150 m*M* in conjunction with 20 m*M* borate buffer (pH 9.3) or phosphate buffer (pH 7.0) represent the most common operating conditions. The mechanism of separation is related to reversed-phase liquid chromatography, at least for neutral solutes. Organic solvents such as 5–20% methanol or acetonitrile are useful to modify selectivity when there is too much "retention" in the system. Alternative surfactants such as bile salts (sodium cholate), cationic surfactants (cetyltrimethylammonium bromide), nonionic surfactants (polyoxyethylene-23-lauryl ether), and alkyl glucosides can be used as well.

Cyclodextrins (CD) are frequently used for chiral recognition [7], although they are quite useful for achiral applications as well. Many classes have been used including native, functionalized, sulfobutylether, and highly sulfated CDs. The latter two are generally most effective for chiral and structural isomer separations. The typical CD concentrations range from 1 to 20 m*M* in 20–50 m*M* of borate (pH 9.3) or phosphate buffer (pH 2.5). Other reagents useful for chiral recognition include macrocyclic antibiotics, bile salts, chiral surfactants, noncyclic oligosaccharides and polysaccharides, and crown ethers.

Additional reagents useful for secondary equilibrium include borate buffer for carbohydrates, chelating agents for transition metals, ion-pair reagents for acids and bases, transition metals for proteins and peptides, silver ion for alkenes, and Mg^{2+} for nucleosides.

Electro-osmotic Flow Control

The control of EOF is critical to the migration time precision of the separation. Among the factors affecting the EOF are buffer pH, buffer concentration, buffer viscosity, temperature, organic modifiers, cationic surfactants or protonated amines, polymer additives, field strength, and the nature of the capillary surface.

At pH 2.5, the EOF is approximately 10^{-5} $cm^2/V/s$ in 50 m*M* buffer. At pH 7, it is an order of magnitude higher. The EOF is inversely proportional to BGE viscosity and is proportional to temperature, up until the point where heat dissipation is inadequate. Organic modifiers such as methanol decrease the EOF because hydro-organic mixtures have higher viscosity compared to water alone. Acetonitrile does not strongly affect the EOF. Polymer additives such as methylcellulose derivatives increase viscosity as well as coat the capillary wall.

Cationic surfactants and protonated polyamines may reverse the direction of the EOF as they impart a positive charge on the capillary wall. This technique is used to prevent wall interactions with cationic proteins. Changing the direction of the EOF is important in anion analysis where comigration of anions and the EOF is required. Otherwise, highly mobile anions such as chloride migrate toward the anode, whereas lower mobility anions are swept by the EOF toward the cathode.

A new series of reagents (CElixir, MicroSOLV, Long Branch, NJ) have been shown to dramatically stabilize the EOF, resulting in highly reproducible run-to-run and capillary-to-capillary migration times [8]. First, a capillary is treated as usual with $0.1N$ sodium hydroxide, followed by a rinse with a polycation solution. Then, a second layer consisting of a polyanion in a buffer at the desired pH is flushed through the capillary. Replicate runs are virtually superimposible, yielding reproducibility seldom found in HPCE. The reagents have been shown to work best for bases below at a pH below the pK_a.

Maintaining Solubility

All solutes and matrix components must remain in solution for an effective separation to occur. In aqueous systems, surfactants and urea are the most useful reagents. Organic solvents can be used as well, but this is less desirable because of evaporation. It can be difficult to separate solutes with widely different solubilities in a single run. In some cases, nonaqueous separations are necessary.

Reducing Wall Effects

Wall effects, or the adherence of material to the bare silica capillary wall, has been a difficult problem since the early days of HPCE, particularly for large molecules such as proteins. Small molecules can have, at most, one point of attachment to the wall and the kinetics of adsorption/desorption are rapid. Large molecules can have multiple points of attachment resulting in slow kinetics. Several solutions have been proposed, including the use of (a) extreme-pH buffers, (b) high-concentration buffers, (c) amine modifiers, (d) dynamically coated capillaries, and (e) treated or functionalized capillaries.

In the first case, it was recognized that if the buffer pH is greater that 2 units above the protein pK_a, the anionic protein would be repelled from the anionic capillary wall [9]. At a pH < 2, the capillary wall is neutral and does not attract the cationic protein. The problem with this approach is that a wide range of pHs are not available for use and separations of similar proteins may not occur. For high-pI proteins such as histones, a buffer pH of 13 is required. The conductivity and UV background of such an electrolyte is too high to be generally applicable.

The use of high-concentration buffers is effective in reducing wall effects. This includes electrolytes containing up to 250 mM added salt. The problem with this approach is the high conductivity of the BGE. This requires a reduction in field strength resulting in lengthy separations. Zwitterionic buffers titrated to their pI can be used as well at concentrations approaching $1M$. At that concentration, it is important to select a reagent with low UV absorption.

The latter three cases are most commonly employed to reduce wall effects. In the third case, amine modifiers such as polyamines are added to the BGE at concentrations ranging from 1 to 60 mM [10]. These reagents coat the free silanols and reduce wall interactions. Now, any pH electrolyte can be employed. Diaminobutane, otherwise known as putrecein, is the preferred reagent because it is less volatile compared to diaminopropane. Monovalent amines such as triethanolamine are not as effective in this regard.

Dynamically coated capillaries (case d) are often used to reduce wall effects [11]. The mechanism of charge reversal is as follows. Ion-pair formation between the cationic head group of the surfactant and the anionic silanol group naturally occurs. The hydrophobic surfactant tail extending into the bulk solution is poorly solvated by water. The molecular need for solvation is satisfied by binding to the tail of another surfactant molecule. The cationic head group of the second surfactant molecule now extends into the bulk solution. The capillary wall becomes positively charged and the EOF is directed toward the anode. Separations are performed using the reversed-polarity mode (inlet side negative). Following this approach, a buffer pH is selected that is below the pI of the target protein. The cationic protein is now repelled from the cationic wall.

When coated capillaries are employed (case e), conventional buffers without additives to reduce wall effects are used. Urea and/or organic solvents can be added to aid solubility. Reagents for secondary equilibrium can be used as well. It is best to operate at a pH below 8 to maximize the stability of the often labile coating material. Coated capillaries are also used simply to eliminate the EOF in some applications.

References

1. P. G. Righetti, C. Gelfi, M. Perego, A. V. Stoyanov, and A. Bossi, *Electrophoresis, 18*: 2145 (1997).
2. D. Burgi and R.-L. Chien, *Anal. Chem. 63*: 2042 (1991).
3. P. Jandik, W. R. Jones, A. Weston, and P. R. Brown, *LC-GC 9*: 634 (1991).
4. M. Macka, P. Andersson, and P. R. Haddad, *Anal. Chem. 70*: 743 (1998).
5. S. A. C. Wren and R. C. Rowe, *J. Chromatogr. 603*: 235 (1992).
6. H. Nishi and S. Terabe, *J. Chromatogr. A 735*: 3 (1996).
7. B. Chankvetadze, *Capillary Electrophoresis in Chiral Analysis*, John Wiley & Sons, Chichester, 1997.
8. R. Weinberger, *Am. Lab. 31*: 59 (1999).
9. H. H. Lauer and D. McManigill, *Anal. Chem. 58*: 166 (1986).
10. J. A. Bullock and L.-C. Yuan, *J. Microcol. Separ. 3*: 241 (1991).
11. J. E. Wiktorowicz and J. C. Colburn, *Electrophoresis 11*: 769 (1990).

Robert Weinberger

Buffer Type and Concentration, Effect on Mobility, Selectivity, and Resolution in Capillary Electrophoresis

Introduction

Resolution in capillary zone electrophoresis (CZE) is, as in elution chromatography, a quantity that describes the extent of the separation of two consecutively migrating compounds, i and j. It is the result of the counterplay of two effects: migration and zone dispersion. The different migration velocities of the two separands lead (at least potentially) to the separation of the sample zones. The simultaneous mixing of the samples with the background electrolyte (BGE), caused by a number of processes, results in zone broadening and counteracts separation. Both effects determine the overall degree of separation. A quantitative measure that describes this degree is the resolution R_{ji}, a dimensionless number. It is expressed by the difference of the apex of the two peaks, on the one hand. It is of advantage not to measure this difference in an absolute scale (e.g., in seconds when the electropherogram is depicted in the time domain). In fact, a relative scale is taken, which is based on the widths of the peaks. We define the resolution as the difference in migration times, t, related to the peak width, taken, for example, by the mean standard deviation of the Gaussian peaks, as the scaling unit:

$$R_{ji} \equiv \frac{t_j - t_i}{2(\sigma_{t,i} + \sigma_{t,j})} \tag{1}$$

Baseline separation is achieved for two peaks with the same area when the resolution is 1.5. For peak area ratios larger than unity, the resolution must be larger.

Selectivity and Efficiency

This definitional equation (1) is not very operative and is, thus, transformed to an expression which more clearly visualizes the dependence of the resolution on sample properties and experimental variables. The migration times are substituted for by the mobilities of the separands, and the standard deviations by the plate height, H, or the plate number, N, respectively. The resulting resolution is then

$$R_{ji} = \frac{1}{4}\frac{\mu_i - \mu_j}{\bar{\mu}}\sqrt{\frac{L}{\bar{H}}} = \frac{1}{4}\frac{\Delta\mu}{\bar{\mu}}\sqrt{\bar{N}} \tag{2}$$

where $\bar{\mu}$, $\bar{H}$, and $\bar{N}$ are the average values; L is the migration distance.

The resolution consists of two terms, the selectivity term, $\Delta\mu/\bar{\mu}$, with the relative difference of the mobilities, and the efficiency term, the square root of the mean plate number. It must be pointed out that the plate height, $\bar{H}$, on which this plate number is based consists of all the contributions to peak broadening.

At this point, a differentiation should be made between two cases: the simple one, where migration is only caused by the electric force on the ionic separands, and the second, where an additional migration due to the occurrence of an electro-osmotic flow (EOF) takes place.

Resolution in Absence of EOF

Two main parameters determine the resolution: the effective mobility and the plate number. The effective mobility of a simple ion (e.g., the anion from a monovalent weak acid) is given by

$$\mu_{\text{eff}} = \frac{\mu_{\text{act}}}{1 + 10^{\text{p}K_a - \text{pH}}} \tag{3}$$

We take, here, only protolysis into consideration and do not discuss such important other equilibria such as complexation or interactions with pseudo-stationary phases. It follows from Eq. (3) that the effective mobility depends on the actual mobility (that of the fully charged particle at the ionic strength of the experiment), on the pK_a value of the analyte, and on the pH of the BGE. It follows that all these properties determine the selectivity term in the resolution.

The actual mobility depends on the following:

1. *The solvent.* There is a more or less pronounced influence of the solvent viscosity, reflected by Walden's rule. However, this rule is obeyed in rare cases; mainly in some mixed aqueous–organic solutions is an acceptable agreement found. On the other hand, in very viscous aqueous solutions of water-soluble polymers, such as poly(ethylene glycol), it was found that the actual mobility is independent of the viscosity.
2. *The size of the solvated ion.* Here, we must note that water is an excellent solvator for anions and cations as well, compared to most organic solvents. Only few exceptions for preferred solvation of the organic solvents are found (e.g., for Ag^+ and acetonitrile).
3. *The ionic strength of the BGE.* The dependence of the mobility on the ionic strength is expressed for simple systems (and simple ions) by the theory of

Debye, Hückel, and Onsager. Without going into detail, we can state that the mobility decreases, in all cases, with increasing ionic strength of the BGE, and the decrease is more pronounced the higher the charge number of the ion.

4. *On the temperature.* In aqueous solutions, the mobility increases with temperature roughly by about 3% per degree. This is a strong effect as, for example, a temperature difference of only 5K between the center and the wall of the separation capillary leads to a mobility difference of about 15%.

The pK_a value is also a function, mainly, of the solvent. Note that the pH scale is strongly dependent on the kind of solvent. Restricting the discussion to protolysis, it can be followed that the pH of the buffer has the most pronounced effect on the effective mobility, because the other effects change the mobility roughly in parallel for all separands. Again, it must be pointed out that other equilibria have an enormous potential to affect the effective mobility (cf. e.g., the use of cyclodextrins to introduce selectivity for the separation of enantiomers).

How is the efficiency influenced by the BGE? Peak broadening is the result of different processes in CZE occurring during migration [in addition, extracolumn effects contribute to peak width (e.g., that stemming from the width and shape of the injection zone, or the aperture of the detector cell)]. If the system behaves linearly, the individual peak variances (the second moments), σ^2_{ind}, are additive according to

$$\sigma^2_{\text{tot}} = \Sigma\,\sigma^2_{\text{ind}} = \sigma^2_{\text{extr}} + \sigma^2_{\text{dif}} + \sigma^2_{\text{Joule}} + \sigma^2_{\text{conc}} + \sigma^2_{\text{ads}} \qquad (4)$$

where the subscripts extr, dif, Joule, conc, and ads indicate the contributions from extracolumn dispersion, longitudinal diffusion, Joule self-heating, concentration overload, and wall adsorption, respectively. All but one effect might be eliminated: The longitudinal diffusion is inevitable. Plate number expressing this contribution is dependent on the voltage, U, applied and on the charge number, z_i, of the analytes according to

$$N_i \approx 20 z_i U \qquad (5)$$

at 20°C. It is obvious that the charge number depends on the pH of the BGE, as it is related to the degree of ionization. It follows that the plate number is a function of the pH as well. Thus, the resolution is influenced by the pH of the BGE twofold: via the selectivity, on the one hand, and via the plate number, on the other hand.

In conclusion, it follows for the limiting case of longitudinal diffusion as the only peak-broadening effect, that the resolution depends on the following:

- Instrumental variables: U and T
- Analyte parameters: μ_{act} and pK_a
- Chemical conditions determining the degree of ionization, α, or charge number z.

Resolution in Presence of EOF

The EOF brings an additional, unspecific velocity vector to the electrophoretic migration of the separands. The total migration velocity of the analyte, i, is then

$$v_{i,\text{tot}} = (\mu_{i,\text{eff}} + \mu_{\text{EOF}})E \qquad (6)$$

Note that the mobilities are taken as signed quantities. By convention, cations have positive electrophoretic mobilities and those of anions are negative. The mobility of the EOF when directed toward the cathode has positive sign, and vice versa.

The effect of the EOF on migration time and selectivity depends on the mutual signs of the mobilities of analytes and EOF, respectively. Concerning the change in separation selectivity, we refer to the expression of the selectivity term in the resolution equation. The difference between the mobilities of the two separands, i and j, will not be influenced by the EOF. However, the mean mobility is larger for the case of comigration. This means that the selectivity term in the expression for the resolution is always reduced in this case. In practice, selectivity is lost for cation separation when the EOF is directed, as is usual in uncoated fused-silica capillaries, toward the cathode. For this reason, cationic additives are applied in the BGE to reverse the EOF direction.

The effect of the EOF on separation selectivity (in comparison with the situation without EOF) can be quantified by the so-called electromigration factor, or reduced mobility, μ_i^*, defined as

$$\mu_i^* = \frac{\mu_{i,\text{eff}}}{\mu_{i,\text{eff}} + \mu_{\text{EOF}}} \qquad (7)$$

The change of the selectivity term in the resolution is directly expressible by μ_j^*. Interestingly, the effect of the EOF on the dispersion effects, expressed by the plate height H, also depends directly on μ^*. For longitudinal diffusion, Joule self-heating, and concentration overload, the variation of the plate height in the presence of the EOF is directly dependent upon this reduced mobility according to

$$H^{\text{EOF}} = H^0 \mu^* \qquad (8)$$

where the superscript 0 indicates the system without EOF. For wall adsorption, the corresponding effect is related to the reciprocal of μ^*.

An analysis of the effect of the EOF on the resolution brings the following result: For comigration of the analyte and EOF, the efficiency always increases and the selectivity term always decreases. As the decrease is directly proportional to μ^* but the gain in plate number is only increasing with the square root of μ^*, the resolution is always worse than without comigrating EOF.

For the case of countermigration, the situation is more complicated, because the overall effect depends on the magnitude of the mobility of analyte and that of the EOF. Roughly, it can be concluded that the resolution is increased for a given pair of analytes when the EOF is counterdirected, and it has a lower mobility than the analytes. Here, efficiency is lost, but selectivity is gained overproportionally. When the EOF mobility reaches a value that is twice as large as the analyte mobility (note that the signs of the mobilities are different), an analogous situation is found as without EOF. At mobilities of the EOF larger than twice the analyte mobility (conditions not impossible for high pH values of the BGE), resolution is worse here than without EOF, but the analysis time is shorter than in all other cases. It should be pointed out that all of these effects can be quantified by the reduced mobility defined in Eq. (7).

Suggested Further Readings

Camillieri, P., *Capillary Electrophoresis, Theory and Practice*, CRC Press, Boca Raton, FL, 1998.

Giddings, J. C., *J. Chromatogr. 480*: 21 (1989).

Guzman, N. A., *Capillary Electrophoresis Technology*, Marcel Dekker, Inc., New York, 1993.

Kenndler, E., *J. Microcol. Separ. 10*: 273 (1998).

Kenndler, E., in *High Performance Capillary Electrophoresis, Theory, Techniques and Applications*, M. G. Khaledi (ed.), John Wiley & Sons, New York, 1998, Vol. 146, pp. 25–76.

Landers, J. P., *Handbook of Capillary Electrophoresis*, 2nd ed. CRC Press, Boca Raton, FL, 1997.

Reijenga, J. C., and E. Kenndler, *J. Chromatogr. A 659*: 403 (1994).

Ernst Kenndler

C

Calibration of GPC–SEC with Narrow Molecular-Weight Distribution Standards

Introduction

In size-exclusion chromatography (SEC), polymer solutions are injected into one or more columns in series, packed with microparticulate porous packings. The packing pores have sizes in the range between ~ 5 and 10^5 nm, and during elution, the polymer molecules may or may not, depending on their size in the chromatographic eluent, penetrate into the pores. Therefore, smaller molecules have access to a larger fraction of pores compared to the larger ones, and the macromolecules elute in a decreasing order of molecular weights. For each type of polymer dissolved in the chromatographic eluent, and eluting through the given set of columns with a pure exclusion mechanism, a precise empirical correlation exists between molecular weights and elution volumes. This relationship constitutes the calibration of the SEC system, which allows the evaluation of average molecular weights (MWs) and molecular-weight distributions (MWDs) of the polymer under examination.

Direct column calibration for a given polymer requires the use of narrow MWD samples of that polymer, with molecular weights covering the whole range of interest. The polydispersity of the calibration standards must be less than 1.05, except for the very low and very high MWs ($< 10^3$ and $> 10^6$), for which polydispersity can reach 1.20. The chromatograms of such standards give narrow peaks and to each standard is associated the retention volume of the peak maximum.

There is a limited number of polymers for which narrow MWD standards are commercially available: polystyrene, poly(methyl methacrylate), poly(α-methyl styrene), polyisoprene, polybutadiene, polyethylene, poly(dimethyl siloxane), polyethyleneoxide, pullulan, dextran, polystyrene sulfonate sodium salt, and globular proteins. In some cases, the standards available cover a limited molecular weight range, so it may be impossible to construct the calibration curve over the complete column pore volume.

Standard methods for calibration of SEC columns with narrow MWD samples have been published by the American Society for Testing and Materials (ASTM D2596-97) and the Deutsches Institut for Normung (DIN 55672-1).

Procedure

Fresh solutions of the standards are prepared in the solvent used as chromatographic eluent. Calibration solutions should be as dilute as possible, in order to avoid any concentration dependence of sample retention volumes. The concentration effect causes an increase of retention volumes with increased sample concentration. As a rule of thumb, when high efficiency microparticulate packings are used, the concentration of narrow standards should be $\leq 0.025\%$ (w/v) for MW over 10^6, $\leq 0.05\%$ for MW between 10^6 and 2×10^5, and $\leq 0.1\%$ for MW down to 10^4. With a lower MW and in the oligomer range, the sample concentration can be higher than the previously suggested values.

Two or more standards may be dissolved and injected together to determine several retention volumes with a single injection. In such a case, the MW difference between the samples in the mixture should be sufficient to give peaks with baseline resolution. A sufficient number of narrow MWD standards, with different MWs, are required for establishing the calibration of a SEC column system. At least two standards per MW decade should be injected, and a minimum of five calibration points should be obtained in the

MW Fractionation Range of the Column Set

The maximum injection volume depends from column size and packing pore volumes, and for high-efficiency 300 × 8-mm columns, it is generally recommended not to exceed 100 μL per column.

The flow rate of the chromatographic apparatus must be extremely stable and reproducible: Flow rate fluctuations about the specified value should be lower than 3%, and long-term drift lower than 1%. Repeatability of flow rate setting is extremely important, as a 1% constant deviation of the actual flow rate from the required value may give 20% differences in calculated MW averages.

The systematic errors introduced by flow rate differences may be avoided by adding to the solutions a minimum amount of a low-molecular-weight internal standard (*o*-dichloro benzene, toluene, acetone, sulfur) which must not interfere with the polymer peaks. Flow rate is monitored in each chromatogram by measuring the retention time of the internal standard, and eventual variations may be corrected accordingly.

The peak retention times for the narrow polymer standards are measured from the chromatograms and transformed into retention volumes according to the real flow rate. For each standard, the logarithm of nominal molecular weight is plotted against its peak elution volume. Often, retention times are directly employed and plotted as the measured variable, and in this case, the condition of equal flow rate elutions for all the standards and for any subsequent sample analysis is achieved by means of the internal standard elution.

The molecular weight of the standards is supplied by the producers, either with a single value which should correspond to that of peak maximum, or with a complete characterization data sheet containing the values of M_n and M_w determined by osmometry and light scattering. In the latter case, the peak molecular weight to be inserted in the calibration plot is the mean value $(M_n M_w)^{1/2}$. A typical calibration curve for a three-column set, 300 × 7.5 mm, packed with a mixture of individual pore sizes is shown in Fig. 1. The calibration curve has a central part which is essentially linear and becomes curved at the two extremes:

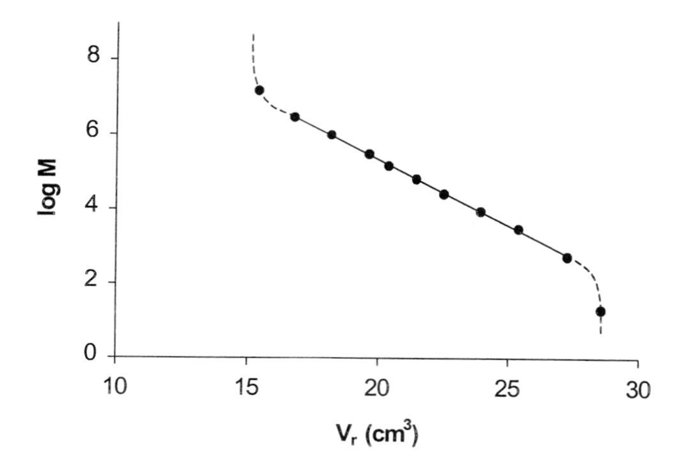

Fig. 1 Example of calibration curve with narrow MWD standards.

on the high-MW side when it approaches the retention value of totally excluded samples; on the low-MW side with a downward curvature until it reaches the retention time of total pore permeation.

The calibration curve, therefore, defines the extremes of retention times (or volumes) for the specific column system, the useful retention interval for sample analysis, and the related MW range. Columns packed with a balanced mixture of different pore sizes are capable of giving linear calibrations over the whole MW range of practical interest, from the oligomer region to more than 10^6.

The plot of log M versus peak retention volumes of narrow standards is represented in the more general form by a nth-order polynomial of the type

$$\log M = A + BV_r + CV_r^2 + DV_r^3 + \cdots$$

the coefficients of which are determined by regression on the experimental data. Most usually, when the linear plot is not sufficient to fit the points, a third-order polynomial will be adequate to represent the curve. Higher-order equations, although improving the fit, should be used with great care, as they can lead to unrealistic oscillations of the function.

The goodness of different equations fitted to the experimental data points is assessed by the results of statistical analysis or by simply considering the standard error of the estimate. It should be also considered that the adequacy of the calibration function for the determination of correct MW values is also dependent on the quality of the narrow MWD standards. Their nominal MWs are determined with independent absolute methods and are affected by experimental errors which may be different between samples with different MWs, or coming from different producers. A check of the quality of the narrow standards may be obtained by calculating the percent MW deviation of each standard from the calibration curve:

$$\Delta M(V_i)\% = \frac{M_{\text{peak}}(V_i) - M_{\text{calc}}(V_i)}{M_{\text{peak}}(V_i)} \times 100$$

A plot of $\Delta M(V_i)\%$ versus log M results in positive and negative values scattered around the MW axis, which allows one to visualize the limits of percent error into which the MW of standards are estimated by the calibration curve. If the MW error of some standard is found to be significantly larger than all the others, it is likely that its nominal MW is incorrect. The point of such sample should be removed from the calibration and the regression recalculated.

The calibration curve should always cover the MW of the samples that must be analyzed. Extrapolation of the calibration outside the range of injected polymer standards should be avoided in MW determinations.

From the calibration curve, the resolution power of the column set may also be evaluated. Resolution between two adjacent peaks, 1 and 2, is defined in terms of their retention volumes, V_r, and peak widths, w:

$$R_S = \frac{2(V_{r2} - V_{r1})}{w_1 + w_2}$$

The calibration is often expressed in the form of ln M versus V_r, and assuming a linear function, it may be written as

$$\ln M = \ln D_1 - D_2 V_r$$

By solving for V_r and substituting into the relationship for R_S, we obtain

$$R_S = \frac{\ln(M_1/M_2)}{wD_2} = \frac{\ln(M_1/M_2)}{4\sigma D_2}$$

valid for peaks of similar width or standard deviation σ, where $w_1 \approx w_2 = w = 4\sigma$. The above equation shows that the MW fractionation of SEC columns is linked to both their useful pore volume (slope D_2 of the calibration curve) and to packing quality (column efficiency or number of plate heights, determining peak widths). Working with columns having linear calibration in their whole fractionation range guarantees equal resolution power over several MW decades.

Suggested Further Readings

ASTM D 5296-97, Standard Test Method for Molecular Weight Averages and Molecular Weight Distribution of Polystyrene by High Performance Size-Exclusion Chromatography (1997).

DIN 55672-1, Gelpermeationschromatographie Teil 1: Tetrahydrofuran als Elutionsmittel (1995–02) (1995).

Janca, J. (ed.), *Steric Exclusion Liquid Chromatography of Polymers*, Marcel Dekker, Inc., New York, 1984.

Mori, S. and H. Barth, *Size Exclusion Chromatography*, Springer-Verlag, Berlin, 1999.

Yau, W. W., J. J. Kirkland, and D. D. Bly, *Modern Size-Exclusion Liquid Chromatography*, John Wiley & Sons, New York, 1979.

Oscar Chiantore

Calibration of GPC–SEC with Universal Calibration Techniques

Direct calibration of GPC–SEC columns requires well-characterized polymer standards of the same type of polymer one has to analyze. However, narrow molecular-weight distribution (MWD) standards are available for a limited number of polymers only, and well-characterized broad MWD standards are not always accessible. The parameter controlling separation in GPC–SEC is the size of solute in the chromatographic eluent. Therefore, if different polymer solutes are eluted in the same chromatographic system with a pure exclusion mechanism, at the same retention volume, molecules with the same size will be found. By plotting the logarithm of solute size versus retention volume, the points of all different polymers will be represented by a unique curve — a universal calibration curve. Thus, by application of the universal calibration, average molecular weights (MWs) and MWDs of any type of polymer may be evaluated from the size-exclusion chromatograms, provided that the relationship between molecular size and polymer molecular weight is known.

Several size parameters can be used to describe the dimensions of polymer molecules: radius of gyration, end-to-end distance, mean external length, and so forth. In the case of SEC analysis, it must be considered that the polymer molecular size is influenced by the interactions of chain segments with the solvent. As a consequence, polymer molecules in solution can be represented as equivalent hydrodynamic spheres [1], to which the Einstein equation for viscosity may be applied:

$$\eta = \eta_0(1 + 2.5\phi_s) \tag{1}$$

η and η_0 are the viscosities of solution and solvent, respectively, and ϕ_s is the volume fraction of solute particles in the solution.

By expressing the solute concentration c in grams per cubic centimeter, the relationship holds:

$$\phi_s = \frac{cN_A V_h}{M} \tag{2}$$

where N_A is Avogadro's number and V_h and M are the hydrodynamic volume and the molecular weight of the solute, respectively. Substituting in Eq. (1) and taking into account that

$$[\eta] = \lim_{c \to 0}\left(\frac{(\eta - \eta_0)/\eta_0}{c}\right) \tag{3}$$

we obtain

$$[\eta]M = 2.5N_A V_h \tag{4}$$

Equation (4) states that the hydrodynamic volume of a polymer molecule is proportional to the product of its intrinsic viscosity times the molecular weight.

The use of $[\eta]M$ as size parameter for GPC–SEC universal calibration was first proposed by Benoit and coworkers [2] and shown to be valid for homopolymers and copolymers with various chemical and geometrical structures. Their data are reported in the semilogarithmic plot of Fig. 1.

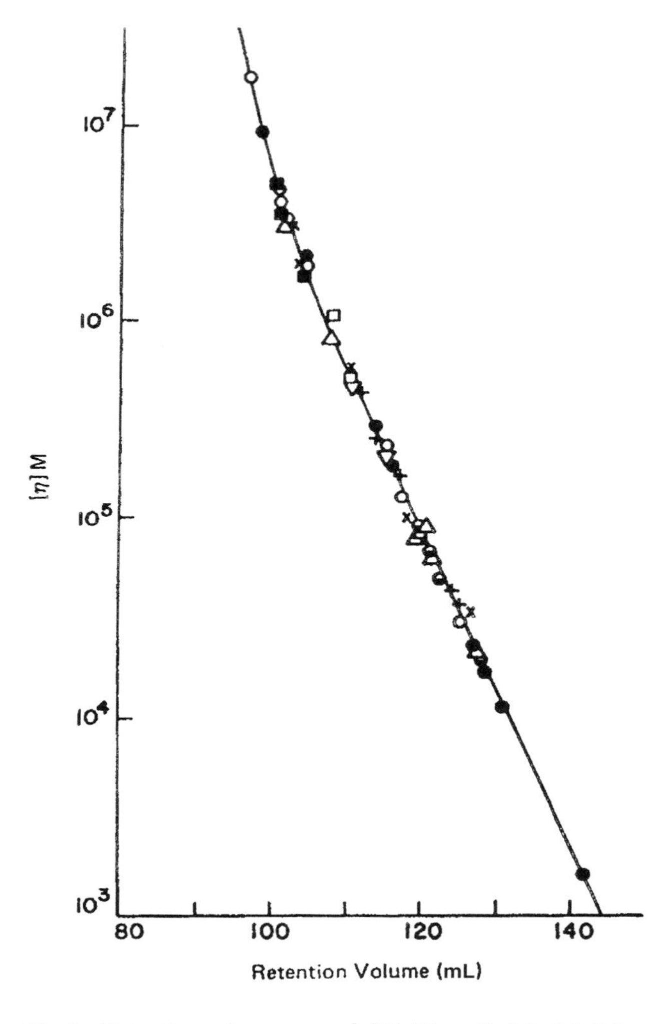

Fig. 1 Retention volume versus $[\eta]M$. (From Ref. 2, © J. Wiley & Sons, Inc.)

The hydrodynamic volume parameter $[\eta]M$ has been proven to be applicable also to the cases of rodlike polymers [3] and to separations in aqueous solvents [4] where, however, secondary nonexclusion mechanisms often superimpose and affect the sample elution behavior. In the latter situation, careful choice of eluent composition must be made in order to avoid any possible polymer-packing interaction.

The application of universal calibration requires a primary column calibration with elution of narrow MWD standards. For SEC in tetrahydrofuran, polystyrene (PS) standards are generally used. Intrinsic viscosities of the standards are either known or calculated from the proper Mark–Houwink equation, so that the plot of $\log[\eta]_{PS}M_{PS}$ values versus retention volumes V_r may be created. The universal calibration equation is obtained by polynomial regression, in the same way described for the calibration with narrow MWD standards.

Average molecular weights and MWDs of any polymer sample eluted on the same columns with pure exclusion mechanism may be calculated by considering that, at any retention volume, the following relationship holds:

$$[\eta]_i M_i = [\eta]_{PS,i} M_{PS,i} \tag{5}$$

from which

$$M_i = \frac{[\eta]_{PS,i} M_{PS,i}}{[\eta]_i} \tag{6}$$

To solve Eq. (6), the denominator must be known. Substituting into the denominator the Mark–Houwink expression $[\eta] = KM^a$ for the investigated polymer and rearranging, we obtain

$$M_i = \left(\frac{[\eta]_{PS,i} M_{PS,i}}{K}\right)^{1/1+a} \tag{7}$$

where K and a are the constants of the viscosimetric equation for that polymer, dissolved in the chromatographic eluent and at the temperature of analysis.

From Eq. (7), the molecular weight of each fraction in the chromatogram is obtained and average molecular weights may be calculated by application of the appropriate summations. The numerator in Eq. (7) is the value of the universal calibration at each retention volume.

The necessary conditions for application of the universal calibration method and for calculation of molecular weights through Eq. (6) is the knowledge of the $[\eta]_i$ values, which are obtained from the Mark–Houwink equations when the pertinent values of K and a constants are known. An alternative way is to make a continuous measurement of $[\eta]_i$ at the different elution volumes with an on-line viscometer detector coupled to the usual concentration detector system.

Methods for application of the universal calibration have been developed also for cases where K and a of the polymer of interest are not known and neither $[\eta]_i$ values are measured. Such methods are based on the availability of two broad MWD standards, having different molecular weights, of the polymer under examination [5].

One important property of the universal calibration concept is that, in the SEC separation of complex polymers (i.e., polymers with different architectures or copolymers with nonconstant chemical composition), at each retention volume, $V_{r,i}$, molecules with same hydrodynamic volume but possibly different molecular weights will elute. It has been demonstrated that, in such a case, the application of the hydrodynamic volume parameter, $[\eta]M$ gives the number-average molecular weight, M_n, of the polymer [6]. In fact, at each retention volume, the intrinsic viscosity of the eluted fraction is given by the weight average over the n different molecular species present:

$$[\eta]_i = w_1[\eta]_1 + w_2[\eta]_2 + \cdots + w_n[\eta]_n \tag{8}$$

Equation (8) may be written as

$$[\eta]_i = \frac{[\eta]_1 M_1 w_1}{M_1} + \frac{[\eta]_2 M_2 w_2}{M_2} + \cdots + \frac{[\eta]_n M_n w_n}{M_n} \tag{9}$$

As the condition holds, at each retention volume

$$[\eta]_1 M_1 = [\eta]_2 M_2 = \cdots = [\eta]_{PS} M_{PS} \tag{10}$$

Equation (9) becomes

$$[\eta]_i = [\eta]_{PS} M_{PS} \sum \left(\frac{w_i}{M_i}\right) = \frac{[\eta]_{PS} M_{PS}}{M_{n,i}} \tag{11}$$

$$[\eta]_i M_{n,i} = [\eta]_{PS} M_{PS} \tag{12}$$

By considering all the fractions of the chromatogram, the M_n value of the whole sample may be then calculated.

Experimental aspects for the determination of molecular weight averages and MWD distributions by GPC–SEC using universal calibration are described in a standard ASTM method [7]. Detailed discussion on the validity and limitations of the method may be also found in Ref. 8.

References

1. P. J. Flory, *Principles of Polymer Chemistry*, Cornell University Press, Ithaca, NY, 1953.
2. Z. Grubisic, P. Rempp, and H. Benoit, *J. Polym. Sci. B 5*:753 (1967).
3. J. V. Dawkins and M. Hemming, *Polymer 16*:554 (1975).
4. P. L. Dubin, *Aqueous Size Exclusion Chromatography*, Elsevier, Amsterdam, 1988.
5. H. Coll and D. K. Gilding, *J. Polym. Sci. A-2, 8*:89 (1970).
6. A. E. Hamielec, A. C. Ouano, and L. L. Nebenzahl, *J. Liquid Chromatogr. 1*:111 (1978).
7. ASTM D 3593-80, Standard Test Method for Molecular Weight Averages and Molecular Weight Distribution of Certain Polymers by Liquid Size-Exclusion Chromatography (Gel Permeation Chromatography — GPC) Using Universal Calibration (1980).
8. J. V. Dawkins, in *Steric Exclusion Liquid Chromatography of Polymers* (J. Janca ed.), Marcel Dekker, Inc., New York, 1984.

Oscar Chiantore

Capacity

The capacity is closely related to the number of active sites of the stationary phase per volume or mass unit. Practically, there are two definitions corresponding to two different approaches to the problem. On the one hand, there is the linear capacity and, on the other, the maximum available capacity.

It is well known that when increasing the injected sample quantity, whether in volume or in concentration, peaks are distorted and/or shifted beyond a certain limit; the column is said to be overloaded. To quantify how much sample can be injected into a column without altering the resolution, it is convenient to define the column linear capacity. It is well known, for small injected quantities, that solute retention times and column efficiency are not affected by the sample size. However, above a critical sample size, a noticeable decrease in retention time and column efficiency are always observed.

Snyder has defined [1] the adsorbent linear capacity as the ratio (weight sample)/(weight stationary phase) giving a value of k' (or V_R) reduced by 10% relative to the constant k' values measured for smaller samples (Fig. 1). In Figure 1, the adsorbent (Silica Davison) has a linear capacity close to 0.5 mg of dibenzyl per gram of silica. When the linear capacity of the column is exceeded, qualitative and quantitative analyses become much more complicated. Retention factors vary according to the injected solute quantity and the column efficiency can be tremendously decreased, entailing a degradation of resolution. Therefore, for analytical separations, it is always preferable to choose operating conditions corresponding to the linear capacity (k' and N values are constant whatever the injected sample sizes).

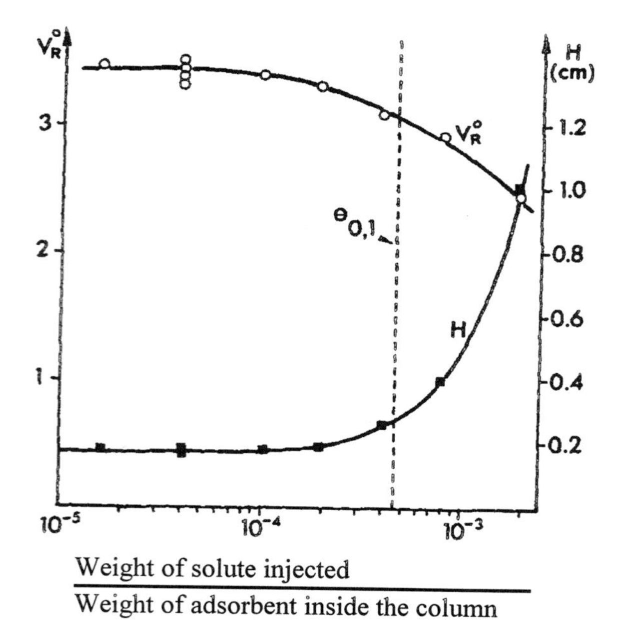

Fig. 1 Variation of the specific retention volume V_R^0 and of the height equivalent to a theoretical plate H as a function of the weight of injected solute (dibenzyl) related to the weight of adsorbent inside the column. (From Ref. 1.)

However, the practical interest of column linear capacity is very limited because its value varies according to various parameters: solute nature and retention and, even for the same quantity of injected solute, both the injected volume and the solute concentration of the injected solution. Thus, although widely accepted, the column linear capacity is misleading because it characterizes not only the thermodynamic nature of the chromatographic system but also the kinetic conditions (in term of column efficiency).

Consequently, it is preferable, according to Gareil et al. [2], to define the concept of maximum available capacity C_D for a stationary phase: mass of solute Q_S entailing the saturation of the mass m of stationary phase contained in the column for given operating conditions:

$$C_D = \frac{Q_S}{m} \tag{1}$$

with

$$k' = \frac{Q_S}{Q_m} = \frac{Q_S}{V_m C_0} \tag{2}$$

The combination of Eqs. (1) and (2) gives

$$C_D = \frac{k' V_m C_0}{m}$$

where k' is the solute retention factor measured for an analytical injection, V_M is the mobile-phase volume contained in the column, and C_0 is the solute concentration in the mobile phase.

Figure 2 shows, for various retention factors, the available capacity variation versus the solution concentration in the mobile phase in reversed-phase chromatography. These curves, called distribution isotherms, can be divided into two parts. In the first part, a linear variation of C_D versus C_0 is observed (bilogarithm scale); in the second part, a plateau is reached. In the first part and for the same retention (k' constant), the available capacity is independent of the solute nature.

The maximum available capacity is defined as the C_D limit value when both C_0 and k' are high ($C_0 \cong 1$ mol/L, $k' \geq 10$). This value does not vary either with C_0, or k', or the solute nature (for the same family).

The maximum available capacity depends on the nature of the stationary phase: specific area for adsorption, the ion-exchange capacity for ion-exchange capacity, and the bonded rate for partition chromatography.

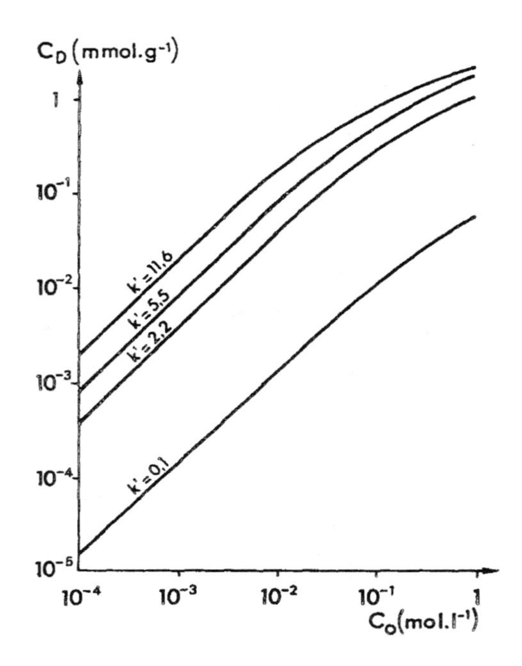

Fig. 2 Variation of the available capacity C_D as a function of the solute concentration in the mobile-phase C_0 (logarithm scales). In the case of reversed-phase chromatography, the stationary phase is *n*-octyl-bonded silica Lichroprep R.P.8 with 11.6% of carbon, the mobile phases are water–methanol mixtures, and the solute is phenol.

As a general rule, the maximum values of available capacity vary from 1.2 mmol/g (silica having a specific area close to 400 m^2/g) to 5 mEq/g for the cation exchanger (sulfonate groups).

References

1. L. R. Snyder, *Anal Chem. 39*:698 (1967).
2. P. Gareil, L. Semerdjian, M. Caude, and R. Rosset, *J. High. Resolut. Chromatogr. Chromatogr. Commun. 7*:123 (1984).

Suggested Further Reading

Knox, J. H. (ed.), *High Performance Liquid Chromatography*, Edinburgh University Press, Edinburgh, 1978, pp. 27–28, 50.
Rosset, R., Caude, M., and Jardy, A., *Chromatographies en phases liquide et supercritique*, Masson, Paris, 1991, pp. 32–37.

M. Caude
A. Jardy

Capillary Electrochromatography: An Introduction

Introduction

In 1998, Dadoo et al. [1] succeeded in achieving near-baseline resolution of five polynuclear aromatic hydrocarbons in less than 5 s by capillary electrochromatography (CEC) (see Fig. 1). These high speeds were obtained from a combination of factors, including a modest column length (6.5 cm), a high column plate number (13,000 plates) associated with 1.5-μm nonporous C_{18} particles, and a high voltage (28 kV). Although this separation is one of the fastest achieved in a liquid phase, higher column plate numbers have been obtained. Smith and Evans [2] report values of greater than 8 million plates per meter for the analysis of tricyclic antidepressants on a 3-μm sulfopropyl-bonded silica. These values are clearly due to a focusing effect within the column, whose reproducibility has not entirely withstood close scrutiny. Dadoo et al. [1] has produced CEC columns which generate plate numbers of about 700,000 per meter when peaks were detected before they passed through the outlet column frit. These results illustrate how closely practical achievements in CEC have now approached predicted theoretical performance maxima. The technique has not always been such a high performer.

History

Pretorius et al. [3] were among the first investigators to carry out packed column liquid chromatography in a tangential electric field (capillary electrochromatography) as a feasible alternative to using pressure. Their 1-mm-i.d. (inner diameter) quartz columns filled with 75–125-μm silica particles gave reduced plate heights of about 3 by

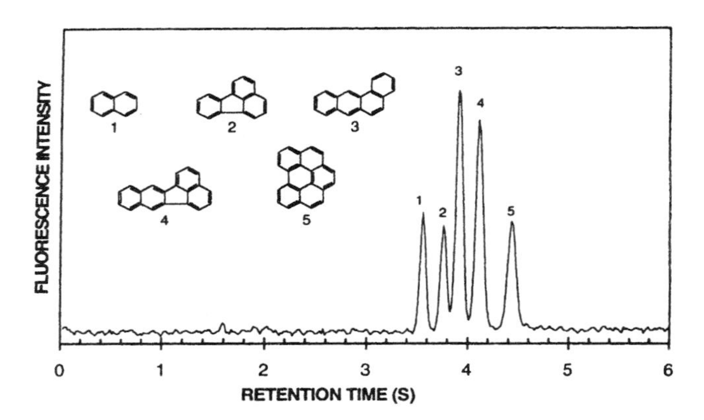

Fig. 1 Electrochromatogram of naphthalene (1), fluoranthene (2), benz[*a*]anthracene (3), benzo[*k*]fluoranthene (4), and benzo[*ghi*]perylene (5), using 1.5-μm nonporous octadecylsilyl bonded (ODS) particles. Column dimensions: 100-μm i.d. × 6.5-cm packed length (10 cm total length). Mobile phase: 70% acetonitrile in a 2-m*M* Tris solution; applied voltage for separation: 28 kV; injection: electrokinetic at 5 kV for 2 s. (Reprinted with permission from Ref. 1.)

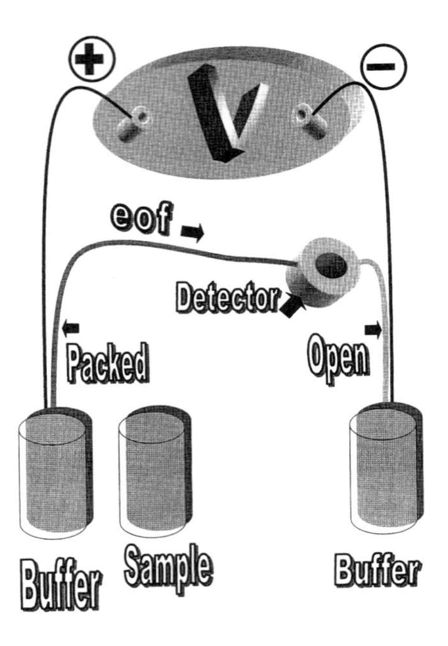

Fig. 2 Schematic of a typical electrochromatograph.

CEC versus the pressure mode values of about 8. This improvement in column efficiency was qualitatively predicted from the fact that the driving force in CEC — electro-osmotic flow — originates from the double electrical layer on the surface of the capillary and sorbent particles and, therefore, generates a relatively flat flow profile across the tube. This has the fundamental effect of producing sharper peaks and, ultimately, higher resolution in a shorter time. Because there are few or no pressure gradients generated within a CEC column, long packed capillaries containing very small particles are possible.

As the mobile phase moves through the capillary containing the sorbent under the effect of this electro-osmotic flow (EOF), sample components partition between the two phases in sorption and diffusive mechanisms characteristic of liquid chromatography. Ions in the sample move both under the influence of EOF and by their added attraction toward the oppositely charged electrode (electrophoresis). Uncharged components, on the other hand, move only under the influence of EOF. Thus, sample components, in general, separate by chromatographic and, sometimes, electrophoretic processes.

The full advantages of electrochromatography were not to be realized until the technology needed to create narrow capillaries (<200 μm i.d.) stable frits and sensitive detection systems had matured. Small-diameter tubes are necessary in order to reduce Joule heating due to the effect of electrical current generated by high voltage.

Thus, in 1981, Jorgenson and Lukacs [4] carried out experiments in CEC using an instrument (see Fig. 2) whose basic design is still used today. In addition to acting as a combined injector, separation medium, and flow cell, their column could also be used in the capillary electrophoresis (CE), CEC, and micro-LC (liquid chromatographic) modes. However, the fused-silica capillaries (170 μm i.d.) drawn in their labs, packed with 10-μm particles and operated in neat acetonitrile, gave rather modest improvements in efficiency over standard liquid chromatographic techniques, with reduced plate heights of no less than 1.9. These disappointing results, coupled with an admission of the technical difficulty of using this technique, led these authors to conclude that CEC would only be useful in wider-bore (several centimeters) preparative scale processes; a suggestion also made by Pretorius et al. [3].

Then, in 1991, Knox and Grant [5], working carefully with 3- and 5-μm particles, showed that it was practical to achieve dimensionless property of less than 1 in the CEC mode. These results confirmed their strongly optimistic view of the future of this technique, and a few years later, interest in CEC rapidly accelerated.

Operational Limits

Knox and Grant [5] have placed a general maximum limit of 200 μm on the inner diameter of capillaries used in CEC in order to avoid problems with excessive internal heating that harms column efficiency in aqueous/organic solutions. In principle, however, wider-bore tubes can be used, provided the current and field strength are kept low or the thermal conductivity of the system is kept high. In general, currents should be kept below 50 μA and field strengths held below 1000 V/cm.

Caillary electrochromatography in nonaqueous mobile phases is possible provided that the electrical double layer is formed with appropriate dissolved salts [3].

The ionic strength of most conventional buffers, such as phosphate, acetate, or borate, needs to be kept within the range 0.002*M* to 0.05*M*, but care needs to be taken with the lower concentration to avoid buffer capacity depletion due to hydrolysis. Zwitterionic buffers such as morpholino ethanesulfonic acid (MES) (whose electrical conductivity is low) can be used in the range 0.010*M* to 0.1*M* without undue heating problems, provided the field strength and aqueous content are kept low and the capillary is cooled.

Unlike high-performance liquid chromatography (HPLC), there is no maximum length for capillaries in CEC, but the longer columns mean slower chromatography and equilibration. Generally, columns in CEC are no longer than 60 cm.

Instrumentation

Creative solutions to practical problems abound in the evolution of instruments designed to carry out CEC. Pretorius and co-workers' [3] graphite electrodes, quartz tubing, glass wool frits, and on-column pressure injection gave way to Jorgenson and Lukacs' [4] fused-silica tubing, sintered frits, and electrokinetic injection (see Fig. 2). Commercially developed automated instruments designed for CE, whose appearance in 1988 followed these last authors' breakthrough research, have been used for most applications in CEC. Modern instruments therefore consist of the column, a cooling system, detector, voltage controller, autosampler, and data processor. In-capillary optical focusing of ultraviolet, visible, and laser radiation has largely solved the problems of detection [1]. In-column (through the packing) detection of appropriate analytes by laser-induced fluorescence has improved the general efficiency by avoiding the deterioration of the peak shape that often occurs as the analyte zone passes through the outlet frit [1].

Columns for modern CEC have been prepared using standard HPLC particles, from 0.5 to 10 μm in diameter, bearing C_{18}, phenyl, C_8, C_6, C_4, CN, amino, sulfo, and other functional groups and a variety of chiral polymers such as proteins and polysaccharides [6]. *In situ* sintered silica-based frits are most often used in these columns [7] which are generally slurry packed at high pressures. Several types of so-called monolithic (single piece) columns have been developed [8] which dispense with frits while generally maintaining high efficiency.

Applications

Most chemical classes have been separated and analyzed by CEC [6]. These include many classes of pharmaceuticals, environmental chemicals, explosives, natural products, drugs of abuse, polypeptides, oligosaccharides, nucleosides, and their bases and polynucleotides. Applications of CEC are readily found in *Analytical Abstracts* for example, a publication of the American Chemical Society, or the indexes of journals such as the *Journal of Chromatography*.

Euerby et al. [9] have systematically investigated the effects of the bonded phase, mobile phase, buffer type, field strength, pH, and temperature on the resolution of specific substituted barbiturates. Critical parameters for the optimization of efficiency of basic drugs by CEC (as for HPLC) include the nature of the sample solvent, pH, and concentration of ion-pair reagents, for example.

Typical buffers include alkali metal and ammonium phosphates and acetates, morpholino ethanesulfonic acid and Tris. Silica-based packings are used in the pH range 2–9. Methanol and acetonitrile are the two most commonly used organic solvents. Ion-pair reagents such as hexylamine and trifluoroacetic acid have been employed for basic compounds.

Problems, Issues, and Future Prospects

A fundamental theory of CEC that will provide a better understanding of mechanisms of separation is being developed. In particular, the work of Rathore and Horváth [10] in elucidating electrical properties of packed is particularly interesting. Technological issues that remain to be addressed include the difficulty of dealing with bubble formation and the fragility of conventional columns due to the aggressive frit-forming methods currently used. Monolithic columns [8,11,12] appear to have advantages in this regard.

Majors [13] has compiled the results of his perspectives survey of 14 leading separation scientists with an interest in CEC. As expected, there is a wide divergence in the opinions of these leaders with regard to current issues and future prospects for CEC. However, few underestimated the current technological difficulties of column manufacture, reproducibility of chromatographic and electroosmotic properties of the packed capillary, and the short-term problems of competing with HPLC or CE, but the majority of scientists interviewed believe that like any new technique, these problems will be overcome and that CEC will become a routine method of analysis in time.

On the other hand, the future of CEC may lie with the exciting developments in microfabrication [14], where capillaries are open channels 1.5 μm wide and 4.5 cm long and can achieve efficiencies of 800,000 plates per meter.

References

1. R. Dadoo, R. N. Zare, C. Yan, and D. S. Anex, *Anal. Chem. 70*:4787 (1998).
2. N. W. Smith and M. B. Evans, *Chromatographia 41*:197 (1995).
3. V. Pretorius, B. J. Hopkins, and J. D. Schieke, *J. Chromatogr. 99*:23 (1974).
4. J. W. Jorgenson and K. D. Lukacs, *J. Chromatogr. 218*:209 (1981).
5. J. H. Knox and I. H. Grant, *Chromatographia 32*:317 (1991).
6. K. D. Altria, N. W. Smith, and C. H. Turnbull, *Chromatographia 46*:664 (1997).
7. M. M. Dittman, G. R. Rozing, G. Ross, T. Adam, and K. K. Unger, *J. Capillary Electrophoresis 5*:201 (1997).
8. E. C. Peters, M. Petro, F. Svec, and J. M. J. Frechét, *Anal. Chem. 70*:2296 (1998).
9. M. R. Euerby, C. M. Johnson, S. F. Smyth, N. Gillot, D. A. Barrett, and P. N. Shaw, *J. Microcolumn Separ. 11*:305 (1999).
10. A. S. Rathore and Cs. Horváth, *Anal. Chem. 70*:3069 (1998).
11. C. K. Ratnayake, C. S. Oh, and M. P. Henry, *J. Chromatogr. A 887*:277 (2000).
12. C. K. Ratnayake, C. S. Oh, and M. P. Henry, *J. High Resol. Chromatogr. 23*:81 (2000).
13. R. E. Majors, *LC–GC 16*:96 (1998).
14. B. He, N. Tait, and F. Regnier, *Anal. Chem. 70*:3790 (1998).

Michael P. Henry
Chitra K. Ratnayake

Capillary Electrophoresis on Chips

Introduction

It is no wonder that capillary electrophoresis (CE) has evolved into one of the premier separation techniques in use today, due to its extremely high efficiencies, fast analysis times, reduced sample and reagent consumption, and vast array of operating modes. The transposition of CE methods from conventional capillaries to channels on planar chip substrates is a more recent phenomenon and has been driven by several factors, including, but not limited to, the need for ever-more sensitive and selective assays, the need to manipulate increasingly smaller samples, and the desire to process many samples in parallel [1]. Perhaps of greater significance to the rapid development of this important field, however, is its amenability to the assimilation of multiple components of an assay — beyond simple separation of analytes — into a single, fully integrated device. The promise of the "lab-on-a-chip," although seemingly ambitious in concept, is clearly attainable, and microchip capillary electrophoresis (μ-chip CE) has quickly established itself as one of the most fundamental constituents of such systems.

One of the first published demonstrations of capillary electrophoresis on a chip appeared in 1992, when Harrison et al. separated a mixture of fluorescein and calcein [2]. Although separation efficiencies and analysis times in this pioneering work did not represent significant improvements over those achievable by way of conventional CE, this work demonstrated the feasibility of miniaturizing a chemical analysis system involving electrokinetic phenomena for sample injection, separation, and solvent pumping. Within 2 years of the appearance of this seminal paper, analysis times on the order of seconds and even milliseconds had been demonstrated with similar μ-chip systems, and efficiencies in excess of 100,000 theoretical plates were routinely obtained. Subsequently, the integration of other functionalities, such as sample manipulations and chemical reactions, alongside the CE separation, has vaulted CE-on-a-chip to new heights.

Chip Fabrication Technology

The evolution of CE on a chip has directly benefited from the tremendous advances in semiconductor microfabrication technologies that have taken place over the past two decades. Although semiconducting substrates are not ideally suited to CE applications due to the high voltages applied for separation and fluid manipulation, many of the established semiconductor microfabrication techniques can be modified for the insulating glass or quartz substrates most commonly encountered in μ-chip CE. Here, the name μ-chip refers to the channel dimensions as opposed to the actual substrate dimensions, which commonly are on the order of 0.5 mm thick and anywhere from 3 to 10 cm in length and width (or diameter for circular substrates). In many cases, standard photolithographic and wet-etching techniques are employed in the manufacture of CE chips. To begin, the clean glass or quartz substrate is uniformly coated with sequential thin layers of chromium/gold and positive photoresist by sputter-coating and spin-coating methods, respectively.

The design for the CE channel structure is then transferred to the substrate by exposure of the photoresist to ultraviolet (UV) light through a photomask of the channel pattern. After photoresist development, a series of wet etches are employed, first to remove the metal etch mask and, second, to etch the channels into the substrate. Channels created in this fashion are trapezoidal in profile due to the isotropic etching of amorphous materials. Typical channel dimensions range from 5 to 40 μm deep and 20 to 100 μm wide (at half the channel depth). Residual photoresist and metal film are stripped from the etched substrate prior to thermal bonding of a cover plate, thereby forming closed channels suitable for electrophoresis. Access to the channels is most commonly gained through holes drilled in the cover plate prior to bonding.

Capillary electrophoresis chips so created are quite rugged due to the monolithic nature of their structure, and they can withstand applied voltages in the same range (up to 30 kV) as those commonly encountered in conventional CE. In addition, they offer greater heat dissipation than conventional CE capillaries, thereby allowing for operation under conditions of higher power. Although quartz substrates have superior optical properties relative to their glass counterparts, both present an optically flat surface for detection schemes, which is a definite advantage over the curvature inherent to conventional capillary walls. As well, the void volumes associated with channel intersections on-chip are virtually nonexistent. Despite their many advantages, these chips are time-consuming and expensive to fabricate. As such, alternative methods for CE chip fabrication are being developed, such as the creation of channels in polymeric materials by casting, molding, and imprinting techniques. The success of these methods will rely, in part, on the concomitant development of suitable surface modification procedures to successfully manage channel wall properties.

Injection on Chips

Clever chip design permits the integration of the sample injector directly on the chip, thereby combining injection and separation functions by default on a single substrate. Most commonly, the injector is fashioned as a simple cross or "double-T" arrangement of etched channels, as shown in Fig. 1. One branch of this cross serves as the sample channel, and the other serves as the separation channel. Fluid flow is manipulated through this cross, just as with all other fluid manipulations on chip, by control of electrokinetic phenomena: electrophoresis and electroosmosis. By first applying the appropriate voltage between the sample and sample waste reservoirs (Fig. 1a), the sample solution crosses the separation channel, filling the double-T intersection. Consequently, injection volumes are defined by the injector geometry. Typical sample plug volumes and lengths are on the order of about 10–100 pL and 50–200 μm, respectively. Provided the injection field strength and time are sufficient to ensure that the least mobile sample component has moved through the channel intersection, this method results in an unbiased injection, with all sample components represented in the intersection volume according to their original proportions. Having thus formed a sample plug, the voltage is switched so as to generate electro-osmotic flow along the separation channel (Fig. 1b), sweeping the sample plug out of the double-T injector and initiating separation along the second branch of the injector. This branch — the separation channel — typically ranges from 1 to 10 cm in length. Further control of sample plug size and shape and prevention of sample leakage can be effected by carefully controlling voltages applied to all four arms of the cross simultaneously during injection and separation phases.

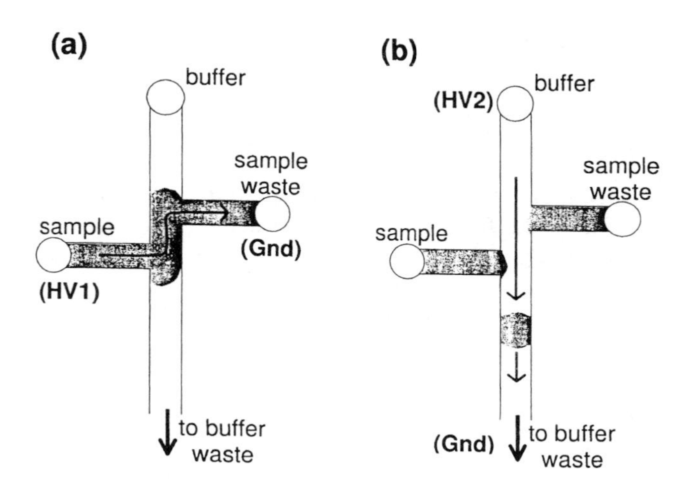

Fig. 1 Illustration of a microchip CE sample injector. Injection consists of (a) sample loading across the separation channel by application of a high voltage (HV1) across the sample and sample waste reservoirs, followed by (b) mobilization of the sample plug along the separation channel by application of HV2 across the buffer and buffer waste reservoirs.

Detection on Chips

It is not surprising that the requirements for detection on chips are very stringent, especially given the extremely small sample sizes discussed earlier. This need for sensitivity, along with the optically flat chip surface, makes laser-induced fluorescence (LIF) detection a natural choice; consequently, LIF detection on chips is the most widespread of all detection types. However, because relatively few analytes are natively fluorescent, LIF detection necessitates the development of selective and sensitive labeling strategies for each assay. Other detection methods, such as UV-vis absorption, chemiluminescence, and electrochemical, are less commonly encountered in chip CE, but they have been successfully demonstrated. Recently, the ability to generate an electrospray from the edge of a CE chip [3] has spawned a flurry of additional work in the area of electrospray ionization–mass spectrometry (ESI–MS) detection for CE chips. This promises to be a particularly powerful and exciting advance, as both quantitative and qualitative information can be provided simultaneously by this method of detection.

Beyond CE: Sample Manipulations

The most fundamental purpose of a CE chip is, of course, the separation and subsequent detection of analytes within a manifold of micromachined channels on a miniaturized substrate. This separation may take place as a result of basic electrokinetic phenomena or it may be enhanced or assisted by implementation of any one of various other separation techniques, such as isotachophoresis, micellar electrokinetic chromatography, isoelectric focusing, or capillary gel electrophoresis, all of which have been successfully demonstrated on-chip. However, the feature that truly distinguishes μ-chip CE from its conventional capillary counterpart is not its separative ability but, rather, its facility to integrate other functions onto the chip alongside the separation. This has already been discussed with respect to the sample injector and it is equally applicable to various sample preparation techniques. For example, controlled sample dilution, achieved by mixing buffer and sample streams directly on-chip, was first shown by Harrison et al. [4]. By increasing the voltage applied to a buffer reservoir while holding the voltage applied to a fluorescein dye sample reservoir constant, a controlled decrease in fluorescence intensity, corresponding to increasingly greater dilutions of the fluorescein, was observed downstream at the detector.

Preconcentration is another commonly encountered sample pretreatment method that has been successfully integrated onto a CE chip. Ramsey and co-workers incorporated a porous membrane structure into a microfabricated injection valve, enabling electrokinetic concentration of DNA samples using homogeneous buffer conditions [5]. Sample preconcentration in nonhomogeneous buffer systems—a technique known as sample stacking—has also been achieved on-chip [6].

Filtration is yet another pretreatment technique commonly encountered in CE. The reduced dimensions of fluid channels on chip substrates make the need for solution filtration all the more critical in this work. Until very recently, filtration was exclusively conducted "off-line" (i.e., before the sample and/or buffer solution was ever introduced to the chip). However, Regnier and co-workers recently micromachined solvent and reagent filters into quartz substrates using deep reactive ion etching [7]. Flow through these microfabricated lateral percolation filters was driven by electro-osmotic flow, thus making them compatible with other fluidic processes in a chip CE system. The on-chip filters were shown to be capable of removing a variety of particulate materials, ranging from dust to cells. Surface fouling and loss of cationic proteins from analyte streams were minimized by applying a polyacrylamide coating to the filter surfaces. Hence, these selected examples of the transposition of some traditional sample pretreatment methods onto chip substrates and their compatibility with on-chip CE separations illustrate the potential for achieving a fully integrated lab-on-a-chip.

Beyond CE: Chemical Reactions on Chip

Full functionality of these chips cannot be realized by the integration of sample pretreatment, injection, and separation methods alone. Additionally, the ability to carry out chemical reactions on-chip must be included in the list of integrated functions in order to extend the utility of these systems. Indeed, a wide variety of on-chip chemical reactions coupled to CE separations have been successfully demonstrated, including fluorescent derivatization, digestion of DNA and proteins, affinity-type reactions, and the polymerase chain reaction (PCR) for DNA amplification. The first of these reactions is necessitated by the laser-induced fluorescence detection schemes commonly used with CE chips. Because few analytes are natively fluorescent, it is often necessary to either (a) react the analyte with a fluorescent tag prior to separation (preseparation or precolumn labeling) or (b) separate the analyte first, followed by reaction of the separated zones with a derivatizing agent (postseparation or postcolumn labeling). The former, although leading to greater sensitivity, often suffers from increased band broadening and reduced separation efficiencies. The latter, although leading to higher efficiencies, offers reduced sensitivity and requires very fast labeling kinetics. Preseparation and postseparation labeling schemes were the first reactions demonstrated in conjunction with CE separations on chip substrates. Although the initial on-chip reactors suffered from inefficient mixing, and therefore inefficient reactions, improvements in channel structures and geometries along with optimization of solution conditions soon led to satisfactory results. The geometry of one such "second-generation" CE reactor chip is shown in Fig. 2, along with the electropherogram generated by postseparation labeling of amino acids with *o*-phthaldialdehyde (OPA). Despite the fact that the amino acids, once separated, had to mix and react with OPA in order to be rendered fluorescent, their corresponding peaks remained as sharp and well-defined as the peak obtained for hydrolyzed dansylchloride (DNS–OH), which did not react with OPA [8].

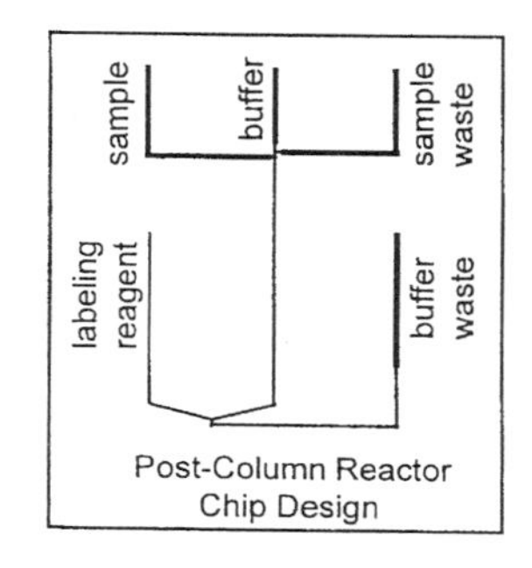

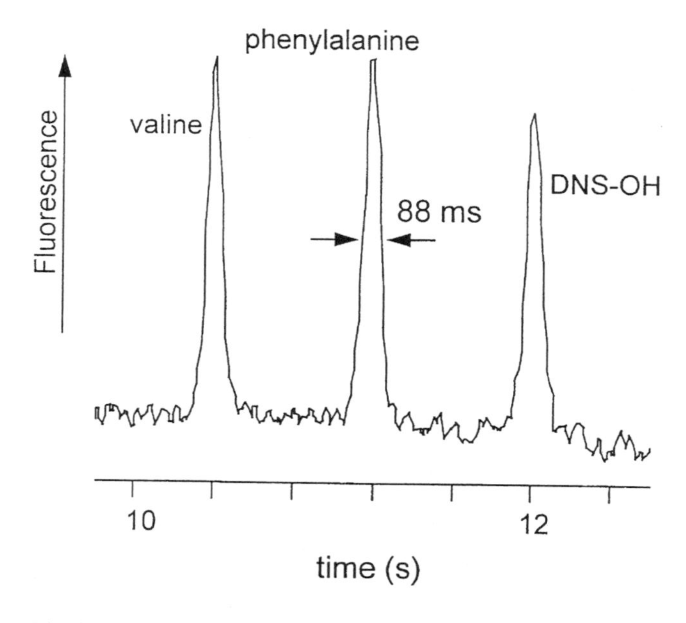

Fig. 2 Electrokinetically driven on-chip reaction in a CE-based system. Postseparation labeling of amino acids with OPA. Sample contained 200 μM each of phenylalanine and valine, and 10 μM of hydrolyzed dansyl chloride. The postcolumn reactor chip design is shown in the inset. [Reproduced from *Electrophoresis 18*:1733 (1997), with permission.]

The products of digestion reactions involving either protein or DNA substrates are conveniently separated and detected by CE on a chip. Improvements in digestion product assays should, therefore, be realized by marrying the digestion reaction and separation on a single chip. Jacobson and Ramsey demonstrated one such marriage by fabricating a chip device capable of both digesting a DNA sample with a restriction enzyme and separating the resulting DNA fragments using electrophoresis in a sieving matrix. Subsequent detection of the DNA restriction fragments was achieved by way of LIF using an intercalating dye that was introduced to the fragments on-chip [9]. In some cases, the products of a digestion reaction may not, in themselves, be of interest, but, rather, they may be used to determine information about the digestion enzyme itself. In one such chip assay, the reaction kinetics for the enzyme β-galactosidase were determined using resorufin β-D-galactopyranoside, a substrate that is hydrolyzed to resorufin, a fluorescent product [10]. Precise concentrations of substrate, enzyme, and inhibitor (phenylethyl β-D-thiogalactoside) were mixed on-chip, and the entire integrated assay was conducted in a 20-min period using only 120 pg of enzyme and 7.5 ng of substrate. Thus, the facility to perform digestion reactions directly on chip prior to separation and detection of digestion products necessarily improves the efficiency of these assay methods and represents a powerful new tool in the area of biochemical analysis.

Affinity-type reactions, which involve an analyte's affinity for a conjugate molecule, such as antibody–antigen interactions, form an important part of biochemical research. CE on a chip provides for the separation of complexes of the analyte with its conjugate from uncomplexed reagents. However, the ability to carry out the reaction between the analyte and its conjugate to form a complex directly on-chip, in conjunction with electrophoretic separation, is an important advance, and many examples of such on-chip affinity reactions exist. In one such example, Chiem and Harrison presented a μ-chip CE device capable of functioning as a complete immunoreactor for the determination of serum theophylline, a therapeutic drug for asthma treatment [11]. In this competitive immunoassay, a serum sample containing theophylline was mixed, directly on the chip, with fluorescently labeled theophylline tracer prior to introducing a limited amount of anti-theophylline antibody, also on-chip. The products of this immunoassay were subsequently separated by electrophoresis and detected by LIF on-chip. As the concentration of theophylline in the serum increased, this competitive assay led to an increase in signal for free, labeled theophylline and a corresponding decrease in signal for the labeled theophylline–antibody complex. Total analysis time, including on-chip reagent mixing, reaction, separation, and detection, was 150 s per sample, demonstrating one of the obvious advantages (along with reduced reagent consumption and increased sensitivity) of being able to conduct reactions on-chip alongside CE separations and other functions.

Another important reaction that has found a place on CE chips is the PCR, which is used to amplify DNA and which is critical to high-throughput genetic analyses. With the demonstrated ability of CE chips to integrate chemical reactions alongside high-speed separations, it is perhaps not surprising that μ-chip devices capable of genetic analysis would be fabricated. For example, a single monolithic chip capable of PCR amplification of up to four DNA samples, followed by product analysis has been demonstrated [12]. Integrated onto this chip are the facilities to thermally lyse cells to release DNA, standard PCR protocols to amplify DNA, gel electrophoresis to separate PCR products, intercalation of a fluorescent dye into PCR products, and detection of labeled products by LIF. The level of sophistication in a device such as this clearly illustrates the reality of the lab-on-a-chip concept: A concept that is founded upon the many advantages of μ-chip CE.

Conclusions and Future Directions

The advantages typically associated with capillary electrophoresis, such as reduced sample and reagent consumption, reduced analysis time, and increased separation efficiency, are augmented when the CE system is transposed to a chip substrate. More importantly, however, μ-chip CE offers the further advantage of integrating analytical processes beyond separation. Sample preparation, injection, reaction, and detection can be seamlessly tied to the electrophoretic separation stage of the analysis. The monolithic CE chips capable of separation coupled to some of these other analytical steps are precursors to the ultimate "lab-on-a-chip," which promises high-throughput sensitive analyses with minimal user intervention. Applications of such devices in biochemical, clinical, forensic, and environmental analyses are seemingly unlimited. However, several challenges remain despite the great promise of these devices. In order to fully realize the advantages offered by the microfluidics regime of the chip, methods of addressing these microvolumes and of interfacing them to the macroscale world beyond the chip must be carefully managed. Although chip fabrication techniques are now well established, they are by no means accessible to the majority of analysts. Fabrication processes that rely less heavily on high-tech processing facilities must be developed or, more realistically, the chips themselves must be made available inexpensively and in a variety of application designs for all potential users. Miniaturization or careful arrangement of the apparatus accompanying the CE chip, including power supplies, detection components, and computer controllers, into a compact and robust system must be considered in order to take full advantage of the chip's inherently small size. Finally, true parallel processing facilities must be routinely developed on single chips in order to increase sample throughput and increase the applicability of these systems to large-scale analytical problems. These challenges, although formidable, are worthy of solutions in order to successfully build labs-on-a-chip around the cornerstone of CE on a chip.

References

1. S. C. Jacobson and J. M. Ramsey, in *High-Performance Capillary Electrophoresis* (M. Khaledi, ed.), John Wiley & Sons, New York, 1998, pp. 613–633.
2. D. J. Harrison, A. Manz, Z. Fan, H. Lüdi, and H. M. Widmer, *Anal. Chem. 64*:1926 (1992).
3. R. S. Ramsey and J. M. Ramsey, *Anal. Chem. 69*:1174 (1997).
4. D. J. Harrison, K. Fluri, K. Seiler, Z. Fan, C. S. Effenhauser, and A. Manz. *Science 261*:895 (1993).
5. J. Khandurina, S. C. Jacobson, L. C. Waters, R. S. Foote, and J. M. Ramsey, *Anal. Chem. 71*:1815 (1999).
6. S. C. Jacobson and J. M. Ramsey, *Electrophoresis 16*:481 (1995).
7. B. He, L. Tan, and F. Regnier, *Anal. Chem. 71*:1464 (1999).
8. K. Fluri, G. Fitzpatrick, N. Chiem, and D. J. Harrison, *Anal. Chem. 68*:4285 (1996).
9. S. C. Jacobson and J. M. Ramsey, *Anal. Chem. 68*:720 (1996).
10. A. G. Hadd, D. E. Raymond, J. W. Halliwell, S. C. Jacobson, and J. M. Ramsey, *Anal. Chem. 69*:3407 (1997).
11. N. H. Chiem and D. J. Harrison, *Clin. Chem. 44*:591 (1998).
12. L. C. Waters, S. C. Jacobson, N. Kroutchinina, J. Khandurina, R. S. Foote, and J. M. Ramsey, *Anal. Chem. 70*:5172 (1998).

Christa L. Colyer

Capillary Electrophoresis: Introduction and Overview

Electrophoresis has been used as a separation technique for decades, particularly by biochemists, in the open-bed format. In this mode, a layer of a gel is formed on a flat-bed support which is in contact with an electrolyte and two electrodes are situated at either end of the open slab. The sample is placed at one end of the separation medium, and when voltage is applied, the molecules migrate through the gel by electrophoresis. The components in the sample are separated based on their differences in electrophoretic mobility. The electrophoretic mobility is controlled by molecular parameters such as charge, size, and shape. After the electric field is turned off, the separation is evaluated by spraying the plate with a dye and the bands of the sample components become visible, similar to the detection format used in paper or thin-layer chromatography.

Although the basic principle was conceived many years ago, the practical development of electrophoresis experiments in a closed or tubular format was only begun a little more than a decade ago. The main problem of the closed system is that the application of high voltages leads to the generation of Joule heat as current flows through the electrolyte solution. The heat generated can often cause sample decomposition or, more frequently, result in a large increase in molecular diffusion, leading to zone broadening that obliterates the separation between adjacent bands. Therefore, for these experiments to work in a tubular format, it is necessary to use capillary tubes with diameters of 100 μm or less in most cases. The answer to overcoming the Joule heat problem is to use fused-silica tubes, similar to those developed for capillary gas chromatography but having a smaller internal diameter. A second advantage of the fused-silica capillary is that it is suitable for direct on-line detection because it is optically transparent to ultraviolet (UV) and visible light. Typical dimensions for the capillary under experimental conditions are an outer diameter (o.d.) of $\sim$ 375 μm, an inner diameter (i.d.) of 50–100 μm, and an overall length of 50–100 cm. To protect the fragile fused-silica tube, the capillary is coated with an external layer of polyimide, allowing it to be flexible and manipulated into a variety of instrumental geometries. A detection window can be made by removing a small amount of the protective coating.

The fused-silica surface also provides another mechanism, electro-osmosis, which drives solutes through the tube under the influence of an electric field. The principle of electro-osmotic flow (EOF) is illustrated in Fig. 1. The inner wall of the capillary contains silanol groups on the surface that become ionized as the pH is raised above about 3.0. This creates an electrical double layer in the presence of an applied electric field so that the positively charged species of the buffer which are surrounded by a hydrated layer carry solvent toward the cathode (negatively charged electrode). This results in a net movement of solvent toward the cathode that will carry solutes in the same direction as if the solvent were pumped through the capillary. This electrically driven solvent pumping mechanism results in a flat flow profile in contrast to the laminar one (parabolic) obtained from mechanical pumps such as those used in high-performance liquid chromatography (HPLC). However, electro-osmotic flow is uniform throughout the capillary and does not depend on any solute properties. All solutes are affected by EOF uniformly and this process does not contribute to the separation mechanism. Therefore, only differences in electrophoretic velocity are responsible for the separation of charged compounds in a fused-silica capillary in the presence of an applied electric field. In fact, EOF is detrimental to the separation process because it moves positively charged species through the capillary faster, thus allowing less time for differences in electrophoretic velocity to separate two species with similar mobilities. This effect can be described mathematically by the following equation:

$$v_{tot} = v_{ep} + v_{EOF}$$

where v_{tot} is the total velocity of the charged species, v_{ep} is the eletrophoretic velocity of that species, and v_{EOF} is the eletro-osmotic velocity. In a typical experiment, sample migration rates through the capillary are as follows: cationic species > neutral compounds > anionic species. Be-

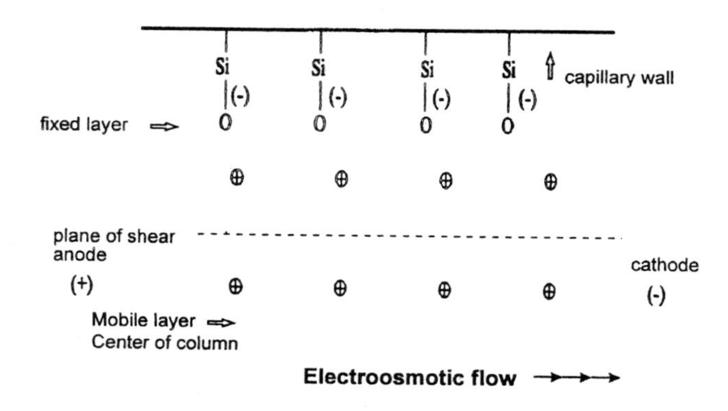

Fig. 1 Principle of electro-osmotic flow.

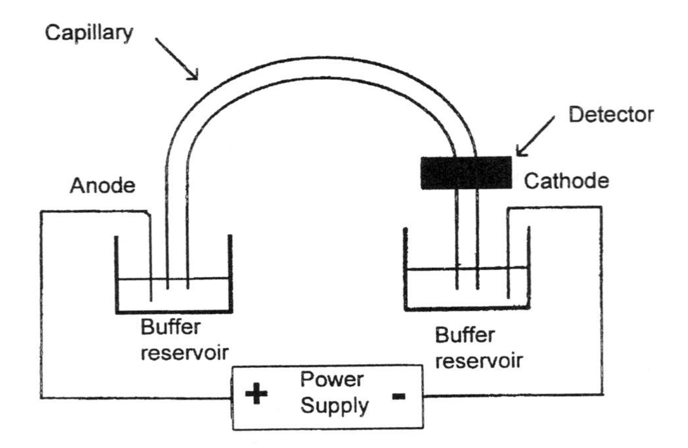

Fig. 2 Basic apparatus for capillary electrophoresis.

cause both cationic and anionic compounds can have different electrophoretic mobilities, they can be separated within the capillary. However, neutral species are carried through the capillary only by electro-osmotic flow, so these compounds will all migrate at the same rate and, therefore, cannot by separated by capillary electrophoresis (CE).

The basic apparatus necessary for a capillary electrophoresis system is shown in Fig. 2. The instrument must have the following components: power supply, electrodes (anode and cathode), vials for electrodes and buffers, separation capillary, detector, and data system or recorder (not shown). The function of each of these components is as follows:

Power Supply. This device supplies the high voltage to the system. Typically, experiments are run at several kilovolts up to 30 kV or more. Most power supplies will also have an ammeter to measure the current flowing through the system.

Electrodes. These components are generally platinum wires which serve as the contact point between the liquid (buffer solutions) in the system and the high-voltage power supply. An inert metal is desirable to avoid an electrochemical reaction or excessive fouling of the electrode surface that would disrupt current flow in the system.

Buffer Reservoirs. These containers hold the buffer solution that provides for a complete electrical circuit through the capillary and connection to the high-voltage supply. Due to electro-osmotic flow as described earlier, the vials also serve as reservoirs to maintain electroneutrality in the system.

Separation Capillary. The fused-silica capillary tube is the focal point in the instrument because sample separation takes place here. The length is typically 50–100 cm and the ends are placed in the buffer reservoirs. The capillary is filled with the running buffer before the analysis begins.

Detector. This device measures a property of the solute in order to determine when each compound has passed through a significant portion of the capillary to the detection window. The detector is not placed at the end of the capillary, as this part of the tube must be in the buffer solution. Solute properties most often used for detection are absorbance and fluorescence, although CE can also be coupled to a mass spectrometer.

Data System/Recorder. The simplest device for output is an ordinary recorder. However, an integrator will provide more information about the peaks (time and area). The most sophisticated apparatus is a computer, which can be used to process and evaluate the data. The computer can also be used to control the operations of the instrument.

The sample can be introduced into the capillary by several methods. The simplest approach is to remove the end of the capillary from the anode buffer reservoir and place it in the sample vial that has been elevated slightly above the level of the cathode buffer container. Gravity flow for several seconds will move some of the sample in the separation capillary. Another approach is to place the anode end of the capillary into the sample vial and apply pressure to the analyte solution. The next method involves placing the anode end of the capillary in the sample vial and applying a vacuum to the cathodic side of the capillary to draw solution into the tube. The previous three means of sample introduction are referred to as hydrodynamic modes of injection. The last method involves placing the anodic end of the capillary in the sample vial and applying a low voltage for several seconds. This approach is referred to as electrokinetic injection.

Information about the analyte can be qualitative and/or quantitative, with the data resembling a chromatogram. The output from the recorder/integrator/data system is in the form of peaks which are indicated by a time (migration time) from the start of the experiment. The migration time is analogous to the retention time in a chromatographic separation and provides qualitative information by comparison to a known compound under identical experimental conditions. Because the majority of detection in CE is by spectroscopic means, the area under the peak is proportional to the concentration. Therefore, quantitative information can be obtained by making a calibration curve from a plot of peak area versus concentration.

Even though capillary electrophoresis is a relatively simple method, several formats exist that allow for analyses of different types of samples or to take advantage of certain solute properties. The primary modes of capillary electrophoresis are as follows:

Capillary Zone Electrophoresis. The most fundamental approach that involves the use of a fused-silica capillary placed between the two buffer vials so that separation of the sample component occurs after an electric field (voltage) is applied to the system. Separation of the analytes is based on differences in electrophoretic mobility. Only charged compounds, both large and small, can be separated in this format.

Capillary Gel Electrohoresis. In this mode, molecules are separated according to size as they migrate through a polymer matrix. The polymer can be in solution, physically coated on the capillary wall, or chemically bonded to the capillary wall. This mode is primarily used for the separation of large molecules like proteins, peptides, and DNA species.

Capillary Isoelectric Focusing. In this approach, the capillary contains a pH gradient. When the sample is introduced and voltage is applied, it migrates to the point in the capillary where it has zero net charge (isoelectric point). The analytes are removed from the capillary by adding a salt to one of the reservoirs and then applying voltage again. The solutes will then migrate past the detector, with the time being related to its position in the capillary.

Capillary Isotachophoresis. In isotachophoresis, the capillary is first filled with a buffer of higher mobility than any of the solutes, then the sample, and, finally, a second buffer with lower mobility than any of the analytes. Separation occurs in the zone formed between the two electrolytes.

Micellar Electrokinetic Capillary Chromatography. Surfactants that form micelles in solution are added to the buffer in the capillary. When the solute is injected, it partitions itself between the buffer and the micelle. Migration of the solute depends on the amount of time it spends in the micelle versus the time it spends in the buffer. Therefore, the separation of analytes occurs due to differences in the partition coefficient between the two phases, much like in a chromatographic process.

Capillary electrophoresis is still an emerging technology. Rapid development is occurring in separation capillaries, detector technology, and applications.

Suggested Further Reading

Altria, K. D., *Capillary Electrophoresis Guidebook: Principles, Operation and Applications*, Humana Press, Totowa, NJ, 1996.

Camilleri, P., *Capillary Electrophoresis: Theory and Practice*, CRC Press, Boca Raton, FL, 1998.

Hjerten, S., *Methods Enzymol. 270*:296 (1996).

Landers, J. P., *Handbook of Capillary Electrophoresis*, 2nd ed., CRC Press, Boca Raton, FL, 1997.

Lunte, S. M. and D. M. Radzik, *Pharmaceutical and Biomedical Applications of Capillary Electrophoresis*, Pergammon, Oxford, 1996.

Parves, H., P. Candy, S. Parvez, and P. Roland-Gosselin, *Capillary Electrophoresis in Biotechnology and Environmental Analysis*, VSP, Utrecht, 1997.

Righetti, P. G., *Capillary Electrophoresis in Analytical Biotechnology*, CRC Press, Boca Raton, FL, 1996.

Joseph J. Pesek
Maria T. Matyska

Capillary Electrophoresis in Nonaqueous Media

Introduction

Organic solvents are used in capillary electrophoresis (CE) for several reasons:

1. To increase the solubility of lipophilic analytes.
2. To affect the actual mobilities of the analytes (those of the fully charged species at the ionic strength of the solution).
3. To change the pK values of the analytes.
4. To influence the magnitude of the electro-osmotic flow.
5. To influence the equilibrium constant of association reactions between analytes and additives (e.g., for the adjustment of the degree of complexation; an important example is the separation of chiral compounds by the use of cyclodextrins).
6. In some rare cases, to allow homoconjunction or heteroconjugation of the analytes with other species present and, thus, enable separation. For such

interactions, a low dielectric constant of the solvent is a prerequisite.

Application of Nonaqueous Solvents

The organic solvents are applied in many cases in order to enhance the separation selectivity by changing the effective mobilities of the analytes. They are either applied as pure solvents, or as nonaqueous mixtures, or as constituents of mixed aqueous–organic systems. Solvents used for CE, as described in the literature, are methanol, ethanol, propanol, acetonitrile, tetrahydrofuran, formamide, *N*-methylformamide, *N*,*N*-dimethylformamide, *N*,*N*-dimethylacetamide, dimethylsulfoxide, acetone, ethylacetate, and 2,2,2-trifluoroethanol.

Organic solvents have relevance in many fields of application: for the separation of inorganic ions, organic anions and cations, pharmaceuticals and drugs, amino acids, peptides, and proteins.

There are some practical restrictions for the use of organic solvents:

1. Many organic solvents have a significant ultraviolet (UV) absorbance in the range of wavelengths that are normally also used for the detection of the analytes. This property leads to a poor signal-to-noise ratio or limits the applicability to solutes with UV absorbances at a higher wavelength.
2. Many electrolytes cannot be used as buffers, due to their low solubilities in organic solvents.
3. The low dielectric constant of solvents suppresses ion dissociation and favors ion-pair formation.
4. Important physicochemical properties (e.g., ionization constants of weak acids and bases) are often not known, which leads to a more or less random experimental approach for the optimization of the resolution.
5. In this context, it should be mentioned that the clear determination of the pH scale in these solvents is not a straightforward task, which may introduce a certain inaccuracy for the description of the experimental conditions. As this aspect is not adequately considered in many articles on CE in nonaqueous solvents, it is discussed here in more detail.

Acidity Scales in Organic Solvents

When investigating the effect of organic solvents on the pK_a of an acid, the significance of the pH scale in this solvent must be questioned. We base such scales on the measurement of the activity of the solvated proton. We define the activity, a_i, of a particle, i, the proton in the case of interest, by the difference between the chemical potential, ω_i, in the given and in a standard state (indicated by superscript 0)

$$\omega_i = \omega_i^0 + RT \ln a_i$$

In practice, we therefore differentiate a number of acidity scales: the standard, the conventional, the operational, and the absolute (thermodynamic) scale.

Standard Acidity Scale

The standard state might be chosen in various ways (e.g., as the state at infinitely diluted solution). The resulting standard acidity scale is characterized by the activity of the proton solvated by the given solvent, HS, according to

$$\text{pH} = -\log a_{\text{SH}_2^+} \quad (1)$$

The range of this scale is defined by the ionic product of the solvent, pK_{HS}.

Measurements in the standard acidity scale are carried out in cells without liquid junctions (e.g., with the following setup: Pt/H_2/HCl in SH/AgCl/Ag). It is assumed, here, that the activities of the solvated proton and the counterion, chloride, are equal. In this case, the electromotive force (emf) of the cell can be expressed by

$$E = E_S^0 - \frac{RT}{F} \ln a_{\text{SH}_2^+} a_{\text{Cl}^-} = E_S^0 - \frac{2RT}{F} \ln(c_{\text{HCl}} \gamma_{\text{HCl}}) \quad (2)$$

where c_{HCl} is the concentration and γ_{HCl} is the mean activity coefficient of HCl. E_S^0 is the standard potential of the silver chloride electrode in the given solvent, S, after extrapolation of the measured emf to zero ionic strength. Rearrangement leads to the expression of the pH in the standard scale:

$$\text{pH} = -\frac{(E - E_S^0)F}{2.3RT} + \log c_{\text{Cl}^-} + \log \gamma_{\text{Cl}^-} \quad (3)$$

Conventional Acidity Scale

The standard acidity scale, although well defined theoretically, has the limitation in practice that only the mean activity coefficient, but not the single-ion activity coefficient, is thermodynamically assessible. The single-ion coefficient depends on the composition of the solution as well. One way to circumvent this problem would be to have a defined value of the activity coefficient for one selected ion. Given that, all other activity coefficients could be obtained from the activity coefficients of the particular electrolytes and that special single-ion coefficient. The value of this selected coefficient could be used, then, as the base of the conventional acidity scale. This single-ion activity coefficient is derived for chloride by the Debye–Hückel theory. This choice is made by convention, initially proposed for aqueous solutions; it is accepted also for other amphiprotic, polar solvents. Note that the measurements of the proton activity are carried out in cells without liquid junction.

Operational Acidity Scale

Due to the disadvantage of working with cells without liquid junctions, in practice the operational scale uses buffer solutions with known conventional pH for the calibration of cells with liquid junction [e.g., the convenient glass electrode (with the calomel or silver electrode, respectively, as reference)]. After calibration of the measuring cell (with a buffer of known conventional pH), the acidities of unknown samples can be measured in the same solvent. It is clear that for the standard buffers used, the conventional and the operational pH are identical. However, we cannot assume such an identity for the unknown samples. This is because the activities and the mobilities of the different ionic species might change the potential on the boundary with all liquid junctions (even without taking effect of the nonelectrolytes into account).

Absolute (Thermodynamic) Scale and Medium Effect

This scale, in fact, would allow comparing the basicities of the different solvents in a general way. It is based on the question of the chemical potential of the proton (as a single-ion species) in water, W, and the organic solvent, S. Taking the hypothetical $1M$ solution as the standard state, the chemical potential is given, according to Eq. (1), as

$$\omega_{H^+} = \omega^0_{H^+} + RT \ln m_{H^+} + RT \ln \gamma_{H^+} \tag{4}$$

where m is the molal concentration. The so-called medium effect on the proton is given by

$$\ln {}_{W}\gamma_{H^+} - \ln {}_{S}\gamma_{H^+} = \ln\left(\frac{{}_{W}\gamma_{H^+}}{{}_{S}\gamma_{H^+}}\right) = \ln {}_{m}\gamma_{H^+} = \frac{{}_{S}\omega^0_{H^+} - {}_{W}\omega^0_{H^+}}{RT} \tag{5}$$

${}_{m}\gamma_{H^+}$ is named the *transfer activity coefficient*. The medium effect is proportional to the reversible work of transfer of 1 mol of protons in water to the solvent, S (in both solutions at infinite dilution). If the medium effect is negative, the proton is more stable in the solvent, S. It is, thus, an unequivocal measure of the basicity of the solvent, compared to water, as it allows us to establish a universal pH scale due to

$$-\log {}_{W}a_{H^+} = -\log {}_{S}a_{H^+} - \log {}_{m}\gamma_{H^+} \tag{6}$$

It is a serious drawback that it is not possible to determine the transfer activity coefficient of the proton (or of any other single-ion species) directly by thermodynamic methods, because only the values for both the proton and its counterion are obtained. Therefore, approximation methods are used to separate the medium effect on the proton. One is based on the simple *sphere-in-continuum* model of Born, calculating the electrostatic contribution of the Gibb's free energy of transfer. This approach is clearly too weak, because it does not consider solvation effects. Different extrathermodynamic approximation methods, unfortunately, lead not only to different values of the medium effect but also to different signs in some cases. Some examples are given in the following: $\log {}_{m}\gamma_{H^+}$ for methanol +1.7 (standard deviation 0.4); ethanol +2.5 (1.8), n-butanol +2.3 (2.0), dimethyl sulfoxide −3.6 (2.0), acetonitrile +4.3 (1.5), formic acid +7.9 (1.7), NH_3 −16. From these data, it can be seen that methanol has about the same basicity as water; the other alcohols are less basic, as is acetonitrile. Dimethyl sulfoxide, on the other hand, is more basic than water. However, the basicity of the solvent is not the only property that is important for the change of the pK values of weak acids in comparison to water. The stabilization of the other particles that are present in the acido-basic equilibrium is decisive as well.

Suggested Further Readings

Bates, R. G., Medium effect and pH in nonaqueous and mixed solvents, in *Determination of pH, Theory and Practice*, John Wiley & Sons, New York, 1973, pp. 211–253.

Covington, A. K. and T. Dickinson, Introduction and solvent properties, in *Physical Chemistry of Organic Solvents Systems* (A. K. Covington and T. Dickinson, eds.), Plenum Press, London, 1973, pp. 1–23.

Kolthoff, I. M. and M. K. Chantooni, General introduction to acid–base equilibria in nonaqueous organic solvents, in *Treatise on Analytical Chemistry, Part I, Theory and Practice* (I. M. Kolthoff and P. J. Elving, eds.), John Wiley & Sons, New York, 1979, pp. 239–301.

Popov, A. P. and H. Caruso, Amphiprotic solvents, in *Treatise on*

Analytical Chemistry, Part I, Theory and Practice (I. M. Kolthoff and P. J. Elving, eds.), John Wiley & Sons, New York, 1979, pp. 303–347.
Sarmini, K. and E. Kenndler, Ionization constants of weak acids and bases in organic solvents, *J. Biophys. Biochem. Methods 38*:123 (1999).
Sarmini, K. and E. Kenndler, Influence of organic solvents on the separation selectivity of capillary electrophoresis, *J. Chromatogr. A 792*:3 (1997).

Ernst Kenndler

Capillary Isoelectric Focusing: An Overview

Introduction

Capillary isoelectric focusing (CIEF) employs a pH gradient developed within the capillary to separate zwitterions, usually proteins and peptides, based on each solute's p*I*. The technique is analogous to slab–gel CIEF [1] with several important differences: (a) Slab–gel IEF is a non-elution process. After running the electrophoretic step, the proteins are detected by staining. CIEF is usually an elution technique. The contents of the capillary are mobilized to pass through the detector region. (b) In the slab–gel, detection is by Commassie or silver staining. In CIEF, detection is by ultraviolet (UV) absorbance at 280 nm. (c) In the capillary format, no gel is required because mechanical stability is provided by the rigid capillary walls. (d) The field strength is at least an order of magnitude higher in the capillary format compared to the slab–gel.

The usual advantages of capillary electrophoresis apply equally to CIEF. Slab–gel IEF is extremely labor intensive and time-consuming. CIEF is simple to run, fully automated, and high speed and provides improved quantitative results compared to slab–gel IEF. This topic has been recently reviewed [2,3], as is usually covered as a chapter in many high-performance capillary electrophoresis textbooks.

pH Gradient Formation

A solution containing 0.5–2.0% carrier ampholytes and 0.1–0.4% methylcellulose (1500 cP for a 2% solution) is filled into the capillary. A coated capillary is used to suppress the electro-osmotic flow in conjunction with the methylcellulose solution. The sample (protein) concentration in the ampholyte blend is usually between 50 and 200 μg/mL. The inlet reservoir (anolyte) is filled with 10 m*M* phosphoric acid in methylcellulose solution. The outlet reservoir (catholyte) contains 20 m*M* sodium hydroxide.

The condition of the capillary immediately upon activation of the voltage is illustrated at the top of Fig. 1. In this case, the capillary is filled with a p*I* 3–10 mixture of ampholytes. Assuming the pH of the solution is 7, charges have been assigned to the individual ampholytes. The ampholyte charge dictates the direction of migration. Should any ampholytes migrate into a reservoir, the extreme pH conditions cause immediate charge reversal. Likewise, as the steady state is approached, should an ampholyte migrate into a more acidic or basic zone, charge reversal occurs as well. The result is the formation of a pH gradient as indicated at the bottom of Fig. 1. As each ampholyte approaches a pH equal to its individual p*I*, migration slows and then ceases. Because overlapping Gaussian zones of each ampholyte are formed, the gradient is smooth and linear.

The conventional pH range for CIEF is pH 3–10. Narrow-range gradients can be created by selecting cus-

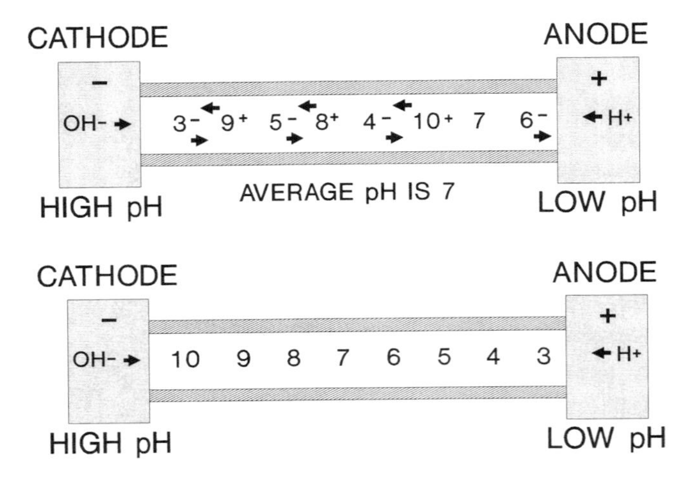

Fig. 1 Illustration of the process of pH gradient formation.

tom ampholyte blends (e.g., pH 6–8). To avoid problems such as a step-gradient or the creation of water zones, the narrow-range ampholyte solution is usually supplemented with 20% pH 3–10 ampholytes. This ensures that there are sufficient ampholyte species to produce Gaussian overlaps.

The focusing step takes 2–5 min. at a field strength of 500–1000 V/cm. Overfocusing causes precipitation of proteins and can damage the capillary as well. Initially, the current is high as ampholytes and proteins are highly charged. As their p*I*'s are approached, the current declines and reaches 1–4 μA when focusing is complete.

The 2:1 ratio of base:acid in the reservoirs is not coincidental. It is selected to minimize drift of the pH gradient. The pH of the anolyte must be lower than that of the most acidic ampholyte; likewise, the pH of the catholyte must be higher than the most basic ampholyte. Otherwise, ampholytes will migrate into the reservoirs and cause gradient drift. If the EOF is not reduced, a form of cathodic drift occurs as well. An exception to this is when one-step mobilization is employed.

Internal standards are always used to calibrate the pH gradient. This is important because the salt content of the sample can compress the gradient. The ideal internal standard absorbs at a wavelength other than 280 nm. It can then be added to the ampholytes and monitored without producing interference. In this case, photodiode array detection is used for multiwavelength monitoring. Methyl red, p*I* 3.8, is ideal in this regard, but other such markers have not been identified. Aminomethylphenyl dyes are often used with monitoring at 400 nm, but they have some absorbance at 280 nm.

To prevent focused zones from reaching the blind side of the capillary past the detector, the ampholytes are either filled just prior to the detector or the reagent TEMED (*N,N,N,N*-tetramethylenediamine) is added to the blend at the appropriate concentration (0.5–2.0%). TEMED, a basic amine, then occupies that space past the detector window.

Resolving Power

The resolving power, ΔpI, of CIEF is described by

$$\Delta pI = 3\sqrt{\frac{D(d\text{pH}/dx)}{E(d\mu/d\text{pH})}} \tag{1}$$

where D is the diffusion coefficient, E is the field strength, μ is the mobility of the protein, and $d\mu/d$pH describes the mobility–pH relationship. The term dpH$/dx$ represents the change in the buffer pH per unit of capillary length. This adjustable parameter is controlled by selecting an appropriate ampholyte pH range as well as the capillary length. Under optimal conditions, a resolution of 0.02 pH units is possible.

Mobilization

There are three ways of mobilizing the contents of the capillary: (a) chemical (salt) mobilization, (b) electroosmotic (one-step) mobilization, and (c) hydrodynamic mobilization. Hydrodynamic mobilization is the simplest and most widely used method. Low pressure is used to evacuate the capillary with the voltage activated. Laminar band broadening is minimized by simultaneous focusing. In one-step mobilization, the EOF is reduced but not absent. If done correctly, focusing occurs prior to any proteins reaching the detector. This is the fastest method but has lower resolution and linearity compared to hydrodynamic mobilization. Salt mobilization is infrequently used today, but it produces the highest resolution at the expense of run time and gradient linearity.

Detection

Because the ampholytes absorb below 250 nm, 280-nm detection is required. It is critical to run ampholyte blanks because the reagents are not checked by the manufacturers for UV absorption at 280 nm. The limit of detection is a few micrograms per milliliter of protein and this is usually limited by the UV reagent background. Proteins that are deficient in aromatic amino acids are poorly detected. Different lots of ampholytes from various manufacturers show variation in the UV background.

The combination of CIEF and mass spectrometry is analogous to two-dimensional electrophoresis [4]. In this case, the mass spectrometer provides the molecular-weight information instead of sodium dodecyl sulfate–polyacrylamide gel electrophoresis. This information can be obtained by on-line CIEF–MS [5] or by using CIEF as a micropreparative technique [6].

For the on-line system, CIEF is performed conventionally in a 20-cm capillary mounted inside an electrospray probe. After focusing, the outlet reservoir (catholyte) is removed and the capillary tip set to 0.5 mm outside of the probe. A sheath liquid of 50% methanol, 49% water, and 1% acetic acid (pH 2.6) pumped with a syringe pump at 3 μL/min produces a stable electrospray. Cathodic mobilization is produced by changing the anolyte to the sheath liquid. The ampholyte ions were observed up to m/z 800 and thus did not interfere with the protein signals.

Additives for Hydrophobic Proteins

The tendency of hydrophobic proteins to aggregate and precipitate is a major problem in IEF whether in the slab–gel or capillary format. The focusing power of CIEF produces an increase in solute concentration by a factor of over 200 [7]. Proteins also readily precipitate as the p*I* is approached because their charge and, thus, electrostatic repulsion approach zero.

Protein precipitation is indicated first by spikes in the electropherogram followed by clogging of the capillary. Additives are required to suppress the aggregation of hydrophobic proteins to keep them in solution. Two excellent review articles describe this in detail [8,9].

Among the reagents used to prevent aggregation are urea, nonionic surfactants such as Brij-35, zwitterionic detergents such as sulfobetains, polyols such as ethylene glycol, glycerol, amd sorbitol or nonreducing sugars. A strategy of mixing polyols and zwitterions is often successful in dealing with solubility problems.

Applications

Capillary isoelectric focusing has been employed for separations of many proteins, recombinant proteins, monoclonal antibodies, and protein glycoforms. The most widely used method employing CIEF is the determination of hemoglobin variants for the screening of genetic disorders, including sickle cell anemia, thalessemisas, and other hemoglobinopathies. Figure 2 illustrates the separation of hemoglobins in a patient with Hb S/β^+ thalassemia.

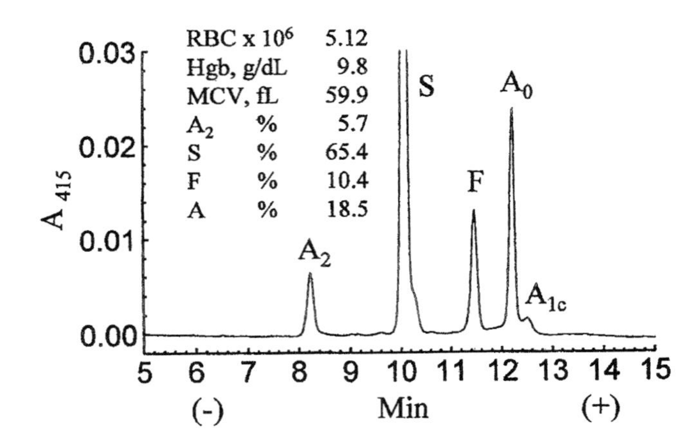

Fig. 2 Separation of hemoglobins by CIEF in blood from a patient suffering from Hb S/β^+ thalessemia. Capillary: 27 cm (20 cm to detector) × 50 μm i.d. DB-1 (J & W Scientific); ampholytes: 2% pH 6–8:10-3 (10:1) Pharmalytes (Pharmaceia Biotech) and 0.375% methylcellulose; catholyte: 20 m*M* sodium hydroxide; anolyte: 100 m*M* phosphoric acid in 0.375% methylcellulose; focusing: 5 min at −30 kV; mobilization: low pressure (0.5 psi) for 10 min at −30 kV; detection: UV, 415 nm. [Reprinted from *Electrophoresis 18*:1785 (1997), copyright (1997) Wiley–VCH.]

References

1. P. G. Righetti, *Isoelectric focusing: Theory, Methodology and Applications*, Elsevier Biomedical Press, Amsterdam, 1983.
2. P. G. Righetti and A. Bossi, *Anal. Chim. Acta 372*:1 (1998).
3. T. J. Pritchett, *Electrophoresis 17*:1195 (1996).
4. Q. Tang, A. K. Harrata, and C. S. Lee, *Anal. Chem. 69*:3177 (1997).
5. Q. Tang, A. Kamel Harrata, and C. S. Lee, *Anal. Chem. 67*:3515 (1995).
6. F. Foret, O. Muller, J. Thorne, W. Gotzinger, and B. L. Karger, *J. Chromatogr. A 716*:157 (1995).
7. G. G. Yowell, S. D. Fazio, and R. V. Vivilecchia, *J. Chromatogr. 652*:215 (1993).
8. T. Rabilloud, *Electrophoresis 17*:813 (1996).
9. R. Rodriguez-Diaz, T. Wehr, and N. Zhu, *Electrophoresis 18*:2134 (1997).

Robert Weinberger

Capillary Isoelectric Focusing of Peptides, Proteins, and Antibodies

Introduction

Capillary isoelectric focusing (CIEF) is a high-resolution technique for protein and peptide separation performed at academic sites and in the biotechnology and pharmaceutical industries for the analysis and characterization of, for example, recombinant antibodies and other recombinant proteins, isoforms of glycoproteins, point mutations in hemoglobin, and peptide mapping. Also, hyphenation to mass spectrometry and chip-based CIEF (microfabrication) have shown promise. CIEF kits and specific recipes/application notes are available from vendors of capillary electrophoresis (CE) equipment, as are a vast amount of publications and handbooks of CE published over recent years.

Capillary isoelectric focusing is a rapid analysis technique with typical run times of 5–30 min, fully automated with on-line detection and real-time data acquisition, and minute sample consumption (a few microliters is enough for repetitive injections). A linear dynamic range over one order of magnitude is achievable, and a detectability down to 5–10 μg/mL. A resolution of ΔpI-0.01 is possible under optimized conditions. Reproducibility of p*I* determination is typically <0.5% (RSD) using internal standards.

Sample Salt Content

For high-performance analysis, it is important that the sample applied has a low salt content [1]. Salt concentrations below 10 m*M* is preferable; concentrations above 40–50 m*M* should be desalted. The salt will compress the pH gradient so that it will occupy only some part of the capillary. Hereby, the resolution will decrease and the risk for protein precipitation will increase (see the section on Focusing, below). Also, focusing time will have to be increased as well as focusing/mobilization time being less reproducible.

Ampholytes

Several commercial ampholytes are available covering broad-range (pH 3–10) and narrow-range pH gradients (e.g., pH 6–8). Broad-range gradients are suitable for analysis of proteins covering a wide spectrum of isoelectric points, whereas narrow-range gradients are preferable for high-resolution separations where only minor differences in protein p*I*'s are expected [2]. Tailor-made gradients can easily be made by mixing ampholytes of different pH ranges or by adding ampholyte spacers [3]. It is preferable, also, to mix ampholytes (with a similar pH interval) from different suppliers so as to create a smoother gradient. Typical concentrations of ampholytes used are 1–5% (v/v); the concentration employed will affect, for example, the protein load, resolution, and focusing/mobilization voltage and time.

Additives

Additives are used in the sample–ampholyte solution either (a) to suppress protein aggregation/precipitation or (b) to decrease the electro-osmotic flow (EOF) velocity and/or protein adsorption to the capillary surface. Protein precipitation is often a major problem in CIEF due to a highly concentrated protein band, and because electrostatic repulsion is minimal at the p*I*, proteins interact strongly by hydrophobic interactions (also hydrogen-bonding) causing irreproducible migration time and peak area quantification and sometimes capillary clogging [1, 2]. Precipitates are often seen as "spikes" in the electropherogram. Strategies to minimize precipitation include reducing the protein concentration and focusing time, applying a lower field strength, and/or using various additives. Because as high a resolution and detection sensitivity as possible are strived for, additives are often the first choice. Many such additives are used to enhance the solubility of proteins like (a) nonionic detergents (e.g., reduced Triton X-100 and Brij-35), (b) zwitterions (e.g., CAPS and sulfobetain), (c) carbohydrates (e.g., sucrose, sorbitol, and cellulose derivatives), (d) urea, and (e) organic modifiers (e.g., ethylene glycol and glycerol). It is also common to mix different additives [3]. There seems to be no universal solubilization recipe covering all sorts of proteins. Instead, tailor-made recipes often must be worked out whenever precipitates are encountered. The commonly used mixture of 8*M* urea and 2% detergents, as employed in the first dimension in two-dimensional gel electrophoresis, often performs well, although attention must be paid to protein modification causing artifacts in the electropherogram [1]. Because urea also denatures proteins (in contrast to the other, more mildly, solubilizers), a shift in p*I* might be expected. See Ref. 3 for new types of solubilizers used in CIEF.

Additives used for decreasing the EOF and/or protein adsorption are often cellulose derivatives [e.g., hydroxypropylmethylcellulose (HPMC)]. The cellulose adsorbs to the capillary surface. Hereby, the viscosity will increase at the capillary surface (more than in bulk solution), causing a reduction in EOF as well as a decrease in protein adsorption to the capillary wall. The tendency for protein precipitation will also decrease by addition of cellulose derivatives.

Focusing

Focusing is the process where ampholytes and proteins migrate to their respective p*I* positions in the capillary. During focusing, current will decrease, as the current carrier ampholytes cease to migrate, and finally reach a constant value when the pH gradient is fully developed and the steady state has been attained (certain gradient drifts are often observed in CIEF whose presence will affect performance and reproducibility [2]). Normally, focusing is considered complete when current has decreased to

10% of its original value; a longer focusing time increases the risk for protein precipitation.

Resolution in CIEF is described by the following formula:

$$\Delta pI = 3\left(\frac{D(dpH/dx)}{E(d\mu/dpH)}\right)^{1/2}$$

where D is the diffusion coefficient, dpH/dx is the change of pH with distance x, E is the electric field strength, and $d\mu/dpH$ is the change of mobility with pH [2]. The importance of the applied voltage is clearly seen, as a higher field strength gives a better resolution and shorter focusing time. Field strengths between 300 and 700 V/cm are normally used. The optimal voltage may be determined experimentally from a series of runs with different voltages. The formula also reveals that proteins attain a higher resolution than peptides because their D values are lower and $d\mu/dpH$ is higher.

Mobilization

Single-Step CIEF

In single-step (electro-osmotic displacement) CIEF, focusing and mobilization take place simultaneously [1,2]. Here, a dynamically coated (or static coated with reasonably high EOF) capillary is used where the EOF is fast enough to sweep all proteins by the detector in a reasonable time frame and slow enough to simultaneously allow the protein zones to focus. Additives are used to manipulate the velocity of the EOF as well as to decrease protein adsorption. The single-step method benefits from short analysis times and an extended capillary lifetime but suffers from lower detectability (due to lower protein load), compromised resolution, and nonlinearity of the p*I* calibration curve [4]. A variant of the single-step CIEF is to use whole-column imaging detection where all the focused bands in the capillary are detected simultaneously without any need for mobilization [5].

Two-Step CIEF

Two-step CIEF is used when EOF in the capillary is strongly reduced. After focusing is complete, a separate mobilization step is applied where the protein bands are transported past the fixed detector. Mobilization can be achieved either by chemical means [for cathodic (anodic) mobilization, the composition of the catholyte (anolyte) is changed; thereby the pH gradient will change], or by applying a positive/negative pressure, or by a mixture of both [1,2]. A pressure will induce zone broadening due to the parabolic flow profile; a modest pressure in combination with high-voltage will preserve the high-resolution pattern 4. Pressure mobilization offers good reproducibility (migration times and p*I* values) and a linear p*I* calibration curve. Chemical mobilization shows the same performance and also exhibits the highest resolution, at the expense of longer migration times. Anodic mobilization affords a better resolution of the acidic region, whereas cathodic mobilization is preferable for the basic region [4].

Detection

Absorbance detection at 280 nm is mostly used. Below 220 nm is preferable for sensitive protein/peptide detection but is not possible because of background-absorbing ampholytes [2]. Proteins with chromophores (e.g., hemoglobins) can be detected at visual wavelengths. However, because CIEF is a concentrating technique (the protein bands will be concentrated about 100-fold during focusing), a fairly good sensitivity for most proteins is still attainable at 280 nm. For high-sensitivity analysis, laser-induced fluorescence detection might be an alternative, but it requires a labeling procedure [5,6]. By labeling, a change in intrinsic p*I* might occur and the same analyte might be subjected to multiple labeling sites showing several peaks in the electropherogram (especially proteins). This might, however, be the only choice for peptides lacking tyrosine and tryptophan. Whole-column imaging detection (no mobilization step needed) with different detection schemes is also possible for proteins and peptides [5].

Capillary isoelectric focusing coupled to mass spectrometry has gained popularity in recent years (by analogy to two-dimensional gel electrophoresis). Additional information obtained from mass spectrometry includes not only precise molecular-weight determination but also the possiblity for peptide sequencing. Analysis of hemoglobin variants, recombinant proteins, and monoclonal antibodies have been demonstrated [1,7,8].

Application of CIEF to Peptides, Proteins, and Antibodies

Peptides

Peptides are not as commonly analyzed by CIEF as are proteins; one reason is their lower resolution, another their lower (or lack of) detectability at 280 nm (the wavelength mostly used). The separation of tryptic digests (peptide mapping) of proteins have been performed by using absorption detection at 280 nm and refractive index gradient imaging detection; no exact correlations were observed between measured and calculated p*I* values [1, 5]. Refractive index detection is a universal detection method (i.e., independent of chromophores like tyrosine and tryptophan) but suffers from low sensitivity. Assays of trypsin activity have also been performed with laser-induced fluorescence detection for enhanced sensitivity, with detectability down to picomolal concentrations [5,6].

Hemoglobins

Analysis of hemoglobin (Hb) variants are of major clinical importance. Many hematological disorders exist where the globin chains have been subjected to alterations (e.g., point mutations or deletion of gene sequence). Four hemoglobin, variants (A, F, S, and C) having very close pI values (from 7 to 7.40) are often employed as standards to demonstrate the high resolving power of CIEF [2]. Several other variants are possible to resolve (e.g., the separation of Hb A from its glycated form Hb A_{1c} whose pI's differ only by 0.01 or less) [1,3]. Quantitative determination of Hb A_{1c} is routinely used for assessing the degree of diabetes. To improve the resolution between Hb A and Hb A_{1c}, spacers (β-alanine and 6-aminocaproic acid) were added to a mixture of pH 3–10 and pH 6–8 ampholytes in a coated capillary [3]. The spacers flatten the pH gradient in the pI region of the Hb's, thus allowing full separation.

Antibodies

Analysis of microheterogeneity in natural and recombinant monoclonal antibodies is a challenge where CIEF seems to be particularly well suited (see Fig. 1 [4]). Minor modifications (e.g., deamidation, improper folding, and a change in glycosylation pattern) arising from protein synthesis or posttranslational modifications are often detected by CIEF [1,2,9]. The use of CIEF and comparison to flat-bed IEF for routine analysis of recombinant immunoglobins have been demonstrated with a coated capillary, methylcellulose, and a two-step mobilization procedure. The performance of the analysis and coating stability were constant for over 150 analyses with qualitatively and quantitatively equivalent immunoglobulin focusing profiles obtained via CIEF and IEF. Intra-assay reproducibility was less than 2% RSD for peak areas and 1% RSD for migration time. Interassay (72 h) reproduciblity was less than 8% RSD for peak areas and 3% for migration time [10].

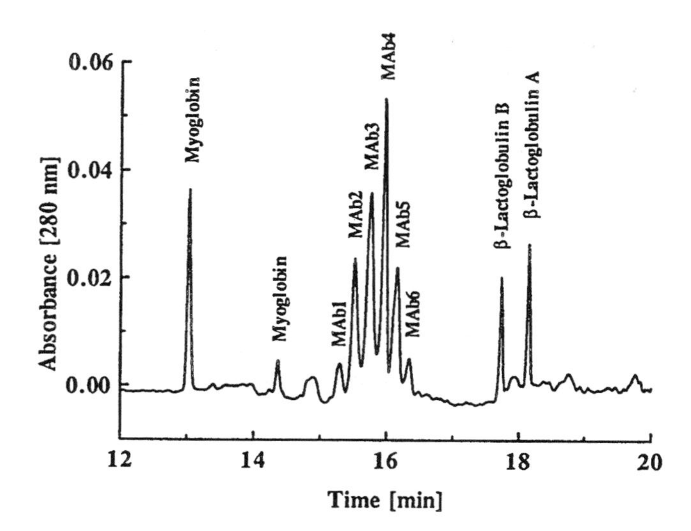

Fig. 1 CIEF of a mouse monoclonal antibody (MAb). Antibody concentration: 0.5 mg/mL; concentration of the marker proteins (myoglobin and β-lactoglobulin): 0.05 mg/mL; capillary: μ-SIL DB-1, 27 cm total length (20 cm to detector); carrier ampholytes: 4% Pharmalyte pH 3–10 including 1% TEMED and 0.8% methylcellulose (MC); anolyte: 10 mM H_3PO_4 in 0.4% MC; catholyte: 20 mM NaOH. Focusing for 2 min at 20 kV, followed by mobilization at low pressure (0.5 psi) at 20 kV. (From Ref. 4.)

Glycoproteins

Applications of CIEF for the separation of isoforms of transferrin have been reported by several groups [1,2,5]. Transferrin contains different number of sialic acid residues, with an additional −1 charge added per residue. Also, transferrin bound to different amount of iron atoms has been separated by CIEF [1,2,5]. Glycoforms of recombinant tissue-type plasminogen activator (rtPA) is another sialic acid containing protein which has been the subject of analysis by CIEF [1,2,11]. A rapid (<10 min) one-step method was developed using a coated capillary, HPMC, and urea in a mixture of pH 3–10 and pH 5–8 ampholytes. Ten species could be detected. Intra-assay precision was less than 5% for peak migration times and 10% for normalized peak areas [11].

References

1. R. Rodriguez-Diaz, T. Wehr, and M. Zhu, *Electrophoresis 18*:2134 (1997).
2. X. Liu, Z. Sosic, and I. S. Krull, *J. Chromatogr. A 735*:165 (1996).
3. P. G. Righetti, A. Bossi, and C. Gelfi, *J. Capillary Electrophoresis 4*(2), 47 (1997).
4. C. Schwer, *Electrophoresis 16*:2121 (1995).
5. X. Fang, C. Tragas, J. Wu, Q. Mao, and J. Pawliszyn, *Electrophoresis 19*:2290 (1998).
6. K. Shimura, H. Matsumoto, and K. Kasai, *Electrophoresis 19*:2296 (1998).
7. J. Wei, C. S. Lee, I. M. Lazar, and M. L. Lee, *J. Microcol. Separ. 11*(3):193 (1999).

8. M-L Hagmann, C. Kionka, M. Schreiner, and C. Schwer, *J. Chromatogr. A 816*:49 (1998).
9. I. S. Krull, X. Liu, J. Dai, C. Gendreau, and G. Li, *J. Pharm. Biomed. Anal. 16*:377 (1997).
10. S. Tang, D. P. Nesta, L.R. Maneri, and K. R. Anumula, *J. Pharm. Biomed. Anal. 19*:569 (1999).
11. K. G. Moorhouse, C. A. Eusebio, G. Hunt, and A. B. Chen, *J. Chromatogr. A 717*:61 (1995).

Anders Palm

Capillary Isotachophoresis

Introduction

Three analytical electrophoretic techniques can be distinguished. They differ in the kind of background electrolyte (BGE) and in its arrangements. Zone electrophoresis has a uniform BGE without a gradient; in isoelectric focusing, the separation is established by the aid of a BGE forming a continuous (linear) pH gradient. In contrast, isotachophoresis (ITP) is an electrophoretic method with a stepwise gradient of the background electrolyte.

Isotachophoresis

In ITP, samples of only one charge type are separated in the same run (i.e., either anions or cations). The BGE in ITP is selected in the way that (the anion or cation of) the leading (L) and the terminating (T) electrolyte will have a higher and a lower mobility, μ, respectively, than the analytes of interest. Thus, the prerequisite for separation by ITP is that $\mu_L > \mu_{analytes} > \mu_T$.

Consider the case that the sample ions are anions (consisting of analytes A^- and B^-), and that the ions exhibit mobilities in the sequence $\mu_L > \mu_A > \mu_B > \mu_T$. The capillary is filled with L^- and T^-, separated by a sharp boundary, and the sample is injected between the two electrolyte zones (for simplicity, it is assumed that the counterions, Q^+, are all the same). After application of an electric field, a certain field strength is established in the zones as depicted in Fig. 1 (it is assumed that the capillary has uniform diameter). Across the individual zone, the field strength is constant, but it increases due to the decreasing mobility (and increasing electrical resistance) from L to T. Across the zone of the mixed sample (A + B) it is constant as well, with strength E_{mix}. In this zone, the two analytes A and B are moving with different migration velocities, ν, according to $\nu_A = \mu_A E_{mix}$ and $\nu_B = \mu_B E_{mix}$. Due to the higher mobility of A, this analyte moves faster here than B: $\mu_A E_{mix} > \mu_B E_{mix}$. This effect leads to the migration of ions A^- out of the mixed zone, forming a separate zone in front with pure A^- (which is always placed behind the zone of the leading electrolyte). For B^-, the analogous situation occurs; it is moving slower in the mixed zone and forms a separate zone at the rear side (but in front of T). The formation of zones of pure analyte ions (obviously with counterions, Q^+, the counter ions of the leading electrolyte) is, therefore, observed. Five zones can be differentiated in this transient state: L^-, pure A^-, mixed $A^- + B^-$, pure B^-, and T^-. Due to the different mobilities, the field strength increases in this sequence. Separation takes place as long as the mixed zone exists. Finally, this zone disappears and the isotachophoretic condition is established: All zones and zone boundaries migrate with the same mean velocity ("isotachophoresis"); consequently,

$$\mu_L E_L = \mu_A E_A = \mu_B E_B = \mu_T E_T \tag{1}$$

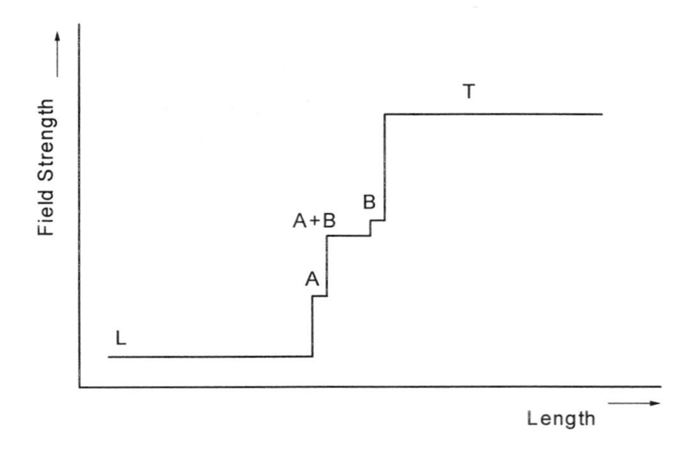

Fig. 1 Electrical field strength in the zones of the leading electrolyte, L, the sample consisting of A and B, and the termination electrolyte, T. For details, see the text.

This is the isotachophoretic condition. The conditions of electroneutrality must be fulfilled as well, which means, in the case considered, that in each zone the concentration of ions and counterions is equal. The third condition is Ohm's law, stating that (for given constant current) the product of electrical conductivity and field strength in each zone is equal. The combination of these conditions leads to the so-called regulation function (Kohlrausch), which reads in a simplified form (for monovalent strong electrolytes):

$$c_A = c_L \frac{\mu_A(\mu_L + \mu_Q)}{\mu_L(\mu_A + \mu_Q)} \tag{2}$$

It relates the concentration of a species in its own zone to the concentration of the species in the subsequent zone and, as a consequence, to the concentration of the first zone, the leading ion. It is also a function of the mobility of the ions involved: the analyte, the leading ion, and the counterion. This function allows two conclusions as follows.

As for given conditions, the mobilities in Eq. (2) are constant under ITP conditions; the concentration of the sample in its zone is constant as well. The concentration distribution is, therefore, given by a rectangular function. It follows that the temperature and the pH within the particular zone is constant, too, and changes stepwise at the boundary to the neighboring zone.

The concentration depends only on that of the leading ion; it is independent of the initial concentration in the sample. Therefore, ITP can act as an enrichment method, analogous to displacement chromatography and in contrast to zone electrophoresis and elution chromatography. The concentration in the steady state is adjusted to the value given in Eq. (2). If the concentration of the analyte species is lower in the initial sample, the higher steady-state concentration is established. This concentration is independent of the migration distance: there is no dilution with a BGE as there is in capillary zone electrophoresis (CZE).

The adjustment of the steady-state concentration to a certain constant value has the consequence that the zone length of an analyte depends on the amount present in the sample. Increasing the amount results in an increase of the zone length under ITP conditions. The length is, therefore, the parameter for quantitative analysis.

The stepwise gradient of the electrical field at the zone boundary is the source of the "self-sharpening" effect in ITP. When, by diffusion, an ion migrates out of its own zone, into a neighboring zone (where the field strength is higher or lower, respectively), the condition given in Eq. (1) is not fulfilled any more. Therefore, the velocity of the considered ion is either accelerated or retarded, and the ion is pushed back into its initial zone. As a consequence, the boundary between the zones remains sharp and its shape is not dependent on the migration distance.

Isotachophoresis might have several advantages compared to zone electrophoresis. The adaptation to a considerably high concentration of the sample components in their own zones in the absence of further BGE favors the use of the electrical conductivity detector. The high concentration and the long sample zone have also some advantage in combining capillary electrophoresis with, for example, mass spectrometry. Also, ITP can be used as an enrichment technique prior to zone electrophoretic separation, a phenomenon that is applied routinely in sodium dodecyl sulfate–polyacrylamide gel electrophoresis using a discontinuous buffer for sample introduction, and a technique called sample stacking in CZE. In fact, both methods rely on an isotachophoretic principle.

Suggested Further Readings

Bocek, P., M. Deml, P. Gebauer, and V. Dolnik, *Analytical Isotachophoresis*. VCH, Weinheim, 1988.

Everaerts, F. M., J. L. Beckers, and T. P. E. M. Verheggen, *Isotachophoresis: Theory, Instrumentation, and Applications*, Elsevier, Amsterdam, 1976.

Mosher, R. A., D. A. Saville, and W. Thormann, *The Dynamics of Electrophoresis*, VCH, Weinheim, 1992.

Ernst Kenndler

Carbohydrates as Affinity Ligands

Since its conception 30 years ago, affinity chromatography has been a powerful technique to separate or purify biological compounds, but also a method to study the interactions between living systems and molecules of biological or therapeutical interest. Among these molecules, oses, polysaccharides, or more complex molecules such as glycosaminoglycans constitute a family of potential ligands.

Introduction

The majority of applications of affinity chromatography has been, for a long time, in the field of protein purification. For example, some of the most rewarding applications of affinity chromatography have been in the area of purification of hormone and drug receptors. Among these drugs, carbohydrate-based structures are well known for their biological activity (e.g., their anticoagulant or antiproliferative properties). Although some of these biological properties are widely used in medical applications, the mechanisms of action at the molecular level is not accurately determined.

An example of the use of carbohydrate-based affinity chromatography is, thus, the separation of proteins which are responsible for the action of bioactive polysaccharides or oses. The strategy consists in the immobilization of these carbohydrates on classical low- or high-pressure affinity phases. We will distinguish two types of ligands: those based on osidic structures and those prepared from glycosaminoglycans or polysaccharides.

Osidic Ligands

All osidic structures which can interact with proteins implicated in biological responses or phenomenons are, a priori, candidates as ligands in affinity chromatography. Here, we give some examples of oses which have been successfully immobilized to make selective chromatographic supports.

Sialic Acid (N-Acetylneuraminic Acid)

Purification of immunoglobulins G is of great interest in biological science. Among other separation methods, affinity chromatography has been used in different ways. Affinity supports were first prepared using protein A or protein G, whose affinity constants take different values according to the IgG subclasses (IgG1, 2, 3, or 4). Among interesting carbohydrates, sialic acid is known to specifically interact with IgGs. Indeed, immunoglobulins are able to react to the presence of tumoral cells. The antigens which are suspected to promote this reaction are gangliosids. Beyond a ceramide molecule, some oses are present in gangliosids, among which are included between 1 and 3 sialic acids.

The hypothesis has been made that sialic acid may lead to the formation of specific interactions with immunoglobulins G. Some workers have, thus, developed affinity supports bearing sialic acid, in order to purify IgGs. Sialic acid can be extracted from swallow nests and coupled on activated coated silica. This support has been found to possess a good specificity for IgGs, in particular for the subclass IgG3 [1].

Supports prepared with sialic acid have also been used in the purification of insulin, as this sugar and *N*-acetylglucosamine have been found to take part in the interaction between insulin and its glycosylated receptor. The affinity between insulin and sialic acid-bearing supports has been found rather strong ($K_a \sim 10^9 M^{-1}$) and the system allows the separation of beef and pig insulins [2], which differ by only two amino acids.

N-Acetylglucosamine

N-Acetylglucosamine also shows a specific interaction with insulin. Affinity chromatography experiments have evidenced results very close to those observed in the case of sialic acid. An interesting result is that the improvement of both affinity and capacity of supports bearing both sugars [3]. These supports are supposed to more accurately mimic the structure of insulin receptor.

Mannose and Derivatives

Mannose has been immobilized in order to separate mannose-binding proteins such as mannose-binding lectin. When grafted onto agarose, it constitutes a purification step of a specific lectin which does not bind to DEAE–cellulose or Affi-gel Blue gel [4]. Another way to obtain mannose-binding lectin is to perform expanded-bed affinity chromatography by immobilizing mannose on a DEAE Streamline support [5]. Affinity agarose supports, grafted with pentamannosyl phosphate, allowed the testing of the functionality (in terms of ligand binding) of truncated and glycosylation-deficient forms of the mannose 6-phosphate receptor from insect cells [6]. Phosphomannan affinity chromatography has been used to purify a human insulin-like growth factor II mannose 6-phosphate receptor [7].

Glycosaminoglycans and Polysaccharides

Certainly, one of the most used glycosaminoglycans is heparin, because of its anticoagulant properties. Other glycosaminoglycans or polysaccharides have shown such properties; among them, sulfated dextran derivatives and naturally sulfated polysaccharides extracted from algae (fucans) will be discussed further.

Heparin

Anticoagulant properties are due to the formation of a complex between heparin and antithrombin (ATIII); heparin increases ATIII activity, inhibiting thrombin, which is responsible for the formation of the clot [8]. Although this complex is already characterized by a weak affinity, the exact mechanism of association between heparin and antithrombin is not exactly known. A multistep protocol of immobilization of heparin on silica beads permitted high-performance chromatographic phases to be obtained. Thus, it has been possible to evidence a slightly stronger affinity of heparin for antithrombin than for thrombin.

ATIII has been also used as a model protein to test a novel affinity chromatographic system: capillary affinity chromatography [9]. Separation quality has been found equivalent to that observed with classical affinity chromatography, whereas the necessary protein amount is strongly reduced to the nanogram level.

Heparin possesses an affinity for many molecules, among which is a phospholipid-binding protein, annexin V. Affinity chromatography evidenced the Ca^{2+} dependence of the binding mechanism [10]; von Willebrand factors with high and low molecular weights have been separated using their different affinities toward heparin [11].

Heparin has been used in enzyme purification such as recombinant human mast cell tryptase. The purified enzyme is fully active [12]. Heparin-based affinity chromatography also permitted the isolation of growth factors such as basic fibroblast growth factor (bFGF). The affinity is lower when bFGF is complexed with acidic gelatin [13]. The elution of synthetic TFPI (tissue factor pathway inhibitor) peptidic fragments on immobilized heparin has allowed one to find the peptidic sequence responsible for the TFPI–heparin interaction [14].

Heparin is able to inhibit smooth-muscle cells (SMCs) proliferation in vitro. SMCs are present in blood vessel walls and may proliferate in the case of an internal injury. The antiproliferative action of heparin is due to its internalization in SMCs, which is probably mediated by membrane receptors. Heparin-based affinity chromatography of SMCs membrane extracts allowed the separation of a few proteins, which could be implicated in the growth inhibition [15].

The different actions taking place in the overall affinity mechanism of immobilized heparin for different biological compounds is not yet elucidated, but the influence of ionic strength demonstrates the important contribution of ionic interactions in this mechanism. However, the large spectrum of biological activities of heparin is also a limit for its specificity.

Other Glycosaminoglycans

Dermatan sulfate is known to specifically catalyze thrombin inhibition by the plasmatic inhibitor heparin cofactor II (HCII). Dermatan sulfate has been immobilized on a dextran- or agarose-coated silica matrix. These systems were tested as high-performance chromatographic supports for the purification of HCII from human plasma. The eluted HCII was obtained with no contamination of ATIII, the other main thrombin inhibitor [16].

Dextran Derivatives

Phosphorylated dextran derivatives, called phosphodextrans, possess a strong affinity for K-vitamin-dependent coagulation factors, like factor II or prothrombin. This property was used to separate them by affinity chromatography on phosphodextrans, which interact in a similar way as phospholipids from the cell membrane [17].

Heparinlike sulfated dextran derivatives, like carboxymethyldextran benzylamide sulfonates (CMDBS), have been immobilized on silica beads. By high-performance liquid affinity chromatography (HPLAC), they allow a good recovery of thrombin, with a yield of 80%. The affinity constant was estimated ($K_a \sim 10^5 M^{-1}$) and was found superior to the value obtained between thrombin and heparin. On the contrary, the affinity of dextran derivatives for ATIII is estimated at a lower value than that of heparin [17].

Fucans

Fucan is a sulfated polysaccharide, naturally present in algae such as *Fucus vesiculosus* or *Ascophyllum nodosum*. Fucan is a general name for a mixture of three polysaccharides, and among them, fucoidan (or homofucan) can be theoretically considered as an homopolymer of α-1,2 L-fucose-4-sulfate and has been studied as a ligand in the same way as fucan himself. Their interaction with two proteins implicated in the coagulation process (thrombin and antithrombin) has been studied and is at least partially ionic. However, the dissociation of the complex fucan–antithrombin seems to include a slower step which could be attributed to a conformation change of the fucan [18].

Fucan was also used as ligand for high-performance liquid affinity chromatography. In the same way as dextran derivatives, a good separation was obtained for thrombin, with a yield of 80%. The affinity constant was estimated in the same order as that obtained for CMDBS ($K_a \sim 10^5 M^{-1}$) and superior to the value obtained for heparin [17].

Conclusion

In this article, we have described several uses of carbohydrate compounds as affinity ligands. These few examples clearly demonstrate the importance of such affinity supports in the separation of biological products. Affinity chromatography is also able to help in the determination of the interaction mechanisms of carbohydrate derivatives in biological reactions. This developing field of research will lead to improved quality and specificity of affinity-chromatographic phases.

References

1. A. Serres, E. Legendre, J. Jozefonvicz, and D. Muller, Affinity of mouse immunoglobulin G subclasses for sialic acid derivatives immobilized on dextran-coated supports, *J. Chromatogr. B 681*:219 (1996).
2. H. Lakhiari, J. Jozefonvicz, and D. Muller, Separation and purification of insulins on coated silica support functionalized with sialic acid by affinity chromatography, *J. Liquid Chromatogr. Related Technol. 19*:2423 (1996).
3. H. Lakhiari, Supports de silice enrobée à ligands biospécifiques: Synthèse, caractérisation et relations structure-propriétés de séparation. Application à la purification de l'insuline et des immunoglobulines G. Thesis, University Paris 13, 1996.
4. L. S. M. Ooi, H. X. Wang, T. B. Ng, and V. E. C. Ooi, Isolation and characterization of a mannose-binding lectin from leaves of the chinese daffodil *Narcissus tazetta*, *Biochem. Cell Biol. 76*:601 (1998).
5. O. Bertrand, S. Cochet, and J. P. Cartron, Expanded bed chromatography for one-step purification of mannose binding lectin from tulip bulbs using mannose immobilized on DEAE Streamline, *J. Chromatogr. A 822*:19 (1998).
6. P. G. Marron-Terada, K. E. Bollinger, and N. M. Dahms, Characterization of truncated and glycosylation-deficient forms of the cation-dependent mannose 6-phosphate receptor expressed in baculovirus-infected insect cells, *Biochemistry 37*:17,223 (1998).
7. M. Costello, R. C. Baxter, and C. D. Scott, Regulation of soluble insulin-like growth factor II mannose 6-phosphate receptor in human serum: Measurement by enzyme-linked immunosorbent assay, *J. Clin. Endocrinol. Metab. 84*:611 (1999).
8. I. Björk, S. T. Olson, and J. D. Shore, Molecular mechanisms of the accelerating effect of heparin on the reactions between antithrombin and clotting proteinases, in *Heparin, Chemical and Biological Properties, Clinical Applications* (D. A. Lane and U. Lindahl, eds.), Edward Arnold, London, 1989, pp. 229–255.
9. X. J. Wu and R. J. Linhardt, Capillary affinity chromatography and affinity capillary electrophoresis of heparin binding proteins, *Electrophoresis 19*:2650 (1998).
10. I. Capila, V. A. Van der Noot, T. R. Mealy, B. A. Seaton, et al., Interaction of heparin with annexin V, *FEBS Lett. 446*:327 (1999).
11. B. E. Fischer, K. B. Thomas, U. Schlokat, and F. Dorner, Selectivity of von Willebrand factor triplet bands towards heparin binding supports structural model, *Eur. J. Haematol. 62*:169 (1999).
12. A. L. Niles, M. Maffit, M. Haak-Frendscho, C. J. Wheeless, et al., Recombinant mast cell tryptase: Stable expression in Pichia pastoris and purification of fully active enzyme, *Biotechnol. Appl. Biochem. 28*:125 (1998).
13. M. Muniruzzaman, Y. Tabata, and Y. Ikada, Complexation of basic fibroblast growth factor with gelatin, *J. Biomater. Sci. Polym. Ed. 9*:459 (1998).
14. Z. Y. Ye, R. Takano, K. Hayashi, T. V. Ta, et al., Structural requirements of human tissue factor pathway inhibitor (TFPI) and heparin for TFPI-heparin interaction, *Throm. Res. 89*:263 (1998).
15. A. S. Clairbois, D. Letourneur, D. Muller, and J. Jozefonvicz, High-performance affinity chromatography for the purification of heparin-binding proteins from detergent-solubilized smooth muscle cell membranes, *J. Chromatogr. B 706*:55 (1998).
16. V. Sinninger, J. Tapon-Bretaudière, F. L. Zhou, A. Bros, et al., Immobilization of dermatan sulphate on a silica matrix and its possible use as an affinity chromatography support for heparin cofactor II purification, *J. Chromatogr. 539*:289 (1991).
17. F. L. Zhou, Supports à base de silice enrobée par des polysaccharides pour chromatographie d'affinité haute performance: Préparation, caractérisation et application dans la purification de protéines, thesis, University Paris 13, 1990.
18. E. Legendre, Etude par chromatographie d'affinité liquide haute performance des interactions entre des protéines de la coagulation et des polysaccharides sulfatés à activité anticoagulante immobilisés sur des supports de silice enrobée, thesis, University Paris 13, 1996.

I. Bataille
D. Muller

Carbohydrates: Analysis by HPLC

Introduction

Carbohydrates are widely distributed molecules in biological systems and pharmaceutical products, not only in free form but also in conjugated form. Because they are present in various forms and there are isomers and analogs, the separation of carbohydrates involves more difficult problems than those of proteins or nucleic acids. Difficulties are also found in detection, especially in biochemical and biomedical analyses due to their low abundance and the fact that photometric and fluorimetric methods cannot be applied directly because of the lack of chromophores and fluorophores.

Analysis of carbohydrates in body fluids by high-performance liquid chromatography (HPLC) using anion-exchange columns was first reported in the 1970s [1–5]. This method has been greatly improved by the use of packing materials of fine, spherical particles and by the development of photometric and fluorimetric postcolumn labeling systems for sensitive detection. Honda et al. established rapid automated methods for microanalysis of aldoses [6], uronic acids [7], and sialic acids using a Hitachi 2633 anion-exchange resin and a photometric and fluorimetric postcolumn labeling system with 2-cyanoacetamide. Alditols [8] were fluorescence labeled by the use of sequential periodate oxidation and the Hantzsch reaction. Microanalysis of aminosugars was successful when their borate complexes were separated in the cation-exchange mode and detected by fluorescence generated either by the reaction with 2-cyanoacetamide [9] or by the Hantzsch reaction [10]. All these methods are suitable for routine analysis of clinical samples because of their speed of analysis and high sensitivity.

The United States Food and Drug Administration and the regulatory agencies in other countries require that pharmaceutical products be tested for composition to verify their identity, strength, quality, and purity. Recently, attention has been given to inactive ingredients as well as active ingredients. Some of these ingredients are nonchromophoric and cannot be detected by absorbance. Some nonchromophoric ingredients, such as carbohydrates, glycols, sugar, alcohols, amines, and sulfur-containing compounds, can be oxidized and, therefore, can be detected using amperometric detection. This detection method is specific for those analytes that can be oxidized at the selected potential, leaving all other nonoxidizable compounds transparent [11]. Amperometric detection is a powerful detection technique with a broad linear range and very low detection limits.

This review outlines current chromatographic methods utilized in the analysis of carbohydrates in biological systems and pharmaceutical products.

Analysis of Carbohydrates by Partition HPLC

Partition HPLC is an important type of chomatography for the analysis of monosaccharides and oligosaccharides. Analysis in this mode has the advantages that it requires a shorter analysis time and gives sharper peaks than anion-exchange chromatography of borate complexes, although it has the drawback of low sensitivity, as detection is usually performed by refractometry. Generally, silica gel whose silanol groups are substituted by alkyl or aminoalkyl groups is used as the stationary phase. HPLC separations using such a stationary phase has been applied successfully to separate oligosaccharides liberated from glycoproteins with hydrazine or borohydride in alkali, permitting quick separation within 60 min [12–14]. Previously, such oligosaccharides were separated and purified by tedious procedures involving gel permeation chromatography on Bio-Gel P-2 or P-4, paper chromatography, and paper electrophoresis. However, modified silica, especially amine-modified silica, has difficulties in durability, being unsuitable for routine analysis.

Analysis of Carbohydrates by Anion-Exchange HPLC

The introduction in the 1980s by Honda and Suzuki [15] of the anion-exchange resin resulted in a significant improvement in the separation of carbohydrates by HPLC using the partition mode. Honda and Suzuki, using this mode, established analytical conditions common to aldoses, amino sugars, and sialic acids. Aldoses in the intact state, amino acids as their *N*-acetates, and sialic acids as *N*-acylmannosamines were separated on a column of a proton-formed, sulfonated styrene–divinylbenzene copolymer and detected by measuring absorption at 280 nm after postcolumn labeling with 2-cyanoacetamide.

Postcolumn labeling is a characteristic feature of carbohydrate analysis in which no direct physical methods are available for sensitive detection. Many labeling methods have hitherto been developed. The methods with phenol in sulfuric acid [16], orcinol in sulfuric acid [17], anthrone in sulfuric acid [18], tetrazolium blue in alkali [19], copper(II)-2-2′-bicinchonitate [20], and 2-cyanoacetamide [21] are used for photometric detection. The methods with 2-cyanoacetamide [6], ethylenediamine [22], ethanolamine [23], taurine [24], and arginine [25] are used for fluorimetric detection. Some labeling methods for electrochemical detection were reported by Honda and Suzuki in 1984 [26,27].

Quantification of Carbohydrates by Anion-Exchange HPLC and Amperometric Detection

Two main columns are used in the analysis of carbohydrates by amperometric detection: the CarboPac™ PA10 and the CarboPac MA1 anion-exchange columns. The CarboPac PA10 column packing consists of a nonporous, highly cross-linked polystyrene–divinylbenzene substrate agglomerated with 460-nm-diameter latex. The MicroBead™ latex is functionalized with quaternary ammonium ions, which create a thin surface-rich anion-exchange site. The packing is specifically designed to have a high selectivity for carbohydrates. The PA10 has an anion-exchange capacity of approximately 100 μEq/column.

The CarboPac MA1 resin is composed of a polystyrene–divinylbenzene polymeric core. The surface is grafted with quaternary ammonium anion-exchange functional groups. Its macroporous structure provides an extremely high anion-exchange capacity of 1450 μEq/column. The CarboPac MA1 column is designed specifically for sugar alcohol and glycol separations. The PA10 but not the MA1 is compatible with eluents containing organic solvents, which can be used to clean these columns.

The equipment used for the analysis of carbohydrates by anion exchange and amperometric detection include a Dionex DX-500 system consisting of a GP40 gradient pump, an ED40 electrochemical detector, a LC30 chromatography oven, and a PeakNet chromatography workstation. A gold electrode is used for both column applications. The flow rate used for the PA10 column is 1.5 mL/min and 0.4 mL/min for the MA1. Injection volumes are typically 10 μL and the oven temperature 30°C. Eluent components include water and 200 m*M* sodium hydroxide for the PA10 column and water and 480 m*M* sodium hydroxide for the MA1 column. Eluent concentration for the PA10 column starts at 91% water/9% 200 m*M* sodium hydroxide for up to 11 min, 100% 200 m*M* sodium hydroxide from 11.1 to 17.6 min, and 91% water/9% 200 m*M* sodium hydroxide from 17.7 to 40.0 min. Eluent concentration for the MA1 column system starts at 52% water/48% 480 m*M* sodium hydroxide and is maintained for up to 60 min [28].

Table 1 shows the separation of alcohols (2,3-butanediol, ethanol, methanol), glycols (glycerol), alditols (erythritol, arabitol, sorbitol, galactitol, mannitol), and carbohydrates (rhamnose, arabinose, glucose, galactose, lactose, sucrose, raffinose, maltose) using a CarboPac MA1 column set with 480 m*M* sodium hydroxide eluent flowing at 0.4 mL/min. The alcohols, sugar alcohols (alditols), glycols, and carbohydrates are well resolved. Maltose elutes at about 60 min.

Table 1 Separation of Carbohydrates, Alditols, Alcohols, and Glycols using a CarboPac MA1 Column and Pulsed Amperometry

Analyte	Retention time (min)
2,3-Butanediol	7.4
Ethanol	7.6
Methanol	7.8
Glycerol	9.0
Erythritol	10.1
Rhamnose	13.4
Arabitol	14.2
Sorbitol	16.3
Galactitol	18.0
Mannitol	19.5
Arabinose	21.8
Glucose	23.3
Galactose	27.4
Lactose	29.7
Ribose	32.0
Sucrose	46.5
Raffinose	52.9
Maltose	61.2

References

1. R. L. Jolley and C. D. Scott, *Clin. Chem. 16*:687 (1970).
2. W. C. Butts and R. L. Jolley, *Clin. Chem. 16*:722 (1970).
3. S. Katz, S. R. Dinsmore, and W. W. Pitt, Jr., *Clin. Chem. 17*:731 (1971).
4. C. D. Scott, D. D. Chilcote, S. Katz, and W. W. Pitt, Jr., *J. Chromatogr. Sci. 11*:96 (1973).
5. D. S. Young, J. A. Epley, and P. Goldman, *Clin. Chem. 17*:765 (1971).
6. S. Honda, M. Takahashi, K. Kakehi, and S. Ganno, *Anal. Biochem. 112*:130 (1981).
7. S. Honda, S. Suzuki, M. Takahashi, K. Kakehi, and S. Ganno, *Anal. Biochem. 134*:34 (1983).
8. S. Honda, M. Takahashi, S. Shimada, K. Kakehi, and S. Ganno, *Anal. Biochem. 128*:429 (1983).
9. S. Honda, T. Konishi, S. Suzuki, M. Takahashi, K. Kakehi, and S. Ganno, *Anal. Biochem. 134*:483 (1983).
10. S. Honda, T. Konishi, S. Suzuki, K. Kakehi, and S. Ganno, *J. Chromatogr. 281*:340 (1983).
11. R. D. Rocklin, *J. Chromatogr. 546*:175 (1991).
12. S. J. Mellis and J. U. Baenziger, *Anal. Biochem. 114*:276 (1981).

13. V. K. Dua and C. A. Bush, *Anal. Biochem. 133*:1 (1983).
14. V. K. Dua and C. A. Bush, *Anal. Biochem. 137*:33 (1984).
15. S. Honda and S. Suzuki, *Anal. Biochem. 142*:167 (1984).
16. M. H. Simatupang, *J. Chromatogr. 180*:177 (1979).
17. D. F. Smith, D. A. Zopf, and V. Ginsburg, *Anal. Biochem. 85*:602 (1978).
18. K. J. Kramer, R. D. Speirs, and C. N. Childs, *Anal. Biochem. 86*:692 (1978).
19. K. Mopper and E. T. Degens, *Anal. Biochem. 45*:147 (1972).
20. K. Mopper and E. M. Gindler, *Anal. Biochem. 56*:440 (1973).
21. S. Honda, Y. Matsuda, M. Takahashi, K. Kakehi, and S. Ganno, *Anal. Chem. 55*:1079 (1980).
22. K. Mopper, R. Dawson, G. Liebezeit, and H. P. Hansen, *Anal. Chem. 52*:2018 (1980).
23. T. Kato and T. Kinoshita, *Anal. Biochem. 106*:238 (1980).
24. T. Kato and T. Kinoshita, *Chem. Pharm. Bull. 26*:1291 (1978).
25. H. Mikami and Y. Ishida, *Bunseki Kagaku 32*:E207 (1983).
26. R. D. Rocklin and C. A. Pohl, *J. Liquid Chromatogr. 6*:1577 (1983).
27. S. Honda, T. Konishi, and S. Suzuki, *J. Chromatogr. 299*:245 (1984).

Juan G. Alvarez

Carbohydrates: Analysis by Capillary Electrophoresis

Carbohydrates play an important role in many research and industrial domains. The huge number of stereoisomers, the immense combination possibilities of carbohydrate monomers in oligosaccharides, and the lack of chromophores are the major problems in the analysis of carbohydrates. Capillary electrophoresis (CE), in its various modes of operation, has been developed as a very useful tool in the analysis of carbohydrate species such as monosaccharides and oligosaccharides, glycoproteins, and glycopeptides.

Some simple sugar mixtures, such as monosaccharides and oligosaccharides, consisting of not more than about 15 monomer units, can be separated in free solution due to their mass-to-charge ratio (m/z). For an increase in selectivity or for analyzing neutral carbohydrates, micellar electrokinetic chromatography (MEKC) can be used for analysis. In this case, charged amphiphilic molecules containing both hydrophilic and hydrophobic regions (e.g., sodium dodecyl sulfate) are used as buffer surfactants.

Higher oligosaccharides or polysaccharides possess unfavorable mass-to-charge ratios, preventing their effective resolution in open tubes. The separation of these carbohydrates is possible with capillary gel electrophoresis (CGE). The analytes are selectively retarded by a sieving network (gel or polymer matrix) due to differences in their sizes and structural conformations.

Complex carbohydrates (in particular, glycoproteins) play an important role in various biological processes and in biotechnological production of glycoproteinaceous pharmaceuticals. To elucidate the relationship between bioactivity and structures of complex carbohydrates, it is necessary to determine the sites of attachment of the oligosaccharide chains to the polypeptide backbone and to characterize the oligosaccharide class (N- or O-linked, high mannose, hybrid, etc.).

For this reason, glycoproteins must first be isolated from the biological matrix by dialysis, preparative chromatography, isoelectric focusing, and so forth or by a combination of several methods. For a structural determination, degradation steps such as a site-specific proteolysis (e.g., with trypsin), removal of oligosaccharides from the polypeptide (by an enzymatic hydrolysis or hydrazine treatment), or chemical hydrolysis, yielding a monosaccharide mixture may be applied. Then, the CE can function as a powerful end method in analytical and structural glycobiology. Due to the complexity of the carbohydrate-dependent microheterogeneity of glycoproteins, several electrophoretic techniques are usually needed, in concert, to characterize the various glycoforms of a given glycoprotein.

For detection of carbohydrates in principle, ultraviolet (UV), laser-induced fluorescence, refractive index, electrochemical, amperometric, and mass spectrometric detection can be used. Mass spectrometry, with its various ionization methods, has traditionally been one of the key techniques for the structural determination of proteins and carbohydrates. Fast-atom bombardment (FAB) and electrospray ionization (ESI) are the two on-line ionization methods used for carbohydrate analysis. The ESI principle has truly revolutionized the modern mass spectrometry of biological molecules, due to its high sensitivity and ability to record large-molecule entities within a relatively small-mass scale.

The refractive index detection (RID), often used in high-performance liquid chromatography, is an interesting detection method in CE with a laser light source and a limit of detection (LOD) in the micromolar range. Electrochemical detection (ECD) and pulsed amperometric detection (PAD) of sugars are common and effective methods used in HPLC. Some recent communications show that the sensitivity of these detection methods in CE have an approximately 1000-fold better LOD than RID.

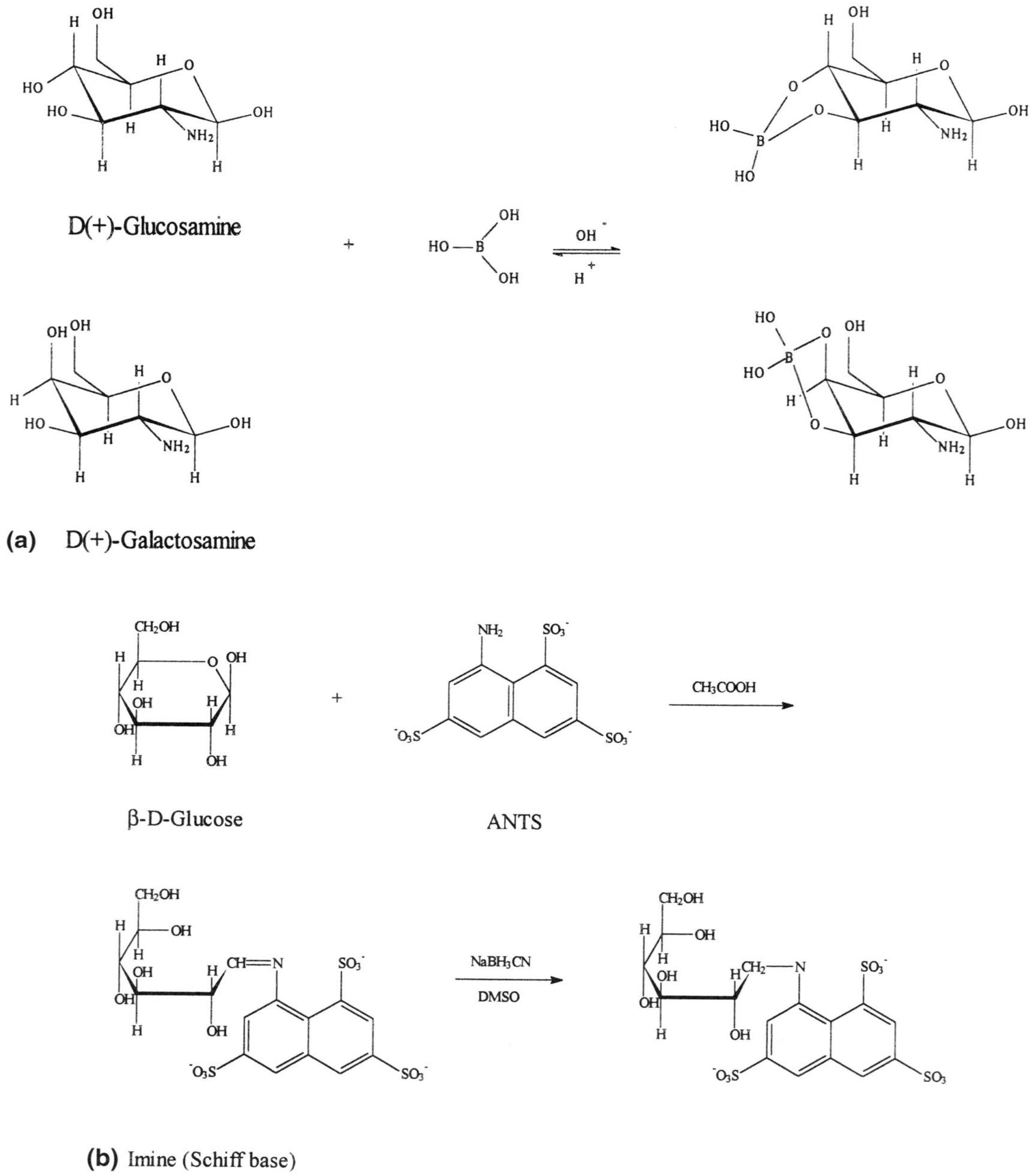
D(+)-Glucosamine
(a) D(+)-Galactosamine
β-D-Glucose
ANTS
CH_3COOH
$NaBH_3CN$
DMSO
(b) Imine (Schiff base)

Fig. 1

Unfortunately, these detectors (RID, ECD, and PAD) are not commercially available for capillary electrophoresis at the moment.

Indirect detection methods are a viable alternative for compounds lacking a chromophoric or a fluorophoric group. An electrolyte containing a chromophore or fluorophore allows the indirect detection of carbohydrates. This method is based on the displacement of the background electrolyte (BGE) by carbohydrates, which are dissociated in strongly alkaline electrolytes. The LOD with indirect LIF detection is in the nanometer range, but the lack of any specificity is a great disadvantage of this detection method, because all sample compounds displace the BGE and the peak identification is only possible by the migration time.

Direct UV detection is the most versatile detection method in CE and is implemented in every commercial CE system. Unfortunately, its use for carbohydrates detection is restricted, because of their lack of conjugated π-electron systems and, consequently, the relatively low extinction coefficients. Despite this fact, it is possible to detect carbohydrates with UV detection without any derivatization at 200 nm. Sensitivity and selectivity can be increased by the use of an alkaline borate buffer as the electrolyte by *in situ* complexation with the tetrahydroxyborate ion rather than the boric acid (Fig. 1a). The LOD is between micromolar and nanomolar. A further advantage of very high pH values (>10) is the negative charge of the carbohydrates, which are repelled by the negatively charged surface. Consequently, the surface problems in high-performance capillary electrophoresis are much lower in carbohydrate analysis than in the analysis of proteins. Therefore, simple carbohydrates are often analyzed in uncoated fused-silica capillaries. Unlike the analysis of simple carbohydrates, for glycopeptides and glycoproteins, the use of coated capillaries such as hydroxypropylcellulose, hydroxyethylmethacrylate, polyether, or other commercially available coated fused-silica capillaries is necessary to achieve high resolution and reproducibility.

In complex matrices, the insufficient specificity at 200 nm and the low sensitivity of direct UV detection make the analysis of carbohydrates more difficult. Therefore, derivatization of carbohydrates with a suitable agents is still a preferred approach for the detection of monosaccharides and oligosaccharides. Derivatization agents like 2-aminopyridine, 8-aminonaphthalein-1,3,6-trisulfonate (ANTS) or 8-aminopyren-1,3,6-trisulfonate (APTS) can be introduced mostly by reductive amination. This reaction is based on imine formation (Schiff base) by the condensation of the aldehyde group in a carbohydrate with the amino group in a primary amine (fluorescent tag), followed by reduction to an N-substituted glycamine with a reductant like sodium cyanoborohydride (see Fig. 1b). Selection of the suitable derivatization reagent is important, because the electrophoretic migration of the carbohydrates and, therefore, the separation power is influenced by the properties of the tags. Fluorescent dyes are better suitable than UV-active derivatization reagents, because CE analysis permits the use of laser-induced-fluorescence (LIF) detection with excellent sensitivity up to the femtomolar-level.

In conclusion, capillary electrophoresis in carbohydrate analysis has advantages in both separation and detection over other techniques of electrophoresis, as well as chromatography. It allows high efficiency (up to a few million plate numbers) and very good sensitivities (up to femtomolar). In addition, CE permits analysis by a variety of separation modes simply by changing the electrolyte (capillary zone electrophoresis, MEKC, CGE).

Suggested Further Reading

El Rassi, Z., *Electrophoresis 18*:2400 (1997).

Grimshow, J., *Electrophoresis 18*:2408 (1997).

Linhardt, R. J. and A. Pervin, *J. Chromatogr. A 720*:323 (1996).

Novotny, M. V., Capillary electrophoresis of carbohydrates, in *High-Performance Capillary Electrophoresis* (M. G. Khaledi, ed.), John Wiley & Sons, New York, 1998, pp. 729–765.

Paulus, A. and A. Klockow, *J. Chromatogr. A 720*:353 (1996).

Suzuki, S. and S. Honda, *Electrophoresis 19*:2539 (1998).

Oliver Schmitz

Carbohydrates: Analysis by TLC—New Visualization

There are many publications and comprehensive handbooks on the thin-layer chromatography (TLC) of carbohydrates (e.g., Refs. 1 and 2). The reason is their great importance in life science and the great diversity of cases: monosaccharide, disaccharide, trisaccharide, oligosaccharide, polysaccharide, aldose, ketose, triose, tetrose, pentose, hexose, as well as reducing and nonreducing sugars. In addition, when extracted from natural products or produced by fermentation, carbohydrates are accompanied by many impurities. That is why separation methods are used predominantly for their analysis.

Carbohydrates are polyhydroxy compounds (i.e., very polar compounds) with low volatility; a gas chromatographic (GC) analysis, therefore, will not be the best choice. GC methods continue to be applied in the cases when low concentrations have to be determined (e.g., in clinical analyses).

Due to high water solubility of monosaccharides, the use of the most routine high-performance liquid chromatography (HPLC) reversed-phase columns is also not suitable for their analysis. Extremely pure solvents have to be used if ultraviolet (UV) detection is applied. If a refractometer is used as the detector (RD), extremely steady chromatographic conditions are necessary. Nevertheless, HPLC is applied in the practice. The modern approach involves the use of propylamino columns (e.g., Refs. 3 and 4).

The best choice remains the TLC method. The impurities can be left on the start or eluted to the front; TLC can be used for qualitative and quantitative analyses, for screening, and so forth. Simultaneous analysis of several samples is possible. TLC affords an opportunity for a broad range of choices (more than 30) of visualizing agents. Most of them are chromogenic agents [1,2] with a mean sensitivity of 500–100 ng. The Stahl reagent—anisaldehid/H_2SO_4—has a limit of detection (LOD) of 50 ng, but this reagent is not suitable for carbohydrate alcohols [3]. In a series of articles [5], Alperin et al. proposed the so-called thermal-UV detection of sugars but, again, with low sensitivity (about 200 ng). The great number of reagents has also one disadvantage—lack of a firm visualization agent which answers the higher daily requirements for LOD.

More than eight types of stationary phases have been mentioned in Ref. 1 for the separation of carbohydrates, but the most popular is pure silica or impregnated with various inorganic ions silica gel G, because the separation is faster and provides more compact spots [2]. A modern stationary phase has become propylamino TLC plates. On both silica and amino plates, the separation is satisfactory (e.g., Refs. 2, 4, 6, and 7).

The remaining problems are (a) sensitivity (kind of visualization) and (b) quantitative analysis. Although the second problem has found its solution through the use of densitometers, the choice of visualization reagent remains a subject of individual personal decision. Klaus et al. [6] proposed a reagent-free visualization by heating propylamino stationary-phase plates. Sugars give fluorescent spots after heating at 120–150°C [8]. The sensitivity remains in the limits of micrograms.

The present article describes how fluorescent spots of carbohydrates can be achieved after heating, using common silica plates, applying the already accepted laboratory mobile phases.

Procedure

The well-known and widely applied TLC silica gel G plates 20 × 20 cm from Merck (Darmstadt, FRG) is used. Stock solutions from the compounds given in Table 1 were prepared. After corresponding dilution, a 1-μL sample is applied and the plate is developed for a distance of 10 cm using *n*-propanol–water, 8:2 (v/v), as the mobile phase. Densitometric evaluation is performed at 365 nm with a Camag TLC Scanner II in absorbance reflection mode. The plate is first air-dried and then heated in an oven for 5 min at 100°C. Immediately after drying, the plate is inserted in a tank saturated with ammonia atmosphere (ammonia solution in a vessel). After 10 min, the plate is pulled out, covered tightly with another plate (which can be preliminarily placed in the tank or can be another spotted plate), and both are heated for 5 min at 160°C. The spots are observed under UV light at 365 nm. The behavior of all studied compounds is presented in Table 1. All studied carbohydrates as well as the sugar alcohols mannitol and sorbitol give orange fluorescence spots. The fluorescence remains stable for more than 1 week (which is less than the stability cited in Ref. 8, but enough for practical use).

The linear range is small—to about 100 ng. For an extended range of application (over the range to 1000 nm), a nonlinear relationship between peak area (A) and the quantity (ng) has been tested. For example, the equation for D-glucose is

$$\ln A = 7.722 - \frac{30.5}{\text{ng}} \qquad (1)$$

with a correlation coefficient 0.993 and maximum error at the 1000-ng level—less than 10%. The area precision (intraplate repeatability) from six peaks of D-glucose at the 500-ng level is 4.1% and the height precision is 1.8%. The calculated (twice the noise of densitogram baseline) limit of detection (LOD) is 10-ng.

The proposed visualization procedure possesses good intraplate repeatability, the limit of detection is satisfactorily low, and it is more ecological than spraying with sulfuric acid-containing sprays (which also have poor repro-

Table 1 Behavior of 10 Carbohydrates, 2 Sugar Alcohols, and 1 Glycoside After Heating the Saturated-with-Ammonia Silica Plate at 160°C

Compound	Fluorescence at ng[a]		
Monosaccharides	50	10	1
Rhamnose	+	−	n
Ribose	−	n	n
Arabinose	+	−	n
Xylose	+	+	−
Glucose	+	−	−
Manose	−	n	n
Galactose	+	−	n
Disaccharides			
Saccharose	+	−	n
Trehalose	+	+	−
Maltose	+	−	n
Lactose	+	−	n
Sorbose	+	−	n
Trisaccharide			
Raphinose	+	−	n
Alcohol			
Mannitol	+	−	n
Sorbitol	+	−	n

[a] +: well-observed spot suitable for quantitation; −: LOD; n: no observation.

ducibility). The disadvantages of the proposed visualization procedure are poor reproducibility from plate to plate (interplate repeatability), dependence of the spot intensity both on structure and the extent of saturation with ammonia and heating, and lack of a wider linear range. The mentioned disadvantages can be overcome with a thorough validation and verification of the particular method. The proposed approach is an easy transferable one and can be applied directly to routine work.

Thus, it is anticipated that this little variation in everyday routine work in laboratories analyzing carbohydrates can be easily adapted and will contribute to a better Good Laboratory Practice (GLP).

References

1. E. Stahl, *Duenschicht-chromatography*, Springer-Verlag, Berlin, 1967.
2. B. Fried and J. Sherma, *Thin-Layer Chromatography. Techniques and Applications*, Marcel Dekker, Inc., New York, 1982; B. Fried and J. Sherma, *Thin-Layer Chromatography*, 4th ed., Marcel Dekker, Inc., New York, 1999.
3. G. W. Hay, B. A. Lewis, and F. Smith, *J. Chromatogr. 11*:479 (1963).
4. M. Okamoto, F. Yamada, and T. Omori, *J. High Resolut. Chromatogr. Chromatogr. Commun. 5*:163 (1982).
5. D. M. Alpein et al., *J. Chromatogr. 242*:299 (1982); *250*:124 (1982); *265*:193 (1983).
6. R. Klaus, W. Fischer, and H. E. Hauck, *Chromatographia 28*:364 (1989).
7. R. Klaus, W. Fischer, and H. E. Hauck, *Chromatographia 39*:97 (1994)
8. R. Klaus, W. Fischer, and H. E. Hauck, *LC–GC Int. 8*:151 (1995).

N. Dimov

CCC Solvent Systems

Countercurrent chromatography has been mainly developed and used for preparative and analytical separations of organic and bio-organic substances [1]. The studies of the last several years have shown that the technique can be applied to analytical and radiochemical separation, preconcentration, and purification of inorganic substances in solutions on a laboratory scale by the use of various two-phase liquid systems [2]. Success in CCC separation depends on choosing a two-phase solvent system that provides the proper partition coefficient values for the compounds to be separated and satisfactory retention of the stationary phase. The number of potentially suitable CCC solvent systems can be so great that it may be difficult to select the most proper one.

Recent studies have made it possible to classify water–organic solvent systems in CCC for separation of organic substances on the basis of the liquid-phase density difference, the solvent polarity, and other parameters from the point of view of stationary-phase retention in a CCC column [1,3–9]. Ito [1] classified some liquid systems as hydrophobic (such as heptane–water or chloroform–water), intermediate (chloroform–acetic acid–water and

n-butanol–water) and hydrophilic (such as n-butanol–acetic acid–water) according to the hydrophobicity of the nonaqueous phase. Thirteen two-phase solvent systems were evaluated for relative polarity by using Reichardt's dye to measure solvachromatic shifts and using the solubility of index compounds [6].

However, the systems for inorganic separations are very different from those for organic separations, as, in most cases, they contain a complexing (extracting) reagent (ligand) in the organic phase and mineral salts and/or acids or bases in the aqueous phase. Thus, the complexation process, its rate, and the mass-transfer rate can play a significant role in the separation process [9]. There are three important criteria for choosing a two-phase liquid system.

First, the systems must be composed of two immiscible phases. Each solvent mixture should be thoroughly equilibrated in a separatory funnel at room temperature and the two phases separated after the clear two phases have been formed. When the nature of the organic sample to be separated is known, one may find a suitable solvent system by searching the literature for solvent systems that have been successfully applied to similar compounds [1, 3–8]. In the case of organic–aqueous two-phase systems, the organic phase consists of one solvent or of a mixture of different solvents. Various nonaqueous–nonaqueous two-phase solvent systems have been used for separation of nonpolar compounds and/or compounds that are unstable in aqueous solutions. Separation of macromolecules and cell particles can be performed with a variety of aqueous–aqueous polymer-phase systems. Among the various polymer-phase systems available, the following two types are the most versatile for performing CCC [1,8]. Poly(ethylene glycol) (PEG)–potassium phosphate systems provide a convenient means of adjusting the partition coefficient of macromolecules by changing the molecular weight of PEG and/or the pH of the phosphate buffer. The PEG 6000–Dextran 500 systems provide a physiological environment, suitable for separation of mammalian cells by optimizing osmolarity and pH with electrolytes.

For preconcentration and separation of inorganic species, a stationary phase containing extracting reagents of different types (cation-exchange, anion-exchange, and neutral) in an organic solvent should be usually applied [2,9–12]. The mobile-phase components should not interfere with the subsequent analysis. Solutions of inorganic acids and their salts are most often used. The mobile phase may also contain specific complexing agents, which can bind one or several elements under separation.

Second, one of the phases (stationary one) must be retained in the rotating column to a required extent. The most important factor, which determines the separation efficiency and peak resolution for both organic and inorganic compounds, is the ratio of the stationary-phase volume retained in a column to the total column volume. The volume of the stationary phase V_s retained in the column depends on various factors, such as the physical properties of the two-phase solvent system, flow rate of the mobile phase, and applied centrifugal force field. In droplet CCC, where the separation is performed in a stationary column, a large density difference between the stationary solvent phases becomes the predominant factor for the retention of the stationary phase. In other CCC schemes, various types of two-phase solvent systems can be used under optimized experimental conditions. The influence of planetary centrifuge parameters and operation conditions on the stationary-phase retention have been well studied for some simple two-phase liquid systems consisting of water and one or two organic solvents [1,3–8].

According to Ito's classifications [1,3], hydrophobic organic phases are easily retained by all types of CCC apparatus. Intermediate solvent systems involve a more hydrophilic organic phase. Their tendency to evolve, after mixing, to a more stable emulsion than the hydrophobic systems decreases the retention of stationary phase. The hydrophilic two-phase systems containing a polar phase are even less retained in the column.

However, the addition of extracting reagents and mineral salts to a two-phase system (in case of inorganic separations) can strongly affect the physicochemical properties of liquid systems and, consequently, their hydrodynamic behavior and S_f value. Varying concentrations of the system constituents used for inorganic separation allows selective changing of a certain physicochemical parameter (interfacial tension γ, density difference between two liquid phases $\Delta\rho$, and viscosity of the organic stationary phase η_{org}). The type of the solvent may often have a great effect on the stationary-phase retention and, consequently, on the chromatographic process. The correlations between the physicochemical parameters of the complex liquid systems under investigation and their behavior in coiled columns are described in detail [10]. The composition and physicochemical properties of the organic phase in inorganic analysis were modified by adding an extracting reagent [e.g., di-2-ethylhexylphosphoric acid (D2EHPA), tri-n-butyl phosphate, trioctylamine] [2, 10]. The density and viscosity of the organic phase were varied by changing the amount of reagents in the stationary phase. For example, a small addition (5%) of D2EHPA in an organic solvent (n-decane, n-hexane, chloroform, and carbon tetrachloride) leads to a considerable increase in the factor S_f in the organic solvent—$(NH_4)_2SO_4$—water systems (from 0 to 0.73 in the case of carbon tetrachloride) [10].

Third, the stationary phase should permit separate elution of the substances into the mobile phase and the selectivity toward samples of interest has to be sufficient to lead to separations with good resolution. The selectivity of solvent systems can be estimated by determination of the partition coefficients for each substance. The batch partition coefficients D^{bat} are calculated as the ratio of the component concentration in the organic phase to that in the aqueous phase. The dynamic partition coefficients of compounds D^{dyn} are determined from an experimental elution curve [7]. Several solvent systems for organic separation were investigated [4–8]. The most efficient evolution usually occurs when the value of the partition coefficient is equal 1. However, in some CCC schemes, the best results are obtained with lower partition coefficient values of 0.3–0.5 [1,4].

In inorganic analysis with the use of CCC, the stationary phase should provide preconcentration of the elements to be determined, if necessary. It should be noted that the element elution depends on the operation conditions for the planetary centrifuge, which influence the quantity of the stationary phase in the column. A chromatographic peak shifts to left and narrows if the volume of the stationary phase lowers (all the other factors being the same) [2]. The reagent concentration in the organic solvent also affects the elution curve shape and, therefore, the dynamic partition coefficient values. An increase of the reagent concentration in the organic phase leads to higher partition coefficients for the elements, and a better separation is achieved. However, a rather large volume of the mobile phase can be required for the elution of elements from the column.

The composition of the mobile phase also has an influence on the partition coefficients of inorganic substances and the separation efficiency. Concentrations of the mobile-phase constituents should provide partition coefficient values needed for the enrichment or separation of components under investigation. If a step-elution mode is used, partition coefficients higher than 10 and less than 0.1 are favorable for the enrichment of components into the stationary phase and their recovery into the mobile phase, respectively. Chemical kinetics factors may also play an important role in the separation of inorganic species by CCC [9]. It has been shown that the values of mass-transfer coefficients determine the type of elution (isocratic or step), which is necessary for the element separation. The data on batch extraction (mass-transfer coefficients and partition coefficients) and parameters of chromatographic peaks (half-widths) can be interrelated by some empirical expressions [9]. The application of CCC in inorganic analysis looks promising because various two-phase liquid systems, providing the separation of a variety of inorganic species, may be used for the separation of trace elements.

References

1. Y. Ito, in *Countercurrent Chromatography. Theory and Practice* (N. B. Mandava and Y. Ito, eds.), Marcel Dekker, Inc., New York, 1988.
2. B. Ya. Spivakov, T. A. Maryutina, P. S. Fedotov, and S. N. Ignatova, in *Metal-Ion Separation and Preconcentration: Progress and Opportunities* (A. N. Bond, M. L. Dietz, and R. D. Rodgers, eds.), American Chemical Society, Washington, DC, 1999, pp. 333–347.
3. W. D. Conway, *Countercurrent Chromatography. Apparatus, Theory and Applications*, VCH, New York, 1990.
4. A. Berthod and N. Schmitt, *Talanta 40*:1489 (1993).
5. J.-M. Menet, D. Thiebaut, R. Rosset, J. E. Wesfreid, and M. Martin, *Anal. Chem. 66*:168 (1994).
6. T. P. Abbott and R. Kleiman, *J. Chromatogr. 538*:109 (1991).
7. S. Drogue, M.-C. Rolet, D. Thiebaut, and R. Rosset, *J. Chromatogr. 593*:363 (1992).
8. A. P. Foucault and L. Chevolot, *J. Chromatogr. A 808*:3 (1998).
9. P. S. Fedotov, T. A. Maryutina, A. A. Pichugin, and B. Ya. Spivakov, *Russ. J. Inorg. Chem. 38*:1878 (1993).
10. T. A. Matyutina, S. N. Ignatova, P. S. Fedotov, B. Ya. Spivakov, and D. Thiebaut, *J. Liquid Chromatogr. Related Technol. 21*:19 (1998).
11. E. Kitazume, M. Bhatnagar, and Y. Ito, *J. Chromatogr. 538*:133 (1991).
12. Yu. A. Zolotov, B. Ya. Spivakov, T. A. Maryutina, V. L. Bashlov, and I. V. Pavlenko, *Fresenius Z. Anal. Chem. 35*: 938 (1989).

T. Maryutina
Boris Ya. Spivakov

CE–MS: Large-Molecule Applications

Introduction

Capillary electrophoresis (CE) is a modern analytical technique which permits rapid and efficient separation of charged components present in small-sample volumes. Separation occurs due to differences in electrophoretic mobilities of ions inside small capillaries. The impetus for CE method developments focused primarily on the separation of larger biopolymers such as polypeptides, proteins, oligonucleotides, DNA, RNA, and oligosaccharides [1]. Mass spectrometry (MS) has long been recognized as the most selective and broadly applicable detector for analytical separations. Currently, electrospray ionization (ESI) serves as the most common interface between CE and MS. Generation of multiply-charged species with an ESI extends the applicability of conventional mass analyzers of limited mass-to-charge (m/z) ranges to molecular mass and structure determination of larger biopolymers. CE–MS combines the advantages of CE and MS so that information on both high efficiency and molecular masses and/or fragmentation can be obtained in one analysis. This article focuses on larger-molecular analysis by on-line CE–MS interfaced via ESI sources [2,3]. However, CE–MS using continuous-flow fast-atom bombardment (CF–FAB) sources employing either "liquid-junction" or "coaxial" interfaces and several off-line CE–MS combination should be noted.

When ESI–MS is employed as detector, the proper choice of a suitable electrolyte system is essential to both a successful CE separation and good quality ESI mass spectra. Even though a wide range of CE buffers were successfully electrosprayed when the liquid-junction and sheath flow CE–MS interfaces were employed since the low CE effluent flow is effectively diluted by a much large volume of sheath liquid [4]; the best detector response is produced by volatile electrolyte systems at the lowest practical concentration and ion strength and by minimizing other nonvolatile and charge-carrying components. Volatile reagents like ammonium acetate (pH 3.5–5.5) or formate (pH 2.5–5; both adjustable to high pH) and ammonium bicarbonate have been proven to be well suited for CE–ESI–MS.

Due to the inherent tendency to adsorb strongly to the inner walls of the fused-silica capillary, the analysis of proteins and peptides by CE has presented unique challenges to the analyst because this phenomenon gives rise to substantial peak broading and loss of separation efficiency. Successful separations of proteins and peptides by CE involve efficient suppression of adsorption to the fused-silica wall. Basically, there are two approaches to prevent protein adsorption: modification of the fused-silica surface by dynamic or static coating or by performing analysis under experimental conditions that minimize adsorption [5]. The static coating capillary is preferred under CE–ESI–MS analysis of large molecules because the CE buffer composition is simplified. This article is meant only to provide the reader with a description of most common approaches taken to analyze large molecules, especially polypeptides and proteins, by CE–ESI–MS.

Large-Molecule Analysis of CE–MS by Neutral Capillary

Because there is no ionizable groups of the coating in the neutral capillary, the interaction between charged molecules with ionic capillary surface is eliminated. Also, the electro-osmotic flow (EOF) of a neutral capillary is eliminated. However, a continuous and adequate flow of the buffer solution toward the CE capillary outlet is an important factor for routine and reproducible CE–ESI–MS analysis; in order to maintain a stable ESI operation, some low pressure applied to the CE capillary inlet is usually needed, especially when the sheathless interface is employed. The disadvantage of the pressure-assisted CE–ESI–MS is the loss of some resolution because the flat flow profile of the EOF is partially replaced by the laminar flow profile of the pressure-driven system. A typical neutral capillary is a LPA (linear polyacrylamide)-treated capillary. Karger and co-workers [6] used mixtures of model proteins, a coaxial sheath flow ESI interface, and a 75-μm-inner diameter (i.d.), 360-μm-outer diameter (o.d.), 50-cm-long LPA-coated capillary to evaluate CE–MS, CITP (capillary isotachophoresis)–MS, and the on-column combination of CITP–CE–MS. In the CE–MS experimental, 0.02M 6-aminohexanoic acid + acetic acid (pH 4.4) was employed and a 18-kV constant voltage was applied during the experiment. Seven model proteins were well resolved. They showed that the sample concentration necessary to obtain a reliable full-scan spectrum was in the range of $10^{-5}M$. However, by proper selection of the running buffers, they demonstrated that the on-column combination of both CITP and capillary zone electrophoresis (CZE) can improve the concentration detection limits for a full-scan CE–MS analysis to approximately $10^{-7}M$.

Large-Molecule Analysis of CE–MS by a Positively Charged Capillary

To help overcome adsorption, positively charged coatings have been employed for the separation of positively charged solutes. In this approach, positively charged proteins are electrostatically repelled from the positively charged capillary inner wall. Two examples of such coatings are aminopropyltrimethoxysilane (APS) and polybrene, a cationic polymer. These coatings reverse the charge at the column–buffer interface and, thus, the direction of the EOF compared to uncoated capillaries.

The CE–MS analysis of the venom of the snake *Dendroaspis polylepis polylepis*, the black mamba, is reported by Tomer and co-workers [7]. A VG 12-250 quadrupole equipped with a Vestec ESI source (coaxial sheath flow interface) was employed for this experiment. The sheath fluid was a 50:50 methanol:3% aqueous acetic acid solution. The CE voltage was set at −30 kV during the analysis and the ESI needle was held at +3 kV. The CE running buffer used was 0.01*M* acetic acid at pH 3.5. The APS column was flushed with buffer solution for 10 min prior to sample analysis. The snake venom was dissolved in water at a concentration of 1 mg/mL and 50 nL of the analyte solution was injected into the column. They demonstrated the existence of at least 70 proteins from this venom.

One interesting example of intact protein analysis was described by Smith and co-workers [8]. They used the high sensitivity and mass accuracy of a Fourier transform ion cyclotron resonance (FTICR) MS detector to analyze hemoglobin α and β in a single human erythrocyte. Human erythrocytes were obtained from the plasma of a healthy adult male. A small drop of blood diluted with saline solution (pH 7.4) was placed on a microscope slide. With the help of a stereomicroscope and a micromanipulator, the etched terminus of the CE capillary was positioned within a few microns of the cell to be injected. Following electroosmotic injection of the cell, the end of the CE capillary was placed in a vial containing the CE running buffer (10 m*M* acetic acid, pH 3.4), and the cell membrane was lysed via osmotic shock from the running buffer and the cellular contents of the cell released for subsequent CE separation and mass analysis. A 1-m APS column and a sheathless interface employing a gold-coated capillary with −30 kV CE separation and +3.8 kV ESI voltage were used for this study. They demonstrated that adequate sensitivity needed to characterize the hemoglobin from a single human erythrocyte (~ 450 μmol) and mass spectra with average mass resolution in excess of 45,000 (full width at half-maximum) were obtained for both the α- and β-chain of hemoglobin. Figure 1 shows the mass spectra obtained from this experiment.

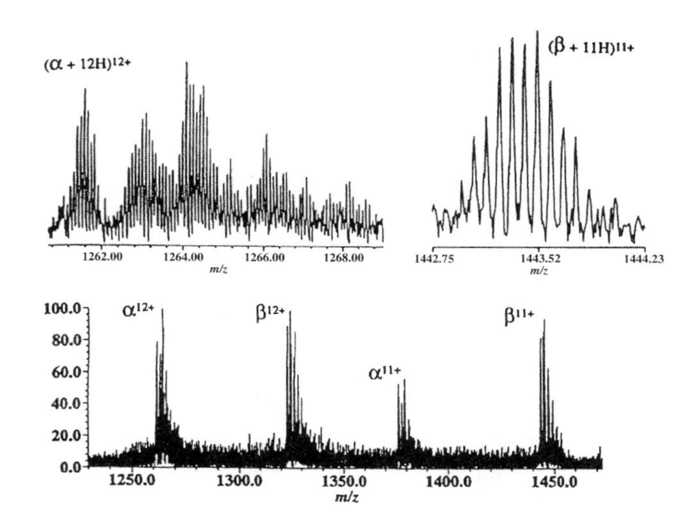

Fig. 1 Mass spectra obtained from CE–MS analysis of a single human erythrocyte using an FTICR mass analyzer. (Reproduced from Ref. 7, with permission).

In order to overcome the bubble formation associated with the sheathless CE–MS interface and quick degradation of the coated capillary, Moini et al. [9] introduced hydroquinone (HQ) as a buffer additive to suppress the bubbles formed due to the electrochemical oxidation of the CE buffer at the outlet electrode. The oxidation of water ($2H_2O\,(l) \leftrightarrow O_2\,(g) + 4H^+ + 4e$) was replaced with that of more easily oxidized HQ (hydroquinone $\leftrightarrow$ *p*-benzoquinone + $2H^+ + 2e$). Formation of *p*-benzoquinone, other than the formation of oxygen gas, effectively suppresses gas bubble formation. The APS-coated capillaries and 10 m*M* acetic acid CE running buffer containing 10 or 20 m*M* HQ were used for the experiments. The CE outlet/ESI electrode was maintained at +2 kV and the CE inlet electrode was held at −30 kV. Tryptic digest of cytochrome-*c* and hemoglobin were used as model proteins. They demonstrated that the combination of the in-capillary electrode sheathless interface using a platinum wire, HQ as a buffer additive, and pressure programming at the CE inlet provides a rugged high-efficiency setup for analysis of peptide mixtures.

Because the concentration limits of detection of CE are often inadequate for most practical applications (approximately $10^{-6}M$), several analyte concentration techniques have been developed, including combining capillary isotachophoresis (CITP) with CE, transient isotachophoresis (tITP) in a single capillary, analyte stacking, and field amplification. Such electrophoretic techniques have extended the applicability of CE for the analysis of dilute analyte solutions. Chromatographic on-line sample concentration has been achieved by using an extraction cartridge which contains a bed of reversed-phase packing [10] or a membrane [11] with reversed-phase properties. Accumulated analyte on the cartridge can be prewashed to remove salts and buffers that are not suited for CE separation or ESI operation. Figeys and Aebersold [12] designed the solid-phase extraction (SPE)–CE–MS–MS system which consists of a small cartridge of C_{18} reverse-phase extraction material immobilized in a Teflon sleeve. Solutions of peptide mixtures typically derived by proteolysis of gel-separated proteins were forced through the capillary by applying positive pressure at the inlet and the peptides were concentrated on the SPE device. After equilibration with an electrophoresis buffer compatible with ESI, eluted peptides were separated by CE and analyzed by ESI–MS. A detection limit of 400 αmol tryptic digest of bovine serum albumin (20 μL of solution at a concentration of 20 αmol/μL was applied) was achieved in the ion trap mass spectrometer-based system. This method was successfully applied to the identification of yeast proteins separated by two-dimensional gel electrophoresis.

Applications of CE–MS to large molecules are progressing rapidly. As biology enters an era of large-scale systematic analysis of biological systems as a consequence of genome sequencing projects, rapid and sensitive identifications of large-scale (proteomewide) proteins that constitute a biological system is essential. CE–MS with its high separation efficiency, rapid separation, and economy of sample size is complementary to microcolumn high-performance liquid chromatography (μHPLC)–MS. In addition, high-resolution, multiple-dimensional separations become increasingly attractive. HPLC–CE–MS, affinity CE–MS, capillary microreactor on line with CE–MS, and microchip-based separations will be used in a broad range of future applications.

References

1. W. G. Kuhr and C. A. Monnig, *Anal. Chem. 64*:389R (1992).
2. R. D. Smith and H. R. Udseth, *Pharmaceutical and Biomedical Applications of Capillary Electrophoresis*, Elsevier Science, New York, 1996, pp. 229–276.
3. J. F. Banks, *Electrophoresis 18*:2255 (1997).
4. R. D. Smith, J. A. Loo, C. G. Edmonds, C. J. Barinaga, and H. R. Udseth, *Anal. Chem. 62*:882 (1992).
5. P. Thibault and N. J. Dovichi, *Capillary Electrophoresis (Theory and Practice)*, 2nd ed., CRC Press, Boca Raton, FL, 1998, pp. 23–90.
6. T. J. Thompson, F. Foret, P. Vouros, and B. L. Karger, *Anal. Chem. 65*:900 (1993).
7. J. R. Perkins, C. E. Parker, and K. B. Tomer, *Electrophoresis 14*:458 (1993).
8. S. A. Hofstadler, J. C. Severs, and R. D. Smith, *Rapid Commun. Mass Spectros. 10*:919 (1996).
9. M. Moini, P. Cao, and A. J. Bard, *Anal. Chem. 71*:1658 (1999).
10. D. Figeys, and R. Aebersold, *Electrophoresis 19*:885 (1998).
11. A. J. Tomlinson, L. M. Benson, N. A. Guzman, and S. Naylor, *J. Chromatogr. 744*:3 (1996).
12. D. Figeys and R. Aebersold, *Electrophoresis 18*:360 (1997).

Ping Cao

Centrifugal Partition Chromatography: An Overview

Introduction

Centrifugal partition chromatography (CPC) belongs to the methods based on counter current chromatography (CCC). Separation is based on the differences in partitioning behavior of components between two immiscible liquids. Like high-performance liquid chromatography (HPLC), the phase retained in the column is called the stationary phase, the other one is called the mobile phase. In CCC, there are two modes to equilibrate the two immiscible phases. They depend on the characteristics of the centrifugal force field which permits retention of the stationary phase inside the column. Devices which equilibrate the phases according to the so-called "hydrodynamic mode" were developed by Ito [1]. They use a centrifugal force which is variable in intensity and direction. Alternated zones of agitation and settling of both phases take place along the column. By contrast, CPC uses a so-called "hydrostatic mode," owing to a centrifugal force which is constant in intensity and direction. Therefore, the mobile phase penetrates the stationary phase by forming either droplets or jets or sprays. The more or less vigorous agitation of both phases depends on the intensity of the centrifugal force, the flow rate of the mobile phase, and the physical properties of the solvent system. Chromatographic separations obtained in hydrostatic mode are less efficient than in the hydrodynamic mode. However, the retention of the stationary phase is less sensitive to the physical properties of solvent systems, such as viscosity, density, and interfacial tension. This justifies the wide application field of CPC.

Apparatus

The CPC column is made of channels engraved in plates of an inert polymer (Fig. 1) and they are connected by narrow ducts. Several plates are put together to form a cartridge. The cartridges are placed in the rotor of a centrifuge and are connected to form the chromatographic column. The mobile phase enters and leaves the column via rotary seals. Because two immiscible liquids are present in the channel, the denser liquid moves away from the axis because of the centrifugal force. The less dense liquid is pushed toward the axis. The mobile phase can be either the lighter or the denser phase. In the latter case, the mobile phase flows through the channels from the axis to the outside of the rotor. This is called the descending mode. The other case, the mobile phase flowing toward the axis, is called the ascending mode.

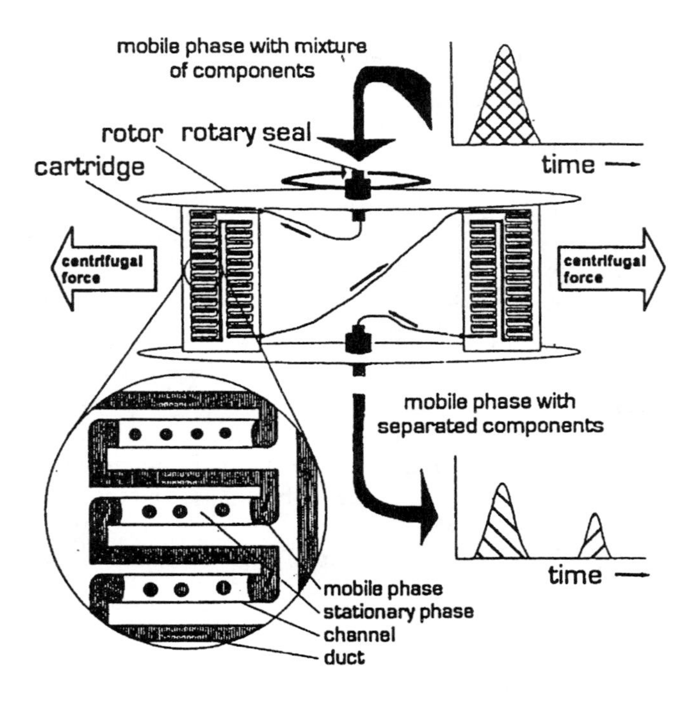

Fig. 1 Schematic representation of the CPC apparatus. (Reproduced with permission from Ref. 5.)

Hydrostatic apparatuses are manufactured by Sanki Engineering Limited (Kyoto, Japan). They include two types of device; the first is designed for analytical or semipreparative scale applications, and the second one for the scale-up at a industrial scale. CPC type LLN was introduced in 1984 but is no longer available since 1992. It can be thermostated from 15°C to 35°C in an ambient temperature of 25°C. Type HPCPC or Series 1000 supersedes type LLN. The HPCPC main frame is a 31 × 47 × 50-cm centrifuge operating in the range 0–2000 rpm; it cannot be thermoregulated. The rotor consist of two packs of six disks each, connected through a 1/16-in. tubing and easily removable. Larger instruments have internal volumes from 1.4 to 30 L and can be used with flow rates ranging from 20 to 700 mL/min and are custom designed for specific separation processes at a small industrial scale.

Retention of Stationary Phase

Before use, the column is first filled with a stationary phase and, afterward, rotated at the desired rotational speed. The mobile phase is then pumped into the cartridge at the desired flow rate and pushes a certain volume of stationary phase out of the column. Hydrostatic equilibrium is reached when the mobile phase is expelled out at the column outlet. The retention of stationary phase, designated S_F, is defined as

$$S_F = \frac{V_s}{V_t}$$

where V_s is the stationary-phase volume in the column after equilibrium and V_t is the total volume of the column.

The value of S_F depends on several parameters, including the hydrodynamic properties of the channels, the centrifugal force (S_F increases to reach a maximum with the centrifugal force), the mobile-phase flow rate (S_F decreases linearly with the mobile-phase flow rate), the physical properties of the solvent system (such as viscosity, density, interfacial tension), the sample volume, the sample concentration, and the tensioactive properties of solutes to separate [2,3]. It is necessary to precisely monitor S_F because various chromatographic parameters depend on it, in particular the efficiency, the retention factor, and the resolution. Foucault proposed an explanation for the variation of S_F with the various parameters previously described. He modeled the mobile phase in a channel as a droplet and applied the Stokes' law which relies on the density difference between the two phases, the viscosity of the stationary phase, and the centrifugal force. Then, he applied the Bond number, derived from the capillary wavelength which was formerly introduced for the hydrodynamic mode [4] and which relies on the density difference between the two phases: the interfacial tension and the centrifugal force [3].

Pressure Drop

Van Buel et al. [5] have proposed a model to explain the considerable pressure drop arising in the column during CPC separation. The overall pressure drop is the sum of the hydrostatic pressure-drop term and of the hydrodynamic pressure-drop term over the individual parts of the system. The hydrostatic contribution is caused by the difference in density between the liquids in the ducts and in the channels ($\Delta P_{\text{stat}} = nl\Delta\rho\omega^2 R$, where n is the number of channels, l is the height of stationary phase in the channel, $\Delta\rho$ is the density difference of the both phases, ω is the rotational speed, and R is the average rotational radius of the cartridge). The hydrodynamic contribution (ΔP_{hydr}) is caused by the friction of the mobile phase with the walls of the channels and ducts. This latter, in a channel and a duct, is proportional to the mobile-phase density, the square of its linear velocity, the lengths of the channel and duct, the reverse of channel, and the duct diameter. Consequently, the overall pressure drop depends on the flow rate and rotational speed (input variables), on the physical properties of the two-phase solvent system (variables), on the geometry of the channels and ducts, on the number of channel–duct combinations (apparatus variables), and on the holdup of the stationary phase in the channel. The maximum pressure is limited by the rotary seals, which can support about 60 bars before leaking. Resolution and efficiency depend on the same variables as the pressure drop. Therefore, it is important to determine which combinations of input variables and liquid two-phase systems can be applied, with respect to the maximum pressure that can be supported by the rotary seal.

Efficiency

The efficiency (N) in CCC can be defined as in HPLC by

$$N = 16\left(\frac{V_r}{\omega}\right)^2$$

where V_r is retention volume of the solute, ω is the peak base width expressed in volume unity as V_r, or for an asymmetrical peak according to the Foley–Dorsey formula

$$N = 41.7\left(\frac{(t_r/\omega_{0.1})^2}{A/B + 1.25}\right)$$

where $\omega_{0.1}$ is the peak width at 10% of the peak height and A/B is the asymmetry factor with $A + B = \omega_{0.1}$.

The efficiency variation shows a minimum when the flow rate of the mobile phase is increased, which is the opposite of the usual HPLC Van Deemter plot. This observation has been modeled by Armonstrong et al. [6]. The mobile phase, when it comes out of the duct, flows very quickly to reach an intermediate emulsified layer and then settles in a third step before being transferred to another channel. In these conditions

$$\ln(1 - E) = \frac{A}{F} - BF^b$$

where

$$E = \frac{C_{m,t} - C_{m,0}}{C_{m,\text{eq}} - C_{m,0}}$$

($C_{m,t}$, $C_{m,0}$, and $C_{m,\text{eq}}$ are the solute concentrations in the mobile phase at a moment t, before equilibrium and after equilibrium, respectively), A depends on S_F, B on the physical properties of the solvent system, and b on solutes and solvent system. This variation is very interesting, because it shows that a high mobile-phase flow rate decreases the retention time without decreasing efficiency. However, it was observed that S_F decreases with the flow rate and the resolution R_s also decreases as described in the following section. The flow rate of the mobile phase may be increased to decrease the separation time, but, at the condition that S_F remain satisfactory to maintain a sufficient R_s [2]. Finally, it has been shown that N increases with the centrifugal force field [2].

Resolution

The resolution (R_s) in CCC can be defined as in HPLC by

$$R_s = 2\left(\frac{V_{r2} - V_{r1}}{\omega_1 + \omega_2}\right) = 2V_s\left(\frac{K_2 - K_1}{\omega_1 + \omega_2}\right)$$

where V_{r1} and V_{r2}, ω_1 and ω_2, and K_1 and K_2 are respectively the retention volumes, the peak base widths expressed in volume unity as V_r, and the partition coefficients of the first and second eluted solutes. R_s is directly proportional to volume V_s of the stationary phase, and, hence, on the flow rate of the mobile phase and the centrifugal force [2,7].

The resolution in CCC as in HPLC is governed by the Purnell relation

$$R_s = \frac{\sqrt{N}}{4}\left(\frac{k_2'}{1 + k_2'}\right)\left(\frac{1 - \alpha}{\alpha}\right)$$

where k_2' is the retention factor of the second solute and α is the separation factor. N is controlled by F, the centrifugal force, S_F, and physical properties of the solvent system, and k' is controlled by the nature of the solvent system (through the partition coefficient K and S_F); α is controlled mainly by the solvent system. This relationship shows that it is essential, in CCC, to control technical parameters and to choose, judiciously, the solvent system to separate the products.

Solvent Systems

The choice of the solvent system is the key parameter to a good separation. On one hand, its physical properties define S_F, N, and R_s; on the other hand, the relative polarities of its two phases define the partition coefficients of the solutes and, as a result, the selectivities and the retention factors. Usually, solvent systems are biphasic and made of three solvents, two of which are immiscible.

We only give basic directions for the choice of a solvent system. If the polarities of the solutes are known, the classification established by Ito [1] can be taken as a first approach. He classified the solvent systems into three groups, according to their suitability for apolar molecules ("apolar" systems), for intermediary polarity molecules ("intermediary" system), and for polar molecules ("polar" system). The molecule must have a high solubility in one of the two immiscible solvents. The addition of a third solvent enables a better adjustment of the partition coefficients. When the polarities of the solutes are not known, Oka's [8] approach uses mixtures of *n*-hexane (HEX), ethyl acetate (EtOAc), *n*-butanol (*n*-ButOH), methanol (MeOH), and water (W) ranging from the HEX–MeOH–W, 2:1:1 (v/v/v) to the *n*-BuOH–W, 1:1 (v/v) systems and mixtures of chloroform, methanol, and water. These solvent series cover a wide range of hydrophobicities from the nonpolar *n*-hexane–methanol–water system to the polar *n*-butanol–water system. Moreover, all these solvent systems are volatile and yield a desirable two-phase volume ratio of about 1. The solvent system leading to partition coefficients close to the unit value will be selected.

Applications

Numerous applications using CPC are described in reference books [1,3]. We will only give key examples extracted from the CPC literature (Table 1).

Polyphenols and Tannins

Open-column chromatography with silica gel and alumina is not applicable to the fractionation of tanins because of their strong binding to these adsorbents, which induces extensive loss of tannins. Such losses do not occur with countercurrent chromatography, as it does not use a solid stationary phase. Such molecules are very polar, so butanol-based solvent systems can be used. Centrifugal partition chromatography is more adequate in this case, as compared to hydrodynamic CCC, because of the good retention of the stationary phase of a such solvent system.

Okuda et al. [9] separated castalagin from vescalagin by using the solvent system *n*-butanol–*n*-propanol–water (4:1:5, v/v/v). They are diastereoisomers which differ only in the configuration of the hydroxyl group of the central carbohydrate moiety. In the same way, they have separated oligomeric hydrolyzable macrocyclic tannins, Oenothein B and Woodfordins by using *n*-butanol–*n*-propanol–water (4:1:5, v/v/v). In spite of a small structural difference (presence or absence of a galloyl group), these dimers showed a considerable difference of the partition coefficients in this solvent system (0.36 for Woodfordin C and 0.19 for Oenothein B).

Preparative Separation of Raw Materials

One of the major applications of CPC is the purification of natural products from vegetal extracts (flowers, roots, etc.) or crude extracts from fermentation broths without previous sample preparation. Hostettman et al. [10] have

Table 1 Applications of CPC

Species [Ref.]	Solvents system
Polyphenols tannins [3]	$CHCl_3$–MeOH–water (7:13:8; v/v/v), $CHCl_3$–MeOH–*n*-ProOH–water (9:12:2:8; v/v/v/v), *n*-ButOH–*n*-ProOH–water (4:1:5; v/v/v)
Triptolide and tripdiolide [3]	Hexane–EtOAc–CH_2Cl_2–MeCN–MeOH–H_2O (12:10:3:10:5:6; v/v/v/v/v/v)
Lanthanoids [3]	Hexane containing bis(2-ethylhexyl) phosphoric/0.1 mol/L (H,Na) Cl_2CHCOO to an appropriate pH
Flavonoids [10]	$CHCl_3$–MeOH–H_2O (5:6:4; v/v/v)
Polyphenols [10]	C_6H_{12}–EtOAc–MeOH–H_2O (7:8:6:6; v/v/v/v)
Tannins [10]	*n*-ButOH–*n*-ProOH–H_2O (2:1:3; v/v/v)
Naphtoquinones [10]	*n*-C_6H_{14}–MeCN–MeOH (8:5:2; v/v/v)
Retinals [10]	C_6H_6–*n*-C_5H_{12}–MeCN–MeOH (500:200:200:11; v/v/v/v)

described many examples of the isolation of natural products by CPC. Some flavonoids are, for instance, purified by using solvent systems containing chloroform, some coumarins by using solvent systems containing *n*-hexane and ethyl acetate, and more polar products such as tannins by butanol based systems. The main interest of this technique lies, however, in the possibility of overloading its column so that all the applications of semipreparative chromatography are available. For instance, Menet and Thiebaut [11] have separated 140 mg of an antibiotic from a crude extract of a fermentation broth. Some fractions up to a 95% purity were collected, whereas the original extract contained only 7% of the molecule of interest. They have also compared the performance of CPC, preparative liquid chromatography, and hydrodynamic mode CCC. They finally showed that the solvent consumption is the lowest for CPC and the enrichment is the best.

Measurement of log K_{oct}*–Water* [12]

Octanol–water partition coefficients (K_{ow}) have been established as the most relevant quantitative physical property correlated with biological activity. CPC, using octanol and water as the two phases, is a useful alternative for providing octanol–water partition coefficients (K_{ow}). It offers the automation advantages as compared to HPLC and the classical shake flask method. Three approaches for determining K_{ow} by CPC have been described. The normal mode consists in equilibrating the CPC column according to a normal equilibrium or an overloading mode by artificially decreasing the volume of the stationary phase. K_{ow} is calculated according to the classic formula $K = k'(V_t - V_s)/V_s$. The second procedure is the dual-mode method, which is based on the exchange of the role of the mobile and stationary phases during experiment. Therefore, the determination range of partition coefficients can be extended. The third procedure, the cocurrent mode, relies on the simultaneous pumping of a ratio of a small flow of octanol and a larger flow of water to elute strongly retained compounds.

References

1. B. N. Mandava and Y. Ito, Principles and instrumentation of counter current chromatography, in *Counter Current Chromatography. Theory and Practice* (B. N. Mandava and Y. Ito, eds.), Chromatographic Science Series Vol. 44 Marcel Dekker, Inc., New York, 1988, pp. 79–442.
2. J.-M. Menet, M.-C. Rolet, D. Thiébaut, R. Rosset, and Y. Ito, *J. Liquid Chromatogr. 15*:2883 (1992).
3. A. P. Foucault, Theory of Centrifugal Partition Chromatography, in *Centrifugal Partition Chromatography* (A. P. Foucault, ed.), Chromatographic Science Series Vol. 68 Marcel Dekker Inc., New York, 1995, pp. 25–50.
4. J.-M. Menet, D. Thiébaut, R. Rosset, J. E. Wesfreid, and M. Martin, *Anal. Chem. 66*:168 (1994).
5. M. J. van Buel, L. A. van der Wielen, and K. Ch. A. M. Luyben, Pressure drop in centrifugal partition chromatography, in *Centrifugal Partition Chromatography* (A. P. Foucault, ed.), Chromatographic Science Series Vol. 68 Marcel Dekker, Inc., New York, 1995, pp. 51–70.
6. D. W. Armstrong, G. L. Bertrand, and A. Berthod, *Anal. Chem. 60*:2513 (1988).
7. W. Murayama, T. Kobayashi., Y. Kosuge, H. Yano, Y. Nunogaki, and K. Nunogaki, *J. Chromatogr. 239*:643 (1982).
8. H. Oka, K.-I. Harada, and Y. Ito, *J. Chromatogr. 812*:35 (1998).
9. T. Okuda, T. Yoshida, and T. Hatano, Fractionation of plant polyphenols, in *Centrifugal Partition Chromatography* (A. P. Foucault, ed.), Chromatographic Science Series Vol. 68 Marcel Dekker, Inc., New York, 1995, pp. 99–132.
10. M. Maillard, A. Marston, and K. Hostettmann, High speed counter current chromatography of natural products, in *High-Speed Counter Current Chromatography* (Y. Ito and W. D. Conway, eds.), Chemical Analysis Series Vol. 32 John Wiley & Sons, New York, 1995, pp. 179–218.
11. M.-C. Menet and D. Thiebaut, *J. Chromatogr. 831*:203 (1999).
12. A. Berthod and K. Talabardon, Operating parameters and partition coefficient determination, in Counter Current Chromatography, in *Counter Current Chromatography* (J.-M. Menet and D. Thiébaut, eds.), Marcel Dekker, Inc., New York, 1999, pp. 121–148.

M.-C. Rolet-Menet

Centrifugal Precipitation Chromatography

For many years, proteins have been fractionated with ammonium sulfate (AS) by stepwise precipitation. In this conventional procedure, an increasing amount of AS is added to the protein solution and the precipitates are removed by centrifugation in each step. Recently, "centrifugal precipitation chromatography" [1,2] has been developed to replace the tedious manual procedure. This novel chromatographic system is capable of internally generating a concentration gradient of AS through a long separation channel under a centrifugal force field. Proteins introduced into the channel are exposed to a gradually increasing AS concentration and finally precipitated at different locations according to their solubility in the AS solution. Then, the AS concentration in the upper channel is gradually reduced so that the AS concentration gradient in the lower channel is proportionally decreased. This manipulation causes the precipitated proteins to be redissolved and eluted out by repeating precipitation and dissolution along the channel. As in liquid chromatography, the effluent is continuously monitored with an ultraviolet (UV) monitor and fractionated into test tubes using a fraction collector.

The principle and unique design of the separation column is shown in Figs. 1a and 1b, respectively. In Fig. 1a, a pair of separation channels is partitioned by a dialysis membrane. A concentrated (C) AS solution is introduced

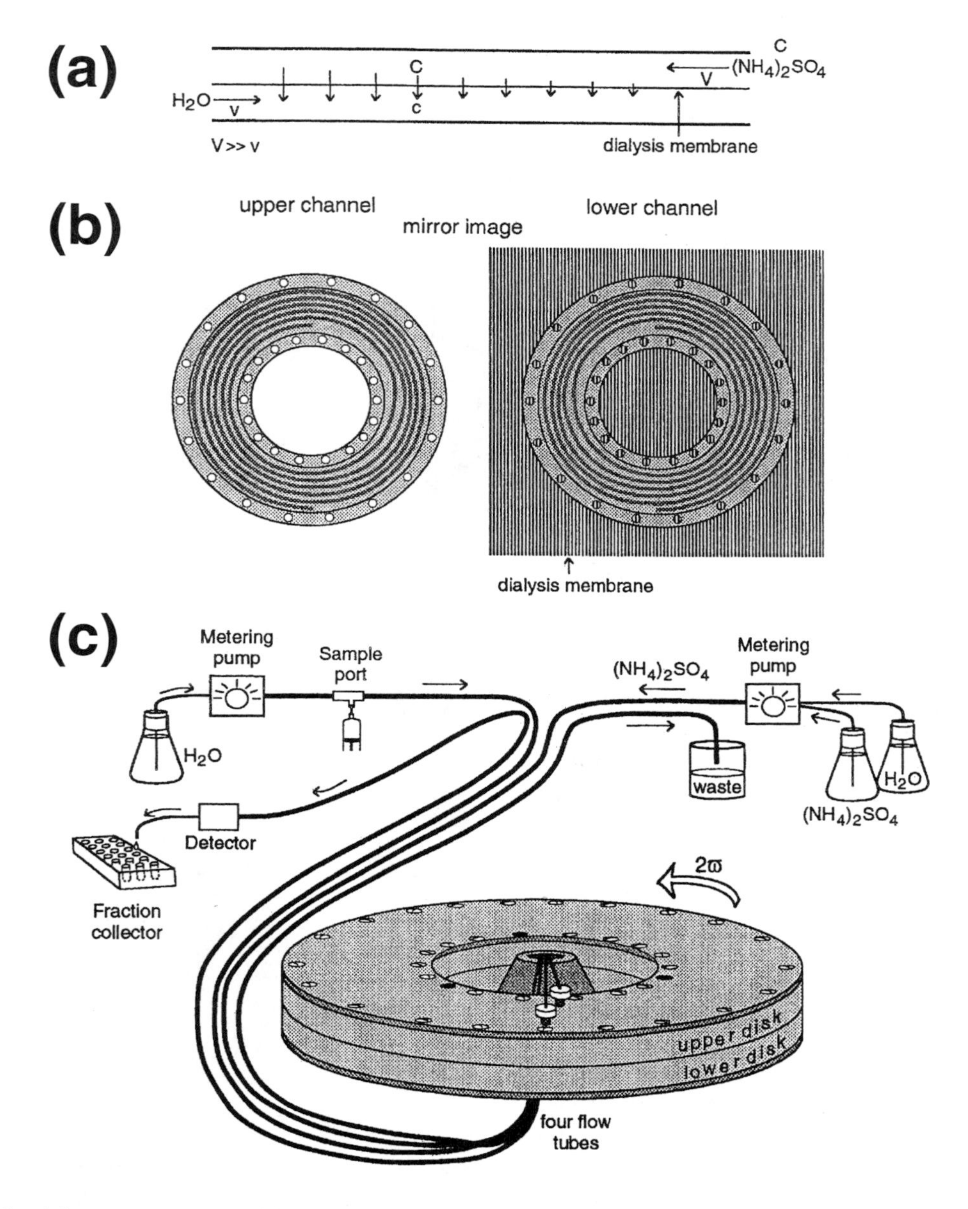

Fig. 1 (a) Principle of the present method; (b) design of the separation column assembly; (c) schematic illustration of the entire elution system of centrifugal precipitation chromatography.

through the upper channel at a high flow rate (V) from the right, whereas water is fed into the lower channel from the left at a much lower rate (v). This countercurrent flow of the two liquids through the channel results in AS transfer from the upper channel to the lower channel at every point, as shown by arrows across the membrane. Because the AS transfer rate through the membrane is proportional to the difference in AS concentration between the two channels, an exponential gradient of AS concentration (c) is formed through the lower channel. The system allows manipulation of the AS concentration in this gradient by modifying the AS concentration in the upper channel, as described earlier. The separation column is fabricated from a pair of plastic disks (high-density polyethylene, 13.5 cm in diameter and 1.5 cm thick) equipped with mutually mirror-imaged spiral grooves (1.5 mm wide, 2 mm deep, and ~ 200 cm long), as shown in Fig. 1b. A regenerated cellulose membrane with a desirable pore size (6000–8000 or 12,000–14,000) is sandwiched between these two disks which are, in turn, tightly pressed between two metal plates with a number of screws. The capacity of each channel is about 5 mL. This column assembly is mounted on the sealless continuous-flow centrifuge that allows continuous elution through the rotating column without the use of a conventional rotary seal device. The principle of this sealless flow centrifuge system is described in the entry Countercurrent Chromatography Instrumentation. Figure 1c schematically illustrates an entire elution system of the present chromatographic system. Two sets of pumps are used, on (upper right) for pumping AS solution at a high rate and the other (upper right) for eluting buffer solution and protein samples at a lower rate. The total of four flow tubes led from these two pumps are bundled together and clamped at the top of the sealless continuous-flow centrifuge to reach the column assembly as illustrated. As mentioned earlier, these flow lines are twist-free as the column rotates around the central axis of the centrifuge. Consequently, the system eliminates various complications such as leakage, clogging, and cross-contamination, often arising from the use of the conventional rotary seal device for multiple flow lines.

A series of experiments was conducted to study the AS transfer rate through the dialysis membrane, pumping a concentrated AS solution into the upper channel at 1 mL/min and water into the lower channel at varied flow rates ranging from 1 to 0.1 mL/min without sample injection. In these experiments, the AS input concentration into the upper channel and the AS output concentration from the lower channel were compared. The rate of the AS transfer rose, as expected, with the decreased flow rate through the water channel, and at a flow rate of 0.1 mL/min, the AS concentration collected through the lower channel reached nearly 100% that of the AS input in the upper channel. Whereas AS diffuses from the upper channel toward the lower channel, water in the lower channel is absorbed into the upper channel. This water transfer rate is estimated by comparing input and output flow rates through the water channel. At an input rate of 0.1 mL/min, the outlet flow was decreased to one-fourth of the input rate, indicating that the separated fractions would be eluted in a highly concentrated state. Generating an AS concentration gradient and concentrating fractions are the two unique capabilities of the present system, which can be effectively utilized for fractionation of proteins.

Figure 2 illustrates serum protein separation by centrifugal precipitation chromatography: the chromatographic tracing of the elution curve in Fig. 2a and so-

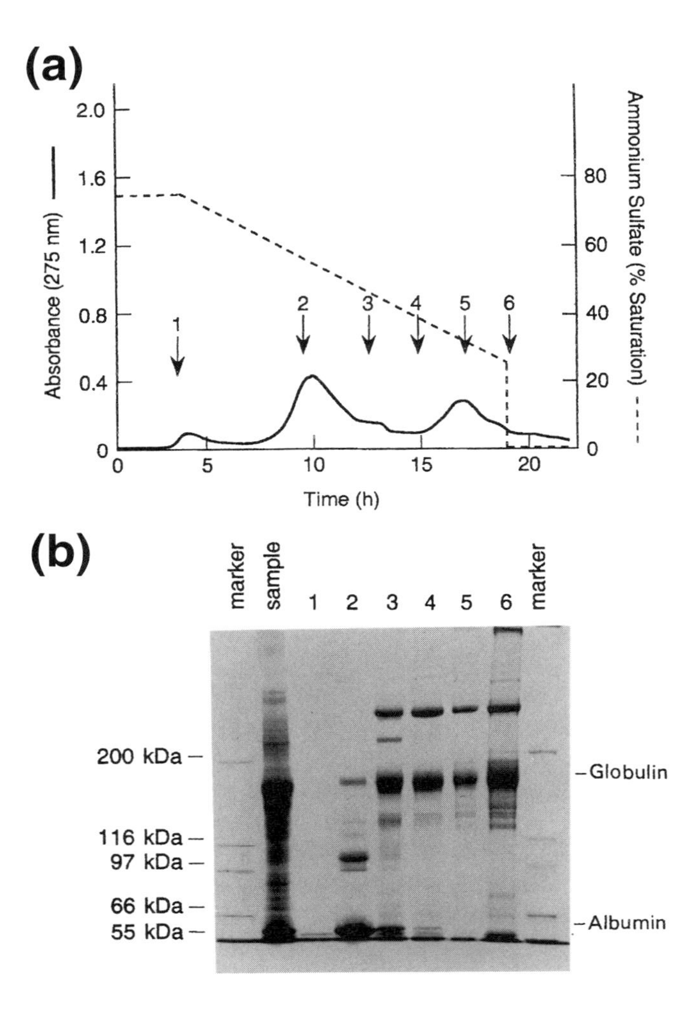

Fig. 2 Separation of human serum proteins by centrifugal precipitation chromatography: (a) Elution curve; (b) SDS-PAGE analysis of separated fractions.

dium dodecyl sulfate–polyacrylamide gel electrophoresis (SDS-PAGE) analysis of separated fractions in Fig. 2b. In this example, 100mL of normal human serum (pooled) was diluted to 1 mL and introduced into the separation channel. The experiment was initiated by filling both upper and lower channel with 75% AS solution followed by sample charge into the lower channel through the sample loop. After the separation column assembly was rotated at 2000 rpm, the upper channel was eluted 75% AS solution at a flow rate of 1 mL/min while the lower channel was eluted with 50 m*M* potassium phosphate at 0.06 mL/min. After 4 h of elution, the AS concentration in the upper channel was linearly decreased down to 25%, as indicated in the chromatogram. The effluent from the lower channel was continuously monitored with an UV monitor (LKB Uvicord S) at 275 nm and fractionated into test tubes using a fraction collector (LKB Ultrorac), and the AS solution eluted from the upper channel was discarded. The chromatogram (Fig. 2a) produced two major peaks, one at AS saturation at 60–50% and the other at 35–30%. The SDS-PAGE analysis of peak fractions (Fig. 2b) revealed that the first peak represents albumin (MW 68,000) and the second peak, γ-globulin (MW 160,000).

Centrifugal precipitation chromatography can produce highly purified protein fractions because the proteins are refined by repetitive precipitation and dissolution. The method has the following advantages over the conventional manual procedure: The method is programmed and automated; the fractions are almost free of small molecules that are dialyzed through the membrane or otherwise quickly eluted out from the channel; noncharged macromolecules such as polysaccharides are washed out, whereas charged biopolymers such as DNA and RNA may also be separated according to their solubility in AS solution; and the method may be amenable for microscale to large-scale fractionation by designing the separation column in suitable dimensions.

The present system has been successfully applied to fractionation of various protein samples, including serum proteins, monoclonal antibodies (IgM against mast cells) from hybridoma culture supernatant, and minor protein components (less than 1% of total proteins) from a crude rabbit reticulocyte lysate containing a large amount of hemoglobin and protein–poly(ethylene glycol) conjugates. One important application of the present method is affinity separation using a ligand that can specifically bind to the target protein to substantially lower its solubility in the AS solution. This ligand–protein complex is then eluted out much later than most of other proteins. This affinity precipitation method has been demonstrated in the purification of recombinant proteins such as ketosteroid isomerase from a crude *Escherichia coli* lysate by adding an affinity ligand (17-estradiol-methyl-polyethylene-glycol-5,000) to the sample solution. The application of the present method may be extended to fractionation of other biopolymers such as DNA and RNA using a pH gradient.

References

1. Y. Ito, Centrifugal precipitation chromatography applied to fractionation of proteins with ammonium sulfate, *J. Liquid Chromatogr. Related Technol. 22*(18):2825 (1999).
2. Y. Ito, Centrifugal precipitation chromatography: Principle, apparatus, and optimization of key parameters for protein fractionation by ammonium sulfate precipitation, *Anal. Biochem. 277*:143 (2000).

Yoichiro Ito

Ceramides: Analysis by Thin-Layer Chromatography

Thin-layer chromatography (TLC) is widely used in all chemical and biochemical disciplines for preparative as well as analytical purposes. TLC has been thoroughly used in lipid biochemistry and the number of applications is still growing, as more and more laboratories are using this method due to its low cost and ease of use and because new developments in TLC techniques allow better performance [1]. Ceramides are molecules of growing interest, because they are key intermediates in sphingolipid metabolism and they are involved in numerous cellular signal transduction processes [2]. Purification or isolation of these molecules is often carried out by liquid chromatography or TLC prior to further analysis by GLC or high-performance liquid chromatography (HPLC) and mass spectrometry. The aim of the present article is to describe the possibilities offered by TLC in the analysis and purification of ceramides on silica gel TLC and high-performance TLC (HPTLC) plates.

Separation of Ceramides Into Group Species

Thin-layer chromatography has been widely used in the separation of free ceramides into different groups. Indeed, ceramides constitute a class of molecules with a large number of molecular species differing in the constituting sphingoid base and in the species of fatty acids linked as amides (see Fig. 1). Despite the great number of molecular species that can be found in a tissue, the different ceramides can be efficiently separated into defined groups of species according to some structural criteria and to their chromatographic behavior on silica gel TLC plates [3]. The presence of hydroxyl groups at various positions in the ceramide molecule can be utilized to separate the species into groups of compounds with structural homology [3,4].

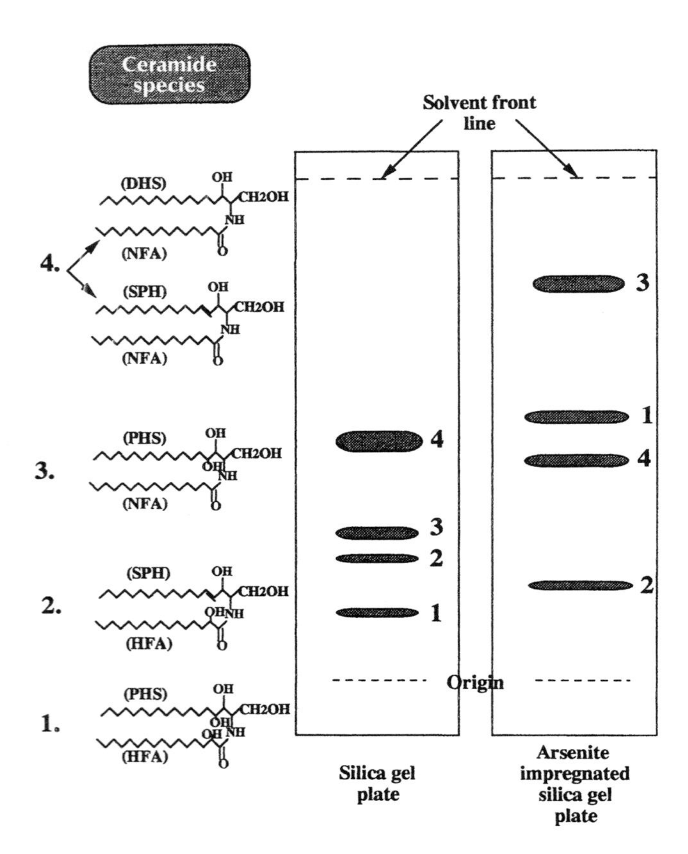

Fig. 1 Schematic representation of ceramide species separation into groups differing in position and number of hydroxyls, after migration onto silica gel or onto an arsenite-impregnated silica gel TLC plate. The plate is developed in chloroform–methanol 50:3.5, (v/v) as the solvent system. The ceramide structure is shown and the numbers refer to those reported close to the ceramide spot on the TLC plates. NFA: normal (nonhydroxy) fatty acid; HFA: hydroxy fatty acid; DHS: dihydrosphingosine; SPH: sphingosine; PHS: phytosphingosine.

Separation of Ceramides into Groups with Respect to the Position and Number of Hydroxyl Groups

A systematic study by Karlsson and Pascher showed the thin-layer chromatographic behavior of ceramides species [3]. In that study, separation of molecular species onto silica gel plates was made possible according to the number and position of hydroxyls on ceramides by using chloroform–methanol, 95:5 (v/v), as the solvent system. The procedure was applied to separate ceramides of sphingomyelin isolated from bovine kidney and intestine for further analysis of their trimethylslyl derivatives by gas chromatography–mass spectrometry (GC–MS) [5]. Separation of ceramide species according to the position and number of hydroxyl groups was also efficiently achieved by Motta et al. using chloroform–methanol–acetic acid, 190:9:1 (v/v) [4]. These authors proposed a designation of free ceramides based on the number and positions of hydroxyl groups in the ceramide molecule (Fig. 1). The migration rate of the molecule will be lowered by the presence of hydroxyl groups. This is due to enhanced interactions between the molecule and the silica gel, because hydrogen-bonding can occur between the hydroxyls of ceramide and silanols. Thus, ceramides containing phytosphingosine, a sphingoid base with three hydroxyl groups, linked to hydroxy fatty acids will have a lower R_f than ceramide containing sphingosine, a dihydroxy long-chain base, linked to a normal nonhydroxy fatty acid [3–7]. Such a separation of ceramide species into defined groups of relative structural homology can be obtained on silica gel plates when using solvent systems consisting of chloroform and methanol mixtures ranging from 50:2 to 50:5 by volume [7].

When the proportion of methanol is enhanced, free ceramides will tend to migrate close to the solvent front line, making the separation from other neutral lipids difficult, particularly with free fatty acids and cholesterol, which tend to tail in these solvent systems as compared to other neutral lipids. This problem can be minimized with prior purification of ceramide fraction by liquid chromatography or preparative TLC. Moreover, with the higher proportion of methanol in chloroform, some groups of ceramide species will tend to overlap. This is particularly true with ceramides containing three hydroxyl groups, such as ceramides containing sphingosine linked to hydroxy fatty acids and ceramides containing phytosphingosine linked to nonhydroxy fatty acids. With a greater proportion of methanol in chloroform (5 volumes of methanol to 50 volumes of chloroform), these last species will mix together, whereas if the methanol proportion in chloroform is lowered (2 volumes of methanol to 50 volumes of chloroform), these two homologous ceramide groups will be efficiently resolved from each other [6].

Little attention has been given to the possible overlapping of some of the separated ceramide groups with monoacylglycerols. In the solvent systems previously mentioned, these latter migrate very close to free ceramides, particularly ceramide containing phytosphingosine linked to normal fatty acids. Although their acyl or alkenyl bond can be removed easily by chemical treatment (alkaline methanolysis and mild acidic hydrolysis), this is not the case if the fatty acid is bound with an ether linkage. One way to prevent such a possible contamination of ceramide spots by monoacylglycerols is to further separate these glycerolipids from ceramides in a second direction perpendicular to the direction followed by the first solvent system.

Better separation of molecules will, indeed, be obtained with multidimensional TLC, because it allows the use of different solvent systems in each direction, so that the resolution of molecules which can overlap each other in one of the solvent systems can be achieved in the second one. This can be done by running the plate in diethyl ether, as monoacylglycerols will tend to migrate in this solvent, whereas ceramides do it poorly. In fact, the resolution of monoacylglycerols from free-ceramide group species can be obtained by a single run of the silica gel plate in chloroform–methanol 50:3.5 (v/v) [7]. This solvent mixture was shown to be optimal for the resolution of monoacylglycerols from ceramides while giving a good separation of ceramide species into groups of homologous compounds. Multidimensional TLC allows the separation of these compounds while preserving the other neutral lipids of interest whose structures can be altered by chemical treatment.

Effect of Carbon Chain Length and Unsaturation

Ceramide species do not only differ by the position and the number of their constituting hydroxyl groups. The carbon chain length and the degree of unsaturation of the sphingoid bases and fatty acids may also affect the chromatographic behavior of ceramides. It has been shown that on silica gel plates, an increase in unsaturation of the fatty acid constituting the ceramide will result in a slight increase in mobility [3]. Conversely, the presence of a 4-5 *trans* double bond in the sphingoid base does not seem to alter the migration of ceramides on a silica gel plate. Hence, ceramides containing sphinganine (dihydrosphingosine) instead of sphingosine linked to normal fatty acids will tend to migrate as a unique spot under the solvent conditions previously cited. The ceramide spot number 2 (see Fig. 1) will be a mixture of ceramides containing sphinganine or sphingosine linked to normal fatty acids.

The chain length of the amide-linked fatty acid must also be considered. This was studied by Karlsson and Pascher, who compared the chromatographic behavior of synthetic ceramides with increasing fatty acid chain length [3]. The general feature is that, with increasing the fatty acid chain length, the ceramides migrate faster. Motta et al. reported the separation of ceramides containing phytosphingosine linked to C_{24}–C_{26} fatty acids from ceramides with phytosphingosine linked to shorter fatty acids ranging from C_{16} to C_{18} [4]. They also succeeded in separating ceramides containing phytosphingosine and alpha-hydroxy fatty acids into two groups according to the chain length of fatty acids. This was possible along with a good resolution of ceramides according to the position and the number of hydroxyl groups [4].

Resolution of Ceramide Group Species onto Arsenite-Impregnated Silica Gel TLC Plates

Separation of ceramides into group species is also possible after impregnation of the silica gel layer with sodium arsenite. This glycol-complexing agent can be added directly into the silica gel mixture when preparing the TLC layer [3] or by spraying a solution of 1% sodium *meta* arsenite in methanol (w/v) onto precoated silica gel plates. Karlsson and Pascher recommended the use of arsenite-impregnated TLC plates for the separation of ceramides because of the better resolution than that which was obtained with plain silica gel plates [3]. When the arsenite-impregnated layer is used, ceramides are also separated according to the number and positions of hydroxyl groups. This separation is efficient using the system chloroform–methanol 95:5 (v/v) as the solvent. However, the chromatographic behavior of ceramide species is different on arsenite-impregnated layers as compared to plain silica gel. Hence, on arsenite silica gel layers, ceramides containing the trihydroxy base phytosphingosine migrate with a higher R_f than ceramides containing dihydroxy bases such as sphingosine and sphinganine linked to non-hydroxy fatty acids. This is not the case with plain silica gel, as shown in Fig. 1. The effect of a long-chain-base unsaturation is negligible when running ceramide mixtures on arsenite silica gel TLC. Hence, homologous ceramides containing either sphinganine or sphingosine will be mixed together at the end of migration.

The use of silica gel TLC plates (impregnated or not with arsenite) allows the separation of ceramide species into different groups of relative structural homology. This is based on the position and numbers of hydroxyl groups on the ceramide molecules. However, TLC allows a possible further fractionation of ceramide species according to the unsaturation of sphingoid bases or fatty acids. This is made possible by the impregnation of the silica gel plates with borate and silver nitrate.

Separation of Ceramides Species on Borate and Silver Nitrate TLC Plates

Ordinary silica gel TLC plates can be impregnated with sodium borate for the separation of ceramides. This is particularly suitable for the separation of ceramides containing saturated sphingoid base (sphinganine) from ceramide containing monounsaturated sphingoid base, such as sphingosine [3]. Resolution of such ceramide species is not really possible on ordinary or arsenite-impregnated silica gel plates. Boration of the TLC plate can circumvent such a challenge, because resolution of ceramides according to the presence or not of the sphingoid 4-5 *trans* double bond is efficiently achieved [3]. Moreover, the effect of borate as a complexing agent on fatty acid unsaturation was shown not to be significant, so that groups of ceramides could be separated with respect to the unsaturation of their sphingoid bases but not to the unsaturation of the N-linked fatty acids. This chromatographic tool has been used to study the bioconversion of dihydroceramide to ceramide [8].

The separation of ceramide species according to fatty acid unsaturation is, however, possible by using silver-ion-containing silica layers. In such conditions, ceramide species will migrate faster when decreasing the fatty acid unsaturation degree. Moreover, the degree of unsaturation of the sphingoid base is of significant importance, because the R_f of ceramide species will decrease by increasing the unsaturation of the sphingoid base. This can result in difficulties for the accurate separation and identification of some ceramide species differing in unsaturation of sphingoid bases or fatty acids [3]. Although such a separation of ceramide molecular species is possible, it is, in fact, seldom used. When a detailed analysis is needed, TLC shows its limits, as other analytical tools give more information. Gas chromatography or HPLC, either alone or coupled with mass spectrometry, are the best tools for the determination of ceramide molecular species. However, TLC represents a reliable chromatographic technique which can be efficiently used for ceramide separation and purification into groups of compounds of relative structural homology. This is often based on the presence of different hydroxyl groups at different locations of the ceramide molecules. The separation scheme can be efficiently performed by running ceramide samples in a single direction on ordinary silica gel or arsenite-impregnated silica gel TLC plates. Potentiation of this separation scheme could be accomplished by running the plate in a second direction, perpendicular to the first one, after boration of the migration field so that further separation of sphinganine-containing ceramides from sphingosine-containing ceramides can simultaneously be achieved with the separation of groups with different hydroxyls. Such preparative and analytical use makes TLC an interesting tool for the investigation of ceramide biochemistry.

References

1. H. Kalasz and M. Bathori, *LC–GC Int.* 440 (1997).
2. D. K. Perry and Y. A. Hannun, *Biochim. Biophys. Acta 1436*: 233 (1998).
3. K. A. Karlsson and I. Pascher, *J. Lipid Res. 12*:466 (1971).
4. S. Motta, M. Monti, S. Sesana, R. Caputo, S. Carelli, and R. Ghidoni, *Biochim. Biophys. Acta 1182*:147 (1993).
5. M. E. Breimer, K. A. Karlsson, and B. E. Samuelsson, *Lipids 10*:17 (1974).
6. J. Bodennec, G. Brichon, O. Koul, M. El Babili, and G. Zwingelstein, *J. Lipid Res. 38*:1702 (1997).
7. J. Bodennec, G. Brichon, J. Portoukalian, and G. Zwingelstein, *J. Liquid Chromatogr. Related Technol. 22*:1493 (1999).
8. L. Geeraert, G. P. Mannaerts, and P. P. Van Veldoven, *Biochem. J. 327*:125 (1997).

Jacques Bodennec
Jacques Portoukalian

Channeling and Column Voids

Channeling can occur when voids that are created in the packing material of a column cause the mobile phase and accompanying solutes to move more rapidly than the average flow velocity. The most common result of channeling is band broadening and, occasionally, elution of peak doublets.

Column voids can develop in a poorly packed column from settling of the packing material or by erosion of the packed bed. In a properly packed column, voids can develop gradually over time or suddenly as the result of pressure surges. A void that forms in the inlet of a column may lead to poor peak shape, including severe band tailing, band fronting, or even peak doubling for every peak in the chromatogram. Filling the column inlet with the same or equivalent column packing can sometimes reduce voids. For this type of repair, the column should be held in a vertical position while the inlet frit is removed. The void will be evident as either settling of the packing material or as holes in the column surface. The new packing should be slurried with an appropriate mobile phase and packed into the column void with a flat spatula. Once the top of the new packing is level with the column end, a new inlet frit can be added and the end fitting reinstalled. The column should then be reconnected to the LC system and conditioned with the mobile phase at a fairly high flow rate to help settle the new column bed. After filling the void, the packing bed will generally be more stable if the repaired column is operated with the direction of flow reversed from the original direction. This repair procedure can be used to extend column life; however, it should be noted that the plate number of the repaired column would be, at best, only 80–90% of the original column. Columns that develop voids over time are often near the end of their useful life spans and in some cases it may be more cost efficient to discard such a column rather than to repair it.

Suggested Further Reading

Dolan, J. W. and L. R. Snyder, *Troubleshooting LC Systems*, Humana Press, Totowa, NJ, 1989.

Majors, R. E., The care and feeding of modern HPLC columns, *LC–GC 16*:900 (1998).

Eileen Kennedy

Characterization of Metalloproteins Using Capillary Electrophoresis

Metalloproteins constitute a distinct subclass of proteins that are characterized by the presence of single or multiple metal ions bound to the protein by interactions with nitrogen, sulfur, or oxygen atoms of available amino acid residues or are complexed by prosthetic groups, such as heme, that are covalently linked to the protein. These metals function either as catalysts for chemical reactions or as stabilizers of the protein tertiary structure. Protein-bound metals may also be labile and, as such, be subject to transport, transient storage, and donation to other molecular sites of requirement within tissues and cells.

Metalloproteins play critical roles in a wide variety of basic cellular functions, including respiration, gene expression, reproduction, and metabolism. Isolation, characterization, and quantification of individual metalloproteins are each necessary and important steps toward understanding their unique biological functions. Alone or in combination, various types of chromatography, electrophoresis, and spectrometric techniques have been employed to study many unique aspects of metalloprotein structure and function. However, no one technique currently offers the ability to isolate, characterize, and quantify individual metalloproteins in a single step from complex matrices such as tissue extracts or physiological fluids. Therefore, there is an ongoing need for new and more capable methodologies. Because of the small-sample volume requirement, high degree of resolution, and advanced instrument automation capabilities, capillary electrophoresis (CE) has gained increasing popularity in the analysis of proteins [1,2]. In fact, many of the CE-based techniques developed for general protein separations are directly applicable to metalloprotein analyses [2,3]. This article will emphasize some recent applications of CE and CE-related methodologies and their utility in providing new

Table 1 Characterization of Metalloproteins Using Capillary Electrophoresis

Characterization
Identification and Purity Assessments
General characteristics (i.e., net charge, molecular weight, isoelectric point, etc.)
Monitoring purification steps
Separation of impurities or degradation products
Detection of unique UV/visible absorbance (chromophore)
Structural Information
Separation of molecular forms (i.e., isoforms, metalloforms, glycoforms, etc.)
Study of macromolecular assembly
Determination of metal-binding sites
Peptide mapping
Stability Determinations
Effects of temperature and pH
Buffer additives (i.e., metals and metal chelators)
Activity Measurements
Enzymatic activity
Electrophoretically mediated microassay (EMMA)
Isozyme profiling
Metal-Binding/Electrophoretic Mobility Shift
Affinity CE (ACE)
Immobilized metal-ion affinity CE (IMACE)
Metal-chelate coated capillaries
REDOX
oxidation state of protein-bound metal
Elemental Analyses
Indirect Detection Methods
Unique UV/visible absorbance spectra (chromophore)
Mass spectrometry (CE-MS)
Direct (Element-Specific) Detection Methods
Inductively coupled plasma–mass spectrometry (CE–ICP–MS)
Proton-induced x-ray emission (PIXE)

insight into the structure and function of a variety of metalloproteins.

Table 1 summarizes some of the ways CE has been applied to metalloprotein characterization. Altering capillary temperature, buffer ionic strength and pH, electric field strength (i.e., voltage), and capillary internal surface coating are but a few of the ways that CE conditions can be varied to influence the efficiency and selectivity of metalloprotein separations. Furthermore, different CE separation modes can be applied to gain new information about a specific metalloprotein [3]. For instance, capillary zone electrophoresis (CZE) can indicate a protein's net charge at a given pH; capillary isoelectric focusing (CIEF) gives a rapid estimate of its isoelectric point (p*I*); capillary gel electrophoresis (CGE), in the presence of sodium dodecyl sulfate (SDS), can be used to estimate its apparent molecular mass; and micellar electrokinetic capillary chromatography (MEKC) can be useful in characterizing its surface hydrophobicity.

Varying CE separation conditions has been shown to be a particularly effective approach for improving the isolation and characterization of metallothioneins, a heterogeneous family of low-molecular-weight, cysteine-rich, heavy-metal-binding proteins [4]. It was found that (a) phosphate and borate buffers enhanced the sensitivity of detection at 200 nm by significantly reducing ultraviolet (UV) absorption of the background electrolyte, (b) the alkaline borate buffer (pH 8.4) gave rapid analysis times with reasonably high resolution, (c) the acidic phosphate buffer (pH 2.5) completely stripped zinc and cadmium from the proteins, yielding higher resolution and more reproducible separations of the apothioneins (metal-free proteins), (d) capillaries coated on their inner surface with a polyamine polymer that reversed electro-osmotic flow (EOF) or polyacrylamide that greatly suppressed EOF significantly improved resolution, (e) the MEKC mode of CE improved separation selectivity, and (f) photodiode array scanning detection to monitor UV absorption spectra of individual protein peaks separated at neutral pH was useful in determining both the presence and the type of metal associated with each.

Capillary electrophoresis is a useful tool for monitoring the purity of metalloproteins isolated from either natural or recombinant sources [3]. CZE was used to follow the purification progress of metallothioneins in samples subjected to gel filtration chromatography and reversed-phase high-performance liquid chromatography (HPLC) [3,4]. Detection of a unique chromophore arising from the interaction of metal ions and specific amino acid residues in the protein or with a prosthetic group attached to the protein can be useful. The selectivity achieved under such conditions can greatly reduce or even eliminate the need to purify metalloproteins prior to their analysis by CE. Two good examples of this are the detection of hemoglobin variants separated from red blood cell lysates by monitoring absorbance at 415 nm and the detection of transferrin in serum at 460 nm [3]. Absorbance at these characteristic wavelengths reflects the presence of iron atoms complexed by the heme moiety (hemoglobin) or by iron-binding sites located at the amino and carboxyl ends of the transferrin protein molecule. CE has also been used to characterize surface metal-binding sites on cytochrome-*c* and myoglobin modified with ruthenium-bis(bipyridine)imidazole, which imparts a strong absorbance at 292 nm to the modified proteins or peptides derived from a tryptic digest of the modified proteins [3].

Capillary electrophoresis has proven to be useful in characterizing different molecular forms of various metalloproteins like metallothionein, transferrin, and conalbumin [2–5]. Molecular forms arise from differences in the amino acid sequence of proteins (isoforms), differences in the amount or type of metal bound (metalloforms), or from differences in the type and amount of carbohydrate side chains linked to the protein (glycoforms). CZE was used to follow the formation of the oligomeric iron core and its incorporation into ferritin, to detect and quantify ferritin species or ferritin subunit proteins in purified or partially purified states, and to study the interaction of different metal ions with ferritin [2,3].

Structural stability of metalloproteins can be quickly assessed by CE under different conditions [2,3]. For example, thermally induced conformational changes in calcium-depleted α-lactalbumin and urea-induced unfolding of serum albumin were studied using CZE. The oxidation state of cysteine sulfhydryl groups in the zinc-containing protein, ribonuclease A, has been assessed using CZE to determine the presence or absence of a disulfide bond. Elevated capillary temperature altered the structure of myoglobin, which, in turn, resulted in reduction of the valence state of the iron atom bound to heme associated with this protein. Similarly, CIEF was used to separate and characterize different heme–iron valence hybrids of hemoglobin [6].

Buffer additives, especially metals and metal chelators, can have dramatic effects on CE-based separations of metalloproteins by causing shifts in their electrophoretic mobility [3]. This observation forms the basis for a unique CE method referred to as affinity capillary electrophoresis or ACE. When a protein forms a complex with a charged metal-ion ligand, there can be a resulting change in electrophoretic mobility of the complexed protein relative to that of the metal-free protein. Scatchard analysis of the change in electrophoretic mobility of the protein as a function of the metal ion concentration in the separation buffer allows for the calculation of a metal-binding constant (K_b). ACE has been used to characterize K_b values for several metalloproteins, including (a) calcium affinity for calmodulin and C-reactive protein [3] and (b) the binding affinity of zinc for two separate sites in a highly basic, zinc-finger protein (NCp7) from the human immunodeficiency virus [7]. Haupt et al. [8] reported the development of an alternative CE affinity method based on immobilized metal affinity chromatography, which they called immobilized metal-ion affinity capillary electrophoresis or IMACE. In IMACE, metal ions (e.g., Cu^{2+}) are fixed to a soluble polyethylene glycol replaceable polymer matrix support added to the CE separation buffer. IMACE was used to study surface-related affinity characteristics (number and accessibility of histidine residues and histidine microenvironment) for particular immobilized metal-ion chelate ligands in such proteins as cytochrome-*c*, ribonucleases A and B, chymotrypsin, and kallikrein [8].

Some of the most promising advances in our understanding of unique characteristics of metalloprotein structure and function come from continuing developments in detection methodologies and from further development and refinement of coupled (hyphenated) systems such as CE–mass spectrometry (CE–MS) and CE–inductively coupled plasma–mass spectrometry (CE–ICP–MS). The major difficulties restraining the routine use of such systems, aside from cost, arise from problems in interfacing the CE instrument with MS and ICP–MS instrumentation, although much progress is being made in this area [9]. CE–MS has been used to characterize metallothionein isoforms and metalloforms, the structures of which were deduced from discrete differences detected in molecular mass of the species separated by CE [10]. Using molecular masses calculated from the amino acid sequence and the type and amount of associated metals, it was possible to unequivocally identify distinct molecular forms.

The most definitive assessment of the metal composition of metalloproteins comes from the application of element-specific detection methods. CE–ICP–MS provides information not only about the type and quantity of individual metals bound to the proteins but also about the isotopes of each element as well [11,12]. Elemental speci-

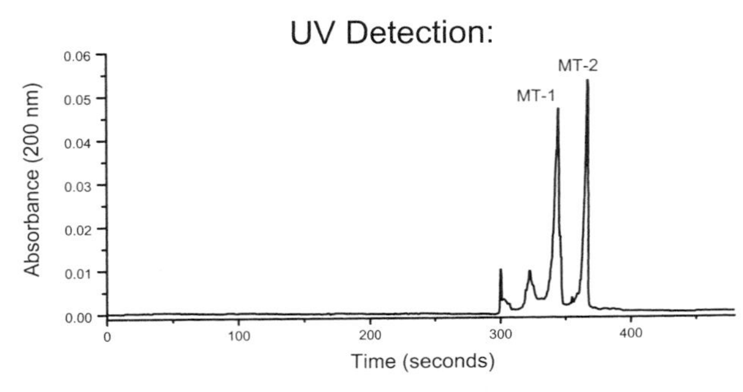

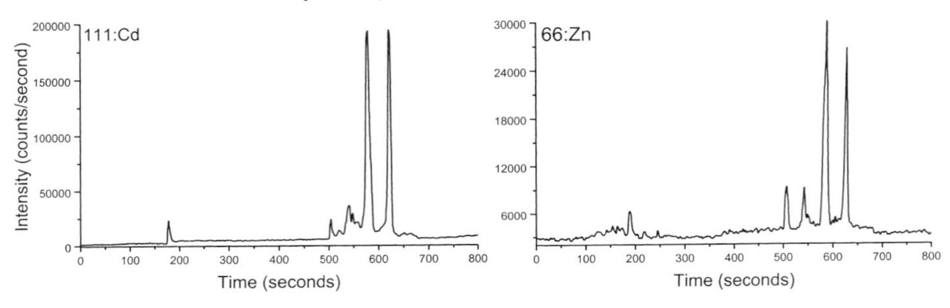

Fig. 1 Separation of rabbit liver metallothionein using CE–ICP–MS. The protein sample (1 mg/mL dissolved in deionized water) was first subjected to CZE with UV detection to optimize CE separation parameters for the major metallothionein isoforms (MT-1 and MT-2) shown in the upper panel. The CE instrument was then coupled to an ICP–MS instrument using a specially modified direct injection nebulizer (CETAC Technologies Inc., Omaha, NB) which enabled the entire capillary effluent from the CE to be directly injected into the ICP plasma torch, thus avoiding postcolumn dilution and band-broadening effects of conventional spray chamber nebulizers. Specific isotopes of cadmium (^{111}Cd) and zinc (^{66}Zn) associated with each isoform peak were monitored as shown by the figures in the lower panel.

ation has become increasingly important to the areas of toxicology and environmental chemistry. Such analytical capability also opens up important possibilities for trace element metabolism studies. Figure 1 depicts the separation of rabbit liver metallothionein containing zinc, copper, and cadmium (the predominant metal) using CE–ICP–MS with a high-sensitivity, direct injection nebulizer (DIN) interface. UV detection (200 nm) was used to monitor the efficiency of the CE separation of the protein isoforms (MT-1 and MT-2), whereas ICP–MS detection made it possible to detect and quantify specific zinc, copper (not shown), and cadmium isotopes associated with the individual isoform peaks.

There are a number of emerging CE-based techniques that will greatly benefit the field of metalloprotein analysis in the near future. Major advances in interfacing instrumentation that will result in more efficient separations and more sensitive detection in coupled systems, especially for CE–MS and CE–ICP–MS, are occurring now [9]. Further development of capillary electrochromatography (CEC), new column packing materials, and commercial systems that allow for gradient elution CEC will have a major impact on improving CE separations of metalloproteins. Moreover, coupling CEC to MS or ICP–MS detectors will offer new and more powerful ways to isolate and characterize metalloproteins. The ability to accurately detect and quantify elemental isotopes offers the promise of being able to conduct isotope dilution experiments involving human and animal subjects in which metal metabolism will be studied and the molecular (metalloprotein) level. Finally, the push toward miniaturization of CE instrumentation (CE on a chip) will find increasing application in the analysis of metalloproteins. This will be especially true in clinical/diagnostic laboratories, where sample size may be severely limited.

References

1. J. P. Landers (ed.), *Handbook of Capillary Electrophoresis*, 2nd ed., CRC Press, Boca Raton, FL, 1997.
2. T. Wehr, R. Rodriguez-Diaz, and M. Zhu, *Capillary Electrophoresis of Proteins* (J. Cazes, ed.), Chromatographic Science Series Vol. 80, Marcel Dekker, Inc., New York, 1999.
3. M. P. Richards and J. H. Beattie, *J. Capillary Electrophoresis 1*:196 (1994).

4. M. P. Richards and J. H. Beattie, *J. Chromatogr. B 669*:27 (1995).
5. M. P. Richards and T.-L. Huang, *J. Chromatogr. B 690*:43 (1997).
6. M. L. Shih and W. D. Korte, *Anal. Biochem. 238*:137 (1996).
7. T. Guszczynski and T. D. Copeland, *Anal. Biochem. 260*:212 (1998).
8. K. Haupt, F. Roy, and M. A. Vijayalakshmi, *Anal. Biochem. 234*:149 (1996).
9. K. L. Sutton and J. A. Caruso, *LC–GC 17*:36 (1999).
10. C. B. Knudsen, I. Bjornsdottir, O. Jons, and S. H. Hansen, *Anal. Biochem. 265*:167 (1998).
11. Q. Lu, S. M. Bird, and R. M. Barnes, *Anal. Chem. 67*:2949 (1995).
12. B. Michalke and P. Schramel, *J. Chromatogr. A 750*:51 (1996).

Mark P. Richards

Chelating Sorbents for Affinity Chromatography (IMAC)

Immobilized metal-ion affinity chromatography (IMAC) is a collective term that includes all kinds of affinity chromatography, where metal atoms or ions immobilized on polymer cause or dominate the interaction at the sorption site. Metal-chelate affinity chromatography was introduced as a specific method for fractionation of proteins by Porath et al. in 1975 [1]. The principle of this type of chromatographic method is that certain amino acid residues, such as histidine, cysteine, lysine, tryptophan, aspartic acid, glutamic acid, or phosphorylated amino acids, which are accessible on the protein surface, can interact through nonbonding lone-pair electron coordination with some metal ions. Metal cations Cu^{2+}, Ni^{2+}, Zn^{2+}, Co^{2+}, Fe^{3+}, Al^{3+}, and Cr^{3+}, which have been chelated to ligands immobilized on support material, have already been used for such specific interactions [2,3]. The most widely used chelating ligands for the isolation of proteins are iminodiacetic acid (IDA) and its analogs, such as tricarboxyethylenediamine (TED). IDA is covalently coupled to an insoluble matrix (e.g., agarose or Sepharose) and forms stable coordinate compounds with a variety of divalent metal ions. These chelates create bases for the above-mentioned specific adsorption. Elution of adsorbed solutes from immobilized metal-ion affinity adsorbents can be provided by changing the pH of the elution buffer or by a specific competing solute, such as histidine, imidazole, or sodium phosphate, depending on the interaction types involved.

Ligands coupled to agarose gels were commonly used at the beginning of IMAC application; however, these sorbents were not suitable for high-performance liquid chromatography (HPLC). Then, Small et al. [4] used a silica-based matrix and demonstrated that such IMA sorbents can be used in HPLC techniques. The metal-chelate adsorbent "TSK gel chelate-5PW," suitable for HPLC, which was prepared by coupling IDA to a hydrophilic resin-based matrix (TSK gel G 5000 PW), later became commercially available (Fig. 1).

Since the introduction of metal-ion affinity sorbents for the fractionation of proteins [1], the method became popular for the purification of a wide variety of biomolecules. Metal-ion affinity sorbents are also widely used for the immobilization of enzymes. At present, IMAC is a powerful method for separation of phosphorylated macromolecules, particularly proteins and peptides. The significance of techniques for separation and characterization of phosphorylated biomolecules is now increasing, because phosphorylation modulates enzyme activities and mediates cell membrane permeability, molecular transport, and secretion. Phosphorylated peptides can be separated from a peptide mixture on IDA–Sepharose with Fe^{3+} ions (Fig. 2). The majority of peptides pass freely through an IMAC column, whereas acidic peptides, including phosphorylated ones, are retained and can be released by a pH gradient.

Acidic peptides are released in the pH range 5.5–6.2 and phoshorylated peptides are eluted in the pH range 6.9–7.5 [5]. Elution of retained peptides can also be per-

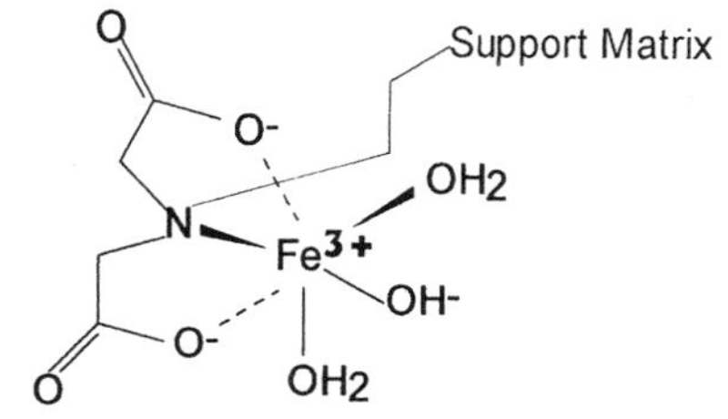

Fig. 1 Complex of water with IDA–Fe^{3+}.

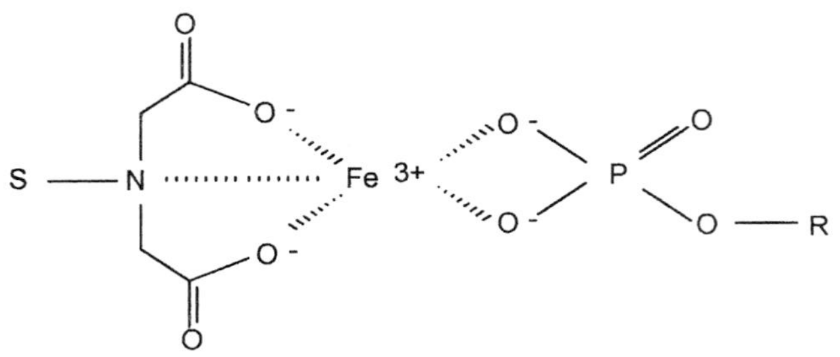

Fig. 2 Interaction of phosphate group with Fe^{3+} ion on IDA–Sepharose.

formed with sodium phosphate. IMAC has been successfully used for the characterization of casein phosphopeptides in cheese extracts [6]. Phosphoproteins can be separated under very similar conditions as phosphopeptides. IMA sorbents were already used for fractionation of proteins according to the number of phosphate groups contained in their molecules.

Separations of biomolecules on IMA sorbents achieved significant advances in the past few years, but detailed analyses of the mechanism of adsorption of molecules should be completed. Various factors, such as support matrix, chelating ligands, buffer composition, temperature, and so forth should be investigated in order to optimize analysis on metal-chelate ion affinity sorbents.

References

1. J. Porath, J. Carlsson, I. Olson, and G. Belfrage, *Nature* *258*:598 (1975).
2. E. Sulkowski, *Trends Biotechnol. 3*:1 (1985).
3. E. S. Hemdan,Y. J. Zhao, E. Sulkowski, and J. Porath, *Proc. Natl. Acad. Sci. USA 86*:1811 (1989).
4. D. A. P. Small, T. Atkinson, and C. R. Lowe, in *Affinity Chromatography and Biological Recognition* (I. M Chaiken, M. Wilchek, and I. Parikh, eds.), Academic Press, New York, 1983, p. 267.
5. G. Muszynska, G. Dobrorowolska, A. Medin, P. Ekman, and J. O. Porath, *J. Chromatogr.* 604:19 (1992).
6. R. Hynek, A. Kozak, V. Dráb, J. Sajdok, and J. Káš, *Adv. Food Sci. (CMTL) 21*(5/6):192 (1999).

Radovan Hynek
Anna Kozak
Jirí Sajdok
Jan Káš

Chiral Chromatography by Subcritical and Supercritical Fluid Chromatography

Introduction

The intrinsic physical properties of supercritical fluids—increased diffusivity and reduced viscosity—when compared to "normal" liquid phases, make subcritical/supercritical fluid chromatography a very attractive technology when short cycle times are required. Chiral subcritical/supercritical fluid chromatography typically is carried out using packed columns (pSFC) that are frequently identical in mechanical construction to the ones used in traditional high-performance liquid chromatography (HPLC). It should be noted, though, that capillary columns coated or packed with a chiral stationary phase have been used for the separation of racemic mixtures. The direct separation of racemic mixtures by chromatographic means can be effected by using chiral stationary-phase or chiral mobile-phase additives. Both techniques have been used successfully in HPLC and pSFC. The use of chiral pSFC is not limited to analytical applications. The relative ease of solvent removal and recycling, typically carbon dioxide modified with a polar organic solvent such as methanol, makes pSFC a very attractive tool for preparative separations. Equipment for laboratory- and industrial-scale pSFC in traditional dicontinuous batch–chromatography mode, as well as in continuous simulated moving-bed (SMB) mode, has been developed and is commercially available. pSFC can be used as an orthogonal method when techniques such as reversed-phase HPLC, capillary electrophoresis, or capillary electrochromatography provide insufficient or ambiguous results.

Characteristics and Advantages of Subcritical and Supercritical Fluids

The advantages of using supercritical mobile phases in chromatography were recognized in the 1950s by Klesper et al. [1], among others. Carbon dioxide is the most frequently used supercritical mobile phase, due to its moderate critical temperature and pressure, almost complete chemical inertness, safety, and cost. Virtually all published chiral pSFC separations have used carbon dioxide as the primary mobile-phase component. Compared to most commonly used organic solvents, it is environmentally friendly. The reduced viscosity of carbon dioxide-based mobile phases, typically one order of magnitude less than water (0.93 cP at 20°C), allows for efficient chromatography at higher flow rates. In addition, diffusion coefficients of compounds dissolved in supercritical mobile phases are about one order of magnitude larger than in traditional aqueous and organic mobile phases [D_M(naphthalene): 0.97×10^{-4} cm^2/s in CO_2 at 25°C, 171 bar, 0.90 g/cm^3]. This directly translates to higher efficiency of the separation due to improved mass transfer.

The first chiral separation using pSFC was published by Caude and co-workers in 1985 [3]. pSFC resembles HPLC. Selectivity in a chromatographic system stems from different interactions of the components of a mixture with the mobile phase and the stationary phase. Characteristics and choice of the stationary phase are described in the method development section. In pSFC, the composition of the mobile phase, especially for chiral separations, is almost always more important than its density for controlling retention and selectivity. Chiral separations are often carried out at $T < T_c$ using liquid-modified carbon dioxide. However, a high linear velocity and a low pressure drop typically associated with supercritical fluids are retained with near-critical liquids. Adjusting pressure and temperature can control the density of the subcritical/supercritical mobile phase. Binary or ternary mobile phases are commonly used. Modifiers, such as alcohols, and additives, such as acids and bases, extend the polarity range available to the practitioner.

A typical pSFC instrument, at first glance, is designed like an HPLC system. The major differences are encountered at the pump, the column oven, and downstream of the column. pSFC is best carried out using pumps in a flow-control mode. A regulator mounted downstream of the column and ultraviolet–visible detector (UV) controls the pressure drop in the chromatographic system. Detection is not limited to UV. If pure carbon dioxide is used as the mobile phase, an easy-to-use, sensitive, and stable universal detector such as the flame ionization detector (FID) can be employed. Other detection techniques are Fourier-transform infrared (FTIR) and evaporative light-scattering detection (ELSD), or hyphenated techniques such as pSFC–MS (mass spectrometry) and pSFC–NMR (nuclear magnetic resonance). Temperature control of the mobile phase and column is achieved by a column oven allowing for operation under cryogenic conditions and/or from ambient temperature to 150°C.

CSFC, although, resembles gas chromatography (GC) at high pressures, with the pressure (density) programming taking the place of temperature programming used in GC. Typical operating temperatures are up to 100°C.

Method Development

Mechanistic considerations (e.g., the extensive work published on brush-type phases) or the practitioner's experience might help to select a chiral stationary phase (CSP) for initial work. Scouting for the best CSP/mobile phase combination can be automated by using automated solvent and column switching. More than 100 different CSPs have been reported in the literature to date. Stationary phases for chiral pSFC have been prepared from the chiral pool by modifying small molecules, like amino acids or alkaloids, by the derivatization of polymers such as carbohydrates, or by bonding of macrocycles. Also, synthetic selectors such as the brush-type ("Pirkle") phases, helical poly(meth)acrylates, polysiloxanes and polysiloxane copolymers, and chiral selectors physically coated onto graphite surfaces have been used as stationary phases.

Generally accepted starting conditions are summarized in Table 1. Typically, 5–15% of alkanol-modified carbon dioxide is used as the mobile phase. Depending on the nature of the analyte, acids or bases can be added to the modifier for controlling ionization of the stationary phase and analyte. If partial selectivity is observed after the first injection, it is advisable to first adjust the modifier concentration. If the peak shape is not satisfactory, then the addition of 0.1% trifluoroacetic acid or acetic acid for acidic compounds or 0.1% diethyl amine or triethyl amine for basic compounds to the modifier can bring an improvement. In case the selectivity cannot be improved by the previous measures, decreasing the operating temperature can result in the desired separation. Although many chiral separations improve as the temperature is reduced, this does not occur in all cases. The temperature

Table 1 Initial Conditions for Chiral Method Development Using Modified Carbon Dioxide as the Mobile Phase

Parameter	Unit	Value
Flow rate	mL/min	2.0
Pressure	bar	200
Temperature	°C	30
Methanol	%	10
Injection volume	μL	5
Sample concentration	mg/mL	1
Detection		Diode array detector, 190–320 nm

dependence of the selectivity does not necessarily follow the van't Hoff equation (ln $\alpha \sim 1/T$), as one might expect, based on experience with other chromatographic techniques. Stringham and Blackwell [7], who have reported several examples of entropically driven separations, studied the effects of temperature in detail. In the temperature range between -10°C, 70°C (T_{iso}), and 190°C, a reversal of elution order for the enantiomers of a chlorophenylamide was observed on a (*S*,*S*)-Whelk-O 1 chiral stationary phase using 10% ethanol in carbon dioxide, at a pressure of 300 bar. The potential for reversing the elution order can be valuable if just one enantiomer of the CSP affecting the separation is available. If all of the above adjustments should fail, a different CSP should be investigated. Due to the low viscosity of carbon dioxide-based mobile phases, multiple columns can be coupled. This provides the opportunity to increase chemical selectivity for the analysis of complex samples by coupling an initial achiral column with a chiral column. Also, the successful coupling of multiple different chiral columns has been reported.

Applications

Analytical applications of chiral pSFC in chemical and pharmaceutical research, development, and manufacturing comprise the screening of combinatorial libraries, monitoring chemical and biological transformations from the laboratory to the process scale, following stereochemical preferences of drug metabolism and pharmacokinetics, and assessing toxicology and stability of drug substance and dosage form. Preparative applications are of considerable interest because of the relative ease with which the mobile phase can be removed and recycled. This is of particular interest in the pharmaceutical environment, because a small amount of the desired product can be obtained, almost free of solvent, quite rapidly. Recent advances in automation and separation technology now allow for a predictable scale-up of the separation from a laboratory to a production scale.

References

1. E. Klesper, A. H. Corwin, and D. A. Turner, *J. Org. Chem. 27*:700 (1960).
2. D. R. Gere, *Science 222*:253 (1983).
3. P. A. Mourier, E. Eliot, M. H. Caude, and R. H. Rosset, *Anal. Chem. 57*:2819 (1985).
4. F. J. Ruffing, J. A. Lux, and G. Schomburg, *Chromatographia 26*:19 (1988).
5. K. Anton, J. Eppinger, L. Fredriksen, E. Francotte, T. A. Berger, and W. H. Wilson, *J. Chromatogr. 666*:395 (1994).
6. G. J. Terfloth, W. H. Pirkle, K. G. Lynam, and E. C. Nicolas, *J. Chromatogr. 705*:185 (1995).
7. R. W. Stringham and J. A. Blackwell, *Anal. Chem. 68*:2179 (1996).
8. K. W. Phinney, L. C. Sander, and S. A. Wise, *Anal. Chem. 70*:2331 (1998).

Suggested Further Reading

Anton, K. and C. Berger, *Supercritical Fluid Chromatography with Packed Columns*, Marcel Dekker, Inc., New York, 1998.

Berger, T. A., *Packed Column SFC*, The Royal Society of Chemistry, Cambridge, 1995.

Chester, T. L., J. D. Pinkston, and D. E. Raynie, *Anal. Chem. 68*:487 (1996).

Gerald J. Terfloth

Chiral Countercurrent Chromatography

Countercurrent chromatography (CCC) can be used for the separation of a variety of enantiomers by adding a chiral selector to the liquid stationary phase [1,2]. The method is free of complications arising from the use of a solid support and also eliminates the procedure of chemically bonding the chiral selector to a solid support as in conventional chiral chromatography.

In the past, various CCC systems, such as droplet CCC, rotation locular CCC (RLCCC), and centrifugal partition chromatography (CPC), have been used for the separation of chiral compounds. None of those techniques, however, is considered satisfactory for preparative purposes in terms of sample size, resolution, and/or separation time. In the early 1980s, the high-speed CCC (HSCCC) technique improved both the partition efficiency and separation time and has been successfully applied to the separation of racemates using a Pirkle-type chiral selector. Both analytical (milligram) and preparative (gram) separations can be performed simply by adjusting the amount of chiral selector in the liquid stationary phase in the standard separation column. A large-scale separation of enantiomers can also be performed by pH-zone-refining CCC, a recently developed preparative CCC technique for the separation of ionized compounds [3]. One of the important advantages of the CCC technique over the conventional chiral chromatography is that the method allows computation of the formation constant of the chiral-selector complex, one of the most important parameters for studies of the mechanism of enantioselectivity [4].

Standard High-Speed CCC Technique in Chiral Separation

The separations are performed using a commercial high-speed CCC centrifuge equipped with a multilayer coil separation column(s). The column is first entirely filled with the stationary phase that contains the desired amount of chiral selector (CS). In order to prevent the contamination of CS in the eluted fractions, some amount of the CS-free stationary phase should be left at the end of the column, typically at about 10% of the total column capacity. After the sample solution is injected through the sample port, the mobile phase is pumped into the column while the column is rotated. Separation can be carried out by the successive injection of samples without renewing the stationary phase containing the chiral selector in the column.

Figure 1 shows the separation of four pairs of DNB–amino acid enantiomers by the standard CCC technique using a two-phase solvent system composed of hexane–ethyl acetate–methanol–10 m*M* HCl and *N*-dodecanoyl-L-3,5-dimethylanilide as a CS which is almost entirely partitioned into the organic stationary phase ($K > 100$) due to its high hydrophobicity. All analytes are well resolved in 1–3 h. The CS used in this separation is similar to the chiral stationary phase which has been introduced by Pirkle et al. for the high-performance liquid chromatography (HPLC) separation of racemic DNB–amino acid *t*-butylamide. A hydrophobic *N*-dodecanoyl group is connected to the CS molecule for retaining the CS in the organic stationary phase.

The effect of CS concentration in the stationary phase was investigated [1]. As the CS concentration is increased, the separation factor and peak resolution are also increased [5]. The result clearly indicates an important technical strategy: The best peak resolution is attained by saturating the CS in the stationary phase in a given column, where the resolution is further improved by using a longer and/or wider-bore coiled column, which can hold greater amounts of CS in the stationary phase.

The preparative capability of the present system is demonstrated in the separation of DNB–leucine enantiomers by varying the CS concentration in the stationary phase. The sample loading capacity is found to be determined mainly by the CS concentration or total amount of CS in the stationary phase; that is, the higher the CS concentration in the stationary phase, the greater the peak resolution and sample loading capacity. Consequently, the standard HSCCC column can be used for both analytical and preparative separations simply by adjusting the amount of CS in the stationary phase.

pH-Zone-Refining CCC for Chiral Separation

pH-Zone-refining CCC is a powerful preparative technique that yields a succession of highly concentrated rect-

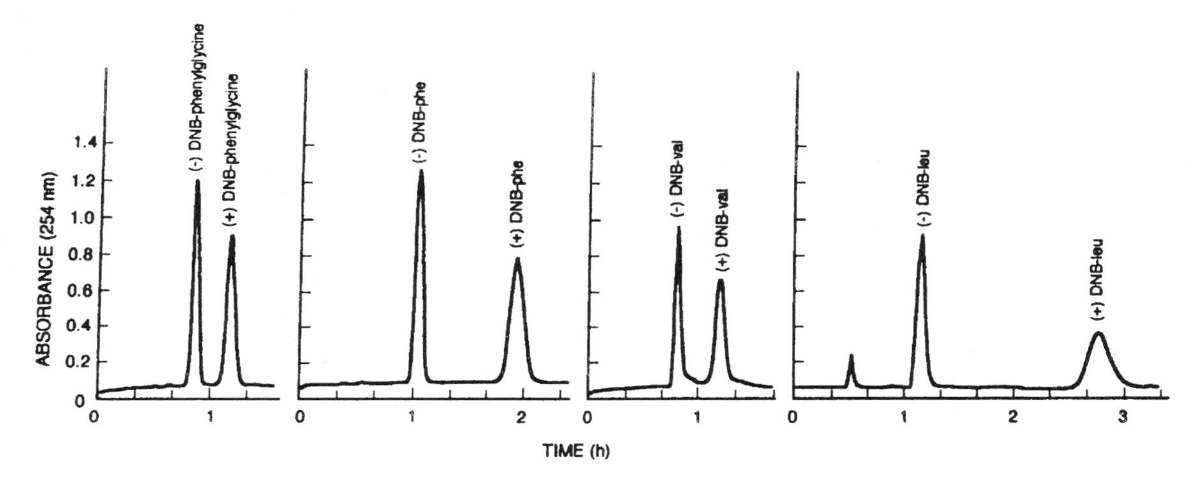

Fig. 1 Separation of four pairs of (±)-DNB-amino acids by the standard analytical HSCCC technique with a CS (*N*-dodecanoyl-L-proline-3,5-dimethylanilide) in the stationary phase.

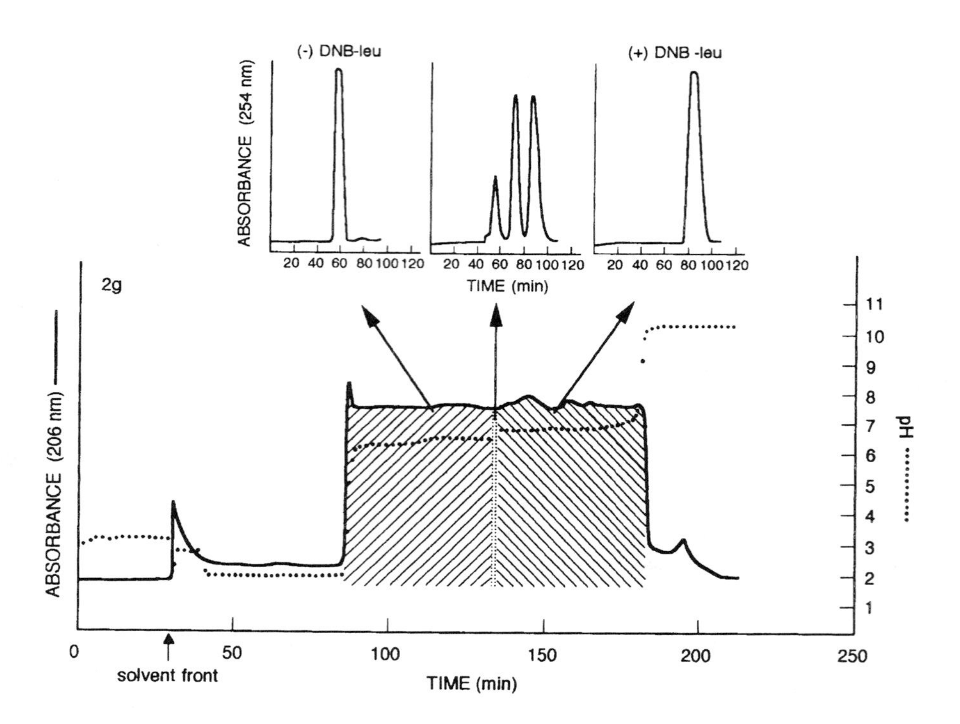

Fig. 2 pH-Zone-refining CCC separation of 2 g (±) DNB–leucine using the same HSCCC centrifuge with a CS (*N*-dodecanoyl-L-proline-3,5-dimethylanilide) in the stationary phase.

angular solute peaks with minimum overlap where impurities are concentrated at the peak boundaries (see the entry pH Peak Focusing pH-Zone-Refining CCC). This technique was applied to the resolution of DNB–amino acid racemates using a binary two-phase solvent system composed of methyl *t*-butyl ether–water where trifluoroacetic acid (retainer) and CS were added to the organic stationary phase and ammonia (eluter) to the aqueous mobile phase. Figure 2 shows a typical chromatogram obtained by pH-zone-refining CCC . The pH of the fraction (dotted line) revealed that the peak was evenly divided into two pH zones, each corresponding to a pure enantiomeric species with a sharp transition. Compared with the standard CCC technique, the pH-zone-refining CCC technique allows separation of large amounts in a shorter elution time.

In both techniques, leakage of the chiral selector into the elute can be completely eliminated by filling the outlet of the column with a proper amount of the CS-free stationary phase so as to absorb the chiral selector leaking into the flowing mobile phase.

Advantages

Countercurrent chromatography can be applied to the separation of enantiomers by adding a suitable chiral selector to the liquid stationary phase by analogy to binding the CS to the solid support in conventional chiral chromatography. The HSCCC technique has the following advantages over the conventional chromatography technique:

1. The method permits repetitive use of the same column for a variety of chiral separations by choosing appropriate CSs.
2. Both analytical and preparative separations can be performed with a standard CCC column by adjusting the amount of CS in the liquid stationary phase, and the method is cost-effective, especially for large-scale preparative separations.
3. The separation factor and peak resolution can be improved simply by increasing the concentration of CS in the stationary phase.
4. The method is very useful for the investigation of the enantioselectivity of CS including determination of the formation constant and separation factor.
5. pH-Zone-refining CCC can be applied for gram-quantity separation in a short elution time.

References

1. Y. Ma and Y. Ito, Chiral separation by high-speed countercurrent chromatography, *Anal. Chem. 67*:3069 (1995).
2. Y. Ma, Y. Ito, and A. Foucault, Resolution of gram quanti-

ties of racemates by high-speed CCC, *J. Chromatogr. A 704*: 75 (1995).
3. Y. Ito and Y. Ma, pH-Zone-refining countercurrent chromatography, *J. Chromatogr. A 753*:1 (1996).
4. Y. Ma, Y. Ito, and A. Berthod, A chromatographic method for measuring K_f of enantiomer–chiral selector complexes, *J. Liquid Chromatogr. 22*(19):2945 (1999).
5. Y. Ma and Y. Ito, Affinity CCC using a ligand in the stationary phase, *Anal. Chem. 68*:1207 (1996).

Ying Ma
Yoichiro Ito

Chiral Separations by Capillary Electrophoresis and Micellar Electrokinetic Chromatography with Cyclodextrins

Introduction

In the area of pharmaceutical research, drug enantiomers are now recognized as potentially different drug entities which may have synergistic, similar, or antagonistic pharmacological properties to those of the desired effect. The pharmaceutical industry and the drug regulatory bodies, therefore, require high resolution, robust, fully validated, and highly selective analytical methodology that can discriminate drug enantiomeric substances for every stage of research and development, from drug discovery to clinical testing [1]. The aim of this brief review is to introduce the use of capillary electrophoresis (CE) and micellar electrokinetic chromatography (MEKC) as chiral separation techniques that employ cyclodextrin (CD) molecules as enantiomeric discriminating agents.

The use of CDs for chiral separations has, to date, been the most common approach when using CE or MEKC, so it would be difficult to discuss and detail every aspect relating to their chemistry, effects on separation, and application in this field. The emphasis will, thus, be placed on a short description of the principle and mechanism of chiral separation, typical method development procedures, and an outline of the influential experimental parameters using CE and MEKC. References to recent published review and research literature will enable the reader to explore this vast area further. It is also beyond the scope of this short introductory review to actually outline the actual CE or MEKC separation principles in detail, but an in-depth discussion can be found in this encyclopedia and references to recent textbooks and can be readily found elsewhere. It must, of course, be pointed out that CDs are not the only useful chiral selectors that can be employed using electrophoretic techniques. The use of chiral surfactants (bile salts), crown ethers, metal-chelation agents, carbohydrates, proteins, and glycopeptides have all been used effectively [2].

Capillary electrophoresis and micellar electrokinetic chromatography are fast, offer high resolution, and are cost-effective separation techniques that have been applied extensively for the discrimination of enantiomers using CDs, resulting in a number of review articles [2–6] and, recently, a book [7]. The separation principle in CE is based on differential migration of ionic solutes according to the mass/charge ratio through an electrolytic solution which is also moving by electro-osmosis under an applied electric field. MEKC is an extension of this technique originally developed for the separation of achiral neutral analytes by the incorporation of hydrophobic micelles to the background electrolyte. This allowed separation according to hydrophobic interactions, in addition to discrimination through differential mobility according to the mass/charge ratio. Fanali [8] and Ueda et al. [9] were among the earlier researchers to report the use of CDs for chiral discrimination by CE and MEKC, respectively. Since that earlier work, a vast number of racemic mixtures have been separated by many groups [2–7], whereas other groups have carried out useful theoretical and practical studies on the chiral mechanism taking place and parameters affecting enantioselectivity in CE [10,11].

Theory

Cyclodextrins are a family of three well-known, nonreducing cyclic oligosaccharides consisting of D-glucopyranose units bonded through α-1,4-linkages. The smallest is α-CD, followed by β-CD and γ-CD, which have six, seven, and eight D-glucopyranose units, respectively. In CDs, the D-glucopyranose units adopt a $^4C_1$ chair confirmation and orient themselves so that the overall molecule forms a toroid/hollow truncated cone structure, much like a doughnut. As a consequence of the CD conformation, all the primary hydroxyl groups are situated at the smaller edge of the truncated cone, whereas the chiral secondary hydroxyl groups are situated at the larger edge of the cone. The primary and secondary hydroxyl groups on the outside of the CD toroid give them a hydrophilic exterior, whereas the electron-rich glucosidic oxygen bridges inside the toroidal cavity result in a hydrophobic interior. These are two aspects that make them attractive chiral-discriminating molecules for CE and MEKC, because they are soluble in aqueous media while being capable of forming strong complexes with analyte molecules typically via hydrophobic insertion of an aromatic ring into the cavity.

Other favorable aspects include ultraviolet (UV) transparency, which allows their incorporation into the electrolyte, and the fact that they are relatively inexpensive. A more complete description of the chemical and physical properties of CDs, together with a schematic of their structure and shape, can be found in Fanali's review [5].

The β-CD molecule contains 35 asymmetric centers and guest racemic compounds can interact with this chiral environment. The interaction is normally governed by the size of the racemic guest molecule and, thus, its ability to form an inclusion complex. The possibility for additional interaction with the chiral secondary hydroxyl groups (at the large opening of the cone) is normally crucial and is considered to occur through hydrogen-bonding. The CD hydroxyl groups, 21 on the β-CD molecule, can also be modified chemically, thereby altering those CDs physical and chemical properties and, thus, the potential nature of interaction with a guest molecule. Derivatives used to date include alkyl-, hydroxyalkyl-, glucosyl-, maltosyl-, methyl-, hydroxyethyl-, hydroxypropyl-, acetyl-, and a range of chargeable functionalities. The aim of this derivatization may be (a) to improve the solubility of the CD, (b) to improve the fit and/or the degree or strength of association between a CD and its guest, or (c) to give the CD and/or the CD–guest complex enhanced or decreased mobility through greater size or by a charged functionality.

The enantioselective mechanism of CDs in CE and MEKC is similar to that in liquid chromatography (LC). Enantioselective interaction is thought to occur through the inclusion complex formation between a hydrophobic group of the analyte and the relatively hydrophobic interior of the CD cavity. As a result of the different three-dimensional spatial arrangement of each enantiomer, they may have a different binding constant with a particular CD. Just as in LC, this is a prerequisite to actually obtaining a chiral separation in CE or MEKC. When a charged enantiomer forms a complex with a CD, its charge/mass ratio and, thus, its mobility decrease. The free or uncomplexed charged enantiomer migrates as it would in the absence of a CD. Therefore, differences in the binding constants determine the ratio of free/complexed enantiomers, and if the binding constants are sufficiently different for the two enantiomers, a chiral separation will occur. Theoretical aspects, including the relevant equations pertaining to the resolution of enantiomers and practical considerations for CE and MEKC when using CDs, can be found in Refs 2–10 and in the references therein.

The application of MEKC for chiral separation is primarily used when the enantiomers of interest are neutral. In conventional CE without micelles, neutral enantiomers will be swept along with the electro-osmotic flow (EOF) because they carry no ionic charge. If a neutral CD is present and forms a complex with these CDs, they will still move with the EOF. Thus, it is necessary for neutral enantiomers to create the potential for differential migration so that the overall complexes and free enantiomers will not just be swept along with the EOF. For this reason, the use of MEKC which utilizes ionic micelles for differential migration (through hydrophobic interaction) modified with CDs for enantioselectivity was applied [9]. The mechanism is as outlined for MEKC; however, because there are hydrophobic micelles inherently present in the electrolyte, additional interactions between the enantiomers and the micelle over those with just a CD may, of course, occur which will normally influence any observed separation.

Another approach may also be adopted for separating neutral enantiomers; the use of CDs that carry an overall anionic or cationic charge, sulfobutylether-, carboxymethyl-, sulfated-, phosphated-, or methylamino- are a few examples. If complexation occurs between neutral enantiomers and an ionic CD, the enantiomer–CD complex will carry an effective charge and will thus migrate to a greater or lesser extent than the EOF, thus creating the opportunity for a chiral separation. In conventional CE, where the EOF is in the cathodic direction, a CD carrying an overall cationic charge will result in the neutral drug enantiomers migrating before the EOF if complexation takes place; conversely, a CD carrying an anionic charge will result in the neutral drug enantiomers migrating after the EOF. Thus, it is not necessary to use a micellar system for the separation of neutral drug enantiomers when utilizing a CD that carries an overall cationic or anionic charge. The results, from theoretical aspects and practical applications of this type of chiral separation, have been reviewed [12].

Obtaining Enantioselectivity in CE and MEKC

A flow diagram outlining possible starting points for obtaining a chiral separation for basic, acidic, and neutral

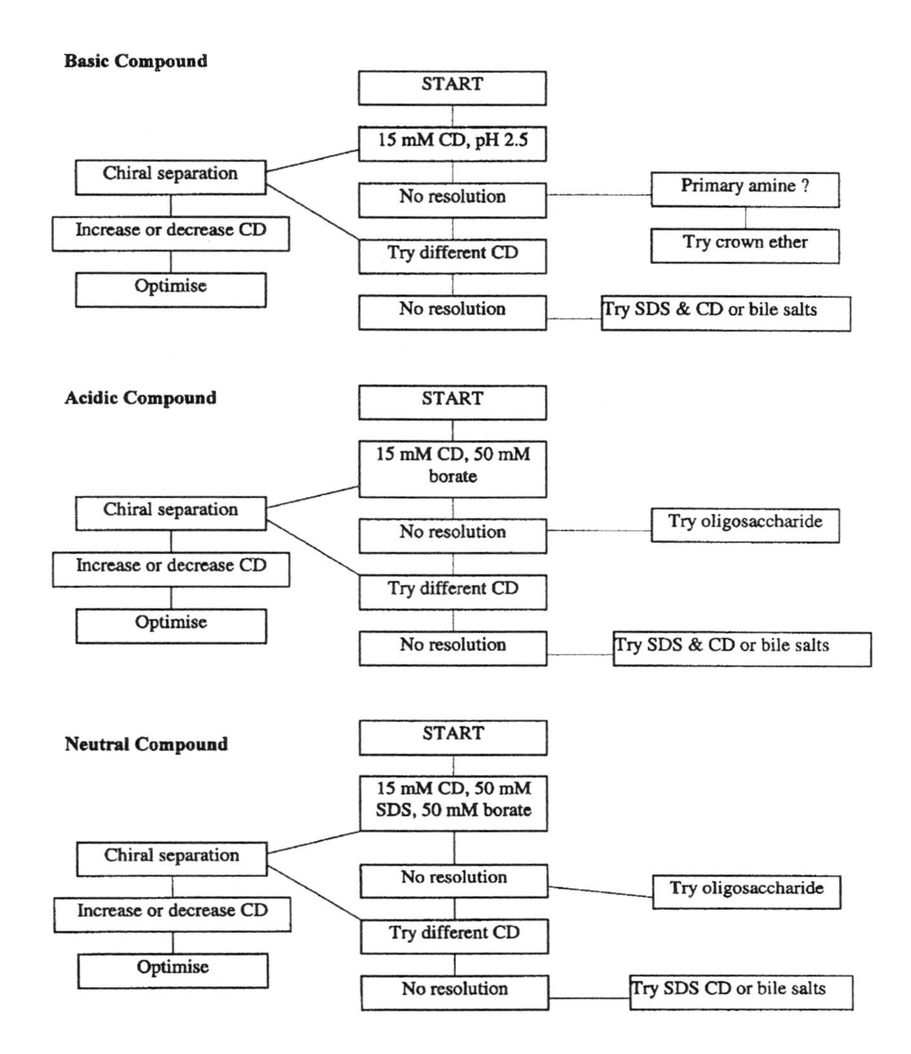

Fig. 1 Flow diagram indicating starting point and optimization procedure for enantiomeric separation of basic, acidic, and neutral compounds. [Adopted with kind permission from M. M. Rogan and K. D. Altria, Introduction to the theory and applications of chiral capillary electrophoresis, Beckman primer Vol. IV, p. 22, Part No. 726388.]

compounds is shown in Fig. 1. A suitable starting point may be a low- and high-pH buffering system for basic and acidic compounds, respectively, with typically 15 m*M* of the medium-sized β-CD present in the electrolyte. It is necessary, as outlined earlier, to use a micellar system for the separation of neutral compounds and the most common surfactant is sodium dodecyl sulfate. In CE or MEKC, it is the background electrolyte that is necessary and responsible for any observed enantioselectivity. As a consequence, the simplest procedure to initially obtain and/or improve any chiral resolution is to modify or control the parameters of that electrolyte. If no resolution is obtained from the initial result or it needs to be improved, many parameters can be modified, including the CD type, CD concentration, operating pH, buffer concentration, temperature, and the use of electrolyte additives like organic solvents or ion-pair reagents. The experimental variables for a chiral separation, including those that are not electrolyte based, and their typical influence are shown in Table 1. The most influential parameters are the nature and concentration of the CD. This is not surprising because these parameters will strongly influence the chiral recognition mechanism, which is based on complexation and interaction. This has been studied, in detail, through theoretical and practical studies by Wren and Rowe [10]. The chiral separations of the basic drug tocainide and re-

Table 1 Table of the Effects of Operating Parameters

Variable	Range	Effect of increasing variable
CD type and size		Large impact on chiral selectivity
CD concentration	1–100 m*M*	Increased viscosity, reduced EOF, increased solute migration if complexation occurs
Voltage	5–30 kV	Reduced analysis time; some loss in resolution
Current	5–250 μA	Reduced analysis time; some loss in resolution
Capillary length	20–100 cm	Increased analysis time; gain in resolution
Capillary bore	25–100 μm	Increased current; some loss in resolution
pH	1.5–11.5	Increased EOF; increased ionization of acids, reduced ionization of bases
Organic solvents	1–30% v/v	Gain or loss of resolution
Urea	1–7*M*	Increased ionization of hydrophobic solutes (and CDs)
Ion-pair reagent	1–20 m*M*	Can reduce or increase resolution
Amine modifiers	1–50 m*M*	Reduced surface charge, reduced peak tailing
Viscosity	Various	Reduced EOF, longer migration times
Electrolyte concentration	5–200 m*M*	Increased resolution, increased current, lower EOF, solute ionization, reduced tailing
Cationic surfactant	1–20 m*M*	Reversal of EOF, increased solubilization, longer migration times
Injection time	1–20 s	Improved signal, some loss in resolution
Cellulose or polymer derivatives	0.1–0.5%	Can improve resolution
Bile salt type	10–50 m*M*	Choice has a large impact on chiral selectivity

Source: Adopted with kind permission from M. M. Rogan and K. D. Altria, Introduction to the theory and applications of chiral capillary electrophoresis, Beckman primer Vol. IV, p. 24 (Part No. 726388).

lated analogs using reasonably typical conditions is shown in Fig. 2, where the choice of CD is shown to strongly influence the enantioselectivity.

In addition to CD-modified MEKC, the use of mixed micelles incorporating CDs, CDs together with other chiral selectors (bile salts, ligand exchangers, crown ethers, etc. [5]), or mixed CD systems, neutral with neutral or neutral with charged CDs can also be extremely useful. A very attractive feature of the latter approach is the ability to control the enantiomeric migration order, an aspect that may be extremely useful if quantitation at low levels is required. Details of this technique are available in the review of Chankvetadze et al. [12].

Conclusions

It is hoped that the aim of this very short review article, to entice the reader into the world of chiral CE and MEKC using CDs, has been fulfilled. A brief description of the importance, historical background, method development, and theoretical and practical aspects concerning this approach is given. Additionally, possible alternative chiral selectors that can be applied in order to achieve enantioselectivity in CE and MEKC are also mentioned.

References

1. W. H. De Camp, The FDA perspective on the development of stereoisomers, *Chirality 1*:2 (1989).
2. M. M. Rogan, K. D. Altria, and D. M. Goodall, Enantioselective separations using capillary electrophoresis, *Chirality 6*:25 (1994).
3. F. Bressolle, M. Audran, T.-N. Pham, and J-J. Vallon, Cyclodextrins and enantiomeric separations of drugs by liquid chromatography and capillary electrophoresis: Basic principles and new developments, *J. Chromatogr. B 687*:303 (1996).
4. H. Nishi and S. Terabe, Micellar electrokinetic chromatography perspectives in drug analysis, *J. Chromatogr. A 735* (1–2):3 (1996).
5. S. Fanali, Controlling enantioselectivity in chiral capillary electrophoresis with inclusion complexation, *J. Chromatogr. A 792*:227 (1997).
6. H. Nishi, Enantioselectivity in chiral capillary electrophoresis with polysaccharides, *J. Chromatogr. A 792*(1–2):327 (1997).
7. B. Chankvetadze, *Capillary Electrophoresis in Chiral Analysis*, John Wiley & Sons, New York, 1997.
8. S. Fanali, Separation of optical isomers by capillary zone electrophoresis based on host–guest complexation with cyclodextrins, *J. Chromatogr. 474*:441 (1989).
9. T. Ueda, F. Kitamura, R. Mitchell, T. Metcalf, T. Kuwana,

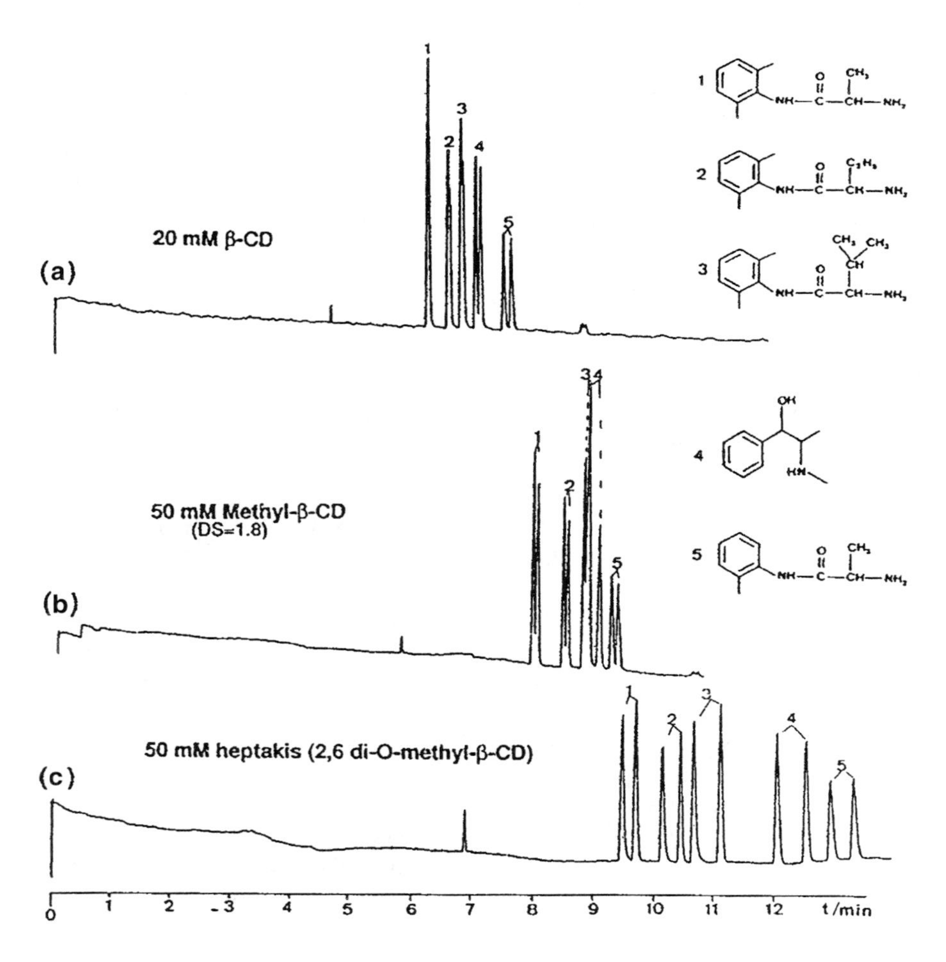

Fig. 2 Influence of cyclodextrin type on the chiral resolution of the basic drug tocainide and related substances using 40 m*M* sodium phosphate (pH 3.0) containing (a) 20 m*M* β-CD and (b) 50 m*M* methyl-β-CD and 50 m*M* *heptakis* (2,6 di-*O*-methyl-β-CD. [Adopted with kind permission from D. Belder and G. Schomburg, Chiral separations of basic and acidic compounds in modified capillaries using cyclodextrin-modified capillary zone electrophoresis, *J. Chromatogr. A 666*:351 (1994).]

and A. Nakamoto, Chiral separation of naphthalene-2,3-dicarboxaldehyde-labeled amino-acid enantiomers by cyclodextrin-modified micellar electrokinetic chromatography with laser-induced fluorescence detection, *Anal. Chem. 63*: 2979 (1991).

10. S. A. C. Wren and R. C. Rowe, Theoretical aspects of chiral separation in capillary electrophoresis. 3. Application to beta-blockers, *J. Chromatogr. 635*:113 (1993).
11. S. M., Branch, U., Holzgrabe, T. M., Jefferies, H. Mallwitz, and M. W., Matchett, Chiral discrimination of phenethylamines with β-cyclodextrin and *heptakis* (2,3-di-*O*-acetyl-β-cyclodextrin by capillary electrophoresis and NMR spectroscopy, *J. Pharm. Biomed. Anal.* 12:1507 (1994).
12. B. Chankvetadze, G. Endresz, and G. Blaschke, Charged cyclodextrin derivatives as chiral selectors in capillary electrophoresis. *Chem. Soc. Rev. 25*:141 (1996).

Paul K. Owens

Chiral Separations by GC

In gas chromatography (GC), chiral selectivity is controlled solely by the choice of the stationary phase and the operating temperature. Thermodynamically, it is achieved by introducing an additional *entropic* component to the standard free energy of distribution. This is accomplished by employing a chiral stationary phase which will have unique spatially oriented groups or atoms that allow one enantiomer to interact more closely with the molecules of the stationary phase than the other. The enantiomer that can approach more closely to the stationary phase molecules will interact more strongly (the dispersive or polar charges being nearer) and, thus, the standard *enthalpy* of distribution of the two enantiomers will also differ. Consequently, the Van't Hoff curves will have different slopes and intersect at a particular temperature (see the entries Thermodynamics of Retention in GC and Van't Hoff Curves). At this temperature, the two enantiomers will co-elute and, hence, temperature is an important variable that must be used to control chiral selectivity. The farther the operating temperature of the column is away from the temperature of co-elution, the greater the separation ratio and the easier will be the separation (less theoretical plates, shorter column, faster analysis).

The first effective chiral stationary phases for GC were the derivatized amino acids [1], which, however, had very limited temperature stability. The first reliable GC stationary phase was introduced by Bayer and co-workers [2], who synthesized a thermally stable, low-volatility polymer by attaching L-valine-*t*-butylamide to the carboxyl group of dimethylsiloxane or (2-carboxypropyl)-methylsiloxane with an amide linkage. This stationary phase was eventually made available commercially as Chirasil-Val and could be used over the temperature range of 30°C to 230°C. OV-225 (a well-established polar GC stationary phase) has also been used for the synthesis of chiral polysiloxanes, which, in this case, possess more polar characteristics than the (2-carboxypropyl)-methylsiloxane derivatives.

Although the polysiloxane phases carrying chiral peptides are still used in contemporary chiral GC, the presently popular phases are based on cyclodextrins. These materials are formed by the partial degradation of starch followed by the enzymatic coupling of the glucose units into crystalline, homogeneous, toroidal structures of different molecular sizes. The best known are the α-, β-, and γ-cyclodextrins which contain six (cyclohexamylose), seven (cycloheptamylose), and eight (cyclooctamylose) glucose units, respectively. The cyclodextrins are torus shaped macromolecules which incorporate the D(+)-glucose residues joined by α-(1-4)glycosidic linkages. The opening at the top of the torus-shaped cyclodextrin molecule has a larger circumference than that at the base. The primary hydroxyl groups are situated at the base of the torus, attached to the C_6 atoms. As they are free to rotate, they partly hinder the entrance to the base opening. The cavity size becomes larger as the number of glucose units increases. The secondary hydroxyl groups can also be derivatized to insert different interactive groups into the stationary phase. Due to the many chiral centers the cyclodextrins contain (e.g., β-cyclodextrin has 35 stereogenic centers), they exhibit high chiral selectivity and, as a consequence, are probably the most effective GC chiral stationary phases presently available.

The α-, β-, or γ-cyclodextrins that have been permethylated do not coat well onto the walls of quartz capillaries and must be dissolved in appropriate polysiloxane mixtures for stable films to be produced. In contrast, underivatized cyclodextrins can be coated directly onto the walls of the column with the usual techniques. The thermal stability of a mixed stationary phase can be improved by including some phenylpolysiloxane in the coating material. Phenylpolysiloxane also significantly inhibits any oxidation that might take place at elevated temperatures. However, unless some methylsiloxane is present the cyclodextrin may not be sufficiently soluble in the polymer matrix for successful coating.

The inherent chiral activity of the cyclodextrins can be strengthened by bonding other chirally active groups to the secondary hydroxyl groups of the cyclodextrin. Certain derivatized cyclodextrins are susceptible to degradation, on contact with water or water vapor. Consequently, all carrier gases must be completely dry and all samples that are placed on the column must also be dry.

Derivatized cyclodextrins can interact with chiral substances in a number of different ways. If, the positions 2 and 6 are alkylated (pentylated), very dispersive (hydrophobic) centers are introduced that can strongly interact with any alkyl chains contained by the solutes. After pentylation of the 2 and 6 positions has been accomplished, the 3-position hydroxyl group can then be trifluoroacetylated. This stationary phase is widely used and it has been found that the derivatized γ-cyclodextrin is more chirally selective than the β material. It has been successfully used for the separation of both very small and very large chiral molecules. The cyclodextrin hydroxyl groups can also be made to react with pure "S" hydroxypropyl groups and then permethylated. As a result, the size selectivity of the stationary phase is reduced, but its interactive character is made more polar (hydrophilic). In general, the α or γ phases have less chiral selectivity than the β material. There are a considerable number of cyclodextrin-based chiral stationary phases commercially available and, without doubt, there will be many more introduced in the future.

References

1. D. Gil-Av, B. Feibush, and R. Charles-Sigler, *Tetrahedron Lett.* 1009 (1988).
2. H. Frank, G. J. Nicholson, and E. Bayer, *J. Chromatogr. Sci. 15*:174 (1974).

Suggested Further Reading

Beesley, T. E. and R. P. W. Scott, *Chiral Chromatography*, John Wiley & Sons, Chichester, 1998.
Scott, R. P. W., *Introduction to Gas Chromatography*, Marcel Dekker, Inc., New York, 1998.
Scott, R. P. W., *Techniques of Chromatography*, Marcel Dekker, Inc., New York, 1995.

Raymond P. W. Scott

Chiral Separations by HPLC

Chirality arises in many molecules from the presence of a tetrahedral carbon with four different substituents. However, the presence of such atoms in a molecule is not a necessary condition for chirality. An object is said to be chiral if it is not superposable with its mirror image and achiral when the object and its mirror image are superposable. A chiral pair can be distinguished through their interaction with other chiral molecules to form either long-lived or transient diastereomers. Diastereomers are molecules containing two or more stereogenic (chiral) centers and having the same chemical composition and bond connectivity. They differ in stereochemistry about one or more of the chiral centers.

Long-Lived Diastereomers

Long-lived diastereomers are generated by chemical derivatization of the enantiomers with a chiral reagent. They may be separated subsequently by achiral means. Their formation energies have no relevance to their chromatographic separation; it is, rather, due to the difference in their solvation energies. Differences in their shape, size, or polarity will affect the energy needed to displace solvent molecules from the stationary phase [1].

There are several characteristics of diastereomeric chiral separations (also known as indirect enantiomeric separations) that are worth mentioning. Achiral phases that are cheaper, more rugged, and widely commercially available are used. The elution order can be controlled by choice of the chirality of the derivatizing agent. This feature is useful for the analysis of trace levels of enantiomers. The separation can be designed such that the minor enantiomer is eluted first, allowing for more accurate quantitation.

Derivatization requires that the species of interest must contain a functional group that can be chemically modified. There should be no enantioselectivity of the rate of the derivatization [2]. There are several disadvantages to an indirect chromatographic chiral separation. The derivatization procedure may be complex and time-consuming and there is always a possibility of racemization during the derivatization procedure. In the case of preparative chromatography of the diastereomeric species, they have to be chemically reversed to the initial enantiomers. Figure 1 shows the main types of derivatives formed from amines, carboxylic acids, and alcohols in reaction with chiral reagents [3].

There are several structural considerations to achieving a diastereomeric separation. The diastereomers should possess a degree of conformational rigidity in order to maximize their physical differences. Large size differences between the groups attached to the chiral center enhance the separation in most cases. The distance between the asymmetric centers should be minimal and ideally less than three bonds. The presence of polar or polarizable groups can enhance hydrogen-bonding, interactions with the stationary phase, resulting in increased resolution.

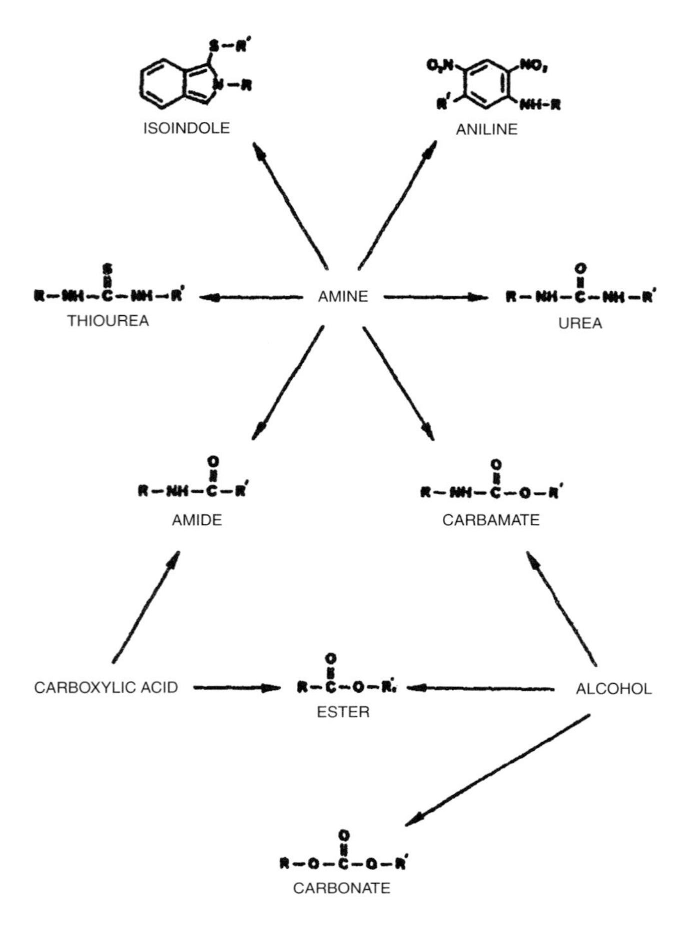

Fig. 1 Main types of derivatives formed from amines, carboxylic acids, and alcohols in reactions with chiral derivatizing reagents. (From Ref. 3.)

Transient Diastereomers

Objects that can distinguish between enantiomers are chiral receptors. Nature gives us plenty of examples of chiral receptors, such as enzymes and nucleic acids. There are also man-made chiral receptors such as chiral phases (CP) used in gas chromatography, high-performance liquid chromatography (HPLC), supercritical fluid chromatography (SFC), and capillary electrophoresis (CE). The operation of a CP involves the formation of transient diastereomeric complexes between the enantiomer (selectand) and the CP (selector). They must be energetically nondegenerate in order to effect a separation. Because of their transient nature, it is usually not possible to isolate them.

There are specific criteria for the interaction between the selectand and the selector which leads to separation on a particular column [4]:

1. Strong interactions, such as $\pi-\pi$ interactions, coordinative bonds, and hydrogen bonds between the selector and selectand
2. Close proximity of the transient bonds to the respective asymmetric carbons
3. Inhibition of free rotation of the transient bonds
4. Minimal noncontributing associative forms that do not bring the respective asymmetric centers to proximity

The diastereomeric associate between selectand and selector is formed through bonds between one or more substituents of the asymmetric carbon. These bonds are the leading selectand–selector interactions. Only when the leading bonds are formed and the asymmetric moieties of the two molecules are brought to close proximity do the secondary interactions (e.g., van der Waals, steric hindrance, dipole–dipole) become effectively involved (Fig. 2). The secondary interactions can affect the conformation and the formation energy of the diastereomeric

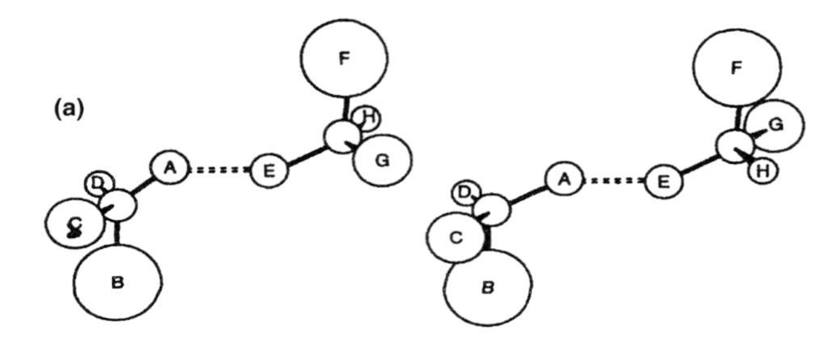

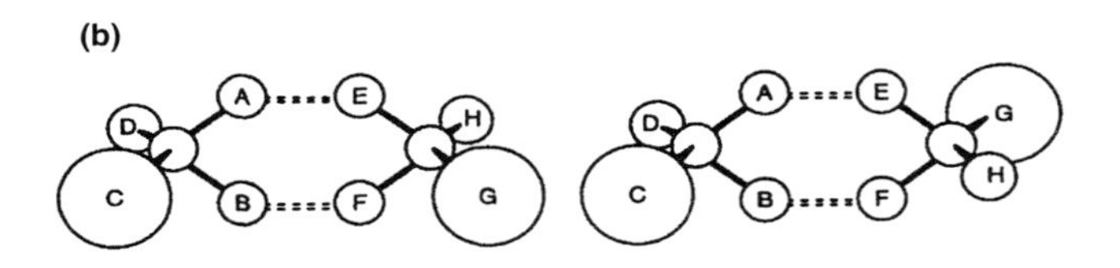

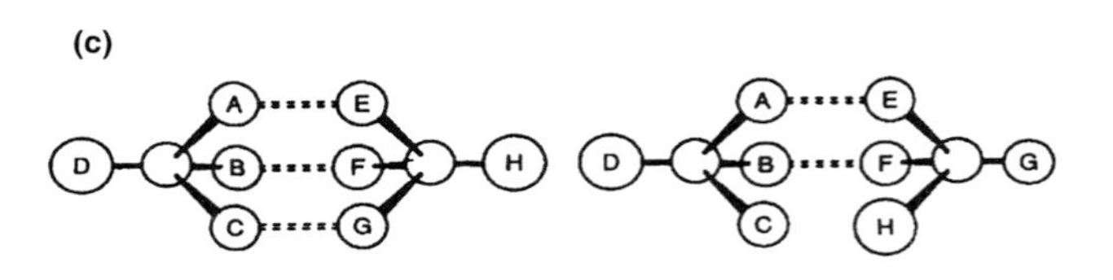

Fig. 2 Schematic representation of selectand–selector association. A dotted line represents a leading interaction between the two molecules. (a) The selectand forms a bond that involves only one substituent of its asymmetric carbon; (b) the selectand binds through two of its substituents; (c) the selectand binds through three substituents. (From Ref. 1.)

associates. In Fig. 2a, the size, shape and polarity of the unbounded B, C, and D substituents of the selectand and their positions to the groups F, G, and H of the selector will determine the enantioselectivity of the system. One particular enantiomer of the selectands will interact more strongly with a particular selector. When the selective associate is formed through interactions A–E and B–F (Fig. 2b), enantioselectivity and elution order are determined by the effective size of unbounded groups C and D their relative positions, syn or anti, to groups G and H of the selector. In most of the cases that include hydrogen-bonding or ligand–metal complexes, the enantiomer with the larger nonbonded groups positioned syn to the selector's larger nonbonded group will elute last from the chiral column. When the selective association is formed through three leading interactions (Fig. 2c), the enantioselectivity is determined by the stereochemistry of the two enantiomers. One enantiomer in one configuration will establish three leading bonds (H bonds or a combination of H bonds and $\pi-\pi$ interactions), whereas the other one will not [1].

In chromatographic systems, the selectors are either added to the mobile phase [chiral mobile phases (CMP)] or are bonded to a stationary phase (e.g., silica gel) as chiral stationary phases (CSP).

Chiral Mobile Phases

In this mode of separation, active compounds that form ion pairs, metal complexes, inclusion complexes, or affinity complexes are added to the mobile phase to induce enantioselectivity to an achiral column. The addition of an active compound into the mobile phase contributes to a specific secondary chemical equilibrium with the target analyte. This affects the overall distribution of the analyte between the stationary and the mobile phases, affecting its retention and separation at the same time. The chiral mobile phase approach utilizes achiral stationary phases for the separation. Table 1 lists several common chiral additives and applications.

Chiral Stationary Phases

Compared to CMP, the mechanism of separation on a chiral stationary phase is easier to predict, due to a much simpler system. Because the ligand is immobilized to a matrix and is not constantly pumped through the system, the detection limits for the enantiomers are much lower. Depending on the ligand immobilized to the matrix, one can have different types of interactions between the selectand and selector: metal complexes, hydrogen-bonding, inclusion, $\pi-\pi$ interactions, and dipole interactions, as well as a combination thereof.

Chiral Separation Where the Leading Interaction Is Established Through Metal Complexes (Ligand Exchange)

Chiral separation using ligand-exchange chromatography involves the reversible complexation of metal ions and chiral complexing agents. The central ion, usually Cu^{2+} or Ni^{2+} forms a bis complex with bidentates ligands. If one of the chelating ligands is anchored to a support, the CSP can form diastereomeric adsorbates with the bidentate selectand. The metal ion is held by the stationary phase through coordination to the bound ligand. If the coordination sphere of the metal is unsaturated or is occupied by weakly bound solvent molecules, it can reversibly attach different solute ligands from the mobile phase. The solute ligands are then resolved according to differences in their binding constants. Ligand exchange is possible only in systems where the interaction of the mobile ligand with the stationary phase is reversible. The coordination bonds must be kinetically labile. If the chelating ligands are amino acids and the metal is copper(II), the amine and carboxylate groups of the ligands are arranged around

Table 1 Main Classes of Chiral Additives and Their Applications

Mechanism	Additive	Application	Mode of separation	Ref.
Ion pair	(+)-10-camphorsulfonic acid	Aminoalcohols, alkaloids	HPLC	5, 6
Ion pair	Quinines	Carboxylic acids	HPLC	7, 8
Inclusion	Dimethyl β-cyclodextrin	Aminoalcohols, carboxylic acids	CE	9, 10
Inclusion	Crown ether	Primary amines	CE	11
Ligand exchange	L-Proline/Cu^{2+}	Amino acids	HPLC	12
Proteins	α_1-Acid glycoprotein	Hexobarbitone	CE	13
Antibiotics	Rifamycin	Amino acids	CE	14

the metal ion in a trans configuration, forming a square planar complex. A third interaction should take place to ensure enantioselectivity. The third interaction may arise through steric hindrance or attractive or repulsive interactions between the selector and the selectand [15,16].

Chiral Separation Where the Leading Interaction Is Established Through Hydrogen-Bonding

A hydrogen bond is formed by the interaction between the partners R–X–H and :Y–R′ according to

$$\text{R–X–H} + \text{:Y–R}' \rightarrow \text{R–X–H}\cdots\text{Y–R}'$$

R–X–H is the proton donor and :Y–R makes an electron pair available for the bridging bond. X and Y are atoms of higher electronegativity than hydrogen (e.g., C, N, P, O, S, F, Cl, Br, I). Hydrogen-bonding acceptors are the oxygen atoms in alcohols, ethers, and carbonyl compounds, as well as nitrogen atoms in amines and N-heterocycles. Hydrogen-bonding donors are hydroxy, carboxyl, and amide protons. Interactions can be modified by changing the elution conditions. The more nonpolar the elution conditions, the stronger the H-bond interactions. Enantioselectivity is determined by the strength of the hydrogen bonds, which is, in turn, affected by secondary interactions such as steric hindrance or attractive or repulsive interactions between the selector and the selectand.

Chiral Separation Through Charge Transfer

Complexes formed by weak interactions of electron donors with electron-acceptor compounds are known as charge-transfer complexes. The necessary condition for the formation of a charge transfer complex is the presence of an occupied molecular orbital of sufficiently high energy in the electron-donor molecule, and the presence of a sufficiently low unoccupied orbital in the electron-acceptor molecule. Small unsaturated hydrocarbons are usually weak donors or weak acceptors. Polynuclear aromatic hydrocarbons are efficient π-donor molecules. Replacement of a hydrogen atom in the parent molecule with an electron-releasing substituent such as alkyl, alkoxy, or amino, increases the capability of the molecule to donate π electrons. Aromatic molecules containing groups such as NO_2, Cl, C≡N are efficient electron acceptors. Carbonyl compounds are acceptors to aromatic hydrocarbons but are donors to bromine.

The overlapping and the orientation of the molecules in the crystal correspond to parallel planes if the bonding occurs only through π orbitals. π-donor–π-donor interactions do not occur in the same fashion because of repulsion between the π clouds. This repulsion leads to edge-to-face interactions, where weakly positive H atoms at the edge of the molecule point toward negatively charged C atoms on the faces of adjacent molecule. The dihedral ring planes are often close to perpendicular. Aromatic rings can act as hydrogen-bond acceptors for the amidic proton [17].

In general, the stability of a charge-transfer complex increases with the increase in the polarity of the solvent. To establish the enantiomeric separation under such conditions, secondary interactions must occur: namely the charge-transfer interactions have to be accompanied by hydrogen bonds and/or steric hindrance. Under these conditions, the mobile-phase conditions should be adjusted such that these interactions are achieved. Figure 3 presents an example of a chiral stationary phase designed by Pirkle's group. This CSP allows for charge-transfer interaction with secondary interactions such as hydrogen-bonding and steric hindrance [18].

Chiral Separation Through Host–Guest Complexation

Cyclodextrins and crown ethers are the main classes of compounds able to undergo host–guest complexes with a particular pair of enantiomers. Cyclodextrins (CD) are natural macrocyclic polymers of glucose that contain 6–12 D-(+)-glucopyranose units which are bound through α-1,4-glucopyranose linkages. The number of glucose units per CD is denoted by a Greek letter: α for six, β for seven, and γ for eight (Fig. 4) [19]. The inherent chirality of the CD renders them useful for chromatographic enantioseparations. In most cases, an inclusion complex is formed between the solute and the cyclodextrin cavity. The host–guest complexation is dependent on the polarity, hydrophobicity, size, and geometry of the guest, as well as the size of the internal cavity of the CD. Enantioselectivity is

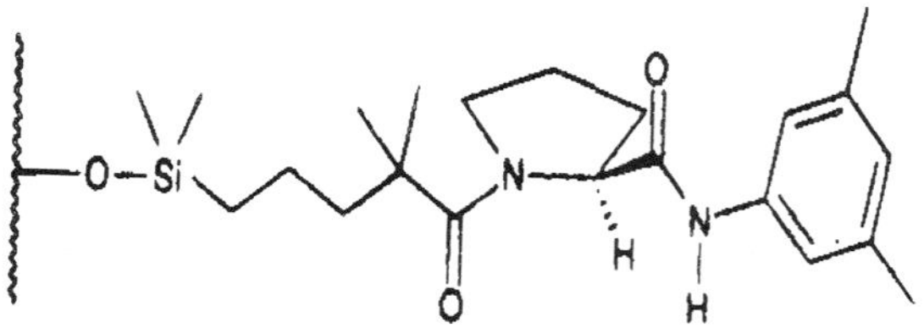

Fig. 3 The structure of the (*S*)-proline derivative chiral stationary phase. (From Ref. 18.)

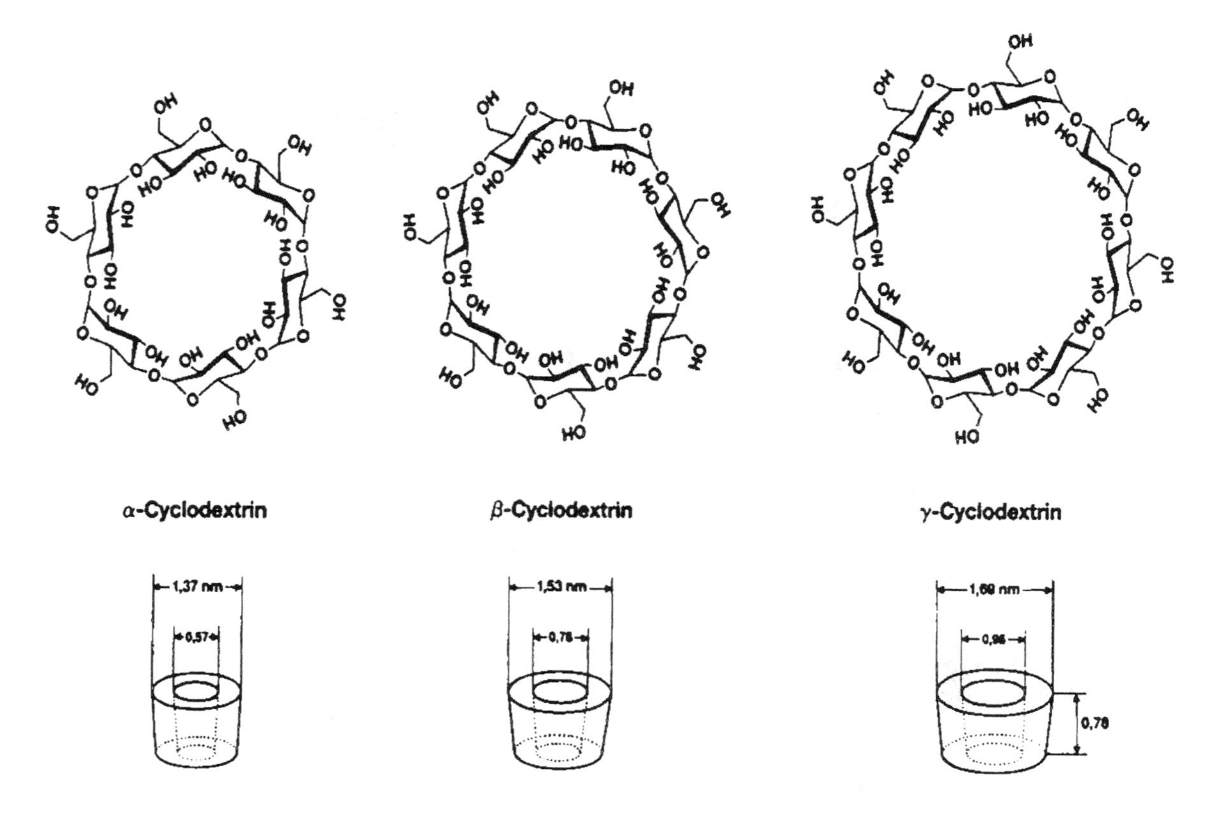

Fig. 4 Schematic representation of α-CD, β-CD, and γ-CD. (From Ref. 19.)

then determined by the fit in the cavity and by the interactions between substituents attached to or near the chiral center of the analyte and the unidirectional secondary hydroxyl groups at the mouth of the cavity. The temperature, pH, and the composition of the mobile phase influence the complexation.

Under reversed-phase conditions, the presence of an organic modifier affects the binding of the guest molecule in the CD's cavity. The inclusion complex is usually strongest in water and decreases upon addition of organic modifiers. The modifier competes with the guest analyte for the cavity. Under normal-phase conditions, apolar solvents such as hexane and chloroform occupy the CD's cavity and cannot be easily displaced by the solute molecules. In these circumstances, the solute is usually restricted to interactions with the exterior of the CD. Chemical modifications of CD has opened new possibilities for enantiorecognition, widening the range of compounds that can be separated into enantiomers [20].

Crown ethers, especially 18-crown-6 ethers, can complex not only inorganic cations but also alkylammonium compounds. The primary interactions occur between the hydrogens of the ammonium group and the oxygens of the crown ether. The introduction of bulky groups such as binaphtyl onto the exterior of the crown ether provides steric barriers and induces enantioselective interactions with the guest molecule.

The rigid binaphthyl units occupy planes that are perpendicular to the plane of the cyclic ether. One of the naphthalene rings forms a wall that extends along the sides and outward from the other face of the cyclic ether. The substituents attached at the 3-position of the naphthalene rings extend along the side or over the face of the cyclic ether. In the presence of a chiral primary amine, it forms a triple hydrogen bond with the primary ammonium cation. The same complex is formed whether the guest approaches from the top or from the bottom of the crown ether, as the crown ether has a C_2 axis of symmetry. In the complex, the large (L), medium (M), and small (S) groups attached to the asymmetric carbon of the guest must adjust themselves into two identical cavities. The L is placed in one cavity and the M and S into the other cavity. M will reside in the pocket with S against the wall for the more stable diastereomeric complex (Fig. 5) [21].

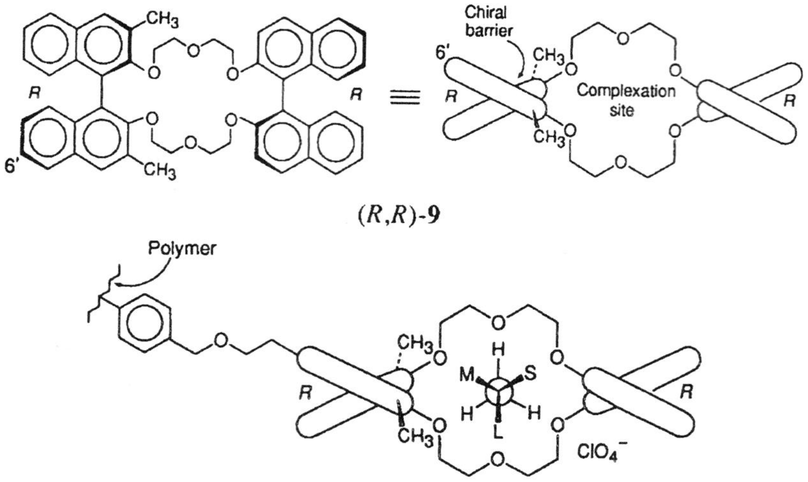

Fig. 5 Structure of the crown ether and the most stable complex. (From Ref. 21.)

Chiral Separation Through Combination of Interactions

Included in this category are stationary phases such as biopolymers (e.g., celluloses and cellulose derivatives, proteins) [22,23], as well as macrocyclic antibiotics [24]. These stationary phases exhibit interactions with a particular enantiomer through hydrogen-bonding, charge transfer, and inclusion interactions. They proved to be very effective in resolving a wide class of racemates encompassing a variety of structures. Describing the mechanism of such separation is very challenging due to the complexity of these stationary phases. Such stationary phases can be operated under reversed-phase conditions (protein phases, cellulose phases, and macrocyclic antibiotics), as well as in the normal-phase conditions (cellulose phases and macrocyclic antibiotics). Conformational changes of biopolymers under the temperature and mobile-phase conditions can occur and they should be controlled such that the separation can be maximized [25,26].

References

1. B. Feisbush, *Chirality 10*:382 (1998).
2. W. Lindner, in *Chromatographic Chiral Separation* (M. Zieff and L. J. Crane, eds.), Marcel Deker, Inc., New York, 1988, p. 91.
3. M. Ahnoff and S. Einarsson, in *Chiral Liquid Chromatography* (W. J. Lough, ed.), Blackie and Son, Glasgow, 1989, p. 39.
4. B. Feibush and N. Grinberg, in *Chromatographic Chiral Separation* (M. Zieff and L. J. Crane, eds.), Marcel Dekker, Inc., New York, 1988, p. 1.
5. C. Pettersson and G. Schill, *J. Chromatogr. 204*:179 (1981).
6. C. Pettersson and G. Schill, *Chromatographia 16*:192 (1982).
7. A. Karlsson and C. Pettersson, *Chirality 4*:323 (1992).
8. C. Pettersson and K. No, *J. Chromatogr. 316*:553 (1984).
9. A. Guttman, *Electrophoresis 16*:1900 (1995).
10. A. Guttman and N. Cooke, *J. Chromatogr. 685*:155 (1994).
11. J.-M. Lin, T. Nakagama, and T. Hobo, *Chromatographia 42*:559 (1996).
12. E. Gil-Av and S. Tishbee, *J. Am. Chem. Soc. 102*:5115 (1980).
13. B. Clar and J. Mame, *J. Pharm. Biomed. Anal. 7*:1883 (1989).
14. D. Armstrong, *Anal. Chem. 66*:1690 (1994).
15. V. A. Davankov, in *Advances in Chromatography* (J. C. Giddings, E. Grushka, J. Cazes, and P. R. Brown, eds.), Marcel Dekker, Inc., New York, 1980, Vol. 18, p. 139.
16. V. A. Davankov, A. A. Kurganov, and A. S. Bochkov, in *Advances in Chromatography* (J. C. Giddings, E. Grushka, J. Cazes, and P. R. Brown, eds.), Marcel Dekker, Inc., New York, 1983, Vol. 22, p. 71.
17. R. Foster, *Organic Charge-Transfer Complexes*, Academic Press, London, 1969, p. 217.
18. W. H. Pirkle and S. R. Selness, *J. Org. Chem. 60*:3252 (1995).
19. W. L. Konig, *Gas Chromatographic Enantiomer Separation with Modified Cyclodextrins*, Hütihig Buch Verlag, Heidelberg, 1992, p. 4.
20. A. M. Stalcup, in *A Practical Approach to Chiral Separations by Liquid Chromatography*, VCH, Weinheim, 1994, p. 1994.
21. D. J. Cram and J. M. Cram, *Container Molecules and Their Guests*, Royal Society of Chemistry, London, 1994, p. 56.

22. Y. Okamoto and Y. Kaida, *J. Chromatogr. 666*:403 (1994).
23. S. G. Allenmark and S. Anderson, *J. Chromatogr. 666*:167 (1994).
24. K. H. Ekborg-Ott, L. Youbang, and D. W. Armstrong, *Chirality 10*:434 (1998).
25. M. Waters, D. R. Sidler, A. J. Simon, C. R. Middaugh, R. Thompson, L. J. August, G. Bicker, H. J. Perpall, and N. Grinberg, *Chirality 11*:224 (1999).
26. T. O'Brien, L. Crocker, R. Thompson, K. Thomson, P. H. Toma, D. A. Conlon, B. Feibush, C. Moeder, G. Bocker, and N. Grinberg, *Anal. Chem. 69*:1999 (1997).

Nelu Grinberg
Richard Thompson

Chiral Separations by Micellar Electrokinetic Chromatography with Chiral Micelles

Since micellar electrokinetic chromatography (MEKC) was first introduced in 1984, it has become one of major separation modes in capillary electrophoresis (CE), especially owing to its applicability to the separation of neutral compounds as well as charged ones. Chiral separation is one of the major objectives of CE, as well as MEKC, and a number of successful reports on enantiomer separations by CE and MEKC has been published. In chiral separations by MEKC, the following two modes are normally employed: (a) MEKC using chiral micelles and (b) cyclodextrin (CD)-modified MEKC (CD–MEKC).

MEKC Using Chiral Micelles

An ionic chiral micelle is used as a pseudo-stationary phase; it works as a chiral selector. When a pair of enantiomers is injected to the MEKC system, each enantiomer is incorporated into the chiral micelle at a certain extent determined by the micellar solubilization equilibrium. The equilibrium constant for each enantiomer is expected to be different more or less among the enantiomeric pair; that is, the degree of solubilization of each enantiomer into the chiral micelle would be different for each. Thus, the difference in the retention factor would be obtained and different migration times would occur.

CD–MEKC

An ionic achiral micelle [e.g., sodium dodecyl sulfate (SDS)] and a neutral CD are typically used as a pseudo-stationary phase and a chiral selector, respectively. When a pair of enantiomers is injected into this system, two major distribution equilibria can be considered for the solutes or enantiomers: (a) the equilibrium between the aqueous phase and the micelle (i.e., micellar solubilization) and (b) the equilibrium between the aqueous phase and CD (i.e., inclusion complex formation). Each enantiomer may have a different equilibrium constant for the inclusion complex formation among the enantiomeric pairs due to the enantioselectivity of the CD. As a result, each enantiomer exists in the aqueous phase at a different time among the enantiomeric pairs; hence, the time spent in the micelle would be varied.

In some cases, an ionic chiral micelle (e.g., a bile salt) is also used as a chiral pseudo-stationary phase with a CD. Moreover, cyclodextrin electrokinetic chromatography (CDEKC), where a CD derivative having an ionizable group is used as a chiral pseudo-stationary phase, has become popular recently since several commercially available ionic CD derivatives have appeared. Although the CDEKC technique is actually beyond the field of MEKC, it is an important method for enantiomer separation by CE.

In this section, chiral separation by MEKC with chiral micelles is mainly treated. The development of novel chiral surfactants adaptable to pseudo-stationary phases in MEKC for enantiomer separation is continuously progressing. It seems somewhat difficult for a researcher to find an appropriate mode of CE when one wants to achieve a specific enantioseparation. However, nowadays, various method development kits for chiral separation have been commercially available and some literature on the topic is also available, so that helpful information may be obtained without difficulty.

MEKC Using Natural Chiral Surfactants

Bile Salts

Bile salts are natural and chiral anionic surfactants which form helical micelles of reversed micelle conformation.

The first report on enantiomer separation by MEKC using bile salts was the enantioseparation of dansylated DL-amino acids (Dns-DL-AAs) and, since then, numerous papers have been available. Nonconjugated bile salts, such as sodium cholate (SC) and sodium deoxycholate (SDC), can be used at pH > 5, whereas taurine-conjugated forms, such as sodium taurocholate (STC) and sodium taurodeoxycholate (STDC), can be used under more acidic conditions (i.e., pH > 3). Several enantiomers, such as diltiazem hydrochloride and related compounds, carboline derivatives, trimetoquinol and related compounds, binaphthyl derivatives, Dns-DL-AAs, mephenytoin and its metabolites, and 3-hydroxy-1,4-benzodiazepins have been successfully separated by MEKC with bile salts. In general, STDC is considered as the the most effective chiral selector among the bile salts used in MEKC.

The use of CDs with bile salt micelles has been also successful for enantiomer separations. For example, Dns-DL-AAs, baclofen and its analogs, mephenytoin and fenoldopam, naphthalene-2,3-dicarboxaldehyde derivatized DL-AAs (CBI-DL-AAs), diclofensine, ephedrine, nadolol, and other β-blockers, and binaphthyl-related compounds were enantioseparated by CD–MEKC with bile salts.

Digitonin and Saponins

Digitonin, which is a glycoside of digitogenin and used for the determination of cholesterol, is a naturally occurring chiral surfactant. By using digitonin with ionic micelles, such as SDS or STDC as pseudo-stationary phases, some phenylthiohydantoin-DL-AAs (PTH-DL-AAs) were enantioseparated.

On the other hand, glycyrrhizic acid (GRA) and β-escin can be employed as chiral pseudo-stationary phases in MEKC. Chiral separations of some Dns-DL-AAs and PTH-DL-AAs were achieved.

MEKC Using Synthetic Chiral Surfactants

N-Alkanoyl-L-Amino Acids

Various *N*-alkanoyl-L-amino acids, such as sodium *N*-dodecanoyl-L-valinate (SDVal), sodium *N*-dodecanoyl-L-alaninate (SDAla), sodium *N*-dodecanoyl-L-glutamate (SDGlu), *N*-dodecanoyl-L-serine (DSer), *N*-dodecanoyl-L-aspartic acid (DAsp), sodium *N*-tetradecanoyl-L-glutamate (STGlu), and sodium *N*-dodecanoyl-L-threoninate (SDThr) have been employed as synthetic chiral micelles in MEKC; several enantiomers have been successfully separated (Fig. 1). In each case, the addition of SDS, urea, and organic modifiers such as methanol or 2-propanol were essential to obtain improved peak shapes and enhanced enantioselectivity.

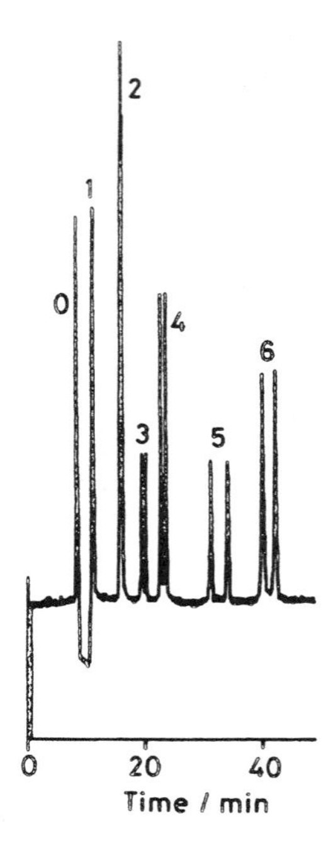

Fig. 1 Chiral separation of six PTH-DL-AAs by MEKC with SDVal. Corresponding AAs: (1) Ser, (2) Aba, (3) Nva, (4) Val, (5) Trp, (6) Nle; (0) acetonitrile. Micellar solution: 50 m*M* SDVal–30 m*M* SDS–0.5*M* urea (pH 9.0) containing 10% (v/v) methanol; separation column: 50 μm inner diameter × 65 cm, 50 cm effective; applied voltage, 20 kV; current, 17 μA; detection wavelength, 260 nm; temperature, ambient. [Reprinted from K. Otsuka et al., *J. Chromatogr.* *559*:209 (1991) with permission.]

N-Dodecoxycarbonyl Amino Acids

Chiral surfactants of amino acid derivatives, such as (*S*)- and (*R*)-*N*-dodecoxycarbonylvaline (DDCV) and *N*-dodecoxycarbonylproline (DDCP) are available for enantiomer separation by MEKC: Several pharmaceutical amines, benzoylated amino acid methyl ester derivatives, piperidine-2,6-dione enantiomers, and aldose enantiomers were successfully resolved. Because both enantiomeric forms of DDCV or (*S*)- and (*R*)-forms are available, we can expect that the migration order of an enantiomeric pair would be reversed.

Alkylglucoside Chiral Surfactants

Anionic alkylglucoside chiral surfactants, such as dodecyl β-D-glucopyranoside monophosphate and monosulfate, and sodium hexadecyl D-glucopyranoside 6-hydrogen sulfate, were used as chiral pseudo-stationary phases in MEKC, where several enantiomers (e.g., PTH-DL-AAs and binaphthol) were resolved.

Several neutral alkylglucoside surfactants, such as heptyl-, octyl-, nonyl-, and decyl-β-D-glucopyranosides and octylmaltopyranoside, were also employed for the enantiomer separation of phenoxy acid herbicides, Dns-DL-AAs, 1,1′-bi-2-naphthyl-2,2′-diyl hydrogen phosphate (BNP), warfarin, bupivacaine, and so forth.

Tartaric Acid-Based Surfactants

A synthesized chiral surfactant based on (*R*,*R*)-tartaric acid was used for the enantiomer separation in MEKC, where enantiomers having fused polyaromatic rings were separated easier than those having only a single aryl group.

Some PTH-DL-AAs and drug enantiomers were successfully resolved by using tartaric acid-based chiral surfactants.

Steroidal Glucoside Surfactants

Neutral steroidal glucoside surfactants, such as *N*,*N*-bis-(3-D-gluconamidopropyl)-cholamide (Big CHAP) and *N*,*N*-bis-(3-D-gluconamidopropyl)-deoxycholamide (Deoxy Big CHAP), which contain a cholic or deoxycholic acid moiety, respectively, have been introduced for use as chiral pseudo-stationary phases in MEKC. By using a borate buffer under basic conditions, these surfactant micelles could be charged via borate complexation. Some binaphthyl enantiomers, Tröger's base, phenoxy acid herbicide, and Dns-DL-AAs were enantioseparated.

MEKC Using High-Molecular-Mass Surfactants

The use of a high-molecular-mass surfactant (HMMS) or polymerized surfactant has been recently investigated as a pseudo-stationary phase in MEKC. Because a HMMS forms a micelle with one molecule, enhanced stability and rigidity of the micelle can be obtained. Also, it is expected that the micellar size is controlled easier than with a conventional low-molecular-mass surfactant (LMMS). The first report on enantiomer separation by MEKC using a chiral HMMS appeared in 1994, where poly(sodium *N*-undecylenyl-L-valinate) [poly(L-SUV)] was used as a chiral micelle and binaphthol and laudanosine were enantioseparated. The optical resolution of 3,5-dinitrobenzoylated amino acid isopropyl esters by MEKC with poly(sodium (10-undecenoyl)-L-valinate) as well as with SDVal, SDAla, and SDThr was also reported.

As for the use of monomeric and polymeric chiral surfactants as pseudo-stationary phases for enantiomer separations in MEKC, a review article has been available.

The use of an achiral HMMS butyl acrylate–butyl methacrylate–methacrylic acid copolymer (BBMA) sodium salt was also investigated for enantiomer separations with CDs or as a CD–MEKC mode. A better enantiomeric resolution of Dns-DL-AAs was obtained by a β-CD–BBMA–MEKC system than an β-CD–SDS–MEKC system.

Polymerized dipeptide surfactants, which are derived from sodium *N*-undecylenyl-L-valine-L-leucine (L-SUVL), sodium *N*-undecylenyl-L-leucine-L-valine (L-SULV), sodium *N*-undecylenyl-L-leucine-L-leucine (L-SULL), and sodium *N*-undecylenyl-L-valine-L-valine (L-SUVV), were employed. Among these dipeptides, poly(L-SULV) showed the best enantioselectivity for the separation of 1,1′-bi-2-naphthol (BN).

Suggested Further Reading

Camilleri, P., Chiral surfactants in micellar electrokinetic capillary chromatography, *Electrophoresis 18*:2332 (1997).

Chankvetadze, B., *Capillary Electrophoresis in Chiral Analysis*, John Wiley & Sons, New York, 1997.

Otsuka, K. and S. Terabe, Micellar electrokinetic chromatography, *Bull. Chem. Soc. Jpn. 71*:2465 (1998).

Otsuka, K. and S. Terabe, Enantiomer separation of drugs by micellar electrokinetic chromatography using chiral surfactants, *J. Chromatogr. A 875*:163 (2000) .

Terabe, S., K. Otsuka, and H. Nishi, Separation of enantiomers by capillary electrophoretic techniques, *J. Chromatogr. A 666*:295 (1994).

Koji Otsuka
Shigeru Terabe

Classification of Organic Solvents in Capillary Electrophoresis

Introduction

In capillary electrophoresis (CE), several criteria can be applied to classify solvents [e.g., for practical purposes based on the solution ability for analytes, on ultraviolet (UV) absorbance (for suitability to the UV detector), toxicity, etc.]. Another parameter could be the viscosity of the solvent, a property that influences the mobilities of analytes and that of the electro-osmotic flow (EOF) and restricts handling of the background electrolyte (BGE). For more fundamental reasons, the dielectric constant (the relative permittivity) is a well-recognized parameter for classification. It was initially considered to interpret the change of ionization constants of acids and bases according to Born's approach. This approach has lost importance in this respect because it is based on too simple assumptions limited to electrostatic interactions. Indeed, a more appropriate concept reflects solvation effects, the ability for H-bonding, or the acido-base property of the solvent.

Classification of Solvents

A first classification, according to the dielectric constant ε (with a somewhat arbitrary value about 20 or 30 to distinguish the two main classes), is still useful, because the dielectric constant reflects the extent of ion association and ion-pairing. In solvents with a high dielectric constant, called polar solvents, ion-pairing is nearly negligible in dilute solutions. Acid strength can be assigned by a numerical value, independent of the base with which it undergoes reaction. Nonpolar solvents, on the other hand (i.e., those with low dielectric constant), support ion-pairing, and the acidity is dependent on the choice of the reference base. Further grouping of the solvents is based on their ability for H-bond formation. Note that this scheme disregards the solvents' own acidity or basicity (this criterion is taken for the latter scheme). This is shown in Table 1. A typical H-bonded solvent, termed protic solvents, is water or methanol. Aprotic solvents (although containing H atoms or being able to accept H bonds) lack the ability to donate hydrogen bonds. The lower alcohols (except methanol) are grouped in the class with low ε. They are protic as well (they are able to donate hydrogen bonds), in contrast to the fourth group, where, for example, dioxane and tetrahydrofurane are found.

The solvents can be grouped according to another classification scheme introduced by Brønsted. It has significance for the evaluation of acido-basic equilibria in the particular solvents, especially for the interpretation of the changes of the pK values of weakly acidic (or basic) analytes and their degree of protolysis. Obviously, it is also of main importance for the evaluation of the pH scales in the different solvents or solvent mixtures. It differentiates, mainly, three groups: amphiprotic, dipolar aprotic, and inert solvents. As the first scheme, it also relates to the "polarity" of the solvent (expressible by the dielectric constant, too). *Amphiprotic* solvents have acidic and basic properties as well. These solvents, HS, are able to form lyonium ions, H_2S^+, with the proton, and stable lyate ions, S^-. Examples for *amphiprotic neutral* solvents (with high ε) are water and methanol. In the classification given in Table 1, they belong to the H-bonded solvents.

Amphiprotic protogenic solvents have higher acidic

Table 1 Classification of Solvents for Potential Use in CE, According to Their Dielectric Constant, ε, and Their Ability for H-Bond Formation

High ε		Low ε	
H-bonded	Non-H-bonded	H-bonded	Non-H-bonded
Water	Acetonitrile	Ethanol	Acetone
Methanol	*N,N*-Dimethylformamide	*n*-Propanol	Dioxane
Ethylene glycol	*N,N*-Dimethylacetamide	*i*-Propanol	Tetrahydrofuran
Formamide	Dimethyl sulfoxide	*n*-Butanol	
Acetamide		*tert*-Butanol	
N-Methylformamide			
N-Methylacetamide			

properties, but lower basic ones (always in comparison to water). Examples are formic and acetic acid. *Amphiprotic protophilic* solvents have lower acidity and higher basicity than water, with formamide or ethanolamine as examples. *Aprotic dipolar* solvents have low acidity and (occasionally) basicity as well, with *N,N*-dimethylformamide and dimethylsulfoxide as examples for protophilic dipolar solvents and acetonitrile for a protophobic dipolar solvent.

Solvents with low ε and without the ability to form H bonds are also classified as dipolar aprotic solvents in some cases. Examples are dioxane and tetrahydrofuran, which have ε values smaller than 7. They have reduced applicability in CE; however, their aqueous mixtures might be of some interest.

Inert solvents (according to the second classification scheme) have insignificant applicability for CE, as dissociation into free ions is reduced (and solubility of electrolytes is very low). Charged analytes form stable ion pairs there. Examples are the halogenated or aliphatic and aromatic hydrocarbons (e.g., octane, benzene).

Suggested Further Reading

Bates, R. G., Medium effect and pH in nonaqueous and mixed solvents, in *Determination of pH, Theory and Practice*, John Wiley & Sons, New York, 1973, pp. 211–253.

Covington, A. K. and T. Dickinson, Introduction and solvent properties, in *Physical Chemistry of Organic Solvents Systems* (A. K. Covington and T. Dickinson, eds.), Plenum Press, London, 1973, pp. 1–23.

Kenndler, E., Organic solvents in capillary electrophoresis, in *Capillary Electrophoresis Technology* (N. A. Guzman, ed.), Basel, Hong Kong: Marcel Dekker, Inc., New York, 1993, pp. 161–186.

King, E. J., Acid-base behaviour, in *Physical Chemistry of Organic Solvent Systems* (A. K. Covington and T. Dickinson, eds.), Plenum Press, London, 1973, pp. 331–403.

Kolthoff, I. M. and M. K. Chantooni, General introduction to acid-base equilibria in nonaqueous organic solvents, in *Treatise on Analytical Chemistry, Part I, Theory and Practice* (I. M. Kolthoff and P. J. Elving, eds.), John Wiley & Sons, New York, 1979, pp. 239–301.

Ernst Kenndler

Cold-Wall Effects in Thermal FFF

In field-flow fractionation (FFF), like chromatography, retention and resolution are affected by temperature. For chromatographic systems, thermostatted ovens are used to precisely control the column temperature, but in thermal FFF, the issue of temperature control is more complex. Under a given set of conditions, the temperature in the channel varies between the hot and cold walls, but more importantly, the average temperature of an eluting component varies with the temperature of the cold wall. As a result, the retention of a given analyte in different channels will be identical only if the cold-wall temperatures (as well as the field strengths) are identical. The retention of polystyrene in ethylbenzene, for example, is reduced by 1% for each 2-degree increase in the cold-wall temperature (T_c), even when the temperature gradient is held constant [1].

The effect of T_c on the retention of polystyrene in ethylbenzene is typical of most polymer–solvent systems. Therefore, fluctuations in T_c of only a couple degrees are generally not a problem. Larger fluctuations can be a significant problem, however, especially when retention is used to monitor small batch-to-batch variations in a quality control situation. The magnitude of T_c depends on several factors. Heat, which is transferred from the hot wall, is typically removed by heat exchange with water flowing beneath the cold wall. The cooling efficiency is affected by the temperature of the incoming water, which can vary by several degrees among different laboratories. The incoming water temperature may also vary by several degrees over time in a given laboratory. The effect of variations in the incoming water temperature can be attenuated with the use of a flow-control valve, because T_c also varies with the flow rate of the water. For example, a lower coolant flow rate can be used in the winter to offset the effect on T_c of the lower water temperature. An alternative to varying the flow rate is to maintain a constant water temperature by running it through a thermostatted heater/chiller.

When T_c is not controlled to within a couple of degrees, the fluctuations ultimately affect the accuracy of molecular-weight determinations that rely on the calibration of thermal FFF retention to molecular weight. For a detailed discussion of the role of T_c in thermal FFF calibrations, see the entry Molecular Weight and Molecular-Weight Distributions by Thermal FFF. In summary, variations in retention with T_c can be handled when characterizing molecular-weight distributions by either matching the T_c between calibration and analysis or by incorporating the T_c into the calibration equation. Matching the T_c can be difficult when data are to be compared among channels in different laboratories. On the other hand, incorporating T_c into the calibration equation requires the systematic accumulation of large amounts of data.

In addition to molecular weight, thermal FFF is used to measure transport coefficients. For example, the measurement of thermodiffusion coefficients is important for obtaining compositional information on polymer blends and copolymers (see the entry Thermal FFF of Polymers and Particles). Thermal FFF is also used in fundamental studies of thermodiffusion because it is a relatively fast and accurate method for obtaining the Soret coefficient, which is used to quantify the concentration of material in a temperature gradient. However, the accuracy of Soret and thermodiffusion coefficients obtained from thermal FFF experiments depends on properly accounting for several factors that involve temperature. In order to understand the effect of temperature on transport coefficients, as well as the effect on thermal FFF calibration equations, a brief outline of retention theory is given next.

In all FFF subtechniques, retention depends on a balance of two opposing motions. The first motion is induced by the applied field and results in the concentration of material at the accumulation wall (typically, the cold wall in thermal FFF). The buildup in concentration induces the opposing motion of diffusion. Both motions are accounted for in the retention parameter λ, which is defined for all FFF subtechniques as

$$\lambda = \frac{D}{Uw} \tag{1}$$

Here, D is the (mass) diffusion coefficient, U is the field-induced velocity of the sample, and w is the channel thickness. In thermal FFF, U is governed by the thermal diffusion coefficient (D_T) and the temperature gradient (dT/dx), which is applied in the same dimension (x) as the channel thickness (x varies in value from 0 at the cold wall to w at the hot wall). Using the dependence of U on D_T and dT/dx, the retention parameter in thermal FFF can be expressed as

$$\lambda = \frac{D}{D_T(dT/dx)w} \cong \frac{D}{D_T \Delta T} \tag{2}$$

where ΔT is the difference in temperature between the hot and cold walls. Because x and w are in the same dimension, dT/dx can be approximated by $\Delta T/w$; therefore, $w(dT/dx)$ is approximated by ΔT on the right-hand side of Eq. (2).

When the parameters in Eq. (1) are constant throughout the channel, the volume V_r of fluid required to flush a sample component through the channel is related to λ by the following equation:

$$V_r = \frac{V^\circ}{6\lambda}\left[\coth\left(\frac{1}{2\lambda} - 2\lambda\right)\right]^{-1} \tag{3}$$

Here, V° is the volume of fluid required to flush a sample that is not affected by the field ($U = 0$). In most FFF subtechniques, the parameters in Eq. (1) are, in fact, constant throughout the channel. In thermal FFF, however, these parameters vary across the channel because they depend on temperature, which varies between the hot and cold walls. As a result, Eq. (3) is only an approximation in thermal FFF. Fortunately, the approximations associated with Eq. (3) are inconsequential for the determination of molecular-weight distributions, so that the only concern is variations in T_c, as outlined earlier. For measuring transport coefficients, on the other hand, the approximations can lead to significant errors.

When calculating transport coefficients from measured values of V_r, the retention parameter λ is first calculated using Eq. (3). Next, the Soret coefficient D_T/D is calculated from λ using Eq. (2). From the Soret coefficient, the value of D_T can be calculated if an independent measure of D is available. The accuracy of the resulting value is compromised by the approximations involved in deriving Eq. (3). First, a parabolic velocity profile is assumed. In reality, the velocity profile is skewed toward the hot wall by a carrier-liquid viscosity (η) that varies across the channel as a result of the temperature gradient. Variations in thermal conductivity (κ) with temperature across the channel must also be considered. Finally, the accuracy of Eq. (3) is compromised by an assumption that the analyte forms an exponential concentration profile, whereas, in reality, the profile is more complicated due to the temperature dependence of the transport coefficients.

The consequences of the temperature dependence of η, κ, D, and D_T have been discussed in several articles [3–6]. Ko et al. [3] demonstrated that the temperature dependence of the Soret coefficient actually increases the resolution of different molecular-weight components. In a theoretical study by van Asten et al. [4], it was shown that the consequence of ignoring the temperature dependence of κ has a nearly negligible effect on the accuracy of D/D_T values calculated using Eqs. (2) and (3). Ignoring the temperature dependence of η and D/D_T, on the other hand, can lead to errors of up to 8% when D/D_T values are calculated from retention data. Several refinements to Eq. (3) have been made over the years [2,5,6]. When these refinements are used, they yield accurate values for the transport coefficients. Although the resulting equations are quite complex, they are not required for the routine analysis of polymers by thermal FFF.

References

1. S. L. Brimhall, M. N. Myers, K. D. Caldwell, and J. C. Giddings, *J. Polym. Sci. Polym. Phys. Ed. 23*:2443 (1985).
2. J. J. Gunderson, K. D. Caldwell, and J. C. Giddings, *Separ. Sci. Technol. 19*:667 (1984).
3. G.-H. Ko, R. Richards, and M. E. Schimpf, *Separ. Sci. Technol. 31*:1035 (1996).
4. A. C. van Asten, H. F. M. Boelens, W. Th. Kok, H. Poppe, P. S. Williams, and J. C. Giddings, *Separ. Sci. Technol. 29*:513 (1994).
5. M. Martin and J. C. Giddings, *J. Phys. Chem. 85*:727 (1981).
6. M. Martin, C. van Batten, and M. Hoyos, *Anal. Chem. 69*: 1339 (1997).

Martin E. Schimpf

Concentration of Dilute Colloidal Samples by Field-Flow Fractionation

Introduction

Many colloidal systems, such as those of natural water, are too dilute to be detected by the available detection systems. Thus, a simple and accurate method for the concentration and analysis of these dilute samples should be of great significance in analytical chemistry. In the present work, two methodologies of the field-flow fractionation (FFF) technique for the concentration and analysis of dilute colloidal samples are presented. Both the conventional and potential barrier methodologies of FFF are based on the "adhesion" of the samples at the beginning of the channel wall, followed by their total removal and analysis. In the conventional sedimentation FFF (SdFFF) concentration procedure, the apparent adhesion of a dilute sample is due to its strong retention, which can be achieved by applying high field strengths and low flow rates. In the potential barrier SdFFF (PBSdFFF) concentration procedure, the true adhesion of a dilute sample is due to its reverse adsorption at the beginning of the column, which can be achieved by the appropriate adjustment of various parameters influencing the interactions between the colloidal particles and the material of the channel wall. The total release of the adherent particles is accomplished either by reducing the field strength and increasing the solvent velocity (conventional SdFFF) or by varying the potential energy of interaction between the particles and the column material—for instance, by changing the ionic strength of the carrier solution (PBSdFFF).

Methodology

Field-flow fractionation is a one-phase chromatographic system in which an external field or gradient replaces the stationary phase. The applied field can be of any type that interacts with the sample components and causes them to move perpendicular to the flow direction in the open channel. The most highly developed of the various FFF subtechniques is sedimentation FFF (SdFFF), in which the separations of suspended particles are performed with a single, continuously flowing mobile phase in a very thin, open channel under the influence of an external centrifugal force field.

In the normal mode of the SdFFF operation, a balance is reached between the external centrifugal field, driving the particles toward the accumulation wall, and the molecular diffusion in the opposite direction. In that case, the retention volume increases with particle diameter until steric effects dominate, at which transition point there is a foldback in elution order.

Potential barrier SdFFF (PBSdFFF), which has been developed recently in our laboratory, is based either on particle size differences or on Hamaker constant, surface potential, and Debye–Hückel reciprocal distance differences.

The retention volume of a component under study, V_r, in the normal SdFFF and the PBSdFFF methodologies is a function of the following parameters:

1. SdFFF:

$$V_r = f(d, G, \Delta\rho) \tag{1}$$

2. PBSdFFF:

$$V_r = f(d, G, \Delta\rho, \psi_1, \psi_2, A, I) \qquad (2)$$

where d is the particle diameter, G is the field strength expressed in acceleration, $\Delta\rho$ is the density difference between solute and solvent, ψ_1 and ψ_2 are the surface potentials of the particle and of the wall, respectively, A is the Hamaker constant, and I is the ionic strength of the carrier solution.

The conventional concentration procedure in SdFFF consists of two steps: the feeding (or concentration) and the separation (or elution) step. In the feeding step, the diluted samples are fed into the column with a small flow velocity while the channel is rotated at a high field strength to ensure the "apparent adhesion" of the total number of the colloidal particles at the beginning of the channel wall as a consequence of the particles' strong retention. In the separation step, the field is reduced and the flow rate is increased to ensure the total release and the consequence elution of the adherent dilute particles.

In PBSdFFF, the concentration step consists of feeding the column with the diluted samples at such experimental conditions, so as to decrease the repulsive component and increase the attractive component of the total potential energy of the particles under study. Because the stability of a colloid varies (increases or decreases) with a number of parameters (surface potential, Hamaker constant and ionic strength of the suspending medium), the proper adjustment of one or more of these parameters can lead not only to the adhesion of the dilute colloidal samples, which leads to their "concentration," but also to the total release of the adherent particles during the elution step.

Applications

Conventional SdFFF

As model samples for the verification of the conventional SdFFF as a concentration methodology monodisperse polystyrene latex beads (Dow Chemical Co.) with nominal diameters of 0.357 μm (PS1) and 0.481 μm (PS2) were used. They were either used as dispersions containing 10% solids or diluted with the carrier solution (triple-distilled water + 0.1% (v/v) detergent FL-70 from Fisher Scientific Co. + 0.02% (w/w) NaN_3) to study sample dilution effects. Diluted samples in which the amount of the polystyrene was held constant (1 μL of the 10% solids) while the volume in which it was contained was varied over a 50,000-fold range (from 1 to 50 mL of carrier solution) were introduced into the SdFFF column. During the feeding step, the flow rate was 5.8 mL/h for the PS1 polystyrene, and 7.6 mL/h for the polystyrene PS2, and the channel was rotated at 1800 rpm for the PS1 sample and at 1400 rpm for the PS2 sample. In the separation (elution) step, the experimental conditions for the two samples were as follows:

PS1: Field strength = 880 rpm, flow rate = 12–53 mL/h
PS2: Field strength = 500 rpm, flow rate = 24–59 mL/h

Figure 1 provides a comparison of fractograms for the 0.357-μm polystyrene injected as a narrow pulse (Fig. 1a) and injected at 10 mL dilution (Fig. 1b) by the conventional SdFFF concentration procedure described previously. Figure 1b shows that the eluted peak from the diluted sample emerges intact and without serious degradation, compared to the peak of Fig. 1a, despite the fact that the sample volume (10 mL) is over twice the channel volume (4.5 mL). The same concentration procedure was also successfully applied to the separation of the two polystyrene samples initially mixed together in a volume of 10 mL, as well as to the concentration of the colloidal particles contained in natural water samples collected from the Colorado, Green, and Price rivers in eastern Utah (U.S.A.).

As a general conclusion, the on-column concentration procedure of the conventional SdFFF method works quite successfully in dealing with highly diluted samples. Optimization, particularly higher field strengths during the concentration step, would allow higher flow rates and increased analysis speed. However, experimental confirmation would be necessary to give assurance that the particle–wall adhesion is not irreversible at higher spin rates.

Potential Barrier SdFFF

As model samples to test the validity of the PBSdFFF as a concentration procedure of diluted samples the monodisperse colloidal particles of α-Fe_2O_3 with nominal diameters of 0.271 μm were used. Diluted samples of α-Fe_2O_3 containing 2 μL of the 10% solid, in which the volume was varied over a 10,000-fold range (from 2 to 20 mL), were introduced into the column with a carrier solution containing 0.5% (v/v) detergent FL-70 + 3 $\times$ $10^{-2}M$ KNO_3 to ensure the total adhesion of the α-Fe_2O_3 particles at the beginning of the SdFFF Hastelloy-C channel wall. In the separation step, the carrier solution was changed to one containing only 0.5% (v/v) detergent FL-70 (without electrolyte) to ensure the total detachment of the adherent particles. In that case, a sample peak appeared (cf. Fig. 1c) as a consequence of the desorption of the α-Fe_2O_3 particles. The mean diameter of the α-Fe_2O_3 particles (0.280 μm) obtained by the proposed PBSdFFF

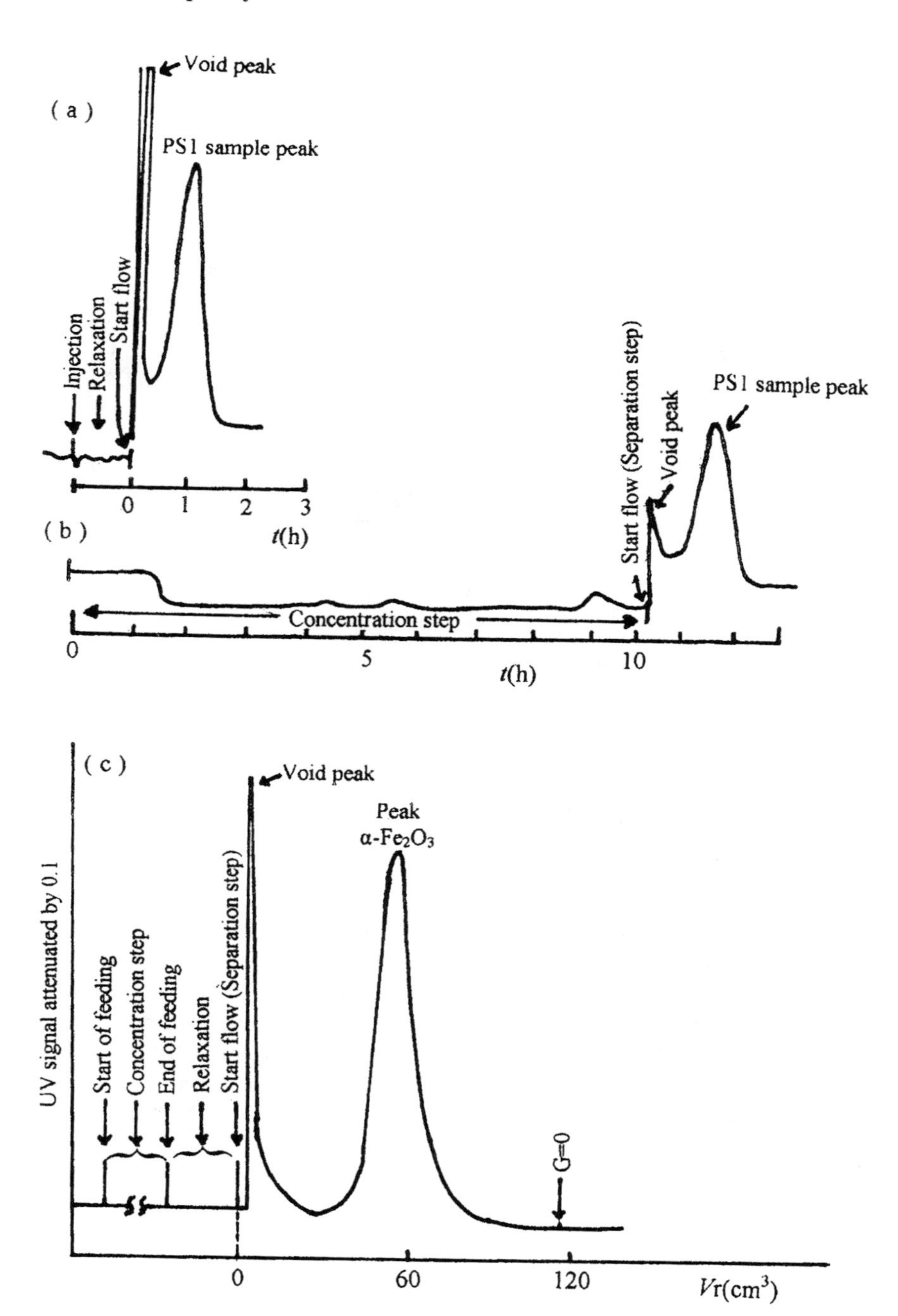

Fig. 1 Fractograms of the polystyrene latex beads of 0.357 μm (PS1) obtained by the direct injection of 1 μL of PS1 (a) and by the concentration procedure of the PS1 sample diluted in 10 mL of the carrier solution (b) using the conventional SdFFF technique, as well as of the α-Fe_2O_3 sample with nominal particle diameter of 0.271 μm diluted in 6 mL of the carrier solution obtained by the PBSdFFF concentration methodology (1c).

methodology for the on-channel concentration procedure of the sample diluted in 8 mL of the carrier solution is very close to that found (0.271 μm) by the direct injection of the same particles into the channel, using a carrier in which no adsorption occurs.

As a general conclusion, one could say that the proposed PBSdFFF concentration procedure works quite successfully in dealing with highly dilute samples, separating them according to size, surface potential, and Hamaker constant. At the same time, as separation occurs, the particle sizes of the colloidal materials of the diluted mixture can be determined. The major advantage of the proposed concentration procedure is that the method can concentrate and analyze dilute mixtures of colloidal particles even of the same size but with different surface potentials and/or Hamaker constants. The method has considerable promise for the separation and characterization, in terms of particle size, of dilute complex colloidal materials, where particles are present in low concentration.

Future Developments

Looking to the future, we believe that the efforts of the researchers will be focused on the extension of the FFF concentration methodologies to the ranges of more dilute and complex colloidal samples, without lengthening the analysis time.

Suggested Further Reading

A. Athanasopoulou, A. Koliadima, and G. Karaiskakis, *Instrum. Sci. Technol. 24*(2):79 (1996).

J. C. Giddings, G. Karaiskakis, and K. D. Caldwell, *Separ. Sci. Technol. 16*(6):725 (1981).

P. C. Hiemenz, *Principles of Colloid and Surface Chemistry*, Marcel Dekker, Inc., New York, 1977.

G. Karaiskakis and J. Cazes (eds.), *J. Liquid Chromatogr. Related Technol. 20*(16 & 17) (1997).

G. Karaiskakis, K. A. Graff, K. D. Caldwell, and J. C. Giddings, *Int. J. Environ. Anal. Chem. 12*:1 (1982).

A. Koliadima and G. Karaiskakis, *J. Liquid Chromatogr. 11*: 2863 (1988).

A. Koliadima and G. Karaiskakis, *J. Chromatogr. 517*:345 (1990).

A. Koliadima and G. Karaiskakis, *Chromatographia 39*:74 (1994).

George Karaiskakis

Concentration Effects on Polymer Separation and Characterization by ThFFF

The Phenomena of the Effects of Sample Concentration

The understanding of the effects of sample concentration (sample mass) in field-flow fractionation (FFF) has being obtained gradually with the improvement of the sensitivity (detection limit) of high-performance liquid chromatography (HPLC) detectors. Overloading, which was used in earlier publications, emphasizes that there is an upper limit of sample amount (or concentration) below which sample retention will not be dependent on sample mass injected into the FFF channels [1]. Recent studies show that such limits may not exist for thermal FFF (may be true for all the FFF techniques in polymer separation), although some of the most sensitive detectors on the market were used [2].

Experimental results indicate that the effects of sample mass include, but not exclusively, the following aspects.

Increased Polymer Retention

Figure 1 shows the fractograms of thermal FFF (ThFFF) to show the concentration effects for poly(methyl methacrylate) (PMMA) in THF, where M_p is the peak average molecular weight, m is the sample mass in micrograms injected into the ThFFF channel, and t^0 is the retention time of a nonretained species such as toluene. When the molecular weight (MW) of a polymer is moderate or higher, say above 300×10^3 g/mol for PMMA in THF, a moderate increase in concentration will result in longer retention. As reported in Ref. 2, the detector limits for the study was 0.09 μg of sample mass for 1000×10^3 g/mol polystyrene using an ultraviolet (UV) detector and 1 μg for 570×10^3

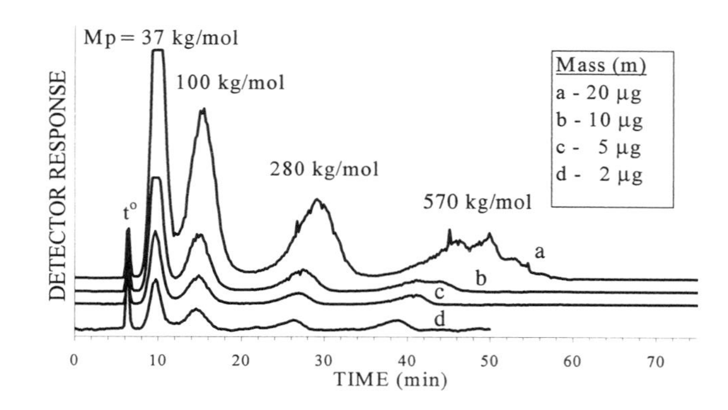

Fig. 1 Fractograms of PMMA in THF showing the effect of sample concentration on retention. Experimental conditions: cold-wall temperature, 25°C, T, 50°C; flow rate, 0.1 mL/min.

g/mol PMMA with an evaporative light-scattering detector. The retention was measured for sample masses ranging from these limits to more than 20 μg and was consistently found to increase with the increase in sample mass. The high limit, below which polymer retention is not dependent on concentration, was not found.

Broader Polymer Peaks

Increased concentration will increase band broadening in all chromatographic techniques, but it seems that the effect of concentration on band broadening is more serious in FFF, due to its concentration enhancement as shown by Fig. 1, and by Figure 5 of Ref. 1. More details will be discussed in the next section.

Distorted Peaks and Double-Topped Peaks (Ghost Peaks)

When the amount of sample mass injected into the ThFFF channel is moderate, say 1–10 μg for a typical channel, the peaks are pretty symmetrical and not much distortion may be observed for small polymers, as shown by Fig. 1. Increased retention may be observed for high-MW polymers [2]. When sample mass is further increased, say more than 20 μg, distorted peaks, even double-topped peaks or ghost peaks, may be observed for high-MW polymers as shown by Fig. 1, Fig. 6 of Ref. 1, and Fig. 3 of Ref. 2. The detailed report and discussion of double-topped peaks can be found in both Refs. 1 and 2.

Enhanced Viscosity Is Blamed for the Sample Concentration Effects

The viscosity of a polymer solution is highly dependent on concentration, temperature, and MW, as discussed below.

Concentration and Viscosity Enhancement in FFF

The amount of sample injected into the FFF channels can affect the retention time, primarily by influencing the viscosity of the solute–solvent mixture in the sample zone [1,2]. Unlike other chromatographic polymer separation methods [e.g., GPC–SEC (size-exclusion chromatography) and TREF, etc.], the viscosity of the fluid is not homogeneous at a given channel (column) cross section. In order for the samples to be retained by FFF, the concentration must be larger near the accumulation wall than that near the depletion wall [3,4]. In chromatography, sample concentration changes only along one dimension (i.e., the flow axis) whereas in FFF, sample concentration varies along two dimensions, one being the flow axis and the other one is across the channel thickness, which is perpendicular to the flow axis. The concentration across the channel thickness varies due to the migration of the molecules under the influence of the temperature gradient across the ThFFF channel [3]. The concentration distribution is approximately exponential as given by

$$c(x) = c_0 \exp\left(\frac{-x}{\lambda w}\right) \tag{1}$$

where $c(x)$ is the concentration at distance x across the channel thickness measured from the accumulation wall, c_0 is the concentration at the accumulation wall, w is the channel thickness, and λ is the retention parameter or reduced mean thickness of the sample zone. Shortly after injection, the sample zone is assumed to broaden into a Gaussian distribution along the z axis, corresponding to the direction of flow down the channel. The two-dimensional concentration becomes [5]

$$c(x, z) = c_{00} \exp\left(\frac{-(z - Z)^2}{2\sigma^2}\right) \exp\left(\frac{-x}{\lambda w}\right) \tag{2}$$

where Z is the distance traveled by the center of the zone down the channel. The concentration at the accumulation wall at the center of the zone, c_{00}, is found from [6]

$$c_{00} \cong \frac{V_{\text{inj}} c_{\text{inj}} L}{(2\pi\sigma^2)^{1/2} V^0 \lambda} \tag{3}$$

where V_{inj} is the volume of sample injected, V^0 is the void volume (channel volume), c_{inj} is the concentration of the injected sample, L is the length of the channel, and σ is the sum of the variances contributing to the zone breadth.

A rough calculation using Eq. (3) indicates that the concentration of c_{00} can be as high as 20 times the concentration of the original polymer solution. The concentration of the sample zone, therefore, can be enhanced dramatically in FFF.

The relationship between viscosity and concentration of polymer solution is very complex. Several empirical equations are necessary to describe the viscosity behavior of a polymer solution's dependence upon concentration. As an example, the following equation can be used for dilute solutions [7]:

$$\eta = 0.54 + 1.3374C + 1.1593C^2 \tag{4}$$

where η is viscosity in centipoise and C is the concentration in grams per deciliter.

For the concentration where a microgel may be formed, the following equation is proposed [8]:

$$\eta = BM^3C^{3.7} \tag{5}$$

where B is a constant and M is the polymer's molecular weight.

Although various empirical equations can be found in the literature, the common aspect is that the viscosity of a polymer solution is highly dependent on concentration and molecular weight.

Temperature Dependence of Viscosity

The effect of temperature on the viscosity of the carrier can be expressed as [5]

$$\frac{1}{\eta} = a_0 + a_1 T + a_2 T^2 + a_3 T^3 \qquad (6)$$

where a_0, a_1, a_2, and a_3, are empirically obtained coefficients.

As Eq. (6) shows, the viscosity of a polymer solution is highly dependent on temperature. The sample zone of a high-MW polymer is pressed much closer to the cold wall in ThFFF. Its viscosity is more enhanced than with a low-MW polymer. The concentration effect, therefore, is more serious for high-MW polymers in ThFFF.

Molecular-Weight Dependence of Viscosity

If the temperature and concentration are kept the same, the viscosity of higher-MW polymer solution is higher, as shown by Eq. (2); thus, more distortion of its peak is expected, as shown by Fig. 1.

Unlike most of the elution separation methods, such as HPLC, GPC/SEC, gas chromatography (GC), and so forth, where the concentration of the sample zone will never be higher than the stock solution before injection, FFF will concentrate the samples, that is to say that sample concentration will be enhanced near the accumulative wall of FFF and the cold wall in ThFFF. The higher the MW of the polymer, the more the concentration will be enhanced and the lower the temperature of the sample zone will be. All three factors, concentration, temperature, and MW, contribute simultaneously to enhance the viscosity of the sample zone of the polymers in ThFFF. The viscosity of the sample zone can reach such extension that there is a tendency for the carrier fluid to flow over the top of the zone, with increased velocity in the region above the sample zone. The moving fluid will go over the sample zone, thus resulting in a longer retention for the sample zone; this is like a sticky slump going slowly on the floor of a river. A longer retention will be observed even if the flow rate of the carrier is constant.

When the MW of the polymer is so large that its zone is compressed close to the cold wall, the temperature of the sample zone becomes, essentially, the temperature of the cold wall, 25°C in many experiments. The viscosity is enhanced so much that the flow velocity of the carrier fluid is further distorted, so that deformed or double-topped peaks will be produced.

For the double-topped peaks, pseudo-gel, formed near the cold wall, is also proposed due to the low temperature and high concentration of the sample zone in ThFFF [2, 9]. The behavior of a pseudo-gel solution is quite different from the polymer solution from which it is formed. The diffusion coefficient of a pseudo-gel is much smaller than that of the original polymer, and the viscosity of the pseudo-gel solution will be much larger than that of the original polymer solution. The pseudo-gel, in theory, will be compressed closer to the cold wall and will elute out of the channel later than the parent molecules. However, as the size of the pseudo-gel cluster increases, hydrodynamic effects will result in an earlier emergence from the channel [3]. If either of these scenarios occurs in the ThFFF channel, double peaks might be observed for a sample of a single peak without "overloading."

Any attempts to obtain the parameters of the chromatograms and the physicochemical constants which are measurable in theory, by FFF, will be affected by the sample mass injected into the FFF channel. All of the concentration effects on the chromatograms discussed in the previous sections will be transferred, in turn, to those measured parameters and the physicochemical constants, such as the mass selectivity (S_m), the common diffusion coefficient (D), the thermal diffusion coefficient (D_T), and so forth. The increased retention of large polymers will result in enhanced mass selectivity in ThFFF. For a long time, this enhanced selectivity, in turn, the enhanced ThFFF universal calibration constant n, has led to confusion concerning the accuracy and repeatability of FFF, because different research groups have reported different data for selectivity and physicochemical constants measured by FFF for a given polymer–solvent combination [2,11]. Recent studies show that the enhanced selectivity and the different values of the physicochemical constants reported by different laboratories, measured by ThFFF, may be caused by different concentrations (sample mass) used by different laboratories.

References

1. K. D. Caldwell, S. L. Brimhall, Y. Gao, and J. C. Giddings, *J. Appl. Polym. Sci. 36*:703 (1988).
2. W. J. Cao, M. N. Marcus, P. S. Williams, and J. C. Giddings, *Int. J. Polym. Anal. Charact. 4*:407 (1998).
3. J. C. Giddings, *Science 260*:1456 (1993).
4. J. C. Giddings, *Anal. Chem. 66*:2783 (1994).
5. J. C. Giddings, F. J. F. Yang, and M. N. Myers, *Anal. Chem. 46*:1917 (1974).
6. K. D. Caldwell, S. L. Brimhall, Y. Gao, and J. C. Giddings, *J. Appl. Polym. Sci. 36*:703 (1988).

7. C. Tanford, *Physical Chemistry of Macromolecules*, John Wiley & Sons, New York, 1961, Chap. 6.
8. P. G. DeGennes, *Macromolecules 9*:594 (1976).
9. H. Tan, A. Moet, A. Hiltner, and E. Baer, *Macromolecules 16*:28 (1983).
10. M. Hoyos and M. Martin, *Anal. Chem. 66*:1718 (1994).
11. R. M. Sisson and J. C. Giddings, *Anal. Chem. 66*:4043 (1994).

Wenjie Cao
Mohan Gownder

Conductivity Detection in Capillary Electrophoresis

Introduction

In contrast to component-specific detectors, such as ultraviolet (UV) absorbance and fluorescence, conductivity detection is a universal detection method. This means that a bulk property (conductivity) of the buffer solution is continuously measured. A migrating ionic component locally changes the conductivity and this change is monitored. As such, conductivity detection is universally sensitive because, in principle, all migrating ionic compounds show detector response, although not to the same extent.

Types of Conductivity Detection

Two kinds of conductivity detector are distinguished: contact detectors and contactless detectors. Both types were originally developed for isotachophoresis in 0.2–0.5-mm-inner diameter (i.d.) PTFE tubes. Contactless detectors are based on the measurement of high-frequency cell resistance and, as such, inversely proportional to the conductivity. The advantage is that electrodes do not make contact with the buffer solution and are, therefore, outside the electric field. As these types of detectors are difficult to miniaturize down to the usual 50–75-μm capillar inner diameter, their actual application in capillary electrophoresis (CE) is limited.

Contact detectors are somewhat easier to miniaturize. There are generally two subtypes: those with twin axially mounted electrodes and those with twin or quadruple radially mounted electrodes. The former can be operated in DC mode or AC mode. In the DC mode, the detector signal directly originates from the field strength between the electrodes and, given the current, is inversely proportional to the detector cell resistance. In the AC mode, both axially and radially mounted electrodes form part of a closed primary circuit of an isolation transformer, the output of which is also inversely proportional to the cell conductivity. Alternatively, the output can be linearized with respect to the conductivity.

Conductivity Detector Response

As mentioned, the detector continually measures the conductivity of the buffer solution in the capillary. If an ionic component enters the detector cell, the local conductivity will change. At first glance, one would expect the conductivity to increase, because of additional ionic material. This is a simplified and incorrect approach, however. Suppose, in a buffer consisting of 0.01M potassium and 0.02M acetate (pH 4.7), a $10^{-4}M$ sodium solution is analyzed. Electroneutrality requires that with an increase of the sodium concentration from zero to, in this case, initially $10^{-4}M$, the potassium and/or charged acetate concentration cannot remain unchanged. This process is governed by the so-called Kohlrausch law. For strong ions, this equation reads

$$\Lambda = \sum_i \frac{c_i}{\mu_i}$$

in which Λ is the so-called Kohlrausch regulating function, c_i is the concentration of component i, and μ_i is the mobility of component i. Generally speaking, potassium will be partly displaced by sodium, whereas acetate will remain approximately (but not, by definition, exactly) constant. In the example given, the conductivity detector will give a negative response (see line A in Fig. 1), because potassium (with a high mobility and, hence, a higher contribution to conductivity) is, to some extent, replaced with sodium which has a ~30% lower mobility. From this example, it automatically follows that a potassium peak in a sodium acetate buffer, by contrast, will yield a positive amplitude. This makes interpretation of conductivity detector signals less straightforward.

Sensitivity of Conductivity Detection

A further example will illustrate aspects related to sensitivity. Suppose a 100 times more concentrated (10 mM)

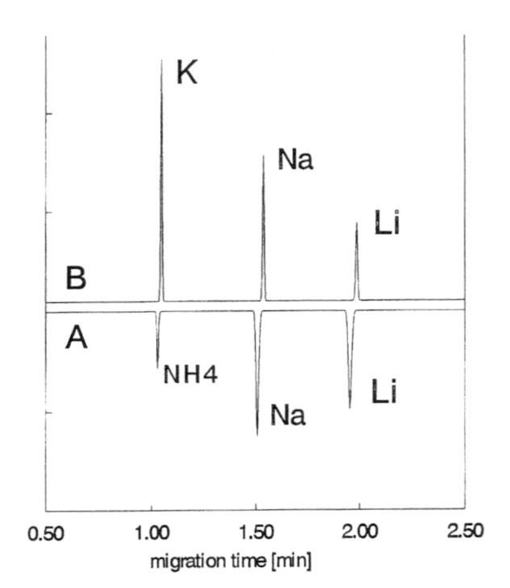

Fig. 1 Relative sensitivities in conductivity detection in CE. Trace A: sample of 10 m*M* NH_4, 0.1 m*M* Na, and 0.005 m*M* Li in a 0.01*M* potassium–acetate buffer; trace B: sample of 0.1 m*M* each of K, Na, and Li in a 10 m*M* Tris–acetate buffer.

solution of ammonium is coseparated in the potassium–acetate system mentioned earlier. Naturally, ammonium will displace potassium, but as the mobilities of potassium and ammonium are almost equal, the resulting change in conductivity is minor. Sensitivity in this example is, consequently, very low (line A in Fig. 1). On the other hand, 0.005 m*M* lithium has a much lower conductivity than sodium and, consequently, shows a higher specific response (line A in Fig. 1).

Generally, one cannot expect a high sensitivity anyhow, as the background signal (originating from the buffer) is generally much higher than the eventual change superimposed upon that background. One might argue that background conductivity can easily be decreased by diluting the buffer. Potential gain with this approach is very limited, because diluting the buffer below an ionic strength of 1 m*M* will lead to unacceptable loss in buffering capacity and, moreover, in severe sample overload. Another possibility to decrease the background conductivity is to use buffer components with lower mobility, such as GOOD buffers. This, however, will sooner lead to nonsymmetric peaks on sample overload (peak triangulation). Using low-mobility Tris as a buffer co-ion will lead to positive peaks for 0.1 m*M* potassium, sodium, and lithium alike (line B in Fig. 1).

Suggested Further Reading

Beckers, J. L., Isotachophoresis, some fundamental aspects, *Thesis*, Eindhoven University of Technology, 1973.

Everaerts, F. M., J. L. Beckers, and Th. P. E. M. Verheggen, *Isotachophoresis: Theory, Instrumentation and Applications*, Elsevier, Amsterdam, 1976.

Hjertén, S., *Chromatogr. Rev. 9*:122 (1967).

Kohlrausch, F., *Ann. Phys. (Leipzig) 62*:209 (1897).

Li, S. F. Y., *Capillary Electrophoresis — Principles, Practice and Applications*, Elsevier, Amsterdam, 1992.

Reijenga, J. C., Th. P. E. M. Verheggen, J. H. P. A. Martens, and F. M. Everaerts, *J. Chromatogr. A 744*:147 (1996).

Jetse C. Reijenga

Conductivity Detection in HPLC

Conductivity detection is used to detect inorganic and organic ionic species in liquid chromatography. As all ionic species are electrically conducting, conductometric detection is a universal detection technique, considered as the mainstay in high-pressure ion chromatography, in the same way as is ultraviolet (UV) detection in high-performance liquid chromatography (HPLC).

The principle of operation of a conductivity detector lies in differential measurement of mobile-phase conductivity prior to and during solute ion elution. The conductivity cell is either placed directly after an analytical column or after a suppression device required to reduce background conductivity, in order to increase the signal-to-noise ratio and, thus, sensitivity. In the first mode, known as *nonsuppressed* or *single-column ion chromatography*, aromatic acid eluents are used, with low-capacity fixed-site ion exchangers and dynamically or permanently coated reversed-phase columns. In the second mode, known as *eluent-suppressed ion chromatography*, the separated ions are detected by conductance after passing through a suppression column or a membrane, to convert the solute ions to higher conducting species (e.g., hydrochloric acid in the case of chloride ions and sodium hydroxide in the case of sodium ions). In the meantime, the eluent ions are converted to a low-residual-conductivity medium such as carbonic acid or water, thus reducing background noise.

Conductance G is the ability of electrolyte solutions in an electric field applied between two electrodes to transport current by ion migration. According to Ohm's law, ohmic resistance R is given by

$$R = \frac{U}{I} \tag{1}$$

where U is the voltage (V) and I is the current intensity (A). The reciprocal of ohmic resistance is the conductance G, where

$$G = \frac{1}{R} \tag{2}$$

expressed in Siemens in the International System of Units (SI), formerly reported in the literature as mho (Ω^{-1}). The measured conductance of a solution is related to the interelectrode distance d (cm) and the microscopic surface area (A) (geometric area × roughness factor) of each electrode (A is assumed identical for the two electrodes) as well as the ionic concentration, given by

$$G = \frac{kA}{d} \tag{3}$$

where k is the specific conductance or conductivity. The ratio d/A is a constant for a particular cell, referred as the cell constant K_c (cm^{-1}) and is determined by calibration. The usual measured variable in conductometry is conductivity k (S/cm):

$$k = GK_c \tag{4}$$

The conductance G (in μS) of a solution is given by

$$G = \frac{(\lambda^+ + \lambda^-)CI}{10^{-3}K_c} \tag{5}$$

where λ^+ and λ^- are limiting molar conductivities of the cation and anion, respectively, and C is the molarity and I the fraction of eluent that is ionized. If the eluent and solute are fully ionized, the conductance change accompanying solute elution is

$$\Delta G = \frac{(\lambda_s - \lambda_e)C_s}{10^{-3}K_c} \tag{6}$$

The specific conductance/conductivity k (S/cm) of salts measured by a conductivity detector is given by

$$k = \frac{(\lambda_{s+} + \lambda_{s-})C_s + (\lambda_{e+} + \lambda_{e-})C_e}{1000} = \frac{\Lambda_s C_s + \Lambda_e C_e}{1000} \tag{7}$$

where C_s and C_e are the concentration (mol/L) of the solute and eluent ions, respectively, and Λ is the molar conductivity of the electrolyte.

The change in conductance when a sample solute band passes through the detector results from replacement of some of the eluent ions by solute ions, although the total ion concentration C_{tot} remains constant:

$$C_{tot} = C_s + C_e \tag{8}$$

The background ion conductivity when $C_s = 0$ is

$$k_1 = \frac{\Lambda_e C_{tot}}{1000} \tag{9}$$

When a solute band is eluted, the ion conductivity k_2 is given by

$$k_2 = \frac{\Lambda_e C_{tot}}{1000} + \frac{(\Lambda_s - \Lambda_e)C_s}{1000} \tag{10}$$

The difference in conductivity is obtained after subtraction of the first equation from the second:

$$\Delta k = k_2 - k_1 = \frac{(\Lambda_s - \Lambda_e)C_s}{1000} \tag{11}$$

From Eq. (11), it is obvious that when a sample band is eluted, the observed difference in conductivity is proportional to the concentration of the sample solute C_s. However, the linear relation holds only for dilute solutions, as Λ is itself dependent on concentration, according to Kohlrausch's law:

$$\Lambda = \Lambda^\circ - A\sqrt{C} \tag{12}$$

where A is a constant and Λ° is the limiting molar conductivity in an infinitely dilute solution, given by the sum

$$\Lambda^\circ = \Lambda^{\circ+} + \Lambda^{\circ-} \tag{13}$$

or

$$\Lambda^\circ = \nu_+\lambda_+^\circ + \nu_-\lambda^\circ \tag{14}$$

where ν_+ and ν_- represent stoichiometric coefficients for the cation and anion, respectively, in the electrolyte.

Eq. (11) shows that the signal observed during solute ion elution is also proportional to the difference in limiting molar ionic conductivities between the eluent and the solute ions.

Values of limiting molar ionic conductivities for a few common ions are shown in Table 1. The data tabulated are referred to 25°C temperature. The term *limiting molar ionic conductivity* is used according to IUPAC recommendation, rather than the formerly used *limiting ionic equivalent conductivity*. The molar and equivalent values are interconvertible through stoichiometric coefficient z.

Table 1 Limiting Molar Ionic Conductivities of some Anions and Cations (S cm^2/mol) at 25°C

Anions	λ^-	Cations	λ^+
OH^-	199.1	H^+	349.6
F^-	55.4	Li^+	38.7
Cl^-	76.4	Na^+	50.1
Br^-	78.1	K^+	73.5
I^-	76.8	NH_4^+	73.5
NO_3^-	71.46	Mg^{2+}	106
NO_2^-	71.8	Cu^{2+}	107.2
SO_4^{2-}	160.0	Ca^{2+}	120
Benzoate$^-$	32.4	Sr^{2+}	118.9
Phthalate^{2-}	76	Ba^{2+}	127.2
Citrate^{3-}	168	Ethylammonium	47.2
CO_3^{2-}	138.6	Diethylammonium	42.0
$C_2O_4^{2-}$	148.2	Triethylammonium	34.3
PO_4^{3-}	207	Tetraethylammonium	32.6
CH_3COO^-	40.9	Trimethylammonium	47.2
$HCOO^-$	54.6	Tetramethylammonium	44.9

Conductivity is measured by applying an alternating voltage to two electrodes of various geometric shapes in a flow-through cell, which results in anion migration, as negatively charged, toward the anode (positive electrode) and cation migration, as positively charged, toward the negative electrode (cathode). An AC potential (frequency 1000–5000 Hz) is required in order to avoid electrode polarization. The cell current is measured and the solution's resistance (or more strictly the impedance) is calculated by Ohm's law. Conductance is further corrected by the conductivity cell constant, thus giving conductivity.

The requirements for a typical conductivity detection cell are small volume (to eliminate dispersion effects), high sensitivity, wide linear range, rapid response, and acceptable stability. The cell generally consists of a small-volume chamber (<5 μL) fitted with two or more electrodes constructed of platinum, stainless steel, or gold.

Most conductivity detectors function according to the Wheatstone Bridge principle. What is actually measured is resistance of the solution. Electronically, the electrodes are arranged in that way to constitute one arm of a Wheatstone Bridge. Eluting ions from chromatographic column subsequently enter the detector cell, leading to a change of electrical resistance and the out-of-balance signal is rectified with a precision rectifier. The DC signal is either digitized and sent to a computer data acquisition system or is passed to a potentiometric recorder, by means of a linearizing amplifier, which modifies the signal so that the output is linearly related to ion concentration. Sometimes, a variable resistance is situated in one of the other arms of the bridge and is used for zero adjustment to compensate for any signal from mobile-phase ions. As mentioned earlier, at constant voltage applied to the cell, the current will be proportional to the conductivity (Fig. 1).

The conductivity k is a characteristic property of the solution rather than a property of the cell used. It contains all the chemical information available from the measurement, such as concentration and mobilities of the ions present. Accordingly, the conductivity detector is a bulk property detector and, as such, it responds to all electrolytes present in the mobile phase as well as the solutes.

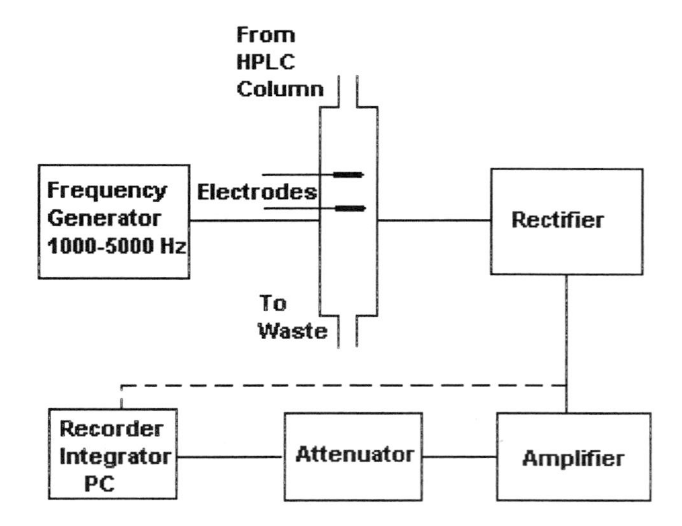

Fig. 1 Block diagram of electrical conductivity detector in HPLC.

Thus, the experimentally determined conductivity is the sum of the contributions from all ions present in the solution. The sensitivity of the conductivity detector depends on the difference between the limiting ionic conductivities of the solute and eluent ions.

The differential mode of detection is mostly effective, provided there is a significant difference in the values of the measured property between the eluent and solute ions. This difference may be either positive or negative. The former case refers to lower conductivity of the eluent ion, described as *direct detection method*, the latter to greater conductivity of the eluent ion, described as *indirect detection method* (Fig. 2).

The thermal stability of a conductivity detector is of great importance. Effective thermostating is highly required, as the temperature greatly affects the mobility of ions and, therefore, conductivity. A 0.5–3% increase of conductivity is usually expected per degree Celsius. Close temperature control is necessary to minimize background noise and maximize sensitivity; this is an especially important issue if nonsuppressed eluents are used.

Typical specifications for an electrical conductivity detector are as follows: sensitivity, 5×10^{-9} g/mL; linear dynamic range, 5×10^{-9} to 1×10^{-6} g/mL; response index, 0.97–1.03.

Conductivity detection in HPLC or, more precisely HPIC, can be applied to ionic species, including all anions and cations of strong acids and bases (e.g., chloride, sulfate, sodium, potassium, etc.). Ions of weaker acids and bases are detected provided that the pH value of the eluent is chosen to maximize the analyte's ionization so as to increase sensitivity. The relatively simple design requirements, accuracy, and low cost contribute to its utility and popularity; thus, it is almost used in over 95% of analyses, where ion-exchange separation procedures are involved.

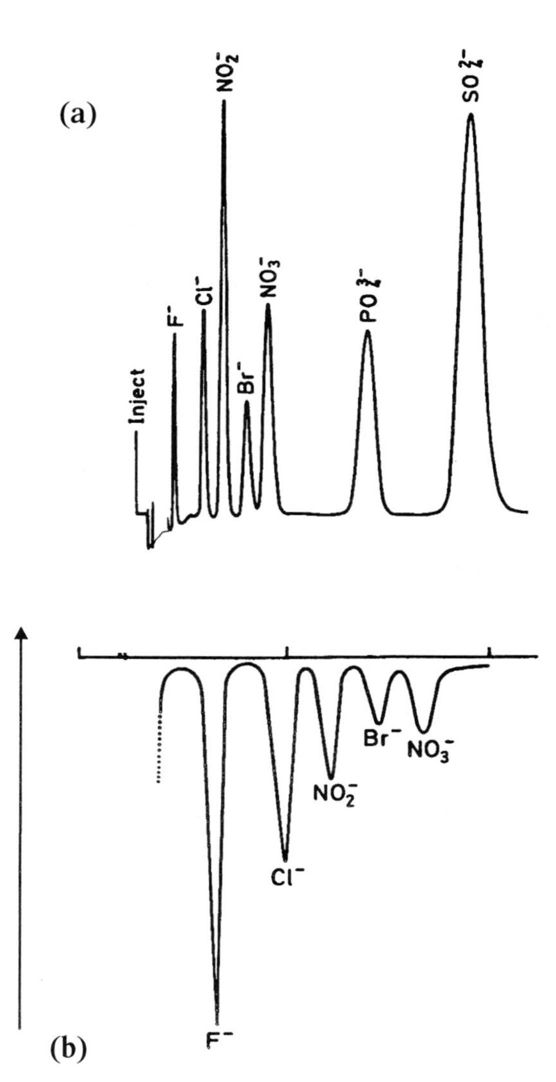

Fig. 2 Conductivity detection of anions in nonsuppressed (single column) ion chromatography using an eluent of (a) low background conductance (direct detection) and (b) high background conductance (indirect detection). The direction of the arrow indicates the increase of conductivity.

Suggested Further Reading

Coury, L., *Curr. Separ. 18*(3):91 (1999).

Papadoyannis, I., V. Samanidou, and A. Zotou, *J. Liquid Chromatogr. 18*(7):1383 (1995).

Parriott, D., *A Practical Guide to HPLC Detection*, Academic Press, San Diego, CA, 1993.

Schaefer, H., M. Laubli, and R. Doerig, *Ion Chromatography*, Metrohm Monograph 50143, Metrohm AG, Herisau, 1996.

Scott, R., *Techniques and Practice of Chromatography*, Marcel Dekker, Inc., New York, 1995.

Scott, R., *Chromatographic Detectors, Design, Function and Operation*, Marcel Dekker, Inc., New York, 1996.

Tarter, J., *Ion Chromatography*, Chromatographic Science Series Vol. 37, Marcel Dekker, Inc., New York, 1987.

Ioannis N. Papadoyannis
Victoria F. Samanidou

Congener-Specific PCB Analysis

Polychlorinated biphenyls (PCBs) are complex mixtures of 209 possible chlorinated biphenyl molecules, referred to as congeners. There are from 3 to 46 isomers at each of the 10 possible levels of chlorination. Isomers of a given chlorination level are referred to as homologs. About 150 of these congeners appear at significant levels in the commercial mixtures. These mixtures, trade named Aroclor (U.S.A.), Clophen (Germany), Kanechlor (Japan), and so forth, found use as electrical insulating fluids in transformers and capacitors and as binders for a wide variety of uncontained applications. Although their manufacture has been largely discontinued, their long-term stability, dispersion into the environment by prior uncontrolled releases, lipophilicity (resulting in biomagnification up food chains), and potential toxicity to humans and biota have sparked extensive research and the requirement for detailed analytical characterization of these mixtures.

This article will not discuss the extensive literature on sample preparation, cleanup, and proper instrumental operation. Adsorption column chromatography and high-performance liquid chromatography (HPLC) procedures find application here, and the book by Erickson [1] provides exhaustive discussions of these and of the history of PCB use and analysis. Methods for measuring total PCB content or measuring and reporting the mixtures by their commercial designation will not be detailed. Congener-specific PCB analysis demands separation and quantitation of either short lists of priority PCB congeners or of the PCB content of all chromatographic PCB peaks that can be separated in particular system(s). This latter mode is referred to as Comprehensive, Quantitative, Congener-Specific Analysis (CQCS). The methods of choice for CQCS PCB analysis employ high-resolution gas chromatography (HRGC) on capillary columns with sensitive and selective detection by electron-capture detectors (ECD), selected ion monitoring–mass spectrometry (MS–SIM), or full-scan, ion-trap MS (ITMS). The most complete discussion of target congeners for specific research applications is in Ref. 2. A descriptive overview of CQCS PCB analysis appears in an *Analytical Chemistry* A-page article [3], and extensive reviews [4–6] provide detailed discussions and large bibliographies.

Figure 1 summarizes PCB congener structure, nomenclature, the Ballschmiter and Zell (BZ) congener numbering system, and the relative abundances in the commercial Aroclor mixtures as a function of single-ring chlorine-substitution patterns. The BZ numbers in the matrix correspond to the chlorine-substitution positions in each ring of the biphenyl structure, which are listed along the top and right sides of the figure matrix. The abbreviated nomenclature (e.g., 234–245 = PCB #138) defines each congener by the substitution pattern in each ring, with the hyphen representing the bond between the two phenyl rings. Rotation about this bond is possible except in congeners with three or four chlorines in the ortho (2 or 6) positions.

In the United States, the commercial mixtures were manufactured until 1977 by Monsanto under the trade name Aroclor. In the four-digit Aroclor designations (e.g., Aroclor 1242), the 12 indicates a biphenyl nucleus and the 42 the weight percentage of chlorine in the mixture. Reference to the matrix in Fig. 1 reveals congeners in black cells which never exceed 0.1 wt% in the mixtures. These "non-Aroclor" congeners are absent due to unfavored or improbable formation in the electrophilic chlorination process employed in the manufacture of Aroclors [7]. Conversely, the ring chlorine-substitution patterns giving rise to the congeners in gray cells are especially favored.

Whereas three or more *ortho*-chlorines block rotation about the ring-connecting bond, congeners with none, or only one, *ortho*-chlorine can relatively easily achieve a planar configuration (colloquially referred to as "coplanars") and may behave as isosteres (compounds with similar shape, functionality, and polarity) to 2,3,7,8 — tetrachlorodibenzodioxin (TCDD). These bind significantly to the "dioxin receptor," and measurement of their concentrations can be combined with their "dioxin-like toxic equivalency factors (TEFs)" to give an estimate of a PCB mixture's "dioxin-like toxic equivalency (TEQ)" [2]. Thus, one "short list" analysis specified by the World Health Organization (WHO) is for 12 such congeners found in commercial mixtures, namely PCBs 77, 81, 105, 114, 118, 123, 126, 156, 157, 167, 169, and 189. Although PCB 126 is generally a trace component, it has such a high TEF that it often dominates the TEQ calculation.

In the United States, the initial regulatory methods were the USEPA 8080 series. In 8080, packed column GC–ECD was recommended and calibration was against Aroclor mixtures and results were reported as Aroclor equivalents. Method 8081 encouraged use of higher-resolution capillary GC columns and MS–SIM detection, and the current version 8082 extends this to suggest measuring some individual congeners against primary standards. An early version of CQCS analysis is EPA Method 680, which employs MS–SIM detection at the molecular ion cluster mass for each homolog level. It does not provide for actually identifying all congeners by determining their elution

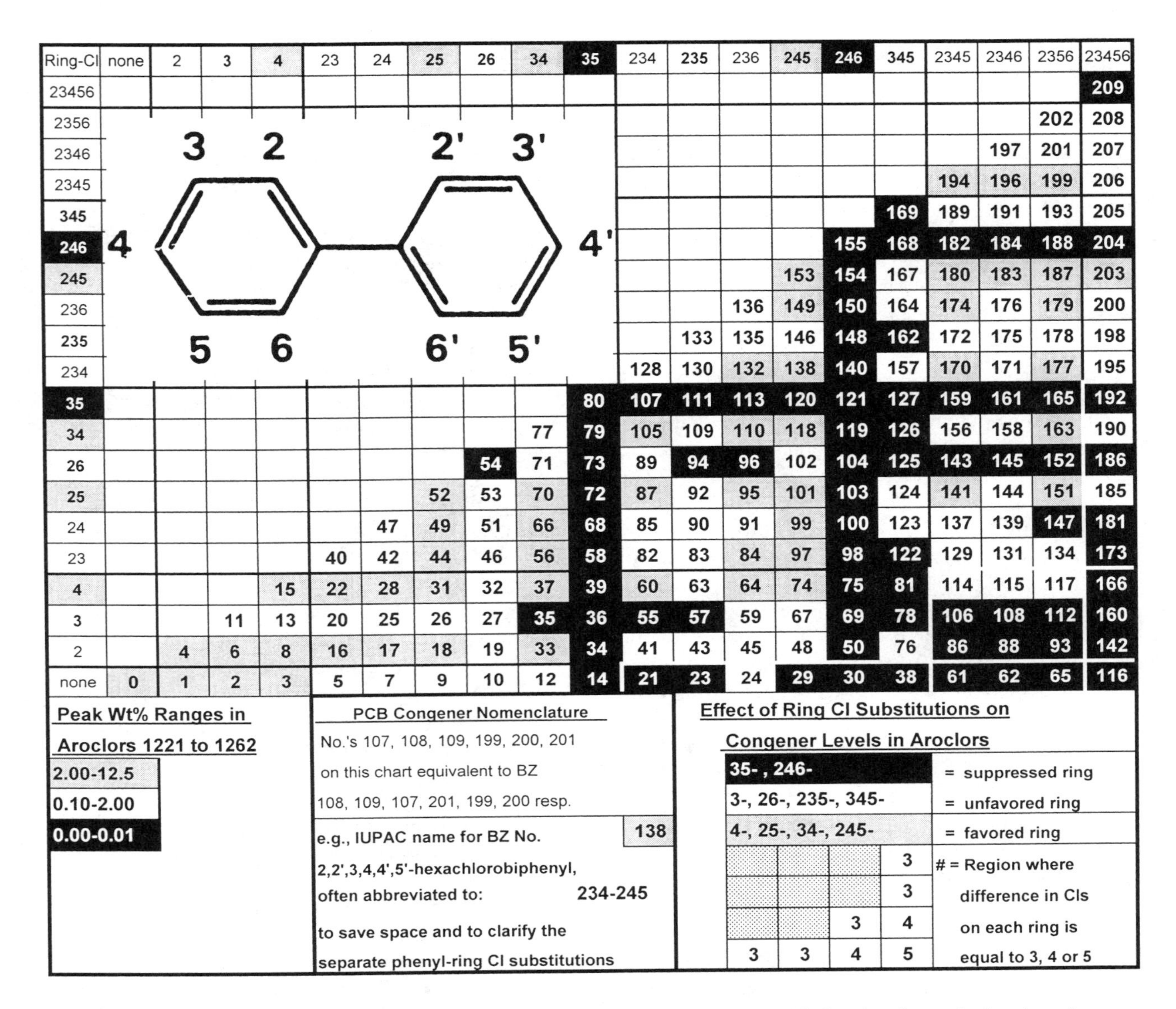

Ring-Cl	none	2	3	4	23	24	25	26	34	35	234	235	236	245	246	345	2345	2346	2356	23456
23456																				209
2356																			202	208
2346																		197	201	207
2345																	194	196	199	206
345																169	189	191	193	205
246															155	168	182	184	188	204
245														153	154	167	180	183	187	203
236													136	149	150	164	174	176	179	200
235												133	135	146	148	162	172	175	178	198
234											128	130	132	138	140	157	170	171	177	195
35										80	107	111	113	120	121	127	159	161	165	192
34									77	79	105	109	110	118	119	126	156	158	163	190
26								54	71	73	89	94	96	102	104	125	143	145	152	186
25							52	53	70	72	87	92	95	101	103	124	141	144	151	185
24						47	49	51	66	68	85	90	91	99	100	123	137	139	147	181
23					40	42	44	46	56	58	82	83	84	97	98	122	129	131	134	173
4				15	22	28	31	32	37	39	60	63	64	74	75	81	114	115	117	166
3			11	13	20	25	26	27	35	36	55	57	59	67	69	78	106	108	112	160
2		4	6	8	16	17	18	19	33	34	41	43	45	48	50	76	86	88	93	142
none	0	1	2	3	5	7	9	10	12	14	21	23	24	29	30	38	61	62	65	116

Fig. 1 PCB congener structure, nomenclature, BZ congener numbering system, and relative abundances in Aroclor mixtures.

times and is, thus, not classified as congener-specific, but rather homolog-class-specific. It is quantitatively calibrated against an average response at each level, resulting in less precise measurement of individual congeners, but it will detect and measure all PCB-containing peaks whether Aroclor derived or not. It is, thus, superior to the 8080 series when a PCB mixture arising from a non-Aroclor source or a substantially altered Aroclor congener distribution is encountered. In Europe, the Community Bureau of Reference (BCR) specifies measurement of a short list of major persistent "indicator congeners," namely PCBs 28, 52, 101, 118, 138, 153, and 180. A number of other congener short lists are detailed in Refs. 2 and 3. A powerful but expensive and difficult-to-implement congener-specific PCB analysis is USEPA Method 1668. The target list is the WHO list of coplanar PCBs with dioxinlike TEQs. The methodology is HRGC with >10,000 resolution HRMS detection and ^{13}C-isotope dilution internal standards for all the analytes. The procedures are modeled on the well-established HRGC–HRMS USEPA Method 1613 for PCDD/Fs. The newer USEPA Method 1668 Revision A (December 1999) describes procedures for extending the analyte list to all 209 congeners which can be resolved on either a SPB-Octyl capillary column or a

DB-1 (100% methyl silicone) capillary. Primary standards for all 209 congeners are distributed among 5 calibration solutions, which avoid any isomer co-elutions on the SPB-Octyl column.

No single column, nor any pair of columns, can completely separate all 209 congeners, or even the 150 or so found in Aroclors. Analysts developing CQCS or even "short list" congener-specific PCB analyses must select GC stationary phases capable of resolving congeners in their target list. Many analysts have employed 5% phenyl-, 95% methyl-substituted silicone polymers (e.g., DB-5) since a very similar phase was the first one for which the relative retention times for all 209 PCB congeners were published [3]. Methyl silicone phases with 50% *n*-octyl or *n*-octadecyl substituents have PCB retention characteristics similar to those of hydrocarbon columns such as Apeizon L or Apolane, but much greater stability and higher temperature limits than the latter. They permit resolution of many pairs of lower homologs which co-elute on the more polar phases. This feature is valuable for characterizing the products of dechlorination by anaerobic bacteria [7]. Phases with arylene or carborane units substituted in the silicone backbone to decrease column bleed (e.g., DB-5MS, DB-XLB, HT-8) have been found to have particularly useful congener-separation capabilities [3,7–9].

A database of relative retention times and co-elutions for all 209 congeners on 20 different stationary phases has been published [8]. For 12 of the most useful of these phases, the elution orders of 9 solutions of all 209 congeners are available from a standard supplier which markets these solutions (AccuStandard, New Haven, CT, U.S.A.). By surveying the database, one can determine the most suitable column(s) for a particular application and can quickly establish a method component table by nine injections of the standard mixtures. This greatly facilitates the development of new CQCS PCB analyses. Tables of the weight percentages of all congeners in each of the numbered Aroclor mixtures, from which the information condensed in the figure matrix were derived, are available [7,8]. These help reduce the number of congeners which a CQCS method is required to separate when one anticipates analyzing only relatively unaltered Aroclor congener mixtures.

Prior to the availability of all 209 congeners in well-designed primary standard mixtures, much effort was expended to use structure–retention relationships on various phases to predict retention for congeners for which standards were not available [5]. In general, PCB retention times increase with chlorination level, and within chlorination levels, with less chlorine substitution in the ortho position (i.e., "coplanar PCBs" are more strongly retained). These relationships are of theoretical interest but are of less use now that accurate retention time assignments are possible with actual standards. The use of commercial mixtures such as Aroclors as quantitative secondary standards for CQCS PCB analysis is now to be discouraged [4], as detailed studies of congener distributions show significantly different proportions among different lots [7]. In the case of Aroclor 1254, there are actually two different mixtures of radically different composition produced by totally different synthetic processes [9].

The other major factor affecting the capability of CQCS PCB analyses is the selection of the GC detector. Initially, the ECD has been most useful for this application. It is selective for halogenated compounds, and its sensitivity is outstanding for the more chlorinated ($Cl \geq 4$) congeners. Its drawbacks are twofold: It has a limited linear range, necessitating multilevel calibration, and the relative response factors vary widely from instrument to instrument and among congeners even at the same chlorination level [3]. For mono- and dichloro-substituted congeners, it is less sensitive than the corresponding MS detectors. Other halogenated compounds such as organochlorine pesticides will produce ECD responsive peaks which may interfere by co-elution with certain PCB congeners. For these reasons, CQCS PCB analyses with ECD detection often employ a procedure of splitting the injected sample to two columns (each with ECD detector) of different polarity and PCB congener elution order [3, 7]. To be reported, a congener must be measured on at least one column without co-elution of PCB or another interfering compound. If separately measurable on each column, the quantities found must match within a preset limit to preclude the possibility of an unexpected co-eluting contaminant on one of the columns. Given the large number of congeners which may need to be measured, the data reduction algorithm for such a procedure is complex and not easily automated.

Another approach to providing a second dimension to CQCS PCB analysis is to employ much more selective mass spectrometric detection [3,6–8]. In EI–MS, the spectra consist of a molecular ion cluster of chlorine isotope MS peaks and similar fragment ion clusters resulting from the successive loss of chlorine atoms. Congeners differing by one chlorine substituent which co-elute on the GC column may often be separately quantitated by MS detection, if the more chlorinated one is not in great excess. This is because the $[M - 1Cl]^+$ fragment which interferes with the lower congener's signal is from a ^{13}C isotope peak and typically has 0.5–12% the signal level of its M^+ peak [9]. In contrast to the ECD, the sensitivity of MS–SIM or full-scan ITMS is greater for the less chlorinated congeners, as their electron affinity is lower and the positive charge of the ions is distributed over a smaller number of fragments. The linearity of the MS detectors is better than that of ECDs, and the ions monitored are more specific for PCBs and less prone to interference from non-PCB compounds. ECDs continue to hold the edge in absolute sensitivity (for the higher chlorinated congeners), and the dual-column/ECD detector systems are slightly less expensive than comparable bench-top, unit-mass-resolution, single-column GC–MS systems. Application to PCB analysis of more advanced (and expensive) MS detection systems, such as high-resolution mass spectrometry (HRMS), MS–MS, and negative-ion MS, is described in several reviews [4,6].

A final refinement of congener-specific PCB analysis arises from the fact that 19 of the congeners actually exist as stable enantiomeric pairs, either component of which can withstand racemization even at the elevated temperatures required to elute them from a capillary GC separation [6]. Some congeners containing either a 236- or 2346-chlorine-substituted ring *and* three or more chlorines in the ortho position exist in two mirror-image forms by virtue of their inability to rotate around the bond between the two rings. These so-called atropisomers do *not* contain asymmetric carbon centers. They are PCB numbers 45, 84, 91, 95, 132, 135, 136, 149, 174, and 176 (containing the 236-ring), as well as PCB numbers 88, 131, 139, 144, 171, 175, 176, 183, 196, and 197 (containing the 2346-ring). They may be separated on chiral GC stationary phases, primarily those employing a family of modified cyclodextrins. A series of 7 such columns has been found, which among them can achieve resolution of all 19 stable PCB atropisomers as well as separation of 11 of them from other possible coeluting PCBs if MS detection is employed [10]. Observation of PCB enatiomeric ratios significantly different from 1 is a certain indication of the action of an enzyme-mediated biological process operating on these congeners.

References

1. M. D. Erickson, *Analytical Chemistry of PCBs*, 2nd ed., Lewis Publishers, New York, 1997.
2. L. G. Hansen, *The* ortho *Side of PCBs: Occurrence and Disposition*, Kluwer Academic, Boston, 1999.
3. G. M. Frame, Congener-specific PCB analysis, *Anal. Chem. 69*:468A (1997).
4. P. Hess, J. de Boer, W. P. Cofino, P. E. G. Leonards, and D. E. Wells, Critical review of the analysis of non- and mono-ortho-chlorobiphenyls, *J. Chromatogr. A 703*:417 (1995).
5. B. R. Larsen, HRGC separation of PCB congeners, *J. High Resolut. Chromatogr., 18*:141 (1995).
6. J. W. Cochran and G. M. Frame, Recent developments in the high resolution gas chromatography of polychlorinated biphenyls, *J. Chromatogr. A 843*:323 (1999).
7. G. M. Frame, J. W. Cochran, and S. S. Bøwadt, Complete PCB congener distributions for 17 Aroclor mixtures determined by 3 HRGC systems optimized for comprehensive, quantitative, congener-specific analysis, *J. High Resolut. Chromatogr. 19*:657 (1996).
8. G. M. Frame, A collaborative study of 209 PCB congeners and 6 Aroclors on 20 different HRGC columns: 1. Retention and coelution database, 2. Semi-quantitative Aroclor distributions, *Fresenius J. Anal. Chem. 357*:701 (1997).
9. G. M. Frame, Improved procedure for single DB-XLB column GC–MS–SIM quantitation of PCB congener distributions and characterization of two different preparations sold as "Aroclor 1254", *J. High Resolut. Chromatogr. 22*: 533 (1999).
10. C. S. Wong and A. W. Garrison, Enantiomer separation of polychlorinated biphenyl atropisomers and polychlorinated biphenyl retention behavior on modified cyclodextrin capillary gas chromatography columns, *J. Chromatogr. A 866*:213 (2000).

George M. Frame II

Copolymer Analysis by LC Methods, Including Two-Dimensional Chromatography

Introduction

Gel permeation chromatography (GPC) is the established method for the determination of molar mass averages and the molar mass distributions of polymers. GPC retention is based on the separation of macromolecules in solution by molecular sizes and, therefore, requires a molar mass calibration to transform elution time or elution volume into molar mass information. This kind of calibration is typically performed with narrow molecular mass distribution polymer standards, universal, or broad calibration methods or molar-mass-sensitive detectors like light-scattering or viscosity detectors.

Copolymer GPC Analysis by Multiple Detection

Conventional GPC data processing is unable to determine other important polymer properties such as copolymer composition or copolymer molar mass. The reason is that the GPC separation is based on hydrodynamic volume rather than the molar mass of the polymer and that molar mass calibration data are only valid for polymers of identical molecular structures. This means that polymer topology (e.g., linear, star-shaped, comb, ring, or branched polymers), copolymer composition, and chain conformation (isomerization, tacticity, etc.) determine the apparent molecular weight. The main problem of copolymer analysis is the calibration of the size-exclusion chromatography (SEC) instrument for copolymers with varying comonomer compositions. However, even if the bulk composition is constant, second-order chemical heterogeneity has to be taken into account (i.e., the composition will vary for a given chain length, in general).

Several attempts have been made to solve the calibration dilemma. Some are based on the universal calibration concept, which has been extended for copolymers. Another approach to copolymer calibration is multiple detection [1]. The advantage of multiple detection can be seen in its flexibility and its ability to yield the composition distribution as well as molar masses for the copolymer under investigation. This method requires a molar mass calibration and an additional detector response calibration to determine chemical composition at each point of the elution profile. No other kind of information, parameters, or special equipment are necessary to do this kind of analysis [2] and to calculate compositional drift, bulk composition, and copolymer molar mass.

Determination of Comonomer Concentration

In order to characterize the composition of a copolymer of k comonomers, the same number of independent detector signals d are necessary in the GPC experiment; that is, in the case of a binary copolymer, two independent detectors (e.g., LUV and RI) are required to calculate the composition distribution $w_k(M)$ and the overall (bulk) composition $\overline{w}_k$. The detector output U_d of each detector d is the superposition of all individual responses from all comonomers present in the detector cell at a given elution volume V. Therefore,

$$U_d(V) = \sum_d f_{dk} c_k(V) \tag{1}$$

with f_{dk} the response factor of comonomer k in detector d, and c_k the true concentration of comonomer k in the detector cell at elution volume V. The detector response factors are determined in the usual way by injecting homopolymers for each comonomer of known concentration and correlating that with the area of the corresponding peak. If no homopolymers are available, model compounds have been used to estimate the detector response factors. In the case of a binary copolymer, the weight fraction, w_A, of comonomer A is then given by

$$W_A(V) = \left(1 + \frac{[U_1(V) - (f_{1B}/f_{2B})U_2(V)][f_{1A} - (f_{1B}/f_{2B})f_{2A}]}{[U_1(V) - (f_{1A}/f_{2A})U_2(V)][f_{1B} - (f_{1A}/f_{2A})f_{2B}]}\right)^{-1} \tag{2}$$

Obviously, the sum of all comonomer weight fractions is unity. The accurate copolymer concentration and the distribution of the comonomers across the chromatogram can be calculated from the apparent chromatogram and the individual comonomer concentrations [3].

The accuracy of the compositional information is not affected by the polymer architecture. Deviations from the true comonomer ratios are only possible if the detected property is dependent on the local environment. This is the case if neighbor-group effects exist. The possibility of electronic interactions causing such deviations is very small, because there are too many chemical bonds between two different monomer units. Other types of interactions, especially those which proceed across space (e.g., charge-transfer interactions), may influence composition accuracy [4].

Determination of Copolymer Molar Mass Averages

The major difficulty in the determination of the copolymer molar mass distribution is the fact that the GPC separation is based on the molecular size of the copolymer chain. Its hydrodynamic radius, however, is dependent on the type of the comonomers incorporated into the macromolecule as well as their placement (sequence distribution).

Consequently, there can be a co-elution of species possessing different chain length *and* chemical composition. The influence of different comonomers copolymerized into the macromolecule on the chain size can be measured by the GPC elution of homopolymer standards of this comonomer. Unfortunately, the influence of the comonomer sequence distribution on the hydrodynamic radius cannot be described explicitly by any theory at present. However, there are limiting cases which can be discussed to evaluate the influence of the comonomer placement in a macromolecular chain.

From a GPC point of view, the most simple copolymer is an alternating copolymer $(AB)_n$, which can be treated exactly like a homopolymer with a repeating unit (AB). The next simple copolymer architecture is a AB block copolymer, where a sequence of comonomer A is followed by a block of B units. The only heterocontact in this chain is the A–B link, which can influence the size of the macromolecule. The A segment and the B segment of the AB block copolymer will hydrodynamically behave like a pure homopolymer of the same chain length. In the case of long A and B segments in the AB block copolymer, only the A–B link acts as a defect position and will not change the overall hydrodynamic behavior of the AB block copolymer chain. Consequently, the molar mass of the copolymer chain can be approximated by the molar masses of the respective segments. Similar considerations are true for ABA, ABC, and other types of block structure and for comb-shaped copolymers with low side-chain densities.

In such cases, the copolymer molar mass M can be determined from the interpolation of homopolymer calibration curves $M_k(V)$ and the weight fractions w_k of the comonomers k according to [1]

$$\log M_c(V) = \sum_k w_k(V) \log M_k(V) \tag{3}$$

The calculation of copolymer molar mass averages $M_{n,c}$, $M_{w,c}$, and so on, and copolymer polydispersity D is done as in conventional GPC calculations using the copolymer molar mass calculated from Eq. (3).

In cases where the number of heterocontacts can no longer be neglected, this simplified reasoning breaks down and copolymer molar masses cannot be measured accurately by GPC alone. This is the case with statistical copolymers, polymers with only short comonomer sequences and high side-chain densities [2]. In such cases, more powerful and universal methods have to be employed (e.g., 2D separations) (see below).

Copolymer Characterization by GPC

Block copolymers are an important class of polymers used in many applications from thermoplastic elastomers to polymer-blend stabilizers. Their synthesis is most often done by ionic polymerization, which is both costly and sometimes difficult to control. However, block copolymer properties strongly depend, for example, on the exact chemical composition, block molar mass, and block yield. These parameters can be evaluated in a single experiment using copolymer GPC with multiple detection.

Figure 1 shows the measured molar mass distribution of a styrene–MMA block copolymer using refractive index (RI) and ultraviolet (UV) detection. The RI responds to the styrene and MMA units, whereas the UV, tuned to 260 nm, predominantly picks up the presence of styrene in the copolymer. After detector calibration, the styrene and MMA content in each fraction can be measured. The MMA content distribution (black solid line) is superimposed on the MWD of the product in Fig. 1. It is obvious that the MMA content is not constant throughout the MWD, but continuously increases with the molar mass. The trimodal MWD itself only shows the presence of three different species. The MMA content information clearly reveals that the copolymerization process was not producing block structures, but that the MMA was added to chains of different styrene molar masses.

Two-Dimensional Chromatography

Complex polymer topologies, polymer blends, and multicomponent formulations require a different approach to perform a proper molecular characterization. In two-dimensional (2D) chromatography, different separation techniques are used to avoid co-elution of species and to measure molar mass and chemical composition in a truly independent way [5].

It is obvious that n independent molecular properties require n-dimensional methods for accurate (independent) characterization of all parameters. Additionally, the separation efficiency of any single separation method is limited by the efficiency and selectivity of this separation mode (i.e., the plate count N of the column and the phase system selected). Adding more columns will not overcome the need to identify more components in a complex sample, due to the limitation of peak capacities, n. The corresponding peak capacity in an n-dimensional separation is substantially higher due to the fact that each dimension contributes to the total peak capacity as a factor and not as an additive term for single-dimension methods:

$$n_{\text{total}} = \prod_{i=1}^{n} n_i \sin^{(i-1)}\theta_i \tag{4}$$

for example, for a 2D system, $n_{2D} = n_1 n_2 \sin\theta$, where n_{total} represents the total peak capacity, n_i is the peak capacity in dimension i, and θ_i is the "angle" between two dimensions; for orthogonal separations, this angle will be 90°

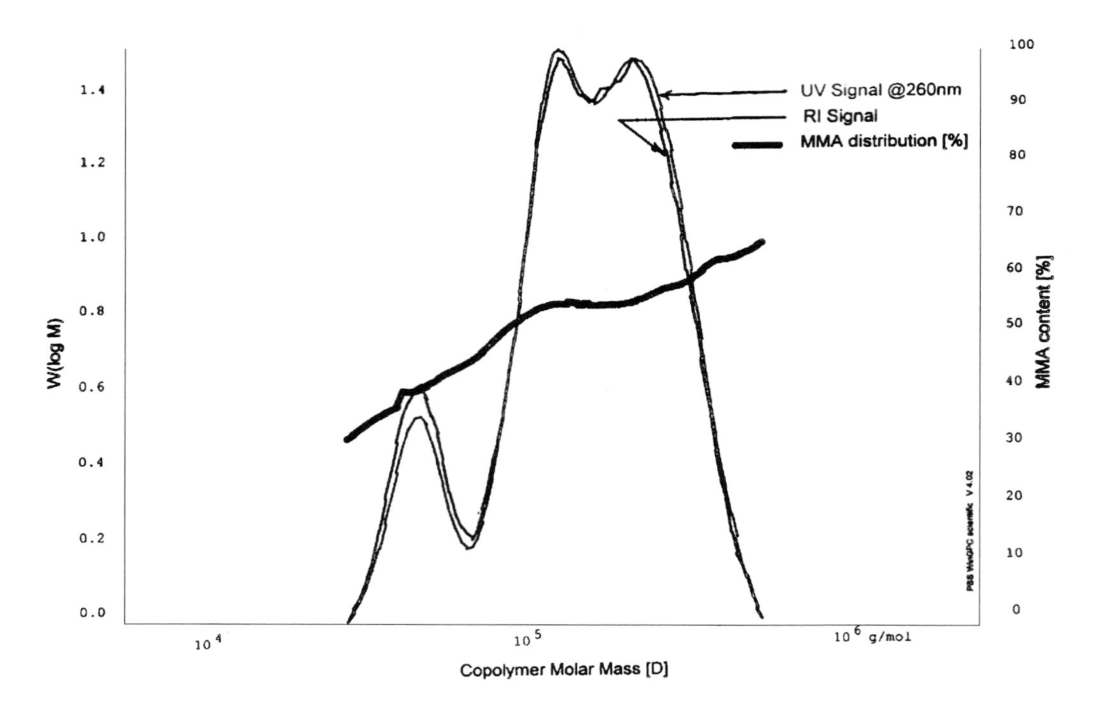

Fig. 1 Molar mass distribution with overlaid chemical composition distribution of a styrene–MMA block copolymer with poor block formation.

and the peak capacity will be maximized. The angle between dimensions is determined by the independence of the methods; a 90° angle is obtained using two methods which are completely independent of each other and will, thus, separate two properties solely on a single parameter without influencing each other.

In 2D chromatography separations, an aliquot from a first column (method) is transferred into the next separation method in a sequential and repetitive manner using automated sample transfer valves which are equipped with one or more sample loops. Alternatively, as a simpler and not quite as useful transfer technique, "heart cuts" from peaks in the first separation mode can be manually injected into the next separation column (second dimension).

The use of different modes of liquid chromatography facilitates the separation of complex samples, selectively, with respect to different properties like hydrodynamic volume, molar mass, chemical composition, or functionality. Using these techniques in combination, multidimensional information on different aspects of molecular heterogeneity can be obtained. If, for example, two different chromatographic techniques are combined in a "cross-fractionation" mode, information on chemical composition distribution and molar mass distribution can be obtained. Reviews on different techniques and applications involving the combination of GPC and various LC methods can be found in the literature [6–8].

Experimental Aspects of Two-Dimensional Separations

Setting up a 2D chromatographic separation system is actually not as difficult as one might first think. As long as well-known separation methods exist for each dimension [8], the experimental aspects can be handled quite easily in most cases. Off-line systems just require a fraction collection device and something or someone who reinjects the fractions into the next chromatographic dimension. In on-line 2D systems, the transfer of fractions is preferentially done by automated injection valves, as was proposed by Kilz et al. [9].

The sequence of the separation methods is important, in order to realize the best resolution and accurate determination of property distributions [10]. It is advisable to use the method with highest selectivity for the separation of one property as the first dimension. This ensures highest purity of eluting fractions being transferred into the subsequent separation. In many cases, interaction chromatography has proven to be the best and most easily adjustable method for the first dimension separation, because (a) more parameters (mobile phase, mobile-phase composition, mobile-phase modifiers, stationary phase, temperature, etc.) can be used to adjust the separation according to the chemical nature of the sample, (b) better fine-tuning in interaction chromatography allows for more homogeneous fractions, and (c) sample load on such columns can be much higher as compared to, for example, GPC columns.

Because of the consecutive dilution of fractions, detectability and sensitivity become important criteria in 2D experimental design [11]. If low-level components have to be detected, only the most sensitive and/or selective detection methods can be used. ELSD detection, despite several drawbacks, has been used mostly due to its high sensitivity for compounds which will not evaporate or sublime under detection conditions. Fluorescence and diode-array UV/VIS are also sensitive detection methods, which can pick up samples at nanogram levels. Mass spectrometers have a high potential in this respect too; however, they are currently not developed to a state where they would be generally usable. Only in special cases, refractive index detection, otherwise very popular in GPC, has been used in multidimensional separations, because of its low sensitivity and strong dependence on mobile-phase composition.

Time consumption is another important aspect of 2D analyses. All early 2D chromatography work was done in either off-line mode or in a stop-flow mode, which are both time-consuming and have poor reproducibility. The first fully automated 2D chromatographic system was developed by Kilz et al. [11]. This system uses 2D chromatography software for data acquisition and processing, which also controls the 2D transfer valve. This setup relieves the operator from all the time-consuming tasks and does all 2D and 3D data processing and reporting.

The recent introduction of high-speed GPC columns reduces time consumption even further and allows a complete 2D analysis in about an hour.

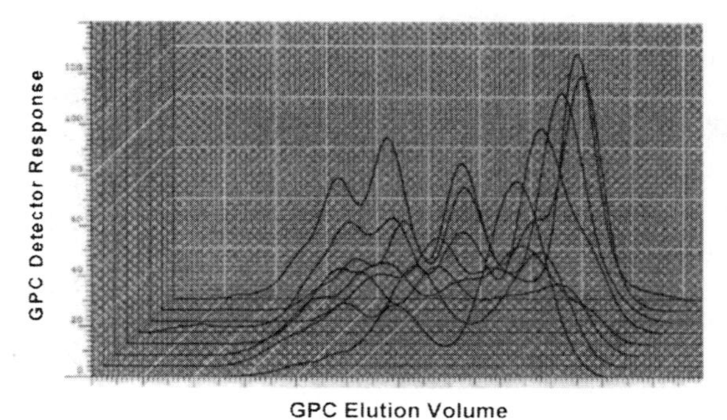

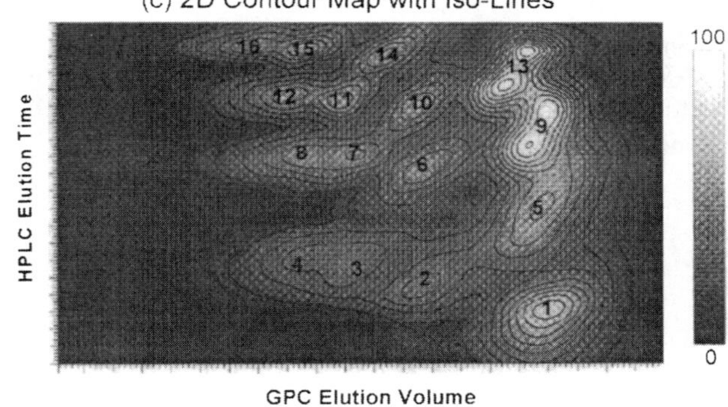

Fig. 2 Various data presentation views in 2D chromatography showing the 2D analysis of a 16-component star block copolymer (see text for details).

2D Analysis of Complex Copolymer Blends

The potential of 2D separations can be explained best by a 16-component blend of a styrene–butadiene star block copolymer. The copolymer mixture consisted of four different molar masses (1*M*, 2*M*, 3*M*, and 4*M* reflecting the 1-, 2-, 3- and 4-arm star molecules) and four styrene compositions (20%, 40%, 60%, and 80%) for each arm. Molar mass and composition were very thoroughly controlled by a special anionic polymerization technique to ensure macromolecular architecture [7]. Gradient HPLC analysis only showed a broad elution profile, whereas high-resolution GPC just separated out the molar masses of the four arms. The GPC result gave no indication of additional peaks with identical molecular weight but different composition hiding behind the detected peaks. The online combination of both techniques, under otherwise identical conditions, allowed for the separation of all 16 species in the mixture. The separation was not completely orthogonal, because the gradient high-performance liquid chromatography (HPLC) separation was partially dependent on molar mass (cf. Fig. 2c).

Further data analysis allows the determination of molar mass, chemical composition, and relative concentration for each component. The 2D chromatography data can be displayed in various forms (cf. Fig. 2): (a) stacked or waterfall presentation showing the individual traces transferred into the second dimension; (b) surface plot allowing to view the 3D surface from different angles; (c) 2D contour maps, which are most useful for data analysis and interpretation. Other data views can be also calculated from the 3D data set: cuts in any direction of properties (e.g., composition or molar mass) and true projections (accumulation) of data to create virtual chromatograms for each separation dimension.

Conclusions

Multidimensional chromatography separations are currently one of the most promising and powerful methods for the fractionation and characterization of complex sample mixtures in different property coordinates. This technique combines extraordinary resolution and peak capacity with flexibility, and it overcomes the limitations of any given single chromatographic method. This is the ideal basis for the identification and quantification of major compounds and by-products, which might adversely affect product properties if not detected in time.

Copolymer GPC is a proven analytical tool to investigate molecular properties of block and other highly segmented copolymers and to correlate structure with bulk properties.

References

1. J. R. Runyon, D. E. Barnes, J. F. Rudel, and L. H. Tung, *J. Appl. Polym. Sci. 13*:2359 (1969).
2. F. Gores and P. Kilz, Copolymer characterization using conventional SEC and molar mass-sensitive detectors, in *Chromatography of Polymers* (T. Provder, ed.), ACS Symposium Series 521, American Chemical Society, Washington, DC, 1993, Chap. 10.
3. B. Trathnigg, S. Feichtenhofer, and M. Kollroser, *J. Chromatogr. A 786*:75 (1997).
4. C. Johann and P. Kilz, *Proc. Int. Conf. Molar Mass Charact.* 1989.
5. H. Pasch, *Adv. Polym. Sci. 150*:1 (2000).
6. G. Glöckner, *Gradient HPLC of Copolymers and Chromatographic Cross-Fractionation*, Springer-Verlag, Berlin, 1991.
7. P. Kilz, R.-P. Krüger, H. Much, and G. Schulz, in *Chromatographic Characterization of Polymers: Hyphenated and Multidimensional Techniques* (T. Provder, M. W. Urban, and H. G. Barth, eds.), Advances in Chemistry Vol. 247, American Chemical Society, Washington, DC, 1995.
8. H. Pasch and B. Trathnigg, *HPLC of Polymers*, Springer-Verlag, Berlin, 1997.
9. P. Kilz, R.-P. Krüger, H. Much, and G. Schulz, *Polym. Mater. Sci. Eng. 69*:114 (1993).
10. M. R. Schure, *Anal. Chem. 71*:1645 (1999).
11. P. Kilz, *Laborpraxis, 6*:628 (1992).

Peter Kilz

Copolymer Composition by GPC–SEC

Determination of the average chemical composition and polymer composition by size-exclusion chromatography (SEC) has been reported in the literature. Two different types of concentration detector or two different absorption wavelengths of an ultraviolet or an infrared detectors are employed; the composition at each retention volume is calculated by measuring peak responses at the identical retention points of the two chromatograms.

However, synthetic copolymers have both molecular-weight and chemical composition distributions and copolymer molecules of the same molecular size, which are eluted at the same retention volume in SEC, may have different molecular weights in addition to different compositions. This is because separation in SEC is achieved according to the sizes of molecules in solution, not according to their molecular weights or chemical compositions.

Molecules that appear at the same retention volume may have different compositions, so that accurate information on chemical heterogeneity cannot be obtained by SEC alone. When the chemical heterogeneity of a copolymer, as a function of molecular weight, is observed, the copolymer is said to have a heterogeneous composition, but, even though it shows constant composition over the entire range of molecular weights, it cannot be concluded that it has a homogeneous composition [1]. Nevertheless, SEC is still extremely useful in copolymer analysis, due to its rapidity, simplicity, and wide applicability.

When one of the constituents, A or B of a copolymer A–B, has an ultraviolet (UV) absorption and the other does not, a UV detector–refractive index (RI) combined detector system can be used for the determination of chemical composition or heterogeneity of the copolymer. A point-to-point composition, with respect to retention volume, is calculated from two chromatograms and a variation of composition is plotted as a function of molecular weight. The response factors of the two components in the two detectors must first be calibrated.

Let A be a constituent that has UV absorption. K_A and K_B are defined as the response factors of an RI detector for the A and B constituents, and K'_A as the response of the UV detector for A. These response factors are calculated by injecting known amounts of homopolymers A and B into the SEC dual-detector system, calculating the areas of the corresponding chromatograms, and dividing the areas by the weights of homopolymers injected as

$$F_A = K_A G_A, \qquad F_B = K_B G_B, \qquad F'_A = K'_A G_A$$

where F_A, F_B, and F'_A are areas of homopolymers A and B in the RI detector and of homopolymer A in the UV detector, and G_A and G_B are the weights of homopolymers A and B injected into the SEC system.

The weight fraction $W_{A,I}$ of constituent A, at each retention volume I of the chromatogram for the copolymer, is given by

$$W_{A,I} = \frac{K_B}{R_I K'_A - K_A + K_B}$$

where $R_I = F_{RI,I}/F_{UV,I}$ for the copolymer at retention volume I. Retention volme I for the RI detector is not equal to the retention volume I for the UV detector. Usually, the UV detector is connected to the column outlet and is followed by an RI detector, and the dead volume between these two detectors must be corrected. The dead volume can normally be measured by injecting a polymer sample having a narrow molecular-weight distribution and by measuring the retention difference between the two peak maxima.

Because the additivity of the RI increments of homopolymers is valid for copolymers, the additivity of the response factors is also valid:

$$K_C = W_A K_A + W_B K_B$$

where K_C is the response factor for the copolymer in the RI detector. If the response factors of one or two homopolymers that comprise a copolymer cannot be measured because of insolubility of the homopolymer(s), then this equation is employed.

Alternatively, the extrapolation of the plot of RI response factors of copolymers of known compositions can be used. An example is that the RI response for polystyrene was 2800 and that for polyacrylonitrile was 2250.

Although the values of these response factors are dependent on several parameters, the ratio of K_A to K_B is almost constant in the same mobile phase.

An infrared detector can be used, at an appropriate wavelength, for detecting one component in copolymers or terpolymers and, thus, expand its range of applicability to copolymers analysis. Information on composition can be obtained by repeating runs, using different wavelengths to monitor different functional groups. A single-detector system is more advantageous than a dual-detector system, such as a combination of UV and RI detectors.

Instead of measuring chromatograms two or three times at different wavelengths for different functional groups, operation in a stop-and-go fashion was introduced for rapid determination of copolymer composition as a function of molecular weight [2].

Pyrolysis gas chromatography has been widely used for copolymer analysis. This technique may offer many advantages over other detection techniques for copolymer analysis by SEC. One obvious advantage is the small sample size required. Another is the capability of application to copolymers which cannot utilize UV or IR detectors [3].

Combination with other liquid chromatographic techniques is also reported by several workers. Orthogonal coupling of an SEC system to another high-performance liquid chromatography (HPLC) system to achieve a desired cross-fractionation was proposed [4]. It was an SEC–SEC mode, using the same polystyrene column, but the mobile phase in the first system was chosen to accomplish only a hydrodynamic volume separation, and the mobile phase in the second system was chosen so as to be a thermodynamically poorer solvent for one of the monomer types in the copolymer, in order to fractionate by composition under adsorption or partition modes as well as size exclusion.

A combination of liquid adsorption chromatography with SEC has recently been developed by several workers. Poly(styrene–methyl methacrylate) copolymers were fractionated according to chemical composition by liquid adsorption chromatography and the molecular weight averages of each fraction were measured by SEC [5,6].

References

1. S. Mori, *J. Chromatogr. 411*:355 (1987).
2. F. M. Mirabella, Jr., E. M. Barrall II, and J. F. Johnson, *J. Appl. Polym. Sci. 19*:2131 (1975).

3. S. Mori, *J. Chromatogr. 194*:163 (1980).
4. S. T. Balke and R. D. Patel, *J. Polym. Sci. Polym. Lett. Ed. 18*:453 (1980).
5. S. Mori, *Anal. Chem. 60*:1125 (1988).
6. S. Mori, *Trends Polym. Sci. 2*:208 (1994).
7. S. Mori and H. G. Barth, *Size Exclusion Chromatography*, Springer-Verlag, New York, 1999, Chap. 12.
8. S. Mori, Copolymer analysis, in *Size Exclusion Chromatography* (B. J. Hunt and S. R. Hodling, eds.), Blackie, Oxford, 1989.

Sadao Mori

Copolymer Molecular Weights by GPC–SEC

It is well known that most copolymers have both molecular weight and composition distributions and that copolymer properties are affected by both distributions. Therefore, we must know average values of molecular weights and composition, and their distributions. These two distributions are inherently independent of each other. However, it is not easy to determine the molecular-weight distribution independently of the composition, even by modern techniques.

Size-exclusion chromatography (SEC) is a rapid technique used to obtain the molecular-weight averages and the molecular-weight distributions of synthetic polymers. The objective of SEC for copolymer analysis must be not only the determination of molecular-weight averages and its distribution but also the measurement of average copolymer composition and its distribution. However, separation by SEC is achieved according to the sizes of molecules in the solution, not according to their molecular weights. Therefore, the retention volume of a copolymer molecule obtained by SEC reflects not the molecular weight, as in the case of a homopolymer, but simply the molecular size.

For example, the elution order of polystyrene (PS), poly(methyl methacrylate) (PMMA), and their copolymers [P(S–MMA)], both random and block, all having the same molecular weight are as follows: random copolymer of P(S–MMA), PS, block copolymer (MMA-S-MMA), and PMMA [1]. Copolymers having the same molecular weight but different composition are different in molecular size and elute at different retention volumes. Therefore, the accurate determination of the values of molecular-weight averages and the molecular-weight distribution for a copolymer by SEC might be limited to the case when the copolymer has the homogeneous composition across the whole range of molecular weights.

A calibration curve for a copolymer consisting of components A and B can be constructed from those for the two homopolymers A and B, if the relationships of the molecular weights and the molecular sizes of the two homopolymers are the same as their copolymer and if the size of the copolymer molecules in the solution is the sum of the sizes of the two homopolymers times the corresponding weight fractions. The molecular weight of the copolymer at retention volume I, $M_{C,I}$, is calculated using

$$\log M_{C,I} = W_{A,I} \log M_{A,I} + W_{B,I} \log M_{B,I}$$

where $M_{A,I}$ and $M_{B,I}$ are the molecular weights of homopolymers A and B, respectively, and $W_{A,I}$ and $W_{B,I}$ are the weight fractions of components A and B, respectively, in the copolymer at retention volume I. This equation was empirically postulated for block copolymers [2].

The use of the so-called "universal calibration" is a theoretically reliable procedure for calibration. For ethylene–propylene (EP) copolymers, Mark–Houwink parameters in *o*-dichlorobenzene at 135°C are calculated as [3]

$$a_{EP} = (a_{PE} a_{PP})^{1/2}$$
$$K_{EP} = W_E K_{PE} + W_P K_{PP} - 2(K_{PE} K_{PP})^{1/2} W_E W_P$$

where W_E and W_P are the weight fractions of the ethylene and propylene units of the copolymer, respectively.

Calculated Mark–Houwink parameters for P(S–MMA) block and statistical copolymers at several compositions in tetrahydrofuran at 25°C are listed in Table 1 [4]. The parameters for PS and PMMA used in the calculation are as follows:

PS: $K = 0.682 \times 10^{-2}$ mL/g, $a = 0.766$
PMMA: $K = 1.28 \times 10^{-2}$ mL/g, $a = 0.69$

If copolymer molecules and PS molecules are eluted at the same retention volume, then

$$[\mu]_C M_C = [\mu]_S M_S$$

where M_C and M_S are the molecular weights of the copolymer and PS, respectively, and $[\mu]_C$ and $[\mu]_S$ are the intrinsic viscosities of the copolymer and PS, respectively. A differential pressure viscometer can measure intrinsic vis-

Table 1 Calculated Mark–Houwink Parameters for P(S–MMA) Block and Statistical Copolymers at Several Compositions in Teterahydrofuran at 25°C

	Block copolymer		Statistical copolymer	
Composition (styrene wt%)	$K \times 10^2$ (mL/g)	a	$K \times 10^2$ (mL/g)	a
20	1.124	0.705	1.044	0.718
30	1.054	0.714	0.953	0.731
40	0.989	0.721	0.879	0.742
50	0.929	0.729	0.821	0.750
60	0.872	0.736	0.779	0.756
70	0.820	0.744	0.747	0.760
80	0.771	0.751	0.722	0.763

cosities for the fractions of the copolymer and PS continuously, followed by the determination of M_C of the copolymer fraction at retention volume i.

The application of a light-scattering detector in SEC does not require the construction of a calibration curve using narrow molecular-weight distribution polymers. However, this method is not generally applicable to copolymers because the intensity of light scattering is a function not only of molecular weight but also of the specific refractive index (the refractive index increment) of the copolymer in the mobile phase. The refractive index increment is also a function of composition. In the case of a styrene–butyl acrylate (30:70) emulsion copolymer, the apparent molecular weight of the copolymer in teterahydrofuran was only 7% lower than true one [5]. A recent study concluded that if refractive index increments of the corresponding homopolymers do not differ widely, SEC measurements combined with light scattering and concentration detectors yield good approximations to molecular weight and its distribution, even if the chemical composition distribution is very broad [6].

References

1. A. Dondos, P. Rempp, and H. Benoit, *Macromol. Chem. 175*:1659 (1984).
2. J. R. Runyon, D. E. Barnes, J. F. Rudd, and L. H. Tung, *J. Appl. Polym. Sci. 13*:2359 (1969).
3. T. Ogawa and T. Inaba, *J. Appl. Polym. Sci. 21*:2979 (1988).
4. J. M. Goldwasser and A. Rudin, *J. Liquid Chromatogr. 6*: 2433 (1983).
5. F. B. Malihi, C. Y. Kuo, and T. Provder, *J. Appl. Polym. Sci. 29*:925 (1984).
6. P. Kratochvil, 8th International Symposium on Polymer Analysis and Characterization (ISPAC-8), 1995, Abstract L14.
7. S. Mori and H. G. Barth, *Size Exclusion Chromatography*, Springer-Verlag, New York, 1999, Chap. 12.
8. S. Mori, Copolymer analysis, in *Size Exclusion Chromatography* (B. J. Hunt and S. R. Holding, eds.), Blackie, Oxford, 1989.

Sadao Mori

Coriolis Force in Countercurrent Chromatography

The Coriolis force acts on a moving object on a rotating body such as the Earth and centrifuge bowl. It was first analyzed by a French engineer and mathematician, Gaspard de Coriolis (1835) [1]. The effect of the Coriolis force produced by the Earth's rotation is weak, whereas that on the rotating centrifuge is strong and easily detected. Fig-

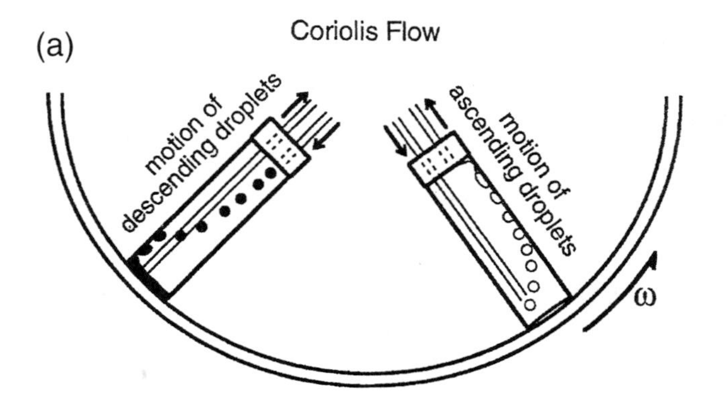

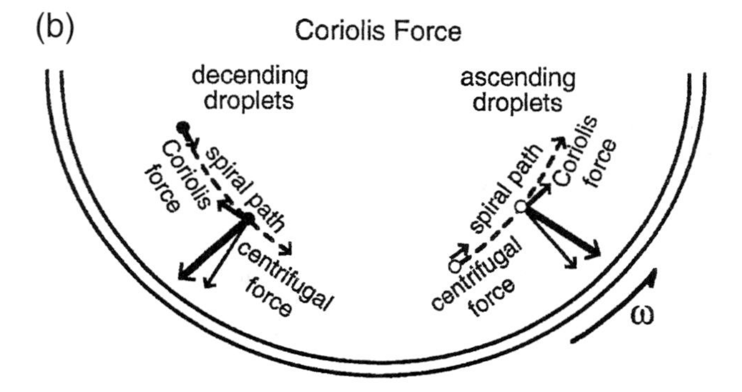

Fig. 1 Effects of Coriolis force on moving droplets in a rotating centrifuge. (a) Motion of the droplets in a flow through cell in a rotating centrifuge; (b) direction of the Coriolis force acting on droplets on the rotating centrifuge bowl.

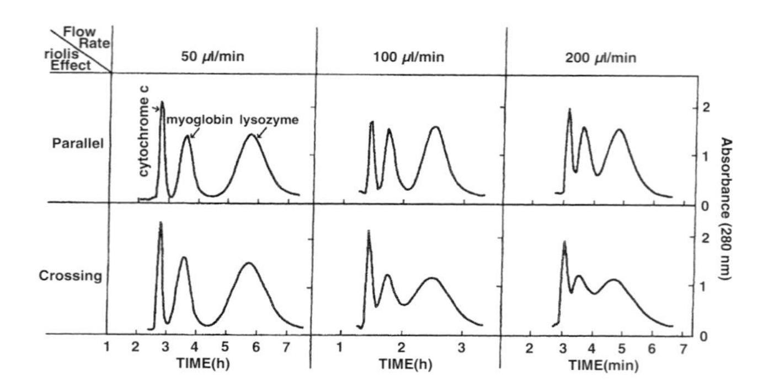

Fig. 2 Effects of Coriolis force on partition efficiency and retention of the stationary phase in protein separation by the toroidal coil centrifuge.

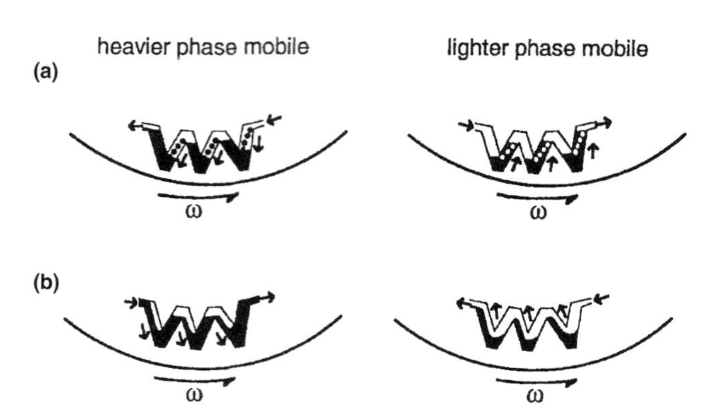

Fig. 3 Effects of Coriolis force on the two-phase flow in the separation coil of the toroidal coil centrifuge: (a) Coriolis force parallel; (b) Coriolis force crossing.

ure 1 illustrates the effect of the Coriolis force on the moving droplets in a rotating centrifuge where the path of the sinking droplets shifts toward the opposite direction to the rotation (left) and this effect is reversed on the floating droplets (right) [2]. The moving droplets on a rotating centrifuge have been photographed under stroboscopic illumination [3,4].

The effects of the Coriolis force on countercurrent chromatography have been demonstrated on the toroidal coil centrifuge which uses a coiled tube mounted around the periphery of the centrifuge bowl [2]. When a protein mixture containing cytochrome-*c*, myoglobin and lysozyme was separated on an aqueous–aqueous polymer-phase system composed of 12.5% (w/w) poly(ethylene glycol) 1000 and 12.5% (w/w) dibasic potassium phosphate, the direction of the elution through the toroidal coil produced substantial effects on peak resolution, as shown in Fig. 2 and Table 1 [2]. Because the toroidal coil separation column has a symmetrical orientation except for the handedness, the above effect is best explained on the basis of the Coriolis force as follows. If the Coriolis force acts parallel to the effective coil segments (parallel orientation), the two phases form multiple droplets which provide a broad interface area to enhance the mass-transfer process, hence improving the partition efficiency (Fig. 3a). When the Coriolis force acts across the effecting coil seg-

Table 1 Effects of Coriolis Force on Partition Efficiencies of Three Stable Proteins in Toroidal Coil CCC

Flow rate (μL/min)	Analyte peak	TP (Para/Cross)	R_s (Para/Cross)	Retention (%) (Para/Cross)
50	Cytochrome-*c*	1860/1490		29.2/32.0
			1.62/1.39	
	Myoglobin	365/266		
			1.66/1.40	
	Lysozyme	156/104		
100	Cytochrome-*c*	1760/2821		30.0/30.3
			1.27/0.86	
	Myoglobin	433/172		
			1.39/0.84	
	Lysozyme	172/63		
200	Cytochrome-*c*	1296/–		22.8/21.3
			0.84/–	
	Myoglobin	330/–		
			0.92/–	
	Lysozyme	123/–		

ments, the two phases form a streaming flow, minimizing the interfacial area for mass transfer and resulting in lower partition efficiency (Fig. 3b).

It is interesting to note that the above effects have not been observed on the separation of low-molecular-weight compounds such as dipeptides [2] and DNP (dinitrophenyl) amino acids [5] on the conventional organic–aqueous two-phase solvent systems, except that at a relatively low revolution speed, the Coriolis force acting across the effective coil segments slightly improves the partition efficiency, probably due to the substantially higher retention of the stationary phase.

References

1. *New Encyclopedia Britannica*, 1995, Vol. 3, p. 632.
2. Y. Ito and Y. Ma, Effect of Coriolis force on countercurrent chromatography, *J. Liquid Chromatogr. 21*:1 (1998).
3. L. Marchal, A. Foucault, G. Patissier, J. M. Rosant, and J. Legrand, *J. Chromatogr. A 869*:339 (2000).
4. A. Morvan, A. Foucault, G. Patissier, J. M. Rosant and J. Legrand, J. Hydrodyn. (in press).
5. Y. Ito, K. Matsuda, Y. Ma, and L. Qi, *J. Chromatogr. A 808*:95 (1998).

Yoichiro Ito

Corrected Retention Time and Corrected Retention Volume

The *corrected retention time* of a solute is the elapsed time between the *dead point* and the peak maximum of the solute. The different properties of the chromatogram are shown in Fig. 1. The *volume of mobile phase* that passes through the column between the *dead point* and the *peak maximum* is called the *corrected retention volume.* If the mobile phase is incompressible, as in liquid chromatography, the retention volume (as so far defined) will be the simple product of the *exit flow-rate* and the *corrected retention time.*

If the mobile phase is compressible, the simple product of the corrected retention time and flow rate will be incorrect, and the corrected retention volume must be taken as the product of the corrected retention time and the *mean* flow rate. The true corrected retention volume has been shown to be given by [1]

$$V_r' = V_{r'}'\frac{3}{2}\left(\frac{\gamma^2 - 1}{\gamma^3 - 1}\right) = Q_0 t_r' \frac{3}{2}\left(\frac{\gamma^2 - 1}{\gamma^3 - 1}\right)$$

where the symbols have the meaning defined in Fig. 1, and $V_{r'}'$ is the corrected retention volume measured at the column exit and γ is the inlet/outlet pressure ratio.

The corrected retention volume, V_r', will be the difference between the retention volume and the dead volume V_0, which, in turn, will include the actual dead volume V_m and the extra column volume V_E. Thus,

$$V_r' = V_r - (V_E + V_m)$$

The retention time can be taken as the product of the distance on the chart between the dead point and the peak maximum and the chart speed, using appropriate units. As in the case of the retention time, it can be more accurately measured with a stopwatch. Again, the most accurate method of measuring V_r' for a noncompressible mo-

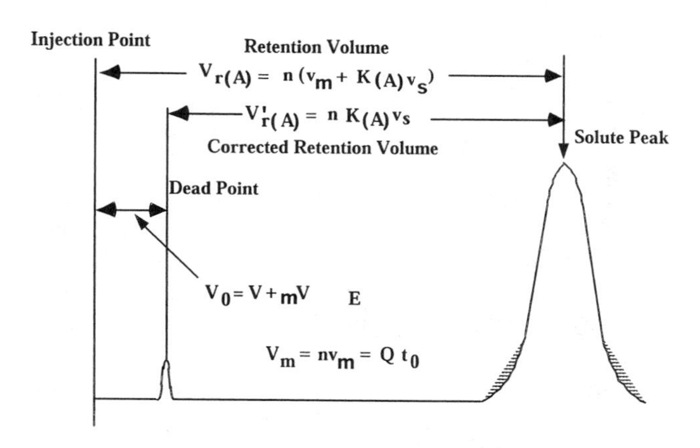

Fig. 1 Diagram depicting the retention volume, corrected retention volume, dead point, dead volume, and dead time of a chromatogram. V_0: total volume passed through the column between the point of injection and the peak maximum of a completely unretained peak; V_m: total volume of mobile phase in the column; $V_{r(A)}$: retention volume of solute A; $V_{r(A)}'$: corrected retention volume of solute A; V_E: extra column volume of mobile phase; v_m: volume of mobile phase, per theoretical plate; v_s: volume of stationary phase per theoretical plate; $K_{(A)}$: distribution coefficient of the solute between the two phases; n: number of theoretical plates in the column; Q: column flow rate measured at the exit.

bile phase, although considered antiquated, is to attach an accurate burette to the detector exit and measure the retention volume in volume units. This is an absolute method of measurement and does not depend on the accurate calibration of the pump, chart speed, or computer acquisition level and processing.

Reference

1. R. P. W. Scott, *Introduction to Analytical Gas Chromatography*, Marcel Dekker, Inc., New York, 1998, p. 77.

Suggested Further Reading

Scott, R. P. W., *Liquid Chromatography Column Theory*, John Wiley & Sons, Chichester, 1992, p. 19.
Scott, R. P. W., *Techniques and Practice of Chromatography*, Marcel Dekker, Inc., New York, 1996.

Raymond P. W. Scott

Coumarins: Analysis by TLC

Coumarins are natural compounds that contain characteristic benzo[α]pyrone (2H-benzopyran-2-one) moiety. They are especially abundant in Umbelliferae, Rutaceae, Leguminosae, and other plant families. Different coumarin derivatives have been isolated. Usually the substituents are at the positions C_5, C_6, C_7, and C_8 [e.g., umbelliferone (7-hydroxycoumarin), hierniarin (7-methoxycoumarin), esculetin (6,7-dihydroxycoumarin), scopoletin (6-methoxy-7-hydroxycoumarin), osthenol (7-hydroxy-8-prenylcoumarin), osthol (7-methoxy-8-prenylcoumarin), and others].

In addition to simple coumarin derivatives, furano- and pyrano-coumarins are also commonly encountered in the Umbelliferae and Rutaceae families. The essential chemical moiety of linear furano-coumarins consists of 2H-furan[3.2-*g*]benzo[*b*]pyran-2-one ring called psoralen (its derivatives; e.g., bergapten, xanthotoxin, isopimpinelin, imperatorin, isoimperatorin, oxypeucedanin, and others). The second type of furano-coumarins (the angular type of angelicin) has a 2H-furan[2.3-*h*]-benzo[*b*]pyran-2-one structure (the angelicin derivatives; e.g., isobergapten, pimpinelin, sphondin). There are also both linear and angular types of pyrano-coumarins. In the linear type, which is named alloxanthiletin, the 2H,8H-pyran[3.2-*g*]benzo[*b*]pyran-2-one ring is characteristic, whereas in the angular type called seselin, the 2H,8H-pyran[2.3-*h*]benzo[*b*]pyran-2-one moiety is typical.

Thin-layer chromatography (TLC) is a very useful method for separation of natural coumarins, furano-coumarins and pyrano-coumarins. Natural coumarins exhibit fluorescence properties which they display in ultraviolet (UV) light (365 nm). Their spots can be easy detected on paper and thin-layer chromatograms without the use of any chromogenic reagents. It is often possible to recognize the structural class of coumarin from the color it displays under UV detection (Table 1). Purple fuorescence generally signifies 7-alkoxycoumarins, whereas 7-hydroxycoumarins and 5,7-dioxygenated coumarins tend to fluorescence blue. In general, furano-coumarins possess a dull yellow or ocher fluorescence, except for psoralen, sphondin, and angelicin. Spot fluorescence is more intense or its color is changed after spraying the TLC chromatogram with ammonia (see Table 1) [1].

Thin-layer chromatograms can also be detected by several nonspecific chromatogenic reactions:

1. 1% aqueous solution of iron(III) chloride
2. 1% aqueous solution of potassium ferricyanide
3. Diazotized sulfonillic acid and diazotized *p*-nitroaniline

None of these reagents is very specific for hydroxycoumarins and its confirmation should be substantiated by other methods. Exposed groups present in many natural coumarins can be detected due to their susceptibility to cleavage by acids and applied over a phosphoric acid spot on a silica TLC plate.

The linear (psoralens) and angular (angelicins) furano-coumarins can be readily differentiated with the Emerson reagent. It is used, also, for detection of pyrano-coumarins (selinidin, pteryxin) on TLC chromatograms.

Thin-Layer Chromatography

Conventional TLC is a well-known technique, used for many years in systematic research on coumarin content of numerous plant species, as well as for chemotaxonomic

Table 1 Chromatographic Methods of Coumarins Identification: Fluorescence Colors of Coumarins Under UV Irradiation (365 nm)

Fluorescence color[a]	Fluorescence color with ammonia	Coumarin or coumarin type
Blue	L. Blue	7-Hydroxycoumarin
B. blue	V. blue	7-Hydroxycoumarins
Blue	B. blue	7-Hydroxy-6-alkoxycoumarins
Blue	Blue	5,7-Dialkoxycoumarins
B. blue	B. blue	6,7-Dialkoxycoumarins
Blue	Blue	6,7,8-Trialkoxycoumarins
W. blue	W. blue	5,6,7-Trialkoxycoumarins
W. blue	B. blue	7-Hydroxy-5,6-dialkoxycoumarins
Blue		Psoralen
Blue	B. blue	6-Methoxyangelicin
Blue	Green	7,8-Dihydroxycoumarin
Pink	Yellow	6-Hydroxy-7-glucosyloxycoumarin
Purple	Purple	8-Hydroxy-5-alkoxypsoralens
W. purple	Pink	6-Hydroxy-7-alkoxycoumarins
Purple	Green	Angelicin, coumestrol
Green		5-Methoxyangelicin
Green		8-Hydroxy-6,7-dimethoxycoumarin
Green	Yellow	7,8-Dihydroxy-6-methoxycoumarin
Yellow		7-Hydroxy-8-methoxycoumarin
Yellow		3,4,5-Trimethoxypsoralen
Yellow	Yellow	6-Hydroxy-5,7-dimethoxycoumarin
Yellow	Yellow	5-Hydroxy-6,7-dimethoxycoumarin
Yellow	Yellow	5-Hydroxypsoralen
Yellow	Yellow	5,6-Dimethoxyangelicin
Yellow	Yellow	8-Alkoxypsoralens
Yellow–green	Yellow–green	5-Alkoxypsoralens
B. yellow	B. yellow	5,8-Dialkoxypsoralens

[a]B. = bright; V. = very bright; L. = light; W. = weak
Source: Ref. 1.

relationships between those species. Great progress in optimization of the TLC–separation process was made by the design of modern, horizontal chambers for TLC. It is a universal design, offering the possibility of developing chromatograms in the space saturated or nonsaturated with mobile-phase vapors; moreover, it is possible to perform gradient elution, stepwise or continuous, or to accomplish micropreparative separation of chemical compound composites (e.g., plant extracts) [2].

Gradient elution in TLC can be obtained in several ways [1]:

1. Multizonal development: the use of multicomponent eluents which are partially separated during development (frontal chromatography), forming an eluent strength gradient along the layer
2. Development with a strong solvent (e.g., acetone) of an adsorbent layer exposed to vapors of a less polar solvent
3. The use of mixed layers of varying surface area, activity (comparison silica and florisil)
4. Delivery of an eluent whose composition is varied in a continuous or stepwise manner by introducing small volumes of more polar eluent fractions (e.g., 0.2 mL).

The possibility of zonal sample dosage in equilibrium conditions (after the front of mobile-phase and continuous-chromatogram development, which is provided by a horizontal "sandwich" chamber) was utilized by Głowniak et al. [3] in preparative chromatography of simple coumarins and furano-coumarins found in *Archangelica* fruits, performed with a short-bed continuous development (SB–CD) technique.

The latter possibility was employed by Wawrzynowicz and Waksmundzka-Hajnos for micropreparative TLC isolation of furano-coumarins from *Archangelica*, *Pastinaca*, and *Heracleum* fruits on silica gel, silanized gel, and florisil.

Superior coumarin compounds separation with use of the described flat "sandwich" chambers was achieved with gradient chromatography on silica gel and stepwise variation of polar modifier concentration in mobile phase, as less polar solvents (hexane, cyclohexane, toluene, or dichloromethane) and polar modifiers (acetonitrile, diisopropyl ether, ethyl acetate) are used.

Complex pyrano-coumarin mixtures can be separated with the TLC technique by alternative use of two different polar adsorbents (silica gel, florisil) and various binary and ternary eluents with different mechanisms of adsorption centers effect on the molecules to be separated [4]. Improved separation can be achieved by high-performance thin-layer chromatography (HPTLC), which employs new highly effective adsorbents of narrow particle size distribution, or with a chemically modified surface. Because of its similarity, HPTLC is applied in designing optimal HPLC systems. Another gradient technique in coumarin compounds separation is programmed multiple development (PMD), also called "reverse gradient" technique, in which chromatograms are developed to increasing distances by a sequence of eluents with decreasing polarity, with eluent evaporation after each stage. Automated multiple development (AMD), providing automatic chromatogram developing and drying, is a novel form of the PMD technique.

Two-dimensional thin-layer chromatography (2D–TLC), is particularly effective in the case of complex extracts when one-dimensional developing yields partial separation. Moreover, it offers the possibility of modifying separation procedures when the development direction is changed.

Overpressured Layer Chromatography

The term "overpressured layer chromatography" (OPLC) was originally introduced by Tyihak et al. [5] in the late seventies. The crucial factor is pressurized mobile-phase flow through the planar media. A short analysis time, low solvent consumption, high resolution, and availability of on-line and off-line modes are the main OPLC advantages in comparison with the classical TLC techniques. OPLC was proved effective in qualitative and quantitative analysis of furano-coumarins by densitometric on-line detection. OPLC can also be performed in two-dimensional mode (2D–OPLC).

Long-distance OPLC is a novel form of OPLC, in which chromatograms are developed over a long distance with optimal (empiric) mobile-phase flow. Used in combination with specialized equipment designs, it produces high performance (70,000 to 80,000 of theoretic plates) and excellent resolution. Botz et al. [6], who initiated long-distance OPLC, proved its efficiency in separation of eight furano-coumarin isomers and in isolation of furanocoumarin complex from *Peucedanum palustre* roots raw extract.

Rotation Planar Chromatography

Rotation planar chromatography (RPC), as with OPLC, is another thin-layer technique with forced eluent flow, employing a centrifugal force of a revolving rotor to move the mobile phase and separate chemical compounds. The RPC equipment can vary in chamber size, operative mode (analytical or preparative), separation type (circular, anticircular, or linear), and detection mode (off-line or on-line). The described technique was applied in analytical and micropreparative separation of coumarin compounds from plant extracts.

References

1. R. D. H. Murray, Nat. Prod. Rep. 591–624 (1989).
2. E. Soczewinski, *Chromatography*, *138*:443 (1977).
3. K. Głowniak, E. Soczewinski, and T. Wawrzynowicz, *Chem. Anal. (Warsaw) 32*:797 (1987).
4. K. Głowniak, *J. Chromatogr. 552*:453 (1991).
5. E. Tyihák, E. Mincsovisc, and H. Kalász, *J. Chromatogr. 174*:75 (1979).
6. L. Botz, S. Nyiredy, and O. Sticher, *J. Planar Chromatogr. 4*:115 (1991).

Kazimierz Głowniak

Countercurrent Chromatography–Mass Spectrometry

Introduction

Countercurrent chromatography (CCC) is a unique liquid–liquid partition technique which does not require the use of a solid support [1–5], hence eliminating various complications associated with conventional liquid chromatography, such as tailing of solute peaks, adsorptive sample loss and deactivation, contamination, and so forth. Since 1970, the CCC technology has advanced in various directions, including preparative and trace analysis, dual CCC, foam CCC, and, more recently, partition of macromolecules with polymer-phase systems. However, most of these methods were only suitable for preparative applications due to relatively long separation times required. In order to fully explore the potential of CCC, efforts have been made to develop analytical high-speed CCC (HSCCC) by designing a miniature multilayer coil planet centrifuge; interfacing analytical HSCCC to a mass spectrometer (HSCCC–MS) began in the late 1980s.

Integration of the high-purity eluate of HSCCC with a low detection limit of MS has led to the identification of a number of natural products, as shown in Table 1 [6–10].

Various HSCCC–MS techniques and their applications are described herein.

Interfacing HSCCC to Thermospray Mass Spectrometry

HSCCC–thermospray (TSP) MS was initiated using an analytical HSCCC apparatus of a 5-cm revolution radius, equipped with a 0.85-mm-inner diameter (i.d.) polytetrafluoroethylene (PTFE) column at 2000 rpm [6–8]. Directly interfacing HSCCC to the MS produced, however, a problem in that the high back-pressure generated by the TSP vaporizer often damaged the HSCCC column. To overcome this problem, an additional high-performance liquid chromatography (HPLC) pump was inserted at the

Table 1 Summary of Previously Reported HSCCC–MS Conditions

Sample	Column	Column capacity (mL)	Revolutional speed (rpm)	Solvent system	Mobile phase	Flow rate (mL/min)	Retention of stationary phase (%)	Ionization	Ref.
Alkaloids	0.85-mm PTFE tube	38	1500	*n*-Hexane–ethanol–water (6:5:5)	Lower phase	0.7	—	Thermospray	6
Triterpenoic acids	0.85-mm PTFE tube	38	1500	*n*-Hexane–ethanol–water (6:5:5)	Lower phase	0.7	—	Thermospray	7
Lignans	0.85-mm PTFE tube	38	1500	*n*-Hexane–ethanol–water (6:5:5)	Lower phase	0.7	—	Thermospray	8
Indole auxins	0.3-mm PTFE tube	7	4000	*n*-Hexane–ethyl acetate–methanol–water (1:1:1:1)	Lower phase	0.2	27.2	Frit–EI	9
Mycinamicins	0.3-mm PTFE tube	7	4000	*n*-Hexane–ethyl acetate–methanol–8% ammonia (1:1:1:1)	Lower phase	0.1	40.4	Frit–CI	9
Colistins	0.55-mm PTFE tube	6	4000	*n*-Butanol–0.04*M* TFA (1:1)	Lower phase	0.16	34.3	Frit–FAB	9
Erythromycins	0.85-mm PTFE tube	17	1200	Ethyl acetate–methanol–water (4:7:4:3)	Lower phase	0.8	—	Electrospray	10
Didemnins	0.85-mm PTFE tube	17	1200	Ethyl acetate–methanol–water (1:4:1:4)	Lower phase	0.8	—	Electrospray	10

interface junction between HSCCC and MS to protect the column against high back-pressures. The effluent from the HSCCC column (0.8 mL/min) was introduced into the HPLC pump through a zero-dead-volume tee fitted with a reservoir supplying extra solvent or venting excess solvent from the HSCCC system. The effluent from the HPLC pump, after being mixed with 0.3*M* ammonium acetate at a rate of 0.3 mL/min, was introduced into the TSP interface. This system has been successfully applied to the analyses of alkaloids [6], triterpenoic acids, [7] and lignans [8] from plant natural products, thereby providing useful structural information. However, a large dead space in the pump at the interface junction adversely affected the resulting chromatogram, as evidenced by loss of a minor peak when HSCCC–UV and HSCCC–TSP–MS total ion current (TIC) chromatograms of plant alkaloids were compared.

In the subsequently developed techniques, the HSCCC effluent is directly introduced into the MS to preserve the peak resolution. Direct HSCCC–MS techniques have many advantages over the HSCCC–TSP method as follows:

1. High enrichment of sample in the ion source
2. High yield of sample reaching the MS
3. No peak broadening
4. High applicability to nonvolatile samples

Various types of HSCCC–MS have been developed using frit fast-atom bombardment (FAB) including continuous flow (CF) FAB, frit electron ionization (EI), frit chemical ionization (CI), TSP, atmospheric pressure chemical ionization (APCI), and electrospray ionization (ESI). Each interface has its specific features. Among those, frit MS and ESI are particularly suitable for directly interfacing to HSCCC, because they generate low back-pressures of approximately 2 kg/cm^2, which is only one-tenth of that produced by TSP.

Interfacing of HSCCC to Frit EI, CI, and FAB–MS

In our laboratory, separations were conducted by newly developed HSCCC-4000 with a 2.5-cm revolution radius, equipped with a 0.3-mm or 0.55-mm-i.d. multilayer coil at a maximum revolution speed of 4000 rpm [9]. The system produced an excellent partition efficiency at a flow rate ranging between 0.1 and 0.2 mL/min, whereas the suitable flow rate for HSCCC–frit MS is between 1 and 5 μL/min. Therefore, the effluent of the HSCCC column was introduced into the MS through a splitting tee which was adjusted to a split ratio of 1:40 to meet the above requirement. A 0.06-mm-i.d. fused-silica tube was led to the MS and a 0.5-mm-i.d. stainless-steel tube was connected to the HSCCC column. The other side of the fused-silica tube extended deeply into the stainless-steel tube to receive a small portion of the effluent from the HSCCC column, and the rest of the effluent was discarded through a 0.1-mm-i.d. PTFE tube. The split ratio of the effluent depended on the flow rate of the effluent and the length of the 0.1-mm-i.d. tube. For adjusting the split ratio at 40:1, a 2-cm length of the 0.1-mm-i.d. tube was needed. Figure 1 shows the HSCCC–MS system, including an HPLC pump, sample injection port, HSCCC-4000, the split tee, and mass spectrometer [9].

In order to demonstrate the potential of HSCCC–frit MS, indole auxins, mycinamicins (macrolide antibiotics), and colistin complex (peptide antibiotics) were analyzed under HSCCC–frit, EI, CI, and FAB–MS conditions.

Three indole auxins, including indole-3-acetamide (IA, MW: 174), indole-3-acetic acid (IAA, MW: 175), and indole-3-butyric acid (IBA, MW: 203) were analyzed under frit EI–MS conditions. In comparison of a TIC with a UV chromatogram, both showed similar chromatographic resolution with excellent theoretical plate numbers ranging from 12,000 to 5500. The results indicate that MS interfacing does not adversely affect chromatographic resolution. In frit EI–MS, the mobile phase behaves like a reagent gas in CI–MS. Both molecular ions and protonated molecules appear in all mass spectra and these data are very useful for the estimation of the molecular weight. Common fragment ions originating from the indole nuclei are found at m/z 116 and 130.

A mixture of mycinamicins was analyzed under HSCCC–frit CI–MS conditions. Mycinamicins consist of six components, mycinamicins I to VI, and isolated mycinamicins IV (MN-IV, MW: 695) and V (MN-V, MW: 711) were used. The structural difference is derived from the hydroxyl group at C-14. These antibiotics were detected under CI conditions, but a reagent gas such as methane, isobutane, or ammonia was not introduced, because the mobile phase behaves like a reagent gas, as described earlier. Both UV and TIC chromatograms showed similar efficiencies, indicating that the MS interfacing does not affect peak resolution, as demonstrated in the analysis of indole auxins. An applicability of this HSCCC–MS system to less volatile compounds was examined under frit FAB–MS conditions.

A peptide antibiotic colistin complex consisting of two major components of colistins A (CL-A, MW: 1168) and B (CL-B, MW: 1154) is difficult to ionize by CI and EI–MS. For HSCCC analysis of these polar compounds, a wider column of 0.55 mm i.d. (instead of 0.3 mm i.d.) was used to achieve satisfactory retention of the stationary phase for a polar n-butanol–trifluoroacetic acid (TFA) solvent system. In addition, for obtaining FAB mass spectra, it is necessary to introduce a sample with an appropriate matrix such as glycerol, thioglycerol, and m-nitrobenzyl alcohol into the FAB–MS ion source. In the present study, glycerol was added as a matrix to the mobile phase at a concentration of 1%. Although a two-phase solvent system containing glycerol was the first trial for a HSCCC study, similarly satisfactory results were obtained in both retention and separation efficiency. Because of the use of a wider column with a viscous n-butanol–aqueous TFA solvent system, the separation was less efficient compared with those obtained from the above two experiments, but

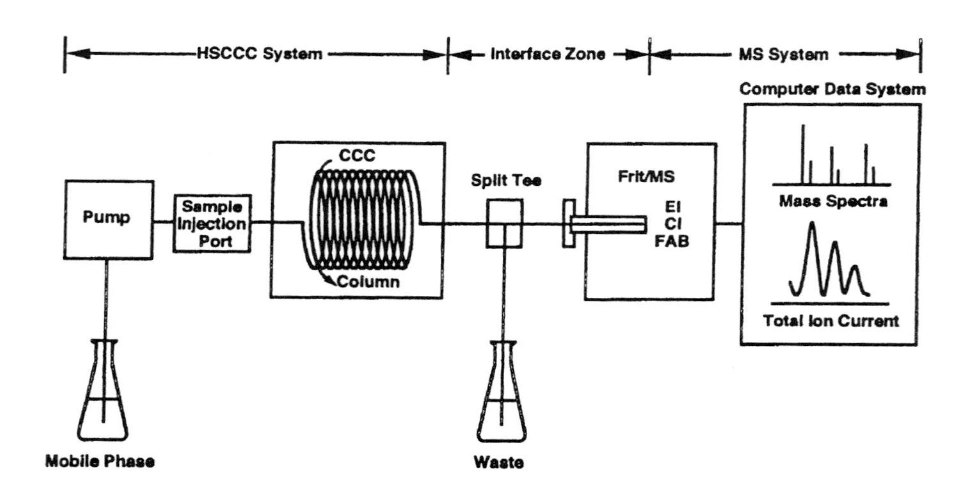

Fig. 1 HSCCC–frit MS system.

the peaks corresponding to CL-A and CL-B were clearly resolved. Mass chromatograms at individual protonated molecules showed symmetrical peaks without a significant loss of peak resolution due to MS interfacing. In all spectra, protonated molecules appeared well above the chemical noise to indicate the molecular weights. These experiments demonstrated that the present HSCCC–frit MS system including EI, CI, and FAB is very potent and is applicable to various analytes having a broad range of polarity. For a nonvolatile, thermally labile and/or polar compound, HSCCC–frit FAB is most suitable, whereas both HSCCC–frit EI and CI can be effectively used for a relatively nonpolar compound.

Interfacing HSCCC to ESI–MS

The experiment was carried out using a small analytical coiled column (17 mL) at 1200 rpm. The effluent from the CCC column at 800 μL/min was split at a 1:7 ratio to introduce the smaller portion of the effluent into ESI–MS using a tee adaptor, as described earlier.

The performance of HSCCC–ESI–MS was evaluated by analyzing erythromycins and didemnins [10]. Because erythromycins (macrolide antibiotics) show weak UV absorbance and cannot be detected easily with a conventional UV detector, mass spectrometric detection is a very useful technique for analysis of these antibiotics. A mixture of erythromycin A (Er-A, MW: 733), erythromycin estolate (Er-E, MW: 789), and erythromycin ethyl succinate (Er-S, MW: 789) was analyzed using HSCCC–ESI–MS with a two-phase solvent system composed of *n*-hexane–ethyl acetate–methanol–water (4:7:4:3). TIC showed, clearly, four peaks corresponding to Er-A, Er-E, Er-S, and an unknown substance. The mass spectra of Er-E and Er-S gave $[M + H]^+$ at *m/z* 862 and 789 and $[M + H - H_2O]^+$ at *m/z* 844 and 772, respectively. In the mass spectrum of Er-A, $[M + H - H_2O]^+$ was observed at *m/z* 761; however, no $[M + H]^+$ was given. The mass spectrum of the unknown peak indicated that it consists of two components with molecular weights of 843 and 772, which correspond to dehydrated Er-S and Er-E, respectively.

Didemnin A (Did-A, MW: 942) is one of the main components of dideminis (cyclic depsipeptides) and is a precursor for synthesis of other dideminins which exhibit antiviral, antitumor, and immunosuppressive activities. Therefore, its purification is very important in the field of pharmaceutical science. However, large-scale purification of Did-A using conventional liquid chromatography is difficult due to the presence of nordidemin A (Nordid-A), which contaminates the target fraction. HSCCC–ESI–MS has been successfully applied to the separation and detection of didemnins. Three peaks were observed on TIC corresponding to dideminins A and B and nordideminin A. Their mass spectra gave only protonated molecules without fragmentation. The first eluted peak was didminin B, which gave $[M + H]^+$ and $[M + Na]^+$ at *m/z* 1112 and 1134, respectively. Did-A appeared as the second peak with $[M + H]^+$ at *m/z* 943 and $[M + Na]^+$ at *m/z* 965. The third peak was Nordid-A, showing $[M + H]^+$ at *m/z* 929 and $[M + Na]^+$ at *m/z* 951. The results indicated that Did-A can be isolated by HSCCC.

Future Prospects

HSCCC–MS has many desirable features for performing the separation and identification of natural and synthetic products, because it eliminates various complications arising from the use of solid support and offers a powerful identification capacity of MS with its low detection limit. We believe that the combination of these two methods, HSCCC–MS, has great a potential for screening, identification, and structural characterization of natural products and will contribute to a rapid advance in natural products chemistry.

References

1. N. B. Mandava and Y. Ito (eds.), *Countercurrent Chromatography: Theory and Practice*, Marcel Dekker, Inc., New York, 1988.
2. W. D. Conway, *Countercurrent Chromatography: Apparatus, Theory and Applications*, VCH, New York, 1990.
3. A. Foucault (ed.), *Centrifugal Partition Chromatography*, Marcel Dekker, Inc., New York, 1995.
4. Y. Ito, *CRC Crit. Rev. Anal. Chem. 17*:65 (1986).
5. Y. Ito and W. D. Conway (eds.), *High-Speed Countercurrent Chromatography*, Wiley–Interscience, New York, 1996.
6. Y.-W. Lee, R. D. Voyksner, Q.-C. Fang, C. E. Cook, and Y. Ito, *J. Liquid Chromatogr. 11*:153 (1988).
7. Y.-W. Lee, T. W. Pack, R. D. Voyksner, Q.-C. Fang, and Y. Ito, *J. Liquid Chromatogr. 13*:2389 (1990).
8. Y.-W. Lee, R. D. Voyksner, T. W. Pack, C. E. Cook, Q.-C. Fang, and Y. Ito, *Anal. Chem. 62*:244 (1990).
9. H. Oka, Y. Ikai, N. Kawamura, J. Hayakawa, K.-I. Harada, H. Murata, and M. Suzuki, *Anal. Chem. 63*:2861 (1991).
10. Z. Kong, K. L. Rinehart, R. M. Milberg, and W. D. Conway, *J. Liquid Chromatogr. 21*:65 (1998).

Hisao Oka
Yoichiro Ito

Cross-Axis Coil Planet Centrifuge for the Separation of Proteins

Introduction

Countercurrent chromatography (CCC) is a form of support-free liquid–liquid partition chromatography in which the stationary phase is retained in the column with the aid of the Earth's gravity or a centrifugal force [1]. Partition of biological samples such as proteins, nucleic acids, and cells has been carried out using various aqueous polymer-phase systems [2]. Among many existing polymer-phase systems, PEG [poly(ethylene glycol)]–dextran and PEG–phosphate systems have been most commonly used for the partition of biological samples. Whereas these polymer-phase systems provide an ideal environment for biopolymers and live cells, high viscosity and low interfacial tension between the two phases tend to cause a detrimental loss of stationary phase from the column in the standard high-speed CCC centrifuge system (known as type J).

The cross-axis CPC (coil planet centrifuge), with column holders at the off-center position on the rotary shaft, enables retention of the stationary phase of aqueous–aqueous polymer-phase systems such as PEG 1000–potassium phosphate and PEG 8000–dextran T500 [3,4]. Since the last decade, various types of cross-axis CPC (types XL, XLL, XLLL, and L) have been developed for performing CCC with highly viscous aqueous polymer-phase systems. The separation and purification of protein samples, including lactic acid dehydrogenase [5], recombinant enzymes, profilin–actin complex, and so on, were achieved using these cross-axis CPCs [6].

Apparatus

The cross-axis CPCs which include types X and L and their hybrids are mainly used for protein separations. These modified versions of the high-speed CCC centrifuge have a unique feature among the CPC systems in that the system provides reliable retention of the stationary phase for viscous polymer-phase systems. Figure 1 presents a photograph of the type XLLL CPC unit and schematically illustrates the orientation and motion of the coil holder in the cross-axis CPC system, where R is a radius of revolution; there are five types of the cross-axis CPC in which the degree of the lateral shift of the coil holder is conventionally expressed by L/R. This parameter for type X, XL, XLL, XLLL, and L CPCs is 0, 1, 2, 3.5, and infinity, respectively. Our studies have shown that the stationary-phase retention is enhanced by laterally shifting the position of the coil holder along the rotary shaft, apparently due to the enhancement of a laterally acting force field across the diameter of the tubing.

The polymer-phase system composed of PEG and potassium phosphate has a relatively large difference in density between the two phases, so that it can be retained well in both XL and XLL column positions which provide efficient mixing of the two phases. On the other hand, the viscous PEG–dextran system, with an extremely low interfacial tension and a small density difference between the two phases, has a high tendency of emulsification under vigorous mixing. Therefore, the use of either the XLLL or L column position, which produces less violent mixing and an enhanced lateral force field, is required to achieve satisfactory phase retention of the PEG–dextran system.

The photograph of the XLLL cross-axis CPC (L/R = 3.5) equipped with a pair of multilayer coil separation columns is shown in fig. 1. The apparatus holds a pair of horizontal rotary shafts symmetrically, one on each side of the rotary frame, at a distance of 3.8 cm from the centrifuge axis. A spool-shaped column holder is mounted on each rotary shaft at an off-center position 13.5 cm away from the midpoint. Each multilayer coil separation column was prepared from a 2.6-mm-inner diameter (i.d.) PTFE (polytetrafluoroethylene) tubing by winding it onto the coil holder hub, forming multiple layers of left-handed coils.

Polymer-Phase Systems for Protein Separation

Countercurrent chromatography utilizes a pair of immiscible solvent phases which have been preequilibrated in a separatory funnel: One phase is used as the stationary phase and the other as the mobile phase. A solvent system composed of 12.5% or 16.0% (w/w) PEG 1000 and 12.5% (w/w) potassium phosphate was usually used for the type XL and XLL cross-axis CPCs. These solutions form two layers; the upper layer is rich in PEG and the lower layer is rich in potassium phosphate. The ratio of monobasic to dibasic potassium phosphates determines the pH of the solvent system; this effect can be used for optimizing the partition coefficient of proteins.

A solvent system composed of 4.4% (w/w) PEG 8000, 7.0% (w/w) dextran T500, and 10 mM potassium phosphate is used with the type XLLL and L cross-axis CPCs for separation of proteins which are not soluble in the PEG–phosphate system. This two-phase solvent system consists of the PEG-rich upper phase and dextran-rich lower phase. The cross-axis CPC may be operated in four different elution modes: P_IHO, P_{II}TO, P_ITI, and P_{II}HI. The parameters P_I and P_{II} indicate the direction of the

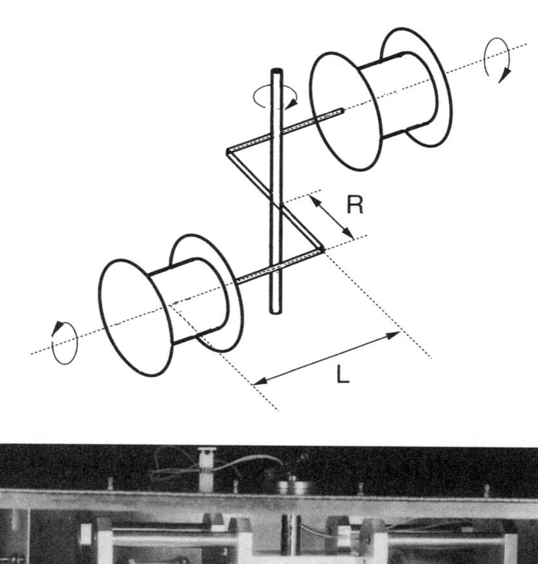

Fig. 1 Diagram of the cross-axis coil planet centrifuge and the photograph of the type XLLL CPC.

planetary motion where P_I indicates counterclockwise and P_{II} clockwise when observed from the top of the centrifuge. H and T indicate the head–tail elution mode, and O and I the inward–outward elution mode along the holder axis. In mode I (inward), the mobile phase is eluted against the laterally acting centrifugal force, and in mode O (outward), this flow direction is reversed. These three parameters yield a total of four combinations for the left-handed coils. Among these elution modes, the inward–outward elution mode plays the most important role in the stationary-phase retention for the polymer-phase systems. To obtain a satisfactory retention of the stationary phase, the lower phase should be eluted outwardly along the direction of the lateral force field (P_IHO and P_{II}TO) or the upper phase in the opposite direction (P_ITI or P_{II}HI).

Application of Cross-Axis CPC for Proteins

Type XL Cross-Axis CPC

The performance of the XL cross-axis CPC, equipped with a pair of columns with a 165-mL capacity, was evaluated for purification of lactic acid dehydrogenase (LDH) from a crude bovine heart filtrate. Successful separation of the LDH fraction was achieved with 16% (w/w) PEG 1000–12.5%(w/w) potassium phosphate at pH 7.3. The separation was performed at 500 rpm at a flow rate of 1.0 mL/min using the potassium phosphate-rich lower phase as a mobile phase. The sodium dodecyl sulfate–polyacrylamide gel electrophoresis (SDS–PAGE) analysis of the LDH fractions showed no detectable contamination by other proteins. The enzymatic activity was also preserved in these fractions.

Type XLL Cross-Axis CPC

The XLL cross-axis CPC, with a 250-mL capacity column, was used for the purification of recombinant enzymes such as purine nucleoside phosphorylase (PNP) and uridine phosphorylase (UrdPase) from a crude *Escherichia coli* lysate. The polymer-phase system used in these separations was 16% (w/w) PEG 1000–12.5% (w/w) potassium phosphate at pH 6.8. The separation was performed at 750 rpm at a flow rate of 0.5 mL/min using the upper phase as a mobile phase. About 1.0 mL of crude lysate, containing PNP in 10 mL of the above solvent system, was loaded into the multilayer coil. Purified PNP was harvested in 45-mL fractions. The SDS-PAGE analysis clearly demonstrated that PNP was highly purified in a one-step elution with the XLL cross-axis CPC.

The capability of the XLL cross-axis CPC was further examined in the purification of a recombinant UrdPase from a crude *E. coli* lysate under the same experimental conditions as described earlier. The majority of the protein mass was eluted immediately after the solvent front (between 105 mL and 165 mL elution volume), whereas the enzyme activity of UrdPase coincided with the fourth protein peak (between 225 mL and 265 mL elution volume). The result indicated that the recombinant UrdPase can be highly purified from a crude *E. coli* lysate within 10 h using the XLL cross-axis CPC.

Type XLLL Cross-Axis CPC

Although PEG–phosphate systems yield a high-efficiency separation, some proteins show a low solubility due to a high salt concentration in the solvent system. In this case, the PEG–dextran polymer-phase system with a low salt concentration can be alternatively used for the separation of such proteins. Because the dextran–PEG system has a high viscosity and an extremely low interfacial tension, it tends to cause emulsification and loss of the stationary phase in the XLL or XL cross-axis CPCs. This problem is minimized using the XLLL cross-axis CPC, which provides a strong lateral centrifugal force to provide a more stable retention of the stationary phase.

Type L Cross-Axis CPC

This cross-axis CPC provides the universal application of protein samples with a dextran–PEG polymer-phase system. Using a prototype of the L cross-axis CPC with a 130-mL column capacity, profilin–actin complex was purified directly from a crude extract of *Acanthamoeba* with the same solvent system as used for the serum protein separation earlier. The sample solution was prepared by adding proper amounts of PEG 8000 and dextran T500 to 2.5 g of the *Acanthamoeba* crude extract to adjust the two-phase composition similar to that of the solvent system used for the separation. The experiment was performed by eluting the upper phase at 0.5 mL/min under a high revolution rate of 1000 rpm. The profilin–actin complex was eluted between 60 mL and 84 mL fractions and well separated from other compounds. The retention of the stationary phase was 69.0% of the total column capacity.

Conclusion

The overall results of our studies indicate that the retention of the stationary phase of polymer-phase systems in the cross-axis CPCs is increased by shifting the column holder laterally along the rotary shaft. Separation of proteins with high solubility in the PEG–phosphate system can be performed with the XL or XLL cross-axis CPC at a high partition efficiency. Proteins with low solubility in PEG–phosphate systems may be separated with a dextran–PEG system using the XLLL or L cross-axis CPC, which provides more stable retention of the stationary phase.

References

1. W. D. Conway, *Countercurrent Chromatography: Apparatus and Applications*, VCH, New York, 1990.
2. P.-Å. Albertsson, *Partition of Cell Particles and Macromolecules*, 3rd ed., Wiley–Interscience, New York, 1986.
3. Y. Ito, E. Kitazume, and M. Bhatnagar, *J. Chromatogr. 538*: 59 (1991).
4. Y. Ito, *J. Chromatogr. 538*:67 (1991).
5. Y. Shibusawa, Y. Eriguchi, and Y. Ito, *J. Chromatogr. B 696*: 25 (1997).
6. Y. Shibusawa, in *High-Speed Countercurrent Chromatography* (Y. Ito and W. D. Conway, eds.), Chemical Analysis Series Vol. 132, Wiley–Interscience, 1996, p. 121.

Yoichi Shibusawa
Yoichiro Ito

D

Dead Point (Volume or Time)

The *injection point* on a chromatogram is that position where the sample is injected. The *dead point* on a chromatogram is the position of the peak maximum of a completely unretained solute. The different attributes of the chromatogram are shown in Fig. 1. The *dead time* is the elapsed time between the injection point and the dead point. The volume that passes through the column between the injection point and the dead point is called the *dead volume*. If the mobile phase is *incompressible*, as in liquid chromatography (LC), the *dead volume* (as so far defined) will be the simple product of the *exit* flow rate and the dead time. However, in LC, where the stationary phase is a porous matrix, the dead volume can be a very ambiguous column property and requires closer inspection and a tighter definition.

If the mobile phase is compressible, the simple product of dead time and flow rate will be incorrect, and the dead volume must be taken as the product of the dead time and the *mean* flow rate. The dead volume has been shown to be given by [1]

$$V_0 = V'_0\frac{3}{2}\left(\frac{\gamma^2 - 1}{\gamma^3 - 1}\right) = Q_0 t_0\frac{3}{2}\left(\frac{\gamma^2 - 1}{\gamma^3 - 1}\right)$$

where the symbols have the meaning defined in Fig. 1, and V'_0 is the dead volume measured at the column exit and γ is the inlet/outlet pressure ratio.

The dead volume will not simply be the total volume of mobile phase in the column system (V_m) but will include extra-column dead volumes (V_E) comprising volumes involved in the sample valve, connecting tubes, and detector. If these volumes are significant, then they must be taken into account when measuring the dead volume.

There are two types of dead volume (i.e., the *dynamic* dead volume and the *thermodynamic* dead volume [2]). The dynamic dead volume is the volume of the *moving phase* in the column and is used in kinetic studies to calculate mobile-phase velocities. In gas chromatography, both the dynamic dead volume and the thermodynamic dead volume can be taken as the difference between the dead volume and the extra-column volume. In LC, however, where a porous packing is involved, some of the mobile phase will be in pores (the *pore volume*) and some between the particles (the *interstitial volume*). In addition, some of the mobile phase in the interstitial volume which is close to the points of contact of the particles will also be stationary. The dynamic dead volume (i.e., the volume of the moving phase) is best taken as the retention volume of a relatively large inorganic salt such as potassiun nitroprusside. This salt will be excluded from the pores of the packing by ionic exclusion and will only explore the moving volumes of the mobile phase [2]. The thermodynamic dead volume will include all the mobile phase that is available to the solute that is under thermodynamic examination. It is best measured as the retention volume of a solvent sample of very similar type to that of the mobile phase and of small molecular weight. If a binary solvent mixture is used (which is the more common situation), then one component of the mobile phase, in pure form, can be used to measure the thermodynamic dead volume. Careful consideration must be given to the measurement of the column dead volume when determining thermo-

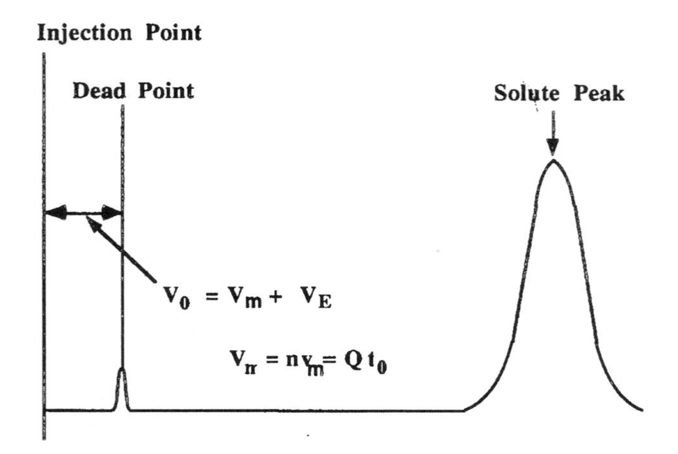

Fig. 1 Diagram depicting the dead point, dead volume, and dead time of a chromatogram. If the mobile phase is not compressible, then V_0 is the total volume passed through the column between the point of injection and the peak maximum of a completely unretained peak, V_m is the total volume of the mobile phase in the column, V_E is the extra column volume of the mobile phase, v_m is the volume of the mobile phase per theoretical plate, t_0 is the time elapsed between the time of injection and the retention time of a completely unretained peak, n is the number of theoretical plates in the column, and Q is the column flow rate measured at the exit.

dynamic data by LC using columns packed with porous materials.

References

1. R. P. W. Scott, *Introduction to Analytical Gas Chromatography*, Marcel Dekker, Inc., New York, 1998, p. 77.
2. A. Alhedai, D. E. Martire, and R. P. W. Scott, *Analyst 114*: 869 (1989).

Suggested Further Reading

Scott, R. P. W., *Liquid Chromatography Column Theory*, John Wiley & Sons, Chichester, 1992, p. 19.

Scott, R. P. W., *Techniques and Practice of Chromatography*, Marcel Dekker, Inc., New York, 1996.

Raymond P. W. Scott

Degassing of Solvents

Degassing, an important step in the preparation of high-performance liquid chromatography (HPLC) mobile phases, is used to remove dissolved atmospheric gases from mobile-phase solvents which may interfere with the normal operation of the HLPC pump and detector. Bubbles that form when these gases come out of solution may result in loss of pump prime and other irregularities, such as inaccurate and irreproducible flow, ultimately resulting in inconsistent retention times and noisy or drifting chromatographic baselines.

A bubble in the detector's flow cell typically manifests itself as a "spike" which may hinder peak detection and integration. Dissolved oxygen poses special problems such as baseline shifts caused by the formation of ultraviolet (UV) absorbing O_2–solvent complexes and diminished fluorescence response (quenching).

Three methods are commonly used to degas HPLC solvents: applying a vacuum, helium sparging, or using an in-line membrane degassing device.

Applying a vacuum for several minutes (alone, or in combination with filtration) to a solvent reservoir is an inexpensive, effective method of degassing. It is typically performed with a side-arm filtration flask and a vacuum aspirator attached to a laboratory faucet. Alternatively, a vacuum pump may be used. Care must be taken when employing vacuum degassing. Because it is an off-line, noncontinuous process, its long-term effectiveness is limited, as air will begin to redissolve as soon as the vacuum is removed. Complete resaturation of the mobile phase will occur within several hours. It is most effective when used on premixed solvents for isocratic separations and is generally unacceptable when performing gradient separations with a low-pressure solvent proportioning device. In addition, vigorous vacuum degassing may alter the composition of mixed solutions as the more volatile organic components (e.g., methanol, acetonitrile, etc.) are selectively evaporated. This may lead to variability in the retention times of separations performed with different batches of mobile phase.

Bubbling helium through a solvent, a technique known as sparging, is a highly effective means of degassing. The rationale behind such an approach is that the helium will displace all other gases present in the solvent, but because the solubility of helium is low in virtually all solvents, the helium-saturated solution is immune from the problems of air-saturated mobile phases. In practice, ultrapure helium is delivered from a gas cylinder and connected to a high-porosity frit via Teflon or similar polymeric tubing. The frit is placed at the bottom of the solvent reservoir and a relatively high flow rate of helium is allowed to bubble through the mobile phase for several minutes. Once the solvent is degassed, the helium flow rate is stopped or, more typically, reduced to a trickle to maintain a continuous deaerated state. Many helium degassing devices are available commercially, including those that allow the solvent reservoirs to be sealed. After the helium flow is reduced to a trickle, the reservoir is sealed, thus blanketing the mobile phase with helium, pressurizing the reservoir, and conserving the sparge gas.

Continuous helium degassing is extremely useful when performing low- or high-pressure solvent proportioning. However, like vacuum degassing, care must be exercised when sparging at high helium flow rates to avoid selective evaporation of the more volatile components of a mixture.

The third common method of degassing performs its function continuously as mobile-phase solvents flow through it. An in-line membrane degasser consists of a vacuum chamber housing coils of gas-permeable tubing, typically composed of thin-walled Teflon. As aerated solvent flows through the tubing, the vacuum removes the dissolved gasses. Typical commercial devices incorporate up to four separate solvent channels to accommodate up to four channels of mobile-phase flow. In-line membrane degassing is an efficient, cost-effective alternative to continuous helium sparging. Selective solvent evaporation is not a problem with these devices, but degassing efficiency does decrease as the mobile-phase flow rate increases.

Whereas degassing is important for all HPLC systems, it is especially important when blending mobile phases automatically with solvent-proportioning devices. The solubility of gases in pure solvents is different than in mixtures, as evidenced by the release of bubbles when two solvents such as methanol and water are mixed together. Because solvent-proportioning devices used to form HPLC gradients continuously blend two or more solvents, it is essential that each solvent be deaerated before meeting in the mixing device. This is particularly true of low-pressure proportioning HPLC systems in which the solvents meet at atmospheric pressure. These systems often require the use of a continuous degassing device such as helium sparging or in-line vacuum. In high-pressure proportioning HPLC systems, the solvents meet under conditions of elevated pressure. This pressure helps to keep the gas in solution and a continuous form of degassing is not always needed.

Although degassing is widely accepted as a necessary step in the preparation of HPLC mobile phases, cases involving the appearance of anomalous peaks have been reported when there is a difference between the amount of dissolved gas in the mobile phase and that of the sample solvent. This is most commonly seen when continuous forms of degassing are used. Although a variety of remedies may be employed, suspension of continuous degassing may be recommended, provided low-pressure solvent proportioning is not being used.

Suggested Further Reading

Bakalyar, S. R., M. P. T. Bradley, and R. Honganen, The role of dissolved gases in high performance liquid chromatography, *J. Chromatogr. 158*: 277 (1978).

Egi, Y. and A. Ueyanagi, Ghost peaks and aerated sample solvent, *LC–GC 16*(2): 112 (1998).

Snyder, L. R. and J. J. Kirkland, *Introduction to Modern Liquid Chromatography*, 2nd ed., John Wiley & Sons, New York, 1979.

Richard DeMuro

Dendrimers and Hyperbranched Polymers: Analysis by GPC–SEC

Dendrimers and hyperbranched polymers are globular macromolecules having a highly branched structure, in which all bonds converge to a focal point or core, and a multiplicity of reactive chain ends. Because of the obvious similarity of their building blocks, many assume that the properties of these two families of dendritic macromolecules are almost identical and that the terms "dendrimer" and "hyperbranched polymer" can be used interchangeably. These assumptions are incorrect because only dendrimers have a precise end-group multiplicity and functionality. Furthermore, they exhibit properties totally unlike that of other families of macromolecules. Highly branched and generally irregular dendritic structures have been known for some time, being found, for example, in polysaccharides, such as amylopectin, glycogen, and some other biopolymers. In the area of synthetic structures, Flory discussed, as early as 1952, the theoretical growth of highly branched polymers obtained by the polycondensation of AB_x structures in which x is at least equal to 2. Such highly branched structures are now known as "hyperbranched polymers."

Today, regular dendrimers can only be prepared using rather tedious, multistep syntheses that require intermediate purifications. In contrast, hyperbranched polymers are easily obtained using a variety of one-pot procedures, some of which mimic, but do not truly achieve, regular dendritic growth [1]. The presence of such a large number of atoms within each dendritic or hyperbranched macromolecule permits an enormous variety of conformations with different shapes and sizes. The distribution of molecular weights focuses on the polydispersity index (M_w/M_n), and the requirements for gelation (or avoidance of gelation) when multimodal monomers are incorporated into the macromolecule. Each of these topics are discussed in Newcome's monograph [2]. Lists of reviews between 1986 and 1996 and Advances series are also given.

Buchard et al. [3] outlined some properties of hyperbranched chains. The dilute solution properties of branched macromolecules are governed by the higher segment density found with linear chains. The dimensions appear to be shrunk when compared with linear chains of the same molar mass and composition. It is shown that the apparent shrinking has an influence also on the intrinsic viscosity and the second virial coefficient. The broad molecular-weight distribution (MWD) has a strong influence on these shrinking factors, which can be defined and used for quantitative determination of the branching density (i.e., the number of branching points in a macromolecule). Here, the branching density can be determined only by size-exclusion chromatography (SEC) in on-line combination with light-scattering and viscosity detectors. The technique and possibilities are discussed in detail.

A dendritic structure generally gives rise to better solubility than the corresponding linear analog. For example, aromatic polyamide dendrimers and hyperbranched polymers are soluble in amide-type solvents and even in tetrahydrofuran. Gel permeation chromatography (GPC) was performed on a Jasco HPLC 880PU fitted with polystyrene–divinylbenzene columns (two Shodex KD806M and KD802) and a Shodex RI-71 refractive index detector in DMF containing 0.01 mol/L of lithium bromide as an eluent. Absolute molecular weights (M_w) of 74,600, 47,800 and 36,800 were determined by light scattering using a MiniDawn apparatus (Wyatt Technology Co.) and a Shimadzu RID-6A refractive index detector. A specific refractive index increment (dn/dc) of the polymer in DMF at 690 nm was measured to be 0.216 mL/g [4].

Standards commonly employed [5] to calibrate SEC columns do not have a well-defined size. Carefully characterized spherical solutes in the appropriate size range are therefore of considerable interest. The chromatographic behavior of carboxylated starburst dendrimers — characterized by quasi-elastic light scattering and viscometry — on a Superose SEC column was explored. Carboxylated starburst dendrimers appear to behave as noninteracting spheres during chromatography in the presence of an appropriate mobile phase. The dependence of the retention time on the solute size seems to coincide with data collected on the same column for Ficoll. Chromatography of the dendrimers yields to a remarkable correlation of the chromatographic partition coefficient with the generation number; this result is, in part, a consequence of the exponential relationship between the generation number and the molecular volume of these dendrimers. All measurements were made in a 9:1 mixture of $NaNO_3$:Na_2HPO_4, pH = 5.5, 0.38M, which has been previously known to minimize electrostatic interactions between a variety of proteins and this stationary phase [4].

The SEC partition coefficient [6] (K_{SEC}) was measured on a Superose 6 column for three sets of well-characterized symmetrical solutes: the compact, densely branched nonionic polysaccharide, Ficoll; the flexible chain nonionic polysaccharide, pullulan; and compact, anionic synthetic polymers, carboxylated starburst dendrimers. All three solutes display a congruent dependence of K_{SEC} on solute radius, R. In accord with a simple geometric model for SEC, all of these data conform to the same linear plot of $K_{SEC}^{1/2}$ versus R. This plot reveals the behavior of noninteracting spheres on this column. The mobile phase for the first two solutes was $0.2M$ NaH_2PO_4–Na_2HPO_4, pH 7.0. In order to ensure the suppression of electrostatic repulsive interactions between the dendrimer and the packing, the ionic strength was increased to $0.30M$ for that solute.

The MWD [7] is derived for polymers generated by self-condensing vinyl polymerization (SCVP) of a monomer having a vinyl and an initiator group ("inimer") in the presence of multifunctional initiator. If the monomer is added slowly to the initiator solution (semibatch process), this leads to hyperbranched polymers with a multifunctional core. If monomer and initiator are mixed simultaneously (batch process), even at vinyl group conversions as high as 99%, the total MWD consists of polymers, which have grown via reactions between inimer molecules (i.e., the normal SCVP process) and those which have reacted with the initiator. Consequently, the weight distribution, $w(M)$, is bimodal. However, the z-distribution, $z(M)$, equivalent to the "GPC distribution," $w(\log M)$ versus log M, is unimodal. Their theoretical studies showed that the hyperbranched polymers generated from an SCVP possess a very wide MWD $M_w/M_n \cong P_n$, where P_n is the number-average degree of polymerization. The evolution of the weight-distribution and z-distribution curves of the total resultant polymer during the SCVP in the presence of the core moiety with $f = 10$ is given. The weight distributions become less bimodal with increasing conversion. In contrast, all z-distributions are unimodal.

Striegel et al. [8] employed SEC with universal calibration to determine the molecular-weight averages, distributions, intrinsic viscosities, and structural parameters of Starburst dendrimers, dextrans, and the starch-degradation polysaccharides (maltodextrins). A comparison has been made in the dilute solution behavior of dendrimers and polysaccharides with equivalent weight-average molecular weights. Intrinsic viscosities decreased in the order $[\eta_{dexstran}] > [\eta_{dextrin}] > [\eta_{dendrimer}]$. A comparison between the molecular radii obtained from SEC data and the radii from molecular dynamics studies show that Starburst dendrimers behave as θ-stars with functionality between 1 and 4. Additionally, electrospray ionization mass spectrometry was employed to determine M_w, M_n, and the PD of Astromol dendrimers.

Size-exclusion chromatography experiments were carried out on a Watters 150CV$^+$ instrument (Waters Associates, Milford, MA) equipped with both differential refractive index single-capillary viscometer detectors. The solvent/mobile phase was H_2O/0.02% NaN_3, at the flow rate of 1.0 mL/min. Pump, solvent, and detector compartments were maintained at 50°C. Separation occurred over a column bank consisting of three analytical columns preceded by a guard column: Shodex KB-G, KS-802, KS-803, and KB-804 (Phenomenex, Torrance, CA). Universal calibration was performed using a series of oligosaccharides (Sigma, St. Louis, MO), and Pullulan Standards (American Polymer Standards, Mentor, OH, and Polymer Laboratories, Amherst, MA).

The solution behavior of several generations of Starburst poly(amido amine) dendrimers, low-molecular-weight ($M_w < 60{,}000$) dextrans, and maltodextrins was also examined by SEC, using the universal calibration. For Starburst and Astramols, supplied M_w values are theoretical average molecular weights. Weight-average molecular weights for the dendrimers determined by SEC with universal calibration using oligosaccharide and polysaccharide narrow standards were slightly, albeit consistently lower than the theoretical averages. In general, the intrinsic viscosity of polymers tends to increase with increasing molecular weight (M), which accompanies an increase in the size of the macromolecule. Exceptions to this are the hyperbranched polymers, in which the Mark–Houwink double logarithmic $[\eta]$ versus M curve passes through a minimum in the low-molecular-weight region before steadily increasing. For solutions of the dendrimers studied in their experiments, it is evident that as M increases, $[\eta]$ decreases. This corresponds to the molecules growing faster in density than in radial growth. Fréchet has pointed out the special situation of this class of polymers, in which their volume increases cubically and their mass increases exponentially [9]. The exponent a in the Mark–Houwink equation for the dendrimers is -0.2 for convergent growth for the generation studied (located in the inverted region of the Mark–Houwink plot). This value for the Starburst dendrimers is comparable to the a value of -0.2 for convergent-growth dendrimers, generations 3–6, studied by Mourey et al. [9]. When the results from SEC are combined with those from computer modeling by comparing the ratios of geometric to hydrodynamic radii for the trifunctional Starbursts to the ratios derived for the other molecular geometries, the dendrimers appear to resemble θ-stars.

Size-exclusion chromatography [9] with a coupled molecular-weight-sensitive detection is a simple convenient method for characterizing dendrimers for which limited sample quantities are available. The polyether dendrimers increase in hydrodynamic radius approximately linearly with generation and have a characteristic maximum in viscosity. These properties distinguish these dendrimers from completely collapsed, globular structures.

The experimental data also indicate that these structures are extended to approximately two-thirds of the theoretical, fully extended length.

Puskas and Grasmüller characterized the synthesized star-branched and hyperbranched polyisobutylenes (PIBs) by SEC–light scattering in tetrahydrofuran (THF), with the *dn/dc* measured as 0.09 mL/g. The radius of gyration gave a slope of 0.3, demonstrating the formation of a star-branched polymer [10].

Gitsov and Fréchet [11] reported the syntheses of novel linear-dendritic triblock copolymers achieved via anionic polymerization of styrene and final quenching with reactive dendrimers. For the characterization of the products in the reaction mixture, SEC with double detection was performed at 45°C on a chromatography line consisting of a 510 pump, a U6K universal injector, three Ultrastyragel columns with pore sizes 100 Å and 500 Å and Linear, a DRI detector M410, and a photodiode array detector M991 (all Millipore Co., Waters Chromatography Division). THF was used as the eluent at a flow rate of 1 mL/min. SEC with coupled PDA detection proves to be particularly useful in the separation and identification of all compounds in the reaction mixture. A detailed discussion can be found in Ref. 11. SEC/VISC studies show that the ABA copolymers are not entangled and undergo a transition from an extended globular form to a statistical coil when the molecular weight of their linear central block exceeds 50,000.

The solution properties of hybrid–linear dendritic polyether copolymers are investigated by SEC with coupled viscometric detection from the same authors [12]. The results obtained show that the block copolymers are able to form monomolecular and multimolecular micelles depending on the dendrimer generation and the concentration in methanol–water (good solvent for the linear blocks).

Large macromolecular assemblies and agglomerates play an important role in living matter and its artificial reproduction. AB and ABA block copolymers are convenient tools used for modeling of these processes. Usually in a specific solvent–nonsolvent system, ABA triblocks form micelles with a core consisting of insoluble B blocks and a surrounding shell of A blocks that extend into the solvent phase. Two Waters/Shodex PROTEIN KW 802.5 and 804 columns were used for the aqueous SEC measurments. The columns were calibrated with 14 PEO and PEG standards. The radius of gyration (R_g) was calculated from the intrinsic viscosity $[\eta]$ and Unical 4.04 software (Viscotek). The calculated values for the Mark–Houwink–Sakurada constant a are 0.583 for PEG ($K = 9.616 \times 10^{-4}$) and 0.776 for PEO ($K = 2.042 \times 10^{-4}$). They are in close agreement with the data reported for the same polymer in other aqueous mixtures (compositions).

The significant decrease in the $[\eta]$ of the copolymer solutions and the parallel decrease in R_g of the hybrid structures containing [G-4] blocks indicate that the block copolymers are undergoing intramolecular micellization. Unimolecular micelles consisting of a small, dense, dendritic core tightly surrounded by a PEO corona are formed. The influence of the size of the dendritic block was investigated with PEO7500. The solution behavior of ABA hybrid copolymers is documented. In general, materials containing more than 30 wt% of dendritic blocks are not soluble in methanol–water. However, it should be emphasized that the solubility of copolymers is also strongly influenced by the size of the dendritic block. Obviously, an optimal balance between the size of the dendrimer and the length of the linear block is required to enable the dissolution of the copolymer in the solvent composition.

Performing SEC with dual detection (DRI and viscometry) permmited application of the concept of universal calibration.

References

1. J. M. J. Fréchet, C. J. Hawker, I. Gitsov, and J. W. Leon, *J. M. S.-Pure Appl. Chem. A33*: 1399 (1996).
2. G. R. Newcome, C. N. Moorefield, and F. Vögtle, *Dendritic Molecules, Concepts, Syntheses, Perspectives*, VCH, Weinheim, 1996.
3. W. Buchard, *Adv. Polym. Sci. 143*: 113 (1999).
4. G. Yang, M. Jikey, and M. Kakimoto, *Macromolecules 32*: 2215 (1999).
5. P. L. Dubin, S. L. Eduards, I. Kaplan, M. S. Mehta, D. Tomalia, and J. Xia, *Anal. Chem. 64*: 2344 (1992).
6. P. L. Dubin, S. L. Edwards, M. S. Mehta, and D. Tomalia, *J. Chromatogr, 635*: 51 (1993).
7. D. Yan, Z. Zhou, and A. Müller, *Macromolecules 32*: 245 (1999).
8. A. M. Strigel, R. D. Plattner, and J. L. Willet, *Anal. Chem. 71*: 978 (1999).
9. T. H. Mourey, S. R. Turner, M. Rubinstein, J. M. J. Fréchet, C. J. Hawker, and K. L. Wooley, *Macromolecules 25*: 2401 (1992).
10. J. E. Puskas and M. Grasmüller, *Macromol. Symp. 132*: 117 (1998).
11. I. Gitsov and J. M. J. Fréchet, *Macromolecules 27*: 7309 (1994).
12. I. Gitsov and J. M. J. Fréchet, *Macromolecules 26*: 6536 (1993).

Nikolay Vladimirov

Derivatization of Acids for GC Analysis

The class of "acids" includes various types of compounds with active hydrogen atoms. The most important group of organic acids is the compounds with the carboxyl fragment —COOH. Some other compounds can be classified not only as O acids [e.g., hydroxamic acids, —CONHOH $\rightleftharpoons$ —C(OH)=NOH], but C—H acids [in the presence of structural fragments $-CH(NO_2)_2$, $-CH(CN)_2$, etc.]. The known substances of this class for gas chromatography (GC) analysis are fatty acids from triglycerides, numerous nonvolatile polyfunctional biogenic compounds (including phenolic carboxylic acids: gallic, vanyllic, syringic, etc.), acidic herbicides (e.g., 2,4-D, 2,4,5-T, MCPB, MCPA, fenoprop, haloxyfop, etc.), and many other substances. Strong inorganic acids (e.g., volatile hydrogen halides HHal or nonvolatile H_2SO_4, H_3PO_4, etc.) can also be the objects of GC analysis.

The simplest monofunctional carboxylic acids have boiling points at atmospheric pressure without decomposition and, hence, can be analyzed directly by GC. However, owing to the high polarities of carboxyl compounds, the typical problem of their GC analysis with standard nonpolar phases is a nonlinear sorption isotherm. As a result, these compounds yield broad, nonsymmetrical peaks that lead to poor detection limits and unsatisfactory reproducibility of their retention indices. The recommended stationary phases for direct analysis of free carboxylic acids are polar polyethylene glycols (Carbowax 20M, DB Wax, SP-1000, FFAP, etc.). However, these phases have less thermal stability compared with polydimethyl siloxanes (approximately 220–230°C versus 300–350°C). This means that the upper limit of GC columns with these polar phases in RI units is not more than 3000–3500 IU. High homologs, even of monocarboxylic acids, cannot be eluted within this RI window (keep in mind that the absence of RI data for palmitic acid $C_{15}H_{31}COOH$ on the mentioned type of polar phases).

Compound	pK_a	T_b(°C)	$RI_{nonpolar}$	RI_{polar}
Acetic acid	4.75	118	638 ± 10[a]	1428 ± 30
Palmitic acid	4.9	351.5	1966 ± 7	No data
Benzoic acid	4.2	250	1201 ± 24	2387 ± 5
Phenylacetic acid	4.2	266	1290 ± 44	No data

[a]RI data with standard deviations are randomized interlaboratory data.

Some of dicarboxylic acids can also be distilled, without decomposition, under reduced pressures. This is at least a theoretical ground for the possibility of their direct GC analysis. Few successful attempts have been described, but these analytes require "on-column" injection of samples and extremely high inertness of chromatographic systems. Many types of polyfunctional carboxylic acids (hydroxy-, mercapto-, amino-, etc.) cannot be analyzed in free, underivatized form, owing to either nonvolatility and/or absence of thermal stability. These features are the principal reasons for the conversion of carboxylic acids, before their GC analysis, into less polar derivatives without active hydrogen atoms.

The general approach for carboxylic acids derivatization is their esterification with formation of alkyl (arylalkyl, halogenated alkyl) or silyl esters:

$$XCO_2H + RY \rightarrow XCO_2R + YH$$

$$XCO_2H + ZSi(CH_3)_3 \rightarrow XCO_2Si(CH_3)_3 + ZH$$

Some of the most widely used reagents for the synthesis of alkyl carboxylates are listed in Table 1. The general recommendations for the silylation of carboxylic acids (TMS and more stable *tert*-butyldimethylsilyl derivatives) are the same as those for other hydroxy-containing compounds (see the entry Derivatization of Hydroxy Compounds for GC Analysis).

The simplest esterification reagents are the corresponding alcohols ROH themselves. Different esters have been used as the analytical derivatives of carboxylic acids: Me, Et, Pr, iso-Pr, isomeric Bu (excluding *tert*-Bu esters, owing to their poorer synthetic yields), and so forth. This method requires the use of excess of dry alcohol and acid catalysis by BCl_3, BF_3, CH_3COCl, $SOCl_2$, and so on. Otherwise, the alcohol being used can be saturated with gaseous HCl, which must then be removed by heating the reaction mixtures after completion of the reaction. The same procedure is used for the synthesis of 2-chloroethyl ($RCO_2CH_2CH_2Cl$), 2,2,2-trichloroethyl ($RCO_2CH_2CCl_3$), and hexafluoroisopropyl esters [$RCO_2CH(CF_3)_2$] for GC analysis with selective detection. Instead of acid catalysis of this reaction, some reagents for the coupling of water were recommended, namely 1,1′-carbonyldiimidazole (**I**) and 1,3-dicyclohexylcarbodiimide (DCC, **II**):

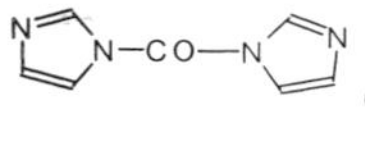

N—CO—N **(I)** —N=C=N— **(II)**

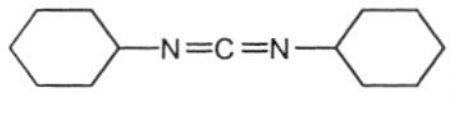

The application of any additional reagents usually leads to the appearance of additional peaks in the chromatograms (including the peaks of by-products; for example, imidazole, $RI_{nonpolar}$ = 1072 ± 17), which must be

Table 1 Physicochemical and Gas Chromatographic Properties of Some Alkylating Derivatization Reagents

Reagent (abbreviation)	MW	T_b(°C)	$RI_{nonpolar}$	By-products ($RI_{nonpolar}$)
Methanol (BCl_3, BF_3, HCl, DCC, etc.)	32	64.6	381 ± 15	—
Diazomethane (in ethyl ether solution)	42	−23	None (unstable)	—
Methyl iodide (DMFA, K_2CO_3)	142	42.8	515 ± 7	CH_3OH (381 ± 15)
Dimethyl sulfate (*tert*-amines)	126	188.5	853 ± 22	CH_3OH (381 ± 15)
1-Iodopropane (DMFA, K_2CO_3)	170	102	711 ± 11	C_3H_7OH (552 ± 13), $(C_3H_7)_2O$ (680 ± 6)
2-Bromopropane (LiH, DMSO)	122	59.4	571 ± 5	$(CH_3)_2CHOH$ (486 ± 9), (iso-Pr)$_2$O (598 ± 5)
Methyl chloroformate	94	71	582 ± 17	CH_3OH (381 ± 15)
Ethyl chloroformate	108	—	640 ± 12	C_2H_5OH (452 ± 18)
Butyl chloroformate	136	—	832 ± 10	C_4H_9OH (658 ± 12)
Pentafluorobenzyl bromide (PFB–Br)	260	174–175	991 ± 11[a]	$C_6F_5CH_2OH$ (934 ± 16)[a]
3,5-bis-(Trifluoromethyl)benzyl bromide (BTBDMA–Br)	306	—	1103 ± 9[a]	$(CF_3)_2C_6H_3CH_2OH$ (1046 ± 15)[a]
Tetramethylammonium hydroxide (TMAH; in 25% aqueous solution)	74	—	Nonvolatile	$(CH_3)_3N$ (418 ± 9)
Trimethylanilinium hydroxide (TMPAH; in 0.2*M* MeOH solution)	136	—	Nonvolatile	$C_6H_5N(CH_3)_2$ (1065 ± 9)
3,5-bis-(Trifluoromethylbenzyl)-dimethylanilinium fluoride	258	—	Nonvolatile	3,5-$(CF_3)_2C_6H_3CH_2N(CH_3)_2$ (no data), $C_6H_5N(CH_3)_2$ (1065 ± 9)
2-Bromoacetophenone (phenacyl bromide)	198	260	1321 ± 4	$C_6H_5COCH_2OH$ (1118)[b]
Silylating reagents	See table in the entry Derivatization of Hydroxy Compounds for GC Analysis			

[a]Estimated RI values.
[b]Single experimental value.

reliably identified and excluded from data interpretation. The by-product from compound **(II)**—1,3-dicyclohexylurea—is nonvolatile for GC analysis.

Other classes of esterification reagents are halogenated compounds (alkyl iodides, substituted benzyl and phenacyl bromides, etc.), which need basic media for their reaction (K_2CO_3 or DMFA are used usually for the neutralization of HBr or HCl as acid by-products). For methylation of carboxylic acids, some tetra-substituted ammonium hydroxides can be used, namely tetramethylammonium hydroxide (in aqueous solutions) or trimethylanilinium hydroxide (in methanolic solutions). The intermediate ammonium carboxylates are thermally unstable and can produce methylalkanoates during subsequent heating of reaction mixtures or even during their introduction into the hot injector of the gas chromatograph (flash methylation):

$$RCO_2H + XNMe_3^+OH^- \rightarrow [RCO_2^-NMe_3^+] \rightarrow RCO_2Me \quad (X = Me, Ph)$$

The possible by-products of these reactions are corresponding amines (Me_3N or $PhNMe_2$). If the appearance of any volatile by-products is undesirable, the methylation of carboxylic acids by diazomethane can be recommended. This reagent (*warning*: highly toxic) is synthesized by alkaline cleavage of *N*-methyl-*N*-nitrosourea **(III)** or *N*-methyl-*N*-nitrosotoluenesulfamide **(IV)** and, owing to its low boiling point (−23°C), can be used only in diethyl ether solutions.

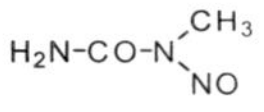

H_2N-CO-N(CH_3)NO **(III)** CH_3–C_6H_4–SO_2–N(CH_3)NO **(IV)**

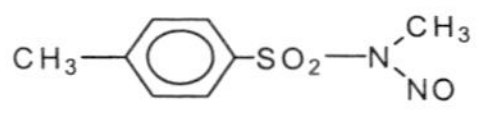

In the absence of acid catalysis, diazomethane reacts only with carboxylic acids (pK_a = 4–5), and phenols (pK_a = 9–10), but has no influence on aliphatic OH groups. Besides CH_2N_2, some more complex diazo compounds (diazoethane, diazotoluene) have been recommended for the synthesis of other esters (ethyl and benzyl, correspondingly). For the synthesis of benzyl (or substituted benzyl) esters, some special reagents have also been proposed [e.g., *N,N'*-dicyclohexyl-*O*-benzylurea **(V)** and 1-(4-methylphenyl)-3-benzyltriazene **(VI)**:

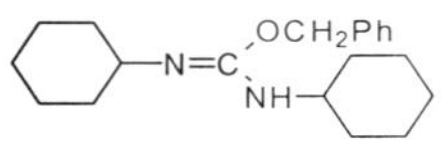

C_6H_{11}–N=C(OCH_2Ph)NH–C_6H_{11} **(V)** CH_3–C_6H_4–N=N-NH-CH_2Ph **(VI)**

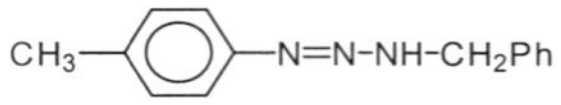

The esterification of carboxylic acids can be provided, also, by synthetic equivalents of alcanols: acetals $RCH(OR')_2$ (with acid catalysis), ortho-esters $RC(OR')_3$ (acid catalysis), and dialkylcarbonates $CO(OR)_2$ (base catalysis). The series of bifunctional reagents of this type [dimethylformamide dialkylacetals $(CH_3)_2N{-}CH(OR)_2$] is commercially available. Besides the esterification of carboxyl groups, these compounds react with primary amino groups and, thus, are used for GC analysis of amino acids (see the entry "Derivatization of Amines, Amino Acids, and Related Compounds for GC Analysis"):

$$R{-}CH(CO_2H)(NH_2) + (CH_3)_2NCH(OCH_3)_2 \rightarrow R{-}CH(CO_2CH_3)(N{=}CHN(CH)_3)_2) + DMFA + CH_3OH$$

A "sandwich" technique (flash methylation) can also be used in this case. This implies the injection the combined sample and reagent in the same syringe into the gas chromatograph (e.g., successively placed 1 mL of derivatization reagent, 1 mL of pyridine with internal standard, and 1 mL of the solution of analytes in the same solvent).

Alkyl chloroformates $ClCO_2R$ (R = Me, Et, Bu) have been recently proposed as convenient alkylating reagents for carboxylic acids:

$$RCO_2H + ClCO_2R' + B \rightarrow RCO_2R' + CO_2 + BH^+Cl^-$$

Two-stage single-pot derivatization of carboxylic acids (with intermediate formation of chloroanhydrides with thionyl chloride, followed by their conversion into amides) was recommended for high-performance liquid chromatography analysis, but the simplest dialkylamides are also volatile enough for GC analysis (the mixture of Ph_3P and CCl_4 can be used in this reaction instead of $SOCl_2$). Moreover, the same procedure is used for the synthesis of diastereomeric derivatives of enantiomeric carboxylic acids (see below):

$$RCO_2H + SOCl_2 \rightarrow RCOCl + SO_2 + HCl$$

$$RCOCl + R'R''NH + B \rightarrow RCONR'R'' + BH^+Cl^-$$

The reactivities of carboxy and OH groups in the polyfunctional hydroxy and phenolic carboxylic acids are different. This indicates the possibility of an independent two-stage derivatization of these compounds, for example,

$$R{-}CH(CO_2H)(OH) \rightarrow 1.\ MeOH/BF_3;\quad 2.\ Ac_2O/Py \rightarrow R{-}CH(CO_2CH_3)(OCOCH_3)$$

$$HO{-}C_6H_4{-}CO_2H \rightarrow 1.\ BuOH/BF_3;\quad 2.\ TFAA \rightarrow CF_3CO_2{-}C_6H_4{-}CO_2C_4H_9$$

If these functional groups are located in vic- (aliphatic series) or ortho-positions (arenecarboxylic acids), methyl or butyl boronic acids are convenient reagents for their one-step derivatization, with the formation of cyclic methyl (butyl) boronates:

$$R{-}CH(CO_2H)(OH) + Me(Bu)B(OH)_2 \rightarrow$$ [cyclic boronate: R–CH–C(=O)–O–B(Me(Bu))–O ring]

A similar method for simultaneous derivatization of two functional groups is the formation of cyclic silylene derivatives for the same types of compounds:

$$C_6H_4(CO_2H)(XH)\ (X{=}O, NR) + Cl_2Si(tert.Bu)_2 \rightarrow$$ [cyclic C_6H_4–C(=O)–O–Si(tert.Bu)$_2$–X ring]

The special type of carbonyl-group derivatization is aimed for gas chromatography–mass spectrometry (GC–MS) determination of double-bond C=C positions in the unsaturated long-chain acids. The analytical derivatives for the solution of this problem are nitrogen-containing heterocycles. These compounds can be synthesized by condensation of carboxylic acids with 2-amino-2-methyl-1-propanol (2-substituted 4,4-dimethyloxazolines), 2-aminophenol (2-substituted benzoxazoles), and so forth.

$$R{-}C(O)OH + H_2N{-}C(CH_3)_2{-}CH_2OH \rightarrow$$ 2-R-4,4-dimethyloxazoline

$$R{-}C(O)OH + H_2N{-}C_6H_4{-}OH \rightarrow$$ 2-R-benzoxazole

Gas chromatographic separation of enanthiomeric carboxylic acids is based on the formation of their esters or amides with optically active alcohols [e.g., (−)-menthol **(VII)**] or amines [α-methylbenzenemethaneamine **(VIII)**], usually through the intermediate chloroanhydrides. These diastereomeric products are not as volatile as other acid derivatives, but, owing to presence of two asymmetric carbon atoms (∗) in the molecule, they can be separated on nonchiral phases:

$$R^*CO_2H \rightarrow [R^*COCl] + \text{(VII)} \rightarrow \text{menthyl–OCOR}^*$$

$$R^*CO_2H \rightarrow [R^*COCl] + C_6H_5C^*H(CH_3)NH_2\text{(VIII)} \rightarrow R^*CONHC^*H(CH_3)C_6H_5$$

A problem closely related to the derivatization of free carboxylic acids is the determination of their composition in the biogenic triglycerides, lipids, and so forth. The sample preparation includes the reesterification (preferably with formation of methyl esters) of these compounds in acid ($MeOH/BF_3$, MeOH/AcCl, etc.) or basic (MeONa, MeOH/KOH, etc.) media. Methyl esters of fatty acids are compounds that are well characterized by both standard mass spectra and GC retention indices on standard phases.

The combination of these analytical parameters provides their reliable identification.

The general method of organic sulfo- (RSO_2OH) and various substituted phosphorus acids [$ROP(O)(OH)_2$, $RP(O)(OH)_2$, etc.] is their silylation. Analogous recommendations have been proposed for the determination of inorganic anions. The values of retention indices on standard nonpolar phases (SE-30) are known for TMS derivatives of most important inorganic acids:

Anion	Volative derivative for GC analysis	$RI_{nonpolar}$	Anion	Volative derivative for GC analysis	$RI_{nonpolar}$
Borate	$B(OTMS)_3$	1010	Arsenite	$As(OTMS)_3$	1149
Carbonate	$CO(OTMS)_2$	1048	Phosphate	$PO(OTMS)_3$	1273
Phosphite	$P(OTMS)_3$	1115	Vanadate	$VO(OTMS)_3$	1301
Sulfate	$SO_2(OTMS)_2$	1148	Arsenate	$AsO(OTMS)_3$	1353

Suggested Further Reading

Blau, K. and J. M. Halket (eds.), *Handbook of Derivatives for Chromatography*, 2nd ed., John Wiley & Sons, New York, 1993.

Brooks, C. J. W. and W. T. Cole, *J. Chromatogr. 441*: 13 (1988).

Burke, D. J. and B. Halpern, *Anal. Chem. 55*: 822 (1983).

Butz, S. and H.-J. Stan, *J. Chromatogr. 643*: 227 (1993).

Drozd, J., *Chemical Derivatization in Gas Chromatography*, Elsevier, Amsterdam, 1981.

Gabelish, C. L., P. Crisp, and R. P. Schneider, *J. Chromatogr. A 749*: 165 (1996).

Gonzalez, G., R. Ventura, A. K. Smith, R. de la Torre, and J. Segura, *J. Chromatogr. A 719*: 251 (1996).

Knapp, D. R., *Handbook of Analytical Derivatization Reactions*, John Wiley & Sons, New York, 1979.

Umeh, E. O., *J. Chromatogr. 51*: 139, 147 (1970).

Zhang, J. T., Q. T. Yu, B. N. Lin, and Z. H. Huang, *Biomed. Environ. Mass Spectrom. 15*: 33 (1988).

Igor G. Zenkevich

Derivatization of Amines, Amino Acids, Amides, and Imides for GC Analysis

The amines are the extensive class of organic compounds of general formulas RNH_2 (primary), $RR'NH$ (secondary), and $RR'R''N$ (tertiary). Their chemical and chromatographic properties are determined by the presence of a basic functional group and active hydrogen atoms in the molecule. Their basicity is strongly different for aliphatic amines ($pK_a = 10.5 \pm 0.8$) and substituted anilines ($pK_a = 4.9 \pm 0.3$) owing to p–π conjugation N–Ar. Amides are the derivatives of carboxylic acids with structural fragments $-CO-N\langle$ or $-SO_2-N\langle$ (including cyclic structures), whereas imides have the fragments $-CO-NH(R)-CO-$.

The simplest members of all amine classes usually are volatile enough for their direct gas chromatography (GC) analysis. When other polar functional groups with active hydrogen atoms are present in the molecule, the derivatization of one or all of them becomes necessary. A typical example of these compounds is amino acids, which exist in the form of inner-molecular salts $RCH(NH_3^+)CO_2^-$ in the solid state.

The principal directions of amino compound derivatization for GC analysis include the following types of chemical reactions:

Acylation: $RR'NH + R''COX + B \rightarrow RR'NCOR'' + BH^+X^-$

Formation of Schiff bases (only for primary amines): $RNH_2 + R'R''CO \rightarrow RN{=}CR'R'' + H_2O$

Alkylation: $RR'NH + R''X \rightarrow RR'NR'' + XH$

Silylation: $RR'NH + XSi(CH_3)_3 \rightarrow RR'N-Si(CH_3)_3 + XH$

The first group of reactions (acylation) includes the greatest number of examples. Numerous recommended reagents are listed in the Table 1; they belong to two classes of reagents: anhydrides ($X{=}OCOR''$) and chloroanhydrides ($X{=}Cl$). Most widely used of them are acetic and trifluoroacetic anhydrides. The by-products of acylation, in all cases, are acids; these reactions need basic media (additives of pyridine or *tert*-amines) to prevent the formation of nonvolatile salts from the analytes. The technique of derivatization is extremely simple: Samples mixtures are allowed to stand with acylating reagents for some minutes prior to analysis.

Table 1 Physicochemical and Gas-Chromatographic Properties of Some Acylation Reagents

Reagent (abbreviation)	MW	T_b(°C) (P)	d_4^{20}	n_D^{20}	$RI_{nonpolar}$	By-products ($RI_{nonpolar}$)
Acetic anhydride	102	139.6	1.08	1.390	706 ± 9	CH_3CO_2H (638 ± 10)
Trifluoroacetic anhydride (TFA)	210	39–40	1.511	1.268	515 ± 6	CF_3CO_2H (744 ± 6)
Pentafluoropropionic anhydride (PFPA)	310	70–72	1.588	—	606 ± 6[a]	$C_2F_5CO_2H$ (781 ± 12)
Heptafluorobutyric anhydride (HFBA)	410	109–111	1.674	—	745 ± 4[a]	$C_3F_7CO_2H$ (863 ± 16)
bis-Trifluoroacetyl methylamine (MBTFA)	223	120–122	1.547	1.346	773 ± 16[a]	$CF_3CONHCH$ (540)
N-Trifluoroacetyl imidazole (TFAI)	164	137–138	1.442	1.424	830 ± 21[a]	Imidazole (1072 ± 17)
Chloroacetic anhydride	170	203	1.550	—	1116 ± 18[a]	$ClCH_2CO_2H$ (864 ± 3)
Dichloroacetic anhydride	238	214–216	—	—	1248 ± 14[a]	Cl_2CHCO_2H (1048 ± 23)
Trichloroacetic anhydride	306	222–224	1.691	1.484	1471 ± 27[a]	CCl_3CO_2H (1270[a])
Benzoyl chloride	140	197–198	1.211	1.553	1046 ± 9	$C_6H_5CO_2H$ (1201 ± 24)
Pentafluorobenzoyl chloride	230	158–159	1.669	1.453	922 ± 14[a]	$C_6F_5CH_2OH$ (934 ± 16[a])
Chlorodifluoroacetic anhydride	242	92–93	—	1.348	679 ± 8[a]	$ClCF_2CO_2H$ (793[a])
Acetyl chloride	78	51.8	1.104	1.389	542 ± 7	CH_3CO_2H (638 ± 10)
Chloroacetyl chloride	112	106.1	1.419	1.454	622 ± 8	$ClCH_2CO_2H$ (864 ± 3)
Dichloroacetyl chloride	146	107–108	1.532	1.460	726 ± 19[a]	Cl_2CHCO_2H (1048 ± 23)
Trichloroacetyl chloride	180	118	1.629	1.470	778 ± 15	CCl_3CO_2H (1270[a])
Pivaloyl anhydride	186	192–193	0.918	1.409	1053 ± 31[a]	$(CH_3)_3CCO_2H$ (804)
Diethylpyrocarbonate	162	93–94 (18)	1.101	1.398	—	C_2H_5OH (452 ± 18)

[a]Estimated RI values.

The anhydrides and chloroanhydrides of chlorinated acetic acids and pentafluorobenzoyl chloride are used for the synthesis of chlorinated amides for GC analysis with selective detectors. Diethylpyrocarbonate converts primary and secondary amines (including NH_3) to N-substituted carbamates:

$$RR'NH + O(CO_2C_2H_5)_2 \rightarrow RR'NCO_2C_2H_5 + CO_2 + C_2H_5OH$$

The next group of derivatization reactions is the formation of Schiff bases from the reaction of primary amines with carbonyl compounds. Some recommended reagents are listed in Table 2. Aromatic aldehydes are much more reactive in this condensation, compared with ketones and aliphatic compounds (from the latter, only low-boiling acetone and cyclohexanone have been used in GC practice). All carbonyl compounds can be taken into reaction with amines in the form of their various synthetic analogs, at first acetals or ketals. So, dimethylformamide dialkylacetals $(CH_3)_2N{-}CH(OCH_3)_2$ react with primary amines with formation of *N*-dimethylaminomethylene derivatives $R{-}N{=}CH{-}N(CH_3)_2$. So far, because these reagents at the same time provide the esterification of carboxyl groups, they have been recommended for single-stage derivatization of amino acids (discussed later).

Table 2 Physicochemical and GC Properties of Some Carbonyl Reagents and Their Analogs for Derivatization of Amino Compounds

Reagent (abbreviation)	MW	T_b(°C)	d_4^{20}	n_D^{20}	$RI_{nonpolar}$
Acetone	58	56.2	0.791	1.359	472 ± 12
Pentafluorobenzaldehyde	196	166–168	1.588	1.450	943 ± 22[a]
Thiophen-2-carboxaldehyde	112	198	1.200	1.590	966 ± 9
Carbon disulfide	76	46.3	1.263	1.628	530 ± 9
Methyl isothiocyanate[b]	73	118	1.069	1.525	689 ± 16
Phenyl isothiocyanate[b]	135	219–221	1.130	1.652	1163 ± 7
Dimethylformamide dimethyl acetal (DMFDMA)[c]	119	106	0.897	1.397	726 ± 4
Dimethylformamide diethyl acetal (DMFDEA)[c]	147	134–136	0.859	1.400	826 ± 5[a]

[a]Estimated RI values.
[b]Only for derivatization of amino acids; with monofunctional amines nonvolatile products can be formed.
[c]Bifunctional reagents; the by-products are MeOH (EtOH) and DMFA (749 ± 16).

Carbon disulfide, as the thio analog of carbonyl compounds, reacts with primary amines with resultant formation of alkyl isothiocyanates, which have lower retention indices than other derivatives of primary amines, including *N*-trimethylsilylated amines. The sole by-product of this reaction is gaseous hydrogen sulfide:

$$RNH_2 + CS_2 \rightarrow RNCS + H_2S$$

R in $RNHSi(CH_3)_3$	$RI_{nonpolar}$	R in R—NCS	$RI_{nonpolar}$
Me	689 ± 21[a]	Me	689 ± 16
Et	756 ± 11	Et	736 ± 5

[a]RI data with standard deviations are randomized interlaboratory data.

The alkylation of amines (including polyamines formed by reduction of polypeptides) was a highly popular method of derivatization in peptide chemistry before the appearance of contemporary mass-spectrometric techniques for analysis of nonvolatile compounds (FFAB, MALDI, etc.). Direct alkylation of amines by alkyl halides (Hoffman reaction) can lead to the final nonvolatile ammonium salts and, hence, other soft reagents must be used. For example, exhaustive methylation can be provided by the mixtures $CH_2O/NaBH_4/H^+$ or CH_2O/formic acid.

Silylation of amines is a relatively rarely used method for their derivatization at present. The problem is the facile hydrolysis of *N*-TMS compounds. It leads to the formation in the reaction mixtures of both mono-TMS (RNHTMS), and bis-TMS $[RN(TMS)_2]$ derivatives. This multiplicity of products from the same precursor creates some difficulties in data interpretation. The recently introduced *N*-(*tert*-butyldimethylsilyl) (DMTBS) derivatives are more stable to the hydrolysis and their formation is more unambiguous (only monosubstituted compounds are formed) owing to steric reasons.

Tertiary amines have no active hydrogen atoms and their derivatization in the generally accepted sense is not required. Only in special cases (GC analysis with selective detectors), the cleavage of $N—CH_3$ bonds by chloroformates can be used:

$$RR'N—CH_3 + CCl_3CH_2OCOCl \rightarrow RR'N—CO_2CH_2CCl_3$$

Mixtures of amino acids are among of the most important substances for chromatographic analysis. Some dozens of methods for derivatization of these compounds for their GC determination have been proposed. The most widely used can be subdivided into two types:

1. Separate derivatization of functional groups $—CO_2H$ and $—NH_2$ by different reagents.
2. Protection of both groups by only one reagent.

The typical derivatives of the first type are various esters (Me, Et, Pr, iso-Pr, Bu, iso-Bu, sec-Bu, Am, iso-Am, etc.) of *N*-acyl (acetyl, trifluoroacetyl, pentafluoropropionyl, heptafluorobutiryl, etc.) amino acids. The butyl esters of *N*-TFA amino acids even have been given the special abbreviation: TAB derivatives. The two-stage process includes the esterification of amino acids by an excess of the corresponding alcohol in the presence of HCl and, after evaporation of volatile compounds, the treatment of the nonvolatile hydrochlorides of alkyl esters by acylating reagents:

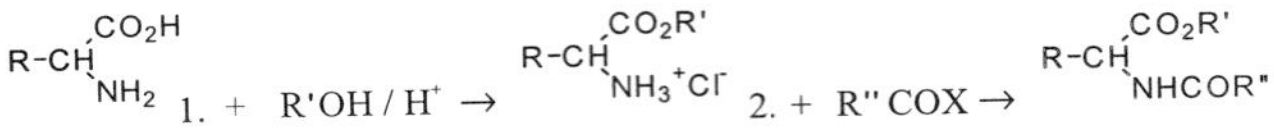

$$R\text{-}CH(CO_2H)(NH_2) \xrightarrow{1.\ +\ R'OH/H^+} R\text{-}CH(CO_2R')(NH_3^+Cl^-) \xrightarrow{2.\ +\ R''COX} R\text{-}CH(CO_2R')(NHCOR'')$$

Some variations of this procedure are known. Instead of N-acylation, the treatment of intermediate esters by CS_2/Et_3N and CH_3OCOCl, with the formation of 2-alkoxycarbonylisothiocyanates **(I)**, or by carbonyl compounds, which leads to the Schiff bases **(II)**, have been proposed. However, the analytical advantages of these derivatives are not so obvious.

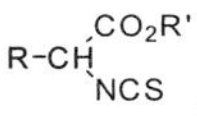

$R\text{-}CH(CO_2R')(NCS)$ **(I)** $R\text{-}CH(CO_2R')(N{=}CHR'')$ **(II)** $CF_3CH_2NH\text{-}CH_2\text{-}CH(CH_2Ph)\text{-}NH\text{-}CH(CH_2Ph)\text{-}CH_2OSi(CH_3)_3$ **(III)**

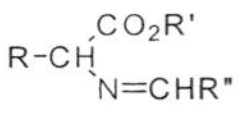

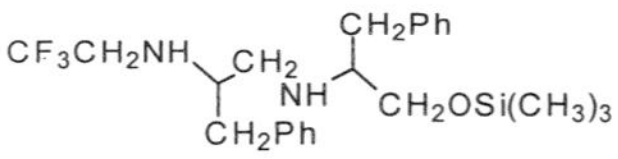

The same types of *N*-acyl-*O*-alkyl derivatives can also be used for GC analysis of the simplest oligopeptides (at least dipeptides and tripeptides). Other, obsolete, recommendations on the GC analysis of these compounds include various sequences for their reduction by $LiAlH_4$ into polyaminoalcohols, followed by N-acylation or permethylation and, finally, silylation of OH groups. For example, *N*-TFA dipeptide Phe-Phe after reduction and silylation gives the compound **(III)** with a retention index of 2390 on semistandard-phase Dexsil-300.

The typical example of one-stage derivatization of amino acids is their treatment by isopropyl bromide in presence LiH with the formation of N-isopropylated isopropyl esters. Unfortunately, this reaction can take place only in high-boiling aprotic bipolar solvents like dimethyl sulfoxide (DMSO, T_b = 189°C, $RI_{nonpolar}$ = 790 ± 18); this is a significant restriction for its application, in practice. A more important method is based on the reaction of amino acids with methyl or phenyl isothiocyanates with the formation of 3-methyl (phenyl) hydantoins:

$R\text{-}CH(CO_2H)NH_2$ + Me(Ph)NCS (pH ≈ 9) → $R\text{-}CH(CO_2H)NHCSNHMe(Ph)$ (pH ≈ 1) → (thiohydantoin, NMe(Ph))

The stable nonvolatile intermediate phenylthiocarbamoyl derivatives are formed in basic media and can be analyzed directly by reverse-phase high-performance liquid chromatography (RP–HPLC). Their cyclization into hydantoins requires acid catalysis. This mode of derivatization is a very important supplement to the Edman's method of N-terminated sequencing of polypeptides. Before GC analysis, any hydantoins can be converted into *N*-trifluoroacetyl or enol-*O*-trimethylsilyl derivatives, which increases the selectivity of their determination in complex matrices.

N-Acylated amino acids, in the presence of water-coupling reagents (dicyclohexylcarbodiimide or the excess of TFAA), form other cyclic derivatives—azlactones (2,4-disubstituted oxazolin-5-ones):

$R\text{-}CH(CO_2H)NHCOR'$ − H_2O → (oxazolin-5-one, R, R')

One of the most popular methods of single-stage amino acid derivatization at present is their conversion to *N,O(S) tert*-butyldimethylsilyl derivatives [the reagent *tert*-butyldimethylsilyl trifluoroacetamide (MTBSTFA) or its *N*-Me analog]. Another way, which was proposed at the beginning of the 1970s is based on amino acid interaction with dimethylformamide dialkylacetals $(CH_3)_2NCH(OR')_2$ (R = Me, Et, Pr, iso-Pr, Bu, Am) with formation of *N*-dimethylaminomethylene derivatives of amino acids esters:

$R\text{-}CH(CO_2H)NH_2$ + $(CH_3)_2NCH(OR')_2$ → $R\text{-}CH(CO_2R')N{=}CHN(CH_3)_2$ + $(CH_3)_2NCHO$ + R'OH

The GC separation of enantiomeric D- and L-amino acids with nonchiral phases requires their conversion to diastereomeric derivatives. The second asymmetric center in the molecule (∗) arises after esterification by optically active alcohols [2-BuOH, 2-AmOH, pinacolol, (−)-menthol, etc.] or NH_2 group acylation by chiral reagents {e.g., α-methoxy-α-trifluoromethylphenylacetyl chloride [MTPAC **(IV)**], *N*-trifluoroacetyl-L-prolyl chloride [N-TFA-L-Pro-Cl **(V)**], *N*-trifluoroacetyl-thiazolidine-4-carbonyl chloride **(VI)**, etc.}:

$C_6H_5\text{-}C^*(OCH_3)(CF_3)\text{-}COCl$ **(IV)**; N-$COCF_3$ pyrrolidine-2-COCl **(V)**; N-$COCF_3$ thiazolidine-4-COCl **(VI)**

The selection of derivatization methods for amides and imides is not as great as for other classes of amino compounds. The active hydrogen atom in the structural fragments —CO—NH— or SO_2—NH— is highly acidic and, hence, sometimes recommended TMS, acetyl, or TFA derivatives of these compounds are unstable during hydrolysis. The best derivatization method is their exhaustive alkylation (preferably methylation), because permethylated amides and imides are volatile enough for their GC analysis. This general statement can be illustrated by retention data for methylated derivatives of urea $CO(NH_2)_2$ as the simplest amide: Both the initial compound and its monomethyl and two dimethyl homologs cannot be analyzed by GC, owing to their nonvolatility. The retention index of trimethyl urea on the standard nonpolar polysimethyl siloxanes is 976 ± 28, whereas, for tetramethyl urea, it is 956 ± 5 (see the high interlaboratory reproducibility of this value compared with the previous one).

The exhaustive methylation of amides can be realized with rather high yields by their reactions with dimethyl sulfate/$EtN(iso\text{-}Pr)_2$, by CH_3I in acetone solution, with CH_3I in the presence of K_2CO_3 or LiH in DMSO, in heterophaseous "water–organic solvent" systems (together with the extraction of derivatives from matrices), and, directly, in the injector of gas chromatograph (so-called flash methylation) by $PhNMe_3^+OH^-$(TMPAH). These modes of derivatization precede the GC determination of numerous diuretics (acetazolamide, ethacrinic acid, clopamide, etc.), some barbiturates and their metabolites, xanthines (theophylline), various urea and carbamate pesticides (monuron, fenuron, linuron and their metabolites), and so forth.

Suggested Further Reading

Avery, M. J. and G. A. Junk, *Anal. Chem. 57*: 790 (1985).

Biermann, C. J., C. M. Kinoshita, J. A. Marlett, and R. D. Steele, *J. Chromatogr. 357*: 330 (1986).

Blau, K. and J. M. Halket (eds), *Handbook of Derivatives for Chromatography*, 2nd ed., John Wiley & Sons, New York, 1993.

Carreras, D., C. Imas, R. Navajas, M. A. Garcia, C. Rodrigues, A. F. Rodrigues, and R. J. Cortes, *J. Chromatogr. A 683*: 195 (1994).

Drozd, J., *Chemical Derivatization in Gas Chromatography*, Elsevier, Amsterdam, 1981.

Horman, I. and F. J. Hesford, *Biomed. Mass Spectrom. 1*: 115 (1974); *Nestle Research News* 100 (1976/77).
Knapp, D. R., *Handbook of Analytical Derivatization Reactions*, John Wiley & Sons, New York, 1979.
Mawhinney, T. P., R. S. R. Robinett, A. Atalay, and M. A. Madson, *J. Chromatogr. 358*: 231 (1986).
Nazareth, A., M. Joppich, A. Panthani, D. Fisher, and R. W. Giese, *J. Chromatogr. 319*: 382 (1985).
Thenot, J. P. and E. C. Horning, *Anal. Lett. 5*: 519 (1972).

Igor G. Zenkevich

Derivatization of Analytes in Chromatography: General Aspects

Any chromatographic analysis is preceded by a priori available information about analytes. Depending on the quantity of this available information, all determinations may be classified as (a) preferably confirmatory (determined components are known) and (b) prospective (any propositions concerning their chemical nature are very approximate). Numerous differences in the design of analytical procedures in these two cases are manifested in the features of all stages of analysis: sampling, sample preparation, the analysis itself, and interpretation of results. Only for procedures being classified as confirmatory, the stage of sample preparation can be supplemented by their chemical treatment by different reagents for the optimization of subsequent chromatographic analysis. The most widely used type of treatment is the synthesis of various chemical derivatives of target analytes, namely derivatization.

Derivatization is a special subgroup of organic reactions used in chromatography for compounds with specific types of functional group. Not all known reactions can be applied as methods for derivatization, because these processes must be in accordance with some specific conditions:

1. The experimental operations must be as simple as possible. The mixing of sample with reagent(s) at ambient temperature, without additional treatment of mixtures, is preferable. In chromatographic practice, the time needed for the completion of the derivatization reaction may be up to 24 h (so-called "stay overnight"). Instead of this long time, the heating of reaction mixtures in ampoules is also permitted. Some processes (including silylation) can be realized by simple injection of reaction mixtures into the hot injector of the GC equipment.
2. The number of stages for every type of functional group must be minimal (one or 2, but no more!). For multistaged processes, the condition "single-pot synthesis" is necessary. Such operations as extraction or reextraction may be used only when the quantities of analytes are not very small or when it is necessary to isolate them from complex matrices. The large excess of derivatization reagent(s) and/or solvents must be easily removable or have no influence on the results of the analysis. The use of high-boiling solvents, typically, is not recommended. The possible by-products of reaction of must have no influence on the results as well.
3. The degree of transformation of initial compounds into products (yield, %) must be maximal and reproducible to provide valid quantitative determination of these compounds by analysis of their derivatives.
4. Mutually unambiguous correspondence between initial analytes and derivatives must be assured. The optimal case is $1 \rightarrow 1$, but some examples of type $1 \rightarrow 2$ are known (e.g., when the derivatives of enantiomers form a pair of diastereomers, *O*-alkyl ethers of oximes exist in syn and anti isomers, etc.). All processes that lead to further uncertainty (chemical multiplication of analytical signals) [e.g., $1 \rightarrow N$ ($N \geq 3$) must be excluded].

In accordance with the last criterion, for example, *N,O*-trimethylsilyl derivatives of amino acids seem not useful in analytical practice, owing to the nonspecific silylation of amino groups or postreaction hydrolysis of the resultant N—Si bonds. Even the simplest compounds of this class $H_2N—CHR—CO_2H$ form three possible products: $H_2N—CHR—CO_2TMS$, TMSNH—CHR—

CO_2TMS, and $(TMS)_2N—CHR—CO_2TMS$ [TMS = $Si(CH_3)_3$] with different GC retention parameters. In the case of diamino monocarboxylic acids with nonequivalent amino groups [e.g., lysine $H_2N(CH_2)_4CH(NH_2)CO_2H$], the number of similar semisilylated derivatives is theoretically increased to nine.

The greatest principal difference between organic reactions in general and those which can be considered as chromatographic derivatization reactions is *de facto* commonly accepted absence of necessity of product structure determinations in the last case. In "classical" organic chemistry, every synthesized compound must be isolated from its reaction mixture and characterized by physicochemical constants or spectral parameters for confirmation or estimation (for new objects) of its structure. However, for the processes that have been classified as derivatization reactions, these operations are not necessary and are really not used in analytical practice. *The reaction itself is considered as the confirmation of structures of derivatives.* Of course, any exceptions from this important rule are very dangerous and must be pronounced as the special warning for applications of every method of derivatization.

The examples of chemical uncertainty (some products of different structures are formed) are the reactions of barbiturates with diazomethane CH_2N_2, which leads to different *N*- and/or *O*-methyl derivatives [1], or the interaction of dimethyl disulfide with conjugate dienes [2]. One of the frequently used derivatization methods for carbonyl compounds RR′CO (including an important group of steroids) implies their one-step treatment by *O*-alkyl hydroxylamines $R''ONH_2$, with formation of alkyl ethers of oximes RR′C=NOR″. However, this reaction has an anomaly for compounds with double bonds C=C conjugated with carbonyl groups, namely the parallel addition of reagent with active hydrogen atoms to the polarized C=C bonds [3]:

$$\begin{array}{ccc} R_2C{=}CR'{-}COR'' + CH_3ONH_2 & \rightarrow & R_2C{=}CR'{-}CR''{=}NOCH_3 \\ \downarrow & & \downarrow \\ CH_3ONHCR_2CHR'{-}COR'' & \rightarrow & CH_3ONHCR_2CHR'{-}CR''{=}NOCH_3 \end{array}$$

This means that, instead of one expected product with molecular weight (MW) = $M_0 + 29$ (with *O*-methyl hydroxylamine as reagent), reaction mixtures may contain two additional compounds with MW = $M_0 + 47$ and $M_0 + 76$. This feature is negligible for analysis of individual compounds, but when samples are mixtures of components of interest, the analysis becomes impossible because of the complexity of the resultant interpretation.

Only if the organic reaction is in accordance with all above-mentioned criteria can it be considered as a method for derivatization. Finding new appropriate processes of this type is complex and often not obvious. Some reagents have been recommended only for synthesis of derivatives for chromatography.

One of the main purposes of derivatization is the transformation of nonvolatile compounds into volatile derivatives. However, it is not the sole purpose of this treatment of analytes. Each chromatographic method [gas chromatography (GC), GC–mass spectroscopy (MS), high-performance liquid chromatography (HPLC), capillary electrophoresis (CE), etc.] being supplemented by derivatization permits us to solve some specific problems. Most principal of them are summarized briefly in Table 1, followed by more detailed comments.

Most monofunctional organic compounds (including alcohols ROH, carboxylic acids RCO_2H, amides $RCONH_2$, etc.) are volatile enough for direct GC analysis. Exceptions are only those compounds with high melting points (sometimes with decomposition) because of strong intermolecular interactions in their condensed phases [e.g., semicarbazones $RR'C{=}N{-}NHCSNH_2$, guanidines $RNH{-}C({=}NH){-}NH_2$, etc.]. Ionic compounds (e.g., quaternary ammonium salts $[R_4N]^+X^-$ are nonvolatile as well. If the compounds contain two or more functional groups with active hydrogen atoms [including the case of inner-molecular ionic structures, such as that seen in amino acids $RCH(NH_3^+)CO_2^-$], their volatility decreases significantly. The purpose of derivatization of all these objects is to substitute active hydrogen atoms by covalently bonded fragments that provide more volatile products. Direct GC analysis of highly reactive compounds (free halogens, hydrogen halides, sulfonic acids RSO_2OH, etc.) is accompanied by their interaction with stationary phases of chromatographic columns; they also require derivatization. If the initial compound A can be analyzed together with derivative B, the comparison of their GC retention indices is an important source of information about the nature of these compounds. The average value of the retention indices $\Delta RI_r = RI(B) - RI(A)$ may be used for identification of both analytes. Some selected ΔRI_r values are presented in Table 2.

Gas chromatography–mass spectrometry analysis completely excludes the second item from the possible aims of derivatization, insofar as the mass spectrometer itself is both a universal and a selective GC detector. At the same time, two new important reasons for derivatization appear in this method. The intensities of molecular ion ($M^{+\cdot}$) signals are small if the compounds have no structural fragments to provide the effective delocalization of a charge and unpaired electron. These fragments are conjugated bonds–atoms systems and isolated het-

Table 1 The Principal Applications of Derivatization in Different Chromatographic Techniques

Aims of derivatization	Typical examples
Gas Chromatography	
1. Transformation of nonvolatile, thermally unstable and/or highly reactive compounds into stable volatile derivatives	$ROH \rightarrow ROSiMe_3$ $ArOH \rightarrow ArOCOCF_3$ $RR'CO \rightarrow RR'C{=}NOCH_3$
2. Synthesis of derivatives for element-specific GC detectors or conversion of nondetectable compounds into suitable products for the minimization of detection limit	$ROH \rightarrow ROCOCCl_3$ (ECD) $RCO_2H \rightarrow RCO_2CH_2CCl_3$ (ECD) $HCO_2H \rightarrow HCO_2CH_2C_6H_5$ (FID)
3. Combination with the stage of sampling (preferably in the environment analyses when derivatization are used as the method of chemosorption)	$RCHO \rightarrow 2,4\text{-}(NO_2)_2C_6H_3{-}NH{-}N{=}CHR$
4. Separation of enantiomeric compounds on non-chiral phases after their conversion into diastereomeric derivatives	$RR'C^*HNH_2 + C_6H_5C^*H(OMe)COCl \rightarrow$ $RR'C^*HNH{-}COC^*H(OMe)C_6H_5$ (C^* = asymmetrical carbon atoms)
Gas chromatography–mass spectrometry	
5. Determination of molecular weights of compounds with $W_M \approx 0$ at the electron impact ionization (synthesis of derivatives with conjugated bonds–atoms systems)	$RR'CO \rightarrow RR'C{=}NNH{-}C_6F_5$ (π–p–π conjugation system) $RNH_2 \rightarrow RN{=}CH{-}NMe_2$ (p–π conjugation system)
6. The increasing of specificity of molecular ion fragmentation for estimation of structure of analytes (an example: the determination of double bonds C=C position in carbon skeleton of the molecules)	"On-site" derivatization: $RCH{=}CHR' \rightarrow R{-}CH(SMe){-}CH(SMe){-}R'$ "Remote-site" derivatization: $RCH{=}CH(CH_2)_nCO_2H + H_2NCH_2CH_2OH \rightarrow$ 2-Substituted oxazoline
High-performance liquid chromatography with UV detection	
7. Synthesis of chromogenic derivatives (with chromophores which provide the adsorption within typical range of UV detection 190–700 nm)	$C_6H_7O(OH)_5 \rightarrow C_6H_7O(OCOC_6H_5)_5$
8. The conversion of hydrophilic analytes into more hydrophobic derivatives	$RCH(NH_2)CO_2H \rightarrow$ $RCH(NHCSNHC_6H_5)CO_2H$
9. The determination of number of functional groups with active hydrogen atoms	$X(OH)_n + [(R_1CO)_2O + (R_2CO)_2O)] \rightarrow$ miscellaneous acyl derivatives

eroatoms with high polarizabilities (S, Se, I). In accordance with this regularity, *O*-TMS derivatives of alcohols indicate no $M^{+\cdot}$ peaks in mass spectra, whereas for the TMS ethers of carbonyl compounds' enols $RCH_2COR' \rightarrow RCH{=}CR'{-}OSiMe_3$ (π–p–d conjugation system), they are very intense.

The determination of double-bond C=C positions in the carbon skeletons of molecules very often is impossible, owing to uncertain charge distribution in molecular ions. The solution of this problem implies the conversion of unsaturated compounds into products whose molecular ions have sufficiently fixed charge localization. There are two concepts to provide this localization: (a) by the addition of heteroatomic reagents to the C=C bond (so-called "on-site" derivatization with the formation of TMS ethers of corresponding diols, adducts with dimethyl disulfide, etc.) and (b) after introduction or formation of nitrogen-containing heterocycles rather far from an unchanged C=C bond ("remote-site" derivatization).

The formation of new chromophores for the optimization of ultraviolet (UV) detection of analytes in HPLC implies the synthesis of derivatives with conjugated systems in the molecule. Compared with GC, there are no restrictions on the volatilities of these derivatives for HPLC analysis. They may be synthesized before analysis (precolumn derivatization) or after chromatographic separation (postcolumn derivatization). The latter technique is rarely used and then only for a few classes of compounds, but it permits us to combine the measurement of retention parameters of initial analytes with the detection of their derivatives.

The range of the most convenient hydrophobicities of organic compounds for reversed-phase HPLC separation may be estimated approximately as $-1 \leq \log P \leq +5$ (log

Table 2 Average Values of Retention Indices Differences ΔRI_r for Different Derivatization Reactions

$\Delta M = M_{deriv} - M_{initial}$	Scheme of reaction (for monofunctional compounds only)	$\Delta RI_r \pm s_{\Delta RI}$ (standard nonpolar polydimethyl siloxanes)
14	$RCO_2H \rightarrow RCO_2Me$	-102 ± 28
14	$ArOH \rightarrow ArOMe$	-62 ± 16
28	$RCO_2H \rightarrow RCO_2Et$	-42 ± 7
28	$RR'CO \rightarrow RR'C{=}NNHMe$	229 ± 23
29	$ROH \rightarrow RONO$	-66 ± 27
42	$ROH \rightarrow ROCOMe$	142 ± 18
42	$ArOH \rightarrow ArOCOMe$	97 ± 20
42	$RNH_2 \rightarrow RNHCOMe$	437 ± 29
42	$ArNH_2 \rightarrow ArNHCOMe$	401 ± 7
42	$RR'CO \rightarrow RR'C{=}NNMe_2$	302 ± 20
72	$ROH \rightarrow ROSiMe_3$	119 ± 18
72	$RCO_2H \rightarrow RCO_2SiMe_3$	76 ± 16
90	$RR'CO \rightarrow RR'C{=}NNHC_6H_5$	858 ± 22
96	$ROH \rightarrow ROCOCF_3$	-85 ± 24
114	$RCO_2H \rightarrow RCO_2SiMe_2$-*tert*-Bu	288 ± 29
144	$ROH \rightarrow ROCOCCl_3$	494 ± 18

P is the logarithm of the partition coefficient of the compound being characterized in the standard solvent system *n*-octanol–water). Highly hydrophilic substances with log $P \leq -1$ need the special choice of analysis condition (e.g., use of ion-pair reagents). Another approach is their conversion to more hydrophobic derivatives by modification of functional groups with active hydrogen atoms.

The examples mentioned here for RP–HPLC analysis of monosaccharides in the form of perbenzoates and amino acids as *N*-phenylthiocarbamoyl derivatives (Table 1) satisfy both principal criteria: introducing the chromophores into molecules of the analytes (C_6H_5CO- and $C_6H_5NH-CS-NH-$) and optimization of their retention parameters.

Sometimes, the generally prohibited multiplication of analytical signals of derivatives may be attained artificially for the solution of partial problems. For example, the treatment of polyhydroxy compounds (phenols, phenolcarboxylic acids, etc.) $X(OH)_n$ by an equimolar mixture of acylation reagents $(R_1CO)_2O + (R_2CO)_2O$ leads to the formation of $n + 1$ miscellaneous acyl derivatives $X(OCOR_1)_n, X(OCOR_1)_{n-1}(OCOR_2), \ldots, X(OCOR_2)_n$. The relative abundances of their chromatographic peaks are similar to the binomial coefficients (i.e., 1:1 at $n = 1$, 1:2:1 at $n = 2$, 1:3:3:1 at $n = 3$, and so forth). The differences between retention indices of these derivatives at $R_1{=}CH_3$ and $R_2{=}C_3H_7$ are small and their average value for phenols is 160 ± 23 [4]. These two regularities permit us to determine the number of hydroxyl groups (n) in analytes.

References

1. D. J. Harvey, J. Nowlin, P. Hickert, C. Butler, O. Gansow, and M. G. Horning, *Biomed. Mass Spectrom. 1*: 340 (1974).
2. M. Vincentini, G. Guglielmetti, G. Gassani, and C. Tonini, *Anal. Chem. 59*: 694 (1987).
3. I. G. Zenkevich, Ju. P. Artsybasheva, and B. V. Ioffe, Zh. *Org. Khim.* (*Russ.*). 25: *487 (1989).*
4. I. G. Zenkevich, *Fresenius' J. Anal. Chem. 365*: 305–309 (1999).

Suggested Further Reading

Blau, K. and G. S. King (eds.), *Handbook of Derivatives for Chromatography*, John Wiley & Sons, New York, 1978.

Blau, K. and J. M. Halket (eds.), *Handbook of Derivatives for Chromatography*, 2nd ed., John Wiley & Sons, New York, 1993.

Drozd, J., *Chemical Derivatization in Gas Chromatography*, Chromatography Library Vol. 19, Elsevier, Amsterdam, 1981.

Heberle, J. and G. Simchen, *Silylating Agents*, 2nd ed., Fluka Chemie AG, 1995.

Knapp, D. R., *Handbook of Analytical Derivatization Reactions*, John Wiley & Sons, New York, 1979.

Igor G. Zenkevich

Derivatization of Carbohydrates for GC Analysis

As a result of the development of special bonded phases, carbohydrates or their derivatives are usually separated by liquid chromatography. However, certain carbohydrate samples are still analyzed by GC due to the inherent high efficiencies obtainable from the technique and to the associated short elution times. In addition, gas chromatography–mass spectrometry (GC–MS) is a particularly powerful analytical technique for carbohydrates, especially for their identification. As a consequence, appropriate derivatives must be formed to render them sufficiently volatile but still easily recognizable from their mass spectra.

It is relatively easy to form the trimethylsilyl derivatives, employing standard silyl reagents such as trimethylchlorsilane or hexamethyltrisilazane, and the reactions normally can be made to proceed to completion. However, there is a major problem associated with the derivatization of natural sugars which arises from the formation of anomers and the pyranose–furanose interconversion. Reducing sugars in solution (e.g., glucose) exist as an equilibrium mixture of anomers. Consequently, each sugar produces five tautomeric forms — two pyranose, two furanose, and one open-chain form. In general, all the anomers can be separated by GC. This autoconversion can be minimized by mild and rapid derivatization. Equilibrium mixtures are to be expected from reducing sugars isolated from natural products.

Mixtures of hexamethyltrisilazane and trimethylchlorsilane are frequently used to derivatize sugars. Pure sugars (e.g., glucose, mannose, and xylose) can be readily derivatized using a mixture of hexamethyltrisilazane:trimethylchlorsilane:pyridine (2:1:10 v/v/v), giving single GC peaks; however, if the proportion of trimethylchlorsilane is doubled, small amounts of the anomeric forins are observed. One disadvantage of this procedure is the formation of an ammonium chloride precipitate, which, if injected directly onto the column, can cause column contamination and eventually column blockage. The formation of ammonium chloride can be avoided by extracting the derivative into hexane or by the use of an alternative derivatizing agent.

N,*O*-Bistrimethylsilyltrifluoroacetamide, usually combined with trimethylchlorosilane (10:1, v/v) is also a popular derivatizing agent for carbohydrates and the reaction mixture can be injected directly onto the column without fear of column contamination. The formation of anomers can also be avoided by preparing the alditols by treating with sodium borohydride. The alditols can then be separated after derivatizing with trimethylchlorosilane, a procedure that is considered preferable to the preparation of their acetates. The rapid preparation of trimethylsilylalidtols using trimethylsilylimidazole in pyridine mixtures at room temperature has the advantage over other methods, which require longer reaction times and higher temperatures. The use of silylaldolnitrile derivatives has been reported for the separation of aldoses. The sugars are reacted with hydroxylamine-*O*-sulfonic acid to form aldonitriles which are then silanated with *N*,*O*-bistrimethylsilyltrifluoroacetamide:pyridine (1:1 v/v). The silylaldonitrile derivatives are readily separated on open tubular columns.

Another silanization procedure for the derivatization of carbohydrates is the formation of trimethylsilyl oximes. The methyloxime is heated with hexarnethyldisilazane:trifluoroacetic anhydride (9:1 v/v) for 1 h at 100°C. Anthrone *O*-glucoside is an important ingredient in skin care cosmetics and can be fully silylated by reaction with *N*,*O*-bistrimethylsilylacetamide:acetonitrile mixture (1:1 v/v) for 1 h at 90°C, and subsequently separated by GC.

Acetylation and the use of trifluoroacetates, originally the more popular derivatives for the separation of carbohydrates by GC, are still used on occasion, but the various silanization methods are, today, the most common and considered the most effective for GC carbohydrate analysis.

Suggested Further Reading

Blau, K. and J. Halket (eds.), *Handbook of Derivatives for Chromatography*, John Wiley & Sons, New York, 1993.

D. W. Grant, *Capillary Gas Chromatography*, John Wiley & Sons, New York, 1996.

R. P. W. Scott, *Techniques of Chromatography*, Marcel Dekker, Inc., New York, 1995.

R. P. W. Scott, *Introduction to Analytical Gas Chromatography*, Marcel Dekker, Inc., New York, 1998.

Raymond P. W. Scott

Derivatization of Carbonyls for GC Analysis

The carbonyl group in aldehydes (RCHO) and ketones (RCOR′) is one of the frequently encountered functionalities in the composition of organic compounds. This group has no active hydrogen atoms, excluding the cases of high content of enols for β-dicarbonyl compounds (β-diketones, esters of β-ketocarboxylic acids, etc.):

Hence, the derivatization, at least of monofunctional carbonyl compounds, is not the obligatory stage of sample preparation before their gas chromatography (GC) analysis. Nevertheless, the objective reasons for their derivatization are the following

1. The simplest aldehydes and ketones are slightly polar, low-boiling substances with small retention indices (RI) both on standard nonpolar and polar stationary phases; for example:

Aldehyde	$RI_{nonpolar}$	RI_{polar}	Ketone	$RI_{nonpolar}$	RI_{polar}
Ethanal	369 ± 7[a]	701 ± 14	Acetone	472 ± 12	820 ± 11
Propanal	479 ± 9	794 ± 9	2-Butanone	578 ± 12	907 ± 14

[a]Here and later, RI values with standard deviations are randomized interlaboratory data.

In the numerous GC analytical procedures, the first part of chromatograms very often can be overloaded by intensive peaks of auxiliary compounds (solvents, by-products, etc.). The optimization of the determination of target carbonyl compounds usually needs the "replacement" of their analytical signals into less "populated" areas of chromatograms (the derivatives with retention parameters which exceed that for initial analytes).

2. The simplest monofunctional compounds with active hydrogens in the functional groups —OH, $—CO_2H$, NH_2, and so forth typically are volatile enough for their GC determinations. However, the presence in the molecules of any extra polar fragments (including C=O) makes the possibilities of GC analysis of these compounds worse; they usually need derivatization of one or (better) of both polar functional groups. Thus, in many cases, the necessity of derivatization of carbonyl fragment in polyfunctional compounds is a secondary procedure, which is caused by the presence of other functional groups.
3. Both aliphatic and aromatic aldehydes are easily oxidized compounds (even by atmospheric oxygen). Hence, one of the aims of their derivatization is to prevent the oxidation of analytes with resultant formation of carboxylic acids.

Some methods of carbonyl compound derivatization have been known in "classical" organic chemistry since the last century. Their predestination was simply the identification by comparison of melting points of purified solid derivatives with reference data. These derivatives are, for example, semicarbazones $RR'C{=}NNHCONH_2$, thiosemicarbazones $RR'C{=}NNHCSNH_2$, 2,4-dinitrophenylhydrazones $RR'C{=}NNHC_6H_3(NO_2)_2$, and so forth. However, most of these derivatives with structural fragments $=N-NH-C(=X)-NH_2$ are not volatile enough for their direct GC analysis, because of the presence of three active hydrogen atoms in these fragments. The convenient derivatives for GC determinations must include not more than one of these atoms (monosubstituted arylhydrazones) or they need to be free of them (*O*-alkyl and *O*-benzyl ethers of oximes). The oximes themselves can be synthesized by the reaction of carbonyl compounds with hydroxylamine, but they are usually used in GC practice in the form of *O*-TMS ethers.

Other types of carbonyl derivatives are acetals and/or ketals, preferably cyclic 1,3-dioxolanes, 1,3-oxathiolanes, or thiazolidines. Some carbonyl compounds (especially steroids) can be analyzed as their enol-TMS derivatives:

$$RR'CO + H_2NNHAr \rightarrow RR'C{=}NNHAr$$

$$RR'CO + H_2N-OR'' \rightarrow RR'C{=}N-OR''$$

$$RR'CO + H_2NOH \rightarrow RR'C{=}NOH \rightarrow \rightarrow RR'C{=}N-OSi(CH_3)_3$$

$$RR'CO + HXCH_2CH_2YH/H^+ \ (X, Y = O, NH, S) \rightarrow \text{[cyclic R,R'-substituted X,Y-heterocycle]}$$

$$RR'CH-COR'' + XSi(CH_3)_3 \rightarrow RR'C{=}CR''-OSi(CH_3)_3 + XH$$

All of these processes (excluding the last one) can be classified as condensation reactions. Hence, in all cases, the target derivatives are theoretically the sole components and no other volatile by-products, excluding water, have been formed. However, some common features of these reactions must be considered:

1. Both O-alkyl hydroxylamines and, especially, arylhydrazines, are slightly oxidized compounds. The presence of any oxidizers in the reaction mixtures must be excluded. Nevertheless, in real practice, these mixtures very often contain some byproducts (e.g., $ArNH_2$, ArOH, ArH, etc.). Usually, there are no problems to reveal their chromatographic peaks, because all of them have lower retention parameters than those for the initial reagents and, moreover, all target derivatives.
2. The condensation reaction of the considered type can be characterized by statistically processed differences of retention indices of products and initial substrates. This mode of additive scheme permits us to estimate these analytical parameters for any new derivatives on standard nonpolar polydimethyl siloxanes. For the simplest reaction scheme, $A + \cdots \rightarrow B + \cdots$, $\Delta MW = MW(B) - MW(A)$ and $\Delta RI_r = RI(B) - RI(A)$:

Scheme of reaction	ΔMW	⟨ΔRI_r⟩ (IU)
$RR'CO \rightarrow RR'{=}NNHCH_3$	28	229 ± 23
$ArRCO \rightarrow ArR{=}NNHCH_3$	28	320 ± 30
$RR'CO \rightarrow RR'C{=}NN(CH_3)_2$	42	302 ± 20
$RR'CO \rightarrow RR'C{=}NNHC_2H_5$	42	345 ± 16
$RR'CO \rightarrow RR'C{=}NNHC_6H_5$	90	858 ± 22
$ArCHO \rightarrow ArCH{=}NNHC_6H_5$	90	996 ± 50
$RR'CO \rightarrow RR'C(OCH_3)_2$	46	189 ± 17
$ROH \rightarrow ROSi(CH_3)_3$ (for the comparison)	72	119 ± 18

This set of ΔRI_r values illustrate that the simplest alkylhydrazones (methyl, ethyl, dimethyl, etc.) have appropriate GC retention parameters and, theoretically, can be recommended as the derivatives for carbonyl compounds. However, in real analytical practice, these hydrazones are not used because of their low yields, especially for the aliphatic ketones.

3. All considered condensation reactions have some anomalies for the α,β-unsaturated carbonyl compounds. Most reagents with active hydrogen atoms can react not only with carbonyl groups but also with polarized conjugated double bonds C=C. As a result of this regularity, three products, instead of the estimated one target derivative, are formed in the reactions of α,β-unsaturated carbonyl compounds with O-alkyl hydroxylamines:

$$>C{=}CR{-}COR' + H_2NOR'' \rightarrow\ >C{=}CR{-}CR'{=}NOR''$$

$$\downarrow \qquad\qquad\qquad\qquad \downarrow$$

$$R''ONHC(C<){-}CHR{-}COR' \rightarrow R''ONHC(C<){-}CHR{-}CR'{=}NOR'' \qquad \textbf{(I)}$$

The relative ratio of these products depends on the excess of derivatization reagent and the pH of the reaction mixtures. The chemical origin of products of the reaction of phenyl hydrazine with unsaturated carbonyl compounds depends on the same factors and, moreover, on the order of the components' mixing (it exerts the influence on the current pH of reaction media). The 2-pyrazolines have resulted, in some cases, as the sole reaction products instead of hydrazones:

$$>C{=}C<_{CHO} + PhNHNH \rightarrow\ >C{=}C<_{CH=NNHPh} + \text{(2-pyrazoline, N-Ph)}$$

The formation of hydrazinohydrazones [similar to compounds **(I)** have been reported for the reaction of unsaturated carbonyl compounds with dialkyl hydrazines. Hence, the existence of different products means that the quantitative yield of each of them is not enough, as is necessary for GC analysis of derivatives. It is quite probable that the same anomalies can take place in the reactions of unsaturated carbonyls with other reagents. For example, unsaturated 2,4-DNPHs seem unstable at the high temperatures of injectors and GC columns. This fact explains the small number of published RI data for the derivatives of these compounds and the necessity to search the new types of derivatization reactions for them.

4. Both hydrazones and O-alkyl oximes of asymmetrical carbonyl compounds exist in two isomeric structures with slightly different GC retention parameters:

CH=N (E)- or *anti-* CH=N (Z)- or *syn-*

Hence, most of these derivatives give two analytical signals (antiisomers typically are more stable and their peaks prevail over the syn derivatives). If the assignment of these two peaks to the isomeric compounds is not obvious, they can be marked by symbols #1 and #2 in the accordance with the usual chromatographic practice. The same duplication of analytical signals takes place, for example, during reversed-phase high-performance liquid chromatography (RP–HPLC) analysis of 2,4-DNPHs.

For the polyfunctional organic compounds, the considered condensation reactions of carbonyls can be combined with different derivatizations of other functional groups. So, the "standard" method of hydroxyketosteroids derivatization for GC analysis is their two-stage treatment with *O*-methyl hydroxylamine $\left(\rangle C{=}O \rightarrow \rangle C{=}NOCH_3\right)$ followed by silylation of OH groups. The resulting MO–TMS derivatives of numerous important compounds of this class are characterized by standard mass spectra and GC retention indices on standard nonpolar phases for their identification. The use of most active silylating reagents permits us to exclude the stage of *O*-methyl oxime formation, so far as ketosteroids can form the enol-TMS ethers. For example, androstenedione **(II)** gives the bis-*O*-TMS derivative:

(II) a). + MSTFA / KOAc → $(CH_3)_3SiO$ … $OSi(CH_3)_3$

b). or + $Me_3SiCH_2CO_2Et$

Because of the presence of p–π–(d) conjugated systems C=C—O—(Si), these derivatives indicate the intensive signals of the molecular ions in their mass spectra.

The formation of acetals from carbonyl compounds requires acid catalysis and (sometimes) the presence of water-coupling reagents (for instance, anhydrous $CuSO_4$). The conversion of aliphatic aldehydes into dimethyl acetals slightly increases the retention parameters of analytes ($\Delta RI_r = 189 \pm 17$). The cyclic ethylene derivatives (1,3-dioxolanes, $\Delta RI_r = 212 \pm 7$, this value is valid only for acyclic carbonyl compounds) are more stable to the hydrolysis and used in GC practice preferably. Their important advantage for gas chromatography–mass spectrometry (GC–MS) analysis is the very specific fragmentation of molecular ions with the loss of substituents R and R′ in the second position of the cycle and the formation of daughter ions $[M-R]^+$ and $[M-R']^+$, that give useful information for the determination of the initial carbonyl compound structure. The use of 2-aminoethanethiol in this reaction instead of ethylene glycol leads to the thiazolidines, which indicate the intense signals of molecular ions.

The GC separation of chiral carbonyl compounds (at first natural terpenoids, for example, camphor, menthone, carvone, etc.) on the nonchiral phases can be carried out in the accordance with the same principles as those for enantiomers of other classes. Their conversion into diastereomeres by the reaction with chiral derivatization reagents is necessary. Examples of them are α-trifluoromethylbenzyl hydrazine **(III)** and optically active 2,3-butanediol **(IV)**:

+ $H_2NNH\text{-}C^*H(CF_3)C_6H_5$ **(III)** → $NNHC^*H(CF_3)C_6H_5$

+ HO OH **(IV)** →

In connection with the problem of carbonyl compounds derivatization, it is expedient to touch upon a question of the application of new proposed reagents which are not yet widely used in GC analytical practice, but seem like promising ones in accordance with different criteria. One of them is low-boiling trimethyltrifluoromethyl silane $(CH_3)_3SiCF_3$ which converts the carbonyl groups into α-trifluoromethyl *O*-TMS carbinol fragments:

$$RR'C{=}O + (CH_3)_3SiCF_3 \rightarrow RR'C(CF_3)OSi(CH_3)_3$$

Moreover, the carbonyl groups are represented in the numerous derivatives of carboxylic acids, namely amides, imides, and so forth. There are no derivatization methods for these compounds at present (excluding reduction), which are based on the reactions of the carbonyl group. Nevertheless, the amides react with $(CH_3)_3SiCF_3$ in a manner similar to other carbonyl compounds. The hydrolysis of intermediate *O*-TMS derivatives followed by dehydration leads to the low boiling α-trifluoromethyl enamines

$$RR'N—C(CF_3){=}CR''R'''$$

These structures are in the mutually unambiguous accordance with the structures of initial amides and, hence, can be considered as new types of derivatives.

$(C_2H_5)_2NCOCH_3$ + $(CH_3)_3SiCF_3$ → $(C_2H_5)_2N(CH_3)C(CF_3)OSi(CH_3)_3$ → $(C_2H_5)_2N\text{-}C(CF_3){=}CH_2$

T_b 185 °C ($RI_{nonpolar}$ 966 ± 13) ... T_b 113.3 °C (RI_{estd} 754 ± 6)

N-methylcaprolactam + $(CH_3)_3SiCF_3$ → (CF₃, OSi(CH₃)₃, N–CH₃) → (N–CH₃, CF₃)

no T_b and RI data ... T_b 152.1 °C (RI_{estd} 897 ± 9)

The retention indices of any members of this new class of derivatives are unknown at present (only the boiling points have been determined). Nevertheless, this infor-

Table 1 Physicochemical and Gas Chromatographic Properties of Some Reagents for the Derivatization of Carbonyl Compounds

Reagent (abbreviation)	MW	T_b (°C)(m.p., °C)	$RI_{nonpolar}$
Phenyl hydrazine	108	244	1157 ± 11
Pentafluorophenyl hydrazine (PFPH)	198	(74–75)	1164 ± 16[a]
2,4,6-Trichlorophenyl hydrazine (TCOH)	210	(141–143)	1654[b]
2,4-Dinitrophenyl hydrazine	198	(200.5–201.5)	—
N-Aminopiperidine	100	146	859 ± 14
O-Methyl hydroxylamine (HCl)	47	47 (148)	
O-Ethyl hydroxylamine (HCl)	61	68 (130–133)	
O-Pentafluorobenzyl hydroxylamine hydroxylamine (HCI) (PFBHA)	213	227 (subl)	1068 ± 12[a]
Hydroxylamine (HCl) (with following silylation of oximes)	33	(151)	—
Methanol/H^+	32	64.6	381 ± 15
Ethylene glycol/H^+, $CuSO_4$	62	197.8	726 ± 28
Trimethyltrifluoromethyl silane	142	45	532 ± 3[a]
Silylating reagents (for synthesis of enols' TMS ethers)	See table in the entry Derivatization of Hydroxy Compounds for GC Analysis		

[a]Estimated RI values.
[b]Single experimental data.

mation is enough for the estimation of retention index (RI) values on the standard nonpolar phases with the following nonlinear equation:

$$\log RI = a \log T_b + b\, MR_D + c$$

where T_b is the boiling point and MR_D is the molar refraction (can be estimated by additive schemes); the coefficients of this equation are calculated by a least squares method with the data for compounds with known RI, T_b, and MR_D. Some results of calculations are presented in Table 1: They indicate that the RIs of α-trifluoromethyl enamines are less than the RIs of initial amides by approximately 200 IU; this is a very important fact for the choice of appropriate derivatization methods for the more complex nonvolatile compounds of this class.

Suggested Further Reading

Biondi, P. A., F. Manca, A. Negri, C. Secchi, and M. Montana, *J. Chromatogr. 411*: 275 (1987).

Blau, K. and J. M. Halket (eds.), *Handbook of Derivatives for Chromatography*, 2nd ed., John Wiley & Sons, New York, 1993.

Drozd, J., *Chemical Derivatization in Gas Chromatography*, Elsevier, Amsterdam, 1981.

Gleispach, H., *J. Chromatogr. 91*: 407 (1974).

Knapp, D. R., *Handbook of Analytical Derivatization Reactions*, John Wiley & Sons, New York, 1979.

Lehmpuhl, D. W. and J. W. Birks, *J. Chromatogr. A 740*: 71 (1996).

Levine, S. P., T. M. Harvey, T. J. Waeghe, and R. H. Shapiro, *Anal. Chem. 53*: 805 (1981).

Nawrocki, J., I. Kalkowska, and A. Dabrowska, *J. Chromatogr. A 749*: 157 (1996).

Zenkevich, I. G., *J. High Resolut. Chromatogr. 21*: 565 (1998).

Zenkevich, I. G., Ju. P. Artsybasheva, and B. V. Ioffe. *Zh. Org. Khim.* (*Russ.*). *25*: 487 (1989).

Igor G. Zenkevich

Derivatization of Hydroxy Compounds for GC Analysis

The hydroxyl group is one of the most propagated functional groups in organic compounds. Important biogenic substances (carbohydrates, flavones, phenolic acids, etc.) belong to the class of hydroxy compounds. One of the principal directions of the metabolism of different ecotoxicants and drugs in vivo is their hydroxylation followed by formation of conjugates with carbohydrates or amino acids. For example, the oxidation of widespread environmental pollutants — polychlorinated biphenyls (PCBs) — by cytochrome P450 leads to hydroxy-PCBs. The determination of OH compounds was one of the important problems of GC analysis during the almost half a century that this method has been in existence.

The simple rule for the prediction of the possibility of GC analysis of organic compounds is based on the reference data of their boiling points. If any compound can be distilled without decomposition at the pressures from atmospheric to 0.01–0.1 torr, it can be subjected to GC analysis, at least on standard nonpolar polydimethylsiloxane stationary phases. In accordance with this rule, most of the monofunctional —OH compounds (alcohols, phenols) and their S analogs (thiols, thiophenols, etc.) may be analyzed directly. The confirmation of chromatographic properties of any analytes must be not only verbal (at the binary level "yes/no") but also based on their GC Kovats retention indices as the most objective criteria; for example:

Compound	T_b (°C)	RI (nonpolar)
1-Tetradecanol	290.8	1664 ± 12[a]
1-Decanethiol	239.2	1320 ± 7
2,6-Di-*tert*-butyl-4-methyl phenol	265	1491 ± 10
2-Methylbenzenethiol	194	1061 ± 11

[a]RIs with standard deviations are randomized interlaboratory data.

Chemical properties of hydroxy compounds depend on the presence of active hydrogen atoms in the molecule. The pK_a values for aliphatic alcohols are comparable to that of water (≈ 16), but phenols are weak acids ($pK_a \approx$ 9–10). An increase in the number of polar functional groups in the molecules leads to an increase of the strength of intermolecular interactions. This is manifested in the rising of melting and boiling points, which can increase the temperature limits of thermal stability of the compounds. For example, some aliphatic diols and triols have boiling points at atmospheric pressure and, hence, are volatile enough for GC analysis. Similar compounds with four or more hydroxyl groups have no boiling points at atmospheric pressure; this means the impossibility of their GC analysis. The same restrictions are valid within the series of polyfunctional phenols:

Compound	T_b (°C)	RI (nonpolar)
Glycerol (three OH groups)	290.5	1196 ± 28
Hydroquinone	287	1338 ± 14
meso-Erythritol (four OH groups)	329–331	1319
Pyrogallol	309	1548
Xylitol (five OH groups)	None	None
1,2,4-Benzenetriol	None	None

Even the simplest bifunctional compounds of these classes being analyzed on nonpolar phases indicate broad nonsymmetrical peaks on chromatograms. This leads to poor detection limits and reproducibility of retention indices (the position of peaks' maxima depends on the quantity of analytes) compared with nonpolar compounds. The general way to avoid these problems is based on the conversion of hydroxy compounds to thermally stable volatile derivatives. This task is a most important purpose of derivatization (see the entry Derivatization of Analytes in Chromatography, General Aspects). This chemical treatment may be used not only for nonvolatile compounds but also for volatile substances. The less polar products typically yield narrower chromatographic peaks that provide the better signal-to-noise ratio and, hence, lower detection limits. Nonpolar derivatives have much better interlaboratory reproducibility of retention indices compared with this parameter for initially polar compounds.

Principal methods of hydroxy compound derivatization may be classified in accordance with types of chemical reactions:

Silylation:
$R(OH)_n + nXSi(CH_3)_3 \rightarrow R(OTMS)_n + nXH$
Acylation:
$R(OH)_n + nR'COX + nB \rightarrow R(OCOR')_n + nBH^+X^-$
Alkylation: $R(OH)_n + nR'X \rightarrow R(OR')_n + nXH$

The large group of silylation reactions (trimethylsilyl derivatives are most widely used) implies the replacement of active hydrogen atoms in molecules of analytes by a silyl group donated from different O-, N-, or C-silylating reagents. The relative order of —OH compounds reactivity is *n*-OH > *sec*-OH > *tert*-OH > Ar—OH > R—SH. The first part of Table 1 includes physicochemical and gas

Table 1 Physicochemical and Gas Chromatographic Properties of Some Silylating Reagents

Reagent (abbreviation)	MW	T_b (°C)(P)	d_4^{20}	n_D^{20}	$RI_{nonpolar}$	By-products ($RI_{nonpolar}$)[a]
Most widely used reagents (in order of increasing silyl donor strength)						
Hexamethyldisilazane (HMDS)	161	126	0.774	1.408	817 ± 29	NH_3 (ND)[b]
Trimethylchlorosilane (TMCS)	108	57.7	0.858	1.338	560 ± 8	HCl (ND)
N-Methyl-*N*-trimethylsilyl acetamide (MSA)	145	159–161	0.904	1.439	947 ± 14	$CH_3CONHCH_3$ (816 ± 22)
N-Trimethylsilyl diethylamine (TMSDEA)	145	125–126	0.767	1.411	817 ± 11[c]	$(C_2H_5)_2NH$ (548 ± 8)
N-Trimethylsilyl dimethylamine (TMSDMA)	117	84	0.732	1.397	660 ± 4[c]	$(CH_3)_2NH$ (425 ± 16)
N-Methyl-*N*-trimethylsilyl trifluoroacetamide (MSTFA)	199	130–132	1.079	1.380	826 ± 3[c]	$CF_3CONHCH_3$ (540)
N,*O*-bis-Trimethylsilyl acetamide (BSA)	203	71–73 (35)	0.832	1.418	1008[c]	CH_3CONH_2 (711 ± 19)
N,*O*-bis-Trimethylsilyl trifluoroacetamide (BSTFA)	257	145–147	0.974	1.384	887[c]	CF_3CONH_2 (675 ± 11)
N-Trimethylsilyl imidazole (TMSI)	140	222–223	0.957	1.476	1176 ± 18[c]	Imidazole (1072 ± 17)
New proposed and special reagents (in order of molecular weights increasing)						
2-(Trimethylsilyloxy)propene (IPOTMS)	130	—	0.780	1.395	675 ± 12	Acetone (472 ± 12)
Chloromethyldimethyl chlorosilane (CMDCS)	142	114	1.086	1.437	755 ± 8[c]	HCl (ND)
N-Trimethylsilyl pyrrolidine (TMSP)	143	139 – 140	0.821	1.433	862 ± 5[c]	Pyrrolidine (686 ± 10)
Dimethyl-*tert*-butyl chlorosilane (DMTBCS, TBDMS-Cl)	150	125	—	—	729 ± 11[c]	HCl (ND)
Ethyl (trimethylsilyl)acetate (ETSA)	160	156–159	0.876	1.415	930 ± 5[c]	$CH_3CO_2C_2H_5$ (602 ± 9)
bis-Trimethylsilyl methylamine (BSMA)	175	144–147	0.799	1.421	903 ± 18[c]	CH_3NH_2 (348 ± 12)
Trimethylsilyl trifluoroacetate (TMSTFA)	186	88–90	1.076	1.336	674 ± 5[c]	CF_3CO_2H (744 ± 6)
Bromomethyldimethyl chlorosilane (BMDCS)	186	—	—	—	842 ± 14[c]	HCl (ND)
bis-Trimethylsilyl formamide (BSFA)	189	158	0.885	1.437	948 ± 14[c]	$HCONH_2$ (637 ± 6)
Bis-Trimethylsilyl urea (BSU)	204	—	—	—	1237 ± 11	$CO(NH_2)_2$ (ND)
N,*O*-bis-Trimethylsilyl carbamic acid	205	77–78 (mp)	—	—	—	H_2NCO_2H, NH_3, CO_2 (ND)
Trimethylsilyl trifluoromethanesulfonate	222	77 (80)	1.228	1.360	—	CF_3SO_2OH
Dimethyl-*tert*-butylsilyl trifluoroacetamide (MTBSTFA)	241	168–170	1.023	1.402	996 ± 13[c]	CF_3CONH_2 (675 ± 11)
Dimethylpentafluorophenyl chlorosilane (in the mixture with dimethylpentafluorosilyl amine, 1:1 v/v)	260	88–90 (10)	1.384	1.447	—	HCl (ND)
N-Methyl-*N*-trimethylsilyl heptafluorobutanamide (MSHFBA)	299	148	1.254	1.353	906 ± 11[c]	$C_3F_7CONH_2$ (750 ± 18)

[a]Common hydrolysis by-products for all reagents are trimethylsilanol $(CH_3)_3SiOH$ (RI = 584 ± 8) or dimethyl-*tert*-butyl silanol (RI = 753 ± 18, estimated value).
[b]ND: By-product is not detected after formation of nonvolatile salts with bases (HCl) or acids (NH_3).
[c]Estimated RI values may be defined more precisely afterward.

chromatographic constants of numerous reagents, listed in the order of increasing silyl-donor strength. The second part of this table presents compounds which have been recently introduced into analytical practice with still unestimated relative silylation activities and some non-TMS reagents. Besides the individual chemicals, their different combinations have been considered as so-called silylating mixtures (e.g., HMDS + TMCS or HMDS + DMFA). The triple mixture BSA + TMSI + TMCS (1:1:1 v/v/v) seems to be the most active currently known silylating agent. It may be used for derivatization of all types of compounds with active hydrogen atoms, including carbonyl's enols and aci forms of aliphatic nitro compounds.

Every reagent in Table 1 is characterized, not only by its own retention index but also by retention index (RI) values of principal by-products of the reaction. This permits us to predict the possible overlapping of their chromatographic peaks with signals of derivatives of target compounds.

The list of recommended silylating compounds is constantly changing. Some older ones have been excluded and replaced by more effective reagents. For example, in one textbook (Knapp, 1979), *N*-trimethylsilyl-*N*-phenyl acetamide $C_6H_5N(COCH_3)SiMe_3$ (Phanalog of MSA) was presented as the silylating reagent. However, no new examples of its application have been recently published. The most principal reason for its being excluded from analytical practice is the inconvenient RIs, both for this reagent itself (1493 ± 28, estimated value) and for the by-product of the reaction (*N*-phenylacetamide, 1362 ± 11). This window of GC retention indices may include peaks of target derivatives and, hence, it must be free from overlap with the initial reagents and by-products.

Standard mass spectra of TMS derivatives of aliphatic hydroxy compounds indicate no peaks of molecular ions $M^{+\cdot}$ but, in all cases, ions $[M-CH_3]^+$ are reliably registered. The same derivatives of phenols and carbonyl compound's enols with $p-\pi$ conjugated systems C=C—O in the molecules indicate the signals of $M^{+\cdot}$ of high intensities. Some typical base peaks in mass spectra of *O*-TMS derivatives are $[Si(CH_3)_3]^+$ (*m/z* 73) and $[Si(CH_3)_2OH]^+$ (*m/z* 75).

The principal disadvantage of TMS derivatives is the ease of postreaction hydrolysis in the presence of water. Most of these compounds cannot exist in aqueous media. Other types of derivatives, namely dimethyl-*tert*-butyl silyl ethers, have approximately 10^3 times lower hydrolysis constants, owing to steric hindrance of Si—O bonds by *tert*-butyl groups. Unfortunately, their mass spectra are not so informative for elucidation of the structures of unknown analytes, because the base peaks for all of them belong to noncharacteristic ions $[C_4H_9]^+$ with *m/z* 57 and $[M-C_4Hg]^+$.

Some halogenated silylating reagents are used in the synthesis of derivatives for GC analysis with selective (element-specific) detectors; for example, $ClCH_2SiMe_2Cl$, $BrCH_2SiMe_2Cl$, and $C_6F_5SiMe_2Cl$.

The second group of hydroxy compound derivatization reactions includes acylation of OH groups with the formation of esters. The most important are listed below (for the table of physicochemical and gas chromatographic constants of acylation reagents refer to the entry Derivatization of Amines, Amino Acids, Amides, and Imides for GC Analysis).

Reagent (abbreviation)	Derivative	ΔMW
Acetic anhydride/(H^+)	$R(Ar)OCOCH_3$	42
Trifluoroacetic anhy dride (TFAA), bis-trifluoroacetyl methylamine (MBTFA), or trifluoroacetylimidazole (TFAI)	$R(Ar)OCOCF_3$	42
Pentafluoropropionic anhydride (PFPA)	$R(Ar)OCOC_2F_5$	146
Heptafluorobutyric anhydride (HFBA)	$R(Ar)OCOC_3F_7$	196
Pentafluorobenzoyl chloride	$R(Ar)OCOC_6F_5$	194
Chloroacetic anhydride	$R(Ar)OCOCH_2Cl$	76
Dichloroacetic anhydride	$R(Ar)OCOCHCl_2$	110
Trichloroacetic anhydride or trichloroacetyl chloride	$R(Ar)OCOCCl_3$	144
N-Trifluoroacetyl-L-prolyl chloride (*N*-TFA-L-Pro-Cl)[a]	*N*-TFA L-prolyl esters	193
Diethyl chlorophosphate	$R(Ar)OP(O)(OC_2H_5)_2$	136
2-Chloro-1,3,2-dioxaphospholane	2-Alkoxy-1,3,2-dioxaphospholanes	90
$NaNO_2/(H^+)$ (special derivatization of simplest C_1–C_5 aliphatic alcohols into volatile alkyl nitrites for headspace analysis)	RONO	29

[a]Used for synthesis of diastereomeric derivatives of enantiomeric alcohols.

It is recommended that these reactions be conducted in the presence of bases without active hydrogen atoms (pyridine, triethylamine, etc.). Exceptions are indicated by the symbol "reagent/(H^+)." These basic media are necessary for the connection of acidic by-products into nonvolatile salts to protect the acid-sensitive analytes from decomposition and avoid the appearance of extra peaks of by-products on the chromatograms. Phenols can be converted into Na salts before acylation.

The third group of derivatization reactions of hydroxy compounds for GC analysis includes formation of their alkyl or substituted benzyl ethers (see also the entry Derivatization of Acids for GC Analysis):

Reagent (abbreviation)	Derivative	ΔMW
Diazomethane (in diethyl ether solutions in presence of HBF_4)	$R(Ar)OCH_3$	14
Methyl iodide/DMFA (dimethylformamide), K_2CO_3	$R(Ar)OCH_3$	14
Pentafluorobenzyl bromide (PFB–Br)	$R(Ar)OCH_2C_6F_5$	180
3,5-bis-(Trifluoromethylbenzyl)dimethylanilinium fluoride (BTBDMA-F) (only for phenols during GC injection)	$ArOCH_2C_6H_3$-3,5-$(CF_3)_2$	226

Methylation by diazomethane is a simple method for derivatization of relatively acidic compounds [e.g., phenols (pK_a = 9–10) or carboxylic acids (pK_a = 4.4 ± 0.2)].

The application of this reagent for methylation of aliphatic alcohols requires additional acid catalysis. Methyl iodide is the most convenient reagent for synthesis of permethylated derivatives of polyols (including carbohydrates) and phenols. Dimethyl sulfate [$(CH_3O)_2SO_2$] can be used in basic aqueous media for methylation of phenols, but the yields of methyl ethers, in this case, are not enough for quantitative determinations of initial compounds by GC.

Some special derivatization methods, which lead to the formation of cyclic products, have been recommended for glycols (triols, tetrols, etc., including carbohydrates) and amino alcohols. An appropriate orientation of two functional groups $(OH)_2$ or (OH) + (NH_2) in 1,2- (*vic*) or (sometimes) in 1,3- and more remote positions, is necessary for their realization:

Initial compound	Reagent (s)	Product	ΔMW
1,2-Diols	Acetone/(H^+)	2,2-Dimethyl-1,3-dioxolanes	40
	Methane- or butaneboronic acid $CH_3B(OH)_2$ or $C_4H_9B(OH)_2$	2-Methyl or 2-butyl 1,3,2-dioxaborolanes (Me or Bu boronates)	24 (Me) 66 (Bu)
2-Amino alcohols	Methane- or butane-boronic acid	2-Methyl or 2-butyl 1,3,2-oxazaborolanes	24 (Me) 66 (Bu)

The number of proposed methods for alcoholic and phenolic S-analog derivatization for GC analysis is significantly less than those for alcohols. The most objective reason for this is the lower frequency of their determinations in real analytical practice. In accordance with general recommendations, thiols and thiophenols may be converted into TFA (PFP, HFB) esters or PFB ethers (*S*-TMS derivatives seems not as stable as *O*-TMS ethers). In addition to the optimization of chromatographic parameters, the derivatization of these compounds is necessary to prevent their oxidation by atmospheric oxygen.

The experimental details of numerous derivatization reactions are presented in specialized textbooks (see below).

Suggested Further Reading

Amijee, M., J. Cheung, and R. J. Wells, *J. Chromatogr. A. 738*: 57–72 (1996).

Blau, K. and G. S. King (eds.), *Handbook of Derivatives for Chromatography*, John Wiley & Sons, New York, 1978.

Blau, K. and J. M. Halket (eds.), *Handbook of Derivatives for Chromatography*, 2nd ed., John Wiley & Sons, New York, 1993.

Drozd, J., *Chemical Derivatization in Gas Chromatography*, Elsevier, Amsterdam, 1981.

Heberle, J. and G. Simchen, *Silylating Agents*, 2nd ed., Fluka Chemie AG, 1995.

Knapp, D. R., *Handbook of Analytical Derivatization Reactions*, John Wiley & Sons, New York, 1979.

Little, J. L., *J. Chromatogr. A. 844*: 1–22 (1999).

Igor G. Zenkevich

Derivatization of Steroids for GC Analysis

Steroids, bile acids, and similar compounds pose certain problems when they require to be derivatized for separation by GC. The hydroxyl groups in the respective structures differ greatly in their reaction rate, which will depend on their nature (whether they are primary, secondary, or tertiary) and also, to a certain extent, on their steric environment. After considerable research, which examined a wide variety of different derivatives, trimethylsilylation has emerged as the procedure of popular choice for steroid and steroidlike compounds. Pure compounds or biological extracts containing 3β-hydroxyl groups can be readily silylated by treatment with *N,O*-bistrimethylsilyltrifluoroacetamide containing 1% trimethylchlorosilane at 60°C for 30 min, with or without added pyridine. *N,O*-Bistrimethylsilyltrifluoroacetamide, under some circumstances, can also be used alone or with a mixture of hexamethyldisilazane and trimethylchlorosilane (10:10:5 v/v/v) at 60°C for 30–60 min. The trimethylsilyl derivatives separate well on capillary columns carrying apolar stationary phases. The derivatives also provide excellent electron-impact mass spectra.

In addition, *N,O*-bistrimethylsilyltrifluoroacetamide has been used very effectively for the silylation of estradiol and catechol estrogens. Employing *N,O*-bistrimethylsilyltrifluoroacetamide:pyridine:trimethylchlorosilane (5:5:1 v/v/v) at 40°C for 8–10 h, tetrahydroaldosterone (11β,18-epoxy-3α,16,21-trihydroxy-5β-pregnene-20-one) and aldosterone (11β,21-dihydroxy3,20-dio-exopregn4-en-18-al) have been derivatized. Employing stable isotope dilution, cortisol has been determined in human plasma by gas chromatography–mass spectrometry (GC–MS) after reacting the dimethoxime cortisol derivative with 50 μL of *N*,*O*-bistrimethylsilyltrifluoroacetamide at 100°C for 2 h.

More sterically hindered steroids and, in particular, the polyhydroxylated compounds were more efficiently derivatized with trimethylsilylimidazole. The ease of silylation of the ecdysteroids (polyhydroxylated anthrapod moulting hormones) tracked the following order: 2, 3, 22, 25 $>$ 20 $\gg$ 14. Those substances containing a 14α-hydroxyl group require very strong reaction conditions. For example, 20-hydroxyecdysone could only be silylated using neat trimethylsilylimidazole at 100°C over a reaction period of 15 h. It has also been established that the addition of 1% trimethylchlorosilane to the trimethylsilylimidazole catalyzed the reaction of the reagent with the 14α-hydroxyl group, reducing the reaction time to 4 h at 100°C. The use of larger quantities of trimethylchlorosilane caused confusing side reactions to occur and should be avoided. However, the addition of potassium acetate also appeared to increase the reaction rate, allowing the reaction time to be reduced to 2–3 h.

The conversion of the enol form of keto steroids to a silyl derivative is somewhat fraught with difficulties, as mixtures of silylated substances can be easily formed. Nevertheless, the quantitative conversion of keto steroids to their trimethylsilyl-enol ethers has been optimized. Silylation of dexamethasone (9α-fluor-11β, 17α,21-trihydroxy-16α-methyl-pregna-1,4-diene-3,20-dione) using *N*,*O*-bistrimethylsilyltrifluoroacetamide in the presence of sodium acetate yielded the pure tetra-trimethylsilyl derivative. In this case, the trimethylsilyl-enol ether of the 20-one moiety was produced, leaving the 3-one group unreacted. An unusual reaction associated with the enolization of keto steroids is the aromatization of the (A) ring of norethynodrel (a 3-keto-5,10-ene-nor-19-methyl steroid) during trimethylsilylation. It has been suggested that, under routine silylation conditions, aromatic derivatives are very likely to form from 3-keto-4,5-epoxides of nor-19-methyl steroids. The yield of aromatic silylated products were greater when the more basic reagents were employed and at higher temperatures.

Formyl derivatives are also popular in situations where several groups have to be blocked, as in steroid analysis, because the formyl group adds little to the molecular weight. To prevent the formation of artifacts, the strength of the formic acid should be kept at 95% and reaction allowed to take place for 30 min at 40°C. Alternatively, sodium formate can be used, an example of which is in the preparation of the enol *tert*-butyldimethylsilyl derivatives of steroids and bile acids. Sodium formate solution (1 mg in 100 μL) is dried under a stream of nitrogen in a 1-mL reaction tube fitted with a Teflon-lined screw cap. The tube is then heated to 270°C for 30 min, cooled, and 10 μg of the steroid in 100 μL of methanol added. The solvent is evaporated under a stream of nitrogen and 20 μL of *t*-butyldimethylsilylimidazole added. The tube is then filled with nitrogen, sealed, and heated at 100°C for 4 h. Twenty microliters of 2-propanol are then added to remove excess reagent, the tube sealed, and then heated again to 100°C for 10 min. One-half milliliter of water is then added to the reaction mixture and then extracted three times with 0.5 mL of hexane. The hexane solution is concentrated under a stream of nitrogen and the concentrated solution is used for analysis.

The literature indicates that *t*-butyldimethylsilylimidazole is probably the most popular reagent for derivatizing steroids, and it is often used in conjunction with *t*-butyldimethylchlorosilane. These reagents form derivatives under relatively mild conditions (room temperature, reaction time, 1 h), reaction goes close to completion, and relatively few side products are generated.

Suggested Further Reading

Blau, K. and J. Halket (eds.), *Handbook of Derivatives for Chromatography*, John Wiley & Sons, New York, 1993.

Grant, D. W., *Capillary Gas Chromatography*, John Wiley & Sons, New York, 1996.

Scott, R. P. W., *Techniques of Chromatography*, Marcel Dekker, Inc., New York, 1995.

Scott, R. P. W., *Introduction to Analytical Gas Chromatography*, Marcel Dekker, Inc., New York, 1998.

Raymond P. W. Scott

Detection in Countercurrent Chromatography

Introduction

Detection of solutes is an essential link in the separation chain. It has to reveal the solutes separation by detecting them in the column effluent and, in some cases, it could permit their characterization. Toward these objectives, it is based on the various physical properties of substances.

Countercurrent chromatography (CCC) is a chromatographic method which separates solutes more or less retained in the column by a stationary phase (liquid in this case) and are eluted at the outlet of column by a mobile phase. Two treatments of column effluent have been used until now in CCC. Either the column outlet is directly connected to a detector commonly used in high-performance liquid chromatography (HPLC) (on-line detection) or fractions of the mobile phase are collected and analyzed by spectrophotometric, electrophoretic, or chromatographic methods (off-line detection).

The first one is more practical and easier to perform. It is commonly used in analytical applications of CCC and also in preparative CCC to analyze effluent continuously and to follow the steps of the separation.

The second one is often tedious, because each fraction must be analyzed individually. However, it is really interesting in preparative applications of CCC, especially to measure the purities of fractions and the biological activities of separated compounds and also to recover a product from one or some selected fractions to resolve its chemical structure.

On-Line Detection

This type of detection can be used as much in preparative CCC to monitor separations, before the fraction collector if any, as in analytical CCC (for instance, during the determination of log $P_{\text{octanol/water}}$).

Several detectors used in HPLC and in supercritical fluid chromatography (SFC) can be connected to a CCC column [1] to detect solutes and, thus, follow separation. They can be, for instance, fluorimetry (very sensitive and used without modifications in CCC), ultraviolet (UV)-visible spectroscopy [1], evaporative light-scattering detection [1], and atomic emission spectroscopy [2]. Some detectors give more information than the detection of the solute, such as structural information of separated components (e.g., infrared spectroscopy [3], mass spectrometry [4], or nuclear magnetic resonance [5]). These detectors are used either on-line with a fraction collector or in parallel if they are destructive detectors.

UV Detection

The UV-visible detector is the universal detector used in analytical and preparative CCC. It does not destroy solutes. It is used to detect organic molecules with a chromophoric moiety or mineral species after formation of a UV-absorbing complex (the rare earth elements with Arsenazo III [6], for instance). Several problems can occur in direct UV detection, as has already been described by Oka and Ito [7]: (a) carryover of the stationary phase due to improper choice of operating conditions with appearance of stationary phase droplets in the effluent of the column, or (b) overloading of the sample, vibrations, or fluctuations of the revolution speed, (c) turbidity of the mobile phase due to the difference in temperature between the column and the detection cell, or (d) gas bubbling after reduction of the effluent pressure. Some of these problems can be solved by optimization of the operating conditions, better control of the temperature of mobile phase, or addition of some length of capillary tubing or a narrow-bore tube at the outlet of the column, before the detector, to stabilize the effluent flow and to prevent bubble formation. The problem of the stationary-phase carryover (especially encountered with hydrodynamic mode CCC devices) can be solved by the addition, between the column outlet and UV detector, of a solvent which is miscible with both stationary and mobile phases and which allows to obtain a monophasic liquid in the cell of detector [1] (a common example is isopropanol).

Evaporative Light-Scattering Detection

Evaporative light-scattering detection (ELSD) involves atomization of the column effluent into a gas stream via a Venturi nebulizer, evaporation of solvents by passing it through a heated tube to yield an aerosol of nonvolatile solutes, and, finally, measurement of the intensity of light scattered by the aerosol. After a suitable evaporation step, in the worst case of segmented or emulsified mobile phase, the column effluent should always be an aerosol of the solutes before reaching the detection cell. It can be used without modifications. For molecules without chromophore or fluorophore groups, or with mobile phases with a high UV cutoff (acetone, ethyl acetate, etc.), ELSD is useful [1]. However, it cannot detect fragile or easy sublimable solutes because nebulizer is heated. Moreover, this detection method does not preserve the solutes. To collect column effluent, a split must be installed at the outlet of the column to allow ELSD in a parallel direction to fraction collection, with a consequent loss of solutes.

Atomic Emission Spectrometry

This detection mode can be used during ion separation. Kitazume et al. [2] used a direct plasma–atomic emission spectrometer (DCP, Spectra-Metrics Model SpectraSpan IIIB system with fixed-wavelength channels) for observation of the elution profile during the separation of nickel, cobalt, magnesium, and copper by CCC. For profile measurement of a single element, an analog recorder signal from the DCP was converted into a digital signal. The digital data was stored in a work station and the elution profile was plotted. For simultaneous multielement measurement, the emission signal for each channel was integrated for 10 s at intervals of 20 s, and the integrated data were printed out.

Infrared Spectrometry

Romanach and de Haseth [3] have used, in CCC, a flow cell for LC–FTIR (liquid chromatography–Fourier transform infrared) spectrometry. The main difficulty is the absorbance of the liquid mobile phase. This problem is exacerbated in LC by low solute-to-solvent ratios in the eluates. On the contrary, CCC leads to a high solute-to-solvent ratio so that it can be used with a very simple interface with a CCC column, without any complex solvent removal procedures. High sample loadings are possible by using the variable path length of the IR detector (from 0.025 to 1.0 mm).

Mass Spectrometry [5]

Several interfaces have been used for CCC–MS (mass spectrometry). The first employed is thermospray (TSP). When a column is directly coupled with TSP MS, the CCC column often breaks due to the high back-pressure generated by the thermospray vaporizer. By contrast, other interfaces, such as fast atom bombardment (FAB), electron ionization (EI), and chemical ionization (CI), have been directly connected to a CCC column without generating high back-pressure. Such interfaces can be applied to analytes with a broad range of polarities. As it is suitable to introduce effluent from the column CCC into MS at a flow rate of only between 1 and 5 μL/min, the effluent is usually introduced into the MS through a splitting tee, which is adjusted to an appropriate ratio.

Nuclear Magnetic Resonance

Nuclear magnetic resonance (NMR) gives maximum structural information and allows the measurement of the relative concentrations of eluted compounds. Spraul et al. [5] experimented with the coupling of pH zone-refining centrifugal partition chromatography (CPC) with NMR by using a biphasic system based on D_2O and an organic solvent. In pH zone refining, solutes are not eluted as separated peaks but as contiguous blocks of constant concentrations, so that it is highly difficult to monitor the separation by means of a UV detector. On-line pH monitoring is generally used, allowing the observation of transitions between solutes, because each zone has its own characteristic pH determined by the pK_a and the solute concentration it contains. The experiment was carried out in stop-flow mode.

Off-Line Detection

The analysis of the mobile-phase fractions collected at the outlet of column is the oldest method used in CCC (droplet countercurrent chromatography and rotation locular countercurrent chromatography) to elucidate the quality of separation and to characterize solutes. With modern CCC, such as CPC, CCC type J, and crossaxis, numerous applications have been described for preconcentration and preparative chromatography.

Table 1 gathers some applications described in the literature [4,8]. The type of detection used with each fraction depends on the isolated solute. They include thin-layer chromatography (TLC) and HPLC, which also enables an estimation of each fraction's purity, for organic solutes, the inductively coupled plasma–atomic emission spectroscopy (ICP–AES) for mineral species, and polyacrylamide gel electrophoresis (PAGE) for biological molecules. If the purity of the compound is satisfactory, a study by direct injection–mass spectrometry and nuclear magnetic resonance allows one to determine its chemical structure. Biochemical tests are also available to verify the biological activity of biomolecules which are often separated and collected in aqueous two-phase CCC solvent systems.

Table 1 Off-Line Detection

Molecules [Ref.]	Fractions analysis
Schisanhenol acetate 5 and 6 of *Schisandra rubriflora* [4]	TLC Purity control by HPLC
Bacitracine complex [4]	Absorbance measure at 234 nm Purity control by HPLC
Dye species [4]	Mass spectrometry
Thyroid hormone derivatives [4]	UV on line at 280 nm; gamma radioactivity measure of fractions Purity control by TLC, HPLC, and UV spectra
Cerium chloride and erbium chloride [4]	Inductively coupled plasma–atomic emission spectroscopy
Recombinant uridine phosphorylase [8]	SDS-PAGE Enzymatic activity by Magni method

References

1. S. Drogue, M.-C. Rolet, D. Thiébaut, and R. Rosset. *J. Chromatogr. 538*: 91–97 (1991).
2. E. Kitazume, N. Sato, and Y. Ito, *Anal. Chem. 65*: 2225–2228 (1993).
3. R. J. Romanach and J. A. de Haseth. *J. Liquid Chromatogr. 11*(1): 133–152 (1988).
4. H. Oka, High-speed counter current chromatography/mass spectrometry, In *High-Speed Counter Current Chromatography* (Y. Ito and W. D. Conway, eds.), John Wiley and Sons, New York, 1995, pp. 73–91.
5. M. Spraul, U. Braumann, J.-H. Renault, P. Thépinier, and J.-M. Nuzillard, *J. Chromatogr. 766*: 255–260 (1997).
6. E. Kitazume, M. Bhatnagar, and Y. Ito, *J. Chromatogr. 538*: 133–140 (1991).
7. H. Oka and Y. Ito, *J. Chromatogr. 475*: 229–235 (1989).
8. Y. W. Lee, Cross-axis counter current chromatography: A versatile technique for biotech purification, in *Counter Current Chromatography* (J.-M. Menet and D. Thiébaut, eds.), Marcel Dekker, Inc., New York, 1999, pp. 149–169.

M.-C. Rolet-Menet

Detection Methods in Field-Flow Fractionation

An analytical separation technique requires a detection method responding to some or all of the components eluting from the separation system. The choice of detector is determined by the demands of the sample and analysis. For Field-Flow Fractionation (FFF) techniques many of the detection systems have evolved from those used in liquid chromatography (LC) techniques.

Detection can be carried out either with an on-line detector coupled to the eluent flow or by the collection and subsequent analysis of discrete fractions. For collected fractions, a range of analytical methods can be used, both quantitative (e.g., radiotracer and metal analysis) and more qualitative (e.g., microscopic techniques). On-line detectors suitable for coupling to the FFF channels include both nondestructive flow through cell systems and destructive analysis systems. It is often desirable to use on-line detection if possible because the total analysis time is much less than for discrete fraction analysis. Regardless of detector type, the dead volumes and flows in the system between the FFF channel and detector or fraction collector must be accurately determined and corrected for.

If the signal from a detector consists of a factor of two properties, it is possible to use another detector on-line to resolve the different properties [e.g., multiangle light scattering (MALS) in combination with the differential refractive index (DRI) or continuous viscometry + DRI].

Alternatively, the two detectors may respond to two different properties of interest; in either case, it is almost as simple to acquire multiple detector signals as a single one. Multiple detectors can be arranged either in series or in parallel. A parallel detector arrangement avoids the band-broadening problem encountered in the serial arrangement, where the first detector may cause significant band-broadening for the second, due to the dead volume in the flow cell. For a serial detector connection, it is best to have the one with the smallest dead volume first, as long as it is not a destructive detector. For parallel coupling, the outflow from the FFF channel needs to be split into two or more detectors, and it is then essential to have control of the individual flows because changes can induce drift in sensitivity and shifts in the dead times between the channel and detector during a run.

When choosing detector and experimental conditions, one needs to consider analyte concentration, detector sensitivity, background level, and detection limits. The maximum amount of analytes that can be injected is usually limited by sample overloading in the FFF channel, which can cause interparticle interactions which disturb the separation. It is necessary to have an analyte detection limit well below the overloading sample concentration to be able to quantify the peaks without a too noisy background. When using multiple detectors on-line, their sensitivity may be quite different either overall or as a function of size range. An example of this is the use of a MALS detector, which has much higher sensitivity for larger particles, together with a DRI detector, which has the opposite sensitivity properties, making the small and large particle ranges difficult to cover.

Optical Detection Systems

Ultraviolet (UV) detectors are the most commonly used detectors for FFF applications because of their availability, simplicity and low cost. The majority of FFF work to date has focused on separation method development, in which the use of a UV detector showing the quality of the separations is sufficient. However, the quantification of the separated particles or macromolecules is not always straightforward because the UV signal is actually a turbidimetric measure, which is a combination of light scattering and absorbance. The absorbance contribution is only dependent on concentration, but there is a more complicated relationship involved in the scattering signal. Large particles scatter light much more effectively than smaller particles, and particles with varying composition and refractive indices give rise to further complications. The correction of the detector signal according to Mie scattering theory is complicated but can often be simplified with appropriate assumptions [1]. For particles larger than 1 μm, efforts have been put into development of an absolute or standard free-quantification method using UV Detection for Gravitational FFF [2].

The DRI detector is very common in LC and records any change in refractive index of the sample stream relative to a reference stream. DRI is a general detector with the advantage of responding to almost all solutes and it is concentration selective. The sensitivity of a DRI detector is not always the best, but new detector models offer different lengths of the optical path, so that the sensitivity can be adjusted to match sample concentration. The DRI detector is not sensitive to changes in flow rate but they are highly sensitive to temperature changes. It is probably the most frequently used detector for FFF applications after UV detectors.

Flow-through fluorescence detectors are very common in liquid chromatography, due to the high selectivity and good signal-to-noise ratio. Only a few articles on FFF with fluorescence detectors are published, but when the analytes have suitable fluorescence properties, this is an excellent choice.

Photon correlation spectroscopy (PCS), also called dynamic light scattering or quasi-elastic light scattering, correlates the frequency of the light-scattering signal to the diffusion coefficients of the sample particles. PCS is a valuable tool in the validation of FFF separations, but it is too slow to be of practical use as an on-line detector. PCS has, for example, been used for verification of the average sizes obtained from FFF theory for discrete fractions of emulsion separated using sedimentation FFF [3]. Improvements in light-scattering theory and instrumentation has been going on for decades, but the development of the MALS instrument from the earlier Low Angle Light Scattering (LALS) technology, presents a breakthrough in particle sizing. Compared to LALS instruments multiangle detection (up to 18 detectors measuring the scattered light at individual angles) allows more physical properties of the particles to be derived from the results. Also, the MALS instrument has a higher sensitivity and is less affected by dust particles in the sample. Light-scattering techniques give average values of the properties of the particle population in the sample and do not describe the property distribution of the sample, but when coupled to a particle sizing technique, such as FFF, distribution of the different properties are derived from each size fraction.

The MALS theory has been thoroughly described in several articles by Wyatt [4] and will only be mentioned briefly here. For each size fraction or batch measurement, the following applies if the assumptions that all diffraction of the light is due to scattering of the particles (i.e., similar refractive index of sample and solvent; no change of light wave front when passing through the sample):

$$\frac{Kc}{R_\theta} = \frac{1}{M_w P(\theta)} + 2A_2 c + \cdots \qquad (1)$$

$$P(\theta) = 1 - a_1[2k \sin(\theta/2)]^2 + a_2[2k \sin(\theta/2)]^4 - \cdots \qquad (2)$$

where K is the light-scattering constant including the refractive index increment and the wavelength of the scattering light and A_2 is the second viral coefficient. If a sample is very diluted, the second term of Eq. (1) can be neglected and the excess Rayleigh ratio, R_θ (net light-scattering contribution from each component at angle θ), becomes directly proportional to $M_w P(\theta)$. On plotting R_θ/Kc against $\sin^2(\theta/2)$, the intercept yields the molecular weight (M_w) at the concentration c, and from the slope the root mean square radius can be derived. One great advantage with the MALS detector is that it does not demand calibration of the channel with reference materials, but the absolute concentration of the analyte is necessary because the signal includes a factor of the concentration. To acquire the sample concentration at each time slice, a concentration-calibrated DRI detector is commonly used on-line with the MALS detector. The FFF–MALS–DRI is receiving much interest and attention and many applications have been developed in recent years, especially in polymer characterization. Thielking and co-workers have published articles on the coupling of FFF–MALS–DRI for analysis of both polystyrene particles and smaller polystyrene sulfonates (PSS) [5]. Figure 1 shows their results of DRI-derived concentration and molecular weight given from MALS data as a function of elution volume for seven PSS standards. However, for small molecules (<10 kDa), the sensitivity of the MALS detector is rather poor.

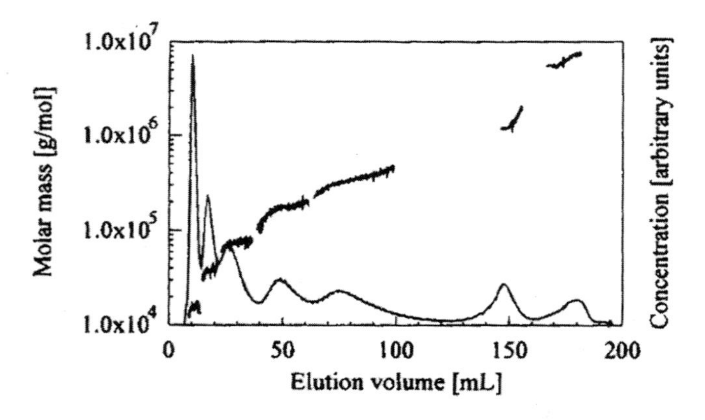

Fig. 1 Results from FlFFF–MALS–DRI analysis of seven PSS standards. The molecular weight is derived from MALS data and the concentration is given by the DRI detector. (From Ref. 5.)

In an evaporative light-scattering detector (ELSD), the sample is nebulized and when the solvent in the resulting droplets is evaporated, their mass content is proportional to the particle's mass in the sample stream. The particles are detected with a laser-light-scattering detector and the signal is related to their size. ELSD has not been extensively used in FFF. Oppenheimer and Mourey [6] showed that it can be a good complement to turbidimetric detection in sedimentation FFF for particles smaller than 0.2 μm. This detector is free from the problems associated with UV detectors when applied to a broad size range or samples with differing extinction coefficients over the size range. Further, it can be used for samples lacking absorbance characteristics. Compton et al. [7] presented a single-particle detector for steric FFF (1–70 μm) based on light scattering of single particles flowing through the laser-light path. Today, there are several other commercial flow-stream particle counters available.

A continuous-viscosity detector has been shown to be a good detection tool for thermal FFF analysis of polymer solutions [8]. Due to the high sample dilution in FFF, the viscosity detector response above the solvent baseline, ΔS, is only dependent on the intrinsic viscosity of every sample point, $[\eta]$, multiplied by the concentration, c, at the corresponding points:

$$\Delta S = [\eta]c \qquad (3)$$

If a concentration-selective detector, such as a DRI, is connected on-line with the viscosity detector, the ratio of the two signals yields the intrinsic viscosity distribution of the polymer sample. In polymer characterization, the intrinsic viscosity can be a property just as important as the molecular-weight distribution. Furthermore, polymer intrinsic viscosity follows the Mark–Houwink relation to the molecular weight, M, where K and a are Mark–Houwink viscosity constants:

$$[\eta] = KM^a \qquad (4)$$

Mass Spectrometric Detection Systems

The mass spectrometry (MS) detection methods covered here are mainly a selection of commonly used LC–MS methods, some of which have been optimized for FFF techniques or could potentially be good detection tools for FFF separations. The issue in the coupling of a liquid-based separation method to a mass spectrometer is the ion source conversion of dissolved analytes to ions in the high vacuum mass analyzer, which, for instance, can be magnetic sectors, quadrupoles, ion traps, or time-of-flight analyzers. Different ion sources give different information depending on the ionization mechanisms and will be discussed for each method.

In most FFF separations, a moderate concentration of dispersion agent, electrolyte, or surfactant is used to improve the separations. A common feature for most MS instruments is that salt in the liquid entering the ion source is deteriorating the performance of the MS by lowering the signal-to-noise ratio and by condensing on surfaces inside the MS, thus continuously increasing the background level.

Today, the most frequently used LC–MS ion source is electrospray ionization (ESI) in which the sample stream ends in a narrow capillary, put on a high voltage (positive or negative). This potential, sometimes together with a sheath gas flow, gives rise to a spray of small charged droplets (~1 μm). When the solvent is evaporated from these droplets, electrostatic repulsion forces smaller droplets (~10 nm) to leave. Before entering the semivacuum region, free analytes with one or more net charges, usually due to proton transfer or ion adducts (e.g., Li^+, Na^+, or NH_4^+), dominate.

Electrospray ionization is a mild ionization source (i.e., almost no fragmentation of the ions occur). It is applicable to all organic compounds applicable to proton exchange or binding to ions in the gas phase, which includes almost all biomolecules and polymers. In ESI–MS, multiple charges occur with a charge distribution for all components. This charge envelope has usually maximum intensity at m/z about 1000 and rarely ranging beyond 2000 in m/z. This has the advantage that large molecules, such as peptides, DNA molecules, or polymers, can be analyzed by all common MS analyzers, but the drawback is that the resulting spectra can be complicated to interpret. For single-mass molecules such as peptides, there are numerical models to deconvolute the single-charge molecular weight from the ESI–MS m/z spectra, but for not completely separated polymer components the overlapping charge distributions for the individual polymer components makes the interpretation complicated. ESI–MS sensitivity is dramatically reduced due to cluster formation in the presence of more than a few mmol/L salt, and surfactants can have a devastating effect on the ESI–MS spectra. Therefore, a volatile buffer should be used if possible (e.g., ammonium acetate, ammonium nitrate). ESI–MS has been used as a detector for FFF analysis of low-molecular-weight ethylene glycol polymers [9], where the effect of different carriers on cluster formation was investigated. ESI–MS has been coupled to size-exclusion chromatography (SEC) in several applications for polymer analysis and other applications where FFF techniques can be successfully used, including proteins, neuropeptides, and DNA molecule segments. Modern ESI–MS has a broad range of flow rates from nanoliters per minute up to milliliters per minute.

Atmospheric pressure chemical ionization (APCI) is a method not yet applied to FFF but could potentially be a good alternative to ESI for semivolatile analytes lacking a natural site for a charge. The analytes are evaporated and exposed to gas-phase molecules ionized by a high-voltage corona discharge electrode. The analytes are subsequently ionized by a charge transfer from the gas molecules. APCI has been shown to be less sensitive to buffer salts than ESI and no fragmentation occurs in the ion source. Primarily, singly charged ions are formed, making APCI less applicable to large molecules, depending on the upper range of the MS analyzer. APCI has a good flow-rate compatibility (0.3–1.5mL/min) with FFF.

Matrix-assisted laser desorption ionization (MALDI) is a frequently used ionization technique, but it is rarely used as an on-line detector. The sample stream is applied to a target plate, and it is allowed to cocrystallize with the matrix, which is subsequently desorbed, ionized with a laser, and analyzed in the MS. There has been attempts of combining FFF and MALDI, and for biomolecules, MALDI is a good ion source, due to the soft ionization with high efficiency and simple mass spectra, even for heavier molecules because the majority carry only single charge.

Inductively coupled plasma–ionization mass spectrometry (ICP–MS) is an ion source for elemental analysis where the analyte stream is introduced into a high-energy plasma with efficient atomization and ionization, producing almost entirely singly-charged elemental ions. ICP–MS has multielement capability with excellent sensitivity and has good flow-rate compatibility (0.3–1.5 mL/min) with FFF techniques. ICP–MS has previously been applied to Sedimentation FFF for determination of major element composition in different size fractions of suspended riverine particles and soil particles in the size range 50–800 nm [10] and recently Flow FFF coupled on-line to ICP–MS has been used to determine elemental size distributions for over 30 elements in freshwater colloidal material (1–50 nm) [11]. Figure 2 shows a selection of elements and the signal from the UV detector, coupled on-line before the ICP–MS, from a river water sample. An interface between the FFF channel and the ICP–MS was used to supply acid, to improve the performance of the nebulizer–spray chamber system, and the internal standard. The interface also serves to dilute and split away some half of the salt content. The salt is necessary for the FFF separation, but harmful to the ICPMS.

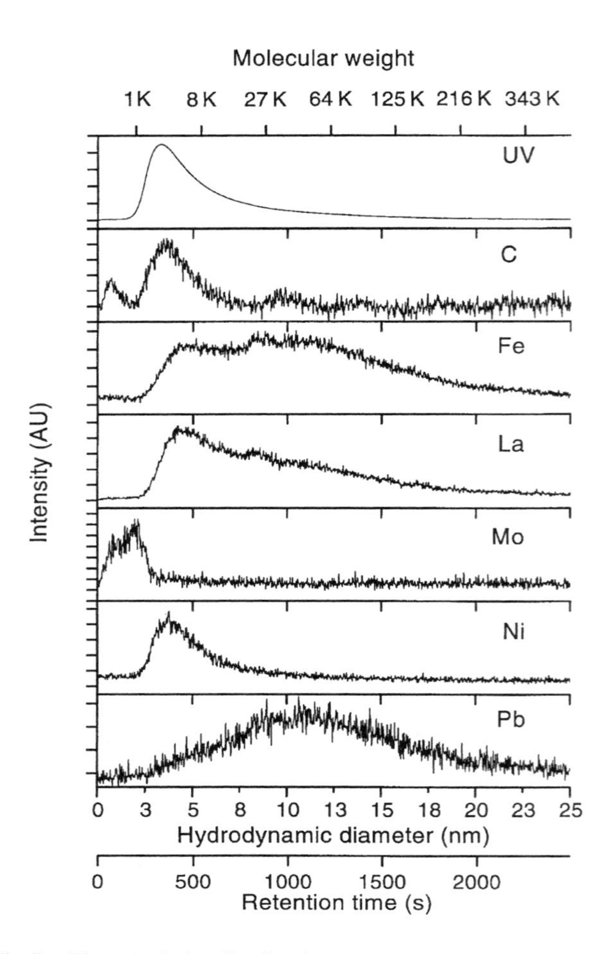

Fig. 2 Elemental size distributions of the colloidal material in a freshwater sample as given from a FIFFF coupled to ICPMS. A UV detector is placed on-line prior to the ICPMS and the UV size distribution is included. The signals are plotted as a function of retention time, hydrodynamic diameter (from FFF theory) and molecular weight (from standardization with PSS standards). (From Ref. 11.)

Density-Based Detection

A continuous-density detector is working on the principles of a liquid flowing through a oscillating U-shaped glass tube where the oscillating frequency is found relating the oscillations to the density of the flowing liquid. A densimetric detector has been evaluated for sedimentation FFF [12], and the conclusions were that it is a universal concentration-selective detector without the need for signal correction or transformation. However, the sensitivity is a limiting factor because the sensitivity is dependent on the density difference between the sample and the carrier liquid. A density difference of 0.2 g/mL is sometimes sufficient, but to achieve higher sensitivity, a difference up to 1.0 is desirable, making the densimetric detection suitable for inorganic particles, but less appropriate for lighter organic analytes. The densimeter detector is sensitive to temperature changes, but insensitive to flow changes, making it most suitable for flow programming applications.

References

1. F.-S. Yang, K. D. Caldwell, and J. C. Giddings, *J. Colloid Interf. Sci. 92*: 81–91 (1983).
2. P. Reschiglian, D. Melucci, A. Zattoni, and T. Giancarlo, *J. Microcol. Separ. 9*: 545–556 (1997).
3. K. D. Caldwell and J. Li, *J. Colloid Interf. Sci. 132*: 256–268 (1989).
4. P. J. Wyatt, *J. Colloid Interf. Sci. 197*: 9–20 (1998).
5. H. Thielking and W.-M. Kulicke, *Anal. Chem. 68*: 1169–1173 (1996).
6. L. E. Oppenheimer and T. H. Mourey, *J. Chromatogr. 298*: 217–224 (1984).
7. B. J. Compton, M. N. Myers, and J. C. Giddings, *Chem. Biomed. Environ. Instrum. 12*: 299–317 (1983).
8. J. J. Kirkland, S. W. Rementer, and W. W. Yau, *J. Appl. Polym. Sci. 38*: 1383–1395 (1989).
9. M. Hassellöv, G. Hulthe, B. Lyvén, and G. Stenhagen, *J. Liquid Chromatogr. Related Technol. 20*: 2843–2856 (1997).
10. H. E. Taylor, J. R. Garbarino, D. M. Hotchin, and R. Beckett, *Anal. Chem. 64*: 2036 (1992).
11. M. Hassellöv, B. Lyvén, C. Haraldsson, and W. Sirinawin, *Anal. Chem. 71*: 3497–3502 (1999).
12. J. J. Kirkland and W. W. Yau, *J. Chromatogr. 550*: 799–809 (1991).

Martin Hassellöv

Detection (Visualization) of TLC Zones

Introduction

After development with the mobile phase, the plate is dried in a fume hood and heated to completely evaporate the mobile phase. Separated compounds are detected on the layer by viewing their natural color (e.g., plant pigments, food colors, dyestuffs), natural fluorescence (aflatoxins, polycyclic aromatic hydrocarbons, riboflavin, quinine), or quenching of fluorescence. Substances that cannot be seen in visible or ultraviolet (UV) light must be visualized with suitable detection reagents to form colored, fluorescent, or UV-absorbing compounds by means of derivatization reactions (postchromatographic derivatization). Although dependent on the particular analyte and the detection method chosen, sensitivity values are generally in the nanogram range for absorbance and picogram range for fluorescence. Detection may be obtained by prechromatographic derivatization, either in solution prior to sample application or directly on the plate at the origin. As an example, dansyl derivatives of amino acids are detected directly by their fluorescence. Prechromatographic derivatization may enhance compound stability or chromatographic selectivity as well as serving for detection.

Direct Detection

Compounds that are naturally colored are viewed directly on the layer in daylight, whereas compounds with native fluorescence are viewed as bright zones on a dark background under UV light. Viewing cabinets or boxes incorporating short-wave (254 nm) and long-wave (366 nm) UV-emitting mercury lamps are available for inspecting chromatograms in an undarkened room.

Fluorescence Quenching

Compounds that absorb around 254 nm, including most compounds with aromatic rings and conjugated double bonds and some unsaturated compounds, can be detected on an "F layer" containing a phosphor or fluorescent indicator (often zinc silicate). When excited with 254-nm UV light, absorbing compounds diminish (quench) the uniform layer fluorescence and are detected as dark violet spots on a bright green background. Detection by natural fluorescence or fluorescence quenching does not modify or destroy the compounds, and the methods are therefore suitable for preparative layer chromatography. Derivatization reactions modify or destroy the structure of the compounds detected, but they are often more sensitive than detection with UV radiation.

Universal Detection Reagents

Universal reactions such as iodine absorption or spraying with sulfuric acid and heat treatment are quite unspecific and are valuable for completely characterizing an unknown sample. Absorption of iodine vapor from crystals in a closed chamber produces brown spots on a yellow background with almost all organic compounds except for some saturated alkanes. Iodine staining is nondestructive and reversible upon evaporation, whereas sulfuric acid charring is destructive. Besides sulfuric acid, 3% copper acetate in 8% phosphoric acid is a widely used charring reagent. The plate, which must contain a sorbent and binder that do not char, is typically heated at 120–130°C for 20–30 min to transform zones containing organic compounds into black to brown zones of carbon on a white background. Some charring reagents initially produce fluorescent zones at a lower temperature before the charring occurs at a higher temperature.

Selective Detection Reagents

Selective reagents form colored or fluorescent compounds on a group- or substance-specific basis and aid in compound identification. They also allow the use of a thin-layer chromatography (TLC) system with less resolution because interfering zones will not be detected. Examples include the formation of red to purple zones by reaction of α-amino acids with ninhydrin, detection of acidic and/or basic analytes with the indicator bromocresol green, and location of aldehydes as orange zones after reaction with 2,4-dinitrophenylhydrazine hydrochloride.

Certain specialized reagents are used to visualize compounds based on their biological activity. Cholinesterase-inhibiting pesticides (e.g., organophosphates, carbamates) are detected sensitively by treating the layer with the enzyme and a suitable substrate, which react to produce a colored product over the entire layer except where colorless pesticide zones are located due to their inhibition of the enzyme–substrate reaction.

Application of Detection Reagents

Chromogenic and fluorogenic liquid detection reagents are applied by spraying or dipping the layer. Various types of aerosol sprayers that connect to air or nitrogen lines are available commercially for manual operation. For safety purposes, spraying is carried out inside a laboratory fume hood or commercial TLC spray cabinet with a blower (fan) and exhaust hose, and protective eyeware and laboratory gloves are worn. The developed, dried plate is placed on a sheet of paper or supported upright inside a cardboard spray box. The spray is applied from a distance of about 15 cm with a uniform up-and-down and side-to-side motion until the layer is completely covered. It is usually better to spray a layer two or three times lightly and evenly with intermediate drying rather than a single, saturating application that might cause zones to become diffuse.

Studies are required with each reagent to determine the optimum total amount of reagent that should be sprayed, but, generally, the layer is sprayed until it begins to become translucent. After visualization, zones should be marked with a soft lead pencil because zones formed with some reagents may fade or change color with time.

Dipping is usually the best method for uniform reagent application and will produce the most reproducible results in quantitative densitometric analysis. The simplest method is to manually dip for a short time (5–10 s) in a glass or metal dip tank. A more uniform dip application of reagents can be achieved by the use of a battery-operated automatic, mechanical chromatogram immersion instrument (Fig. 1), which provides selectable, consistent vertical immersion and withdrawal speeds between 30 and 50 mm/s and immersion times between 1 and 8 s for plates with 10- or 20-cm heights. The immersion device can also be used for impregnation of layers with detection reagents prior to initial zone application and development and for postdevelopment impregnation of chromatograms containing fluorescent zones with a fluorescence enhancememt and stabilization reagent such as paraffin. Dip application to the layer cannot be used when two or more aqueous reagents must be used in sequence without intermediate drying, such as the detection of primary aromatic amines as pink to violet zones by diazotization and coupling with the Bratton–Marshall reagent composed of sodium nitrite and *N*-(1-naphthyl)ethylenediamine dihydrochloride aqueous spray solutions. Dip reagents must be prepared in a solvent that does not cause the layer to be removed from the plate or the zones to be dissolved from the layer or to become diffuse.

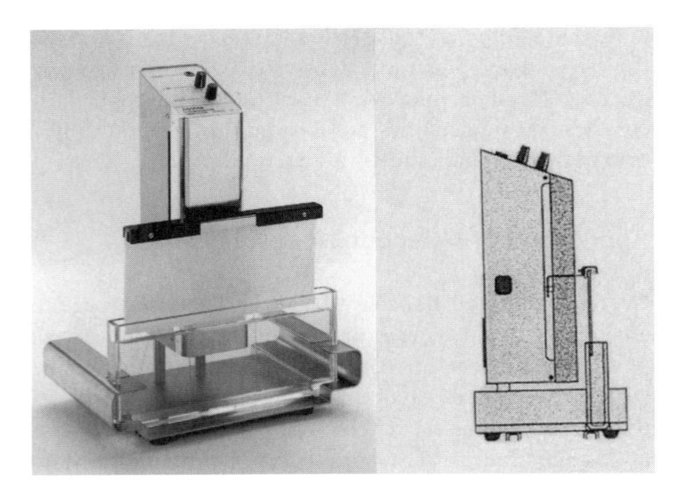

Fig. 1 Camag chromatogram immersion device set for a 10-cm dipping depth with high-performance TLC plates. Vertical dipping and removal rates and the residence time in the reagent can be preselected. [Photograph (left) supplied by D. Jaenchen, Camag, Muttenz, Switzerland; schematic diagram (right) reprinted from *Handbook of Thin Layer Chromatography* (J. Sherma and B. Freid, eds.), p. 146, with permission of Marcel Dekker, Inc.]

Heating the Layer

Layers often require heating after applying the reagent in order to complete the reaction upon which detection is based and ensure optimum color development. Typical conditions are 10–15 min at 105–110°C. If a laboratory oven is used, the plate should be supported on a solid metal tray to help ensure uniform heat distribution. A plate heater (Fig. 2), which contains a 20 × 20-cm flat, evenly heated surface, a grid to facilitate proper positioning of TLC and high-performance TLC plates, programmable temperature between 25°C and 200°C, and digital display of the programmed and actual temperatures, will usually provide more consistent heating conditions than an oven. Prolonged heating time or excessive temperature can cause decomposition of the analytes and darkening of the layer background and should be avoided.

Some reagents can be impregnated into the layer before spotting of samples if the selectivity of the separation is not affected. Detection takes place only upon heating after development. This method has been used for the detection of lipids as blue spots on a yellow background on silica gel layers preimpregnated with phosphomolybdic acid. A few detection reagents (HCl, sulfuryl chloride, iodine) can be transferred uniformly to the layer as vapors in a closed chamber rather than as solutions.

Fig. 2 Camag TLC plate heater. (Photograph supplied by D. Jaenchen, Camag, Muttenz, Switzerland.)

Zone Identification and Confirmation

The identity of the detected TLC zones is obtained initially by comparison of characteristic R_f values between samples and reference standards chromatographed on the same plate; R_f equals the migration distance of the center of the zone from the start (origin) divided by the migration distance of the mobile phase front from the start. Identity is more certain if a selective chromogenic detection reagent yields the same characteristic color for sample and standard zones. Because the chromatogram is stored on the layer, multiple compatible, specific detection reagents can be applied in sequence to confirm the identity of unknown zones. As an example, almost all lipids are detected as light green fluorescent zones by use of 2,7-dichlorofluorescein reagent, and absorption of iodine vapor differentiates between saturated and unsaturated lipids or lipids containing nitrogen. The identity of zones is confirmed further by recording UV or visible absorption spectra directly on the layer using a densitometer (*in situ* spectra) or by direct or indirect (after scraping and elution) measurement of Fourier transform infrared, Raman, or mass spectra.

Suggested Further Reading

Details of procedures, results, and applications for many hundreds of reagents for detection of all classes of compounds and ions are available in the following literature references:

Bauer, K., L. Gros, and W. Sauer, *Thin Layer Chromatography: An Introduction*, EM Science, Darmstadt, 1991.

Fried, B. and J. Sherma, *Thin Layer Chromatography: Techniques and Applications*, 4th ed., Marcel Dekker, Inc., New York, 1999, pp. 145–175.

Jork, H., W. Funk, W. Fischer, and H. Wimmer, *Thin Layer Chromatography, Volume 1a, Physical and Chemical Detection Methods*, VCH, Weinheim, 1990.

Jork, H., W. Funk, W. Fischer, and H. Wimmer, *Thin Layer Chromatography, Volume 1b, Reagents and Detection Methods*, VCH, Weinheim, 1994.

Kovar, K.-A. and G. E. Morlock, Detection, identification, and documentation, in *Handbook of Thin Layer Chromatography* 2nd ed., (J. Sherma and B. Fried, eds.), Marcel Dekker, Inc., New York, 1996, pp. 205–239.

Macherey-Nagel GmbH & Co. KG, *TLC Applications*, Dueren, pp. 75–78.

Merck, KGaA, *Dyeing Reagents for Thin Layer Chromatography and Paper Chromatography*, Darmstadt.

Stahl, E., *Thin Layer Chromatography: A Laboratory Handbook*, Academic Press, San Diego, CA, 1965, pp. 485–502.

Touchstone, J. C., *Practice of Thin Layer Chromatography*, 3rd ed., Wiley–Interscience, New York, 1992, pp. 139–183.

Zweig, G., and J. Sherma, *Handbook of Chromatography, Volume II*, CRC Press, Boca Raton, FL, 1972, pp. 103–189.

Joseph Sherma

Detector Linear Dynamic Range

The linearity of most detectors deteriorates at high concentrations and, thus, the *linear dynamic range* of a detector will always be less than its dynamic range. The symbol for the linear dynamic range is usually taken as (D_{LR}). As an example, the linear dynamic range of a flame ionization detector might be specified as

$$D_{LR} = 2 \times 10^5 \quad \text{for } 0.98 < r < 1.02$$

where r is the response index of the detector.

Alternatively, according to the ASTM E19 committee report on detector linearity, the linear range may also be defined as that concentration range over which the response of the detector is constant to within 5%, as determined from a linearity plot. This definition is significantly looser than that using the response index.

The lowest concentration in the linear dynamic range is usually taken as equal to the *minimum detectable concentration* or the *sensitivity* of the detector. The largest concentration in the linear dynamic range would be that where the response factor (r) falls outside the range specified, or the deviation from linearity exceeds 5% depending on how the linearity is defined. Unfortunately, many manufacturers do not differentiate between the dynamic range of the detector (D_R) and the linear dynamic range (D_{LR}) and do not quote a range for the response index (r). Some manufacturers do mark the least sensitive setting on a detector as N/L (nonlinear), which, in effect, accepts that there is a difference between the linear dynamic range and the dynamic range.

Suggested Further Reading

Fowlis, I. A. and R. P. W. Scott, *J. Chromatogr.*, *11*: 1 (1963).
Scott, R. P. W., *Chromatographic Detectors*, Marcel Dekker, Inc., New York, 1996.

Raymond P. W. Scott

Detector Linearity and Response Index

It is essential that any detector that is to be used directly for quantitative analysis has a linear response. A detector is said to be truly linear if the detector output (V) can be described by the simple linear function

$$V = Ac$$

where A is a constant and c is the concentration of the solute in the mobile phase (carrier gas) passing through it.

As a result of the imperfections inherent in all electromechanical and electrical devices, true linearity is a hypothetical concept, and practical detectors can only approach this ideal response. Consequently, it is essential for the analyst to have some measure of detector linearity that can be given in numerical terms. Such a specification would allow quantitative comparison between detectors and indicate how close the response of the detector was to true linearity. Fowlis and Scott [1] proposed a simple method for measuring detector linearity. They assumed that for an approximately linear detector, the response can be described by the power function

$$V = Ac^r$$

where r is defined as the response index of the detector.

For a truly linear detector, $r = 1$, and the proximity of r to unity will indicate the extent to which the response of the detector deviates from true linearity. The response of some detectors having different values for r are shown as curves relating the detector output (V) to solute concentration (c) in Fig. 1. It is seen that the individual curves appear as straight lines but the errors that occur in assuming true linearity can be quite large. The errors actually involved are shown in the following, which is an analysis of a binary mixture employing detectors with different response indices:

Solute	$r = 0.94$	$r = 0.97$	$r = 1.00$	$r = 1.03$	$r = 1.05$
1	11.25%	10.60%	10.00%	9.42%	9.05%
2	88.75%	89.40%	90.00%	90.58%	90.95%

It is clear that the magnitude of the error for the lower-level components can be as great as 12.5% (1.25% absolute) for $r = 0.94$ and 9.5% (0.95% absolute) for $r = 1.05$. In general analytical work, if reasonable linearity is assumed, then $0.98 < r < 1.03$. The basic advantage of defining linearity in this way is that if the detector is not perfectly linear, but the value for r is known, then a correction can be applied to accommodate the nonlinearity.

There are alternative methods for defining linearity which, in the author's opinion, are somewhat less precise and less useful. The recommendations of the ASTM E19 committee on linearity measurement are as follows:

> The linear range of a detector is that concentration range of the test substance over which the response of the detector is constant to within 5% as de-

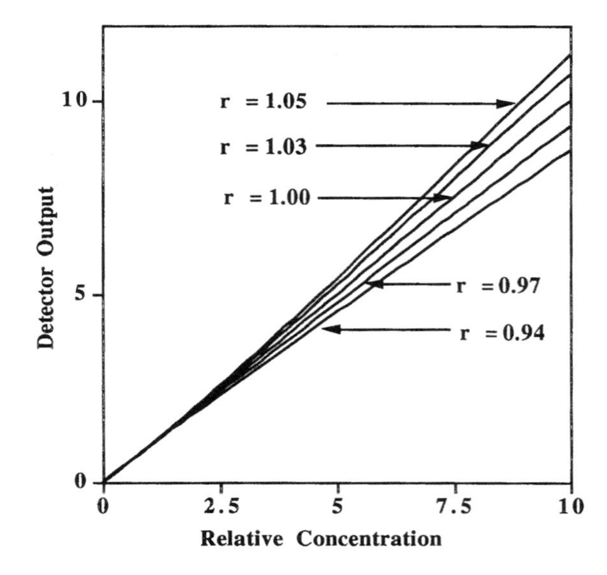

Fig. 1 Graph of detector output against solute concentration for detectors having different response indices.

termined from a linearity plot,—the linear range should be expressed as the ratio of the highest concentration on the linearity scale to the minimum detectable concentration.

This method for defining detector linearity is satisfactory up to a point and ensures a minimum linearity from the detector and, consequently, an acceptable quantitative accuracy. However, the specification is significantly "looser" than that given above, and it is not possible to correct for any nonlinearity that may exist, as there is no correction factor provided that is equivalent to the response index. It is strongly advised that the response index should be determined for any detector that is to be used for quantitative analysis. In most cases, r need only be measured once, unless the detector undergoes some catastrophic event that is liable to distort its response, in which case, r may need to be checked again.

There are two methods that can be used to measure the response index of a detector: the *incremental method* of measurement and the *logarithmic dilution method* of measurement [2]. The former requires no special apparatus, but the latter requires a log-dilution vessel, which, fortunately, is relatively easy to fabricate. The incremental method of measurement is the one recommended for general use.

The apparatus necessary is the detector itself with its associated electronics and recorder or computer system, a mobile-phase supply, pump, sample valve, and virtually any kind of column. In practice, the chromatograph to be used for the subsequent analyses is normally employed. The solute is chosen as typical of the type of substances that will be analyzed and a mobile phase is chosen that will elute the solute from the column in a reasonable time. Initial sample concentrations are chosen to be appropriate for the detector under examination.

Duplicate samples are placed on the column, the sample solution is diluted by a factor of 3 and duplicate samples are again placed on the column. This procedure is repeated, increasing the detector sensitivity setting where necessary until the height of the eluted peak is commensurate with the noise level. If the detector has no data acquisition and processing facilities, then the peaks from the chart recorder can be used. The width of each peak at 0.607 of the peak height is measured and the peak volume can be calculated from the chart speed and the mobile-phase flow rate. Now, the concentration at the peak maximum will be twice the average peak concentration, which can be calculated from

$$c_p = \frac{ms}{wQ}$$

where c_p is the concentration of solute in the mobile phase at the peak height (g/mL), m is the mass of solute injected, w is the peak width at 0.6067 of the peak height, s is the chart speed of the recorder or printer, and Q is the flow rate (mL/min).

The logarithm of the peak height y (where y is the peak height in millivolts) is then plotted against the log of the solute concentration at the peak maximum (c_p). Now,

$$\log(V) = \log(A) + (r)\log(c_p)$$

Thus, the slope of the $\log(V)/\log(c_p)$ curve will give the value of the response index (r). If the detector is truly linear, $r = 1$ (i.e., the slope of the curve will be $\sin \pi/4 = 1$). Alternatively, if suitable software is available, the data can be curved fitted to a power function and the value of r extracted directly from the curve-fitting analysis. The same data can be employed to determine the linear range as defined by the ASTM E19 committee. In this case, however, a linear plot of detector output against solute concentration at the peak maximum should be used and the point where the line deviates from 45° by 5% determines the limit of the linear dynamic range.

References

1. I. A. Fowlis and R. P. W. Scott, *J. Chromatogr.*, *11*: 1 (1963).
2. R. P. W. Scott, *Chromatographic Detectors*, Marcel Dekker, Inc., New York, 1996.

Raymond P. W. Scott

Detector Noise

Detector noise is the term given to any perturbation on the detector output that is not related to the presence of an eluted solute. As the minimum detectable concentration, or detector sensitivity, is defined as that concentration of solute that provides a signal equivalent to twice the noise, the detector noise determines the ultimate perfor-

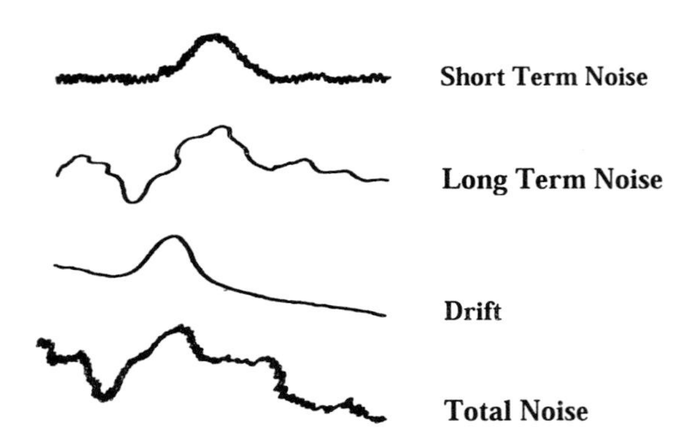

Fig. 1 Different types of detector noise.

mance of the detector. Detector noise has been arbitrarily divided into three types, *short-term noise*, *long-term noise*, and *drift*, all three of which are depicted in Fig. 1.

Short-Term Noise

Short-term noise consists of baseline perturbations that have a frequency that is significantly higher than that of the eluted peak. Short-term detector noise is usually not a serious problem in practice, as it can be easily removed by appropriate electronic noise filters that do not significantly affect the profiles of the peaks. The source of this noise is usually electronic, originating from either the detector sensor system or the amplifier electronics.

Long-Term Noise

Baseline perturbations that have a frequency that is similar to that of the eluted peak are termed long-term noise. This type of detector noise is the most significant and damaging, as it is often indiscernible from very small peaks in the chromatogram. Long-term noise cannot be removed by electronic filtering without affecting the profiles of the eluted peaks and, thus, destroying the integrity of the chromatogram. It is clear in Fig. 1 that the peak profile can easily be discerned above the high-frequency noise, but it is lost in the long-term noise. Long-term noise usually arises from temperature, pressure, or flow-rate changes in the sensing cell. Long-term noise can be controlled by careful detector cell design, the rigorous stabilization of operating variables such as sensor temperature, flow rate, and sensor pressure. Long-term noise is the primary factor that ultimately limits the detector *sensitivity* or the *minimum detectable concentration*.

Drift

Baseline perturbations that have a frequency significantly larger than that of the eluted peak are called drift. In gas chromatography (GC), drift is almost always due to either changes in detector temperature, changes in carrier gas flow rate, or column bleed. As a consequence, with certain detectors, baseline drift can become very significant at high column temperatures. Drift is easily constrained by choosing operating parameters that are within detector and column specifications.

A combination of all three sources of noise is shown by the trace at the bottom of Fig. 1. In general, the sensitivity of the detector (i.e., in most cases, the amplifier setting) should never be set above the level where the combined noise exceeds 2% of the FSD (full scale deflection) of the recorder (if one is used), or appears as more than 2% FSD of the computer simulation of the chromatogram.

Measurement of Detector Noise

The detector noise is defined as the maximum amplitude of the combined short- and long-term noise measured over a period of 10 min (the ASTM E19 committee recommends a period of 15 min). The detector must be connected to a column and carrier gas passed through it during measurement. The detector noise is obtained by constructing parallel lines embracing the maximum excursions of the recorder trace over the defined time period. The distance between the parallel lines, measured in millivolts, is taken as the measured noise (v_n), and the noise level (N_D) is calculated in the following manner:

$$N_D = v_n A = \frac{v_n}{B}$$

where v_n is the noise measured in volts from the recorder trace, A is the attenuation factor, and B is the alternative amplification factor.

Note: Attenuation is the reciprocal of amplification; manufacturers may use either function as a control of detector sensitivity.

The noise levels of detectors that are particularly susceptible to variations in column pressure or flow rate (e.g., the katharometer) are sometimes measured under static conditions (i.e., no flow of carrier gas). Such specifications are not really useful, as the analyst can never use the detector without a column flow. It could be argued that the manufacturer of the detector should not be held responsible for the precise control of the mobile phase, whether it may be a gas flow controller or pressure controller. However, all carrier supply systems show some variation in flow rates (and, consequently, pressure) and it is the responsibility of the detector manufacturer to design devices that are as insensitive to pressure and flow changes as possible.

At the high-sensitivity-range settings of some detectors, electronic filter circuits are automatically introduced to reduce the noise. Under such circumstances, the noise level should be determined at the lowest attenuation (or highest amplification) that does not include noise-filtering devices (or, at best, the lowest attenuation with the fastest response time) and then corrected to an attenuation of unity.

Suggested Further Reading

Scott, R. P. W., *Chromatographic Detectors*, Marcel Dekker, Inc., New York, 1996.

Scott, R. P. W., *Introduction to Gas Chromatography*, Marcel Dekker, Inc., New York, 1998.

Raymond P. W. Scott

Displacement Chromatography

Displacement chromatography is one of the three basic modes of chromatographic operation, the other two being frontal analysis and elution chromatography. Displacement chromatography is rarely, if ever, used for analytical separations, but it is useful for preparative separations. It has also been used for trace enrichment. Many retentive chromatographic methods have been performed in the displacement mode, including normal-phase, reversed-phase, ion-exchange, and metal affinity chromatographies. Much of the recent work has focused on the use of ion-exchange displacement chromatography for the preparative purification of the products of biotechnological products. Solutes purified by displacement chromatography have included metal cations, small organic molecules, antibiotics, sugars, peptides, proteins, and nucleic acids.

Tswett recognized the difference between elution and displacement development, although Tiselius was the first to clearly define these differences. Although displacement was popular in the 1940s, that popularity waned in the 1950s. In the 1980s, there was a resurgence of interest in displacement operation due to the efficient utilization of the stationary phase possible in that mode. Frenz and Horvath have published a comprehensive review of the history and applications of displacement chromatography.

Displacement chromatography is characterized by the introduction of a discrete volume of sample into the chromatographic column that has been previously equilibrated with a weak mobile phase, termed the carrier. This carrier is chosen so that the individual components of the sample (the solutes) are significantly retained by the stationary phase. The displacement is accomplished by following the sample with a new mobile phase containing some concentration of the displacer, a molecule with a higher affinity for the stationary phase than that of any of the solutes. The solutes are displaced from the stationary phase by the higher-affinity displacer and move further down the column, readsorbing. That solute with the highest affinity for the stationary phase moves the least before readsorbing and that solute with the lowest affinity for the stationary phase moves the most. This process is repeated as the displacer solution moves further down the column until a series of separated, but adjacent, bands is formed, termed the isotachic train. Each component of the train moves at the same velocity as the velocity of the displacer front.

Following elution of the isotachic train and the displacer solution from the column, the column must be regenerated and reequilibrated with the carrier before any subsequent displacement separation. This reequilibration step can be lengthy and is frequently considered a major limitation to efficient displacement operation. Displacement chromatography requires the competitive isotherms of the solutes and the displacer to be convex upward and to not intersect each other. (See the entry Distribution Coefficient for related information.)

The isotherm of the displacer must have a higher saturation capacity than any of the solutes. This is shown in Fig. 1a. The operating line is the line that connects the origin with the concentration of the displacer used: 9 g/L in the figure. The points of intersection between the operating line and the isotherms of each of the solutes determine the concentrations of each solute in the isotachic train, roughly 3.4, 4.5, and 6.8 g/L as shown in Fig. 1b. The width,

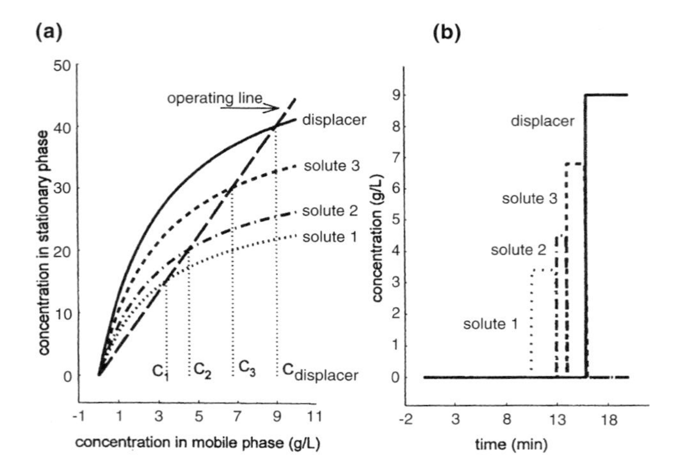

Fig. 1 Relationship between solute and displacer adsorption isotherms and the resultant isotachic train. (a) Nested set of upward isotherms showing the operating line at a displacer concentration of 9 g/L. (b) Resultant isotachic train showing that each solute elutes at the concentration determined by the intersection of the operating line with that solute's isotherm.

not the height, of the solute band within the train varies as the amount of solute in the sample varies. The concentration of the eluted solutes can be greater than their concentrations in the sample in displacement chromatography, unlike isocratic elution chromatography, wherein dilution necessarily occurs. Note that an actual displacement chromatogram would not have the appearance of that shown in the figure unless all solutes and the displacer had equal response factors for the detector employed.

The choice of the displacer concentration, which determines the operating line for a given displacement system, is critical for successful displacement. If the displacer concentration is increased (i.e., the slope of the operating line is decreased), then the solute concentrations in the isotachic train also increase. If the displacer concentration is decreased to the point that the operating line does not intersect the isotherms of the solutes, then displacement does not occur and the solutes elute as overloaded peaks in the elution mode. Rhee and Amundson have shown that there is a critical displacer concentration below which displacement cannot occur. This concentration depends primarily on the saturation capacities of the solutes and displacer.

When the solute isotherms cross one another, the situation becomes more complex. It then becomes possible to experience selectivity reversal; that is, at one displacer concentration, the solutes elute in the order A first, then B, whereas at another displacer concentration, the order is B first, then A. In a study of this problem, Antia and Horvath showed the existence of the separation gap. This is a region in the isotherm plane, the position of which depends on the ratio of the saturation capacities of the solutes in question. If the operating line is outside the separation gap, displacement occurs in the normal fashion. The elution order of the solutes then depends on the position of the operating line relative to the separation gap. However, if the operating line is within the separation gap, displacement operation does not separate the displaced solutes, but results in the elution of a mixture of the solutes.

In addition to appropriate isotherm behavior and displacer concentration, other factors are important in determining the effectiveness of a displacement chromatographic method. Highly efficient columns and fast-mass-transfer kinetics are necessary to achieve sharp boundaries between the adjacent solute bands in the isotachic train. Diffuse boundaries mean significant regions of overlap between adjacent solute bands, and thus a low recovery of purified material.

Successful displacement requires the establishment of the isotachic train before elution of the solutes from the column. As might be expected, the column length is an important parameter in displacement chromatography. The column should be sufficiently long (or sufficiently efficient) to allow complete formation of the isotachic train while lengths beyond that minimum do not improve the separation and increase the separation time. An inadequate length results in the elution of an incompletely formed isotachic train with inadequately resolved solute bands, again reducing the recovery yield.

Similarly, the sample size, column length, and displacer concentration jointly influence the establishment of the isotachic train and, thus, the effectiveness of the displacement separation. For a given displacer concentration and column length, increasing amounts of sample result in increasingly diffuse boundaries, and in sufficiently large samples, significant deterioration of the isotachic train. Likewise, for a given sample size and column length, increasingly high concentrations of displacer cause increasingly diffuse boundaries — termed overdisplacement.

Displacement chromatography has the attractive benefit of concentrating the solute. If the conditions are selected appropriately, large injection volumes of low-concentration samples can result in isotachic trains having high solute concentrations, essentially identical to those obtained for narrow pulses of high-concentration samples. This is one of the features that has caused the increased interest in displacement as a preparative mode. However, detailed comparisons of the production rates of displacement versus overloaded elution operation are limited (see Guiochon et al., pp. 641–648 and references therein). The limited experimental studies suggest that displacement operation is superior, although regeneration time was not included in the production rate calculation. Alternately, extensive theoretical studies indicate that for solutes having Langmuirian behavior, optimized overloaded elution chromatography is superior. Resolution of this issue currently awaits further studies.

Suggested Further Reading

Anita, F. D. and Cs. Horvath, Displacement chromatography of peptides and proteins, in *HPLC of Peptides and Proteins: Separation, Analysis, and Conformation* (C. Mant and R. Hodges, eds.), CRC Press, Boca Raton, FL, 1990, pp. 809–821.

Antia, F. and Cs. Horvath, *J. Chromatogr. 556*: 119–143 (1991).

Cramer, S. M. and G. Subramanian, *Separ. Purif. Methods 19*(1): 31–91 (1990).

Freitag, R., Displacement chromatography: application to downstream processing in biotechnology, in *Bioseparation Bioprocess* (G. Subramanian, ed.), Wiley–VCH, Weinheim, 1998, Vol. 1, pp. 89–112.

Frenz, J. and Cs. Horvath, High-performance displacement chromatography, in *High-Performance Liquid Chromatography: Advances and Perspectives* (Cs. Horvath, ed.), Academic Press, San Diego, CA, 1988, Vol. 5, pp. 211–314.

Guiochon, G. and S. G. Shirazi, and A. M. Katti, *Fundamentals of Preparative and Nonlinear Chromatography*, Academic Press, Boston, 1994.

Rhee, H.-K. and N. R. Amundson, *AIChE J. 28*: 423–433 (1982).

John C. Ford

Distribution Coefficient

Principle

In high-performance liquid chromatography (HPLC), as in many chromatographic techniques, separations result from the great number of repetitions of the analyte distribution between the mobile and stationary phases that are linked. At each elementary step, the distribution is governed by the distribution equilibrium

$$X_M \rightleftharpoons X_S$$

where X stands for the solute and the subscripts M and S for the mobile phase and the stationary phase, respectively. Conventionally, this equilibrium is characterized by the distribution coefficient

$$K = \frac{C_S}{C_M} \qquad (1)$$

where C_S and C_M are the molar concentrations of the solute X in the two phases. The distribution coefficient is also sometimes defined in terms of mole fractions of the solute in both phases. In elution analytical chromatography, concentrations are low enough to be comparable to the activities as an approximation. Although K should be dimensionless, as any thermodynamic constant, some authors use different units for the concentrations in both phases. For example, in LSC chromatography, concentrations are given in moles per square meter for the adsorbent. In such a case, K should be considered as an "apparent" constant, the dimension of which is m. However, in practice, the product KV_S in Eq. (2) must have the dimension of a volume.

K is interrelated with the retention factor k' by

$$k' = Kq$$

where q is the phase ratio, defined as the ratio of stationary phase volume V_S to the mobile phase volume V_M. Under linear conditions, K is linked to the solute retention volume (first-order moment of the elution peak) through

$$V_R = V_M + KV_S \qquad (2)$$

The distribution coefficient is the reflection of the ternary interactions schematically represented by

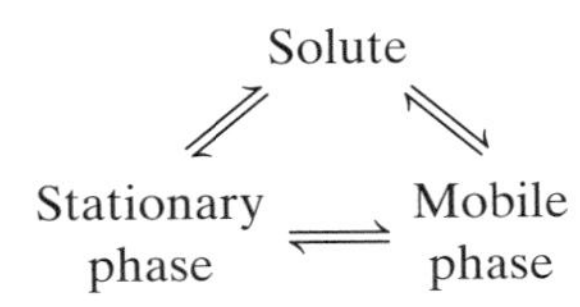

These interactions are as follows:

- Solute ↔ stationary phase for the retention
- Solute ↔ mobile phase for the solubilization
- Mobile phase ↔ stationary phase because of the competition between at least one constituent of the mobile phase (the strongest, also called the modifier) and the solute toward the active sites.

For an isobaric and isothermal process, the equilibrium constant K is given by

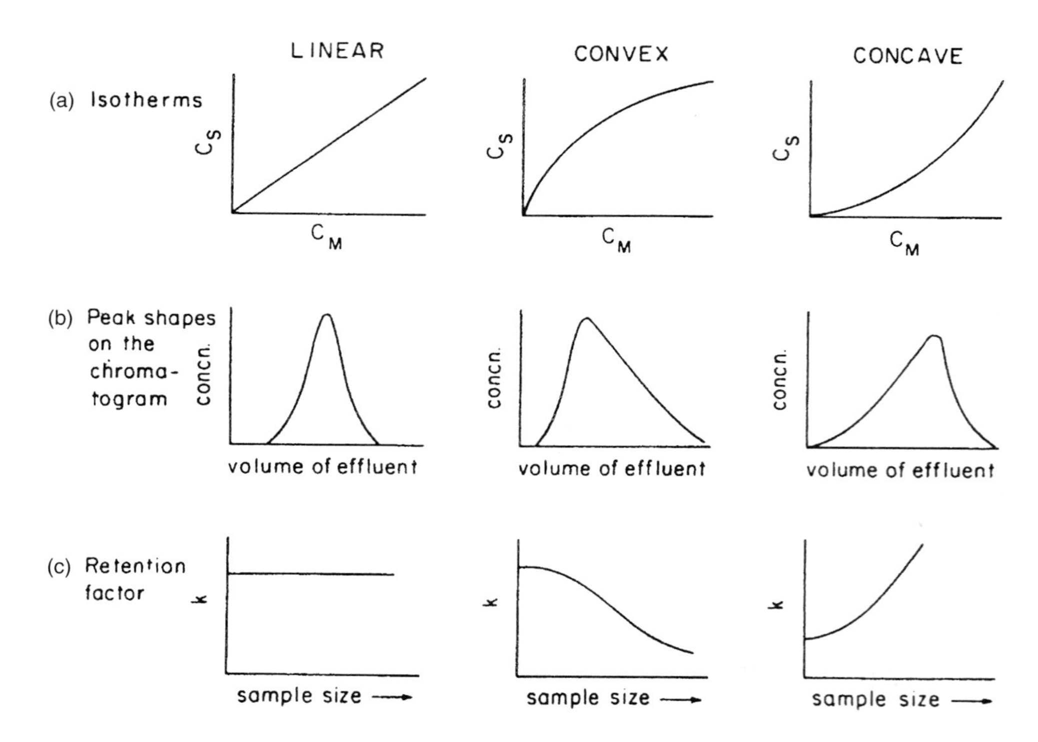

Fig. 1 Effect of isotherm shape on certain chromatographic properties. (a) Three different shapes of sorption isotherms encountered in chromatography; (b) peak shapes resulting from these isotherms; (c) dependence of the retention factor on the amount of solute injected. (From Karger, B. L., L. R. Snyder, and C. Horvath, *An Introduction to Separation Science*, John Wiley & Sons, New York, 1973.)

$$\ln K = -\frac{\Delta G^\circ}{RT} \tag{3}$$

where ΔG° is the difference in standard Gibbs free energy linked with the solute transfer from the mobile phase to the stationary phase:

$$\Delta G^\circ = \Delta H^\circ - T\Delta S^\circ \tag{4}$$

where ΔH° and ΔS° are the enthalpy and entropy, respectively.

In many cases, the value of ΔH° is independent of temperature so that linear Van't Hoff plots ($\ln K$ versus $1/T$) are observed. However, irregular retention behaviors are often observed, in practice, due to the dependence of ΔH° on the temperature.

Relationship (3) shows, clearly, the obtaining of reproducible results needed to thermoregulate the chromatographic apparatus, especially the column and mobile phase. Similarly, Eq. (3) explains why temperature changes implemented in order to increase efficiency can affect, drastically, the selectivity.

From Eq. (1), K can be related to the sorption isotherm, as it corresponds to the chord slope at each point. Therefore, Eq. (2) is valid only if K is constant; that is, the sorption isotherm is linear or, at least, is in a region where it becomes linear (i.e., if the dilution is great enough). The effect of the isotherm shape is shown in Fig. 1.

For the low concentrations used in liquid chromatography, the elution peak is Gaussian in shape and its retention factor is independent of the sample size. When isotherms are nonlinear (convex or concave, as illustrated in Fig. 1), an asymmetric elution peak is obtained and the retention factor measured at the peak apex is dependant on the sample size. This peak asymmetry is due to the dependence of the solute migration velocity versus the slope of the isotherm, which varies the solution concentration in the mobile phase.

Suggested Further Reading

Katz, E., R. Eksteen, P. Schoenmakers, and N. Miller, *Handbook of HPLC*, Marcel Dekker, Inc., New York, 1998, Chap. 1.

Karger, B. L., L. R. Snyder, and C. Horvath, *An Introduction to Separation Science*, John Wiley & Sons, New York, 1973, pp. 12–33.

Rosset, R., M. Caude, and A. Jardy, *Chromatographies en phases liquide et supercritique*, Masson, Paris, 1991, pp. 729–730.

M. Caude
A. Jardy

DNA Sequencing Studies by CE

Human Genome Sequencing

Development and commercialization of capillary array electrophoresis has a great impact on the Human Genome Project. The commercial 96-capillary array instrument has been achieved high-throughput DNA sequencing in the Human Genome Project and a 384-capillary array version has been developing to obtain a fourfold increase in throughput. Accordingly, the original key deadline for sequencing the 3 billion bases of the human genome by the end of 2005 has been brought forward significantly by use of capillary array electrophoresis. June 26, 2000, President Clinton announced that scientists in the international sequencing effort completed a working draft, which covers at least 90% of the human genome.

Because the read length for each capillary of the commercial instrument is limited to less than 500–600 bases with >99% accuracy, long read sequencing more than 1000 bases for each capillary with >99% accuracy is required to achieve much higher throughput for DNA sequencing. Unfortunately, sequence read length degrades as the electric field and sequencing speed increases. Several investigations to improve the read length for DNA sequencing have been reported and elevated capillary temperature, development of new separation matrices, use of lower electric fields, use of a gradient of electric field strength, and a change in denaturing agent proved to be effective in achieving long reading for DNA sequencing. The longest sequencing read lengths have been obtained at modest electric fields, high temperature, and with low-concentration, non-cross-linked polymers. In parallel, the second track of DNA sequencing development is the design of large-scale capillary instruments, wherein hundreds of DNA samples can be sequenced in parallel.

Four-color LIF Detector

To develop a four-color detectable laser-induced fluorescence (LIF) detection system for capillary electrophoresis is the first challenge toward accurate DNA sequencing, because the sequencing DNA fragment is fluorescently labeled with four different dyes on every base. The first high-speed, four-color capillary electrophoresis system for DNA sequencing was reported in 1990. The DNA sequencing fragments, which are fluorescently labeled with four different dyes on every base, were separated in a single capillary. A multiline argon ion laser was used to illuminate the narrow capillary simultaneously with light at 488 and 514 nm. Fluorescence was collected at right angles from the capillary. A set of beam splitters was used to direct the fluorescence to a set of four photomultiplier tubes, each equipped with a bandpass spectral filter, and fluorescence was recorded simultaneously in the four spectral channels. At an electric field of 300 V/cm, 80 min were required to separate fragments up to 360 bases in length. Some other detection schemes for DNA sequencing using a single capillary have been developed, including time-multiplexed fluorescence detection, two-spectral channel direct reading fluorescence spectrometer, and four-level peak-height encoded sequencing technique.

Capillary Array

Capillary electrophoersis (CE) provides rapid separation of long DNA sequencing fragments. However, the use of capillary arrays, instead of a single capillary, allows capillary DNA sequencing instrumentation to function as a high-throughput system. Array instruments typically use about 100 capillaries. A number of capillary array instruments have been reported. The systems fall into two classes: systems built around a scanning optical detector and systems that employ a detector array. Capillary array instruments put special requirements on the detection system. The first instrument used a confocal fluorescence scanner. In another approach, a multiple-capillary DNA sequencer was reported, in which a ribbon of capillaries was illuminated with a line-focused laser beam. Fluorescence was collected at right angles and imaged onto a CCD camera. The use of the CCD camera ensured that all capillaries were monitored simultaneously. To eliminate light scattering, a sheath-flow cuvette for multiple capillaries was developed (a multiple-sheath flow method). Fluorescence was collected at right angles and imaged onto a CCD camera. Optical waveguides have been proposed to collect fluorescence in the detection cell. A postelectrophoresis capillary scanning method has been developed to detect DNA sequencing fragments. The confocal fluorescence-scanning method is used in the first commercial system for DNA sequencing by capillary array electrophoresis, the MegaBACE 1000 by Molecular Dynamics, and the multiple-sheath flow detection method produces the second commercially system, PE-Biosystems PRISM 3700.

Microfabricated Chip

Substantial effort is being directed toward microfabrication of a device on a chip for DNA sequencing by capil-

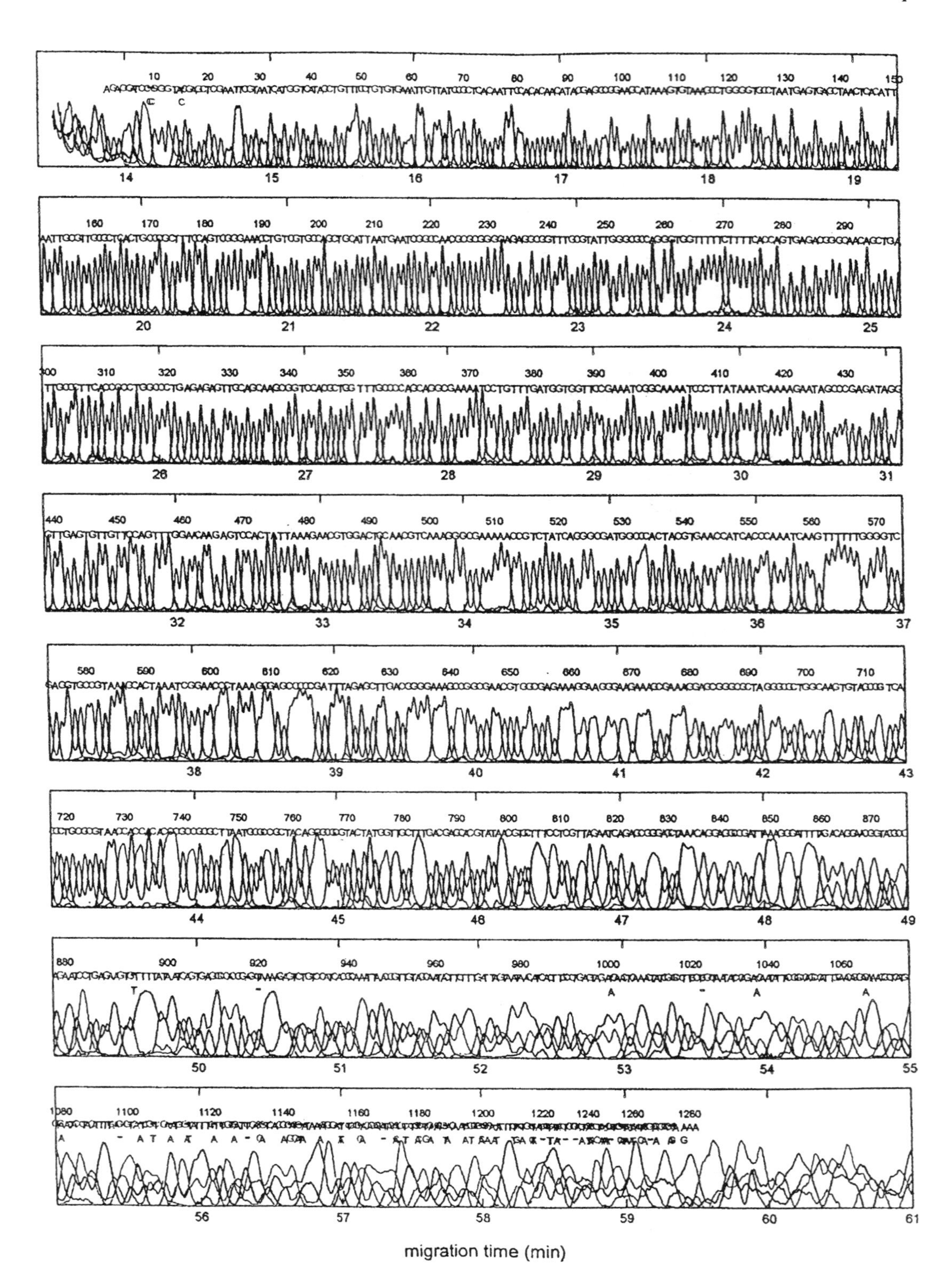

Fig. 1 Rapid 1000-base DNA sequencing by capillary electrophoresis.

lary electrophoresis. Several laser-confocal scanners have been developed to be applicable to the detection of DNA sequencing fragments migrating in the microchannel array on a glass chip. An efficient microfabricated 96-microchannel (capillary) array chip was developed for DNA sequencing and achieved 570 bases sequencing within only 20 min. Capillary array electrophoresis on a chip brings new potentials into DNA sequencing and, after solving some technical problems, it may become a starting point for an instrument of future.

Separation Matrix

A key characteristic of an electrophoresis system is the sequencing read length. The number of primers required for directed sequencing strategy and the number of templates generated and sequenced in shotgun sequencing are inversely proportional to the read length. A long read length minimizes sample preparation. More subtly, long reads minimize the computational effort required to assemble shotgun-generated data into finished sequence. Finally, a long read length allows sequencing of difficult templates that contain long stretches of repeated sequence.

Among several factors affecting the read length for DNA sequencing, the separation matrix is the most important factor to gain long sequencing read length, even 1000 bases sequencing. The first experiments with DNA sequencing by capillary electrophoresis used cross-linked polyacrylamide gels. However, the cross-linked polyacrylamide capillary had a rather limited lifetime. The use of cross-linked polymers in capillary electrophoresis leads to a serious limitation: The entire capillary must be replaced when the separation medium has degraded. The replacement can be quite tedious because of alignment constraints of the optical system with the narrow-diameter capillaries. Instead, it is attractive to use low-viscosity polymers for DNA sequencing by capillary electrophoresis. Low-viscosity polymers may be pumped from the capillary and replaced with a fresh matrix without replacement of the capillary or realignment of the optical system. Replaceable sieving matrices of linear polymers were introduced with the use of a linear polyacylamide. With some exceptions, replaceable sieving matrices are now used in DNA sequencing, exclusively.

Several polymers have been used successfully in DNA sequencing: linear polyacrylamide (PAA), poly(ethylene oxide) (PEO), hydroxyethyl cellulose (HEC), polydimethyacrylamide (PDMA), polyvinylpyrrolidone (PVP), poly(ethylene glycol) (PEG) with a fluorocarbon tail, polyacryloylaminopropanol, and a copolymer of acrylamide and allylglucopyranose. It is not well understood why some polymers are better sieving matrices than others. It has been observed that a larger molecular mass of the polymer means a longer read length but results in poorer separation of short fragments. Linear polymers are much more suitable for separation of long DNA sequencing fragments than are branched polymers.

Linear polyacrylamide represents the best replaceable sieving polymer today in terms of read length and speed, although formation of a dynamic wall coating by PDMA and PEO, which allows DNA sequencing in bare fused-silica capillaries, sometimes makes these latter polymers the preferred choice. DNA sequencing performance, expressed as the read length generated per unit time, is the more important criterion. The read length itself depends on a number of factors, namely temperature, voltage, and the molecular mass of the polymer. It is not clear why polyacrylamide currently provides the best DNA sequencing data. It may be because of inherently superior properties of linear polyacrylamide or it may simply be because acrylamide is available in much higher purity than other monomers. Karger and his colleagues showed that linear polyacrylamide can generate DNA-sequencing read lengths beyond 1000 bases with an accuracy of almost 99% in 53 min (Fig. 1). This accuracy is good enough, because the value generally accepted for routine DNA sequencing is 98.5%. With linear polyacrylamide, the run-to-run base calling accuracy was 99.2% for the first 800 bases and 98.1% for the first 900 bases. Poly(ethylene oxide) is another extensively studied sieving polymer. A mixture of low-molecular-mass populations of PEO was possible to separate DNA sequencing fragments over 1000 bases, but the separation time exceeded 7 h.

Suggested Further Reading

Dolnik, V., *J. Biochem. Biophys. Methods 41*: 103–119 (1999).

Dovichi, N. J., *Electrophoresis 18*: 2393–2399 (1997).

Kambara, H. and S. Takahashi, *Nature 361*: 565–566 (1993).

Karger, B. L., *Nature 339*: 641–642 (1989).

Mathies, R. A., *Anal. Chem. 71*: 31A–37A (1999).

Mathies, R. A. and X. C. Hunag, *Nature 359*: 167–169 (1992).

Salas-Solano, E. Carrilho, L. Kotler, A. W. Miller, W. Goetinger, Z. Sosic, and B. L. Karger, *Anal. Chem. 70*: 3996–4003 (1998).

Simpson, P. C., D. Roach, A. R. Woolley, T. Thorsen, R. Johanston, G. F. Sensbauch, and R. A. Mathies, *Proc. Natl. Acad. Sci. USA 95*: 2256–2261 (1998).

Ueno, K. and E. S. Yeung, *Anal. Chem. 66*: 1424–1431 (1994).

Zhang, J., Y. Fang, J. Y. Hou, H. J. Ren, R. Jiang, P. Roos, and N. J. Dovichi, *Anal. Chem. 67*: 4589–4593 (1995).

Yoshinobu Baba

Dry-Column Chromatography

Dry-column chromatography (DCC) is a modern chromatographic technique that allows easy and rapid transfer of the operating parameters of analytical thin-layer chromatography (TLC) to preparative column chromatography (CC). The dry-column technique bridges the gap between preparative column chromatography and analytical thin-layer chromatography.

Thin-layer chromatography has become an important technique in laboratory work, because it permits the rapid determination of the composition of complex mixtures. TLC allows the isolation of substances in micro amounts. If, however, milligrams or even grams of substance are required, column chromatography (CC) has to be applied, as TLC would involve a high cost and excessive time. In many cases, even the so-called thick layer or prep layer is but a poor choice because of time, cost, and sometimes inadequate transferability of the parameters of the analytical technique. In addition, the transfer from TLC to column chromatography, however, often proves to be difficult because the column chromatographic adsorbent is not usually analogous to the TLC adsorbent.

It is imperative that when transferring conditions of TLC separations to preparative columns, the conditions responsible for the TLC separation be meticulously transferred. Both CC and TLC use the same principle of separation. For normal operating conditions, a TLC layer has a chromatographic activity of II–III of the Brockmann and Schodder scale. Therefore, the sorbent used for DCC has to be brought to the same grade of activity. TLC layers often contain a fluorescent indicator in which case the DCC sorbent has to contain the same phosphor.

In TLC, the silica or alumina layer is "dry" before it is used and contacts the solvent only after it has been placed into the developing chamber. This is why, in DCC, the dry column is charged with the sample. Contrary to the normal CC, DCC is a nonelution technique. Therefore, only a limited amount of eluent is used in DCC to merely fill the interstitial volume between the adsorbent particles.

Scientific Adsorbents, Inc. DCC adsorbents, which are commercially available from Scientific Adsorbents, Inc. (Atlanta, GA, U.S.A.) are adjusted to meet the physical–chemical properties of TLC as closely as possible. These adjustments are made during the manufacturing cycle, and the material is packaged ready to use. With similar physical–chemical properties, the R_f values obtained for the substances under investigation from TLC are practically identical to those obtained with dry-column chromatography.

Using these especially adjusted adsorbents for DCC, one can use the same sorbent and the same solvent for the column work and can transfer the TLC results to a preparative scale column operation rapidly, saving time and money. DCC materials are available corresponding with the most common thin layers: silica DCC and alumina DCC.

These DCC sorbents have found wide use when it is necessary to scale up TLC separations in order to prepare sufficient quantities of compounds for further chemical reactions and/or analytical processes. DCC can be practically used for every separation achievable by TLC (Fig. 1).

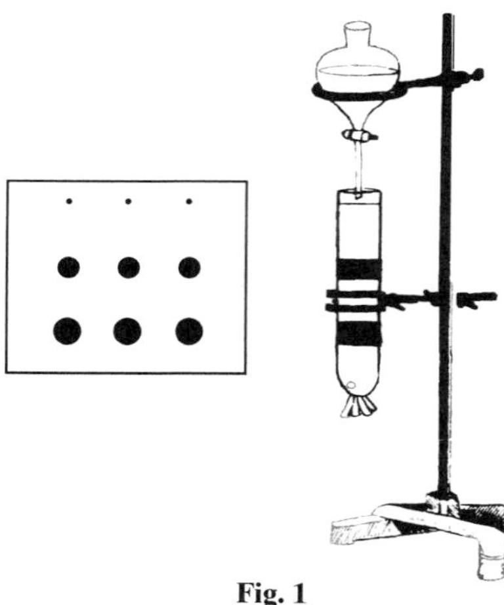

Fig. 1

Simplified Procedure

Preparation

1. Use the same solvent system that was developed on a TLC plate.
2. Cut a Nylon tube to the desired length. To isolate 1 g of material, use approximately 300 g of sorbent in a 1-m × 740-mm tube (Fig. 2).

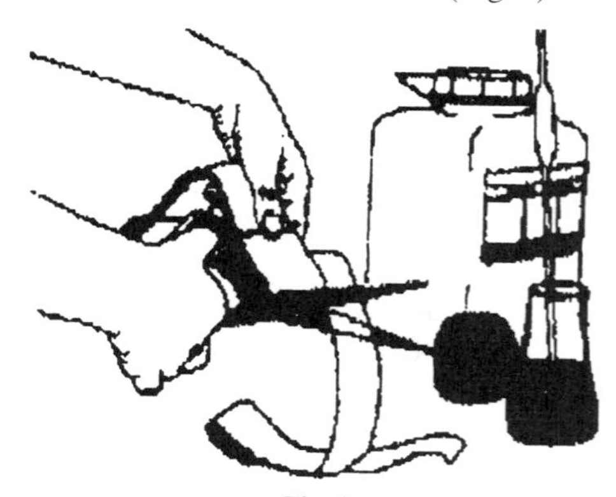

Fig. 2

3. Close the tube by rolling one end and securing it by a seal or a clip/staple.
4. Insert a small pad or wad of glass wool at the bottom of the column; pierce holes at the bottom with a needle.
5. Dry fill the column to three-fourths of its length (Fig. 3).

Fig. 3

6. The sample to be separated should be combined with at least 10 times its weight of the same sorbent in a conical test tube.
7. Add an additional centimeter of sorbent on top of the sample, followed by a small pad of glass wool (Fig. 4).

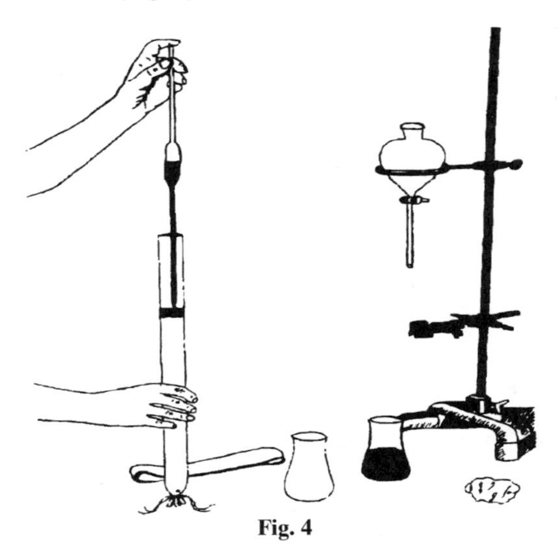

Fig. 4

8. Fasten the tube to a clamp on a stand.
9. Open the stopcock of the solvent reservoir and add solvent until it reaches the bottom of the column. Stop. Elapsed time: approximately 30 min. (Fig. 5).

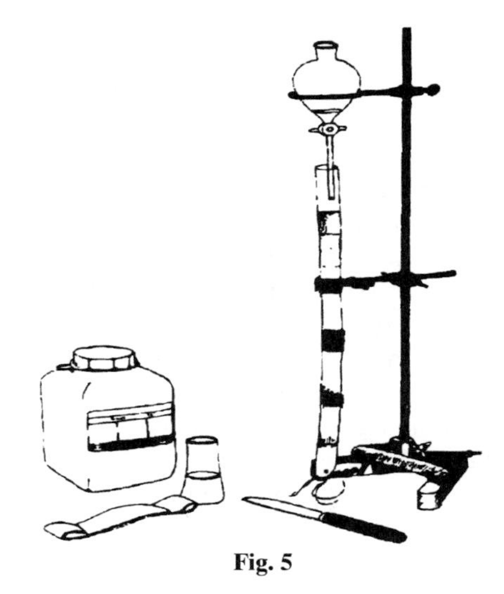

Fig. 5

10. Find the locations of the separated bands by visible, ultraviolet (UV), or UV quenching. Alternatively, cut a 1/16-in. vertical slice off the tube. Spray the exposed area with an appropriate visualization reagent and align with the untreated column to identify (mark) the separated bands.
11. Mark the location of the bands on the Nylon tube.
12. Remove the column from the clamp.
13. Slice the column into the desired sections (Fig. 6).
14. Elute the pure compounds from the sliced sections with polar solvents.

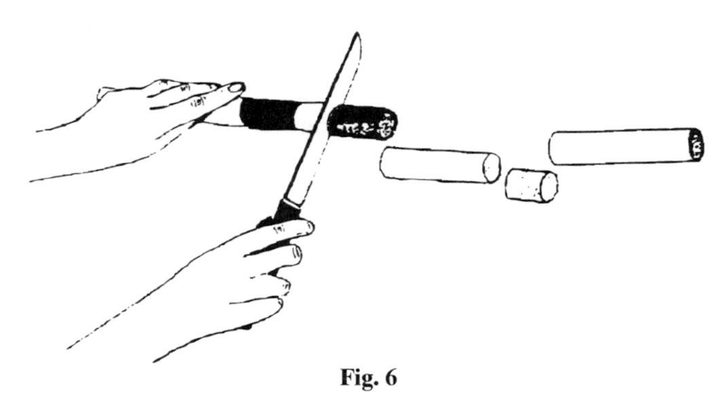

Fig. 6

Suggested Further Reading

Love, B. and M. M. Goodman, *Chem. Ind.* (*London*) 2026 (1967).
Love, B. and K. M. Snyder, *Chem. Ind.* (*London*) 15 (1965).

Mark Moskovitz

Dual Countercurrent Chromatography

Dual countercurrent chromatography (DuCCC) is a powerful separation method, which allows the performance of classic countercurrent distribution in a highly efficient manner. The system consists of a multilayer coiled column integrated with two inlet and two outlet flow tubes for a nonmiscible two-phase solvent system and a sample feed line, which is connected to the middle of the coiled column. Subjecting the system to a particular combination of centrifugal and planetary motions produces a unique hydrodynamic effect, which allows two immiscible liquids to flow countercurrently through the coiled column. The sample solution is fed at the middle portion of the column and eluted simultaneously through the column in opposite directions by the two solvents. This distinct feature of maintaining constant fresh two mobile phases within the coiled column permits a rich domain of applications. The principles of DuCCC and its applications in the purification of natural products and synthetic peptides are reviewed.

Introduction

The development, in the 1980s, of modem high-speed countercurrent chromatography (HSCCC) based on the fundamental principles of liquid–liquid partition has caused a resurgence of interest in the separation sciences. The advantages of applying continuous liquid–liquid extraction, a process for separating of a multicomponent mixture according to the differential solubility of each component in two immiscible solvents, have long been recognized. For instance, the countercurrent distribution method, which prevailed in the 1950s and 1960s, was applied successfully to fractionate commercial insulin into two subfractions, which differed only by one amide group in a molecular weight of 6000 [1]. In recent years, significant improvements have been made to enhance the performance and efficiency of liquid–liquid partitioning [2–8]. The high-speed centrifugal partition chromatographic (CPC) technique utilizes a particular combination of coil orientation and planetary motion to produce a unique hydrodynamic, unilateral phase distribution of two immiscible solvents in a coiled column. The hydrodynamic properties can effectively be applied to perform a variety of liquid–liquid partition chromatographies including HSCCC [2], foam countercurrent chromatography [8,9] and DuCCC [10,11]. In most cases, for the two-phase solvent system selected for HSCCC, one liquid phase serves as a stationary phase and the second phase is used as a mobile phase. An efficient separation can be achieved by continuous partitioning of a mixture between the stationary phase and the mobile phase.

By definition, this mode of separation should be called high-speed liquid–liquid partition chromatography or centrifugal partition chromatography, because only one solvent phase is mobile. In the case of DuCCC, for the two-phase solvents countercrossing each other inside the coiled column from opposite directions, both phases are mobile and there is no stationary phase involved.

The name "dual" countercurrent chromatography is redundant; however, it is useful to distinguish it from ordinary HSCCC. DuCCC shares several common advantages with other types of liquid–liquid partition chromatography. For instance, there are an unlimited number of two-phase solvent systems which can be employed, and there are no sample losses from irreversible adsorption or decomposition on the solid support. In addition, DuCCC is extremely powerful in separating crude natural products, which usually consist of multicomponents with an extremely wide range of polarities. In a standard operation, the crude sample is fed through the middle portion of the column. The extreme polar and nonpolar components are readily eluted from the opposite ends of the column followed by components with decreasing orders of polarity in one phase and increasing order of polarity in the other phase. A component with a partition coefficient equal to 1.0 will remain inside the coiled column. Essentially, the DuCCC resembles a highly efficient performance of classic countercurrent distribution. They differ in that CCC is a dynamic process, whereas CCD is an equilibrium process. The principles, instrumentation of DuCCC, and its capabilities in natural products isolation are illustrated in the remainder of the article.

Principles and Mechanism

The fundamental principle of separation for modem DuCCC is identical to classic countercurrent distribution. It is based on the differential partitions of a multicomponent mixture between two countercrossing and immiscible solvents. The separation of a particular component within a complex mixture is based on the selection of a two-phase solvent system, which provides an optimized partition coefficient difference between the desired component and the impurities. In other words, DuCCC and HSCCC cannot be expected to resolve all the components with one particular two-phase solvent system. Nevertheless, it is always possible to select a two-phase solvent system, which will separate the desired component. In general, the crude sample is applied to the middle of the coiled column through the sample inlet, and the extreme polar and nonpolar components are readily eluted by two immiscible solvents to opposite outlets of the column.

Contrary to the classic countercurrent distribution method, modern DuCCC allows the entire operation to be carried out in a continuous and highly efficient manner. DuCCC is based on the ingenious design of Ito [8]. A cylindrical coil holder is equipped with a planetary gear, which is coupled to an identical stationary sun gear (shaded) placed around the central axis of the centrifuge. This gear arrangement produces an epicyclic motion; the holder rotates about its own axis relative to the rotating frame and simultaneously revolves around the central axis of the centrifuge at the same angular velocity as indicated by the pair of arrows. The epicyclic rotation of the holder is necessary to unwind the twist of the five flow tubes caused by the revolution, eliminating the use of rotary seals to connect each flow tube.

As shown in Fig. 1, this unique design enables the performance of DuCCC using five flow channels connected directly to the column without using a rotational seal. When a column with a particular coil orientation is subjected to an epicyclic rotation, it produces a unique hydrodynamic phenomenon in the coiled column in which one phase entirely occupies the head side and the other phase occupies the tail side of the coil column. This unilateral phase distribution enables the performance of DuCCC in an efficient manner. A theoretical calculation of the hydrodynamic forces resulting from such an epicyclic rotation is very complicated and has not been elucidated.

Methods and Apparatus

The DuCCC experiments are performed with a tabletop (type J) high-speed plant centrifuge equipped with a multilayer coiled column connected to five flow channels. The multilayer coiled column is prepared from 2.6-mm-inner diameter PTFE tubing by winding it coaxilially onto the holder to a total volume capacity of 400 mL. The multilayer coiled column is subjected to an epicyclic rotation at 500–800 rpm. The fractions are collected simultaneously from both ends of the column and analyzed by thin-layer chromatography (TLC) or high-performance liquid chromatography (HPLC) [8,12].

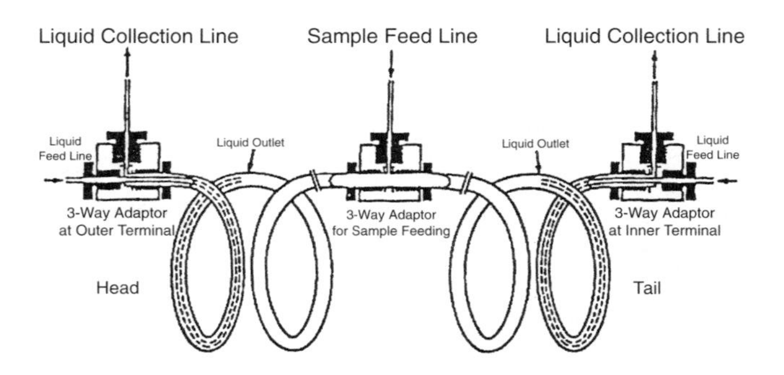

Fig. 1 Column design for DuCCC.

Applications

In the past decade, the rapid development of sophisticated spectroscopic techniques, including various two-dimensional nuclear magnetic resonance (2D NMR) methods, automated instrumentation and routine availability of x-ray crystallography has greatly simplified structural elucidation in natural product investigations. Consequently, the challenge to today's chemists has shifted to one's capability of isolating the bioactive components from crude extracts of either plants or animals. The extract of crude natural products usually is comprised of hundreds of components over a wide range of polarities. In isolating these natural products, it is essential to preserve the biological activity while performing chromatographic purifications. DuCCC represents one of the most efficient methods for isolation of the desired compound from a complex mixture.

Dual CCC has several advantages over HSCCC or CPC [13] in dealing with crude natural products. One distinct feature of DuCCC is the capability of performing normal-phase and reversed-phase elusions simultaneously. This provides a highly efficient and unique method for separation of crude natural products. In many instances, fractions eluted from DuCCC are pure enough for recrystallization or structural study. For example, an HPLC trace of the crude ethanol extract of *Schisandra rubriflora* shows that the major bioactive lignan, schisanhenol, is closely eluted with its acetate, it has been a major problem to isolate the pure schisanhenol. The fractions collected from DuCCC after injection of a crude ethanol extract of *Schisandra rubriflora* (125 mg) were analyzed by TLC and reversed-phase HPLC. The solvent system employed for DuCCC was hexane:ethyl acetate:methanol:water (10:5:5:1).

The upper phase, being less polar than the lower phase, results in a sequence of elution similar to normal-phase chromatography, whereas the lower phase provides a sequence of elution resembling reversed-phase chromatography. The bioactive components, schisanhenol acetate and schisanhenol, were eluted in the lower phase. Reversed-phase HPLC analyses of fractions 36–40 accounted for 32 mg of almost pure schisanhenol [6]. A total of 4 mg of schisanhenol acetate was also obtained from fractions 50–57. As evidenced by this experiment, DuCCC offers an excellent method for semipreparative isolation of bioactive components from very crude natural products [11]. The isolation of the topoisomerase inhibitor boswellic acid acetate from its triterpenoic acid mixture has also been accomplished by DuCCC [12]. As shown in Fig. 2, when an isomeric mixture of triterpenoic acids (400 mg) was subjected to DuCCC, using a hexane: ethanol: water (6:5:1) as the solvent system, 215 mg of the boswellic acid acetates and 135 mg of the corresponding boswellic acid were obtained. Some highly polar impurities were eluted immediately in the solvent front, from fraction 1 to 4. The isomeric boswellic acid was eluted in the lower-phase solvent and the less polar acetates were eluted simultaneously in the upper-phase solvent. Although the isomers were only partially resolved by DuCCC, this experiment demonstrates that DuCCC is a highly efficient system for preparative purification.

The conformationally restricted cyclic, disulfide-containing, enkephalin analog (D-Pen, D-Pen) enkephalin (DPDPE) was synthesized by solid-phase methods. Its purification was accomplished previously by partition on Sephadex G-25 block polymerizate using the solvent system (1-butanol: acetic acid: water: 4:1:5), followed by gel filtration on Sephadex G-15 with 30% acetic acid as the eluent [14]. DuCCC demonstrated a highly efficient and one step method for the purification of DPDPE. The crude DPDPE (500 mg), which contained impurities and salts, was purified by DuCCC with a two-phase solvent system consisting of 1-butanol containing 0.1% TFA and water also containing 0.1% TFA in a 1:1 (v/v) ratio. The desired DPDPE was eluted from the upper phase in fractions 15–19. The purity of each fraction collected was monitored by HPLC. A total of 24 mg pure DPDPE was obtained within 2 h. As evidenced, DuCCC can be a highly cost-effective procedure for the purification of polypeptides.

Conclusion

The capability and efficiency of DuCCC in performing classic countercurrent distribution has been demonstrated in the isolation of bioactive lignans and triter-

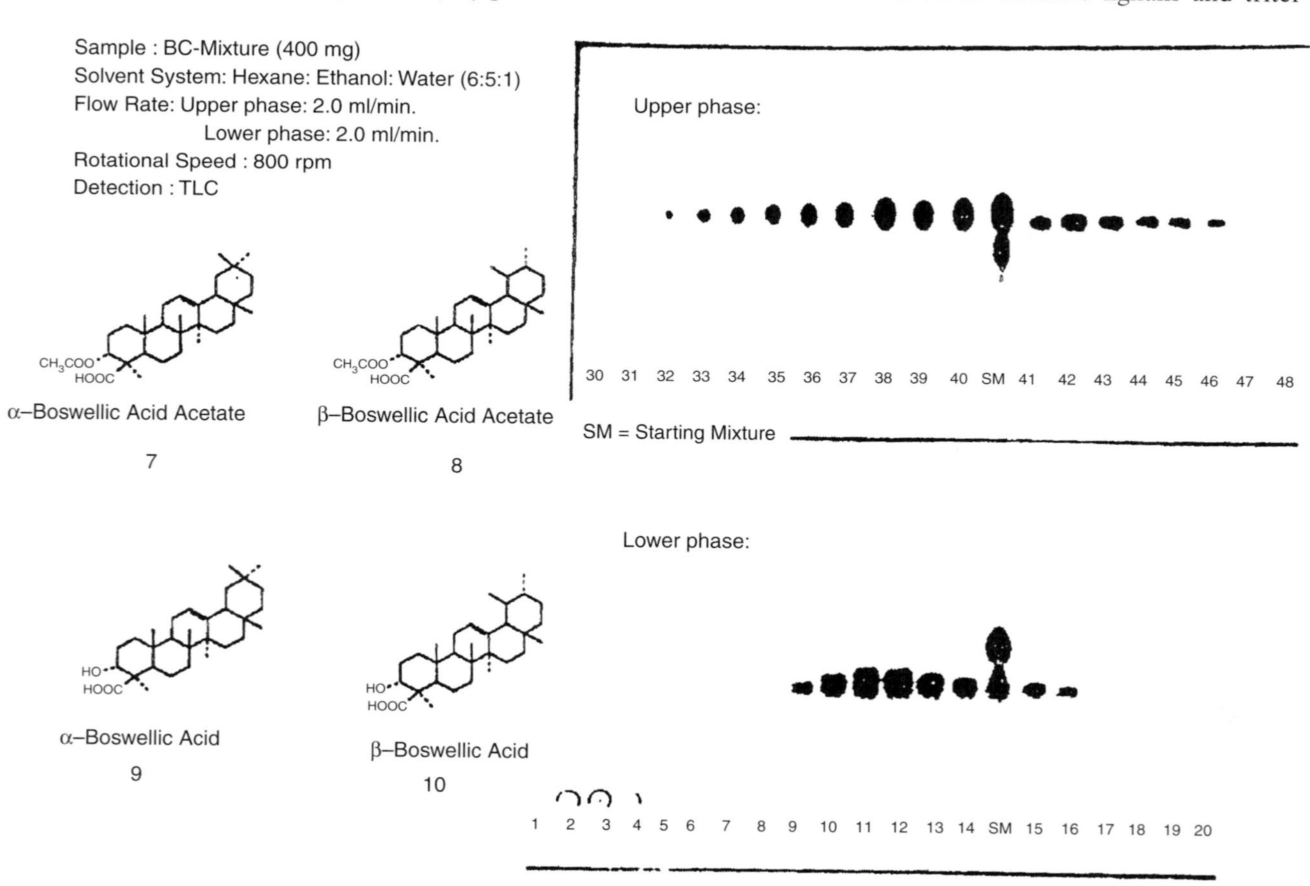

Fig. 2 DuCCC of triterpenoic acids.

penoic acids from crude natural products and in the purification of synthetic polypeptides. DuCCC provides excellent resolution and sample loading capacity. It offers a unique feature of elution of the nonpolar components in the upper-phase solvent (assuming the upper phase is less polar than the lower phase) and concomitant elution of the polar components in the lower phase. This capability results in an efficient and convenient preparative method for purification of the crude complex mixture. The capability of DuCCC has not yet been fully explored. For instance, a particular solvent system can be selected to give the desired bioactive component a partition coefficient of 1. This will allow the "stripping" of the crude extract with DuCCC to remove the impurities or inactive components. Consequently, the bioactive component will be concentrated inside the column for subsequent collection. This strategy can also be applied to extract and concentrate certain metabolites in biological fluids such as urine or plasma. Because there is no saturation of the stationary phase, a large amount of sample can also be processed by DuCCC. In addition, the system can be easily automated with computer-assisted sample injection and fractionation.

References

1. L. C. Craig, W. Hausmann, P. Ahrens, and E. J. Harfenist, *Anal. Chem. 23*: 1326 (1951).
2. Y. Ito, *CRC Crit. Rev. Anal. Chem. 17*: 65–143 (1986).
3. Y. W. Lee, Y. Ito, Q. C. Fang, and C. E. Cook, *J. Liquid Chromatogr. 11*(1): 75–89 (1988).
4. T. Y. Zhang, X. Hua, R. Xiao, and S. Kong, *J. Liquid Chromatogr. 11*(1): 233–244 (1988).
5. Y. W. Lee, C. E. Cook. Q. C. Fang, and Y. Ito, *J. Chromatogr. 477*: 434–438 (1989).
6. G. M. Brill, J. B. McAlpine, and E. J. Hochlowski, *J. Liquid Chromatogr. 8*: 2259 (1985).
7. D. G. Martin, R. E. Peltonen, and J. W. Nielsen, *J. Antibiot. 39*: 721 (1986).
8. Y. Ito, *J. Liquid Chromatogr. 8*(12): 2131 (1985).
9. H. Oka, K.-L. Harada, M. Suzuki, H. Nakazawa, and Y. Ito, *J. Chromatogr. 482*: 197 (1989).
10. Y. W. Lee, C. E. Cook, and Y. Ito, *J. Liquid Chromatogr. 11*(1): 37–53 (1988).
11. Y. W. Lee, Q. C. Fang, Y. Ito, and C. E. Cook, *J. Nat. Products 52*(1): 706–710 (1989).
12. W. Lee, unpublished data.
13. W. Murayarna, Y. Kosuge, N. Nakaya, Y. Nunogaki, N. Nunogaki, J. Cazes, and H. Nunogaki, *J. Liquid Chromatogr. 11*(1): 283–300 (1988).
14. H. J. Mosberg, R. Hurst, V. J. Hruby, K. Gee, H. I. Yamamura, J. J. Galligan, and T. F. Burks, *Proc. Natl. Acad. Sci. USA 80*: 5871–5874 (1983).

David Y. W. Lee

Dyes: Separation by Countercurrent Chromatography

Countercurrent chromatography (CCC) is a liquid–liquid partition technique that does not involve use of a solid support. In high-speed countercurrent chromatography (HSCCC), one of the liquid phases (the stationary phase) is retained in an Ito multilayered coil column by centrifugal force while the other liquid phase (the mobile phase) is pumped through the column. A variation of HSCCC was recently developed and is known as pH-zone-refining CCC. Both conventional HSCCC and pH-zone-refining CCC have been applied to the separation of dyes. The principles of these techniques, the instrumentation involved, the basis for selecting the two-phase solvent systems, and the separation procedure itself have all been discussed by Ito in the countercurrent chromatography entries of the present volume.

Conventional HSCCC

Conventional HSCCC has been used for dye separations since the mid-1980s. At that time, Fales et al. [1] separated various methylated homologs and contaminants from a 6-mg portion of the triphenylmethane dye Methyl Violet 2B (CI 42535), and Freeman and Willard [2] purified up to 520 mg of azo textile and ink dyes. Later documented applications have been for the colors Sulforhodamine B (CI 45100) [3] and Gardenia Yellow [4].

There are two notable advantages of this technique for the separation of dyes. One is its applicability for the separation of both the ionic and nonionic components of a dye mixture. Second, it has been shown to be one of the best methods for successfully separating small quantities

(20 mg or less) of dye mixtures. On the other hand, several hundred milligrams of dye mixtures represent the upper limit beyond which conventional HSCCC cannot be applied as the method of separation when the common 1.6-mm-inner diameter, 325-mL-volume column is used.

pH-Zone-Refining CCC

pH-Zone-refining CCC is a relatively new preparative method of separation [5,6] that allows separation of organic acids and bases according to their pK_a values and hydrophobicities. In contrast to conventional HSCCC, it permits the separation or purification of multigram quantities of dye mixtures, as long as those mixtures contain ionic or ionizable components. It has been applied to the separation of dyes containing various functional groups, such as $-COOH$, $-OH$, $-NH_2$, and $-SO_3H$. Details on its application to dye separation were recently described [7]. Briefly, pH-zone-refining CCC of dyes can be divided into two categories:

1. Standard separations, for carboxylic acid- or amine-containing dyes
2. Affinity-ligand separations, for sulfonated dyes

Standard pH-Zone-Refining CCC

For separation by standard pH-zone-refining CCC, the sample components should be stable over a pH range of 1–10 during the separation period and the minimum quantity of each targeted component should be no less than 0.1 mmol and preferably over 1 mmol. The requirements for the selection of a solvent system are different from those needed to select a solvent system for conventional HSCCC separations. Carboxylic acid-containing dyes (e.g., xanthene dyes or their lactone analogs, fluoran dyes) may be separated by using an organic acid (e.g., trifluoroacetic acid) as a retainer in the organic stationary phase and a base (e.g., ammonia) as an eluter in the aqueous mobile phase. Dyes containing amino groups may be separated by using an organic base as a retainer (e.g., triethylamine) and an inorganic acid (e.g., HCl) as an eluter. The recommended steps in selecting an appropriate solvent system for the separation of dyes and the separation procedure have been described in detail elsewhere [7].

Standard pH-zone-refining CCC was applied to the separation of pure components, sometimes multigram quantities, from various halogenated fluorescein (F) dyes such as 4,5,6,7-tetrachloroF, D&C Orange No. 5 (mainly 4′,5′-dibromoF), D&C Red No. 22 (Eosin Y, mainly the disodium salt of 2′,4′,5′,7′-tetrabromoF), D&C Orange No. 10 (mainly 4′,5′-diiodoF), FD&C Red No. 3 (Erythrosine, mainly the disodium salt of 2′,4′,5′,7′-tetraiodoF), D&C Red No. 28 (Phloxine B, mainly the disodium salt of 2′,4′,5′,7′-tetrabromo-4,5,6,7-tetrachloroF), and Rose Bengal (mainly the disodium salt of 2′,4′,5′,7′-tetraiodo-4,5,6,7-tetrachloroF). These dyes are used as biological stains and, except for Rose Bengal, for coloring food (FD&C Red No. 3), drugs, and cosmetics in the United States. Standard pH-zone-refining CCC was also applied to the separation of contaminants from these colors and their intermediates. The purified contaminants were further used as reference materials for the development of analytical methods [high-performance liquid chromatography (HPLC), thin-layer chromatography (TLC), etc.]. Furthermore, the capability of this method to purify multigram quantities of dyes used as biological stains may facilitate standardization of the biological stains themselves, a well-documented goal. A detailed review [8] describes the separation of the above-mentioned dyes by standard pH-zone-refining CCC.

Affinity-Ligand pH-Zone-Refining CCC

Sulfonated dyes have not been amenable to separation by standard pH-zone-refining CCC because of their very low pK_a values, which prevent their partitioning into the organic phase of a conventional two-phase solvent system. The addition of a ligand, an ion-exchange reagent (e.g., dodecylamine) or an ion-pairing reagent (tetrabutylammonium hydroxide), and a retainer acid (e.g., HCl, H_2SO_4) will enhance the partitioning of the sulfonated dye components into the organic stationary phase, thus enabling their separation by pH-zone-refining CCC. The more stringent conditions require that the components of a sulfonated dye mixture be stable over a pH range of 0.5–13.5 during the time of the experiment. The details and conditions required for the separation of sulfonated dyes by affinity-ligand pH-zone-refining CCC in the ion-exchange mode and in the ion-pairing mode can be found elsewhere [7]. An example [9] of such a separation is described below. Figure 1 shows the separation by affinity-ligand pH-zone-refining CCC in the ion-exchange mode of the main components from a 1.8-g portion of U.S. certified color additive D&C Yellow No. 10 (Quinoline Yellow, CI 47005). The two-phase solvent system used consisted of iso-amyl alcohol:methyl *tert*-butyl ether:acetonitrile:water (3:1:1:5). The ligand (dodecylamine) at a concentration of 5%, and sulfuric acid were added to the organic stationary phase as the retainer. The separation obtained for the two monosulfonated positional isomers 6SA (0.6 g) and 8SA (0.18 g) is shown by the associated HPLC chromatograms in Fig. 1.

Using a modified procedure, the applications of this technique have been extended to the separation of the highly polar disulfonated and trisulfonated components

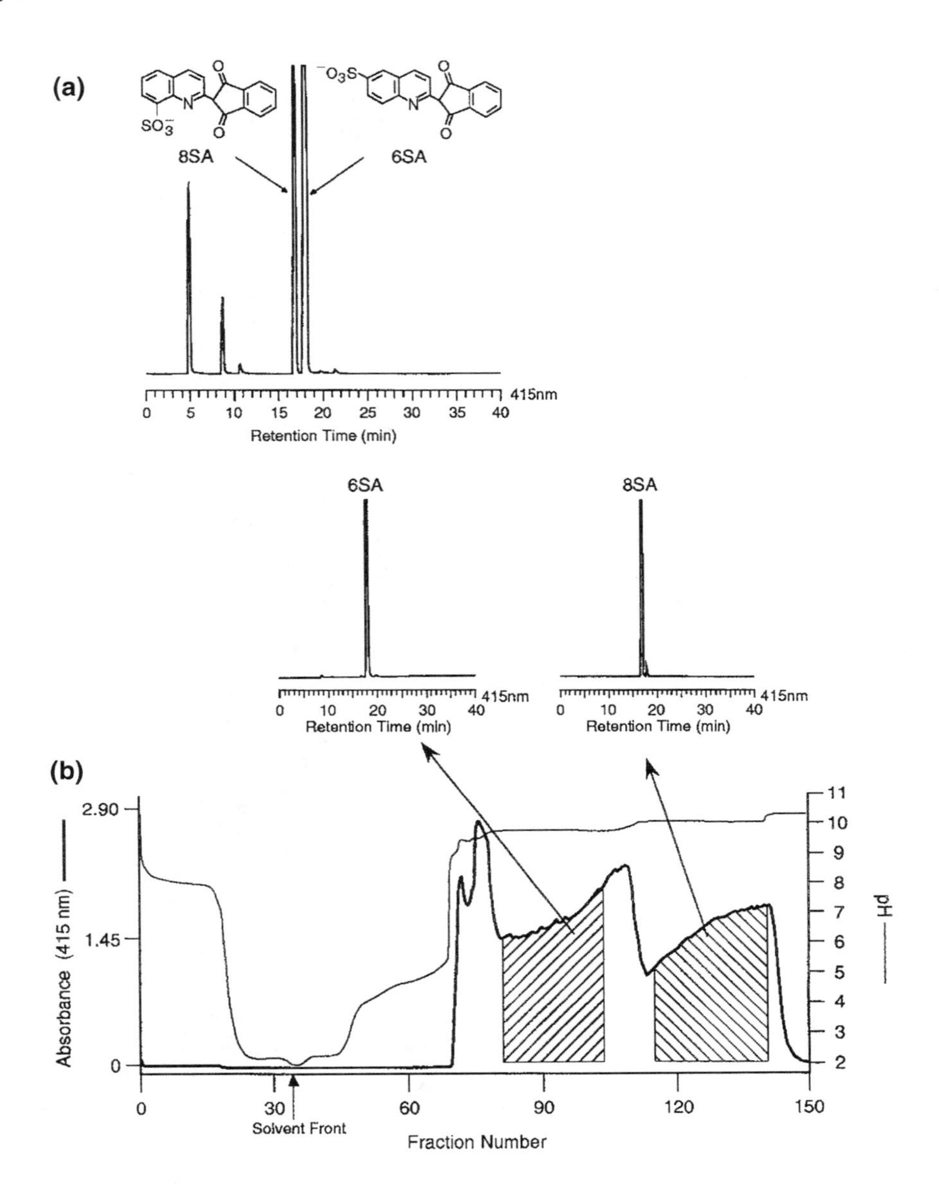

Fig. 1 Separation by affinity-ligand pH-zone-refining CCC in the ion-exchange mode of the main components from a sample of D&C Yellow No. 10 (Quinoline Yellow, CI 47005). (a) HPLC analysis of the original mixture; (b) pH-zone-refining CCC elution profile and HPLC analyses of the combined fractions 81–103 and 114–138, respectively. For experimental conditions, see text and Ref. 9.

of D&C Yellow No. 10 and Yellow No. 203 (the Japanese Quinoline Yellow, CI 47005) (7) and of other sulfonated dyes such as FD&C Yellow No. 6 (Sunset Yellow, CI 15985), D&C Green No. 8 (Pyranine, CI 59040), and FD&C Red No. 40 (Allura Red, CI 16035).

References

1. H. M. Fales, L. K. Pannell, E. A. Sokoloski, and P. Carmeci, Separation of methyl violet 2B by high-speed countercurrent chromatography and identification by Californium-252 plasma desorption mass spectrometry, *Anal. Chem. 57*: 376–378 (1985).
2. H. S. Freeman and C. S. Williard, Purification procedures for synthetic dyes: 2. Countercurrent chromatography, *Dyes Pigments 7*: 407–417 (1986).
3. H. Oka, Y. Ikai, N. Kawamura, J. Hayakawa, M. Yamada, K-I. Harada, H. Murata, M. Suzuki, H. Nakazawa, S. Suzuki, T. Sakita, M. Fujita, Y. Maeda, and Y. Ito, Purification of food color Red No. 106 (Acid Red) using high-speed counter-current chromatography, *J. Chromatogr. 538*: 149–156 (1991).

4. H. Oka, Y. Ikai, S. Yamada, J. Hayakawa, K-I. Harada, M. Suzuki, H. Nakazawa, and Y. Ito, Separation of Gardenia Yellow components by high-speed countercurrent chromatography, in *Modern Countercurrent Chromatography* (W. D. Conway and R. J. Petroski, eds.), American Chemical Society, Washington, DC, ACS Symposium Series, Vol. 593 1995, pp. 92–106.
5. A. Weisz, A. L. Scher, K. Shinomiya, H. M. Fales, and Y. Ito, A new preparative-scale purification technique: pH-Zone-refining countercurrent chromatography, *J. Am Chem Soc 116*: 704–708 (1994).
6. Y. Ito and Y. Ma, Review: pH-Zone-refining countercurrent chromatography, *J. Chromatogr. A, 753*: 1–36 (1996).
7. A. Weisz and Y. Ito, High-Speed countercurrent chromatography. Dyes, in *Encyclopedia of Separation Science*, Academic Press, London 2000 vol. 6 pp. 2588–2602.
8. A. Weisz, Separation and purification of dyes by conventional high-speed countercurrent chromatography and pH-zone-refining countercurrent chromatography, in *High-Speed Countercurrent Chromatography* (Y. Ito and W. D. Conway, eds.), John Wiley & Sons, New York, 1996, pp. 337–384.
9. A. Weisz, E. P. Mazzola, J. E. Matusik, and Y. Ito, Separation of components of the color additive D&C Yellow No. 10 (Quinoline Yellow) by pH-zone-refining countercurrent chromatography, *J. Chromatogr.* (in press).

Adrian Weisz

E

Eddy Diffusion in Liquid Chromatography

Among the causes of widening of peaks corresponding to components of a mixture undergoing separation by liquid chromatography (LC) is the phenomenon known as eddy diffusion. This results from molecules of a solute traversing a packed bed of a column through different pathways, in and around the stationary phase. Some molecules travel more rapidly through the column through more open, shorter pathways, whereas others will encounter longer, restricted areas and lag behind. The result is a solute band that passes through the column with a Gaussian distribution around its center [1].

The degree of band broadening of any chromatographic peak may be described in terms of the height equivalent to a theoretical plate, H, given by

$$H = \frac{L}{N} \tag{1}$$

where L is the length of the column (usually measured in cm) and N is the number of theoretical plates, which can be calculated from Eq. (2), where t_R and W are the retention time and width of the peak of interest, respectively:

$$N = 16\left(\frac{t_R}{W}\right)^2 \tag{2}$$

Because higher values of N correspond to lower degrees of band broadening and narrower peaks, the opposite is true for H. Therefore, the goal of any chromatographic separation is to obtain the lowest possible values for H.

The contribution of eddy diffusion and other factors to band broadening in liquid chromatography can be quantitatively described by the following equation, which relates the column plate height H to the linear velocity of the solute, μ:

$$H = A\mu^{0.33} + \frac{B}{\mu} + C\mu + D\mu \tag{3}$$

where A, B, C, and D are constants for a given column [2]. The linear velocity μ is related to the mobile-phase flow rate and is determined by

$$\mu = \frac{L}{t_0} \tag{4}$$

where t_0 (the so-called "dead time") is determined from the retention time of a solute which is known not to interact with the stationary phase of the column. The first term in Eq. (4), $A\mu^{0.33}$, includes the contribution of eddy diffusion to chromatographic band broadening. This term, which is dependent on the cube root of the linear velocity, is less dependent on mobile-phase flow rate than the other terms in the equation, which are either directly or inversely proportional to linear velocity.

Minimizing eddy diffusion in an LC column results in a lower $A\mu^{0.33}$ term in Eq. (3), which minimizes band spreading and gives narrower chromatographic peaks. The most common methods used for this purpose, in LC, are the following: (a) using a column of the smallest practical diameter; this obviously reduces the number of alternate pathways which a solute can take through the column; (b) using a stationary phase of smallest practical particle size; Giddings [3] and others have shown that the effects of eddy diffusion are directly proportional to the average diameter of stationary-phase particles; thus, smaller stationary phase particles give narrower peaks; (c) making sure the column is uniformly packed; again, this limits open space in the column, thus minimizing the number of pathways.

Those who prepare and/or manufacture LC columns must use the above methods to limit the effects of eddy diffusion on the chromatographic separations. However, there are practical limitations. Column and stationary-phase particle diameters can only be reduced to points that are compatible with the pressure limitations of the pumps used in chromatographic instruments and the required sample capacities of the columns. The degree of training and experience of those who pack the columns may also limit the quality of the procedure used in packing the column. Nevertheless, most commercial manufacturers of LC columns have adopted column designs and packing procedures which generally reduce the effects of eddy diffusion on modern LC separations to an inconsequential level. Still, these effects may increase as a column ages, and practicing chromatographers should be on the watch for them.

References

1. C. F. Poole and S. K. Poole, *Chromatography Today*, Elsevier, New York, 1991, Chap. 1.
2. L. R. Snyder and J. J. Kirkland, *Introduction to Modern Liquid Chromatography*, 2nd ed., John Wiley & Sons, New York, 1979, pp. 15–37.
3. J. C. Giddings, *Dynamics of Chromatography*, Marcel Dekker, Inc., New York, 1965, pp. 35–36.

J. E. Haky

Effect of Organic Solvents on Ion Mobility

Introduction

The ionic mobility, μ_i, of a species, i, is the velocity, v, of a particle that moves under the influence of an electric field, E, of unit strength:

$$\mu_{\mathrm{i}} = \frac{v_{\mathrm{i}}}{E} \qquad (1)$$

The dimension of the mobility is square meter per volt per second. The values of the mobilities range from more than 300 units for the proton in water to about 5 units for large organic ions in solvents with high viscosities. The mobility depends on the size and shape of the solvated ion, on its charge, and on the viscosity and temperature of the solution. Thus, it is clear that the mobility is a function of the solvent.

Three kinds of mobility can be distinguished: the absolute mobility, μ^0, in an infinitely diluted solution, the actual mobility of the fully charged ion at the ionic strength, I, of the solution, and the effective mobility, μ^{eff}, which depends on the degree of ionization, α.

Ionic Mobilities of Weak Electrolytes

For higher ionic strengths, no quantitative theoretical description of the dependence of the mobility on the ionic strength exists. Even for lower ionic strengths, theory is directed, rather, to spherical ions of 1:1 electrolytes with low charge (Debye–Hückel–Onsager theory, see below) than to multiple charged ions with irregular geometry. An expression derived empirically relates the logarithm of the correction factor for the mobility on the square root of the ionic strength and the charge number of the analyte.

The degree of dissociation, on the other hand, depends, in a well-defined manner, on the pK of the analyte and the pH of the solution, according to the Henderson–Hasselbalch equation. α, is a function of pK_a and pH: for example, for monobasic neutral acids

$$\alpha = \frac{1}{1 + 10^{\mathrm{p}K_a - \mathrm{pH}}} \qquad (2)$$

For this type of acid, the total mobility can be expressed as

$$\mu_i^{\mathrm{tot}} = \mu_i^{\mathrm{eff}} + \mu_{\mathrm{EOF}} = \frac{\mu_i^0 f}{1 + 10^{\mathrm{p}K_{a,i} - \mathrm{pH}}} + \mu_{\mathrm{EOF}} \qquad (3)$$

Here, μ_i^0 is the absolute mobility, that of the ion at infinite dilution, f is the correction factor that takes into account the deviation from ideal behavior. It can be seen that an additional parameter occurs in this equation: the mobility of the electro-osmotic flow, μ_{EOF}, which occurs in many cases in the separation systems and leads to an additional velocity vector of the solutes.

Influence of Organic Solvents

Equation (3) is the key expression that enables understanding of the influence of the solvent on the mobility. From this expression, it follows that solvents may affect the following:

- the absolute mobility
- the correction factor, f
- the ionization constants, K_a, of the analytes and the buffering electrolytes
- the mobility of the EOF

Absolute Mobility

A first approach to take into account the solvent's effect on the absolute mobility of an ion was made by Walden. It is based on the Stokes' law of frictional resistance. Walden's rule states that the product of absolute mobility and solvent viscosity is constant. It is clear that the serious limitation of this model is that it does not consider specific solvation effects, because it is based on the *sphere-in-continuum* model. However, it delivers an appropriate explanation for the fact that, within a given solvent, the mobility depends on temperature to the same extent as the viscosity (in water, for example, the mobility increases by about 2.5% per degree Kelvin). The mobilities do not deviate too much from Walden's rule in some binary mixtures of water with organic solvents. This model is, on the other hand, not appropriate for forecasting or explaining the effect of the solvent on the mobility in a more general manner (see Table 1).

Table 1 Absolute Mobilities and Walden Products of Some Anions in Pure Solvents

	Water		Methanol		Acetonitrile	
Ion	μ_{abs}	$\eta\lambda^0$	μ_{abs}	$\eta\lambda^0$	μ_{abs}	$\eta\lambda^0$
Li^+	40.0	0.34	41.0	0.22	71.8	0.24
K^+	76.2	0.65	54.5	0.29	86.6	0.29
$(CH_3)_4N^+$	46.5	0.40	71.2	0.33	98.0	0.33
$(C_4H_9)_4N^+$	20.2	0.17	40.4	0.21	63.6	0.21
Cl^-	79.2	0.68	54.3	0.28	92.2	0.31
J^-	79.6	0.68	65.1	0.34	105.7	0.35
CH_3COO^-	42.4	0.36	40.8	0.21	110.9	0.37

Note: μ_{abs} is measured in 10^{-5} cm^2/Vs; $\eta\lambda^0$ is measured in $Pcm^2/\Omega mol$ where λ^0 is the single ion conductance at infinite dilution.

Taking specific solvation effects into account makes it clear that the hydrodynamic radius, r_h, of an ion depends on the solvent, due to the different solvation shell established:

$$r_{h,i} = \frac{z_i e_0}{6\pi\eta\mu_i} \tag{4}$$

where e_0 is the electron charge and z_i is the charge number. This expression is valid for a spherical ion. It is obvious that by a simple geometrical argument, this radius changes for an ion with crystal radius, r_{cryst}, according to solvation by n_h molecules of solvent, S, with radius, r_S, in the solvation shell:

$$r_h^3 = r_{cryst}^3 + n_h r_S^3 \tag{5}$$

Because n_h and r_S may change, the hydrodynamic radius and, therefore, the absolute mobility change as well.

Correction Factor

The correction factor, f, relates the actual mobility of a fully charged particle at the ionic strength under the experimental conditions to the absolute mobility. It takes ionic interactions into account and is derived for not-too-concentrated solutions by the theory of Debye–Hückel–Onsager using the model of an ionic cloud around a given central ion. It depends, in a complex way on the mean ionic activity coefficient. The resulting equation contains the solvent viscosity and dielectric constant in the denominator. In all cases, the factor is <1. The actual mobility is always smaller than the absolute mobility.

Ionization Constant

Organic solvents influence the ionization constants of weak acids or bases in several ways (note that they influence the analytes and the buffer as well). Concerning ionization equilibria, an important solvent property is the basicity (in comparison to water), which reflects the interaction with the proton. From the most common solvents, the lower alcohols and acetonitrile are less basic than water. Dimethyl sulfoxide is clearly more basic. However, stabilization of all particles involved in the acido-basic equilibrium is decisive for the pK_a shift as well. For neutral acids of type HA, the particles are the free, molecular acid, and the anion, A^-. In the equilibrium of bases, B, stabilization of B and its conjugated acid, HB^+, takes place. As most solvents have a lower stabilization ability toward anions (compared to water), they shift the pK_a values of acids of type HA to higher values in general. No such clear direction of the change is found for the pK_a values of bases; however, they undergo less pronounced shifts.

Mobility of the EOF

According to the Smoluchowsky equation, the mobility of the EOF is

$$\mu_{\mathrm{EOF}} = -\frac{\varepsilon\varepsilon_0\zeta}{\eta} \quad (6)$$

where ε and ε_0 are the permittivity of the medium and the vacuum, respectively, η is the solvent viscosity, and ζ is the zeta potential of the electric double layer. Roughly, the mobility of the EOF should depend on the ratio of relative permittivity to viscosity of the solvent and on the conditions of the electric double layer of the surface of the capillary in capillary electrophoresis (CE). As in most cases, fused-silica capillaries are used in CE, and its ζ potential is a function of the buffer pH, because the silanol groups at the surface are weak acids, with pK values around 5–6. It follows that the solvent affects the pK of the silanol as well, in a manner similar to that for the ionization constants of the analytes and the buffers. Nearly all solvents shift the pK values of the silanol groups toward a higher, pure acetonitrile and to even to very high values.

Suggested Further Reading

Conway, B. E., *Ion Hydration in Chemistry and Biophysics*. Elsevier, Amsterdam, 1981.

Fernandez-Prini, R. and M. Spiro, Conductance and transference numbers, in *Physical Chemistry of Organic Solvent Systems* (A. K. Covington and T. Dickinson, eds.), Plenum, London, 1973, pp. 525–679.

Kenndler, E., Organic solvents in capillary electrophoresis, in *Capillary Electrophoresis Technology* (N. A. Guzman, ed.), Marcel Dekker, Inc., New York, 1993, pp. 161–86.

Schwer, C. and E. Kenndler, Electrophoresis in fused-silica capillaries: The influence of organic solvent on the electroosmotic velocity and the zeta potential, *Anal. Chem. 63*: 1801–1807 (1991).

Ernst Kenndler

Efficiency in Chromatography

One of the most important characteristics of a chromatographic system is the efficiency or the number of theoretical plates, N. The number of theoretical plates can be defined from a chromatogram of a single band as

$$N = \left(\frac{t_R}{\sigma_t}\right)^2 = \frac{L^2}{\sigma_t^2} \quad (1)$$

where, for a Gaussian shaped peak, t_R is the time for elution of the band center, σ_t is the band variance in time units, and L is the column length [1]. N is a dimensionless quantity; it can also be expressed as a function of the band elution volume and variance in volume units:

$$N = \left(\frac{V_R}{\sigma_V}\right)^2 = 5.56\left(\frac{t_R}{W_{1/2}}\right)^2 \quad (2)$$

In a chromatographic system, it is desirable to have a high column plate number. The column plate number increases with several factors [2]:

- Well-packed column
- Longer columns
- Smaller column packing particles
- Lower mobile-phase viscosity and higher temperature
- Smaller sample molecules
- Minimum extracolumn effects

Dr. Rosario LoBrutto coauthored the article ("Efficiency in Chromatography") at Merck Research Laboratories, Rahway, New Jersey.

In an open-bed system, N can be measured from the distance passed by a zone along the bed:

$$N = \left(\frac{d_R}{\sigma_d}\right)^2 \quad (3)$$

where d_R is the distance from the point of sample application to the point of the band center and σ_d is the variance of the band in distance units [1].

In fact, the plate theory describes the movement of a particular zone through the chromatographic bed. As the zone is washed through the first several plates, a highly discontinuous concentration profile is obtained, with the solute being distributed in plates following the Poisson distribution [3]. At an intermediate stage (approximately 30–50 plates), much of the abrupt discontinuity disappears due to a similar concentration of the analyte in the neighboring plates. As the process continues (after 100 plates), the concentration profile is smooth and, even though the distribution is still Poisson, it can be approximated by a Gaussian curve. The standard deviation, σ, of the Gaussian curve, which is a direct measure of the zone spreading, is found to be

$$\sigma = \sqrt{HL} \tag{4}$$

where H is the plate height and L is the distance migrated by the center of the zone. In practice, the plate height is used to describe the zone spreading, including both nonequilibrium and longitudinal effects. In a uniform column, free from concentration and velocity gradients, the plate height is defined as

$$H = \frac{\sigma^2}{L} \tag{5}$$

In a nonuniform column, the zone spreading varies from point to point and its local value is

$$H = \frac{d\sigma^2}{dL} \tag{6}$$

which represents the increment of plate height in the variance σ^2 per unit length of migration. In practice, the smaller the value of H, the smaller the magnitude of band spreading per unit length of the column. The determination of H does not require the measurement of σ^2, as long as N is known. Thus, combining Eqs. (1) and (6) yields [4]

$$H = L\left(\frac{\sigma}{L}\right)^2 = \frac{L}{N} \tag{7}$$

In practice, because the separation in a particular chromatographic column is linked to the time spent by the analyte in the stationary phase and the time spent by the analyte in the mobile phase is irrelevant for the separation, a new parameter is defined (i.e., *effective plate number*, N_{eff}). The effective plate number is related to the separation factor k' and N by

$$N_{\text{eff}} = N\left(\frac{k'}{1 + k'}\right)^2 \tag{8}$$

Similarly, an expression for H_{eff} can be written

$$H_{\text{eff}} = H\left(\frac{1 + k'}{k'}\right)^2 \tag{9}$$

The effective parameters are more meaningful when comparing different columns [4].

There are several major contributions that will influence the band broadening and, consequently, H [5]: eddy diffusion, mobile-phase mass transfer, longitudinal diffusion, stagnant mobile-phase mass transfer, and stationary-phase mass transfer. The effect of each process on the band broadening and, consequently, on the plate height is related to all the experimental variables: mobile-phase velocity, u; particle diameter, d_p; sample diffusion coefficient in the mobile phase, D_m; the thickness of the stationary-phase layer, d_f; and the sample diffusion coefficient in the stationary phase, D_s. In general, H will vary with the velocity of the mobile phase, u, as it travels through the column. In a gas chromatography (GC) system, a plot of u versus H will lead to a curve which has a hyperbolic shape [6], characterized by the equation

$$H = A + \frac{B}{u} + Cu \tag{10}$$

Equation (10) is known as the van Deemter equation, and no correction was made for gas compressibility. Using the reduced parameters $h = H/d_p$ and $v = ud_p/D_m$, Eq. (10) becomes

$$h = a + \frac{b}{v} + cv \tag{11}$$

where A, B, C, a, b, and c are constants for a particular sample compound and set of experimental conditions as the flow rate varies. The B term in Eq. (10) relates to band broadening occurring by diffusion in the gas phase in the longitudinal direction of the column. According to Einstein's equation for diffusion,

$$\sigma^2 = 2D_m t_0 = \frac{2D_m L}{u} \tag{12}$$

Because $H = \sigma^2/L$, the B term becomes

$$B = 2\frac{D_m}{u} \tag{13}$$

The inverse velocity term in Eq. (13) becomes important at low velocities. Because the D_m in liquids is 10^5 times smaller than in gases, the longitudinal term plays no practical role in band broadening in liquid chromatography.

The A term in Eq. (10) describes the nonhomogeneous flow, also called eddy diffusion. In this case,

$$\frac{\sigma^2}{L} = 2\lambda d_p = A \tag{14}$$

where λ is a packing correction factor of $\sim$0.5. In classical GC, the A term is a constant, representing a lower limit on column efficiency, equivalent to $H = d_p$ or $h = 1$.

At velocities above H_{min}, the C term controls H and relates to nonequilibrium resulting from resistance to mass transfer in the stationary and mobile phases [6].

In high-performance liquid chromatography (HPLC), the van Deemter equation still holds. However, Giddings [7] argued that the equation is too simplistic because it ignores the coupling that exists between the flow velocity and the radial diffusion in the void space of the packing around the particles. He suggested replacing the term A by a term $a/(1 + bu^{-1})$ to account for the flow velocity, because both the eddy diffusion and the radial diffusion are responsible for the transfer of the molecules between the different flow paths of unequal velocity. To include the coupling between the laminar flow and the molecular diffusion in porous media, Horvath and Lin [8] introduced a new parameter, δ, which is the thickness of the stagnant film surrounding each stationary-phase particle. However, at high velocities required in HPLC, Horvath and Lin's model reduces to the Knox equation, which is a variation of the van Deemter equation [9]:

$$h = av^{0.33} + \frac{b}{v} + cv \qquad (15)$$

where a, b, and c are empirical parameters related to the analyte and the experimental flow rate conditions.

References

1. B. L. Karger, L. R. Snyder, and Cs. Horvath, *An Introduction to Separation Science*, John Wiley & Sons, New York, 1973, p. 136.
2. L. R. Snyder, J. J. Kirkland, and J. L. Glajch, *Practical HPLC Method Development*, John Wiley & Sons, New York, 1997, p. 42.
3. J. C. Giddings, *Dynamic of Chromatography, Part I, Principles and Theory*, Marcel Dekker, Inc., New York, 1965, p. 23.
4. Cs. Horvath and W. R. Melander, in *Chromatography, Fundamentals and Applications of Chromatographic and Electrophoretic Methods, Part A: Fundamentals and Techniques* (E. Heftmann, ed.), Elsevier Scientific, Amsterdam, 1983, p. A45.
5. L. R. Snyder and J. J. Kirkland, *Introduction to Modern Liquid Chromatography*, 2nd ed., John Wiley & Sons, New York, p. 168.
6. B. L. Karger, *Modern Practice of Liquid Chromatography* (J. J. Kirkland, ed.), Wiley–Interscience, New York, 1971, p. 23.
7. J. C. Giddings, *Dynamics of Chromatography, Part I, Principles and Theory*, Marcel Dekker, Inc., New York, 1965, p. 61.
8. Cs. Horvath and H. J. Lin, *J. Chromatogr., 149*:43 (1978).
9. G. Guiochon, S. G. Shirazi, and A. M. Katti, *Fundamentals of Preparative and Nonlinear Chromatography*, Academic Press, Boston, 1994, p. 201.

Nelu Grinberg
Rosario LoBrutto

Electrochemical Detection

Introduction

With respect to chromatography, "electrochemical detection" means amperometric detection. Amperometry is the measurement of electrolysis current versus time at a controlled electrode potential. It has a relationship to voltammetry similar to the relationship of an ultraviolet (UV) detector to spectroscopy. Whereas conductometric detection is used in ion chromatography, potentiometric detection is never used in routine practice. Electrochemical detection has even been used in gas chromatography in a few unusual circumstances. It has even been attempted with thin-layer chromatography (TLC). Its practical success has only been with liquid chromatography (LC) and that will be the focus here.

Most chemists remember electrochemistry as a difficult subject they heard about in physical chemistry courses and they regard it as having something to do with batteries. Both of these impressions are true! What is important here is to understand that (a) redox reactions can be made to occur at surfaces (electrodes) and (b) amazingly enough, such reactions are not just the fate of metals ($Fe^{3+} \rightarrow Fe^{2+}$) but actually occur quite widely among organic compounds of interest such as drugs, pesticides, explosives, food additives, neurotransmitters, DNA, and so forth. There are good references for the novice wishing to understand the analytical electrochemistry of organic substances [1]. A few common examples are presented for both oxidations (electrons are lost, the process is anodic) and reductions (electrons are gained by the analyte, the process is cathodic):

HO, HO (catechol ring with R) ⟶ O, O (quinone ring with R) + 2H⁺ + 2e

$$2RSH \longrightarrow RSSR + 2H^+ + 2e$$

NO_2 (ring with R) $+ 4e + 4H^+ \longrightarrow$ HNOH (ring) $+ H_2O$

O, O (quinone ring with R) $+ 2e + 2H^+ \longrightarrow$ OH, OH (hydroquinone ring)

Although we all remember (or try not to) the confusing math associated with electrochemistry and thermodynamics, all we need here is an appreciation of the fact that the current, i, is proportional to the moles, N, of analyte reacted per unit time. The latter is proportional to concentration at a constant flow rate through a detector cell. The key equation is

$$i = \frac{dQ}{dt} = nF\frac{dN}{dt}$$

where Q is the amount of electricity (charge in coulombs), n is the number of electrons, and F is the Faraday constant. As one can see from the above examples, most organic analytes are involved in reactions where $n = 1, 2,$ or 4. To use an electrochemical detector, it is very important to know that i is proportional to the concentration and the amount injected, just as UV absorbance is proportional to the concentration or the amount injected.

Liquid chromatography–electrochemistry (LC–EC) is now over 30 years old [1]. In recent years, an emphasis has been placed on miniaturizing the technology to accommodate the study of smaller biological samples, often with a total available volume of only a few microliters. Both liquid chromatography and electrochemistry are largely controlled by surface science. Considering this fact, both technologies benefit from reducing the distance from the bulk of the solution phase to the surface. In LC, this is accomplished by using smaller-diameter stationary-phase particles. In electrochemistry, it is accomplished by using packed-bed or porous electrodes and/or thin-layer cells with greatly restricted diffusion pathways.

For analytical purposes, there is no loss in the concentration detection limit by reducing the total surface area available in both methodologies. In LC, this reduction is accomplished by using smaller-diameter columns, and in EC, by using smaller electrodes. With LC column diameters of 0.1–1.0 mm and radial flow thin-layer cells with dead volumes of a few tens of nanoliters, it is possible to build analytical instruments capable of routine use by neuroscientists, drug metabolism groups, and pharmacokinetics experts. LC–EC has been used for foods, industrial chemicals, and environmental work. Nevertheless, biomedical applications have dominated. It shows no potential for preparative chromatography and is generally used when nanograms or picograms hold some appeal.

Detector Cells

A wide variety of detector cells has been used for LC–EC [1]. The choice can be baffling to a nonexpert. These all "work" to some degree. The key issues are as follows:

1. An electrode (the "working electrode") exposed to the mobile phase in a dead volume (small) appropriate to the column diameter chosen
2. A place to locate at least one other electrode (a "counter electrode") or preferentially two (an auxiliary electrode and a reference electrode)
3. The possibility for a choice of different working electrode materials (see the following section)
4. The possibility of multiple channels in series or parallel

Figure 1 is representative of one choice that meets these criteria. Such a cell is normally described as a "thin-layer

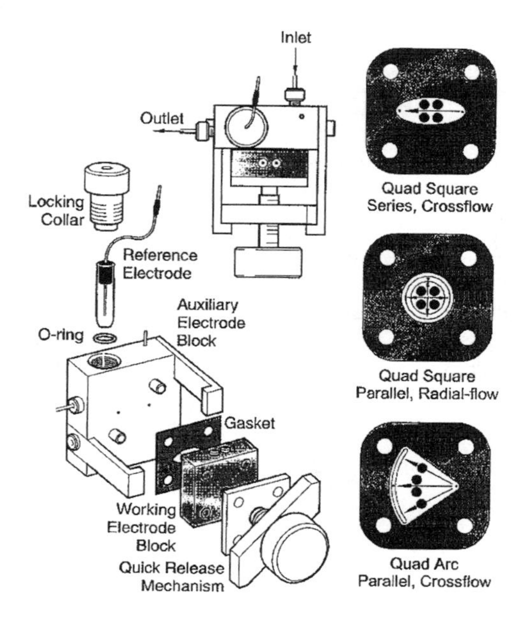

Fig. 1 One example of a sandwich-type thin-layer LC–EC detector with adjustable dead volume, flow pattern, and up to four channels. [Reprinted with permission from *Current Separations* 18: 114 (2000).]

sandwich configuration." The working electrode(s) is in the form of an interchangeable block. Electrodes of different sizes, shapes, or materials can be accommodated with a flow pattern established by a gasket shape and thickness. Such cells can easily be adapted for LC flow rates of from 5 to 5000 μL/min. Different designs are used for capillary separation tools such as capillary electrophoresis (CE).

Electrode Materials

The most common electrode material used in LC–EC is carbon, either as solid "glassy carbon" disks in thin-layer cells, or as a high-surface-area porous matrix through which the mobile phase can flow. Gold electrodes are useful to support a mercury film and these are primarily used to determine thiols and disulfides, and also for carbohydrates using pulsed electrochemical detection (PED) with high-pH mobile phases. Platinum electrodes are occasionally useful for specific analytes, but they are most frequently employed to determine hydrogen peroxide following an oxidase immobilized enzyme reactor (IMER). More recently, copper electrodes have begun to attract serious interest for the determination of carbohydrates in basic mobile phases. Glassy carbon is the overwhelming favorite choice because of its wide range of applicable potentials and its rugged convenience. Bulk glassy carbon is difficult to use in geometries other than disks and plates. There are a number of other geometries that have practical interest for multiple-electrode detectors. One of the more valuable recent contributions to LC–EC derives from the ability to deposit conducting vitreous carbon films on silicon or quartz substrates using lithography techniques. The lithography technology makes it possible to lay down a variety of electrode geometries, which could not possibly be manufactured in small sizes by traditional machining. Although such electrodes are still at the research stage, they show considerable promise. Detector cells with 2, 4, or even 16 electrodes are commercially available. There are obvious parallels with diode-array detection (DAD). When two electrodes are used in series, there are similarities to fluorescence or mass spectrometry–mass spectrometry (MS–MS) in the way that selectivity is often enhanced.

Pulsed Electrochemical Detection

There are many substances which would appear to be good candidates for LC–EC from a thermodynamic point of view but which do not behave well due to kinetic limitations. Johnson and co-workers at Iowa State University used some fundamental ideas about electrocatalysis to revolutionize the determination of carbohydrates, nearly intractable substances which do not readily lend themselves to ultraviolet absorption (LC–UV), fluorescence (LC–F), or traditional DC amperometry (LC–EC) [2]. At the time that this work began, the LC of carbohydrates was more or less relegated to refractive index detection (LC–RI) of microgram amounts. The importance of polysaccharides and glycoproteins, as well as traditional sugars, has focused a lot of attention on pulsed electrochemical detection (PED) methodology. The detection limits are not competitive with DC amperometry of more easily oxidized substances such as phenols and aromatic amines; however, they are far superior to optical detection approaches.

Postcolumn Reactions

Electrochemical detection is inherently a "chemical" rather than a "physical" technique (such as ultraviolet, infrared, fluorescence, or refractive index). It is, therefore, not surprising to find that many imaginative postcolumn reactions have been coupled to LC–EC. These include photochemical reactions, enzymatic reactions, halogenation reactions, and Biuret reactions. In each case, the purpose is to enhance selectivity and therefore improve limits of detection. While simplicity is sacrificed with such schemes, there are many published methods that have been quite successful.

Capillary Electrophoresis and Capillary Electrochromatography

Because there is LC–EC, it is only logical that there should be CE–EC and capillary electrochromatography (CEC)–EC. This area was pioneered by Andrew Ewing at the Pennsylvania State University. Richard Zare (Stanford University) and Susan Lunte (Kansas University) have explored this idea in a number of unique ways. The basic technology has been recently reviewed [3]. There are several fundamental problems that do not occur with LC–EC. First, the capillaries must be of small diameter to properly dissipate resistive heating. Thus, the electrodes used in CE–EC are normally carbon fibers or metallic wires placed in or at the capillary end. Second, the electrical current through the capillary which establishes the electro-osmotic pumping is much larger than the electrolysis current measured in determining analytes of interest. The ionic and electrolytic currents need to be "decoupled" in some way. A third concern is that the flow rate in CE or CEC is not independent of the choice of "mobile phase" or even the sample, whereas in LC, it is easily predetermined and maintained by a volume displacement pump. In spite of these concerns, CE is very attractive because of its high resolution per unit time and the small sample volumes required. In the case of CEEC, the concentration detection limits are frequently superior to optical detectors for suitable analytes. This is because electrochemical detection is a surface (not volume)-dependent technique. In the grand scheme of things, at this writing, CE and CEC are very rarely used instead of LC and, therefore, CE–EC and CEC–EC must be considered academic curiosities until this situation changes.

Conclusions

Electrochemical detection has matured considerably in recent years and is routinely used by many laboratories, often for a very specific biomedical application. The most popular applications include acetylcholine, serotonin, catecholamines, thiols and disulfides, phenols, aromatic amines, macrocyclic antibiotics, ascorbic acid, nitro compounds, hydroxylamines, and carbohydrates. As the last century concluded, it is fair to say that many applications for which LC–EC would be an obvious choice are now pursued with LC–MS–MS. This only became practical in the 1990s and is clearly a more general method applicable to a wider variety of substances. In a similar fashion, LC–MS–MS has also largely supplanted LC–F for new bioanalytical methods. Nevertheless, there remain a number of key applications for these more traditional detectors known for their selectivity (and therefore excellent detection limits).

References

1. P. T. Kissinger and W. R. Heineman (eds.), *Laboratory Techniques in Electroanalytical Chemistry*, 2nd ed., Marcel Dekker, Inc., New York, 1996.
2. W. R. LaCourse, *Pulsed Electrochemical Detection in High-Performance Liquid Chromatography*, John Wiley & Sons, New York, 1997.
3. L. A. Holland and S. M. Lunte, Capillary electrophoresis coupled to electrochemical detection: A review of recent advances, *Anal. Commun. 35*: 1H–4H (1998).

Peter T. Kissinger

Electrochemical Detection in CE

Capillary electrophoresis (CE) is a powerful separation tool which has its primary strength in the high separation efficiency and short analysis times. By decreasing the internal diameter (i.d.) of the capillaries used, the situation can be further improved due to the possibility of using higher separation voltages. Such a miniaturization, however, often involves a challenge regarding how the detection is to be made in the narrow capillaries for sample volumes in the nanoliter to sub-picoliter range. Electrochemical (EC) methods, usually based on the use of microelectrodes, are relatively inexpensive and are readily miniaturized and adapted to such low volumes and capillary sizes without loss of performance. Electrochemical detection is based on the monitoring of changes in an electrical signal due to a chemical system at an electrode surface, usually as a result of an imposed potential or current. The principles, advantages, and drawbacks of currently used EC methods will be discussed briefly below.

In a solution, the equilibrium concentrations of the reduced and oxidized forms of a redox couple are linked to the potential (E) via the Nernst equation

$$E = E^{0'} + \frac{RT}{nF} \ln\left(\frac{c(\text{ox})}{c(\text{red})}\right) \quad (1)$$

with $E^{0'}$ the standard potential and $c(\text{ox})$ and $c(\text{red})$ the concentration of the oxidized and reduced forms, respectively; the other symbols have their usual meaning.

In electrochemical detection, the potential of a working electrode can be measured versus a reference electrode, usually while no net current is flowing between the electrodes. This type of detection is referred to as "potentiometry." Alternatively, a potential is applied to the working electrode with respect to the reference electrode while the generated oxidation or reduction current is measured. This technique is referred to as "amperometry."

When applying a negative potential to the working electrode, the energy of the electrons in the electrode is increased and, eventually, an electron can be transferred to the lowest unoccupied level of a species in the nearby solution. This species is thus reduced; vice versa, species can be oxidized by applying a sufficiently high positive potential. In both cases, the generated current (i) can be expressed by

$$i = -aFDnc\delta_N^{-1} \tag{2}$$

with a being the electrode area, D the diffusion coefficient, δ_N the thickness of diffusion layer; the other symbols have their usual meaning. For each redox couple, there exists a potential, the standard potential E^0, for which the reduced and oxidized forms are present in equal concentrations. By applying a potential more positive than E^0, the concentration of the reduced form is forced to decrease at the electrode surface while the concentration of the oxidized form increases. This process is the cause of the current measured in amperometric techniques. By choosing the applied potential, it is also possible to discriminate between different analytes. The range of potentials that can be applied in amperometric detection is, however, generally limited by redox processes involving the solvent [e.g., the oxidative and reductive evolution of oxygen ($2H_2O \rightarrow O_2 + 4H^+ + 4e^-$) and hydrogen ($H_2O + e^- \rightarrow 0.5\ H_2 + OH^-$), respectively, in water]. A wide range of physiologically and pharmacologically important substances, as well as many heavy metals, transition metals, and their complexes, exhibit standard potentials within this accessible potential range. In fact, many metabolic pathways involve redox processes taking place in aqueous systems. Neurotransmitters of the catechol type (*o*-dihydroxy benzene derivatives) were consequently among the first reported analytes for electrochemical detection in CE (CE–EC). Detection limits down to 10^{-9} mol/L can be achieved in this way.

The choice of working electrode material is an important factor in amperometric detection. For catechols and similar substances, such as phenolic acids, electrodes made of glassy carbon have shown good performances. Other good detectable and biological important substances include thiols and disulfides (e.g., cysteine, glutathione, and their disulfides which are best detected on an Au/Hg amalgam electrode); amino acids and peptides, which can be detected using Cu electrodes, and carbohydrates, glycopeptides, and nucleotides, detected on Au, Cu, or Ni electrodes.

Commonly in amperometric detection, a fiber or disk microelectrode is used where the electrode is positioned in or close to the outlet of the capillary (see Fig. 1). A complication when working with EC in CE is the need for careful alignment of the electrode(s) and capillary outlet. This alignment, which mostly is carried out with micromanipulators under a microscope, is essential to ensure both a good sensitivity and reproducibility and, hence, constitutes the main challenge while adapting EC for routine CE. Another complication in CEEC involves the interference of the high-voltage (HV) separation field on the EC detection. In the first combination of EC and CE, it was assumed that the HV field had to be totally removed from the detection area. This was done by various kinds of decouplers, which unfortunately also introduced additional band broadening and decreased sensitivities. Furthermore, the manufacturing of the decouplers requires considerable labor-intensive experience and skill.

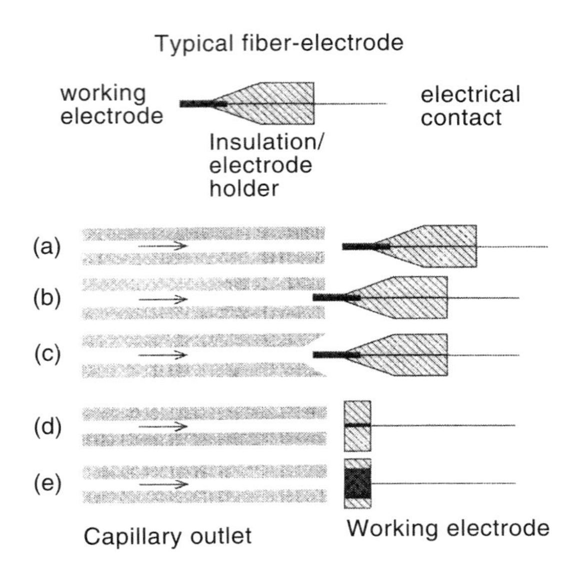

Fig. 1 Electrode setups for CE–EC with fiber microelectrodes: (a) end column, (b) on-column, (c) improved on-column, (d) wall tube, and (e) wall-jet detection.

A later approach is based on the utilization of small-inner-diameter (<25 μm) capillaries to reduce the influence of the HV field.

In potentiometry, all ions present in the solution principally contribute to the potential of the working electrode. As the ratio between the analyte concentration and that of other species in the solution generally is rather low, the analyte contribution to the detector signal is often low, which results in relatively poor detection limits. To circumvent this problem, ion-selective membranes (ISM), which permit only some ions to pass through the membranes, are commonly employed. In this way, detection limits down to 10^{-7} mol/L can be achieved. The ISM also reduces the influence from matrix components, which allows measurements in complex matrices such as blood or serum without interferences. The long-term stability of these electrode may, however, be a problem, as the electrodes might have to be replaced after a few hours or days. Common analytes are inorganic anions and cations, especially alkali and alkaline earth metals ions. A further application is the indirect detection of amino acids, where the complexing of amino acids with Cu^+ ions selectively alters the potential of a copper electrode.

In conductometry, two working electrodes placed either in or at the end of the capillary, along the capillary axis, are commonly employed. A high-frequency AC potential is applied between the working electrodes and the conductance (L) of the solution is continually monitored. In this way, the passing of any zone deviating in its ion composition from the background is detected. As the ion mobilities contribute to the magnitude of the signal as seen in Eq. (3), slowly moving large molecules and low charged biomolecules are less straightforwardly detected, whereas detection limits down to some hundred parts per thousand have been reported for small inorganic and organic ions:

$$L = \frac{FA\Sigma i|z_i|u_i c_i}{l} \tag{3}$$

with A as the cross-sectional area perpendicular to the AC field, l the electrode distance, and u the mobility. Applications have been described for ions up to a size of sulfate or Cd^+ and MES or benzylamine, respectively.

Suggested Further Reading

The field of electrochemical detection in CE have been extensively reviewed in Refs. 1–3. Instructive applications can be found for amperometry in Ref. 4, for potentiometry in Ref. 5, and for conductometry in Ref. 6. An example of miniaturized on-chip EC–CE is given in Ref. 7. The theoretical aspects of electrochemical detection have been discussed in detail in Ref. 8.

References

1. P. D. Voegel and R. P. Baldwin, *Electrophoresis, 18*(12–13): 2267–2278 (1997).
2. S. M. Lunte, et al., *Pharmaceut. Res., 14*(4): 372–387 (1997).
3. T. Kappes and P. C. Hauser, *J. Chromatogr. A 834*(1–2): 89–101 (1999).
4. L. A. Holland and S. M. Lunte, *Anal. Chem. 71*(2): 407 (1999).
5. T. Kappes and P. C. Hauser, *Anal. Chim. Acta 354*: 129–134 (1997).
6. C. Haber, et al., *J. Capillary Electrophoresis 3*(1): 1–11 (1996).
7. A. T. Woolley, K. Lao, A. N. Glazer, and R. A. Mathies, *Anal. Chem. 70*(4): 684 (1998).
8. C. M. A. Brett and A. M. Oliveira Brett, *Electrochemistry*, Oxford University Press, Oxford, 1993.

Oliver Klett

Electrokinetic Chromatography Including Micellar Electrokinetic Chromatography

Introduction

Separation science technology has provided the analyst with numerous methods for quantitative determinations, the more established being high-performance liquid chromatography (HPLC) and gas chromatography (GC). With huge advancements in computer technology, extremely sensitive methods have been made available to the user such as tandem mass spectrometry (MS–MS), liquid chromatography–mass spectrometry (LC–MS), and gas chromatography–mass spectrometry (GC–MS). However, an electrophoretic technique which has been developed through joint efforts from a number of scientific disciplines is rapidly generating interest for its wide applicability and highly sensitive assays. The field of capillary electrophoresis (CE) borrows principles from conventional electrophoresis, liquid chromatography, and gas chromatography. High-performance capillary electrophoresis (HPCE) refers to all techniques that have been developed on the subject. In this article, the electrokinetic chromatographic (EKC) analysis method will be discussed, with emphasis on micellar electrokinetic chromatography (MEKC). Before doing so, some fundamental principles of capillary electrophoresis will be discussed.

Electrophoretic and Electro-osmotic Migration

Capillary zone electrophoresis (CZE) [1] is a basic mode of HPCE and serves as a good starting point for laying the

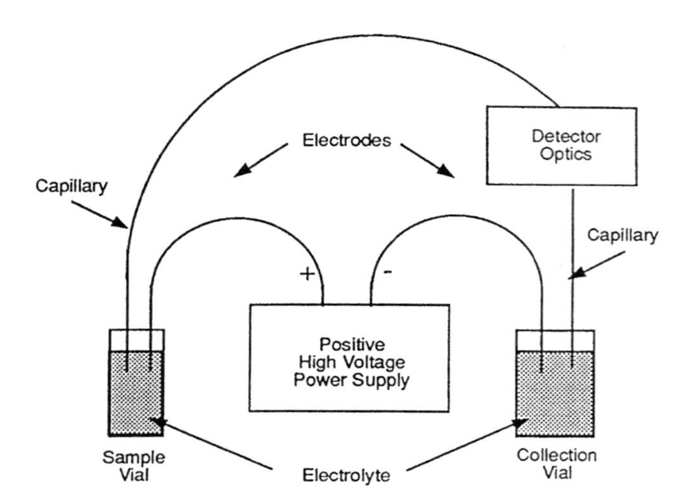

Fig. 1 Diagram of a CE system. [Reprinted from Waters Quanta 4000E Capillary Electrophoresis System Operator's Manual, Waters Corp., 1993.]

background information on EKC and MEKC. Only ionic or charged compounds are separated, based on their differential eletrophoretic mobilities. Figure 1 illustrates the setup for a CE system. In summary, electrolyte buffer solutions and electrodes are present at both ends of the open-tube fused-silica capillary. A positive high-voltage power supply is the source of current. A sample is injected hydrostatically or electrokinetically at the positive end of the capillary (capillary head) and, in the presence of an applied voltage potential, moves toward the negative electrode, where it is detected by an ultraviolet (UV) absorbance instrument. The signal produced is recorded as a chromatogram where sample components are identified as chromatographic peaks according to their retention times, and peak areas or heights are calculated for quantitative purposes.

During analyte migration, electrophoresis and electro-osmosis are taking place. The velocity (v_s, cm/s) and mobility (μ_s, cm^2/V s) of the solute are defined by the following equations:

$$v_s = v_{eo} + v_{ep} \quad (1)$$

$$\mu_s = \mu_{eo} + \mu_{ep} \quad (2)$$

where v_{eo} and μ_{eo} are electro-osmotic velocity and mobility, respectively, and v_{ep} and μ_{ep} are electrophoretic velocity and mobility, respectively. The relationship between velocity and mobility is given by

$$v = \mu E \quad (3)$$

where E (V/cm) is the electric field strength. Because E is constant for all solutes in a separation analysis, the solute velocities are differentiated by their mobilities.

Electrophoresis is an electrokinetic phenomenon whereby charged compounds in an electric field move through a continuous medium and separate by preferentially obtaining different electrophoretic mobilities according to their charges and sizes. Cations move toward the negative electrode (cathode) and anions move toward the positive electrode (anode).

Electro-osmosis is created by the electric double-layer effect [2]. A fused-silica capillary at neutral pH attains fixed negative charges at the inner wall surface as its silanol groups undergo ionization. A layer of hydrated cations will form adjacent to the inner wall to counter the fixed negative charges. An applied voltage potential will cause the positively charged layer to migrate toward the cathode with a flat velocity profile, simultaneously dragging the bulk solution inside the capillary and transporting charged compounds at the electro-osmotic flow velocity. It is assumed that v_{eo} is faster than v_{ep} and determines the direction of solute migration. Hence, a cation will have its sum total [Eq. (1)] greater than the individual component velocities ($v_s > v_{eo}$) and the anion will migrate slower than the electro-osmotic flow ($v_s < v_{eo}$).

Electro-osmosis need not be present in open-tube CE. Coatings exist that can be applied to the capillary surface to eliminate the electrical double layer. However, electro-osmotic flow can be used to reduce solute retention times, which can be advantageous for certain analyses.

As previously mentioned, electrophoretic separations using open-tube capillaries are based on solute differential mobility, which is a function of charge and molecular size. A different approach is required for separating neutral or uncharged compounds. Because charge is absent, electrophoretic mobility is zero. Electro-osmotic flow would allow them to migrate, but their velocities would be equal. Separation would not be possible with the above method.

Electrokinetic Chromatography

Terabe [3] developed a method that separates neutral or uncharged compounds; he named it electrokinetic chromatography. The experimental design is that of capillary zone electrophoresis (Fig. 1). The difference lies in the separation principle. In liquid chromatography, a solute freely distributes itself between two phases [i.e., a mobile phase usually made of a mixture of aqueous and organic solvents and a stationary phase (a solid material packed in a steel housing known as a chromatographic column)]. Under high pressure, the mobile phase is delivered by a liquid chromatographic pump and continuously solvates the stationary phase, thereby transporting nonvolatile compounds of interest that are introduced into the system via chromatographic injection. Separation is based on their phase-distribution profiles. EKC follows the above principle but uses electro-osmosis and electrophoresis to displace analytes and "chromatographic phases" in capillaries.

The electrolyte buffer solution is analogous to the mobile phase. A charged substance, referred to as the carrier, is dissolved in the electrolyte buffer. The neutral solute present in the separation medium will partition itself between the carrier (incorporated form) and the surrounding solution (free form). The carrier ("chromatographic phase") corresponds to the stationary phase in conventional chromatography with modifications, in that it is not a fixed support and exists homogeneously in solution. For this reason, the carrier is called the "pseudo-stationary phase." As discussed in the previous section, the charged carrier will transport the incorporated solute electrophoretically (here, v_{ep} is the carrier velocity) at a slower velocity than the free solute migrating with the electro-osmotic flow velocity (v_{eo}) in the opposite direction. The point to keep in mind is that the carrier migrates with a different velocity than the bulk solution. The variation of the ratio of the amount of incorporated solute to the amount of total solute between separands in a sample mixture will lead to sample component separation.

Micelles in Electrokinetic Chromatography

Different types of EKC have been developed. Cyclodextrins (CDEKC) have been used to form inclusion complexes with solutes to effect their separation. Other examples of EKC include microemulsion electrokinetic chromatography (MEEKC). The MEKC technique (for a detailed treatise, the reader is referred to Ref. 4) utilizes the presence of micelles in the electrolyte buffer solution to influence the migration time of solutes. In this case, the separation carrier is the micelle [5].

Surfactants produce micelles. Their amphophilic nature classifies them as detergents, surface-active agents that are composed of a hydrophilic group and a hydrophobic hydrocarbon chain. In addition to what is known as the critical micelle concentration (CMC), individual surfactant molecules (monomers) interact with each other to form aggregates or micelles, establishing a state of equilibrium between a constant monomer concentration and a rapidly increasing micelle concentration.

As shown in Fig. 2a, micelles are depicted as "round-like" structures with their polar moieties exteriorly located in the vicinity of the aqueous medium and their hydrophobic tails oriented inward forming a cavity. The sizes and shapes of the structures formed when monomer units aggregate is affected by electrolyte concentration, pH, temperature, and hydrocarbon chain length. During aggregation, interactions occur not only between the aggregates but also among the monomer units within the aggregate structure. Monomer unit distribution (the number of surfactant molecules in an aggregate) is characteristic of the surfactant used. Figure 2 displays several forms of aggregates that exist in solution and Table 1 gives examples of surfactants.

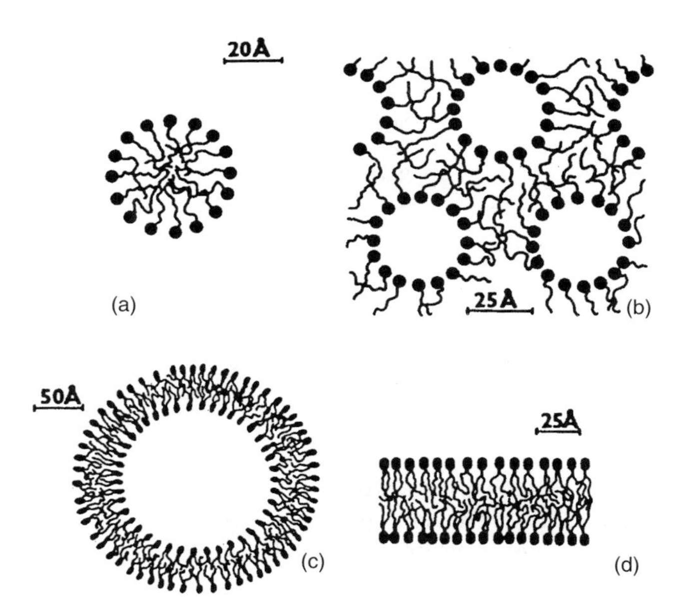

Fig. 2 Examples of aggregate structures of surfactants in solution: (a) micelle; (b) inverted micelles; (c) bilayer vesicle; (d) bilayer. (From Ref. 7.)

Basically, MEKC is an EKC application with the micelle as the designated carrier. A surfactant at a concentration above the CMC is added to the running buffer and initiates micelle formation. Because the separation principle has already been dealt with and the flow scheme in Fig. 3 is an illustrative summary notated for MEKC, it is clear that a neutral analyte residing in the hydrophobic interior of a micelle (depicted as a sphere in Fig. 3) will be transported with the micelle's velocity (v_{mc}). The free analyte will migrate with the electro-osmotic flow velocity (v_{eo}).

The chromatographic aspect (solute partitioning) of the separation [6] can be explained in terms of a commonly used parameter in chromatography, the retention or capacity factor (k'). We begin with the following equation:

$$k' = \frac{n_{mc}}{n_{aq}} \tag{4}$$

Table 1 Some Common Surface-Active Agents

Anionic	
Sodium stearate	$CH_3(CH_2)_{16}COO^-Na^+$
Sodium oleate	$CH_3(CH_2)_7CH{=}CH(CH_2)_7COO^-Na^+$
Sodium dodecyl sulfate	$CH_3(CH_2)_{11}SO_4^-Na^+$
Sodium dodecyl benzene sulfonate	$CH_3(CH_2)_{11} \cdot C_6H_4 \cdot SO_3^-Na^+$
Cationic	
Laurylamine hydrochloride	$CH_3(CH_2)_{11}NH_3^+Cl^-$
Cetyltrimethylammonium bromide	$CH_3(CH_2)_{15}N(CH_3)_3^+Br^-$
Nonionic	
Polyethylene oxides	$CH_3(CH_2)_7 \cdot C_6H_4 \cdot (O \cdot CH_2 \cdot CH_2)_8OH$

Source: Ref. 9.

where n_{mc} and n_{aq} are the mole amounts of the analyte in the micellar and aqueous phases, respectively. The corresponding mole fractions are given by

$$\frac{n_{mc}}{n_{mc} + n_{aq}} \quad \text{and} \quad \frac{n_{aq}}{n_{mc} + n_{aq}}$$

where $n_{mc} + n_{aq}$ is the total amount of analyte present in the electrolyte buffer. The relationship in Eq. (4) and appropriate substitutions transform the above ratios into $k'/(1 + k')$ for the micelle analyte mole fraction and $1/(1 + k')$ for the aqueous analyte mole fraction. The total analyte velocity (v_s) takes the form

$$v_s = \frac{1}{1 + k'}v_{eo} + \frac{k'}{1 + k'}v_{mc} \tag{5}$$

where $[1/(1 + k')]v_{eo}$ represents the velocity of the analyte mole fraction in the aqueous phase and $[k'/(1 + k')]v_{mc}$ is the velocity of the analyte mole fraction in the micellar phase.

Because velocity is a function of length (the capillary length from the point of injection to the detector cell) over time ($v = l/t$), we can substitute and rearrange the terms in Eq. (5) to obtain a relationship between the migration time (t_R) of the analyte and k':

$$T_R = \frac{1 + k'}{1 + (t_0/t_{mc})k'}t_0 \tag{6}$$

$$k' = \frac{t_R - t_0}{(1 - t_R/t_{mc})t_0}\left[\left(\frac{1 - t_R}{t_{mc}}\right)t_0\right]^{-1} \tag{7}$$

Figure 4a is a snapshot of the capillary tube following a sample injection at its positive end (inj.). The micelle, neutral solute, and aqueous solution (water) migrate toward the negative electrode (det.), establishing zones depicted by vertical bands, as they separate inside the capillary. The corresponding chromatogram in Fig.4b shows the migration order where the t_R value for a neutral analyte is range bound between the migration times of the micelle (t_{mc}) and water (t_0). This limitation is reflected in the denominator of Eq. (6). When t_{mc} approaches infinity, the micelle

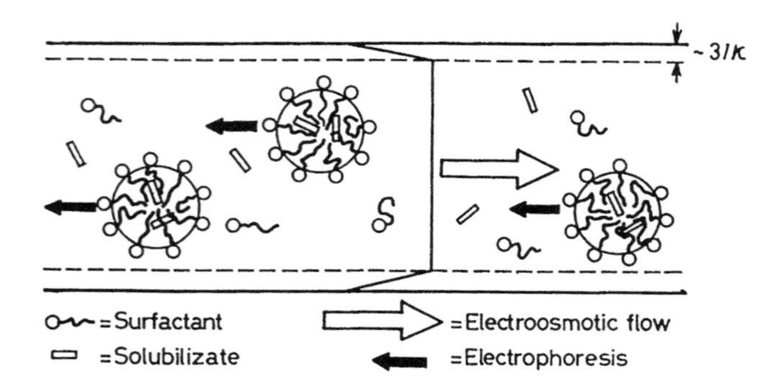

Fig. 3 Schematics of the separation principle of MEKC. (From Ref. 6.)

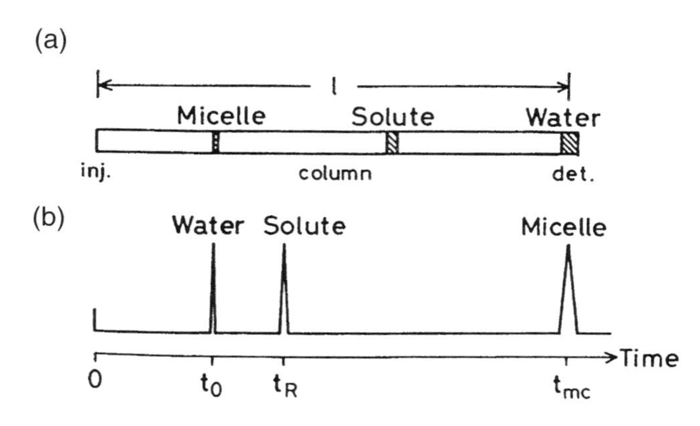

Fig. 4 (a) Representation of zone migration inside the capillary tube and (b) the corresponding chromatogram. (From Ref. 8.)

is assumed to be stationary. The ratio t_0/t_{mc} will become zero and Eq. (6) turns into

$$t_R = 1 + k' \quad (8)$$

In this case, a neutral solute completely solubilized within the micelle will have a t_R value approaching infinity as its k' value does the same. Solute elution is assumed not to occur. Similarly, the free neutral analyte is unretained and its k' is equal to zero. Therefore, the t_R value will be equal to t_0 [Eq. (8)]. This explains why the neutral solute can migrate no slower than the micelle (t_{mc}) and no faster than the aqueous solution (t_0).

Although the above discussion focuses on neutral analyte separation, MEKC can be applied to ionic species which have their own electrophoretic mobilities and a broader migration time range.

Conclusion

The purpose of this article was to provide the reader with a basic understanding of capillary electrophoresis and to describe how a technique such as MEKC uses basic principles of chromatography to perform separations which are not possible electrophoretically. As the applications for electrokinetic chromatography rapidly expand, the future direction will develop on two fronts:

1. The development of novel separation carriers that will broaden the species range of separable analytes. EKC is suitable for separating small molecules, considering the size of the cavities of the established carriers.
2. The scope for further partition mechanisms, as new separation carriers are discovered, is promising. The separation principle is basic chromatography and with research efforts introducing new carriers in the pipeline, modified versions of the separation mechanism are possible.

Its rapid analysis time, low sample and solvent volume requirements, high resolution, and selectivity will continue to attract researchers who are involved in separation analysis.

References

1. B. J. Radola (ed.), *Capillary Zone Electrophoresis*, VCH, Weinheim, 1993.
2. B. L. Karger and F. Foret, Capillary electrophoresis: Introduction and assessment, in *Capillary Electrophoresis Technology* (N. A. Guzman, ed.), Marcel Dekker, Inc., New York, 1993, pp. 3–64.
3. S. Terabe, *Trends Anal. Chem. 8*: 129 (1989).
4. P. Muijselaar, Micellar electrokinetic chromatography: Fundamentals and applications, Ph.D. thesis, Eindhoven University of Technology, Eindhoven, The Netherlands, 1996.
5. F. Foret, L. Kivánková, and P. Boek, Principles of capillary electrophoretic techniques: Micellar electrokinetic chromatography, in *Capillary Zone Electrophoresis* (B. J. Radola, ed.), VCH, Weinheim, 1993, pp. 67–74.
6. S. Terabe, Micellar electrokinetic chromatography, in *Capillary Electrophoresis Technology* (N. A. Guzman, ed.), Marcel Dekker, Inc., New York, 1993, pp. 65–87.
7. J. N. Israelchvilli, in *Physics of Amphiphiles: Micelles, Vesicles and Microemulsions* (V. Degiorgio and M. Corti, eds.), Proceedings of the International School of Physics "Enrico Fermi," CourseXC, North-Holland, Amsterdam, 1985, pp. 24–37.
8. S. Terabe, K. Otsuka, and T. Ando, *Anal. Chem. 57*: 834 (1985).
9. D. J. Shaw, *Introduction to Colloid and Surface Chemistry*, Butterworths, London, 1966, pp. 57–72.

Hassan Y. Aboul-Enein
Vince Serignese

Electron-Capture Detector

The electron-capture detector (ECD) is probably the most sensitive GC detector presently available. However, like most high-sensitivity detectors, it is also very specific and will only sense those substances that are electron capturing (e.g., *halogenated* substances, particularly fluorinated materials). The ECD detector was invented by Lovelock [1] and functions on an entirely different principle from that of the argon detector. A low-energy β-ray source is used in the sensor to produce electrons and ions. The first source to be used was tritium absorbed onto a silver foil, but, due to its relative instability at high temperatures, this was replaced by the far more thermally stable ^{63}Ni source.

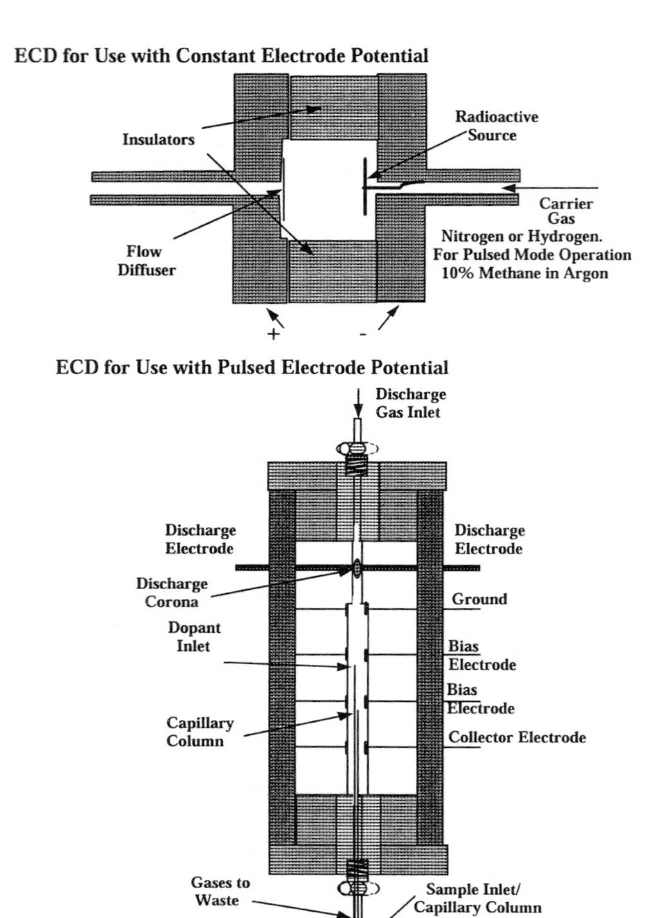

Fig. 1 The two types of electron capture detector. (Courtesy of Valco Instruments Company, Inc.)

The detector can be made to function in two ways: either a constant potential is applied across the sensor electrodes (the DC mode) or a pulsed potential is used (the pulsed mode).

A diagram of the ECD is shown in Fig. 1. In the DC mode, a constant electrode potential (a few volts) is employed that is just sufficient to collect all the electrons that are produced and provide a small standing current. If an electron-capturing molecule (e.g., a molecule containing a halogen atom which has only seven electrons in its outer shell) enters the sensor, the electrons are captured by the molecules and the molecules become charged. The mobility of the captured electrons are much reduced compared with the free electrons and, furthermore, are more likely to be neutralized by collision with any positive ions that are also generated. As a consequence, the electrode current falls dramatically. In the pulsed mode of operation, which is usually the preferred mode, a mixture of methane in argon is usually employed as the carrier gas. Pure argon cannot be used very effectively as the carrier gas, as the diffusion rate of electrons in argon is 10 times less than that in a 10% methane–90% argon mixture. The period of the pulsed potential is adjusted such that relatively few of the slow negatively charged molecules reach the anode, but the faster moving electrons are all collected. During the "off-period," the electrons reestablish equilibrium with the gas. In general use, the pulse width is set at about 1 μs and the frequency of the pulses at about 1 kHz. This allows about 1 ms for the sensor to reestablish equilibrium in the cell before the next electron collection occurs. The peak potential of each pulse is usually about 30 V but will depend on the geometry of the sensor and the strength of the radioactive source. The average current resulting from the electrons collected at each pulse is about 1×10^{-8} A and usually has an associated noise level of about 5×10^{-12} A. Both the standing current and the noise will also vary with the strength of the radioactive source that is used.

The sensor consists of a small chamber, 1 or 2 mL in volume, with metal ends separated by a suitable insulator. The metal ends act both as electrodes and as fluid conduits for the carrier gas to enter and leave the cell. The cell contains the radioactive source, electrically connected to the conduit through which the carrier gas enters and to the negative side of the power supply. A gauze "diffuser" is connected to the exit of the cell and to the positive side of the power supply. In the pulsed mode, the sensor operates with oxygen-free nitrogen or argon–methane mixtures. The active source is ^{63}Ni, which is stable up to 450°C. The sensor is thermostatted in a separate oven which can be operated at temperatures ranging from 100°C to 350°C. The column is connected to the sensor at the base and makeup gas can be introduced into the base of the detector. If open tubular columns are employed, the columns are operated with hydrogen or helium as the carrier gas. The electron-capture detector is extremely sensitive (i.e., minimum detectable concentration $\sim 1 \times 10^{-13}$ g/mL) and is widely used in trace analysis of halogenated compounds — in particular, pesticides.

In the DC mode, the linear dynamic range is relatively small, perhaps two orders of magnitude, with the response index lying between 0.97 and 1.03. The pulsed mode has a much wider linear dynamic range and values up to five orders of magnitude have been reported. The linear dynamic range will also depend on the strength of the radioactive source and the detector geometry. The values reported will also rest on how the linearity is measured and defined. If a response index lying between 0.98 and 1.02 is assumed, then a linear dynamic range of at least three orders of magnitude should be obtainable from most pulsed-mode electron-capture detectors.

Pulsed-Discharge Electron-Capture Detector

The pulsed-discharge electron-capture detector is a variant of the pulsed ECD detector, a diagram of which is shown in the lower part of Fig. 1. The detector functions in exactly the same way as that of the traditional electron-capture detector but differs in the method of electron production. The sensor consists of two sections: the upper section, where the discharge takes place, has a small diameter and the lower section where the column eluent is sensed and the electron capturing occurs, has a wider diameter. The potential across the discharge electrodes is pulsed at about 3 kHz with a discharge pulse width of about 45 μs for optimum performance. The discharge produces electrons and high-energy photons (which can also produce electrons) and some metastable helium atoms. The helium doped with propane enters just below the second electrode, metastable atoms are removed, and electrons are generated both by the decay of the metastable atoms and by the photons. The electrons are collected by appropriate potentials applied to each electrode in the section between the third and fourth electrode and, finally, collected at the fourth electrode. The collector electrode potential (the potential between the third and fourth electrodes) is pulsed at about 3 kHz with a pulse width of about 23 μs and a pulse height of 30 V.

The device functions in the same way as the conventional electron-capture detector with a radioactive source. The column eluent enters just below the third electrode, any electron-capturing substance present removes some of the free electrons, and the current collected by the fourth electrode falls. The sensitivity claimed for the detector is 0.2–1.0 ng, but this is not very informative as its significance depends on the characteristics of the column used and on the k' of the solute peak on which the measurements were made. The sensitivity should be given as that solute *concentration* that produces a signal equivalent to twice the noise. Such data allow a rational comparison between detectors. The sensitivity or minimum detectable concentration of this detector is probably similar to the conventional pulsed ECD (viz. 1×10^{-13} g/mL). The linear dynamic range appears to be at least three orders of magnitude for a response index of r, where $0.97 < r < 1.03$, but this is an estimate from the published data. The modified form of the electron-capture detector, devoid of a radioactive source, is obviously an attractive alternative to the conventional device and appears to have similar, if not better, performance characteristics.

The high sensitivity of the electron-capture detector makes it very popular for use in forensic and environmental chemistry. It is very simple to use and is one of the less expensive, high-sensitivity selective detectors available.

Reference

1. J. E. Lovelock and S. R. Lipsky, *J. Am. Chem. Soc. 82*: 431 (1960).

Suggested Further Reading

R. P. W. Scott, *Chromatographic Detectors*, Marcel Dekker, Inc., New York, 1996.

R. P. W. Scott, *Introduction to Gas Chromatography*, Marcel Dekker, Inc., New York, 1998.

Raymond P. W. Scott

Electro-osmotic Flow

Electro-osmosis refers to the movement of the liquid adjacent to a charged surface, in contact with a polar liquid, under the influence of an electric field applied parallel to the solid–liquid interface. The bulk fluid of liquid originated by this electrokinetic process is termed electro-osmotic flow (EOF). It may be produced both in open and in packed capillary tubes, as well as in planar electrophoretic systems employing a variety of supports, such as paper or hydrophilic polymers. The formation of an electric double layer at the interfacial region between the charged surface and the surrounding liquid is of key importance in the generation of the electro-osmotic flow [1–3]. Most solid surfaces acquire a superficial charge when are brought into contact with a polar liquid. The acquired charge may result from dissociation of ionizable groups on the surface, adsorption of ions from solution, or by virtue of unequal dissolution of oppositely charged ions of which the surface is composed. This superficial charge causes a variation in the distribution of ions near the solid–liquid interface. Ions of opposite charge (counterions) are attracted toward the surface, whereas ions of the same charge (co-ions) are repulsed away from the surface. This, in combination with the mixing tendency of thermal motion, leads to the generation of an electric double layer formed by the charged surface and a neutralizing excess of counterion over co-ions distributed in a diffuse manner in the polar liquid. Part of the counterions are firmly held in the region of the double layer closer to the surface (the compact or Stern layer) and are believed to be less hydrated than those in the diffuse region of the double layer where ions are distributed according to the influence of electrical forces and random thermal motion. A plane (the Stern plane) located at about one ion radius from the surface separates these two regions of the electric double layer.

Certain counterions may be held in the compact region of the double layer by forces additional to those of purely electrostatic origin, resulting in their adsorption in the Stern layer. Specifically, adsorbed ions are attracted to the surface by electrostatic and/or van der Waals forces strongly enough to overcome the thermal agitation. Usually, the specific adsorption of counterions predominates over co-ion adsorption.

The variation of the electric potential in the electric double layer with the distance from the charged surface is depicted in Fig. 1. The potential at the surface (ψ_0) linearly decreases in the Stern layer with respect to the value of the zeta potential (ζ). This is the electric potential at the plane of shear between the Stern layer (plus that part of the double layer occupied by the molecules of solvent associated with the adsorbed ions) and the diffuse part of the double layer. The zeta potential decays exponentially from ζ to zero with the distance from the plane of shear between the Stern layer and the diffuse part of the double layer. The location of the plane of shear, a small distance further out from the surface than the Stern plane, renders the zeta potential marginally smaller in magnitude than the potential at the Stern plane (ψ_δ). However, in order to simplify the mathematical models describing the electric double layer, it is customary to assume identity of ψ_δ and ζ, and the bulk experimental evidence indicates that errors introduced through this approximation are usually small.

According to the Gouy–Chapman–Stern–Grahame (GCSG) model of the electric double layer [4], the surface density of the charge in the Stern layer is related to the adsorption of the counterions, which is described by a Langmuir-type adsorption model, modified by the incorporation of a Boltzman factor. Considering only the adsorption of counterions, the surface change density σ_S of

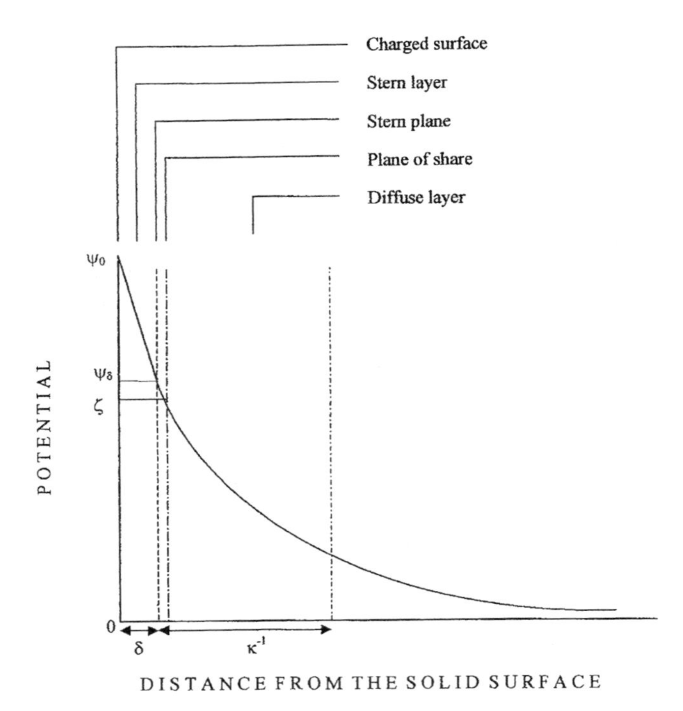

Fig. 1 Schematic representation of the electric double layer at a solid–liquid interface and variation of potential with the distance from the solid surface: ψ_0, surface potential; ψ_δ, potential at the Stern plane; ζ, potential at the plane of share (zeta potential); δ, distance of the Stern plane from the surface (thickness of the Stern layer); κ^{-1}, thickness of the diffuse region of the double layer.

the Stern layer is related to the ion concentration C in the bulk solution by the following equation:

$$\sigma_S = zen_0 \frac{C}{V_m} \exp\left(\frac{ze\xi + \Phi}{kT}\right) \times \left[1 + \frac{C}{V_m} \exp\left(\frac{ze\xi + \Phi}{kT}\right)\right]^{-1} \tag{1}$$

where e is the elementary charge, z is the valence of the ion, k is the Boltzman constant, T is the temperature, n_0 is the number of accessible sites, V_m is the molar volume of the solvent, and Φ is the specific adsorption potential of counterions.

The surface charge density of the diffuse part of the double layer is given by the Gouy–Chapman equation

$$\sigma_G = (8\varepsilon kTc_0) \sinh\left(\frac{ze\xi}{2kT}\right) \tag{2}$$

where ε is the permittivity of the electrolyte solution and c_0 is the bulk concentration of each ionic species in the electrolyte solution.

At low potentials, Eq. (2) reduces to

$$\sigma_G = \frac{\varepsilon\xi}{\kappa^{-1}} \tag{3}$$

where κ^{-1} is the reciprocal Debye–Hückel parameter, which is defined as the "thickness" of the electric double layer. This quantity has the dimension of length and is given by

$$\kappa^{-1} = \left(\frac{\varepsilon kT}{2e^2 I}\right)^{1/2} \tag{4}$$

in which I is the ionic strength of the electrolyte solution.

Equation (3) is identical to the equation that relates the charge density, voltage difference, and distance of separation of a parallel-plate capacitor. This result indicates that a diffuse double layer at low potentials behaves like a parallel capacitor, in which the separation distance between the plates is given by κ^{-1}. This explains why κ^{-1} is called the double-layer thickness.

Equation (2) can be written in the form

$$\xi = \frac{\sigma_G \kappa^{-1}}{\varepsilon} \tag{5}$$

which indicates that the zeta potential can change due to variations in the density of the electric charge, in the permittivity of the electrolyte solution, and in the thickness of the electric double layer, which depends, throughout the ionic strength [see Eq. (4)], on the concentration and valence of the ions in solution.

The dependence of the velocity of the electro-osmotic flow (v_{eo}) on the zeta potential is expressed by the Helmholtz–von Smoluchowski equation

$$v_{eo} = -\frac{\varepsilon_0 \varepsilon \zeta}{\eta} E \tag{6}$$

where E is the applied electric field, ε_0 is the permittivity of vacuum, and ε and η are the dielectric constant and the viscosity of the electrolyte solution, respectively. This expression assumes that the dielectric constant and viscosity of the electrolyte solution are the same in the electric double layer as in the bulk solution.

The Helmholtz–von Smoluchowski equation indicates that under constant composition of the electrolyte solution, the electro-osmotic flow depends on the magnitude of the zeta potential, which is determined by the different factors influencing the formation of the electric double layer, as discussed earlier. Each of these factors depends on several variables, such as pH, specific adsorption of ionic species in the compact region of the double layer, ionic strength, and temperature.

The specific adsorption of counterions at the interface between the surface and the electrolyte solution results in a drastic variation of the charge density in the Stern layer, which reduces the zeta potential and, hence, the electro-osmotic flow. If the charge density of the adsorbed counterions exceeds the charge density on the surface, the zeta potential changes sign and the direction of the electro-osmotic flow is reversed.

The ratio of the velocity of the electro-osmotic flow to the applied electric field, which expresses the velocity per unit field, is defined as electro-osmotic coefficient or, more properly, electro-osmotic mobility (μ_{eo}).

$$\frac{v_{eo}}{E} = \mu_{eo} = \frac{\varepsilon_0 \varepsilon \zeta}{\eta} \tag{7}$$

Using SI units, the velocity of the electro-osmotic flow is expressed in meters per second (m/s) and the electric field in volts per meter (V/m). Consequently, in analogy to the electrophoretic mobility, the electro-osmotic mobility has the dimension square meters per volt per second. Because electro-osmotic and electrophoretic mobilities are converse manifestations of the same underlying phenomenon, the Helmholtz–von Smoluchowski equation applies to electro-osmosis as well as to electrophoresis. In fact, when an electric field is applied to an ion, this moves relative to the electrolyte solution, whereas in the case of electro-osmosis, it is the mobile diffuse layer that moves under an applied electric field, carrying the electrolyte solution with it.

According to Eq. (6), the velocity of the electro-osmotic flow is directly proportional to the intensity of the applied electric field. However, in practice, the nonlinear dependence of the electro-osmotic flow on the applied electric field is obtained as a result of Joule heat production, which causes an increase of the electrolyte temperature with a consequent decrease of viscosity and variation of all other temperature-dependent parameters (protonic equilibrium, ion distribution in the double layer, etc.). The electro-osmotic flow can also be altered during a run by variations of the protonic and hydroxylic concentration in the anodic and cathodic electrolyte solutions as a result of electrolysis. This effect can be minimized by using electrolyte solutions with a high buffering capacity and electrolyte reservoirs of relatively large volume and by frequent replacement of the electrolyte in the electrode compartments with fresh solution.

The magnitude and direction of the electro-osmotic flow depend also on the composition, pH, and ionic strength of the electrolyte solution [5–7]. Both the pH and ionic strength influence the protonic equilibrium of fixed-charged groups on the surface and of ionogenic substances in the electrolyte solution which affect the charge density in the electric double layer and, consequently, the zeta potential. In addition, the ionic strength influences the thickness of the double layer (κ^{-1}). According to Eq. (4), increasing the ionic strength causes a decrease in κ^{-1}, which is currently referred to as the compression of the double layer that results in lowering the zeta potential. Consequently, increasing the ionic strength results in decreasing the electro-osmotic flow.

The charge density in the electric double layer and, hence, the electro-osmotic flow are also influenced by the adsorption of potential-determining ions in the Stern region of the electric double layer. A variety of additives can be incorporated into the electrolyte solution with the purpose of controlling the electro-osmotic flow by modifying the solid surface dynamically. These include simple and complex ionic compounds, ionic and zwitterionic surfactants, and neutral and charged polymers. The incorporation of these additives into the electrolyte solution may result either in increasing or in reducing the electro-osmotic flow, or even in reversing its direction. The impact of these additives on the electro-osmotic flow is generally concentration dependent. Such behavior is in accordance to the Langmuir-like adsorption model describing the variation of the charge density in the Stern layer on the concentration of adsorbing ions in the electrolyte solution [see Eq. (1)].

The proper control of the electro-osmotic flow can be also obtained by adding organic solvents to the electrolyte solution. The influence of organic solvents on the electro-osmotic flow may result from a multiplicity of mechanisms. Organic solvents are expected to influence both the dielectric constant and viscosity of the bulk electrolyte solution. Generally, this leads to the variation of the ratio of the dielectric constant to the viscosity of the electrolyte solution, to which the electro-osmotic flow depends according to Eq. (6). In addition, the local viscosity within the electric double layer [8] can be varied by the adsorption of the organic–solvent molecules in the Stern layer, which may also influence the adsorption of counterions, depending on the different solvation properties of the organic solvent. Organic solvents may also influence the zeta potential by affecting the ionization of potential-determining ions on the surface.

Different methods can be employed to measure the magnitude of the electro-osmotic flow [9]. One possibility involves measuring the velocity of the electro-osmotic flow by measuring the change in weight or in volume in one of the electrolyte solution reservoirs. The addition of an electrically neutral dye to one electrode reservoir and its detection in the other where the electro-osmotic flow is directed is another possible method. Other methods based on monitoring electric current while an electrolyte solution of different conductivity is drawn into the system by electro-osmosis or determining the zeta potential from streaming potential measurements are less popular and accurate. More common is the method of calculating the electro-osmotic velocity from the migration time of an electrically neutral marker substance incorporated into the sample solution. The selected compound must be soluble in the electrolyte solution, neutral in a wide pH range, and easily detectable. In addition, it should neither become partially charged by complexation with the components of the electrolyte solution nor interact with the capillary tube, the chromatographic stationary phase, or the slab gel employed in capillary electrophoresis, capillary electrochromatography, and planar electrophoresis, respectively. This method has the advantage of simplicity and can be used to monitor the electro-osmotic flow during analysis in any of the above techniques, provided that the analytes and the electro-osmotic flow are directed toward the same electrode.

References

1. P. C. Hiemenz, *Principles of Colloid and Surface Chemistry*, 2nd ed., Marcel Dekker, Inc., New York, 1986, pp. 677–735.
2. A. W. Adamson, *Physical Chemistry of Surfaces*, 5th ed., John Wiley & Sons, New York, 1990, pp. 203–257.
3. D. J. Shaw, *Introduction to Colloid and Surface Chemistry*, 3th ed., Butterworths, London, 1980, pp. 148–182.
4. D. C. Grahame, *Chem. Rev. 41*: 441–501 (1947).
5. K. D. Lukacs and J. W. Jorgenson, *J. High Resolut. Chromatogr. Chromatogr. Commun. 8*: 407–411 (1985).
6. K. D. Altria and C. F. Simpson, *Chromatographia 24*: 527–532 (1987).
7. M. G. Cikalo, K. D. Bartle, and P. Myers, *J. Chromatogr. A 836*: 35–51 (1999).
8. S. Hjerten, *Chromatogr. Rev. 9*: 122–219 (1967).
9. A. A. A. M. Van de Goor, B. J. Wanders, and F. M. Everaerts, *J. Chromatogr. 470*: 95–104 (1989).

Danilo Corradini

Electro-osmotic Flow in Capillary Tubes

The electro-osmotic flow in open capillary tubes is generated by the effect of the applied electric field across the tube on the uneven distribution of ions in the electric double layer at the interface between the capillary wall and the electrolyte solution. In bare fused-silica capillaries, ionizable silanol groups are present at the surface of the capillary wall, which is exposed to the electrolyte solution. In this case, the electric double layer is the result of the excess of cations in the solution in contact with the capillary tube to balance the negative charges on the wall arising from the ionization of the silanol groups. Part of the excess cations are firmly held in the region of the double layer closer to the capillary wall (the compact or Stern layer) and are believed to be less hydrated than those in the diffuse region of the double layer [1]. When an electric field is applied across the capillary, the remaining excess cations in the diffuse part of the electric double layer move toward the cathode, dragging their hydration spheres with them. Because the molecules of water associated with the cations are in direct contact with the bulk solvent, all the electrolyte solution moves toward the cathode, producing a pluglike flow having a flat velocity distribution across the capillary diameter [2].

The flow of liquid caused by electro-osmosis displays a pluglike profile because the driving force is uniformly distributed along the capillary tube. Consequently, a uniform flow velocity vector occurs across the capillary. The flow velocity approaches zero only in the region of the double layer very close to the capillary surface. Therefore, no peak broadening is caused by sample transport carried out by the electro-osmotic flow. This is in contrast to the laminar or parabolic flow profile generated in a pressure-driven system, where there is a strong pressure drop across the capillary caused by frictional forces at the liquid–solid boundary. A schematic representation of the flow profile due to electro-osmosis in comparison to that obtained in the same capillary column in a pressure-driven system, such as a capillary high-performance liquid chromatography (HPLC), is displayed in Fig. 1.

The dependence of the velocity of the electro-osmotic flow (v_{eo}) on the applied electric field (E) is expressed by the Helmholtz–von Smoluchowski equation

$$v_{eo} = -\frac{\varepsilon_0 \varepsilon_r \zeta}{\eta} E \tag{1}$$

where ζ is the zeta potential, ε_0 is the permittivity of vacuum, ε_r is the dielectric constant, and η is the viscosity of the electrolyte solution. This expression assumes that the dielectric constant and viscosity of the electrolyte solution are the same in the electric double layer as in the bulk solution. The term $-\varepsilon_0 \varepsilon_r \xi / \eta$ is the defined electro-osmotic coefficient or, more properly, electro-osmotic mobility (μ_{eo}) and expresses the velocity of the electro-osmotic flow per unit field. Accordingly, the Helmholtz–von Smoluchowski equation can be written

$$\mu_{eo} = \frac{v_{eo}}{E} \tag{2}$$

In a capillary tube, the applied electric field E is expressed by the ratio V/L_T, where V is the potential difference in volts across the capillary tube of length L_T (in meters). The velocity of the electro-osmotic flow, v_{eo} (in meters per second), can be evaluated from the migration time t_{eof} (in seconds) of an electrically neutral marker substance and the distance L_D (in meters) from the end of the capillary where the samples are introduced to the detection windows (effective length of the capillary). This indicates that, experimentally, the electro-osmotic mobility can be easily calculated using the Helmholtz–von Smoluchowski equation in the following form:

$$\mu_{eo} = \frac{L_T L_D}{V t_{eo}} \quad (\mathrm{m^2/V/s}) \tag{3}$$

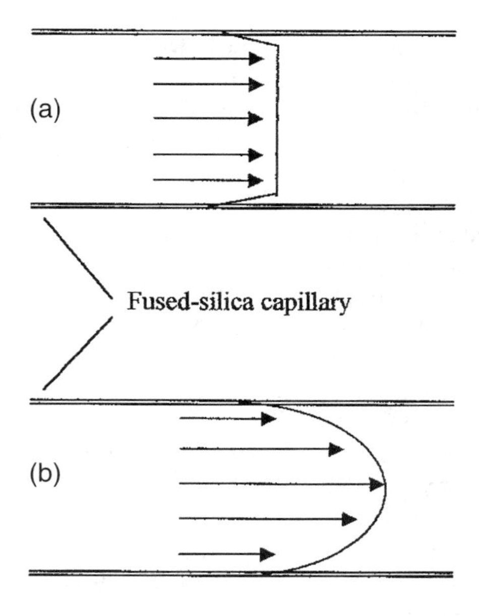

Fig. 1 Schematic representation of the flow profiles obtained with the same capillary column connected to an electric-driven system (a) and to a pressure-driven system (b). Arrows indicate flow velocity vectors.

which demonstrates that, by analogy to the electrophoretic mobility, the electro-osmotic mobility has the dimension of square meters per volt per second.

The electrically neutral marker substance employed to measure the velocity of the electro-osmotic flow has to fulfill the following requirements. The compound must be soluble in the electrolyte solution and neutral in a wide pH range and no interaction with the capillary wall must occur. In addition, the electrically neutral marker substance should be easily detectable in order to allow a small amount to be injected. If the electrically neutral marker interacts with the capillary wall or becomes partially charged by complexation with the components of the electrolyte solution, the measured electro-osmotic velocity may appear slower or faster than the real flow. Some compounds that adequately serve as electrically neutral markers include benzyl alcohol, riboflavin, acetone, dimethyl-formamide, dimethyl sulfoxide, and mesityl oxide.

Alternatively, the velocity of the electro-osmotic flow can be measured by weighing the volume of the electrolyte solution displayed by electro-osmosis from the anodic to the cathodic reservoir. When detection is performed by ultraviolet (UV) absorbance, a "solvent dip" equal to the electro-osmotic flow appears in the electropherogram after any sample injection. In most cases, the sample solvent has a lower UV absorbance than the electrolyte solution, resulting in a negative UV signal. On the other hand, if the UV absorbance of the sample solvent is higher than that of the electrolyte solution, a positive system peak can be observed at the time corresponding to the velocity of the electro-osmotic flow. The time at which the "solvent dip" appears in the electropherogram can be used to measure the velocity of the electro-osmotic flow in a very simple but less accurate way than those using an electrically neutral marker substance or the weight of the displaced liquid.

The Helmholtz–von Smoluchowski equation indicates that under constant composition of the electrolyte solution, the electro-osmotic flow depends on the magnitude of the zeta potential which is determined by many different factors, the most important being the dissociation of the silanol groups on the capillary wall, the charge density in the Stern layer, and the thickness of the diffuse layer. Each of these factors depends on several variables, such as pH, specific adsorption of ionic species in the compact region of the electric double layer, ionic strength, viscosity, and temperature.

Secondary equilibrium in solution, generation of Joule heat, and variation of protonic and hydroxylic concentration due to electrolysis may alter the hydrogen ion concentration in the capillary tube when electrolyte solutions having low buffering capacities are employed. A change in the protonic equilibrium directly influences the zeta potential through the variation of the charge density on the capillary wall resulting from the deprotonation of the surface silanol groups, which increases with increasing pH. The shape of a curve describing the dependence of the zeta potential on the electrolyte pH resembles a titration curve, the inflection point of which may be interpreted as the pK value of the surface silanol groups. At acidic pH, the ionization of the surface silanol groups is suppressed and the zeta potential approaches zero, determining the virtual annihilation of the electro-osmotic flow. Under alkaline conditions, the silanol groups are fully charged and the zeta potential reaches its maximum value, which corresponds to a plateau value of the electro-osmotic flow. Between these extreme conditions, the zeta potential rapidly increases with increasing pH up to the complete dissociation of the silanol groups, determining the well-known sigmoidal pH dependence of the electro-osmotic flow. The concentration and ionic strength of the electrolyte solution also have a strong impact on the electro-osmotic flow. The ionic strength influences the thickness of the diffuse part of the electric double layer to which the zeta potential is directly proportional. Because the thickness of the diffuse part of the electric double layer is inversely proportional to the square root of the ionic strength, the electro-osmotic flow decreases with the concentration of the electrolyte solution according to the following relationship [3]:

$$\mu_{eo} \approx \frac{e}{3 \times 10^7 |z| \eta \sqrt{C}} \tag{4}$$

where, e, z, η, and C are the total charge per unit surface area, the electron valence of the electrolyte, the viscosity, and the concentration of the electrolyte in the bulk solution, respectively.

Another model that accounts for the decrease of the electro-osmotic flow with increasing the electrolyte concentration relates the electro-osmotic mobility to the concentration of a monovalent counterion, introduced with the buffer, according to the following relationship [4]:

$$\mu_{eo} = \frac{Q_0}{\eta(1 + K_{wall}[M^+])}\left(d_0 + \frac{1}{K'\sqrt{[M^+]}}\right) \tag{5}$$

where η is the viscosity of the electrolyte solution and K' is a constant that, for a dilute aqueous solution at 25°C, is equal to $3 \times 10^9/\mathrm{m(mol/L)}^{-1/2}$. The first term on the right side of Eq. (5) is related to the dependence of the surface charge on the concentration of the monovalent cation in the electrolyte solution. The model postulates that the initial charge per unit area at the surface of the silica capillary wall (Q_0) is reduced by the factor $1/1 + K_{wall}[M^+]$ upon incorporating a monovalent buffer of concentration $[M^+]$ into the electrolyte solution. This is a result of the neutralization of the free silanol groups on the capillary surface caused by the adsorption of the monovalent cations. The constant K_{wall} is defined as the equilibrium constant between the cations in the buffer solution and adsorption sites on the capillary wall. The second term describes the influence of the concentration of the monovalent cation on the thickness of the mobile region of the electric double layer. This is postulated to be composed of a fixed thickness (d_0) and the Debye–Hückel thickness $\delta = 1/K'[M^+]^{1/2})$, which is inversely proportional to the square root of the concentration of the monovalent ion. According to this model, increasing the concentration of the monovalent buffer cation in the bulk solution influences the electro-osmotic mobility by reducing the Debye–Hückel thickness of the diffuse double layer and by neutralizing the negative charges on the capillary wall resulting from the ionization of the silanol groups.

Certain counterions, such as polycationic species, cationic surfactants, and several amino compounds can be firmly held in the compact region of the electric double layer by forces additional to those of simple Coulombic origin. The specific adsorption of counterions at the interface between the capillary wall and the electrolyte solution results in a drastic variation of the positive charge density in the Stern layer, which reduces the zeta potential and, hence, the electro-osmotic flow. If the positive charge density of the adsorbed counterions exceeds the negative charge density on the capillary wall resulting from the ionization of silanol groups, the zeta potential becomes positive and the concomitant electro-osmotic flow is reversed from cathodic to anodic.

The dependence of the electro-osmotic flow on the specific adsorption of counterions in the electric double layer can be described by a model which correlates the electro-osmotic mobility to the charge density in the Stern part of the electric double layer (arising from the adsorption of counterions) and the charge density at the capillary wall (resulting from the ionization of silanol groups) [5]. According to this model, the dependence of the electro-osmotic mobility on the concentration of the adsorbing ions (C) in the electrolyte solution is expressed as

$$\mu_{\text{eo}} = \frac{\kappa^{-1}}{\eta}\left\{zen_0 \frac{C}{V_m}\exp\left(\frac{ze\psi_d + \Phi}{kT}\right) \times \left[1 + \frac{C}{V_m}\exp\left(\frac{ze\psi_\delta + \Phi}{kT}\right)\right]^{-1} - \left(\frac{\gamma}{1 + [H^+]/K_a}\right)\right\} \tag{6}$$

where κ^{-1} is the Debye–Hückel thickness of the diffuse double layer, η is the viscosity of the electrolyte solution, e is the elementary charge, z is the valence of the adsorbing ion, k is the Boltzman constant, T is the absolute temperature, n_0 is the number of accessible sites in the Stern layer, V_m is the molar volume of the solvent, Φ is the specific adsorption potential of counterions, γ is the sum of the ionized and protonated surface silanol groups, $[H^+]$ is the bulk electrolyte hydrogen ion concentration, and K_a is the silanol dissociation constant. According to this equation, at constant ionic strength, viscosity, and pH, the electro-osmotic mobility depends mainly on the surface density of the adsorbed counterions in the Stern region of the electric double layer, which follow a Langmuir-type adsorption model.

The reversal of the direction of the electro-osmotic flow by the adsorption onto the capillary wall of alkylammonium surfactants and polymeric ion-pair agents incorporated into the electrolyte solution is widely employed in capillary zone electrophoresis (CZE) of organic acids, amino acids, and metal ions. The dependence of the electro-osmotic mobility on the concentration of these additives has been interpreted on the basis of the model proposed by Fuerstenau [6] to explain the adsorption of alkylammonium salts on quartz. According to this model, the adsorption in the Stern layer as individual ions of surfactant molecules in dilute solution results from the electrostatic attraction between the head groups of the surfactant and the ionized silanol groups at the surface of the capillary wall. As the concentration of the surfactant in the solution is increased, the concentration of the adsorbed alkylammonium ions increases too and reaches a critical concentration at which the van der Waals attraction forces between the hydrocarbon chains of adsorbed and free-surfactant molecules in solution cause their association into hemimicelles (i.e., pairs of surfactant molecules with one cationic group directed toward the capillary wall and the other directed out into the solution).

Lowering the velocity or reversing the direction of the electro-osmotic flow may have a beneficial effect of on the resolution of two adjacent peaks, as evidenced by the following expression for resolution in electrophoresis elaborated by Giddings [7]:

$$R_s = \frac{\sqrt{N}}{4}\left(\frac{\Delta\mu}{\mu_{\text{av}} + \mu_{\text{eo}}}\right) \tag{7}$$

where N is the number of theoretical plates, $\Delta\mu$ and μ_{av} are the difference and the average value of the electrophoretic mobilities of two adjacent peaks, respectively, and μ_{eo} is the electro-osmotic mobility. According to this equation, the highest resolution is obtained when the electro-osmotic mobility has the same value but opposite direction of the average electrophoretic mobility of the two adjacent peaks.

Neutral polymeric molecules, such as polysaccharides and synthetic polymers, may also adsorb onto the Stern layer, causing a variation of viscosity in the double layer with distance from the capillary wall, which affects the electro-osmotic mobility according to the following relationship [2]:

$$\mu_{eo} = \frac{\varepsilon_r}{4\pi} \int_0^{\zeta} \frac{1}{\eta} \, d\psi \tag{8}$$

where ε_r is the dielectric constant, ζ is the zeta potential, η is the viscosity, and ψ is the electric potential. The value of the integral in this expression will approach zero when the viscosity in the double layer approaches infinity. Accordingly, the electro-osmotic flow is drastically reduced when the local viscosity of the double layer is increased as a result of the adsorption of a neutral polymer onto the Stern layer. It is worth noting that at constant value of the viscosity in the electric double layer, Eq. (8) is equivalent to the Helmholtz–von Smoluchowski expression for the electro-osmotic flow.

The incorporation of an organic solvent into the aqueous electrolyte solution also leads to a variation of the electro-osmotic flow [8]. The general trend is that the electro-osmotic flow decreases steadily with increasing concentration of the organic solvent in the hydro-organic electrolyte solution. This effect can be attributed, to some extent, to the increasing viscosity and decreasing dielectric constant of most hydro-organic electrolyte solutions with increasing concentration of organic solvent. However, in most cases, the decrease of the electro-osmotic flow is also observed at organic solvent concentrations greater than 50–60% (v/v), at which the ratio of the dielectric constant and the viscosity, ε_r/η, is generally increasing. This indicates that the variation of the electro-osmotic flow caused by the incorporation of an organic solvent into the electrolyte solution cannot be solely related to the changes of the ratio ε_r/η.

Similar to the neutral polymers, organic solvents can adsorb at the interface between the capillary wall and the electrolyte solution, through hydrogen-bonding or dipole interaction, thus increasing the local viscosity within the electric double layer. Organic solvents may also influence the zeta potential by affecting the ionization of the silanol groups at the capillary surface, whose pK_a has been found to be shifted toward higher values with increasing the content of organic solvents in the electrolyte solution. The dependence of the zeta potential on the fraction of an organic solvent incorporated into the electrolyte solution may be also related to the variation of both the dielectric constant and the adsorption of counterions in the Stern layer. In practice, introducing a neutral polymer or an organic solvent into the electrolyte solution results in multiple changes, generally involving the viscosity and the dielectric constant of the bulk solution, the ionization of the silanol groups on the capillary wall, and the charge density in the Stern layer, as well as the local viscosity and the dielectric constant of the electric double layer.

References

1. P. C. Hiemenz, *Principles of Colloid and Surface Chemistry*, 2nd ed., Marcel Dekker, Inc., New York, 1986, pp. 677–735.
2. S. Hjertén, *Chromatogr. Rev. 9*: 122–219 (1967).
3. T. Tsuda, K. Nomura, and G. Nakagawa, *J. Chromatogr. 248*: 241–247 (1982).
4. K. Salomon, D. S. Burgi, and J. C. Helmer, *J. Chromatogr. 559*: 69–80 (1991).
5. D. Corradini, A. Rhomberg, and C. Corradini, *J. Chromatogr. A 661*: 305–313 (1994).
6. D. W. Fuerstenau, *J. Phys. Chem. 60*: 981–985 (1956).
7. J. C. Giddings, *Separ. Sci. 4*: 181–189 (1969).
8. C. Schwer and E. Kenndler, *Anal. Chem. 63*: 1801–1807 (1991).

Danilo Corradini

Electro-osmotic Flow Nonuniformity: Influence on Efficiency of Capillary Electrophoresis

There are two types of nonuniformities of electro-osmotic flow (EOF) that can contribute significantly to the solute peak broadening and are important for capillary electrophoresis (CE). The first is the transversal nonuniformity of the usual EOF in the capillary with the zeta potential of the walls and longitudinal electric field strength constant and independent of coordinates. The second one is the nonuniformity of EOF caused by the dependence of the zeta potential of the walls or electric field strength on coordinates.

The first type of EOF nonuniformity was described in the classical article by Rice and Whitehead [1] written much earlier than the first works on CE. The equation for the EOF velocity profile in the infinitely long tube with radius a was given by

$$V(r) = \frac{\zeta \varepsilon \varepsilon_0 E}{\eta}\left(1 - \frac{I_0(\kappa r)}{I_0(\kappa a)}\right) \tag{1}$$

where ζ is the zeta potential of the wall, ε and η are the dielectric constant and viscosity of the buffer, respectively, ε_0 is the permitivity of the free space, $\kappa^{-1} = (\varepsilon \varepsilon_0 k_B T/2ne^2)^{1/2}$ is the Debye layer thickness, k_B is the Boltzmann constant, T is the temperature, n is the number of ions per unit volume (proportional to the concentration of buffer C_0), e is the proton charge, and $I_0(x)$ is the modified Bessel function. It is evident from Eq. (1) that the nonuniformity of the EOF profile can be substantial only if the capillary radius and Debye length are commensurate. In fact, for the case of $\kappa a \approx 1$, the profile of EOF according to Eq. (1) is very close to parabolic. Luckily, it is not the case of usual capillaries for CE with $a \geq 25$ μm because, even for distilled water, $\kappa^{-1} \approx 0.1$ μm. Another important result of Ref. 1 is the prediction of the EOF profile in the long capillary with closed ends. In such a capillary, liquid moves in one direction in the vicinity of the walls and in the opposite direction near the axis of the capillary, thus making the total flow through the cross section equal to zero. With this result, it is quite evident that CE must be realized in a capillary with open ends; otherwise, nonuniformity of EOF would ruin the separation.

Results of Ref. 1 were produced by employing a linear approximation of the exponential terms in the Poisson–Boltzmann equation for electrical potential and charge distribution. Strictly speaking, this linearization is valid only for $|\zeta| \ll kT/e \approx 0.03$ V, whereas the range of the values of the zeta potential is $|\zeta| \leq 0.1$ V. In Ref. 2, the Poisson–Boltzmann equation and the Navier–Stokes equations for EOF velocity profile were solved numerically without linearization. The dependence of buffer viscosity on temperature and the existence of temperature gradients due to Joule heating were also taken into consideration. Calculated EOF profiles were compared with and predicted by Eq. (1), showing that the difference in flow profiles for $|\zeta| = 0.1$ V could be significant, especially for thin capillaries and low buffer concentrations ($\kappa a \approx 10$). Calculated flow profiles were used to predict the stationary value of HETP by using the results of generalized dispersion theory. It was shown that for low buffer concentrations ($C_0 \leq 10^{-4}M$), the contribution of electroosmotic flow nonuniformity to the HETP value could be larger than the contribution of the thermal effects and molecular diffusion.

A similar approach was used in Ref. 3, where the contributions of EOF nonuniformity (H_{eo}) and molecular diffusion (H_{diff}) to HETP were compared for different values of solute diffusion coefficients. It was shown [3] that for the typical CE velocities of EOF (1–2 mm/s) and rather high buffer concentration ($C_0 = 10^{-2}M$), $H_{eo}/H_{diff} \geq 1$ for $D \leq 2 \times 10^{-12}$/m^2/s. For lower buffer concentrations, the influence of EOF nonuniformity is substantial for smaller molecules also ($H_{eo} = 1.3 \times 10^{-8}$ m, $H_{diff} = 1.6 \times 10^{-8}$ m for $D = 2.4 \times 10^{-11}$ m^2/s, corresponding to α_2-macroglobulin).

The joint effect of EOF nonuniformity and particle–wall electrostatic interactions was studied in Ref. 4. Two types of solute particles were examined: one with the charge of the same sign as the zeta potential of the wall, and the other of the opposite sign. The particles of the first type are moving electrophoretically in the direction opposite to the direction of EOF and are electrostaticaly subtracted by the wall, whereas the particles of the second type are attracted by the wall and are moving electrophoretically in the same direction as EOF. Particles of the second type spend a large portion of time in the vicinity of the capillary wall and, so, EOF nonuniformity contributes significantly to peak broadening, whereas for the particles subtracted by the wall, the influence of EOF nonuniformity is negligible because their residence time in the vicinity of the wall is close to zero. For example, for the particles with $D = 5 \times 10^{-11}$ m^2/s, in the capillary with $a = 10$ μm, $\zeta = -0.1$ V, and $E = 40$ kV/m, filled with diluted buffer ($C_0 = 10^{-5}$ M), one has HETP ≈ 10 μm for particles attracted by the wall, whereas for particles subtracted by the wall, HETP ≈ 0.1 μm. For neutral particles not interacting with the walls, HETP ≈ 0.2 μm was predicted. The difference was much less dramatic for the case of the higher buffer concentrations and the lower zeta potential. For example, for $\zeta = -0.02$ V, $C_0 = 10^{-3}M$, and the rest of parameters being the same as described earlier, HETP is determined mainly by molecular diffusion and is close to 0.1 μm.

The influence of EOF nonuniformity on efficiency of CE in the capillary with the zeta potential of the wall being the function $\zeta(x)$ of the longitudinal coordinate x was studied in Ref. 5. To calculate the EOF velocity profile, an important approximation was justified by the fact that usually $\kappa a \gg 1$ in CE. Thus, the double-layer region was neglected and the following boundary condition was formulated:

$$V_x(x, a, t) = \frac{\varepsilon \varepsilon_0 E}{\eta}\zeta(x) \tag{2}$$

With this boundary condition, the Navier–Stokes equations for longitudinal V_x and radial V_r components of EOF velocity were solved numerically, and the calculated EOF profiles were used to simulate the solute peak shapes. The situation where the part of the capillary length was modified to the zero value of the zeta potential and the part of capillary was not modified ($\zeta \neq 0$) was studied. It was shown that the radial component of the velocity is nonzero only in the rather short transition region between the uncovered and covered parts of the capillary. At a distance of a few capillary diameters from the transient region, the radial flows are negligible and the axial component of the velocity in the covered section of the capillary has an almost parabolic profile. Peak shapes and peak variances were studied, and the general conclusion of Ref. 5 was that the main contribution to the peak width was from the parabolic velocity profiles, the contribution of the radial flow in the transient regions being less significant. Based on this result, the mathematical model of CE in the capillary made of several sections with various nonequal values of the zeta potentials and radii was developed [6]. For each of the sections, the total flow was considered to be the sum of EOF caused by electrical potential differences along the section and the Poiseuille flow, caused by the pressure drop along the section. The values of the pressure drops and potentials differences were determined by the solution of the set of $2N$ algebraic equations, where N is the number of sections in the capillary. These equations reflect the fact that the total flow of liquid and total current are constant along the capillary, and the sums of pressure drops and potential differences at the sections are equal to the total pressure drop and total potential difference at the whole capillary, respectively. The lengths of the sections were considered to be much larger than the capillary radius, so the results of the model are valid everywhere except the immediate vicinity of the points where the radius of capillary or the zeta potential of the wall change their values. When calculating the values of HETP in such a capillary, particle–wall electrostatic interactions were taken into consideration, and it was shown that HETP values are considerably larger for particles attracted by the wall. It was also shown that differences in the values of the zeta potential contributes to HETP much more significantly than the differences in radii values. The situation that might happen in the case of a bubble-cell detector was modeled and considerable growth of HETP was predicted.

In Ref. 7, the case was studied in which the zeta potential of the wall was the linear function of the longitudinal coordinate. This situation may happen when the value of the zeta potential is controlled by the external electrical potential applied to the wall. Electrical potential value inside the capillary is naturally a linear function of the longitudinal coordinate x; therefore, if the electrical potential applied to the outer boundary of the capillary wall is constant, then the potential difference across the wall is a linear function of x. The theoretical approach used in Ref. 7 is similar to the one in Ref. 5. Secondary parabolic flow was shown to be generated, leading to the increase of HETP. It was predicted theoretically, and verified experimentally, that a pressure profile superimposed on the capillary can, in some cases, compensate for the disturbed profile and reduce the HETP value.

In Ref. 8, the mathematical model of capillary electrophoresis in rectangular channels with nonequal values of the zeta potentials of the walls was developed. This model may be of interest for the case of CE on a microchip, where the microgroves are produced by wet chemical etching and, so, the walls of the groove can have different values of zeta potential than the cover plate that is not etched. Flow profiles for the channels with different aspect ratios and different combinations of the zeta potential values were examined. It was shown, for example, that a 10% difference in the values of the zeta potentials of upper and lower walls can cause a sixfold growth of the HETP value.

The above-mentioned examples show that EOF nonuniformities may occur in different situations and must be given considerable attention, as they can reduce the CE efficiency dramatically.

References

1. C. L. Rice and R. Whitehead, Electrokinetic flow in a narrow cylindrical capillary, *J. Phys. Chem. 69*: 4017 (1965).
2. V. P. Andreev and E. E. Lisin, Investigation of the electroosmotic flow effect on the efficiency of capillary electrophoresis, *Electrophoresis 13*: 832 (1992).
3. B. Gas, M. Stedry, and E. Kenndler, Contribution of the electroosmotic flow to peak broadening in capillary zone electrophoresis with uniform zeta potential, *J. Chromatogr. A 709*: 63 (1995).
4. V. P. Andreev and E. E. Lisin, On the mathematical model of capillary electrophoresis, *Chromatographia 37*: 202 (1993).
5. B. Potocek, B. Gas, E. Kenndler, and M. Stedry, Electroosmosis in capillary zone electrophoresis with non-uniform zeta potential, *J. Chromatogr. A 709*: 51 (1995).
6. V. P. Andreev and N. V. Shirokih, Electroosmotic flow profile in the capillary made of several sections, 20th Int. Symp. on Capillary Chromatography, Proceedings on CD, 1998, paper H 11.
7. C. A. Keely, T. A. A. M. van de Goor, and D. McManigill, Modeling flow profiles and dispersion in capillary electrophoresis with nonuniform zeta potential, *Anal. Chem. 66*: 4236 (1994).
8. V. P. Andreev, S. G. Dubrovsky, and Y. V. Stepanov, Mathematical modeling of capillary electrophoresis in rectangular channels, *J. Microcol. Separ. 9*: 443 (1997).

Victor P. Andreev

Electrospray Ionization Interface for CE–MS

The development of the electrospray ionization (ESI) source for mass spectrometry provided an ideal means of detection for capillary electrophoretic (CE) separations. The ESI source is currently the preferred interface for CE–MS, due to the fact that it can produce ions directly from liquids at atmospheric pressure and with high sensitivity and selectivity for a wide range of analytes.

Electrospray ionization is initiated by generating a high potential difference between the spray capillary tip and a counterelectrode [1]. This electric field leads to the production of micron-sized droplets with an uneven charge distribution, generally accepted to be due to an electrophoretic mechanism acting on electrolytes in the solvent [2]. This mechanism, combined with a shrinkage of the droplets due to solvent evaporation (aided by heat and an applied gas flow into the source), leads to electrostatic repulsion overcoming surface tension in the droplet. The "Rayleigh" limit is reached, a "Taylor cone" is formed, and smaller highly charged droplets are emitted, eventually leading to the production of gas-phase ions [1–3]. These ions are accelerated through a skimmer into successive vacuum stages of the mass analyzer. The ESI source has been demonstrated to act as an electrolytic cell, generating electrochemical oxidation and reduction [2]. The exact ionization mechanism will vary with experimental conditions and is still an area of continuing in-depth research and discussion [3].

Electrospray ionization is classified as a "soft" ionization technique. It produces molecular-weight information and very little, if any, fragmentation of the analyte ion, unless induced in the vacuum region of the mass analyzer. The number of charges accumulated by an analyte ion is proportional to its number of basic or acidic sites. The spray polarity and conditions, solution pH and nature, as well as solute concentration will all effect the charge state distribution observed in the mass spectrum. Multiple charging of an analyte ion encourages the release of very high-molecular-weight ions. It is mainly due to this fact that ESI has gained such enormous interest, especially among biochemists. Employing only small, relatively inexpensive mass analyzers, spectrometrists are able to obtain high-sensitivity information on analytes with molecular weights of up to 200 kDa. The multiple-charging phenomenon means that the mass-to-charge (m/z) range of the analyzer does not generally need to exceed 3000. A deconvolution algorithm [4], generally built nowadays into the instrument software, can be applied to the series of multiply-charged, molecular-ion peaks, and a single peak, representing the molecular weight, is then displayed on a "true mass" scale. The m/z scale is calibrated with standards of known exact mass. Whereas ESI–MS (mass spectrometry) has made the largest impact on large biomolecule analysis, CE–ESI–MS has also been applied with great success to the analysis of many small-molecule applications.

The development of the first CE–MS was prompted by the early reports on electrospray ionization (ESI–MS) by Fenn and co-workers in the mid-1980s [1], when it was recognized that CE would provide an optimal flow rate of polar and ionic species to the ESI source. In this initial CE–MS report, a metal coating on the tip of the CE capillary made contact with a metal sheath capillary to which the ESI voltage was applied [5]. In this way, the sheath capillary acted as both the CE cathode, closing the CE electrical circuit, and the ESI source (emitter). Ideally, the interface between CE and MS should maintain separation efficiency and resolution, be sensitive, precise, linear in response, maintain electrical continuity across the separation capillary so as to define the CE field gradient, be able to cope with all eluents presented by the CE separation step, and be able to provide efficient ionization from low flow rates for mass analysis.

Several research groups have presented work on the development of CE–ESI–MS interfaces. The interfaces developed can be categorized into three main groups: coaxial sheath flow, liquid junction, and sheathless interfaces. A schematic of the sheath-flow interface first developed for CE–ESI–MS by Smith et al. [6] is illustrated in Fig. 1a. A sheath liquid, with an electrolytic content, is infused into the ESI source at a constant rate, through the coaxial sheath capillary which surrounds the end of the separation capillary and terminates near the end of the separation capillary. This sheath liquid mixes with the separation buffer as it elutes from the tip of the CE capillary, thus providing the necessary electrical contact between the ESI needle and the CE buffer, and closing the CE circuit. Because the CE terminus and ESI source are at the same voltage, if the ESI source requires a high voltage (2–5 kV) (rather than ground potential), then the ESI voltage chosen also directly affects the potential difference across the separation capillary. To date, the sheath–liquid interface has been the most widely used and accepted system, being the simplest to construct, with numerous results published employing sheath liquids typically containing 60–80% organic solvent, modified with 1–3% acid in water and typically introduced at flow rates of 1–4 μL/min. The composition of the sheath liquid should be optimized for the specific systems under investigation. Recent

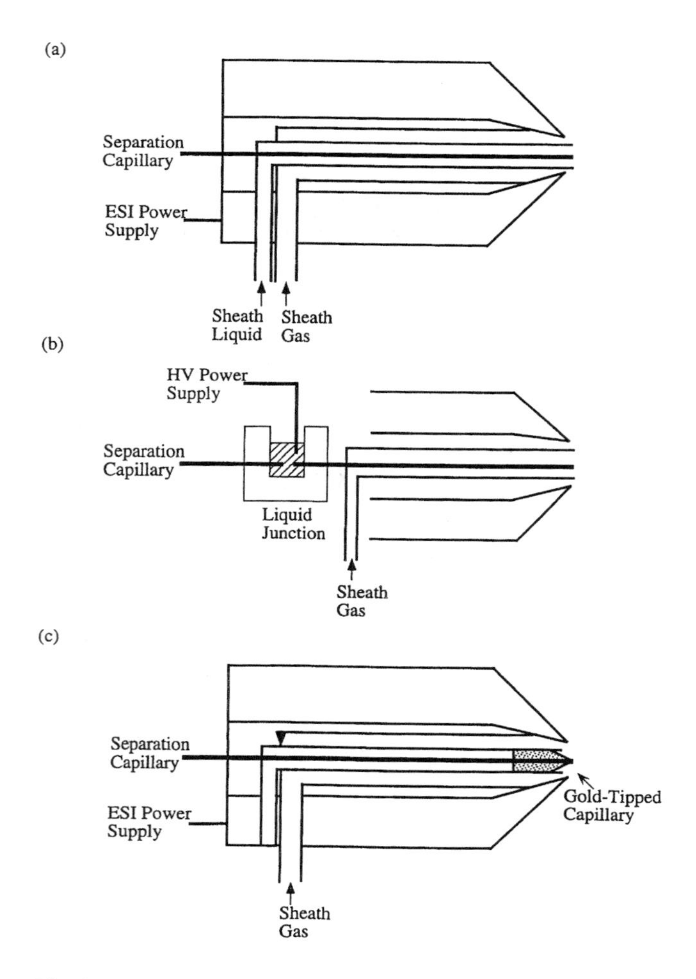

Fig. 1 Schematic illustration of CE–MS interfaces to an ESI source: (a) a coaxial sheath-flow interface; (b) a liquid-junction interface; (c) a sheathless interface.

reports have confirmed that the relative dimensions and positioning of the separation and sheath capillaries also influence sensitivity and stability.

Although the additional flow of an organic-containing electrolyte into the ESI source moderately extends the range of CE buffer systems that can be used, the CE buffer composition still has a dramatic effect on the ESI signal, minimizing the buffer choice for best sensitivity to volatile solutions. Reports have also highlighted the need for a considered selection of sheath-liquid composition due to the possibility of formation of moving ionic boundaries inside the capillary [7]. The possibility of these effects occurring should be considered and minimized when transferring a CE method from an alternative detection system to MS. It should be noted, however, that these effects are minimized or eliminated when there is a sufficiently strong flow toward the CE terminus.

A "liquid-junction interface" has also been suggested and applied for CE–ESI–MS [8]. Electrical contact with this interface is established through the liquid reservoir which surrounds the junction of the separation capillary and a transfer capillary, as shown in Fig. 1b. The gap between the two capillaries is approximately 10–20 μm, allowing sufficient makeup liquid from the reservoir to be drawn into the transfer capillary while avoiding analyte loss. The flow of makeup liquid into the transfer capillary is induced by a combination of gravity and the Venturi effect of the nebulizing gas at the capillary tip [8].

In comparisons of coaxial sheath-flow and liquid-junction interfaces, it has been noted that although both provide efficient coupling, the former is generally easier to operate. One of the major disadvantages in employing the liquid-junction interface is in establishing a reproducible connection inside the tee piece. Also, the use of a transfer capillary, which has no potential difference applied across it, can lead to peak broadening. Advantages of this interface, however, include the possibility of combining different outer-diameter capillaries through the junction and the extra mixing time provided for the makeup liquid and CE eluant.

The problem with both interfaces described so far is that they depend on the addition of excess electrolyte to the ESI source to maintain the circuit, generally leading to a decrease in analyte sensitivity. As previously mentioned, the first CE–MS interface reported made electrical connection between the separation buffer and the ESI needle via a metal coating on the tip of the CE capillary [5], as represented in Fig. 1c. Although femtomole detection limits and separation efficiencies of up to half a million theoretical plates were achieved, problems included a high dependence on the buffer system used and the need to regularly replace the metal coating on the capillary tip.

The further development of interfaces which do not rely on an additional liquid flow are currently underway. Generally, they have employed metal deposition on the CE terminus that is tapered (by chemical etching or mechanical pulling) to provide an increased electric field at the capillary tip. These so-called "microspray" and "nanospray" approaches, with more effective ionization mechanisms, have been adopted recently by several groups for interfacing infusion systems, LC and CE to ESI–MS, and in all cases, significant gains in sensitivity and sample usage have been observed [9]. Attomole level detection limits from nanoliter sample volumes can now be attained, and the ability to form an electrospray from a purely aqueous solution is now possible. Alternative sheathless interfaces have also been briefly investigated [10]. Stability problems still need consideration in most cases. An interface which does not use an additional makeup flow can, as well as aiding sensitivity, also avoid such problems as charge state distribution shifts in the mass spectrum. In addition, the ability to electrospray purely aqueous systems is often advantageous for looking at fragile biological and noncovalently bound analytes. In some cases, however, a makeup liquid may be found necessary. For example, for certain separations, the EOF may need to be minimized or eliminated in the CE capillary, and thus flow rates into the source will not be sufficiently high as to maintain a stable electrospray. If a capillary needs to be coated to avoid analyte interaction with the capillary wall, then a cationic coating, which reverses the EOF rather than eliminating it, should preferably be chosen if a sheathless system is to be employed. Also, it may be found that a makeup liquid is necessary to increase the volatility of a specific CE electrolyte system.

Another disadvantage at present in using the sheathless interface is the time dispensed in preparing the tapered, coated tips. Although the coatings now employed are more stable than those initially used, the tips do not regularly survive more than a day or two of use. This can, however, be due to the tip "plugging" rather than the metal coating deteriorating. Filtering of electrolyte and analytes and rinsing of the capillary can, therefore, often prolong the capillary lifetime.

An instrumental attribute which aids the development and interfacing of CE to ESI–MS is the ability to pressurize the CE capillary, at low pressure for sample injection and higher pressures for capillary content elution. Balancing of the heights of the capillary termini is also an important consideration in order to avoid syphoning effects. In all cases of CE–ESI–MS application, safety, with respect to the electrical circuits, should be considered. It should be verified that all circuits have a common ground, and the addition of a resistor in the ESI power supply line when interfaced to CE is a wise precaution.

An incompatability that does need to be considered in CE–MS method development is the use of certain CE buffer systems and additives which are detrimental to the ESI process. For example, although sample concentration can be increased by the use of more conductive buffers, this approach is not advantageous for ESI–MS detection. These characteristics result in a significant demand upon ESI interface efficiency [11]. Ideally, the chosen CE buffer should be volatile, such as ammonium acetate or formate. The use of pure acids or bases rather than a true buffer has also been shown to be advantageous for certain molecules. Nonaqueous buffer systems are also being employed more widely.

Capillary electrokinetic chromatography (CEKC) with ESI–MS requires either the use of additives that do not significantly impact the ESI process or a method for their removal prior to the electrospray. Although this problem has not yet been completely solved, recent reports have suggested that considered choices of surfactant type and reduction of electro-osmotic flow (EOF) and surfactant in the capillary can decrease problems. Because most analytes that benefit from the CEKC mode of operation can be effectively addressed by the interface of other separations methods with MS, more emphasis has until now been placed upon interfacing with other CE modes. For "small-molecule" CE analysis, in which micellar and inclusion complex systems are commonly used, atmospheric pressure chemical ionization (APCI) may provide a useful alternative to ESI, as it is not as greatly affected by involatile salts and additives.

The efficiency of the ESI detection process for CE–MS can be considered in terms of the simple model of Kebarle and Tang [3] and has been discussed in great detail by Smith and co-workers [11]. These considerations indicate that analyte sensitivity in CE–ESI–MS may be increased by reducing the mass flow rate of the background components. This decrease in background flow rates can be experimentally accomplished by decreasing the electric field or employing smaller-diameter capillaries, and this predicted increase in analyte sensitivity is now well supported by experimental studies [11].

To reduce the elution speed of the analyte ions into the source, the electrophoretic voltage can be decreased just prior to elution of the first analyte of interest, minimizing the experimental analysis time while allowing more scans to be recorded without a significant loss in ion intensity [11]. Alternatively, the use of smaller-diameter capillaries than conventionally used for CE also increases sensitivity [11]. A capillary diameter should, ideally, be commercially available, amenable to alternative detection methods, provide the necessary detector sensitivity, and be free from clogging. Capillary internal diameters of between 20 and 40 μm have been shown to be optimal and are compatible with "microspray" techniques.

The further development of microscale preconcentration and cleanup techniques and the resulting improvements in CE–MS concentration detection limits are likely to expand the use of this analytical technique. The more common use of small-diameter capillaries and even tiny etched microplate devices [10], along with the improvements in ESI spray techniques are pushing research along. Further investigations into improving interface design, durability, reproducibility and sensitivity are still necessary. The availability of improved, less expensive, and smaller mass spectrometers will almost certainly lead to increased use of CE–MS. However, the sensitivity and selectivity already demonstrated by CE–MS systems, in combination with the minute analyte volumes sampled, already make this a highly powerful technique.

References

1. C. M. Whitehouse, R. N. Dreyer, M. Yamashita, and J. B. Fenn, *Anal. Chem. 57*: 675–679 (1985).
2. P. Kebarle and L. Tang, *Anal. Chem. 65*: 972A (1993).
3. M. G. Ikonomou, A. T. Blades, and P. Kebarle, *Anal. Chem. 63*: 1989–1998 (1991).
4. M. Mann, C. K. Meng, and J. B. Fenn, *Anal. Chem. 61*: 1702–1708 (1989).
5. J. A. Olivares, N. T. Nguyen, C. R. Yonker, and R. D. Smith, *Anal. Chem. 59*: 1230–1232 (1987).
6. R. D. Smith, J. A. Olivares, N. T. Nguyen, and H. R. Udseth, *Anal. Chem. 60*: 436–441 (1988).
7. F. Foret, T. J. Thompson, P. Vouros, B. L. Karger, P. Gebauer, and P. Bocek, *Anal. Chem. 66*: 4450–4458 (1994).
8. E. D. Lee, W. Mück, J. D. Henion, and T. R. Covey, *Biomed. Environ. Mass Spectrom. 18*: 844–850 (1989).
9. S. K. Chowdhury and B. T. Chait, *Anal. Chem. 63*: 1660–1664 (1991); M. Wilm and M. Mann, *Anal. Chem. 68*: 1–8 (1996).
10. D. Figeys and R. Aebersold, *Electrophoresis 19*: 885–892 (1998) and references therein.
11. J. H. Wahl, D. R. Goodlett, H. R. Udseth, and R. D. Smith, *Electrophoresis 14*: 448–457 (1993); J. P. Landers, Capillary electrophoresis–mass spectrometry, in *Handbook of Capillary Electrophoresis*, CRC Press, Boca Raton, FL, 1997, and references therein.

Joanne Severs

Elution Chromatography

Elution chromatography is the one of the three basic modes of chromatographic operation, the other two being frontal analysis and displacement chromatography. All three modes were known to Tswett in the early 1900s, although a systematic definition was not made until 1943. Elution chromatography is, by far, the most common chromatographic mode and is virtually the only mode used for analytical separations. Most theoretical work has been directed at the elution mode, although, frequently, the results are applicable to other modes as well.

Elution chromatography is characterized by the introduction of a discrete volume of sample into the chromatographic column that has been previously equilibrated with the mobile phase. Typically, the volume of the sample is small compared to the volume of the column. The individual components of the sample (the solutes) move through the column at different average velocities, each less than the velocity of the mobile phase. The differences in velocities are caused by differences in the interactions of the solutes with the stationary and mobile phases. Assuming essentially equivalent interactions with the mobile phase, solutes which interact strongly with the stationary phase spend less time on the average in the mobile phase and, consequently, have a lower average velocity than components which interact weakly with the stationary phase. If the difference in the average velocities of two solutes is sufficiently large, the dispersive transport within the column is sufficiently small, and the column is sufficiently long, the solute bands will be resolved from one another by the time they exit the column.

Elution chromatography can be performed with a constant mobile-phase composition (isocratic elution) or with a mobile-phase composition that changes during the elution process (gradient elution). The following discussion focuses on isocratic operation (see the entry Gradient Elution, Overview for further information on gradient operation). Further, each of the mechanistic categories of chromatography can be performed in the elution mode and additional information on elution chromatography can be obtained by reference to the appropriate entries of this encyclopedia.

Elution chromatography is categorized as being linear or nonlinear, depending on the distribution isotherm, and as being ideal or nonideal, with ideal behavior requiring both infinite mass-transfer kinetics and negligible axial dispersion. Although truly linear distribution isotherms are rare, at low solute concentrations or over small ranges of solute concentration, sufficient linearity may exist to approximate linear elution. Linear, ideal elution would result in band profiles that are identical to the injection profiles — an unrealistic situation. Under linear, nonideal elution conditions, thermodynamic factors control band retention and kinetic factors such as mass-transfer resistances control the band shape.

Figure 1 shows the effect of isotherm nonlinearity on band shape. Figure 1a corresponds to linear elution; the

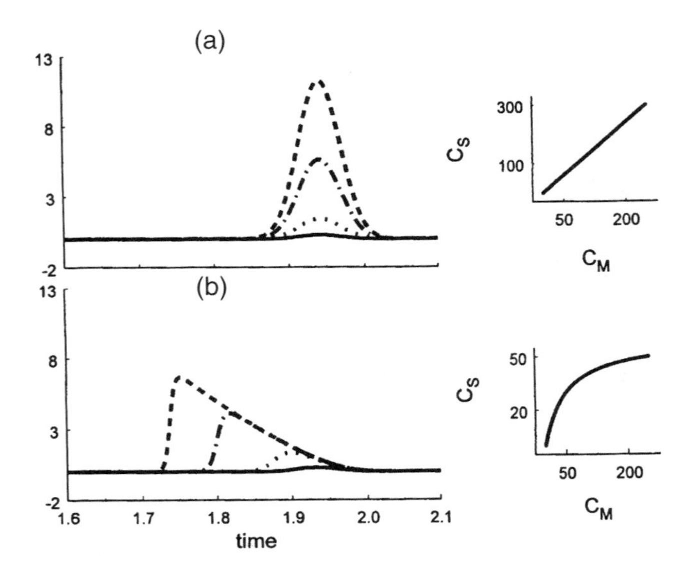

Fig. 1 Effect of distribution isotherm curvature on peak shape in elution chromatography. (a) A linear isotherm resulting in linear elution; the band shape is independent of the solute concentration; (b) a nonlinear isotherm results in asymmetric peaks when increasing amounts of solute are injected. In the example shown, the isotherm results in a diffuse rear edge and a sharp leading edge (i.e., tailing). If the isotherm were curved in the other direction, the leading edge would be diffuse and the read edge sharp (fronting).

band shape is independent of the solute concentration because the isotherm is linear. In Fig. 1b, increasing amounts of solute result in a diffuse rear edge and a sharp leading edge (i.e., tailing). The peak maximum moves forward with increasing sample load. If the isotherm were curved in the other direction, the leading edge would be diffuse and the read edge sharp (fronting) and the peak maximum would move back with increasing sample load. However, isotherm nonlinearity is not the only cause of asymmetric peak shape.

The retention of a solute in elution chromatography is usually expressed as the retention factor, k (capacity factor or k'), given by $k = (t_R - t_M)/t_M$, where t_R is the retention time of the solute and t_M is the holdup time (void time or t_0). The holdup time is the time required to elute a component that is not retained at all by the stationary phase. [See the entry Dead Point (Volume or Time).] One can relate k to the distribution coefficient, K, by $k = K\beta$, where β is the phase ratio, the ratio of the stationary-phase volume to the mobile-phase volume. Rearranging the definition of retention factor, we find that $t_R = t_M(1 + k) = t_M(1 + K\beta)$. Because it is usually reasonable to assume that t_M and β are the same for different solutes, the retention time differences are due to distribution coefficient differences. Under the appropriate conditions, the distribution coefficient can be related to the thermodynamic distribution constant, and the elution chromatographic measurements can be used for physicochemical determinations of thermodynamic parameters.

Differences in solute retention are usually expressed as the separation factor (selectivity coefficient or α), given by $\alpha = k_b/k_a$, where k_a and k_b are the retention factors of the two solutes in question. By convention, k_b is the more retained solute and $\alpha > 1$, although this is not always followed. Again, because it is reasonable to assume that β is the same for different solutes, $\alpha = K_b/K_a$, where K_a and K_b are the distribution coefficients of the two solutes, and, again, retention time differences are due to distribution coefficient differences. If two solutes have the same distribution coefficient (i.e., $\alpha = 1$) in a particular combination of mobile and stationary phases, they cannot be separated by elution chromatography in that system. However, $\alpha \neq 1$ is a necessary, but not sufficient, condition for a successful separation.

As a solute moves through the column, it undergoes dispersive transport as well as separative transport. Under typical elution chromatographic conditions, the dispersive transport is caused by axial diffusion and mass-transfer considerations, such as slow adsorption–desorption kinetics. This dispersive transport results in band spreading (see the entry Band Spreading, Mechanism), which can prevent adequate separation of different solutes. The plate number (plate count, number of theoretical plates, theoretical plate number, or N), defined as $N = t_R^2/\sigma_t^2$, where σ_t^2 is the variance of the band in time units, is a measure of the column efficiency (i.e., the ratio of separative to dispersive transport). Several alternate forms of this equation are commonly used, usually based on the assumption of a Gaussian peak shape. The effective plate number, N_{eff}, is a combination of the plate number and the capacity factor {i.e., $N_{\text{eff}} = N[k/(1 + k]^2$} and is generally more useful than N for comparing the resolving power of different columns. Another common measure of column efficiency is the plate height (HETP, height equivalent to a theoretical plate, or H), defined by $H = L/N$, where L is the length of the column, usually in centimeters. This is frequently presented as the reduced plate height h, the ratio of the plate height to the diameter of the packing material. A "good" column has a high plate count (a low plate height, $2 < h < 5$). Further discussion can be found in the Further Readings section.

Efficiency in High-Performance Liquid Chromatography

The overall quality of the separation of two solutes is measured by their resolution (R_S), a combination of the thermodynamic factors causing separative transport and the kinetic factors causing dispersive transport and is an index of the effectiveness of the separation. Defined by $R_S = (t_{r,b} - t_{r,a})/\frac{1}{2}(w_{t,b} + w_{t,a})$, where a and b refer to the two solutes, $t_{r,x}$ is the retention time of solute x, and $w_{t,x}$ is the peak width at the base of solute x in units of time, it is frequently estimated by use of the fundamental resolution equation,

$$R_S = \left(\frac{\sqrt{N}}{4}\right)\left(\frac{\alpha - 1}{\alpha}\right)\left(\frac{k_b}{1 + k_b}\right)$$

where k_b is the retention factor of the more retained solute, α is the separation factor of the solute pair under consideration, and N is the plate count. This equation assumes that the peak shapes are Gaussian and that the peak widths are equivalent.

Easy recognition of the two peaks over a wide range of relative concentrations is possible for $R_s = 1$ and this is essentially the practical minimum resolution desirable. It is usually stated that $R_s = 1$ corresponds to a peak purity of about 98%; however, this is correct only for equal concentrations of the two solutes. As the ratio of relative concentrations of the two solutes deviates from 1, the recovery of the lower concentration solute at a given level of purity becomes poorer.

Examination of the fundamental resolution equation shows that improvements in resolution can be obtained by the following:

Increasing the column efficiency. The dependence of R_S on $\sqrt{N}$, rather than N, means that this method is most effective when the column efficiency is initially low. In other words, when using efficient columns to develop a separation, major improvements in R_S are not generally obtained by increasing N.

Increasing α. If α is close to 1.0, the greatest increase in R_S can be obtained by changing those parameters which influence α (i.e., the mobile-phase composition, the choice of stationary phase, the temperature, or, less frequently, the pressure). Increasing α from 1.1 to 1.2 increases R_S by more than 80%. However, as α increases, the amount of increase in R_S decreases, so that increasing α from 2.1 to 2.2 increases R_S by only about 4%.

Increasing k. If k_b (and thus k_a) < 1, R_S can be significantly increased by changing the mobile-phase composition to increase k_b. As for α, the amount of increase decreases as k_b increases, so that changing k_b from 0.5 to 1.5 improves R_S by about 80%, increasing k_b from 1.5 to 2.5 increases R_S by about 20%. Moreover, increasing k_b increases the analysis time; thus, this approach is also of limited practicality.

To summarize, the most successful approach to obtaining an adequate R_S is usually to increase α by varying the mobile-phase composition [e.g., choice of solvent(s), pH, or temperature] or by varying the stationary phase. Increasing R_S by increasing N or k_b works in selected instances, but it is not as generally applicable.

Developing an elution separation method to be used for the analysis of numerous samples requires more than obtaining the minimal resolution of the solutes of interest. A successful method should not only achieve the desired separation but should also do so in a cost-effective and robust manner. High-performance liquid chromatography (HPLC) method development has its own, extensive literature, reflecting the importance of HPLC as an analytical technique.

Snyder *et al.* state the goals of the HPLC method development as follows: (a) precise and rugged quantitative analysis requires that R_S be greater than 1.5; (b) a separation time of $<$5–10 min; (c) $\leq$2% RSD for quantitation in assays ($\leq$5% for less demanding analyses and $\leq$15% for trace analysis); (d) a pressure drop of $<$150 bar; (e) narrow peaks to give large signal-to-noise ratios; and (f) minimal mobile-phase consumption per run. Additionally, thorough testing of the robustness of proposed methods is recommended.

Suggested Further Reading

Bidlingmeyer, B. A., *Practical HPLC Methodology and Applications*, John Wiley & Sons, New York, 1992.

Giddings, J. C., *Unified Separation Science*, John Wiley & Sons, New York, 1991.

Guiochon, G., S. G. Shirazi, and A. M. Katti, *Fundamentals of Preparative and Nonlinear Chromatography*, Academic Press, Boston, 1994.

Karger, B. L., L. R. Snyder, and Cs. Horvath, *An Introduction to Separation Science*, John Wiley & Sons, New York, 1973, pp. 11–167.

Meyer, V. R., *Practical High-Performance Liquid Chromatography*, 3rd ed., John Wiley & Sons, New York, 1998.

Rizzi, A., Retention and selectivity, in *Handbook of HPLC* (E. Katz, R. Eksteen, P. Schoenmakers, and N. Miller, eds.), Marcel Dekker, Inc., New York, 1998, pp. 1–54.

Snyder, L. R. and J. J. Kirkland, *Introduction to Modern Liquid Chromatography*, 2nd ed., John Wiley & Sons, New York, 1979.

Snyder, L. R., J. J. Kirkland, and J. L. Glajch, *Practical HPLC Method Development*, 2nd ed., John Wiley & Sons, New York, 1997.

John C. Ford

Elution Modes in Field-Flow Fractionation

Field-flow fractionation is, in principle, based on the coupled action of a nonuniform flow velocity profile of a carrier liquid with a nonuniform transverse concentration profile of the analyte caused by an external field applied perpendicularly to the direction of the flow. Based on the magnitude of the acting field, on the properties of the analyte, and, in some cases, on the flow rate of the carrier liquid, different elution modes are observed. They basically differ in the type of the concentration profiles of the analyte. Three types of the concentration profile can be derived by the same procedure from the general transport equation. The differences among them arise from the course and magnitude of the resulting force acting on the analyte (in comparison to the effect of diffusion of the analyte). Based on these concentration profiles, three elution modes are described.

Introduction

Field-flow fractionation (FFF) represents a family of versatile elution techniques suited for the separation and characterization of macromolecules and particles. Separation results from the combination of a nonuniform flow velocity profile of a carrier liquid and a nonuniform transverse concentration profile of an analyte caused by the action of a force field. The field, oriented perpendicularly to the direction of the flow, forms a specific concentration distribution of the analyte inside the channel. Because of the flow velocity profile, different analytes are displaced along the channel with different mean velocities, and, thus, their separation is achieved.

According to the original concept [1], the field drives the analytes to the accumulation wall of the channel. This concentrating effect is opposed by diffusion, driven by Brownian motion of the analytes, which causes a steady state when the convective flux is exactly balanced by the diffusive flux. The concentration profile is exponential and the corresponding elution mode is referred to as the normal mode. Recently, it has been called the Brownian elution mode [2].

During last two decades, new elution modes were described [3–5] that were not suggested in the original concept [1]. Basically, they differ in the type of the analyte concentration profile.

In 1978, Giddings and Myers described another elution mode for large particles under conditions when diffusion effects can be neglected [3]. The particles form a layer on the channel wall and, under the influence of the carrier liquid flow, they roll on the channel bottom to the channel outlet. This elution mode is referred as the steric mode [3].

The above-mentioned elution modes apply to the situation when the resulting force acting on the analytes does not change its orientation inside the channel. However, there exist conditions when the resulting force acting on analytes may change its orientation inside the channel [e.g., two counteracting forces, a gradient of a property (pH, density) of the carrier liquid, influence of hydrodynamic lift forces]. Under such conditions, the analytes form narrow zones at the positions where the resulting forces acting on them equal zero. The resulting force is changing its sign below and above this position in such a way that the analyte is focused into this equilibrium position. The concept of formation of narrow zones of analytes inside the FFF channel was first described in 1977 by Giddings [6]. In 1982, a technique utilizing sedimentation–flotation equilibrium and centrifugal field was suggested [7]. However, this technique has not been yet verified experimentally. Later, the general features of this elution mode were described and several techniques were implemented; for a review, see Ref. 8. The mode was called either the hyperlayer [4] or focusing [5] elution mode.

Some other elution modes have been described. They are induced by various factors — cyclical field, secondary chemical equilibria, adhesion chromatography, asymmetrical electro-osmotic flow; for a review, see Ref. 2. However, the number of their implementations is rather limited, and for this reason, these modes are not discussed here.

Theory

Field-flow fractionation experiments are mainly performed in a thin ribbonlike channel with tapered inlet and outlet ends (see Fig. 1). This simple geometry is advantageous for the exact and simple calculation of separation characteristics in FFF. Theories of infinite parallel plates are often used to describe the behavior of analytes because the cross-sectional aspect ratio of the channel is usually large and, thus, the end effects can be neglected. This means that the flow velocity and concentration profiles are not dependent on the coordinate y. It has been shown that, under suitable conditions, the analytes move along the channel as steady-state zones. Then, equilibrium concentration profiles of analytes can be easily calculated.

Generally, the concentration profile of analytes in FFF can be obtained from the solution of the general trans-

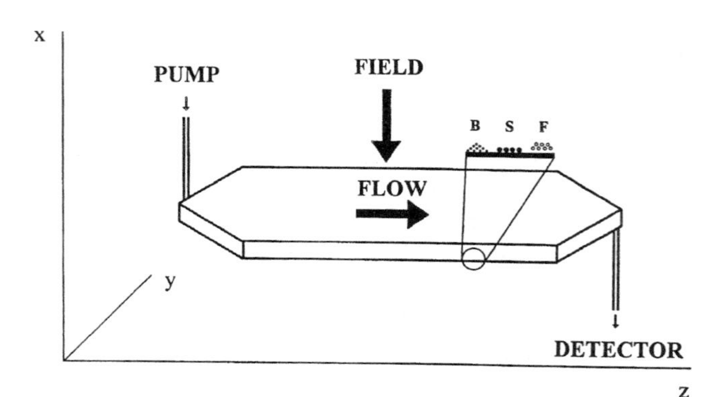

Fig. 1 The orientation of the field and flow in the given coordinate system. The zoomed inset shows the schematic representation of the zone shapes in particular elution modes (B for Brownian, S for steric, and F for focusing).

port equation. For the sake of simplicity, the concentration profile of the steady-state zone of the analyte along the axis of the applied field is calculated from the one-dimensional transport equation:

$$J_x = W_x c(x) - D\frac{\partial c}{\partial x} \tag{1}$$

where J_x and W_x are the components of the flux density of the analyte and of the transport velocity of the analyte along the axis of the applied field, $c(x)$ is the analyte concentration distribution along the direction of the applied field, and D is the total effective diffusion coefficient. The term $W_x c(x)$ corresponds to the x component of the convective flux of the analyte and the term $D(\partial c/\partial x)$ corresponds to the x component of the diffusive flux of the analyte. W_x equals the sum of the x components of the transport velocity of the analyte induced by the external field applied U_x and the transport velocity of the analyte induced by the carrier liquid flow v_x ($W_x = U_x + v_x$). Because of the direction of the carrier liquid flow inside the FFF channel, the component v_x equals zero (the x axis is perpendicular to the direction of the flow) and, thus, W_x equals U_x.

Following the treatment given by Giddings [9], imposing for the condition of the steady-state zone of the analyte, which is characterized by the null flux density, and applying the equation of continuity, the general solution of the analyte concentration profile can be expressed in the form

$$c(x) = c_0 \exp\left[\int_0^x \left(\frac{U_x}{D}\right) dx\right] \tag{2}$$

The integration limit $x = 0$ corresponds to the accumulation wall boundary. The particular solutions for the concentration profile are dependent on the course of the force field inducing the transport of the analyte, and the ratio of U_x and D.

The equation of the field-induced transport velocity was derived by Giddings [10]:

$$U_x = -ax^n \tag{3}$$

where a is constant and n equals 0 or 1. If $n = 0$, then U_x is constant; if $n = 1$, then U_x is dependent on the position inside the channel.

Discussion

Brownian Elution Mode

The field-induced velocity of the analyte in the separation channel is constant and comparable with its diffusive motion (U_x = constant, $U_x t \approx \sqrt{2Dt}$, where t is time). The resulting concentration profile of the analyte is given by the exponential relationship [9]

$$c(x) = c_0 \exp\left(-\frac{|U_x|}{D}x\right) \tag{4}$$

where c_0 is the maximum concentration at the accumulation channel wall. The elution mode with the exponential concentration profile is called Brownian [2].

It is known that there are two main factors influencing the behavior of analytes in this elution mode: the properties of the analytes (characterized by the so-called analyte–field interaction parameter [11] and the diffusion coefficient) and the strength of the field applied.

In Brownian elution mode, the retention ratio R is indirectly dependent on both the applied force F and the thickness of the channel w, and independent on the flow rate [9]. It can be expressed in an approximate form:

$$R = \frac{6kT}{Fw} \tag{5}$$

where k is the Boltzmann constant and T is the absolute temperature.

Steric Elution Mode

The velocity of transport induced by the force field in the separation channel is constant and much higher than the velocity caused by the diffusive motion of the analyte (U_x = constant, $U_x t \gg \sqrt{2Dt}$). In this case, the analyte forms a layer on the accumulation channel wall and its concentration in any other position inside the channel equals zero. The particle radius r_p describes the distance of the particle center from the accumulation wall:

$$c(r_p) = c_0 \quad \text{and} \quad c(x \neq r_p) = 0 \tag{6}$$

The elution mode is called steric [3]. The retention ratio can be expressed in the form

$$R = \frac{6r_p}{w} \tag{7}$$

This shows that R is independent of both the field applied and the flow rate, and it is dependent only on the particle radius and the channel thickness. In fact, the retention ratio values corresponding to the pure steric elution mode have been seldom observed experimentally [12]. The observed values often correspond to the focusing elution mode as a result of the action of some additional forces influencing retention behavior of analytes.

Focusing Elution Mode

In this elution mode, the velocity of analyte transport induced by a force field in the separation channel is dependent on the position across the channel ($U_x \neq$ constant). Based on Eq. (3), the nonconstant transport velocity can be, in the simplest case, described as

$$U_x = -a(x - s) \tag{8}$$

where s is the distance of the center of the focused zone from the channel wall (i.e., the position where the resulting force acting on the analyte equals zero). Combining this equation with Eq. (2), we obtain a relation for the resulting concentration profile of the analyte:

$$c(x) = c_0 \exp\left(-\frac{a}{2D}(x - s)^2\right) \tag{9}$$

where c_0 is the maximum concentration at the center of the focused zone at the position s. The concentration profile of the analyte across the channel thickness, in this simplest case, is Gaussian. In other cases, where other secondary effects act on the retention in the focusing elution mode, the observed concentration profile is more complex. However, even in these cases, the main feature remains the same; that is, the maximum concentration of the analyte is at the equilibrium position, where the resulting force acting on the analyte is zero, and not on the channel wall as in the case of the Brownian and steric elution modes. The elution mode is called focusing [5] or hyperlayer [4]. In the focusing elution mode, the retention ratio can be expressed in a form formally similar to the expression given for the steric elution mode [see Eq. (7)]:

$$R = \frac{6s}{w} \tag{10}$$

At least two counteracting forces are necessary for the formation of the focused zone of the analyte. The center of the zone is located at the position s where the resulting force is zero. Changing of both forces can control the resulting position of the particle zone.

Conclusions

In the majority of FFF techniques, the retention ratio is dependent on the analyte size. This dependence for Brownian and steric mode is described by Eq. (11), derived by Giddings [13]:

$$R = 6(\alpha - \alpha^2) + 6\lambda(1 - 2\alpha)\left[\coth\left(\frac{1 - 2\alpha}{2\lambda}\right) - \frac{2\lambda}{1 - 2\alpha}\right] \tag{11}$$

where $\alpha = d/2w$, $\lambda = kT/Fw$, and d is the analyte diameter. The curve describing this dependence is shown in Fig. 2. The values of R for the focusing mode lie above the curve. This complex situation shows that determination of the elution mode is very important for evaluation of the measured retention data because different elution modes can act on particular analytes in the same experiment.

Acknowledgment

This work was supported by a grant No. A4031805 from the Grant Agency of Academy of Sciences of the Czech Republic.

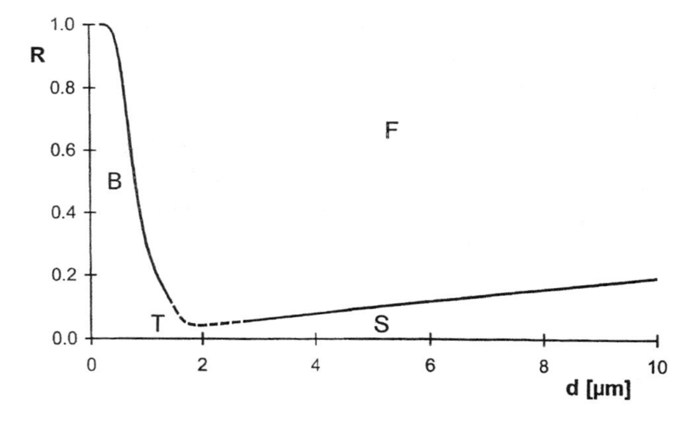

Fig. 2 Schematic representation of the dependence of the retention ratio R on the analyte size. Curve B corresponds to Brownian mode and the line S to the steric mode. The dashed part T denotes the transition between these two modes. The area F shows the range of applicability of the focusing mode.

References

1. J. C. Giddings, *Separ. Sci. 1*: 123 (1966).
2. M. Martin, in *Advances in Chromatography, Vol. 39* (P. R. Brown and E. Grushka, eds.), Marcel Dekker, Inc., New York, 1998, pp. 1–138.
3. J. C. Giddings and M. N. Myers, *Separ. Sci. Technol. 13*: 637 (1978).
4. J. C. Giddings, *Separ. Sci. Technol. 18*: 765 (1983).
5. J. Janća and J. Chmelík, *Anal. Chem. 56*: 2481 (1984).
6. J. C. Giddings, *Am. Lab. 24*: 20D (1992).
7. J. Janća, *Makromol. Chem. Rapid Commun. 3*: 887 (1982).
8. J. Janća, J. Chmelík, V. Jahnová, N. Nováková, and E. Urbánková, *Chem. Anal. 36*: 657 (1991).
9. J. C. Giddings, *Unified Separation Science*, John Wiley & Sons, New York, 1991.
10. J. C. Giddings, *J. Chem. Phys. 49*: 81 (1968).
11. J. C. Giddings and K. D. Caldwell, in *Physical Methods of Chemistry, Vol. 3B* (B. W. Rossiter and J. F. Hamilton, eds.), John Wiley & Sons, New York, 1989, p. 867.
12. J. Pazourek, K.-G. Wahlund, and J. Chmelík, *J. Microcol. Separ. 8*: 331 (1996).
13. J. C. Giddings, *Separ. Sci. Technol. 13*: 241 (1978); *14*: 869 (1979).

Josef Chmelík

Enantiomer Separations by TLC

Enantiomers are compounds which have the same chemical structure but different conformations, whose molecular structures are not superimposable on their mirror images, and, because of their molecular asymmetry, these compounds are optically active. The most common cause of optical activity is the presence of one or more chiral centers which, in organic chemistry, are usually related to tetrahedral structures formed by four different groups around carbon, silicon, tin, nitrogen, phosphorus, or sulfur.

Many molecules are chiral, even in the absence of stereogenic centers; that is, molecules containing adjacent π-systems, which cannot adopt a coplanar conformation because of rotational restrictions due to steric hindrance, can exist in two mirror forms (atropisomers). This is the case for some dienes or olefins, for some nonplanar amides, and for the biphenyl or binaphthyl types of compounds.

Designation of optical isomers can be by the symbols D and L, which are used to indicate the relationship between configurations based on D (+)-glyceraldehyde as an arbitrary standard. If such relationship is unknown, the symbols (+) and (−) are used to indicate the sign of rotation of plane polarized light (i.e., dextrorotatory and levorotatory). In 1956, Cahn et al. [1] presented a new system, the (*R*) and (*S*) absolute configurations of compounds.

Many enantiomers show different physiological behaviors and it is, therefore, desirable to have reliable methods for the resolution of racemates and the determination of the enantiomeric purity. To this end, thin-layer chromatography (TLC) is a simple, sensitive, economic, and fast method which allows easy control of a synthetic process and can be used for preparative separations.

TLC Separation of Enantiomers by Use of Diastereomeric Derivatives

Because the stationary phases originally used in liquid chromatography were achiral, much research was devoted to the separation of enantiomers as diastereomeric derivatives produced by reaction of the enantiomers with an optically pure reagent (A_R). The resultant diastereomers could, because of their different physicochemical properties, then be separated on conventional stationary phases:

$$\begin{matrix} E_R & & & & A_R E_R \\ & + & A_R & \Leftrightarrow & \\ E_S & & & & A_R E_S \end{matrix}$$

Enantiomers (similar properties)	Diastereomers (different properties)

In addition, a significant increase in the sensitivity of detection and the location on the layers of some compounds not otherwise identifiable can be achieved by this method. There are, however, some disadvantages: (a) It is very important to use derivatization reagents with 100% optical purity; (b) quantitation is founded on the assumption that the reaction is complete and not associated with racemization; (c) the distance between the two chiral centers should be as close as possible to each other in order to maximize the difference in chromatographic properties.

Many chiral derivatization reactions were used and the compounds examined are mostly amphetamines, β-blocking agents, amino acids, and anti-inflammatory drugs. Silica gel and, to a lesser extent, silanized silica were used as stationary phases. The ΔR_f values obtained for the diastereomeric pairs were not usually very high (0.04–0.07), with the exception of amino alcohol and amino acid diastereomers obtained with Marphey's reagent, a derivative of L-alanine amide (0.06 — 0.22).

TLC Separations of Enantiomers by Chiral Chromatography

In chiral chromatography, the two diastereomeric adducts A_RE_R and A_RE_R are formed during elution, rather than synthetically, prior to chromatography. The adducts differ in their stability using chiral stationary phases (CSP) or chiral coated phases (CCP) and/or in their interphase distribution ratio adding a chiral selector to the mobile phase (CMP). The difference between the interactions of the chiral environment with the two enantiomers is called enantioselectivity.

According to Dalgliesh [2], three active positions on the selector must interact simultaneously with the active positions of the enantiomer to reveal differences between optical antipodes. This is a sufficient condition for resolution to occur, but it is not necessary. Chiral discrimination may happen as a result of hydrogen-bonding and steric interactions, making only one attractive force necessary in this type of chromatography. Moreover, the creation of specific chiral cavities in a polymer network (as in "molecular imprinting" techniques) could make it possible to base enantiomeric separations entirely on steric fit.

The most important technique for enantiomeric separation in TLC is chiral ligand-exchange chromatography (LEC). LEC is based on the copper(II) complex formation of a chiral selector and the respective optical antipodes. Differences in the retention of the enantiomers are caused by dissimilar stabilities of their diastereomeric metal complexes. The requirement of sufficient stability of the ternary complex involves five-membered ring formation, and compounds such as α-amino and α-hydroxy-acids are the most suitable.

Chiral Stationary Phases and Chiral Coated Phases

Few chiral phases are used in TLC; one of the main reasons for this is that stationary phases with a very high ultraviolet (UV) background can be used only with fluorescent or colored solutes. For example, amino-modified ready-to-use layers bonded or coated with Pirkle-type selectors [3], such as *N*-(3,5-dinitrobenzoyl)-L-leucine or *R*(−)-α-phenylglycine, are pale yellow and strongly adsorb UV radiation.

Another reason is the high price of most CSPs. In spite of this, Pirkle-type CSPs, based on a combination of aromatic π–π bonding interactions (charge-transfer complexation), hydrogen-bonding and dipole interactions, allows the resolution of racemic mixtures of 2,2,2-trifluoro-1-(9-anthryl)ethanol,1,1′-bi-2-naphthol, benzodiazepines, hexobarbital, and β-blocking agents derivatized with achiral 1-isocyanatonaphthalene. However, the most widely used CSPs or CCPs are polysaccharides and their derivatives (cellulose, cellulose triacetate, and triphenylcarbamate) and silanized silica gel impregnated with an optically active copper(II) complex of (2S,4*R*, 2′*RS*)-*N*-(2′-hydroxydodecyl)-4-hydroxyproline (ChiralPlate, Macherey-Nagel, and HPTLC-Chir, Merck, Germany) for ligand-exchange chromatography. The chiral layer on the latter plates is combined with a so-called "concentrating zone."

The resolution of optical antipodes on polysaccharides is mainly governed by the shape and size of solutes (inclusion phenomena) and only to a minor extent by other interactions involving the functional groups of the molecules. In the case of microcrystalline cellulose triacetate (MCTA), the type and composition of the aqueous–organic eluent affect the separation because these result in different swelling of MCTA.

The use of silica gel impregnated with chiral polar selector, such as D-galacturonic acid, (+)-tartaric acid, (−)-brucine, L-aspartic acid, or a complex of copper(II) with L-proline, should also be mentioned.

In CSPs, owing to the nature of the polymeric structure, the simultaneous partecipation of several chiral sites or several polymeric chains is conceivable. In CCPs, the chiral sites are distributed at the surface or in the network of the achiral matrix relatively far away from each other and only a bimolecular interaction is generally possible with the optical antipodes.

A survey of optically active substance classes separated with ChiralPlate and HPTLC Chir layers and with MCTA plates is shown in Fig. 1.

Chiral Mobile Phases

Chiral mobile phases enable the use of conventional stationary phases and show minor detection problems compared to CSPs or CCPs. However, high cost chiral selectors (i.e., γ-cyclodextrin) are certainly not advisable for TLC. Enantiomer separations can be achieved using chiral mobile phases in both normal- and reversed-phase

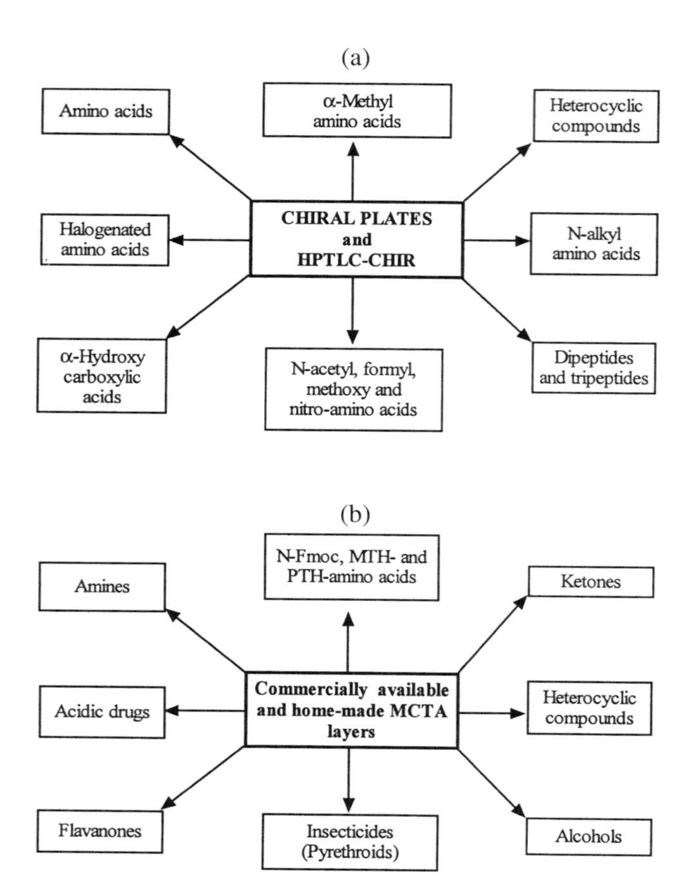

Fig. 1 Classes of chiral organic compounds resolved (a) by ligand-exchange chromatography on Chiral and HPTLC Chir plates and (b) on MCTA layers.

chromatography. The first technique uses silica gel and, mostly, diol F_{254} HPTLC plates (Merck) and, as chiral selectors, D-galacturonic acid for ephedrine, *N*-carbobenzoxy (CBZ)-L-amino acids or peptides and 1*R*-(−)-ammonium-10-camphorsulfonate for several drugs and 2-*O*-[(R)-2-hydroxypropyl)]-β-cyclodextrin for underivatized amino acids. Extremely high ΔR_f values (0.05–0.25) were observed for the various pairs of enantiomers, proving the strong enantioselectivity of this system.

Most separations have been obtained by reversed-phase chromatography on hydrophobic silica gel (RP-18W/UV_{254} and Sil C_{18}-50/UV_{254} from Macherey-Nagel, Germany; KC2F, KC18F and chemically bonded diphenyl-F from Whatman, USA, and RP-18W/F_{254} from Merck, Germany) as stationary phase and β-cyclodextrin and its derivatives, bovine serum albumin (BSA), and the macrocyclic antibiotic vancomycin as chiral agents. Enantiomers which interact selectively with β-cyclodextrin cavities are generally N-derivatized amino acids, whereas the use of BSA as chiral selector is able to resolve many N-derivatized amino acids, tryptophan and its derivatives, derivatized lactic acid, and unusual optical antipodes such as binaphthols.

Quantitative Analysis of TLC-Separated Enantiomers

Although TLC–MS (mass spectrometry) has been shown to be technically feasible and applicable to a variety of problems, thin-layer chromatography is generally coupled with spectrophotometric methods for quantitative analysis of enantiomers. Optical quantitation can be achieved by *in situ* densitometry by measurement of UV-vis absorption, fluorescence or fluorescence quenching, or after exctraction of solutes from the scraped layer. The evaluation of detection limits for separated enantiomers is essential because precise determinations of trace levels of a D- or L-enantiomer in an excess of the other become more and more important. Detection limits as low as 0.1% of an enantiomer in the other have been obtained.

References

1. R. S. Cahn, C. K. Ingold, and V. Prelog, Specification of asymmetric configuration in organic chemistry, *Experentia 12*: 81–94 (1956).
2. C. E. Dalgliesh, The optical resolution of aromatic amino acids on paper chomatograms, *J. Chem. Soc. 3*: 3940–3943 (1952).
3. W. H. Pirkle and J. M. Finn, Chiral high-pressure liquid chromatographic stationary phases. General resolution of arylalkylcarbinols, *J. Org. Chem. 46*: 2935–2938 (1981).

Suggested Further Reading

Gunther, R. and K. Möller, Enantiomer separations, in *Handbook of Thin Layer Chromatography* (J. Sherma and B. Fried, eds.), Marcel Dekker, Inc., New York, 1996, pp. 621–682.

Lepri, L., Enantiomer separation by TLC, *J. Planar Chromatogr. Mod. TLC 10*: 320–331 (1997).

Prosek, M. and M. Puki, Basic principles of optical quantitation in TLC, in *Handbook of Thin Layer Chromatography* (J. Sherma and B. Fried, eds.), Marcel Dekker, Inc., New York, 1996, pp. 273–306.

Luciano Lepri
Alessandra Cincinelli

Enantioseparation by Capillary Electrochromatography

Introduction

Capillary electrochromatography (CEC) is considered to be a hybrid technique that combines the features of both capillary high-performance liquid chromatography (HPLC) and capillary electrophoresis (CE). In CEC, a mobile phase is driven through a packed or an open tubular coating capillary column by electro-osmotic flow [1,2] and/or pressurized flow [3]. The first electrochromatographic experiments were done in early 1974 by Pretorius et al. [4], who applied an electric field across a packed column. This allows the analyte to partition between the mobile and stationary phases. As a high voltage is applied, electrophoretic mobility should also contribute to the chromatographic separation for charged analyses. The ability of CEC to combine electrophoretic mobility with partitioning mechanisms is one of its strongest advantages. For electro-osmotically driven capillary electrochromatography (ED–CEC), the resulting flow profile is almost pluglike; thus, a high column efficiency, comparable to that in CE, can be obtained. For pressure-driven capillary electrochromatography (PD–CEC), although dispersion caused by flow velocity differences causes zone broadening, plate numbers are higher than in capillary HPLC due to the contribution of the electric field to total flow rate. Unlike ED–CEC, the use of an HPLC pump provides stable flow conditions and, thus, offers improvements in retention reproducibility, in sample introduction (e.g., split injection), in suppression of bubble formation, and in gradient elution. More importantly, because the solvent can be mainly driven by pressurized flow, the change of the direction of electric field is no longer limited, and the separation of mixtures of cationic, anionic and neutral compounds becomes possible in a single run. Additionally, neutral molecules can be separated without micelles or other organic additives; this makes CEC more amenable to coupling with mass spectrometry.

Chiral separation in capillary electrophoresis is usually achieved by the addition of chiral complexing agents to form *in situ* diastereometric complexes between the enantiomers and the chiral complexing agent. Many of the chiral selectors successfully used in HPLC [5] can also be applied in CE, and thus the experience from both HPLC and CE can be transferred to CEC. During the last few years, interest in CEC has increased due to the improvement in the preparation of capillary columns [6,7] and in the stability and efficiency of separations [6–9]. A limited but dramatically increasing number of chiral separations in CEC have been reported so far. This review will be mainly devoted to recent developments and applications. We are also interested in exploring the potential advantages offered by capillary electrochromatography and, in particular, its practical utility for enantioseparation.

Enantioselectivity in CEC

Capillary electrochromatography is a more complicated system than CE and HPLC due to the combination of both electrophoretic and chromatographic transport mechanisms. It is difficult to define an effective selectivity (separation factor) as in the case of general chromatography or general electrophoresis. To better illustrate the interactions that control selectivity, we defined a relative selectivity ($\alpha_r = \Delta t_e/t_{e2}$), and postulated a model that illustrates the effect of separation parameters on the enantioselectivity [10].

For enantioseparation chiral stationary phases (CSPs), an expression of the relative selectivity is obtained:

$$\alpha_r = \frac{\phi(K_{f2} - K_{f1})}{1 + \phi K_2 + \phi K_{f2}} \tag{1}$$

Interestingly, this equation indicates that the electrophoresis mechanism does not influence the enantioselectivity and the electric field only plays a role in driving the mobile phase.

For enantioseparation with chiral additives in CEC, we derived another expression:

$$\alpha_r = \frac{(K_{f1} - K_{f2})[\phi K v_c + (\mu_c - \mu_f)E][\mathrm{C}]_m}{(1 + \phi K + K_{f2}[\mathrm{C}]_m)(v_f + v_c K_{f1}[\mathrm{C}]_m)} \tag{2}$$

where v_f and v_c are the apparent flow velocity of the free analyte and the complexed analyte, respectively. Both Eqs. (1) and (2) show that the enantioselectivity is not only dependent on the difference in formation constants (K_f) between a pair of enantiomers with the chiral agents but also is influenced by some experimental factors. Substantially, chiral recognition of enantiomers is the direct result of the transient formation of diastereomeric complex between enantiomeric analytes and the chiral complexing agent (i.e., the difference in formation constants). However, the importance of experimental factors lies in the fact that they can convert the intrinsic difference into the apparent difference in migration velocity along the column. Therefore, the overall selectivity in chiral separation can be considered to be made up of two contributing factors: the intrinsic difference (intrinsic selectivity) in formation constants of a pair of enantiomers, and the conversion efficiency (exogenous selectivity) of the intrinsic difference into the apparent difference in the migration velocity. According to Eq. (2), these experimental factors may include the equilibrium concentration of a chiral selector, the electric field strength, and the properties of the stationary phase.

In CEC with chiral additives, Eq. (2) shows that there exists a maximum selectivity at the optimal concentration of chiral selector. The optimal concentration is not only dependent on the formation constants (K_{f1}, K_{f2}) but also on properties of the column (ϕ and K) {i.e., $[C]_{opt} = \sqrt{(1 + \phi K)/K_{f1}K_{f2}}$}.

Unlike in the case of a chiral column, the selectivity in CEC with chiral additives is determined by both partition and electrophoresis, and the electric field either increases or decreases the selectivity. Table 1 summarizes the relationship between the direction of field strength and the electrophoretic mobility of the free and complexed analytes. For PD–CEC, the solvent is mainly driven by pressurized flow; thus, there is no limitation to change the direction of electric field.

For enantioseparation on CSPs in CEC, nonstereospecific interactions, expressed as ϕK, contribute only to the denominator as shown in Eq. (1), indicating that any nonstereospecific interaction with the stationary phase is detrimental to the chiral separation. This conclusion is identical to that obtained from most theoretical models in HPLC. However, for separation with a chiral mobile phase, ϕK appears in both the numerator and denominator [Eq. (2)]. A suitable ϕK is advantageous to the improvement of enantioselectivity in this separation mode. It is interesting to compare the enantioselectivity in conventional capillary electrophoresis with that in CEC. For the chiral separation of salsolinols using β-CyD as a chiral selector in conventional capillary electrophoresis, a plate number of 178,464 is required for a resolution of 1.5. With CEC (i.e., $\phi K = 10$), the required plate number is only 5976 for the same resolution [10]. For PD–CEC, the column plate number is sacrificed due to the introduction of hydrodynamic flow, but the increased selectivity markedly reduces the requirement for the column efficiency.

Table 1 Relationship Between the Field Strength and the Electrophoretic Mobility for Getting High Enantioselectivity in CEC

Direction of μ_{ep}		Size relationship (absolute value)	Direction of E
μ_f	μ_c		
+	+	$\mu_f < \mu_c$	+
+	+	$\mu_f > \mu_c$	−
+	−	[a]	−
−	−	$\mu_f < \mu_c$	−
−	−	$\mu_f > \mu_c$	+
−	+	[a]	+

[a]The selection of direction of electric field is not influenced by size relationship in absolute values between the electrophoretic mobility of the free and complexed analytes.

Chiral Separation in CEC

There are different ways of performing chiral separation by CEC. Mayer and Schurig immobilized the chiral selectors by coating or chemically binding them to the wall of the capillary [11,12]. Permethylated β- or γ-CyD was attached via an octamethylene spacer to dimethylpolysiloxane (Chirasil-Dex) as the stationary phase. A high efficiency (~250,000/M) was obtained for the separation of 1,1′-dinaphthyl-2,2′-diyl hydrogenphosphate. An alternative coating approach was developed by Sezeman and Ganzler. Linear acrylamide was coated on the capillary wall, and after polymerization, CyD derivatives were bound to the polymer [13].

Chiral separation can also be performed with packed capillaries. β-CyD-bonded CSPs that are most frequently used in HPLC and CE were successfully applied in CEC. The separation of a variety of chiral compounds, such as some amino acid derivatives benzoin and hexobarbital was achieved by using CSPs bonded with different CyD derivatives [14,15]. Proteins are not ideal for use as buffer additives in CE because of their large detector response; however, CEC may be a good way to use this type of chiral selectors. Lloyd et al. have performed CEC enantioseparation by using commercially available protein CSPs, such as AGP and HAS [16,17]. The resolution obtained on protein CSPs was good; the efficiency, however, was rather poor. Another HPLC–CSP based on cellulose derivatives has been also reported for enantioseparation by CEC [18]. CSPs modified by covalent attachment of poly-N-acryloyl-L-phenylalanineethylester or by coating with cellulose tris(3,5-dimethylphenylcarbamate) can be performed in the reversed-phase mode. Acetonitrile as organic modifier was found to be advantageous for this type of CSP. An anion-exchange-type CSP was recently developed for the separation of N-derivatized amino acids [19]. The new chiral sorbent was modified with a basic *tert*-butyl carbamoyl quinine. Enantioselectivity obtained in CEC was as high as in HPLC and efficiency was typically a factor of 2–3 higher than in HPLC. A recent innovative approach is the use of imprinted polymers as CSPs in CEC [20,21]. Imprinted polymers possess a permanent memory for the imprinted species, and, thus, their enantioselectivity is predetermined by the enantiomeric form of the templating ligand. The use of imprint-based CSPs in HPLC is hampered by their poor chromatographic performance. CEC, however, was found to greatly improve the efficiency of the imprint-based separation. The most successful approach is the use of capillary columns filled with a monolithic, superporous imprinted polymer obtained by an *in situ* photo-initiated polymerization process. This technique enables imprint-based column to be operational within 3 h from the start of preparation. Generally, the imprint-based CSPs show high enantioselectivity but somewhat low efficiency and are limited to the separation of very closely related compounds.

Enantioseparation can be achieved on a conventional achiral stationary phase by the inclusion of an appropriate chiral additive into the mobile phase. It is theoretically predicted that the enantioselectivity in CEC with a chiral additive may be higher than that using a chiral column with the same chiral selector [10]. Lelievre et al. compared an HP-β-CyD column and HP-β-CyD as an additive in the mobile phase with an achiral phase (ODS) to resolve chlortalidone by CEC [22]. It was demonstrated that resolution on ODS with the chiral additive was superior on the CSP; however, efficiency was low. With an increasing amount of acetonitrile, the peak shape was improved and the migration time was decreased. We achieved the separation of salsolinol by the use of CEC with β-CyD as a chiral additive in the mobile phase containing sodium 1-heptanesulfonate, as shown in Fig. 1. Salsolinol is a hydrophilic amine and is difficult to enantioseparate due to

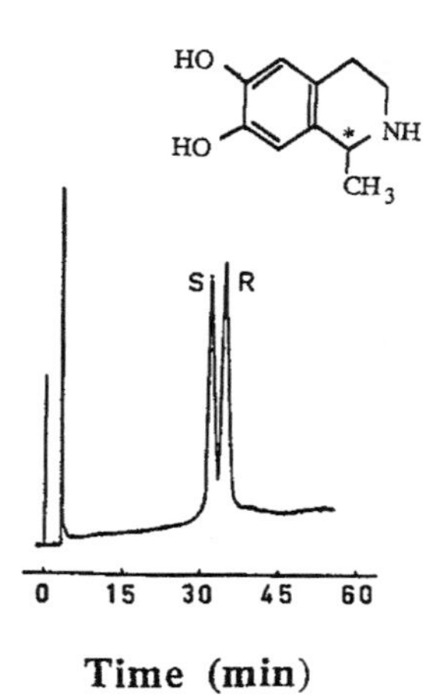

Fig. 1 Electrochromatogram of salsolinol enantiomers on a packed capillary column. Column: ODS-C18, 29 cm (23 cm effective length) × 75 μm ID; applied electric field strength: ~ 250 V/cm; mobile phase: 20 mM sodium phosphate buffer (pH 3.0) containing 12 mM β-cyclodextrin and 5 mM sodium 1-heptanesulfonate. The pump was set at the constant pressure of 100 kg/cm^2.

the small k' values on the reversed stationary phases. Sodium 1-heptanesulfonate was used as a counterion to improve the retention.

In conclusion, CEC has great potential in separation technology. Our theoretical model as well as many published practices in CEC show clearly that the benefit of combining electrophoresis and partitioning mechanisms in CEC is the increase in selectivity for the separation. The intrinsic difference in formation constants is critical, but the experimental factors, such as electric field or the stationary and mobile phases, can also contribute to the improvement of the overall enantioselectivity via increasing the conversion efficiency. However, only when both electrophoretic and partitioning mechanisms act in the positive effects, can high overall enantioselectivity in CEC be obtained.

References

1. T. Tsuda, K. Nomura, and G. Nagakawa, *J. Chromatogr. 248*: 241 (1982).
2. J. W. Jorgenson and K. D. Lukacs, *J. Chromatogr. 218*: 209 (1981).
3. T. Tsuda, *LC–GC Int.* 5:26 (1992).
4. V. Pretorius, B. J. Hopkins, and J. D. Schieke, *J. Chromatogr. 99*: 23 (1974).
5. Y. Deng, W. Maruyama, M. Kawai, P. Dostert, and M. Naoi, *Progress in HPLC and HPCE*, VSP, Utrecht, 1997, Vol. 6, pp. 301.
6. R. J. Boughtflower, T. Underwood, and C. J. Paterson, *Chromatographia 40*: 329 (1995).
7. C. Yan, U.S. Patent 5453163 (1995).
8. M. R. Taloy, P. Teale, S. A. Westwood, and D. Perrett, *Anal. Chem. 69*: 2554 (1997).
9. J. Eimer, K. K. Unger, and T. Tsuda, *Fresenius J. Anal. Chem. 352*: 649 (1995).
10. Y. Deng, J. Zhang, T. Tsuda, P. H. Yu, A. A. Boulton, and R. M. Cassidy, *Anal. Chem. 70*: 4586 (1998).
11. S. Mayer and V. Schurig, *J. High Resolut. Chromatogr. 15*: 129 (1992).
12. S. Mayer and V. Schurig, *J. Liquid Chromatogr. 16*: 915 (1993).
13. J. Sezemam and K. Ganzler, *J. Chromatogr. A 668*: 509 (1994).
14. S. Li and D. K. Lloyd, *J. Chromatogr. A 666*: 321 (1994).
15. D. Wistuba, H. Czesla, M. Roeder, and V. Schurig, *J. Chromatogr. A 815*: 183 (1998).
16. S. Li and D. K. Lloyd, *Anal. Chem. 65*: 3684 (1993).
17. D. K. Lloyd, S. Li, and P. Ryan, *J. Chromatogr. A 694*: 285 (1995).
18. K. Krause, M. Girod, B. Chankvetadze, and G. Blasehk, *J. Chromatogr. A 837*: 51 (1999).
19. M. Lammerhofer and W. Lindner, *J. Chromatogr. A 829*: 115 (1998).

20. L. Schweitz, L. I. Andersson, and S. Nilsson, *Anal. Chem. 69*: 1179 (1997).
21. L. Schweitz, L. I. Andersson, and S. Nilsson, *J. Chromatogr. A 817*: 5 (1998).
22. F. Lelievre, C. Yan, R. N. Zare, and P. Gareil, *J. Chromatogr. A 723*: 145 (1996).

Yulin Deng

End Capping

A typical stationary phase for chromatography, especially liquid chromatography (LC), is a chemically alkyl (C_{18})-bonded phase on silica gel particles. For the preparation of this type of bonded phase, alkylsilane is used to react with the silica gel surface by a silane-coupling reaction. In order to perform this synthesis, the silica gel to be bonded is treated to remove heavy metals and to prepare the surface for better bonding. Generally, only one of the functional groups bonds to form a Si — O — Si bond. Less often, two of the functional groups react to form adjacent Si — O — Si bonds. The remaining functional groups on each reagent molecule hydrolyze to form Si — O — H groups during workup, following the initial reaction. These groups, however, which form with the di- and trifunctional reagents, can cross-link with one another near the surface of the silica gel support. Thus, bonded phases made with any di- or tri- functional reagents are termed "polymeric" phases. A monofunctional silane reagent can only bond to the silanols and any excess is washed free as the ether resulting from hydrolysis of the reagent. Any packing made with a monofunctional silane reagent is referred to as a "monomeric" bonded phase. These schemes are summarized in Fig. 1a(i) and (ii). Other chemically bonded phases, such as cyano-, amino-, and shorter or longer alkyl phases are synthesized by similar bonding chemistries.

The products made by the above synthetic processes still have large numbers of residual silanols, which lead to poor peak shapes or irreversible adsorption, because chemically bonded groups on the silica gel surface have large, bulky molecular sizes and, after the bonding, the functionalized silane cannot react with the silanols around the bonded ligands. Because such alkyl-bonded phases are used for reversed-phase separations, especially for chromatography of polar molecules, any silanol groups that remain accessible to solutes after the bonding are likely to make an important contribution to the chromatography of such solutes; this is generally detrimental to the typical reversed-phase LC separations. It is a common fact that the residual silanols produce peak tailing for highly polar compounds which will interact with these silanol groups with deleterious effects. Therefore, the attempt to reduce the number of residual silanols on the silica gel is a common procedure in the preparation of chemically bonded stationary phases, where the surface of a reversed-phase material is ensured to be uniformly hydrophobic, for example, by blocking residual silanol groups with some functional groups. This process is the so-called "end capping." The end-capping process is possible with a smaller molecule than alkylchlorosilanes, such as a trimethyl-substituted silane (from trimethylchlorosilane or hexamethyldisilazane) as seen in Fig. 1a(iii). Because the molecular weights of these reagents are small, they do not add much to the total percent carbon, compared with the initial bonded phase. It must be known that all chemically bonded phases on silica gel cannot be end-capped by this process, because the above reagents can react with diol and amino phases, and not only with silanol groups on the surface. To block, end cap, and then unblock these phases would be very time-consuming and too expensive to be practical. If the final bonded phase is, in fact, a diol, this silane-bonding reagent is made from glycerol and has the structure Si — O — CHOH — CH_2OH. The cyano or amino phases are most often attached with a propyl group between the silicon atom and the CN or NH_2 group.

Often, when various bonded phases are studied for suitability for a particular separation, the question arises as to which is bonded most completely. This is a common question, because all phases, no matter how they are bonded, will have some residual silanols, even after an end-capping process. It is impossible for the bulkier bonding reagents to reach any but the most sterically accessible silanols. It is much easier for the smaller solutes to reach the silanols, however, and be affected by them. The final surface of the silica gel has three different structures, as demonstrated in Fig. 1b(i), (ii), and (iii), for a monomeric C_{18}, end capped by trimethylchlorosilane and residual silanols, respectively.

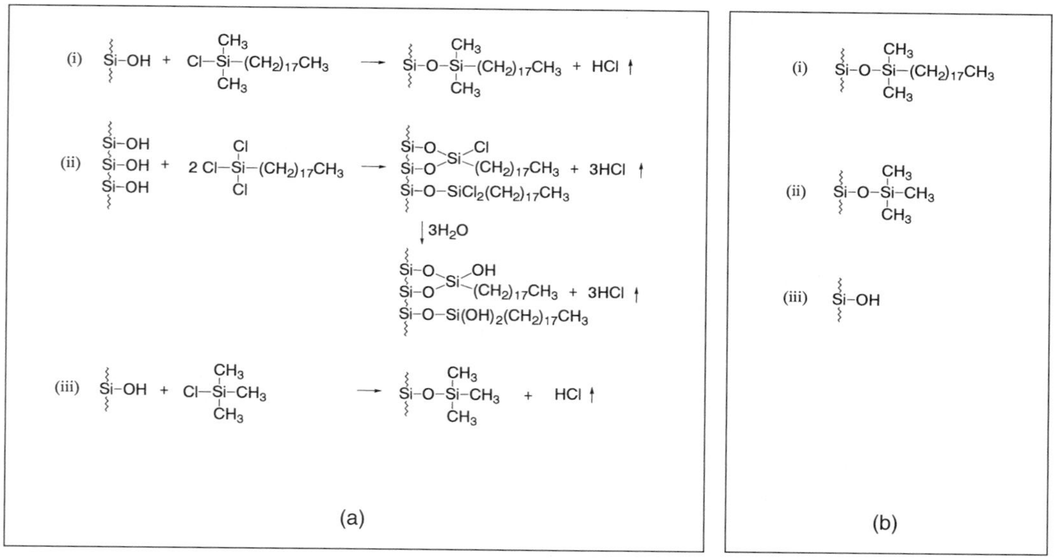

Fig. 1 (a) Scheme of bonding chemistry for chemically bonded C_{18} silica phase: (i) synthesis of monomeric C_{18}; (ii) synthesis of polymeric C_{18}; (iii) end-capping process. (b) Surface structure of a monomeric C_{18} phase: (i) monomeric C_{18} ligand; (ii) end-capped trimethyl ligand; (iii) residual silanol.

The presence of residual silanol groups can be detected most readily by using Methyl Red indicator [1], which turns red in the presence of acidic silanol groups, but a more sensitive test is to chromatograph a polar solute on the reversed-phase material.

To test, chromatographically, any phase for residual silanols, the column has to be conditioned with heptane or hexane (which has been dried overnight with spherical 4A molecular sieves). The series of solvents to use if the column has been used with water or a water–organic mobile phase, such as water → ethanol → acetone → ethyl acetate → chloroform → heptane. Once activated, a sample of nitrobenzene or nitrotoluene is injected, eluted with heptane or hexane, and detected at 254 nm. The degree of retention is then a sensitive guide to the presence or absence of residual silanols; if the solute is essentially unretained, the absence of silanols may be assumed. The better the bonding, the faster the polar compound will be eluted from the column. A well-bonded and end-capped phase will have a retention factor of between 0 and 1. Less well-covered silicas can have retention factors greater than 10. This is a comparative test, but it can also be useful for examining a phase to see if the end-capping reagent or primary phase has been cleaved by the mobile phase used over a period of time. Other methods to measure the silanol content of silica and bonded silica have been discussed by Unger [2]. Solid-state nuclear magnetic resonance spectrometry is the most powerful method to identify the species of residual silanol groups on the silica gel surface [3].

In order to avoid the contribution of the residual silanols to solute retention, many packing materials that should not have silanols have been developed [4]. They are polymer-based materials and also polymer-coated silica phases. These polymer-based or polymer-coated phases can be recommended as very useful and stable stationary phases in LC separations of polar compounds; they also offer much better stability for use at higher pH alkaline conditions.

References

1. K. Karch, I. Sebestian, and I. Halasz, *J. Chromatogr. 122*: 3 (1976).
2. K. K. Unger (ed.), *Packings and Stationary Phases in Chromatographic Techniques*, Marcel Dekker, Inc., New York, 1990.
3. M. Pursch, L. C. Sander, and K. Albert, *Anal. Chem. 71*: 733A (1999).
4. K. K. Unger and E. Weber, *A Guide to Practical HPLC*, GIT Verlag, Darmstadt, 1999.

Kiyokatsu Jinno

Environmental Applications of SFC

Because supercritical fluids have liquidlike solvating power and gaslike mass-transfer properties, supercritical fluid chromatography (SFC) is considered to be the bridge between gas chromatography (GC) and liquid chromatography (LC) and possesses several advantages over GC and LC, as summarized in Table 1. For example, SFC can separate nonvolatile, thermally labile, and high-molecular-weight compounds in short analysis times. Another advantage of SFC is its compatibility with both GC and high-performance liquid chromatographic (HPLC) detectors. Because of these advantages of SFC, there is a large number of SFC applications in environmental analysis. However, only selected recent works are reviewed here. Although sample preparation is often required before SFC analysis to remove the analytes from environmental matrices and to enrich them, sample preparation is not intensively discussed in this review. To facilitate the discussion, the environmental pollutants are classified and reviewed separately in this article.

Pesticides and Herbicides

The analysis of pesticides and herbicides has mainly been done either by GC with selective detectors or by HPLC with ultraviolet (UV) detection. As summarized in Table 1, GC is limited to thermally stable volatile compounds, whereas the HPLC with UV can only detect compounds with chromophores. These limitations of GC and HPLC led to the use of SFC in the analysis of pesticides and herbicides. Among the SFC works in environmental analysis, one-third of the works concerns the analysis of pesticides and herbicides.

Many detectors have been used to detect pesticides and herbicides in SFC. Among these detectors, the flame ionization detector (FID) is most commonly used for detection of a wide range of pesticides and herbicides, with a detection limit ranging from 1 ppm (for carbonfuran) to 80 ppm (for Karmex, Harmony, Glean, and Oust herbicides). The UV detector has frequently been used for the detection of compounds with chromophores. The detection limit was as low as 10 ppt when solid-phase extraction (SPE) was on-line coupled to SFC. The mass spectrometric detector (MSD) has also been used in many applications as a universal detector. The MSD detection limit reached 10 ppb with on-line SFE (supercritical fluid extraction)–SFC. Selective detection of chlorinated pesticides and herbicides has been achieved by an electron-capture detector (ECD). The limit of detection for triazole fungicide metabolite was reported to be 35 ppb. Other detectors used for detection of pesticides and herbicides include thermoionic, infrared, photometric, and atomic emission detectors.

A variety of both packed and open tubular columns have been used for separation of pesticides and herbicides. The columns were either used separately or coupled in series to achieve better separations. Although environmental water samples were mostly analyzed by SFC, analyses of pesticides and herbicides from soil, foods, and other samples were also reported.

Polychlorinated Biphenyls

Since 1929, polychlorinated biphenyls (PCBs) have been produced and used as heat-transfer, hydraulic, and di-

Table 1 Comparison of Characteristics of GC, SFC, and LC

	GC[a]	SFC	LC[b]
Suitability for polar and thermolabile compounds	Low	High	High
Size of analyte molecule	Small–Medium	Small–Large	Small–Large
Sample capacity	Low	High (packed column)	High
Possibility of introducing selectivity in the mobile phase	Low	High	Medium
Toxicity and disposal cost of the mobile phase	No	No (with pure CO_2) Low (with modifier)	High
Efficiency	High	Medium–High	Low
Use of gas-phase detectors	Yes	Yes	No
Analysis time	Medium	Medium	Long

[a]Only capillary GC is used for these evaluations. Fast GC and packed column GC are not included here.
[b]Capillary HPLC is not included.

electric fluids. Because of their chemical and physical stability, PCBs have been found in many environmental samples. Generally, PCBs have been analyzed by GC with electron-capture detection. There are many reports on subcritical and supercritical fluid extraction of PCBs, but only a few on supercritical fluid separation of PCBs.

Among the works of supercritical fluid separations of PCBs, UV has been the most popular detector. A Microbore C_{18} column was used to separate individual PCB congeners in Aroclor mixtures. Density and temperature programming was also utilized for separation of PCBs. Both packed (with phenyl and C_{18}) and capillary (Sphery-5 cyanopropyl) columns were used in this work. Carbon dioxide, nitrous oxide, and sulfur hexafluoride were tested as mobile phases for the separation of PCBs.

A flame ionization detector and MSD were also used for detection of PCBs in SFC. Capillary columns packed with aminosilane-bonded silica and open tubular columns coated with polysiloxane were employed for PCB separation in these works.

Polycyclic Aromatic Hydrocarbons

Polycyclic aromatic hydrocarbons (PAHs) have routinely been analyzed by GC and LC. However, both techniques have limitations in terms of analyte molecular weight and analysis time. The greater molecular-weight range of SFC with respect to GC makes it better suited for determining a wide range of PAHs. SFC also has advantages over HPLC for the analysis of PAHs when the same kind of columns is used. Supercritical fluid has similar solvating power as a liquid does, and the solute diffusion coefficients are much greater than those found in liquids. Therefore, comparable efficiencies to HPLC can be obtained by SFC in shorter analysis time. Because of these characteristics of SFC, the separation of PAHs by SFC with different kinds of packed and capillary columns is a well-investigated and established method.

The most popular detector for PAHs is the UV detector. The detection limit was 0.2–2.5 ppb for 16 PAHs. A diode-array detector was also used for PAHs in SFC, and the detection limit was reported to be as low as 0.4 ppb. Other detections used for PAHs include mass spetrometric, thermoionic, infrared, photoionization, sulfur chemiluminescence, and fluorescence detectors.

Although CO_2 has mainly been used as the mobile phase in SFC, modifiers have often been added to CO_2 to increase the solvating power of the mobile phase. Although the most frequently used modifier has been methanol, many other modifiers were also tested. The modifier effect on retention is discussed separately in this encyclopedia. Because organic modifiers are incompatible with FID, flame ionization detection was rarely used for PAHs in SFC.

Fast separations of 16 PAHs were achieved within 6–7 min using packed columns. A comparison study of the PAH molecular shape recognition properties of liquid-crystal-bonded phases in packed-column SFC and HPLC found that the selectivity was enhanced in SFC. The result of an interlaborotory round-robin evaluation of SFC for the determination of PAHs also shows that SFC possesses distinct advantages over GC–mass spectrometry (MS) and nuclear magnetic resonance (NMR) including speed, cost, and applicability.

Polar Pollutants

Because carbon dioxide is nonpolar, the separation of polar compounds by supercritical carbon dioxide is difficult. Thus, polar modifiers are often used for the separation of phenols and amines. Derivatization has also been employed to obtain nonpolar analytes in some applications. The UV detector has mainly been used for the detection of polar compounds. Oxidative and reductive amperometric detection was also utilized with a detection limit of 250 pg for oxidative detection of 2,6-dimethylphenol. The detection of amines has generally been achieved by FID. Other detectors used for the detection of polar analytes include Fourier transform infrared (FTIR), photodiode array, and flame photometry.

It should be pointed out that separation of more than one class of organic compounds can be achieved by SFC. For example, Fig. 1 shows the chromatogram of 35 PAHs,

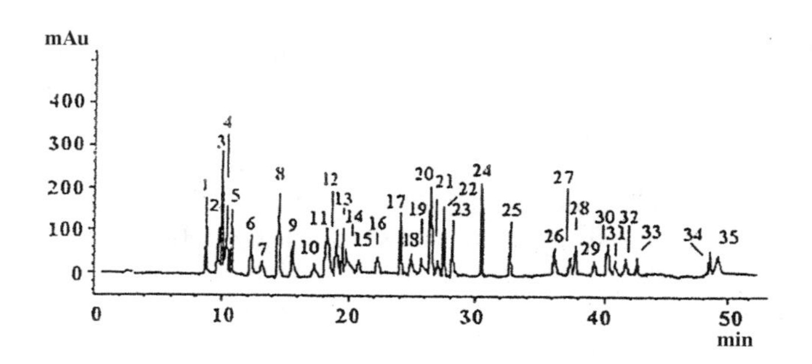

Fig. 1 Chromatogram of PAHs, herbicides, and phenols obtained by supercritical carbon dioxide modified with methanol. [Reprinted from L. Toribio, M. J. del Nozal, J. L. Bernal, J. J. Jimenez, and M. L. Serna, Packed-column supercritical fluid chromatography coupled with solid-phase extraction for the determination of organic microcontaminants in water, *J. Chromatogr. A 823*: 164 (1998). Copyright 1998, with permission from Elsevier Science.]

herbicides, and phenols from a contaminated water sample. Solid-phase extraction was used for sample preparation. Five Hypersil silica columns were coupled in series for separation of these contaminants. The percentage of methanol (as modifier) was varied from 2% (5 min) to 10% (29 min) at 0.5%/min. A pressure program was also applied. A diode-array detector was used in this work.

Organotin, Mercury, and Other Inorganic Pollutants

Organotin compounds are used extensively as biocides and in marine antifouling paints. These compounds accumulate in sediments, marine organisms, and water, as they are continuously released into the marine environment. Many of these organotin compounds are toxic to aquatic life. Most organotin separation techniques have been based on the GC resolution of volatile derivatives and coupled to elemental detection techniques that are often not sensitive enough to detect trace organotin compounds. However, the separation of organotin compounds was achieved by capillary columns (SB-Biphenyl-30 or SE-52) with pure CO_2 as the mobile phase. Inductively coupled plasma–mass spectrometry (ICP–MS) was used in most of the applications to improve the sensitivity for detecting trace organotin species. The reported detection limits range from 0.2 to 0.8 pg for tetrabultin chloride, tributyltin chloride, triphenyltin chloride, and tetraphenyltin. However, the detection limits obtained by FID are 15- to 45-fold higher than those obtained by ICP–MS for the above-mentioned organotin compounds. Flame photometric detector was also used to detect organotin species with a detection limit of 40 pg for tribultin chloride.

The separation of organomercury was conducted by using a SB-methyl-100 capillary column and pure CO_2 as the mobile phase. FID and atomic fluorescence were used for detection. The same column was also used for separation of mercury, arsenic, and antimony species using carbon dioxide as the mobile phase. A chelating reagent, bis-(trifluoroethyl)dithiocarbamate, was used in this case to convert the metal ions to organometallic compounds before the separation. The detection limit of FID was 7 and 11 pg for arsenic and antimony, respectively.

Figure 2 shows an example of separating organomercury using supercritical CO_2. A 10-m × 50-μm-inner diameter SB-Methyl 100 column was used for the separation. Due to their poor solubility in supercritical carbon dioxide, monoorganomercury compounds were derivatized by diethyldithiocarbamate. An interface for a system consisting of SFC and atomic fluorescence spectrometry was developed for the detection of organomercurials.

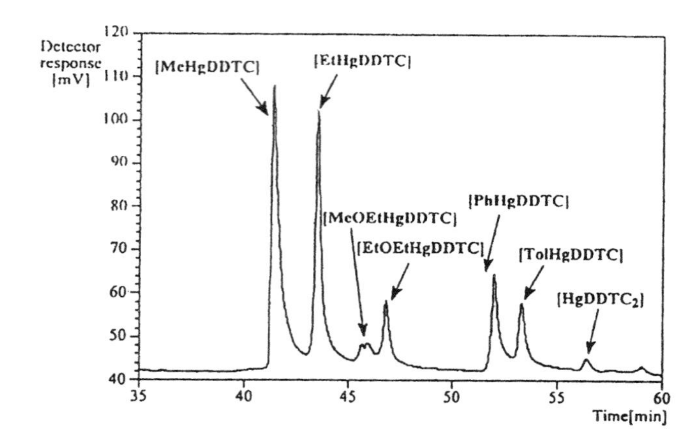

Fig. 2 Chromatogram of a standard mixture after complexation with sodium diethyldithiocarbamate. Composition of the standard: mercury dichloride, methylmercury chloride, ethylmercury chloride, methoxyethylmercury chloride, ethoxyethylmercury chloride, phenylmercury chloride, and tolymercury chloride. [Reprinted from A. Knochel and H. Potgeter, Interfacing supercritical fluid chromatography with atomic fluorescence spectrometry for the determination of organomercury compounds, *J. Chromatogr. A 786*: 192 (1997). Copyright 1997, with permission from Elsevier Science.]

In closing, supercritical fluid chromatography is a promising technique for the analysis of environmental pollutants. The analytes range from inorganic species to polar and nonpolar organic compounds. The sample matrices cover water, soil, sediments, sludge, and air particulate matters. The sample preparation has been done by solid-phase extraction, supercritical fluid extraction, or traditional solvent extraction. Modifiers are often used to enhance the solubility of analytes and to yield a better separation for polar and high-molecular-weight analytes. Packed columns are preferred for trace analysis because of their high sample capacity. Both gas-phase and liquid-phase detectors have been used in SFC to detect a wide range of environmental pollutants.

Suggested Further Reading

J. M. Bayona and Y. Cai, *Trends Anal. Chem. 13*: 327–332 (1994).

T. A. Berger, *J. Chromatogr. A 785*: 3–33 (1997).

T. L. Chester, J. D. Pinkston, and D. E. Raynie, *Anal. Chem. 70*: 301R–319R (1998).

S. F. Dressman, A. M. Simeone, and A. C. Michael, *Anal. Chem. 68*: 3121–3127 (1996).

Z. Juvancz, K. M. Payne, K. E. Markides, and M. L. Lee, *Anal. Chem. 62*: 1384–1388 (1990).

A. Knochel and H. Potgeter, *J. Chromatogr. A 786*: 188–193 (1997).

K. E. Laintz, G. M. Shieh, and C. M. Wai, *J. Chromatogr. Sci. 30*: 120–123 (1992).
D. R. Luffer and M. Novotny, *J. Chromatogr. 517*: 477–489 (1990).
A. Medvedovici, A. Kot, F. David, and P. Sandra, The use of supercritical fluids in environmental analysis, in *Supercritical Fluid Chromatography with Packed Columns* (K. Anton and C. Berger, eds.), Marcel Dekker, Inc., New York, 1998, pp. 369–401.
A. Medvedovici, F. David, G. Desmet, and P. Sandra, *J. Microcol. Separ. 10*: 89–97 (1998).
E. Moyano, M. McCullagh, M. T. Galceran, and D. E. Games, *J. Chromatogr. A 777*: 167–176 (1997).
L. J. Mulcahey, C. L. Rankin, and M. E. P. McNally, Environmental applications of supercritical fluid chromatography, in *Advances in Chromatography Vol. 34*, 1994, pp. 251–308.
S. Shan, M. Ashraf-Khorassani, and L. T. Taylor, *J. Chromatogr. 505*: 293–298 (1990).
R. M. Smith and D. A. Briggs, *J. Chromatogr. A 688*: 261–271 (1994).
L. Toribio, M. J. del Nozal, J. L. Bernal, J. J. Jimenez, and M. L. Serna, *J. Chromatogr. A 823*: 163–170 (1998).

Yu Yang

Exclusion Limit in GPC–SEC

A given gel permeation chromatography–size-exclusion chromatography (GPC–SEC) column can analyze the molecular weight (MW) of a polymer only over a limited range of MWs. Figure 1 illustrates a typical calibration curve for the column. The logarithm of the MW is plotted as a function of the retention time t_R. At low and high ends of MW, t_R barely depends on MW, effectively limiting the range of analysis to $M_1 < MW < M_2$. The exclusion limit refers to M_2.

The sharp slope of the calibration curve at the high-MW end of the calibration curve is caused by a drastic decline of the partition coefficient as the chain dimension increases beyond the accessible pore size of the column packing material. Polymer chains of $MW > M_2$ have a molecular size dimension which is much greater than the pore size. It is virtually impossible for these chains to enter the stationary-phase pores. Thus, at almost every plate in the column, they are partitioned to the mobile phase, thus eluting with little separation at around the dead time (volume) of the column. By contrast, polymer chains smaller or comparable to the available pore sizes can pentrate the pores to be partitioned to the stationary phase with a partition coefficient which depends on their chain dimensions. The dependence of the partition coefficient on the chain dimension allows polymer chains of different MWs to be separated and elute at different times. Columns packed with porous materials of a larger pore size have greater M_1 and M_2.

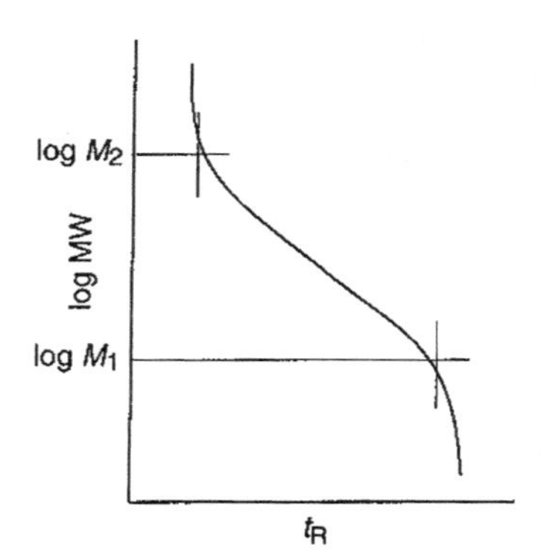

Fig. 1 Calibration curve of GPC–SEC column. Logarithm of the molecular weight, M, is plotted as a function of the retention time (volume) t_R (V_R).

For the high-MW chains that are excluded by the pores, there is a small dependence of t_R (V_R) on MW. The latter is mostly caused by the velocity gradient of the mobile phase and the population gradient of the polymer near the stationary-phase particles' surface. The mobile phase flows more slowly near the particles' surface because of the no-slip boundary condition of the fluid at the particles' surface. Among sufficiently long polymer chains to be excluded by the pore, those with a smaller dimension can more easily approach the particles' surface, compared with those of a greater dimension. Therefore, shorter chains flow more slowly. Longer chains stay away from the particles and flow along the fastest-flowing mobile phase, eluting earlier than other components. This mode of separation is called "hydrodynamic chromatography."

Iwao Teraoka

Extra-Column Dispersion

In addition to the dispersion that takes place during the normal function of the column, dispersion can also occur in connecting tubes, injection system, and detector sensing volume, and as a result of injecting a finite sample mass and sample volume onto the column. The major sources of extra column dispersion are as follows:

1. Dispersion due to the sample volume (σ_S^2)
2. Dispersion occurring in valve-column and column-detector connecting tubes (σ_T^2)
3. Dispersion in the sensor volume from Newtonian flow (σ_{CF}^2)
4. Dispersion in the sensor volume from peak merging (σ_{CM}^2)
5. Dispersion from the sensor and electronics time constant (σ_t^2)

The sum of the variances will give the overall variance for the extra-column dispersion (σ_E^2). Thus,

$$\sigma_E^2 = \sigma_S^2 + \sigma_T^2 + \sigma_{CF}^2 + \sigma_{CM}^2 + \sigma_t^2 \qquad (1)$$

Equation (1) shows how the various contributions to extra-column dispersion can be combined. According to Klinkenberg [1], the total extra-column dispersion must not exceed 10% of the column variance if the resolution of the column is not to be seriously denigrated; that is,

$$\sigma_E^2 = \sigma_S^2 + \sigma_T^2 + \sigma_{CF}^2 + \sigma_{CM}^2 + \sigma_t^2 = 0.1\sigma_c^2$$

In practice, σ_T^2, σ_{CF}^2, σ_{CM}^2, and σ_t^2 are all kept to a minimum to allow the largest contribution to extra-column dispersion to come from σ_S^2. This will allow the largest possible sample to be placed on the column, if so desired, to aid in trace analysis. Each extra-column dispersion process can be examined theoretically and two examples will be the evaluation of σ_S^2 and σ_T^2.

Maximum Sample Volume

Consider the injection of a sample volume (V_i) that forms a rectangular distribution of solute at the front of the column. The variance of the final peak will be the sum of the variance of the sample volume plus the normal variance from a peak for a small sample. Now, the variance of a rectangular distribution of sample volume (V_1) is $V_i^2/12$, and assuming the peak width is increased by 5% due to the dispersing effect of the sample volume (a 5% increase in standard deviation is approximately equivalent to a 10% increase in peak variance), then by summing the variances,

$$\frac{V_i^2}{12} + [\sqrt{n}(v_m + Kv_s)]^2 = [1.05\sqrt{n}(v_m + Kv_s)]^2$$

where the dispersion due to the column alone is $[\sqrt{n}(v_m + Kv_s)]^2$ (see the entry Plate Theory). Simplifying and rearranging,

$$V_i^2 = n(v_m + Kv_s)^2(1.22)$$

Bearing in mind that

$$V_r = n(v_m + Kv_s)$$

then

$$V_i = \frac{1.1V_r}{\sqrt{n}}$$

Thus, the maximum sample volume that can be tolerated can be calculated from the retention volume of the solute concerned and the efficiency of the column. A knowledge of the maximum sample volume can be important when the column efficiency available is only just adequate, and the compounds of interest are minor components that are only partly resolved.

Dispersion in Connecting Tubes

The column variance is given by V_r^2/n, and for a peak eluted at the dead volume, the variance will be V_0^2/n (see the entry Plate Theory). Thus, for a connecting tube of radius r_t and length l_t, the dead volume (V_0) (i.e., the volume of the tube) is

$$V_0 = \pi r_t^2 l_t$$

Thus,

$$\sigma_E^2 = \frac{0.1(\pi r_t^2 l_t)^2}{n}$$

Now, for the dead volume peak from an open tube, $n = 1/0.6r_t$ (see the entry Golay Dispersion Equation for Open-Tubular Columns). Thus,

$$\sigma_E^2 = 0.06\pi^2 r_t^5 l_t$$

However, when assessing the length of tube that can be tolerated, it must be remembered that the 10% increase in variance that can be tolerated before resolution is seriously denigrated involves *all* sources of extra-column dispersion, not just for a connecting tube. In practice, the connecting tube should be made as short as possible and the radius as small as possible commensurate with reasonable pressures and the possibility that if the radius is too small, the tube may become blocked. The different sources of extra-column dispersion have been examined in Refs. 2 and 3.

References

1. A. Klinkenberg, *Gas Chromatography 1960* (R. P. W. Scott, ed.), Butterworths, London, 1960, p. 194.
2. R. P. W. Scott, *Liquid Chromatography Column Theory*, John Wiley & Sons, New York, 1992, p. 19.
3. R. P. W. Scott, *Introduction to Gas Chromatography*, Marcel Dekker, Inc., New York, 1998.

Suggested Further Reading

Scott, R. P. W., *Chromatographic Detectors*, Marcel Dekker, Inc., New York, 1998.

Raymond P. W. Scott

F

Field-Flow Fractionation Data Treatment

Introduction

Field-flow fractionation (FFF) methods are classified into two main categories [1–3]: *polarization* FFF and *focusing* FFF. Their basic characterization is given in the entry Field-Flow Fractionation Fundamentals. Whereas the polarization FFF methods allow to fractionate the samples on the basis of the differences in the extensive properties (such as the molar mass or particle size, etc.) of the individual species, the focusing FFF methods discriminate among the species, according to their intensive property differences (such as the charge or density, etc.). This article deals with the data treatment of the experimental results from polarization FFF, thus with the quantitative characterization of the extensive properties. However, a principally identical approach can be applied to the intensive properties data treatment of the results obtained from the focusing FFF experiments.

In general, the methodology of the data treatment, concerning the separations of the macromolecular or particulate samples, does not depend on the particular separation method or technique. The basics of this methodology were elaborated in parallel with the development of size-exclusion chromatography (SEC) [4] and of the techniques of particle size analysis [5], but they originate at the very beginning [6,7] of liquid chromatography of macromolecules and remain substantially unchanged until today.

Macromolecular or particulate samples fractionated by the FFF are usually not uniform but exhibit a distribution of the concerned extensive or intensive parameter [8] or, in other words, a polydispersity. Molar mass distribution (MMD), sometimes called molecular weight distribution (MWD), or particle size distribution (PSD) describes the relative proportion of each molar mass (molecular weight), M, or particle size (diameter), d_p, species composing the sample. This proportion can be expressed as a number of the macromolecules or particles of a given molar mass or diameter, respectively, relative to the number of all macromolecules or particles in the sample:

$$N(M) = \frac{n_i(M)}{\sum_{i=1}^{\infty} n_i} \qquad N(d_p) = \frac{n_i(d_p)}{\sum_{i=1}^{\infty} n_i} \tag{1}$$

or as a mass (weight) of the macromolecules or particles of a given molar mass or diameter relative to the total mass of the sample:

$$W(M) = \frac{m_i(M)}{\sum_{i=1}^{\infty} m_i} \qquad W(d_p) = \frac{m_i(d_p)}{\sum_{i=1}^{\infty} m_i} \tag{2}$$

Accordingly, the MMD (MWD) and PSD are called number or mass (weight) MMD or PSD, respectively. FFF provides a fractogram which has to be treated to obtain the required MMD or PSD. These distributions can be used to calculate various average molar masses or particle sizes and polydispersity indices.

Average Molar Masses, Particle Sizes, and Polydispersities

As mentioned, in addition to the MMD and PSD, various average molar masses, particle sizes, and polydispersity indexes can be calculated from the FFF fractograms. If the detector response, h, is proportional to the mass of the macromolecules or particles, the mass-average molar mass or mass average particle diameter can be calculated from

$$\overline{M_m} = \overline{M_w} = \frac{\sum_{i=1}^{\infty} M_i h_i}{\sum_{i=1}^{\infty} h_i}$$
$$\overline{d_m} = \overline{d_w} = \frac{\sum_{i=1}^{\infty} d_i h_i}{\sum_{i=1}^{\infty} h_i} \tag{3}$$

and the corresponding number average values are calculated from

$$\overline{M_n} = \frac{\sum_{i=1}^{\infty} M_i n_i}{\sum_{i=1}^{\infty} n_i} = \frac{\sum_{i=1}^{\infty} h_i}{\sum_{i=1}^{\infty} (h_i/M_i)}$$
$$\overline{d_n} = \frac{\sum_{i=1}^{\infty} d_i n_i}{\sum_{i=1}^{\infty} n_i} = \frac{\sum_{i=1}^{\infty} h_i}{\sum_{i=1}^{\infty} (h_i/d_i)} \tag{4}$$

The width of the MMD or PSD (polydispersity) can be characterized by the index of polydispersity:

$$I_{\text{MMD}} = \frac{\overline{M_m}}{\overline{M_n}}$$
$$I_{\text{PSD}} = \frac{\overline{d_m}}{\overline{d_n}} \tag{5}$$

Practical Data Treatment

Provided that the correction for the zone broadening should not be applied, the first step in the data treatment is to convert the retention volumes (or the retention ratios R) into the corresponding molecular or particulate parameter, characterizing the fractionated species. Whenever the zone broadening correction procedure has to be applied, the data treatment protocol is modified, as described in the entry Zone Dispersion in Field-Flow Fractionation.

The dependences of the retention ratio R on the size of the fractionated species (molar mass for the macromolecules or particle diameter for the particulate matter) are presented for various polarization FFF methods in the entry Field-Flow Fractionation Fundamentals. The raw, digitized fractogram, which is a record of the detector response as a function of the retention volume, is represented by a differential distribution function $h(V)$. It can be processed to obtain a series of the height values h_i corresponding to the retention volumes V_i, as shown in Fig. 1.

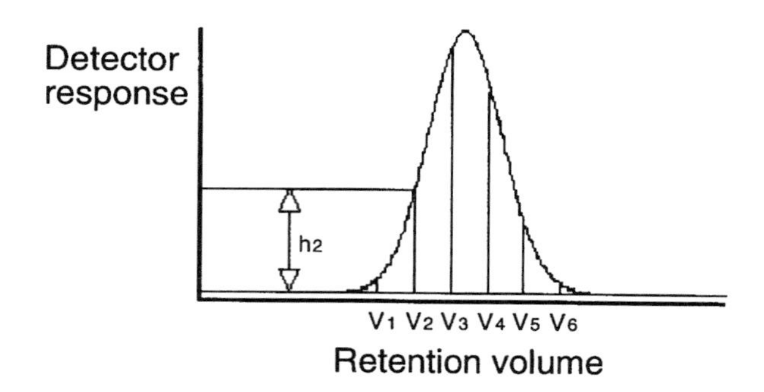

Fig. 1 Treatment of an experimental FFF fractogram of a polydisperse sample.

Subsequently, the retention volumes V_i, are converted into the retention ratios R_i:

$$R_i = \frac{V_0}{V_i} \tag{6}$$

The retention ratio R in polarization FFF is related to the retention parameter λ (see the entry Field-Flow Fractionation Fundamentals) by

$$R = 6\lambda\left[\coth\left(\frac{1}{2\lambda}\right) - 2\lambda\right] \tag{7}$$

or by an approximate relationship

$$(\lim R)_{\lambda \to 0} = 6\lambda \tag{8}$$

and the parameter λ is directly related to the molecular or particulate parameters by the general relationships

$$\lambda = f(M^{-n}) \quad \text{or} \quad \lambda = f(d_p^{-n}) \tag{9}$$

where the exponent $n = 1, 2$, or 3. As concerns the focusing FFF methods, similar relationships exist between the retention ratio R and the intensive properties of the fractionated species.

Having the V_i values converted into the R_i values by using Eq. (6), the corresponding molar mass M_i or the particle diameter d_i values are calculated by applying Eqs. (7)–(9). The difficulty is that Eq. (7) is a transcendental function $R = f(\lambda)$ for which the inversion function $\lambda = f'(R)$ does not exist. As a result, Eq. (8) can be used as a first approximation to estimate the λ_i values from the experimental R_i data, and by applying a rapidly converging iteration procedure, the accurate λ_i values can be calculated. The subsequent attribution of the corresponding M_i or d_i values to the calculated λ_i values, by using the appropriate relationship, Eq. (9), is not mathematically complicated.

In order to obtain an accurate result, the regular segmentation ΔV_i of the raw fractogram must be converted into the ΔR_i and, thereafter, into the appropriate increment of the molar mass ΔM_i or of the particle diameter Δd_i. The corresponding conversions of the raw experimental fractograms into the MMD or PSD can be carried out according to the following protocol. Equations (3) and (4) can be rewritten in integral form:

$$\overline{M_m} = \frac{\int_0^\infty W(M)M\,dM}{\int_0^\infty W(M)\,dM} \qquad \overline{d_m} = \frac{\int_0^\infty d_p W(d_p)\,dd_p}{\int_0^\infty W(d_p)\,dd_p} \tag{10}$$

and

$$\overline{M_n} = \frac{\int_0^\infty N(M)M\,dM}{\int_0^\infty N(M)\,dM} \qquad \overline{dn} = \frac{\int_0^\infty d_p N(d_p)\,dd_p}{\int_0^\infty N(d_p)\,dd_p} \tag{11}$$

where it holds for the normalized MMD or PSD:

$$\int_0^\infty W(M)\,dM = \int_0^\infty W(d_p)\,dd_p = \int_0^\infty N(M)\,dM = \int_0^\infty N(d_p)\,dd_p = 1 \tag{12}$$

By considering all of the above-mentioned transformations, Eqs. (10)–(12) give

$$\overline{M_m} = \int_0^\infty W(M)M\left(\frac{\partial M}{\partial \lambda}\right)\left(\frac{\partial \lambda}{\partial R}\right)\left(\frac{\partial R}{\partial V}\right) dV \left(\int_0^\infty W(M)\,dV\right)^{-1}$$

$$\overline{d_m} = \int_0^\infty W(d_p)\,d_p\left(\frac{\partial d_p}{\partial \lambda}\right)\left(\frac{\partial \lambda}{\partial R}\right)\left(\frac{\partial R}{\partial V}\right) dV \left(\int_0^\infty W(d_p)\,dV\right)^{-1}$$

$$\overline{M_n} = \int_0^\infty N(M)M\left(\frac{\partial M}{\partial \lambda}\right)\left(\frac{\partial \lambda}{\partial R}\right)\left(\frac{\partial R}{\partial V}\right) dV \left(\int_0^\infty N(M)\,dV\right)^{-1}$$

$$\overline{d_n} = \int_0^\infty N(d_p)\,d_p\left(\frac{\partial d_p}{\partial \lambda}\right)\left(\frac{\partial \lambda}{\partial R}\right)\left(\frac{\partial R}{\partial V}\right) dV \left(\int_0^\infty N(d_p)\,dV\right)^{-1} \tag{13}$$

Any of Eqs. (13) can further be rewritten in a numerical form of Eqs. (3) and (4), which are convenient for the data treatment and calculations using the discrete M_i or d_i and h_i values. The acquisition of the experimental data and the treatment of the fractogram is easily performed by a computer connected on-line to the separation system.

References

1. J. Janča, *Field-Flow Fractionation: Analysis of Macromolecules and Particles*, Marcel Dekker, Inc., New York, 1988.
2. J. Janča, *J. Liquid Chromatogr. Related Technol. 20*: 2555 (1997).
3. H. Cölfen and M. Antonietti, *Adv. Polym. Sci. 150*: 67 (2000).
4. C. Quivoron, in *Steric Exclusion Chromatography of Polymers* (J. Janča, ed.), Marcel Dekker, Inc., New York, 1984.
5. H. G. Barth (ed.), *Modern Methods of Particle Size Analysis*, John Wiley & Sons, New York, 1984.
6. J. Cazes, *J. Chem. Educ. 43*: A567 (1966).
7. J. Cazes, *J. Chem. Educ. 43*: A625 (1966).
8. J. V. Dawkins, in *Comprehensive Polymer Science, Vol. 1* (C. Booth and C. Price, eds.), Pergamon Press, Oxford, 1989.

Josef Janča

Field-Flow Fractionation with Electro-osmotic Flow

It is well known that the essence of field-flow fractionation (FFF) is in the interaction between the distribution of the sample particles in the transversal field and the nonuniformity of the longitudinal flow profile. The classical FFF is realized in the channel with the flow driven by the pressure drop. The flow, in this case, is called Poiseuille flow and its profile is parabolic.

Electro-osmotic flow (EOF) is widely used for the propulsion of liquid in modern chromatographic methods, so it was natural to study the possibility of FFF with

EOF, generated by applying an electric field, E, along a channel or a tube with charged (having the nonzero zeta-potentials) walls. The usual EOF is very close to uniform. For the cylindrical tube of radius a, the EOF velocity profile is described by

$$V(r) = \frac{\zeta\varepsilon\varepsilon_0 E}{\eta}\left(1 - \frac{I_0(\kappa r)}{I_0(\kappa a)}\right) \quad (1)$$

where ζ is the zeta-potential of the wall, ε and η are the dielectric constant and viscosity of the buffer, respectively, ε_0 is the permitivity of the free space, $\kappa^{-1} = (\varepsilon\varepsilon_0 k_B T/2ne^2)^{1/2}$ is the Debye layer thickness [k_B is the Boltzman constant, T is the temperature, n is the number of ions per unit volume (proportional to the concentration of buffer C_0), e is the proton charge], and $I_0(x)$ is the modified Bessel function. As can be seen from Eq. (1), the velocity profile of the EOF in the tube is very close to uniform everywhere except the Debye layer vicinity of the wall. Thus, it is hard to exploit such a profile for FFF unless the concentration of buffer is very low.

That is why it was proposed [1,2] to realize the asymmetrical FFF in the flat channel by making its walls of different materials or chemically modifying them. If the channel walls have nonequal values of the zeta-potentials, then the shape of the EOF profile can be quite different from uniform. The flow profiles that can be generated in the FFF channel with the applied electric field E and pressure drop Δp are presented in Fig. 1. These profiles can be described by

$$V(r) = \frac{\zeta_2\varepsilon\varepsilon_0 E}{\eta} \cdot \left[(\zeta_R - 1)\frac{\sinh kY}{\sinh k} + (\zeta_R + 1)\frac{\cosh kY}{\cosh k} + (1 - \zeta_R)Y - (1 + \zeta_R)\right] + V_0(1 - Y^2) \quad (2)$$

where $\zeta_R = \zeta_1/\zeta_2$ is the ratio of the zeta-potential of the accumulation wall to the zeta-potential of the depletion wall, $k = \kappa w/2$, w is the channel depth, $Y = 1 - 2y/w$, and $V_0 = \Delta p/2\eta L$.

For large values of k, the first two terms in the square brackets are substantially nonzero only in the Debye layer vicinity of the walls, whereas everywhere else the EOF profile is dominated by the last two linear terms in the square brackets. Therefore, the asymmetric EOF profile can be close to trapezoidal or close to triangular depending on the exact values of the zeta-potentials of the walls. If the signs of the zeta-potentials of the walls are different,

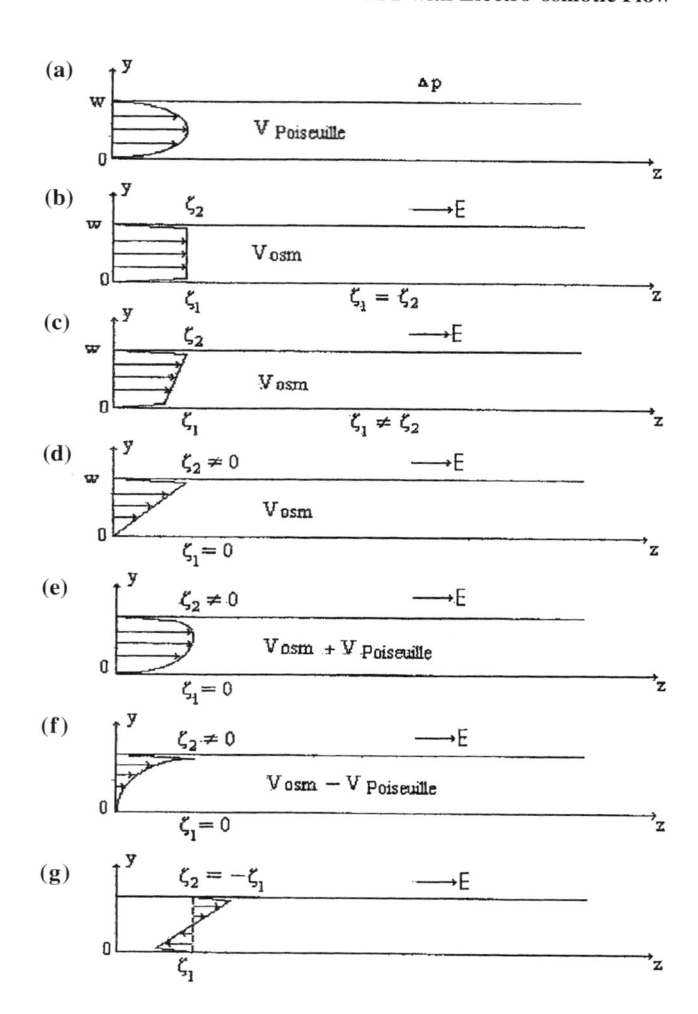

Fig. 1 Outline of the flow profiles in the FFF channels: (a) Poiseuille flow; (b) EOF (the equal zeta-potentials of the walls; (c) trapezoidal EOF (the nonequal zeta-potentials of the walls); (d) triangular EOF (the zero zeta-potential of the accumulating wall); (e) codirected triangular EOF and Poiseuille flow; (f) counterdirected triangular EOF and Poiseuille flow; (g) antisymmetric EOF (different signs of the zeta-potentials of the walls).

then the liquid moves in one direction near one wall and in the opposite direction near another wall (this case can be interesting for the preseparation of the particles having different densities).The last term in Eq. (2) corresponds to the pressure-driven Poiseuille flow.

Having Eq. (2) for flow profile enables one to calculate the retention ratio R and χ coefficient describing the Taylor dispersion part of the theoretical plate height H for arbitrary flow profile, according to [3]

$$H = \frac{2D}{R\langle V\rangle} + \chi\frac{w^2\langle V\rangle}{D} \quad (3)$$

where $\langle V \rangle$ is the average velocity of the flow and D is the diffusion coefficient of sample molecules. Usually, in FFF, the second term of Eq. (3) is much larger than the first one.

Comparison of R and χ values for the flow profiles presented in Fig. 1 for the case of the FFF parameter $\lambda \ll 1$ gives $R = 6\lambda$ and $\chi = 24\lambda^3$ for classical FFF with Poiseuille flow and $R = 2\lambda$ and $\chi = 8\lambda^3$ for FFF with a triangular EOF ($\zeta_1 = 0$). The most interesting result corresponds to the case of FFF with a combined triangular EOF and counterdirected Poiseuille flow (with $V_0 = \zeta_2 \varepsilon \varepsilon_0 E/4\eta$ leading to $dV/dY = 0$ for $Y = 0$). In this case, $R = 6\lambda^2$ and $\chi = 24\lambda^4$. Thus, the selectivity $S = d \ln R/d \ln \lambda = 2$ and is twice as large as in the case of classical FFF and FFF with a triangular EOF. The χ coefficient is very small for $\lambda \ll 1$, so that the Taylor dispersion is very low and efficiency is high. The function $F = S/\sqrt{\chi}$ (fractionating power), characterizing the resolution for the given value of $\langle V \rangle$, is proportional to λ^{-2} in this case; in the rest of the cases, it is proportional to $\lambda^{-3/2}$. The situation with this kind of combined flow is very similar to the one described in Ref. 4 for the case of Poiseuille flow combined with the natural convection flow in the thermogravitational FFF channel.

High selectivity, efficiency, and fractionating power makes FFF with combined EOF and Poiseuille flows very interesting, as it can, at least theoretically, lead to finer separations for given values of λ. Experimental realization of FFF with asymmetrical EOF have not yet been reported due to some technical problems. However, considerable progress in this field is accomplished by a Finnish group (Riekkola, Vastamaki, and Jussila) working on the experimental realization of thermal FFF with asymmetrical EOF [5] and a Russian group (Andreev, Stepanov, and Tihomolov) working on gravitational FFF with asymmetrical EOF [6].

The situation is more complicated, even theoretically, when the sample particles are charged. In this case, they are not only moving with the longitudinal flow (here, asymmetrical EOF) but are also forced by the longitudinal electric field to move along the channel electrophoretically. If the electrophoretic mobilities of the particles are different, then there are two types of separations combined: The FFF type due to the difference in λ values and the capillary zone electrophoresis (CZE) type due to the difference in electrophoretic mobilities. A great variety of variants of FFF and CZE combinations in the FFF channel with asymmetrical EOF could be imagined, depending on various factors such as the ratio of the zeta-potentials of the channel walls, the sign and the value of the ratio of eletrophoretic and electro-osmotic velocities, and the type of the transversal field. Some of these combinations are examined in Ref. 6. They could lead both to the new possibilities of the method and to some new complications in the interpretation of the experimental results.

Another possibility for realizing FFF with EOF is to reduce the concentration of the buffer, thus making the Debye length commensurate, if not with the depth of the channel, then with the thickness $l = \lambda w$ of the layer of sample particles compressed to the accumulating wall of FFF channel (for $C_0 = 10^{-5} M$ Debye length, $\kappa^{-1} = 0.1\ \mu m$). In this case, EOF will be nonuniform enough to realize FFF in a channel with equal zeta-potentials of the walls.

In Ref. 7 the mathematical model of CZE, taking into consideration EOF nonuniformity and particle–wall electrostatic interactions, was developed. It was shown that for the particles electrostatically attracted by the wall of the capillary, two mechanisms of separation exist. The first is the usual CZE mechanism and it dominates for the case of high buffer concentrations; the second is the FFF accompanying the CZE mechanism and it dominates for low buffer concentrations. As is usual in CZE, the total velocity of the particle is the sum of its electrophoretic velocity and electro-osmotic velocity of the flow. The larger the electrical charge of sample particles, the stronger they are attracted to the wall and the higher is their concentration in the Debye layer vicinity of the wall, where the EOF is substantially nonuniform. Thus, for the particles with a higher charge, the mean velocity of movement with EOF will be lower than for the particles with the lower charge. Especially interesting with this type of FFF is for the particles with equal electrophoretic mobilities but different charges. Such types of particles (e.g., DNA fragments) cannot be fractionated by usual CZE, but can be fractionated by FFF accompanying CZE, where the separation is due to the difference of electrical charges, not the difference of mobilities. Note that for this type of FFF, there is no need for any external transversal field, because the particles are attracted to the walls by the field of the electrical double layer. As is usual in FFF, there is the transition point from normal diffusional FFF to steric FFF mode, taking place when the size of the particle is commensurable with λw. In Ref. 8, it was theoretically predicted that steric FFF accompanying the CZE mode can be realized for the separation of DNA fragments in the range of 20–3000 bases with high resolution and speed. To realize this type of separation, one needs to develop a modified capillary with the positive value of the zeta-potential of the wall and without the irreversible sorption of DNA fragments on the walls. Such an attempt seems to be worthy because, unlike DNA separation by CZE in gel or polymer solution, in the case of FFF–CZE there is the possibility of on-line coupling with a mass spectrometer without the risk of gel particles going inside the spectrometer.

References

1. V. P. Andreev, M. E. Miller, and J. C. Giddings, Field-flow fractionation with asymmetrical electroosmotic flow, *5th Int. Symp on FFF*, 1995.
2. V. P. Andreev, Y. V. Stepanov, and J. C. Giddings. Field-flow fractionation with asymmetrical electroosmotic flow. I. Uncharged particles, *J. Microcol. Separ. 9*: 163 (1997).
3. M. Martin and J. C. Giddings, Retention and nonequilibrium peak broadening for generalized flow profile in FFF, *J. Phys. Chem. 85*: 727 (1981).
4. J. C. Giddings, M. Martin, and M. N. Myers, Thermogravitational FFF: An elution thermogravitational column, *Separ. Sci. Technol. 14*: 611 (1979).
5. P. Vastamaki, M. Jussila, and M.-L. Riekkola, The effect of electrically nonconductive wall coating on retention in ThFFF, *7th Int. Symp. on FFF*, 1998.
6. V. P. Andreev and Y. V. Stepanov, Field-flow fractionation with asymmetrical electroosmotic flow. II. Charged particles, *J. Liquid Chromatogr. Related Technol. 20*: 2873 (1997).
7. V. P. Andreev and E. E. Lisin, On the mathematical model of capillary electrophoresis, *Chromatographia 37*: 202 (1993).
8. V. P. Andreev and Y. V. Stepanov, Steric FFF accompanying capillary electrophoresis, *5th Int. Symp on FFF*, 1995.

Victor P. Andreev

Field-Flow Fractionation Fundamentals

Introduction

Field-flow fractionation (FFF) is a separation methodology suitable for the analysis and characterization of the macromolecules and particles. The separation is based on the interaction of the effective physical or chemical forces (e.g., temperature gradient; electric, magnetic, gravitational, or centrifugal forces; chemical potential gradient; etc.) with the separated species. The field acting across a separation channel concentrates them at a given position inside the channel. The formed concentration gradient induces an opposite diffusion flux. This leads to a steady-state distribution of the sample components across the channel. The velocity of the longitudinal flow of the carrier liquid also varies across the channel. A flow velocity profile is established inside the channel. As a result, the components of the separated sample are transported in the longitudinal direction at different velocities depending on their transversal positions within the flow of the carrier liquid. This general principle of the FFF is demonstrated in Fig. 1.

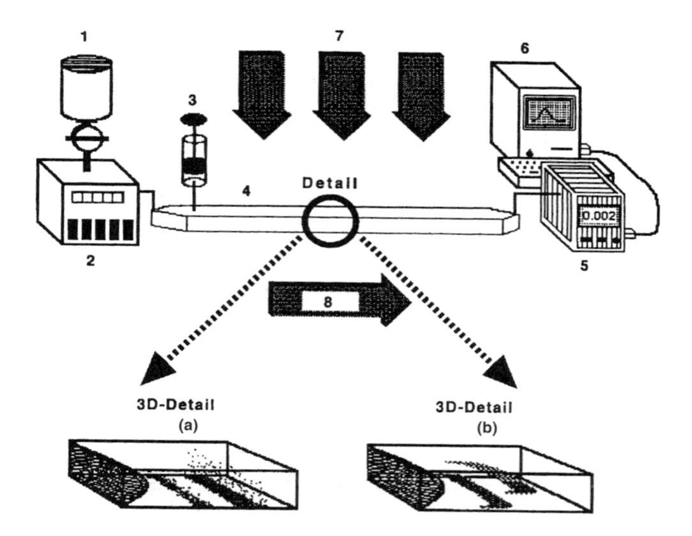

Fig. 1 Principle of field-flow fractionation: (1) solvent reservoir, (2) carrier liquid pump, (3) injection of the sample, (4) separation channel, (5) detector, (6) computer for data acquisition, (7) transversal effective field forces, (8) longitudinal flow of the carrier liquid. (a) Section of the channel demonstrating the principle of polarization FFF with two distinct zones compressed differently at the accumulation wall and the parabolic flow velocity profile. (b) Section of the channel demonstrating the principle of focusing FFF with two distinct zones focused at different positions and the parabolic flow velocity profile.

Two mechanisms, *polarization* [1] and *focusing* [2], lead to the formation of different concentration distributions across the fractionation channnel. The components of the fractionated sample are either compressed to the accumulation wall of the channel or focused at different positions, as shown in Fig. 1. Steady state inside the channel is reached in a short time due to a small channel thickness. The strength of the field can be controlled within a wide range in order to manipulate the retention conveniently. Many operational variables in FFF can be manipulated during the experiment by a suitable programming.

The *polarization FFF* methods are classified according to the nature of the applied field, whereas the *focusing FFF* methods are classified by considering the combination of various gradients and fields emphasizing the focusing processes. Polarization FFF methods make use of the formation of an exponential concentration distribution of each sample component across the channel with the maximum concentration at the accumulation wall, which is a consequence of constant and position-independent velocity of transversal migration of the affected species due to the field forces. This concentration distribution is combined with the velocity profile formed in the flowing liquid. Focusing FFF methods make use of transversal migration of each sample component under the effect of driving forces that vary across the channel. As a result, the sample components are focused at the positions where the intensity of the effective forces is zero and are transported longitudinally with different velocities according to the established flow velocity profile. The concentration distribution within a zone of a focused sample component can be described by Gaussian or similar distribution function.

Principle and Theory

The carrier liquid flows in the direction of the channel longitudinal axis, whereas the field forces act perpendicularly across the channel. The driving forces can be generated by a single field or by the coupled action of two or more different fields. Polarizing and focusing forces can operate simultaneously, resulting in a complex mechanism of separation. The field force F and, consequently, the velocity U are independent of the position in the direction of the x axis in polarization FFF:

$$F \neq 0 \quad \text{and} \quad U \neq 0 \quad \text{for } 0 < x < w \tag{1}$$

where w is the distance between the main channel walls in the direction of the x axis with $x = 0$ at the accumulation wall. On the other hand, it holds for the x-axis-dependent direction of the field force in focusing FFF:

$$F = F(x) \quad \text{and} \quad U = U(x) \quad \text{within } 0 < x < w \tag{2}$$

$$F(x) = 0 \quad \text{and} \quad U(x) = 0 \quad \text{for } x = x_{max} \text{ with } 0 < x_{max} < w \tag{3}$$

$$F(x) > 0 \quad \text{and} \quad U(x) > 0 \quad \text{for } x < x_{max} \tag{4}$$

$$F(x) < 0 \quad \text{and} \quad U(x) < 0 \quad \text{for } x > x_{max} \tag{5}$$

where the coordinate x_{max} corresponds to the position at which the concentration distribution of a sample component across the channel attains its maximal value.

Polarization FFF

The equilibrium concentration distribution in the direction of the x axis across the channel of a given component of the sample can be calculated from the continuity equation

$$-D\frac{\partial c}{\partial x} - Uc = 0 \tag{6}$$

where D is the diffusion coefficient and c is the concentration. The solution of Eq. (6) gives the exponential concentration distribution of the sample component across the channel [3]:

$$c(x) = c(0)\exp\left(-\frac{xU}{D}\right) \tag{7}$$

By defining the mean layer thickness, $l = D/U$, Eq. (7) can be rewritten

$$c(x) = c(0)\exp\left(-\frac{x}{l}\right) \tag{8}$$

The mean layer thickness is practically equal to the center of gravity of the concentration distribution.

Focusing FFF

It holds for a focused species at equilibrium that

$$D\frac{\partial c}{\partial x} - U(x)c = 0 \tag{9}$$

The force $F(x)$, acting on one particle undergoing the focusing, can be written as

$$F(x) = U(x)f \tag{10}$$

with the friction coefficient defined by

$$f = \frac{kT}{D} \tag{11}$$

where k is the Boltzmann constant and T is the absolute temperature. Then, it holds that

$$\frac{dc}{dx} = \frac{F(x)c}{kT} \tag{12}$$

The focusing force can be approximated by [4]

$$F(x) = -\left|\left(\frac{dF(x)}{dx}\right)_{x \approx x_{max}}\right|(x - x_{max}) \tag{13}$$

where $(dF(x)/dx)_{x=x_{max}}$ is the gradient of the driving force. The solution is

$$c(x) = c_{max} \exp\left[-\frac{1}{2kT}\left|\left(\frac{dF(x)}{dx}\right)_{x \approx x_{max}}\right|(x - x_{max})^2 \right] \quad (14)$$

which is a Gaussian concentration profile of a single focused component. A more accurate approach [4] is based on the real gradient of the focusing forces and results in a concentration distribution of the focused species, which is not Gaussian.

Flow Velocity Profiles

The separation is usually carried out in a belt-shaped narrow channel of constant thickness. The cross section of the channel is rectangular. Velocity distribution in such a channel (provided that the flow is isoviscous) is parabolic:

$$v(x) = \frac{\Delta P x(w - x)}{2L\mu} \quad (15)$$

where $v(x)$ is the longitudinal velocity at the x coordinate, ΔP is the pressure drop along the channel of the length L, and μ is the viscosity of the carrier liquid. The average velocity is

$$\langle v(x) \rangle = \frac{\Delta P w^2}{12L\mu} \quad (16)$$

Other shapes of the flow velocity profiles can be formed in channels whose cross section is not rectangular but, for example, trapezoidal. The use of such nonparabolic flow velocity profiles can be advantageous, especially in focusing FFF.

Separation

The separation is due to the coupled action of the concentration and flow velocity distributions. The concentration distribution across the channel of each sample component is established and the sample components are eluted along the channel with different velocities, depending on the distance of their centers of gravity from the accumulation wall. The average velocity of the zone of a retained sample component is

$$\langle v \rangle = \frac{\langle c(x)v(x) \rangle}{\langle c(x) \rangle} \quad (17)$$

The retention ratio R is defined as the average velocity of a retained sample component to the average velocity of the carrier liquid:

$$R = \frac{\int_0^w c(x)v(x)\,dx \int_0^w dx}{\int_0^w c(x)\,dx \int_0^w v(x)\,dx} \quad (18)$$

where $v(x)$ and $c(x)$ are the local velocity and concentration, respectively, of the retained species. From the practical point of view, the retention ratio R can be expressed as the ratio of the experimental retention time t_0 or the retention volume V_0 of an unretained sample component to the retention time t_r or the retention volume V_r of the retained sample component. Provided that the relationship between the position of the center of gravity of the zone and the molecular parameters of the sample component exists, these parameters can be calculated from the retention data without a calibration.

The retention ratio in polarization FFF is thus given by [5]

$$R = 6\lambda\left[\coth\left(\frac{1}{2\lambda}\right) - 2\lambda\right] \quad (19)$$

where $\lambda = l/w$. If λ is small, the following approximations hold:

$$(\lim R)_{\lambda \to 0} = 6(\lambda - 2\lambda^2) \quad \text{or} \quad (\lim R)_{\lambda \to 0} = 6\lambda \quad (20)$$

The retention parameter λ is the ratio of thermal energy to the effective of the field on the retained species:

$$\lambda = \frac{kT}{Fw} \quad (21)$$

When the size of the separated species is commensurable with the thickness of the channel, the limit retention ratio in this mode of steric FFF is [6]

$$\lim_{\alpha \to 0} R = 6\alpha \quad (22)$$

where $\alpha = r/w$ and r is the particle radius.

The retention ratio in focusing FFF carried out in rectangular cross-section channel is given by the approximate relationship [7]

$$R = 6(\Gamma_{max} - \Gamma_{max}^2) \quad (23)$$

where $\Gamma_{max} = x_{max}/w$ is the dimensionless coordinate of the maximal concentration of the focused zone.

Methods and Applications

The retention is related to the size, charge, diffusion coefficient, thermal diffusion factor, and so forth of the separated species in polarization FFF. As concerns the focusing FFF, the retention is usually related with the intensive properties of the fractionated species. Consequently, the FFF can be used to characterize the properties related to the retention. Because the entry Focusing FFF of Particles and Macromolecules is fully devoted to the focusing FFF, only the polarization FFF methods will be described here.

The particular methods of polarization FFF are denominated by the nature of the applied field. The most important of them are described in the following subsections.

Sedimentation FFF

The sedimentation FFF is shown schematically in Fig. 2a. The separation channel is situated inside a centrifuge rotor and the centrifugal forces are applied radially [8]. The method can be used for the analysis and characterization

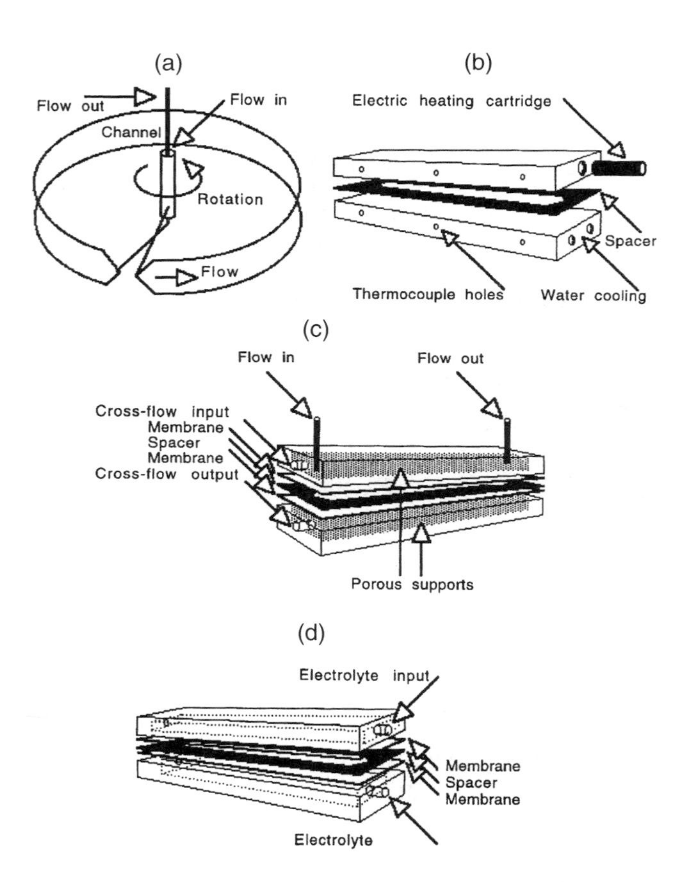

Fig. 2 Methods of polarization field-flow fractionation: (a) sedimentation FFF; (b) thermal FFF; (c) flow FFF; (d) electric FFF.

of various latexes, inorganic particles, emulsions, biological cells, and so forth. The retention parameter λ depends on the effective mass of the particles:

$$\lambda = \frac{6kT}{\pi d^3 g w \Delta \rho} \tag{24}$$

where g is the gravitational or centrifugal acceleration and $\Delta \rho$ is the density difference between the particles and the carrier liquid. The calculation of the particle size distribution is possible directly from the retention data.

Thermal FFF

Thermal FFF was the first experimentally implemented method [9]. It is used mostly for the fractionation of macromolecules. The temperature difference between two metallic bars, forming the channel walls with highly polished surfaces and separated by a spacer in which the channel proper is cut, produces the flux of the sample components, usually toward the cold wall. The channel for thermal FFF is shown in Fig. 2b. The relation between λ and the operational variables is given by

$$\lambda = \frac{D}{w D_T (dT/dx)} \tag{25}$$

where D_T is the coefficient of the thermal diffusion which depends on the chemical composition and structure of the fractionated species but not on their size. On the other hand, the diffusion coefficient D depends on the size. As a result, the differences in thermal diffusion coefficients allow the fractionation according to differences in chemical composition and structure, whereas different diffusion coefficients allow the fractionation based on the size differences. The performances favor thermal FFF over the competitive methods.

Flow FFF

Flow FFF is the most universal method because the cross-flow field acts on all fractionated species in the same manner and the separation is due to the differences in diffusion coefficients [10]. The channel, schematically demonstrated in Fig. 2c, is formed between two parallel semipermeable membranes. The carrier liquid can permeate through the membranes but not the separated species. The retention parameter λ is related to the diameter d_p of the separated species by

$$\lambda = \frac{kTV_0}{3\pi \mu V_c w^2 d_p} \tag{26}$$

where V_0 is the void volume of the channel, μ is the viscosity of the carrier liquid, and V_c is the volumetric velocity of the cross-flow. The separations of various kinds of particles such as proteins, biological cells, colloidal silica, polymer latexes, and so forth, were described as well as of the soluble macromolecules.

Electric FFF

Electric FFF uses the electric potential across the channel to generate the transversal flux of the charged species [11]. The walls of the channel can be formed by semipermeable membranes that allow the passage of small ions but not of the separated species. The channel is shown in Fig. 2d. The dependence of the retention parameter λ on the electrophoretic mobility μ_e, and the diffusion coefficient of the charged particles is given by

$$\lambda = \frac{D}{\mu_e E w} \tag{27}$$

where E is the electric field strength. As a result, the ratio of the diffusion coefficient to the electrophoretic mobility determines the retention. The species exhibiting only small differences in electrophoretic mobilities but important differences in diffusion coefficients can be separated. Electric FFF is especially suited for the separations of the biological cells as well as for charged polymer latexes and other colloidal particles and charged macromolecules.

Other Polarization FFF Methods

Other polarization FFF methods have recently been proposed theoretically and some of them implemented experimentally. Their use in current laboratory practice needs further development in methodology and instrumentation. One of the most recent review articles brings an excellent review of the state of the art of the polarization FFF methods [12].

References

1. J. C. Giddings, *Separ. Sci. Technol. 18*: 765 (1983).
2. J. Janča, *Makromol. Chem. Rapid Commun. 3*: 887 (1982).
3. J. Janča, *Field-Flow Fractionation: Analysis of Macromolecules and Particles*, Marcel Dekker, Inc., New York, 1988.
4. J. Janča, in *Chromatographic Characterization of Polymers, Hyphenated and Multidimensional Techniques* (T. Provder, H. G. Barth, and M. W. Urban, eds.), Advances in Chemistry Series Vol. 247, American Chemical Society, Washington, DC, 1995.
5. M. E. Hovingh, G. H. Thompson, and J. C. Giddings, *Anal. Chem. 42*: 195 (1970).
6. J. C. Giddings and M. N. Myers, *Separ. Sci. Technol. 13*: 637 (1978).
7. J. Janča and J. Chmelík, *Anal. Chem. 56*: 2481 (1984).
8. J. C. Giddings, M. N. Myers, M. H. Moon, and B. N. Barman, in *Particle Size Distribution* (T. Provder, ed.), ACS Symposium Series Vol. 472, American Chemical Society, Washington, DC, 1991.
9. S. J. Jeon and M. E. Schimpf, in *Particle Size Distribution III: Assessment and Characterization* (T. Provder, ed.), American Chemical Society, Washington, DC, 1998.
10. S. K. Ratanathanawongs and J. C. Giddings, in *Chromatography of Polymers: Characterization by SEC and FFF* (T. Provder, ed.), ACS Symposium Series Vol. 521, American Chemical Society, Washington, DC, 1993.
11. M. E. Schimpf and K. D. Caldwell, *Am. Lab. 27*: 64 (1995).
12. H. Cölfen and M. Antonietti, *Adv. Polym. Sci. 150*: 67 (2000).

Josef Janča

Flame Ionization Detector for GC

The flame ionization detector (FID) is, by far, the most commonly used detector in gas chromatography (GC) and is probably the most important. It is a little uncertain as to who was the first to invent the FID; some gave the credit to Harley and Pretorius [1], others to McWilliams and Dewer [2]. In any event, it would appear that both contenders developed the device at about the same time, and independently of one another; the controversy had more patent significance than historical interest. The FID is an extension of the flame thermocouple detector and is physically very similar, the fundamentally important difference being that the ions produced in the flame are measured, as opposed to the heat generated.

The principle of detection is as follows. Hydrogen is mixed with the column eluent and burned at a small jet. Surrounding the flame is a cylindrical electrode and a rela-

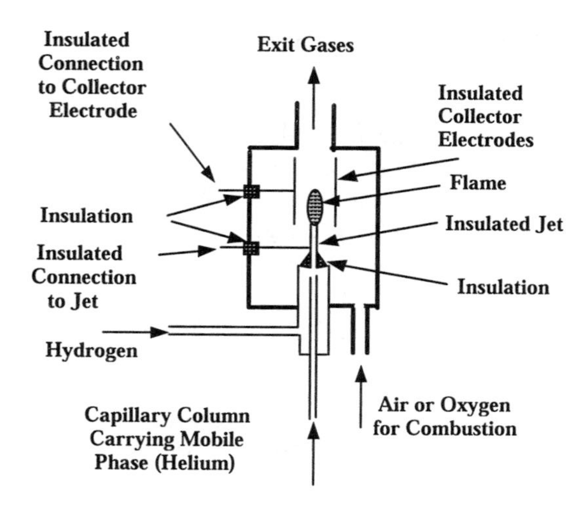

Fig. 1 The FID sensor.

tively high voltage is applied between the jet and the electrode to collect the ions that are formed in the flame. The resulting current is amplified by a high-impedance amplifier and the output fed to a data acquisition system or a potentiometric recorder.

A detailed diagram of the FID sensor is shown in Fig. 1. The body and the cylindrical electrode is usually made of stainless steel and stainless-steel fittings connect the detector to the appropriate gas supplies. The jet and the electrodes are insulated from the main body of the sensor with appropriate high-temperature insulators. Some care must be taken in selecting appropriate insulators as many glasses (with the exception of fused quartz) and some ceramic materials become conducting at high temperatures (200–300°C) [3].

As a result of the relatively high voltages used in conjunction with the very small ionic currents being measured, all connections to the jet or electrode must be well insulated and electrically screened. In addition, the screening and insulating materials must be stable at the elevated temperature of the detector oven. In order to accommodate the high temperatures that exist at the jet tip, the jet is usually constructed of a metal that is not easily oxidized, such as stainless steel, platinum, or platinum–rhodium. The detector electronics consist of a high-voltage power supply and a high-impedance amplifier. The jet and electrode can be connected to the power supply and amplifier in basically two configurations. The floating jet configuration is the most commonly used and in this arrangement, +250 to +400 V is applied to the cylindrical electrodes and the jet is connected to a ground by a very high resistance. The signal developed across the resistance is amplified, modified, and passed to a recorder of the data acquisition system. In the second alternative, the jet is grounded and the high-voltage power supply is electrically floated. Then, +250 to +400 V is applied to the cylindrical electrodes and the negative terminal of the power supply is connected to a ground by a very high resistance. The signal that is developed across the resistance is again amplified, modified, and passed to a recorder of the data acquisition system.

Response Mechanism of the FID

The flame ionization detector has a very wide dynamic range, has a high sensitivity, and, with the exception of about half a dozen low-molecular-weight compounds, will detect all substances that contain carbon. The response mechanism of the FID has been carefully investigated by a number workers. It was originally thought that the ionization mechanism in the FID flame is similar to the ionization process in a hydrocarbon flame, but it quickly became apparent that ionization in the hydrogen flame is many times higher than could be accounted for by thermal ionization alone. It would appear that the ionization potentials of organic materials become much lower when they enter the flame.

The generally accepted explanation of this effect is that the ions are not formed by thermal ionization but by thermal emission from small carbon particles that are formed during the combustion process. Consequently, the dominating factor in the ionization of organic material is not their ionization potential but the work function of the carbon that is transiently formed during their combustion. The flame plasma contains both positive ions and electrons which are collected on either the jet or the plate, depending on the polarity of the applied voltage. Initially, the current increases with applied voltage, the magnitude of which depend on the electrode spacing. The current continues to increase with the applied voltage and eventually reaches a plateau at which the current remains sensibly constant. The voltage at which this plateau is reached also depends on the electrode distances.

As soon as the electron–ion pair is produced, recombination starts to take place. The longer the ions take to reach the electrode and be collected, the more the recombination takes place. Thus, the greater the distance between the electrodes and the lower the voltage, the greater the recombination. As a result, initially the current increases with the applied voltage and then eventually flattens out, and at this point, it would appear that all the ion–electron pairs were being collected. In practice, the applied voltage would be adjusted to suit the electrode geometry and ensure that the detector operates under conditions where all electrons and ions are collected.

It was also shown that the airflow should be at least six times that of the hydrogen flow for stable conditions and complete combustion. The base current from the hydrogen flow depends strongly on the purity of the hydrogen. Traces of hydrocarbons significantly increase the base current, as would be expected. Consequently, very pure hydrogen should be employed with the FID if maximum sensitivity is required. Employing purified hydrogen, Desty et al. reported a base current of 1.45×10^{-12} A for a hydrogen flow of 20 mL/min. This was equivalent to 1×10^{-7} C/mol. The sensitivity reported for n-heptane, assuming a noise level equivalent to the base current from hydrogen of $\sim 2 \times 10^{-14}$ A (a fairly generous assumption), was 5×10^{-12} g/mL at a flow rate of 20 mL/min. It follows that although the sensitivity is amazingly high, the ionization efficiency is still very small ($\sim$0.0015%). The general response of the FID to substances of different type varies very significantly from one to another. For a given homologous series, the response appears to increase linearly with carbon number, but there is a large difference in response between different homologous series (e.g., hydrocarbons and alcohols).

The linear dynamic range of the FID covers at least four to five orders of magnitude for $0.98 < r < 1.02$. This is a remarkably wide range that also helps explain the popularity of the detector. Examination of the different commercially available detectors shows considerable difference in electrode geometry and operating electrode voltages, yet they all have very similar performance specifications.

Operation of the FID

The FID is one of the simplest and most reliable detectors to operate. Generally, the appropriate flow rates for the different gases are given in the detector manual. The hydrogen flow usually ranges between 20 and 30 mL/min and the airflow is about six times that of the hydrogen flow (e.g., 120–200 mL/min. The column flow that can be tolerated is usually about 20–25 mL/min, depending on the chosen hydrogen flow. However, if a capillary column is used, the flow rate may be less than 1 mL/min for very small-diameter columns. The mobile phase can be any inert gas — helium, nitrogen, argon, and so forth. To some extent, the detector is self-cleaning and rarely becomes fouled. However, this depends a little on the substances being analyzed. If silane derivatives are continuously injected on the column, then silica is deposited both on the jet and on the electrodes and may need to be regularly cleaned. In a similar way, the regular analysis of phosphate-containing compounds may eventually contaminate the electrode system. Electrode cleaning is best carried out by the qualified instrument service engineer.

Apparently, the sole disadvantage of the FID as a general detector is that it normally requires three separate gas supplies, together with their precision flow regulators. The need for three gas supplies is a decided inconvenience but is readily tolerated in order to take advantage of the many other attributes of the FID. The detector is normally thermostatted in a separate oven; this is not because the response of the FID is particularly temperature sensitive but to ensure that no solutes condense in the connecting tubes.

The FID has an extremely wide field of application and is used in the analysis of hydrocarbons, solvents, essential oils, flavors, drugs, and their metabolites — in fact, any mixture of volatile substances that contain carbon.

References

1. J. Harley, W. Nel, and V. Pretorius, *Nature (London)*, *181*: 177 (1958).
2. I. G. McWilliams and R. A. Dewer, *Gas Chromatography 1958* (D. H. Desty, ed.), Butterworths, London, 1957, p. 142.
3. S. A. Beres, C. D. Halfmann, E. D. Katz, and R. P. W. Scott, *Analyst 112*: 91 (1987).

Suggested Further Reading

Scott, R. P. W., *Chromatographic Detectors*, Marcel Dekker, Inc., New York, 1996.

Scott, R. P. W., *Introduction to Analytical Gas Chromatography*, Marcel Dekker, Inc., New York, 1998.

Raymond P. W. Scott

Flow Field-Flow Fractionation: Introduction

Principles

Flow field-flow fractionation (flow FFF or FlFFF) is one of the FFF subtechniques in which particles and macromolecules are separated in a thin channel by aqueous flow under a field force generated by a secondary flow. As with other FFF techniques, separation in FlFFF is based on the applied force directed across the axis of separation flow. In FlFFF, this force is generated by cross-flow of liquid delivered across the channel walls. In order to maintain the uniformity of cross-flow moving in a typical rectangular channel, two ceramic permeable frits are used as channel walls and the flow stream enters and exits through these walls. The force applied in FlFFF is a Stokes force that depends only on the sizes of sample components.

In FlFFF, particles or macromolecules entering the channel are driven toward an accumulation wall by the cross-flow. Normally, a sheet of semipermeable membrane is placed at the accumulation wall in order to keep sample materials from being lost by the wall. While sample components are being transported close to the accumulation wall, they are projected against the wall by Brownian diffusion. The diffusive transport against the wall leads the sample components to be differentially distributed against the wall, according to their sizes: The larger particles, having a small diffusion coefficient, are placed at an equilibrium position closer to the vicinity of accumulation wall than the smaller ones. Thus, small particles, which are located further from the wall, will be exposed to the fast streamline of a parabolic flow profile, and they will be eluted earlier than the larger ones. This is the typical elution profile that can be observed in the normal operating mode of FFF (denoted as Fl/Nl FFF). Retention time in Fl/Nl FFF is inversely proportional to the diffusion coefficient of the sample; it is represented as

$$t_r = \frac{w^2}{6D}\frac{\dot{V}_c}{\dot{V}} \quad \left(\text{where } D = \frac{kT}{3\pi\eta d_s}\right) \tag{1}$$

where w is the channel thickness, D is the diffusion coefficient, $\dot{V}_c$ is the cross-flow rate, and $\dot{V}$ is the channel flow rate. Because the diffusion coefficient D is inversely proportional to the viscosity of carrier solution η and hydrodynamic radius d_s, the retention time can be simply predicted provided the particle diameter or the diffusion coefficient is known. Conversely, the particle diameter of an unknown sample can be calculated from experimental retention time by rearranging Eq. 1.

As the particle size becomes large at or above 1 μm, the diffusional process of particles becomes less dominant in FFF. In this regime, a particle's retention is largely governed by the particle size itself, in which the center of large particles is located at a higher position than small ones. Thus, large particles meet the faster streamlines and they elute earlier than the small ones; the elution order is reversed. However, it is known, from experimental results, that particles migrate at certain positions elevated from the wall due to the existence of hydrodynamic lift forces that act in the opposite direction to the field. This is described as the steric/hyperlayer operating mode of separation in flow FFF and is denoted by Fl/Hy FFF. Whereas the theoretical expectation of particle retention in Fl/Nl FFF is clearly understood, retention in Fl/Hy FFF is not predictable because the hydrodynamic lift forces are not yet completely understood. Therefore, the particle size calculation in Fl/Hy FFF relies on the calibration process in which a set of standard latex particles of known diameter is run beforehand as

$$\log t_r = -S_d \log d_s + \log t_{r1} \tag{2}$$

where S_d is the diameter-based selectivity and t_{r1} is the interpolated intercept representing the retention time of a unit diameter. The S_d values found experimentally are about 1.5 in Fl/Hy FFF. By using Eq. (2), the particle diameters of unknown samples can be calculated once the calibration parameters S_d and t_{r1} are provided.

Types of Channel in FlFFF

There are two main categories of flow FFF channel systems, depending on the use of frit wall. The above-described flow FFF system has a frit on both walls; this is classified as a symmetrical channel, as shown in Fig. 1a. An asymmetrical channel system is being widely studied in which only one permeable frit wall is used, at the accumulation wall, and the depletion wall is replaced with a glass plate (Fig. 1b). In an asymmetrical channel, part of the flow entering the channel is lost by the accumulation wall and this acts as a field force to retain the sample components in the channel, as does the cross-flow in a symmetrical channel.

The separation efficiency of an asymmetrical flow FFF system has been known to be higher than that of a conventional symmetrical channel. Because an asymmetrical channel utilizes only one frit, nonuniformity of flow that could arise from the imperfection of frits can be reduced. In addition, the initial sample band can be kept narrower in an asymmetrical channel, due to the focusing–relaxation procedure, which is an essential process in an asym-

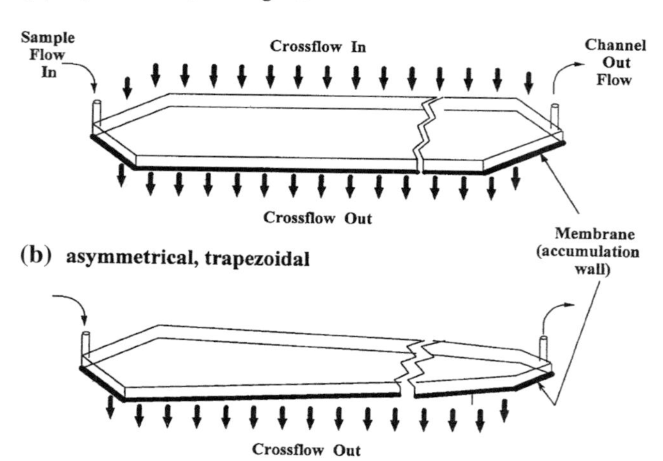

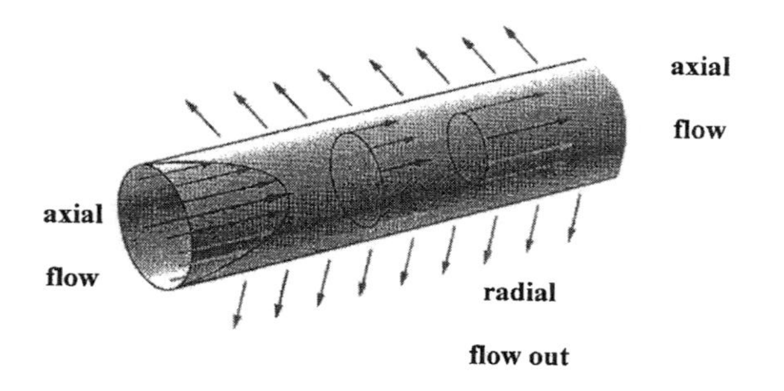

Fig. 1 Types of channel in FFF.

metrical channel. The relaxation processes, which provide an equilibrium status for sample components, are necessary in both symmetrical and asymmetrical channels for a period of time prior to the separation. For a symmetrical channel, this is normally achieved by stopping channel flow immediately after sample injection, while the cross-flow is applied.

During the relaxation process, sample components seek their equilibrium positions where the drag of the cross-flow is counterbalanced with diffusive transports (or lift forces) against the walls. After relaxation, flow is resumed and separation begins. However, in an asymmetrical channel, the relaxation process is achieved by two convergent focusing flow streams originating at the channel inlet and outlet (focusing–relaxation). Thus, injected sample can be focused at a certain position near the inlet end and the broadening of the initial sample band can be better minimized. This will lead to a decrease in band broadening of an eluted peak in an asymmetrical channel.

In asymmetrical flow FFF, two channel designs are utilized: rectangular and trapezoidal. Because flow velocity decreases along the axis of migration, a trapezoidal channel in which the channel breadth decreases toward the outlet is known to be more efficient in eluting low-retaining materials such as high-molecular-weight proteins. Retention in an asymmetrical flow FFF system follows the basic FFF principle and the retention time is calculated as

$$t_r = \frac{w^2}{6D} \ln\left(1 + \frac{\dot{V}_c}{\dot{V}_{\text{out}}}\right) \tag{3}$$

where $\dot{V}_c$ is the cross-flow rate and $\dot{V}_{\text{out}}$ is the outlet flow rate.

In addition to the rectangular channels in FlFFF described thus far, a cylindrical channel system has been developed with the use of hollow fibers in which the fiber wall is made of a porous membrane, as shown in Fig. 1c. It also requires a focusing–relaxation process, as does an asymmetrical channel. Retention in hollow-fiber flow FFF (HF-FlFFF) is controlled by the radial flow, which effectively acts as the cross-flow of a conventional flow FFF system, and the retention in a hollow fiber resembles that of an asymmetrical channel system.

However, the retention ratio in HF-FlFFF is approximately 4λ for a sufficiently retained component, which is somewhat different from that of a conventional channel system ($R \cong 6\lambda$). The retention time in a hollow fiber is calculated as

$$t_r = \frac{r_f^2}{8D} \ln\left(1 + \frac{\dot{V}_{\text{rad}}}{\dot{V}_{\text{out}}}\right) \tag{4}$$

where r_f^2 is the radius of the fiber and $\dot{V}_{\text{rad}}$ is the radial flow rate. Although a number of experiments have indicated a great potential of hollow fibers as an alternative for a flow FFF channel, a great deal of study related to their performance and optimization is needed.

Suggested Further Readings

Giddings, J. C., *Science 260*: 1456 (1993).

Giddings, J. C., F. J. Yang, and M. N. Myers, *Anal. Chem. 48*: 1126 (1976).

Jönsson, J. A., and A. Carlshaf, *Anal. Chem. 61*: 11 (1989).

Litzén, A. and K.-G. Wahlund, *J. Chromatogr. 476*: 413 (1989).

Litzén, A. and K.-G. Wahlund, *Anal. Chem. 63*: 1001 (1991).

Moon, M. H., Y. H. Kim, and I. Park, *J. Chromatogr. 813*: 91 (1998).

Ratanathanawongs, S. K. and J. C. Giddings, in *Chromatography of Polymers: Characterization by SEC and FFF* (T. Provder, ed.), ACS Symposium Series Vol. 521, American Chemical Society, Washington, DC, 1993, pp. 13–29.

Myeong Hee Moon

Fluorescence Detection in Capillary Electrophoresis

One cannot overestimate the importance of fluorescence detection in high-performance capillary electrophoresis (HPCE) [1]. The success of the human genome project along with the forthcoming revolutions in forensic testing and genetic analysis might not have occurred without the sensitivity and selectivity of laser-induced fluorescence (LIF) detection.

Basic Concepts

The stunning sensitivity of fluorescence detection arises from two areas: (a) detection is performed against a very dark background and (b) the use of the laser as an excitation source provides a high photon flux. The combination of the two can yield single-molecule detection in exceptional circumstances, although picomolar ($10^{-12}M$) is typically obtained. Under conditions that are easy to replicate, LIF detection is often 10^6 times more sensitive compared to ultraviolet (UV) absorption detection.

Most molecules absorb light in the ultraviolet or visible portion of the spectrum, but only few produce significant fluorescence. This provides for the extreme selectivity of the technique. Molecular fluorescence is usually quenched through vibronic or collosional events resulting in a radiationless decay of excited singlet-state energy to the ground state. In aromatic structurally rigid molecules, quenching is less significant and the quantum yield increases.

The selectivity of fluorescence is to itself a problem because the technique is applicable to fewer separations. Sophisticated derivatization schemes have been developed for these applications to take advantage of the attributes contributed by fluorescence detection. Because there are two instrumental parameters to adjust, the excitation and emission wavelengths, the inherent selectivity of the method is further enhanced.

The fundamental equation governing fluorescence is

$$I_f = \Phi_f I_0 abc E_x E_c E_m E_{\text{pmt}}$$

where I_f is the measured fluorescence intensity, Φ_f is the quantum yield (photons emitted/photons absorbed), I_0 is the excitation power of the light source, a, b, and c are the Beer's Law terms, and the E terms are the efficiencies of the excitation monochromator or filter, the optical portion of the capillary, the emission monochromator or filter, and the detector (photomultiplier or charge-coupled device), respectively. It is no wonder why optimization of fluorescence detection is difficult for the uninitiated.

Excitation Sources

The optimal excitation wavelength is usually a combination of the power of the light source and the molar absorptivity of the solute at the selected wavelength. The argon-ion laser is used for most DNA applications since the primers, intercalators, and dye terminators have been optimized for 488-nm excitation. For other applications, particularly for small molecules, where native fluorescence is measured, a tunable light source is desirable. The deuterium lamp is useful for low-UV excitation and the xenon arc is superior in the near-UV to visible region. With a 75-W xenon arc, the limit of detection (LOD) is 2 ng/mL ($6 \times 10^{-9}M$) for fluorescein using fiber-optic collection of the fluorescence emission [2]. This is a 100-fold improvement compared to absorption detection. By using a microscope objective to focus the light along with a sheath-flow cuvette (to reduce scattering, see below) and lens to collect the light, the LOD is reduced to $8 \times 10^{-11}M$ [3]. Nevertheless, the LOD using conventional tunable sources will never be superior to that found with the laser.

It is possible to select lasers other than the argon-ion laser for LIF detection. A 625-nm diode laser is available on a commercial unit (Beckman P/ACE and MDQ). Tunable dye lasers would be desirable but cost and reliability has precluded widespread use. The KrF laser is particularly useful because it emits in the UV at 248 nm. If fiber optics are employed to direct the laser light, then a UV transparent fiber optic must be used. A table of lasers and their wavelengths of emission is given in Table 1.

Low-power lasers are often used in HPCE. Because scattered light is the factor that often limits detectibility, raising the power level is ineffective. At high laser power, photobleaching becomes more likely to occur as well.

Methods for Collecting Fluorescent Emission

The goal here is to minimize the collection of scattered radiation and optimize the collection of emitted fluorescence. Scattered radiation comes from two sources: Rayleigh scattering and Raman scattering. Rayleigh scattering occurs at the wavelength of excitation. To optimize

Table 1 Laser Light Sources for LIF Detection

Laser	Available wavelengths (nm)
Ar ion (air-cooled)	457, 472, 476, 488, 496, 501, 514
Ar ion (full frame)	275, 300, 305, 333, 351, 364, 385, 457, 472, 476, 488, 496, 501, 514
Ar ion (full frame, frequency doubled)	229, 238, 244, 248, 257
ArKr	350–360, 457, 472, 476, 488, 496, 501, 514, 521, 514, 521, 531, 568, 647, 752
HeNe	543, 594, 604, 612, 633
Excimer	
XeCl (pulsed)	308
KrF (pulsed)	248
Nitrogen (pulsed)	337
Nitrogen-pumped dye (tunable)	360–950
Solid state	
YAG (frequency doubled)	532
YAG (frequency quadrupled)	266
Diode lasers	
Frequency doubled ($LiNbO_3$)	415
Frequency doubled (KTP)	424
Frequency tripled (Nd-doped YLiF)	349

Source: Data from Ref. 13.

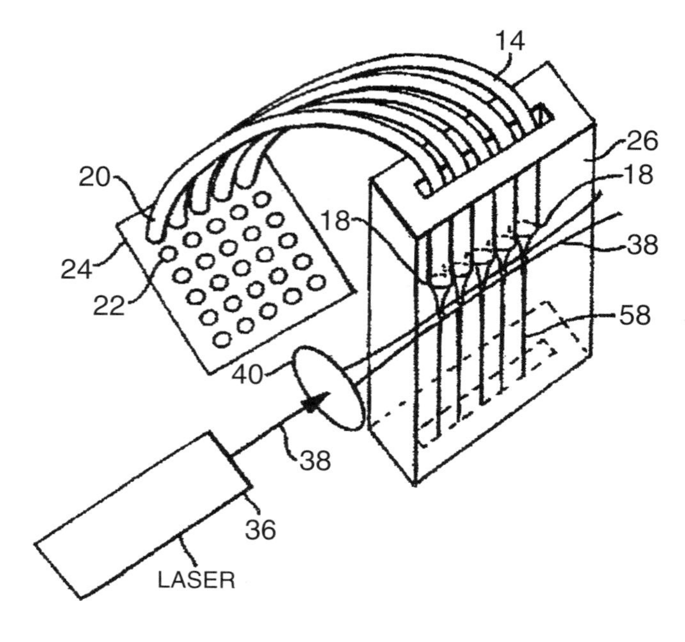

Fig. 1 Multiple-capillary instrument employing the sheath-flow technique. Key: 14, capillary; 18, capillary outlet; 20, capillary inlet; 22, buffer well; 24, microtiter plate; 26, quartz chamber; 36, laser; 38, laser beam; 40, lens; 58, fluidic stream. The electrodes are not shown nor is the device for delivering the sheath fluid. (Reprinted in part from U.S. Patent No. 5,741,412, Figure 1.)

the LOD, virtually all of this radiation must be excluded from detection. Raman scattering is observed at longer wavelengths than Rayleigh scattering and it is 10^6 times less intense. Despite the weakness of Raman scattering, this effect can significantly elevate the background if left unchecked. Bandpass and/or cutoff filters are often used to reduce the impact of scattering. It is important to ascertain that the selected filter does not fluoresce as well.

Fiber optics held at right angles to the capillary can be employed to route emitted light toward the photomultiplier tube (PMT) [2]. The Beckman LIF detector employs a collecting mirror to increase the amount of collected emission. One problem with both of these approaches is the failure to prevent small amounts of scattered light from reaching the PMT. Cutoff and/or bandpass filters are not 100% efficient in this regard. This is particularly important when lasers are used because of the intense scattering of light.

The sheath-flow design is an important advance in reducing scattering because detection occurs after the solutes have exited the capillary [4]. Scattering occurs whenever a refractive index (RI) change occurs in the optical path. These RI changes include the air–capillary interface and the buffer–capillary interface. Eliminating the capillary from the optical path effectively removes four scattering surfaces. This becomes most important in multiple-capillary systems such as the DNA sequencer because many surfaces are now involved. The sheath-flow device patented in 1998 [5] is illustrated in Fig. 1 for a five-capillary system. In actual practice, 96 capillaries are employed. The laser beam is sufficiently strong that attenuation is not significant or at least can be compensated for in the software. Fluorescence from each capillary is then imaged onto a charge-coupled device (CCD) camera.

For single-capillary systems, a conventional PMT is used for detection. Light is routed to that PMT either with fiber optics, a collecting mirror, epi-illumination microscopy, or a microscope objective. For multiple-capillary systems, the system must be scanned [6] or the light imaged onto a CCD camera.

Derivatization

Derivatization is important in capillary electrophoresis to enhance the detectibility of solutes that are nonfluorescent [7]. The chemistry can occur precapillary, on capillary, postcapillary. Typically, the solutes are amino acids, catecholamines, peptides, or proteins, all of which contain primary or secondary amine groups.

Reagents such as *ortho*-phthaldehyde (OPA), naphthalenedialdehyde (NDA), 3-(4-carboxy-benzoyl)-2-quinoline carboxaldehyde (CBQCA), fluorescein, and fluorenylmethyl chloroformate (FMOC) are all useful for precapillary derivatization, the most common of the three techniques. For carbohydrates, reagents such as amino-pyrene naphthalene sulfonate (APTS) are used for precapillary derivatization. For chiral recognition, prederivatization with optically pure fluorenylethyl chloroformate (FLEC) provides for both enantioseparation by micellar electrokinetic capillary chromatography (MECC or MEKC) and a tag that absorbs at 260 nm and emits above 305 nm. Reagents for derivatizing carbonyl, hydroxyl, and other functional groups are also available.

For on-capillary and postcapillary derivatization, the reagent must not fluoresce until reacted with the solute. For these purposes, NDA and OPA are the best choices. With on-capillary derivatization, it is possible to use a reagent that fluoresces, but its removal prior to solute detection can be difficult.

The advantage of precapillary and on-capillary derivatization is the lack of the need for additional instrumentation beyond the basic HPCE instrumentation. The disadvantage of precapillary derivatization is the need for extra sample-handling steps. For postcapillary derivatization, the need for additional miniaturized instrumentation is the principle disadvantage. This problem may be overcome when dedicated microfabricated systems become available.

Important non-DNA application areas for precapillary derivatiation with LIF detection include the determination of amino acids and amines in cerebrospinal fluid to distinguish disease states such as Alzheimer's disease and leukemia from the normal population. In vivo monitoring of microdialysates from the brain of living animals has been employed for the determination neuropeptides, amphetamine, neurotransmitters, and amino acids. The contents of single neurons and red blood cells have been studied as well.

A variant of postcapillary derivatization is chemiluminescence (CL) detection [8]. In this case, the chemical reaction replaces the light source for excitation. The detector is a PMT run at high voltage. Solutes can be tagged with CL reagents such as luminol or directly excited via the peroxyoxalate reaction. The latter works best for aminoaromatic hydrocarbons such as dansylated amines. The LODs using CL detection approach laser levels because of the low background. However, the need for specialized apparatus has limited the applicability of CL detection.

Fluorescence Detection for Microfabricated Systems

The so-called micro-total analytical systems (μTAS) can integrate sample handling, separation, and detection on a single chip [9]. Postcapillary reaction detectors can be incorporated as well [10]. Fluorescence detection is the most common method employed for these chip-based systems. A commercial instrument (Agilent 2100 Bioanalyzer) is available for DNA and RNA separations on disposable chips using a diode laser for LIF detection. In research laboratories, polymerase chain reaction (PCR) has been integrated into a chip that provides size separation and LIF detection [11].

Indirect Fluorescence Detection

When detecting solutes that neither absorb nor fluoresce, indirect detection can be employed. With this technique, a reagent is added to the background electrolyte that absorbs or fluoresces and is of the same charge for the solute being separated. This reagent elevates the baseline. When solute ions are present, they displace the additive as required by the principle of electroneutrality. As the separated ions migrate past the detector window, they are measured as negative peaks relative to the high baseline. The advantage of indirect fluorescence compared to indirect absorption is an improved LOD.

The sensitivity of indirect detection is given by the following equation [12]:

$$C_{\text{LOD}} = \frac{C_R}{(\text{DR})(\text{TR})}$$

where the CLOD is the concentration limit of detection, C_R is the concentration of the reagent, DR is the dynamic reserve, and TR is the transfer ratio. Thus, the lowest CLOD occurs when the reagent concentration is minimized.

With 100 μm fluorescein, a mass limit of detection of 20 μM was measured for lactate and pyruvate in single red blood cells. Fluorescein is a good reagent because it absorbs at 488 nm and thus matches the argon-ion laser emission wavelength. In indirect absorption detection, the additive concentration is usually 5–10 mM. Band broadening due to electrodispersion is less unimportant in indirect fluorescence detection because the solute concentration is so low. At higher solute concentrations, the system will be less useful because of electrodispersion. The concentration of the indirect reagent could be increased, but then indirect absorption detection becomes applicable.

With the advent of microfabricated systems that employ LIF detection, it is expected that indirect fluorescence will gain importance as a general-purpose detection scheme.

References

1. C. E. MacTaylor and A. G. Ewing, *Electrophoresis 18*: 2279 (1997).
2. M. Albin, R. Weinberger, E. Sapp, and S. Moring, *Anal. Chem. 63*: 417 (1991).
3. E. Arriaga, D. Y. Chen, X. L. Cheng, and N. J. Dovichi, *J. Chromatogr. 652*: 347 (1993).
4. Y. F. Cheng and N. J. Dovichi, *SPIE 910*: 111 (1988).
5. N. J. Dovichi and J. Z. Zhang, U.S. Patent 5,741,412 (April 21, 1998).
6. X. C. Huang, M. A. Quesada, and R. A. Mathies, *Anal. Chem. 64*: 967 (1992).
7. H. A. Bardelmeijer, et al., *Electrophoresis 18*: 2214 (1997).
8. T. D. Staller and M. J. Sepaniak, *Electrophoresis 18*: 2291 (1997).
9. A. Manz, et al., *Analusis 22*: M25 (1994).
10. S. C. Jacobson, L. B. Koutny, R. Hergenroeder, and A. W. Moore, Jr., *Anal. Chem. 66*: 4372 (1994).
11. L. C. Waters, et al., *Anal. Chem. 70*: 158 (1998).
12. E. S. Yeung and W. G. Kuhr, *Anal. Chem. 63*: 275 (1991).
13. H. E. Schwartz, K. J. Ulfelder, F.-T. A. Chen, and J. Pentoney. *J. Capillary Electrophoresis 1*: 36 (1994).

Robert Weinberger

Fluorescence Detection in HPLC

Detection based on analyte fluorescence can be extremely sensitive and selective, making it ideal for trace analysis and complex matrices. Fluorescence has allowed liquid chromatography to expand into a high-performance technique. High-performance liquid chromatography (HPLC) procedures with fluorescence detection are used in routine analysis for assays in the low nanogram per milliliter range and concentrations as low as picogram per milliliter often can be measured. The linearity range for these detectors is similar to that of ultraviolet (UV) detectors (i.e., 10^3-10^4).

One major advantage of fluorescence detection is the possibility of obtaining three orders of magnitude increased sensitivity over absorbance detection and its ability to discriminate analyte from interference or background peaks. Contrary to absorbance, fluorescence is a "low-background" technique. In an absorbance detector, the signal measured is related to the difference in light intensity in the presence of the sample versus the signal in the absence of the sample. For traces of analyte, this difference becomes extremely small and the noise level of the detector increases significantly. In a fluorescence detector, however, the light emitted from the analyte is measured against a very low-light (dark) background and, thus, against a very low noise level. The result is a much lower detection limit, which is limited by the electronic noise of the instrument and the dark current of the photomultiplier tube.

Another major advantage of fluorescence detection is selectivity. The increased selectivity of fluorescence versus absorbance is mainly due to the following reasons: (a) Most organic molecules will absorb UV/visible light but not all will fluoresce. (b) Fluorescence makes use of two different wavelengths (excitation and emission) as opposed to one in absorbance, thus decreasing the chance of detecting interfering chromatographic peaks.

Quantitative analysis can be performed with fluorescence detection even when poor column resolution occurs, provided there is enough detection selectivity to resolve the peaks.

One of the weak points of fluorescence is that relatively few compounds fluoresce in a practical range of wavelengths. However, chemical derivatization allows many nonfluorescent molecules containing derivatizable functional groups to be detected, thus expanding the number of applications. Fluorescence derivatization can be accomplished either via precolumn or postcolumn methods.

Theoretical Background of Fluorescence Detection

Fluorescence is a specific type of luminescence. When a molecule is excited by absorbing electromagnetic radiation (a photon) supplied by an external source (i.e., an incandescent lamp or a laser), an excited electronic singlet state is created. Eventually, the molecule will attempt to lower its energy state, either by reemitting energy (heat or light) by internal rearrangement or by transferring the energy to another molecule through a molecular collision. This process distinguishes fluorescence from chemiluminescence, in which the excited state is created by a chemical reaction. If the release of electromagnetic energy is immediate or stops upon the removal of the excitation source, the substance is said to be fluorescent.

In fluorescence, the excited state exists for a finite time (1–10 ns). If, however, the release of energy is delayed or persists after the removal of the exciting radiation, then the substance is said to be phosphorescent.

Once a photon of energy $h\nu_{ecx}$ excites an electron to a higher singlet (absorbance) state (1 fs), emission of the photon $h\nu_{em}$ occurs at longer wavelengths. This is due to the competing nonradiative processes (such as heat or bond breakage) occurring during energy deactivation. The difference in energy or wavelength represented by $h\nu_{em} - h\nu_{exc}$ is called the *Stokes shift.*

The fluorescence signal, I_f, is given by

$$I_f = \varphi I_0(1 - e^{-kcl})$$

where φ is the quantum yield (the ratio of the number of photons emitted to the number of photons absorbed), I_0 is the intensity of the incident light, c is the concentration of the analyte, k is the molar absorbance, and l is the path length of the cell.

With few exceptions, the *fluorescence excitation spectrum* of a single fluorophore in dilute solution *is identical to its absorption spectrum*. Under the same conditions, the *fluorescence emission spectrum is independent of the excitation wavelength*, due to the partial dissipation of the excitation energy during the excited lifetime. The emission intensity is proportional to the amplitude of the fluorescence excitation spectrum at the excitation wavelength.

Deactivation Pathways in Fluorescence

The excited state exists for a finite time (1–10 ns) during which the fluorophore undergoes conformational changes and is also subject to several interactions with its molecular environment. The processes which deactivate the excited state may be radiational or nonradiational (see Fig. 1) and are the following.

Internal Conversion

A transition from a higher (S_3, S_2) to the first singlet excited energy state (S_1) occurs through an internal conversion (in 1 ps). Internal conversion is increased with increasing solvent polarity.

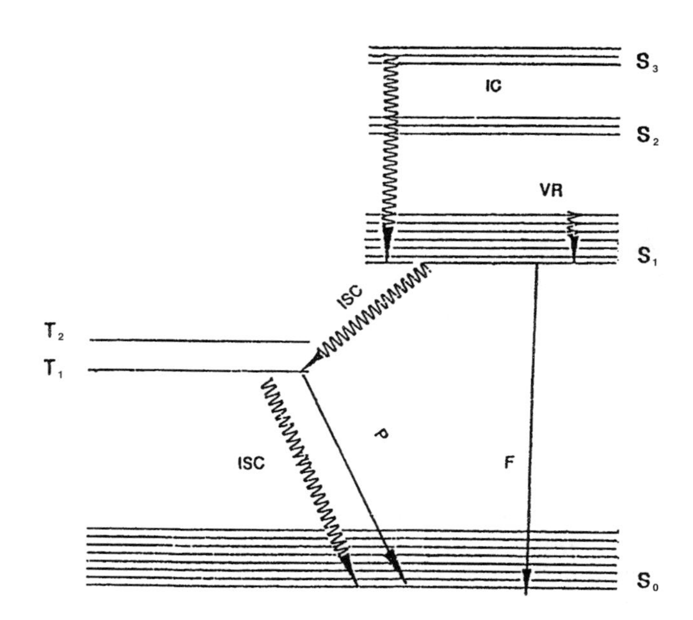

Fig. 1 Deactivation pathways in fluorescence; S_1, S_2, and S_3 are singlet excited states; S_0 is the ground state; T_1 and T_2 are triplet excited states; VR is vibrational relaxation, IC is internal conversion, ISC is intersystem crossing, P is phosphorescence, and F is fluorescence.

External Conversion (Quenching)

This is a chemical or matrix effect and can be defined as a bimolecular process that reduce the fluorescence quantum yield without changing the emission spectrum. Fluorescence radiation is transferred to foreign molecules after collisions.

Vibrational Relaxation

The energy of the first excited singlet state is partially dissipated through vibrations, yielding a relaxed singlet excited state. *Increased vibrations lower the fluorescence intensity*, due to the fact that they occur much faster (1 ps) than the fluorescence event. The molecular structure itself will determine the amount of vibrations. Rigid and planar molecules usually do not favor vibrations and they are prone to fluoresce.

Intersystem Crossing (Photobleaching)

This is a nonradiational process under high-intensity illumination conditions and in the same timescale as fluorescence (1–10 ns). It is defined as a transition from the first excited singlet (S_1) to the excited triplet (T_1) state. This is a "forbidden" transfer and necessitates the change of electron spin. *The quantum yield of fluorescence is reduced and phosphorescence also occurs.*

Phosphorescence

This event occurs due to a radiational relaxation to the ground singlet (S_1) state and in the 0.1 ms to 10 s time frame. Therefore, the emission is at even longer wavelengths than in fluorescence. Energy addition to the molecule in the form of heat or collisions of two triplet-state molecules can cause delayed fluorescence.

Factors Affecting Fluorescence

Molecular structure and environmental factors such as acidity, solvent polarity, and temperature variations exert significant influence on fluorescence intensity. Also, variations in mobile-phase composition will cause excitation and emission-wavelength changes in the fluorophore.

Molecular Structure

Common fluorophores possess aromaticity and electron-donating substituents on the ring. Only compounds with a high degree of conjugation will fluoresce. The possible molecular transitions resulting in fluorescence are $\sigma \rightarrow \sigma^*$, occurring only on alkanes in the vacuum UV region, and $\pi \rightarrow \pi^*$ with very high extinction coefficients, occurring in alkenes, carbonyls, alkynes, and azo compounds. The majority of strong fluorophores undergo this transition and the excited state is more polar than the ground state.

Solvent Polarity

Polar solvents affect the excited state differently in $\pi \rightarrow \pi^*$ and $n \rightarrow \pi^*$ transitions. The excited state in $\pi \rightarrow \pi^*$ transition is stabilized. A reduction in the energy gap will occur and the emission will be shifted to a longer wavelength (red shift). Therefore, the difference between excitation and emission wavelengths will be greater in polar solvents.

Temperature

A rise in temperature increases the rate of vibrations and collisions, resulting in increased intersystem crossing, internal and external conversion. Consequently, the fluorescence intensity is inversely proportional to the temperature increase. Additionally, an increased temperature causes a red shift of the emission wavelength.

Acidity

Acidity can drastically affect the fluorescence intensity. The pK_a of concern is the pK_a of the excited state. Because protonation is faster than fluorescence, the pK_a can be quite different than it is for the molecule in the ground state. Therefore, a pH optimization versus fluorescence intensity is needed for molecules that are particularly prone to pH changes.

Fluorescence Detector Instrumentation

Fluorescence detectors for HPLC use come in many designs from the manufacturer. Differences in detector design can lead to markedly different results during interlaboratory comparisons.

Fluorescence detectors are based either on the straight-path design (similar to UV photometers) or on the more often encountered right-angle design. The common excitation source lamps used are continuous deuterium, xenon, xenon–mercury, and pulsed xenon. Recently, the use of high-power light sources for excitation, such as laser sources, allows the development of much smaller volume flow cells with less scatter (noise), resulting in improved efficiency. Photomultiplier tubes are commonly used as the photodetectors (photocells) versus photodiodes in UV detectors. They convert a light signal to an electronic signal.

Detector flow cells are the link between the chromatographic system and the detector system. The cell cuvettes are made of quartz, with either cylindrical or square shapes and volumes between 5 and 20 μL. The sensitivity is directly proportional to the volume. However, resolution decreases with increasing volume. Fluorescence is normally measured at an angle perpendicular to the incident light. An angle of 90° has the lowest scatter of incident light. However, fluorescence from the flow cell is isotropic and can be collected from the entire 360°.

With the straight-path design, a standard UV cell can be used, but the filters must be selected so as to prevent stray light from reaching the photodetector. The right-angle design often uses a cylindrical cell. This design is less efficient than the straight-path cell because light-scattering problems result in a lower light intensity reaching the photodetector. However, this design is less susceptible to interference from stray light from the lamp, because the photodetector is not in line with the lamp.

With respect to monochromator type, three general detector designs are available: filter–filter, grating–filter, and grating–grating, where either a filter or monochromator grating is used to select the correct excitation and emission wavelengths. Gratings allow a choice of any desired wavelength, whereas filters are limited to a single wavelength.

Fluorescence detectors that use *filters* to select excitation and emission wavelengths are called *filter fluorometers*. This type of detector is the most sensitive, yet the simplest and least expensive. A diagram of this simple form of fluorescence detector is shown in Fig. 2. Usually, in order to enhance the fluorescence collected from the flow cell, lenses are employed along with filters. The lenses are positioned before the excitation filter and after the flow cell to focus and collect the light.

The ultimate in fluorescence detection is a detector that uses a diffraction grating to select the excitation wavelength and a second grating to select the wavelength of the fluorescent light. These dual monochromatic *grating–grating* fluorescence detectors are called *spectrofluorometers*. If the gratings are used in the scanning mode, the detector is a *scanning spectrofluorometer*. A fluorescence or excitation spectrum can be provided by arresting the flow (stop-flow technique) of the mobile phase when the solute resides in the detecting cell or by scanning the excitation or fluorescent light, respectively. In this way, it is possible to obtain excitation spectra at any chosen fluorescent wavelength or fluorescence spectra at any chosen excitation wavelength.

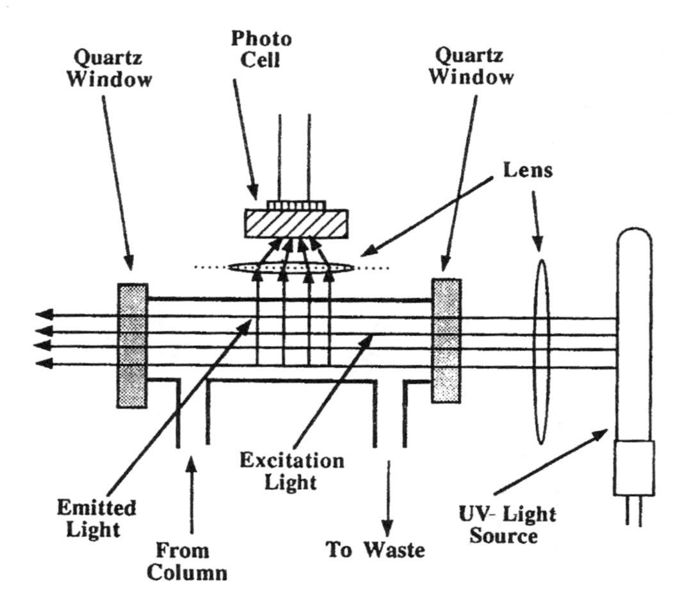

Fig. 2 Schematic of a single-wavelength excitation fluorescence detector.

The *grating–filter* detector is a hybrid between the filter–filter and the grating–grating types. Both high sensitivity and intermediate selectivity are achieved. The use of a filter in combination with gratings is ideal for lowering the background.

Grating–grating fluorometers are convenient for method development, because they permit selection of any excitation or emission wavelength. Filter–filter instruments, on the other hand, are simpler, easier in use, less expensive, more sensitive, and better suited for transferring an HPLC method between laboratories.

With the vast development of technology, fluorescence detectors have become programmable. Optimization of wavelength-pair maxima for each analyte can be time programmed during the chromatographic run.

The proper use of fluorescence detectors necessitates knowledge and understanding of noise sources. Dual monochromatic detectors have stray light leakage. When the wavelength pair is close, the background noise can significantly limit the detection limit. The *stray light*, along with *reflection* and *scattering*, increases the blank signal, resulting in reduced signal-to-noise ratio. Reflection occurs at interfaces that have a difference in the refractive index. Scattering can be of *Rayleigh* or *Raman* type.

In Rayleigh scatter, the wavelength of the absorbed and emitted photons are the same. Ultraviolet wavelengths scatter more than visible. Rayleigh scatter can be a significant problem when the wavelength pair overlaps (less than 50 nm) and instruments do not have filter accommodations and adjustable slits.

Raman scatter can also be troublesome. Depending on the wavelength pair of the sample, Raman scatter from the mobile phase can overlap the fluorescence signal and, thus, can be misdiagnosed as the fluorescence signal itself. This problem arises during increasing instrument sensitivity. However, satisfactory separation can be achieved by changing the excitation wavelength because emission is independent of the excitation wavelength.

To summarize, in terms of instrumental operation, the following practices should be followed: proper zeroing of the blank and nontampering with the gain during serial dilutions. Increased sensitivity should be accomplished by varying the full-scale range.

The basic sequence in instrumental adjustments is to select the minimum gain necessary to allow a full-scale deflection, at the least sensitive scale. When linear curves are prepared, the gain need not be adjusted. Amplification should always be done using the range control. Any small changes in the gain during calibration will cause nonlinearity. Once the gain has been set, the zero can be set. To ensure reproducibility, zeroing the detector from time to time during the day is recommended, because the dark current can change during the day.

Suggested Further Reading

Dolan, J. W. and L. R. Snyder, *Troubleshooting LC Systems*, Humana Press, Clifton, NJ, 1989, pp. 337–339.

Gilbert, M. T., *High Performance Liquid Chromatography*, IOP Publishing, Bristol, U.K., 1987, pp. 34–35.

Hancock, W. S. and J. T. Sparrow, *HPLC Analysis of Biological Compounds, A Laboratory Guide*, Marcel Dekker, Inc., New York, 1984, pp. 166–169.

Haugland, R. P., *Handbook of Fluorescent Probes and Research Chemicals*, 6th ed., Molecular Probes, Inc., Eugene, OR, 1996, pp. 1–4.

O'Flaherty, B., *Fluorescence Detection*, in *A Practical Guide to HPLC Detection* (D. Parriott, ed.), Academic Press, San Diego, CA, 1993, pp. 111–139.

Papadoyannis, I. N., *HPLC in Clinical Chemistry*, Marcel Dekker, Inc., New York, 1990, pp. 74–75.

Scott, R. P. W., *Techniques and Practice of Chromatography*, Marcel Dekker, Inc., New York, 1995, pp. 288–292.

Scott, R. P. W., *Chromatographic Detectors, Design, Function and Operation*, Marcel Dekker, Inc., New York, 1996, pp. 199–211.

Snyder, L. R. and J. J. Kirkland, *Introduction to Modern Liquid Chromatography*, 2nd ed., John Wiley & Sons, New York, 1979, pp. 145–147.

Snyder, R. L., J. J. Kirkland, and J. L. Glajch, *Practical HPLC Method Development*, John Wiley & Sons, New York, 1997, pp. 81–84.

Ioannis N. Papadoyannis
Anastasia Zotou

Foam Countercurrent Chromatography

Introduction

When a foam moves through a liquid, it carries particles caught at its interface, resulting in the accumulation of these particles at the surface. For many years, this phenomenon has been utilized for the separation of minerals and metal ions. Because the method only employs an inert gas and an aqueous solution, it should have a great potential for the separation of biological samples. This idea has been realized using the high-speed countercurrent chromatographic (CCC) system. In this foam CCC method, foam and liquid undergo rapid countercurrent movement through a long, fine Teflon tubing [2.6 mm inner diameter (i.d.) × 10 m] under a centrifugal force field. This foam CCC technology has been applied to the separation of variety of samples.

Apparatus for Foam CCC

Figure 1a illustrates a cross-sectional view of the foam CCC apparatus. The rotary frame holds a coiled separation column and a counterweight symmetrically at a distance of 20 cm from the central axis of the centrifuge. When the motor drives the rotary frame, a set of gears and pulleys produces synchronous planetary motion of the coiled column in such a manner that the column revolves around the central axis of the centrifuge while it rotates about its own axis at the same angular velocity, in the same direction. The rotating force field resulting from this planetary motion induces a countercurrent movement between foam and its mother liquid through a long, narrow coiled tube. Introduction of a sample mixture into the coil results in the separation of sample components. Foam active components are quickly carried with the foaming stream and are collected from one end of the coil while the rest moves with the liquid stream in the opposite direction and is collected from the other end of the coil.

Figure 1b illustrates the column design for foam CCC. The coiled column consists of a 10-m-long, 2.6-mm-i.d. Teflon tube with a 50-mL capacity. The column is equipped with five flow channels. The liquid is fed from the liquid feed line at the tail and collected from the liquid collection line at the head. Nitrogen gas is fed from the gas feed line at the head and discharged through the foam collection line at the tail while the sample solution is introduced through the sample feed line at the middle portion of the coil. The head–tail relationship of the rotating coil is conventionally defined by an Archimedean screw force, where all objects in different density are driven toward the head. The liquid feed rate and sample injection rate are each separately regulated with a needle valve and the foam collection line is left open.

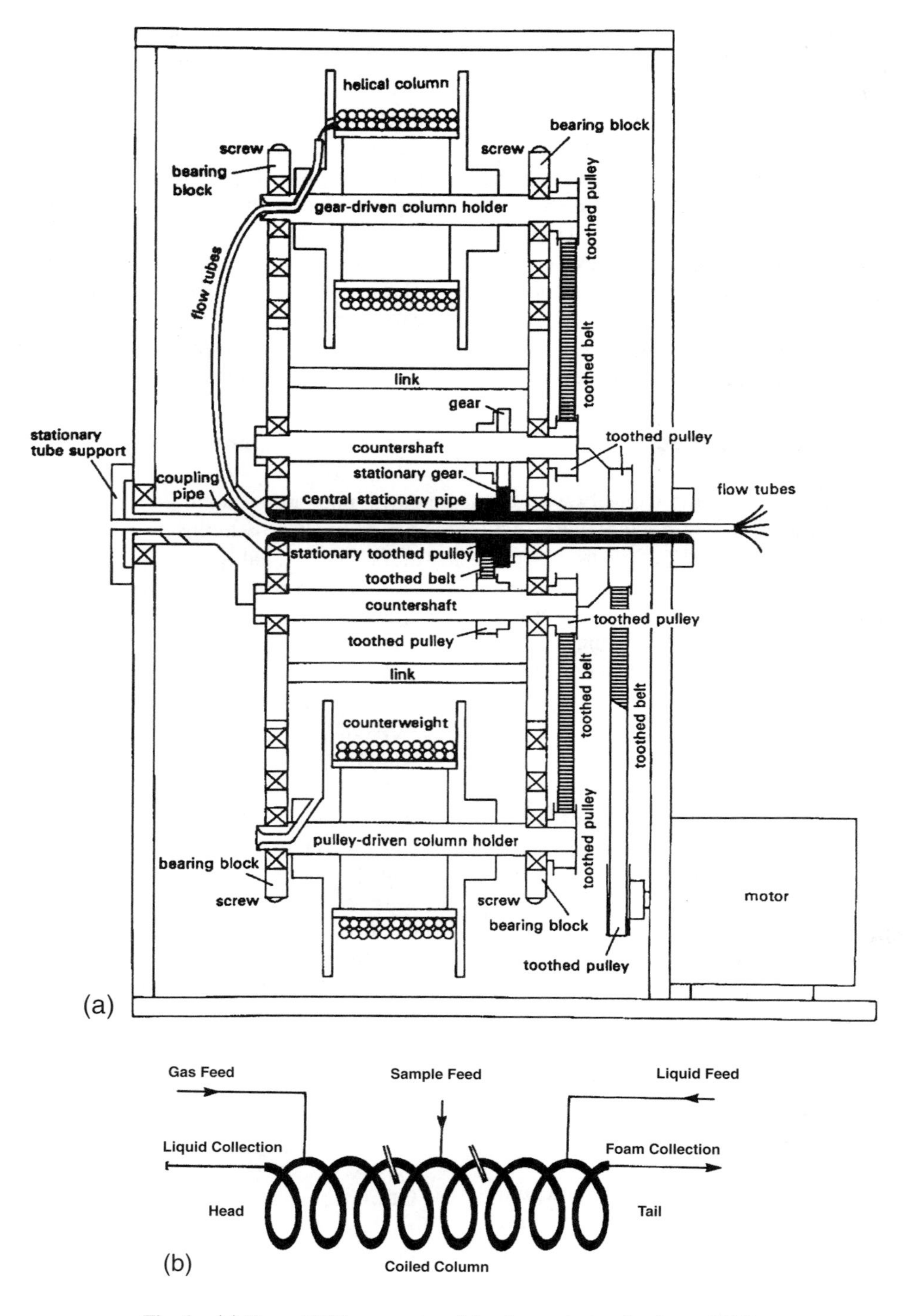

Fig. 1 (a) Foam CCC apparatus; (b) column design for foam CCC.

Application

Foam CCC can be applied to two types of samples: with (a) the affinity to the foam-producing carrier and (b) the direct affinity to the gas–liquid interface.

Foam Separation Using Surfactants

This technique was demonstrated by the separation of methylene blue and DNP-leucine, having affinity to the foam-producing carrier. Sodium dodecyl sulfate (SDS) and cetyl pyridinium chloride (CPC) were used as carriers to study their effects of electric charges on the foam affinities of various compounds. When the sample mixture was introduced with the anionic SDS surfactant, the positively charged Methylene Blue was adsorbed onto the foam and quickly eluted through the foam collection line; the negatively charged DNP-leucine was carried with the liquid stream in the opposite direction and eluted through the liquid collection line. Similarly, when the same sample mixture was introduced with the cationic CPC surfactant, the negatively charged DNP-leucine was totally eluted through the foam collection line, and positively charged Methylene Blue through the liquid collection line.

Foam Separation Without a Surfactant

Many natural products have foaming capacity, so foam CCC may be performed without a surfactant. This technique was demonstrated using the bacitracin complex (BC) as a test sample because of its strong foaming capacity. The foam CCC experiment for separation and enrichment of BC components was conducted using nitrogen gas and distilled water, which was entirely free of surfactant or other additives.

Batch Sample Loading

Bacitracin is a basic cyclic peptide antibiotic consisting of more than 20 components, but, except for the major components BC-A and BC-F, the chemical structures of other components are still unknown.

Foam CCC of BC components was initiated by simultaneously introducing distilled water from the liquid feed line at the tail and nitrogen gas from the gas feed line at the head into the rotating column while the needle valve at the liquid collection line was fully open. After a steady-state hydrodynamic equilibrium was reached, the pump was stopped and the sample solution was injected through the sample feed line at the middle of the column. After a lapse of the predetermined standing time, the needle valve opening was adjusted to the desired level and pumping was resumed. Effluents were collected at 15-s intervals. The bacitracin components were separated in the order of hydrophobicities of the molecules in the foam fractions, with the most hydrophobic compounds being eluted first. This method can also be applied to continuous sample feeding, as described in the next section.

Continuous Sample Feeding

The experiment was initiated by introducing nitrogen gas from the gas feed line at the head of the rotating column. Then, a 2.5-L volume of the BC solution was continuously introduced into the coil from the sample feed line at 1.5 mL/min. The hydrophobic components produced a thick foam which was carried with the gas stream and collected from the foam collection line at the tail; other components stayed in the liquid stream and eluted from the liquid collection line at the head. High-performance liquid chromatographic analysis of the foam fraction revealed that the degree of enrichment increased with the hydrophobicity of the components. These results clearly indicate that the present method will be quite effective for the detection and isolation of small amounts of natural products present in a large volume of aqueous solution.

Foaming Parameters

For application of foam CCC to various natural products, it is desirable to establish a set of physicochemical parameters which reliably indicate their applicability to foam CCC. Two parameters were selected for this purpose (i.e., "foaming power" and "foam stability") which can be simultaneously determined by the following simple procedure. In each test, the sample solution (20 mL) is delivered into a 100-mL graduated cylinder with a ground-glass stopper, and the cylinder vigorously shaken for 10 s. The foaming power is expressed by the volume ratio of the resultant foam to the remaining solution; the foam stability is expressed by the duration of the foam.

In order to correlate the foaming parameters to the foam productivity in foam CCC, the following five samples were selected because of their strong foaming capacities: bacitracin, gardenia yellow, rose bengal, phloxine B, and senega methanol extract. The results of our studies indicated that a sample having the foaming power greater than 1.0 and a foam stability for over 250 min could be effectively enriched by foam CCC. These minimum requirements of foaming parameters, derived from the bacitracin experiment, were found to be applicable to other four samples.

Conclusions

Foam CCC can be applied successfully to a variety of samples having foam affinity, with or without surfactants. The method offers important advantages over the conventional foam separation methods by allowing the efficient chromatographic separation of sample in both batch loading and continuous feeding. The foam CCC technique has a great potential in enrichment, stripping, and isolation of foam active components from various natural and synthetic products in both research laboratories and industrial plants.

Suggested Further Reading

Bhatnagar, M. and Ito, Y., Foam countercurrent chromatography on various test samples and the effects of additives on foam affinity, *J. Liquid Chromatogr. 11*: 21 (1988).

Ito, Y., Foam countercurrent chromatography: New foam separation technique with flow-through coil planet centrifuge, *Separ. Sci. 11*: 201 (1976).

Ito, Y., Foam countercurrent chromatography based on dual countercurrent system, *J. Liquid Chromatogr. 8*: 2131 (1985).

Ito, Y., Foam countercurrent chromatography with the cross-axis synchronous flow-through coil planet centrifuge, *J. Chromatogr. 403*: 77 (1987).

Oka, H., Foam countercurrent chromatography of bacitracin complex, in *High-Speed Countercurrent Chromatography* (Y. Ito and W. D. Conway, eds.), John Wiley & Sons, New York, 1996, pp. 107–120.

Oka, H., Harada, K.-I., Suzuki, M., Nakazawa, H., and Ito, Y., Foam countercurrent chromatography of bacitracin with nitrogen and additive-free water, *Anal. Chem. 61*: 1998 (1989).

Oka, H., Harada, K.-I., Suzuki, M., Nakazawa, H., and Ito, Y., Foam countercurrent chromatography of bacitracin I. Batch separation with nitrogen and water free of additives, *J. Chromatogr. 482*: 197 (1989).

Oka, H., Harada, K.-I., Suzuki, M., Nakazawa, H., and Ito, Y., Foam countercurrent chromatography of bacitracin II. Continuous removal and concentration of hydrophobic components with nitrogen gas and distilled water free of surfactants or other additives, *J. Chromatogr. 538*: 213 (1991).

Oka, H., Iwaya, M., Harada, K.-I., Muarata, H., Suzuki, M., Ikai, Y., Hayakawa, J., and Ito, Y., Effect of foaming power and foam stability on continuoius concentration with faom countercurrent chromatography, *J. Chromatogr. A 791*: 53 (1997).

Hisao Oka
Yoichiro Ito

Focusing Field-Flow Fractionation of Particles and Macromolecules

Introduction

The original idea of the focusing field-flow fractionation (focusing FFF) [1] appeared in 1982. Giddings [2] proposed an identical principle in 1983 under the name hyperlayer FFF. More detailed methodology of focusing FFF was developed, subsequently, by exploiting various separation mechanisms [3]. Recently, emerging isoperichoric focusing FFF represents the generalization of the original concept [4].

The principle of the focusing FFF is different compared with that of polarization FFF. The crucial difference between the focusing and polarization mechanisms is that the intensity and direction of the driving field force must be dependent on the position across the channel and converging in focusing FFF, whereas it is position independent in polarization FFF. The sample components are focused at different altitudes across the channel and, consequently, eluted at different velocities corresponding to their positions within the flow velocity profile in focusing FFF, as shown in Fig. 1a. Although focusing FFF is still in a stage of fundamental investigation, some applications concerning the fractionation of the macromolecular and particulate species were published.

Methods and Techniques

The focusing can appear only if a gradient of the effective forces exists, and the magnitude of these converging forces is position dependent and is zero at the focusing point. Various combinations of the fields and gradients determining the focusing FFF methods and techniques can be exploited, as demonstrated in the following subsections.

Effective Property Gradient of the Carrier Liquid, Combined with a Field Action

The gradient of an effective property of the carrier liquid, combined with the action of a field, can lead to the focusing of the macromolecules or particles; for example, a density gradient combined with the gravitational or centrifugal field generates the focusing of the species at their isopycnic positions, the amphoteric species focus in the pH gradient combined with the electrical field at their isoelectric points, and so forth. All of these phenomena are called by the general term *isoperichoric focusing*, introduced by Kolin [5].

Usually, the same primary field forces which produce

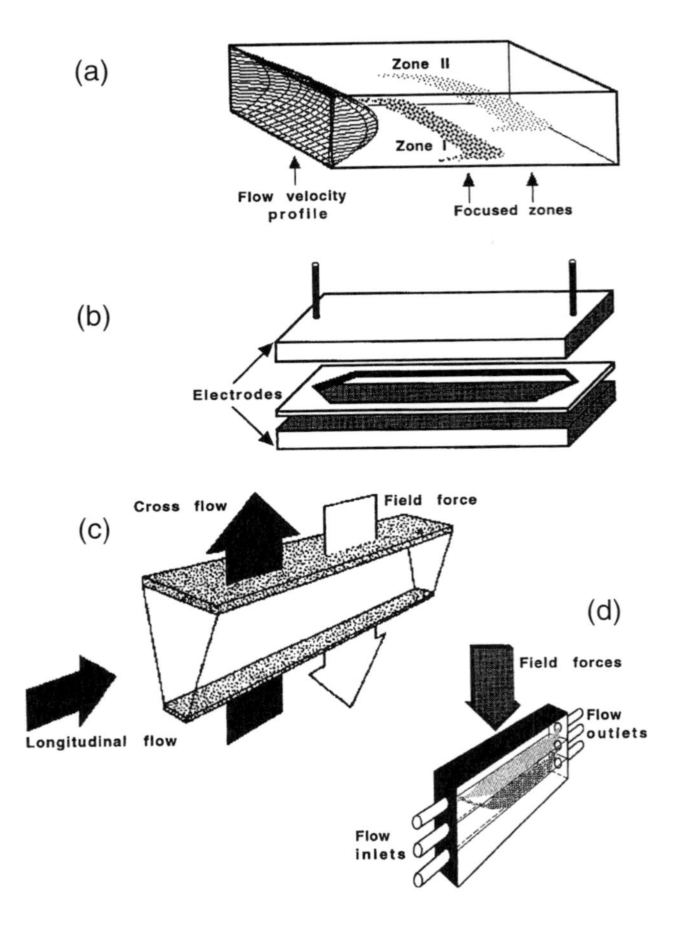

Fig. 1 Principle of focusing field-flow fractionation. (a) Section of the channel demonstrating the principle of focusing FFF with two distinct zones focused at different positions and the parabolic flow velocity profile. (b) Design of the channel for dynamic focusing FFF in coupled electrical and gravitational fields. (c) Schematic representation of the trapezoidal cross-section channel for elutriation focusing FFF. (d) Separation channel for continuous preparative focusing FFF operating in natural gravitational field with three inlet capillaries allowing one to preform the step density gradient by pumping three liquids of different densities and with three outlet capillaries to collect the separated fractions.

the effective property gradient are used to generate the focusing. However, the use of the secondary field forces of a different nature to generate the focusing phenomenon within the corresponding gradient established by the primary field is possible; for example, the isopycnic focusing of the large-sized uncharged particles due to the weak gravitational field force was found effective under dynamic focusing FFF conditions, whereas the density gradient was generated by the electrical field acting on small charged colloidal particles suspended in the carrier liquid. The construction of the fractionation channel was extremely simple, as shown in Fig. 1b. This principle, applied under static or dynamic FFF conditions, is promising for high-performance analytical and micro-preparative separations.

Although the focusing under the static conditions, without the action of perpendicularly (with respect to the focusing axis) applied bulk flow, can lead to a good separation of the focused species, the theoretical calculations, as well as the experimental tests, have shown the increase of the resolution under the dynamic conditions of focusing FFF.

Cross-Flow Velocity Gradient Combined with a Field Action

The velocity gradient of the carrier liquid across the fractionation channel, generated by the transversal flow through the semipermeable walls, which opposes the action of an external field, can produce the focusing phenomenon. The longitudinal flow is applied simultaneously. This method is called elutriation focusing FFF and was used to separate model mixtures of polystyrene latex particles and silica particles in a trapezoidal cross-section channel. The principle of such a fractionation channel is demonstrated in Fig. 1c. A similar focusing FFF principle can be exploited in a rectangular cross-section channel with two opposite semipermeable walls if the flow rates through the walls are different.

Lift Forces Combined with a Field Action

The hydrodynamic lift forces appearing at high flow rates of the carrier liquid, combined with the field forces, are able to concentrate the suspended particles into the focused layers. While the field forces in polarization FFF concentrate the retained species at the accumulation wall, the lift forces, becoming operational at high flow rates, pull the particles away from this wall. As a result, the transition from polarization to focusing FFF appears first, followed by the proper focusing effect. For example, Wahlund and Litzen [6] observed the interference of the lift forces in polarization flow FFF carried out in an asymmetrical channel with one semipermeable wall.

Moreover, the retention of the particles can vary with the nature of the field forces. Consequently, the fractionation data interpretation concerning the species eluting within the transition range must be performed with care.

Shear Stress Combined with a Field Action

A high-shear gradient can lead to the deformation of the macromolecular coils. The entropy gradient thus generated produces the driving forces that displace the macromolecules into a low-shear zone. The reversed elution order of high-molecular-weight polystyrenes in thermal FFF at high flow rates could be attributed to this phenomenon [7], but another possibility to explain the reversed elution order cannot be neglected [8].

Gradient of a Nonhomogeneous Field Action

The use of a high-gradient magnetic field was proposed to separate the paramagnetic and diamagnetic species by a mechanism of focusing FFF [9]. Various aspects of focusing FFF carried out under these conditions were discussed, but no experimental results appeared until now.

Preparative Fractionation

No principal difference distinguishes the analytical and preparative uses of focusing FFF. Both types of fractionations can be carried out under conditions of continuous operation [10], which represents the high-performance experimental arrangement for preparative FFF. The fractionation channel, equipped with several outlet capillaries at various positions (and occasionally with several inlets to preform a stepwise gradient in the direction of the focusing), allows one to fractionate the sample which is introduced continuously into the channel and to collect the focused layers eluting by the individual outlets. A schematic representation of such a fractionation channel is shown in Fig. 1d.

The experimental demonstration of this technique was given by the fractionations of various samples of silica particles, by applying natural gravitation and a counteracting cross-flow gradient. The silica particles were separated according to size. The isopycnic or isoelectric focusing FFF already performed on an analytical scale can easily be transformed into such continuous large-scale separations.

Applications

Focusing FFF represents an important contribution to the science and technology of separation and analysis of the macromolecules of synthetic or natural origin. The range of molar masses and of the sizes of the particles in submicron and micron ranges, of the supramolecular structures, and of the organized biological species, such as the cells, microorganisms, and so forth, which can be fractionated by focusing FFF, is very large.

The molecules which do not interact sufficiently with the imposed fields, such as low-molar-mass species, and which, consequently, do not exhibit the focusing effect can still be separated. The condition is that an equilibrium between them and the effectively focused species is established. As a result, the species which originally do not undergo the separation processes can be transported and thus fractionated with the "carrier" focused species.

The most important field of potential applications of focusing FFF is in the research and technologies related to the life sciences and to macromolecular chemistry. The problems related to trace analysis, which have an enormous importance in the protection of the environment and many other scientific and technological activities, have already stimulated the development of new analytical separation methods. Focusing FFF is one of them, representing an alternative of choice whenever macromolecular or particulate species are concerned.

The newest achievements in focusing FFF clearly indicate that most of the experimental implementations have been obtained with model systems. Practical applications for daily laboratory use, elaborated to the minutest details, were not yet described. However, the most significant advantages of these methods, mentioned earlier, are evident. Some of these advantages are inherently related to the separation principle of the focusing FFF, such as the absence of a large surface area within the separation channel, which is of crucial importance for sensitive biological materials that can be denatured in contact with the active surfaces.

The operational variables, such as the strength of the field, the flow rate, and so forth, can be continuously manipulated within a very wide range. Another advantage is that although the specific FFF apparatuses are already produced, the commercially available instrumentation for liquid chromatography can be adapted easily for use with focusing FFF methodology. All particular components of the complete focusing FFF apparatus are identical to those for liquid chromatography, except the separation channel, which, in most cases, is not difficult to build in the laboratory. Certainly, focusing FFF represents a large field of challenges, soliciting the creativity and invention in the theory, methodology, and practical applications.

References

1. J. Janča, *Makromol. Chem. Rapid Commun. 3*: 887 (1982).
2. J. C. Giddings, *Separ. Sci. Technol. 18*: 765 (1983).
3. J. Janča, *Field-Flow Fractionation: Analysis of Macromolecules and Particles*, Marcel Dekker, Inc., New York, 1988.
4. J. Janča, *J. Liquid Chromatogr. Related Technol. 20*: 2555 (1997).
5. A. Kolin, in *Electrofocusing and Isotachophoresis* (B. J. Radola and D. Graesslin, eds.), de Gruyter, Berlin, 1977.

6. K. G. Wahlund and A. Litzen, *J. Chromatogr. 461*: 73 (1989).
7. J. C. Giddings, S. Li, P. S. Williams, and M. E. Schimpf, *Makromol. Chem. Rapid Commun. 9*: 817 (1988).
8. J. Janča and M. Martin, *Chromatographia 34*: 125 (1992).
9. S. N. Semyonov, A. A. Kuznetsov, and P. P. Zolotaryov, *J. Chromatogr. 364*: 389 (1986).
10. J. Janča and J. Chmelik, *Anal. Chem. 56*: 2481 (1984).

Josef Janča

Forensic Applications of Capillary Electrophoresis

Introduction

During the last decade, capillary electrophoresis (CE) has developed into a widely applied method for the analysis of pharmaceuticals (both for the evaluation of pharmaceutical formulations and metabolites). These applications established the basis for introducing CE into the forensic field also. Today, capillary electrophoresis can be applied to a number of analytical problems in forensic science, including the analysis of gunshot residues, explosives, inks, dusts, soils, and, of course, illicit drugs, diverse toxicants, DNA fingerprinting, protein analysis, and so forth (for reviews, see Refs. 1 and 2).

Several features of capillary electrophoresis are particularly interesting for forensic scientists, namely high separation efficiency, sensitivity, and small amount of samples (nanoliters) and solvents (a few milliliters per day). Regarding different operational modes, all of them are applied, although to a different extent, depending on the type of compounds to be assayed.

Forensic Toxicology

Capillary electrophoresis has a particularly wide potential for the analysis of illicit drugs [3]. For this category of applications, mainly capillary zone electrophoresis (CZE) and micellar electrokinetic chromatography (MEKC) are used; however, other modifications of this approach, such as separation in buffers containing a high proportion (up to 20%) of an organic modifier in the background electrolyte, can be used also. More recently, some hints appeared indicating the possibility of applying electrochromatographic techniques for this purpose as well, although, admittedly, this latter approach has not reached the stage of maturity of other techniques. It is to be foreseen that particularly electrokinetic separations exploiting the properties of reversed-phase packings, will be used for these purposes in the near future.

Regarding tissues and body fluids to be analyzed, blood and urine serve most frequently as source material, although the interest of forensic analytical chemists can be easily extended to other specimens (e.g., saliva, bile, or vitreous humor). A tissue that has attracted a lot of interest in the course of recent years is hair [4]; at the root end, the tissue is penetrated by a number of drugs during hair matrix formation. Because the hair stalk is basically devoid of metabolic processes, the drugs, once sorbed, remain in the tissue (depository effects). If one considers the average hair growth of about 1 cm/month the analysis of hair sections may yield information about past exposure of the individual to the toxicants; it has to be kept in mind, however, that external contamination may contribute also to the amount of the toxicant recovered.

As the analytes to be assayed are nearly always present in minute amounts, preconcentration steps are almost always necessary. The techniques used for this purpose have been mostly adopted from analytical procedures using gas or liquid chromatography as the separation step [5]. This fact is emphasized here because the appropriate adjustment of the sample preparation is necessary because the original procedures do not respect the fact that the sample volume injected into a CE system represents a few nanoliters only and requires a relatively high concentration of analytes assayed. However, in selected types of analysis, direct sample application is also possible (for a review, see Ref. 3).

There are three points emphasized in the quoted review which limit the use of direct injection: (a) protein-bound drugs display a different mobility in CZE in comparison to unbound species, (b) high conductivity of the untreated biological samples may cause undesirable peak broadening, and (c) selectivity of the analysis may be negatively influenced by using a nonselective wavelength for detection (usually 200 nm), which is needed to reveal low concentrations of the analytes of interest. Consequently, desalting and deproteinization (at last partial) is frequently done by adding different proportions of organic solvents to the sample ($\sim 1{:}1 \rightarrow 1{:}4$). If MEKC is to be used, it is recommended to remove the organic solvent prior to analysis (for the first application of MEKC for forensic purposes, see Ref. 6; for additional information, see Ref. 7). Standard preconcentration conditions, such as solid-phase and liquid–solid extractions, are widely used.

Some idea about actual conditions applicable for the separation of drugs of forensic interest can be obtained from Table 1. Extensive information about the MEKC of drugs is offered in a review of Nishi and Terabe [8].

As indicated in Table 1, ultraviolet (UV) light at short wavelengths (190–220 nm) is routinely used for detection. In these cases, laser-induced fluorescence appears to be the method of choice. Unfortunately, the commercially available laser units emit at wavelengths not suitable for direct drug analysis, which limits the practical applicability of this approach. Nevertheless, where applicable, laser-induced fluorescence can easily improve the detection limit by a factor of 1000, in comparison with UV detection; Ar-ion lasers (emitting at 488 nm wavelength) or He–Cd lasers (emitting at 325 nm wavelength) are commonly used for this purpose. Typically, with enzymatically hydrolyzed urine, the detection limit can be about 2 ng/mL of the assayed compound (zolpidem).

If the investigated drug does not possess a suitable fluorophore, derivatization may be required [fluorescein isothiocyanate for compounds possessing a free amino group (e.g., amphetamines) may serve as a typical example; an Ar-ion laser emitting at 488 nm was used for this purpose]. Generally, detection limits achieved with laser-induced fluorescence after derivatization can be around 3 mM in concentration terms (or 3 μmol in terms of absolute mass detection). This is about three orders of magnitude less than what can be achieved with gas chromatography–mass spectrometry.

Another possibility is to use xenon-arc lamp irradiation for the same purpose, which extends the possibilities of excitation wavelengths to the 272–382-nm range. Exploiting competitive binding of trace amounts of (misused) drugs with fluorescence-labeled immunotracers can be spotted in the literature; however, this is not a widely used approach at the moment.

Amperometric detection, which generally offers quite high sensitivity, has not been used so far for forensic (toxicological) applications. The reason probably reflects some problems with commercially available coupling of the high-performance amperometric cell with the capillary electrophoresis device.

Surprisingly, not very many methods using the capillary electrophoresis–mass spectrometry (CE–MS) combination (mainly electrospray ionization) are in use in the area of forensic drug analysis. The first application of CE–MS for forensic purposes was described by Johansson et al. in 1991 [10] for the analysis of sulfonamides and benzodiazepines in urine. The main problem faced in this case is the need of improving the concentration sensitivity of the CE–MS combination. In order to improve the concentration of analytes, a special method of in-capillary coupling of isotachophoresis with CZE–MS has been proposed. CE–MS instrumentation and its application was reviewed by Cai and Henion [11].

Forensic Biology

One of the major areas of forensic science is DNA fingerprinting, which is used for personal identification and paternity testing. Most of these analyses are based on the polymerase chain reaction (PCR) of individual loci, followed by analysis of differences in length or sequence. Capillary electrophoresis, with laser-induced fluorescence detection, is becoming an alternative method to polyacrylamide gel electrophoresis (PAGE) and agarose slab gel electrophoresis. Fully automated CE-based instruments are now available for DNA sizing, quantitation, screening, and sequencing. Detailed information about the separation of nucleic acids is a matter of a specialized entry; for detailed information, specialized reviews are available (see, e.g., Refs. 12 and 13).

Using the separation of DNA fragments in media with cross-linked polyacrylamide or agarose gels makes it possible to achieve high efficiency (tens of millions of theoretical plates); however, from the practical point of view, replaceable, entangled polymers (e.g., derivatives of cellulose, linear polyacrylamide) are preferred. The main advantage of replaceable media is the ease of renewing the gel media in the capillary with every single run.

Ultraviolet light is routinely used for detection; however, laser-induced fluorescence of labeled DNA offers better results. For example, a PCR-amplified DNA fragment comprised of 120–400 base pairs can be separated with a resolution up to four base pairs using 1% hydroxyethylcellulose and DB-17 capillary (60 cm effective length $\times$ 0.1 mm inner diameter with 0.1 μm phase thickness). Laser-induced fluorescence detection can yield a sensitivity of about 500 pg/mL of DNA (after staining with fluorescent intercalating dye YO-PRO-1) [14].

Table 1 Examples of the Capillary Electrophoretic Separations of Misused Drugs, Their Enantiomers, and Metabolites

Analytes (remarks)	Detection	Separation
17 Basic drugs (amphetamine, lidocaine, codeine, diazepam, methaqualone, etc.; extracted by chloroform–2-propanol, 9:1 v/v)	UV at 214 nm	CZE, 50 m*M* phosphate buffer pH 2.35
Abused drugs (includes heroin, heroin impurities, *cis*- and *trans*-cinnamoyl cocaine)	UV at 210 nm	MEKC, 85 m*M* SDS, 8.5 m*M* borate, pH 8.5; containing 15% acetonitrile
Abused drugs and metabolites (includes benzoylecgonine, morphine, heroin, methamphetamine, codeine, amphetamine, cocaine, methadone, benzodiazepines)	Fast scanning UV	MEKC, borate–phosphate buffer pH 9.1, 75 m*M* SDS
Amphetamines (enantiomers; LLE[a] or SPE[b])	UV at 200 nm	CZE, 20 m*M* (2-hydroxy)-propyl-β-cyclodextrin in 200 m*M* phosphate pH 2.5
Amphetamines and ephedrine (enantiomers; LLE)	UV at 200 nm	CZE, 20 m*M* β-cyclodextrin in 150 m*M* phosphate pH 2.5
Barbiturates (after LLE)	UV at 214 nm	MEKC, 100 m*M* SDS in 10 m*M* borate, 10 m*M* phosphate, pH 8.5, 15% acetonitrile added (by volume)
Barbiturates (phenobarbital can be assayed without sample pretreatment)	Multiwavelength detection (195 and 320 nm)	MEKC, 50 m*M* SDS, phosphate–borate buffer, pH 7.8
Caffeine metabolites (direct injection, LLE)	UV at 254 nm	MEKC, 70 m*M* SDS, phosphate–borate, pH 8.43
Caffeine metabolites (LLE)	Scanning UV 195–320 nm	MEKC, 70 m*M* SDS, 16.2 m*M* phosphate, pH 8.6
Cannabis constituents (alkaline hydrolysis of urine followed by SPE[b])	Fast scanning UV	MEKC, 75 m*M* SDS in phosphate–borate buffer pH 9.1)
Dextromethorphan and dextrophan (direct injection)	UV at 200 nm	CZE, 175 m*M* borate pH 9.3
Dihydrocodeine metabolites and O-demethylation (hydrolysis, direct injection, SPE)	UV at 213 nm, scanning UV 195–320 nm	MEKC, 75 m*M* SDS, 6 m*M* borate, 10 m*M* phosphate, pH 9.2
Flurazepam metabolites, sulfonamides (hydrolysis, LLE)	UV at 254 nm; MS	CZE, 15 or 0.2 m*M* ammonium acetate pH 2.5 or 1.3 adjusted with TFA, 15% methanol
Haloperidol metabolites (SPE)	UV at 214 nm, scanning UV 195–320 nm, MS	CZE, 50 m*M* ammonium acetate, 10% methanol, 1% acetic acid, pH 4
Mephenytoin and dextromethorphan metabolites (hydrolysis)	Scanning UV 195–320 nm	MEKC, 75 m*M* SDS, 6 m*M* borate, 10 m*M* phosphate pH 9.2–9.3 or CZE, 140 m*M* borate pH 9.4
Morphine and cocaine in hair (hydrolysis by 0.25 *M* HCl at 45°C followed by LLE)	UV at 200 nm (230 nm for cocaine, 214 nm for morphine)	CZE, 50 m*M* borate pH 9.2
Nitrazepam and metabolites (SPE)	UV at 220 nm	MEKC, 60 m*M* SDS in 6 m*M* phosphate–borate buffer pH 8.5, 15% methanol (by volume)
Opiates (heroin, morphine, and metabolites; SPE)	Spectral UV analysis	CZE, 12 m*M* borate, 20 m*M* phosphate, pH 9.8 or MEKC, 75 m*M* SDS in phosphate borate buffer, pH 9.2
Opiates (morphine, heroin, codeine, etc., SPE)	UV at 200 nm	CZE, 100 m*M* phosphate pH 6
Opiates (morphine, heroin, codeine, amphetamine, caffeine)	UV at 200 nm	MEKC, 50 m*M* SDS, 50 m*M* glycine, pH 10.5
Purines, substituted (direct injection, LLE, SPE)	Scanning UV 195–320 nm	MEKC, 75 m*M* SDS, 6 m*M* borate, 10 m*M* phosphate pH 9
Racemethorphan, racemorphan (optical isomers)	UV at 200 nm	MEKC, 60 m*M* β-cyclodextrin in 50 m*M* borate pH 9.05, 50 m*M* SDS, 20% 1-propanol
Theophylline metabolites (SPE)	Scanning UV 195–320 nm	MEKC, 200 m*M* SDS in 100 m*M* borate, 100 m*M* phosphate, pH 8.5 (ration 12:7, final pH 6.5)

Note: For detailed specifications of individual procedures, see Ref. 2.
[a]Liquid–liquid extraction.
[b]Solid-phase extraction.

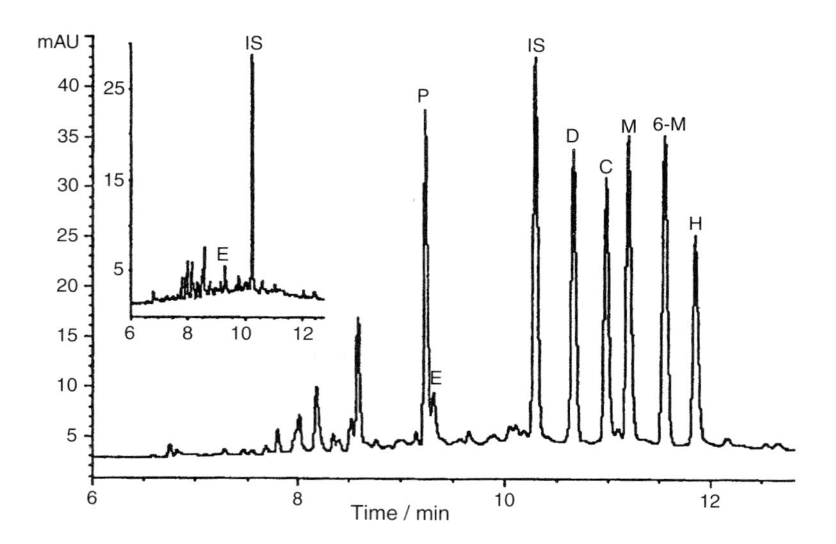

Fig. 1 Capillary electrophoretic separation of opiates. Running conditions: 100 mM phosphate buffer pH 6.0. Electrokinetic injection with field-amplified sample stacking after solid-phase extraction of spiked urine using "double mechanism" cartridges. Precision of migration times 1.2% R.S.D. (relative standard deviation), resolution >2 with all peaks shown. Within the day and day-to-day repeatability 1–4% R.S.D., respectively; detection by UV at 200 nm. Peak identification: pholcodine (P), MAM (6-M), heroin (H), codeine (C), morphine (M), dihydrocodeine (D), and levallorphan (I.S.); E represents an unidentified endogenous compound present in urine (see inset). (From Ref. 9 with permission.)

Capillary electrophoresis can be also applied for DNA sequencing. For this purpose, multicapillary (array) instruments with laser-induced fluorescence detection are being developed. Detailed descriptions of these methods is beyond the scope of this entry.

Proteins and enzymes are also of interest in forensic science. In this context, it is possible to mention acetaldehyde–protein adducts, which can be used as potential markers of alcoholism. Another application is the determination of globins, saliva, and semen proteins. Capillary electrophoresis of proteins is a broad and complex area of analytical chemistry which, like the separation of nucleic acids, is beyond the scope of this entry (for a review, see, e.g., Ref. 15 and many others). In principle, it is possible to use different operational modes such as CZE in acid or alkaline media, capillary isotachophoresis, capillary isoelectric focusing, capillary gel electrophoresis, and, recently, MEKC. A considerable problem in protein–enzyme separations in untreated capillaries is sticking of these analytes to the inner capillary wall, which can be eliminated (at least in part) by running the separation at very high or very low pH values, by adding some modifiers (as salts, etc.) to the buffer, or by appropriate modification of the capillary surface.

Conclusion

In conclusion, CE is a valuable analytical tool that offers a number of possibilities for the analysis of a wide spectrum of forensically interesting compounds. Practically all compounds which have been traditionally analyzed by GC, high-performance liquid chromatography, thin-layer chromatography, or slab–gel electrophoresis, can be assayed by capillary electrophoretic procedures. All methods of capillary electrophoresis can be validated and can meet the demands of good laboratory practice.

References

1. F. Tagliaro, F. P. Smith, L. Tadeschi, F. Castagna, M. Dobosz, I. Boschi, and V. Pascali, Toxicological and forensic applications, in *Advanced Chromatographic and Electromigration Methods in BioSciences* (Z. Deyl, I. Mikšík, F. Tagliaro, E. Tesarová, eds.), Journal of Chromatography Library Vol. 60, Elsevier, Amsterdam, 1998, pp. 917–961.
2. F. Tagliaro, Z. Deyl, and I. Mikšík, Applications of HPLC/HPCE in forensic, in *HPLC in Enzymatic Analysis* (E. F. Rossomando, ed.), Methods in Biochemical Analysis Vol. 38, John Wiley & Sons, New York, 1998, pp. 164–206.

3. F. Tagliaro, S. Turrina, P. Pisi, F. P. Smith, and M. Marigo, *J. Chromatogr. B 713*: 27 (1998) (and reviews therein).
4. F. Tagliaro, W. P. Smyth, S. Turrina, Z. Deyl, and M. Marigo, *Forensic Sci. Int. 70*: 93 (1995).
5. D. K. Lloyd, *J. Chromatogr. A 735*: 29 (1996).
6. R. Weinberger and I. S. Lurie, *Anal. Chem. 63*: 823 (1991).
7. P. Wernly and W. Thormann, *Anal. Chem. 63*: 2878 (1991).
8. H. Nishi and S. Terabe, *J. Chromatogr. A 735*: 3 (1996).
9. R. B. Taylor, A. S. Low, and R. G. Reid, *J. Chromatogr. B 675*: 213 (1996).
10. I. M. Johansson, R. Pavelka, and J. D. Henion, *J. Chromatogr. 559*: 515 (1991).
11. J. Cai and J. Henion, *J. Anal. Toxicol. 20*: 27 (1996).
12. A. Guttman and K. J. Ulfelder, *Adv. Chromatogr. 38*: 301 (1998).
13. P. G. Righetti and C. Gelfi, *Forensic Sci. Int. 92*: 239 (1998).
14. B. R. McCord, D. L. McClure, and J. M. Jung, *J. Chromatogr. A 652*: 75 (1993).
15. J. F. Banks, Protein analysis, in *Advanced Chromatographic and Electromigration Methods in BioSciences* (Z. Deyl, I. Mikšík, F. Tagliaro, E. Tesarová, eds.), Journal of Chromaography Library Vol. 60, Elsevier, Amsterdam, 1998, pp. 525–573.

Ivan Mikšík

Forskolin Purification Using an Immunoaffinity Column Combined with an Anti-Forskolin Monoclonal Antibody

Introduction

Forskolin, a labdane diterpenoid, was isolated from the tuberous roots of *Coleus forskohlii* Briq. (Lamiaceae) [1]. *C. forskohlii* has been used as an important folk medicine in India. Forskolin was found to be an activator of adenylate cyclase [2], leading to an increase of c-AMP, and now a medicine in India, Germany, and Japan. The production of forskolin is completely dependent on the commercial collection of wild and cultivated plants in India. We have already set up the production of monoclonal antibodies (MAbs) against forskolin [3]. The practical application of enzyme-linked immunosorbent assay (ELISA) for the distribution of forskolin contained in clonally propagated plant organs and the quantitative fluctuation of forskolin depend on the age of *C. forskohlii* [4,5]. As an extension of this approach, we present the production of the immunoaffinity column using anti-forskolin MAb and its application [6].

Materials and Methods

Chemicals

Bovine serum albumin (BSA) was provided by Pierce (Rockford, U.S.A.). Forskolin and 7-deacetyl forskolin were isolated from the tuberous root of *C. forskohlii*, as previously reported [1]. 1-Deoxyforskolin, 1,9-dideoxyforskolin, and 6-acetyl-7-deacetylforskolin were purchased from Sigma Chemical Company (St. Louis, MO, U.S.A.). The mixture (approximately 20 μg) of forskolin and 7-deacetyl forskolin, purified by the immunoaffinity column, was acetylated with pyridine and acetic anhydride mixture (each 100 ml) at 4°C for 2 h to give pure forskolin.

Preparation of Immunoaffinity Column Using Anti-Forskolin Monoclonal Antibody [6]

Purified IgG (10 mg) in PBS was added to a slurry of CNBr-activated Sepharose 4B (600 mg; Pharmacia Biotech) in coupling buffer (0.1*M* $NaHCO_3$ containing 0.5*M* NaCl). The slurry was stirred for 2 h at room temperature and then treated with 0.2*M* glycine at pH 8.0 for blocking of activated groups. The affinity gel was washed four times with 0.1*M* $NaHCO_3$ containing 0.5*M* NaCl and 0.1*M* acetate buffer (pH 4.0). Finally, the affinity gel was centrifuged and the supernatant was removed. The immunoaffinity gel was washed with phosphate buffer solution (PBS) and packed into a plastic mini-column in volumes of 2.5 mL. Columns were washed until the absorption at 280 nm was equal to the background absorption. The columns were stored at 4°C in PBS containing 0.01% sodium azide.

Direct Isolation of Forskolin from Crude Extractives of Tuberous Roots and Callus Culture of C. forskohlii *by Immunoaffinity Column*

The dried powder (10 mg dry weight) of tuberous root was extracted five times with diethyl ether (5 mL). After evaporation of the solvent, the residue was redissolved in MeOH and diluted with PBS (1:16), and then filtered by Millex-HV filter (0.45-μm filter unit; Millipore Products, Bedford, MA, U.S.A.) to remove insoluble portions. The filtrate was loaded onto the immunoaffinity column and allowed to stand for 90 min at 4°C. The column was washed with the washing buffer solution (10 mL). After forskolin disappeared, the column was eluted with PBSM (45%) at a flow rate of 0.1 mL/min. The fraction containing forskolin was lyophilized and extracted with diethyl ether. Forskolin was determined by thin-layer chromatography (TLC) developed with C_6H_6–EtOAc (85:15) [Rf; forskolin (0.21), 7-deacetyl forskolin (0.16)] and ELISA.

Results and Discussion

We established a simple and reproducible purification method for forskolin using an immunoaffinity column chromatography method. Because forskolin is almost insoluble in water, various buffer solutions were tested for the solubilization of forskolin. It became evident that 6% MeOH in PBS was necessary for the solubilization of forskolin [3,5]. Next, the elution system for the immunoaffinity column was investigated by using various elution buffers based on PBS. Only 9% of bound forskolin can be recovered by the PBS supplemented with 10% of MeOH. The forskolin concentrations eluted increased rapidly from 20% of MeOH, and reached the optimum at 45% of MeOH.

To assess the capacity and the recovery of forskolin from the affinity column, 30 μg of forskolin was added and passed through the column (2.5 mL of gel), and the forskolin content was analyzed by ELISA. After washing with 5 column volumes of PBST, 22.5 μg of forskolin remained bound and was then completely eluted with the PBS containing 45% of MeOH. Therefore, the capacity of affinity column chromatography was determined to be 9.4 μg/mL.

The crude diethyl ether extracts of the tuberous root of *C. forskohlii* were loaded onto the immunoaffinity column chromatography system, washed five times with PBS containing 6% of MeOH, and eluted with the PBS containing 45% of MeOH. Figure 1 shows a chromatogram detected by ELISA. Fractions 2–8 contained 45 μg of forskolin that were over the column capacity, together with the related compounds 1-deoxyforskolin, 1,9-dideoxyforskolin, 7-deacetylforskolin and 6-acetyl-7-deacetylforskolin, and other unknown compounds which were detected by TLC, as indicated in Fig. 2. The peak of fractions 26–30 shows the elution of forskolin (21 μg) eluted with the PBS containing 45% of MeOH. Forskolin eluted

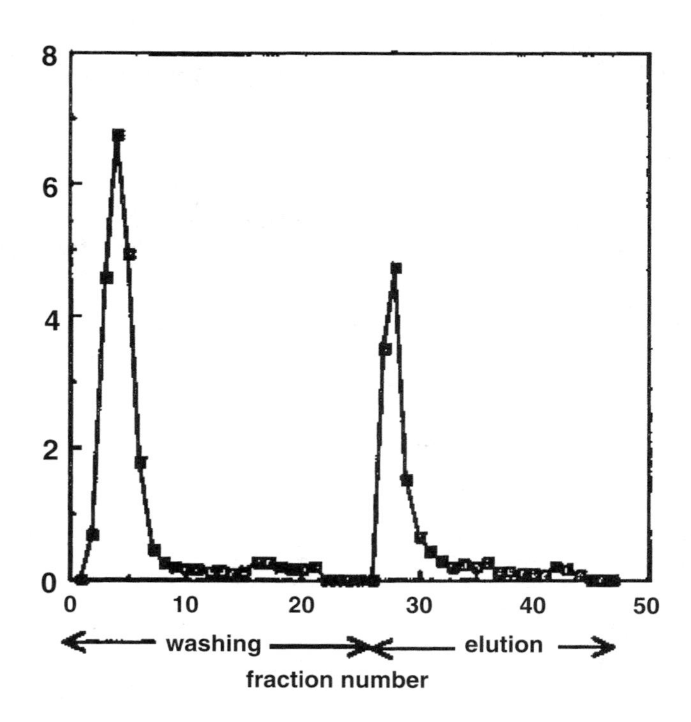

Fig. 1 Elution profile of forskolin in the tuberous root of *C. forskohlii* by purification on immunoaffinity column chromatography. The column was washed with PBSM, then eluted by PBS containing 45% of methanol after the forskolin disappeared. Individual fractions were assayed by ELISA.

by washing solution (fractions 2–8) was repeatedly loaded and finally isolated. However, forskolin purified by the immunoaffinity column chromatography was still contaminated with a small amount of 7-deacetyl forskolin (Fig. 2) because this compound has a 5.5% cross-reactivity against Mab, as previously indicated [3]. Therefore, the mixture was treated with pyridine and acetic anhydride at 4°C for 2 h to give pure forskolin. In our case, the stability of antibody against PBS containing 45% MeOH is also quite high, because the immunoaffinity column has been used over 10 times, under the same conditions, without any substantial loss of capacity. Therefore, we concluded that the PBS supplemented with 45% MeOH can be routinely used as an elution buffer solution.

References

1. S. V. Bhat, B. S. Bajwa, H. Dornauer, N. J. de Sousa, and H. W. Fehlhaber, *Tetrahedron Lett.* 1669–1672 (1977).
2. H. Metzger and E. Lindner, *Drug Res. 31*: 1248–1250 (1981).
3. R. Sakata, Y. Shoyama, and H. Murakami, *Cytotechnology 16*: 101–108 (1994).

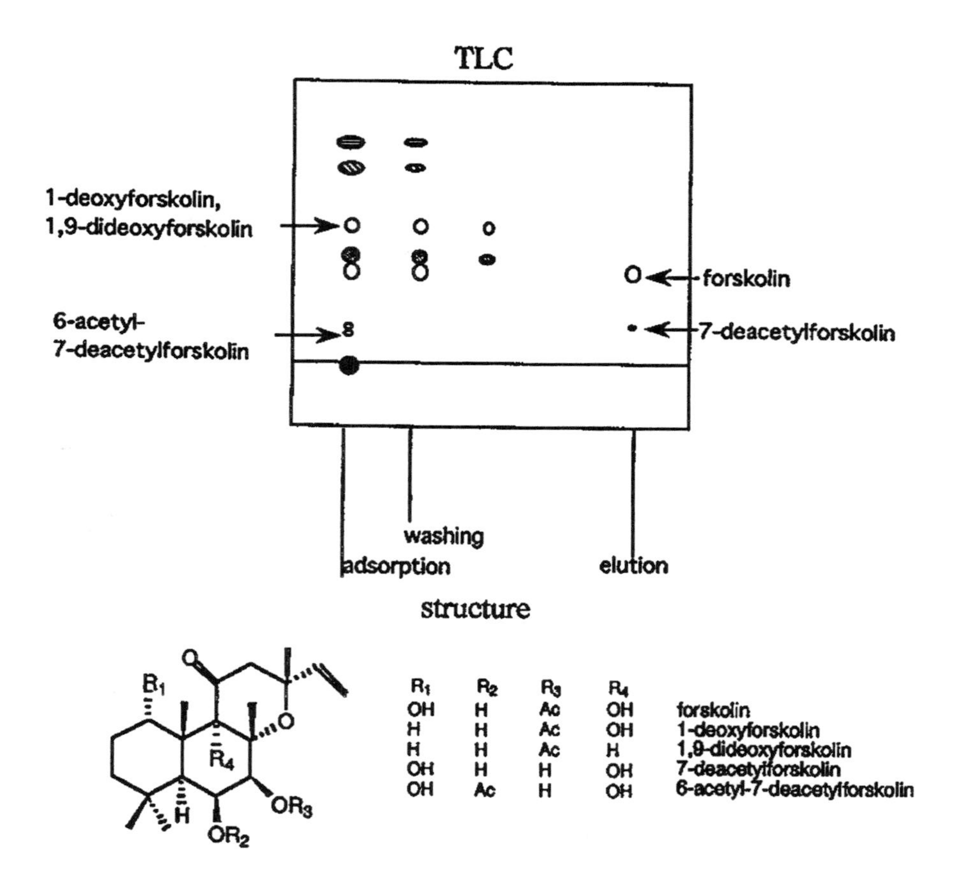

Fig. 2 TLC of adsorption, washing, and elution solutions, and structures of forskolin and the related compounds.

4. H. Yanagihara, R. Sakata, Y. Shoyama, and H. Murakami, *Biotronics 24*: 1–6 (1995).
5. H. Yanagihara, R. Sakata, Y. Shoyama, and H. Murakami, *Planta Med. 62*: 169–172 (1996).
6. H. Yanagihara, H. Minami, H. Tanaka, Y. Shoyama, and H. Murakami, *Anal. Chim. Acta 335*: 63–70 (1996).

Hiroyuki Tanaka
Yukihiro Shoyama

Fraction Collection Devices

Purposes and Uses of Fraction Collectors

The growth of high-performance liquid chromatography (HPLC) has prompted significant interest in preparative liquid chromatography as a tool for purifying organic compounds in complex liquid mixtures. On any scale, preparative LC implies separation of target compounds and subsequent collection of the column eluent, containing the compounds, in appropriate vessels for their isolation. This frequently occurs in complex mixtures containing both major and minor peaks. Characterization of the isolated compound is then usually performed by complementary techniques such as mass spectroscopy (MS), nuclear magnetic resonance (NMR), or infrared (IR) spectroscopy.

Fraction collectors are especially useful in applications involving unattended, overnight, and automated chromatographic purification schemes. If only several fractions are to be collected from the liquid chromatograph and only a few samples are to be purified, then the eluent from the column can sometimes be collected manually. However, a fraction collection device is recommended if the analyst has many samples to process and/or if there are many fractions to collect.

The goals of preparative HPLC are different compared to analytical and trace analyses. In analytical applications, the goal is to obtain quantitative and qualitative information about the sample. With preparative HPLC, however, the goal is to isolate and purify compounds from complex mixtures. Preparative liquid chromatography traditionally meant large flow rates, large-inner-diameter (i.d.) columns and relatively large column particles to isolate and purify milligram to gram quantities of compounds of interest. In recent years, however, preparative liquid chromatography has been "scaled down" to analytical (4.6 mm i.d.) and narrow-bore ($\leq$2 mm i.d.) column proportions as biopharmaceutical and biomedical research focuses on trace analytes at the picogram level or lower.

Preparative liquid chromatography and fraction collectors are widely used to purify target compounds for applications found in many disciplines:

Pharmaceuticals. Determination of drug candidate/metabolite structure in discovery and development studies; end product and intermediate purification

Organic chemistry. Preparation of standards; purification of starting materials; identification of impurities in end products; determination of intermediate structures during organic synthesis

Biotechnology/biomedical. Identification of molecular structures in genetic engineering applications; end product and intermediate purification; isolation and purification of enzymes, proteins, nucleic acids, carbohydrates, lipids, and other biomolecules; polymerase chain reaction (PCR) product and monoclonal antibody purifications

Food/beverage. Additive purification; determination of intermediate structures during fermentation processes

Operational Considerations of Fraction Collectors in Liquid Chromatography

The functional requirements of fraction collectors have changed as liquid chromatography evolved from low-pressure (low-resolution) to high-pressure (high-resolution) techniques. This evolution was led by developments in column technologies and packing materials, and followed by adaptations to pumps, detectors, injectors, and fraction collectors. The following subsections briefly describe some of the key operational considerations of fraction collection devices.

Low-Pressure (Low-Resolution) Techniques

Classical low-pressure LC is a relatively simple technique developed in the 1950s, but still commonly used today by organic chemists and biochemists. It is very effective as a crude purification step, especially before preparative HPLC, to remove contaminants and interferences, even though it is slow and offers low resolution. For many laboratory studies using classical low-pressure LC, requirements are generally for simple time and drop mode fraction collectors. The entire chromatographic eluent is automatically collected by selecting equal fractions based on time per tube or drops per tube, as shown in Fig. 1.

If the elution time(s) of peaks of interest are known and reproducible, the user can program most microprocessor-based fraction collectors to automatically collect only the peaks of interest and discard the between-peak eluent or peaks of noninterest. This is a variable time window programming mode, also known as Time Program plus Time (or Drop) mode, and involves a sequence of time-based collection and drain steps, as shown in Fig. 2. Each collection step is commonly referred to as a collection time window. It conveniently allows the user to define time intervals during which the column eluent is either collected into fractions or discarded into waste. Each selected peak will be subfractionated by equal slices based on time counting or drop counting. Column void volumes, equilibration volumes, and peaks of no interest are discarded.

High-Pressure (High-Resolution) Techniques

Advances in column technology and small-particle packing materials ($\leq$15 μm) have enabled small-scale preparative chromatography using "analytical scale" HPLC columns, pumps, injectors, and detectors. However, the resultant high-resolution chromatograms are frequently very complex and require fraction collectors capable of "cutting" pure peaks of less than 20 s duration.

This has created a need for a new generation of flexible fraction collectors capable of sophisticated collection based on the detector output signal that corresponds to the target compound(s). This is sometimes known as peak detection and enables isolation of the purest part of each peak into a single vessel. When very high-purity levels are required, peaks of interest containing target compounds should not be contaminated by neighboring peaks or diluted with mobile phase, which would be the case if the collector were to collect equal fractions throughout the run based simply on time or drop counts (see Fig. 1). These

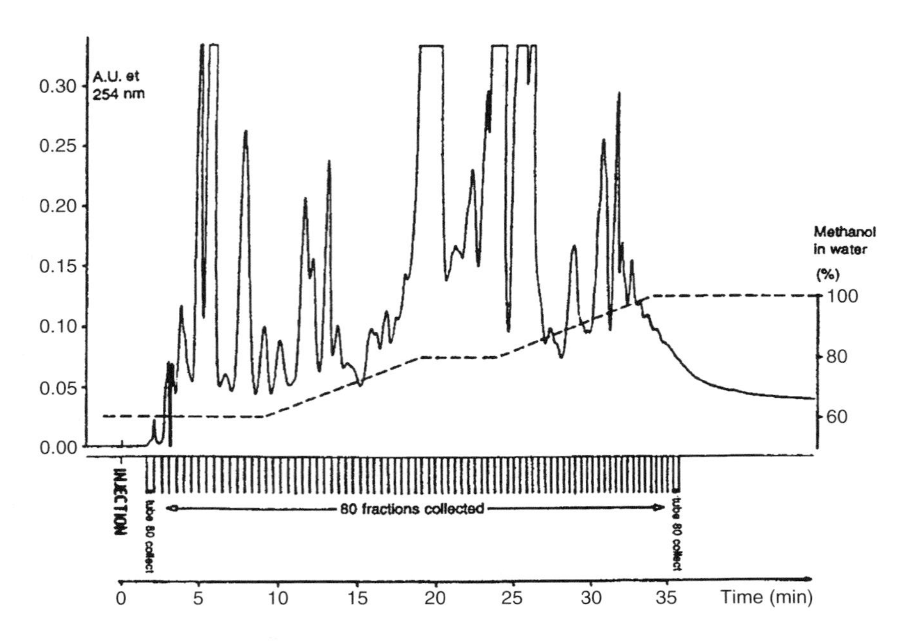

Fig. 1 Collection in Drop mode. The entire run is collected by selecting equal fractions based on drops per tube.

high-purity requirements are usually found in small-scale laboratory studies where optimum purity of fractions (sometimes ≥99.9%) is critical.

Detection of chromatographic peaks by the fraction collector is determined by either calculation of a minimum baseline slope or by a user-specified millivolt threshold level of the detector's output. In the former case, sophisticated algorithms are used to calculate the baseline slope based on user input of peak width and peak sensitivity. An example of fraction collection based on the detector sig-

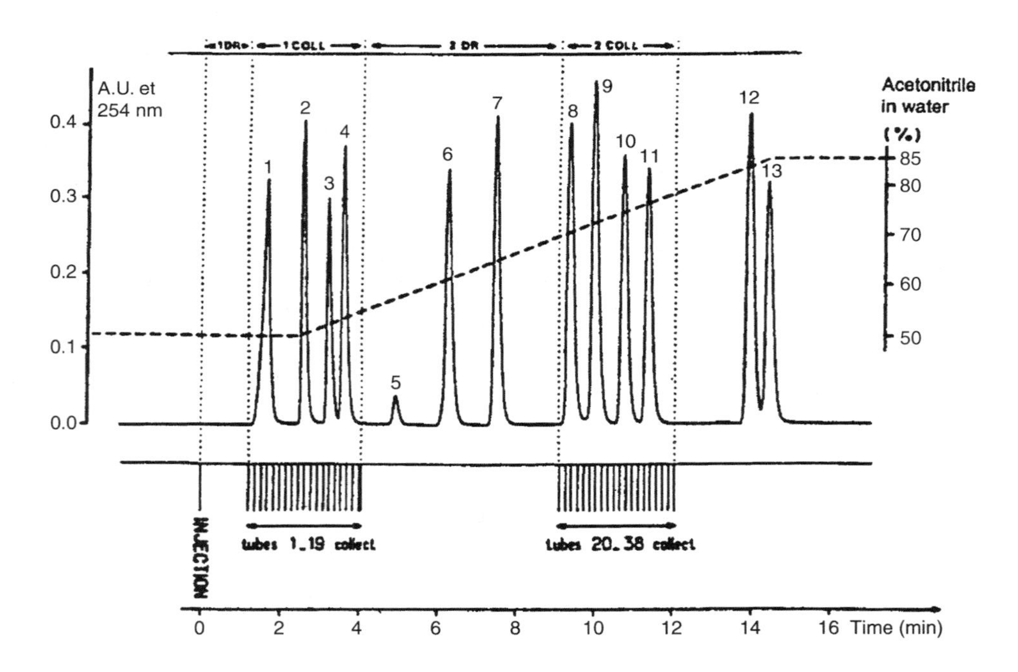

Fig. 2 Collection in Time Program plus Time mode. Two collection time windows are subfractionated into equivalent fractions based on time per tube.

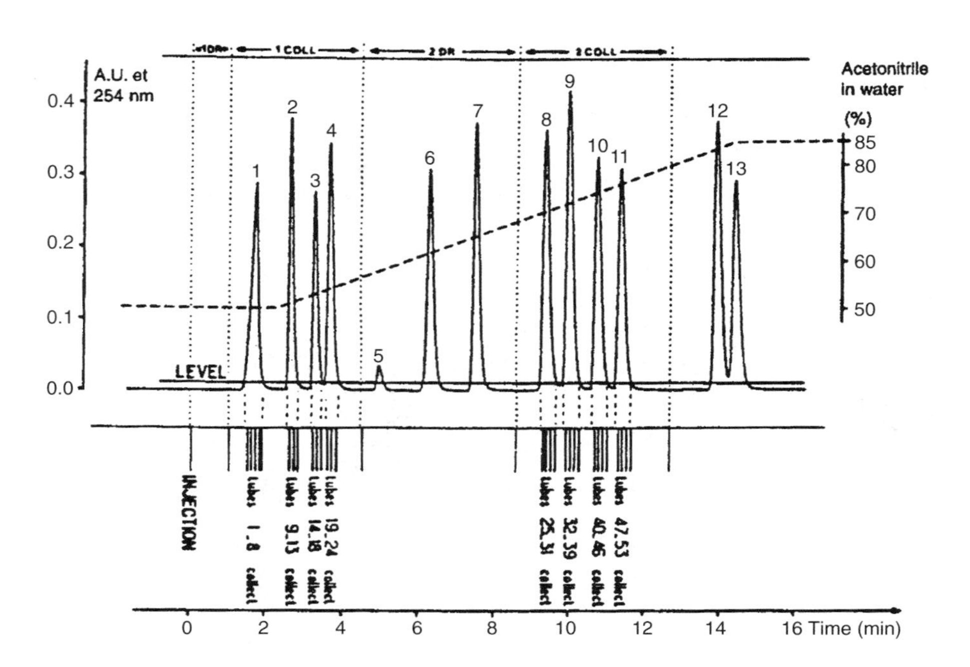

Fig. 3 Collection in Time Program plus Peak plus Time mode. This multimode protocol involves two collection time windows, peak detection (based on slope) within those time windows, and subfractionation of the peaks by time per tube.

nal is shown in Fig. 3. Each fraction corresponds to a peak detected by the collector. Unwanted peaks, between-peak eluent, column void volumes, and equilibration volumes are discarded to waste. Because the chromatographer is not collecting the entire column effluent from beginning to end, there is no need to stop the system and change tubes or racks, thereby enabling automated, unattended operation.

Applications

In preparative HPLC, fraction collection is likely to be employed for two extreme applications: (a) the purification of one or a few major components or (b) the isolation of trace components or impurities in the presence of main components. The first problem is generally solved by the millivolt level (threshold) collection mode with time or drop "subfractionation" of each peak. A slope-detecting peak collection mode with equal time slices is often the best solution to the second problem.

Suggested Further Reading

Katz, A., Eksteen, R., Schoenmakers, P., and Miller, N., Collection devices, in *Handbook of HPLC*, Marcel Dekker, Inc., New York, 1998, Chap. 19.

Gordon S. Hunter

Frit-Inlet Asymmetrical Flow Field-Flow Fractionation

Frit-inlet asymmetrical flow field-flow fractionation (FIA–FlFFF) [1–3] utilizes the frit-inlet injection technique, with an asymmetrical flow FFF channel which has one porous wall at the bottom and an upper wall that is replaced by a glass plate. In an asymmetrical flow FFF channel, channel flow is divided into two parts: axial flow for driving sample components toward a detector, and the cross-flow, which penetrates through the bottom of the channel wall [4,5]. Thus, the field (driving force of separation) is created by the movement of cross-flow, which is constantly lost through the porous wall of the channel bottom. FIA–FlFFF has been developed to utilize the stopless sample injection technique with the conventional asymmetrical channel by implementing an inlet frit nearby the channel inlet end and to reduce possible flow imperfections caused by the porous walls.

The asymmetrical channel design in flow FFF has been shown to offer high-speed and more efficient separation for proteins and macromolecues than the conventional symmetrical channel. However, an asymmetrical channel requires a focusing–relaxation procedure for sample components to reach their equilibrium states before the separation begins. The focusing–relaxation procedure is achieved by two counterdirecting flow streams from both the channel inlet and outlet to a certain point slightly apart from the channel inlet end for a period of time. This is a necessary step equivalent to the stop-flow procedure as is normally used in a conventional symmetrical channel system. Although the stop-flow and the focusing–relaxation procedures are essential in each technique (symmetrical and asymmetrical channels, respectively), they are basically cumbersome in system operation due to the stoppage of flow with valve operations. In addition, they often cause baseline shifts during the conversion of flow. For these reasons, the frit-inlet injection technique, which can be an alternative to bypass those flow-halting processes, is adapted to an asymmetrical flow FFF channel in order to take advantage of hydrodynamic relaxation of sample components.

The frit-inlet injection device was originally applied to the conventional symmetrical channel in order to bypass the stop-flow procedure [6]. However, the lowest axial flow rate that can be manipulated in a frit-inlet symmetrical system is limited, because the total axial flow rate becomes the sum of the injection flow rate and frit flow rate, and the incoming cross-flow penetrates through the bottom wall at the same rate. The relatively high axial flow rate in a symmetrical system needs a very high cross-flow rate in order to separate relatively low-retaining materials, such as proteins or low-molecular-weight components.

Compared to the limited choice in the selection of flow rate conditions, application of the frit-inlet injection technique to an asymmetrical flow FFF channel can be more flexible in allowing the selection of a low axial flow rate condition which is suitable for low-retaining materials without the need of using a very high cross-flow rates and for the reduction of injection amount resulting from the concentration effect.

In FIA–FlFFF, sample materials entering the channel are quickly driven toward the accumulation wall and are transported to their equilibrium positions by the com-

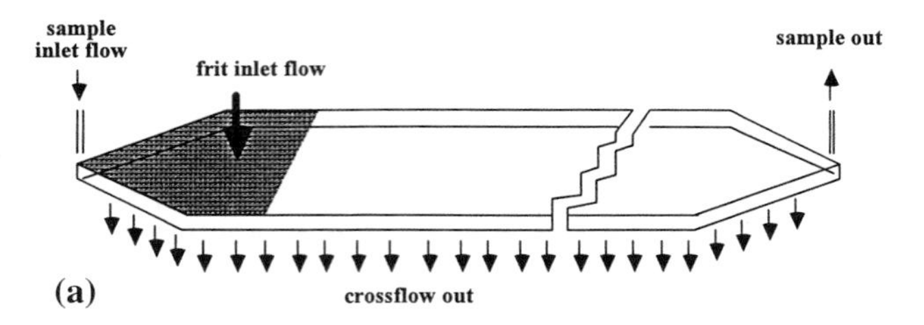

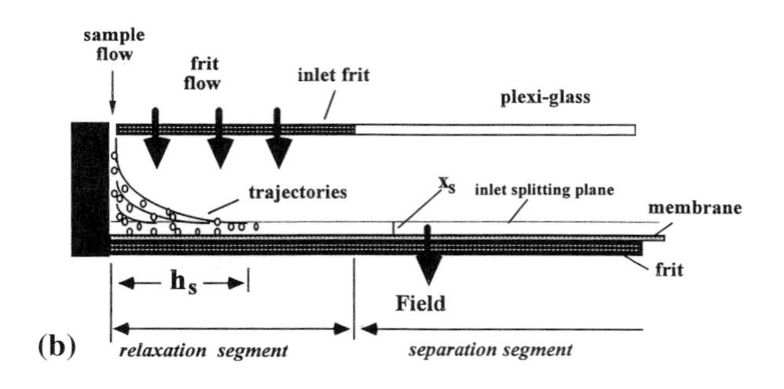

Fig. 1 Schematic view of an FIA–FlFFF channel.

pressing action of a rapidly flowing frit flow entering through the inlet frit. The schematic view of an FIA–FlFFF channel is shown in Fig. 1. In the relaxation segment of a FIA–FlFFF channel, the frit flow stream and the sample stream of relatively low speed will merge smoothly. During this process, sample materials are expected to be pushed below the inlet splitting plane formed by the compressing effect of frit flow, as illustrated in Fig. 1b. Thus, sample relaxation is achieved hydrodynamically in the relaxation segment (under the inlet frit region), and the sample components are continuously carried to the separation segment where the separation of sample components takes place. System operation requires only a simple one-step injection procedure, with no need for valve switching or interruption of flow. This is far simpler and more convenient than the operation of the conventional relaxation techniques, such as stop-flow and focusing–relaxation procedures.

In the first experimental work on FIA–FlFFF [1], the system efficiency was studied by examining the effect of the ratio of injection flow rate to frit flow rate on hydrodynamic relaxation; the initial tests showed a possibility of using hydrodynamic relaxation in asymmetrical flow FFF with a number of polystyrene latex standards, in both normal and steric/hyperlayer modes of FFF. Normally, relaxational band broadening under hydrodynamic relaxation arises from a broadened starting band. The length of an initial sample band during hydrodynamic relaxation is dependent on flow rates as

$$h_s = \frac{\dot{V}_s}{\dot{V}_f}\frac{\dot{V}}{\dot{V}_c}L \tag{1}$$

where L is the channel length; $\dot{V}_s$, $\dot{V}_f$, $\dot{V}$, and $\dot{V}_c$ represent the flow rates of the sample stream, frit stream, effective channel flow, and cross-flow, respectively. Equation (1) suggests that a small ratio of sample flow rate to frit flow rate, with a combined high cross-flow rate, is preferable in reducing h_s, leading to minimized relaxational band broadening.

Experimentally, the optimum ratio of $\dot{V}_s/\dot{V}_f$ has been found to be about 0.03–0.05 for the separation of latex beads and for proteins.

Retention in the separation segment of the FIA–FlFFF channel is expected to be equivalent to that observed in a conventional asymmetrical channel system, if complete hydrodynamic relaxation can be obtained. It will follow basic principles, as shown by the retention ratio, R, given by

$$R = \frac{t^0}{t_r} = 6\lambda\left[\coth\left(\frac{1}{2\lambda} - 2\lambda\right)\right] \quad \left(\text{where } \lambda = \frac{D}{w^2}\frac{V^0}{\dot{V}_c}\right) \tag{2}$$

where t^0 is the void time, t_r is the retention time, λ is the retention parameter, D is the diffusion coefficient, w is the channel thickness, and V^0 is the channel void volume. The void time in an FIA–FlFFF channel system is complicated to calculate, because sample flow and frit flow enter the channel simultaneously, and part of the merged flow exits through the accumulation wall. For this reason, channel flow velocity varies along the axial direction of channel. By considering these, the determination of void time can be represented as

$$t^0 = \frac{V^0 A_f/A_c}{\dot{V}_f - \dot{V}_c A_f/A_c}\ln\left(\frac{\dot{V}_s + \dot{V}_f - \dot{V}_c A_f/A_c}{\dot{V}_s}\right) + \frac{V^0}{\dot{V}_c}\ln\left(\frac{\dot{V}_s + \dot{V}_f - \dot{V}_c A_f/A_c}{\dot{V}_{\text{out}}}\right) \tag{3}$$

where $\dot{V}_{\text{out}}$ is the channel outflow rate and A_f and A_c are the area of the inlet frit and the accumulation wall, respectively. Equation (3) represents the void time calculation in terms of volumetric flow rate and channel dimensions only; it is valid for any channel geometry, such as rectangular, trapezoidal, and even exponential design. Retention in FIA–FlFFF has been shown to follow the general principles of FFF with the confirmation of experimental work. It has also been found that the trapezoidal channel design provides a better resolving power for the separation of protein mixtures than a rectangular channel in FIA–FlFFF.

References

1. M. H. Moon, H. S. Kwon, and I. Park, *Anal. Chem. 69*: 1436 (1997).
2. M. H. Moon, H. S. Kwon, and I. Park, *J. Liquid Chromatogr. Related Technol. 20*: 2803 (1997).
3. M. H. Moon, P. Stephen Williams, and H. S. Kwon, *Anal. Chem. 71*: 2657 (1999).
4. A. Litzén and K.-G. Wahlund, *Anal. Chem. 63*: 1001 (1991).
5. A. Litzén, *Anal. Chem. 65*: 461 (1993).
6. J. C. Giddings, *Anal. Chem. 57*: 945 (1985).

Myeong Hee Moon

Frontal Chromatography

Introduction

Frontal chromatography is a mode of chromatography in which the sample is introduced continuously into the column. The sample components migrate through the column at different velocities and eventually break through as a series of fronts. Only the least retained component exits the column in pure form and can, therefore, be isolated; all other sample components exit the column as mixed zones. The resulting chromatogram of a frontal chromatography experiment is generally referred to as a breakthrough curve, although the expression *frontalgram* has also been used in the literature [1].

The exact shape of a breakthrough curve is mainly determined by the functional form of the underlying equilibrium isotherms of the sample components, but secondary factors such as diffusion and mass-transfer kinetics also have influence. The capacity of the column is an important parameter in frontal chromatography, because it determines when the column is saturated with the sample components and, therefore, is no longer able to adsorb more sample. The mixture then flows through the column with its original composition.

The Use of Frontal Chromatography

Frontal chromatography can also be called *adsorptive filtration* because it can be used for the purpose of filtration. The purification of gases and solvents are two classical applications of frontal chromatography. Another important use is the purification of proteins, where a frontal chromatography step is used in the initial purification procedure [2,3].

One of the most important applications of frontal chromatography is the determination of equilibrium adsorption isotherms. It was introduced for this purpose by Shay and Szekely and by James and Phillips [4,5]. The simplicity as well as the accuracy and precision of this method are reasons why the method is so popular today and why it is often preferred over other chromatographic methods {e.g., elution by characteristic points (ECP) or frontal analysis by characteristic points (FACP) [6,7]}. Frontal chromatography as a tool for the determination of single-component adsorption isotherms will be discussed in the following section.

Frontal Chromatography for the Determination of Isotherms

Theory

First, the column is filled only with sample at concentration C_n; then, a step injection is performed (i.e., sample with the concentration C_{n+1} is introduced into the column). This results in a breakthrough curve, as shown in Fig. 1. The amount adsorbed at the stationary phase Q_{n+1} can be calculated by

$$Q_{n+1} = q_{n+1}V_s = (C_{n+1} - C_n)(V_{R,n+1} - V_0) + q_nV_s$$

where q_n and q_{n+1} are the initial and final sample concentrations, respectively, in the stationary phase and C_n and C_{n+1} are the initial and final sample concentrations in the mobile phase, respectively. V_s is the volume of adsorbent in the column, V_0 is the holdup volume, and $V_{R,n+1}$ is the retention volume of the breakthrough curve. The retention volume is calculated from the area over the breakthrough curve:

$$V_R = \int_0^{\infty} \frac{(C_{n+1} - C)\,dV}{C_{n+1} - C_n}$$

The retention volume defined by the area method always gives the theoretically correct result for the amount

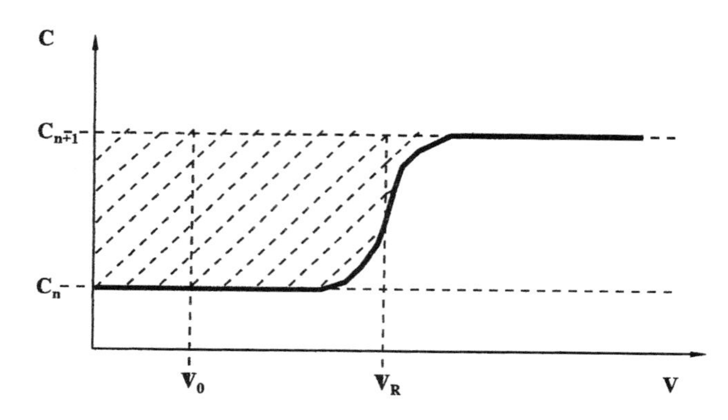

Fig. 1 Example of a frontal chromatography experiment; breakthough curve of a single component.

adsorbed. In practice, it is, however, often easier and better to use the retention volume from half-height [i.e., at the concentration $(C_{n+1} + C_n)/2$] or the retention volume derived from the inflection point of the breakthrough curve. The reason for this is that the calculation of the area incorporates signal noise and it is very dependent on the integration limits. This is often a problem, especially when the mass transfer is slow, because, in this case, the plateau concentration C_{n+1} is only reached slowly and, therefore, systematic errors in the calculated area occur. It has been shown that the use of the retention volumes derived from the inflection point or the half-height gives satisfactory results. The half-height method is, however, easier to use and slightly more accurate than the inflection-point method [5].

It has to be noted that the half-height and inflection-point methods do not give reliable results if the isotherm is concave upward and ascending concentration steps are performed. The same is true for a convex upward isotherm and descending concentration steps. The reason for this is that, in these cases, a diffuse breakthrough profile is obtained and, consequently, errors are made in the accurate determination of the retention volumes when they are derived from the half-height or the inflection point. The diffuse profile can, however, be used for the determination of isotherms by the frontal analysis by characteristic points method (FACP).

Modes of Frontal Analysis

There are two possibilities for performing a frontal chromatography experiment for the purpose of the determination of equilibrium isotherms. The step-series method uses a series of steps starting from $C_n = 0$ to C_{n+1}. After each experiment, the column has to be reequilibrated and a new step injection with a different end concentration

C_{n+1} can be performed. In the staircase method, a series of steps is performed in a single run with concentration steps from 0 to C_1, C_1 to C_2, . . . , C_n to C_{n+1}. The column does not have to be reequilibrated after each step and, therefore, the staircase method is faster than the step-series method. Both modes of frontal analysis give very accurate isotherm results.

Determination of Multicomponent Isotherms by Frontal Analysis

It is possible to extend the frontal chromatography method for the measurement of binary and multicomponent isotherms. In this case, the profiles are characterized by successive elution of several steep fronts. The use of these profiles for the determination of competitive isotherms in the binary case has been developed by Jacobsen et al. [8].

Combination of Frontal Analysis with Chromatographic Models

Frontal chromatography can be used in combination with chromatographic models to study mass-transfer and dispersion processes (e.g., the equilibrium dispersive or the transport model of chromatography [7]).

Constant Pattern, Self-Sharpening Effect, Shock-Layer Theory

Frontal chromatography generally requires the adsorption isotherm to be convex upward if the step injection is performed with ascending concentration (i.e., $C_{n+1} > C_n$) because, in this case, the profile of a breakthrough curve tends asymptotically toward a limit. After this constant profile has been reached, the profile migrates along the column without changing its shape. This state is called *constant pattern* [9]. This phenomenon arises because the self-sharpening effect associated with a convex isotherm is balanced by the dispersive effect of axial dispersion and a finite rate of mass-transfer kinetics. If the equilibrium adsorption isotherm is linear or concave upward, no constant pattern behavior is observed and the breakthrough curve spreads constantly during its migration through the column. This case is unfavorable. If the adsorption isotherm is concave upward, then a descending concentration step (i.e., $C_{n+1} < C_n$) leads to the formation of a constant pattern.

A very detailed study of the combined effects of axial dispersion and mass-transfer resistance under a constant pattern behavior has been conducted by Rhee and Amundson [10]. They used the *shock-layer* theory. The shock layer is defined as a zone of a breakthrough curve where a specific concentration change occurs (i.e., a concentration change from 10% to 90%). The study of the shock-layer thickness is a new approach to the study of column performance in nonlinear chromatography. The optimum velocity for minimum shock-layer thickness (SLT) can be quite different from the optimum velocity for the height equivalent to a theoretical plate (HETP) [9].

Instrumentation

There are many possibilities for performing frontal chromatography experiments. In general, standard chromatographic equipment can be used. The preparation of a series of solutions of known concentration can be easily accomplished by using a chromatograph with a gradient delivery system applied as a mobile-phase mixer. If this system is not available, then the solutions have to be prepared manually. Two pumps can be used to perform the step injections or a single pump with a gradient delivery system. An injector having a sufficient large loop can also be used. Even a single pump without gradient delivery system can be used. In this case, the step injection has to be made by manually switching the solvent inlet line to the prepared sample reservoir. The choice of the system is dependent on the application. For fast and accurate measurements of adsorption isotherms, a multisolvent gradient system with two pumps and a high-pressure mixer is a very good choice.

References

1. J. Parcher, *Adv. Chromatogr. 16*: 151 (1978).
2. F. Antia and Cs. Horváth, *Ber. Bunsenges. Phys. Chem. 93*: 968 (1989).
3. A. Lee, Aliao, and Cs. Horváth, *J. Chromatogr. 443*: 31 (1988).
4. D. James and C. Phillips, *J. Chem. Soc.*, 1066 (1954)
5. G. Shay and G. Szekely, *Acta Chim. Hung. 5*: 167 (1954).
6. H. Guan, B. Stanley, and G. Guiochon, *J. Chromatogr. A 659*: 27 (1994).
7. P. Sajonz, Ph.D. thesis, University Saarbrücken, Germany, 1996.
8. J. Jacobsen, J. Frenz, and Cs. Horváth, *Ind. Eng. Chem. Res. 26*: 43 (1987).
9. G. Guiochon, S. Golshan-Shirazi, and A. Katti, *Fundamentals of Preparative and Nonlinear Chromatography*, Academic Press, Boston, 1994.
10. H. Rhee and N. Amundson, *Chem. Engng. Sci. 27*: 199 (1972).

Peter Sajonz

Fronting of Chromatographic Peaks: Causes

Peaks with strange shapes represent one of the most vexing problems that can arise in a chromatographic laboratory. Fronting of peaks is a condition in which the front of a peak is less steep than the rear relative to the baseline. This condition results from nonideal equilibria in the chromatographic process.

Fronting peaks, as well as tailing or other misshaped peaks, can be hard to quantitate. Some data systems have difficulty in measuring peak size accurately. As a result, the precision and/or reliability of assay methods involving fronting or other misshaped peaks is often poor when compared to good chromatography. There are a number of different causes of peak fronting, and discovering why peaks are thus misshaped and then fixing the problem can be a difficult undertaking. Fortunately, there is a systematic approach based on logical analysis plus practical fixes that have now been documented in numerous laboratories.

Fronting peaks are less commonly encountered in liquid chromatography (LC), but they are readily distinguished from other peak-shape problems. Fronting peaks are the opposite of tailing peaks. Whereas tailing peaks suggest that sample retention decreases with increasing sample size or concentration, fronting peaks suggest the opposite: retention increases with larger samples. In both cases, a decrease in sample size may eliminate peak distortion. However, this is often not practical, because some minimum sample size is required for good detectability. In the case of tailing peaks, it is believed that peak distortion often arises because large samples use up some part of the stationary phase. However, the cause of fronting peaks is seldom fully understood.

Ion-pair chromatography (IPC) is more susceptible to peak fronting than other modes in LC. *Column temperature* problems can cause fronting peaks in IPC. Figure 1 shows the separation of an antibiotic amine at ambient temperature. Repeating the separation at 45°C eliminated the fronting problem. Some studies have shown peak fronting in IPC that can be corrected by operating at a higher column temperature, whereas some other separations are best carried out at lower temperatures. The reason for this peculiar peak-shape behavior is unclear, but it may be related to the presence of reagent micelles in the mobile phase for some experimental IPC conditions. Generally, it is good practice to run ion-pair separations under thermostatted conditions, because relative retention tends to vary with temperature in IPC. Usually, narrower bands and better separation results when temperatures of 40–50°C are used for IPC.

The use of a *sample solvent other than the mobile phase* is another cause of fronting peaks in IPC. In this case, the sample should only be injected as a solution in the mobile phase. No more than 25–50 μl of sample should be injected, if possible.

Silanol effects can adversely alter peak shape in IPC, just as in reversed-phase separations. Therefore, when separating basic (cationic) compounds, the column and mobile phase should be chosen bearing this in mind. When ion-pair reagents are used, however, silanol effects are often less important. The reason is that an anionic (acidic) reagent confers an additional negative charge on the column packing and this reduces the relative importance of sample retention by ion exchange with silanol groups. Similarly, cationic (basic) reagents are quite effective at blocking silanols because of the strong interaction between reagent and ionized silanol groups.

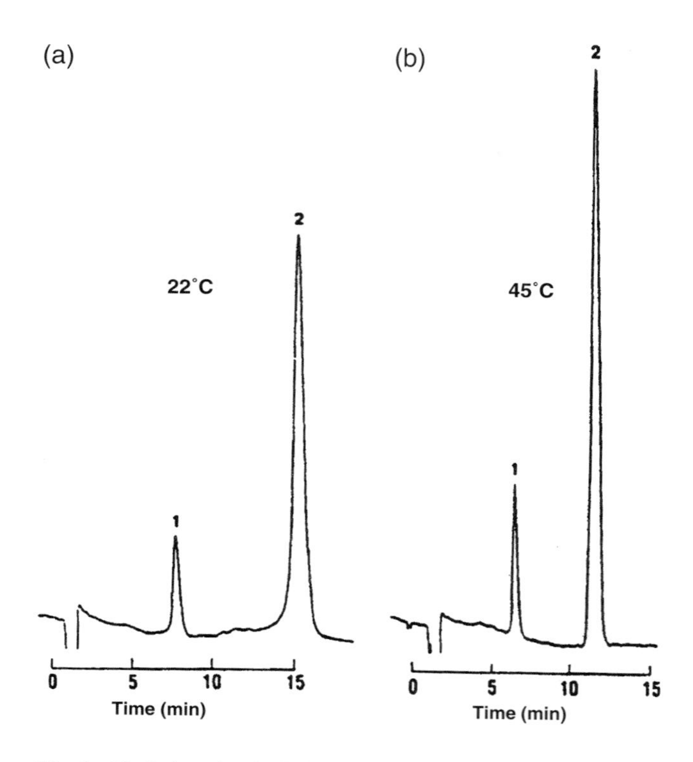

Fig. 1 Peak fronting in IPC as a function of separation temperature. Column: Zorbax C_8; mobile phase: 10 m*M* sodium dodecyl sulfate and 150 m*M* ammonium phosphate in 33% acetonitrile; pH: 6.0; flow rate: 2.0 mL/min; temperature: (a) = 22°C and (b) = 45°C. Peaks: 1 = lincomycin B; 2 = lincomycin A. (Reprinted from Ref. 1 with permission from Elsevier Science.)

Still another cause of peak fronting is for the case of *anionic (acidic) sample molecules separated with higher-pH mobile phases*. For silica-based packings, the packing has an increasingly negative charge as the pH increases, and this results in the repulsion of anionic sample molecules from the pores of the packing. With larger sample sizes, however, this effect is overcome by the corresponding increase in ionic strength, caused by the sample. A remedy for this problem is to increase the ionic strength of the mobile phase, by increasing the mobile-phase buffer concentration to the range of 25–100 m*M*. It should be mentioned here that ionic or ionizable samples should never be separated with unbuffered mobile phases.

Finally, *column voids* and *blocked frits* can also cause peak fronting.

Suggested Further Reading

Asmus, P. A., J. B. Landis, and C. L. Vila, *J. Chromatogr. 264*(2): 241 (1983).

Bidlingmeyer, B. A., *Practical HPLC Methodology and Applications*, John Wiley & Sons, New York, 1992, p. 20.

Dolan, J. W. and L. R. Snyder. *Troubleshooting LC Systems*, Humana Press, Totowa, NJ, 1989, pp. 400–401.

Sadek, P. C., P. W. Carr, and L. D. Bowers, *LC, Liq. Chromatogr. HPLC Mag. 3*: 590 (1985).

Ioannis N. Papadoyannis
Anastasia Zotou

G

Gas Chromatography–Mass Spectrometry Systems

Despite the speed and accuracy of contemporary analytical techniques, the use of more than one, separately and in sequence, is still very time-consuming. To reduce the analysis time, many techniques are operated concurrently, so that two or more analytical procedures can be carried out simultaneously. The tandem use of two different instruments can increase the analytical efficiency, but due to unpredictable interactions between one technique and the other, the combination can be quite difficult in practice. These difficulties become exacerbated if optimum performance is required from both instruments. The mass spectrometer was a natural choice for the early tandem systems to be developed with the gas chromatograph, as it could easily accept samples present as a vapor in a permanent gas.

The first gas chromatography–mass spectrometry (GC–MS) system was reported by Holmes and Morrell in 1957, only 4 years after the first description of GC by James and Martin in 1953. The column eluent was split and passed directly to the mass spectrometer. Initially, only packed GC columns were available and thus the major problem encountered was the disposal of the relatively high flow of carrier gas from the chromatograph (~ 25 mL/min or more). These high flow rates were in direct conflict with the relatively low pumping rate of the MS vacuum system. This problem was solved either by the use of an eluent split system or by employing a vapor concentrator. A number of concentrating devices were developed (e.g., the jet concentrator invented by Ryhage and the helium diffuser developed by Biemann).

The jet concentrator consisted of a succession of jets that were aligned in series but separated from each other by carefully adjusted gaps. The helium diffused away in the gap between the jets and was removed by appropriate vacuum pumps. In contrast, the solute vapor, having greater momentum, continued into the next jet and, finally, into the mass spectrometer. The concentration factor was about an order of magnitude and the sample recovery could be in excess of 25%.

The Biemann concentrator consisted of a heated glass jacket surrounding a sintered glass tube. The eluent from the chromatograph passed directly through the sintered glass tube and the helium diffused radially through the porous walls and was continuously pumped away. The helium stream enriched with solute vapor passed into the mass spectrometer. Solute concentration and sample recovery were similar to the Ryhage device, but the apparatus was bulkier although somewhat easier to operate. An alternative system employed a length of porous polytetrafluorethylene (PTFE) tube, as opposed to one of sintered glass, but otherwise functioned in the same manner.

The introduction of the open-tubular columns eliminated the need for concentrating devices as the mass spectrometer pumping system could cope with the entire column eluent. Consequently, the column eluent could be passed directly into the mass spectrometer and the total sample can enter the ionization source. The first mass spectrometer used in a GC–MS tandem system was a rapid-scanning magnetic sector instrument that easily provided a resolution of one mass unit. Contemporary mass spectrometers have vastly improved resolution and the most advanced system (involving the triple quadrupole mass spectrometer) gives high in-line sensitivity, selectivity, and resolution.

Ionization Techniques for GC–MS

There are a number of ionization processes that are used, probably the most important being electron-impact ionization. Electron-impact ionization is a harsh method of ionization and produces a range of molecular fragments that can help to elucidate the structure of the molecule. Nevertheless, although molecular ions are usually produced that are important for structure elucidation, sometimes only small fragments of the molecule are observed, with no molecular ion invoking the use of alternative ionizing procedures. A diagram showing the configuration of an electron-impact ion source is shown in Fig. 1. Electrons, generated by a heated filament, pass across the ion source to an anode trap. The sample vapor is introduced in the center of the source and the solute molecules drift, by diffusion, into the path of the electron beam. Collision with the electrons produce molecular ions and ionized molecular fragments, the size of which is determined by the energy of the electrons. The electrons are generated by thermal emission from a heated tungsten or rhenium filament and accelerated by an appropriate potential to the anode trap. The magnitude of the collection potential may range from 5 to 100 V, depending on the electrode geometry and the ionization potential of the substances being ionized. The ions that are produced are driven by a potential applied to the ion-repeller electrode into the accelerating region of the mass spectrometer.

Unfortunately, with electron-impact ionization, there is a frequent absence of a molecular ion in the mass spectrum, which makes identification uncertain and complicates structure elucidation. One solution is to employ chemical ionization. If an excess of an appropriate reagent gas is fed into an electron-impact source, an entirely different type of ionization takes place. As the reagent gas is in excess, the reagent molecules are preferentially ionized and the reagent ions then collide with the sample molecules and produce sample + reagent ions or, in some cases, protonated ions. In this type of ionization, very little fragmentation takes place and parent ions + a proton or + a molecule of the reagent gas are produced. Little modification to the normal electron impact source is required and an additional conduit to supply the reagent gas is all that is necessary.

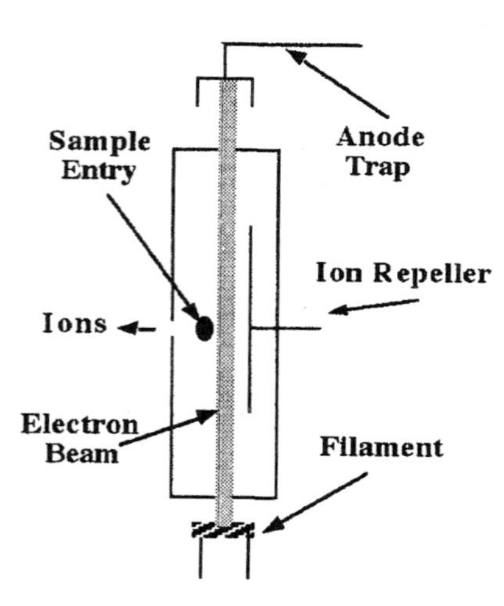

Fig. 1 An electron-impact ionization source: (a) Reagent gas methane; (b) reagent gas isobutane.

Chemical ionization was first observed by Munson and Field, who introduced it as an ionization procedure in 1966. A common reagent gas is methane and the partial pressure of the reagent gas is arranged to be about two orders of magnitude greater than that of the sample. The process is gentle and the energy of the most reactive reagent ions never exceeds 5 eV. Consequently, there is little fragmentation, and the most abundant ion usually has a *m/z* value close to that of the singly-charged molecular ion. The spectrum produced depends strongly on the nature of the reagent ion; thus, different structural information can be obtained by choosing different reagent gases. This adds another degree of freedom in the operation of the mass spectrometer. Using methane as the reagent ion, the following reagent ions can be produced:

$$CH_4 \rightarrow CH_4^+, CH_3^+, CH_2^+$$
$$CH_4^+ + CH_4 \rightarrow CH_5^+ + CH_3$$
$$CH_3^+ + CH_4 \rightarrow C_2H_4^+ + H_2$$

Other reactions can occur that are not useful for ionization but, in general, these are in the minority. The interaction of positively charged ions with the uncharged sample molecules can also occur in a number of ways, and the four most common are as follows:

1. Proton transfer between the sample molecule and the reagent ion

$$M + BH^+ \rightarrow MH^+ + B$$

2. Exchange of charge between the sample molecule and the reagent ion

$$M + X^+ \rightarrow M^+ + X$$

3. Simple addition of the sample molecule to the reagent ion

$$M + X^+ \rightarrow MX^+$$

4. anion extraction

$$AB + X^+ \rightarrow B^+ + AX$$

As an example, CH_5^+ ions, which are formed when methane is used as the reagent gas, will react with a sample molecule largely by proton transfer; that is,

$$M + CH_5^+ \rightarrow MX^+ + CH_4$$

Some reagent gases produce more reactive ions than others and will produce more fragmentation. For example, methane produces more aggressive reagent ions than isobutane. Consequently, whereas methane ions produce a number of fragments by protonation, isobutane, by a similar protonation process, will produce almost exclusively the protonated molecular ion. This is shown in the mass spectra of methyl stearate in Fig. 2. Spectrum (a) was produced using methane as the reagent gas and exhibits fragments other than the protonated parent ion. In contrast, spectrum b obtained with butane as the reagent gas, exhibits the protonated molecular ion only. Continuous use of a chemical ionization source causes significant source contamination, which impairs the performance of the spectrometer and thus the source requires cleaning by baking-out fairly frequently. Retention data on two-phase systems coupled with matching electron-impact mass spectra or confirmation of the molecular weight from chemical ionization spectra are usually sufficient to establish the identity of a solute.

The inductively coupled plasma (ICP) source is used largely for specific element identification and evolved from the ICP atomic emission spectrometer; it is probably more commonly employed in liquid chromatography (LC–MS) than GC–MS. In GC–MS, the ICP ion source is used in the assay of organometallic materials and in metal speciation analyses. The ICP ion source is very similar to the volatilizing unit of the ICP atomic emission spectrometer, and a diagram of the device is shown in Fig. 3. The argon plasma is an electrodeless discharge, often initiated by a Tesla coil spark, and maintained by radio-frequency (rf) energy, inductively coupled to the inside of the torch by an external coil, wrapped around the torch stem. The plasma is maintained at atmospheric pressure and at an average temperature of about 8000 K. The ICP torch consists of three concentric tubes made from fused silica. The center tube carries the nebulizing gas, or the column eluent, from the gas chromatograph. Argon is used as the carrier gas, and the next tube carries an auxiliary supply of argon to help maintain the plasma and also to prevent the hot plasma from reaching the tip of the sample inlet tube. The outer tube also carries another supply of argon at a very high flow rate that cools the two inner tubes and prevents them from melting at the plasma temperature. The coupling coil consists of two to four turns of water cooled copper tubing, situated a few millimeters behind the mouth of the torch. The rf generator produces about 1300 W of rf at 27 or 40 MHz, which induces a fluctuating magnetic field along the axis of the torch. Temperature in the induction region of the torch can reach 10,000 K, but in the ionizing region, close to the mouth of the sample tube, the temperature is 7000–9000 K.

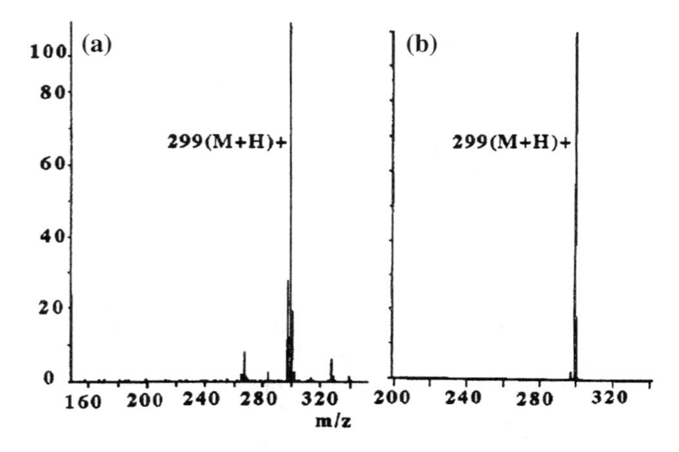

Fig. 2 Mass spectrum of methyl stearate produced by chemical ionization.

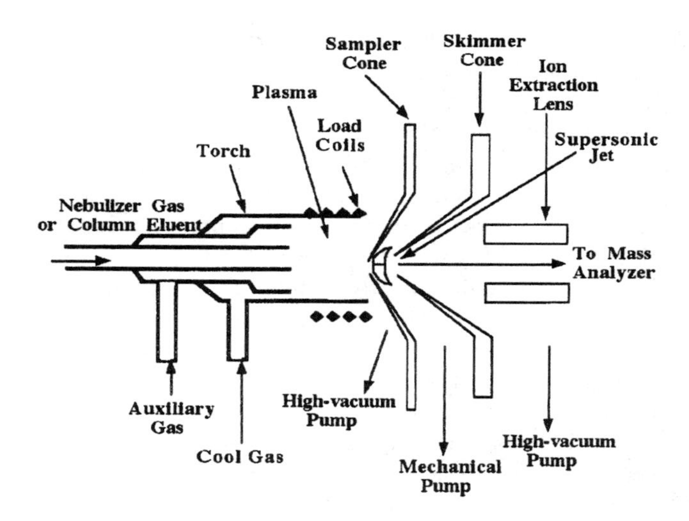

Fig. 3 ICP mass spectrometer ion source.

The sample atoms account for less than 10^{-6} of the total number of atoms present in the plasma region; thus, there is little or no self-quenching. At the plasma temperature, over 50% of most elements are ionized. The ions, once formed, pass through the apertures in the apex of two cones. The first has an aperture about 1 mm inner diameter (i.d.) and ions pass through it to the second skimmer cone. The space in front of the first cone is evacuated by a high-vacuum pump. The region between the first cone and the second skimmer cone is evacuated by a mechanical pump to about 2 mbar and, as the sample expands into this region, a supersonic jet is formed. This jet of gas and ions flows through a slightly smaller orifice into the apex of the second cone. The emerging ions are extracted by negatively charged electrodes (−100 to −600 V) into the focusing region of the spectrometer, and then into the mass analyzer.

The ICP ion source has the advantages that the sample is introduced at atmospheric pressure, the degree of ionization is relatively uniform for all elements, and singly-charged ions are the principal ion product. Furthermore, sample dissociation is extremely efficient and few, if any, molecular fragments of the original sample remain to pass into the mass spectrometer. High ion populations of trace components in the sample are produced, making the system extremely sensitive. Nevertheless, there are some disadvantages: the high gas temperature and pressure evoke an interface design that is not very efficient and only about 1% of the ions that pass the sample orifice pass through the skimmer orifice. Furthermore, some molecular ion formation does occur in the plasma, the most troublesome being molecular ions formed with oxygen. These can only be reduced by adjusting the position of the cones, so that only those portions of the plasma where the oxygen population is low are sampled.

Although the detection limit of an ICP–MS is about 1 part in a trillion, as already stated, the device is rather inefficient in the transport of the ions from the plasma to the analyzer. Only about 1% pass through the sample and skimming cones and only about 10^{-6} ions will eventually reach the detector. One reason for ion loss is the diverging nature of the beam, but a second is due to space-charge effects, which, in simple terms, is the mutual repulsion of the positive ions away from each other. Mutual ion repulsion could also be responsible for some non-spectroscopic interelement interference (i.e., matrix effects). The heavier ions having greater momentum suffer less dispersion than the lighter elements, thus causing a preferential loss of the lighter elements.

Mass Spectrometers for MS–GC Tandem Operation

The most common mass spectrometer used in GC–MS systems is the quadrupole mass spectrometer, either as a single quadrupole or as a triple quadrupole, which can also provide MS–MS spectra. A diagram of a quadrupole mass spectrometer is shown in Fig. 4. The operation of the quadrupole mass spectrometer is quite different from that of the sector instrument. The instrument consists of four rods which must be precisely straight and parallel and so arranged that the beam of ions is directed axially between them. Theoretically, the rods should have a hyperbolic cross section, but in practice, less expensive cylindrical rods are nearly as satisfactory. A voltage comprising a DC component (U) and a rf component ($V_0 \cos wt$) is applied between adjacent rods, opposite rods being electrically connected. Ions are accelerated into the center, between the rods, by a potential ranging from 10 to 20 V. Once inside the quadrupole, the ions oscillate in the x and y dimensions induced by the high-frequency electric field. The mass range is scanned by changing U and V_0 while keeping the ratio U/V_0 constant. The quadrupole mass spectrometer is compact, rugged, and easy to operate, but its mass range does not extend to very high values. However, under certain circumstances, multiply-charged ions can be generated and identified by the mass spectrometer. This, in effect, increases the mass range of the device proportionally to the number of charges on the ion.

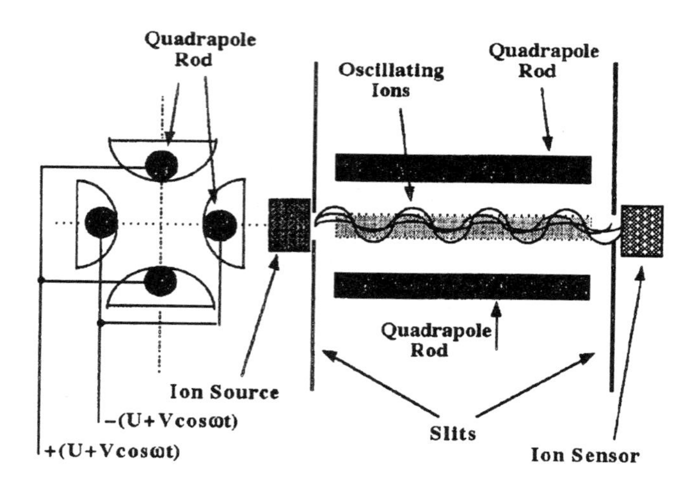

Fig. 4 Quadrupole mass spectrometer.

The quadrupole mass spectrometer can also be constructed to provide MS–MS spectra by combining three quadrupole units in series. A diagram of a triple quadrupole mass spectrometer is shown in Fig. 5. The sample enters the ion source and is usually fragmented by either an electron-impact or chemical ionization process. In the first analyzer, the various charged fragments are separated in the usual way, which then pass into the second quadrupole section, sometimes called the collision cell. The first quadrupole behaves as a straightforward mass spectrometer. Instead of the ions passing to a sensor, the ions pass

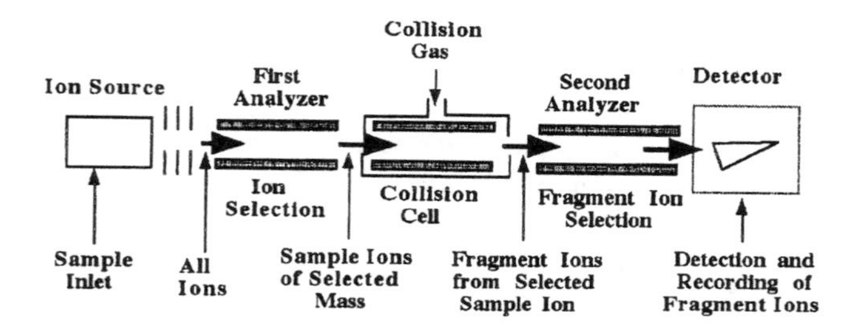

Fig. 5 Triple quadrupole mass spectrometer.

into a second mass spectrometer and a specific ion can be selected for further study. In the center quadrupole section, the selected ion is further fragmented by collision ionization and the new fragments pass into the third quadrupole, which functions as a second analyzer. The second analyzer resolves the new fragments into their individual masses producing the mass spectrum. Thus, the exclusive mass spectrum of a particular molecular or fragment ion can be obtained from the myriad of ions that may be produced from the sample in the first analyzer. This is an extremely powerful analytical system that can handle exceedingly complex mixtures and very involved molecular structures.

Another form of the quadrupole mass spectrometer is the ion trap detector, which has been designed more specifically as a chromatography detector than for use as a tandem instrument. The electrode orientation of the quadrupole ion trap mass spectrometer is shown in Fig. 6. The ion trap mass spectrometer has an electrode arrangement that consists of three cylindrically symmetrical electrodes comprised of two end caps and a ring. The device is small, the opposite internal electrode faces being only 2 cm apart. Each electrode has accurately machined hyperbolic internal faces. An rf voltage together with an additional DC voltage is applied to the ring, and the end caps are grounded. The rf voltage causes rapid reversals of field direction, so any ions are alternately accelerated and decelerated in the axial direction and vice versa in the radial direction. At a given voltage, ions of a specific mass range are held oscillating in the trap. Initially, the electron beam is used to produce ions, and after a given time, the beam is turned off. All the ions, except those selected by the magnitude of the applied rf voltage, are lost to the walls of the trap, and the remainder continue oscillating in the trap. The potential of the applied rf voltage is then increased, and the ions sequentially assume unstable trajectories and leave the trap via the aperture to the sensor. The ions exit the trap in order of their increasing m/z values. The first ion trap mass spectrometers were not very efficient, but it was found that the introduction of traces of helium to the ion trap significantly improved the quality of the spectra. The improvement appeared to result from ion–helium collisions that reduced the energy of the ions and allow them to concentrate in the center of the trap. The spectra produced are quite satisfactory for solute identification by comparison with reference spectra. However, the spectrum produced for a given substance will probably differ considerably from that produced by the normal quadrupole mass spectrometer.

The time-of-flight mass spectrometer was invented many years ago, but the performance of the modern version is greatly improved. A diagram of the time-of-flight mass spectrometer is shown in Fig. 7. In a time-of-flight mass spectrometer, the following relationship holds:

$$t = \left(\frac{m}{2zeV}\right)^{1/2} L$$

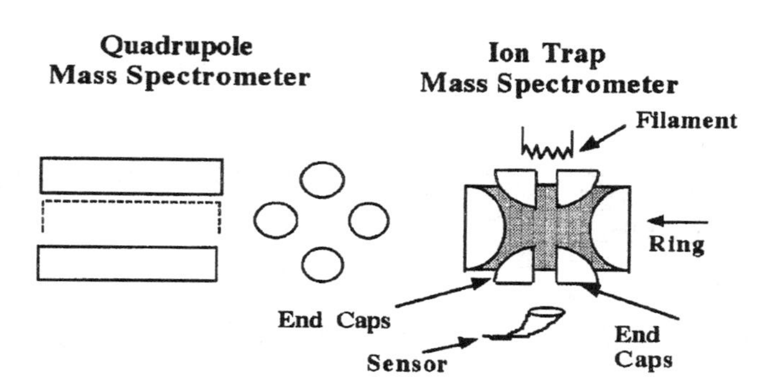

Fig. 6 Pole arrangement for the quadrupole and ion trap mass spectrometers.

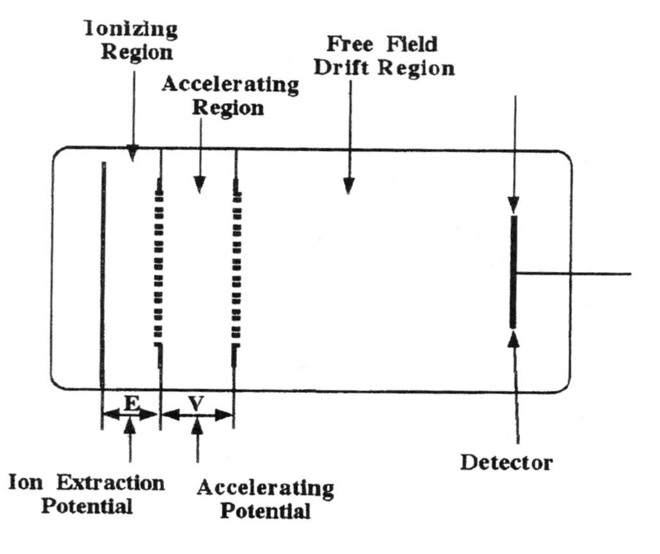

Fig. 7 The time-of-flight mass spectrometer. (Courtesy of VG Organic Inc.)

where t is the time taken for the ion to travel a distance L, V is the accelerating voltage applied to the ion, and L is the distance traveled by the ion to the ion sensor.

The mass of the ion is directly proportional to the square of the transit time to the sensor. The sample is volatilized into the space between the first and second electrodes and a microsecond burst of electrons is allowed to produce ions. An extraction voltage is then applied for another short time period, which, as those further from the second electrode will experience a greater force than those closer to the second electrode, will focus the ions. After focusing, the accelerating potential (V) is applied for about 100 ns so that all the ions in the source are accelerated almost simultaneously. The ions then pass through the third electrode into the drift zone and are then collected by the sensor electrode. The particular advantage of the time-of-flight mass spectrometer is that it is directly compatible with surface desorption procedures. Consequently, it can be employed with laser-desorption and plasma-desorption techniques. An excellent discussion on general organic mass spectrometry is given in *Practical Organic Mass Spectrometry* edited by Chapman [1].

The combination of the gas chromatograph with the single quadrupole mass spectrometer or with the triple quadrupole mass spectrometer are the most commonly used tandem systems. They are used extensively in forensic chemistry, in pollution monitoring and control, and in metabolism studies. The quadrupole mass spectrometers provide both high sensitivity and good mass spectrometric resolution. They can be readily used with open-tubular columns, and an example of the use of the single quadrupole monitoring a separation from an open-tubular column is shown in Fig. 8. The column was 30 m long with a 0.25-mm i.d. and carried a 0.5-mm film of stationary phase. A 1-mL sample was used and the column was programmed from 50°C to 300°C at 10°C/min.

An elegant example of the use of GC–MS in the analysis of pesticides in river water is given by Vreuls et al. [2]. A 1-mL sample was collected in an LC sample loop and the internal standard added. The sample was then displaced through a short column 1 cm long with a 2-mm i.d. packed with 10-mm particles of a proprietary PLRP-S adsorbent (styrene–divinylbenzene copolymer) by a stream of pure water. The extraction column was then dried with nitrogen and the adsorbed materials displaced into a gas chromatograph with 180 mL of ethyl acetate. The sample was passed through a short retention gap column and then to a retaining column. The GC oven was maintained at 70°C so that the ethyl acetate passed through the retaining column and was vented to waste. The solutes of interest were held in the retaining column at this temperature during the removal of the ethyl acetate. The temperature

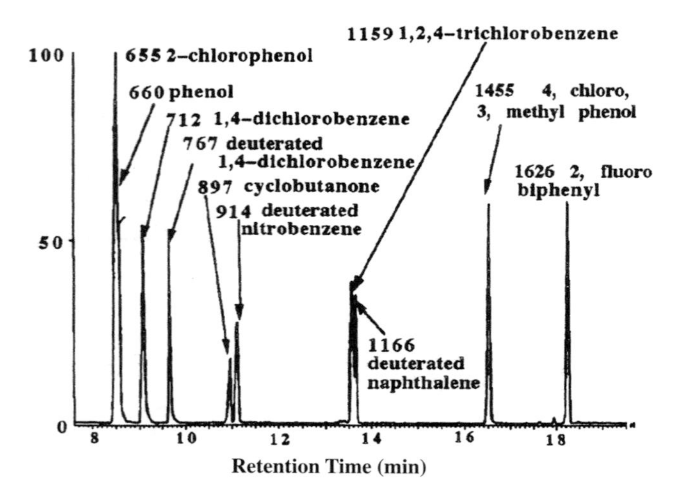

Fig. 8 A separation from an open-tubular column monitored by a single quadrupole mass spectrometer.

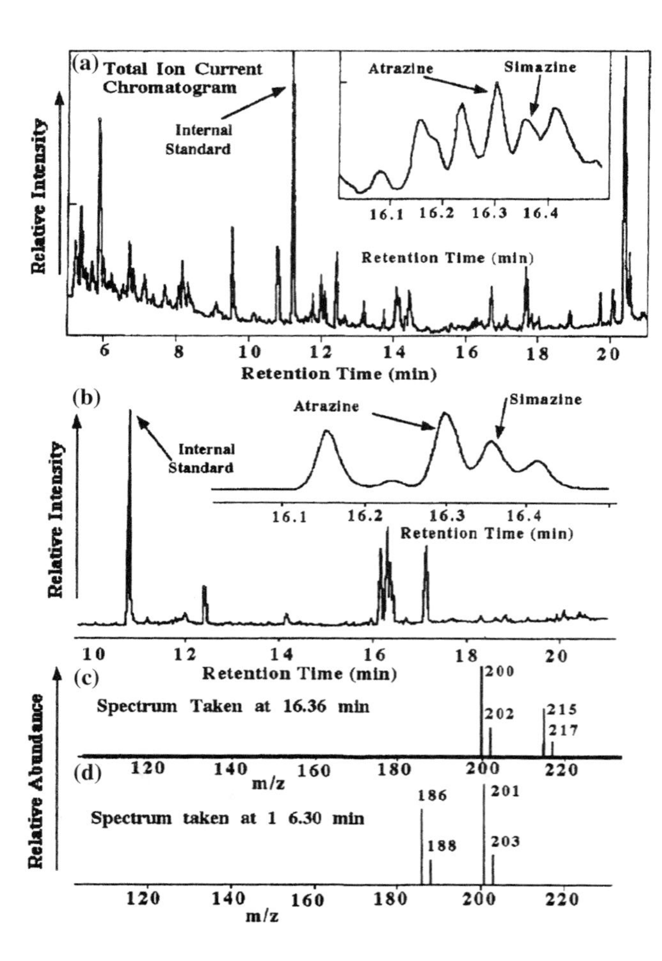

Fig. 9 Chromatogram and spectra from a sample of river water containing 200 ppt of atrazine and simazine. (From Ref. 2.)

was then increased and the residual material separated on an analytical column using an appropriate temperature program. The eluents from the analytical column passed to a quadrupole mass spectrometer. An example of the chromatograms and spectra obtained are shown in Fig. 9. Figure 9a shows the total ion current chromatogram from a sample of Rhine River water containing 200 ppt of the herbicides atrazine and simazine. The pertinent peaks are shown enlarged in the inset. Figure 9b shows a section of the same chromatogram presented in the selected ion mode. It is seen that the herbicide peaks are clearly and unambiguously revealed. In Figs. 9c and 9d, the individual mass spectra of atrazine (eluted at 16.30 min) and simazine (eluted at 16.36 min) are shown. The spectra are clear and more than adequate to confirm the identity of the two herbicides.

References

1. J. R. Chapman (ed.), *Practical Organic Mass Spectrometry*, John Wiley & Sons, New York, 1994.
2. J. J. Vreuls, A.-J. Bulterman, R. T. Ghijsen, and U. Th. Brinkman, *Analyst 117*: 1701 (1992).

Suggested Further Reading

Message, G. M., *Practical Aspects of GC/MS*, John Wiley & Sons, New York, 1984.

Scott, R. P. W., *Tandem Techniques*, John Wiley & Sons, New York, 1984.

Raymond P. W. Scott

Golay Dispersion Equation for Open-Tubular Columns

The open-tubular column or capillary column is the one most commonly used in gas chromatography (GC) today. The equation that describes dispersion in open tubes was developed by Golay [1], who employed a modified form of the rate theory, and is similar in form to that for packed columns. However, as there is no packing, there can be no multipath term and, thus, the equation only describes two types of dispersion. One function describes the longitudinal diffusion effect and two others describe the combined resistance to mass-transfer terms for the mobile and stationary phases. The Golay equation takes the following form:

$$H = \frac{2D_m}{u} + \frac{f_1(k')r^2}{D_m}u + \frac{f_2(k')r^2}{K^2 D_s}u \tag{1}$$

where H is the height of a theoretical plate or the variance/unit length, D_m is the diffusivity of the solute in the mobile phase, D_s is the diffusivity of the solute in the stationary phase, r is the column radius, k' is the capacity ratio of the solute, K is the distribution coefficient of the solute, and u is the mobile-phase linear velocity.

Open-tubular columns behave in exactly the same way as packed columns with respect to pressure. The same mathematical arguments can be educed which results in the modified form of the equation shown in Eq. (2). As the column is geometrically simple, the respective functions of k' can also be explicitly developed.

$$H = \frac{2D_m}{u_0} + \frac{(1 + 6k' + 11k'2)r^2}{24(1 + k')^2 D_{m(0)}}u_0 + \frac{2k'\,df^2}{3(1 + k')^2 D_s(\gamma + 1)}u_0 \tag{2}$$

where u_0 is the exit velocity of the mobile phase and $D_{m(0)}$ is the diffusivity of the solute measured at the exit pressure. As the film is thin, $r \gg df$; then,

$$\frac{(1 + 6k' + 11k'^2)r^2}{24(1 + k')^2 D_{m(0)}} \gg \frac{2k'\,df^2}{3(1 + k')^2 D_s(\gamma + 1)}$$

and, thus,

$$H = \frac{2D_{m(0)}}{u} + \frac{(1 + 6k' + 11k'^2)r^2}{24(1 + k')^2 D_{m(0)}}u_0 \tag{3}$$

By differentiating Eq. (3) and equating it to zero, expressions can be obtained for u_{opt} and H_{min} in a manner similar to the method used for a packed column:

$$u_{0(opt)} = 2\frac{D_{m(0)}}{r}\left(\frac{12(1 + k')^2}{1 + 6k' + 11k'^2}\right)^{1/2} \tag{4}$$

$$H_{min} = \frac{r}{2}\left(\frac{1 + 6k' + 11k'^2}{3(1 + k')^2}\right)^{1/2} \tag{5}$$

The approximate efficiency of a capillary column operated at its optimum velocity (assuming the inlet/outlet pressure ratio is small) can be simply calculated. If only the dead volume is considered (i.e., $k' = 0$), Eq. (3) reduces to

$$H = \frac{2D_m}{u} + \frac{1}{24}\frac{r^2}{D_m}u \quad (6)$$

Differentiating and equating to zero,

$$\frac{dH}{du} = -\frac{2D_m}{u^2} + \frac{1}{24}\frac{r^2}{D_m} = 0 \quad \text{or} \quad u = \frac{\sqrt{48}D_m}{r}$$

Substituting for u in Eq. (6) and simplifying,

$$H = \frac{2D_m r}{\sqrt{48}D_m} + \frac{1}{24}\frac{r^2}{D_m}\frac{\sqrt{48}D_m}{r}$$
$$= 0.289r + 0.289r = 0.577r$$

Thus, the efficiency of a capillary column of length (l) can be assessed as

$$n = \frac{l}{0.6r} \quad (7)$$

The column efficiency will be inversely proportional to the column radius and the analysis time will directly proportional to the column radius and inversely proportional to the diffusivity of the solute in the mobile phase.

Reference

1. M. J. E. Golay, *Gas Chromatography. 1958* (D. H. Desty, ed.), Butterworths, London, 1958, p. 36.

Suggested Further Reading

Scott, R. P. W., *Techniques and Practice of Chromatography*, Marcel Dekker, Inc., New York, 1996.
Scott, R. P. W., *Introduction to Analytical Gas Chromatography*, Marcel Dekker, Inc., New York, 1998.

Raymond P. W. Scott

GPC–SEC: Introduction and Principles

Introduction

The basic principle of chromatography involves the introduction of the sample into a stream of mobile phase that flows through a bed of a stationary phase. The sample molecules will distribute so that each spends some time in each phase. *Size-exclusion chromatography* (SEC) is a liquid column chromatographic technique which separates molecules on the basis of their sizes or hydrodynamic volumes with respect to the average pore size of the packing. The stationary phase consists of small polymeric or silica-based particles that are porous and semirigid to rigid. Sample molecules that are smaller than the pore size can enter the stationary-phase particles and, therefore, have a longer path and longer retention time than larger molecules that cannot enter the pore structure. Very small molecules can enter virtually every pore they encounter and, therefore, elute last. The sizes, and sometimes the shapes, of the mid-size molecules regulate the extent to which they can enter the pores. Larger molecules are excluded and, therefore, are rapidly carried through the system. The porosity of the packing material can be adjusted to exclude all molecules above a certain size. SEC is generally used to separate biological macromolecules and to determine molecular-weight distributions of polymers.

History

It is not obvious who was the first to use SEC. However, the first effective separation of polymers based on *gel filtration chromatography* (GFC) appears to be that reported by Porath and Flodin [1]. Porath and Flodin employed insoluble cross-linked polydextran gels, swollen in aqueous medium, to separate various water-soluble macromolecules. GFC generally employs aqueous solvents and hydrophilic column packings, which swell heavily in water. Moreover, at high flow rates and pressures, these lightly cross-linked soft gels have low mechanical stability and collapse. Therefore, GFC stationary phases are generally used with low flow rates to minimize high-back pressures. GFC is mainly used for biomolecule separations at low pressure [2].

Moore described an improved separation technique relative to GFC and introduced the term *gel permeation chromatography* (GPC) in 1964 [3]. GPC performs the same separation as GFC, but it utilizes organic solvents and hydrophobic packings. Moore developed rigid polystyrene gels, cross-linked with divinylbenzene, for separating synthetic polymers soluble in organic media. These extensively cross-linked gels are mechanically stable enough to withstand high pressures and flow rates. The more rugged GPC quickly flourished in industrial laboratories where polymer characterization and quality control are of primary concern. Since its introduction in the 1960s, the understanding and utility of GPC has substantially evolved. GPC has been widely used for the determination of molecular weight (MW) and molecular-weight distribution (MWD) for numerous synthetic polymers [4].

Other names such as gel chromatography, exclusion chromatography, molecular sieve chromatography, gel exclusion chromatography, size separation chromatography, steric exclusion chromatography, and restricted diffusion chromatography have been utilized to reflect the principal mechanism for the separation. The fundamental mechanism of this chromatographic method is complex and certainly will not be readily incorporated into one term. Strong arguments have been made for many of the above-listed titles [5]. In an attempt to minimize the dispute over the proper name, the term SEC will be used in this entry, as it appears to be the most widely used.

Mechanism

Size-exclusion chromatography is a liquid chromatography technique in which a polymer sample, dissolved in a solvent, is injected into a packed column (or a series of packed columns) and flows through the column(s) and its concentration as a function of time is determined by a suitable detector. The column packing material differentiates SEC from other liquid chromatography techniques where sample components primarily separate by differential adsorption and desorption. The SEC packing consists of a polymer, generally polystyrene, which is chemically cross-linked so that varying size pores are created. Several models are discussed by Barth et al. [6] to illustrate SEC separation theory. A rather simplified separation mechanism is described here. A polymer sample dissolved in the SEC mobile phase is injected in the chromatographic system. The column eluent is monitored by a mass-sensitive detector, which responds to the weight concentration of polymer in the mobile phase. The most common detector for SEC is a differential refractometer. The raw data in SEC consists of a trace of detector response proportional to the amount of polymer in solution and the corresponding retention volume. A typical SEC sample chromatogram is depicted in Fig. 1. An SEC chromatogram generally is a broad peak representing the entire range of molecular weights in the sample. For synthetic polymers, this can extend from a few hundred mass units up to a million or more. The average molecular weight can be calculated in a number of ways. Both natural and synthetic polymers are molecules containing a distribution of molecular weights. The most commonly calculated molecular-weight averages using SEC are the weight-average molecular weight (M_w) and number-average molecular weight (M_n). These terms have been well defined by Cazes [7].

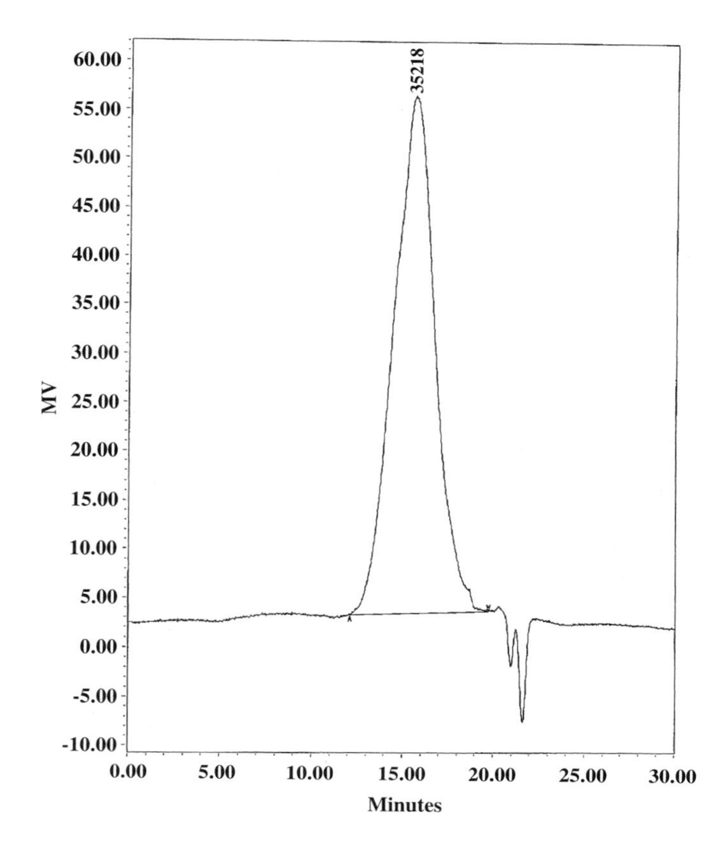

Fig. 1 Typical size-exclusion chromatogram of a polymer sample. SampleName: 6B Vial: 15 Inj: 1 Ch: 410 Type: Broad Unknown.

The weight-average molecular weight is defined as

$$\overline{M}_w = \frac{\sum_{i=1}^{\infty} W_i M_i}{\sum_{i=1}^{\infty} W_i} \tag{1}$$

and the number-average molecular weight is defined as

$$\overline{M}_n = \frac{W}{\sum_{i=1}^{\infty} N_i} = \frac{\sum_{i=1}^{\infty} M_i N_i}{\sum_{i=1}^{\infty} N_i} \tag{2}$$

where W is the total weight of the polymer, W_i is the weight fraction of a given molecule i, N_i is the number of moles of each species i, and M_i is the molecular weight of each species i.

M_w is generally greater than or equal to M_n. The samples in which all of the molecules have a single molecular weight ($M_w = M_n$) are called monodisperse polymers. The degree of polydispersity (i.e., the ratio of M_w to M_n) describes the spread of the molecular-weight-distribution curve. The broader the SEC curve, the larger the polydispersity.

The detector response on the SEC chromatogram is proportional to the weight fraction of total polymer, and suitable calibration permits the translation of the retention volume axis into a logarithmic molecular-weight scale. Calibration of SEC is perhaps the most difficult aspect of the technique because polymer molecules are separated by size rather than by molecular weight. Size, in turn, is most directly proportional to the lengths of the polymer molecules in solution. A length, however, is proportional to molecular weight only within a single polymer type. An absolute SEC calibration would require the use of narrow molecular-weight range standards of the same polymer that is being analyzed. This is not always practical because a wide range of polymer types needs to be evaluated. SEC calibration is often achieved using the "universal calibration" technique, which assumes hydrodynamic volume is the sole determinate of retention time or volume [8]. A series of commercially available monodisperse molecular-weight polystyrenes are the most commonly used SEC calibration standards. If polystyrene standards are used to calibrate the analyses of any other type of polymer, the molecular weights obtained for a polymer sample are actually "polystyrene-equivalent" molecular weights. Numeric conversion factors are available for correlating "molecular weight per polystyrene length" to that of other polymers, but this approach only produces marginally better estimates of the absolute molecular weights. In addition, approaches such as these are usually invalid because the calibration curve for the polymer being analyzed does not often have the same shape as the curve generated with the polystyrene standards.

Size-exclusion chromatograms of narrow-distribution polystyrene standards along with a typical polystyrene calibration curve are shown in Fig. 2. The peak retention volume and corresponding molecular weights produce a calibration curve. With a calibration curve, it is possible to determine M_w and M_n for a polymer. The SEC curve of a polymer sample is divided into vertical segments of equal retention volume. The height or area of each segment and the corresponding average molecular weight, calculated from the calibration curve, are then used for M_w and M_n calculations. There are several commercially available software packages that simplify the calculation process for molecular-weight determinations.

Application

Until the mid-1960s, molecular-weight averages were determined only by techniques such as dilute solution viscosity, osmometry, and light scattering. Most of these techniques work best for polymers with a narrow MWD. None of these techniques, either alone or in combination, could readily identify the range of molecular weights in a given sample. SEC was introduced in the mid-1960s to determine MWDs and other properties of polymers. During the first two decades of SEC acceptance, the emphasis was on improving the fundamental aspects of chromatography, such as column technology, optimizing solvents, and the precision of analysis. Over the past 10 years, there has been an increasing demand for deriving more information from SEC, driven by the need to characterize, more fully, an increasingly complex array of new polymers. Significant developments in SEC detection systems include light scattering, viscometry, and matrix-assisted laser desorption ionization time-of-flight (MALDI–TOF) mass spectrometry and, most recently, nuclear magnetic resonance (NMR) detection in conjunction with SEC for determining MW and chemical composition of polymers. The use of SEC for measuring physiological properties of polymers, especially biopolymers, has become an important area of research.

Finally, SEC is merely a separation technique based on differences in hydrodynamic volumes of molecules. No direct measurement of molecular weight is made. SEC itself does not render absolute information on molecular weights and their distribution or on the structure of the polymers studied without the use of more specialized detectors (e.g., viscometry and light scattering). With these detectors, a "self-calibration" may be achieved for each polymer sample while it is being analyzed by SEC. However, it is possible to calibrate the elution time in relation to molecular weight of known standards. With proper column calibration, or by the use of molecular-weight-sensitive detectors such as light scattering, viscometry, or mass spectrometry, MWD and average molecular weights can be obtained readily [6]. The combined use of concentration sensitive and molecular-weight-sensitive detectors has greatly improved the accuracy and precision of SEC measurements. Thus, SEC has become an essential technique that provides valuable molecular-weight informa-

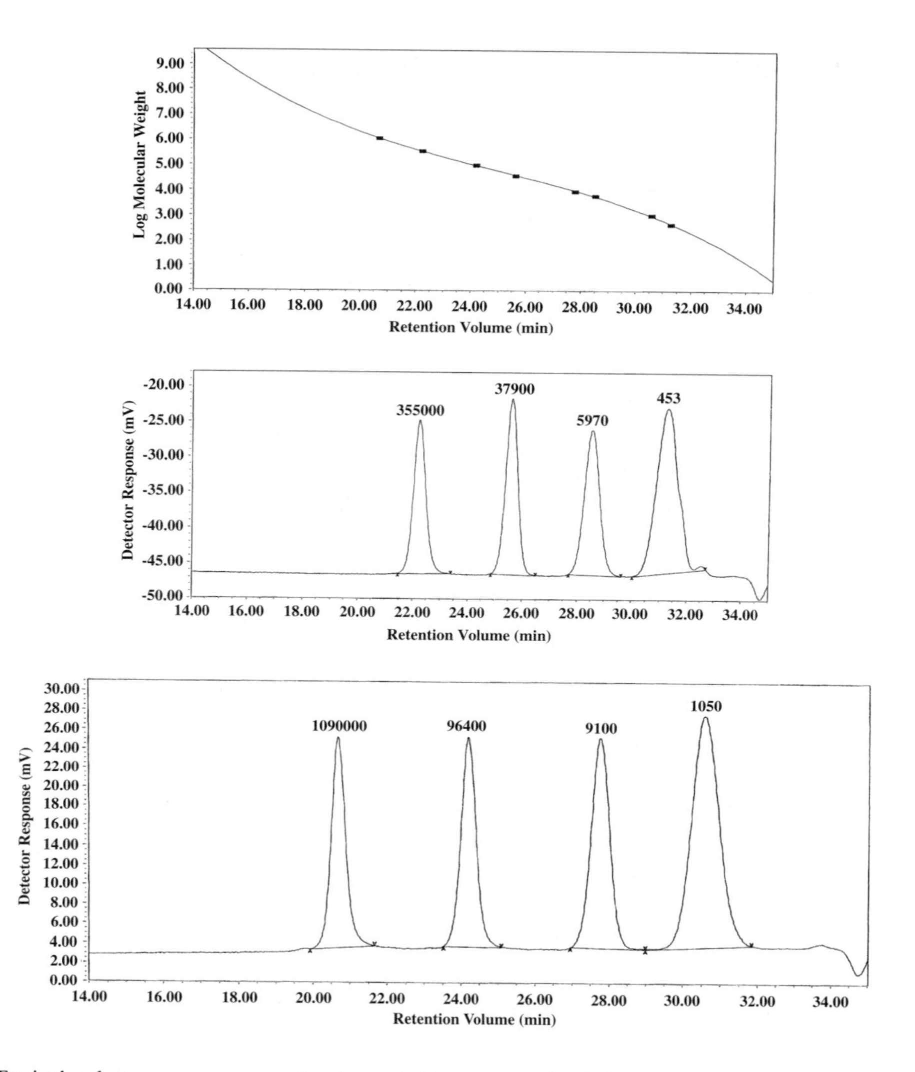

Fig. 2 Typical polystyrene narrow molecular-weight range standard chromatograms and calibration curve.

tion, which can be related to polymer physical properties, chemical resistance, and processability.

References

1. J. Porath and P. Flodin, *Nature 183*: 1657 (1959).
2. A. V. Danilov, I. V. Vagenina, L. G. Mustaeva, S. A. Moshnikov, E. Y. Gorbunova, V. V. Cherskii, and M. B. Baru, *J. Chromatogr. A 773*: 103 (1997).
3. J. C. Moore, *J. Polym. Sci., Part A 2*: 835 (1964).
4. V. S. Lafita, Y. Tian, D. Stephens, J. Deng, M. Meisters, L. Li, B. Mattern, and P. Reiter, *Proc. Int. GPC Symp. 1998*, Waters Corp.; Milford, MA, 1998, pp. 474–490.
5. J. Johnson, R. Porter, and M. Cantow, *J. Macromol. Chem., Part C 1*: 393 (1966).
6. H. G. Barth, B. E. Boyes, and C. Jackson, *Anal. Chem. 70*: 251R (1998).
7. J. Cazes, *J. Chem. Educ., 43*: A567 (1966).
8. R. H. Boyd, R. R. Chance, and G. Ver Strate, *Macromolecules 29*: 1182 (1996).

Vaishali Soneji Lafita

GPC–SEC Analysis of Nonionic Surfactants

Nonionic surfactants are one of the most important and largest surfactant groups. They are amphiphilic molecules composed, in most cases, of poly(oxyethylene) blocks (PEO) as the water-soluble fragment and fatty alcohols, fatty acids, alkylated phenol derivatives, or various synthetic polymers as the hydrophobic part [1]. This class of surfactants is widely used as surface wetting agents, emulsifiers, detergents, phase-transfer agents, and solubilizers for diverse industrial and biomedical applications [2]. Several of them have been used for many years under different trade names: Brij (ethoxylated fatty alcohols), Synperonic (PEO copolymers) and Tween (ethoxylated sorbitan esters) by ICI Surfactants; Igepal (PEO copolymers) by Rhone-Poulenc, Rhodia; Pluronic [poly(oxyethylene)-*block*-poly(oxypropylene) copolymers] by BASF; Triton DF (ethoxylated fatty alcohols) and Triton X (ethoxylated octylphenols) by Union Carbide; and others. The exploitation characteristics of nonionic surfactants depend on the oligomer distribution, the molecular-weight characteristics of the constituent blocks, and the hydrophilic/hydrophobic ratio of their chemical composition. Therefore, the quantitative determination of these factors is of primary importance for their performance evaluation. Several high-performance liquid chromatography (HPLC) separation techniques have been used in combination with different detection methods to characterize poly(ethylene glycol)s and their amphiphilic derivatives [3]. Size-exclusion chromatography (SEC) is a particularly attractive analytical tool for the investigation of nonionic surfactants because it can provide information for their composition, molecular weight, and molecular-weight distribution along with their micellization in selective solvents. This entry will survey briefly both applications with major emphasis on the choice of the most appropriate eluent and stationary phase.

Molecular-weight determinations are performed in good solvents for both blocks of the PEO copolymers. The most widely used analysis conditions are as follows: eluents — tetrahydrofuran (THF) and chloroform (CHL); flow rate — 1.0 mL/min; detection — differential refractive index (dRI) detector. The temperature interval is between 20°C and 40°C. The stationary phase is typically a polystyrene–divinylbenzene cross-linked matrix supplied by different vendors: Phenogel (Phenomenex, USA); PL Gel (Polymer Laboratories, UK), PSS Gel (Polymer Standards Service, Germany), TSKgel (tosoHaas, U.S.A.); UltraStyragel (Waters Corporation, U.S.A.); and others. The pore size range of the column set should be adjusted to the molecular-weight range of the investigated materials. The SEC analysis in THF requires low-dRI response correction factors (1.66 for M_n = 106 and 1.00 for M_n = 20,000 [4]), whereas the low-molecular-weight PEO derivatives are almost invisible in CHL. On the other side, the solubility of PEO in THF decreases with the molecular weight, as evidenced by steeper calibration curves and broadening of the peaks in the eluograms (Fig. 1). This complicates the precise molecular weight calculations of PEO copolymers and the choice of calibration standards becomes crucial [5]. The increasing content of PEO in the copolymer results in lower hydrodynamic volumes in THF and, consequently, yields lower apparent molecular weights, regardless of the macromolecular architecture of the analytes [6]. The problems with calibration mismatch can be avoided to some extent by using the universal calibration approach with on-line differential viscome-

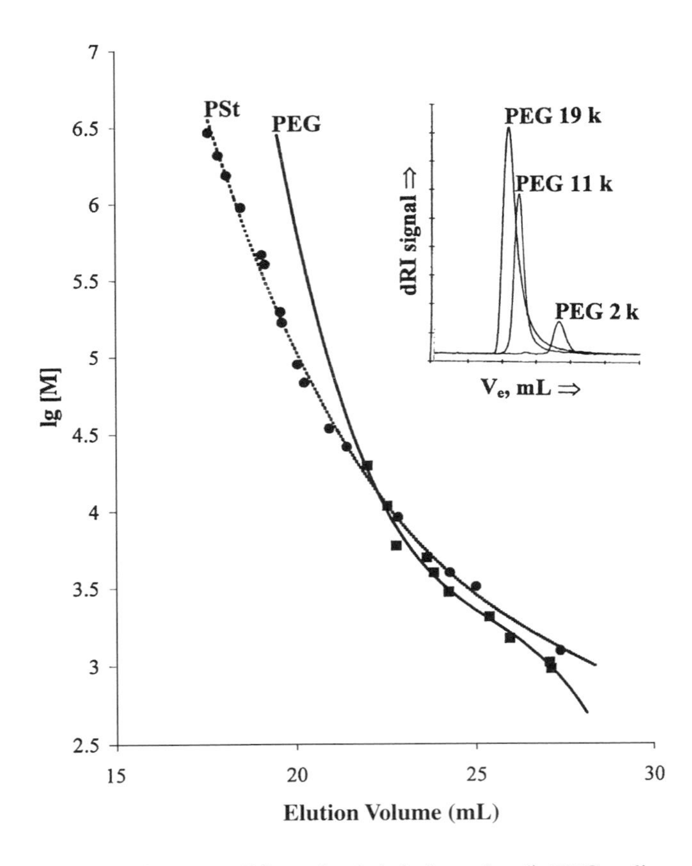

Fig. 1 Polystyrene, PSt, and poly(ethylene glycol), PEG, calibration curves obtained on PL Gel column set (Mixed C, 10^3 Å, 5×10^2 Å, and 10^2 Å). Eluent: THF; flow rate: 1 mL/min; temperature: 40°C. Inset: SEC elution profiles of poly(ethylene glycol)s (PEGs) at the same analysis conditions.

try. This method provides accurate molecular-weight information for most linear and comb-graft PEO copolymers but has been proved less precise for copolymers with complex linear-dendritic or hyperbranched architecture. On-line laser-light-scattering detection eliminates the need for a calibration curve and yields correct M_w values for macromolecules with molecular weights higher than 500 g/mol. The precision of the method depends largely on accurate *dn*/*dc* values that need to be measured for each copolymer investigated.

The self-assembly process of nonionic surfactants in aqueous media differs in several aspects from the micellization of amphiphilic copolymers:

1. Micelles constructed of low-molecular-weight surfactants have much lower molecular weights than that of polymeric ones.
2. The critical micelle concentration (CMC) is much lower for polymer surfactants.
3. The vast majority of polymer micelles have spherical shape in dilute or semidilute solutions, whereas low-molecular-weight surfactants form structures that are strongly concentration dependent: lamellae, sheets, rods, and spheres.
4. The kinetics of micelle formation and the dynamics of the micelle–unimer equilibrium are considerably slower for polymeric surfactants.

The last factor is particularly important for the successful utilization of aqueous SEC in the investigation of the micellization process. In order to provide a realistic picture for the micellization equilibrium, the chromatographic system needs to meet several strict requirements: The packing material and the eluent should be appropriately chosen to prevent the occurrence of non-size-exclusion phenomena and adsorption of micelles in the stationary phase. In all cases, mass balance of the material injected and recovered from the columns must be performed in order to verify the absence copolymer entrapment. Cross-linked copolymers containing either poly(vinyl alcohol) or poly(glycidyl methacrylate) as the hydrophilic component are the most widely used column packing materials for aqueous SEC. Both Shodex Protein KW (Showa Denko, Japan) and Micropak TSK-gel PW (Toyo Soda, Japan) columns have been reported to afford good information on the micellization behavior of PEO copolymers without interference of side effects [7]. The mobile phase is methanol–water (1:1, v/v) or pure water eluting at 1.0 mL/min. The analysis temperature is between 22°C and 40°C. Historically, the major concern for the use of SEC in the investigation of micellar systems has always been the lack of suitable calibration and, consequently, the inability of the method to furnish accurate information for the size (hydrodynamic volume) of the micelles and their molecular weight. The incorporation of light-scattering and viscosity detectors for on-line measurement seems to eliminate this problem. With no solute–column interaction present and slow unimer–micelle equilibrium, a multiangle light–scattering detector (DAWN-DSP, Wyatt Technology, U.S.A.) provides accurate information for the molecular weight and radius of gyration of the micelle using Zimm's formalism:

$$\frac{R_\theta}{K^*c} = M_w P(\theta) - 2A_2 c M_w^2 P^2(\theta)$$

where R_θ is the excess Rayleigh ratio, K^* is an optical constant that includes the differential refractive index increment (*dn*/*dc*) of the solvent–solute mixture, c is the concentration of the solute molecules in the analyzed solution, M_w is the weight-average molecular weight, $P(\theta)$ is a form factor, and A_2 is the second virial coefficient. If the *dn*/*dc* value for the copolymer above CMC is known, the extrapolation to zero concentration and zero angle will yield the Z-average of the radius of gyration and A_2, respectively. The double extrapolation to zero angle and zero concentration will afford M_w. The hydrodynamic radii can be calculated using an on-line viscometric detector (Viskotek, U.S.A., and Waters) and the following relationship:

$$R_\eta^3 = \frac{3[\eta]M}{10\pi N_A}$$

where the value of $[\eta]M$ could be extracted directly from the universal calibration and N_A is Avogadro's number (6.022×10^{23}). It should be pointed out, however, that this formula is strictly valid only for spherical structures. The hydrodynamic radius can also be measured by a combination of SEC and dynamic light scattering (Precision Detectors, U.S.A.) or NMR spectroscopy [8]. However, the same assumption for a spherical shape of the investigated macromolecules has to be made.

In conclusion, modern SEC is a versatile technique for the investigation of nonionic surfactants in aqueous and organic media. In combination with different spectroscopic and viscometric detectors it will provide useful information for the molecular weight characteristics, chemical composition and solution behavior of this important class of materials.

References

1. B. Jönsson, B. Lindman, K. Holmberg, and B. Kronberg, *Surfactants and Polymers in Aqueous Solution*, John Wiley & Sons, New York, 1998.
2. J. E. Glass (ed.), *Hydrophilic Polymers: Performance with*

Environmental Acceptability, Advances in Chemistry Series Vol. 248, American Chemical Society, Washington, DC, 1996.
3. K. Rissler, *J. Chromatogr. A 742*: 1 (1996).
4. S. Mori, *Anal. Chem. 50*: 1639 (1978).
5. B. Trathnigg, S. Feichtenhofer, and M. Kollroser, *J. Chromatogr. A 786*: 75 (1997), and references therein.
6. D. Taton, E. Cloutet, and Y. Gnanou, *Macromol. Chem. Phys. 199*: 2501 (1998); I. V. Berlinova, I. V. Dimitrov, and I. Gitsov, *J. Polym. Sci., Part A: Polym. Chem. 35*: 673 (1997); I. V. Berlinova, A. Amzil, and N. G. Vladimirov, *J. Polym. Sci., Part A: Polym. Chem. 33*: 1751 (1995).
7. I. V. Berlinova, N. G. Vladimirov, and I. M. Panayotov, *Makromol. Chem. Rapid Commun. 10*: 163 (1989); R. Xu, Y. Hu, M. A. Winnik, G. Riess, and M. D. Crocher, *J. Chromatogr. 547*: 434 (1991).
8. G. R. Newkome, J. K. Young, G. R. Baker, R. L. Potter, L. Audoly, D. Cooper, C. D. Weis, K. Morris, and C. S. Johnson, Jr., *Macromolecules 26*: 2394 (1993).

Ivan Gitsov

GPC–SEC: Effect of Experimental Conditions

In order to calculate the molecular-weight averages of a polymer from the SEC chromatogram, the relationship between the molecular weight and the retention volume (called the "calibration curve") needs to be known, unless a molecular-weight-sensitive detector is used. The retention volume of a polymer changes with changing experimental conditions; therefore, when molecular-weight averages of the polymer are calculated using the calibration curve, care must be taken with the effect of experimental conditions [1].

Sample concentration is one of the most important operating variables in SEC, because the retention volumes of polymers increase with increased concentration of the sample solution. The concentration dependence of the retention volume is a well-known phenomenon and the magnitude of the peak shift to higher retention volume is more pronounced for polymers with a higher molecular weights than for those with lower molecular weights. This phenomenon is almost improbable for polymers with a molecular weight lower than 10^4 and is observed ever at a low concentration, such as 0.01%, although the peak shift is smaller than that at a higher concentration.

In this sense, this concentration dependence of the retention volume should be called the "concentration effect," not "overload effect" or "viscosity effect." If a large volume of a sample solution is injected, an appreciable shift in retention volume is observed, even for low-molecular-weight polymers; this is called the "overload effect."

The retention volume increases with increasing concentration of the sample solution and the magnitude of the increase is related to the increasing molecular weight of the sample polymers [2]. The reason for the increase in retention volume with increasing polymer concentration is considered to result from the decrease in the hydrodynamic volume of the polymer molecules in the solution

Molecular-weight averages calculated with calibration curves of varying concentrations may differ in value. As the influence of the sample concentration on the retention volume is based on the essential nature of the hydrodynamic volume of the polymer in solution, it is necessary to select experimental conditions that will reduce the errors produced by the concentration effect.

By rule of thumb, the preferred sample concentrations, if two SEC columns of 8 mm inner diameter (i.d.) $\times$ 25 cm in length are used, are as follows. The sample concentrations should be as low as possible and no more than 0.2%. For high-molecular-weight polymers, concentrations less than 0.1% are often required, and for low-molecular-weight polymers, concentrations of more than 0.2% are possible. The concentrations of polystyrene standards for calibration should be one-half of the unknown sample concentration. For polystyrene standards with a molecular weight over 10^6, it is preferable that they are one-eighth to one-tenth and for those with a molecular weights between 5×10^5 and 10^6, a quarter to one-fifth of the sample concentration.

The retention volume of a polymer sample increases as the injection volume increases [3]. In some cases, the increase in the retention volume from an injection volume increase from 0.1 to 0.25 mL was 0.65 mL, whereas that from 0.25 to 0.5 mL was only 0.05 mL, suggesting that a precise or constant injection is required even if the injection volume is as small as 0.1 or 0.05 mL. In view of the significant effect of the injection volume on the retention volume, it is important to use the same injection volume

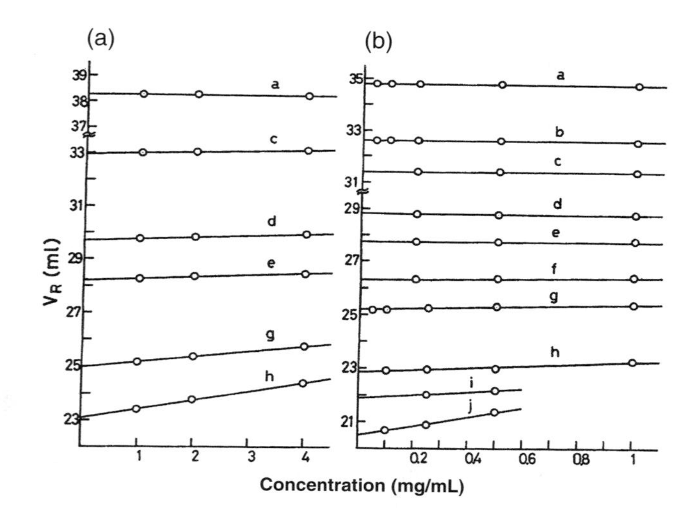

Fig. 1 Concentration dependence of retention volume for polystyrene in good solvents on polystyrene gel columns: (a) in toluene on microstyragel columns (3/8 in. × 1 ft × 4) (10^6, 10^5, 10^4, and 10^3 nominal porosity) at a flow rate 2 mL/min and injected volume 0.25mL; (b) in Tetrahydrofuran on Shodex A 80*M* columns (8 mm × 50 cm × 2) (mixed polystyrene gels of several nominal porosities) at a flow rate 1.5 mL/min and injected volume 0.25 mL. Molecular weight of polystyrene standards: (a) 2100; (b) 10,000; (c) 20,400; (d) 97,200; (e) 180,000; (f) 411,000; (g) 670,000; (h) 1,800,000; (i) 3,800,000; (j) 8,500,000.

for the sample under examination as that used when constructing the calibration curve. The use of a loop injector is essential, and the same injection volume must be employed for all sample solutions including calibration standards, regardless of their molecular-weight values. The increase in the injection volume results in a decrease in the number of theoretical plates, due to band broadening, which means that the calculated values of the molecular-weight averages and distribution deviate from the true values (Fig. 1).

The retention volume in SEC increases with increasing flow rate [3]. This is attributed to nonequilibrium effects, because polymer diffusion between the intrapores and extrapores of gels is sufficiently slow that equilibrium cannot be attained at each point in the column. With a decreasing flow rate, the efficiency and the resolution are increased. Bimodal distribution of a PS standard (NBS706) with a narrow molecular weight distribution was clearly observed at the lower flow rate.

Separation of molecules in SEC is governed, mainly, by the entropy change of the molecules between the mobile phase and the stationary phase, and the temperature independence of peak retention can be predicted. However, an increase in retention volume with increasing column temperature is often observed. A temperature difference of 10°C results in a 1% increase in the retention volume, which corresponds to a 10–15% change in molecular weight [4].

Two main factors that cause retention-volume variations with column temperature are assumed: an expansion or a contraction of the mobile phase in the column and the secondary effects of the solute to the stationary phase. When the column temperature is 10°C higher than room temperature, the mobile phase (temperature of the mobile phase is supposed to be the same as room temperature in this case) will expand about 1% from when it entered the columns, resulting in an increase in the real flow rate in the column due to the expansion of the mobile phase and the decrease in the retention volume. The magnitude of the retention-volume dependence on the solvent expansion is evaluated to be about one-half of the total change in the retention volume. The residual contribution to the change in retention volume is assumed to be that due to gel–solute interactions such as adsorption.

In order to obtain accurate and precise molecular-weight averages, the column temperature, as well as the difference of both temperatures, the solvent reservoir and the column oven, must be maintained.

Other factors affecting retention volume are the viscosity of the mobile phase, the sizes of gel pores, and the effective size of the solute molecules. Of these, the former two can be ignored, because they exhibit either no effect or only a small effect. The effective size of a solute molecule may also change with changing column temperature. The dependence of intrinsic viscosity on column temperature for PS in chloroform, tetrahydrofuran, and cyclohexane were tested [5]. The temperature dependence of intrinsic viscosity of PS solutions was observed over a range of temperatures. The intrinsic viscosity of PS in tetrahydrofuran is almost unchanged from 20°C up to 55°C, whereas the intrinsic viscosity in chloroform decreased from 30°C to 40°C. Cyclohexane is a theta solvent for PS at around 35°C and intrinsic viscosity in cyclohexane increased with increasing column temperature.

Because the hydrodynamic volume is proportional to the molecular size, the intrinsic viscosity can be used as a measure of the molecular size and optimum column temperatures and solvents must be those where no changes in intrinsic viscosity are observed.

References

1. S. Mori and H. G. Barth, *Size Exclusion Chromatography*, Springer-Verlag, New York, 1999, Chap 5.

2. S. Mori, Effect of experimental conditions, in *Steric Exclusion Liquid Chromatography of Polymers* (J. Janča, ed.), Marcel Dekker, Inc., New York, 1984.
3. S. Mori, *J. Appl. Polym. Sci. 21*: 1921 (1977).
4. S. Mori and T. Suzuki, *Anal. Chem. 52*: 1625 (1980).
5. S. Mori and M. Suzuki, *J. Liquid Chromatogr. 7*: 1841 (1984).

Sadao Mori

GPC–SEC–HPLC Without Calibration: Multiangle Light Scattering Techniques

Traditional size-exclusion chromatography [SEC or gel permeation chromatography (GPC)] as used to obtain molar masses and their distributions has been described elsewhere in this volume. The method suffers from three shortcomings:

1. The calibration standards generally differ from the unknown sample;
2. The results are sensitive to fluctuations in chromatography conditions (*e. g.* temperature, pump speed fluctuations, *etc.*); and
3. Calibration must be repeated frequently.

By adding a multiangle light-scattering [1] detector directly into the separation line, as shown schematically in Fig. 1, the eluting molar masses are determined *absolutely*, thus obviating the need for calibration and elimination of all of the three shortcomings listed. Figure 1 illustrates also two most important elements associated with making quality light-scattering measurements: an in-line degasser and an in-line filter. The in-line degasser is essential to minimize dissolved gases and, thereby, prevent the production of bubbles during the measurement process. Scattering from such bubbles can overwhelm the signals from the solute molecules or particles. Perhaps even more importantly, the system requires that the mobile phase be dust-free. The filter illustrated is placed between the pump and the injector. Usually, this filter station is comprised of two holders, holding, respectively, a 0.20-μm filter followed by a 0.02- or 0.01-μm filter. Although providing for such pristine operating conditions may seem bothersome, it has been shown that so-called "dirty" solvents, although rarely affecting the refractive index detector (RID) signal, do actually contribute significantly to the degradation of high-performance liquid chromatography (HPLC) and SEC columns as well as resulting in the more frequent need to rebuild pumps. The additional solvent cleanup effort is well worth it!

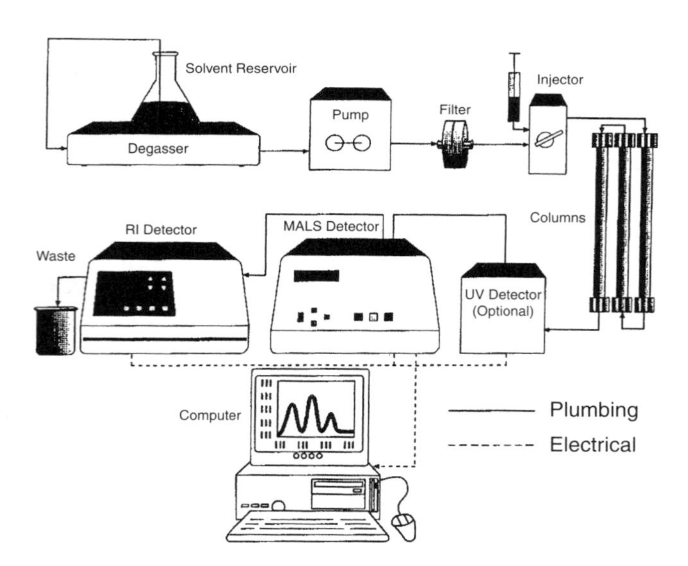

Fig. 1 Schematic diagram showing elements of traditional chromatograph with added MALS detector and dust- and bubble-reducing elements.

An "absolute" light-scattering (LS) measurement is one that is independent of calibration standards which have "known" molar masses to which the unknown is compared. A LS measurement requires the chromatographer to determine the fundamental properties of the solution (refractive index, *dn*/*dc* value) and the detector response (field of view, sensitivity, solid angle subtended at the scattering volume). In addition, other factors must be determined, such as the light wavelength and polarization, geometry of the scattering cell, refractive index of all regions through which the scattered and incident light will pass, and the ratio of the scattered light to the incident light. Generally, these determinations are made in conjunction with appropriate multiangle light-scattering (MALS) software. The importance of light scattering's independence of a set of reference molar masses to determine the molar mass of an unknown cannot be overemphasized.

Theory

As described in detail in Refs. 1–3, the fundamental equation relating the quantities measured during a MALS detection and the quantities derived is, in the limit of "... vanishingly low concentrations...," [2], given by

$$\frac{K^*c}{R(\theta)} \approx \frac{1}{MP(\theta)} + 2A_2c \tag{1}$$

where $K^* = 4\pi^2(dn/dc)^2n_0^2(N_A\lambda_0^4)^{-1}$, M is the weight-average molar mass, N_A is Avogadro's number, dn/dc is the refractive index increment, λ_0 is the vacuum wavelength, θ is the angle between the incident beam and the scattered light, and n_0 is the refractive index of the solvent. The refractive index increment, dn/dc, is measured off-line (or looked up in the literature) by means of a differential refractive index (DRI) operating at the same wavelength as the one used for the MALS measurements. It represents the incremental refractive index change dn of the solution (solvent plus solute) for an incremental change dc of the concentration in the limit of vanishingly small concentration. Most importantly, the excess Rayleigh ratio, $R(\theta)$, and form factor $P(\theta)$ are defined respectively by

$$R(\theta) = \frac{f(\theta)_{\text{geom}}[I(\theta) - I_S(\theta)]}{I_0} \tag{2}$$

$$P(\theta) = 1 - \alpha_1 \sin^2(\theta/2) + \alpha_2 \sin^2(\theta/2) - \cdots \tag{3}$$

where

$$\alpha_1 = \frac{1}{3}\left(\frac{4\pi n_0}{\lambda_0}\right)\langle r_g^2\rangle \tag{4}$$

I_0 is the incident light intensity (ergs/cm^2 s), $f(\theta)_{\text{geom}}$ is a geometrical calibration constant that is a function of the solvent and scattering cell's refractive index and geometry, and $I(\theta)$ and $I_S(\theta)$ are the normalized intensities respectively of light scattered by the solution and by the solvent per solid angle. The mean square radius is given by Eq. (5), where the distances r_i are measured from the molecule's center of mass to the mass element m_i:

$$\langle r_g^2\rangle = \frac{\sum_i r_i^2 m_i}{\sum_i m_i} = \frac{1}{M}\int r^2\, dm \tag{5}$$

For MALS measurement following GPC separation, the sample concentration at the LS detector is usually diluted sufficiently that the term $2A_2c$ often may be safely dropped from Eq. (1). In some applications involving very high molar masses, it is often worthwhile to perform an off-line determination of the second virial coefficient from a Zimm plot [1,2] to confirm its negligible effect on the derived molar mass of Eq. (1).

Basic Principles

In the limit of vanishingly small concentrations, and the extrapolation of Eq. (3) to very small angles, the two basic principles of light scattering are evident:

1. The amount of light scattered (in excess of that scattered by the mobile phase) at $\theta \approx 0$ is directly proportional to the product of the weight-average molar mass and the concentration (*ergo*, measure the concentration and derive the mass!).
2. The angular variation of the scattered light at $\theta \approx 0$ is directly proportional to the molecule's mean square radius (i.e., size).

The successful application of absolute MALS measurements requires a sufficient number of resolved scattering angles to permit an accurate extrapolation to $\theta \approx 0$. Again, all required calculations are performed by the software. Whenever the mobile phase is changed, its corresponding refractive index must be entered into the software program, which should correct automatically for the resultant change of scattering geometry. Figure 2 shows the normalized light-scattering signals at each scattering angle (detector) as a function of elution volume for a relatively broad sample. Also indicated is the corresponding concentration detector signal.

In conventional SEC measurements, it is necessary to calibrate the mass detector [DRI or ultraviolet (UV)] so that its response yields concentration directly. For example, a DRI detector, following calibration, should produce a response proportional to the refractive index change (Δn) detected. This is related to the concentration change Δc by the simple result $\Delta c = \Delta n/(dn/dc)$. Implicit in the use of a DRI detector, therefore, is that measurement of the concentration of the unknown requires that

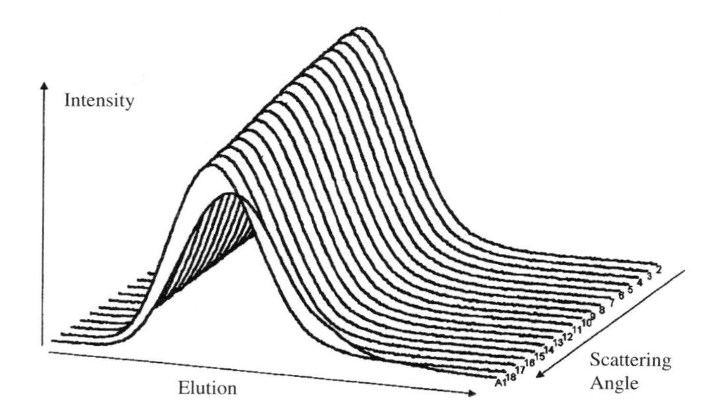

Fig. 2 Light-scattering and DRI signals from MALS setup shown in Fig. 1.

its differential refractive index, *dn/dc*, be measured, or otherwise determined.

Combining SEC with MALS to produce absolute molar mass data without molecular calibration standards also requires prior calibration of the concentration detector as well as calibration of the MALS detector itself. The latter calibration involves the determination of all geometrical contributions such that the MALS detector measures the Rayleigh excess ratio at each scattering angle. This is most easily achieved by using a turbidity standard such as toluene. Details are found in Ref. 2. Once the refractive index of the mobile phase is entered, the software [4] performs the required calibration.

Derived Mass, Size, and Conformation

The MALS detector produces the absolute molar mass and mean square radius $\langle r_g^2 \rangle$ at each eluting slice. The *root mean square* (rms) *radius* $r_g = \langle r_g^2 \rangle^{1/2}$ is often referred to by the misnomer "radius of gyration." There is a lower limit to its determination, which is generally about 8–10 nm. Below this value, MALS cannot generally produce a reliable value. Nevertheless, whenever both r_g and molar mass M are determined by MALS over a range of fractions present in an unknown sample, the sample's so-called conformation may be determined by plotting the logarithm of the rms radius versus the logarithm of the corresponding molar mass. A resultant slope of unity indicates a rodlike structure, a slope of 0.5–0.6 corresponds to a random coil, and a slope of $\frac{1}{3}$ would indicate a sphere. Values below $\frac{1}{3}$ generally suggest a highly branched molecular conformation.

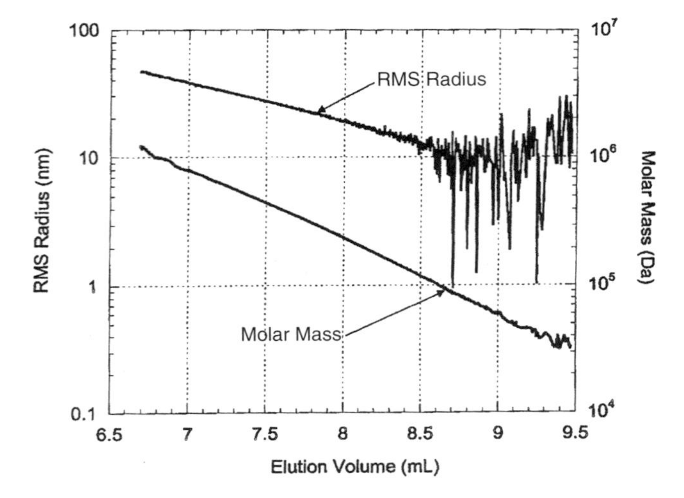

Fig. 3 Molar mass and rms radius generated from data of Fig. 2 as a function of elution volume.

Figure 3 shows the MALS-derived molar mass and rms radius as a function of elution volume for a broad polystyrene sample. Measurements were made in toluene at 690 nm. The value of *dn/dc* chosen was 0.11. Also shown superimposed on these curves is the DRI signal proportional to the eluting sample concentration. From Fig. 3, it should be noted that the radius data begins to deteriorate around 10 nm, whereas the mass data extends to its detection limits. From the mass and radius data of Fig. 3, a conformation plot is easily generated with a slope of about 0.57 (i.e., corresponding to a random coil). These same data can also be used immediately to calculate the mass and size moments of the sample as well as its polydispersity as shown in the next section.

Mass and Size Moments

If we assume that the molecules in each slice, i, following separation by SEC, are monodisperse, the mass moments of each sample peak selected are calculated from the conventional definitions [3,5] by

$$M_n = \frac{\sum_i n_i M_i}{\sum_i n_i} = \frac{\sum_i c_i}{\sum_i c_i / M_i} \tag{6}$$

for the *number*-average molar mass, where n_i is the number of molecules of mass M_i in slice i and the summations are over all the slices present in the peak; the concentration c_i of the ith species, therefore, is proportional to $M_i n_i$;

$$M_w = \frac{\sum_i c_i M_i}{\sum_i c_i} = \frac{\sum_i n_i M_i^2}{\sum_i n_i M_i} \tag{7}$$

for the *weight*-average molar mass; and

$$M_z = \frac{\sum_i n_i M_i^3}{\sum_i n_i M_i^2} = \frac{\sum_i c_i M_i^2}{\sum_i c_i M_i} \tag{8}$$

for the z-average ("centrifuge") molar mass. Note how these "moments" are defined. In particular, the z-average moment corresponds to "the next higher weighting" of both numerator and denominator by the factor M_i. Equations (6) and (7), of course, have a simple physical interpretation in terms of molecular numbers and concentration. From Eq. (8), it is a simple matter to write down expressions for the $z + 1, z + 2, \ldots$ moments.

A similar set of expressions may be written down for the so-called size number and weight moments by replacing the mass terms M_i by the mean square radius values at each slice $\langle r_g^2 \rangle_i$ in Eqs. (6) and (7). A z-average term, on the other hand, takes on a more convoluted form [5]. For a random coil conformation under so-called theta conditions, the molar mass is directly proportional to $\langle r_g^2 \rangle$, and an expression that looks identical to Eq. (8) with one of the M_i of the numerator sum replaced by $\langle r_g^2 \rangle_i$ is obtained. However, in general, this "equivalence" is not the case and the "light-scattering" value LS is a better description, namely

$$\langle r_g^2 \rangle_{LS} = \frac{\sum_i c_i M_i \langle r_g^2 \rangle_i}{\sum_i M_i c_i} \tag{9}$$

Despite the non-random-coil-at-theta conditions, Eq. (9) is commonly referred to as the z-average mean square radius. The cross-term $M_i \langle r_g^2 \rangle_i c_i$ of Eq. (9) is a quantity measured directly by light scattering, at small $\sin^2(\theta/2)$, as clearly may be seen by expanding the term $1/P(\theta)$ in Eq. (1) using the expansion of $P(\theta)$ of Eq. (3).

Polydispersity

Within the peak selected, the sample polydispersity is simply the ratio of the weight to number average (viz. M_w/M_n) obtained from Eqs. (7) and (6), respectively.

Differential Mass Weight Fraction Distribution

The MALS measurements illustrated by Fig. 2 also may be used directly to calculate the differential mass weight fraction distribution, $x(M) = dW(M)/d(\log_{10}M)$ by using the measured $\log_{10}M$ as a function of the elution volume V. Thus, if the concentration detector's baseline subtracted response is $h(V)$, then $dW/dV = \pm h(V)/\int h(V)\, dV$, the integral representing the sum over all contributing concentrations to the peak. It is then easily shown [6] that

$$x(M) = -\frac{h(V)/\int h(V)\, dV}{d(\log_{10}M)/dV} \tag{10}$$

Note that for so-called "linear" column separations, the denominator $d(\log_{10}M)/dV$ is just a constant and, therefore, the differential weight fraction distribution will appear as a reflection (small mass first, from left to right) of the DRI signal. In general, column separations are not linear, so the DRI signal is not a good representation of the mass-elution distribution.

Branching

The MALS measurements which eliminate the need for column calibration and all of its subsequent aberrations also permit the direct evaluation of branching phenomena in macromolecules because the basic quantitation of branching may only be achieved from such measurements as shown in the article by Zimm and Stockmayer [7]. Empirical approaches to quantitate branching, using such techniques as viscometry, have been shown to yield consistently erroneous results especially when long-chain branching becomes dominant.

Reversed-Phase and Other Separation Techniques

Because MALS determinations are independent of the separation mechanism, they may be applied to many types of HPLC. Reversed-phase separations are of particular significance because they cannot be calibrated, as sequential elutions do not occur in a monotonic or otherwise predictable manner. Again, as with all MALS chromatography measurements,* all that is required is that the concentration and MALS's signals be available at each elution volume (slice).

Another separation technique of particular application for proteins, high-molar-mass molecules, and particles is the general class known as field-flow fractionation (FFF) in its various forms† (cross-flow, sedimentation, thermal, and electrical). Once again, MALS detection permits mass and size determinations in an absolute sense without calibration. For homogeneous particles of relatively simple structure, a concentration detector is not required to calculate size and differential size and mass fraction distributions. Capillary hydrodynamic fractionation (CHDF) is another particle separation technique that may be used successfully with MALS detection.

References

1. P. J. Wyatt, *Anal. Chim. Acta 272*: 1 (1993).
2. B. H. Zimm, *J. Chem. Phys. 16*: 1093 (1948); *16*, 1099 (1948).
3. N. C. Billingham, *Molar Mass Measurements in Polymer Science*, John Wiley & Sons, New York, 1977.
4. ASTRA® software, Wyatt Technology, Santa Barbara, CA, 1999.
5. P. J. Wyatt, in *Analytical and Preparative Separation Methods of Biomacromolecules* (H. Y. Aboul-Enein, ed.), Marcel Dekker, Inc., New York, 1999.
6. D. W. Shortt, *J. Liquid Chrom.*, 16, 3371 ± 3391 (1993).
7. B. H. Zimm and W. H. Stockmayer, *J. Chem Phys. 17*: 1301 (1949).

Suggested Further Reading

Huglin, M. B. (ed.) *Light scattering from Polymer Solutions*, Academic, London, 1972.

Philip J. Wyatt

*Current MALS light-scattering bibliography and applications may be found at www.wyatt.com/Bibliography.html.

†Current FFF bibliography and applications may be found at www.rohmhaas.com/fff/fff.html.

GPC–SEC Viscometry from Multiangle Light Scattering

Intrinsic Viscosity and Universal Calibration

Viscometric techniques have long been used in combination with GPC–SEC separations since the early discovery [1] that the elution of many classes of divers polymers follows a so-called "universal calibration" curve. A plot of the logarithm of the hydrodynamic volume, $M[\eta]$, where M is the molar mass and $[\eta]$ the intrinsic (or "limiting") viscosity, against the elution volume V yields a common curve (differing for each mobile phase, operating temperature, and column set) along which polymers of greatly differing conformation appear to lie. Neglecting the fact that the errors of such fits can be quite large (the results are usually presented on a logarithmic scale), the concept of universal calibration (UC) allows one to estimate (from the UC curve) the molar mass of an eluting fraction by measuring only the intrinsic viscosity, $[\eta]$, and the corresponding elution volume (time). Key to the measurement of $[\eta]$ is the determination of the specific, η_{sp}, or relative, η_{rel}, viscosity and the concentration c, both in the limit as $c \rightarrow 0$. These viscosities are defined by

$$\eta_{sp} = \frac{\eta \pm \eta_0}{\eta_0} \tag{1}$$

and

$$\eta_{rel} = \frac{\eta}{\eta_0} \tag{2}$$

where η is the solution viscosity and η_0 is the viscosity of the pure solvent. Because $\eta_{rel} = \eta_{sp} + 1$, it is easily shown for η_{sp} small compared to unity that

$$\lim_{c \rightarrow 0} \frac{\ln(\eta_{rel})}{c} = \lim_{c \rightarrow 0} \frac{\eta_{sp}}{c} = [\eta] \tag{3}$$

For the case of GPC–SEC elutions, the concentration c following separation is generally so small that Eq. (3) is assumed to be valid.

The Mark–Houwink–Sakurada Equation

Even without the use of a UC curve (one must be generated for each series of measurements), measurement of $[\eta_0]$ is believed by some to yield an intrinsic viscosity-weighted molar mass [2]. Most importantly, there is a historic interest in the relation of $[\eta]$ to molar mass and/or size. Indeed, the study and explanation of UC has occupied the theorists for some time and, accordingly, there are various formulations describing such relationships [2]. For linear polymers, the most popular empirical relationship between $[\eta]$ and molar mass is the Mark–Houwink–Sakurada (MHS) equation

$$[\eta] = KM^a \tag{4}$$

where K and a are the MHS coefficients. For many polymer–solvent combinations, a plot of $\log([\eta])$ versus $\log(M)$ is linear over a wide range of molar masses. In other words, both K and a are constant throughout the range. Thus, the equation may be used for such polymer–solvent combinations to determine molar mass by measuring $[\eta]$.

Unfortunately, for some solvent–polymer combinations, even for nearly ideal random coils such as polystyrene, the coefficients are not constant but vary with molar mass.

The Flory–Fox Equation

In the various theoretical attempts to explain the relation between $[\eta]$ and the molar mass M, a relation derived by Flory and Fox for random coil molecules is often applied to interpret viscometric measurements for even more general polymer structures. Although applicable to a broader range of polymers than the MHS equation, the Flory–Fox relation has its own shortcomings. Nevertheless, its frequent use and good correlation with experimental data over a wide range of polymer types confirms its potential for combination with light-scattering measurements to eliminate the need for separate viscometric determinations. In its most general form, the Flory–Fox equation is given by

$$M[\eta] = \Phi(\sqrt{6}r_s)^3 \tag{5}$$

where r_g is the root mean square radius (or "radius of gyration"). The excluded volume effect is taken into account by representing the Flory–Fox coefficient as $\Phi = \Phi_0(1 \pm 2.63\varepsilon + 2.86\varepsilon^2)$. The constant $\Phi_0 = 2.87 \times 10^{23}$ and ε is related to the MHS coefficient a by the relation $2a = 1 + 3\varepsilon$. Thus, ε ranges from 0 at the theta point to 0.2 for a good solvent. Equation (5) is of particular interest because multiangle light-scattering (MALS) measurements [3] determine M and r_g directly. Thus, if a polymer–

solvent combination is well characterized by Eq. (5), then this equation may be used directly to calculate the IV without need for a viscometer.

Viscometry Without a Viscometer

As we have seen earlier, $[\eta]$ may be calculated directly from the (absolute) MALS measurements of M and r_g using Eq. (5). For linear polymers spanning a relatively broad molecular range (an order of magnitude or more), the measurement of M and r_g permits the determination of the molecular conformation defined by

$$r_g = kM^{\alpha} \tag{6}$$

where k and α are constants generally calculated from the intercept and slope of the least-squares fitted plot of $\log(M)$ against $\log(r_g)$. Combining Eqs. (4) and (5), we obtain

$$KM^{\alpha-1} = (\Phi\sqrt{6}r_g)^3 \tag{7}$$

Solving for r_g and substituting into Eq. (6) yields

$$\frac{K^{1/3}M^{(a-1)/3}}{\Phi\sqrt{6}} = kM^{\alpha} \tag{8}$$

Therefore, we have the following relations between the coefficients:

$$a = 3\alpha - 1 \quad \text{and} \quad K = \Phi(\sqrt{6}k)^3 \tag{9}$$

Equations (9) show that we can obtain the MHS coefficients a and K directly from a MALS measurement and a determination from such measurements of the molecular conformation parameters α and k. Note that when long-chain branching becomes significant and the molecular conformation becomes more compact such that $\alpha \rightarrow \frac{1}{3}$, the MHS equation, Eq. (4), shows that the intrinsic viscosity no longer varies with molar mass, but becomes constant. This condition also represents a failure of the Flory–Fox equation and the concepts associated with the use of intrinsic viscosity as a means (through UC, for example) to determine molar mass. For linear polymers for which MALS measurements yield values for r_g and M at each eluting slice, all of the important viscometric parameters may be derived directly from the Flory–Fox relation and the MHS equation, as has been shown. For more complex molecular structures or solvent–solute interactions, the MHS coefficients are no longer constants and the empirical theory itself begins to fail.

It is well known [3], however, that the MALS measurements begin to fail in the determination of r_g once r_g falls below about 8–10 nm, even though the M values generated still remain precise. This lack of precision is due to the limitations of the laser "ruler" to resolve a size much below about one-twentieth of the incident wavelength. The trouble with empirical relations, such as the relation between intrinsic viscosity and molar mass, is that they too are often limited to regions where such concepts are applicable. For very small molar masses, the conformation of a polymer molecule may be poorly described by the same theory applied for the larger constituents of a sample. Although MALS conformation measurements may be extrapolated in r_g for the case of linear polymers, such extrapolations must be used with great caution. Similar remarks apply, of course, to the use of viscometric measurements for characterizing complex molecules whose conformations (and, therefore, MHS coefficients) are changing with M.

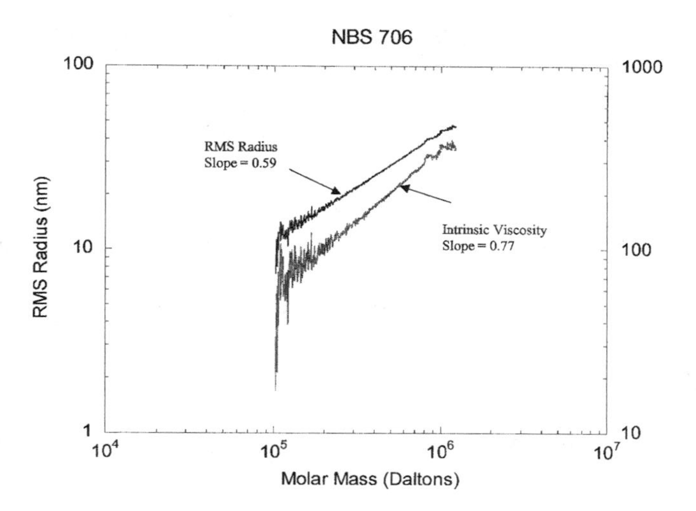

Fig. 1 Conformation plot for $\log(r_g)$ versus $\log(M)$.

Figure 1 presents the conformation plot for $\log(r_g)$ versus $\log(M)$ as obtained from a MALS measurement for the polystyrene broad linear standard NIST706 in toluene. Superimposed thereon is a plot of the calculated $\log[\eta]$ as a function of $\log(M)$ for the same sample. From the latter plot, the MHS coefficients may be deduced by inspection of the slope and intercept to yield $a = 0.77$ and $K \approx 0.008$.

References

1. H. Benoit, Z. Grubisic, and R. Rempp, *J. Polym. Sci. B5, 753*: (1967).

2. K. Kamide and M. Saito, in *Determination of Molecular Weight* (A. R. Cooper, ed.), John Wiley & Sons, New York, 1989.
3. P. J. Wyatt, *Anal. Chim. Acta 272*: 1 (1993).

Suggested Further Reading

Billingham, N. C., *Molar Mass Measurements in Polymer Science*, John Wiley & Sons, New York, 1977.
Zimm, B. H., *J. Chem. Phys. 16*: 1093 (1948); *16*: 1099 (1948).

Philip J. Wyatt
Ron Myers

Gradient Development in Thin-Layer Chromatography

The main tasks of analytical thin-layer chromatography (TLC) are the separation of sample components and the measurement of peak heights or areas for quantitative analysis. In the final effect, the peaks should be narrow and symmetrical. Two problems are related to analysis: choice of suitable conditions of elution and full separation. In practice, two modes of development are applied (i.e., isocratic and gradient elution). In isocratic elution, the band migrates under constant conditions; in gradient elution, the changes in migration conditions are consciously programmed. Both in column and in thin-layer chromatography, the isocratic mode is preferred unless the "general elution problem" is encountered. Its solution may consist in gradient elution (stepwise or continuous), gradient of stationary phase, development with a mixed eluent composed of solvents of different polarity (polyzonal TLC), or temperature programming [1]. The gradient of the mobile phase is both simple and practical in application. In gradient and isocratic development, we consider the properties and composition of the eluent delivered to the adsorbent bed; if these are constant, we have isocratic analysis, if varied — usually stepwise — it is the gradient of the mobile phase. In a simple gradient, the eluent strength is increased from the beginning to the end of development process. One of advantages of gradient TLC is the feasibility of application of both simple and reversed gradients (decreasing modifier concentration) and a complex gradient — combination of both types of gradient. The reversed gradient can be applied in the case of multiple development.

The gradient is defined by the variation of composition of the mobile phase: by the percent contents of the weaker component A and the stronger component B, called the modifier. The gradient is also characterized by its steepness and shape. The steepness is defined by the concentration of the modifier of the first and last fractions of the eluent delivered to the adsorbent layer. When the differences of modifier concentrations among all steps are constant, the gradient is called linear. When these differences are large in the beginning and then decrease in the consecutive steps, it is known as a convex gradient program, and in the opposite case, a concave gradient.

To illustrate the "general elution problem" and its solution, let us consider the following situation. Consider a 20-component mixture with capacity factors k of the components forming a geometrical progression and exponentially dependent on the modifier concentration (molar or volume fraction c), in accordance with the Snyder–Soczewinski model of adsorption [2]. The log k versus log c plots of the 20 solutes are given in Fig. 1, which has a parallel R_f axis subordinated to the right-hand-side log k axis. It can be seen that no isocratic eluent can separate all the components. A pure modifier [$c = 1.0$ (100%)] separates well solutes 1–7, and the less polar solutes are accumulated near the solvent front; for $c = 0.1$ (10%), solutes 8–14 are well separated, the remaining ones being accumulated either near the start line or the front line; for $c = 0.01$ (1%), solutes 1–14 are accumulated on the start line. Thus, only less than half of the components can be satisfactorily separated by isocratic elution.

Mobile-Phase Gradients

The use of a stepwise gradient of the mobile phase is well described by theoretical models [1,3,4]. The migration of zones is shown in Fig. 2a. In the ideal situation, the spots of solutes overtaken by the consecutive fronts of increased modifier concentrations accelerate their migration so that the strongly retained solutes also start to migrate. De-

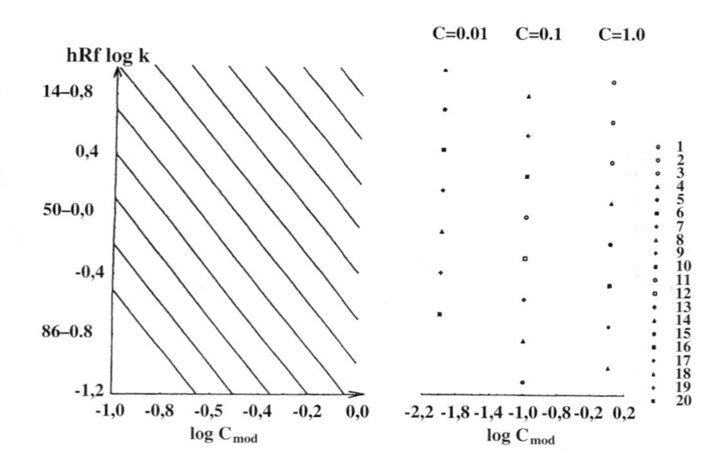

Fig. 1 Family of log k versus log C_{mod} plots for hypothetical solutes 1–20 with capacity factors forming a geometrical progression according to Snyder–Soczewinski equation. For isocratic elution at $C_{mod} = 1.0$, 0.1, and 0.01. Only seven to eight solutes give R_f values in the range 0.09–0.9 (left-hand ordinate).

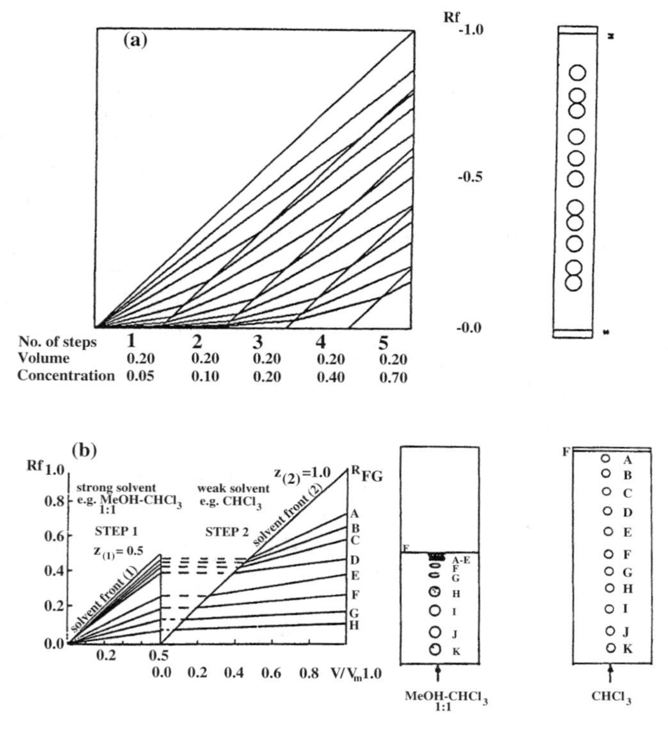

Fig. 2 Separation of hypothetical solutes A–K with a wide range of polarity: (a) five-step simple gradient of mobile phase (concentration of modifier C_{mod} in the range 0.05–0.7), volume of solvent expressed in void volume units (dimensionless); (b) two-step development (solid lines). In the first step, polar solutes E–H are separated; the less polar solutes A–D, poorly separated during the first step, are separated during the second step.

pending on the polarities of the solutes, they migrate all the time in the first concentration zones or are overtaken by the consecutive zones of higher concentration. It can be seen that both weakly polar (A–E) and strongly polar (F–K) compounds are well separated in the final chromatogram. The migration of the components is given by the following equation [3]:

$$R_f = \sum_{i=1}^{h-1} \frac{V(i) R_{f(j,i)}}{1 - R_{f(j,i)}} + R_{f(j,h)} \left(1 - \sum_{i=1}^{h-1} \frac{V(i)}{1 - R_{f(j,i)}} \right) \quad (1)$$

(For detailed derivation and discussion, see Ref. 3.)

This equation could be applied to formulate computer programs that calculate the final R_f values for a given gradient program and retention–eluent composition relationships. In the equation, it is assumed that the stagnant mobile phase in the pores of adsorbent is rapidly displaced. In reality, demixing takes place (especially in the first fractions of slow concentrations of B) and the exchange of the stagnant solvent in the pores with the mobile phase is low; therefore, the boundaries of the concentration zones are not sharp but somewhat diffuse and the solutes migrate in zones of intermediate properties. Planning the gradient program, it is necessary to choose conditions under which the weakly retained components do not migrate with the front of the mobile phase and the strongly retained ones do not remain on the start line. The following series of solvents can be applied to TLC on silica: heptane, trichloroethylene, dichloromethane, diisopropyl ether, ethyl acetate, and isopropanol. The application of the stepwise mobile-phase gradient greatly improves the separation of complex mixtures (e.g., plant extracts); in many cases, twice as many spots were obtained in comparison to isocratic elution. Densitograms of lanatosides obtained by isocratic and gradient elution, for example, indicated that the polar lanatosides were separated on silica using a rapid nonaqueous eluent, avoiding the frequently used cellulose powder and viscous aqueous eluents [5].

Gradient Multiple Development

In all techniques of multiple development, the plate is repeatedly developed in the same direction, with intermittent evaporation of the mobile phase between the consecutive developments (Fig. 2b). If the layer is developed many times to the same distance with the same eluent, the technique is called unidimensional multiple chromatography. A variation of this technique, called incremental multiple development, consists in the stepwise change of the development distance that is the shortest in the first step and is then increased, usually by a constant increment (equal distance or time); the last development step corresponds to the maximum development distance. If, in the process of multiple development, the solvent strength of the mobile phase is varied, the technique is then called gradient multiple development in either the unidimensional or the incremental version. The change in the mobile phase may concern several or all steps.

Depending on the properties of the components of the mixture to be analyzed, an increasing or decreasing gradient of solvent strength can be applied. The process of multiple development with any variation of the mobile-phase composition can be described by a model and equations reported, modified to take into account the intermittent evaporations of solvents:

$$R_{f(n)} = S_{(n-1)} + (Z_{(n)} - S_{(n-1)})R_{(n)} \qquad (2)$$

where $S_{(n-1)}$ is the position of the zone in the $n-1$ development step and $Z_{(n)}$ is the final development distance [1,6]. In the simplest case of a decreasing stepwise gradient, the layer is developed to half the distance with a polar eluent that separates the most polar components in the lower part of the chromatogram; the less polar components are accumulated in the front area. Their separation occurs in the second stage when the layer is developed to the full distance with a less polar eluent (Fig. 2b). The incremental, multistep version of this technique, with programmed, automated development and evaporation steps, is called automated multiple development (AMD) [7] and the method is considered to be the most effective and versatile TLC technique [8].

Mechanism of Compression of Chromatographic Zones

One of the advantages of gradient elution is the compression of zones. Each passage of the front of the increased concentration of the mobile phase through the spot leads to compression of the spot in the direction of development. This is due to the fact that the front of the increased eluent concentration first reaches the lower edge of the spot so that the solute molecules in this fragment start to move (multiple development) or accelerate their migration (gradient) earlier than the molecules in the farther part of the spot. When the front of the mobile phase or of the concentration zone overtakes the whole spot, the compressed spot continues to migrate and gradually becomes more diffuse as in isocratic elution. If the two mechanisms, compression and diffusion, become counterbalanced, the spot may migrate through considerable distances without any marked broadening (Fig. 3). Incremental multiple development provides a superior separation compared to multiple chromatography, in this case, by minimizing zone broadening and enhancing the zone center separation by migrations of the sample components over a longer distance while maintaining a mobile-phase flow rate range closer to the best value for the separation. This variant can also be achieved by the change of the point of delivery of the eluent to the layer.

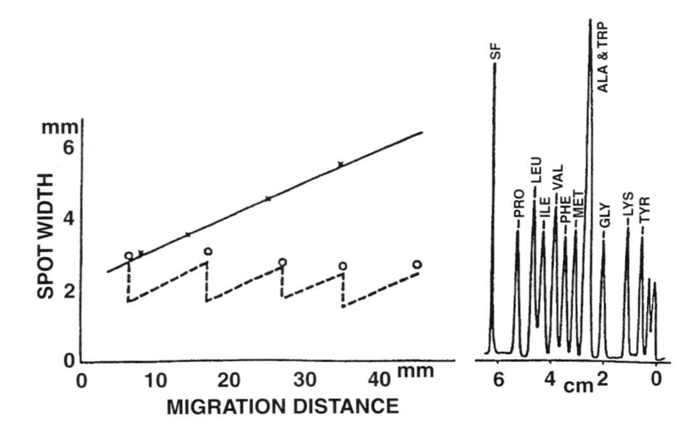

Fig. 3 An illustration of the zone refocusing mechanism (left) and its application to the separation of a mixture of phenylthiohydantoin–amino acids (right). The broken line on the left-hand side represents the change in spot size due to the expansion and contraction stages in multiple development and the solid line depicts the expected zone width for a zone migrating the same distance in a single development.

Not all compounds are suitable for separation by multiple development. Compounds with significant vapor pressure may be lost during the repeated solvent evaporation steps. Certain solvents of low volatility and/or high polarity, such as acetic acid, triethylamine, dimethyl sulfoxide, and so forth, are unsuitable selections for mobile phases because of the difficulty of removing them from the layer by vacuum evaporation between development steps. Water can be used, but the drying steps will then be lengthy. Solvent residues remaining after the drying step can modify the selectivity of mobile phases used in later steps, resulting in irreproducible separations. Although precautions can be taken to minimize the production of artifact peaks in multiple development, the separation of light and/or air-sensitive compounds is probably better handled by other techniques such as simple gradient development.

Polyzonal TLC

Polyzonal TLC is the simplest method for formation of gradients in TLC. The main effect utilized is solvent demixing, which occurs in the case of the application of binary and polycomponent solvents, especially those of differentiated polarity. For an n-component mixed eluent, $n - 1$ solvent fronts are formed, ordered in the sequence of polarity of the components. A gradient of eluent strength is thus formed along the layer; the solutes migrate in various zones, and the passage of fronts leads to compression of TLC spots.

Equipment for Gradient TLC

Depending on the type of the gradient, various apparatuses are applied for its generation. Numerous gradient generators have been described [1]. The gradient of the mobile phase can be formed in some types of horizontal chamber (e.g., Camag, Muttenz, Switzerland; Chromdes, Lublin, Poland). The generation of stepwise gradients is simple for sandwich chambers with distributors, which allow for complete absorption of the eluent fractions from the reservoir. For both chambers, the eluent fractions of increasing eluent strength (increasing concentrations of the polar modifier) are introduced under the distributor. After absorption of the preceding eluent fraction by the adsorbent layer, the next fraction of eluent of changed strength is delivered; the total volume of the eluent fractions corresponds to the development distance. Any gradient program, including continuous or multiple-component gradients, can be generated in this way. The process of multiple development can be fully automated (AMD chamber, Camag Scientific [7]). An apparatus comprises an N-type chamber with connections for adding and removing solvents and gas phases. AMD involves the use of a stepwise gradient of different mobile phases with decreasing strength in 10–30 successive developments increasing in length about 1–5 mm. The initial solvent, which is the strongest, focuses the zones during the first short run, and the solvent is changed for each, or most, of the following cycles. The mobile phase is removed from the chamber, the plate dried and activated by vacuum evaporation, and the layer conditioned with a controlled atmosphere of vapors prior to the next development. High resolution and improved detection limits are achieved because zones are reconcentrated during each development stage. Widths of the separated zones are approximately constant at 2–3 mm, and separation capacity for baseline-resolved peaks is 25–40. Zones migrate different distances according to their polarity. The reproducibility of R_f values is 1–2% (CV) for multiple spots on the same plate or different plates from the same batch. A typical universal gradient for a silica gel layer involves 25 steps, with methanol, dichloromethane, or *tert*-butyl ether, and hexane as the solvents.

Discontinuous gradient of the stationary phase can be obtained easily using an ordinary spreader. The trough is divided into separate chambers filled with suspensions of mixtures of adsorbents. The carrier plates are covered in the usual way [1]. Another method of formation of gradients of stationary-phase activity is the use of a Vario-KS chamber, which permits adsorption of various vapors on the adsorbent surface or to control the activity of adsorbent.

To sum up, gradient elution may be applied to the separation of samples composed of solutes of differentiated chromatographic behavior, for increasing throughput in preparative TLC, for removal of ballast (nonpolar) matrix from the analytes in the first gradient step, for decreasing the detection limit by the compression of spots, for the preliminary estimation of the polarity of sample components, and for acceleration of the choice of optimal conditions of chromatographic analysis.

References

1. W. Golkiewicz, Gradient development in thin-layer chromatography, in *Handbook of Thin-layer Chromatography* (J. Sherma and B. Fried, eds.), Marcel Dekker, Inc., New York, 1997, pp. 135–154.
2. E. Soczewinski, *Anal. Chem. 41*: 179 (1969).
3. E. Soczewinski and W. Markowski, *J. Chromatogr. 370*: 63 (1986).
4. E. Soczewinski, *J. Chromatogr. 369*: 11 (1986).
5. G. Matysik and E. Soczewinski, *J. Planar Chromatogr. 9*: 404 (1996).
6. W. Markowski, *J. Chromatogr. 485*: 517 (1989).
7. C. F. Poole and S. K. Poole, *J. Chromatogr. 703*: 573 (1995).
8. C. F. Poole and M. T. Belay, *J. Planar Chromatogr. 4*: 345 (1991).

Wojciech Markowski

Gradient Elution: Overview

Gradient elution is the elution method in which the mobile-phase composition changes during time. It may be considered as an analogy to the temperature programming in gas chromatography.

The main purpose of gradient elution is to move strongly retained components of a mixture faster, while having the least retained components well resolved. At the beginning of the analysis, the solvent used is appropriate to elute some of the components, but is "weak" in terms of its ability to remove other compounds from the column and separate them from one another.

Gradient elution operates on the principle that, under the initial mobile-phase conditions, many of the components have a k' (capacity factor) value of essentially infinity, in that these components are stopped in a narrow band near the head of the column. As the solvent composition is changed and its solvent strength is increased, sample components dissolve at a characteristic solvent strength and then migrate down the column, leaving the remaining components behind. Changes in the mobile-phase composition may be "continuous" with a predetermined set of conditions or may be done in "steps" of substantial solvent composition changes.

The typical gradients used in reversed-phase chromatography are linear or binary (i.e., involving two mobile phases). Convex and concave gradients are used occasionally for analytical purposes, particularly when dealing with multicomponent samples requiring extra resolution either at the beginning or at the end of the gradient (Fig. 1).

The concentration of the organic solvent is lower in the initial mobile phase (mobile phase A) than it is in the final mobile phase (mobile phase B). The gradient then, regardless of the absolute change in percent organic modifier, always proceeds from a condition of high polarity (high aqueous content, low concentration of organic modifier) to low polarity (higher concentration of organic modifier, lower aqueous content). Reversed-phase separations can be achieved using either a stepwise or a continuous gradient to elute sample components. Step gradients (i.e., a series of isocratic elutions at different percentages of B) are useful for applications such as desalting, but for separations requiring high resolution, a linear continuous gradient is required.

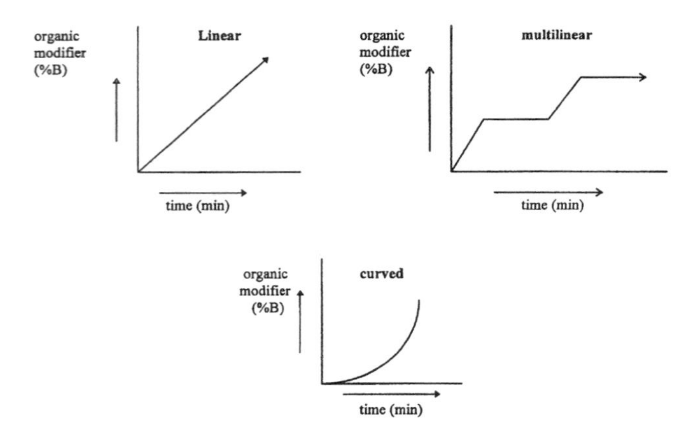

Fig. 1 Various gradient shapes.

Step gradients are also ideal when performing process scale applications providing the desired resolution can be obtained; less complex instrumentation is required to generate step gradients. Additionally, step gradients can be generated more reproducibly than linear gradients.

Gradient shape (combination of linear gradient and isocratic conditions), gradient slope, and gradient volume are all important considerations in reversed-phase chromatography. Typically, when first performing a reversed-phase separation of a complex sample, a broad gradient is used for initial screening in order to determine the optimum gradient shape.

The ideal gradient shape and volume are empirically determined for a particular separation. Generally, the sample is chromatographed using a broad-range linear gradient to determine where the molecules of interest will elute. The initial conditions usually consist of mobile phase A containing 10% or less organic modifier and mobile phase B containing 90% or more organic modifier. The initial gradient runs from 0% B to 100% B over 10–30 column volumes. A blank gradient is usually run prior to injecting the sample, in order to detect any baseline disturbances resulting from the column or impurities originating in the mobile phase.

After the initial screening is completed, the gradient shape may be adjusted to optimize the separation of the desired components. This is usually accomplished by decreasing the gradient slope, where the desired components elute, and increasing it before and after. The choice of gradient slope will depend on how closely the contaminants elute to the target molecule. Generally, decreasing the gradient slope increases the resolution. However, the peak volume and retention time increase with decreasing gradient slope. Shallow gradients with short columns are generally optimal for high-molecular-weight biomolecules.

Gradient slopes are generally reported as change in percent B per unit time (% B/min) or per unit volume

(% B/mL). When programming a chromatography system in the time mode, it is important to remember that changes in flow rate will affect gradient slope and, therefore, resolution.

Resolution is also affected by the total gradient volume (gradient volume × flow rate). Although the optimum value must be determined empirically, a good rule of thumb is to begin with a gradient volume that is approximately 10–20 times the column volume. The slope can then be increased or decreased in order to optimize the resolution.

Except for optimizing gradient elution methods, which is a very important parameter in chromatographic analysis, another parameter of great significance is the mixing of the mobile-phase components.

There are two primary methods of mixing the mobile-phase components, known as "low-pressure mixing" and "high-pressure mixing." The first method employs electrically actuated solenoid valves located ahead of a single-solvent delivery system (pump). The precision of the gradient depends on the ability of the solenoid to reproducibly dispense solvents in segments of variable size (volume), depending on the composition desired. If reproducible retention times and stable detector baselines are to be obtained, these "segments of solvent plugs" must be well mixed into a homogenous mobile-phase stream before entering the chromatographic column.

The second method uses a separate solvent delivery device for each solvent, with each being capable of delivering smooth, precise flow rates of as low as a few microliters per minute. Gradients are formed by varying the delivery speeds and simply blending the concurrent solvent streams on the high-pressure side of the pumps.

Each method of gradient formation has advantages and disadvantages. The low-pressure gradient formation is preferred most of the time, because it uses only one pump, whereas the high-pressure method uses two pumps which might go wrong during use. Because low-pressure gradient systems have only one pump, they are, of course, less expensive. These systems require extensive degassing and have a large lag time (delay volume) in starting the gradient, whereas, in high-pressure systems, degassing is desired but not essential.

Moreover, low-pressure systems often use three or four solvents. This multiple-solvent blending might also be useful for the optimization of both isocratic and gradient elution methods. This is an advantage that the high-pressure system also has; when not using a gradient system, the operator has two independent isocratic pumps.

Gradient elution is ideal for separating certain kinds of sample which cannot be easily handled by isocratic methods, because of their wide k' range. Nevertheless, there is a strong bias against the use of gradient elution in many laboratories.

Some of the reasons for not preferring gradient elution are as follows: Gradient equipment is not available in some laboratories, because of its higher cost, and gradient elution is more complicated and makes both method development and routine analysis more difficult; the most important issue is that it is not compatible with some high-performance liquid chromatographic (HPLC) detectors (e.g., refractive index detectors). Furthermore, gradient runs take longer, because of the need of column equilibration after each run. Baseline problems are more common with gradient elution and the solvent must be of high purity.

Although the disadvantages of gradient elution must be taken into serious consideration, many separations are only possible using gradient elution. The use of gradient elution for routine applications is suggested for the following kinds of sample:

1. Samples with a wide range of k'
2. Samples composed of large molecules > 1000 in molecular weight
3. Samples containing late eluting interferences that can either foul the column or overlap subsequent chromatograms

Using gradient elution to develop HPLC methods has many advantages compared with using isocratic experiments. First, errors in solvent strength can be adjusted when changing from one solvent to another. Second, the ability to increase resolution during early exploratory runs is a distinct advantage when doing solvent mapping. Early bands often are severely overlapped in isocratic separations, so that it may not be clear how resolution is changing as separation conditions are varied. Gradient elution opens up the front of the chromatogram, allowing a better view of what is happening as conditions are varied (Fig. 2).

Third, using gradient elution runs during the initial stages of method development makes it easier to locate compounds that elute either very early or very late in the chromatogram. With isocratic separation, early-eluting compounds are often lost in the solvent front, whereas late-eluting compounds disappear into the baseline or overlap the next sample. Finally, gradient elution method development works for either gradient or isocratic elution.

In conclusion, gradient elution is not preferred as a quantitative technique because it is more complex than isocratic elution and, hence, more things can potentially go wrong. However, with proper control of operating parameters and good instrumentation, it is possible to obtain a separation with excellent quantitative results. This requires that the operator understand the hardware and

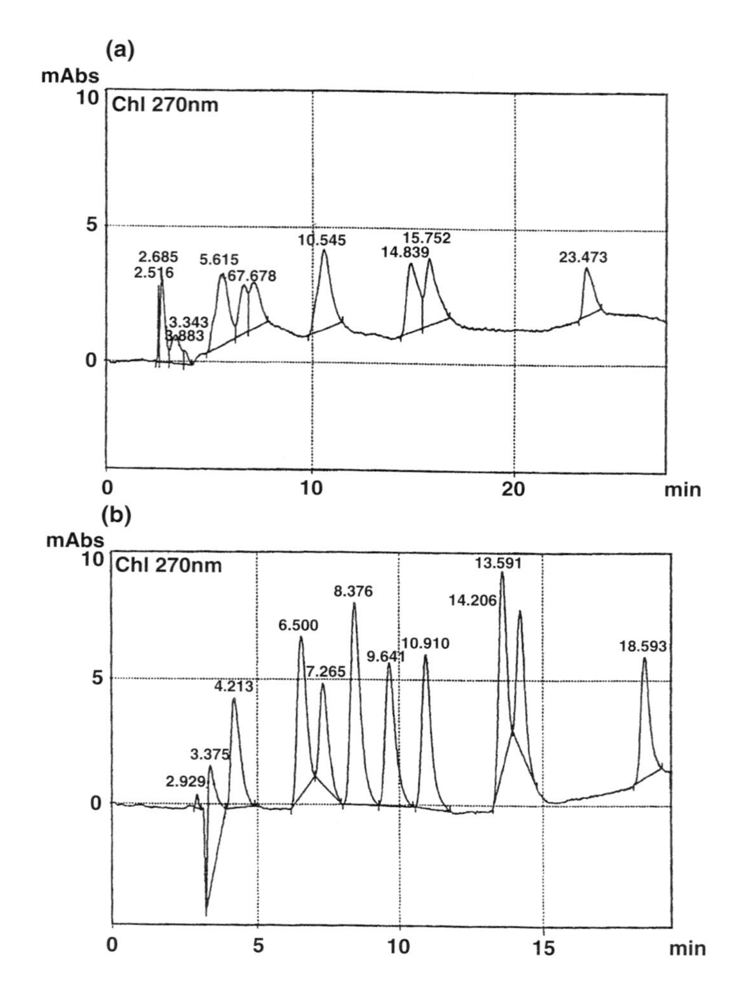

Fig. 2 HPLC analysis of eight methylxanthines: (a) with isocratic elution; (b) with gradient elution.

determine that it is working correctly before attempting a separation. The ideal gradient system should be easy to operate, reproducible to provide consistent retention times, versatile to provide capability of generating various concave, convex, and linear gradient shapes, and convenient to provide a rapid turnaround time to initial eluent conditions (equilibration) for fast throughput from analysis to analysis.

Suggested Further Reading

1. Bidlingmeyer, B. A., *Practical HPLC Methodology and Applications*, John Wiley and Sons, Inc., New York, 1992.
2. Papadoyannis, I., V. Samanidou, and K. Georga, *J. Liq. Chromatogr. 19*(16): 2559 (1996).
3. Pharmacia Biotech, *Reversed Phase Chromatography*, Pharmacia Biotech, Uppsala, Sweden, 1996.
4. Snyder, L. R., J. L. Glajch, and J. J. Kirkland, *Practical HPLC Method Development*, John Wiley and Sons, Inc., New York, 1988.
5. Snyder, L. R., J. J. Kirkland and J. L. Glajch, *Practical HPLC Method Development*, 2nd ed., New York, 1997.

Ioannis N. Papadoyannis
Kalliopi A. Georga

Gradient Elution

The term *gradient elution* refers to a systematic, programmed increase in the elution strength of the mobile phase during the chromatographic run. Of all the techniques used to provide quality separations among complex mixtures, gradient elution offers the greatest potential [1]. Basically, the composition of the mobile phase is varied throughout the separation so as to provide a continual increase in solvent strength and, thereby, a more convenient elution time and sharper peaks for all sample components [2]. What makes this method so useful is the ability to choose from a variety of different eluents. Although most instruments permit gradients to be automatically prepared from various concentrations of only a two-eluent mixture, sample mixtures of a wide range of polarities can be separated efficiently.

The process of mixing eluents is a sensitive one. When two solvents with a large difference in their elution strengths are used, even a small increase in the polar component produces a sharp rise in elution strength. Such an effect is undesirable because the components are almost always eluted at the beginning of the analysis and displacement effects may result from demixing of eluent mixtures [1]. According to Poole et al. [3], the most frequently used gradients are binary solvent systems with a linear, convex, or concave increase in the percent volume fraction of the stronger solvent, as depicted in the following equations:

Linear gradient

$$\theta_B = \frac{t}{t_G} \tag{1}$$

Convex gradient

$$\theta_B = 1 - \left(1 - \frac{t}{t_G}\right)^n \tag{2}$$

Concave gradient

$$\theta_B = \left(\frac{t}{t_B}\right)^n \tag{3}$$

In these equations, θ_B is the volume fraction of the stronger eluting solvent, t is the time after the gradient begins, t_G is the total gradient time, and n is an integer controlling gradient steepness.

Complex gradients can be constructed by combining several gradient segments (i.e., rates of increase of strong solvent composition) to form the complete gradient program [3]. Linear gradients are most commonly used, with convex and concave gradients employed only when necessary to optimize more complex separations. In a linear-solvent-strength gradient, the logarithm of the capacity factor for each sample component, k', decreases linearly with time, according to Eq. (4):

$$\log k = \log k_0 - b\frac{[t]}{[t_m]} \tag{4}$$

In this equation, k_0 is the value of k determined isocratically in the starting solvent, b is the gradient steepness parameter, t is the time after the start of gradient and sample injection, and t_m is the column dead time. Ideally, this equation shows that a linear-solvent-strength gradient should result in equal resolution and bandwidths of all components. Unfortunately, this is not always possible. There are certain cases where linear-solvent-strength gradient is not the ideal method. In some cases, for example, b actually increases regularly with solute retention, which reduces the separation of late-eluting components. Such an effect is observed in the separation of polycyclic aromatic hydrocarbons [3].

There are three things to consider when finding a suitable gradient for a separation: (a) the initial and final mobile-phase compositions, (b) the gradient shape, and (c) the gradient steepness [3]. A convex gradient leads to the elution of bands with a lower average capacity factor and a shorter total analysis time. In other words, the later-eluting bands appear wider and better resolved than the early eluting bands. A concave gradient resolves the early bands to a greater degree than the later bands.

Solvent selection is one of the most important facets of gradient elution. The choice of the first solvent influences the separation of the initial bands, whereas the strength of the final solvent influences the selectivity of the separation and the retention times and peak shapes of later-eluting bands. If solvent B is too weak, the analysis time may become very long and the later-eluting bands might broaden excessively; thus, a stronger solvent B may be required [3].

Abbott et al. [4] devised a method designed to predict the retention times in gradient elution under the assumption that the retention factor as determined under isocratic conditions is a log-linear function of solvent composition according to Eq. (5), where k_w is the retention factor obtained in water, φ_0 refers to the volume fraction of the organic component, and S refers to the solvent strength for which the values can be obtained as the negative slope of plots of log k versus volume fraction:

$$\log k = \log k_w - S\varphi_0 \tag{5}$$

Engelhardt and Elgass [5] found that if the gradient volume is held constant and the initial and final compositions of the eluent are fixed, each component of a sample is eluted at a given solvent composition. Snyder et al. [6] derived a simple relationship between the elution time of a solute and the rate of change of solvent composition in gradient elution. Utilizing Eq. (6), they found that the elution time t_e is related to column dead time, t_0, and an experimental parameter b whereby k_0 is the retention factor that would be obtained in isocratic elution with mobile-phase composition used at the beginning of the gradient. [1]:

$$t_e = \left(\frac{t_0}{b}\right) \log(2.31\ k_0 b + 1) + t_0 \tag{6}$$

The parameter b is defined as

$$b = \frac{\Phi S t_0}{100} \tag{7}$$

where (Φ is the rate of increase in the concentration of the solvent component having eluent strength S and given as volume percent of organic solvent component per minute [1].

Many technical problems can occur with gradient elution, some of which can be avoided through various methods. To begin, gradient elution relies upon the purity of the solvents used. The high-performance liquid chromatography (HPLC) column can collect impurities, in the mobile phase, which may or may not elute as sharp peaks at a certain eluent composition. These can be mistaken for sample components. Such peaks are called "ghost peaks" and can result in inaccurate data. Water presents its own set of problems. Contaminated water can also result in ghost peaks. Even deionization of water by ion exchangers can leach out organics from the resin [1]. For this reason, it is advisable to run a gradient first without injecting the sample and use commercially available, purified solvents, including the water, to determine if they result in the elution of ghost peaks.

Another thing to consider with gradient elution is changes in the eluent viscosity. When gradient elution with a hydro-organic mobile phase is used (e.g., methanol–water), systematic variations in the flow rate are expected under conditions of constant-pressure operation, and systematic variations in the operating pressure will be found when a constant flow rate is used [1]. The compressibility of the solvent is species-specific.

In summary, gradient elution is a powerful method for the separation and analysis of complex mixtures containing components with a wide variety of polarities and hydrophobicities. It can also be used to help establish an isocratic mobile phase for the analysis of simpler mixtures. In either case, utmost care must be taken in the selection and use of solvents of high purity and selectivity.

References

1. C. Horvath, *High Performance Liquid Chromatography: Advances and Perspectives*, Vol. 2, Academic Press, New York, 1980.
2. J. J. Kirkland and J. L. Glajch, *J. Chromatogr. 255*: 27 (1983).
3. C. F. Poole and S. A. Schuette, *Contemporary Practice of Chromatography*, Elsevier, Amsterdam, 1984.
4. S. R. Abbott, J. R. Berg, P. Achener, and R. L. Stevenson, *J. Chromatogr. 126*: 421 (1976).
5. H. Englehardt, and H. Elgass, *J. Chromatogr. 158*: 249 (1978).
6. L. R. Snyder, J. W. Dolan, and J. R. Gant, *J. Chromatogr. 165*: 3 (1979).

J. E. Haky
D. A. Teifer

Gradient Elution in Capillary Electrophoresis

Gradient elution is routinely used in high-performance liquid chromatography (HPLC) to achieve the complete resolution of a mixture which could not be resolved using isocratic elution. Unlike isocratic elution, where the mobile-phase composition remains constant throughout the experiment, in gradient elution the mobile-phase composition changes with time. The change could be continuous or stepwise, known as the *step-gradient*. In the continuous gradient, the analyst can pick one of three general shapes: linear, concave, or convex. Gradient elution in HPLC is achieved using two pumps, two different solvent reservoirs, and a solvent mixer. In capillary electrophoresis (CE), electro-osmotic flow controls the flow of the mobile phase, which is, in most cases, an aqueous buffer and is used in place of a mechanical pump.

A manual step-gradient was used by Balchunas and Sepaniak [1] to separate a mixture of amines by micellar electrokinetic chromatography (MEKC). Stepwise gradients were produced by pipetting aliquots of a gradient solvent to the inlet reservoir which was filled with 2.5 mL of running buffer. A small magnetic stirring bar was used to ensure thorough mixing of the added gradient solvent with the starting mobile phase. The gradient elution solvent was manually added, in four 0.5 mL increments, spaced 5 min apart, 5 min after start of the experiment.

Bocek and his group [2] developed a method for controlling the composition of the operational electrolyte directly in the separation capillary in isotachophoresis (ITP) and capillary zone electrophoresis (CZE). The method is based on feeding the capillary with two different ionic species from two separate electrode chambers by simultaneous electromigration. The composition and pH of the electrolyte in the separation capillary is thus controlled by setting the ratio of two electric currents. This procedure can be used, in addition to generating the mobile-phase gradient, for generating pH gradients [3,4]. Sepaniak et al. [5–7] produced continuous gradients of different shapes (linear, concave, or convex) by using a negative-polarity configuration in which the inlet reservoir is at ground potential and the outlet reservoir at a very high negative potential. This configuration allows two syringe pumps to pump solutions into and out of the inlet reservoir. Tsuda [8] used a solvent-program delivery system, similar to that used in HPLC, to generate pH gradients in CZE. A pH gradient derived from temperature changes has also been reported [9]. Chang and Yeung [10] used two different techniques (i.e., the dynamic pH gradient and electro-osmotic flow gradient) to control selectivity in CZE. A dynamic pH gradient from pH 3.0 to 5.2 was generated by a HPLC gradient pump. An electro-osmotic flow gradient was produced by changing the reservoirs containing different concentrations of cetylammonium bromide for injection and running.

Capillary electrochromatography (CEC) is a separation technique which combines the advantages of micro-HPLC and CE. In CEC, the HPLC pump is replaced by electro-osmotic flow. Behnke and Bayer [11] developed a micro-bore system for gradient elution using 50- and 100-μm fused-silica capillaries, packed with 5 μm octadecyl reversed phase silica gel and voltage gradients, up to 30,000 V, across the length of the capillary. A modular CE system was combined with a gradient HPLC system to generate gradient CEC. Enhanced column efficiency and resolution were realized. Zare and his co-workers [12] used two high-voltage power supplies and a packed fused-silica capillary to generate an electro-osmotically driven gradient flow in an automated manner. The separation of 16 polycyclic aromatic hydrocarbons was resolved in the gradient mode; these compounds were not separated when the isocratic mode was employed. Others [13–16] used gradient elution in combination with CEC to resolve various mixtures.

Multiple, intersecting narrow channels can be formed on a glass chip to form a manifold of flow channels in which CE can be used to resolve a mixture of solutes in seconds. Harrison and co-workers [17] showed that judicious application of voltages to multiple channels within a manifold can be used to control the mixing of solutions and to direct the flow at the intersection of channels. The authors concluded that such a system, in which the applied voltages can be used to control the flow, can be used for sample dilution, pH adjustment, derivatization, complexation, or masking of interferences. Ramsey and co-workers [18] used a microchip device with electrokinetically controlled solvent mixing for isocratic and gradient elution in MEKC. Isocratic and gradient conditions are controlled by proper setting of voltages applied to the buffer reservoirs of the microchip. The precision of such control was successfully tested for gradients of various shapes (linear, concave, or convex) by mixing pure buffer and buffer doped with a fluorescent dye. By making use of the electro-osmotic flow and employing computer control, very precise manipulation of the solvent was possible and allowed fast and efficient optimization of separation problems.

Acknowledgment

This project has been funded in whole or in part with federal funds from the National Cancer Institute, National Institutes of Health, under Contract No. NO1-CO-56000.

By acceptance of this article, the publisher or recipient acknowledges the right of the U.S. government to retain nonexclusive, royalty-free license to any copyright covering the article.

The content of this publication does not necessarily reflect the views of the Department of Health and Human Services, nor does the mention of trade names, commercial products, or organizations imply endorsement by the U.S. government.

References

1. A. T. Balachunas and M. J. Sepaniak, *Anal. Chem. 60*: 617 (1988).
2. J. Popsichal, M. Deml, P. Gebauer, and P. Bocek, *J. Chromatogr. 470*: 43 (1989).
3. P. Bocek, M. Deml, J. Popsichal, and J. Sudor, *J. Chromatogr. 470*: 309 (1989).
4. V. Sustacek, F. Foret, and P. Bocek, *J. Chromatogr. 480*: 271 (1989).
5. M. J. Sepaniak, D. F. Swaile, and A. C. Powell, *J. Chromatogr. 480*: 185 (1989).

6. A. C. Powell and M. J. Sepaniak, *J. Microcol. Separ.* *2*: 278 (1990).
7. A. C. Powell and M. J. Sepaniak, *Anal. Instrum.* *21*: 25 (1993).
8. T. Tsuda, *Anal. Chem.* *64*: 386 (1992).
9. C. W. Wang and E. S. Yeung, *Anal. Chem.* *64*: 502 (1992).
10. H-T. Chang and E. S. Yeung, *J. Chromatogr.* *608*: 65 (1992).
11. B. Behnke and E. Bayer, *J. Chromatogr.* *680*: 93 (1994).
12. C. Yan, R. Dadoo, R. N. Zare, D. J. Rakestraw, and D. S. Anex, *Anal. Chem.* *68*: 2726 (1996).
13. K. Schmeer, B. Behnke, and E. Bayer, *Anal. Chem.* *67*: 3656 (1995).
14. M. R. Taylor, P. Teale, S. A. Westwood, and D. Perrett, *Anal. Chem.* *69*: 2554, (1997).
15. M. R. Taylor and P. Teale, *J. Chromatogr. A* *768*: 89 (1997).
16. P. Gfrorer, J. Schewitz, K. Psecker, L-H. Tseng., K. Albert, and E. Bayer, *Electrophoresis* *20*: 3 (1999).
17. K. Seller, Z. H. Fan, K. Fluri and J. Harrison, *Anal. Chem.* *66*: 3485 (1994).
18. J. P. Kutter, S. J. Jacobson, and J. M. Ramsey, *Anal. Chem.* *69*: 5165 (1997).

Haleem J. Issaq

Gradient Generation Devices and Methods

Introduction

Chromatography represents a large number of principles, methods, and approaches and can be divided into many different groups. One of the classifications is based on the "continuity" of conditions during the separation. Using such a classification, we can recognize two types of chromatographic elution or separation (i.e., isocratic and gradient).

Isocratic elution is the term used when the sample is introduced into the separation unit and eluted from it under a constant set of conditions. Isocratic separation is suitable for those applications in which the sample components have similar retention behavior and are eluted rapidly, one after the other.

Gradient elution involves a change in chromatographic conditions during a chromatographic run in order to achieve separation of sample components of widely varying affinities for the stationary phase or of different solubilities in the mobile phase.

There is a good correlation between the retention behavior of solutes under isocratic conditions and the retention observed under conditions of gradient elution [1]. Migration of the solute bands through a chromatographic column under both isocratic and gradient conditions is described in Fig. 1. It demonstrates that the gradient elution cannot improve the resolution, but it is a convenient tool for controlling the speed of the solute migration through the column and, as a result of this, the position of the solute peaks on a chromatogram.

From the practical point of view, it is not important which mechanism controls the chromatographic process during a gradient elution, but how the changes in experimental conditions affect the resulting retention times of separated compounds. Typically, conditions that support a strong retention are applied at the start of the run, and conditions enhancing an elution are applied more and more over the course of the separation. This allows for sufficient resolution of the early eluted, weakly retained solutes, while ensuring that the elution time of the later peaks is not excessively long. Changes in experimental conditions that affect elution can be either continuous or happening in discrete steps. The continuous gradient can be of any shape. The most popular is a linear gradient, but convex, concave, or a combination of several profiles are also used. Step-elution involves a sudden change in the

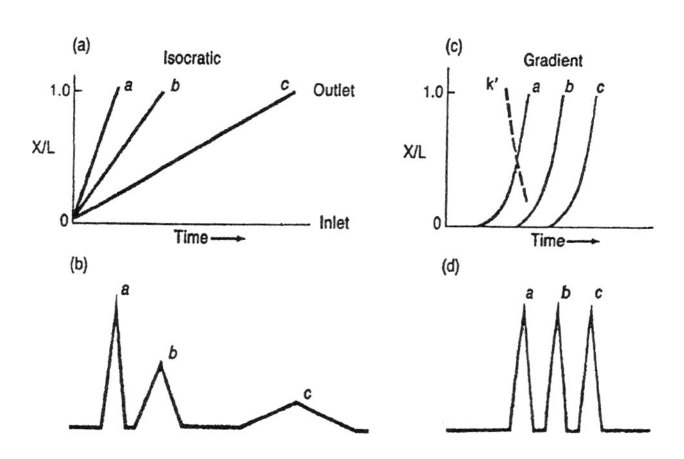

Fig. 1 Fractional migration (X/L) of solute bands along the column and resulting chromatograms for isocratic versus gradient elution. (Reprinted with permission from L. R. Snyder, M. A. Stadalius, and M. A. Quarry, *Analytical Chemistry*, Vol. 55, No. 14, 1983, pp. 1413A–1430A. Copyright 1983 American Chemical Society.)

composition of the mobile phase, followed by a period where the mobile phase is held constant. This process may be repeated several times during an analysis. Optimization of the gradient profile may lead to a successful separation of a large variety of the solutes in a single run.

Theoretical principles, approaches, applications, and advantages of gradient elution are described in detail in a large variety of books and review articles. We recommend reading some of those issued in the last 2 years and the papers referenced therein (see Ref. 2 for some examples). In this entry, we will briefly discuss various types of gradient elution and compare different ways of gradient generation.

Discussion

The devices and methods for gradient generation can be sorted out, for example, according to parameters that are being intentionally changed during the chromatographic run. From those parameters, composition of the mobile phase, flow rate, and temperature are the most important factors influencing the separation in a desirable way.

Mobile-Phase Composition Gradient

The mobile-phase composition gradient involves a change in solvent ratio or concentration of an additive. The properties that affect the separation conditions are, for example, solvent strength, polarity, ionic strength, and pH. Some gradients can be simply generated by an injection of a liquid having an elution strength different from that of the original mobile phase [3]. In a vast majority of cases, the gradient is created by a controlled blending of several eluents together before entering the injection port. A binary gradient refers to a gradient employing two different eluents, a ternary gradient refers to a system in which three different eluents are used, and a quaternary gradient where four different eluents are used.

Gradient Pumps

Gradient pumping (solvent delivery) systems are those that can accurately mix and deliver more than one solvent during an analysis. There is a large variety of gradient pumps for liquid chromatography on the market [4] and many different ways of using them [5]. The major parameters for judging the quality of an isocratic pump are the flow rate precision and accuracy, robustness, and back-pressure capability. In addition to this, accuracy and smoothness of the mobile-phase mixing are the main parameters describing the quality of a gradient pump. The blending of the solvents can occur in one of two ways (i.e., high-pressure mixing or low-pressure mixing).

High-Pressure Mixing

High-pressure mixing means that the eluents are blended on the "high-pressure" side of the pump. Two or more isocratic pumps are required, one for each solvent used to generate the gradient. The output fluid lines of the pumps are joined with a mixing device. Then, the mobile-phase composition during the gradient is easily controlled by controlling the ratio of the flow rates of the pumps. The schematic of such a setup is shown in Fig. 2a. More than two solvents can also be mixed in this way, by using pumps connected to the mixing device.

For some "high-speed" or "low-volume" applications, it is very important to minimize the time interval needed for the mobile phase to travel from the mixing point to the injection port. High-pressure mixing has an advantage of very low delay volumes, which makes it attractive for rapid analyses, micro high performance liquid chromatography (HPLC), mass spectrometry (MS), and many other applications. The gradient solvent delivery systems can be

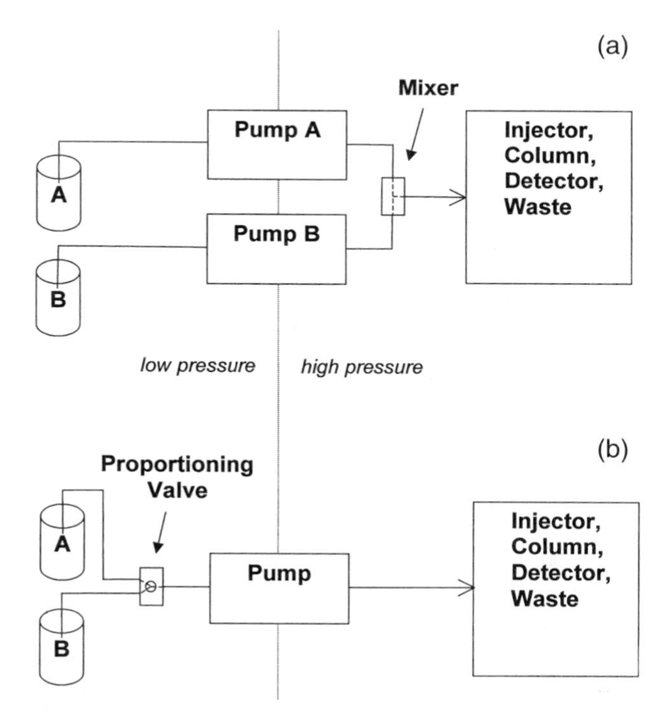

Fig. 2 Schematic of the mobile-phase gradient generation device based on high-pressure (a) and low-pressure (b) mixing of two liquids.

configured with delay volumes of only a few microliters. The disadvantage of the high-pressure mixing is that piston pumps have limited precision at the extremes of the flow rate ranges, which becomes worse with increasing volume of the piston and decreasing volume of the mixing device. In addition, blending of solvents, compressibility, or combining volumes which are nonadditive may result in unwanted fluctuations of the flow rate at the injector and through the column.

Typically, a constant flow rate (i.e., the sum of the flow rates of all of the isocratic pumps) is desired at any point of the run. However, in certain cases, a combination of the mobile-phase composition and flow rate gradients might be useful.

Low-Pressure Mixing

In contrast to the high-pressure mixing that had prevailed in the early years of gradient liquid chromatography, low-pressure mixing became a common configuration for most of the modern HPLC systems. In the low-pressure mixing arrangement, the solvents are blended at atmospheric pressure, ahead of the pump, and a single high-pressure pump is used to deliver the mixture to injector and column. Figure 2b shows the schematic of such a configuration. It is easier to control mixing of several solvents at the low-pressure side of the pump, as compared to a high-pressure mixing. Therefore, the precision of the blending is good, even at extremes of mobile-phase compositions. Typically, the low-pressure gradient pumps are equipped with a valve that can mix up to four solvents. After the mixing, the blend of the solvents travels through the pump head, pressure transducer, and pulse damper before entering the injection port. This results in a slightly higher delay volume than that offered by a high-pressure gradient system. The delay volume was reduced below 1 mL only for the most advanced low-pressure gradient systems, recommended for separations requiring short and narrow columns or for connection to a low flow rate requiring detection systems such as mass spectrometer.

Electro-osmotic Mixing

The mobile phase can be driven through the chromatographic column not only by a mechanical pressure but also by applying voltage on both ends of the column that generates an electro-osmotic flow. The technique, which uses electro-osmotic flow for driving a mobile phase through a column to achieve a chromatographic separation, is called electrochromatography. Various approaches for gradient elution in electrochromatography have been explored.

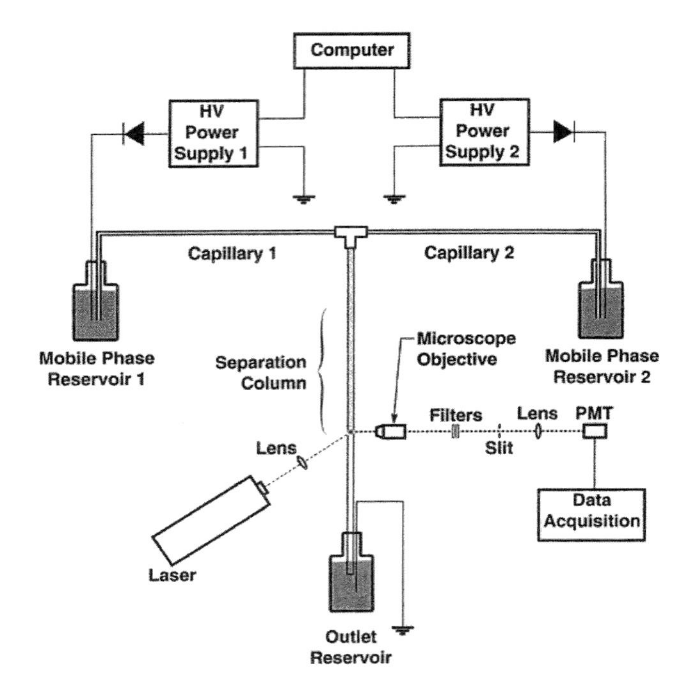

Fig. 3 Schematic of the mobile-phase gradient generation device based on electro-osmotic mixing of two liquids. (Reprinted with permission from C. Yan, R. Dadoo, R. N. Zare, D. J. Rakestraw, and D. S. Anex, *Analytical Chemistry*, Vol. 68, No. 17, 1996, pp. 2726–2730. Copyright 1996 American Chemical Society.)

Simple combination of pressure-driven mixing with electro-osmotic delivery of the resulting blend has been used [6]; however, a more perspective approach seems to be the one that includes merging of two or more electro-osmotic flows, regulated by computer-controlled voltages [7]. An example of such a system is shown in Fig. 3. The mobile-phase composition can be easily changed in time via changes of the applied voltages. One of the main advantages of this approach is that the electro-osmotically driven gradient devices can be miniaturized to the extreme level and micromachined to a high precision to be a part of the new "lab-on-a-chip" technology [8].

On-Line Gradient Generation

Various gradient generation devices have been used for transforming deionized water into solutions of varying pH or ionic strength. The on-line generator allows one to apply mobile-phase gradients without the necessity of blending several starting eluents [9]. In addition, those systems offer very short gradient delays and prevent the corrosive acidic, basic, or salt solutions from contacting sensitive parts of the pump.

Flow Rate Gradient

The main purpose of gradient elution is to achieve separation of sample components which differ widely in retention properties, in a single run. This means that, after eluting all of the fast-moving components of the sample, we want to speed up the elution of the more retained ones. It is obvious that an increase in the flow rate during the run would serve well for this purpose. Any isocratic pump that allows one to program flow rate versus time can be used for the flow rate gradient elution. In contrast to the mobile-phase composition changes, the speed of the sample band migration through a column increases only linearly with a linear change in the flow rate. Also, the back-pressure increases proportionally with the flow rate increase. These factors severely limit applicability of the flow rate gradients.

Temperature Gradient

Because viscosity of the mobile-phase decreases and the speed of diffusion of the separated molecules increases with an increase in temperature, the temperature gradient represents another way of speeding up elution during a chromatographic run. In addition to these effects, the stationary-phase surface or some of the separated molecules may go through phase-transition changes during the temperature gradient, leading to a strong effect on the separation. Unfortunately, temperature changes, especially during the cooling period, often a require significant time for equilibration; this limits the applicability of the temperature-gradient elution in liquid chromatography.

On the other hand, optimization of the column temperature may significantly improve the overall performance of the chromatographic system. Several studies describe a combination of temperature control with the mobile-phase composition changes and the effect of the combined parameters on reversed-phase separations [10].

References

1. L. R. Snyder, M. A. Stadalius, and M. A. Quarry, *Anal. Chem. 55*: 1413A (1983).
2. L. R. Snyder and J. W. Dolan, *Adv. Chromatogr. (NY) 38*: 115 (1998); J. B. Li, J. Morawski, *LC–GC 16*: 468 (1998); P. Schoenmakers, Programmed analysis, in *Handbook of HPLC* (E. Katz, R. Eksteen, P. Schoenmakers, and N. Miller, eds.), Chromatography Science Series Vol. 78, Marcell Dekker, Inc., New York, 1998, pp. 193–231; A. Weston and P. Brown, *HPLC and CE: Principles and Practice*, Academic Press, San Diego, CA, 1997, pp. 1–130.
3. B. Streel, A. Ceccato, P. Chiap, Ph. Hubert, and J. Crommen, *Biomed. Chromatogr. 9*: 254 (1995); V. Berry, *J. Liq. Chromatogr. 13*: 1529 (1990).
4. See for example, http://www.lcgcmag.com/lcgcbg/prod42.htm
5. R. L. Stevenson, Mobile-phase delivery systems for HPLC, in *Handbook of HPLC* (E. Katz, R. Eksteen, P. Schoenmakers, and N. Miller, eds.), Chromatography Science Series Vol. 78, Marcel Dekker, Inc., New York, 1998.
6. C. G. Huber, G. Choudhary, and C. Horvath, *Anal. Chem. 69*: 4429 (1997).
7. C. Yan, R. Dadoo, R. N. Zare, D. J. Rakestraw, and D. S. Anex, *Anal. Chem. 68*: 2726 (1996).
8. D. Figeys and R. Aebersold, *Anal. Chem. 70*: 3721 (1998); J. P. Kutter, S. C. Jacobson, and J. M. Ramsey, *Anal. Chem. 69*: 5165 (1997).
9. Y. Liu, N. Avdalovic, C. Pohl, R. Matt, H. Dhillon, and R. Kiser, *Am. Lab. 30 (22)*: 48 (1998).
10. J. W. Dolan, L. R. Snyder, N. M. Djordjevic, D. W. Hill, D. L. Saunders, L. Van Heukelem, and T. J. Waeghe, *J. Chromatogr. A 803*: 1 (1998); M. H. Chen and C. Horvath, *J. Chromatogr. A 788*: 51 (1997).

Miroslav Petro

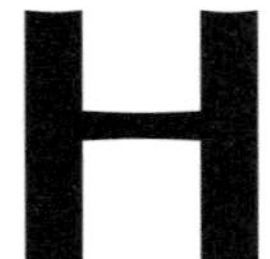

Headspace Sampling

Headspace sampling is usually employed to identify the volatile constituents of a complex matrix without actually taking a sample of the material itself. There are three variations of the technique: (a) static headspace sampling, (b) dynamic headspace sampling, and (c) purge and trapping.

The first technique, commonly used to monitor the condition of foodstuffs, particularly for detecting food deterioration (food deterioration is often accompanied by the characteristic generation of volatile products such as low-molecular-weight organic acids, alcohols, and ketones, etc.), involves first placing the sample in a flask or some other appropriate container and warming to about 40°C. Raising the temperature increases the distribution of the volatile substances of interest in the gas phase. A defined volume of the air above the material is withdrawn through an adsorption tube by means of a gas syringe. Graphitized carbon is often used as the adsorbing material, although other substances such as porous polymers can also be employed. Carbon adsorbents having relatively large surface areas ($\sim$ 100 m^2/g) are used for adsorbing low-molecular-weight materials, whereas for large molecules, adsorbents of lower surface areas are used ($\sim$ 5 m^2/g). After sampling, the adsorption trap is placed in an oven and connected to the chromatograph. The column is maintained at a low temperature (50°C or less) to allow the desorbed solutes to concentrate at the beginning of the column. The trap is then heated rapidly to about 300°C and a stream of carrier gas sweeps the desorbed solutes onto the column. When desorption is complete, the temperature of the column is programmed up to an appropriate temperature and the components of the headspace sample are separated and quantitatively assayed. The proportions of each component in the gas phase will not be the same as that in the sample, as they are modified by the distribution coefficient. Thus, analyses will be comparative or relative, but not absolute.

The second analytical procedure is somewhat similar, but a continuous stream of gas is passed over the sample and through the trap. This produces a much larger sample of the volatile substances of interest and, thus, can often detect trace materials. The adsorbed components are desorbed by heat in the same manner and passed directly onto a gas chromatography (GC) column. The results are still determined by the distribution coefficient of each solute between the sample matrix and the air and, thus, the quantitative results remain comparative or relative, but not absolute.

The third method (purge and trap) is used for liquids and, in particular, for testing for water pollution by volatile solvents. In this method, air or nitrogen is bubbled through the water sample and then through the adsorbent tube. In this way, the substances of interest can be completely leached from the water; the results will give the total quantity of each solute in the original water sample. Thus, with this method, the results can be actual and not relative or comparative. The solutes are desorbed by heat in exactly the same way as the previous two methods, but provision is usually made to remove the water that is also collected before developing the separation.

A good example of the use of headspace analysis is in the quality control of tobacco. Despite the health concern in the United States, tobacco is an extremely valuable export and its quality needs to be carefully moni-

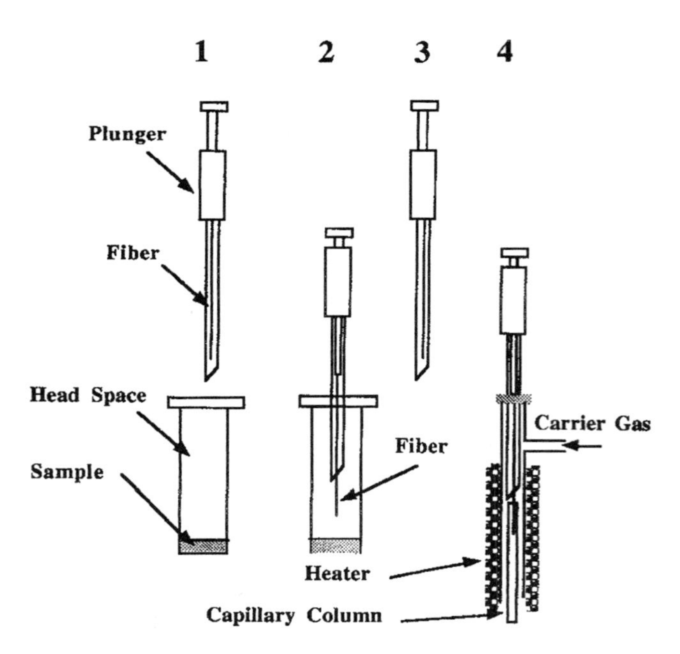

Fig. 1 The SPME apparatus.

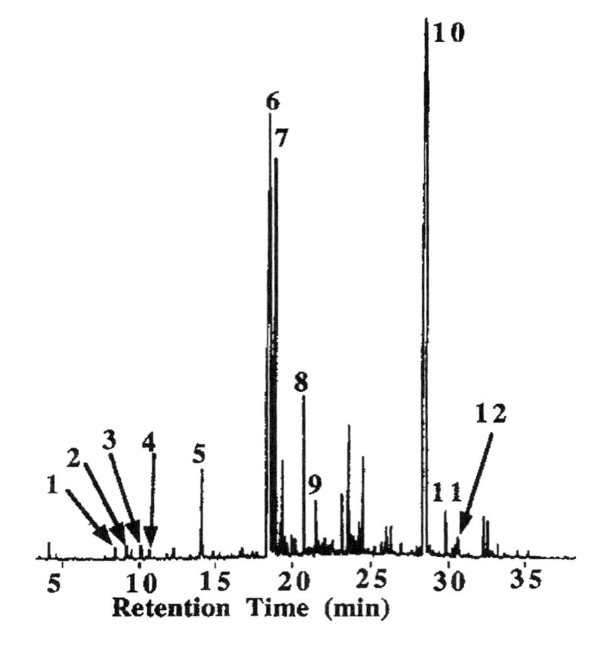

Fig. 2 A chromatogram of tobacco headspace. 1: Benzaldehyde; 2: 6-methyl-5-heptene-2-one; 3: phenylacetaldehyde; 4: ninanal; 5: menthol; 6: nicotine; 7: solanone; 8: geranyl acetone; 9: β-nicotyrine; 10: neophytadiene; 11: famesylacetone; 12: cembrene.

tored. Tobacco can be flue cured, air cured, fire cured, or sun cured, but the quality of the product can often be monitored by analyzing the vapors in the headspace above the tobacco.

The headspace over tobacco can be sampled and analyzed using a solid-phase micro-extraction (SPME) technique. The apparatus used for SPME is shown in Fig. 1. The basic extraction device consists of a length of fused-silica fiber, coated with a suitable polymeric adsorbent, which is attached to the steel plunger contained in a protective holder. The steps that are taken to sample a vapor are depicted in Fig. 1. The sample is first placed in a small headspace vial and allowed to come to equilibrium with the air in the vial (1). The needle of the syringe containing the fiber is then made to pierce the cap, and the plunger pressed to expose the fiber to the headspace vapor. The fiber is left in contact with air above the sample for periods that can range from 3 to 60 min, depending on the nature of the sample (2).

The fiber is then removed from the vial (3) and then passed through the septum of the injection system of the gas chromatograph into the region surrounded by a heater (4). The plunger is again depressed and the fiber, now protruding into the heater, is rapidly heated to desorb the sample onto the GC column. In most cases, the column is kept cool so the components concentrate on the front of the column. When desorption is complete (a few seconds), the column can then be appropriately temperature programmed to separate the components of the sample. A chromatogram of the headspace sample, taken over tobacco, is shown in Fig. 2. The actual experimental details were as follows. One gram of tobacco (12% moisture) is placed in a 20-mL headspace vial and 3.0 mL of 3*M* potassium chloride solution is added. The fiber is coated with polydimethyl siloxane (a highly dispersive adsorbent) as a 100-μm film. The vial is heated to 95°C and the fiber is left in contact with the headspace for 30 min. The sample is then desorbed from the fiber for 1 min at 259°C. The separation can be carried out on a column 30 cm long with a 250-μm inner diameter, carrying a 0.25-μm-thick film of 5% phenylmethylsiloxane. The stationary phase is predominantly dispersive, with a slight capability of polar interactions with strong polarizing solute groups by the polarized aromatic nuclei of the phenyl groups. Helium can be used as the carrier gas, at 30 cm/s. The column is held isothermally at 40°C for 1 min, then programmed to 250°C at 6°C/min and held at 250°C for 2 min. It is seen that a clean separation of the components of the tobacco headspace is obtained and the resolution is quite adequate to compare tobaccos from different sources, tobaccos with different histories, and tobaccos of different quality.

References

1. D. W. Grant, *Capillary Gas Chromatography* (R. P. W. Scott, C. F. Simpson, and E. D. Katz, eds.), John Wiley & Sons, Chichester, 1996.
2. R. P. W. Scott, *Introduction to Analytical Gas Chromatography*, Marcel Dekker, Inc., New York, 1998.
3. R. P. W. Scott, *Techniques of Chromatography*, Marcel Dekker, Inc., New York, 1995.

Raymond P. W. Scott

Helium Detector

The outer group of electrons in the noble gases is complete, and as a consequence, collisions between noble gas atoms and electrons are perfectly elastic. It follows that if a high potential is set up between two electrodes in a noble gas and ionization is initiated by a suitable radioactive source, electrons will be accelerated toward the anode and will not be impeded by energy absorbed from collisions with the noble gas atoms. However, if the potential of the anode is high enough, the electrons will develop sufficient kinetic energy that, on collision with a the noble gas atom, energy can be absorbed and a *metastable* atom can be produced. A metastable atom carries *no* charge, but adsorbs energy from collision with a high-energy electron by displacing an orbiting electron to an outer orbit. Metastable helium atoms have an energy of 19.8 and 20.6 eV and thus can ionize and, consequently, detect all permanent gas molecules and, in fact, the molecules of all other volatile substances. A collision between a metastable atom and an organic molecule will result in the outer electron of the metastable atom collapsing back to its original orbit, followed by the expulsion of an electron from the organic molecule. The electrons produced by this process are collected at the anode and produce a large increase in anode current. However, when an ion is produced by collision between a metastable atom and an organic molecule, the electron, simultaneously produced, is also immediately accelerated toward the anode. This results in a further increase in metastable atoms and a consequent increase in the ionization of other organic molecules.

This cascade effect, unless controlled, results in an exponential increase in ion current. It is clear that the helium must be extremely pure or the production of metastable helium atoms would be quenched by traces of any other permanent gases that may be present.

Originally, a very complicated helium-purifying chain was necessary to ensure the helium detector's optimum operation. However, with high-purity helium becoming generally available, the helium detector is now a more practical system.

The metastable atoms that must be produced in the argon and helium detectors need not necessarily be generated from electrons induced by radioactive decay. Electrons can be generated by electric discharge or photometrically, which can then be accelerated in an inert gas atmosphere under an appropriate electrical potential to produce metastable atoms. This procedure is the basis of a highly sensitive helium detector that is depicted on the left-hand side of Fig. 1. The detector does not depend solely on metastable helium atoms for ionization and, for this reason, is called the helium discharge ionization detector (HDID).

The sensor consists of two cavities, one carrying a pair of electrodes across which a potential of about 550 V is applied. In the presence of helium, this potential initiates a gas discharge across the electrodes. The discharge gas passes into a second chamber that acts as the ionization chamber and any ions formed are collected by two plate electrodes having a potential difference of about 160 V. The column eluent enters the top of the ionization chamber and mixes with the helium from the discharge chamber and exits at the base of the ionization chamber.

In this particular detector, ionization probably occurs as a result of a number of processes. The electric discharge produces both electrons and photons. The electrons can be accelerated to produce metastable helium atoms which, in turn, can ionize the components in the column eluent. However, the photons generated in the discharge have, themselves, sufficient energy to ionize many eluent components and so ions will probably be produced by both mechanisms. It is possible that other ionization processes may also be involved, but the two mentioned are likely to account for the majority of the ions produced. The response of the detector is largely controlled by the collect-

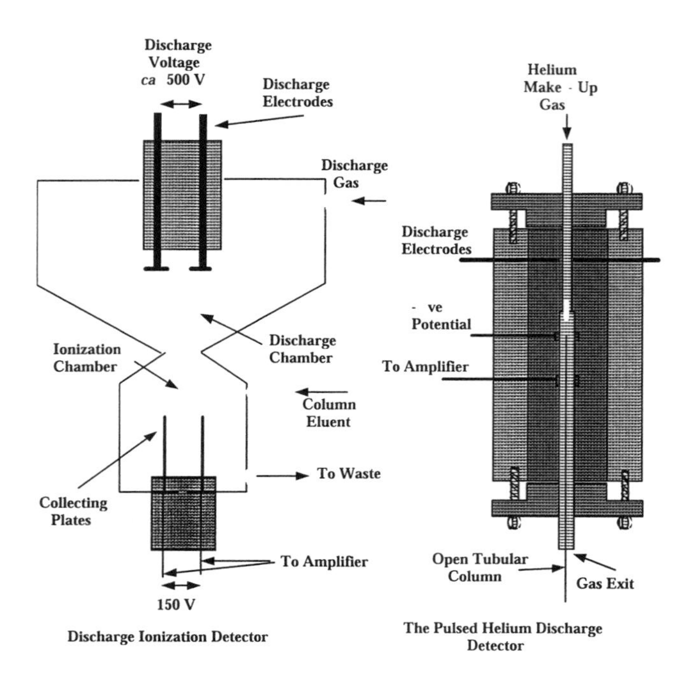

Fig. 1 The discharge ionization detector (courtesy of GOW-MAC Instruments) and the pulsed helium discharge detector (courtesy of Valco Instruments).

ing voltage and is very sensitive to traces of inert gases in the carrier gas. Peak reversal is often experienced at high collecting voltages, which may also indicate that some form of electron capturing may take place between the collecting electrodes. This peak reversal appears to be significantly reduced by the introduction of traces of neon in the helium carrier gas.

The helium discharge ionization detector has a high sensitivity toward the permanent gases and has been used very successfully for the analysis of trace components in ultrapure gases. It would appear that the detector response is linear over at least two, and possibly three, orders of magnitude, with a response index probably lying between 0.97 and 1.03. In any event, any slight nonlinearity of the sensor can be corrected by an appropriate signal-modifying amplifier. The potential sensitivity of the detector to organic vapors appears to be about 1×10^{-13} g/mL.

The Pulsed Helium Discharge Detector

The pulsed helium discharge detector [1,2] is an extension of the helium detector, a diagram of which is shown on the right-hand side of Fig. 1. The detector has two sections: the upper section consisting of a tube 1.6 mm i.d. (where the discharge takes place) and the lower section, 3 mm i.d. (where reaction with metastable helium atoms and photons takes place). Helium makeup gas enters the top of the sensor and passes into the discharge section. The potential (about 20 V) applied across the discharge electrodes and for optimum performance is pulsed at about 3 kHz with a discharge pulse width of about 45 μs. The discharge produces electrons and high-energy photons (that can also produce electrons), and probably some metastable helium atoms. The photons and metastable helium atoms enter the reaction zone where they meet the eluent from the capillary column. The solute molecules are ionized and the electrons produced are collected at the lower electrode and measured by an appropriate high-impedance amplifier. The distance between the collecting electrodes is about 1.5 mm. The helium must be 99.9995 pure, otherwise permanent gas impurities quench the production of metastable atoms. The base current ranges from 1×10^{-9} to 5×10^{-9} A, the noise level is about 1.2×10^{-13} A, and the ionization efficiency is about 0.07%. It is claimed to be about 10 times more sensitive than the flame ionization detector and to have a linear dynamic range of 10^5. The pulsed helium discharge detector appears to be an attractive alternative to the flame ionization detector and would eliminate the need for three different gas supplies. It does, however, require equipment to provide specially purified helium, which diminishes the advantage of using a single gas.

References

1. W. E. Wentworth, S. V. Vasnin, S. D. Stearns, and C. J. Meyer, *Chromatographia 34*: 219 (1992).
2. W. E. Wentworth, H. Cai, and S. D. Stearns, *J. Chromatogr. 688*: 135 (1994).

Suggested Further Reading

Scott, R. P. W., *Chromatographic Detectors*, Marcel Dekker, Inc., New York, 1996.
Scott, R. P. W., *Introduction to Analytical Gas Chromatography*, Marcel Dekker, Inc., New York, 1998.

Raymond P. W. Scott

High-Temperature High-Resolution Gas Chromatography

Introduction

Gas chromatography (GC), in its early days, used packed columns with chemically inert solid supports coated with stationary phases. These columns presented low efficiency due to the wide range of particle sizes used, causing inhomogeneity in the packed bed and, consequently, high instability due to a poor deactivation and thermal instability at high-temperature operations [1]. This characteristic limited the use of the GC to only volatile and low-mass molecular compounds. The later development of columns with a stationary phase coated on the inner wall of the capillary provided a more inert environment. In this form, columns with higher thermal stability and more efficiency (higher N) were produced, allowing the analysis of semivolatile and medium molecular mass compounds. This technique was named high-resolution gas chromatography (HRGC) [1]. The possibility of using thermally stable, highly efficient columns, stimulated scientists to search for new stationary phases and chemical manufacturing processes to produce capillary columns with high thermal stabilities, capable of operating at higher temperatures [2] (to 360°C).

Lipsky and McMurray [3] suggested, in their pioneering work on high-temperature high-resolution gas chromatography (HT-HRGC), the use of column temperatures equal to, or higher than, 360°C. However, other column temperature values have also been reported for this technique [4].

The thermal stability of the high-temperature capillary columns allowed the analysis of higher molecular masses (more than 600 Das) and nonvolatile compounds never before directly analyzed by gas chromatography [2].

Instrumentation for HT-HRGC

The instrumentation used for HT-HRGC is the same as used for conventional GC, with only minor modifications.

Columns

The columns utilized in HT-HRGC are short (usually equal to, or shorter than, 10 m) coated with thin films (~ 0.1 μm or less) and having an inner diameter (i.d.) around 0.2 mm [5].

A smaller inner diameter (e.g., 0.1 mm) can also be used, but with the inconvenience of limiting the work to more diluted samples in order to avoid column overload. On the other hand, this type of column permits carrier gas speeds higher than with columns of inner diameters in the range 0.2–0.3 mm. Columns with inner diameters equal to 0.1 mm exhibit fewer plates with the increment of the carrier gas speed, in contrast to the columns with equivalent characteristics, but of 0.3 mm i.d. [5]. The increase of the carrier gas speed in smaller-i.d. columns performs an analysis in a shorter time, without undermining the efficiency of separation [6].

Capillary columns, to be suitable to HT-HRGC, must be extremely robust and must be coated with a thin film of the stationary phase with the purpose of reducing the retention of the less volatile compounds and preventing stationary-phase bleed at high temperatures [7].

Using such proper columns, elution of substances with carbon numbers in excess of n-C_{130} has been reported, at column temperatures of up to 430°C [8].

Tubing Material for HT-HRGC Columns

There are four major types of materials being utilized to prepare columns for high-temperature capillary columns [2]:

1. Glass (borosilicate)
2. Polyimide-clad fused silica
3. Aluminum-clad fused silica
4. Metal-clad fused silica

Columns of aluminum-clad fused silica [2,4] and metal-clad fused silica support temperatures up to 500°C, representing an advantage in comparison with borosilicate glass columns, with a temperature limit to 450°C, and columns of polyimide-clad fused silica for high temperature [2,9], limited temperature to 400–420°C. On the other hand, aluminum-clad fused silica columns present leakage, principally in the connections, after a short time of use [2,9]. Polyimide-clad fused-silica capillaries, after prolonged exposure to temperatures above 380°C, tend to break spontaneously at many points, thus losing the polyimide coating [9]. Borosilicate columns are inexpensive, being an alternative to fused silica for high-temperature applications. However, these columns have been reported to leak when coupled with retention gap and to mass spectrometry detectors [2]. An important alternative for HT-HRGC are HT metal-clad fused-silica columns which resist temperatures above 500°C for long-term exposure [9].

Stationary Phases

The first results on HT-HRGC [3,10] were published in 1983, dealing with stationary-phase immobilization (polysiloxane —OH terminated). Due to the column instability, when submitted to high temperature, stationary-phase loss was common at that time. These works can be considered to be the precursor of high-temperature gas chromatography, because the phase immobilization process developed resulted in a series of OH-terminated polysiloxane phases compatible with the inner surfaces of borosilicate glass and fused-silica tubing. These phases are thermally stable and capable of withstanding elevated temperatures [11] used in HT-HRGC. After this report, many other articles dealing with the ideal stationary phase for high-temperature gas chromatography appeared. Nonpolar stationary phases of the carborane–siloxane-type bonded phase (temperature range >480°C) and siloxane–silarylene copolymers suitable for HTGC were developed [7] around 1988.

A medium-polarity stationary phase based on fluoralkyl–phenyl substitution, which is thermally stable up to 400°C, was reported [12], and a CH_3O-terminated polydimethyl siloxane, diphenyl-substituted stationary phase made possible the analysis of complex high-molecular-mass mixtures such as free-base porphyrins and triglycerides using narrow-bore capillary columns [5]. Since these developments, a variety of stationary phases for analysis of specific analytes by HT-HRGC were found [2].

Sample Introduction

The sampling and elution of such high-molecular-weight materials requires careful attention in order to avoid quantitative sample losses during the sample introduction step. In general, "cold" injection techniques are required for accurate nondiscriminative sample transfer into the column. Cold on-column and programmed temperature (PT) split/splitless injection have been used with success for a large number of HT-HRGC analyses. In certain cases, however, significant losses of compounds above n-C_{60} have been observed with PT splitless injection [13]. This effect was identified as a time-based discrimination process caused by purging the PT inlet too soon after injection, resulting in incomplete sample vaporization [14].

Actually, same articles show the possibility of use split injection [8] in HT-HRGC analyses of substances up to C_{78}. However, volatile materials from the septum accumulate at the head of the column during the cool-down portion of the temperature program. When the columns are reheated to analyze the next sample, these accumulated volatiles are eluted, producing peaks, a baseline rise, or both. This difficulty can be solved using commercial septa already available for HT-HRGC, which exhibit very low bleed levels.

Detectors

High-temperature high-resolution GC is a technique similar to conventional GC; however, it presents high column bleeding due to the high temperature to which the column is submitted. Selective detectors, when used in HT-HRGC, require special attention. As an example, the electron-capture detector (ECD) is a very sensitive detector and should not be used in HT-HRGC because of its ability to detect column bleeding. This fact limits the detectors used to a few, such as the flame ionization detector (FID), alkali-flame ionization detector (AFID), and mass spectrometry detector (MS). In HT-HRGC, these detectors usually need small adjustments; for example, the MS detector requires a special interface when used for HT-HRGC [2].

HT-HRGC Application

High-temperature high-resolution GC has opened to many scientists the opportunity to analyze compounds of high molecular mass (600 Da or more) with similar efficiency to conventional high-resolution gas chromatography (HRGC). Actually, HT-HRGC has been applied to the analyses of compounds from several different areas [15–18]. As a general rule, this will avoid the time-consuming and usually expensive step of derivatization. In natural products, underivatized triterpenic compounds found in medicinal plants can be analyzed by this technique. The HT-HRGC analysis of triterpenes in aqueous alcoholic extracts of *Maytenus ilicifolia* and *M. aquifolium* leaves clearly allows the detection of the presence of friedelan-3-ol and friedelin and, therefore, allows distinguishing between the two varieties [15]; this differentiation is very important in pharmacological studies, because they present different biological activities.

Cyclopeptidic alkaloids (molecular mass ~ 600 Da), a class of important alkaloids which present biological activity, were analyzed by HT-HRGC without derivatization [16]. Figure 1 illustrates the separation of cyclopeptidic alkaloids in the chloroform fraction. The following selected compounds were identified: (1) Franganine, (2) Miriantine-A, (3) Discarine-C, and (4) Discarine-D.

Triacylglycerides from animal and vegetable sources have been separated and identified by HT-HRGC and

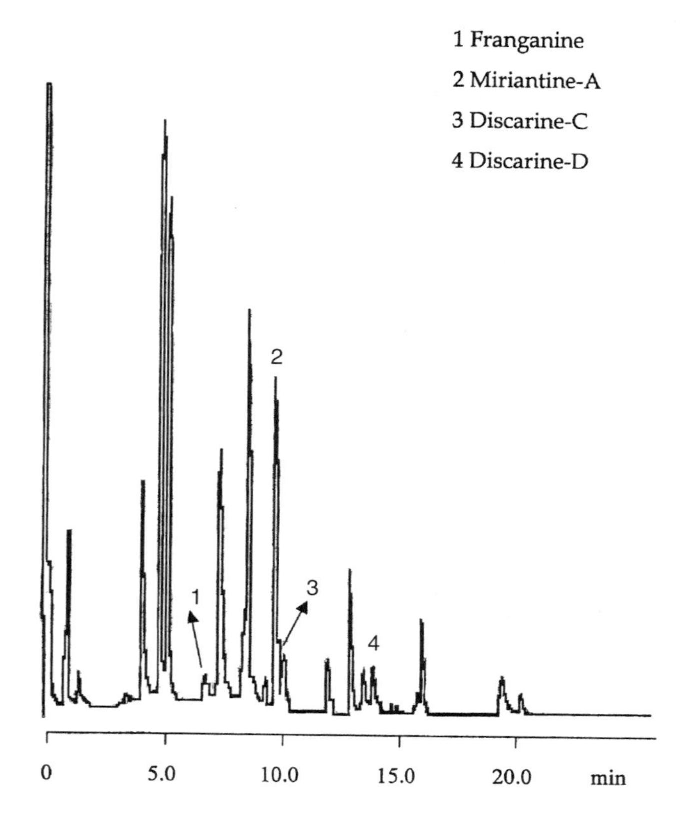

Fig. 1 Analysis of underivatized cyclopeptidic alkaloids in chloroform extract using HT-HRGC. Condition: fused-silica capillary column (6 m × 0.25 mm × 0.08 μm) coated with a LM-5 (5% phenyl, 95% polymethylsiloxane immobilized bonded phase) stationary phase. Temperature condition: column at 200°C (1 min), increased by 4°C/min, then 300°C (5 min); inlet: 250°C; FID detector: 310°C.

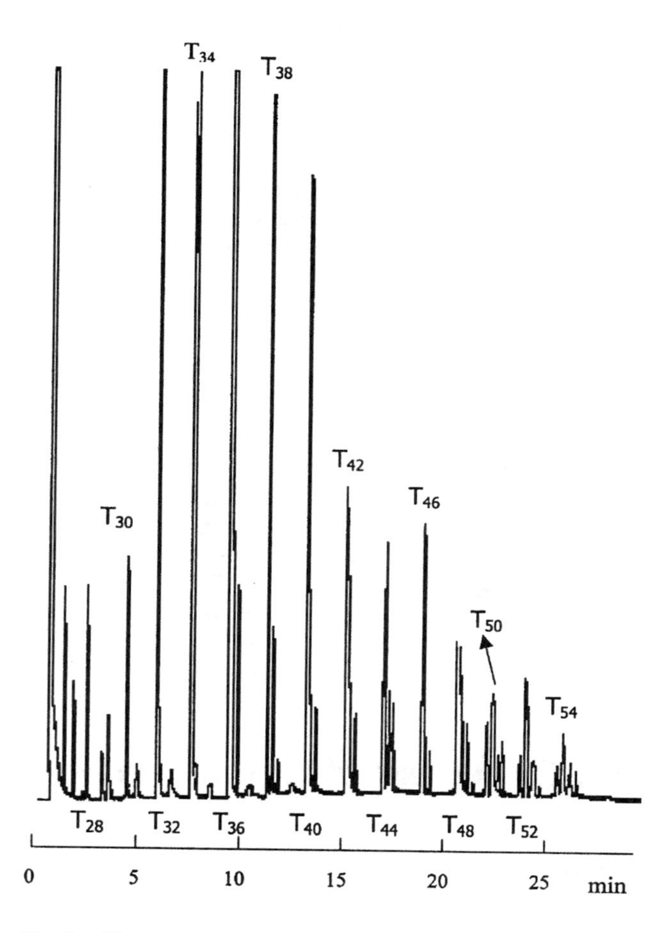

Fig. 2 Chromatogram of underivatized Palmist Oil (*Elaesis guineensis* L.) triacylglyceridic fraction using HT-HRGC. Condition: fused-silica capillary column (25 m × 0.25 mm × 0.1 μm) with the stationary phase OV-17-OH (50% phenyl, 50% methylpolysiloxane immobilized phase). Temperature condition: column at 350°C isothermic; injector: 360°C; FID detector: 380°C. T is the number of the underivatized triacilglyceride (e.g., T_{50} means a triacylglyceride having 50 carbon atoms).

high-temperature gas chromatography coupled to mass spectrometry (HT-HRGC/MS). Figure 2 shows the chromatographic profile of palm oil (*Elaeis guineensis* L.) by HT-HRGC, and the triacylglyceride compounds identification [17].

The HT-HRGC/MS technique was also used as an important tool to identify and quantify cholesterol present in the total lipid extracts of archeological bones and teeth, constituents of a new source of paleodietary information [19].

The detection of vanadium, nickel, and porphyrins in crude oils were analyzed by high-temperature gas chromatography–atomic emission spectroscopy (HT-GC–AES), presenting characteristic metal distributions of oils from different sources [18]. Other related applications of HT-HRGC, including the analysis of α, β, and γ cyclodextrins, antioxidants, and oligosaccharides [2].

Considering that HT-HRGC is still a young separation technique and that it presents several attractive features, including the analysis of higher-molecular-weight compounds within short analysis times, without the necessity of sample derivatization, we can envisage a bright future for this technique, with many new applications being developed in the near future.

References

1. I. A. Fowlis, *Gas Chromatography*, 2nd ed., John Wiley & Sons, New York, 1994, pp. 1–11.

2. W. Blum and R. Aichholtz, *Hochtemperatur Gas-Chromatographie*, Hüthing, Germany, 1991, pp. 26–114.
3. S. R. Lipsky and W. J. McMurray, *J. Chromatogr. 279*: 59 (1983).
4. F. M. Lanças and M. S. Galhiane, *J. High Resolut. Chromatogr. Chromatogr. Commun. 13*: 654 (1990).
5. L. M. P. Damasceno, J. N. Cardoso, and R. B. Coelho, *J. High Resolut. Chromatogr. Chromatogr. Commun. 15*: 256 (1992).
6. K. Grob and R. Tschuor, *J. High Resolut. Chromatogr. Chromatogr. Commun. 13*: 193 (1990).
7. J. Hubball, *LC–GC 8*: 12 (1990).
8. J. V. Hinshaw and L. S. Ettre, *J. High Resolut. Chromatogr. Chromatogr. Commun. 12*: 251 (1989).
9. W. Blum and L. Damasceno, *J. High Resolut. Chromatogr. Chromatogr. Commun. 10*: 472 (1987).
10. M. Verzele, F. David, M. van Roelenbosch, G. Diricks, and P. Sandra, *J. Chromatogr. 270*: 99 (1983).
11. S. R. Lipsky and M. L. Duffy, *J. High Resolut. Chromatogr. Chromatogr. Commun. 9*: 376 (1986).
12. R. Aichholz and E. Lorbeer, *J. Microcol. Separ. 8*: 553 (1996).
13. S. Trestianu, G. Zilioli, A. Sironi, C. Saravelle, F. Munari, M. Galli, G. Gaspar, J. Colin, and J. L. Jovelin, *J. High Resolut. Chromatogr. Chromatogr. Commun. 8*: 771 (1985).
14. J. V. Hinshaw, *J Chromatogr. Sci. 25*: 49 (1987).
15. F. M. Lanças, J. H. Y. Vilegas, and N. R. Antoniosi Filho, *Chromatographia 40*: 341 (1995).
16. F. M. Lanças and J. J. S. Moreira, High temperature gas chromatography (HT-GC) analysis of underivatized cyclopeptidic alkaloids, Proc. of the 23rd Int. Symp. Capill. Chromatogr., 2000.
17. N. R. Antoniosi Filho, Analysis of the vegetable oils and fats using high resolution gas chromatography and computational methods, Ph.D. thesis, University of São Paulo, Institute of Chemistry at São Carlos, Brazil, 1995, pp. 140–152.
18. Y. Zeng and P. C. Uden, *J. High Resolut. Chromatogr. Chromatogr. Commun. 17*: 223 (1994).
19. A. W. Stott and R. P. Evershed, *Anal. Chem. 68*: 4402 (1996).

Fernando M. Lanças
J. J. S. Moreira

HPLC Analysis of Amino Acids

Amino acids are small organic molecules that posses both an amino and a carboxyl group. Amino acids occur in nature in a multitude of biological forms, either free or conjugated to various types of compounds, or as the building blocks of proteins. The amino acids that occur in proteins are named α-amino acids and have the empirical formula $RCH(NH_2)COOH$. Only 20 amino acids are used in nature for the biosynthesis of the proteins, because only 20 amino acids are coded by the nucleic acids. Amino acids show acid–base properties, which are strongly dependent on the varying R groups present in each molecule. The varying R groups of individual amino acids are responsible for specific properties: polarity, hydrophilicity–hydrophobicity [1,2]. Hence, the 20 α-amino acids could be categorized in the 4 distinct groups listed in Table 1.

Their dipolar (zwitterionic) behavior is a fundamental factor in any separation approach. At low pH, amino acids exist in their cationic form with both amino and carboxyl groups protonated. The ampholyte form appears at a pH of 6–7, whereas at higher values, amino acids are in their anionic form (carboxyl group dissociated). Another important parameter is that all α-amino acids (with the exception of Gly) are asymmetrical molecules exhibiting optical isomerization (L being the isomer found in nature). As can be seen in Table 1, amino acids are actually small [molecular weight (MW) ranging from 75 to 204] molecules, exhibiting pronounced differences in polarity and a few chromophoric moieties.

The determination of amino acids in various samples is a usual task in many research, industrial, quality control, and service laboratories. Hence, there is a substantial interest in the high-performance liquid chromatography (HPLC) analysis of amino acids from many diverse areas like biochemistry, biotechnology, food quality control, diagnostic services, neuro-chemistry/biology, and so forth. As a result, the separation of amino acids is probably the most extensively studied and best developed chromatographic separation in biological sciences. The most known system is the separation on a cation-exchange column and postcolumn derivatization with ninhydrin, which was described in 1951 by Moore and Stein. With this approach, a sulfonated polystyrene column achieved a separation of the 20 naturally occurring amino acids within approximately 6 h; modifications of the original protocol enhanced color stabilization of the derivatives and enabled the application of the method in various real samples. Since then, immense developments in instrumentation, column technologies, and automation established HPLC as the dominant separation technique in chemical analysis. Numerous published reports described the

Table 1 Amino Acids Found in Proteins

	Amino acid	Structure at pH 6-7	MW	pKa
Hydrophobic Aminoacids (Nonpolar R)	Alanine Ala	$CH_3-CH(NH_3^+)-COO^-$	89	2.35, 9.69
	Leucine Leu	$CH(Me)_2CH_2-CH(NH_3^+)-COO^-$	131	2.36, 9.60
	Isoleucine Ile	$C_2H_5CH(CH_3)-CH(NH_3^+)-COO^-$	131	2.36, 9.68
	Valine Val	$CH(Me)_2-CH(NH_3^+)-COO^-$	117	2.36, 9.68
	Proline Pro	COO^-, NH_2^+	115	1.99, 10.60
	Methionine Met	$CH_3SC_2H_5-CH(NH_3^+)-COO^-$	149	2.28, 9.21
	Phenylanine Phe	$-CH_2-CH(NH_3^+)-COO^-$	165	1.83, 9.13
	Tryptophan Trp	$-CH_2-C(NH_3^+)-COO^-$ (N)	204	2.38, 9.39
Hydrophilic Aminoacids (Not Charged R)	Clycine Gly	$H-CH(NH_3^+)-COO^-$	75	2.34, 9.6
	Serine Ser	$CH_2OH-CH(NH_3^+)-COO^-$	105	2.21, 9.15
	Threonine Thr	$CH_3CHOH-CH(NH_3^+)-COO^-$	119	2.63, 10.43
	Cysteine Cys	$HSCH_2-CH(NH_3^+)-COO^-$	121	1.71, 10.78
	Tyrosine Tyr	$HO-\ -CH_2-CH(NH_3^+)-COO^-$	181	2.20, 9.11
	Glutamine Gln	$NH_2COC_2H_5-CH(NH_3^+)-COO^-$	146	2.17, 9.13
	Asparagine Asn	$NH_2COCH_2-CH(NH_3^+)-COO^-$	132	2.02, 8.8
Acidic Aminoacids	Aspartic Acid Asp	$^-OOCCH_2-CH(NH_3^+)-COO^-$	133	2.09, 3.86, 9.86
	Glutamic Acid Glu	$^-OOCC_2H_5-CH(NH_3^+)-COO^-$	147	2.19, 4.25, 9.67

(*continued*)

Table 1 Continued

	Amino acid	Structure at pH 6-7	MW	pKa
Basic Aminoacids	Lysine Lys	$^{+}H_3NC_4H_8—CH(NH_3^+)—COO^-$	133	2.18, 8.95, 10.53
	Arginine Arg	$H_2NC(=^{+}NH_2)NHC_3H_6—CH(NH_3^+)—COO^-$	133	2.17, 9.04, 12.48
	Histidine His	(imidazolium ring)$—CH_2—CH(NH_3^+)—COO^-$	155	1.82, 8.95, 10.53

HPLC analysis of amino acids in a great variety of samples. To no surprise, a two-volume handbook is entirely devoted to HPLC for the separation of amino acids, peptides and proteins [3]. Many of the initial reports employed soft resins or ion exchangers such as polystyrene or cellulose as stationary phases. These materials show some disadvantages (e.g., compaction under pressure, reduced porosity, and wide particle size distribution). The last decades' developments in manufacturing silica-based materials resulted in the domination of reversed-phase (RP) silica-based packing in liquid chromatography. As a result, RP–HPLC is, at present, widely used for the separation of amino acids, because it offers high resolution, short analysis time, ease in handling combined with low cost, and environmental impact per analysis circle.

In ligand-exchange chromatography (LEC), the separation of analytes is due to the exchange of ligands from the mobile phase with other ligands coordinated to metal ions immobilized on a stationary phase. LEC has been used successfully for the resolution of free amino acids, amino acid derivatives, and for enantiomeric resolution of racemic mixtures [3].

Apart from ninhydrin, many other derivatization reagents have been used; both precolumn and postcolumn derivatization modes have been extensively employed [3–5]. Derivatization procedures offer significant advantages in both separation and detection aspects and, thus, will be discussed in further detail. The rest of the entry will be divided into two sections: separation of underivatized amino acids, where the determination of free amino acids and postcolumn derivatization procedures are described; and separation of derivatized amino acids, where precolumn derivatization approaches are discussed.

Separation of Underivatized Amino Acids

The differing solubilities, polarities, and acid–base properties of free amino acids have been exploited in their separation by partition chromatography, ion-exchange chromatography, and electrophoresis. For example, the elution order obtained from a polystyrene ion-exchange resin with an acidic mobile phase corresponds to the amino acid classification depicted in Table 1: Acidic amino acids are eluted early, neutral between, and basic amino acids later. In this case, ionic interactions between the sample and the stationary phase are the driving force for the separation of the groups. However, hydrophobic van der Waals and π–π aromatic interactions are responsible for the separation of amino acids within the groups [3].

The dominant stationary phase in HPLC is modified silica and, to be more specific, octadecyl silica (ODS). It should be pointed out that there could be great differences between various types of ODS materials or even between different batches of the same material. Carbon load, free silanol content, endcapping, type of silica, and coupling chemistry to the C_{18} moiety, not to mention the several physical characteristics of the packing material all involve the behavior of an ODS column. However, a rather safe generalization is that, in such material, hydrophobic interactions are a dominant mechanism of separation [3–7].

In typical ODS materials, polar amino acids are very weakly retained on column; thus, they are insufficiently resolved. In contrast, nonpolar amino acids are stronger retained and adequately separated. To overcome the poor resolution of polar amino acids, two strategies are the most promising:

1. Derivatization (as discussed in the next section)
2. Modification of the mobile phase with the addition of ion-pairing reagents

Alkyl sulfates/sulfonates added to the mobile phase form a micellar layer interacting with both the stationary phase and the amino acids (which under these conditions are protonated). A mixed mechanism (ion-pairing and dynamic ion exchange) is observed. Furthermore, the ion-pairing reagent masks underivatized silanols of the ODS material, reducing nonspecific unwanted interactions. Sodium dodecyl sulfate (SDS) is the most often used ion-pairing reagent. Gradient elution is often required to achieve reasonable analysis time for nonpolar amino acids. Despite the above-mentioned advantages, ion-pairing shows some disadvantages, such as irreproducibility (especially in gradient runs), long equilibration times, and difficulties in ultraviolet (UV) detection.

Another possibility is the use of alternative stationary phases. A strong trend of the last decade is the employment of specialty phases in challenging and complex separations. Thus, newer C_8, NH_2, CN, mixed-mode phases (materials incorporating both ion exchange and reversed-phase moieties), new polymeric phases, and zirconia-based materials offer attractive stationary-phase selectivities.

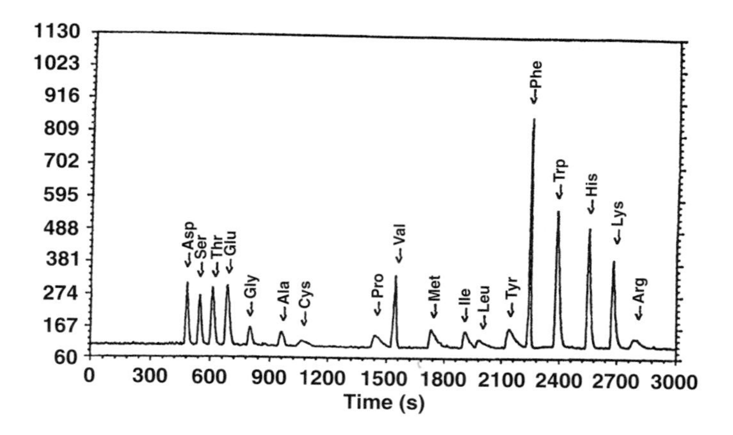

Fig. 1 HPLC analysis of 18 common amino acids. Conditions: stationary phase: CS-10 cation exchange; mobile phase: gradient of aqueous 0.01% TFA and ammonium acetate; detection at the ELSD. (From Ref. 11.)

Postcolumn Derivatization

The nonderivatized amino acids, following their chromatographic separation, can either be directly detected as free amino acids, on-line derivatized, or by postcolumn derivatization. Derivatization with ninhydrin, the classical amino acid analysis, was the first reported postcolumn derivatization method. Modern postcolumn derivatization protocols employ sophisticated instrumentation and achieve high resolution and sensitivity. In such configurations, derivatization occurs in a reaction coil placed between the analytical column and the detector. Additional pumps and valves are required; thus, such systems typically run fully automated and controlled by a computer. The major disadvantages of postcolumn derivatization are the need for sophisticated and complex instrumentation and the band broadening occuring in the reactor. *Ortho*-phthaldialdehyde (OPT) is the most common reagent in postcolumn derivatization. OPT reacts with primary amino acids under basic conditions, forming a fluorescent derivative (OPA derivative) that allows detection at femtomole levels. Disadvantages of OPA derivatization are the instability of the resultant derivatives and the fact that secondary amino acids are not detected.

If no derivatization takes place, detection is preferably accomplished by UV at a low wavelength (200–210 nm) in order to enhance detection sensitivity. However, detection selectivity is sacrificed at such low wavelengths. Electrochemical detection, when applied to the analysis of free amino acids, offers higher selectivity but suffers from a small linearity range. Furthermore, most amino acids (with the exception of tryptophan, tyrosine, and cysteine) are not intrinsically electrochemically active within the current useful potential range [5]. Lately, the development of the evaporative light-scattering detector (ELSD) offers an attractive alternative for the determination of nonderivatized amino acids (see Fig. 1).

Separation of Derivatized Amino Acids

Precolumn derivatization is the generally accepted approach for the determination of amino acids, because it offers significant advantages: increased detection sensitivity, enhanced selectivity, enhanced resolution, and limited needs for sophisticated instrumentation (in contrast with postcolumn derivatization techniques).

In modern instrument configurations, derivatization can take place in a conventional autosampler; the resultant derivatives are separated on the analytical column. Detection limits at the femtomole level are achieved, and the resolution of polar amino acids is greatly enhanced.

The most common derivatization reagents are as follows:

- Dimethylamino azobenzene isothiocyanate (DABITC)
- 4-(Dimethylamino)azobenzene-4-sulfonyl chloride (dabsyl chloride or DABS-Cl) [dabsyl derivatives]
- 1-*N*-*N*-Dimethylaminonaphthalene-5-sulfonyl chloride [dansyl derivatives]
- Fluorodinitrobenzene (DNP derivatives)
- Fluorescamine
- 9-Fluorenylmethyl chloroformate (FMOC-Cl)
- 4-Chloro-7-nitro-2,1,3-benzoxadiazole (NBD-Cl)

Phenylisothiocyanate (PITC) [phenylthiohydantoin (PTH) derivatives]
Methylisothiocyanate (MITC) [methylthiohydantoin (MTH) derivatives]

Figure 2 illustrates the structure of the product resulting from the derivatization of an amino acid with the above-mentioned reagents. Mixtures of derivatives with the most commonly used reagents (dansyl, DNP, PTH) are readily provided in kits, to be directly used as reference standards in HPLC analysis.

Phenylthiohydantoin derivatization offers a special value because it is actually performed during Edman degradation, the sequencing technique mostly used for the determination of the primary structure of proteins and peptides. PTH derivatives are separated in many different stationary phases, in either normal- or reversed-phase mode and are mostly detected at 254 nm [8,9]. Using radiolabeled proteins, sequencing of proteins down to the 1–100-pmol range can be achieved. The formed derivatives are basic and thus interact strongly with base silica materials. RP separations are mostly carried out in acidic conditions with the addition of appropriate buffers (sodium acetate mostly, but also phosphate, perchlorate, etc). Failings of PTH derivatization are the lengthy procedure and the higher detection limits obtained (compared to fluo-

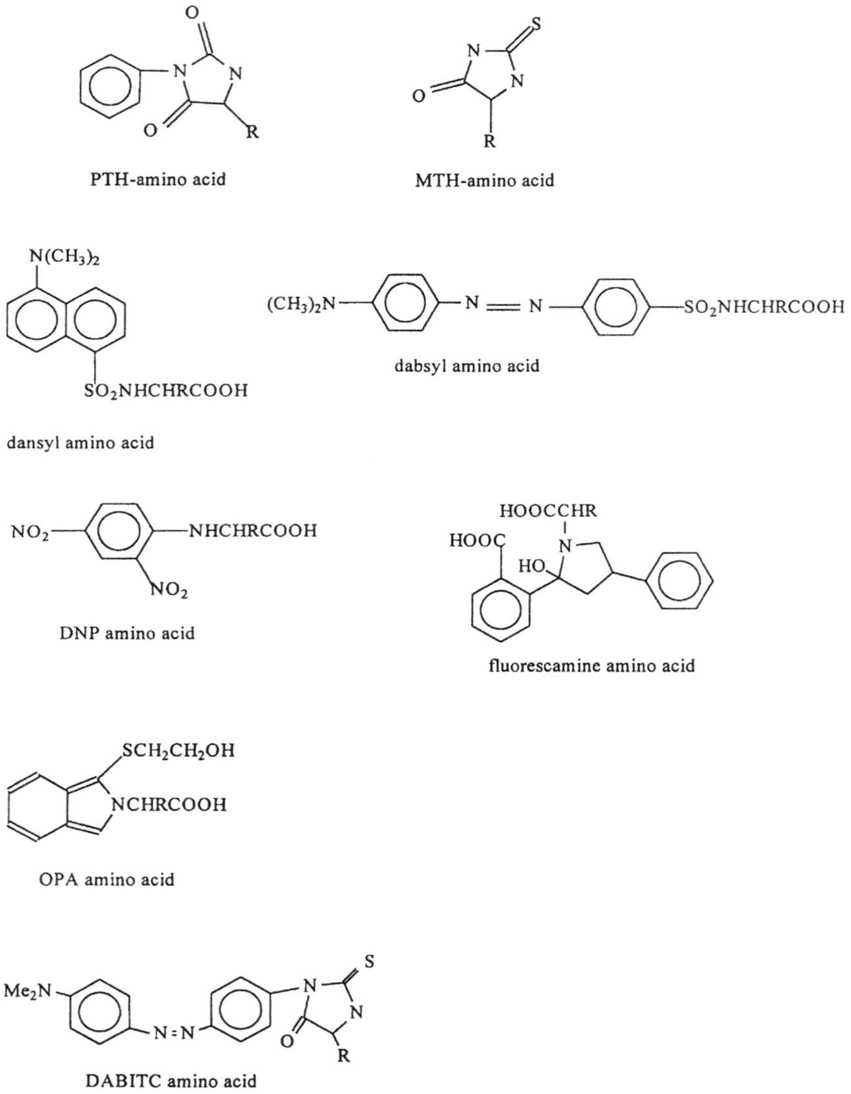

PTH-amino acid

MTH-amino acid

dansyl amino acid

dabsyl amino acid

DNP amino acid

fluorescamine amino acid

OPA amino acid

DABITC amino acid

Fig. 2 Structures of the most common amino acid derivatives.

rescent derivatives). Potent advantages of the method are its robustness and reproducibility, and the extensive research literature that covers any possible requirement. An alternative to PTH is MTH derivatization, a method well suited for solid-phase sequencing [3].

Dimethylamino azobenzene isothiocyanate microsequencing results in red–orange derivatives, which exhibit their absorbance maximum at 420 nm with $\varepsilon = 47.000$, in other words offering threefold higher sensitivity compared to PTH derivatives. DABITC derivatives are separated in C_8 or C_{18} columns in acidic environment, within 20 min.

Dabsyl chloride is an alternative to DABITC as a derivatization reagent to be used for manual sequencing. Dabsyl chloride reacts with primary and secondary amino acids forming red–orange derivatives that are stable for months. The method offers excellent sensitivity, ease, and speed of preparation and high-resolution capabilities. However, it suffers from interferences with ammonia present in biological samples. Furthermore, it results in a relatively reduced column lifetime due to the utilization of excess of Dabsyl chloride [9].

The dansyl derivatization has been extensively studied to label α- or ε-amino groups. DNS derivatives are formed within 2 min and are detected by either UV or fluorescence. A typical example of a separation of dansyl amino acids is illustrated in Fig. 3.

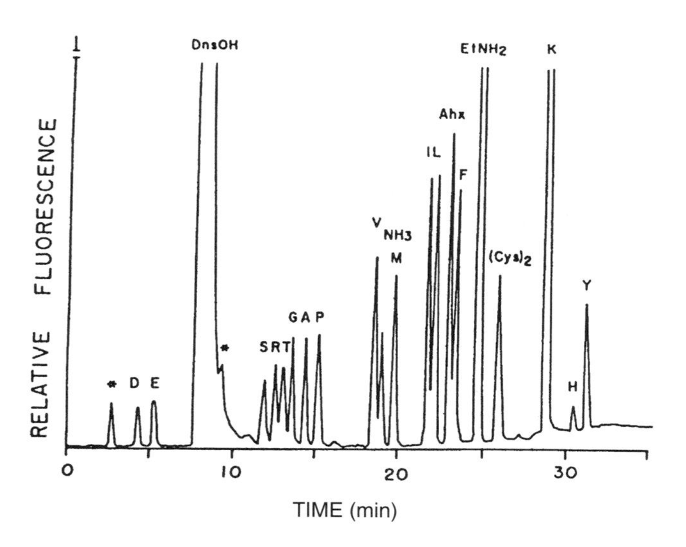

Fig. 3 RP–HPLC analysis of a mixture of dansyl amino acids. Conditions: stationary phase: 4 μm Nova Pak C_{18}; mobile phase: gradient of methanol and tetrahydrofuran versus aqueous phosphate buffer; detection in a fluorescent detector; excitation 338 nm, emission 455 nm. Amino acids are abbreviated by the one-letter system. (From Ref. 12.)

The FMOC derivatization offers high fluorescent detection sensitivity, but it requires an extraction step to remove unreacted FMOC and by-products. This step is a potential cause of analyte losses. Furthermore, it not suitable for Trp and Cys, because the corresponding derivatives exhibit a lower response due to intramolecular quenching of fluorescence.

The DNP derivatives are analyzed either in normal or reversed phase. Disadvantages of this method are the lower detection sensitivity (60 times less sensitive compared to dabsyl detection) and the lower separation resolution. However, this approach has proven useful for the determination of lysine in food materials.

Finally, the incorporation of an electroactive functionality into a chromatographic label is an attractive alternative for the HPLC of amino acids. Reagents like π–*N* and *N*-dimethylaminosothiocyanate have been used to facilitate amperometric detection of the derivatives.

Chiral Separation of Amino Acids

The importance of chirality has rapidly evolved the last decade. Both analytical and preparative separations are needed for biochemical, pharmaceutical, and alimentary purposes. Amino acids are asymmetrical molecules. L is the form appearing in proteins; however the D form is also present in nature. Enantiomeric separation of amino acids has been achieved in various stationary phases, such as polystyrene and polyacrylamide to which chiral ligands were covalently bound. Metal ions, in conjunction with chiral ligands, have also been utilized in the mobile phase in the reversed-phase and ligand-exchange mode. Novel stationary chiral phases developed for enantiomeric analysis incorporate chiral ligands (e.g., cyclodextrins or even amino acids) immobilized on silica. Generally, L-amino acid-bonded phases retain L-amino acids stronger than the D species [3,5,10].

Recently, a strong trend in molecular recognition is the development of molecular imprinting polymers (MIP). MIPs have been used as synthetic antibodies in immunoassays and biosensors, but also as catalysts and separation media (employed both in analysis and extraction). One of the first applications of MIPs in separations was the enantiomeric separation of amino acids derivatives.

Conclusions

High-performance liquid chromatography, when compared to other instrumental methods [thin-layer chromatography (TLC), GC, automated amino acid analyzer], offers significant advantages in the analysis of amino acids:

high resolution, high sensitivity, low cost, time saving (one-third of the analysis time of an amino acid analyzer), and a multivariate optimization scheme offering versatility and flexibility. Furthermore, optimization of HPLC determination enables the practitioner to overcome typical problems of other methods (e.g., the well-known interferences of ammonia in amino acid analyzer). An additional advantage of HPLC is its direct compatibility with mass spectrometry. The widespread use of LC–MS in proteomic analysis, which at present utilizes state of the art mass spectrometers (e.g., matrix-assisted laser desorption ionization–time-of-flight — mass spectrometry), is seen as a potent future trend.

The variety of instrumentation and experimental conditions (columns, buffers, organic modifiers, derivatization procedures, etc.) reported in the vast literature may hinder the novice from pinpointing the best method to use. The choice of the appropriate method depends on the specific needs of each analytical problem and the nature of the sample to be analyzed. Aspects such as specificity and speed of the derivatization reaction should always be considered. Furthermore, in such multivariate dynamic systems, precision, accuracy, and linearity of the chosen method is a very important factor. The practitioner should carefully follow the developed protocol; the use of automated systems, especially in derivatization procedures, could greatly enhance the reproducibility of the method.

References

1. L. M. Silverman and R. H. Christenson, Amino Acids and Proteins, in *Fundamentals of Clinical Chemistry*, 4th ed. (C. A. Burtis and E. R. Ashwood, eds.), Saunders, Philadelphia, 1996.
2. C. K. Matthews and K. E. van Holde, Biochemistry, 2nd ed., Benjamin-Cummings, Menlo Park, CA, 1995, pp. 129–214.
3. W. S. Hancock (ed.), *CRC Handbook of HPLC for the Separation of Amino Acids, Peptides and Proteins*, CRC Press, Boca Raton, FL, 1984.
4. W. S. Hancock and J. T. Sparrow, *HPLC of Biological Compounds*, Marcel Dekker, Inc., New York, 1984, pp. 187–207.
5. I. N. Papadoyannis, HPLC in the analysis of amino acids, in *HPLC in Clinical Chemistry*, Marcel Dekker, Inc., New York, 1990, pp. 97–154.
6. R. M. Kamp, High sensitivity Amino acid analysis, in *Protein Structure Analysis* (R. M. Kamp, T. Choli-Papadopoulou, and B. Wittman-Liebold, eds.), Springer-Verlag, Berlin, 1997.
7. F. Lottspeich and A. Hernschen, Amino acids, peptides, proteins, in *HPLC in Biochemistry* (A. Hernschen, K. P Hupe, F. Lottspeich, and W. Voelter, eds.), VCH Weinheim, 1985.
8. M. D. Waterfield, G. Scrace, and N. Totty, Analysis of phenylthiohydantoin amino acids, in *Practical Protein Chemistry — A Handbook* (A. Darbre, ed.), John Wiley & Sons, Chichester, 1986.
9. T. Bergman, M. Carlquist, H. Jornvall, Amino acid analysis by high performance liquid chromatography of phenylthiocarbamyl derivatives, and amino acid analysis using DABS-Cl precolumn derivatization method, R. Knecht and J. Y Chang, in *Advanced Methods in Protein Microsequence Analysis* (B. Wittmann-Liebold, J. Salnikow, and V. A. Erdmann, eds.), Springer-Verlag, Berlin, 1986.
10. D. Vollenbroich and K. Krause, Quantitative Analysis of D- and L-amino acids by HPLC, in *Protein Structure Analysis* (R. M. Kamp, T. Choli-Papadopoulou, and B. Wittman-Liebold, eds.), Springer-Verlag, Berlin, 1997.
11. J. Petterson, L. J. Lorenz, D. S. Risley, and B. J. Sanmann, *J. Liquid Chromatogr. Related Technol. 22*: 1009 (1999).
12. A. R. Martins and A. F. Padovan, *J. Liquid Chromatogr. Related Technol. 19*: 467 (1999).

Ioannis N. Papadoyannis
Georgios A. Theodoridis

Hydrodynamic Equilibrium in CCC

In all cases, countercurrent chromatography (CCC) utilizes a hydrodynamic behavior of two immiscible liquid phases through a tubular column space which is free of a solid support matrix. The most versatile form of CCC, called the hydrodynamic equilibrium system, applies a rotating coil in an acceleration field (either in the unit gravity or in the centrifuge force field). Two immiscible liquid phases confined in such a coil distribute themselves along the length of the coil to form various patterns of hydrodynamic equilibrium [1].

According to the hypothesis proposed by Ito [2], the multitude of hydrodynamic phenomena observed in the rotating coils can be attributed to the following types of liquid distribution.

1. The basic hydrodynamic equilibrium (the two liquid phases are evenly distributed from one end of the coil, called the head, and any excess of either phase is accumulated at the other end, called the tail). Here, the tail–head relationship of the rotating coil is defined by the direction of the Archimedean screw force which drives all objects toward the head of the coil.
2. The unilateral hydrodynamic equilibrium [the two solvent phases are unilaterally distributed along the length of the coil, one phase (head phase) entirely occupying the head side and the other phase (tail phase) the tail side of the coil]. The head phase can be the lighter or the heavier phase and also can be the aqueous or the nonaqueous phase, depending on the physical properties of the liquid system and the applied experimental conditions. This type of equilibrium may also be called bilateral, indicating the distribution of the one phase on the head side and the other phase on the tail side [3].

To illustrate the process of establishing the hydrodynamic equilibrium, it is worthwhile to begin with the distribution of two immiscible solvent phases in the "closed" coil, simply rotated around the horizontal axis in the unit gravitational field. The coil is filled with equal volumes of the lighter and heavier phase and then sealed at both ends. At a slow rotation of 10–20 revolutions per minute (rpm), two liquid phases are evenly distributed in the coil (basic hydrodynamic equilibrium) due to the Archimedean screw force. As the rotational speed increases, the heavier phase quickly occupies more space on the head side of the coil and, at the critical speed range of 60–100 rpm, the two phases are completely separated along the length of the coil, with the heavier phase on the head side and the lighter phase on the tail side (unilateral hydrodynamic equilibrium).

After this critical speed range, the amount of the heavier phase on the head side decreases sharply, reaching substantially below the 50% level at about 160 rpm. Further increase of the rotational speed again distributes the two phases fairly evenly throughout the coil, apparently due to the strong radial centrifugal force field produced by the rotation of the coil. The phase distribution described can be observed in many solvent systems [chloroform–acetic acid–water (2/2/1), hexane–methanol, *n*-butanol–water, etc.], glass coils [10–20 mm inner diameter (i.d.)] with different helical diameters (5–20 cm) being applied.

As a first approximation, the complex hydrodynamic phenomenon taking place in the rotating coil may be explained by the interplay between two force components acting on the fluid. The tangential force component (F_t) generates the Archimedean screw effect to move two phases toward the head of the coil, and the radial force component (F_r) which acts against the Archimedean force. The critical speed range is the most interesting. An increase of the rotational speed up to 60–100 rpm alters the balance of the hydrodynamic equilibrium by an enhanced radial centrifugal force field that increases the net force field acting at the bottom of the coil and decreases that acting at the top. Under this asymmetrical force distribution, the movement of the heavier phase toward the head is accelerated, whereas the movement of the lighter phase toward the head is retarded. This results in a unilateral hydrodynamic phase distribution in the rotating coil.

The hydrodynamic equilibrium condition may be used for performing CCC as follows. First, the coil is completely filled with the stationary phase, either the lighter or the heavier phase, and the other phase is introduced from the head end of the coil while the coil is rotated around its axis. Then, the two liquid phases establish equilibrium in each turn of the coil and the mobile phase finally emerges from the tail end of the coil, leaving some amount of the stationary phase permanently in the coil. Solutes locally introduced at the head of the coil are subjected to a partition process between two phases and eluted in order of their partition coefficients. In general, higher retention of the stationary phase significantly improves the peak resolution. Consequently, the unilateral hydrodynamic equilibrium condition provides a great advantage in performing CCC, because the system permits retention of a large amount of stationary phase in the coil if the lighter phase is eluted in a normal mode (head-to-tail direction) or the heavier phase in a reversed mode (tail-to-head direction).

In general, the retention of the stationary phase in the coil rotated in the unit gravity field entirely relies on relatively weak Archimedean screw force. In this situation, application of a high flow rate of the mobile phase would cause a depletion of the stationary phase from the column. This problem can be solved by the utilization of synchronous planetary centrifuges, free of rotary seals, which enable one to increase the rotational speed and, consequently, enhance the Archimedean screw force. The seal-free principle can be applied to various types of synchronous planetary motion. In all cases, the holder revolves around the centrifuge axis and simultaneously rotates about its own axis at the same angular velocity ω.

When the coil is mounted coaxially around the holder, which revolves around the central axis of the device and counterrotates about its own axis, two axis being parallel (Type I), two solvent phases are distributed along the length of the coil according to the basic hydrodynamic equilibrium. It does not favor the stationary-phase retention. Another, similar planetary motion, except that the holder revolves around the central axis of the centrifuge and rotates about its own axis in the same direction (Type J), produces, regardless of the rotational speed, a totally different phase-distribution pattern which is typical for the unilateral hydrodynamic equilibrium. The unilateral distribution can also be attained in the coaxially mounted coils in cross-axis planetary centrifuges [4]. It is important to note that all the planetary motions providing the unilateral distribution form an asymmetrical centrifuge force field that closely resembles that observed in the coil rotating at the critical speed in the unit gravity.

The unilateral hydrodynamic equilibrium conditions provide the basis for high-speed CCC (HSCCC, ω = 800 rpm or more) which has mainly gained acceptance for CCC separations. The stroboscopic observation on two-phase flow through the running spiral column of a Type J system reveals the following pattern. When the lower phase (chloroform) is eluted through the stationary lighter phase (water) from the head toward the tail of the spiral column, a large volume of the stationary phase is retained in the column and the spiral column is divided into two distinct zones: the mixing zone in about one-fourth of the area near the center of the centrifuge and the settling zone showing a linear interface between the two phases in the rest of the area. The mixing zone is always fixed at the vicinity of the central axis of the centrifuge while the spiral column undergoes the planetary motion. In other words, the mixing zone in each loop is traveling through the spiral column toward the head at a rate equal to the column rotation. Consequently, at any portion of the column, the two liquid phases are subjected to a typical partition process of repetitive mixing and settling at a high frequency, over 13 times per second at 800 rpm of column revolution, while the mobile phase is being continuously pumped through the stationary phase [3].

At a first approximation, the hydrodynamic phenomenon observed also may be explained by the interplay between two force components acting on the fluid. At the distal portion of the coil, both the strong radial force field and the reduced relative flow of the two phases establish a clear and stable interface between the two liquid phases. At the proximal portion of the spiral column, where the strength of the radial-force component is minimized, the effect of the Archimedean screw force becomes visualized as agitation at the interface caused by the relative movement of two liquid layers [2].

It should be noted that the centrifuge force field acts on the fluid in the rotated coil in parallel with other forces of different nature [5]:

F_A, buoyancy force due to the difference between the stationary and mobile phases
F_i, inertial force caused by coil motion, comprises components of centrifugal force field
F_η, viscosity force due to the overflow of the stationary phase along the coil tube walls
F_γ, interfacial tension force
F_W, adhesion force
F_h, hydraulic resistance force caused by moving of two immiscible phases relative to each other

The following balance of these forces of a different nature is considered:

$$F_i = F_A + F_\eta + F_\gamma + F_W + F_h$$

From this, the basic equation of the stationary-phase retention process can be derived, a number of assumptions and complex theoretical treatments being required. Taking as example the planetary centrifuge of Type J, the average cross-sectional area of a stationary-phase layer has been estimated for hydrophobic liquid systems, which are characterized by high values of interfacial tension γ, low values of viscosity η, and low hydrodynamic equilibrium settling times:

$$\left(\sqrt{\frac{S_c}{S}} - 1\right)(S_c - S) \approx \frac{v_m^{1/2} r^{1/2} \eta_s}{\rho_m \Delta\rho R \omega^{3/2}}$$

where S and S_c are the cross-sectional areas of the stationary-phase layer and the spiral column, respectively, v_m is the linear speed of the mobile phase flow, η_s is the viscosity of the stationary phase, and r and R are rotation and revolution radii, respectively; ρ_m is the density of the mobile phase and $\Delta\rho$ is the density difference between two phases. After a few assumptions, it can be rewritten as

$$\frac{S}{S_c} \approx 1 - k_1 \frac{\beta^{1/4}}{\omega^{3/4} R^{1/4}} \approx 1 - k_2 \frac{1}{\omega^{3/4}}$$

where $\beta = r/R$, k_1 is a proportional coefficient characterizing peculiarities of the liquid system (it is dependent on the interfacial tension, viscosity of the stationary phase, and density difference between two phases); $k_2 = k_1(r^{1/4}/R^{1/2})$.

The ratio of the cross-sectional area of the stationary-phase layer to that of the coil tube (S/S_c) governs the volume of the stationary phase retained in the column. The theoretical dependence of S/S_c on the rotation speed ω and the experimental dependencies of the S_f value (ratio of the volume of the stationary phase retained in the column to the total column volume) on ω for n-decane–water and chloroform–water liquid systems are in good agreement (Fig. 1).

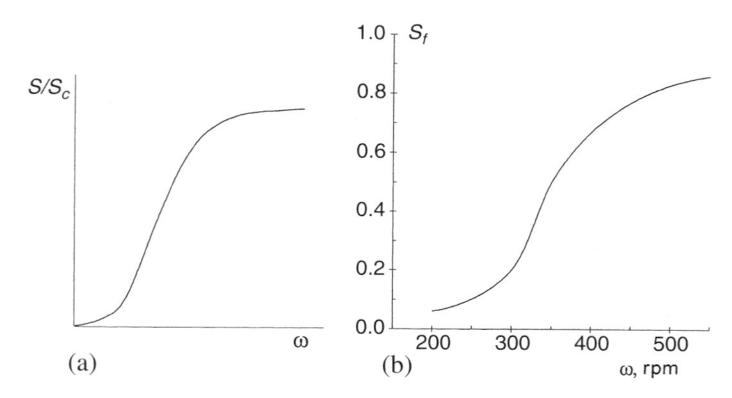

Fig. 1 (a) Theoretical ω-dependence of S/S_c; (b) Experimental ω dependence of S_f for the n-decane–water system. Planetary centrifuge of Type J; $\beta = 0.37$; flow rate = 1 mL/min.

Hence, an approach based on considering the balance of forces of a different nature acting on the fluid in the rotating coil may give some correlation among the peculiarities of the liquid system, operation conditions, design parameters of the planetary centrifuge, and the stationary-phase retention. However, any rigorous mathematical model describing the complex hydrodynamic equilibrium of two liquid phases in the rotating coiled column has not been yet elaborated. This issue remains open.

References

1. W. D. Conway, *Countercurrent Chromatography. Apparatus, Theory and Application*, VCH, New York, 1990.
2. Y. Ito, *J. Liquid Chromatogr. 15*: 2639 (1992).
3. Y. Ito, Principle, apparatus, and methodology of high-speed countercurrent chromatography, in *High-Speed Countercurrent Chromatography* (Y. Ito and W. D. Conway, eds.), John Wiley & Sons, New York, 1996, pp. 3–44.
4. J.-M. Menet, K. Shimomiya, and Y. Ito, *J. Chromatogr. 644*: 239 (1993).
5. P. S. Fedotov, V. A. Kronrod, T. A. Maryutina, and B. Ya. Spivakov, *J. Liquid Chromatogr. Related Technol. 19*: 3237 (1996).

Petr S. Fedotov
Boris Ya. Spivakov

Hydrophobic Interaction Chromatography

Mechanism

Hydrophobic interaction chromatography (HIC) is a mode of separation in which molecules in a high-salt environment interact hydrophobically with a nonpolar bonded phase. HIC has been predominantly used to analyze proteins, nucleic acids, and other biological macromolecules by a hydrophobic mechanism when maintenance of the three-dimensional structure is a primary concern [1–4]. The main applications of HIC have been in the area of protein purification because the recovery is frequently quantitative in terms of both mass and biological activity.

In HIC, a high-salt environment causes the association of hydrophobic patches on the surface of an analyte with the nonpolar ligands of the bonded phase. Elution is generally effected by an "inverse" gradient to lower salt concentration. This is considered "inverse" because it is the opposite of gradients used for ion-exchange chromatography. Effective salts for HIC are those which are "antichaotropic"; that is, they promote the ordering of water molecules at interfaces. Because interaction is only with the surface of a macromolecule such as a protein, the number of amino acids involved in the chromatography is relatively small, and changes in surface structure can cause differential binding and, hence, separation.

Reversed-phase chromatography (RPC) and HIC are both based on interactions between hydrophobic moieties, but the operational aspects of the techniques render selectivities totally different. The physical properties and selectivities of the two methods are contrasted in Fig. 1. The bonded phase of HIC supports consists of a hydrophilic matrix into which hydrophobic chains are inserted, generally in low density. This can be contrasted with the higher-density organosilane chemistry used in RPC. The chromatograms illustrate that both the selectivity and the number of peaks obtained for a protein mixture vary between the two modes. Cytochrome c is not retained at all by HIC and myoglobin is split into two peaks by RPC. A primary reason for the vast difference is the mobile-phase environment for each method. The organic solvents and generally acidic conditions used in RPC cause dena-

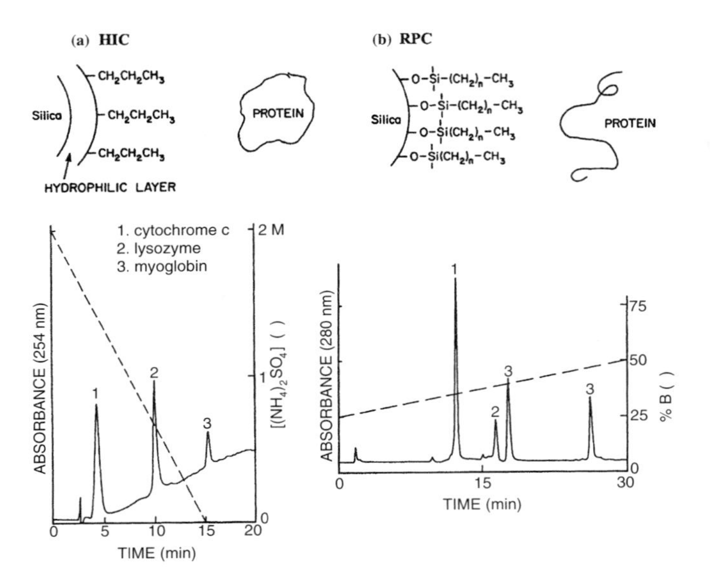

Fig. 1 (a) SynChropak Propyl; 15 min gradient from $2M$–$0M$ $(NH_4)_2SO_4$ in $0.1M$ potassium phosphate, pH 6.8. (b) SynChropak RPP (C_{18}); 30-min gradient from 25% to 50% ACN with 0.1% TFA. (Used with permission of Eichrom.)

turation of most proteins and even splitting into subunits, whereas the high-salt concentrations at neutral pH used in HIC result in stabilization of globular or three-dimensional structures for biological macromolecules. The hydrophobic amino acid residues of globular proteins are generally folded inside the structure or located in a few patches on the surface. As a protein is denatured, the buried amino acids are exposed, yielding more sites for hydrophobic binding. The hydrophobic interaction system thus encounters primarily surface amino acids — far fewer hydrophobic residues than the reversed phase.

Supports

Bonded phases for HIC consist of a hydrophilic polymeric layer into which hydrophobic ligands are inserted. The hydrophilic layer totally covers the silica or polymer matrix, providing a wettable and noninteractive surface which is neutral to the protein. In HIC, even short ligands cause substantial binding and there is a definite relationship between ligand chain length and retention, contrary to the minimal effect of chain length observed in the RPC of proteins and peptides. The ligand chains are postulated either to interact with hydrophobic surface patches on the proteins or to be inserted into their hydrophobic pockets; it is the latter interaction which is strengthened by and related to chain length. The strength of the binding causes some proteins to bind irreversibly if the ligand is too long; therefore, most ligands are either aromatic or 1–3 carbon alkyl chains.

Because HIC supports are designed for macromolecules, they either possess pore diameters of at least 300 Å to allow inclusion or are nonporous. Both silica and polymer matrices are used because the hydrophilic polymeric coating minimizes or eliminates most matrix-based effects. The absolute retention and selectivity of an HIC support may be affected by the specific composition of the bonded phase, as well as the ligand. For example, protein mixtures have shown distinct selectivity on different HIC columns which have propyl functional groups [5].

Operation

Mobile Phase

In HIC, the concept of weak and strong solvents is different than in other modes because the weak solvent, or the one which promotes binding, is that containing high-salt concentration. The strong solvent, or one which causes elution, is that with low-salt concentration.

Salt

The most important variable in HIC retention, other than the ligand chain, is the composition of the salt used to promote binding. The effectiveness is based on the molal surface tension increment, which is parallel to the Hofmeister salting-out series for precipitation of proteins. The strength of HIC binding for some commonly used salts is K_3citrate > Na_2SO_4 > $(NH_4)_2SO_4$ > Na_2HPO_4 > NaCl.

Although potassium citrate and sodium sulfate cause stronger retention, ammonium sulfate is probably the most popular choice for HIC. Besides being effective for retention, it is highly soluble, stabilizing for enzymes, and resistant to microbial growth. Ammonium sulfate is available in high purity because of its use for salt fractionation. Sodium sulfate is less soluble and may precipitate under conditions of high concentration. The initial concentration of salt must be at a level high enough to cause binding of all the proteins to the bonded phase to avoid variable retention of early eluting peaks, which may also be broad [6]. Most proteins will bind when $2M$ ammonium sulfate is used. In HIC, the concentration of antichaotropic salt is proportional to log k, as has been shown for conalbumin in four different salts [7]. The exact relationship varies for each salt, as well as for the specific protein.

pH

In HIC, the mobile phase should be buffered to provide control of ionization because amino acids which are not ionized are more hydrophobic than those which are charged. The effect of pH on hydrophobicity produces some variation of retention with pH; however, it is not directly related to the p*I* of the analyte because only surface amino acids interact with the ligands. In a study of the effect of pH on retention by HIC for a series of lysozymes from different bird species, those containing histidine residues in the hydrophobic contact region exhibited deviation for pH values of 6–8, which is near the p*K* of histidine (pK = 6) [7].

Additives

Because HIC is based on surface-tension phenomena, changing those characteristics by the addition of surfactants affects retention. In a study of the effects of surfactants on retention of proteins by HIC, the addition of CHAPS {3-[(3-cholamidopropyl) dimethylammonio]-1-propane sulfonate} to the mobile phase resulted in shortened retention, improvement of peak shape, and a change in peak order for enolase and bovine pancreatic trypsin inhibitor [8]. The effects were dependent on the concentration of the surfactant. Surfactants can usually be washed easily from hydrophobic interaction columns because the bonded phases are neither highly hydrophobic nor ionic.

The hydrophobic basis of HIC means that alcohols may reduce interaction with supports; however, disruption of protein conformation may also occur. Because of the high salt concentrations used in HIC, organic solvents should only be added after compatibility with the mobile phase has been tested to ensure that precipitation will not take place. Generally, no more than 10% organic is added. Other additives that increase the stability of a given protein can often be included in the mobile phase for HIC without adversely changing the separation.

Flow Rate and Gradient

Almost all HIC separations are performed in the gradient mode because proteins bind with multipoint interactions. The flow rate and gradient have an effect on retention in HIC because HIC follows the linear solvent strength model [9]. The time of the gradient is another determinant in improving resolution in that longer gradients provide increased resolution. Generally, a 20–60-min gradient from 2*M*–0*M* ammonium sulfate in 0.02*M* buffer at neutral pH, with a moderate flow rate (1 mL/min for a 4.6-mm inner diameter), will provide a satisfactory starting point for an HIC analysis [1].

Temperature

Hydrophobic interaction chromatography is different than other modes of chromatography in that it is an entropy-driven process, characterized by increased retention with increased temperature. This is a major benefit when subambient temperatures must be used to preserve the structure and biological activity of labile proteins. Retention is usually decreased rather than increased as temperatures are lowered. In one study, the retention of lysozyme was relatively unchanged throughout a temperature range 0–45°C, whereas bovine serum albumin exhibited two peaks which changed in proportion with temperature, as well as increased in retention [10]. Some of the increase in retention with elevated temperatures, in this or other studies, can be attributed to protein unfolding and the increased exposure of internal hydrophobic residues, especially when peak broadening also occurs.

Loading

Loading capacities for proteins on HIC columns are quite high because proteins retain their globular forms during the procedure [1,4]. High loading is generally accompanied by high recoveries of biological activity. Dynamic and absolute loading capacities of HIC supports are in the range of 10 mg/mL and 30 mg/mL, respectively. Loading is also related to the relative sizes of the pore diameter and the solute, with 300 Å giving maximum capacity for many proteins.

Applications

The primary application for HIC has been in protein analysis and purification due to the good selectivity and preservation of biological activity [1–4]. Because of the major differences in selectivity, HIC can be used as an orthogonal technique to RPC, as well as to ion-exchange and size-exclusion chromatography.

Although the best HPLC method for peptide analysis is RPC, HIC offers a different selectivity for those peptides possessing three-dimensional conformations under high-salt conditions. When the separations of peptide mixtures by HIC and RPC have been compared, peaks were generally narrower on RPC due to the organic mobile phase. In a study of calcitonin variants, it was seen that peptides with certain amino acid substitutions could not be re-

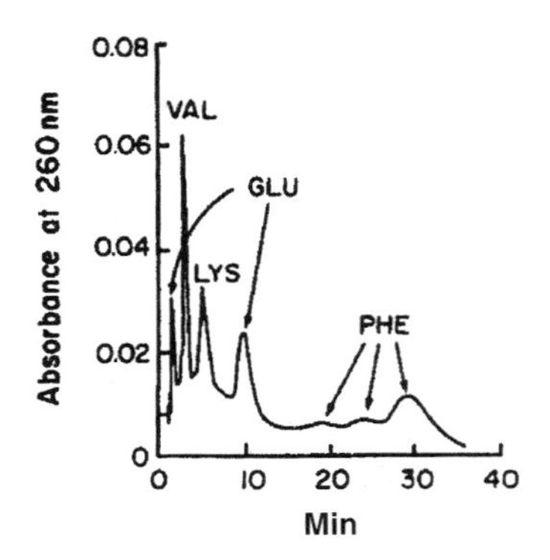

Fig. 2 Column: Polyol HIC, 100 nm; mobile phase: 0.7*M* disodium hydrogen phosphate, pH 6.3. [Reprinted from El Rassi and Horvath, *J. Chromatogr.* 326 (1985) p. 79 with kind permission of Elsevier Science NL, Sara Burgerhartstraat 25, 1055 Amsterdam, The Netherlands.]

solved by RPC, but were separated by HIC [11]. The main utility of HIC for peptide separations seems to lie in applications for extremely hydrophilic or hydrophobic peptides, or those with three-dimensional structures stable in high salt.

The separation of nucleic acids, particularly t-RNA, has been another useful application of HIC for biological macromolecules. The tertiary structure of t-RNA has made analysis under the gentle conditions of HIC very feasible [12]. Figure 2 shows an example of the purification of t-RNA molecules specific for different amino acids on a 100-nm polyol HIC column. Separation of t-RNA molecules has also been accomplished successfully by using HIC conditions on supports with alkylamino ligands, which are functionally similar to those traditionally used to separate nucleic acids [1].

Conclusions

Hydrophobic interaction chromatography is a mode of chromatography particularly effective for the analysis of proteins and other macromolecules. The hydrophobic interactions are primarily with nonpolar groups on the surface of the analytes due to maintenance of the tertiary structure. High loading and recovery of both mass and biological activity are achieved.

References

1. R. L. Cunico, K. M. Gooding, and T. Wehr, Hydrophobic interaction chromatography, in *Basic HPLC and CE of Biomolecules*, Bay Bioanalytical Laboratories, Richmond, CA, 1998.
2. R. E. Shansky, S.-L. Wu, A. Figueroa, and B. L. Karger, Hydrophobic interaction chromatography of proteins, in *HPLC of Biological Macromolecules* (K. M. Gooding and F. E. Regnier, eds.), Marcel Dekker, Inc., New York, 1990, p. 95.
3. M. I. Aguilar and M. T. W. Hearn, Reversed-phase and hydrophobic-interaction chromatography of proteins, in *HPLC of Proteins, Peptides and Polynucleotides* (M. T. W. Hearn, ed.), VCH, New York, 1991, p. 247.
4. R. H. Ingraham, Hydrophobic interaction chromatography of proteins, in *High-Performance Liquid Chromatography of Peptides and Proteins* (C. T. Mant and R. S. Hodges, eds.), CRC Press, Boca Raton, FL, 1991, p. 425.
5. A. J. Alpert, *J. Chromatogr. 359*: 85 (1986).
6. Y. Kato, T. Kitamura, S. Nakatani, and T. Hashimoto, *J. Chromatogr. 483*: 401 (1989).
7. J. L. Fausnaugh and F. E. Regnier, *J. Chromatogr. 359*: 131 (1986).
8. D. B. Wetlaufer and M. R. Koenigbauer, *J. Chromatogr. 359*: 55 (1986).
9. L. R. Snyder, Gradient elution separation of large biomolecules, in *HPLC of Biological Macromolecules* (K. M. Gooding and F. E. Regnier, eds.), Marcel Dekker, Inc., New York, 1990, p. 95.
10. S. C. Goheen and S. C. Engelhorn, *J. Chromatogr. 317*: 55 (1984).
11. M. L. Heinitz, E. Flanigin, R. C. Orlowski, and F. E. Regnier, *J. Chromatogr. 443*: 229 (1988).
12. Z. El Rassi and Cs. Horvath, *J. Chromatogr. 326*: 79 (1985).

Karen M. Gooding

Immobilized Metal Affinity Chromatography

The foundations for immobilized metal affinity chromatography (IMAC) were first laid in 1961 when Helferich introduced "ligand-exchange chromatography" [1]. The modern-day usage of this technique and its practical applications as a purification tool did not emerge, however, until 1975 and the seminal work by Porath et al. [2].

Among the many new protein purification approaches introduced in recent years, IMAC stands out for its ease of use and widespread applicability. This highly versatile and efficient technique is based on the interaction between biological molecules and covalently bound chelating ligands immobilized on a chromatographic support. Indeed, because the popularization of the Qiagen Qiaexpress® bacterial expression and one-step purification system [3], the use of IMAC has become nearly ubiquitous as tool for molecular biologists.

The principle behind IMAC lies in the fact that many transition metal ions [i.e., Ni(II) and Cu(II)] can coordinate to the amino acids histidine, cysteine, and tryptophan via electron-donor groups on the amino acid side chains.

An IMAC column may be loaded with a given metal-ion by perfusing the column with a metal-ion solution until equilibrium is reached between the metal chelated to the stationary phase and the metal ion in solution. The solid support (typically agarose, cross-linked dextran, or silica) is covalently linked to a metal-chelating ligand. The two most common ligands are iminodiacetic acid (IDA) and nitrilotriacetic acid (NTA) [4]. All major chromatography suppliers now offer their own brands of IMAC supports, with IDA typically the ligand of choice. The IDA residue is very suitable as an immobilized chelating agent because a bidentate chelating moiety remains free after immobilization, to which a metal ion can be coordinated. The NTA ligand contains an additional chelating site for metal ions, which can minimize metal leakage on the column. Free coordination sites of the metal ion are then used to bind different proteins and peptides.

Pearson systematized metal ions into three categories according to their reactivity toward nucleophiles: hard, intermediate, and soft [5]. Hard metal ions, such as Fe(III), prefer oxygen, whereas soft metal ions prefer sulfur. Intermediate types of ions such as Cu(II), Zn(II), Ni(II), and Co(II) coordinate nitrogen but also oxygen and sulfur. All the metals mentioned have been successfully employed for use in IMAC [6]. The immobilized metal-ion adsorbents may be prepared by charging the chelating gels with a slightly acid solution of the metal salt (pH 3–5). Charging the gel under acidic conditions is essential in the case of Fe(III) to avoid the formation of ferric hydroxide particles in solution. The use of colored Ni(II) or Cu(II) ions facilitates checking of leakage and the possible presence of metal ions bound to the protein eluate.

In his pioneering contribution, Porath postulated that the histidine, cysteine, and tryptophan residues of a protein were most likely to form stable coordination bonds with chelated metal ions at near neutral pH [2]. To date, an analysis of several protein models [7] lend full to his original theory. Having said that, histidine, by far and away, plays the most prominent role in IMAC binding. In a very real sense, IMAC has subtly become synonymous as a histidine affinity technique. The absence of a histidine residue on a protein surface correlates with the lack of retention of that protein on any IDA–metal column. The presence of even a single histidine on a protein surface, available for coordination, results in retention of that protein on an IDA–Cu(II) column. Also, a protein needs to display at least two histidine residues on its surface to be retained on an IDA–Ni(II) column. Thus, beyond its role as a purification technique, IMAC has been used as a tool to probe the surface topography of proteins [6].

The Qiaexpress® system is based on the selectivity of Ni–NTA for proteins with an affinity tag of six consecutive histidine residues: the 6x His tag. The 6x His tag is much smaller than such affinity tags as glutathione *S*-transferase, protein A, and maltose-binding protein and is uncharged at physiological pH. It has been shown to rarely contribute to a protein's immunogenicity, interfere with protein structure, function, or affect secretion from its expression system [3].

As in any chromatography technique, one can break down the separation process to its two most fundamental aspects: adsorption and elution. On a more practical level, the execution of an IMAC experiment involves five discrete steps which can be readily automated: column equilibration (charging of the gel), sample loading, removal of unbound material (washing), elution, and regeneration.

Adsorption of a protein to an IMAC column has to be performed at a pH at which an electron-donor group(s) on the protein's surface is at least partially unprotonated. Because the pK_a value of histidine groups (which supply the strongest metal interactions) lies in the neutral range, the binding of protein samples to the column should normally occur at a pH value of approximately 7. However, the actual pK_a value of an individual amino acid varies strongly depending on the neighboring amino acid value. Various experiments show that depending on the protein structure, the pK_a value of an amino acid can deviate from the theoretical value up to one pH unit [4]. Therefore, an application buffer of pH 8 often achieves improved binding. In order to eliminate any nonspecific electrostatic interactions, it is common to include salt in the equilibrating buffer. Typically, sodium chloride is used in concentrations between 0.1*M* and 1.0*M* [8].

The buffer itself should not effectively compete with a protein for coordination to the metal ligand. Sodium phosphate or sodium acetate are recommended buffers (depending on the pH choice) and the presence of EDTA or sodium citrate should be avoided. The presence of detergents (Triton X-100, Tween-20, urea, etc.) in the buffer does not normally affect the adsorption of proteins [4].

Elution of proteins can be achieved by one of three methods: protonation, ligand exchange, or column stripping. Protonation is the most common method and probably the simplest. The pH is reduced by either a linear gradient or step-gradient in the range of pH 8 to 3 or 4, reflecting the titration of the histidyl residues. Most proteins elute between pH 6 and 4. Again, sodium phosphate or sodium acetate are the buffers of choice. Competitive elution with ammonium chloride (0*M*–2.0*M*), imidazole (0*M*–0.5*M*) or its analogs histidine (0*M*–0.05*M*) and histamine yield similar selectivity [8]. Competitive elution with a linear gradient or step-gradient is best run at a constant nearly neutral pH. The final method of elution is to use chelating agents such as EDTA or EGTA (0.05*M* solutions) which will strip the metal ions from the gel and cause the proteins to elute. Unless the protein of interest is the only one still bound on the column, this method will result only in recovery and not in purification or resolution [3]. Another undesirable feature of this protocol is that the eluate will contain a high concentration of free metal ion.

Most resin manufacturers recommend that to maintain reproducibility and consistency, IMAC columns should be stripped of their metal ligands after each use and subsequently recharged with metal before the next run [6].

Recently, Fe(III)–IMAC has found specific application in the separation of phosphorylated macromolecules and other biological substances [9]. Unlike Cu(II)–IDA complexes which have no formal charge, the metal–ligand complex Fe(III)–IDA has a net positive charge. In terms of use, the highest protein capacity is reached at low pH (< 6) rather than at or above neutrality and at low ionic strength rather than at high salt concentrations. Electrostatic interactions for Fe(III) complexes play an important in protein binding [2]. However, Fe(III)–IMAC systems do not interact with phosphoproteins in the same way as ordinary ion-exchange resins. Fe(III)–IMAC can be employed to resolve proteins with a wide range of isoelectric points (p*I* 4–11) something that is not generally possible in a simple, single ion-exchange chromatographic step.

Immobilized metal affinity chromatography has been shown to be effective for isolating proteins from crude mixtures, as well as for selective separations of closely related proteins [2]. With respect to separation efficiency, IMAC compares well with biospecific affinity chromatography and the immobilized metal-ion complexes are much more robust than antibodies or enzymes. These factors make IMAC particularly well suited for scale-up to process scale chromatography. The main scale-up points to be aware of are the degree to which the column is metal saturated, the chelating agent content of the sample, and the potential of leached metal (or its interactions) within the product eluate.

Leakage of metals from the column during elution can be the most significant problem due to their toxicity, but there are several ways to avoid this pitfall. Some references suggest the precaution of underloading IMAC columns (by as much as 20%) with the metal ion to begin with [8]. Another precaution is to add EDTA with imidazole, histidine, or histamine to the column fractions. EDTA competitively blocks formation of coordination complexes between protein carboxyl clusters and divalent metal cations, whereas the imidazolium groups block histidyl–metal complexation. For best reproducibility and general ease of use, a two-column format is preferred for process scale. A second scavenging column with 5% of the volume of the metal saturated purification column is simply placed in line.

Besides accommodating raw feedstreams, the relative independence of protein binding from salt concentration offers a great deal of flexibility for process sequencing. IMAC can follow virtually any other technique without the requirement for buffer exchange. The main exception is hydrophilic interaction chromatography (HIC) with ammonium sulfate. IMAC can also be used as the initial capture step enabling purification up to a 1000-fold [4] and subsequent preequilibration for downstream low-ionic-strength methods. Although high salt loading improves IMAC-binding specificity [5], its concentration can be reduced after the major contaminants are washed through the column. Even with sodium chloride concentrations of up to $0.1M$, just a minor dilution can allow for the following charge-based chromatography method. This flexibility reveals IMAC as a valuable tool for streamlining the overall process design.

References

1. F. Helferich, *Nature 189*: 1001 (1961).
2. J. Porath, J. Carlsson, I. Olsson, and G. Belfrage, *Nature 258*: 598 (1975).
3. *The Qiaexpressionist*, 2nd ed., Qiagen Inc., Chatsworth, CA, 1992.
4. J. Porath, *Protein Express. Purif. 3*: 263 (1992).
5. E. Sulkowski, *Trends Biotechnol. 3*: 170 (1985).
6. E. Sulkowski, *BioEssays 10*: 170 (1989).
7. R. D. Johnson and F. H. Arnold, *Biotechnol. Bioeng. 48*: 437 (1995).
8. J. Porath and B. Olin, *Biochemistry 22*: 1621 (1983).
9. L. D. Holmes and M. R. Schiller, *J. Liquid Chromography Related Technol. 20*: 123 (1997).

R. Musil

Immunoaffinity Chromatography

Introduction

Immunoaffinity chromatography (IAC) refers to any chromatographic method in which the stationary phase consists of antibodies or antibody-related binding agents. *Antibodies*, or *immunoglobulins*, are a diverse class of glycoproteins that are produced by the body in response to a foreign agent, or *antigen*. The high selectivity of antibodies in their interactions with other molecules and the ability to produce antibodies against a wide range of substances has made IAC a popular purification tool for the isolation of hormones, peptides, enzymes, proteins, receptors, viruses, and subcellular components. The high selectivity of IAC has also made it appealing as a means for developing a variety of specific analytical methods.

Antibody Structure

The key component of any IAC method is the antibody preparation that is used as the stationary phase. The basic structure of a typical antibody (i.e., immunoglublin G or an IgG-class antibody) consists of four polypeptides that are linked by disulfide bonds to form a Y- or T-shaped structure. The two upper arms of this structure are called the *Fab fragments* and contain two identical antigen-binding regions. The lower stem region is known as the *Fc fragment* and has a structure which is highly conserved between antibodies that belong to the same class. Other classes of antibodies (e.g., IgA, IgM, IgD, and IgE) have the same basic structure as IgG but may contain multiple units that are cross-linked through the presence of additional peptide chains. The amino acid composition within the Fab fragments is highly variable from one type of antibody to the next. It is this variability that allows the body to produce antibodies that have a large variety of affinities and binding specificities for foreign agents.

Typical antigens in nature include bacteria, viruses, and foreign proteins from animals or plants. All of these agents are fairly large compared to the binding sites on an antibody. As a result, these antigens usually have many different locations on their surfaces to which an antibody can bind; each of these locations is called an *epitope*. Smaller antigens (i.e., those with molecular masses below several thousand Daltons) are too small to produce an immune response by themselves. However, these can be made to give rise to antibody production if they are first coupled to a larger species, such as a *carrier protein*. The agent that is coupled to the carrier protein is then called a *hapten*.

Antibody Production

One way to produce antibodies to a given compound is to inject the corresponding antigen or hapten-carrier protein conjugate into a suitable laboratory animal, such as a mouse or rabbit. Samples of the animal's blood are then taken at specified intervals to collect any antibodies that have been produced to the foreign agent. This method results in a heterogeneous mixture of antibodies that bind with a variety of strengths and to various epitopes on the antigen or hapten–carrier protein conjugate. These antibodies are called *polyclonal antibodies*, because they are produced by several different immune system cell lines within the body. Techniques have also been developed that allow for the isolation of single antibody-producing cells and the subsequent hybridization of these with cancer cells to produce new cell lines that are stable and relatively easy to grow over long periods of time. These combined immune system/cancer cells are known as *hybridomas*, and their product is a single type of well-defined antibody called a *monoclonal antibody*. Both polyclonal and monoclonal antibodies are commonly used in IAC methods.

Immunoaffinity Supports

The support material is another important item to consider in the development of a successful IAC method. In the past, most IAC applications have been based on low-performance supports (i.e., nonrigid media that can be operated under gravity or in the presence of peristaltic flow or a small applied vacuum). The supports used in this situation have typically been carbohydrate-related materials, like agarose and cellulose, or synthetic organic supports, like acrylamide-based polymers. However, IAC can also be used in high-performance liquid chromatography (HPLC) if more rigid, pressure-resistant, and higher efficiency materials are employed. Some examples of HPLC supports that have been used for IAC include derivatized glass, silica, polystyrene-based perfusion media, and azalactone beads. When these types of supports are used in IAC, the resulting method is often referred to as *high-performance immunoaffinity chromatography* (*HPIAC*). Of these two approaches, low-performance IAC is the method most often used for the purification of solutes or in sample pretreatment prior to analysis by other techniques. HPIAC can also be used for sample pretreatment or compound isolation, but it is more commonly employed as an analytical tool for the measurement of specific chemicals in complex mixtures.

Antibody Immobilization

One common approach to antibody immobilization involves direct, covalent attachment between the support and free amine groups on the antibodies. Examples include reductive amination (i.e., the Schiff base method) or the reaction of antibodies with supports that have been activated with reagents such as carbonyldiimidazole or *N*-hydroxysuccinimide. Antibodies or antibody fragments can also be covalently immobilized through more site-selective methods. For instance, free sulfhydryl groups that are generated during the production of Fab fragments can be used to couple these fragments to thiol-activated supports. Another example involves the mild oxidation of the carbohydrate residues which occur in the Fc region of antibodies, followed by the reaction of these oxidized residues with amine- or hydrazide-activated materials. The main advantage of site-selective immobilization is that it produces immobilized antibodies or antibody fragments that have fairly well-defined points of attachment and greater accessibility of their binding regions to analytes, thus giving rise to higher binding activities than are obtained by more general coupling methods.

Noncovalent immobilization can also be used for the site-selective coupling of antibodies to supports. One common approach for this involves absorbing the antibody to a secondary ligand such as protein A or protein G, which both bind to the Fc region of many antibody classes. This binding is quite strong under physiological conditions but can be easily disrupted by decreasing the pH of the surrounding solution. This method is useful when high antibody activity is needed or when it is desirable to have frequent replacement of antibodies in the IAC column.

Application and Elution Conditions

The mobile phases that are used in IAC are another group of factors that need to be considered when using this method. The *application buffer* that is used during sample injection should facilitate quick and efficient binding of the analyte to immobilized antibodies. This mobile phase is usually selected so that it mimics the natural surroundings of the antibody (i.e., physiological pH and ionic strength). The association equilibrium constants for antibody–antigen interactions under such conditions are often in the range of $10^6 M^{-1}$–$10^{12} M^{-1}$. This results in extremely strong binding between the analytes and the immunoaffinity column during sample application.

Although it is possible to use isocratic elution for IAC columns that contain low-affinity antibodies (i.e., those with association constants below $10^6 M^{-1}$), this is not practical for higher-affinity antibodies. The only way that solutes can be quickly eluted from these antibodies is to change the column conditions to lower the effective strength of the antibody–analyte interaction. This is done by applying an *elution buffer* to the column. Usually, an acidic buffer (pH 1–3) or one that contains a chaotropic agent such as sodium thiocyanate is used for analyte elution, but, occasionally, a competing agent, an organic modifier, a temperature change, or a denaturing agent is employed. The elution buffer is typically applied in a step-gradient; however, more gradual linear or nonlinear gradients can also be used.

Traditional IAC

There a variety of formats in which IAC can be performed. Some examples of these are shown in Fig. 1. The simplest format is the *on–off mode* or *direct-detection mode* of IAC. In this technique, the sample is first injected onto the IAC column in the presence of the application buffer. As the analyte is being retained, other solutes present in the sample pass through nonretained and are washed from the column. After these nonretained solutes have been removed, the elution buffer is applied. The analyte is then collected or detected as it elutes from the column. Afterward, the initial application buffer is reapplied and the antibodies are allowed to regenerate before the next sample is injected. This particular format is the one most commonly used in IAC for the purification of compounds. The on–off mode is also used in analytical applications that involve analytes which are labeled or occur at sufficiently high levels to allow direct detection as they elute from the IAC column.

Immunoextraction Methods

Another set of IAC methods are those that involve *immunoextraction*. This refers to the use of IAC for the removal of a specific solute or group of solutes from a sample prior to determination by a second analytical method. *Off-line immunoextraction* is generally the easiest way for combining IAC with other techniques. For this method, antibodies are typically immobilized onto a low-performance support that is packed into a small disposable syringe or solid-phase extraction cartridge. After sample application and the washing away of undesired sample components, an elution buffer is applied and the analyte is collected. In most situations, the collected fraction is dried down and reconstituted in a solvent that is more suited for a subsequent analysis (e.g., a volatile solvent for compound quantitation by gas chromatography). This approach has been used to analyze substances in a variety of samples, ranging from plasma and urine to food, water, and soil extracts.

The relative ease with which IAC can be directly coupled to an HPLC system makes *on-line immunoextraction* appealing as a means for automating and reducing the time required for sample pretreatment in HPLC. Although IAC has been directly coupled with both size-exclusion and ion-exchange chromatography, the vast majority of on-line immunoextraction has involved coupling IAC with reversed-phase liquid chromatography. Part of the reason for this is the popularity of reversed-phase HPLC in routine chemical separations. Another reason is the fact that the elution buffer for an IAC column is an aqueous solvent with little or no organic modifier, making this act as a weak mobile phase for reversed-phase columns. On-line immunoextraction coupled with reversed-phase HPLC has been used to quantitate compounds in such samples as food extracts, bodily fluids, enzyme digests, cell extracts, and environmental samples.

Chromatographic Immunoassays

Another important technique in IAC is the use of immobilized antibody columns to perform *chromatographic* (or *flow-injection*) *immunoassays*. One way this can be done is in a competitive binding format. The simplest approach to a competitive binding scheme is to mix the sample and a labeled analyte analog (the label) and apply these simultaneously to the IAC column; this is a method known as a *simultaneous injection competitive binding immunoassay*. If the sample is applied to the IAC column and followed later by a separate injection of the label, then the technique is called a *sequential injection competitive binding immunoassay*. In both formats, an indirect measure of the sample analyte is obtained by examining the amount of label that elutes in either the nonretained or retained IAC fractions.

An alternative format is the *displacement competitive binding immunoassay*. Here, the IAC column is first saturated with the labeled analog, followed by application of sample to the column. As the sample travels through the column, it is able to bind to any antibody-binding regions that are momentarily unoccupied by the label as this undergoes local dissociation and reassociation with the immobilized antibodies. This results in displacement of the

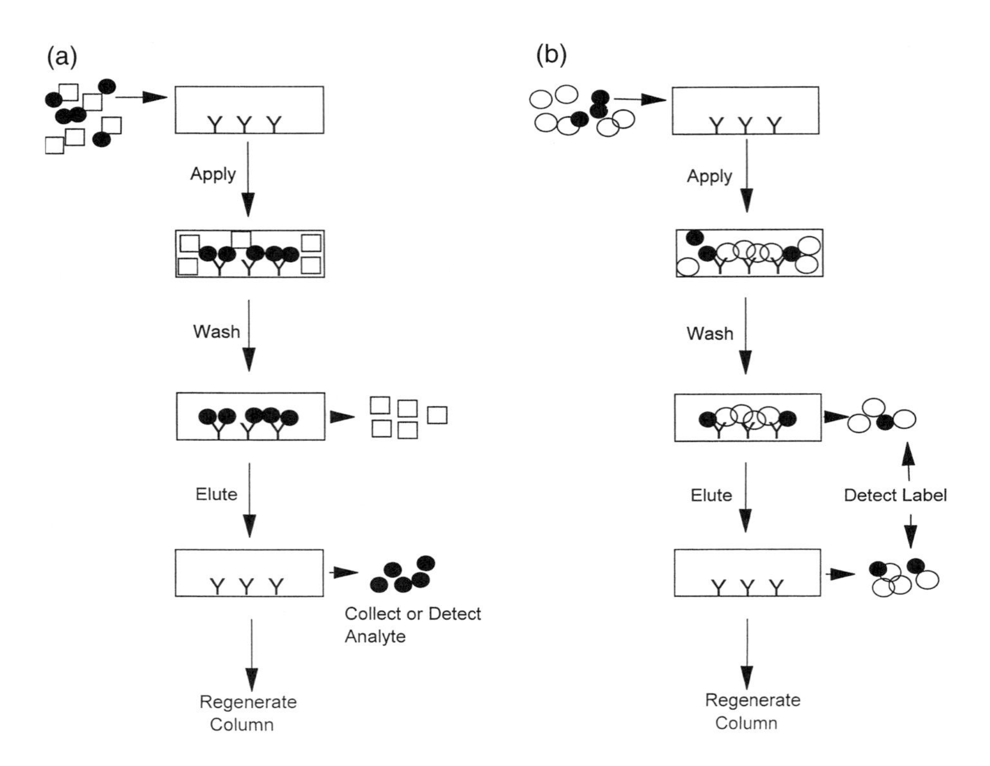

Fig. 1 Typical schemes for (a) the on–off mode of immunoaffinity chromatography and (b) a chromatography-based competitive binding immunoassay. The solid circles represent the analyte, the open circles represent a labeled analog of the analyte, and the open squares represent nonretained sample components.

label from the column, with the degree of this displacement being directly proportional to the amount of applied analyte.

A *sandwich immunoassay* in IAC involves the use of two different types of antibody that each bind to the analyte of interest. The first type is attached to a solid-phase support and is used for extraction of the analyte from samples. The second contains an easily measured label and is added in solution to the analyte either before or after extraction. This label allows a substance to be quantitated by providing a signal that is directly proportional to the amount of analyte that is present in the IAC column.

In a *one-site immunometric assay*, the sample is first incubated with a known excess of labeled antibodies (or Fab fragments) that are specific for the analyte of interest. After this binding has occurred, the mixture is applied to a column that contains an immobilized analog of the analyte; this is done to extract any antibodies that are not bound to the analyte. Those antibodies that are bound to the analyte will pass through the column in the nonretained peak. Detection is performed by either looking at the nonretained labeled antibodies or by monitoring the amount of excess antibodies that later dissociate from the column during the elution step.

Postcolumn Immunodetection

The technique of *postcolumn immunodetection* involves the use of an IAC column that is attached to the exit of an analytical HPLC system. The IAC column in this approach serves to collect and retain a specific analyte from the HPLC column eluent for later detection. Both the on–off mode and immunoassay formats of IAC have been used as strategies for postcolumn immunodetection. The most common of these approaches is the one-site immunometric assay. One reason for this is that the immobilized analog columns in one-site immunometric assays often have a much more flexible selection of elution conditions than immobilized antibody columns. Another reason is that one-site immunometric columns can usually be used for many sample injections before they are eluted and regenerated, which helps to decrease the overall analysis time associated with their use in postcolumn detection.

Suggested Further Reading

Calton, G., *Methods Enzymol. 104*: 381 (1984).
de Frutos, M. and F. E. Regnier, *Anal. Chem. 65*: 17A (1993).
Hage, D. S., *J. Clin. Ligand Assay 20*: 293 (1997).
Hage, D. S., D. H. Thomas, and M. S. Beck, *Anal. Chem. 65*: 1622 (1993).
Irth, H., A. J. Oosterkamp, U. R. Tjaden, and J. van der Greef, *Trends Anal. Chem. 14*: 355 (1995).
Oosterkamp, A. J., H. Irth, U. R. Tjaden, and J. van der Greef, *Anal. Chem. 66*: 4295 (1994).
Wilchek, M., T. Miron, and J. Kohn, *Methods Enzymol. 104*: 3 (1984).

David S. Hage
John Austin

Immunodetection

Introduction

Monitoring a liquid chromatographic effluent by means of an immunoassay provides sensitive and selective detection in combination with the separation of cross-reactive compounds [1,2]. When implementing the immunoassay as a postcolumn reaction detection system after liquid chromatography, it is frequently referred to as immunodetection [3,4]. Automation and assay speed are the main advantages of immunodetection over off-line coupling of immunoassays to liquid chromatography by means of fraction collection [5,6].

The typical setup of immunodetection is illustrated in Fig. 1 [5,6]. The column effluent is mixed with labeled antibodies which will bind selectively to the analytes while passing through a reaction coil. This binding is based on the affinity between analyte and antibody and is characterized by the association and dissociation rate constants of the affinity reaction. Whereas the association rate constant (k_{+1}) is diffusion controlled, the dissociation rate constant (k_{-1}) depends on the interactions between the antibody and its antigen. Generally, k_{+1} lies in the range of 10^7–10^8 L/ml s, whereas k_{-1} is comparably slow (10^3–10^{-5} s^{-1}). The volume of the reaction coil and the flow rates used determine the reaction time during which the labeled antibodies can bind to analyte molecules. Typical reaction times lie in the range of a few minutes and thus allow the fast association reaction to take place, whereas the dissociation reaction can practically be neglected. Quantification of the analyte concentration is then possible by distinguishing labeled antibody which has bound to analyte from free antibody. For that purpose, the free and the bound antibody needs to be separated (e.g., by means of an affinity column which traps free antibody), whereas the analyte–antibody complex passes the affinity column unretained for detection in a conventional flow through the detector. Using this setup, both analyte recognition and quantification occurs through the labeled antibody.

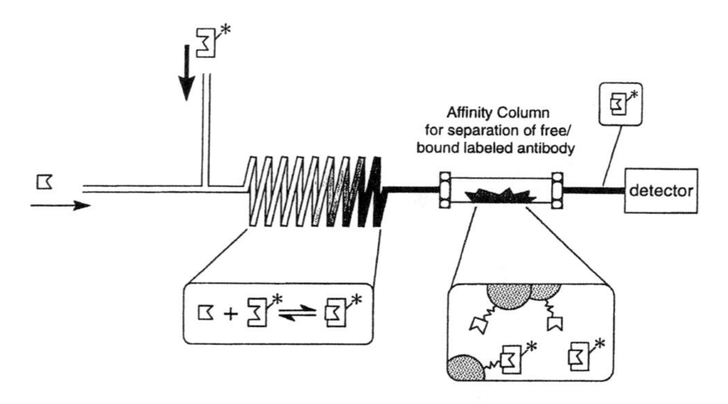

Fig. 1 Scheme of the immunodetection system employing labeled antibodies (*) and an affinity column for separating labelled antibodies which have reacted with an analyte (*) from free-labeled antibodies.

Alternatively, it is possible to use untreated antibodies in combination with a labeled antigen; see Fig. 2 [5,6]. Under these circumstances, a two-step reaction is performed after the analytical separation. First, the column effluent is mixed with antibodies to allow the recognition of analyte(s). In the second step, the labeled antigen is added to saturate the fraction of free antibodies and allow quantification. When binding of labeled antigen to the antibody causes a change in detection properties, the reaction mixture can be monitored directly for quantification (homogeneous assay). However, generally the labeled antigen which has reacted with the antibody needs to be separated from the free labeled antigen to allow quantification. Again, affinity columns can be used for this purpose. Other forms of separating free and bound labeled anti-

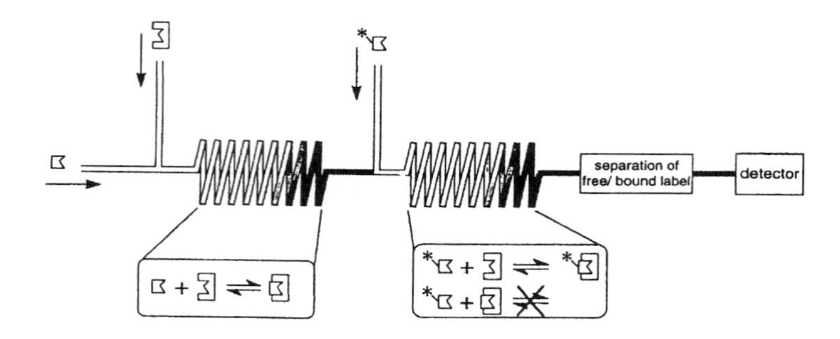

Fig. 2 Scheme of the immunodetection system employing untreated antibodies ( ) and a labeled antigen (*).

gens comprise restricted access columns, free-flow electrophoresis, and cross-flow filtration.

Reagents

So far, primarily antibodies and their Fab fragments have been implemented in immunodetection for analyte recognition. Antibodies can be raised against virtually any compound of interest; accordingly, their implementation into detection for liquid chromatography provides a general approach. Antibody affinity and selectivity can be modulated by appropriate design of the hapten, by adequate screening of the antibodies, and by site-directed mutagenesis. Because the chemical structure and properties of antibodies against different antigens is comparatively homogenous, immobilization, stabilization, calibration, and storage procedures can be standardized.

The approach of implementing a biological assay as a postcolumn reaction detection system after liquid chromatography can not only be applied to antibody-based assays (immunoassays) but also to assays employing other affinity interactions with high association and low dissociation rate constants, such as receptors. Information obtained from such a detection system not only provides quantitative results but also indicates the biological activity of the detected compound.

Requirements with respect to the label used to mark one of the immunoreagents are comparable to those in other postcolumn reaction detection systems [4]. The label should preferably allow sensitive and rapid detection and be nontoxic, stable, and commercially available. So far, mainly fluorescence labels have been employed (e.g., fluorescein), although, in principle, also liposomes, time-resolved fluorescence, and electrochemical or enzymatic labels are feasible. On the other hand, labels providing a slow response, including radioactive isotopes and glow-type chemiluminescence, are less suitable for immunodetection. When attaching the label to the immunoreagent, care has to be taken not to affect the affinity reaction between the antibody and its antigens and thus deteriorate assay performance.

A concern in recent research involving immunodetection has been availability, quality, and cost of reagents, especially of antibody and receptor preparations. In the future, this concern may be overcome with novel cloning techniques providing possibilities to drastically reduce the cost of producing proteins as well as to develop proteins for specific applications.

Interfacing Liquid Chromatography–Immunodetection

The attractiveness of immunodetection consists in its on-line coupling to a separation step, such as liquid chromatography. Parameters to consider are band broadening caused by the postcolumn reaction and interference of the liquid chromatographic mobile phase with the immunoreaction [4,7].

In conventional immunoassays with long incubation times, the environment in which the affinity reaction is taking place needs to be strictly controlled with respect to, for example, pH, salt, and organic modifier content. In contrast, immunodetection takes place within a few minutes, entailing less stringent requirements with respect to reaction conditions. Nevertheless, the mobile phase needs to be consistent with the affinity reaction; that is, the mobile phase should not denature the immunoreagents or compete with the analyte for the available binding sites. Mobile-phase compatibility has mainly been evaluated with reversed-phase liquid chromatography, as it is a frequently used analytical separation technique and constitutes the greatest challenges in interfacing to biological assays. The crucial consideration is the organic-modifier content in the reversed-phase liquid-chromatography mobile phase. Investigations have shown that up to 15–25% (v/v) of organic modifier can be used in immunodetection without affecting the antibody–antigen interaction [5]. These results are in concurrence with immunodetection systems which have been coupled to reversed-phase liquid chromatography. At higher concentrations of organic modifier, the affinity reaction can be hampered seriously, which typically is overcome by dilution of the column effluent [8,9].

However, many interesting analytes (e.g., peptides and proteins) are commonly separated by means of a gradient. The challenge of coupling immunodetection to gradient liquid chromatography is twofold: On the one hand, the affinity interaction will be affected; on the other hand, the detection properties of the label will vary. For example, using a gradient of organic modifier affects the conformation of the antibody and, thus, its affinity characteristics, as well as the detection properties of a fluorescence label. When acceptance of an increasing baseline [10,11] is out of the question, additional interfacing between the separation step and the immunodetection is required. One approach is to introduce a buffer-exchange step after the separation (e.g., with on-line dialysis [12] or an ion-exchange column [7]). However, this will introduce extra band broadening as well as affect robustness with yet another part in the system.

Applications

Feasibility of liquid chromatography–immunodetection has been shown for quantitative analysis and for screening for biological activity, as summarized in Table 1. For analytical purposes, immunodetection in combination with liquid chromatography is most promising in those cases when conventional immunoassays or conventional liquid chromatographic methods by themselves do not suffice for accurate analytical determinations. Being an approach offering high selectivity and sensitivity, applications are directed toward measurement of trace levels of compounds which lack appropriate detection properties in complex matrices. This is illustrated, for example, by measuring endogenous levels of the protein granulocyte colony-stimulating factor (GCSF) in biological matrices [10]. Affinity chromatography for sample preparation introduces high selectivity into the system but does not provide a means to improve detection properties of the protein. By combining the affinity chromatography for sample preparation with an analytical separation by means of reversed-phase liquid chromatography and immunodetection, levels of GCSF were determined in plasma.

In other bioanalytical applications, the emphasis lies more on overcoming cross-reactivity. For example, the heart glycoside digoxin is cross-reactive with several of its metabolites as well as with plasma constituents, thus hampering approaches solely based on immunoassays. By treating plasma samples with solid-phase extraction on a C_{18}-restricted access column, coupled on-line to reversed-phase liquid chromatography–immunodetection, digoxin and two of its cross-reactive metabolites were analyzed in patients treated with the heart glycoside [8].

A key to successful drug development is the identification of new lead compounds. Lead compounds can be identified through receptor assays, where the receptor–ligand interaction reflects the biomolecular mechanism associated with a disorder. By implementing the receptor interactions into postcolumn reaction detection systems, biologically active compounds can be separated and detected with high sensitivity and selectivity. This concept has been described for the analysis of estrogens using a recombinant steroid-binding domain of the human estrogen receptor for analyte recognition and coumestrol, a fluorescent estrogen, as reporter molecule [13]. Prior to detection, samples were treated on-line and automated by reversed-phase solid-phase extraction and reversed-phase liquid chromatography. Selectivity of this system is demonstrated for analysis in urine samples, which shows the feasibility of using this method in the determination of the abuse of steroid hormones in performance doping or cattle breeding. When performing both liquid chromatography–immunodetection and mass spectrometry, information on biological activity is combined with structure elucidation [11].

Recently, also direct recognition of active ligands attached to bead surfaces has been achieved with immunodetection [14]. This provides a rapid and automated screening tool which is compatible with solid-phase bound compounds originating from solid-phase chemistry in combinatorial chemistry. However, this approach has so far only been published for a model system.

Conclusions

Immunodetection coupled on-line to liquid chromatography as a tool for quantitative analysis has been developed for model compounds as well as been applied in relevant applications. The approach is particularly appealing for trace analysis in complicated matrices and for identifying ligands for certain receptors in drug discovery. However, each application still requires a fair amount of method development and optimization, and obtaining the desired, pure immunoreagents still is a concern, although advances

Table 1 Review of Immunodetection Applications

Application area	Compound	Matrix	Immunodetection type	Detection limit	Ref.
Protein bioanalysis	Granulocyte colony-stimulating factor (GCSF)	Plasma	Fluorescence-labeled antibodies, affinity chromatography	0.6 nmol/L	10
	Growth-hormone-releasing factor (GHRF)	Plasma	Fluorescence- and enzyme-labeled antibodies, affinity chromatography, interface for dilution between liquid chromatography and immunodetection for compatibility	0.2 ng/mL	9
	Urokinase	Plasma	Fluorescence-labeled receptor, affinity column	40 nmol/L	11
	Interleukine 4	?	fluorescence-labeled antibodies, affinity chromatography	2 fmol	16
Drug bioanalysis	Diogoxin	Plasma	Fluorescence-labeled antibodies, affinity chromatography	0.2 nmol/L	8
Biomarker analysis	Sulfodipeptide Leukotrienes	Urine, human cell culture extract	Untreated antibodies and fluorescence-labeled ligand, reversed-phase restricted-access chromatography	0.4 nmol/L	15

in recombinant protein production are providing us with an increased choice and availability of affinity reagents at reduced cost.

In parallel with miniaturization of liquid chromatography, immunodetection will be downscaled, thus lowering reagent consumption and, consequently, cost. However, increased challenges with respect to band broadening in a postcolumn reaction detection system and nonspecific binding to capillary walls will need to be addressed. The trend toward miniaturization will simultaneously give the opportunity to couple immunodetection to other analytical separation methods, such as capillary electrophoresis. Combination of immunodetection with mass spectrometry enables the combination of information of biological activity with structure elucidation. Other developments in immunodetection concern the separation of free and bound labels, enabling the implementation of suspended materials in detection liquid chromatography, including, for example, suspended membrane receptors, whole cells, and molecularly imprinted polymers serving as artificial receptors. Consequently, immunodetection potentially conquer new application areas, such as the evaluation of absorption profiles of drugs and the investigation of drug metabolism on a cellular level.

References

1. B. Mattiasson, M. Nilsson, P. Berdén, and H. Håkansson, *Tr. A. C. 9*: 317 (1990).
2. M. De Frutos and F. E. Regnier, *Anal. Chem. 65*: 17 (1992).
3. D. S. Hage, *J. Chromatogr. B 715*: 3 (1998).
4. I. S. Krull, B.-Y. Cho, R. Strong, and M. Vanderlaan, *LC–GC Int.* 278 (May 1997).
5. H. Irth and A. J. Oosterkamp, *Tr. A. C. 14*: 355 (1995).
6. E. S. M. Lutz, A. J. Oosterkamp, and H. Irth, *Chim. Oggi 15*: 11 (1997).
7. K. Shahdeo, C. March, and H. T. Karnes, *Anal. Chem. 69*: 4278 (1997).
8. A. J. Oosterkamp, H. Irth, M. Beth, K. K. Unger, U. R. Tjaden, and J. van der Greef, *J. Chromatogr. B 653*: 55 (1994).
9. B.-Y. Cho, H. Zou, R. Strong, D. H. Fisher, J. Nappier, and I. S. Krull, *J. Chromatogr. A 743*: 181 (1996).
10. K. J. Miller and A. C. Herman, *Anal. Chem. 68*: 3077 (1996).
11. A. J. Oosterkamp, R. van der Hoeven, W. Glässgen, B. König, U. R. Tjaden, J. van der Greef, and H. Irth, *J. Chromatogr. B 715*: 331 (1998).
12. M. Kaufmann, T. Schwarz, and P. Batholmes, *J. Chromatogr. A 639*: 33 (1993).
13. A. J. Oosterkamp, M. T. Villaverde Herraiz, H. Irth, U. R. Tjaden, and J. van der Greef, *Anal. Chem. 68*: 1201 (1996).
14. E. S. M. Lutz, H. Irth, U. R. Tjaden, and J. van der Greef, *Anal. Chem. 69*: 4878 (1997).
15. A. J. Oosterkamp, H. Irth, L. Heintz, G. Marko-Varga, U. R. Tjaden, and J. van der Greef, *Anal. Chem. 68*: 4101 (1996).
16. T. Schenk, H. Irth, L. Heintz, G. Marko-Varga, U. R. Tjaden, and J. van der Greef, submitted.

E. S. M. Lutz

Industrial Applications of CCC

Introduction

In human activity, industry has to produce material products and goods. The chemical industry produces millions of metric tons of basic chemicals such as soda, ethylene, sulfuric acid, or urea, and a few kilograms or less of fine and/or complicated chemicals such as chiral drugs, catalysts, antibiotics, or delicate perfumes. Countercurrent chromatography (CCC) is useful in the production of the latter class of chemicals. This entry explains the role that CCC can play in industrial processes, revealing concepts and ideas rather than detailing examples that can be found elsewhere. At the moment, only a handful of chemical companies are using CCC in commercial processes. Often, they are, apparently, very successful with the technique, because they purchase more CCC systems and CCC becomes part of the production process. The problem is the companies do not make nor want their chemical competitors to know that CCC works.

The Liquid Stationary Phase

It is the liquid nature of the stationary phase in CCC that renders it useful, for three reasons:

1. The solutes can access the whole volume of the stationary phase, not only the surface of a solid stationary phase as in most other chromatographic techniques. Large amounts of substance can be processed in a single run. CCC is truly a preparative technique.
2. The retention mechanism is very simple. The only physicochemical parameter responsible for solute retention is the liquid–liquid partition coefficient, P. The retention volume, V_R, is simply

$$V_R = V_M + PV_S = V_C + (P - 1)V_S \quad (1)$$

   where V_M and V_S are the volumes of the mobile and stationary phases, respectively, inside the CCC system. The sum of these two volumes is the CCC system volume, V_C.
3. It is possible to switch the roles of the phases during a run. The mobile phase becomes the stationary phase and vice versa (see the entry Measurement of $P_{O/W}$ by CCC). The combination of these three points produces the following advantages.

A study and optimization of a separation of a complex mixture can be accomplished with a low-volume CCC system, V_C, injecting small quantities of the sample. The biphasic liquid system composition is *optimized rapidly* because the V_R volumes are reduced.

If the retention volumes of some constituents of the sample are too large, the dual mode is used. The retained constituents are eluted in the reversed mode. It is absolutely certain that *no part of the sample can be trapped* inside the CCC apparatus.

Once the separation is optimized, the partition coefficient of each constituent is then calculated [Eq. (1)]. The very same liquid system is used in a large-volume CCC system. So, the *retention volumes can be predicted* because the partition coefficients, which depend entirely upon the liquid–liquid system, are the same as for the small-volume system. The scaling-up is straightforward [1].

Large-Scale Separation or Purification

Classical Use of the Technique

In a recent work, the fractionation of a tannin sample was studied. The separation was optimized with a 150-mL CCC apparatus. The butanol–ethyl acetate–water (pH 2.8) system (3.5:46.5:50% v/v) was found to be efficient for the separation. The partition coefficients of the 12 peaks were calculated. Then, it was possible to fractionate 26 g of the tannin sample in one run with the same liquid system and a 2-L CCC system [1]. The dual-mode approach was used for this separation.

The high loading capability of large-volume CCC systems is commonly used in industry. Large amounts (gram to kilogram scale) of natural products with high added values are separated by CCC. Alkaloids, antibiotics, enzymes, macrolides, peptides, rare fatty acids, saponins, tannins, taxoids and/or precursors of Taxol®, and other fine chemicals have been isolated, separated, and/or purified by preparative CCC [1].

What makes CCC preferred to classical prep liquid chromatography (LC) are as follows:

- An original or unique selectivity obtained with a subtle polarity difference between the two liquid phases
- The possibility of injecting heavily concentrated or even polyphasic samples (e.g., fermentation broths)
- The gentle interactions during the separation process that preserve delicate molecules (e.g., proteins) from denaturation.

Displacement Chromatography

In displacement chromatography, the sample to be purified is injected in a large volume, or even continuously, into the CCC machine. The sample components have different affinities for the stationary phase in an exclusive way: The component with a higher affinity for the stationary phase displaces another one with a lower affinity. Bands of pure components form. This method of using CCC offers the maximum throughput capability [1].

pH Zone Refining

pH Zone refining sorts compounds by their ionization constants, K_a, using a stationary phase with a different pH from the mobile phase [2]. For example, the stationary phase will be an acidic organic phase [e.g., methyl *tert*-butyl ether (MTBE) with 1% acetic acid] and the mobile phase is an aqueous basic buffer (e.g., 0.015M NH_3 solution, pH = 10). Up to 60% of the CCC system volume of a mixture of organic acids can be injected into the apparatus (in the ammonium salt form at pH = 10) [2]. The injected organic salts are protonated by the acidic stationary MTBE. The protonated molecular forms stay in the organic phase, and the ionized basic forms prefer the aqueous phase. The weaker acid is protonated first, so it is the most retained. Bands of pure organic acids form in the order of decreasing pK_a values.

Complexation

Complexation can be used to separate metallic ions on an industrial scale. Figure 1 illustrates the process in the case of the separation of nickel and cobalt ions [3]. A complexing agent (e.g., diethyl hexyl phosphoric acid) is added to a heptane stationary phase. A large volume (up to 20 times the CCC machine volume V_C) of the ionic solution is injected. The nickel ions are displaced in the aqueous phase. The cobalt ions can be collected in the stationary phase. More than two ions can be separated in bands of increasing complexation constants order [3]. Because no ions can stay trapped inside the CCC machine, it could be a very potent tool in the separation of radionuclides in the processing of nuclear wastes.

Extraction

Countercurrent chromatography can be used to extract and to concentrate, in a low volume of stationary phase, a component present in large volumes of mobile phase. It was shown that a 60-mL CCC instrument was able to extract 285 mg of a nonionic surfactant contained in 20 L of water (at 16.5 ppm or mg/L) and to concentrate it into 30 mL of ethyl acetate (at 9500 ppm or 9.5 g/L) [4].

Step 1: aqueous mobile phase, heptane + HA stationary phase

Step 2: the CCC machine works like a deionizer, bands form

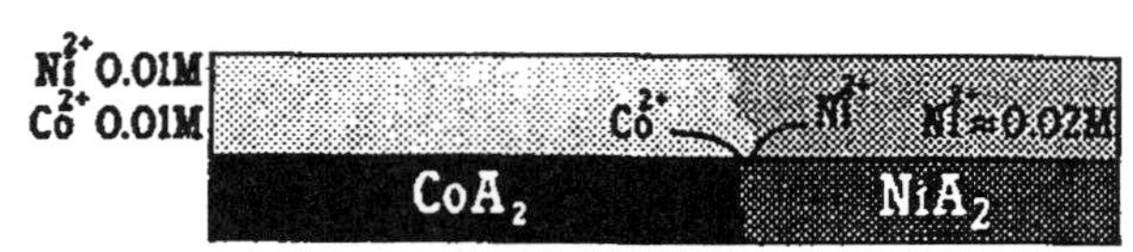

Step 3: only nickel ions are eluted more concentrated

Step 4: cobalt ions are in the liquid stationary phase

Fig. 1 Removal and separation of cobalt and nickel ions by CCC. Stationary phase: heptane + diethyl hexyl phosphoric acid (HA 0.5M); mobile phase: aqueous solution of cobalt and nickel acetate (0.01M each). Step 1: The CCC machine is equilibrated with water. Step 2: The ionic solution is introduced in the machine, the ions are extracted into the stationary phase, and the cobalt complex displaces the nickel one less stable. Step 3: The stationary phase is saturated in nickel ions. The greenish effluent leaving the machine contains only nickel ions two times more concentrated than the entering solution. Cobalt ions are still extracted, displacing nickel ions. Step 4: End of the process — the stationary phase is saturated in cobalt ions. The machine is stopped, the dark blue stationary phase is collected, and cobalt ions are recovered by an acid wash. (From Ref. 3.)

A Continuous Plug-Flow Reactor

Recently, the use of a CCC system as an original and powerful plug-flow liquid–liquid reactor for a biphasic catalytic reaction was demonstrated [5]. Benzaldehyde (BZA) can be reduced to benzyl alcohol (BZOH) by sodium formate in the aqueous phase, at room temperature, when a rhuthenium phosphine complex is used. BZA is

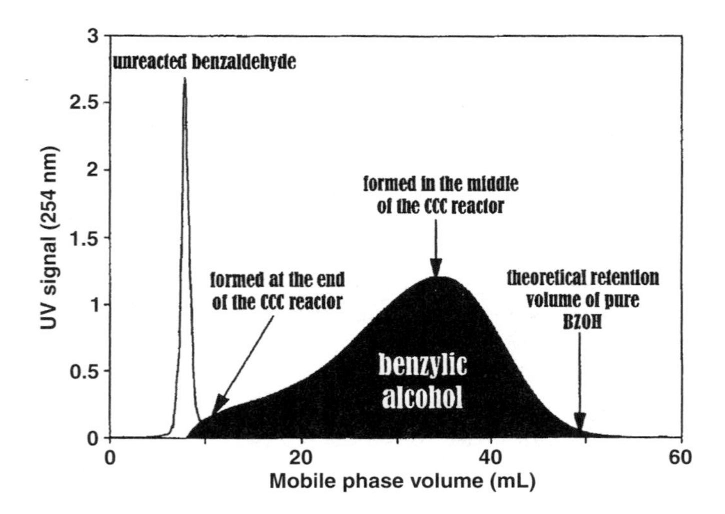

Fig. 2 Fast evaluation of a new catalyst capability: reduction of benzaldehyde (BZA) in benzylic alcohol (BZOH) by sodium formate catalyzed by a ruthenium–triphenylphosphine trisulfonated sodium complex. Twenty Tmoles BZA were injected. Seventeen Tmoles BZOH were formed (85% conversion). A 59-mL CCC machine, V_S = 52 mL of 5*M* sodium formate aqueous solution, V_M = 7 mL cyclohexane at 1.5 mL/min in the ascending tail to head direction, 750 rpm. (From Ref. 5.)

located in the cyclohexane mobile phase. Sodium formate, the complex, and BZOH are located in the aqueous stationary phase. A 79% conversion of BZA to BZOH was obtained at 30°C [5]. Even more interesting, injecting a 200-TL plug of BZA into the CCC apparatus with the aqueous phase containing the catalyst and sodium formate (5*M*), the result shown in Figure 2 was obtained. The elution band profile of the BZOH formed should allow one to model the kinetic behavior of the catalyst complex used. This study would need numerous batch experiments with different experimental conditions using, each time, some amount of an expensive and rare new catalyst. This can be done in one run, for one temperature, with the CCC reactor, saving time and catalyst.

At the moment, CCC is scarcely used in industry. The reasons are that the technique is not well known and few small-instrument companies market good CCC systems. The capabilities of the technique could be of great help in many industrial processes such as classical: extraction, purification, and separation of fragile compounds, as well as novel: use of a CCC system as a powerful liquid–liquid reactor.

References

1. A. Berthod and B. Billardello, *Adv. Chromatogr.* (1999).
2. Y. Ito, K. Shinomiya, H. M. Fales, and A. Weisz, in *Modern Countercurrent Chromatography* (W. D. Conway and R. J. Petroski, eds.), ACS Symposium Series No. 368, American Chemical Society, Washington, DC, (1995), pp. 156–183.
3. A. Berthod, J. Xiang, S. Alex, and C. Collet-Gonnet, *Can. J. Chem. 74*: 277–286 (1996).
4. A. Berthod, C. D. Chang, and D. W. Armstrong, in *Centrifugal Partition Chromatography* (A. P. Foucault, ed.), Chromatography Science Series No. 68, 1995, pp. 1–23.
5. A. Berthod, K. Talabardon, S. Caravieilhes, and C. De Bellefon, *J. Chromatogr. 828*: 523–530 (1998).

Alain Berthod
Serge Alex
Sylvain Caravieilhes
Claude De Bellefson

Influence of Organic Solvents on pK_a

Introduction

The influence of solvents on the ionization equilibrium is related to their electrostatic and their solvation properties. The value of the ionization constant of an analyte is closely determined, in practice, by the pH scale in the particular solvent. It is clear that it is most desirable to have a universal scale which is able to describe acidity (and basicity) in a way that is generally valid for all solvents. It is, in principle, not the definition of an acidity scale in theory which complicates the problem; it is the difficulty of approximating the measured values in practice to the specifications of the definition. The pH scale, as is common in water, is applicable only to some organic solvents (i.e., mainly those for which the solvated proton activity is compatible with the Brønsted theory of acidity). The applicability of an analog to the pH scale in water decreases with decreasing relative permittivity of the solvents and with their increasing aprotic character.

A scale that would enable us to compare the acidity in

all solvents could be based on the transfer activity coefficient on the proton (see the entry Capillary Electrophoresis in Nonaqueous Media). The effect of the solvent of any species can be expressed in the same way as for the proton by this concept and applied to all particles involved in the thermodynamic equilibrium.

Medium Effect and Ionization Constants of Weak Acids and Bases

The ionization of a weak neutral acid, HA, is described according to

$$HA = H^+ + A^-$$

For the weak base, B, for formal reasons, it is favourably expressed for its conjugated cation acid:

$$HB^+, \quad \text{by } HB^+ = H^+ + B$$

The corresponding changes of the ionization constants of these weak acids can be expressed by the particular transfer activity coefficients, ${}_m\gamma_i$, on the single species, i, according to

$$\Delta pK_{a,HA} = {}_S pK_{a,HA} - {}_W pK_{a,HA} = \log\left(\frac{{}_m\gamma_{H^+}\,{}_m\gamma_{A^-}}{{}_m\gamma_{HA}}\right) \quad (1)$$

$$\Delta pK_{a,HB^+} = {}_S pK_{a,HB^+} - {}_W pK_{a,HB^+} = \log\left(\frac{{}_m\gamma_{H^+}\,{}_m\gamma_{B}}{{}_m\gamma_{HB^+}}\right) \quad (2)$$

S and W indicate organic solvent and water, respectively. The transfer activity coefficients, ${}_m\gamma_i$, is the ratio of the activity coefficients in the particular solvents: ${}_m\gamma_i = {}_W\gamma_i / {}_S\gamma_i$.

Equations (1) and (2) enable the interpretation of the changes in pK_a values in the different organic solvents, compared to water. One must take into account not only the stabilization of the proton (this is given by the mutual basicity of the solvents) but also the different ability to stabilize the other individual particles. Besides the neutral particles HA and B, oppositely charged ions H^+ and A^- take part in the equilibrium in case of the neutral acids, and equally charged HB^+ and B^+ ions in case of the cation acid. This occurrence leads to a different change of the pK_a values for these two different types of weak electrolytes, depending on the ability of the solvent to stabilize anions or cations.

Nearly all organic solvents stabilize anions worse than water. For this reason, the pK_a values of neutral acids are larger in organic solvents (it is obvious that strongly basic solvents like amines may level out this effect). In lower alcohols, for example, the pK_a values increase by several units. In acetonitrile, which has generally even less cation stabilization ability in addition, the pK_a values may increase by 16 units. The effect of many solvents on weak bases (expressed by the pK_a of the corresponding cation acid) is much lower. pK_a values change only by few units. An exception is acetonitrile, due to the reason mentioned earlier.

It should be noted that traces of water have a great influence on the shift of pK_a values in all solvents, because there is a steep change of the pK_a with increasing water content of the organic solvent. It should be also pointed

Table 1 Ionization Constants of Neutral and Cation Acids of type HA and HB^+, respectively, in Water, Amphiprotic, and Dipolar Aprotic Solvents

	pK_A						
Acid	W	MeOH	EtOH	*t*-BuOH	ACN	DMSO	DMF
Acetic	4.73	9.7	10.3	14.2	22.3		13.3
Chloro acetic	2.81	7.8	8.3	12.2	18.8		10.1
Benzoic	4.21	9.4	10.1	15.1	20.7	11.0	12.3
3,4-Dimethyl benzoic	4.4	9.7		15.4	21.2	11.4	13.0
3-Bromo benzoic	3.81	8.8	9.4	13.5	20.3	9.7	11.3
4-Nitro benzoic	3.45	8.3	8.9	12.0	18.7	9.0	10.6
2,4,6-Trinitro phenol	0.3	3.7	4.1	4.8	11.0		
Ammonium	9.2				16.5	10.5	
Ethylammonium	10.6				18.4	11.0	
Triethylammonium	10.7	10.9			18.5	9.0	
Anilinium	4.6		5.7		10.6	3.6	
Pyridinium	5.2	5.2			12.3	3.4	

out that the pK_a of the silanol groups at the surface of the commonly used fused-silica material is affected by the choice of the solvent, too. It is also shifted to higher values (in water the pK_a is around 5–6).

Suggested Further Reading

Bates, R. G., Medium effect and pH in nonaqueous and mixed solvents, in *Determination of pH, Theory and Practice*. John Wiley & Sons, New York, 1973, pp. 211–253.

Kenndler, E., Organic solvents in capillary electrophoresis, in *Capillary Electrophoresis Technology* (N. A. Guzman, ed.), Marcel Dekker, Inc., New York, 1993, pp. 161–186.

Kolthoff, I. M. and M. K. Chantooni, General introduction to acid–base equilibria in nonaqueous organic solvents, in *Treatise on Analytical Chemistry* (I. M. Kolthoff and P. J. Elving, eds.), John Wiley & Sons, New York, 1979, pp. 239–301.

Sarmini, K. and E. Kenndler, Ionization constants of weak acids and bases in organic solvents, *J. Biophys. Biochem. Methods* *38*: 123–137 (1999).

Ernst Kenndler

Injection Techniques for Capillary Electrophoresis

Introduction

In liquid chromatography, a loop containing a defined volume is used to introduce the sample into the flowing mobile phase. Injection in high-performance capillary electrophoresis (HPCE) differs in two ways: (a) the injection volume is not as well defined and (b) injection is performed with the electric field turned off. Both of these features can contribute to quantitative errors of analysis. In addition, the length of the injection plug must be kept quite small to maintain the efficiency of the electrophoretic process [1]. The use of stacking electrolytes permits large injections to be made [2]. This is necessary to achieve acceptable limits of detection.

There are two modes of injection in capillary electrophoresis: hydrodynamic injection and electrokinetic injection. In hydrodynamic injection, pressure or vacuum are placed on the inlet sample vial or the outlet waste vial, respectively. For electrokinetic injection, the voltage is activated for a short time with the capillary and electrode immersed in the sample.

The general process of performing an injection and run is as follows:

1. The capillary is rinsed with 0.1N sodium hydroxide or 0.1N phosphoric acid for 1–2 min.
2. A second rinse with background electrolyte (BGE) is performed for 2–3 min.
3. The inlet side of the capillary is immersed in the sample.
4. Injection is performed for 1–30 s.
5. The voltage is ramped up (15 s) to the designated value and the separation is performed.
6. The process repeats for the next sample.

Volumetric Constraints on Injection Size

Because the entire internal volume of a 50-cm × 50-μm-inner diameter (i.d.) capillary is only 981 nL, the injection volume must be kept quite small. The contribution to band broadening (variance) from a plug injection is given by

$$\sigma_{\text{inj}}^2 = \frac{l_{\text{inj}}^2}{12} \quad (1)$$

where l is the length of the injection plug. To calculate the band broadening from the injection process, the diffusion-limiting case can be considered using the Einstein equation:

$$\sigma_{\text{diff}}^2 = 2D_m t \quad (2)$$

Because the squares of the variances are additive, the contributions to band broadening from injection and diffusion can be inserted into the theoretical plate equation:

$$N = \left(\frac{L_d}{\sigma_{\text{tot}}}\right)^2 \quad (3)$$

For a 50-cm capillary and a solute migration time of 600 s, the impact of the injection size for a small molecule ($D_m = 10^{-5}$ cm^2/s) and large molecule ($D_m = 10^{-6}$ cm^2/s) is shown in Fig. 1.

As illustrated in Fig. 1, injection of 1% (0.5 cm) of the capillary volume with sample produces a 92% loss of efficiency for a large molecule and an 8% loss of efficiency for a small molecule. Because diffusion is a limiting cause of efficiency, the large molecule provides a higher number of theoretical plates. The more efficient the separation process, the more difficult it is to maintain that inherent efficiency.

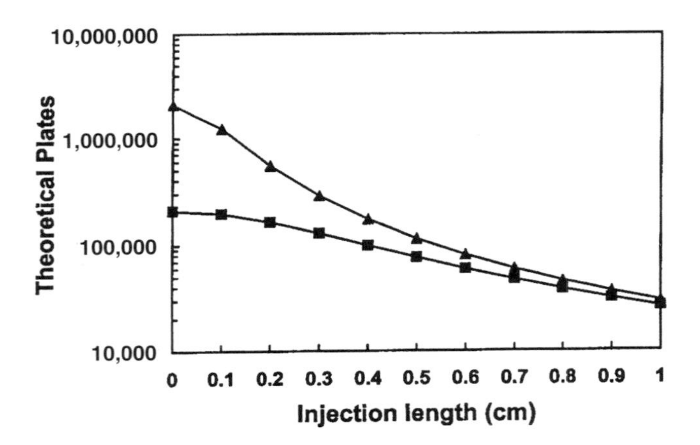

Fig. 1 Effect of the injection zone length on the number of theoretical plates for a small molecule (■, $D_m = 10^{-5}$ cm²/s) and large molecule (▲, $D_m = 10^{-6}$ cm²/s) as solved by Eq. (3). Conditions: capillary length = 50 cm to detector; migration time = 600 s.

This model assumes that the sample is dissolved in BGE. Through the use of a low-ionic-strength solution as the sample diluent, sample stacking permits large-volume injections to be made.

In a well-controlled separation, injection can be the greatest source of band broadening [1]. This is one of the reasons that micromachined systems may become important for high-resolution DNA separations. In this case, it is possible to inject minute amounts of sample and use shortened separation channels [3]. Sensitivity does not suffer because laser-induced detection is employed.

Hydrodynamic Injection

The volume of material injected per unit time (V_t, nL/s) is determined by the Poiseuille equation.

$$V_t = \frac{\Delta P D^4 \pi}{128 \eta L} \tag{4}$$

where ΔP equals the pressure drop, D is the capillary internal diameter, η is the viscosity, and L is the length of the capillary. On some instruments, the pressure is generated by raising the capillary inlet side (siphoning).

The problems generated using an open-ended injection system as shown by the Poiselle equation dictate that changes in the experimental conditions will result in variations of the amount of material injected. Internal standards are best used to compensate for some of the experimental variables.

Pressure-driven systems are preferred compared to vacuum-driven systems for two reasons: (a) Generation of pressures over 1 atm is important when viscous polymer networks are used for size separations and (b) interface to the mass spectrometer is simpler.

Electrokinetic Injection

The quantity (Q) of a solute injected is given by

$$Q = (\mu_{ep} + \mu_{eo})\pi r^2 ECt \tag{5}$$

where μ_{ep} and μ_{eo} are the electrophoretic and electroosmotic mobilities, respectively, r is the capillary radius, E is the field strength, t is the time of injection, and C is the concentration of each solute.

As illustrated in Fig. 2, solutes with high mobility are preferably injected compared to those with low mobility [4]. Note the smaller peak heights for lithium and arginine compared to rubidium when electrokinetic injection is employed. Solutes that have identical mobility in free solution show no such bias (e.g., oligonucleotides and DNA

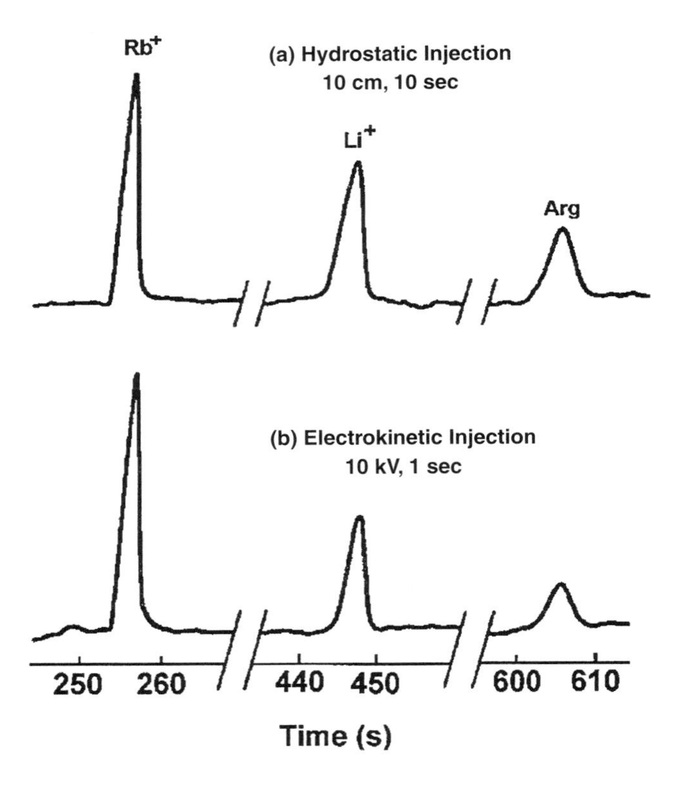

Fig. 2 Hydrostatic versus electrokinetic injection. Buffer: 20 mM MES adjusted with histidine to pH 6.0; solutes: Rb^+, Li^+, and arginine, $5 \times 10^{-5}M$; injection: (top) hydrostatic, Δh = 10 cm, t = 10 s; (bottom) electrokinetic, 1 s at 10 kV; detection: conductivity. [Reprinted with permission from *Anal. Chem. 60*: 375 (1988), copyright © 1988 American Chemical Society.]

fragments). This is fortunate because it is often necessary to use electrokinetic injection with gel-filled capillaries or when high-viscosity polymer networks are employed.

The problem with electrokinetic injection is that the field strength at the point of injection is inversely proportional to the sample conductivity. Calibration curves for ionic solutes show negative deviations from linearity because of this. Internal standards are necessary unless it is certain that the sample conductivities are identical. Low-conductivity samples are preferable because they stack.

The advantage of electrokinetic injection is that extreme trace enrichment is possible [5]. If the electroosmotic flow approaches zero, it is possible to inject only solute ions, omitting the sample diluent.

"Short-End" Injection

The section of capillary between the outlet vial and the detector can be used for high-speed separations if sufficient selectivity is designed into the separation [6]. The process is as follows. (1) Equilibrate the capillary in BGE as usual. (2) Place the sample at the capillary outlet. (3) Inject by pressuring the outlet vial or with electrokinetic injection using negative polarity (inlet-side negative). (4) Set the power supply to negative polarity and perform the usual voltage ramp. Because the capillary length is short, the injection should also be kept small and stacking buffers should be used. Be sure to set the detector time constant to 10–20% of the peak width to minimize that form of band broadening. Depending on the instrument, the short-end of the capillary usually ranges from 6 to 10 cm.

Injection Artifacts, Problems, and Solutions

No Injection

A plugged capillary is the usual culprit. Cut a few millimeters of the inlet or pressurize the outlet with a syringe to unplug it. When plugged, the observed current is usually zero. No injection can also occur if an empty or incorrect sample vial is used, if an incorrect vial is called for in the method, if the vial cap is missing or badly leaking, or if the external pressure source (if required) is not activated. It is possible that the capillary is broken. Breaks usually occur at the detection window. Check that the voltage polarity is correctly set.

Peak Tailing

Peak tailing can result from a poorly cut capillary inlet. If the capillary is not cut squarely, a concentration gradient can occur upon injection [7].

Peak Splitting

Artifactual-injection-related peak splitting can occur under certain conditions. When the sample diluent contains organic solvents and micellar electrokinetic capillary electrophoresis or cyclodextrin containing electrolytes are employed, splitting can occur due to the distribution of the solute between two phases moving at different speeds at the point of injection [8]. The problem is solved by dissolving the solute in aqueous media. If the sample is insoluble in totally aqueous solvents, 6*M* urea can be added both to the BGE and the sample diluent.

A fracture near the capillary inlet can also cause peak splitting. The break can occur if the capillary hits a vial wall or seal. The polyimide coating keeps the cracked portion intact. During injection, the sample moves into the capillary from both the open end of the tube and through the crack. The split peak is usually smaller than the main component and always has a migration time that is a little shorter. This is confirmed by examining the capillary inlet. If fractured, a small piece often detaches and the peak splitting is resolved.

References

1. X. Huang, W. F. Coleman, and R. N. Zare, *J. Chromatogr. 480*: 95 (1989).
2. D. Burgi and R.-L. Chien, *Anal. Chem. 63*: 2042 (1991).
3. S. C. Jacobson, R. Hergenroder, L. B. Koutny, R. J. Warmack, and M. J. Ramsey, *Anal. Chem. 66*: 1107 (1994).
4. X. Huang, M. J. Gordon, and R. N. Zare, *Anal. Chem. 60*: 375 (1988).
5. C.-X. Zhang and W. Thormann, *Anal. Chem. 70*: 540 (1998).
6. M. R. Euerby, C. M. Johnson, M. Cikalo, and K. D. Bartle, *Chromatographia 47*: 135 (1998).
7. A. Guttman and H. E. Schwartz, *Anal. Chem. 67*: 2279 (1995).
8. R. Weinberger, *Am. Lab. 29*: 24 (1997).

Robert Weinberger

Instrumentation of Countercurrent Chromatography

Countercurrent chromatography (CCC) is a support-free liquid–liquid partition system in which solutes are partitioned between the mobile and stationary phases in an open-column space. The instrumentation, therefore, requires a unique approach for achieving both retention of the stationary phase and high partition efficiency in the absence of a solid support. A variety of existing CCC systems may be divided into two classes [1] (i.e., hydrostatic and hydrodynamic equilibrium systems). The principle of each system may be illustrated by a simple coil as shown in Fig. 1.

Two Basic CCC Systems

The basic hydrostatic equilibrium system (Fig. 1, left) utilizes a stationary coil. The mobile phase is introduced into the inlet of the coil which has been filled with the stationary phase. The mobile phase then displaces the stationary phase completely on one side of the coil (dead space), but only partially displaces it on the other side of the coil due to the effect of gravity. This process continues until the mobile phase elutes from the coil. Once this hydrostatic equilibrium state is established throughout the column, the mobile phase only displaces the same phase, leaving the stationary phase permanently in the coil. Consequently, the solutes locally introduced at the inlet of the coil is subjected to a continuous partition process between two phases in each helical turn and separated according to their partition coefficient in the absence of a solid support.

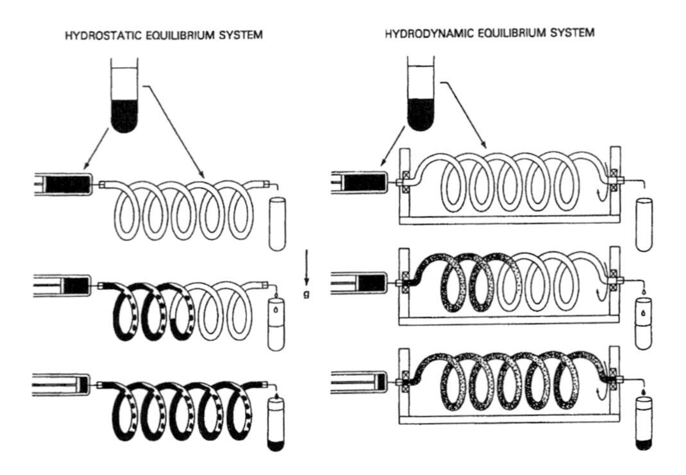

Fig. 1 Two basic CCC systems.

The basic hydrodynamic equilibrium system (Fig. 1, right) uses a rotating coil which generates an Archimedean screw effect where all objects in different density present in the coil are driven toward one end, conventionally called the "head." The mobile phase introduced through the head of the coil is mixed with the stationary phase to establish a hydrodynamic equilibrium, where a portion of the stationary phase is retained in each turn of the coil. This process continues until the mobile phase elutes from the tail of the coil. After the hydrodynamic equilibrium is established throughout the coil, the mobile phase displaces only the same phase, leaving the other phase stationary in the coil. Consequently, solutes introduced locally at the head of the coil is subjected to an efficient partition process between the two phases and separated according their partition coefficients.

Each basic system has its specific advantages as well as disadvantages. The hydrostatic system provides stable retention of the stationary phase while it yields relatively low partition efficiency due to a limited degree of mixing. The hydrodynamic system, on the other hand, produces a high partition efficiency in a short elution time while the retention of the stationary phase tends to become unstable due to violent mixing, often resulting in emulsification and extensive carryover of the stationary phase.

Development of Hydrostatic CCC Systems

In the early 1970s, the hydrostatic system was quickly developed into several efficient CCC schemes, as shown in Fig. 2 [2]. The development has been made by utilizing the unit gravity (Fig. 2, top) or the centrifugal force (Fig. 2, bottom).

In droplet CCC, which utilizes unit gravity, one side of the coil (Fig. 1, left) entirely occupied by the mobile phase is reduced to a fine-flow tube and the other side of the coil is replaced by a straight tubular column. The column is first filled with the stationary phase, and the mobile phase is introduced into the column in a proper direction so that it forms a parade of droplets in the stationary phase by the effect of gravity. The system necessitates the formation of droplets, which limits the choice of the solvent system. In order to allow a more universal application of solvent systems, a locular column was devised by inserting centrally perforated disks into the tube at regular intervals to form a number of compartments called locules. The locular column is held in an angle and rotated along its axis to mix the two phases in each locule. As in droplet

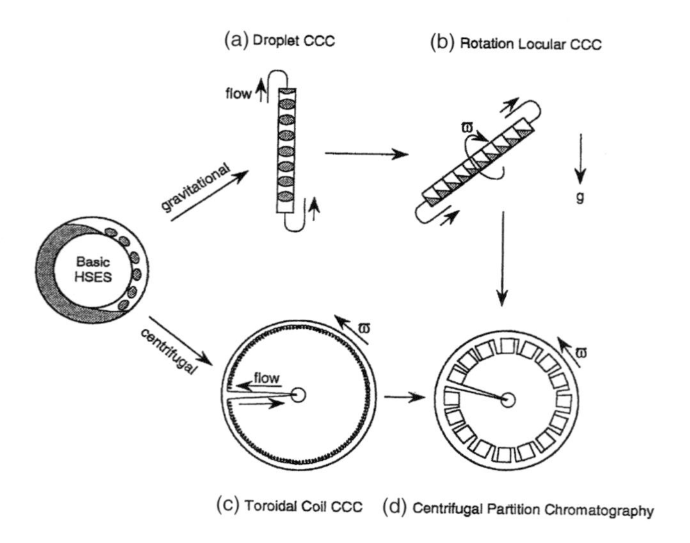

Fig. 2 Development of hydrostatic CCC systems.

CCC, the lower phase is eluted from the upper end of the locular column, and the upper phase is eluted from the lower end for better retention of the stationary phase.

In the toroidal coil CCC (helix CCC) system operated under a centrifugal force, the dimensions of the coil is reduced (Fig. 2c). The coil is mounted around the periphery of the centrifuge bowl so that the stable radially acting centrifugal force field retains the stationary phase, either upper or lower phase, in one side of the coil as in the basic hydrostatic system (Fig. 1, left). The effective column capacity and retention of the stationary phase can be increased by replacing the coil with a locular column arrangement (centrifugal partition chromatography).

Development of Hydrodynamic CCC Systems

The performance of the hydrodynamic CCC system is remarkably improved by rotating the coil in the centrifugal force field, but this requires a planetary motion of the coil. During the 1970s, a series of flow-through centrifuge schemes has been developed for performing CCC. In these centrifuge systems, the use of the conventional rotary seal devise is eliminated because it would produce various complications such as leakage, clogging, and cross-contamination. These sealless flow-through centrifuge schemes are divided into three classes: synchronous, nonplanetary, and nonsynchronous according to their mode of planetary motion (Fig. 3).

In type I synchronous planetary motion (Fig. 3, upper left), a vertical holder revolves around the center of the centrifuge while it counterrotates around its own axis at the same angular velocity. This counterrotation of the holder unwinds the twist of the tube bundle caused by revolution, thus eliminating the need for the rotary seal. This principle works well for the rest of the synchronous schemes with tilted (types I-L and I-X), horizontal (types L and X), inversely tilted (types J-L and J-X), and even inverted orientation (type J) of the holder. When the holder of type I is moved to the center of the centrifuge, the counterrotation of the holder cancels out the revolution effect, resulting in no rotation (Fig. 3, upper center). On the contrary, when this shift is applied to the type J planetary motion, the rotation of the holder is added to the revolution, resulting in the rotation of the holder at a doubled speed while the tube bundle revolves around the holder to unwind the twisting (Fig. 3, bottom center). This nonplanetary scheme is a transitional form to the nonsynchronous planetary motions. On the base of the nonplanetary scheme, the holder is again shifted toward the periphery to undergo a synchronous planetary motion. Because the net revolution speed of the coil is the sum of the nonplanetary and the synchronous planetary motions, the ratio of the rotation and revolution becomes freely adjustable.

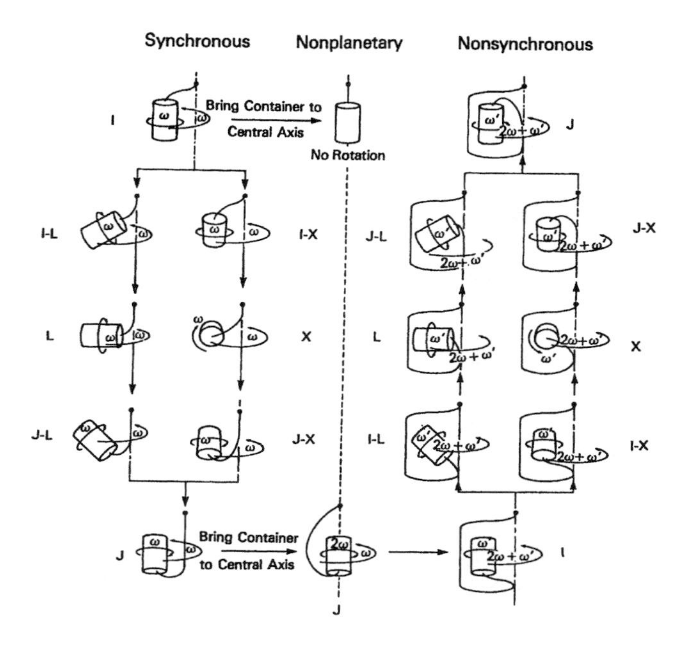

Fig. 3 A series of sealless flow-through centrifuge systems for performing CCC.

Several useful CCC systems have been developed from these centrifuge schemes. The nonplanetary scheme has been used for toroidal coil CCC [3,4], centrifugal precipitation chromatography [5], and on-line apheresis in the blood bank [6,7]. The nonsynchronous scheme has been applied to partition of cells with polymer phase systems and also to cell elutriation with physiological solutions [8, 9]. The type J synchronous scheme is further developed into a highly efficient CCC system called high-speed CCC (HSCCC) [10].

Development of HSCCC

The development of HSCCC was initiated by the discovery that when the type J planetary motion is applied to an end-closed coil coaxially mounted on the holder, the two solvent phases are completely separated in such a way that one phase occupies the head side and the other phase the tail side of the coil. This bilateral hydrodynamic distribution can be utilized for performing CCC, as illustrated in Fig. 4, where all coils are each schematically shown as a straight tube to indicate the overall distribution of the two phases.

Figure 4a shows the bilateral distribution of the two phases as mentioned earlier, where the white phase occupies the head side and the black phase the tail side. This hydrodynamic distribution of the two phases can be utilized for performing CCC.

In Fig. 4b, the upper coil is filled with the white phase and the black phase is introduced from the head end. The mobile black phase then travels rapidly through the coil, leaving a large volume of the white phase stationary in the coil. Similarly, the lower coil is filled with the black phase and the white phase is introduced from the tail end. The mobile white phase then travels through the coil, leaving a large volume of the black phase stationary in the coil. In either case, solutes locally injected at the inlet of the coil are efficiently partitioned between the two phases and quickly eluted from the coil in the order of their partition coefficient, thus yielding a high partition efficiency in a short elution time.

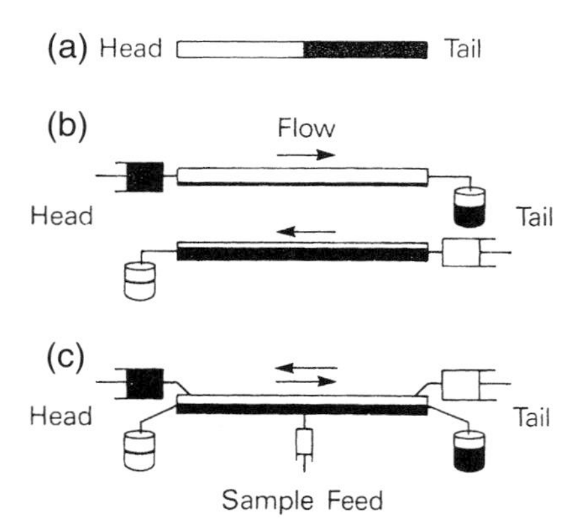

Fig. 4 Mechanism of high-speed CCC : (a) bilateral phase distribution in an end-closed coil; (b) two elution procedures for high-speed CCC; (c) dual CCC system.

The present system also permits simultaneous introduction of the two phases through the respective terminals, as illustrated in Fig. 4c. This dual countercurrent operation requires an additional flow tube at each terminal to collect the effluent, and if desired, a sample injection port is made in the middle portion of the coil. This system has been effectively applied to foam CCC [11,12] and dual CCC [13].

References

1. Y. Ito, Minireview: Countercurrent chromatography, *J. Biophys. Biochem. Methods 3*: 77–87 (1980).
2. Y. Ito, Review: Recent advances in countercurrent chromatography, *J. Chromatogr. 538*: 3–25 (1991).
3. Y. Ito and R. L. Bowman, Countercurrent chromatography with flow-through centrifuge without rotating seals, *Anal. Biochem. 85*: 614–617 (1978).
4. K. Matsuda, S. Matsuda, and Y. Ito, Toroidal coil countercurrent chromatography. Achievement of high resolution by optimizing flow-rate, rotation speed, sample volume and tube length, *J. Chromatogr. A 808*: 95–104 (1998).
5. Y. Ito, Centrifugal precipitation chromatography applied to fractionation of proteins with ammonium sulfate, *J. Biophys. Biochem. Methods*, submitted.
6. Y. Ito, J. Suaudeau, and R. L. Bowman, New flow-through centrifuge without rotating seals applied to plasmapheresis, *Science 189*: 999–1000 (1975).
7. Y. Ito, Sealless continuous flow centrifuge, in *Apheresis: Principles and Practice* (B. McLeod, T. H. Price, and M. J. Drew, eds.), AABB Press, Bethesda, MD, 1997, pp. 9–13.
8. Y. Ito, G. T. Blamblett, R. Bhatnagar, M. Huberman, L. Leive, L. M. Cullinane, and W. Groves, Improved nonsynchronous flow-through coil planet centrifuge without rotating seals. Principle and application, *Separ. Sci. Technol. 18*: 33–48 (1983).
9. T. Okada, D. D. Metcalf, and Y. Ito, Purification of mast cells with an improved nonsynchronous flow-through coil planet centrifuge, *Int. Arch. Allergy Immunol. 109*: 376–382 (1996).
10. Y. Ito and W. D. Conway (eds.), *High-speed Countercurrent Chromatography*, Wiley–Interscience, New York, 1996.
11. Y. Ito, Foam countercurrent chromatography based on dual countercurrent system, *J. Liquid Chromatogr. 8*: 2131–2152 (1985).
12. H. Oka, Foam countercurrent chromatography, in *High-Speed Countercurrent Chromatography* (Y. Ito and W. D. Conway, eds.), Wiley–Interscience, New York, 1996, pp. 107–120.
13. Y. W. Lee, Dual countercurrent chromatography, in *High-Speed Countercurrent Chromatography* (Y. Ito and W. D. Conway, eds.), Wiley–Interscience, New York, 1996, pp. 93–104.

Yoichiro Ito

Intrinsic Viscosity of Polymers: Determination by GPC

The intrinsic viscosity is a widely used measure of molecular weight, M, and size (dimensions) of macromolecules in dilute solution. Important information about macromolecular architecture and conformations can be obtained from the molecular-weight dependence of intrinsic viscosity for a homologous series of polymers. Size-exclusion chromatography (SEC) provides a unique opportunity to measure this dependence in a single chromatographic run. Another striking coincidence is that the dimensions of a macromolecule associated with its frictional properties in dilute solution (i.e., with the intrinsic viscosity) determine the elution time (volume) in size-exclusion separation. This allows one to use the intrinsic viscosity measurement as a crucial intermediate step in the determination of molecular weights and molecular-weight distributions of polymers. From the above, it might be assumed that on-line intrinsic viscosity measurements represent an important aspect of contemporary gel permeation chromatography (GPC).

Solution Viscosity

There are several dilute solution viscosity quantities used in the determination of the intrinsic viscosity. The size of macromolecules in solution is associated with an increase in viscosity of the solvent brought about by the presence of these molecules. Relative viscosity is a dimensionless quantity representing a solution/solvent viscosity ratio, $\eta_r = \eta/\eta_0$, where η and η_0 are the solution and solvent viscosities, respectively. The specific viscosity $\eta_{sp} = \eta_r - 1$ is the fractional increase in viscosity between the solution and solvent. The effect of the concentration can be normalized by division of η_{sp} by the concentration C, expressed in grams per deciliter (g/dL). This concentration-normalized viscosity is termed the reduced specific viscosity or η_{sp}/C (the IUPAC preferred term is viscosity number). Another related term, the inherent viscosity, is expressed as $\eta_{inh} = (\ln \eta_r)/C$ (the IUPAC preferred term is logarithmic viscosity number).

In order to relate viscosity to molecular weight, the value of reduced (or inherent) viscosity is extrapolated to zero concentration. This parameter is called the intrinsic viscosity, $[\eta]$, and is usually expressed in deciliters per gram (dL/g) (the IUPAC preferred term is limiting viscosity number, mL/g):

$$[\eta] = \lim(\eta_{sp}/C) = \lim \eta_{inh} \quad (C \rightarrow 0) \tag{1}$$

Practically, the limit in Eq. (1) is achieved when the concentration is so low that the frictional interactions between an individual macromolecule and a solvent are not affected by the presence of other macromolecules in the same solution. Under these conditions, which are typical for the GPC–SEC experiments, the difference between the reduced and intrinsic viscosities is negligible.

Calculation of Intrinsic Viscosity in GPC

The opportunity to measure the dilute polymer solution viscosity in GPC came with the continuous capillary-type viscometers (single capillary or differential multicapillary detectors) coupled to the traditional chromatographic system before or after a concentration detector in series (see the entry Viscometric Detection in GPC–SEC). Because liquid continuously flows through the capillary tube, the detected pressure drop across the capillary provides the measure for the fluid viscosity according to the Poiseuille's equation for laminar flow of incompressible liquids [1]. Most commercial on-line viscometers provide either relative or specific viscosities measured continuously across the entire polymer peak. These measurements produce a viscometry elution profile (chromatogram). Combined with a concentration-detector chromatogram (the concentration versus retention volume elution curve), this profile allows one to calculate the instantaneous intrinsic viscosity $[\eta]_i$ of a polymer solution at each data point i (time slice) of a polymer distribution. Thus, if the differential refractometer is used as a concentration detector, then for each sample slice i,

$$[\eta]_i \approx \frac{\eta_{sp,i}}{C_i} = \frac{\nu \eta_{sp,i}}{\Delta n_i} \tag{2}$$

where Δn is the refractive index change due to the polymer in solution, detected by the refractometer and $\nu = dn/dc$ is the refractive index increment of the polymer. The quantity calculated from Eq. (2) is often designated as "observed" intrinsic viscosity, $[\eta]_{obs}$.

As can be seen from Eq. (2), the observed intrinsic viscosity for each slice is proportional to the ratio of two detectors' responses. It follows that detector noise, which is an irreducible component of the measurement process, introduces noise in the intrinsic viscosity that depends on this ratio. However, two detectors have different sensitivities at the tails of polymer distribution: the concentration detector is less sensitive to the high-molecular-weight end and the viscometer is less sensitive to the opposite end. Thus, the noise increases dramatically on both tails of the distribution, where the ratio (2) does not produce physically meaningful values. For example, the logarithm of in-

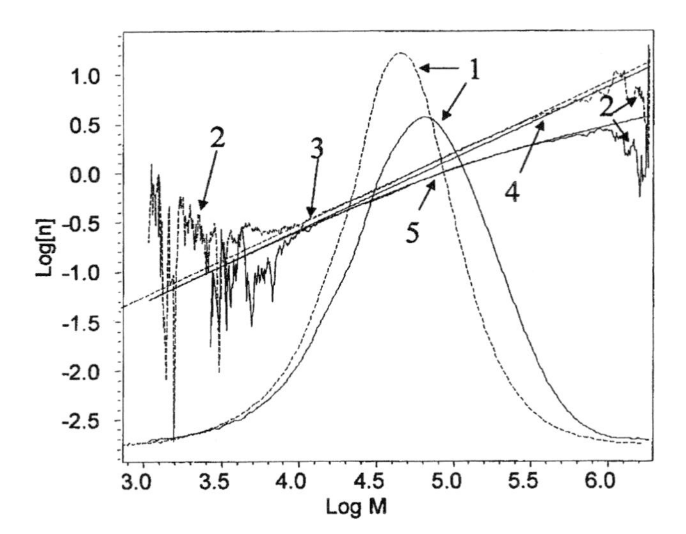

Fig. 1 Molecular-weight-distribution (MWD) and viscosity-law plots for NIST PE1475 high-density polyethylene (dashed lines) and NIST PE1476 low-density polyethylene (solid lines). The curves are (1) MWD, (2) observed viscosity, (3) fitted viscosity for linear polyethylene, (4) extrapolated fitted viscosity for polyethylene with short-chain branches only, and (5) fitted viscosity for branched polyethylene.

trinsic viscosity computed from the slice ratios (2) sometimes does not increase monotonically with molecular weight (i.e., with decreasing elution volume V) even for the flexible coillike polymers (curves 2 in Fig. 1).

Fitting a smooth, multivariate model to a time series of noisy data is an effective way to produce a more precise estimate of the measured quantity at each sample time. Typically, the logarithm of intrinsic viscosity is modeled as a low-order polynomial in elution volume V using a least-squares fitting to the experimental data [2–4]. The intrinsic viscosity calibration curve ($\log[\eta]_{\mathrm{fit}}$ versus V) obtained this way depends on properties of the polymer as well as that of the chromatographic system (e.g., columns). It can be used for diagnostic information concerning the GPC–viscometry system [5] and also to refine such system parameters as interdetector volume (see the entry Interdetector Delay Volume) and band broadening (see the entry Axial Dispersion Correction in GPC–SEC).

Application of Intrinsic Viscosity in GPC to Polymer Characterization

Molecular Weight and Molecular-Weight Distribution Determination

The most important feature that has been added to conventional GPC by the viscometer detector through the intrinsic viscosity calculation is the ability to determine the "absolute" MWD without any additional assumptions about the polymer chemical structure. This goal is accomplished by applying the universal calibration concept, which establishes the hydrodynamic volume $H = [\eta]M$ as a universal parameter governing the size-exclusion separation. The MWD can be determined in three steps. First, the set of narrow polydispersity polymers with known molecular weights (narrow standards) covering the entire region of column size-exclusion separation is selected to construct the universal (or hydrodynamic volume) calibration curve, $\log H$ versus elution volume V (see the entry Calibration of GPC–SEC with Universal Calibration Techniques). This curve is then used to calculate the molecular-weight calibration curve via the relationship $\log M = \log H - \log[\eta]_{\mathrm{fit}}$, where these quantities are obtained from the (smooth) hydrodynamic and intrinsic viscosity calibration curves. Finally, the MWD is constructed by plotting the concentration C as a function of $\log M$ across the polymer distribution (curves 1 in Fig. 1). Different statistical moments of this distribution (i.e., average molecular weights, including viscosity-average molecular weight M_v) can be calculated by appropriate summation over the slice data and compared with the values obtained by bulk measurements [6].

Intrinsic Viscosity Distribution

The intrinsic viscosity is a fundamental property of the polymer sample in solution, and thus the intrinsic viscosity distribution (IVD) (C versus $\log[\eta]$) with associated statistical moments may be used to characterize polymers without converting this distribution into a MWD [7]. The IVD can be determined in GPC–viscometry directly, without resorting to universal calibration. This distribution depends not only on the polymer sample itself but also on the solvent and the temperature, and hence does not possess the versatility of the MWD. Nevertheless, the IVD measurement in GPC–viscometry is much less sensitive to experimental conditions than any calibration curve and, hence, can be successfully used in industry (e.g., for quality control of polymers in production).

Size and Molecular Structure of Polymers

The GPC–viscometry with universal calibration provides the unique opportunity to measure the intrinsic viscosity as a function of molecular weight (viscosity law, $\log[\eta]_{\mathrm{fit}}$ versus $\log M$) across the polymer distribution (curves 3 and 4 in Fig. 1). This dependence is an important source of information about the macromolecule architecture and conformations in a dilute solution. Thus, the Mark–Houwink equation usually describes this law for linear polymers: $\log[\eta] = \log K + \alpha \log M$ (see the entry Mark–Houwink Relationship). The value of the exponent α is affected by the macromolecule conformations: Flexible coils have the values between 0.5 and 0.8, the higher values are typical for stiff anisotropic ("rod"-like) molecules, and much lower (even negative) values are associated with dense spherical conformations.

The determination of the viscosity law in GPC–viscometry is even more important for branched polymers. Branches reduce the sizes of a macromolecule, including its hydrodynamic volume H. This size reduction is reflected by the changes in the shape and position of the viscosity law plot for a branched polymer. Short-chain branches usually do not change the linearity and slope of the Mark–Houwink plot and just decrease the value of parameter K, whereas the long-chain branches cause bending of the corresponding plot.

These features of the viscosity-law plots for branched polymers are demonstrated in Fig. 1 with two NIST (National Institute of Standards and Technology, U.S.A.) polyethylene standards as examples: high-density linear polyethylene PE1475 and low-density branched polyethylene PE1476. This last one contains both short- and long-chain branches. Dashed straight line 3 represents the Mark–Houwink plot for linear polyethylene, parallel solid line 4 takes into account the short-chain branches, and polyethylene with both types of branches (PE1476) is described by solid curve 5 (see the entry Long-Chain Polymer Branching, Determination by GPC–SEC for further discussion).

For further information on the intrinsic viscosity determination in GPC, including the use of the light-scattering detector, see Ref. 8 and the entry GPC–SEC Viscometry from Multiangle Light Scattering.

References

1. J. W. Mays and N. Hadjichristidis, Polymer characterization using dilute solution viscometry, in *Modern Methods of Polymer Characterization* (H. G. Barth and J. W. Mays, eds.), John Wiley & Sons, New York, 1991, pp. 227–269.
2. C.-Y. Kuo, T. Provder, M. E. Koehler, and A. F. Kah, Use of a viscometric detector for size exclusion chromatography, in *Detection and Data Analysis in Size Exclusion Chromatography* (T. Provder, ed.), American Chemical Society, Washington, DC, 1987, pp. 130–154.
3. R. Lew, P. Cheung, S. T. Balke, and T. H. Mourey, *J. Appl. Polym. Sci. 47*: 1685–1700 (1993).
4. Y. Brun, R. Nielson, M. Gorenstein, and N. Hay, New results in polymer characterization using multidetector GPC, in *Proceedings, International GPC Symposium*, 1998, pp. 48–67.
5. S. T. Balke, P. Cheung, R. Lew, and T. H. Mourey, *J. Liquid Chromatogr. 13*: 2929–2955 (1990).
6. J. Lesec, Problems encountered in the determination of average molecular weights by GPC viscometry, in *Liquid Chromatography of Polymers and Related Materials II* (J. Cazes and X. Delamare, eds.), Chromatographic Science Series, Vol. 13, Marcel Dekker, Inc., New York, 1980, pp. 1–17.
7. W. W. Yau and S. W. Rementer, *J. Liquid Chromatogr. 13*: 627–675 (1990).
8. C. Jackson and H. G. Barth, Molecular weight-sensitive detectors for size exclusion chromatography, in *Handbook of Size Exclusion Chromatography* (C. Wu, ed.), Chromatographic Science Series, Vol. 69, Marcel Dekker, Inc., New York, 1995, pp. 103–145.

Yefim Brun

Ion Chromatography Principles, Suppressed and Nonsuppressed

Ion chromatography (IC) is a mode of high-performance liquid chromatography (HPLC) in which ionic analyte species are separated on cationic or anionic sites of the stationary phase. The separation mechanisms can be broadly compared to ion exchange, using fixed-site exchange resins of various composition and ion-interaction methods, using a variety of columns as substrates to support dynamically exchanged or permanently bonded ionic groups. Alternative approaches of minor significance also exist. The mobile phase is an aqueous buffer solution. The rate of migration of the ion (inorganic ions and organic acids and bases) through the column is directly dependent on the type and concentration of eluent ions. Retention is based on the affinity of different ions for the ion-exchange sites and on the competition between eluent buffer ions and analyte ions, which is dependent on the ionic strength of the buffer and can be adjusted by altering the pH of the mobile phase or the concentration of any organic modifier in it.

Ion chromatography operates at pressures ranging from several hundred to several thousand pounds per square inch. In most cases, the same chromatographic components (pumps, injectors, etc.) can be employed in both HPLC and IC. Most of the chromatographic principles developed in HPLC stand for IC also, with possible minor modifications. Injection volumes in ion chromatography are generally somewhat larger than those normally in HPLC, typically in the range 50–100 μL, in contrast to HPLC, where 5–20 μL are injected.

It was in 1975 when Small and his co-workers introduced the high-pressure operation mode of ion chromatography. In their original paper, they described a novel system for the chromatographic determination of inorganic ions, in which a resin was used for the separation and a second ion-exchange column was combined to chemically suppress the background conductance of the eluent, thus improving detection limits for eluted ions. Since 1975, IC has grown rapidly and ion chromatographic methods for ions are currently among the best available and have been applied to a wide range of inorganic species. This can be attributed to concurrent advances in separation technology and detection methods.

Detection techniques can be subdivided into three broad categories:

1. Electrochemical detection (using conductivity, amperometry, or potentiometry).
2. Spectroscopic detection (using ultraviolet/visible (UV/vis) absorbance, refractive index, fluorescence, atomic absorption or atomic emission).
3. Techniques based on postcolumn reactions.

Conductivity detectors provide the advantage of universal detection, as all ions are electrochemically conducting. Thus, the majority of ion chromatography detectors rely upon conductivity measurements.

The principle of conductivity-detector operation is the differential measurement of conductance of the eluent, prior to and during elution of the analyte ion. The detector response depends on analyte concentration, the degree of ionization of both eluent and analyte (governed by the eluent pH), and limiting equivalent conductances of the eluent cation and of the eluent and analyte anions (where an anion-exchange system is considered). If the eluent and analyte are fully ionized, the signal is proportional to the analyte concentration and to the difference (positive or negative) in limiting equivalent conductances which determines sensitivity.

Conductivity detection provides a sensitive measure of ion concentrations in solution, but its measurement is hampered by high conductivity of the eluent, as ion exchange requires a competing electrolyte to displace the analytes from the column. In order to eliminate background conductivity and thus to improve the analyte signal, H, Smith et al. proposed the use of a second ion-exchange column. In this way, two different ion chromatography techniques are distinguished: eluent suppressed and nonsuppressed (also called single-column ion chromatography) using different packing materials and different eluents, leading to specific advantages and disadvantages for each technique.

Eluent Suppressed Ion Chromatography

Various schemes have been devised to improve the signal-to-noise ratio (S/N) by decreasing the background signal of the eluent/displacer or increasing the conductance of the analyte, or both.

The principle of conductivity suppression is the reduction of background conductivity by converting the eluent to a less conductive medium (H_2O) through acid–base neutralization while the analyte ions' conductivity is increased, by converting them to a more conductive medium: Anions are converted to their acid forms and cations to their hydroxide forms. These reactions lead to higher S/N ratios, thus significantly improving baseline stability and detection limits.

Suppressor devices include packed column suppressors, hollow-fiber membrane suppressors, micromembrane suppressors, suspension postcolumn reaction suppressors, autoregenerated electrochemical suppressors, and so forth.

The packed column suppressor, originally introduced by Small et al., suffers from a number of drawbacks, such as time shifts due to Donnan exclusion effects, band broadening (due to a large dead volume and high dispersion), and oxidation of nitrite, which is easily oxidized to nityrate, due to the formation of nitrous acid in the suppressor. Because of these limitations, they were only practical for isocratic elution. However, the main disadvantage of the method is the necessity for periodical regeneration of the suppressor (also called stripper) to restore its ion-exchange capacity.

For anion analysis, the regenerant must supply a source of hydrogen ions to convert the eluent anions to a less conductive form. The most common regenerant is dilute sulfuric acid, whereas for cation analysis, the most common regenerant is hydroxide (sodium, potassium, or tetramethylammonium hydroxide).

The preferred eluents for anions are dilute carbonate–bicarbonate mixture, sodium hydroxide and, for common alkali metals and simple amines, dilute mineral acids (HCl, HNO_3, $BaCl_2$, $AgNO_3$, amino acids, alkyl and aryl sulfonic acids). The most common choice is HCl, but in the case of divalent ions, an eluent of much higher affinity for the ion-exchange resin, such as $AgNO_3$, must be used.

Typical neutralization reactions for chemical suppressors are as follows:

Anion-exchange chromatography:
Eluent reaction: NaOH + resin–$SO_3^-H^+ \rightarrow$ Resin–$SO_3^-Na^+$ + H_2O
Analyte reaction: NaX + resin–$SO_3^-H^+ \rightarrow$ Resin–$SO_3^-Na^+$ + HX, where X = anions (Cl^-, Br^-, NO_2^-, etc.).
Cation-exchange chromatography:
Eluent reaction: HCl + resin–$NR_3^+OH^- \rightarrow$ Resin–$NR_3^+Cl^-$ + H_2O
Analyte reaction: MCl + resin–$NR_3^+OH^- \rightarrow$ Resin–$NR_3^+Cl^-$ + MOH, where M = cations (Na^+, K^+, etc.).

The reaction, in the case of bicarbonate, yields the largely undissociated carbonic acid that does not contribute significantly to the conductivity.

Without chemical suppression, the contribution to the total measured conductivity from the eluent is many orders of magnitude higher than that from the analyte, leading to low sensitivity (Fig. 1).

Some of the drawbacks that packed column suppressors have were eliminated when hollow-fiber membrane suppressors were introduced in 1981. These were found to be even more convenient and efficient, with low dead volume and high capacity, and they are dynamically regenerated. Eluent passes through the core of the fiber and regenerant washes the outside. However, they have also limited suppression capacity and are restricted only to isocratic operation.

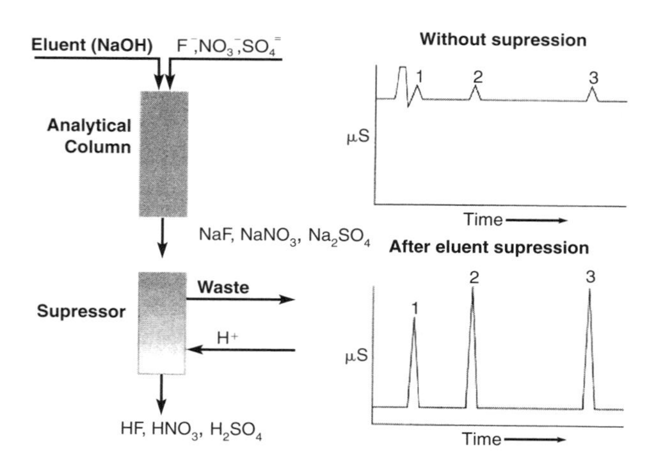

Fig. 1 The effect of the background conductivity suppression on the monitored signal of the analyte anions, after separation by means of ion chromatography. Peaks: 1 = fluoride, 2 = nitrate, 3 = sulfate.

Micromembrane suppressors introduced in 1985 use thin, flat ion-exchange membranes to enhance ion transport while maintaining a very low dead volume, providing a high suppression capacity, with low dispersion.

Later, electrochemically regenerated suppression modules were introduced, where an electrochemical process is used to regenerate a solid-phase chemical suppressor for continuous reagent-free operation. Self-regenerating suppressors are similar to micromembrane suppressors, except that regenerant hydronium and hydroxide ions are produced, *in situ*, by electrolysis of water supplied by recycle or an external source. This is achieved by incorporating electrodes inside the regenerant chambers; thus, external acid or base supply are unnecessary. The two electrolysis reactions taking place are

Anode: $2H_2O \rightarrow 4H^+ + O_2 + 4e$
Cathode: $2H_2O + 2e \rightarrow H_2 + 2OH^-$

Another technique of improving the S/N ratio is the one that uses postcolumn addition of a solid-phase reagent (SPR), which is a colloidal suspension of ultrafine ion-exchange particles. The SPR reacts with the analyte to increase its conductivity. Additionally, the SPR has a low electrophoretic mobility and, hence, conductance. This technique avoids the dead time due to suppressor column and also eliminates the regeneration cycle.

Nonsuppressed Single-Column Ion Chromatography

Another approach of ion chromatography is the nonsuppressed single column, in which no suppressor device is used. In this case, the only method for improving the sensitivity is to maximize the difference between mobile-phase conductivity and analyte conductivity.

Nonsuppressed single-column ion chromatography (SCIC) was introduced in 1979 by Gierde and co-workers, based on a two-principal innovation:

1. The use of a special anion-exchange resin of very low capacity (0.007–0.007 mEq/g).
2. The adoption of an eluent having a very low conductivity, which can be passed directly through the conductometric detector. Typical eluents used are benzoate, phthalate, or other aromatic acid salts, with low limiting equivalent conductances (leading to direct detection) or potassium hydroxide eluent, with high conductivity for anions or dilute nitric acid for cations, leading to indirect detection mode (decrease of conductivity as the analyte is eluting).

The major limitation of nonsuppressed conductivity detection is that gradient systems cannot be used; thus, the background conductivity remains constant.

Virtually every type of high-performance liquid chromatography (HPLC) detector can be combined with SCIC: refractive index, UV absorbance (direct and indirect), electrochemical, and so forth.

A typical nonsuppressed SCIC separation obtained with a low-capacity resin-based strong anion exchanger (PRP-X100 Hamilton) used as the analytical column is illustrated in Fig. 2 for the simultaneous determination of eight inorganic anions (F^-, Br^-, NO_2^-, Cl^-, NO_3^-, PO_4^{3-}, SO_4^{2-}, CO_3^{2-}), with conductometric detection, using a mixture of 2.0 m*M* sodium benzoate and 2.5 m*M* *p*-hydroxybenzoic acid (pH 9.0 adjusted with 1*N* NaOH) as eluent, with the organic modifier methanol 8% v/v, at a flow rate of 0.7 mL/min. The detection limits (S/N = 3) were 100 μg/L for carbonate and 50 μg/L for the rest of the cited anions, when 50 μL of the samples were injected onto the analytical column.

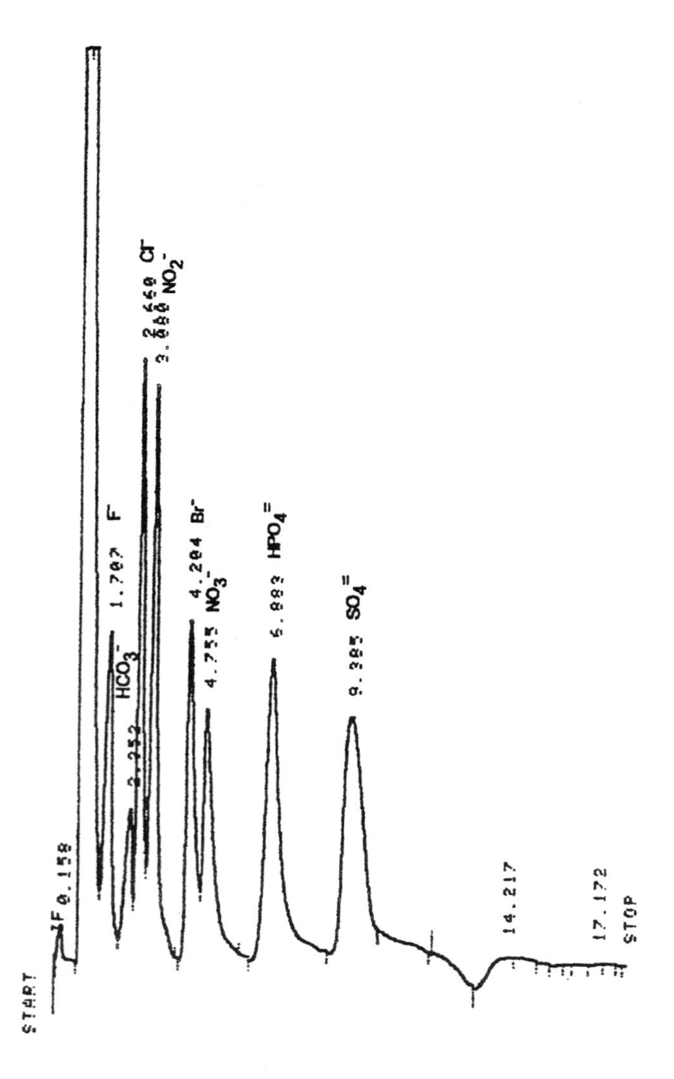

Fig. 2 Nonsuppressed SCIC determination of eight inorganic anions.

Columns

Two types of packing materials are commonly used for ion chromatography: silica-based and polymer-based ion exchangers. The polymer-based ion exchangers typically contain a PSDVB (polystyrenedivinylbenzene) backbone, lightly sulfonated (cation exchanger) or lightly aminated (anion exchanger), whereas the silica-based ion exchangers use a porous silica bead, chemically prepared to form the anion or cation exchanger. The resins have the advantage that they can be used over the entire pH range, whereas silica based materials can be used in a narrow working pH range (2–6.5).

Detection limits for ions vary with the sensitivity of the detector, with the volume of sample injected, and with the identity, concentration, and pH of the eluent, as well as with chromatographic factors, such as column efficiency and so forth.

Comparison of ESIC and SCIC

The main advantages of eluent suppressed ion chromatography (ESIC) are that a wide range of eluents and columns can be used, the wide dynamic range, and the higher sensitivity; the main disadvantage is the periodical necessity for suppressor-column regeneration.

On the other hand, SCIC is rapid, sensitive, with easy sample preparation, and simple instrumentation; however, it requires a significant difference in conductance between eluent and analyte ions and the temperature stability is crucial. The answer to the question of which IC technique is most efficient is dependent on several considerations, such as the nature of sample analytes, the concentration of the solute ions, the sensitivity required, the equipment available, and so forth.

Applications

Ion chromatography, suppressed and nonsuppressed, can be applied both to anion and cation analysis. The current situation is that the methods for anion determination have far outnumbered those for cation analysis, for the reason that there are available methods for the latter, which are rapid and sensitive (e.g., AAS, ICP, ASV). It is difficult to mention all the ionic species detectable by this analytical technique. Practically, any compound that can be converted to an ionic form is amenable to analysis by IC. Among the inorganic ions determined are (F^-, Br^-, NO_2^-, Cl^-, NO_3^-, PO_4^{3-}, SO_4^{2-}, CO_3^{2-}, CrO_4^{2-}, I^-, IO_3^-, $C_2O_4^{2-}$, BrO_3^-, SCN^-, Na^+, K^+, Mg^{2+}, Ca^{2+}, NH_4^+, at the ppm or ppb levels, in drinking water, food samples, food additives, beverages, environmental samples (soil extracts, rain water, surface water or groundwater), cosmetics, pharmaceuticals, biomedical, plating bath analysis, biological fluids, industrial process products, wastewater, and so forth. IC is also capable of speciation analysis of polyvalent anions or transition metal ions with multiple oxidation states, at levels lower than those possible with ICP or AAS. Organic species of biological and biochemical interest can also be determined.

Suggested Further Reading

Dasgupta, P., *Anal. Chem. 64*(15): 775A–783A (1992).

Gierde, D., J. Fritz, and G. Schmuckler, *J. Chromatogr. 186*: 509–519 (1979).

Gierde, D., G. Schmuckler, and J. Fritz, *J. Chromatogr. 187*: 35–45 (1980).

Haddad, P., and A. Heckenberg, *J. Chromatogr. 300*: 357–394 (1984).

Henderson, I., R. Saari-Nordhaus, and J. Anderson, Jr., *J. Chromatogr. 546*: 61–71 (1991).

Henshall, A., S. Rabin, J. Statler, and J. Stilian, *Int. Chromatogr. Lab. 12*: 7–14 (1993).

Papadoyannis, I., V. Samanidou, and K. Moutsis, *J. Liquid Chromatogr. 21*(3): 361–379 (1998).

Papadoyannis, I., V. Samanidou, and A. Zotou, *J. Liquid Chromatogr. 18*(7): 1383–1403 (1995).

Pietrzyk, D., Z. Iskandarani, and G. Schmitt, *J. Liquid Chromatogr. 9*(12): 2633–2659 (1986).

Saari-Nordhaus, R. and J. Anderson, Jr., *Int. Chromatogr. Lab. 18*: 4–10 (1994).

Schmuckler, G., *J. Chromatogr. 313*: 47–57 (1984).

Small, H., T. Stevens, and W. Bauman, *Anal. Chem. 47*(11): 1801–1809 (1975).

Tarter, J., *Ion Chromatography*, Chromatographic Science Series Vol. 37. Marcel Dekker Inc., New York, 1987.

Walker, T., N. Akbari, and T. Ho, *J. Liquid Chromatogr. 14*(4): 619–641 (1991).

Ioannis N. Papadoyannis
Victoria F. Samanidou

Ion-Exchange Buffers

Ion-exchange chromatography is a separation method based on the exchanging of ions in a solution with ions of the same charge present in a porous insoluble solid. The method is used for the deionization of water [1,2]. It is often employed for the separation and identification of the rare earth and transuranium elements [2]. Additionally, ion-exchange chromatography is also used in clinical laboratories for the automated separation and analysis of amino acids and other physiologically important amines used for pharmaceutical purposes [3].

In ion-exchange chromatography, ions are separated on the basis of their differences in relative affinity for ionic functional groups on the stationary phase. Anionic and cationic functional groups are covalently attached to the stationary phase, usually resins, which are amorphous particles of organic material [1–3]. Sulfonated styrene-based polymers are the most widely used cation-exchange resin, and similar polymers containing quaternary ammonium groups are the most widely used anion exchangers [4]. Oppositely charged solute ions are attracted to ionic functional groups on the stationary phase by electrostatic forces. Retention is based on the attraction between solute ions and charged sites bound to the stationary phase [4,5].

Due to the desirable solvent and ionizing properties of water, most ion-exchange chromatographic separations are carried out in aqueous media. Once the selection of the column type has been made, the resolution of components in the sample can be optimized by adjusting ionic strength, temperature, flow rate, and, most importantly, the pH and concentration of buffer or organic modifier in the mobile phase.

Solvent strength, which is defined as the ability of the solvent to elute a given solute from the stationary phase, increases with increased ionic strength of the mobile phase. Selectivity is generally not affected by changes in ionic strength, except for samples containing solutes with different valence charges. With increased temperature, the rate of solute exchange between the stationary phases and mobile phases increases, and the viscosity of the mobile phase decreases, resulting in increased solvent strength. Solvent strength also increases with the volume percent of organic modifier for hydrophobic solutes. However, most ion-exchange chromatography is performed in totally aqueous mobile phases, due to the hydrophilic nature of most ionic solutes. Flow rates of the mobile phase can change resolution in ion-exchange chromatography, but the effects are often minimal [5].

Increases in mobile-phase pH cause decreases in solute retention in cation-exchange chromatography and increases in retention in anion-exchange chromatography. Separation selectivity can also be greatly influenced by small changes in pH. In ion-exchange chromatography with aqueous mobile phases, buffers are used to maintain the pH in the mobile phase. A buffered solution can resist the changes in pH when an acid or base is added or when dilution is occurring. The pH of a buffer is given by the Henderson–Hasselbalch equation:

$$\mathrm{pH} = \mathrm{p}K_a + 109\frac{[\mathrm{A}^-]}{[\mathrm{HA}]} \qquad (1)$$

where pK_a refers to the acid dissociation constant of the species in the denominator, HA, and A refers to the conjugate base of the acid HA. Buffer capacity, the measure of how well a solution resists changes in pH when a strong acid or base is added, increases as the concentration of the buffer increases. However, the pH of a buffer solution is virtually independent of dilution. When the pH = pK_a, the maximum buffer capacity is met and a good working range of the buffer is approximately when the pH = pK_a = 1 ± 1 [1]. A buffer is very easy to make. For example, to prepare 1.00 L of buffer containing 0.100*M* tris(hydroxymethyl)aminomethane hydrochloride at pH of 7.4, simply weigh out 0.100 mol of its hydrochloride salt and dissolve it in a beaker containing about 900 mL of water. Then, add a base (e.g., NaOH), until the pH is exactly 7.4. Then, quantitatively transfer the solution to a volumetric flask. Finally, dilute to the volumetric mark and mix [1].

By increasing the buffer concentration, the concentration of the counterions are increased in the mobile phase and stronger competition is provided between the sample components and the counterions for the exchangeable ionic centers, resulting in reduced solute retention [5]. As stated earlier, selectivity and retention can also be adjusted by changing the pH of the mobile phase. This occurs because such a change in pH modifies the character of both the ion-exchange medium and the acid–base equilibrium as well as the degree of ionization of the sample [3]. A pH gradient in which the pH of the mobile phase is changed during the chromatographic analysis can also be used to control the solvent strength and retention of ionic solutes. Such gradients can also be used to control selectivity [6].

The working pH range for a separation can be estimated from the pK_a values of the sample components. If such pK_a values are not available, they can often be estimated by considering the number and types of functional groups present and the molecular structures of the components in the sample [1]. In order to ensure that solutes are ionized and retained by the ion exchanger, the optimum buffer pH of the mobile phase should be 1 or 2 pH units above the pK_a of acids and 1 or 2 pH units below the pK_a of bases [3].

Table 1 Typical Buffers for Ion-Exchange Chromatography

Buffer salt	pH Range
Ammonia	8.2–10.2
Ammonium acetate	8.6–9.8
Ammonium phosphate	2.2–6.5
Citric acid	2.0–6.0
Disodium hydrogen citrate	2.6–6.5
Potassium dihydrogen phosphate	2.0–8.0 / 9.0–13
Potassium hydrogen phthalate	2.2–6.5
Sodium acetate	4.2–5.4
Sodium borate	8.0–9.8
Sodium dihydrogen phosphate	2.0–6.0 / 8.0–12
Sodium formate	3.0–4.4
Sodium perchlorate	8.0–9.8
Sodium nitrate	8.0–10.0
Triethanolamine	6.7–8.7

Two criteria should be met when choosing the components of the buffer. First, the buffer must be able to maintain the operating pH for the separation to be performed. Second, the exchangeable buffer counterion must yield the desired eluent strength [3]. Some common buffer salts used in ion-exchange chromatography and their usable pH ranges are summarized in Table 1. Examples of their use includes the chromatography of amino acids, polymeric, cation exchanger using various combinations of citrate and borate buffer [3]. Additionally, carbohydrates can be separated by anion-exchange chromatography using an aqueous solution of sodium hydroxide–sodium acetate as the eluent [3]. Being weak acids, the ion-exchange behavior of such compounds is significantly affected by the pH of the mobile phase. Similar separations of ionizable compounds through the use of ion-exchange chromatography with these and other buffers have been reported [1–7].

References

1. D. C. Harris, *Quantitative Chemical Analysis*, 5th ed., W. H. Freeman, New York, 1998, pp. 755–766.
2. H. F. Walton, *Ion-Exchange Chromatography*, Hutchinson and Ross, Dowden, U.K., 1976.
3. C. F. Poole and S. K. Poole, *Chromatography Today*, Elsevier, New York, 1991, pp. 422–439.

4. H. Small, *Ion Chromatography*, Plenum, New York.
5. D. T. Gjerde and J. S. Fritz, *Ion Chromatography*, 2nd ed., Huthig, New York, 1987.
6. L. R. Snyder and J. J. Kirkland, *Introduction to Modern Liquid Chromatography*, 2nd ed., John Wiley & Sons, New York, 1979, pp. 410–452.
7. W. Rieman and H. F. Walton, *Ion Exchange in Analytical Chemistry*, Pergamon Press, New York, 1976.

J. E. Haky
H. Seegulum

Ion Exchange: Mechanism and Factors Affecting Separation

Introduction

Ion-exchange chromatography (IEC) is a technique in which ionic solutes bind to charged functional groups on the bonded phase. The power and versatility of IEC as an analytical and preparative technique is due in large part to the ability to drastically change the selectivity through manipulation of the mobile phase. Although it is obvious that the pH determines the charge on the support and the analytes, the nature of the salt is an equally important parameter. The constituent ions of the salt associate with the support functional groups and/or those of the solute, yielding distinct ionic interactions. Mobile-phase additives, temperature, and gradient conditions also contribute to the separation in IEC.

Mobile Phase

pH

Adjustment of the pH is a critical factor in IEC because the pH dictates the charge of both the solutes and the ion exchanger, thus controlling their affinity for one another or their ability to release from a bound state. The essential nature of pH in the process necessitates its exact control; therefore, any mobile phase used for IEC should contain an effective buffer ($0.02M$–$0.1M$) within its optimum pH range. Some common buffers which cover much of the range of pH used in IEC are phosphate, citrate, acetate, and tris(hydroxymethyl)aminomethane (Tris) [1,2]. The pH should be selected to yield ionization of the functional groups on the support as well as those on the analytes. For molecules with a single charge, the pH should be at least two units from the pK in the direction of ionization. The guideline for zwitterions is that the pH be at least two units from the isoelectric point (pI). A pH near neutrality is often effective for complex mixtures of diverse substances. Even carbohydrates, whose hydroxyl groups do not ionize until the pH is greater than 12, can be separated by IEC when the pH is adjusted to a high enough value with low concentrations of base as the mobile phase [3].

The choice of pH for IEC of proteins or other macromolecules is not as simplistic as it is for small molecules. Although using the pI as a guide frequently yields an adequate separation, the pI encompasses all the charged groups in the molecule, whereas, because of their defined tertiary structures, only the surface amino acids of proteins are actually involved in the binding. Under denaturing conditions, more amino acids are likely to be exposed to the bonded phase.

Salt Concentration

Ion-exchange chromatography is a very predictable technique because the mechanism is well defined. The capacity factor (k) for the binding of an ionic solute to an ion-exchange functional group in IEC is directly related to the concentration (c) of salt in the mobile phase:

$$\log k = \log K_0 + Z_c \log\left(\frac{1}{c}\right)$$

where K_0 is the distribution coefficient and Z_c is an experimentally determined parameter that reflects the apparent number of ionic charges associated with the process of a specific solute with a specific surface [4]. For isocratic separations of simple molecules with up to several charges, the analysis time can be optimized along with resolution by adjustment of the salt concentration. For more complex analytes or mixtures, salt gradients are often necessary to achieve acceptable separations. Generally, a gradient from $0M$ to $1M$ salt in a buffer at a suitable pH will yield a preliminary separation.

An opposite mechanism to ion exchange occurs when the ionic strength is too low. Ion exclusion is a phenomenon in which a charged analyte is repelled by the like charges within a pore. This is very likely to occur if water is used alone as the mobile phase with ion-exchange supports or with other modes of silica-based columns. Adding buffer and salt usually eliminates the problem.

Salt Composition

Elution with increased concentrations of salt is the most common and readily controlled method of achieving displacement of molecules which are strongly bound by an ion exchanger. The salt counterions competitively displace solute ions from the charged sites on the stationary phase. Smaller, more highly charged ions are most effective at this displacement. Specifically, the strength of displacement for cations is

$$Mg^{2+} > Ca^{2+} > NH_4^+ > Na^+ > K^+$$

and for anions, it is

$$SO_4^{2-} > HPO_4^{2-} > Cl^- > CH_3COO^-$$

The strength of the ions for displacement is not necessarily related to optimum selectivity or resolution. Selectivity is dictated by the effect of the salt on both the solute and the bonded phase. Besides displacing the solute from the support, either of the ions of the salt can complex with the ion-exchange functional group or the solute, alter the tertiary structure of the solute, or enhance hydrophobic properties. It is this combination of effects which results in selectivity. For example, when a mixture of proteins was run on a polyethyleneimine (PEI) weak anion-exchange column with gradients formed with 1.0*N* salt, substitution of sodium acetate for sodium phosphate produced not only longer retention but also much better resolution of the proteins [5]. Sodium phosphate produced narrower peaks with less tailing, but the peaks had only slight differences in retention. In this case, the short retention was proven to be due to a special affinity of phosphate for PEI, which did not occur with anion-exchange supports having quaternary (Q) or diethylaminoethanol (DEAE) functional groups. The salt effects on selectivity encompass anions and cations in both anion-exchange and cation-exchange chromatography, as illustrated in Fig. 1, implying that the selectivity occurs because of ionic interactions with the functional groups of both the support and the solute. In the case of adenosine 5′-diphosphate (ADP), divalent ions like calcium can bridge between the oxygens in the phosphate and thus reduce the ionic properties. Phosphate salts reduce the retention of ADP on PEI supports due to the phosphate–PEI affinity discussed earlier. Another example of ion-based selectivity is the excellent resolution obtained for sugars when a calcium salt is used with a cation-exchange resin. This ability to change selectivity so dramatically by varying the salt significantly broadens the utility of IEC.

The only restrictions on the choice of salt are those involving analyte solubility or stability. Volatile salts such as ammonium acetate even allow IEC to be interfaced with mass spectrometry or evaporative light-scattering detection. It is very important that a given salt be totally stripped from a support before changing to other ions to avoid mixed ion effects. An acid such as trifluoroacetic acid is often effective as a bridge/washing solvent for this purpose.

Surfactants and Organic Solvents

Secondary separation which may be present in IEC is generally size exclusion or hydrophobicity. Size exclusion will occur if macromolecules are larger than the pores in a support. Hydrophobic interactions are most often observed under conditions of high salt for solutes with significant nonpolar characteristics, such as certain peptides. The hydrophobicity of an ion-exchange support is due to either the matrix or the cross-linking agents which were employed in the synthesis of the bonded phase. Any hydrophobic interactions are fundamentally undesirable and can be minimized by the addition of 1–10% of an organic solvent, such as methanol, ethanol, or acetonitrile, to the running buffer. The solubility of the salt in the organic mobile phase should always be verified to avoid precipitation.

Nonionic detergents may also reduce hydrophobic interactions with a column. These detergents, such as CHAPS or urea, can also be added to ion-exchange mobile phases to aid in the solubilization of membrane or other insoluble proteins. Such detergents are easy to equilibrate and remove from ion-exchange columns; however, ionic detergents should be avoided because of their very strong binding to the column or the solutes.

Flow Rate and Gradient

Small molecules can often be effectively separated isocratically by IEC; however, due to multipoint interactions, isocratic IEC of proteins and most biological macromolecules is not usually feasible, yielding no resolution and extreme tailing. Such complex molecules are generally separated by gradient elution.

As a salt gradient proceeds to higher levels in IEC, molecules elute at a specific salt concentration, generally without binding from secondary effects. The relationship of gradient conditions to elution (k^*) can be described by

$$k^* = 0.87 t_G \frac{F}{V_M}\left(\log \frac{C_2}{C_1}\right) Z$$

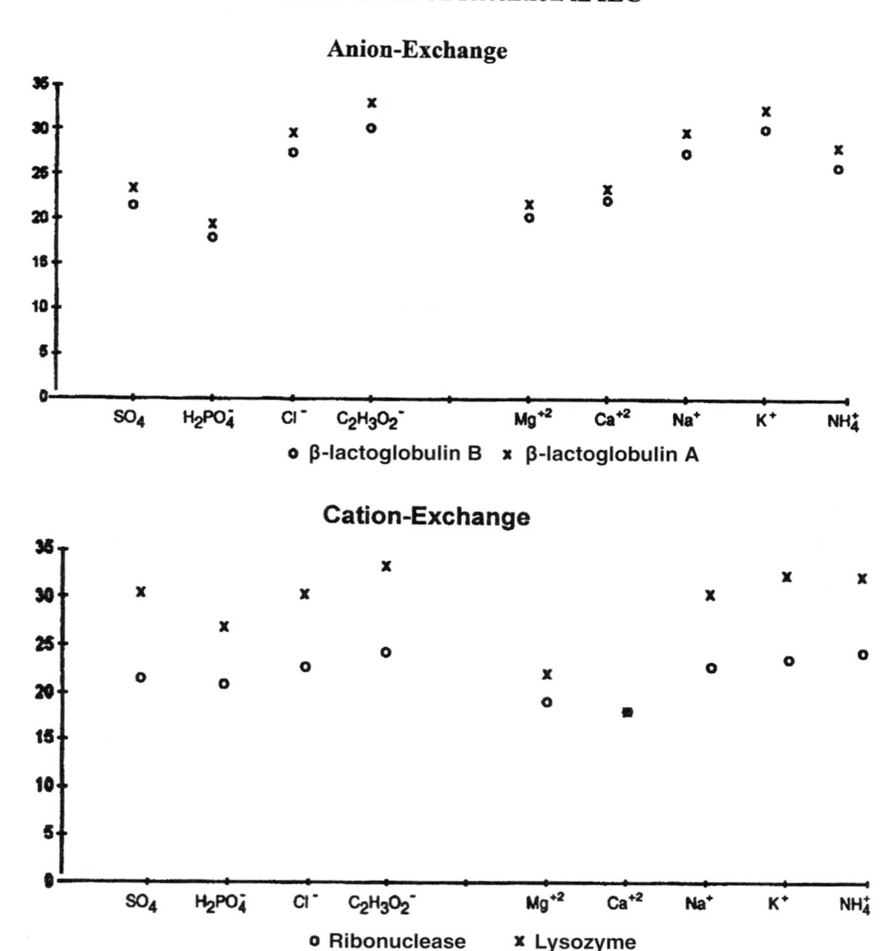

Fig. 1 Anion-exchange chromatography (AEX): SynChropak AX300 (polyethyleneimine, 300 Å, 6 μm); cation-exchange chromatography (CEX): SynChropak CM300 (carboxymethyl, 300 Å, 6 μm); 30-min gradient (0–1N) of sodium or chloride salts in 0.02M Tris, pH 7. (Reprinted with permission of MICRA Scientific.)

where C_1 and C_2 are the total salt concentrations (salt plus buffer) at the beginning and the end of the gradient, respectively, Z is the effective charge on the solute molecule, F is the flow rate; V_M is the total mobile-phase volume, and t_G is the gradient time [6]. The Z number will vary with solute and pH. An initial ion-exchange protocol of a 20–30-min linear gradient from 0M–1M salt in a buffer at a suitable pH will usually yield a separation which can be later optimized, if necessary. For shortest analysis times, a gradient should begin at the highest salt concentration where the analytes are bound and it should end at the lowest ionic strength that causes elution. The pH gradients may also be used to elicit elution during IEC, although this has been a less popular strategy than salt gradients. Ion-exchange columns can be effectively washed with a mobile phase of higher ionic strength than the upper gradient limit or with low pH. For gradients, intermediate flow rates of 1mL/min for a 4.6-mm-inner diameter column are usually satisfactory.

Temperature

The use of elevated temperature in IEC reduces the mobile-phase diffusion coefficient and concomitantly decreases band spreading. Most mobile phases in IEC are composed of water with salts and thus produce efficiencies which are less than those obtained in modes using organic solvents. Because increased temperatures decrease retention, they may permit the use of lower salt concen-

Amino Acid Analysis by Cation-Exchange Chromatography

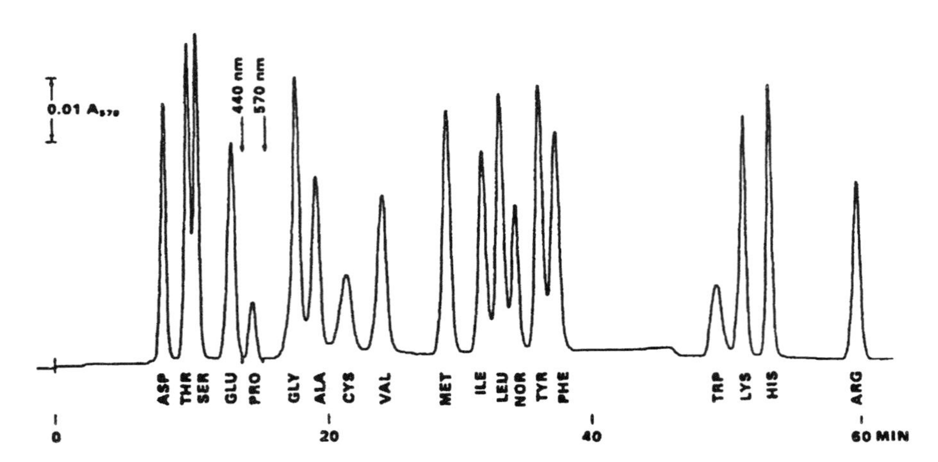

Fig. 2 Column: Micropak AA (sulfonated polystyrene); solvent A: 0.2*M* sodium citrate, pH 3.25; solvent B: 1*M* sodium citrate, pH 7.40. Gradient: 5 min 100% A; 100–75% A in 20 min; 75–70% A in 5 min; 70–35% A in 5 min; 10 min 35%; 35–0% A in 1 min. $T = 50°C$ for 25 min, then 90°C. Detection after ninhydrin postcolumn reaction. (Reprinted from Amino acid analysis with ninhydrin postcolumn derivatization, *LC at Work*, Varian Associates with permission.)

trations. Elevated temperatures have been especially effective in amino acid analyses by cation-exchange chromatography, as illustrated in Fig. 2.

Conclusions

The effectiveness of ion-exchange chromatography as a method for separating charged species is enhanced by the ability of many operational factors to change the selectivity and resolution. Salt concentration, salt composition, and pH are the most important operational parameters which strengthen the versatility of the technique.

References

1. R. L. Cunico, K. M. Gooding, and T. Wehr, Ion-exchange chromatography, in *Basic HPLC and CE of Biomolecules*, Bay Bioanalytical Laboratories, Richmond, VA, 1998.
2. *Ion-Exchange Chromatography, Principles and Methods*, Pharmacia Biotech, Sweden.
3. R. R. Townsend, High-pH anion exchange chromatography of recombinant glycoprotein glycans, in *High Performance Liquid Chromatography: Principles and Methods in Biotechnology* (E. D. Katz, ed.), John Wiley & Sons, New York, 1996.
4. M. I. Aguilar, A. N. Hodder, and M. T. W. Hearn, HPIEC of proteins, in *HPLC of Proteins, Peptides and Polynucleotides* (M. T. W. Hearn, ed.), VCH, New York, 1991, p. 199.
5. M. P. Nowlan and K. M. Gooding, HPIEC of proteins, in *High-Performance Liquid Chromatography of Peptides and Proteins* (C. T. Mant and R. S. Hodges, eds.), CRC Press, Boca Raton, FL, 1991.
6. L. R. Snyder, Gradient elution separation of large biomolecules, in *HPLC of Biological Macromolecules: Methods and Applications* (K. M. Gooding and F. E. Regnier, eds.), Marcel Dekker, Inc., New York, 1990.

Karen M. Gooding

Ion-Exchange Stationary Phases

Introduction

In ion-exchange chromatography (IEC), molecules bind by the reversible attraction of electrostatic charges located on the outer surface of a solute molecule with dense clusters of groups with an opposite charge on an ion-exchange support. To maintain electrical neutrality, the charges on both the analytes and the matrix are associated with ions of opposite charge, termed counterions, which are either provided by preequilibration with the mobile phase or during manufacturing. Because a solute must displace the counterions on the matrix during attachment, the technique is termed "ion exchange." If the support possesses a positive charge, it is used for anion-exchange chromatography, whereas if it carries a negative charge, it is for cation exchange. Generally, the molecule of interest will have a charge that is opposite (positive or negative) to that on the support and the same as the competitively displaced counterions.

There are several major variables which distinguish ion-exchange packings and determine their utility for specific classes of solutes and for analytical or preparative applications. Those variables are as follows:

1. Structure of the bonded phase, including the chemistry of the functional group, its pK, and the properties of the spacer arm and/or bonded phase layer
2. Charge density and related nominal capacity
3. Properties of the support matrix, including composition and pore diameter

Bonded Phase

Functional Groups

The functional groups of an ion-exchange bonded phase are ionizable under specific pH conditions. The extent of their charge dependence on pH is the basis for distinguishing two types of ion exchangers — strong and weak. These designations do not refer to the strength of binding or to the capacity of the gel, but simply to the pK of the ionizable ligand group, similar to the designations for acids and bases. The structures and approximate pK and pH ranges of some typical strong and weak ion-exchange groups are shown in Table 1 [1–5].

Generally, strong ion-exchange groups retain their charge over a wide range of pH, with binding capacity dropping off at the extremes. For example, quaternary ammonium (Q) resins are strong anion-exchange groups which are effective throughout the pH range of about 2–12. Similarly, sulfonyl groups are strong cation-exchange groups that remain negatively charged until acidic pH levels are used. Strong ion-exchange groups can be considered to possess a permanent positive or negative charge.

The diminished ionization of weak ion-exchange groups near neutral pH result in less predictable separations if operation in this range is necessary for analyte stability, as in the case of many proteins. In these cases, the use of a strong ion exchanger allows the pH of the mobile phase to be manipulated to protonate or deprotonate the analytes without changing the ionic properties of the packing. For example, certain amino acids are most highly charged at pH less than 4, where a weak cation-exchange support would not be fully charged, but a strong cation-exchange group would.

Because weak ion-exchange groups are not fully charged in certain pH ranges, column equilibration may require more mobile phase or time under those conditions. Conversely, highly bound molecules may release more easily from supports which are not totally ionized. Clearly, careful consideration of the titration curves for an ion-exchange support is an essential aspect of designing appropriate conditions for a separation. A complete description of the charged group of an ion exchanger is necessary to understand its pH characteristics because they are dependent on the exact chemical composition of the bonded phase and the matrix. Convenient descriptions such as "strong," "S," "stable weak ion-exchange," and so forth do not sufficiently describe the ionic characteristics of the packing. The exact pK and functional pH range are also affected by the chemistry of the remainder of the bonded phase and of the matrix. For example, a silica ma-

Table 1 Properties of Ion-Exchange Groups

	Functional group	Type	pK	pH Range (approximate)
Anion exchange				
DEAE (diethylaminoethyl)	$-O-CH_2-CH_2-N^+H(CH_2CH_3)_2$	Weak	5–9	2–9
PEI (polyethyleneimine)	$(-NHCH_2CH_2)_n-N(CH_2CH_2-)_{n'}CH_2CH_2NH_2$	Weak	5–9	2–9
Q (quaternary ammonium)	$-CHOH-CH_2-N^+(CH_3)_3$	Strong	>13	2–12
Cation exchange				
CM (carboxymethyl)	$-O-CH_2-COO^-$	Weak	4–6	6–10
SP (sulfopropyl)	$-CH_2-CH_2-CH_2SO_3^-$	Strong	<1	4–13
S (sulfonate)	$-R-CH_2SO_3^-$ (R may be methyl with hydroxyl or amide groups)	Strong	<1	3–11

trix may ion-pair with cationic functional groups or a polymeric layer with amines may ion-pair with anionic functional groups. The actual titration curves, pK, and/or pH range for a given support should always be consulted.

Hydrophobic Spacer Arms

Ion-exchange functional groups are chemically bonded to the support, often through a polymeric layer which totally covers the matrix. The chemical nature of this coupling chemistry and its spatial characteristics can affect the chromatographic properties. Hydrophobic linkages may impart a nonpolar aspect to the separations. Spacer arms make the functional groups more accessible by distancing them from the support surface. Tentacle IEC bonded phases are a spacer design incorporating a hydrophilic ligand arm [6].

Charge Density

The number of charges, as measured by titration, defines the nominal capacity of a support. the charge density of an ion-exchange support is determined by the number of ionic groups divided by the surface area or the volume. Typical values range from 3 to 370 μEq/mL of support. The lower values are generally found in nonporous supports. High loading capacities are associated with IEC, especially for porous supports. Weak ion-exchange groups only have maximum capacity in the pH range where they maintain charge — pH less than 9 for DEAE supports and pH greater than 6 for CM.

Counterions

In certain cases, ion-exchange columns are preequilibrated with distinct counterions by the manufacturer. These ions, such as calcium for amino acids, impart a specific selectivity (see the entry Ion-Exchange, Mechanism and Factors Affecting Separation). Alternatively, a layer of counterions is applied by the user by conditioning a column with the salt of interest. An intermediate step of washing with a weak acid may accelerate the equilibration process.

Matrix

Composition

Ion-exchange supports based on derivatized cellulose and agarose have been popular since the 1960s, particularly for protein analysis. For high-performance liquid chromatography (HPLC), less compressible supports, such as silica and cross-linked polymers, are most commonly used.

Carbohydrate Matrix

Carbohydrate supports such as dextran or agarose are very hydrophilic and easily derivatized with ionic functional groups. They have been very popular for analysis and purification of biological molecules like proteins. One major drawback to these supports is that their volume changes with mobile-phase composition. This has been alleviated in part by higher cross-linking.

Silica Matrix

In silica-based ion exchangers, the silica is bonded through a polymeric layer to a charged ligand group. Operating pH is generally limited to pH 2–8 due to the silica backbone. Although some small-pore silica-based ion exchangers have been synthesized with silane bonding, large-pore supports (≥300 Å) designed for protein analysis have polymeric layers containing ionic functional groups which are very stable and even protect the silica matrix from erosion. Silica columns have several advantages:

1. High mechanical stability
2. Minimal shrinkage or swelling with changes in counterions
3. Stability to organic modifiers (with the restriction of salt solubility)
4. High capacity
5. Good mass transfer
6. Large variety of particle and pore sizes

Polymeric Matrix

Polymeric matrices are also widely available for IEC. Polystyrene cross-linked with divinylbenzene (PSDVB) is one such polymer, typically available with pore diameters of at least 1000 Å. The repetitive structure of polystyrene permits reproducible coupling of both strong and weak ion-exchange groups; cross-linking adds the rigidity required for high-pressure applications. These polymeric supports have most of the same advantages as silica for IEC. Methacrylate copolymers, which are also used as matrices in IEC, are more hydrophilic than PSDVB.

Pellicular Matrix

A third group of ion-exchange supports are pellicular, consisting of a solid inert core made of PSDVB agglomerated with 350 nm functionalized latex. The quaternary amine groups are closely and uniformly bound on the microbeads, improving flow and reducing nonspecific retention. These pellicular supports are primarily used for carbohydrate analysis [7].

Pore Diameter

Pore diameter is a major determinant in ion-exchange capacity because as the pore diameter decreases, there is a tremendous increase in surface area. Nominal loading capacity is directly related to the surface area and the ligand density; consequently, matrices with the smallest pores exhibit the highest ion-exchange capacities for small, totally included solutes.

The ion-exchange capacities of picric acid correlate with surface area. For example, that of a 100-Å pore was seen to be 1415 μmol/g, whereas that of a 300-Å pore was only 656 μmol/g [8]. The capacities for macromolecules such as proteins do not relate directly to surface area because they are excluded by size from portions of small pores and are effectively prevented from reaching all the reactive exchange sites [5,8]. Consequently, larger pores exhibit maximum capacity for macromolecules. For example, a 300-Å pore exhibited maximum capacities of 98 and 130 mg/g for ovalbumin (45,000 MW) and bovine serum albumin (65,000 MW) respectively, because they were able to permeate and bind to the optimum available surface area [5,8].

References

1. R. L. Cunico, K. M. Gooding, and T. Wehr, Ion-exchange chromatography, in *Basic HPLC and CE of Biomolecules*, Bay Bioanalytical Laboratories, Richmond, VA, 1998.
2. *Ion-Exchange Chromatography, Principles and Methods*, Pharmacia Biotech, Sweden.
3. E. D. Katz (ed.), *High Performance Liquid Chromatography: Principles and Methods in Biotechnology*, John Wiley & Sons, New York, 1996.
4. M. I. Aguilar, A. N. Hodder, and M. T. W. Hearn, HPIEC of Proteins, in *HPLC of Proteins, Peptides and Polynucleotides* (M. T. W. Hearn, ed.), VCH, New York, 1991, p. 199.
5. C. T. Mant and R. S. Hodges (eds.), *High-Performance Liquid Chromatography of Peptides and Proteins*, CRC Press, Boca Raton, FL, 1991.
6. W. Muller, *J. Chromatogr. 510*: 133 (1990).
7. Analysis of Carbohydrates by HPAE-PAD, Technical Note 20, Dionex, 1993.
8. G. Vanecek and F. E. Regnier, *Anal. Biochem. 109*: 345 (1980).
9. M. P. Nowlan and K. M. Gooding, HPIEC of proteins, in *High-Performance Liquid Chromatography of Peptides and Proteins* (C. T. Mant and R. S. Hodges, eds.), CRC Press, Boca Raton, FL, 1991, p. 203.

Karen M. Gooding

Ion-Exclusion Chromatography

Ion exclusion is the term that describes the mechanism by which ion-exchange resins are used for the fractionation of neutral and ionic species. Ionic compounds are rejected by the resin, due to Donnan exclusion, and they are eluted in the void volume of the column. Nonionic or weakly ionic substances penetrate into the pores of the packing, they are retained and, thus, separation is achieved, as they partition between the liquid inside and outside the resin particles.

Ion-exclusion Chromatography is a mode of high-performance liquid chromatography (HPLC) and, thus, the same equipment can be used, with the proper eluent, column, and detection technique. The technique is mostly used for the analysis of organic acids, sugars, alcohols, phenols, and organic bases. It provides a convenient way to separate molecular acids from highly ionized substances. Ionized acids pass rapidly through the column while molecular acids are held up to varying degrees. A conductivity detector is commonly used. Carboxylic acids can be separated by using water, a dilute mineral acid, or a dilute benzoic or succinic acid as eluent.

As neutral species, rather than ions, are being separated, ion-exclusion chromatography cannot be considered as a form of ion chromatography; although ion-exchange polymers are used, ion-exchange mechanisms are not involved.

Anions, most commonly simple carboxylic acids (e.g., tartaric, malic, citric, lactic, acetic, succinic, formic, propionic, butyric, etc.), are separated on cation-exchange resins in acidic form. Salts of weak acids can also be analyzed, as they are converted to the corresponding acid by the hydrogen ions in the exchanger. Cations (weak bases and their salts) are separated on anion-exchange resins in the hydroxide form.

In order to understand the mechanism of ion-exclusion chromatography, the behavior of the resin, in an aquatic medium, must be taken into account. In this case, three parts can be distinguished:

1. The resin network
2. The liquid inside the resin particles
3. The liquid between the resin particles

The first acts as a semipermeable membrane between the stationary liquid phase within the resin and the mobile liquid phase between the resin beads.

Ionic groups are fixed on the resin and movement of ions across the membrane takes place as predicted by Donnan theory. The ion-exclusion mechanism involves interaction between partially ionized species and fully ionized polymer matrix. Electrostatic repulsive forces, between strong electrolytes (e.g., chloride, in the case of using HCl as eluent) and the ionic groups fixed on the resin (e.g., sulfonate), prevent them from entering into the resin, due to high ionic concentration inside the resin. Because ionized analytes are not retained, they are excluded from the polymer, migrate rapidly through the column, and are eluted at the column void volume. Partially ionized and neutral species (e.g., the undissociated forms of the analyte acids), as they penetrate into the pores of the resin, are distributed between the mobile phase in the column and the immobilized liquid in the pores of the packing. Separation is accomplished by differences in acid strength, size, and hydrophobicity.

The degree of retardation increases with the decrease of the ionization degree and, additionally, depends on polar attractions between analyte and fixed functional groups and on different van der Waals forces between an analyte and the hydrocarbon part of the resin. Elution order is related to pK_a values for ionic species and to the molecular size for neutral compounds.

Members of a homologous series, such as formic, acetic, and propionic acids, elute in the order of increasing pK_a (decreasing acid strength). Dibasic acids elute sooner than monobasic acids of the same carbon number. Isoacids elute earlier than normal acids. Double bonds retard elution, whereas keto groups increase elution rate.

Microporous polystyrene divinylbenzene resins are used, operating at pressures sometimes exceeding 3000 psi; unlike silica-based packings, they are stable from pH 0 to 14. For ion-exclusion separation of organic acids and weakly acidic compounds, strongly acidic, high-capacity, sulfonated styrene divinylbenzene in the hydrogen form are used. For organic bases, separation columns are packed with strongly basic copolymer with a quaternary ammonium functional group. The degree of cross-linking (the percentage of divinylbenzene in the copolymer) affects the retention of weakly ionized species; the lower the degree of cross-linking, the longer the retention time of acid, either strong or weak. This is due to the fact that as cross-linking decreases, ions more readily penetrate the resin, where they are held up.

As aforementioned, a large number of organic and weak inorganic acids can be eluted from the hydrogen form of cation-exchange resin using water as the eluent. However, the addition of mineral acid to the water eluent suppresses the ionization of strong and moderately strong organic acids, allowing them to partition into the resin phase and, thus, improve selectivity, as retention times on the resin are increased. The addition of inorganic salts, such as $(NH_4)_2SO_4$ or organic modifiers (acetonitrile, isopropanol, ethanol, methanol), to the eluent may improve separation. Acetonitrile, for example, decreases the retention time of relatively nonpolar compounds.

Ion-exclusion chromatography can couple to ion chromatography to improve the chromatographic resolution of inorganic anions and organic acids in complex matrices. The dual system can be either in the order IEC/IC or IC/IEC.

Various detection systems can be used in ion-exclusion chromatography, among them ultraviolet (UV)/vis spectrophotometry, conductivity, electrochemistry, fluorometry, refractive index (RI) measurement, are the most common techniques. Additionally, combined detection systems (e.g., UV/amperometry, UV/RI) may be used, leading to enhanced selectivity.

Ultraviolet detection is useful, especially when water or sulfuric acid, which do not absorb in the UV region, are used. Detection for most nonaromatic carboxylic acids is accomplished at 210 nm.

Conductivity detection is preferred when water is used as eluent; then ionizable analytes are readily detected. However, in the case where HCl is used as the eluent, the analytical column is followed by a suppressor column, packed with a cation-exchange resin in the silver form. The hydrogen ions of the eluent are exchanged for silver ions, which then precipitate chloride ions, thus removing the ions contributed by the eluent and enhancing the analyte's signal.

Electrochemical detectors (coulometric and amperometric) are used when the analytes are electrochemically active or capable of being coupled to an electrochemical reaction.

Refractive index monitors are used in food analysis, for detecting carbohydrates, alcohols, and other substances with weak or no UV absorption.

With the combination of RI and UV, simultaneous detection of organic acids, carbohydrates, and alcohols with one sample injection can be achieved. Postcolumn reactions can be used for fluorometric detection of amino acids, with excellent sensitivity and selectivity.

Ion-exclusion chromatography finds numerous applications for identification and determination of acidic spe-

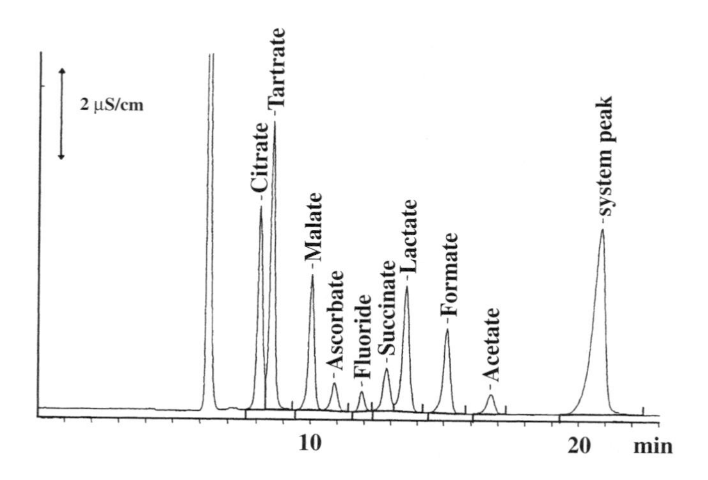

Fig. 1 Determination of organic acids and fluoride using ion-exclusion chromatography with direct conductivity detection, using mmol/L H_2SO_4, and 10% acetone as eluent. (From Metrohm Ltd., with permission.)

cies in complex matrix materials, such as dairy products, coffee, wine, beer, fruit juice, and other commercial products which can be quickly analyzed with minimal sample preparation before injection (usually only filtration, dilution, or centrifugation). Organic acid determination is also of great importance in biomedical research (e.g., physiological samples, in which most of the Krebs cycle acids (tricarboxylic acid cycle) are present).

Organic acids can be detected in the parts per billion range. With preconcentration, this limit can be further decreased. A typical ion-exclusion chromatogram of organic acids separation is presented in Fig. 1.

Suggested Further Reading

Gierde, D. and J. Fritz, *Ion Chromatography*, 2nd ed., Alfred Huethig Verlag, New York, 1987.

Gierde, D. and H. Mehra, *Advances in Ion Chromatography Volume 1* (P. Jandik and R. Cassidy, eds.), Century International, Franklin, MA, 1989.

Haddad, P. and P. Jackson, *Ion Chromatography, Principles and Application*, Elsevier, Amsterdam, 1990.

Kaine, L., J. Crowe, and K. Wolnic, *J. Chromatogr. 602*: 141–247 (1992).

Metrohm IC Application Note No. O-5, Application Notes, Metrohm, Herisau (1996).

Small, H., *Ion Chromatography*, Plenum Press, New York, 1989.

Tanaka, K. and J. Fritz, *Anal. Chem. 59*: 708–712 (1987).

Tarter, J., *Ion Chromatography*, Chromatographic Science Series Vol. 37, Marcel Dekker, Inc., New York, 1987.

Togami, D., L. Treat-Clemons, and D. Hometchko, *Int. Lab. 2*: 29–33 (1990).

Ioannis N. Papadoyannis
Victoria F. Samanidou

Ion-Interaction Chromatography

Introduction

Under reversed-phase high-performance liquid chromatography (HPLC) conditions, ionic compounds are weakly retained. On the contrary, when an ion-interaction reagent (IIR), which is a large lipophilic ion, is added to the mobile phase, ionized species of opposite charge are separated on reversed-phase columns with adequate retention. This is the chromatographic approach of reversed-phase ion-interaction chromatography (IIC). It has become a widely used separation mode in analytical HPLC because it provides a useful and flexible alternative to ion-exchange chromatography. Better selectivity, enhanced resolution, and retention are usually gained by this separation strategy.

According to the qualitative retention model of Bidlingmeyer, the lipophilic IIR, flowing under isocratic conditions, dynamically adsorbs onto the alkyl-bonded apolar surface of the stationary phase, forming a primary charged ion layer. The corresponding counterions are found in the diffuse outer region to form an electrical double layer. This charged stationary phase can then more strongly retain analyte ions of the opposite charge.

Unlike conventional ion exchange, IIC can be used to separate nonionic and ionic or ionizable compounds in the same sample, because retention of an analyte involves its transfer trough the electrical double layer and depends on both electrostatic interactions and adsorptive (reversed-phase) effects.

In recent years, many examples of applications of IIC have been reported. They essentially concern the separation of organic and inorganic ions in the environmental, pharmaceutical, food, and clinical fields.

Retention Mechanism

The larger number of names (e.g., ion-pair chromatography, dynamic ion-exchange chromatography, hetaeric chromatography, soap chromatography) which have been given to the IIC mode sheds light on the uncertainty concerning the retention mechanism. A majority of the proposed models are stoichiometric. They suggest that the oppositely charged analyte and IIR form a complex, according to a clear reaction scheme, either in the mobile-phase (ion-pair model) or at the stationary-phase surface (dynamic ion-exchange model). According to the first theory, the uncharged ion pair between oppositely charged analyte and IIR, which is formed in the mobile phase, is then more strongly retained by the stationary phase. The second theory presumes that solute ions undergo an ion-exchange process, at exchange sites dynamically generated by the adsorption of the IIR at the stationary phase. Knox and Hartwick demonstrated that both models lead to identical retention equation.

These models, although of practical and intuitive value, are not well founded in physical chemistry. The pioneeristic, even if qualitative, work of Bidlingmayer demonstrated that IIRs adsorb onto the stationary phase. It follows that stoichiometric equilibrium constants, which depend on the change in free energy of adsorption of the analyte, cannot be considered constant if the IIR concentration in the mobile phase increases, because the stationary phase surface properties (including its charge density) are modified. The multibody interactions and long-term forces involved in IIC can better be described by a thermodynamic approach.

A quantitative nonstoichiometric model was developed by Ståhlberg and coworkers. The model applies the Gouy–Chapman electrostatic theory to describe the interactions between charged species and it does not assume the formation of any chemical complexes: The adsorption of the IIR onto the stationary phase establishes a certain electrostatic surface potential, because its counterion has a lower adsorption tendency. An electrical double layer develops and a difference in electrostatic potential is created between the electroneutral bulk of the mobile phase and the net charged surface. The intuitive view of the effect of the IIR on retention is an electrostatic repulsion or attraction of the analyte to the charged stationary-phase surface, according to the analyte and IIR charge status. However, the adsorptive (reversed phase) effects are also considered, to evaluate the total free energy of adsorption of the solute: The latter is partitioned into a "chemical" and an electrostatic free energy. This is a first approximation: The "chemical" part depends on the concentration of the IIR, as it determines a dynamic modification of the stationary-phase properties. This electrostatic theory of IIC has been implemented by taking into account the competition between IIR and analyte for a limited surface area, and the different surface area requirements of analyte and IIR (multisite occupancy model). However, the main drawback of this powerful electrostatic theory is the complex algebraic form of the resulting equations; hence, a series of approximations has to be made to obtain a relationship between the analyte capacity factor and mobile-phase concentration of IIR, which is of interest for practical work.

Cantwell and co-workers proposed a surface adsorption, diffuse-layer ion-exchange double-layer model in which they underlined the role of the diffuse part of the double layer by assigning a stoichiometric constant for the exchange of ions.

Stranahan and Deming proposed a thermodynamic model for IIC in which the distribution a sample between the mobile and the stationary phase is discussed in terms of chemical potentials in both phases.

Additional peaks relative to the number of components injected are often obtained in IIC. These so-called "system peaks" confirm the proposed mechanism of dynamic functionalization of the stationary phase. They can be explained by taking into account that IIR ions are locally adsorbed onto (desorbed from) the stationary phase by injection of adsorbophilic solute ion of the opposite charge (of same charge). This change in the eluent composition, created by the sample injection, migrates along the column and give a signal if at least one of the eluent components can be detected. The same rationale provides the explanation for the indirect ultraviolet (UV) visualization (or amplification) of otherwise non-UV-absorbing samples, when a UV-absorbing lipophilic ion is added to the eluent.

Influence of Experimental Parameters on Retention

The optimization of separations performed with IIC and the rationalization of analytes retention behavior are not easy tasks because they are influenced by many interdependent factors. This allows a fine modulation of their effects to achieve tailor-made separations.

Experimental design can be very helpful, and a number of chemometric optimization methods are present in the literature. Neural network models provided a good prediction power and a great versatility, without the need to develop any equations.

The following presents the effect of varying some individual factors on analyte retention.

Ion-Interaction Reagent

Type

The hydrophobic character of the IIR increases with increasing its chain length. More lipophilic reagents have higher adsorption constants, hence the effect of increasing chain length is qualitatively similar to the effect of increasing IIR concentration (see below) with regard to the degree of stationary-phase coverage. The use of multiply-charged IIRs allows the chromatographer to obtain larger changes of analyte retention. If chiral compounds are used as the IIR, the separation of the enantiomeric forms of the analyte may be achieved. The most popular IIRs are listed in Table 1.

Concentration

If the eluent concentration of the IIR increases, the amount of the adsorbed IIR also increases, according to its adsorption isotherm. This induces a higher surface potential on the stationary phase but also adsorption competes between analyte and IIR for the available stationary phase sites. Therefore, the following hold:

1. If the charge status of analyte and the IIR is the same, a decrease in retention is observed because of electrostatic repulsion between solute and charged stationary phase, and because of adsorption competition.
2. If the charge status of analyte and the IIR is the opposite, an increase in retention is expected because of electrostatic attraction between solute and charged stationary phase. A parabolalike dependence of analyte capacity factors on IIR concentration is observed if the investigated concentration range is broad. For narrower ranges, a linear increase may hold. Some authors have emphasized that analyte retention passes through a maximum because if the ionic strength is not kept constant when increasing the IIR concentration, there is a competition between analyte ion and IIR counterion; this competition counteracts the retention increase. However, a foldover of the plot may still occur even if the ionic strength is kept constant, because there is a critical value of the IIR concentration at which the positive effect of the electrostatic attraction is balanced by the negative effect of adsorption competition for the available stationary-phase surface area.
3. If the analyte is uncharged, a very weak decrease in retention is usually observed, primarily because of adsorption competition for the stationary phase.

Table 1 Commonly Used Ion-Interaction Reagents

Cationic IIRs	Anionic IIRs
Tetramethylammonium	Butanesulfonate
Tetraethylammonium	Pentanesulfonate
Tetrabutylammonium	Hexanesulfonate
Cetyltrimethylammonium	Octanesulfonate
Octylammonium (from octylamine)	Dodecanesulfonate

Increasing the IIR above its critical micelle concentration leads into the field of micellar chromatography in which analyte may partition between the mobile phase and both the stationary phase and the micelle.

Mobile-Phase Composition

Organic-Modifier Concentration

In IIC, the logarithm of the analyte capacity factor is described as a linear function of the organic-modifier concentration in the mobile phase. When the sample ion is in the same charge status as the IIR, the slope of the linear relationship, if compared to the original reversed-phase slope, becomes steeper (the contrary is observed for oppositely charged combinations). This can be explained by taking into account that the organic modifier, through desorption effects, decreases the retention of ionic solutes via the simultaneous decrease of the free energy of adsorption of both the analyte and IIR.

Ionic Strength

An increase in salt concentration in the bulk mobile phase provides those counterions which are able to reduce, according to the Gouy–Chapman electrostatic theory, the electrostatic stationary-phase surface potential. Hence, the adsorption of the IIR may increase, even if its concentration in the eluent is the same, because of lower electrostatic "self"-repulsion. However, the net effect is a reduced surface potential: The ion interactions decrease, and analyte retention may be modulated.

From an intuitive point of view, the inorganic ions are eluting agents because they limit the interaction of oppositely charged analyte and IIR, via a competing equilibrium for adsorbed lipophilic ions. This view gives the rationale for the use of mobile-phase additives, such as sodium carbonate, to avoid the unnecessarily high resolution which may be obtained between analytes of different charge.

It has to be emphasized that the nature of the electrolyte ions influences the surface potential value because the effective surface charge concentration is reduced if slight hydrophobic, adsorbophilic electrolytic counterions are included in the eluent.

The influence of moderate increase of ionic strength on the "chemical" part of the free energy relative to the analyte transfer from the mobile to the stationary phase has been usually neglected.

Mobile-Phase pH

The eluent pH value affects the degree of ionization of the species involved in ion interaction. Hence, the greatest retention is obtained for completely dissociated species. This is the opposite of what is observed in reverse-phase chromatography.

Unexpected pH dependencies were explained by (a) competition between negative analyte ions and OH^- ions for interaction with the electrical double layer and (b) a mixed retention mechanism in which reverse-phase partition or interaction with unreacted silanols from the stationary-phase base may play a significant role.

Reversed-Phase Stationary Phase

A number of different packings were used in IIC, including the newly developed graphitized carbon column, which has excellent chemical and physical resistance. The use of polymeric material has the drawback of poor physical resistance. However, a wider pH range is investigable and the affinity for certain IIR is higher, by comparison with the silica-based reversed-phase columns. However, discordant results are present in literature reports with regard to the chromatographic efficiency.

With regard to the silica-based reversed-phase stationary phase, unreacted residual silanol groups may play a significant role in IIC because it was shown that they are ion-exchange sites not only for analyte cations but also for alkylammonium IIR. The higher retentions that were noticed for the silica-based stationary phase if compared to end-capped or polymer-based packings supports this.

The reproducibility of results obtained with silica-based reversed phases, of the same declared characteristics but from different manufacturers, was sometimes poor, probably because of the properties of the silica used and the different reaction conditions in the alkylation of the support.

Stationary phases with higher hydrophobicities and adsorption capacities show increased retention of both solute and IIR. Hence, an increased capacity factor value should be expected, even if anomalies can be due to direct competition of solute and IIR for the available stationary phase.

Temperature

Temperature control is very important for obtaining reproducible separations. Indeed, the adsorption of the IIR onto the stationary phase follows an adsorption isotherm; hence, an increase of the column temperature leads to a decreased amount of the adsorbed IIR, even if its concentration in the mobile phase is constant. This, in turn, determines a decreased absolute surface potential and a modification of the solutes' capacity factors. Usually, a temperature increase results in an improved resolution and faster separation, even if a reversal of the elution sequence of the components of a mixture may sometimes be observed, because of the interplay of electrostatic and reversed-phase interaction which are characterized by different enthalpies.

Suggested Further Reading

Bartha, A. and J. Ståhlberg, *J. Chromatogr. A 668*: 255–284 (1994).

Bidlingmeyer, B. A., *J. Chromatogr. Sci. 18*: 525–539 (1980).

Chen, J. C., S. G. Weber, L. L. Glavina, and F. F. Cantwell, *J. Chromatogr. 656*: 549–576 (1993).

Gennaro, M. C., *Adv. Chromatogr. 35*: 343–381 (1995).

Knox, J. H. and R. A. Hartwick, *J. Chromatogr. 204*: 3–21 (1981).

Okamoto, T., A. Isozaki, and H. Nagashima, *J. Chromatogr. A 800*: 239–245 (1998).

Pietrzy, D. J., *Chromatogr. Sci. 78*: 413–462 (1998).

Sacchero, G., M. C. Bruzzoniti, C. Sarzanini, E. Mentasti, H. J. Metting, and P. M. J. Coenegracht, *J. Chromatogr. A 799*: 35–45 (1998).

Stranahan, J. J. and S. N. Deming, *Anal. Chem. 54*: 2251–2256 (1982).

Weiss, J., *Ion Chromatography*, 2nd ed. VCH, Weinheim, 1995, pp. 239–289.

Teresa Cecchi

Ion-Pairing Techniques

Ion-pair chromatography (IPC) is of relatively recent origin, being first applied in the mid-1970s. Much of the development work in both theory and practice was performed by Schill and co-workers. At various times, IPC has also been called extraction chromatography, chromatography with a liquid ion exchanger, soap chromatography, paired-ion chromatography and ion-pair partition chromatography.

When solute ions (A^-) are added to a chromatographic system containing pairing ions (B^+) and associated counter ions (C^-), the degree of retention of (A^-) depends on the following equilibrium:

$$A_{aq}^- + B_{aq}^+ = AB_{org} \tag{1}$$

with an extraction constant

$$E_{AB} = \frac{[AB_{org}]}{[A_{aq}^-][B_{aq}^+]} \tag{2}$$

In the simplest case of IPC, it can be assumed that the sample and counterions are soluble only in the aqueous mobile phase and the ion pair formed is soluble only in the organic stationary phase.

Assuming that the concentration of the pairing ion in the aqueous phase is high compared to that of the solute ion, the *distribution coefficient* of A^-, D_{A^-}, is given by

$$D_{A^-} = \frac{[AB_{org}]}{[A_{aq}^-]} = E_{AB}[B_{aq}^+] \tag{3}$$

The *capacity factor* k' is related to E_{AB} as follows (in the reversed-phase mode):

$$k' = \frac{V_S}{V_m}\left(\frac{[AB_{org}]}{[A_{aq}^-]}\right) \tag{4}$$

$$= \frac{V_S}{V_m}(E_{AB}[B_{aq}^+]) \tag{5}$$

or

$$k' = D_{A^-}\frac{V_S}{V_m} \tag{6}$$

Because the capacity factor k' is proportional to $1/D$ in normal-phase chromatography and to D in reversed-phase chromatography, it follows that k' in IPC is inversely proportional to the pairing ion concentration in the normal-phase situation but directly proportional in the reversed-phase case.

When the pairing ion is very hydrophobic, B^+ will be extracted into the organic phase with its normal counterion C^-, according to

$$C_{aq}^- + B_{aq}^+ = CB_{org} \text{ (ion pair)} \tag{7}$$

Substracting it from Eq. (1) gives

$$A_{aq}^- + CB_{org} = AB_{org} + C_{aq}^- \tag{8}$$

This is very similar to ion-exchange chromatography with an equilibrium constant:

$$K_{IE} = \frac{[AB_{org}][C_{aq}^-]}{[CB_{org}][A_{aq}^-]} \tag{9}$$

This gives

$$D_{A^-} = \frac{[AB_{org}]}{[A_{aq}^-]} = K_{IE}\frac{[CB_{org}]}{[C_{aq}^-]} \tag{10}$$

from which it follows that k' is inversely proportional to the concentration of the counterion in the aqueous phase.

The latter situation is usual in the reversed-phase mode, where the hydrophobic ion is adsorbed onto the bonded hydrocarbon of the packing material. Thus, we can distinguish three different techniques:

1. Normal-phase IPC, where the support is coated with an aqueous stationary phase containing the pairing ion and the ion pairs are partitioned between the stationary phase and an organic mobile phase
2. Reversed-phase IPC, where the liquid stationary phase is organic and the pairing ion is introduced in the aqueous mobile phase
3. Reversed-phase IPC, using a chemically bonded stationary phase and a hydrophobic pairing ion in the aqueous mobile phase

The use of bonded-phase partition systems is generally preferred over mechanically held stationary phases; this gives advantage to technique 3.

Normal-Phase Ion-Pair Chromatography

The support is loaded with the aqueous stationary phase containing the pairing ion by one of the following three methods:

1. The stationary phase or a concentrated solution of the stationary phase in acetone is pumped through the packed column bed. The excess is then

removed by passing eluent or hexane, followed by eluent saturated with stationary phase, until equilibrium is reached. Equilibrium is normally achieved when stable k' values are obtained for a series of representative solutes. This usually requires passage of several hundred milliliters of eluent.
2. The stationary phase can be loaded onto the column in several large plugs (0.1–1.0 mL) using a stopped-flow technique and equilibrium is achieved in the same way as previously.
3. The eluent, which has been preequilibrated with the stationary phase, is pumped through the column until stable k' values are obtained. The stationary phase is adsorbed onto the support surface, but at equilibrium, the pores of the support are not as completely filled as they are in the columns obtained by the first two methods. The equilibration can be a very time-consuming procedure, but the columns thus obtained are stable and reproducible.

Because the columns are in an equilibrium situation, it is obvious that gradient elution is not possible. In the normal-phase situation, the k' value of a solute is inversely proportional to the pairing ion concentration. Because the pairing ion is in the stationary phase, this concentration is not readily changed, and for this reason, retention is normally controlled by modification of the eluent. Hydrocarbon or chlorinated hydrocarbon solvents are usually employed with a small percentage of alcohol as a modifier. Varying the concentration or nature of the alcohol can produce the required changes in retention or selectivity.

Very high efficiencies have not usually been achieved with normal-phase IPC; the advantage of the reversed-phase mode, where the pairing ion concentration can be easily altered, has led to almost total takeover in the ion-pair field. The normal phase has two advantages compared to the reversed-phase mode:

1. The use of ultraviolet (UV) absorbing or fluorescent ions to enhance or enable the detection of nonabsorbing solutes
2. The possibility of varying selectivity by varying the organic-phase composition.

Reversed-Phase Ion-Pair Chromatography

Most often, pentanol or butyronitrile is used as the stationary phase loaded onto a hydrophobic support such as silanized silica.

The equilibration time depends on the hydrophobicity of the support (the coating of pentanol on a hydrocarbon-bonded silica takes no longer than 2 h at a flow rate of approximately 1 mL/min).

The retention of analytes can be regulated by varying the following factors:

1. The capacity factor increases with the hydrophobicity of the pairing ion. For hydrophilic solutes, hydrophobic pairing ions are chosen, and vice versa.
2. The capacity factor increases linearly with pairing ion concentration. Alternatively, gradient elution can be performed by decreasing the pairing ion concentration.
3. The choice of the organic phase affects the selectivity of the system.

Reversed-Phase Ion-Pair Chromatography Using a Chemically Bonded Stationary Phase

This is, by far, the most commonly used form of reversed-phase IPC. This technique has been also called *soap chromatography* although, in soap chromatography, the use of detergents as counterions is introduced.

Here, the columns (with C_8 or C_{18} packings) are prepared by equilibrating the stationary bonded phase with the mobile phase containing the pairing ion. The ion-pair reagent is attracted to the stationary phase because of its hydrophobic alkyl group and the charge carried by the reagent thereby attaches to the stationary phase.

The surface of a C_8 or C_{18} column packing is shown in Fig. 1 as a rectangle covered by sorbed molecules of

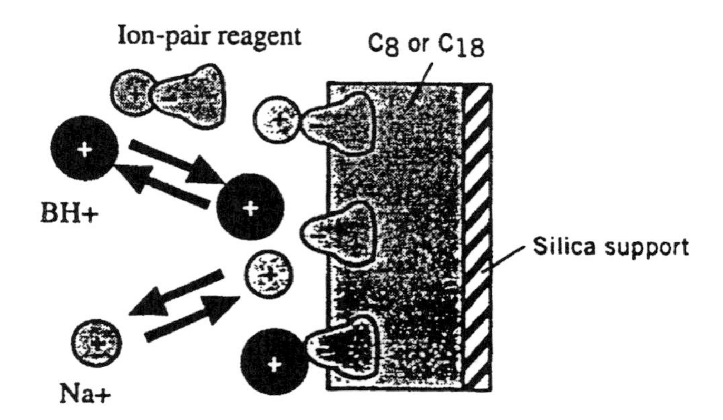

Fig. 1 Pictorial representation of IPC retention of a protonated base (BH^+); Na^+ is the mobile-phase cation; the IPC reagent is hexane sulfonate. (Reprinted from L. R. Snyder, J. J. Kirkland, and J. L. Glajch, *Practical HPLC Method Development*, 2nd ed., 1997, by permission of John Wiley & Sons, Inc.)

a negative ion-pair reagent (e.g., hexane sulfonate, $C_6\text{-}SO_3^-$). The negative charge on the stationary phase is balanced by the positive ions (Na^+) from the reagent and/or buffer. A positively charged sample ion (protonated base BH^+) can exchange with a Na^+ ion as shown (arrows), resulting in the retention of the sample ion by an ion-exchange process.

For each ion-pair reagent, the column uptake increases for a higher reagent concentration in the mobile phase, but then levels off as the column becomes saturated with the reagent. The more hydrophobic reagents are retained more strongly and saturate the column at a lower mobile-phase reagent concentration ($10^{-5}M$), but equilibration may take several hours. Less hydrophobic ion-pair reagents are added at a slightly higher concentration ($10^{-4}M$–$10^{-3}M$) and equilibrium is reached much faster (1–2 h at a flow rate of 1mL/min). This is shown in Fig. 2a, where the concentration of reagent in the stationary phase $(P^-)_S$ is plotted versus the concentration of reagent in the mobile phase $(P^-)_m$ for two reagents of different hydrophobicity.

The change in sample retention, as the ion-pair reagent concentration increases, is shown in Fig. 2b for a hydrophilic sample compound BH^+. Once the column becomes saturated with the reagent, the sample retention levels off. Because IPC retention involves an ion-exchange process, further increases in reagent concentration lead to an increase in the counterion concentration (Na^+), which competes with the retention of the sample ion on the column.

In practice, when very hydrophobic pairing ions are used, the columns are irreversibly altered, because the ions can never be completely removed. Once equilibrium is reached, the columns are stable and can be used for several months. The columns should be stored in the mobile phase because of the lengthy equilibration times. Only if the column is not used for an extended period of time should one consider storing the column in an organic solvent.

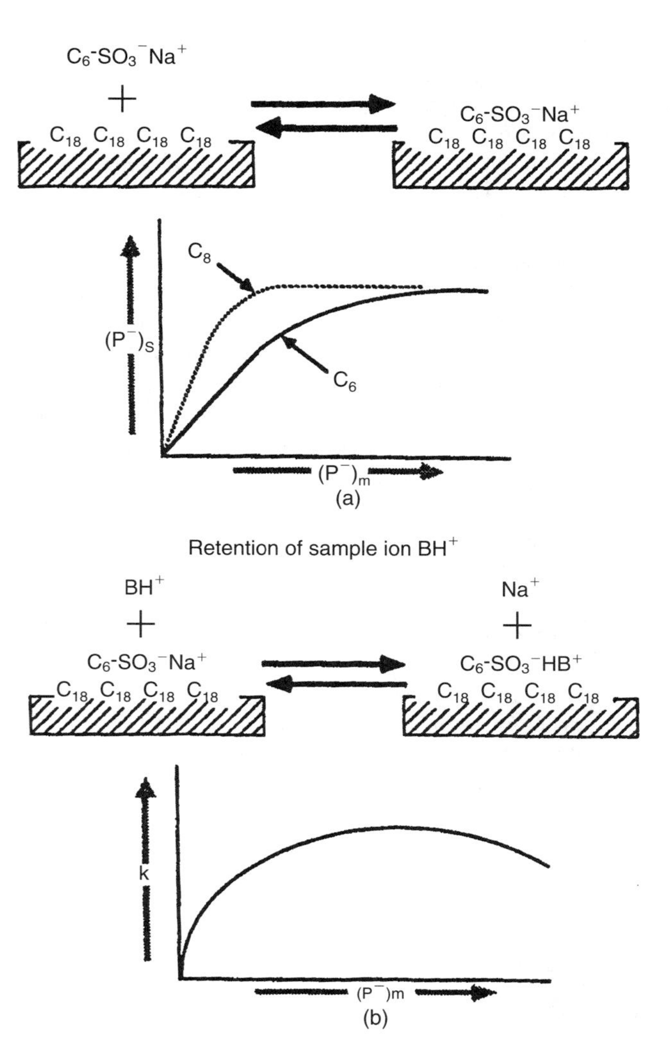

Fig. 2 Effect of ion-pair reagent concentration on separation. (a) Sorption of the ion-pair reagent as a function of concentration for reagents of different hydrophobicity (C_6- and C_8-sulfonates); (b) retention as a function of reagent concentration. (Reprinted from L. R. Snyder, J. J. Kirkland, and J. L. Glajch, *Practical HPLC Method Development*, 2nd ed., 1997, by permission of John Wiley & Sons.)

Design of an Ion-Pair Separation

Unless there is a specific reason to choose a normal-phase system, IPC should be carried on in the reversed-phase mode, using chemically bonded stationary phases.

The best counterion and pH depend on the kind of sample to be separated. Most ion-pair reagents used today are either alkyl sulfonates or tetraalkyl ammonium salts, either of which allow UV detection above 210 nm. The IPC aqueous phase must be adequately buffered with respect to both pH and concentration of the counterion.

Inadequate buffering of the aqueous phase is a source of band tailing in IPC. Conventional buffers, such as citrate and phosphate, have been used and, in some cases, the counterion itself is an adequate buffer. For separations at low pH, $0.1M$–$0.2M$ solutions of a strong acid provide adequate buffering. Inadequate buffering of the aqueous phase is a source of band tailing in IPC.

In reversed-phase IPC, maximum k' values are obtained at intermediate values of pH, where the sample compounds are completely ionized and ion-pair formation is at a maximum. As the pH of the mobile phase is lowered, sample anions A^- begin to form the un-ionized acids HA, leading to a smaller number of sample ion pairs in the stationary phase. Acids are usually separated at a pH of 7–9, whereas bases are separated at a pH of 1–6.

In reversed-phase systems, the solvent strength is readily varied by changing the counterion or its concentration. When all sample ions are fully ionized, a change in solvent strength via a change in counterion concentration leads to minimal changes in separation selectivity. The concentration of the counterion is usually $0.005M$–$0.05M$, except for perchlorate ($0.5M$–$1M$) or the detergents used in soap chromatography (e.g., 1 wt% of counterion). Buffer concentrations are similar to those used in ion-exchange chromatography ($0.001M$–$0.5M$).

An increase in the alkyl chain length of the counterion increases retention in reversed-phase IPC by up to 2.5 times per added $-CH_2-$ group in the counterion.

Apart from an increase in the counterion concentration, an increase in ionic strength of the aqueous phase generally reduces the formation of ion pairs, as a result of the competition of secondary ions in forming ion pairs with the counterion. One study showed a twofold to threefold change in k' for each doubling of ionic strength.

For reproducible separations by IPC, it is important to thermostat the column. Temperature effects in IPC are more important than in some other liquid chromatography methods.

Suggested Further Reading

Bidlingmeyer, B. A., *Practical HPLC Methodology and Applications*, John Wiley & Sons, New York, 1992, pp. 157–165.

Gilbert, M. T., *High Performance Liquid Chromatography*. IOP Publishing, Wright, Bristol, U.K., 1987, pp. 227–253.

Snyder, L. R. and J. J. Kirkland, *Introduction to Modern Liquid Chromatography*, 2nd ed. John Wiley & Sons, New York, 1979, pp. 454–482.

Snyder, L. R., J. J. Kirkland, and J. L. Glajch, *Practical HPLC Method Development*, 2nd ed. John Wiley & Sons, New York, 1997, pp. 317–341.

Su, S. C., A. V. Hartkopf, and B. L. Karger, *J. Chromatogr. 199*: 523 (1976).

Ioannis N. Papadoyannis
Anastasia Zotou

K

Katharometer Detector for Gas Chromatography

The katharometer detector [sometimes spelled "catherometer" and often referred to as the *thermal conductivity detector* (TCD) or the *hot-wire detector* (HWD)] is the oldest commercially available gas chromatographic (GC) detector still in common use. Compared with other GC detectors, it is a relatively insensitive detector and has survived largely as a result of its almost universal response. In particular, it is sensitive to the permanent gases to which few other detectors have a significant response. Despite its relatively low sensitivity, the frequent need for permanent gas analysis in many industries probably accounts for it still being the fourth most commonly used GC detector. It is simple in design and requires minimal electronic support and, as a consequence, is also relatively inexpensive compared with other detectors.

In the late 1940s and early 1950s, the katharometer was developed for measuring the amount of carbon dioxide in flue gases. However, with the advent of GC, its use as a detector was investigated by Ray [1]. It was soon established as a very effective GC detector and was found to be simpler to fabricate than the gas density bridge, but had about the same sensitivity and linearity. For a while, it was the only detector that was commercially available. At the time, its mode of action was the subject of some controversy, as it was not clear whether it responded to changes in the *thermal conductivity* or the *specific heat* of the column eluent. The response of the detector was examined in detail by Mellor [2] and Harvey and Morgan [3] in 1956 and it would appear that no such detailed studies have been carried out since that time. It was concluded that the katharometer responded to both changes in thermal conductivity *and* to changes in the specific heat of its surroundings. In any particular system, depending on the operating conditions employed, one or the other property may dominate in controlling the response of the detector. The relationship, however, is not simple and it was not found possible to accurately predict the response of the detector from a knowledge of the specific heat and thermal conductivity of the gases or vapors involved.

The basic design of a katharometer is as follows. A filament carrying a current is situated in the column eluent. Under equilibrium conditions, the heat generated in the filament will equal the heat lost by conduction, convection, and radiation and the filament will assume a constant temperature. The filament is constructed from a metal, such as platinum, that has a high temperature coefficient of resistance, and at the equilibrium temperature, the resistance of the filament and, thus, the potential across it will be constant. The heat lost from the filament will depend on the thermal conductivity of the gas, its specific heat, and the thermal emissivity of the filament surface. Both the thermal conductivity and the specific heat of the gas will change in the presence of a different gas or solute vapor. As a result, the temperature of the filament will change, causing a change in potential across the filament. This potential change is amplified and either fed to a suitable recorder or regularly sampled by an appropriate data acquisition system.

As the device responds to the heat lost from the filament, the katharometer detector is extremely *flow* and *pressure* sensitive. Consequently, all katharometer detectors must be carefully thermostatted and must be fitted with reference cells to help compensate for changes in pressure or flow rate. There are two basic katharometer

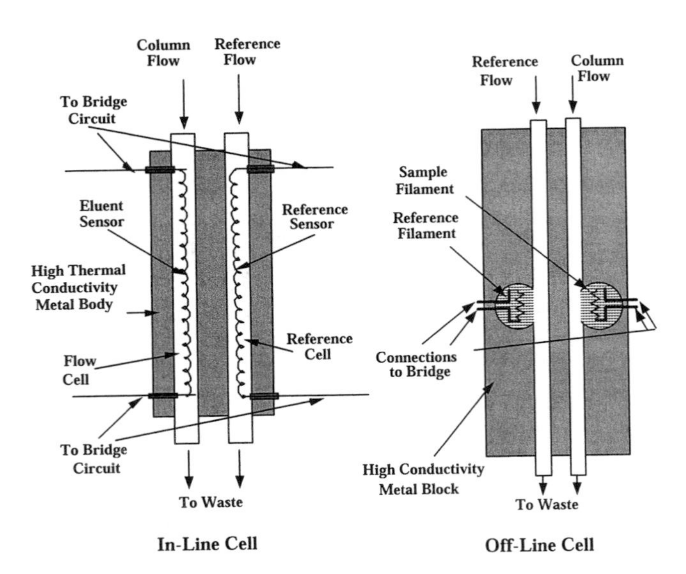

Fig. 1 Katharometer cells.

designs: the "in-line" cell, where the column eluent actually passes directly over the filament, and the "off-line" cell, where the filaments are situated away from the main carrier gas stream and the gases or vapors only reach the sensing element by diffusion. Due to the high diffusivity of vapors in gases, the diffusion process can be considered as almost instantaneous. Diagrams of the two katharometer designs are shown in Fig. 1.

The sensitivity of the katharometer is only about 10^{-6} g/mL (probably the least sensitive of all GC detectors) and has a linear dynamic range of about 500 (the response index lying between 0.98 and 1.02). It is, however, a general detector and will sense all permanent gases and vapors other than the gas that is used as the carrier gas. Its universal response is one reason for its survival as a GC detector, despite its very limited sensitivity. Although the least glamorous, this detector can be used in most GC analyses that utilize packed columns and where there is no limitation to sample availability. Although small-volume katharometers have been designed for use with capillary columns, the katharometer is rarely used with such columns, again due to its relatively low sensitivity. The device is simple, reliable, rugged, and, as already stated, comparatively inexpensive. As a consequence, it has found use in less than ideal environments where GC is employed for process monitoring and process control. If constructed of platinum and Teflon, the katharometer is one of the few detectors that can be used for detecting very corrosive materials such as the halogens, uranium hexafluoride, volatile inorganic acids, and so forth.

References

1. N. H. Ray, *J. Appl. Chem. 4*: 21 (1954).
2. N. Mellor, in *Vapor Phase Chromatography* (D. H. Desty and C. L. A. Harbourn, eds.), Butterworths, London, 1957, p. 63.
3. D. Harvey and G. O. Morgan, in *Vapor Phase Chromatography* (D. H. Desty and C. L. A. Harbourn, eds.), Butterworths, London, 1957, p. 74.

Suggested Further Reading

Scott, R. P. W., *Chromatographic Detectors*, Marcel Dekker, Inc., New York, 1996.

Scott, R. P. W., *Introduction to Gas Chromatography*, Marcel Dekker, Inc., New York, 1998.

R. P. W. Scott

Kovats Retention Index System

In "classical" chromatographic methods (excluding hyphenated techniques), analytical signals are two dimensional. Every chromatographic peak may be characterized by two parameters: area, which is proportional to the quantity of substance being eluted from column, and position on the chromatogram (retention time, t_R), which reflects the interaction between the sorbate (analyte) and sorbent (stationary phase). This interaction is the sole source of information about the chemical nature and structure of the analytes. However, the "raw" retention times by themselves are not useful for any chemical interpretation, owing to their dependence on numerical conditions of analysis. In gas chromatography, for example, these conditions include oven temperature, type of stationary phase, its content on the support of packed columns or film thickness in the capillary columns, length of the column, carrier gas flow, and the pressure gradient between the inlet and outlet of the chromatographic system.

The characterization of analytes in chromatography by net retention times may be compared to the measurement of temperature by thermometers with different arbitrary scales, or even without any scales at all. Thus, the standardization of chromatographic retention parameters is a problem of extremely high importance.

One possible solution is based on the complete interlaboratory standardization of all of the above-mentioned experimental conditions. The realization of the so-called RTL concept (retention time locking) became possible only during the middle of the 1990s. The second most widely used solution requires the recalculation of the data being measured at different conditions into the interlaboratory comparable scale. The mathematical method suitable for this recalculation — linear interpolation — has been known long before the appearance of chromatography. Given pairs of function (y) and argument (x) values $[y_1(x_1), y_2(x_2), \ldots, y_n(x_n)]$ that are connected by a known or unknown functional dependence $y = f(x)$, we can estimate the unknown values y_i, which are located between the known values $y_k < y_i < y_{k+1}$ from the value x_i ($x_k < x_i < x_{k+1}$) by the following relationship:

$$y_i = y_k + \frac{(y_{k+1} - y_k)(x_i - x_k)}{x_{k+1} - x_k} \tag{1}$$

In the case of a nonlinear or unknown dependence $y = f(x)$, the application of simple linear interpolation leads to some uncertainty in the results. However, if the equation for nonlinear dependence is known, a transformation of Eq. (1) into Eq. (2) can be applied to remove the nonlinearity. This approach gives the results of the same precision as that of direct calculation with the equation $y = f(x)$:

$$y_i = y_k + (y_{k+1} - y_k)\left(\frac{f(x_1) - f(x_k)}{f(x_{k+1}) - f(x_k)}\right) \tag{2}$$

The application of this concept in chromatography requires the introduction of some extra components with previously known (postulated) retention indices [$RI = f(t_R)$] into the samples being analyzed. Their peaks form a "mobile" coordinate system for the recalculation of t_R data of the target analytes. Hence, the establishment of any retention index system needs the following:

1. The choice of set of reference compounds (most often they are members of the same homologous series which differ by a homologous difference CH_2)
2. The attribution of standard (conventional) RI values for these compounds
3. The choice of a formula for the calculation of RI values for all other analytes

The first RI system was proposed by Kovats in 1958 [1] for isothermal conditions of gas chromatographic (GC) analysis. The easily accepted n-alkanes n-C_nH_{2n+2} with postulated RI values of $100n_C$ [e.g., methane (CH_4) — 100; n-nonane (C_9H_{20}) — 900, n-hentriacontane ($C_{31}H_{64}$) — 3100, etc.] was recommended as reference compounds. Insofar as, at the isothermal conditions of GC analysis, the linear dependence of logarithms of corrected retention times $t'_R = t_R - t_0$ (t_0 is the dead time of chromatographic system) versus the number of carbon atoms in the molecule of the homolog is observed, $\log t'_R = an_C + b$, and (by definition) $100n_C = RI$; this means the existence of following linear dependence:

$$RI = a \log t'_R + b \tag{3}$$

If we consider the last relationship as the function $y = f(x)$ in Eq. (2), we come to the final equation of the Kovats Retention Index System:

$$RI_x = RI_k + (RI_{k+1} - RI_k)\left(\frac{\log(t'_{R,x}) - \log(t'_{R,k})}{\log(t'_{R,k+1}) - \log(t'_{R,k})}\right) \tag{4}$$

where $t'_{R,x} < t'_{R,k} < t'_{R,k+1}$ are the corrected retention times of reference n-alkanes with number of carbon atoms k and $k + 1$ being eluted immediately before ($t'_{R,k}$) and after ($t'_{R,k+1}$) the target compound ($t'_{R,x}$).

As far as the basis of formula (4) is linear dependence [Eq. (3)], it is possible to use retention times of reference n-alkanes which differ by not one but by a greater number of carbon atoms {i.e., RI_{k+m} and RI_k with $t_{R,k+m}$ and $t_{R,k}$ (in the original publication of Kovats [1] just the difference $m = 2$ was used)}. When $m = 1$, Eq. (3) may be simplified to the various visually different but the same relationship formulas; for example,

$$RI_x = 100\left(k + \frac{\log(t'_{R,x}/t'_{R,k})}{\log(t'_{R,k+1}/t'_{R,k})}\right) \tag{5}$$

where k is the number of carbon atoms in the n-alkane that elutes before the compound undergoes characterization.

The proposed form of data presentation became highly popular and opportune in gas chromatography. Up to the present, some thousand references to the Kovat's work [1] have been known. The RI values are proportional to the free energies of sorption; this is their thermodynamic interpretation. Further development of the RI concept was aimed at its application to nonisothermal conditions of gas chromatographic (GC) analysis. For linear temperature programming regimes (which are characterized by two variables: initial temperature, T_0, and rate of its increase, r, deg $\times$ min^{-1}), the linear relationship (3) does not hold. In some partial cases, other linear dependence seems more precise for the retention time approximation:

$$RI \approx at_R + b \tag{6}$$

This is a reason to change the formula for the RI calculation (the set of reference compounds and their attributed RI values remain the same). This version of RIs developed especially for a linear temperature programming regime (linear retention indices) have been proposed by Van den Dool and Kratz in 1963 [2]:

$$\mathrm{RI}_x = \mathrm{RI}_k + (\mathrm{RI}_{k+1} - \mathrm{RI}_k)\frac{t_{R,x} - t_{R,k}}{t_{R,k+1} - t_{R,k}} \tag{7}$$

or, after the same simplifications as those which were used for Eq. (4),

$$\mathrm{RI}_x = 100\left(k + \frac{t_{R,x} - t_{R,k}}{t_{R,k+1} - t_{R,k}}\right) \tag{8}$$

Owing to the approximate character of dependence (6), Eqs. (7) and (8) give less comparable results for the same compounds in different temperature programming regimes. As a consequence, the replacement of the $k, k+1$ pair of reference n-alkanes on the $k, k+m$ $(m > 1)$ also increases the errors to an unpredictable extent and is usually not recommended.

Complex dependencies $\mathrm{RI} = f(t_R)$ in nonisothermal conditions of GC analysis may be described by polynomials of different degrees (up to 13 have been tested) or splines (cubic splines seem most convenient). However, the calculation of coefficients of an N-degree polynomial needs t_R data for at least $N + 1$ reference compounds instead of only two t_R values as in "classical" RI systems. In connection with this fact, it is interesting to mention the combined lin-log RI system, which was proposed in 1984 (3,4). If both the dependencies (3) and (6) are nonlinear at temperature programming, every local window of retention times for reference compounds may be precisely approximated by linear and logarithmic addends in variable proportion:

$$\mathrm{RI} = a(t_R + q \log t'_R) + c \tag{9}$$

and

$$\mathrm{RI}_x = \mathrm{RI}_k + (\mathrm{RI}_{k+m} - \mathrm{RI}_k)\left(\frac{f(t'_{R,x}) - f(t'_{R,k})}{f(t'_{R,k+m}) - f(t'_{R,k})}\right) \tag{10}$$

where $f(t'_R) = t_R + q \log t'_R$. The variable parameter q may be calculated in different manners, but in the simplest case, it needs t_R data only for three successive reference compounds with retention times $t_{R,k-1}$, t_R, and $t_{R,k+1}$:

$$q = \frac{2t_{R,k} - t_{R,k-1} - t_{R,k+1}}{\log t'_{R,k-1} + \log t'_{R,k+1} - 2\log t'_{R,k}} \tag{11}$$

The most convenient advantage of a lin-log RI system is the possibility of its application in any temperature regime of GC analysis without special choice of formulas for calculations. Under isothermal conditions, the logarithmic contribution to the total dependence $\mathrm{RI} = f(t_R)$ exceeds the linear one by many times, which automatically reflects on the value of q $(|q| \to \infty)$. Only the lin-log RI system provides the most comparable results for different temperature conditions of analysis.

The maximal influence on RI values is the nature of stationary phase in the chromatographic column. Use of these parameters as the constants of chemical compounds (similar to other known physicochemical constants like boiling points, T_b, refractive index, n_D^{20}, density, d_4^{20}, etc.) requires the choice of standard phases for their determination. In accordance with the criteria of the most often used application in practice, two types of phases may be classified as standards:

Nonpolar polydimethyl siloxanes: $[-\mathrm{Si(CH_3)_2}-\mathrm{O}-]_n$; maximal temperature of application ≈ 300°C
Polar polyethylene glycols: $[-\mathrm{CH_2CH_2}-\mathrm{O}-]_n$; maximal temperature of aplication ≈ 225°C

Each of these groups of phases includes the numerous items of various trade names, different average molecular weights, viscosity, thermal stability, and so forth, but all of them are very close to each other by polarity. Up to the middle of the 1970s, the preferred nonpolar phase was squalane (isoprenoid alkane $C_{30}H_{62}$). This phase is no longer used because of its low thermal stability (only about 110°C). However, this obsolete phase maintains its importance as a nonpolar standard in gas chromatography. Other phases may be characterized by differences of RI values of specially selected test compounds between the phase under consideration and squalane, for example:

	Test compounds				
	Benzene	1-Butanol	2-Pentanone	Nitro propane	Pyridine
$\Delta\mathrm{RI} = \mathrm{RI}_{\text{polydimethyl siloxanes}} - \mathrm{RI}_{\text{squalane}}$	16 ± 1	53 ± 2	44 ± 1	65 ± 2	42 ± 1

The comparison of RI values of the same compounds measured with the same stationary phase but at different conditions of analysis indicates some deviations. From some objective reasons for these deviations, the temperature dependence of retention indices seems like the most important contribution (coefficients β are the measure of this dependence, typically $\beta > 0$):

$$\beta = \frac{\mathrm{dRI}}{\mathrm{dT}} \approx \frac{\mathrm{RI}(T_2) - \mathrm{RI}(T_1)}{T_2 - T_1} \tag{12}$$

Among the multitude of organic compounds, there are objects with $\beta \approx 0$ (all types of noncyclic compounds most topologically relevant to n-alkanes). The increase in the number of cycles in the molecules leads to the increase of β up to 0.3–0.5 (cycloalkanes, arenes, etc.), 0.5–0.8 (naphthalenes, biphenyl, etc.), and 1.0 and more retention index units per degree (i.u. $\times$ deg^{-1}) for tricyclic and polycyclic structures. Hence, it is not surprising that RI data for isoalkanes, ethers, esters, and so forth being measured at different conditions are in good agreement with each other (standard deviations of randomized interlaboratory values are not more than 1–3 i.u.). The same statistical characteristic for substituted benzenes is about 8 i.u., and for naphthalenes it may exceed 10–15 i.u.

The choice of n-alkanes as a reference set of compounds for the determination of RI values is accepted up to the present. Meanwhile, numerous other homologous series have been recommended for different specific applications. For example, GC analysis with electron-capture detectors (selective to halogenated compounds) requires other reference compounds of similar chemical origin. These series are alkyl trichloroacetates $CCl_3CO_2C_nH_{2n+1}$, alkyl methyl phosphonofluoridates $CH_3P(O)(F)OC_nH_{2n+1}$ (so-called P series), O-alkyl bis(trifluoromethyl)phosphinothionates $(CF_3)_2P(S)OC_nH_{2n+1}$ (A series), alkyl bis(trifluoromethyl)thiophosphines $(CF_3)_2P(S)C_nH_{2n+1}$ (M series), and others. The last two series seem most universal for different types of GC detectors, owing to the simultaneous presence of various elements in the molecule (Hal, P, S, CH). A special set of reference objects have been proposed by Lee [5] for the analysis of polycyclic aromatic compounds to eliminate the strong temperature dependence of their RIs in the n-alkane scale. This system is based on benzene (postulated RI value 100 = 100 $\times$ number of cycles), naphthalene (200), phenanthrene (300), chrysene (400), picene (500), benzo[b]picene (600), and dinaphtho[2,1-a:2,1-h]anthracene (700) (i.e., condensed aromatic hydrocarbons). Of course, RI values being measured with different reference series are not directly comparable to each other but, if necessary, may be recalculated.

Since the 1980s, some applications of the RI system have been reported for reversed-phase high-performance liquid chromatography (RP–HPLC). The principal requirement for the reference compounds in this method, with UV detection, is the presence of chromophores in the molecules. The most widely accepted RI system in HPLC is based on homologous alkyl phenyl ketones $PhCOC_nH_{2n+1}$ (so-called Smith's system of retention indices; by analogy with GC, the RI values, attributed for reference compounds, are $100n_C$). Other RI scales imply the use of homologous monoalkylbenzenes PhC_nH_{2n+1}, 1-nitroalkanes $C_nH_{2n+1}NO_2$, glycerol 1-(4-acylphenyl) ethers 4-$C_nH_{2n+1}CO–C_6H_4–OCH_2–CH(OH)–CH_2OH$, and so forth. The last set of substances is convenient for the determination of RIs of most hydrophilic organic compounds; this set is eluted before simplest reference components of other series. In the isocratic regimes of HPLC separation, which are analogous to isothermal conditions in GC, linear dependence (3) is correct and RIs must be calculated with formulas (4) and (5). With gradient elution, by analogy with temperature programming, formulas (7) and (8), which are based on dependence (6), are preferable. Of course, lin-log RI system may be used in any regimes of RP–HPLC, as well as in gas chromatography.

It is interesting to note that by analogy with chromatographic retention parameters, the values of some other properties of organic compounds may be presented in the linear interpolated form relative to the set of reference compounds. These equivalent to indices forms are known for boiling points [6], molecular weights [7], and molar refractions, $MR_D = (MW/d)(n^2 - 1)/(n^2 + 2)$, where MW is the molecular weight, n_D^{20} is the refractive index, and d_4^{20} is the density [8]. For example,

$$I(T_b) = 100\left(k + \frac{\log(T_{b,x}/T_{b,k})}{\log(T_{b,k+1}/T_{b,k})} \right) \tag{13}$$

where $T_{b,k} < T_{b,x} < T_{b,k+1}$ are the boiling points of n-alkanes with k and $k + 1$ carbon atoms in the molecule and the target compound. For nonpolar compounds, the values of $I(T_b)$ are close to the experimental RI data being measured with nonpolar stationary phases.

The most significant feature of retention indices, as the constants of organic compounds, seems to be the possibility of their precalculation both from other physicochemical parameters and by different additive schemes [9]; this is impossible for values of net retention times themselves. The methods of RI precalculation unite as complex algorithms, as very simple but useful rules. For instance, it is interesting to note that even in the first publication of Kovats [1], the rule for precalculation of RIs for compounds of general type A–B by arithmetical averaging of data for compounds A–A and B–B were recommended [i.e., RI(A–B) = [RI(A–A) + RI(B–B)]/2. It is very surprising that any attempts to use the same rule in mathematically transformed form, namely RI(B–B) = 2RI(A–B) − RI(A–A) were unknown during the past 40 years. Nevertheless, if B is a more complex structural fragment of molecule than A, this is simplest way to precalculate RIs of high-molecular-weight compounds from data for more simple precursors. This general statement may be illustrated by the following example: The estimation of unknown RI value of 1,1,1,3,3,3-hexachloropropane $CCl_3—CH_2—CCl_3$ needs the data for 1,1,1-trichloroethane and propane, so far as $2 \times CCl_3—CH_2—CH_3 - C_3H_8 = CCl_3—CH_2—CCl_3$, namely $2(736 \pm 3) - 300 = 1172 \pm 4$.

One of the most practical contemporary problems of GC RI application seems to be the formation of available and representative databases by analogy with well-organized databases in mass spectrometry.

References

1. E. Kovats, *Helv. Chim. Acta 41*: 1915 (1958).
2. H. Van den Dool and P. Kratz, *J. Chromatogr. 11*: 463 (1963).
3. I. G. Zenkevich, *Zh. Anal. Khim.* (*Russ.*) *42*: 1297 (1984).
4. I. G. Zenkevich and B. V. Ioffe, *J. Chromatogr. 439*: 185 (1988).
5. M. L. Lee, M. V. Novotny, and K. D. Bartle, *Analytical Chemistry of Polynuclear Aromatic Hydrocarbons*, Academic Press. New York, 1981.
6. P. G. Pobinson and A. L. Odell, *J. Chromatogr. 57*: 1 (1971).
7. M. B. Evans, J. K. Haken, and T. Toth, *J. Chromatogr. 351*: 155 (1986).
8. I. G. Zenkevich and L. M. Kuznetsova, *Collect. Czech Chem. Commun. 56*: 2042 (1991).
9. R. Kaliszan, *Quantitative Structure—Chromatographic Retention Relationships*, John Wiley & Sons, New York, 1987.

Igor G. Zenkevich

L

Large-Volume Injection for Gas Chromatography

Trace or ultratrace analyses for environmental and biological samples require sensitive methods, including sample preparation and detection techniques, to be used. The enhancement of the detectability of analytical procedure is often brought about by some form of preconcentration, such as Soxhlet extraction, liquid–liquid extraction, and solid–phase extraction. Capillary gas chromatography (GC) coupled with different detectors is one of the most frequently employed techniques for trace analysis of micropollutants. However, because of the limited sample capacity of the capillary column, only a small portion (1–5 μL) of the final sample extract is introduced into the gas chromatographic system. This means that after careful workup, which often comprises analyte isolation by extraction and changing to a GC-compatible solvent, a maximum few percent could be injected and reach the detector [1]. The concentration detection limit of the method can be improved by increasing the concentration factor of the extraction and cleanup procedure. This is generally achieved by using a larger amount of sample or/and by reducing the volume of final extract through evaporation. Both methods have limitations for practical application. Extraction of a large amount of sample is very time consuming and requires a large volume of toxic organic solvents. Sampling a large amount of sample is sometimes difficult or even impossible. Although solvent evaporation is frequently used, it is a rather critical step in which the more volatile compounds may be lost from the sample because of the coevaporation with the solvent [1,2]. Recently, there has been increased interest in the introduction of large sample volumes in capillary GC [1–6]. Several hundred microliters of sample can be introduced by using these techniques. Among the several techniques for the introduction of a large volume of sample, on-column injection and programmed-temperature vaporization (PTV) injection are best developed.

On-Column Injection

The on-column injection techniques, in which the solvent is generally vaporized in a few meters of uncoated deactivated capillary (retention gap) and vented via an early vapor exit valve, are of good accuracy and reproducibility [3–5]. A schematic diagram for on-column injection is shown in Fig. 1. The on-column large-volume injection system generally consists of a regular gas chromatograph equipped with an on-column injector, a retention gap, a retaining precolumn, an analytical column, and a heated early solvent vapor exit.

The introduction of a large volume of sample can be performed using a technique called partially concurrent solvent evaporation (PCSE). With PCSE, some 90% of the solvent injected is evaporated during the injection through the early solvent vapor exit [1]. The injection rate of the sample is controlled based on the length of the retention gap and the boiling points of the analytes. If the sample volume exceeds the capacity of the uncoated retention gap to retain liquid (it is true for most large-volume injection applications), the injection speed should be adjusted to result in a sufficiently large proportion of concurrent evaporation to prevent liquid solvent from spreading into the analytical column [2]. Most of the vapor escapes through the open solvent vapor exit. After large-

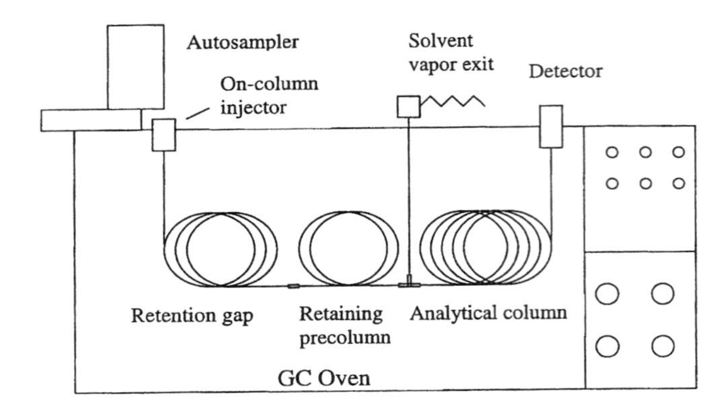

Fig. 1 Schematic diagram of GC with on-column large-volume injection.

volume on-column injection, the analytes are spread out over several meters of the uncoated retention gap. Solvent evaporation continues, removing solvent from the rear to the front of the sample film. Relatively volatile analytes evaporate and are reconcentrated by the liquid ahead. This process is described as the solvent-trapping effect. Less volatile components do not evaporate with the solvent and remain on the dry retention gap surface. The vapor exit valve is closed shortly before the end of solvent evaporation when the residual liquid still remains the volatile components. The rest of the solvent is discharged through the analytical column. Reconcentrations of the less volatile compounds are carried out by phase-ratio focusing; that is, the difference in migration speed in the retention gap and in the coated column causes the rear end of the zone to catch up with the front end at an appropriately increased oven temperature [1,2]. As soon as the front end of the zone reaches the stationary phase, its migration speed reduces dramatically, while the remaining part, which is still in the retention gap, continues to migrate at a higher speed.

With on-column large-volume injection, the selection of appropriate experimental conditions is complicated [2]. Generally, the following parameters need to be carefully optimized.

1. The sample must be introduced at a rate slightly exceeding the evaporation rate. Slower injection causes the loss of solvent trapping because all solvent evaporate concurrently, whereas a too high introduction rate results in flooding of the solvent into the retaining precolumn and, eventually, the analytical column and the vapor exit.
2. The temperature of the column oven during injection should be set slightly below the pressure-corrected boiling point of the solvent used. At too high temperatures, the solvent starts to boil and causes the backflush of solvent into the injector.
3. The vapor exit valve must be closed at a right time. Early closing of the valve may cause the remaining solvent in the retention gap to exceed the capacity of the analytical column. However, delayed closure will result in the loss of volatile components.

On-column large-volume injection has wide applications in terms of volatility and thermostability of the analytes [4]. If the solvent exit valve is operated carefully, compounds with boiling points only slightly above that of the solvent can be recovered quantitatively. However, the technique is generally not very suitable for dirty samples. Frequent analysis of samples with high contents of matrix compounds can rapidly decrease the column performance [3–6].

Programmed-Temperature Vaporization

Vogt and co-workers in 1979 described an injector that allowed the injection of up to 250 μL into a cold glass insert filled with glass wool [7,8]. This technique has been modified and refined in the recent years [3–6,9,10]. In fact, the large-volume sampling technique using PTV injectors is modified from a conventional split/splitless injector [1]. The main differences between the conventional split/splitless and the PTV injectors are the temperature control of the injector and the solvent venting capacity of the split/splitless valve. In PTV injectors, the injection port should be heated or cooled rapidly, and the split/splitless valve should be large enough to be able to vent the solvent vapor produced during injection. Before a large volume of sample is introduced, the temperature of the injection port is reduced to below the boiling point of the solvent. The sample is injected into the liner of the injector at a controlled rate. Upon introduction, the solvent is selectively eliminated and solvent vapor is vented via the split/splitless valve. The less volatile components are retained in the cold liner. When solvent elimination is finished, the components retained in the liner are transferred to the analytical column by rapid temperature-programmed heating of the injector.

In order in obtain a quantitative analysis for the analytes with a wide range of volatility using the PTV injection technique, the following parameters are of great importance: injection speed, liner packing material, solvent venting temperature, and solvent venting time. The speed of sample introduction should roughly equal the rate of solvent elimination [1]. If the sample is injected at a rate exceeding the evaporation rate, the sample will accumulate in the liner, which eventually will result in overloading of the liner. This will cause flooding of the analytical column and severe losses of both volatile and nonvolatile components via the split exit. On the other hand, a too slow sample introduction rate will cause losses of volatile sample components. The use of liners packed with an absorbent is an efficient means to retain the liquid sample and minimize losses of volatile compounds [1,10]. A number of packing materials, such as glass wool, quartz wool, cup liner, Tenax TA, PTFE wool, have been investigated in terms of their interaction with different types of analytes [10]. An unsuitable choice of the packing material can cause degradation of the analyte in the liner. The solvent venting temperature and solvent venting time must be carefully optimized. The liner temperature is held below the boiling point of the solvent during the solvent venting time. The split/splitless valve must be closed at the right time. Early closing will cause accumulation of the solvent, which may exceed the capacity of the column, whereas delayed closure can result in losses of the volatile sample components.

It has been shown that the PTV injector is a very useful technique for large-volume injection, especially for the analysis of a dirty sample. Because the vaporization of the solvent is carried out at a low temperature, nonvolatile matrix constituents remaining in the liner will not contaminate the GC column. However, the PTV injection technique is less suited when analyzing volatile compounds because only components with volatility significantly below that of the solvent are trapped in the cold liner, unless liners packed with a selective adsorbent is used [4].

References

1. H. G. J. Mol, H. G. M. Janssen, C. A. Cramers, J. J. Vreuls, and U. A. Th. Brinkman, *J. Chromatogr. A 703*: 277 (1995).
2. K. Grab, *J. Chromatogr. A 703*: 265 (1995).
3. H. G. J. Mol, M. Althuizen, H. G. Janssen, C. A. Cramers, and U. A. Th. Brinkman, *J. High Resolut. Chromatogr. 19*: 69 (1996).
4. J. C. Bosboom, H. G. Janssen, H. G. J. Mol, and C. A. Cramers, *J. Chromatogr. A 724*: 384 (1996).
5. H. J. Stan and M. Linkerhagner, *J. Chromatogr. A 727*: 275 (1996).
6. F. Munari, P. A. Colombo, P. Magni, G. Zilioli, S. Trestianu, and K. Grab, *J. Microcol. Separ. 7*: 403 (1995).
7. W. Vogt, K. Jacob, and H. W. Obwexer, *J. Chromatogr. 174*: 437 (1979).
8. W. Vogt, K. Jacob, A. B. Ohnesorge, and H. W. Obwexer, *J. Chromatogr. 186*: 197 (1979).
9. S. Ramalho, T. Hankemeier, M. de Jong, U. A. Th. Brinkman, and R. J. J. Vreuls, *J. Microcol. Separ. 7*: 383 (1995).
10. H. G. J. Mol, P. J. M. Hendriks, H. G. Janssen, C. A. Cramers, and U. A. Th. Brinkman, *J. High Resol. Chromatogr. 18*: 124 (1995).

Yong Cai

Large-Volume Sample Injection in FFF

Field-flow fractionation (FFF) techniques are used for a range of sample types, including polymers, macrobiomolecules, glass beads, silica, and other inorganic particles due to their good separation characteristics (e.g., high-molecular-weight separation selectivity and a large dynamic operating range). These analytes are dissolved or suspended in an appropriate liquid and small aliquots of the suspensions (typically 2–20 μL) are injected into the FFF channel. For these applications, it has usually been possible to adjust the analyte concentration to ensure an acceptable signal-to-noise ratio given detector sensitivity and the dilution in the FFF system. However, there are a number of applications where the solutes are very dilute and any preconcentration may cause perturbation of the size distribution or other changes of sample characteristics. Such applications could include most of the groups of sample types mentioned, and especially in environmental particles and colloids, which are often present in low concentrations and are easily disturbed.

Schure has recently evaluated dilution factors and detection limits for FFF in comparison with liquid chromatography and capillary electrophoresis [1]. It is obvious that samples are subject to much higher dilution in FFF channels (~200–1000 times) compared with most liquid chromatography techniques (~3–25) due to the rather low efficiency and large elution volumes. A narrower and thinner channel than one with more conventional dimensions would improve the dilution factors, but there are practical problems involved in the production and maintenance of such a channel. There is the option of using any of the outlet flow splitting devices which have been developed. These are different approaches having in common removal of some of the overlying liquid at the end of the channel and thus reducing the dilution. The principles builds on horizontal flow splitters, capillaries exiting both above and below the channel, or a second porous frit section in the overlying channel wall letting out liquid. There have been difficulties in accurate production, maintenance, and precision of some of these technical solutions, but there is now one system commercially available using the frit-outlet system in flow FFF (FlFFF) for removal of excess liquid [2], with which at least a 10-fold reduction in dilution is achieved.

The injection volume of the conventional stop-flow sample injection technique is limited typically to 2–20 μL. This procedure is carried out using a sample loop and injecting the sample plug just onto the channel carried with the channel flow before stopping that flow and letting the sample attain its equilibrium distribution relative to the accumulation wall (relaxation). Sedimentation FFF (SdFFF) is the only technique where the field is not applied during the complete stop-flow procedure, but the centrifuge is started as soon as the sample is injected and the channel flow switched away for practical reasons.

Kirkland et al. [3] introduced an injection technique for SdFFF, where the sample is slowly injected with the centrifuge spinning at a high field and thereby retaining the analytes in a narrow band at the entrance of the channel. When the sample is injected, the flow is stopped, the centrifuge speed is decreased to the initial elution speed, and an appropriate relaxation time is allowed. This technique has been further refined and developed by Giddings and co-workers into an on-channel preconcentration step for concentrating dilute particulate samples in SdFFF [4].

Another injection technique, which involves a sample injection port a small distance downstream of the channel inlet, where the sample was injected and relaxed between two focusing flows with a reversed flow from the channel outlet, was first employed for asymmetrical FlFFF [5]. The theoretical basis of the horizontal transport was derived and fundamental practical aspects were discussed. This procedure is now standard for asymmetrical FlFFF, partly due to the difficulties of adapting the stop-flow injection mode and was further optimized by Wahlund and Litzén [6] to include injection of up to 5 mL of sample in a 0.8-mL channel.

This approach has subsequently been used in symmetrical FlFFF [7,8], where the sample is injected in either the forward or backward focusing flow streams (Fig. 1). Lyvén et al. [7] showed a virtually quantitative recovery for preconcentration of 1.7 to over 100 mL of low-molecular-weight polymers and natural water colloids (Fig. 2) with a molecular weight of the same order as the membrane cutoff by optimizing injection flows, membrane, and carrier.

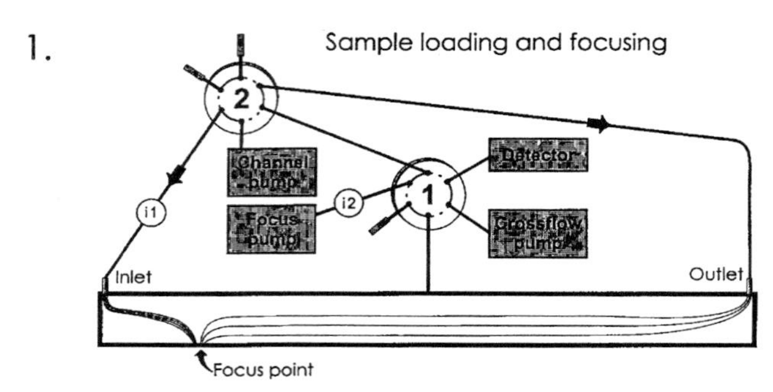

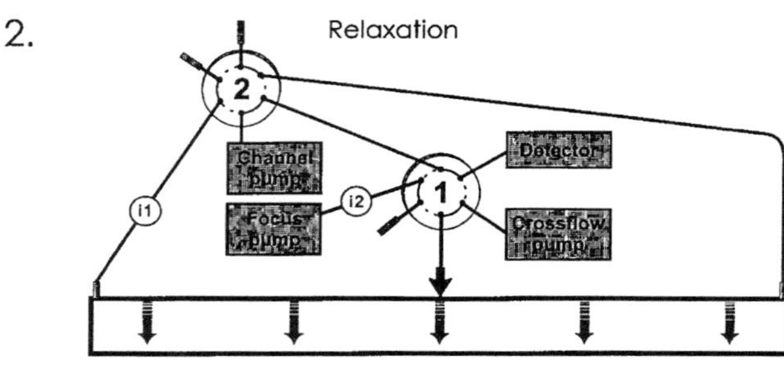

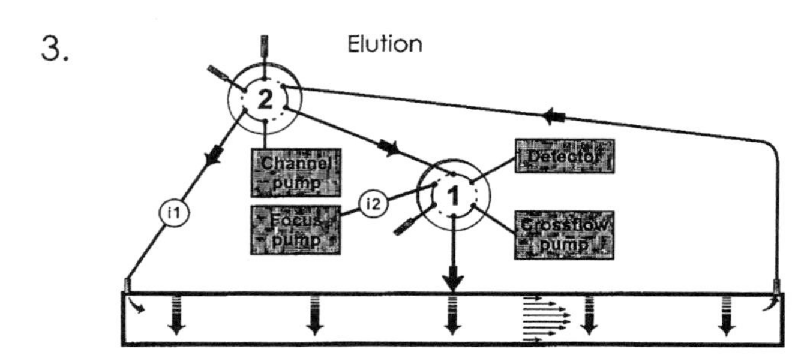

Fig. 1 Instrumental schematics of FlFFF with on-channel preconcentration showing the three different procedure steps. The first involves emptying of the sample loop into either the forward or backward flows and subsequent focusing of the sample material at the focusing point. During the next step (2), the sample is allowed relaxation to the equilibrium position by applying cross-flow only, and then the channel flow is switched on and elution is commenced.

Injection in both the forward and backward focusing flow stream was evaluated. An experimental approach was used to determine the minimum time for sample loading and focusing. In order to ensure sufficient focusing time for all colloids, the focusing time was increased until a stabile retention time was attained for the largest reference standard. This on-channel preconcentration method has later been used in a FlFFF coupling with inductively coupled plasma-mass spectrometry for trace element size distributions for natural water colloids [9]. The colloids and the associated metals would have been impossible to study without the preconcentration procedure due to the

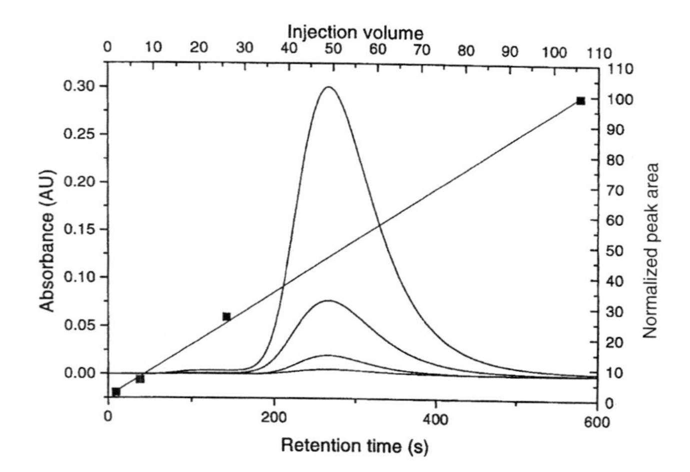

Fig. 2 Results showing preconcentration of 1.7–106 mL of a natural freshwater sample using the procedure in Fig. 1. The solid lines show the ultraviolet detector response, and the squares with the regression line (slope 0.94 and $R^2 = 0.9993$) represent peak area versus injection volume. The sample was first diluted in order not to exceed the overloading point. (From Ref. 7.)

very low concentrations of both colloids and trace metals often found in natural waters. Lee et al. [8] presented a method for preconcentrating 10-mL samples of dilute polystyrene latex beads, river sediment particles, and proteins, which are injected in the channel inlet stream with an opposed focusing flow from the channel outlet. Results of recoveries are shown for different membrane materials. On-channel preconcentration for FlFFF and SdFFF are rather simple to incorporate in a conventional instrument, just with the addition of a second switching valve and a liquid pump. The sample is exposed to the same media as in the separation step and the preconcentration step is carried out just prior to analysis, which is not always the case for external preconcentration techniques such as ultrafiltration or centrifugation.

Precautions must be taken in all FFF analysis when injecting a large amount of material with any injection technique. During the sample relaxation, local concentrations of analytes can become quite high and, at some point, the sample components start affecting each other. Then, conformational changes or intermolecular repulsion can occur, preventing the analytes from attaining a true equilibrium distribution in the channel. This phenomenon is often referred to as overloading and is usually indicated by a later eluting peak with a fronting peak shape.

Overloading in polymer analysis using flow and thermal FFF is thoroughly described by Caldwell et al. [10], where different overloading mechanisms during relaxation and elution are theoretically and experimentally evaluated. Usually, for FFF separations, the detector sensitivity is sufficiently high to detect the separated components at sample concentrations well below the overloading concentration. For further characterization of the particulate material by other detection systems (e.g. mass spectrometry) or subsequent analysis of a collected fractions, it is often necessary to inject large amounts of sample material to exceed the detection limits and then it is essential to be observant of overloading effects. The best way of investigating the elution behavior in order to rule out overloading is to inject various amounts of sample and follow the retention time as well as the shape of the peak. The applied field in stop-flow relaxation or focusing flows in on-channel preconcentration could also have an affect on nonideal behavior; that is, the sample components are compressed during high fields, causing both higher local concentrations and potential interaction with the accumulation wall. When working with sample amounts near the overloading point, varying the field strength acting on the analytes can be used to find conditions where overloading starts and thereby avoid it.

References

1. M. R. Schure, *Anal. Chem. 71*: 1645–1657 (1999).
2. P. Li, M. Hansen, and J. C. Giddings, *J. Microcol. Separ. 10* (1998).
3. J. J. Kirkland, W. W. Yau, and W. A. Doerner, *Anal. Chem. 52*: 1944–1954 (1980).
4. J. C. Giddings, G. Karaiskakis, and K. D. Caldwell, *Separ. Sci. Technol. 16*: 725–744 (1981).
5. K.-G. Wahlund and J. C. Giddings, *Anal. Chem. 59*: 1332–1339 (1987).
6. K.-G. Wahlund and A. Litzén, *J. Chromatogr. 461*: 73–87 (1989).
7. B. Lyvén, M. Hassellöv, C. Haraldsson, and D. R. Turner, *Anal. Chim. Acta 357*: 187–196 (1997).
8. H. Lee, S. K. Ratanathanawongs Williams, and J. C. Giddings, *Anal. Chem. 70*: 2495–2503 (1998).
9. M. Hassellöv, B. Lyvén, C. Haraldsson, and W. Sirinawin, *Anal. Chem.* (in press).
10. K. C. Caldwell, L. B. Steven, Y. Gao, and J. C. Giddings, *J. Appl. Polym. Sci. 36*: 703–719 (1988).

Martin Hassellöv

Lipid Analysis by HPLC

Introduction

Lipids are mixtures of fatty molecules that contain polar and nonpolar groups. Polar lipids include phospholipids and glycolipids, whereas nonpolar lipids include fatty acids and their esters, cholesterol and its esters, essential oils, wax esters, and squalene. Lipids are important ingredients in the production of modern food and play key roles in physiological systems. The role of lipids in certain cardiovascular and dermotological conditions has been established. Understanding the roles of lipids in cellular and physiological systems have been the driving force behind recent developments in lipid analysis. Historically, thin-layer chromatography (TLC) and gas chromatography (GC) have been used for the analysis of lipids. Due to the limitations of these techniques and the advances in high-performance liquid chromatography (HPLC) detection technology, HPLC is gaining popularity for lipid analysis. The high temperatures used in the GC causes degradation of some molecules, whereas many fat molecules are not volatile enough to go through the GC. On the other hand, detection of fats on TLC is somewhat cumbersome.

Today, HPLC is the dominant analytical technique used for the analysis of most classes of compounds. The analyses can be carried out at room temperature and the collection of fractions for reanalysis or further manipulation is straightforward. The main reason for the slow acceptance of the HPLC technique for lipid analysis has been the detection system. Traditionally, HPLC used ultraviolet/visible (UV/vis) detection, which requires the presence of a chromophore in the analyte. Most lipid molecules do not contain chromophores and therefore would not be detected by UV/vis. Modern HPLC detection techniques, such as the use of a mass spectrometer as the detector, derivatization techniques to introduce chromophores, and the availability of pure solvents to reduce interference, have allowed HPLC to compete with and/or complement GC and other traditional methods of lipid analysis. In addition to analytical HPLC, preparative HPLC has been used extensively to collect pure samples of the lipids for the derivatization or synthesis of new compounds.

Modes of Separation of Lipids by HPLC

The mode of separation in the HPLC depends on the selection of the stationary and mobile phases. In HPLC of lipids, normal- and reversed-phase modes are primarily used, with the reverse phase being more common than the normal phase. Separation in the reversed-phase mode is mainly by partition chromatography, whereas separation in the normal phase mode is primarily by adsorption chromatography. Normal-phase HPLC is used for the separation of the lipids into classes of lipids [1,F]. Reversed-phase HPLC (RP–HPLC), on the other hand, is mainly used to separate each lipid class into individual species [2,B1]. For example, several triglycerides were separated from each other via nonaqueous reversed-phase HPLC, involving an octadecyl (ODS) column and a nonpolar (non-aqueous) mobile phase. RP–HPLC alone can be used to separate the fat molecules into classes and species [2,B1].

In reversed-phase chromatography of lipids [B2], the lipophilic interaction between the octadecyl chains of the column and the fatty molecules of the analyte determines the retention. For this reason, the retention times of the lipid molecules depend on the stationary-phase chain length and number of double bonds in each lipid structure. In general, the retention time increases as the number of double bonds decrease. Glycerides with the shortest acyl chain length and highest number of double bonds elute first. This general rule applies if the column contains a C_8 or C_{18} stationary phase. Using such a general correlation and knowing the relative retention time of a reference compound such as a phospholipid on ODS columns, the retention time of other phospholipids can be calculated [B1].

In addition to normal- and reversed-phase chromatography, silver ion HPLC, RP ion-pair HPLC, chiral separation, and supercritical fluid chromatography have been used for analysis of lipids [3]. In silver-ion HPLC, the counterion of an ion-exchange column such as sulfonate is exchanged with silver ions. Only the degree of unsaturation in the lipid molecule determines retention. In RP ion-pair HPLC, different ion-pairing agents, such as alkylamonium phosphates, are added to the ODS column for molecular species separation. In supercritical fluid chromatography (SFC), which is used in combination with flame ionization detection (FID), solubility of lipids in carbon dioxide allows the separation to be performed. This mode of analysis is not widely used.

Selection of the Stationary Phase

The type of packing in the HPLC column is the most important factor affecting the HPLC separation. Silica is the most common packing material for normal-phase chromatography because it is stable under pressure and has a neutral pH. Silica-bonded C_{18} and C_8, on the other hand, are the most common packing materials for RP chromatography because they are stable to pressure and low pH. Diol-bonded phases are also used as polar column packing [4]. In cases where gradient elution is necessary, diol-bonded stationary phases are more suitable than silica. For lipid classes found in plants, chemically bonded phases have produced better resolution than silica gel [5]. Cyanopropyl, amino, and other stationary phases have all been used. Aminopropyl phases are suitable for acidic and nonpolar lipids. These columns are compatible with commonly used solvents and modifiers. Florisil is another stationary phase used for the separation of lipids; however, irreversible retention may occur.

The method of stationary-phase preparation has a major effect on the resolution, column stability, retention time, reproducibility, and peak shape. When preparing C_{18} or C_8, for example, it is important that the residual silanol groups are capped to prevent peak tailing. The extent of capping must be consistently maintained between different batches of the stationary phase for reproducible results.

In terms of particle size and column dimensions, stationary phases with particle sizes of 5–10 μm in 250 × 4.6-mm analytical columns are most common. HPLC columns with smaller particle sizes (2–3 μm) have also been prepared and used. Such columns have the advantages of higher efficiency, better resolution, and shorter analysis times. For example, the triglycerides of cocoa butter were baseline separated with columns that contained 2–3-μm particles. Research and development of columns with smaller particles is very active.

Selection of the Mobile Phase

The mobile phase consists of one or more solvents that are pumped through the chromatographic system, resulting in the separation of analytes. Mobile phases may also contain modifiers. Examples of frequently used solvents include hexane, methanol, 2-propanol, acetonitrile (ACN), and water. Examples of modifiers include trifluoroacetic acid, acetic acid, or formic acid. In general, the composition of the mobile phase should be kept simple. Factors that influence the choice of mobile phase include the solubility of the sample in the mobile phase, the polarity of the mobile phase, ultraviolet (UV) absorption wavelength, refractive index, and viscosity of the solvents. The purity of the solvents in the mobile phase is also important because the region of UV that is used for the detection of lipids (200–215 nm) must be free of interferences. For phospholipids, the most popular solvent systems are transparent to UV in the range of 200–215 nm; they include hexane–2-propanol–water and acetonitrile–methanol–water. Variations include hexane–methanol–isopropanol or water–methanol. The ratios of these solvents in the mobile phase depends on the nature of the lipid substances being separated. Several examples of mobile phases commonly used in the analysis of lipids are listed in the general references.

In order to separate the molecular species of lipids that have very close characteristics, small differences such as the number of double bonds in each molecule can occasionally be used as a guide to devise a solvent system to separate the molecules. This method is not always successful and may have to be further optimized for a good separation.

Detection

No single HPLC detector has all the characteristics of a good detector, which include sensitivity, specificity, detectability, linearity, repeatability, and dependability. Detection by UV/vis is widely used for the analysis of lipids; it is simple, concentration sensitive, and nondestructive. However, the analyte to be monitored by UV/vis absorption must contain a chromophore, and because many fat molecules do not contain a chromophore, this detection system cannot be used in many cases. Fortunately, in cases where no chromophore is present in the molecule, a chromophore can be introduced through derivatization. If the derivatization is done before the analyte enters the column, it is called precolumn derivatization, whereas if it is done after the elution of analyte, it is called postcolumn derivatization.

For the detection of phospholipids, UV/vis at 205 nm is often used. In cases where the molecules contain several lipid groups, UV detection in the range of 205–215 nm has been used. Vitamin E, α-, β-, δ-, and γ-tocopherols have been analyzed by RP–HPLC using UV at 215 [B1] and 280 nm. In other cases, UV diode-array detection at 190–350 nm or 200–400 nm has been used. Tocopherols and triglycerides from vegetable oils and human lipoproteins were analyzed by RP–HPLC using a UV/vis diode-array detector. With diode-array detection, UV spectra of the peaks can be collected during the analysis. In many analyses, the UV absorption of the lipid analyte is due to the carbonyl group C=O in the acyl group of the molecules.

More recently, a less sensitive detection technique, refractive index (RI) detection, has been used. The disadvantage of RI is that one is limited to isocratic elution, because RI detection is affected by changes in the pressure and temperature of the mobile phase. On the other hand, infrared (IR) detection can be used when the solvents in the mobile phase do not absorb infrared light, which can create interference. Because of these limitations, RI and IR have not been used as widely as UV detection.

In contrast, newer techniques such as FID and evaporative light-scattering detection (ELSD, "Universal Detector") are more sensitive and have been used for the analysis of many lipids, including the polar and neutral lipid classes [6]. For preparative scale separations, if the UV absorbance is more than what the detector can handle, an evaporative light-scattering detector is used [B2].

Fluorescence (FL) detection is another selective and sensitive technique. Vitamin E, α-, β-, δ-, and γ-tocopherols have been analyzed by RP–HPLC using FL as well as UV detection. In addition, a newer technique, electrochemical detection, has been reported in the analysis of phenolic compounds of olive oil [7].

Finally, the detection by liquid chromatograph–mass spectrometer (LC–MS), which has been largely dependent on the price, mode of ionization, and ease of operation of the mass spectrometer, is becoming popular [8]. It is predicted that LC–MS will become the method of choice for lipid analysis in coming years.

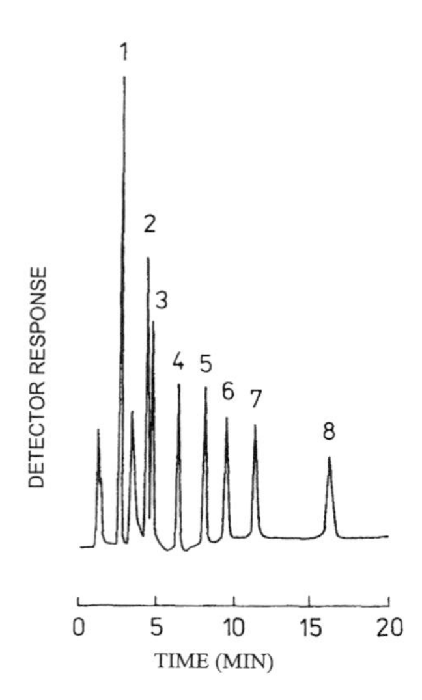

Fig. 1 Separation of free fatty acids by HPLC with conductivity detection. Mobile phase is methanol–5 m*M* tetrabutylammonium salt (75:25, v/v; pH 7.5) at a flow rate of 0.8 mL/min. Peaks: 1, lauric (12:0); 2, myristic (14:0); 3, linolenic (18:3); 4, linoleic (18:2); 5, palmitic (16:0); 6, oleic (18:1); 7, margaric (17:0); 8, stearic acids (18:0). (Reproduced by kind permission of the authors and of *Journal of Chromatography*).

Derivatization

Samples of lipids may be derivatized either before they are injected into the HPLC or after they emerge from the HPLC column. The purpose of derivatization is either to make the lipid molecules detectable or to improve peak shape during the analysis. The addition of functional groups, such as phenyl, 2,4-dinitrophenyl, or anisyl to the molecule enhances the detectability by changing the absorption wavelength of a lipid molecule. On the other hand, making derivatives such as acetates or methyl esters help to improve the peak shape. To make the lipid molecules fluorescent, groups such as anthroyl, 7-methoxycoumaryl, naproxen, and naphthyl [B1] are added to the lipid molecules. The reactions are carried out postcolumn and the derivatized molecules can be detected by both excitation and emission. For the detection of fatty acids, they are derivatized by reacting with either *p*-bromophenacylbromide, 9-anthryldiazomethane, or bromomethylmethoxy-cumarin. These derivatizations also make the molecules UV absorbers. Fatty acids have also been analyzed directly without derivatization using conductivity for detection (Fig. 1). Many examples of derivatization techniques have been tabulated in Ref. B1. Phosphatidilic acids, which result from the enzymatic reaction of phospholipase D with phospholipids, are derivatized to the methyl ester by treatment with diazomethane. In the reaction, the phosphate salt is converted to a neutral organophosphate, thereby changing its chromatographic behavior without affecting UV absorption of the lipid. Several commercially available phosphatidylethanolamines (PEs) have been derivatized for HPLC analysis as dansyl, pyrenesulphonyl, and fluoresceinthiocarbomoyl derivatives. Diacylglycerol acetates, from bovine brain ethanolamineglycerophospholipid (EGP), were separated into molecular species using an ODS column. The chromatograms of derivatized phospholipids showed better resolution when compared with the chromatograms of underivatized phospholipids.

Sample Cleanup

Before analysis, lipids are extracted from biological matrices with either a single organic solvent or a mixture of several organic solvents, such as chloroform–methanol or 2-propanol–hexane. Some lipids that contain several double bonds are easily oxidized by air during extraction. To prevent oxidation of such lipids, it is best to perform the extraction in an atmosphere of nitrogen and/or at low temperature. Prior to extraction, an antioxidant, such as BHT, may be added to inhibit oxidation. After the extraction, the extracts are cleaned up by solid-phase extraction (e.g., aminopropyl), column chromatography, or TLC [9]. Then, one-dimensional or two-dimensional TLC is used to separate the lipids. The separated lipids are eluted from the silica gel by scraping the zones of interest from the plate and extracting the lipids with a solvent.

In many cases, HPLC can replace TLC in this step. HPLC has the advantage that, during the analysis, oxygen-sensitive lipids are not exposed to air and autoxidation is limited. After the cleanup steps, samples are analyzed by GC or HPLC. Analysis of triglycerides often involves saponification before the HPLC analysis. Because samples of lipids may originate from a variety of different sources, the specific cleanup steps prior to analysis depend largely on the sample and the original sample matrix. For analysis of cholesterol in blood plasma, a sample of plasma was mixed with isopropanol and centrifuged. After centrifugation, the supernatant was directly analyzed by HPLC [9]. For analysis of monoglycerides in emulsifiers, the sample was dissolved in hexane–2-propanol (90:10), diluted, and injected into the HPLC. For analysis of vegetable oils, a dilute solution of oil in hexane–2-propanol is prepared and analyzed.

For quantitation of cholesterol and its derivatives in muscle and liver tissues, the extracts of the tissue homogenates are evaporated, dissolved in a mobile phase such as hexane–isopropanol, and injected onto a normal-phase column. For analysis of soybean oil by reversed-phase HPLC, after extraction with chloroform–methanol (9:1), the neutral lipids, chlorophyls, and the phospholipids are separated by TLC. The lipids recovered from the TLC are analyzed by HPLC.

Examples of Lipid Analysis by HPLC

For examples of separation of lipids see general references. For tabulated examples, see Ref. B1 for separation of molecular species of phospholipids by HPLC, Ref. I for separation of lipids in food by HPLC, Ref. H for HPLC of phosphatidic acid, and Ref. B2 for preparative HPLC of lipids.

For separation of intact polar lipids by HPLC and detection by mass spectrometer, see Ref. 8. For triglycerides and fatty acids, see Ref. B4. For separation of neutral lipids into classes and species, see Ref. 2.

Acknowledgments

The author wishes to thank Sue Hammer for preparation of the manuscript, Andrea Wong for reviewing the manuscript, and Peter Austin for library searches.

References

General References

For important references in lipid analysis see Appendices in *Advances in Lipid Methodology, Volumes 1–4* (W. W. Christie, ed.), The Oily Press, Dundee (1990, 1992, 1994, 1997).

For more detailed description of the subjects consult the following:

The following chapters in *Advances in Lipid Methodology, Volume 4*: (B1) Chapter 2, Separation of molecular species of phospholipids by high-performance liquid chromatography, M. V. Bell; (B2) Chapter 3, Preparative high-performance liquid chromatography of lipids, P. Van der Meeren and J. Vanderdeelen; (B3) Chapter 4: Structural analysis of fatty acids, W. W. Christie; (B4) Chapter 6: Reverse-phase high-performance liquid chromatography: General principles and application to the analysis of fatty acids and triglycerols, B. Nikolova-Damyanova.

The following chapters in *Advances in Lipid Methodology, Volume 3*: (C1) Chapter 3, Separation of phospholipid classes by high-performance liquid chromatography. W. W. Christie; (C2) Chapter 6, Plant glycolipids: Structure, isolation and analysis, E. Heinz.

The following chapters in *Advances in Lipid Methodology, Volume 2*: (D1) Chapter 2, Preparation of ester derivatives of fatty acids for chromatographic analysis, W. W. Christie; (D2) Chapter 3, Size exclusion chromatography in the analysis of lipids, M. C. Dobarganes and G. Marquez-Ruiz; (D3) Chapter 5, Capillary isotachophoresis in the analysis of lipoproteins, G. Schmitz, G. Nowicka, and C. Mollers; (D4) Chapter 6, Preparation of lipid extracts from tissues, W. W. Christie.

The following chapters in *Advances in Lipid Methodology, Volume 1*: (E1) Chapter 1, Solid-phase extraction columns in the analysis of lipids, W. W. Christie; (E2) Chapter 3, Supercritical fluid chromatography of lipids, P. Laakso; (E3) Chapter 4, The chromatographic resolution of chiral lipids, W. W. Christie; (E4) Chapter 6, Silver ion chromatography and lipids, B. Nikolova-Damyanova; (E5) Chapter 7, Detectors for high-performance liquid chromatography of lipids with special reference to evaporative light-scattering detection, W. W. Christie.

Also, consult the following:

Christie, W. W., Lipid class separations using high-performance liquid chromatography, in *New Trends in Lipid and Lipoprotein Analyses* (J. L. Sebedio and E. G Perkins, eds.), AOCS Press, Champaign, IL, 1995, p. 1934.

Shulka, High performance liquid chromatography, normal phase, reverse phase detection methodology, in *New Trends in Lipid and Lipoprotein Analyses* (J. L. Sebedio and E. G Perkins, eds.), AOCS Press, Champaign, IL, 1995, p. 38.

Abidi, S. L. and T. L. Mounts, High-performance liquid chromatography of phosphatidic acid, *J. Chromatogr. B 671*: 281–297 (1995).

Berg, K. A. and C. E. Canessa, HPLC applications in food and nutritional analysis, in *Chromatographic Science, Volume 78* (L. M. L. Nollet, ed.), Marcel Dekker, Inc., New York, 1998, pp. 753–787.

Additional References

1. J. G. Hamilton and K. Comai, Separation of neutral lipid, free fatty acid and phospholipid classes by normal phase HPLC, *Lipids 23*: 1150–1158 (1988).
2. Antonopoulou, N. K. Andrikopolos, and C. A. Demopoulos, Separation of the main neutral lipids into classes and species by RP–HPLC and UV detection, *J. Liquid Chromatogr. 17*(3): 633–648 (1994).
3. W. W. Christie, Some recent advances in the chromatographic analysis of lipids, *ANALUSIS Mag. 26*(3): M34–M40 (1998).
4. I. Elfman-Borjesson and M. Harrod, Analysis of non-polar lipids by HPLC on a diol column, *J. High Resolut. Chromatogr. 20*: 516 (1997).
5. W. W. Christie and R. Anne Urwin, Separation of lipid classes from plant tissues by HPLC on chemically bonded stationary phases, *J. High Resolut. Chromatogr. 18*: 97 (1995).
6. Liu, T. Lee, E. Bobik, Jr., M. Guzman-Harty, and C. Hastilow, Quantitative determination of monoglycerides and diglycerides by HPLC and evaporative light-scattering detection, *J. Am. Oil Chem. Soc. 70*: 343 (1993).
7. K. Seta, H. Nakamura, and T. Okuyama, Determination of α-tocopherol, free cholesterol, esterified cholesterols and triacylglycerols in human lipoproteins by HPLC, *J. Chromatogr. 515*: 585–595 (1990).
8. A. A. Karlsson, Analysis of intact polar lipids by HPLC-mass spectrometry/tandem mass spectrometry with use of thermospray or atmospheric pressure ionization, in *Lipid Anal. Oils Fats* (R. J. Hamilton, ed.), Blackie, London, 1998, pp. 290–316.
9. H. P. Nissen and H. W. Kreysel, The use of HPLC for determination of lipids in biological samples, *Chromatographia 30*(11/12): 686 (1990).
10. Y. Tsuyama, T. Uchida, and T. Goto, Analysis of underivatized C12–C18 fatty acids by reverse phase ion-pair high-performance liquid chromatography with conductivity detection, *J. Chromatogr. 596*: 181–184 (1992).

Jahangir Emrani

Lipid Classes: Purification by Solid-Phase Extraction

In recent years, solid-phase extraction (SPE) has emerged as an important tool for the purification of compounds in various fields of chemistry and biochemistry. This technique is taking place with others such as thin-layer chromatography (TLC) and high-performance liquid chromatography (HPLC) in the wide range of preparative tools available to the analyst. Solid-phase extraction technology is relatively new, as the first applications were published during the past two decades, and its real development occurred only in the nineties. The advantages of SPE, as compared to other methods, makes it an attractive method of choice for purification of molecules. Among these advantages are less time and lower solvent consumption when compared to classical liquid–liquid extraction, low cost, and excellent reproducibility of SPE products. Moreover, SPE is easily automated, so that numerous samples can be processed at the same time. SPE is frequently used in lipid chemistry and biochemistry for the isolation of particular compounds or groups of molecules for analytical or preparative purposes [1–3]. Here, we will focus on the applications of SPE in the field of lipid biochemistry, and, primarily, on the use of the aminopropyl-bonded silica gel matrix for lipid fractionation.

General Principles and Methodology of SPE

Solid-phase extraction is a technique which is used for concentration and purification of analytes from solution by adsorption onto a disposable solid-phase-containing cartridge, disk, or syringe barrel, followed by elution of analytes with an appropriate solvent. From a general point of view, the principles of SPE are the same as conventional liquid chromatography and HPLC, so that the retention and elution depend on interactions of the analyte with the stationary solid and mobile liquid phases. Retention mechanisms include normal and reversed phases and ion exchange. The SPE procedure begins by sorbent conditioning to remove cartridge impurities and to wet the functional groups at the surface of the matrix. Then, the sample is loaded onto the SPE matrix and the sorbent is rinsed to remove the components which are not desired. In a final step, the analytes are eluted from the SPE matrix and recovered in an appropriate solvent for further analysis. A wide variety of solid phases are available. Most applications are performed on the silica gel solid phase, which can be bonded with the same substituents used for HPLC or conventional liquid chromatography columns. SPE principles are described in detail in ad hoc reviews or books [1,4]. This method (as with all chromatographic methods) requires an accurate determination of the conditions of column preparation, sample loading, and stepwise elution. Hence, incomplete elution or analyte breakthrough can occur if optimal conditioning or elution parameters are not well defined. A good choice of solid sorbent must be made according to the nature of the analytes that are to be recovered. The choice of solvents and their volumes, as well as flow rates, must also be carefully determined to ensure optimal recovery of analytes. An understanding of the interactions between sorbent and analytes is needed to develop and optimize SPE procedures [4].

Its advantages and ease of use have made SPE an appropriate method for concentration and fractionation of lipid classes [2,3]. Many different solid-phase types have been used in lipid fractionation on silica, alumina, porous carbon, and ion-exchange resins. Various bonded phases have been used, particularly with the silica gel matrix.

Separation of Lipid Classes by Reversed-Phase SPE

The reversed-phase procedure has been applied in some cases for the isolation of lipid classes on SPE columns. This procedure is mainly suitable for the isolation of lipids dissolved in polar solvents, such as aqueous samples. The mechanism involves partitioning of organic solutes from the polar mobile phase to a nonpolar sorbent phase, which can be C_2, C_4, C_8, and C_{18} aliphatic chains, or cyclohexyl and phenyl groups. Elution of analytes is accomplished by choosing a solvent that will disrupt the van der Waals forces retaining the molecules on the matrix. Methanol, ethyl acetate, and acetonitrile are often used for this purpose, as they can overcome van der Waals forces and they succeed in bonding to free silanol groups of the column and dissolve residual water coming from the aqueous matrix sample. Reversed-phase separations are not as common as normal-phase separation of lipids by SPE. C_{18} SPE tubes have been often used in the isolation of phosphatidylcholine, cerebrosides, sulfatides, and gangliosides from water-soluble compounds (Figlewicz et al., 1985., quoted in Ref. 2). C_{18} has been shown to be an efficient tool for the purification of these lipids and is easier to use than other liquid–liquid methods. Reversed-phase SPE columns have also been used for the isolation of short- and long-chain fatty acids from distilled water and seawater (Pempkoviak. J. 1983, quoted in Ref. 2). For more information on the use of SPE in the reverse phase, references cited at the end of this entry should be consulted.

Separation of Lipid Classes by Normal-Phase SPE

Most of applications deal with silica gel sorbent and with aminopropyl-bonded silica sorbents. Silica is a sorbent of choice for lipid fractionation, because these analytes are first recovered in nonpolar solvents as chloroform. The polar silanol groups present at the surface of the silica sorbent will more strongly adsorb polar compounds such as phospholipids, rather than neutral lipids such as sterols and triglycerides. The retention of analytes and their elution will depend on solvent polarity. The mildly acidic nature of silica can also interfere with the separation mechanism by ion-exchange effects. Silica matrix solid-phase extraction has been used to purify some individual lipid classes, such as phosphatidylcholine (PC) or phosphatidylethanolamine (PE) from total lipids. More complex separations are also possible with silica SPE columns. Hence, a procedure for isolating cholesterol esters, triglycerides, free fatty acids (FFAs), cholesterol, acidic phospholipids such as phosphatidylinositol (PI) and phosphatidylethanolamine (PE) from a neutral phospholipid fraction containing PC, lyso-PC, and sphingomyelin (SPH) has been described [5]. This procedure was modified and adapted for further isolation of specific classes of molecules such as the choline phospholipid platelet-activating factor (PAF).

Numerous chemically bonded stationary phases can be formed by reacting the silanol groups of silica with various organic reagents. These can be successfully used for the isolation of specific lipid classes, because the bonded phases will have greater bonding potential for some specific molecules according to their functional groups. This is the case for aminopropyl-bonded phases whose primary amine group can develop a strong interaction by hydrogen-bonding with molecules having functional groups such as hydroxyl. The amino group can also be used to separate molecules on the basis of an ion-exchange mechanism, because it can also have weak anion-exchange properties [2].

As with silica gel sorbent, neutral lipids will be poorly retained as compared to polar lipids such as phospholipids or glycosphingolipids. Aminopropyl-bonded silica gel SPE sorbent can be used according to its chemical properties, compared to an unbonded silica matrix. Surprisingly, relatively few procedures have been described for lipid fractionation and purification using aminopropyl-bonded silica SPE columns. Kaluzny et al. [6] first described a detailed procedure which allows the fractionation of neutral lipids into different classes. In this procedure, all neutral lipids were eluted from the column with a mixture of chloroform–methanol (2:1, v/v) and transferred for further fractionation onto a second column to obtain cholesterol esters, triacylglycerols, cholesterol, diacylglycerols, and monoacylglycerols to separate fractions, with good yields. With such elution, FFAs and phospholipids still bind to the column. They are then eluted in a stepwise manner by washing this column with 2% acetic acid in diethyl ether (FFA) followed by methanol to elute neutral phospholipids such as PC and SPH. Aminopropyl columns show, here, some advantage compared to un-bonded silica matrix to isolate FFA, since these molecules are firmly bound to the amino group of the column and with relative selectivity compared to other neutral lipids. In the case of an unbonded silica matrix, FFAs would have been eluted before monoacylglycerols, with the other neutral lipids.

Elution of FFAs from the aminopropyl-bonded column requires a change in pH of the elution solvent to weaken the interaction between the FFA carboxyl and the column's amino groups. This is achieved by washing the column with diethyl ether containing 2% acetic acid. This example shows the advantage which can be realized by adapting the nature of the solid phase to the kind of molecule to be recovered. This is particularly useful with molecules which tend to elute together on an unbonded silica sorbent. This is often the case with free ceramides and FFAs, which will tend to migrate close to each other on a silica gel matrix; thus, it can sometimes be difficult to individually purify these compounds, particularly when high levels of FFAs are present in the sample. By choosing aminopropyl SPE tubes, this problem can be avoided, as FFA will be firmly retained, allowing further efficient elution and purification of these compounds.

However, Hamilton and Comai described a procedure which allows good recovery and separation of the neutral lipids from FFA on a silica SPE matrix [5]. Separation of the different phospholipids can be also achieved on an aminopropyl column. These polar compounds are tightly bound to the matrix, so they are retained while eluting neutral lipids and FFAs. Then, the different phospholipids can be washed from the column by changing the pH and the ionic strength of the solvent. Stepwise elution of phospholipids was obtained (PC, PE, PS, and PI) by increasing, progressively, the polarity of the solvent and its pH [7].

Suzuki et al. also separated phospholipids by using a combination of aminopropyl-bonded silica and an unbonded silica SPE to recover PC, PE, cardiolipin, phosphatidylglycerol, and phosphatidylserine. Phospholipids were recovered with good purity and high yields [8]. The present procedure allows the optimized use of the advantages of each kind of column. Such a combination of two SPE tubes with different matrices was already successfully used by Prietto et al. (1992, in Ref. 2). Steryl esters, TGs, FFAs, diglycerides, monoglycerides, monogalactosylmonoglycerides, and monogalactosyldiglycerides and their digalactosyl derivatives were separated from PC and lysoPC onto a silica gel SPE tube by different solvents of increasing polarity. Then, phosphatidylethanolamine and its lyso derivatives, which were co-eluting from the silica tube, were separated on an aminopropyl-bonded SPE.

Recovery of phospholipids, as well as the various neutral lipids, is obtained with good yields on aminopropyl tubes. Hence, Kaluzny's procedure was shown to recover up to 100% of all the lipid classes studied, giving better results than the TLC procedure [6]. This latter procedure has been extensively used by many workers to separate lipid mixtures from different origins. It has been slightly modified and adapted for particular use with poorer fractionation of lipid samples [2]. This SPE procedure has been used to achieve partial fractionation of phopholipids by various researchers [2] and, particularly, to recover acidic phospholipids, which can only be eluted from an aminopropyl matrix by changing the ionic strength of the solvent.

Surprisingly, the development of SPE procedures for fractionation of lipid samples has attracted relatively little attention as compared to conventional methods, such as TLC. However, because of its qualities and advantages, it promises to be applied much more in the future. The development of such separation methods is still in progress, as new sorbents and enhanced quality of packing will become available in the near future, from the more than 50 manufacturers currently making SPE products [9].

References

1. W. W. Christie, Solid-phase extraction columns in the analysis of lipids, in *Advances in Lipid Methodology, Volume 1* (W. W. Christie, ed.), The Oily Press, Arly, Scotland, 1992, pp. 1–17.
2. S. E. Ebeler and T. Shibamoto, Overview and recent developments in solid-phase extraction for separation of lipid classes, in *Lipid Chromatographic Analysis* (T. Shibamoto, ed.), Marcel Dekker, Inc., New York, 1994, pp. 1–49.
3. S. E. Ebeler and J. D. Ebeler, *INFORM 7*: 1094–1103 (1996).
4. E. M. Thurman and M. S. Mills, in *Solid-Phase Extraction*.

Principles and Practice (J. D. Winefordner, ed.), John Wiley & Sons, New York, 1998.
5. J. G. Hamilton and K. Comai, *Lipids 23*: 1146–1149 (1988).
6. M. A. Kaluzny, L. A. Duncan, M. V. Merritt, and D. E. Epps, *J. Lipid Res. 26*: 135–140 (1985).
7. A. Pietsch and R. L. Lorenz, *Lipids 28*: 945–947 (1993).
8. E. Suzuki, A. Sano, T. Kuriki, and T. Miki, *Biol. Pharm. Bull. 20*: 299–304 (1997).
9. R. E. Majors, Current trends and developments in sample preparation, *LC–GC Int. 11*(Suppl.): 8–16 (1998).

Jacques Bodennec
Jacques Portoukalian

Lipids Analysis by Thin-Layer Chromatography

Introduction

The term *lipids* in this entry is restricted to esters and amides of the long-chain aliphatic monocarboxylic acids, the fatty acids, and to their biosynthetically or functionally related compounds. The most abundant lipids are the esters of fatty acids with glycerol (1,2,3-trihydroxypropane), denoted as glycerolipids. Lipids are classified according to the number of hydrolytic products per mole. *Simple* (or *neutral*) lipids release two types of products (e.g., fatty acids and glycerol). The most abundant simple lipids are the triacylglycerols. *Complex* (or *polar*) lipids give three or more products, such as fatty acids, glycerol, phosphoric acid, and an organic base. Typical complex lipids are the glycerophospholipids (phospholipids), glycoglycerolipids (galactolipids), and the sphingolipids. A natural lipid mixture comprises different types (denoted as lipid classes) of simple and complex lipids.

Lipids are important constituents of all living organisms. Thus, for example, the triacylglycerols serve as an energy reserve, whereas complex lipids are structural components of the cell membranes with a substantial role in cell functions. Lipids are also important components of the human and animal diet. Disturbances in the lipid metabolism of the organism lead to various disorders and malfunctions. In humans, these are unambiguously related to the development of cardiovascular disease.

The complexity of natural lipids requires relevant methods for examination. Adequate approaches for separation and isolation of lipid components are mandatory, and among these, the chromatography techniques are of primary importance. Among the different chromatographic methods, thin-layer chromatography (TLC) has its special place, being one of the first separation methods applied in lipid analysis. Most of the present basic knowledge on the structure and biological role of lipids has been achieved by using various TLC techniques.

The first description of TLC goes back to 1938, but Kirchner (1951) and Stahl (1965) were those who converted the idea into a full-scale analytical technique. In the early 1960s, Kaufmann and co-workers in Germany and the group of Privet in the United States introduced TLC in the lipid analysis.

Thin-layer chromatography is a powerful tool in the analysis of lipids. It is easy to perform, versatile, and relatively cheap, and it allows for direct quantitative measurements of the separated compounds by means of scanning densitometry. An important feature is that the analyst gets a full picture of the examined sample.

The General TLC Technique for Lipids

Thin-layer chromatography is a separation technique in which the components of a lipid mixture are differently distributed between a solid stationary phase, spread as a thin layer on a plate made of inert material, and a solvent mobile phase. Depending on their type, the components are retained with different strengths on the layer to give distinctive spots or bands. The migration of a band is presented quantitatively by the corresponding R_f value. The stronger the retention, the lower is the R_f value.

Three modifications of TLC are in use in lipid analysis: (a) separation on unmodified silica gel layer, silica gel TLC; (b) separation on a layer impregnated with silver ions, silver-ion TLC (Ag TLC), and (c) separation on a layer modified with silanes or long-chain hydrocarbons to give a nonpolar stationary phase, reverse-phase TLC (RP–TLC).

The universal TLC facilities are utilized: plates, adsorbents, microcapillars or micropipettes for sample application, development tanks, detection spray reagents and devices for spraying, and densitometers for quantification. Plates are either commercially precoated or handmade.

Silica gel G (G, for gypsum as a binding substance), silica gel H (no binding substance), and, rarely, alumina and kieselguhr form the thin-layer stationary phases. Complete sets of devices necessary for the preparation of handmade plates are commercially available. After the silica gel slurry is spread on the plates, they are left to dry in the air for at least 24 h and briefly in an oven at 110°C. The plates are then ready for either direct use or for modification of the layer. From the great variety of precoated plates which are commercially available and preferred nowadays, silica gel plates and plates with layers modified with carbon chains from C_2 to C_{18} are of interest in lipid analysis. Understandably, precoated plates for Ag TLC are not commercially available. Preparative TLC is performed mostly on 20-cm × 20-cm plates with a layer thickness of 1 mm. Analytical separations are usually performed on 5-cm × 20-cm plates with a layer thickness of 0.2 mm.

The preparation of a lipid sample includes extraction of the lipid material from the examined object (seeds, tissues, food, etc.), choosing among the several widely accepted procedures. The extraction with chloroform–methanol (2:1) (the Folch extraction) is the most popular. A solution of known concentration in hexane or dichloroethane is prepared and a suitable aliquot is applied on the plate as a small spot or, better, as a narrow band. Two, three, or more solvents, mixed in different proportions, give the mobile phase. Development is performed in common tanks (Desaga type, for example), in the ascending mode. For fine separations, cylindrical or sandwich-type tanks provide better results.

In preparative TLC, detection is performed by spraying the plate with 2,7-dichlorofluoroescein and viewing the plate under ultraviolet (UV) light. The spraying reagent does not affect the lipid and can be easily removed during the isolation process, if necessary. Nonspecific destructive reagents are used in analytical TLC. Those most widely used are the alcoholic solutions of sulfuric (up to 50%) or phosphomolybdic (5%) acid, applied as spraying reagents. They are equally suitable for nonmodified as well as for modified layers. Reliable results are also obtained by saturating the layer with vapors of sulfurylchloride (for silica gel TLC and Ag TLC only). To visualize the separated components, the plate is heated at 180–220°C. The substances carbonize to give intensively stained spots, contrasting well with the background. The concentration of the lipid substance in the spots can be measured directly on the plate by using scanning densitometry.

Separation by Silica Gel TLC

Silica gel TLC is used for the identification of lipid classes in the sample (analytical TLC) and for isolation of a given lipid class for further examination (preparative TLC). Silica gel TLC is a good aid in checking both the identity and purity of individual components and different derivatives.

The separation is based on the interaction (hydrogen-bonding, van der Waals' forces, and ionic bonding) between the lipid molecule and the silica gel. Lipid classes with free hydroxyl, keto, and carboxyl groups are held stronger than those which contain only fatty acid residues.

Mobile phases of hexane or light petroleum ether as main components and acetone or diethyl ether as polar modifiers are used for the separation of simple lipids. Acetic or formic acid is often added to keep the free fatty acids in the fully protonated form. The retention of simple lipids increases in the order waxes, sterol esters, methyl esters, triacylglycerols, free fatty acid, sterols, diacylglyerols, and monoacylglycerols (see Fig. 1a). If no acid is present

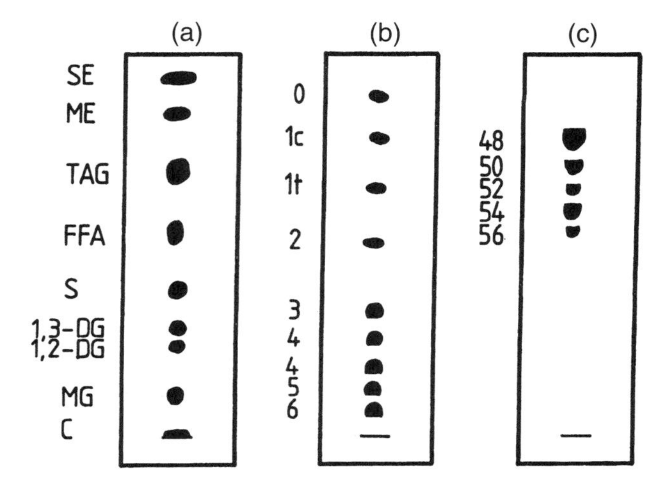

Fig. 1 Schematic presentation of lipid separation by TLC: (a) Reference mixture of simple lipid classes, silica gel TLC; mobile phase: hexane–ethyl ether–acetic acid, 80:20:2 (v/v/v); SE = sterol esters, ME = fatty acid methyl esters, TG = triacylglycerols, FFA = free fatty acids, S = sterols, 1,3-DG = 1,3-diacylglycerols, 1,2-DG = 1,2-diacylglycerols, MG = monoacylglycerols, C = complex lipids; detection by spraying with 5% ethanolic phosphomolibdic acid and heating for several minutes at 180°C; (b) reference mixture of fatty acid methyl esters, Ag TLC on silica gel layer impregnated with 0.5% methanolic solution of silver nitrate; mobile phase: 5 mL of light petroleum ether–acetone–formic acid, 92:2:1 (v/v/v), development in open cylindrical tank; detection by treatment for 30 min with sulfurylchloride vapors and heating for 5 min at 180°C; the numbers denote the number of double bonds, *t*, for a trans double bond, *c* for a cis double bond; (c) triacylglycerols of the 001 class (0 for zero and 1 for one double bond in the molecule), reversed-phase TLC on kieselguhr layer silanized by treatment with dimethyldichlorosilane vapors; mobile phase: acetone–acetonitrile–water, 70:30:12 (v/v/v), detection by spraying with 50% ethanolic sulfuric acid and heating at 220°C for 10 min; the numbers denote the partition number.

in the mobile phase, the free fatty acids migrate between diacylglyerols and monoacylglycerols. The complex lipids remain on the start. The retention of phospholipids depends on the nature of the polar "head" and increases in the order cardiolipin, phosphathidylethanolamin, phosphathidylserine, phosphatidylinositol, phosphatidylcholine, [silica gel H, mobile-phase chloroform–methanol–acetic acid–water, 25:15:4:2 (v/v/v/v)]. Galactolipids can be separated with the same solvent system after changing the solvent proportions to increase the polarity. Predictably, monogalactosyl diacylglycerols migrate ahead digalactosyl diacylglycerols.

Identification of the lipid components is performed easily by applying a standard lipid mixture on the same plate prior to the development.

Separation by Ag TLC

Silver-ion TLC is the modification with the most important impact on the development of lipid chemistry and has been of immense importance for the understanding of lipid structure. It is used to resolve the molecular species of a single lipid class. The separation is based on the ability of unsaturated fatty acid moieties in lipid molecules to form weak reversible charge-transfer complexes with silver ions. The complexation includes the formation of a σ-type bond between the occupied $2p$ orbitals of the olefinic double bond in the fatty acid (FA) moiety and the free $5s$ and $5p$ orbitals of the silver ion, and a (probably weaker) π-acceptor backbond between the occupied $4d$ orbitals of the silver ion and the free antibonding $2p\pi^*$ orbitals of the olefinic bond. Thus, Ag TLC separates lipid classes into molecular types depending on the number, configuration, and, occasionally, the position of the double bond in the fatty acid moieties.

The impregnation of the layer is performed by immersing the plate in a solution of the silver salt in methanol, acetone or acetonitrile or by spraying the plate with one of these solutions. Preparative plates are usually treated with 1–20% silver nitrate solutions. For analytical Ag TLC, the concentration of silver nitrate varies in the range 0.5–10%. The impregnation procedures must be standardized to provide reproducible separation. Plates are left in the air for the solvent to evaporate and are usually activated prior to use (between 5 min and 1 h depending on the purpose) by heating at 110°C.

The separation is affected by the dimensions of the tank, the volume of the mobile phase, the development mode (covered or "open" tanks, with or without saturation of the atmosphere, respectively), the atmospheric humidity, and the temperature. The Ag TLC plates are normally developed at ambient temperature.

Hexane or light petroleum ether, chloroform, benzene, and toluene are most often the major components of the mobile phase, whereas smaller proportions of diethyl ether, acetone, methanol, ethanol, or acetic acid are added as modifiers. Chloroform–methanol and hexane–acetone mobile phases reportedly provide very good separations. Often more than one development is required for reliable resolution. The separation starts with the most polar phase and proceeds, after drying between runs, with mobile phases of gradually decreasing polarity. Highly unsaturated components are resolved first and do not move further with subsequent developments when the more saturated components are separated.

In general, the migration order of any lipid class is determined by the overall number of double bonds in the molecule. Thus, the retention of common fatty acids (chain lengths of 16–22 carbon atoms, methylene-interrupted double bonds) increases with increasing number of double bonds from zero to six (Fig. 1b). For the triacylglycerols which contain the above type of acyl moieties, the order of increasing retention is 000, 001, 011, 002, 111, 012, 112, 003, 112, 013, 113, 222, 023, 123, 223, 133, 233, and 333 (the numbers indicate the number of double bonds in the fatty acid residue but not the position in the glycerol backbone). The same order of retention is valid for complex lipids, but because of different technical difficulties, Ag TLC is only rarely applied in the analysis of these lipids. Species with cis double bonds are held stronger than those with trans double bonds and this differentiation is of great practical importance. Ag TLC is capable, under specific conditions, to differentiate fatty acids and triacylglycerols according to the position of the double bond in the carbon chain.

Both handmade and precoated plates provide reliable separation of fatty acids. Successful separation of triacylglycerols has been achieved on handmade plates only.

Silver ion TLC offers an effective means of fractionation of lipid mixtures into distinct fractions differing in the number of double bonds. It is often used to simplify the further examination with gas chromatography, GC–mass spectrometry, Fourier transform infrared, and so forth. Ag TLC serves also as an enrichment procedure for minor components and allows for more accurate estimation of their content and identity. Quantitative procedures have been developed for the determination of fatty acids and triacylglycerols by using Ag TLC and densitometry.

Separation by RP–TLC

Reversed-phase (RP) TLC is less popular and has been applied so far only for the resolution of fatty acids and triacylglycerols. It is based on the distribution of lipid molecules between a nonpolar stationary phase and a relatively polar mobile phase. Lipids are, therefore, separated according to their overall polarity, expressed by the partition number (PN). PN relates the migration of a component to the total number of carbon atoms, CN (in the acyl residues only), and the total number of double bonds, n, so that PN = CN − $2n$. The higher the PN, the stronger is the component retained in the nonpolar layer (the lower the R_f value).

Reversed-phase TLC also uses the common supporting facilities. In the laboratory, the nonpolar stationary phase can be produced by impregnating the layer (kieselguhr G or silica gel G) with long-chain hydrocarbons or liquid paraffin or by treatment with dimethyldichlorosilane (DMCS). Although commercial RP–TLC plates are available, so far it has been experimented with C_{18} plates only.

Mobile phases which provide good separation for triacylglycerols are (a) acetone–acetonitrile–water, 70:30: X (v/v/v, the water proportion, X, increases with the increasing unsaturation of the lipid class) which is suitable for handmade kieselguhr layers, treated with DMCS and (b) acetonitrile–2-butanone–chloroform, 50:35:15 (by volume) which is suitable for precoated C_{18} plates.

At present, RP–TLC finds application only as a quantitative technique complementary to the analytical and preparative Ag TLC of triacylglycerols. The triacylglycerol mixture is first fractionated into classes according to the unsaturation and then each class is subjected to RP–TLC to give a series of species with different PNs. An example is shown on Fig. 1c.

Suggested Further Reading

Ackman, R. G., Application of thin-layer chromatography to lipid separation, in *Analysis of Fats, Oils and Lipoproteins* (G. E. Perkins, ed.), American Oil Chemical Society, Champaign, IL, 1991, pp. 60–82.

Christie, W. W., *Lipid Analysis*, 2nd ed., Pergamon Press, Oxford, 1982.

Fried, B. and J. Sherma, *Thin-Layer Chromatography, Techniques and Applications*, Marcel Dekker, Inc., New York, 1994.

Hamilton, R. J., Thin-layer chromatography and high-performance liquid chromatography, in *Analysis of Oils and Fats* (R. J. Hamilton and J. B. Rossel, eds.), Elsevier, London, 1986, pp. 243–311.

Kuksis, A., Lipids, in *Chromatography, Part B. Application*, 5th ed. (E. Heftman, ed.), Elsevier, Amsterdam, 1992, pp. B171–B227.

Nikolova-Damyanova, B., Silver ion chromatography of lipids, in *Advances of Lipid Methodology, Volume 1* (W. W. Christie, ed.), The Oily Press Ltd., Ayr, U.K., 1992, pp. 181–237.

Boryana Nikolova-Damyanova

Lipophilicity Determination of Organic Substances by Reversed-Phase Thin-Layer Chromatography

Introduction

Lipophilicity seems to be an important physicochemical parameter for organic substances, influencing their biological activity. Usually, lipophilicity is defined by the partition coefficient, P, of an organic compound between an organic phase and water:

$$P = \frac{C_o}{C_a} \quad (1)$$

where C_o and C_a are the compound concentrations in the organic and the aqueous phases, respectively, when equilibrium is established. The organic phase can be a hydrocarbon, oil, or, most frequently, n-octanol. From a qualitative point of view, the partition coefficient measures the preference of the compound for the nonpolar phase, and from the thermodynamic perspective, it gives information about the intermolecular forces affecting a compound in solution.

Discussion

The partition coefficient between n-octanol and water, $P_{o/w}$, can be measured experimentally using the "shake-flask" method or can be calculated from structural fragments. The direct measurement of log $P_{o/w}$ values by equilibration between n-octanol and water faces some difficulties, such as the necessary high purity of the substance that must be available in an adequate quantity, and the method is time-consuming too. In addition, it is not applicable to very lipophilic or very hydrophilic compounds.

Reversed-phase liquid chromatography (RP–LC) has been proposed as an alternative method for log $P_{o/w}$ determination, simulating the partition of a specific compound between the lipid layer and the biological membrane of a cell. This is the process which rules the biological activity of substances, such as the absorption of drugs and their metabolites, the bioaccumulation of organic compounds, and so forth.

Reversed-phase thin-layer chromatography (RP–TLC) is one of the liquid chromatographic techniques showing distinct advantages compared to conventional methods for log $P_{o/w}$ determination. It is a rapid and reliable method, having a good reproducibility, and it needs only a small amount of sample that does not have to be of high purity. In a RP–TLC experiment, several samples can be run simultaneously. In TLC, the retention parameter, R_F, is defined as the ratio between the migration distances of the compound and of the mobile phase [Eq. (2)]; The term R_M was introduced for a linear relationship between the log P and R_M values [Eq. (3)]:

$$R_F = \frac{z_x}{z_f} \qquad (2)$$

where z_x and z_f are the migration distances of the component and of the mobile phase between the start and the front line, and

$$R_M = \log\left(\frac{1}{R_F} - 1\right) \qquad (3)$$

The stationary phase in RP–TLC can be silica-C_8 or silica-C_{18}; however, better correlations between the R_M values and log P have been obtained for silica-C_{18}, probably due to the strong interactions which appear between the polar groups of organic compounds and the more accessible free silanol groups on silica-C_8. Impregnated silica with paraffin oil or silicone oil was also used in order to improve the detectability of different organic compounds by spraying with derivatisation reagents. The measurements of R_M values on silica layers impregnated with oils have a lower reproducibility than the measurements on silica C_8- or C_{18}-bonded phases.

The mobile phase used in the lipophilicity measurements by RP–TLC is usually a binary mixture between an organic solvent and water. The organic solvent can be methanol, acetonitrile, or acetone. The first two can also be used in reversed-phase high-performance liquid chromatography (RP–HPLC) measurements, but acetone is restricted by its absorption in ultraviolet (UV) light.

When a compound contains one or more dissociable polar substituents, the pH and the ionic strength of the mobile phase influence the apparent lipophilicity of the compound. Usually, the mobile phase is buffered at a pH which ensures the neutral form of the studied compound.

The lipophilicity determination of organic substances by RP–TLC is based on the linear relationship established between the R_M values and the concentration of organic modifier in the mobile phase [Eq. (4)]:

$$R_M = a_0 + a_1\varphi \qquad (4)$$

where φ is the concentration of the organic modifier in the mobile phase, a_0 and a_1 are the intercept and the slope of the linear relationship (4). The significance of a_0 is the R_{M_w} value obtained for pure water as the mobile phase. Experimentally, the R_{M_w} value can be measured only for very hydrophilic compounds. Practically, the R_{M_w} value is extrapolated from the linear relationship (4) to 0% organic modifier in the mobile phase. The extrapolation method was confirmed by the good correlations obtained between the experimental and the extrapolated R_{M_w} values. From a theoretical point of view, the nature of the organic modifier in the aqueous mobile phase should lead to the same R_{M_w} value. Practically, these values are slightly different, but they do not affect the lipophilicity measurements. Sometimes, a positive or negative deviation of R_M values from the linear relationship (4) was observed, especially at a low concentration of organic modifier in the mobile phase (Fig. 1).

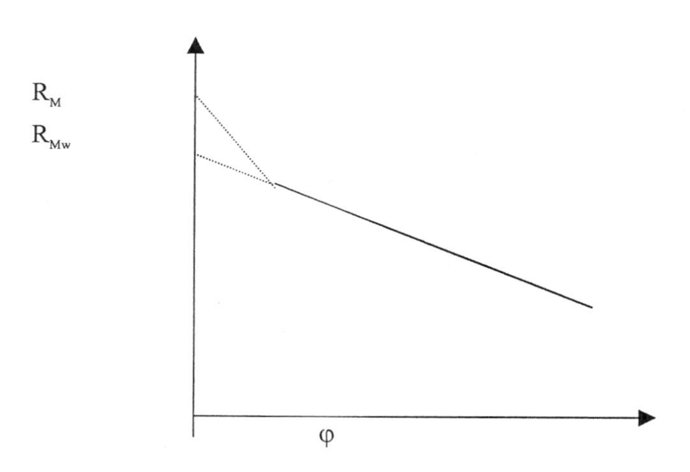

Fig. 1 The relationship between the R_M values and the organic-modifier concentration (φ) in the mobile phase for RP–TLC lipophilicity measurements.

The use of the R_{M_w} value as a measure of a compound lipophilicity is based on the linear relationship with log $P_{o/w}$ values [Eq. (5)]:

$$R_M = b_0 + b_1 \log P_{o/w} \quad (5)$$

where b_0 and b_1 are the coefficients, the intercept, and the slope, respectively. Equation (5) is similar to the linear relationships obtained in RP–HPLC between log k and log P values.

The extreme values for R_F are 0 and 1. As a consequence, the obtained R_M values can be situated between $-\infty$ and $+\infty$. Practically, the R_F values can be measured in the range 0.03–0.97, which means that $-1.5 \leq R_M \leq +1.5$. A particular case corresponds to very hydrophilic compounds, which migrates in the mobile-phase front even at a low concentration of the organic modifier.

For hydrophilic compounds, the R_{M_w} values can be obtained experimentally, or by extrapolation. The relationship obtained between the direct experimental and the extrapolated R_{M_w} is linear, having and intercept value $m \cong 0$ and the slope n of approximately 1 [Eq. (6)]:

$$R_{M_w,\text{experimental}} = m + nR_{M_w,\text{extrapolated}} \quad (6)$$

Equation (6) proves the validity of the extrapolation method for obtaining the R_{M_w} values in cases when the direct measurement is not possible, due to the lipophilic character of the substance.

The R_{M_w} values obtained in different solvent systems are similar for acetone or methanol aqueous mobile phases, but for *N,N*-dimethylformamide, they are significantly lower. This observation can be explained by the structure of *N,N*-dimethylformamide, which is very different from water.

Irrespective of the organic modifier used in the aqueous mobile phase, a linear relationship can be obtained between the intercept and the slope of Eq. (4). This means that the intercept value, $a_0 = R_{M_w}$, is correlated with the slope a_1 [Eq. (7)]:

$$R_{M_w} = a + ba_1 \quad (7)$$

The linear correlation shown in Eq. (7) is not influenced by the value of the slope a_1. For a methanol–water mobile phase, an increase of methanol concentration will produce a more rapid decrease of R_M values for the lipophilic compounds than for the hydrophilic ones. In other words, the lipophilic substances are more sensitive to variations in the mobile-phase composition.

The linear relationship between a_0 and a_1 values [Eq. (4)] is an important characteristic of lipophilicity measurements. The intercept, $a_0 = R_{M_w}$, can be considered a measure for the partition of a compound between a polar mobile phase and a nonpolar stationary phase. The physicochemical significance of the slope, a_1, is not completely established. It was demonstrated experimentally that the slope shows the rate of increasing the compound solubility in the mobile phase. The solubility increases as a consequence of the decreasing of the mobile-phase polarity. The slope, a_1, depends on the nonpolar surface of the organic compound (i.e., the part of the compound which interacts with the nonpolar stationary phase).

The correct measurement of the R_F values is very important because it affects the calculation of the R_{Mw} values and the further correlations. The manual measurement of the R_F values will include a error of ± 1 mm, comparing with the precision of a densitometer, which is ± 0.01 mm. The start and the front lines should be also exactly established, in order to reduce the measurement error. The "front line" in TLC is approximately equivalent with the "dead time" in HPLC. The marker substance used in HPLC (such as inorganic salts) can be applied to estimate the exact front line in TLC. The measurement error for the front line is about 0.03 mm for a migration distance of 80 mm, for any concentration of the organic modifier in the aqueous mobile phase.

Suggested Further Reading

G. L. Biagi, A. M. Barbaro, and M. Recanatini, *J. Chromatogr. A 678*: 127–137 (1994).

G. L. Biagi, A. M. Barbaro, A. Sapone, and M. Recanatini, *J. Chromatogr. A 669*: 246–253 (1994).

G. L. Biagi, A. M. Barbaro, A. Sapone, and M. Recanatini, *J. Chromatogr. A 662*: 341–361 (1994).

G. Cimpan, C. Bota, M. Coman, N. Grinberg, and S. Gocan, *J. Liquid Chromatogr. Related Technol. 22*: 29–40 (1999).

R. Kaliszan, *Quantitative Structure–Chromatographic Retention Relationships*, John Wiley & Sons, New York, 1987.

Lambert, W. J., *J. Chromatogr. A 656*: 469–484 (1993).

Leo, A. and C. Hansch, *Chem. Rev. 71*: 525–616 (1971).

Tomlinson, E., *J. Chromatogr. 113*: 1–45 (1975).

Gabriela Cimpan

Lipophilic Vitamins by Thin-Layer Chromatography

Vitamins are defined as biologically active, organic compounds, controlling agents which are essential for an organism's normal health and growth, not synthesized within the organism, available in the diet in small amounts, and carried in the circulatory system in low concentrations to act on target organs or tissues. Vitamins are classified according to their solubility in water and in fats. Lipophilic vitamins are the vitamins of the groups A, D, E, and K. In chromatography of vitamins, the following problems must be solved: identification and the determination of vitamins in pharmaceutical preparations, identification and determination of vitamins and related substances in natural materials and foodstuffs, and chemical and biochemical determination of vitamins and their metabolites in fats and tissues. The isolation of the vitamins, their metabolites, and related substances from natural material is the most difficult task [1–4].

Vitamin A

There are different physiological forms known as vitamins A, namely retinol (vitamin A_1) and esters, 3-dehydroretinol (vitamin A_2) and esters, retinal (retinene, vitamin A aldehyde), 3-dehydroretinal (retine-2), retinoic acid, neovitamin A, and neo-b-vitamin A_1. There are active analogs and related compounds known as vitamins A, namely α-, β-, and γ-carotene, neo-β-carotene B, cryptoxanthine, myxoxanthine, torularhodin, aphanicin, and echinenone [3]. Vitamin A is susceptible to oxidation and degradation. Therefore, the control of the vitamin A level is recommended.

On silica gel thin-layer chromatography (TLC) plates eluted with a mixture of cyclohexane–ethanol (97:3, by volume), vitamin A_1 alcohol and *retro*-vitamin A_1 alcohol (R_F values 0.10 and 0.09, respectively) are separated from the pair vitamin A_1 acetate/*retro*-vitamin A_1 acetate (R_F values 0.48 and 0.45, respectively) and vitamin A_1 palmitate (R_F value 0.78) and anhydrovitamin A_1 (R_F value 0.87) [5]. Geometric isomers of retinol are separated on silica gel plates using hexane–ether (50:50, by volume) mixture as a mobile phase. Syn and anti forms of retinal oximes are separated from each other using cyclohexane–toluene–ethyl acetate (50:30:20, by volume) as a mobile phase [1,2]. On magnesium hydroxide, plates eluted with a benzene retinol (R_F value 0.29) is separated from retinal (R_F value 0.61) and retinyl acetate (R_F value 0.75). On magnesium oxide layers, and eluted with petrol ether (boiling point 90–110°C)–benzene (50:50, by volume) as the solvent, the carotenes are separated: ε-carotene, α-carotene, β-carotene, δ-carotene, and γ-carotene (R_F values 0.70, 0.66, 0.49, 0.20, and 0.11, respectively) [4]. In partition TLC on kieselguhr G, plates impregnated with paraffin oil (8% paraffin oil in petrol ether), and eluted with acetone–methanol–water (50:47:3, by volume), cryptoxanthin, echinenone, torularhodin methyl ester, and β-carotene (R_F values 0.91, 0.69, 0.57, and 0.22, respectively) are separated [4].

Very small amounts of the intensely colored pigments can be detected in ultraviolet (UV) light and also in visible light; the limit of detection is 0.01 μg of the carotenoids; 0.02–0.03 μg retinal can be detected after spraying of the rhodanine (orange-red color of the spot of retinal). Many vitamin A compounds fluoresce yellow–green in light of 365 nm wavelength (limit of detection 0.05 μg). Vitamin A compounds can be detected with antimony(III) and antimony(V)-chlorides (blue color of spot), with concentrated sulfuric acid (blue color of spot), with molybdophosphoric acid (green–blue color of spot), and with potassium dichromate in sulfuric acid (limits of detection 0.1–0.3 μg) [1,2,4].

Vitamin D

There are various physiological forms known as vitamins D, namely vitamin D_2 (calciferol, ergocalciferol), vitamin D_3 (cholecalciferol), phosphate esters of D_2, D_3, 25-hydroxycholecalciferol, 1,25-dihydroxycholecalciferol, and 5,25-dihydroxycholecalciferol. There are active anologs and related compounds known as vitamins D, namely 22-dihydroergosterol (vitamin D_4), 2-dehydrostigmasterol (vitamin D_6), and 7-dehydrositosterol (vitamin D_5) [3].

On silica gel plates, using a mixture of cyclohexane–dichloroethane–diethyl ether (5:3:2, by volume) as a mobile phase, provitamin D_3, tachysterol, lumisterol, and previtamin D_3 (R_F values 0.18, 0.23, 0.27, and 0.31, respectively) are separated [1,2]. The vitamins D_2 and D_3 (R_F values 0.50 and 0.41, respectively) are separated on kieselguhr F_{254} plates impregnated with a 10% solution of paraffin oil in benzene and eluted with binary mixtures of acetonitrile–water (9.5:0.5, by volume) [6]. Previtamins D_2 and D_3 (R_F value 0.50) are separated from vitamins D_2 and D_3 (R_F value 0.40) on silica gel GF_{254} plates using a mixture of cyclohexane–ether (50:50, by volume) as solvent. Vitamins D_2 and D_3 (R_F values 0.79 and 0.57, respectively) are separated from cholesterol, ergosterol, and 7-dehydrocholestrol (R_F values 0.41, 0.30, and 0.13, respectively) on silica gel G impregnated with silver nitrate and eluted with acetic acid–acetonitile (25:75, by volume) as solvent [4]. On silica gel high-performance thin-layer chromatography (HP–TLC) plates, eluted with a mixture of hexane–isopropanol (85:15, by volume), vitamin D_3 (R_F value 0.66) is separated from hydroxylated metabolites of vitamin D_3: $1,24,25(OH)_3D_3$, $1,25\text{-}(OH)_2D_3$, $25,26(OH)_2D_3$, $24,25(OH)_2D_3$, $23,25(OH)_2D_3$, and $25(OH)D_3$ (R_F values 0.14, 0.26, 0.35, 0.46, 0.51, and 0.59, respectively) [1,2].

Spots of vitamin D derivatives can be inspected in short-wavelength ultraviolet (UV) light (limit of detection 0.025–0.5 μg). Vitamins D_2 and D_3 can be detected with antimony(III) and antimony(V) chlorides (gray–blue and orange–red color of spots, respectively; limit of detection 0.025–0.3 μg), with concentrated sulfuric acid (brown and green color of spots respectively of vitamins D_2 and D_3; limit of detection 30 μg), with tungstophosphoric acid (gray–brown color of spots; limit of detection 0.2 μg), with molybdophosphoric acid (gray–blue color of spots; limit of detection 0.3 μg), and with trichloroacetic and trifluoroacetic acids (limit of detection 0.1–0.2 μg) [4].

Vitamin E

There are different physiological forms known as vitamins E, namely *d*-α-tocopherol, tocopheronolactone, and their phosphate esters. There are active analogs and related compounds knows as vitamins E, namely *dl*-α-tocopherol, *l*-α-tocopherol, esters (succinate, acetate, phosphate), and β-, ζ_1-, ζ_2-tocopherols [3].

α-, β-, γ-, and δ-Tocopherols are separated (R_F values 0.26, 0.48, 0.50, and 0.65, respectively) on kieselguhr G plates impregnated with a 10% solution paraffin oil in hexane and using a mixture of metanol–water (9.5:0.5, by volume) as solvent [7]. Tocotrienols are separated on silica gel G plates using a mixture of methanol–benzene (1:99, by volume); the R_F values for ζ_1-, ε-, and η-tocopherol and δ-T-3 are 0.55, 0.41, 0.39, and 0.29, respectively [4].

Vitamin E compounds can be detected (about 20 μg) as dark spots in UV light. They appear violet and detection is appreciably more sensitive (0.02 μg) on layers that contain 0.02% Na-fluorescein. Moreover, these are visible in daylight as reddish spots (limit of detection 2 μg). The same effect is produced by spraying with fluorescein or dichlorofluorescein reagent. Nonspecific visualization procedures for tocopherols and tocotrienols are based on spraying with sulfuric acid, molybdophosphoric acid, antimony(V) chloride, dipyridyl-iron reagent, nitric acid, and copper(II) sulfate–phosphoric acid [1–4].

Vitamin K

There are different physiological forms known as vitamins K, namely vitamin K_1 (phylloquinone, phytonadione) and vitamin K_2 (farnoquinone). There are active analogs and related compounds knows as vitamins K, namely menadiol diphosphate, menadione (vitamin K_3), menadione bisulfite, phthiocol, synkayvite, menadiol (vitamin K_4), menaquinone-*n* (MK-*n*), ubiquinone (Q-*n*), and plastoquinone (PQ-*n*) [3].

Selected K vitamins and related quinones are separated on silica gel G impregnated with silver nitrate using a diisopropyl ether as a solvent; the R_F values for vita-

Table 1 hR_F Values of Lipophilic Vitamins Using Various Separation Systems and Detection of Lipophilic Vitamins

					Detection in		
Layer[a] (~0.25 mm):	S_1	S_1	S_1	S_2	UV	UV	
Solvent[b]:	F_1	F_2	F_3	F_2	(254 nm)	(365 nm)	Daylight
β-Carotene[a]	84	100	100	78	Dark[d]	Dark[d]	Orange
Vitamin A–alcohol[c]	10	8	22	8	Dark[d]	Yellow–green[e]	—
Vitamin A–acetate[c]	45	41	69	62	Dark[d]	—	—
Vitamin A–palmitate[c]	72	75	94	74	Dark[d]	—	—
Vitamin D_2 or D_3	15	9	14	17	Dark[d]	—	—
α-Tocopherol	32	35	56	37	Dark[d]	—	—
α-Tocopherol–acetate	40	40	76	60	Dark[d]	—	—
Vitamin K_1[c]	61	67	81	73	Dark[d]	Dark[d]	Yellow
Vitamin K_3[c]	38	29	49	63	—	—	—

[a] S_1 = silica gel G; S_2 = alumina G.
[b] F_1 = cyclohexane–ether (80:20, by volume); F_2 = benzene; F_3 = chloroform.
[c] All-trans compounds.
[d] Absorption.
[e] Fluorescence.

min K (trans K-4), MK-3, MK-4, MK-5, MK-6, MK-7, and menadione are 0.72, 0.58, 0.49, 0.41 0.29, 0.21, and 0.45, respectively [4].

All lipoquinones in amounts of 0.5 μg and more are visible as dark spots on layers containing inorganic fluorescent material when illuminated with UV light, by adding Na-fluorescein or rhodamine B or 6G to the adsorbent or by spraying the chromatographed layer with fluorescein or dichlorofluorescein reagents. These compounds can be detected in daylight and, with high sensitivity, in UV light. Vitamin K compounds can be detected with iodine vapor (brown color of spots), with concentrated sulfuric acid followed by heating (violet color of spots, limit of detection 3 μg), with molybdophosphoric acid (gray–blue color of spots, limit of detection 0.5 μg) [1–4].

Lipophilic vitamins A, D_2, and E (α-tocopherol) (R_F values 0.34, 0.44, and 0.62, respectively) are separated on silica gel plates and a mixture of benzene–chloroform–acetone (88.5:8.8:2.7, by volume) applied [8]. Table 1 gives information about detection and the hR_F values of lipohilic vitamins in mixtures, using the various layers and solvents.

The assay of vitamins A, D, E, and K in food, feedstuffs, pharmaceutical preparations, organs, blood, and vegetable matter is described in Refs. 1–4. The application of TLC techniques, together with a densitometric scanning apparatus, now permit precise and sensitive quantification of vitamins A, D, E, and K compounds on TLC plates.

References

1. A. P. de Leenheer, W. E. Lambert, and H. J. Nelis, Lipophilic vitamins, in *Handbook of Thin-Layer Chromatography* (J. Sherma and B. Fried, eds.), Chromatographic Science Series Vol. 55, Marcel Dekker, Inc., New York, 1991, pp. 993–1019.
2. A. P. de Leenheer and W. E. Lambert, Lipophilic vitamins, in *Handbook of Thin-Layer Chromatography* (J. Sherma and B. Fried, eds.), Chromatographic Science Series Vol. 71, Marcel Dekker, Inc., New York, 1996, pp. 1055–1077.
3. R. Strohecker and H. M. Henning, *Vitamin Assay, Tested Methods*, Verlag Chemie, Weinheim, 1966.
4. E. Stahl (ed.), *Thin-Layer Chromatography, A Laboratory Handbook*, Springer-Verlag, Berlin, 1969.
5. J. Kahan, *J. Chromatogr. 30*: 506–513 (1967).
6. B. Kocjan and J. •liwiok, *J. Planar Chromatogr. Mod. TLC 7*: 327–328 (1994).
7. J. •liwiok and B. Kocjan, *Fat. Sci. Technol. 94*: 157–159 (1992).
8. M. Ranný, *Thin-Layer Chromatography with Flame Ionization Detection*, Prague, 1987.

Alina Pyka

Liquid Chromatography–Mass Spectrometry

The coupling of liquid chromatography and mass spectrometry (LC–MS) was initially considered impossible and actually is still characterized as a hyphenated technique. This can be seen, at first glance, to be due to the incompatibilities of the two methods. In order to introduce a conventional LC flow (0.5–1.5 mL/min) into the mass spectrometer, one should evaporate it, producing a vapor of a volume that is far beyond the capacity of the pump systems of a mass spectrometer. This means that an interface is required to couple the instruments, thus increasing price, complexity, and prestige of any such system. Until recently, most LC–MS systems were state-of-the-art, room-filling machines, reachable only by specialists. The last years' evolution in technology and electronics, together with stronger cooperation of LC with MS manufacturers, brought to the market bench-top, low-price, dedicated systems that require less handling and can even be used for routine analysis.

The advantages of an on-line LC–MS approach are many. Both techniques show high separation power and their combination on-line is a powerful tool for identification purposes as well as quantitative studies. Many detectors are available for high-performance liquid chromatography (HPLC): ultraviolet (UV), conductivity, electrochemical, fluorescence, refractometer, and so forth. Unfortunately, most of them lack specificity, selectivity, and sensitivity. Hence, identification of unknown compounds is actually impossible.

An ideal detector for HPLC should combine optimum sensitivity with maximum identification capability. It should also respond to all solutes and increase its signal linearly with the amount of the solute but, at the same time, be unaffected by changes in the mobile phase. Finally, it should also not contribute to extracolumn peak broadening and should be reliable and robust. Of the detectors used with HPLC, the mass spectrometer is the closest to the above requirements.

A mass spectrometer offers many attractive features as a chromatographic detector. Low detection limits down to the picogram range can be accomplished in optimized configurations. Mass spectrometry offers high specificity and mass distinction according to nominal mass (low-resolution MS). The specificity is even further increased in high-resolution MS, where mass distinction is made according to accurate mass. This enables the resolution of compounds with the same nominal mass but with different elemental composition (e.g., isomers). Mass spectrometry provides compound-specific information by fragmentation obtained either by hard ionization or by tandem mass spectrometry. Fragmentation patterns greatly increase identification power in the analysis of unknown compounds in complex matrices. Finally, a mass spectrometer is potentially a universal detector.

A diagram of a typical mass spectrometer scheme is depicted in Fig. 1. The simplest form of an MS system should perform the following fundamental tasks:

1. Vaporize compounds of varying volatility. This is accomplished in the inlet system. Introduction of the sample is done by direct insertion probe, reservoir inlet, or following a chromatographic separation (GC, HPLC, and CE). As mentioned earlier, to introduce the LC flow to the mass spectrometer on-line, we need an appropriate interface. Development of appropriate interfaces was the utmost for evolution of the LC–MS coupling.
2. Produce ions from neutral molecules in the vapor phase. This takes place in the ion source. As can be seen in Fig. 1, there are several modes of ion sources. Of those, the most used in LC–MS are electron spray ionization (ESI), chemical ionization (CI), thermospray ionization (TSP), fast-atom bombardment (FAB), and electron impact (EI).
3. Separate the formed ions according to their mass-to-charge ratio (*m*/*z* ratio). This takes place in the mass analyzer. By far, the most used analyzer is the quadropole, either as single or as triple quadropole, combined with a soft ionization technique. In the latter configuration, tandem mass spectrometry can be accomplished. The magnetic sector is also coupled to LC when there is a need for higher-resolution power or higher sensitivity.
4. Detect and record the separated ions. Multipliers are the most common detectors used in LC–MS instruments. Proper fully computerized data manipulation systems are required to handle the massive information flux from the detector.

Interfacing LC and MS benefits both techniques. An obvious question could be: Why do you need such a seemingly unorthodox coupling when other robust and well-behaving techniques (GC–MS) are already available? The great potential of the LC–MS approach is its applicability in the analysis of thermolabile or nonvolatile compounds;

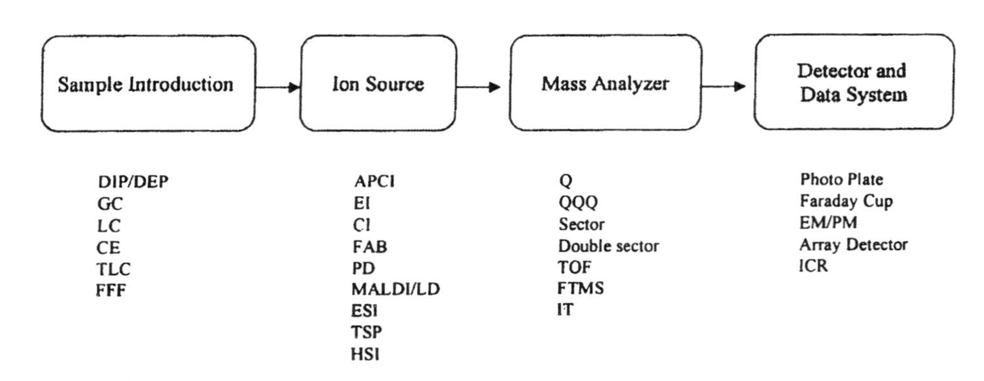

Fig. 1 Schematic representation of a mass spectrometer depicting its main components and the different modes used. Abbreviations: DIP: direct insertion probe; DEP: direct exposure probe; GC: gas chromatography; LC: liquid chromatography; CE: capillary chromatography; TLC: thin-layer chromatography; FFF: field-flow fractionation; APCI: atmospheric pressure ionization; EI: electron impact; CI: chemical ionization; FAB: fast-atom bombardment; PD: plasma desorption; MALDI: matrix-assisted laser desorption ionization; LD: laser desorption; TSP: thermospray; ESI: electron spray ionization; HSI: hypherthermal surface ionization; Q: quadropole; QQQ: triple quadropole; TOF: time-of-flight; FTMS: Fourier transform mass spectrometer; IT: ion trap; EM: electrom multiplier; PM: photomultiplier; ICR: ion cyclotron resonance.

this is of great interest in bioanalysis. Despite its high-resolving power, GC is actually suitable only for the analysis of volatile compounds of low molecular mass. LC–MS analysis gives a real boost to bioanalysis, enabling the positive identification and low-level detection of peptides, proteins, carbohydrates, polar analytes, and many other types of synthetic or naturally occurring compounds.

Interfaces

Numerous types of LC–MS interface have been developed. Early attempts were focused on methods of overcoming the incompatibility of the liquid flow rate and maintenance of the MS high vacuum. Recent approaches pay more respect to the practical use of ionization techniques that do not require sample volatilization. The main approaches to overcome incompatibilities are as follows:

1. Splitting of the flow. A high proportion of the sample is wasted this way.
2. Increasing the source housing pump speed, usually with the addition of a cryopump.
3. Miniaturizing the LC step.
4. Removal of the solvent prior to the introduction in the MS. This approach is used in transport interfaces (MB) and PB.
5. Attaching additional pump to the ion source, which can then accept higher flow rates. This approach is used in TSP.
6. Atmospheric pressure ionization (ESI, ISP).

The main features of LC–MS interfaces are illustrated in Table 1.

The first commercially available LC–MS system (Finnigan, 1976) used a moving-belt interface. This was a straightforward approach based on the 1960s instruments design of moving wires and cords. An endless continuous moving belt transports the analytes from the LC system to the MS. The LC flow is deposited on a polyimide belt and the mobile phase is removed by heating and subsequent evaporation in two consecutive vacuum chambers. Next, the analyte is desorbed from the belt and introduced into the ion source. On the return path, the belt passes over a cleanup heater to remove residual solvent and sample and, finally, through a water bath to remove any nonvolatile material.

More robust, sensitive, and user-friendly devices soon replaced the first types of interfaces (MB and DLI). During the 1980s, much effort was put on the development of TSP and PB. In the 1990s, ESI is by far the most used interface for LC–MS.

In ESI, the column effluent is passed through a small jet maintained at a high voltage (kilovolt range). Due to electronic charging, the liquid is dispersed into very small droplets. Desolvation of the droplets increases the electric field strength at the surface and leads to the ejection of charged compounds by ion evaporation. ESI demonstrates three major advantages: It is a gentle technique and leads to less thermal degradation compared to other techniques; because it is an atmospheric pressure ionization technique, up to 100% ionization can be achieved, thus enhancing method sensitivity; and multiple ions (sometimes up to 70–100 charges) are often formed, thus lowering the m/z ratio and permitting the analysis of high-molecular-weight compounds.

However, despite the great advantages it offers, LC–MS coupling has some limitations, apart from the interfacing need: Incompatibility with some of the nonvolatile buffers and other mobile phase additives. Hence, phosphates, ion-pairing agents, and amine modifiers are re-

Table 1 General Features and Applicability of Various LC–MS Interfaces

Type	Solvent	Flow rate (μL/min)	Ionization	Analyte type
MB	NP	<2000	EI, CI, FAB	Non/slightly polar
	RP (<50% H_2O)	<500	EI, CI, FAB	RV
DLI	RP (<50% H_2O)	<50	Solvent-CI	Non/medium polar, MV
PB	RP, NP	100–500	EI, CI, FAB	Non/medium polar, MV
TSP	RP, NP	500–2000	TSP, solvent-CI	Nonpolar to polar V to RIV
ESI	RP, NP	1–1000	IE, API	Polar, IV
ISP	RP, NP	10–1000	IE, API	Polar, IV
HNI	RP, NP	1–2000	APCI	Non/medium polar, RV
cfFAB	RP, NP	<20	FAB	Polar, IV

Note: MB: moving belt; DLI: direct liquid introduction; ISP: ion spray; HNI: heated nebuliser interface; cfFAB: continuous-flow FAB; RP: reversed phase; NP: normal phase; IE: ion evaporation; API: atmospheric pressure ionization; RV: relatively volatile; RIV: relatively involatile; IV: involatile. The rest of the abbreviations as in Fig. 1.

Table 2 Characteristics of Various LC Modes

LC mode	Inner diameter	Injection volume	Flow rate
Conventional	3–6 mm	10–50 μL	0.5–2 mL/min
Microbore	0.5–2 mm	0.5–2 μL	10–300 μL/min
Packed micro-capillary	0.2–0.5 mm	0.1 μL	1–10 μL/min
Drawned packed capillary	50–200 μm	10 nl	<0.1 μL/min
Open tubular	5–25 μm	1 nl	<0.1 μL/min

placed by ammonium acetate, ammonium formate, and so forth.

Care has to be taken that any connection will not introduce peak broadening; Interfaces are not compatible with all the ionization modes, thus although the MS is a universal detector, the fact remains that not all compounds can be analyzed with any instrumental configuration. The price still remains much higher compared to GC–MS.

The application area of LC–MS is rapidly growing. LC–MS is now regularly used for the analysis of many different types of compound: drugs and metabolites, herbicides–pesticides and metabolites, surfactants, dyes, saccharides, lipids–phospholipids, steroids, and many others. In our opinion, the area that profits more from the development of LC–MS is bioanalysis: natural products, proteins, peptides, nucleosides, and metabolic studies. Despite the current trends toward immunoassays–biospecific assays and capillary electrophoresis, LC–MS is an extremely powerful analytical technique that is considered complementary to the above mentioned, rather than competitive.

Finally, an aspect that should not be left out of consideration is the off-line combination of LC with MS. A great advantage of on-line techniques is the integration of the evaluation systems used. In on-line LC–MS systems, identification of incompletely resolved peaks is easily and unambiguously accomplished, taking advantage of the mass spectrometer separation power. In contrast, with off-line LC–MS, complete resolution of the peaks is essential. Moreover, fraction collection, evaporation, and sample introduction may result in sample loss or contamination. However, off-line LC–MS can still be of great value, for preparative purposes or in cases where a special mass spectrometry technique is required. For example, characterization of biopolymers or proteins by MALDI–TOF cannot be performed on-line with a chromatographic separation method, due to the absence of an appropriate interface. In this case, off-line LC–MS is the method of choice, enabling separation and analysis of very high molecular compounds (more than 1,000,000 Da).

Suggested Further Reading

P. J. Arpino, Combined liquid chromatography/mass spectrometry. A review, in *Mass Spectrometry in Biological Sciences: A Tutorial* (M. L. Gross, ed.), Kluwer, Dordrecht, 1992.

J. R. Chapman, Mass spectrometry as an LC detection technique, in *A Practical Guide to HPLC Detection* (D. Parriott, ed.), Academic Press, San Diego, CA, 1993.

C. Heeremans, Thermospray liquid chromatography tandem mass spectrometry, Ph.D. thesis, Leiden University, 1990.

P. S. Kokkonen, Continuous-flow fast atom bombardment liquid chromatography mass spectrometry in bioanalysis, Ph.D. thesis, Oulou University, 1991.

W. M. A. Niessen and J. Van der Greef, *Liquid Chromatography–Mass Spectrometry*, Marcel Dekker, Inc., New York, 1991.

A. P. Tinke, Surface ionization. A new approach in liquid chromatography–mass spectrometry, Ph.D. thesis, Leiden University, 1996.

K. B. Tomer, HPLC detection by mass spectrometry, in *HPLC Detection, Newer Methods* (G. Patonay, ed.), VCH, New York, 1992.

E. R. Verheij, Strategies for compatibility enhancement in liquid chromatography–mass spectrometry, Ph.D. thesis, Leiden University, 1993.

A. L. Yergey, C. G. Edmonds, I. A. S. Lewis, and M. L. Vestal, *Liquid Chromatography–Mass Spectrometry Techniques and Applications*, Plenum Press, New York, 1990.

Ioannis N. Papadoyannis
Georgios A. Theodoridis

Liquid–Liquid Partition Chromatography

Introduction

Liquid–liquid partition chromatography (LLPC), employing an organic–aqueous two-phase solvent system and a silica support to separate amino acids, was pioneered by Martin and Synge [1], for which they received the Nobel Prize. Its invention presented a significant advancement over conventional adsorption-based chromatographic processes that were rather unfavorable, due to the energetic heterogeneity of the adsorbent surface. By replacing the solid surface with a layer of an immiscible liquid to retain the solutes, much higher separation efficiencies could be obtained with LLPC. Its scope was later expanded to resolve water-insoluble sample mixtures as well as to isolate biopolymers by introducing organic–organic and aqueous–aqueous two-phase systems, respectively, and by extending the range of supports to include cellulose, dextran, and polymer-based supports.

The rapid development of hydrocarbonaceous bonded phases (nonpolar stationary phases with hydrocarbon chains covalently attached to the silica support) in the 1970s and 1980s for the popular reversed-phase chromatographic (RPC) technique has, however, eclipsed this technique somewhat, lately. Nevertheless, LLPC has become one of the most powerful separation techniques for the isolation of natural products and biopolymers.

Fundamental Principles and Theoretical Background

Martin and Synge provided the first theoretical treatment of LLPC by adapting the concept of theoretical plates which had been developed, mainly, for distillation and countercurrent extraction. According to the theory, a chromatographic column is considered to consist of a number of theoretical plates, within each of which perfect equilibrium occurs between the mobile and the stationary phases. Unlike in RPC employing hydrocarbonaceous bonded stationary phases where the retention mechanism is still the subject of controversy [2], three distribution processes contributing to the total retention of the solute may be clearly identified in LLPC systems. They are (a) partitioning between the two bulk liquid phases, (b) adsorption at the surface of the support, and (c) adsorption at the liquid–liquid interface. Assuming an inert support with large pores that minimize solute adsorption at the support surface and weak adsorption at the liquid–liquid interface, solute retention associated with undesirable adsorption effects may be neglected. Thus, separation in LLPC may be visualized as a multistep partition process, similar to countercurrent distribution, in which the differential migrations of various sample constituents are governed by their partition coefficients and by the volume ratio of the two phases. Then, the magnitude of solute retention in LLPC, measured under isocratic condition by the retention factor, k', is given by

$$k' = \frac{KV_s}{V_m} \tag{1}$$

where K is the partition coefficient of the solute in the two-phase system employed, V_m is the volume of the mobile phase, measured by an inert tracer and V_s is the volume of the stationary phase. The retention factor is evaluated directly from the chromatogram as

$$k' = \frac{V_r - V_m}{V_m} \tag{2}$$

where V_r is the retention volume of the solute.

Shake-flask measurements are often employed to design a suitable LLPC system for a given sample mixture by assisting in the selection of the two phases in which the compound of interest shows a partition coefficient sufficiently different from those of the impurities. One of the two phases is then immobilized on a suitable support that is packed into the column, and the second phase is used as the mobile phase. In general, partition coefficients of solutes obtained from static experiments compare favorably with those obtained from chromatographic experiments [3]. A comprehensive thermodynamic treatment of LLPC can be found in Ref. 4, and the prediction and control of zone migration is discussed extensively in Ref. 5.

Supports for Stationary Phases

Liquid–liquid partition chromatography systems have been used with or without stationary-phase support. Support-free LLPC has been classified as countercurrent chromatography (CCC). High-speed CCC techniques, based on planetary motion and coaxial orientation of the coiled column, have been developed that achieve both high partition efficiency and excellent retention of the stationary phase, thus circumventing the need for stationary-phase supports [6].

Liquid–liquid partition chromatography systems employing stationary-phase supports have been applied in both paper and column chromatographic modes. Paper strips impregnated with the liquid stationary phase are employed as supports for paper chromatography [5]. Column supports for LLPC bind the liquid stationary phase selectively enough to immobilize it and thus prevent its bleeding by the mobile phase. The support surface is homogeneously coated with the stationary phase and is sufficiently large to generate an interface for solute partitioning. Given these considerations, macroporous supports of average pore diameter greater than 50 nm are most suitable as column packings. Supports with smaller pores might contribute to retention by adsorption of solutes. In principle, both organic- and inorganic-based column packings can be employed in LLPC. Soft packings, such as cellulose, dextran, polymers, and diatomaceous earth, which do not withstand high pressures, are preferably used as larger particles at low pressures. On the other hand, microparticulate and bonded silicas are the supports of choice in the high-performance mode.

In the separation of large biopolymers, such as proteins and nucleic acids, specific supports characterized by large pores and bonded structure have been developed for LLPC employing poly(ethylene glycol) (PEG)–dextran two-phase aqueous–aqueous system. Bonded silica [e.g., LiChrospher® Diol (typically 1000 Å pore size and 10 μm particle size)] or a hydrophilic methacrylate polymer (e.g., Fractogel®) are chemically modified to form polyacrylamide chains of defined lengths that facilitate the selective binding of a dextran-rich aqueous phase. Polymer-based materials serve frequently as phase supports insensitive to high-pH systems.

Columns for LLPC are prepared by using solvent evaporation, direct coating, precipitation, or a dynamic coating technique [3]. In the solvent-evaporation technique, the nonvolatile liquid stationary phase is dissolved in a volatile solvent and mixed with the dry support. Then, the solvent is slowly removed by rotating the suspension in a rotary evaporator until the coated support remains as a dry, free-flowing powder. In the direct coating technique, the liquid stationary phase is pumped through the column, and then the excess is displaced by elution with a mobile phase saturated with the stationary phase. In the precipitation technique, the liquid stationary phase precipitates in the support pores when a solvent, which is not miscible with the liquid stationary phase, is pumped through the column. Finally, the liquid mobile phase saturated with a liquid stationary phase is pumped through the prepacked column in the dynamic coating technique. The liquid stationary phase is adsorbed at the surface of the support until it builds a multilayer and completely fills the pore volume.

Liquid–Liquid Systems

Binary and sometimes ternary systems employing phases of different polarity, which is characterized by Hildebrand's solubility parameter [7] or Snyder's polarity index [8], are used for LLPC separations. Applying the solubility parameter concept, various organic–organic, organic–aqueous, and aqueous–aqueous LLPC systems have been utilized to modulate solute retention. Based on polarity differences, LLPC can be operated in both normal phase (polar stationary phase, nonpolar mobile phase) and reversed-phase (polar mobile phase and nonpolar stationary phase) modes. Mobile phases typically used in the normal-phase mode are aliphatic hydrocarbons such as hexane, heptane and isooctane, alcohols, tetrahydrofuran (THF), methylene chloride, dioxane, and aromatic solvents. Stationary phases in the normal-phase mode include water, formamide, glycerol, and glycols. In the reversed-phase mode, hydro-organic mixtures such as acetonitrile–water and methanol–water are employed as mobile phases, whereas nonpolar solvents such as squalane, octanol, and aliphatic hydrocarbons are used as stationary phases. For the separation of large biopolymers based on aqueous–aqueous LLPC systems, a PEG-rich aqueous phase forms the mobile phase, and the dextran-rich aqueous phase is immobilized on the polyacrylamide support to form the stationary phase.

Applications

A number of advantages of current LLPC systems make them attractive for widespread application in the separation and isolation of low-molecular-weight compounds as well as macromolecules. Among them are simple preparation of bulk liquid phases, high reproducibility of retention and selectivity under isothermal conditions and lesser contamination, higher efficiency, and longer lifetime compared to adsorbent columns. Some of the most common applications of LLPC are now described.

Small Molecules

Liquid–liquid partition chromatography employing organic–organic, organic–aqueous, and aqueous–organic binary systems with column support have been used to separate small molecules of wide polarity [3]. A majority of applications involve the use of LLPC systems in the normal-phase mode (e.g., the separation of polyaromatic hydrocarbons, alcohols, esters, alkylated and chlorinated aromatic derivatives, pesticides, herbicides, and steroids). Several nitrogen-containing compounds, radionuclides, derivatives of aromatic carboxylic acids, benzodiazepines, aromatic compounds, and naphthalenesulfonic acids have successfully been separated in the reversed-phase mode. Many ternary LLPC systems in both the normal-phase and reversed-phase modes have also been employed in the separation of polyaromatic hydrocarbons, benzodiazepines, steroids, metal chelates, glycosides, metabolites, dansyl amino acids, herbicides, aliphatic carboxylic and dicarboxylic acids, sulfonic acids, nucleosides, barbiturates, chlorinated phenols, and alkylbenzenes. Support-free high-speed CCC has been employed in the separation of dyes, rare earth and certain inorganic elements, as well as in the preparative and analytical applications of medicinal herbs and other natural products, such as alkaloids, tannins, flavonoids, marine compounds, and anthraquinones [6].

Biopolymers

Classical LLPC using aqueous–aqueous polymer systems based on Albertsson's [9] PEG–dextran system has provided a versatile tool for the separation of proteins and nucleic acids, thus increasing the arsenal of biopolymer purification methods currently dominated by gel filtration, ion-exchange chromatography, and affinity chromatography RPC. The technique operates on the basis of the biopolymers partitioning between the top PEG-rich aqueous mobile phase and the bottom dextran-rich aqueous stationary [10]. Factors that depend on the nature of the protein, such as size, surface area, and hydrophobicity, determine their partition in the PEG–dextran systems. Other factors associated with the nature of the two-phase system that strongly influence protein retention in LLPC are the size of the phase-forming polymers, the pH of the system, and the presence of salts, zwitterions, detergents and organic solvents.

Liquid–liquid partition chromatography techniques based on aqueous–aqueous systems have successfully been employed in the fractionation of crude human serum, purification of steroid hormone-binding proteins from human serum, isolation of basic proteins from crude bacterial extracts, purification of immunoglobulins and monoclonal antibodies, DNA fractionations by size, topology and base sequence, as well as the isolation of soluble and ribosomal RNAs in preparative amounts from bulky mixtures [10]. High-speed CCC using PEG–dextran system has also been employed in the separation of proteins [6].

References

1. A. J. P. Martin and R. L. M. Synge, *Biochem. J. 35*: 1358–1368 (1941).
2. A. Vailaya and Cs. Horváth, *J. Chromatogr. A. 829*: 1–27 (1998).
3. J. C. Kraak and J. P. Crombeen, in *Practice of High Performance Liquid Chromatography* (H. Engelhardt, ed.), Springer-Verlag, Berlin, 1986, pp. 182–194.
4. D. C. Locke, *Adv. Chromatogr. 8*: 47–89 (1969).
5. E. Soczewinski, *Adv. Chromatogr. 5*: 3–78 (1968).
6. Y. Ito and W. D. Conway, *High Speed Countercurrent Chromatography*, Wiley–Interscience, New York, 1996.
7. Cs. Horváth, L. R. Synder, and B. L. Karger, *Introduction to Separation Science*, Wiley–Interscience, New York, 1974, pp. 49–55, 268–276.
8. L. R. Snyder, *J. Chromatogr. Sci. 16*: 223–234 (1978).
9. P. Å. Albertsson, *Partition of Cell Particles and Macromolecules*, 2nd ed., Wiley–Interscience, New York, 1971.
10. W. Müller, *Bioseparation 1*: 265–282 (1990).

Anant Vailaya

Long-Chain Polymer Branching: Determination by GPC–SEC

Introduction

Polymer branching has long been recognized as a main influencer of macromolecular properties, both chemical and physical. The increase in the number (and, possibly, variety) of end groups of a branched polymer, with respect to its linear counterpart, along with the concomitant increase in the number of branch points, has the potential to greatly alter the chemical reactivity of the polymer, the polymer's ability to crystallize, and so forth. The effects of branching, however, tend to manifest themselves most greatly in changes in the physical or space-filling properties of the molecule, both in the melt and in solution. Properties such as the viscosity and elasticity of melts, as well as the intrinsic viscosity, angular distribution of scattered radiation, and sedimentation behavior of dilute solutions are all affected by branching. As a branched molecule will possess a more compact structure in solution than a linear molecule composed of the same type and number of repeat units, its elution behavior as determined via size-exclusion chromatography [SEC, also known as gel permeation chromatography (GPC)] will be likewise altered, the branched molecule taking a longer time to elute from the column under identical chromatographic conditions. At this point, two factors should be mentioned. First, for the sake of the present discussion, it is assumed that only homopolymers are being analyzed. Nonetheless, when confronted by two SEC peaks with different retention volumes, the possibility presents itself that these peaks could either be different with respect to molecular weight, to chain branching, or to both. The synergistic nature of combining various SEC detectors, which are sensitive to dissimilar polymer properties, has now become a favored way to address this concern. These combinations of detectors have also proven to be fundamental in determining and quantifying polymer branching by SEC. Second, this entry deals exclusively with long-chain branching (LCB) in polymers, not with short-chain branching (SCB). Although a clear difference between these two types of branching is yet to be established, LCB is usually thought of as when the length of the side chains is comparable to that of the main chain or, in the absence of the latter, when random dendritic branching occurs.

Polymer Branching by SEC

Virtually every method of quantifying branching of dilute polymer solutions harkens back to a single publication by Zimm and Stockmayer from 1949 [1]. Here, the authors derived the root-mean-square radius of a polymer ($\langle R_g^2 \rangle^{1/2}$) and showed how the ratio of the mean-square radius of a branched polymer ($\langle R_g^2 \rangle_B$) to that of a linear polymer ($\langle R_g^2 \rangle_L$) may be used as the basis for branching calculations. This ratio, termed the ratio of the mean-square radii, has been given the symbol g and is defined by

$$g = \left[\frac{\langle R_g^2 \rangle_B}{\langle R_g^2 \rangle_L} \right]_M \tag{1}$$

with the subscript M referring to values obtained for the same molecular weight. (It should be noted that the term "radius of gyration" is often used in the literature both indiscriminately and mistakenly to describe, variously, the root-mean-square radius, the viscosity radius, the Stokes radius, and the hydrodynamic radius of a molecule. Caution is recommended when encountering this term undefined.) The assumptions inherent in the branching calculations were well recognized at the time: Only materials of equal chemistry and molecular weight should be compared, and the type (functionality, f) of the branch points should remain uniform throughout the molecule. The calculations were shown to be invariant to changes in the length of the branches throughout the molecular-weight distribution (MWD) of the polymer. In the case of trifunctional ($f = 3$) branching, the number- and weight-average number of branches per molecule (B_{3n} and B_{3w}, respectively) are given by Eqs. (2) and (3),

$$g = [(1 + B_{3n}/7)^{0.5} + 4B_{3n}/9\pi]^{-0.5} \tag{2}$$

$$g = \frac{6}{B_{3w}} \left[0.5 \left(\frac{2 + B_{3w}}{B_{3w}} \right)^{-0.5} \cdot \ln\left(\frac{(2 + B_{3w})^{0.5} + B_{3w}^{0.5}}{(2 + B_{3w})^{0.5} - B_{3w}^{0.5}} \right) - 1 \right] \tag{3}$$

whereas for tetrafunctional branching ($f = 4$), these same parameters are defined via Eqs. (4) and (5):

$$g = \left[\left(1 + \frac{B_{4n}}{6} \right)^{0.5} + \frac{4B_{4n}}{3\pi} \right]^{-0.5} \tag{4}$$

$$g = \frac{\ln(1 + B_{4w})}{B_{4w}} \tag{5}$$

(Many of the symbols used here are those which have found favor in the literature over the years and which are presently being used. As such, they may differ from the symbols used in the original publications.) From the number of branches, the branching frequency (λ) of the molecule may be calculated by Eq. (6),

$$\lambda = \frac{RB}{M} \tag{6}$$

where B is defined by Eqs. (2)–(5), R is the molecular weight of the repeat unit and M is the molecular weight of the branched material for the SEC slice under consideration. For f-functional stars, Stockmayer and Fixman [2] computed the number of arms (f) from g using

$$g = \frac{3f - 2}{f^2} \tag{7}$$

$\langle R_g^2 \rangle^{1/2}$ averages and the $\langle R_g^2 \rangle^{1/2}$ distribution with respect to the molecular weight of the polymer are best calculated by SEC using either a variable-angle or, more commonly and conveniently, a multiangle laser light-scattering (MALLS) detector, in conjunction with a concentration-sensitive detector (e.g., a differential refractometer). Various MALLS detectors are available commercially, and their data acquisition and manipulation software has the ability to perform the desired branching calculations. For a truly quantitative measure of the branching number and branching frequency of a polymer, it should be compared to a linear standard with identical chemical composition. Moreover, the MWD of the linear standard should span that of the unknown, so that at each molecular weight slice, the $\langle R_g^2 \rangle$ of the latter may be compared to that of the former. Also, the functionality of the polymer should be known, so that the proper branching equations (either those for B_3 or those for B_4) may be used.

If a viscometer is used instead of a MALLS detector in SEC (again in conjunction with a concentration-sensitive detector), either by applying universal calibration or the newly emerging technique of SEC [3], the ratio (g') of the intrinsic viscosities of the branched molecule and the linear standard at the same molecular weight may be used for the branching calculations (Eq. (8) [1]):

$$g' = \left(\frac{[\eta]_B}{[\eta]_L} \right)_M \tag{8}$$

(the same requirements for a linear standard as mentioned earlier continue to apply). The relationship between g and g' is given by

$$g' = g^{\varepsilon} \tag{9}$$

where ε is known as the viscosity shielding ratio. This ratio was defined by Debye and Bueche [3] as the distance within the hydrodynamic sphere occupied by the molecule in solution over which the flow of solvent decreases by a factor $1/e$ of that in the free solution, divided by the radius of said hydrodynamic sphere. (It should be noted that even though this definition applied strictly to the Debye–Bueche linear-density pearl-string model of a polymer in solution, the authors demonstrated its suitability to the more accurate Kirkwood–Riseman Gaussian-density polymer model.) The value of ε is dependent on a number of factors, including solvent, temperature, and branching, and may be considered indicative of how "draining" a polymer coil is in dilute solution. ε has been found to fall in the range 0.5–1.5, with the smaller value determined by Zimm and Kilb [4] to correspond to non-free-draining, high-molecular-weight, regular stars, and the larger value determined by Berry [5] for comb-shaped polystyrenes in good solvents. It has been noted that the majority of values for ε fall in the range 0.7–0.8 [6], although it is recommended that an exact value be either sought in the literature for the appropriate polymer under the given solvent/temperature conditions of the experiment or determined by a comparison of MALLS and viscometry data or by other experimental techniques (for a table of ε values, see Ref. 6).

An alternative method, also using viscometry, for determining g' without recourse to a linear standard is that implemented in the Waters Millennium software for SEC. In this method, the lower-molecular-weight portion of the MWD is assumed to be nonbranched, as branching normally does not express itself markedly at low molecular weights. The linear double-logarithmic intrinsic viscosity–molecular weight relationship (also known as the Mark–Houwink plot) is extrapolated from this portion, and deviations from this relationship at high molecular weights are used to calculate g'. Although this method has proven effective in corroborating the linearity of natural polymers and should be useful for calculating g' in certain types of LCB, doubts exist as to its applicability to extremely highly branched molecules such as hyperbranched polymers and dendrimers that have no linear portion in their MWDs from which to perform the requisite extrapolation [7]. One may alternatively calculate g' by what is known as the "mass method" [8], by comparing the molecular weights of the unknown (M_B) and of the linear standard (M_L) at the same elution volume (V), via

$$g' = \left(\frac{M_L}{M_B}\right)_V^{(a+1)/\varepsilon} \tag{10}$$

with a being the exponent of the Mark–Houwink relationship ($[\eta] = KM^a$) and ε the viscosity shielding ratio.

In 1953, Stockmayer and Fixman [2] observed that the ratio of hydrodynamic (Stokes) radii (R_H), given by the symbol h, could also be used as a measure of branching. Although they presented their relationship as a ratio of intrinsic viscosities [see Eq. (11)], the relatively recent introduction of on-line dynamic light scattering (DLS) detectors

$$h^3 = \frac{[\eta]_B}{[\eta]_L} \tag{11}$$

permits direct determination of h by

$$h = \frac{(R_H)_B}{(R_H)_L} \tag{12}$$

In certain cases, such as with lightly branched polymers, the h^3 rule appears more suitable than either the $g^{0.5}$ or $g^{1.5}$ rule. Future research on polymer branching using SEC with on-line DLS detection may shed new light on this area of polymer characterization.

Applications

For a detailed list of applications of SEC to determining LCB in polymers through the mid-1970s, see Ref. 9. For comprehensive, up-to-date reviews of the solution and conformational properties of branched molecules, see Ref. 10.

Conclusions

Long-chain polymer branching, which can have a great effect on the physical and chemical properties of macromolecules, may be determined using multidetector SEC by a variety of means; calculations will likely be aided by the rapid advances in computer modeling of polymer conformations and properties [7]. Whether one employs static or dynamic light scattering, viscometry, or a combination of viscometry, refractometry, and light scattering, stringent requirements on the nature of the linear standard must be met in order to assure accurate, quantitative results.

References

1. B. H. Zimm and W. H. Stockmayer, *J. Chem. Phys. 17*: 1301–1314 (1949).
2. W. H. Stockmayer and M. Fixman, *Ann. NY Acad. Sci. 57*: 334–352 (1953).
3. P. Debye and A. M. Bueche, *J. Chem. Phys. 16*: 573–579 (1948).
4. B. H. Zimm and R. W. Kilb, *J. Polym. Sci. 37*: 19–42 (1959).

5. G. C. Berry, *J. Polym. Sci. A-2 9*: 687–715 (1971).
6. J. Roovers, *Encyclopedia of Polymer Science and Engineering, Volume 2*, John Wiley & Sons, New York, 1985, pp. 478–499.
7. A. M. Striegel, R. D. Plattner, and J. L. Willett, *Anal. Chem. 71*: 978–986 (1999).
8. L.-P. Yu and J. E. Rollings, *J. Appl. Polym. Sci. 33*: 1909–1921 (1987).
9. E. E. Drott, *Liquid Chromatography of Polymers and Related Materials* (J. Cazes, ed.), Marcel Dekker, Inc., New York, 1977, pp. 161–167.
10. J. Roovers (ed.), *Advances in Polymer Science, Volume 143*, Springer-Verlag, Berlin, 1999.

André M. Striegel

Longitudinal Diffusion in Liquid Chromatography

Longitudinal diffusion refers to the natural spreading of a solute band from regions of high concentration to those of lower concentration as it passes through a chromatographic system [1]. It is a simple process which is dependent on the time that the solute spends on the chromatographic system, which, in turn, is related to the flow rate of the mobile phase.

As an extreme illustration of the longitudinal diffusion process, consider a situation in which a solute is put onto a liquid chromatographic column and the flow of the mobile phase is shut off prior to any of the solute eluting from the column. The column is then sealed. Under these conditions, the solute molecules in the column would begin to diffuse from a concentrated band outward to regions of lower concentrations in the column. Given enough time, this diffusion would continue until the solute concentration throughout the mobile phase in the sealed column is constant. Although this is an extreme example of band broadening caused by this phenomenon, such longitudinal diffusion occurs to a lesser extent even when the mobile phase is flowing.

The degree of band broadening of any chromatographic peak may be described in terms of the height equivalent to a theoretical plate, H, given by

$$H = \frac{L}{N} \tag{1}$$

where L is the length of the column (usually in cm) and N is the number of theoretical plates, which can be calculated from Eq. (2),

$$N = 16\left(\frac{t_R}{W}\right)^2 \tag{2}$$

where t_R and W are the retention time and width of the peak of interest, respectively. Because higher values of N correspond to lower degrees of band broadening and narrower peaks, the opposite is true for H. Therefore, the goal of any chromatographic separation is to obtain the lowest possible values for H.

The contribution of longitudinal diffusion and other factors to band broadening in liquid chromatography can be quantitatively described by the following equation, which relates the column plate height H to the linear velocity of the solute, μ:

$$H = A\mu^{0.33} + \frac{B}{\mu} + C\mu + D\mu \tag{3}$$

In this equation, A, B, C, and D are constants for a given column [2]. The linear velocity μ is related to the mobile-phase flow rate and is determined by

$$\mu = \frac{L}{t_0} \tag{4}$$

where t_0 (the so-called "dead time") is determined from the retention time of a solute which is known not to interact with the stationary phase of the column.

The second term in Eq. (3), B/μ, describes the contribution of longitudinal diffusion to band broadening of the solute as it passes through the chromatographic system. This is the only term in the equation inversely proportional to the linear velocity of the mobile phase; the other terms increase in value as the linear velocity increases. Giddings and others have also shown that this term is also directly proportional to the diffusion coefficient D_m of the solute in the mobile phase according to the following equation, where C_d is a constant [3]:

$$\frac{B}{\mu} = \frac{C_d D_m}{\mu}$$

Fortunately, the mobile-phase flow rates used in most modern liquid chromatographic separations are sufficiently high enough to minimize the effects of longitudinal diffusion on chromatographic band broadening under usual conditions. Other factors, such as solute mass transfer, are usually more important contributors to band broadening in LC. However, longitudinal diffusion can be an important factor in situations where the flow rate of the mobile phase is either stopped or significantly reduced during a chromatographic separation. Additionally, because it occurs throughout the path of a solute passing through a chromatographic system, efforts should be made to minimize longitudinal diffusion effects by minimizing open space ("extracolumn volume") in the chromatographic system before and after the column.

References

1. D. C. Harris, *Quantitative Chemical Analysis*, W. H. Freeman, New York, 1999, p. 662.
2. L. R. Snyder and J. J. Kirkland, *Introduction to Modern Liquid Chromatography*, 2nd ed., John Wiley & Sons, New York, 1979, pp. 168–173.
3. J. C. Giddings, *Dynamics of Chromatography*, Marcel Dekker, Inc., New York, 1965, pp. 47–61.

J. E. Haky

M

Magnetic FFF and Magnetic SPLITT

Magnetic field-flow fractionation (FFF) employs static or quasi-static magnetic fields and excludes electromagnetic fields. Electromagnetic fields having frequencies in the kilohertz to megahertz range are used in dielectrophoretic FFF. Static electric fields are used in electrical FFF (see the entry Field-Flow Fractionation Fundamentals).

Magnetophoretic Mobility and Magnetic Field Energy Density Gradient

The magnetic field is described by two vectors: magnetic field strength, H, and magnetic flux density (also called magnetic induction or, simply, magnetic field), B, described by Maxwell's equations. The magnetic field is solenoidal; that is, its sources are electric currents or magnetic dipoles (but not "magnetic charges"). The magnetic property of matter is described by its magnetic permeability, μ, which for an isotropic medium is a scalar. The difference between the external, applied field H and the induced field B is the magnetization of the matter, which is a vector quantity M. An external magnetic field always induces magnetization in matter and, in this sense, all matter is magnetic. (The microscopic basis of magnetization is provided by quantum physics and is beyond the scope of this entry.) For a large class of materials, magnetization is a linear function of the applied, external magnetic field strength, and it always consists of a diamagnetic contribution and, less often, a paramagnetic contribution. In the International System of Units, we have

$$B = \mu\mu_0 H, \qquad M = \frac{1}{\mu_0}B - H, \qquad \chi = \frac{|M|}{|H|} = \frac{M}{H}$$

where B is measured in tesla (T), H is measured in A/m, $\mu_0 = 4\pi \times 10^{-7}$ T m/A is a constant called the magnetic permeability of a vacuum, χ is the volumetric susceptibility of the material (dimensionless), and the symbol $|\cdot|$ denotes the magnitude of the vector. Note that $\mu = 1 + \chi$. Diamagnetic magnetization is antiparallel to the field strength ($\chi < 0$), and paramagnetic magnetization is parallel to the field strength ($\chi > 0$). Diamagnetic effects are very weak (with the notable exceptions of bismuth and graphite) and are dominated by much stronger paramagnetic effects.

Typical fluid media encountered in magnetic separation are diamagnetic (for water, $\chi = 9.05 \times 10^{-6}$). For certain metals, their alloys, and oxides, the coordinated behavior of the neighboring atoms or molecules (the totality of which is referred to as a magnetic domain) gives rise to very strong magnetic effects known as ferromagnetism and ferrimagnetism. For such material, magnetization increases very rapidly with the external field strength H, and the value of the magnetization depends on the history of the material (hysteresis effect). In particular, the magnetization does not disappear with the removal of the external field (residual magnetization) and requires application of an opposite field to be brought down to zero (coercive force). For convenience, diamagnetics and paramagnetics are characterized by their magnetic susceptibility, χ (typically, $\chi = 10^{-2}$ to 10^{-5}), whereas ferromagnetics are characterized by their magnetic permeability, μ (typically, $\mu = 10^5$). For particles having a size not larger than that of the magnetic domain, the hysteresis curve reduces to a straight line, and such particles are referred to as superparamagnetic particles. Paramagnetic and superparamagnetic particles and colloids are used to tag (or label) weakly magnetic particles and biological cells for their selective separation.

The fluid dynamic properties of continuous magnetic media, including liquid ferromagnetic media (ferrofluids), are described by the Navier–Stokes equations with the addition of terms describing the interaction with the magnetic field and are the subject of a considerable number of studies (for a review, see Ref. 1). The motion of discrete magnetic particles in a very weakly magnetic ("nonmagnetic") medium can be described by a combination of long-range, magnetic body forces and local shear stresses due to the resulting motion through the viscous medium. In the case of colloidal particles for which inertial effects are much lower than the viscous effects (low-particle Reynolds number), and for laminar flows (low-channel Reynolds number), the equation of motion reduces to

$$0 = 6\pi\eta R\nu - \Delta\chi V \nabla\left(\frac{B^2}{2\mu_0}\right)$$

from which

$$|\nu| = \nu = \frac{\Delta\chi V}{6\pi\eta R}\left|\nabla\left(\frac{B^2}{2\mu_0}\right)\right|$$

where η is the viscosity of the medium, R is the particle radius, V is the particle volume, ν is the particle linear velocity vector relative to the medium, and $\Delta\chi$ is the difference between the particle magnetic susceptibility and that of the medium. The operand acted on by the nabla operator, ∇ (resulting in a gradient of scalar quantity), is the external magnetic field energy density, $B^2/2\mu_0$, in a nonmagnetic medium. The expression for the field-induced velocity can be conveniently presented as a product of the particle magnetophoretic mobility, m, and the magnetic field energy density gradient, S_m,

$$\nu = mS_m$$

where

$$m = \frac{\Delta\chi V}{6\pi\eta R} \quad \text{and} \quad S_m = \left|\nabla\left(\frac{B^2}{2\mu_0}\right)\right|$$

For paramagnetic (and diamagnetic) particles in diamagnetic media, $\Delta\chi$ is independent of S_m and the magnetophoretic mobility depends entirely on the intrinsic properties of the particle and those of the medium. For a ferromagnetic particle, $\Delta\chi$ is, in general, a function of H, and therefore the magnetophoretic mobility depends on the properties of the particle, the medium, and the field strength. Particle magnetophoretic mobility provides the basis of magnetic separation, and its role is analogous to the electric and sedimentation mobilities encountered in the electrical and sedimentation split-flow thin-channel (SPLITT) separations, respectively.

The order-of-magnitude analysis of the magnetophoretic mobility of weakly paramagnetic particles, 10 μm in diameter, acted on by the body forces available for the magnetic separation, returns the following numerical values: $\Delta\chi = 10^{-3}$, $V = 524\ \mu\text{m}^3$, $\eta = 10^{-3}$ kg/m s, $R = 5\ \mu$m, $B = 1$ T, gradient of $B = 200$ T/m, $m = 5.6 \times 10^{-15}$ (m/s)/(N/m^3), and $S_m = 1.6 \times 10^8$ N/m^3. It follows, that the particle velocity induced by the magnetic field, $\nu = 9 \times 10^{-7}$ m/s, or about 1 μm/s.

Advantages of the magnetostatic or quasi-static field as compared to the oscillating electromagnetic fields in application to separation include the following: no interaction with the solvent, which typically is diamagnetic or very weakly paramagnetic, and therefore no Joule heating, no convective effects, and no need for a cooling system; no electro-osmotic effects; long range of the magnetic interactions (as compared to dielectrophoresis); the possibility of using permanent magnets for the magnetic field and thus no requirement for power supply; high specificity in targeting the separands by using magnetic labels as ligands to specific receptors; and no demonstrated biological effects in the practical range of magnetic fields and gradients. The limitations of the magnetic separation include a requirement for magnetic labels for nonmagnetic separands and, generally, a complex geometry of the magnetic force field, which tends to make it difficult to control the separation process.

Magnetic FFF

There is a limited number of publications concerning magnetic field-flow fractionation (magnetic FFF or MgFFF), which, characteristically, describe different technical approaches to achieve a uniformly high magnetic field and gradient over the entire channel volume. In this sense, magnetic FFF has not achieved the level of maturity of other FFF techniques for which the field and flow geometries are well established. Interestingly, there is a large volume of work on "magnetic chromatography," or "magnetic affinity chromatography," or "magnetic columns," which does not make reference to FFF and does not use its analytical methods, although it is based on differential retention of separands in the magnetic field [2]. The most successful of those devices is the "high-gradient magnetic separation" (HGMS) column, which is used for separation of small particulate matter (such as cells using magnetic ligands as labels) in biological and clinical laboratories [3].

The earliest, explicit reference to magnetic FFF was made by Vickrey and Garcia-Ramirez [4], although their methods and conclusions were questioned by Semenov and Kuznetsov [5]. Those early efforts pointed to the difficulties of achieving a high, uniform magnetic field energy gradient over a substantial part of a thin flow channel. (Vickrey and Garcia-Ramirez proposed an electromagnet with an iron core, and Semenov and Kuznetsov a ferromagnetic fiber exposed to an external magnetic field, as a source for S_m). There were no experimental data in the Semenov and Kuznetsov study however. In a more rigorous work by Schunk et al. [6], an electromagnet with an iron core and a variable current was used to generate a well-characterized magnetic field. Due to the low field gradient, and thus low S_m, these investigations were limited to particulate materials of high magnetic permeability (iron oxide, Fe_2O_3, 0.8-μm particles used in recording media). They showed that their magnetic FFF method was able to characterize flocculation of those particles in suspension in acetonitrile (singlets from dimers). In a more recent work, Ohara discussed theoretically the separation of small radioactive waste particles (1–50 nm in diameter) from nuclear fuel (not specified) using magnetic FFF consisting of a very thin flow channel (50 μm) lined with a regular grid of 20-μm ferromagnetic wires in the field of a superconducting magnet (2.0–28.0 T) and concluded that magnetic FFF would be superior to HGMS for such separations (no experimental data to support this conclusion were given) [7].

Magnetic SPLITT

The requirement of a highly regular field and flow geometries is even more stringent in split-flow thin channel (SPLITT) separation than it is in FFF separation and this may be the reason for a relatively late appearance of the magnetic SPLITT technique. The solenoidal character of the magnetic field requires a radical departure from the usual rectangular geometry of the SPLITT channel. This has led to the design of an annular channel around an axisymmetric magnetic quadrupole field (similar to that used in quadrupole mass spectroscopy) [8]. Quadrupole magnetic SPLITT offers the advantages of a highly regular S_m, which is a linear function of the distance from the axis of symmetry and directed along the radial position vector, and of an edgeless flow channel. The theory of the quadrupole magnetic SPLITT separation has been developed and its experimental verification in biological applications is ongoing [9]. It appears that it has advantages over a more traditional approach utilizing a rectangular channel.

A SPLITT system based on a quadrupole magnetic field combined with annular channel geometry possesses an axial symmetry that not only lends itself to the separation but also to the tractability of the mathematical description of the separation process. Given the magnetophoretic mobility of a paramagnetic particle, m, the magnetic field energy gradient in the annular channel, the system dimensions, and flow rates, it is possible to predict particle trajectory through the system as a function of starting position. It follows that it is possible to calculate the flow rate conditions required to obtain any desired cutoff in mobility. All particles having mobilities higher than a specified critical value will be collected at the outer channel outlet, and those having mobilities lower than another critical mobility will be collected at the inner channel outlet; see Fig. 1. The resolution of the fractionation is related to how close these two critical values are and the volumetric throughput is inversely related to this resolution.

The availability of immunomagnetic colloidal labels opens up the possibility of purifying or isolating certain specific biological cell types using magnetic SPLITT. These procedures are useful for cancer treatment and various cell therapies. Unlabeled biological cells tend to be only very weakly influenced by magnetic fields. Flow conditions may be selected so that when a mixture of labeled and unlabeled cells is passed through the magnetic SPLITT system, the labeled cells are selectively drawn toward the outer channel wall and are collected at the outer channel outlet. An example of this type of separation is shown in Fig. 1.

In general, the magnetophoretic mobility of a weakly magnetic particle, or a cell, labeled with a magnetic ligand, is directly proportional to the number of ligand molecules, or moieties, attached to the target particle or cell. An important question in magnetic SPLITT is this: Given the upper and lower limit of the number of magnetic receptors per particle (or cell), and an upper and lower limit of the particle diameter, into how many different fractions may one separate the sample? If one assumes a Normal distribution in size for the particle (or cell), the theoretical number of subpopulations that a magnetically labeled particle (or cell) population can be divided into is given by

$$n = \ln\left(\frac{\alpha_n}{\alpha_0}\right)\left[\ln\left(\frac{1+z\sigma}{1-z\sigma}\right)\right]^{-1}, \quad 0 < |z\sigma| < 1$$

where α_n and α_0 are the high and low values, respectively, of the magnetic receptor density and σ is the relative standard deviation in cell diameter. The number z is chosen according to the desired level of certainty of including cells of a given receptor density in a sorted fraction. For

(a)

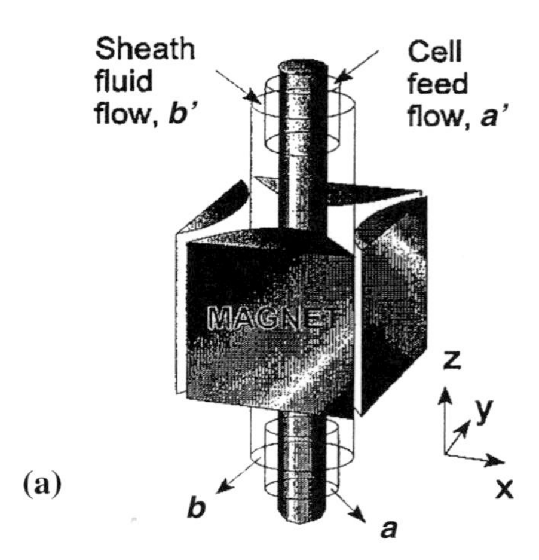

(b)

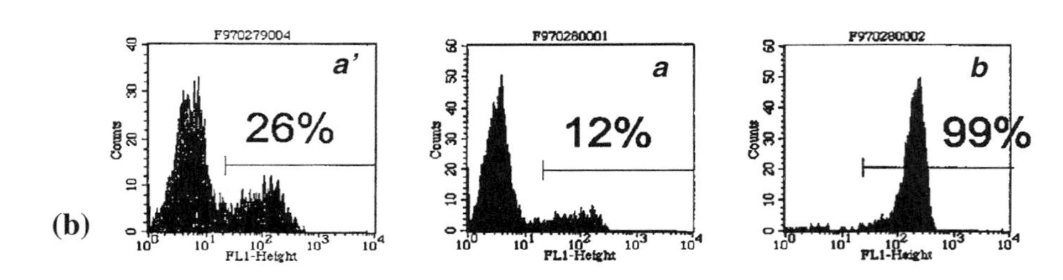

Fig. 1 (a) Schematic view of magnetic SPLITT based on quadrupole magnetic field and annular flow channel. Note the positions of short cylindrical flow splitters, concentric with the solid cylindrical core. Note direction of the feed flow, a' sheath fluid flow, b', and sorted fraction flows (a and b) as indicated by arrows. (b) Composition of human peripheral lymphocytes labeled with anti-CD8 antibody and magnetic colloid, as analyzed by flow cytometry, before and after magnetic SPLITT sorting: a' = original sample, a = nonmagnetic fraction, b = magnetic fraction. The percentages indicate fractional composition of the magnetically labeled cells in the sample. Note enrichment of the positive cells in b fraction as compared to the original cell sample, a'.

example, for $z = 2$, the percentage of cells of a given receptor density in the sorted fraction is 95.5%. A more complete discussion of this relationship can be found in Ref. 10.

References

1. R. E. Rosensweig, *Ferrohydrodynamics*, Cambridge University Press, New York, 1985.
2. C. H. Evans, A. P. Russell, and V. C. Westcott, Demonstration of the principle of paramagnetic chromatography for resolving mixtures of particles, *J. Chromatogr. 351*: 409–415 (1986).
3. A. Radbruch, B. Mechtold, A. Thiel, S. Miltenyi, and E. Pflueger, High-gradient magnetic sorting, *Methods Cell Biol. 42*: 387–403 (1994).
4. T. M. Vickrey and J. A. Garcia-Ramirez, Magnetic field-flow fractionation: Theoretical basis, *Separ. Sci. Technol. 15*: 1297–1304 (1980).
5. S. N. Semenov and A. A. Kuznetsov, Flow fractionation in a transverse high-gradient magnetic field, *Russ. J. Phys. Chem. 60*: 424–428 (1986) (translated from *Zh. Fizi. Khim.*).
6. T. C. Schunk, J. Gorse, and M. F. Burke, Parameters affecting magnetic field-flow fractionation of metal oxide particles, *Separ. Sci. Technol. 19*: 653–666 (1984).
7. T. Ohara, Feasibility of using magnetic chromatography for ultra-fine particle separation, in *High Magnetic Fields: Applications, Generation, Materials* (H. J. Schneider-Muntau, ed.), World Scientific, Singapore, 1997.
8. M. Zborowski, P. S. Williams, L. Sun, L. R. Moore, and J. J. Chalmers, Cylindrical SPLITT and quadrupole magnetic field in application to continuous-flow magnetic cell

sorting, *J. Liquid Chromatogr. Related Technol. 20*: 2887–2905 (1997).
9. P. S. Williams, M. Zborowski, and J. J. Chalmers, Flow rate optimization for the quadrupole magnetic cell sorter, *Anal. Chem.* (1999).
10. J. J. Chalmers, M. Zborowski, L. R. Moore, S. Mandal, B. Fang, and L. Sun, Theoretical analysis of cell separation based on cell surface marker density, *Biotechnol. Bioeng. 59*: 10–20 (1998).

Maciej Zborowski
Jeffrey J. Chalmers
P. Stephen Williams

Mark–Houwink Relationship

The molecular weight of polymer molecules can be determined by the measurement of the viscosity of dilute polymer solutions [1]. The relationship used is the so-called Mark–Houwink (MH) empirical equation:

$$[\eta] = KM^{a} \tag{1}$$

where the intrinsic viscosity $[\eta]$, also called the limiting viscosity number, is proportional to the polymer molecular weight, M, through the constants K and a, valid for each polymer–solvent system at a given temperature. The constants of Eq. (1) are obtained by measuring the intrinsic viscosities, in the solvent and at the temperature of choice, of a series of polymer samples having different and known molecular weights.

For flexible macromolecules, the exponent a takes values between 0.5 and 0.8, whereas, for rigid chains, it can reach values higher than 1, up to 2.

The intrinsic viscosity of a polymer is obtained from the viscosities η and η_0 of solution and solvent, respectively, through the following transformations. The relative viscosity is the ratio $\eta_{rel} = \eta/\eta_0$. By assuming that the viscosity η of a dilute solution is given by the sum of viscosities from solvent and solute molecules, the specific viscosity, η_{sp}, represents the polymer contribution to viscosity:

$$\eta_{sp} = \frac{\eta - \eta_0}{\eta_0} = \eta_{rel} - 1 \tag{2}$$

and dividing by the concentration, c, the reduced viscosity η_{sp}/c is obtained. The intrinsic viscosity is the value of reduced viscosity at infinite dilution:

$$[\eta] = \lim_{c \to 0} \frac{\eta_{sp}}{c} \tag{3}$$

Experimentally, the viscosity of dilute polymer solutions is, in most cases, determined with glass capillary viscometers, making application of the Hagen–Poiseuille's law for laminar flow of liquids. The time required for a specific volume of a liquid to flow through a capillary of defined length and radius is proportional to the ratio of the viscosity by the density of the liquid itself. As the density of a dilute solution may be considered practically equal to that of the pure solvent, the ratio of efflux time of the solution, t, to that of solvent t_0, gives the relative viscosity:

$$\eta_{rel} = \frac{t}{t_0} = \frac{\eta}{\eta_0} \tag{4}$$

The relative viscosities of polymer solutions are measured at different concentrations and a plot of the reduced viscosity versus concentration is made, in order to extrapolate to zero concentration. The concentration dependence of the viscosity of polymer solutions, in the dilute regime, may be expressed by several linear equations. For practical extrapolation to zero concentration, the most commonly employed are the Huggins equation:

$$\frac{\eta_{sp}}{c} = [\eta] + k_H[\eta]^2 c \tag{5}$$

and the Kraemer equation, where a new quantity, the inherent viscosity η_{inh}, is defined:

$$\eta_{inh} = \ln\left(\frac{\eta_{rel}}{c}\right) = [\eta] + k_K[\eta]^2 c \tag{6}$$

The constants of the two equations are connected by the relationships $k_K = k_H - 0.5$. Given that, for flexible polymers, the k_H values vary between 0.3 and 0.5, the slope of Kraemer equation is generally negative, with absolute values lower than the Huggins slope. This helps in the extrapolation procedure which is conveniently made, in order to reduce experimental uncertainties, by plotting in the same graph the viscosity data according to Eqs. (5) and (6), as shown in the example of Fig. 1.

The combination of Eqs. (5) and (6) with the assumption that $k_H + k_K = 0.5$ leads to the Solomon–Ciuta equation:

$$[\eta] = \frac{[2(\eta_{sp} - \ln \eta_{rel})]^{0.5}}{c} \tag{7}$$

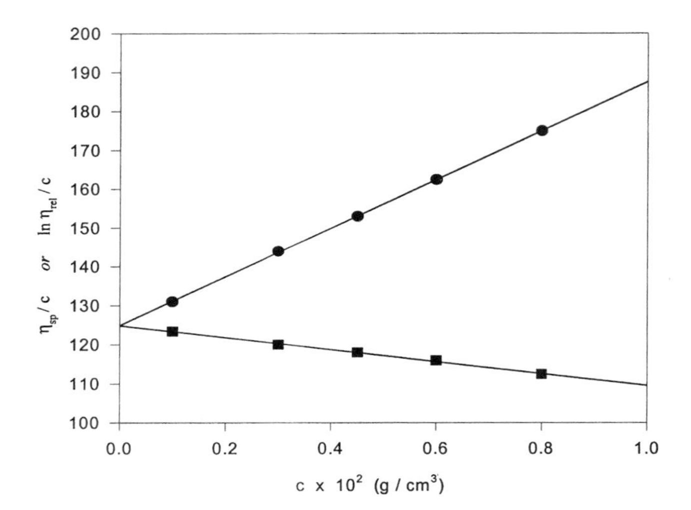

Fig. 1 Double extrapolation of viscometric data according to the Huggins and Kraemer equations: ●: reduced viscosity; ■: inherent viscosity.

which may be used to obtain intrinsic viscosity by a single measurement performed at reasonably low concentration. Equation (7) finds application also in the evaluation of data from viscometer absolute detectors in gel permeation–size-exclusion permeation (GPC–SEC) [2].

In the case of polymers with different chain lengths, the molecular weight obtained from the Mark–Houwink equation is a viscosity-average molecular weight, M_v, whereas the intrinsic viscosity is the weight average. In fact, at infinite dilution, one may write

$$[\eta] = \Sigma\left(\frac{\eta_{sp,i}}{c_i}\right) = \Sigma\left(\frac{c_i[\eta]_i}{c_i}\right) = \Sigma\, w_i[\eta]_i = [\eta]_w \tag{8}$$

where $w_i = c_i/\Sigma\, c_i$ is the weight fraction of the ith component.

The expression for M_v is immediately derived by combining Eq. (8) with the Mark–Houwink relationship:

$$[\eta]_w = \Sigma\, w_i[\eta]_i = K\, \Sigma\, w_i M_i^a = KM_v^a \tag{9}$$

$$M_v = (\Sigma\, w_i M_i^a)^{1/a} \tag{10}$$

The viscosity-average molecular weight is located between the number- and weight-average values but is, in any case, closer to the weight average. From Eq. (10), it may be seen that viscosity and weight averages coincide for $a = 1$. It is also worth noting that the M_v of a polymer is not a unique definite value because, depending on the exponent a, it varies with the solvent, where the polymer is dissolved.

Mark–Houwink relationships are also important for the application of the universal calibration procedure in GPC–SEC of polymer molecules, where the calibration curve is expressed in terms of the size of fractionated molecules against retention volumes. The intrinsic viscosity of a polymer (expressed in cm^3/g) is, in practice, a measure of the volume occupied by a unit weight of the macromolecules in the solution. From the Flory–Fox equation for viscosity of flexible chain molecules and from the Einstein relationship for viscosity of a dispersion of spheres [3], it turns out that the product $[\eta]M$ can be used to represent the dimension of polymer molecules in the solution:

$$[\eta]M = \Phi'\alpha^3\langle s_0^2\rangle^{3/2} = 2.5N_A V_h \tag{11}$$

Φ' is the so-called Flory's constant, α is the expansion factor of the polymer molecule, which depends from the thermodynamic quality of the solvent ($\alpha = 1$ in ideal solvent), $\langle s_0^2\rangle$ is the mean-square radius of gyration, N_A is Avogadro's number, and V_h is the volume of the equivalent hydrodynamic sphere.

In GPC–SEC with universal calibration, at each retention volume i of the chromatogram, a value of $[\eta]_i M_i$ is read, and for molecular-weight and molecular-weight-distribution calculations the values of $[\eta]_i$ are needed. These are obtained, whenever the equation is known, from the Mark–Houwink constants of the polymer in the solvent and at the temperature of chromatographic elution.

A selection of literature values for Mark–Houwink constants K and a in tetrahydrofuran, which is the eluent most commonly used in size-exclusion chromatography, is collected in Table 1 for the principal polymer structures. An accurate, more extensive compilation of the same constants for homopolymers and copolymers may be found in Ref. 22.

References

1. N. C. Billingham, *Molar Mass Measurements in Polymer Science*, John Wiley & Sons, New York, 1977.
2. W. W. Yau, S. D. Abbot, G. A. Smith, and M. Y. Keating, *ACS Symp. Ser. 352*: 80 (1987).
3. P. J. Flory, *Principles of Polymer Chemistry*, Cornell University Press, Ithaca, NY, 1953.
4. M. Kolinsky and J. Janca, *J. Polym. Sci. A-1 12*: 1181 (1974).
5. U. K. O. Schroder and K. H. Ebert, *Makromol. Chem. 188*: 1415 (1987).
6. D. Goedhar and A. Opshoor, *J. Polym. Sci. A-2 8*: 1227 (1970)
7. Z. Grubisic, P. Rempp, and H. Benoit, *J. Polym. Sci. B 5*: 753 (1967).
8. Y-J. Chen, J. Li, N. Hadjichristidis, and J. W. Mays, *Polym. Bull. 30*: 575 (1993).
9. G. Samay, M. Kubin, and J. Podesva, *Angew. Makromol. Chem. 72*: 185 (1978).
10. M. Szesztay and F. Tuedoes, *Polym. Bull. 5*: 429 (1981).
11. J. Xie, *Polymer 35*: 2385 (1994).
12. J. M. Evans, *Polym. Eng. Sci. 13*: 401 (1973).
13. X. Zhongde, S. Minghsi, N. Hadjichristidis, and L. J. Fetters, *Macromolecules 14*: 1591 (1981).

Table 1 Literature Values for Mark–Houwink Constants K and a in Tetrahydrofuran

Polymer	Temp. (°C)	$K \times 10^2$ (cm^3/g)	a	Ref.
Polystyrene	23	1.11	0.723	4
Poly(methyl styrene)	25	4.2	0.608	5
Poly(vinyl chloride)	25	1.50	0.77	4
Poly(vinyl acetate)	25	3.50	0.63	6
Poly(methyl methacrylate)	23	0.93	0.69	7
	25	1.08	0.702	8
Poly(ethyl methacrylate)	25	1.549	0.679	9
Poly(butyl methacrylate)	25	0.503	0.758	9
Poly(methyl acrylate)	25	0.388	0.82	10
Polyisobutylene)	40	5.79	0.593	11
Polycarbonate	25	3.99	0.77	12
Polybutadiene	30	2.56	0.74	13
Polyisoprene	25	1.77	0.753	14
Butyl rubber	25	0.85	0.75	12
Poly(vinyl butyral)	25	1.4	0.80	15
Poly(2-vinyl pyridine)	25	2.23	0.66	16
Poly(dimethyl siloxane)	25	0.65	0.77	17
Cellulose nitrate	25	25.0	1.00	18
Poly(DL-lactic acid)	31.15	5.49	0.639	19
Poly(ethylene-*co*-vinyl acetate) (27–29% VA)	20	9.7	0.62	20
Poly(ethylene-*co*-propylene-*co*-ethylidene norbornene) (EPDM: 27% PP, 11.5%: ENB)	35	27.4	0.54	21

14. C. Kraus and C. J. Stacy, *J. Polym. Sci. A-2 10*: 657 (1972).
15. L. Mrkvickova, J. Danhelka, and S. Pokorny, *J. Appl. Polym. Sci. 29*: 803 (1984).
16. C. Hugelin and A. Dondos, *Makromol. Chem. 126*: 207 (1969).
17. L. Mrkvickova and N. Radhakrisnan, *Eur. Polym. J. 33*: 1403 (1997).
18. A. Rudin and H. W. Hoegy, *J. Polym. Sci. A-1 10*: 217 (1972).
19. J. A. P. P. van Dijk, J. A. M. Smit, F. E. Kohn, and J. Feijten, *J. Polym. Sci., Polym. Chem. Ed. 21*: 197 (1983).
20. J. Echarri, J. J. Iruin, G. M. Guzman, and J. Amsorena, *Makromol. Chem. 180*: 2749 (1979).
21. O. Chiantore, P. Cinquina, and M. Guaita, *Eur. Polym. J. 30*: 1043 (1994).
22. S. Mori and H. Barth, *Size Exclusion Chromatography*, Springer-Verlag, Berlin, 1999.

Oscar Chiantore

Mass Transfer Between Phases

Chromatography is a separation method which involves two phases — one stationary and one mobile. A mixture is introduced into the mobile phase and is carried through the system by it. At some point, the mobile phase passes over and through the stationary phase. The components of the mixture partition between the two phases, resulting in different migration rates through the system [1]. At any given point, an analyte molecule will either be moving along the mobile phase or be held immobile in the stationary phase. A separation results as the molecules emerge from the bed at different times, which are called retention times.

The retention time of a solute is partially controlled by how effectively it interacts with the stationary phase as it passes through the column. As a solute molecule moves through a chromatographic system, it is carried through the mobile phase to a new solution site in the stationary phase. Simultaneously, other solute molecules are moving from the stationary phase and are being conducted through the column by the mobile phase. In any separation of components of a mixture by liquid chromatography, the rate at which this repeated transfer of solutes between the stationary phase and the mobile phase is an important factor affecting retention, peak shape, and resolution.

Mass transfer in both the stationary and mobile phases are not instantaneous and, consequently, complete equilibrium is not established under normal separation conditions [2]. The result is that the solute concentration profile in the stationary phase is always displaced slightly behind the equilibrium position and the mobile-phase profile is slightly in advance of the equilibrium position [2]. A high degree of displacement will lead to wider peaks and reduced resolution. In fact, the largest problem associated with mass transport in the packed column revolves around moving the solute from the stationary phase to the mobile phase.

The degree of band broadening of any chromatographic peak may be described in terms of the height equivalent to a theoretical plate, H, given by

$$H = \frac{L}{N} \tag{1}$$

where L is the length of the column (usually in cm) and N is the number of theoretical plates, which can be calculated from

$$N = 16\left(\frac{t_R}{W}\right)^2 \tag{2}$$

where t_R and W are the retention time and width of the peak of interest, respectively.

Because higher values of N correspond to lower degrees of band broadening and narrower peaks, the opposite is true for H. Therefore, the goal of any chromatographic separation is to obtain the lowest possible values for H.

The contribution of mass transfer and other factors to band broadening in liquid chromatography can be quantitatively described by the following equation, which relates the column plate height H to the linear velocity of the solute, μ:

$$H = A\mu^{0.33} + \frac{B}{\mu} + C\mu + D\mu \tag{3}$$

In this equation, A, B, C, and D are constants for a given column [3]. The linear velocity μ is related to the mobile-phase flow rate and is determined by

$$\mu = \frac{L}{t_0} \tag{4}$$

where t_0 (the so-called "dead time") is determined from the retention time of a solute, which is known not to interact with the stationary phase of the column.

The first term in Eq. (3), $A\mu^{0.33}$, includes the contribution of eddy diffusion to band broadening as well as that of mass transfer of the solute through the mobile phase. This contribution of this mobile-phase mass transfer to this term, H_i, increases with the square of the stationary-phase particle diameter d_p. It is also inversely proportional to the diffusion coefficient of the solute in the mobile phase, D_m, according to

$$H_i = \frac{C_m d_p^2 \mu}{D_m} \tag{5}$$

where C_m is a constant.

The third and fourth terms of Eq. (3) also relate to mass transfer. The third term, $C\mu$, describes the contribution of mass transfer of solutes to and from areas in the column where the mobile phase is stagnant (e.g., within the pores of the packing). The size of this term is related to stationary-phase particle diameter and solute diffusion coefficient according to

$$C\mu = \frac{C_{sm} d_p^2 \mu}{D_m} \tag{6}$$

where C_{sm} is a constant.

Finally, the fourth term in Eq. (3), $D\mu$, describes the contribution of mass transfer of solutes to and from the stationary phases. This term is related to the thickness d_f of the phase that coats the stationary phase and the diffusion coefficient D_s of the solute in the stationary phase according to

$$D\mu = \frac{C_s d_s^2 \mu}{D_s} \tag{6}$$

where C_s is a constant.

To minimize band broadening in liquid chromatography, conditions must be established to minimize each of the terms described by Eqs. (5)–(7). Because each of these terms is directly proportional to mobile-phase linear velocity, employing the lowest possible mobile-phase flow rates would seem to serve this purpose. However, use of extremely low flow rates [<0.5 mL/min for a standard high-performance liquid chromatography (HPLC) column] can increase solute retention times to impractical levels and may actually increase band broadening due to increased longitudinal diffusion [described by the second term in Eq. (3)]. For this reason, other factors are usually adjusted to minimize these mass transfer terms. Such adjustments include (a) using monomerically bonded stationary phases with small particle diameters [this reduces the size of d_f and d_p terms in Eqs. (5)–(7), which, in turn, has an exponential effect on reducing the size of the mass transfer terms]; (b) employing mobile phases of low viscosity at high temperatures (this increases the sizes of the diffusion coefficients D_s and D_m in the equations, resulting in fast mass transfer and narrow chromatographic bands [4].)

Column manufacturers and researchers have optimized the above parameters to produce LC columns of remarkably high selectivities and efficiencies. However, there are practical limitations to adjustments of these parameters. For example, stationary phases with particle diameters below 3 μm generally cannot be routinely used, owing to excessively high back-pressures and short column lifetimes. Additionally, for obvious reasons, operating temperatures must be kept below the boiling points of the components of the mobile phase. Use of nonporous pellicular column packings has also been attempted in an effort to eliminate any areas of stagnant mobile phase in the chromatographic system, thus reducing the size of the third term in Eq. (3) to zero. However, such pellicular stationary phases have very low sample capacities, and diffusion coefficients of many solutes on their polymeric coatings are often low, which, of course, results in increased band broadening [3].

At commonly used mobile-phase flow rates, mass transfer of solutes through and between the stationary phase and the mobile phase is generally the most important factor controlling the widths of chromatographic bands in liquid chromatography. Although manufactures have designed packings and columns to minimize their effects, consideration of solute mass-transfer effects are extremely important in the development of any chromatographic method.

References

1. J. M. Miller, *Chromatography: Concepts and Contrasts*, John Wiley & Sons, New York, 1988, Chap. 2.
2. J. C. Giddings, *Dynamics of Chromatography*, Marcel Dekker, Inc., New York, 1965, pp. 95–118.
3. L. R. Snyder and J. J. Kirkland, *Introduction to Modern Liquid Chromatography*, 2nd ed., John Wiley & Sons, New York, 1979, Chap. 5.
4. L. R. Snyder, J. J. Kirkland, and J. L. Glajch, *Practical HPLC Method Development*, 2nd ed., John Wiley & Sons, New York, 1997, Chap. 2.

J. E. Haky
D. A. Teifer

Metal-Ion Enrichment by Countercurrent Chromatography

Introduction

Countercurrent chromatography (CCC) is a useful method for separating metal ions as well as organic compounds. In addition, highly efficient chromatographic separation has been achieved using a multilayer coil system and strong force field by over several hundreds of revolutions per minute. However, there have been no applications to the enrichment of inorganic elements until quite recently.

The latest study has revealed that CCC has a great potential in the ultratrace determination of metals, because it can concentrate minute amounts of metal prior to the instrumental multielement analysis, such as atomic absorption spectrometry (AAS), inductively coupled plasma–atomic emission spectrometry (ICP–AES), and inductively coupled plasma–mass spectrometry (ICP–MS).

Enrichment of the desired trace elements would not only allow highly sensitive determination of the trace elements but also alleviate various problems including interferences, high risk of exposure to toxic chemicals and radiation from radioactive samples, and so forth.

Extraction of Metal Ions in the Stationary Phase

The existence of the extracting reagent in the stationary phase is one of the essential factors in the enrichment of inorganic elements as well as in the separation itself. However, the values of the distribution ratios, determined by batch extraction measurements in the two-phase system, is sometimes considerably different from that of the dynamic distribution ratios calculated from the elution curve. Further theoretical and basic investigations are necessarily concerned with extraction kinetics, as well as hydrodynamics behavior of two phases in the high-speed CCC (HSCCC) column [1].

Organophosphorus extractants such as di(2-ethylhexyl) phosphoric acid (DEHPA), 2-ethylhexylphosphonic acid mono-2-ethylhexyl ester (EHPA), *N*-benzoyl-*N*-phenylhydroxylamine (BPHA), and tetraoctylethylenediamine (TOEDA) are often used due to their solubility properties in the stationary organic phase [1–3].

Enrichment in the Effluent by Conventional Elution

After metal ions are enriched in the organic stationary phase including the extracting reagent in the CCC column, they can be eluted simultaneously or chromatographically into the eluent stream by a conventional elution mode.

It was demonstrated that systems with DEHPA can extract and preconcentrate Zr(IV), Hf(IV), and Nb(V) into the organic stationary phase and can separate them from the majority of other elements [1]. The preconcentration of Zr(IV) and Hf(IV) and subsequent back-extraction, as well as the selective extraction of Zr(IV), Hf(IV), Nb(V), and Ta(V) into a 4-mL volume of the organic phase can be performed with a mixture of DEHPA and BPHA.

A 1000-mL sample solution containing $5 \times 10^{-7} M$ of each rare earth element were effectively enriched onto the CCC column head using carboxylic acid–toluene including the EHPA system [2]. Then, each element concentrated on the column head was eluted chromatographically by the mobile phase with a stepwise pH gradient.

On the other hand, the capability of sample preconcentration for instruments such as AAS, ICP–AES, ICP–MS, and so forth was studied [3]. After metal ions were enriched, they were eluted almost simultaneously by inorganic acid at low pH, because of their diffusion in the column is at a disadvantage for improvement of the detection limits. It has been demonstrated that metal ions such as Ca, Cd, Mg, Mn, Pb, and Zn were enriched with a good recovery at a concentration of 10 ppb each in 500 mL of the sample solution. However, the final enriched sample volume eluted from the CCC column was as large as several milliliters, due to longitudinal diffusion of the sample band in the retained stationary phase [1,3]. Additional band spreading occurred in the flow tube when the concentrated solution was eluted with an acid solution for subsequent analysis.

Also, the mechanism of the separation, based on displacement chromatography followed by the enrichment in the CCC column, was studied and a 10-fold enrichment of the transition metal ions was observed in the stationary phase [4]. However, the diffusion of the sample is an inherent process in chromatographic elution, and spreading of the sample bands in the column is unavoidable in CCC as it is in conventional liquid chromatography.

Therefore, it may be difficult to obtain highly concentrated metal ions in an extremely small volume of effluent such as under-mL-order, even if a small-bore column was used with higher concentration efficiency.

Enrichment in the Effluent by pH-Peak Focusing Technique

Recently, pH-zone-refining countercurrent chromatography (pH-ZRCCC), which is a quite unique technique based on neutralization reaction between mobile and stationary phases, has been developed for preparative-scale separation and enrichment of various organic compounds [5]. The pH-ZRCCC can realize chemical reaction in a quite limited thin area (i.e., the interface between organic and aqueous phase, where there is spreading over wide direction in a small-bore tube). If the pH between two phases in the column is reversed, the stationary phase will be continuously neutralized with the mobile phase. Therefore, the pH border, where neutralization has just finished, moves to the tail (outlet) from the head (inlet) of the column. The moving rate of the pH border in the column can be controlled by adjusting the pH in each phase. This means that "another flow rate" concerned with pH, different from the "real flow rate of the eluent," can be realized in the column. Impurities in the sample solution can be quantitatively trapped and enriched in the pH border at a specific condition of the moving rate. This enrichment method for trace organic impurities has been called pH-peak-focusing countercurrent chromatography (pH-PFCCC).

It has been shown that the pH-PFCCC could be successfully applied to enrichment of inorganic trace elements in solution [6–8]. It has great potential for on-line enrichment and subsequent analysis, when CCC is combined with another analytical instrument for solution. The peak intensities for a 10-mL standard sample in the effluent stream was increased over 100-fold, compared with the conventional plasma atomic emission spectrometry. In this method, Ca, Cd, Cu, Mg, Mn, and Zn are chromatographically extracted, in a basic organic stationary phase containing a complex-forming reagent such as DEHPA, by introducing the sample solution into the column rotating at 1200 rpm. When the column is eluted with the acidified mobile phase, metal ions are trapped and concentrated around the sharp pH border formed between the acidic and the basic zones, moving toward the outlet

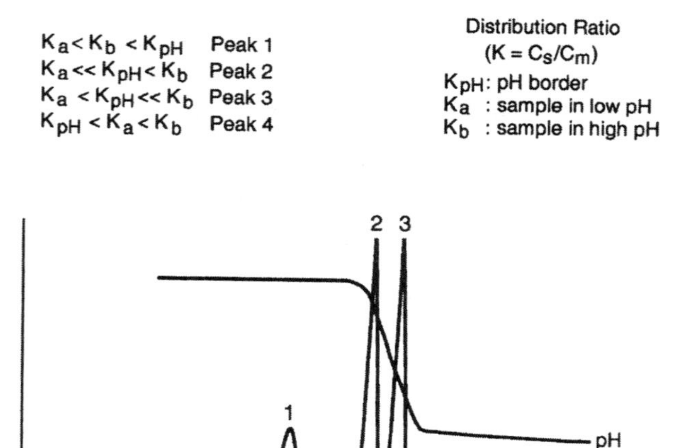

Fig. 1 Conditions of distribution ratio between two-phase solvent for pH-peak focusing.

of the column. Enriched metal ions are finally eluted with the sharp pH border as a highly concentrated peak into less than a 100-μL volume.

The conditions of distribution ratio (K) between the two-phase solvent for pH-PFCCC are shown in Fig. 1. In a relatively basic environment at the first stage of the sample injection, the metal ion present in an aqueous phase forms a complex with the ligand and it partitions into the organic phase with a distribution ratio of K_b, whereas in an acidic environment, the metal ion is mainly distributed into the aqueous phase with distribution ratio of K_a.

As the elution proceeds, acid (e.g., HCl) present in the mobile phase steadily neutralizes the base (e.g., NH_3) in the stationary phase, forming a narrow pH border between the basic front zone and the acidic rear zone. The traveling rate of this sharp pH border through the column is determined mainly by the molar ratio between the base in the stationary phase and acid in the mobile phase, but it is substantially lower than the flow rate of the mobile phase. The metal ion present in the acidic zone quickly moves with the mobile phase passing through the pH border into the basic zone, where it forms a metal–ligand complex and is transferred into the stationary phase. As the pH border moves forward, the complex is exposed to a lower pH where the metal is displaced by a proton (H^+) and released into the aqueous phase as its ionic form (M^+) to repeat the above cycle. Consequently, the metal element is always confined to a narrow region around the sharp pH border and, finally, eluted as a highly concentrated sharp peaks (peaks 2 and 3) in the pH slope, as shown in Fig. 1. Thus, the system eliminates longitudinal spreading of the sample band due to the separation process, which is inherent in other liquid chromatographic methods.

In above method, trapping the metal element by the sharp pH border is essential, and this requires a certain relationship between the distribution ratios of the metal element and the traveling rate of the pH border through the column, as illustrated in Fig. 1. The distribution ratio is shown as K_a and K_b in the acidic and the basic zones, respectively. V_{pH} shows the volume of the eluent when the pH border comes out on the chromatogram. K_{pH} is assumed as a temporary distribution ratio when solute peak was appeared at the retention volume of V_{pH}. K_{pH} is adjustable by changing the molarity between the basic phase and the acidic phase. If K_{pH} is greater than K_a and K_b, the metal ion will elute earlier than the sharp pH border (peak 1), and if K_{pH} is smaller than K_a and K_b, the metal ion will elute after the sharp pH border (peak 4). The peak trapping takes place only when K_{pH} falls between K_a and K_b (peaks 2 and 3). The two metal peaks (peaks 2 and 3) may be resolved within a narrow range if they have a substantial difference in their K_a and K_b values, as shown in Fig. 1.

Conclusion

In contrast to HPLC, CCC has the unique feature that there is no solid support in the column. Because the distribution abilities, including the capacity of the stationary phase, are easy to control, CCC can be applied to the various inorganic treatment, such as enrichment as well as separation and purification, over a wide range in concentration. Moreover, the reproducibility for the enrichment operation has a substantial advantage over other enrichment systems, such as ion exchange, because the absorber system in CCC is always "fresh" in each operation. In particular, enrichment of trace elements using pH-PFCCC will be an ideal preconcentration method for subsequent inorganic determination of modern instrumental analytical methods. It can be combined directly with the flow injection technique, so there would be great potentials for a new enrichment (preconcentration) system of a desired trace element prior to the determination. On-line enrichment and subsequent analysis may take the place of conventional sample preparation using a beaker and separatory funnel, in the future investigation in this field.

References

1. P. S. Fedotov, T. A. Maryutina, O. N. Grebneva, N. M. Kuz'min, and B. Ya. Spivakov, *J. Anal. Chem. 52*: 1034 (1997).
2. S. Nakamura, H Hashimoto, and K. Akiba, *J. Liquid Chromatogr. A 789*: 381 (1997).
3. E. Kitazume, N. Sato, and Y. Ito, *J. Liquid Chromatogr. Related Technol. 21*: 251 (1998).
4. K. Talabardon, M. Gagean, J. M. Marmet, and A. Berthod, *J. Liquid Chromatogr. Related Technol. 21*: 231 (1998).
5. Y. Ito and Y. Ma, *J. Chromatogr. A 753*: 1 (1996).
6. E. Kitazume, N. Sato, and Y. Ito, A new preconcentration-detection method for trace metals by pH-zone-refining countercurrent chromatography, 1995 Pittsburgh Conference and Exposition on Analytical Chemistry and Applied Spectroscopy, 1995.
7. E. Kitazume, N. Sato, and Y. Ito, Effective concentration method for trace metals in solution by pH-zone-refining countercurrent chromatography, 1997 Pittsburgh Conference and Exposition on Analytical Chemistry and Applied Spectroscopy, 1997.
8. E. Kitazume, T. Higashiyama, N. Sato, M. Kanetomo, T. Tajima, S. Kobayashi, and Y. Ito, A novel on-line micro extraction system of metal traces for their subsequent determination by plasma atomic emission spectrometry using pH-zone refining high speed countercurrent chromatography, 1999 Pittsburgh Conference and Exposition on Analytical Chemistry and Applied Spectroscopy, 1999.

Eiichi Kitazume

Metal-Ion Separation by Micellar High Performance Liquid Chromatography

Introduction

The separation of target metal ions from a complex mixture is an extremely important area of research for the purpose of recovery of metal values from wastes and for environmental remediation and restoration. Conventional approaches to metal-ion separation and recovery fall into two broad classes, namely (a) solid–liquid and (b) liquid–liquid separations.

Solid–liquid methods include ion-exchange resins, which involve electrostatic interaction between the positively charged metal ions in solution and a negatively charged functional group on a polymer backbone such as SO_3^-, chelating polymers containing complexing ligands such as iminodiacetic acid, and membrane-mediated separations using solid membranes.

Liquid–liquid methods include solvent extraction with immiscible liquid–liquid systems in which a suitable ligand is dissolved in an organic phase and contacted with a metal ion containing an aqueous phase and liquid membranes. Separations can also be achieved with pseudo-phase systems such as micelles, microemulsions, and vesicles. Such separations can be solid–liquid or liquid–liquid and include separations with normal- and reversed-phase silica, and polymeric supports where the mobile phase contains the organized molecular assembly (OMA) of micelles, microemulsions, or vesicles. Separation of metal ions using the pseudo-phase systems is still in its infancy and a brief account will be provided here.

Organized molecular assemblies (OMAs) are inherently capable of providing greater selectivities in separations of both organic compounds and metal ions mainly due to the ability of the organized microenvironments to discriminate among analytes with similar properties. Selectivities in metal-ion separation are best achieved through complexation rather than through ion-pair formation, and, in general, chelating ligands provide higher selectivities than monodendate ligands. The selectivities of the chelating ligands are limited in the conventional approaches using chelating resins and solvent extraction because the ligands are present in random macroenvironments. The incorporation of these ligands into organized microsturcutres such as micelles can provide dramatic improvements in their metal-ion selectivities. The factors that influence metal-ion selectivities are the stability constants of their complexes, the geometry and coordination number of the complexes, and the equilibrium and kinetics of metal complex formation and dissociation reactions, especially in the interfacial region. It is evident that these factors can be better exploited to achieve metal-ion selectivities in an organized microenvironment compared to a random macroenvironment. High metal-ion selectivities are needed for the separation of target metal ions from complex matrices such as spent catalysts and electrochemical baths, geothermal brines, and nuclear wastes. Organized microenvironments will be key to achieving the selectivities demanded by such complex and difficult matrices.

Micelles

Separations of metal ions mediated by the OMA micelles will be the focus of this entry. Micelles are formed from surfactants in aqueous solutions above a certain concentration of the surfactants called the critical micelle concentration (CMC) [1]. These surfactants can be neutral, such as Triton X-100 and Brij 35, which have the general structure $R(OCH_2CH_2)_nOH$, where R is a long-chain alkyl group (C_{12} and above) and n the number of oxyethylene groups ($n = 9$ for Triton X-100 and 23 for Brij 35), or anionic such as sodium dodecyl sulfate (SDS; $C_{12}H_{25}SO_4^-Na$), or cationic such as cetyltrimethylammonium bromide [CTAB; $C_{16}H_{33}N(CH_3)_3^+Br^-$]. The CMC values of Triton X-100, SDS, and CTAB at an ionic strength of 0.1 are $2 \times 10^{-4}M$, $2 \times 10^{-3}M$, and $1.8 \times 10^{-4}M$, respectively, and their aggregation numbers are 100, 40, and 60, respectively. The neutral micelles in addition to the CMC are also characterized by the cloud point, which is the temperature at which a solution of the neutral micelle separates into two phases, namely the surfactant-rich phase and the water-rich phase. This property has been exploited to achieve cloud point separations analogous to separations employing two immiscible liquid phases. When the micelles are formed in the aqueous phase, the polar head groups of the surfactants, namely OH, SO_4^-, and $C_{16}H_{33}N(CH_3)_3^+$, are present on the surface of the micelles, and the interior of the micelle is hydrophobic like an organic phase in the case of the anionic and cationic micelles. In the case of the neutral micelles, the interior is composed of the oxyethylene groups terminating in an alkyl chain which form the hydrophobic interior called the corona. Neutral surfactants form reverse micelles in nonpolar solvents where the OH group is in the interior and the R group is on the surface. The presence of polar groups on the surface of the normal micelles in the aqueous phase results in an electrical double layer and an interfacial potential, which have significant influence on metal-ion selectivities. This potential is close to zero for neutral micelles, negative for anionic micelles, and positive for cationic micelles.

Adsorption of Surfactants on Silica and Reversed-Phase Silica

The neutral, anionic, and cationic surfactants will adsorb on silica and reversed-phase silica such as octadecylsilanized silica (ODS) from an aqueous solution to form a monolayer of the surfactant on the surface. The adsorption of surfactants on silica surface is more complex than their adsorption on ODS surface. The SiO_2 surface is polar due to the presence of surface silanol groups, and, as a result, the surfactants will adsorb with the polar group on the surface, namely the OH, SO_4^-, and $C_{16}H_{33}N(CH_3)_3^+$, with the alkyl groups pointed away from the surface. The adsorption of the surfactants on SiO_2 is pH dependent as the surface charge changes with pH. This adsorption is also dependent on the concentration of the surfactant with monolayers being formed at low surfactant concentrations and bilayers being formed at high surfactant concentrations. The bilayers are formed when a second layer of the surfactant adsorbs on the initial monolayer through its alkyl chains and the bilayer has the polar head groups pointing toward the bulk aqueous phase. Due to the complex adsorption phenomenon associated with adsorption of surfactants on SiO_2 surface, it is not a useful stationary phase for performing micellar chromatography.

The adsorption of surfactants onto ODS is more straightforward and well understood. Here, the formation of the monolayer of the surfactant on the ODS surface proceeds through the adsorption of the surfactant through its alkyl chain with the polar head group pointing into the bulk aqueous phase. Such an adsorption only results in a monolayer coverage, and the chromatographic behavior of these adsorbed monolayers can be discerned with established principles. The absorbed monolayers of surfactants on reversed-phase silica have been characterized by several spectroscopic methods [2] and will not be discussed here.

Micellar Chromatographic Separations of Metal Ions

The adsorbed monolayers of surfactants on reversed-phase silica can be utilized as stationary phases and an aqueous solution of the same surfactant above its CMC as the mobile phase to separate both organic and inorganic analyte mixtures [3]. In the case of organic analytes, the selectivity between analyte components can be improved dramatically by employing micellar chromatography. The mechanism of separation, as in reversed-phase chromatography, is the distribution of the analyte between the micellar pseudo-phase and the stationary phase containing the adsorbed monolayer of the surfactant. Because the partitioning of the analyte now occurs between the organized microstructure of the micellar pseudo-phase and the adsorbed monolayer, the selectivity between closely related analytes is significantly improved. The separation of metal ions requires the addition of a complexing ligand

$-O-Si(CH_3)_2-$~~~CH_3 CH_3~~~$SO_4^- M^{2+} + 2\,HL \rightleftharpoons ML_2 + 2\,H^+$

ODS Adsorbed SDS monolayer Aqueous Mobile Phase

Fig. 1 Mechanism of separation of metal ions by micellar chromatography employing sodium dodecyl sulfate micelles in the mobile phase and octadecylsilanized silica as the stationary phase.

in order to discriminate among the various metal ions. The SDS adsorbed on ODS as shown in Fig. 1 can function as an ion-exchange column but will not possess any selectivity between metal ions with similar charges analogous to a cation-exchange resin. Metal-ion separations can be achieved, in principle, on neutral, anionic, and cationic surfactants adsorbed on ODS columns, but only the separation with anionic surfactants adsorbed on ODS columns has been studied so far because of its complementary nature to conventional cation-exchange resins. The separations with neutral and cationic surfactants adsorbed are more complex and their behavior needs to understood before they will be useful for the separation of metal ions.

Metal-Ion Separations on the ODS–SDS Column

The separation on the ODS–SDS column is achieved by the addition of a suitable complexing ligand to the mobile phase to discriminate between the various metal ions. The selectivity between two metal ions is achieved through two competing equilibria, namely the complexation equilibrium of the free metal ion in the micellar pseudo-phase and the complexation equilibrium of the metal ion adsorbed on the SDS monolayer and this is shown in Fig. 1. It is evident from this equilibrium that the distribution ratio (capacity factor) of the metal ion M^{2+} is given by Eq. (1), where the subscripts a and s represent the concentration of the free metal ion in the bulk aqueous phase and the adsorbed monolayer, respectively:

$$D = \frac{[M^{2+}]_s}{[M^{2+}]_s} \tag{1}$$

The associated equilibrium constant is defined in Eq. (2), where HL represents the ligand, and the relationship between log D and log[HL] and pH is given in Eq. (3):

$$K_{eq} = \frac{D[HL]^2}{[H^+]^2} \tag{2}$$

$$\log D = \log K_{eq} - 2\log[HL] - 2\text{pH} \tag{3}$$

As is evident from Eq. (3), a plot of log D as a function of log[HL] at constant pH and as a function of pH at constant [HL] will yield the stoichiometry of the equilibrium involved in the separation of the metal ion and the associated equilibrium constants.

The above-mentioned separation principle can be understood by the separation of Co(II), Ni(II), Cu(II), and Zn(II) by micellar chromatography employing SDS above its CMC in the mobile phase, ODS as the stationary phase, and 8-quinolinol-5-sulfonic acid ($pK_1 = 3.84$; $pK_2 = 8.35$) as the ligand [4]:

SO_3^- / N / OH

The logarithm of the stability constant for the 1:2 metal:ligand complex of this ligand with Co(II), Ni(II), Cu(II), and Zn(II) are 16.1, 16.77, 21.9, and 14.3, respectively. This separation employing a pH gradient is shown in Fig. 2, from which it is evident that the metal ions are eluted in the reverse order of their stability constants with 8-quinolinol-5-sulfonic acid. This elution order is readily understood from the equilibrium in Fig. 2, which indicates that the metal ion with the larger stability constant will be

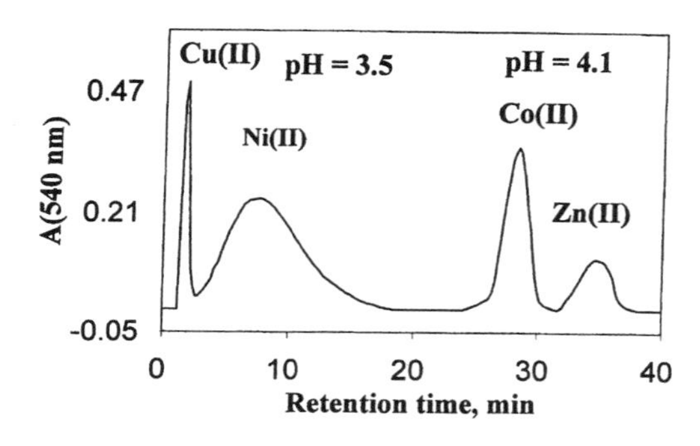

Fig. 2 Separation of Co^{2+}, Ni^{2+}, Cu^{2+}, and Zn^{2+} by micellar chromatography employing a pH gradient. Experimental conditions: mobile phase: [SDS] = 0.004M, [8-quinolinol-5-sulfonic acid] = 0.002M, ionic strength = 0.1; ODS stationary phase. 10 μL of the metal-ion mixture containing 0.002M of each ion was injected at a mobile-phase flow rate of 1 mL/min. Cu^{2+} and Ni^{2+} elute at pH = 3.5 and Co^{2+} and Zn^{2+} elute at pH = 4.1.

stripped from the adsorbed SDS monolayer more easily compared to the metal ion with a smaller stability constant. It may also be noted that the chromatographic bandwidths are much larger than those encountered with organic analytes and this is a direct consequence of slow metal complex formation and dissociation kinetics. Information on such kinetics can be obtained by studying the dependence of HETP on the concentrations of ligand and pH. This indicates that in the case of Co(II), Cu(II), and Zn(II), the dissociation of the metal complex is mainly responsible for the band broadening, whereas in the case of Ni(II), both metal complex formation and dissociation affect the efficiency.

Several simple water-soluble chelating ligands such iminodiacetic acid have been employed for the separation of a variety of metal ions. Okada demonstrated the separation of several divalent metal ions (Mn^{2+}, Co^{2+}, Ni^{2+}, Cu^{2+}, Zn^{2+}, Cd^{2+}) on ODS column with SDS micelles in the mobile phase and tartaric acid as the ligand [5]. The selectivity between the various divalent metal ions is much smaller than that observed with 8-quinolinol-5-sulfonic acid, which forms complexes with much larger stability constants than does tartaric acid. The stability constants of the metal complexes with a given ligand as mentioned earlier is one of the factors that affect selectivity. Karcher and Krull have published a review of the use of simple water-soluble ligands such as acetic acid, oxalic acid, hydroxyisobutyric acid, citric acid, tartaric acid, oxalic acid, and EDTA in micellar chromatography and with simple ion-exchange columns [6]. The selectivities obtained with these simple ligands is not as large as with ligands of the 8-quinolinol, acylpyrazolone, and aromatic oxime families, which are yet to be extensively investigated. Inorganic anions such as nitrate, nitrite, and phosphate can be separated by micellar chromatography using cationic micelles in the mobile phase and ODS as the stationary phase, as shown by Cassidy and Elchuk [7].

Chelating Micelles

The studies that have been conducted so far have involved the addition of a suitable ligand to the micellar pseudo-phase in order to separate the mixture of metal ions. The location of the ligand in this pseudo-phase is a function of the hydrophilic and hydrophobic nature of the ligand. If the ligand is hydrophilic simple ones such as tartaric acid, then they will be present predominantly in the bulk aqueous phase, and if the ligand is hydrophobic such as 8-quinolinols, then they will distribute into the micellar pseudo-phase. The location of the ligands in the micellar pseudo-phase is a function of the hydrophobicity of the ligand. Thus, the distribution of the ligand between the bulk aqueous phase and the micellar pseudo-phase introduces uncertainty in the location of the ligand and does not fully exploit the organized microenvironment. The aqueous–micelle interfacial region has a very high interfacial area and possesses an electrical double layer with an interfacial potential. These can be effectively exploited to achieve even higher selectivities in metal-ion separations if the ligand can be specifically located at the interfacial region. We have recently achieved this by chemically derivatizing neutral surfactants such as Brij 35 such that a chelating ligand like 8-quinolinol or acylpyrazolone is the head group instead of OH. Such chelating surfactants form chelating micelles at CMC values much lower than Brij 35. The charge of the micelle is pH and ligand dependent and can be neutral, anionic, or cationic. In these chelating micelles, the ligand is exclusively present on the surface of the micelle and its chelating properties and metal-ion selectivities are strongly influenced by the interfacial region. We have synthesized chelating surfactants containing 2-methyl-8-quinolinol and 1-phenyl-3-methyl-4-benzoyl-5-pyrazolone as the head groups [8]. We have been able to obtain very large selectivities in the separation of divalent transition metal ions and trivalent lanthanide ions using these chelating micelles. The selectivities that we have achieved in the separation of the trivalent lanthanide metal ions are unprecedented and clearly indicate the tremendous gains in selectivities that can be achieved by the incorporation of ligands into organized microstructures. We have also shown that the mixed system of chelating surfactants with the 8-quinolinol and acylpyrazolone head groups form vesicles which also exhibit unique selectivities in the separation of transition and lanthanide metal ions on an ODS stationary phase. The chelating micelles are also interesting candidates for the separation of metal ions by electrokinetic chromatography, where the interfacial properties of the chelating micelles can be further exploited to achieve very high metal-ion selectivities.

References

1. M. J. Rosen, *Surfactants and Interfacial Phenomena*, John Wiley & Sons, New York, 1989.
2. D. A. Piasecki and M. A. Wirth, Spectroscopic probing of the interfacial roughness of sodium dodecyl sulfate adsorbed to a hydrocarbon surface, *Langmuir 10*: 1913 (1994).
3. W. L. Hinze and D. W. Armstrong, eds., *Ordered Media in Chemical Separations* American Chemical Society. Washington, DC, 1987.
4. S. Muralidharan, unpublished results.
5. T. Okada, Interpretation of retention behavior of transition-metal cations in micellar chromatography using an ion-exchange model, *Anal. Chem. 64*: 589 (1992).

6. B. D. Karcher and I. S. Krull, The use of complexing eluents for the high performance liquid chromatographic determination of metal species, in *Trace Metal Analysis and Speciation* (I. S. Krull, ed.), Journal of Chromatography Library Series Vol. 47, Elsevier, Amsterdam, 1991, pp. 123–166.
7. R. M. Cassidy and S. Elchuk, Dynamically coated columns for the separation of metal ions and anions by ion chromatography, *Anal. Chem. 54*: 1558 (1982).
8. S. Muralidharan, unpublished results.

S. Muralidharan

Metals and Organometallics: Gas Chromatography for Speciation and Analysis

Speciation analysis is usually defined as the determination of the concentrations of the individual physicochemical forms of the element in a sample that together constitute its total concentration. Recently, there has been increasing interest in speciation information of elements present in environmental and biological samples because the toxicological and biological importance of many metals and metalloids greatly depends on their chemical forms [1–3]. The determination of the total amount of an element is important, but it is not sufficient to assess its toxicity. Information about concentrations of the individual species of an element, including its organic derivatives, is particularly crucial. Frequently, the lack of the speciation information is the major limitation to our understanding of the biogeochemical cycling of the element.

The identification of the chemical forms of an element has become an important and challenging research area in environmental and biomedical studies. Two complementary techniques are necessary for trace element speciation. One provides an efficient and reliable separation procedure, and the other provides adequate detection and quantitation [4]. In its various analytical manifestations, chromatography is a powerful tool for the separation of a vast variety of chemical species. Some popular chromatographic detectors, such flame ionization (FID) and thermal conductivity (TCD) detectors are bulk-property detectors, responding to changes produced by eluates in a characteristic mobile-phase physical property [5]. These detectors are effectively "universal," but they provide little specific information about the nature of the separated chemical species. Atomic spectroscopy offers the possibility of selectively detecting a wide rang of metals and nonmetals. The use of detectors responsive only to selected elements in a multicomponent mixture drastically reduces the constraints placed on the separation step, as only those components in the mixture which contain the element of interest will be detected [6]. It is not surprising that the coupling of chromatographic techniques [gas chromatography (GC) and high-performance liquid chromatography (HPLC)] with a highly sensitive and selective atomic spectrometry detector has been widely exploited and accepted for the speciation of metals and organometals. GC has enjoyed particular attention because of its high sensitivity and simplicity of coupling. Details about GC coupled to atomic absorption spectrometry (AAS), MIP–AES, and inductively coupled plasma–mass spectrometry (ICP–MS) are discussed here.

Sample Preparation

The native species of most metals and metalloids, such as mercury, lead, tin, arsenic, and selenium, are generally present as ionic forms in sample matrices. For GC-based coupling techniques, these compounds need to be extracted from the sample matrix and to be converted to volatile and thermally stable derivatives. Frequently, the derivatives are then concentrated by cryotrapping or extracting into an organic solvent prior to injection onto a GC column [1].

Grignard Reaction

The Grignard reaction is one of the most widely used derivatization techniques for the speciation of a number of elements [7,8]. The main advantage of this reaction is that different alkyl groups can be chosen to make fully alkylated species. Reactions for mercury analysis can be described as follows:

$$\mathrm{RMgX} + \mathrm{R'Hg^{+}} \rightarrow \mathrm{RHgR'} \quad (1)$$

$$\mathrm{RMgX} + \mathrm{Hg^{2+}} \rightarrow \mathrm{R_2Hg} \quad (2)$$

where R = propyl, butyl, and pentyl groups and R′ = methyl and ethyl groups. However, Grignard reagents are very sensitivity to water. As a consequence, metal and organometallic compounds have to be extracted prior to derivatization into organic solvent with assistance of complexing reagents, such as dithiocarbamates and tropolone. The whole sample preparation can be tedious and time-consuming.

Hydride Generation

Several elements (Hg, Ge, Sn, Pb, Se, As, Te, Sb, Bi, and Cd) can be transformed into volatile hydrides, providing the basis of their analysis [1,3]. Sodium borohydride ($NaBH_4$) is commonly used as a derivatization reagent. The reaction for methylarsenicals is

$$Me_nAsO(OH)_{3-n} + NaBH_4 \rightarrow Me_nAsH_{3-n} \quad (n = 1\text{–}3) \tag{3}$$

The usefulness of this procedure for speciation analysis, however, is restricted by either the thermodynamic inability of hydride formation for some species or considerable kinetic limitation to hydride formation. Nevertheless, the technique is still essential for some classes of compounds [1].

Aqueous Derivatization with Tetraalkyl(aryl)borates

The restricted versatility can, to a certain degree, be overcome by replacing $NaBH_4$ by alkyl or arylborates [1]. The most common derivatization procedure relies on ethylation with sodium tetraethylborate, which was initiated for determining methyllead ionic compounds [9]:

$$Me_nPb^{(4-n)+} + NaBEt_4 \rightarrow Me_nPbEt_{(4-n)} \quad (n = 1\text{–}4) \tag{4}$$

$NaBEt_4$ acts as an aqueous-phase ethylation reagent, quantitatively transferring Et^- ions to ionic metals and organometals. Direct aqueous-phase ethylation with $NaBEt_4$ has been used for the speciation of a variety of metals, such as lead, mercury, and tin in different environmental and biological samples. This reaction has significant advantages because the derivatization reaction is performed in the aqueous phase, subsequently reducing the analysis time and eliminating the need for organic solvent extraction. However, this method cannot be used for the speciation of the natively occurring ethylated species, such as ethyllead and ethylmercury [8].

Several efforts have been carried out to develop a new aqueous derivatization reagent. Phenylation with sodium tetraphenylborate is a promising procedure for the speciation of several metals. Its application for organomercury analysis has been comprehensively studied [10]. Sodium tetrapropylborate is another reagent that has been investigated for determining organolead, tin, and mercury compounds [1]. However, its application is limited because it is not commercially available.

Coupling GC with Atomic Spectrometric Detection

GC–AAS

Gas chromatography coupled with AAS has been the most popular hyphenated technique for the speciation of metals and organometals. Among different atomizer designs employed in coupling GC with AAS, electrothermal atomization, especially the electrothermally heated quartz tube, is preferred because of its high sensitivity, simple operation, and low cost [6]. The quartz tube (~ 100 × 10 mm inner diameter) is usually constructed in house and wrapped with Nichrome resistance wire and ceramic isolation material. A thermocouple is attached to the surface of the quartz burner for temperature control (Fig. 1). If necessary, hydrogen and air can be supplied from side arms of the burner. The furnace can be heated to more than 1000°C by means of a variable transformer.

Two types of sample introduction–interface coupling GC and quartz furnace have been generally used. Figure 1 shows a schematic representation of the first design, usually called purge and trap. The instrumental setup consists of a reaction vessel where the sample is brought in contact with a reagent, such as $NaBH_4$ and $NaBEt_4$, and a U-tube filled with a GC sorbent. During the reaction, the U-tube is maintained in liquid nitrogen and the produced derivatives are purged with helium, then trapped on the sorbent. Once the reaction is finished, the liquid nitrogen is removed and the U-tube is heated electrically. The trapped compounds are then separated according to

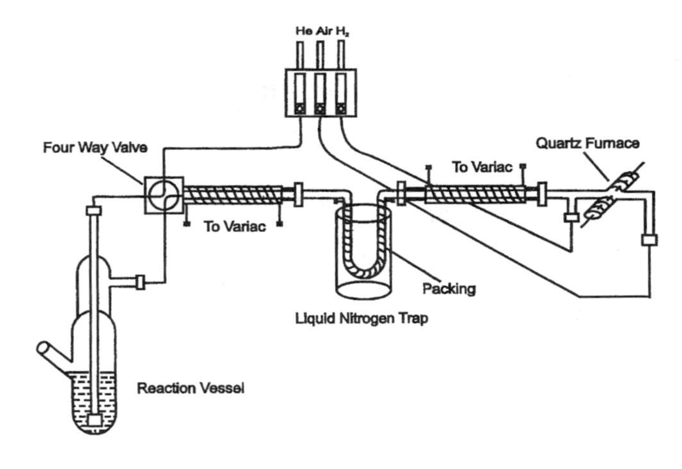

Fig. 1 Schematic diagram of a purge-and-trap GC–AAS system.

their volatility and measured by AAS. The principal advantages of this setup are its very high sensitivity and on-line aqueous derivatization. However, it suffers from a water-condensation problem in the U-trap, which limits the length of the purge time. A water trap is often installed upstream from the sorbent tube to avoid its blocking with ice crystals [1].

The second design is actually modified from a regular GC device equipped with either a packed or a capillary column. The outlet of the GC is directly connected to the entrance of the quartz furnace through a piece of stainless-steel tubing, which is electrically heated to avoid condensation of the compounds of interest. To be able to use this technique, the target compounds have to be extracted from the sample matrix, derivatized with the appropriate reagent, and back-extracted into organic solvent before injection. Absolute detection limits obtained with this system are similar to those using purge and trap because they use same detection technique. However, the concentration detection limits can be different as large as several orders of magnitude. With the purge-and-trap system, all target compounds in the sample are on-line preconcentrated and analyzed at once, whereas with regular GC, only a small portion of the sample extract is injected for analysis. However, the regular GC method offers a high versatility because of the possibility of using different Grignard reagents.

GC/MIP–AES

MIP–AES has two basic characteristics that can be utilized when coupling to a GC instrument [6]. The low gas temperature of the MIP allows small amounts of sample, compatible with those of GC solutes, to be introduced without extinguishing the plasma. In addition, sample introduction is easily facilitated because the carrier and plasma gases are the same (helium). These advantages have made the coupled GC/MIP–AES a popular technique and many applications have been reported [5,6,10]. A commercial instrument is currently available.

Helium plasma is maintained within a "cavity" which serves to focus power from a microwave source (usually operated at 2.45 GHz) into a discharge cell (usually a quartz capillary tube). The cavity, which has been most used for GC detection, has been the atmospheric-pressure TM_{010} cylindrical resonance cavity developed by Beenakker [11]. The effluent from a GC is connected directly to the discharge tube via a heated transfer line to prevent analyte condensation. The GC/MIP–AES requires a consistent supply of high-purity gases for optimal performance. Hydrogen and oxygen are often used as plasma reagent gases for metal and metalloid analysis. In addition, the AED spectrometer requires a continuous nitrogen purge.

Advantages of MIP–AES are low cost, simple operation, and high sensitivity. Its sensitivity for mercury analysis is compatible with that using GC coupled with atomic fluorescence detection [10], which is currently recognized as the most sensitive method for determining mercury. The main drawback is the low tolerance for organic solvents. A venting system has to be used to prevent solvent from getting into plasma tube.

GC/ICP–MS

Over the last decade, ICP–MS has proven to be a highly sensitive and selective technique for the determination of trace and ultratrace amounts of metals in various samples. It allows multielement detection in a single run and offers isotopic information of the elements of interest [12–14]. Solvent venting to prevent plasma instability is unnecessary, unlike in GC/MIP–AES, and no carbon accumulates, as it does on the MIP discharge tube. These features make the hyphenation GC/ICP–MS unique.

Coupling GC to ICP–MS is easily accomplished by connecting the column to the inner tube of torch using a transfer line between the GC oven and the plasma torch (Fig. 2). The transfer line usually consists of an electrically heated stainless-steel tube through which a piece of deactivated fused silica is passed. The transfer line capillary ends at the tip of the ICP injector. Generally, the stainless-steel tubing is maintained at a temperature that prevents the condensation of the GC effluent in the transfer line. Fluctuation in the transfer line temperature can affect GC peak shape and resolution.

The ICP–MS instrument requires a carrier gas flow rate of approximately 1 L/min, whereas the capillary GC effluent is less than 5 mL/min. It is, therefore, necessary

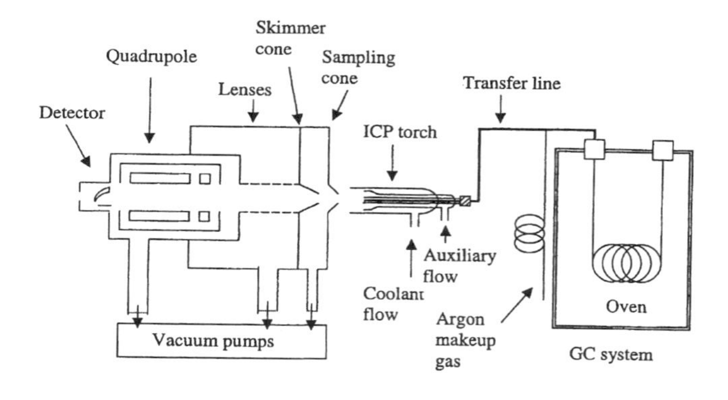

Fig. 2 Schematic diagram of a GC/ICP–MS system.

to introduce argon as a makeup gas. The makeup gas produces a central channel in the plasma and helps to carry analytes from the GC column to the plasma. The makeup gas must be heated to avoid condensation of the column effluent and this has been done by passing the argon gas through a heated stainless-steel tubing and added it at the beginning of the transfer line [12]. However, a high flow rate of argon makeup gas results in an undesirable dilution effect.

This hyphenated technique has been successfully applied to the speciation of a number of metals, including lead, mercury, and tin. The advantages of determining multielements simultaneously and the wide dynamic liner range are obvious [13].

References

1. R. Lobinski and F. C. Adams, *Spectrochim. Acta B 52*: 1865 (1997).
2. K. Sutton, R. M. C. Sutton, and J. A. Caruso, *J. Chromatogr. A 789*: 85 (1997).
3. Y. Cai, *Trend Anal. Chem.* (2000).
4. N. P. Vela, L. K. Olson, and J. A. Caruso, *Anal. Chem. 65*: 585A (1993).
5. P. Uden, in *Element-Specific Chromatographic Detection by Atomic Emission Spectroscopy* (P. Uden, ed.), ACS Symposium Series Vol. 479, American Chemical Society, Washington, DC, 1990.
6. L. Ebdon, S. Hill, and R. W. Ward, *Analyst 111*: 1113 (1986).
7. Y. K. Chau, P. T. S. Wong, G. A. Bengert, and J. L. Dunn, *Anal. Chem. 56*: 271 (1984).
8. Y. Cai, R. Jaffe, and R. D. Jones, *Environ. Sci. Technol. 31*: 302 (1997).
9. S. Rapsomanikis, O. F. X. Donard, and J. H. Weber, *Anal. Chem. 58*: 38 (1986).
10. S. Monsalud, M.S. thesis, Florida International University, 1999.
11. C. I. M. Beenakker, *Spectrochim. Acta 32B*: 173 (1977).
12. T. De Smaele, L. Moens, R. Dams, and P. Sandra, *LC–GC 14*: 876 (1996).
13. T. De Smaele, J. Vercauteren, L. Moens, R. Dams, and P. Sandra, *Hewlett-Packard Peak*, N. 2, 10 (1999).
14. F. A. Byrdy and J. A. Caruso, *Environ. Sci. Technol. 28*: 529A (1994).

Yong Cai
Weihua Zhang

Migration Behavior: Reproducibility in Capillary Electrophoresis

Introduction

Identification of sample components based solely on migration time in capillary electrophoresis (CE) requires reproducibilities not normally obtained. These are caused, mainly, by two effects: temperature effects and electro-osmotic affects. Migration times in CE are determined by the electro-osmotic velocity v_{EOF} and effective electrophoretic migration velocity v_{EFF}; the net migration velocity v_i is the vector sum of both velocities:

$$v_i = v_{\mathrm{EFF}} + v_{\mathrm{EOF}}$$

In terms of mobilities, this relationship becomes

$$v_i = \left(-\frac{\zeta\varepsilon}{\eta} + \mu_i \right) \cdot E$$

in which ζ is the zeta-potential of the capillary, ε is the dielectric constant of the liquid, η is the viscosity of the liquid, μ_{eff} is the effective mobility of the sample component, and E is the electric field strength. From this relationship, it is obvious that migration time reproducibility depends on the reproducibility of a number of parameters. In addition to the ones mentioned, sample matrix effects can also play a significant role.

Electrophoretic Effects

The reproducibility of the effective mobility is governed by temperature and pH effects. An acceptable run-to-run pH reproducibility can usually be assured, at least in equipment where the effects of possible electrode reactions can be avoided. Again, temperature plays a leading role: The effective mobility has a temperature coefficient of 2–3% per degree. Most buffer solutions, incidentally, have a temperature-dependent pH value, so that changing the temperature in the capillary will, for weak ions, even lead to changing degrees of dissociation and, hence, effective mobility. These effects are, by definition, different for different buffers and different sample components.

Sample Matrix and Injection Effects

When injecting very dilute sample solutions, a relatively large proportion of the total voltage drop across the capillary will take place over this sample plug, resulting in a proportionally lower field strength in the remainder of the separation capillary. This will lead to systematically longer migration times. In this case, also, identification on the basis of effective mobilities is better than on migration times alone.

Another effect takes place when injecting different sample volumes in the capillary: This may systematically change the migration length and, thus, migration times. A constant injection volume is, therefore, advised in cases of required high migration time reproducibility.

Electro-osmotic Effects

In coated capillaries, the $\zeta\varepsilon E/\eta$ term is small, possibly close to zero, compared to the $\mu_{\mathrm{EFF}} \cdot E$ term. If, however, one works with unsuppressed electro-osmosis, the reproducibility of the $\zeta\varepsilon E/\eta$ term plays an important role. Both ζ and η depend on temperature, and in some equipment, thermostating is not perfect, at least not in all parts of the capillary. The viscosity alone accounts for an $\sim 2\%$ per degree dependence. Another effect that might take place is the gradual change in ζ-potential due to adsorption of sample matrix components on the innner capillary surface. The good news is that run-to-run change in electroosmotic velocity is directly obvious from the EOF marker peak in the detection signal, so that changes in EOF can be accounted for in cases where they cannot be avoided. However, this also means that measures have to be taken to assure a clear EOF marker peak [e.g., by addition to the sample of a neutral ultraviolet (UV)-absorbing component, such as mesityloxide]. In complicated samples, such an EOF marker peak may not be possible. For reasons of EOF reproducibility, identification is preferentially based not on migration time but on effective mobility calculation by means of

$$\mu_{\mathrm{EFF}} = \frac{L_d}{t_i E} - \mu_{\mathrm{EOF}}$$

in which L_d is the capillary length to the detector and t_i is the migration time of the component i.

For quantitation under conditions of changing migration velocity, peak areas are usually divided by corresponding migration times to make them more independent of run-to-run differences in EOF. Migration times are especially susceptible to small EOF variations in the case of opposite signs for EOF and sample mobility (i.e., usually for anions). This is illustrated in Fig. 1, a separation of three cations before and eight anions after the EOF marker (not indicated in the figure). The traces given are for a ζ-potential of -46 and -45 mV, respectively. This 2% EOF change corresponds to the equivalent of only 1 degree temperature effect.

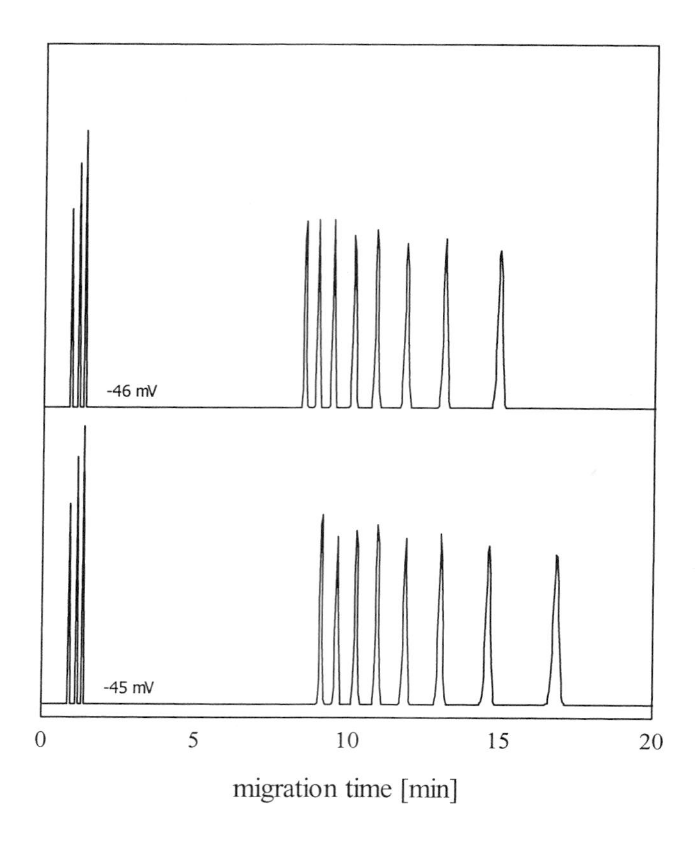

Fig. 1 Effect of capillary ζ-potential on the analysis of a mixture of anions in an uncoated fused-silica capillary in a 10 m*M* Tris/acetate buffer of pH 8. Normal ζ-potentials are around 50 mV at this pH.

References

1. J. W. Jorgenson and K. D. Lucaks, *Science 222*: 266 (1983).
2. S. F. Y. Li, *Capillary Electrophoresis—Principles, Practice and Applications*, Elsevier, Amsterdam, 1992.
3. E. Kenndler, *J. Capillary Electrophor. 3*(4): 191–198 (1996).

Jetse C. Reijenga

Minimum Detectable Concentration (Sensitivity)

Detector sensitivity or the minimum detectable concentration (MDC) is defined as the minimum concentration of solute passing through the detector that can be unambiguously distinguished from the noise. The size of the signal that will be distinct from the noise (the signal-to-noise ratio) will be an arbitrary choice. In electronic measuring systems, it is generally accepted that a signal can be differentiated from the noise when the signal-to-noise ratio is 2. This criterion has been generally adopted for physical–chemical measurements and is used to define detector sensitivity in chromatography.

Thus, for a concentration-sensitive detector, the detector sensitivity (X_D) or MDC is given by

$$X_D = \frac{2N_D}{R_c} \quad \left(\frac{\text{g}}{\text{mL}}\right)$$

where R_c is the response of the detector (i.e., the voltage output for unit change in concentration (V/g/mL) and N_D is the detector noise level (V).

The sensitivity or MDC of a detector is *not* the same as the minimum mass that can be detected. This would be the *system mass sensitivity*, which will depend on the characteristics of the column and the chromatographic properties of the solute, as well as the detector specifications. In all chromatographic systems, the peak becomes broader as the retention increases. Consequently, a given mass may be detected if eluted as a narrow peak early in the chromatogram, but if eluted later, its peak height may be reduced to such an extent that it is impossible to discern it from the noise. Thus, detector sensitivity quoted as the "minimum mass detectable" must be carefully examined and related to the chromatographic system and, particularly, the column with which it is to be used. If the data to do this are not available, then the sensitivity must be calculated from the detector response and the noise level in the manner described earlier.

Some manufacturers have taken the minimum detectable concentration and multiplied it by the sensor volume and defined the product as the minimum detectable mass. This gives values that are very misleading. For example, a detector having a true sensitivity of 10^{-6} g/mL and a sensor volume of 10 μL would be attributed to a mass sensitivity of 10^{-8} g. This is grossly incorrect, as it is the *peak volume* that controls the mass sensitivity, not the *sensor volume*. Conversely, if the peak volume does approach that of the sensor, then a very serious peak distortion occurs with loss of resolution; thus, this way of specifying sensitivity remains meaningless.

Suggested Further Reading

Scott, R. P. W., *Chromatographic Detectors*, Marcel Dekker, Inc., New York, 1996.

Scott, R. P. W., *Introduction to Gas Chromatography*, Marcel Dekker, Inc., New York, 1998.

R. P. W. Scott

Mixed Stationary Phases in GC

The desired interactive character of a GC stationary phase is usually obtained by choosing a thermally stable substance that contains the appropriate polar and dispersive groups that can provide the necessary molecular interactions for sample resolution.

However, Purnell and co-workers [1] pioneered an alternative approach to stationary-phase polarity control for GC. These workers demonstrated that for a wide range of stationary phases made up of binary mixtures, the corrected retention volume of a solute was linearly related to the volume fraction of either one of the two phases. At the time of discovery, this was quite an unexpected relationship, as it was generally accepted that the expression for the retention volume would take the form of the exponent of the stationary-phase composition. The results of Purnell and his co-workers can be summarized by

$$V'_{r(A)} = K_A \alpha V_S + K_B(1 - \alpha)V_S$$

where $V'_{r(AB)}$ is the retention volume of the solute on a mixture of stationary phases A and B, K_A is the distribution coefficient of the solute with respect to the pure stationary phase A, K_B is the distribution coefficient of the solute with respect to the pure stationary phase B, V_S is the total volume of the stationary phase in the column, and α is the volume fraction of phase A in the stationary-phase mixture; that is,

$$V'_{r(AB)} = \alpha V'_A + (1 - \alpha)V'_B \qquad (1)$$

where V'_A is the retention volume of the solute on the same volume of pure phase A and V'_B is the retention volume of the solute on the same volume of pure phase B.

Rearranging Eq. (1),

$$V'_{AB} = \alpha(V'_A - V'_B) + V'_B \qquad (2)$$

This remarkably simple relationship is depicted in Fig. 1.

Purnell carried out three experiments to examine the effect of the composition of binary mixtures of stationary phases on solute retention. In the first experiment, the two fractions were mixed, coated on some support, and packed into the column. In the second experiment, each of the two fractions was coated on separate aliquots of support and the coated supports mixed and packed in a column. In the third experiment, each fraction was coated on a support and each support packed into separate columns and the columns joined in series. Purnell demonstrated that all three columns gave exactly the same corrected retention volume for a given solute.

It was found, however, that where strong association occurred between the two phases, there were exceptions to this relationship. If a strong association took place, the blend would no longer be a simple binary mixture but would contain the associate of the two phases as a third component. It is clear that for a ternary mixture, the simple linear relationship obtained for a binary mixture would not hold. The simple linear relationship for the binary mixture is not surprising. The distribution coefficient of the solute with either pure component is a constant. It follows that the volume fraction of each phase will determine the probability that a solute molecule will interact with a molecule of that phase. This is analogous to the partial pressure of solute determining the probability that a solute molecule will collide with a gas molecule.

Doubling the volume fraction of one phase doubles the probability of solute interaction and, consequently, doubles its contribution to retention. There is another interesting outcome from the results of Purnell and his coworkers. Where a linear relationship existed between the retention volume and the volume fraction of the stationary phase, the linear functions of the distribution coefficients could be summed directly, but their logarithms could *not*. In many classical thermodynamic descriptions of the effect of the stationary-phase composition on solute retention, the stationary-phase composition is often taken into account by including an extra term in the expression for the standard free energy of distribution. The results of Purnell indicate that this is not acceptable, as the solute retention or distribution coefficient is *linearly* not exponentially related to the stationary-phase composition. The stationary phases of intermediate polarities can easily be constructed from binary mixtures of a strongly dispersive stationary phase and one that is strongly polar. This procedure is not commonly used for commercial columns, although it is a simple and economic method for constructing columns having intermediate polarities. Mixed phases are always worth considering as a flexible alternative to the use of a specific proprietary material.

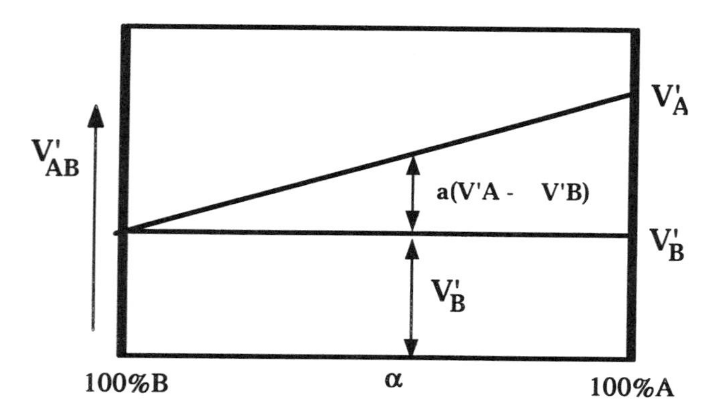

Fig. 1 Graph of corrected retention volume versus volume fraction of the stationary phase.

References

1. M. McCann, J. H. Purnell, and C. A. Wellington, Proceedings of the Faraday Symposium, Chemical Society, 1980, p. 83.

Suggested Further Reading

Scott, R. P. W., *Techniques of Chromatography*, Marcel Dekker, Inc., New York, 1995.

Scott, R. P. W., *Introduction to Gas Chromatography*, Marcel Dekker, Inc., New York, 1998.

R. P. W. Scott

Mobile Phase Modifiers for SFC: Influence on Retention

Carbon dioxide has been the most common mobile phase in supercritical fluid chromatography (SFC) due to its low critical point, nontoxicity, and wide availability in pure form. However, the polarity and solvating power of carbon dioxide are fairly low, which prevents polar and high-molecular-weight compounds to elute by using pure CO_2 as the mobile phase. Therefore, a small amount of organic or inorganic solvents are often added to CO_2 to enhance the solvating power of the mobile phase, so that polar and high-molecular-weight solutes can be eluted. Both polar and nonpolar modifiers have been added to carbon dioxide to alter the chromatographic retention and selectivity. The modifiers and CO_2 can be either premixed in a compressed cylinder, mixed using a saturator column (device), or mixed by using two delivery pumps. Among many modifiers tested in SFC, methanol has been the most popular additive for carbon dioxide. Although binary systems have been used in many applications, ternary systems have also been investigated and found to be very successful in separating organic acids and bases. The modifier influence on chromatographic retention and the mechanism of modifier effect in SFC are also discussed in this entry.

Preparation of the Modified Mobile Phase

There are a number of ways to mix modifiers with CO_2. A mixture of modifier and CO_2 can be purchased as premixed cylinders. Because the modifier is premixed with CO_2 in a cylinder, a modifier gradient cannot be carried out by this approach. Another drawback of this premixing method is that the concentration of the modifier may be increased as the cylinder is evacuated. Modifiers can also be directly introduced into an empty syringe pump that is then filled with CO_2. Again, similar limitations exist in this method. Another way to mix the modifier with CO_2 is to use a saturator column, usually a silica column saturated with polar modifiers. However, because the modifier-holding capacity of the silica column is limited, the amount of modifier dissolved in the mobile phase varies as the mobile phase passes through the saturator column. An improved mixing device in which highly porous stainless-steel filters were used to generate a modified CO_2 mobile phase has been developed to substitute for the saturator column. Although this mixing device can maintain the amount of modifiers dissolved in supercritical CO_2 for a long time, the performance of the modifier gradient remains impossible with this system. The best and most effective (also the most expensive) way of delivering a modifier in SFC is by using two supply pumps. One pump is used for CO_2 delivery and the other for modifier delivery. The modifier and CO_2 meet inside a mixing chamber that mixes the modifier with CO_2 and equilibrates the mixture in a thermostated zone, and then the mixed fluid is delivered to the SFC injection port and column.

Methanol as the Modifier

Among the published articles regarding modifiers in SFC, methanol has been involved in at least two-thirds of the works. Therefore, methanol has been the most popular modifier for CO_2 in SFC. The mole fraction of methanol in the mobile phase is generally low, ranging from 1% to 10%. However, methanol concentrations up to 45% have also been tested. Due to the high cost of dual-pump systems, a constant concentration of methanol has normally been used to modify carbon dioxide. Once a dual-pump system is available, a programmed concentration of the modifier is preferred, as better separations can be achieved by using the modifier gradient technique. The methanol-modified mobile phase has been used to successfully separate phenols and their derivatives, amines and their derivatives, carboxylic acids, carbohydrates and their derivatives, herbicides and pesticides, alkylbenzenes, chlorobenzenes, chlorophenyls, higher-molecular-weight *n*-alkanes, polycyclic aromatic hydrocarbons (PAHs) and their derivatives, polychlorinated biphenyls (PCBs), pharmaceutical compounds, active ingredients in natural products, explosives, and other solutes.

Ternary Systems

Because methanol is not very polar, the elution of strong organic acids and bases requires a mobile phase with even greater polarity. This has normally been done by adding a very low concentration of acids or bases into methanol, and then the modified methanol is mixed with CO_2 for separation. Citric, acetic, and chlorinated acetic acids have been used as acidic secondary modifiers, whereas isopropylamine, triethylamine, and tetrabutylammonium hydroxide have been served as basic secondary modifiers. A good example with ternary systems is the separation of benzylamines, as shown in Fig. 1. None of the three tested amines were effectively eluted by pure CO_2 (Fig. 1a); however, some of these amines were eluted with very poor peak shapes when methanol was added to the CO_2

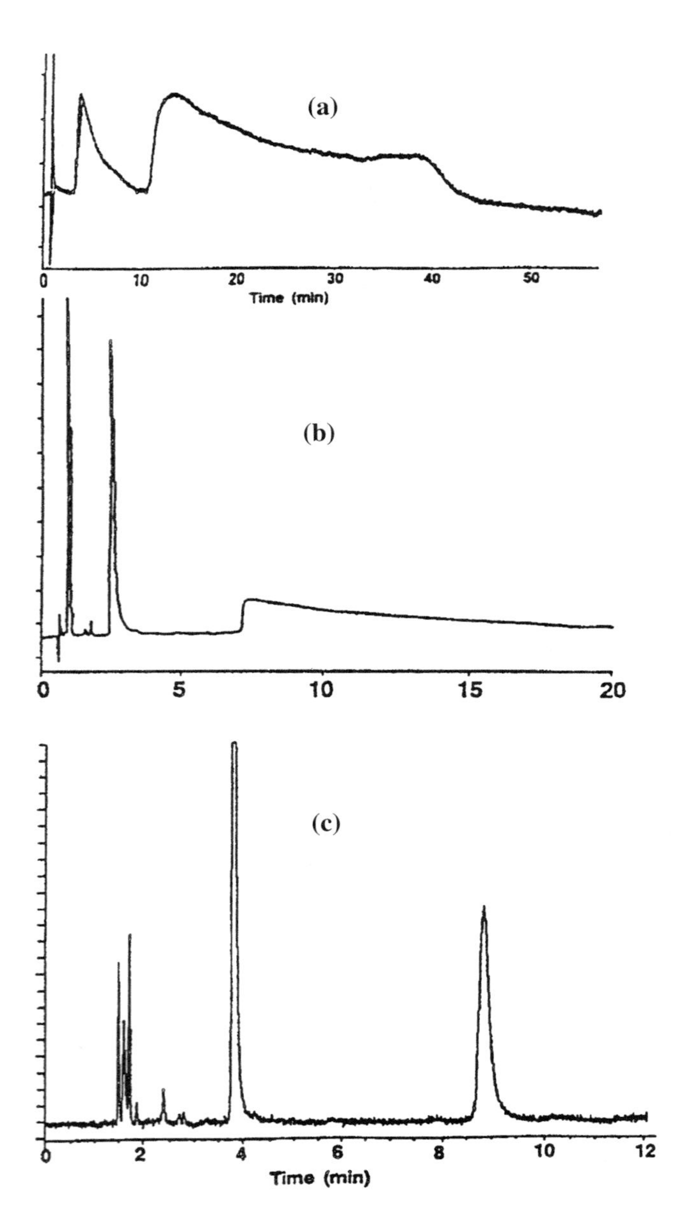

Fig. 1 Separation of benzylamines: (a) On a Deltabond Octyl column (100 × 2 mm, 5 μm) using pure carbon dioxide (0.5 mL/min) at 40°C and 180 bar; (b) on a Diol column (100 × 2 mm, 7 μm) using 5% methanol in carbon dioxide (0.5 mL/min) at 40°C and 182 bar; (c) on a Diol column (250 × 4.6 mm, 5 μm) using 10% methanol (containing 0.6% isopropylamine) in carbon dioxide (2 mL/min) at 40°C and 200 bar. Reprinted from *J. Chromatogr. A* Vol. 785, T. A. Berger, Separation of polar solutes by packed column supercritical fluid chromatography, p. 9. Copyright ©1997, with permission from Elsevier Science.)

(Fig. 1b). However, the addition of isopropylamine to the methanol-modified mobile phase effectively eluted all of the three benzylamines and dramatically improved the peak shapes, as shown in Fig. 1c.

Other Modifiers

Besides methanol, many other polar and nonpolar modifiers have also been used to successfully improve the separation in SFC. These modifiers include acetone, acetonitrile, acetic acid, butane, butanol, *n*-butyl chloride, carbon tetrachloride, dioxane, ethanol, formic acid, heptane, hexane, *n*-hexylamine, methylene chloride, nitromethane, propanol, proprionitrile, tetrahydrofuran, toluene, triethanolamine, trifluoroacetic acid, trifluoroethanol, trimethyl phosphate, and water.

FID-Compatible Modifiers

One of the major advantages of SFC over HPLC (high-performance liquid chromatography) is its compatibility with the flame ionization detector (FID), a universal and sensitive detector for carbon compounds. Unfortunately, most modifiers used in SFC are incompatible with FID. Therefore, a search for polar modifiers that have less response to FID led to the use of water, formic acid, and formamide. These modifiers produce acceptably low background noise and enable the use of FID. Because both water and formic acid have poor solubilities in carbon dioxide, they have been used as modifiers at very low concentrations. However, the modifier effect is significant even at this low level. For example, when water or formic acid was used as modifiers, the resolution of free fatty acids was significantly improved.

Very recently, the separation of polar analytes has also been performed by using pure water under subcritical conditions. Subcritical water has several unique characteristics. For example, the dielectric constant, surface tension, and viscosity of water are dramatically decreased by raising the water temperature while a moderate pressure is applied to keep water in the liquid state. At 200–250°C, the values of these physical properties are similar to those of pure methanol or acetonitrile at ambient conditions. Therefore, subcritical water may be a potential mobile phase for polar analytes. SFC mobile phases other than CO_2 are reviewed separately in this encyclopedia.

Modifier Effect on Retention

It is very clear that chromatographic retention is changed by adding modifiers into carbon dioxide. However, modi-

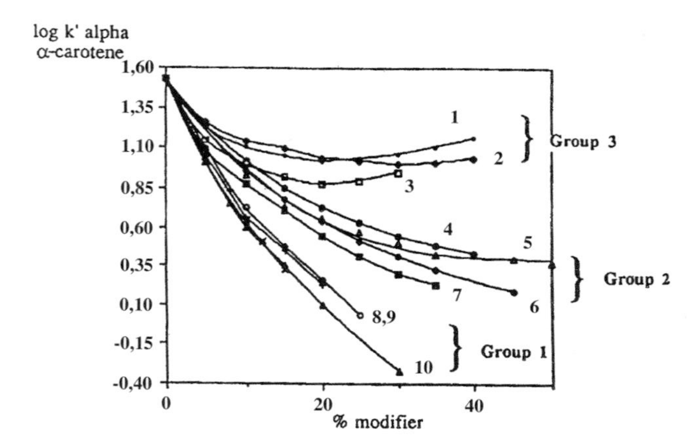

Fig. 2 Variation of the log k' for α-carotene versus the percentage of modifier in carbon dioxide. Temperature: 25°C (subcritical); outlet pressure: 150 bar; flow rate: 3 mL/min; UV at 450 nm; column: UB 225 (250 × 4.6 nm; 5 μm). 1. Acetonitrile; 2. methanol; 3. nitromethane; 4. ethanol; 5. proprionitrile; 6. acetone; 7. 1-propanol; 8. heptane; 9. tetrahydrofuran; 10. methylene chloride. [Reprinted from *Chromatographia, 36*: 275 (1993), E. Lesellier, A. M. Krstulovic, and A. Tchapla, with permission from Friedr Vieweg und Sohn Verlagsgesellschaft mbH.]

fiers used in SFC can be classified in three groups, depending on the retention behavior of the solutes. Figure 2 shows a good example of the modifier effect on the retention of α-carotene. For separations using heptane-, tetrahydrofuran-, and methylene chloride-modified CO_2 (group 1), the retention of α-carotene was decreased rapidly and almost inversely proportional to the modifier concentration, as shown in Fig. 2. The second group of modifiers include ethanol, proprionitrile, acetone, and 1-propanol. A more gradual decrease in the retention of α-carotene was obtained by using modifiers in this group. Methanol, acetonitrile, and nitromethane are in the third group. Even though the retention decreased significantly with increasing modifier percentage up to ~ 10% for all three modifiers in this group, the retention reached a minimum with acetonitrile- and nitromethane-modified carbon dioxide. For separation using the methanol-modified CO_2, the retention remained almost unchanged at higher methanol concentrations. A similar behavior was observed for separations of many other solutes with methanol as the modifier.

Mechanism of the Modifier Effect

Although the exact mechanism is not very clear, the following factors may contribute to the modifier effect on chromatographic retention. Polar modifiers may cover the active sites of the stationary phase (deactivation) so that solute retention is reduced. This can be explained by the differences in retention change between packed and open-tubular columns when small amounts of modifiers were used. Open-tubular columns normally do not show the drastic changes in retention or efficiency upon the addition of small amounts (<2%) of modifier as most packed columns do. These less drastic differences were caused by the differences in the degree of deactivation of the packed column stationary phase as compared with the open-tubular-column stationary phase. An open-tubular column has fewer active sites present and, thus, fewer active sites are present for the modifier to deactivate.

Modifiers may also swell the stationary phase, causing retention change of solutes. The density of the mobile phase can be increased by most polar and nonpolar modifiers so that the solvating power of the mobile phase is enhanced. The polarity of the mobile phase is definitely increased by adding polar modifiers; thus, the retention of polar analytes is reduced. Specific intermolecular interactions between the solute and the modifier in the mobile phase may be additional factors for the modifier effect.

In conclusion, both polar and nonpolar modifiers can be added to the SFC mobile phase to increase the solvent strength. Unlike pure carbon dioxide, the modified CO_2 can elute polar and high-molecular-weight solutes due to its enhanced solvating power. The retention factors are reduced and peak shapes greatly improved by using binary or ternary mobile phases. Although ultraviolet detection can be applied for separations with many modifiers, only water, formic acid, and formamide are compatible with FID.

Suggested Further Reading

Berger, T. A. and Deye, J. F., *Chromatographia 31*: 529–534 (1991).

Berger, T. A., *J. Chromatogr. A 785*: 3–33 (1997).

Cantrell, G. O., Stringham, R. W., Blackwell, J. A., Weckwerth, J. D., and Carr, P. W., *Anal. Chem. 68*: 3645–3650 (1996).

Francis, E. S., Lee, M. L., and Richter, B. E., *J. Microcol. Separ. 6*: 449–457 (1994).

Janssen, H.-G., Schoenmakers, P. J., and Cramers, C. A., *J. Chromatogr. 552*: 527–537 (1991).

Kuepper, S., Grosse-Ophoff, M., and Klesper, E., *J. Chromatogr. 629*: 345–359 (1993).

Lesellier, E., Krstulovic, A. M., and Tchapla, A., *Chromatographia 36*: 275 (1993).

Lesellier, E. and Tchapla, A., Supercritical fluid chromatography with organic modifiers on octadecyl packed columns: Recent developments for the analysis of high molecular organic compounds, in *Supercritical Fluid Chromatography*

with Packed Columns, (K. Anton and C. Berger, eds.), Marcel Dekker, Inc., New York, 1998, pp. 195–221.
Morrissey, M. A., Giorgetti, A., Polasek, M., Pericles, N., and Widmer, H. M., *J. Chromatogr. Sci. 29*: 237–242 (1991).
Pyo, D. and Ju, D., *Anal. Sci. 10*: 171–174 (1994).
Pyo, D., Li, W., Lee, M. L., Weckwerth, J. D., and Carr, P. W., *J. Chromatogr. A 753*: 291–298 (1996).
Roth, M., *J. Phys. Chem. 100*: 2372–2375 (1996).
Smith, R. M. and Briggs, D. A., *J. Chromatogr. A 688*: 261–271 (1994).
Taylor, L. T., *J. Chromatogr. Sci. 35*: 374–381 (1997).
Upnmoor, D. and Brunner, G., *Chromatographia 33*: 261–266 (1992).

Yu Yang

Molecular Interactions in GC

The retention of a solute is directly proportional to the magnitude of its distribution coefficient (K) between the mobile phase (gas) and the stationary phase. The magnitude of K depends on the relative affinity of the solute for the two phases; thus, the stronger the forces between the solute molecule and the molecules of the stationary phase, the larger the distribution coefficient and the more the solute is retained. It follows that the stationary phase must interact strongly with the solutes to be retained and to achieve a separation. Molecular interaction results from *intermolecular forces*, of which there are only two types effective in gas chromatography (GC).

In total, there are three different basic types of molecular force, all of which are electrical in nature. These forces are called *dispersion forces*, *polar forces*, and *ionic forces*. Despite there being many different terms used to describe molecular interactions (e.g., hydrophobic forces, π–π interactions, hydrogen-bonding, etc.), all interactions between molecules are the result of composites of these three different types of molecular force.

Dispersion Forces

Dispersion forces [1] arise from charge fluctuations throughout a molecule, resulting from electron/nuclei vibrations. Glasstone [2] described them in the following way:

> Although the physical significance probably cannot be clearly defined, it may be imagined that an instantaneous picture of a molecule would show various arrangements of nuclei and electrons having dipole moments. These rapidly varying dipoles, when averaged over a large number of configurations, would give a resultant of zero. However, at any instant, they would offer electrical interactions with another molecule, resulting in interactive forces.

Dispersion forces are those that occur between hydrocarbons and other substances that have either no permanent dipoles or can have no dipoles induced in them. In biotechnology and biochemistry, dispersive interactions are often referred to as "hydrophobic" or "lyophobic" interactions, apparently because dispersive substance such as the aliphatic hydrocarbons do not dissolve readily in water. To a first approximation, the interaction energy (U_D) involved with dispersive forces has been deduced to be

$$U_D = \frac{3h\nu_0\alpha^2}{4r^6}$$

where α is the polarizability of the molecule, ν_0 is a characteristic frequency of the molecule, h is Planck's constant, and r is the distance between the molecules. The dominant factor that controls dispersive force is the polarizability (α) of the molecule, which for substances that have no dipoles is given by

$$\frac{D-1}{D+2} = \frac{4}{3}\pi n\alpha$$

where D is the dielectric constant of the material and n is the number of molecules per unit volume.

If ρ is the density of the medium and M is the molecular weight, then the number of molecules per unit volume is N_ρ/M, where N is Avogadro's number. Thus,

$$\frac{4}{3}\pi N\alpha = \frac{D-1}{D+2}\frac{M}{\rho} = P$$

where P is called the molar polarizability. It is seen that the molar polarizability is proportional to M/ρ, the molar volume.

Polar Forces

Polar interactions arise from electrical forces between localized charges such as permanent or induced dipoles. Polar forces are always accompanied by dispersive interactions and may also be combined with ionic interactions. Polar interactions can be very strong and produce molecular associations that approach, in energy, that of a weak chemical bond (e.g., "hydrogen-bonding").

Dipole–Dipole Interactions

The interaction energy (U_P) between two dipoles molecules is given, to a first approximation, by

$$U_P = \frac{2\alpha\mu^2}{r^6}$$

where α is the polarizability of the molecule, μ is the dipole moment of the molecule, and r is the distance between the molecules. The energy depends on the square of the dipole moment, which can vary widely in strength. The numerical value of the dipole moment does not always give an indication of the strength of any polar interactions, as there is often internal electric field compensation when more than one dipole is present in the molecule. Although, the *polarizability* of a substance containing no dipoles may give an indication of the *strength of the dispersive* interactions, due to possible self-association or internal compensation, the *dipole moment* of a substance will *not* always give an indication of the *strength of any polar interaction* that might take place with another molecule.

Dipole–Induced-Dipole Interactions

Compounds, such as those containing the aromatic nucleus and thus π electrons, are polarizable. When such molecules are in close proximity to a molecule with a permanent dipole, the electric field from the dipole induces a counterdipole in the polarizable molecule. This induced dipole acts in the same manner as a permanent dipole and, thus, polar interactions occur between the molecules. Induced-dipole interactions are, as with polar interactions, always accompanied by dispersive interactions. Aromatic hydrocarbons can be retained and separated in GC purely by dispersive interactions when using a hydrocarbon stationary phase or they can be retained and separated by combined induced-polar and dispersive interactions using a poly(ethylene glycol) stationary phase. Molecules can possess different types of polarity, phenyl ethanol, for example, will possess both a permanent dipole as a result of the hydroxyl group and also be polarizable due to the aromatic ring. More complex molecules can have many different interactive groups.

Ionic Forces

Ionic interactions arise from permanent negative or positive charges on the molecule and, thus, usually occur between ions. Ionic interactions are exploited in ion-exchange chromatography, where the counterions to the ions being separated are situated in the stationary phase; ionic interactions are rarely active in GC separations.

To achieve the necessary retention and selectivity between the solutes for complete resolution, it is necessary to select a stationary phase that will provide the optimum balance of dispersive, polar, and induced-polar interactions between the solute molecules and those of the mobile phase.

References

1. F. London, *Phys. Z 60*: 245 (1930).
2. S. Glasstone, *Textbook of Physical Chemistry*, D. Van Nostrand, New York, 1946, pp. 298 and 534.

Suggested Further Reading

Scott, R. P. W., *Techniques of Chromatography*, Marcel Dekker, Inc., New York, 1995.

Scott, R. P. W., *Introduction to Analytsical Gas Chromatography*, Marcel Dekker, Inc., New York, 1998.

R. P. W. Scott

Molecular Weight and Molecular-Weight Distributions by Thermal FFF

In field-flow fractionation (FFF), retention can be related through a well-defined equation to the applied field and governing physicochemical parameters of the analyte. Therefore, in principle, FFF is a primary measurement technique that does not require calibration, but only if the governing physiochemical parameters are either the analyte parameters of interest or their relationship to the parameter of interest (such a molecular weight) is well defined.

In thermal FFF, the applied field is a temperature drop (ΔT) across the channel, and the physicochemical parameter that governs retention is the Soret coefficient, which is the ratio of the thermodiffusion coefficient (D_T) to the ordinary (mass) diffusion coefficient (D). Because ΔT is set by the user, retention in a thermal FFF channel can be used to calculate the Soret coefficient of a polymer–solvent system. However, in order to calculate the molecular weight (M) or molecular-weight distribution (MWD) of the polymer, the dependence of the Soret coefficient on M must be known. Because D_T is virtually independent of M, at least for random coil polymers, the dependence of retention on M reduces to the dependence of D on M. The separation of molecular-weight components by D (or hydrodynamic volume, which scales directly with D) is a feature that thermal FFF shares with size-exclusion chromatography (SEC). In the latter technique, the dependence of retention on D forms the basis for universal calibration, as D scales directly with the product $[\eta]M$, where $[\eta]$ is the intrinsic viscosity. Thus, a single calibration plot prepared in terms of $\log([\eta]M)$ versus retention volume (V_r) can be used to measure M for different polymer compositions, provided an independent measure of $[\eta]$ is available. In thermal FFF, a single calibration plot can only be used for multiple polymers when the values of D_T for each polymer–solvent system of interest are known. However, a single calibration plot can be used with multiple channels. In summary, calibration plots in SEC are specific to each column but can be made universal with respect to their application to a variety of different polymer compositions. Calibration plots in thermal FFF are specific to each polymer–solvent system but universal with respect to different thermal FFF channels. The following discussion outlines the various forms that can be used to calibrate a thermal FFF channel and how they are used to obtain average values of M and MWDs.

The simplest calibration plot takes the following form:

$$\log V_r = a + S_m \log M \tag{1}$$

where a and S_m are calibration constants. The parameter S_m is referred to as the mass-based selectivity in the FFF literature and is very close but not identical in value to the exponent (b) which defines the dependence of D on M:

$$D = AM^b \tag{2}$$

The difference between S_m and b can be explained by the fact that components of different M experience different temperatures as they separate in the thermal FFF channel [1].

An alternate form of Eq. (1) allows for the use of low levels of retention (associated with components of low molecular weight) without losing linearity in the calibration plot [2]:

$$\log(V_r - V^\circ) = a + S_m \log M \tag{3}$$

Here, V° is the geometric (void) volume of the FFF channel.

Because ΔT can be varied to optimize the separation of different samples, it is convenient to incorporate the dependence of retention on ΔT into the calibration plot:

$$\log(\lambda \Delta T) = \phi + n \log M \tag{4}$$

Here, n and ϕ are calibration constants and λ is the FFF retention parameter, which is calculated from V_r using

$$V_r = \frac{V^\circ}{6\lambda[\coth(2\lambda)^{-1} - 2\lambda]} \tag{5}$$

When ΔT is incorporated into the calibration plot, retention parameter λ is used in place of $V_r - V^\circ$ in order to maintain linearity in the plot over a wide range of applied fields and retention levels.

Once the calibration constants ϕ and n have been determined for a given polymer–solvent system, Eq. (4) can be used for all thermal FFF channels, provided the temperature of the cold wall (T_c) is held constant. The cold-wall temperature affects the calibration plot because the Soret coefficient (D_T/D) and, therefore, ϕ varies with T_c. For a detailed discussion of temperature effects; see the entry Cold-Wall Effects in Thermal FFF. In a thorough study of temperature effects, Myers and co-workers [3] demonstrated that the dependence of the Soret coefficient on T_c can be accurately modeled by

$$\frac{D}{D_T} = \lambda \Delta T = a' T_c^m \tag{6}$$

The validity of this model is illustrated in Fig. 1 by the linearity in plots of $\log(\lambda \Delta T)$ versus $\log(T_c/298)$. Based on

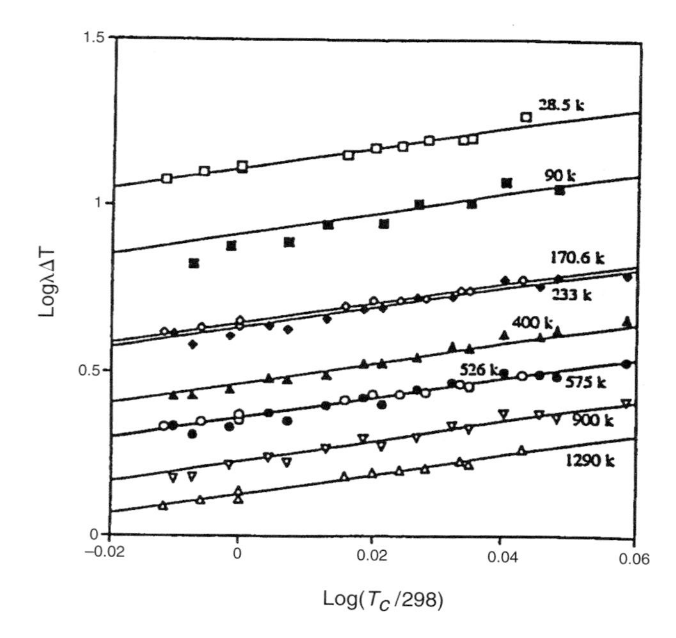

Fig. 1 Plots of log($\lambda\Delta T$) versus log(T_c/298) for polystyrene in tetrahydrofuran. The data were gathered using a variety of different thermal FFF channels. Values of ΔT ranged from 30 to 70 K.

Eq. (6), the following universal calibration equation has been proposed:

$$\log(\lambda\Delta T) = \phi + m\log(T_c/298) - n\log\left(\frac{M}{10^6}\right) \quad (7)$$

In order to utilize Eq. (7), a set of linear plots of log($\lambda\Delta T$) versus log($M/10^6$) is established for a given polymer–solvent system, with one plot being generated for each of several cold-wall temperatures. Such plots run parallel to one another with a slope equal to n and with intercepts that equal $\phi + m\log(T_c/298)$. To obtain the values of ϕ and m, linear regression is performed on the intercept values as a function of $T_c/298$. In Eq. (7), T_c and M are divided by 298 and 10^6, respectively, to avoid large extrapolations in obtaining the various intercept values by regression.

When a mass-sensitive detector such as a refractometer or photometer is used, the weight-average molecular weight (M_w) of a polymer sample is calculated from the calibration equation using the value of V_r that corresponds to the center of gravity of the elution profile. The center of gravity is defined by placing a vertical line through the elution profile such that the line bisects the profile into two halves of equal area. The intersection of this line with the elution–volume axis defines the weight-average V_r, which is converted to M_w through the calibration equation. If the elution profile has a Gaussian or other symmetrical shape, then the weight-average V_r is the value of V_r that corresponds to the peak of the elution profile.

In thermal FFF, the shape of the elution profile is not significantly affected by band broadening for polydisperse ($\mu > 1.2$) polymers, provided that low flow rates (<0.2 mL/min) are used. In that case, accurate information on the MWD can be obtained directly from the elution profile. For example, if a mass-sensitive detector is used, the signal (s) is linearly related to the mass of polymer in the eluting stream, and the number-average molecular weight (M_n) of the sample is calculated as [2,4]

$$M_n = \frac{\sum s_i}{\sum s_i/M_i} \quad (8)$$

Here, the summation extends over small equal elements of elution volume from the beginning to the end of the elution profile. Thus, s_i is the detector signal of the ith digitized increment and M_i is calculated from the associated value of V_r using the calibration equation. The weight-average molecular weight (M_w) is calculated as

$$M_w = \frac{\sum s_i M_i}{\sum s_i} \quad (9)$$

The polydispersity, μ, defined as the ratio M_w/M_n, thus becomes

$$\mu = \frac{(\sum s_i M_i)(\sum s_i/M_i)}{(\sum s_i)^2} \quad (10)$$

Thermal FFF elution profiles can also be converted into a detailed MWD. With a mass-sensitive detector, the elution profile is essentially a plot of the polymer concentration c (with units of mass/volume) in the eluting stream versus V_r. In order to obtain the mass-based MWD, $m(M)$, we need to transform this profile using

$$m(M) = c(V_r)\frac{dV_r}{dM} \quad (11)$$

The normalized and digitized form of this equation is

$$m_i = \frac{s_i}{\sum s_i}\frac{\Delta V_r}{\Delta M_i} \quad (12)$$

where ΔV_r is the fixed elution volume element corresponding to one digitized interval. In the case of linear calibrations, as defined by Eqs. (1) and (3), $d(\log V_r)/d(\log M)$ is a constant equal to S_m and Eq. (12) is more conveniently expressed as

$$m_i = \frac{s_i}{\sum s_i}\frac{M_i}{V_{r,i}}S_m \quad (13)$$

If needed, the differential number MWD can be obtained by dividing $m(M)$ by M:

$$n(M) = \frac{m(M)}{M} \tag{14}$$

For polymers with a low polydispersity ($\mu < 1.2$), a detailed MWD is not generally needed, and simply calculating M_n and M_w from Eqs. (8) and (9) is generally adequate. However, the elution profile of such a narrow MWD is affected significantly by band broadening, which must be accounted for in order to obtain the most accurate possible values of M_n and μ [5]. Even the elution profile of more polydisperse samples ($\mu > 1.2$) can be affected by band broadening if high flow rates (>0.2 mL/min) are used. Fortunately, band broadening is well defined in thermal FFF and can be removed by one of two methods. For samples of low polydispersity, μ is well approximated by [5]

$$\mu = 1 + \frac{H_P}{LS_m^2} \tag{15}$$

where L is the channel length and H_P is the polydispersity contribution to plate height. H_P is obtained by plotting the experimental plate height as a function of flow rate. With the proper technique [5], such plots are linear and the y intercept is equivalent to H_P. Alternatively, H_P can be obtained by subtracting the nonequilibrium band-broadening contribution to plate height (H_N) from the experimentally measured value. Methods for calculating H_N from well-established models of band broadening in thermal FFF can be found in the literature [2]. Because the elution profile of samples with low polydispersity are generally symmetrical, M_w is calculated using the peak value of V_r in the calibration equation, Eq. (15) is used to calculate μ. Finally, M_n is calculated from $\mu = M_w/M_n$.

Samples with higher polydispersity do not generally yield symmetrical elution profiles. Moreover, a detailed MWD [either $m(M)$ or $n(M)$] is often desired. As mentioned earlier, band broadening does not significantly affect the elution profile when low flow rates are used. Therefore, the elution profile can be directly converted into a MWD by the procedures outlined. If, on the other hand, fast flow rates are used to shorten the analysis time, the observed elution profile must be adjusted to account for the effects of band broadening. A deconvolution algorithm that filters out the band-broadening contribution to the elution profile is described in the literature [2]. Figure 2 illustrates the elution profiles before and after the effects of band broadening are removed for samples of both low and high polydispersity.

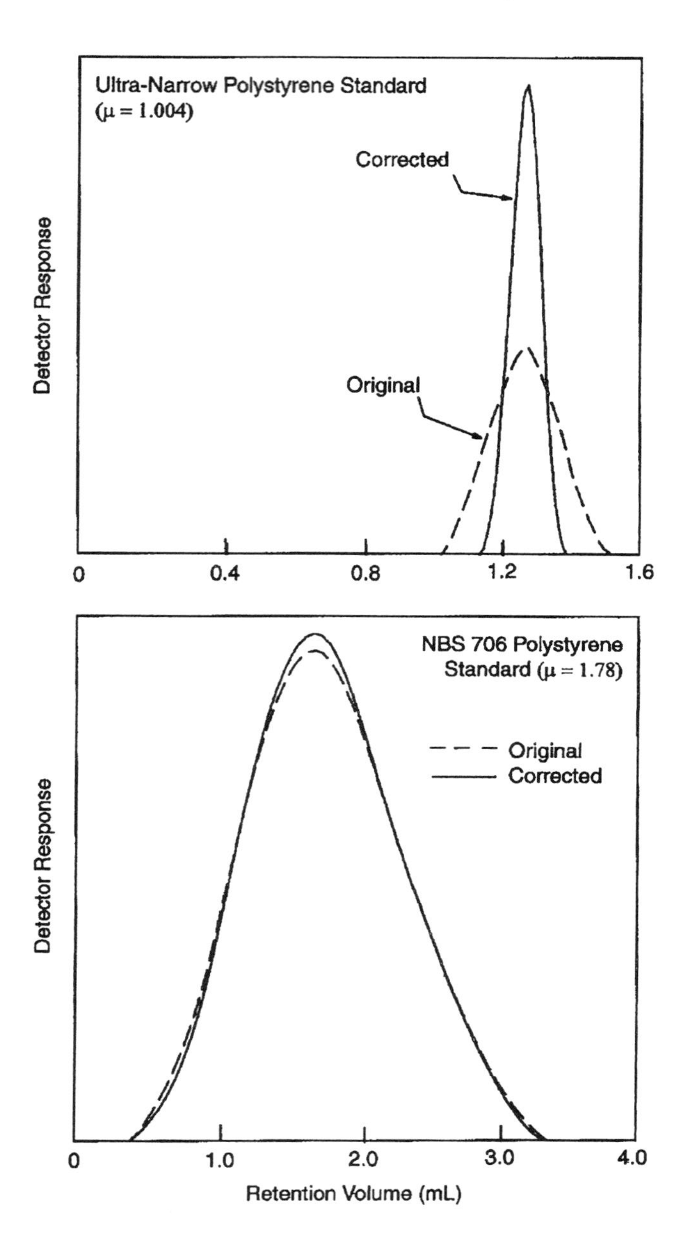

Fig. 2 Thermal FFF elution profiles before (original) and after (corrected) removing the effects of band broadening. With the polydisperse sample (NBS 706), which was analyzed at a flow rate of 0.4 mL/min, the effect of band broadening on the elution profile is minimal. The polydispersity values listed were determined using thermal FFF.

References

1. G. H. Ko, R. Richards, and M. E. Schimpf, *Separ. Sci. Technol. 31*: 1035 (1996).
2. M. E. Schimpf, P. S. Williams, and J. C. Giddings, *J. Appl. Polym. Sci. 37*: 2059 (1989).

3. M. N. Myers, W. Cao, C.-I. Chen, V. Kumar, and J. C. Giddings, *J. Liquid Chromatography Related Technol. 20*(16 & 17): 2757 (1997).
4. L. H. Tung, in *Polymer Fractionation* (M. J. R. Cantow, ed.), Academic, New York, 1967.
5. M. E. Schimpf, M. N. Myers, and J. C. Giddings, *J. Appl. Polym. Sci. 33*: 117 (1987).

Martin E. Schimpf

Multidimensional TLC

Introduction

Since the introduction of thin-layer chromatography (TLC), there has been great interest regarding the increase of separation capacity (spot capacity). This interest was motivated by the separation of complex mixtures that contain a great number of compounds. Some of them are difficult to separate because they have similar properties.

Two-Dimensional Development

Two-dimensional thin-layer chromatography (2D TLC) involves the application of a single sample to one corner of a TLC plate, which is subjected to two development processes. The TLC plate is developed with the first mobile phase, dried, and redeveloped in an orthogonal (the plate rotated through 90°) direction with a second mobile phase having different selectivity characteristics. Thus, the components migrate from the point of application into a two-dimensional thin layer, ensuring more space for resolution compared to one-dimensional separation.

Two-dimensional TLC is an analytical separation technique recommended for the separation of sample of compounds that are difficult to separate in a single dimension. This technique has been mostly used for qualitative clinical and biochemical analysis, where high selectivity separation is required.

In 2D TLC, any spot can be defined by a pair of x and y coordinates; the quality of a separation can be established by comparing the distance between all pairs of spots in the chromatogram. High resolution will be obtained when the selectivity between the two directions will be significantly different. In practice, several methods have been used to achieve this purpose. The potential methods for obtaining two different separation mechanisms in orthogonal directions are the following:

1. Two eluent systems of different selectivities for the sample components are used and a single sorbent thin layer for two dimensions is developed.
2. Two sorbent layers (e.g., silica gel and reversed phase) as adjacent zones with different selectivities can be used. The sorbent layer for the first development is a narrow strip (reversed phase), and for the second development, it is a large surface (normal phase). A suitable eluent system has to be used for each sorbent layer.

There are others possibilities, but all of them have the same principle (i.e., the selectivities to be different in the two orthogonal developments). If the selectivities are close to each other, the 2D separation effect will be a diagonal arrangement of the spots. A separation will be considered better when the spots will be uniformly spread along the entire surface of the plate.

Two-dimensional TLC is used for a great number of compounds. When spot capacity in the first direction is n_1 and is n_2 in the orthogonal direction, then the total spot capacity will be equal to n_1n_2. References 1 and 2 should also be consulted for routine procedures.

Multiple Unidimensional Development

Multiple unidimensional development (MUD) is the simplest approach for enhancement of the separation capacity in TLC [2]. In this approach, the TLC plate is developed for a selected distance, then the plate is withdrawn from the developing chamber and the adsorbed solvent is evaporated before repeating the development process. MUD is a very versatile strategy for the separation of complex mixtures. The main feature of MUD is that it leads to an increase in the spot reconcentration mechanism. There is an optimum number of developments that provide maximum separation.

Programmed Multiple Development

The term *programmed multiple development* (PMD) was introduced by Perry et al. [3] for a TLC developing technique and defined as follows: A TLC plate is repeatedly developed in the same direction with the same solvent. Each development run is longer than the previous development step. Following each run, the layer is dried by radiant heat, optionally assisted by a flow of inert gas. The lower edge of the layer remains, at all times, in the contact with the solvent in the reservoir, which is shielded from the radiant heat. The solvent migration distance is controlled, programmed via the lengths of intervals between the heating cycles.

Automated Multiple Development

Automated multiple development (AMD) combines the advantages of MUD and mobile-phase gradient elution. This multiple development approach was improved by Burger [4], who maintained the basic idea of development. The chromatogram is developed in the same direction, stepwise, over an increasing solvent migration distance, but changes all other characteristics. The characteristics of the AMD system, according to Burger, are as follows: An HPTLC plate is developed several times in the same direction with eluents differing in elution power. In general, the polarity of the eluent is decreased step by step over the solvent migration distance. Between each run, the plate is completely dried by vacuum.

Gradient elution in AMD starts with the most polar eluent and is varied toward decreasing polarity. Figure 1a shows a typical universal elution gradient, made up of the three solvents: methanol, dichloromethane, and hexane. Figure 1b illustrates the increasing duration of the development cycles. Time increments are chosen to obtain uniform increases of the running distance of ~3 mm/step.

Automated multiple development causes sample fractions to become concentrated into narrow bands. The eluent flows over the lower part of a sample before reaching the upper part, concentrating the sample in the top to bottom direction. This is due to the fact that molecules in the lower part of the sample zone start their upward movement earlier than those in the upper part of that sample zone each time the eluent front passes through that area. A strong solvent used at the beginning of AMD causes concentration effects similar to a plate with a concentration zone. The separation power is increased by factor of 3, as compared with regular high-pressure TLC (HP–TLC). The combination of the focusing effect and gradient elution results in extremely narrow bands. Their typical peak width is about 1 mm. This means that with the available separation distance of 80 mm, up to 40 components can be completely resolved (i.e., with a resolution greater than 1).

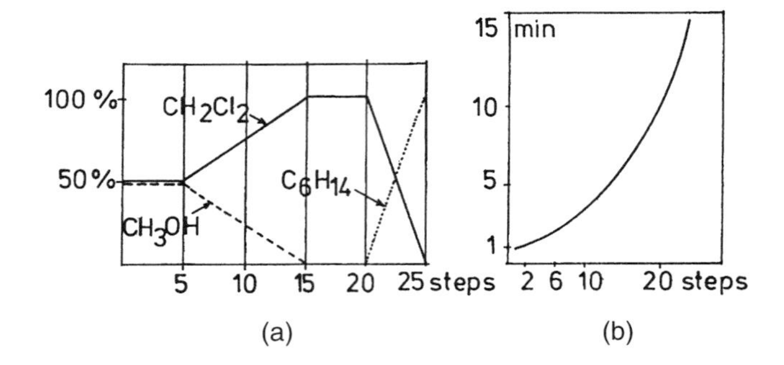

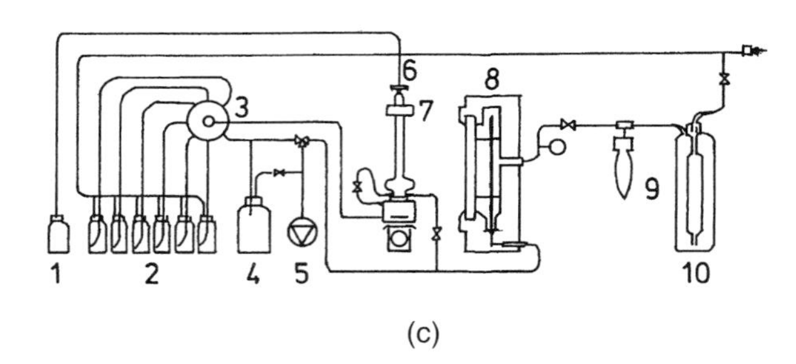

Fig. 1 (a) Typical universal elution gradient; (b) time of development versus number of steps; (c) AMD system flow diagram.

Design and Operation of an AMD System [5]

The Camag AMD system consists of two main components: the AMD developing unit (Fig. 1c) and the microprocessor-based controller. This system provides an AMD under reproducible conditions. For the AMD microprocessor-based controller, the following parameters may be chosen: the eluent composition, by selecting the number of solvent reservoir; the number of developing steps; the developing time for each step; the number of preconditions; the option of emptying the mixer after a selected step.

Functional Principle of the AMD Device

During the development process, the system is controlled by computer. The main component is an enclosed developing chamber (8), which is connected for introducing and withdrawing the developing solvent and for pumping the gas phase in and out. The six reservoir bottles (2) contain the individual solvent components. The gradient mixer (6) is connected, via motor valve (3), to reservoir bottles (2). The gas phase is made up externally by passing nitrogen through the wash bottle (10) into the gas-phase reservoir (9). At the same time, the mixer is filled with solvent from the solvent reservoir bottle (1). Then, the chamber is equilibrated for 15 s with gas from the reservoir (9). The first step of development can start after filling the chamber (8) with the contents (8 mL) of the upper part of the gradient mixer (6), which is controlled by a light barrier (7).

Migration of the solvent in the layer starts immediately. At the end of the programmed time (determining the migration distance), the liquid solvent is removed by vacuum from the chamber (8) into the waste-collecting container (4). Then, after all liquid solvent has been removed, vacuum is applied by pump (5), thus drying the layer. The drying period is time programmed. Before the next developing cycle is started, the chromatographic thin layer is reconditioned by feeding the gas phase from the blender (9) into the chamber. After reconditioning the chamber, the next development step can start. While the drying phase is carried out, the solvent for the next step is made up in the gradient mixer (6).

Application

When the sample components differ widely in their polarities, they can be separated by using a universal elution gradient. With AMD, it is possible to simultaneously analyze 12 samples. AMD–HP-TLC has been applied for screening of pesticides from groundwater or drinking water and soil. A universal gradient based on dichloromethane was used to detect the presence of pesticides from different classes such as organochlorine, organophosphorus, carbamate, phenylureas, triazines, phenoxycarboxylic acids, and others. In this way, 283 pesticides were analyzed using this universal gradient [6,7].

Plant extracts have a widespread application in the drug and cosmetic industries. For the separation of plant extracts, the AMD–HP-TLC method is the most suitable because it has a higher separating power. Isocratic and AMD development are shown in Fig. 2 [8].

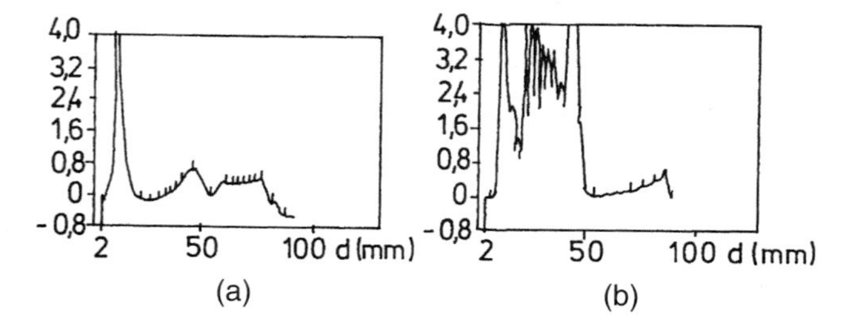

Fig. 2 Densitograms: (a) isocratic development TLC; (b) AMD–TLC.

On-Line Coupling of HPLC–AMD–TLC

Combining different separation methods, governed by different separation mechanisms, to multidimensional methods is suitable for multiplying the potential of the individual techniques. Reversed-phase chromatography high-performance columns (RP–HPLC) can be coupled with normal phase TLC [9,10].

Principle

The system for mass transfer HPLC–TLC consists of a column filled with RP-C_{18} sorbent and a modified Linomat (Camag) sample spray unit for TLC. The effluent is transferred, via a capillary tubing, to a TLC applicator unit. While the effluent of the HPLC column is sprayed onto the thin layer with nitrogen, the plate is moved. The separation of 56 pesticide residues in drinking water in a single sample using RP–HPLC coupled with TLC and AMD has been demonstrated. For example, by AMD, a chromatogram of RP–HPLC cut no. 13 was separated and over 10 pesticides were detected through multiwavelength scanning by absorbance between 200 and 300 nm, [11].

References

1. N. Grinberg (ed.), *Modern Thin-layer Chromatography*, Marcel Dekker, Inc., New York, 1990.
2. C. F. Poole and S. K. Poole, *Chromatography Today*, Elsevier, Amsterdam, 1991.
3. J. A. Perry, K. W. Haag, and L. J. Glunz, *J. Chromatogr. Sci. 11*: 447 (1973).
4. K. Burger, *Z. Anal. Chem. 318*: 228 (1984).
5. D. E. Jaenchen, in *Proceedings 3rd International Symp. on Instrumental HPTLC* (R. E. Kaiser (ed.), Institute for Chromatography, Bad Duerkheim, Germany, 1985, p. 71.
6. S. Butz and H. J. Stan, *Anal. Chem. 67*: 620 (1995).
7. R. Koeber and R. Niessner, *Fresenius J. Anal. Chem. 354* (1996).
8. S. Gocan, G. Cimpan, and L. Muresan, *J. Pharm. Biomed. Anal. 14*: 1221 (1986).
9. D. E. Jaenchen, in *Proceedings 4th International Symposium on Instrumental High Performance Thin-Layer Chromatography*, H. Traitler, A. Studer, and Kaiser (eds.), Institute for Chromatography, Bad Duerkheim, Germany, 1987, p. 185.
10. J. W. Hofstraat, M. Engelsman, R. J. Van De Nesse, C. Gooijer, N. H. Velthorst, and U. A. Th. Brinkman, *Anal. Chim. Acta 186*: 247 (1986).
11. B. Protze, Diploma thesis, Fachhochschule Niederrnein Krefeld, Germany, 1986.

Simion Gocan

N

Natural Products Analysis by CE

Natural products chemistry is a science that investigates the identification, metabolism, biosynthesis, distribution, and biological activity of various organic compounds derived from living things. It concerns, essentially, the separation, purification, and structural determination of each compound. It was not until the late 1700s that, in plants or animals, the separation of natural products was carried out; for example, tartaric acid was extracted from grapes, citric acid from lemons, and malic acid from the apple by Scheele, a pharmacist in Sweden. In the nineteenth century, most of the compounds, alkaloid, terpenoid, and glycoside, were first isolated in pure form and elucidated structurally, since morphine was extracted from opium by Serturner. In the twentieth century, natural products chemistry achieved a great promotion through development of micro element analysis and column chromatography, which allowed physiological activity materials in crude form to be isolated with ease. Natural products are the organic and inorganic compounds found in nature: in plants (leaves, roots, barks, rhizoma, flowers, and seeds), in marine organisms (plants, animals, and microbes), in the fungi found in highly diverse and sometimes extreme environments, and in soil [1].

There are two major classes of natural products: primary and secondary metabolites. Primary metabolites are compounds that exist in all organisms and are involved in basal and vital metabolism (e.g., glucose, fatty acids, amino acids, etc.). Secondary metabolites, alkaloids, terpenoids, and flavonoids, are unique to a particular species and vary in their basic structures. Secondary metabolites are usually accumulated, as most of their end metabolites are in plants, but are excreted in animals and microorganisms and some of these are proven to have pharmacological and ecological significance.

The pharmaceutical industry, in its drug discovery efforts, has relied heavily on natural products. Not all natural products are bioactive; in fact, most have little or no measurable activity at all. Simple sugars, lipids, amino acids, flavonoids, and so forth are nature's essential building blocks, and measurements of their quantities are needed for the study of metabolism, disease processes, and aging. In today's public maketplace, there is a growing interest in the so-called crude drugs, where an increasing number of herbal products claimed as health aids, bodybuilding supplements, nutrients, vitamin supplements, cosmetics, and so forth are being sold.

In Asia, plants have been applied in various diseases and studied with their components for many years. For research of their components, at first, fresh plants are dried at low temperature, for short times, in well-ventilated places so as to minimize chemical change of components, and then they are extracted. They are generally finely ground and extracted first with an organic solvent, then by water, to remove the widest possible range of compounds, from the hydrophilic to the most hydrophobic. The solvent is then removed and the crude extracts are tested for biological activity. If they are found to be active, fractionation by a variety of chromatographic techniques is initiated. At each step, the fractions are reevaluated, and only those fractions in which bioactivity is increasing are fractionated further. Finally, a pure chemical and structural identity of the molecule is determined by a combination of nuclear magnetic resonance (NMR), infrared (IR), ultraviolet (UV), mass spectrometry (MS), x-ray crystallography, and other analytical tools. Once the structure and chemical properties of an active compound are known, quantitative and qualitative analytical methods are employed to detect and quantify it. Recently, the pharmacological significance of the crude drugs is highlighted in terms of the suitability of their medicinal value and low number of side effects. Most crude drugs are comprised of more than one component and their number is determined through the analysis of major components.

Although high-performance liquid chromatography (HPLC) and gas chromatography (GC) have been mainly applied in quantitative method of analysis, they have difficulty with simultaneous quantitative analysis of crude drugs and have problems concerning the instability of experimental conditions, poor reproducibility because of pretreatment of crude drugs, efficacy of columns, and delay of separation time. GC has a limitation that components must be volatile. HPLC requires longer times for simultaneous analysis of samples, although gradient could be used advantageously for shortening analysis times.

With capillary electrophoresis (CE), it is possible to carry out simultaneous quantitative analysis with small sample volumes (1–10 nL); the amount of solvent waste generated is in the order of 1–2 mL/day and requires much less analysis time, with outstanding resolution. CE may be able to resolve a component of interest more quickly and with less effort invested in sample preparation than alternative techniques, because it can resolve neutral as well as ionic compounds using the same column in the same analytical run. Although CE is an excellent micro-analytical technique, its use in a preparative format is limited at the present time. CE in its capillary zone electrophoresis (CZE) and micellar electrokinetic chromatography (MEKC) formats allows the analyst to resolve both ionic and neutral compounds on the same column using simple buffers, with or without an organic modifier, a micelle, a cyclodextrin, or a mixture of all of these. The reproducibility of migration time is the factor requiring improvement with CE, as the longer the migration time is, the more the peak area increases.

The velocity of a solute in the capillary is determined by its electrophoretic mobility and electro-osmotic flow (EOF), which are affected by temperature, and which is influenced by the diameter and length of the capillary, its contents, concentration, and pH of running buffer, applied voltage, current, viscosity, and zeta-potential. The subtle variation of EOF is also a main factor in maintaining high reproducibility if CE is automated by the constant temperature of the capillary and running buffer with proper buffering capacity. However, it is difficult to keep the temperature constant, as EOF depends on the condition of the fused silica.

The zeta-potential present on the silica surface is changed by the cleaning, preconditioning, storage method, and use time. Therefore, it is important to keep the surface condition reproducible. Usually, herb drugs are made of several crude drugs, as extracts, powders, or pills. The components of herb drugs are so varied that their quality control is applied to just one or two of their index components. CE has been applied to the analytical study of various components ever since it was applied to individual components in glycyrrhizin of an herb drug by Iwagami in 1991 [2]. Presently, new analysis methods that are combinations of MEKC have been developed for the distribution of samples in an interfacially active agent by Terabe. CZE is able to analyze components with and/or without electric charge [3].

Micellar electrokinetic chromatography was originally developed for the separation of electrically neutral substances by capillary electrophoresis and has proven to be a highly efficient separation method for various kinds of analyte. Although various ionic substances can be separated by CE alone, the separation of many components in complex mixtures is not always succesful. For example, a number of drugs are ionic, but some of them are not easily separated by CE. MEKC has been shown to be a powerful technique for the separation of complex drug mixtures. Although most of these drugs can be separated by HPLC, MEKC usually gives a better resolution in shorter analysis times.

Compared to HPLC, CE is simpler, faster, more convenient, and has a higher resolution power, which is a necessity when analyzing complex multicomponent mixtures.

A literature search reveals that CE has been used for the analysis of widely different natural compounds from extracts of leaves and needles, bark, roots, marine organisms, moss, soils, and so forth. To illustrate the problems associated with analyzing natural product extracts, especially from plant materials, by chromatography and to demonstrate how CE can overcome some of these problems as a consequence of its high resolving power, we compared the separation of an extract from the radix of the *Scutellaria baicalensis* Georgi by HPLC and CE. *Scutellaria radix* is the root of *Scutellaria baicalensis* Georgi. It is well known and is frequently used in oriental pharmaceutical preparations as a remedy for inflammation, suppurative dermatitis, allergic diseases, hyperlipidemia, and arteriosclerosis [4].

Flavonoids are major components of *Scutellaria radix*, and about 40 kinds of flavonoids have been identified in it so far [5]. These flavonoids are known to have a broad range of physiological activities. The activities of baicalin,

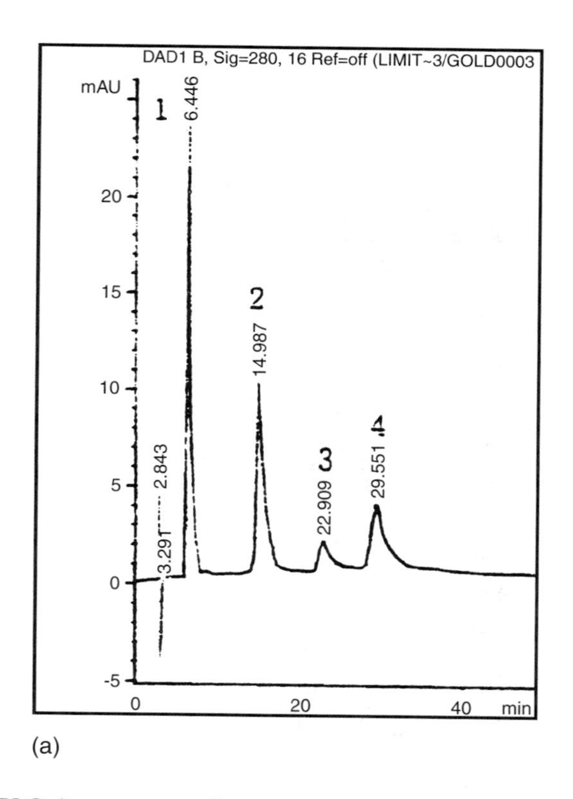

(a)

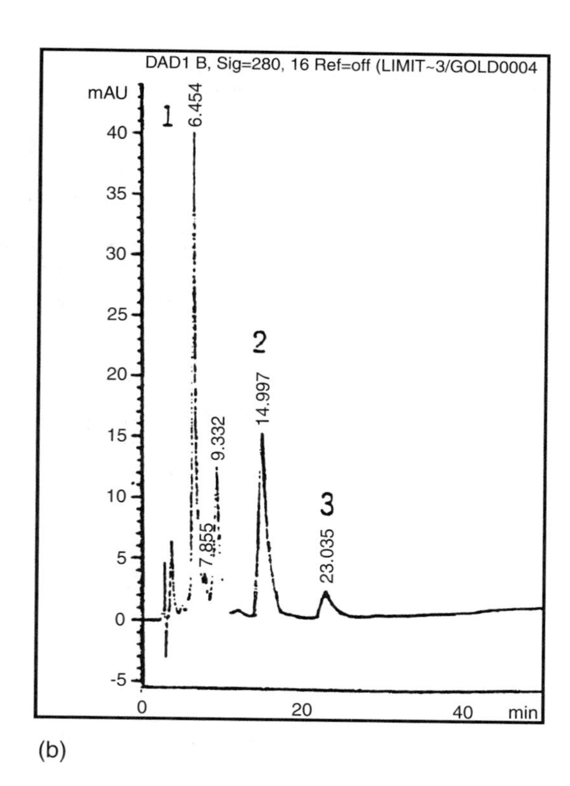

(b)

Fig. 1 HPLC chromatogram of standard mixtures (a) and a *Scutellaria radix* extract (b). Key: 1: baicalin; 2: baicalein; 3: wogonin; 4: chrysin.

which is the main component, baicalein, and wogonin, in particular, have been investigated. From the results, it has been concluded that the most potent antiallergic material is baicalein, and the other flavonoids have low activities. Therefore, the quantitative analysis of individual flavonoids is important for evaluating the quality of *Scutellaria radix*. Figures 1 and 2 show the HPLC chromatogram and CE electropherogram of a crude drug radix extract (*Scutellaria baicalensis* Georgi). These samples were obtained by first extracting with 50% ethanol. In the ethanol extract, many compounds are left. It is clear from Fig. 1 that HPLC is not the method of choice in this case. Important characteristics of an analytical procedure are sufficiently high precision and accuracy and also the time of analysis. Therefore, often it is not relevant in method development to simply maximize the resolution of the sample components without fulfilling the demands on analysis time. Decreasing the pH of the buffer leads to an increase in the degree of dissociation of baicalin, baicalein, wogonin, and chrysin. However, crude drugs have many components. As *Scutellaiae radix* has about 40 kinds of flavonoid, it is very difficult to simultaneously separate baicalin, baicalein, wogonin, chrysin, and other flavonoids in a short time. Based on the dependence of the migration times and resolution on pH and phosphate buffer concentration (Fig. 2), it can be concluded that the most favorable electrolyte system is that with pH 7.0, 35 m*M* phosphate buffer. This system was applied for the determination of baicalin, baicalein, wogonin, and chrysin in the extracts of *Scutellaiae radix*. A typical HPLC chromatogram obtained from the extract of *Scutellaiae radix* with a KFDA regulation is shown in Fig. 1. Baicalin is not clearly distinguished from other components and the analysis time is greater than 35 min.

By comparing the records of Figs. 1 and 2 for the results of HPLC and CE analysis, it is seen that (a) CE provides higher sensitivity than HPLC and (b) the CE analysis time is shorter than HPLC.

Conclusions

The proposed method is well suited for the rapid and simultaneous determination of baicalin, baicalein, wogonin, and chrysin. The data presented in this report indicates that CE has application for crude drugs. The separation

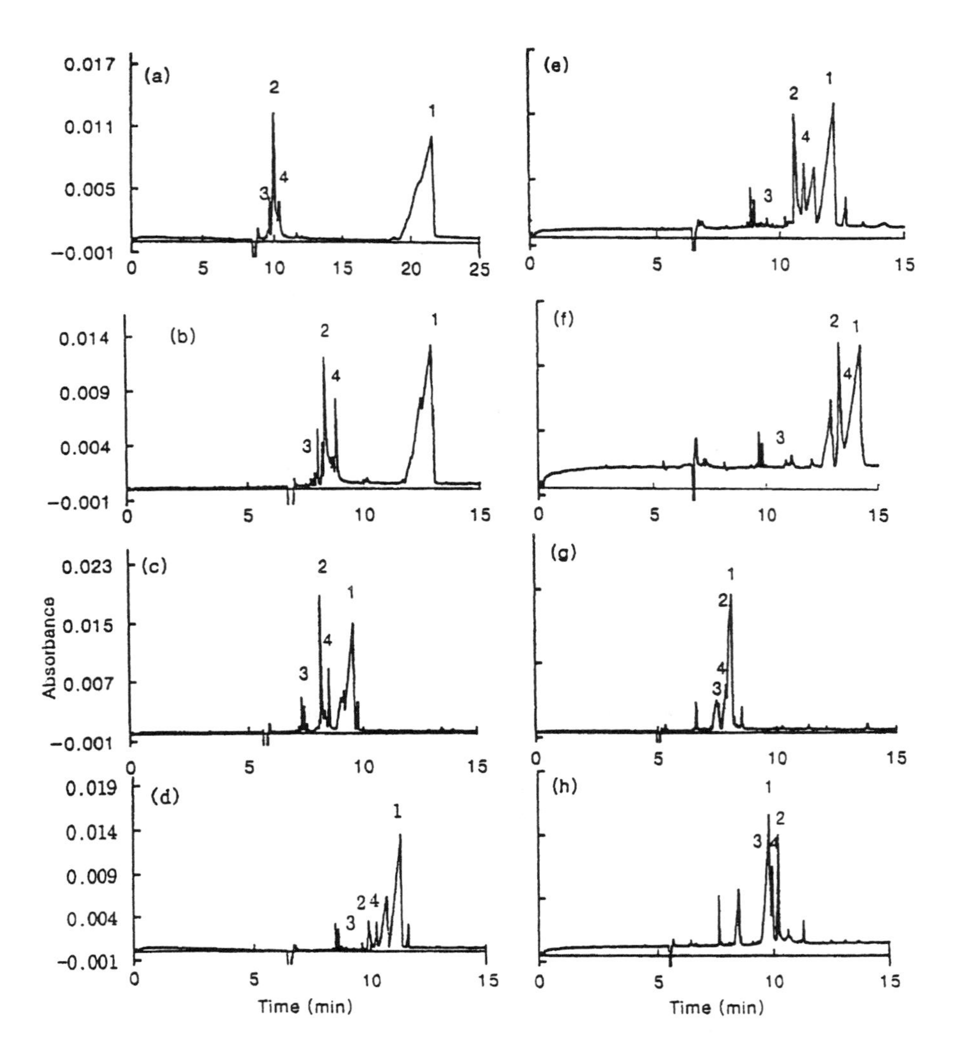

Fig. 2 Effect of running buffer pH and phosphate concentration on the separation of four components in the *Scutellaria radix* CE (1: baicalin; 2: bicalein; 3: chrysin; 4: wogonin). (a) pH 6.0 sodium citrate buffer (20 m*M*), (b) pH 6.5 sodium phosphate buffer (20 m*M*), (c) pH 7.0 sodium phosphate buffer (20 m*M*), (d) pH 7.0 sodium phosphate buffer (35 m*M*), (e) pH 7.0 sodium phosphate buffer (50 m*M*), (f) pH 7.0 sodium phosphate buffer (100 m*M*), (g) pH 7.5 sodium phosphate buffer (50 m*M*), (h) pH 8.0 sodium phosphate buffer (35 m*M*).

by CE is completed within 15 min and is much faster than HPLC. A comparision of the analysis time for both techniques is made. Resolution, recovery, and reproducibility for four flavonoids in *Scutellaiae radix*, separated by CE, are greatly improved in crude drug analysis.

References

1. H. J. Issaq, Capillary electrophoresis of natural products, *Electrophoresis 18*: 2438 (1997).
2. S. Iwagami, Y. Sawaba, and T. Nakagawa, Micellar electrokinetic chromatography for the analysis of crude drugs. Determination of glycyrrhizin in oriental pharmaceutical preparations, *Shoykugaku Zasshi 45*(3): 233 (1991).
3. S. Terabe, K. Otsuka, K. Ichikawa, A. Tsuchiya, and T. Ando, Electrokinetic separations with micellar solutions and opentublar capillaries, *Anal. Chem 56*: 111 (1984).
4. S. I. Lee and D. G. An, Natural herb products sciences, Young Lim Co., Seoul, p. 178 (1992).
5. H. S. Yoon, Flavonoid components in plants of the Genus Scutellarin. *Korean J. Pharmacogen. 23*: 201 (1992).

Noh-Hong Myoung

Natural Rubber: GPC–SEC Analysis

Introduction

Natural rubber, produced from *Hevea brasiliensis*, a very high-molar-mass polymer, differs from most of its synthetic counterparts through its more complex microstructure, due to the interactions of nonrubber compounds with the polyisoprene chains. This "associative" microstructure is gradually destroyed and, part when the polyisoprene is dissolved in a conventional solvent. However, in very many cases, a proportion of the natural rubber remains insoluble in such solvents; this fraction is commonly called the gel or macrogel phase [1,2] and is usually eliminated and quantified by centrifugation. The soluble fraction contains the polyisoprene macromolecules and a variable quantity of microaggregates, between 1 and 15 μm in diameter [3], forming the microgel.

Numerous *Hevea* varieties, referred to as "clones" in the professional jargon, can be found in estates. Subramaniam [4] was the first to study natural rubber from various clones by size-exclusion chromatography (SEC), to study the native molar mass distribution.

This article describes the different stages of natural rubber analysis by SEC, concentrating on the points that distinguish natural rubber from other more conventional polymers.

Dissolving and Eluant

In most cases, tetrahydrofuran (THF) has been used as the mobile phase, except by Bartels et al. [5], who used cyclohexane. However, using THF to analyze natural rubber by SEC can raise certain problems [6].

As a general rule in our laboratory, the samples of natural rubber are dissolved in cyclohexane (HPLC grade) stabilized with 2,6-di-*tert*-butyl-4-methylphenol (internal standard) at a rate of 120 mg for 30 mL of solvent. The solutions, stored at 30°C, are gently stirred for 1 h, periodically, for 14 days.

Equipment

The equipment used consists of a conventional SEC system [gas extractor, a pumping system, an injector, column(s), and detector(s)]. It is important to use a column oven. At room temperature, injecting natural rubber solutions into an SEC system can block the columns. The oven temperature used depends on the solvent used as the mobile phase. With cyclohexane, the oven temperature must be 65°C to overcome adsorption phenomena due to the very low polarity of this solvent. When using THF, the oven is heated to 50°C.

Using columns with a porosity of 20 μm is also recommended, in order to minimize shearing the long chains of the polyisoprene. As molar mass distribution (MMD) in natural rubber is quite broad ($12 < I < 3$), the columns need to offer a considerable separation range.

Because cyclohexane does not require an added stabilizer in the mobile phase, this means that ultraviolet (UV) detection at 220 nm can be used. This is important, as this type of detector is more sensitive than a refractometer. In fact, as natural rubber is a polymer with a very high molecular weight, it is recommended that low-concentration solutions at around 0.2 mg/mL are injected, so as to overcome viscosity effects and avoid excessive shearing of the macromolecules [7]. Of course, a light-scattering detector, or viscometer, can be added to the system to access the branching rate [8,9], which is an important parameter for natural rubber.

Calibration

Cyclohexane as the mobile phase requires the use of polyisoprene standards, as polystyrene standards are not soluble in this solvent. It should be noted that calibration by polystyrenes results in an overestimation of molar masses by a factor of around 2, compared to the use of polyisoprene standards [10]. It is, therefore, necessary to carry out universal calibration or to convert molar masses using the Mark–Houvink coefficients relative to synthetic or natural polyisoprenes [4,5,8,11,12].

Determining the Quantity of Macrogel

Once solubilization is complete, the solution is centrifuged (35,000*g* for 1 h at 17°C). The quantity of macrogel is determined by weighing the centrifugation residue, after drying.

Filtering the Solutions Prior to Injection

The solution obtained after centrifugation is diluted to 0.2 mg/mL. It is allowed to stand for 24 h and is filtered through a disposable filter (glass fiber) with a porosity of

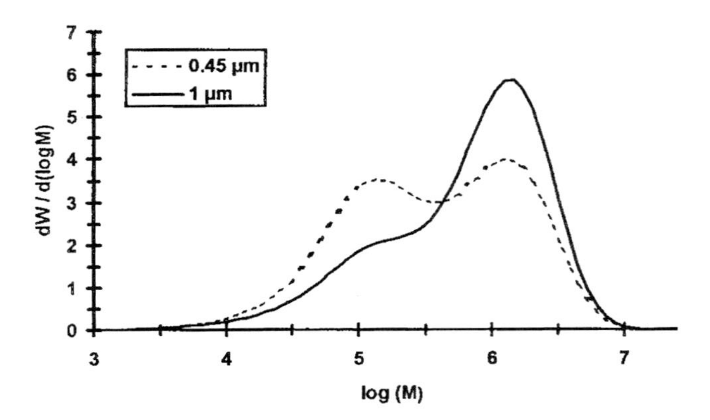

Fig. 1 Influence of the filter porosity on the molar mass distribution.

1 μm to eliminate the microgel. The filter porosity is very important, as the MMD observed after filtration through 0.45 μm or 1 μm are very different (Fig. 1). By filtering through 0.45 μm, shearing is considerable in the case of natural rubber.

References

1. P. W. Allen and G. M. Bristow, *J. Appl. Polym. Sci. 7*: 603 (1963).
2. Y. Tanaka, J. Tangpakdee, and S. Kawahara, *Kautsch. Gummi Kunstst. 50*: 6 (1997).
3. A. P. Voznyakovskii, I. P. Dmitrieva, V. V. Klyubin, and S. A. Tumanova, *Polym. Sci. Series A 38*(10): 1153 (1996).
4. A. Subramaniam, *Rubber Chem. Technol. 45*: 346 (1972).
5. H. Bartels, M. L. Hallensleben, G. Pampus, and G. Schulz, *Angnew. Makromol. Chem. 180*: 73 (1990).
6. F. Bonfils, A. K. Achi, J. Sainte Beuve, S. Sylla, A. Allet Don, and J. C. Laigneau, *J. Nat. Rubber Res. 10*(3): 143 (1995).
7. A. D. Edwards, in *Natural Rubber Science and Technology* (A. D. Roberts, ed.), Oxford University Press, Oxford, 1988, pp. 995–997.
8. J. L. Angulo-Sanchez and P. Caballero-Mata, *Rubber Chem. Technol. 54*: 34 (1981).
9. J. Tangpakdee and Y. Tanaka, *J. Nat. Rubber Res. 1*(1): 14 (1998).
10. C. L. Swanson, M. E. Carr, and H. C. Nielsen, *J. Polym. Mater. 3*: 211 (1986).
11. A. K. Bhowmick, J. Cho, A. MacArthur, and D. MacIntyre, *Polymer 27*: 1889 (1986).
12. K. N. G. Fuller and W. S. Fulton, *Polymer 31*: 609 (1990).

Frederic Bonfils

Neuropeptides and Neuroproteins by Capillary Electrophoresis

Introduction

Neuropeptides comprise peptide neurotransmitters and neuromodulators occurring primarily in the central nervous system (CNS), but may also be found in peripheral tissues. In order to study their physiological effects and their role in a variety of physiological functions (e.g., sensory information, food intake control, regulation of hormone secretion, and control of the sleep–waking cycle), methods for accurate determination of neuropeptides in biological tissues and fluids are required. Typically, immunoassays are used for this purpose, but cross-reactivity, especially with unknown fragments and matrix components, compromises quantitative analysis. In addition, immunoassays are generally designed for a single analyte, whereas neuropeptide research regularly aims to separately determine several neuropeptides in the same sample. One approach to inducing selectivity and allowing multiple neuropeptide analysis is to employ high-performance separations such as capillary electrophoresis (CE) in combination with appropriate sample handling and sensitive detection.

Capillary Electrophoresis of Neuropeptides and Neuroproteins

Because peptides and proteins are zwitterionic substances whose charge can be controlled through the pH, CE is well suited to provide high-speed separations with high efficiency, good mass sensitivity, low sample consumption, and high resolution. Differences relative to CE of low-molecular-mass compounds are due to the larger size of peptides and proteins, their secondary and tertiary structures, and their tendency to interact with the wall of bare fused-silica capillaries. This is valid for peptides and proteins occurring primarily in the CNS and for other peptide and protein analytes. For further reading, see the entry on Capillary Isoelectric Focusing of Peptides, Proteins, and Antibodies as well as to more extensive reviews (e.g., CE on opioid peptides by Lee et al. [1]).

Briefly, CE of peptides and proteins is often performed in uncoated fused-silica capillaries. In order to reduce interactions with the wall and to ensure that all peptides and proteins are charged, a buffer with a low pH (e.g., 50 mmol/L phosphate pH 2.5) is commonly used. Consistent rinsing protocols are required between runs, such as 0.1 mol/L NaOH followed by rinsing with the running buffer in order to remove adsorbed contaminants. When such rinsing procedures are not in compliance with the total analytical system or when adsorption of the analyte and/or matrix components hampers analysis, coating of the capillaries may prove useful. Adsorption phenomena can also be influenced by pH adjustment, temperature control, or by adding ionic and nonionic surfactants, organic solvents, or ion-pairing reagents to the running buffer. Both capillary coating and use of buffer additives affect the electro-osmotic flow leading to a reduction or a reversal of the flow, which, in turn, will affect the time required for separation.

Concentration sensitivities achieved in CE typically lie in the micromoles per liter or nanograms per liter range and are limited by sample loadability and detector path length (often 20–100 μm). Whether such concentration sensitivity is sufficient for a certain biochemical or pharmacological investigation depends primarily on the application. Complexity of the sample, concentration range of the analyte, and sample volume are all factors determining whether CE can be applied to analyze neuropeptides and neuroproteins in real-life samples. Plasma concentrations often are in the picomoles per liter range and can, accordingly, only be analyzed by CE after rigorous sample concentration. For instance, average concentrations of β-endorphin, methionin and leucine–enkephalin in 20 human plasma samples were determined to be 20 ng/ml, 2 ng/ml, and 2 ng/ml, respectively [2]. On the other hand, sampling of discrete areas in the brain can provide small samples with considerably higher concentrations of neuropeptides, and clear differences in concentrations may exist in the healthy and diseased states. For example, base levels of vasoactive intestinal peptide, substance P, and neuropeptide Y in nasal tissue of normal individuals are 28, 16, and 12 pg/μg protein, whereas patients suffering from allergy may have concentrations of 63, 165, and 98 pg/μg protein in perivascular lesion areas in the nasal tissue [3].

Even though free-solution CE is most commonly used for neuropeptides and neuroproteins, other forms of CE have also been employed. For instance, as an alternative to conventional slab-gel electrophoresis, a method using sodium dodecyl sulfate (SDS) capillary gel electrophoresis was developed. It was applied to low-molecular-mass proteins (β-trace protein, β_2-microglobulin, φ-trace protein, and myelin basic protein) in cerebrospinal fluid [4]. Advantageous features of capillary gel electrophoresis over slab-gel electrophoresis are compatibility with small sample volumes, shorter analysis times, and more accurate quantification of the analytes.

Generally, integration of sample collection, preparation, and introduction with the analytical separation and detection is decisive for successful application of CE to neuropeptide analysis; why both sample handling and detection will be discussed in this context.

Sample Handling

Sample collection and preparation are crucial issues for any bioanalytical application in order to address the complexity of samples originating from biological tissues and fluids. It is necessary to cope with the lack in concentration sensitivity typical for capillary separation techniques, to avoid interference from matrix components as well as to ensure analyte stability. In peptide analysis, a strong focus exists on handling small-volume samples and on selective concentration of the analyte in order to overcome limitations with respect to loadability. In addition, loss of analyte frequently occurs due to degradation by proteases and due to adsorption to surfaces, which accordingly needs to be minimized.

In brain research, microdialysis sampling employing a miniaturized dialysis unit (probe) containing a dialysis membrane of a few millimeters length has become popular. The probe is implanted into the tissue or organ of the test animal and is infused with an isotonic solution (typically at 0.5–25 μL/min). A steady-state osmotic flux across the membrane removes molecules with a mass below the cutoff of the membrane from the extracellular matrix. Microdialysis yields relatively clean samples of volumes in the range 20–100 μL. However, the recovery of neuropeptides can be as low as 0.5–15%, leading to a low neuropeptide concentration in the samples [5,6].

Problems with recovery are avoided when sampling the extracellular brain tissue by means of a push–pull cannula. A push–pull cannula consists of two coaxial assembled hollow needles (cannulae) which are implanted into the brain. Artificial cerebrospinal fluid is infused through the inner cannula and withdrawn through the outer cannula [5–8].

In addition, a whole-tissue sample can be taken by means of routine pathological biopsy. Tissue samples are typically homogenized under denaturating conditions (0.1 mol/L HCl), centrifuged, and the supernatant is then used for further analysis [5]. When of interest, even morphologically defined areas can be investigated after being isolated by means of microdissection prior to further sample treatment. For example, nasal tissue biopsies were taken and perivascular lesion tissue or normal tissue 5 mm from the lesion were isolated by means of microdissection preceding further analysis [3].

As discussed in-depth elsewhere in this encyclopedia, sample composition is important with respect to peak efficiencies achieved in CE. When the sample has a lower conductivity than the running buffer, focusing occurs, which often is referred to as sample stacking. This is a convenient way of increasing concentration sensitivity 5–10-fold. However, when the sample has a higher conductivity than the running buffer, uneven migration of the analyte(s) and zone spreading will occur, resulting in lower concentration sensitivities. Biological samples therefore need to be desalted prior to submitting them to capillary electrophoretic separation.

Desalting of biological samples can be achieved with traditional sample preparation methods, including ultrafiltration, liquid–liquid extraction, and solid-phase extraction (SPE). Sample preparation, especially of plasma samples, involves, most often, protein precipitation by means of an acid or an organic solvent (e.g., acetonitrile), followed by SPE. After evaporation of the eluate and reconstitution in the running buffer used for CE, [D-Pen^{2,5}]enkephalin was analyzed in rat serum in this way [9]. These traditional sample preparation methods usually include deproteination, which reduces the protease content of the sample and improves analyte stability. Another option to reduce peptide degradation during sample handling is to add a protease inhibitor (e.g., 1 mmol/L leupeptin) [3].

Even under ideal circumstances, samples that are ready for CE analysis are in the microliter range, whereas typical injection volumes in CE are a few nanoliters. In order to take full advantage of the sample, increased injection volumes without column overload are desirable and can be achieved by coupling (transient) isotachophoresis (ITP) to CE. During ITP, the concentration of the sample introduced into the capillary is adapted to the concentration of the leading buffer, thus focusing the sample in a discrete zone. A discontinuous buffer is used during ITP, and after a change in conditions, separation by capillary zone electrophoresis continues. On-column ITP enables focusing of 10–100 times higher injection volumes (100–1000 nL), but it is limited to the total volume of the analytical system.

Another approach to increase the loadability is to enrich the analyte at the capillary inlet by means of an adsorptive phase, as reviewed in detail in Ref. 10. Reversed-phase materials such as octadecyl-bonded silica have regularly been used in the form of membranes or column materials, thus, in principle, performing miniaturized SPE (μSPE) in-line with CE, allowing injections of 10–15 μL. More selective sample concentration is obtained when using antibodies or Fab fragments for coating the inner wall of the capillary inlet [3,5].

Often a combination of techniques is used to achieve a sufficient concentration of the sample in combination with selective cleanup—for instance, a combination of microdialysis and immunoaffinity CE [11], microdissection and SPE [7], or microdialysis and μSPE [12].

Detection

Detection of neuropeptides in CE is usually performed by ultraviolet (UV) absorbance, fluorescence and mass spectrometry (MS). UV absorbance is widely used for detection in CE, often at 214 nm, because it is inexpensive, robust, and widely available on commercial instruments. Typically, concentration sensitivities lie in the micromoles per liter range, as shown in an application starting from 2.5-mL lumbar cerebrospinal fluid, where detection limits of neuroproteins with UV absorbance detection at 214 nm between 5 and 10 μg/mL, corresponding to 0.1–1 μmol/L, were reported [4]. In order to address the lack of sensitivity achieved with UV absorbance detection, path lengths in the detection window have been increased threefold by means of the bubble cell and 10-fold by applying the so-called *z*-cell.

Tryptophan and tyrosine are intrinsic fluorophores that are present in many peptides, which then can be identified with fluorescence detection. However, most peptides have no native fluorescence, thus making derivatization a prerequisite for fluorescence detection. Derivatization has been described with naphthalene-2,3-dicarboxaldehyde-β-mercaptoethanol for determination of substance P and its metabolites [6], fluorescamine [5], and 5-carboxyfluorescein succinimidyl ester [8] for luteinizing hormone-releasing hormone (LHRH), neuropeptide Y, and β-endorphin. Kostel and Lunte [6] compared various postcolumn reactor designs, whereas Advis et al. [5,8] employed precolumn derivatization, among others. In order to improve sensitivity, laser light is frequently employed for exciting the fluorescent molecules referred to as laser-induced fluorescence (LIF) and provides a 500–1000 times improved sensitivity compared to UV detection.

Mass spectrometry is another detection technique widely used in neuropeptide analysis. Concentration sensitivities in CE–MS do not reach those obtained by CE–LIF; nevertheless, tedious derivatization procedures are avoided. In addition, CE–MS has proven to be a powerful tool for structure elucidation as illustrated by the investigation of the in vivo metabolic fate of peptide E by Caprioli's group [12]. After microdialysis and in-line SPE, neuropeptides migrating out of the electrophoresis capillary were deposited directly onto a precoated cellulose target used in matrix-assisted laser desorption–time of flight (MALDI–TOF) MS subsequently. Structural information is then obtained along with the mass of the peptide(s).

More extensive structural information is obtained when using MS–MS, where information on the peptide sequence can be extracted from the fragmentation pattern. So far, CE–MS–MS has predominantly been performed with electrospray ionization (ESI)–triple quadrupole and ESI–ion trap instruments, as reviewed in Ref. 13. The advent of ESI–TOF and ESI–quadrupole–TOF instruments is believed to have a strong impact on CE–MS. TOF instruments require an extremely short time to produce a full mass spectrum and are especially attractive as a detection device for a separation technique producing sharp peaks, as illustrated by the separation of three enkephalins in a time window of 6 s with detection by means of ESI–TOF [14].

Conclusions

In this entry, an overview is given over the implementation of CE in neuropeptide and neuroprotein analysis. So far, relatively few applications in biological matrices, including cerebrospinal fluid, plasma, and neural tissue, have been published; nevertheless, the potential of this approach has been demonstrated over the past years and the number of publications is growing steadily. Accordingly, it is expected that more research groups will be enticed to use CE for neuropeptide and neuroprotein analysis.

References

1. H. G. Lee and D. M. Desiderio, *Anal. Chim. Acta 383*: 79 (1999).
2. E. Ban, D. Kim, E. A. Yoo, and Y. S. Yoo, *Anal. Sci. 13*: 489 (1997).
3. T. M. Phillips, *Anal. Chim. Acta 372*: 209 (1998).
4. A. Hiraoka, T. Arato, I. Tominaga, N. Eguchi, H. Oda, and Y. Urade, *J. Chromatogr. B 697*: 141 (1997).
5. J. P. Advis, K. Iqbal, A. W. Malick, and N. A. Guzman, *Handbook of Endocrine Research Techniques*, Academic Press, San Diego, CA, 1993, p. 127.
6. K. L. Kostel and S. M. Lunte, *J. Chromatogr. B 695*: 27 (1997).
7. J. P. Advis, L. Hernandez, and N. A. Guzman, *Peptide Res. 2*: 389 (1989).
8. S. P. SungAe, W.-L. Hung, D. E. Schaufelberger, N. A. Guzman, and J. P. Advis, *Methods in Molecular Biology, Neuropeptide Protocols* (G. B. Irvine and C. H. Williams, eds.), Humana Press, Totowa, NJ, 1997, Vol. 73, p. 101.
9. C. Chen, D. Jeffery, J. W. Jorgenson, M. A. Moseley, and G. M. Pollack, *J. Chromatogr. B 697*: 149 (1997).
10. A. J. Tomlinson, L. M. Bensson, N. A. Guzman, and S. Naylor, *J. Chromatogr. A 744*: 3 (1996).
11. T. M. Phillips, L. M. Kennedy, and E. C. De Fabo, *J. Chromatogr. B 697*: 101 (1997).
12. H. Zhang, M. Stoeckli, P. E. Andren, and R. M. Caprioli, *J. Mass Spectrom. 34*: 377 (1999).
13. D. Figeys and R. Aebersold, *Electrophoresis 19*: 885 (1998).
14. I. M. Lazar, E. D. Lee, A. L. Rockwood, and M. L. Lee, *J. Chromatogr. A 829*: 1 (1998).

E. S. M. Lutz

Neurotransmitter and Hormone Receptors: Purification by Affinity

The isolation of membrane receptors has great application in a number of fields ranging from structural biochemistry to pharmacology and the medicinal sciences. Although many biochemical approaches to receptor purification are available, the application of affinity techniques holds great promise. The use of selective receptor ligands promises isolation not only of a specific receptor but also possibly in a bioactive form. The use of immobilized receptor ligands often ensures that the integrity of the receptor structure is maintained during the isolation process due to protection of the receptor-binding domains by the ligand itself. Additionally, receptors can be isolated either via a chromatographic system or by batch technology. The application of ligand-coated magnetic beads offers a quick and relatively simple approach to receptor isolation, although its use is only just becoming evident. The application of biotinylated receptor substrates is another approach, incubating the labeled substrate with the receptors prior to isolation on an avidin-coated support. In such cases, biotinylation with a cleavable biotinylation reagent such as Sulfo-NHS-SS-biotin or NHS-Iminobiotin would be essential for recovery of the isolated receptor. Alternatively, the receptor could be recovered by substrate competition. Perhaps one of the major drawbacks to the application of affinity techniques is the relative low molecular weight or small size of the receptor substrates, making them difficult ligands to immobilize. However, affinity procedures have been applied to the purification of a number of different receptors although relatively little work has been reported on those involved in the processing of neurotransmittors, neuropeptides, and hormones [1,2].

Ligands used to isolate neurotransmittor and hormone receptors have ranged from immobilized receptor substrates to immobilized antireceptor antibodies, the latter being used extensively in recent years. This marks the increased availability of specific monoclonal antibodies to specific receptors and/or receptor domains and the increasing popularity of using solid-phase extraction techniques over more physicochemical ones. The use of antibodies also circumvents problems arising from the immobilization chemistry required not only to maintain the integrity of the ligand but also its correct orientation. The rise of molecular biological techniques has opened a new approach to studying receptor structure and function. Cloning and expression of recombinant receptors in bacterial, yeast, or insect cells allows the investigator to engineer the receptor structure as well as the quantity of materials available for study. This situation has also increased the use of antibodies for the isolation of these recombinant receptors from culture medium.

Cosgrove et al. [3] described studies that isolated a recombinant ectodomain of the human insulin receptor using immobilized insulin as an affinity ligand. The ligand was immobilized to a chromatographic support and the receptor domain was isolated by affinity chromatography. The efficiency of the system was shown by the similarities between the isolated receptor domain and the native form isolated by physicochemical techniques. The process also allowed for the receptor to be isolated in a bioactive form as shown by its ability to bind insulin with a K_d of approximately 2×10^{-9} M. Feng et al. [4] approached the isolation of a recombinant human estrogen α-receptor in a different manner. They used a sequential heparin and 17-β-estradiol-17-hemisuccinate–bovine serum albumin affinity chromatography system to first remove the majority of the contaminating materials in the culture medium, followed by selective binding of the receptor to its substrate. This approach easily isolated the receptor yielding a 100-fold purification of a bioactive receptor. Ohtaki et al. [5] employed a biotinylated substrate to isolate a recombinant human pituitary adenylate cyclase-activating polypeptide (PACAP) receptor expressed in both insect cells and Chinese hamster ovary cells. The biotinylated ligand was immobilized on an avidin-coated support, allowing the receptor to be isolated by affinity chromatography.

The use of immobilized antibodies is perhaps the most popular approach to isolating these receptors. Andersen and Stevens [6] employed a combination of immobilized metal affinity and immunoaffinity chromatography to isolate a functional recombinant human D1A dopamine receptor expressed in *Saccharomyces cerevisiae*. This was achieved by engineering both a FLAG and a His6 tag to the C-terminus of the receptor. The histamine tag was used to perform a primary selection via metal affinity chromatography followed by a second purification step involving immobilized antibodies directed against the FLAG tag. In this way, a bioactive recombinant receptor could be isolated in a relatively pure form. Recombinant human β-1 thyroid hormone receptors can be isolated using a one-step immunoaffinity chromatography procedure [7]. Antibodies directed against the receptor are immobilized to a suitable chromatographic support, enabling the recovery of a bioactive receptor capable of binding 3,3′,5-triiodo-L-thyronine with a $K_a = 2 \times 10^{-9}$ M.

Eckard et al. [8] isolated a series of rhodopsin-like, G-protein-coupled neuropeptide Y receptors (Y1, Y2, Y4, and Y5) using immunoaffinity chromatography. Antibodies were raised against synthetic fragments of the second (E2) and third (E3) extracellular loops of the Y receptors and used to recover the receptors of interest. In a similar manner, murine glucocorticoid receptors have been isolated from a lymphoma cell line by immunoaffinity chromatography using an immobilized monoclonal antibody [9]. The antibody called BUGR-2 was coupled to protein A-coated Sepharose 4B beads and used in a chromatography system. Recovery of the bound receptors was achieved by epitope competition. This procedure released not only multiple receptors but also the heat-shock proteins 70 and 90, suggesting that the murine receptor interacts with these proteins, which may act as chaperones under physiological conditions. Immunoaffinity chromatography was also used by Repa et al. [10] for the isolation of members of the steroid/thyroid hormone receptor superfamily. Isolation of recombinant retinoic acid receptors expressed in an insect cell line was achieved by immunoaffinity chromatography using a monoclonal antibody to the human γ-retinoic acid receptor. The immunoaffinity-purified receptor was found to be biochemically greater than 90% pure as revealed by silver-stained electrophoretic gels. The isolated receptor was also shown to be functional, binding its ligand with a K_d of approximately 2 nM.

References

1. A. Azzi, U. Brodbeck, and P. Zahler, *Membrane Proteins. A Laboratory Manual*, Springer-Verlag, New York, 1981.
2. J. C. Venter, *Receptor Biochem. Methodol. 4*: 117–139 (1984).
3. L. Cosgrove, G. O. Lovrecz, A. Verkuylen, L. Cavaleri, L. A. Black, J. D. Bentley, G. J. Howlett, P. P. Gray, C. W. Ward, and N. M. McKern, *Protein Express. Purif. 6*: 789–798 (1995).
4. W. Feng, K. Graumann, R. Hahn, and A. Jungbauer, *J. Chromatogr. A 852*: 161–173 (1999).
5. T. Ohtaki, K. Ogi, Y. Masuda, K. Mitsuoka, Y. Fujiyoshi, C. Kitada, H. Sawada, H. Onda, and M. Fujino, *J. Biol. Chem. 273*: 15,464–15,473 (1998).
6. B. Andersen and R. C. Stevens, *Protein Express. Purif. 13*: 111–119 (1998).
7. J. B. Park, K. Ashizawa, C. Parkison, and S. Y. Cheng, *J. Biochem. Biophys. Methods 27*: 95–103 (1993).
8. C. P. Eckard, A. G. Beck-Sickinger, and H. A. Wieland, *J. Recept. Signal Transduct. Res. 19*: 379–394 (1999).
9. C. E. Powell, C. S. Watson, and B. Gametchu, *Endocrine 10*: 271–280 (1999).
10. J. J. Repa, J. A. Berg, M. E. Kaiser, K. K. Hanson, S. A. Strugnell, and M. Clagett-Dame, *Protein Express. Purif. 9*: 319–330 (1997).

Terry M. Phillips

Nitrogen/Phosphorus Detector

The nitrogen/phosphorus detector (NPD) is extremely sensitive [more so than the flame ionization detector (FID)] but is also highly selective. As its name suggests, it responds strongly to substances containing nitrogen and/or phosphorus. Physically, the design appears very similar to that of the FID but, in fact, operates on an entirely different principle. A diagram of an NPD detector is shown in Fig. 1.

The essential change that differentiates the NPD sensor from that of the FID is a rubidium or cesium bead contained inside a heater coil and situated close to the hydrogen flame. The bead, heated by a current through the coil, is situated above a jet, through which passes the helium carrier gas from the column mixed with hydrogen from a separate supply. If the detector is to respond to both nitrogen and phosphorus, then the hydrogen flow is arranged to be minimal so that the gas does not ignite at the jet. If the detector is to respond to phosphorus only, a large flow of hydrogen can be used and the mixture burned at the jet.

The detector functions in the following manner. The heated alkali bead emits electrons by thermionic emission, which are collected at the anode, providing a base current across the electrode system. When a solute that contains nitrogen or phosphorus is eluted, the partially combusted nitrogen and phosphorus materials are adsorbed on the surface of the bead. This adsorbed material *reduces* the work function of the surface and, as a consequence, the emission of electrons is increased, which *increases* the current collected at the anode. The sensitivity of the NPD is very high and only about an order of magnitude less than that of the electron capture detec-

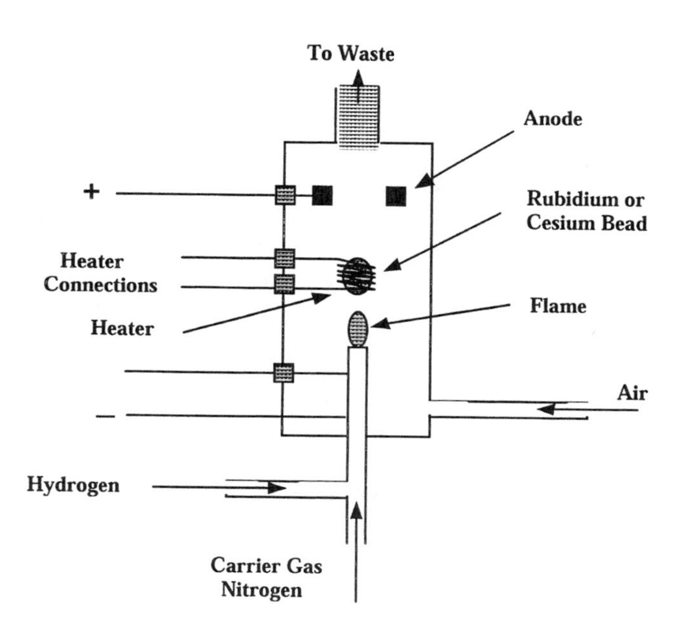

Fig. 1 The nitrogen/phosphorus detector.

tor (~10^{-12} g/mL for phosphorus and 10^{-11} g/mL for nitrogen).

A significant disadvantage of this type of detector is that its performance deteriorates gradually with time and eventually does not function at all. Reese [1] examined the performance of the NPD in great detail. The alkali salt employed as the bead is usually a silicate and Reese demonstrated that the loss in response was due to water vapor from the burning hydrogen converting the alkali silicate to the hydroxide and free silica.

Unfortunately, at the normal operating temperature of the bead, the alkali hydroxide has a significant vapor pressure and, consequently, the rubidium or cesium is continually lost during the operation of the detector. Eventually, all the alkali is evaporated, leaving a bead of inactive silica. This is an inherent problem with all NPDs and, as a result, the bead needs to be replaced regularly if the detector is in continuous use. The detector can be made "linear" over three orders of magnitude, although no values for the response index appear to have been reported. Like the FID, it is relatively insensitive to pressure, flow rate, and temperature changes but is usually thermostatted at 260°C or above.

The specific nature of the NPD response to nitrogen and phosphorus, coupled with its relatively high sensitivity, makes it especially useful for the analysis of many pharmaceuticals and, in particular, in environmental analyses of samples containing herbicides. Employing appropriate column systems, traces of herbicides at the 500-pg level can easily be determined. Virtually all the basic drugs presently employed in medicine contain nitrogen. Consequently, the specific detection of the NPD allows these drugs to be selectively monitored and quantitatively assayed, even when they are eluted among a large number of other unresolved compounds not containing nitrogen.

Reference

1. C. H. Reese, Ph.D. thesis, University of London (Birkbeck College), 1992.

Suggested Further Reading

Scott, R. P. W., *Chromatographic Detectors*, Marcel Dekker, Inc., New York, 1996.

Scott, R. P. W., *Introduction to Analytical Gas Chromatography*, Marcel Dekker, Inc., New York, 1998.

Raymond P. W. Scott

Nonionic Surfactants: GPC–SEC Analysis

Nonionic surfactants are one of the most important and largest surfactant groups. They are amphiphilic molecules composed, in most cases, of poly(ethylene oxide) (PEO) blocks as the water-soluble fragment and fatty alcohols, fatty acids, alkylated phenol derivatives, or various synthetic polymers as the hydrophobic part [1]. This class of surfactants is widely used as surface wetting agents, emulsifiers, detergents, phase-transfer agents, and solubilizers for diverse industrial and biomedical applications [2].

Several of them have been used for many years under different trade names: Brij (ethoxylated fatty alcohols), Synperonic (PEO copolymers) and Tween (ethoxylated sorbitan esters) by ICI Surfactants; Igepal (PEO copolymers) by Rhone-Poulenc, Rhodia; Pluronic [poly(oxyethylene)-*block*-poly(oxypropylene) copolymers] by BASF; Triton DF (ethoxylated fatty alcohols), and Triton X (ethoxylated octylphenols) by Union Carbide; and others. The exploitation characteristics of nonionic surfactants depend on the oligomer distribution, the molecular-weight characteristics of the constituent blocks, and the hydrophilic/hydrophobic ratio of their chemical composition. Therefore, the quantitative determination of these factors is of primary importance for their performance evaluation. Several high-performance liquid chromatography (HPLC) separation techniques have been used in combination with different detection methods to characterize poly(ethylene glycol)s and their amphiphilic derivatives [3]. Size-exclusion chromatography (SEC) is a particularly attractive analytical tool for the investigation of nonionic surfactants because it can provide information for their composition, molecular weight and molecular-weight distribution, and their micellization in selective solvents. This entry will survey briefly both applications, with major emphasis on the choice of the most appropriate eluent and stationary phase.

Molecular-weight determinations are performed in good solvents for both blocks of the PEO copolymers. The most widely used analysis conditions are eluents [tetrahydrofuran (THF) and chloroform (CHL)], flow rate (1.0 mL/min), and detection [differential refractive index (dRI) detector]. The temperature interval is between 20°C and 40°C. The stationary phase is typically a polystyrene/divinylbenzene cross-linked matrix supplied by different vendors: Phenogel (Phenomenex, U.S.A.); PL Gel (Polymer Laboratories, U.K.), PSS Gel (Polymer Standards Service, Germany), TSKgel (tosoHaas, U.S.A.), Ultra-Styragel (Waters Corporation, U.S.A.), and others. The pore size range of the column set should be adjusted to the molecular-weight range of the investigated materials. The SEC analysis in THF requires low dRI response correction factors, between 1.66 for $M_n = 106$ and 1.00 for $M_n = 20{,}000$ [4], whereas the low-molecular-weight PEO derivatives are almost invisible in CHL. On the other hand, the solubility of PEO in THF decreases with the molecular weight, as evidenced by steeper calibration curves and broadening of the peaks in the eluograms; see Fig. 1. This complicates the precise molecular-weight calculations of PEO copolymers and the choice of calibration standards becomes crucial [5]. Increasing the content of PEO in the copolymer results in lower hydrodynamic volumes in THF and, consequently, yields lower apparent molecular weights, regardless of the macromolecular architecture of the analytes [6]. The problems with calibration mismatch can be avoided to some extent by using the universal calibration approach with on-line differential viscometry. This method provides accurate molecular-weight information for most linear and comb-graft PEO copolymers, but has been proved less precise for copolymers with complex linear-dendritic or hyperbranched architecture. On-line laser-light-scattering detection eliminates the need for a calibration curve and yields correct M_w values for macromolecules with molecular weights higher than 500 g/mol. The precision of the method depends largely on accurate *dn/dc* values that need to be measured for each copolymer investigated.

The self-assembly process of nonionic surfactants in aqueous media differs in several aspects from the micellization of amphiphilic copolymers:

1. Micelles constructed of low-molecular-weight surfactants have much lower molecular weights than that of polymeric ones.
2. The critical micelle concentration (CMC) is much lower for polymer surfactants.
3. The vast majority of polymer micelles have spherical shape in dilute or semidilute solutions, whereas low-molecular-weight surfactants form structures that are strongly concentration dependent—lamellae, sheets, rods, and spheres.
4. The kinetics of micelle formation and the dynamics of the micelle–unimer equilibrium are considerably slower for polymeric surfactants.

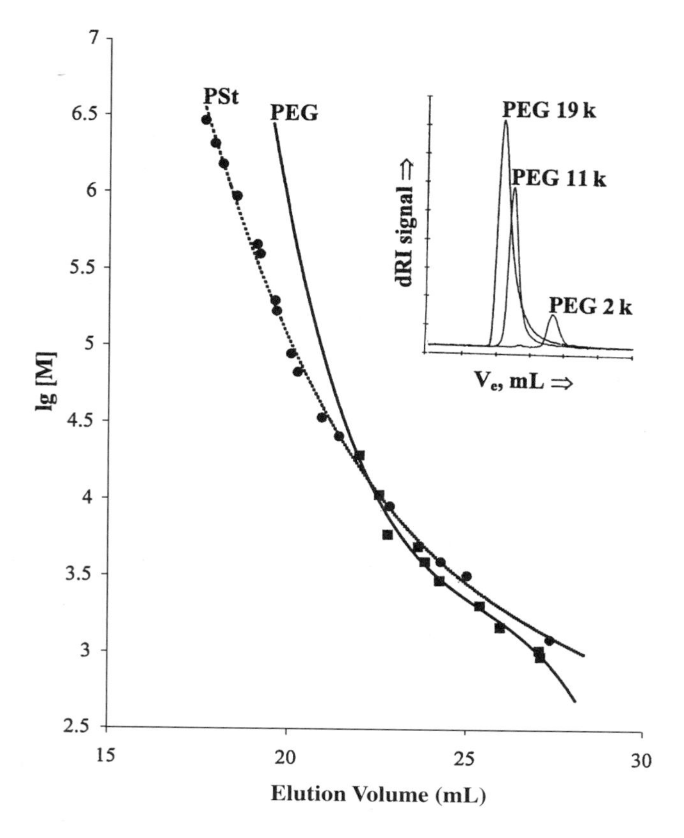

Fig. 1 Poly(styrene) (PSt) and poly(ethylene glycol) (PEG) calibration curves obtained on PL Gel column set (Mixed C, 10^3 Å, 5×10^2 Å, and 10^2 Å). Eluent: THF; flow rate: 1 mL/min; temperature: 40°C. Inset: SEC elution profiles of PEGs at the same analysis conditions.

The last factor is particularly important for the successful utilization of aqueous SEC in the investigation of the micellization process. In order to provide a realistic picture for the micellization equilibrium, the chromatographic system needs to meet several strict requirements: The packing material and the eluent should be appropriately chosen to prevent the occurrence of non-size-exclusion phenomena and adsorption of micelles in the stationary phase. In all cases, mass balance of the material injected and recovered form the columns must be performed in order to verify the absence of copolymer entrapment. Cross-linked copolymers containing either poly(vinyl alcohol) or poly(glycidyl methacrylate) as the hydrophilic component are the most widely used column packing materials for aqueous SEC. Both Shodex Protein KW (Showa Denko, Japan) and Micropak TSK-gel PW (Toyo Soda, Japan) columns have been reported to afford good information on the micellization behavior of PEO copolymers without the interference of side effects [7]. The mobile phase is methanol–water (1:1, v/v) or pure water eluting at 1.0 mL/min. The analysis temperature is between 22°C and 40°C.

Historically, the major concern for the use of SEC in the investigation of micellar systems has always been the lack of suitable calibration and, consequently, the inability of the method to furnish accurate information for the size (hydrodynamic volume) of the micelles and their molecular weight. The incorporation of light-scattering and viscosity detectors for on-line measurement seems to eliminate this problem. With no solute–column interaction present and slow unimer–micelle equilibrium, a multiangle light-scattering detector (DAWN-DSP, Wyatt Technology, U.S.A.) provides accurate information for the molecular weight and radius of gyration of the micelle using Zimm's formalism:

$$\frac{R_\theta}{K^*c} = M_w P(\theta) - 2A_2 c M_w^2 P^2(\theta)$$

where R_θ is the excess Rayleigh ratio, K^* is an optical constant that includes the differential refractive index increment (dn/dc) of the solvent–solute mixture, c is the concentration of the solute molecules in the analyzed solution, M_w is the weight-average molecular weight, $P(\theta)$ is a form factor, and A_2 is the second virial coefficient. If the dn/dc value for the copolymer above the CMC is known, the extrapolation to zero concentration and zero angle will yield the Z-average of the radius of gyration and A_2, respectively. The double extrapolation to zero angle and zero concentration will afford M_w. The hydrodynamic radii can be calculated using on-line viscometric detector (Viskotek, U.S.A., and Waters) and the following relationship:

$$R_\eta^3 = \frac{3[\eta]M}{10\pi N_A}$$

where the value of $[\eta]M$ could be extracted directly from the universal calibration and N_A is Avogadro's number (6.022×10^{22}). It should be pointed out, however, that this formula is strictly valid only for spherical structures. The hydrodynamic radius can also be measured by a combination of SEC and dynamic light scattering (Precision Detectors, U.S.A.) or nuclear magnetic resonance spectroscopy [8]. However, the same assumption for a spherical shape of the investigated macromolecules has to be made.

In conclusion, modern SEC is a versatile technique for the investigation of nonionic surfactants in aqueous and organic media. In combination with different spectroscopic and viscometric detectors, it will provide useful information for the molecular-weight characteristics, chemical composition, and solution behavior of this important class of materials.

References

1. B. Jönsson, B. Lindman, K. Holmberg, and B. Kronberg, *Surfactants and Polymers in Aqueous Solution*, John Wiley & Sons, New York, 1998.
2. J. E. Glass (ed.), *Hydrophilic Polymers: Performance with Environmental Acceptability*, Advances in Chemistry Series Vol. 248, American Chemical Society, Washington, DC, 1996.
3. K. Rissler, *J. Chromatogr. A 742*: 1–54 (1996).
4. S. Mori, *Anal. Chem. 50*: 1639–1643 (1978).
5. B. Trathnigg, S. Feichtenhofer, and M. Kollroser, *J. Chromatogr. A 786*: 75–84 (1997), and references therein.
6. D. Taton, E. Cloutet, and Y. Gnanou, *Macromol. Chem. Phys. 199*: 2501–2510 (1998); I. V. Berlinova, I. V. Dimitrov, and I. Gitsov, *J. Polym. Sci., Part A: Polym. Chem. 35*: 673–679 (1997); I. V. Berlinova, A. Amzil, and N. G. Vladimirov, *J. Polym. Sci., Part A: Polym. Chem. 33*: 1751–1758 (1995).
7. I. V. Berlinova, N. G. Vladimirov, and I. M. Panayotov, *Makromol. Chem., Rapid Commun. 10*: 163–166 (1989); R. Xu, Y. Hu, M. A. Winnik, G. Riess, and M. D. Crocher, *J. Chromatogr. 547*: 434–438 (1991).
8. G. R. Newkome, J. K. Young, G. R. Baker, R. L. Potter, L. Audoly, D. Cooper, C. D. Weis, K. Morris, and C. S. Johnson, Jr., *Macromolecules 26*: 2394–2396 (1993).

Ivan Gitsov

Normal-Phase Chromatography

There are a number of modes or mechanisms into which chromatography is divided. These include adsorption, normal-phase partition, reversed-phase partition, and ion exchange. Often, the term "partition" is deleted from the discussions of the differences and similarities of these modes. The word "partition" initially arose when supports had to be coated with a liquid phase (and the mobile phases saturated with them) to accomplish separations with these two modes. Today, bonded-phase versions of these liquid phases are available, making them easier to use with greater reproducibility. Perhaps it has been the use of these bonded supports that have enabled the name of the mode to be simplified.

The reasons for distinguishing between these modes are as follows:

1. To better understand the operating parameters of each mode
2. To understand what type of separations (isomers, equal polarity, different polarity) each is able to do better than another mode

Often the modes can be complementary, allowing various types of selectivity. This is advantageous when doing separations of unknown substances, as different mechanisms will allow unique selectivities, thereby guaranteeing more complete separation. Applying different modes of chromatography to a single-separation problem is often called "multidimensional" chromatography.

Before bonded phases became available, all normal-phase partition — and reversed-phase partition — separations were done on silica and other supports which were only "coated" with different polar and nonpolar phases or oils. Obviously, their use presented many problems, because it was absolutely necessary to keep both the mobile phase and the stationary phases saturated with these phases. The laboratory work of the analytical chemist was made infinitely simpler with the introduction of bonded phases, which this topic addresses.

Normal-phase chromatography is a close parallel to adsorption chromatography. Briefly, adsorption chromatography most often uses polar silica gel as the stationary phase and a mobile phase that is predominately nonpolar, possibly with some polar modifier. An example of such a mobile phase would be hexane with 2% ethanol. When increasing the percentage of the polar modifier, the elution times decrease.

Normal-phase (NP) chromatography generally uses the same types of mobile phase as for the adsorption mode. The difference, however, is the nature of the stationary phase. In NP chromatography, the packing is silica gel that has been bonded with a polar phase. The usual polar phases widely available from many manufacturers include cyano, amino, nitro, and diol phases. These are illustrated in Fig. 1. The first three often have a propyl group between the Si — O — and the X groups. The diol phase is derived from glycerol bonded to the silane reagent. Of course, each manufacturer uses a different silica, different reagents, different reaction conditions, and a different final workup; thus, any given polar bonded phase from the various suppliers can differ in relative selectivity and/or observed retention times.

Cyano - $Si-O-SiR_2-(CH_2)_3-CN$

Amino - $Si-O-SiR_2-(CH_2)_3-NH_2$

Nitro - $Si-O-SiR_2-(CH_2)_3-NO_2$

Diol – $Si-O-Si-O-CH_2OH-CHOH$

Fig. 1 The usual structures of polar bonded phases. The R groups may be $-OCH_3$, $-OCH_2CH_3$, $-H$, $-CH_3$, $-CH_2CH_3$, or others. Combinations of reagents or active groups are also possible. One manufacturer, for instance, offers a PAC (polar aminocyano) bonded phase.

There are two major differences between using silica gel (for adsorption) and a polar bonded phase (for normal phase) for a separation. First, the polar bonded phase adds a different surface to the silica gel, which can impart unique selectivities (i.e., separation characteristics) to an analysis. This is because the stationary-phase surface is no longer as polar as it was before bonding. Many of the silanols are replaced with the organo-polar groups. As with any bonded phase, perhaps only 50% of the available silanols are actually bonded, because of their incomplete accessibility to any reagent. It is possible to also endcap (a second-step bonding reaction, with a trimethylsilyl reagent) more of the remaining silanols. This is not generally done; so, there is often some definite silanol interaction and, also, with amino and diol phases, they would, themselves, react with the reagent, silanizing them. This would defeat the purpose of trying to produce a different selectivity packing.

The second difference comes from the fact that if there are fewer silanols available on the surface of the packing, it will be less sensitive to any moisture that gets into the mobile phase. One of the biggest problems observed by users of silica gel separations in the adsorption mode is the control of the separation's reproducibility. Unless one is very careful to use dry solvents, or "controlled water content" mobile phases, the retention times of their compounds can change dramatically from analysis to analysis. If more water gets onto the silanols, they are deactivated, and sample components elute more rapidly. Removing water with an exhaustively dried solvent combination, for instance, removes moisture from the silanol groups and components are then retained longer. As a result, with less silanols available for deactivation by water, the polar bonded phases are more reproducible and often preferred by chromatographers.

If variable retention times become a problem in the adsorption mode, a switch to a polar bonded-phase column is often the simplest solution. When using the same solvent mixture as was used on the bare, unbonded silica gel column, it is often found that the elution order is often the same or very similar. Of course, a switch in mode would necessitate identifying the sample components, to guarantee that their elution order is, indeed, changed or not. With fewer silanols, however, the components of the mixture will elute faster, so readjusting the mobile phase is required to ensure adequate resolution. As mentioned earlier, the polar bonded-phase separation will be much less sensitive to small moisture variations in the mobile phase. Moisture in the mobile phase should, nevertheless, be controlled, but will present few, if any, problems over the time the samples in a study are being run.

As with adsorption chromatography, the normal-phase mode separates molecules on the basis of solubility in the mobile phase (so it dissolves relatively nonpolar compounds) and the differences in the polarities of the components in the mixture and their attraction to the solvated polar bonded phase. Relatively nonpolar components elute first; more polar components elute later. These relative polarities and elution orders are summarized in Table 1.

At the nonpolar end of the elution scheme, the elution of alkanes to ethers with perhaps only hexane or heptane would be expected. For elution of moderate to polar compounds, a polar modifier has to be added to the major nonpolar component of the mobile phase. Much of the guesswork as to which solvent combinations to use has been simplified by the work of Kirkland and Snyder [1]. Their work has grouped solvents of similar selectivities [e.g., alcohols give identical (or similar) elution sequences; only the elution times change from one alcohol to another]. As a result of their studies, the recommended solvents to be mixed with hexane or heptane to effect the greatest possible selectivity differences are diethyl ether, chloroform, and methylene dichloride.

It is possible to separate more polar components by introducing a gradient from a low concentration (0–10%) of a polar modifier to a higher concentration of the same polar solvent (to 90%, for example). The important consideration is the solubility limit of the polar component. Thus, methanol is only soluble in hexane to about 5%, but ethanol can be brought up to 20% before reaching maximum solubility. If more polar samples are to be separated, then hexane can be replaced by ethyl acetate or chloroform.

Table 1 Relative Group Polarity and Elution Order in Normal-Phase Partition Chromatography

$RH < RX < RNO_2 < ROR < RCOOR < RCONHR < RNH_2 < ROH$

$RCRO$ $\quad$ R_2NH

$RCHO$ $\quad$ R_3N

Increasing polarity of organic structures/groups → →

Increasing retention of compounds containing these structures/groups → →

Notes: (a) The more functional groups in the compound, the greater its polarity; (b) for polyfunctional cpds, the *most polar* group determines the retention of the compound.

Using such an approach does not limit the chromatographer to the use of only relatively nonpolar mobile phases. Much work with these bonded phases is also done with polar solvent mixtures, such as methanol–water or acetonitrile–water. They might appear to be reversed-phase separations, but this may not be the case. Often, when a CN (cyanopropyl) phase is used, it is used as a slightly more polar bonded phase than would be a C4 bonded phase. Then, the polar organic–water combinations used almost always are invoking the reversed-phase mode.

One widely used application using NP chromatography in which the mobile-phase composition is deceptive in the actual mode being used is the separation of sugars (carbohydrates) and oligosaccharides. The packing used is an NH_2 (aminopropyl, most often) bonded phase and the mobile phase is acetonitrile–water in the ratio of 20–35% water. This mobile phase might lead one to suspect a reversed-phase mechanism at work, but, in fact, adding more water (the polar component) to the mobile phase decreases the retention times of the sugars, proving it is, indeed, a normal-phase partition mode being used (if it were a reversed phase, the addition of more water would increase the retention of the compounds). *Note*: If this column is supplied by the manufacturer in heptane or hexane, an intermediate solvent such as ethyl acetate should be used, on going to the acetonitrile–water mobile phase, to prevent immisibility of one solvent in another.

It is also possible to also use an NH_2 bonded phase as an ion exchanger, because it will form the quaternary ion, NH_3^+ in buffers between pH's of 2 to 6 (i.e., a weak cationic exchanger). This can also be a problem when attempting the sugar separation described earlier, if acidic components in the mixture inadvertently transform the NH_2 to the NH_3^+ form. The separation will not work as well, if at all. This, fortunately, is reversible by taking the column to pure water, then passing through 10 column volumes ($V_M \times 10$) of 0.01*M* NH_4OH, then pure water, to regenerate the free base.

Other possible problems with this reactive bonded phase is that it can form Schiff bases with aldehydes and ketones, so samples containing these should be avoided. The NH_2 group also can be easily oxidized, so peroxides in any easily oxidizable solvents (e.g., diethyl ether, dioxane and tetrahydrofuran) should be avoided [2].

The types of compounds that can be separated in the normal-phase mode is as vast as for the adsorption mode.

Likewise, because the range of solvents which can be used in this mode is virtually unlimited, it will lend itself to even more sample types. Unfortunately, not as many references exist for normal-phase separations as do for reversed-phase separations. As a guide to the types of solvents to use in the normal-phase mode, refer to the many references on silica gel. As mentioned, usually only minor changes, if any, need be made to the mobile phase when adapting to the normal bonded-phase column.

References

1. L. R. Snyder and J. J. Kirkland, *Introduction to Modern Liquid Chromatography*, 2nd ed., John Wiley & Sons, New York 1979, pp. 247–264.
2. V. R. Meyer, *Practical High Performance Liquid Chromatography*, 2nd ed., John Wiley & Sons, New York, 1994, p. 166.

Fred M. Rabel

Normal-Phase Stationary Packings*

Introduction

Normal-phase chromatography is a mode of liquid chromatography employing polar stationary phases and nonpolar eluents. Retention is predominately governed by hydrogen-bonding, electrostatic interactions, and, more specifically, dipole–dipole interactions. The stationary phases most typically used for the normal phase are silica, alumina, and chemically bonded phases such as aminoisopropyl, cyanopropyl, nitrophenyl, and diol. Other phases designed for particular types of analytes have also proved to be successful. These include modified alumna, titania, and zirconia, modified silica gels, impregnated silica gels, and nitrophases.

Silica

Silica is the most widely used commercially available normal-phase packing in various forms, having standard particle sizes ranging from 3 to 10 μm and surface areas from 200 to 800 m^2/g [1]. It is classified as an acidic adsorbent because its surface consists of acidic hydroxyl groups that are covalently bound to the Si atoms (i.e., silanol groups). Silica contains weak Brønsted acidic sites and does not contain any Lewis-acid sites.

The silica gel surface is heterogeneous and consists of siloxane and silanol groups. The active silanol groups on the surface are electron acceptors and hydrogen-bond with polar or unsaturated molecules. They exist in several forms: single (isolated or free), geminal, bound, and reactive surface hydroxyls (vicinal). Free silanols contain an Si atom that has three bonds in the bulk structure and the fourth bond is attached to a single OH group. Geminal silanols contain two hydroxyl groups attached to one silicon atom that are unable to hydrogen-bond. A bound hydroxyl is denoted as a hydrogen-bound surface hydroxyl. The reactive surface hydroxyls have two hydroxyl groups attached to different silicon atoms that are able to hydrogen-bond with each other. The reactive surface hydroxyls are believed to have the greatest relative strength as adsorption sites, among all hydroxyls formed. At higher temperatures (200–400°C), surface siloxane groups are formed from the condensation of reactive and geminal silanols and decomposition of free silanols.

The geometric properties of silica also play a role on the retention of analytes, because they vary the surface activity. Because the silica surface is heterogeneous, different silica gels have varied surface hydroxyls and, therefore, their concentrations can be different in pores of different dimensions. The smaller the pore diameter, the higher the surface area of the corresponding silica. It also has been shown that higher concentrations of reactive and bound hydroxyls are present in smaller-pore silicas, typically with a pore diameter less than 100 Å, whereas free hydroxyls predominate on large-pore silicas having pore diameters greater than 150 Å [1].

Silica may have different amounts of adsorbed water on the surface which decrease the activity of the silica, because the water blocks the underlying surface. If the silica is heated between 150°C and 250°C, the water may then be removed without the loss of surface hydroxyls and, therefore, the silica retains maximum activity [2]. Water is selectively adsorbed onto reactive hydroxyls, and because the concentration of the reactive hydroxyls is greater in small-pore silicas, there is a large deactivation effect, consequently leaving a surface of bound hydroxyls. Deactiva-

*This article was prepared while the author was affiliated with Seton Hall University, South Orange, New Jersey.

tion of the large-pore silicas generally leaves, mostly, free hydroxyls. Therefore, heavily deactivated large-pore silicas, when compared to small-pore silicas, have a higher surface activity due to the presence of the free hydroxyls versus bound hydroxyls.

The silica may also be contaminated by metallic impurities such as aluminum, nickel, and iron, depending on the synthesis of the silica or the manufacturing process. These metals may be present either in the form of oxides and hydrous oxides or through oxygen bonds attached to an Si atom [3]. The metal impurities may also have an effect on the chromatography, causing peak tailing due to complexation with the trace metal impurities. The acidity of the surface silanols is increased with the presence of these metal impurities. Depending on the pH of the silica, metal ions can exist in either nonhydrated or hydrated forms.

The surface of the silica can be modified with the use of a buffer and this may be an effective alternative method for the separation of polar analytes. This may lead to the enhancement and change in selectivity of ionic samples while exhibiting no effect on the behavior of nonionic samples. The pH of the buffer, concentration, and the type of buffer used have a significant effect on the retention and peak shape of the analytes [4–6]. However, the most influential parameter is the pH of the buffer, where the pH of the buffer solution should be lower than the pK_a of the acidic analytes and be higher than the pK_a of the basic analytes.

The surface of the silica may be coated with heavy metals and the selectivities observed can be attributed to the complexes formed between the metal ions and analyte species. The use of silver-impregnated silica has been used for the analysis of saturated and unsaturated fatty acid methyl esters (FAME) and triacylglycerols (TAG) [7]. The retention of the unsaturated FAME and TAG can be attributed to the stability of the complex that is formed between the π electrons of the carbon–carbon double bonds and the silver ions. The predominant interaction for saturated analytes is with the polar silanol groups. The secondary interactions are those of the silver ions with the unpaired electrons of the carbonyl oxygens of the analytes. The amount of silver adsorbed onto the silica and the pH have been determined to have an effect on the retention and resolution of certain acidic and basic compounds and fatty acids [8].

Alumina

Alumina is an inorganic oxide just as silica, but it is less widely used. Alumina may exist in many forms containing several hydroxides and oxide–hydroxides which are stable only at low temperatures. γ-Alumina, a low-temperature crystalline form, is the type that is most commercially available, having surface areas of 50–200 m^2/g, specific pore volumes ranging up to 0.6 mL/g, pore structures consisting of cylindrical micropores, and larger-diameter irregularly shaped pores [3]. The characteristics of low-temperature aluminas include different surface hydration and imperfect crystal structure, in which different crystal planes may be formed. The alumina surface contains weak Brønsted-acid sites and, upon calcination treatment, these sites are transformed into aprotic Lewis-acid sites (Al^{3+} atoms) that lie in one plane and Lewis-base sites (oxide ions, O^{2-}), which form the surface layer [1]. The pretreatment of the alumnina determines the acidity, basicity, or neutrality of the packing; the basic alumina has cation-exchange properties and the acidic alumina behaves as an anion exchanger.

The surface of alumina is covered by five distinct types of surface hydroxyls in their coordination to the aluminum. The total hydroxyl groups of the γ-alumnina is about 3 $\mu mol/m^2$. Upon heating to temperatures above 200°C, there is a consequent loss of these surface hydroxyls. Even though there is a loss of surface hydroxyls that may participate as weak Brønsted sites, the activity of the alumina increases with increase of hydroxyl loss, because they are converted into Lewis acidic and basic sites which may act as stronger adsorption sites. The activity of the five types of hydroxyl sites on the alumina is dependent on the amount of water present on the surface. Furthermore, the highest surface activity would be obtained with lesser amounts of physically sorbed water. The presence of Na_2O, a common impurity of the γ-alumina, is known to affect the pH of the γ-alumnina to a more basic alumina [3].

γ-Alumina is generally more polar and, therefore, more retentive than silica. It is typically used for the analysis of organic compounds that have carbon–carbon double bonds, such as olefinic hydrocarbons, weak acids, and electron-rich aromatic compounds such as polynuclear aromatic hydrocarbons. The aromatic π electrons can interact with the Lewis acid.

Some drawbacks of this packing include a strong interaction with polar analytes such as organic acids, leading to peak tailing or maybe even irreversible adsorption, and decomposition of the analyte due to chemisorbtion. On the other hand, if acid-treated alumninas are employed for the analysis of strongly acidic samples, then some of the aforementioned deleterious effects may be avoided. The analysis of basic compounds may also be used on acid-treated and neutral alumninas.

Zirconia and Titania

Zirconia and titania both contain Lewis-acid and Lewis-base sites, with the latter having stronger adsorption properties. The titania phase also has strong Brønsted acidic sites. Basic compounds are less retained on zirconia and titania phases, due to their basic nature. Neutral compounds such as polyaromatic hydrocarbons (PAH), due to their π-electron system, behave as Lewis bases and the interactions with Lewis acid sites on the zirconia and titania packing materials become dominant for retention.

Chemically Bonded Phases

Polar bonded phases for normal-phase separations have recently gained popularity. These include the dihydroxypropyl propyl ether (diol), aminopropyl, cyanopropyl, and nitrophenyl bonded silicas. These phases are advantageous to silica because they are less active and, yet, produce similar interactions, require shorter equilibration times, and are influenced less by the water content of the mobile phase. The retention of most analytes upon the diol and amino phases is similar to that of the parent silica and alumnia, whereas the cyanopropyl and nitrophenyl phases generally show less retention.

Amino Propyl Silica

The aminopropyl phase is generally prepared from trimethoxy or triethoxy aminopropysilanes. The amino phase acts as a strong proton donor. It generally retains acids longer than bases. For example, phenol elutes after aniline in various mobile-phase solvents of different solvent strengths such as chloroform and methyl-tert-butyl ether (MTBE) [9]. The amino phase generally shows more retention for samples of acidic nature than do silica, diol, and cyanopropyl phases. Alcohols and phenols were shown to be preferentially retained on amino, compared to diol and cyano, in pentane–diethyl ether mobile phases [10]. It also has been shown that steroids with phenolic groups generally show a higher retentivity on the amino phase [11,12]. The separation of a mixture of saturated hydrocarbons, olefinic hydrocarbons, and aromatic compounds have been shown to be separated on the amino and alumnina phases, but not on the silica [13].

Basic solutes such as amines, ethers, esters, and ketones are preferentially retained on amino and diol columns when compared to cyano columns [14,15]. The amino phase is a good alternative to the cyano column for a change in selectivity. When analyzing ketones and aldehydes, the aminopropyl phase should be carefully used due to the possible reactivity of the amino group with these substances. This may lead to the formation of imines and the bonded phase may be easily oxidized [13].

Cyano

The cyano column may be used in the normal- or reversed-phase mode and can be regarded as the first column of choice for method development when both modes are under consideration. Typically, it is prepared from mono-, di-, or trifunctional silane. A more reproducible packing is obtained when monofunctional silanes are used for synthesis.

Dipolar compounds such as those with chloro, nitro, and nitrile substituents are more strongly retained on cyano columns, compared to amino or diol [14]. Also, cyanopropyl silica can exhibit acidic or basic character, depending on the mobile phase used. It was shown that a complete reversal of elution order was obtained for phenol and aniline when MTBE and chloroform were used as the mobile phases, because phenol eluted first in the MTBE solvent and second in the chloroform solvent [9].

The cyano phase may physically collapse in solvents of intermediate polarity and it is recommended to be used solely with nonpolar or polar solvents. If solvents of intermediate polarity are to be used, the flow should not be changed or stopped, because the back-pressure holds the bonded phase in place and this helps prevent the collapse of the packed bed [13].

Diol Phases

The diol phases are usually prepared from trimethoxyglycidoxypropyl or triethoxyglycidoxypropyl silane, followed by hydrolysis of the epoxy group to form the diol functionality. The most widely used diol phase is the 1,2-dihydroxypropyl propyl ether phase.

Esters and ethers are preferentially retained on diol silica when compared to amino or cyano in systems when employing a pentane–diethyl ether mobile phase. The diol groups can form hydrogen bonds with esters and may interact strongly with sp^3-hybridized oxygen of ether bridges, thus leading to an increased adsorption [10]. For a change in selectivity, the diol phase is a good alternative to the cyano column, but it is less stable [14]. The interaction of basic and acidic samples is equivalent, because the diol phase is neutral and possesses no ionizable groups.

Nitrophenyl Phases

The phase that is most commercially available contains a weak π-electron acceptor mononitrophenyl bonded to the silica and usually contains a propyl spacer arm. This packing interacts with analytes through π–donor interactions. Other nitrophases exist and, as the number of nitro groups increases on these acceptor-type phases, an increased retention of aromatic donor compounds occurs. The phases of 3-(2,4-dinitroanilino)-propyl and (2,4,6-trinitronanilino)-propyl have shown preferential adsorption of PAHs, alkyl aromatic hydrocarbons, and polyarylalkanes, in comparison to aminopropyl silica [3].

Retention and Adsorption in Normal Phase

Retention in normal-phase chromatography increases as the polarity of the mobile phase decreases. The selectivity of the analytes may arise from the differences in solvent strengths (ε_0), acidity, basicity, and dipolar nature of the mobile phase. Furthermore, solvent localization of the mobile phase plays a major role in the retention of the analytes [15,16]. These solvent strengths have been shown to be different when used with varied stationary-phase packings such as alumina, diol, and silica [3,17].

The retention mechanism in the normal phase is often referred to as adsorption chromatography. It is described as the competition between analyte molecules and mobile-phase molecules on the surface of the stationary phase. It is assumed that the adsorbing analyte displaces an approximate equivalent amount of the adsorbed solvent molecules from the monolayer on the surface of the packing throughout the retention process [18]. The solvent molecules that cover the surface of the adsorbent may or may not interact with the adsorption sites, depending on the properties of the solvent. This retention model, proposed by Snyder, was originally used to describe retention with silica and alumnina adsorbents, but several other studies have shown that this model may also be used for polar bonded phases, such as diol, cyano, and amino bonded silica [10,19].

References

1. L. L. Synder, *Principles of Adsorption Chromatography*, Marcel Dekker, Inc., New York, 1968.
2. E. Heftmann, *Chromatography*, 2nd ed., Reinhold Publishing, New York, 1961.
3. K. K. Unger, *Packings and Stationary Phases in Chromatographic Techniques*, Marcel Dekker, Inc., New York, 1990.
4. R. Schwarzenbach, *J. Liquid Chromatogr. 2*(2): 205–216 (1979).
5. R. Schwarzenbach, *J. Liquid Chromatogr. 334*: 35 (1985).
6. S. H. Hansen, P. Helboe, and M. Thomsen, *J. Chromatogr. 368*: 39 (1986).
7. O. R. Adlof, *J. Chromatogr. A. 764*: 337–340 (1997).
8. M. Okamoto, H. Kakamu, K. Nobuhara, and D. Ishii, *J. Chromatogr. A 722*: 81–85 (1996).
9. P. L. Smith and W. T. Cooper, *J. Chromatogr. 410*: 249 (1987).
10. M. Lubke, J. L. le Quere, and D. Barron, *J. Chromatogr. A 690*: 41 (1995).
11. S. Hara and S. Ohnishi, *J. Liquid Chromatogr. 7*(1): 59–68 (1984).
12. S. Hara and S. Ohnishis, *J. Liquid Chromatogr. 7*(1): 69–82 (1984).
13. U. D. Neue, *HPLC Columns*, Wiley–VCH, New York, 1997.
14. L. R. Synder, J. J. Kirkland, and J. L. Glach, *Practical HPLC Method Development*, 2nd ed., Wiley–Interscience, New York, 1997.
15. L. R. Synder, J. L. Glajch, and J. J. Kirkland, *J. Chromatogr. 218*: 299 (1981).
16. J. J. Kirkland, J. L. Glajch, and L. R. Snyder, *J. Chromatogr. 238*: 269 (1982).
17. L. R. Synder, *High-Performance Liquid Chromatography Advances and Perspectives*, C. Horvath, ed., Academic Press, San Diego, CA, 1983, Vol. 3, p. 157.
18. L. R. Synder and H. Poppe, *J. Chromatogr. 184*: 363–413 (1980).
19. T. C. Schunk and M. F. Burke, *Int. J. Environ. Anal. Chem. 25*: 81 (1986).

Rosario LoBrutto

Nucleic Acids, Oligonucleotides, and DNA: Capillary Electrophoresis

Introduction

Rapid progress in the Human Genome Project has stimulated investigations for gene therapy and DNA diagnosis of human diseases through mutation or polymorphism analysis of disease-causing genes. The recent development of capillary electrophoresis (CE) technologies has facilitated the application of CE to the analysis of polymorphism and mutations on human genome toward DNA diagnosis and gene therapy for human diseases.

A CE system is a very simple analytical separation instrumentation consisting of only a high-voltage power supply, detector, and capillary. Separations are performed in fused-silica capillaries which are supplied with a thin outer coating of polyimide to make them strong and flexible. Coated capillaries with an internal diameter of 50–100 μm were usually filled with gel or polymer solution for DNA separation.

The advantage of electrophoresis using a capillary is to be able to apply a high voltage, compared with gel electrophoresis, that has enabled us to analyze DNA with high resolution and with high speed. This is because a small-diameter capillary leads to efficient heat dissipation, because the fused-silica wall acts as a heat sink, absorbing heat generated inside the capillary by Joule heating and dissipates it from the relatively large surface area of the outer wall of the capillary. In addition, DNA separation by CE is performed in a buffer including a polymer such as non-cross-linked polyacrylamide, cellulose derivatives, or poly(ethylene glycol), not only in gel. The polymer solution is set automatically into the capillary, which is much easier than with gel electrophoresis. DNA is detected with PDA (photodiode array) or LIF (laser-induced fluorescence).

Theory for DNA Analysis by CE

In a polymer solution such as methylcellulose, polymer molecules start to overlap to form dynamic pores. The pores do not have a definite size, but the average mesh size depends on the polymer concentration. DNA separation is achieved by a molecular sieving effect during electrophoresis in polymer solution, as well as in gel. DNA is separated depending on its size while migrating in the entanglement polymer solution. Polymer solution can be separated into three different concentration regimes; dilute, semidilute, and entangled. When the polymer concentration is low enough, polymer chains are hydrodynamically isolated from one another in solution. As the concentration of polymer is increased, the polymer chains begin to overlap. With higher concentration of polymer, each polymer molecule strongly interacts with other polymer molecules in the solution, forming physical networks. The transition between dilute and entangled solution is called a semidilute regime, which occurs at concentrations near their entanglement threshold. Thus, we need the information of such an entanglement threshold for testing polymer solutions to use the molecular sieving effect. The entanglement threshold for the polymer, c^*, is

$$c^* = \left(\frac{1.5}{K}\right)M_w^{-a} = IM_w^{-a} \tag{1}$$

where M_w is the molecular mass of the polymer, and K, I, and a are all molecular parameters which are different for each polymer. For example, the molecular mass of methylcellulose is about 400,000, so the value of c^*, in this case, is estimated at 0.4%.

An entangled solution can be characterized as a network with an average mesh size, ξ, similar to the pore size of the gel. Mesh sizes for various polymers are expressed as a function of the polymer concentration:

$$\xi = 1.43\left[\frac{1.5^{1+1/a}}{2.5N_A}\right]^{1/3} K^{-1/3a}\, c^{-(a+1)/3a} = Jc^{-\beta} \tag{2}$$

where N_A is Avogadro number, and J and b are parameters which differ for each polymer. Each mesh size for various concentrations of methylcellulose was estimated as $\xi = 48$ nm (0.4%), $\xi = 39$ nm (0.5%), $\xi = 33$ nm (0.6%), $\xi = 28$ nm (0.7%), $\xi = 25$ nm (0.8%), $\xi = 22$ nm (0.9%), and $\xi = 20$ nm (1.0%).

To explain electrophoretic behavior of DNA, the Ogston model and theory are well used, in general. On the basis of the Ogston model, a separation medium, such as gel or polymer, is treated as a random network of fibers with limited lengths. A sample DNA molecule is assumed to migrate through the network as a sphere with radius R. The sample molecule migrates as a whole toward the direction of the electric field, diffusing laterally until it encounters a pore large enough to allow its passage. The network of the polymer matrix is assumed to be a round post with radius r; the probability that a sample does not go through those regions is equal to the probability that a sample passes through polymer matrices. In the Ogston model, the mobility, μ, or the migration time, t, of DNA is expressed as follows:

$$\log\left(\frac{\mu}{\mu_0}\right) = -\log\left(\frac{t}{t_0}\right) = -\pi l'(r + R)^2 T \times 10^{-16} \tag{3}$$

where l' (cm) is the total fiber length produced by 1 g of dry polymer, T is the concentration of polymer (or gel), μ and t are the mobility and the migration time of DNA in the polymer solution, respectively, and μ_0 and t_0 are those in free solution, respectively.

DNA molecules, however, are not actually spheres; rather, they are considered random coil polymers with some degree of flexibility. Thus, DNA molecules with total length L (nm) as a random coil polymer can be assumed to adopt a spherical shape with radius of gyration, R_g, given by

$$R_g^2 = \frac{1}{3}pL\left[1 - \frac{p}{L} + \left(\frac{p}{L}\right)\exp\left(\frac{-L}{p}\right)\right] \tag{4}$$

In this equation, p (nm) is the persistence length, a measure of the flexibility of the DNA chain. Because single-stranded DNA is extremely flexible, the p/L terms are assumed to be negligible for smaller DNA molecules, which allows us to express R_g^2 with a simpler formula:

$$R_g^2 = \frac{1}{3}pL = \lambda N \tag{5}$$

where N is the length of the chain, expressed in nucleotide units, and λ is a constant.

Using the radius of gyration, R_g, instead of R in Eq. (3), R_g is much larger than the polymer strand radius, we obtain the relationship $(r + R)^2 = R_g^2$. Further, using Eq. (3) allows us to obtain Eq. (6) as follows:

$$\ln(\mu) = \ln(\mu_0) - CTN \tag{6}$$

where $C\ (= \pi/\lambda \times 10^{-16})$ is a constant. This formula predicts the linear relationship between the logarithm of mobility and the DNA size, N, or polmyer concentration, T, with the Ogston model. Although Eq. (6) is successfully used in the analysis of DNA mobilities in both gel and polymer solution, the migrating molecule is treated as a spherical coil in this theory, and it is generally applied only for the molecules that can be treated as small spherical coils, typically with a radius of gyration R_g much smaller than the average pore size ξ.

The Ogston model has been successful in fitting experimental data for small DNA fragments with the limitation of low electric fields. However, DNA with radii of gyration much larger than the average gel pore size will still migrate rapidly through the gel during electrophoresis, implying that the assumptions inherent in the Ogston model are invalid for larger DNA.

Because a DNA molecule is a flexible polymer, the assumption in the Ogston model that DNA migrates as an undeformable particle loses validity with increasing size of DNA. The Ogston model predicts that the mobility of DNA with R_g larger than the pore size of the gel will approach 0. However, it actually was observed experimentally that DNA with its R_g larger than the gel pore size keeps migrating.

The reptation theory has been established to explain the movement of polyelectrolyte during the electrophoresis in neutral polymer network, and the reptation theory was applied to biomolecules. In the reptation theory, a polymer chain moves through a network, not as an undeformable particle, but rather snakelike through "tubes" in the polymer network. DNA chains move through the "tube" and the friction coefficient, ζ, is propotional to chain length, $\zeta = \zeta_0 L$, where ζ_0 is the friction coefficient per unit length.

According to the reptation theory, the polyelectrolyte (it can be DNA in this case) is assumed to be something which consists of N units (length $= a$, the number of bases in case of DNA). The total length of DNA is equal to Na. In the reptation theory, only the force along the DNA chain (longitudinal force) is taken into account in the electric field. The mobility of DNA, μ, therefore, is then expressed by

$$\mu = \frac{Q\langle hx^2\rangle}{\zeta_0 a L^2} \tag{7}$$

where Q is the total charge of DNA and hx is the end-to-end distance of the molecule in the direction of the electric field.

When the electric field is low enough, the DNA molecule can be considered as a random coil polymer with some degree of flexibility. When the DNA size is very large, electrophoretic behavior is expressed as

$$\mu \propto L^{-1} \propto N^{-1} \tag{8}$$

On the other hand, when the electric field is strong, DNA is no longer considered to move as a random coil polymer, but moves with extended rodlike conformation. If DNA is completely extended and moves straight to the field direction, $\langle hx^2\rangle = L^2$, showing that the mobility in Eq. (7) is independent of DNA size:

$$\mu \propto L^0 \propto N^0 \tag{9}$$

This can be well applied to the electrophoretic behavior for extremely large sizes of DNA as well as the case of strong electric field.

The regime for Eq. (8) obtained from reptation theory is termed reptation-without-stretching regime; the one for Eq. (9) is termed reptation with stretching regime. The mobility decreases exponentially in accordance with DNA size when it is small enough (Ogston regime); then, as the size becomes larger, the mobility is proportional to the size (reptation-without-stretching regime).

Separation Matrix and Capillary Detection

Separation matrices mainly used for the separation of DNA by capillary electrophoresis are gel or polymer solutions as listed in Table 1. The separation of single-strand oligonucleotide is achieved by using cross-linked polyacrylamide gel, in which the concentration varies from 3% to 8% T and the degree of cross-linking varies from 3% to 5% C. High-concentration linear polyacrylamide (8–10%) has been successfully applied to the separation of oligonucleotides, too. Because linear polyacrylamide is replaceable from the capillary, gels are now increasingly being replaced by polymer solution. Electric fields between 200 and 500 V/cm have been used.

Table 1 Proposed Concentration Range of the Polymer-Type Separation Matrices for the Separation of DNA Fragments

Effective DNA size range of separation (bp)	Concentration range of polymer (%, w/v)		
	PAA	HEC, MC	PEG, PEO
1–100	8.0–12	1.0–3.0	6.0–8.0
100–300	7.0–8.0	0.7–1.0	3.0–6.0
300–1,000	5.0–7.0	0.5–0.7	2.0–3.0
1,000–10,000	3.0–5.0	0.3–0.5	0.5–2.0
10,000–30,000	2.0–3.0	0.01–0.3	

Note: PAA = polyacrylamide; MC = methylcellulose; HEC = hydroxyethylcellulose; PEG = poly(ethylene glycol); PEO = poly(ethylene oxide).

For the separation of double-stranded DNA fragments, several polymer networks are used primarily, as well as cross-linked polyacrylamide gel. The matrices currently used are methyl cellulose (MC), hydroxyethyl cellulose (HEC), hydroxypropyl cellullose, hydroxypropylmethyl cellulose (HPMC), poly(ethylene glycol) (PEG), poly(ethylene oxide) (PEO), poly(vinyl alcohol) (PVA), and agarose. For the separation of polymerase chain reaction (PCR) products and restricted DNA fragments, less concentrated solutions of polymers can be used. Typical polymer concentrations are about 0.1–1.0%, as listed in Table 1, and most commercially available capillary electrophoresis instruments are now able to fill and empty the capillaries with such solutions automatically. For the separation of DNA fragments ranging from 300 to 5000 bp, 0.5% cellulose derivative solution would be selected. Concentrations of cellulose derivative solution should be higher (0.7–1.0%) for the separation of shorter DNA fragments less than 300 bp. The application of dilute polymer solution (less than 0.3%) will be appropriate for the separation of larger DNA fragments ranging from 5000 to 50,000 bp. When low-concentration polymer solutions are applied, inner-wall-coated capillaries are usually used for eliminating electro-osmotic flow. Such capillaries are prepared by the chemical attachment of linear polyacrylamide to the capillary inner wall or some commercially available coated capillaries are used (e.g., J&W DB-17). The electric fields between 50 and 800 V/cm have been used for the separation of double-stranded DNA. For the optimization of the separation of double-stranded DNA fragments, some systematic studies on the separation conditions have been published. We can easily reach the optimum conditions for our own separations by choosing appropriate length, diameter, and coating for the capillary, electric field, temperature, pH of the buffer, type of polymer network, and its concentration.

DNA fragments separated by polymer networks are detected by ultraviolet (UV) detector or laser-induced fluorescence (LIF) detection. UV detection of DNA fragments is based on the UV absorption of the DNA bases; that is, the wavelength and the molar absorption coefficient for the UV absorption maxima of DNA bases are 260 nm for *A*, 254 nm for *G*, 267 nm for *T*, and 271 nm for *C*, respectively.

The LIF detection of DNA fragments requires some fluorescent dye (e.g., intercalating dyes and chemically labeling dyes). The intercalating dyes are currently used for the LIF detection of DNA fragments. Ethidium bromide has been the most widely used as an intercalating dye for capillary electrophoresis, but it has some disadvantages (e.g., high background and low sensitivity). More recently, some new, efficient intercalating dyes have been developed. Some monomeric dye including thiazole orange (TO), TO-PRO-1, and oxazole yellow (YO), YO-PRO-1, and dimeric dyes including TOTO-1, YOYO-1, and YOYO-3 are applied to the highly sensitive detection of double-stranded DNA fragments. These dyes have several advantages, including low background and high sensitivity. Monomeric dyes, TO, TO6, and YO-PRO-1 are especially better in detection sensitivity than dimeric dyes. Recently, another monomeric dye, SYBR Green I, has been developed as a fluorescent dye well suited for efficient separation and quantitative, sensitive, and precise determination of double-stranded DNA using capillary electrophoresis. Most monomeric dyes, including ethidium bromide, acridine orange, TO, YO-PRO-1, TO PRO-1, and SYBR Green I, are optimally excited by the argon ion laser (488 nm and/or 514 nm). Some dimeric dyes, YOYO-1 and TOTO-1, are suitable for the excitation by argon ion laser too, and others, YOYO-3 and POPO-3, are excited by the He-Ne laser (543 nm and/or 633 nm). YOYO-1 is a benzoxazolium-4-quinolinium dimer that has one carbon atom bridging the aromatic rings of the unsymmetrical cyanines. YOYO-3, which differs from YOYO-1 only in the number of bridging carbon atoms (three), has longer wavelength spectral properties. The addition of intercalating dye also improves the resolution of DNA fragments.

Another LIF detection scheme for DNA involves direct labeling of the analyte with a suitable fluorophore, Fluorescently labeled probes and primers are used in many molecular biology applications involving hybridization, PCR, DNA sequencing, and multicolor detection for accurate SSCP analysis. Mostly used fluorescent labeling agents are so-called ABI dyes, including FAM, JOE, TAMRA, and ROX. Recently, new types of labeling agents called energy-transfer (ET) primers have been developed. DNA primers and probes are usually synthesized with a fluorescent label attached to the 5′ end of the molecule.

Suggested Further Reading

Baba, Y., *Mol. Biotech. 6*: 143–153 (1996).
Baba, Y., *J. Chromatogr. B 687*: 271–302 (1996).
Dolnik, V., *J. Biochem. Biophys. Methods 41*: 103–119 (1999).
Heller, C. (ed.), *Analysis of Nucleic Acids by Capillary Electrophoresis*, Vieweg, Weisbaden, 1997.
Heller, M. and A. Guttman (eds.), *Integrated Microfabricated Device Technology: Advances in Genomic, Drug Discovery, and Clinical Diagnosis*, Marcel Dekker, Inc., New York, 2001.
Mitchelson, K. and J. Cheng (eds.), *Capillary Electrophoresis of Nucleic Acids*, Humana Press, Totowa, NJ, 2001.
Mitnik, L., L. Salome, J. L. Viovy, and C. Heller, *J. Chromatogr. A 710*: 309–321 (1995).
Righetti, P. G. (ed.), *Capillary Electrophoresis in Analytical Biotechnology*, CRC Press, Boca Raton, FL, 1996.
Righetti, P. G. and C. Gelfi, *J. Chromatogr. A 806*: 97–112 (1998).
Sunada, W. M. and H. W. Blanch, *Electrophoresis 18*: 2243–2254 (1997).

Yuriko Kiba
Yoshinobu Baba

Octanol–Water Partition Coefficients by CCC

Introduction

Hydrophobicity, from the greek *hydro* water and *phobia* aversion, is a term referring to the way a molecule "likes" or "does not like" water. A compound with a high hydrophobicity will not be water soluble. It is apolar. Conversely, a compound with a low hydrophobicity is said to be hydrophilic or polar. It is likely to be water soluble. In between the two extremes, the hydrophobicity varies. A scale is needed. The problem is that the hydrophobicity, or the polarity of a compound, depends on several parameters such as the dipole moment, the dielectric constant, the polarizability, the proton donor or acceptor character, or even the boiling point to molecular mass ratio. Since the end of the nineteenth century, the *octanol–water partition coefficient*, $P_{o/w}$, was used with success as a measure of hydrophobicity. The log $P_{o/w}$ is the convenient scale. Compounds with a positive log $P_{o/w}$ value are more and more hydrophobic or apolar as the value increases. Compounds with a negative log $P_{o/w}$ value are hydrophilic or polar [1].

It is of paramount importance to be able to measure, accurately, the $P_{o/w}$ and log $P_{o/w}$ value of a compound, because it is the accepted parameter used by the Food and Drug Administration (FDA) and the Environmental Protection Agency (EPA) and many other international drug and environmental agencies to estimate the tendency of an organic chemical to bioconcentrate into living cells. A new drug cannot be accepted by the FDA and EPA without the $P_{o/w}$ parameter.

The most extensive and useful sets of $P_{o/w}$ data were obtained by simply shaking a solute with the two immiscible octanol and water phases and then analyzing the solute concentration in one or both phases. For many solutes, repeated inversion (say ~100) of a 25-mL tube with ~0.01 M solute and the two phases establishes equilibrium in ~15 min. Very vigorous shaking can produce troublesome emulsions. The solute can be analyzed in only one phase and the concentration in the other can be obtained by the difference. The phase analysis is most often done by gas chromatography, liquid chromatography, or ultraviolet (UV)-visible spectroscopy. The shake flask method gives reliable results over the wide 10^{-4}–10^{4} $P_{o/w}$ range. However, it requires highly pure solutes and is very sensitive to the smallest contamination.

Reversed-phase liquid chromatography (RPLC), capillary electrophoresis (CE), micellar liquid chromatography (MLC), and electrochromatography (EC) can be used to estimate values of log $P_{o/w}$ from the corresponding log k values; k is the retention factor, directly related to the retention parameter of the solute of interest. Good correlations are generally found between log k and log $P_{o/w}$ for structural congeners. Unfortunately, the correlations are much poorer with dissimilar compounds. Trace amounts of octanol were added in the mobile phase to enhance log k– log $P_{o/w}$ correlations with a wide variety of solutes. The $P_{o/w}$ range is 1–$10^{5.5}$. The advantages of the RPLC method are its relative simplicity and the fact that it does not need highly pure solutes. At the moment, the correlation remains the main drawback.

Direct $P_{o/w}$ Measurement by CCC

The decisive advantage of countercurrent chromatography (CCC) in $P_{o/w}$ measurement is that there is no correlation at all. Water, saturated with octanol, is the mobile phase. Octanol saturated with water is the stationary phase. The octanol–water partition coefficient of a given solute is the only physicochemical parameter responsible for the solute retention. If the solute is not highly pure, it is likely that the impurities will have differing $P_{o/w}$ values. This means that if the impurities have differing retention volumes, they are separated during the measurement from the solute of interest. The $P_{o/w}$ value is easily derived from the CCC retention equation:

$$V_R = V_M + PV_S \tag{1}$$

using

$$P_{o/w} = \frac{V_R - V_M}{V_S} = 1 + \frac{V_R - V_C}{V_S} \tag{2}$$

If octanol is the stationary phase and water the mobile phase, $P_{o/w}$ is the octanol–water partition coefficient without any assumption. Correlations of the $P_{o/w}$ or log $P_{o/w}$ values obtained with the same liquid system by the shake flask method and by CCC produce straight lines with a slope unity and a negligible intercept. The validity and solidity of the method was assessed by Gluck and Martin for $P_{o/w}$ coefficients [2]. The $P_{o/w}$ range that can be obtained directly by CCC is 0.05–200 [1]. It is limited on the high side by the experiment duration. A $P_{o/w}$ value of 200 corresponds to a V_R retention volume of 6 L with a V_S value of only 30 mL [Eq. (1)]. This is 1200 min or 20 h with a 5-mL/min flow rate. The lower-side limitation is due to experimental precision. The difference between the retention volume V_R and the dead volume V_M is equal to PV_S [Eq. (1)]. With a 30-mL V_S volume, the $P_{o/w}$ value of 0.05 corresponds to a $V_R - V_M$ value of only 1.5 mL. Such a low value may be difficult to evaluate with an acceptable accuracy. To increase the measurable $P_{o/w}$ range, the fact that the CCC stationary phase is a liquid can be used. This led to the dual-mode use of CCC and the cocurrent operation.

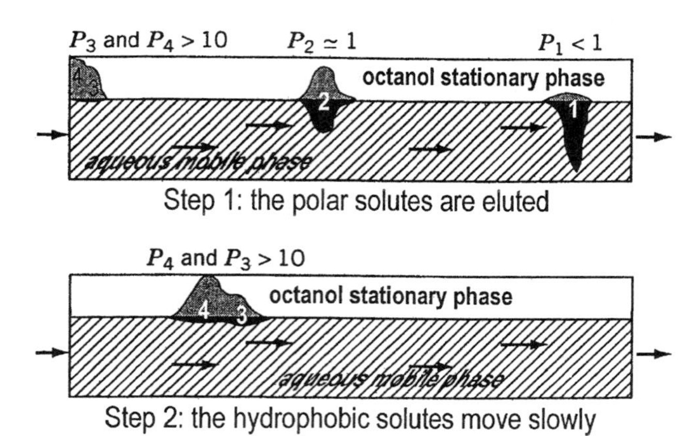

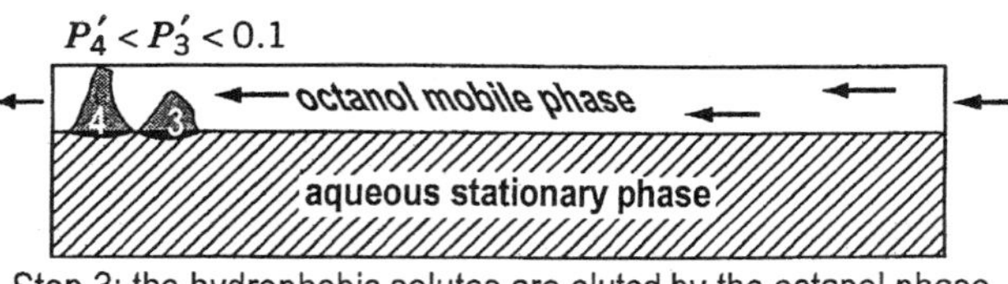

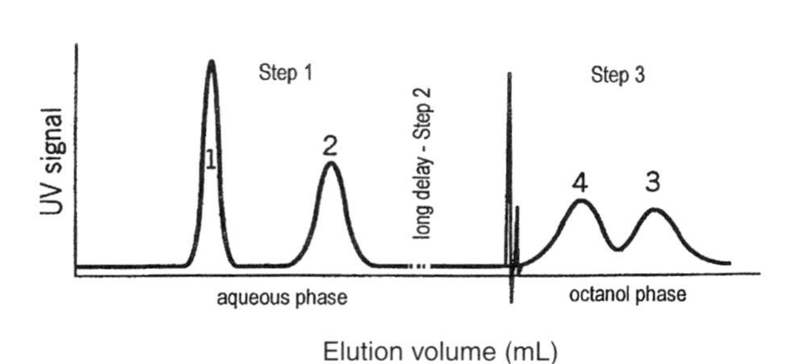

Fig. 1 Dual-mode CCC.

Dual-Mode CCC

The idea is simple: Solutes with very high $P_{o/w}$ values move very slowly in the octanol phase; they need too long a time to emerge from the apparatus. To force them out of the CCC apparatus, the role of the aqueous and octanol phases and their flow directions are reversed after some reasonable flowing time in the normal direction. The dual-mode operation is illustrated in Fig. 1. It was demonstrated that the $P_{o/w}$ value can be simply expressed by

$$P_{o/w} = V_{aq}/V_{oct} \tag{3}$$

in which V_{aq} is the aqueous-phase volume passed in the normal way (descending or head to tail, Steps 1 and 2) and V_{oct} is the octanol-phase volume passed in the reversed way (ascending or tail to head, Step 3). The highest $P_{o/w}$ value that can be measured is again limited by the lowest V_{oct} volume that can be accurately determined. Due to band broadening, the practical minimum V_{oct} value is about 3 mL [1]. Then, with a 6000-mL V_{aq} volume, the corresponding $P_{o/w}$ value is 2000. Table 1 shows the experimental conditions corresponding to the dual-mode measurement of some $P_{o/w}$ values in the 40–3000 range. It was shown that the error on the $P_{o/w}$ determination was minimized when the octanol flow rate in the reversed mode was very low.

Table 1 Some Practical Examples of $P_{o/w}$ Measurements by CCC

Solute Direct measurement		V_R (mL)	t_R (h)	$P_{o/w}$	log $P_{o/w}$ CCC	log $P_{o/w}$ Literature
Benzamide		240	0.8	4.4	0.643	0.64
Acetophenone		1,020	3.4	40	1.60	1.6
2-Chlorobenzoic acid		2,280	7.6	97	1.99	2.0
2-Chlorophenol		3,270	10.9	145	2.16	2.15
Dual-Mode CCC	V_{aq} (mL)	V_{oct} (mL)	t_R (h)	$P_{o/w}$	log $P_{o/w}$	log $P_{o/w}$ (lit.)
Benzoic acid	178	2.36	0.8	75.4	1.88	1.87
2-Chlorophenol	487	2.71	2.2	180	2.25	2.15
Toluene	2,630	5.5	9.0	480	2.68	2.71
Biphenyl	19,600	1.0	65	19,600	4.29	3.80
Cocurrent CCC	V_{oct} (mL)	V_R (mL)	t_R (h)	$P_{o/w}$	log $P_{o/w}$	log $P_{o/w}$ (lit.)
Benzene	29.7	2,530	10.5	107	2.03	2.14
Toluene	20.2	6,520	12	500	2.70	2.71
Naphthalene	20.2	15,500	28.6	5,100	3.7	3.2
Phenanthrene	22.2	9,790	18.1	20,000	4.3	4.4

Source: Data from Refs. 1–4.

Cocurrent CCC

The cocurrent CCC operation takes advantage of the liquid nature of the stationary phase. If a lipophilic solute stays too long inside the CCC apparatus, why not push it out, pushing the liquid stationary phase, slowly, in the same direction as the mobile phase? The theoretical treatment [3] and the practical promise of the method were established [4]. Three pumps are needed. Pump 1 allows the adjustment of the aqueous-phase flow rate in the few milliliter per minute range. Pump 2 governs the octanol phase flow rate in the microliter per minute range. Pump 3 is used to add a clarifying agent to the phase mixture leaving the CCC apparatus. The clarifying agent can be 2-propanol; it solubilizes the trace amounts of octanol present in the aqueous phase. The interest of the method is that there is no abrupt change; that is, it is continuous. The octanol volume retained in the CCC system is very stable, more stable than with other methods because there is a constant input of octanol. The octanol volume changes, due to dissolution that was noted in the direct method or due to phase reversal as noted in the back-flushing method, do not exist with the cocurrent CCC method. The V_{oct} volume was determined using a test solute (2-chlorophenol, $P_{o/w} = 147$ in Ref. 4). Another very important effect that was experimentally observed is the increased peak efficiency due to the octanol flow rate. The measurable $P_{o/w}$ range was extended up to 20,000 (log $P_{o/w} = 4.3$). Table 1 lists the actual conditions of some $P_{o/w}$ measurements by the cocurrent CCC method. The CCC methods presented here were extensively used to measure the $P_{o/w}$ value of molecular compounds. They were recently adapted to measure the $P_{o/w}$ values of ionizable compounds using buffered octanol-saturated aqueous phases [5].

References

1. A. Berthod, Liquid–liquid partition coefficients, in *Centrifugal Partition Chromatography* (A. P. Foucault, ed.), Chromatographic Science Series Vol. 68, Marcel Dekker Inc., New York, 1995, pp. 167–198.
2. S. J. Gluck and E. J. Martin, *J. Liquid Chromatogr. 13*: 2529–2551 (1990).
3. A. Berthod, *Analusis 18*: 352–358 (1990).
4. A. Berthod, R. A. Menges, and D. W. Armstrong, *J. Liquid Chromatogr. 15*: 2769–2785 (1992).
5. A. Berthod, S. Carda-Broch, and M. C. G. Alvarez-Coque, *Anal. Chem. 71*: 879–888 (1999).

Alain Berthod

Open-Tubular (Capillary) Columns

Open-tubular columns were discovered by Golay [1] in the late 1950s and the first commercial columns were introduced in the early 1960s. The first capillary columns were fabricated from copper tubing 0.01 in. inner diameter but, due to their somewhat variable geometry, were quickly replaced with the more rigid cupronickel tubing and, subsequently, by stainless-steel tubing. Metal capillary columns need to be cleaned to remove traces of extrusion lubricants by washing them with methylene dichloride, methanol, and then water. They should also be washed with dilute acid to remove any metal oxides or corrosion products that remain adhering to the walls. The acid is removed with water and the tubing is again washed with methanol and methylene dichloride and dried in a stream of hot nitrogen.

Metal columns provide the expected high efficiencies and were used successfully for the analysis of low-polarity materials such as petroleum and fuel oils and, today, they are still extensively used for the analysis of hydrocarbons. Metal columns, however, although easily coated with dispersive stationary phases (e.g., squalane, Apiezon grease, etc.), do not coat well with the more polar stationary phases such as Carbowax®. In addition, the hot metal surface can cause decomposition and molecular rearrangement of many thermally labile materials that are being separated (e.g., the terpenes in essential oils). Metal can also react directly with some solutes by chelation and, as a result of surface adsorption, produce asymmetric and tailing peaks. Nevertheless, metal columns are rugged, easy to handle, and easy to remove and replace in the chromatograph, so their use has persisted in many applications despite the introduction of fused-silica columns.

In an attempt to eliminate surface activity, Desty et al. [2] introduced the first silica-based columns and invented an extremely clever device for drawing soft glass capillary columns. Desty produced both rigid soft glass and rigid Pyrex capillary columns, although their permanent circular shape rendered them a little difficult to connect to the injector and detector. It was found that, with special surface treatment, the rigid glass tubes could be coated with polar stationary phases. The demand for special surface processing evoked a large number of proprietary methods for column treatment. Fortunately, the frenetic interest in the surface deactivation of soft glass capillary tubes was curtailed by the introduction of the flexible fused-silica capillary columns by Dandenau and Zenner [3].

The quartz fiber drawing technique used in the manufacture of data transmission lines was used to produce flexible *fused-silica* tubing. Basically, the solid quartz rod used in quartz fiber drawing was replaced by a quartz tube. In a similar manner to that used in the quartz fiber production, the quartz tubes were coated with polyimide to prevent moisture from attacking the surface and producing stress corrosion. Soft glass capillaries can be produced by the same technique at much lower temperatures [4], but the tubes are not as mechanically strong or as inert as quartz capillaries. Flexibility was the main advantage to quartz capillaries, as it greatly facilitated the installation of the columns in the chromatograph. However, surface treatment is still necessary with a fused-quartz column to reduce adsorption and catalytic activity and render the surface wettable for efficient coating. The treatment may involve washing with acid, silanization, and other types of chemical treatment, including the use of surfactants.

Deactivation procedures used for commercial columns also tend to be highly proprietary. A deactivation program for silica and soft glass columns that is suitable for most applications would first entail an acid wash. The column is filled with 10% (w/w) hydrochloric acid, the ends sealed, and the column then heated to 100°C for 1 h. The column is then washed free of acid with distilled water and dried. This procedure is believed to remove traces of heavy metal ions that can cause adsorption and peak tailing. The column is then filled with a solution of hexamethyldisilazane, sealed, and heated to the boiling point of the solvent for 1 h. This procedure blocks any hydroxyl groups on the surface that were generated during the acid wash. A polar or semipolar silane reagent might be preferable to facilitate coating if a polar stationary phase is to be used. The column is then washed with the pure solvent, dried at an elevated temperature in a stream of pure nitrogen, and is ready for coating.

Open-tubular columns can be coated internally with a liquid stationary phase or with polymeric materials that are subsequently polymerized to form a relatively rigid polymer coating. The two methods of coating are the *dynamic method of coating* and the *static method of coating*. In the dynamic coating procedure, a plug of solvent containing the stationary phase is placed at the beginning of the column. The strength of the solution, among other factors, determines the thickness of the stationary-phase film. In general, the film thickness of an open-tubular column ranges from 0.25 μm to about 1.5 μm. As an estimate, a 5% (w/w) solution of stationary phase will provide a stationary film thickness of about 0.5 μm. After the plug has been run into the front of the column (sufficient solution should be added to fill about 10% of the column length), a gas pressure is used to force the plug through the column at about 2–4 mm/s. When the plug has passed through the column, the gas flow is continued for about 1 h. The gas flow should not be increased too soon, as ripples of stationary phase solution will form on the walls of the tube, which produces a very uneven film. After 1 h, the flow rate is increased and the column stripped of solvent. The last traces of the solvent are removed by heating the column above the boiling point of the solvent at an increased gas flow rate.

In static coating, the entire column is filled with a solution of the stationary phase and one end connected to a vacuum. As the solvent evaporates, it retreats back down the tube, leaving a coating on the walls. The optimum concentration will depend on the stationary phase, the solvent, the temperature, and the condition of the wall surface. This process is very time-consuming but can proceed without attention and is often carried out overnight. This procedure is more repeatable than the dynamic method of coating, but, in general, it produces columns having a similar performance to those dynamically coated.

However well the column may be coated, the stability of the column depends on the stability of the stationary phase film, and thus on the constant nature of the surface tension forces holding it to the column wall. These surface tension forces can change with temperature or be effected by the samples used for analysis. As a consequence, the surface tension can be suddenly reduced and the film break up. It follows that the stationary phase should be bonded in some way to the column walls or polymerized *in situ*. Such coatings are called immobilized stationary phases and cannot be removed by solvent washing.

Some stationary phases that are polymeric in nature can sometimes be formed by coating the monomers or dimers on the walls and then initiating polymerization either by heat or a suitable catalyst. This locks the stationary phase to the column wall and is thus completely immobilized. Polymer coatings can be formed in the same way using dynamic coating. Techniques used for immobilizing the stationary phases are highly proprietary and little is known of the methods used. In any event, most chromatographers do not want to go to the trouble of coating their own columns and are usually content to purchase proprietary columns.

Porous-Layer Open-Tubular Columns

There are two basic disadvantages to the coated capillary column. First, the limited solute retention that results from the small quantity of stationary phase in the column. Second, if a thick film is coated on the column to compensate for this low retention, the film becomes unstable resulting in rapid column deterioration. Initially, attempts were made to increase the stationary-phase loading by increasing the internal surface area of the column. Attempts were first made to etch the internal column surface, which produced very little increase in surface area and very scant improvement. Attempts were then made to coat the internal surface with diatomaceous earth, to form a hybrid between a packed column and coated capillary. None of the techniques were particularly successful and the work was suddenly eclipsed by the production of immobilize films firmly attached to the tube walls. This solved both the problem of loading, because thick films could be immobilized on the tube surface, and that of phase stability. As a consequence, porous-layer open-tubular (PLOT) columns are not extensively used. The PLOT column, however, has been found to be an attractive alternative to the packed column for gas–solid chromatography (GSC) and effective methods for depositing adsorbents on the tube surface have been developed.

The open-tubular column is, by far, the most popular type of GC column in use today. As a result of its small internal cross section, however, extracolumn dispersion can become a serious problem. This means that open-tubular columns must be used with special types of injector and reduced volume connectors, and certain detectors must have specially designed sensor cells to avoid impairing column performance.

References

1. M. J. E. Golay, *Gas Chromatography. 1958* (D. H. Desty, ed.), Butterworths, London, 1958, p. 36.
2. D. H. Desty, A. Goldup, and B. F. Wyman, *J. Inst. Petrol. 45*: 287 (1959).
3. R. D. Dandenau and E. M. Zenner, *J. High Resolut. Chromatogr. 2*: 351 (1979).
4. K. L. Ogan, C. Reese, and R. P. W. Scott, *J. Chromatogr. Sci. 20*: 425 (1982).

Suggested Further Reading

Scott, R. P. W., *Techniques and Practice of Chromatography*, Marcel Dekker, Inc., New York, 1996.
Scott, R. P. W., *Introduction to Analytical Gas Chromatography*, Marcel Dekker, Inc., New York, 1998.

Raymond P. W. Scott

Open-Tubular and Micropacked Columns for Supercritical Fluid Chromatography

Introduction

Supercritical fluid chromatography (SFC) with open-tubular columns was first demonstrated in 1981 by Novotny and co-workers [1]. This technique, known as capillary SFC, was made available to the analytical community through the introduction of several commercial instruments in 1986. Initially difficult to use, improvements in instrumentation and hardware, coupled with a wider array of columns and restrictor options designed specifically for the technique, becoming available, have led to a general acceptance of the method in many laboratories. Not only useful as a research tool, capillary SFC is firmly established as an essential analytical method for production support and quality control in many industries. Some of these include chemical and petroleum manufacturing, pharmaceuticals, polymers, and environmental monitoring.

Packed columns have also been used in SFC for many years, predating capillaries by nearly 20 years. Many columns originally developed for liquid chromatography have found utility in SFC and have varied in internal diameter from smaller than 50 μm to very large-preparative-scale sizes. Definitions vary, but for purposes here, micropacked columns are considered to have internal diameters less than 2 mm. These smaller-diameter columns are also in wide use and offer significant benefits with regard to mobile-phase consumption and detector compatibility than their large-bore counterparts. The selectivity and performance of micropacked columns are complimentary to those of capillaries, and instrumentation is available that is compatible with both separation techniques, allowing for the separation of a wide range of analytes and rapid switchover between methods. Several reviews have been published [2–4].

Pressure Drop Effects

Elution of a particular compound in SFC is a function of its extent of interaction with the column stationary phase and the solvating strength of the mobile phase, with the latter being a direct function of density. The density is affected by temperature and pressure and, in the case of separations with capillary columns that are inherently open and exhibit little pressure drop across their length, it is essentially constant throughout. By contrast, packed columns exhibit much more resistance to mobile-phase flow and can experience a considerable density drop during SFC analysis, producing a potentially significant loss in separation efficiency. Commercial packed columns, tested only by high-performance liquid chromatography (HPLC), may not show these deficiencies in their test reports. The only reliable gauge of suitability of a column for SFC is a performance test in the SFC mode. Columns tested under SFC conditions and tested for suitability for a specific SFC method have been commercially available for some time.

The pressure drop effect limits the usable length of packed columns to approximately 25 cm, although micropacked columns prepared specifically for SFC can be used to longer lengths [5]. The particle size also plays a role, with packing materials smaller than 5 μm producing the highest pressure drops. Whereas short columns dominate in packed column SFC, typical parameters for capillary columns are 3–10 m in length, 50 μm in inner diameter, and a stationary-phase film thickness of 0.25 μm, which give the best compromise in loadability, analysis speed, and efficiency.

Calculated practical efficiencies for a compound with a capacity factor of 2 and a CO_2 mobile phase are shown for each type of column in Table 1. It is clear that capillary columns are capable of delivering high efficiency separations in SFC, but at the expense of analysis time when compared to packed columns.

Activity

Silica surfaces are the chief source of activity in columns for SFC and, even though many of the columns are well deactivated, the residual silanol sites can lead to tailing or adsorption of analytes. The low surface area of capillary columns is responsible for much higher levels of inertness than their packed counterparts based on silica particles. Capillary columns have been used successfully in the analysis of active compounds, including isocyanates, acid halides, organic acids, amines, peroxides, azo compounds, and many others. The low temperatures required for elution make analysis of active and labile compounds viable.

Silica particles have high surface areas and usually contain a large number of exposed residual silanol groups after derivatization. These groups impart a significant degree of polarity to packed columns and can be used to advantage, for example, in the determination of aromatics

Table 1 Calculated Practical Efficiencies for Compound with a Capacity Factor of 2 and CO_2 Mobile Phase

Column type	Particle diameter, internal diameter (μm)	Length (m)	Plates at low density (100 atm, 100°C)	Plates at high density (400 atm, 100°C)	Linear velocity (cm/s) at low density	Linear velocity (cm/s) at high density
Packed	5	0.1	5,200	9,100	0.6	2.1
Capillary	50	10	102,000	19,000	2.5	5.8

Source: Data from Ref. 2.

in fuels [6]. For more active solutes, modifiers are used to reduce tailing and improve quantitation.

Modifiers

The addition of cosolvents to the mobile phase can be effective in adjusting selectivity and improving sample solubility. As the most dramatic effect with polar modifiers is seen in the interaction with the surface silanol groups, even small amounts of cosolvents change the elution characteristics of packed columns. With capillary columns, the effect is related more to solvent strength of the mobile phase than surface modification, and higher modifier levels are required to produce significant changes in retention.

One of the drawbacks of using modifiers is their response in some of the detectors. The flame ionization detector (FID) is very popular with capillary and micropacked columns in SFC because of its near-universal response and high sensitivity and the lack of response of CO_2 as the most popular mobile phase. The low mass flow rate of the mobile phase in small columns allows for a direct interfacing of the column to the FID and other detectors without flow splitting or back-pressure regulation.

Sample Introduction

The small internal volume and low mobile-phase mass flow rates in capillary and micropacked column SFC place significant demands on the injection system and connections. The injector must deliver a small, narrow band of material onto the head of the column and must not contain any void volume or unswept area in the flow path. Several methods of injection are in common use, including the following:

1. Split, where the column is placed in the injector such that it intercepts a portion of the sample stream with the excess carried past and out of the system through a flow restrictor. This method gives the highest efficiencies, but it can produce some sample discrimination.
2. Timed-split, where the sample loop is placed in the flow stream for short periods, and the time in the inject position determines the amount on column. This is the most popular injection method and gives good efficiency and reproducibility. It requires fast actuation and an internal sample loop.
3. Split-splitless, which is performed with a split assembly and a split vent shutoff valve. This method enables larger volumes to be admitted onto the columns and the split activates to reduce tailing by sweeping residual amounts of material out of the system.

The use of a retention gap can allow for higher efficiencies and larger injection volumes on capillary columns [7]. The retention gap is a section of uncoated tubing placed between the column and the injector, which allows the analytes to refocus into a narrow band at the head of the column. This uncoated section can be built right into the capillary column such that no additional connections are required.

Restrictors

Restrictors are required at the ends of SFC columns to maintain supercritical conditions throughout the column and to limit overall flow. Several options exist, with frit restrictors being the most popular, followed by integral and linear formats. The frit restrictor is made by casting a porous ceramic material inside fused-silica tubing with the flow rate dependent on length and pore size. These restrictors are robust and are easily tuned to the desired flow rate by trimming small sections off of the frit end. The multiple flow paths are also resistant to plugging. Frit restrictors are supplied in varied porosities in the end of deactivated 50-μm-inner diameter tubing and are attached to the end of the column using low dead-volume connectors. Integral restrictors are made by heating fused-silica tubing to its melting point and allowing it to collapse to a single orifice of very small diameter. The end can be ground to form a larger opening, but this process requires considerable patience. This type of restrictor can be fabricated in the end of the column such that no connectors are required, but the single orifice is more susceptible to plugging with stray particles than are other types. Linear restrictors are made from short lengths of fused-silica tubing with narrow internal diameters. These are interfaced to the column with low-dead-volume connectors, but the long pressure drop across the tubing length can cause some analytes to precipitate prematurely and produce detector spiking.

Stationary Phases

A wide variety of stationary phases and bonded-phase particles for SFC are available. Capillary columns are coated with substituted and cross-linked polysiloxanes, which exhibit good inertness, efficiency, and stability. There are three main classes of capillary column stationary phases for SFC: apolar, polarizable, and polar.

Apolar

Methyl silicone, 5% phenyl-substituted silicone, and 50% octyl-substituted silicone separate generally on the basis of solute volatility. The most significant interactions are inherently weak van der Waal's. These phases have the highest diffusion properties and give the highest efficiencies. Highly polar materials overload easily on these columns and produce wedge-shaped peaks.

Polarizable

The 50% phenyl-substituted silicone and 30% biphenyl-substituted silicone stationary phases are moderately polar and contain polarizable aromatic rings that exhibit induced dipoles in the presence of dipolar solutes such as alcohols, phenols, amines, nitriles, ketones, and so forth. They give selectivity without extended retention of polar solutes because the dipole-induced–dipole interaction is relatively weak. Temperature affects the extent of this polarization and can be used as a variable in optimizing separations.

Polar

The 25% and 50% cyanopropyl phases exhibit permanent dipoles that interact strongly with polar solutes. Because this translates into longer retention times for polar solutes, only lower-molecular-weight materials of this type can be eluted. Polarizable (aromatic and unsaturated hydrocarbons) and weakly dipolar solutes are good candidates for analysis with these phases. Aliphatic hydrocarbons overload easily but elute rapidly.

Micropacked columns are available with most of the bonded-phase packings used in high-performance liquid chromatography. Porous and nonporous silica particles are optionally functionalized with covalently bound silanes or other strongly adsorbed materials. Alkyl-bonded silicas produce separations, generally based on solute volatility, but with the potential for selectivity differences based on interaction with silanol groups. Underivatized silica is popular for petroleum separations of aliphatic and aromatic hydrocarbons. Silver-ion-containing silica columns are selective for olefin separations. Fluoroalkyl-bonded silicas produce unique selectivities and show good sample capacities for fluorocarbons. Polybutadiene-derivatized zirconia particles have also been used in SFC as have particles based on cross-linked organic polymers. These latter types show different selectivities because of the absence of surface silanol groups. Chiral-bonded phases capable of resolving enantiomers are seeing wide use, particularly in the pharmaceutical market.

Recent Advances

Recent developments in capillary and micropacked column SFC have centered on making the technique easier and more reliable to use. Columns are available that are fitted with restrictors, performance tested by SFC, and are ready to install. Dead-volume issues have been resolved with low-mass couplers and auto-depth-adjusting finger-tight fittings suitable for high-pressure use. Packed columns have been developed and optimized for SFC use that have low pressure drops and high stabilities. The future should see a continuation of this trend, with more column options and formats becoming available and additional methods utilizing them seeing wide acceptance.

References

1. M. Novotny, S. R. Springston, P. A. Peaden, J. C. Fjeldstead, and M. L. Lee, *Anal. Chem. 53*: 407A (1981).
2. M. L. Lee and K. E. Markides (eds.), *Analytical Supercritical Fluid Chromatography and Extraction*, Chromatography Conferences, Provo, UT, 1990.
3. M. Caude and D. Thiebaut (eds.), *Practical Supercritical Fluid Chromatography and Extraction*, Harwood, Amsterdam, 1999.
4. L. G. Blomberg, M. Demirbueker, I. Haegglund, and P. E. Andersson, *Trends Anal. Chem. 13*(3): 126–137 (1994).

5. W. Li, A. Malik, and M. L. Lee, *J. Microcol. Separ.* 6: 557–563 (1994).
6. W. Li, A. Malik, M. L. Lee, B. A. Jones, N. L. Porter, and B. E. Richter, *Anal. Chem.* 67(3): 647–654 (1995).
7. T. L. Chester and D. P. Innis, *Anal. Chem.* 67(17): 3057–3063 (1995).

Brian Jones

Optical Activity Detectors

Optical activity detectors are capable of specifically detecting chiral compounds, taking advantage of their unique interactions with polarized light. Much of the work on the development of prisms and other devices for the production of polarized light was done in the early part of the nineteenth century. However, the measurement of optical activity is often used for enantiomeric purity determination of chiral compounds, which by definition have either a center or plane of asymmetry. Enantiomers rotate the plane of polarized light in opposite directions, although in equal amounts. The isomer that rotates the plane to the left (counterclockwise) is called the levo isomer and is designated (−), whereas the one that rotates the plane to the right (clockwise) is called the dextro isomer and is designated (+). Questions of optical activity are of extreme importance in the field of asymmetric chemical synthesis and in the pharmaceutical industry.

Detection Principle

Figure 1 shows the basic optimal system of the optical rotation detector, which is based on the nonmodulated polarized beam-splitting method. The light radiated from the light source is straightened by the plane polarizer, then to the lens for beam formation and concentration, and then to the flow cell.

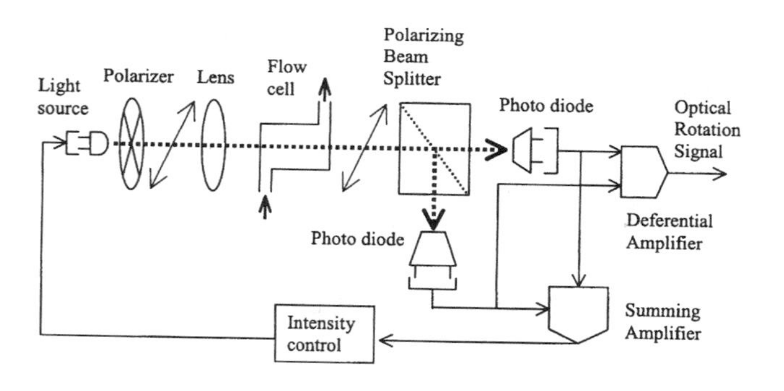

Fig. 1 Optical rotation detector.

The plane-polarized light which goes through the flow cell is rotated by optically active substances (chiral compounds) according to their specific optical rotations and concentrations. The light then enters the polarized beam splitter and is divided into two beams according to the polarized beam directions. These beams are detected by two photodiodes as shown.

The angle of the plane polarizer is adjusted so that the two photodiodes may receive the same beam intensity when no optically active substance is present in the flow cell. When optically active substances are present in the flow cell, the difference between the beam intensities received by the two photodiodes is not zero. Therefore, the difference has a linear relation with specific optical rotation and concentration of the optically active substance and can be expressed by

$$V_0 = K[\alpha]C$$

where V_0 is the difference of beam intensities received by the two photodiodes (i.e., output of signal level), K is a constant determined by cell structure and light intensity of the light source, $[\alpha]$ is the specific optical rotation of the chiral compound, and C is the concentration of the chiral compound.

Polarimetry Theory

Most forms of optical spectroscopy are usually concerned with the measurement of the absorption or emission of electromagnetic radiation. Ordinary, natural, unreflected light behaves as though it consists of a large number of electromagnetic waves vibrating in all possible orientations around the direction of propagation. If, by some means, we sort out from the natural conglomeration only those rays vibrating in one particular plane, we say that we have plane-polarized light. Of course, because a light wave consists of an electric and a magnetic component vibrating at right angles to each other, the term "plane" may not be quite descriptive, but the ray can be considered planar if we restrict ourselves to noting the direction of the electrical component. Circular polarized light represents a wave in which the electrical component (and, therefore, the magnetic component also) spirals around the direction of propagation of the ray, either clockwise ("right-handed" or dextrorotatory) or counterclockwise ("left-handed" or levorotatory). If, following the passage of the plane-polarized ray through some material, one of the circularly polarized components, say the left circularly polarized ray, has been slowed down, then the resultant would be a plane-polarized ray rotated somewhat to the right from its original position. In addition, lasers have been incorporated into two optical rotation methods to date: polarimetry and circular dichroism.

Optical Rotation and Optical Rotatory Dispersion

A polarimeter measures the direction of rotation of plane-polarized light caused by an optically active substance. The specific optical activity of an asymmetrical molecule varies with the wavelength of the light used for its determination. This variation is called optical rotatory dispersion (ORD). In ORD, rotations are measured over a range of wavelengths rather than at a single wavelength, usually covering the ultraviolet (UV) as well as the visible region.

Circular Dichroism

In this technique, the molecular extinction coefficients of a compound are measured with both left and right circularly polarized light, and the difference between these values is plotted against the wavelength of the light used. The phase angle between the projections of the two circularly polarized components is altered by passage through the chiral medium, but their amplitudes will be modified by the degree of absorption experienced by each component. This differential absorption of left- and right-circularly polarized light is termed circular dichroism (CD). So, circular dichroism measurements provide both absorbance and optical rotation information simultaneously.

Circularly Polarized Luminescence Spectroscopy

Circularly polarized luminescence spectroscopy (CPLS) is a measure of the chirality of a luminescent excited state. The excitation source can be either a laser or an arc lamp, but it is important that the source of excitation be unpolarized to avoid possible photoselection artifacts. The CPLS experiment produces two measurable quantities, which are obtained in arbitrary units and related to the circular polarization condition of the luminescence. It is appropriate to consider CPLS spectroscopy as a technique that combines the selectivity of CD with the sensitivity of luminescence. The major limitation associated with CPLS spectroscopy is that it is confined to emissive molecules only.

Vibrational Optical Activity

The optical activity of vibrational transitions has been conducted. The infrared (IR) bands of a small molecule can easily be assigned with the performance of a normal coordinate analysis, and these can usually be well resolved. One of the problems associated with vibrational optical activity is the weakness of the effect. Instrumental limitations of infrared sources and detectors create additional experimental constraints on the signal-to-noise ratios.

Two methods suitable for the study of vibrational optical activity have been developed:

Vibrational Circular Dichroism: Vibrational circular dichroism (VCD) could be measured at good signal-to-noise levels. Vibrational optical activity is observed in the classic method of Grosjean and Legrand.

Raman Optical Activity: The Raman optical activity (ROA) effect is the differential scattering of left- or right-circularly polarized light by a chiral substrate where chirality is studied through Raman spectroscopy.

Fluorescence-Detected Circular Dichroism

Fluorescence-detected circular dichroism (FDCD) is a chiroptical technique in which the spectrum is obtained by measuring the difference in total luminescence obtained after the sample is excited by left- and right-circularly polarized light. For the FDCD spectrum of a given molecular species to match its CD spectrum, the luminescence excitation spectrum must be identical to the absorption spectrum.

Factors Affecting the Measurement of Optical Rotation

The rotation exhibited by an optically active substance depends on the thickness of the layer traversed by the light, the wavelength of the light used for the measurement, and the temperature of the system. In addition, if the substance being measured is a solution, then the concentration of the optically active material is also involved and the nature of the solvent may also be important. There are certain substances that change their rotation with time. Some are substances that change from one structure to another with a different rotatory power and are said to show mutarotation. Mutarotation is common among the sugars. Other substances, owing to enolization within the molecules, may rotate so as to become symmetrical and, thus, lose their rotatory power. These substances are said to show racemization. Mutarotation and racemization are influenced not only by time, but also by pH, temperature, and other factors. Of course, rotations that determined for the same compound under the same conditions are identical. Therefore, in expressing the results of any polarimetric measurement, it is, therefore, very important to include all experimental conditions.

Temperature

Temperature changes have several effects on the rotation of a solution or liquid. An increase in temperature increases the length of the tube; it also decreases the density, thus reducing the number of molecules involved in the measurement. It causes changes in the rotatory power of the molecules themselves, due to association or dissociation and increased mobility of the atoms, and affects other properties. In addition, temperature changes cause expansion and contraction of the liquid and a consequent change in the number of active molecules in the path of the light.

The unique ability of the optical rotation detector to respond to the sign of rotation allows precise enantiomeric purity determination even if the enantiomers are only partially resolved. The sign of rotation is also useful in establishing enantiomer elution order.

Because the optical rotation detectors only respond to optically active compounds, enantiomeric purity determination to precisions of better than 0.5% can be achieved and is possible in even the complex mixtures. The detection can also be used as part of a flow injection analysis system to determine amount and enantiomeric purity of a drug in dosage form.

The applications using optical rotation detectors include the following:

1. Qualitative analysis of chiral compounds, including drugs, pesticides, carbohydrates, amino acids, liquid crystals, and other biochemicals
2. Determination of enantiomeric purity of chiral compounds
3. Monitoring an enzymatic reaction
4. Qualitative analysis of proteins
5. Use as a conventional polarimeter

However, the disadvantages of optical rotation detectors may be limited by shot or flicker noise, which are dependent on the optical and mechanical properties of the system or by noise in the detector electronics. Generally, the usefulness of this technique has been limited by the lack of sensitivity of commercially available instruments.

Suggested Further Reading

Allenmark, S., Techniques used for studies of optically active compounds, in *Chromatographic Enantioseparation: Methods and Application*, 2nd ed., Ellis Horwood Ltd., London, 1991.

Beesley, T. E. and R. P. W. Scott, An introduction to chiral chromatography, in *Chiral Chromatography*, John Wiley & Sons, Inc., New York, 1998, pp. 1–11.

Dodziuk, H., Physical methods as a source of information on the spatial structure of organic molecules, in *Modern Conformational Analysis, Elucidating Novel Exciting Molecular Structures*, VCH, New York, 1995, pp. 48–54.

Edkins, T. J. and D. C. Shelly, Measurement concepts and laser-based detection in high-performance micro separation, in *HPLC Detection: Newer Methods* (G. Patonay, ed.), VCH, New York, 1992, pp. 1–15.

Goodall, D. M. and D. K. Lloyd, A note on an optical rotation detector for high-performance liquid chromatography, in *Chiral Separations* (D. Stevenson and D. Wilson, eds.), Plenum Press, New York, 1988, pp. 131–133.

Sheldon, R. A., Introduction to optical isomersion, in *Chirotechnology: Industrial Synthesis of Optically Active Compounds*, Marcel Dekker, Inc., New York, 1993, pp. 25–27.

Weston, A. and P. R. Brown, *HPLC and CE Principles and Practice*, Academic Press, San Diego, CA, 1997.

Yeung, E. S., Polarimetric detectors, in *Detectors for Liquid Chromatography* (E. S. Yeung, ed.), John Wiley & Sons, New York, 1986, pp. 204–228.

Hassan Y. Aboul-Enein
Ibrahim A. Al-Duraibi

Optical Quantification (Densitometry) in TLC

Nondensitometric Quantification

Quantitative evaluation of thin-layer chromatograms can be performed by direct, *in situ* visual, and indirect elution techniques. Visual evaluation involves comparison of the sizes and intensities of color or fluorescence between sample and standard zones spotted, developed, and detected on the same layer. The series of standards is chosen to have concentrations or weights that bracket those of the sample zones. After matching a sample with its closest standard, accuracy and precision are improved by respotting a more restricted series of bracketing standards with a separate sample spot between each of two standard zones. Accuracy no greater than 5–10% is possible for trained personnel using visual evaluation. The determination of mycotoxins in food samples is an example of a practical application of visual comparison of fluorescent zones.

The elution method involves scraping off the separated zones of samples and standards and elution of the substances from the layer material with a strong, volatile solvent. The eluates are concentrated and analyzed by use of a sensitive spectrometric method, gas or liquid column chromatography, or electroanalysis. Scraping and elution must be performed manually because the only commercial automatic micropreparative elution instrument has been discontinued by its manufacturer. The elution method is tedious and time-consuming and prone to errors caused by the incorrect choice of the sizes of the areas to scrape, incomplete collection of sorbent, and incomplete or inconsistent elution recovery of the analyte from the sorbent. However, the elution method is being rather widely used (e.g., some assay methods for pharmaceuticals and drugs in the USP Pharmacopoeia).

Introduction to Densitometry

In order to achieve the optimum accuracy, precision, and sensitivity, most quantitative analyses are performed by using high-performance thin-layer chromatography (TLC) plates and direct quantification by means of a modern optical densitometic scanner with a fixed sample light beam in the form of a rectangular slit that is variable in height (e.g., 0.4–10 mm) and width (20 μm to 2 mm). Densitometers measure the difference in absorbance or fluorescence signal between a TLC zone and the empty plate background and relate the measured signals from a series of standards to those of unknown samples through a calibration curve. Modern computer-controlled densitometers can produce linear or polynomial calibration curves relating absorbance or fluorescence versus weight or concentration of the standards and determine bracketed unknowns by automatic interpolation from the curve. Samples and standards are best applied using an automated instrument such as the one shown in Fig. 1. Use of manual spotting and less efficient TLC plates results in greater errors and poorer reproducibility in quantitative results.

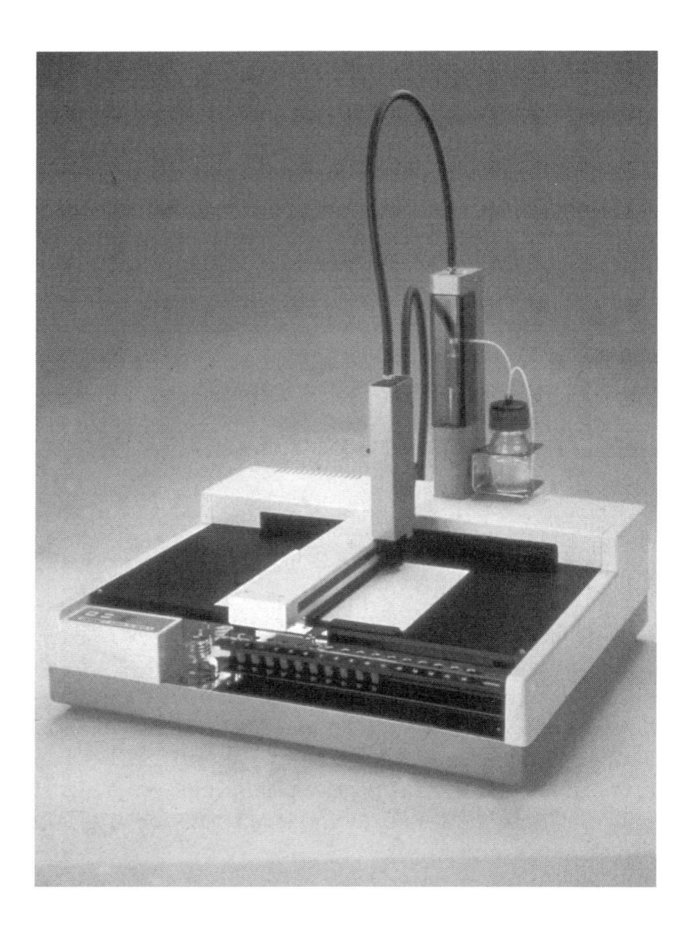

Fig. 1 Automatic TLC sampler (ATS 3) used for computer-controlled application of precisely controlled volumes of samples and standards between 10 nL and 50 μL from a rack of vials as spots or bands to preselected origins on a plate. (Courtesy of Camag Scientific Inc., Wilmington, NC.)

Instrumental Design and Scanning Modes

A commercial densitometer and a schematic diagram of the light-path arrangement used in scanning are shown in Fig. 2. The plate is mounted on a moveable stage controlled by a stepping motor drive that allows each chromatogram track to be scanned in or against the direction of development. A tungsten or halogen lamp is used as the source for scanning colored zones in the 400–800-nm range (visible absorption) and a deuterium lamp for scanning ultraviolet (UV)-absorbing zones directly or as quenched zones on phosphor-containing layers (F-layers) in the 190–450-nm range. The monochromator used with these continuous-wavelength sources can be a quartz prism or, more often, a grating. The detector is a photomultiplier or photodiode placed above the layer to measure reflected radiation. [Some scanners (e.g., Fig. 2) make use of a reference photomultiplier in addition to the measuring photomultiplier in the single-beam mode; the reference photomultiplier puts out a constant signal that is compared to the signal from the measuring photomultiplier to produce a difference signal that is more accurate than a direct signal from a single measuring photomultiplier would be.]

For normal fluorescence scanning, a high-intensity xenon continuum source or a mercury vapor line source is used, and a cutoff filter is placed between the plate and detector to block the exciting UV radiation and transmit the visible emitted fluorescence. For fluorescence measurement in the reversed-beam mode, a monochromatic filter is placed between the source and plate and the monochromator between the plate and detector. In this mode, the monochromator selects the emission wavelength, rather than the excitation wavelength as in the normal mode.

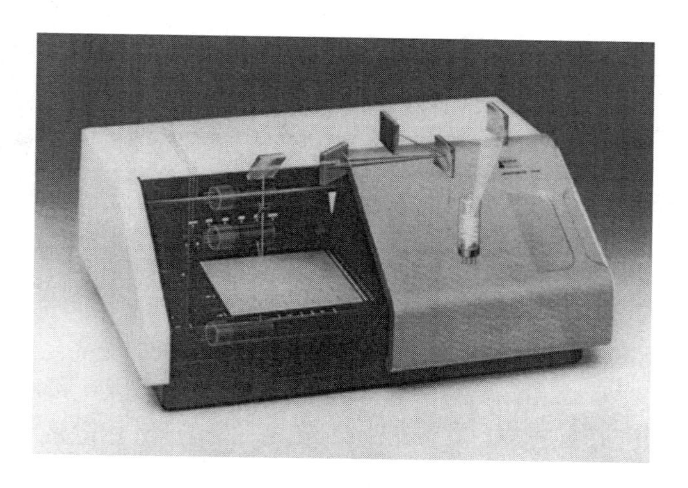

Fig. 2 Photograph of the DESAGA Densitometer CD 60 with a superimposed schematic diagram of the light path including (right to left) the source lamp, two mirrors, grating monochromator, mirror, beam splitter, plate with chromatograms to be scanned, reference and measuring detectors (reflection) above the plate and detector (transmission) below the plate. (Courtesy of DESAGA GmbH, Weisloch, Germany.)

Simultaneous measurement of reflection and transmission, or transmission alone, can be carried out by means of a detector positioned on the opposite side of the plate (Fig. 2). Ratio-recording double-beam densitometers, which can correct for background disturbances and drift caused by fluctuations in the source and detector, were designed earlier with two photomultiplier detectors simultaneously recording the two beams (double beam in space), but, today, such densiotmeters are equipped with a chopper and one detector (double beam in time). For dual-wavelength, single-beam scanning, which will correct for scattering of the absorbed light by subtracting out the (presumably equal) scattering at a nonabsorbed wavelength, a light beam is selected by a mirror and passes through two separate monochromators to isolate the two different wavelengths. The two beams are alternated by a chopper and recombined into a single beam representing a difference signal at the detector. Zigzag or meandering scanning with a small point or spot of light is possible with densitometers having two independent stepping motors to move the plate in the X and Y axes. Computer algorithms integrate the maximum absorbance measurements from each swing to produce a distribution profile of zones having any shape. The potential advantages of scanning with a moving light spot are offset by problems with lower spatial resolution and errors in data processing, and the method is not as widely used as conventional scanning of chromatographic tracks with a fixed slit. Some densitometers have the ability to rotate the plate while scanning for measurement of circular and anticircular chromatograms.

Single-wavelength, single-beam, fixed-slit scanning is most often used and can produce excellent results when high-quality plates and analytical techniques are employed.

Spectral Measurement

Many modern scanners have a computer-controlled motor-driven monochromator that allows automatic recording of *in situ* absorption and fluorescence excitation spectra. These spectra can aid compound identification by comparison with stored standard spectra, test for identity by superimposition of spectra from different zones on a plate, and check zone purity by superimposition of spectra from different areas of a single zone. The spectral maximum determined from the *in situ* spectrum is usually the optimal wavelength for scanning standard and sample areas for quantitative analysis.

Data Handling

The densitometer is connected to a recorder, integrator, or computer. A personal computer with software designed specifically for TLC is most common for data handling and automation of the scanning process in modern instruments. With a fully automated system, the computer can carry out the following functions: data acquisition by scanning a complete plate following a preselected geometric pattern with control of all scanning parameters; automated peak searching and optimization of scanning for each fraction located; multiple-wavelength scanning to find, if possible, a common wavelength for all substances to be quantified, to optically resolve fractions incompletely separated by TLC, and to identify fractions by comparison of spectra with standards cochromatographed on the same plate or stored in a spectrum library through pattern recognition techniques; baseline location and correction; computation of peak areas and/or heights of samples and codeveloped standards and processing of the analog raw data to quantitative digital results, including calculation of calibration curves by linear or polynomial regression, interpolation of sample concentrations, statistical analysis of reproducibility, and presentation of a complete analysis report; and storage of raw data on disk for later reintegration, calibration, and evaluation with different parameters.

Calibration Curves

Densitometric calibration curves relating absorption signal and concentration or weight of standards on the layer are usually nonlinear, especially for higher amounts of standards, and do not pass through the origin. Fluorescence calibration curves are generally linear and pass through the origin, and analyses based on fluorescence are more specific and 10–1000 times more sensitive. The advantages of fluorescence measurement may be realized for nonfluorescent compounds by prechromatographic or postchromatographic derivatization reactions with suitable fluorogenic reagents.

Because the incident monochromatic light is absorbed, reflected, and scattered by the opaque layer material, the theoretical relationship between amount of absorption and amount of substance does not follow the simple Beer–Lambert law that is valid for solutions. The Kubelka–Munk equation is the most accepted theoretical relationship for TLC, but its use is not necessary because of the ability of densitometer software to handle empirical nonlinear regression functions.

Image Analysis (Videodensitometry)

Video camera systems are available from several manufacturers for documentation and densitometric quantification of TLC plates. As an example, the Camag VideoScan instrument consists of a lighting module with short- and long-wave UV and visible sources upon which the layer is placed, a charge-coupled device (CCD) camera with zoom and long-time integration capability, and a PC under MS-Windows control with frame grabber, monitor, and printer. The available software for quantitative evaluation allows the display of the tracks of the chromatogram image acquired with the video camera as analog curves and calculation of their peak properties (R_f, height, area, height percent, and area percent). For quantification, the computer creates a standard curve from the areas or heights of the standards and interpolates unknown values from the curve.

Video scanners have potential advantages, including rapid data collection, simple design with virtually no moving parts, and ability to quantify two-dimensional chromatograms, but they have not yet been shown to have the required capabilities, such as sufficient spectral discrimination or the ability to illuminate the plate uniformly with monochromatic light of selected wavelength, to replace slit-scanning densitometers. Current video scanners can measure spots in the visible range in transmittance, reflectance, or fluorescence modes, but they cannot perform spectral analysis.

Applications and Practical Aspects of Densitometry

Densitometric quantification has been applied to virtually every type of analyte and sample. For example, the greatest number of applications is for the analysis of drug and pharmaceutical compounds, most of which have structures including chromophores that cause strong UV absorption. These compounds are readily quantified in the fluorescence quenching mode on F-layers or in the direct UV absorption mode on unimpregnated layers. Lipids are compounds that are not easily analyzed by gas chromatography (GC) or high-performance liquid chromatography (HPLC) because they lack volatility and the presence of a chromophore leading to UV absorption. The most successful way to quantify lipids is by densitometry after separation and detection on the layer with a chromogenic reagent, most notably phosphomolybdic acid. The quantification of amino acids after detection with ninhydrin is another example of densitometry in the visible absorption mode. Fluorescence densitometry has been applied to the determination of naturally fluorescent compounds (e.g., quinine in tonic water) or compounds derivatized with a fluorogenic reagent pre-TLC or post-TLC (e.g., amino acids reacted with fluorescamine, or carbamate pesticides with dansyl chloride after hydrolysis).

The steps in a typical densitometric quantitative analysis, regardless of analyte type, are the following:

1. Prepare a standard reference solution.
2. Prepare a sample solution in which the analyte is completely dissolved and impurities have been reduced to a level at which they do not interfere with scanning of the analyte.
3. Choose a layer and mobile-phase combination that will separate the analyte as a compact zone with an R_f value in the range 0.2–0.8.
4. Apply the standard and sample aliquots to the layer using an instrument (Fig. 1) or manually with a micropipette, onto preadsorbent, laned plates. Generally, three or four standard zones are applied in constant volumes from a series of standard solutions with increasing concentrations, or in a series of increasing volumes from a single standard solution. The sample volume applied must provide an amount of analyte zone with a weight or concentration that is bracketed by the standard amounts.
5. Develop the plate in an appropriate chamber and dry the mobile phase under in a fume hood or oven.
6. Apply a detection reagent, if necessary, by spraying or dipping. The reagent should produce a stable colored, UV-absorbing, or fluorescent zone having high contrast with the layer background.
7. Scan the natural or induced absorption or fluorescence of the standard and sample zones on the plate using a densitometer with optimized parameters.
8. Generate a calibration curve by linear or polynomial regression of the scan areas and weights of the standards and interpolate the weights in the sample zones from the curve.
9. Calculate the concentration of analyte in the sample from the original weight of the sample, the original total volume of the sample test solution, the aliquot volume of the test solution that is spotted, the interpolated analyte weight in that spotted volume from the calibration curve, and any numerical factor required because of dilution or concentration steps needed for the test solution to produce a bracketed scan area for the analyte zone in the sample chromatogram.
10. Validate the precision of the TLC analysis by replicated determination of the sample and accuracy by comparison of the results to those obtained from analysis of the same sample by an established independent method or calculation of recovery from analysis of a spiked preanalyzed sample or spiked blank sample.

The following are some advantages of TLC densitometry compared to HPLC:

1. The simultaneous analysis of multiple samples on a single plate leads to higher sample throughput (lower analysis time) and less cost per sample. Up to 36 tracks are available for samples and standards on a 10-cm × 20-cm high-performance TLC plate.
2. The ability to generate a unique calibration curve using standards developed under the same conditions as samples on each plate (in-system calibration) leads to statistical improvement in data handling and better analytical precision and accuracy and eliminates the need for an internal standard for most analyses.
3. Detection is versatile and flexible because the mobile phase is removed prior to detection. Because the detection process is static (the zones are stored on the layer), multiple, complementary detection methods can be used.
4. Storage of the chromatogram also allows scanning to be repeated with various parameters without time constraints and assures that the entire sample is available for detection and scanning.
5. Less sample cleanup is often required because plates are not reused. Every sample is analyzed on a fresh layer without sample carryover or cross-contamination.
6. Solvent use is very low for TLC, both on an absolute and per-sample basis, leading to reduced purchase and disposal costs and safety concerns.

Suggested Further Reading

Fried, B. and J. Sherma, *Thin Layer Chromatography—Techniques and Applications*, 4th ed., Marcel Dekker, Inc., New York, 1999, pp. 197–222.

Jaenchen, D. E., Instrumental thin layer chromatography, in *Handbook of Thin Layer Chromatography*, 2nd ed. (J. Sherma and B. Fried, eds.), Marcel Dekker, Inc., New York, 1996, pp. 129–148.

Petrovic, M., M. Kastelan-Macan, K. Lazaric, and S. Babic, Validation of thin layer chromatography quantitation with CCD camera and slit-scanning densitometer, *J. AOAC Int. 82*: 25–39 (1999).
Pollak, V. A., Theoretical foundations of optical quantitation, in *Handbook of Thin Layer Chromatography* (J. Sherma and B. Fried, eds.), Marcel Dekker, Inc., New York, 1991, pp. 249–281.
Poole, C. F. and S. K. Poole, *Chromatography Today*, Elsevier, New York, 1991, pp. 649–734.
Prosek, M. and M. Pukl, Basic principles of optical quantitation in TLC, in *Handbook of Thin Layer Chromatography*, 2nd ed. (J. Sherma and B. Fried, eds.), Marcel Dekker, Inc., New York, 1996, pp. 273–306.
Robards, K., P. R. Haddad, and P. E. Jackson, *Principles and Practice of Modern Chromatographic Methods*, Academic Press, San Diego, CA, 1994, pp. 180–226.

Joseph Sherma

Optimization of Thin-Layer Chromatography

The principal task of chromatography is the separation of mixtures of substances. By "optimization" of the chromatographic process, we mean enhancement of the quality of the separation by changing one or more parameters of the chromatographic system. An ability to foresee, correctly, the direction and scope of these changes is the most important goal of each optimization procedure.

Use of chemometrics to devise procedures suitable for the most crucial stage of optimization, optimization of selectivity, is generally performed in three steps:

1. Selection of the experimental method which best suits the analytical problem considered. At this stage, a chromatographic technique is chosen that ensures that the best possible range of retention parameters is obtained for each individual component of the separated mixture.
2. Establishing the experimental conditions that enable quantification of the influence of the optimized parameters of a chromatographic system on solute retention.
3. Fixing the experimental conditions at values that provide the optimum separation selectivity.

Chemometric optimization of the chromatographic system consists, in fact, in predicting local maxima in multiparametric space and, then, in further deciding which of these parameters is global with regard to the overall efficiency of a given chromatographic system.

Quality of Chromatographic Separations

Elementary Criteria

The simplest way of quantifying the separation of two chromatographic bands, 1 and 2, is to calculate the difference between their respective retention parameters; that is, the difference between their R_F values,

$$\Delta R_F = R_{F_2} - R_{F_1} \tag{1}$$

or between their R_M values,

$$\Delta R_M = R_{M_1} - R_{M_2} = \log \frac{k_1}{k_2} = \log \alpha \tag{2}$$

where k_1 and k_2 are the capacity (retention) factors of the chromatographic bands and α is the separation factor.

The terms most frequently used to characterize the separation of two chromatographic bands are the separation factor, α,

$$\alpha = \frac{k_1}{k_2} \tag{3}$$

where $k_1 > k_2$, and the resolution, R_S [1],

$$R_S = \frac{2(z_2 - z_1)}{w_1 + w_2} = \frac{2l\Delta R_F}{w_1 + w_2} \tag{4}$$

where z_1 and z_2 are the distances of the geometric centers of two chromatographic bands, 1 and 2, from the origin, l is the distance from the origin to the mobile phase front, and w_1 and w_2 are the diameters of the two chromatographic bands, measured in the direction of eluent flow.

Other elementary criteria include the separation factor, S [2],

$$S = \frac{k_2 - k_1}{k_1 + k_2 + 2} \tag{5}$$

the peak-to-valley ratio of the bands, P [3],

$$P = \frac{f}{g} \tag{6}$$

(where f and g are, respectively, the average peak height and valley depth, characteristic of a given pair of neighboring solutes on a chromatogram), the fractional peak overlap, FO [4],

$$\mathrm{FO} = \frac{A_n - A_{n,n-1} - A_{n,n+1}}{A_n} \tag{7}$$

(where A_n is the surface area of the part of the band originating from the pure single compound, $A_{n,n-1}$ is the surface area of the fractional overlap of the nth and $(n-1)$th bands, and $A_{n,n+1}$ is the surface area of the fractional overlap of the nth and $(n+1)$th bands), and the selectivity parameter, R_R [5],

$$R_R = \frac{R_{F_1}}{R_{F_2}} \tag{8}$$

where $R_{F_1} > R_{F_2}$.

Criteria for the Quality of Chromatograms

One method which can be used to establish the optimum conditions for the separation of a complex mixture (i.e., not only a pair) of compounds consists in searching for the maximum of a function denoted the chromatogram quality criterion. The evaluation of separation selectivity can be conducted with the aid of different criteria of chromatogram quality such as the sum of resolution, ΣR_S [6], the sum of separation factors, ΣS [2], and other sums and products of elementary criteria, selected examples of which are the resolution product, ΠR_S [7],

$$\prod R_S = \exp\left(\sum \ln R_S\right), \tag{9}$$

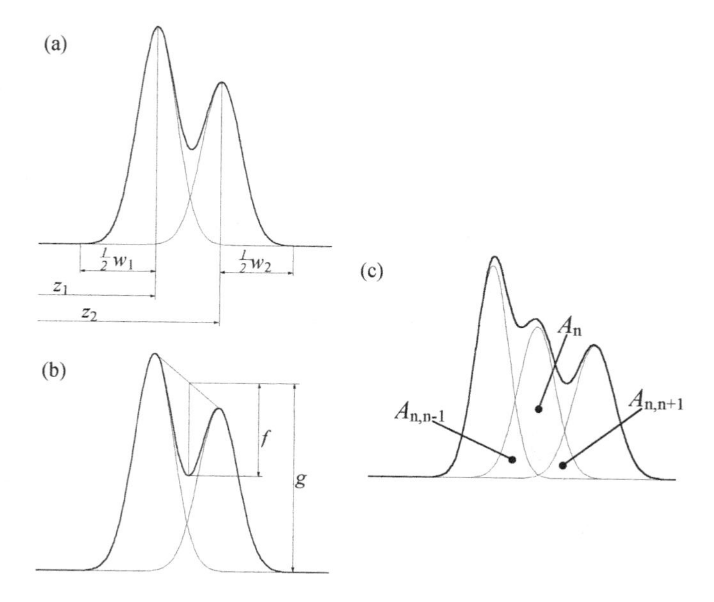

Fig. 1 Graphical interpretation of the selected elementary criteria: (a) resolution, R_s; (b) the peak-to-valley ratio, P; (c) the fractional peak overlap, FO.

the product of the separation factors, ΠS [8], the product of the fractional peak overlap, Π FO [9], and the product of the peak-to-valley ratio of the bands, ΠP [10].

There are also other, more complex criteria, including the normalized resolution product, r [11],

$$r = \prod_{i=1}^{n-1} \frac{R_{S_{i,i+1}}}{\overline{R_S}} = \prod_{i=1}^{n-1} \frac{S_{i,i+1}}{\overline{S}} \tag{10}$$

where n is the number of the chromatographic bands,

$$\overline{R_S} = \frac{1}{n-1} \sum_{i=1}^{n-1} R_{S_{i,i+1}}$$

and

$$\overline{S} = \frac{1}{n-1} \sum_{i=1}^{n-1} S_{i,i+1}$$

and the minimum R_S [8],

$$R_{S,\min} \geq x \quad \text{or} \quad \max R_{S,\min} \tag{11}$$

The minimum of α is used as a criterion of the quality of chromatograms in liquid chromatography [12].

Other criteria are the total peak overlap, φ [13],

$$\varphi = \sum \exp(-2R_S), \tag{12}$$

the informing power, $P_{\inf}$ [14],

$$P_{\inf} = \sum_{i=1}^{n} \log_2 S_i \tag{13}$$

and the chromatographic response function, CRF [10],

$$\mathrm{CRF} = \sum_{i=1}^{n} \ln P_i \tag{14}$$

(where P_i is the peak-to-valley ratio for the ith pair of chromatographic bands).

Performance of the Chromatographic System

One measure of the performance of a given chromatographic system is the number of the theoretical plates per chromatographic band (N). In its simplest form, this can be defined as

$$N = \frac{l}{H} \tag{15}$$

where l is the distance from the origin to the eluent front and H is the height equivalent to one theoretical plate (H is sometimes also denoted HETP).

The average height equivalent to one theoretical plate ($\overline{H}$) can be calculated from the relationship [15]

$$\overline{H} = \frac{(\sigma_i)^2}{(z_r - z_0)R_F} = \frac{(w_i)^2}{16 z_x} \tag{16}$$

where σ_i is the standard deviation, which characterizes the width of the chromatographic spot, or the band width on the densitogram, w_i is the spot width (or the width of the peak base on the densitogram), $z_r - z_0$ is the distance from the origin to the eluent front, and z_x is the distance from the origin to the geometric center of the chromatographic spot.

The relationship between the height equivalent to one theoretical plate (H) and the velocity of the mobile-phase flow is given by the simplified van Deemter equation [16]

$$H = A + \frac{B}{u} + Cu \tag{17}$$

where u is the linear velocity of the mobile phase, A is a constant characterizing eddy diffusion, B is a constant characterizing molecular diffusion, and C is a constant characterizing resistance to interphase mass transfer. This particular issue is of considerable significance in planar chromatographic separations, during the course of which the velocity of the mobile phase changes.

The concept of separation number (SN) in planar chromatography is a practical approach to the task of quantification of chromatographic system performance. According to this concept, such performance can simply be evaluated by calculating how many components of the separated mixture can be comfortably accommodated (i.e., without any overlap of adjacent components) along the direction of migration of the eluent. A convenient relationship proposed in Ref. 17 enables easy calculation of the numerical value of SN:

$$\mathrm{SN} = \left(\log\frac{b_0}{b_1}\right)\left(\log\frac{1 - b_1 + b_0}{1 + b_1 - b_0}\right)^{-1},$$

or, simplified,

$$\mathrm{SN} \approx \frac{1}{b_0 + b_1} \tag{18}$$

where b_0 is the width at half-height of a spot at the origin and b_1 is the width at half-height of a spot at $R_F = 1$ (extrapolated) (b_0 and b_1 are in R_F units).

Semiempirical Optimization Strategies

Strategies used for optimization of selectivity can basically be divided into three separate groups: (a) the simultaneous strategy, (b) the sequential strategy, and (c) the interpretative strategy.

Simultaneous Strategy

In this strategy, one must accomplish all the experiments according to a plan devised earlier. All the results obtained must then be carefully evaluated, the optimum experimental conditions being chosen on this basis.

Sequential Strategy

In this strategy, the optimum experimental conditions are approached in a series of consecutive steps. The choice of any step results strictly from the outcome of all those accomplished previously. One example of a relevant algorithm is the simplex method [18]; the PRISMA [19] geometrical method is a suitable example of the overall optimization approach.

Interpretative Strategy

This method enables prediction of the quality of a separation on the basis of a relatively limited number of the experimental data, collected in previous experiments. According to this approach, the chromatographic results are interpreted in terms of the retention functions, valid for each individual solute separately. Some good examples of the interpretative strategy are the so-called "window diagrams" approach [20] and the search for the extremum of the multiparameter response function with the aid of the genetical algorithm [21].

References

1. T. Kowalska, Theory and mechanism of thin-layer chromatography, in *Handbook of Thin-Layer Chromatography* (J. Sherma and B. Fried, eds.), Chromatographic Science Series. Vol. 55, Marcel Dekker, Inc., New York, 1991, p. 50.
2. P. Jones and C. A. Wellington, *J. Chromatogr. 213*: 357–361 (1981).
3. R. Kaiser, *Gas-Chromatographie*, Geest und Portig, Leipzig, 1960.
4. P. J. Schoenmakers, *Optimization of Chromatographic Selectivity. A Guide to Method Development*, Journal of Chromatography Library, Vol. 35, Elsevier, Amsterdam, 1986, pp. 123–125.
5. W. Prus and T. Kowalska, *J. Planar Chromatogr. 8*: 205–215 (1995).
6. J. C. Berridge, *J. Chromatogr. 244*: 1–14 (1982).
7. J. L. Glajch, J. J. Kirkland, K. M. Squire, and J. M. Minor, *J. Chromatogr. 199*: 57–79 (1980).
8. P. J. Schoenmakers, A. C. J. H. Drouen, H. A. H. Billiet, and L. de Galan, *Chromatographia 15*: 688–696 (1982).
9. R. Smits, C. Vanroelen, and D. L. Massart, *Fresenius Zeitschr. Anal. Chem. 273*: 1–5 (1975).
10. S. L. Morgan and S. N. Deming, *J. Chromatogr. 112*: 267–285 (1975).
11. A. C. J. H. Drouen, P. J. Schoenmakers, H. A. H. Billiet, and L. de Galan, *Chromatographia 16*: 48–52 (1982).

12. S. N. Deming and M. L. H. Turoff, *Anal. Chem. 50*: 546–548 (1978).
13. J. C. Giddings, *Anal. Chem. 32*: 1707–1711 (1960).
14. D. L. Massart and R. Smits, *Anal. Chem. 46*: 283–286 (1974).
15. G. Guiochon and A. M. Siouffi, *J. Chromatogr. 245*: 1–20 (1982).
16. J. J. Van Deemter, F. J. Zuiderweg, and A. Klinkenberg, *Chem. Eng. Sci. 5*: 271 (1965).
17. F. Geiss, *Fundamentals of TLC (Planar Chromatography)*, Hüthig, Heidelberg, 1987.
18. W. Spendley, G. R. Hext, and F. R. Hinsworth, *Technometrics 4*: 441 (1962).
19. Sz. Nyiredy, B. Meier, C. A. J. Erdelmeier, and O. Sticher, *J. High Resolut. Chromatogr. Chromatogr. Commun. 8*: 186–188 (1985).
20. R. J. Laub and J. H. Purnell, *J. Chromatogr. 112*: 71–79 (1975).
21. J. H. Holland, *Adaptation in Natural and Artificial Systems*, University of Michigan Press, Ann Arbor, 1975.

Wojciech Prus
Teresa Kowalska

Overpressured Layer Chromatography

Introduction and Theory

Conventional thin-layer chromatography (TLC) in our experience, known under the name planar chromatography, uses horizontal or vertical glass or Teflon chambers for the development of chromatograms. As stationary phases, commonly known adsorbents or supports based on silica gel, aluminium oxide, magnesium silica, cellulose, and so forth are used; particle sizes are about 20 μm. The migration of the mobile phase is based on the phenomenon of capillary forces. This chromatographic method is described, in detail, in other sections of this volume.

This method is characterized by many limitations which either can cause unsatisfactory separation of a mixture of substances or lead to long development times (even up to several hours) or, sometimes, makes use of solvents of high viscosity impossible. The efficiencies of such chromatographic systems are also rather low.

In conventional TLC, the velocity of chromatogram development depends on the dimension of stationary-phase particles, viscosity of the mobile phase, distance from the start line of the mobile phase, and other parameters. Therefore, there is no possibility of regulation of resolution by change of migration velocity of mobile-phase flow; the distance between the solvent reservoir and the solvent front (z_f) varies with time (t) according to

$$z_f^2 = kt$$

where k is a constant that depends on the chromatographic system (mobile phase and adsorbent) and the size of sorbent particles constituting the layer and presents a parabolic relationship.

As the most popular planar liquid chromatographic technique, TLC uses a vapor phase of solvent above the sorbent layer, which has an important influence on the resolving power.

Many of the inconveniences of TLC are avoided in overpressured layer chromatography (OPLC), which is a logical extension of the theory and practice developed in high-pressure liquid chromatography (HPLC) which can now be used in the field of planar liquid chromatography. This extension offers some exceptional advantages to a chromatographer. OPLC is, in practical terms, a planar HPLC technique. OPLC integrates many of the benefits of TLC, high-performance TLC, and HPLC. This technique corresponds to an HPLC column having a relatively thin, wide cross section and using a pressurized ultramicrochamber with standard chromatoplates. Eluent is forced into the sorbent layer by the means of a pump which enables development of chromatograms with forced flow of the mobile phase (more precise penetration into micropores). The eluent migrates against the sorbent resistance imposed by external pressure on the sorbent surface, and the vapor phase is excluded. In the case of OPLC, there is a linear relationship between the distance (z_f) of the solvent front from the starting point and the migration time (t):

$$z_f = kt$$

where k is a constant that depends on the rate of solvent flow, on the externally applied pressure, and on the size of the particles constituting the layer. In principle, k is constant throughout the development and independent of the rate of solvent migration. In OPLC, the parameter R_F

is also used to describe the position of the separated analyte and, in this case, the R_F values do not depend on the starting distance.

The parameters characterizing chromatographic systems in TLC, such as average plate height (H), reduced plate height (h), and theoretical plate number (N), are calculated in a similar way in OPLC, but, practically, they do not depend on either average particle diameter (d_p) or the start distance (s_0). In OPLC, the start distance has no influence on the efficiency of separation, and the average plate hight is nearly constant on a layer of exceptionally fine particles, even over a longer development distance. Thus, the major advantage of OPLC over other planar techniques lies in this fact. We can say that OPLC permits relatively large plate numbers to be obtained and can be applied more favorably in the case of smaller particles.

Basic Instrumentation for OPLC

As far as OPLC is concerned, the method, in principle, differs from conventional TLC in the design of the equipment that is used. The first attempts at construction of chromatographic pressure chambers were made in the beginnings of the 1960s. However, only at the end of the 1970s, Tyihák, Mincsovics, and Kalász were successful in construction of a well-operating OPLC chromatograph called Chrompress 10 (maximum pressure permitted in this chamber was 1.0 MPa) and, later, in the 1980s, Chrompress 25 (2.5 MPa) (Labor MIM, Hungary) and the most modern Personal OPLC BS-50 (OPLC-NIT, Budapest, Hungary) (5.0 MPa). In Poland, during 1980–1990, a pressure thin-layer chromatograph was constructed (Cobrabid, Warsaw); however, due to its narrow range of operating pressures (0.8 MPa), it was not widely used.

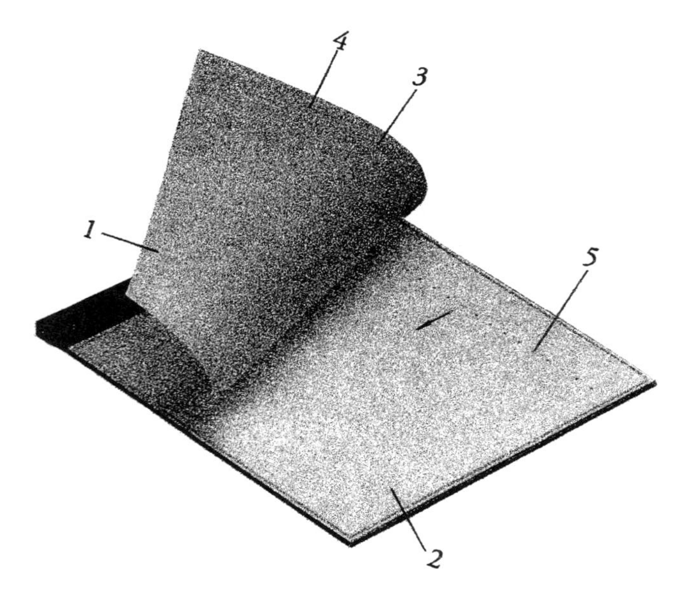

Fig. 1 Cassette of foil-backed layer for linear development: 1 = cover sheet, 2 = sorbent layer, 3 = eluent puncture, 4 = eluent trough, 5 = sample application site.

We will describe only the most up-to-date OPLC system. This fully automatic OPLC system allows separation of mixtures on an analytical and on a semipreparative scale. The fundamental separation process occurs on a chromatographic plate (constructed of glass or aluminum foil) with sorbent (Fig. 1) covered and compressed by a special polyethylene or Teflon foil, pressured by water. In this way, a flat, thin chromatographic column is created. This technique also requires a special chromatoplate which is sealed at the edges, which prevents the eluent from flowing off the chromatoplate in an unwanted direction. According to the technique of chromatogram development used (unidirectional, bidirectional, circular, on-line, off-line, parallel coupled multilayer, serial coupled multilayer), all four margins of a plate, three margins, two opposite margins can be impregnated, or they can be left uncoated.

In linear development of a chromatogram, unidirectional or bidirectional developments of the chromatogram are possible. Similarly, as in liquid column chromatography, there are possible, in this case, either on-line or off-line techniques of sample application, separation, and detection, as well as various modifications (e.g., partly off-line method). Bidirectional development can also be vertical. Using vertical bidimensional development, applying different eluents, components of complex, difficult mixtures can be separated. The separation of such mixtures is also possible by means of this technique using multiple automatic development of chromatogram.

In OPLC, the changes in composition of the eluent give good possibilities for special separation techniques such as isocratic and gradient separation. The choice of mobile phase can be effectively and quickly determined using the optimization model Prisma according to Nyiredy. This is a three-dimensional model that correlates the solvent strength and the proportion of eluent constituent, which determines the selectivity of the mobile phases according to Snyder's solvent classification.

The newest apparatus for overpressured layer chromatography, "Personal OPLC BS-50" manufactured in Hungary, is shown in Fig. 2. Generally speaking, it consists of the separation chamber and a liquid delivery system. The separation chamber contains the following units: (a) holding unit, (b) hydraulic unit, (c) troy layer cassette, and

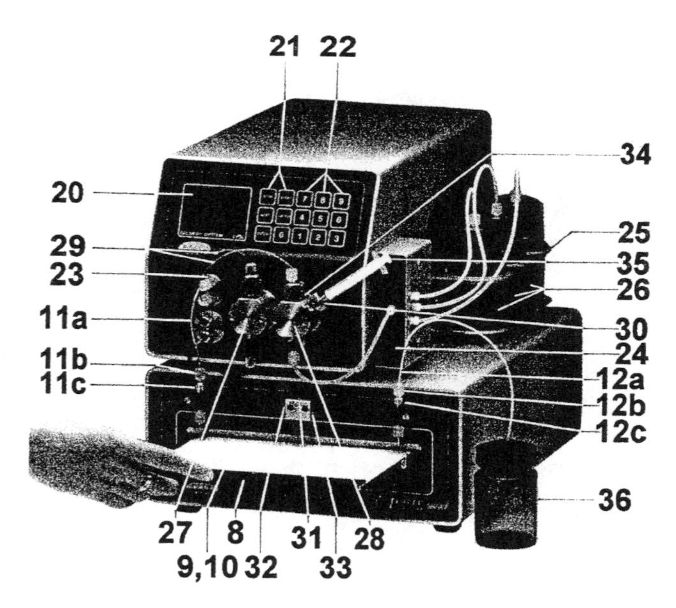

Fig. 2 Personal OPLC BS-50 apparatus: 9,10 = Teflon cover sheet of cassette, upper and lower, respectively; 11a = tube driving eluent from pressure gauge to chamber inlet connector; 11b = end connector for tube 11a; 11c = chamber–eluent–inlet connector; 12a = tube for eluent outlet of the chamber; 12b = end connector for tube 12a; 12c = chamber–eluent–outlet connector; 20 = display; 21 = function keys; 22 = numeric keys; 23 = pressure gauge for eluent; 24 = eluent switching valve; 25 = tank holder; 26 = eluent tanks A, B, and C; 27 = pump head for hydraulic liquid delivery; 28 = pump head for eluent delivery; 29 = hydraulic liquid and eluent connecting tubes; 30 = connecting stub for syringe to fill up eluent pump; 31 = middle hole of T distributor for fitting tubes in case of two-directional development; 32 = left hole of T distributor for fitting tubes in case of two-directional development; 33 = right hole of T distributor for fitting tubes in case of two-directional development; 34 = hole for piston rinsing against deposition; 35 = syringe.

(d) drain valve. The apparatus also has a pumping system for eluent delivery and for the hydraulic liquid delivery. The entire apparatus and the total chromatogram development process are controlled by a computer system. The apparatus and the method of chromatogram development is characterized by high reproducibility of results and chromatographic parameters. External pressure (5 MPa), eluent flow rate and its volume, and development time can be automatically programmed. The OPLC-BS-50 chamber works in off-line and on-line systems. Using this equipment, it is possible to separate 70–100 or even more samples at the same time, depending, of course, on the chosen technique of the OPLC process.

This OPLC method is charaterized by high precision for determination of retention parameters and reproducibility of results.

Advantages

- Separation of components which the former TLC techniques failed to achieve
- Smaller attainable plate height over longer migration distances
- Rapid separation and high resolution for industrial control
- Optimization of resolution as a function of solvent velocity, development distance, and temperature
- Possibility of using high-viscosity eluents and poorly wettable stationary phases
- Possibility of both quantitative evaluation and preparative applications
- Efficient separation of multicomponent samples
- Different development modes: unidirectional, bidimensional, continuous on-line, and off-line
- Long migration distances on fine-particle layers with short development times
- No air interactions
- Minute consumption of developing solvent
- Programmable operating system.

Some applications include analytical and preparative analyses in all types of biological, biochemical, pharmaceutical, clinical, forensic, food, and environmental laboratories.

Suggested Further Reading

Kaiser, R. E., *Einfuerung in die HPLC*, Huethig, Heidelberg, 1987.

Kaiser, R. E. and R. I. Rieder, in *Planar Chromatography, Vol. 1* (R. E. Kaiser, ed.), Huethig, Heidelberg, 1986, p. 165.

Mincsovics, E., K. Ferenczi-Fodor, and E. Tyihák, in *Handbook Thin-Layer Chromatography* (J. Sherma and B. Fried, eds.), Marcel Dekker, Inc., New York, 1996, p. 173.

Nyiredy, Sz., C. A. J. Erdelmeier, and O. Sticher, in *Proc. Int. Symp. TLC with Special Emphasis on Overpressured Layer Chromatography (OPLC)* (E. Tyihak, ed.), LABOR MIM, Budapest, 1986, p. 222.

Nyiredy, Sz., S. Y. Meszaros, K. Dallenbach-Toelke, K. Nyiredy-Mikita, and O. Sticher, *J. High Resolut. Chromatogr. Chromatogr. Commun. 10*: 352 (1987).

Różyło, T. K., R. Siembida, and E. Tyihak, *Biomed. Chromatogr. 13*: 1 (1999).

Ruoff, A. D. and J. C. Giddings, *J. Chromatogr. 3*: 438 (1960).

Tyihák, E., E. Mincsovics, and H. Kalász, *J. Chromatogr. 174*: 75 (1979).
Tyihák, E., E. Mincsovics, H. Kalász, and J. Nagy, *J. Chromatogr. 211*: 45 (1981).
Tyihák, E., E. Mincsovics, P. Tetenyi, I. Zambo, and H. Kalász, *Acta Horticult. 96*: 113 (1980).

Jan K. Różyło

P

Packed Capillary Liquid Chromatography

Introduction

Liquid chromatography (LC) was the first chromatographic mode to be developed in the beginning of the twentieth century. For almost 70 years, it was employed without major modifications until the end of the 1960s when an instrumental version of liquid chromatography was finally produced. Before this milestone, LC was performed mainly in large-bore glass tubing packed with large-diameter solid particles. To differentiate the instrumental version developed in the late 1960s from the noninstrumental, usually referred as the "classical" version, the former was named high-pressure liquid chromatography and, later, high-performance liquid chromatography (HPLC). Because HPLC used smaller particles as the stationary phase, the columns had to be packed at higher pressures in order to obtain a more stable bed required by the higher pressures used in these techniques. Altogether, HPLC offered a much higher efficiency (number of plates) than "conventional LC" and, as a consequence, higher resolution (separation power) as well. The standard HPLC columns used in the 1970s consisted of particles of 5–10 μm packed in stainless-steel tubing of 4.0–4.6 mm inner diameter (i.d.) and 15–25 cm long. Typical flow rates under these conditions are ~1–2 mL/min. In a typical quality control laboratory (8 h a day; 5 days a week, 20 days a month; 12 months a year), more than 100 L of chromatographic solvent are generated in a 1-year period. Most of these solvents are highly toxic to man and the environment, requiring special waste storage, transportation, and final disposal. As a consequence, a miniaturization of the HPLC techniques using less solvent became important immediately after its development.

A major step in the miniaturization of HPLC columns was done early in 1967 by Horváth and co-workers [1, 2], when investigating the parameters that influence the separation of nucleotides in a 1-mm-i.d. column. These columns were then named microbore columns. A further step in the miniaturization process was done in 1973, by Ishii and co-workers, by separating polynuclear aromatic hydrocarbons (PAHs) in a 0.5-mm-i.d. PTFE column. The term micro-LC was then introduced to differentiate this technique from HPLC, which uses larger-bore columns [3–5]. Shortly after, Scott and Kucera published several articles dealing with microbore (1-mm-i.d. columns) LC [6,7].

In spite of the fast development in its early days (late 1960s and early 1970s), the miniaturization of HPLC followed a slow progress until recently, with the development of LC–mass spectrometry (MS) using electrospray-type interfaces.

Capillary Liquid Chromatography

Capillary liquid chromatography (CLC) is a mode of HPLC that deals with columns having internal diameters equal to, or smaller than, 0.5 mm. This number is limited by the internal diameter of the fused-silica tubing commercially available, which is the most popular tubing used in this area. The CLC columns are usually 15–60 cm long, having internal diameters <0.5 mm and being either coated or packed with the stationary phase. Due to the small inner diameter of the CLC columns, this technique is more demanding in instrumentation than HPLC, particularly with respect to the solvent delivery, sample introduction, and detection systems.

Sample Introduction

Because the column inner diameter is small, the amount of stationary phase is also very small and, as a consequence, the amount of sample that can be introduced into the column without overloading is very small (typically a few nanoliters). In most cases, the preferred sample introduction system consists of an injection valve containing an internal loop smaller than 0.1 μL.

Pumping System

Because the eluent flow rate is relatively small (typically a few microliters per minute), the pump used to deliver it to the column is critical. There are two major approaches being used: pumps capable of delivering flow rates in the range of few microliters per minute (usually syringe-type pumps) or reciprocating pumps using a flow splitter. In both cases, reproducible flow rates are hard to obtain using commercially available pumps.

Detectors

Almost all detectors currently used in HPLC have been evaluated to be used in CLC. The major modification required, in most cases, is a decrease in the detector cell volume in order to accommodate the small sample volume without considerable peak broadening. Ultraviolet-visible (UV-vis), fluorescence, electrochemical, mass spectrometric, and several other detectors have been successfully used with CLC.

Columns

Capillary LC columns can be generally made in two different ways: wall-coated open tubular (WCOT) or packed columns. Coating the internal wall of the tubing (usually fused silica) with a thin film of a solvent-resistant polymer makes WCOT columns and is the same technology as used for capillary GC columns. Usually, cross-linked or immobilized phases are preferred in order to avoid stationary-phase removal by the eluent. The major drawback of these columns is that they have to be made with an internal diameter smaller than 20 μm in order to be highly efficient for complex separations, thus justifying their use instead of the packed capillary columns [8]. This places great demands on the instrumentation, the eluent quality, the sample preparation step, and so forth, thus making it impractical at this moment. As an alternative, the packed capillary columns using the technology already available to prepare HPLC columns is less instrument demanding and have been gaining more acceptance, every day becoming the preferred form of CLC.

Advantages of Capillary Liquid Chromatography

The advantages of CLC are consequences of its miniaturization [9]. Due to its miniaturized size, it requires much less stationary phase than does ordinary HPLC and, as a consequence, more expensive phases can be used to prepare the columns. This includes chiral phases, experimental new materials, expensive biocompounds, and so forth. In the same way, the amount of mobile phase is very small, thus leading to a savings in buying, storing, and discarding the solvent, allowing the use of expensive eluents such as deuterated solvents and chiral modifiers such as cyclodextrins, transition metals, and so forth. In many cases, the total amount of mobile phase in one separation is just 10 μL (1 μL/min; 10-min run); this explains why this technique is sometimes referred to as "one-drop chromatography." The amount of sample injected is also very small, so it becomes an important technique when the sample size is critical, such as in biomedical studies (brain, spine liquid, newborn tests, etc.), forensic chemistry (fire debris, explosives, blood residues), environmental analysis, and several other application fields.

Other advantages of CLC, when compared to HPLC, includes its higher permeability [10], chemical inertia, easier coupling to other separation and identification systems such as mass spectrometry, gas chromatography, and nuclear magnetic resonance, and the possibility of making longer columns, thus achieving more plates (efficiency) and resolution.

Figure 1 shows a chromatogram of a separation of PAHs using a packed capillary column. As can be veri-

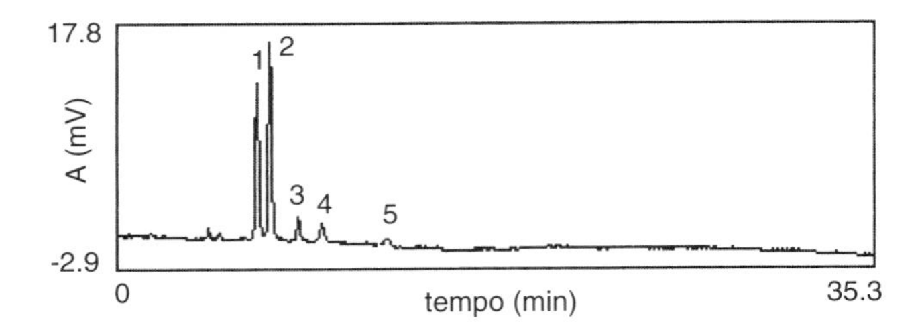

Fig. 1 Capillary LC chromatogram of a river water solid-phase extract containing PAHs. Column: fused silica (20 cm × 0.25 mm) H_2O home-packed with RP-18 (5 μm). Mobile phase: ACN/H2O (75:25), flow rate 4 μL/min, UV detection at 254 nm. Compound identity: 1 = fenanthrene, 2 = anthracene, 3 = fluoranthene, 4 = pyrene, 5 = crysene.

fied, a good separation is obtained with minute amounts of stationary phase, mobile phase, and sample.

Limitations of Capillary Liquid Chromatography

In spite of the several advantages over HPLC, CLC has not yet achieved its maturity as a separation technique to be used worldwide, particularly as a routine technique for quality control laboratories. Among the limitations still hindering the further development of CLC, one of the most critical ones is the very limited availability of commercial equipment dedicated to this technique. Even so, most systems are simple adaptations of parts already used for HPLC, by just decreasing their sizes and volumes without specifically having CLC in consideration. Therefore, in order to become a routine technique as its counterpart in gas chromatography, capillary liquid chromatography still has to have a broader interest for the instrument manufacturing companies in the technique before it will spread out beyond the academic environment. Those who have worked with packed columns in gas chromatography in the 1960s have already seen this same history.

Acknowledgments

Professor F. Lanças wishes to thank FAPESP (Fundação de Apoio à Pesquisa do Estado de São Paulo) and CNPq (Conselho Nacional de Desenvolvimento Científico e Tecnológico) for financial support to his laboratory.

References

1. C. G. Horváth, B. A Preiss, and S. R. Lipsky, *Anal. Chem. 39*: 1422 (1967).
2. C. G. Horváth and S. R. Lipsky, *Anal. Chem. 41*: 1227 (1969).
3. D. Ishii and K. Sakurai, in *Tokyo Conference of Applied Spectrometry*, 1973, Abstract 1B05, p. 73.
4. D. Ishii, K. Asai, K. Hibi, T. Jonokuchi, and M. Nagaya, *J. Chromatogr. 144*: 157 (1977).
5. D. Ishii, K. Asai, and T. Jonokuchi, *J. Chromatogr. 151*: 147 (1978).
6. R. P. W. Scott and P. Kucera, *J. Chromatogr. 125*: 251 (1976).
7. R. P. W. Scott and P. Kucera, *J. Chromatogr. 169*: 51 (1979).
8. H. Menet, P. Gareil, and R. Rosset, *Anal. Chem. 56*: 2990 (1984).
9. M. Verzele and C. Dewaele, *J. High Resolut. Chromatogr. 10*: 280 (1987).
10. D. Shelly, J. Gluckman, and M. Novotny, *Anal. Chem. 56*: 2990 (1984).

Fernando M. Lanças

Particle Size Determination by Gravitational FFF

Introduction

Particle Size Distribution: A Key Property of Particulate Samples

Particle size distribution analysis was considered, in one of the latest Pittsburgh Conferences, as one of the most outstanding trends in analytical science. This is not an overstatement, as most of the real samples of analytical interest occur either in dispersed form or in dispersed matrices. Just for argument's sake, in industrial applications the characterization of the size of sample particles is routine and is an essential part of the overall quality control procedures. In the medical field, for particles used to carrier drugs, the size is a critical performance factor (e.g., liposomes). In the food industry, the alcoholic yield from fermentation of starch, and even the taste of chocolate, depends on the size of particles of which these samples are composed.

Before discussing our method for determining particle size, it is necessary to briefly review the definition of size distribution. If all particles of a given system were spherical in shape, the only size parameter would be the diameter. In most real cases of irregular particles, however, the size is usually expressed in terms of a sphere *equivalent* to the particle with regard to some property. Particles of a dispersed system are never of either perfectly identical size or shape: A spread around the mean (*distribution*) is found. Such a spread is often described in terms of standard deviation. However, a frequency function, or its integrated (cumulative) distribution function, more properly defines not only the spread but also the shape of such a spread around the mean value. This is commonly referred to as the *particle size distribution* (PSD) profile of the dispersed sample.

An examination of technical literature and trade publications indicates that a wide variety of instruments are commercially available for PSD analysis [1]. The classical methods are based on either electrical properties (e.g., the Coulter Counter® principle) or optical properties (e.g., laser scattering) of the analyte. However, none of these techniques are separation methods. Because particulate dispersions are often highly complex in terms of the polydispersity index, multimodal size distribution, and density, it is hardly possible, without the use of separative methods, to obtain an accurate determination of their size distribution. Among separative chromatography-like methods, one can consider hydrodynamic chromatography and field-flow fractionation (FFF). The application to PSD of a subset of the latter family of methods is the topic of this article.

PSD by FFF

Field-flow fractionation is a broad family of liquid chromatographic-like techniques which have been shown, over more than 20 years, to be able to fractionate and characterize high-molecular-weight species in a size range spanning five orders of magnitude, from macromolecules to micron-size particles [2]. FFF has been demonstrated to be a rapid method for the determination of the mean diameters and polydispersities of particulate samples. When compared to standard methods for PSD analysis, the main advantage of FFF lies in the fact that FFF is a separation method which has some common features with liquid chromatography. The output from an FFF experiment is a function of the detector signal versus the retention time of the analyte. Whereas, in liquid chromatography, such an analytical response is referred to as the *chromatogram*, in FFF it is commonly defined as the *fractogram*. However, when it is compared to classical liquid chromatography, the existence of a direct relationship between retention and some physical properties of the analyte, such as the size, is a fundamental feature of most FFF techniques. The theory of FFF retention has been fully explained elsewhere, [2, and references therein] as well as in other entries of this encyclopedia. What is important to focus on here is that, in FFF, particle size determination of the analyte can be obtained by means of a direct numerical conversion of the retention scale, whereas the relative amount of separated analyte is, as in the case of chromatograms, in some way proportional to the signal intensity. The basic procedures for the two conversions is the topic of this entry.

Gravitational FFF: An Economical Device for PSD Analysis of Micron-Size Dispersions

Here, we treat the case of PSD analysis of particulate systems of micron-size range (i.e., with a size distribution extending above 1 μm). Since 1994, in our laboratories, this topic has been dealt with by means of a low-cost subset of sedimentation FFF (SdFFF), the gravitational field-flow fractionation (GrFFF) technique [3]. GrFFF had already been applied to the fractionation of a variety of micron-size dispersion, either inorganic as commercial chromatographic supports [4] or biological, as cells and parasites [5]. In no cases, however, had PSD been performed through GrFFF. As an SdFFF subset, GrFFF requires the application of a sedimentation field that, in this case, is simply Earth's gravity applied perpendicularly to a very thin, empty channel with a rectangular cross section. The big advantage of GrFFF, compared to other techniques for the characterization of particulate matter, lies in its very low cost ($\sim$\$50 for a homemade channel) and easy implementation in a standard high-pressure liquid chromatography (HPLC) system (the channel can simply replace the standard HPLC column). The GrFFF channel employed here can be easily built as described elsewhere [3–9]. It is basically a ribbonlike capillary channel which consists of two mirror-polished plates, of either glass or plastic material (e.g., polycarbonate) which are clamped together over a thin sheet of either Teflon or Mylar from which the channel volume has been removed. Simplicity and economy of use make it possible for laboratories that are not specialized in PSD analysis to perform dimensional characterization of supermicron particles dispersions with limited effort and cost.

We shall show here that GrFFF is capable of performing reliable, quantitative PSD analysis of particulate matter. Some basics of the overall procedure will be overviewed and the relevant questions presented. In fact, in order to obtain a PSD by means of GrFFF, the conversion of the retention time axis into the analyte size axis is necessary. For the same reason, the detector signal axis must be converted into mass (or concentration) of the fractionated analyte.

Procedure and Discussion

From a GrFFF Fractogram to a PSD

The diameter scale can be obtained from retention coordinates (i.e., the retention time axis) by applying to the fractogram the well-known, approximate expression that is valid for highly retained samples in GrFFF [3]:

$$d_i = \frac{wV_0}{3\gamma} \frac{1}{V_{r,i}} \tag{1}$$

where d_i (cm) is the diameter value corresponding to the ith data point of the fractogram, the retention volume of which is $V_{r,i}$ (cm^3). V_0 (cm^3) is the void volume of the GrFFF channel, w (cm) is the channel thickness and γ is the so-called hydrodynamic correction factor, the knowledge of which is, therefore, required for PSD analysis.

In practice, PSD curves can be obtained directly from the experimental, digitized peak (fractogram), $y_i(V_{r,i})$ once it is converted to a function of particle diameter d_i with the use of Eq. (1). The frequency function of particle size f_m is expressed as [3]

$$f_{m,i} = \frac{\delta m_i}{\delta d_i} = \frac{\delta m_i}{\delta V_{r,i}} \frac{\delta V_{r,i}}{\delta d_i} \tag{2}$$

where $\delta m_i/\delta V_{r,i}$ (g/cm^3) is the mass concentration of the analyte at the ith digitized point and $\delta V_{r,i}$ and δd_i are the differences in retention volume and particle diameter between the ith and the $(i-1)$th digitized points, respectively. The incremental quantity δd_i can be calculated for any given $\delta V_{r,i}$ by Eq. (1).

As far as the conversion of the analytical response y_i is concerned, the most used detectors in GrFFF have been, until now, conventional ultraviolet (UV) detectors commonly used for HPLC. With this type of detector, the amount of particles with diameter d_i is proportional to the detector response at the ith point. With particulate samples, in fact, because of UV detector optics, the response is a *turbidity* signal read within an angle between the incident light and the photosensor (i.e., usually smaller than ~10°) rather than the *absorbance*. This turbidity signal can be assumed to be directly proportional to the sum of all cross-sectional areas of the particulate sample components at any time. The validity of the above assumption, in the case of particles which are about 10-fold larger than the incident wavelength, is discussed elsewhere [6]. The mass frequency function can thus be expressed as [7]

$$f_{m,i} \propto (\text{UV signal})_i(d_i)\frac{\delta V_{r,i}}{\delta d_i} \tag{3}$$

For the reader's convenience, a scheme of the required conversion is represented in Fig. 1. It is evident that it is rather straightforward to derive a PSD from a GrFFF experiment [3]. An example of GrFFF–PSD analysis for a sample of silica particles commonly used as the stationary phase in HPLC (5 μm LiChrospher, Merck) is reported in Fig. 2a. For the sake of comparison, the PSD of the same sample obtained by laser diffraction is reported in Fig. 2b. We can observe that PSD resolution is higher in GrFFF than in laser diffraction, where just a histogram is obtained. On the other hand, accuracy is comparable when the experimental distribution moments (i.e., the percentiles indicated as d_{10}, d_{50}, d_{90}) are compared to the nominal values given by the manufacturer. We must point out that differences in distribution moment values as high as 10% are commonly reported when different, uncorrelated techniques for PSD studies are compared [1].

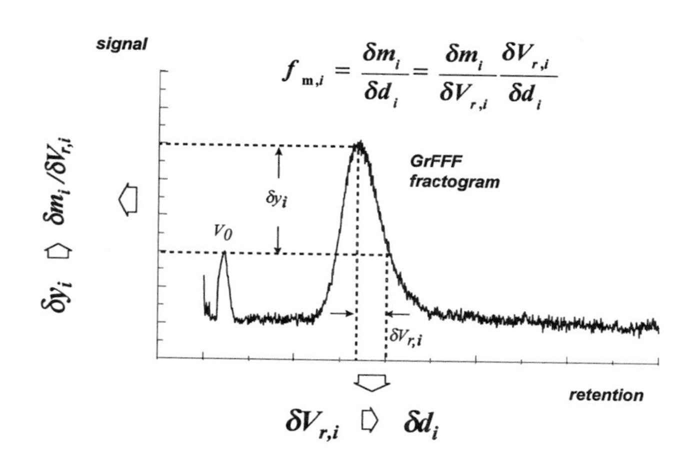

Fig. 1 Scheme for the conversion from a fractogram to a frequency function of particle size.

Quantitative Particle Size and Sample Amount Distribution in GrFFF

We have derived an original method by which quantitative particle size and sample amount distribution (PSAD) in GrFFF can be obtained by applying to Eq. (2) a derivation of the Lambert–Beer law in flow-through systems [7]. If compared to standard PSD, a PSAD thus represents a distribution of the *real* mass of the analyte as a function of size, rather than a functional expression only *proportional* to mass.

For particle dispersions in the micron-size interval, which is the typical application range of GrFFF, it has been demonstrated that the sample amount exiting the detector cell, $N_0(g)$, can be expressed as

$$\frac{\overline{A}F}{Kb} = N_0 \tag{4}$$

Where $\overline{A}$ (min) is the peak area, F (cm^3/min) is the flow rate, b (cm) is the cell thickness, and K (cm^2/g) is the total extinction coefficient of the particulate sample. If the extinction coefficient can be assumed to be approximately constant, as in the case of particles whose size is at least

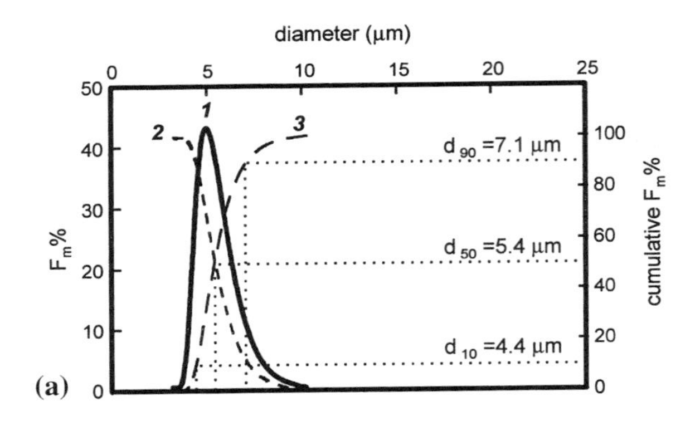

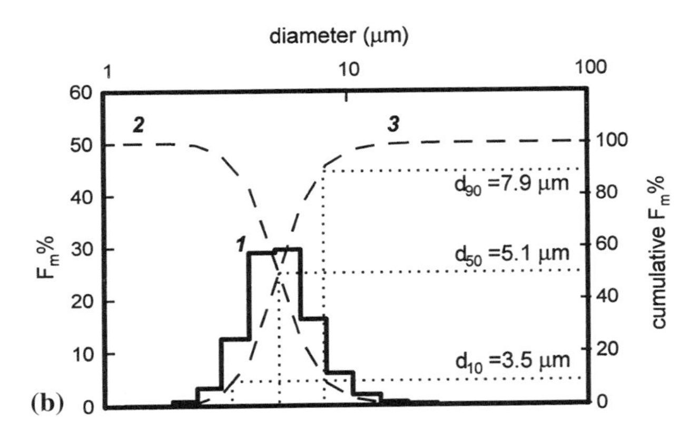

Fig. 2 Comparison between normalized, frequency functions of size (F_m %) of silica particles for HPLC packing (5 μm; LiChrospher, Merck, Darmstad); nominal distribution percentiles: $d_{10} = 3.7\ \mu$m, $d_{50} = 5.0\ \mu$m, $d_{90} = 6.8\ \mu$m); curve 1 (full line): F_m %; curve 2 (dashed line): F_{cum} %; curve 3 (dashed line): $100 - F_{cum}$ %. (a) GrFFF/PSD: Sample load: 100 μg; channel: $90 \times 2 \times 0.020$ cm; mobile phase: Milli-Q water/Triton X-100, 0.1% (v/v)/NaN_3 0.02% (w/v); flow rate: $1.010 \pm 0.004\ cm^3$/min; UV detection: 330 nm; experimental $\gamma = 0.70$. (b) Laser diffraction PSD (Malvern MASTERSIZER®, Malvern Instruments Ltd., UK).

10 times higher than the incident wavelength [6], the detector reading, expressed as "absorbance" at the ith point (A_i), is related to the real turbidity signal from the detector (τ_i) by the equation [7]

$$A_i = Kc_ib = \frac{\tau_i}{2.303}b \tag{5}$$

where c_i (g/cm^3) is the analyte concentration at the ith point of the fractogram; that is,

$$c_i = \frac{\delta m_i}{\delta V_{r,i}}$$

Therefore, the turbidity signal can be expressed as

$$\tau_i = 2.303K\frac{\delta m_i}{\delta V_{r,i}} \tag{6}$$

and the PSD expression for the frequency function, which is, in Eq. (3), just proportional to particle mass, can be transformed into a real function in mass directly from Eq. (6), thus giving [7]

$$f_{m,i} = \frac{\delta m_i}{\delta d_i} = \frac{\tau_i}{2.303K}\frac{\delta V_{r,i}}{\delta d_i} \tag{7}$$

Integration of Eq. (7) yields the cumulative distribution which gives, plotted as percent distribution, the size distribution percentiles (d_{10}, d_{50}, d_{90}). Moreover, once it is related to the injected sample amount, it gives the cumulative distribution of analyte mass as a function of size, with its asymptotic value giving the total sample recovery. Some examples of GrFFF/PSAD of silica samples used as HPLC column packing are reported in Ref. 7.

Direct Conversion of Retention to Size; Secondary Effects

As shown earlier, the direct conversion from retention time to particle diameter values [Eq. (1)] requires that the correction factor γ is predicted or experimentally estimated. It is known that, in GrFFF, γ can be influenced by either hydrodynamic or other effects as those due to the mobile phase [8] and the channel walls' nature [9]. All of these effects can influence particle size determination by GrFFF.

We have been developing an approach to the evaluation of the second-order effects, which act on GrFFF retention, and to the prediction of the correction factor γ. Among these effects, prominent are those due to hydrodynamic forces which lift the analyte particles away from the accumulation wall during their elution. GrFFF really shows a significant dependence of retention (and, thus, of the parameter γ) on the flow rate: The higher the flow rate, the higher the lift and, therefore, the lower the retention. In order to evaluate particle lift and, thus, particle retention, the semiempirical model given by Williams et al. has been applied [10–12]. This model is known to predict particle elevation from the accumulation wall in sedimentation field-flow fractionation (SdFFF), of which GrFFF it is just a subset, as noted earlier. A description of the hydrodynamic and other secondary effects on GrFFF retention is far above the introductory nature of this entry. However, just to introduce the reader to the possibility of obtaining a direct conversion of retention to size by predicting γ, the above-mentioned model can be used for relating retention volume to particle mean elevation during elution as follows [10]:

$$V_{r,i} = V_0\left[6f\left(\frac{2\delta}{d_i}\right)\frac{x_i}{w}\left(1 - \frac{x_i}{w}\right)\right]^{-1} \tag{8}$$

where x_i (cm) is the distance of the center of the particles from the accumulation wall ($x_i = \delta + d_i/2$) and $f(2\delta/d_i)$ is an empirical function. It was shown that under optimized experimental conditions, in a properly designed GrFFF system, a balance between secondary effects of forces other than hydrodynamic forces can give negligible effects [9]. In this way, particle elevation is predictable. In this case, also, the value of γ can be estimated, thus allowing for the direct conversion of retention time to analyte size without previous calibrations (*standardless*). This possibility of calculating γ is a task still in progress and it will open more promising uses of GrFFF for dimensional analysis of suspended particulate matter, because PSD can be obtained in the "single-run" mode (i.e., without previous calibration). This could be a significant enhancement in the future evolution of GrFFF/PSD. In fact, in the GrFFF/PSD example in Fig. 2a, the conversion from retention to size was performed only by means of an experimental evaluation of the parameter γ, with a calibration plot formerly obtained with standards [3].

Acknowledgments

Giancarlo Torsi, Dora Melucci, Andrea Zattoni, and Gabriele Berardi of the Department of Chemistry "G. Ciamician," Bologna, Italy, are duly acknowledged.

References

1. H. G. Barth (ed.), *Modern Methods of Particle Size Analysis*, John Wiley & Sons, New York, 1984.
2. J. C. Giddings, *Science 260*: 1456 (1993).
3. P. Reschiglian and G. Torsi, *Chromatographia 40*: 467 (1995).
4. J. Pazourek and J. Chmelík, *J. Microcol. Separ. 9*: 611 (1997).
5. A. Bernard, B. Paulet, V. Colin, and Ph. J. P. Cardot, *Trends Anal. Chem. 14*: 266 (1995).
6. P. Reschiglian, D. Melucci, and G. Torsi, *Chromatographia 44*: 172 (1997).
7. P. Reschiglian, D. Melucci, A. Zattoni, and G. Torsi, *J. Microcol. Separ. 9*: 545 (1997).
8. P. Reschiglian, D. Melucci, and G. Torsi, *J. Chromatogr. A 740*: 245 (1996).
9. D. Melucci, G. Gianni, A. Torsi, A. Zattoni, and P. Reschiglian, *J. Liquid Chromatogr. Related Technol. 20*: 2615 (1997).
10. P. S. Williams, T. Koch, and J. C. Giddings, *Chem. Eng. Commun. 111*: 121 (1992).
11. P. S. Williams, S. Lee, and J. C. Giddings, *Chem. Eng. Commun. 130*: 143 (1994).
12. P. S. Williams, M. H. Moon, Y. Xu, and J. C. Giddings, *Chem. Eng. Sci. 51*: 4477 (1996).

Pierluigi Reschiglian

Peak Skimming for Overlapping Peaks

Any discussion of quantitating overlapped peaks should be prefaced by stating that baseline resolution of peaks is the only means of absolutely assuring the accuracy of their integration. All deconvolution methods involve assumptions that can affect accuracy. These methods are particularly inaccurate in cases of small peaks eluting on the tails of much larger peaks where percent errors are measured by two to three orders of magnitude [1,2]. In any case, the accuracy of all baseline methods generally increases with increasing resolution between the overlapping pair of peaks.

This discussion is confined to single-channel chromatographic analyses such as high-performance liquid chromatography (HPLC) with single-wavelength ultraviolet (UV) detection or gas chomatography with flame ionization or thermal conductivity detection. Three-dimensional (multichannel) techniques such as photodiode-array, multiwavelength detection may allow deconvolution based on component characteristics such as absorptivity at multiple wavelengths rather than peak shape. These techniques and/or the required information about the components are not always available, however, and many routine analyses are still conducted with single-channel detection because of their lower cost and relative simplicity.

In the overlap region between two unresolved chromatographic peaks, the detector response is a function of the response due the first peak and the response due to the second peak. For a single-channel detector, this is mathematically equivalent to a single equation with two unknowns, which is impossible to solve. Detection and integration schemes, in such cases, must make assumptions regarding the shapes of the chromatographic peaks in order to attribute the response in the overlap region to one peak or the other. In some cases, the assumption is based on the ease of determining the baseline and, in other cases, on theoretical models of chromatographic peaks such as Gaussian or exponentially modified Gaussian distributions.

Chromatographic data analysis systems generally employ three methods for determining baselines in overlapping peaks: *perpendicular drop*, *linear tangential skim*, and *exponential skim* (see Fig. 1). In order to choose the most appropriate method, the analyst must understand the assumptions and weaknesses of the three methods.

The perpendicular drop method produces accurate peak areas for symmetrical overlapped peaks of similar height and width. In this case, the portion of the second peak attributed to the first peak is offset by the portion of the first peak attributed to the second peak and vice versa. For overlapped peaks of significantly different size, the perpendicular drop method always overestimates the peak area of the smaller peak as the smaller peak gains more area from the larger peak than it "loses" to the larger peak. Quantitation errors are further exacerbated in the case of a smaller peak imposed on the tail of a much larger tailing peak. This method tends to show little injection-to-injection variability because of its simplicity.

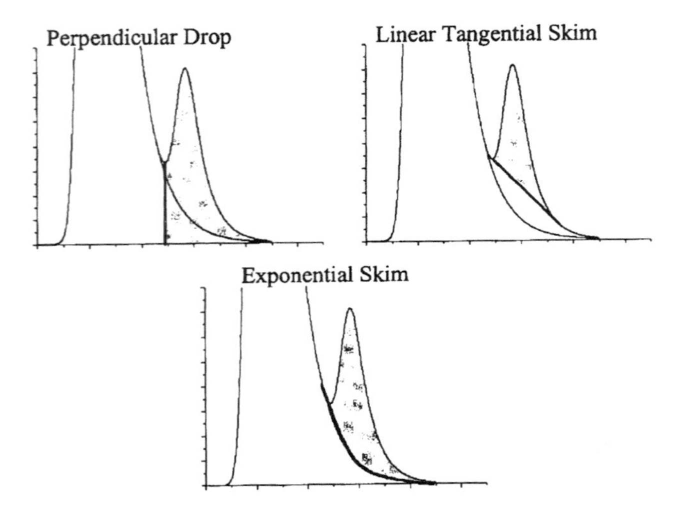

Fig. 1 Common methods for determining the baselines of overlapping peaks.

Linear tangential skimming, on the other hand, consistently underestimates the area of the smaller peak by neglecting that portion of the peak area under the tangential baseline. As seen in Fig. 1, it is easy to visually underestimate the size of a peak that is unresolved from a significantly larger one, giving the appearance that the linear tangential skim method is much more accurate than it is. The relative simplicity of this method makes it rugged as well.

Many commercial software packages offer a nonlinear exponential skimming method as well. As it has been estimated that 90% of all chromatographic peaks can be modeled by an exponentially modified Gaussian function (EMG) [3], this method should be the most accurate for tailing overlapped peaks. The algorithms for calculating such baselines are considerably more complicated, however, and the reproducibility of the technique may be problematic for some separations. The calculation of necessary parameters may limit its usefulness in extreme cases of tailing and peak width. In addition, the ability of exponential skimming to mimic symmetrical (Gaussian) peaks may also be an issue in some software. Examination of the constructed baselines is crucial in determining the proper use of this technique.

The peak area errors for the two most studied deconvolution methods (i.e., perpendicular drop and linear tangential skim) are dependent on a complex combination of resolution, relative peak width, relative peak height, and asymmetry ratio [1]. Exponential skimming assumes that the tailing of the first peak can be described by an exponential decay and that the peaks are sufficiently resolved to determine the decay parameters. Nonetheless, some broad generalizations can be made:

1. For symmetrical overlapped peaks of equal size and width, the perpendicular drop method provides accurate peak areas. Peak heights may still be overestimated.
2. In general, exponential skimming should be used when the first peak tails significantly or when the peak widths are significantly different. Care should be given that the area calculated by this method does not exceed the area by the perpendicular drop method, which already overestimates peak area.
3. For cases when exponential skimming is not possible or appropriate, due to extremely poor resolution or tailing, the tangential skim method should be used when the peak widths of the unresolved peaks vary significantly (>2:1) [2].
4. Quantitation by peak height should also be considered. It has been shown that peak height is

more accurate in cases where the sizes of the overlapping peaks vary widely (>100:1). Quantitation is also generally more accurate for peak pairs of disparate size if the smaller peak elutes first [1].

Although only doublets have been considered here, the arguments can be extended to multiplets by considering the effects on the front and tail of surrounded peaks.

References

1. V. R. Meyer, *Chromatographia 40*(1): 15–22 (1995).
2. J. P. Foley, *Anal. Chem. 59*: 1984–1987 (1987).
3. A. N. Papas and T. P. Tougas, *Anal. Chem. 62*: 234–239 (1990).

Wes Schafer

Pellicular Supports for HPLC

In current practice, high-performance liquid chromatography (HPLC) supports consist of siliceous or polymer-based materials having pore sizes in the range of 80–120 Å for chromatography of small molecules and of 300–1000 Å for large biological molecules. Diffusional resistances in the stagnant mobile phase in the retentive material can have a significant influence on the efficiency of separation, particularly for large molecules of biological origin. Furthermore, pore size distribution of many mesoporous particles is rather wide and poor mass recovery is frequently encountered due to entrapment of macromolecules in the porous interior. One approach to minimize mass-transfer resistance in a stagnant mobile phase employs specially designed particles with a bimodal network of pores. The larger pores (≥1000 Å) facilitate convective transport of the mobile phase inside the particles, whereas the small pores (≤500 Å) are explored by the sample components by diffusion only and provide the necessary surface area for adequate sorption capacity.

Another approach consists in eliminating the pore structure, using pellicular column packing materials. These HPLC supports are mechanically stable, fluid-impervious microspheres with a thin retentive layer on the surface [1,2]. Such a stationary-phase configuration facilitates the interactions of the analytes with the active moieties of the stationary phase, which are completely exposed to the mobile-phase steam in the interstitial space of the column packing material. Because the diffusional path length in the retentive surface layer of the pellicular stationary phase is very short, the plate height contribution to the C term in the van Deemter equation is relatively small. It should be recalled that the C term in the van Deemter equation estimates the contribution to the plate height of the resistance to mass transfer. As a result, the absence of intraparticulate diffusional resistance and the fast mass transfer between the stationary and the mobile phases due to the small particle size allows high column efficiency, even at relatively high flow rates. In addition, the lack of internal pore structure eliminates the undesirable steric effects encountered in HPLC of large biological molecules with mesoporous stationary phases, resulting in good sample recovery.

With porous materials, the largest contribution to the total surface area is due to the area contained within particles; it is related to the pore volume, surface area, and pore diameter. As a result, there is little variation in the total surface area of packing materials with different diameters, but identical pore size. On the other hand, with pellicular particles, the total surface area within a column is a function of particle size. Consequently, stationary phases having the pellicular configuration have an adsorption capacity lower than the conventional mesoporous particles. This appears to be particularly so with micropellicular materials in HPLC of small molecules. With large molecules such as proteins, the loading capacity of columns packed with traditional mesoporous particles may be only three or four times higher than that of columns packed with the micropellicular stationary phases [3]. However, this somewhat low loading capacity is still adequate in the analytical mode and is more than compensated for by the high analytical speed and efficiency obtained with pellicular stationary phases.

Conventionally, the phase ratio is the volume of the active moieties of the stationary phase divided by the volume of the mobile phase in the column. From the particular configuration of pellicular stationary phases described earlier, it follows that they have a relatively low phase ratio with respect to that of the conventional packing materials employed in HPLC.

A variety of micropellicular packing materials has been

developed for the analysis of both small and large molecules by various HPLC modes, including ion exchange (IEC), metal interaction (MIC), reversed phase (RPC) [4], and affinity chromatography (AC) [5]. Besides analytical applications, other possible utilization of micropellicular stationary phases includes fundamental kinetic and thermodynamic studies of the retention mechanisms on a well-defined surface. Nevertheless, a relatively limited variety of micropellicular columns are commercially available. They are mainly restricted to ion-exchange and reversed-phase stationary phases. This may reflect certain practical disadvantages of micropellicular sorbents.

Columns packed with pellicular stationary phases of small particle size have low permeability and, therefore, cannot be operated at relatively high flow rates due to the pressure limitation of commercial HPLC instruments. However, due to the nonporous structure, micropellicular particles are generally more stable at higher temperature than conventional porous materials. Consequently, in the absence of limitation due to the thermal stability of either the sample or the bonded stationary phase, a column packed with nonporous materials can be operated at elevated temperatures with practical advantages from column permeability and sorption kinetics. This appears to be particularly so with several micropellicular materials employed in reversed-phase HPLC. By increasing the column temperature, the viscosity of the mobile phase decreases with concomitant increases in column permeability, which allows operating the column at high flow rates. In addition, with increasing temperature, sample diffusivity also increases and, in many cases, the sorption kinetic improves.

The elevated mechanical and thermal stabilities of pellicular stationary phases having a solid, fluid-impervious core have favored the development of nonporous particles tailored for the rapid HPLC analysis of peptides, proteins, and other biopolymers. Most of these applications are in RPC.

Although governed by the same separation mechanism, RPC of proteins differs significantly from that of small molecules because of high molecular weight and complex tertiary structure of these macromolecules. A large molecular size is associated with low diffusivity and molecular complexity is associated with multipoint interactions of proteins with the hydrophobic stationary phase, which complicate the dynamics of the separation process. In addition, the elution of certain proteins may be further complicated by conformational changes associated with the retention mechanism. Hence, RPC of proteins and other complex macromolecules is generally performed under the gradient elution mode, which reduces analysis time and improves the performance of separation. In the reversed-phase gradient elution mode, proteins are believed to be retained at the column inlet until, at some point in the gradient, corresponding to the proper mobile-phase composition, they are desorbed completely. Proteins then move through the column without apparent further interaction with the hydrophobic stationary phase.

In most cases, RPC of peptides and proteins are performed under denaturing conditions using hydro-organic mobile phases containing acidic additives such as trifluoroacetic or phosphoric acid. Because the elution strength of the mobile phase increases by decreasing its polarity, the appropriate gradient is produced by increasing the concentration of the organic solvent in the hydro-organic mobile phase during the chromatographic run. The variation of the eluent composition as a function of time gives the gradient shape, which is steep in the case of fast analysis performed with micropellicular reversed phase (i.e., high rate of change of mobile-phase composition during gradient elution).

Steep gradients require high flow rates in order to maintain satisfactory differences in retention of the analytes. According to Snyder and Standalius [6], the influence of the gradient parameters on retention is given by

$$k = \frac{t_g F}{1.15 \Delta\phi V_m S} \tag{1}$$

where k is the effective capacity factor for gradient elution, which is the value of the corresponding capacity factor under isocratic conditions (k') for the peak when it reaches the column midpoint, t_g is the gradient time (time from the beginning to the end of the gradient), F is the flow rate, $\Delta\phi$ is the variation of the fraction of organic solvent in the mobile phase during gradient, V_m is the column void volume, and S is the slope of the plot of the logarithmic retention factor (k') against mobile-phase composition under isocratic conditions. The value of the parameter S is related to the magnitude of the hydrophobic contact area and the number of the interaction sites established between the solutes and the stationary phase during the separation process. In comparison to small molecules, biopolymers have a larger contact area with the stationary-phase surface and, as a result, the retention is very sensitive to the content of the organic solvent in the hydro-organic mobile phase (i.e., the value of the parameter S is relatively large).

Equation (1) clearly shows that in order to maintain K constant while decreasing t_g, the flow rate must be proportionally increased. Because of the favorable mass-transfer properties of pellicular stationary phases, columns packed with such material can be employed at a high flow rate without loss in resolution. Moreover, pellicular column packings have a negligible intraparticulate void volume and, after the gradient, can be rapidly reequilibrated to the starting conditions. Finally, the thermal stability generally exhibited by these stationary phases allows running the gradient at relatively elevated temperature with the beneficial effect of reducing the viscosity of the mobile phase, which reflects on reducing the column inlet pressure. In addition, temperature influences the retention behavior of the analytes. Generally, the chromatographic retention decreases with increasing temperature, with concomitant improvement in column efficiency due to increased solute diffusivity and faster mass transfer. However, with proteins, the retention may increase and efficiency decrease when temperature promotes further unfolding of the protein.

Pellicular packings may also consist of a fully functionalized layer encapsulating solid particles, where there is no physical attachment of the active layer to the core particles [7] or as colloidal particles bearing charged moieties (latex) electrostatically bound to a solid core functionalized with groups of opposite charge [8] (Fig. 1). Most of the anion-exchange columns in use today for either carbohydrate or ion chromatography are packed with electrostatically latex-coated pellicular ion exchangers. These sorbents consist of three parts: (a) a highly cross-linked polymeric nonporous core, (b) a sulfonated layer at the outer surface of the solid core, and (c) a monolayer coating of anion-exchange latex particles functionalized with quaternary ammonium compounds. The solid support and the latex particles are manufactured separately and brought together after independent quality control performed to remove particle agglomerates. This allows easy control of phase thickness, which results from the monodisperse latex particles, attached to the solid core by electrostatic forces. As a result, loading capacity increases with increasing the diameter of the latex particles, which generally ranges from 50 to 500 μm. For a 250-mm × 4-mm-inner diameter column, the corresponding loading capacity ranges from 5 to 150 μEq/column.

The highly cross-linked polymeric nonporous core may consist of either a polystyrene–divinylbenzene or an ethylvinylbenzene–divinylbenzene substrate. The latex coatings are generally made from vinylbenzenylchloride polymer cross linked with divinylbenzene and fully functionalized with an appropriate quaternary amine for introducing anion-exchange properties.

Pellicular anion-exchange sorbents may also consist of quaternized latex hydrophobically coated onto the surface of an unsulfonated polystyrene solid core. However, using hydro-organic mobile phases can easily wash off the latex particles held onto the particle surface by hydrophobic interactions.

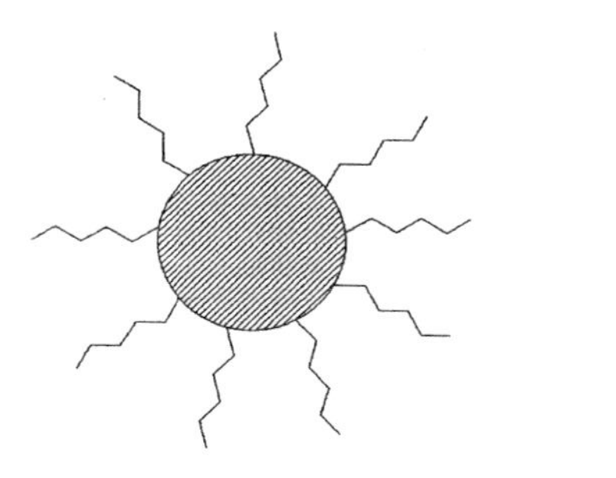

(a) Chemically coated

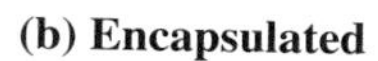

(b) Encapsulated

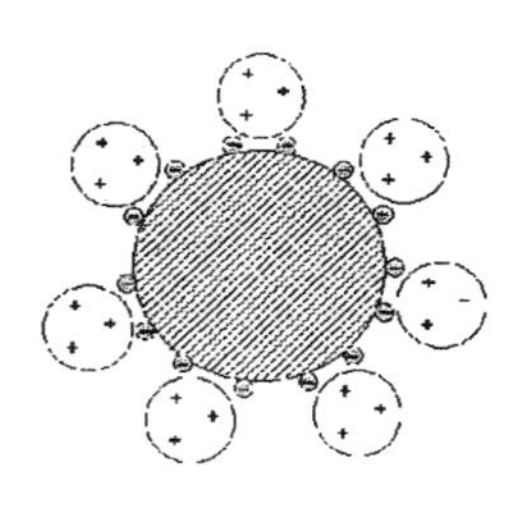

(c) Latex coated

Fig. 1 Schematic representation of pellicular stationary phases consisting of covalently bonded functional groups at the surface (a), functionalized layer-encapsulating solid particles (b), and functionalized latex particles electrostatically bound to a solid core bearing groups of opposite charge (c).

References

1. Cs. Horváth, B. A. Preiss, and S. R. Lipsky, *Anal. Chem. 39*: 1422–1428 (1967).
2. K. K., Unger, G. Jilge, J. K. Kinkel, and M. T. W. Hearn, *J. Chromatogr. 359*: 61–72 (1986).
3. L. Varady, K. Kalghatgi, and Cs. Horváth, *J. Chromatogr. 458*, 207–215 (1988).
4. K. Kalghatgi, *J. Chromatogr. 499*: 267–278 (1990).
5. Q. M. Mao, A. Johnston, I. G. Prince, and M. T. W. Hearn, *J. Chromatogr. 548*: 147–163 (1991).
6. L. R. Snyder and M. A. Standalius, in *High-Performance Liquid Chromatography: Advances and Perspectives* (Cs. Horváth, ed.), Academic Press, New York, 1984, p. 195–309.
7. H. Giddings, U.S. Patent 3,488,922 (1970).
8. H. Small, T. S. Stevens, and W. C. Bauman, *Anal. Chem. 47*: 1801–1809 (1975).

Danilo Corradini

Peptide Analysis by HPLC

Introduction

The rapid advancement in peptide research over the past 25 years must be attributed, in part, to the effectiveness of high-performance liquid chromatography (HPLC), particularly reversed-phase chromatography, in the separation and analysis of peptides. The resolution and selectivity of this technique allows peptides to be effectively isolated and purified from closely related substances. It also separates most or all of the components of complex biological mixtures such as tryptic digests of proteins.

Modes of HPLC

Peptides, which are composed of amino acids linked by amide bonds, are often found in random coil to semi-defined conformations, depending on their lengths and structures. As such, most of the composite amino acids are available to interact with the bonded phase of an HPLC support. Although the variety of amino acid characteristics, such as charge, polarity, and hydrophobicity, would suggest that multiple modes of HPLC would be effective for separation, reversed-phase chromatography has shown the ultimate success in selectivity and resolution of peptides. Hydrophilic interaction chromatography is a good alternative for hydrophilic peptides or other mixtures that are not separated well by reversed-phase chromatography, whereas cation-exchange chromatography can be effective for highly cationic species. Size-exclusion chromatography is a difficult technique for analysis of peptides, due to their varying solubilities and high degrees of hydrophobicity and charge; however, it is invaluable for resolution from dimers and aggregates.

Reversed-Phase Chromatography

Reversed-phase chromatography (RPC) is a method in which molecules are bound hydrophobically to nonpolar ligands in the presence of a polar solvent. Solutes are generally bound in an acidic mobile phase with elution occurring during a gradient to an organic solvent. Molecules will tend to be unfolded due to the combination of acidic and organic mobile phases and the hydrophobic bonded phase. Consequently, binding involves most of the amino acids, depending on the conformation. RPC can differentiate peptides which vary in the positions of their amino acids, as well as in their identities, thus making it a powerful analysis tool. Very high selectivity is attained even for complex biological mixtures, such as the tryptic digest of human growth hormone shown in Fig. 1.

There are many ligands used for RPC, but the most popular for peptide analysis are octadecyl (C_{18}) and octyl (C_8) chains. Little difference in selectivity for peptides is observed with ligand chain-length variation; however, mass recovery of very hydrophobic species may be enhanced on the shorter chains. Many RPC bonded phases are synthesized by silane bonding, with or without end-capping, which results in a hydrophobic layer on a silanol surface. Recently, more hydrophilic–hydrophobic phases have been developed whereby amide linkages are embedded in the nonpolar layer to provide a wettable area which may be more compatible with some biological molecules. Due to their higher efficiencies, silica-based supports are generally used for peptide analysis; however, polymeric supports are effective for operation at high pH. The silica matrix is sometimes a factor in retention because silanols on reversed-phase supports are rarely totally eliminated. Most peptide analyses use supports with

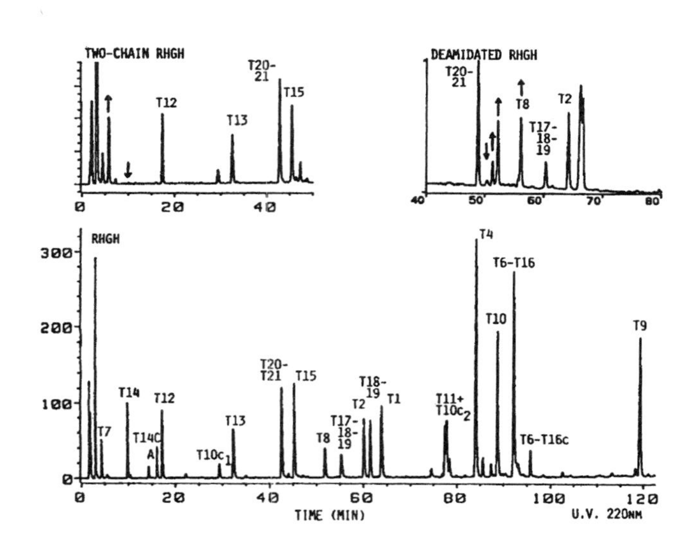

Fig. 1 Tryptic maps of the intact and (insets) degraded forms of recombinant human growth hormone (RHGH). The separation was achieved on a Nucleosil C-18 column (150 × 3.9 mm inner diameter) with a 120-min mobile-phase gradient from 10 m*M* potassium phosphate (pH 2.85) to 60% acetonitrile in the starting buffer. The flow rate was 1 mL/min and detection was at 214 nm. [Reprinted from J. Frenz, W.S. Hancock, W. J. Henzel and Cs. Horvath, in *HPLC of Biological Macromolecules: Methods and Applications* (K. M. Gooding and F. E. Regnier, eds.), Marcel Dekker, Inc., New York, 1990, p. 145.]

pore diameters of at least 300 Å to allow access to the bonded phase; nonporous supports offer a high-resolution option with significantly lower capacities. Dynamic loading capacities of peptides on porous analytical reversed-phase columns (250 × 4.6 mm inner diameter) are usually 100–500 μg.

Due to the multiple functional groups found in peptides, selectivity can be vastly changed by operational factors. Variable mobile-phase parameters include organic modifiers, pH, ion-pair agents, and gradient rates. Acetonitrile is the most popular organic solvent for RPC of peptides due to its transparency at low wavelengths (<210 nm) and its tendency to yield narrow peak widths. The strength of the organic solvent which causes elution increases from methanol to acetonitrile to isopropanol. Acidic pH is generally utilized to minimize silanol interactions; however, distinct pH conditions yield different selectivities. Ion-pairing agents, such as trifluoroacetic acid, are often added to the mobile phase to change the ionic or hydrophobic properties of either or both of the solute and the bonded phase. Amounts of 0.1% added to both the aqueous and the organic mobile phases are usually adequate, but optimal concentrations can be determined experimentally. Trifluoroacetic acid (TFA) is the most popular ion-pairing agent for peptides, imparting some additional hydrophobicity to amine groups and neutralizing cationic charges.

Most RPC peptide separations utilize gradient rather than isocratic conditions. Small peptides may be resolved well isocratically; however, due to multipoint interactions, larger ones yield broad peaks and tailing unless gradients are used. Shallow gradients generally improve resolution in an analytical mode or purity in preparative applications. For complex mixtures, 0.5–1% organic/min is usually a satisfactory rate. Increased temperatures can be implemented to reduce both retention times and the pressure generated by small-particle-diameter supports.

Hydrophilic Interaction Chromatography

Hydrophilic interaction chromatography (HILIC) is a variation of normal-phase chromatography in which solutes are retained on a polar bonded phase under high concentrations (80–90%) of organic solvent and released during a gradient to a more aqueous solvent. If the polar bonded phase contains a charge, a salt gradient can also be implemented, yielding separation by both charge and polarity. HILIC is most effective for polar peptides which are often poorly retained by RPC. Figure 2 compares the differences in selectivity and resolution of a cation-exchange (CEX) procedure on a HILIC column with that of RPC for a peptide mixture.

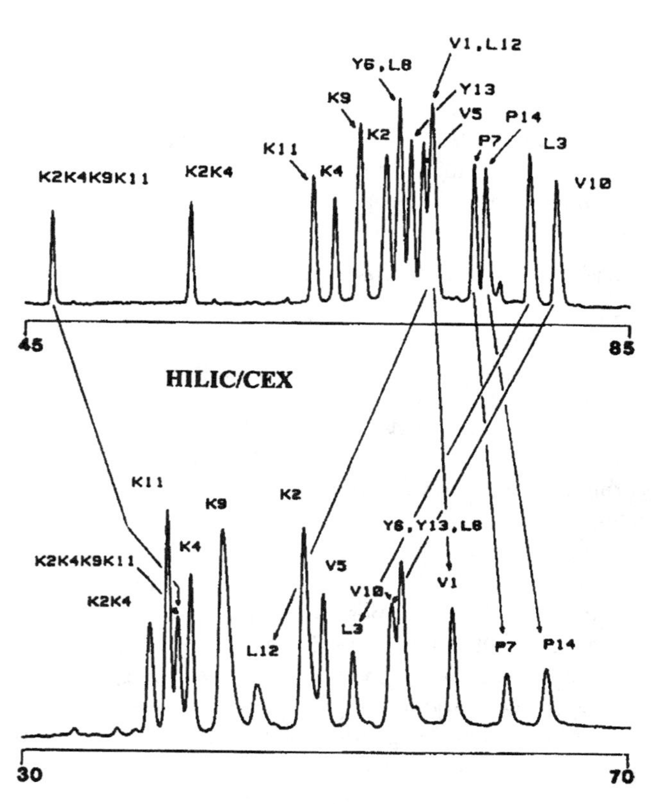

Fig. 2 Comparison of RPLC and HILIC/CEX elution profiles of cyclic peptides. RPC: Zorbax 300XDB-C_8 column; linear gradient (0.5% acetonitrile/min) from 0.05% aqueous trifluoroacetic acid (TFA) to 0.05% TFA in acetonitrile at 1 mL/min; at 70°C and detection at 210 nm. HILIC/CEX: Polysulfoethyl A column; 5 min isocratic elution with buffer A (20 m*M* aqueous triethylammonium phosphate with 90% acetonitrile), followed by linear gradient (2.5 m*M* sodium perchlorate/min) from buffer A to buffer B (buffer A containing 400 m*M* sodium perchlorate with 80% acetonitrile); at 30°C and detection at 210 nm. [Reprinted from C. T. Mant, L. H. Kondejewski, and R. S. Hodges, *J. Chromatogr.* 816 (1998), p. 79, with permission of Elsevier Science.]

The HILIC bonded phases are hydrophilic, consisting of amide and/or polyhydroxy functionalities. Pore diameters are at least 300 Å to allow penetration of peptides. Supports can be based on either silica or polymer because the matrix is not exposed to the solutes.

Mobile-phase manipulation causes variations in selectivity in HILIC. Although a buffer is used for pH control, it may also serve to change the hydrophobicity of the solutes by ion-pairing. More hydrophobic ions such as acetate will decrease retention. The pH is a less important factor than the identity of the salt in HILIC selectivity. The organic modifier and its initial concentration directly affect retention and resolution. At least 80–90% organic is usually required to achieve adequate retention of hydrophilic peptides. For HILIC/cation exchange (HILIC/CEX), a salt gradient is implemented to separate the hydrophilic peptides by charge, as can be seen in Fig. 2.

Ion-Exchange Chromatography

Ion-exchange chromatography (IEC) separates peptides by ionic attraction of their composite amino acids to charges on the stationary phase. Selectivity is dependent on the number and the identity of the amino acids, as well as their spatial arrangement. A peptide with charges grouped together will bind differently than one whose charges are dispersed. IEC has been most successful for cationic peptides that cannot always be analyzed effectively by RPC.

Supports for IEC possess either anion- or cation-exchange functionalities which are positively or negatively charged, respectively. They are also classified as weak or strong to correspond to their titration curves, similar to acid and base designations. Generally, strong cation-exchange supports are used for peptide analysis. The pore diameter is important in that it must be large enough to allow access of the peptides so that they are effectively retained and have optimum loading capacities.

In IEC, solutes are bound in a low-ionic-strength buffer ($0.02M$–$0.05M$) at an appropriate pH (often 1–2 pH units from the pI). Elution occurs when the ionic strength is increased during a concentration gradient of a salt in the same buffer. Isocratic elution may successfully separate small peptides with one or two charged groups, but larger peptides with more charged groups will require gradients for good peak shapes, resolution, and reproducibility. The nature of the salt in IEC has a major effect on selectivity due to interaction of the composite ions with either the stationary phase or the solutes. The hydrophobicity of many peptides may necessitate the addition of 5–10% organic solvent to improve peak shape. This can usually be included in the mobile phase without deleterious effects on the separation.

Size-Exclusion Chromatography

Size-exclusion chromatography (SEC) is a method in which molecules are separated by size due to differential permeation into a porous support. It requires complete solubility of the analytes in the mobile phase and elimination of all interactions with the bonded phase. In these respects, SEC is not as useful for the separation of peptides as it is for proteins because peptides vary drastically in solubility, charge, and hydrophobicity. Peak capacity in SEC is fairly low compared to other HPLC methods because all separations must occur in the internal volume (V_i) of the support, which is generally less than half the volume of mobile phase in the column. Despite these deficiencies, SEC can be very effective for separating peptides from dimers, aggregates, small molecules, proteins, and other molecules which differ by size.

Supports for SEC of proteins are designed to be neutral and very hydrophilic to avoid interaction of the solutes with the support by ionic or hydrophobic mechanisms. The base matrix can be either silica or polymer; efforts are made to totally mask its properties with a carbohydratelike stationary phase. The pore structure is critical to successful SEC. Not only must the total pore volume (V_i) be adequate for separation, the pore diameter must be consistent and nearly homogeneous for attainment of maximum resolution between molecules with relatively small differences in molecular size (radius of gyration or molecular weight). A twofold difference in size is usually required for separation by SEC. Retention in SEC is directly related to the logarithm of the radius of gyration or the molecular weight. Because peptides do not all have the same shapes, their molecular volumes may not correspond uniformly to their molecular weights unless they are in the presence of sodium dodecyl sulfate (SDS) or another denaturing agent.

The mobile phase is a critical factor in SEC because it must eliminate all solute–support interactions. This may require adjustment to low pH or to an ionic strength which eliminates ionic interactions ($0.05M$–$0.2M$) and/or the addition of 5–10% organic solvent to remove hydrophobic binding. Due to their variations in physical properties, it is sometimes difficult to eliminate all the ionic and hydrophobic interactions of a mixture of peptides using a single mobile phase.

Detection

The detectability of peptides varies greatly, depending on their amino acid composition. The aromatic amino acids (tyrosine, tryptophan, and phenylalanine) offer selective detection by ultraviolet (UV) light at 280 or 254 nm or by fluorescence detection. In the absence of these amino acids, low wavelengths (<220 nm), which also detect many other substances, including components of the mobile phase, must be used. TFA and acetonitrile are compatible with low wavelengths, making this system a popular one for peptide analysis by RPC. Alternatively, the amine functionalities of peptides can be derivatized using precolumn or postcolumn techniques with compounds such as fluorescamine. In this way, selective and sensitive detection methods like fluorescence can be implemented. Mass spectrometry also provides a viable, albeit, expensive means of detection of peptides.

Conclusions

High-performance liquid chromatography provides a rapid and effective means for the analysis and purification of peptides. RPC is the primary and most universally successful mode, with HILIC and IEC offering alternative methods for hydrophilic peptides or others where RPC is ineffective. HILIC and IEC are also orthogonal modes to RPC for preparative or identification purposes. The ability to change selectivity using the mobile phase in HPLC methods makes them versatile techniques which can be rapidly optimized and implemented.

Suggested Further Reading

Cunico, R. L., K. M. Gooding, and T. Wehr, *Basic HPLC and CE of Biomolecules*, Bay Bioanalytical Laboratories, Richmond, CA, 1998.

Gooding, K. M., and F. E. Regnier (eds.), *HPLC of Biological Macromolecules: Methods and Applications*, Marcel Dekker, Inc., New York, 1990.

Hancock, W. S. (ed.), *High Performance Liquid Chromatography in Biotechnology*, John Wiley & Sons, New York, 1990.

Hearn, M. T. W. (ed.), *HPLC of Proteins, Peptides and Polynucleotides*, VCH, New York, 1991.

Katz, E. D. (ed.), *High Performance Liquid Chromatography: Principles and Methods in Biotechnology*, John Wiley & Sons, New York, 1996.

Mant, C. T. and R. S. Hodges (eds.), *High-Performance Liquid Chromatography of Peptides and Proteins*, CRC Press, Boca Raton, FL, 1991.

Mant, C. T., L. H. Kondejewski, and R. S. Hodges, *J. Chromatogr. 816*: 79 (1998).

Karen M. Gooding

Pesticide Analysis by Gas Chromatography

Introduction

In spite of the worldwide controversy which has surrounded the use of pesticides for many years, there can be little doubt that they provide one of the most effective contributions to increasing crop production and they have helped the farmer to improve the quality and variety of our foodstuffs. However, even when used correctly, these compounds can cause ecological consequences, public health problems, and the occurrence of toxic residues in foodstuffs. These problems make necessary the development of analytical methodologies that allow the appropriate monitoring of pesticides residues.

One of the most important analytical techniques used today is high-resolution gas chromatography (HRGC). The pesticide residues can be analyzed by specific or multiresidue methods. When crops are treated with several pesticides, the use of a multiresidue method is preferable due the reduced cost and time of analysis. The methods of pesticides residue analysis usually present an initial step of extraction with a solvent which is not miscible with water, followed by a cleanup step and the analyte determination by gas chromatography.

Extraction Methods

Sample preparation represents a formidable challenge in the chemical analysis of the "real-world" samples. Not only is the majority of total analysis time spent in sample preparation, but also it is the most error-prone, least glamorous, and the most labor-intensive task in the laboratory. The components to be separated from the matrix are usually taken up with an auxiliary substance such as a carrier gas, an organic solvent, or an adsorbent. These separation processes can be regarded as extraction procedures (i.e., liquid–liquid extraction, liquid–solid extraction, Soxhlet extraction, solid-phase extraction, supercritical fluid extraction, solid-phase microextraction, etc.).

Soxhlet extraction (SE) has been the standard extraction method of the analyst for nearly 90 years. The complete extraction produces a high-volume diluted solution that usually needs to be concentrated prior to analysis. The choice of solvent determines the solvating power as well as the temperature of the extraction. Perhaps the greatest disadvantage of using the Soxhlet method is the utilization of expensive, high-purity organic solvents such as acetone and methylene chloride.

Liquid–liquid extraction (LLE) uses two immiscible liquids as the two phases. The sample is dissolved in one of the liquids (refinate) which comes in contact with the other liquid (extractant) into a separatory funnel, under agitation, to increase the contact area among the phases. Some mixing time is usually necessary for efficient phase exchange. Multiple extractions are also mandatory if quantitative extraction is desired. Sample transfer can become a problem, especially if phase emulsions are produced.

Solid–liquid extraction (SLE), normally performed at room temperature, is a simpler version of Soxhlet extraction (see Fig. 1). The extraction of sample components is performed by blending the sample with a solvent. The choice of the extraction solvent can be determined by the analyst's experience, the equipment available in the laboratory, the type of sample to be analyzed, and the range of target analytes. Ethyl acetate is usually more powerful than acetone for extraction of more polar compounds. As regards selectivity, acetone is preferable because the amount of polar coextracted matrix interference will be less.

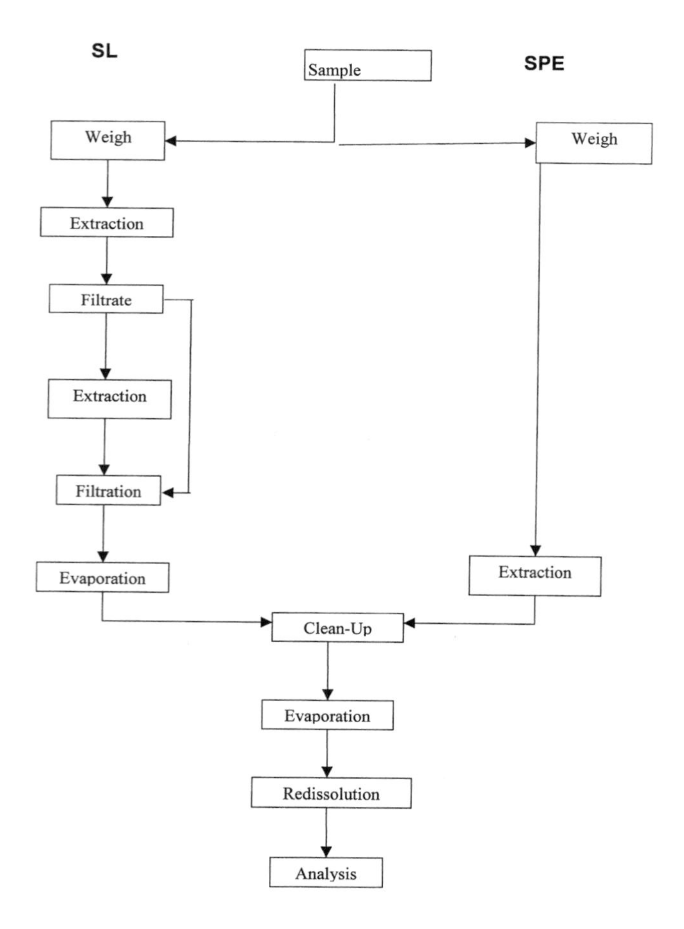

Fig. 1 Schematic diagram of the main steps of the solid–liquid extraction (SLE) and supercritical fluid extraction (SFE) methodologies for pesticide residue analysis.

Solid-phase extraction (SPE) is based on low-pressure liquid chromatography, where a short column is filled with an adsorbent. The separation mechanisms are based on the intermolecular interactions among analyte molecules and functional groups of sorbent. The choice of eluent is made by the relationship between the eleutropic value ($\Sigma°$) and the analyte polarity. SPE is fast, selective, and economical if compared with the extraction methods described previously. It can be applicable to both nonpolar and polar analytes, but both matrix and analyte must be in the liquid state.

Supercritical fluid extraction (SFE) provides, for the first time, a viable alternative to other traditional sample preparation techniques which are slow, composed of several steps, and make use of organic solvents. A supercritical fluid can be defined as any substance that is above its critical temperature and critical pressure. A supercritical fluid exhibits physicochemical properties intermediate between those of liquids and gases. Specifically, its relatively high (liquidlike) density gives good solvating power, whereas its relatively low viscosity and high diffusivity (gaslike) values provide appreciable penetrating power into the matrix. These latter two properties have been shown to give rise to higher rates of solute mass transfer into a supercritical fluid than into a liquid.

Supercritical CO_2 is the fluid with more applications at the present time, due to its readily amenable critical conditions (T_c = 31.3°C; P_c = 72.9 atm) and high volatility and diffusivity, and it also is inert, inexpensive, nonflammable, nontoxic, and miscible with most solvents. It has been used to extract compounds ranging from low polarity to moderate polarity. The extraction of polar compounds can be performed by supercritical CO_2 modified by the addition of a small volume of a polar liquid solvent.

The major parameters that influence the supercritical fluid extraction are temperature, pressure/density, fluid composition, particle size, matrix type, and extract collection system.

Solid-phase microextraction (SPME) is based on the adsorption of an analyte in a fused-silica fiber externally coated with a stationary phase and following a thermal desorption in the injector of a gas chromatograph. The fiber is introduced into the aqueous sample. In SPME, usually equilibrium among the aqueous samples and the stationary organic phase occurs instead of an exhaustive extraction. The pH, ionic strength, temperature, time, and agitation of the sample can exert an influence on the quality of the extract. The SPME process can integrate sampling, extraction, concentration, and sample introduction in just one step.

Cleanup Procedures

No single cleanup method is able to cover the entire matrix range. The need for a cleanup procedure prior to gas chromatography analysis is largely dependent on the complexity of the matrix and the range of analytes to be analyzed.

The most important cleanup procedures are liquid–liquid partition, liquid chromatography in a column of silica, florisil, and/or alumina, gel permeation chromatography, and solid-phase extraction.

Gel permeation chromatography (GPC) is the most universally applicable cleanup method for the removal of high-molecular-weight compounds. It is most favorable towards the multimatrix aspect and includes most pesticides. GPC has its limitations in the analyses of samples with a high load of coextractives and does not offer selectivity with respect to interference with low molecular weights. A selectivity gain can be obtained by the application of an additional cleanup using a small-scale chromatographic separation on silica, florisil, or alumina.

Gas Chromatographic Analysis

For the set up a GC-based multiresidue method for a specific pesticide–matrix combination, information is needed on GC retention and detectability of the analytes; also, the need for a cleanup procedure must be evaluated.

In general, the nature of the analyte determines the choice of stationary phase. For example, for the separation of organochlorine and pyrethroid pesticides, a nonpolar stationary phase such as DB-1 (or OV-1) is recommended. For the separation of somewhat more polar compounds, such as organophosphorus compounds, OV-17 (or DB-1701) can be applied. In addition, for confirmation purposes, the use of two columns with distinct stationary-phase polarities (e.g., DB-1 and DB-1701) is certainly required. A polar stationary phase (e.g., DB-wax) is suitable for the more polar compounds such as methamidofos, but its application to some detection modes is limited due to stationary-phase bleeding.

Due its robustness, particularly toward uncleaned samples, on-column, splitless injection is the most widely applied technique for sample introduction.

The conventional sensitive and specific GC detection such as electron-capture detector (ECD) (see Fig. 2),

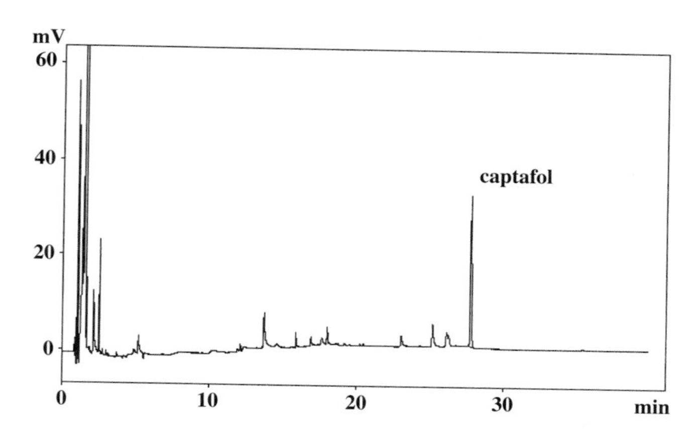

Fig. 2 Gas chromatogram (ECD) resultant from the analysis of captafol residues in tomato, after supercritical fluid extraction with neat CO_2 without further cleanup.

flame thermionic detector (FTD), and flame photometric detector (FPD) are still widely used in pesticide residue analysis. In recent years, mass spectrometric detection is becoming more and more important. Although other types of mass analyzers are commercially available, the equipment used in modern residue laboratories is based on two major types: the classical quadrupole mass analyzers and those based on the ion trap (also called tridimensional quadrupole).

For most compounds, the information on the m/z fragments were obtained with a quadrupole instrument. It should be mentioned, however, that the sensitivity of quadrupole detectors must be enhanced by means of limited mass range scanning or by selected ion monitoring, whereas ion-trap-based instruments offer a fair sensitivity with simultaneous monitoring of the complete m/z range.

Acknowledgments

Professor Lanças wishes to express his appreciation to the Brazilian Agencies FAPESP (Fundação de Amparo à Pesquisa do Estado de São Paulo) and CNPq (Conselho Nacional de Desenvolvimento Científico e Tecnológico) for their financial support for our research programs.

Suggested Further Reading

Font, G., J. Mañes, J. C. Moltó, and Y. Picó, Solid-phase extraction in multi-residue pesticides analysis of water, *J. Chromatogr. 642*: 135–161 (1993).

General Inspectorate for Health Protection, Ministery of Public

Health, Welfare and Sport, *Analytical Methods for Pesticides Residues in Foodstuffs*, 6th ed., The Netherlands, 1996.
Hedrick, J. L., L. J. Mulcahey, and L. T. Taylor, Supercritical fluid extraction, *Microchim. Acta 108*: 115–132 (1992).
Hennion, M. C., C. Call Dit-Coumes, and V. Pichon, Traces analysis of polar organic pollutants in aqueous samples: tools for the rapid prediction and optimisation of the SPE parameters, *J. Chromatogr. A 823*: 147–161 (1998).
Ling, Y. C., H. C. Teng, and C. Castwright, Supercritical fluid extraction and cleanup of organochlorine pesticides in chinese herbal medicine, *J. Chromatogr. A 835* (1-2), 145–157 (1999).
Mol, H. G. J., H. G. M. Janssen, C. A. Cramers, J. J. Vreuls, and U. A. T. Brinkman, Trace level analysis of micropollutants in aqueous samples using gas chromatography with on-line sample enrichment and large volume injection, *J. Chromatogr. A 703*: 277–307 (1995).
Peñalver, A., F. Pocurull, and R. M. Marcé, Trends in solid-phase microextraction for the determining organic pollutants in environmental samples, *Trends Anal. Chem. 18*(8): 557–568 (1999).
van der Hoff, G. R. and P. van Zoonen, Traces analysis of pesticides by gas chromatography, *J. Chromatogr. A 843*: 301–322 (1999).

Fernando M. Lanças
M. A. Barbirato

Pesticide Analysis by Thin-Layer Chromatography

Introduction

Thin-layer chromatography (TLC) is complementary to gas chromatography (GC), high-performance liquid chromatography (HPLC), capillary electrophoresis, and immunochemical methods for the determination of single residues and multiresidues of many classes of pesticides in a variety of food, feed, biological, and environmental samples of importance in maintaining human health. The off-line, development arrangement of TLC allows significant advantages in many pesticide analyses, including high sample throughput, because many samples can be chromatographed simultaneously on a single plate.

Materials and Techniques

Sample Preparation

Trace pesticide analysis involves extraction and cleanup steps prior to TLC. Extraction is carried out with a solvent of appropriate polarity by Soxhlet, ultrasonication, homogenization, or shaking, followed by liquid–liquid partitioning and/or column adsorption or gel permeation chromatography (GPC) cleanup. Because each plate is used only once, another advantage of TLC is the possibility of analyzing cruder samples than could be injected into a GC or HPLC column without ruining the analysis, thereby reducing the number of sample preparation steps. These conventional procedures, which are slow and consume large volumes of solvents, have been superseded by solid-phase extraction (SPE) using small, disposable cartridges, columns, or disks for isolation and cleanup of pesticides from water and other samples prior to TLC analysis, especially using octadecyl (C_{18}) bonded silica gel phases. Microwave-assisted extraction is a time- and solvent-saving method for removing residues from samples such as soils. Supercritical fluid extraction (SFE) has been used for sample preparation in the screening of pesticide-contaminated soil by automated multiple development.

Stationary and Mobile Phases

Most pesticide analyses have been performed by normal- or straight-phase (NP) TLC using commercial plates precoated with silica gel and a less polar mobile phase containing combinations of two or more solvents such as acetone, methanol, chloroform, hexane, and toluene. Lipophilic bonded-phase silica gel, mainly C_{18}, and a more polar mobile phase, such as methanol or acetonitrile mixed with water, have been used for reversed-phase (RP) TLC. Other precoated layers used less often for pesticide analysis include aluminum oxide, magnesium silicate (Florisil), polyamide, cellulose, acetylated cellulose (reversed phase), and polar-modified silica gel layers containing bonded amino, cyano, diol, and thiol groups. Mixtures of sorbents or layers impregnated with various reagents have been used to prepare layers with special selectivity properties. High-performance TLC (HPTLC) plates with smaller particles of sorbent provide improved resolution, shorter analysis time, higher detection sensitivity, and more precise and accurate *in situ* quantification compared to conventional TLC plates. Plates with a preadsorbent or concentrating zone may provide cleanup by retaining some interfering substances from impure samples, and they allow the application of larger sample amounts for quantification of very low pesticide concentrations without loss of zone resolution.

The mobile phase for a particular separation is usually selected empirically using prior personal experience and literature reports of similar separations as a guide or by use of a systematic mobile-phase optimization scheme, usually the PRISMA model. Typical mobile phases that have been used for separations of many classes of pesticides on silica gel have been mixtures of hexane–acetone, toluene–acetone, chloroform–diethyl ether, and toluene–methanol, whereas mobile phases for RPTLC analyses on C_{18} layers are usually methanol–water and acetonitrile–water mixtures.

Application of Samples and Standards

Initial zones in the form of round spots are applied manually to the origins of the layer using a glass micropipette such as a 10- or 25-μL digital microdispenser. In addition, partly or fully automated instruments are available for the application of solutions as spots or bands in the microliter to nanoliter range. Compact bands are also produced when samples are applied manually with a micropipette as diffuse vertical streaks to plates containing a preadsorbent zone. Band application is advantageous for obtaining tighter zones, high-resolution separations, and precise quantitative results by scanning densitometry.

Chromatogram Development Techniques

Pesticide analyses have usually been carried out in paper-lined, vapor-saturated glass N-chambers using a single ascending development with the mobile phase. Increased resolution is obtained by overpressured layer chromatography (OPLC) and automated multiple development (AMD) with gradient elution. An AMD separation of a complex pesticide mixture, with multiple wavelength scanning to provide confirmation of identity, is shown in Fig. 1.

Zone Detection

Pesticides are detected after development as colored, ultraviolet (UV)-absorbing, or fluorescent zones after reac-

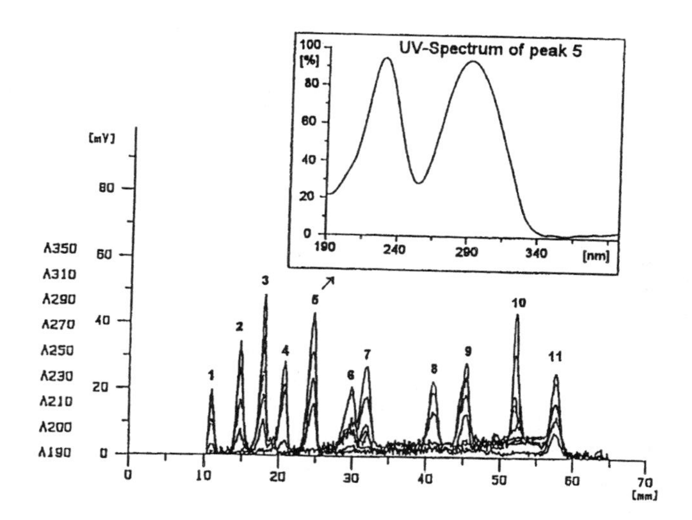

Fig. 1 Multiwave densitogram of a mixture of 11 pesticides (50 ng/compound) and *in situ* ultraviolet spectrum of peak 5. Compounds represented by peaks: 1, clopyralid acid; 2, triclopyr acid; 3, bitertanol; 4, atraton; 5, chloridazon; 6, sethoxydim; 7, atrazine; 8, iprodione; 9, desmedipham; 10, ethofumesate; 11, pendimethalin.

tion with a more or less selective reagent applied to the layer by spraying or dipping. Silver nitrate with UV irradiation is an example of a chromogenic detection reagent used to visualize chlorinated pesticides, whereas arsenic trichloride–sulfuric acid has been used to detect various pesticide classes. Phenolic pesticides and metabolites are detected by use of 7-chloro-nitrobenzo-2-oxa-1,3-diazole to produce fluorescent 4-nitrobenzofuran derivatives or by means of Pauly's reagent for those compounds that can form intensely colored azo dyes. Compounds that absorb short-wavelength (254-nm) UV light, particularly those with aromatic rings and/or conjugated double bonds, can be detected by fluorescence quenching on a layer containing a fluorescent indicator (phosphor). Another important advantage of the off-line operation of TLC is the versatility achieved by use of multiple detection methods. For example, the layer can be viewed under long- and short-wave UV light, followed by one or more chromogenic, fluorogenic, or biological detection methods. Cholinesterase inhibition has been used widely for detection of certain pesticide classes, such as organophosphates, with very low detection limits.

Quantitative Analysis

Quantitative pesticide analyses are performed by measuring the areas of sample and standard zones using a slit-scanning or video densitometer. Calibration curves are prepared for each analyte, and sample amounts are interpolated from the curves. Excellent accuracy and precision can be obtained because samples and standards are chromatographed and measured in parallel under the same conditions on a single TLC plate.

Identification and Confirmation of Zones: TLC Combined with Spectrometry

The identification of unknown pesticide zones is initially based on the comparison of the migration of sample zones relative to standards developed on the same layer and colors obtained with selective chromogenic and fluorogenic detection reagents. Many densitometers can record *in situ* UV and visible absorption and fluorescence excitation spectra to confirm compound identification by the comparison of unknown spectra with stored standard spectra obtained under identical conditions or spectra of standards measured on the same plate. Additional confirmation methods include off-line and on-line combination of TLC with infrared, Raman, or mass spectrometry or with gas or column liquid chromatography. Identification of pesticides by multiwavelength UV scanning is demonstrated in Fig. 1.

Special Techniques

Thin-layer radiochromatography (radio-TLC) is widely applied for a variety of environmental studies involving radiolabeled pesticides, such as plant uptake from soil, bioaccumulation in fish, dissipation from soil, metabolism in soil, plants, and fish, and environmental fate. The determination of the lipophilicity of pesticides is important because their bioaccumulation and tendency for degradation and biotransformation are related to lipid solubility. TLC has advantages for lipohilicity studies compared to traditional partition coefficient measurement in an octanol–water system.

Examples of Pesticide TLC Analysis Applications

Determination of Carbaryl and Propoxur in Water

Sample preparation: Chloroform extraction; grain samples were also analyzed.
Layer: Silica gel G.
Mobile phase: Acetone–hexane (1:4) or butanol–acetic acid–water (3:1:6, butanol layer).
Detection: Spray with diazotized *p*-nitroaniline followed by aqueous NaOH to produce blue and purple spots, respectively.
Quantification: Colored sample and standard zones scraped and eluted with 6*M* NaOH and the solutions measured by spectrometry.

Determination of 24 Pesticides in Water

Sample preparation: One thousand-milliliter samples of ground, surface, and drinking water were extracted using a C_{18} SPE cartridge; the cartridge was eluted with 3 mL of acetonitrile, and the eluate was used directly for TLC or cleaned up on a silica column eluted with acetonitrile.
Layer: HPTLC silica gel 60F-254; samples applied by the spray-on technique using a Linomat IV.
Mobile phase: AMD with 20-step universal gradients based on methylene chloride containing 0.1% acetic acid, methanol, and hexane (screening gradient) and *t*-butyl methyl ether containing 0.1% acetic acid, acetonitrile, and hexane (confirmatory gradient).
Detection and identification: Comparison of sample and standard chromatograms scanned at six different wavelengths with a densitometer.

Determination of Carbamate Insecticides in Water

Sample preparation: Pesticides recovered from pond water by SPE on a C_{18} column eluted with ethyl acetate.
Layer: Channeled preadsorbent HPTLC silica gel.
Mobile phase: Toluene–acetone (4:1) (for carbaryl, carbofuran, methiocarb) or hexane–acetone–chloroform (75:5:10) (propoxur).
Detection: Plate dipped into *p*-nitrobenzenediazonium fluoborate chromogenic reagent.
Quantification: Densitometric scanning of sample and standard zones at 610 nm (carbaryl), 550 nm (carbofuran, propoxur), or 510 nm (methiocarb).

Determination of Atrazine and Its Deethyl, Deisopropyl, and Hydroxy Metabolites in Surfacesoils and Subsoils

Sample preparation: Soil was extracted with methanol, the suspension was filtered, and the filtrate was concentrated.
Layer: HPTLC reversed-phase plates; standards and samples were applied with a Nanomat III.
Mobile phase: Methanol-water (7:3), development in a horizontal chamber.
Quantification: Zones were scanned at 222 nm with a

dual-wavelength flying-spot densitometer; standard curves were prepared by quadratic regression analysis, which gave a higher correlation coefficient than linear regression.

Determination of Abate in Environmental Water

Sample preparation: Acidified water was extracted with chloroform and the extract dried by filtering through Whatman PS paper.
Layer: Channeled preadsorbent C_{18} chemically bonded silica gel; paper-lined glass chamber, 10 cm development.
Mobile phase: Acetonitrile–water (8:2).
Detection: Spraying with 5% magnesium chloride in methanol, air-drying, spraying with 0.3% *N*,2,6-trichlorobenzoquinoneimine (TCQ) in hexane, heating for 5–10 min at 110°C; bright red-orange zone, 200 ng sensitivity.
Quantification: Scanning densitometry.

Determination of Chlorpyrifos and Its Metabolite 3,5,6-Trichloro-2-Pyridinol in Bananas and Tap Water

Sample preparation: Water was extracted with hexane for chlorpyrifos (a) and benzene for 3,5,6-trichloro-2-pyridinol (TCP) (b); banana samples were prepared using standard Food and Drug Administration procedures based on extraction, solvent partitioning, and silica gel or alumina column chromatography.
Layer: Channeled preadsorbent silica gel G.
Mobile phase: (a) hexane–chloroform (8:2); (b) hexane–acetone–methanol–acetic acid (60:30:10:0.2); paper-lined glass N-chamber, 10 cm development.
Detection: Dipping into silver nitrate reagent followed by exposure to UV light; dark brown spots on a white background, sensitivity 25–100 ng.
Quantification: Scanning densitometry at 440 nm.

Determination of Cymiazole and Pentachlorophenol in River Water and Honey

Sample preparation: The pesticides were extracted from water and honey using C_{18} SPE.
Layer: Channeled preadsorbent high-performance silica gel.
Mobile phase: Toluene–methanol (9:1) for pentachlorophenol (PCP); hexane–acetone–methanol–glacial acetic acid (35:10:5:0.1) for cymiazole; paper-lined twin-trough chamber.
Detection: Fluorescence quenching under 254-nm UV light; 200 ng sensitivity.
Quantification: Scanning densitometry at 215 nm for PCP and 265 nm for cymiazole.

Determination of Diflubenzuron in Water

Sample preparation: Extraction on a C_{18} SPE column eluted with acetonitrile.
Layer: Channeled preadsorbent high-performance silica gel.
Mobile phase: Ethyl acetate–toluene (1:3); paper-lined glass HPTLC chamber.
Detection: Spraying with 6*M* HCl, heating for 10 min at 180°C, and spraying with Bratton–Marshall reagent [sodium nitrite and *N*-(1-naphthyl)ethylenediamine dihydrochloride]; purple-blue band; 100 ng sensitivity.
Quantification: Scanning densitometry at 550 nm.

Suggested Further Reading

The following references contain detailed information on the procedures and instrumentation of TLC and applications to pesticide residue analysis.

Fodor-Csorba, K., Pesticides, in *Handbook of Thin Layer Chromatography*, 1st ed. (J. Sherma and B. Fried, eds.), Marcel Dekker, Inc., New York, 1991, pp. 663–715.
Fodor-Csorba, K., Pesticides, in *Handbook of Thin Layer Chromatography*, 2nd ed. (J. Sherma and B. Fried, eds.), Marcel Dekker, Inc., New York, 1996, pp. 753–817.
Follweiler, J., and J. Sherma, *Handbook of Chromatography—Pesticides*, CRC Press, Boca Raton, FL, 1984.
Fried, B. and J. Sherma, *Practical Thin Layer Chromatography—A Multidisciplinary Approach*, CRC Press, Boca Raton, FL, 1996.
Fried, B. and J. Sherma, *Thin Layer Chromatography*, 4th ed., Marcel Dekker, Inc., New York, 1999.
Sherma, J., Pesticide analysis by thin layer chromatography, in *Analytical Methods for Pesticides and Plant Growth Regulators* (G. Zweig and J. Sherma, eds.), Academic Press, San Diego, CA, 1973, Vol. VII, pp. 3–87; 1980, Vol. XV, pp. 79–122; 1986, Vol. XIV, pp. 1–39.
Sherma, J., Thin layer chromatography of pesticides—A review of recent techniques and applications, *J. Liquid Chromatogr. 5*: 1013–1032 (1982).
Sherma, J., Thin layer chromatography of pesticides, *J. Planar Chromatogr.—Mod. TLC 4*: 7–14 (1991).
Sherma, J., Determination of pesticides by thin layer chromatography, *J. Planar Chromatogr.—Mod. TLC 7*: 265–272 (1994).
Sherma, J., Thin layer chromatography in environmental analysis, *Rev. Anal. Chem. 14*(2): 75–142 (1995).
Sherma, J., Review: Determination of pesticides by thin layer chromatography, *J. Planar Chromatogr.—Mod. TLC 10*: 80–89 (1997).

Sherma, J., Planar chromatography, *Anal. Chem. 72*: 9R–25R (2000); see also reviews of TLC each even numbered year in this journal beginning in 1970.
Sherma, J., Recent advances in thin layer chromatography of pesticides, *J. AOAC Int. 82*: 48–53 (1999).
Sherma, J. and B. Fried, *Handbook of Thin Layer Chromatography*, 2nd ed., Marcel Dekker, Inc., New York, 1996.
Singh, K. K. and M. S. Shekhawat, Thin layer chromatographic methods for analysis of pesticides residues in environmental samples, *J. Planar Chromatogr.—Mod. TLC 11*: 164–185 (1998).

Joseph Sherma

pH, Effect on MEKC Separation

In micellar electrokinetic chromatography (MEKC), the effect of the constituents of the buffer is not significant, whereas the pH is critical, especially for ionizable analytes, as well as to the electrokinetic velocities. The change in the buffer pH, especially in the lower-pH region, causes a significant change in the velocity of the electro-osmotic flow (EOF).

Electrokinetic Velocities

When sodium dodecyl sulfate (SDS) is employed as an ionic micelle or pseudo-stationary-phase in MEKC, the relationship between the velocity of the EOF, v_{EOF}, and the migration velocity of the SDS micelle, v_{mc}, is given as

$$v_{mc} = v_{EOF} + v_{ep} \tag{1}$$

where v_{ep} is the electrophoretic velocity of the SDS micelle. The sign of each velocity is defined as plus when the migration is toward the cathode and as minus when toward the anode.

The dependence of these electrokinetic velocities on pH is shown in Fig. 1. In the case of capillary zone electrophoresis (CZE), with a bare fused-silica capillary, the pH greatly affects the EOF velocity (i.e., v_{EOF} significantly decreases with the decrease in pH from 8 to 3). In MEKC, however, the dependence of v_{EOF} on pH is different from that in CZE, especially under weakly acidic conditions (pH 7.0–5.5). In the range of pH between 7.0 and 5.5, v_{EOF} slightly decreases with the decrease in pH, due to the adsorption of the SDS molecule or monomer on the inside wall of the capillary. On the other hand, v_{EOF} rapidly decreases with the decrease in the pH below 5.5. The decrease of v_{EOF} is mainly caused by the decrease in the zeta-potential of the inside wall of the capillary, because the dissociation of silanol groups on the capillary wall is more suppressed as the solution becomes more acidic.

The electrophoretic velocity of the SDS micelle (v_{ep}) (i.e., $v_{mc} - v_{EOF}$) is almost constant over the pH range in Fig. 1; that is, the charge of the SDS micelle is almost constant in this pH range.

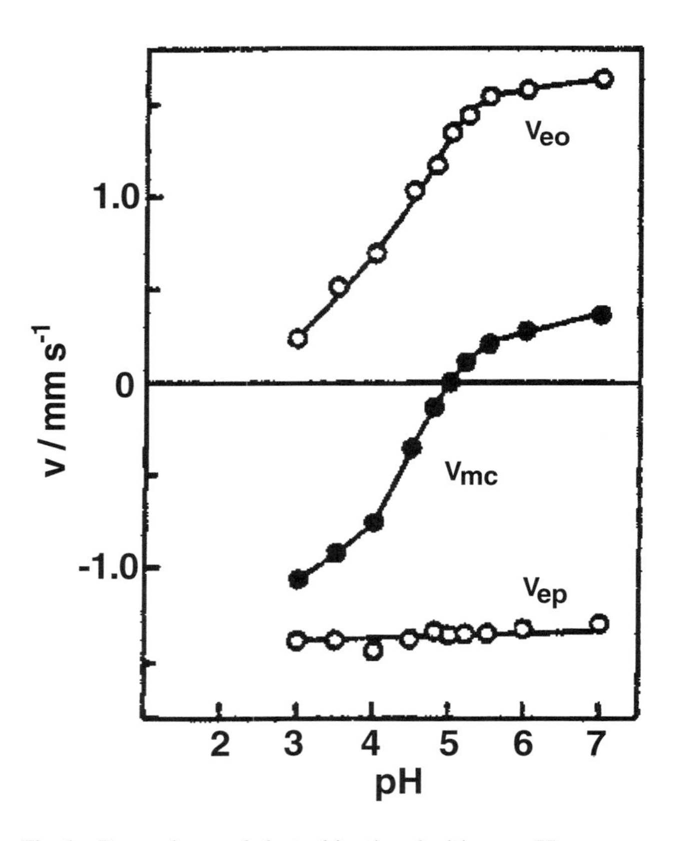

Fig. 1 Dependence of electrokinetic velocities on pH: v_{eo} = velocity of the EOF, v_{mc} = migration velocity of the micelle, v_{ep} = electrophoretic velocity of the micelle. (Reprinted from K. Otsuka and S. Terabe, *J. Microcol. Separ.*, 1989, 1, 150 with permission.)

The migration velocity of the SDS micelle (v_{mc}) changes from positive to negative at a pH below 5.0, which means that the SDS micelle migrates toward the anode; thus, the migrating direction of the SDS micelle is opposite that of the EOF. One should note that the reproducibility of retention times was low in acidic solutions, especially below pH 5.0, compared with that in neutral SDS solutions.

Migration Time

The migration time, t_R, of a neutral solute in MEKC is represented by

$$t_R = \left(\frac{1 + k}{1 + (t_0/t_{mc})k}\right)t_0 \tag{2}$$

where t_0 is the migration time of an unretained solute or insolubilized solute by the micelle at all, t_{mc} is that of the micelle, and k is the retention factor of the solute. Here, we define the sign of the migration time as positive when the solute migrates toward the cathode or the velocity of the solute, v_s, is positive, and vice versa. When neutral SDS solutions are employed, the t_{mc} is positive and the t_R of any neutral solute is always positive and limited to between t_0 and t_{mc}. Under acidic condition, or typically pH below 5.0, the neutral solute totally solubilized by the SDS micelle, such as Sudan III, migrates toward the anode and, hence, t_{mc} becomes negative, whereas the solute insolubilized by the micelle (e.g., methanol) migrates toward the cathode. By considering Eq. (2), it is apparent that the migration time of the solute whose capacity factor is equal to $-(t_{mc}/t_0)$ becomes infinity when $t_{mc} < 0$ and the solute never migrates in the column, whereas the solute of $k > -(t_{mc}/t_0)$ migrates toward the same directions as the micelle.

Resolution

In MEKC, the resolution, R_s, of two peaks of which the retention factors are k_1 and k_2 is described as (see the entry Retention Factor, Effect on MEKC Separation)

$$R_s = \frac{N^{1/2}}{4}\left(\frac{\alpha - 1}{\alpha}\right)\left(\frac{k_2}{1 + k_2}\right)\left(\frac{1 - t_0/t_{mc}}{1 + (t_0/t_{mc})k_1}\right) \tag{3}$$

For the closely migrating peaks, we can assume that $k_1 = k_2 = k$. Then, the fourth term of the right-hand side of Eq. (3) will become infinity when k is equal to $-(t_{mc}/t_0)$; thus, the resolution will become maximum or infinity. The function $f(k)$, the product of the last two terms in Eq. (3), is written as

$$f(k) = \left(\frac{k_2}{1 + k_2}\right)\left(\frac{1 - t_0/t_{mc}}{1 + (t_0/t_{mc})k_1}\right) \tag{4}$$

The value of $f(k)$ approaches infinity, as in the case of t_R, as k becomes close to $-t_{mc}/t_0$, and, consequently, R_s becomes quite large. Hence, a considerably large R_s will be obtained for a solute having k close to $-t_{mc}/t_0$, although a quite long t_R is required for such a solute.

Migration Window

The parameter t_0/t_{mc} is related to the migration window. As long as t_{mc} is positive, a smaller value of t_0/t_{mc} gives a wider migration window. Some attempts to extend the migration window have been made by (a) the addition of organic modifiers (e.g., methanol, acetonitrile, and 2-propanol) to the micellar solution and (b) the use of capillaries of which the inside walls were chemically modified to reduce the silanol-group concentration. In each case, the extension was mainly achieved through decreasing the EOF owing to a decrease of the zeta-potential of capillaries. As mentioned earlier, the migration window is no longer limited when acidic micellar solutions (i.e., pH below 5.0) are employed. The value of zero for t_0/t_{mc} corresponds to the case of conventional chromatography, in which the elution range is infinity; in other words, the solute of $k = \infty$ (e.g., Sudan III) never comes out from the capillary. The case of $t_0/t_{mc} < 0$ also causes the infinite migration window.

Ionizable Solutes

If the ionized form of the solute has the same charge as the micelle, it will be incorporated into the micelle less than with its neutral form. By contrast, the ionized form of the solute will be bound to the micelle more strongly than its neutral form if the ionized solute has the opposite charge to that of the micelle. The dependence of the apparent retention factor, k_{app}, on the buffer pH for chlorinated phenols has been investigated in an SDS/MEKC, where both the ionizable solutes or chlorinated phenols and SDS have negative charge. The apparent retention factor was calculated by the usual equation for the retention factor in MEKC:

$$k_{app} = \frac{t_R - t_0}{[1 - (t_R/t_{mc})]t_0} \tag{5}$$

regardless of whether the solutes were ionized or not. For acidic compounds, the increase in pH will promote ionization; then, the distribution coefficient to the SDS or an anionic micelle will be decreased. For example, the apparent retention factor for 2,3,4,5-tetrachlorophenol (pK_a = 5.6) was dramatically changed from ~100 at pH 6.0 to 4 at pH 9.0.

It is often essential to find the optimum pH for the separation of ionizable solutes, where closely spaced peaks are obtained.

Suggested Further Reading

Foret, F., L. Kriváková, and P. Bocek, *Capillary Zone Electrophoresis*, VCH, Weinheim, 1993, pp. 67–74.

Lukacs, K. D. and J. W. Jorgenson, Capillary zone electrophoresis: Effect of physical parameters on separation efficiency and quantitation, *J. High Resolut. Chromatogr. Chromatogr. Commun. 8*: 405–411 (1985).

Otsuka, K. and S. Terabe, Effects of pH on electrokinetic velocities in micellar electrokinetic chromatography. *J. Microcol. Separ. 1*: 150–154 (1989).

Otsuka, K. and S. Terabe, Micellar electrokinetic chromatography, *Bull. Chem. Soc. Jpn. 71*: 2465–2481 (1998).

Otsuka, K., S. Terabe, and T. Ando, Electrokinetic chromatography with micellar solutions: Retention behaviour and separation of chlorinated phenols, *J. Chromatogr. 348*: 39–47 (1985).

Koji Otsuka
Shigeru Terabe

pH-Peak-Focusing and pH-Zone-Refining Countercurrent Chromatography

These two countercurrent chromatography (CCC) techniques are mutually related: pH-peak-focusing CCC is for analytical-scale separations and pH-zone-refining CCC is for preparative-scale separations. The pH-peak-focusing CCC technique was developed from an accidental finding that a thyroxine analog produced an unusually sharp elution peak [1,2]. The cause was found that an acid present in the sample solution affected the retention time of the analyte (Fig. 1a). The mechanism of this peak-sharpening effect is shown in Fig. 1b, where a portion of the separation column contains the organic stationary phase in the upper half and the aqueous mobile phase in the lower half. Because of its nonlinear isotherm, the acid in the sample solution forms a sharp trailing border which traps the analyte in the following manner: When the analyte molecule is present in the mobile phase (position 1), it is protonated by low pH and transferred into the organic stationary phase (position 2). As the sharp acid border moves forward, the analyte molecule is exposed to high pH (position 3), deprotonated, and transferred back to the aqueous mobile phase (position 4), where it quickly migrates through the acid border to repeat the above cycle. Consequently, the analyte molecules are trapped with the sharp acid border and eluted as a sharp peak, together with the acid border. A similar effect can be produced by introducing an organic acid in the stationary phase.

Figure 2a shows the separation of DNP-amino acids using three spacer acids (i.e., acetic acid, propionic acid, and *n*-butyric acid) in the stationary phase [2]. Hydrophilic DNP-glutamic acid is eluted between acetic and propionic acids; DNP-alanine between propionic and *n*-butyric acids, and hydrophobic DNP-leucine after *n*-butyric acid. The method can be effectively applied for the separation and concentration of a small amount of organic ions present in a large volume of the sample solution. However, the most useful application has been found in the preparative-scale separation. When the sample size of the above DNP-amino acids is increased each from 6 mg to 600 mg, a strange chromatogram was produced as shown in Fig. 2b, where all peaks became rectangular in shape, each associated with its own specific pH. The elimination of the spacer acids resulted in the fusion of these peaks while maintaining their own pH, suggesting that the rectangular shape of each peak is well preserved (Fig. 2c) [2–4].

The mechanism of this pH-zone-refining CCC (for the separation of acidic compounds) is illustrated by the following model experiment [2,4,5]. Figure 3a shows the preparation of the solvent phases. Ether and water are equilibrated in a separatory funnel and separated. A suitable amount (usually 10–40 m*M*) of TFA (trifluoroacetic acid) ("retainer") is added to the lighter organic phase, which is then used as the stationary phase. Ammonia (10–40 m*M*) ("eluter") is added to the heavier aqueous phase, which is used as the mobile phase. The experiment is initiated by filling the entire column with the stationary

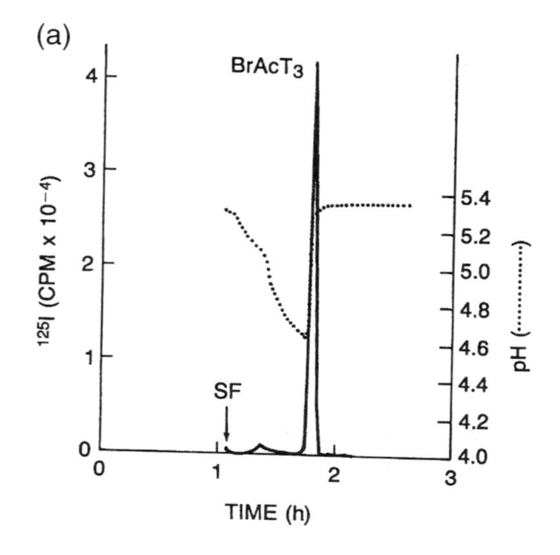

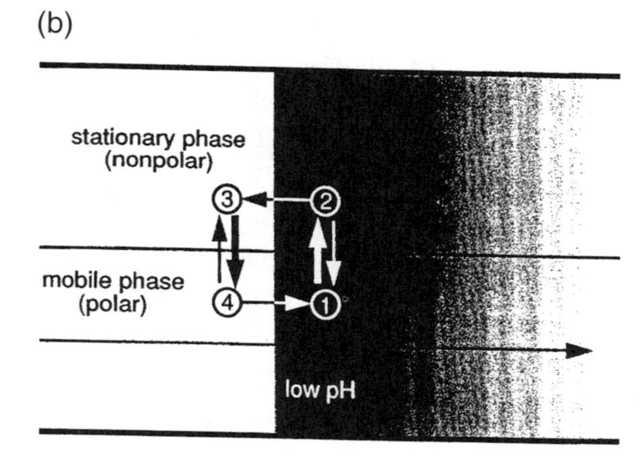

Fig. 1 (a) Sharp analyte peak produced by an acid in the sample solution; (b) mechanism of sharp peak formation.

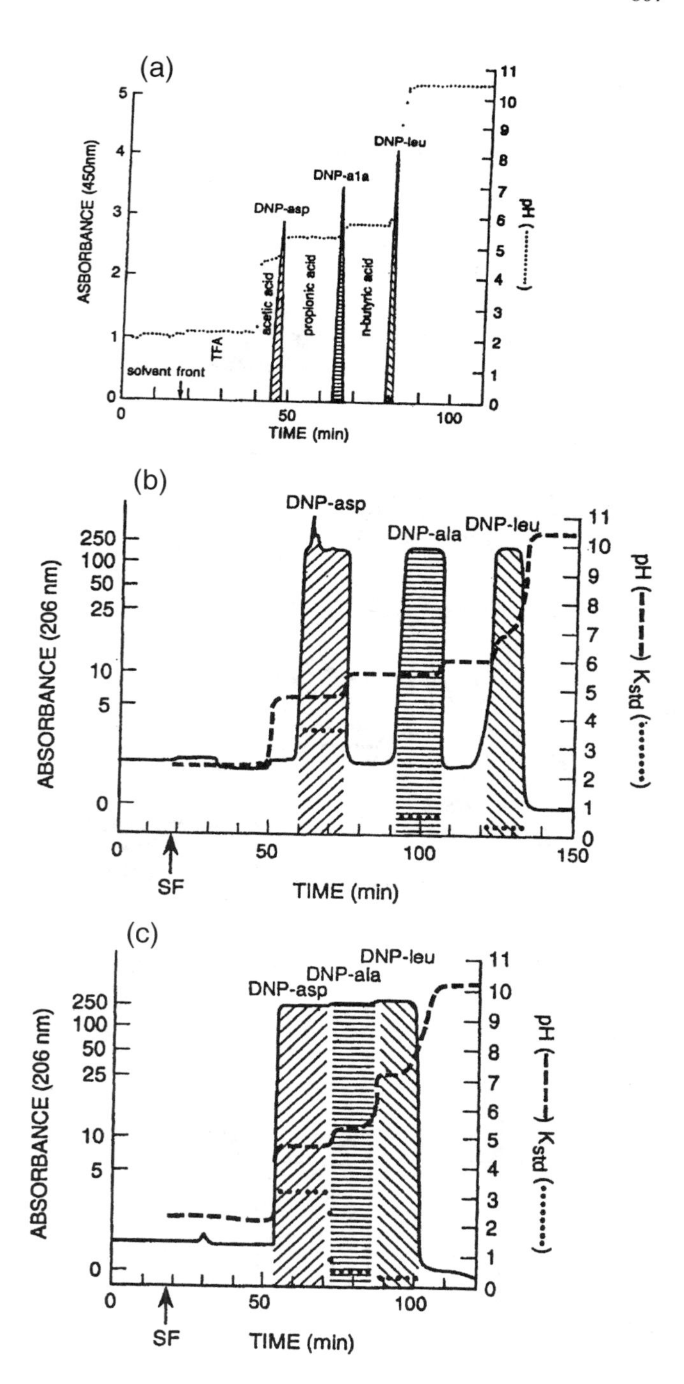

Fig. 2 Separation of three DNP-amino acids showing a transition from pH-peak-focusing CCC to pH-zone-refining CCC. (a) Analytical separation by pH-peak-focusing CCC (6 mg each component); (b) formation of rectangular peaks by increasing sample size (600 mg each component); (c) elimination of three spacer acids to form fused rectangular peaks.

phase, followed by injection of a sample solution containing three major acidic analytes (S_1, S_2, and S_3). Then, the column is eluted with the mobile phase while the apparatus is rotated at a desired g-force. Figure 3b shows steady-state chemohydrodynamic equilibrium established in the separation column. The retainer acid TFA forms a sharp trailing border which travels through the column at a constant rate substantially lower than that of the mobile phase.

Three analytes S_1, S_2, and S_3 each form a discrete pH zone behind the sharp retainer border in the order of their pK_a's and hydrophobicities. The proton transfer takes place at each zone boundary according to the difference in pH between the neighboring zones, causing solute ex-

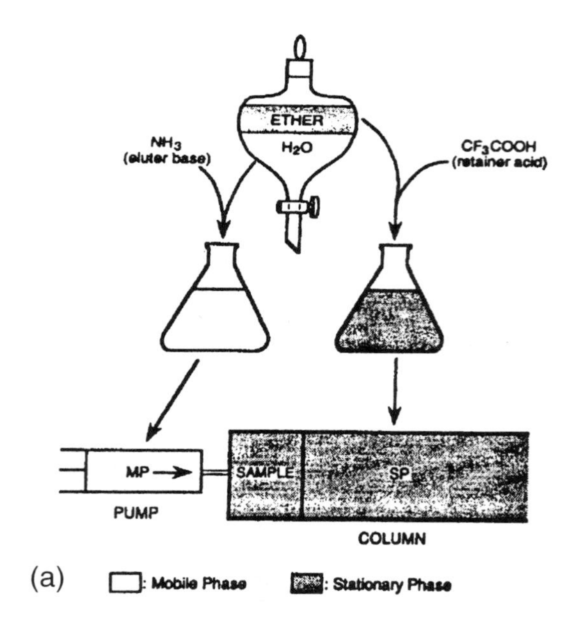

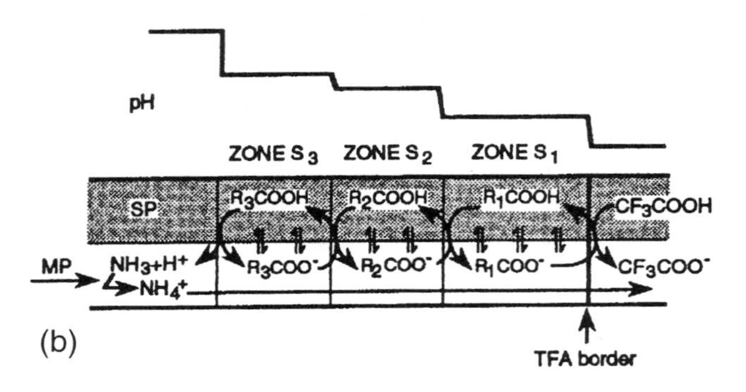

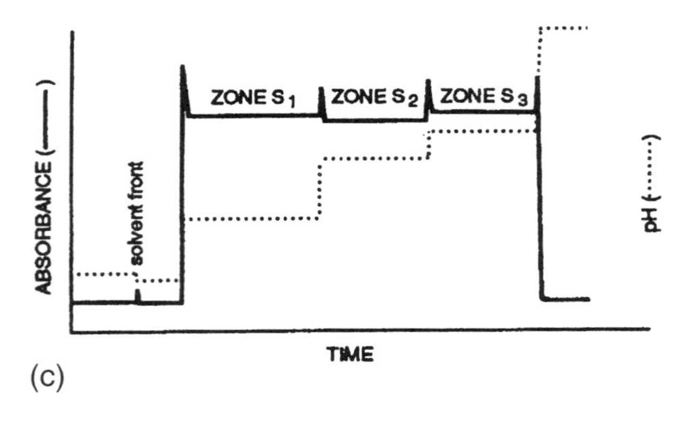

Fig. 3 Mechanism of pH-zone-refining CCC. (a) Preparation for the model experiment; (b) chemohydrodynamic equilibrium in the separation column; (c) elution profile of three major analytes.

change between the two phases, as indicated by curved arrows. Once the equilibrium is established, all solute zones move at a same rate determined by that of the sharp retainer border. Charged impurities present in each zone are eliminated either forward or backward according to their pK_a's and hydrophobicities, and accumulated at the zone boundaries. Consequently, the analytes are eluted as fused rectangular peaks with minimum overlap associated with sharp impurity peaks at their boundaries (Fig. 3c).

The relationship between the zone pH (pH_{zone}) and pK_a/hydrophobicity in the present method is given by the following equation [2,4,5]:

$$pH_{zone} = pK_a + \log\left(\frac{K_D}{K} - 1\right) \tag{1}$$

where K_D and K indicate the distribution coefficient (an indicator for hydrophobicity) and distribution ratio of the analytes, respectively. When the pK_a and K_D of the analyte is known, the zone pH can be computed from the K value.

The pH-zone-refining CCC technique shares many unique features with displacement chromatography [6] and has several important advantages over the standard CCC technique such as (a) large sample-loading capacity, (b) highly concentrated fractions, (c) concentration and detection of minor impurities, and (d) monitoring the effluent by pH. The method has been successfully applied to the separation of various organic acids and bases, including the derivatives of amino acids [4,7] and peptides [8], alkaloids [4,9], hydroxyxanthene dyes [3,4,10], anti-human immunodeficiency virus lignans [11], indole auxins [4], structural and stereoisomers [4], and so forth. (Table 1). By being analogous to affinity chromatography, the method allows the use of an affinity ligand dissolved in the liquid stationary phase for separations of special analytes, including highly polar compounds such as catecholamines [4,12] and sulfonated dyes [4], enantiomers [4], and zwitterions such as free peptides [4,13] (Table 2). Figures 4a–4d illustrate a few examples of these applications.

References

1. Y. Ito, Y. Shibusawa, H. M. Fales, and H. J. Cahnmann, *J. Chromatogr. 625*: 177–181 (1992).
2. Y. Ito, pH-Peak-focusing and pH-zone-refining countercurrent chromatography, in *High-Speed Countercurrent Chromatography* (Y. Ito and W. D. Conway, eds.), Chemical Analysis Series Vol. 132, Wiley–Interscience, New York, 1996, pp. 121–175.
3. A. Weisz, A. L. Scher, K. Shinomiya, H. M. Fales, and Y. Ito, *J. Am. Chem. Soc. 116*: 704–708 (1994).
4. Y. Ito and Y. Ma, *J. Chromatogr. A 753*: 1–36 (1996).

Table 1 Samples and Solvent Systems Applied to Standard pH-Zone-Refining CCC

Sample[a]	Solvent system[b] (vol. ratio)	Pair of acid–base reagents[c]	
		Retainer	Eluter
DNP-Amino acids	MBE/AcN/H_2O (4:1:5)	TFA (200 μL/SS)	NH_3 (0.1%/MP)
Proline (OBzl) (1 g)	MBE/H_2O	TEA (10 m*M*/SP)	HCl (10 m*M*/MP)
Amino acid (OBzl) (0.7 g)	MBE/H_2O	TEA (10 m*M*/SP)	HCl (10 m*M*/MP)
Amino acid (OBzl)) (10 g)	MBE/H_2O	TEA (5 m*M*/SP)	HCl (20 m*M*/MP)
CBZ dipeptides (0.8 g)	MBE/AcN/H_2O (2:2:3)	TFA (16 m*M*/SP)	NH_3 (5.5 m*M*/MP)
CBZ dipeptides (3 g)	MBE/AcN/H_2O (2:2:3)	TFA (16 m*M*/SP)	NH_3 (5.5 m*M*/MP)
CBZ tripeptides (0.8 g)	BuOH/MBE/AcN/H_2O (2:2:1:5)	TFA (16 m*M*/SP)	NH_3 (2.7 m*M*/MP)
Dipeptide-βNA (0.3 g)	MBE/AcN/H_2O (2:2:3)	TEA (5 m*M*/SP)	HCl (5 m*M*/MP)
Indole auxins (1.6 g)	MBE/H_2O	TFA (0.04%/SP)	NH_3 (0.1%/MP)
TCF (0.01–1 g)	DEE/AcN/10 m*M* $AcONH_4$ (4:1:5)	TFA (200 μL/SS)	MP
Red #3 (0.5 g)	DEE/AcN/10 m*M* $AcONH_4$ (4:1:5)	TFA (200 μL/SS)	MP
Orange #5 (0.01–5 g)	DEE/AcN/10 m*M* $AcONH_4$ (4:1:5)	TFA (200 μL/SS	MP
Orange #10 (0.35 g)	DEE/AcN/10 m*M* $AcONH_4$ (4:1:5)	TFA (200 μL/SS	MP
Red #28 (0.1–6 g)	DEE/AcN/10 m*M* $AcONH_4$ (4:1:5)	TFA (200 μL/SS	MP
Eosin YS (0.3 g)	DEE/AcN/10 m*M* $AcONH_4$ (4:1:5)	TFA (200 μL/SS	MP
Amaryllis alkaloids (3 g)	MBE/H_2O	TEA (5 m*M*/SP)	HCl (5 m*M*/MP)
	MBE/H_2O (DPCCC)	HCl (10 m*M*/SP)	TEA (10 m*M*/MP)
Vinca alkaloids (0.3 g)	MBE/H_2O (DPCCC)	HCl (5 m*M*/SP)	TEA (5 m*M*/MP)
Structural isomers (15 g)	MBE/AcN/H_2O (4:1:5)	TFA (0.32%/SP)	NH_3 (0.8%/MP)
Stereoisomers (0.4 g)	Hex/EtOAc/MeOH/H_2O (1:1:1:1)	TFA, octanoic acid	NH_3 (0.025%/MP)
Fish oil (0.5 g)	Hex/EtOH/H_2O (4:1:5)	TFA (10 m*M*/SP)	NH_3 (0.1%/MP)
NDGA derivatives (10 g)	MBE/H_2O	TFA (25 m*M*/SP)	NaOH (100 m*M*/MP)

[a]DNP: dinitrophenyl; CBZ: carbobenzoxy; OBzl: benzylesters; βNA: naphthyl amide; TCF: tetrachlorofluorescein; amaryllis alkaloids: crinine, powelline, and crinamidine; vinca alkaloids: vincamine and vincine; structural isomers: 2- and 6-nitro-3-acetamido-4-chlorobenzoic acid; stereoisomers: 4-methoxymethyl-1-methyl-cyclohexane carboxylic acid; fish oil: mixture of docosahexaenoic acid and eicosapentaenoic acid; NDGA: nordihydroguaiaretic acid.

[b]The upper organic phase was used as the stationary phase (SP) and the lower aqueous phase as the mobile phase (MP), except in DPCCC, where the relationship is reversed. MBE: methyl-*t*-butyl ether; AcN: acetonitrile; BuOH: *n*-butanol; Hex: hexane; EtOAc: ethyl acetate; MeOH: methanol; $AcONH_4$ ammonium acetate; DEE: diethyl ether; DPCCC: displacement mode.

[c]TFA: trifluoroacetic acid; AcOH: acetic acid; SP: in stationary phase; MP: in mobile phase; SS: in sample solution; TEA: triethylamine.

Table 2 Samples and Solvent Systems Applied to Affinity pH-Zone-Refining CCC

Sample[a]	Solvent systems[b] (vol. ratio)	Set of key reagents[c]		
		Retainer	Eluter	Ligand
(±)-DNB-leucine (2 g)	MBE/H_2O	TFA (40 m*M*/SP)	NH_3 (20 m*M*/MP)	DPA (40 m*M*/SP)
(±)-DNB-valine (2 g)	MBE/H_2O	TFA (40 m*M*/SP)	NH_3 (20 m*M*/MP)	DPA (40 m*M*/SP)
Catecholamines (3 g)	MBE/H_2O	NH_4OAc (200 m*M*/SP)	HCl (50 m*M*/MP)	DEHPA (20%/SP)
Dipeptides (1 g) (hydrophobic)	MBE/AcN/50 m*M* HCl 4:1:5 (SP)	TEA (20 m*M*/SP)		DEHPA (10%/SP)
	MBE/AcN/H_2O (4:1:5) (MP)		HCl (20 m*M*/MP)	
Dipeptides (1 g) (hydrophilic)	MBE/BuOH/AcN/50 m*M* HCl (2:2:1:5) (SP)	TEA (20 m*M*/SP)		DEHPA (30%/SP)
	MBE/BuOH/AcN/H_2O (2:2:1:5) (MP)		HCl (20 m*M*/MP)	
Bacitracins (5 g)	MBE/50 m*M* HCl (1:1) (SP)	TEA (40 m*M*/SP)		DEHPA (10%/SP)
	MBE/H_2O (MP)		HCl (20 m*M*/MP)	
FD&C Yellow #6 (2 g)	MBE/AcN/H_2O (2:2:3)	H_2SO_4 (0.2%/SP)	NH_3 (0.4%/MP)	TDA (5%/SP)

[a]DNB: dinitrobenzoyl.

[b]MBE: methyl *t*-butyl ether; AcN: acetonitrile; BuOH: *n*-butanol

[c]TFA: trifluoroacetic acid; NH_4OAc: ammonium acetate; TEA: triethylamine; DPA: *N*-dodecanoyl-L-proline-3,5-dimethylanilide; DEHPA: di-(2-ethylhexyl) phosphoric acid; TDA: tridodecylamine; SP: organic stationary phase; MP: aqueous mobile phase.

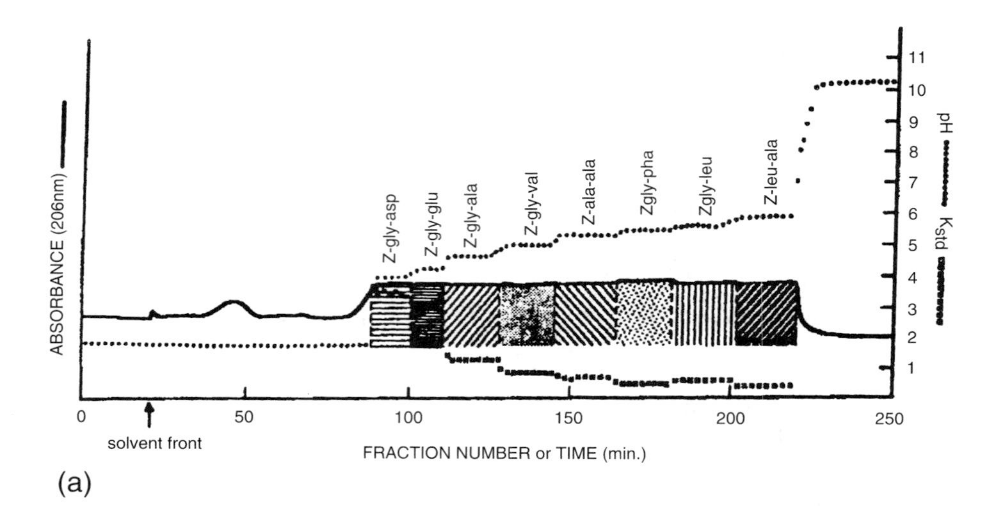

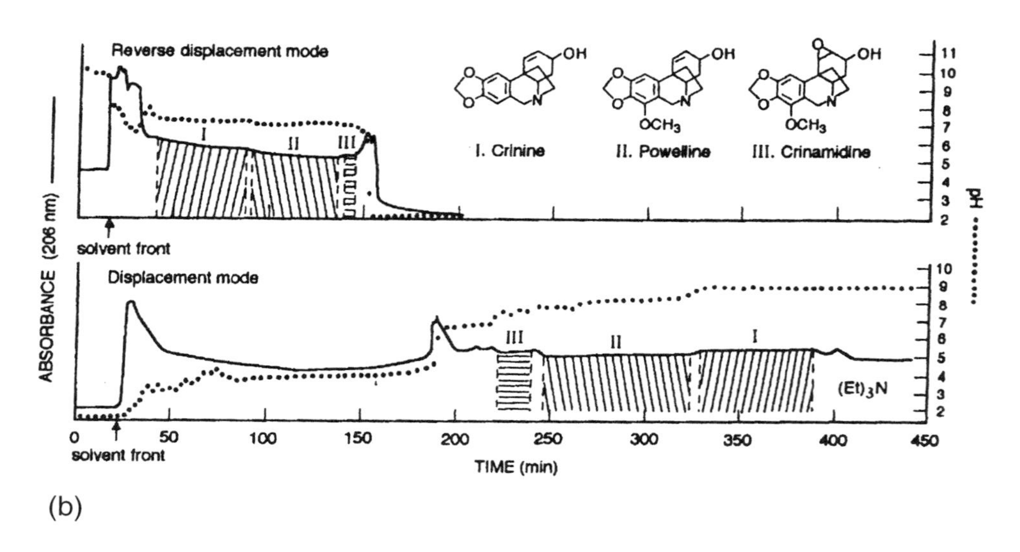

Fig. 4 Some applications of pH-zone-refining CCC. (a) Separation of eight CBZ dipeptides (see Table 1) [4,8]; (b) separation of amaryllis alkaloids using both the lower phase (upper chromatogram) and upper phase (lower chromatogram) as the mobile phase (see Table 1) [4,9]; (c) separation of catecholamines using a ligand (see Table 2) [4,12]; (d) separation of two groups of dipeptide each using an affinity ligand [4, 13] (see Table 2).

5. Y. Ito, K. Shinomiya, H. M. Fales, A. Weisz, and A. L. Scher, *pH-Zone-Refining Countercurrent Chromatography: A New Technique for Preparative Separation*, (W. D. Conway and R. J. Petroski, eds.), ACS Monograph on Modern Countercurrent Chromatography, 1995, pp. 154–183.
6. C. Horvath, A. Nahum, and J. H. Frens, *J. Chromatogr. 218*: 365 (1981).
7. Y. Ma and Y. Ito, *J. Chromatogr. A 678*: 233–240 (1994).
8. Y. Ma and Y. Ito, *J. Chromatogr. A 702*: 197–206 (1995).
9. Y. Ma and Y. Ito, E. Sokoloski, and H. M. Fales, *J. Chromatogr. A 685*: 259–262 (1994).
10. A. Weisz, Separation and purification of dyes by conventional countercurrent chromatography and pH-zone-refining countercurrent chromatography, in *Countercurrent Chromatography* (Y. Ito and W. D. Conway, eds.), Chemical Analysis Series Vol. 132, Wiley–Interscience, New York, 1996, pp. 337–384.
11. Y. Ma, L. Qi, J. N. Gnabre, R. C. C. Huang, F. E. Chou, and Y. Ito, *J. Liquid Chromatogr. 21*: 171–181 (1998).
12. Y. Ma, E. Socoloski, and Y. Ito, *J. Chromatogr. A 724*: 348–353 (1996).
13. Y. Ma and Y. Ito, *J. Chromatogr. A 771*: 81–88 (1997).

Yoichiro Ito

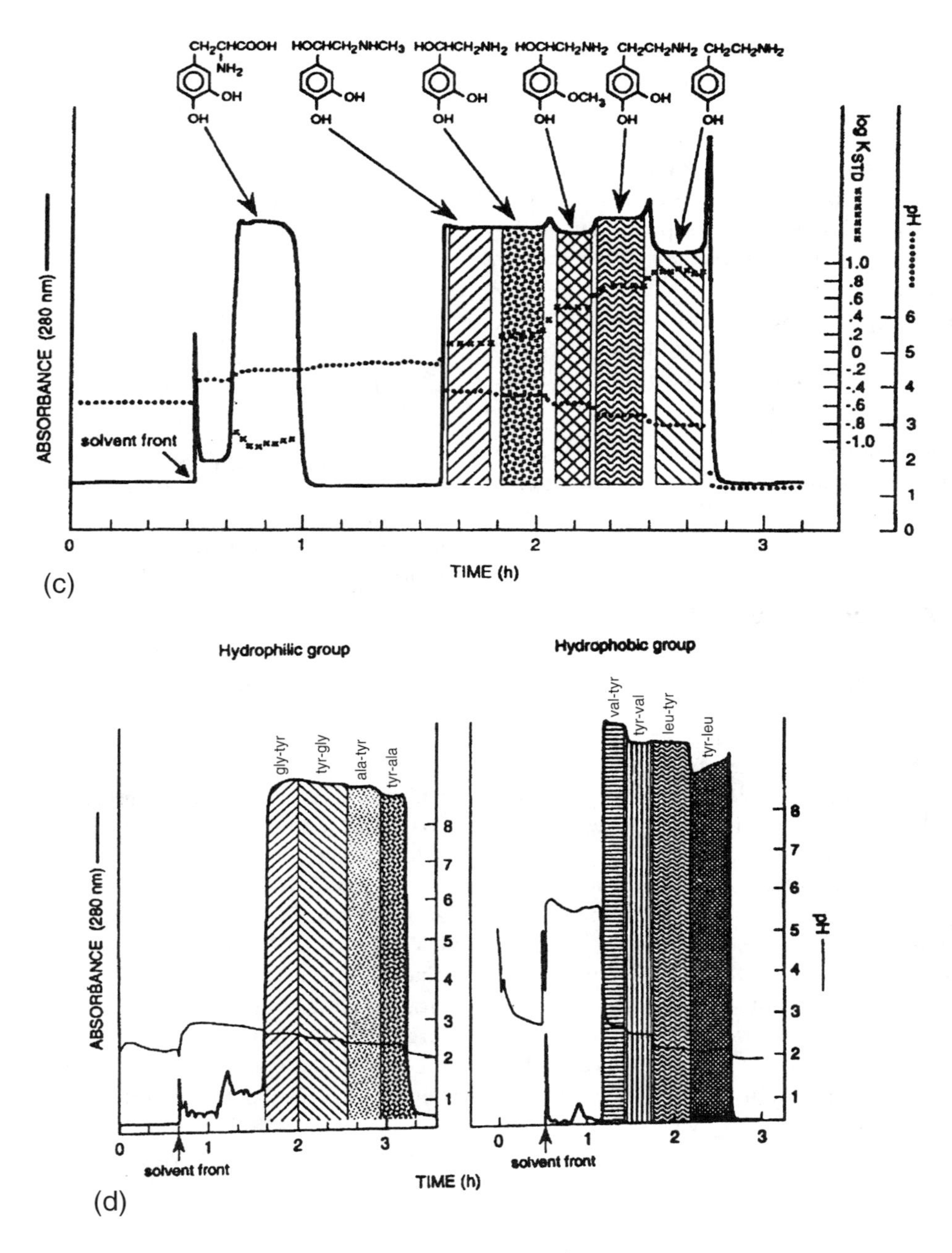

Fig. 4 *Continued*

Pharmaceuticals: Analysis by TLC

Introduction

The pharmaceutical industry has been producing hundreds of new compounds on the market during the last decades. This evolution would certainly not have been possible without the potentialities of analytical techniques. Indeed, analysis appears necessary in at least three fields of the pharmaceutical area:

1. In the development of new drugs, for the identification of interesting compounds and their metabolites or derivatives.
2. In the whole manufacturing process, where it is essential to guarantee that the product obtained is both efficient and safe. Part of this process relies on the analytical determination of the purity and quality of the active ingredient(s), their by-products, and the excipients.
3. In human beings, as well as in other animals, analytical methods may be useful to confirm a poisoning case or to monitor undesired residues in edible tissues of food-producing animals.

In this entry, we will briefly overview these fields and see how thin-layer chromatography (TLC) or high-performance thin-layer chromatography (HPTLC) can be of value.

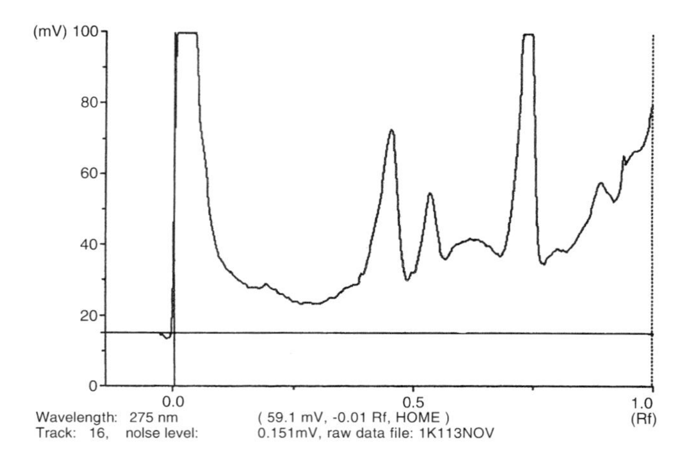

Fig. 1 Analysis of vitamin K_1 administered in dogs by the intravenous or intrarectal route (P. Berny, S. Viallet, F. Buronfosse, and G. Lorgue, European Association of Veterinary Pharmacology and Toxicology, Madrid 6–10 July, 1997).

Research and Development of New Drugs

In the development of new drugs, HPTLC may offer very interesting features. Contrary to other chromatographic techniques, all the constituents of a mixture are spotted on the plate and can be analyzed, even if they are not identified easily. Therefore, it is possible to determine if and how a given product is broken down or metabolized. For example, nimesulide, a common nonsteroidal anti-inflammatory drug (NSAID), was detected in equine blood and urine samples following a race [1]. An unidentified spot was also detected on the same plate. This spot was further identified as a nitro derivative of the active compound. This derivative appeared to be unique to horses. Other published data mention the use of HPTLC alone or in combination with other analytical techniques such as high-performance liquid chromatography (HPLC) or gas chromatography–mass spectrometry (GC–MS), to determine the presence and amount of new metabolites of drugs in various biological fluids and tissues. The use of TLC–HPTLC offers numerous advantages over other techniques (e.g., its simplicity and rapidity, and the potential for convenient repeated analysis), which is interesting when several tissues have to be analyzed in order to identify an unknown metabolite. Kinetic studies may also be performed with the help of TLC.

Indeed, after appropriate sample preparation, standards and samples can be spotted or sprayed onto the same plate, thereby providing a rapid and cost-effective analytical procedure to evaluate all sampling times. If an ultraviolet (UV) scanner is used, it is also possible to determine the amount of substance for each sample. Furthermore, the UV scanner sensitivity may be adapted specifically to analyze highly concentrated samples and poorly concentrated ones on the same plate, provided the calibration curve includes the values. As an example, Fig. 1 presents some kinetic data obtained for vitamin K_1, which is used as an antidote in dogs suffering from anticoagulant rodenticide poisoning [2]. The method was developed on C_{18}-coated silica gel plates. Elution was based on acetone–acetonitrile–chloroform (4:5:1) and took about 15 min for each plate. Detection was performed by UV scanning at 275 nm, followed by integration and solid-phase UV spectrum evaluation. Under such conditions, vitamin K_1 has an R_F value of 0. There were no interferences from plasma and limit of detection was 0.23 μg/mL, which was consistent with the therapeutic values obtained. The results also indicated that vitamin K_1 had a bioavailability of 82% by this route.

Standard approaches with HPTLC in research and development are usually based on the former TLC methods developed for older products. Unfortunately, these official methods (pharmacopoeias or associations of analytical chemistry) were based on TLC plates and material. The development of HPTLC material (100- or 200-μm-thick plates, microspheres of 5 μm, etc.) should improve many of these methods as far as duration of elution, resolution, and quantification are concerned. Therefore, older methods may have to be completely reviewed before being fully adapted to newer techniques.

A new step has been implemented with the development of automatic developing chambers and polarity gradients. With these techniques, development of a plate over 5–7 cm may be sufficient to separate up to 50 compounds and identify them. Universal gradient systems based on methanol, dichloromethane, and hexane, for instance (see *Camag Bibliography Service* for more information), have been developed as starting points for method development.

An example of the use of gradient techniques is given with plant extracts of the *Artemisia* genus [3]. In this entry, four related plant extracts were analyzed by isocratic and gradient techniques; the authors concluded that the gradient technique offered greater resolution and enabled the separation of more active ingredients in these plant extracts.

Drug Manufacturing Processes

Many methods have been published, based on TLC–HPTLC, for the qualitative analysis of drug or medical products. Numerous examples of quality control procedures could be presented here. It is interesting to look at quality control procedures for medicinal plants. Indeed, in this case, the only way to control the quality of a medicinal product is to make sure that the raw materials are correct (i.e., that they contain the desired active ingredients and the other constituents at the "usual" standard concentrations). The reader should refer to a review article such as in Ref. 4. HPTLC offers unbeatable characteristics for such uses: its simplicity, rapidity, and the possibility of analyzing several samples over a very brief period of time. Identification of the active ingredients and by-products is achieved, as well as the quantitative analysis of the crude extract or the final product. Densitometry may prove to be a useful tool to establish calibration curves, but quantitation may be achieved by means of derivatization and further densitometry.

A good example of the use of TLC in quality control procedures is given by Dhanesar [5]. The author used reversed-phase HPTLC (C_{18}-coated silica gel) to quantify ceftriaxone, a novel cephalosporin-derived antibiotic. In this example, direct spotting and quantitation was used based on UV densitometry, as there was no need for separation. Using such a technique, quantitative evaluation of the product was easy and rapid. As the author concluded, neither GC nor HPLC could be used that way. Results reported by another group of researchers [6] also indicated that HPTLC analysis for cephalexin and cefaclor gave satisfactory results with regards to the limit of detection and the limit of quantitation, with precision around 3% and recovery close to 100%. Even before final products are quantified, the fermentation processes in antibiotic production can be monitored by means of HPTLC. Indeed, there is no need for extraction and a mere dilution of the fermentation matrix with methanol provides rapid and reliable results to check that the fermentation process is functioning properly. The high polarity of the silica gel plate used retains many of the matrix components and the specificity of the chromatographic procedure may be increased by the use of postderivatization techniques [7].

Many derivatization techniques, based on chemical reactions or even heat and chemical reactions, can be applied to TLC or HPTLC to specifically determine some compounds or groups of compounds; this is specific to TLC techniques. For instance, exposure to fluorescamine in acetone will convert sulfonamide antibiotics into fluorescent derivatives which can be visualized under UV radiation [8]. Other techniques are reported and, as suggested earlier, the reader should refer to the latest edition of the *Official Methods of Analysis* published by the Association of Official Analytical Chemists (Arlington VA, U.S.A.) for complete details about all the available techniques for the routine analysis of drugs.

Detection of Residues and Monitoring of Poisoning Cases

More and more often, animal products are checked for the presence of undesirable residues of veterinary drugs. This is an important issue in public health, because residues may prove either toxic or allergenic and must be kept below their maximum residual limit (MRL). For this kind of analysis, it is necessary to analyze many types of biological samples [tissues (e.g., meat) or fluids (e.g., milk)]. The analytical techniques applied need to be very sensitive (some residues are monitored below the nanogram per gram threshold) and robust to give similar results when applied in different laboratories. It may also be of interest to analyze several compounds with only a single technique. For instance, several sulfonamide residues can be determined simultaneously in eggs by TLC with fluorescamine and densitometric analysis of the fluorescent derivatives. The limit of detection was 3 $\mu g/kg$, which is compatible with most MRLs. A similar technique, with a solid-phase extraction technique, was applied to milk samples to determine residues of these drugs as well [8]. In both cases, the techniques were validated and had mean percent recoveries as well as linearity and coefficients of variation compatible with recommended values of official agencies (Committee of Veterinary Medicinal Products in the European Union, or Food and Drug Administration in the United States). This regulatory requirement has to be carefully considered when a method is developed for analyzing residues in food. The analytical techniques developed for antibiotics appear to be valuable, but they are usually applied after microbiological evaluation of biological samples.

Other products can only be determined by chromatographic or other analytical methods. A major area of investigation concerns growth promoters (anabolic steroids or β-agonists). The use of these hormones is restricted in many parts of the world, or even prohibited (European Union). Some compounds are of special interest because they may have carcinogenic effects. This is true for diethylstilbestrol, a synthetic estrogen. Several methods have been published to detect these products [9]. Although we cannot offer any specific reference here, TLC techniques would appear suitable to control doping in sports. However, many sports authorities usually require mass spectrometry to identify the potential doping agents. It is our opinion, however, considering their sensitivity and selectivity, that HPTLC techniques would certainly provide a quick and cost-effective way of screening samples.

Our final comment concerns clinical toxicology. Thin-layer chromatographic procedures can be especially useful to screen samples for the presence of various drugs and chemicals involved in poisoning cases in human beings and in animals as well. A recent review article considered using TLC on biological samples to look for benzodiazepines [10], which are commonly involved in suicidal attempts. Another example involves meprobamate. This product is a common antiepileptic drug, and it is difficult to detect with standard HPLC–UV methods. A screening method with TLC and visualization with Erlich's reagent results in yellow spots, typical of meprobamate. This screening technique for neutral drugs appears fairly simple and is suggested prior to HPLC analysis [11]. Finally, a method has been developed for salicylates and their metabolites in urine samples. This method [12] involves acidic extraction of urine samples and analysis on silica gel plates with ferric chloride added as a chromogen. This technique allows for the determination of salicylic acid, methyl salicylate, and *p*-aminosalicylate at the same time, with detection limits below 1 μg/mL for each compound. Analysis of the parent drug, together with its metabolites, is interesting for diagnostic purposes, because it might give some hints as to when the product was taken (approximately) or what prognosis should be expected.

Conclusion

This entry is a brief introduction to the world of TLC in pharmaceutical research and should be considered merely as such. TLC and HPTLC techniques are numerous and have been developed for several specific needs. With the introduction of gradient development, newer techniques can be expected, which should be able to detect more compounds on a single plate than commonly performed today.

References

1. P. Sarkar, J. M. Mcintosh, R. Leavitt, and H. J. Gouthro, *Anal. Toxicol. 21*: 197–202 (1997).
2. P. Berny, S. Viallet, F. Buronfosse, and G. Lorgue, *J. Vet. Pharmacol. Therapeu., 20* (Suppl. 1): 270–271 (1997).
3. N. K. Olah, L. Muresan, G. Compan, and S. Gocan, *J. Planar Chromatogr. 11*: 361–364 (1998).
4. F. Li, S. Sun, J. Wang, and D. Wang, *Biomed. Chromatogr. 12*: 78–85 (1998).
5. S. C. Dhanesar, *J. Planar Chromatogr. 11*: 258–262 (1998).
6. D. Agbaba, S. Eric, D. Zivanov, S. Stakic, and S. Vladimirov, *Biomed. Chromatogr. 12*: 133–135 (1998).
7. Anon., *Camag Bibliography Service 81*: 10–13 (1998).
8. J. Unruh, E. Piotrovski, D. Schwartz, and R. Barford, *J. Chromatogr. 519*: 179–187 (1990).
9. G. Garcia, R. Saelzer, and M. Vega, *J. Planar Chromatogr. 4*: 223–225 (1991).
10. O. H. Drummer, *J. Chromatogr. B 713*: 201–225 (1998).
11. W. E. Lambert, *Clin. Toxicol. 30*: 683–684 (1992).
12. R. L. Kincaid, M. M. McMullin, D. Sanders, and F. Rieders, *J. Anal. Toxicol. 15*: 270–271 (1991).

Philippe J. Berny

Phenols and Acids: Analysis by TLC

Phenols are stronger acids than alcohols, because the oxygen atom acquires a positive charge by resonance and, thus, proton release is facilitated:

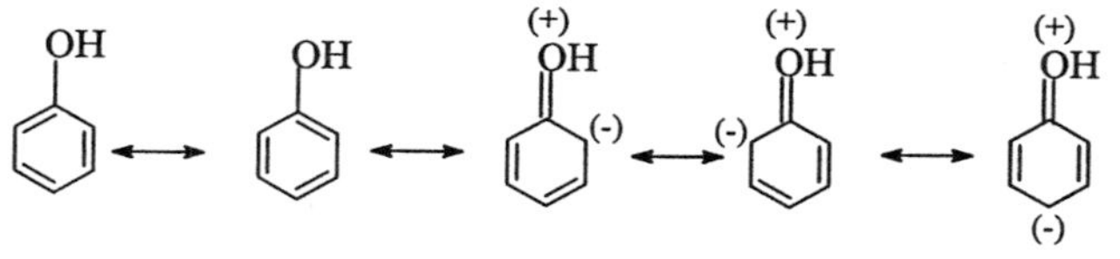

Phenol is a weak acid ($pK_a = 9.98$) and the effect of a ring substituent on the acid strength depends on whether the group is electron withdrawing or releasing, its position, and its ability to give resonating structures (i.e., the methyl group is electron releasing and decreases the acid strength from all ring positions).

The phenolic group occurs in a large number of natural and industrial products, extending from phenolic resins, herbicides, surfactants, alkaloids, steroids, and glycosides to numerous other groups.

A comprehensive review of phenolic compounds in biochemical, environmental, industrial, and consumer products, as well as of sample preparation prior to thin-layer chromatography (TLC) has been effected by Tyman [1].

Chromatographic Behavior of Phenols

Substituted monocyclic phenols have been widely studied on several stationary phases (alumina, silica gel, cellulose, polyamide, silanized silica gel) and also on chemically modified adsorbents (cyano- and amino-silica plates), ion-exchange layers, and impregnated plates.

Phenols can be detected with diazotized orthanilic acid or dianisidine by spraying an ammoniacal silver nitrate solution, followed by exposure to ultraviolet light, or with a modified ferric ferricyanide reagent, and also by exposing the wet layer successively to nitrogen dioxide and ammonia vapors.

Silica Gel, Alumina, Cellulose, and Polyamide

Layers of silica gel and alumina have been employed for the separation and identification of 126 monocyclic phenols eluted with 3 solvents of increasing polarity (benzene, diisopropylether, and ethanol) [2]. Alumina is more basic than silica gel and strongly retains phenolic compounds, particularly those with more acidic properties such as chlorophenols and nitrophenols, even when eluting with a medium-polarity solvent.

Table 1 reports the retention data relative to some substituted monocyclic phenols examined on silica gel in benzene and diisopropylether.

The presence of a 2-substituent or 2,6-substituents results in an increase in the R_f value; this is likely due to hydrogen-bonding or steric effects. However, 2,6-dinitrophenol is more retained in benzene than the 2,4 and 2,5 isomers. The sequence of their R_f values ($R_{f2,6} = 0.04 < R_{f2,4} = 0.07 < R_{f2,5} = 0.38$) is in agreement with that of the corresponding pK_a values ($pK_{a_{2,6}} = 3.71 < pK_{a_{2,4}} = 4.09 < pK_{a_{2,5}} = 5.22$) and, therefore, these two facts can be closely bound up with one another.

Diisopropyl ether allows the separation of nitrophenol isomers and of several nitroalkyl and nitrochlorophenols, as well as of dihydroxy and trihydroxybenzenes. On alumina plates, dimethyl and trimethylphenols are better resolved than on silica gel when eluting with the above-mentioned solvent, as shown by the following R_f value sequences:

Dimethylphenols

$$R_{f3,4} = 0.22 < R_{f3,5} = 0.26 < R_{f2,4} = 0.28 < R_{f2,3} = 0.39 < R_{f2,5} = 0.44 < R_{f2,6} = 0.52.$$

Trimethylphenols

$$R_{f3,4,5} = 0.16 < R_{f2,4,5} = 0.23 < R_{f2,3,5} = 0.46 < R_{f2,4,6} = 0.51$$

Hydrogen-bonding, steric effects, and acid–base properties of phenols are involved in their retention on silica gel and alumina with benzene and isopropylether as eluents.

Only a limited number of researches have been focused on cellulose plates, microcrystalline cellulose being the most used stationary phase with solvents such as ethylacetate–*n*-propanol–25% ammonia (3:5:2), water–formic acid (98:2), *n*-amyl alcohol–acetic acid–water (10:6:5), and benzene–propionic acid–water (4:9:3).

Polyamide is an especially useful adsorbent for the separation of phenols owing to the formation of hydrogen bonds between the phenolic compounds and the amide group of the polymer. Organic solvents of increasing polarity and aqueous–organic solutions have been used as eluents: benzene, chloroform, ethylacetate, water–methanol, water–acetone, water–acetic acid, and cyclohexane–acetic acid (93:7) mixtures.

Water–propanol–27% ammonia (1:8:1), *n*-butanol–

Table 1 R_f Values of Substituted Phenols in Different Chromatographic Conditions

	Silica gel		Cellulose + ethyl oleate[a]	Silanized silica + 4% DBS[b]		RP-18 + 4% HDBS[c]		
Compound	A	B	C	D	E	F	G	pK_a (25°C)
Phenol	16	74	79	35	62	—	—	10.02
2-Methylphenol	24	78	62	20	35	—	—	10.32
3-Methylphenol	16	75	67	20	40	—	—	10.09
4-Methylphenol	15	74	66	19	36	—	—	10.27
2,3-Dimethylphenol	25	80	45	11	20	—	—	10.54
2,4-Dimethylphenol	28	83	42	11	19	—	—	10.60
2,5-Dimethylphenol	25	77	47	11	21	—	—	10.41
2,6-Dimethylphenol	40	76	40	12	20	—	—	10.63
3,4-Dimethylphenol	15	75	53	12	24	—	—	10.36
3,5-Dimethylphenol	15	73	51	12	25	—	—	10.19
2-Ethylphenol	27	77	40	10	21	—	—	10.2*
3-Ethylphenol	17	72	48	11	28	—	—	9.9*
4-Ethylphenol	15	69	45	11	24	—	—	10.0*
2-Chlorophenol	42	75	53	24	89	48	77	8.48
3-Chlorophenol	20	74	43	14	64	34	50	9.02
4-Chlorophenol	16	69	46	14	50	34	37	9.38
2,3-Dichlorophenol	38	63	28	22	85	35	78	7.45**
2,4-Dichlorophenol	38	65	22	16	78	27	69	7.75**
2,5-Dichlorophenol	41	77	20	28	85	41	81	7.35**
2,6-Dichlorophenol	56	83	31	55	88	59	86	6.79**
3,4-Dichlorophenol	15	63	21	7	57	18	48	8.39**
3,5-Dichlorophenol	—	—	14	9	64	16	55	7.93**
2,3,4-Trichlorophenol	—	—	—	18	68	29	66	7.59
2,3,5-Trichlorophenol	—	—	—	25	69	36	63	7.23
2,3,6-Trichlorophenol	49	76	—	45	82	52	74	6.12
2,4,5-Trichlorophenol	32	67	7	22	70	42	63	7.33
2,4,6-Trichlorophenol	49	76	15	37	73	43	63	6.42
3,4,5-Trichlorophenol	—	—	—	9	52	16	50	7.74
2,3,4,5-Tetrachlorophenol	—	—	—	14	46	35	44	6.96
2,3,4,6-Tetrachlorophenol	36	50	4	—	—	—	—	—
2,3,5,6-Tetrachlorophenol	—	—	—	25	50	38	43	5.44
Pentachlorophenol	20	21	2	13	33	—	—	5.26
2-Nitrophenol	69	79	—	61	92	—	—	7.23
3-Nitrophenol	7	62	—	29	92	—	—	8.40
4-Nitrophenol	4	46	—	66	92	—	—	7.15
Catechol (1,2)	2	49	—	—	—	—	—	—
Resorcinol (1,3)	0	39	—	56	87	—	—	9.81
Hydroquinone	0	43	—	67	e.s.	—	—	10.35

Eluents: A = benzene; B = isopropylether; C = 25% aqueous ethanol; D = 0.1 M NH_3 + 0.1 M NH_4Cl in 30% methanol (pH = 9.02); E = 1 M NH_3 in 30% methanol (pH = 11.30); F = 0.1 M NH_3 + 0.1 M NH_4Cl in 60% methanol; G = 1 M NH_3 in 40% methanol.

[a] 15 g cellulose impregnated with a solution of ethyl oleate in ether (70 ml of a 0.75% v/v solution);

[b] 20 g silanized silica gel 60 HF (C_2) mixed with a 4% triethanolamine dodecylbenzensulphonate (DBS) solution in 95% ethanol;

[c] RP-18 ready-to-use plates dipped in a 4% dodecylbenzensulphonic acid solution in 95% ethanol;

* pK_a values at 28°C;

** pK_a values at 29°C.

$5M$ ammonia (100:33), and n-butanol–ethanol–ammonia (5:1:1) have also been employed for nitrophenols.

Ion-Exchange Resins and Impregnated Plates

A wide study of the chromatographic behavior of alkyl, halogenated phenols, and phenols containing alkyl and halogeno groups by reversed-phase TLC has been performed by Bork and Graham [3] on cellulose impregnated with ethyl oleate eluting with aqueous ethanol (see Table 1).

The phenols can be removed by the stationary phase as a result of solvation of the phenolic group by the proton acceptor eluent (water or ethanol), which may be influenced by steric factors and by altering the polarity of the phenolic grouping.

Long-chain alkylphenols present in natural cashew nut shell liquid have been chromatographed on argentated silica gel G [10% (w/w) silver nitrate] with diethylether–light petroleum–formic acid (30:70:1) as eluent for the separation of unsaturated constituents.

Alkylphenols, nitrophenols, halogenophenols and polyhydroxybenzenes have been extensively studied on a thin layer of anion and cation exchangers with cellulose, paraffin, and polystyrene matrices and on silanized silica gel impregnated with anionic and cationic surfactants. The best results have been obtained by using cation exchangers and anionic surfactants as impregnating agents [4,5].

The parameters that determine the retention of phenols on layers of silanized silica gel, untreated and impregnated with anionic surfactants, are the same that affect retention on cation exchangers (i.e., the organic modifier percentage, the ionic strength, and, particularly, the pH of the eluent).

With regard to the influence of pH, the protonated form of the phenols exhibits a higher affinity toward the stationary phase than the deprotonated form. From the relationship

$$K_d = \frac{[\mathrm{HA}]_R + [\mathrm{A}^-]_R}{[\mathrm{HA}]_S + [\mathrm{A}^-]_S} \frac{V}{W} \tag{1}$$

(where K_d is the distribution coefficient, $[\mathrm{HA}]_R$, $[\mathrm{A}^-]_R$, $[\mathrm{HA}]_S$, and $[\mathrm{A}^-]_S$ are the concentrations of the protonated and deprotonated form of the phenol in the resin and in the solution, V is the volume of the solution, and W is the weight of the resin) and introducing the K_a value into Eq. (1), combined with the Martin–Synge equation for partition thin-layer chromatography, Lepri et al. [4] obtained the following relationship:

$$\left(\frac{1}{R_f} - 1\right) = \left(\frac{1}{R_{f\,\mathrm{ac}}} - 1\right)\frac{[\mathrm{H}^+]}{K_a + [\mathrm{H}^+]} + \left(\frac{1}{R_{f\,\mathrm{alk}}} - 1\right)\frac{K_a}{K_a + [\mathrm{H}^+]} \tag{2}$$

where $R_{f\,\mathrm{ac}}$ and $R_{f\,\mathrm{alk}}$ are the R_f values of the protonated and deprotonated form of the phenol achieved by eluting with strong acidic and alkaline solutions, respectively.

Although the major change in $(1/R_f) - 1$ with pH occurs at pH $=$ pK_a, differentiating R_f twice with respect to $\log[H^+]$ and equating to zero, the following relation is obtained:

$$[\mathrm{H}^+] = K_a \frac{R_{f\,\mathrm{ac}}}{R_{f\,\mathrm{alk}}} \tag{3}$$

On the basis of Eq. (3), we can predict that the mean R_f value of $R_{f\,\mathrm{ac}}$ and $R_{f\,\mathrm{alk}}$ will be shifted more with respect to the pH $=$ pK_a value the lower the $R_{f\,\mathrm{ac}}/R_{f\,\mathrm{alk}}$ ratio is. Many nitrophenols, chlorophenols, and bromophenols can be easily separated by this technique eluting with aqueous–organic solutions at different pH values (see Table 1).

Chromatographic Behavior of Phenolic Acids and Their Derivatives

In general, silica gel has been more widely used than cellulose, polyamide, and silanized silica gel for the separation of phenolic acids and their derivatives.

Selected eluents for silica gel are chloroform–ethylacetate–formic acid (5:4:1), n-hexane–ethylacetate–formic acid (15:9:2), chloroform–acetic acid–water (2:1:1), toluene–dichloromethane–formic acid (40:50:10), and dichloromethane–acetic acid–water (100:50:50, lower phase).

Recently, two-dimensional TLC on cellulose plates has been used for the separation of 14 phenolic acids eluting with methanol–acetonitrile–benzene–acetic acid mixtures in the first direction and sodium formate–formic acid–water (10:1:200, v/v/v) in the second direction.

Phenolic aldehydes and ketones have been chromatographed on silica gel G and cellulose plates as their phenylhydrazones formed *in situ*. Toluene–chloroform–acetone (5:3:2), chloroform–acetone (8:2), anisole–methanol (8:2), and anisole–chloroform–acetone (5:3:2) are the eluents used for silica gel, whereas the layers of cellulose have been eluted with 2% formic acid, 20% potassium chloride, 10% acetic acid, or isopropanol–ammonia–water (8:1:1).

Long-chain alkylphenolic acids and their derivatives have been separated on silica gel G eluting with light petroleum (60–80°C)–ethylether–dimethylformamide–acetic acid (75:85:5:1).

Reversed-phase TLC on silanized silica gel layers (OPTI-UPC_{12}, $SilC_{18}$-50 and RP-18), untreated and impregnated with anionic and cationic surfactants, has been used for the separation of catecholamines, phenolic acids, and glycols excreted in the urine. Many interesting separations have been achieved on OPTI-UPC_{12} plates eluting with 1*M* hydrochloric acid +3% KCl in water (biogenic amines) and with 1*M* sodium acetate in water (urinary phenolic acids and glycols).

Quantitative Determinations

Phenols occurring in water have been quantified by *in situ* densitometry after coupling with diazotized *p*-nitroaniline or by using vanadium pentoxide and dichlorofluorescein.

A recent article reported the analysis of phenols included in the list of priority pollutants of the Environmental Protection Agency on silica gel G F_{254} HPTLC plates and polyamide plates after solid-phase extraction from water [6].

In situ quantitation has been performed by absorption ultraviolet or visible light measurements of the color developed with Würsters salts.

Planar chromatography has also been used to separate and quantify several common phenols in contaminated land leachates after derivatization with 3-methyl-2-benzothiazolinone hydrazone (MBTH) and extraction of the resulting azo dyes.

References

1. J. H. P. Tyman, Phenols, aromatic carboxylic acids, and indoles, in *Handbook of Thin-Layer Chromatography* (J. Sherma and B. Fried, eds.), Marcel Dekker, Inc., New York, 1996, pp. 877–920.
2. F. Dietz, J. Trandy, P. Koppe, and C. Rubelt, Detection and photometric determination of 126 phenolic compounds in water using for group-specific reagents, *Chromatographia 9*: 380–396 (1976).
3. L. S. Bark and J. T. Graham, Studies in the relation ship between molecular structure and chromatographic behaviour, *J. Chromatogr. 23*: 417–442 (1966); *J. Chromatogr. 25*: 357–366 (1966).
4. L. Lepri, P. G. Desideri, M. Landini, and G. Tanturli, Chromatographic behaviour of phenols on thin-layers of cation and anion exchangers, *J. Chromatogr. 109*: 365–376 (1975); *J. Chromatogr. 129*: 239–248 (1976).
5. L. Lepri, P. G. Desideri, and D. Heimler, Reversed-phase and soap thin-layer chromatography of phenols, *J. Chromatogr. 195*: 339–348 (1980).
6. J. Bladek, A. Rostkowshi, and S. Neffe, The application of TLC to the determination of phenol residues in water, *J. Planar Chromatogr.—Mod. TLC 11*: 330–335 (1998).

Luciano Lepri
Alessandra Cincinelli

Photodiode-Array Detection

Introduction

Analysis is an integral part of research, clinical, and industrial laboratory methodology. The determination of the components of a substance or the sample in question can be qualitative, quantitative, or both. Techniques that are available to the analyst for such determinations are abundant. In absorption spectroscopy, the molecular absorption properties of the analyte are measured with laboratory instruments that function as detectors. Those that provide absorbance readings over the ultraviolet-visible (UV-vis) light spectrum are commonly used in high-performance liquid chromatography (HPLC). The above method is sufficiently sensitive for quantitative analysis and it has a broader application than other modes of detection.

The most advanced UV-vis detector that is used in HPLC is the diode array. Compared to its predecessors, the diode array generates a large amount of spectral information without compromising sensitivity or wavelength resolution. In order to discuss the properties of the diode-array detector, a brief historical description of its technological development is presented.

Historical Background and Schematics

Early on, UV-vis detectors produced data at one wavelength only (fixed-wavelength detector, FWD). Because compounds of interest do not absorb light at a fixed wavelength with equal efficiencies, the next step was to develop a detector with an adjustable wavelength range. Modifications were required in the light source. The high-energy-emitting mercury lamp in the FWD was replaced with a deuterium and/or tungsten lamp. The latter light source gave detectable energies over a wider wavelength range. The addition of a wavelength selector (monochromator) provided accurate wavelength adjustments from 190 to 800 (UV-vis) nm. Hence, the variable-wavelength detector (VWD) was established.

At this point, single-wavelength data acquisition remained a bottleneck, restricting the analyst to one basic characteristic tool to identify the analyte: the amount of time the analyte is retained on the chromatographic column (i.e., the retention time). If there are four compounds in the sample to be analyzed, this would require a minimum of four injections to identify them. Additional work would be needed for cross-identification and so on. The concept of multiple-wavelength detection was introduced.

The diode-array detector collects data with a maximum wavelength bandwidth of 190–800 nm. All wavelengths in that range are accounted for simultaneously. As shown in Fig. 1, the upper portion represents a single detector element with three components: the photodiode, a capacitor, and a FET switch known as the field effect transistor switch. Shining light onto a single detector element produces a signal which is processed and then expressed as a digital absorbance reading. This response is for one wavelength only. The FWDs and VWDs are equipped with a single detector element. The bottom portion of Fig. 1 displays the schematics of a linear diode-array detector. Many detector elements, all of the same composition as described above, are in a linear arrangement. Each detector element is dedicated to a particular wavelength bandwidth such that there are enough of them to cover the UV-vis light spectrum. This unit produces signals that cover the entire UV-vis light spectrum. Simultaneous wavelength detection was now possible with the diode array.

Figure 2 displays the optical components of the diode-array detector. A deuterium lamp will emit light onto the flow cell, which has liquid continuously flowing through it. A shutter is provided between the light source and the flow cell, which can either be fully open or closed. The beam of light will then travel to the diffraction grating, where it will be separated into wavelengths ranging from 190 to 800 nm. The separated light will finally reach the diode array and a signal will be produced that is proportional to the amount of light received. For a detailed account of this topic, the reader is referred to the selected references cited.

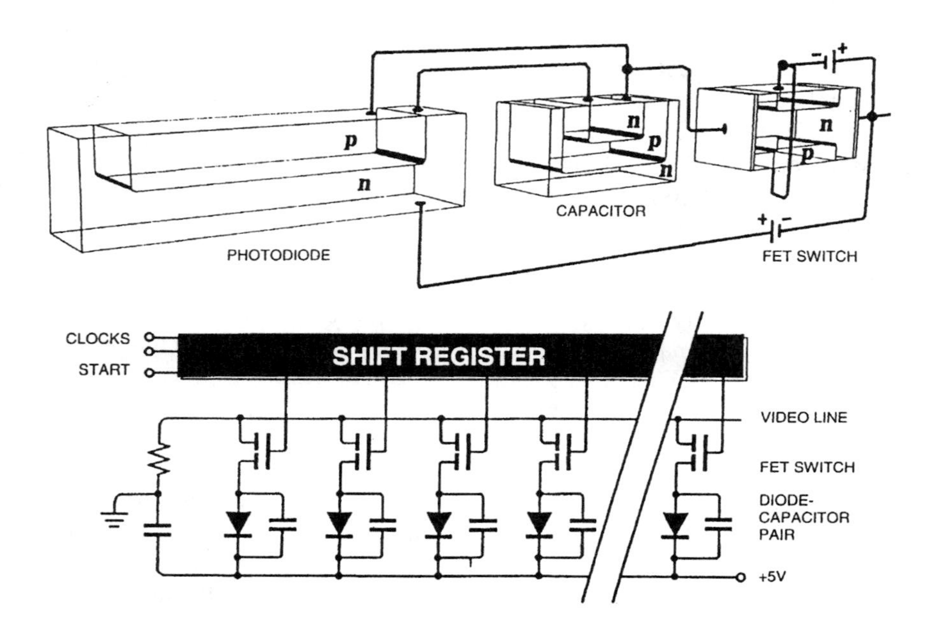

Fig. 1 Diagram of a single photodiode-capacitor switch (top) and an electronic schematic of the linear array (bottom).

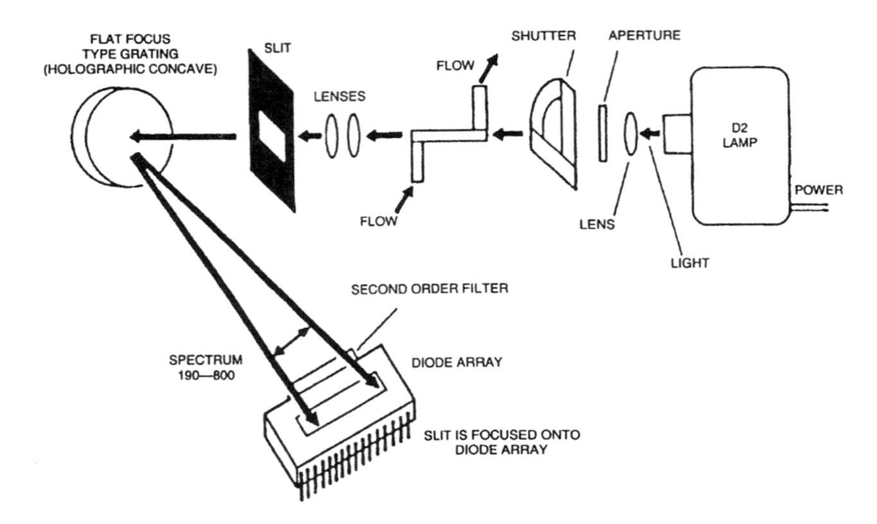

Fig. 2 Optics unit components. (Reprinted from *Waters 990+ Photodiode Array Detector Operator's Manual 1989*; courtesy of Waters Corporation, Milford, MA.)

Applications

Thus far, we have given a brief account on the technological development of the diode-array detector. Its applications offer several advantages to the user, which have been made possible primarily by the current computer hardware and software that are available on the market. Data acquisition and analysis are computer driven. The information acquired from the diode array can be analyzed to tailor the user's preferences. This area represents an emerging field and further advancements in analytical software tools will ultimately determine the true potential of diode-array detection.

As stated previously, a major improvement achieved with the diode array is that fewer attempts are required to identify the components of a sample mixture. A single injection can be sufficient for sample identification when spectral information in addition to peak retention time is part of the collected data. Through software capabilities, spectra can be extracted from the chromatogram of each individual peak. The spectral information combined with the retention times can be used to identify the chromatographic peaks. Furthermore, the spectra can be compared to spectral libraries compiled from the literature for purposes of spectral matching. Many proprietary software programs have been developed for the diode-array instrument. It is up to the user's discretion to determine how he can best utilize these resources.

At first, one may show skepticism at the usefulness of diode-array detection because other analytical systems are more sensitive and offer similar features such as peak identification and purity checks. For these alternatives, one would have to refer to gas chromatography–mass spectrometry (GC–MS), liquid chromatography–mass spectrometry (LC–MS), and tandem mass spectrometry (MS–MS). However, one must keep in mind the significantly higher costs as a trade-off for enhanced sensitivity.

When assessing the merits of the diode array, the user should consider the various steps that are involved in methods development. It will become clear that for every hurdle one encounters, diode-array detection renders the task less laborious. It is most often than not that in developing new procedures you are confronted with drawbacks such as contaminants, peak overlap, artifacts, sample cleanup, and baseline noise. For instance, the presence of a shoulder on a major chromatographic peak raises questions as to its origin. With the aid of the diode array, a spectral profile of the shoulder peak will quickly determine its existence. A flat line of low absorbance throughout its spectra indicates that it is an artifact. The user can then go back to his procedure and ascertain if the root of the problem is systematic. The direction problem-solving takes is crucial and if gone astray, vast amounts of time are wasted. The same shoulder obtained with a FWD could have led the user to perform a time-dependent study. In this case, it was incorrectly assumed that the shoulder peak was a degradation product and additional injections at increased time intervals would show the shoulder peak increasing in size, indicating the compound of interest was degrading. Even though the analyst would eventually achieve the same result in the latter situation, the former approach is less time-consuming.

Diode-array detection has found its way into the pharmaceutical industry and the clinical laboratory. The widespread use of HPLC analysis in the product development cycle of pharmaceuticals provided an open door for the diode array. Throughout the different phases of drug development, the identification of a peak and its homogeneity must be established. The diode array confirms peak identity and compound purity by spectral analysis. Peak spectra are overlaid with spectra from standards (that are stored on the computer hard drive) for identification and spectra from the upslope, apex, and downslope of a chromatographic peak are overlaid to test for purity. Because the diode array performs continuous-flow spectral acquisition in the HPLC mode, the method can be automated to screen for thousands of compounds where only one will have the desired pharmacological action. Drug screening is common in drug development procedures, which goes hand-in-hand with drug metabolism and quantitative analysis. Diode-array detection is effective in all of these areas.

Peak purity is of utmost importance in the quality control of pharmaceutical products. The contents of the product must be known with great certainty before it is sold to the consumer. Adjunct laboratories test the stability of the product in order to determine its expiry date. All by-products are accounted for and the amount of active ingredient is referenced. The diode array in HPLC analysis is used to determine all of the above information.

In the clinical laboratory setting, methods of analysis are used to diagnose disease and for therapeutic drug monitoring. Diode array–HPLC techniques will determine the drug concentrations and its metabolite levels in biological matrices. From these results, dosing regimens can be outlined that are specifically tailored to patients with respect to therapeutic drug concentration and drug dose, thereby avoiding drug toxicity. These procedures will ensure the safety and efficacy of clinically administered drugs.

Conclusion

The photodiode-array detector is a powerful analytical instrument that has provided enhanced detection capabilities with the addition of detailed spectral information via its multisignal detection technology. Its applications are HPLC based and can be found in basic research, automated analysis, pharmaceutical product development, and the clinical laboratory environment. Through spectral acquisition and analysis, a wealth of information can be obtained about the identity and purity of a compound. Combined with high selectivity and sensitivity, this mode of detection is essential for qualitative and quantitative HPLC analysis.

Selected Further Reading

Green, R. B., Absorption detectors for high-performance liquid chromatography, in *Detectors for Liquid Chromatography* (E. S. Yeung, ed.), John Wiley & Sons, New York, 1986, pp. 42–46.

Huber, L. and S. A. George (eds.), *Diode Array Detection in HPLC*, Marcel Dekker, Inc., New York, 1993.

Lloyd, D. K., Instrumentation: detectors and integrators, in *High Performance Liquid Chromatography* (W. J. Lough and I. W. Wainer, eds.), Blackie Academic & Professional/Chapman & Hall, London, 1996, pp. 120–125.

McMaster, M. C., Hardware specifics, in *HPLC: A Practical User's Guide*, VCH, Weinheim, 1994, pp. 112–114.

Scott, R. P. W., Liquid chromatography detectors, in *Techniques and Practice of Chromatography*, Marcel Dekker, Inc., New York, 1995, pp. 284–287.

Hassan Y. Aboul-Enein
Vince Serignese

Photophoretic Effects in FFF of Particles

High versatility and remarkable instrumental power of the field-flow fractionation (FFF) family of techniques are based substantially on the large variety of driving forces used [1]. Thus, the further development of FFF is associated, considerably, with the search for new types and combinations of physical and physicochemical agents suitable for the particles' and molecules' redistribution across the flow. Among the possible new force agents, light interaction with particles leading to their motion (photophoresis) deserves special attention. This is due both to the universal character of this interaction and to a large variety of particular interaction mechanisms. These mechanisms form two families.

The "direct photophoretic" mechanisms include light pressure and gradient force, both associated with the direct momentum transfer from the incident light to a particle or a molecule during the absorption or scattering of photons (see Refs. 2–4 for details). The "gradient force" [3] arises from the spatial nonhomogeneity of electromagnetic field (if any) and has the same nature as the force moving electric dipoles in a static divergent field. It is characteristic of the focused light beams, evanescent waves near the refractive index boundaries, and so forth.

The "indirect photophoretic" mechanisms involve, as a primary light action, either nonuniform heating of particles or establishing a temperature gradient across the channel due to the light absorption in the carrier fluid. Such heating leads to conventional thermophoresis of particles due to tangential gradients of interfacial energy arising near the particle surface. The whole phenomenon can be referred to as photothermophoresis [5].

From the physical picture, it is evident that photophoretic effects in FFF are influenced by a remarkable diversity of particles material properties, including geometrical, optical, and thermophysical characteristics. The use of these effects offers a unique possibility of including the optical properties of particles and molecules into the nomenclature of separation parameters. It promises to exploit the high spectral selectivity, in particular, the resonant character of light pressure exerted upon some solid-state and molecular structures [6,7], to separate the species identical in their density and geometry parameters.

Such selectivity can be achieved in several ways; first, by exploiting the intrinsic optical properties of the species and, second, using special preparative procedures, such as selective staining of species, or their binding to special carrier particles (molecules), having appropriately designed optical characteristics and serving as a kind of light tugs.

Let us consider, briefly, the basic theoretical principles underlying the photophoretic effects and their possible use in FFF. For further reading, see Ref. 5 and the references therein. The calculations of the direct photophoretic force, as well as of the spatial distribution of absorbed energy in the case of indirect mechanisms, require the solution of the optical problem of light scattering by a particle. This can be done using either the wave optics theory [4] or the geometrical (ray) optics description [3, 5], depending on the particle size parameter $\rho = 2\pi a/\lambda$. Here, λ is the wavelength of light and a is the characteristic particle dimension (radius for the spherical particles). In practice, the boundary between the "geometrically small" and "large" particles is $a \sim 10\ \mu$m for the visible light.

The direct photophoretic force F_{ph} for "geometrically small" particles can be written as a sum of the radiation pressure force F_{pr} and the gradient force F_{∇}:

$$\begin{aligned} \vec{F}_{ph} &= \vec{F}_{pr} + \vec{F}_{\nabla} \\ \vec{F}_{pr} &= \varepsilon_m \frac{\pi a^2}{c} Q_{pr}(\varepsilon, \rho) \cdot \vec{J}_0, \\ \vec{F}_{\nabla} &= \varepsilon_m \frac{2\pi a^3}{c} \left(\frac{\varepsilon' - 1}{\varepsilon' + 2} \right) \cdot \nabla \vec{J}_0 \end{aligned} \tag{1}$$

Here J_0 is the incident light irradiance (measured in units of energy per unit time per unit area), c is the light velocity, Q_{pr} is the dimensionless efficiency factor of the radiation pressure [4], $\varepsilon = \varepsilon' - i\varepsilon''$ is the complex dielectric permittivity of particle material related to the permittivity ε_m of surrounding medium. In general, the efficiency factor can be calculated numerically using the electromagnetic theory [4]. It can be obtained analytically as an instructive example of a spherical particle placed in a wide slightly divergent light beam with effective radius $w_0 \gg a$ and angular divergence $|\Delta\theta| \ll 1$, $|\Delta\theta| \geq a/w_0$ [5]. Taking $\Delta\theta > 0$ for divergent beam, $\Delta\theta < 0$ for convergent beam, considering the transverse and the longitudinal gradients of beam irradiance to be $\nabla_{\perp}(J_0) \approx J_0/w_0$, $\nabla_{\|}(J_0) \approx -\Delta\theta(J_0/w_0)$, and regarding $\varepsilon' + 2 \approx 3$ because $|\varepsilon - 1| \ll 1$, we get [5]

$$\begin{aligned} F_{ph\|} &\equiv F_{pr} + F_{\nabla\|} \approx \varepsilon_m \frac{\lambda^2 \rho^2}{4\pi c} \\ &\quad \cdot \left(\frac{1}{4}(\varepsilon' - 1)^2 I_1(\rho) + \frac{4}{3}\varepsilon'' \rho \right) J_0 - \Delta\theta F_{\nabla\perp} \\ F_{ph\perp} &\equiv F_{\nabla\perp} \approx \varepsilon_m \frac{\lambda^3 \rho^3}{12\pi^2 c w_0} (\varepsilon' - 1) J_0 \end{aligned} \tag{2}$$

Here $I_1(\rho)$ is an increasing function of ρ, with the small-amplitude oscillations superimposed [5]. For small particles ($\rho \ll 1$), $I_1(\rho) \approx (32/27)\rho^4$, whereas for large ones ($\rho \gg 1$), $I_1(\rho) \approx 2[\ln(4\rho) + 0.5772] - 3$. The analysis [5] of Eq. (2) shows that the longitudinal component $F_{ph\|}$ of direct photophoretic force can be dominated both by the light pressure term and by the gradient term, depending on the particle optical parameters and the beam nonhomogeneity: $|F_{\nabla\|}| > |F_{pr}|$ if $w_0 < \lambda|\Delta\theta||\varepsilon' - 1|/4\pi\varepsilon''$. The direction of $F_{\nabla\|}$ is determined by the sign of $\Delta\theta$ (beam's convergence or divergence). Hence, in general, both the positive (away from the light source) and the negative photophoresis are possible. The direction of $F_{\nabla\perp}$ depends on the sign of $\varepsilon' - 1$. The particles with $\varepsilon' > 1$ are pulled into the maximum of the radial distribution of field intensity, whereas the particles with $\varepsilon' < 1$ are pushed either out of the beam or into the minimum of the field intensity. The direct photophoretic force F_{ph} has a strong dependence on the particle size: the term $F_{pr} \propto a^6$ for small particles and

$F_{pr} \propto a^2$ for large particles. It depends also on their optical properties: $F_{pr} \propto |\varepsilon' - 1|^2$ for weakly absorbing particles and $F_{pr} \propto \varepsilon''$ for particles with a strong light absorption. The direct photophoretic force acting on very large particles ($\rho \gg 1$) is better for calculations using the geometrical optics picture of reflected and refracted light rays, which exert the pressure on the particle surface at their points of incidence. The total force is obtained by the integration of this pressure over the particle surface [3].

Turning to the indirect photophoretic mechanisms, or photothermophoresis, consider the problem of light-absorbing particle surrounded by transparent liquid. The opposite extreme, a transparent particle placed into a channel flow of light-absorbing fluid, reduces to an ordinary thermophoresis, however, with unusual temperature profile in the channel. The photothermophoresis problem consists of three independent stages [5]. The first is the calculations of the optical field inside the particle, in order to obtain the source function $B(\vec{r}) = E_i^2(\vec{r})/E_0^2$, which describes the heat production due to the light absorption. Here, $E_i(\vec{r})$ is the electric field strength at the point $\vec{r}$ inside the particle and E_0 is the incident wave amplitude. The second stage is the determination of the temperature field inside and outside the particle generated by the known distribution of heat sources. The third is the calculation of the photothermophoretic velocity u_{phth} of a particle in the temperature field thus obtained on the basis of specific physicochemical models for the particle, surrounding medium, and their interfacial region. The result has the form [5]

$$\begin{aligned} u_{phth} &= b_{th} g_{an}(n_r, \kappa\rho) \frac{T_0}{a} \\ \frac{T_0}{a} &= 2\kappa\rho \frac{n_r J_0}{m_0 k_i} \\ g_{an}(n_r, \kappa\rho) &= -\frac{1}{4}\left(2 + \frac{k_i}{k_e}\right) \int_0^{\pi} \frac{d\tau(a,\theta)}{d\theta} \sin^2\theta \, d\theta \end{aligned} \quad (3)$$

Here, b_{th} is the ordinary thermophoretic mobility of a particle, k_i and k_e are the thermal conductivity of a particle and fluid, respectively, T_0 is the characteristic temperature of particle heating due to the light absorption, T_0/a is the characteristic temperature gradient across the particle, $\tau(a, \theta)$ is the dimensionless temperature distribution over the particle surface in units of T_0, the polar angle θ is measured from the shadow pole of illuminated particle, $n(\lambda)$ is the refraction index, $\kappa(\lambda)$ is the absorption index of particle material, m_0 is the refractive index of fluid, and $n_r = n(\lambda)/m_0$. The function $g_{an}(n_r, \kappa\rho)$ describes the anisotropy of particle heating by light. Together with b_{th}, it defines the magnitude and sign of photothermophoretic velocity. The analysis for a solid particle in electrolyte solution shows that u_{phth} is negative for $\kappa\rho \ll 1$, goes through the zero in the range $\kappa\rho \sim 0.1$–0.3, then stays positive and increases monotonically for $\kappa\rho > 0.5$, saturating at very high $\kappa\rho$ values ($\sim$40) [5].

The practical implementation of photophoretic FFF encounters two major difficulties. First, the generation of substantial photophoretic force over the full length of separation channel requires very powerful light sources and special optical delivery systems, including transparent channel walls. Second, the transparency of the channel wall should not degrade substantially with time despite the permanent contact with suspension flow. The feasibility studies of photophoretic effects in FFF were made with the focused Ar^+-ion laser beam propagating along the axis of the round metallic capillary along the flow of water suspension of carbon-black particles [5]. Their results evidence that the practical transverse geometry Photophoretic FFF technique can be developed also. The most promising schemes are highly convergent light sheet normal to the channel wall and enabling the gradient trapping of particles, and the use of photophoretic forces in combination with some counterbalancing force, such as the gravity force.

References

1. J. C. Giddings, *Chem. Eng. News 66*: 34–45 (1988).
2. A. Askin, *Science 210*: 1081–1088 (1980).
3. A. Askin, *Biophys. J. 61*: 569–582 (1992).
4. C. F. Bohren and D. R. Huffman, *Absorption and Scattering of Light by Small Particles*, Interscience, New York, 1983.
5. V. L. Kononenko, J. K. Shimkus, J. C. Giddings, and M. N. Myers, *J. Liquid Chromatogr. Related Technol. 20*: 2907–2929 (1997).
6. S. Arnold, Spectroscopy of single levitated micron sized particles, in *Optical Effects Associated with Small Particles* (P. W Barber and R. K. Chang, eds.), World Scientific, Singapore, 1988, pp. 65–137.
7. V. G. Minogin and V. S. Letokhov, *Laser Light Pressure on Atoms*, Gordon & Breach, New York, 1987.

Vadim L. Kononenko

Plant Extracts: Analysis by TLC

Introduction

Thin-layer chromatography (TLC) is a powerful method for separating mixtures of compounds of very different polarity. Plant extracts are usually hydroalcoholic solutions, containing complex mixtures of compounds, many of them being still unidentified. Although plant extracts are widely used in homeopathic medicine, for the treatment of different diseases, the control of these drugs is often performed by qualitative analysis. TLC can provide a chromatographic "fingerprint" of a plant extract, very useful for identification purposes, and usually a photograph is attached to the analysis certificate [1]. The colors of the separated spots and their positions relative to standard substances are important characteristics for the plant extract identification. Quantitative analysis is seldom applied in this field, and it is usually performed by spectrophotometry, which is included in pharmacopoeias.

Discussion

TLC Analysis of Plant Extracts in Pharmacopoeias

The TLC separation of plant extracts is described as a method of analysis in different pharmacopoeias [2–5] and is usually performed on silica layers and sometimes on a silica hydrocarbon (C_8, C_{18}) bonded phase. However, alumina or other stationary phases are not excluded.

The plant extract samples can be applied directly onto the plate, or a specific class of compounds is extracted in a suitable solvent before TLC. The polarity of the solvent used for extraction should be similar to that of the desired compounds. The samples can be applied onto the plates manually, by using calibrated micropipettes or automated applicators. The applied samples can be spots or bands, and the mobile-phase migration distances vary between 8 and 15 cm.

Normal presaturated chambers are used for development. The mobile phases used for the TLC development are characteristic of different classes of compounds. The classification of medicinal plants takes into account the presence of different classes of compounds which are separated by using the so-called TLC fingerprint method. The analyzed compounds can be alkaloids, anthracene derivatives, bitter drugs, cardiac glycoside drugs, coumarin drugs, drugs containing essential oils, flavonoids, saponin drugs, drugs containing triterpenes, and so forth. For example, mixtures containing chloroform and diethylamine are used for the TLC separation of plant extracts containing alkaloids and toluene–ethyl acetate is used for drugs containing essential oils [1].

After development, the plates are dried in a gentle airstream, at room temperature, and are examined in ultraviolet or visible light, with or without a derivatization, depending on the chemical nature of the separated compounds. The examination in ultraviolet light is performed at 254 nm by quenching the thin-layer fluorescence or at 366 nm when the natural fluorescence of compounds is observed. Several alkaloids (quinine, quinidine, cinchonine, cinchonidine, noscapine, berberine), anthraglycosides, coumarins derivatives (scopoletin, aesculetin), or flavonoids (chlorogenic acid, rutin) have natural fluorescence and can be visualized at 366 nm. The natural fluorescence of compounds can be enhanced by spraying with different reagents, leading to low detection limits. Flavonoids show a yellow–green fluorescence after consecutively spraying with alcoholic solutions of diphenylboryloxyethylamine and poly(ethylene glycol) 4000.

The compounds which cannot be derivatized for fluorescence at 366 nm are sprayed with specific reagents and examined in visible light. The derivatization reactions can take place at room temperature or the TLC plate should be heated on a thermostated plate, at a specific temperature, for approximately 10 min. In both situations, the chromatographic plate should be examined immediately and eventually photographed. Ninhydrine is a good detecting reagent for amino acids and other biocompounds containing the primary amino group, and the compounds can be observed as blue–violet spots after heating at 100°C. Terpenoids, bitter principles, or saponins can be visualized as red–violet spots by spraying the plate with an acidic solution of anisaldehyde and heating at 100°C.

The pharmacopoeias describe the obtained colored spots after a TLC separation, taking into account their color, order, and position on the plate in the presence of reference substances. The description should match the TLC separation of the studied plant extract. The correspondence of the spots is important for the qualitative analysis of the medicinal plant. The chosen reference substances mentioned in pharmacopoeias can also be present in the analyzed plant extract or can be used for a comparison with the R_F values of the separated compounds in the plant extract.

Actual Trends in the TLC Analysis of Plant Extracts

The exact chemical composition of a plant extract is not always completely known. Many articles published in recent years attempt to identify the compounds structure by coupling chromatography with spectrometric methods. Modern densitometers are able to record the *in situ* ultraviolet-visible (UV-vis) spectra of a separated substance on a TLC plate [6]. Thin-layer chromatography can be also coupled with other methods in order to enhance the identification of compounds, such as mass spectrometry (MS) or nuclear magnetic resonance (NMR). There are devices able to record the *in situ* spectra on the TLC plate, or the separated substance is removed from the plate together with the layer, then extracted in a small volume of an adequate solvent, and the sample can be used for obtaining the spectra [6,7].

It is well known that the concentration of active substances in medicinal plants can vary widely in different parts of the plant, and it depends on the harvesting time. The quantitative determination of active substances is then very useful for a plant extracts analysis. Densitometry is the method for the evaluation of the separated substances on a TLC plate. The quantitative determination can use a calibration curve of a reference substance or the internal or external standard method. Figure 1 shows the densitograms of two medicinal plant extracts, *Uva ursi* and *Vaccinium vitis-idaea*, both containing arbutin as one of the bioactive substances. The TLC plates are scanned in UV or vis light, usually in the reflectance mode, linear or zigzag. In the zigzag mode, the absorbance of a band around the separated sample is measured and it is useful for the samples that do not migrate in a straight line due to the layer imperfections.

Automated multiple development (AMD) has been successfully applied for the separation of compounds from

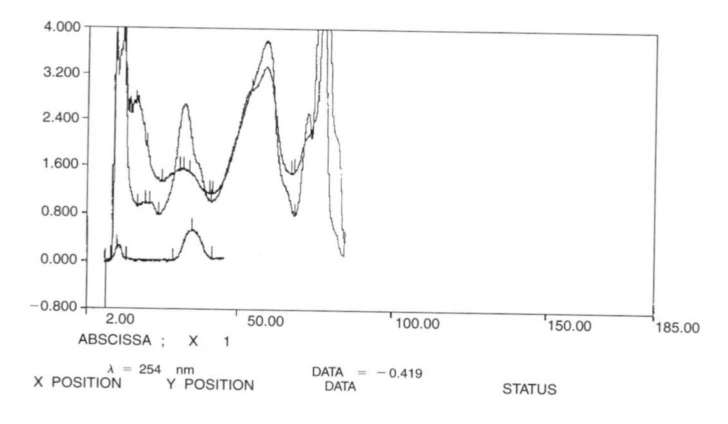

Fig. 1 The densitometry of *Uva ursi* and *Vaccinium vitis-idaea*, two alcoholic plant extracts containing arbutin. Arbutin (reference substance) is the last one, and *Vaccinium vitis-idaea* contains less arbutin than *Uva ursi*; silica gel plate, $\lambda = 254$ nm.

plant extracts. AMD is a technique using the concentration gradient to separate substances differing widely in polarity. Usually, the gradient is started from a polar composition decreasing in steps to a medium polar mobile phase and ending with a nonpolar mixtures of solvents or to a single nonpolar solvent. In this way, all the compounds from a plant extract can be separated, from very polar to nonpolar, between the start and the front line. Usually, the migration distance is 8 cm, and the number of steps can be a maximum of 25 in order to achieve a reasonable development time. The development distances increase as the solvent polarity is decreased. As the total number of compounds in a plant extract is very difficult to estimate, the AMD technique always yields a greater number of separated spots than a monodimensional or a two-dimensional isocratic TLC development [8]. Most of the reference substances used in plant extract analysis have a natural origin. As a consequence, they are not always very pure or can lead to decomposition compounds. The separation power of an AMD method can show a greater number of compounds in a reference substance than an isocratic TLC method.

Glycerinic plant extracts have various application in the naturist medicine, but they cannot be analyzed directly by TLC due to the presence of glycerin which should be removed first from the sample [9]. Solid-phase extraction (SPE) is a fast and convenient method for the separation of glycerin. The organic compounds from the analyzed plant extract are retained in the SPE cartridge on a nonpolar stationary phase (usually silica–C_{18}); the glycerin and the nonselectively retained compounds are eluted with a polar mobile phase (can be a diluted solution of methanol in water). The retained compounds are eluted from the cartridge with a stronger mobile phase, usually acidified with a mineral acid, and the obtained solution can be analyzed by TLC directly or after concentration. The SPE recovery is very good, around 100%, and the method is successfully applied for the plant extracts cleanup before TLC.

References

1. H. Wagner and S. Bladt, *Plant Drug Analysis*, Springer-Verlag, Berlin, 1996.
2. *Homöopathisches Arzneibuch*, Deutscher Apotheker Verlag, Stuttgart/Govi-Verlag, Frankfurt, 1978 (Vol. 1) and 1991 (Vol. 2).
3. *La Pharmacopée Française, Xe Édition*, Assocoation pour la Recherche Appliqué à la Pharmacopée, Paris, 1989.
4. *DAB 10, Deutsches Arzneibuch 10, Ausgabe 1991*, Deutscher Apotheker Verlag, Stuttgart, 1991.
5. *The United States Pharmacopoeia, XXIII Revision*, United States Pharmacopoeial Convention, Rockville, MD, 1995.

6. S. Gocan and G. Cimpan, *Rev. Anal. Chem. 16*: 1–24 (1997).
7. V. Ossipov, K. Nurmi, J. Loponen, E. Haukioja, and K. Pihlaja, *J. Chromatogr. A 721*: 59–68 (1996).
8. N.-K. Olah, L. Muresan, G. Cimpan, and S. Gocan, *J. Planar Chromatogr.-Mod. TLC 11*: 361–364 (1998).
9. S. Cobzac, G. Cimpan, N. Olah, and S. Gocan, *J. Planar Chromatogr.—Mod. TLC 12*: 26–29 (1999).

Gabriela Cimpan

Plate Number, Effective

The concept of the effective plate number was introduced in the late 1950s by Purnell and Bohemen [1] and Desty and Golup [2]. Its introduction arose directly as a result of the development of the capillary column. It was noted that the very high efficiencies were only realized from open-tubular columns for solutes eluted close to the column dead volume (i.e., for solutes eluted at very low k' values). In addition, the high efficiencies in no way reflected the increase in resolving power that would be expected from a packed column with much higher stationary-phase loading.

This poor performance, relative to the high efficiencies produced, results from the high phase ratio inherent with open-tubular columns. The high phase ratio is due to there being very little stationary phase in the capillary column (the film is very thin). The corrected retention volume of a solute is directly proportional to the amount of stationary phase in the column, and solutes, in general, are eluted from a capillary column at relatively low k' values relative to the magnitude of their distribution coefficient.

To compensate for what appeared to be very misleading efficiency values, the *effective plate number* was introduced. The effective plate number uses the *corrected retention distance*, as opposed to the *total retention distance* to calculate the efficiency. Thus, the effective plate number is significantly smaller than the true plate number for solutes eluted at low k' values. At high k' values, the two measures of efficiency converge. In this way, the effective plate number appears to more nearly correspond to the column resolving power. The efficiency of the column (n) in number of theoretical plates has been shown to be given by

$$n = 4\frac{y^2}{x^2}$$

where y is the retention distance, and x is the peak width.

Now, the number of "effective plates" (N), by definition, is given by

$$N = 4\frac{(y - y_0)^2}{x^2} \tag{1}$$

where y_0 is the retention distance of an unretained solute (the position of the dead point). Now, from the plate theory,

$$\frac{y}{x} = \frac{n(v_m + Kv_s)}{2\sqrt{n}(v_m + Kv_s)}$$

thus,

$$\frac{y - y_0}{x} = \frac{n(v_m + Kv_s) - nv_m}{2\sqrt{n}(v_m + Kv_s)}$$

By dividing through by v_m and noting that

$$\frac{Kv_s}{v_m} = k'$$

then,

$$\frac{y - y_0}{x} = \frac{\sqrt{n}k'}{2(1 + k')}$$

Consequently,

$$4\left(\frac{y - y_0}{x}\right)^2 = n\left(\frac{k'}{1 + k'}\right)^2 = N \tag{2}$$

Equation (2) describes the relationship between the efficiency of a column in theoretical plates and the efficiency given in "effective plates." It is also seen that the calculation of the number of "effective plates" in a column does not provide an arbitrary measure of the column performance, but is directly related to the number of theoretical plates in the column as defined by the plate theory. It should be noted that as k' becomes large, n and N converge to the same value.

The effective plate number has an interesting relationship to the function for the resolution of a column that was suggested by Giddings [3]. Giddings proposed that the function $k'/\Delta k'$ could be a means of defining the resolving power R of a column. He employed this function in an analogous manner to the function used in spectroscopy to define resolution (i.e., $\lambda/\Delta\lambda$, where $\Delta\lambda$ is the minimum wavelength increment that can be differentiated at a wavelength λ). The value taken by Giddings for $\Delta k'$ was the bandwidth at the base of the eluted peak which is equivalent to twice the peak width or 4σ. Thus, from the plate theory,

$$R = \frac{k'}{\Delta k'} = \frac{nkv_s}{4\sqrt{n}(v_m + Kv_s)}$$

Again, dividing through by v_m and noting that

$$\frac{Kv_s}{v_m} = k'$$

then,

$$R = \frac{\sqrt{n}k'}{4(1 + k')} = \frac{\sqrt{N}}{4} \quad (3)$$

It is seen from Eq. (3) that the resolving power of the column, as defined by Giddings, will be directly proportional to the square root of the number of effective plates. As a consequence, R can be used by the chromatographer to directly compare the resolving power of columns of any size or type. However, the value of R will vary with the value of k' for the solute, and so comparison between columns must be made using solutes that have the same k' value.

References

1. J. H. Purnell and J. Bohemen, *J. Chem. Soc.* 2030 (1961).
2. D. H. Desty and A. Goldup, *Gas Chromatography 1960* (R. P. W. Scott, ed.), Butterworths, London, 1960, p. 162.
3. J. C. Giddings, *The Dynamics of Chromatography*, Marcel Dekker, Inc., New York, 1965, p. 265.

Suggested Further Reading

Scott, R. P. W., *Liquid Chromatography Column Theory*, John Wiley & Sons, Chichester, 1992, p. 19.

Scott, R. P. W., *Introduction to Analytical Gas Chromatography*, Marcel Dekker, Inc., New York, 1998.

Raymond P. W. Scott

Plate Theory

Originally derived by Martin and Singh [1] and extended by Said [2], the plate theory provides an equation that describes the elution curve (the chromatogram) of a solute. By differentiating the elution curve equation and equating to zero, an expression for the retention volume of a solute can be obtained. By equating the second differential to zero, an equation for the variance and standard deviation (the peak width) can be obtained, and from these equations, methods for calculating the column efficiency can be derived, together with the numerous equations that describe resolution.

The plate theory assumes that the solute is in equilibrium with the mobile and stationary phases. Due to the continuous exchange of solute between the two phases as it progresses down the column, equilibrium between the phases can *never* actually be achieved. To accommodate this nonequilibrium condition, a technique originally introduced in distillation theory is adopted, where the column is considered to be divided into a number of cells or plates. Each cell is allotted a finite length and, thus, the solute spends a finite time in each cell. The size of the cell is such that the solute is considered to have sufficient residence time to achieve equilibrium with the two phases. Thus, the smaller the plate, the more efficient the solute exchange between the two phases and, consequently, the more plates there are in the column. As a result, the number of theoretical plates contained by a column has been termed the *column efficiency*. The *plate theory* shows that the peak width (the dispersion or peak spreading) is inversely proportional to the square root of the efficiency and, thus, the higher the efficiency, the narrower the peak. Consider the equilibrium that is assumed to exist in each plate; then

$$X_s = KX_m \quad (1)$$

where X_m is the concentration of solute in the mobile phase, X_s is the concentration of solute in the stationary phase, and K is the distribution coefficient of the solute between the two phases.

It should be noted that K is defined with reference to the stationary phase (i.e., $K = X_s/X_m$), thus the larger the distribution coefficient, the more the solute is distributed in the stationary phase. Differentiating Eq. (1),

$$dX_s = KdX_m \quad (2)$$

Consider three consecutive plates in a column, the $p - 1$, the p, and the $p + 1$ plates and let there be a total of n plates in the column. The three plates are depicted in Fig. 1.

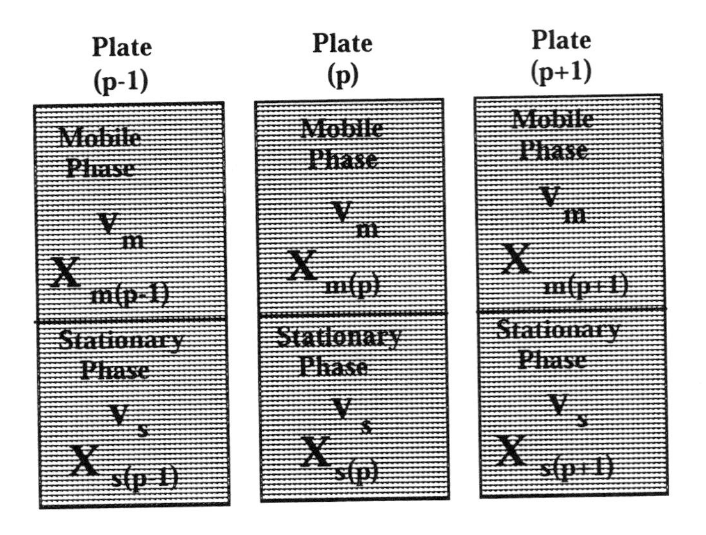

Fig. 1 The three consecutive plates in a column.

Let the volumes of mobile phase and stationary phase in each plate be v_m and v_s, respectively, and the concentrations of solute in the mobile and stationary phase in each plate be $X_{m(p-1)}$, $X_{s(p-1)}$, $X_{m(p)}$, $X_{s(p)}$, $X_{m(p+1)}$, and $X_{s(p+1)}$, respectively. Let a volume of mobile phase, dV, pass from plate $p - 1$ into plate p, at the same time, displacing the same volume of mobile phase from plate p to plate $p + 1$. In doing so, there will be a change of mass (dm) of solute in plate p that will be equal to the difference in the mass entering plate p from plate $p - 1$ and the mass of solute leaving plate p and entering plate $p + 1$. A simple mass balance procedure can be applied to plate p. Thus, bearing in mind that mass is the product of concentration and volume, the change of mass of solute in plate p will be

$$dm = (X_{m(p-1)} - X_{m(p)})\, dV \tag{3}$$

Now, if equilibrium is to be maintained in the plate p, the mass (dm) will distribute itself between the two phases, which will result in a change of solute concentration in the mobile phase of $dX_{m(p)}$ and in the stationary phase of $dX_{s(p)}$. Then,

$$d_m = v_s\, dX_{s(p)} + v_m\, dX_{m(p)} \tag{4}$$

Substituting for $dX_{s(p)}$ from Eq. (2),

$$d_m = (v_m + Kv_s)\, dX_{m(p)} \tag{5}$$

Equating Eqs. (3) and (5) and rearranging,

$$\frac{dX_{m(p)}}{dV} = \frac{X_{m(p-1)} - X_{(p)}}{v_m + Kv_s} \tag{6}$$

The volume flow of the mobile phase will now be measured in units of $v_m + Kv_s$, instead of milliliters. Thus, the new variable (v) can be defined as

$$v = \frac{V}{v_m + Kv_s} \tag{7}$$

The function $v_m + Kv_s$ is termed the "plate volume" and, thus, the flow of mobile phase will be measured in "plate volumes" instead of milliliters. The "plate volume" can be defined as that volume of mobile phase that would contain all the solute that is in the plate at the equilibrium concentration of the solute in the mobile phase.

Differentiating Eq. (7),

$$dv = \frac{dV}{v_m + Kv_s} \tag{8}$$

Substituting for dV from Eq. (8) in Eq. (6),

$$\frac{dX_{m(p)}}{dV} = X_{m(p-1)} - X_{(p)} \tag{9}$$

Equation (9) is the basic differential equation that describes the rate of change of concentration of solute in the mobile phase in plate p with the volume flow of mobile phase through it. Thus, the integration of Eq. (9) will provide the equation for the elution curve of a solute eluted from any plate in the column. A simple algebraic solution to Eq. (9) is given in Ref. 3 and the resulting elution curve equation for plate n of a column of n plates is shown to be

$$X_{m(n)} = \frac{X_0 e^{-v} v^n}{n!} \tag{10}$$

Equation (10) is the basic elution curve equation; it is a Poisson function, but when n is large, the function approximates to a normal error function or Gaussian function. In practical chromatography systems, n is always greater than 100 and, thus, all chromatographic peaks will be Gaussian or nearly Gaussian in shape.

The Retention Volume of a Solute

The retention volume of a solute is that volume of mobile phase that passes through the column between the injection point and the peak maximum. It is, therefore, now possible to determine that volume by differentiating Eq. (10) and equating to zero and solving for v. Restating Eq. (10),

$$X_{m(n)} = X_0 \frac{e^{-v} v^n}{n!}$$

$$\frac{dX_{m(n)}}{dv} = X_0 \frac{-e^{-v} v^n + e^{-v} n v^{n-1}}{n!}$$

$$= X_0 \frac{-e^{-v} v^{n-1}}{n!} (n - v)$$

Equating to zero and solving for v,

$$n - v = 0 \quad \text{or} \quad v = n$$

Thus, n plate volumes of mobile phase have passed through the column (remembering that the volume flow is measured in "plate volumes" and not in milliliters). Thus, the volume passed through the column (in ml) will be

$$V_r = n(v_m + Kv_s)$$
$$= nv_m + nKV_s \qquad (11)$$

Now, the total volume of mobile phase and stationary phase in the column (V_m and V_s, respectively) will be the volume of mobile phase and stationary phase per plate multiplied by the number of plates (i.e., nv_m and nv_s). Thus,

$$V_r = V_m + KV_s \qquad (12)$$

Returning to Eq. (13), it is also possible to derive an equation for the adjusted retention volume, V'_r. Now,

$$V'_r = V_r - V_m \qquad (13)$$

Thus, from Eqs. (11) and (13),

$$V'_r = V_m + KV_s - V_m = KV_s \qquad (14)$$

For the use of the plate theory to determine peak widths and column efficiency, see the entry Resolution.

References

1. A. J. P. Martin and R. L. M. Synge, *Biochem. J. 35*: 1358 (1941).
2. A. S. Said, *Am. Inst. Chem. Eng. J. 2*: 477 (1956).
3. R. P. W. Scott, *Liquid Chromatography Column Theory*, John Wiley & Sons, Chichester, 1992, p. 19.

Suggested Further Reading

Scott, R. P. W., *Techniques and Practice of Chromatography*, Marcel Dekker, Inc., New York, 1996.

Scott, R. P. W., *Introduction to Analytical Gas Chromatography*, Marcel Dekker, Inc., New York, 1998.

Raymond P. W. Scott

Pollutant–Colloid Association by Field-Flow Fractionation

Instrumental Methods

The association of pollutants such as trace metals, nutrients, and toxic organic molecules to colloids is intimately connected to the health of natural waters. Colloids, with their large specific surface area, play a dominant role in the transportation and eventual deposition of these pollutants. Of particular interest is the size speciation data. It is important to know not only the total amount of pollutant present but also where it is distributed. It has been inherently difficult to study pollutant–colloid interactions because of the lack of methods for particle size determination and fractionation as well as the low concentrations of pollutants present in many systems. This entry outlines a new approach using field-flow fractionation (FFF).

Field-flow fractionation is a separation and elution technique similar to chromatography [1]. It is based on the application of a field perpendicular to the flow of the axis of a thin (100–500 μm) channel. An externally applied field drives unlike particles to different positions across the thin channel, where they are caught up in different flow velocities. For small particles, typically less than 1 μm, the elution time depends on the particle's interaction with the field and its diffusivity. Separations in this mode, termed the *normal* mode, have the smaller particles eluting ahead of the larger particles. Larger particles tend to stay near the channel wall and move through the channel with the lower flow velocities. An alternate mode, termed the *steric* or *hyperlayer* mode, is for particles greater than about 1 μm. A reversal of the elution order is observed because they necessarily protrude into the higher flow velocities (due to their physical size). Utilizing these two modes, it is possible to probe a mass range spanning 15 orders of magnitude, starting from molecules of 1000 Da molecular weight up to particles 50 μm in diameter.

Further, it is possible to utilize different fields to yield the various FFF subtechniques. The two most common fields are centrifugal and fluid cross-flow, which give rise to the sedimentation and flow FFF subtechniques. Other fields currently in use include thermal, electrical, and magnetic fields. In the normal mode, it is possible to extract physical parameters from retention data. For example, sedimentation FFF using a centrifugal force gives information about the buoyant mass, and flow FFF gives information about the sample's diffusivity or hydrodynamic diameter.

Environmental samples often have a broad size distribution and are often heterogenous in density and chemical composition, and to add to the complexity, natural particles often have an irregular shape [2]. Thus, it is easy to appreciate that FFF with its wide and flexible range of operating parameters is the ideal tool to divide these broad distributions into discrete, roughly monodisperse fractions for subsequent analysis. Sedimentation and flow FFF, in particular, have been used to measure the size distribution of environmental samples. Karaiskakis et al. first demonstrated that it is possible to correlate chemical content (major elements such as Al, Ca, Fe, Si, and S found in bulk minerals) with particle size using sedimentation FFF with energy dispersive x-ray analysis (EDXA) [3]. However, due to the low sensitivity and long analysis time of FFF–EDXA, the technique was abandoned for the analysis of pollutants.

Much more sensitive and less time-consuming techniques such as mass spectrometry, atomic emission, and atomic absorption are needed for the analysis of pollutants. Detectors such as graphite furnace–atomic absorption spectrometer (GF–AAS), inductively coupled plasma–mass spectrometer (ICP–MS), or inductively coupled plasma–atomic emission spectrometer (ICP–AES) seem to be ideal candidates for the analysis of trace metals because of their very low detection limits. The high temperatures used avoid the need for tedious digestions in many samples. FFF–gas chromatography–mass spectrometry could perhaps be used in the analysis of particular organic molecules. Another extremely sensitive technique applied in the study of adsorption behavior of pollutants is to add radiolabeled adsorbates (such as $^{33}PO_4^{3-}$, ^{14}C-atrazine, and ^{14}C-glyphosate) to study the distribution of the pollutant as a function of size.

Contaminant Speciation Data

In recent years, the direct coupling of the FFF channel to GF–AAS, high-resolution ICP–MS, and ICP–AES has been implemented. It has enabled high resolution size-based speciation data for pollutants to be collected. In the first instance, the data acquired by these methods is a fractogram [an instrumental signal, usually but not necessarily from an ultraviolet (UV) detector, representing the mass of eluted sample as a function of the retention time]. The retention time is rigorously but not linearly related to the particle size (i.e., diameter) and may be easily calculated from FFF theory. Suitable algorithms can be used to generate a mass-based size distribution. Figures 1a and 1b show an example of recent work conducted in our laboratory demonstrating the distribution of copper in contaminated soils represented as a fractogram and a size distribution.

Element concentrations in the eluent are also recorded if a suitable detector is used. This can be processed in a similar fashion to the UV signal to yield an eluent-based size distribution. If the element detector signal is divided by the mass detector signal, we obtain a quantity which is proportional to the concentration in the sample particles. When this quantity is plotted against particle diameter, we obtain the element concentration distribution for the sample (Fig. 1c). It is often useful to plot the element atomic ratio distributions for elements of interest. This graph is particularly useful for deducing size-based speciation data for trace elements.

The main feature of this experiment is that although the copper content in the soil roughly follows the fractogram (Fig. 1a) and the size distribution (Fig. 1b), the concentration of copper is higher for the smaller particles (Fig. 1c). However, a comparison of the copper/aluminium ratio shows that there is no change across the size distribution for this sample. Figure 1d shows that the surface-coating density of copper is increasing with particle size, suggesting a denser or thicker coating of copper as the particle size increases.

Contado et al. [4] coupled sedimentation FFF indirectly to GF–AAS as well as directly to ICP–MS to produce element composition data across the size distribution. The high levels of Cu, Pb, Cr, and Cd found were associated with colloidal particles taken from a river situated in a highly industrialized site. The two methods give comparable results, with on-line coupling of ICP–MS having a higher resolution, but ICP–AES yields data for some elements (such as potassium and calcium) where ICP–MS produces interferences.

Adsorption Behavior of Pollutants

The adsorption behavior of colloidal material onto river particles can play a vital role in the transport and fate of pollutants. FFF methods provide a means to evaluate the

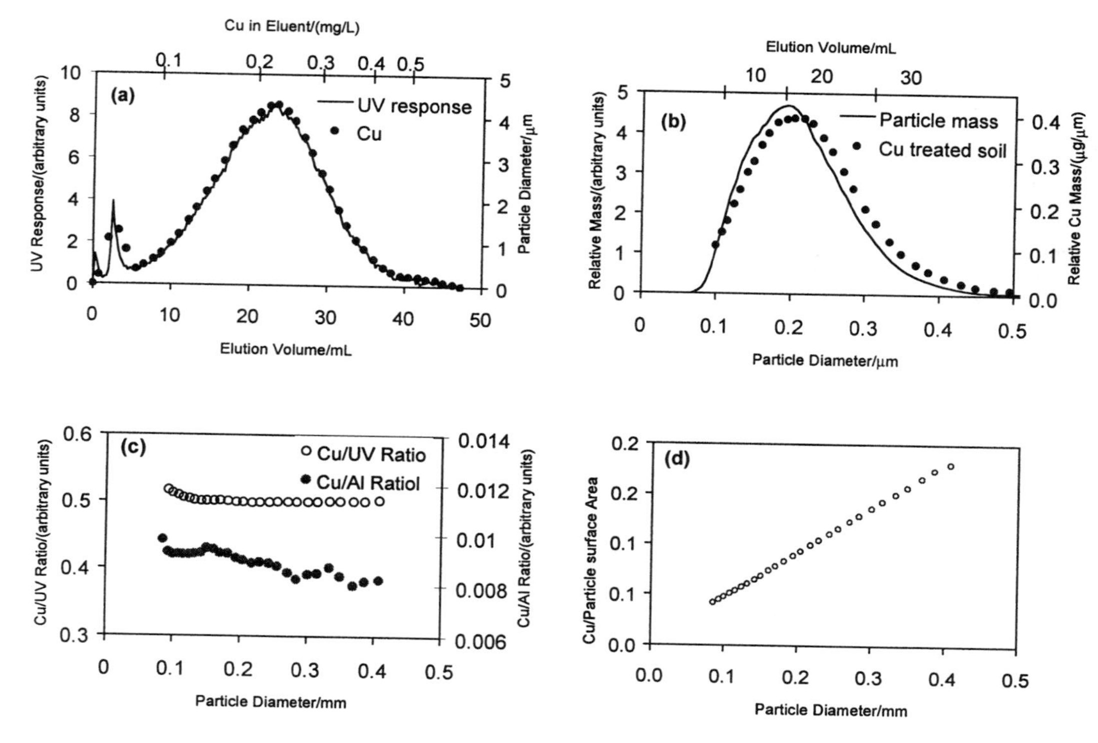

Fig. 1 Graphs showing (a) fractogram and copper concentration in eluent, (b) corresponding size distribution, (c) copper concentration distribution and element ratio distribution of copper in soil sample, and (d) copper per unit surface area distribution.

relative importance of different fractions in adsorption of contaminants in soils and sediments.

One particularly sensitive method employs the adsorption of radioactive material to natural particles [5]. Radiolabeled ^{32}P (as H_3PO_4) and ^{14}C (as glyphosate and atrazine) were adsorbed onto the river water colloid samples for several hours, then separated using sedimentation FFF and specific size fractions were collected. The fractions were subsequently analyzed for their β-activity. These data yield adsorbate (activity)-based fractograms and size distribution (Fig. 2a and b).

Dividing the adsorbate size distribution by the mass size distribution gives an adsorbate concentration (i.e., the amount adsorbed per mass of particles) distribution as outlined previously. Figure 2b shows that the smaller particles contain the highest pollution content. This is consistent with the concept that smaller particles have a higher specific surface area, but changes in geochemistry of the particle with size could also be involved.

If the adsorption were uniform, we would expect the adsorbate concentration to increase with decreasing size. This effect can be eliminated if the amount of adsorbate per unit area of particles is estimated. Assuming that the particles are spherical and have constant density, it is possible to calculate the relative amount adsorbed per unit surface area as a function of particle diameter. This plot is known as a surface adsorption density distribution (SADD). Figure 2c shows the SADD plot for ^{14}C-glyphosate adsorbed onto a river suspended-colloid sample. This trend could be attributed to changes of particle size (as these calculations are based on a spherical particle), changes in mineralogy, and coating density across the size range.

The method outlined is also applicable to trace metal data collected by ICP–MS and ICP–AES. Recently, Hassellov et al. demonstrated that it is possible to measure major elements as well as a range of trace metals, including Cs, Cd, Cu, Pb, Zn, and La [6]. Further, they showed that it was possible to obtain speciation data across the size range.

Field-flow fractionation separations combined with other high-sensitivity analytical techniques are capable of yielding more detailed information than has been possible with existing methods. Although, at this stage, there are still many uncertainties to the interpretation of the trends observed, this method is certain to provide further insights into the nature of pollutant–colloid interactions in natural waters.

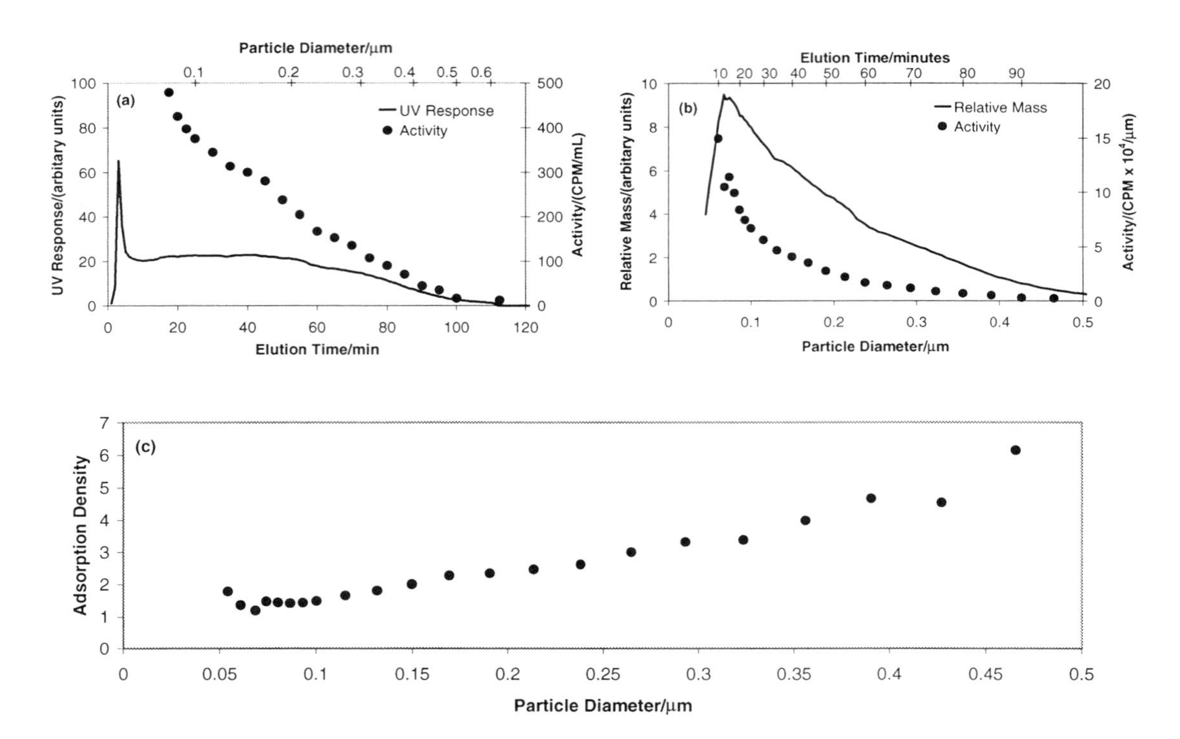

Fig. 2 Graphs showing ^{14}C-glyphosate on river suspended-particulate matter. (a) Particle mass and adsorbate fractograms; (b) particle size distribution and pollutant adsorption distribution; (c) surface-adsorption density distribution.

References

1. J. C. Giddings, F. J. Yang, and M. N. Myers, *Science 193*: 1244–2145 (1976).
2. R. Beckett, *Environ. Tech. Lett. 8*: 339–354 (1987).
3. G. Karaiskakis, K. A. Graff, K. D. Caldwell, and J. C. Giddings, *Int. J. Environ. Anal. Chem. 12*: 1–15 (1982).
4. C. Contado, G. Blo, F. Fagioli, F. Dondi, and R. Beckett, *Colloids Surf. A: Physicochem. Eng. Aspects 120*: 47–59 (1997).
5. R. Beckett, D. M. Hotchin, and B. T. Hart, *J. Chromatogr. 517*: 435–447 (1990).
6. M. Hassellov, B. Lyven, and R. Beckett, *Environ. Sci. Technol. 33*: 4528–4531 (1999).

Ronald Beckett
Bailin Chen
Niem Tri

Pollutants in Water by HPLC

Introduction

Over the last decade, the problem of diffuse pollution caused by industry, agricultural, and human activities, accidental spills, and waste discharges has resulted in directives to control the sources of pollution with the aim of protecting the water quality, contribute to the protection of human salubrity, and guarantee the utilization of natural resources. The European Union 76/464/CEE directive and the Environmental Protection Agency of the United States (US EPA) have listed the more toxic and persistent pollutants, including some of their degradation products, and the maximum permissible levels in surface waters [1]. The need for water monitoring has led to the development of a substantial number of analytical procedures based on an efficient sample-treatment technique and chromatographic determination. Because pollutants of different physicochemical properties are present at extremely low concentrations in complex environmental water samples, the analytical procedure should provide both a sensitive and selective detection and should generate accurate and precise data.

Traditionally, gas chromatography (GC) was the preferred approach for the analysis of pollutants in water, due to the high sensitivity and selectivity achieved, thanks to its selective detectors such as the nitrogen–phosphorus (NPD), the flame photometric detector (FPD), and electron-capture detector (ECD), and to the ease of coupling to mass spectrometry (MS). However, high-performance liquid chromatography (HPLC or LC) is the most powerful approach for the determination of polar, nonvolatile, and thermolabile compounds (i.e., those which are not GC amenable).

The purpose of this entry is to provide an overview of the main HPLC techniques to determine priority organic pollutants in water. The recent developments in detection techniques (including diode-array detection, fluorescence detection, electrochemical detection) and especially the ability to connect a HPLC to MS has increased substantially the use of this technique for environmental monitoring. The methods described here are group-specific and cover the analysis of polycyclic aromatic hydrocarbons, phenols, pesticides, herbicides, aniline derivatives, surfactants, and explosives.

Table 1 classifies the different types of pollutants according to their chemical structure and summarizes the main HPLC techniques used for their determination. The objective is to suggest a suitable HPLC technique for each type of organic compound and to provide information on quality parameters and applicability in real environmental problems. All methods proposed are actually being accepted by the US EPA and permit one to achieve limits of detection at the low ppb–ppt level and reproducibilities below 10%.

Pesticides

Much concern is being given to the determination of pesticides in various environmental matrices, due to the increased number of pesticides detected and to the severe rules imposed by the legislation which aims to protect natural resources. Pesticides are classified as insecticides, herbicides, fungicides, and so forth, depending on the type of organism targeted. From an analytical point of view, pesticides cannot be treated as a group due to the great number of compounds with different chemical structures and physicochemical properties [2]. Many groups of pesticides, including their degradation products, are water soluble, nonvolatile, and polar, which leads to LC as the preferred approach. LC with ultraviolet (UV) or diode-array detection (DAD) are especially used for the analysis of organophosphorus and phenoxyacid pesticides, triazines and phenylurea herbicides, and quat herbicides and, in general, for compounds presenting a suitable chromophore (e.g., an aromatic moiety).

The main advantages of LC–DAD is related to its robustness and ease of use and to the fact that it offers an

Table 1 Families of Priority Pollutants, Principal Method of Analysis Proposed, and Main Sources

Compound class	Extraction	LC method	Main sources
Alcohol ethoxylates	SPE	MS	Leather, textile industry
Anilines and chloroanilines	SPE, LLE	ECD, DAD, MS	Dye industry
BS and NPS	SPE	MS	Textile, tannery industry
Benzidines	SPE, LLE	ECD, DAD, MS	Industry
Carbamates	SPE, LLE	FLD, MS	Agriculture
Chlorophenols	SPE, LLE	DAD, ECD, CD, MS	Paper, pulp, plastic industry
Glyphosate	SPE	FLD, MS	Agriculture
Linear alkylbenzenesulfonates	SPE	MS	Household, chemical industry
Nonylphenol ethoxylates	SPE	MS	Leather, textile industry
Organophosphorus pesticides	SPE, LLE	DAD, MS	Agriculture, household
Phenoxyacid pesticides	SPE, LLE	DAD, MS	Agriculture
Phenylurea compounds	SPE, LLE	DAD, MS	Agriculture
Polyaromatic hydrocarbons	SPE, LLE	FLD, DAD, MS	Natural, antropogenic sources
Quats	SPE	DAD, MS	Agriculture
Triazines	SPE, LLE	DAD, MS	Agriculture, nonagricultural

SPE = solid-phase extraction; LLE = liquid–liquid extraction; DAD = diode-array detector; EC = coulimetric detector; ECD = electrochemical detector; FLD = fluorescence detector; MS = mass spectrometry.

absorbance spectra that can be used to identify pesticides through spectral comparison. UV–DAD can best be combined with acetonitrile–water mixtures because of their absorbance cutoff. Interferences caused by UV-absorbing compounds (e.g., humic and fulvic acids) can seriously affect the quantification of target analytes. LC with fluorescence detection (LC–FLD) is more selective than LC–DAD and provides better sensitivity for naturally fluorescent compounds, as well as for many other compounds, such as glyphosates or carbamates, which are precolumn or postcolumn derivatized to yield fluorescent reaction products.

An alternative to LC–FLD for the analysis of aryl *N*-methylcarbamates and *N*-phenylcarbamates is LC with electrochemical detection (LC–ECD) because these compounds can be easily oxidized at 1.1 V. However, one should be aware of the problems associated with these techniques for water monitoring. Complex water samples, such as wastewater or industrial effluents might (a) increase the detection limits, (b) inhibit analyte detectability, or (c) interfere or co-elute with the analytes. In conclusion, these detection techniques are suitable for target determination of pesticides only if combined with a preconcentration and adequate cleanup step.

The matrix effect can be avoided by using MS. This technique is characterized by being highly selective and sensitive and, in addition, it offers spectral information which permits the unequivocal identification of target compounds. LC–MS with thermospray (TSP) and particle beam (PB) interfaces have been widely used for pesticide monitoring and produce molecular information and electron-impact mass spectra, respectively. High detection limits and poor reproducibility are the main problems related to these techniques. Modern atmospheric-pressure chemical ionization (APCI) and electrospray (ESI) sources for LC–MS coupling are currently applied for the analysis of pesticides from environmental waters and limits of detection (LODs) of a few nanograms per liter and the linear response range over two orders of magnitude have been reported [3].

Polycyclic Aromatic Hydrocarbons

Polycyclic aromatic hydrocarbons (PAHs) are considered as priority pollutants due to their mutagenic and carcinogenic properties. PAHs are introduced in the environment from natural sources [e.g., incomplete combustion of organic matter from natural processes (volcanic eruptions, fires)] or anthropogenic, such as burning of fossil fuels, waste incineration, traffic, and so forth.

Polycyclic aromatic hydrocarbons have been encountered in different environmental matrices (e.g., soil, atmosphere, water, biota, etc.). Because PAHs have natural fluorescent properties, LC–FLD is advantageous over GC–MS because of its ability to measure the different PAH isomers and comparable detection limits are obtained. LC–DAD and LC–APCI–MS can also be applied and have the advantage that hydroxylated PAH transformation products can be identified [4].

Phenolic Compounds

Phenol and its derivatives are generated in a wide variety of industrial processes (e.g., manufacture of plastics, dyes, drugs, glues) and in the pulp industry, but can also be formed as degradation products of pesticides. Chlorophenols are toxic to aquatic organisms and to man, and even at low concentrations in water, they can be the cause of unpleasant taste and odor.

Current official analytical methods include GC with flame ionization detection (FID), ECD, or MS and involve a derivatization step. However, there is a general trend to switch to LC procedures to avoid sample manipulation and because derivatization of phenols is not straightforward [5]. Common detection techniques are UV detection at 280 nm and 310 nm for nitrophenols and pentachlorophenol. LC–DAD is recommended because spectral libraries can be used for confirmation purposes. In terms of sensitivity, LC with electrochemical (ECD) or coulometric detectors (CD) produce better response than UV or DAD. However, electrochemical detection for LC has never become so popular as its inherent sensitivity and selectivity seem to indicate due to problems of electrode fouling and poor reproducibility. Moreover, electro-active matrix components can increase the background current and make ECD less robust for water monitoring. Therefore, this technique can only be recommended for monitoring clean-water samples. An attempt to use LC with direct or indirect FLD for phenols indicated acceptable detection limits, but problems related to the derivatization step made the technique less attractive. Similar to other organic pollutants, phenolic compounds can be detected by LC–APCI–MS and LC–ESI–MS and produce, in general, the protonated molecule as the base peak.

The advantages of these techniques is that an increase of the extraction voltage enhances fragmentation via collision-induced dissociation (CID) and provide structural information, useful for confirmation and identification purposes. Moreover, phenols are good candidates for capillary electrophoresis (CE) and capillary zone electrophoresis (CZE) because they permit a fast separation of polar and ionic compounds, have a high resolution, and are compatible with most LC detectors, including MS. In addition, trace enrichment can be performed in the capillary using isotachophoresis, making the technique suitable for water analysis.

Aniline Derivatives

In the recent years, the occurrence of anilines, chloroanilines, and benzidines in environmental waters has started to be of concern due to their widespread use in various industrial processes. Detection of these compounds is generally carried out with LC–ECD working in the amperometric mode using a glassy-carbon electrode. Problems encountered are similar to those for phenolic compounds and, therefore, confirmation is always necessary by UV, DAD, or, preferably, MS detection [6]. These compounds are also CE amenable and this approach is really promising in the sense of sensitivity and selectivity because humic and fulvic material can be separated from the analytes due to their different migration kinetics.

Surfactants

Linear alkylbenzenesulfonates (LAS), alcohol ethoxylates (AEO), and nonylphenol ethoxylates (NPEO) are synthetic surfactants used in the formulation of detergents and other cleaning products and are widely applied in the dye and leather industry and other industrial processes. These compounds, considered as estrogenic, have aroused considerable interest due to the large quantities produced globally. Their low volatility and anionic form make LC-based methods the preferred approach [7]. Due to the presence of different positional isomers, to the biodegradation intermediates, and to the lack of reference standards, LC–MS, and in particular with ESI, is the only technique which enables their identification and quantification in environmental waters.

Aromatic Sulfonates

Benzene (BS) and naphthalene sulfonates (NPS) are commonly used in the textile industry as dye bath auxiliaries and in the tannery industry as dispersants and wetting agents. After application, these compounds are discharged into surface waters and their presence in industrial effluents was not reported due to the lack of an appropriate analytical technique. Recently, ion-pair chromatography–ESI–MS and LC–ESI–MS was developed to determine BS and NPS with the final goal of determining polar, ionic, and water-soluble pollutants in wastewater [7].

Conclusions

Many organic pollutants in water can be analyzed with HPLC techniques at trace levels. From the different HPLC methods discussed in this entry, several remarks can be made: (a) For routine water monitoring, LC–DAD is the most common detection device used to analyze polar, thermolabile, and nonvolatile compounds due to its robustness, high sample throughput, and the possibility of obtaining an UV spectrum that can be used for analyte confirmation; (b) LC–FD and ECD are more selective and sensitive than DAD and are especially useful for the monitoring of PAHs, carbamates, phenols, and anilines, respectively; (c) LC–MS with API sources have become highly robust techniques and the preferred option for the identification, confirmation, and quantification of organic pollutants in water. In addition, all HPLC techniques described can be coupled, on-line, with solid-phase extraction, so that limits of detection can be decreased and their total automation permit to achieve high precision.

References

1. D. Barceló, *J. Chromatogr. 643*: 117–143 (1993).
2. D. Barceló and M. C. Hennion, *Trace Determination of Pesticides and Their Degradation Products in Water*, Elsevier, Amsterdam, 1997.
3. J. Abián, *J. Mass Spectrom., 34*: 157–168 (1999).
4. R. Koeber, R. Niessner, and J. M. Bayona, *Fresenius J. Anal. Chem. 359*: 267–273 (1997).
5. D. Puig and D. Barceló, *Trends Anal. Chem. 15*(8): 362–376 (1996).
6. S. Lacorte, M. C. Perrot, D. Fraisse, and D. Barceló, *J. Chromatogr. A 833*: 181–194 (1999).
7. M. Castillo, M. C. Alonso, J. Riu, and D. Barceló, *Environ. Sci. Technol. 33*: 1300–1306 (1999).

Silvia Lacorte
Damià Barceló

Polyamide Analysis by GPC–SEC

Introduction

Synthetic polyamides (PA) are among the mostly widely used engineering thermoplastics, owing to their high strength and toughness, stiffness, abrasion resistance, and retention of physical and mechanical properties over wide temperature ranges. The material's outstanding properties are, to a large extent, due to their semicrystalline morphology and the cooperative intermolecular hydrogen-bonding of the amide groups. The strength of these dipolar interactions are reflected in the melting points, which are much higher in polyamides than in polyesters having comparable structures.

Proteins, including wool and silk fibers, are made up by the linkage of α-amino acids and constitute the most widespread source of natural polyamides.

Synthetic PAs are produced by polycondensation of bifunctional monomers or by cationic and anionic ring-opening polymerization of lactams. Polymers obtained with the first technique are linear, whereas chain branching may occur with anionic polymerization. Based on their chemical structure, synthetic polyamides may be classified into two categories [1]:

1. Aliphatic polyamides, also known under the generic name nylon. Industrial nylons with the general structure $[-RNHCO-]_n$, such as PA-6, PA-11, and PA-12, are called monadic; those with the general structure $[-NHR_1NHCOR_2CO-]_n$ are called dyadic (PA-4,6, PA-6,6, PA-6,9, PA-6,10, and PA-6,12).
2. Aromatic polyamides or aramids. The best known in the specialty fiber market are Nomex™ (poly-*m*-phenyleneisophthalamide) and Kevlar™ (poly-*p*-phenyleneterephthalamide).

A number of polymers with structures related to those of polyamides have been developed, some of which have achieved commercial importance, such as

—CO—NH—NH— (polyhydrazide),
—O—OC—NR— (polyurethane), and
—CO—NR—CO— (polyimide).

SEC Eluants for Polyamides

High cohesive strength confers an exceptional solvent resistance to PAs, which can be dissolved only under rather drastic conditions (i.e., in highly corrosive solvents at temperatures close to the polymer melting point). This limited solubility leads to serious difficulties in the solution characterization of these polymers, such as in the determination of molecular-weight distribution (MWD). Selection of a solvent medium which is good for the polymer, chemically inert, and compatible with the stationary phase is crucial to a successful size-exclusion chromatographic (SEC) analysis [2]. These criteria are particularly demanding for many aramids, which are also rigid-rod polymers. Poly-*p*-phenyleneterephthalamide, for instance, has been analyzed in 97% sulfuric acid on silica columns.

Recently, a less corrosive room-temperature eluant consisting of methane sulfonic acid + 5% methane sulfonic anhydride + 0.1*M* sodium methane sulfonate has been reported for Kevlar and Technora fibers. Separation was performed on Hastelloy C columns packed with 4000 Å SAX 10-μm particles (Polymer Laboratories), using ultraviolet (UV) detection and poly(benzoxazole) for calibration [3]. Work on the SEC characterization of aramids is extremely limited and generally lacks details of molecular-weight (MW) accuracy. In the following section, we will focus exclusively on aliphatic polyamides.

High-Temperature Solvents

The first mention of SEC analysis of PA-6 and PA-6,6 is by Goebel (1967), using *m*-cresol at 135°C with Styragel® columns. It was soon realized, however, that PA degrades extensively within a few hours under the above experimental conditions; alternative solvents have been proposed, most of which can be used only at elevated temperatures, due to solubility or viscosity problems [4]:

- *o*-Chlorophenol at 100°C
- Hexamethylphosphoramide (85°C)
- Dimethylacetamide + 2.5% LiCl (100–130°C)
- Benzyl alcohol (130°C).

In addition to the necessity for high-temperature SEC, each of the above solvents has its own drawbacks, such as adsorptive effects with phenol derivatives, column blocking, and carcinogenicity with hexamethylphosphoramide and corrosion of stainless steel by the dimethylacetamide–LiCl mixture.

Benzyl alcohol has been tested at 100°C on silica columns. To prevent potential solute interactions with silica stationary phase, it is recommended to prefer cross-linked PS columns with benzyl alcohol at 130°C and polytetrahydrofurans as calibration standards (polystyrene and polyethylene oxide standards interact with the stationary phase under the analytical conditions) [5].

Low-Temperature Eluants

Fluorinated Alcohols

Most solvents which dissolve nylon at room temperature are either too toxic (nitrobenzene) or too corrosive [strong acids: sulfuric, formic, dichloroacetic; concentrated alcoholic salt solutions (e.g. methanol + $CaCl_2$)] to be safely used as an SEC eluant. Fluorinated alcohols constitute an exception: It has been known for a long time that trifluoroethanol (TFE), tetrafluoropropanol (TFP), or hexafluoroisopropanol (HFIP), owing to the presence of strong electron-withdrawal groups ($—CF_2—$ and $—CF_3$), have the propensity to dissolve a large number of polymeric materials which possess receptive sites for H-bonding formation (proteins, stereoregular polyesters, and, more recently, polyaniline).

The application of fluoroalcohols for dissolving polyethylene terephthalate (PET) and polyamides dates back from the early 1960s. Dissolution tests show that the solubility of polyamides improves with the acidity of the fluoroalcohol and decreases with increasing steric hindrance of substituents adjacent to the hydroxyl group. The use of HFIP for the SEC of PA-6,6 and PET on porous glass and μ-Styragel columns was described by Drott in 1971. TFE was tested in the following year for the SEC separation of Nylon 6, using fractionated PMMA for calibration. Fluoro-alcohols are hygroscopic and should be dried before use to avoid formation of corrosive products. Polymer adsorption was observed when the water content in the eluent is >0.03%. Moisture in TFE and HFIP can induce the formation of carboxylate end groups in nylon. Electrostatic repulsion between these groups and residual carboxyl ions present in cross-linked PS packings will lead to ion exclusion, which can be suppressed with the addition of a salt (sodium acetate, sodium trifluoroacetate, or tetraethyl ammonium chloride) in the $10^{-2}M$ concentration range. At present, HFIP is the preferred room-temperature eluant for the SEC characterization of nylons and polyesters. TFE, although four times less expensive than HFIP, is more toxic and has limited dissolution capability for polyesters and for nylons with long alkyl sequences. In addition to high cost, the use of HFIP entails a number of difficulties, such as high volatility, health hazards, incompatibility with a PS–DVB-based stationary phase (particularly with columns of pore sizes <1000 Å), insolubility of PS standards, and the presence of nonsteric exclusion effects. Most of these problems have now been largely solved: Calibration can be done with commercially available anionic poly(methyl methacrylate) (PMMA) standards or with "absolute" detection, incompatibility with the stationary phase is minimized with columns specially designed for HFIP (Polymer Laboratories, Shodex, Waters), and closed-loop distillation and the use of narrow-bore columns substantially reduces solvent consumption.

Despite this progress, SEC with HFIP remains an expensive procedure due to extra investment in fluorinated solvent and special columns. Finally, it should be noted that HFIP with a pK_a of 9.3 is sufficiently acidic to react with the nitrogen in aromatic polyimides, resulting in ring-opening solvolysis.

Mixed Eluants

In search of less expensive, less toxic, and lower viscosity eluants, a few authors have proposed diluting the active ingredient with a common SEC eluant such as toluene, dichloromethane, or chloroform. To lower the operating temperature and minimize polymer degradation, mixtures of *m*-cresol with chlorobenzene (50:50, v/v, 43°C), dichloromethane (50:50, room temperature), and chloroform have been used, with 0.25 wt% benzoic acid added to prevent adsorption. In the same vein, *o*-chlorophenol has been diluted with chloroform (25:75) and used at 20°C. The main disadvantage in this latter solvent was a small dn/dc for the polymer, which rendered refractive index measurements difficult. In addition, careful purification of the phenol is required to obtain a detection signal. Dichloroacetic acid diluted to 20 vol% with dichloromethane has been proposed as the mobile phase. However, even at this concentration, PA tends to degrade at room temperature.

Among the different solvent combinations, it appears that mixtures based on fluorinated alcohols give the best results in terms of stability of the polymer and compatibility with the stationary phase (conventional silica or PS–DVB columns can both be used). As with pure fluoroalcohols, a salt should be added to suppress polyelectrolyte effects. Toluene, with 20 vol% HFIP, has been used for the SEC fractionation of PA-12. However, based on thermodynamic excess properties, this solvent combination should be less efficient in terms of solvation power than mixtures of HFIP with chloroalkanes.

Mixtures of HFIP or TFE with $CHCl_3$ and CH_2Cl_2 have been successfully tested as SEC eluants. HFIP–CH_2Cl_2 mixtures have the advantage of being more stable, of low toxicity, and UV transparent up to 200 nm (an assess for UV-absorption detection) and are better solvents for PA than the other mixtures [6]. On the negative side, HFIP–CH_2Cl_2 forms a 30:70 (v/v) azeotrope at 30°C, thus limiting its use to room temperature. HFIP–$CHCl_3$ does not have an azeotrope, whereas TFE–$CHCl_3$ possesses a lower azeotrope at 55°C.

Problems Associated with Mixed Eluants

The use of mixed eluants in SEC analysis was proposed long ago to take advantage of the desirable properties of both solvent systems. Possible consequences of preferential solvation of the stationary phase and the polymer by one of the eluant components must, however, be critically evaluated: changes in mixture composition with polymer hydrodynamic volume, differences in composition of the mobile phase and of the solvent in the pores, dependence of the refractive index signal on elution volume, and the possibility of interference of the system peak with the oligomer signal in refractive index detection. Solvent recycling is more difficult and the original composition must be readjusted after distillation.

N-Trifluoroacetylation

N-Trifluoroacetylation remains a common technique to solubilize aliphatic PA for SEC analysis [7]:

$$\text{—CONH—} + (CF_3CO)_2O \rightarrow \text{—CON(COCF}_3\text{)—} + CF_3COOH \qquad (1)$$

With a twofold to threefold excess in trifluoroacetic anhydride, reaction (1) is quantitative within 1 day at room temperature. The *N*-trifluoroacetylated polyamides (NTFA–PA) are soluble in many ordinary organic solvents, such as acetone, methylene chloride, chloroform, and tetrahydrofuran. In addition to the change in solubility, another remarkable feature of *N*-trifluoroacetylation is a near-two-orders-of-magnitude increase in the UV-absorption coefficient of the amide band, accompanied by a large bathochromic shift. Trifluoroacetylated polyamides have a rather small *dn*/*dc* in CH_2Cl_2 and detection was better achieved by UV absorption. Once prepared, the NTFA–PA solutions are unstable in the presence of atmospheric humidity and should be used immediately to avoid reverse hydrolysis, which converts the fluorinated polymer back into the original polyamide. It has been reported that SEC separation of NTFA–PA does not follow the universal calibration in dichloromethane, whereas better agreement was obtained in tetrahydrofuran. Comparative SEC analyses of NTFA–PA in both solvents on Styragel columns show, nevertheless, identical results [6].

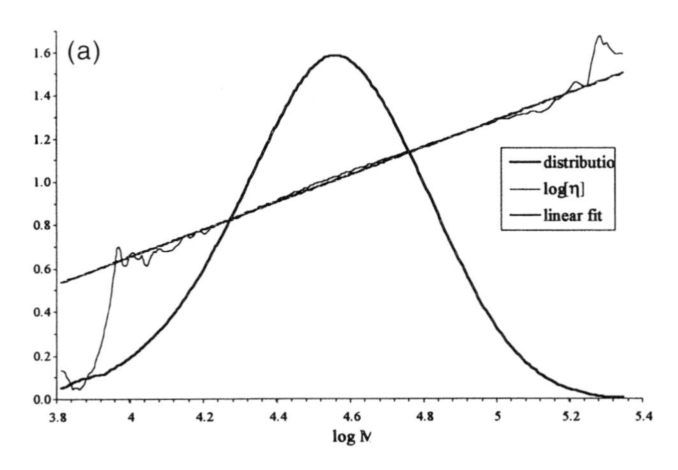

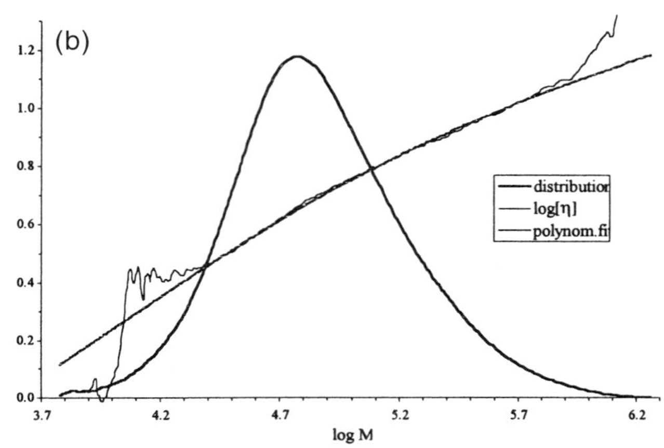

Fig. 1 SEC–viscometry characterization of a linear (a) and a branched (b) PA-12 synthesized by anionic polymerization, with different activators. Analysis conditions: mixed eluant 5:95 v/v; HFIP/CH_2Cl_2 at 30°C; Styragel–HR2/3/4/5 columns.

Conclusions

Size-exclusion chromatography analysis of PAs remains a difficult enterprise in comparison to other amorphous thermoplastics which can be readily dissolved in common solvents at room temperature. To improve MW accuracy, calibration should be performed with absolute detection (on-line viscometer, light scattering) or with a broad MWD polymer standard of identical chemical composition to the analyzed sample. Among the high-temperature eluants, benzyl alcohol is the less corrosive and seems to give the most reliable results. In low-temperature SEC, HFIP remains the best eluant, provided the necessary extra investment is not a problem. A mixed-eluant based on HFIP is a reasonable alternative to pure HFIP for long alkyl nylons (PA-12, PA-11). For shorter alkyl sequences, the high percentage of HFIP (>10 vol%) required for dissolution of the polymer may interfere with the size-exclusion mechanism of separation [6]. *N*-Trifluoroacetylation of the PA remains an interesting alternative, provided due care is given to prevent contact of the derivatized compound with reagents (water, alcohols, etc.) able to cause scission of the $-N-COCF_3$ bond.

References

1. M. S. M. Alger, *Polymer Science Dictionary*, Elsevier Applied Science, London, 1989, p. 288.
2. P. J. Wang, in *Handbook of Size Exclusion Chromatography* (C. S. Wu, ed.), Chromatography Science Series Vol. 69, Marcel Dekker, Inc., New York, 1995, p. 161.
3. R. M. Nelson and M. W. Warren, *International GPC Symposium '96*, Waters Corporation, Milford, MA, 1996, p. 453.
4. D. J. Goedhart, J. B. Hussem, and B. P. M. Smeets, in *Liquid Chromatography of Polymers and Related Materials II* J. Cazes and X. Delamare, eds.), Chromatography Science Series 13, Marcel Dekker, Inc., New York, 1980, p. 203.
5. G. Marot and J. Lesec, *J. Liquid Chromatogr. 11*: 3305 (1988).
6. T. Q. Nguyen, *International GPC Symposium '98*, Waters Corporation, Milford, MA, 1998, p. 135.
7. E. Jacobi, H. Schuttenberg, and R.C. Schulz, *Makromol. Chem. Rapid Commun. 1*: 397 (1980).

Tuan Q. Nguyen

Polycarbonates: GPC–SEC Analysis

Polycarbonate (PC) polymers are amorphous thermoplastics with excellent toughness, clarity, and high heat-deflection temperatures. They are one of the better thermoplastics where strength and elastic moduli are concerned, offering both excellent impact strength and rigidity. In general, PCs are linear aromatic polyesters of carbonic acid. Polycarbonates can be described as polymeric combinations of bifunctional phenols, or bisphenols, linked together through carbonate linkages.

Robertson et al. used gel permeation chromatography (GPC) to examine PCs produced by different synthetic routes. They concluded that this method is superior to end-group analysis. Hoore and Hillman carried out fractionations of PCs by GPC to procure narrow molecular-weight-distribution (MWD) fractions which were used to calibrate the columns of a GPC system and an experimental "*Q* factor" of 23.8 was found. The calibration was further confirmed by membrane osmometry and light-scattering measurements [1]. The "*Q* factor" method is the simplest conversion method for molecular-weight calculations that provides data closer to the real molecular weights while still using polystyrene (PS) standards, but it lacks accuracy because it assumes the extended chain length of the polymer in solution. Most flexible polymers are not extended, but, rather, they are randomly coiled in solution. The more reliable method is to use the concept of the "hydrodynamic volume" [2]. A comprehensive listing of Mark–Houwink coefficients K and a for different solvents and temperatures are published in the *Polymer Handbook* [3]. PC standards for calibration have been commercially available for the past several years from Polymer Standard Service Co. (Mainz, Germany) and American Polymer Standards Co. (Mentor, OH, U.S.A.).

The polymerization of trimethylene carbonate (1,3-dioxan-2-one) with a complexation catalyst was studied by Kriecheldorf et al. [4]. On the basis of GPC measurements employing a universal calibration, weight-average molecular weights up to approximately 22,000 were found. A combination of four Ultrastyragel columns was used, with nominal pore sizes of 100 Å, 500 Å, 10^3 Å, and 10^4 Å. The detector was a Waters M410 differential refractometer. The molecular weights were estimated from GPC results by means of intrinsic viscosities determined in dichloromethane at 25°C and an universal calibration curve, based on commercial PS standards.

Wang and Gonsalves [5] investigated the enzyme-catalyzed synthesis of poly(ethyl phenol) (PEP), which was modified by copolymerization with PCs. GPC was carried out on a system equipped with a differential refractive index detector (DRI) and four Ultrastyragel columns in tetrahydrofuran (THF) with a flow rate of 1 mL/min. A calibration plot, constructed with PS standards, was used to determine the molecular weights. Based on the results, a slight difference between the high- and low-temperature products was observed for the pristine PC. The copolymers PC-*co*-PEP had higher molecular weights when compared to the pristine material, due to the linkage between PC and PEP, which effectively doubled the molecular weights.

Gel permeation chromatography was used for the analysis of fullerene-functionalized polycarbonates, achieved by direct reaction of PC and fullerenes with aluminum chloride as a catalyst. THF was used as the eluent and the system was calibrated with PS standards. The working wavelength for the ultraviolet (UV) detector was 340 nm, where the starting PC was hardly detectable, but the reaction product gave two UV peaks, a strong peak in the "normal" molecular-weight (MW) region and a weak peak in the very high MW region. Because THF is a nonsolvent for fullerenes, the solubility of the polymers in THF and the GPC data served as evidence for the existence of a covalent bonding between the fullerene and PC [6].

The influence of molecular structure on the degradation mechanism of degradable polymers of poly(trimethylene carbonate), poly(trimethylene carbonate-*co*-caprolactone), and poly(adipic anhydride) was explored by Albertsson and Eklund [7]. Measuring the change of the mass and the molecular weight for the polymeric samples during degradation is a good indication of the degradation rate and an initial measure of material deterioration. The value of the molecular weight is not absolute, but, rather, it is relative to PS standards. This is sufficient for this study, which is simply concerned with the change in molecular weight relative to the starting value in order to evaluate the degradation. While the molecular weight decreased, an increase in the polydispersity was observed for some of the samples.

Studies of the thermal decomposition of the products under a nitrogen atmosphere were carried out by combining liquid chromatography under critical conditions of adsorption (LACCC) and size-exclusion chromatographic (SEC) methods with matrix-assisted laser desorption/ionization–time-of-flight spectrometry (MALDI–TOF) and postcolumn derivatization. The critical conditions of polymer adsorption in the liquid chromatography and an optimal matrix system for MALDI–TOF are reported. The changes of molar mass distribution and chemical heterogeneity are said to be due to the simultaneous processes of degradation and recombination [8].

Branching by reactive end groups, syntheses, and thermal branching of 4-hydroxybenzocyclobutene/*p-tert*-butylphenol co-terminated bisphenol PC was investigated by Marks et al. [9]. Reactive end groups on bisphenol A PCs allow for significant structural changes in this condensation polymer that are not accessible by direct synthetic routes. Commercially offered branched PCs have improved melt rheological properties, compared to their linear analogues; they are manufactured by incorporation of a trifunctional comonomer in the PC polycondensation process. A GPC system, equipped with a UV detector (GPC–UV), and liquid chromatography (LC) were used for the analyses. A Hewlett-Packard 1047 DRI detector and a LALLS (CMX-100) (low-angle light scattering) photometer were connected in series. Values of *dn*/*dc* were estimated from the DRI responses and sample concentrations. Five TSK MicroPak H-series (3 GMH6, 1 G500H6, and 1 G400H8) columns were used at 30°C with THF as the eluent at 1 mL/min. Heating to 300°C for 20 min causes the BCB-OH/PTBT BA PCs to branch or cross-link. The molecular weight and polydispersity of the branched polymers increase to a value at which the onset of the gel formation is observed. GPC–LALLS (M_w 135,100 and 310,700) was performed on two of the branched samples; it shows, as expected, a significantly higher molecular weight, particularly M_w, than indicated by GPC–UV (M_w 70,600 and 64,600, respectively).

The M_w by GPC–UV of each series of BCB-OH/PTBT BA PCs increases exponentially up to the gel point. The apparent decrease in M_w after the gel point reflects the values of the soluble fraction only, which are, therefore, not representative of the entire sample. LC was applied to investigate the types and distribution of BCB reaction products formed.

Unfortunately, none of the hyperbranched polymers studied to date has demonstrated good mechanical properties. A hyperbranched PC is also expected to be a brittle material, but such a structure may prove interesting as a highly functionalized prepolymer for composites, coatings, and other applications. Hyperbranched PCs were synthesized and characterized by Bolton and Wooley [10]. The products were prepared by the polymerization of an A_2B monomer derived from 1,1,1-*tris*(4′-hydroxyphenyl)ethane. Silylation of the phenol terminated material with *tert*-butyldimethylsilyl chloride, followed by degradation of the carbonate linkages by reaction with lithium aluminum hydride and analysis of the products by high-performance liquid chromatography (HPLC) allowed for the determination of the degree of branching, which was found to be 53%. The molecular weights of the carbonyl imidazolide-, phenol-, and *tert*-butyldimethylsilyl ether-terminated hyperbranched polycarbonates were 16,000, 77,000, and 82,000, respectively, from GPC, based on polystyrene standards, 23,000, 180,000, and 83,000, respectively, from GPC with LALLS, and 24,000, 160,000, and 88,000, respectively, from GPC with SEC software. SEC was conducted on a Hewlett-Packard series 1050 HPLC with a Hewlett-Packard 1047A refractive index detector, a Wyatt MiniDawn laser-light-scattering detector, and a Viscotek Model 110 differential viscometer. Two 5-μm Polymer Laboratories PL gel columns

(300 × 7.7 mm), connected in series, were used with THF as an eluent.

Bailly et al. investigated the separation of polycarbonate oligomers by SEC and reversed-phase LC [11]. The separation of PC oligomers was achieved by SEC with styrene–divinylbenzene microparticle gel as a stationary phase and dichloromethane as a mobile phase. Oligomers were separated up to 10 repeating monomer units. SEC is useful for the analyses of the low-molecular-weight content of commercial PC samples. Retention times are significantly influenced by the nature of end groups (O— or OH—O—) {O=phenyl}, indicating that adsorption occurs with hydroxy-terminated oligomers. Also, SEC results were compared with reversed-phase HPLC data.

Although the molecular weight of the bisphenol A polycarbonate repeat unit is relatively high, the separation of PC oligomers by SEC is difficult and has, thus, received little attention up to now. The problem is complicated by the existence of three oligomer families resulting from the addition of chain modifiers (monofunctional phenols) during the synthesis of the polymer by the usual interfacial method. The SEC results are compared with those obtained by reversed-phase HPLC for the same systems. Adsorption of phenolic compounds is known to occur on the surface of PS–DVB gels. When several types of oligomers are simultaneously present, interpretation of chromatograms becomes difficult and should be done with caution. This is particularly true when analyzing the oligomer content of commercial PC samples. The resolution obtained by reversed-phase HPLC remains superior, but, in most cases, quantitative result cannot be obtained by this technique because of solubility problems. In contrast with previous claims, the nature of the end groups was observed to significantly influence the retention time of PC oligomers by SEC. The presence of the hydroxy terminal groups induces a reversible adsorption of the compounds onto the columns. This effect is very important for oligomers having two hydroxy end groups.

As a consequence, the precise determination of the number-average molecular weights of PC samples by SEC is possible only if the nature of the oligomers present is taken into account.

References

1. T. R. Crompton, *The Analysis of Plastics*, Pergamon Press, New York, 1984, p. 354.
2. S. Mori, HPLC application to polymer analysis, in *Handbook of HPLC* (E. Katz, R. Eksteen, P. Schooenmakers, and N. Miller, eds.), Marcel Dekker, Inc., New York, 1998, p. 836.
3. J. Brandrup and E. H. Immergut (eds.), *Polymer Handbook*, 3rd ed., John Wiley & Sons, New York, 1989, p. VII/23.
4. H. R. Kricheldorf, J. Jensen, and I. Kreiser-Saunders, *Makromol. Chem. 192*: 2391–2399.
5. J. Wang and K. E. Gonsalves, *J. Polym. Sci. Part A: Polym. Chem. 37*: 169–178 (1999).
6. B. Z. Tnag, H. Peng, M. Leung, Ch. Song, M. Takashi, Kai Su et al., *Macromolecules 31*: 103–108 (1998).
7. A.-C. Albertsson and M. Eklund, *J. Appl. Polym. Sci. 57*: 87–103 (1995).
8. O. Wachen, K. H. Reichert, R. P. Kruger, H. Much, and G. Schulz, *Polym. Degrad. Stabil. 55*(2): 225–231 (1997).
9. M. J. Marks, J. Newton, D. C. Scott, and S. E. Bales, *Macromolecules 31*: 8781–8788 (1998).
10. D. H. Bolton and K. L. Wooley, *Macromolecules 30*: 1890–1896 (1997).
11. Ch. Bailly, D. Daoust, R. Legras, and J. P. Mercier, *Polymer 27*: 776–782 (1986).

Nikolay Vladimirov

Polyester Analysis by GPC–SEC

Gel permeation chromatography (GPC) also known as size-exclusion chromatography (SEC) has proven to be an extremely useful analytical technique for characterizing the molecular-weight distribution (MWD) of polyesters. The MWD of a polymer can provide valuable information regarding the molecular composition. For example, along with relative molecular-weight averages, polydispersity, intrinsic viscosity, and branching information can be obtained.

Thermoplastic polyesters are popular engineering polymers because they offer excellent chemical and heat resistance, along with desirable electrical properties. When reinforced with glass or mineral fillers, they can be used to replace materials such as metals, ceramics, composites, and other less suitable plastics. In addition, thermoplastic polyesters offer excellent mechanical properties while retaining good processability [1]. Unfortunately, the polyester molecule is readily susceptible to hydrolysis. Hydrolytic degradation of the polymer causes random chain scissions to occur, normally at the ester linkages (R—CO—O—R′), which causes a reduction in molecular weight and, in turn, a reduction in mechanical integrity [2,3]. GPC can be used to monitor the hydrolytic process and predict product performance [4].

Aromatic polyesters, because of their crystalline structure and polar nature, require the use of aggressive solvents and/or elevated temperatures to dissolve the polymer [5]. Various solvent blends can also be incorporated to aid in the dissolution process [6,7].

Early work involving high-temperature GPC of polyesters utilized solvents such as meta-cresol and orthochlorophenol as mobile-phase solvents. These solvents are very viscous and require system temperatures of between 140°C and 150°C. Both solvents are also very dangerous and difficult to handle [8,9].

More recently, hexafluoroisopropanol (HFIP) has been used successfully as a mobile phase for both polyesters and polyamides [4,9,10]. Unfortunately, HFIP is very expensive and dangerous to handle. HFIP requires special care and attention due to its extreme corrosiveness to eyes and skin. *Handling should be conducted only by trained professionals equipped with proper splash goggles and face shield for eye/face protection, and appropriate impervious gloves, apron and other protective equipment as noted on the Material Safety Data Sheet for HFIP.* Mobile-phase distillation is necessary, due to the extreme cost of the solvent. Some laboratories have reported using blends of HFIP and methylene chloride or HFIP and chloroform successfully in an effort to reduce solvent costs [7,10–13].

Hexafluoroisopropanol is able to dissolve most polyesters and polyamides (nylons) at room temperature in about 4–8 h. Sodium trifluoroacetate (NATFAT) is typically added to suppress any polyelectrolyte effects that could occur in HFIP [9]. GPC columns made from cross-linked polystyrene–divinylbenzene are typically used to perform the separation [14]. Calibration is generally performed using poly(methyl methacrylate) standards instead of polystyrene standards, due to solubility constraints [5,15].

Typical polyesters characterized by GPC include polyethylene terephthalate (PET), polybutylene terephthalate (PBT), and polycylohexylenedimethylene terephthalate (PCT). Polyphthalamides (PPA) and polyamides are also commonly analyzed.

Polyester detection is normally accomplished by using refractive index detection in conjunction with either viscometry or low-angle laser light scattering (LALLS).

The viscosity data are a quantitative gauge for monitoring the loss of high-molecular-weight chains within the polymer. Because the viscometer detector operates by measuring the pressure drop across a capillary tube, the intrinsic viscosity at a constant flow rate can be measured as follows [11,16]:

$$P = \left(\frac{8}{\pi}\right)\left(\frac{L}{r^4}\right)\eta F$$

where P is the pressure drop across the capillary, L is the capillary length, r is the internal radius of the capillary, η is the viscosity of fluid (GPC column effluent), and F is the flow rate.

Universal calibration incorporates intrinsic viscosity data with molecular-weight information provided by polymer standards of differing composition. Absolute molecular-weight determinations have proven to be inaccurate in HFIP due to non-size-exclusion interactions occurring within the process [6,17]. Typically, *relative* molecular-weight comparisons are reported when using this procedure. However, if *absolute* molecular-weight values are required, light-scattering techniques have been utilized with greater success [18–20].

Data derived from GPC analysis is commonly used to monitor the hydrolytic stability of various thermoplastic polyesters used under conditions requiring exposure to elevated temperatures and humidity. Polyester hydrolysis results in shifts in MWD toward lower molecular weight with a corresponding lowering of the intrinsic viscosity. A reduction in mechanical properties also is evident when hydrolysis occurs. Changes in mechanical properties can also be caused by moisture absorption and by weakening of the glass–fiber bond. GPC can distinguish between chemical and physical changes occurring to the material.

Figure 1 compares the changes that occur to PBT due to aging of the polymer at 60C and 100% relative humidity for a total of 25 weeks. Samples were removed at 5-week intervals, and the molecular-weight distributions were compared. Polymer hydrolysis was evident by regular shifts in molecular-weight distribution toward lower molecular weight [4].

Gel permeation chromatographic analysis will serve as a valuable technique for evaluating the newly developed hydrolysis-resistant polyesters and for monitoring the weatherability of polyesters in outdoor applications.

References

1. E. Yokley, Thermoplastics for electrical applications, *Electrical Manuf.*, 11–13 (November 1991).
2. P. G. Kellenher, G. H. Bebbington, D. R. Falcone, J. T. Ryan, and R. P. Wentz, Thermal and hydrolytic stability of poly(butylene terephthalate), in *37th Antec SPE*, 1979, pp. 527–531.
3. R. J. Gardner and J. R. Martin, Effect of relative humidity

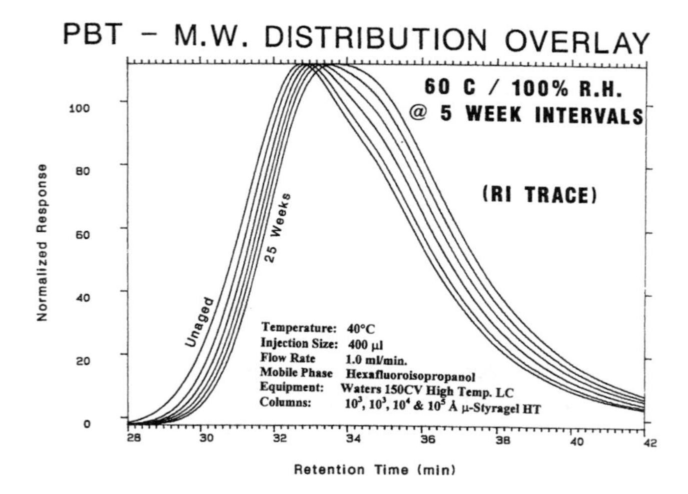

Fig. 1 Degradation of PBT at 60°C and 100% relative humidity; samples for analysis were taken at 5-week intervals.

on the mechanical properties of thermoplastic polyesters, in *37th Antec SPE*, 1979, pp. 831–834.

4. S. J. Ferrito, An analytical approach toward monitoring degradation in engineering thermoplastic materials used for electrical applications, in *International GPC Symposium Proceedings*, 1994, pp. 675–684.
5. S. Mori, Size exclusion chromatography of poly(ethylene terephthalate) using hexafluoro-2-propanol as the mobile phase, *Anal. Chem. 61*: 1321–1325 (1989).
6. A. Moroni and T. Havard, Characterization of polyesters and polyamides through SEC and light scattering using 1,1,1,3,3,3-hexafluoro-2-propanol as eluent, in *International GPC Symposium '96*, 1996, pp. 229–238.
7. C-s. Wu (ed.), *Handbook of Size Exclusion Chromatography*, Chromatographic Science Series Vol. 69, pp. 169–172.
8. Waters Division of Millipore, Solubilizing polyester (PET) for high temperature GPC analysis, *Polym. Notes. 3*(1) (April 1988).
9. W. W. Yau, J. J. Kirkland, and D. D. Bly, *Modern Size-Exclusion Liquid Chromatography*, John Wiley & Sons, Inc., New York, 1979, pp. 390–394.
10. Q. Nguyen, GPC–viscometry characterization of polyamides and polyethylene terephthalate in HFIP/CH_2CL_2 mixtures, *International GPC Symposium '98*, 1998, pp. 135–153.
11. R. L. Miller, R. W. Seymour, and L. W. Branscome, The application of size-exclusion chromatography to study molecular-weight changes of flame-retardant fiber-reinforced polyesters, *Polym. Eng. Sci. 592*: 31–38 (1991).
12. M. R. Milana, M. Denaro, L. Arrivabene, A. Maggio, and L. Gramiccioni, Gel permeation chromatography (GPC) of repeatedly extruded polyethylene terephthalate (PET), in *Food Addit. Contamin. 15*(3): 355–361 (1998).
13. S. Weidner, G. Kuehn, B. Werthmann, H. Schroeder, U. Just, R. Borowski, R. Decker, B. Schwarz, I. Schmueck-ing, and I. Seifert, *A New Approach of Characterizing the Hydrolytic Degradation of Poly(ethylene terephthalate) by MALDI–MS*, John Wiley & Sons, New York, 1997, pp. 2183–2192.
14. E. Meehan, S. Oakley, and F. Warner, *The Application of a Novel Particle Technology for GPC Using HFIP as Eluent*, Polymer Laboratories Ltd., Shropshire, UK, pp. 353–356.
15. *The Gel Permeation Chromatography of Poly(tetramethylene terephthalate)*, John Wiley & Sons, New York, 1977, pp. 2293–2295.
16. J. L. Ekmanis, GPC analysis of polymers with an on-line viscometer detector, in *International GPC Symposium Proceeding*, 1989.
17. M. Szesztay, Z. S. Laszlo-Hedvig, and F. Tudos, Gel permeation chromatography identification of polyester oligomers with different endgroups, *J. Appl. Polym. Sci. 48*: 227 (1991).
18. J. P. Sibilia (ed.), *A Guide to Materials Characterization and Chemical Analysis*, 2nd ed., VCH Publishers, Inc., New York, 1996.
19. A. V. Pavlov, V. V. Gur'yanova, O. M. Karan'yan, A. G. Morozov, and T. N Prudskova, Modern methods for studying the molecular characteristics of polymers, *Int. Polym. Sci. Technol. 20*(12): T/41–T/44 (1993).
20. S. Berkowitz, Viscosity–molecular weight relationships for poly(ethylene terephthalate) in hexafluoroisopropanol–pentafluorophenol using SEC–LALLS, *J. Appl. Polym. Sci. 29*: 4353–4361 (1984).

Sam J. Ferrito

Polymer Additives: Analysis by Chromatographic Techniques

Introduction

The need for developing methods of analysis for additives in plastics materials is increasing day by day because of several aspects. First, as toxicological studies are developed, there is the presence of some substances which have been banned in specific products. Second, in production, it is necessary to carry out a quality control for additive levels. It is also necessary to determine the stability of the additives when the polymer is processed. In this case, some additional compounds, often undesirable, can be generated and their control becomes necessary. Finally, the analysis will let us know the composition of a plastic product.

It is possible to use techniques in which the additives can be determined by direct analysis of the sample, such as nuclear magnetic resonance (NMR) spectrometry, ultraviolet (UV) spectrometry, and UV desorption–mass spectrometry. These techniques are very useful when the concentrations of additives in the polymer are high. However, when additives are present in trace levels, it is necessary to carry out a preliminary extraction/concentration step before analysis. Some of the most common additives used in plastics materials are presented in Table 1.

Sample Preparation

As was previously mentioned, a preliminary extraction step is necessary for samples in which a low level of organic compounds of interest must be determined. The extraction presents advantages when compared to the direct plastics analysis solution. In this case, the polymer is dissolved directly in an organic solvent and then precipitated. This procedure is not selective and the polymer residues can affect the analysis of the additives. When an extraction is carried out, higher selectivities can be obtained when the appropriate extraction conditions are used.

Traditional methods of extraction, such as Soxhlet, have been replaced by modern techniques as supercritical fluid extraction (SFE), microwave-assisted extraction (MAE), ultrasonic extraction, and accelerated solvent extraction (ASE) during recent years. The application of specific methods to these kinds of samples has permitted the development of a great number of other extraction methods. In the following list, a brief description is given:

SFE: This method uses pressure and temperature for fluids above their critical points. Under these conditions, the density and diffusivity of a supercritical fluid are between those of liquids and gases, resulting in an increase of the solvent power.

MAE: In this case, the sample and solvent are placed in a vessel and microwave radiation is applied. The solvent is heated under pressure above its normal boiling point so it remains in the liquid state.

ASE: Common solvents are used, at elevated temperatures and pressures, to increase the speed and efficiency of the extraction process.

In general, each of these procedures presents advantages: They are faster, an extract concentration step is not

Table 1 Common Additives Used in Plastic Products

Additive type	Chemical name	Commercial name
Antioxidants	Di-*t*-butyl-*p*-cresol	Bisphenol A
	Dioctadecyl (3,5-di-t-butyl-4-hydroxybenzyl)phosphate	Irganox 1093
	2,2′-Methylene bis (4-methyl-6-*tert*-butylphenol)	Irganox 2246
	Tris(2-methyl-4-hydroxy-5-*t*-butylphenyl)-butane	Topanol CA
	4,4′-Thio-bis (6-*t*-butyl-*m*-cresol)	Santonox
Light stabilizers	2-(2′-hydroxy-3,3,5-di-*tert*-amylphenyl)benzotriazole	Tinuvin 328
	Bis(2,2,6,6-tetramethylpiperidin-4-yl)sebacate	Tinuvin 770
	2-(2-Hydroxy-5-methylphenyl)-2*H*-benzotriazole	Tinuvin P
	2-Hydroxy-4-*n*-octoxybenzophenone	Chimassorb 81
Plasticizers	Di(2-ethylhexyl) phthalate	DEHP
	Disononyl phthalate	DINP
	Di(2-ethylhexyl) adipate	DOA
	Dipropylene glycol dibenzoate	Benzoflex 9–88
	Tributyl-*O*-acetylcitrate	Citroflex A
	Tris(2-ethylhexyl) phosphate	TOP
	Trioctyl trimellitate	TOTM
Colorants	Monoazo benzimidazole pigment	
	Pigment Red 176	
	Solvent Blue 97	
Flame retardants	Tetrabromobisphenol A	TBBA
	Hexabromocyclododecane	HBCD

necessary, a minimal amount of solvents is used, so they are environmentally friendly, and cleaner extracts are obtained. Thus, when the extract is obtained, a separation technique such as chromatography, can be applied in order to analyze the sample qualitatively and quantitatively.

There are many publications showing the applications of SFE to the determination of polymer additives. Antioxidants such as Irganox 2246, BHT and others, as well as UV stabilizers such as Tinuvin P, have been effectively extracted with supercritical CO_2. Extraction conditions varied from 15 to 25 MPa at 60°C and with a total time of 30 min [1]. If microwaves are applied to extract these compounds, a mixture of solvents can be used (acetone–heptane, for example) and time is considerably reduced (from 3 to 6 min).

Phthalate plasticizers are extracted with the same supercritical fluid at a pressure of 48 MPa at 95°C for the same amount of time [2]. Other phthalate plasticizers, such as TOP and TOTM are extracted from poly(vinyl chloride) (PVC) by ASE with ether at 100°C for 20 min.

Characterization of Polymer Additives with Chromatographic Techniques

There are several techniques that can be used for the determination of polymer additives after extraction; for example, high-performance liquid chromatography (HPLC), supercritical fluid chromatography (SFC), and high-temperature capillary gas chromatography (GC) are the most widely used. Capillary gas chromatography is especially useful because it combines high resolution with the possibility of coupling the instrument with some sophisticated detection techniques, such as mass spectrometry (MS), Fourier transform infrared (FTIR), or atomic emission detection (AED). MS is the most suitable detector for GC; it is very easy to prepare a personal spectral library for identification of separated species. There are also commercially available spectral libraries. Apart from the fact that the versatility of GC is due to its detectors, the injection systems provide a higher number of applications; for instance: on-column, split/splitless, and head-space techniques, among others. Today, almost every organic substance can be analyzed by gas chromatography, because of the development of high-temperature injectors. The combination of SFC with MS is also useful. In some cases, an on-line extraction is used, such as SFE–SFC [3].

The case of HPLC is quite different, because of the peak broadening or, sometimes, when the additives cannot be eluted from the stationary phase. However, it is a very useful technique when analyzing high-molecular-weight additives and, in combination with diode-array detection, provides an important tool for qualitative analysis. An important application is the determination of organic colorants in cosmetics [4]. HPLC is also applicable for the determination of linear and cyclic derivatives from poly(ethylene terephthalate), extracts of which were obtained using supercritical CO_2 [5].

Other Detection Systems

The analysis of extracts is also possible by using other techniques, such as infrared spectrometry. This kind of detector is very useful because it provides a total spectrum for the analyte. For example, a method for analyzing poly(dimethylsiloxane) oil from polymer samples was developed by Kirschner et al. [6]. In this case, the extractor is directly coupled to the spectrometer. However, it is also possible to carry out off-line analyses with high recoveries. This combination was also used by Raynor et al. [7] in order to analyze some UV stabilizers such as Tinuvin 770, Irganox 1010, and Topanol OC. In this case, a preliminary separation was carried out by supercritical fluid chromatography.

References

1. P. J. Arpino, D. Dilettato, K. Nguyen, and A. Bruchet, Investigation of antioxidants and UV stabilizers from plastics. Part I: Comparison of HPLC and SFC; preliminary SFC/MS study, *J. High-Resolut. Chromatogr. 13*: 5–12 (1990).
2. M. L. Marín, A. Jiménez, V. Berenguer, and J. López, Optimization of variables on the supercritical fluid extraction of phthalate plasticizers, *J. Supercrit. Fluids 12*: 271–277 (1998).
3. H. Daimon and Y. Hirata, Directly coupled supercritical fluid extraction/capillary supercritical-fluid chromatography of polymer additives, *Chromatographia 32*: 549–554 (1991).
4. S. C. Rastogi, V. J. Barwick, and S. V. Carter, Identification of organic colorants in cosmetics by HPLC–diode array detection, *Chromatographia 45*: 215–228 (1997).
5. St. Küppers, The use of temperature variation in supercritical fluid extraction of polymers for the selective extraction of low molecular weight components from poly(ethylene terephthalate), *Chromatographia 33*: 434–440 (1992).
6. C. H. Kirschner, S. L. Jordan, L. T. Taylor, and P. D. Seemuth, Feasibility of extraction and quantification of fiber finishes via on-line SFE/FT-IR, *Anal. Chem. 66*: 882–887 (1994).
7. M. W. Raynor, K. D. Bartle, I. L. Davies, A. Williams, A. A. Clifford, J. M. Chalmers, and B. W. Cook, Polymer additive characterization by capillary supercritical fluid chromatography/Fourier transform infrared microspectrometry, *Anal. Chem. 60*: 427–433 (1988).

M. L. Marín
Alfonso Jiménez Migallon

Polymer Degradation in GPC–SEC Columns

Size-exclusion chromatography (SEC) is a well-established method for the determination of the molar mass distribution (MMD) of polymers. However, the determination of the MMD by SEC substantially excludes ultra-high-molar-mass (UHMM) polymers. Actually, it is well accepted that UHMM polymeric samples degrade, by shearing or elongational forces, in the SEC columns. The upper limit of the molar mass for a successful SEC fractionation without degradation of the sample depends on the broadness of the MMD of the sample, from the SEC columns used and, obviously, from the experimental conditions. Successful SEC fractionations of narrow MMD standards up to 1×10^7 g/mol of molar mass have been reported. Instead, when the MMD of the sample is broad, rarely does the molar mass of the polymeric samples exceed the upper limit of 1×10^6 g/mol.

Whenever SEC fractionation has been applied to UHMM polymers, severe problems, strongly interrelated, have been invariably reported. The degradation of UHMM polystyrene during SEC fractionation in a good solvent has been often reported. The degradation of some other UHMM synthetic polymers [polyethylene, polyisoprene, polyacrylamide, polyisobutylene, poly(methyl methacrylate), etc.] has been reported. Besides, the degradation of some natural polysaccharides such as hyaluronic acid has been reported. The possible degradation of the polymeric sample is not the only problem of the SEC fractionation of UHMM polymers. The degradation of the UHMM sample in the SEC columns is often accompanied by some other important problems, such as (a) concentration effects, (b) anomalous flow, (c) poor column resolution (band broadening), and (d) very poor reproducibility.

Generally, the practical difficulties in the SEC fractionation exponentially increase with the polymer molecule's dimension, hydrodynamic volume, hence with the molar mass and the conformation, stiffness, of the polymer. For many years, SEC has been considered inadequate for the fractionation of UHMM polymers. In the past, there have been many theoretical and experimental studies [1,2] to elucidate a better knowledge of the degradation of UHMM polymers. Furthermore, there have been notable improvements in the SEC column's performance. At this time, there are commercially available SEC columns specifically designed for the fractionation of UHMM polymers.

Evidence of the degradation of the polymeric sample may be found by using qualitative and quantitative methods. From a qualitative point of view, an accurate analysis of the chromatogram of a series of narrow MMD standards with increasing molar mass, under identical analytical conditions, would be self-evident. The resolution of the SEC columns, almost in the low-molar-mass range, generally is very good. In the high-molar-mass range, $M > 1 \times 10^6$ g/mol, the resolution of the SEC columns quickly decreases. In the presence of the degradation of the sample, the resolution of the SEC columns is poor or absent. Furthermore, the peak shapes of degraded UHMM narrow standards is abnormally broad with a severe tailing. Obviously, we assume the absence of nonsteric fractionation. To find evidence of the degradation of broad MMD samples is more difficult. The analysis of the peak shape of a broad MMD sample requires experience. Generally speaking, a chromatogram that presents too steep slopes, shoulders, excessive tailing, or multipeaks is suspect from the point of view of the degradation of the sample. Obviously, the origin of these anomalous shapes of the chromatogram could be different. Concentration effects and anomalous flow often accompany the degradation of the sample. The qualitative analysis is used only to identify the problem.

A quantitative analysis of the degradation of the sample substantially requires the use of an on-line absolute detector such as light scattering (SEC–LS). In this case, the on-line LS detector directly measures, without calibration, the molar mass of the sample. The characterization of a series of narrow MMD standards with increasing and known molar mass evidences the degradation of the sample. The simple direct comparison of the known molar mass of the standards and the molar mass value obtained from the LS detector provides an estimate of the extent of the degradation under the experimental conditions used. Furthermore, if the dispersity D of the standards is very narrow (i.e., $D \leq 1.05$), the central part of the $M = f(V)$ plot, where M is the molar mass and V the elution volume, of each standard from the LS detector could be substantially flat. In fact, the central part of the peaks of the standards could be considered homogeneous in molar mass and the residual broadness of the peak is substantially due to the band broadening of the system. In the presence of degradation of the sample, the $M = f(V)$ plot of a narrow standards is not flat (M = constant) but steep. The method could be used to set up the experimental condition (columns set, flow rate, sample concentration) to avoid the degradation of the sample.

Figure 1 shows the experimental $M = f(V)$ plot of a narrow MMD polystyrene (PS) standard ($M = 1.15 \times 10^7$ g/mol, $D = 1.03$) from an on-line LS detector. The

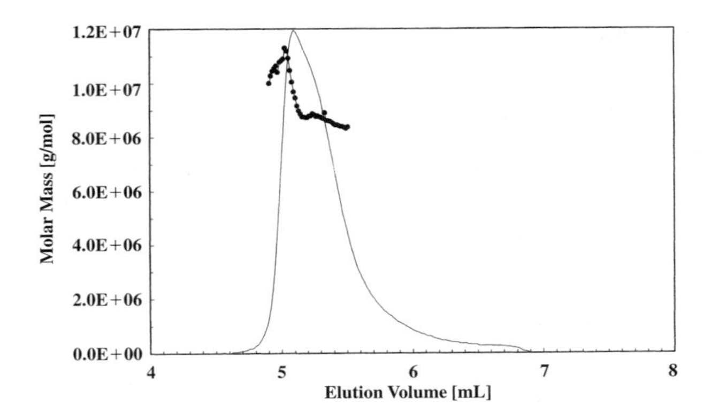

Fig. 1 Experimental $M = f(V)$ plot from an on-line LS detector.

experimental conditions were as follows: one column, packed with 16-μm spherical particles of porous silica; tetrahydrofuran as the mobile phase; 0.8-mL/min flow rate; 0.1 mg/mL of sample concentration; 100-μL injection volume. Figure 1 evidences the quick decrease of the molar mass of the sample due to the degradation of the UHMM PS sample in the column. A fractionation with the identical experimental conditions, column, and concentration and a 0.2-mL/min flow rate shows a constant $M = f(V)$ plot (i.e., the absence of degradation).

The origin of the degradation of the sample could be in the SEC columns and, for UHMM polymers, also in some other critical part (detector cell, long narrow capillary tube, injector, etc.) of the fluidic system. The extent of degradation also depends on the experimental conditions, particularly the flow rate and the concentrations of the polymeric solutions. In an SEC column, there are three parameters that have to be optimized for a successful fractionation of UHMM polymers. Two parameters are related to the packing of the column: particle size and pore size. The last parameter, often incorrectly ignored, is the dimension of the pores of the inlet and outlet frits of the column. It is well accepted that for a successful fractionation of UHMM polymers, one must use SEC columns with the diameter of the particles larger than 10 μm. In organic solvents, 20-μm particle columns are customarily used. However, in aqueous solvent, there are commercially available 15-μm nominal, maximum 17-μm particle columns. Obviously, the efficiency of the column quickly decreases with the increase of particle size. As usual, one needs to find a reasonable compromise between discordant effects.

The influence of the pore size on the degradation of the sample is not generally accepted. However, to obtain a meaningful fractionation on the basis of the dimension of the macromolecules, it is necessary that the pore size be higher than the dimensions of the molecules. The maximum dimension of UHMM macromolecules could be of several hundreds of nanometers. Polymer degradation strongly decreases with the large pore size of the frits of the columns. The commercially available 20-μm-particle SEC columns commonly use 10-μm inlet and outlet frits. Finally, it is very important to remember that partially obstructed frits could cause strong polymer degradation.

Also, the experimental conditions have to be optimized to avoid the degradation of UHMM samples. Particularly, the flow rate and concentration of the sample are two critical parameters. Polymer degradation strongly increases with the increase of the flow rate. For a successful fractionation of UHMM macromolecules, it is necessary to use a very low flow rate. With UHMM macromolecules, it is not unusual to use a flow rate of 0.2 mL/min or less.

The influence of the polymer concentration is well known in SEC. Theoretically, SEC assumes independence of the elution volume with regard to the concentration of the molecular species. Practically, it is well known that a high concentration of the sample solution leads to distortion of the polymer peak and dependence of the elution volume on the concentration of the sample. The "concentration effect" [3] depends on the difference in molecular mobility between the pure solvent and the viscous polymeric solution. This difference causes nonuniform flow, often called "viscous fingering," and, as a result, multiple peaks. In addition, it is well known that a high sample concentration decreases the hydrodynamic volume of the macromolecules. SEC fractionation is based on the hydrodynamic volume of the macromolecules; hence, this effect causes the dependence of the elution volume on the concentration, at least for high-molar-mass polymers.

The concentration, with UHMM macromolecules, also influences the degradation of the sample. Hence, in the fractionation of UHMM polymers, the concentration is very critical and has to be as low as possible. A sample concentration of 0.1 mg/mL or less is not unusual with UHMM polymers. In this case, it is necessary to use the minimal sample concentration that is consistent with the sensitivity of the concentration detector.

To avoid the degradation of the sample, some authors suggest the use of an ideal "theta" solvent as the mobile phase [4]. It is well known that the dimension of the macromolecules depends on the expansion factor α (excluded volume). For an ideal theta solvent, $\alpha = 1$ and the dimension of the macromolecules is minimal. For many polymers, an ideal theta solvent is known. For example, cyclohexane at 34.0°C is an ideal solvent for polystyrene polymers. For a polyelectrolyte polymer, ideal solvent means infinite ionic strength. Obviously, this condition is impracticable. Nevertheless, for example, the dimension of the macromolecules of an hyaluronic acid, anionic polymer, sample ($M = 7.4 \times 10^6$ g/mol) approximately decreases 20% when the solvent is changed from 0.15M to 0.5M NaCl. The use of ideal solvents to avoid the degradation of the sample is not generally accepted. In fact, using ideal solvents as the mobile phase could cause absorption of the polymer onto the column's packing. Also, some authors hypothesize that the degradation of the sample could also increase with ideal solvents.

At equilibrium conditions, at very low concentration, the elution volume of a macromolecule should be independent of the flow rate. However, with increasing molar mass in the UHMM range, in the absence of degradation, the elution volume strongly depends on the molar mass of the sample. This result does not depend on the concentration of the sample. This "retardation" effect occurs also at very low concentrations below the overlapping concentration c^*. The "retardation" of UHMM macromolecules has been studied by several workers [5,6]; it is a very complex effect and substantially still not well understood. The "retardation" effect is particularly meaningful in proximity to the exclusion limit of the columns and when the pore size approximately equals or is lower than the sizes of the macromolecules. A trivial conclusion is that for a successful fractionation of UHMM macromolecules without "retardation" effects, one must use SEC columns with ultralarge pore sizes.

Degradation of the sample and related problems, such as the "concentration effect" and anomalous flow, are the more important problems in the fractionation of UHMM polymers. The critical point in the characterization of the UHMM polymers is the fractionation in the SEC columns. For a successful fractionation of UHMM macromolecules, one must use specifically designed SEC columns with large particle sizes and ultralarge pore sizes. Furthermore, many aspects of the experimental protocol, such as flow rate and sample concentration, which is not critical in the usual molar mass range, become determining with UHMM polymers. A successful characterization of UHMM polymers requires optimization of the experimental protocol. Each step of the experimental protocol should be performed methodically to achieve the absence of sample degradation and reliable results.

References

1. F. J. Bueche, *J. Appl. Polym. Sci. 4*: 101 (1960).
2. J. C. Giddings, *Adv. Chromatogr. 20*: 217 (1982).
3. A. Rudin and R. A. Wagner, *J. Appl. Polym. Sci. 20*: 1483 (1976).
4. L. Soltes, D. Berek, and D. Mikulasova, *Colloid & Polym. Sci. 258*: 702 (1980).
5. E. V. Chubarova and V. V. Nesterov, *ACS Symp. Ser. 635*: 127 (1996).
6. W. Cheng and D. J. Hollis, *J. Chromatogr. 40*: 9 (1987).

Raniero Mendichi

Polymerase Chain Reaction Products: Analysis Using Capillary Electrophoresis

It is now possible to routinely analyze very small amounts of DNA using a procedure known as the polymerase chain reaction (PCR) [1]. PCR involves repeatedly subjecting a buffered salt solution containing deoxyribonucleotide triphosphates (dNTPs), two strand-specific oligonucleotide primers, a thermal-stable DNA polymerase enzyme (Taq), and a small amount of the DNA to be analyzed to a three-step temperature cycle. Using PCR, discrete regions of the source DNA molecules are copied and amplified by the repetitive cycling to readily detectable levels. It is also possible to analyze RNA sequences with PCR after an initial DNA strand (cDNA) and one complementary to it are produced from the original RNA template by the reverse transcription (RT) reaction [2]. The combined process, RT–PCR (also referred to as RNA–PCR), has been effectively used to study small amounts of RNA, such as individual messenger RNAs (mRNAs) or viral RNA, present in tissues and physiological fluids. Both PCR and RT–PCR produce double-stranded DNA (dsDNA) fragments of various sizes which are subsequently isolated and characterized by a number of qualitative and quantitative methodologies.

The most common method of analyzing PCR and RT–PCR products involves separating a portion of the reaction mixture by agarose slab gel electrophoresis with ethidium bromide staining to detect the presence of the amplified dsDNA fragments [3]. There are several disadvantages associated with slab-gel techniques, including the following:

1. Gel casting and handling are costly, labor intensive, and not readily automated.
2. A significant portion of the PCR or RT–PCR sample is typically consumed by this mode of analysis.
3. Buffer and reagent consumption, and hazardous waste generated from the use of radioactive probes and ethidium bromide stain, can be considerable.
4. Quantitation requires additional steps and instrumentation for gel imaging and analysis.

During the past decade, capillary electrophoresis (CE)-based techniques have been developed and refined for the analysis of dsDNA products of PCR and RT–PCR [4–7]. CE has several advantages over conventional slab-gel separation techniques, including the following:

1. Capillary electrophoresis instrumentation is fully automated with respect to sample injection, separation, on-capillary detection, and postrun data analysis.
2. Because the separation is conducted in a narrow-bore capillary that facilitates Joule heat dissipation, higher field strengths can be used, resulting in enhanced resolution and shorter run times.
3. Very small amounts (nL) of sample are required for the analysis, thus preserving more of the original sample for subsequent procedures, such as cloning or sequencing.

Considering the expanding base of CE applications, it is becoming clear that the majority, if not all, of conventional slab-gel separation techniques can be readily adapted to capillary format. Analysis of PCR products by CE is becoming more routine and it will likely become one of the primary applications of CE in the area of DNA separations [6]. This entry describes the use of CE-based techniques for the analysis of dsDNA products from PCR and RT–PCR.

Table 1 summarizes selected parameters that are important for establishing a robust and reproducible technique for the separation and quantification of dsDNA

Table 1 Analysis of PCR Products by CE: Selected Technique Parameters

Capillaries
– Untreated (bare fused silica): not frequently used
– Coated: polyacrylamide, polysiloxane (e.g., DB-1, DB-17), polyvinyl alcohol
Separation Matrix
– Buffer: 89–100 m*M* Tris-boric acid, 2 m*M* EDTA, pH 8.2–8.5 (1X TBE)
Sieving gel
– Chemical (fixed) gels: cross-linked polyacrylamide, bonded to capillary wall
– Replaceable Gels (entangled polymer networks): linear polyacrylamide, methyl cellulose, hydroxypropylmethyl cellulose, hydroxyethylcellulose, polyethylene oxide, polyvinyl alcohol, agarose
– Intercalating dyes: 9-aminoacridine (nonfluorescent), ethidium bromide, TOTO, YOYO, YO-PRO-1, TO-PRO-1, TO-PRO-3, SYBR Green I, EnhanCE™
Sample injection
– Hydrodynamic: high reproducibility; useful for quantitative analyses; direct injection of untreated samples possible
– Electrokinetic: affected by sample salt concentration (i.e., Cl^-); prior dialysis or dilution of the sample required
Detectors
– Ultraviolet (UV): 254–260 nm, least sensitive
– Laser-induced fluorescence (LIF): up to $1000\times$ more sensitive than ultraviolet
Data analysis
– Qualitative: optimization of PCR conditions and product characterization
– Quantitative:
– Relative: ratio of product (target) to "housekeeping" gene or other heterologous dsDNA internal standard
– Competitive: ratio of product (target) to homologous, modified internal standard (competitor); most accurate estimate
Applications
– Clinical/diagnostic: screening for genetic abnormalities and diseases
– Forensic: human identity testing
– Biotechnology: genetic analysis, gene expression, genotyping

products of PCR and RT–PCR using CE. Important advances have been made in a number of areas, including capillary coatings, sieving gel matrices, and high-sensitivity detection methods. Because of a nearly identical linear negative charge density at neutral pH and above, dsDNA molecules exhibit an electrophoretic mobility in free solution that is independent of molecular size [3]. Therefore, a gel or sieving matrix is required to effect a separation based on molecular size and, for that reason, capillary gel electrophoresis (CGE) has become the specific separation mode most often used for PCR product analysis. Because of the negative charge on DNA molecules, uncoated (bare fused silica) capillaries, which above pH 7 exhibit a strong electro-osmotic flow (EOF) in the direction of the cathode, are rarely if ever used. Instead, capillaries treated with a specific interior surface coating to greatly reduce or completely eliminate EOF are routinely employed in the separation of DNA, a process that is conducted in reversed polarity mode (i.e., cathode at the capillary inlet side). Capillary surface coatings can either be covalently bound to the surface or dynamically adsorbed to the wall. Examples of typical surface coatings include polyacrylamide, polysiloxanes (dimethyl and phenyl–methyl), cellulose derivatives, and polyvinyl alcohol [4–7].

Early CGE separations of dsDNA made use of capillaries in which a polyacrylamide gel was polymerized in and linked to the wall of the capillary, producing what has been referred to as a fixed or chemical gel [4–7]. Although such gels are capable of extremely high resolution due to a well-controlled pore size, they are not commonly used for PCR product analysis because of the problems related to air bubble formation and limited useful lifetime [5]. The development of replaceable sieving gels (also referred to as entangled polymer networks) that can be flushed from the capillary after the separation is complete has been one of the major factors in establishing CGE as a routine method for PCR product analysis. With this system of replaceable gels, a "new" gel is used for each separation. Some of the most widely used replaceable gel compounds include linear polyacrylamide, alkylcelluloses, polyethylene oxide, agarose, and polyvinyl alcohol [4–7]. These polymer compounds are employed to produce viscous buffered solutions that enable the separation of dsDNA in a capillary based on molecular size [5]. In general, the pore size and, hence, the resolving capacity for dsDNA molecules are controlled by simply manipulating the gel concentration.

One factor that initially hampered the direct analysis of PCR samples by CGE was the presence of high levels of salt, especially chloride ions, in the samples injected into the capillary. Electrokinetic injection, which is the requisite loading method when using fixed gels, is severely affected by the presence of high salt concentration because it impairs the loading of dsDNA into the capillary. Although PCR samples can be injected directly into capillaries using replaceable gels, the presence of salts adversely affects the quality of the results obtained. Also, the presence of other components in the PCR sample (dNTPs, primer oligonucleotides, etc.) not only affect the separation, but they can also obscure peaks when ultraviolet (UV) detection is employed. Fortunately, two relatively simple cleanup methods have been devised to counteract the effects of the PCR sample matrix. The two most common methods are sample microdialysis (float dialysis) and dilution of the sample (20–100-fold) with deionized water prior to CGE [6]. Both methods are effective in reducing the adverse effects of salts, but sample dilution necessitates the use of high-sensitivity detection to compensate for the reduction in the dsDNA concentration.

An important development in CE technology that has helped to promote the analysis of PCR products by CGE is the introduction of laser-induced fluorescence (LIF) detection [4–7]. Because LIF can increase the sensitivity of detection for dsDNA by more than 400-fold over UV detection, it has become the method of choice for the vast majority of dsDNA separations [5]. A practical illustration of the advantage of LIF detection is that typical separations of PCR products by slab-gel electrophoresis with ethidium bromide staining require approximately 5 ng of DNA per band for adequate detection, whereas, with CGE–LIF, subpicogram levels of DNA are readily detected [6].

In order to employ LIF detection, it is necessary to label the dsDNA molecules with a fluorescent compound prior to and/or during their separation by CGE. Two approaches have been employed. The first, and most common, involves the incorporation of an intercalating dye into the separation gel buffer (and, in some cases, into the sample loading buffer), which is highly fluorescent only when bound to dsDNA. Table 1 lists a number of commonly used intercalating dyes. Both monomeric and dimeric dyes have been developed and used to detect PCR products [8]. They offer unique advantages in that they enhance detection sensitivity two to three orders of magnitude over UV detection, and separation resolution and selectivity are often improved with their inclusion [4]. A second approach involves labeling of the primers used in PCR with a fluorophore, such as fluorescein, to produce 5′-end-labeled dsDNA products that can be separated and detected by CGE–LIF. The former approach offers the highest sensitivity with the amount of intercalating dye bound proportional to the size of the dsDNA fragment (i.e., the larger the fragment, the more dye will be bound). Not only do intercalating dyes label the dsDNA for detection, but they also can enhance the selectivity and resolution of dsDNA fragments of similar size [4,6]. Intercalating dyes can also produce anomalous effects on peak shape, depending on such factors as their concentration in the separation or sample buffer and certain sequence-dependent properties of the dsDNA (e.g., %GC composition). These effects result from the binding and retention of differing amounts of dye molecules by the dsDNA fragments during CGE. Therefore, care must be taken to carefully evaluate the use of a specific intercalating dye with a particular PCR product in order to generate reproducible results.

Capillary gel electrophoresis–LIF analysis of PCR and RT–PCR products has been applied to the areas of clinical diagnostics, forensics, and biotechnology. Screening of patients for genetic and infectious diseases, human identity testing using PCR-amplified DNA fragments from specific polymorphic genomic regions (loci) defined by a variable number of tandem repeats (VNTRs) or short tandem repeats (STRs), analysis of mitochondrial DNA, genotyping, and gene expression studies are only a few examples of these applications [4–7]. The types of results that can be gained from CGE analysis of PCR products are twofold. First, CGE is useful in a qualitative evaluation of PCR by separating target DNA from nonspecific products and demonstrating that a single dsDNA fragment resulted from the amplification. CGE is also a rapid method of evaluating various PCR parameters (e.g., cycle number, temperature, [Mg^{2+}], [dNTPs], etc.) for optimizing the efficiency of the reaction. Another use for CGE is to accurately determine the size of the PCR products. This approach has been applied to DNA profiling in human subjects by an assessment of PCR amplified alleles resulting from VNTRs and STRs [6].

The ability to do on-capillary detection and to calculate integrated peak areas from the collected data makes CGE–LIF very useful for the quantification of PCR- and RT–PCR-generated dsDNA products. The quantity of the amplified product can be indicative of the efficiency of PCR and this information can be used to optimize the reaction. Such information is also useful in determining the amount of a specific DNA or RNA present in the analyzed sample. Quantitation can be achieved by relative or absolute estimates. A ratio of target DNA peak area to the peak area of an added dsDNA internal standard gives an estimate of the relative amount of target DNA generated by PCR. Internal-standard dsDNAs can derive from genes that remain at constant levels in the sample (are unaffected by experimental treatments), such as the so-called "housekeeping" genes or they can be added amounts of a known quantity of a purified and well-characterized dsDNA, such as restriction enzyme digest fragments of genomic DNA. For example, the digestion of ϕX174 bacteriophage DNA with *Hae*III produces 11 distinct dsDNA fragments ranging in size from 72 to 1353 bp [6]. The latter type of standardization also affords the opportunity to accurately determine the size of the PCR product in addition to estimating its quantity when appropriate standards are chosen. Such standards can be obtained from commercial sources.

The most accurate means for quantitation of PCR products, especially those of low copy number, involves a method known as competitive PCR. In quantitative–competitive PCR (QC–PCR), known amounts of an internal standard (competitor) are co-amplified along with an unknown amount of target DNA. The competitor's sequence is chosen to be nearly identical to that of the target except for a small addition or deletion of sequence. The competitor is designed to use the same set of primers as the target so that a competition for them develops. Because the target and competitor are exposed to identical PCR conditions, the ratio of the two products should remain constant even after the reaction has reached its plateau phase. Thus, by plotting the different competitor/target peak area ratios against the amount of added competitor and extrapolating from the point at which the ratio is equal to 1, the amount of target DNA in the original sample can be determined in absolute terms. For RT–PCR, a competitor RNA is used to correct for variable conditions in both the RT and PCR steps [9,10].

Figure 1 depicts a scheme for the estimation of leptin mRNA (encoded by the obese gene, *ob*) contained in total RNA samples isolated from liver and adipose tissues of chickens. Using QC–RT–PCR with CGE–LIF, it was possible to derive absolute estimates (in attomoles) of leptin mRNA. Others have demonstrated a further extension of QC–RT–PCR with CGE–LIF called multiplexing, in which more than one competitor–target pair is subjected to co-amplification and subsequent analysis by CGE–LIF [9,11]. With multitarget QC–RT–PCR, it is possible to monitor PCR-amplified dsDNA corresponding to several genes simultaneously in a single sample assuming PCR conditions have been optimized for each product formed [11].

Future applications of CE to PCR product analysis will arise from improvements in CE instrumentation. Advances in miniaturization of CE devices by producing glass chips with etched channels of <1 cm in length have already been demonstrated as a feasible method for ultrafast (<45 s) separations of dsDNA [4–7]. The use of multiple capillaries or capillary arrays has proven to be useful in dedicated devices for DNA sequencing [4–7]. Recently, it has been possible to integrate PCR amplification and CE separation in a single device using an array of

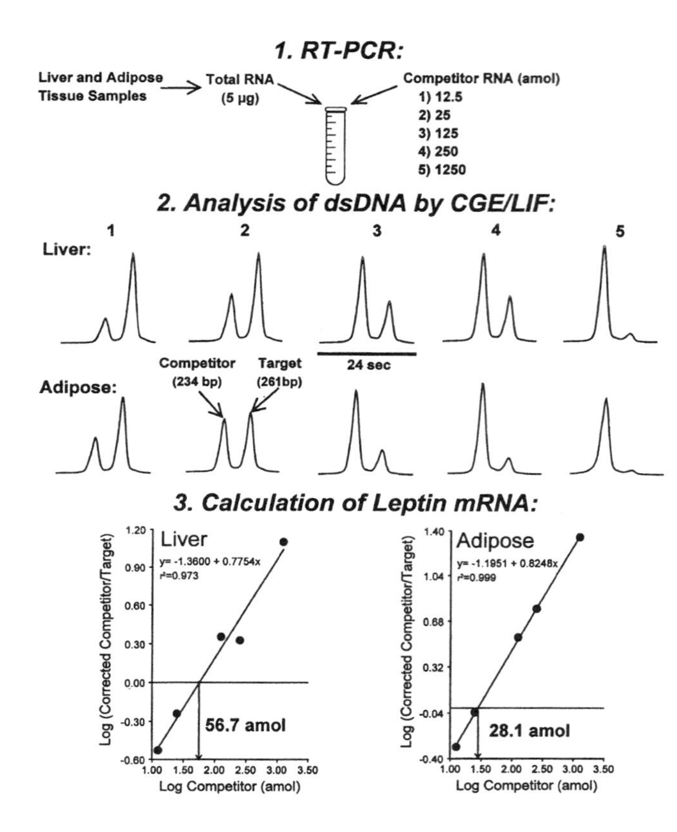

Fig. 1 Analysis of leptin gene expression in chicken liver and adipose tissue by QC–RT–PCR using capillary gel electrophoresis with laser-induced fluorescence detection (CGE–LIF). Target (261 bp) and competitor (234 bp) dsDNA amplicons were separated on a DB-1-coated capillary (27 cm × 100 μm inner diameter) at a field strength of 300 V/cm in a replaceable sieving matrix consisting of 0.5% HPMC in 1X TBE buffer with 0.5 μg/mL EnhanCE™ intercalating dye. RT–PCR samples (1–2 μL) were diluted 1:100 with deionized water and introduced into the capillary by electrokinetic injection. Separations were completed in under 5 min. A portion (4.4–4.8 min) of each separation shows the changes in the competitor and target peaks. CGE–LIF was more sensitive in detecting both amplicons than agarose slab gel electrophoresis with ethidium bromide staining. The integrated peak area ratio of competitor/target for a series of five individual samples (to which increasing amounts of a synthetic competitor RNA were added prior to RT–PCR) is used to calculate the amount of leptin mRNA (amol) in total RNA isolated from liver and adipose tissue by linear regression analysis.

eight capillaries for high sample throughput [12]. This technology will undoubtedly produce dedicated CE instruments for PCR product analysis that will feature rapid run time and high throughput. New detection methods such as mass spectrometry offer the promise of increases in selectivity, detection sensitivity, and more accurate quantification of PCR products. It is now clear that, in the future, CE-based analyses of PCR products will continue to get faster and more reliable while achieving wider acceptance by biomedical and biotechnology laboratories.

References

1. K. B. Mullis and F. A. Faloona, *Methods Enzymol. 155*: 335–350 (1987).
2. J. Chelly and A. Kahn, RT–PCR and mRNA quantitation, in *The Polymerase Chain Reaction* (K. B. Mullis, F. Ferre, and R. A. Gibbs, eds.), Birkhauser, Boston, 1994, pp. 97–109.
3. A. E. Barron and H. W. Blanch, *Separ. Purif. Methods 24*: 1–118 (1995).
4. A. Guttman and H. E. Schwartz, Separation of DNA, in *Capillary Electrophoresis Theory and Practice*, 2nd ed. (P. Camilleri, ed.), CRC Press, Boca Raton, FL, 1998, pp. 397–439.
5. K. J. Ulfelder and B. R. McCord, Separation of DNA by capillary electrophoresis, in *Handbook of Capillary Electrophoresis*, 2nd ed. (J. P. Landers, ed.), CRC Press, Boca Raton, FL, 1997, pp. 347–378.
6. J. M. Butler, Separation of DNA restriction fragments and PCR products, in *Analysis of Nucleic Acids by Capillary Electrophoresis* (C. Heller, ed.), Verlag Vieweg, Wiesbaden, 1997, pp. 195–217.
7. P. G. Righetti and C. Gelfi, Capillary electrophoresis of DNA, in *Capillary Electrophoresis in Analytical Biotechnology* (P. G. Righetti, ed.), CRC Press, Boca Raton, FL, 1996, pp. 431–476.
8. J. Skeidsvoll and M. Ueland, *Anal. Biochem. 231*: 359–365 (1995).
9. M. Fasco, C. P. Treanor, S. Spivack, H. L. Figge, and L. S. Kaminsky, *Anal. Biochem. 224*: 140–147 (1995).
10. N. D. Borson, M. A. Strausbauch, P. J. Wettstein, R. P. Oda, S. L. Johnston, and J. P. Landers, *Biotechniques 25*: 130–137 (1998).
11. W. Lu, D. S. Han, J. Yuan, and J. M. Andrieu, *Nature 368*: 269–271 (1994).
12. N. Zhang, H. Tan, and E. S. Yeung, *Anal. Chem. 71*: 1138–1145 (1999).

Mark P. Richards

Potential Barrier Field-Flow Fractionation

Introduction

Potential barrier field-flow fractionation (PBFFF), developed by Karaiskakis, is a combination of potential barrier chromatography and field-flow fractionation. It can be applied to separate colloidal particles and is based on differences in size or in any physicochemical parameters involved in the potential energy of interaction between the particles and the material of the FFF channel wall. Of the various quantities which affect the total potential energy (surface potential, Hamaker constant, and Debye–Huckel reciprocal distance), none is as accessible to empirical adjustment as the ionic strength of the suspending medium. This quantity depends on both the concentration and the cationic or anionic charge of the indifferent electrolyte added to the carrier solution. In its simplest form, the PBFFF technique consists in changing the ionic strength of the carrier solution from a high value, where only one of the colloidal components of the mixture subject to separation is totally adhered at the beginning of the FFF channel wall, to a lower value, where the total number of adhered particles is released.

The method has been applied to the separation of hematite and titanium dioxide submicron spherical particles, to the separation of submicron hematite spherical particles with different sizes, to the separation of various mixed sulfides with supramicron polydisperse irregular particles, and to the concentration of dilute colloidal samples, in both the normal and the steric modes of operation of sedimentation FFF.

Principle of PBFFF

Field-flow fractionation is a family of high-resolution techniques capable of separating and characterizing colloids and macromolecules. In normal FFF, the particles form a Brownian-motion cloud that extends a short distance into the channel. Separation is possible because the solvent flows at different velocities at various points within the channel. The smaller particles, whose cloud protrudes out into the faster laminae, are transported more rapidly than the larger particles, so that the two populations are soon separated. In the steric mode of operation, which happens when the protrusion of particles into the flow stream is determined by their physical bulk instead of by diffusion, the larger particles are elute earlier than the smaller ones.

In normal sedimentation FFF (SdFFF) the retention volume, V_r, of a spherical particle is immediately related to its diameter, d, via

$$V_r = \frac{\pi G w \Delta\rho V_0}{36kT} d^3 \tag{1}$$

where G is the sedimentation field strength expressed in acceleration, w is the channel thickness, $\Delta\rho = |\rho_s - \rho|$ is the density difference between the particle (ρ_s) and the carrier liquid (ρ), V_0 is the void volume of the channel, k is Boltzmann's constant, and T is the absolute temperature.

In the sedimentation steric FFF (Sd/StFFF), the retention volume of a spherical particle is immediately related to the diameter via

$$V_r = \frac{wV_0}{3\gamma}\frac{1}{d} \tag{2}$$

where γ is a dimensionless factor that accounts for the drag-induced reduction in velocity, as well as for the increase in velocity due to the activity of lift forces. To find γ, it is necessary to have a linear calibration curve of $\log V_r$ versus $\log d$ for standard particles of known size and nature (e.g. polystyrene latex beads). The easier way of working in the steric mode of FFF is that using the Earth's gravity as the external field (gravitational FFF = GFFF).

In the normal mode of the SdFFF operation, the potential energy, $\varphi(x)$, of a spherical particle is given by

$$\varphi(x) = \frac{\pi d^3}{6}\Delta\rho G x \tag{3}$$

where x is the coordinate position of the center of particle mass.

When the colloidal particles interact with the SdFFF channel wall, the potential energy given by Eq. (3) must be corrected, so as to include the potential energy of interaction, $\varphi(h)$. The latter can be estimated by the sum of the contributions of the van der Waals, $\varphi_6(h)$, and double-layer, $\varphi_{DL}(h)$, forces, and the total potential energy, φ_{tot}, of a spherical particle in PBSdFFF is given by the relation

$$\begin{aligned}\varphi_{tot} &= \varphi(x) + \varphi(h) \\ &= \varphi(x) + \varphi_6(h) + \varphi_{DL}(h) \\ &= \frac{4}{3}\pi a^3 \Delta\rho G x + \frac{A_{132}}{6}\left[\ln\left(\frac{h+2a}{h}\right) - \frac{2a(h+a)}{h(h+a)}\right] \\ &\quad + 16\varepsilon a\left(\frac{kT}{e}\right)^2 \tan h\left(\frac{e\psi_1}{4kT}\right)\tan h\left(\frac{e\psi_2}{4kT}\right)e^{-\kappa h}\end{aligned} \tag{4}$$

where h is the separation distance between the sphere and the channel wall, a is the particle radius, A is the effective Hamaker constant, ε is the dielectric constant of the liquid phase, e is the electronic charge, ψ_1 and ψ_2 are the surface potentials of the particle and the wall, respectively, and κ is the reciprocal Debye length.

The last equation shows that the energy φ_{tot} in PBSdFFF is a function of the size and of the surface potential of the particle, of the Hamaker constant, and of the ionic strength of the carrier solution, as the reciprocal double-layer thickness is immediately related to the ionic strength of the suspending medium. Thus, selectivity in PBSdFFF results from differences in particle size or chemical composition of the particles and of the suspending medium, where the latter will affect the surface potential and the Hamaker constant of the particle, as well as the medium's ionic strength.

Applications of PBFFF

As model samples for the verification of the PBFFF as a separation technique, colloidal samples of hematite and titanium dioxide with submicron monodisperse spherical particles were used. In the first example, the fractionation of titanium dioxide [TiO_2 with the nominal diameter obtained by a transmission electron microscope (TEM) of 0.298 μm] and hematite-I [α-Fe_2O_3(I) with nominal diameter obtained by TEM of 0.148 μm] spherical particles was succeeded by the PBSdFFF technique. Figure 1a shows the fractionation of the TiO_2 and α-Fe_2O_3(I) particles by the normal SdFFF mode of operation, and Fig. 1b shows the fractionation of the same particles by the potential barrier mode of SdFFF. The latter is based on the difference of the total potential energy of interaction between the colloidal particles and the channel wall due to the variation of the ionic strength of the suspending medium. In the PBSdFFF technique, the mixture was introduced into the channel with a carrier solution containing 0.5% (v/v) detergent FL-70, 0.02% (w/w) NaN_3, and $3 \times 10^{-2} M$ KNO_3. At this high electrolyte concentration, all of the TiO_2 colloidal particles adhered at the beginning of the SdFFF Hastelloy-C channel wall, whereas all of the α-Fe_2O_3(I) particles were eluted from the channel. The average diameter of the eluted α-Fe_2O_3(I) particles determined by Eq. (1) was found to be 0.143 μm, in good agreement with that obtained by TEM (0.148 μm) or determined by normal SdFFF (0.150 μm). Changing the carrier solution to one containing only 0.5% (v/v) detergent FL-70, whose ionic strength is $1 \times 10^{-3} M$, released all the adhered TiO_2 particles and gave a particle diameter (0.302 μm) in good agreement with that obtained by TEM (0.298 μm).

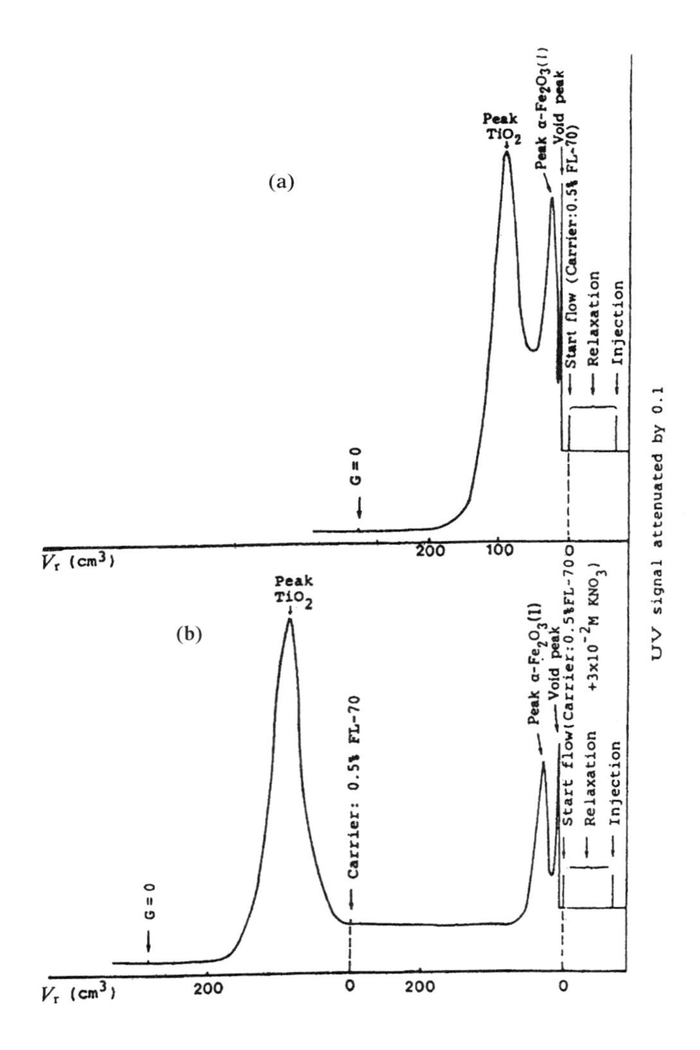

Fig. 1 Fractionation of α-Fe_2O_3(I) (with nominal diameter 0.148 μm) and TiO_2 (with nominal diameter 0.298 μm) colloidal particles by (a) the normal SdFFF and (b) the PBSdFFF technique. The experimental conditions, except for those given in the scheme, were as follows: field strength = 15.5 g, flow rate = 150 cm^3/h, void volume of the channel = 2.06 cm^3. [Reproduced with permission from G. Karaiskakis et al. (1990) *J. Chromatogr. 517*: 345; copyright Elsevier Science Publishers B.V.]

In order to show whether the size of the particles and not their nature is responsible for the variation of the total interaction energy between the colloids and the channel wall, a second example of fractionation was performed by using two samples of hematite with different particle diameters [α-Fe_2O_3(I) with nominal diameter 0.148 μm and α-Fe_2O_3(II) with nominal diameter 0.248 μm]. In the present case, the separation is based only on the particle size difference, as the Hamaker constants and the surface potentials of the two samples are identical. That obtained by the PBFFF technique fractogram had the same form as that of Fig. 1b, except for the fact that the peak of α-Fe_2O_3(II) was in the position of TiO_2. The critical electrolyte (KNO_3) concentration for the adhesion of α-Fe_2O_3(II) particles at the beginning of the SdFFF channel wall was found to be $3 \times 10^{-2}M$, whereas at this high concentration of KNO_3, the whole number of α-Fe_2O_3(I) particles are eluted from the channel. Variation of the carrier solution to one containing only 0.5% (v/v) detergent FL-70 and 0.02% (w/w) NaN_3 without any amount of KNO_3 released the total number of adherent α-Fe_2O_3(II) particles. The particle diameters obtained by the PBFFF technique [0.151 μm for α-Fe_2O_3(I) and 0.244 μm for α-Fe_2O_3(II)] are in excellent agreement with those found by normal SdFFF [0.145 μm for α-Fe_2O_3(I) and 0.237 μm for α-Fe_2O_3(II)] or determined by TEM [0.148 μm for α-Fe_2O_3(I) and 0.248 μm for α-Fe_2O_3(II)].

In PBFFF, the variation of the potential energy of interaction between the colloidal particles and the channel wall can be succeeded, except for the variation of the ionic strength, by changing also the pH and the nature of the suspending medium. Polydisperse, irregular supramicron colloidal particles of the mixed sulfides $Cu_xZn_{1-x}S$ ($0 < x < 1$) were used as model samples to verify the applicability of the potential barrier gravitational field-flow fractionation (PBGFFF), based on the variation of the above parameters, to fractionate colloidal particles.

A general methodology for the analysis of a colloidal mixture by PBSdFFF consists in injecting into the column the mixture with a carrier solution in which the ionic strength is too high to assure total adhesion of all the components of the mixture, except for the one with the lower attractive force with the channel wall. Then, a programmed decrease of the ionic strength of the carrier solution is applied to release, *in time*, the adherent particles according to their size and/or surface characteristics. As the PBSdFFF technique is based on particle–wall interactions, its applications can be extended by using different from Hastelloy-C materials, such as stainless steel, Teflon, and polyimide. PBSdFFF is also a convenient and accurate method for the *concentration* and analysis of *dilute* colloidal samples. The major advantage of the proposed concentration procedure is that the method can concentrate particles even of the same size but with different surface potentials and/or Hamaker constants. The method has considerable promise for the separation and characterization, in terms of particle size, of dilute complex colloidal materials, such as those of natural water, where particles are present in low concentration.

Future Developments

Looking to the future, it is reasonable to expect continuous efforts to improve the theoretical predictions and to expand the applications of the potential barrier methodology of FFF to more complex and dilute colloidal samples. The latter can be easily succeeded by constructing FFF channels from various materials of well-defined composition.

Suggested Further Reading

A. Athanasopoulou and G. Karaiskakis, *Chromatographia 40*: 734 (1995).

A. Athanasopoulou and G. Karaiskakis, *Chromatographia 43*: 369 (1996).

J. C. Giddings, *Separ. Sci. 1*: 123 (1966).

M. E. Hansen and J. C. Giddings, *Anal.Chem. 61*: 811 (1989).

M. E. Hansen, J. C. Giddings, and R. Beckett, *J. Colloid. Interf. Sci. 132*(2): 300 (1989).

G. Karaiskakis and J. Cazes (eds.), *J. Liquid Chromatogr. Related Technol. 20*(16 & 17) (1997) (special issue).

G. Karaiskakis and A. Koliadima, *Chromatographia 28*: 31 (1989).

A. Koliadima and G. Karaiskakis, *J. Chromatogr. 517*: 345 (1990).

A. Koliadima and G. Karaiskakis, *Chromatographia 39*: 74 (1994).

A. Koliadima, D. Gavril, and G. Karaiskakis, *J. Liquid Chromatogr. Related Technol. 22*(18): 2779 (1999).

George Karaiskakis

Preparative HPLC Optimization

The approach used for preparative high-performance liquid chromatography (PHPLC) optimization depends on the goals and scale of a particular laboratory or industrial chromatographic separation. Parameters such as quantities (from milligrams to tons), required purity and recovery of the final material, cost of separation, and availability of the equipment need to be considered. The separation expenses including time, cost of solvents, stationary phase, and isolation of the product should be compared to the cost of the starting material and the value of the product being purified. Overall production cost is a function of purification costs and production rate [1]. In multiton production of a selected material, it is worthwhile to invest a substantial effort on trial experiments and mathematical modeling of the nonlinear chromatographic process. In contrast, in pharmaceutical research and development, time is the most crucial parameter for the delivery of milligram to kilogram quantities of new compounds for initial biological testing. Other applications that require fast separations on a smaller scale are the preparation of pure standards and isolation of byproducts for structural determination.

Preparative HPLC optimization goals which ultimately lead to a product with a given minimum purity may include the maximum amount of the purified material per weight unit of stationary phase per time unit (g/kg/day), the maximum amount of the purified material per mobile phase unit per time unit (g/L/day), the maximum production rate (g/day), the lowest cost ($/kg), the maximum recovery (%), and the maximum production rate with maximum recovery. Regardless of the differences in application, it is important to be aware of the following parameters that may affect the purity and recovery of the product as well as the time and cost required for the separation:

Selectivity

The separation of the material of interest and closely retained impurities is greatly affected by changes in selectivity (α) when the relative distribution of the sample components between the stationary and mobile phases changes. A practical value for the selectivity (separation factor), described as $\alpha = k_2/k_1$, where k_1 and k_2 are the retention factors for adjacent peaks, is greater than 1.2. However, separation can be improved by increasing the selectivity value up to 2 or 3 whenever possible. In normal-phase chromatography, selectivity is altered by varying the organic solvents in the mobile phase. In a reversed-phase separation, changing both the type of organic solvent and water/organic ratio in the mobile phase is also used to optimize selectivity [2]. Selectivity is also dependent on the temperature and stationary phase being used, especially with enantioselective (chiral) separations. The newest addition to the stationary-phase materials designed to improve selectivity includes custom-tailored molecular-imprinted polymers [3].

Retention

It is advantageous to use the minimum retention (k) necessary for separating the product and providing the desired purity, because the cycle time is decreased and the production rate is increased. In both normal-phase and reversed-phase modes, the concentration of the product in collected fractions increases as the retention decreases. In displacement chromatography, longer sample retention provides a more concentrated product fraction. As retention increases, column efficiency increases, but the cycle time and solvent consumption are increased as well. The optimum retention is in the range of $k = 1.2$ to 2.0 for isocratic separations and $k = 3$ to 4 for gradient separations. Poorly retained impurities can be eluted from the column in the void volume. Impurities that are retained significantly longer than the product need to be washed from the column at the end of the run cycle using a step-gradient with a strong eluent; this step may include flow reversal. Although regeneration is a regular procedure in reversed-phase chromatography, in normal-phase chromatography polar impurities tend to strongly bind to the silica surface and full regeneration is difficult. However, because of the low price of the silica, the stationary phase is often discarded after one or several runs. Methods for use with modern flash chromatography employing bare silica are easy to develop using preliminary thin-layer chromatography (TLC) results followed by scale-up with disposable cartridges. In this flash chromatography technique, an increase of the difference in column volumes of elution between the product and a closely retained impurity (ΔC_v) increases production rate of the separation:

$$\Delta C_v = \frac{1}{R_{f1}} - \frac{1}{R_{f2}}$$

where R_{f1} and R_{f2} are mobility coefficients of the separated compounds using TLC.

Stationary Phase

The high loading capacity of the stationary phase provides increased production rate. Column loadability can be increased with an increase in accessible surface area of the stationary phase by using a smaller particle size. However, this results in higher back-pressure during separation. If the stationary phase contains particles of different sizes, the fraction of larger particles controls the efficiency and the fraction of smaller particles controls the back-pressure. The use of high-quality packing with a narrow particle size distribution allows for increased production rates with less back-pressure and higher flow rates. Current average particle sizes for reversed-phase HPLC are in the range of 10–20 μm.

In preparative liquid-carbon-dioxide-based supercritical flow chromatography (SFC), smaller particles in the 5–10 μm range are used due to the decreased viscosity of the mobile phase. The pore size of the particles should be large enough to allow the molecules to readily diffuse into and out of the pores. In reversed-phase HPLC, longer alkyl chains provide better loadability because of the higher volume of interaction between the separated compounds and the stationary phase but lower resolution because of the slower kinetic rates of mass transfer [4]. The production rate (g/day) in batch chromatography increases with the increased size of the column. Columns of different length, packed with particles of different sizes of the same material, may have a similar production rate provided that the ratio of d_p^2/L is the same, where d_p is the diameter of the particles and L is the length of the column. Spherical particles have better mechanical stability and, thus, longer life than irregular particles.

Axial compression columns allow for efficient in-house packing of large columns with stationary phases. Systems utilizing radially compressed, preloaded, disposable cartridges are commercially available. In reversed-phase PHPLC, the choice between stationary phases with longer or shorter alkyl chains (C_3, C_8, C_{18}) depends mainly on the lipophilicities of the compounds to be separated. In order to assure that retention times are not too long for highly lipophilic compounds, a shorter alkyl chain length is preferred. This allows for faster separation to be achieved and, consequently, less organic solvent in the mobile phase, thereby cutting down on solvent costs. However, if evaporation of the collected fractions is critical, then using a longer-alkyl-chain-length stationary phase with a more volatile mobile phase may be advantageous.

Mobile Phase

The viscosity of the mobile phase influences the back-pressure and efficiency of the separation. The higher the back-pressure, the lower the flow rate that can be used and, consequently, the longer the run time. Lowering the viscosity increases mass transfer and, hence, the efficiency of the separation is improved (the peaks are narrower). Although the maximum flow rate does not provide the maximum separation, it does maximize the production rate.

The addition of nonvolatile additives to the mobile phase complicates isolation of the product after separation in reversed-phase PHPLC. The use of trifluoroacetic acid (TFA) simplifies workup due to its high volatility; it is common to add TFA to water–organic mobile phases as an ion-pairing reagent. It is important, however, to maintain a sufficient concentration of TFA, especially during loading of the sample, in order to ensure an adequate concentration of the counterion. Any residual TFA can be effectively removed from the isolated product using anion-exchange chromatography.

The solubility of the product in the mobile phase is an important parameter. Low solubility of the crude material in a mobile phase may require alterations in the method to prevent on-column precipitation of the material. If a number of different mobile-phase mixtures provides similar selectivity, the one in which the product is most soluble should be selected. For example, in normal-phase PHPLC, a mixture of 15% tetrahydrofuran in heptane may provide the same selectivity but better solubility than a mixture of 5% methanol in heptane. Separation is much more reproducible when the ratio of the mobile-phase components is 50/50 rather than 99/1 because there is less chance that a small change in the mobile-phase composition will change the separation.

Some enantioselective stationary phases may be used only with a limited number of solvents. Because the cost of the mobile phase often represents more than half of the total cost of the separation (except lower cost of carbon dioxide in SFC), it is worthwhile to minimize the solvent consumption and also to increase the concentration of the product in the eluted fractions. Regeneration of the mobile phase is an economical choice at high solvent-consumption rates. Regeneration may include distillation of the used mobile phase followed by the adjustment of the mobile-phase composition if necessary. In normal-phase PHPLC, nonaqueous (nonhalogenated) solvents are easily removed from the product by evaporation and are less expensive to dispose of than water-containing mobile phases. At the same time, the use of high volumes of low-boiling flammable organic solvents may represent a safety concern. Accordingly, a less volatile solvent, heptane, for example, rather than hexane, should be used.

Flow Rate

The production rate increases at higher flow rates; however, some decrease in separation efficiency occurs. The upper limit of the flow rate depends on the ability of the stationary phase and hardware of the PHPLC system to withstand the higher back-pressure. Decreasing the flow rate may improve efficiency, and this effect is more distinctive with a reversed stationary phase containing long alkyl chains due to the more pronounced effect of the mass-transfer kinetics. Many noncovalently bonded enantioselective stationary phases may not be able to withstand higher flow rates and pressure.

Temperature

Increased temperature usually improves loadability and solubility and decreases the viscosity of the mobile phase. Consequently, the production rate increases, provided that the selectivity of separation or stability of the separated compounds and stationary phases are not compromised. However, resolution in enantioselective chromatography often improves at lower temperatures.

Gradient Program and Elution Order

Gradient elution provides a better production rate and higher recovery than using isocratic conditions if isocratic separation requires a regeneration step. The gradient profile in reversed-phase PHPLC is usually shallow, which decreases peak tailing compared to isocratic elution. After the sample injection, the gradient elution starts with a slightly weaker mobile phase (lower percentage of organic) than that required to elute the product with a desired retention, and then increases slowly in order to decrease the tailing. The gradient used is shallower when the target compound is the late-retained peak of the two peaks separated. The slope of the gradient should be steep enough to avoid dilution of the product fraction. Too shallow a gradient decreases the production rate and complicates isolation of the product. After most of the product has eluted, the gradient is raised stepwise in order to elute all late-retained impurities from the column. The column is equilibrated again with at least one column volume before the cycle is repeated.

If a compound is the major component in a mixture, the production rate increases if the impurities are eluted first. However, it is preferable for the product to be eluted prior to any closely retained impurity if a compound is not the main component in the mixture. Under self-displacement conditions, the product can actually be separated in a more concentrated solution than in a touching-band separation, especially if a later-retained impurity is in high concentration. The production rate in this case can be improved by an increase in loading, despite some decrease in recovery.

Sample Loading

Preferably, the sample is dissolved in a solvent mixture weaker than the mobile phase and introduced (concentrated) on the column in a narrow zone. Although a higher concentration of the sample contributes to a narrower zone after injection, it is not recommended to use a solution of more than twice the viscosity of the mobile phase. Column hardware should provide an even distribution of the injected solution throughout the cross section of the column.

The "dry injection" technique is often used in normal-phase flash chromatography when the solubility of the compound in a mobile phase is low. In this technique, a suspension of silica in the sample solution is concentrated to dryness and placed in a separate precolumn introduced before the actual column. In the reversed-phase mode, when there is limited sample solubility, an injection of a larger volume of the less concentrated feed solution is preferable to an injection of a more concentrated material in a solution with a higher percentage of the organic solvent. Simple flash chromatographic prepurification of the crude material is always recommended for all types of PHPLC.

Direct scale-up of an analytical HPLC method at touching-band conditions is feasible only when there is a need for small (mg) amounts of the isolated material and the mobile phase does not contain nonvolatile additives. Using a touching-band optimization technique, the amount of material loaded may be increased until visible peak broadening occurs and the first sign of overlapping between the product peak and closely retained impurities is observed. The amount of product that may be loaded onto a column under touching-band conditions can be estimated as [5]

$$w \approx \frac{1}{9} w_s \left(\frac{\alpha - 1}{\alpha} \right)^2$$

where w_s is a column capacity that may be experimentally found by frontal analysis or less exactly estimated as w_s (mg) ≈ 0.4 (column surface area, m^2). The sample capacity of an analytical 25×0.46-cm column is estimated to be 150–400 mg for small-molecule (<1000 Da) compounds.

Overloading of the column provides a much better production rate than the touching-band technique. Unlike analytical HPLC, Gaussian peak shape and baseline resolution are not a requirement in PHPLC. Overloading causes interaction between the compounds during chromatography that creates a sample self-displacement effect. This reduces peak tailing of the earlier eluting peak of the product and leads to increased purity and concentration of the eluted band. In contrast, if the concentration of the closely retained, earlier eluted compound is higher than the product, the tag-along effect compromises separation. The total capacity of the column may be found using frontal analysis [6]. A loading factor that represents the percentage of the loaded amount versus the total capacity of the column is one of the major parameters of optimization as well as efficiency (plate number) of the column. The loading factor and concentration of the product collected are usually higher in gradient and displacement chromatography than with isocratic elution.

Collection and Isolation of Separated Product

The composition of the mobile phase controls the product isolation procedure. Isolation from reversed-phase PHPLC fractions requires evaporation, lyophilization, or dialysis of the water-containing mobile phase.

Another approach is to concentrate the fraction by diluting it with water and injecting the product on a separate column, followed by elution using a pure organic solvent. This method provides a more concentrated solution of the product in an organic phase, but it generates more waste solvents. In addition, automated collection at optimized cut points can contribute significantly to the overall optimization process.

Design of the Optimization Experiment

Nonlinearity of the Langmuir adsorption isotherms is observed even in noncompetitive chromatographic processes. Individual adsorption isotherms can be found experimentally using frontal analysis at overload conditions; however, the adsorption isotherms in the separation of mixtures are different because of the interference of other compounds in the mixture. In PHPLC method development, it is necessary to optimize separation conditions and column loading experimentally.

For optimization purposes, it is convenient to use analytical 0.46×25-cm columns filled with the stationary phase available in bulk quantities for PHPLC. Primary evaluation and selection of the stationary and mobile phases, based on the selectivity of separation, column efficiency, and product retention, should be done initially on an analytical scale, including experiments under overload conditions on an analytical column. A particular set of chromatographic conditions includes the column, programmed mobile phase, flow rate, sample solution composition and amount injected, temperature, and collection points. These conditions are evaluated to obtain optimization of the maximum recovery of the desired material with required purity, or the maximum available purity for the required yield based on the HPLC analysis of collected fractions.

Not all of the separation parameters can be improved simultaneously and the optimization process often requires multiple experiments. In addition, separation problems, such as the sample "bleeding" through the column during loading or the product not completely eluting from the column, should be identified and corrected. Steps of the HPLC optimization experiment include (a) preliminary recording of the peak area–concentration plot for a pure product standard, (b) calculation of the amount of the target material in the solution prepared for separation, (c) running the designed separation and collecting all fractions by time, starting after the eluted void volume and ending after the column is washed with the strong mobile phase for regeneration (*Note*: During the time in which the product and closely retained impurities elute, samples have to be taken more frequently), (d) all collected fractions should be analyzed to determine the purity (A) and the recovery of the product (C) in each fraction, and (e) the material balance of the product amount before and after separation should be measured.

Calculated product recovery $C_{\Sigma i}$ (%) if i fractions combined:

$$C_{\Sigma i} = C_1 + C_2 + \cdots + C_i$$

Calculated product purity $A_{\Sigma i}$ (%) if i fractions combined:

$$A_{\Sigma i} = \frac{A_1C_1 + A_2C_2 + \cdots + A_iC_i}{C_1 + C_2 + \cdots + C_i}$$

where $C_1, C_2, \ldots, C_i$ is the product recovery (%) in fractions $1, 2, \ldots, i$ and is calculated as a part of the total desired product presented in the injected crude material, and $A_1, A_2, \ldots, A_i$ is the product purity (%) in fractions $1, 2, \ldots, i$.

Comparison of the optimization results on recovery and purity assists in the selection of the optimized PHPLC conditions for scale-up. After optimum conditions are determined, a scale-up factor (Y) can be calculated, based on the results obtained on an analytical column, using the following equation:

$$Y = \frac{L_1D_1^2}{L_2D_2^2}$$

where L_1 and L_2 specify column lengths for the scale-up column and analytical column respectively and D_1 and D_2 are respective internal column diameters.

Continuous Process and Automation

Several techniques, such as continuous simulated moving bed (SMB) as well as closed-loop and closed-loop steady-state recycling chromatographic techniques, are designed to improve the production rate and cost by recycling partially separated product, thus decreasing solvent consumption compared to conventional batch PHPLC. As a result, products are isolated in a more concentrated solution. These binary techniques are mostly used in the separation of pairs of stereoisomers and they require special hardware. If more than one impurity is present, the target material should be either the first or the last band to elute.

Automation allows batch chromatography to be run as a continuous process. Multiple injections using a separate pump and fraction collection provide an opportunity for continuous unattended operation. In isocratic separations, sample injection is often made before previously injected product elutes from the column, thus reducing cycle time and solvent consumption. Continuous and automated processes are always used with smaller columns and lower amounts of expensive enantioselective stationary phases. One of the future goals for modern PHPLC optimization would be the creation of software that would allow computer simulation modeling of nonlinear effects in preparative chromatography.

References

1. G. Ganetsos and P. E. Barker (eds.), *Preparative and Production Scale Chromatography*, 1993, Marcel Dekker, Inc., New York, 1993.
2. L. R. Snyder, *J. Chromatogr. B 689*: 105 (1997).
3. R. A. Bartsch and M. Maeda (eds.), *Molecular and Ionic Recognition with Imprinted Polymers*, American Chemical Society, Washington, DC, 1998.
4. D.-R. Wu and H. C. Greenblatt, *J. Chromatogr. A 702*: 157 (1995).
5. L. R. Snyder, J. J. Kirkland, and J. L. Glajch, *Practical HPLC Method Development* 2nd ed., John Wiley & Sons, New York, 1997.
6. G. Guiochon, S. G. Shirazi, and A. M. Katti, *Fundamentals of Preparative and Nonliner Chromatography*, Academic Press, New York, 1994.

Michael Breslav

Preparative TLC

Thin-layer chromatography (TLC) is mostly used as a separation technique combined with qualitative (identification) and quantitative analytical methods; the lower limit of the sample size depends on the sensitivity of detection. For ultraviolet (UV) and chemical detection, it is usually in the microgram or nanogram range. The separated spots can also be isolated for further investigations (chemical derivatization, mass spectrometry, or as chromatographic standards). To increase the scale, the dimensions and thickness of the layer are increased so that it is no longer "thin"; Nyiredy [1] proposed denoting preparative planar chromatography with the abbreviation PLC (preparative layer chromatography) (1 mg to 1 g). The flow of mobile phase during development may be due to capillary forces, pressure [forced-flow planar chromatography (FFPLC), overpressured layer chromatography (OPLC)], or centrifugal force (rotation planar chromatography, RPC).

In analytical TLC, linear adsorption isotherms and compact spots are obtained for the loading of sample below 10^{-3} g mixture/1 g adsorbent. To increase throughput, PLC is operated under overloaded conditions, at 1 mg/1 g adsorbent or more. The overloading can be of concentration or volume type; the former is more advantageous [2].

The main goal of PLC is not the maximal peak (spot) capacity, but the maximal yield of separation.

The increased sample size is obtained by multiple spotting along the start line; when larger volumes are spotted, a series of microcircular chromatograms is obtained as the starting band; streaking from a syringe moving along the start line (e.g., as aerosol — using a programmed applicator); especially in OPLC, on-line zonal application across the plate is possible: The sample is injected into the continuous stream of eluent, distributed across the plate in a narrow channel parallel to the edge, and collected at the farther end of the layer so that the layer acts as a flat column [1]. Similar on-line injection and collection is possible in centrifugal RPC.

The on-line application of sample, even of large volumes, is possible with certain horizontal chambers provided with distributors (Camag linear chamber, DS.-chamber [3,4]). When the mixture contains few compo-

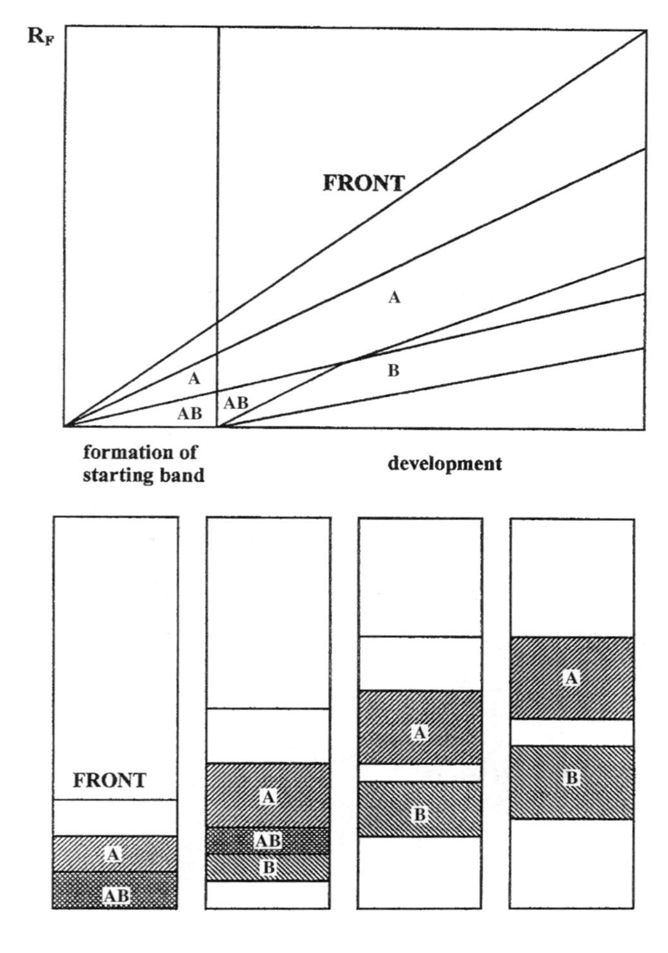

Fig. 1 Mutual displacements effects.

nents and the selectivity of the system is high, even large volumes of sample can be introduced from the edge of the layer so that the components are partly separated already in the application stage (frontal chromatography) and, because of mutual displacement effects, may become fully separated during development (Fig. 1), with high yield. To form a compressed starting band, the sample should be dissolved in a solvent weaker than the eluent.

Wide starting bands can also be formed by putting the edge of the plate in the sample solution. To avoid contamination of the eluent in the chamber during development, its bottom is covered by a strip of paper wetted with the eluent; more eluent is introduced when the starting band leaves the edge of the plate.

For thicker layers, the attainment of equilibrium in the gas–mobile phase–adsorbent system to avoid complicating effects (solvent demixing, preadsorption) is more difficult. The solutes migrate in a nonequilibrated layer with differentiated velocity — more rapidly in the surface layer (because of the evaporation of solvent) and less rapidly closer to the carrier plate.

Two stages of PLC determine the success of separation: application of the sample and development. In the former stage, three different situations are possible: (a) sample dissolved in the eluent; (b) sample dissolved in diluted eluent; (c) sample dissolved in a solvent of different quantitative and qualitative compositions.

In case (a), minimal disturbances can be expected at the start of development; however, a wide starting band may be obtained. To obtain a narrow starting band, it is advantageous to dissolve the sample in a weaker solvent, {i.e., of lower content of modifier [case (b)]}. The solubility of the sample is the limiting factor here; the rule is to dissolve a large sample in a small volume of solvent. A good solvent of low eluent strength is chloroform or dichloromethane (e.g., for alkaloids), toluene, or ethers. The application of sample solvent different from the eluent [case (c)] may lead to precipitation of the solutes at the beginning of development; the gradual dissolution of the precipitate in the mobile phase is reflected by comblike tailing of the starting band. It should also be taken into account that, in TLC, the ratio of volumes of the sample solvent (e.g., 0.5 mL) and eluent is greater than in analytical and preparative column chromatography and may have a more significant effect on the variation of eluent strength of the mobile phase. Evaporation of the sample solvent before development may cause precipitation of the solutes; their delayed dissolution in the mobile phase leads to tailing, which is detrimental to separation. Therefore, on-line application of the sample is advantageous.

The choice of mobile phase in PLC is also determined by the subsequent recovery of the separated solutes. Less volatile components (water, acetic acid, butanol) should be avoided, as well as nonvolatile components — buffer solutions and ion association reagents. Normal-phase chromatography and nonaqueous eluents should, therefore, be preferred. The rules of choice of eluents are otherwise similar to those for analytical chromatography (i.e., basing on eluotropic and isoeluotropic series) depending on the properties of the separated solutes. Usually, eluents giving R_F values in the lower range (0.1–0.5) are chosen, because the application of larger sample volumes leads to wide starting zones and increase of R_F values. For polar adsorbents, ethyl acetate belongs to the recommended modifiers, because of the good solubility of many nonpolar and moderately polar solutes, rapid equilibrium, and easy evaporation of the separated fractions.

In PLC, sorbents applied in TLC are usually used: silica, alumina, Florisil, cellulose, and silanized silica. Binding agents such as gypsum and somewhat lower amounts of water are recommended (e.g., for 1-mm silica layers, the weight ratio of water and adsorbent is 1:1.5 or 1:2). Plates with thicker layers should be dried in air in a horizontal position for a longer time (e.g., for 1-mm layers, ~ 1 h; for 2-mm layers, 2–4 h) and then dried and activated in an oven in gradually increasing temperatures. Equilibration in the chamber should also be prolonged. Because chemically bonded adsorbents are relatively expensive and, moreover, reversed phase (RP) sorbents are poorly wettable by aqueous eluents, polar adsorbents and nonaqueous eluents of low viscosity are usually applied in PLC. Special precoated plates of 0.5–2-mm layers are commercially available (e.g., silica gel 60 $F_{254+366}$). Such plates can be applied directly without activation; however, drying at 80°C for 2 h is recommended. Although the capacities of layers increase with their thickness, the separation efficiency decreases for thickness above 1.5 mm, so that optimal for PLC are layers of 0.5–1 mm thickness.

The development distance in PLC should not exceed 20 cm, because of the decreasing flow rate for longer distances and increasing diffusion of zones. A suitable system should have a resolution $R_s \geq 1.5$ in the analytical scale.

Marked improvement of separation efficiency in the separation of complex samples may be obtained by stepwise gradient elution [5] because of enhanced mutual displacement of the components in the concentrated starting band. A simple stepwise gradient of four to five steps is frequently sufficient; the generation of stepwise gradients is possible in some types of horizontal chambers [6] by consecutive delivery of eluent fractions of increasing concentrations of modifier.

Preparative layer chromatography can also be used as a pilot technique for column preparative chromatography in the same solvent–adsorbent system.

References

1. S. Nyiredy, in *Handbook of TLC*, 2nd ed. (J. Sherma and B. Fried, eds.), Marcel Dekker, Inc., New York, 1996.
2. L. R. Snyder and J. W. Dolan, *Adv. Chromatogr. 38*: 115 (1998).
3. E. Soczewinski, in *Planar Chromatography, Volume 1* (R. E. Kaiser, ed.) Hüthig, Heidelberg, 1986, pp. 79–117.
4. T. H. Dzido, G. Matysik, and E. Soczewinski, *J. Planar Chromatogr. 4*: 161 (1991).
5. E. Soczewinski, G. Matysik, and B. Polak, *Chromatographia 39*: 497 (1994).
6. E. Soczewinski, K. Czapinska, and T. Wawrzynowicz, *Sep. Sci. Technol. 22*: 2101 (1987).

Edward Soczewinski
Teresa Wawrzynowicz

Programmed Flow Gas Chromatography

There are three methods that can be used to accelerate the elution of strongly retained peaks during chromatographic development. Flow programming, where the flow of mobile phase is continuously increased during the development of the separation, temperature programming, and gradient elution, the latter being exclusively used in liquid chromatography. Flow programming is not as effective in reducing the elution time of well-retained components and tends to cause increased band dispersion; it is, however, more gentle than temperature programming and would be chosen when separating thermally labile materials. The complexity of the theoretical treatment depends on whether the mobile phase is compressible or not. In gas chromatography, the mobile phase is compressible, and this must be taken into account in the first theoretical treatment.

We shall assume that under flow programming conditions, the mass flow rate will be increased linearly with time (i.e., $Q_{0(t)} = (Q'_0 + \alpha t)$, where Q'_0 is the *initial exit flow rate*, $Q_{0(t)}$ is the exit flow rate after time t and α is the program rate. These conditions are usual for modern gas flow programming devices that utilize mass flow controllers which are computer operated. Now, if ΔV_0 is an increment of exit flow, measured at atmospheric pressure, then employing the usual pressure correction factor, the corrected gas flow (ΔV_r) will be

$$\Delta V_r = \frac{3}{2}\Delta V_{r(0)}\left(\frac{\gamma^2 - 1}{\gamma^3 - 1}\right)$$

where γ is the inlet/outlet pressure ratio of the column. Then, under the above-defined programming conditions,

$$\Delta V_{r(t)} = \frac{3}{2}\left(\frac{\gamma_t^2 - 1}{\gamma_t^3 - 1}\right)(Q_0 + \alpha t)\Delta t \tag{1}$$

where (γ_t) is the inlet/outlet pressure ratio at time t and $\Delta V_{r(t)}$ is the increment of volume flow at time t.

Now, as the flow is increased, the inlet pressure will also increase and, thus, the inlet/outlet pressure ratio (γ) will change progressively during the program; the *mean* flow rate will be reduced according to the pressure correction function and the decrease in elution rate will not be that which would be expected. Consider an open-tubular column; from Poiseuille's equation,

$$P_0 Q_{(0)t} = \frac{(P_t^2 - P_0^2)\pi a^4}{16\eta l}$$

or

$$P_0(Q_0 + \alpha t) = \frac{(P_t^2 - P_0^2)\pi a^4}{16\eta l} \tag{2}$$

where P_t is the inlet pressure at time t, P_0 is the outlet pressure (atmospheric), η is the viscosity of the gas at the column temperature, l is the length of the open-tubular column, and a is the radius of the open-tubular column.

A similar equation would be used for a packed column, except the constant $(\pi/16)$ would be replaced by the D'Arcy constant for a packed bed. Rearranging,

$$\frac{P_0(Q_0 + \alpha t)16\eta l}{\pi a^4} + P_0^2 = P_t^2$$

thus,

$$\gamma_t = \frac{P_t}{P_0} = \left(\frac{(Q_0 + \alpha t)16\eta l}{P_0 \pi a^4} + 1\right)^{0.5} \tag{3}$$

Assuming the column dimensions are 320 μm in inner diameter (radius $a = 0.0160$ cm) and 30 m in length, and it is operated at 120°C using nitrogen as the carrier gas which, at that temperature, has a viscosity of 129×10^{-6} P, then, by using Eq. (3), the change (γ) can be calculated for different flow rates. The relationship between flow rate (as measured at the column exit and at atmospheric pressure) and the column inlet/outlet pressure ratio is shown in Fig. 1. The inlet/outlet pressure ratio changes significantly during a mass flow rate program, which will attenuate the elution rate as shown by the pressure correction factor. It is seen that the curve is a close fit to a second-order polynomial function, but this relationship is fortuitous, although it might be used empirically to predict inlet/outlet pressure ratios.

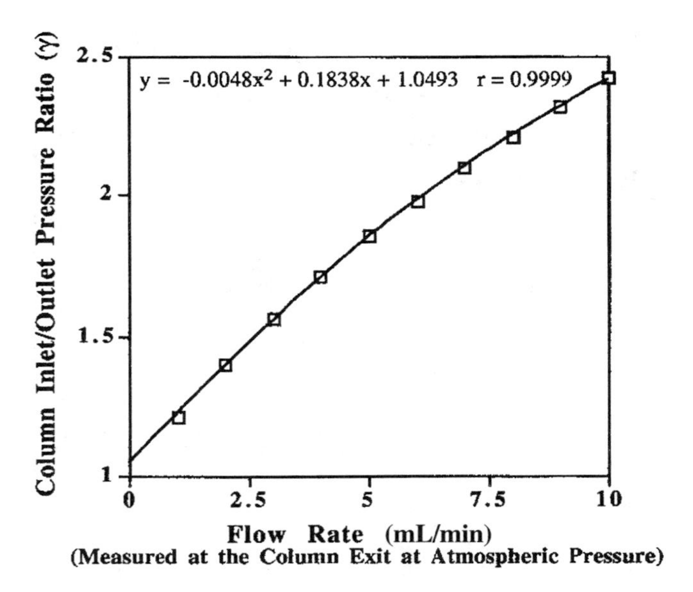

Fig. 1 The relationship between the inlet/outlet pressure ratio and exit flow rate for an open-tubular column.

It is now possible to use the values for γ_t to calculate when the solute is eluted at the retention time t_r for different flow program rates. Employing Eq. (1), when the sum of all the increments of ΔV_r is equal to the retention volume V_r, then t will be t_r, the retention time:

$$V_r = \sum_{t=0}^{t=t_r} \frac{3}{2}\left(\frac{\gamma_t^2 - 1}{\gamma_t^3 - 1}\right)(Q_0 + \alpha t)$$

With a simple computer program, the retention time t_r can be calculated for a range of different program rates (α) and for solutes having retention times of 10, 50, and 250 mL, respectively. Ther column was again assumed to be 320 mm in inner diameter and 30 in length and operated at 120°C using nitrogen as the carrier gas which, at that temperature, has a viscosity of 129×10^{-6} P. The results are shown in Fig. 2.

The effect of program rate on retention time is much as would be expected. For any individual solute, the retention time is related to some power of the solute retention volume, but the indices vary significantly with the retention volume of the solute. This relationship does not have a theoretical explanation at this time but might be useful for predicting retention times from experimental data. Despite the attenuating effect of the pressure correction factor, the use of flow programming is effective in reducing the retention time of strongly retained solutes. However, unless the diffusivities of the solutes in the mobile phase are high, (i.e., mobile phases such as hydrogen or

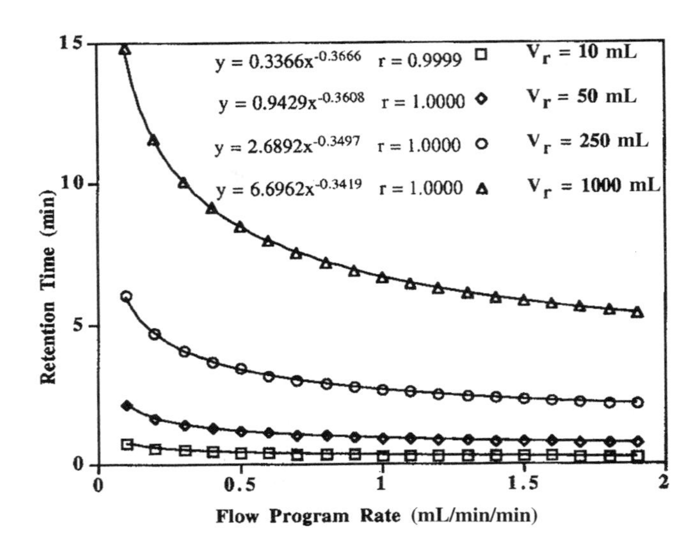

Fig. 2 Curves relating elution time to flow program rate for solutes having different retention volumes. (From Ref. 1.)

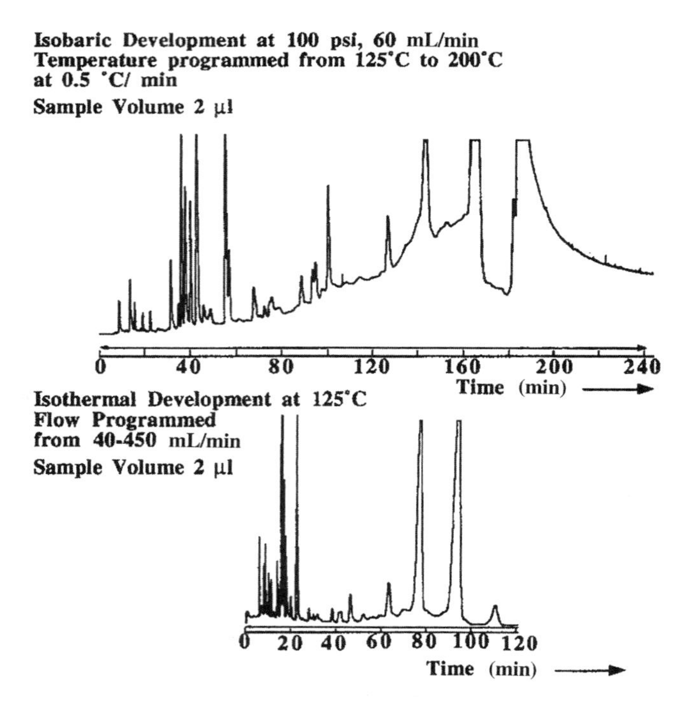

Fig. 3 The separation of lemon grass oil by temperature programming and by flow programming.

helium are employed), there will be significant peak dispersion at the higher velocities, the column efficiency will be reduced, and resolution will be lost.

An excellent example of the use of flow programming was given in a very early article describing the technique [1]. Lemon grass oil contains two substances that are very thermally labile, and these are eluted very late in the chromatogram. If temperature programming is employed, the substances decompose at the higher temperatures, producing a sloping baseline as shown in the upper chromatogram in Fig. 3. The baseline does not return to its normal level until the thermally unstable compounds have been completely eluted. If the separation is carried out isothermally at a temperature where the decomposition is minimal, and the mixture is developed by flow programming, all the solutes are eluted on a relatively stable baseline. The elution times are high, due to long packed columns being employed, as opposed to capillary columns.

Reference

1. R. P. W. Scott, *Gas Chromatography 1964* (A. Goldup, ed.), The Institute of Petroleum, 1964, p.25.

Suggested Further Reading

Scott, R. P. W., *Introduction to Analytical Gas Chromatography,* Marcel Dekker, Inc., New York, 1998.

Scott, R. P. W., *Techniques of Chromatography,* Marcel Dekker, Inc., New York, 1995.

Raymond P. W. Scott

Programmed Temperature Gas Chromatography

Temperature programming becomes necessary when the sample contains components that have polarities and/or molecular weights that extend over a wide range. Such samples, if separated isothermally, may well result in the less retained solutes being adequately resolved and eluted in a reasonable time. However, the more polar or higher-molecular-weight solutes may be held on the column for an inordinately long period, and the solute peaks, when they are eluted, are likely to be wide and flat and difficult to evaluate quantitatively. To avoid this situation, the column temperature can be progressively increased during development so that the late eluting peaks are accelerated through the column and are still sharp and eluted in a reasonable time. This procedure is called temperature programming.

The corrected retention volume of a solute (V'_r) is given by (see the entry Rate Theory in Gas Chromatography)

$$V'_r = KV_s$$

where K is the distribution coefficient of the solute with respect to the stationary phase and V_s is the volume of stationary phase in the column.

The value of K varies with temperature in the following manner (see the entry Thermodynamics of Retention in Gas Chromatography)

$$\ln(K) = -\left(\frac{\Delta H}{RT} - \frac{\Delta S}{R}\right)$$

$$\text{or} \quad K = \exp\left[-\left(\frac{\Delta H}{RT} - \frac{\Delta S}{R}\right)\right]$$

where ΔH is the standard enthalpy of distribution, ΔS is the standard entropy of distribution, R is the gas constant, and T is the absolute temperature. Thus,

$$V'_r = \exp\left[-\left(\frac{\Delta H}{RT} - \frac{\Delta S}{R}\right)\right]V_s$$

It is seen that as T increases, K decreases. It follows that K and, consequently, the retention volume can be progressively decreased by increasing the column temperature. In practice, this is usually achieved by situating the column in an oven, the temperature of which is controlled by appropriate electronic circuitry. Theoretically, the temperature of the oven can be increased as any function of time, but almost all program profiles are linear in form.

Consequently, programming the column from temperatures T_1 for a period of t, using a linear program (i.e., $T = T_1 + \alpha t$, where t is the elapsed time), the mean value of K will be given by the following function:

$$K = t_1^{-1} \sum_{t=0}^{t=t_1} \exp\left[-\left(\frac{\Delta H}{R(T_1 + \alpha t)} - \frac{\Delta S}{R}\right)\right]$$

where t_1 is the time period of the program. [Preferably t_1, for maximum accuracy, should be defined in small units (e.g., seconds and not minutes)]. Similarly, the program rate α must be defined in degrees Celcius per second.

Now, when the solute is eluted, V'_r will equal the product of the mean flow rate Q_m and t_1, which will be the retention time; that is,

$$V'_r = Q_m t_1$$

The mean flow rate through a gas chromatography column is given by [1]

$$Q_m = Q_0 \frac{3(\gamma^2 - 1)}{2(\gamma^3 - 1)}$$

where γ is the inlet/outlet pressure ratio and Q_0 is the exit flow rate.

Thus, when the solute elutes,

$$V'_r = V_s K \sum_{t=0}^{t=t_1} \exp\left[-\left(\frac{\Delta H}{R(T_1 + \alpha t)} - \frac{\Delta S}{R}\right)t_1\right] = Q_0 \frac{3(\gamma^2 - 1)}{2(\gamma^3 - 1)} t_1$$

where V_s is the volume of stationary phase in the column.

Solving for t in the classical manner is a cumbersome mathematical procedure and it is easier to employ a numerical method to calculate t. To demonstrate the effect of temperature programming on solute elution, a computer program was used to calculate the retention time of a given solute eluted at different programming rates. The data used were that of Liao and Martire [1] for 3-methyl hexane chromatographed on *n*-octadecane at temperatures of 30°C, 40°C, 50°C, and 60°C.

By curve-fitting the retention data to the reciprocal of the absolute temperature, the numeric form of the distribution coefficient was found to be

$$K = \exp\left(\frac{1737.777}{T} - 2.75115\right)$$

A simple computer program provided values for the retention time, calculated for different linear program rates, and the results are shown as curves relating retention time to program rate in Fig. 1. The inlet/outlet pressure ratio (γ) was assumed to be 2 and the flow rate at the column exit was 20 mL/min (0.3333 mL/s). The logic of the program was based on a search for that value of t where the value of (V'_r) equaled the product of t and the mean column flow rate. It should be noted that the data were reported as the corrected retention volume per gram of stationary phase; thus, for simplicity, the curves are calculated on the assumption that the column contains 1 g of stationary phase. Retention times for columns containing more stationary phase would be appropriately longer. Due to the change in both the density and viscosity of a gas with temperature, the mean flow rate will change slightly during the program, but the error will be small, provided that a *mass flow controller* is employed and not a pressure controller. For the calculations, the inlet pressure was assumed to remain constant throughout the program.

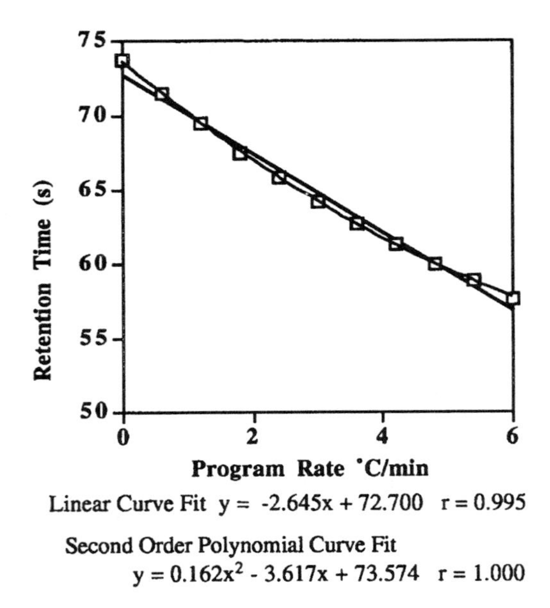

Fig. 1 Graph of retention time versus program rate.

This procedure is a general method for calculating the effect of the program rate on retention time. Basically, the corrected retention volume must be measured for each solute of interest at two different temperatures to provide the thermodynamic constants. The above equations will then allow the effect of different linear temperature programs on the corrected retention volume to be calculated for each solute. The effect of program rate on resolution can also be observed if some solutes elute close together. In fact, the equation can be used for program functions other than linear, but these are rarely employed. The relationship between retention time and program rate is approximately linear and can be assumed so for most practical purposes. For more accurate work, a second-order polynomial should be used to describe the retention time as a function of the program rate, which can give very accurate values.

Reference

1. H. L. Liao and D. E. Martire, *J. Am. Chem. Soc. 94*(10): 2058 (1972).

Suggested Further Reading

Scott, R. P. W., *Introduction to Analytical Gas Chromatography*, Marcel Dekker, Inc., New York, 1998.

Scott, R. P. W., *Techniques of Chromatography,* Marcel Dekker, Inc., New York, 1995.

Raymond P. W. Scott

Prostaglandins: Analysis by HPLC

Introduction: Physiology and Chemistry [1]

Prostaglandins (PG) are a class of substances representing natural metabolites of three 20 carbon fatty acids differing in their number of double bonds: arachidonic acid (AA, 5,8,11,14-eicosatetraenoic acid), 8,11,14-eicosatrienoic acid, and 5,8,11,14,17-eicosapentanoic acid. In the following, only the bisenoic PGs (produced from AA), which dominate most biological systems, will be discussed. Free AA acts as a substrate for the cyclooxygenase enzyme complex and consecutively several spezific enzymes, including isomerases, reductases, and synthases, produce thromboxanes (TX) and further PGs as demonstrated in Fig. 1. Their weak ultraviolet (UV) absorption maxima are found at 192 nm and at 217 nm for PGA_2, 228 nm for 15-oxo-PGE_2, and 278 nm for PGB_2 arising from delocalized electron systems of conjugated double bonds and oxo groups. Prostanoids are known for their high and widespread physiological potency acting as chemical messengers capable of regulating cellular behavior in mammalian tissues. The influences on platelet aggregation, contractory or dilatory effects on muscles and vessels, mediating inflammatory diseases, and pain are only a few examples which make these compounds of great interest for diagnostic and therapeutic aims. Therefore, analytical methods are needed which allow the trace-level determination of PGs from different complex biological sources like serum, plasma, seminal fluid, urine, or culture medium.

In most cases, PGs are measured by highly specific immunoassays, although not enabling simultaneous determination of multiple components in a single experiment.

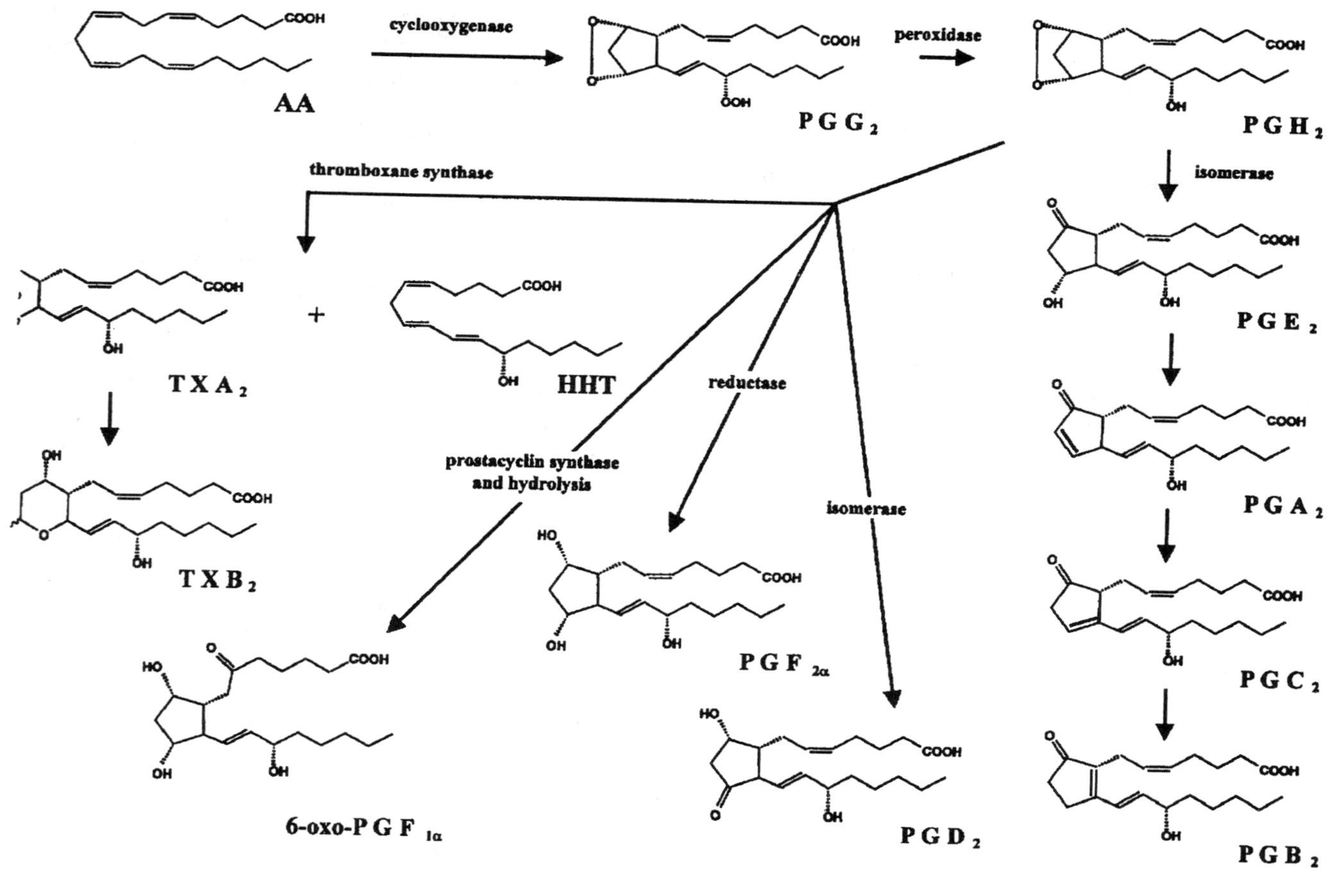

Fig. 1 Metabolism of arachidonic acid (AA) to prostaglandins (PG) and thromboxanes (TX).

Beside high-performance liquid chromatography (HPLC) techniques, GC methods are often used in combination with mass spectrometry (MS), flame ionization (FID), or electron-capture detectors (ECD). However, HPLC is still a very useful, versatile, and widespread tool for analytical or preparative separation and detection under mild conditions, allowing qualitative and quantitative measurement of oxygen-sensitive prostanoids.

Sample Preparation [2–4]

Prostanoids only occur in trace levels in complex biological samples. Therefore, the removal of substances impairing chromatographic resolution or detection and concentration of the analytes is required. Most of the samples are extracted using liquid–liquid (LLE) or solid-phase (SPE) methods without longer time of storage, thus avoiding the nonenzymatic production of isoprostanes by autoxidation of AA [5]. Due to the free carboxylic groups of prostanoids, samples are acidified to pH 3–4, usually by acetic (HOAc), formic, citric, or hydrochloric acid. Acidification to lower pH values has to be avoided due to chemical instability and degradation of some prostaglandins: PGE_2 dehydrates irreversible to PGA_2, isomerizing to PGB_2; PGD_2 might be dehydrated to an inverted version of PGA_2. Epimerization at C_{15} and cis–trans isomerization may occur, as recently shown for the thromboxane synthase product 12-*S*-hydroxyheptadecatrienoic acid (HHT) [6]. LLE is carried out by the addition of a onefold to fivefold volume of nonpolar solvents like diethyl ether, chloroform, or ethyl acetate, followed by shaking or stirring for a few minutes. Consecutive centrifuging enables efficient phase separation and pellets the particulate matter. Recoveries of 80–98% (dependent on the prostaglandin) are achieved.

Repetition of this procedure is seldom done and does not improve the recovery significantly. The organic layers are evaporated to dryness under reduced pressure or a gentle stream of nitrogen to concentrate the prostanoids prior to further preparations or chromatography. Extraction of biological samples using petrol ether or hexane at neutral pH values prior to extraction of acidified matrix removes nonpolar fatty acids and lipids, preventing their possible negative interferences.

Using SPE methods, adsorption of prostanoids is routinely and efficiently done by straight-phase material like silic acid or by reversed-phase C_{18} material. Cleanup and washing steps using eluents of rising polarity for silica-based columns (e.g., increasing percentages of methanol in an ether–hexane–toluene mixture) or decreasing polarity for octadecyl columns [e.g., increasing percentages of acetonitrile (ACN), ethanol, or ethyl acetate in an aqueous mixture] enable suffcient extraction. These stepwise and fractionating procedures allow consecutive elution of different substance classes like polar compounds (polar lipids), fatty acids and their monohydroxy derivatives, PGs and TXs, and hydrophilic components. Recoveries for all prostanoids are in the range 90–98%. Extracts are evaporated to dryness prior to further procedure. Avoiding these manual and time-consuming extraction methods, the use of packed precolumns coupled to reversed-phase (RP)–HPLC equipped with a conventional six-port injector is a suitable alternative especially for small sample volumes.

Mobile and Stationary Phases [4,7,8]

In most cases, PGs or their derivatives are separated by partition chromatography on reversed-phase columns using gradients of organic solvents and water with increasing organic content during elution. ACN enables better separation than MeOH when combined with acidified aqueous media using HOAc, trifluoric acid, or phosphoric acid. Separated by rising hydrophobicity, pure PGs are eluted typically in the following order of retention times: 6-oxo-$PGF_{1\alpha} < TXB_2 < PGF_{2\alpha} < PGE_2 < PGD_2 <$ 13,14-dihydro-15-oxo-$PGF_{2\alpha} <$ 13,14-dihydro-15-oxo-$PGE_2 < PGA_2 < PGB_2$. Monohydroxy metabolites of AA and other fatty acids often present in PG-containing samples are eluted at much higher percentages of organic solvents, demanding a wide-range gradient.

Early PG analysis using HPLC techniques was carried out as adsorption chromatography on normal-phase (NP) columns packed with silica or alumina. The nonpolar mobile phase comprizing of organic solvents (hexane, toluene, ethyl acetate, and HOAc) allows separation of PGs which are unstable in aqueous media (e.g., PGH_2 on cyano- or phenyl-bonded phases). Usually, the injection medium must be fairly polar to dissolve the PGs. This is achieved by the addition of small volumes of isopropanol resulting in much poorer resolution of monohydroxy acids simultaneously analyzed. Furthermore, the NP chromatography enables excellent separation of geometrical isomers like PGE_2 and PGD_2. The lock–key-type steric fitting of solute molecules with the discrete adsorption sites of the silica surface is responsible for this effect. However, adsorption chromatography requires a strict temperature control influencing retention time and, therefore, reproducibility and ruggedness.

Besides the dominating NP and RP techniques, silver-ion-loaded cation- and strong-anion-exchange chromatographies have been introduced as well. The silver-ion method based on interactions between Ag^+ and double bonds, and polar–polar attractive forces between the stationary phase and the solute is carried out using polar solvents containing low concentrations of ACN. The separation of cis–trans isomers can be achieved by this chromatographic modification. Anion-exchange chromatography has been described for underivatized PGs using tromethamine acetate–ACN mixtures at neutral pH as the mobile phase and a pellicular strong anion-exchange column as the stationary phase.

However, NP or RP methods represent the most powerful and common chromatographic techniques allowing PG analysis with very good reproducibility, sensitivity, wide linear ranges for quantitative analysis, and a broad spectrum of detectors.

Detection [4,8,9]

Investigating metabolism or stability of prostanoids, radiolabeled precursors or analytes are often used. The tritiated or ^{14}C-labeled compounds can easily be detected without any further derivatization using on-line or off-line liquid scintillation, which is not impaired by any interferences derived from matrix components. Efficient but less sensitive PG analysis is possible by UV detection (190–210 nm) of underivatized substances demanding the remove of interfering contaminants or simple sample matrices like buffers or some cell supernatants.

Derivatization [8,10]

Due to the low physiological concentrations in the femtomolar to picomolar range and the very weak UV absorptivity of prostanoids, derivatization is often required when analyzing by HPLC procedures. Improving detection sensitivity and selectivity, many methods have been developed allowing the quantitative and qualitative determination of these eicosanoids. Most of these techniques represent a selective and rapid precolumn derivatization of the extracted sample. Many procedures, especially for esterification of the carboxylic function, have been established. In the following, the stationary phases used for the separation of derivatized PGs are indicated in parentheses. Fluorescent derivatives with exceedingly high UV absorptivities can be obtained in almost quantitative yields by *p*-(9-anthroyloxy)-phenacyl bromide, also known as panacylbromide (LiChrosorb 100 Diol, or μ-Bondapak C18, or Radial Pak B). Panacyl esters are produced within 0.5–2 h using mild reaction temperatures between 20°C and 45°C. This reaction requires alkaline reagents like KOH or K_2CO_3 as catalysts in an organic solvent and, additionally, crown ethers as phase-transfer agents.

Ultraviolet detection is possible at 254 nm combined with fluorescence detection at about 450 nm after excitation at about 360 nm. Interferences of native fluorescent compounds are weak due to their emission wavelength range from 300 to 400 nm. Comparable phenacyl esters are synthezised with *p*-bromophenacyl bromide or *p*-nitrophenacyl bromide with a similar spectroscopic behavior (both μ-Bondapak C18). Useful derivatization by tagging the COOH— group can also be done with α-bromo-2′-acetonaphthone resulting in well-soluble and strongly UV-absorbing (254 nm) naphthanyl esters after a reaction time of only 10 min at room temperature (LiChrosorb Si-100). Furthermore, coumarin derivatives (substituted at position 7 for higher fluorescence quantum yields) are predestinated to fluorescent ester preparation ($\lambda_{ex} \approx 330$–390 nm and $\lambda_{em} \approx 410$–470 nm) using 4-bromomethyl-7-methoxycoumarin (LiChrosorb 100 Diol or Varian CN-10 Micropak) or 4-bromomethyl-7-acetoxycoumarin (BrMAC) (LiChrosorb RP18). Luminarin-4 (a labeling reagent with a quinolizinocoumarin structure) enables chemiluminescence detection in combination with peroxyoxalate (Ultraspher ODS-2 or Spherisorb ODS-2). These reactions are carried out at more drastical conditions at about 70°C in aprotic solvents, requiring catalysis and phase-transfer agents.

When using BrMAC, postcolumn hydrolysis of separated PG esters is neccessary for on-line detection of the fluorophore, resulting in detection limits of about 10 fmol. Hydrazone or amide derivatives are also suitable for fluorescence detection as well as 3-bromomethyl-6,7-dimethoxy-1-methyl-2(1*H*) quinoxalinone (YMC Pack C8) or 9-anthroyldiazomethane (ADAM) characterized by its low stability (Nucleosil ODS silica). Electrochemical detection of active compounds readily reduced can be carried out with low applied potential due to low background current and noise level in the oxidative mode after COOH derivatization using compounds with aromatic structures like 2,4-dimethoxyaniline (Nucleosil C18), 2-bromo-2′-nitroacetophenone, *p*-nitrobenzyloxyamine, or 2,4-dinitrophenylhydrazine. Thermospray (TSP)–MS detection has been used after esterification with diazomethane followed by oxime synthesis using trimethylanilinium hydroxide (TMAH) and separation on Zorbax ODS.

Additional to the mentioned COOH-derivatization PGs can also be modified using their hydroxyl functions. Oxidation with pyridinium dichromate to 15-oxo-PGs enables UV detection at 230 nm due to the conjugated oxo and $C_{13}=C_{14}$ double bonds (Microbore C18). TSP–MS analysis in the positive ion mode is possible after overnight reaction at 5°C with acetic anhydride in pyridine to form acetyl derivatives capable to better chemical ionization. About 20 PGs have been determined using the selected ion monitoring (SIM) method without a gradient system in the range from 0.5 to 10 pmol (Nucleosil 100-5C18) [9]. Combined derivatization to panacyl esters prior to methoximation using methoxamine hydrochloride enables good chromatographic separations, especially for the hemiacetalic TXB_2 and $PGF_{2\alpha}$ (Ultraspher ODS C18 or LiChrocart Superspher 100-RP-18). Furthermore, oxo functions present in most PGs can be modified to *p*-nitrobenzyloximes at 40°C within 2 h after methyl ester formation by diazomethane allowing UV detection at 254 nm (μ-Bondapak C18).

In general, quantitative analyses are carried out as usual, using internal or external standards and calibration curves. Detection limits can be achieved in the lower picogram range dependent on the selected derivatization procedure.

Special Features of HPLC in Prostaglandin Analysis [3,4,6–8,11]

Besides quantitative determinations of endogenous PGs, HPLC techniques have been used for some special features in clinical, chemical, or pharmaceutical research, including investigations of stability, metabolization and enantiomeric purity, thermodynamic characterizations of chemical equilibria, validating immunological methods by matrix separations, or preparative purifications.

References

1. M. Hamberg and B. Samuelsson, Prostaglandin endoperoxides. Novel transformations of arachidonic acid in human platelets, *Proc. Natl. Acad. Sci. USA 71*: 3400–3404 (1974).
2. W. S. Powell, Rapid extraction of arachidonic acid metabolites from biological samples using octadecylsilyl silica, *Methods Enzymol. 86*: 467–477 (1982).
3. W. S. Powell and D. F. Colin, Metabolism of arachidonic acid and other polyunsaturated fatty acids by blood vessels, *Prog. Lipid Res. 26*: 183–210 (1987).

4. K. Green, M. Hamberg, B. Samuelsson, and J. C. Frölich, Extraction and chromatographic procedures for purification of prostaglandins, thromboxanes, prostacyclin, and their metabolites, in *Advances in Prostaglandin and Thromboxane Research* (J. C. Frölich, ed.), Raven Press, New York, 1978, Vol. 5, pp. 15–38.
5. J. A. Lawson, H. Li, J. Rokach, M. Adiyaman, S.-W. Hwang, S. P. Khanapures, and G. A. FitzGerald, Identification of two major F_2 isoprostanes, 8,12-iso- and 5-epi-8,12-iso-isoprostane $F_{2\alpha}$-VI, in human urine, *J. Biol. Chem. 273*: 29,295–29,301 (1998).
6. H. John and W. Schlegel, Thermodynamic and structural characterization of cis-trans isomerization of 12-S-hydroxy-(5Z,8E,10E)-heptadecatrienoic acid by high-performance liquid chromatography and gaschromatograpy-mass spectrometry, *Chem. Phys. Lipids 95*: 181–188 (1998); *97*: 195–196 (1999).
7. J. G. Hamilton and R. J. Karol, High performance liquid chromatography (HPLC) of arachidonic acid metabolites, *Prog. Lipid. Res. 21*: 155–170 (1982).
8. T. Toyo'oka, Use of derivatization to improve the chromatographic properties and detection selectivity of physiologically important carboxylic acids, *J. Chromatogr. B 671*: 91–112 (1995).
9. M. Yamane and A. Abe, High-performance liquid chromatography–thermospray mass spectrometry of prostaglandin and thromboxane acetyl derivatives, *J. Chromatogr. 568*: 11–24 (1991).
10. K. Blau and J. Halket, *Handbook of Derivatives for Chromatography*, 2nd ed. John Wiley & Sons, New York, 1993.
11. H. John and W. Schlegel. Reversed-phase high-performance liquid chromatographic method for the determination of the 11-hydroxythromboxane B_2 anomers equilibrium, *J. Chromatogr. B 698*: 9–15 (1997).

Harald John

Protein Analysis by HPLC

Introduction

Proteins were first separated chromatographically in the 1950s when carbohydrate gels were found to be effective as matrices for liquid chromatography. When high-performance liquid chromatography (HPLC) gained popularity, it was assumed that its higher flow rates and concomitantly higher pressures would destroy the biological activity of proteins by disrupting their three-dimensional (tertiary) structures. This was proven incorrect in the mid-1970s, when several groups, including those of Fred Regnier at Purdue university and Jiri Coupek of Czechoslovakia, demonstrated that high recoveries accompanied the high resolution of HPLC methods for protein analysis.

Modes of HPLC

The importance of tertiary structure on the biological activity of proteins implies that the most successful HPLC methods are those employing mobile and stationary phases, which cause minimal disruption of these features. Size-exclusion and ion-exchange chromatography are the gentlest methods in this regard and are generally compatible with any additives needed to enhance stability. Hydrophobic interaction and affinity chromatography have somewhat higher risk of denaturation due to either their mobile phase [high salt for hydrophobic interaction chromatography (HIC)] or stationary phase (affinity). Reversed-phase and hydrophilic interaction chromatography cause greater disruption of tertiary structure due to the presence of organic solvents, acidic pH, and/or hydrophobic stationary phases. Nonetheless, reversed-phase chromatography has shown great utility in the separation of many proteins.

Size-Exclusion Chromatography

Size-exclusion chromatography (SEC) is a method in which molecules are separated by size due to differential permeation into a porous support. This technique is especially useful for the separation of proteins because they are macromolecules frequently found in the presence of smaller and larger species. The peak capacity in SEC is fairly low compared to other HPLC methods because all separation must occur in the internal volume (V_i) of the support, which is generally less than half the volume of the mobile phase in the column. Despite this deficiency, SEC is very effective for separating proteins from small molecules, polymeric forms, and other molecules which differ by size.

Supports for SEC of proteins are designed to be neutral and very hydrophilic to avoid disruption of protein structure and interaction of the solutes with the support by ionic or hydrophobic mechanisms. The base matrix can be either silica or polymer; efforts are made to totally mask its properties with a carbohydratelike stationary phase. The pore structure is critical to successful SEC. Not only must the total pore volume (V_i) be adequate for separation, the pore diameter must be consistent and nearly homogeneous for attainment of maximum resolution between molecules with relatively small differences in molecular size (radius of gyration or molecular weight). A twofold difference in size is usually required for separation by SEC. Pore homogeneity can be assessed from the slope of the calibration curve of the logarithm of the molecular weight versus the retention time or the partition coefficient (K_D): $K_D = (V_R - V_0)/V_i$, where V_R is the retention volume and V_0 is the void or excluded volume. A less steep slope results in the separation of more closely related molecular weights. Figure 1 shows both an analysis of a protein mixture on an SEC column and its calibration curve. The second calibration curve is for a column with a smaller pore diameter, showing its effectiveness for separating smaller solutes.

The mobile phase is a critical factor in SEC because it must eliminate all solute–support interactions. This is effected by adjustment of pH (usually to neutrality), ionic strength (0.05M–0.2M), and/or addition of 5–10% organic solvent or stabilizing agents. The ionic and hydrophobic properties of proteins and their attraction to the stationary phase must be totally removed.

Ion-Exchange Chromatography

Ion-exchange chromatography (IEC) separates proteins by ionic interaction of their surface amino acids with charges on the stationary phase. Selectivity is dependent on the number and the identity of the amino acids, as well as their spatial arrangement. A protein with charges grouped in a patch on its surface will bind differently than one whose charges are dispersed throughout the surface. IEC is so selective that it can resolve isoforms and variants differing by only one amino acid. Loading on porous IEC supports is high (100 mg/g), making IEC invaluable in purification techniques.

Supports for IEC possess either anion- or cation-exchange functionalities which are positively or negatively charged, respectively. They are also classified as weak or strong to correspond to their titration curves, similar to acid and base designations. The stationary phase totally covers the silica or polymer support matrix. The pore di-

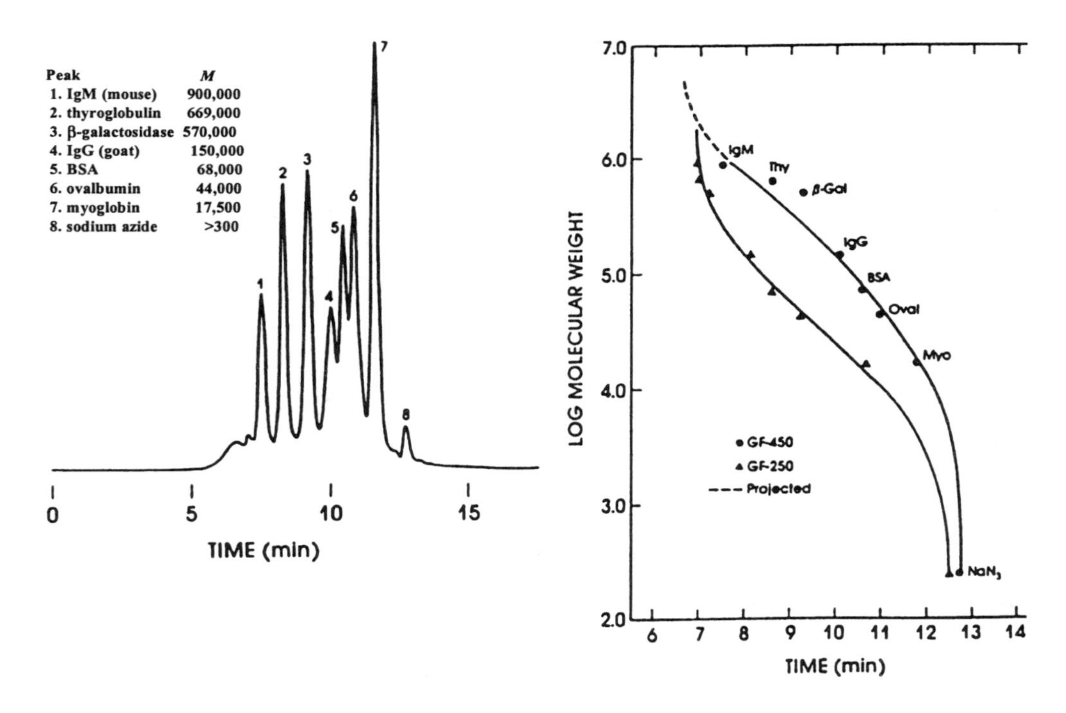

Fig. 1 Analysis of a mixture of proteins on ZORBAX® GF450. Calibration curves of proteins on ZORBAX GF250 and GF450 columns. Mobile phase: 0.2M sodium phosphate, pH 7.5; UV detection at 280 nm. (Printed with permission of Hewlett-Packard.)

ameter is important in that it must be large enough to allow access of the protein. This affects not only retention but also loading capacity.

In IEC, proteins are bound in a low-ionic-strength buffer ($0.02M$–$0.05M$) at an appropriate pH (often 1–2 pH units from the pI). Elution occurs when the ionic strength is increased during a concentration gradient of a salt in the same buffer. Because proteins bind via multipoint interactions, gradients are necessary for good peak shapes, resolution, and reproducibility. The pH gradients are less commonly used but can also be effective. The nature of the salt in IEC has a major effect on selectivity due to interaction of the composite ions with either the stationary phase or the solutes. Figure 2 illustrates the differences in retention of several proteins when various salts are used as the mobile phase for anion-exchange and cation-exchange chromatography. Additives which increase protein stability or improve peak shape can usually be in-

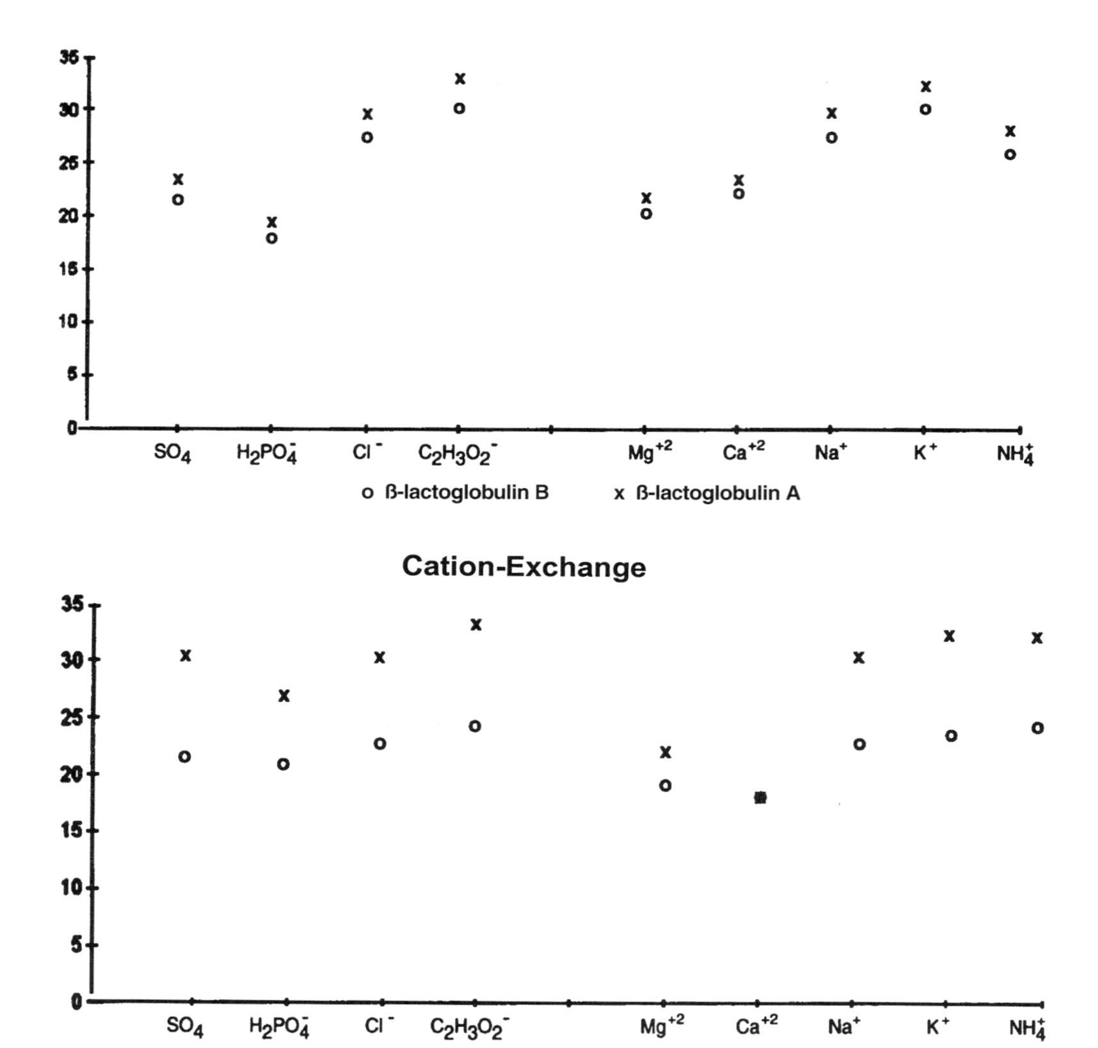

Fig. 2 Effect of salt on protein retention. Columns: SynChropak AX300 and SynChropak CM300; 30 min gradient from 0–1N salt in $0.02M$ Tris, pH 7. (Reprinted with permission from *Basic HPLC and CE of Biomolecules*, Bay Bioanalytical Laboratories, Richmond, VA, 1998.)

cluded in the mobile phase without deleterious effects on the separation. Lower temperatures can be used to preserve biological activity, but they also result in higher retention.

Hydrophobic Interaction Chromatography

Hydrophobic interaction chromatography (HIC) is a method in which proteins in a high salt environment interact hydrophobically with nonpolar ligands. Effective salts are antichaotropic, meaning that they promote the ordering of water molecules at surfaces. Selectivity is based on the hydrophobic amino acids and patches on the surface of the proteins. Many proteins, which generally remain in their native states in this technique, retain their biological activities when separated by HIC.

Supports for HIC have short alkyl chains or phenyl functionalities, the length of which is related to retention. The matrix can be either silica or polymer, as it is totally covered by the bonded phase and, thus, not exposed to the solute. Pore diameters are at least 300 Å to allow access by proteins. Loading capacities are high and similar to those of ion-exchange supports.

In HIC, molecules are bound with a high concentration of salt, usually ammonium or sodium sulfate ($1M$–$2M$) in a buffer ($0.02M$–$0.05M$). Elution is attained by a gradient to a lower concentration of salt in the buffer. The pH is controlled and is usually in the range of 6–8, but it is not a critical factor in selectivity. Additives to enhance protein stability are generally compatible with the process. Contrary to its effect on other modes of chromatography, reducing the temperature decreases the retention in HIC due to its being an entropy-driven technique.

Affinity Chromatography

Affinity chromatography (AFC) is a method in which biomolecules are attracted to ligands due to biospecificity. AFC is a very specific and selective technique in the aspect that many ligands are used, being customized to the analyte of interest. It is generally unlike other HPLC processes because particle diameters are larger and retention tends to be bind-release rather than partitioning, thus separating one protein or class of proteins from everything else in a mixture. Recoveries of biological activity are often high.

The variety of ligands for AFC includes proteins, cofactors, and any other molecules for which a solute has special affinity. Some common ligands are protein A, antibodies, and concanavalin A. Elution is attained by either a change in pH or ionic strength conditions or the addition of a competitor for the binding. Conditions are as varied as the number of ligands.

Reversed-Phase Chromatography

Reversed-phase chromatography (RPC) is a method in which molecules are bound hydrophobically to nonpolar ligands in the presence of a polar solvent. Solutes are generally bound in an acidic mobile phase with elution occurring during a gradient to an organic solvent. The combination of acidic and organic mobile phases and a hydrophobic bonded phase usually results in tertiary structure disruption, which may or may not be reversible. Binding involves internal as well as surface amino acids depending on the extent of unfolding. The utility of RPC for protein analysis is limited by its generally denaturing characteristics; however, it can be a good method for nonpreparative techniques where preservation of biological activity is unimportant. Some enzymes, such as trypsin and chymotrypsin, can be renatured after RPC, regaining biological activity.

There are many ligands used for RPC, but the most popular for protein analysis are butyl (C_4) and octyl (C_8). Little difference in selectivity for proteins is observed with ligand-chain-length variation, but mass recovery is often enhanced on the shorter chains. Due to their higher efficiencies and wettability, silica-based supports are generally used for protein analysis. The silica matrix is sometimes a factor in retention because reversed-phase bonded phases often do not totally eliminate silanols on the support. Pore diameters used for proteins are at least 300 Å; nonporous supports offer a high-resolution option with lower capacities. Loading capacities of RPC are at least 50% lower than those of IEC or HIC.

Many operational factors can change selectivity in RPC. The strength of the organic solvent which causes elution increases from methanol to acetonitrile to isopropanol. Acetonitrile is the most popular solvent due to its transparency at low wavelengths (<210 nm) and its tendency to yield narrow peak widths. The transparency is not critical for proteins, which can usually be detected at 254 or 280 nm. Ion-pairing agents, such as trifluoroacetic acid, can be added to the mobile phase to change the ionic or hydrophobic properties of either or both of the bonded phase or the solute. Generally, acidic pH is utilized to minimize silanol interactions; however, distinct pH conditions yield different selectivities. Increased temperatures result in shorter retention times and are especially effective in reducing the pressure generated by small-particle-diameter supports.

Hydrophilic Interaction Chromatography

Hydrophilic interaction chromatography (HILIC) is a variation of normal-phase chromatography in which solutes are retained on a polar bonded phase under high concentrations (80–90%) of organic solvent and released during a gradient to a more aqueous solvent. The organic mobile phase usually causes at least partial denaturation of proteins.

The HILIC bonded phases are hydrophilic, including amide and/or polyhydroxy functionalities. Pore diameters are at least 300 Å to allow penetration of proteins. Supports can be based on either silica or polymer because the matrix is not exposed to the solutes.

The mobile phase can offer some differences in selectivity in HILIC. A buffer is usually used to control pH. It may also serve to change the hydrophobicity of the solutes by ion-pairing; thus, it may greatly effect retention. For this reason, the pH is less important than the identity of the salt. The denaturing aspects of HILIC diminish its utility in protein analyses, particularly for preparative purification.

Conclusions

High-performance liquid chromatography provides a number of rapid and effective methods for analysis and purification of proteins and enzymes. In modes such as SEC and IEC, quantitative yields are often obtained with full preservation of biological activity. The ability to change selectivity using the mobile phase has resulted in versatile techniques which can be rapidly optimized and implemented.

Suggested Further Reading

Cunico, R. L., K. M. Gooding, and T. Wehr, *Basic HPLC and CE of Biomolecules*, Bay Bioanalytical Laboratories, Richmond, CA, 1998.

Gooding, K. M. and F. E. Regnier (eds.), *HPLC of Biological Macromolecules: Methods and Applications*, Marcel Dekker, Inc., New York, 1990.

Hancock, W. S. (ed.), *High Performance Liquid Chromatography in Biotechnology*, John Wiley & Sons, New York, 1990.

Hearn, M. T. W. (ed.), *HPLC of Proteins, Peptides and Polynucleotides*, VCH, New York, 1991.

Horvath, Cs. and J. G. Nikelly (eds.), *Analytical Biotechnology: Capillary Electrophoresis and Chromatography*, American Chemical Society, Washington, DC, 1990.

Katz, E. D. (ed.), *High Performance Liquid Chromatography: Principles and Methods in Biotechnology*, John Wiley & Sons, New York, 1996.

Mant, C. T. and R. S. Hodges (eds.), *High-Performance Liquid Chromatography of Peptides and Proteins*, CRC Press, Boca Raton, FL, 1991.

Karen M. Gooding

Protein Immobilization

Introduction

Affinity chromatography is a powerful high-resolution separation technique for biomolecules. It is based on a highly specific and unique stereochemical interactions between biomolecules. Examples of these interactions are the binding of enzymes to coenzymes and inhibitors and the interactions between antigens (e.g., protein A) and antibodies. In affinity chromatography, a ligand is covalently bound to a solid matrix which is packed into a chromatography column. A mixture of components is then applied to the column. The unbound contaminants, which have no affinity for the ligand, are washed through the column, leaving the desired component (protein, peptide, etc.) bound to the matrix. Elution is accomplished by changing the pH and/or salt concentration or by applying organic solvents or a molecule which competes for the bound ligand. The purpose of this entry is to review methods for immobilization of protein ligands for affinity chromatography.

Matrix Material (Carrier)

Immobilized ligand is prepared by covalently attaching the biospecific ligand to a chromatographic bed support material, the matrix. A variety of insoluble support materials are used as matrices in affinity chromatography. These include agarose, cellulose, cross-linked dextran, polystyrene, polyacrylamide gels, and porous silica gels.

The majority of matrices used in affinity chromatography are either agarose based or polyacrylamide based. The available reactive groups on these matrices are hydroxyl and amide nitrogen groups, respectively. By far the most popular support materials in use are beaded derivatives of agarose due to their ideal physical and chemical characteristics that are necessary for ligand immobilization. For instance, Sepharose, a bead-formed agarose gel manufactured by Amersham Pharmacia Biotech, has virtually all the features required of a successful matrix for immobilizing biologically active molecules. The hydroxyl groups on the sugar residues can easily be used to covalently attach a ligand. Sepharose 4B (Amersham Pharmacia Biotech) is the most popular and widely used matrix. Its advantages include open-pore structure, molecular exclusion limit of 20×10^6 Da, easy and good binding capacities, and low nonspecific attachment. A spacer arm, which is interposed between the matrix and ligand, is sometimes needed to facilitate the binding particularly when coupling small molecules (e.g., enzyme cofactors). Spacer arms are also useful for coupling ligands containing free carboxyl or amine groups and for further chemical reactions to permit the attachment of phenolic groups and diazonium derivatives to agarose matrix.

Protein Ligands

Protein ligands most commonly used in affinity chromatography are as follows: (a) protein with very high specific affinity for moieties on the desired protein; a wide range of plant lectins with a variety of sugar specificities are used as ligands to purify glycoproteins and enzymes having sugar moieties in their structures; (2) immobilized antibodies, which represent the ideal affinity separation because of the precise specificity of many antibodies, particularly monoclonial antibodies, to their specific antigens; (3) proteins or enzymes with unique affinity for other proteins or peptides, which are used as affinity ligands for the particular proteins or peptides.

Preparation of Immobilized Proteins

The immobilization of proteins and enzymes through covalent bond formation between the protein or enzyme and an activated insoluble carrier entail two processes of activation and coupling. These procedures are classified on the basis of the type of reaction that facilitates the bonding between the protein and the insoluble carrier. Mostly, with matrices containing readily available hydroxyl groups such as agarose derivatives, a variety of activation procedures are used, after which the amino groups of protein ligands are readily incorporated into the matrix (coupling reaction). After coupling of the protein ligands to the matrix, unreacted activated coupling groups are usually hydrolyzed to inactive derivatives during washing of the coupling resin at lower pH. Alternatively, these groups can be blocked by adding an excess of an inert blocking reagent. The following methods are most commonly used for activation and coupling of protein ligands to agaraose and other matrices.

Cyanogen Bromide Activation Method

The aim of the reaction of cyanogine bromide with hydroxyl groups on agarose matrix is to produce reactive carrier to which proteins can be coupled through amino groups. Cyanogen bromide activation method is the most used method for the preparation of affinity gels because of its simplicity and its mild reaction conditions, particularly for immobilizing sensitive proteins such as enzymes and antibodies. Cyanogen bromide reacts with these hydroxyl groups and converts them to imidocarbonate groups, as shown in Fig. 1. Agarose activated with cyanogen bromide have been used to bind a variety of proteins and enzymes. The activated groups react with primary amino groups of a protein ligand to form isourea linkages according to the reactions in Fig. 1.

Acylation Reactions

In this type of protein immobilization, reactions involve the acylation of an NH_2 group on a protein or an enzyme by pendent groups of the carrier such as azide, acid anhydride, carbodiimide, sulfonyl chloride, and hydroxysuccinimide esters. Copolymers of acrylamide and maleic anhydride have been useful for enzyme immobilization through the acid anhydride reaction with the enzyme. This system of protein immobilization involves the use of a reagent that contains an acylating or alkylating agent and a group which can form a covalent link with a carrier polymer. For instance, few protein immobilization methods are based on condensation by carbodimides (e.g., dicyclohexylcarbodimide). Mixing the carbodiimide reagent with the matrix and a protein ligand, stable amide bonds can be formed between an amino group on the ligand and a carboxyl group of matrix (or vise versa) in a one-step procedure of activation and coupling as follows:

$$\underset{\text{Soluble protein}}{R{-}COOH} + \underset{\text{Carbomiide}}{R_1N{=}C{=}NR_2} \rightarrow \underset{\text{Immobilized protein}}{R{-}C{-}O{-}\overset{\overset{NHR_1}{|}}{C}{=}NR_2}$$

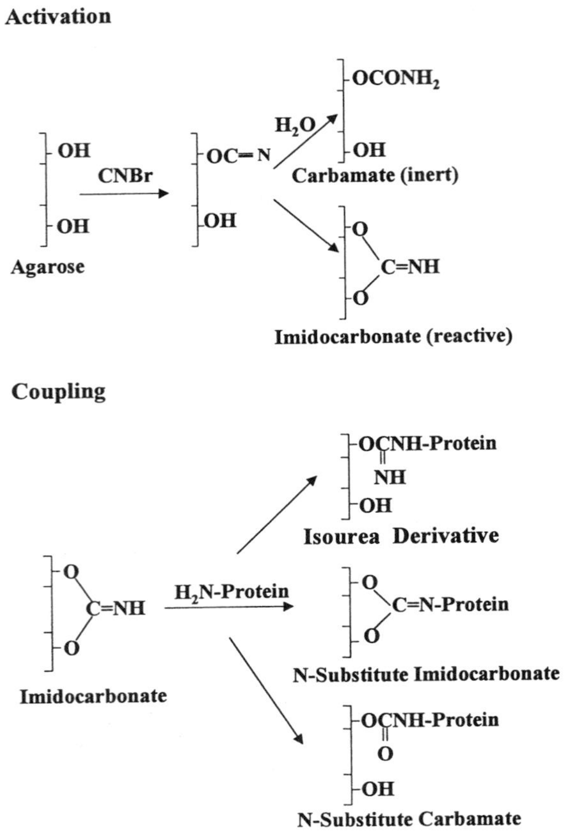

Fig. 1 Immobilization of protein ligands on agarose (activation and coupling) by cyanogen bromide.

Arylation Reactions

Arylation or alkylation are used for activation of support and linking of a protein ligand. In such reactions, the functional group on the carrier combines with an NH_2 group of the protein (e.g., by 3-fluoro-4,6-dinitrophenyl group or 2–4 dichloro-*S*-triazine). In this approach, a chloro-*S*-triazine is coupled to an arylazide. The azide is converted to a nitrene, which upon activation by light reacts with a polymer carrier to form a covalent bond. The chlorotriazine then couples with the protein.

Alkylation Reactions by Bisepoxirane, Epichlorohydrin, and Divinylsulfone

Bisepoxirans activations allows the coupling of ligands containing —OH, —NH_2, and —SH groups with agarose matrix. Here, the ligands are provided automatically with a hydrophilic spacer arm.

Organic Sulfonyl Chlorides, Tosyl and Tresylchloride Methods

These methods enable the activation and the coupling of ligands containing —NH_2 and —SH groups with matrices such as agarose, cellulose, or silica derivatives.

Methods for Polyacrylamide Matrices

The glutaraldehyde and hydrazine reactions are used for matrices having amide groups such as polyacrylamide. Glutaraldehyde and hydrazine react with the polymer, and the enzyme or protein is readily bound to the treated polymer. The mechanism of the reactions involved in the activation and coupling are not well understood. In general, insoluble supports such as polyacrylamide, which tend to swell in water, form immobilized proteins or enzymes when mixed with a solution and treated with glutaraldehyde. Polyacrylamide support is activated by treatment with hydrazine (heated) followed by sodium nitrite in hydrochloric acid. The amino groups of protein ligands can then be coupled to the activated matrix via stable amide bonds.

Affinity Supports: Activated and Ready-to-Use Media

Chromatography suppliers offer affinity gels for a wide range of biomolecular applications. Availability of activated supports allows the users to make their own affinity matrix of choice by coupling a ligand such as an antibody, enzyme, antigen, or receptor. Ready-activated supports or matrices are less expensive than the ready-to-use matrices usually offered by the same manufacturer. These activated matrices are dependable and provide flexibility and convenience in affinity chromatography. With this particular option, the chromatographer has the choice of preparing affinity matrix by activating a support, then coupling or only coupling using ready-to-use activated affinity supports. Table 1 shows the activated supports marketed by two major soft-gel chromatography suppliers that are suitable for coupling of affinity protein ligands.

On the other hand, ready-to-use matrices have a specific ligand already coupled to an affinity support. Amersham Pharmacia Biotech has developed a wide range of ready-to-use affinity gels with specific protein ligands coupled to the matrix. In specified bulletins, Amersham Pharmacia Biotech details the preparation procedures of these affinity gel columns containing immobilized proteins. Applications include monoclonal and polyclonal

Table 1 Activated Affinity Matrices Marketed by Two Major Soft-Gel Chromatography Suppliers for Coupling of Protein Ligands

Matrix	Ligand specificity	Activator and functional groups	Manufacturer
CNBr-activated Sepharose 4B (or Fast Flow)	$-NH_2$	CNBr	Amersham Pharmacia Biotech
EAH Sepharose 4B	$-COOH$	Carbodiimide coupling via 6-carbon spacer arm	Amersham Pharmacia Biotech
ECH Sepharose 4B	$-NH_2$	Carbodiimide coupling via 6-carbon spacer arm	Amersham Pharmacia Biotech
Epoxy-activated Sepharose 6B	$-NH_2$, $-OH$, $-SH$	1,4-bis (2,3-epoxypropoxy-butane)	Amersham Pharmacia Biotech
Activated Thiol Sepharose 4B	$-SH$, $-C{=}O$, $-CNH$	Glutathione-2-pyridyl disulfide	Amersham Pharmacia Biotech
Affi-Gel 10 gel (or Affi-Prep 10, pressure stable)	$-NH_2$	*N*-Hydroxysuccinimide ester via 10-atom spacer arm	Bio-Rad Laboratories
Affi-Gel 15 gel	$-NH_2$	*N*-Hydroxysuccinimide ester via 15-atom spacer arm	Bio-Rad Laboratories
Affi-Gel 102 gel	$-COOH$	Carbodiimide coupling	Bio-Rad Laboratories
CM Bio Gel A	$-NH_2$	Carboxymethyl, carbodiimide coupling	Bio-Rad Laboratories

antibodies, fusion proteins, glycoproteins, enzymes, cells, and other proteins such as fibronectin, and membrane proteins. Likewise, Bio-Rad Laboratories offer two types of supports. One type such as Affi-Gel supports, based on agarose or polyacrylamide, are available as low-pressure gels suitable for most laboratory-scale affinity purification with a peristaltic pump or gravity flow elution. Affi-Prep supports, based on a pressure-stable macroporous polymer, are suitable for preparative and process scale applications. The immobilized protein ligands in these ready-prepared matrices include lectins, protein A, gelatin, avidin, and calmodulin. Affi-Gel and Affi-Prep protein A supports are used to produce highly purified immunoglobulins (IgG), to selectively remove IgG prior to analysis of other immunoglobulin classes, or to adsorb immune complexes to purify antigens. Protein A from *Staphylococcus aureus* binds to the F_c region of immunoglobulins, especially IgG from mammalian species. Affi-Prep supports offer linear flow rates up to 2000 cm/h, pressure stability up to 1000 psi (70 bar), and high chemical stability, according to the manufacturer.

Conclusion

In affinity chromatography, a ligand is covalently bound to a solid matrix by matrix activation and ligand coupling. In protein immobilization, a covalent bond is formed between the protein ligand and an insoluble solid matrix or carrier. Most of protein immobilization steps entail the formation of the insoluble immobilized proteins on cross-linked agarose gels. Most common reactions used in making immobilized proteins and enzymes are discussed. These processes are classified on the basis of the type of reaction which forms the covalent bond between the protein and the insoluble carrier. Also, a wide range of commercial ready-to-use supports with specific protein ligands already coupled to a base matrix are discussed.

Suggested Further Reading

Bell, J. E. and E. T. Bell, Protein purification: affinity chromatography, in *Proteins and Enzymes*, Prentice-Hall, Englewood Cliffs, NJ, 1988, pp. 45–63.

Carlsson, J., J-C. Janson, and M. Sparrma, Affinity chromatography, in *Protein Purification: Principles, High Resolution Methods, and Applications* (J. C. Janson and L. Ryden, eds.), VCH, New York, 1989, pp. 275–329.

Dunlab, R. B., *Immobilized Biochemicals and Affinity Chromatography*, Plenum Press, New York, 1974.

Ganttsos, G. and P. E. Barker, *Preparative and Production Scale Chromatography* (J. Cazes, ed.), Chromatographic Science Series, Marcel Dekker, Inc., New York, 1993.

Kennedy, J. F. and C. A. White, Principle of immobilization of enzymes, in *Handbook of Enzyme Biotechnology* (A. Wiseman, ed.), Ellis Horwood, Chichester, 1986, pp. 147–207.

Low, C. R. L., *Affinity Chromatography*, John Wiley International, London, 1974.
Pharmacia Fine Chemicals, *Affinity Chromatography: Principles & Methods*, Pharmacia Fine Chemicals, Uppsala, 1978.
Schott, H., *Affinity Chromatography*, Chromatographic Science Series Vol. 27, Marcel Dekker, Inc., New York, 1984.
Scouten, W. H., *Affinity Chromatography*, John Wiley & Sons, New York, 1981.
Sofer, G. K., *Bio/Technol. 4*: 712–715 (1987).

Jamel S. Hamada

Proteins as Affinity Ligands

Affinity chromatography using proteins as ligands explores the three-dimensional complementary molecular interactions between the ligand and its target. This noncovalent and reversible binding are the results of ionic interaction, hydrophobic interaction, hydrogen-bonding, and van der Waals–London forces. The binding constant between the immobilized ligand and the target molecule ranges from $10^4 M^{-1}$ to $10^8 M^{-1}$. The multiplicity of the interaction forces offers the superior specificity to protein-based affinity chromatography for biomolecule separation among all the chromatographic techniques.

After the protein ligand is immobilized on a matrix support, a crude sample mixture, including the desired substance, is loaded into the affinity column under appropriate conditions. If the desired molecule forms a stable and reversible complex with the protein ligand, it will be retained and all the other contaminants are washed away. Then, the eluting buffer disrupts the binding between the desired substance and the protein ligand to recover the substance. When the interaction between the desired molecule and the ligand is too weak to form a stable complex with the ligand, it is eluted later than all the other substances in the crude sample and an eluting buffer is not required.

Protein-based affinity chromatography has found wide applications in both academic and industrial settings. The affinity chromatography using proteins as the ligand is an indispensable technique for recombinant protein purification. This review will touch the general aspects of matrix, immobilization of protein ligands, and the application of protein-based affinity chromatography. It is by no means an exhaustive review of the current technologies in protein-based affinity chromatography and only serves as an introduction to this topic. Interested readers should explore those articles listed in Suggested Further Reading for more information.

Matrix

One of the key elements for protein-based affinity chromatography is the matrix, on which the protein ligand is immobilized. The ideal matrix material should have the following characteristics (Porath, 1974): insolubility, sufficient permeability, large surface area, strong mechanical strength, zero adsorption capacity, resistance to microbial and enzyme attack, chemical reactivity for coupling protein ligand, and stability under the operating conditions. Based on their physical strength, the matrices can be divided into two categories: low pressure and high pressure. For example, agarose, dextran, polyacrlyamide, and methacrylate are typical low-pressure matrices and silica and polystyrene–divinylbenzene are high-pressure matrices.

When a matrix is chosen in affinity chromatography applications, its advantages and disadvantages should be taken into account to assess its potentials and limitations. Agarose is hydrophilic and, therefore, its matrix surface has low nonspecific binding. The presence of the hydroxyl group from agarose provides a convenient way for immobilizing the protein ligand. The drawbacks of agarose are its solubility in hot water and nonaqueous solutions and its susceptibility to microbial degradation. The lack of mechanical strength for agarose limits its application only in low-pressure operation and the separation process is slow. Silica and synthetic polymers are the alternative choices of matrix support. The major advantages for those matrix supports over agarose are their rigidity, reduced diffusion rate, and well-defined pore size (i.e., 5–20 μm in diameter). The above factors decide that those matrices can be operated in fast linear flow rate under high pressure to achieve rapid separation. Nonspecific binding in silica-based matrices is a concern because of the negative-charged silanol groups and because also silica has increased solubility at alkaline pH.

Immobilization

The unique specificity of affinity chromatography is the result of the complementary interaction between the protein ligand and the target in the separation process. It is therefore essential to maintain the structure integrity of the protein ligand during immobilization of protein ligand. It is critical to consider the following in the immobilization step: (a) The binding site in the protein ligand should be accessible after the immobilization; (b) the amino acids located at the binding site should not be modified during the immobilization step; (c) the three-dimensional structure of the binding site should be maintained after the coupling step.

Usually, the immobilization of protein ligand is a two-step process: (1) activation and functionalization of the matrix and (2) coupling of the protein ligand to the modified matrix. To avoid the possible reactions between the rest of the activated sites and the protein components from the sample, a third step is often required to quench those activated sites after the protein ligand is coupled. The most exploited function groups for coupling from the protein ligand are the C-terminal carboxyl group, the carboxyl groups of glutamic and aspartic acids, the N-terminal α-amino group, and the ε-amino group of lysine. Sometimes, phenolic hydroxyl groups of tyrosine, imidazole anion of histidine, or the —SH group of cysteine residues may be the candidates for coupling. The ideal linkage between the matrix and the protein ligand from Step 1 should be fairly hydrophilic and neutral to avoid the introduction of a nonspecific interaction. Carbonyldiimidazole and 2-fluoro-1-methylpyridinium toluene-4-sulfonate are two of the popular reagents for activation of the hydroxyl-containing matrix and coupling with an amine-containing protein ligand.

Because there are many options of immobilization chemistries, different immobilization strategies should be evaluated for a particular protein ligand. After immobilization, the binding site should be affected in a minimal way so that the binding function is maintained. The protein ligand could have multiple sites for coupling with the activated matrix surface. The immobilization could be a random process if those sites are located in a similar steric environment. Another undesired scenario is immobilization by multiple-site attachment, which could destroy the native structure of the protein ligand and loss of its binding ability. If feasible, site-specific immobilization is preferred because it would give better reproducibility in both surface chemistry and column performance. Carbohydrate moieties from glycoproteins (i.e., antibodies and enzymes) are often particularly taken advantage of as the coupling sites because, generally, the complementary binding site in the protein ligand is away from the oligosaccharide chains. Immunoglobins (IgG) is often immobilized through the carbohydrates on the F_c region, and, as a result, the binding sites are readily available located on the F_{ab} region of IgG.

The leaching of the protein ligand presents a significant problem for introducing contaminants into the purified product. The solvolysis of the bonds between either the matrix–activator or activator–protein is the major source of leaching. It is the reason that an extra chromatography step is often required to remove protein ligand contaminant after affinity chromatography purification.

Applications

Adsorption and desorption are involved in the operation of affinity chromatography utilizing proteins as affinity ligands. One of the major differences for protein-based affinity chromatography from other modes of chromatographic techniques (i.e., ion-exchange chromatography, reversed-phase chromatography, and hydrophobic interaction chromatography) is that the protein ligand could be denatured or attacked in other ways and lose its affinity consequently. Extreme pH, heat, and organic solvents are worrisome for protein-based affinity chromatography. If any one of them is employed, its effect on the column life should be evaluated.

For method development in protein-based affinity chromatography, the theme is to promote the specific interaction between the protein ligand and its counterpart and minimize the nonspecific interaction between the matrix and the sample components. Because ionic and hydrophobic interactions contribute to the specific binding between the ligand and its counterparts, ionic strength and pH are probably the two simple and convenient parameters to be evaluated for separation method development. The variation of pH can affect the charge distribution on both the protein ligand and its target, leading to a different binding constant between them. Changing the ionic strength of the solution could modulate the hydrophobic interaction. Another very important parameter is the binding kinetics between the immobilized protein ligand and its target. The flow rate study could provide some clue on how fast the binding reaches the equilibrium. The strategy for minimizing the nonspecific interaction depends on the surface chemistry of the matrix. For example, a running buffer of high ionic strength ($0.2M$) can suppress the nonspecific ionic interaction between the sample components and the negative-charged silanol group on the silica matrix.

Many proteins have been utilized as affinity ligands and

there are literally thousands of references using this technique. The following are a few samples of proteins used as affinity ligands. Antibodies or antigens are used as immunoaffinity ligand to separate the complementary counterparts. Often, only a single purification step is needed because of the superior selectivity of immunoaffinity column. Protein A and protein G are the perfect ligands for antibody purification because of their specificities and robustness.

In summary, affinity chromatography using proteins as affinity ligands provides a unique scheme for purification. Currently, a great number of matrices are available and the chemistries for immobilization of protein ligands are significantly improved. The specific three-dimensional interaction between the protein ligand and its counterpart distinguishes this technique from other chromatography methods. The key issue is maintaining the structure integrity of the binding site in the protein ligand throughout the immobilization and column operation.

Suggested Further Reading

Dorsey, J. G., W. T. Cooper, B. A. Siles, J. P. Foley, and H. G. Barth, *Anal. Chem. 70*: 591R–644R (1998).
Lowe, C. R., *Adv. Mol. Cell Biol. 15B*: 513–522 (1996).
Ohlson, S., M. Bergstorm, P. Pahlsson, and A. Lundblad, *J. Chromatogr. 758*: 199 (1997).
Porath, J., *Methods Enzymol. 34*: 13–30 (1974).
Pommerening, *Affinity Chromatography: Practical and Theoretical Aspects*, Marcel Dekker, Inc., New York, 1985.
Turkova, J., *Bioaffinity Chromatography*, Elsevier, Amsterdam, 1993.

Ji-Feng Zhang

Protein Separations by Flow Field-Flow Fractionation

Flow field-flow fractionation (flow FFF) is a separation method that is applicable to macromolecules and particles [1]. Sample species possessing hydrodynamic diameters from several nanometers to tens of microns can be analyzed using the same FFF channel, albeit by different separation mechanisms. For macromolecules and submicron particles, the normal-mode mechanism dominates and separation occurs according to differences in diffusion coefficients. Flow FFF's wide range of applicability has made it the most extensively used technique of the FFF family.

The characteristic feature of flow FFF is the superimposition of a second stream of liquid perpendicular to the axis of separation. This cross-flow drives the injected sample plug toward a semipermeable membrane that acts as the accumulation wall. The cross-flow liquid permeates across the membrane and exits the channel, whereas the sample is retained inside the channel in the vicinity of the membrane surface. Sample displacement by the cross-flow is countered by diffusion away from the membrane wall. At equilibrium, the net flux is zero and sample clouds of various thicknesses are formed for different sample species. As with other FFF techniques, a larger diffusion coefficient D leads to a thicker equilibrium sample cloud that, on average, occupies a faster streamline of the parabolic flow profile and subsequently elutes at a shorter retention time t_r. For well-retained samples analyzed by flow FFF, t_r can be related to D and the hydrodynamic diameter d by

$$\frac{t^0}{t_r} = \frac{6D}{Uw} = \frac{2kTV^0}{\dot{V}_c w^2 \pi \eta d} \tag{1}$$

where t^0 is the void time, U is the field-induced velocity, w is the channel thickness, k is the Boltzmann constant, T is the temperature, V^0 is the channel void volume, $\dot{V}_c$ is the cross-flow rate, and η is the viscosity of the carrier liquid. Equation (1) pertains to normal-mode separations.

It is apparent, from Eq. (1), that the primary sample property measured by flow FFF is the diffusion coefficient. Secondary information includes the hydrodynamic diameter which can be obtained via the Stokes–Einstein equation and the molecular weight if the molecule shape factor is constant. Unlike other FFF techniques, the retention time in flow FFF is determined solely by the diffusion coefficient rather than a combination of sample properties. As a consequence, flow FFF is well suited for analyses of complex sample mixtures and the transformation of the fractogram to a diffusion or size distribution is straightforward. In addition, flow FFF is applicable to a wide range of samples regardless of their charge, size, density, and so forth.

Flow FFF was first introduced as a method for protein separation and characterization by Giddings et al. in 1977 [2]. This first publication discusses the advantages of flow

FFF over other protein separation methods that were used at the time (e.g., polyacrylamide gel electrophoresis and size-exclusion chromatography (SEC)]. Flow FFF permits the calculation of diffusion coefficients and hydrodynamic diameter of proteins in different solution environments directly from retention data using straightforward analytical relationships. No other technique provides simultaneous separation and measurement of D and d for each sample component without the need for additional information such as charge and density. Moreover, the protein molecular weights (MWs) calculated from FFF diffusion coefficient data are on a sounder theoretical basis than that derived from electrophoretic mobility in gels or chromatographic retention data. The absence of packing material in a channel makes flow FFF suitable for the analysis of fragile high-MW proteins and protein complexes in comparison with SEC where shear forces and interfacial interactions can cause changes in the structure and activity of molecules. Flow FFF can be used to separate and characterize mixtures of proteins, protein aggregates, and protein complexes with a single FFF analysis covering a 500–1000-fold size difference. A series of SEC columns with different exclusion limits would be needed to span a similar size range. The cross-flow rate is programmed (i.e., decreased with respect to time) to maintain separation times of 5–20 min. The open FFF channel structure permits the analysis of protein samples suspected of containing precipitated material without special pretreatment. This is not possible by SEC and gel electrophoresis.

The main disadvantage of flow FFF is the lack of commercially available membranes with flat nonadsorbing surfaces, uniform permeability, and batch-to-batch reproducibility. These features may affect retention time, separation efficiency, and sample recovery. Numerous studies and technical improvements have been done to ensure optimum performance.

The traditional flow-FFF method, also called symmetrical flow FFF, utilizes a channel with permeable depletion and accumulation walls. The cross-flow liquid enters the channel through a microfiltration frit that acts as the depletion wall and exits through an ultrafiltration membrane that acts as the accumulation wall [1–3]. Membranes with nominal MW cutoffs in the range of 5000–30,000 Da are used to ensure that the protein sample remains in the FFF channel. Hydrophilic ultrafiltration membranes made from regenerated cellulose, cellulose acetate, and modified polyethersulfone have been used.

The sample capacity in a flow FFF channel with standard dimensions is several hundred micrograms. Sample overloading causes band broadening and a shift in retention time. The experimental procedure involves sample injection followed by a stop-flow period when the axial or channel flow is stopped and the sample is transported to equilibrium positions at the accumulation wall under action of the cross-flow. At the end of the stop-flow period, channel flow is resumed and the sample is fractionated while being swept toward the outlet. The elution of proteins is monitored with an ultraviolet (UV) spectrophotometric detector set at 280 or 210 nm wavelength. Other on-line detectors that are being used or evaluated include multiangle light scattering (MALS) which yields MW and photon correlation spectroscopy (PCS), which yields D. Flow FFF provides the narrow polydispersity sample fractions required for accurate MW and D determinations by the light-scattering detectors. These detectors provide a second independent measure of MW and D for comparison with values calculated from FFF retention times. In addition, MALS can be used to distinguish the polydispersity of the sample from band broadening. The main disadvantages of MALS detection for protein flow-FFF applications are the low sensitivity and particulate-free solvent requirement.

Symmetrical flow FFF has been successfully used to analyze numerous purified proteins and their dimers, protein conjugates, sodium dodecyl sulfate (SDS)–protein complexes, including precipitate [1–3] and other biological samples, as summarized in Table 1. In these applications, sample pretreatment was not required even when the proteins of interest were present in complex matrices such as plasma, dairy products, and wheat flour (in contrast to SEC and electrophoresis). Ultracentrifugation, which is commonly used to fractionate these samples, requires over 24 h in comparison with the 5–40 min for flow FFF. The short analysis time gives flow FFF an advantage in analyzing the large number of samples encountered in the medical diagnostics, pharmaceutical, and food industries. Separation efficiencies are usually of the order of several hundred theoretical plates for a typical channel length of 25–30 cm.

A later modification of flow FFF, called asymmetrical flow FFF, has generated up to 2600 theoretical plates in the separation of viruses [4]. In the asymmetrical channel, the permeable frit or depletion wall is replaced with an impermeable glass plate. The flow of carrier liquid entering the inlet of the channel is the source of both the cross-flow and channel flow. The sample is injected through an additional port located about 2–3 cm downstream from the channel inlet. For asymmetrical flow FFF, the retention mechanism is the same as for the traditional symmetrical mode. The observed higher efficiency may be based on the different relaxation procedures. In the asymmetrical channel, a so-called focusing procedure is used. The injected sample is focused to a narrow zone at the accumulation wall under action of two opposing flows (intro-

Table 1 Examples of Flow FFF Protein Analyses

Application	Flow FFF type	Summary
Ribosomes from *Escherichia coli*	Asymmetrical	Separation of ribosomes, their subunits, and t-RNA/low-MW protein mixture in samples collected at different protein production phases and in the presence of antibiotics, specific genes, and proteins; calculation of a ribosome number per cell and a ribosome fraction using peak area [6]
Yeast acid phosphatase (APase)	Asymmetrical	Separation of APase in cultivation medium; identification of APase peak by enzymatic activity measurements [7]
Proteins from wheat flour	Asymmetrical	Fractionation of wheat proteins in the size range from 5 to 45 nm; separation of gliadins from glutenin [K.-G. Wahlund, M. Gustavson, F. MacRitchie, T. Nylander, and L. Wannerberger, *J. Cereal Sci. 23*: 113–119 (1996)]
Wheat protein fractions	Symmetrical	Size distribution of wheat protein fractions (albumins and globulins, gliadins, glutenins) prepared by extraction; influence of oxidation on size distribution of high-MW glutenin [S. G. Stevenson, T. Ueno, and K. R. Preston, *Anal. Chem. 71*: 8–14 (1999)]
Monoclonal antibodies (Mab) from hybridoma cell culture	Asymmetrical	Separation of five Mab aggregates (immunoglobulins) in half the time needed for GPC; only three partially resolved peaks were obtained by GPC; separation of samples containing precipitated material [A. Litzen, J. K. Walter, H. Krischollek, and K.-G. Wahlund, *Anal. Biochem. 212*: 469–480 (1993)]
Proteins in homogenized dairy products	Symmetrical	Separation of whey proteins from fat globules; fractionation of whey proteins and casein micelles; effect of carrier ionic strength, cross-flow rate, pH, and membrane type on retention and size distribution of micelles [M. A. Jussila, G. Yohannes, and M.-L. Riekkola, *J. Microcol. Separ. 9*: 601–609 (1997)]
Reconstituted skim milk casein	Symmetrical	Fractionation of colloidal Ca^{2+}–caseinate complexes in milk samples in size range 10–50 nm after the preliminary fractionation of 10–600-nm colloids with sedimentation FFF [P. Udabage, R. Sharma, D. Murphy, I. McKinnon, and R. Beckett, *J. Microbiol. Separ. 9*: 557–563 (1997)]
Plasma protein interactions with polymer colloids	Symmetrical	Measurement of changes in particle (PS latex) size due to aggregation and adsorption of proteins; effect of Pluronic™ surfactants on aggregation and protein adsorption [J.-T. Li and K. D. Caldwell, *Colloids Surf. B: Biointerfaces 7*: 9–22 (1996)]
Lipoproteins in human plasma	Symmetrical	Separation of low-MW proteins, HDL , LDL, and VLDL; Determination of subspecies in the lipoprotein fractions with linear programmed cross-flow rate; observation of lipoprotein profiles for different individuals [10]
Drug/plasma protein interactions	Asymmetrical	Separation of albumin, HDL, α-macroglobulin, and LDL; Determination of drug distribution in FFF fractions using a fluorimetric detector [M. Madorin, P. van Hoogevest, R. Hilfiker, B. Langwost, G.M. Kresbach, M. Ehrat, and H. Leuenberger, *Pharm. Res. 14*: 1706–1712 (1997)]
Viruses (purified)	Asymmetrical	Separation of five aggregates of the Satellite tobacco necrosis virus; isolation of the Cow pea mosaic virus [5]

duced simultaneously through the inlet and outlet of the channel). However, this focusing procedure has some potential drawbacks. It may promote chain entanglement of high-MW macromolecules and increase adsorption on the membrane. Lower sample capacity (several micrograms) and recovery have been observed in comparison with the symmetrical flow FFF.

Asymmetrical flow FFF has been successfully applied to several practical problems, such as the separation of monoclonal antibodies and their aggregates in downstream processing, ribosomes, and plasma protein studies (see Table 1). For example, flow FFF separations of ribosomes and their subunits have led to an alternative method of optimizing the fermentation process without having to measure protein yields [5]. Flow FFF is the technique uniquely suited for separating highly glycosylated proteins such as 10^6 Da acid phosphatase (Apase) in cultivation media [6]. Buffers commonly used in chromatographic methods cause changes in the enzymatic activity of APase.

A second variant to the traditional symmetrical flow FFF channel utilizes a hollow-fiber membrane. Even though its feasibility in separating proteins has been demonstrated [7], its development has been hindered by the lack of membranes suitable for use as an FFF channel.

Several studies involving FFF channel modifications and new experimental procedures have been aimed at increasing detectability. For example, frit–outlet flow FFF utilizes a section of the frit depletion wall near the channel outlet to remove sample-free carrier and, thus, concentrate the separated sample just prior to its reaching the detector. A 10-fold increase in the detector response of purified proteins was achieved without any effect on retention time or resolution [8]. Frit–inlet flow FFF involves a modification at the channel inlet that enables the transport of sample to the accumulation wall without using stop-flow relaxation [8,9]. As a result, the experimental operation is simpler and the detector signal is free of fluctuations caused by stopping and resuming axial flow through the channel. Enhanced detectability can also be realized by programming the cross-flow rate during the run to accelerate elution of highly retained components [9]. Another experimental approach involves using the ultrafiltration membrane that serves as the accumulation wall to remove low-MW interferences. A nominal 30,000-Da MW cutoff membrane was used to selectively remove low-MW proteins from a blood plasma sample to allow determinations of high-, low-, and very low-density lipoproteins (HDL, LDL, and VLDL, respectively).

A fundamental study was performed to demonstrate that flow FFF is a good alternative technique for the rapid measurement of protein diffusion coefficients [10]. The results obtained for 15 proteins were in good agreement (within 4%) with the literature data based on classical methods and a group of modern methods such as photon correlation spectrometry (PCS), laminar flow analysis, a chromatographic relaxation method, and analytical split-flow thin-cell (SPLITT) fractionation. The advantages of flow FFF are the high-speed separations and the calculation of D values directly from retention data.

Acknowledgments

G. Kassalainen was supported by grant from ACTR/ ACCELS with funds provided by the U.S. Information Agency and S. K. R. Williams by a Colorado School of Mines start-up grant.

References

1. J. C. Giddings, *Science 260*: 1456–1465 (1993).
2. J. C. Giddings, F. J. Yang, and M. N. Myers, *Anal. Biochem. 81*: 395–407 (1977).
3. J. C. Giddings, M. A. Benincasa, M.-K. Liu, and P. Li, *J. Liquid Chromatogr. 15*: 1729–1747 (1992).
4. A. Litzen and K.-G. Wahlund, *Anal. Chem. 63*: 1001–1007 (1991).
5. M. Nilsson, L. Bulow, and K.-G. Wahlund, *Biotechnol. Bioeng. 54*: 461–467 (1997).
6. A. Litzen, M. B. Garn, and H. M. Widmer, *J. Biotechnol. 37*: 291–295 (1994).
7. J. E. G. J. Wijnhoven, J.-P. Koorn, H. Poppe, and W. Th. Kok, *J. Chromatogr. A 732*: 307–315 (1996).
8. P. Li, M. Hansen, and J. C. Giddings, *J. Microcol. Separ. 10*: 7–18 (1997).
9. P. Li, M. Hansen, and J. C. Giddings, *J. Liquid Chromatogr. Related Technol. 20*: 2777–2802 (1997).
10. M.-K. Liu, P. Li, and J. C. Giddings, *Protein Sci. 2*: 1520–1531 (1993).

Galina Kassalainen
S. Kim Ratanathanawongs Williams

Purification of Peptides with Immobilized Enzymes

Introduction

Affinity chromatography is a powerful separation technique that exploits specific binding properties among biomolecules. For protein purification, the enrichment obtainable in single-step affinity chromatography sometimes exceeds 1000-fold. This high selectivity is due to the vast variety of specific interactions that characterize the functional properties of protein molecules. Like any other affinity chromatography technique, purification of peptides by immobilized enzymes captures the unique interactions between biomolecules (see Refs. 1–3). For example, many enzymes reversibly bind to their organic cofactor molecules and inhibitors. The separation of such enzymes from other proteins can be easily achieved by using a chromatography column containing immobilized inhibitors. Inversely, the enzyme inhibitors can be isolated and purified to a high degree of purity by affinity chromatography using immobilized enzymes. Because affinity chromatography is a most selective technique, it may be used alone as a single chromatography step for peptide purification. However, as many input-material contaminants as possible must be removed prior to using affinity chromatography (e.g., before the affinity chromatography step). For instance, ultrafiltration or gel permeation chromatography is sometimes used concurrently to separate the high- and low-molecular-weight forms while exchanging buffers. In this entry, the focus will be on the use of affinity chromatography to purify peptides when the immobilized ligand is an enzyme.

General Principle of Peptide Purification by Immobilized Enzymes

The first step in this type of chromatography is the preparation of the column by enzyme immobilization on activated support. Enzymes are immobilized by covalent bonding to a packing of a chromatography column (i.e., a support solid matrix such as a cross-linked agarose bead). A mixture of components containing the peptide of interest is applied to the column. Peptides are bound with a high degree of specificity to the immobilized enzyme that functions as an ionic, hydrophobic, aromatic, or stoically active binding site, depending on the particular circumstances. The unbound contaminants, which have no affinity for the ligand, are washed through the column, leaving the desired peptide bound to the matrix. Bound peptides are released by manipulating the composition of the eluant buffers. Generally, the elution of the peptide is accomplished by changing the pH and/or salt concentration or by applying organic solvents or a molecule that competes for the bound ligand (Fig. 1).

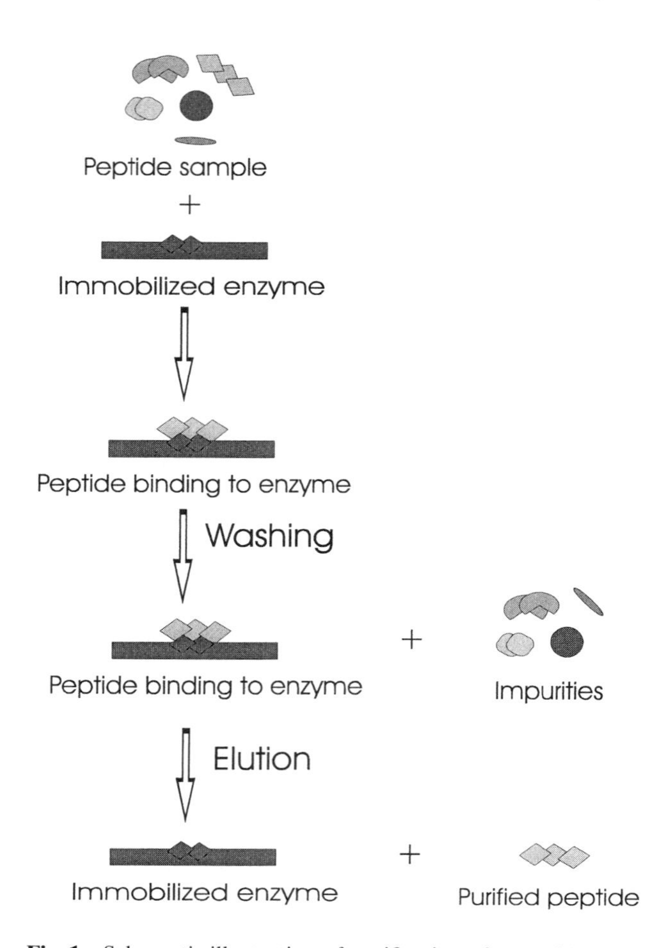

Fig. 1 Schematic illustration of purification of a peptide by affinity chromatography on an immobilized enzyme.

Examples of Purification of Peptides by Immobilized Enzymes

Purification of Protein Hydrolysates

Proteases possess special characters, including their ability to cleave at certain amino acid sites such as at the hydrophobic or neutral residues. This cleavage specificity produces protein hydrolysates that vary significantly in their molecular weights, charge characteristics, functional profiles, and, thus, their potential specific interactions. Using immobilized proteases for peptide purification is a typical application of this type of affinity chromatography. Trypsin is a serine protease; that is, it contains an essential serine at its active site and cleaves the peptide bond next to a basic amino acid residue. Anhydrotrypsin is an inert derivative of trypsin and, thus, has no catalytic activity. It contains dehydroalaninein in place of the active serine residue site. The modified enzyme has 20-fold more affinity for product peptides produced by tryptic cleavage than substrate peptides. C-Terminal arginine peptides adhere more tightly than COOH-terminal lysine peptides. Accordingly, immobilized anhydrotrypsin has been used as an affinity ligand for the purification of these peptides. Because of widely varied peptide structure, further specificity may be obtained during elution, which usually involves decreasing pH gradients.

Purification of Pawpaw Glutamine Cyclotransferase

Pawpaw glutamine cyclotransferase (PQC) was purified 279-fold to near homogeneity by a combination of ion-exchange chromatography on SP-Sepharose, hydrophobic interaction chromatography on Fractogel TSK Butyl-650, and affinity chromatography on immobilized trypsin [4]. Trypsin was immobilized on Affigel-10 (Bio Rad Laboratories) according to procedures recommended by the manufacturer. After the concentration of pooled fractions from the last purification step (Fractogel column) by ultrafiltration, affinity chromatography was used. Partially purified PQC was injected into a 12.5 × 1.6-cm-inner diamater column of immobilized trypsin gel preequilibrated in the eluting buffer (20 m*M* $CaCl_2$ in 50 m*M* Tris-HCl buffer, pH 8.2). Affinity purification was accomplished because PQC contains a unique and highly basic polypeptide chain containing covalently attached phosphate groups without any disulfide bonds.

Purification of Enzyme Inhibitors

Enzyme inhibition may be reversible or irreversible with different inhibitors. Irreversible inhibitors usually form covalent bonds and, thus, are not useful for this type of affinity chromatography. Reversible inhibitors work by a variety of mechanisms, but usually competitive inhibitors structurally resemble the peptide substrates and bind the active center. *Trichosanthes kirilowii* trypsin inhibitor analog (Ala-6-TTI) is a trypsin inhibitor in which methionine at position 6 is replaced by alanine. Affinity chromatography was used to purify expressed fusion protein containing large proteins and (Ala-6)-TTI, with methionine as a connecting residue [5]. After cyanogen bromide cleavage and reduction of the fusion protein, followed by affinity chromatography separation on trypsin-Sepharose 4B as a matrix and on the immobilized trypsin, the fully active (Ala-6)-TTI was obtained in a high purity and high yield. The trypsin inhibitory activity and amino acid composition of the recombinant (Ala-6-TTI) were consistent with those of the natural one.

An essentially pure extracellular glycoprotein proteinase inhibitor was isolated from the latex of green fruits of papaya by a single affinity chromatography purification step [6]. An immobilized trypsin-Sepharose CL 4B column was prepared according to the manufacturer (Amersham Pharmacia Biotech), which provided elaborated bulletins for the preparation procedures and applications of these affinity gel columns. Latex extract was applied to the column after equilibration with 20 m*M* Tris-HCl, pH 8.0, containing 0.5*M* NaCl and 0.05*M* $CaCl_2$. The column was extensively washed with the same buffer until ultraviolet (UV) absorbance became undetectible. The bound trypsin inhibitor was eluted with 0.02*M* HCl and recovered by lyophilization after dialysis against water.

Purification of Dipeptidyl Peptidase IV

Dipeptidyl peptidase IV (E.C. 3.4.14.5) cleaves off N-terminal dipeptides from peptides when a proline or alanine is at the penultimate position. The enzyme was purified from human seminal plasma and prostasomes by a two-step scheme of ion-exchange chromatography on DEAE–Sepharose, followed by affinity chromatography on adenosine deaminase (E.C. 3.5.4.4)–Sepharose [7]. This scheme resulted in a pure, native protein with an overall yield ranging from 35% to 55%. The preparation obtained was free of contaminating aminopeptidase activity.

Purification of Recombinant Proteins via Peptide Accession

Purification of peptides by immobilized enzymes is a relatively recent approach that may have a potential use in the purification of genetically engineered proteins. Purification of recombinant proteins is often difficult and involves many purification steps for a very low yield. Smith et al. [8] found that a specific metal-chelating peptide on the NH_2 terminus of a protein can be used to purify that protein with immobilized ion affinity chromatography. The nucleotide sequence that codes for the expressed protein is extended to include codons for the chelating peptide. This concept of amino acid addition to recombinant proteins can also be beneficial in protein purification by affinity chromatography with immobilized enzymes. This will have an impact on the cloning, expression, and the purification of proteins.

Conclusion

Affinity purification of peptides by immobilized enzymes relies on the highly specific interaction of an ionic, hydrophobic, aromatic, or stoically active binding site on the desired peptide to the immobilized enzyme. The unbound contaminants are washed through and the desired peptide is released by buffer elution. This technique is beneficial in the purification of many peptides and can also be used for affinity chromatography of recombinant proteins after cloning and expression.

References

1. J. E. Bell and E. T. Bell, Protein purification: Affinity chromatography, in *Proteins and Enzymes*, Prentice-Hall, Englewood Cliffs, NJ, 1988, pp. 45–63.
2. J. Carlsson, J-C. Janson, and M. Sparrma, Affinity chromatography, in *Protein Purification: Principles, High Resolution Methods, and Applications* (J. C. Janson and L. Ryden, eds.), VCH, New York, 1989, pp. 275–329.
3 H. Schott, *Affinity Chromatography*, Chromatographic Science Series Vol. 27, Marcel Dekker, Inc., New York, 1984.
4. S. Zerhouni, A. Amrani, M. Nijs, N. Smolders, M. Azarkan,

J. Vincentelli, and Y. Looze, *Biochem. Biophys. Acta 1387*: 275–290 (1998).
5. X. M. Chen, Y. W. Qian, C. W. Chi, K. D. Gan, M. F. Zhang, and C. Q. Chen, *J. Biochem. (Tokyo) 112*: 45–51 (1992).
6. S. Odani, Y. Yokokawa, H. Takeda, S. Abe, and S. Odani, *Eur. J. Biochem. 241*: 77–82 (1996).
7. I. De-Meester, G. Vanhoof, A. M. Lambeir, and S. Scharpe, *J. Immunol. Methods 189*: 99–105 (1996).
8. M. C. Smith, T. C. Furman, T. D. Ingolia, and C. Pidgeon, *J. Biol. Chem. 263*: 7211–7215 (1988).

Jamel S. Hamada

Pyrolysis–Gas Chromatography–Mass Spectrometry Techniques for Polymer Degradation Studies

Introduction

The characterization of relatively complex polymers is usually carried out by means of coupled techniques because sometimes a single technique is not enough to elucidate their structures. Pyrolysis of polymers is an old technique used many years ago to identify materials by their vaporized decomposition products. The coupling of this simple method with a powerful identification technique, such as infrared (IR) spectroscopy or, often, mass spectrometry (MS), has demonstrated its utility for the analysis of polymeric materials and, mainly, for the characterization of their degradation products.

The pyrolysis of a sample can be defined as the conversion of a substance into other(s) with lower molecular weights by the action of heat. In the case of organic polymers, the most interesting data we can obtain from the fragments of pyrolysis are the composition of the volatiles produced and of the nonvolatile residue, changes in molecular weight, as a function of the temperature, and evolution of the thermal decomposition process, kinetics, and activation energies of the degradation reactions.

In order to obtain these data, the most often applied approach has been coupling of gas chromatography (GC) with MS. The pyrolysis products are first separated in the column and then the separated fragments are directly analyzed in the mass spectrometer. Therefore, it is possible to obtain reliable and reproducible results with a single run and in a relatively short time of analysis.

Pyrolysis (Pyr)–GC–MS, however, presents some limitations associated with many factors, such as the complexity of the chemical reactions involved in pyrolysis, the incomplete separation of the degradation products, and, sometimes, poor peak identification and interpretation of chromatograms. However, some recent developments have permitted the application of this technique to a broader range of samples. Those developments are principally the use of highly specific pyrolysis devices, highly efficient capillary columns, and MS libraries with a wider range of spectra. This fact has permitted the use of the technique not only for polymer identification but also for structural characterization of polymers and blends.

An important aspect of analytical pyrolysis is the production and detection of thermal fragments containing essential structural information, in order to provide insight, at least partially, into the original structures of the compounds present in the chemical matrix. Larger fragmentation products are the more informative among all fragments. Therefore, there is a current tendency in analytical Pyr–GC–MS to preserve and detect higher-molecular-weight fragments. This has led to developments in the instrumentation, such as improvement of the direct transfer of high-molecular-weight and polar products to the ion source of the mass spectrometer, the measurement of these compounds over extended mass ranges, and the use of soft ionization conditions, such as field ionization (FI) and chemical ionization (CI) instead of electron impact (EI). Methods such as FI and CI are useful due to the difficulties arising from EI, such as the variation of fragmentation depending on instrumental conditions and the fact that only low-mass ions are observed. However, one problem with the soft ionization methods is the higher cost of instrumentation.

Instrumentation

A typical Pyr–GC–MS instrument for the use in polymer analysis consists of a pyrolyzer coupled to a gas chromatograph and a mass spectrometer. A pyrolyzer, through which the carrier gas flows (usually He or N_2), is directly coupled to a gas chromatograph with a high-resolution capillary column.

Pyrolyzer

The most commonly used pyrolyzers can be classified into three groups:

1. Resistively heated electrical filaments
2. Radio-frequency induction heated wires (Curie-point pyrolyzers)
3. Microfurnace type

Each of these designs has its particular characteristics. Thus, the filament-type permits multistep pyrolysis, which enables the discriminative analysis of formulations in a compounded material. The Curie-point type offers the most precisely controlled equilibrium temperature, but the heating conditions depend greatly on the shape of the sample holder. On the other hand, a precise temperature regulation of the microfurnace type is not easy, but it is most suited for thermally labile compounds such as biopolymers.

Gas Chromatograph

The use of highly effective capillary columns is essential for high-resolution Pyr–GC–MS, because pyrolysis products of polymers are generally very complex mixtures. The bonded-phase fused-silica columns are especially effective for Pyr–GC–MS because of their low level of stationary-phase bleed at elevated column temperatures.

The split mode is usually favored for Pyr–GC–MS, because the very low velocity of the carrier gas for the splitless mode often causes undesirable secondary reactions of the degradation products in the pyrolyzer. In the split mode, however, the product composition entering the capillary column sometimes differs from the original, depending on the volatility of each component and the split temperature. This can cause problems in reproducibility and precision of results because the degradation products of polymers usually consist of complex mixtures with a broad range of volatilities. In order to overcome this problem, the splitter is modified and the dead volume is packed with an ordinary chromatographic packing material. The splitter is then maintained, independently, at the maximum temperature of the column. This arrangement permits the obtaining of reproducible results and protection of the instrument.

Mass Spectrometer

The advances in the specific determinations for peaks by GC–MS were significant for the development of the technique. In particular, a computer-controlled GC–MS system readily leads to the rapid and accurate identification of peaks.

Methods of Analysis

A very important aspect to be controlled in Pyr–GC–MS is the reproducibility of results. This is important not only for the determination of structure but also for the reactions between fragments in the GC. In order to obtain high reproducibility, two different approaches can be used for a particular sample:

1. Isothermal (flash pyrolysis). The temperature of the sample is suddenly increased (10–100 ms) to reach the thermal decomposition level (500–800°C). This process can be carried out by means of a platinum or platinum–rhodium filament heated by an electrical current directly coupled to the injector port of the GC. Some pyrolysis fragments are obtained in a very short time and can be directly sent to the column and detector. In spite of this short time for the pyrolysis, it is possible to indicate three different phases: (a) heating (10^{-2}–10^{-1} s), (b) stabilization of the maximum temperature, and (c) cooling. However, the main drawback of this technique is the lack of equilibrium between temperatures with the pyrolyzer.
2. Programmed temperature pyrolysis. The sample is heated at a given heating rate in a manner similar to the linear heating used in thermogravimetry or differential scanning calorimetry. The desirable conditions of this method are their uniform heating, a minimum dead volume, and a device to introduce samples. Pyrolysis fragments are then analyzed as a function of temperature.

Direct Inlet Mass Spectrometry

Pyrolysis-direct chemical ionization mass spectrometry (Pyr–DCI–MS) was recently introduced as a pyrolysis technique for the characterization of complex macromolecular samples and for the analysis of biopolymers. This technique does not require special pyrolysis equipment and can be performed with an instrument which is equipped with a chemical ionization source and a standard

DCI probe, which consists of an extended wire used to introduce the sample material directly into the chemical ionization plasma. An important characteristic of this technique is the pyrolysis in the ion plasma, using a filament that is resistively heated by a current programming device. This arrangement guarantees an optimum transfer for large and polar molecules. Moreover, temperature-resolved pyrolysis data can be obtained which can provide additional chemical information on a variety of classes of compounds with different thermal stabilities and desorption characteristics.

Direct inlet mass spectrometry (DIMS) is one of the best techniques to elucidate the structure of low-molecular-weight organic and inorganic compounds evolved from polymers. In this technique, polymer samples are introduced into the ion source of a mass spectrometer by means of a direct inlet for solid samples, and thermal degradation is carried out by a linear temperature program. The volatile products are ionized and detected immediately by repeated mass scans. This technique presents some advantages, such as the possibility to avoid manipulation of samples, because it can be used as an on-line method. Some additional advantages are the fast in-vacuum ionization, which reduces secondary reactions, the possibility of obtaining the molecular weight from the molecular ion, and the relatively low cost of the equipment.

Table 1 Relative Intensities in the Dehydrochlorination Reaction in a PVC Resin

m/z	Intensity	m/z	Intensity
15.00	0.16	43.00	0.20
16.00	0.13	44.00	0.17
17.00	0.95	49.00	0.15
18.00	4.08	50.00	0.84
19.00	0.24	51.00	0.96
26.00	0.32	52.00	0.84
27.00	0.47	55.00	0.11
28.00	0.44	63.00	0.22
29.00	0.23	74.00	0.17
35.00	22.20	76.00	0.14
36.00	100.00	77.00	0.69
37.00	6.86	78.00	2.34
38.00	31.30	79.00	0.19
39.00	0.84	91.00	0.13
41.00	0.20	128.00	0.10

Applications

Linear Polymers

The pyrolysis of linear polymers results in a series of fragments that are introduced into the GC–MS instrument. Polymers are completely degraded (depending on the ionization conditions) giving rise to the characteristic fragments which permit the identification of the degradation products. For instance, in Table 1, the most probable structure of the degradation of a poly(vinyl chloride) (PVC) resin is presented.

As can be observed, the maximum m/z ratio is presented at 36, which corresponds to HCl, the main degradation product of PVC. Other important products are those with m/z 35 and 38 also corresponding to HCl. Other significant peaks are those at 18 (corresponding to water) and 78 (benzene). This peak is especially important, as it can be considered as the result of a reaction between double-bond fragments leading to an aromatic product.

Analysis of Copolymers

The Pyr–GC–MS analysis of a copolymer presents a complicated fragmentation pattern, resulting in a wide variety of products and highly complicated spectra. This is the case particularly when EI is used as the ionization method. CI and FI are better selections, and it is possible to obtain only the most significant peaks, corresponding to each component of the copolymer.

Analysis of Thermosets

This is a new and promising area of research and its interest is based on the possibility of identification of particular bonds or functions responsible of the scission of the chain during the first steps of the degradation. As these polymers are becoming increasingly used in industry, Pyr–GC–MS is an important technique for the analysis.

Natural Polymers

The chromatograms of natural polymers, such as proteins and polysaccharides, are so complex that they have been mostly used for general identification. The potential of Pyr–GC–MS has been greatly enhanced by the use of high-resolution capillary columns combined with computer-assisted techniques.

Suggested Further Reading

Bate, D. M. and R. S. Lehrle, *Polym. Degrad. Stabil. 64*: 75 (1999).

Dadvand, N., R. S. Lehrle, I. W. Parsons, M. Rollinson, I. M. Horn, and A. R. Skinner, AR, *Polym. Degrad. Stabil. 67*: 407 (2000).

Dadvand, N., R. S. Lehrle, I. W. Parsons, and M. Rollinson, *Polym. Degrad. Stabil. 66*: 247 (1999).

Erdogan, M., T. Yalçin, T. Tinçer, and S. Suzer, *Eur. Polym. J. 27*: 413 (1991).

Haken, J. K., *J. Chromatogr. A 825*: 171 (1999).

Irwin, W. J., *Analytical Pyrolysis. A Comprehensive Guide*, Chromatographic Science Series Vol. 22, Marcel Dekker, Inc., New York, 1982.

Miranda, R., H. Pakdel, C. Roy, H. Darmstadt, and C. Vasile, *Polym. Degrad. Stabil. 66*: 107 (1999).

Plage, R. and H. R. Schulten, *J. Anal. Appl. Pyrolysis 19*: 285 (1991).

Pouwels, A. D., G. B. Eijkel, and J. J. Boon, *J. Anal. Appl. Pyrolysis 14*: 237 (1989).

Statheropoulos, M., *J. Anal. Appl. Pyrolysis 10*: 89 (1986).

Statheropoulos, M. and S. A. Kyriakou, *Anal. Chim. Acta 409*: 203 (2000).

Tas, A. C., A. Kerkenaar, G. F. LaVos, and J. Van der Greef, *J. Anal. Appl. Pyrolysis 15*: 55 (1989).

Wang , F. C., *J. Chromatogr. A 833*: 111 (1999).

Wilkie, C. A., *Polym. Degrad. Stabil. 66*: 301 (1999).

Alfonso Jiménez Migallon
M. L. Marín

Q

Quantitation by External Standard

An important feature of modern high-performance liquid chromatography (HPLC) is its excellent quantitation capability. HPLC can be used to quantify the major components in a purified sample, the components of a reaction mixture, and trace impurities in a complex sample matrix. The quantitation is based on the detector response with respect to the concentration or mass of the analyte. In order to perform the quantitation, a standard is usually needed to calibrate the instrument. The calibration techniques include an *external standard method*, an internal standard method, and a standard addition method. For cases in which a standard is not available, a method using normalized peak area can be used to estimate the relative amounts of small impurities in a purified sample.

In this entry, only the external standard method is discussed. Detailed discussions of other quantitation methods can be found in other entries of this encyclopedia or in Ref. 1.

The external standard method is the most general method for determination of the concentration of an analyte in an unknown sample. It involves the construction of a calibration plot using external standards of the analyte, as shown in Fig. 1. A fixed volume of each standard solution of known concentration is injected into the HPLC and the peak response of each injection is plotted versus the concentration of the standard solution. The standards used are called "external standards" because they are prepared and analyzed separately from the unknown sample(s). After constructing the calibration plot, the unknown sample is prepared, injected, and analyzed in exactly the same manner. The concentration of the analyte in the unknown sample is then determined from the calibration plot or from the response factor of the unknown sample versus that of the standard.

If the calibration plot is linear, the concentration of the analyte in the unknown sample can be determined based on the linear equation of the calibration plot:

$$Y = a + bX \tag{1}$$

where Y is the peak response of the analyte, X is the concentration of the analyte, a is the intercept, and b is the slope. The concentration X is

$$X = \frac{Y - a}{b} \tag{2}$$

In the example shown in Fig. 1, where $a = 0$ and $b = 2 \times 10^6$, the concentration (X) of the unknown sample (U), which shows a peak response of $Y = 500{,}000$, is determined to be 0.25 mg/mL using Eq. (2).

Alternatively, after the linear calibration range has been established, the concentration of an unknown sample can be determined using the *response factor*. A response factor (RF), sometimes called a *sensitivity factor*, is determined from a standard within the linear calibration range as:

$$\text{RF} = \frac{\text{Standard peak response}}{\text{Standard concentration}} \tag{3}$$

If two or more standards of different concentrations within the linear range are measured, the RF value can be taken as the average value of the response factors for all these standards to minimize the uncertainty in determining the RF. In the example shown in Fig. 1, the RF is the slope of the calibration plot and equals 2×10^6 for all three stan-

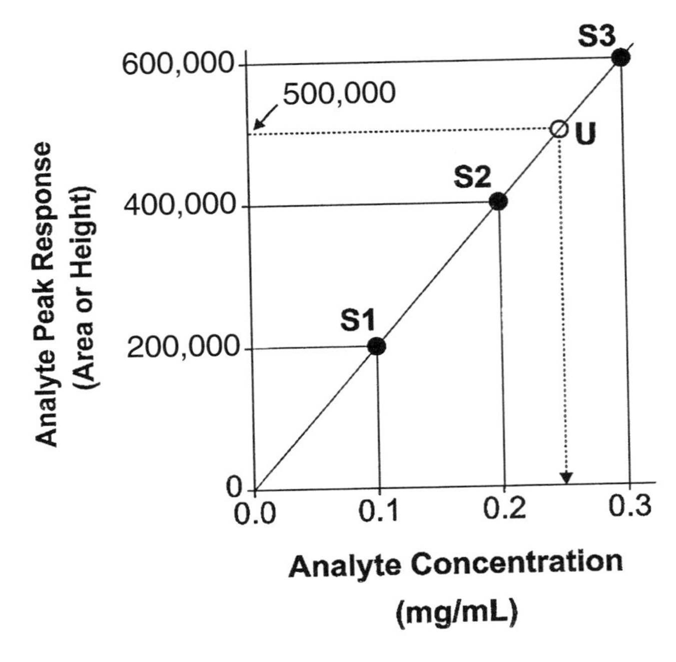

Fig. 1 Linear calibration plot for external standard method. S1, S2, and S3: external standards for calibration; U: unknown sample.

dards because the plot is linear with a zero intercept. The concentration (X) of the analyte in the unknown sample (U) can be calculated as

$$X = \frac{\text{Sample peak response}}{\text{RF}} \tag{4}$$

which is $500{,}000/(2 \times 10^6) = 0.25$ mg/mL. When RF is used for quantitation, the concentration(s) of the standard(s) selected should be similar to the expected concentration of the unknown sample.

In the aforementioned techniques, it is important that the calibration plot be linear over certain concentration range and the concentration of the sample fall within this linear range (as shown in Fig. 1) so that an interpolation can provide accurate measurement of the sample concentration. If the concentration of the sample falls outside the established linear range, extrapolation of the calibration plot should be used with caution. To ensure accuracy, it is recommended that dilution be carried out to bring the concentration of the sample into the linear range prior to the analysis.

In unusual cases, where the calibration plot is not linear, the concentration of an unknown sample can be determined by interpolation of results between standards or by fitting the results of the standards into a nonlinear equation and using the nonlinear equation to calculate the concentration of the unknown sample. When interpolation is used, the concentrations of the two standards used to bracket the sample should be as close to the concentration of the sample as possible to enhance the accuracy of the result. When a nonlinear equation is used, a large enough number of standards is needed in order to more accurately determine the nonlinear equation. Obviously, quantitation in a nonlinear range involves more labor and presents a higher potential of error. Therefore, this approach is used only when there are no other alternatives.

In the external standard method, it is critical that the injections are precise. With modern instruments which employ autosamplers, adequate precision (typically $\leq 0.5\%$) can be achieved using full-loop injection. Poorer injection precision is normally associated with manual injections or with partial-loop injections by autosamplers.

It is also critical to keep the chromatographic conditions (such as flow rate and column temperature) constant during the analysis of all standards and samples. Fluctuations of the chromatographic conditions during the analysis can cause inconsistent peak responses, which, in turn, will cause quantitation error. It is common that the HPLC system is tested using standards to ensure that the system is performing properly and reliably prior to the analysis of samples. This procedure of validating system performance is referred to as "system suitability test."

Detailed discussions on sources of error related to the external standard method, including sampling and sample preparation, chromatographic effects, and data system effects, can be found in Ref. 1. Reference 2 presented detailed discussions on the precision in HPLC.

The peak response used for quantitation can be either peak height or peak area. Peak height is usually used when incomplete resolution of the analyte peak is encountered, because the peak height measurement is subject to less interference from the adjacent overlapping peaks. On the other hand, peak area is less influenced by changes in instrumental or chromatographic parameters. The choice of peak height or peak area for quantitation requires the understanding of the effects of chromatographic parameters on the precision of each approach. The influence of certain chromatographic parameters on the precision of peak height and peak area methods, as well as the preferred method among the two when various chromatographic parameters are subject to change, are discussed in detail in Refs. 1 and 3.

More thorough discussions on external standard method can be found in Refs. 1, 3, and 4.

References

1. L. R. Snyder, J. J. Kirkland, and J. L. Glajch, *Practical HPLC Method Development*, 2nd ed., John Wiley & Sons, New York, 1997, pp. 643–684.

2. E. Grushka and I. Zamir, in *Chemical Analysis*, Vol. 98, *High Performance Liquid Chromatography*. (P. R. Brown and R. A. Hartwick, eds.), John Wiley & Sons, New York, 1989, pp. 529–561.
3. L. R. Snyder and J. J. Kirkland, *Introduction to Modern Liquid Chromatography*, 2nd ed., John Wiley & Sons, New York, 1979, pp. 541–574.
4. R. P. W. Scott, in *Quantitative Analysis using Chromatographic Techniques* (E. Katz, ed.), John Wiley & Sons, New York, 1987, pp. 63–98.

Tao Wang

Quantitation by Internal Standard

Principle

The principle involved is the addition of a known quantity of a foreign substance (internal standard) to the analyzed sample, the response coefficient of which is known or arbitrarily fixed. Quantitation by the internal standard method enables one to compensate for errors in the injected volume [1].

The same quantity of constituent I (internal standard) is added both to the reference solution and to the solution to be analyzed. I is supposed to interfere with none of the constituents present in the sample. This methodology is based on the constancy of the ratios between the proportionality coefficient observed on both chromatograms (determination and calibration).

Assay Chromatogram

The assay solution includes precisely and accurately known weights of the product to be determined and of the internal standard. An injection of an approximately known volume of this solution is made. From the resulting chromatogram, areas of peaks corresponding to the internal standard and to the product(s) to be analyzed are measured. Let M_E be the weight of the sample including solute D, the assay of which τ_D is to be determined, M_I be the weight of the internal standard I, A_D be the peak area of solute D, A_I be the peak area of the internal standard, K_D be the response coefficient of the product D, and K_I be the response coefficient of the internal standard. It is possible to establish the following relationships:

$$M_E \tau_D = K_D A_D \tag{1}$$

$$M_I = K_I A_I \tag{2}$$

and

$$\tau_D = \frac{K_D}{K_I} \frac{A_D}{A_I} \frac{M_I}{M_E} \tag{3}$$

Calibration Chromatogram

The protocol is similar to the calibration solution, including accurately and precisely known weights of reference and internal standard (dilutions must be analogous to those of the assay chromatogram). An injection of an approximately known volume of this calibration solution is made. From the resulting chromatogram, areas of peaks corresponding to the internal standard and to the reference material are measured. Let M_R be the weight of the reference material, the assay of which τ_R is known, M'_I be the weight of the internal standard I, A_R be the peak area of the reference material, and A'_I be the peak area of the internal standard. Thus, it is possible to establish the following relations:

$$M_R \tau_R = K_D A_R \tag{4}$$

$$M'_I = K_I A'_I \tag{5}$$

and

$$\frac{K_D}{K_I} = \frac{M_R}{M'_I} \tau_R \frac{A'_I}{A_R} \tag{6}$$

Calculation of τ_D, the Assay of Product D in the Sample

Combining relation (3) and relation (6), it is possible to determine τ_D, the purity of the product D, by

$$\tau_D = \frac{A_D}{A_I} \frac{A'_I}{A_R} \frac{M_I}{M_E} \frac{M_R}{M'_I} \tau_R \tag{7}$$

Some conditions are required for Eq. (7) to be valid: Areas and weights must be expressed in the same unit system both for analysis and calibration; because precision and accuracy of this method only depend on the precision and accuracy of weighings, they depend neither on the precision and accuracy of the dilutions nor on the injected volume (unlike the external standard method). It requires no preliminary determination of the proportionality coefficients. However, if some points of the sample handling are fully corrected by the use of an internal standard, other difficulties still remain [2].

Moreover, this method can quickly become laborious because an internal standard elution which is compatible with the analysis conditions must be found. Conditions that must be fulfilled by the internal standard are its purity must be known and it must be chemically inert toward solutes and mobile phase; on the one hand, its retention time must be different from those of all the constituents present in the sample and, on the other hand, it should be as close as possible as the retention time(s) of the product(s) to be determined. It has also been demonstrated that a necessary correlation was required between chromatographic behaviors of the internal standard and the product to quantify [3]. Otherwise, the use of an internal standard can even degrade the precision of the results. A comparison of the precision of internal and external standard has also been carried out through a liquid chromatography collaborative study [4].

The internal standard's coefficient of response for the detector used must be of the same order of magnitude as the one of the product to be determined; in no way can it be present as an impurity in the sample; it must be added at a concentration level that gives a peak area more or less equivalent to the one of the product to be determined. Homologues of the product to be analyzed may be used as internal standards.

The chromatogram given Fig. 1 illustrates the internal standard methodology. Here, methomyl was quantified using benzanilide as the internal standard. Using the calibration curve, the unknown methomyl assay in insecticides is deduced from the area ratio, after the addition to the sample of the same quantity of internal standard as in the calibration steps.

The internal standard method is less often used in liquid chromatography than in gas chromatography because injection of repeatable volumes has been made easier by the use of precise and reliable injection systems (loop valves). More generally, gradually, the internal standard method is being abandoned. The external standard method is, nowadays, the most common method and the use of an internal standard seems to be restricted to very specific applications; for example, when preliminary to the chromatographic analysis, the solute of interest must be extracted by means of a complex protocol.

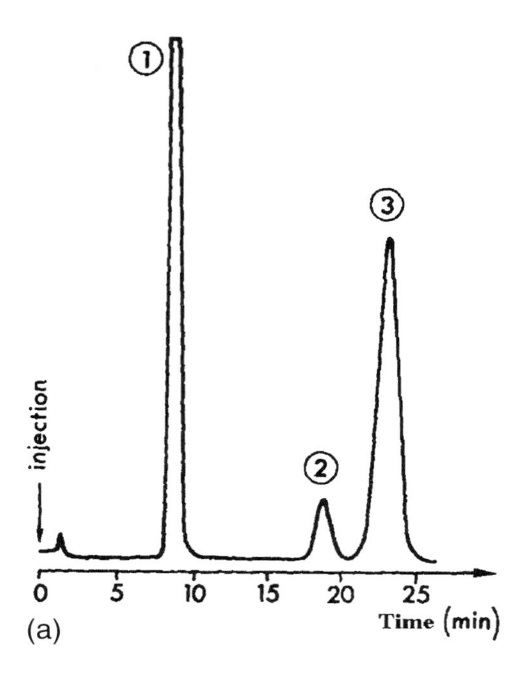

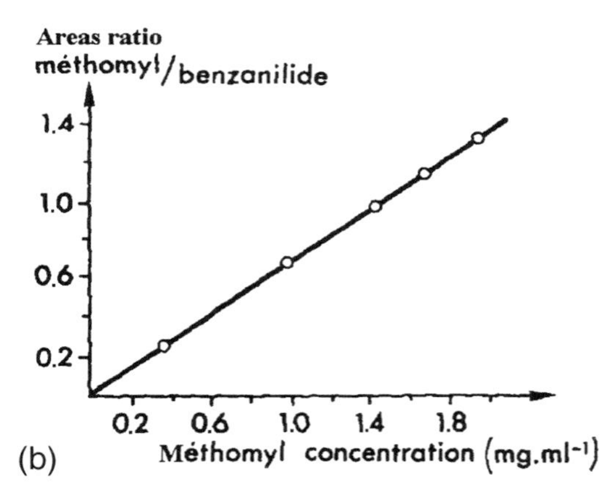

Fig. 1 Insecticide analysis by the internal standard method: (a) chromatogram; (b) calibration curve. 1: Benzanilide (internal standard); 2: methyl-*N*-hydroxythioacetimidate; 3: methomyl. (From Ref. 5.)

References

1. R. Rosset, M. Caude, and A. Jardy, *Chromatographies en phases liquide et supercritique*, Masson, Paris, 1991, pp 731–733.
2. L.R. Snyder and S. Van der Wal, *Anal. Chem. 53*: 877 (1981).
3. P. Haefelfinger, *J. Chromatogr. 218*: 73 (1981).
4. R. E. Pauls and R. W. McCoy, *J. High-Resolut. Chromatogr. 9*: 600 (1986).
5. R. E. Leitch, *J. Chromatogr. Sci. 9*: 531 (1971).

Suggested Further Reading

Snyder, L. R. and J. J. Kirkland, *Introduction to Modern Liquid Chromatography*, 2nd ed., John Wiley & Sons, New York, 1974, pp. 552–556.
Snyder, L. R., J. J. Kirkland, and J. L. Glajch, *Practical HPLC Method Development*, 2nd ed., John Wiley & Sons, New York, 1997, pp. 657–660.

J. Vial
A. Jardy

Quantitation by Normalization

Principle

The principle of this method is quite simple. Provided that for each solute, i, the analytical signal lies within the linearity range, the peak area is proportional to the weight of solute having passed through the detector cell, thus, that was present in the injected volume,

$$m_i = K_i A_i \quad (1)$$

where A_i is the peak area of solute i and K_i is the response coefficient. Therefore, the percentage in weight of each analyte is given by

$$\%_i = \frac{K_i A_i}{\sum_i K_i A_i} \times 100 \quad (2)$$

To apply this method in high-performance liquid chromatography (HPLC), several conditions have to be fullfilled: All the analytes present in the sample to be analyzed must elute from the column (no irreversible retention), with enough resolution and, furthermore, have to be detected. All the response coefficients have to be known, or at least attainable experimentally, which, in turn, implies that all the solutes are available separately in a high degree of purity. This method is unable to determine the percentage of any constituent of the mixture if a response coefficient is missing.

On the other hand, an advantage lies in the fact that there is no need to accurately know the amount of sample injected.

Practically, quantitation by the normalization method is not in as common use in HPLC as it is in GC. It is highly recommended to avoid its implementation in the case of samples for which the qualitative composition is not known exactly. It is only convenient in routine analysis, as in quality control, when the qualitative composition does not vary.

Moreover, note that, except in very particular cases, the approximation to consider all the K_i equal [Eq. (3)] is highly hazardous in HPLC, unlike in GC:

$$\%_i = \frac{A_i}{\sum_i A_i} \times 100 \quad (3)$$

A few examples of where it is possible are the following: Use of a refractive index detector, trace analysis of related substances (impurities) in pharmaceutical products, with UV detection at very low wavelength, when the accuracy of the result is of little interest [1] (e.g., when chromatography is used to monitor a chemical reaction), and when a relative value is enough. In this situation, the simplified relationship (3) is sufficient, but the analyst must not forget that the proportions found in this way are not the true proportions.

In conclusion, its efficiency and simplicity make quantitation by normalization a very attractive method that generally requires few injections. Nonetheless, it must be kept in mind that it requires the knowledge of the response coefficients for all the constituents in the mixture. The a priori hypothesis of equality for all response coefficients seldom corresponds to reality and can lead to hazardous results.

Reference

1. J. Tranchant, *Manuel pratique de chromatographie en phase gazeuse*, Masson, Paris, 1995, pp. 620–623.

Suggested Further Reading

Parris, N. A., *Instrumental Liquid Chromatography*, Elsevier, Amsterdam, 1976, p. 243.
Rosset, R., M. Caude, and A. Jardy, *Chromatographies en phases liquide et supercritique*, Masson, Paris, 1991, pp. 729–730.
Snyder, L. R., J. J. Kirkland, and J. L. Glajch, *Practical HPLC Method Development*, 2nd ed., John Wiley & Sons, New York, 1997, pp. 654–655.

J. Vial
A. Jardy

Quantitation by Standard Addition

Principle

The principle involved is that, provided the analytical signal is proportional to concentration, the initial analyte content is determined through measurement of this signal before and after the addition of a known amount of the analyte to the analyzed sample. The method of standard addition, also denoted as "spiking," is used when an analyte is to be quantified inside a matrix, the effects of which are likely to affect the chromatographic peak behavior. In this case, the sample itself is used as the calibration matrix.

Commonly, the unknown concentration is deduced from the increase in the peak height or the peak area resulting from the addition of known amounts of pure analyte to the sample (at least one or two additions). A comparison of peak height versus peak area for quantitation is given in Ref. 1. A more convenient protocol, which gives information both on the precision and the accuracy of the results, is the following. After each addition, the area of the chromatographic peak is measured. Then, the parameters of the regression line are computed using a least-squares regression method. The analyte quantity in the sample corresponds to the intercept-to-slope ratio, denoted x_0. When the dilution produced by additions cannot be neglected, it is necessary to take it into account. Let A_0 be the peak area of the analyte to be determined on the chromatogram obtained with the unknown sample, A_i be the peak area of the same analyte on the chromatogram obtained after the ith addition, x_i be total quantity of analyte added after the ith addition, and f_i be the dilution factor after the ith addition. Then, a line of the areas corrected by the dilution factor versus the amount of analyte added is plotted (Fig. 1). The expression of the corrected area A_i^* is

$$A_i^* = A_0 + (A_i - f_i A_0) \tag{1}$$

If, on the chromatogram, there is another peak other than that of the analyte, well resolved, but close enough, it can be used as a tracer to evaluate f_i. Let A_t be the peak area of the tracer on the chromatogram obtained with the unknown sample and A_{ti} the peak area of tracer on the chromatogram obtained after the ith addition of pure analyte:

$$f_i = \frac{A_{ti}}{A_t} \tag{2}$$

Otherwise, the dilution factor f_i must be evaluated in another way.

Then, the parameters of the calibration curve are computed by mean of a linear least-squares regression:

$$A_i^* = a + bx \tag{3}$$

The result is given by

$$x_0 = \left(\frac{a}{b}\right) \tag{4}$$

The method of standard additions would give the true result if there were no experimental errors. Practically, this is never the case and, so, a weakness of the method appears concerning the reliability of the provided result (unknown concentration). Effectively, the use of extrapolation of a calibration curve is always less conformable and reliable than interpolation, especially concerning the errors on predicted values [2–4]. The problem is even more explicit when the result is expressed along with its confidence interval. Confidence curves of the regression line must, thus, be used. Their expression is

$$A_i^* = a + bx \pm \hat{\sigma}_{y/x} t_{\alpha,(n-2)} \left(\frac{1}{n} + \frac{(x - \bar{x})^2}{\sum_i (x_i - \bar{x})^2}\right)^{1/2} \tag{5}$$

In Eq. (5), $\hat{\sigma}_{y/x}$ represents the estimated residual variance of the regression, $t_{\alpha,(n-2)}$ is the limit value that a Student's variable with $n - 2$ degrees of freedom has α chances out of 100 not to exceed in module, and n is the number of points used. Confidence limits of the result correspond to the zero of Eq. (5). No exact analytical solutions are available, but either it is possible to solve it numerically or approximate solutions can be used. An approximate expression of the standard deviation of the result is

$$\hat{\sigma}_{x_0} \cong \frac{\hat{\sigma}_{y/x}}{b} \left(\frac{1}{n} + \frac{\bar{y}^2}{b^2 \sum_i (x_i - \bar{x})^2}\right)^{1/2} \tag{6}$$

This leads to the following confidence interval (CI) for the unknown concentration:

$$\text{CI} = \lfloor x_0 - t_{\alpha,(n-2)} \hat{\sigma}_{x_0}, x_0 + t_{\alpha,(n-2)} \hat{\sigma}_{x_0} \rfloor \tag{7}$$

To avoid having an excessively large interval, it is strongly advised to take enough points in a relatively widespread range. A compromise must be found for the spacing out of the points. It must be maximum to decrease the error on predicted values, but all the points must, nevertheless, belong to the linear range. An example of an application describing the analysis of 5-hydroxyindolacetic acid in human cerebrospinal fluid is given in Ref. 1.

However, the method of standard addition is not only useful to quantify an analyte present in a matrix, it can

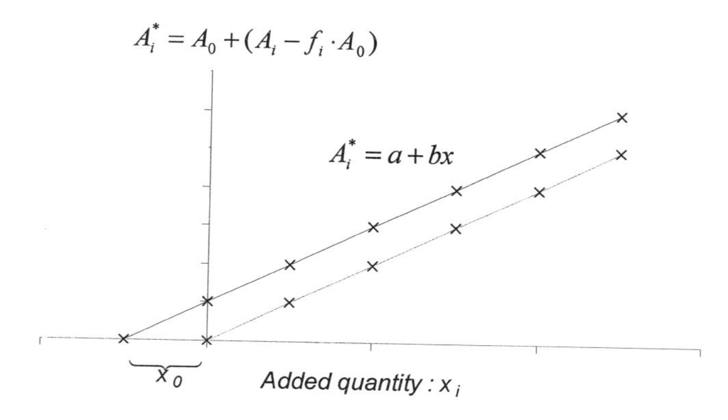

Fig. 1 Visualization of the standard addition methodology. The dotted line represents the calibration line obtained when the analyte of interest is not present in the matrix.

also be employed to check if the matrix introduces any proportional error. Similar additions of analyte are made to a blank matrix (the dotted line in Fig. 1) and if there is no proportional error, then the slopes of both regression lines are equal. Therefore, the slope equality has to be tested through a convenient statistical test.

To conclude, the method of standard additions is a powerful method that enables to quantify an analyte present in a matrix susceptible to modify its behavior. Nevertheless, this method is somewhat tedious, because it requires many preparations and injections to obtain enough points for a sufficient reliability.

References

1. L. R. Snyder, and J. J. Kirkland, J. L. Glajch, *Practical HPLC Method Development*, 2nd ed., John Wiley & Sons, New York, 1997. Chap. 14.
2. Commissariat à l'énergie atomique, *Statistique appliquée à l'exploitation des mesures, Tome I & Tome II*, Masson, Paris, 1978, pp. 139–141.
3. D. L. Massart, B. G. M. Vandeginste, S. N. Deming, Y. Michotte, and L. Kaufman, *Chemometrics, a Textbook*, Elsevier, Amsterdam, 1988, pp. 34, 117, 119.
4. J. C. Miller and J. N. Miller, *Statistics for Analytical Chemistry*, 2nd ed., Ellis Horwood, Chichester, 1988, pp. 103, 117–120.

Suggested Further Reading

Snyder, L. R. and J. J. Kirkland, *Introduction to Modern Liquid Chromatography*, 2nd ed., John Wiley & Sons, New York, 1974, p. 571.

J. Vial
A. Jardy

Quantitative Structure–Retention Relationship in Thin-Layer Chromatography

The retention behavior of a compound in a chromatographic system is governed by three global factors: stationary phase–mobile phase combination, experimental conditions, and the compound's structure. In thin-layer chromatography (TLC), there is a huge choice of combinations: between stationary phases, a much greater choice of mobile-phase components, and less complex experimental conditions. It follows that an optimization strategy is necessary in order to come closer to the wanted separation. Keeping in mind that, for example, with an experimental design, 2^n experiments are necessary, where n is the number of factors, and one has to accept several factors a priori before starting the optimization. Depending on the result achieved, the separation is accepted as satisfactory or the optimization efforts continue. Once the analytical method is established (validated), it is available for application. The new problem became the reference substances necessary for system suitability verification and for identification. Although two difficult-to-separate compounds are enough for the verification, many reference materials are necessary for the identification of all compounds of interest. One of the steps helping to overcome this problem is the so-called quantitative structure–retention relationship (QSRR). It is assumed, in QSRR, that the first two global factors (stationary phase–mobile phase combination and experimental conditions) are already established and the obtained retention depends only on the compound's structure.

There is a great number of publications on QSRR in TLC (e.g., Refs. 1–3). The protocol to be followed in

QSRR calculations has, in general, the following steps: (1) composition of the experimental data set, (2) molecular structure entry, (3) structure descriptor calculation, (4) regression between experimental data and descriptors.

The compound's structure is entered using, typically, computer graphics. Each structure is optimized by molecular mechanics, which is followed by molecular orbital calculations. Topological indices, electronic parameters, physicochemical properties, indicator variables, and so forth are used as molecular structure descriptors.

The existing QSRR calculation methods are limited to the ideal case of an isolated molecule; that is, in all cases, the descriptor values are calculated at minimum molecular energy. Thus, it is accepted a priori that the structure of the analyte molecule is rigid enough to maintain its three-dimensional (3D) structure after contacts with the stationary phase. Although such assumption is disputable, this practice continues, because of the lack of knowledge of how the solute molecule is located on the stationary phase, lack of identity of chromatographic contact regions, and the impossibility of calculating the interaction forces developed between them.

To move aside from this impediment, the following have been realized:

- The interaction forces between the solute molecule and chromatographic phases during the chromatographic process release energy higher than the rotational barrier of solute bonds.
- Due to this excess of energy, the solute 3D structure undergoes changes that allow its better location on stationary phase.
- Predicting the exact 3D structure, as it is sorbed on the surface of the stationary phase, is impossible, but presuming several energetically possible structures for one and the same structure allows us to enter into Step 2 several descriptor values for one and the same analyte and use them further in Step 3. It is assumed that one of the calculated descriptor's values could give the "searched best fit" between the experimental and calculated retention values.

If the descriptors used in the most accurate predictive equation are calculated at the minimum heat of formation (CHF), then the analyte has been sorbed as a rigid molecule. If greater accuracy is achieved at higher CHF, the starting 3D structure has been changed during the chromatographic process.

This assumption will be tested first with the retention data for R_m of 9 benzene derivatives (predominantly rigid structures) and, next, for the R_f of 15 benzodiazepines (versatile structures with many σ bonds) separated on silica gel. The data for benzene derivatives were taken from

Rotamer "a" or Rotamer "b"

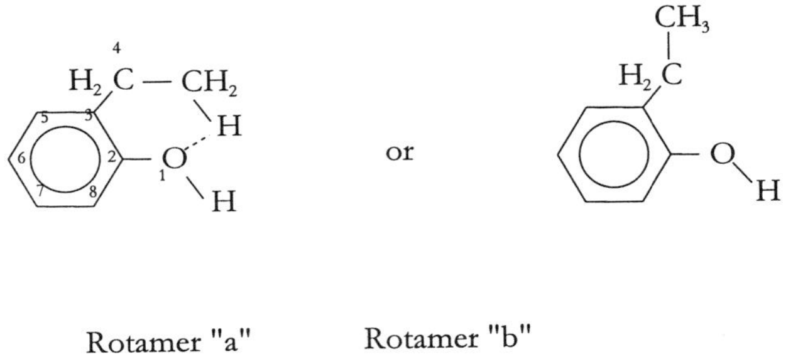

Fig. 1 Rotamer structures.

the literature [4], where several mobile phases had been used. Two mobile phases with different polarities have been chosen.

In the first case, only two from all nine studied benzene derivatives have rotamers (see Fig. 1). From all tested combinations among the calculated descriptors, only one answers both to the accuracy and statistical requirements:

$$R_M = -5.50 \pm 0.22 + (24.57 \pm 0.88)\mathrm{DD}[1] + (4.19 \pm 0.19)Q[6] + (1.26 \pm 0.19)Q[5] \quad (1)$$

for mobile phase I with correlation coefficient $r = 0.9966$, variance $v = 1.96 \times 10^{-4}$, and $F = 483$, and

$$R_M = -7.04 \pm 0.52 + (29.84 \pm 2.08)\mathrm{DD}[1] + (4.51 \pm 0.45)Q[6] + (1.88 \pm 0.44)Q[5] \quad (2)$$

for mobile phase II with correlation coefficient $r = 0.9870$, variance $v = 1.1 \times 10^{-3}$, and $F = 127$. DD[1] stands for the donor delocalization energy of the —O— atom, and $Q[5]$ and $Q[6]$ stand for the charges of the fifth and sixth atoms, respectively. The increased influence of the —O— atom with the most polar mobile phase suggests checking another quantum-chemical local descriptor for the —O— atom. Equation (3) gives a statistically better prediction of the experimental values of R_M:

$$R_M = -0.93 \pm 0.09 + (10.73 \pm 0.55)\mathrm{AD}[1] + (3.00 \pm 0.31)Q[6] + (1.65 \pm 0.29)Q[5] \quad (3)$$

with $r = 0.9913$, variance $v = 6.4 \times 10^{-4}$, and $F = 267$. AD[1] stands for the acceptor delocalization energy of the —O— atom. The better-fitted rotamers from *o*-ethylphenol and *o*-chlorophenol are with the higher heat of formation. In the above-given final regression, only these rotamers have been included.

The calculated retention is compared with the experimental value in Table 1. From a statistical point of view, the improvement when using the rotamer with better fit is insignificant ($F_{9,9} = 1.26$ and 1.07) and the assumption could be considered as disputable. This can be explained

Table 1 Comparison of Experimentally Obtained Data with Two Different Mobile Phases R_M with R_M Values Calculated According to Eqs. (1) or (2), Respectively

No.	Compound	R_M exp., phase I[a]	R_M calc., Eq. (1) phase I	R_M exp., phase II[b]	R_M calc., Eq. (2) phase II	R_M calc., Eq. (3) phase II
1	*o*-Ethylphenol rotamer a	−0.31	−0.31	−0.50	−0.50	−0.48
2	*o*-Ethylphenol rotamer b	−0.31	−0.33	−0.50	−0.52	−0.47
3	*m*-Ethylphenol	−0.16	−0.18	−0.35	−0.40	−0.38
4	*p*-Ethylphenol	−0.14	−0.13	−0.31	−0.31	−0.30
5	*o*-Chlorophenol rotamer a	−0.23	−0.23	−0.43	−0.41	−0.44
6	*o*-Chlorophenol rotamer b	−0.23	−0.21	−0.43	−0.40	−0.46
7	*m*-Chlorophenol	−0.12	−0.11	−0.37	−0.32	−0.35
8	*p*-Chlorophenol	−0.07	−0.07	−0.25	−0.25	−0.25
9	*o*-Toluidine	0.07	0.07	−0.03	−0.03	−0.03
10	*m*-Toluidine	0.16	0.16	0.05	0.04	−0.05
11	*p*-Toluidine	0.27	0.26	0.16	0.17	−0.16

[a] Silica gel plates 60 F_{254} with n-C_7H_{16}–C_6H_6–diethylether = 1:1:1.
[b] Same plate with n-C_7H_{16}–C_6H_6–diethylether = 1:2:2.

by the small number of compounds that are able to rotate around the σ-bond in the total matrix. Independent of the reliability of the assumption, a beneficial conclusion can be drawn from Eqs. (1)–(3). They demonstrate that the retentions of the studied benzene derivatives are governed in the studied cases only by local descriptors and predominately by the local descriptors for the —O— atom in the hydroxyl group. The connectivity index used in many studies (e.g., Ref. 5) showed, in the studied case, a correlation coefficient of only 0.285.

The assumption is undoubtedly verified with the retention data for R_f of 15 benzodiazepines separated on silica. The data were taken from the literature [6] because the authors had corrected, graphically, the R_f values to those obtained with reference materials and, hence, more exact values can be expected.

Three statistically equivalent equations have been obtained. One is presented as follows:

$$R_f = -234.9 \pm 43.2 + (47.27 \pm 4.50)\text{EN} + (2918 \pm 350)\text{DD}[5] + (10.7 \pm 1.8)\ \text{InVNH} \qquad (4)$$

with correlation coefficient $r = 0.949$, variance $v = 13$, and $F = 69$. EN represents the electronegativity, DD[5] is the donor delocalization energy of atom 5 (see Fig. 2), and InVNH is an arbitrarily chosen indicator variable, accounting for the presence (value 1) or absence (value 0) of a substituent at the azepine N atom. All parameters have statistically insignificant interrelations, r_i. The R_f values calculated with Eq. (4) are given in column 4 of Table 2 and can be compared with the experimental results given in column 3. Although some of the chosen rotamers are identical to the conformers at the lowest molecular energy, the variance obtained when the descriptors were calculated for all molecules at their minimum calculated heats of formation (CHF_{min}) are statistically higher (32.7). Their R_f values are given in column 7 of Table 2. Taking into account the criterion relevant to acceptance of the hypothesis, it is apparent that the accuracy, if a flexible structure approach is assumed, is higher.

Another conclusion from Eq. (4) is that benzodiazepine's retention, again, is governed by local descriptors (about 65% contribution of DD[5]) rather than by global molecular properties (about 35% from EN). The general conclusion is that it seems possible to increase the accuracy of QSRR calculations in TLC, assuming that flexible 3D analyte structures undergo some changes during the

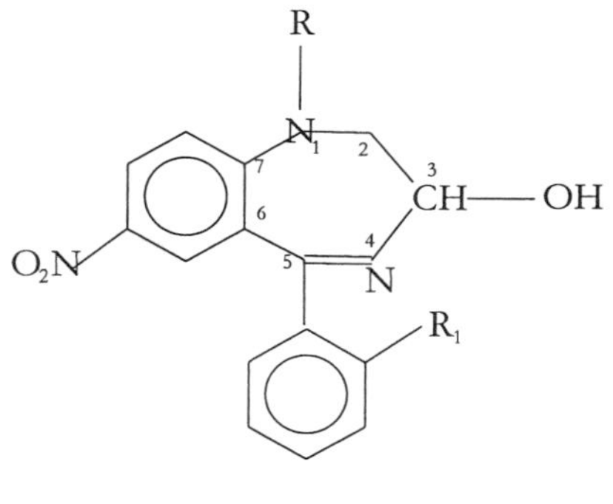

R

N 1 2 3 CH—OH 7 6 5 4 N O_2N R_1

R = H or substituent

Fig. 2 Diazepine structures.

Table 2 Experimental and Calculated R_f Values Obtained with Two Mobile Phases: Cyclohexane–Toluene–Diethylamine (m.ph.I) and Chloroform–Methanol (m.ph.II)

		m.ph. I		m.ph. II		R_f	CHF_{min}
No. (1)	-azepam (2)	$R_{f\,exp}$ (3)	$R_{f\,calc}$ (4)	$R_{f\,exp}$ (5)	$R_{f\,calc}$ (6)	m.ph. I (7)	m.ph. II (8)
1	Nitra-	0	0.1	36	39.5	0.1	41.9
2	Oxa-	0	−0.7	40	40.0	−2.8	43.4
3	Lor-	1	−0.7	36	34.5	−2.7	37.8
4	Nord-	4	10.4	55	51.8	14.0	63.8
5	Lormet-	6	6.6	67	64.3	11.2	75.7
6	Tem-	8	6.2	59	63.2	13.4	67.0
7	Flunitra-	10	11.5	72	66.0	11.7	66.0
8	Nimet-	12	12.7	77	73.4	12.9	61.7
9	Cam-	13	13.2	79	80.3	10.3	74.8
10	Hal-	18	17.9	76	79.2	17.3	79.6
11	Flur-	30	34.0	48	49.8	34.6	46.2
12	Pin-	31	27.5	79	79.3	24.4	72.3
13	Tetra-	34	33.5	75	73.8	32.7	66.3
14	Pra-	36	28.5	74	69.4	26.8	79.3
15	Med-	40	42.4	74	79.1	42.8	79.1

process of adsorption onto silica gel. As a result, a better fit between the calculated and experimental retention data can be achieved if several energetically possible structures for one and the same analyte are presumed.

References

1. L. B. Kier, L. H. Hall, M. J. Murrey, and M. Randic, *J. Pharm. Sci. 64*: 19–25 (1975).
2. J. Sherma and B. Fried, *Handbook of Thin-Layer Chromatography*, Marcel Dekker, Inc., New York, 1989.
3. W. Kiridena, *J. Chromatogr. 802*: 335–347 (1998).
4. W. Wardas and A. Pyka, *J. Planar Chromatogr. 3*: 425–428 (1990).
5. A. Pyka, *J. Planar Chromatogr. 9*: 181–184 (1996).
6. M. Japp, K. Garthwaite, A. Geeson, and M. Osselton, *J. Chromatogr. 439*: 317–324 (1988).

N. Dimov

R

R_f

The R_f value is the fundamental parameter in planar chromatography which describes, numerically, the position of a spot on the developed chromatogram.

The R_f Value in Linear Chromatography

The method for determining R_f values is based on the measurement of two lengths in a thin-layer chromatogram and the calculation of their ratio:

$$R_f = \frac{\text{Distance of spot center from start}}{\text{Distance of solvent front from start}}$$

where "start" is the sample application point.

R_f values are between 0 and 1 (solute remains at start or runs with the solvent front, respectively) and the maximum number of significant figures after the decimal point is currently two. R_f values are often multiplied by a factor of 100 (hR_f).

R_f values can be disturbed by side effects or demixing of the multicomponent solvent used. In order to obtain reproducible R_f values, much attention must be placed on the reproducibility of the system.

The R_f Value in Circular Chromatography

Linear R_f values can be transferred to the circular technique with the equation

$$R_{f\text{lin}} = R_{f\text{circ}}^2$$

This equation holds only if the starting line is close to the center of the layer. Circular R_f values are higher than the linear ones, with the exception of $R_f = 0$ and $R_f = 1$. The increase in the circular R_f values is greater in the lower range.

Definition of Thermodynamic R_f Value

According to the Martin–Synge model of partition chromatography [1,2], the thermodynamic R_f value (R_f'), based on the chromatographic equilibrium process of solute distribution between the mobile and stationary phase, can be expressed as the fraction of the relative time spent by a solute molecule in the mobile phase (A) or fraction of solute molecules in the mobile phase (B):

$$R_f' = \underset{\text{(A)}}{\frac{t_m}{t_m + t_s}} = \underset{\text{(B)}}{\frac{n_m}{n_m + n_s}}$$

where the subscripts m and s refer to mobile and stationary phase, respectively.

Because the fractions of solute molecules and respective mole numbers are identical, it is possible to achieve the fundamental relation (1) connecting R_f' value with the distribution coefficient $K_d = C_s/C_m$ and the phase ratio V_s/V_m:

$$R_f' = \frac{m_m}{m_m + m_s} = \frac{C_m V_m}{C_m V_m + C_s V_s} = \frac{1}{1 + K_d(V_s/V_m)} \qquad (1)$$

In the Martin–Synge relationship (1), C_m and C_s are molar concentration of the solute in the mobile and stationary phase, respectively, and V_s and V_m are the volumes of these two phases. V_s/V_m is numerically equal to A_s/A_m, the ratio of the phase cross section normal to the direction of the solvent flow, which better describes the local conditions in thin-layer chromatography. The validity of the equation is limited because the amount of solvent on the layer decrease going toward the solvent front and, therefore, the phase ratio changes.

R'_f values are generally higher than R_f values and can be related to each other by the following experimental relationship

$$R'_f = \xi R_f$$

where ξ is the disturbing factor ($1 \leq \xi \leq 1.6$) [3]. This relationship usually holds in the R_f region up to 0.7 because the greatest changes of the solvent front are observed at the end of the chromatogram.

The R_{st} Number

Relative R_f values, generally called R_{st} or R_x values, can also be used but are inadequate to render R_f values independent of uncontrolled parameters, because they are dependent of the phase ratio:

$$R_{st} = \frac{(R_f)_i}{(R_f)_{st}} = \frac{\text{Migration distance of substance } i}{\text{Migration distance of reference substance, st}}$$

These values can be greater than 1.

References

1. A. J. P. Martin and R. L. M. Synge, *Biochem. J. 35*: 1358 (1941).
2. A. J. P. Martin, *Biochem. J. 50*: 679 (1952).
3. F. Geiss, Foundamentals of Thin-Layer Chromatography (Planar Chromatography), Alfred Hüthig Verlag, Heidelberg, 1987, pp. 87–114.

Luciano Lepri
Alessandra Cincinelli

Radiochemical Detection

Radiochemical detectors are devices that allow the measurement of low-energy γ-rays or β particles emitted by radioisotopes. Substances called scintillators absorb the energy that is produced by the radioactive decay and transform it into light. The light from the scintillator, the intensity of which is directly proportional to the energy of the radioisotope emission, can be detected by a photomultiplier to provide an electrically countable pulse. Radiochemical detectors have been developed for various chromatographic techniques, the most common of which is high-performance liquid chromatography (HPLC). In HPLC, the detection of radioisotopes in a column eluate has allowed quantitation of discrete radiolabeled peaks in real time and in a generally nondestructively manner.

Gamma-emitters, such as iodine-125, which give off energy in the form of photons rather than particles, are more strongly penetrating than β-emitters. A radiochemical detector for γ-rays is generally composed of a cylindrical block of specially activated sodium iodide enclosed in a thin aluminum shell. One face of the block is optically coupled to a single photomultiplier. The eluate from the HPLC column passes through a coil or U-tube, which is placed in the well of the block. As the eluate passes through the coil, γ-radiation is stopped by the very dense sodium iodide, producing light that can be measured by the photomultiplier.

Beta-emitters such as tritium (H-3) or carbon-14 (C-14) are the most commonly used radioisotopes in drug metabolism, agricultural metabolism, and toxicology studies. In HPLC, the radiochemical detector can be off-line or on-line. Off-line detection requires coupling the chromatograph to a fraction collector. The collected fractions are combined with a suitable liquid scintillation cocktail and then counted by a liquid scintillation counter. This method allows for the control of parameters such as counting time to improve sensitivity, but it suffers from being labor and time intensive. On-line or flow-through radiochemical detectors can be homogeneous or heterogeneous. In homogeneous flow through detectors, a scintillation cocktail is added to the column eluate prior to detection in a liquid flow cell. The energy of decay from the β radionuclide is transferred to the scintillation cocktail via a solvent molecule. The liquid flow cell offers the highest sensitivity in on-line detectors but precludes recovery of the entire sample and requires an additional pump for delivery of the scintillation cocktail. In heterogeneous counting, the column eluate passes directly through a flow cell packed with a suitable solid scintillator such as yttrium silicate, calcium fluoride, or lithium glass. Use of the solid flow cell involves no additional costs for the scintillation cocktail and simplifies waste disposal. However, the solid cells have a lower counting efficiency, particularly with tritium, and can suffer from high backgrounds due to contamination. In both homogeneous and heterogeneous counting, light energy is detected by a pair of photomultiplier tubes located on either side of the flow cell, which is most often composed of a flat coil of thin-wall tubing situated in a transparent case. The photomultiplier tubes are located in fairly close proximity to each other (typically less than $\frac{1}{2}$ in. apart). Most commercially available radiochemical HPLC detectors use coincidence electronics in order to reduce noise pulses. With a coincidence counter, events are only recorded when both photomultiplier tubes are stimulated by light to give a pulse output within 20–50 ns of each other. A multichannel pulse-height analyzer sorts the outputs from the coincidence counter. A computer is then generally used to process data from the radiochemical detector as well as other on-line detectors and the results are presented either on a monitor or printer.

Unlike other chromatographic measurements such as ultraviolet (UV) absorption, the measurement and quantitation of radioactivity are based on time. Although a known percentage of a radioisotope will decay over a relatively long time, within that time period, the instantaneous rate of decay is not known. Because radioactive decay is a statistical process, the quality of a radioactivity measurement improves as a function of time. In other words, the longer the sample is counted, the more accurate (or statistically relevant) the measurement of the radioactivity. With off-line radiochemical detection, even very small peaks can be counted for a sufficiently long time to allow for a statistically relevant measurement. With on-line detection, the total flow rate and the cell volume fix the counting time. On-line detection of low-level peaks can be improved by either a slower flow rate or by a larger flow cell. For solid flow cells, the total flow rate is equal to the mobile-phase flow rate. For liquid flow cells, the total flow rate is equal to the mobile-phase flow rate plus the flow rate of the liquid scintillator. The latter is frequently expressed as a ratio of scintillator to mobile phase. As compared to other HPLC detectors, the volume of the flow cell is relatively large in the radiochemical detector. Flow cells typically range from 250 to 500 μL for solid cells and from 1000 to 2000 μL for liquid cells. Upward limits on the size of the flow cell are governed by the possibility of more than one peak being present in the cell at any one given time. However, assuming sufficient resolution, a larger flow cell should allow for a longer counting time and, hence, a more accurate measurement of radioactivity in a given peak. The actual amount of time that the column eluate will remain in the vicinity of the photomultiplier tubes is defined as the residence time. The residence time (R), in seconds, can be calculated as follows:

$$R = \frac{V}{F} \times 60$$

where V is the cell volume in microliters and F is the total flow rate in milliliters per minute.

The counts per minute (cpm) can then be calculated for each peak in the sample as follows:

$$\text{cpm} = \frac{N}{R}$$

where N is the total net counts observed in the peak after background subtraction and R is the residence time in minutes.

Disintegrations per minute (dpm) are equal to cpm in the rare cases where the counting efficiency is 100%. Otherwise, dpm can be calculated as follows:

$$\text{dpm} = \frac{\text{cpm}}{E}$$

where E is the counting efficiency expressed as a percentage.

The efficiency of a radiochemical detector can be defined as the number of counts that are detected in the peak after background subtraction divided by the total number of radioactive events that actually occurred in the sample during the counting period. The approximate counting efficiencies that can be achieved with solid flow cells are fairly low for tritium (up to 10%) and significantly better for C-14 (up to 90%). With liquid flow cells, the counting efficiencies are greater than 55% for tritium and greater than 95% for C-14. Under isocratic conditions, the counting efficiency remains constant and can be readily determined by injecting a known amount of activity into an HPLC system, collecting the resultant peak as a single entity and counting it in a liquid scintillation counter. The number of disintegrations in the recovered peak is then compared to the calibrated activity of the standard. With gradient elution, the counting efficiency will generally change as the mobile-phase composition changes. In this situation, an efficiency calibration or quench curve can be obtained by injecting a constant known level of radioactivity during an otherwise blank gradient run. The calibration curve is then used to correct subsequent HPLC runs.

The minimum amount of radioactivity that can be detected by a flow-through radiochemical detector is a subject of continuing debate. It is generally accepted that a fairly sharp peak that contains counts that are at least twice background can be detected. One formula for calculating the minimum detectable activity (MDA) is given by

$$\mathrm{MDA} = \frac{BW}{RE}$$

where B is the background count in cpm, W is the base width of the peak in minutes, R is the residence time in minutes, and E is the counting efficiency expressed as a percentage.

In practical terms, for a 500-dpm peak with a base width of 20 s, a counting efficiency of 70% and a residence time of 12 s, we would observe 70 counts, which would be spread out over the peak width, plus the residence time (a total of 32 s). For a detector that has a 10-cpm background, we would observe 5.3 counts over the same time period of 32 s. The ratio of the observed peak counts to background would be 13:1 and the peak should be clearly visible. However, under the same conditions, a 50-dpm peak would have a count to background ratio of only 1.3:1 and it is doubtful that this peak could be observed in on-line counting. In the latter situation, off-line radiochemical detection may be used to clarify the presence or absence of such a low-level peak.

Suggested Further Reading

INUS Home Page. http://www.inus.com (last accessed August 1999).

Parvez, H., A. R. Reich, S. Lucas-Reich, and S. Parvez (eds.), *Progress in HPLC Volume 3: Flow Through Radioactivity Detection in HPLC*. VSP, Ultrecht, 1988.

Scott, R. P. W., *Chromatographic Detectors*, Marcel Dekker, Inc., New York, 1996, pp. 315–327.

Eileen Kennedy

Radius of Gyration Measurement by GPC–SEC

Fundamental properties of macromolecules, such as viscoelasticity and flow behavior, primarily depend on the dimensions and the conformations of macromolecules. Primary biological functions substantially depend on the dimensions of natural macromolecules such as proteins and enzymes. Hence, a primary method to understand the physical properties of macromolecules, synthetic and natural, involves determination of the dimension as a function of the molar mass. A convenient method to fractionate macromolecules is the size-exclusion chromatography (SEC) technique. SEC fractionation is rapid and efficient and requires small amounts of the polymeric sample. A method to measure the dimension of the macromolecules as a function of the molar mass is an on-line technique to a SEC chromatographic system.

The simplest parameter to quantify the dimension of a macromolecule is the end-to-end distance. The end-to-end distance is the spatial distance between the end groups of a linear chain. Unfortunately, there are no experimental direct methods to measure the end-to-end distance of a macromolecule. Furthermore, the end-to-end distance has no meaning for a chain without end groups (ring) and with many end groups (branched). Instead, the dimension of every type of macromolecule can be always quantified by the gyration radius $\langle s^2\rangle^{1/2}$.

There are several experimental methods to measure $\langle s^2\rangle^{1/2}$. Macromolecules can be represented as N segments (monomers, groups) of mass m_i and distance r_i from the center of gravity. The gyration radius $\langle s^2\rangle^{1/2}$ is defined as the mass average of r_i:

$$\langle s^2\rangle^{1/2} = \left(\frac{\langle \sum m_i r_i^2\rangle}{\sum m_i}\right)^{1/2} \tag{1}$$

Actually, the term "gyration radius" is a misnomer; it originates from the kinematics, but is very popular. More correctly, it should be called "root mean square radius." Nonrigid macromolecules possess conformational mobility; hence, their dimension requires an additional average, $\langle\,\rangle$, over all the conformations. The dimension of a macromolecule also depends on its long-range interactions, excluded volume, and the polymer–solvent interactions. Therefore, $\langle s_0^2\rangle^{1/2}$ denotes the dimension of the macromolecules in the unperturbed ideal state and $\langle s^2\rangle^{1/2}$ denotes the expanse dimension as a consequence of the excluded volume. Finally, macromolecules are generally polydisperse and, by analogy to the molar mass, $\langle s^2\rangle_n^{1/2}$ denotes the numeric average, $\langle s^2\rangle_w^{1/2}$ denotes the weight average, and $\langle s^2\rangle_z^{1/2}$ denotes the z average [1].

Size-exclusion chromatography fractionation is steric, that is, dimensional. In theory, a SEC system could be calibrated by means of some appropriate standards of known dimensions and, in this way, to measure $\langle s^2\rangle^{1/2}$. SEC fractionation depends on the hydrodynamic radius R_H of the macromolecules: $R_H \propto M[\eta]$ where $[\eta]$ is the intrinsic viscosity. $\langle s^2\rangle^{1/2}$ and R_H are two different parameters. $\langle s^2\rangle^{1/2}$ is an equilibrium parameter; R_H is a dynamic parameter and depends on the method by which it is obtained. R_H becomes the Stokes radius R_D in diffusion measurements and the Einstein radius R_η in viscosity measurements. Because SEC fractionation depends on R_H, the method is not appropriate for a direct measure of $\langle s^2\rangle^{1/2}$. Convenient experimental methods to measure $\langle s^2\rangle^{1/2}$ are scattering techniques.

The $\langle s^2\rangle^{1/2}$ value of macromolecules usually ranges from 2 to 3 nm (globular proteins) to several hundred nanometers (particles). The range of $\langle s^2\rangle^{1/2}$ values determines the more appropriate scattering technique. Smaller dimensions of the molecules require shorter wavelengths of the radiation. Light-scattering (LS) wavelengths range from approximately 400 to 700 nm. The minimum $\langle s^2\rangle^{1/2}$ value that could be measured by LS is 8–10 nm and 5–6 nm in the more favorable case (shorter wavelength). If the dimension of the molecules is small, we need to use other scattering techniques such as x-ray or neutrons.

A modern LS photometer uses coherent light (i.e., laser) with vertical polarization. In our specific case, LS concerns the interaction of the light with matter with a solution of macromolecules. In an LS experiment, we measure the intensity of the scattering. Furthermore, to measure the $\langle s^2\rangle^{1/2}$ value, we need the angular variation of the scattering. Because we want to measure $\langle s^2\rangle^{1/2}$ on-line to a SEC system, after the fractionation in the columns, we need to measure the angular variation of the scattering instantaneously (i.e., simultaneously) over a wide range of angles. In this case, we need an on-line multiangle light-scattering (MALS) detector.

Following Zimm [2,3], the intensity of the scattering of a solution of macromolecules is in relation to the molar mass of the sample by

$$\frac{Kc}{\Delta R(\theta)} = \frac{1}{MP(\theta)} + 2A_2c + \cdots \tag{2}$$

where $\Delta R(\theta)$ is the scattering excess (Rayleigh factor) at angle θ of the solution with regard to the pure solvent, θ is the angle between the incident light and the detector, M is the molar mass, c is the concentration, A_2 is the second virial coefficient, and K is an optical constant [$K = (4\pi^2 n_0^2 (dn/dc)^2)/(N_a\lambda_0^4)$, where n_0 is the refractive index of the solvent, dn/dc is the refractive index increment of the polymer, λ_0 is the wavelength of the light *in vacuo*, and N_a is Avogadro's number].

The parameter of interest for the measurement of $\langle s^2\rangle^{1/2}$ is the form factor $P(\theta)$. A macromolecule may not be considered as a single point of scattering. In this case, the light scattered from two different points of the same macromolecule will not be in phase: destructive interference. The intensity of the scattering in the presence of the destructive interference is lower for large molecules. The interference depends on the angle of measure of the intensity of the scattering. The interference is absent at 0° angle and highest at 180°. The interference depends on the shape and on the dimension of the molecules. Therefore, there has been introduced a form factor $P(\theta)$ which quantifies the interference. $P(\theta)$ is defined as the ratio between the Rayleigh factor in the presence of interference ($\theta > 0°$) and the Rayleigh factor in the absence of interference ($\theta = 0°$). Thus, by definition,

$$P(\theta) \equiv \frac{R(\theta)}{R(\theta = 0°)} \tag{3}$$

Equation (3) states that $P(\theta) = 1$ for $\theta = 0°$, independent of the dimension of the molecules, and $P(\theta) < 1$ for $\theta > 0°$ when the dimension of the molecules is comparable with the wavelength λ. Fortunately, $P(\theta)$ is a useful method for measuring the dimension of the molecules $\langle s^2\rangle^{1/2}$. Debye [4] showed that the $P(\theta)$ could be expressed independently of the shape and conformation of the macromolecules. Considering the reciprocal of $P(\theta)$ $[P(\theta)^{-1}]$, Debye obtained the following equation:

$$P(\theta)^{-1} = 1 + \tfrac{1}{3}\mu^2\langle s^2\rangle \tag{4}$$

where $\mu = 4\pi/\lambda \sin(\theta/2)$, $\lambda = \lambda_0/n_0$ is the wavelength of the light in the medium. Equation (4) is valid in the limit $\mu^2\langle s^2\rangle/3 \ll 1$. From the initial slope of the $P(\theta)$ versus $\sin^2(\theta/2)$ plot, we can estimate the gyration radius value $\langle s^2\rangle^{1/2}$.

To estimate M and $\langle s^2\rangle^{1/2}$ values, Eqs. (2) and (4), we need to extrapolate to infinite dilution: $c = 0$. When the MALS detector is used as an on-line detector in a SEC system to estimate the $\langle s^2\rangle^{1/2}$ of the molecules, we have only a concentration. In a SEC fractionation, it is generally assumed that each instant, slice, contains molecules homogeneous in molar mass and in dimension. In this case, considering the very low concentrations of the macromolecules that elute from the SEC columns, the term $2A_2c$ of Eq. (2) is ignored. However, the exact concentration of each slice has to be known. Thus, the measure of M and $\langle s^2\rangle^{1/2}$ values by a MALS detector requires a concentration detector also, usually a differential refractometer or an ultraviolet (UV) photometer. Combining Eq. (2), deprived of the $2A_2c$ term, with Eq. (4), we obtain

$$\frac{Kc}{\Delta R(\theta)} = \frac{1}{M} + \frac{16\pi^2\langle s^2\rangle}{3\lambda^2 M}\sin^2\left(\frac{\theta}{2}\right) \tag{5}$$

Using an on-line dual-detector SEC system, MALS, and concentration from a linear regression of $Kc/\Delta R(\theta)$ on $\sin^2(\theta/2)$, we obtain the molar mass M from the intercept and the dimension of the molecules $\langle s^2\rangle^{1/2}$ from the slope.

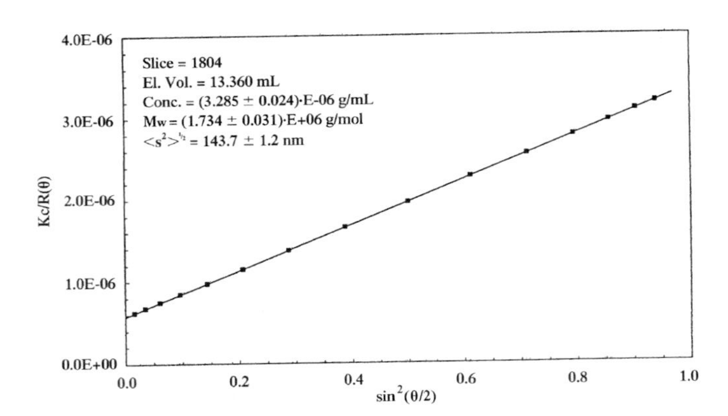

Fig. 1 $Kc/\Delta R(\theta)$ versus $\sin^2(\theta/2)$ plot of a single slice from a hyaluronic acid sample.

If the macromolecules are polydisperse, we obtain [3] the averages M_w and $\langle s^2\rangle_z^{1/2}$. Figure 1 shows the $Kc/\Delta R(\theta)$ versus $\sin^2(\theta/2)$ plot of a single slice from a polysaccharide, hyaluronic acid, sample. Figure 1 shows the classical three parameters that could be obtained at each elution volume: c_i from the concentration detector, M_i and $\langle s^2\rangle_i^{1/2}$ from the MALS detector. An accurate measure of $\langle s^2\rangle_i^{1/2}$ is not simple. The measurement requires a good signal-to-noise ratio, especially at low angles. In addition, if the dimension of the macromolecules is small, the measure of the slope of the $Kc/\Delta R(\theta)$ versus $\sin^2(\theta/2)$ plot is inaccurate. Conversely, if the dimension of the macromolecules is too large, the angular variation of the intensity of the scattering could be not linear. In this case, it is quite difficult to estimate an accurate value of the initial slope of the plot.

Because, for each fraction of the sample, the on-line MALS detector measures both M_i and $\langle s^2\rangle_i^{1/2}$ from a single sample, it is possible to obtain the $\langle s^2\rangle^{1/2} = f(M)$ power law of the polymer. The power law $\langle s^2\rangle^{1/2} = f(M)$ is a very important function for understanding the conformation (flexible coils, compact spheres, rigid rods) of the macromolecules. In fact, if the molar mass distribution of the sample is adequately broad, it is possible, from a linear regression of $\log(\langle s^2\rangle^{1/2})$ on $\log(M)$, to estimate the constants K and α of the power law $\langle s^2\rangle^{1/2} = KM^{\alpha}$. Specifically, the slope α of the power law contains fundamental information of the conformation of the macromolecule in solution.

There are several commercially available on-line MALS detectors. Wyatt Instruments (Santa Barbara, CA, U.S.A.) commercializes two MALS detectors: Dawn-DSP (18 angles) and mini-Dawn (3 angles: 45°, 90°, and 135°). Precision Detectors (Franklin, MA, U.S.A.) commercializes a double-angle LS: 15° and 90°. SEC on-line measurement of $\langle s^2\rangle^{1/2}$ could be performed by other techniques. Viscotek Co. (Houston, TX, U.S.A.) commercializes an alternative three-detector SEC system, the SEC³. The SEC³ system consists of (1) an on-line viscometer (VISC), (2) an on-line right-angle LS detector (RALS), and (3) a concentration detector. This three-detector SEC system also measures $\langle s^2\rangle^{1/2}$. More exactly, $\langle s^2\rangle^{1/2}$ is calculated, using an iterative algorithm, from the intrinsic viscosity, by VISC, and the molar mass obtained by the RALS detector. For each instant i, the algorithm is divided into three steps:

$[\eta]_i$ is measured from VISC; the molar mass M_i is calculated by RALS using Eq. (2) and $P(\theta) = 1$. In this first step, M is generally underestimated as a consequence of the destructive interference.

$\langle s^2\rangle^{1/2}$ is calculated by the Flory–Fox equation [Eq. (6)]. The Ptitsyn–Eizner equation [Eq. (7)] considers the expansion of the macromolecules due to the excluded volume. In Eqs. (6) and (7), $\Phi_0 = 2.86 \times 10^{23}$ and $\varepsilon = (2a - 1)/3$, where a is the slope of the Mark-Houwink–Sakurada equation.

Finally, $P(\theta)$ is calculated by the Debye equation [Eq. (4)]:

$$\langle s^2\rangle^{1/2} = \frac{1}{\sqrt{6}}\left(\frac{M[\eta]}{\phi(\varepsilon)}\right)^{1/3} \tag{6}$$

$$\phi(\varepsilon) = \phi_0(1 - 2.63\varepsilon + 2.86\varepsilon^2) \tag{7}$$

At this point, the algorithm restarts the first step, with $P(\theta)$ equal to the new calculated value from the previous iteration. The algorithm converges very quickly. A detailed description of the SEC³ algorithm can be found in Ref. 5.

To measure the dimension of macromolecules is a very delicate task. However, the dimension is a primary parameter needed to understand the physical properties of a macromolecule. We have described two experimental methods, MALS and SEC³, to measure $\langle s^2\rangle^{1/2}$ on-line to a SEC system.

References

1. P. J. Wyatt, *Anal. Chim. Acta 272*: 1 (1993).
2. B. H. Zimm, *J. Chem. Phys. 16*: 1093 (1948).
3. P. Kratochvil, *Classical Light Scattering from Polymer Solutions*, Elsevier, Amsterdam, 1987.
4. P. Debye, *J. Phys. Colloid Chem. 51*: 18 (1947).
5. W. W. Yau and S. W. Rementer, *J. Liquid Chromatogr. Related Technol. 13*: 627 (1990).

Raniero Mendichi

Rate Theory in Gas Chromatography

The rate theory examines the kinetics of exchange that takes place in a chromatographic system and identifies the factors that control band dispersion. The first explicit height equivalent to a theoretical plate (HETP) equation was developed by Van Deemter et al. in 1956 [1] for a packed gas chromatography (GC) column. Van Deemter et al. considered that four spreading processes were responsible for peak dispersion, namely *multi-path dispersion*, *longitudinal diffusion*, *resistance to mass transfer in the mobile phase*, and *resistance to mass transfer in the stationary phase*.

The Multipath Effect

In a packed column, the individual solute molecules will describe a tortuous path through the interstices between the particles, and some will randomly travel shorter routes than the average and some will travel longer routes. Consequently, those molecules taking the shorter paths will move ahead of the mean and those that take the longer paths lag behind the mean which will result in band dispersion. Van Deemter et al. derived the following function for the multipath variance contribution (σ_M^2) to the overall variance per unit length of the column (σ^2):

$$\sigma_M^2 = 2\lambda d_p \tag{1}$$

where d_p is the particle diameter of the packing and λ is a constant that depends on the quality of the packing.

Longitudinal Diffusion

Driven by the concentration gradient, solutes naturally diffuse when contained in a fluid. Thus, a discrete solute band will diffuse in a gas or liquid, and because the diffusion process is random, it will produce a concentration curve that is Gaussian in form. This diffusion effect occurs in the mobile phase of both packed GC and liquid chromatography (LC) columns. The longer the solute band remains in the column, the greater will be the extent of diffusion. Because the residence time of the solute in the column is inversely proportional to the mobile-phase velocity, the dispersion will also do the same. Van Deemter et al. derived the following expression for the variance contribution by longitudinal diffusion (σ_L^2) to the overall variance per unit length of the column (σ^2):

$$\sigma_L^2 = \frac{2\gamma D_m}{u} \tag{2}$$

where D_m is the diffusivity of the solute in the mobile phase, u is the linear velocity of the mobile phase, and γ is a constant that depends on the quality of the packing.

Resistance to Mass Transfer in the Mobile Phase

During migration through the column, the solute molecules are continually transferring from the mobile phase to the stationary phase and back again. This transfer process is not instantaneous; a finite time is required for the molecules to traverse (by diffusion) through the mobile phase in order to reach the interface and enter the stationary phase. Thus, those molecules close to the stationary phase enter it immediately, whereas those molecules some distance away will find their way to it some time later. However, because the mobile phase is moving, during this time interval, those molecules that remain in the mobile phase will be swept along the column and dispersed away from those molecules that were close and entered the stationary phase immediately. This dispersion is called the resistance to mass transfer in the mobile phase.

Van Deemter derived the following expression for the variance contribution by the resistance to mass transfer in the mobile phase (σ_{RM}^2) to the overall variance per unit length of the column (σ^2):

$$\sigma_{\mathrm{RM}}^2 = \frac{f_1(k')d_p^2}{D_m}u \tag{3}$$

where k' is the capacity ratio of the solute and the other symbols have the meaning previously ascribed to them.

Resistance to Mass Transfer in the Stationary Phase

Dispersion caused by the resistance to mass transfer in the stationary phase is exactly analogous to that in the mobile phase. Solute molecules close to the surface will leave the stationary phase and enter the mobile phase before those that have diffused further into the stationary phase and have a longer distance to diffuse back to the surface. Thus, as those molecules that were close to the surface will be swept along in the moving phase, they will be dispersed from those molecules still diffusing to the surface.

Van Deemter derived an expression for the variance from the resistance to mass transfer in the stationary phase (σ_{RS}^2), which is as follows:

$$\sigma_{\mathrm{RS}}^2 = \frac{f_2(k')d_f^2}{D_S}u \tag{4}$$

where k' is the capacity ratio of the solute, d_f is the effective film thickness of the stationary phase, D_S is the diffusivity of the solute in the stationary phase, and the other symbols have the meaning previously ascribed to them.

As all the dispersion processes are random, the individual variances can be added to arrive at the total variance of the peak leaving the column:

$$\sigma^2 = \sigma_M^2 + \sigma_L^2 + \sigma_{RM}^2 + \sigma_{RS}^2 \quad (5)$$

where σ^2 is the total variance/unit length of the column.

Thus, substituting for σ_M^2, σ_L^2, σ_{RM}^2, and σ_{RS}^2 from Eqs. (1)–(4), respectively,

$$\sigma^2 = 2\lambda d_p + \frac{2\gamma D_m}{u} + \frac{f_1(k')d_p^2}{D_m}u + \frac{f_2(k')d_f^2}{D_S}u \quad (6)$$

Now, the variance per unit length of a column is numerically equivalent to ratio of the column length to the column efficiency [2] [i.e., the height of the theoretical plate (H)]; thus,

$$H = 2\gamma d_p + \frac{2\gamma D_m}{u} + \frac{f_1(k')d_p^2}{D_m}u + \frac{f_2(k')d_f^2}{D_S}u \quad (7)$$

hence the term "HETP equation" for the equation for the variance per unit length of a column. Unfortunately, due to the compressibility of the gaseous mobile phase, neither the linear velocity nor the pressure is constant along the column, and as the diffusivity (D_m) is a function of pressure, the above form of the equation is only approximate. The Van Deemter equation was modified to take into account the compressibility of the carrier gas by Ogan and Scott [3]. The complete HETP equation for a GC column that takes into account the compressibility of the carrier gas will be

$$H = 2\lambda d_p + \frac{2\gamma D_m(o)}{u_o} + \frac{f_1(k')d_p^2}{D_{m(o)}}u_o + 2\frac{f_2(k')d_f^2}{D_S(\gamma + 1)}u_o \quad (8)$$

where u_o is the linear gas velocity at the column outlet, D_m is the diffusivity of the solute measured at the column outlet pressure, and γ is the inlet/outlet column ratio.

It is seen that Eq. (8) is very similar to Eq. (7) except that the velocity used is the *outlet* velocity, *not* the *average* velocity, and that the diffusivity of the solute in the gas phase is taken as that measured at the column outlet pressure (i.e., atmospheric). The shape of the H versus u curve is hyperbolic; it has a minimum value of H_{min} at the optimum velocity u_{opt} (i.e., at the optimum velocity, the column will have a maximum efficiency). Expressions for H_{min} and u_{opt} can be obtained by differentiating Eq. (8) with respect to u and equating to zero, solving for u_{opt} and substituting u_{opt} for u in Eq. (8) to obtain H_{min}.

References

1. J. J. Van Deemter, F. J. Zuiderweg, and A. Klinkenberg, *Chem. Eng. Sci.* 271 (1956).
2. R. P. W. Scott, *Liquid Chromatography Column Theory*, John Wiley & Sons, Chichester, 1992, p. 97.
3. K. Ogan and R. P. W. Scott, *J. High Resolut. Chromatogr.* 7: 382 (July 1984).

Suggested Further Reading

Scott, R. P. W., *Techniques of Chromatography*, Marcel Dekker, Inc., New York, 1995.

Scott, R. P. W., *Introduction to Analytical Gas Chromatography*, Marcel Dekker, Inc., New York, 1998.

Raymond P. W. Scott

Refractive Index Detector

The first practical refractive index detector was described by Tiselius and Claesson [1] in 1942 and, despite its limited sensitivity and its use being restricted to separations that are isocratically developed, it is still probably the fifth most popular detector in use today. Its survival has depended on its response, as it can be used to detect any substance that has a refractive index that differs from that of the mobile phase. It follows that it has value for monitoring the separation of such substances as aliphatic alcohols, acids, carbohydrates, and the many substances of biological origin that do not have ultraviolet (UV) chromophores, do not fluoresce, and are nonionic. When a monochromatic ray of light passes from one isotropic medium (A) to another (B), it changes its wave velocity and direction. The change in direction is called refraction and the relationship between the angle of incidence and the angle of refraction is given by Snell's law of refraction, namely

$$n'_B = \frac{n_B}{n_A} = \frac{\sin(i)}{\sin(r)}$$

where i is the angle of incident light in medium A, r is the angle of refractive light in medium B, n_A is the refractive index of medium A, n_B is the refractive index of medium B, and n'_B is the refractive index of medium B relative to that of medium A.

The refractive index of a substance is a dimensionless constant that normally decreases with increasing temperature; values are taken at 20°C or 25°C using the mean value taken for the two sodium lines of the spectrum. The optical systems that are used to exploit the refractive index for detection purposes are many and varied. One procedure is to construct a cell in the form of a hollow prism through which the mobile phase can flow. A ray of light is passed through the prism, which will be deviated from its original path, and is then focused onto a photocell. As the refractive index of the mobile phase changes, due to the presence of a solute, the angle of deviation of the transmitted light will also alter and the amount of light falling on the photocell will change. This method of refractive index monitoring is used by many manufacturers in their refractive index detector designs.

Another method evolved from the work of Fresnel. The relationship between the reflectance from an interface between two transparent media and their respective refractive indices is given by Fresnel's equation:

$$R = \frac{1}{2}\left(\frac{\sin^2(i-r)}{\sin^2(i+r)} + \frac{\tan^2(i-r)}{\tan^2(i+r)}\right)$$

where R is the ratio of the intensity of the reflected light to that of the incident light and the other symbols have the meanings previously assigned to them.

Now,

$$\frac{\sin(i)}{\sin(r)} = \frac{n_1}{n_2}$$

where n_1 is the refractive index of medium 1, and n_2 is the refractive index of medium 2.

Consequently, if medium 2 represents the liquid eluted from the column, then any change in n_2 will result in a change in R, and, thus, the measurement of R could determine changes in n_2 resulting from the presence of a solute. An example of a refractive index detector that functions on the Fresnel principle is shown in Fig. 1. Light from a tungsten lamp is directed through an infrared (IR) filter (to prevent heating the cell) to a magnifying assembly

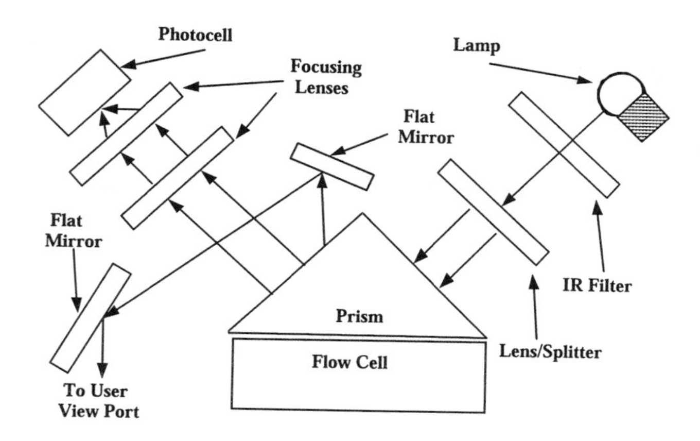

Fig. 1 A diagram of the optical system of a refractive index detector operating on the Fresnel method. (Courtesy of the Perkin Elmer Corporation.)

that splits the beam into two beams. The two beams are focused through the sample and reference cells, respectively. Light refracted from the mobile-phase/prism surface passes through the prism assembly and is then focused on two photocells. The prism assembly also reflects light to a user port where the surface of the prism can be observed. The output from the two photocells is electronically processed and either passed to a potentiometric recorder or to a computer data acquisition system. The range of refractive index covered by the instrument for a given prism is limited and, consequently, three different prisms are usually made available to cover the refractive index ranges of 1.35–1.4, 1.31–1.44, and 1.40–1.55, respectively.

Another variant on the refractive index detector arose from the work of Christiansen on crystal filters [2]. If a cell is packed with particulate material having the same refractive index as the mobile phase passing through it, light will pass through the cell with little or no refraction or scattering. If, however, the refractive index of the mobile phase changes, there will be a refractive index difference between the mobile phase and that of the packing. This difference results in light being refracted away from the incident beam, reducing the intensity of the transmitted light. If the transmitted light is focused onto a photocell and the refractive index of the packing and mobile phase initially matched, then any change in refractive index resulting from the elution of a peak will cause light scattering and a reduction in light falling on the sample photocell and, thus, provide a differential output.

In practice, because the optical dispersions of the media are likely to differ, the refractive index will only match at one particular wavelength and, thus, the fully transmitted light will be largely monochromatic. Light of other wavelengths will be proportionally dispersed depending on the their difference from the wavelength at which the two media have the same optical dispersion. It follows that a change in refractive index of the mobile phase will change both the intensity of the transmitted light and its wavelength. This device has been manufactured but was not a commercial success due to its limited sensitivity and the need for different packing for different applications.

Another detector that functions on the change in refractive index of the column eluent is the interferometer detector, which was first developed by Bakken and Stenberg [3] in 1971. The detector responds to the change in the effective path length of a beam of light passing through a cell when the refractive index of its contents changes due to the presence of an eluted solute. If the light transmitted through the cell is focused on a photocell coincident with a reference beam of light from the same source, interference fringes will be produced; the fringes will change as the path length of one light beam changes with reference to the other, and, consequently, as the concentration of solute increases in the sensor cell, a series of electrical pulses will be generated as each fringe passes the photocell. The effective optical path length (d) depends on the change in refractive index (Δn) and the path length (l) of the sensor cell as follows:

$$d = \Delta n l$$

Further, it is possible to calculate the number of fringes (N) (sensitivity) which move past a given point (or the number of cyclic changes of the central portion of the fringe pattern) in relation to the change in refractive index by the equation

$$N = \frac{2\Delta n l}{\lambda}$$

where λ is the wavelength of the light employed.

The larger the value of N for a given Δn, the more sensitive the detector will be. Therefore, l needs to be made as large as possible but will be limited by the dead volume of the column and the dispersion that can be tolerated before chromatographic resolution is impaired. The smallest cell (1.4 μL) (a cell volume that would be suitable for use with microbore columns) is reported to give a sensitivity of about 2×10^{-7} RI units at a signal-to-noise ratio of 2. Consequently, for benzene (RI = 1.501) sensed as a solute in n-heptane (RI = 1.388), this sensitivity would represent a minimum detectable concentration of 5.6×10^{-5} g/mL. The alternative 7-μL cell would decrease the minimum detectable concentration to about 1×10^{-6} g/mL, similar to that obtained for other refractive index detectors. However, the cell volume is a little large for modern high-efficiency columns. This type of RI detector has also been made commercially available but is somewhat more expensive with little gain in sensitivity over that obtained from simpler devices.

Another detector based on refractive index change is the thermal lens detector. When a laser is focused on an absorbing substance, the refractive index may be affected in such a way that the medium behaves as a lens. This effect was first reported by Gorden et al. [4]. Thermal lens formation results from the absorption of laser light, which may be extremely weak. The excited-state molecules subsequently decay back to ground state and, as a result, localized temperature increases occur in the sample. Because the refractive index of the medium depends on the temperature, the resulting spatial variation of refractive index produces an effect which appears equivalent to the formation of a lens within the medium. The temperature coefficient of refractive index is, for most liquids, negative; consequently, the insertion of a liquid in the laser beam produces a concave lens that results in beam divergence.

The thermal lens effect has been used for LC detection with a small-volume sensor cell. Basically, it consists of a *heating* laser, the light from which is passed directly through the sample, and another laser which passes light through the cell in the opposite direction. When an absorbing solute arrives in the cell, a thermal lens is produced that causes the probe light to diverge and, consequently, the intensity of the light falling on a photocell is reduced. The cell can be made a few microliters in volume and would thus be suitable for use with microbore columns. A sensitivity of 10^{-6} AU is claimed with a linear dynamic range of about three orders of magnitude. The use of two lasers adds significantly to the cost of the device. Basically, as the thermal lens detector is a special form of the refractive index detector, it can be considered as a type of universal detector. However, like other RI detectors, it cannot be used with gradient elution or flow programming and its sensitivity is no better than, if as good as, other refractive index detectors.

The refractive index detectors are very versatile in that they can detect all substances that have a different refractive index than that of the mobile phase. However, they are also one of the least sensitive detectors ($\sim 1 \times 10^{-6}$ g/mL); they have a linear dynamic range of about two to three orders of magnitude. They are extremely sensitive to flow rate, temperature, and pressure changes and cannot be used with gradient elution. Nevertheless, they are very popular for the detection of certain classes of compounds.

References

1. A. Tiselius and D. Claesson, *Ark. Kem. Mineral. Geol. 15B*(18) (1942).

2. C. Christiansen, *Ann. Phys. Chem. 3*: 298 (1884).
3. M. Bakken and V. J. Stenberg, *J. Chromatogr. Sci. 9*: 603 (1971).
4. J. P. Gorden, R. C. C. Leite, R. S. Moore, S. P. S. Posto, and J. R. Whinnery, *Bull. Am. Phys. Soc.* 9(2): 501 (1964).

Suggested Further Reading

Scott, R. P. W., *Liquid Chromatography for the Analyst*, Marcel Dekker, Inc., New York, 1994.
Scott, R. P. W., *Chromatographic Detectors*, Marcel Dekker, Inc., New York, 1996.

Raymond P. W. Scott

Resolution in HPLC: Selectivity, Efficiency, and Capacity

Liquid chromatography involves the analysis of mixtures. The goal of such an analysis is to achieve the greatest possible separation of the components in the mixture in the least amount of time. If performed successfully, the resulting chromatogram can be employed to obtain precise and accurate data describing the concentrations of the components in the mixture being analyzed.

The degree of separation of components of a mixture by a liquid chromatographic method is reflected in the resulting chromatogram. For best analytical results, peaks in the chromatogram must be completely resolved from each other, with little or no overlap. The degree of separation of between adjacent chromatographic peaks is a function of the distance between peak maxima and their corresponding peak widths. For Gaussian peaks, this is adequately described by the peak resolution, R_s, defined as the the ratio between the difference in the retention times t_1 and t_2, of two peaks and the average of the widths, W_1 and W_2, of the two peaks at their baselines, as shown by

$$R_s = \frac{t_2 - t_1}{0.5(W_1 + W_2)} \qquad (1)$$

A resolution of $R_s = 1$ corresponds to about a 4% overlap of two adjacent peaks and is adequate for many chromatographic analyses. Baseline resolution occurs at R_s values of 1.5 or higher [1].

As shown by Eq. (1), the resolution of components in a liquid chromatographic separation is dependent on (1) their relative retention on a particular chromatographic system and (2) their peak widths. To optimize these parameters for maximum resolution, a clear understanding of their nature and the factors that affect them is necessary. Although the retention time of a component adequately describes the amount of time a particular solute takes to elute from a chromatographic system, a more useful parameter describing chromatographic retention is the capacity factor k. This parameter is defined as the ratio of time spent by a solute in the stationary phase to the time it spends in the mobile phase. It can be calculated by Eq. (2), where t_R is the retention time of the peak of interest and t_0 (the "dead-volume time") is the retention time of a solute that is known not to interact with the stationary phase:

$$k = \frac{t_R - t_0}{t_0} \qquad (2)$$

Additionally, the relative retention of two peaks in a chromatogram may be defined by the selectivity factor, α, defined by Eq. (3), where k_1 and k_2 are capacity factors of the early- and late-eluting peaks, respectively:

$$\alpha = \frac{k_2}{k_1} \qquad (3)$$

The widths of chromatographic peaks are dependent on the degree to which a band of solute molecules spreads out, over the time it spends passing through a chromatographic system. This band spreading is best defined in terms of theoretical plates, N, which can be calculated from Eq. (4), where t_R and W are the retention time and width of the peak of interest, respectively:

$$N = 16\left(\frac{t_R}{W}\right)^2 \qquad (4)$$

Higher values of N correspond to lower degrees of band broadening and narrower peaks. On this basis, theoretical plates can be described as a measure of the efficiency of a given chromatographic system [2].

Chromatographic resolution of any two components in a mixture is dependent on three factors: (1) the overall efficiency of the chromatographic system, as described by the number of theoretical plates N, (2) the inherent selectivity of the system, described by the selectivity factor α, and (3) the degree of retention of each of the components, described by their capacity factors k. For two peaks having approximately equal widths, capacity factors of k_1 and k_2, and a mean theoretical plate number N, the following quantitative relationship can be derived between chromatographic resolution and these parameters [3]:

$$R_s = \left(\frac{N^{1/2}}{4}\right)\left(\frac{\alpha - 1}{\alpha}\right)\left(\frac{k_2}{1 + k_2}\right) \quad (5)$$

To a first approximation, each of the terms in Eq. (5) can be treated as independent of each other. Therefore, in the development of a liquid chromatographic method for analysis or isolation of the components of any mixture, experimental conditions can be varied to modify each of these three terms to maximize resolution.

Analysis of Eq. (5) indicates that resolution increases with the square root of the number of theoretical plates in the chromatographic system. Thus, a fourfold increase in N is required to increase the resolution by a factor of 2. Experimentally, N may be increased most directly by (1) increasing the length of the column, (2) reducing the size of the particles of the stationary phase, or (3) using a mobile phase of lower viscosity. However, in modern analytical liquid chromatography, most of these options are impractical for the following reasons: (1) Increasing the length of the column can increase system back-pressures and the time required to complete the separation to unworkably high levels. (2) Stationary phases with particle sizes below 3 μm are not generally available, owing to excessively high system back-pressures they can cause and their short lifetimes. (3) Because most recommended mobile phases for liquid chromatographic separations are those with low viscosity, there are few alternative mobile phases with lower viscosity available.

For these reasons, attempting to improve resolution by increasing the number of theoretical plates in the chromatographic system is not generally recommended as a first choice in liquid chromatographic method development. Nevertheless, the effect of theoretical plates on overall resolution is important to consider in certain situations, such as in the scaling up of an analytical method to a preparative method, when stationary phases of larger particle sizes are often employed, resulting in lower values of N and poorer resolutions [1].

Adjusting the selectivity of the chromatographic system, as measured by α, is often a useful technique in improving separations in liquid chromatography. Such adjustments need to be made with consideration of the second term in Eq. (5), which is $(\alpha - 1)/\alpha$. When $\alpha = 1$, the term is equal to zero, resulting in no resolution. This indicates that the chromatographic system must exhibit some selectivity toward the components of the mixture before any separation is possible. The term rapidly increases as α increases up to about $\alpha = 2$, beyond which only small increases are exhibited [2]. Thus, varying chromatographic conditions to obtain a selectivity factor equal or greater than 2 will often give the best resolutions.

In liquid chromatography, system selectivity can be modified by a number of techniques, including (1) changing the composition or pH of the mobile phase, (2) changing the column temperature, and (3) changing the type of stationary phase that is employed. Of these, the first method is the easiest to accomplish and is most often the first to be utilized in method development for improving selectivity.

Resolution can often be improved in liquid chromatographic separations simply by changing the retention of the components, which corresponds to changing the capacity factor k of the components to be separated. Rapid increases in the third term in Eq. (5), $k_2/(1 + k_2)$, occur as the capacity factor increases, up to $k_2 = 5$. The term increases only slowly beyond this value [2]. Therefore, in most separations by liquid chromatography, optimal resolution of mixtures occurs when each component has capacity factor values between 2 and about 10. Increasing k values to higher values results in longer analysis times with minimal improvements in resolution [2].

The effect of the capacity factor on resolution can be further exhibited by manipulation of Eq. (5) to predict the number of required theoretical plates N_{req} for a given resolution. Equation (6) enables such predictions:

$$N_{req} = 16R_S^2\left(\frac{\alpha}{\alpha - 1}\right)^2\left(\frac{k_2 + 1}{k_2}\right)^2 \quad (6)$$

Assuming a difficult separation of two components with $\alpha = 1.05$ and a reasonable goal of obtaining a resolution R_s of 1.0, the number of theoretical plates required at various capacity factor values can be calculated. For a capacity factor of 1.0, a minimum of 28,200 plates is required. For a capacity factor of 2.0, the required number of plates drops to 15,880. Because many liquid chromatographic systems usually do not exhibit efficiencies greater than 25,000 theoretical plates, it is evident from these calculations that developing methods for difficult separations should be performed when the capacity factors of the components to be separated are adjusted so they are all greater than 2.

In liquid chromatography, adjusting capacity factors is most often accomplished by changing the composition of the mobile phase. In reversed-phase chromatography, for example, decreasing the amount of organic component in the mobile phase (e.g., from 80% methanol, 20% water to 50% methanol, 50% water) will generally result in lower capacity factors for all components of the mixture. Empirical models and computer software have been developed which allow the user to predict mobile-phase compositions, which will result in capacity factors of all components in a mixture to be in the optimal range of 2–10. Such programs utilize data obtained from an analysis performed using gradient elution (i.e. continuously varying mobile-phase composition throughout the separation process) [3]. In many cases, however, experimental variation of mobile-phase compositions through a trial-and-error process can be equally effective in obtaining a composition giving an optimal capacity factor range for all components.

In summary, resolution in liquid chromatography is dependent on three factors: (1) the efficiency of the chromatographic system, measured by the theoretical plate value N; (2) the selectivity of the chromatographic system, measured by the selectivity factor α; and (3) the degree of retention of the components on the chromatographic system, measured by the capacity factor k. Although each of these can be independently varied to obtain acceptable separations, most method development in liquid chromatography is successfully performed through the initial choosing of the proper liquid chromatographic technique (e.g., normal phase, reversed phase, ion exchange) followed by optimizing capacity factors and system selectivity by varying the composition of the mobile phase.

References

1. L. R. Snyder and J. J. Kirkland, *Introduction to Modern Liquid Chromatography*, 2nd ed., John Wiley & Sons, New York, 1979.
2. C. F. Poole and S. K. Poole, *Chromatography Today*, Elsevier, New York, 1991, pp. 1–50.
3. L. R. Snyder, J. J. Kirkland, and J. L. Glajch, *Practical HPLC Method Development*, 2nd ed., John Wiley & Sons, New York, 1997.

J. E. Haky

Resolving Power of a Column

Two solutes will be resolved if their peaks are moved apart in the column and maintained sufficiently narrow to permit them to be eluted as discrete peaks. Resolution is usually defined as the ratio of the distance between the peaks to the peak width at the points of inflection. It is generally accepted that a separation of 4σ is adequate for accurate quantitative analysis, particularly when employing peak heights measurements. It is, therefore, necessary to derive an expression for the peak width in order to equate to the peak separation. The plate theory gives an expression for the elution curve of a solute as

$$X_{m(n)} = \frac{X_0 e^{-v} v^n}{n!} \tag{1}$$

where $X_{m(n)}$ is the concentration of solute in the nth plate on elution, X_0 is the concentration placed on the first plate on injection, n is the number of plates in the column, and v is the flow of mobile phase in plate volumes.

By differentiating and equating Eq. (1) to zero gives the following expression for the retention volume of a solute:

$$V_r = n(v_m + Kv_s)$$

Now, by equating the second differential of the elution equation to zero and solving for v, an expression for the peak width at the points of inflexion can obtained:

$$\frac{d_2\left(X_0 \dfrac{e^{-v} v^n}{n!}\right)}{dv^2} = X_0 \frac{e^{-v} v^n - e^{-v} n v^{n-1} - e^{-v} n v^{n-1} + e^{-v} n(n-1) v^{n-2}}{n!}$$

Thus,

$$\frac{d_2(X_0(e^{-v} v^n / n!))}{dv^2} = X_0 \frac{e^{-v} v^{n-2}(v^2 - 2nv + n(n-1))}{n!} \tag{2}$$

Now, at the points of inflexion,

$$\frac{d_2(X_0(e^{-v} v^n / n!))}{dv^2} = 0$$

Hence,

$$v^2 - 2nv + n(n-1) = 0$$

and

$$v = \frac{2n \pm \sqrt{4n^2 - 4n(n-1)}}{2}$$
$$= \frac{2n \pm \sqrt{4n}}{2}$$
$$= n \pm \sqrt{n}$$

It is seen that the points of inflexion occur after $n - \sqrt{n}$ and $n + \sqrt{n}$ plate volumes of mobile phase have passed through the column. Thus, the volume of the mobile phase that has passed through the column *between* the inflexion points will be

$$n + \sqrt{n} - n + \sqrt{n} = 2\sqrt{n} \tag{3}$$

Thus, the peak width at the points of inflexion of the elution curve will be $2\sqrt{n}$ plate volumes which, in milliliters of mobile phase, will be obtained by multiplying by the *plate volume*; that is,

$$\text{Peak width} = 2\sqrt{n}(v_m + Kv_s) \tag{4}$$

The peak width at the points of inflexion of the elution curve is twice the standard deviation, and, thus, from Eq. (4), it is seen that the variance (the square of the standard deviation) is equal to n, the total number of plates in the column. Consequently, the variance of the band (σ^2) in milliliters of mobile phase is given by

$$\sigma^2 = n(v_m + Kv_s)^2$$

Now,

$$V_r = n(v_m + Kv_s)$$

Thus,

$$\sigma^2 = \frac{V_r^2}{n}$$

Let the distance between the injection point and the peak maximum (the retention distance on the chromatogram) be y cm and the peak width at the points of inflexion be x cm. If the chromatographic data are computer processed, then the equivalent retention times can be used. Then, as the retention volume is $n(v_m + Kv_s)$ and twice the peak standard deviation at the points of inflexion is $2\sqrt{n}(v_m + Kv_s)$, then

$$\frac{\text{Ret. distance}}{\text{Peak width}} = \frac{y}{x} = \frac{n(v_m + Kv_s)}{2\sqrt{n}(v_m + Kv_s)} = \frac{\sqrt{n}}{2}$$

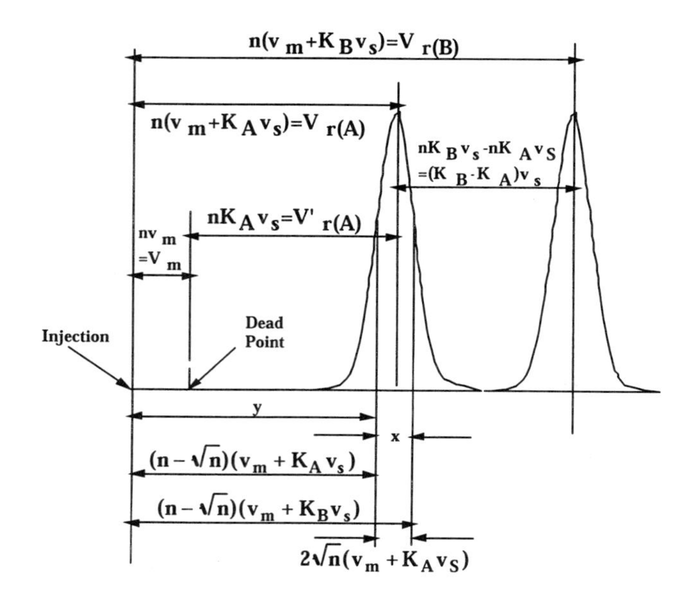

Fig. 1 A chromatogram showing two resolved solute peaks.

Thus,

$$n = 4\left(\frac{y}{x}\right)^2 \tag{5}$$

Equation (5) allows the efficiency of any solute peak, from any column, to be calculated from measurements taken directly from the chromatogram.

Consider the two peaks depicted in Fig. 1. The difference between the two peaks, for solutes A and B (see the entry Plate Theory), measured in volume flow of mobile phase, will be

$$n(v_m + K_B v_s) - n(v_m + K_A v_s) = n(K_B + K_A)v_s \tag{6}$$

Assuming the widths of the two peaks are the same, then the peak width in volume flow of mobile phase will be

$$2\sigma = 2\sqrt{n}(v_m + K_A v_s) \tag{7}$$

where K_A is the distribution coefficient of the first of the eluted pair of solutes between the two phases. Taking the already discussed criterion that resolution is achieved when the peak maxima of the pair of solutes are 4σ apart, then

$$4\sqrt{n}(v_m + K_A v_s) = n(K_B + K_A)v_s$$

Rearranging,

$$\sqrt{n} = \frac{4(v_m + K_A v_s)}{(K_B - K_A)v_s}$$

dividing through by v_m,

$$\sqrt{n} = \frac{4(1 + k'_A)}{(k'_B - k'_A)v_s}$$

Now, as α, the separation ratio between the two solutes, has been defined as

$$\alpha = \frac{k'_B}{k'_A}$$

then

$$\sqrt{n} = \frac{4(1 + k'_A)}{k'_A(\alpha - 1)}$$

and

$$n = \left(\frac{4(1 + k'_A)}{k'_A(\alpha - 1)}\right)^2 = 16\frac{(1 + k'_A)^2}{k'^2_A(\alpha - 1)^2} \tag{8}$$

Equation (8) is extremely important and was first developed by Purnell [1] in 1959. It allows the necessary efficiency to achieve a given separation to be calculated from a knowledge of the capacity factor of the first eluted peak of the pair and their separation ratio.

Reference

1. J. H. Purnell, *Nature (London) 184* (Suppl. 26): 2009 (1959).

Suggested Further Reading

Scott, R. P. W., *Chromatographic Detectors*, Marcel Dekker, Inc., New York, 1996.

Scott, R. P. W., *Introduction to Analytical Gas Chromatography*, Marcel Dekker, Inc., New York, 1998.

Raymond P. W. Scott

Retention Factor: Effect on MEKC Separation

In micellar electrokinetic chromatography (MEKC), an ionic surfactant micelle, such as sodium dodecyl sulfate (SDS), is used as a pseudo-stationary phase that corresponds to the stationary phase in liquid chromatography (LC). Here, the separation principle of MEKC with an anionic micelle (e.g., SDS) under a neutral condition is briefly considered. When high voltage is applied across the whole capillary, the entire solution migrates toward the cathode by electro-osmotic flow (EOF) while the SDS micelle is forced toward the anode by electrophoresis. The EOF is stronger than the electrophoretic migration of the SDS micelle and, hence, the micelle migrates toward the cathode at a more retarded velocity than the EOF.

When a neutral analyte is injected into the micellar solution at the anodic end of the capillary, it will be distributed between the micelle and the surrounding aqueous phase. The analyte, which is not incorporated into the micelle at all, migrates toward the cathode at the same velocity as the EOF. The analyte totally incorporated into the micelle migrates at the lowest velocity, or at the same velocity as the micelle, toward the cathode. The more the analyte is incorporated into the micelle, the slower the analyte will migrate. A neutral analyte always migrates at a velocity between the two extremes (i.e., the velocities of the EOF and micelle). The analytes are detected in an increasing order of the distribution coefficients by a detector located at the cathodic end of the capillary. The migration time of the electrically neutral analyte is limited between the two extremes: the migration time of a solute that is not incorporated into the micelle at all, t_0, and that of the micelle, t_{mc}.

Under an acidic condition, however, the absolute value of the velocity of the EOF becomes lower than that of the electrophoretic velocity of the SDS micelle and, therefore, the micelle migrates toward the anode. By contrast, when a cationic surfactant is employed instead of SDS, the direction of the EOF will be reversed or toward the anode, due to the adsorption of the surfactant molecule on the inside wall of the capillary and changing the surface charges.

Retention Factor

In MEKC, the retention factor, k, for a neutral compound can be defined as n_{mc}/n_{aq}, where n_{mc} and n_{aq} are the number of the analyte incorporated into the micelle and in the surrounding aqueous solution, respectively. The retention factor can be related to the migration time of the solute, t_R, as

$$k = \frac{t_R - t_0}{t_0(1 - t_R/t_{mc})} \tag{1}$$

or

$$t_R = \left(\frac{1 + k}{1 + (t_0/t_{mc})k}\right)t_0 \tag{2}$$

The reciprocal of t_0/t_{mc}, or t_{mc}/t_0, is a parameter representing the migration time window. One should note that when the migration time of the micelle is infinite or the micelle does not migrate in the capillary at all, the value t_0/t_{mc} will be zero; then, Eqs. (1) and (2) become identical to those for conventional LC.

In MEKC, $k = \infty$ means that t_R becomes equal to t_{mc} and the solute migrates at the same velocity as the micelle.

When $t_0 = 0$ or the EOF is completely suppressed, Eq. (2) becomes

$$t_R = \left(1 + \frac{1}{k}\right)t_{mc} \tag{3}$$

Here, the surrounding aqueous phase does not move at all in the capillary and only the micelle migrates toward the anode if an anionic micelle is employed. Note that the EOF is not essential in MEKC.

When the solute has an electrophoretic mobility Eq. (1) will be more complicated, that is, the migration of the ionic solute includes a portion generated by the micelle when the solute is incorporated into the micelle and also the other portion generated by the electrophoresis of the solute itself.

Resolution

Resolution, R_s, in MEKC is given as

$$R_s = \left(\frac{N^{1/2}}{4}\right)\left(\frac{\alpha - 1}{\alpha}\right)\left(\frac{k_2}{1 + k_2}\right)\left(\frac{1 - (t_0/t_{mc})}{1 + (t_0/t_{mc})k_1}\right) \tag{4}$$

where N is the theoretical plate number, α is the separation factor equal to k_2/k_1, and k_1 and k_2 are the retention factors of analytes 1 and 2, respectively. When $t_0/t_{mc} = 0$, Eq. (4) will be identical to that for conventional LC.

The separation factor α is altered by the combination of the structure of the micelle as the pseudo-stationary phase and the aqueous phase as a solvent of the micelle. Because k is included in the last term of the right-hand side of Eq. (4), the effect of k on R_s in MEKC is different from that in conventional chromatography. The last two terms in Eq. (4) are defined by the function $f(k)$ as

$$f(k) = \left(\frac{k_2}{1 + k_2}\right)\left(\frac{1 - (t_0/t_{mc})}{1 + (t_0/t_{mc})k_1}\right) \tag{5}$$

Then, we can calculate the optimum value of the retention factor, k_{opt}, for accomplishing the maximum R_s by differentiating Eq. (5), that is,

$$k_{opt} = \left(\frac{t_{mc}}{t_0}\right)^{1/2} \tag{6}$$

Under a neutral condition, k_{opt} is close to 2 for SDS micelles as the pseudo-stationary phase, as shown in Fig. 1. Practically, the recommended range of k is between 0.5 and 10.

Retention Factor and Distribution Coefficient

The retention factor can be related to the distribution coefficient, K, between the micelle and aqueous phase by

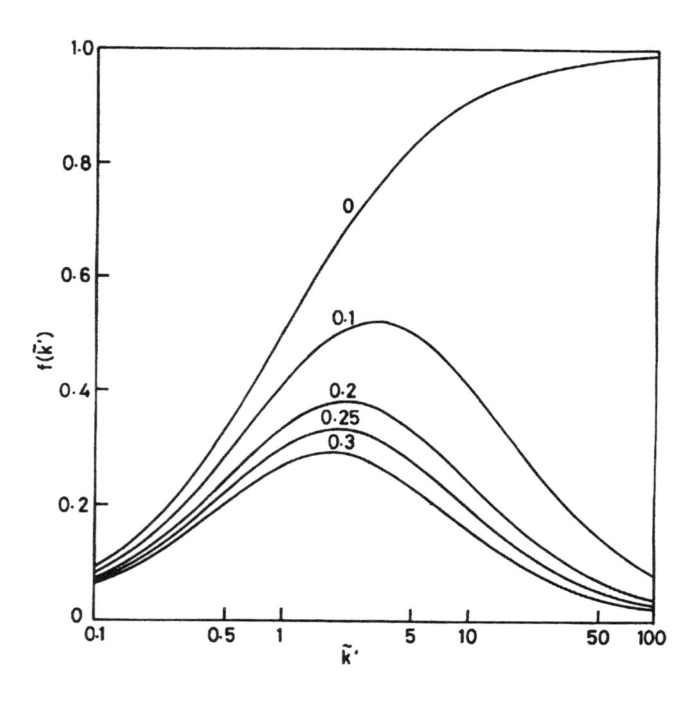

Fig. 1 Dependence of $f(k)$ on capacity factor k. The values of t_0/t_{mc} are given on each line. [Reprinted from S. Terabe, K. Otsuka, and T. Ando, *Anal. Chem. 57*: 834–841 (1985) with permission from the American Chemical Society.]

$$k = K\left(\frac{V_{mc}}{V_{aq}}\right) \quad (7)$$

where V_{mc} and V_{aq} are the volumes of the micelle and aqueous phase, respectively. The value V_{mc}/V_{aq}, or the phase ratio, can be written as

$$\frac{V_{mc}}{V_{aq}} = \frac{\bar{\nu}(C_{sf} - \text{CMC})}{1 - \bar{\nu}(C_{sf} - \text{CMC})} \quad (8)$$

where C_{sf}, $\bar{\nu}$, and CMC are the concentration of the surfactant, partial specific volume of the micelle, and critical micelle concentration, respectively. At a low micellar concentration, we can arrange Eq. (8) as

$$k \simeq K\bar{\nu}(C_{sf} - \text{CMC}) \quad (9)$$

Thus, k can be adjusted by manipulating C_{sf}. Equation (9) shows that k increases linearly with C_{sf}, and we can calculate K from the slope of this relationship. Also, K remains constant regardless of C_{sf}. The applicability of Eq. (9) under various conditions and for various surfactants has been examined in a number of reports.

Suggested Further Reading

K. Otsuka and S. Terabe, Micellar electrokinetic chromatography, *Bull. Chem. Soc. Jpn. 71*: 2465–2481 (1998).

J. P. Quirino and S. Terabe, Electrokinetic chromatography, *J. Chromatogr. A 856*: 465–482 (1999).

S. Terabe and Z. Deyl, Micelles as separation media in chromatography and electrophoresis, *J. Chromatogr. A 780*: (1997).

S. Terabe, K. Otsuka, and T. Ando, Electrokinetic chromatography with micellar solution and open-tubular capillary. *Anal. Chem. 57*: 834–841 (1985).

J. Vindevogel and P. Sandra, *Introduction to Micellar Electrokinetic Chromatography*, Hüthig, Heidelberg, 1992.

Koji Otsuka
Shigeru Terabe

Retention Gap Injection Method

In gas chromatographic analysis employing capillary columns, split injections are usually necessary to ensure that a very small sharp sample is placed on the column. This is important for maintaining column efficiency by not overloading the column. However, split injections generally result in an unrepresentative sample being placed on a capillary column (see the entry Split/Splitless Injector), and because of this, on-column injection is usually preferred for accurate quantitative analysis. On-column injection requires a relatively large-diameter capillary column to be used to permit the penetration of the injection syringe needle into the column. However, although this procedure ensures that a representative sample is placed onto the column, other problems can arise. On injection, the sample readily separates into droplets that act as separate, individual injections. These separate sample sources can cause widely dispersed peaks and serious loss of resolution and, in the extreme, double or multiple peaks. Grob [1] suggested a solution to this problem which he termed the *retention gap method of injection*.

This procedure (Fig. 1) involves removing the internal coating of stationary phase from the first few centimeters of the column. This can be done by heating and volatilizing or burning off the phase. Alternatively, if the stationary phase is sufficiently soluble, it can be removed by a suitable solvent. The sample is then injected into the uncoated section of the column, and although the sample will probably split into droplets, the solvent will still vaporize in the normal way. As there is no stationary phase present, all the components of the mixture will travel at the speed of the mobile phase down the uncoated length of column until they reach a coated section. At this point, they will be absorbed into the stationary phase and all the components of the mixture will accumulate and form a compact sample at the start of the coated portion of the column.

This technique is usually practiced in conjunction with temperature programming, the program being started at a fairly low temperature. The relatively low temperature facilitates the accumulation of all the solutes at one point in the column (i.e., where the stationary-phase coating begins). The temperature program is then started, and the solutes are eluted through the column in the normal way. The success of this method depends on there being a significant difference between the boiling points of the sample solvent and those of the components of the sample.

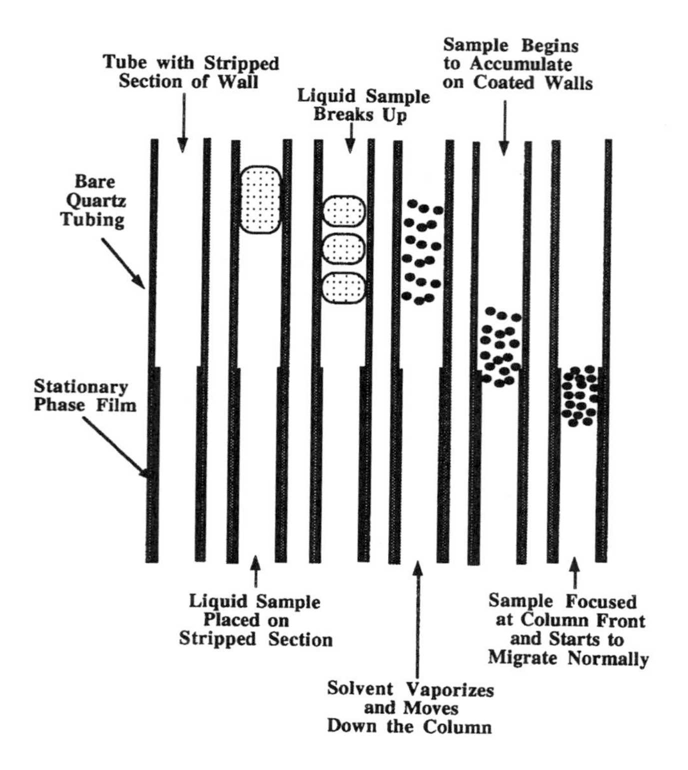

Fig. 1 The retention gap method of injection.

In general, however, this procedure does significantly improve the quality of the separation and allows accurate quantitative results to be obtained.

Reference

1. K. Grob, *Classical Split and Splitless Injection in Capillary Gas Chromatography*, Huethig, Heidlburg, 1987.

Suggested Further Reading

Grant, D. W., *Capillary Gas Chromatography*, John Wiley & Sons, Chichester, 1995.
Scott, R. P. W., *Techniques and Practice of Chromatography*, Marcel Dekker, Inc., New York, 1996.
Scott, R. P. W., *Introduction to Analytical Gas Chromatography*, Marcel Dekker, Inc., New York, 1998.

Raymond P. W. Scott

Retention Time and Retention Volume

The *retention time* of a solute is the elapsed time between the *injection point* and the peak maximum of the solute. The different properties of the chromatogram are shown in Fig. 1. The *volume of mobile phase* that passes through the column between the injection point and the peak maximum is called the *retention volume*. If the mobile phase is incompressible, as in LC, the retention volume (as so far defined) will be the simple product of the ***exit** flow rate* and the *retention time*.

If the mobile phase is compressible, the simple product of retention time and flow rate will be incorrect, and the *retention volume* must be taken as the product of the *retention time* and the ***mean** flow rate*. The true *retention volume* has been shown to be given by [1]

$$V_r = V_{r'}\frac{3}{2}\left(\frac{\gamma^2 - 1}{\gamma^3 - 1}\right) = Q_0 t_r \frac{3}{2}\left(\frac{\gamma^2 - 1}{\gamma^3 - 1}\right)$$

where the symbols have the meanings defined in Fig. 1, and $V_{r'}$ is the retention volume measured at the column exit and γ is the inlet/outlet pressure ratio.

The retention volume V_r will include the dead volume V_0, which, in turn, will include the actual dead volume V_m and the extracolumn volume V_E.

Thus,

$$V_r = V_E + V_m + V'_r$$

The retention time can be taken as the product of the distance on the chart between the injection point and the peak maximum and the chart speed, using appropriate units. More accurately, it can be measured with a stopwatch. The most accurate method of measuring V_r for a noncompressible mobile phase, although considered antiquated, is to attach an accurate burette to the detector exit and measure the retention volume in volume units. This is

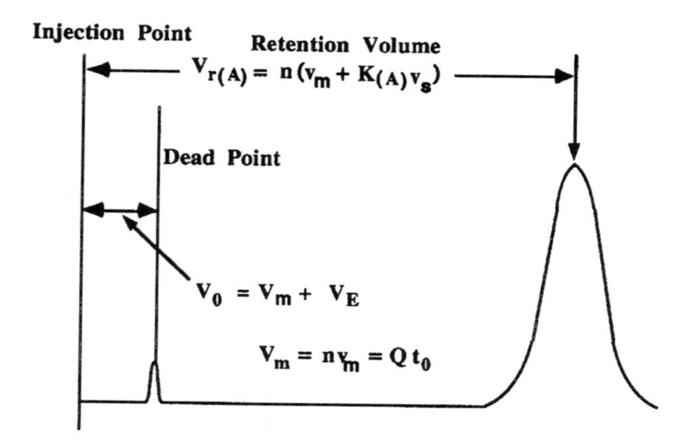

Fig. 1 Diagram depicting the dead point, dead volume, and dead time and retention volume of a chromatogram. V_0 is the total volume passed through the column between the point of injection and the peak maximum of a completely unretained peak, V_m is the total volume of mobile phase in the column, $V_{r(A)}$ is the retention volume of solute A, V_E is the extra column volume of mobile phase, v_m is the volume of the mobile phase per theoretical plate, v_s is the volume of the stationary phase per theoretical plate, K_A is the distribution coefficient of the solute between the two phases, n is the number of theoretical plates in the column, and Q is the column flow rate measured at the exit.

an absolute method of measurement and does not depend on the accurate calibration of the pump, chart speed, or computer acquisition level and processing.

Reference

1. R. P. W. Scott, *Introduction to Analytical Gas Chromatography*, Marcel Dekker, Inc., New York, 1998, p. 77.

Suggested Further Reading

Scott, R. P. W., *Liquid Chromatography Column Theory*, John Wiley & Sons, Chichester, 1992, p. 19.

Scott, R. P. W., *Techniques and Practice of Chromatography*, Marcel Dekker, Inc., New York, 1996.

Raymond P. W. Scott

Reversed-Phase Chromatography: Description and Applications

Introduction

Classical liquid chromatography is typically practiced in what is referred to as the normal-phase mode; that is, the stationary phase is usually a polar sorbent such as silica and alumina and the mobile phase consists of a nonpolar constituent such as hexane modified with a somewhat more polar solvent such as chloroform or ethyl acetate. In this mode, the more polar compounds are preferentially retained. The reversed-phase (RP) mode utilizes the opposite approach for the separation of nonpolar analytes or compounds that have some hydrophobic character. In this case, the stationary phase must consist of sorbent that is nonpolar in nature and the mobile phase is composed of a primary polar solvent, usually water, that is modified by a more nonpolar constituent such as methanol, acetonitrile, or tetrahydrofuran.

In order to make RP chromatography a rapid and efficient method, it is necessary to force the mobile phase through the stationary phase using high pressure. Therefore, the stationary phase must be a mechanically stable entity possessing the desired nonpolar properties for reversed-phase operation. This result is accomplished typically by using particulate silica, which is stable under high pressure, and modifying the surface with a nonpolar organic moiety. The modification takes place by reacting the silanol (Si—OH) groups on the silica surface with a suitable reagent, most often an organosilane compound (X_3Si—R or XR′R′Si—R, where X is a reactive group such as Cl or methoxy, R′ is a small organic group such as methyl, and R is another organic moiety, most often octyl (C_8) or octadecyl (C_{18}), in the RP mode. These silica modifications result in a primarily hydrophobic surface that can preferentially retain the more nonpolar compounds in a mixture. The degree of hydrophobicity is controlled by both the length of the alkyl chain and the density of bonded groups on the surface, usually expressed in terms of micromoles per square meter. Due to the fact that original silica material has a high surface area (typically 100–300 m^2/g), the amount of hydrophobic material in the chromatographic column is considerable (from a few to as much as 20% by weight), leading to substantial interactions between solutes and the stationary phase. RP stationary phases are available with a variety of hydrophobic groups on the surface and bonding densities on silica particles of different diameters, surface areas, and pore sizes. In addition to the RP separation materials consisting of silica, some commercial products are also fabricated on other oxides such as alumina or zirconia or consist of polymeric matrices.

The second major component in modern high-performance liquid chromatography (HPLC) is the mobile phase. Since the stationary phase is a nonpolar entity, the mobile phase must be more polar to allow retention of the analytes. The most polar solvent for RP-HPLC is water, but the overall polarity of the mobile phase can be adjusted by introducing variable amounts of any of a number of organic solvents. In liquid chromatography, retention of solutes is a result of its relative affinity for the stationary and mobile phases. This can be described mathematically by the equation

$$k' = \frac{(\text{Amount of analyte})_{SP}}{(\text{Amount of analyte})_{MP}}$$

where k' is the equilibrium constant referred to as the capacity factor that relates the amounts of the analyte in the stationary phase (SP) and the mobile phase (MP). Therefore, the mobile phase exerts considerable influence on the retention and, hence, the separation of solutes. This factor makes HPLC a very powerful separation technique in that the mobile phase can be adjusted to accommodate a wide variety of solutes (from large biomolecules to small organic and inorganic compounds) having a range of chemical properties. Simultaneously, the selection of the mobile-phase composition will determine the degree of interaction between the solute and the stationary phase.

Most RP-HPLC separations are done in the isocratic mode (i.e., where the composition of the mobile phase is held constant during the analysis). This approach is suitable when the sample consists of analytes having similar properties or where their hydrophobicities encompass a small or moderate range. Under these conditions, all solutes in the sample will be eluted over a reasonable time span (i.e., not too short to prevent resolution of individual analytes and not too long to result in an inconvenient analysis period). Therefore, proper selection of the mobile-phase composition is essential in the development of any reversed-phase separation method. Fortunately, due to the decades of long practice of RP-HPLC, there exists in the literature and from commercial sources, a wealth of information on suitable mobile-phase compositions for particular types of sample, especially for the C_{18} stationary phase. In addition, the retention of solutes on hydrophobic phases has been modeled mathematically and there exist computer programs for assisting in the optimization of mobile-phase composition in the solution of various separation problems.

A single mobile composition is often not suitable for samples that contain a wide range of chemical properties or hydrophobicities. Under these conditions, an isocratic method may leave the early eluting components unresolved and the analytes having strong retention with inconveniently long elution times. The solution to this problem is to change the mobile-phase composition in a systematic way during the course of the separation. This approach is referred to as gradient elution. In gradient elution, the mobile-phase composition initially is weak (with a large percentage of the most polar component) and becomes increasingly stronger (containing greater amounts of the less polar modifier) as the separation process continues. With this approach, the retention of the less hydrophobic compounds is increased at the beginning of the separation, whereas the retention of the more hydrophobic compounds is diminished at the end of the elution period. The simplest approach to gradient elution is to vary the mobile-phase composition linearly from the beginning to the end of the analysis period. In addition to the rate of change of the mobile-phase composition, the initial and final amounts of the two solvents are also variables that can be changed to improve resolution within the shortest analysis times. Besides linear gradients, other formats have been developed to optimize separations. These gradient methods include a constant composition at the beginning and/or the end of the analysis as well as concave, convex, or step profiles. The main disadvantage of the gradient method is the time required for the column to reequilibrate to the initial mobile-phase conditions. This reequilibration time can be from several minutes up to a half-hour or longer. However, modern instrumentation (pumps and pump controllers) has made reproducible gradients relatively easy to achieve.

Another means of controlling eluent strength is the use of ternary or quaternary solvent mixtures instead of the more common binary approach. Each solvent has its own unique properties that can be used to improve the separation of difficult-to-resolve analytes or to shorten the analysis time without sacrificing resolution. Although gradients and more complex solvent matrices are more difficult to model than binary isocratic systems, software exists for such purposes and can assist in method development.

The basic equipment for the RP mode is similar to most other types of HPLC. It consists of solvent reservoirs (one to four), a high-pressure pump, a mixing device that can create any combination of binary solvents or higher order as well as gradients (optional), an injection device, the column, and a detector connected to a data processing device. Ultraviolet (UV) detection is most often used in the reversed-phase mode, but fluorescence, refractive index, and electrochemical properties, as well as coupling to a mass spectrometer are also possible. Qualitative information is obtained by comparing the retention times of unknown compounds to those of known standards, whereas quantitative information comes from calibration curves of the peak area versus concentration. The coupling of liquid chromatographs to mass spectrometers and nuclear magnetic resonance (NMR) spectrometers is becoming more common, which makes positive identification of unknown compounds in a mixture much easier.

Applications

One of the primary factors responsible for the development of HPLC was the need to separate mixtures containing hydrophobic compounds that were not sufficiently volatile for analyzing by gas chromatography (GC) or were thermally unstable after volatilization. Although some compounds that are normally nonvolatile can be made volatile by derivatization, this process adds an extra step to the analytical method. However, under any circumstances, a large majority of chemical species, perhaps as much as 70%, cannot be analyzed by GC. Among the most significant of these compounds are ionic species, both organic and inorganic, as well as most biomolecules. With greater demand for the analysis of biologically related samples for medical, pharmaceutical, and biotechnological purposes, the need for reliable HPLC reversed-phase methods continues to increase. Although it is impossible to review all types of sample amenable to RP-HPLC analysis, a few examples will be given to illustrate the breadth of applications possible by this technique.

Because the mechanism of separation is primarily based on differences in hydrophobicity, a simple mixture of aromatic hydrocarbons can be used to illustrate the operation of the reversed-phase method. A chromatogram of such a separation is shown in Fig. 1, where the elution times are benzene < toluene < ethylbenzene < isopropylbenzene < *t*-butylbenzene < anthracene. When the reversed-phase mechanism is functioning, compounds are eluted in order of increasing hydrophobicity, as illustrated in Fig. 1. By increasing the degree of hydrophobicity either through longer alkyl chains [more saturated or unsaturated (aromatic) hydrocarbon groups] or more alkyl chains (higher bonding density), retention times (larger k' values) become longer under constant mobile-phase conditions. This principle applies to a wide variety of organic compounds. The organic molecules can also have a polar functional group such as an alcohol, ether, amine, or cyano, for example, but the RP method can still be used. In this case, the polar groups may diminish the overall hydrophobicity of the compound, but there will still be some retention on a typical RP stationary phase such as octadecyl (C_{18}). A simple example is benzene and phenol. The addition of a hydroxyl group makes phenol less hydrophobic than benzene, so it will be eluted first.

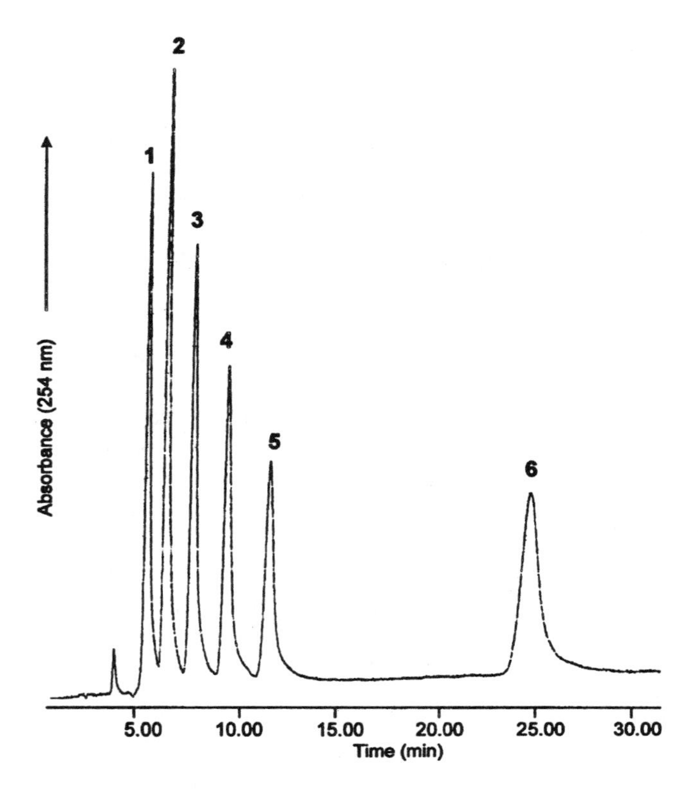

Fig. 1 Separation of reversed-phase test mixture on a C_{22} bonded phase. Mobile phase: 50:50 acetonitrile–water. Solutes: 1 = benzene; 2 = toluene; 3 = ethylbenzene; 4 = isopropylbenzene; 5 = *t*-butylbenzene; 6 = anthracene.

The above example illustrates the principle of relative retention (i.e., that benzene is retained more strongly than phenol). In order to determine absolute retention, the k' values of each compound must be measured as follows:

$$k' = \frac{t_R - t_0}{t_0}$$

where t_R = the retention time of compound and t_0 is the time to elute an unretained compound. In HPLC, the t_0 is equivalent to measuring the elution time for air in GC. Therefore, selection of a suitable compound that will not be retained is crucial to accurate measurement of k' values. Because retention is based on hydrophobicity, the t_0 marker should be very hydrophilic (i.e., very polar or ionic). Two compounds often selected for this determination are KNO_3 and uracil. They both fulfill the requirement for hydrophilic properties and also have absorbance in the UV, which facilitates detection.

Whereas the overall hydrophobic nature of the stationary phase is the most important factor in determining retention, bonded-phase structure can also influence k' values. This effect can be observed in the separation of polycyclic aromatic hydrocarbons (PAHs). For stationary phases with a high bonding density and/or a high degree of association between adjacent bonded organic moieties, molecules that are more planar are preferentially retained. The National Institute of Standards and Technology (NIST) has developed reference mixtures to measure this effect.

In addition to a wide range of polar and nonpolar hydrocarbons that can be analyzed by RP-HPLC, it is also possible to separate ionic species. Because water is used as part of almost all mobile phases, those species which are acids and bases can be neutralized by control of pH. In cases where neutralization is not possible, then the addition of a counterion into the mobile phase so that the analyte will form a neutral complex can be used to enhance RP retention. The same principle can be applied to inorganic species by forming a neutral complex that results in reversed-phase retention.

Large biomolecules, while being charged under most aqueous mobile-phase conditions, still have significant hydrophobic portions that interact with the stationary phase. In many complex mixtures of proteins and peptides, the degrees of interaction with the stationary phase (k' values) vary over a broad range. Therefore, gradient elution methods are often required. An example of such a gradient method for the separation of a biochemical mixture is shown in Fig. 2. Finally, although water is used almost exclusively as the weak solvent in RP methods, a few types of sample require the use of other mobile-phase components. For example, the separation of triglycerides and fatty acids often utilize acetone as the weak solvent in the RP mode.

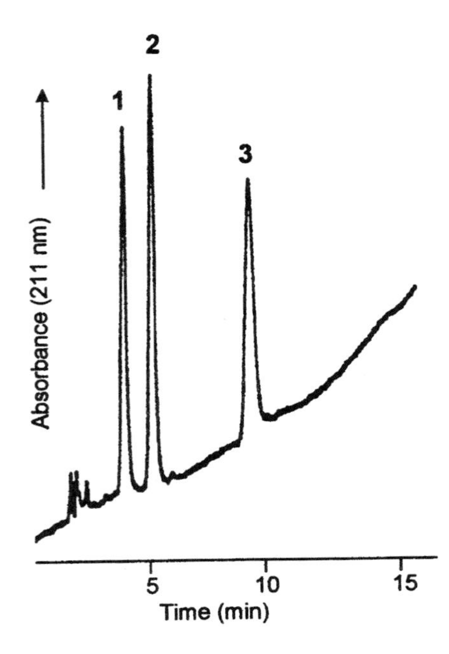

Fig. 2 Gradient separation of peptide mixture on a C_{30} bonded phase. Mobile phase, linear gradient from 25% to 45% A in 15 min. A = 0.1% trifluoroacetic acid (TFA) in 75:25 acetonitrile–water and B = 0.1% TFA in water. Solutes: 1 = bradykinin; 2 = angiotensin III; 3 = angiotensin I.

Suggested Further Reading

Cunico, R. L., K. M. Gooding, and T. Wehr, *Basics HPLC and CE of Biomolecules*, Bay Bioanalytical Laboratory, Richmond, CA, 1998.

Kirkland, J. J., *LC–GC* (May 1997).

Mant, C. T. and R. S. Hodges (eds.), *High Performance Liquid Chromatography of Peptides and Proteins*, CRC Press, Boca Raton, FL, 1991.

Poppe, H., Column liquid chromatography, in *Chromatography*, 5th ed. (E. Heftmann, ed.), Elsevier, Amsterdam, 1992.

Snyder, L. R., Theory of chromatography, in *Chromatography*, 5th ed. (E. Heftmann, ed.), Elsevier, Amsterdam, 1992.

Vansant, E. F., P. Van Der Voort, and K. C. Vrancken, *Characterization and Chemical Modification of Silica*, Elsevier, Amsterdam, 1995.

Joseph J. Pesek
Maria T. Matyska

Reversed-Phase Stationary Phases

The primary purpose for the developement of chemically modified stationary phases was to provide a separation medium that was suited to the type(s) of solute present in the mixture to be analyzed. Historically, silica gel was the most common material used in the early development of column liquid chromatography (LC). However, silica is a polar material that contains hydroxyl groups (silanols) that are both acidic and strongly hydrogen-bonding in character. These properties make it unsuitable as a stationary phase for many typical organic molecules that are predominantly hydrophobic compounds. In addition, the silanols interact strongly with basic compounds leading to poor chromatographic results. In order to overcome these undesirable effects of silica and to have a medium more suitable for the separation of a large variety of organic compounds, modification of the surface is necessary to provide a more nonpolar (hydrophobic) material. It is advantageous to retain silica as the primary material in the column because it possess physical and mechanical properties that make it particularly useful for modern liquid chromatography [i.e., the use of high-pressure liquid chromatography (HPLC) to force the mobile phase and sample through the system at a reasonable flow rate].

The desirable characteristics of silica are as follows: high mechanical strength, a narrow range of particle diameters, a variety of pore sizes, a broad range of surface areas, and the ability to be modified either chemically or physically by adsorption.

It is the latter property (i.e., the ability for modification) that makes silica particularly useful as a separation medium in chromatography. Although physical adsorption has been used occasionally to modify silica surfaces for chromatographic purposes, its usefulness is limited because of the nature of modern HPLC. The use of high pressure creates shear forces at the stationary phase–mobile phase interface so that the absorbed moiety is removed from the silica surface even though the coating may be insoluble in the liquid being pumped through the column. Therefore, chemical modification is the only practical approach to modifying the silica surface in order to create a stationary phase that is compatible with the types of solutes to be separated. The most common method for modifying silica in order to produce a hydrophobic surface is organosilanization. Two types of reaction are available by this method. The first alternative is referred to as a monomeric approach:

$$\equiv \mathrm{Si{-}OH} + \mathrm{X{-}SiR'_2R} \rightarrow \equiv \mathrm{Si{-}O{-}SiR'_2R} + \mathrm{HX}$$

Here, the organosilane reagent is composed of a reactive group, X, which can be a halide, usually chloride, methoxy, or ethoxy; R′ is one of two small organic groups usually methyl, and R is the main group that gives the surface its hydrophobic properties. The end result is a single point of attachment between the organosilane reagent and the surface. The second alternative shown in the following reaction is referred to as the polymeric approach for bonding.

$$\begin{array}{ccccccccc}
| & & & & | & & | & & \\
\mathrm{O} & & & & \mathrm{O} & & \mathrm{O} & & \\
| & & & & | & & | & & \\
-\mathrm{Si} & -\mathrm{OH} + \mathrm{X_3Si{-}R} & \rightarrow & - & \mathrm{Si} & -\mathrm{O}- & \mathrm{Si} & -\mathrm{R} + 3\mathrm{HX} & \\
| & & & & | & & | & & \\
\mathrm{O} & & & & \mathrm{O} & & \mathrm{O} & & \\
| & & & & | & & | & &
\end{array}$$

In this case, the organosilane reagent is composed of three reactive groups; X is as described earlier and R is the main organic group that provides the hydrophobic properties to the surface. Here, the end result is that the bonded phase is attached to the surface at one point and cross-linked to neighboring bonded organosilanes through a siloxane linkage. Both of these synthetic routes are used in the production of commercially available stationary phases for HPLC. The monomeric approach generally is more reproducible from batch to batch, whereas the polymeric approach leads to higher bonding densities (more R groups per unit surface area) and some additional stability due to the multiple sites of attachment.

In both types of reactions, it is the R group that determines the overall characteristics of the surface if there is a resonable bonding denisity as measured in terms of micromoles per square meter. There must be a significant number of organic moieties per unit surface area so that most of the silica is covered by the R groups and relatively few of the siloxane and silanols are accessible. Under these conditions, when the organic moiety is hydrophobic, nonpolar solutes will be selectively retained by the stationary phase. Even in the case where the bonding density is reasonably high, there is still the possibility that some silanols may be accessible to solutes. This is mainly a problem when the analytes are strongly basic compounds. In order to diminish the effect of unreacted silanols on the surface or those that can be created in the polymeric reaction process when complete cross-linking does not take place, a secondary reaction involving a small reactive organosilane can be used. Typically, this reagent is trimethylchlorosilane, a compound with one reactive group and three small organic moieties. This compound is small enough to fit into the larger spaces between bonded hydrophobic groups so that access to the surface will be even more limited for typical solutes. The process of bonding a small moiety to diminish the number of accessible silanols is referred to as "endcapping." Many commercial sources will often designate whether or not a particular bonded phase has been endcapped. The presence or absence of endcapping will determine the nature of the stationary phase surface and, hence, its retention characteristics.

However, it is still the main R group that controls the overall degree of hydrophobicity of the surface. Within this context, the predominant factors in determining the hydophobicity are the length of the alkyl chain or the total number of carbon atoms as well as the bonding density. Some examples of various alkyl groups that have been used as reversed-phase (RP) materials are shown in Table 1. The most common types of these phases are designated by C_n, where n is the number of carbon atoms for bonded linear alkyl hydrocarbon moieties. The simplest case is where $n = 1$ for the methyl-bonded phase (C_1). This material has the lowest degree of hydrophobicity and provides limited retention for most small organic molecules. However, for large biomolecules such as proteins and peptides that can have extensive hydrophobic regions as part of their three-dimensional structure, these phases can prove useful in limiting the strong interactions, leading to excessively long retention times for these compounds. As the degree of hydrophobicity decreases for these large species (i.e., the macromolecule has larger hydrophilic regions or the hydrophobic areas are buried within the three-dimensional structure), the stationary phase will have to become more nonpolar. This is accomplished by extending the chain length of the bonded alkyl group. Hence, the C_2 and C_4 phases have been developed to accomplish this purpose. In general, the bonded phases C_1, C_2, and C_4 have been used for separations of large molecules. In order to develop more hydrophobic interactions, the next most common phase utilizes the octyl-bonded moiety (C_8). At relatively high bonding densities (3–4 μmol/m^2), a wide range of compounds can be separated in the reversed-phase mode with this bonded moiety. Although applications involving large molecules are readily found in the literature, the predominant use involves the separation of typical small [molecular weight (MW) < 500] organic compounds. The most common reversed-phased material contains the octadecyl moiety ($n = 18$) as the bonded group. Although there are reports of phases in the literature with n values between 8 and 18, these are relatively uncommon and have not found widespread use or commercial development. The C_{18}-bonded phase was the separation material used in most of the early development of HPLC; therefore, there are several decades of applications documented in the literature. It is still by far the most often used bonded material in reversed-phase HPLC and is available in a wide variety of forms (type of silica, pore size, surface area, monomeric, polymeric, endcapped, nonendcapped, etc.) from more than 100 commercial sources. Although small organic molecules account for the majority of applications, its early commercial availability and its role in the development of HPLC has lead to examples of separations involving a broad range of compounds, including ionic species, polar compounds, biomolecules, fatty acids, and diastereomers. Because most laboratories with HPLC equipment will have a C_{18} column available, and sometimes the only one on hand, it is the first choice for initial experiments. In addition, with the broad range of applications accessilble in the literature or from commercial sources, it is often easy to find a separation that is similar, allowing for selection of mobile-phase conditions that are likely to be suitable for solving a particular analytical problem.

Table 1 Bonded Hydrophobic Groups

Methyl	$-CH_3$
Ethyl	$-CH_2CH_3$
Butyl	$-CH_2-(CH_2)_2-CH_3$
Octyl	$-CH_2-(CH_2)_6-CH_3$
Octadecyl	$-CH_2-(CH_2)_{16}-CH_3$
Triacontyl	$-CH_2-(CH_2)_{28}-CH_3$
Phenyl	$-CH_2-(CH_2)_x-C_6H_5$
Perfluoro	$-CH_2-(CF_2)_x-CF_3$

As shown in Table 1, a number of other bonded groups have also found use in reversed-phase HPLC. Theoretically, there is no limit to the value of n for bonded alkyl groups. However, until recently, there has been little interest in phases longer than 18 carbons. Some recent studies have demonstrated interesting applications for the C_{30} phase so that its use as well as materials with alkyl chain lengths between 18 and 30 might become more common. A phenyl-bonded group (with alkyl chains attaching it to the surface of various lengths) can also function in the reversed-phase mode. The possibility of utilizing $\pi-\pi$ interactions or charge-transfer effects with the phenyl phase leads to a different selectivity than the solely hydrophobic intereactions that are available from the common alkyl-bonded materials. A similar reasoning can be applied for the phases where F is substituted for H in the bonded organic group.

Although the vast majority of stationary phases for RP-HPLC are based on chemically modified silica, there are a few other supports that have been investigated and some which are available commercially. Although silica has many advantages, its main limitation is the pH range over which it is stable. Depending on the type of silica, bonding method, and surface coverage, most chemically modified silicas are useful from pH 2 to 8. Outside of this range, most materials will experience some type of accelerated degradation. One solution to this problem is to substitute an oxide with a greater pH stability than silica. Some possibilities include alumina, zirconia, and titania, which can all be fabricated in particles with properties similar to those of silica (size, porosity, and surface area) as well as having hydroxide groups on the surface that can be used for chemical modification. Another approach is to use polymeric materials as supports in RP-HPLC. Polymers can be formed into beads similar to oxide particles, can be chemically modified to contain various organic functional groups to control their chromatographic properties, and can possess pH stability in strong acids and bases. If such modification or the basic structure of the polymer is hydrophobic, then these materials can be used in the reversed-phase mode. The main disadvantage to many polymeric materials is that they often expand or contract in various mobile-phase compositions, leading to nonreproducible chromatographic performance. Despite the potential pH advantages of these alternative supports, they have not been extensively exploited because of the long-term use of silica in the development of chemically bonded stationary phases and the limited number of applications where either very acidic or basic eluents are an absolute necessity.

The structure of the alkyl-bonded moiety on the support surface has been the subject of many investigations. A variety of spectroscopic and chromatographic methods have been employed to determine the configuration of various bonded organic groups (although the vast majority of studies have been on C_{18}) in order to understand the mechanism of separation for typical solutes. There are many variables to be considered in these investigations, which include type of bonded group, bonding density, and the nature of the support surface. Some studies involve the presence of solvents to mimic the mobile phase, whereas others utilize the bonded material in the absence of any liquids. Despite these differences, some generalizations can be made about the structure of typical bonded phases in the presence of water–organic solvents, as illustrated in Fig. 1. At low concentrations of an organic constituent (A), the environment around the bonded moiety is polar and the hydrophobic chains tend to collapse on each other in order to minimize their exposure to the surronding solvent. As the percent of organic in the liquid around the bonded group increases (B $\rightarrow$ C), the medium is less polar and the groups are no longer strongly associated with each other. Although reversed-phase bonded materials have been availabe for many years, there is continued development to improve their chromatographic performance and to develop new phases for specialized applications.

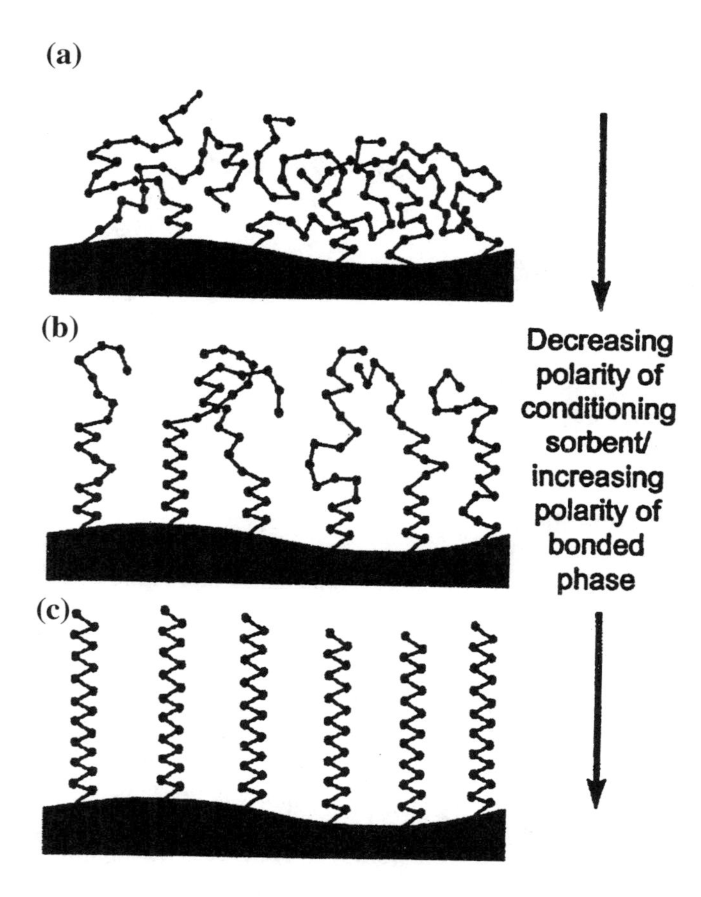

Fig. 1 Structure of a bonded phase as a function of polarity of a mobile phase: (a) highly polar mobile phase; (b) intermediate polarity mobile phase; (c) low-polarity mobile phase.

Suggested Further Reading

Iler, R. K., *The Chemistry of Silica*, John Wiley & Sons, New York, 1979.

Marciniec, B., *Comprehensive Handbook on Hydrosilylation*, Pergamon Press, Oxford, 1992.

Nawrocki, J., *Chromatographia 31*: 177 (1991).
Nawrocki, J., *Chromatographia 31*: 193 (1991).
Pesek, J. J. and M. T. Matyska, *Interf. Sci. 5*: 103 (1997).
Pesek, J. J., M. T. Matyska, J. E. Sandoval, and E. J. Williamsen, *J. Liquid Chromatogr. Related Technol. 19*: 2843 (1996).
Unger, K. K., *Porous Silica*, Elsevier, Amsterdam, 1979.
Vansant, E. F., P. Van Der Voort, and K. C. Vrancken, *Characterization and Chemical Modification of Silica*, Elsevier, Amsterdam, 1995.

Joseph J. Pesek
Maria T. Matyska

Rotation Locular Countercurrent Chromatography

Introduction

Rotation locular countercurrent chromatography (RLCCC) was introduced in the early 1970s [1,2] as a preparative CCC system. In general, the existing CCC systems may be classified into two groups according to the mode of solute partitioning. One is called the hydrostatic equilibrium system (HSES) and the other is called the hydrodynamic equilibrium system (HDES). RLCCC belongs to HSES as does droplet CCC, whereas the high-speed CCC is the most advanced form of HDES, which has been widely used for the separation and purification of natural products.

Although RLCCC is less efficient than high-speed CCC, in terms of resolution and separation times, it has advantages of a large-sample loading capacity and universal application of two-phase solvent systems. Retention of the stationary phase is accomplished simply by adjusting the column rotation speed and flow rate according to physical properties of the solvent system. In addition, RLCCC can be effectively performed with a short column by alternatingly eluting the column with the two solvent phases. This "alternating CCC" method [3,4] is described later in some detail.

Rotation locular countercurrent chromatography is particularly suitable for the preparative separation of natural products, and the apparatus is commercially available through Tokyo Rikakikai Co., Ltd., Tokyo, Japan.

Apparatus

Rotation locular countercurrent chromatography uses a separation column containing a series of cylindrical partition units called "locules." This locular column is made by inserting multiple centrally perforated disks into a PTFE (polytetrafluoroethylene) or glass tubing at regular intervals. Multiple column units are connected in series with PTFE tubing and mounted in parallel around the rotary shaft of the apparatus. The column assembly is held at a constant angle from the horizontal plane and rotated at a moderate rate (60–80 rpm). Figure 1 schematically illustrates the RLCCC apparatus. In each locule, the two phases form a horizontal interface and efficient stirring of each phase is produced by rotation of the column assembly. The system provides the choice of the mobile phase, where the upper phase is eluted in an ascending mode and the lower phase in a descending mode through the inclined column. The solutes present in the sample solution are subjected to an efficient partition process between the two phases, in each locule, and, finally, eluted according to their partition coefficients.

In the early prototype instrument [1], the columns were fabricated from relatively large-bore PTFE tubing of 4.6 mm inner diameter (i.d.) with PTFE disk inserts having 0.8-mm-diameter holes. These disks were spaced in 3-mm intervals to form 47 locules in each unit. A number of column units were connected in series to provide 5000 locules with a total capacity of 100 mL. The capability of the system was demonstrated with the separation of DNP (dinitrophenyl)–amino acids using a two-phase solvent system composed of chloroform–acetic acid–0.1*M* HCl at a 2:2:1 volume ratio. In this system, nine DNP–amino acids were resolved within 70 h at about 3000 theoretical plates.

A commercial RLCCC instrument is equipped with a set of 16 locular column units of 16-mm i.d. and 61 cm in length, containing 37 locules in each unit. The column assembly consists of 592 locules with an 800-mL capacity. At a flow rate of 15–25 mL/h, the system can yield 250–400 theoretical plates, which corresponds to 2.3–1.5 locules/plate [5].

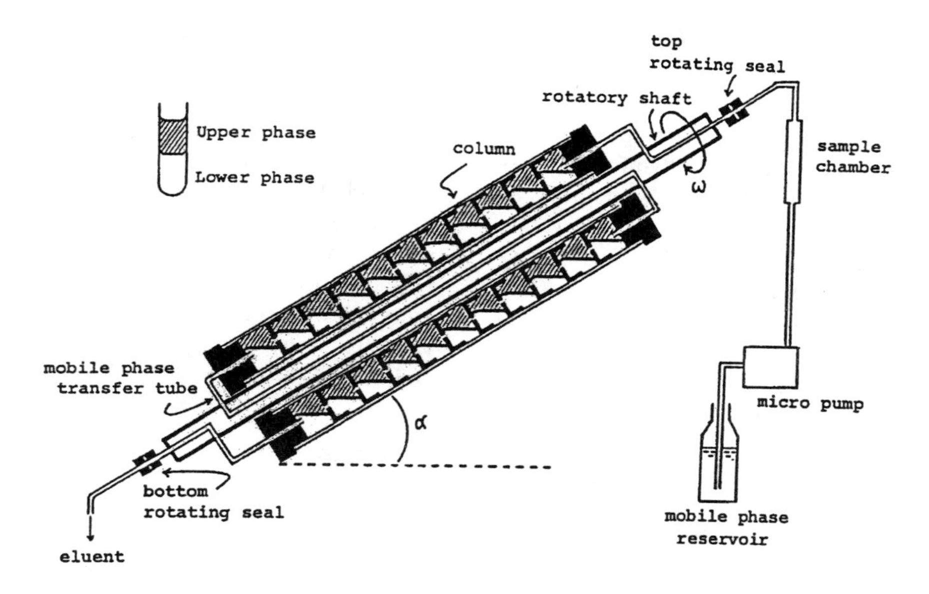

Fig. 1 Rotation locular countercurrent chromatography apparatus.

Separation Procedure

Each separation is initiated by filling the column with either the upper or lower phase of an equilibrated two-phase solvent system. In order to avoid trapping air bubbles in the column, the solvent should be introduced through the bottom of each column, which is kept in a vertical position. Then, the column assembly is tilted at a desired angle (25°–30°) from the horizontal plane. After the sample solution is introduced into the column, the mobile phase is eluted from the column while the apparatus is rotated at a desired rate (60–80 rpm). In order to retain a large volume of the stationary phase, the lower phase is eluted downward from the upper terminus and the upper phase upward from the lower terminus of the column assembly. The effluent from the outlet of the column is continuously monitored with an ultraviolet (UV) monitor and collected into test tubes using a fraction collector.

Applications

Because the apparatus became commercially available in the late 1970s, RLCCC has been applied mainly to the preparative separation of natural products, due to its large sample loading capacity. As with other CCC systems, the partition efficiency of the RLCCC system highly depends on the choice of the suitable two-phase solvent system which gives a partition coefficient close to unity ($K \approx 1$) for the targeted compound. The K value can be obtained from a simple spectrophotometric measurement, thin-layer chromatography (TLC), or high-performance liquid chromatography (HPLC), whichever is appropriate.

Separation of Natural Products

Two-phase solvent systems composed of chloroform–methanol–water at various volume ratios are frequently used for the separation of moderately hydrophobic compounds, including flavone aglycones, phenylpropanoids, iridoid glycosides, and so forth [6].

The separation of more polar compounds, such as glycosides, can be achieved using a polar solvent system composed of ethyl acetate–water with a suitable modifier. Flavonoid glycosides were separated with ethyl acetate–1-propanol–water (2:1:2) and saponins with ethyl acetate–ethanol–water (2:1:2).

Chiral Separation

The separation of (±)-norephedrine was first performed by RLCCC using a solvent system composed of 1,2-dichloroethane and 0.5M aqueous sodium hexafluorophosphate (pH 4) containing chiral tartaric acid ester (di-non-5-yl tartrate) [7]. This method produced an efficient resolution of enantiomers at purities of over 95% from 200 mg of racemate.

Rotation locular countercurrent chromatography can be applied to the chiral separation with an aqueous–aqueous polymer phase system using bovine serum albumin (BSA) as a chiral selector. In our laboratory, the RLCCC separation of D- and L-enantiomers of kynurenine was achieved from 200 mg of D,L-kynurenine using a solvent system composed of 10% (w/w) polyethylene glycol 8000 and 5% (w/w) disodium hydrogen phosphate containing 6% (w/w) BSA [8]. Because of a long settling time of the polymer phase system under unit gravity, the method required a discontinuous operation as used in the conventional countercurrent distribution apparatus, which consisted of 3 min for mixing, 10 min for settling, and 1 min for transfer of the mobile phase to the next locule at a flow rate of 1.0 mL/min. Using the lower mobile phase, L-kynurenine was eluted first, followed by D-kynurenine, and the separation was completed in 60 h.

Alternating CCC Method

In this modified method, upper and lower phases are alternatingly used as the mobile phase by eluting the lower phase in the descending mode and the upper phase in the ascending mode through the respective terminus of a short locular column assembly.

Each separation is initiated by filling the entire column with the upper phase of the equilibrated two-phase solvent system. Following the injection of the sample solution, the column is eluted with the lower phase while the apparatus is rotated at 60–70 rpm. After a desired period of elution, when the target compound is about to elute, the mobile phase is switched to the upper phase, which is eluted at the same flow rate but in an ascending mode in the opposite direction. This alternating elution process with the upper and the lower phases is repeated until the desired component is well resolved.

In our laboratory, this method was applied to the purification of food mono-azo dyes [3]. Amaranth, New Coccine, and Sunset Yellow FCF were purified at 99.7%, 99.5%, and 99.3%, respectively, from 1–2.5 g of commercial dyes. Continued research has led to the purification of impurities present in commercial Sunset Yellow FCF that include RS-SA (trisodium salt of 3-hydroxy-4-[sulfophenyl]azo-2,7-naphthalene disulfonic acid), GS-SA (1-[4-sulfophenyl]azo)-2-naphthol-6,8-disulfonic acid), DONS (disodium salt of 6,6′-oxybis-2-naphthalene sulfonic acid), and 2N-SA (sodium salt of 4-[(2-hydroxy-1-naphthalenyl)azo]benzenesulfonic acid). The method successfully isolated GS-SA from Sunset Yellow FCF [4].

References

1. Y. Ito and R. L. Bowman, *J. Chromatogr. Sci. 8*: 315–323 (1970).
2. Y. Ito and R. L. Bowman, *Anal. Chem. 43*: 69A–75A (1971).
3. Y. Kabasawa, T. Tanimura, H. Nakazawa, and K. Shinomiya, *Anal. Sci. 8*: 351–353 (1992).
4. N. Ogura, Y. Nakamura, K. Shinomiya, and Y. Kabasawa, *Anal. Sci. 11*: 759–763 (1995).
   J. K. Snyder, K. Nakanishi, K. Hostettmann, and M. Hostettmann, *J. Liquid Chromatogr. 7*: 243–256 (1984).
   I. Kubo, G. T. Marshall, and F. J. Hanke, *Countercurrent Chromatography: Theory and Practice* (N. B. Mandava and Y. Ito, eds.), Marcel Dekker, Inc., New York, 1988, pp. 493–507.
5. W. D. Conway, *Countercurrent Chromatography: Apparatus, Theory, and Applications*, VCH, New York, 1990.
6. K. Hostettmann, M. Hostettmann, and A. Marston, *Preparative Chromatography Techniques: Applications in Natural Product Isolation*, Springer-Verlag, Berlin, 1986.
7. B. Domon, K. Hostettmann, K. Kovacevic, and V. Prelog, *J. Chromatogr. 250*: 149–151 (1982).
8. Y. Sato, K. Shinomiya, and Y. Kabasawa, *J. Chem. Soc. Japan* 1067–1071 (1994).

Kazufusa Shinomiya

S

Sample Preparation

Before any sample can be subjected to chromatography, some type of sample preparation is required, which can be as simple as filtration or an involved solid-phase extraction protocol. Sample preparation is that activity or those activities necessary to prepare a sample for analysis. The ultimate goal of sample preparation is to provide the component of interest in solution, free from interferences and at a concentration appropriate for detection. This entry will briefly discuss seven topic areas included in sample preparation: standard methods, solid-phase extraction (SPE), matrix solid-phase dispersion (MSPD), solid-phase microextraction (SPME), microdialysis, ultrafiltration (UF), and automated systems.

Should the sample be solid, then the first step would involve the extraction of the sample with an appropriate solution that would solubilize the compound of interest and remove as few interfering compounds as possible. This operation is sometimes conducted using a blender or other mixer to provide as homogeneous extract as possible.

Alternatively, one could use a Soxhlet or similar apparatus to extract the sample. Other methods to prepare a sample for analysis through extraction include supercritical fluid extraction (SFE), pressurized fluid extraction, and microwave-assisted solvent extraction. There will continue to be additions to this list, as techniques evolve and modifications of the more standard techniques are made.

The standard methods that one would use in sample preparation for chromatography include filtration, sedimentation, centrifugation, liquid–liquid extraction (LLE) open-column chromatography, and concentration/evaporation. Filtration for sample preparation may be performed on numerous occasions in a sample preparation protocol, with the first filtration being used to separate large-particulate matter from solvent. The final filtration before chromatography likely uses a 0.45-μm or smaller disposable filter unit to prevent small-particulate matter from contaminating the chromatographic system.

One of the techniques that has become increasingly used and replaced more of the traditional methods of sample preparation is SPE. SPE introduced in the early 1970s offered the possibility of if not eliminating but at least reducing the tedium in sample preparation. SPE has been called "digital chromatography," where samples can be introduced onto a device, interferences removed, and the analyte of interest eluted in a small amount of solvent. Conversely, the SPE device can be used as a flow-through cleanup device. SPE can be used on many occasions as a substitute for LLE.

There is a wide diversity not only in SPE packing types but also SPE formats. The packing types parallel those in used in open-column chromatography and HPLC. The addition of a second-generation-type support such as the Water's OASIS and the Varian NEXUS that are seen as somewhat universal supports for many sample types with other similar supports unique for SPE in continual development. Although the early SPE devices were cartridge or syringe barrel based, the formats have evolved to support the growing demands in drug discovery through the development of microtiter plate formats to support high-throughput screening (HTS) activities. In addition to columns there has also been a growth in disk-based SPE devices. Figure 1 provides an overview of SPE phase and solvent selection.

Solid-phase microextraction (SPME) was developed

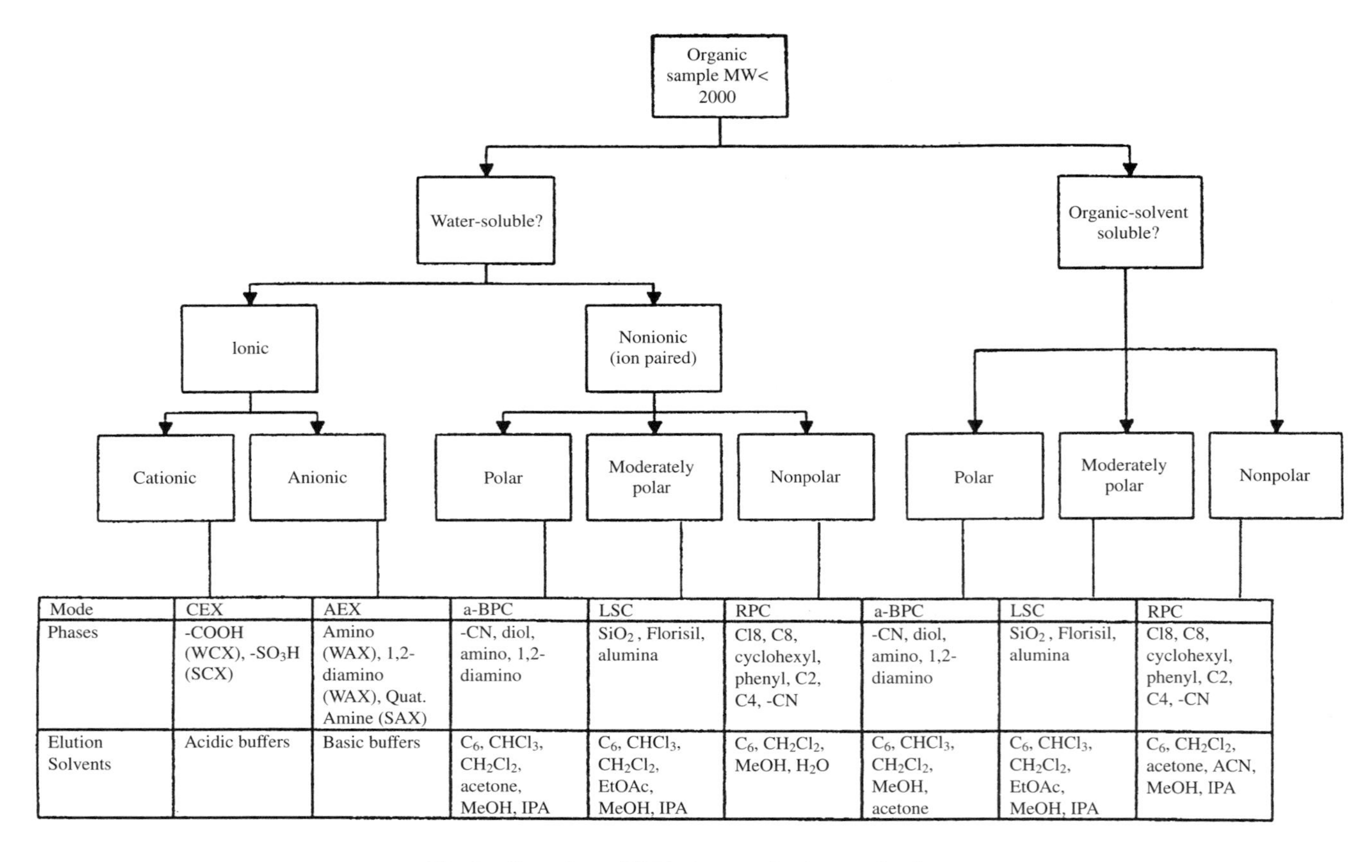

Mode	CEX	AEX	a-BPC	LSC	RPC	a-BPC	LSC	RPC
Phases	-COOH (WCX), -SO_3H (SCX)	Amino (WAX), 1,2-diamino (WAX), Quat. Amine (SAX)	-CN, diol, amino, 1,2-diamino	SiO_2, Florisil, alumina	C18, C8, cyclohexyl, phenyl, C2, C4, -CN	-CN, diol, amino, 1,2-diamino	SiO_2, Florisil, alumina	C18, C8, cyclohexyl, phenyl, C2, C4, -CN
Elution Solvents	Acidic buffers	Basic buffers	C_6, $CHCl_3$, CH_2Cl_2, acetone, MeOH, IPA	C_6, $CHCl_3$, CH_2Cl_2, EtOAc, MeOH, IPA	C_6, CH_2Cl_2, MeOH, H_2O	C_6, $CHCl_3$, CH_2Cl_2, MeOH, acetone	C_6, $CHCl_3$, CH_2Cl_2, EtOAc, MeOH, IPA	C_6, CH_2Cl_2, acetone, ACN, MeOH, IPA

Fig. 1 Overview of SPE phase and solvent selection.

as an alternative to many other sample preparation methods because it uses virtually no solvents or complicated equipment. It is an adsorption/desorption device where the compounds of interest are adsorbed onto a fused-silica fiber. After a given time, the fiber is placed into a gas chromatograph (GC), where the compounds are thermally desorbed. SPME has recently been adapted for use in HPLC, where compounds that are adsorbed are desorbed using an appropriate solvent.

Matrix solid-phase dispersion (MSPD) is an SPE variant where samples are ground and mixed with a C_{18} support. This is placed in a disposable column previously packed with Florisil that traps the fat from the sample, allowing the compounds of interest to be eluted. This has successfully been applied to the determination of lipophilic pesticides from both fatty and nonfatty matrixes.

When supercritical fluid extraction (SFE) was initially introduced, it was thought that it might be the panacea for sample extraction because it used a very innocuous solvent, CO_2. The operator varied pressure, temperature, flow rate, and extraction time, with some extraction protocols requiring the use of small amounts of polar modifiers. All of these variables affected the solvating power of the carbon dioxide. In addition to the carbon dioxide, other supercritical fluids have been used. The technique continues to evolve with increasing numbers of applications being published but has not developed as once might have been predicted.

In microwave-assisted solvent extraction (MASE), the sample and solvent are heated directly rather than in more conventional schemes where the vessel is heated to extract the sample. The sample solvents are placed into a closed vessel that does absorb microwaves. This facilitates the extraction of the samples of interest and has been applied to a variety of sample types and matrixes.

The accelerated solvent extraction (ASE) system introduced by Dionex uses standard solvents at elevated temperatures and pressures to increase extraction efficiency. Samples are placed in stainless-steel extraction vessels that are loaded into the ASE that has been programmed for the extraction protocol. The instrument allows for the unattended extraction of 24 samples. The initial units allowed for solvent blending by premixing solvents before they were placed into the ASE. Recent modifications allow solvent blending to be accomplished in-line. There has been some controversy about the use of the name ASE because it points to instrumentation from one company and other companies have introduced competing products. It has been proposed that the ASE technology be more correctly referred to as pressurized fluid extraction because ASE denotes a commercial device.

Two additional techniques that are used in sample preparation protocols are ultrafiltration and microdialysis. In ultrafiltration, pressure is applied to a membrane and those molecules smaller than the molecular-weight cutoff can pass through while molecules larger are retained. This technique can be used as a way of sample concentration or as a way to eliminate higher-molecular-weight compounds from an analytical scheme. Membranes are available with cutoffs ranging from 300 to 300,000. Microdialysis differs from the other techniques because it is in vivo sampling and has been applied to the determination of drugs and other biomolecules from tissues, organs, and biological fluids. In microdialysis, molecules can diffuse across a membrane resulting in either direct or reverse dialysis.

Finally, no discussion of this topic would be complete without the mention of automation in sample preparation and its impact on this activity. Laboratory robotics' initial focus was on the automation of sample preparation and is used in that way in many laboratories, but "islands of automation" have developed within certain organizations where certain portions of sample preparation such as SPE are automated.

Sample preparation is an extremely broad subject because there are techniques that are more likely used by those in different industry segments. This entry has provided an overview of some of the methods of sample preparation that are widely used. It likely has missed some of the more esoteric ways of sample preparation. The reader is referred to the Suggested Further Readings for additional reading.

Suggested Further Reading

Baker, S. A., Preparation of milk samples for immunoassay and liquid chromatographic screening using matrix solid-phase dispersion, *JAOAC Int. 77*: 848 (1994).

Current Trends and Developments in Sample Preparation, Supplement to *LC-GC*, Advanstar Communications Eugene, OR, 1998.

Current Trends and Developments in Sample Preparation, Supplement to *LC-GC*, Advanstar Communications Eugene, OR, 1999.

Henion, J., E. Brewer, and G. Rule, Sample preparation for LC/MS/MS, *Anal. Chem*, *70*: 650A–656A (1998).

Snyder, L. R, J. J. Kirkland, and J. L. Glajch, *Practical HPLC Method Development*, 2nd ed., John Wiley & Sons, New York, 1997.

VanHorne, K. C., *Handbook of Sorbent Extraction Technology*, Varian Sample Preparation Products, Harbor City, CA, 1994.

Zang Z., and J. Pawliszyn, Headspace solid phase microextraction, *Anal. Chem 65*: 1843 (1993).

W. Jeffrey Hurst

Sample Preparation and Stacking for Capillary Electrophoresis

Introduction

Unlike high-performance liquid chromatography (HPLC), the composition (matrix) of the sample itself in capillary electrophoresis (CE) affects, greatly, the separation, quantification, detection, and precision. Although the sample, in most instances, constitutes a very small portion of the overall volume in the capillary once injected (<1%), the matrix of the sample has profound effects in CE. As the sample size increases, these effects become much more important because the current conductance is affected by any change of the ionic strength along its path and, in turn, affects the sample migration.

Samples obtained from clean sources do not require much, or any, preparation, whereas samples obtained from biological fluids, food, and industrial sources often have a complex matrix, which necessitates manipulation before the CE step. This also depends on the concentration of the analytes relative to the contaminants. Complicating matters further, CE has a relatively low sensitivity, due to the short light path of the capillary, creating a need for clever manipulation of the sample to enhance the detection.

Unlike HPLC, sample preparation in CE requires careful thinking and strategy to obtain a good analysis. There is a relationship between the sample matrix and the separation buffer. Based on how the sample is prepared and how the separation buffer is selected, sample matrix effects can be both favorable and detrimental to the analysis. Matrix effects are different in capillary zone electrophoresis (CZE) from those observed in micellar electrokinetic capillary chromatography (MEKC). Understanding sample matrix effects is the first step in sample preparation in order to obtain a good separation by CE [1].

Effect of Sample Matrix in Capillary Zone Electrophoresis

Sample Ionic Strength

High ionic strength is important for solubility of many proteins. However, there is an inverse relationship between electrophoretic mobility and ionic concentration. An excess of ions in the sample, especially with large sample volumes, ruins the separation in CE. In the electrokinetic injection mode, the excess ions in the sample decreases the migration or transfer of the analytes into the capillary, leading to a diminished detector signal. Pure standards prepared in water compared to those added to serum or saline solutions show a large difference in detector signal, causing the quantification to be difficult. In the hydrodynamic injection mode, a high ionic strength in the sample leads to band broadening or multiple peaks.

Proteins

At low concentrations and under appropriate conditions, proteins have little effect on separation in CE. However, at high concentrations in the sample, the cationic proteins especially tend to adsorb preferentially to the inlet side of untreated capillary walls, changing the zeta-potential, which, in turn, affects the electroosmotic flow (EOF). Slight variations in the EOF along the capillary length cause an increase in band spreading and a decrease in peak symmetry, leading to a poor reproducibility.

pH

The ionization and net charge of the sample components are affected greatly by changes in the pH. As a result, the migration rate, the solubility, the theoretical plate number, and peak height could be all affected.

Matrix Effects in Micellar Electrokinetic Capillary Chromatography

Micellar electrokinetic capillary chromatography (MEKC) is used, often, for separating neutral and hydrophobic molecules. The surfactants in MEKC have the added advantage of solubilizing proteins. This can eliminate the need for extraction or deproteinization, allowing direct sample injection. The effect of sample matrix in MEKC is less dramatic than that in CZE. A high SDS concentration in the running buffer causes an increase in peak height. However, at the same time, a high sodium dodecyl sulfate (SDS) concentration produces excessive current generation and long migration time.

Surfactants in the Sample

In general, a high concentration in the sample of the same surfactant used in the electrophoresis buffer decreases the peak height. Hence, a surfactant in the sample different from that of the running buffer is preferred. The peaks with higher k' are more affected by the surfactants in the buffer.

Organic Solvents in the Sample

Micelle electrokinetic capillary chromatography is often employed to separate nonpolar compounds. The addition of organic solvent to the sample, especially with large-sample injection, decreases the peak height as well as the resolution.

Sample Volume and Stacking

Because of the relatively low sensitivity of detection in CE, a high sample volume is very desirable to increase the signal [1–4]. Unfortunately, sample overloading can occur easily in CE. The theoretical plate number in CE depends greatly on the sample (volume). For example, the theoretical plate numbers dropped from 800,000 to less than 10,000 by increasing the injection time from 0.2 to 15 s [5]. A low sample volume offers very high theoretical plate numbers, but, at the same time, it yields small detector signals. The rule of thumb is to keep the sample plug, under nonstacking conditions, less than 1% of the capillary length [1].

Stacking in CZE

Sample concentration on the capillary is called stacking. Under these conditions a large sample volume is injected and the two edges of the sample are induced to migrate at different rates towards each other before entering the electrophoresis buffer. This mechanism is very useful in CE as it leads to enhanced sensitivity, higher plate numbers, and better separations. Stacking is very simple and can be brought about by several manipulations of the sample:

Low Ionic Strength in the Sample

The sample is dissolved in the same electrophoresis buffer but at a 10 times lower concentration. This causes the sample resistance and the field strength (V/cm) in the sample plug to increase. In turn, this causes the ions to migrate rapidly and stack as a sharp band at the boundary between the sample plug and the electrophoresis buffer with the positive ions lining up in front of the negative ones before entering the electrophoresis buffer. As a result of this simple manipulation, the sample can be concentrated up to 10-fold [6]. The stacking effect can be utilized with both pressure and electrokinetic injections (field amplification injection). Similar stacking can be accomplished also by injecting a very short plug of water into the capillary before injecting the sample.

pH Adjustment

Stacking based on adjusting the pH of the sample has been described. Peptides can be concentrated by dissolving the sample in a buffer 2 units above the net *pI*, so the peptides are negatively charged. As the potential is turned on, the peptides initially migrate toward the anode until they are stopped by the interface of the electrophoresis buffer, where they concentrate. After the short pH gradient of the sample dissipates in the electrophoresis buffer, the peptides become positively charged. Thus, they migrate toward the cathode as a sharp zone. Catecholamines can be concentrated based on the same principle at acidic conditions in the sample.

Acetonitrile–Salt Mixture

When acetonitrile is present in the sample at about 66% together with a low concentration of sodium chloride, another unique type of sample stacking occurs. In practice, biological samples (1v) are deproteinized with 2 volumes of acetonitrile and injected directly into the CE. The acetonitrile is used mainly to remove proteins; however, it has additional advantages: (1) It reverses the harmful effects of ions and allows larger volumes of sample (about one-third of the capillary volume) to be injected. The overall effect is an increased sensitivity of about 20-fold. Cationic compounds stack better in acetonitrile using a high concentration of zwitterionic buffers. The mechanism behind this stacking is not clear but it is thought to involve the low conductivity of the acetonitrile with a transient-isotacophoretic-like effect brought about by the salts [2–4].

Stacking in MEKC

Stacking in MEKC [7] is more difficult than in CZE. In general it depends on solubilizing the neutral molecules in micelles under low conductivity to accelerate their migration at a negative polarity. Another approach is using reversed-migration micelles for stacking. In a third technique, the pH of the separation buffer used is acidic so the micelles have a higher electrophoretic velocity than the EOF.

Sample Preparation

Dilution and Direct Sample Injection

One of the main advantages of CE for routine analysis in industrial and clinical settings is the simplicity of sample introduction. If the compound of interest has a strong absorptivity and/or is present in a high concentration relative to the interfering compounds, it may be injected directly without any sample treatment [1,4,8]. In order to tolerate a higher ion concentration in the sample, the ionic strength of the buffer should be as high as possible. Several drugs have been successfully analyzed with direct serum or urine injection especially by the MEKC, where the micelles solubilize the proteins.

Extraction

If the compound is present at a low concentration in the presence of many interfering compounds, cleanup and concentration steps become necessary [8]. Solvent extraction procedures are used often for sample preparation for drugs and many other small molecules in the chromatographic technique. Solid-phase methods are very popular because of the wide choice of the packing material. Sample extraction methods have two other main advantages in CE, namely sample concentration and elimination of both sample ions and proteins. Double-solvent extraction is very useful for electrokinetic injection, especially for basic compounds. It eliminates variability due to matrix effects in electrokinetic injection.

Filtration and Dialysis

Large molecules can be separated from the small ones through special dialysis and filtration membranes which are available with different molecular weight cutoff points.

However, the use of small commercial dialysis cells or blocks is more suitable for CE than the traditional bags. Both sides of the dialysis chambers can be used for CE analysis based on whether the compound of interest has a high or low molecular weight. Small volumes can also be filtered rapidly through special filtration devices in a microfuge at 15,000*g*. Both of these techniques can be applied to the cleanup of urine proteins for analysis by CE.

Organic Solvent Deproteinization

In HPLC, alcohols and especially acetonitrile are often added to the sample to remove serum proteins. In CE, in addition to removing proteins, the presence of acetonitrile in the sample leads to stacking and indirectly improves precision of quantification because of the ability to increase the sample volume. We have analyzed several drugs and other natural compounds by CE after acetonitrile deproteinization. Protein removal can also be accomplished by alcohols such as ethanol and by acids such as perchloric. Precipitation with acids is less desirable in CE than with organic solvents because it increases the salt load. Following precipitation, proteins could also be dissolved in the appropriate buffers and assayed by CE.

Desalting

Desalting is a difficult procedure to perform in routine assays. There are several methods for desalting, such as (1) dialysis as described earlier, (2) ion exchangers (e.g., Chelax 100 and AG 50X2), and (3) reversing the polarity during CE. Organic solvent extraction is another simple method for eliminating ions. In general, the use of high-ionic-strength running buffers enables the direct analysis of samples containing a relatively high salt concentration, eliminating the need for desalting in many instances.

References

1. Z. K. Shihabi, Effect of sample matrix on capillary electrophoresis, in *Handbook of Electrophoresis*, 2nd ed. (J. P. Landers, ed.), CRC Press, Boca Raton, FL, 1997, pp. 457–477.
2. Z. K. Shihabi, Sample stacking by acetonitrile-salt mixtures, *J. Capillary Electrophoresis 2*: 267–271 (1995).
3. M. A. Friedberg, M. Hinsdale, and Z. K. Shihabi, Effect of pH and ions in the sample on stacking in capillary electrophoresis, *J. Chromatogr. A 781*: 35–42 (1997).
4. Z. K. Shihabi, Review: Therapeutic drug monitoring by capillary electrophoresis, *J. Chromatogr. A 807*: 27–36 (1998).
5. A. Vinther and H. Soeberg, Mathematical model describing dispersion in free solution capillary electrophoresis under stacking conditions, *J. Chromatogr. 559*: 3 (1991).
6. D. S. Burgi R-L. Chien, Optimization in sample stacking for high-performance capillary electrophoresis, *Anal. Chem. 63*: 2042–2047 (1991).
7. J. P. Quirino and S. Terabe, On-line concentration of neutral analytes for micellar electrokinetic chromatography. 3. Stacking with reverse migration micelles. *Anal. Chem. 70*: 149–157 (1998).
8. J. Caslavska, S. Lienhard, and W. Thormann, Comparative use of three elektrokinetic capillary methods for the determination of drugs in body fluids. Prospects for rapid determination of intoxications, *J. Chromatogr. 638*: 335–342 (1993).

Zak K. Shihabi

Scale-up of Countercurrent Chromatography

Introduction

There are few processes that can be predictably scaled up from laboratory to production scale without difficulties. Preparative high-performance liquid chromatography (HPLC), for example, is not a linear scale-up; it is expensive and uses large volumes of solvents. The product can become hydrolyzed by or react with the column, which can induce chemical/steric/chiral conformation changes and often requires significant prepurification with further risk of degradation.

Countercurrent chromatography (CCC) [1,2] is a process that avoids these difficulties. It is a form of liquid–liquid chromatography without a solid support, which separates soluble natural product substances on their partition, or differential solubility, between two immiscible solvents. The principle of separation (partition) is the same in both the laboratory and the production plant and is generic in that it can be applied to an extremely broad range of purification problems in many industries. Furthermore, because there is no solid support, there is 100% sample recovery and no need for any pre-purification.

A recent review on CCC as a preparative tool [3] described an extremely useful comparison of four different CCC approaches and concluded that "the real future belongs to the new generation of centrifugal instruments." They concluded that more reliable designs were required, that there was a need to accommodate higher loads on the 100-g to 1-kg scale, and that truly preparative instruments needed to be developed. They called for a better understanding of the mechanisms of separation in order to achieve this.

Ito's work [4] on pH zone refining makes a valuable contribution to the scale-up scenario. It offers a method of operating existing instruments preparatively when purifying ionizable compounds with the ability of achieving sample loadings two orders of magnitude higher than normal. Sutherland et al. [5] demonstrate that preparative gram-quantity separations of crude plant extracts use one-tenth the volume of solvents compared to the equivalent prep-HPLC. Sandlin and Ito [6] have shown that CCC is feasible using a "J"-type coil planet centrifuge with tubing bore up to 5.5-mm internal diameter and have successfully demonstrated fractionations in 750-mL coils, but at relatively low flow, speed, and β value. They have also investigated the effect on resolution of increasing sample volume and sample concentration [7]. Ito and colleagues [7–9] have described unit-gravity (noncentrifugal) slowly rotating coil devices, which would be suitable for large-scale CCC separations.

Parameters Affecting Scale-up

The main parameters affecting scale-up have been analyzed in detail by Sutherland et al. [10]. For scale-up of CCC to be successful, they recommended that two measures or responses had to be maintained as the process was scaled up: retention of the stationary phase and resolution of the sample components. It was emphasized that even if it was possible to retain phases as the tubing bore increased, it was possible that the hydrodynamics of the mixing and settling zones may not work as well, as the bulk volume to surface area ratio increased.

They studied three "J"-type coil planet centrifuges with different coil sizes: analytical (d = 0.76 mm), lab prep (d = 1.6 mm), and process (d = 3.68 mm). By constructing the coils from stainless steel, they were able to increase flow considerably without risk of bursting the tubing or causing the tubing to work loose under the action of high cyclic forces, which can be a common problem (see later). They first showed that there was no difference in retention between coils made with stainless steel or polytetrafluoroethylene (PTFE). This was an important experiment that showed that retention was a hydrodynamic process and not governed by the surface properties of the tubing-wall material.

The Effect of These Scale-up Parameters on Retention

Figure 1 shows Sutherland et al.'s plot of retention against flow for the three CCC units with different bore sizes. It clearly shows how increasing the tubing bore not only allows higher throughput but also shows that retention with larger bores is far more tolerant or stable when flow is increased, a very important discovery for industrial scale-up. They went on to demonstrate that increasing speed allowed even higher retention and linear flows of the mobile phase and that the mean Reynold's numbers of the mobile-phase flow were still well within the laminar flow region.

They concluded that increasing two of the three major variables affecting scale-up, speed and tubing bore, actually improved retention. The third, flow, decreased retention as flow increased but less so as the bore increased. Tubing material and retention of the stationary phase therefore are no barrier to the industrial scale-up of CCC.

The Retention Behavior of Different Phase Systems

Du et al. [11] have shown that there is a linear relationship between retention and the square root of flow. The negative gradient of this line gives an indication of the stabil-

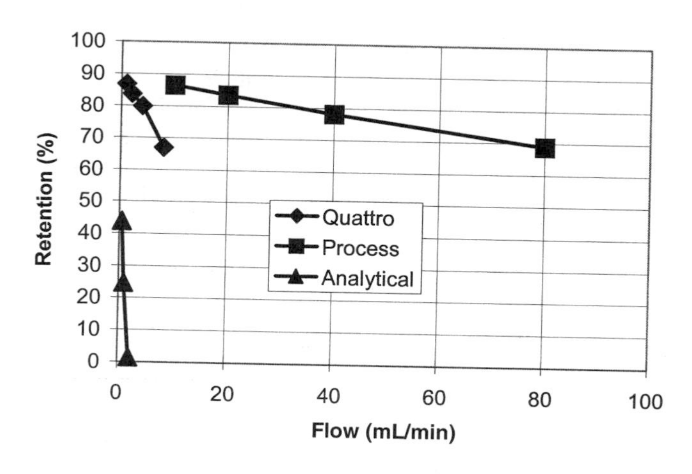

Fig. 1 Variation of retention with flow for different tubing bores: an analytical CCC (0.76 mm), a Bunel CCC (1.6 mm), and a process CCC (3.68 mm).

Table 1 Regression Analysis between Retention (S_f) and the Square Root of Flow ($\sqrt{F}$) for Phase Systems Tested on the Process CCC at 800 rpm

Solvent system	Linear regression	Correlation
Heptane–ethyl acetate–methanol–water (1.4:0.1:0.5:1.0)	$S_f = 97.27 - 3.1341\sqrt{F}$	0.9936
Heptane–ethyl acetate–methanol–water (1.4:0.6:1.0:1.0)	$S_f = 95.73 - 3.3658\sqrt{F}$	0.9988
Heptane–ethyl acetate–methanol–water (1.4:2.0:2.0:1.0)	$S_f = 102.02 - 6.0416\sqrt{F}$	0.9950
Iso-hexane–acetonitrile (1:1)	$S_f = 106.74 - 6.9389\sqrt{F}$	0.9994

Table 2 Regression Analysis between Retention (S_f) and the Square Root of Flow ($\sqrt{F}$) for Phase Systems Tested on the Process CCC at 1200 rpm

Solvent system	Linear regression	Correlation
Heptane–ethyl acetate–methanol–water (1.4:0.1:0.5:1.0)	$S_f = 97.724 - 2.3506\sqrt{F}$	0.9991
Heptane–ethyl acetate–methanol–water (1.4:0.6:1.0:1.0)	$S_f = 100.41 - 3.107\sqrt{F}$	0.9877
Heptane–ethyl acetate–methanol–water (1.4:2.0:2.0:1.0)	$S_f = 103.63 - 5.2379\sqrt{F}$	0.9996
Iso-hexane–acetonitrile (1:1)	$S_f = 105.84 - 5.2228\sqrt{F}$	0.9996
Butyl alcohol–acetic acid–water (4:1:5)	$S_f = 100.65 - 11.484\sqrt{F}$	1.0

ity of the retention process for a given phase system: The shallower the negative gradient, the more stable the process and the higher the flow possible for a given retention. Tables 1 and 2 give the linear regressions for the phase systems tested on the process-scale CCC [10] at 800 and 1200 rpm, respectively. In all cases, the lower phase is the mobile phase pumping from head (center) to tail (periphery). Retention was measured at four different flow rates: 10, 20, 40, and 80 mL/min, except for the butyl alcohol–acetic acid–water (4:1:5) phase system where only flows of 10 and 20 mL/min were tested before retention dropped below 50%. A range of two-phase solvent systems are listed across the polarity range. In the case of hydrophilic low interfacial tension phase systems like butyl alcohol–acetic acid–water (4:1:5), a high speed of rotation was required to achieve a reasonably high retention.

The Effect of These Scale-up Parameters on Resolution

There has not been any significant change in resolution detected [10] as the bore size increases, provided the sample volume injected maintains the same ratio of coil volume. This is a significant finding, as in most chromatography processes, resolution reduces as the process is scaled up. Resolution was found to increase with increasing speed of rotation as would be expected due to the increased number of mixing and settling cycles per unit time.

The effect of flow on resolution is shown in Fig. 2 for the process CCC [10] running at 1200 rpm. The resolu-

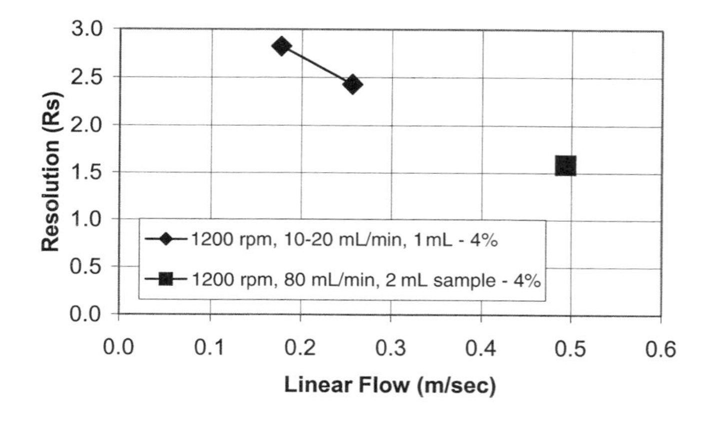

Fig. 2 Variation of resolution with flow for benzyl alcohol and phenyl ethanol in a heptane–ethyl acetate–methanol–water (1.4:0.1:0.5:1.0) phase system.

tion is between benzyl alcohol and phenyl ethanol resolved using the heptane–ethyl acetate–methanol–water (1.4:0.1:0.5:1.0) phase system. Resolution drops off with increasing flow as would be expected, as the sample will have experienced fewer mixing and settling steps before it elutes and the retention is lower. However, the increased flow appears to improve mixing, as it can be seen that this drop off is only gradual. Doubling flow does not halve resolution and so it would appear advantageous to increase flow as much as possible in the scale-up process.

Engineering Challenges of Scale-up

Sutherland et al. [10] and, earlier, Sandlin and Ito [6] have shown that scale-up is feasible. It can be seen that over 60% retention has been achieved for a broad range of phase systems with flows of 0.1 L/min in a 1-L capacity coil. This leads to the solvent front ($k = 0$) eluting in 4 min and the $k = 1$ point in 10 min with sample volumes of at least 0.1 L possible. All this adds up to sample process throughputs of up to 1 L/h or in weight terms as much as 1 kg/day.

However, before this can be realized, the engineering of the coil planet centrifuge will have to be made more reliable. The cyclical forces that produce the unique mixing and settling zones within the coiled tubes can cause them to shake apart and loosen. Janaway et al. [12] have solved this problem by developing new techniques for winding coils. As the scale increases, the volumes of samples being pumped through become extremely high; therefore, designing flying leads that can be guaranteed to not leak becomes paramount. The coil planet centrifuge is a rotating piece of equipment and bearings can wear out. With such high cyclical forces, the reliable engineering of larger CCC units will not be trivial.

So far, Sutherland et al. [10] have only been working with 110-mm-radius coil planet centrifuge (CPC) rotors with a capacity of 1 L, which can be operated in a conventional laboratory. Sandlin and Ito [6] have gone as high as 150 mm with capacities of 0.75 L but with a lower speed and β value. The engineering challenge will be to build the next generation of process units at larger rotor radius with capacities of tens or hundreds of liters, but they would need to be installed in hazards plants using intrinsically safe manufacturing practices.

Conclusions

The chromatographic scale-up of countercurrent chromatography appears feasible, but there are engineering challenges ahead which will need to be solved before this promising new technology can be realized.

Acknowledgments

Some of the work presented was undertaken as part of a BBSRC/DTI LINK Consortium study on the "Industrial Scale up of Countercurrent Chromatography." The author would like thank both the BBSRC and the DTI for their financial support and the members of the consortium [10] who have also contributed toward progressing the scale-up of countercurrent chromatography near to reality.

References

1. W. D. Conway, *Countercurrent Chromatography: Apparatus, Theory and Applications*, VCH, New York, 1990.
2. Y. Ito, Principle, apparatus, and methodology of high-speed countercurrent chromatography, in *High Speed Countercurrent Chromatography*, (Y. Ito and W. D. Conway, eds.), Chemical Analysis Series Vol. 132, John Wiley & Sons, New York, 1996, pp. 3–44.
3. A. Marston and K. Hostettmann, Countercurrent chromatography as a preparative tool — Applications and perspectives, *J. Chromatogr. 658*: 315–341 (1994).
4. Y. Ito, pH-Peak-focusing and pH-zone-refining countercurrent chromatography, in *High Speed Countercurrent Chromatography* (Y. Ito and W. D. Conway, eds.), Chemical Analysis Series Vol. 132, John Wiley & Sons, New York, 1996, pp. 121–175.
5. I. A. Sutherland, L. Brown, S. Forbes, D. Games, D. Hawes, K. Hostettmann, E. H. McKerrell, A. Marston, D. Wheatley, and P. Wood, Countercurrent chromatography (CCC) and its versatile application as an industrial purification & production process, *J. Liquid Chromatogr. 21*(3): 279–298 (1998).
6. J. L. Sandlin and Y. Ito, Gram quantity separation of DNP (dinitrophenyl) amino acids with multi-layer coil countercurrent chromatography (CCC), *J. Liquid Chromatogr. 7*(2): 323–340 (1984).
7. J. L. Sandlin and Y. Ito, Large-scale preparative countercurrent chromatography with a coil planet centrifuge, *J. Liquid Chromatogr. 8*(12): 2153–2171 (1985).
8. Y. Ito and R. Bhatnagar, Improved scheme for preparative CCC with a rotating coil assembly, *J. Chromatogr. 207*: 171–180 (1981).
9. Q. Du, P. Wu, and Y. Ito, Low-speed rotary countercurrent chromatography using a convoluted multilayer helical tube for industrial separation, *Anal. Chem. 72*: 3363–3365 (2000).
10. I. A. Sutherland, A. Booth, L. Brown, S. Forbes, D. E. Games, A. S. Graham, D. Hawes, M. A Hayes, S. Jackson, L. Janaway, B. Kemp, H. Kidwell, G. Lye, P. Massey,

C. Preston, P. Shering, T. Shoulder, C. Strawson, F. Veraitch, R. Whiteside, H. Wolff, and P. Wood, Industrial scale-up of countercurrent chromatography, *J. Liquid Chromatogr.*, *24,* in press (2001).
11. Q. Du, C. Wu, G. Qian, P. Wu, and Y. Ito, Relationship between the flow-rate of the mobile phase and retention of the stationary phase in counter-current chromatography, *J. Chromatogr. A 835*: 231–235 (1999).
12. L. Janaway, D. Hawes, I. A. Sutherland, and P. Wood, Chromatography Apparatus (coil winding process and winding tubing into a coil) UK Patent Application No 0015486.4 filed 23 June 2000.

Ian A. Sutherland

SEC with On-Line Triple Detection: Light Scattering, Viscometry, and Refractive Index

Introduction

During the 1980s, accurate molecular weights (M) and molecular-weight distributions (MWD) could be obtained by SEC in conjunction with multiangle light scattering (LS), size-exclusion chromatography (SEC) using conventional calibration, or SEC in combination with viscometry (VISC) using universal calibration (UC). In addition to generating M and MWD, the viscometer with UC yields conformational and branching information. The impetus to combine the two advanced detector technologies of LS and VISC into a single, efficient, and accurate SEC method has been fueled by a growing interest to characterize both natural polymers and the increasing array of synthetic polymers [1,2].

New electronics and improved computer data acquisition capabilities have permitted the development of SEC with on-line triple detection using LS, VISC, and refractometry. On-line triple detection is known as size-exclusion chromatography cubed (SEC^3) with the three dimensions being defined by the three detectors [3]. The use of SEC^3 eliminates the requirement for column calibration, unlike conventional and universal calibration, where a premium is put on control of variables such as flow rate, temperature, and column resolution. SEC^3 can offer advantages in polymer production quality control as well as in research and development of new polymers.

Theory

The schematic for one possible configuration of SEC^3 hardware is shown in Fig. 1. When polymer molecules exit from the SEC column(s), they are simultaneously monitored in real time by three on-line detectors: right-angle laser LS [4], VISC, and refractive index (RI). The following simplified equations illustrate the variables that relate to the responses of the three detectors:

$$(M)\left(\frac{dn}{dc}\right)^2(C) \rightarrow \text{LS} \tag{1}$$

$$([\eta])(C) \rightarrow \text{VISC} \tag{2}$$

$$\left(\frac{dn}{dc}\right)(C) \rightarrow \text{RI} \tag{3}$$

The term dn/dc refers to the change in RI of a polymer relative to its concentration. The LS detector responds to M, the VISC detector responds to the intrinsic viscosity ($[\eta]$), which is inversely proportional to molecular density, and the RI detector monitors concentration (C). A single narrow standard is used to determine the offset constants related to the interdetector volume for a given three-detector system [5]. Either C or dn/dc of a polymer sample must be known a priori in order to calculate the other variable using the RI detector [Eq. (3)]. Once both dn/dc

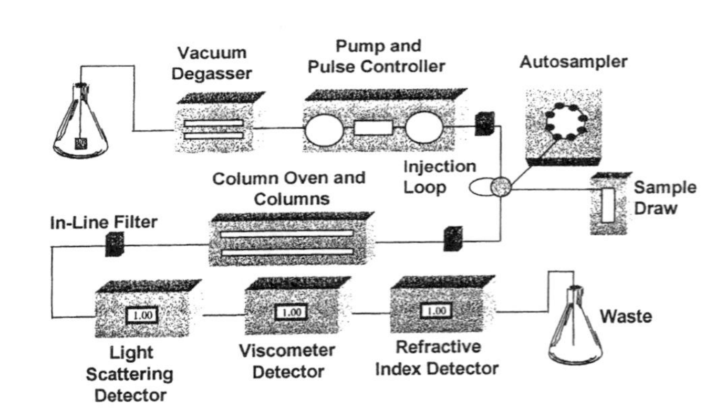

Fig. 1 Hardware schematic for a typical SEC^3 triple on-line detector configuration.

and C are known, the LS and VISC [Eqs. (1) and (2)] can be solved to determine M and $[\eta]$, respectively, for a polymer sample [3]. Structural information, such as chain flexibility, branching, and intramolecular interactions are all related to $[\eta]$. Several key polymer properties related to $[\eta]$ are as follows:

Chain Length. As the chain length increases, $[\eta]$ increases and the density decreases. This behavior can be fitted to the well-known Mark–Houwink (M-H) equation [Eq. (4)] relating M (approximate chain length) to $[\eta]$. The M-H constant a is the slope of the double-logarithmic plot of $[\eta]$ versus M, and log K is its intercept.

$$[\eta] = KM^a \tag{4}$$

Conformation. If a polymer molecule is folded onto itself, instead of keeping the fully extended chain, the density will be higher resulting in a lower $[\eta]$. This can be induced either by strong intramolecular attractions (e.g., hydrogen-bonding) or by a poor solvent. The Flory–Fox equation [3] calculates $[\eta]$ for a linear flexible coil molecule in solution, relating $[\eta]$ to radius of gyration, R_g. Equation (5) shows this linear flexible coil example, where ϕ is the Flory-Fox constant.

$$[\eta]M = 6^{2/3}\phi R_g^3 \tag{5}$$

Chain Flexibility. If two polymers have the same M, the stiff chain one will produce a coil of lesser density and greater $[\eta]$, compared with its flexible coil counterpart.

Chain Branching. A branched molecule is more compact, having greater density and lower $[\eta]$ than its linear counterpart. The Zimm–Stockmayer theory defines the g factor for a polymer as the ratio of $[\eta]$ for the branched polymer to $[\eta]$ of the linear polymer, at the same molecular weight, with ε being the shape factor (~0.75).

$$g = \left(\frac{[\eta]_{\text{branched}}}{[\eta]_{\text{linear}}}\right)^{1/\varepsilon} \tag{6}$$

Once g is determined, the branching number B_n (number of branches per molecule), the branching frequency λ (number of branches per arbitrarily selected repeat unit of molecular weight), and f (number of arms for a star) can be calculated. Determinations of B_n, λ and f require equations specific to the type of branching for that polymer [6]. (Consult the entry Long-Chain Polymer Branching: Determination by GPC–SEC an explanation of the related equations.)

Aggregation. Colloidal suspension particles are aggregates, which are formed due to poorly dissolved molecules. Aggregates are more dense and have a lower $[\eta]$ than their nonaggregated counterparts. The LS detector responds strongly to such aggregates. When a low VISC response is coupled with a high LS response, the presence of an aggregate is confirmed.

Thus, the SEC^3 data obtained from its LS detector determine the MWD, whereas the VISC detector characterizes conformation and branching. The efficiency of SEC^3 is a consequence of no column calibration requirement for the determination of M and MWD. The precision of the system is limited only by the signal-to-noise ratios of the LS and RI detectors, not by chromatographic variables such as flow rate and column retention. Sophisticated software is required to display the SEC^3 picture of molecular structure.

Applications

Four examples of polymer characterization by SEC^3 will be discussed: a dextran sample with branching transitions, a pair of brominated polystyrene (PS) samples, aggregation in chitosan, and a PS star polymer. SEC^3 numerical results for dextran, chitosan, and star-branced PS are listed in the corresponding figure captions.

Dextran is a randomly branched polysaccharide with both long- and short-chain branching. The overlay of the traces generated by the three detectors in Fig. 2 shows a large shift toward a higher M for the LS detector, compared to the other two detectors. This indicates polydispersity within the sample, especially in the high-molecular-weight region of the MWD. Because long-chain branching decreases $[\eta]$ more than short-chain branching, the M-H plot indicates a transition from short- to long-chain branching at $\log(M) = 5.5$, where the plot deviates from linearity. The branching frequency plot is another visual presentation of the transition from short- to long-chain branching within dextran's MWD. Again, the slope of the curve change indicates a branching transition.

An overlay of the MWDs of two samples of brominated PS is shown in Fig. 3. From the MWD overlay, it is not clear if the MWD difference is a result of two PS samples of the same M_w brominated at different levels, or two PS samples with different M_w brominated at the same level. The M-H plot, which also includes linear PS without bromination, shows that the plots of the two samples in question lie on top of each other. The superposition of the two graphs shows that two PS samples of different M_w

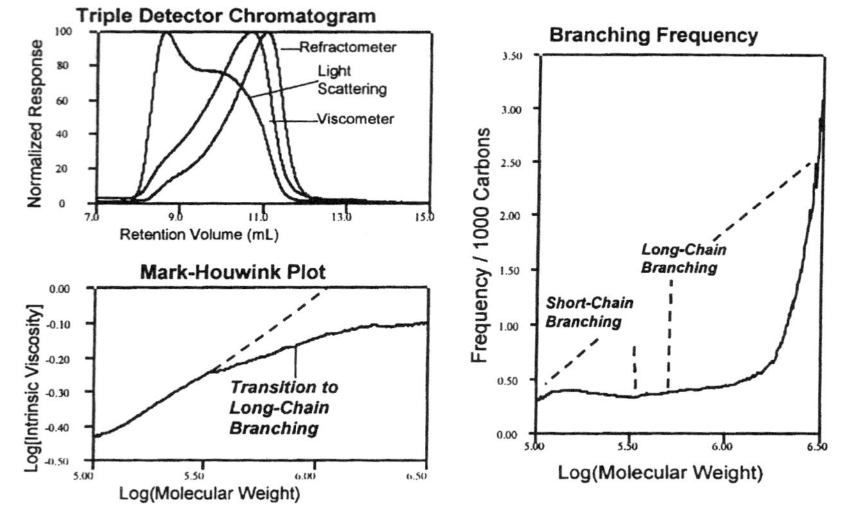

Fig. 2 In the triple-detector overlay of dextran, the shift of the LS detector toward a higher M indicates polydispersity. Both the M-H and branching frequency plots show a randomly branched polysaccharide, with both short- and long-chain branching. $M_n = 230{,}000$, $M_w = 540{,}000$, $M_z = 1{,}160{,}000$, $[\eta] = 0.54$ dL/g, $B_n = 26.6$, $\lambda = 0.43$, $dn/dc = 0.142$, $a = 0.287$, $\log K = -1.852$.

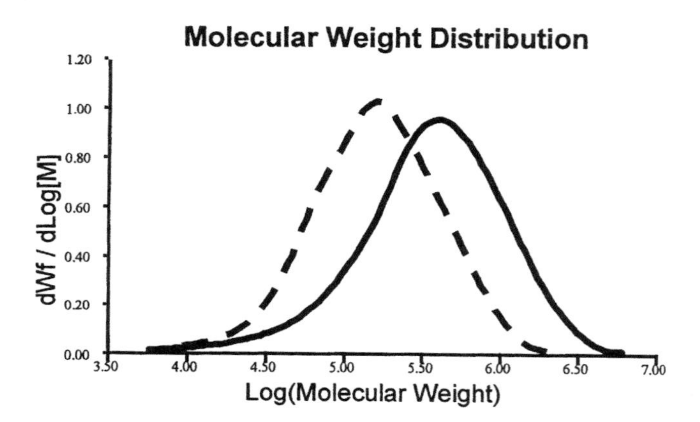

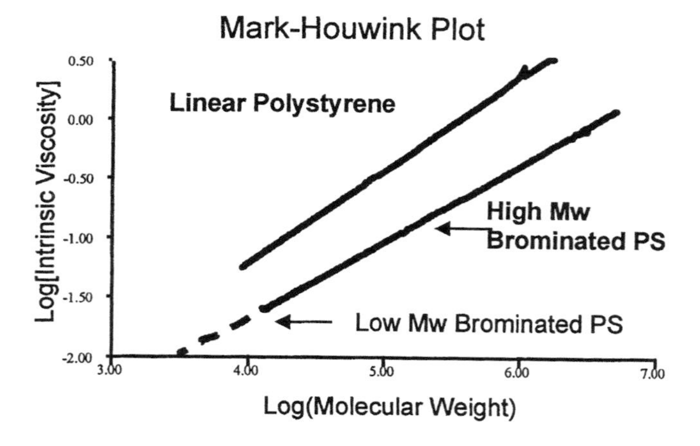

Fig. 3 Two polystyrene (PS) samples with different MWDs are compared in the upper chromatogram. In the lower graph, M-H plots of these same two samples lie on top of each other, indicating they have the same $[\eta]$ across their MWDs. It can be concluded that two PS samples of different M_w were brominated at the same level. A linear PS sample is included in the M-H plot for the purpose of comparison.

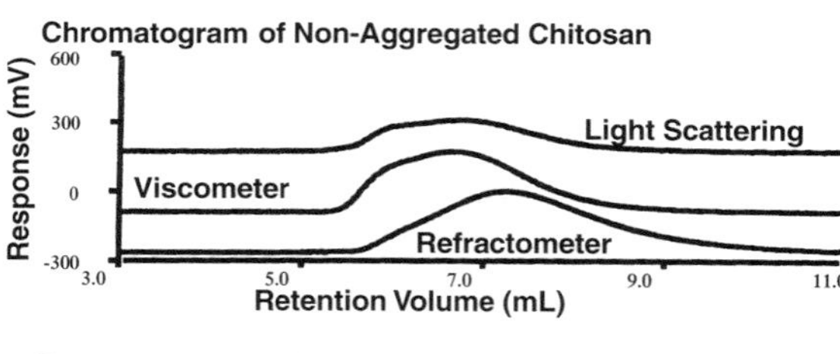

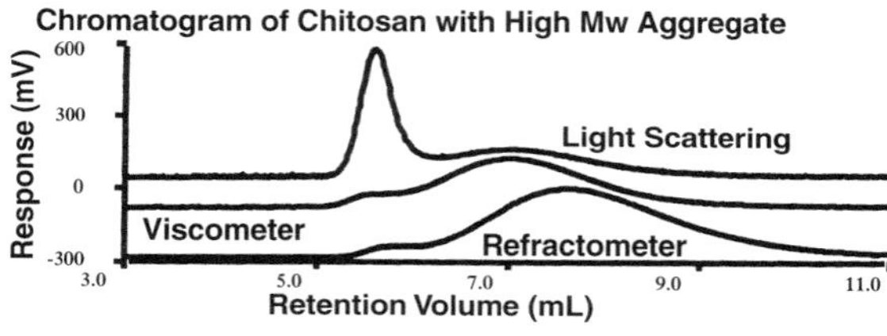

Fig. 4 Triple-detector chromatograms of nonaggregated and aggregated chitosan are compared. For the nonaggregated sample, both the LS and VISC detectors respond similarly. Aggregation in chitosan is indicated in the lower chromatogram, where the VISC response is low and the LS response is high. Nonaggregated chitosan: $M_n = 75{,}000$, $M_w = 260{,}000$, $M_z = 1{,}100{,}000$, $[\eta] = 7.9$ dL/g. Aggregated chitosan: $M_n = 90{,}000$, $M_w = 780{,}000$, $M_z = 3{,}000{,}000$, $[\eta] = 6.8$ dL/g.

were brominated to the same level. The M-H plots of these two samples would be parallel to each other if PS samples of the same M_w were brominated at different levels.

Chitosan is a stiff-chain polysaccharide that has a tendency to aggregate in aqueous solution. Aggregation is indicated when a low VISC response is coupled with a high response from the LS detector. Examples of chitosan with and without aggregation are shown in Fig. 4. Note the close similarity between the nonaggregated chromatograms of the VISC and LS detectors' responses, respectively. For

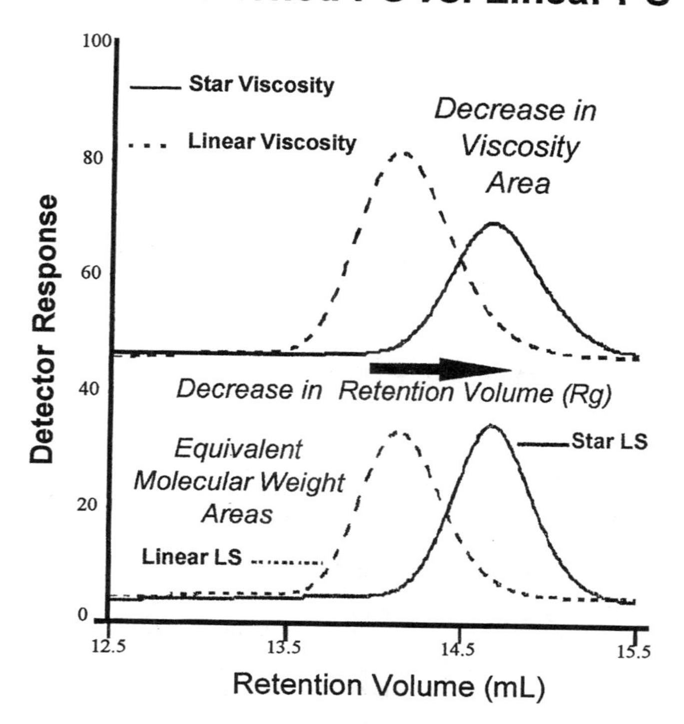

Fig. 5 The VISC traces (upper chromatogram) and the LS traces (lower chromatogram) are overlayed for a linear and a star-branched polymer with the same M. Star-branching creates a denser polymer with lower $[\eta]$ and a smaller R_g. Linear PS: $M_w = 100{,}000$, $[\eta] = 0.495$ dL/g, $R_g = 14$ nm. PS-star: $M_w = 100{,}000$, $[\eta] = 0.318$ dL/g, $R_g = 10$ nm, $f = 5$.

the aggregated sample, the LS response is much greater than that for the viscometer.

Star polymers are created when long-polymer chains are grouped covalently to a center core. The resulting polymer is denser with decreased $[\eta]$, compared with a linear polymer of the same M. Figure 5 compares VISC and LS detector chromatograms obtained for linear and star-branched PS, both with the same M. The difference in $[\eta]$ is demonstrated by the area difference between the viscometer peaks of the two samples. The areas of the LS detector response are the same for both samples, but the delay elution of the star polymer confirms the star's increased density and smaller R_g, in accordance with Eq. (5).

Summary

The graphics associated with SEC^3 capture its ability to reveal qualitative structural information in a visual format. The applications discussed show how peak displacement of the triple-detector chromatograms reflects polymer polydispersity (dextran), how detector response can relate to aggregation (chitosan), and how peak area differences can indicate a change in polymer chemical composition (PS versus PS-star). The M-H plots (dextran, brominated PS) give information about polymer conformational changes, structural differences, and branching distributions [7].

The SEC^3 technique has been in existence for 10 years. It is a relative newcomer to the analytical arena. The amount of information (molecular weight, conformational, and branching) produced, given the ease with which it can be generated, makes SEC^3 a very attractive technique. Recently, the triple detector system has been used in conjunction with temperature rising elution fractionation (TREF) to expand fundamental understanding of polymer structure–property relationships [8].

Acknowledgments

The author thanks Dr. Max A. Haney (Viscotek Corp.) and Dr. Tze-Chi Jao (Ethyl Petroleum Additives, Inc.) for reviewing the manuscript, Dr. Wei Sen Wong (Viscotek Corp.) for contributing figures and related information, and Dr. André M. Striegel (Solutia Inc.) for helpful discussions of SEC theory.

References

1. W. W. Yau, New polymer characterization capabilities using SEC with on-line MW-specific detectors, *Chemtracts–Macromol. Chem. 1*: 1–36 (1990).
2. C. Jackson, H. G. Barth, and W. W. Yau, Polymer characterization by SEC with simultaneous viscometry and laser light scattering measurements, *Waters International GPC Symposium Proceedings*, 1991, pp. 751–764 (www.waters.com).
3. M. A. Haney, T. Gillespie, and W. W. Yau, Viewing polymer structures through the triple "Lens" of SEC^3, *Today's Chemist Work*, *3*(11): 39–43 (December 1994).
4. M. A. Haney, C. Jackson, and W. W. Yau, SEC–viscometry–right angle light scattering, *Waters International GPC Symposium Proceedings*, 1991, pp. 49–63 (www.waters.com).
5. P. Cheung, S. T. Balke, and T. H. Mourey, Data interpretation for coupled molecular weight sensitive detectors in SEC: Interdetector transport time, *J. Liquid Chromatogr. 15*(1): 39–69 (1992).
6. D. T. Gillespie, H. K. Hammons, and S. R. Bryan, Branching and polymer modification analysis through SEC^3, *Mol-Mass International Conference Proceedings*, 1996 (www.chem.leeds.ac.uk/molmass 99).
7. L. J. Rose and F. Beer, Characterization of long chain branching in LDPE's using SEC with on-line viscosity and

light scattering detectors, *MolMass International Conference Proceedings*, 1999 (www.chem.leeds.ac.uk/molmass 99).
8. W. W. Yau and D. T. Gillespie, Triple-detector TREF instrument for polyolefin research, *Waters International GPC Symposium Proceedings*, 1998, pp. 252–256 (www.waters.com).

Suggested Further Reading

Brandrup, J. and E. H. Immergut (eds). *Polymer Handbook*, 4th ed., John Wiley & Sons, New York, 1999.

Burchard, W., Solution properties of branched macromolecules, *Adv. Polym. Sci. 143*, 113–194 (1999).

Lovell, P. A., Dilute solution viscometry, in *Comprehensive Polymer Science* (C. Booth and C. Price, eds.), Pergamon Press, New York, 1989, pp. 173–197.

Susan V. Greene

Sedimentation Field-Flow Fractionation of Living Cells

Introduction

Living cell analysis and sorting in suspension is a major aim in life sciences. A wide panel of techniques and methodologies are available and they can be divided into three main groups. The first is based on physical criteria such as cell size, density, and shape (centrifugation or elutriation). The second group is linked to the cell surface characteristics driving to selective absorption and migration (immunodependent and electrophoretic methods). The third one corresponds to flow cytometry technologies which take advantage of cell biochemical and biophysical properties.

In their fundamental principle, field-flow fractionation (FFF) methods exploit the cell physical characteristics by means of their selective elution in a parallelepipedic channel laminarily flowed by a carrier phase under the effect of an external field applied perpendicularly to the great surface of the channel and, by consequence, perpendicularly to the flow direction. In sedimentation FFF, the external field is gravitational (G-FFF) or multigravitational (Sd-FFF).

Living cells are, in large majority, micron-sized species. Their elution mode is known to follow, roughly, the "steric-hyperlayer" model. However, the cell separation, isolation, and purification requires some specific methodological and technological features, caused by their unique multidimensional polydispersity and the necessity to preserve their viability at a reasonable rate of recovery. The objectives of this entry are, first, to drive the reader to a qualitative understanding of the FFF principle applied to micron-sized species such as living cells. Then, to focus on some specific limitations imposed by the respect of the biological interests of these separations and, finally, to monitor the reader's entrance into that exciting domain.

FFF Principle and Cell Multidimensional Polydispersity

Field-flow fractionations are chromatographlike methods, operated in the elution mode, in which the separator (column) is a capillary ribbonlike channel whose critical dimension is its thickness. With a length/thickness ratio of over 50, the carrier phase in motion along the separator in a laminar mode creates, in the channel thickness, a parabolic flow profile. The general setup of a sedimentation FFF system is schematically described in Fig. 1a. Therefore, species focused in different positions in the channel thickness are carried out by flow streams of different velocities, creating separation. The unique feature of FFF operated in the "steric-hyperlayer" mode is that the focusing process in the channel thickness is provoked by the balance between hydrodynamic lifting forces and the external field generated ones, as schematically described in Fig. 1b(1–3). In sedimentation FFF (Sd-FFF), the force generated by the external field, applied to a single particle, is a function of the particle hydrodynamic radius and of its differential density (particle/carrier phase). The particle in motion in the channel is submitted to lifting forces of hydrodynamic nature whose origin is still under discussion (lubrification, inertia). Whatever its origin, lifting-force intensity depends on the particle size and its position into the channel thickness, as described in Fig. 1b(1) and legend. This balance drives to a flow-rate-dependent retention ratio of the eluted species. In the "steric-hyperlayer" elution mode, this balance focuses the particles into specific flow stream layers where the lifting-force module equalizes the external field generated one, as shown by the c′ particle of Fig. 1b(1). The consequences are (1) that species of the same density are eluted according to their size, the bigger being eluted first, as shown in

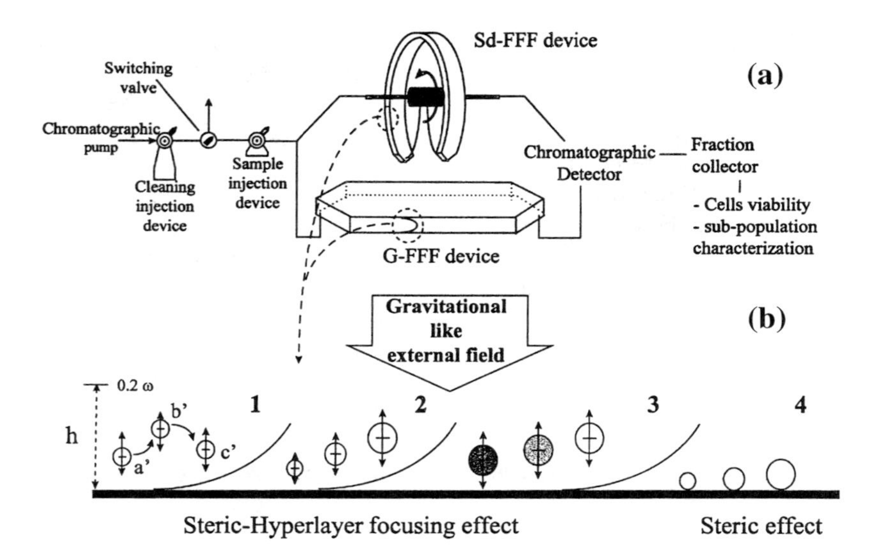

Fig. 1 (a) Schematic representation of the two sedimentation FFF systems (upper: multigravitational device or Sd-FFF; lower: gravitational device or G-FFF). Injection instrumentation is detailed with the cleaning (and decontamination) injection device (1- or 2-mL loop), the switching valve for stop-flow relaxation procedure, and the sample injection device (5–100-μL loop). (b) The "steric-hyperlayer" mechanism: (1) Dynamics of the balance of the lifting force and of the external field created one. Under the predominant effect of the lifting force, the particle translate vertically away from the accumulation wall while in motion along the channel (a′). Under the predominant effect of the sedimentation force, the particle is driven toward the accumulation wall (b′). Finally, after a given time, an equilibrium position is reached (c′). (2) Equilibrium position and size-dependent elution order of spherical particles of identical density. (3) Equilibrium position and density-dependent elution order of spherical particles of identical size. (4) Schematic representation of the size-dependent steric elution mode.

Fig. 1b(2), (2) that, at equivalent size, the denser are eluted last, as shown in Fig. 1b(3), (3) if the lifting force is overpowered by the sedimentation force, the elution mode becomes flow rate independent and only size related, the bigger particles being eluted first, as described in Fig. 1b(4). This case, described as "steric" elution mode, is often associated with very low recovery, reproducibility, and cell viability, linked to intense particles–accumulation wall interaction leading to, reversible or not, sticking of cells. It is obvious that separating cells only according to their size, at low recovery, reduces the separation power of Sd-FFF. The "steric-hyperlayer" mode is, therefore, the most versatile methodology to separate particle populations according to both their size and density, among other parameters, nonformulated so far, such as shape and rigidity.

The polydispersity (P0%) concept developed for polymer analysis and characterization expressed in percentage is defined by the equation

$$\text{P0\%} = \left(\frac{\sigma}{\overline{m}}\right) \times 100 \tag{1}$$

where σ is the standard deviation of the parameter under concern and $\overline{m}$ its mean value. In the case of micron-sized spherical particles, a size polydispersity can occur as well as a density one. In terms of cell polydispersity, this concept can be much more complex, involving numerous cell characteristics. If size expressed by the hydrodynamic radius can be evaluated as well as the density, cell shape, rigidity, and surface characteristics are more difficult to assess whatever the considerable bibliographic sources. Therefore, the cell polydispersity characteristic is a multidimensional one, described as "multipolydisperse." It is not possible, so far, to define, precisely, the exact dimension of this multipolydispersity matrix. If the cells are considered as nonspherical particles such as red blood cells (RBCs), for example, the multipolydispersity matrix is at least of dimension 3, with a polydispersity in size, in density, and in shape. In that case, a cell population to isolate or to purify is characterized by a 3×2 matrix of average values and polydispersity in size, density, and shape (sphericity index).

In the "steric-hyperlayer" elution condition, the cell characteristic differences will drive every subpopulation in specific flow streams of different intensity, leading to a large separation potential described in Fig. 1b(2,3) even among cell populations claimed as morphologically homogenous.

Elution Methodologies in Sd-FFF

Sedimentation FFF techniques encompass two different instrumentations; one is very simple and uses the Earth's gravity (G-FFF), the second, much more complex, uses centrifugational field created by rotation (Sd-FFF). In any case, common features must be described.

Numerous injection procedures have been used in Sd-FFF, in two different instrumental configurations, for cell retention. In terms of instrumentation, cells can be introduced either through the upper wall or the accumulation wall. Methodologically, cells can be injected into the flowing stream or with a time-controlled stop-flow procedure. The stop-flow procedure is known to lead to an increase of retention time of the cells, but it is often associated with (1) reduced recovery and (2) poisoning of the channel. Flow injection through the accumulation wall was demonstrated to produce an optimized response between retention loss and recovery increase. In some cases, time-controlled low flow injection is also possible. Stop-flow injection procedures are not recommended because cells are driven toward the accumulation wall in the absence of lifting forces (no flow). Particles–particles or particles–wall interactions will occur, leading to a biased retention pattern or modified viability characteristics.

To be retained, cell population retention ratio (R) (i.e., species velocity to average flow velocity ratio) must be <1. As already described, cells should not be eluted according to a pure "steric" elution mode whose R, for a monodisperse population of size a, is

$$R_{\text{steric}} = \frac{6a}{\omega} \tag{2}$$

where a is the hydrodynamic radius of the species and ω is the channel thickness. Therefore, the R domain available for cell separation is

$$\frac{6a'}{\omega} < R < 1 \tag{3}$$

where a' is the size of the smallest micron-sized particle involved.

For a given cell population, by increasing the external field and using low flow rates, species are eluted at decreased retention ratio (i.e., closer to the accumulation wall). The eluted cells and wall proximity may drive to particles–wall interactions leading to particles sticking (reversible or not), to cell activation (biophysical and biochemical modifications), and even to cell death or destruction (decrease in recovery and viability). A price is to be paid in using a limited R domain, leading to decreased separation power. However, depending on the populations under concern, two instrumental strategies are possible, predicted by the "steric-hyperlayer" model: (1) If the population to analyze is very polydisperse, a large R domain is expected and relatively high channel thickness is chosen (250–300 μm); (2) if slight differences in cell characteristics must be emphasized, thinner channels are to be used (50–175 μm) to increase cell subpopulation separation.

The major characteristic of carrier phase in FFF is to respect the cell integrity, viability, and, if possible, an enhanced recovery. Particle integrity is respected by means of the use of an iso-osmotic buffer of a given pH and ionic force. Most FFF studies are performed using an iso-osmotic (310 mOsm) phosphate buffer saline at pH 7.4. Particles–particles interactions are limited by means of a biocompatible surfactant such as bovine albumin (0.1% m/v) added to the buffer. In this case, a typical fractogram is obtained as shown in Fig. 2a. However, cell suspensions are complex mixtures. Some of their components (cell membrane, cytoplasmic fractions, etc.) will absorb onto the channel wall, leading to the modification of its surface characteristics, and often described as "channel poisoning," as shown in Fig. 2b (after 47 injections). Such effect limits reproducibility, viability, and recovery and can be limited or eliminated by two complementary methodological procedures. The most described is a systematic postelution channel washing or cleaning procedure. In Sd-FFF and in the absence of external field, the carrier phase is flushed at a high flow rate to release the reversibly adsorbed cells in their integrity. Then, an osmotic shock is provoked by means of a void-volume equivalent injection of doubly-distilled water. The remaining adsorbed proteins are released by means of an equivalent large volume injection of a "protein cleaning agent." Finally, a second distilled water flushing completes the procedure, as shown in Fig. 2c, signing reproducible elutions.

Fractograms of Figs. 2a–2c were obtained with polycarbonate channel walls and RBCs as the cell population. When more fragile nucleated cells (e.g., neuroblasts) are eluted with this procedure, no elution signal is recorded and collected fraction analysis shows a very low cell recovery (<3%). By simply replacing the polycarbonate plates by polystyrene plates, the neuroblast (NB) fractogram of Fig. 2d is obtained. Such an experimental result is confirmed by the elution of an artificial mixture of neuroblast and RBCs as shown in Fig. 2e. Considerable studies on cells–material interactions, confirmed by the FFF experimentation, allowed to define empirical rules: To avoid or reduce interactions, the channel material should be as "hydrophobic" as possible if the cell surface is "hydrophilic." However, cells–material-specific interactions could be used for specific cell subpopulation purification by depletion. If these methodological requirements, associated to soft elution conditions ("hyperlayer"

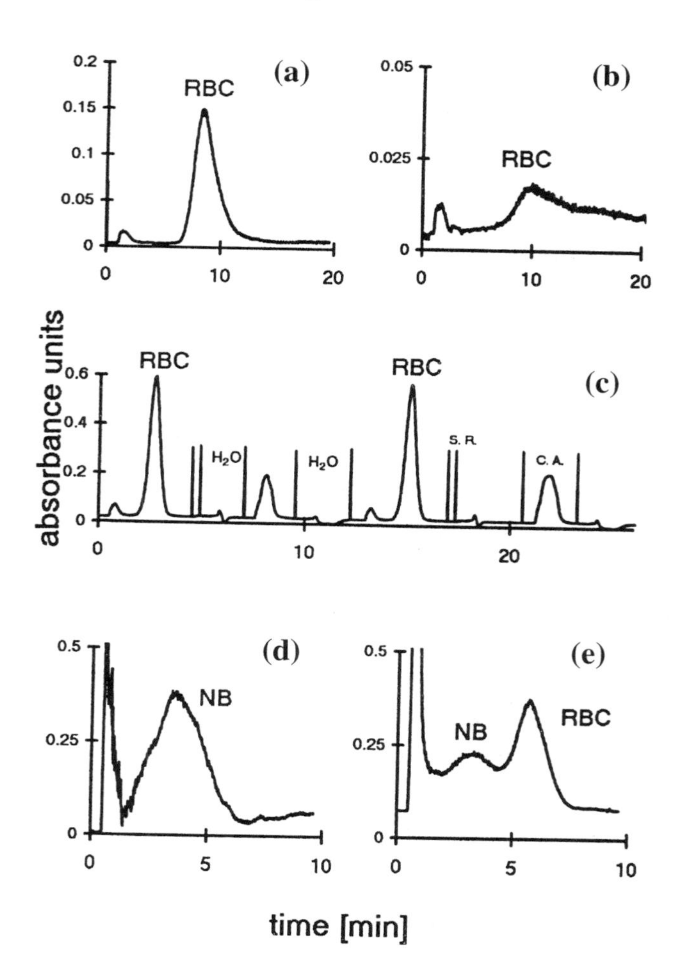

Fig. 2 Fractograms were obtained with a Sd-FFF apparatus (77 × 1 × 0.0125 cm). (a) RBC elution profile on a new or properly washed FFF channel. Elution conditions: flow injection of 5×10^6 RBCs (1/20 dilution of total blood in phosphate buffer saline pH 7.4/0.1 % of bovine albumin); external field 9.45*g* (1 *g* = 9.81 cm/s^2); flow rate: 0.7 mL/min, photometric detection at $\lambda = 313$ nm. (b) Channel poisoning effect observed after 47 identical injections [described in (A)]. (c) Two sequences of RBC elution and channel cleaning procedure. Each sequence is RBC fractogram (flow injection of 5×10^6 RBCs (1/20 dilution of total blood in phosphate buffer saline pH 7.4), external field 25.7*g*, flow rate of 1.02 mL/min, photometric detection at $\lambda = 313$ nm), external field stopped (S.R.), hypo-osmolar shock with doubly distilled water, cleaning agent (C.A.) signal, second water washing. (d) Example of fragile nucleated cells eluted in Sd-FFF: neuroblasts (NB) case. Elution conditions: flow injection of 1.5×10^6 neuroblasts in phosphate buffer saline pH 7.4/0.1 % of bovine albumin), external field 60.0*g*, flow rate of 1.25 mL/min, photometric detection at $\lambda = 254$ nm. (e) Separation of components from an artificial mixture of neuroblasts and RBC. Elution conditions: flow injection of 1.5×10^6 neuroblasts and 5×10^6 RBC in phosphate buffer saline pH 7.4/0.1 % of bovine albumin, external field 50.0*g*, flow rate of 1.25 mL/min, photometric detection at $\lambda = 254$ nm.

mode), are fulfilled, recovery over 80 % is obtained, associated with higher reproducibility, allowing cumulated fraction collection.

Usually, Sd-FFF separator decontamination and sterilization are needed to prevent bacterial contamination; this is done by means of a 0.01% (m/v) sodium azide solution added to the carrier phase. This compound is a cellular toxic and must absolutely avoided for cell viability and operator safety. Safe and effective decontamination process consists of an additional step to the cleaning procedure. The whole FFF system is flooded with a hypochlorite solution (3–6 Chlorometric degrees) whose volume corresponding to three to four times the Sd-FFF void volume. Prior to cell injection, the channel is flushed with ethanol and rinsed with the mobile phase. It is noted that this decontamination procedure is also a sterilization procedure if the hypochlorite solution is at least 5 Chlorometric degrees and if ethanol is over 70°. This sterilized situation is maintained by the use of sterilized mobile phase and daily bacteriological control is performed.

Cell Detection, Viability, and Recovery

Separation analysis at the outlet of the detector must respect three major conditions. The first is the cell integrity (i.e., the diagnosis of the particle). This can be operated on line, by means of classical photometric devices operated in the light-scattering mode (opacimetry) at 254 nm. Off-line methods, after fraction collection, are possible and recommended, by microscopy and granulometric analysis (Coulter® counting). The second objective is to analyze cell viability. Off-line methods after fraction collection are equally possible. The blue trypan exclusion test, motility measurements, or specific enzymatic activities (esterase) can be performed on an aliquot of the collected fraction.

Cell integrity and viability characteristics are of major importance for assessing the elution recovery by comparison with noneluted samples. Recovery can be described in terms of the percentage of cells in their integrity and in terms of cells alive. The third parameter under concern is the cell characteristic specificity, which signifies the biological interest of the separation. Membrane antigenic characteristics, cell surface receptor analysis, and specific protein composition and activities can be tested. In the early twenty-first century, the cell analysis and sorting reference system are flow cytometry technologies. Aliquots of FFF fractions can be characterized by means of this sophisticated method. Nevertheless, FFF separations and flow cytometry analyses are off-line techniques, and an exiting instrumental development domain is opened to realize the on-line coupling of the both techniques.

Conclusion

The number of FFF cell separation reports is increasing. Cell retention and separation rules are qualitatively defined with the help of the "steric-hyperlayer" mode. Separation proofs can be assessed by an arsenal of detection and viability techniques. However, the extent of this family of methods is somehow limited by the lack of commercially available and life-science-adapted separators. This text has described instrumental and methodological basic rules to adapt FFF technologies to life science. The development of FFF in cell separation is an interfacial process between instrumental separation sciences and biological skills. Unique cell separation selectivity, unique separation speed, and unique practicability were observed with these methods, and the development of continuous FFF-based separation devices opens a field to "massive" separation opportunities whose interest in transplantation or biotechnologies will rapidly emerge.

Acknowledgment

Special thanks to Frédéric Bodeau, who performed all the fractograms shown in this text.

Appendix

How to begin with FFF cell separations: gravitational separation of RBC. The general FFF setup, is analogous to classical HPLC systems. The FFF separator will simply replace the chromatographic column. To begin in cell separations with gravitational FFF, an outdated high-performance liquid chromatography (HPLC) chromatograph may be used. Basic steps are (1) construction of the separator, (2) initial filling of the FFF channel, (3) sample preparation, injection, and detection.

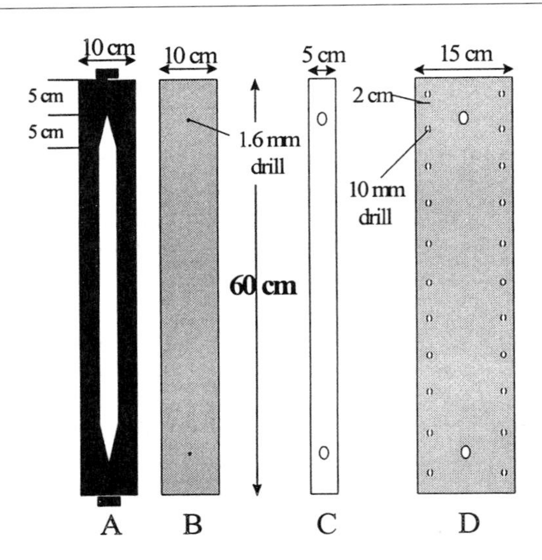

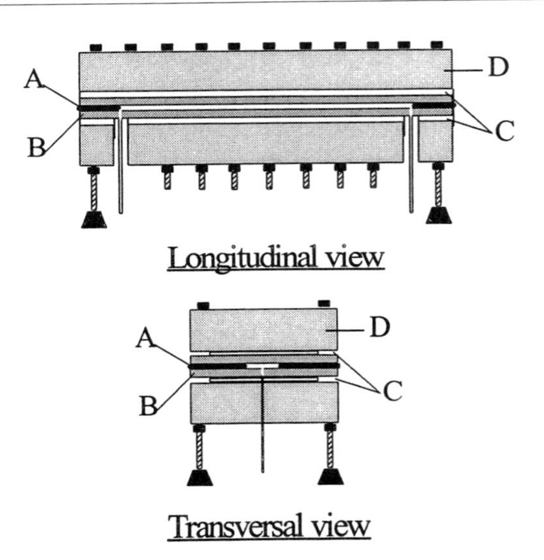

Channel construction: The channel (50 × 2 × 0.025 cm) is carefully cut in a Mylar sheet (60 × 10 × 0.025 cm, Scheme A) and sandwiched between two polycarbonate plates (Lexan) (60 × 10 × 1 cm) . One of them being (1.6 mm) drilled to glue Peek® inlet and outlet tubings (scheme B). The channel is then sandwiched between two Rhodoid (60 × 5 × 0.3 cm, Scheme C and transversal view) bands whose purpose is to avoid channel shape deformation. This double-sandwich setup is sealed between two Plexiglas blocks (Scheme D) by means of a series of screw. During the sealing process special attention must be taken to fit channel tapered ends with the connection tubings. The sealing begin by the middle of the separator toward the ends with a dynanometric device , 10–20-N m forces assume complete watertightness. Special attention must be taken to ensure a complete flatness of the channel by means of equalizing stands as shown on transversal and longitudinal views. *Initial use*: The channel is flowed with 100 mL methanol at 0.1 mL/min and cleaned with doubly-distilled water. Remaining bubbles are eliminated (patiently!!) with a series of surfactant solutions of decreased concentration and the mobile carrier phase finally applied. For cell separations, a simple NaCl (9 g/L) solution can be used. *Sample preparation, elution, detection*: 10 μL (injection loop volume) of 50-fold dilution of a whole-blood suspension on EDTA crystal in the carrier phase is performed at 0.1 mL/min and photometric detection recorded at 254 or 313 nm. *Additional materials*:

22 (0.8 × 20 cm) screws with locknuts, 50-cm Peek® tubing (1/16-in. external diameter and 0.0256-cm internal diameter), Araldite® glue (low-speed polymerization quality), four adjustable stands (4 × 0.8 cm dimension), and dynanometric screw sealing device.

Suggested Further Reading

Bernard, A., B. Paulet, V. Colin, and Ph. J. P. Cardot, Red blood cell separations by gravitational-field-flow fractionation: Instrumentation and applications, *Tr. A. C. 14*(6), 266–273 (1995).

Caldwell, K. D., Z. Q. Cheng, and P. Hradecky, Separation of human and animal cells by steric field-flow fractionation, *Cell Biophys. 6*: 233–251 (1984).

Giddings, J. C., Field-flow fractionation analysis of macromolecular, colloidal and particulate materials, *Science 260*: 1456–1465 (1993).

Hoffstetter-Khun, S., T. Rosler, M. Ehrat, and H. M. Widmer, Characterization of yeast cultivations by steric field-flow fractionation, *Anal. Biochem. 2*: 300–308 (1992).

Martin M. and T. S. William, Theoretical basis of field-flow fractionation, in *Theoretical Advancement in Chromatography and Related Separation Techniques* (F. Dandy and G. Guiochon, eds.), NATO ASI Series C, Vol. 383: Mathematical and Physical Sciences, Kluwer, Dordrecht, 1992, pp. 513–580.

Metreau, J. M., S. Gallet, Ph. J. P. Cardot, V. Lemaire, F. Dumas, A. Hernvann, and S. Loric, Sedimentation field-flow fractionation of cellular species. *Anal. Biochem. 251*: 178–186 (1997).

Yue, V., R. Kowal, L. Neargarder, L. Bond, A. Muetterties, and R. Parson, Miniature field-flow fractionation system for analysis of blood cells., *Clin. Chem. 40*(9), 1810–1814 (1994).

Philippe Cardot
T. Chianea
S. Battu

Selectivity

The selectivity *a*, also known as the relative retention, the separation factor, or chemistry factor, of a chromatographic column is a function of thermodynamic of the mass-transfer process and can be measured in terms of the relative separation of the peaks:

$$\alpha = \frac{t_{R2} - t_0}{t_{R1} - t_0} = \frac{k'_2}{k'_1} = \frac{K_2}{K_1}$$

where t_{R2} and t_{Ri} are the retention times of compounds 2 and 1, respectively, t_0 is the retention time of unretained compounds, k'_2 and k'_1 are the capacity factors of compounds 2 and 1, respectively, and K_2 and K_1 correspond to the distribution coefficients of compounds 2 and 1, respectively.

So, the selectivity of the chromatographic system is a measure of the difference in retention times (or volume) between two given peaks and describes how effectively a chromatographic system can separate two compounds with slight variations in structure or molecular weight. For compounds with the same molecular weight, the structure difference may involve no more than compounds that are mirror images (i.e., optical isomers resulting from the presence of one or more asymmetric atoms). Therefore, when components interact with a column and are retained, they will be separated if their degrees of retention are not identical. Two components with identical retentions would have $\alpha = 1$, or no separation. For effective separation, an $\alpha = 1.5$ is desired.

Optimizing the Selectivity

Separation problems become substantially more difficult as the number of components increases much above 10. Such complexity is often characteristic of environmental and biological samples. Different chromatographic modes offer potentially unlimited selectivity, but the conditions for optimal selectivity are correspondingly more difficult to find. A systematic basis for the combining of independent selectivity mechanism can provide a major boost to the overall selectivity. The overall effect is multiplicative, based on the separating power, or peak capacity, of each of the steps. The serial implementation of multiple origins of selectivity is the most practical approach at present.

The net retention of a particular solute depends on all the solute–solute, solute–mobile phase, solute–stationary phase, and stationary phase–mobile phase interactions that contribute to the retention, which, consequently, affect the selectivity.

The selectivity is dependent on the temperature and the chemistry of the components that make up the chromatographic system (i.e., column, solvent, and the sample). So, it is necessary to understand the physicochemical basis of retention and the retention mechanism involved in high-performance liquid chromatography (HPLC) separation.

The Sample or Solute

The basic structure and number of functional groups in the solute molecule largely determine chromatographic retention. The functional group must be able to interact with the stationary-phase surface. Moreover, the strength of the retention is increased by the introduction of a second functional group. In addition, the type of functional group determines the elution order. Therefore, changing the chemical nature of the sample compounds or altering the functional group by chemical derivatization of an analyte should lead to compounds that can be will separated with higher α, because it will alter the chromatographic properties and the solubility of the analytes. Furthermore, the quantity of injected sample could affect k' values and column efficiency.

Stationary Phase

Selectivity enhancement through choice of stationary phase can be a simplifying approach for difficult separations, where interest is in certain critical pairs and where practical capacity is deemed important.

The nature of the stationary phase plays an important role in the improvement of the selectivity of a chromatographic system. Therefore, changing the chemical composition of the column from very nonpolar to a higher polarity will cause the nonpolar compounds to elute faster. Also, some phases have an affinity toward some compounds; therefore, selecting the ideal phase will improve the selectivity. For instance, the chromatographic selectivity in electron-acceptor and electron-donor stationary phase depends on the ability of the stationary phase to form complexes with solutes. The selectivity depends not only on the number and mutual position of electron-accepting substituents attached to aromatic skeleton but also on the nature of the spacer connecting the ligand with the silica surface. Also, in liquid–solid chromatography (LSQ), the sample retention is governed by adsorption to the stationary phase. For retention to occur, a sample molecule must displace one or more solvents from the stationary phase. In addition to this displacement effect, polar solvent or sample molecules can exhibit very strong interactions with particular sites on the stationary phase. The separation of enantiomers using a chiral mobile phase is possible only if transient diasteromeric complexes are formed in the stationary phase. For this to happen, the stationary phase must be chiral.

Another way to improve selectivity is by using column switching, which is, in its simplest form, the use of a number, N, of different chromatographic mechanisms in sequence, which will expand the overall selectivity of a liquid chromatography system by the Nth power of that obtained from a single selectivity mechanism. Column switching can create tremendous separating power, but it is a requirement that each one in the sequence of selectivity mechanism not be redundant.

Temperature

Temperature is the first of the variables affecting selectivity. Increased temperature decreases retention time on the column, sharpens peaks, and produces a change in selectivity. However, temperature is generally limited, by solvent vapor pressures, to an effective range of 20–60°C; also important is the effect temperature has on the column packing.

Temperature affects sample solubility, solute diffusion, and mobile-phase viscosity in liquid chromatography. With increasing temperature, the solute diffusion coefficient tends to increase while the mobile phase viscosity decreases, producing a favorable influence on the selectivity.

The change in selectivity with temperature appears more pronounced in ion-pair chromatography than other HPLC methods. Therefore, temperature may be an important variable for optimizing selectivity in certain applications of ion-pair chromatography.

Mobile Phase

The most powerful approach to increase α is to change the composition of the mobile phase. If changing the concentration of the components in the mobile phase provides insufficient change, altering the chemical nature of one of the components will often be sufficient. Also, we can produce other α changes by adding mobile-phase modifiers to the mobile phase. The shifts in selectivity under certain circumstances have been attributed to the change in mobile-phase composition rather than to the stationary phase. Also, selectivity arises from the combined action of mobile phase and stationary phase.

Change in mobile phase can result in significant differences in selectivity for various sample analytes which can be obtained when the relative importance of the various intermolecular interactions between solvent and solute molecules is markedly changed. It is frequently preferable to use mixtures of solvents, rather than a single, pure solvent, as the mobile phase. However, in many cases, selecting a mobile phase is still a trial-and-error procedure. Moreover, pH has a prominence as a tool to affect the separation of some compound solutes. Likewise, an impressive separation of optically active compounds has been demonstrated through the use of chiral reagents that induce a ligand-exchange mechanism. Therefore, it should be recognized that the harnessing of liquid-phase composition to control HPLC selectivity provides a major corridor for achieving separation in an increasingly systematic manner.

The difficulty in eliminating the silanol groups from the silica substrate make it necessary to neutralize them using additives in the mobile phase. Also, solvent strength generally increases with the volume percent of organic modifier. Its effect is most important when hydrophobic mechanisms contribute significantly to retention. In this case, changing the organic modifier can be used to adjust solvent selectivity, as normally practiced in reversed-phase chromatography. Mobile-phase additives (in normal phase), which are very polar, influence the adsorption of substances strikingly, even in the very low concentration range, because they are adsorbed preferentially. On the other hand, the stationary-phase selectivity can be altered for some phases by the addition of some compounds or metallic complexes to the mobile phase.

High-performance liquid chromatography offers options to control selectivity through the mobile phase. Therefore, it is important to improve the practical understanding of liquid-phase compositions needed to achieve chemical selectivity.

Suggested Further Reading

Ahuja, S., *Selectivity and Detectability Optimization in HPLC*, John Wiley & Sons, New York, 1989.

Freeman, D., Advances in liquid chromatographic selectivity, in *Ultrahigh Resolution Chromatography* (S. Ahuja, ed.), American Chemical Society, Washington, DC, 1984.

Lochmiller, C. H., Approaches to ultrahigh resolution chromatography: Interaction between relative peak (N), relative retention (a), and absolute retention (k'), in *Ultrahigh Resolution Chromatography* (S. Ahuja, ed.), American Chemical Society, Washington, DC, 1984.

Meyer, V. R., *Practical High-Performance Liquid Chromatography*, 2nd ed., John Wiley & Sons, New York, 1993.

Poole, C. F. and S. K. Poole, *Chromatography Today*, Elsevier Science, Amsterdam, 1991.

Riley, C. M., Efficiency, retention, selectivity and resolution in chromatography, in *High Performance Liquid Chromatography, Fundamental Principles and Practice* (W. J. Laugh and I. W. Wainer, eds.), Blackie Academic and Professional, Glasgow, 1996, pp. 29–35.

Snyder, L. R., J. L. Glajch, and J. J. Kirkland, *Practical HPLC Method Development*, John Wiley & Sons, New York, 1988.

Weston, A. and P. R. Brown, *HPLC and CE Principles and Practice*, Academic Press, San Diego, CA, 1997.

Hassan Y. Aboul-Enein
Ibrahim A. Al-Duraibi

Selectivity: Factors Affecting, in Supercritical Fluid Chromatography

Retention and selectivity in supercritical fluid chromatography (SFC) are a complex function of many experimental variables and are not as easily rationalized as in the case of gas and liquid chromatography. Retention in SFC is dependent on temperature, density (and pressure drop), stationary-phase composition, and the mobile-phase composition. Many of these variables are interactive and do not change in a simple or easily predicted manner [1].

Effect of Temperature

Changes in retention at constant density are predictable from van't Hoff plots. The logarithm of the capacity factors is a linear function of the reciprocal of the column temperature, even down to subcritical conditions [1]. Analysis of the thermodynamics of the temperature-driven selectivity shifts in capillary SFC, at a constant mobile-phase fluid density, demonstrates the importance of stationary-phase polymer swelling. The other thermodynamic derivative contributing to temperature-driven selectivity shifts is the thermal pressure coefficient of the mobile-phase fluid [2]. Usually, temperature programming in SFC is done by increasing the temperature during a pressure, density, or eluent program, although negative temperature programs can also be employed to increase density. Although density conditions are the same either by decreasing temperature at constant pressure or by increasing pressure at constant temperature, the latter is preferable, as the higher diffusion coefficients at the higher constant temperature provide more favorable mass-transport properties.

Effect of Pressure Drop (Density Drop)

Selectivity is almost independent of pressure in high-performance liquid chromatography (HPLC) and gas chromatography (GC), whereas pressure (and corresponding density) is a very important parameter controlling selectivity in SFC, particularly if a significant pressure or density drop occurs along the column. In general, pressure drops are low when open-tubular columns are used, but they are significantly higher with packed columns and, therefore, have a significant effect on chromatographic resolution with packed column systems [3]. The observed selectivity, α_{obs}, can be described by $\alpha_{\text{obs}} = e^{(B-mD)} \times (e^{\text{bwL}} - 1)/mwL$, where the values for the constants B, m, and b will vary depending on the compound types being separated, the mobile phase, the stationary phase, and the temperature. D is the density of the mobile phase at the head of the column, w is the rate at which the density changes along the column, and L is the total column length. Because wL is simply the density change, $\Delta\rho$, across the column, the net result is that observed SFC selectivity changes caused by a linear density change along a column are only dependent on the total density drop which occurs. Therefore, in order to maintain constant selectivity as the density drop is increased, the density at the head of the column must be increased. Alternatively, if the density at the head of the column is kept constant while $\Delta\rho$ is increased, both selectivity and retention will increase [4]. Figure 1 demonstrates the effects of the density drop across a column for n-alkanes in carbon dioxide (data from Ref. 4). Some general conclusions include that, on average, α_{obs} changes by 0.001 per 10% density drop up to a 30% density drop. Also, selectivity decreases more rapidly as density drops become larger and that selectivity increases at larger k' values where the mobile-phase density is lower.

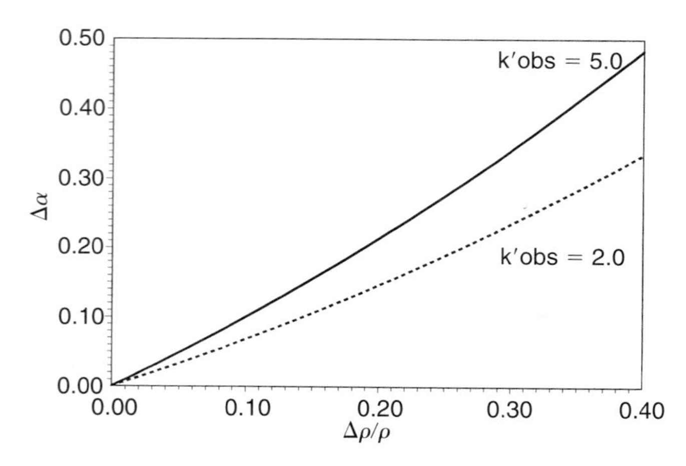

Fig. 1 Density drop effects on α at different k's for n-alkanes in carbon dioxide.

Effect of Overall Density (Pressure at Constant Temperature)

The overall density of the mobile phase is one of the most important parameters used to optimize separations in SFC with density programming as common in SFC as temperature programming in GC and eluent composition in HPLC [5]. Capacity ratios, k', decrease roughly linearly at higher densities with different slopes for different classes of compounds, thereby affording changes in selectivity [5]. A similar effect is seen for the supercritical fluid elution of analytes from octadecylsilica sorbents, as seen in Fig. 2 [6].

Effect of Stationary Phase

In SFC, both packed and capillary columns are used, each with their specific advantages and disadvantages. Packed columns in SFC are very similar to those used in HPLC, with the most often used stationary phases being modified silicas. Column selectivity follows the same rules as it does in HPLC with aromatic hydrocarbons (e.g., more retained on octadecyl silica column than on bare silica) [5]. A great variety of different selective stationary phases have been used in packed column SFC. For example, ra-

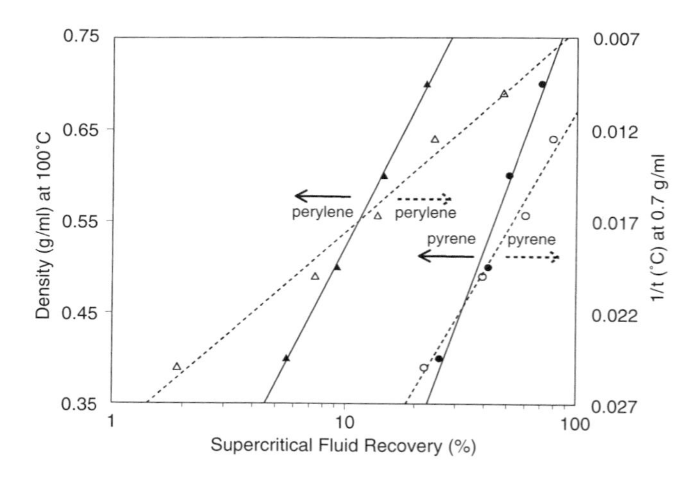

Fig. 2 Supercritical fluid recoveries of polycyclic aromatic hydrocarbons as a function of density and temperature. (From Ref. 6.)

cemic *N*-acetylamino acid *t*-butyl esters have been resolved on chiral (*N*-formyl-L-valylamino)propylsilica using methanol-modified carbon dioxide [5]. The most often used stationary phases for open-tubular SFC are immobilized films of polymeric materials — most commonly, polysiloxanes common to GC. Selecting a suitable stationary phase follows the same rules as in GC or HPLC, bearing in mind the frequently used carbon dioxide is a relatively nonpolar eluent. For example, nonpolar substrates such as hydrocarbons are strongly retained on a dimethyl column, whereas free carboxylic acids are more retained on a cyanopropyl column.

Effect of Mobile-Phase Composition (Polarity Modifiers)

Carbon dioxide is the most commonly used mobile phase in SFC, due to its low cost, low expense, low toxicity, and low critical temperature and pressure. However, using the classification scheme of eluents by Snyder, carbon dioxide shows a polarity similar to that of hexane [5]. Therefore, the solvent power of eluents used in SFC is generally enhanced by adding small amounts of a second eluent modifier. Selection of the optimum solvents can be achieved in much the same way that selections are made for HPLC solvents, namely utilizing a solvent polarity/selectivity scheme. To be useful, a solvent characterization scheme must efficiently determine the solvent strength or polarity and the solvent selectivity. The polarity of nonelectolytes is the capacity of the solvent for all intermolecular interactions (primarily dispersion, induction, orientation, and proton donor–acceptor interactions). Solvent selectivity is a measure of the relative capacity to enter into each specific interaction. The three primary specific interactions evaluated in all solvent characterization schemes are orientation (dipolar interaction), proton-donor (acidity), and proton-acceptor (basicity) interactions. One of the most widely used schemes is the solvent triangle introduced by Snyder and reevaluated over the years [7]. In Snyder's approach, solvent selectivity factors x_n (using nitromethane), x_e (using ethanol), and x_d (using dioxane) are used to characterize the relative importance of orientation, proton acceptor (basicity) and proton donor (acidity), respectively. When these three terms are graphed against one another for the common solvents, a so-called selectivity triangle is generated where solvents with similar selectivities are clustered into eight major selectivity "groups." Additionally, a solvent polarity index, P', is calculated to provide a measure of the relative polarity of each solvent. Values for the various polarity/selectivity terms and critical constants are summarized in Table 1 for some common polarity/selectivity modifier solvents. (data from Refs. 7 and 8). The most recent data for Snyder's terms from Ref. 3 have been included where available. Popular alternative schemes have utilized solvatochromic parameters based on the concept of linear solvation energy relationships to quantitatively probe-specific chemical interactions such as polarizability, hydrophobicity, and hydrogen-bonding interactions.

Solvatochromic descriptors described by Abraham are summarized in Table 1, including the overall or summation hydrogen-bond acidity scale ($\Sigma\,\alpha_2^H$), basicity scale ($\Sigma\,\beta_2^H$), and dipolarity/polarizability descriptor ($\Sigma\,\pi_2^H$) [9]. It has been shown that solvatochromic parameters may be successfully used to predict retention near the

Table 1 Critical Constants and Polarity and Selectivity Parameters for Common Organic Modifiers

Solvent	P_c (psi)	T_c (°C)	P'	χ_d	χ_e	χ_n	Group	$\Sigma\,\alpha_2^H$	$\Sigma\,\beta_2^H$	$\Sigma\,\pi_2^H$
Carbon dioxide	1070.4	31.1						0.00	0.10	0.42
n-Hexane	436.6	234.4						0.00	0.00	0.00
Triethylamine	439.5	262.0	2.19	0.08	0.66	0.26		0.00	0.79	0.15
Diethyl ether	527.9	193.7	3.15	0.13	0.53	0.34	I	0.00	0.45	0.25
Ethylene chloride	735.3	250.0	3.5	0.21	0.30	0.49	V	0.10	0.11	0.64
Isopropanol	690.4	235.3	3.92	0.17	0.57	0.26	II	0.33	0.56	0.36
Ethyl acetate	555.5	250.2	4.24	0.22	0.36	0.42	VIa	0.00	0.45	0.62
Tetrahydro-furan	752.7	267.1	4.28	0.19	0.41	0.40	III	0.00	0.48	0.52
Methylene chloride	913.7	237.0	4.29	0.33	0.27	0.40	VII	0.10	0.05	0.57
Chloroform	778.9	263.4	4.31	0.35	0.31	0.34	VIII	0.15	0.02	0.49
Acetone	681.7	235.1	5.40	0.24	0.36	0.40	VIa	0.04	0.49	0.70
Pyridine	816.6	347.0	5.53	0.22	0.42	0.36	III	0.00	0.52	0.84
Acetonitrile	700.5	272.5	5.64	0.25	0.33	0.42	VIb	0.07	0.32	0.90
Acetic acid	839.8	319.7	6.13	0.30	0.41	0.30	IV	0.61	0.44	0.65
Methanol	1173.4	239.6	6.60	0.19	0.51	0.30	II	0.43	0.47	0.44
Water	3208.2	374.3	10.2	0.37	0.37	0.25	VIII	0.82	0.35	0.45

critical point in packed column SFC and may be useful in controlling selectivity of chiral separations [10]. Supercritical fluid selectivities are comparable to subcritical selectivities with minor differences attributable to the physical nature of modifier behavior under near-critical conditions where binary mobile phases may exhibit gross compositional hererogeneity at interfaces [10]. When organic modifiers are increasingly added to the mobile phase at constant pressure (density) and temperature, the retention of analytes increases or decreases, depending on whether the supercritical analytes are more or less soluble in the modifier compared to the supercritical fluid, provided that the column activity remains the same [1].

Miscellaneous and Combined Effects

Temperature, pressure, and density may also influence SFC selectivity in other ways. For example, water solubility in supercritical fluids generally increases with temperature, causing a shift the equilibrium of the number of water-deactivated silanol groups to carbon-dioxide-deactivated groups [1]. Therefore, the solubility of analytes in the mobile phases increases but so does retention for polar analytes due to increased stationary-phase activity.

References

1. C. F. Poole and S. K. Poole, *Chromatography Today*, Elsevier, Amsterdam, 1991, pp. 601–643.
2. M. Roth, *J. Chromatogr. 718*: 147 (1995).
3. X. Lou, H.-G. Janssen, H. Snijder, and C. A. Cramers, *J. Chromatogr. 718*: 147 (1995).
4. P. A. Peaden and M. L. Lee, *J. Chromatogr. 259*: 1 (1983).
5. R. M. Smith (ed.), *Supercritical Fluid Chromatography*, Royal Society of Chemistry, London, 1988.
6. K. G. Furton and J. Rein, in *Supercritical Fluids Technology: Theoretical and Applied Approaches in Analytical Chemistry* (F. V. Bright and M. E. P. McNally, eds.), ACS Symposium Series Vol. 488, American Chemical Society, Washington, DC, 1992, pp. 237–250.
7. S. C. Rutan, P. W. Carr, W. J. Cheong, J. H. Park, and L. R. Snyder, 463: 21 (1989).
8. *Isco Tables*, 9th Edition, Isco, Inc., 1987, p. 10.
9. M. H. Abraham, G. S. Whiting, R. M. Doherty, and W. J. Shuely, *J. Chromatogr. 587*: 213 (1991).
10. G. O. Cantrell, R. W. Stringham, J. A. Blackwell, J. D. Weckwerth, and P. W. Carr, *Anal. Chem. 68*: 3645 (1996).

Kenneth G. Furton

Separation of Antibiotics by Countercurrent Chromatography

Introduction

An antibiotic is a chemical compound made either by a living organism or by chemical synthesis. It has the property to inhibit, at low concentrations, some vital processes of viruses, microorganisms (such as bacteria, fungi, etc.) and some cells of a pluri-cellular individual (cancerous cells, parasites cells, etc.). The development of antibiotics made by microorganisms requires isolation and purification of a desired compound from a complicated matrix such as a fermentation broth. These bioactive microbial metabolites are often produced in very small quantities and have to be removed from other secondary metabolites and nonmetabolized media ingredients. Antibiotics are normally biosynthesized as mixtures of closely related congeners and many are labile molecules, thus requiring mild separation techniques with a high-resolution capability. Although recent advances in the HPLC technology, using sophisticated equipment and refined adsorbents, greatly facilitate the isolation of antibiotics, some drawbacks remain, which are related to various complications arising from the use of a solid support, such as adsorptive loss, deactivation of sample components, contamination, and so forth. Moreover, high-performance liquid chromatography (HPLC) purification always requires sample preparation, prepurification, concentration, and so forth. Liquid–liquid partition techniques, particularly countercurrent chromatography (CCC), are suitable for the separation of antibiotics because it utilizes a separation column which is free of a solid support matrix and which is made of inert Teflon® channels or tubes.

Raw material can be injected into the column without any previous sample treatment, which simplifies the purification procedure. Oka et al. [1] have gathered antibiotics purification by CCC from crude extract and fermentation broth. They have shown that CCC has been successfully applied to the separation of macrolides and of various antibiotics, including various peptide antibiotics which are generally strongly adsorbed to silanol groups on silica gel used in the stationary phase in HPLC. Several CCC types are used, such as DCCC (droplet countercurrent chromatography) [2] and the more recent, *X*-axis CCC, foam CCC, CPC (centrifugal partition chromatography), and HSCCC (high-speed countercurrent chromatography).

Antibiotics

Antibiotics have a more or less high polarity. They are synthesized by various living materials, such as bacterial strains (such as *Streptomyces* [3], *Bacillus*), marine sponges, and so forth. We will focus our discussion on the separation of macrolides and polypeptide antibiotics by CCC.

Several separations of macrolide and polypeptidic antibiotics by CCC are reported in the literature. Macrolides are heterosides of which aglycone is a cyclic macrolactone with at least 14 atoms. They act by stopping protein syntheses. Polypeptide antibiotics are frequently cyclic molecules. They act by disorganizing proteinic structure of the bacterial membrane. Figures 1–3 show the structures of several molecules whose purification will be described later. Sporaviridines are produced by *Kutzneria viridigrisea*. They are very polar molecules and are water-soluble basic glycoside antibiotics (Fig. 1). They consist of six components, each having a 34-membered lactone and seven monosaccharide units, a pentasaccharide (viridopentaose), and two monosaccharides. They are active against Gram-positive bacteria, acid-fast bacteria, and trichophyton.

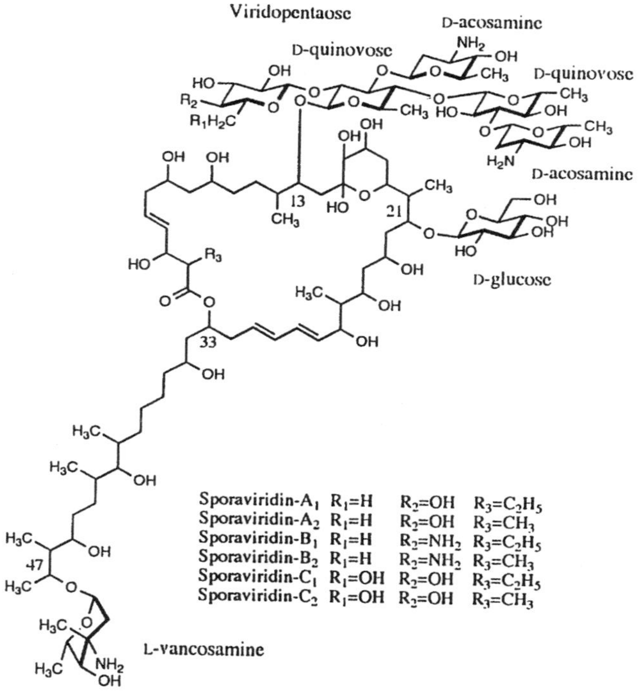

Sporaviridin-A_1 R_1=H R_2=OH R_3=C_2H_5
Sporaviridin-A_2 R_1=H R_2=OH R_3=CH_3
Sporaviridin-B_1 R_1=H R_2=NH_2 R_3=C_2H_5
Sporaviridin-B_2 R_1=H R_2=NH_2 R_3=CH_3
Sporaviridin-C_1 R_1=OH R_2=OH R_3=C_2H_5
Sporaviridin-C_2 R_1=OH R_2=OH R_3=CH_3

Fig. 1 Chemical structure of sporaviridins.

Ivermectins B1 are derived from the avermectines B1, the natural fermentation products of *Streptomyces avermitilis*. The avermectins B1 have double bonds between carbon atoms at 22 and 23, whereas the ivermectins B1 have single bonds in these positions (Fig. 2a). They have an intermediate polarity. The ivermectins B1 are a mixture of two major homologs, ivermectins B1a (>80%) and ivermectins B1b (<20%), but a crude ivermectin complex also contains various minor components. Ivermectins B1 are broad-spectrum antiparasitic agents used against *Onchocerca volvulus* in human medicine and for food production animals such as cattle, swine, and horses.

The bryostatins have been isolated from marine materials, bryozoan *Bugula neritina* [4] (Fig. 2b). They are macrolides with intermediate polarities. They show significant activity against lymphocytic leukemia in vitro.

Finally, bacitracins are peptide antibiotics produced by *Bacillus subtilis* and *B. licheniformis*. Over 20 components are contained in the bacitracin complex medium, among which the major active components are bacitracin A and F (Fig. 3). They exhibit an inhibitory activity against Gram-positive bacteria and are among the most commonly used antibiotics as animal feed additives.

Solvent Systems

The polarities of these molecules are, more or less, high according to the oses number contained in the chemical formula. Several procedures for choosing a solvent system are described in the literature. Usual solvent systems are biphasic and consist of three solvents, two of which are immiscible. If the polarities of the solutes are known, the classification etablished by Ito and co-workers [1] can be taken as a first approach. They classified the solvent systems into three groups, according to their suitability for apolar molecules ("apolar" systems, based on *n*-hexane), for intermediate polarity molecules ("intermediary" system, based on chloroform), and for polar molecules ("polar" system, based on *n*-butanol). The molecule must have a high solubility in one of two immiscible solvents. The addition of a third solvent enables a better adjustment of the partition coefficients (K).

Oka et al. [5] proposed a choice of various solvent systems to purify antibiotics. They have to fulfill various criteria. The settling time of the solvent system should be shorter than 30 s to ensure the satisfactory retention of the stationary phase. The partition coefficient of the tar-

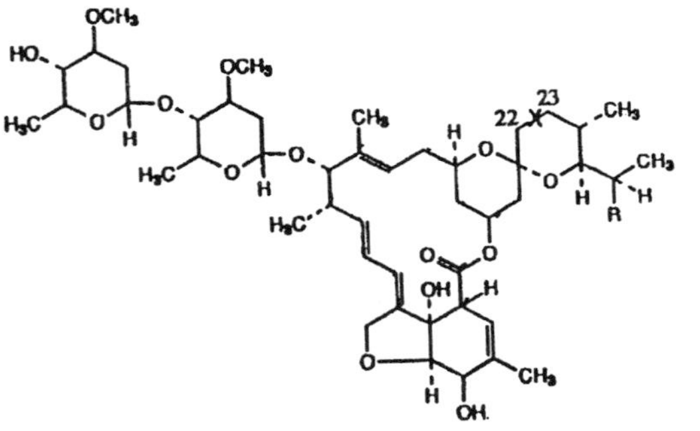

	R	$-C_{22}$-X-C_{23}-
Ivermectin B1a	C_2H_5	$-CH_2$-CH_2-
Ivermectin B1b	CH_3	$-CH_2$-CH_2-
Avermectin B1a	C_2H_5	-CH=CH-
Avermectin B1b	CH_3	-CH=CH-

(a)

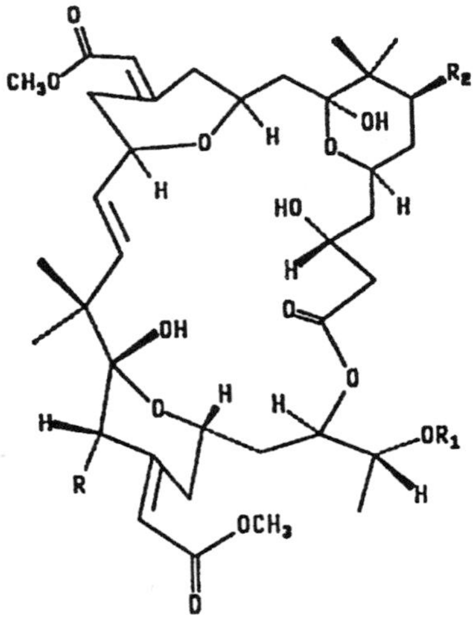

STRUCTURE	R	R_1	R_2	
1	B	H	A	Bryostatin 1
2	B	H	OH	Bryostatin 2
4	D	H	C	Bryostatin 4
5	A	H	C	Bryostatin 5
6	A	H	D	Bryostatin 6
7	A	H	A	Bryostatin 7
8	D	H	D	Bryostatin 8
9	D	H	A	Bryostatin 9
10	H	H	C	Bryostatin 10
11	H	H	A	Bryostatin 11
12	B	H	D	Bryostatin 12
13	H	H	D	Bryostatin 13
14	OH	H	C	Bryostatin 14
14a	A	A	C	
15	E	H	A	Bryostatin 15

(b) A, B, C, D, E,

Fig. 2 Chemical structure of ivermectins (a) and bryostatins (b).

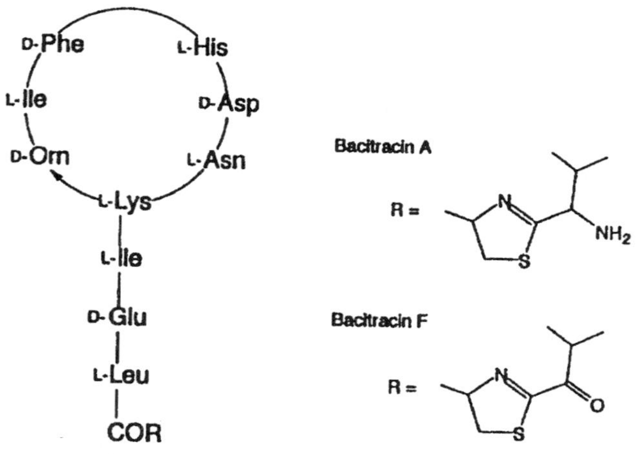

Fig. 3 Chemical structure of bacitracins.

get compounds should be close to 1, and the separation factor (α) between the compounds must be larger than 1.5. Two series of solvent systems can provide an ideal range of the K values for a variety of samples: n-hexane–ethyl acetate–n-butanol–methanol–water and chloroform–methanol–water. These solvent series cover a wide range of hydrophobicity, continuously, from the nonpolar n-hexane–methanol–water system to a more polar n-butanol–water system.

CCC for Purification of Antibiotics

Several CCC devices are commonly used to purify antibiotics, such as rotating coil instruments particularly used in HSCCC and cartridge instruments used in CPC. An entry of this encyclopedia is entirely devoted to the various CCC devices, so that we only give some indications about performances of CCC as compared to preparative HPLC.

Menet et al. [6] have compared performances of CCC and preparative HPLC owing to a separation of two antibiotics X and Y. The CCC apparatus used was a centrifugal partition chromatograph (CPC, Sanki* LLN) of 250 mL internal volume. For this purpose, classical parameters of preparative-scale chromatography were calculated: experimental duration, including the sample preparation and the separation time, solvent consumption, including the volume of the mobile phase, the stationary phase and the injection solvent, and purity of the purest fraction in Y. The parameter "purity in Y" was chosen because Y is the solute most difficult to purify because of its physical properties (particularly hydrophobicity) which are close to those of the main impurities. The hourly yield (g/h) is defined as the ratio of the recovered quantity to the experimental duration. The volumic yield (g/L) is defined as the ratio of the recovered quantity to the solvent consumption. Table 1 summarizes the results of separations of Y by CCC and preparative HPLC. The solvent volume consumption is the volume of the stationary and mobile phases in CCC or the volume of the mobile phase used in HPLC and the samples. The injected sample in CCC was not prepurified to concentrate it in Y from 7% to 25%. So the injected quantity in Y in CCC is lower (0.28 g against 1.59 g in preparative HPLC). For similar volumic yields (i.e., 0.20 g/L in CCC and 0.15 g/L in preparative HPLC), the enrichment in Y is higher with CCC than with preparative HPLC. Indeed, starting from a crude extract at 7% in Y with CCC or from a 25% in Y extract with preparative HPLC leads to the same 95% highest purity. These results demonstrate the interest of CCC in directly purifying crude extracts. Moreover, no preliminary purification of the extract is required, contrary to preparative HPLC, which requires a 1-day enrichment of the crude extract from 7% to 25% in Y.

Table 1 Comparison of CCC and HPLC Performance

	CCC	HPLC
Crude extract purity in Y (%)	7	25
Injected quantity of Y (g)	0.28	1.59
Experimental duration (h)	6.2	2.2[a]
Solvent volume consumption (L)	1.4	10.8
Purity of the purest fraction in Y	>95%	>95%
Hourly yield (g/h)	0.035	0.72
Volumic yield (g/L)	0.20	0.15

[a]One hour for the column equilibration at the 90 mL/min flow rate + 1 h for the separation.

Examples of Purification

Sporaviridins Separation [7]

Chemical structures of sporaviridins are described in Fig. 1. They are only soluble in polar solvents such as water, methanol, and n-butanol. Therefore, a two-phase solvent system containing n-butanol was examined. A nonpolar solvent such as diethyl ether has been added to the n-butanol–water system to decrease the solubility of molecules in n-butanol and to obtain partition coefficients close to 1. The partition coefficients, K, are defined as the ratio of the solute concentration in the upper phase (butanol rich) to its concentration in the lower one (water rich). A two-phase solvent system of n-butanol–diethyl ether–water (10:4:12, v/v/v) was selected because it allows one to obtain the almost equally dispersed partition coefficients among six components (C2, B2, A2, C1, B1,

A1). The preparative separation of six components from sporaviridin complex by HSCCC was performed in 3.5 h (500 mL of elution volume). The six components were eluted in the order of their partition coefficients, yielding pure components A1 (1.4 mg), A2 (0.6 mg), B1 (0.7 mg), B2 (0.5 mg), C1 (1.1 mg), and C2 (1.4 mg) from 15 mg of the sporaviridin complex.

Ivermectins Separation [8]

These molecules have an intermediary polarity (Fig. 2a). A two-phase solvent system, composed of *n*-hexane, ethyl acetate, methanol, and water, has been selected. In this case, the partition coefficients, K, are defined as the ratio of the solute concentration in the upper phase to its concentration in the lower one. A solvent mixture of *n*-hexane–ethyl acetate–methanol–water (19:1:10:10, v/v/v/v) yielded the best K values from 0–2.83. 25 mg of crude ivermectin separated in 4.0 h. This separation yielded 18.7 mg of 99.0% pure ivermectin B1a, 1.0 mg of 96.0 % pure ivermectin B1b, and 0.3 mg of 98.0% pure avermectin B1a.

Bryostatins Separation [4]

By extraction from 1000 kg of *Bugula neritina* 906.5 g of lymphocytic leukemia cell line active fraction was obtained. Further purification was performed with HSCCC. Bryostatins have intermediate polarity, so that *n*-hexane–ethyl acetate–methanol–water (3:7:5:5, v/v/v/v) was employed with the upper layer as the mobile phase and the lower layer as the stationary phase. By this technique, from 300 mg to 3 mg of seven bryostatins have been isolated, including a new molecule called bryostatin 14 (Fig. 2b).

Bacitracins Separation [9]

The bacitracin complex (Fig. 3) was purified by foam CCC. The column design for foam CCC consists of a Teflon tube. Simultaneous introduction of N_2 and the liquid phase through the respective flow tube produces a countercurrent between the gas and the liquid phase through the coil. The sample mixture injected through the middle portion of the column is separated according to the foaming capability: The foam-active components travel through the coil with the gas phase and elute through the foam collection line, whereas the rest of components move with the liquid phase and elute through the liquid collection line.

After the experiment, fractions from the foam and liquid outlets are collected and analyzed. The elution curve of bacitracin components from the foam outlet shows three major peaks, and the one from the liquid outlet. The HPLC analysis of fractions clearly indicate that the bacitracin components are separated in the order of hydrophobicity of the molecule in the foam fractions and in increasing order of their hydrophilicity in the liquid fractions.

References

1. H. Oka, K.-I. Harada, Y. Ito, and Y. Ito, *J. Chromatogr. 812*: 35–52 (1998).
2. K. Hostettman, C. Appolonia, B. Domon, and M. Hostettmann. *J. Liquid Chromatogr.* 7: 231–242 (1984).
3. G. M. Brill, J. B. McAlpine, and J. E. Hochlowski, *J. Liquid Chromatogr. 8*(12): 2259–2280 (1985).
4. G. R. Pettit, F. Gao, D. Sengupta, J. C. Coll, C. L. Herald, D. L. Doubek, J. M. Schimdt, J. R. Van Camp, J. Rudloe, and R. A. Nieman, *Tetrahedron 47*(22): 3601–3610 (1991).
5. F. Oka, H. Oka, and Y. Ito, *J. Chromatogr. 538*: 99–108 (1991).
6. M.-C. Menet and D. Thiébaut, *J. Chromatogr. 831*: 203–216 (1999).
7. K.-I. Harada, I. Kimura, A. Yoshikawa, M. Suzuki, H. Nakazawa, S. Hattori, and Y. Ito, *J. Liquid Chromatogr. 13*: 2373–2388 (1990).
8. H. Oka, Y. Ikai, J. Hayakawa, K.-I. Harada, M. Suzuki, A. Shimizu, T. Hayashi, K. Takeba, H. Nakazawa, and Y. Ito, *J. Chromatogr. 723*: 61–68 (1996).
9. H. Oka, K.-I. Harada, M. Suzuki, N. Nakazawa, and Y. Ito, *J. Chromatogr 482*: 197–205 (1989).

M.-C. Rolet-Menet

Separation of Chiral Compounds by CE and MEKC with Cyclodextrins

Cyclodextrins (CD) and their derivatives represent a unique group of chiral selectors which can be used for enantioseparations in almost all instrumental separation techniques, such as gas chromatography (GC), high-performance liquid chromatography (HPLC), super/subcritical fluid chromatography (SFC), and capillary electrophoresis (CE). CDs are nonreducing, cyclic oligosaccharides produced enzymatically from starch. The most widely applied native α-, β-, and γ-CD are constructed from six, seven, and eight glucose units bonded through 1,4-α-linkages. The inner cavity of the CD, which is lined with hydrogen atoms and glycosidic oxygen bridges, is hydrophobic, which favors hydrophobic interactions between a guest and the CD host. The outer CD rims are formed by the secondary 2- and 3- and the primary 6-hydroxyl groups. The location of the polar hydroxyl groups on the outer rim determines the solubility of the CDs in aqueous solutions. The hydrogen-bonding between the secondary C(2) and C(3) hydroxyl groups of adjacent D-glucopyranosyl residues stabilizes the shape and structure of the CD macrocycle.

The ability of CDs to form intermolecular complexes with other molecules was already known in the early twentieth century. Another important property of CDs is that each glucose molecule in this macrocycle contains five chiral carbon atoms, which results in a chiral recognition ability in complex formation. This property of CDs was first evidenced by Cramer [1]. The relative easy availability from regenerable natural sources, the existence in various sizes, the stable structure, the localized hydrophobic area, the solubility in the hydrophilic solvents, the ability of intermolecular complex formation and the chiral recognition ability together with their nontoxicity, ultraviolet (UV) transparency, feasibility of their modification, and so forth contributed greatly to the establishment of CDs as a major chiral selectors in CE and CD-modified MEKC separations of enantiomers.

The first applications of CDs as chiral selectors in CE were reported in capillary isotachophoresis (CITP) [2] and capillary gel electrophoresis (CGE) [3]. Soon thereafter, Fanali described the application of CDs as chiral selectors in free-solution CE [4] and Terabe used the charged CD derivative for enantioseparations in the capillary electrokinetic chromatography (CEKC) mode [5]. It seems important to note that although the experiment in the CITP, CGE, CE, and CEKC is different, the enantiomers in all of these techniques are resolved based on the same (chromatographic) principle, which is a stereoselective distribution of enantiomers between two (pseudo) phases with different mobilities. Thus, enantioseparations in CE are commonly based on an electrophoretic migration principle and on a chromatographic separation principle [6].

Chiral separations in CE are characterized by some peculiarities compared to true electrophoretic separations based on the different charge-to-mass (size) ratios of analytes. These peculiarities need to be realized when looking at the differences between enantioseparations using charged and neutral chiral selectors, between CZE and MEKC separations, for the evaluation of the role of the electro-osmotic force (EOF) in chiral CE separations, and so forth.

The separation of enantiomers in CE means that they reach a detection window in a different period of time after their simultaneous injection at the capillary inlet. Thus, the enantiomers must migrate with a different velocity along the longitudinal axis of a separation capillary. For a species possessing a different charge-to-mass (size) ratio, this occurs automatically after the application of a voltage gradient at the ends of the separation capillary. However, enantiomers do not differ from each other in terms of their effective charge-to-mass ratio in achiral medium. Therefore, in order to achieve enantioseparations, additional chiral substances are required that are able to transform an achiral medium into a chiral one. These chiral substances are called chiral selectors. A chiral selector will interact with enantiomers of an analyte stereoselectively and this secondary equilibrium can generate a mobility difference between enantiomers ($\Delta\mu$) that can be calculated according to the equation [7]:

$$\Delta\mu = \frac{\mu_f + \mu_{C,R}K_R[C]}{1 + K_R[C]} - \frac{\mu_f + \mu_{C,R}K_S[C]}{1 + K_S[C]} \tag{1}$$

where μ_f is the mobility of the free enantiomers and $\mu_{C,R}$ and $\mu_{C,S}$ are the mobilities of complexed R and S enantiomers. K_R and K_S are the binding constants between the chiral selector and the R and S enantiomers, and [C] is the concentration of a chiral selector.

Analyzing Eq. (1), one realizes that when $\mu_{C,R} \neq \mu_{C,S}$, the requirement $K_R \neq K_S$ is not necessary for obtaining $\Delta\mu \neq 0$. This means that in CE, a chiral recognition in the classical meaning of this definition ($K_R \neq K_S$) is not always necessary for a chiral separation. This is theoretically feasible but has not been undoubtedly evidenced yet experimentally. More common is the case when $\mu_{C,R} = \mu_{C,S}$ and then $K_R \neq K_S$ is the necessary requirement for enantioseparation. A combined contribution of both a stereoselective binding of the enantiomers to a chiral selector ($K_R \neq K_S$) and a mobility difference between the transient diastereomeric complexes ($\mu_{C,R} \neq \mu_{C,S}$) is also possible.

In the case when $\mu_{C,R} = \mu_{C,S}$ (the most common case), Eq. (1) can be simplified as follows [7]:

$$\Delta\mu = \frac{C(\mu_f - \mu_C)(K_R - K_S)}{1 + C[K_R + K_S] + C^2K_RK_S} \tag{2}$$

This simplified equation allows one to calculate the optimal concentration of a chiral selector [7], a better design of chiral separation systems containing multiple chiral selectors [6], and the design of the enantiomer migration order [6,8] and indicates why micellar additives are sometimes required in chiral CE with CDs.

As shown in Eq. (2) together with the chiral recognition ($K_R \neq K_S$), the other necessary requirement for enantioseparations in CE is a mobility difference between the free and the complexed analyte ($\mu_f - \mu_c \neq 0$). Otherwise, it will be impossible to transfer a chiral recognition into a chiral separation. This requirement does not hold when neutral analytes are analyzed with neutral chiral selectors. In such a case, an additional buffer component is required that will assist in generating a difference between the mobilities of an analyte in its free and complexed forms with a chiral selector. This is achieved by an achiral micellar phase in cyclodextrin-modified micellar electrokinetic chromatography (CD–MEKC) [9]. However, a charged CD or a chiral micellar phase can combine the above-mentioned functions of both a neutral CD and a micellar phase [5,6,8,10–12].

The EOF contributes significantly to the mobility of analytes in CE. The EOF is considered to be a nonselective mobility. However, for enantiomers, both the EOF and the electrophoretic mobility of the analyte are inherently nonenantioselective. The stereoselective analyte–selector interactions may turn both of these mobilities into a selective transport with equal success. This is the principal difference between the roles of the EOF in true electrophoretic separations and in chiral CE separations.

The principal mechanism of enantioseparations is a stereoselective interaction of enantiomers with CDs, which is the same in CE and MEKC as well as in CITP, CGE, and even in capillary isoelectric focusing (CIEF). One of the significant advantages of chiral CE compared to other instrumental enantioseparation techniques together with the extremely high peak efficiency, miniaturized size, low costs, less environmental problems, and so forth [8] is the high flexibility. The separation in different modes can be performed using the same instrumental setup and it takes just a few minutes to change from one chiral selector to another one, to combine two or more chiral selectors, to vary the concentration of a chiral selector, and so forth. All of these variations are very difficult and time-consuming in a chromatographic system.

There are much more variables that can be applied for the optimization of a separation in CE than in a pressure-driven system. These include the type and concentration of a chiral selector, the pH of the background electrolyte, the concentration and type of the buffer, the achiral buffer additives, the capillary dimensions and the nature of the inner surface, the EOF, the temperature, and so forth.

Cyclodextrins are commercially available in various sizes (α, β, γ), carrying different functionalities, charge, and so forth. At present, β-CD and its neutral and ionic derivatives are considered to be the most suitable chiral selectors in CE. However, α- and γ-CD sometimes offer a complementary chiral separation ability [8]. Among the native CDs, β-CD is characterized by the lowest solubility in aqueous solutions. Therefore, the neutral derivatives carrying alkyl and hydroxyalkyl substituents which possess a higher solubility in aqueous buffers and sometimes offer complementary chiral recognition properties are widely used as chiral selectors in CE. Among the neutral CD derivatives, single-component heptakis-(2,6-di-*O*-methyl)-β-CD and heptakis-(2,3,6-tri-*O*-methyl)-β-CD represent special interests. The latter exhibits the opposite binding affinity toward many chiral compounds compared to native β-CD. Elucidation of the molecular mechanisms of this phenomenon can contribute markedly to a better understanding of the nature of the forces determining the binding and chiral recognition properties of CDs.

Ionic derivatives of CDs represent another group of effective chiral selectors in CE [5,6,8,10–12]. They offer the following advantages for enantioseparations: (a) enhanced solubility in the aqueous buffers; (b) self-electrophoretic mobility enabling their application also for enantioseparations of uncharged chiral compounds; (c) additional groups for more effective intermolecular interactions; (d) use as chiral carriers; (e) higher separation power toward oppositely charged enantiomers which are not only due to more tight interactions but also to the countercurrent mobility of an analyte and a selector; (f) easier on-line coupling of chiral CE with mass spectrometry (MS), nuclear magnetic resonance (NMR), and so forth [8].

Charged CD derivatives are commercially available with cationic or anionic groups. Among cationic CDs, randomly substituted 2-hydroxypropyltrimethylammonium salts of β-CD and 6-monoamino-6-deoxy-β-CD have been intensively studied [6,8,10]. Cationic CD derivatives tend to be adsorbed to the negatively charged inner surface of a fused-silica capillary and revert the direction of the EOF from the cathode to the anode. Therefore, these derivatives need to be used with capillaries having neutral or positive inner-wall coatings. CD derivatives with positively charged amino, alkylamino, and ammonium groups in a desired position can be synthesized relatively easily from a native CD. In spite of this, there are just two of these derivatives commercially available at present. This seems to be the reason for their incomplete evaluation as chiral selectors in CE [5,6,8,10].

Among anionic CDs, randomly substituted carboxyalkyl, sulfoalkyl, and sulfate derivatives are commercially available and play an important role in the development of chiral CE [6,8,10–12]. However, all randomly substituted neutral and charged derivatives of CDs suffer from the disadvantage to be a multicomponent mixture. The individual components of these mixtures may exhibit markedly different chiral-resolving properties. In rare cases, even the opposite migration order of the enantiomers has been reported depending on the degree of substitution of a charged CD [6]. Thus, it is extremely difficult to optimize and validate a chiral CE separation using randomly substituted derivatives of CD. Recently, Vigh and coworkers developed single-isomer CD polysulfates which are currently commercially available [10]. These derivatives represent a significant interest not only for the development of reproducible, validated methods in chiral CE but also for mechanistic studies.

One important advantage of charged CD derivatives is their use as chiral carriers [5,6,8,10–12]. This enables one to mobilize a neutral analyte even in the absence of the EOF, a charged analyte in the opposite direction to its electrophoretic mobility, and to suppress a mobility of an analyte in the uncomplexed form. The last offers a significant advantage for the improvement of a separation selectivity.

Two unique advantages of chiral CE, using the selector in a double function for the enantioseparation and transport of the resolved enantiomers to the detector as well as the feasibility of a combination of two chiral selectors, are illustrated in Fig. 1. The chiral compounds are the well-known former sedative drug thalidomide (TD) and its hydroxylated metabolites recently observed in human blood plasma and liver. The drug had been withdrawn from clinical practice several decades ago due to severe toxic effects probably residing in one enantiomer, but it has been reapproved again by Food and Drug Administration in 1998 as a result of its recently discovered antileprosy, anti-HIV (human immunodeficiency virus), anti-inflammatory, and antireumatoid activity. The compounds are uncharged and, therefore, lack an electrophoretic mobility. This makes a separation difficult. In the separation system described in Fig. 1, the analytes are stereoselectively accelerated with one of the CDs (sulfobutyl-β-CD) and also stereoselectively decelerated with another CD (β-CD). In

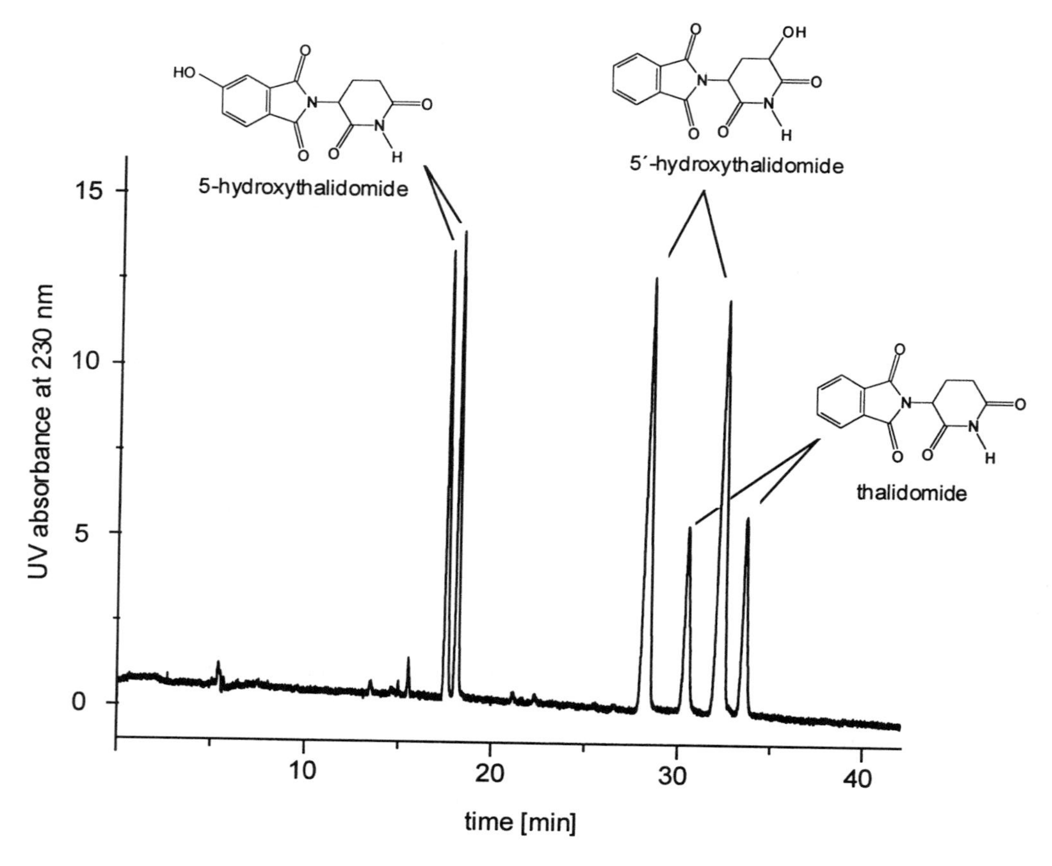

Fig. 1 Simultaneous separation and enantioseparation of thalidomide, 5-hydroxythalidomide, and 5′-hydroxythalidomide in CE using polyacrylamide-coated capillary and a mixture of 15 mg/mL sulfobutyl (4.0)-β-CD and 10 mg/mL β-CD as the chiral carrier.

combination with an easy variation of the CD concentration, this allows one to adjust simultaneous baseline separations and enantioseparations of all components. This example also illustrates the application of chiral CE separations for solving of practical biomedical problems. This seems to be one of the major application areas of chiral CE [8,12].

Together with the high application potential of chiral CE, due to the high separation efficiency and versatility, this technique allows to study very fine nuances of binding and chiral recognition mechanisms by CDs.

Thus, native CDs and their derivatives are definitely established as a major group of chiral selectors in CE, CEKC, and MEKC. Perhaps, in the near future, the main emphasis will be directed to the use of a well-characterized, single isomer native and derivatized (neutral and charged) CDs as chiral selectors and to the use of chiral CE for solving problems in chemical, agrochemical, food, environmental, and, mainly, in biomedical fields. Mechanistic studies for a better understanding of the nature of intermolecular interactions governing the binding and chiral recognition properties of CDs seems to be an interesting topic also.

References

1. F. Cramer, *Angew. Chem. 64*: 136 (1952).
2. J. Snopek, I. Jelinek, and E. Smolkova-Keulemansova, *J. Chromatogr. 438*: 211–218 (1988).
3. A. Guttman, A. Paulus, A. S. Cohen, N. Grinberg, and B. L. Karger, *J. Chromatogr. 448*: 41–53 (1988).
4. S. Fanali, *J. Chromatogr. 474*: 441–446 (1989).
5. S. Terabe, *Trends Anal. Chem. 8*: 129–134 (1989).
6. B. Chankvetadze, *J. Chromatogr. A 792*: 269–295 (1997).
7. S. A. C. Wren and R. C. Rowe, *J. Chromatogr. 603*: 235–241 (1992).
8. B. Chankvetadze, *Capillary Electrophoresis in Chiral Analysis*, John Wiley & Sons, Chichester, 1997.
9. S. Terabe, Y. Miyashita, O. Shibata, E. R. Barnhart, L. R. Alexander, D. G. Patterson, B. L. Karger, K. Hosoya, and N. Tanaka, *J. Chromatogr. 516*: 23–31 (1990).
10. Gy. Vigh and A. D. Sokolowski, *Electrophoresis 18*: 2331–2342 (1997).
11. A. M. Stalcup and K.-H. Gahm, *Anal. Chem. 68*: 1360–1368 (1996).
12. M. Fillet and P. Hubert, *J. Crommen. Electrophoresis 19*: 2834–2840 (1998).

Bezhan Chankvetadze

Separation of Metal Ions by Centrifugal Partition Chromatography

Introduction

Centrifugal partition chromatography (CPC), a multistage countercurrent liquid–liquid distribution technique employing discrete stages and two immiscible bulk liquid phases, is ideally suited for the detailed examination through separation factors and efficiencies, the influence of bulk aqueous and liquid–liquid interfacial equilibria and kinetics, on the separations of metal ions. This has been demonstrated by separations of transition metals, platinum group metals, and trivalent lanthanides; the significant findings are as follows: (a) Separation efficiencies are mainly limited by back-extraction kinetics which occur in the bulk aqueous phase and at the organic–aqueous interface as indicated by a direct linear correlation between the half-lives ($t_{1/2}$) of the dissociation reactions and the reduced plate height. (b) The interfacial areas calculated through this correlation are much larger in many cases than those generated in highly stirred two-phase mixtures. (c) The addition of surfactants and interfacial catalysis of the formation and dissociation of the complexes dramatically improve efficiencies. Examples of the separation of platinum group metal and trivalent lanthanides and the kinetic information that can be derived from their chromatograms are discussed.

Centrifugal Partition Chromatography

The CPC apparatus, manufactured by Sanki Engineering Company, Japan [1], consists of a series of cartridges, with each cartridge containing 40–400 channels, depending on the desired internal volume. These channels serve as stages in the separation experiment and the total number of channels is sometimes 400–4800, depending on the number of cartridges employed. The cartridges are arranged in a rotor which is rotated at 700–1200 rpm and the centrifugal force generated keeps one of the two phases (usually the organic phase) stationary while the other phase (usually the aqueous phase) is moved through it at a constant flow rate. The injected analyte mixture is carried by the aqueous mobile phase into the cartridges where they are extracted into the organic stationary phase by simple distribution if they are organic or by complexation with a suitable ligand if they are metals. When the mobile phase is depleted of the analytes, further flow of the mobile phase of the same (isocratic elution) or different (gradient elution) composition causes the back-extraction of the analytes, which can be detected by a suitable method. If the analytes are completely separated, they appear as discrete peaks similar to conventional chromatographic methods such as high-performance liquid chromatography (HPLC), and, hence, it is called centrifugal partition chromatography. CPC has these unique features: a large number of discrete stages (400–4800 depending on the operational volume chosen); high loading capacity for extractants and analytes; negligible loss of stationary phase due to "bleeding"; flexible organic–aqueous phase volume ratios; high stationary phase to mobile phase ratio; and ready adaptability to the process scale.

The basic parameters employed in the analysis of the CPC chromatograms are the retention volume V_r, which is related to the stationary phase and mobile phase volumes V_s and V_m, respectively, and the distribution ratio of the analyte D [Eq. (1)], the chromatographic efficiency, as measured by the number of theoretical plates N, which is calculated from the retention volume V_r and the width of the chromatogram w, [Eq. (2)], the chromatographic inefficiency represented by the channel equivalent of a theoretical plate (CETP), which is analogous to reduced plate height and is the ratio of the number of channels (CH) (2400 in our experiments) to N [Eq. (3)], and the selectivity I achieved in the separation of two analytes (1 and 2), which is the ratio of their distribution ratios D_1 and D_2 [Eq. (4) [2]:

$$V_r = V_m + DV_s \qquad (1)$$

$$N = 16\left(\frac{V_r}{w}\right)^2 \qquad (2)$$

$$\mathrm{CETP}_{\mathrm{obs}} = \frac{\mathrm{CH}}{N} \qquad (3)$$

$$\alpha = \frac{D_2}{D_1} \quad (D_2 > D_1) \qquad (4)$$

Separation of Platinum Group Metals

Separation, extraction, and purification of the platinum group metals (PGM) Pt, Pd, Ir, and Rh in their various oxidation states continues to be challenging and represent interesting areas of research. The separation of PGM from chloride media by solvent extraction can be achieved either by complexation with a suitable ligand or through ion-pair formation with a large cation. Complexation with a ligand is more selective but generally suffers from slow complex formation and dissociation kinetics. By contrast, ion-pair formation is diffusion controlled and not very selective, but it is necessary to separate kinetically inert species such as $PtCl_6^{2-}$ and $IrCl_6^{2-}$.

Trioctylphosphine oxide (TOPO) is an organophosphorus compound and is a stable and inexpensive extractant. TOPO, as we have shown [3–5], is unique in that it can function as a monodentate ligand and as a cation for ion-pair extraction when protonated and the extraction equilibrium for the neutral ligand is shown in Eq. (5), indicating the extraction of the neutral complex $MCl_2(TOPO)_2$ [M = Pd(II), Pt(II); n = 2–4]:

$$MCl_n^{(n-2)-} + 2TOPO_0 \underset{}{\overset{K_{ex}}{\rightleftharpoons}} MCL_2(TOPO)_{2,0} + (n-2)Cl^- \qquad (5)$$

The K_{ex} values for $PdCl_2$, $PdCl_3^-$ and $PdCl_4^{2-}$ are 794.3 M^{-2}, 2.75 M^{-1}, and 0.14 respectively. A single peak was observed in the CPC chromatogram of Pd(II) at any concentration of Cl^-, as its hydrolytic equilibria are rapid. The corresponding values for the three Pt(II) chloro species are 48 M^{-2}, 0.047 M^{-1} and 0.018, clearly indicating the better extractibility of Pd(II) over Pt(II). The difference in the K_{ex4} values for the MCl_4^{2-} species can be exploited to obtain an efficient separation of Pt(II) and Pd(II) from Rh(III) and Ir(III).

Formation of $HTOPO^+$, at HCl concentrations of 0.1 M, resulted in the extraction of $(HTOPO)_2MCl_4$ (M = Pt or Pd):

$$MCl_4^{2} + 2H^+ + 2TOPO_0 \overset{K_{ex}}{\rightleftharpoons} (HTOPO)_2MCl_{4,0} \qquad (6)$$

The chromatogram of the separation of $RhCl_6^{3-}$, $PdCl_4^{2-}$, and $PtCl_4^{2-}$ by $HTOPO^+$ is shown in Fig. 1. The K_{ex} values of Pd(II) and Pt(II) are 93.3 M^{-4} and 1961 M^{-4}, respectively, indicating that Pd(II) elutes ahead of Pt(II) in the ion-pair separation, whereas the opposite is true in the separation by complexation. While the chromatogram of Pt(II) involves only the formation of $(HTOPO)_2PtCl_4$, the chromatogram of Pd(II) also involves the formation of $(HTOPO)PdCl_3$. In fact, under the experimental conditions employed in these separations, this is the major Pd ion pair that is extracted. The extraction equilibrium constant for $(HTOPO)PdCl_3$ is 18.25 M^{-1}. Similarly, Pt(IV)

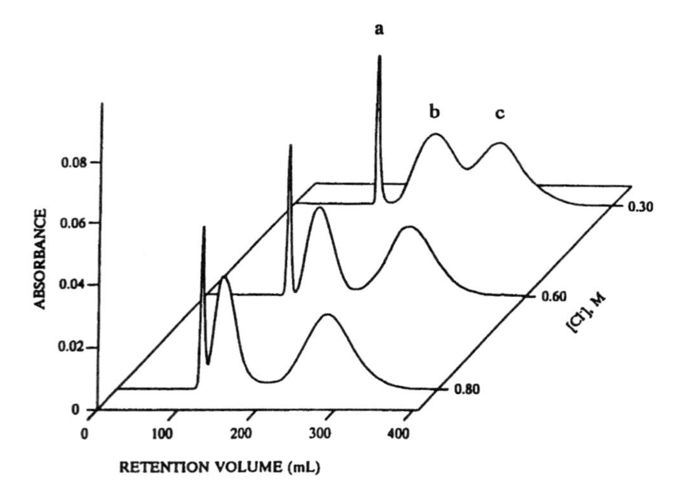

Fig. 1 Separation of 10^{-4} M $IrCl_6^{2-}$, $PtCl_4^{2-}$, and 10^{-3} M $PdCl_4^{2-}$ as their ion pairs with $HTOPO^+$ as a function of $[Cl^-]$ with 0.5 M TOPO at 0.1 M HCl and 4.0 mL/min flow rate. Eluting species are (a) $IrCl_6^{2-}$, (b) $PdCl_4^{2-}$, and (c) $PtCl_4^{2-}$.

and Ir(IV) could be separated by $HTOPO^+$ by ion-pair formation with their MCl_6^{2-} species. The K_{ex} values for the Pt(IV) and Ir(IV) species are 1576 M^{-4} and 8035 M^{-4}, respectively.

Separation Efficiencies of Platinum Group Metals

We observed, early on, that CETP values are significantly larger for metal-ion separations than those for simple organic analytes under the same conditions. They are far larger than could be explained in terms of mass-transfer and diffusion factors. Moreover, they increase more rapidly with increasing flow rate than those of organic analytes, indicating a chemical kinetic component affecting the CETP. The CETP values observed with metal ions, after correction for mass transport and diffusion (achieved using an organic analyte with similar distribution characteristics), reflect the half-lives of chemical reactions causing the added inefficiencies. Metal-complex formation and dissociation reactions with half-lives of milliseconds (i.e., rapid enough that in batch experiments they reach equilibrium "instantaneously") will lower the efficiencies of CPC chromatograms. Conversely, CETP values can be used to study rapid reaction kinetics if this relationship is found to be generally valid. Thus, CPC is a useful tool not only for uncovering kinetics of metals separations but also for obtaining detailed mechanisms of those reactions responsible for inefficiencies in multistage metals separations. This demonstrates the utility of CPC for examining the kinetics of metal-complex formation and dissociation reactions in two-phase systems that are too rapid for the automated membrane extraction system (AMES).

It was evident, from the separations of PGM, that their experimental CETP values were much larger compared to that for an organic analyte at identical distribution ratios. These results indicated that factors other than mass transfer and diffusion were responsible for the additional bandwidths in the case of the metal ions. The most likely factor is the slow kinetics of back-extraction of the metal ions as the forward extraction reactions are usually rapid. To test this hypothesis, 3-picoline was used as the model compound for the determination of the CPC bandwidth due to mass transfer and diffusion ($CETP_{dif}$), and the CETP value due to slow chemical kinetics ($CETP_{ck}$) was derived by expressing the experimental CETP ($CETP_{obs}$) as a sum of $CETP_{dif}$ and $CETP_{ck}$:

$$CETP_{obs} = CETP_{dif} + CETP_{ck} \tag{7}$$

The $CETP_{ck}$ values determined by varying the concentrations of the species in the aqueous and organic phases clearly showed that the slow back-extraction kinetics of the metal complexes were indeed responsible for the broad bands in the CPC chromatograms. On the basis of these results, a mechanism of the dissociation step could be deduced, which indicated that the dissociation of $MCl_2(TOPO)$ is rate limiting in the back-extraction of $MCl_2(TOPO)_2$ (M = Pd, Pt). A plot of the $CETP_{ck}$ values for $MCl_2(TOPO)_2$ against the $t_{1/2}$ values yielded a straight line. Further, these points and those for the Pd(II) system fall on a single line, indicating a general correlation for the separation of these two metals using TOPO in the heptane–H_2O phase pair.

Dissociation reactions with half-lives ranging from milliseconds to seconds can adversely affect the CPC efficiencies. It is important to realize the consequence of these findings: Extraction and back-extraction reactions that appear to be rapid in single-stage equilibrations may still be slow enough to reduce the efficiencies of multistage separations. A further significant finding of this work is that a direct linear correlation exists between $CETP_{ck}$ and $t_{1/2}$ for the Pd(II)–TOPO system and several other systems. Because the $CETP_{ck}$ values are a measure of the half-lives of the slow dissociation steps in metal-complex dissociation reactions, CPC is a useful tool for examining the kinetic and the equilibrium aspects of such reactions.

Separation of Trivalent Lanthanides

The trivalent lanthanides have been separated using acidic organophosphorous ligands and the acylpyrazolones [6–8]. The phosphinic acid bis(2,4,4-trimethylpentyl)phosphinic acid (Cyanex 272) in the heptane–water phase pair

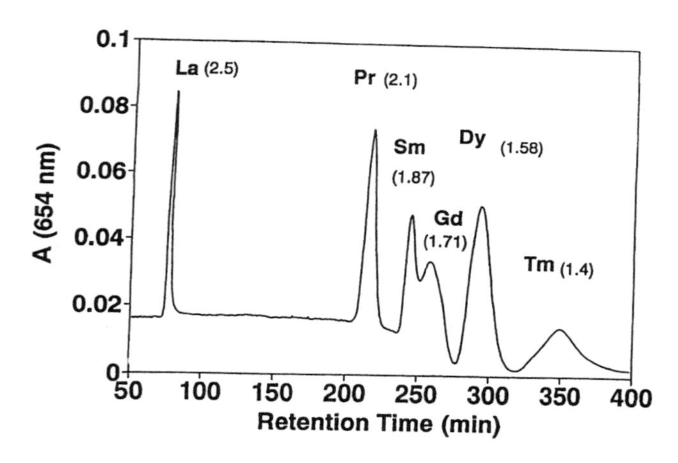

Fig. 2 Separation of lanthanides by use of a pH gradient with 0.1 *M* Cyanex 272 at $V_s/V_m = 0.18$ and flow rate of 1 mL/min. The concentrations and pH of elution are as follows: La (2 ppm; 2.5), Pr (6 ppm; 2.1), Sm (4 ppm; 1.87), Gd (4 ppm; 1.71), Dy (10 ppm; 1.58), and Tm (8 ppm; 1.4).

is dimeric and provides excellent separations of the adjacent light lanthanides at a fixed pH and a mixture of light and heavy lanthanides using a pH gradient (Fig. 2). Cyanex 272 is a chelating ligand that extracts the trivalent lanthanides by chelating them in its dimeric form [Eq. (8)]. The acylpyrazolones, 1-phenyl-3-methyl-4-benzoyl-5-pyrazolone (HPMBP) and 1-phenyl-3-methyl-4-capryloyl-5-pyrazolone (HPMCP, see structure below) have also been used in the toluene–H_2O phase pair for the extraction and separation of the trivalent lanthanides. The extraction equilibrium constant for HPMBP and HPMCP are given in Eq. (9).

$$M^{3+} + 3(HL)_{2(0)} \underset{}{\overset{K_{ex}}{\rightleftarrows}} M(HL_2)_{3(0)} + 3H^+ \tag{8}$$

$$M^{3+} + 3HL_0 \overset{K_{ex}}{\rightleftarrows} ML_{3(0)} + 3H^+ \tag{9}$$

Significant differences are seen between Cyanex 272 and the acylpyrazolones. The extractibility of the trivalent lanthanides is higher with Cyanex 272 than with the acylpyrazolones, where HPMBP shows better extraction than HPMCP. The stability constants of the lanthanides increases from the light to heavy, and the values for the Cyanex 272 and HPMCP complexes are larger than those of HPMBP. The separation factor (or selectivity) for a pair of lanthanides is much better with Cyanex 272 than with HPMBP or HPMCP, which have similar separation factors.

The lanthanide complexes lack distinct ultraviolet–visible spectra and, hence, kinetic information on their complex formation and dissociation reactions was obtained indirectly by the metallochromic indicator method [9]. These studies indicate that in the case of the Cyanex 272 complexes, the CPC efficiencies are mainly limited by the slow dissociation of the $M(HL_2)HL^+$ complex at the heptane–H_2O interface. In the case of the complexes of the acylpyrazolones, the CPC efficiencies are again limited by the dissociation of the lanthanide–pyrazolone complexes at the organic–aqueous interface with the rate-limiting step being the dissociation of the ML^{2+} complex. It was also shown that because the dissociation reactions are interfacial separations, efficiencies can be dramatically improved by the addition of surfactants like Triton X-100 to the organic phase and by interfacial catalysis by the addition of a aqueous soluble metallochromic indicator, which formed highly interfacially active lanthanide–indicator complex. The adsorption of this complex at the organic–aqueous interface catalyzed the metal-complex formation and dissociation reactions leading to high efficiencies in CPC separations [10].

Two significant results have emerged from the CPC separations of metals ions, namely (a) separation efficiencies are significantly reduced by slow metal-complex formation and dissociation kinetics and (b) CPC separations can be entirely interfacially driven analogous to conventional liquid chromatography.

References

1. A. P. Foucault (ed.), *Centrifugal Partition Chromatography*, Marcel Dekker, Inc., New York, 1994.
2. A. Berthod and D. W. Armstrong, Centrifugal partition chromatography II. Selectivity and efficiency, *J. Liquid Chromatogr. 11*: 567 (1988).
3. Y. Surakitbanharn, S. Muralidharan, and H. Freiser, Separation of palladium(II) from platinum(II), iridium(III) and rhodium(III) using centrifugal partition chromatography, *Solvent Extract. Ion Exchange 9*: 45 (1991).
4. Y. Surakitbanharn, S. Muralidharan, and H. Freiser, Centrifugal partition chromatography of palladium(II) and the influence of chemical kinetic factors on separation efficiency, *Anal. Chem. 63*: 2642 (1991).
5. Y. Surakitbanharn, H. Freiser, and S. Muralidharan, Centrifugal partition chromatographic separations of platinum group metals by complexation and ion pair formation, *Anal. Chem. 68*: 3934 (1996).
6. R. Cai, S. Muralidharan, and H. Freiser, Improved separation of closely related metal ions by centrifugal partition chromatography, *J. Liquid Chromatogr. 13*: 3651 (1990).
7. K. Inaba, H. Freiser, and S. Muralidharan, Effect of kinetic factors on the efficiencies of centrifugal partition chromatographic separations of tervalent lanthanides with bis(2,4,4,-trimethylpentyl)phosphinic acid as extractant, *Solvent Extract. Res. Dev. Japan 1*: 13 (1994).
8. G. Ma, H. Freiser, and S. Muralidharan, Centrifugal partition chromatographic separation of tervalent lanthanides

using acylpyrazolone extractants, *Anal. Chem. 69*: 2835 (1997).

9. K. Inaba, S. Muralidharan, and H. Freiser, Simultaneous characterization of extraction equilibria and back-extraction kinetics: Use of arsenazo III to characterize lanthanide–bis(2,4,4-trimethylpentyl)phosphinic acid complexes in surfactant micelles, *Anal. Chem. 65*: 1510 (1993).
10. G. Ma, H. Freiser, and S. Muralidharan, Interfacial catalysis of formation and dissociation of tervalent lanthanide complexes in two phase systems, *Anal. Chem. 69*: 827 (1977).

S. Muralidharan

Separation Ratio

The *separation ratio* between two solutes has two uses. The first is to help identify a solute or confirm its identity. The second is to help calculate the minimum efficiency required to achieve a given separation (this aspect of the separation ratio will be discussed under resolution). The first chromatographic parameter to be used for solute identification, other than the corrected retention volume, is the capacity ratio of the solute.

The capacity ratio of a solute (k') was defined as the ratio of the distribution coefficient (K) of the solute to the phase ratio (a) of the column. In turn, the phase ratio of the column was defined as the ratio of the volume of mobile phase in the column (V_m) to the volume of stationary phase in the column (V_s); that is,

$$a = \frac{V_m}{V_s}$$

and, as

$$V'_r = KV_s$$

Thus,

$$k' = \frac{K}{a} = \frac{KV_s}{V_m} \quad \text{and} \quad k' = \frac{V'_r}{V_m}$$

where V'_r is the corrected retention volume of the solute.

As the measurement of k' does not depend on flow rate, it is unaffected by flow changes and is, thus, a more reliable measurement than corrected retention volume for solute identification.

Unfortunately, both V_m and V_s will vary between different columns and, due to the partial exclusion that can occur with porous supports and stationary phases, may vary between different solutes. For this reason, the separation ratio (α) was introduced as an identification parameter. For two solutes, (A and B), the separation ratio is defined as

$$\alpha_{A/B} = \frac{V'_{r(A)}}{V'_{r(B)}} = \frac{K_A V_s}{K_B V_s} = \frac{K_A}{K_B}$$

It is seen that the separation ratio is independent of all column parameters and depends only on the nature of the two phases and the temperature. Thus, comparing data from two different columns, providing that the same phase system is used in each, and the columns operated at the same temperature, then any two solutes will have the same separation ratio on both systems. Thus, the separation ratio will be *independent of the phase ratios* of the two columns and the *flow rates*. It follows that the separation ratio of a solute can be used more reliably as a means of solute identification.

In practice, a standard substance is often added to a mixture and the separation ratio of the substance of interest to the standard is used for identification purposes. The separation ratio is taken as the ratio of the distances in centimeters between the dead point and the maximum of each peak, or if data processing is employed and the flow rate is constant, chart distances can be replaced by the corresponding retention times.

Suggested Further Reading

Scott, R. P. W., *Liquid Chromatography Column Theory*, John Wiley & Sons, Chichester, 1992, p. 26.

Scott, R. P. W., *Chromatographic Detectors*, Marcel Dekker, Inc., New York, 1996.

Scott, R. P. W., *Introduction to Analytical Gas Chromatography*, Marcel Dekker, Inc., New York, 1998.

Raymond P. W. Scott

Sequential Injection Analysis in HPLC

All of the samples analyzed using a chromatographic technique need special preparation before they are introduced into the column. This process is laborious, not reliable enough, and often expensive. There are several steps involved in sample preparation: dilaysis, dilution, extraction (selective extraction or concentration), and derivatization. Sometimes, the derivatization step is part of the extraction process. The expenses refer to the reagents and solvents of chromatographic purity grade. The sample preparation will become easier and not so expensive by automation.

Sequential injection analysis (SIA), introduced in 1990 [1,2], is a technique with a high potential for on-line process measurements. It is simple and convenient because sample manipulation can be automated. Furthermore, it consumes low volumes of reagents and solvents. Up to now, all the necessary steps done manually before introduction of the sample into the high-performance liquid chromatogram (HPLC) were separately introduced in SIA systems. By including SIA in sample preparation, it became faster, more accurate, and precise. Contamination is reduced substantially and the objectivity of the analysis increased. The main advantages of the coupling of SIA with HPLC are high precision of sample injection into the column, low contamination, low consumption of sample, reagents, and solvents, and the short time of preparation that decreases the time of analysis.

Dialysis

For on-line measurements, the dialysis step is very important because, through dialysis, the solid particles can be retained, the solution can be purified, and also some of the interference can be eliminated. The main disadvantages of dialysis are the slow speed involved and the low recovery of analyte. These parameters can be improved by introducing the dialyser into the conduits of a SIA system [3].

With the incorporation of a passive, neutral, semipermeable dialysis membrane into the conduits of the sequential injection system, the contact time of the sample zone with the membrane had much influence on the quantity of analyte that dialyzed through the membrane. It is necessary to determine, first, the time necessary to propel the entire sample zone over the membrane. After the propelling of the sample zone over the membrane, the flow direction is reversed and the sample zone is drawn back into the holding coil for fixed periods of time; usually these periods of time can vary from 2 to 60 s. By increasing of the dialysis time, the sensitivity of dialysis and, finally, the sensitivity of the analytical information are increased. A long dialysis time is also not good to consider, because this increases the dispersion of the dialyzed sample (situated below the membrane) due to a longer time delay before it is drawn in the specific holding coil for analysis.

To increase the percentage of dialysis, as well as the dialysis time, multiple flow reversals with a time of 20 s between each flow reversal is selected. Similar results are obtained by using a sequential injection system with the stopped-flow period around 150 s. The advantage of utilizing the stopped-flow mode over multiple flow reversals in the sequential injection analysis systems is that it needs less programming and, also, it reduces the strain on the pump.

Dilution

It is well known that HPLC techniques are performed at low concentration ranges. Sometimes, the sample is too concentrated in the analyte to be determined, and a dilution step is absolutely necessary. When a SIA system containing a dialyzer is utilized for sampling before a HPLC, the sample is already diluted. If the dilution is still not enough, a special step must be adopted in the program of the SIA system. The next step, when the analyte is extracted into a solvent, can also be considered a dilution.

There are two methods that can be adopted for a dilution in SIA: by using a dilution coil and by using a dilution step [4,5]. The easier dilution technique in SIA is by using a dilution step which can also be accomplished in a shorter time. There are three types of volumes that can control the dilution: the sample volume (the volume of sample or standard that is drawn into the holding coil via the sample port), the transfer volume (the volume of sample plus accompanying wash in the holding coil and tubing that is transferred into the dilution conduit from the holding coil), and the analysis volume (the volume taken from the dilution conduit to the holding coil) [4].

Concentration

The concentration step is very easy to implement in sequential injection analysis. The system is very simple and the results are reproducible. Most of the time, this step is not necessary for HPLC. When it is necessary, it can easily be done in the same time with the extraction step.

Extraction

There are two types of extraction that can be used in sampling. The first one involves a chemical reaction before the extraction, and the other one is just a simple extraction of analyte(s) from the solution. When a chemical reaction is involved, the derivatization step that may be necessary is included in the extraction step.

Extraction techniques that involve a chemical reaction can be classified as nonselective extraction or concentration, when more than one analyte is extracted from the solution by using the organic collectors (e.g., 8-hydroxyquinoline and dithizone derivatives) and selective extraction or separation. The first step in such an extraction technique is the formation of the complex by adding the reagent(s) to the solution of analyte, and after the extraction of the complex in an organic solvent. The problem that can arise in a SIA system with these types of extraction is the precipitate that is formed, and this can contaminate the other sample and also can block the tubing. To avoid these problems, it is necessary either to dilute the sample in such a way that the precipitation equilibria will not be reached and that all the complex will remain dissolved in the solution, or by derivatization of the ligands to make the complexes soluble in the aqueous solution by introducing hydrophilic groups into their structures (e.g., in the place of 8-hydroxyquinoline, the 7-iodo, 8-hydroxyquinoline, 5-sulfonic acid can be used).

Three types of SIA system were proposed in coupling with the extraction technique: The first one is based on the introduction of bubbles into the system [6], the second system is based on wetting film that is formed on a Teflon tube wall [7], and the third one is based on solid-phase extraction [8]. The most utilized system is the one based on bubbles. The most reliable is the one based on a Teflon tube wall, and the principle of functioning of this system is as follows. The aqueous sample is propelled through the segment of organic solvent whose flow is impeded due to hydrophobic interactions with the walls of Teflon extraction coil. This wall drag allows the faster moving aqueous sample to penetrate through and ultimately separate from the organic solvent. These steps are repeated with a re-extraction into a second aqueous segment that is collected and which is going to the analyzer [9].

Derivatization

This step can be included in the extraction techniques because, in most of the cases of extraction, the analytes are being transformed. This step is only necessary for analytes that cannot be determined directly in the form that they already exist in solution. In the SIA system, the derivatization process can be assimilated with a reaction between analyte(s) and reagent with the optimum parameters for both the reaction and SIA system [8,10]; the difference is that the product of the reaction is not channeled to the detector, but is channeled to the chromatograph.

SIA–HPLC Systems

Sequential injection is the perfect vehicle for HPLC, which, in turn, enhances sequential injection by eliminating the problem of dialysis, dilution, or concentration, extraction, and mixing reactants during the loading process. HPLC can be carried out in different modes: affinity chromatography, ion chromatography, extraction chromatography, and so forth.

Most of the SIA–HPLC systems have been applied for the separation and assay of radionuclides. The reason for selecting such a system is the potential radiation and contamination of an operator during the sampling process. By using SIA–HPLC systems, all steps are automated and the contact of the operator with them is minimal. Grate and Egorov [11] reviewed the radiometric separation and gave to SIA–HPLC systems the main place between automatic analytical separation in radiochemistry. They found that the type of chromatography suitable for coupling with SIA is extraction chromatography. For radiochemical separation, a wide-bore holding coil, in combination with air segmentation and sequential loading and delivery of solutions, instead of zone stacking in the holding coil, is proposed.

Enzymes and antibody–antigen systems have been used to measure a large number of analytes in relation to SIA–HPLC systems [12]. A very interesting application of these systems is given when the HPLC is carried out in the bead injection (BI) form [13]. As BI is presently restricted to relatively short columns, it is focused on separations based on mobile-phase changes, rather than relying on the separating power provided by a large number of theoretical plates. The SIA–BI system has also been applied with very good results for the separation of radionuclides.

An automated sequential trace enrichment dialyzer and gradient HPLC system is proposed for pharmacokinetic studies of drugs and their metabolites [14]. The dialyzer is essential in the determination of pharmaceutical compounds from tablets and biological fluids (e.g., blood). By its incorporation into the conduits of a SIA system and coupling with the HPLC, the objectivity and reproducibility of the measurements were increased.

Features for SIA–HPLC Systems

The ideal system for sample preparation in a chromatographic method is that the operation of all the steps be-

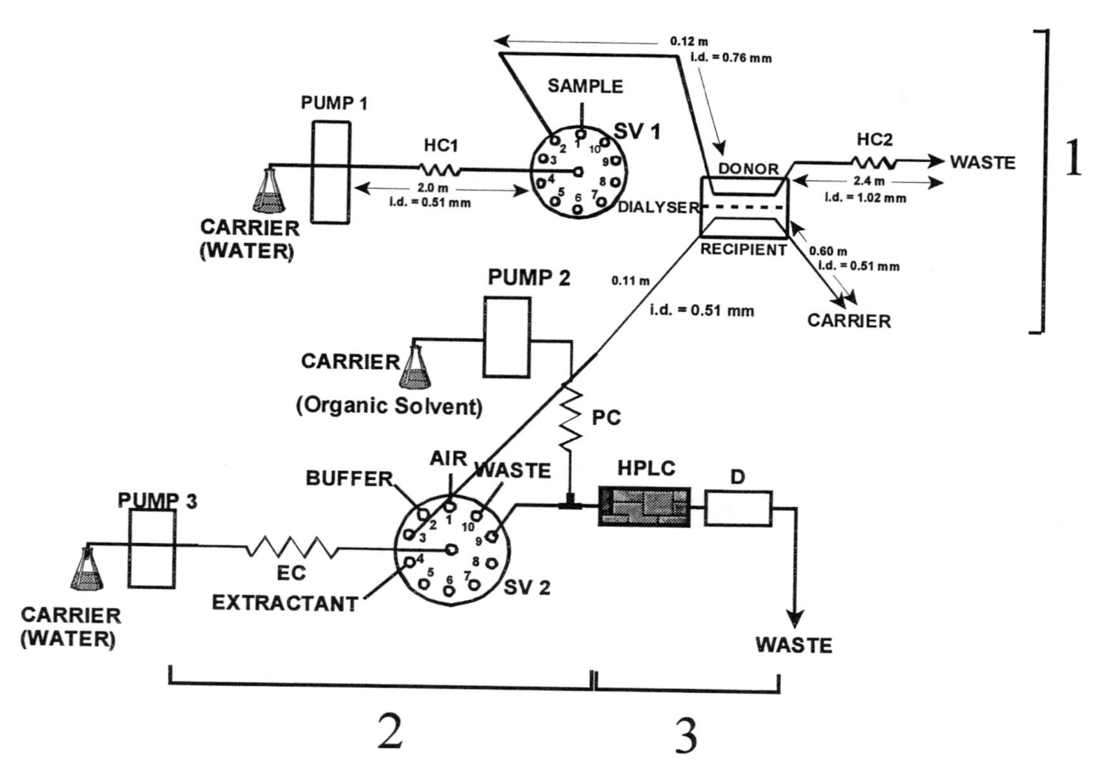

Fig. 1 Schematic representation of SIA–HPLC–detector system: 1: the dialysis unit; 2: the extraction (dilution, concentration, derivatization) unit; 3: the HPLC–D unit. SV is the selection valve; EC is the extraction coil; D is the detector.

tween sample dissolution and chromatography is done through a SIA technique (Fig.1). The first part of the system will consist of a sample dialysis (unit 1) and the outlet will be channeled into the second unit consisting of concentration or dilution steps and extraction of the analyte. It is always assumed that the concentration, dilution, and derivatization steps can be done by extraction — in most cases, it is absolutely necessary.

The proposed SIA–HPLC (Fig. 1) system operates by a well-programmed computer which will be able to analyze all types of sample: from environment, from the food industry, from the pharmaceutical industry, and also biological samples.

Conclusions

The utilization of sequential injection analysis coupled with HPLC systems increases the reliability of an HPLC analysis considerably because the primary factor that contributes to the increasing uncertainty is the sample preparation. It is always necessary to look to the most reliable methods for sample preparation, because only these methods will give the best results after the automation by using sequential injection analysis. The best coupling must be concerned with the selectivity and sensitivity assured by a sequential analysis system and by the selectivity and sensitivity of the HPLC technique. The introduction of bead injection considerably improves the reliability of the discussed system.

References

1. J. Rüzicka and G. D. Marshall, *Anal. Chim. Acta 237*: 329 (1990).
2. J. Rüzicka, G. D. Marshall, and G. D. Christian, *Anal. Chem. 62*: 1861 (1990).
3. J. F. van Staden, H. du Plessis, and R. E. Taljaard, *Anal. Chim. Acta 357*: 141 (1997).
4. M. Boron, J. Guzman, J. Rüzicka, and G. D. Christian, *Analyst 117*: 1839 (1992).
5. J. F. van Staden and R. E. Taljard, *Fresenius J. Anal. Chem. 357*: 577 (1997).
6. Y. Luo, R. Al-Othman, J. Rüzicka, and G. D. Christian, *Analyst 121*: 601 (1996).
7. J. W. Grate and R. H. Taylor, *Field Anal. Chem. Technol. 1*: 39 (1996).
8. J. F. van Staden and R. E. Taljard, *Anal. Chim. Acta 344*: 281 (1997).

9. R. E. Taljaard and J. F. van Staden, *S. Afr. J. Chem. 52*: 36 (1999).
10. J. F. van Staden and R. E. Taljard, *Anal. Chim. Acta 331*: 271 (1996).
11. J. W. Grate and O. B. Egorov, *Anal. Chem. 70*: 779A (1998).
12. J. Emneus and G. Marko-Varga, *J. Chromatogr. A 703*: 191 (1995).
13. J. Rüzicka and L. Scampavia, *Anal. Chem. 71*: 257A (1999).
14. J. D. H. Cooper, N. J. Shearsby, J. E. Taylor, and C. T. C. Fook-Sheung, *J. Chromatogr. B: Biomed. Appl. 702*: 227 (1997).

Raluca-Ioana Stefan
Jacobus F. van Staden
Hassan Y. Aboul-Enein

Settling Time of Two-Phase Solvent Systems in Countercurrent Chromatography

Introduction

Countercurrent chromatography (CCC) is based on the use of two-phase liquid solvent systems. One of the characteristics of a two-phase system is its settling time, which is the time required for the mixture of both phases to be completely separated into two layers, usually in the Earth's gravitational field. The measurement of the settling time itself is helpful in preparing the experiment, for instance, when preparing the mobile and stationary phases in the same vessel. However, it is also intrinsically linked to the hydrodynamic behavior of the two-phase system and, therefore, to its physical parameters such as densities, viscosities, and interfacial tension. It was, therefore, used as a parameter for predicting the hydrodynamic behavior of various solvent systems in J-type CCC devices.

How to Measure the Settling Time?

The settling time depends not only on the nature of the solvent system but also on the experimental environment. As its definition is the time needed for the two phases to completely separate into two layers, the worker has to precisely choose the experimental conditions for the measurement. For instance, Ito [1] chose the following:

- Equilibrate the two-phase solvent system in a separatory funnel at room temperature.
- Separate the two phases.
- Take an aliquot of each phase.
- Put each aliquot in a graduated cylinder, which is then sealed with a glass stopper.
- Gently mix the two-phase system by inverting the cylinder five times.
- Place the cylinder immediately on a flat table in an upright position.
- Measure the time required for the two phases to settle.
- Repeat the experiment several times and take the mean time.

It is necessary to perform several experiments, as significant differences in the values are observed. A second set of tests may be performed by vigorously shaking the graduated cylinder five times and then averaging the times measured on several experiments.

The described experimental environment can be summarized as a 1 : 1 ratio of both phases, the choice of a graduated cylinder, room temperature, five-time inversion of the cylinder in a gentle or a vigorous manner, and the Earth's gravitational field. This should be only considered as an example of measuring the settling time, but what is important is to remain with a single definition in order to then use the resultant values in a comparative way.

Values of Settling Times for Usual CCC Solvent Systems

Ito has carried out the experiments, as described earlier, for 15 solvent systems. All mean values for gentle mixing and for vigorous mixing are gathered in Table 1 in the T and T' columns, respectively, [1].

The T and T' values are close; thus, they both can be considered as reliable measurements of the settling time for each solvent system. The values range from a few seconds (i.e., 3–5 s) to about 1 minute.

Other types of solvent system can be used in CCC. For instance, aqueous two-phase polymer systems (discussed

Table 1 Values of the Settling Times for 15 Solvent Systems

Two-phase solvent system	Volume ratio	Settling time	
		T (s)	T' (s)
Hexane–water	1:1	<1	8
Ethyl acetate–water	1:1	15.5	21
Chloroform–water	1:1	3.5	5.5
Hexane–methanol		5.5	6
Ethyl acetate–acetic acid–water	4:1:4	15	16
Chloroform–acetic acid–water	2:2:1	29	27.5
Butanol-1–water	1:1	18	14
Butanol-1–0.1*M* NaCl	1:1	16	14.5
Butanol-1–1*M* NaCl	1:1	23.5	21.5
Butanol-1–acetic acid–water	4:1:5	38.5	37.5
Butanol-1–acetic acid–0.1*M* NaCl	4:1:5	32	30.5
Butanol-1–acetic acid–1*M* NaCl	4:1:5	26.5	24.5
Butanol-2–water	1:1	57	58
Butanol-2–0.1*M* NaCl	1:1	46.5	49.5
Butanol-2–1*M* NaCl	1:1	34	33.5

Note: T is the average value after five gentle mixings, and T' stands for five vigorous mixings.
Source: Ref. 1.

elsewhere in this volume) are made of two aqueous liquid phases containing various polymers. Such systems are gentle toward biological materials and they can be used for the partition of biomolecules, membrane vesicles, cellular organites, and whole cells. They are characterized by a high content of water in each phase, very close densities, very low interfacial tension, and high viscosities of the phases. Settling times are particularly long and may last up to 1 h. We refer the reader to the ATPS entry (see entry Aqueous Two-Phase Solvent Systems for Countercurrent Chromatography) for further information on these systems.

Hydrodynamic Behavior and Settling Time

The true purpose of Ito's study was to build a classification among the 15 solvent systems, based on their hydrodynamic behavior inside the column, measured through the retention of the stationary phase for various experimental conditions, on a J-type CCC device.

The first three solvent systems of Table 1 define the "hydrophobic" group, because they require the same given combination of two experimental parameters [choice of a lighter or heavier phase as stationary one, and choice of the pumping direction (i.e., from the head to the tail of the column or vice versa)] to achieve a good retention of the stationary phase and because their organic phase is hydrophobic. The last six solvent systems define the "hydrophilic" group, based on the same combination of two experimental conditions, which are reversed as compared to "hydrophobic" systems, because their organic phase can be considered as hydrophilic. There remains six solvent systems whose hydrodynamic behavior is more complicated, as it depends on the ratio of the radius of the coil on the rotation radius of the apparatus. They define the "intermediate" group.

Table 1 reveals that the classification may be defined from the values of the settling times. Indeed, the settling times of the hydrophobic solvent systems are the shortest ones, ranging from 1 to 20 s, whereas those of hydrophilic systems are in the 25–60-s range. Solvent systems belonging to the intermediate group exhibit moderate settling times, ranging from 6 to 30 s.

Consequently, the measurement of the settling time for a solvent system which is different from the 15 studied is a simple way of roughly classifying it among the 3 groups defined by Ito and then to know the best experimental combinations to obtain the highest retention of the stationary phase on a J-type device.

Increasing the temperature or the concentration of a salt in the aqueous phase are known ways of reducing the settling time. This is interesting, as systems with low settling times are easier to handle and are characterized by a "hydrophobic" hydrodynamic behavior in J-type CCC devices, which leads to high values of the retention of the stationary phase.

Hydrodynamics and Settling Time

Three theoretical parameters were introduced by Menet et al. in order to better understand the hydrodynamic behaviors of two-phase solvent systems [2]. We will not discuss, here, the capillary wavelength, as it only enables the description of the formation of droplets of one liquid in another liquid. The two other parameters were introduced because it appeared interesting to introduce other "theoretical" parameters to better describe the dynamic phenomena occurring inside a CCC column (i.e., after the formation of the droplet described by the capillary wavelength). Two of these are presented here, namely V_{low} for the fall of a droplet of the heavier liquid phase (lower) in the continuous lighter one (upper) and V_{up} for the rise of a droplet of the lighter liquid phase in the continuous heavier one, and are defined as follows:

$$V_{\text{low}} = \gamma\left(\eta_{\text{up}}\frac{2 + 3(\eta_{\text{low}}/\eta_{\text{up}})}{3 + 3(\eta_{\text{low}}/\eta_{\text{up}})}\right)^{-1} \quad \text{and}$$

$$V_{\text{up}} = \gamma\left(\eta_{\text{low}}\frac{2 + 3(\eta_{\text{up}}/\eta_{\text{low}})}{3 + 3(\eta_{\text{up}}/\eta_{\text{low}})}\right)^{-1}$$

with γ the interfacial tension between the two liquid phases, η_{up} the dynamic viscosity of the lighter phase, and η_{low} the dynamic viscosity of the heavier phase.

It is interesting to note that as neither V_{up} nor V_{low} depend on g, these velocities do not depend on the selected angular velocity of rotation. This is because the field intensity influences in the same way the size of the capillary wavelength and the sedimentation velocity of the droplet.

Menet et al. have computed the V_{low} and V_{up} values for the 15 solvent systems studied by Ito, along with 6 not previously studied [i.e., dimethyl sulfoxide (DMSO)–heptane (1:1, v/v), dimethyl formamide (DMF)–heptane (1:1, v/v), toluene–water (1:1, v/v), *o*-xylene–water (1:1, v/v), heptane–acetic acid–methanol (1:1:1, v/v) and chloroform–ethyl acetate–water–methanol (2:2:2:3, v/v)]. All the results have allowed them to define ranges of settling velocities linked to the hydrophobic, intermediate, or hydrophilic behavior of the solvent system. They have demonstrated the most reliable scale was that based on V_{up}, which is the description of the rise of a droplet of the lighter phase in a continuous heavier one.

It was further demonstrated that this theoretical parameter was the right one to use, as Menet et al. have carried out experiments on the evolution of the hydrodynamic behavior of the butanol-1–water system with the temperature [3]. They showed that the observed change in behavior was explained by an increase in the V_{up} settling velocity with the temperature, and, thus, a decrease of the settling time, explaining the "hydrophobic" behavior of the solvent system.

As a conclusion, the research worker can use the settling time for both purposes. One is the knowledge of the solvent system, mainly its ease of use, as long settling times complicate the use of the solvent system. The other one is to roughly predict the hydrodynamic behavior inside a J-type CCC device, in order to know the best combinations of some experimental conditions to obtain the highest retention of the stationary phase. For this purpose, the research worker may also use the theoretical settling velocity of a droplet of a lighter phase inside the heavier continuous liquid phase. It was demonstrated to be the better way of predicting the hydrodynamic behavior and, consequently, the best combination of experimental parameters for the highest retention of the stationary phase.

References

1. Y. Ito, Principles and instrumentation of CCC, in *Countercurrent Chromatography — Theory and Practice* (N. B. Mandava and Y. Ito, eds.), Marcel Dekker, Inc., New York, 1988.
2. J. M. Menet, D. Thiébaut, R. Rosset, J.-E. Wesfreid, and M. Martin, *Anal. Chem. 66*: 168–176 (1994).
3. J.-M. Menet and M. C. Rolet-Menet, Characterization of the solvent systems used in countercurrent chromatography, in *Countercurrent Chromatography*, (J. M. Menet and D. Thiébaut, eds.), Marcel Dekker, Inc., New York, 1999.

Jean-Michel Menet

Silica Capillaries: Chemical Derivatization

Capillaries for capillary electrophoresis (CE) are made of fused silica that has been drawn to precise internal and external diameters. Virtually all fused-silica capillaries used in CE, whether for home-made instruments or for commercial systems, have an external diameter of approximately 375 μm. Internal diameters vary over a wider range but generally lie between 50 and 100 μm. Smaller-diameter capillaries generally lead to detection problems, especially if spectroscopic methods are used, because the optical path length becomes too short. Larger-diameter capillaries dissipate heat inefficiently and can lead to band-broadening at higher applied voltages. All capillaries are covered externally with a polyimide coating for protection against breakage and to provide flexibility in fitting the typical column (50–100 cm) into the instrument.

Fused silica has surface properties that are similar to the porous particulate matter used as a support material in high-performance liquid chromatography (HPLC) packings. The most important features are the presence of silanol (Si — OH) groups that are polar and ionizable and siloxane linkages (Si — O — Si) that have a hydrophobic character. It is generally recognized that the silanols are the most influential in determining the surface properties of silica. For capillary electrophoresis, the Si — OH moieties contribute in at least three ways to the overall performance of the electrophoretic experiment. The presence of silanols on silica surfaces can be considered as a result of the formation of a polymer during condensation of silicic acid. When the polymer cross-links, all four bond sites on each silicon atom do not form siloxane linkages, leaving a hydroxyl group in one position. Because the silanols are acidic groups, they can dissociate in the presence of aqueous solution and behave as any weak acid. The pK_a of this moiety is near 5 but can vary depending on the purity of the silica material. Therefore, when the pH of the solution in contact with the inner wall of the capillary is approximately 3 or less, the sites will be fully protonated and the surface will be polar. If the pH is 7 or greater, then the silanols will be fully ionized. The acidic nature of the silanols leads to two of the salient features of the fused-silica surface with respect to electrophoretic experiments. Upon ionization of at least some of the silanols, a double layer is created at the surface when a voltage is applied to a capillary filled with aqueous buffer. This double layer is responsible for electro-osmotic flow (EOF), the movement of solvent toward the cathode. Because of the acidic nature of the silanols, there also is a strong tendency to adsorb basic compounds when these groups are ionized. Finally, the silanol moiety can be considered a reactive group on the surface and it functions as the site for chemical modification of the inner wall. This property will be discussed in more detail later.

Because the first two properties of the silanol, creation of EOF and strong affinity for bases, can often be regarded as undesirable, the third property, the possibility of chemical modification, is used to eliminate these unwanted effects. The presence of EOF diminishes the ability of the CE experiment to separate solutes with very similar electrophoretic mobilities. For basic solutes, the acidic nature of the silanol can result in irreversible adsorption on the surface, which leads to either a complete loss of or greatly reduced detectability. When the silanol group has been modified with an organic moiety, the EOF is greatly diminished and the strong affinity for bases is significantly reduced or eliminated. In addition to chemical derivatization of the surface through a reaction at the silanol, it is also possible to modify the inner wall by adsorption of various compounds that masks the effect of the Si—OH group.

Wall Coating Through Chemical Modification

Chemical modification of the inner wall of fused-silica capillaries and the surfaces of porous silica supports for HPLC utilize the same reactions. The most common method is based on organosilanization. Within this general reaction scheme, there are two possible approaches, as shown in Table 1. The first possibility involves the use of an organosilane reagent (RR′R′SiX) with only a single reactive group (X). The substituents on the silicon atom are as follows: X = halide most often Cl; R = the organic moiety giving the surface the desired properties (i.e., hydrophobic, hydrophilic, ionic, etc.); R′ = a small organic group typically methyl. This reaction leads to a single siloxane bond between the reagent and the surface. Because of the single point of attachment of the reagent, the resulting bonded material is referred to as a monomeric phase. The second approach to organosilanization involves a reagent with the general formula $RSiX_3$. The substituents on the silicon atom in this reagent are defined earlier. The basic difference between the approaches as shown in Table 1 is that the reagent with three reactive groups results in bonding to the surface as well as cross-linking among adjacent bonded moieties and is referred to as a polymeric phase. This cross-linking effect provides extra stability to the bonded moiety but is less reproducible than the monomeric method. The one-step organosilanization procedure is relatively easy and the modification of the surface can be done by forcing the reagent continuously through the capillary or simply filling the capillary with the organosilane solution. The capillary is heated for about 1–2 h and then rinsed with a solvent such as toluene to remove the excess reagent. As is the case with the production of stationary phases for HPLC, organosilanization accounts for virtually all of the commercially available chemically modified capillaries.

A second modification scheme that has been reported for the modification of capillaries is based on a chlorination/organometalation two-step reaction sequence. This process is also depicted in Table 1. In the first step, the silanols on the surface are converted to chlorides via a reaction with thionyl chloride. This step must be done under extremely dry conditions because the presence of any water results in the reversal of the reaction with hydroxyl replacing the chloride (Si—Cl), resulting in the regeneration of silanols (Si—OH). If the chlorinated surface can be preserved, then an organic group can be attached to the surface via a Grignard reaction or an organolithium reaction. The main advantage of this process is that it results in a very stable silicon–carbon linkage at the surface. However, the stringent reaction conditions for the first step and the possibility of forming salts as by-products in the second reaction have resulted in relatively little commercial use of this process.

The third method reported for the modification of capillary inner walls involves first silanization of the silica surface followed by attachment of the organic group through a hydrosilation reaction. The process is also depicted in Table 1. In the first step, the use of triethoxysilane under

Table 1 Types of Reactions for Modifying Capillary Walls

Reaction type	Reaction	Surface linkages
Organosilane	a) $Si—OH + X—SiR'_2R \rightarrow Si—O—SiR'_2R + HX$ or b) 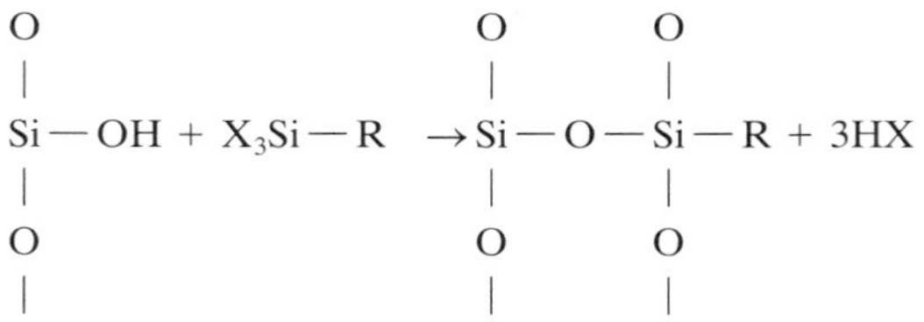 O	Si—OH + $X_3Si—R \rightarrow$ Si—O—Si—R + 3HX	O	(surface Si bonded to O above and below; in the product, both Si atoms bonded to O above and below)	Si—O—Si—C
Chlorination followed by reaction of Grignard reagents or organolithium compounds	$Si—OH + SOCl_2 \xrightarrow{\text{Toluene}} Si—Cl + SO_2 + HCl$ a) $Si—Cl + BrMgR \rightarrow Si—R + MgClBr$ or b) $Si—Cl + Li—R \rightarrow Si—R + LiCl$	Si—C
TES silanization		O	Si—OH	O	Si—OH $\rightarrow$	O	Si—OH	O
O	Si—O—Si—H	O	Si—O—Si—H	O	Si—O—Si—H	O		Si—H monolayer
Hydrosilation	$Si—H + CH_2{=}CH—R \xrightarrow{\text{Catalyst}} Si—CH_2—CH_2—R$	Si—C

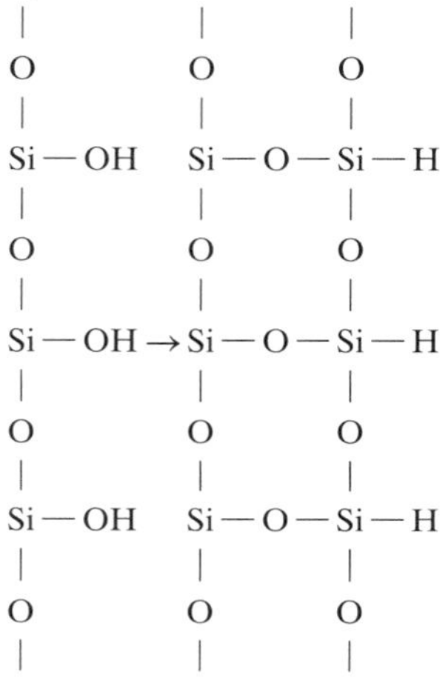

controlled conditions results in a monolayer of the cross-linked reagent being deposited on the surface. This reaction results in the replacement of hydroxides by hydrides. In the second step, an organic moiety is attached to the surface via the hydride moiety in a catalyzed hydrosilation reaction. The catalyst is usually hexachloroplatinic acid (Speier's catalyst) but can be other transition metal complexes or a free-radical initiator. This process also results in a silicon–carbon bond at the surface, does not required dry conditions (water is required as a catalyst in the first step), and is applicable to a variety of unsaturated functional groups in the hydrosilation reaction, although terminal olefins are the most common. The silanization/hydrosilation method also has seen limited commercial utilization to date.

The result of all three of the chemical modification schemes described here is to eliminate or drastically reduce EOF. In some cases, the EOF can be reversed by the attachment of a positively charged group, such as $R—NH_3^+$, to the surface. Whether the EOF is diminished, eliminated, or reversed, separation is improved because electrophoretic mobility differences are enhanced. The replacement of silanols by various organic moieties also has beneficial effects with respect to the separation of basic compounds. An example of the difference in peak shapes seen for bare and modified capillaries is shown in Fig. 1. In some cases, the tailing observed for highly basic compounds is more severe than shown in the figure and in the worst cases irreversible adsorption results in the complete absence of a peak in the electropherogram.

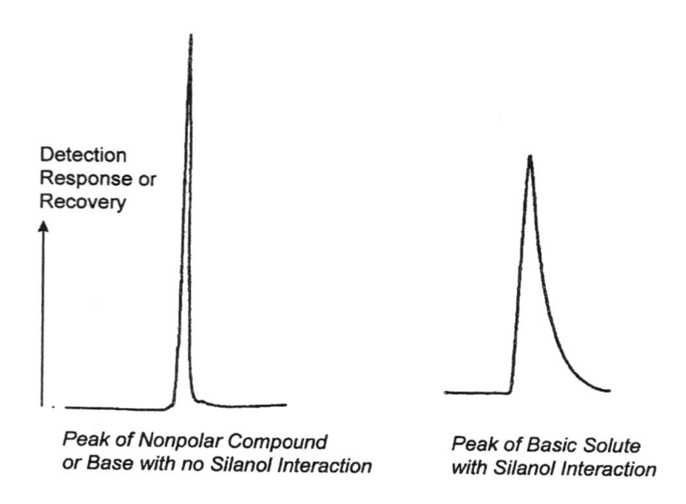

Fig. 1 Comparison of peak shapes for modified and bare capillaries.

Once a reaction scheme has been selected, then a choice must be made as to the type of surface that is suitable for a particular separation. The surface properties are controlled by the "R" group of the reactions shown in Table 1. Hydrophobic coatings can be achieved by using organic moieties that are common in reversed-phase HPLC. The most common would be either octadecyl (C_{18}) or octyl (C_8). In general, hydrophobic coatings are used for small molecule separations where the main concern is the suppression of EOF and the elimination of possible adsorption at the silanol sites. However, for the separation of proteins, peptides and DNA related species that have considerable hydrophobic character, it is more desirable to have a hydrophilic coating. The presence of any neutral bonded group will reduce or eliminate the EOF and a hydrophilic species will prevent strong interactions with the surface that would occur if the coating was hydrophobic. Because of the importance of CE in the separation of biomolecules, considerable effort has been devoted to the development of hydrophilic wall coatings. By far, the most extensively used hydrophilic coating is polyacrylamide. The surface is first modified with a linker such as 3-methacryloxypropyltriethoxysilane having a double bond available as a site for acrylamide to attach and polymerize. Derivatives of polyacrylamide have also been bonded which result in increased stability at high pH. Other polymers such as polyethylene glycol, cellulose, and poly (vinyl alcohol) have also been used in order to achieve a hydrophilic surface. The presence of any polymer on the capillary usually results in a reasonably thick layer that shields the surface and drastically lowers or eliminates EOF. Polymers containing cationic or anionic species can also be useful in preventing adsorption of hydrophobic compounds on the surface as well as controlling EOF. The type charge on the polymer controls the direction of the EOF. For negatively charged groups (sulfonic acid), the direction is cathodic, and for positively charged groups (quaternary amine), the direction is anodic.

Wall Coating Through Buffer Additives

Another approach to reducing the EOF as well as wall adsorption is to add a compound to the running buffer that will compete with the solute for the silanol sites on the surface. These materials must have some affinity for the charged or polar sites on the inner wall and so they must, themselves, be hydrophilic or charged. Nonionic surfactants are hydrophilic to prevent solute adsorption on the wall and block the silanols in order to reduce the EOF. The use of cationic polymers in the buffer results in a reversal of EOF to the anodic direction, whereas the use of anionic polymers preserves the cationic direction but tends to stabilize the flow in comparison to a bare capillary. Other additives such as diaminoalkanes and polyvinylalcohol result in reduced EOF and less solute adsorption on the wall.

Column technology is one of the most rapidly developing areas of CE. The capillary is the key to separation, so it is likely that numerous column formats will be established to meet specific separation needs similar to stationary-phase development in HPLC.

Suggested Further Reading

Altria, K. D., *Capillary Electrophoresis Guidebook: Principles, Operation and Applications*, Humana Press, Totowa, NJ, 1996.

Camilleri, P., *Capillary Electrophoresis*, CRC Press, Boca Raton, FL, 1998.

Landers, J. P., *Handbook of Capillary Electrophoresis*, 2nd ed., CRC Press, Boca Raton, FL, 1997.

Pesek, J. J. and M. T. Matyska, *Electrophoresis 18*: 2228 (1997).

Vansant, E. F., P. Van Der Voort, and K. C. Vrancken, *Characterization and Chemical Modification of the Silica Surface*, Elsevier, Amsterdam, 1995

Joseph J. Pesek
Maria T. Matyska

Silica Capillaries: Epoxy Coating

Introduction

Capillary electrophoresis (CE) has proven to be one of the best possible techniques available for the separation of proteins. However, CE separation of proteins is often complicated by the interaction between the analytes and the silanol groups on the inner surface of the capillary. This interaction often results in broad asymmetrical peaks, poor efficiency, altered electro-osmotic flow (EOF), reduced protein recovery, and poor reproducibility. Attempts have been made to reduce this interaction by chemically modifying the surface with a layer of chemical coating. Capillary coating has the advantage of allowing analysts to freely modify the composition of the buffer to optimize the separation. Different coatings have been made and each of them has unique characteristics. For protein separations, the most commonly used capillary coatings are hydrophilic coatings [1–6]. For example, epoxy coating is one of these hydrophilic coatings; various protein samples have been separated successfully on this coating.

Preparation of Epoxy Coating

A typical procedure for preparing epoxy coating can be found in the literature [7,8]. Basically, a fused-silica capillary is activated with 1.0*M* NaOH solution for 10–20 min and then washed with dilute HCl and water for another 20 min each. The washed capillary is then heated in an oven for 3 h at 120°C with N_2 slowly passing through. A γ-glycidoxypropyltrimethoxysilane (GOX) in CH_2Cl_2 is pushed into the pretreated capillary and heated for 3 h. Next, a solution of ethyleneglycol diglycidyl ether (EGDE) and 1,4-diazabicyclo-[2.2.2]-octane (DABCO) is forced through the column and allowed to react at 120°C for at least 3 h. Finally, the capillary is washed with methanol at room temperature.

Separation of Proteins on the Epoxy Coating

Epoxy coating is a hydrophilic coating due to the high content of ether and hydroxyl groups. It is expected that this hydrophilic surface will reduce protein adsorption and, thus, is suited for protein separations.

Separation of Model Proteins at Neutral pH

The usefulness of this epoxy coating for protein separation can be demonstrated with the separation of various model proteins using 10–50 m*M* phosphate buffer near pH 7. The model proteins are lysozyme (p*I* 11), cytochrome-*c* (p*I* 10.2), ribonuclease-A (p*I* 9.3), α-chymotrypsinogen (p*I* 8.8), trypsinogen (p*I* 8.7), α-chymotrypsin (p*I* 8.4, 8.8), and myoglobin (horse heart, p*I* 7.3). They are all positively charged at pH 7 and have high tendency to adsorb onto the negatively charged walls of uncoated capillaries. Therefore, very poor separation is seen with an uncoated capillary. However, a good separation of these proteins can be achieved on the epoxy-treated surface. Figure 1 shows that all five proteins are baseline resolved with a high separation efficiency.

Coating the inner surface of capillary reduces the amount of surface silanol groups and, thus, the EOF in the capillary. The epoxy coating retains about one-third of the EOF of an uncoated capillary at pH 7. This EOF is critical for the separation and detection of samples containing neutral to slightly negatively charged proteins. For example, α-chymotrypsin, myoglobin (whale, p*I* 6.9), conalbumin (p*I* 6.3), carbonic anhydrase (p*I* 6.1), and α-amylase (p*I* 5.9) can also be separated at pH 7.

Separation of Model Proteins at Other pHs

One of the major advantages of coating the inner surface of the capillary is that it allows the buffer pH to be freely adjustable to achieve the best separation. Coating reduces the effect of pH on EOF, electrophoretic mobility, and protein separation. For example, the EOF in the coated capillary has less than half (from 0 to 3×10^{-4} cm^2/VS) the variation as seen in the uncoated capillary (from 1×10^{-4} to 8×10^{-4} cm^2/VS) between pH 3 and 11.

The five positive model proteins as shown in Fig. 1 can also be separated at various other pHs (pH 4–10) [7]. However, the migration times of these individual proteins vary with pH change differently. Although the electrophoretic mobility of these proteins changes accordingly with pH, the overall effect of pH on both EOF and electrophoretic mobility varies at different pH values. This variation provides a basis for optimizing pH for the separation of these proteins.

The theoretical plate number for the separation of these model proteins has a maximal value at pH 7. How-

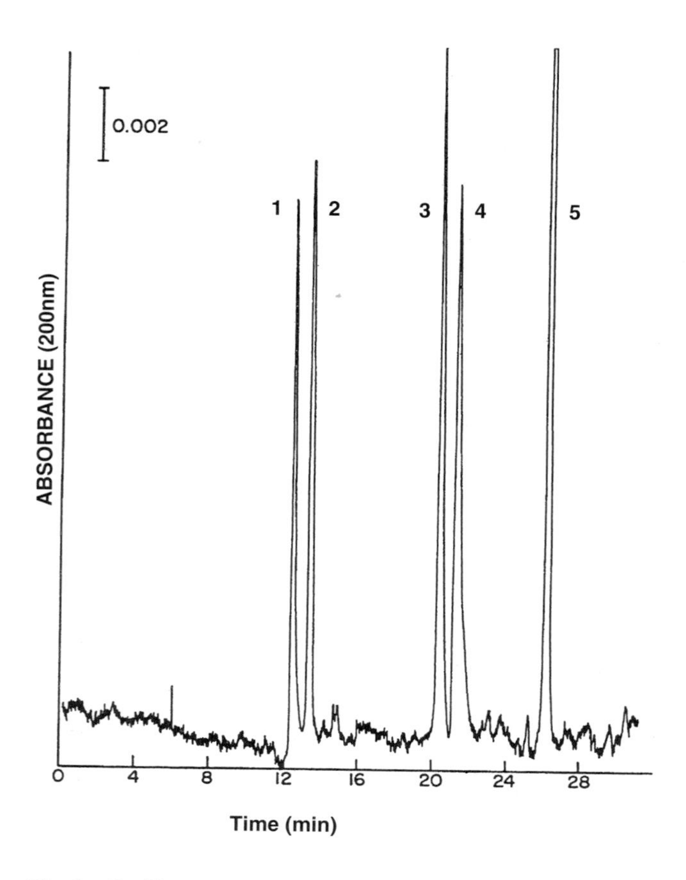

Fig. 1 Capillary electrophoretic separation of five basic model proteins in an epoxy-coated capillary. Experimental condition: 65 cm separation length, inner diameter/outer diameter 50/363 μm, 0.01*M* phosphate buffer, pH 7, 300 V/cm, 17 μA. Peaks: (1) lysozyme, (2) cytochrome-*c*, (3) ribonuclease A, (4) α-chymotrypsonagen, and (5) myoglobin (horse heart).

ever, the peak capacity increases with pH because the separation window between lysozyme and myoglobin increases with pH, whereas the peak width does not change significantly. The resolution between the two pairs of proteins, ribonuclease A and α-chymotrypsinogen, and α-chymotrypsinogen and myoglobin, increases with the pH. However, the resolution between lysozyme and cytochrome-*c* increases slightly from pH 5 to 7, drops to zero at pH 8 (no separation at all), and then increases again as the pH is increased.

Protein Recoveries

In general, the efficiency of protein separation in CE should be directly proportional to the separation length (i.e., $N = L/H$), where N is the theoretical plate number, L is the separation length, and H is the height-equivalent theoretical plate. Results obtained by locating two detectors at 20 and 65 cm from the injection end show that longer separation length gave much better separation. Much higher than expected separation efficiency is obtained from the 65-cm separation length segment than that from the 20-cm separation length. Therefore, a long capillary should be used when better resolution and higher efficiency are desired. However, no significant improvement in separation efficiency can be achieved when the capillary length exceeds 1 m. On the other hand, a short capillary is preferred when separation time is of major concern.

The adsorption of positive proteins onto the epoxy capillary surface can be quantitatively evaluated by using the two on-line detector design. From the responses of the two detectors, it is possible to determine the adsorption of proteins on the surface between the two detectors. Zero-percent recoveries have been reported on an uncoated capillary at pH 7 for lysozyme, cytochrome-*c*, ribonuclease A, and α-chymotrypsinogen. However, most of these proteins had recoveries between 84.4% and 95%, except for lysozyme (55.5%), on an epoxy-coated capillary. Therefore, coating does reduce protein adsorption significantly.

Reproducibility and Stability

The reproducibility of protein separation from run to run, day to day, segment to segment, column to column, and chemist to chemist is very high. From run to run, the % relative standard deviation (RSD) ($n = 5$) of migration times for neutral marker (MO) and lysozyme are 0.94 and 0.71, respectively. The %RSD in EOF for day to day ($n = 5$) and segment to segment ($n = 6$) are 2% and 3.5%, respectively. For column to column, the EOF at pH 7 varied from 0.5×10^{-4} to 2×10^{-4} cm^2/VS with an average of 1.14×10^{-4} cm^2/VS for nine columns coated over a 3-month period.

The major challenge to capillary coatings, especially hydrophilic coatings, is the stability. Adding a hydrophobic moiety into the coating may increase coating stability but interfere with protein separations. Epoxy coating provides a balance between hydrophilicity and stability. Epoxy is well known for its chemical and mechanical stability. The epoxy is covalently bound to the silica surface and is cross-linked to generate a stable surface. This coating has shown to be suitable for protein separations at pH 4–10. When stored at room temperature, this column is stable for at least several months.

Separation of Cytochrome C Variants

The separation of cytochrome-*c* variants was challenging because of their similarity in structure. Each of them differs by only a few amino acids in their sequences. When five cytochrome-*c* variants were injected, only four peaks were obtained, with the horse and dog variants migrating together. Varying the experimental conditions still did not help the separation of the horse and dog variants. Very reproducible separations were obtained at pH 6–8 [8].

Separation of Recombinant Proteins

BMP-2 is a recombinant protein which has the potential to develop into a biotech drug. Previous experiments with an uncoated capillary and a commercially available hydrophilic-coated column (Celect P150) failed to separate the BMP-2 components with reproducible results. By using the epoxy coating, the various BMP-2 components were separated. IL-11 is another recombinant protein under investigation. Using this epoxy coating, a CE method was developed to monitor the stability of IL-11 at different storage conditions. The results showed that there were some significant differences in the CE profiles of the three IL-11 samples stored at different temperatures for different periods of time [8].

Conclusions

Epoxy coating is well suited for the separation of proteins in CE. This coating is easy to prepare and can be used for a broad range of pHs. High recoveries of proteins prove that this epoxy modified surface has reduced the protein adsorption significantly.

References

1. S. Hjerten, *J. Chromatogr. 347*, 191 (1985).
2. R. M. McCormick, *Anal. Chem. 60*: 2322 (1988).
3. G. J. M. Bruin, R. Huisden, J. C. Kraak, and H. Poppe, *J. Chromatogr. 480*: 339 (1989).
4. W. Nashabeh and Z. El Rassi, *J. Chromatogr. 559*: 367 (1991).
5. J. T. Smith and Z. El Rassi, *J. High Resolut. Chromatogr. 15*: 573 (1992).
6. X. Ren, Y. Shen, and M. L. Lee, *J. Chromatogr. A 741*: 115 (1996).
7. J. K. Towns, J. Bao, and F. E. Regnier, *J. Chromatogr. 599*: 227 (1992).
8. J. J. Bao, *J. Liquid Chromatogr. Related Technol. 23*: 61 (2000).

James J. Bao

Silica Capillaries: Polymeric Coating for Capillary Electrophoresis

The performance of capillary electrophoresis with an unmodified fused-silica capillary is dependent on the chemical properties of the silica surface. The residual Si—OH groups on the surface lead to *electro-osmotic flow* (EOF), which contributes to solute migration. For cationic solutes moving toward the negative electrode, the EOF will diminish the resolution, because it contributes toward migration in this direction and reduces the overall migration time. For anionic solutes, EOF is necessary for migration toward the negative electrode or it retards the movement toward the positive electrode and enhances resolution. Therefore, when cationic solutes are the analytes, it is often desirable to diminish the EOF in order to enhance resolution and, therefore, a bare capillary may not be the best choice for a column in this situation.

Another drawback of the unmodified capillary is that the silanols are potential sites for adsorption of certain solutes, especially high-molecular-weight proteins, which leads to poor recovery of the analyte and variable migration times due to the decrease in EOF.

Capillary columns modified with polymeric coatings are a desirable approach for controlling EOF and adsorption of solutes on the wall. In general, for a coating to be successful in CE, the polymers should be able to (a) modify or suppress the electro-osmotic flow, (b) be stable for a long period (many injections) in the presence of aqueous buffer solutions so that migration times remain constant and good quantitative determinations are possible, and (c) suppress strong, or even irreversible, adsorption of analyte molecules (e.g., proteins).

According to the way polymers are attached to the column surfaces, the polymeric coatings can be differentiated between those substances that are covalently attached to the capillary surface and those coatings that are not covalently attached but are adsorbed to the surface by physical or ionic forces. Comparatively, adsorbed coatings are simpler to prepare, whereas covalent bonded coatings require elaborate chemical reactions. With regard to the mechanism by which prevention of adsorption of proteins occurs and the properties that they render to the coated surfaces, the compounds currently used as adsorbed coatings belong mainly to two categories: aminated or cationic polymers and hydroxylic or neutral polymers. Aminated polymers, such as polybrene, are adsorbed to the silica wall by Coulombic attraction.

The mechanism for which wall interactions are minimized is primarily due to the ionic repulsion of proteins and peptides at a pH below their p*I*. These polymers generate a positively charged layer at the surface of the silica wall and, therefore, lead to an anodal EOF. These polymer coatings are very stable and are useful over a wide pH range. Hydroxylic or neutral polymers are attached onto the silica wall by weaker interactions such as hydrogen-bonding. The mechanism for which wall interactions are minimized is described as a shielding of the silanol groups. Because these polymers are not charged, the EOF is, in most cases, suppressed. The working pH range is narrower and is limited to the acidic pH regime. Coating procedures may be varied to generate the coating thickness and homogeneity. Most commonly, the reagent is simply passed through the capillary in a suitable buffer. In the case of hydroxylic polymers, thermal immobilization is usually required. Prior to electrophoresis, the unbonded reagent is flushed from the capillary. The main limitation of adsorbed coatings is their instability under basic conditions.

A polymeric coating which is covalently bonded to the surface of capillary wall is the most significant approach among the surface deactivation methods. This method provides a more flexible approach in preventing adsorption of analytes and, at the same time, permits manipulation of separation parameters to optimize selectivity and efficiency. The silanization of the silica surface is an elegant method which enables the production of a large variety of polymer coatings that are chemically bonded to the capillary column surface. In general, polymers are attached to the silica surface by an Si — O — Si linkage or Si — C linkage. Many different mono-, di-, or tri-functional silanization reagents are commercially available or can be easily synthesized. The capillary coated with polymers via Si — C linkage show better stability under alkaline conditions and improved reproducibility of the separation than that of the Si — O — Si linkage.

Among the bonded materials described in the literature, polyacrylamide is one of the most popular and successful for achieving good protein separations. Initial studies were based on the bonding of the linker 3-methacryloxyptopyltriethoxysilane between the surface and the polymerized, and in some cases cross-linked, acrylamide. The main drawback of this type of wall coating is its long-term stability. In order to overcome this effect, the surface silanization agent was replaced with 7-oct-1-enyltrimethoxysilane. Improvement in stability, as well as efficiency and reproducibility of migration times, in comparison to the columns with linear polyacrylamide bonded by the conventional method, was achieved. However, at the high pH range, the amido bonds in polyacrylamide are possibly hydrolyzed, leading to degradation of the attached polymer and a loss of column performance. In order to overcome this effect, N-substituted acrylamide monomers can be used which provide steric protection to the amido bond. Poly-(acryoylaminoethoxyethanol) was shown to have dramatically improved stability over polyacrylamide at a high pH. Poly-(*N*-(acroylaminoethoxy)-ethyl-β-D-glucopyranose was demonstrated to be even more stable. In addition to the improved stability of the amido bond, the linkage of polymer to the capillary surface via a Si — C bond made through a silanization/hydrosilation process contributed greatly toward the length of coating's service.

Various types of polyethers and diol moieties are also effective hydrophilic coating, which often function in a manner similar to polyacrylamide. Different combinations of polyether with triethoxysilane groups or anchored poly(ethylene glycol)s have resulted in low electro-osmotic flow and low adsorptive coatings. It has also been possible to combine the characteristics of the two most common hydrophilic coatings by linking a polyvinylmethylsiloxanediol to linear polyacrylamide. As combining cellulose, functionalized polyethyleneimine and polyether in various proportions thus allows the control of EOF that is independent of pH. The mixed phase from ethylene glycol diglycidyl ether and glycidol also exhibits low electro-osmotoc flow and protein adsorption. Polysiloxanes are another type of capillary coating used in the formation of mixed materials to produce surfaces more conducive to electrophoretic conditions.

Other types of polymers have also been attached to the inner walls of fused-silica capillaries through different linking agents. A common approach involves the bonding of a methacryl group via standard organosilane chemistry. Then, bonding of another species to the linker takes place by a double-bond reaction and polymerization. For example, in the case of a cellulose coating, the species attached to the linker can be allyl methylcellulose.

Other polymers include cross-linked dextrin, triblock poly(ethylene oxide)–poly(propylene oxide) and poly-(vinyl alcohol).

Charged polymeric coatings are also bonded to the column surface to control the adsorption and EOF, especially in the analysis of oppositely charged species in a single run, or for fast separation of analytes with sufficient differences in electromobilities. One effective approach is the cryptand-containing polymer coating, which generates a switchable EOF depending on the pH of the running buffer. Other charged coatings can be synthesized with sulfonic acid or quaternary ammonia groups that provide more stable EOF over a fixed buffer pH range.

Capillaries with chiral polymer coatings have been applied in CE for resolution of enantiomers. Possibly because of its inclusive effect, cyclodextrin seems to be an effective chiral selective agent when bonded to a fused-silica capillary surface. In this case, the purpose of the modification is to induce interactions with the chiral material on the surface. Certainly, the cyclodextrin moiety lowers EOF like other wall modifications because it diminishes the number of silanols. The lower EOF allows for slower migration of the solute through the column and, hence, more time for interaction with the chiral selector. The diminished number of silanols also results in less nonspecific interactions with the fused-silica surface, which would tend to degrade the enantiomeric separation.

Capillaries bonded with polymeric coatings are also applied to capillary electrochromatography (CEC) for separation of neutral molecules. In this case, the polymeric coating participate to solute–bonded phases interaction in a manner similar to open-tubular LC. Most polymeric coating preparations have followed the procedures typically used in open-tubular LC and GC.

Polymer-coated columns offer advantages in the respect that the surface is generally well shielded from the solutes as they migrate through the system. Most polymer procedures involve a multistep process that can often be time-consuming and/or experimentally difficult. Recently, it has been demonstrated that polymer coatings can be produced in a single step by mixing the polymer, a surface-derivatizing agent, and a cross-linking agent in a solvent that is then placed in the capillary. The coated capillary is then heat-treated, which removes the solvent and immobilizes the polymer film on the surface. A number of polymers were tested and higher efficiencies obtained for basic protein separation. The formation of organic–inorganic polymeric coatings by the sol–gel technique has also been reported to be a simpler way.

Further development in the preparation of polymeric coatings with more reproducible reaction conditions and better reagents will make the capillary electrophoresis a more challenging separation technique.

Suggested Further Reading

Chiari, M., M. Neri, and P. G. Righetti, in *Capillary Electrophoresis in Analytical Biotechnology* (P. G. Righetti, ed.), CRC Series in Analytical Biotechnology, CRC Press, Boca Raton, FL, 1996.

Heiger, D. N. and R. E. Majors, *LC–GC 13*: 13–23 (1995).

Li, S. F. Y., Capillary electrophoresis, principles practice and applications, *J. Chromatogr. Lib.* (1993).

Schomburg, G., *Trends Anal. Chem. 10*: 163–169 (1991).

Xi-Chun Zhou
Lifeng Zhang

Size Separations by Capillary Electrophoresis

Introduction

Electrophoretic separation based on molecular size is the predominant technique for the separation of large biomolecules [1]. Using the traditional slab-gel, a rigid anticonvective gel is required to provide mechanical stability for the separation. The gels are designed for a single use at relatively low electric field strengths.

During the early days of high-performance capillary electrophoresis (HPCE), rigid gels were polymerized *in situ* within the capillary. These gel-filled capillaries were used for multiple runs at high field strengths. The capillaries were prone to failure and proved too unreliable for routine use. It was soon discovered that low-viscosity pumpable media was capable of defining molecular pores required for a size separation. In the capillary format, the walls of the tube provide the requisite mechanical stability, so high viscosity gels are unnecessary. These solutions are known as polymer networks, entangled polymers, or physical gels. When polymer networks are used for size separations, a fresh matrix is employed for each run. Through the use of high-pressure pumping systems, polymer networks suitable for DNA sequencing can be pumped in and out of the capillary.

This technology has facilitated the development of instruments containing arrays of 96 capillaries for high-throughput applications such as DNA sequencing. Other high-throughput DNA applications that will eventually be incorporated include genetic analysis and human identification. Ultimately, microfabricated devices may be used for many of these applications.

Separation Mechanism

Several mechanisms for the migration of macromolecules through polymer networks have been described. The Ogsten model considers the molecule as a nondeformable sphere. The speed of migration is based on the mobility of the solute in free solution modified by the probability of an encounter with a restricting pore. This mechanism is operative when the radius of gyration of the macromolecule is less than or equal to the average pore size of the polymer network. Separation of sodium docecyl sulfate (SDS)–proteins is thought to occur following this mechanism.

Large biopolymers, such as DNA and oligosaccharides, do not follow the Ogston model. These molecules can deform during transit through the porous network. Instead, a strand of DNA can move through the polymer matrix in a snakelike manner known as reptation. It is also known that fragment resolution decreases as the length of macromolecule increases. The molecules align with the electric field in a size-dependent manner. This process is known as baised reptation. The effect limits the size of DNA molecules that can be separated using conventional slab-gel techniques. The high electric field strength used in CE further limits the separation. Beyond 20,000 base pairs (bp), separations become poor and pulsed-field techniques must be employed [2]. This equipment is not available for commercial capillary electrophoresis instrumentation.

Further masking a full understanding of the separation mechanism is the interaction of the macromolecule with the polymer network reagents. Separations have been reported in polymer concentrations far below what is required to define pores [3].

Denaturation of Macromolecules

Because separations occur based of molecular size, the macromolecule must be denatured to ensure that all solutes have the same charge-to-mass ratio. DNA and RNA are denatured by heating in formamide at 90°C for a few min. This improves both the separation and sizing accuracy.

To denature proteins, the disulfide bonds must be reduced and the molecule must be unfolded. Heating to 90–95°C for a few minutes using a solution composed of 0.1% SDS and a reducing agent, β-mercaptoethanol or dithiothreitol (DTT), is sufficient to denature most proteins. When the molecular weight of a protein is less than 10 kDa, the SDS binding stoichiometry may change resulting in errors when calculating molecular weight [4].

Materials for Polymer Networks

A wide variety of polymeric materials can be employed for size separations. It appears that the molecular weight and concentration of the polymer is more important that the polymer type itself. Once the polymer concentration is greater than the overlap threshold, the porous matrix is defined and reproducible. It is best if the polymer does not interact with the macromolecules being separated.

Among the materials used as polymer network reagents are linear polyacrylamide, dimethylpolacrylamide, methylcellulose derivatives, poly(ethylene oxide) and others. The appropriate molecular weight of the polymer is important. Sometimes, blends of different molecular weights are used. For example, the mixture of 2% linear polyacrylamide (LPA), molecular weight (MW) = 9 mDa and 0.5% LPA, MW = 50 kDa is used to separate DNA sequencing reaction products of up to 1000 bases in less than 1 h, as shown in Fig. 1 [5]. The viscosity of this polymer is 30,000 cps. The solution exhibits non-Newtonian properties as the viscosity drops upon the initiation of flow. The use of 2% 16 mDa LPA and 0.5% 250 kDa at 125 V/cm extends the read length to 1300 bases in 2 h [6].

Injection

Electrokinetic injection provides the highest-efficiency separations. If the salt concentration of the sample is greater than 50 mM, hydrodynamic injection gives better results. It is better to desalt the sample by dialysis, precipitation, or ultracentrifugation.

Detection

For SDS–proteins, low-ultraviolet (UV) detection at 200 nm or 220 nm is used depending on the UV transparency of the polymer network. For oligonucleotides, UV detection at 260 nm is employed. When performing DNA sequencing, short tandem repeat or genetic analysis, laser-induced fluorescence (LIF) is the method of choice. Intercalating dyes such as YOYO or YO-PRO fluorescence when complexed in between the DNA strands are

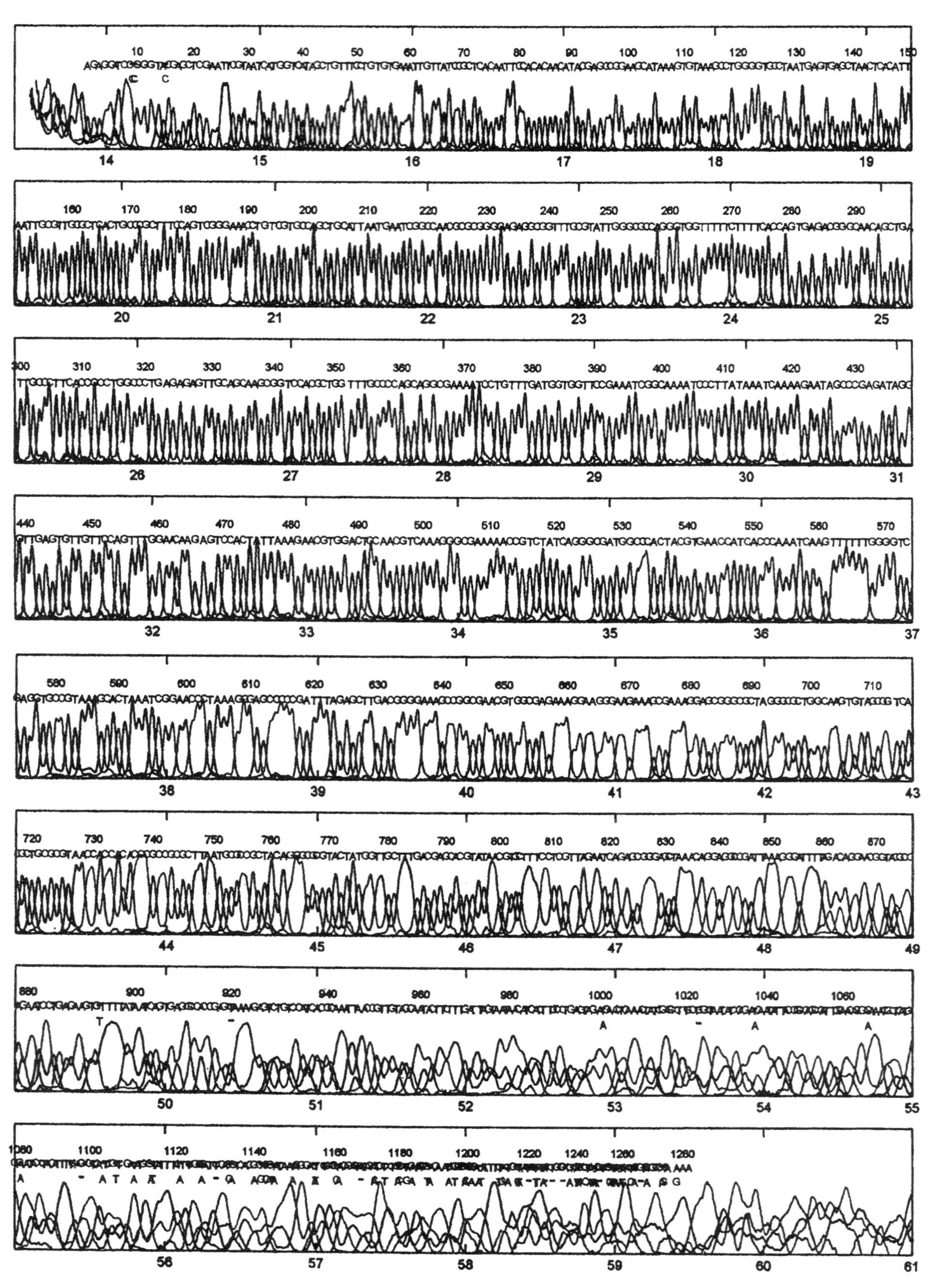

migration time (min)

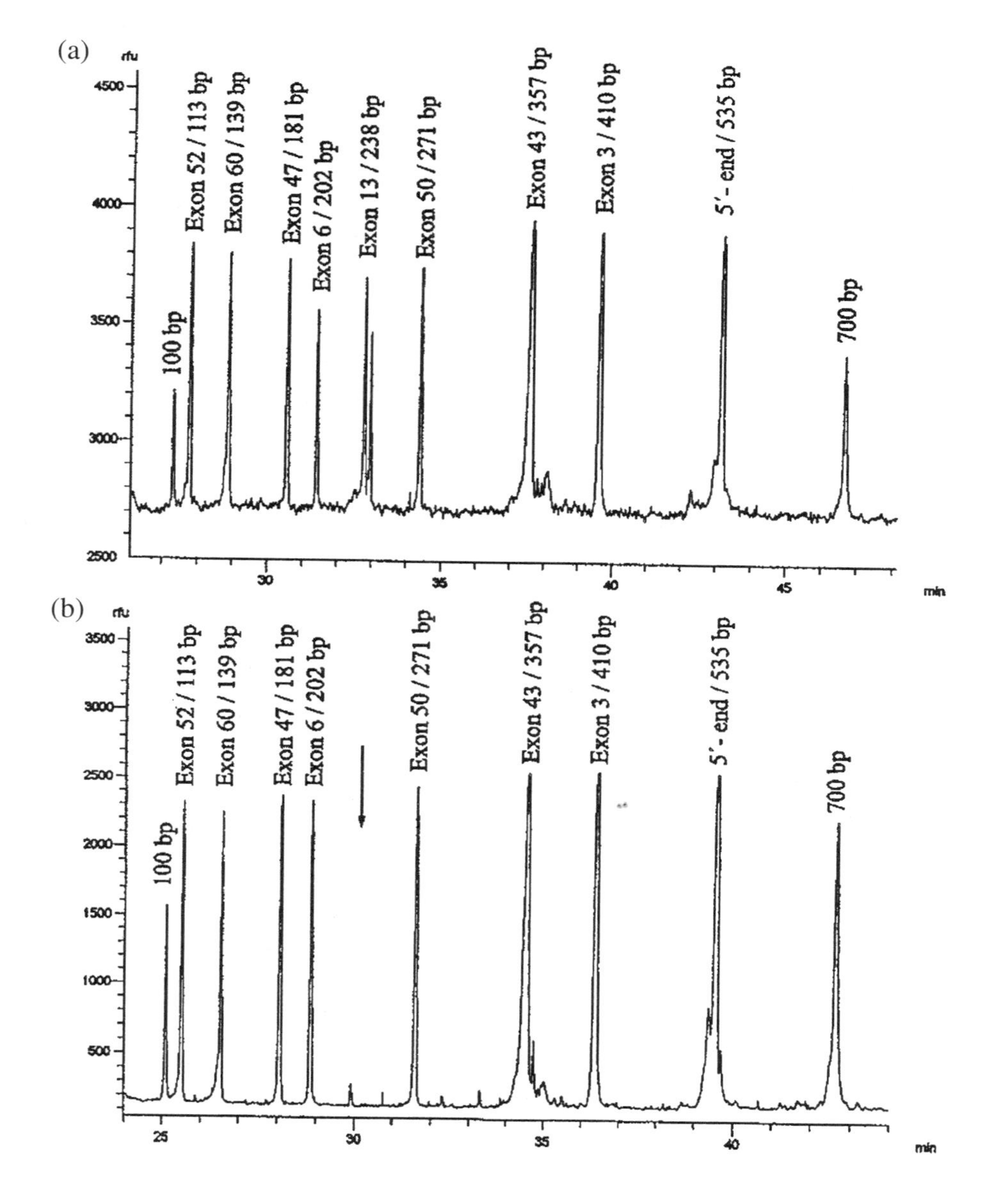

Fig. 2 Multiplex PCR profile of exons in Duchenne or Becker muscular dystrophy genes combined with two flanking standards. Capillary: 40 cm (65 cm total length) × 75 μm polyacrylamide coated; background electrolyte: 0.5% poly(ethylene oxide), 1 MDa in 1X Tris–borate–EDTA, 10μ*M* aminoacridine, 2 n*M* Vistra Green; field strength: 108 V/cm; detection: LIF, 488 nm. [Reprinted with permission from *J. Chromatogr. A 781*: 295 (1997), copyright 1997, Elsevier Science Publishers.]

Fig. 1 Separation of DNA sequencing fragments. Conditions: capillary: 30 cm (45 cm total length) × 75 μm inner diameter coated with poly(vinyl alcohol); background electrolyte: 2% linear polyacrylamide (LPA), 9 mDa and 0.5% LPA, 50 kDa in 7*M* urea and 500 m*M* Tris/500 m*M* TAPS/20 m*M* EDTA; buffer: 50 m*M* Tris/50 m*M* TAPS/2 m*M* EDTA with urea in catholyte; injection: 25 V/cm for 10 s; field strength: 200 V/cm; temperature: 60°C; detection: laser-induced fluorescence. [Reprinted with permission from *Anal. Chem. 70*: 3996 (1998), copyright 1998, American Chemical Society.]

sometimes added to the background electrolyte when monitoring polymerase chain reaction (PCR) products and restriction digests and for genetic analysis. For DNA sequencing, the fluorescent tag is incorporated in the chain-terminating dideoxynucleotide reagent. When separating short tandem repeats for human identification, the PCR primers can incorporate the fluorescent tag.

When separating oligosaccharides in polymer networks with LIF detection, a good tagging reagent is 9-aminopyrene-1,4,6-trisulfonic acid (APTS). This reagent is used in conjunction with the argon-ion laser.

Applications

DNA applications, in particular DNA sequencing, human identification, and genetic analysis dominate the field. Other DNA applications including oligonucleotides, antisense DNA, restriction fragments, plasmids, PCR products, hydridization (DNA probe), and RNA have also been demonstrated [7,8].

DNA testing is rapidly becoming the predominant technique for human identification. The restriction fragment-length polymorphism (RFLP) method requires large amounts of DNA (20–100 ng) and is extremely time-consuming and labor-intensive.

A PCR method employing short tandem repeats (STR) is now the method of choice. STRs are sequences where two to seven nucleotides of DNA are constantly repeated. Unlike the DNA of a gene, STRs are prone to DNA replication errors. The lengths of these fragments vary from one person to the next, thereby providing the potential for DNA fingerprinting. The use of PCR and LIF detection provides high sensitivity, so very little DNA is required. The need for higher specificity is addressed with multiplex PCR, where several dye-labeled primers simultaneously amplify multiple locations throughout the genome.

The most widely used genetic screening technique, PCR–RFLP detects a mutation at a specific restriction endonuclease cleavage site at the mutation locus [9]. The products from other techniques such as ARMS (amplification refractory mutation system), SSCP (single-strand conformational polymorphism), HPA (heteroduplex polymorphism), and CDCE (constant denaturant capillary electrophoresis) and PCR are usually separable using polymer networks.

Polymerase chain reaction is particularly useful for genetic analysis because both amplification and primer specific isolation of gene fragments occur simultaneously. Allele-specific amplification can be employed to detect a single base-pair mutation through the use of a specially designed primer which is complementary to the mutated DNA [10]. PCR amplification only takes place if the mutation is present. Figure 2 illustrates the separation of multiplex PCR fragments in this case, searching for specific deletions in the dystrophin gene thought to result in Duchenne muscular dystrophy. The deletion is indicated by the arrow.

References

1. R. Westheimer, N. Barnes, Gronau-Czybalka, and C. Habeck, *Electrophoresis in Practice: A Guide to Methods and Applications of DNA and Protein Separations*, 2nd ed., John Wiley & Sons, New York, 1997.
2. Y. Kim and M. D. Morris, *Anal. Chem. 66*: 3081 (1994).
3. A. E. Barron, H. W. Blanch, and D. S. Soane, *Electrophoresis 15*: 597 (1994).
4. T. Takagi, *Electrophoresis 18*: 2239 (1997).
5. O. Salas-Solano et al., *Anal. Chem. 70*: 3996 (1998).
6. A. W. Miller, Z. Sosic, B. Buckholz, A. E. Barron, L. Kotler, B. L. Karger, *Anal. Chem. 72*: 1045 (2000).
7. P. G. Righetti (ed.), *Capillary Electrophoresis in Analytical Biotechnology*, CRC Press, Boca Raton, FL, 1996.
8. C. Heller (ed.), *Analysis of Nucleic Acids by Capillary Electrophoresis*, Vieweg, Weinheim, 1997.
9. K. R. Mitchelson and J. Cheng, *J. Capillary Electrophoresis 2*: 137 (1995).
10. C. Barta, M. Sasvari-Szekely, and A. Guttman, *J. Chromatogr. A 817*: 281 (1998).

Robert Weinberger

Solute Focusing Injection Method

As discussed elsewhere (see the entries Split/Splitless Injector and Retention Gap Injection Method), in capillary-column gas chromatography, split injections are necessary to ensure that a very small, compact sample is placed on the column. However, split injections generally result in an unrepresentative sample being placed on a capillary column; thus, on-column injection is usually preferred for accurate quantitative analysis. This demands the use of large-diameter tubes to permit the penetration of the injection syringe needle into the column. However, this procedure also causes other problems to arise. On injection, the sample readily separates into droplets which act as separate, and individual, injections that cause widely dispensed peaks and serious loss of resolution. In the extreme, double or multiple peaks are formed. Grob [1] suggested two solutions to this problem: the retention gap method of injection and the solute focusing method.

The solute focusing method is claimed to be more effective than the retention gap method, but the technique requires more complicated equipment. In the solute focusing method, the injector is designed so that there are two consecutive, independently heated and cooled column zones, located at the beginning of the column. A diagram of the solute focusing system and its mode of action is shown in Fig. 1. Initially, both zones are cooled and the sample is injected onto the first zone, where immediate sample splitting almost inevitably occurs. The carrier gas is then allowed to preferentially remove the solvent by eluting it through the column, leaving the contents of the sample dispersed along the cooled section of the tube. The selective removal of the solvent occurs because the solvent components are significantly more volatile than the components of the sample, even at the reduced temperature. The first zone is then heated and the second zone is continued to be kept cool. The solutes in the first zone progressively elute through the zone at the higher temperature until they meet the cooled zone. The movement of all the components is now significantly slowed down and they begin to accumulate at the beginning of the cooled second zone. The net effect is that the entire sample is now focused at the beginning of the cooled portion of the column. The temperature of the second zone is now programmed to the appropriate rate and the separation developed in the usual manner. This technique has more flexibility than the retention gap method, but the apparatus and the procedure is more complex and expensive. It should be pointed out that sample splitting does not occur in packed columns. It follows that if the sample is amenable to separation on such columns, then the packed column may be the column of choice if high accuracy and precision are required.

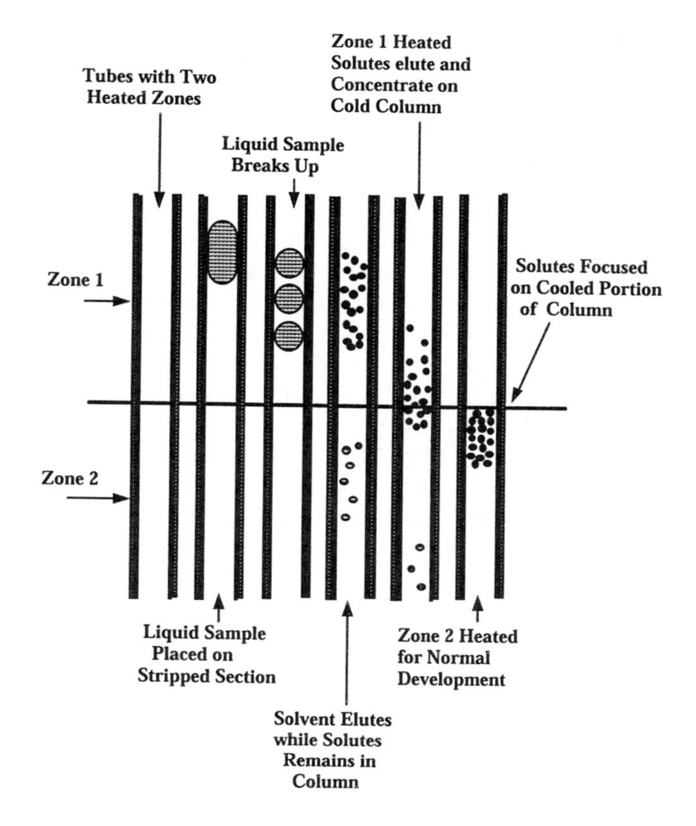

Fig. 1 The solute focusing method.

Reference

1. K. Grob, *Classical Split and Splitless Injection in Capillary Gas Chromatography*, Huethig, Heidlburg, 1987.

Suggested Further Reading

Grant, D. W., *Capillary Gas Chromatography*, John Wiley & Sons, Chichester, 1995.
Scott, R. P. W., *Techniques and Practice of Chromatography*, Marcel Dekker, Inc., New York, 1996.
Scott, R. P. W., *Introduction to Analytical Gas Chromatography*, Marcel Dekker, Inc., New York, 1998.

Raymond P. W. Scott

Solvent Effects on Polymer Separation by ThFFF

The early research of Myers et al. [1,2] shows that polymer thermal field-flow fractionation (ThFFF) retention and thermal diffusion are solvent dependent. Recently, Sisson and Giddings [3] indicated that polymer ThFFF retention could be increased by mixing solvents. Rue and Schimpf [4] extended the molecular-weight range that can be retained by ThFFF to much lower molecular weights (<10 kDa) by using solvent mixtures without using extreme experimental conditions. There are several other reports on the effect of solvents on polymer retention, selectivity, and the universal calibration in FFF in last few years [5].

The dissolving of a polymer into a solvent is governed by the free energy of mixing [6],

$$\Delta G_m = \Delta H_m - T\Delta S_m \tag{1}$$

where ΔG_m is the free-energy change of the solution, ΔH_m is the enthalpy change of the solution, and ΔS_m is the entropy change of the solution.

The enthalpy change is given by

$$\Delta H_m = V\left(\left(\frac{\Delta E_1^v}{V_1}\right)^{1/2} - \left(\frac{\Delta E_2^v}{V_2}\right)^{1/2}\right)^2 \phi_1\phi_2 \tag{2}$$

where V is the volume of the mixture, ΔE_i^v is the energy of vaporization for species i, V_i is the molar volume of species i, and ϕ_i is the volume fraction i in the solution.

The solubility parameter, δ_i, has been defined [6] as the square root of the cohesive energy density (CED), $\Delta E_i^v/V_i$, and describes the strength of attraction between molecules:

$$\delta_i = \left(\frac{\Delta E_i^v}{V_i}\right)^{1/2} \tag{3}$$

Equation (1) may be rewritten as

$$\Delta G_m = V(\delta_1 - \delta_2)^2\phi_1\phi_2 - T\Delta S_m \tag{4}$$

Because the second term, the entropy term, is always positive for the process of polymer dissolution, the deciding factor in determining the sign of the Gibbs energy change in Eq. (4) is the first term, which contains the solubility parameters of the polymer and the solvent. In general, $(\delta_1 - \delta_2)^2$ must be small for the polymer and the solvent to be miscible. A polymer dissolves most easily in a solvent whose solubility parameter matches it or is close to its own. This is consistent with the "like dissolves like" rule.

If a polymer is easily soluble in a solvent, by convention, the solvent is called a good solvent, and the converse, it is a poor solvent [6]. Therefore, a solvent whose δ value is close to the δ value of a polymer family is a good solvent for this polymer family. As examples, the δ value of polystyrene (PS) is about 18, which is closer to the δ value of 18.6 of tetrahydrofuran (THF) than the δ value of 16.8 of cyclohexane (CH). Therefore, THF is expected to be a better solvent for PS than CH. For polyisoprene (PIP), the situation is reversed; CH is a better solvent than THF.

Polymer molecules tend to spread out to a larger hydrodynamic size in a good solvent than they do in a poor solvent, which results in a reduced diffusion coefficient D as the Stoke–Einstein equation shows:

$$D = \frac{kT}{6\pi\eta R_h} \tag{5}$$

where k is the Boltzmann constant, T is the absolute temperature, η is the solvent viscosity, and R_h is the hydrodynamic radius of the polymer molecules in a solvent.

The retention of polymer molecules in thermal field–flow fractionation is determined by the diffusion coefficient and the thermal diffusion coefficient D_T, illustrated approximately by [8]

$$\frac{t_r}{t^0} \approx \frac{1}{6\lambda} = \frac{\Delta T}{6}\frac{D_T}{D} \tag{6}$$

where λ is the reduced mean layer thickness of a sample zone in the ThFFF channel, t_r is the polymer retention time, t^0 is the elution time of the nonretained species, and ΔT is the temperature difference across the channel thickness.

From the point of view of diffusion, the reduced D, obtained by replacing a poor solvent with a good solvent, will result in smaller λ if ΔT and other experimental conditions are kept unchanged. The reduced λ (sample zone is closer to the accumulation wall) will cause the polymer molecules to elute out of the ThFFF channel later because of the parabolic flow inside a ThFFF channel [8]. Higher retention, therefore, should be expected for polymers eluted by good solvents in ThFFF.

D_T for the polymer solution is not well understood. There are several theories of polymer thermal diffusion in liquids. Extensive reviews have been done by Schimpf and colleagues in several publications [4,9]. As an example, the radiation pressure theory of Gaeta and Scala shows that D_T is proportional to the cross-sectional area of the solute molecules [10]:

$$D_T = Z_A\left(\frac{\kappa_s}{U_s} - \frac{\kappa_p}{U_p}\right)\frac{\sigma}{f} \tag{7}$$

where Z_A is the acoustical impedance, a quantity related to the density and velocity of sound waves in the polymer and solvent, κ_p and κ_s are the thermal conductivity of the polymer and solvent, respectively, U_p and U_s are the corresponding velocities of sound in the two media, $f (f \propto \eta R_h)$ is the frictional coefficient, and σ ($\sigma = \pi R_h^2$) is the cross-sectional area of the polymer molecules.

As discussed earlier, the sizes, or cross-sectional areas, of polymer molecules dissolved in a good solvent are larger than that in a poor solvent. Therefore, the D_T of a polymer in a good solvent is predicted to be larger than that in a poor solvent according to Eq. (7).

From the point of view of both diffusion and thermal diffusion (assuming Gaeta's theory is correct in this case), ThFFF retention of a polymer should be larger in a good solvent than in a poor solvent.

The study of Sisson and Giddings [3] shows that polymer retention in some binary solvents can be enhanced while retention is relatively unaffected or even diminished in other cases. In the case of a mixture of 30% dodecane and 70% THF, polymer retention was enhanced by 35% relative to pure THF. The low end of the molecular-weight range, which can be analyzed by ThFFF, therefore, can be expanded further down by using binary carriers.

Rue and Schimpf [4] investigated component partition of a solvent mixture under thermal gradient. It is indicated that when one solvent is a significantly better solvent for a polymer, then the gradient of this solvent will cause the polymer to migrate in the same direction as the thermophoretic motion of this solvent. This is called solubility-based migration as a partitioning effect. When the better solvent partitions to the cold wall of the ThFFF channel, polymer retention is enhanced; when the better solvent partitions to the hot wall, retention is diminished. Five polystyrene samples, ranging in molecular weight from 2500 to 179,000, were separated in a mixture carrier containing 45 vol% THF in n-dodecane. This separation was the first time that polymers of molecular weight below 2500 become separated from the void peak without using extreme experimental conditions such as channel pressurization, ultrahigh ΔT, and so forth.

If the solvent effect is the same for both small and large polymer homologs, the selectivity will not be changed by switching solvents. This might not be the case, because small polymer molecules tend to be spread out to the maximum length of the flexible chain, even in a relatively poor solvent, whereas larger polymer molecules may need better solvents to spread out to their maximum. A different solvent effect might, therefore, be expected when a poor solvent is replaced by a good solvent for a homologous polymer series, and vice versa. It is possible, therefore,

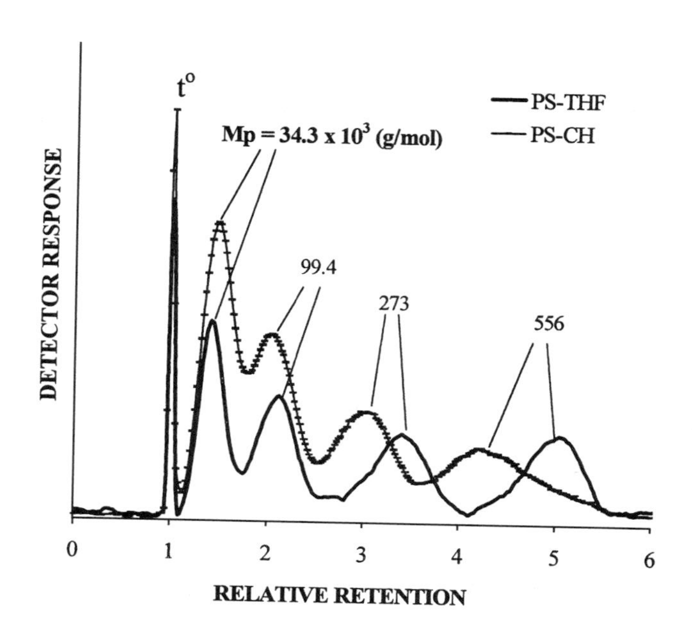

Fig. 1 Fractograms of PS in THF and CH. $\Delta T = 50°C$; $T_c = 25°C$. Flow rate is 0.1 mL/min and sample mass is 1 μg.

to manipulate the selectivity of polymer separation by ThFFF through solvent selection. Any dependence of D_T on molecular weight may be different for different solvents, and this may also contribute to changes in selectivity. Better retention and, probably, better selectivity are expected for PS in THF and PIP in CH among the possible four polymer–solvent pairs, as an example.

Figure 1 shows the overlaid fractograms of PS eluted by THF and CH under the same experimental conditions of cold-wall temperature, temperature drop, flow rate, and sample mass (1 μg) [5]. The fractograms show that a small PS of molecular weight 34,300 is retained approximately the same in both THF and CH, whereas the retentions of larger polymers show significant differences in the two solvents, with the larger PS retained longer in THF. This indicates that using better solvents may enhance both retention and selectivity.

References

1. M. N. Myers, K. D. Caldwell, and J. C. Giddings, *Separ. Sci. Technol. 9*: 47–70 (1974).
2. J. J. Kirkland, L. S. Boone, and W. W. Yau, *J. Chromatogr. 517*: 377–393 (1990).
3. R. M. Sisson and J. C. Giddings, *Anal. Chem. 66*: 4043–4053 (1994).

4. C. A. Rue and M. E. Schimpf, *Anal. Chem. 66*: 4054–4062 (1994).
5. W. J. Cao, P. S. Williams, and M. N. Myers, Solvent effects on polymer retention and universal calibration in ThFFF, in preparation.
6. J. Brandrup and E. H. Immergut, *Polymer Handbook*, 3rd ed., John Wiley & Sons, New York, 1989.
7. P. G. de Gennes, *Scaling Concepts in Polymer Physics*, Cornell University Press, Ithaca, NY, 1979.
8. J. C. Giddings, *Science 260*: 1456–1465 (1993).
9. M. E. Schimpf and J. C. Giddings, *J. Polym. Sci. Part B: Polym. Phys. 27*: 1317–1332 (1989).
10. F. S. Gaeta and G. Scala, *J. Polym. Sci.: Polym. Phys. Ed. 13*: 177 (1975).

Wenjie Cao
Mohan Gownder

Split/Splitless Injector

The split/splitless detector has been designed for use with open-tubular columns or solid-coated open-tubular (SCOT) columns. Due to the small dimensions of such columns, they have very limited sample load capacity and, thus, for their effective use, require sample sizes that are practically impossible to inject directly. The split injector allows a relatively large sample (a sample size that is practical to inject with modern injection syringes) to be volatilized, and by means of a split-flow arrangement, a proportion of the sample is passed to the column while the remainder is passed to waste. A diagram of a split/splitless injector is shown in Fig. 1.

The body of the injector is heated to ensure the sample is volatilized and inside is an inert glass liner. This glass liner helps minimize any sample decomposition that might occur when thermally labile materials come in contact with hot-metal surfaces. The carrier gas enters behind the glass liner and is thus preheated. The sample is injected into the stream of carrier gas that passes down the center of the tube, a portion passes down the capillary column, and the remainder passes out of the system through a needle valve. The needle valve is used to adjust the relative flow rates to the column and to waste and, thus, controls the amount of sample that is placed on the column.

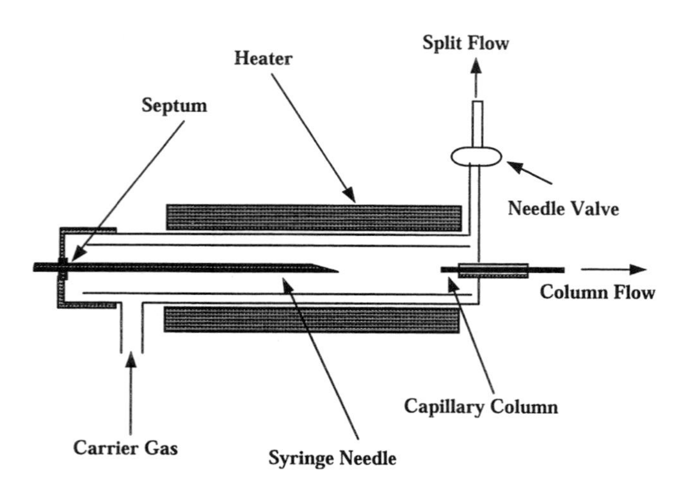

Fig. 1 Split/splitless injector.

This process of sample injection has certain disadvantages. Due to the range of solute types present in most mixtures for analysis, the components will have different volatilities and their vapors will have different diffusivities in the carrier gas. Differential volatilization and diffusion rates will result in the sample that enters the capillary tube, being unrepresentative of the original mixture. For example, the more rapidly diffusing solutes will be more dispersed and, consequently, more diluted in the carrier gas than those of lesser diffusivity. This results in the slower-diffusing substances having a higher concentration in the sample entering the column than those of higher diffusivity. As a consequence, the sample will be proportionally unrepresentative of the original mixture. The differential sampling that results from split injection systems can become a serious problem in quantitative analysis with capillary columns.

An alternative approach is to use a splitless injection system. If the valve in Fig. 1 is closed, then all the sample passes into the column and there is no split; *ipso facto*, the device is a *splitless* injector. When used in the splitless mode, however, it is usual to employ a somewhat wider capillary column, which will allow the penetration of a small-diameter injection syringe and thus permit *on-column* injection. Under these circumstances, there can be no differential sampling of the form described. This procedure, however, introduces other injection problems that can affect both resolution and quantitative accuracy that need to be addressed (See the entries Retention Gap Injection Method and Solute Focusing Injector Method).

Suggested Further Reading

Grant, D. W., *Capillary Gas Chromatography*, John Wiley & Sons, Chichester, 1995.

Scott, R. P. W., *Techniques and Practice of Chromatography*, Marcel Dekker, Inc., New York, 1996.

Scott, R. P. W., *Introduction to Analytical Gas Chromatography*, Marcel Dekker, Inc., New York, 1998.

Raymond P. W. Scott

Stationary Phases for Packed Column Supercritical Fluid Chromatography

Introduction

The name, packed column supercritical fluid chromatography (pSFC), has been applied to separations performed on particulate stationary phases, using mobile phases pressurized to supercritical, near-critical, or subcritical conditions [1,2]. Typically, the stationary phases used for pSFC are commercially available high-performance liquid chromatography (HPLC) columns, consisting of a porous solid support, usually with a covalently linked bonded phase to provide desired chromatographic interactions. Various column configurations have been used. Separate consideration of supercritical, near-critical, and subcritical pSFC applications provides a convenient, although arbitrary, means of subgrouping a discussion of the types of stationary phase used. More specific information on the stationary phases discussed in this overview can be found in Refs. 1–4.

Supercritical Applications

Supercritical pSFC applications can be defined as those in which the mobile phase is a single substance heated and pressurized above its critical point. Carbon dioxide has overwhelmingly been the compound of choice for these mobile phases. Stationary phases typically used for these applications have been polymeric materials or polymer-coated porous silica. Chromatography on uncoated silica-based stationary phases with CO_2 has, in general, been unsuccessful.

Polystyrene–divinyl benzene beads are the most common polymer material phases reported. The successful use of these phases has been mainly limited, however, to relatively hydrophobic compounds. Also, problems associated with bead physical instability, such as shrinking and swelling, affecting chromatographic reproducibility, have also been encountered. Stationary phases consisting of porous silica coated with a covalently bonded polysiloxane layer, containing cyano, phenyl, or alkyl functional groups, have been used with more success. The polymer coating is applied to the silica to mask excessive analyte–stationary phase polar interactions that generate poor peak shapes and yield incomplete analyte recovery with the neat CO_2 mobile phase. As with the polymer materials, these phases work best for hydrophobic compounds such as petroleum-based compounds, lipids, and plastics additives. The apparent degree of silica deactivation obtained with the polymer coating is insufficient to provide quality CO_2 chromatography for polar compounds. An example of plastic additives chromatography obtained with CO_2 on a polymer-coated stationary phase is shown in Fig. 1 [5].

Recently, interest in polymer-coated silica phases has been renewed, with investigators (Chen and Lee [2]) exploring the use of more efficient deactivation techniques and more polar polymers to coat silica particles for neat CO_2 chromatography. Polyethyleneimine-coated silica and amino-terminated polyethylene oxide-coated silica appear promising for pSFC of moderately polar basic compounds. Similarly, hydroxy-terminated polyethylene-oxide-coated silica has been used successfully for pSFC of alcohols and acids. Optimization and commercial production of these stationary phases could significantly extend the polarity range of compounds that can be chromatographed with neat supercritical CO_2.

Near-Critical Applications

Near-critical pSFC applications can be described as those where the mobile phase is solvent-modified CO_2, pressur-

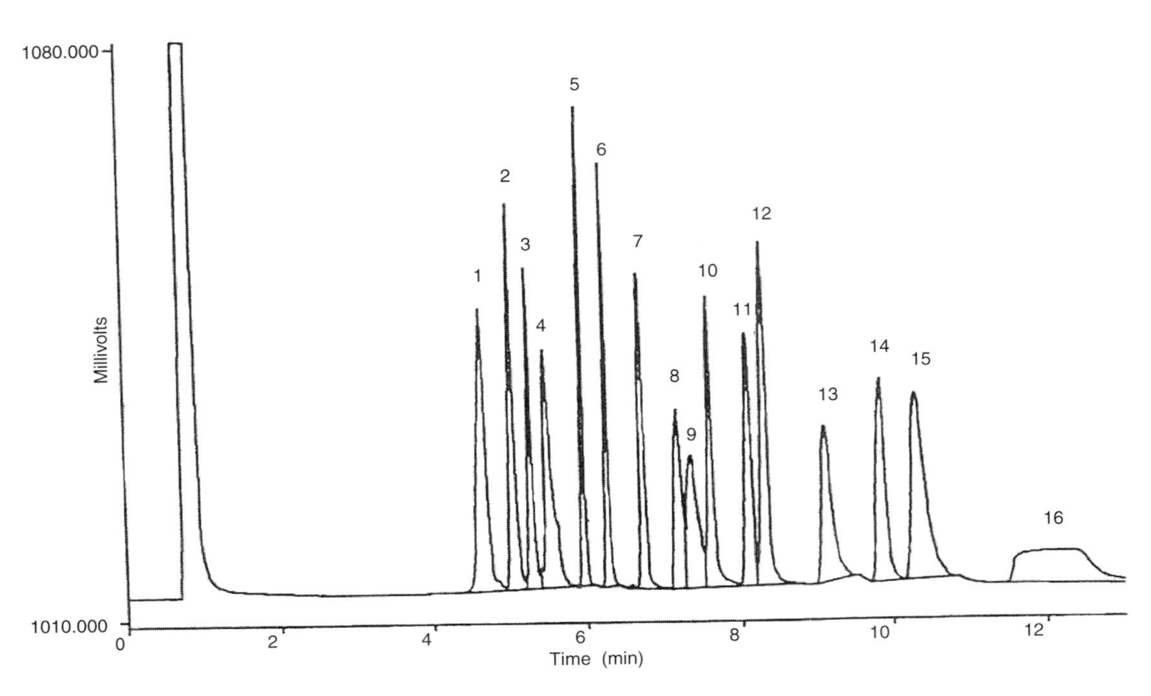

Fig. 1 Use of polymer-coated silica stationary phase with a neat carbon dioxide mobile phase. Column: Deltabond SFC Methyl (150 × 2.0 mm); mobile phase: CO_2; pressure gradient: 75 bar initial for 2 min, then 50 bar/min ramp to 180 bar, then 15 bar/min ramp to 300 bar; flow: 0.5 mL/min; temperature: 100°C; injection: 5 μL; detection: flame ionization detector; sample solvent: methylene chloride. Peak key: 1: BHT; 2: dimethyl azelate; 3: triethyl citrate; 4: tributyl phosphate; 5: methyl palmitate; 6: methyl stearate; 7: diethylhexyl phthalate; 8: Tinuvin 327; 9: Spectra-Sorb UV531; 10: tri(2- ethylhexyl) trimellitate; 11: dilauryl thiodipropionate; 12: Irganox 1076; 13: 1,3-diolein; 14: distearyl thiodipropionate; 15: Ionox 330; 16: Irganox 1010.

ized only enough to maintain a single phase, with temperatures near (typically less than) the critical temperature. Many commercially available HPLC bonded silica phases have been used with modified-CO_2 mobile phases to achieve normal-phase separations, the choice of stationary phase being dictated by sample polarity. The modifiers added to CO_2 acceptably overcome the unwanted analyte–silica interactions obeserved with neat CO_2 mobile phases. For structural separation of polar compounds such as pharmaceuticals [typically weak acids or bases of molecular weight (MW) <1000], polar phases such as diol-, amino-, and cyano-bonded silica (or bare silica) are used. Numerous applications for pharmaceutical, natural product, environmental and other compound classes have been reported in the recent literature (reviewed in Refs. 1 and 2). For structural separation of higher-molecular-weight, less polar compounds, octyl- or octadecyl silane (ODS)-bonded phases are used (Berger [1], Lesellier and Tchapla [2]). The reported applications include stationary-phase columns obtained from many different commercial manufacturers, covering almost the complete range of packing particle size and pore size.

Modified-CO_2 mobile phases excel at stereochemical separations, more often than not outperforming traditional HPLC mobile phases. For the separation of diastereomers, silica, diol-bonded silica, graphitic carbon, and chiral stationary phases have all been successfully employed. For enantiomer separations, the derivatized polysaccharide, silica-based Chiralcel and Chiralpak chiral stationary phases (CSPs) have been most used, with many applications, particularly in pharmaceutical analysis, readily found in the recent literature (reviewed in Refs. 1 and 2). To a lesser extent, applications employing Pirkle brush-type, cyclodextrin and antibiotic CSPs have also been described. In addition, the use of silica and graphitic carbon stationary phases with chiral modifiers added to the CO_2 mobile phase has been reported.

A major advantage of modified-CO_2 pSFC is that due to the low pressure and similarity of the mobile phases used for structural and stereochemical separations, multiple stationary-phase columns can be connected in series, generating separations not achievable by other chromatographic methods. Multiple columns of the same phase have been serially connected to provide significantly am-

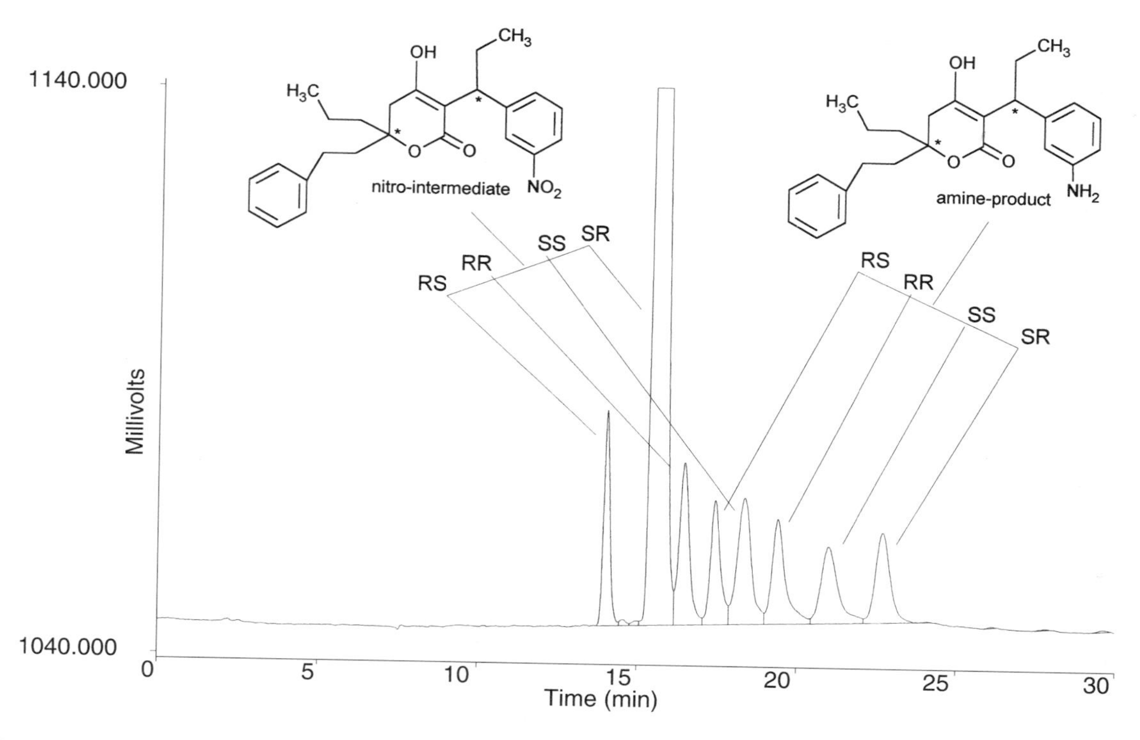

Fig. 2 Use of multiple columns (same stationary phase) in series to improve chromatographic efficiency. Columns: four Chiralpak AD (5 μm, 250 × 4.6 mm) in series; mobile phase: CO_2 with modifier gradient; modifier: ethanol–methanol–isopropylamine (50:50:0.5); gradient: 20% modifier for 2 min, then 1%/min ramp to 35%; flow: 2.0 mL/min; pressure: 150 bar; temperature: 35°C; injection: 5 μL; detection: UV 210 nm; sample solvent: methanol.

plified chromatographic efficiencies [1]. A recent example chromatogram obtained by connecting four 250 × 4.6-mm Chiralpak AD columns in series is shown in Fig. 2 [6]. With the four columns in series, concurrent separation of the four stereoisomers of each of two structurally different compounds was achieved.

Serial connection of different stationary phases provides some very interesting separations. The combination of different CSPs provides systems capable of resolving a wide range of enantiomers (Sandra et al. [2], Gyllenhaal [2]). Recent applications combining normal-phase bonded-silica (diol, cyano, amino) columns with CSPs in series, to provide concurrent structural and stereochemical separations, have been described (Phinney [4], Kline and Matuszewski [4], Williams et al. [4]). An example of chromatography obtained for a pharmaceutical compound and its degradants by connecting a Zorbax SB cyano column and three Chiralpak AD columns in series is shown in Fig. 3 [7]. The cyano column provides the structural separation of the parent compound and the two degradants, and the Chiralpak columns provide concurrent enantiomer separations for the two degradants. The ability to combine different or multiple traditional HPLC stationary phases to generate unique separations is a hallmark of modified-CO_2 pSFC.

Subcritical Applications

Recent work describing the use of subcritical water as a chromatographic mobile phase has been reported. Water heated to 100–200°C, pressurized to 20–50 bar, can be used as a reversed-phase chromatography eluant. This application exists somewhere in the boundary region between pSFC and HPLC. Stationary phases that have been used successfully for subcritical (or superheated) water chromatography include polystyrene–divinyl benzene beads [8,9], ODS silica [8,10], porous graphitic carbon [11], and polybutadiene-coated zirconia [11]. Of these phases, relatively rapid performance deterioration was reported for the ODS silica materials [11], presumably due to silica solubility. As research in this area increases, undoubtedly so will the number of identified suitable stationary phases.

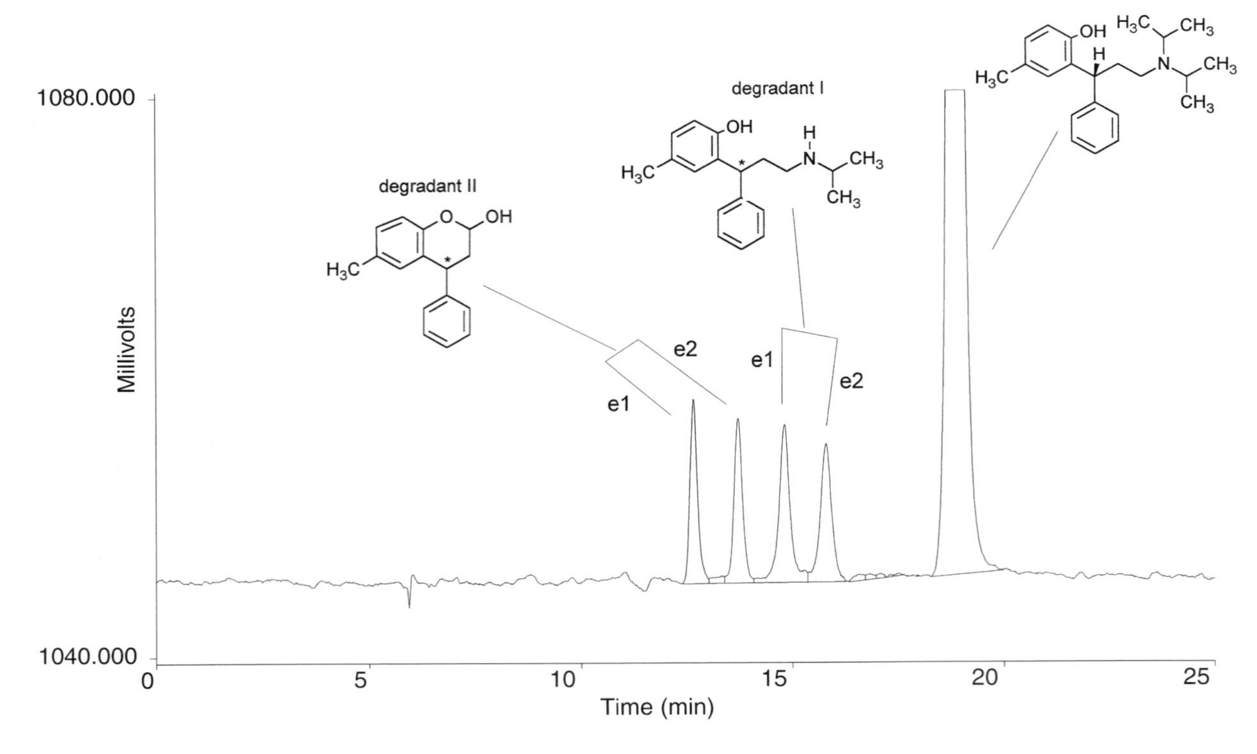

Fig. 3 Use of multiple columns in series (achiral and chiral stationary phases) to provide mixed-mode selectivity. Columns: Zorbax SB CN (5 μm, 250 × 4.6 mm) plus three Chiralpak AD (5 μm, 250 × 4.6 mm) in series; mobile phase: CO_2–methanol (containing 0.5% isopropylamine) (85:15); flow: 2.0 mL/min; pressure: 150 bar; temperature: 35°C; injection: 5 μL; detection: UV 220 nm; sample solvent: methanol.

Conclusions

Packed column SFC stationary phases are very similar or identical to those used for HPLC. With neat CO_2 mobile phases, polymer or polymer-coated silica stationary phases have typically been used. With modified-CO_2 mobile phases, bonded-phase silica columns are typically used. For structural separations, diol, amino, or cyano stationary phases are most often used. For stereochemical separations, derivatized polysaccharide-bonded silica columns are most often the stationary phases of choice. A powerful feature of modified-CO_2 pSFC is the ability to serially connect different stationary phases to obtain enhanced or multiple mechanism separations. With subcritical (superheated) water mobile phases, the use of polymer, porous graphitic carbon, and polymer-coated zirconia stationary phases has been described.

References

1. T. A. Berger, *Packed Column SFC*, Royal Society of Chemistry, Cambridge, 1995.
2. K. Anton and C. Berger (eds.), *Supercritical Fluid Chromatography with Packed Columns*, Marcel Dekker, Inc., New York, 1998.
3. 7th International Symposium on Supercritical Fluid Chromatography and Extraction, 1996.
4. 8th International Symposium on Supercritical Fluid Chromatography and Extraction, 1998.
5. S. L. Secreast, unpublished application. Packed-column supercritical fluid chromatography of plastics additives. Pharmacia Study Report, 1996.
6. S. L. Secreast and L. K. Wade, 8th International Symposium on Supercritical Fluid Chromatography and Extraction, 1998.
7. S. L. Secreast, American Association of Pharmaceutical Scientists Annual Meeting, 1999.
8. R. M. Smith and R. J. Burgess, *J. Chromatogr. A 785*: 49–55 (1997).
9. D. J. Miller and S. B. Hawthorne, *Anal. Chem. 69*: 623–627 (1997).
10. Y. Yang, M. Belghazi, A. Lagadec, D. J. Miller, and S. B. Hawthorne, *J. Chromatogr. 810*: 149–159 (1998).
11. R. M. Smith, R. J. Burgess, O. Chienthavorn, and J. Rose, *LC-GC 17*(10): 938–945 (1999).

Stephen L. Secreast

Stationary-Phase Retention in Countercurrent Chromatography

Introduction

The retention of the stationary phase is a key parameter for countercurrent chromatography (CCC), as it influences all the chromatographic parameters describing a separation. First, it is important to closely monitor its value, commonly named SF; we give the reader three methods to determine this value. Then, the best conditions for obtaining the highest SF value are described for the three main CCC devices, which include (a) a Sanki centrifugal partition chromatograph, (b) a type J high-speed countercurrent chromatograph, and (c) a cross-axis countercurrent chromatograph. Finally, some theoretical approaches are introduced in order to estimate the value before any experimental work is performed.

How to Measure SF?

Figure 1 indicates the principle of use of any CCC device for the equilibrium of two liquid, nonmiscible phases. In this case, the stationary phase which is chosen is the lighter phase of the solvent system (dark gray in Fig. 1), whereas the mobile phase is indicated in white. For simplification, the coil is considered as an empty cylinder and the phenomena which occur inside the column are highly schematized as a stack of disks of mobile and stationary phases. This allows us to visualize the progression of the mobile phase inside the column. After the solvent system has reached equilibrium (complete settling of the two phases), the phase chosen as the stationary phase is pumped into the apparatus. The latter is considered as filled as soon as droplets of this stationary phase are expelled out of the column; this is Step 1 of Fig. 1.

The apparatus is then started, and when the desired rotational speed is reached, the mobile phase is pumped into the apparatus. A graduated cylinder is then put at the outlet of the apparatus. The two phases undergo a hydrodynamic or hydrostatic equilibrium inside the column while the mobile phase progresses toward the outlet of the column; this is Step 2. After a certain time, the mobile phase has reached the end of the column and then the first droplet of the mobile phase falls into the graduated cylinder; this Step 3. The experimenter then reads the volume, V_1, of the stationary phase which has been expelled from the column. The experiment is continued until the desired total volume is reached in the graduated cylinder. The experimenter can read the respective volumes of the stationary phase, named V_2, and the mobile one; this is Step 4.

Finally, the apparatus can be emptied (for instance, by pushing with nitrogen gas) and the liquids collected in another graduated cylinder; the volume of the stationary phase is V_3. For simplification purposes, the extracolumn

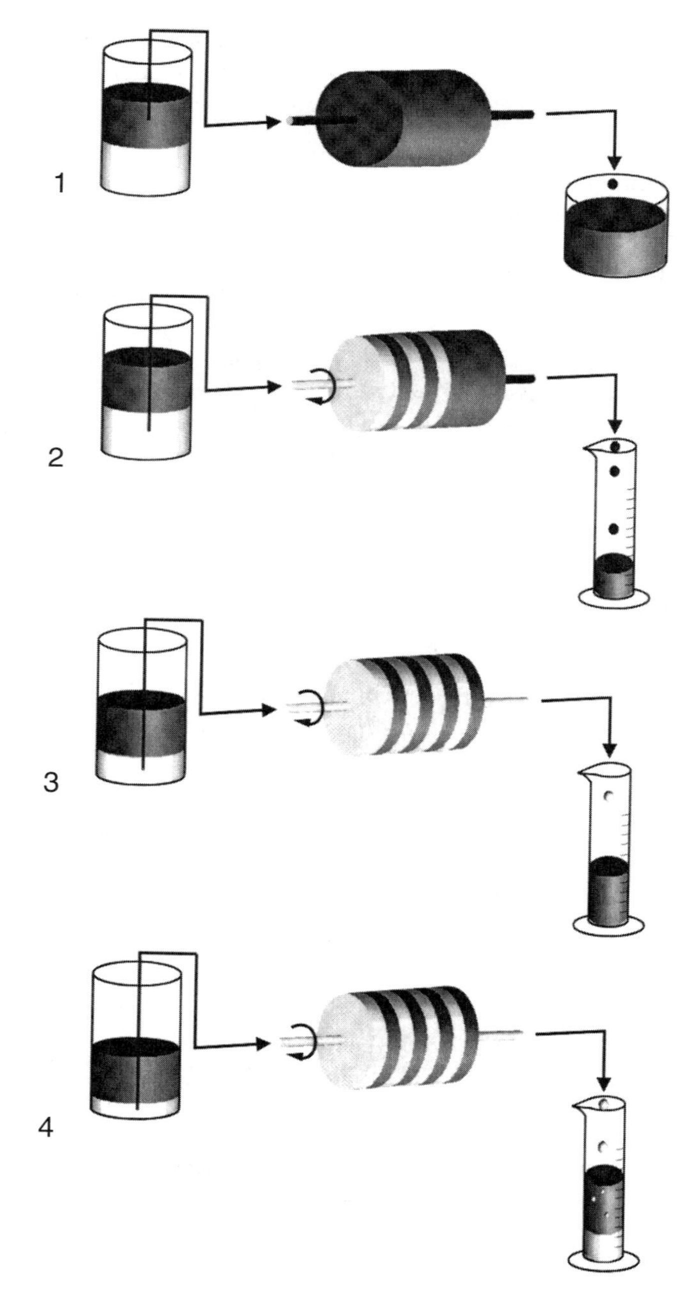

Fig. 1 Principle of the two-phase equilibrium inside a CCC column.

volumes are neglected. Three measurements of the stationary phase retention are available:

- One just after the equilibrium inside the column: $SF_1 = (V_{column} - V_1)/V_{column}$
- One after a certain amount of time: $SF_2 = (V_{column} - V_2)/V_{column}$
- One by emptying the apparatus: $SF_3 = V_3/V_{column}$

If the equilibrium of the two phases was stable and not disturbed by any external event (change in rotational speed, flow rate, etc.), the three values of SF should be similar by a few percent of precision.

Best Conditions for SF for the Three Main CCC Devices

The best combinations of experimental parameters (e.g., choice of lighter or heavier phase, choice of the inlet to pump the mobile phase into, etc.) in order to retain the maximum amount of stationary phase inside the column are related to complex hydrodynamic phenomena which are based on the behavior of the solvent system inside the column of a given CCC apparatus. Many experiments have been carried out on the three main types of CCC devices by varying the experimental parameters and the solvent systems, in order to gather solvent systems by groups characterized by a combination of experimental parameters.

Sanki-Type Apparatus

The principle of this instrument has been precisely described in the entry (see entry Centrifugal Partition Chromatography: An Overview) devoted to this topic in this volume. The apparatus is a centrifuge, in which cartridges or plates are installed. Two rotating seals are required to allow the flow of the liquid phase: One stays at the top of the centrifuge, the other one at the bottom (solvent inlet and outlet).

Whatever the solvent system may be, the optimization for the best retention of the stationary phase is quite simple. Among the four possibilities, only two of them lead to a good retention of the stationary phase inside the cartridges or the plates; they are based on the combination of the lighter mobile phase pumped from the bottom to the top seal, also called the "ascending" mode, and the heavier mobile phase pumped from the top to the bottom seal, also called the "descending" mode.

Type J Apparatus

The principle of this apparatus has been precisely described in an entry (see entry Dual Countercurrent Chromatography) devoted to this topic in this volume. Two main parameters for using a two-phase solvent system with this apparatus are the choice of the heavier or the lighter mobile phase and the pumping mode [i.e., from the tail to the head or the head to the tail of the column (as a rotating coil defines a tail and a head, carrying, for instance, a small solid from the tail to the head)].

The designer of this type of device [Dr. Yoichiro Ito (National Institutes of Health, Washington, DC)], has tried various solvent systems in order to ascertain the best combinations of the two main parameters [1]. He observed that, among the four possibilities, only two led to the best retention of the stationary phase. However, the two optimal conditions were dependent on the nature of the solvent system and, for some solvent systems, on the geometrical dimension of the apparatus.

Ito has, consequently, decided to carry out a systematic study with 15 solvents [1]. His first conclusion was that only one condition was optimal among the two pumping modes for a given phase (i.e., lighter or heavier phase). The second conclusion was that the pumping modes to be used are reversed if the liquid phase is chosen as lighter instead of heavier, or vice versa.

Two groups of solvent systems have, consequently, been defined. One gathers solvent systems for which the two best combinations are the pumping of the lighter phase from the tail to the head of the column and the pumping of the heavier phase from the head to the tail of the column. The organic phases of such solvent systems are hydrophobic; thus, this group is called "hydrophobic" (e.g., hexane–water). The other group gathers solvent systems for which the two best combinations are the pumping of the lighter phase from the head to the tail of the column and of the heavier phase from the tail to the head of the column. The organic phases of such solvent systems are quite hydrophilic; thus, this group is called "hydrophilic" (e.g., butanol-2–water). The best combinations are reversed between the two groups.

However, there was a need to define a third group to take into account the behaviors of some solvent systems for which optimal combinations depend on the geometric dimensions of the apparatus. The discriminating parameter was found to be the β ratio of the coil radius on the distance between the two axis of rotation. For β values smaller than 0.3, solvent systems belonging to this third group behave like the solvent systems of the "hydrophilic" group. Conversely, for β values greater than 0.3, they behave like solvent systems of the "hydrophobic" group. The group was named "intermediate," which was also consistent with the mild hydrophobic (or hydrophilic) character of the organic phase (e.g., chloroform–methanol–water).

The main drawback of this classification comes from the experimental determination of the three groups. For a solvent system not previously studied, either by Ito or cited in the literature, the experimenter has to carry out four experiments in order to determine the two best combinations and, consequently, the group to which it belongs.

Cross-Axis-Type Apparatus

The general principle of this type of apparatus is described in the corresponding entry (see entry Cross-Axis Coil Planet Centrifuge for the Separation of Proteins) of this volume. Contrary to the two previous CCC devices, four main parameters have to be considered here. Two of them are common to the other types of CCC units (i.e., choice of a lighter or a heavier phase and pumping mode from tail to head or from head to tail). Two additional parameters intervene: the pumping direction, from the inside to the outside of the core or reverse, and the rotation direction, clockwise or counterclockwise.

The same designer of this type of apparatus as for type J device, Ito, has applied the same procedure to classify various solvent systems by varying the main running parameters [2,3]. However, it has proven difficult to draw clear and precise conclusions from the results, because of the number of operating parameters. The methodology of experimental design has, consequently, emerged as the rational method to use for this purpose; it is easy to use and it elucidates the effects of the parameters and their interactions. A thorough, but global, analysis based on the experimental design methodology applied to the cross-axis-type device, was reported by Goupy et al. [4]. The overall analysis carried out by the use of experimental design with a coil mounted in the L position simplifies the operation of the cross-axis apparatus, as the best combination does not depend on the solvent system. Indeed, the retention of the stationary phase is mainly related to the choice of the mobile phase and of the elution direction. A heavier mobile phase requires the outward elution direction, whereas a lighter mobile phase requires an inward elution direction.

Consequently, no classification among solvent systems may be built from their behaviors inside a cross-axis device.

Theoretical Approaches to Correlate Physical Parameters and Observed Behavior Inside a CCC Column

The capillary wavelength and settling velocities have been retained as interesting parameters to step forward the description of the behavior of solvent system inside a CCC column. Moreover, they have also enabled a simple prediction of the effect of the temperature on the behavior of solvent systems, thus, on "Ito's classification."

The "theoretical" parameter capillary wavelength, λ_{cap}, has been precisely described by Menet et al. [4,5]. The capillary wavelength is a means of describing the microscopic behavior at the interface between two immiscible liquids. It stands for the wavelength of the deformations which may occur at the interface of the two liquids or represents the mean diameter of drops of a liquid in another one. For common liquids, its average value is 1 cm in the Earth's gravitational field. For CCC devices, it is in the order of 1 mm, because of the generated centrifugal force field.

As the capillary wavelength only enables the description of the formation of droplets of one liquid in another one, it seemed interesting to introduce other "theoretical" parameters to better describe the dynamic phenomena occurring inside a CCC column. Two of these are presented here, namely V_{low} for the fall of a droplet of the heavier liquid phase (lower) in the continuous lighter one (upper) and V_{up} for the rise of a droplet of the lighter liquid phase in the continuous heavier one.

In order to determine if a correlation exists between "Ito's classification" and the values of the previous "theoretical" parameters, the 15 solvent systems used for the design of the classification have been studied. The values of interfacial tension, of densities, and of dynamic viscosities of the solvent systems were used to compute the values of the capillary wavelength and the settling velocities. The values of the three theoretical parameters have allowed to set ranges within which a solvent system is named "hydrophobic," "intermediate," or "hydrophilic." The main interest is that the knowledge of some physical parameters of the solvent systems allows one to know, before any experiments, the best combinations for the greatest stationary-phase retenion.

The values of the capillary wavelengths are smaller for "hydrophilic" solvent systems than those for "hydrophobic" ones: The first family of solvent systems tends to form small droplets of one phase in the other one, hence leading to a more stable emulsion. Their stationary phase is consequently less retained in the CCC column than the one of "hydrophobic" solvent systems; this phenomenon is well known in CCC [1].

From additional studies of six new solvent systems and by studying the influence of the temperature, Menet et al. have shown the best classification is that based on V_{up} [4].

References

1. N. B. Mandava and Y. Ito, *Countercurrent Chromatography – Theory and Practice*, Marcel Dekker, Inc., New York, 1988.
2. Y. Ito and T.-Y. Zhang, *J. Chromatogr. 449*: 135–151 (1988).
3. Y. Ito, *J. Chromatogr. 538*: 67 (1991).
4. J. Goupy, J. M. Menet, and D. Thiébaut, Experimental de-

signs applied to countercurrent chromatography: Definitions, concepts and applications, in *Countercurrent Chromatography* (J. M. Menet and D. Thiébaut, eds.), Marcel Dekker, Inc., New York, 1999.

5. J. M. Menet, D. Thiébaut, R. Rosset, J.-E. Wesfreid, and M. Martin, *Anal. Chem. 66*: 168–176 (1994).

Jean-Michel Menet

Steroid Analysis by TLC

Introduction

Steroids, which are a class of compounds that occur in nature and in synthetic products, have a cyclopentanoperhydrophenanthrene skeleton. The carbon atoms and rings are labeled according to the scheme shown in Fig. 1. The following classes of compounds belongs to steroids: sterols, bile acids, cardenolides, androgens, estrogens, corticosteroids, steroid sapogenins, steroid alkaloids, ecdysteroids, and vitamin D.

In the naturally occurring steroids, the fusion of ring B and C is always trans and that of the ring C and D is usually trans (cis in the cardenolides and bufadienolides). Rings A and B are fused in cis and trans configurations with about equal frequency. The configuration of the substituents is referred to that of the 19-methyl group on C_{10}. Every substituent that is in a configuration identical to the methyl group is indicated by the β-position; substituents of opposite configuration are termed α-substituents (dotted lines). If the substituents lie in the planes of the rings, they are termed equatorial (e), and if perpendicular to the rings, they are called axial (a). Thus, for example, in an A/B *cis*-steroid, a 3β-hydroxy group is axial and equatorial in an A/B *trans*-steroid (see Fig. 1). In general, equatorial substituents are more stable than axial substituents. In various reactions, the formation of the former is favored.

In the field of steroid analysis, thin-layer chromatography (TLC) is still the method of choice, especially when many simultaneous analyses have to be carried out; hundreds of analyses can be performed in a short time and with small demands on equipment and space. Samples can be analyzed with minimal cleanup, and analyzing a sample by the use of multiple separation steps and static postchromatographic detection procedure is also possible because all sample components are stored on the layer without the chance of loss. The time required in TLC analysis is about 10–60 min. As little as 0.01 μg of steroids/spot can be detected by TLC. Using a TLC plate with thicker adsorbent layers (0.5–2 mm), several grams of substance can be isolated (preparative TLC).

Sorbents

The TLC of steroids has been tried on a great variety of sorbents, but various forms of silica gels are most frequently used. In order to achieve adequate stability, adhesion, and resistance to abrasion of the (precoated) layers, the sorbents usually contain a binder. Either they do not contain any additive at all, designated by "H" in the article designation, or gypsum ("G") as binder. The sorbents designated by "P" are suitable for preparative purposes. An "F" designates the addition of flourescent indicators and, if applicable, the number that follows gives the excitation wavelenght and an "s" following this indicates the acid stability of the indicators. A number placed immediately after the name of the sorbent indicates the pore diameter of the sorbents (in Å). In high-performance thin-layer chromatography (HPTLC), the particle diameter is about 3–10 μm, whereas in TLC, it is about 4–25 μm. The plates can be easily prepared in the laboratory or readily purchased from commercial sources as precoated plates or sheets, but it is recommended to use ready-made (precoated) plates or sheets because they are more convenient and more uniform than those manually prepared in the laboratory.

An addition of 3–10% of silver nitrate to silica gel or kieselguhr is considerably helpful in the separation of sterols and steroids that differ only through an unconjugated double bond. For C_{27} steroid sapogenins and alkaloids, it was recommended to use a higher concentration of silver nitrate (15%). Precoated TLC–HPTLC plates can be dipped in a 20% silver nitrate solution for 15–20 min, then, in the absence of light, drying the plate in air. Finally, it is activated in a drying oven. Impregnating silica gel with 10–25% formamide in acetone can be used for separating digitalis glycosides, estrogens, and equilin. Impregnation with boric acid could be used, also, for the delicate separation of cardiac glycosides on silica gel layers. The silica gel can also be modified into nonpolar reversed phases (RP) such as C_{18} (octadecyl function), C_8 (octyl function), CN (cyanopropyl function), NH_2 (aminopropyl function), and diol (vicinal hydroxyl function on C chains). Proges-

Cyclopentanoperhydrophenanthrene skeleton

Cholesterol (Sterol)

A/B trans

Testosterone (Androgen)

Aglycone of digitalis-glycosides

Estron (Estrogen)

Cortisone (Corticosteroid)

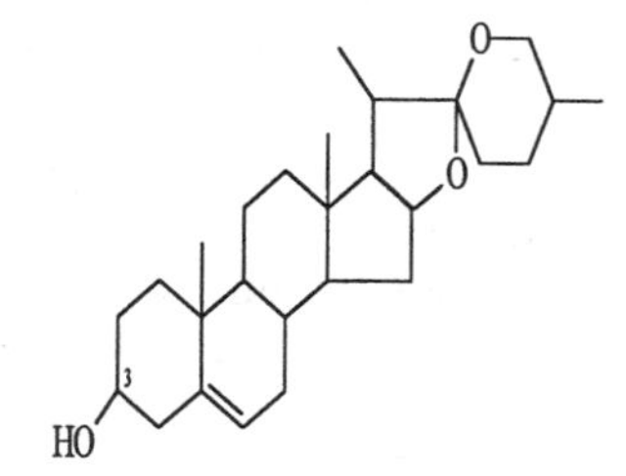

Diosgenin (Sapogenin steroid)

Fig. 1 The structure of some important steroids.

terone steroids can be well separated using CN F_{254} and RP plates. Reversed phases C_{18}, C_8, and C_2 could be used for analyzing the ecdysone steroids and RP C_{18} for estrogen conjugates.

Other sorbents that have also been used for separating steroids in TLC are cellulose, kieselguhr, alumina, polyamide, magnesium oxide, and celite. Androstanes can be analyzed using cellulose layers impregnated with 1,2 propanediol. Layers consisting of alumina and magnesium oxide can be used to separate some sterols and sterol acetates.

Layers of alumina that are deactivated with 2.5% acetic acid could be employed for separating some corticosteroids. An advantage of using alumina rather than silica gel in TLC of Δ^4-3-ketosteroids is that they form fluorescent oxidation products when the plate is heated to 150–180°C. As little as a nanogram of steroids can be detected by this method. By using a layer that consists of a mixture of $MgO–Al_2O_3–CaSO_4$(15:5:1), complex mixtures of sterols as their 3,5 dinitrobenzoate derivatives can be resolved using hexane–ethyl acetate (9:1, v/v) as the mobile phase.

Although most steroids are stable on the silica gel layers, some steroids (e.g., estrogens and vitamin D) can decomposed readily. To avoid the decomposition, a preliminary treatment of the sorbent with a solution of ascorbic acid (antioxidants) in ethanol is recommended.

Development

Simple development will usually be enough for the TLC of steroids. In some cases, multidevelopment in the same or different solvent system or a two-dimensional technique is required to achieve better separation. For ensuring reproducibility of the R_f value, Neher [1] recommended the use of the same number of plates and always using fresh solvent for each run. A system containing a similar volatility maintains a more stable composition and, thus, has more reproducible R_f values during several runs. Our experiences shown that, by weighing the solvent components, the R_f values of testosterone (0.37) and estradiol (0.52) on silica gel plates were unchanged after four runs by using cyclohexane–ethyl acetate (1:1, w/w) as the mobile phase.

Detection

Destructive Detection Methods

Most of the reagents used for detecting steroid spots by *in situ* reactions (destructive reagents) contain sulfuric acid. Sulfuric acid, without any additive, can also produce characteristic colors and fluorescence response, as well as permanent black zones after heating. The initial color, the color after heating for 10 min, and the color in ultraviolet (UV) (366 nm) of various classes of steroids (141 compounds) were presented by Heftmann [2]. Other destructive reagents that are most generally applicable are antimony trichloride (Carr–Price's reaction; for vitamin D, cardenolides, bufadienolides, triterpenoids), aromatic aldehyde–acids (for sapogenin steroids, steroid alkaloids, ketosteroids), molybdophosphoric acid (reducing and unsaturated steroids, cholesterol ester, bile acids; blue on yellow background), chlorosulphonic acid–acetic acid (cardenolides, green, blue violet fluorescence), and phosphoric acid (color similar to sulfuric acid).

For the detection of ketosteroids, an *m*-dinitrobenzene solution can be used (17-keto-steroids, violet; 3-keto-Δ^4 groups, blue). Δ^4-3-Oxo steroids and $\Delta^{4,6}$-3-oxo steroids can be distinguished using a phthalic acid–*p*-phenylenediamine reagent, which gives yellow and orange-brown colors, respectively. Detailed discussion of the destructive detection methods of steroid spots is described in the books of Macek [3] and Touchstone [4]. Thus, by spraying and heating the plates, various class compounds of steroids can be deduced; to confirm this, cochromatography with authentic standards is needed. If the standard is unavailable, isolating the substance and then identification by spectroscopic analysis is recommended.

Nondestructive Detection Methods

Steroids containing an α,β-unsaturated structure, such as Δ^4-3-oxo, Δ^7-6-oxo, Δ^5-7-oxo and Δ^{16}-20-oxo groups, can be visualized under UV light (254 nm). For this purpose, the samples should be spotted on plates that contain a fluorescent indicator (GF_{254} or F_{254}). The steroid spots will appear as dark zones on the layer. Sorbent layers without the indicator must be sprayed with a dilute solution of fluorescein or morin before exposure to UV light. This method has a great advantage over the destructive methods because the unchanged steroids can be eluted or isolated from the sorbents for further analysis.

Other nondestructive detection methods use iodine vapors or iodine–potassium iodide reagents (Mylius's reaction). Yellow, orange, or brown zones will appear on the layer. The zones will have to be marked, because the iodine will eventually evaporate. The sensitivity of the iodine test can be greatly increased by the use of layers containing rhodamine 6G. Most of the steroids are recovered unchanged after exposure for 30 min with iodine vapor, except for estrogen and Vitamin D.

Formation of Derivatives (Microreaction)

It is well known that steroid derivatives such as acetates, benzoates, propionates, and trifluoracetates can be better separated in TLC–HPTLC than the free substances themselves. These derivatization reactions can be performed directly on the TLC layers. Acetylation can be accomplished by treating 10–100 μg of steroids with 0.1 mL acetic anhydride and 0.1 mL pyridine for 8–16 h at room temperature. Following the reaction, blowing with nitrogen at 60°C can remove the reagent. Benzoylation is carried out in the same way, but by using 0.1 mL benzoyl chloride. Propionylation is performed by dissolving the steroids in 0.3 mL of warm propionyl chloride and allowing the mixture to stand 10 min at 20°C. Extraction with hexane and washing with water and sodium carbonate solution can purify the ester. Trifluoroacetylation is performed by mixing the steroids with a small excess of trifluoroacetic anhydride in hexane.

The steroid critical pairs (such as saturated or unsaturated with one double bond) can be separated better by spotting with 0.1% bromine solution in chloroform directly at the starting spot. A similar differentiation is accomplished by spotting with a solution of *m*-chlorperbenzoic acid in chloroform; but, by this method, epoxidation of the double bond occurs.

Quantitative Analysis

Spot Elution Technique

After elution from the chromatoplates, the separated steroids can be analyzed by using various methods such as UV–vis spectroscopy or fluorometry. Although this method is very simple, at the present time its application is significantly diminished due to some disadvantages (e.g., it is difficult to locate the spot position accurately, nearly quantitative elution of the spots is required, the loss originating from irreversible adsorption during chromatography must be minimized, etc.). When the UV method is used, special care should be taken because the eluate of silica gel with some semipolar solvents (e.g., ethanol, methanol) exhibit absorption in the UV region. It is recommended to use a blank containing the same amount of adsorbents from an empty part of the plate.

In Situ *TLC Technique (Densitometry)*

In situ quantification of steroids on the chromatoplate can be performed by using UV (for UV-active steroids), vis (usually using destructive reagents), and fluorometric methods. For UV-active steroids, such as corticosteroids, it is common to scan the steroid spots on the basis of fluorescence quenching at 254 nm using F_{254} precoated layers. For steroids that use destructive reagents for their visualization, a reflectance–absorbance method in the vis range (370–700 nm) is used. The steroid sapogenins (diosgenin, hecogenin, manogenin, etc.), steroid alkaloids (solasodine, tomatidine), and total sterols may be assayed using an absorbance–reflectance densitometry method (in the vis region) after treatment of the steroid spots with anisaldehyde–sulfuric acid reagent. It was found that this densitometric method was faster, simpler, and less expensive when compared to HPLC or gas liquid chromatography (GLC). Three methods are available for *in situ* fluorometric measurement. The first is to produce active derivatives on the layer by spraying with a suitable reagent. The second is to induce fluorescence on the plate, and the third is to prepare fluorescent derivatives prior to the analysis. The DL (detection limit) of the fluorometric method is much lower compared to the UV and vis evaluation methods.

Validation of the Method

Before the assay methods can be used for routine application (e.g., in a quality control laboratory), it must first be validated. The parameters for the validation methods are specificity, linearity, accuracy, precision (repeatability and intermediate precision), detection limit (DL), quantitation limit (QL), and applicable range. A detailed discussion is provided in Ref. 5.

Applications

Many publications dealing with TLC–HPTLC steroid analysis have appeared every year. The publications can be summarized into categories as follows: analytical control of steroid formulations (drug preparations), determination of steroids in biological media and natural resources, and analytical control of the production of steroids (including raw material, syntheses, and biotransformation). A cumulative database of thousands of TLC methods (including steroids) is provided in compact-disk (CD) format by Camag [6].

References

1. R. Neher, *TLC of Steroids and Related Compounds*, in *Thin Layer Chromatography* (E. Stahl, ed.), Springer International Student Edition, Springer-Verlag, Berlin, 1969, pp. 311–357.
2. E. Heftmann, *Chromatography of Steroids*, Journal of Chromatography Library Vol. 8, Elsevier Scientific, Amsterdam, 1976, pp. 14–27.
3. K. Macek, *Pharmaceutical Application of Thin-Layer and Paper Chromatography*, Elsevier, Amsterdam, 1972, pp. 275–348.
4. J. C. Touchstone, *CRC Handbook of Chromatography of Steroids*, CRC Press, Boca Raton, FL, 1986, pp. 27–40.
5. B. Renger, H. Jehle, M. Fisher, and W. Funk, *J. Planar Chromatogr. 8*: 269–278 (1995).
6. Camag Bibliography Service, *Thin-Layer Chromatography, Cummulative CD version 1.00*, Camag, Muttenz, 1997.
7. S. Görög and G. Szasz, *Analysis of Steroid Hormone Drugs*, Elsevier Scientific, Amsterdam, 1978.
8. S. Görög, *Steroid Analysis in the Pharmaceutical Industry*, Ellis Horwood, Chichester, 1989.

Muhammad Mulja
Gunawan Indrayanto

Supercritical Fluid Chromatography: An Overview

Supercritical Fluids: Definition and Basic Properties

A phase diagram, as shown in Fig. 1, can describe the physical stage of a substance of fixed composition. In this pressure–temperature diagram for CO_2, there are three lines describing the sublimation, melting, and boiling processes. These lines also define the regions corresponding to the gas, liquid, and solid states. Points along the lines (between the phases) define the equilibrium between two of the phases. The vapor pressure (boiling) starts at the triple point (Tp) and ends at the critical point (Cp). The critical region has its origin at the critical point. At this point, we can define a supercritical fluid (SF) as any sub-

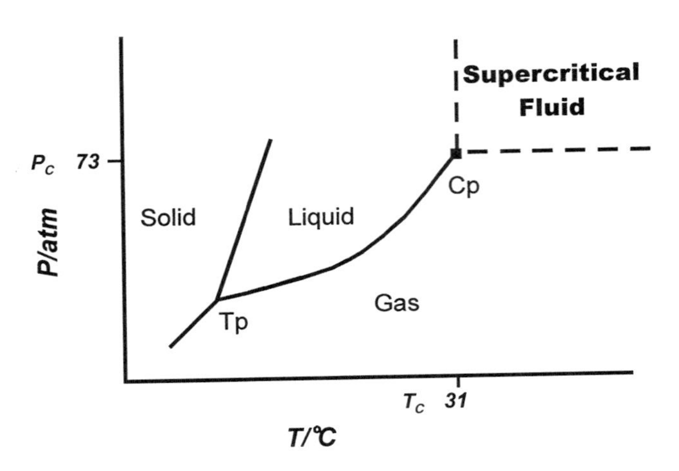

Fig. 1 Phase diagram for CO_2. P_c = critical pressure; T_c = critical temperature; Cp = critical point; Tp = triple point.

stance that is above its critical temperature (T_c) and critical pressure (P_c). The critical temperature is, therefore, the highest temperature at which a gas can be converted to a liquid by an increase in pressure. The critical pressure is the highest pressure at which a liquid can be converted to a traditional gas by an increase in the liquid temperature. In the so-called critical region, there is only one phase and it possesses some of the properties of both a gas and liquid. Subcritical (liquid) CO_2 is found in the triangular region formed by the melting curve, the boiling curve, and the line that defines the critical pressure [1].

Supercritical fluids begin to exhibit significant solvent strength when they are compressed to liquidlike densities. This makes physical sense intuitively because it is known that gases are not considered as good solvents.

The density of a pure solvent changes in the region of its critical point. For a reduced temperature ($T_r = T/T_c$) in the range 0.9–1.2°C, the reduced solvent density ($\rho_r = \rho/\rho_c$) can increase from gaslike values of 0.1 to liquidlike values of 2.5 as the reduced pressure ($P_r = P/P_c$) is increased to values higher than ~1.0 atm. However, as T_r is increased to 1.55, the supercritical fluid becomes more expanded and reduced pressures greater than 10 are needed to obtain liquidlike densities. By operating in the critical region, the pressure and the temperature can be used to regulate density, which regulates the solvent power of a supercritical fluid [2].

The viscosity changes rapidly in the critical region; even at the high-pressure levels of 300–400 bar, it is only about 0.09 cP, an order of magnitude below typical viscosities of liquid organic solvents.

The properties of gaslike diffusivity and viscosity, zero surface tension, coupled with liquidlike density, combined with the pressure-dependent solvating power of SF have provided the impetus for applying SF technology to analytical separation problems.

Supercritical Fluid Chromatography: An Introduction

The first reported observation of the occurrence of a supercritical phase was made by Baron Cagniard de la Tour in 1822 [3]. He noted visually that the gas/liquid boundary disappeared when heating each of them in a closed glass container increased the temperature of certain materials. From these early experiments, the critical point of a substance was first discovered. The first workers to demonstrate the solvating power of supercritical fluids for solids were Hannay and Hogarth in 1879 [4]. They studied the solubility of cobalt(II) chloride, iron(III) chloride, potassium bromide, and potassium iodide in supercritical ethanol (T_c = 243°C, P_c = 63 atm).

Klesper et al. first demonstrated, in 1962, SFC by the separation of nickel porphyrins using supercritical chlorofluoromethanes as mobile phases [5]. Sie and Rijnders [6] and Giddings [7], in 1966, developed the technique further, both practically and theoretically, as well as many applications. A few years later, Gouw and Jentof reviewed the general aspects of SFC, including different mobile phases, solute retention, selectivity, and applications [8].

Until the beginning of the 1980s, SFC was characterized by the utilization of packed columns, in the so-called "LC-like SFC" [9]. The introduction in 1981 of capillary open columns with small internal diameters and immobilized stationary phases has opened perspectives to the "GC-like SFC," or c-SFC (capillary supercritical fluid chromatography), with the great advantage of the high-resolution power of capillary columns. The combination of these columns with detectors traditionally utilized in gas chromatography (GC) allows the analysis of compounds with lower volatility and/or higher molecular weight than those in GC [10].

Instrumentation in SFC

A schematic drawing of the main parts of the SFC system is shown in Fig. 2. It consists of a high-pressure pump for pressurizing and delivering the solvent, usually CO_2, connected to an oven, generally a modified gas chromatograph used as the temperature controller for the SFC column. The injector should introduce small sample volumes into the column and a restrictor is placed between the end of the column and the detector to maintain the mobile phase in the supercritical state. A detailed description of each part of the system follows.

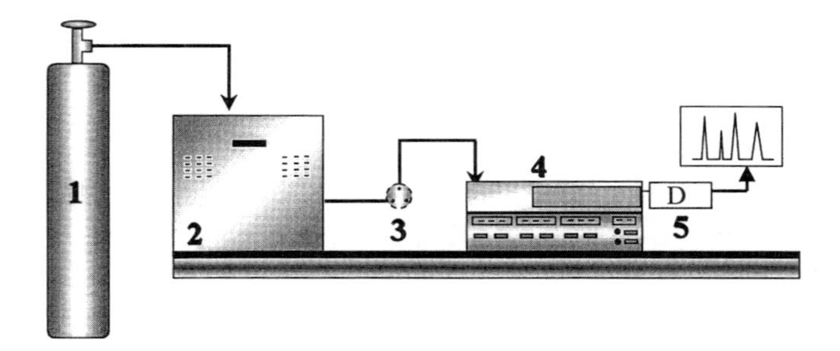

Fig. 2 Schematic drawing of a SFC apparatus. 1: CO_2 tank, 2: high-pressure pump, 3: injection valve, 4: oven (containing the column and restrictor), 5: detector (D).

Mobile Phase

Pure CO_2 has been the preferred solvent due to its favorable properties. The CO_2 is used in siphoned cylinders to assist the transference of the solvent to the pump. CO_2 passes through a cooling system to increase its density before being inserted in the heating system. When required, a vessel containing a modifier can be added to the system in a way similar to that already well known in supercritical fluid extraction (SFE).

Pump

Although several high-pressure pumps have been used in SFC, the syringe-type pump has been the preferred to deliver CO_2 into the system. This choice is made due to the absence of pulses of syringe pumps and the possibility of flow rate and pressure control.

Sample Introduction

Samples are usually injected through a high-pressure injection valve fitted with a small internal loop.

Temperature Control

To control and maintain the critical temperature of the mobile phase (CO_2), the column is installed in an oven, similar to those used for GC or high-performance liquid chromatography (HPLC), depending on the type of column used (Fig. 2).

Columns

Two types of columns are used in SFC: packed columns containing solid particles of small inner diameter or wall-coated open-tubular columns (WCOT), usually called just capillary columns. Packed columns have been preferred when capacity is the most relevant issue; capillary columns are selected when efficiency is the goal.

Restrictor

In order to maintain the desired SF mobile-phase conditions, the end of the column is connected to a restrictor. Although several types of restrictor are available [11], the most popular is the linear restrictor, which consists of a small piece ($\sim$10 cm) of a fused silica or metal tube of small inner diameter (50 μm or less).

Detectors

One of the attractions of SFC is that it can use both GC- and LC-like detectors, including the almost universal flame ionization detector (FID) for nonvolatile and volatile analytes after separation on either capillary or packed columns. Selective responses could be also obtained from a number of detectors as NPD, ECD, FPD, ultraviolet, Fourier transform infrared, nuclear magnetic resonance, and mass spectrometry.

Application

An important field of application of analytical chemistry involves the isolation, identification, and quantification of components in complex samples. Chromatography is one of the most used techniques, because modern chromatographic methods have an excellent separation power, are versatile, and can be used with several detection techniques.

During the last 15 years, the applications of supercritical fluids (SFC and SFE) have shown a fast advance; among others, from a historical perspective, SFC was developed after GC was well established and when HPLC was starting. The interest for SFC has grown with the GC and HPLC development and technological innovations that had occurred independently of SFC research, but surely allowed that commercial SFC instruments could be introduced in the 1980s.

Supercritical fluid chromatography has been applied to environmental analyses, chemical foods, polymers, pharmaceutical, and agro-industry research. This process generates quite complex products, which has been analyzed by different chromatographic approaches, including SFC. Considering the complexity of these samples, a high-resolution technique is required. Even considering that packed column SFC has some advantages in certain cases, capillary columns coated with polymeric phases presents more efficiency (N) per column, being more adequate for complex samples.

As an example, in natural product analysis, SFC offers perspectives in the analysis of several classes of compounds that present difficulties in either conventional LC or GC. In this area, it is very common that the analytes do not have chromophore groups, thus making difficult the detection through UV–vis, the most popular HPLC detector. At the same time, several of them are not volatile enough to be analyzed by GC. In this case, the use of SFC with capillary columns and FID detection is a valuable tool. Figure 3 shows a chromatogram of a mixture of triterpenes containing a —COOH functional group (betulinic acid, oleanolic acid, ursolic acid, and polpunonic acid) [12]. These compounds are a good example of a class of compounds that presents biological activities and are difficult to be analyzed by either GC or HPLC without an additional derivatization step.

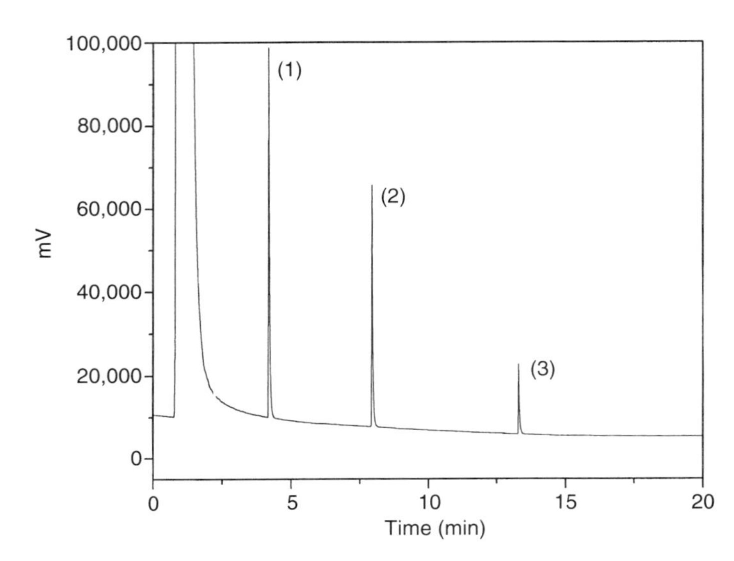

Fig. 3 c-SFC chromatogram of a mixture of triterpenic acids: (1) oleanolic acid, (2) ursolic acid, (3) polpulnonic acid. Column: 20 m × 100 μm × 0.20 μm (5% phenyl, 95% methyl polysiloxane cross-linked); $T = 80°C$; $P = 120$ atm.

Conclusion

Supercritical fluid chromatography is a very important chromatographic technique still underestimated and underutilized. It presents characteristics similar to both GC and HPLC, although having its own characteristics. Whereas the column temperature control is the way to achieve a good separation in GC and the solvating power of the mobile phase is controlling factor in HPLC, in SFC the density of the fluid is the major factor to be optimized. Both packed (LC-like) and capillary (GC-like) columns have been used in this technique, which has found applications in practically all areas in which GC or HPLC has shown to be the selected separation technique.

References

1. L. T. Taylor, *Supercritical Fluid Extraction*, John Wiley & Sons, New York, 1996, pp. 1–30.
2. M. A. McHugh and V. J. Kruponis, *Supercritical Fluid Extraction. Principles and Practice*, 2nd ed., Butterworth–Heinemann, Boston, 1994, pp. 1–26.
3. C. Cagniard de la Tour, *Ann. Chim. Phys. 21*(2): 127, 178 (1822).
4. J. B. Hannay and J. Hogarth, *Proc. Roy. Soc. (London), 29*: 324–326 (1879).
5. E. Klesper, A. H. Corwin, and D. A. Turner, *J. Org. Chem. 27*: 700–701 (1962).
6. S. T. Sie and G. W. A. Rijnders, *Separ. Sci. 1*: 459–490 (1966); *2*: 699–727, 729–753, 755–777 (1967).
7. J. C. Giddings, *Separ. Sci. 13*: 73–80 (1966).
8. T. H. Gouw and R. E. Jentoft, *J. Chromatogr. 68*: 303–323 (1972).
9. D. R. Gere, R. Board, and D. McManigill, *Anal. Chem. 54*: 736–740 (1982).
10. M. Novotny, S. R. Springston, P. A. Peaden, J. C. Fjeldted, and M. L. Lee, *Anal. Chem. 53*: 407A–414A (1981).
11. R. D. Smith, J. L. Fulton, R. C. Petersen, A. J. Kopriva, and B .W. Wright, *Anal. Chem. 58*: 2057–2064 (1986).
12. M. C. H. Tavares, J. H. Y. Vilegas, and F. M. Lanças, *Phytochem. Anal.* (in press).

Fernando M. Lanças
M. C. H. Tavares

Supercritical Fluid Chromatography with Evaporative Light-Scattering Detection

The evaporative light-scattering detector (ELSD) was originally developed for use with high-performance liquid chromatography (HPLC) to detect nonvolatile compounds by mass rather than ultraviolet (UV) absorbance detection [1]. The response is dependent on the light scattered from particles of the solute remaining after the mobile phase has evaporated and is proportional to the total amount of the solute. Because no chromophore is necessary, a response can be measured for any solute less volatile than the mobile phase.

Although ELSD is considered a universal detector for HPLC [2], there are additional advantages obtained from coupling ELSD to packed column supercritical fluid chromatography (SFC). SFC provides better selectivity and faster analysis times over HPLC as a result of the low-viscosity and high solute diffusion coefficients characteristic of supercritical fluids [3]. Detection limits for some solutes are improved using ELSD with SFC relative to HPLC [4]. In order to increase solvating power and improve peak shape, CO_2 is often modified with a polar organic solvent such as methanol [5,6]. Using this binary fluid allows for improved separation efficiency of compounds having a wide range of polarities that may otherwise require a buffer.

When compared to other mass-sensitive detectors such as flame ionization (FID), refractive index (RI), and mass spectrometry (MS), the ELSD can detect analytes without interference from organic modifiers and additives. The use of organic solvents in FID limits usefulness due to an increase in baseline noise, and FID cannot be used with HPLC. RI detectors, in general, are less sensitive than other detectors and are incompatible with gradient elution. Although the MS can be used with modifier gradients, ionization efficiencies can vary over orders of magnitude depending on the solute and mode of ionization. The ELSD is more practical than conventional UV detectors because solutes lacking in UV-absorbing chromophores can be directly detected without any sample derivatization or pretreatment. Baseline disturbances due to absorption of the mobile-phase solvents are not observed with ELSD. However, solvents containing trace levels of impurities and columns with low bleed characteristics must be employed for high-sensitivity work. The ELSD can be optimized to generate a narrow range of response factors to components within a structural class, and the use of appropriate standards would allow for quantitative analysis of these compounds.

Because organic solvent gradients do not interfere with ELSD, the detector is an ideal choice for coupling with SFC. A wide range of SFC–ELSD biomedical and pharmaceutical applications have demonstrated higher sensitivity, shorter analysis times, and better separation efficiencies with SFC than HPLC. Compounds without UV chromophores, such as carbohydrates and ginkgolide extracts, have been reported by Lafosse et al. [2], Carraud et al. [4], and Strode et al. [7] using SFC modifier gradients. More efficient baseline separations of these compounds were achieved by SFC–ELSD than with HPLC, and no time-consuming derivatization steps were necessary. The analysis of triglycerides using SFC–ELSD, which required a polar organic modifier to elute, yielded a significant increase in sensitivity over HPLC–ELSD [4]. Underivatized amino acids were also effectively separated by SFC–ELSD [3]. A more complete review of the various SFC–ELSD interfaces and applications was recently published by Lafosse [2].

Evaporative light-scattering detection response was found to have an exponential relationship to the mass of the solute by the equation:

$$A = am^b \tag{1}$$

where A is the peak area of the ELSD signal, m is the solute mass, and a and b are constants which depend on the nature of the mobile phase and of the solutes [8]. Because the peak area response is proportional to the amount of solute, a linear response would be more desirable if quantitation of sample components is required. Linearity can be achieved by plotting calibration curves on a log–log scale as in Eq. (2):

$$\log A = b \log m + \log a \tag{2}$$

The ELSD detector response is influenced by the functions of its three main units: the nebulizer, drift tube, and light-scattering cell. As the mobile phase passes through the nebulizer, it becomes dispersed by a flow of carrier gas such as nitrogen and forms an aerosol. The resultant droplets vary in size depending on factors including the flow rate of the nebulizer gas and the geometry of the nebulizer [6]. The droplets then travel through a heated drift tube where the mobile phase is evaporated, leaving behind only unsolvated particles. Upon exiting the drift tube, the solute particles enter a detection chamber and pass through a beam of light from either a polychromatic (tungsten lamp) or a monochromatic (laser) source. The

(a)

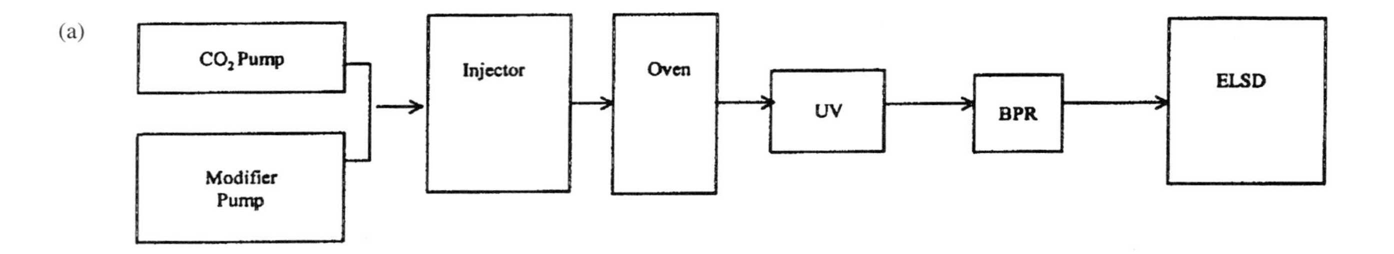

(b)

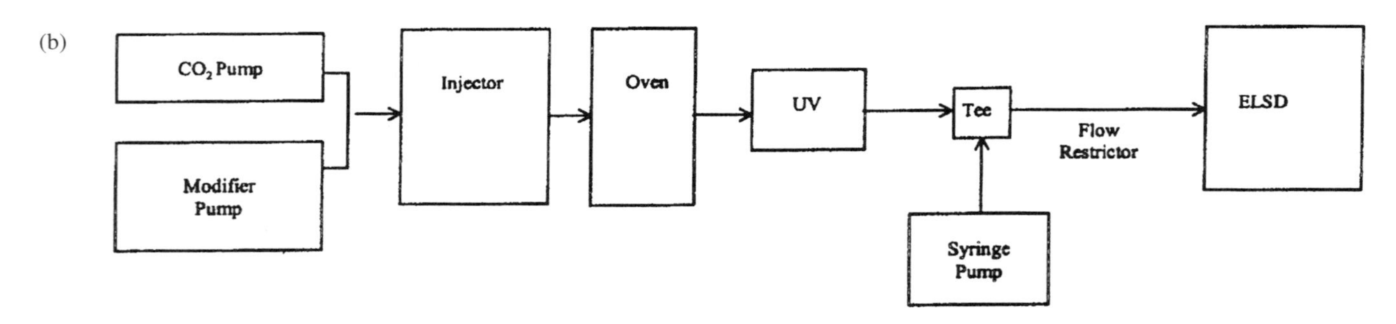

Fig. 1 Two common arrangements for SFC–ELSD coupling: (a) shows pressure control by a back-pressure regulator (BPR), and in (b), the pressure is regulated by a makeup fluid delivered by a pressure-controlled syringe pump. (From Ref. 12.)

light is scattered and a photomultiplier or photodiode detector, which measures the light intensity, produces a chromatographic signal. Refer to Ref. 1 for a more detailed discussion of the principles of light scattering.

When coupling a low-pressure detector such as the ELSD with SFC, detection takes place at atmospheric pressure, usually downstream of the back-pressure regulator [2]. Figure 1a shows a common SFC–ELSD interface with downstream pressure control. Factors affecting ELSD response in this configuration include nebulizer design, evaporation conditions, carrier gas flow rate, and the use of makeup fluid.

Most commercial ELSDs employ a standard or modified HPLC nebulizer (Venturi flow type). It was believed that this nebulizer was not necessary for SFC because nebulization of the SFC mobile phase is accomplished by gas expansion in a restrictor which controls pressure and mobile-phase flow rates. To counter the cooling effects of CO_2 decompression in the linear fused-silica restrictor and improve heat transfer, Nizery et al., using a Cunow Clichy Model DDL 10 detector, placed the restrictor tip into a heated brass ring and applied heat to a small section of tubing between the restrictor and the drift tube [9]. They found that baseline noise resulting from the formation of ice crystals decreased and the performance of the ELSD was unaffected.

The droplet sizes formed and the flow rate of the particles in the drift tube are influenced by the design of the nebulizer. In order to maintain a constant nebulization, droplet sizes should not be too large making them difficult to evaporate or too small where solute vaporization could occur. This requires sufficient liquid and a carrier gas (usually nitrogen). Carraud et al. replaced a conventional ELSD nebulizer (Cunow Clichy Model DDL 10) with a short fused-silica restrictor and determined that the ELSD signal response was dependent on the CO_2 flow rate, although this was later disproved [4].

Larger particle sizes produce higher intensities of scattered light. In order to obtain maximal ELSD sensitivity, evaporator temperatures must be sufficient to allow for the formation of appropriately sized particles. A loss in sensitivity is observed if temperatures are too high because smaller particles may result from sublimation of some compounds. Upnmoor and Bruner [10] studied the effects of varying evaporator temperature on ELSD sensitivity and found that the optimal range of temperatures was between 40°C and 70°C. At lower temperatures, longer residence times required in the drift tube produced peak broadening as well as an increase in baseline noise.

The flow rate of the carrier gas (usually N_2) influences the residence time of the sample in the light-scattering chamber. Low gas flows may allow solute bands to broaden as they travel in the drift tube to the detector. Strode and Taylor [11] observed a decrease in ELSD signal with an increase in the carrier gas flow rate. However, the increase in gas flow improved peak width compared to that observed with the UV detector. It was later found that the total flow of gas (carrier gas plus the decompressed CO_2) through the detector influenced the signal response and the peak width [11].

Most SFC–ELSD instruments employ a direct connection of the outlet of a back-pressure regulator to the detector inlet, as outlined in Fig. 1a. By operating in this manner, peak broadening in the transfer line between the back-pressure regulator and the detector may occur. Additionally, the pressure decrease in the transfer line may affect the strength of the mobile phase and, thus, the ability of the solutes to become completely solubilized. Pinkston bypassed the back-pressure regulator with a postcolumn tee that introduced a makeup fluid such as methanol from a high-pressure syringe pump under pressure control [12]. A fused-silica linear restrictor at the ELSD inlet maintained the pressure and was regulated by the flow of the makeup fluid, as shown in Fig. 1b. Pinkston theorized that this method of pressure control would prevent mass-transfer problems that diminish detector sensitivity and decrease the dependence of the ELSD response on mobile-phase composition [12]. The flow of makeup solvent enhanced the solubility of the analytes in the mobile phase. Additionally, the efficiency of forming appropriately sized particles in the ELSD was improved, generating better peak shapes and higher signal-to-noise ratios.

In conclusion, an ELSD with SFC provides a sensitive analytical tool for qualitative and quantitative analysis of solutes. Detection depends only on the solute being less volatile than the least volatile mobile-phase component. Detection is independent of the basicity or presence of a chromophore for a given solute. The detector response is a logarithmic function of the mass of the solute. The SFC–ELSD combination should be considered whenever a universal high-throughput analysis is needed.

References

1. J. T. B. Strode, L. T. Taylor, K. Anton, M. Bach, and N. Pericles, *Supercritical Fluid Chromatography with Packed Columns: Techniques and Applications*, Marcel Dekker, Inc., New York, 1997, pp. 97–123.
2. M. Lafosse, Evaporative light scattering detection in SFC, *Chromatogr. Princ. Pract.* 201–218 (1999).
3. M. Lafosse, C. Elfakir, L. Morin-Allory, and M. Dreux, *J. High Resolut. Chromatogr. 15*: 312–318 (1992).
4. P. Carraud, D. Thiebaut, M. Caude, R. Rosset, M. Lafosse, and M. Dreux, *J. Chromatogr. Sci. 25*: 395–398 (1987).
5. T. A. Berger, *Packed Column SFC*, The Royal Society of Chemistry, Cambridge, 1995.
6. M. Dreux and M. Lafosse, *LC-GC Int. 10*: 382–390 (1997).
7. J. T. B. Strode, L. T. Taylor, and T. A. van Beek, *J. Chromatogr. A 738*: 115–122 (1996).
8. M. Dreux, M. Lafosse, and L. Morin-Allory, *LC-GC Int. 9*: 148–153 (1996).
9. D. Nizery, D. Thiebaut, M. Caude, R. Rosset, M. Lafosse, and M. Dreux, *J. Chromatogr. 467*: 49–60 (1989).
10. D. Upnmoor and G. Brunner, *Chromatographia 33*: 255–260 (1992).
11. J. T. B. Strode III and L. T. Taylor, *J. Chromatogr. Sci. 54*: 261–270 (1996).
12. T. L. Chester and J. D. Pinkston, *J. Chromatogr. A 807*: 265–273 (1998).

Christine M. Aurigemma
William P. Farrell

Supercritical Fluid Chromatography with Mass Spectrometric Detection

In recent years, supercritical fluid chromatography (SFC) has been exploited as an alternative to high-performance liquid chromatography (HPLC) because of its superior speed and enhanced selectivity for most organic compounds, with the exception of highly polar species. The higher diffusion coefficient and lower viscosity of the SFC mobile phase, primarily consisting of condensed CO_2, permit faster run times than HPLC with longer columns and, thus, higher plate counts. These properties, combined with the advantages of mass spectrometry (MS) instrumentation used as chromatographic detectors, give rise to interest in coupling SFC with MS. When one considers the chromatographic and mass spectrometric advantages inherent with such an interface, SFC–MS is an attractive alternative to LC–MS for many applications.

Considerations of primary importance in developing an interface for SFC–MS are (a) preserving performance of the ionization source and mass spectrometer, (b) preserving integrity of the chromatography, (c) minimizing thermal degradation, and (d) a rugged, simple design. The "guiding hand" behind the design of the interface is the type of mass spectrometer ionization source employed.

Vacuum ionization sources such as electron-impact (EI) or chemical ionization (CI), commonly used for gas chromatography–mass spectrometry (GC–MS), were initially used when the first SFC–MS interfaces were assembled. By EI, a high degree fragmentation occurs, usually obscuring the analyte molecular ion, but potentially elucidating some structural information. EI spectra for many compounds are well known and are stored in the National Bureau of Standards library. In CI, a gas-phase charge transfer with a reagent gas in the source ionizes solutes in a softer fashion than EI, usually producing significant levels of MH^+, MH^-, or M+ charged adduct ions. In either EI or CI, more than a minimal gas load on the source is not tolerable. For SFC–MS with EI or CI, low source pressure is maintained using a low flow rate or split-flow direct fluid introduction (DFI), sometimes termed DLI (direct liquid introduction) [1]. Alternatively, differential pumping of the effluent from a restrictive nozzle is feasible as used with the molecular beam interface employed in some early SFC–MS experiments [2]. The DFI interface includes an isothermal union of a transfer line and column, attached to a heated fixed restrictor to counteract cooling from the adiabatic expansion of CO_2 prior to introduction of the effluent into the ionization source. Performance of such systems is compromised by adverse effects of gradient conditions on ionization and ion transmission. Increased pressure in the ion source resulting from high flow rates reduces ionization efficiency by EI and CI, thus limiting such experiments to low-polarity species, which do not require high levels of polar organic modifier to elute by SFC.

Complex interfaces were also designed to eliminate the gas load on the mass spectrometer from the SFC effluent. Particle beam and moving-belt interfaces originally used for LC–MS were employed with some success, due to their capacity to limit the effect of the mobile-phase on the MS analysis. The particle beam interface generally employs two momentum separators to remove volatilized mobile-phase molecules from the solute beam entering the EI or CI source [3]. Although spectral interferences and chemical noise levels are reduced, the interface suffers from reduced efficiency due to losses of more volatile components in the momentum separator and nonlinear response factors, especially for more nonvolatile compounds. The high-flow-rate moving-belt interface functions by transporting adsorbed solutes on the belt while eliminating more volatile mobile-phase molecules that do not adhere or thermally desorb from the belt prior to introduction to the ionization region [4]. Problems associated with moving-belt interfaces include losses of volatile or thermally labile solutes, and for some solutes, poor sensitivity and carryover due to inefficient desorption from the belt. In both cases, sensitivity is compromised due to inefficient mass transport and thermal degradation.

Thermospray interfaces, originally designed for constant vapor pressure conditions, have nonetheless been utilized for SFC–MS by several researchers [5]. The column effluent is vaporized in a heated fixed restrictor probe in which solutes are ionized by mobile-phase-mediated chemical ionization (e.g., with ammonium acetate), producing CI-like spectra (prevalent in MH^+ or MH^- ions). The ionization process is sometimes assisted by an electron current beam across the probe outlet into the source vacuum region, producing EI-like spectra. The resulting beam of ions is then deflected by an electric potential into the mass analyzer region. Recently, themospray usage for SFC–MS has declined in prominence due to low sensitivity from mass-transport inefficiency to the analyzer or thermal degradation, and it suffers from poor reproducibility in general. The recent alternative of atmospheric pressure ionization (API) sources has slowed further improvements to the thermospray interface.

The recent success of SFC–MS using API sources have contributed to a general decline in the use of many interface designs described earlier. API sources include electrospray ionization (ESI) and atmospheric pressure chemical ionization (APCI). A general schematic illustrates the basic form of the interface of SFC to an API source (Fig. 1). Along with providing usually unambiguous mass spectral identification (MH^+ or MH^- signals predominate), API sources for SFC–MS provide attractive advantages for the analytical chemist. Many commercially available interfaces originally designed for LC–MS require minimal modification to effectively couple an SFC to API–MS systems. Another advantage of such interfaces is the "self-volatilizing effect" provided by CO_2 in nebulization of the mobile-phase solutes.

Electrospray ionization occurs in the solution phase prior to vaporization at atmospheric pressure in the mass spectrometer source. Ions are desolvated in the strong

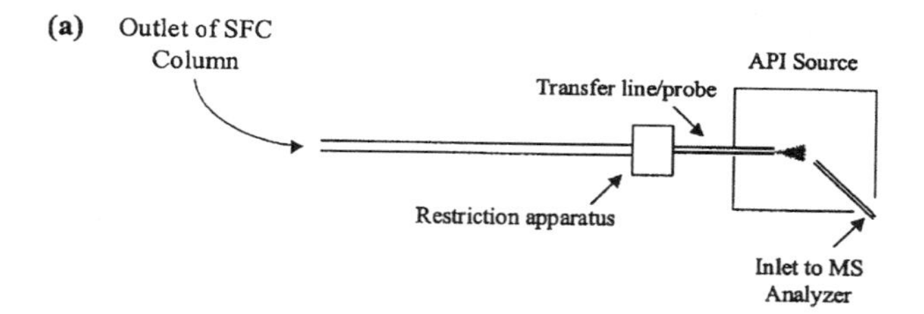

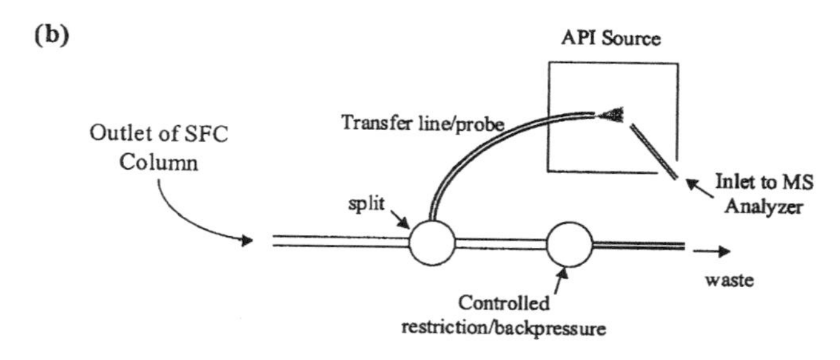

Fig. 1 SFC–MS interfaces.

electric potential gradient from the needle tip to the inlet of the mass analyzer. In the early period of SFC–ESI-MS development, Sadoun et al. [6] constructed an interface using a 25–30-cm-long fused-silica restrictor [25 or 60 μm inner diameter (i.d.)] with one end attached to the outlet of a packed column and the other end to the source (see Fig. 1a). The last 10 cm of silica was coated with a layer of conductive nickel paint to simulate an electrospray needle. Employing a gradient with methanol–water as the polar modifier, detection limits in the low picogram range were reported; however, ionization efficiency was significantly affected by the mobile-phase composition and severe tailing for low-volatility compounds was observed. Some of these deficiencies were addressed by Pinkston and Baker [7] using a sheath liquid of methanol–water–ammonium acetate with an ion spray (or pneumatically assisted electrospray) interface for open-tubular or packed column SFC–MS. In the latter application, a syringe pump supplied methanol through a tee between the mobile-phase outlet and the transfer line to provide back-pressure regulation without a variable restrictor valve (Fig. 1a). Advantages of this interface included reduced tailing and lowered modifier composition requirements in the gradient, allowing for higher flow rates and shorter retention times. Sjöberg and Markides described a similar sheath liquid-type ESI interface also convertible to an APCI interface [8].

Atmospheric pressure chemical ionization is similar to ESI in that both ionization processes occur at atmospheric pressure. Ionization by APCI is fundamentally different, however, in that nebulized solute molecules encounter a plasma of protons, ions, and electrons generated by the corona discharge ionization of background N_2, H_2O, or methanol, and ionize either by proton transfer or charge exchange. The proton-transfer agents, usually water or methanol, are supplied by saturating the N_2 nebulizing gas or using the SFC mobile-phase polar modifier, again usually methanol. The first report of packed column SFC with APCI by Huang et al. [9] described an interface in which the back-pressure regulator was bypassed and the restrictor was a 20-μm-i.d. stainless-steel pinhole diaphragm. The effectiveness of this interface was limited by inadequate heating in the nebulization region, resulting in chromatographic tailing for more involatile substances. This problem was addressed by the interface designed by Tyrefors et al. [10]. This interface was constructed to maintain the temperature of the mobile phase in the column up to the tip of the restrictor and provide consistent intense vaporization conditions somewhat independent of flow rate. The outer tube was actively insulated with a heating coil while a coaxial nebulizing gas flowed around the restrictor tube into which a fine heating wire was inserted.

An SFC–APCI interface style employed by Morgan et al. [11] was constructed and tested independently on three commercial mass spectrometers equipped with API inlet systems. Comparison studies were performed to test the ideal dimensions (length and i.d.) of the PEEK™ transfer tube as well as the fused-silica used in the APCI source inlet. Significantly, neither the PEEK tubing length nor its diameter below 1 mm had any chromatographic implications. The fused-silica diameter impacted chromatographic integrity as diameters above 100 μm gave rise to extreme peak deformation and tailing. The optimal dimensions of the tubing were shown to be independent of the MS source on test. Another APCI interface design reported by Ventura et al. [12] employs a postcolumn (post-UV detector) split upstream of the variable restrictor in the style of some DFI interfaces (Fig. 1b). A 50-μm-i.d. fused-silica capillary carries the split flow from the tee to the exit tip of the APCI needle housing, whereupon expansion and nebulization occur. This method reduces delay between UV and MS signal detection while maintaining chromatographic integrity for MS peaks. An orthogonal analyzer inlet allows flow rates up to 6 mL/min to be used routinely with methanol modifier compositions as high as 60%. The interface is effective for applications in which trace detection is not required.

Further information may be obtained from reviews on SFC–MS, including those by Pinkston and Chester [13], Arpino and Haas [14], and Combs et al. [15].

Two general styles for interfacing SFC to API sources are depicted in Fig. 1. In Fig. 1a, the SFC column effluent is directed entirely into the API source. In some systems, the restriction apparatus is a simple transfer tube restriction used for low-flow experiments. In others, a variable restrictor or back-pressure regulator is used between the outlet and the source. In the configuration used by Pinkston and Baker [7], the restriction apparatus is a series of tees to add a coaxial flow of nebulizing gas, electrospray buffer sheath flow, and another to introduce liquid from a syringe pump to regulate mobile-phase pressure. In Fig. 1b, a direct interface is shown in which a split directs a fraction of the effluent flow toward the API source while the remainder is sent to waste through a back-pressure regulator or some controlled restriction to maintain system pressure.

References

1. L. G. Randall and A. L. Wahrahaftig, *Anal. Chem. 50*: 1703 (1978).
2. L. G. Randall and A. L. Wahrahaftig, *Rev. Sci. Intrum. 52*: 1283 (1981).
3. P. O. Edlund and J. D. Henion, *J. Chromatogr. Sci. 27*: 274 (1989).
4. A. J. Berry, D. E. Games, and J. R. Perkins, *J. Chromatogr. 363*: 147 (1986).
5. J. Via and L. T. Taylor, *Anal. Chem. 66*: 1385 (1994).
6. F. Sadoun, H. Virelizier, and P. J. Arpino, *J. Chromatogr. 647*: 351 (1993).
7. J. D. Pinkston and T. R. Baker, *J. Am. Soc. Mass Spectrom. 9*: 498 (1998).
8. P. J. R. Sjöberg and K. E. Markides, *J. Chromatogr. A 785*: 101 (1997).
9. E. Huang, J. D. Henion, and T. R. Covey, *J. Chromatogr. 511*: 257 (1990).
10. L. N. Tyrefors, R. X. Moulder, and K. E. Markides, *Anal. Chem. 65*: 2835 (1993).
11. D. G. Morgan, K. L. Harbol, and N. P. Kitrinos, Jr., *J. Chromatogr. A 800*: 39 (1998).
12. M. C. Ventura, W. P. Farrell, C. M. Aurigemma, and M. J. Greig, *Anal. Chem. 71*: 4223 (1999).
13. J. D. Pinkston and T. L. Chester, *Anal. Chem. 67*: 650A (1995).
14. P. J. Arpino and P. Haas, *J. Chromatogr. A 703*: 479 (1995).
15. M. T. Combs, M. Ashraf-Khorassani, and L. T. Taylor, *J. Chromatogr. A 785*: 85 (1997).

Manuel C. Ventura

Supercritical Fluid Chromatography with Nitrogen Chemiluminescence Detection

Supercritical fluid chromatography (SFC) with nitrogen chemiluminescence detection (NCD or CLND) is a chromatographic technique that can be used to obtain quantitative results for nitrogen-containing analytes without the use of matching standards. For industries where sample throughput is an issue, such as combinatorial chemistry, SFC–NCD could reduce the need to isolate large numbers of standards and improve the speed of chromatographic characterization of synthetic reaction products. This entry will discuss the fundamental parameters that need to be considered to make SFC and NCD interfacing successful. Two other forms of nitrogen detection used with SFC will also be mentioned, but they fall outside the scope of this entry.

Supercritical fluid chromatography utilizes a compressed or dense gas as a mobile phase (i.e., carbon dioxide) and has been shown to be a powerful separation technique. The low viscosity and high diffusivity of supercritical fluids allow for the use of longer columns than in high pressure liquid chromatography (HPLC) without increasing runtimes. The increased number of theoretical plates available, combined with optimum linear velocities (μ_0) three to five times greater than in LC, can provide more separations per unit time. Berger has written a more detailed description of the functionality of packed column SFC [1].

Supercritical fluid chromatography has some of the same characteristics of both HPLC and gas chromatography (GC). Packed column SFC uses the same column technology as HPLC, and when used with binary or tertiary solvents, has a broad range of applicability [1]. This range is much broader than GC, because compounds need not be volatile or thermally stable. As in GC, SFC can be coupled to most modern chromatographic detectors, such as element-specific detectors. These detectors are often very selective for the element under study, with detection limits below parts per billion [1]. Coupled to an SFC, an element-specific detector can become a powerful tool that offers selectivity, sensitivity, and speed.

Two other types of element-specific detector for nitrogen currently in use coupled to SFCs are the nitrogen phosphorus detector (NPD) and the thermal energy analyzer (TEA). The NPD uses a hot, catalytically active solid surface immersed in a layer of dissociated H_2 and O_2 to form electronegative N and P ions which are detected on a nearby electrode [2]. NPD has been shown to have broad application in SFC, especially in the agrochemical industry [3]. The TEA, as described by Fine et al. [4], uses low-temperature pyrolysis, followed by ozone-induced chemiluminescence, for the detection of compounds containing NO_2 groups. The TEA has been used for the determination of tobacco-specific nitrosamines and explosives [5]. Both of these detectors require specific standards of the analytes of interest for quantitation

The use of NCD, especially in pharmaceutical applications, have focused on exploiting the chemiluminescence properties of nitrogen dioxide (NO_2) formed from the high-temperature combustion of nitrogenous compounds [6–8]. This type of detector, originally developed for GC by Parks and Marietta [9], forms nitric oxide followed by ozone-induced chemiluminescence. The reaction is characterized as follows:

$$\begin{aligned} R{-}N + O_2 &\rightarrow NO + \text{or other compounds} \\ NO + O_3 &\rightarrow NO_2^* + O_2 \\ NO_2^* &\rightarrow NO_2 + \text{light } (600\text{–}3200 \text{ nm}) \end{aligned} \quad (1)$$

Excited-state nitrogen-dioxide-mediated chemiluminescence occurs on a 1:1 molar ratio with solute nitrogen, provided that the combustion products form NO (and not N_2). With complete combustion of the solute, the NCD can directly provide a quantitative measure of the amount of nitrogen in a chromatographic peak. This means that a single calibration curve using a stable nitrogen-containing compound can be used to quantify nitrogen content of any unknown peak in a chromatogram [6]. With a linear range of at least three orders of magnitude and picogram detection limits, SFC–CLND can provide an efficient technique for the quantification of nitrogen-containing compounds [6].

The basic setup of a NCD system is shown in Fig. 1. In almost all cases, only a portion of the entire SFC mobile phase is diverted to the CLND detector using a fused-silica capillary or restrictor [6–8,10,11]. Use of a restrictor minimizes the effects of the SFC decompressed carbon dioxide (CO_2) and solvent composition on the pyrolysis reaction and chemiluminescence. The CO_2 flow rate, dictated largely by the SFC outlet pressure, can affect the residence time of the solute in the pyrolysis chamber and the efficiency of the ozone reaction. The addition of modifiers to the SFC mobile phase (e.g., methanol) can compete for the available pyrolysis oxygen and can reduce signal response. High concentrations of modifier can dramatically affect the detection limits of some compounds [7]. Moreover, the sample concentration also appears to be limited by competition for oxygen, resulting in incomplete combustion.

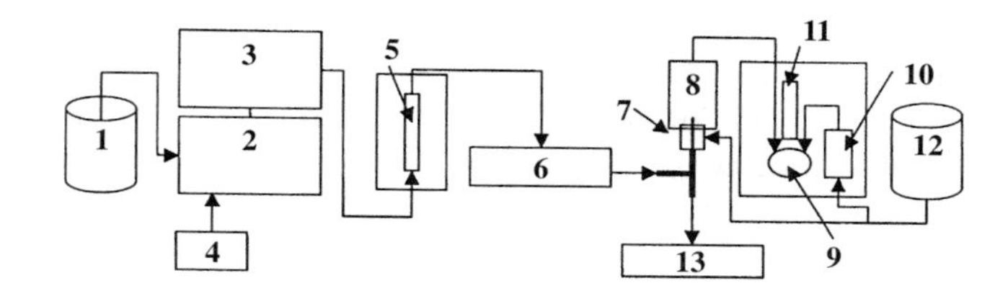

Fig. 1 Schematic diagram of SFC–CLND system: (1) CO_2 tank, (2) SFC pumping system, (3) autosampler, (4) modifier pump, (5) column, (6) UV detector, (7) pyrolysis inlet with linear restrictor, (8) pyrolysis chamber, (9) reaction cell, (10) ozone generator, (11) photomultiplier tube, (12) O_2 tank, (13) back-pressure regulator.

The parameters that need special consideration for the successful coupling on an SFC to a CLND are restrictor position, ozone generation, combustion efficiency (temperature and oxygen flow rate), and quenching. Each of these will be discussed in detail.

Restrictor Position

An important aspect for maintaining peak integrity using any split flow into a detector is the position of the restrictor tip relative to the detector source. Because SFC utilizes a gas under pressure, the rapid release of that pressure causes localized adiabatic cooling at the restrictor tip. On the other hand, at high temperatures, the mobile phase expands while the fluid density drops dramatically, resulting in precipitation of solutes at the end of the restrictor [1]. Careful attention to the restrictor placement should eliminate solute precipitation, maintain peak integrity, and provide sufficient mixing with oxygen for combustion. Several articles have discussed optimized placement of the restrictor for the Antek 705D CLND [6, 7,10]. Inserting the restrictor beyond the inlet oxygen port to the pyrolysis chamber appears to provide both adequate solute nebulization and mixing prior to introduction into the furnace [7].

Ozone Generation

Because ozone is very expensive to obtain in pure form and is quite unstable over long periods of time, CLND detectors utilize an ozone generator. Some of the oxygen flow is diverted over an electric arc to form ozone, which is then fed directly into the reaction cell. Because chemiluminescence occurs on a 1:1 molar ratio with NO, as described in Eq. (1), there must be excess ozone for chemiluminescence of NO independent of the other combustion products. Inlet oxygen flows of 5–10 mL/min appear to be adequate for most of the SFC–CLND experiments reported [6,10,12].

Combustion Efficiency

Because detection is based on the formation of NO, the combustion conditions must be sufficient to completely convert the nitrogenous analytes to NO in the presence of mobile-phase modifiers. Temperatures ranging from 1050°C to 1100°C appear to provide the best environment for combustion. The amount of available oxygen for combustion is also highly dependent on the application. Table 1 outlines some of the conditions that have been reported. These conditions cover a wide range of mobile-phase flow rates (as decompressed CO_2) and modifier percentages. Shi et al. reported that too high an oxygen flow may reduce the residence time of the solute in the pyrolysis chamber and can negate any increase in combustion efficiency [7].

Quenching

Nitrogen chemiluminescence takes place in a reaction cell, under vacuum, where the emitted light is measured using a photomultiplier tube. The primary combustion products from a methanol-modified SFC mobile phase are CO_2 and H_2O. Excessive CO_2 and water can increase the total number of colliding molecules in the reaction cell and reduce the signal. The addition of a membrane drier to selectively remove water has been shown to increase the detection limit of sulfamethazine by an order of magnitude [11].

Careful attention to the above-described parameters should allow for the successful interfacing of a CLND detector to an SFC system. Combining the separation efficiency of SFC with the specificity and sensitivity of CLND results in a powerful analytical technique for the quantification of nitrogenous solutes.

Table 1 Reported Oxygen Flow Rates for Several SFC–CLND Applications

CO_2 flow rate (mL/min)	Restrictor inner diameter (mm)	Modifier (%)	Oxygen flow (mL/min)	Ref.
3	0.025[a]	0	48	5
1200	0.025	8	185–200	11
600	0.075	10	50	3
150	0.075	15	50	4

[a]Open-tubular SFC.

References

1. T. A. Berger, *Packed Column SFC*, Royal Society of Chemistry, Cambridge, 1995.
2. P. Patterson, *Detectors for Capillary Chromatography*, John Wiley & Sons, New York, 1992, Chap. 7.
3. T. A. Berger, *Chromatographia 41*(7/8), (October 1995).
4. D. H. Fine, F. Rufeh, and B. Gunther, *Anal. Lett. 6*: 731 (1973).
5. E. S. Francis, D. J. Eatough, and M. L. Lee, *J. Microcol. Separ. 6*: 395 (1994).
6. H. Shi, J. T. B. Strode III, L. T. Taylor, and E. M. Fujinari, *J. Chromatography A 757*: 183–191 (1997).
7. H. Shi, J. T. B. Strode III, L. T. Taylor, and E. M. Fujinari, *J. Chromatography A 734*: 303–310 (1996).
8. H. Shi, J. T. B. Strode III, L. T. Taylor, and E. M. Fujinari, *Instrumental Methods in Food and Beverage Analysis*, Elsevier, Amsterdam, 1998.
9. R. E. Parks and R. L. Marietta, U.S. Patent 4,018,562 (24 October 1975).
10. J. T. B. Strode III, T. P. Loughlin, T. M. Dowling, and G. R. Bicker, *J. Chromatogr. Sci. 36*: 511–515 (1998).
11. M. T. Combs, M. Ashraf-Khorassani, and L. T. Taylor, *Anal. Chem. 69*: 3044–3048 (1997).
12. H. Shi and L. T. Taylor, *J. High Resolut. Chromatogr. 19*: 213–216 (1996).

William P. Farrell

Surface Phenomena in Sedimentation FFF

Introduction

Field-flow fractionation (FFF) presents a unique method where particles move in a liquid flow, maintaining a quasi-equilibrium Boltzmann transverse concentration distribution in an FFF channel [1]. It allows one to obtain, from experiments, the transverse Peclet number Pe defining the thickness of the layer, where particles are accumulated, and the retention of the FFF process Ret:

$$\mathrm{Pe} = \frac{bW}{D} \tag{1}$$

where b is the generalized mobility of a particle, a droplet, or a macromolecule (the velocity in the unit gradient of the transverse potential), W is half of the potential drop across the FFF channel (assuming that the transverse particle velocity over the FFF channel is constant), and D is the particle diffusion coefficient. If the transverse Peclet number is higher than about 5, we can write [1]

$$\mathrm{Ret} = \tfrac{1}{3}\mathrm{Pe} \tag{2}$$

One can predict the parameter Pe by measurement of the diffusion coefficient D or by deriving it from the Stokes expression [2–4] (for the spherical or ellipsoidal particles), derivation of the b value, and measure or calculate the transverse potential drop. In sedimentation FFF (SdFFF), the common way to derive the transverse Peclet number Pe is the substitution of the gravity or sedimentation force acting on the particle,

$$F = \Delta\rho v G \tag{3}$$

where $\Delta\rho$ is the density difference between the particle and the carrier liquid, v is the particle volume, and G is the acceleration of the centrifugal or gravitational force, into the standard expression for the transverse Peclet number. However, for small particles being the objects of SdFFF with sizes ranging from 10^{-6} to 10^{-4} cm [1], this approach may fail, because it does not take into account phenomena occurring in the surface layer of the particle and causing a liquid flow there. In electrolytes, a so-called sedimentation electrostatic potential arises as a consequence of these surface phenomena [2]. To have the complete picture of the particle sedimentation, one should to derive the particle mobility in a force field, accounting for the phenomena near the particle surface.

How Surface Effects Affect the Particle Sedimentation

As the actual theory shows, the "surface-driven" movement may be caused by the flow of a carrier liquid in the surface layer, where the surfactant ions or molecules are accumulated due to action of the surface potential. The surfactant accumulation leads to an increase of the osmotic pressure in the surface layer. If the gradient of such an osmotic pressure arises, it causes a liquid flow in the surface layer. This "slipping" flow in a surface layer can cause either phoretic (surface-driven) movement of a particle or the osmotic flow of a liquid (when the solid surface is fixed) [2,3]. Gradients of the osmotic pressure in the particle surface layer may be due to the macroscopic gradients of surfactant concentration or temperature established in the FFF channel. If the particle movement is due to the "body" force, similar to the centrifugal one, a gradient of the osmotic pressure also may arise. This may be caused by the intensive transport of surfactant molecules or ions along the particle surface, which is due to its sedimentation movement and should be compensated by the exchange of these molecules or ions between the surface layer and the outer liquid. As a consequence, the longitudinal gradient of the concentration may arise, which leads to the osmotic pressure gradient and, in turn, to the "slipping" of the liquid in the surface layer. This mechanism may add surface-driven movement to the "body-driven" sedimentation movement of the particle.

Slip-Flow Velocity in the Surface Layer

The theory of the surface kinetic phenomena is based on the Navier–Stokes equations for the liquid in the surface layer [2,3]:

$$\eta\frac{\partial^2 u_z}{\partial y^2} = \frac{\partial p_0}{\partial z} + \frac{\partial p_{\mathrm{osm}}}{\partial z} \tag{4}$$

where $u_z(y)$ is the tangential flow velocity profile in the surface layer at a distance y from the particle surface, η is the liquid velocity, p_0 is the pressure excess due to sedimentation movement of particle and the corresponding hydrodynamic viscous stresses, p_{osm} is the osmotic pressure, and z is the longitudinal coordinate on the particle surface. For a spherical particle $z = R\vartheta$, where ϑ is the angle between the direction of the vector $\vec{F}$ given by Eq. (3) and the radius vector $\vec{r}$, and R is the particle radius. On the outer surface of the surface layer, the surfactant concentration is $c_{\mathrm{os}}(y, z)$, which may differ from the surfactant concentration in a liquid far from the particle c_0 due to concentration polarization. The surfactant concentration distribution in the surface layer is the Boltzmann distribution, which allows us to write the pressure distribution in the surface layer as

$$p = p_0 + kTc_{\mathrm{os}}\left[\exp\left(-\frac{\Phi}{kT}\right) - 1\right] \tag{5}$$

The first boundary condition to Eq. (4) is the absence of the liquid slipping on the particle surface:

$$u_z = 0 \quad \text{at } y = 0 \tag{6}$$

The second boundary condition is related to the transformation of the flow profile in the surface layer into the flow profile for the Stokes problem on the flow around a sphere outside the thin surface layer near the particle surface [4] and can be written as

$$\left.\frac{\partial u_z}{\partial y}\right|_{r=\infty} = \left.\frac{\partial u_\varphi^e}{\partial r}\right|_{r=R} \tag{7}$$

where u_φ^e is the tangential flow velocity in the conventional Stokes problem. Using Eqs. (5)–(7) and the standard expression for the outer pressure p_0 [4], one can solve Eq. (4) and have the boundary conditions for the "outer" Stokes problem, where surface phenomena are accounted for:

$$u_\varphi^e(r = R) = u_s \tag{8}$$

$$u_r^e(r = R) = 0 \tag{9}$$

where u_r^e is the radial flow velocity in the standard Stokes problem and

$$u_s = \frac{kT}{\eta R}\frac{\partial c_{os}(R,\varphi)}{\partial \varphi}\int_0^\infty dy \int_\infty^y dy'\left[\exp\left(-\frac{\Phi(y')}{kT}\right) - 1\right] \tag{10}$$

is the slip flow velocity in the surface layer.

Equations (8)–(10) allow the determination of the velocity U of the particle, which is driven by a "body" force compensated by the hydrodynamic friction force established in consequence [5].

Concentration Polarization of Moving Particle

Surface phenomena affect the particle sedimentation when a tangential surfactant concentration gradient exists near the particle surface. This situation is described by the stationary diffusion equation for the surfactant concentration outside the surface layer in the form $\Delta c_{os} = 0$ [2, 3], where the boundary condition, which reflects the substance conservation in a very thin surface layer, may be written as

$$-D_s\left.\frac{\partial c_{os}}{\partial r}\right|_{r=R} = \int_0^\infty \left[c_0 u_z(y,z) - D_s\left.\frac{\partial c_{os}}{\partial z}\right|_{r=R}\right] \cdot \left[\exp\left(-\frac{\Phi(y)}{kT}\right) - 1\right] dy \tag{11}$$

where D_s is the diffusion coefficient of the surfactant [2]. This diffusion equation is solved using the assumption on the dipole form of the surfactant concentration distribution [2]. For the deep surface potential well ($\varepsilon_0 \geq 5$, where ε_0 is the depth of the surface potential well in kT units), the concentration polarization should be most significant, due to the intensive transport of the surfactant which is highly accumulated in such a well. In this case, we can assume that

$$\frac{\Phi}{kT} \approx -\varepsilon_0\left(1 - \frac{y}{h}\right) \tag{12}$$

where h is the characteristic width of this well. The physicochemical nature of these parameters is discussed in the next section. The general expression for the slip velocity can be written compactly as

$$u_s = \left(\frac{a'}{R} - \frac{3b'}{R^3}\right)\frac{h}{\varepsilon_0 R} f(\delta, \varphi, \mathrm{Rel}) \tag{13}$$

where φ is the volume fraction of the surfactant, $\delta = hR/\varepsilon_0 d^2$ is the reduced parameter characterizing the relationship between the parameters of the surface potential well, the particle, and the surfactant (d is the surfactant molecule diameter), the criterion $\mathrm{Rel} = he^{\varepsilon_0}/\varepsilon_0 R$ characterizes the degree of the concentration polarization, and

$$f(\delta, \varphi, \mathrm{Rel}) = 4\varphi\delta\mathrm{Rel}^2/(2 + \mathrm{Rel} + 4\varphi\delta\mathrm{Rel}^2) \tag{14}$$

is the function characterizing the concentration polarization. The parameters a and b are unknown constants of the Stokes problem [4,5]. This function ranges from zero (at $\mathrm{Rel} = 0$) to one (at very large Rel values). Using Eq. (14) and boundary conditions [Eqs. (8) and (9)], we have, for the particle sedimentation velocity,

$$U = U_0\left(1 - h\frac{f}{\varepsilon_0 R}\right) \tag{15}$$

where $U_0 = F/6\pi\eta R$ is the standard value of the sedimentation velocity. The maximum relative change of the particle sedimentation velocity at very large criterion Rel values is

$$\frac{U - U_0}{U_0} = -\frac{h}{\varepsilon_0 R} \tag{16}$$

Parameters of the Surface Potential Well and Criterion Rel

There are several main mechanisms to accumulate the ions or molecules near the surface [5]: dispersion or van der Waals interaction with the potential $\Phi(y) = -Ad^3/y^3$, where A is the Hamaker constant; Coulomb electrostatic interaction in electrolytes with the potential $\Phi(y) = -q\zeta \exp(-y/\lambda)$, where q is the electric charge of the surfactant ion, ζ is the particle electrokinetic potential, and λ is the Debye length [2]; and the adsorption and structure forces due to structural changes in the surface layer, which have no analytical dependence of the surface potential on the distance but have the parameters ε_0 and h derived from the experimental data (see Table 1).

Thus, Eq. (16), together with the data from Table 1, predict an observable contribution of surface phenomena to the particles' sedimentation velocity for all the types of surface interaction, except the van der Waals interaction. The possible values of criterion Rel range from about

Table 1

Surface potential	Analytical expressions for $\Phi(y)$, ε_0 and h			Ranges of values for ε_0 and h	
	$\Phi(y)$	ε_0	h	ε_0	h
van der Waals forces	$-Ad^3/y^3$	A/kT	$d/3$	5–50[a]	$\approx 10^{-8}$ cm (low-molecular surfactant)
Coulomb electrostatic forces	$-q\zeta e^{y/\lambda}$	$q\zeta/kT$	λ	0–10	$10^{-7} \div 10^{-4}$ cm (aqueous electrolytes)[b]
Adsorption forces	None	None	None	0–10	$\approx 10^{-7}$ cm
Structure forces	None	None	None	0–10	$\approx 10^{-5}$ cm

[a]The maximum values of Hamaker constant are characteristic for metals.
[b]The maximum value of the Debye length is reached in the pure water, where only H^+ and OH^- ions exist.

1000 ($h \approx 10^{-6}$ cm and $\varepsilon_0 = 5$) to about 2000 ($h \approx 10^{-6}$ cm and $\varepsilon_0 = 10$).

The volume fraction of surfactant φ may be estimated, taking into account that the molar concentration of surfactant is usually about 10^{-3}–10^{-2} m/L; that is, the numeric concentration of the surfactant molecules c_0 is about 10^{18}–10^{19} cm^3. For a typical surfactant molecule radius $d = 3 \times 10^{-8}$ cm, we have $\varphi \approx 10^{-4}$–10^{-3}. The values of parameter δ at $\varepsilon_0 = 10$ range from 10 (at $h = 10^{-7}$ cm and $R = 10^{-6}$ cm) to 10^4 (at $h = 10^{-6}$ cm and $R = 10^{-4}$ cm). As the data of Table 1 show, the most of the particles that are the object of FFF should have an apparent degree of polarization. The small particles seem to be strongly polarized, and Eq. (16) should describe, adequately, the contribution of the surface phenomena in the particle sedimentation. For the smaller particles, where the saturation may be reached even at $\varepsilon_0 = 5$, the related velocity change may reach 0.2. Such a change of the sedimentation velocity must be measurable in the real FFF conditions. This difference can be useful in the determination of the role of surface phenomena in sedimentation, and the decrease of the sedimentation velocity with the surfactant concentration also may indicate the significant role of the surface phenomena at moderate values of the criterion Rel.

References

1. M. Martin and P. S. Williams, Theoretical basis of field-flow fractionation, in *Theoretical Advancement in Chromatography and Related Separation Techniques* (F. Dondi and G. Guiochon, eds.), Kluwer, Dordrecht, 1992, pp. 513–580.
2. V. G. Levich, *Physicochemical Hydrodynamics*, Prentice-Hall, Englewood Clifts, NJ, 1962.
3. J. L. Anderson, *Anal. Rev. Fluid Mech. 21*: 61–97 (1989).
4. L. D. Landau and E. M. Lifshits, *Mechanics of Continuous Media*, State Publishing of Technical and Theoretical Literature, Moscow, 1954 (in Russian).
5. S. N. Semenov, *J. Liquid Chromatogr. Related Technol. 20*: 2669–2685 (1997).
6. B. V. Derjagin, *Theory of Stability of Colloids and Thin Films*, Nauka, Moscow, 1986 (in Russian).

S. N. Semenov

Surfactants: Analysis by HPLC

Introduction

Surfactants are widely used for a variety of reasons, including surface wetting agents, detergents, emulsifiers, lubricants, gasoline additives, and enhanced oil-recovery agents. The type of surfactants selected for a particular application often depends on the chemical and physical properties required and on economics or other considerations such as environmental concerns. To meet these requirements, a typical surfactant formulation may contain blends of a variety of commercial products, which could include ionic and nonionic ethoxylated surfactants, alkylsulfonates, and alkylarylsulfonates, and petroleum sulfonates.

Commercial surfactants contain mixtures of isomers and homologs and may also contain variable amounts of unreacted starting material or extraneous oil that is added as a diluent or thinning agent. Variable amounts of water and inorganic salts are usually present in these products. In order to maintain quality assurance, considerable effort must be put in the development of accurate quantitative techniques for the characterization of components present in these surfactants. Several publications and reviews are available that describe techniques developed for surfactant analysis [1–6].

Problems are often found in many analytical methods due to the complex nature of the mixture and the lack of adequate detection means, thus leading to poor quantitation techniques. For the routine separation of a broad range of surfactants, high-performance liquid chromatography (HPLC) appears to be the most cost-effective [7–18]. Ultraviolet (UV) and fluorescence detectors are commonly used in HPLC analysis of surfactants because of their compatibility with separation techniques requiring gradient elution. However, these detectors have two inherent limitations: (a) the detector response is dependent on molecular structure (i.e., degree of aromaticity and type of substitution) and (b) only species with a chromophore can be detected. To overcome those limitations, postcolumn reaction detectors, based on extraction of fluorescent ion pairs, were introduced for on-line detection of alkylsulfonates in HPLC [19–22]. However, the ion-pair formation and extraction efficiency were still dependent on the molecular structure and could not easily be used for quantitation.

The recent introduction of evaporative light-scattering (ELS) detectors, also known as a mass detector and as a universal detector for nonvolatile compounds, has significantly changed the landscape in the analysis of surfactants. The ELS detector measures light refracted by the nonvolatile particles after the effluent from the HPLC column is nebulized and the carrier solvent is evaporated. The amount of refracted light is proportional to the concentration of the analyte species. Based on the above, this entry will focus on the separation and quantitation of different surfactants by HPLC by means of the ELS detector for universal detection based on the experience of the author, thus enough details can be provided concerning the methodology used.

Methodology

High-performance liquid chromatography is performed using a Hewlett-Packard 1090 chromatograph equipped with a ternary-solvent delivery system, an autoinjector with a 0–20-μL injection loop, an oven compartment, and a diode-array UV detector. An ELS detector (Alltech Associates, Deerfield, IL) is connected in series to the UV detector. Hexane, 2-propanol, and water were used for the analysis of nonionic surfactants. Water and tetrahydrofuran (THF) are used for the analysis of anionic surfactants. No preliminary sample preparation is used other than dilution. The nonionic surfactants are diluted 1:40 (v/v) with hexane. The anionic surfactants (alkyl ether sulfates and synthetic and petroleum sulfonates) are diluted 1:20 (v/v) with water–THF (50:50). The calcium sulfonate surfactants were diluted 1:20 (v/v) with a THF–38% hydrochloric acid solution of pH 1. Hydrochloric acid is required to prevent salt precipitation by converting any excess water-insoluble calcium carbonate into water-soluble calcium chloride. All diluted samples are filtered through a 0.2-μm filter (Gelman Acrodisc CR) directly into the injector vials.

The nonionic ethoxylates are separated according to the number of ethylene oxide (EO) groups (n) using normal-phase chromatography. The separation is achieved on an amino column [DuPont Zorbax NH2, 25 cm × 4.6 cm inner diameter (i.d.), 5 μm particle size]. A precolumn (Zorbax BP NH2, 2.5 cm × 0.2 cm i.d.) is connected to the analytical column. The solvent system is a gradient of hexane, 2-propanol, and water as follows: 100% hexane at time 0, and 37% hexane, 60% 2-propanol, and 3% water at 55 min. Components of the alkyl ether sulfate surfactants are separated into inorganic salt, sulfates, and unreacted alcohol using reversed-phase chromatography. The column used for this separation is a 2.5-cm × 0.2-cm-i.d. column packed with 10 μm C_{18}. The solvent system consists of a 4-min gradient program of water and THF, as summarized in Table 1. The synthetic and petroleum sulfonate components are separated into inorganic salt, sulfonates, and unreacted oil by the same reversed-phase chromatographic method.

The diode-array UV and ELS detectors are connected in series. The UV signals are monitored at 230 and 254 nm. The operating conditions of the ELS detector are optimized for maximum detector response and stable baseline. Surfactants with UV absorbance are detected by both

Table 1 Gradient Elution Program for Reversed-Phase HPLC of Alkyl Ether Sulfate Synthetic and Petroleum Sulfonate Surfactants

Time (min)	Water (%)	THF (%)	Mode of operation
0.0	90	10	Normal flow
0.5	90	10	Normal flow
1.0	40	60	Normal flow
2.5	40	60	Normal flow
2.6	0	100	Backflush
4.0	90	10	Backflush

Table 2 Quantitative Analysis of NP11 Oligomers Using UV and ELS Detection

No. of EO	UV (%)	ELS (%)	Elution time (min)
3	0.83	0.01	10.0
4	2.30	0.71	12.3
5	4.16	2.16	14.6
6	6.24	4.25	16.9
7	8.54	7.27	19.2
8	10.51	10.65	20.5
9	11.76	12.78	23.0
10	11.91	13.72	24.4
11	11.05	13.07	26.4
12	9.46	11.30	27.7
13	7.66	8.95	29.2
14	5.84	6.19	30.8
15	4.02	3.92	31.9
16	2.60	2.39	34.6
17	1.60	1.38	35.7
18	0.95	0.80	37.6
19	0.56	0.40	40.0

detectors, whereas the UV-transparent surfactants could only be detected by the ELS detector.

Separation of Surfactants by HPLC

Analysis of Nonionic Ethoxylates

Aliphatic and aromatic nonionic ethoxylated surfactants, $RO(CH_2CH_2O)_nH$, were used in order to determine the distribution of the ethoxylate oligomers. Oligomers with different numbers of EO groups are separated by normal-phase HPLC, as summarized in Table 2. The separated components are monitored by both the ELS and UV detectors. Table 2 shows the high resolution of components in a nonylphenolethoxylated alcohol, NP11, revealing a range of oligomers from $n = 3$ to 20.

Analysis of Alkyl Ether Sulfates

Anionic alkyl ether sulfate surfactants are produced by sulfating nonionic alcohol polyalkyloxylates such as the ethoxylated surfactants discussed earlier. The sulfated products generally contain variable amounts of unconverted alcohols and inorganic salts as reaction by-products. Determination of the ratio of anionic to nonionic components in surfactant mixtures is desired for quality control and performance evaluation. Separation of the

Table 3 HPLC Analysis of Inorganic Salt, Sulfated Surfactant, and Unreacted Alcohol in Alkyl Ether Sulfate Surfactants

	Retention time (min)		
Surfactant	Inorganic salt	Sulfated surfactant	Unreacted alcohol
BU-6B2ECOS	9.2	10.0	12.5

ionic sulfate and nonionic alcohol components is achieved by reversed-phase chromatography. The separation of four alkyl ether sulfate surfactants is shown in Table 3.

The first component is inorganic salt and it is eluted with 90% water and 10% THF. As the THF concentration increases to 60%, the ionic sulfate surfactant components are eluted. After elution of these ionics, the nonionic components are backflushed with 100% THF. The analysis time is 4 min per sample.

Analysis of Synthetic and Petroleum Sulfonates

Synthetic and petroleum sulfonates are analyzed by the same reversed-phase chromatographic system used for the analysis of alkyl ether sulfate surfactants. Similar to alkyl ether sulfates, the sulfonate mixtures are separated into three fractions: inorganic salt, sulfonates, and unreacted oil. The analysis of two petroleum sulfonates, NaPS-1 and NaPS-2, is shown in Table 4. Good separation was obtained between the inorganic salt and the sulfonated components. The oil present in NaPS-1 and NaPS-2 surfactants consisted of low-molecular-weight components, which were totally volatile under the detector operating conditions and, therefore, could not be detected. These two sulfonates are considerably different in molecular structure distribution. However, their elution characteristics were the same as those observed for the synthetic single-component sulfonates.

Table 4 HPLC Analysis of Inorganic Salt, Sulfated Surfactant, and Unreacted Alcohol in Alkyl Ether Sulfate Surfactants

	Retention time (min)	
Surfactant	Inorganic salt	Sulfated surfactant
NaPS-1	3.0	4.0
NaPS-2	2.5	3.5

References

1. D. Hummel, *Identification and Analysis of Surface-Active Agents*, Interscience, New York, 1962.
2. M. J. Rosen and H. A. Goldsmith, *Systematic Analysis of Surface-Active Agents*, Wiley–Interscence, New York, 1972.
3. G. F. Longman, *The Analysis of Detergents and Detergent Products*, John Wiley & Sons, New York, 1976.
4. M. Kuo and H. A. Mottola, *CRC Crit. Rev. Anal. Chem. 9*: 297 (1980).
5. R. A. Llenado and R. A. Jamieson, *Anal. Chem. 55*: 174R (1981).
6. R. A. Llenado and T. A. Neubecker, *Anal. Chem. 55*: 93R (1983).
7. P. Jandera and J. Churacek, *J. Chromatogr. 197*: 181 (1980).
8. A. Nakae, K. Tsuji, and M. Yamanaka, *Anal. Chem. 53*: 1818 (1981)
9. G. R. Bear, C. W. Lawley, and R. M. Riddle, *J. Chromatogr. 302*: 65 (1984).
10. M. Ahel and W. Giger, *Anal. Chem. 57*: 2584 (1985).
11. K. Levsen, W. Wagner-Redeker, K. H. Shafer, and P. Dobberstein, *J. Chromatogr. 323*: 135 (1985).
12. I. Zeman, *J. Chormatogr. 363*: 223 (1986).
13. M. S. Holt, E. H. McKerrell, J. Perry, and R. J. Watkinson, *J. Chormatogr. 362*: 419 (1986).
14. G. R. Bear, *J. Chromatogr. 371*: 387 (1986).
15. A. Marcomini, S. Capri, and W. Giger, *J. Chromatogr. 403*: 243 (1987).
16. A. Marcomini and W. Giger, *Anal. Chem. 59*: 1709 (1987).
17. J. A. Pilc and P. A. Sermon, *J. Chromatogr. 398*: 375 (1987).
18. R. H. Schreuder and A. Martijn, *J. Chromatogr. 435*: 73 (1988).
19. J. F. Lawrence, U. A. Th. Brinkman, and R. W. Frei, *J. Chromatogr. 185*: 473 (1979).
20. W. M. A. Niessen, J. F. Lawrence, C. F. Werkhoen-Goewie, U. A. Th. Brinkman, and R. W. Frei, *Int. J. Environ. Anal. Chem. 9*: 45 (1981).
21. F. Smedes, J. C. Kraak, C. F. Werkhoven-Goewie, U. A. Th. Brinkman, and R. W. Frei, *J. Chromatogr. 247*: 123 (!982)
22. Y. Hirai and K. Tomokumi, *Anal. Chim. Acta 167*: 409 (1985).

Juan G. Alvarez

T

Taxoids Analysis by TLC

Taxoids are diterpenoid compounds isolated from different yew species and possessing strong anticancer activity [1]. There are different chemical groups of natural taxoids, but among the most important taxoids is a group derived from 10-deacetylbaccatin III (10-DAB III), a diterpenoid compound occurring in high concentration in the European yew, *Taxus baccata* L. It possesses a scheme of structure of four skeletons (6/8/6/4-membered) which was named taxan and 10-DAB III is a derivative of hexahydroxy-11-taxen-9-one. Baccatin III, which is another compound from this group, has, additionally, an acetyl group estrificated with the β-OH group at position C-10. Paclitaxel and cephalomannine are less polar taxoids because they possess amide-acid side chains at position C-13. In the case of paclitaxel, this is (2*R*,3*S*)-*N*-benzoyl-3-phenylsoserine and cephalomannine (2*R*,3*S*)-*N*-tigloyl-3-phenylsoserine side chains. There are also compounds in this group which have an epimer OH group at position C-7 and other substituents are also met [1]. Thus, there are many compounds usually with similar polarity and small differences in structure which are, thus, difficult to separate.

Thin-layer chromatography (TLC) is mainly applied in micropreparative taxoids separation [2–4]. Silica gel $60F_{254}$ preparative plates are usually applied for this purpose. The problem of taxoids separation involves not only their similar chemical structure (e.g., paclitaxel versus cephalomannine) but also, due to different coextracted compounds usually encountered in crude yew extracts (polar compounds such as phenolics and nonpolar ones such as chlorophylls and biflavones), the separation is very difficult. The common band of paclitaxel and cephalomannine was satisfactorily resolved from an extraneous fraction in isocratic elution with ethyl acetate as a polar modifier [4] and *n*-heptane–dichloromethane as the solvent mixture and it was of suitable purity for high-performance liquid chromatography (HPLC) quantitative determination.

The combination of micropreparative TLC separation of callus extract of *Taxus baccata* with *n*-hexane–ethyl acetate (2:3) mobile phase and preparative HPLC on a Lichrosorb RP-18 column enabled 10-deacetylbaccatin III isolation [3].

Systematic studies on the selection of the best mobile phases to assure the best micropreparative separation of analyzed taxoids, especially of 10-DAB III, as well as its less polar derivatives: baccatin III, paclitaxel, and cephalomannine obtained from the extracts of fresh and dried needles and stems of *Taxus baccata* L. by Głowniak and Mroczek have been undertaken [2]. The TLC investigation on silica gel included solvent systems with one and two polar modifiers, multicomponent mobile phases, as well as some multiple development techniques and gradient elution. As polar modifiers, methanol, acetone, dioxane, ethyl acetate and ethylmethylketone, as well as their mixtures, have been reinvestigated, but dichloromethane, chloroform, benzene, toluene, heptane, and their mixtures were used as solvents.

Using binary mobile phases containing 25–30% of one polar electron-donor modifier (acetone, dioxane) in dichloromethane or chloroform, high values of separation factor α (10-DAB III/paclitaxel) are observed as well as low elution of polar extraneous compounds. Such chromatographic systems can be applied to separation by column chromatography of polar from less polar taxoids and their polar coextracted compounds in preliminary CC investigations on taxoids before further detailed studies.

The addition (already about 15%) of a small amount of π-electron solvents such as benzene improves the separation of the band of less polar taxoids (paclitaxel, cephalomannine) and closely eluted chlorophylls.

The mobile phases with two polar modifiers (acetone–methanol, ethylmethylketone–methanol) assure relatively high values of the α factor and the separation in the area of less polar taxoids is better in comparison with the separation obtained by single electron-donor mobile phases.

Because of the complexity of the composition of yew extracts, different gradient elution programs can be considered as further steps of detailed CC or TLC separations of different taxoids on silica gel.

The multicomponent solvent system consisted of benzene–chloroform–acetone–methanol (20:92.5:15:7.5), developed over a distance of 15 cm (2*x*; in some cases, a third development was necessary) enabled very good separation of coextracted compounds, both with 10-DAB III and paclitaxel as well as the separation of the whole yew extract (*Taxus baccata*), which was also very good. The separation of taxoids was better in the extracts obtained from fresh needles and stems than from the dried plant material. The extracts from the dried needles contained the highest concentrations of extraneous compounds which interfered with the bands of analyzed taxoids. The common band of paclitaxel, cephalomannine, and baccatin III was satisfactorily purified from coextractives; the 10-DAB III band could be easily isolated and further HPLC determination in two isocratic mobile phases have been applied. The investigated taxoids eluted in the following order: paclitaxel + cephalomannine, baccatin III, and 10-DAB III. Interesting was the considerable slope of baccatin III retention on silica gel in comparison with retention of its 10-deacethyl derivative (10-DAB III) (the separation factor α of 10-DAB III/baccatin III amounted to 2.8) [2]. This indicates an increase of hydrophobic properties in the molecule of baccatin III. Such chromatographic behavior of baccatin III can be explained, first, by the absence of a free OH group at position C-10, capable of effectively competing with free OH groups on the surface of silica gel (hydrogen-bonding interactions) but, on the other hand, the presence of an acetyl (ester) group at position C-10 can impede adsorption of baccatin III on silica gel caused by hydrogen-bonding interactions between the carbonyl group at position C-9 and OH groups on the surface of silica gel (a type of spherical hindrance, observed in computer modeling of spatial structures of taxoids by HYPERCHEM) because a role of this carbonyl group in polar properties of taxoids was confirmed by analysis of ^{1}H-NMR (nuclear magnetic resonance) spectra. This can be responsible for the decrease of baccatin III retention on silica gel, which is similar to the retention of paclitaxel and cephalomannine; both possess the acetyl group at position C-10 and low-polarity side chain at position C-13. Because of a very slight difference in the structure of the amide substituent of the acyl side chain at position C-13 (*N*-benzoyl versus *N*-tigloyl) and low polarities, these two compounds have the same retention on silica gel.

Similar solvent systems can be applied for analytical TLC of taxoids in screening of, for example, callus cultures for taxoids presence [5].

The reversed-phase TLC system [stationary phase: RP-18F254s; mobile phase: water–methanol–tetrahydrofuran 5:2:3 (v/v/v) applied two times] was used in analysis of taxoids fractions obtained after high-speed countercurrent chromatography isolation and was suitable for paclitaxel and cephalomannine qualitative determination [6].

The TLC–densitometric paclitaxel determination in different yew samples with two-stage gradient development has been also attempted [7], but, due to the high concentration of coextractives, it is difficult to estimate the precision of this method.

Stasko et al. [8] elaborated two interesting procedures of multimodal TLC for the separation of taxol and related compounds from *Taxus brevifolia*. For the first procedure, a cyano-modified silica gel plate was developed, in the first dimension, in dichloromethane–hexane–acetic acid (9:10:1) and, in the second dimension, in water–acetonitrile–methanol–tetrahydrofuran (8:5:7:0.1). For the second procedure, a diphenyl-modified silica gel plate was developed, in the first dimension, in hexane–isopropanol–acetone (15:2:3) and, in the second dimension, in methanol–water (7:3). These two methods enabled paclitaxel resolution from cephalomannine, which is impossible on silica gel plates, and at least another 20 compounds.

Besides typical ultraviolet detection on plates with a flourescent agent (F_{254}), different visualization methods can be applied:

1. Anisaldehyde spray reagent [8] (76% methanol, 19% *o*-phosphoric acid, and 5% *p*-anisaldehyde followed by heating at 110°C for 5 min. Paclitaxel appeared as a gray–brown spot.
2. A 3% sulfuric acid methanolic solution followed by heating at 115°C for 5 min (visualization: UV at 366 nm and VIS) [6].
3. A 3% sulfuric acid ethanolic solution, then a 1.5% solution of vanillin in ethanol, followed by heating at 105°C for 10 min [5]. DAB III appeared as a gray spot.

Summing up this entry, it should be stated that because of differences in taxoids' and other compounds' compositions in different yew and yew-derived materials, a suitable approach for TLC taxoids separation ought to include problems involving both taxoids separation and appropriate separation of coextracted compounds from taxoids which usually are in higher concentration. The separation method should also be elaborated precisely for each type of extract.

References

1. M. Suffness and G. A. Cordell, *Taxus* alkaloids, in *The Alkaloids (Chemistry and Pharmacology)* (A. Brossi, eds.), Academic Press, New York, 1985, Vol. 25, pp. 6–18.
2. K. Głowniak and T. Mroczek, *J. Liquid Chromatogr. 22*(16): 2483 (1999).
3. Zhiri, M. Jaziri, Y. Guo, R. Vanhaelen-Fastre, M. Vanhaelen, J. Homes, K. Yoshimatsu, and K. Shimomura, *Biol. Chem. Hoppe–Seyler 376*: 583–586 (1995).
4. K. Głowniak, G. Zgórka, A. Józefczyk, and M. Furmanowa, *J. Pharm. Biomed. Anal. 14*: 1215 (1996).
5. Y. Gou, M. Jaziri, B. Diallo, R. Vanhaelen-Fastre, A. Zhiri, M. Vanhaelen, J. Homes, and E. Bombardelli, *Biol. Chem. Hoppe–Seyler 375*: 281 (1994).
6. R. Vanhaelen-Fastre, B. Diallo, M. Jaziri, M. L. Faes, J. Homes, and M. Vanhaelen, *J. Liquid Chromatogr. 15*(4): 697 (1992).
7. G. Matysik, K. Głowniak, A. Józefczyk, and M. Furmanowa, *Chromatographia 78*(41): 485 (1995).
8. M. W. Stasko, K. M. Witherup, T. J. Ghiorzi, T. G. Mc Cloud, S. Look, G. M. Muschik, and H. J. Issaq, *J. Liquid Chromatogr. 12*(11): 2133 (1989).

Tomasz Mroczek
Kazimierz Głowniak

Temperature: Effect on MEKC Separation

The distribution coefficient, K, of a solute for an equilibrium between the aqueous phase and micelle, or the micellar solubilization, depends on temperature; generally, the distribution coefficient decreases with an increase in temperature. This means that the migration time of a solute, t_R, will be reduced when the temperature is elevated under typical micellar electrokinetic chromatography (MEKC) conditions, where, for example, sodium dodecyl sulfate (SDS) is employed as a pseudo-stationary phase at a neutral condition (i.e., pH 7). Also, the velocity of the electro-osmotic flow (EOF), u_{EOF}, and the electrophoretic velocity of the micelle, u_{ep} (mc), will be increased by an increase in temperature because of a reduced viscosity of the micellar solution employed in a MEKC system.

One should note that the increase in temperature during a separation is observed due to Joule heating: Ideally, the retention factor, k, should be independent of the applied voltage or current, but, actually, k does depend on the applied voltage. Although the temperature of the separation capillary is controlled with a liquid coolant or a circulating airstream in most commercially available capillary electrophoresis (CE) instruments, the retention factor usually decreases, almost linearly, with an increase in the velocity of the EOF or an increase in the current. However, the dependence is much less than that observed in a CE system without forced-cooling apparatus.

The critical micelle concentration (CMC) and the partial specific volume, $\overline{v}$, depend on the temperature. The relationship between k and K is represented as

$$k = K\left(\frac{V_{\mathrm{mc}}}{V_{\mathrm{aq}}}\right) \tag{1}$$

where V_{mc} and V_{aq} are the volumes of the micelle and the remaining aqueous phase, respectively. The phase ratio, $V_{\mathrm{mc}}/V_{\mathrm{aq}}$, is described as

$$\frac{V_{\mathrm{mc}}}{V_{\mathrm{aq}}} = \frac{\overline{v}(C_{\mathrm{sf}} - \mathrm{CMC})}{1 - \overline{v}(C_{\mathrm{sf}} - \mathrm{CMC})} \tag{2}$$

where C_{sf} is the surfactant concentration. Approximately, the volume of the micelle is negligible and, hence, Eq. (1) can be rewritten as

$$k = K\overline{v}(C_{\mathrm{sf}} - \mathrm{CMC}) \tag{3}$$

For example, the CMC of SDS and $\overline{v}$ of the SDS micelle depend on temperature as presented in Table 1, where a 100-mM borate–50-mM phosphate buffer (pH 7.0) was used to prepare SDS solutions. It should be noted that the CMC value observed in a buffer solution or electrolyte solution is smaller than that in pure water (e.g., 8.1 mM at 20°C).

In MEKC, an ionic surfactant is used as a pseudo-stationary phase, and the Krafft point is also an important temperature. At a temperatures lower than the Krafft

Table 1 CMC of SDS and $\overline{v}$ of SDS Micelle Dependence on Temperature

Temp. (°C)	CMC (m*M*)	$\overline{v}$ (mL/g)
20	—	0.8562
22	2.8	—
25	2.9	0.8610
30	2.5	0.8686
35	2.6	0.8710
40	3.0	0.8758

point, C_{sf} does not exceed the CMC, due to reduced solubility and, therefore, no micelle is formed. At the Krafft point, C_{sf} reaches the CMC and then the formation of the micelle is begun. The Krafft point of SDS is ~16°C in a pure water, whereas it is ~31°C for potassium dodecyl sulfate in pure water. Thus, a potassium salt is not an adequate buffer component for the SDS–MEKC system.

Figure 1 shows the dependence of the distribution coefficients on temperature for several solutes in MEKC, where SDS is employed as a pseudo-stationary phase. Different dependencies are observed among the solutes; that is, temperature affects the selectivity.

As mentioned, temperature seriously affects the migration time, whereas its effect on selectivity is not remarkable. It is important to maintain temperature precisely to obtain reproducible results.

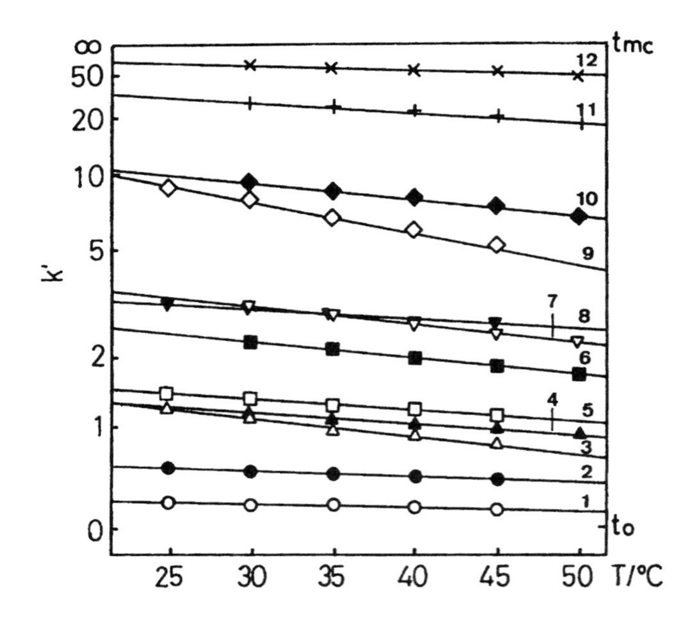

Fig. 1 Dependence of the retention factor, k', on temperature, T. The solutes are (1) resorcinol, (2) phenol, (3) *p*-nitroaniline, (4) *o*-cresol, (5) nitrobenzene, (6) 2,6-xylenol, (7) 2,4-xylenol, (8) toluene, (9) 2-naphthol, (10) *p*-propylphenol, (11) *p*-butylphenol, and (12) *p*-amylphenol. Capillary: 50 μm inner diameter × 570 mm (effective length, 500 mm); separation solution: 50 m*M* SDS in 100 m*M* borate–50 m*M* phosphate buffer (pH 7.0); applied voltage: 15 kV; detection wavelength: 214 nm. Reprinted from Ref. 3, with permission.

References

1. S. Terabe, K. Otsuka, and T. Ando, Electrokinetic chromatography with micellar solution and open-tubular capillary, *Anal. Chem. 57*: 834 (1985).
2. S. Terabe, T. Katsura, Y. Okada, Y. Ishihama, and K. Otsuka, Measurement of thermodynamic quantities of micellar solubilization by micellar electrokinetic chromatography with sodium dodecyl sulfate, *J. Microcol. Separ. 5*: 23 (1993).
3. S. Terabe, Micellar electrokinetic chromatography, in *Capillary Electrophoresis Technology* (N. A. Guzman, ed.), Chromatographic Science Series Vol. 64, Marcel Dekker, Inc., New York, 1993; Chap. 2.
4. K. Otsuka and S. Terabe, Micellar electrokinetic chromatography, *Bull. Chem. Soc. Jpn. 71*: 2465 (1998).

Koji Otsuka
Shigeru Terabe

Temperature: Effects on Mobility, Selectivity, and Resolution in Capillary Electrophoresis

Introduction

Although not a major separation parameter, as in gas chromatography, the operating temperature clearly affects migration behavior in capillary electrophoresis (CE). This alone should be reason to work under uniform, well-thermostated conditions. The temperature affects bulk viscosity, electro-osmotic flow, ionic mobilities, even p*K* values and buffer pH.

Buffer Behavior

A buffer for electrophoresis is usually prepared at room temperature by adding ingredients and measuring the final pH. When this buffer is used for CE at a different temperature, a different pH might result. For example, when a buffer consists of Tris-acetate at pH 5, Tris is fully charged and the temperature dependence of p*K* of acetic acid is the one involved. This value is small so that the pH of the buffer is virtually independent of temperature. This is not the case if the same buffer ingredients are used to prepare a buffer around pH 8; in that case, acetate is fully ionized and the p*K* of Tris is involved. This p*K* has a temperature coefficient of −0.031 per degree; a 10°C increase leads to a 0.31 decrease in buffer pH (Table 1).

In case of single-buffering background electrolyte solutions, the temperature coefficient of the buffer pH equals the temperature coefficient of the p*K* of the buffering ion. In case of a double-buffering system, p*K*'s of both buffer anion and buffer cation are involved, which obviously makes the situation potentially more complicated.

In some situations, the pH of the buffer can be very critical. An example of such a critical pair of components is given in Fig. 1; a mixture of benzoic and 2,3-dimethoxy benzoic acid, analyzed around pH 4.67. A pH change of as little as 0.04 pH units will change the migration order.

Another aspect involved is buffer conductivity; on average, ionic conductivities have a temperature coefficient of around 2.5%/degree. A higher temperature, therefore (at constant voltage), leads to a higher current. This can have a significant effect on power dissipation and, through the corresponding plate-height terms, on separation efficiency (see the entry Band Broadening in Capillary Electrophoresis).

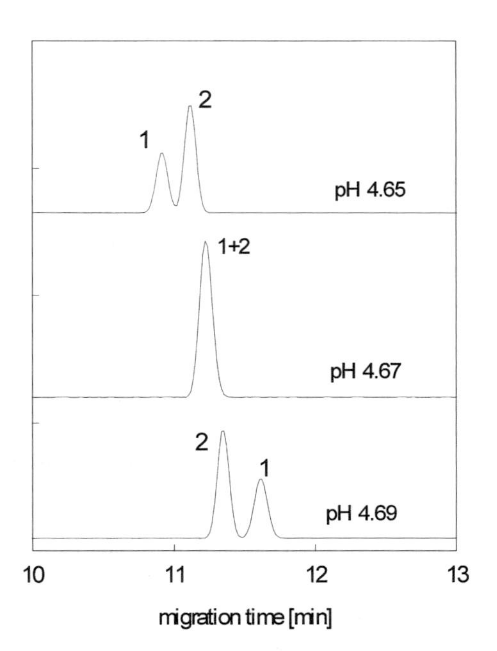

Fig. 1 Separation of benzoic acid (1) and 2,3-methoxy benzoic acid in 10 m*M* Tris-acetic acid buffer of three different pH values.

Table 1 p*K* Values and Their Temperature Dependence of Some Buffer Components

Name	p*K* (20°C)	d*pK*/*dT*
Phosphoric acid, pK_1	2.12	
Citric acid, pK_1	3.13	
Formic acid	3.75	−0.000
Glycine, pK_1	3.75	−0.000
Acetic acid	4.76	−0.000
Histidine	6.12	−0.045
MES	6.15	−0.011
ACES	6.90	−0.020
Imidazole	6.95	−0.018
MOPS	7.20	−0.006
HEPES	7.55	−0.014
Tris	8.30	−0.031
Boric acid	9.24	−0.007
CHES	9.50	−0.009
Glycine, pK_2	9.77	
CAPS	10.40	−0.009

Sample Ion Behavior

Ionic mobilities generally have a temperature coefficient of around +2.5%/degree, but it is not exactly the same for all ions. Incidentally, water viscosity has a temperature coefficient of approximately −2.5%/degree. This sometimes leads to the erroneous conclusion that ionic mobility is, by definition, inversely proportional to liquid viscosity, which is an oversimplification, originating also from the following relationship:

$$\mu = \frac{q}{6\pi\eta r}$$

in which q is the effective charge, η is the liquid viscosity, and r is the radius. The misunderstanding is that, in reality, the above relationship is valid only for rigid spherical particles, not necessarily for individual solvated ions.

For the electro-osmotic flow, temperature-induced viscosity change at the plane of shear for the electric double layer is more straightforward. Analysis times will generally shorten with increasing temperature.

Suggested Further Reading

Boček, P., M. Deml, P. Gebauer, and V. Dolník, *Analytical Isotachophoresis*, VCH, Weinheim, 1988.

Everaerts, F. M., J. L. Beckers, and Th. P. E. M. Verheggen, *Isotachophoresis: Theory, Instrumentation and Applications*, Elsevier, Amsterdam, 1976.

Friedl, W., J. C. Reijenga, and E. Kenndler, *J. Chromatogr. A 709*: 163 (1995).

Mohan, C., Buffers, in *A Guide for the Preparation and Use of Buffers in Biological Systems*, Calbiochem, San Diego, CA, 1997.

Reijenga, J. C. and E. Kenndler, *J. Chromatogr. A 659*: 403 (1994).

Reijenga, J. C. and E. Kenndler, *J. Chromatogr. A 659*: 417 (1994).

Reijenga, J. C., Th. P. E. M. Verheggen, J. H. P. A. Martens, and F. M. Everaerts, *J. Chromatogr. A 744*: 147 (1996).

Jetse C. Reijenga

Temperature Program: Anatomy

The principle of temperature programming in gas chromatography (GC) is based on the thermodynamic explanation of retention (see the entry Thermodynamics of Retention in Gas Chromatography) and is employed for samples that have components that cover a wide range of polarities and/or molecular weights. Its purpose is to accelerate the late eluting peaks through the column to reduce the analysis time and, at the same time, to maintain symmetrical elution profiles that are amenable to accurate quantitative assessment. As the temperature is increased, the magnitude of the distribution coefficient with respect to the stationary phase is reduced and, as a consequence, the retention volume and retention time is reduced. The programming rate must be chosen such that the integrity of the separation is maintained and the peak shapes are not distorted.

The programming procedure usually involves three stages. An initial isocratic period is introduced to efficiently separate the early eluting peaks with adequate resolution. The isocratic period is followed by a linear increase in column temperature with time, which accelerates the well-retained peaks so that they also elute in a reasonable time and are adequately resolved. The effect of linear programming can be calculated employing appropriate equations and the retention times of each solute predicted for different flow rates (see the entry Programmed Temperature Gas Chromatography). To do this, some basic retention data must be measured at two temperatures and the results are then employed in the retention calculations. The temperature program often ends with a final isothermal period. This is usually introduced either because the upper temperature limit of the stationary phase has been reached and so a higher temperature will damage the column, or to purge any remaining, strongly retained solutes from the column. The upper temperature limit of the stationary phase should not be exceeded, as not only will the column be spoiled but the performance of the detector is often impaired and deposits are formed that can permanently increase the detector noise.

Temperature programming is employed in most GC separations and the optimum programming conditions are usually easy to identify from a few preliminary separations using different gradients. If, however, there is a pair of solutes that elute close together (e.g., a pair of enantiomers) and there is a temperature of coelution, then some considerable effort may be necessary to identify the optimum temperature program [1] to achieve a satisfactory separation in a reasonable time. Such situations, however, are not common, unless the applications involves complex samples such as multicomponent chiral mixtures.

Reference

1. T. E. Beesley and R. P. W. Scott, *Chiral Chromatography*, John Wiley & Sons, Chichester, 1998, p. 39.

Suggested Further Reading

Scott, R. P. W., *Techniques of Chromatography*, Marcel Dekker, Inc., New York, 1995.

Scott, R. P. W., *Introduction to Analytical Gas Chromatography*, Marcel Dekker, Inc., New York, 1998.

Raymond P. W. Scott

Theory and Mechanism of Thin-Layer Chromatography

Modes of Thin-Layer Chromatography

General classification of the modes of thin-layer chromatography (TLC) is based on the chemical nature of the stationary and mobile phases. The following three types of thin-layer chromatography given are widely recognized as different modes:

Adsorption TLC

In this mode, active inorganic adsorbents (e.g., silica, alumina, or Florisil) are usually employed as stationary phases and, hence, the overall mechanism of retention is governed predominantly by the specific intermolecular interactions between the functionalities of the solutes, on the one hand, and active sites on the adsorbent surface, on the other. In adsorption TLC, aqueous mobile phases are never used, and stationary-phase activity prevails over the polarity of the mobile phase employed.

Normal-Phase TLC

This mode of chromatography usually involves organic chemically bonded stationary phases with polar (e.g., 3-cyanopropyl) ligands. This particular mode is characterized by a mixed mechanism of solute retention: Solute molecules interact specifically with the polar functionalities of the organic ligand and with the residual active sites of the silica matrix, whereas their interactions with the hydrocarbon moiety of the organic ligands are entirely nonspecific in nature. Again, aqueous mobile phases are never employed in normal-phase (NP) TLC, and stationary phase activity prevails over the polarity of the mobile phase employed.

Reversed-Phase TLC

This chromatographic mode usually involves aliphatic chemically bonded stationary phases with, for example, octyl, octadecyl, or phenyl ligands. The mode of chromatography also is characterized by a mixed mechanism of solute retention: Solute molecules interact specifically with the residual active sites of the silica matrix, whereas their interactions with the aliphatic ligands are nonspecific (and predominantly hydrophobic) in nature. Reversed-phase (RP) TLC is usually performed with aqueous mobile phases containing organic modifiers [such as, e.g., methanol, acetonitrile (ACN), tetrahydrofuran (THF), etc.], and in this case, the activity of the stationary phase — as an exception — is less than that of the high-polarity mobile phase employed.

TLC Parameters of Solute Retention

The parameter R_f is the quantity most commonly used to express the position of a solute in the developed chromatogram. It is calculated as a ratio:

$$R_f = \frac{\text{Distance of chromatographic spot center from origin}}{\text{Distance of solvent front from origin}} \tag{1}$$

Using symbols from Fig. 1, R_f can be given as

$$R_f = \frac{z}{l} \tag{2}$$

R_f values vary between 0 (solute remains at the origin) and 0.999 (solute migrates with the mobile-phase front). From a practical standpoint, the most reliable analytical

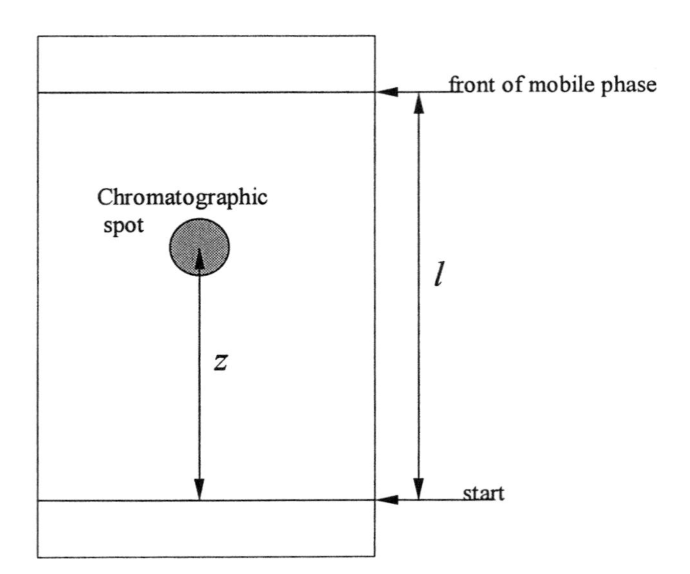

Fig. 1 The thin-layer parameters used to calculate the retention parameter R_f.

results are achieved when the parameter R_f ranges from 0.20 to 0.80.

In the theory and practice of chromatography, another parameter of solute retention is also employed, the so-called R_M value. This quantity was defined by Bate-Smith and Westall [1] as

$$R_M = \log\left(\frac{1 - R_f}{R_f}\right) \tag{3}$$

Selected Models of the Chromatographic Process

The adsorption and partition mechanisms of solute retention are the two most universal mechanisms of chromatographic separation, both operating on a physical principle. In fact, almost all solutes can be adsorbed on a microporous solid surface or be partitioned between two immiscible liquids. It is the main aim of semiempirical chromatographic models to couple the empirical parameters of retention with the established thermodynamic quantities generally used in physical chemistry. The validity of these models in chromatographic practice can hardly be overestimated, because when incorporated in separation selectivity-oriented optimization strategies, they often successfully help overcome the old trial-and-error approach used to optimize analyses. In the forthcoming sections, we will discuss a selection of the most popular and best-performing models and concepts of solute retention.

Martin and Synge Model of Partition Chromatography

Partition chromatography was the first chromatographic technique to be given a thermodynamic foundation, by the pioneering work of Martin and Synge [2], the 1952 Nobel Prize winners for chemistry. The Martin and Synge model describes the idealized parameter R_f (i.e., the parameter R'_f) as

$$R'_f = \underset{\text{(I)}}{\frac{t_m}{t_m + t_s}} = \underset{\text{(II)}}{\frac{n_m}{n_m + n_s}} = \underset{\text{(III)}}{\frac{m_m}{m_m + m_s}} \tag{4}$$

where t_m and t_s denote the time spent by a solute molecule in the mobile and stationary phases, respectively, n_m and n_s are numbers of solute molecules present in the mobile and stationary phases at equilibrium, and m_m and m_s are the respective numbers of moles.

Equation (4) can be further transformed as follows:

$$R'_f = \frac{m_m}{m_m + m_s} = \frac{c_m V_m}{c_m V_m + c_s V_s} = \frac{1}{1 + c_s V_s / c_m V_m} \tag{5}$$

where c_m and c_s are the molar concentrations of the solute in the mobile and stationary phases, respectively, and V_m and V_s are volumes of these phases, respectively.

Assuming that

$$K = \frac{c_s}{c_m} \tag{6}$$

where K is the thermodynamic equilibrium constant for solute partitioning, we obtain

$$R_f = \frac{1}{1 + K\Phi} \tag{7}$$

where Φ is the so-called phase ratio (i.e., $\Phi = V_s/V_m$).

Equation (7) unites the empirical retention parameter of the solute, R'_f, with the established thermodynamic (i.e., theoretical) quantity K, expressed as

$$\ln K = \frac{\Delta\mu_p^0}{RT} \tag{8}$$

where $\Delta\mu_p^0$ is the standard chemical potential for partition. Hence, the retention model given by Eq. (8) can rightfully be called semiempirical.

Snyder and Soczewinski Model of Adsorption Chromatography

The semiempirical model of adsorption chromatography, analogous to that of Martin and Synge, was established only in the late 1960s by Snyder [3] and Soczewinski [4] independently, and it is often referred to as the displacement model of solute retention. The crucial assumption of this model is that the mechanism of retention consists in competition among the solute and solvent molecules for the active sites of the adsorbent and, hence, in a virtually endless process of the solvent molecules displacing those of the solute from the solid surface (and vice versa). Further, the authors assumed that some of the mobile phase remains adsorbed and stagnant on an adsorbent surface. This adsorbed mobile phase formally resembles the liquid stationary phase in partition chromatography. Thus — utilizing with imagination the main concept of the Martin and Synge model of partition chromatography — Snyder and Soczewinski managed to define the R_f' parameter valid for adsorption chromatography as

$$R_f = \frac{t_m}{t_m + t_a} \equiv \frac{n_m}{n_m + n_a} = \frac{m_m}{m_m + m_a} = \frac{c_m(V_m - V_a W_a)}{c_m(V_m - V_a W_a) + c_a V_a W_a} \quad (9)$$

where t_m and t_a denote the time spent by a solute molecule in the mobile phase and on the adsorbent surface, respectively, n_m and n_a are numbers of solute molecules contained in the mobile phase and on the adsorbent surface at equilibrium, respectively, m_m and m_a are the numbers of moles of solute molecules contained in the nonadsorbed and adsorbed moieties of the mobile phase, respectively, c_m and c_a are molar concentrations of solute in the nonadsorbed and the adsorbed moieties of mobile phase, respectively, V_m is the total volume of mobile phase, V_a is the volume of the adsorbed mobile phase per unit mass of adsorbent, and W_a is the mass of adsorbent considered.

Transformation of Eq. (9) results in the relationship

$$R_f = \frac{1}{1 + K_{\text{th}}[V_a W_a/(V_m - V_a W_a)]} \quad (10)$$

where $K_{\text{th}} = c_a/c_m$, K_{th} being the thermodynamic equilibrium constant of adsorption, and $\Phi = V_a W_a/(V_m - V_a W_a)$.

Simplified Relationships Derived from the Snyder and Soczewinski Model

Two very simple relationships have been derived from the general framework of the Snyder and Soczewinski model of adsorption chromatography; these have proved useful for rapid prediction of solute retention in chromatographic systems employing binary mobile phases. One (known as the Soczewinski equation) proved successful for adsorption and normal-phase TLC; the other (known as the Snyder equation) proved similarly successful in reversed-phase TLC.

Soczewinski Equation

The Soczewinski equation [5] [Eq. (11)] is a simple linear relationship with respect to log X_s, linking the retention parameter (i.e., R_m) of a given solute with the quantitative composition of the binary mobile phase used:

$$R_M = C - n \log X_s \quad (11)$$

where C is, in the first instance, the equation constant (although with clear physicochemical significance), X_s is the molar fraction of the stronger solvent in the nonaqueous mobile phase, and n is the number of active sites on the surface of the adsorbent.

Apart from enabling rapid prediction of solute retention, the Soczewinski equation enables molecular-level scrutiny of solute–stationary phase interactions. Thus, a numeric value of the parameter n of Eq. (11) of approximately unity ($n \approx 1$) implies one-point attachment of the solute molecule to the stationary-phase surface. Numerical values of n higher than unity indicate that in a given chromatographic system, solute molecules interact with the stationary phase at more than one point (so-called multipoint attachment).

Snyder Equation

The Snyder equation [6] [Eq. (12)] is another simple linear relationship with respect to φ, which links the retention parameter (i.e., ln k) of a given solute with the volume fraction of the organic modifier in the aqueous binary mobile phase (φ):

$$\ln k = \ln k_w - S\varphi \quad (12)$$

where k is the retention coefficient of the solute [$k = (1 - R_f)/R_f$], k_w is the retention coefficient extrapolated for pure water as the mobile phase, and S is a constant characteristic of a given stationary phase.

Chromatographic Activity of Adsorbents and Elution Strength of Solvents [7–9]

Consequences of the Snyder and Soczewinski model are manifold, and they are of significant practical importance. The most spectacular conclusions of this model are (a) the possibility of quantifying the activity of an adsorbent and (b) the possibility of defining and quantifying the "chromatographic polarity" of solvents (known as their elution strength). These two conclusions could be drawn only upon the assumption of the displacement mechanism of solute retention. An obvious necessity in this model was to quantify the effect of displacement, which resulted in the relationship given by Eq. (13) for the thermodynamic equilibrium constant of adsorption, K_{th}, for an active chromatographic adsorbent and a monocomponent mobile phase:

$$\log K_{th} = \log V_a + \alpha(S^0 - A_S\varepsilon^0) \qquad (13)$$

where α is a function of the adsorbent surface energy and is independent of the properties of the solute (it is known as the activity coefficient of the adsorbent; practical determination of its numerical values can be regarded as quantification of adsorbent activity), S^0 is the adsorption energy of a solute chromatographed on an active adsorbent with n-pentane as the mono-component mobile phase, A_S is the surface area of the adsorbent occupied by an adsorbed solute molecule, and ε^0 is the parameter usually referred to as the solvent elution strength, or simply solvent strength (it is the energy of adsorption of solvent per unit surface area of adsorbent).

Assuming that the adsorbent surface occupied by an adsorbed solute molecule (A_S) and that occupied by a stronger solvent (n_B) are equal, the eluent strength of a binary mobile phase, ε_{AB}, has the following dependence on its quantitative composition:

$$\varepsilon_{AB} = \varepsilon_A + \frac{\log(x_B \times 10^{\alpha n_B(\varepsilon_B - \varepsilon_A)} + 1 - x_B)}{\alpha n_B} \qquad (14)$$

where ε_A is the eluent strength of the weaker component (A) of a given binary mobile phase, ε_B is the eluent strength of the stronger component (B) of the same mobile phase, and x_B is the molar volume of the component B.

Combining Eqs. (13) and (14) gives the following relationship, which expresses the dependence of the retention parameter of a solute, R_M (= log k), on the quantitative composition of a given binary mobile phase:

$$\log k = \log V_a + \alpha[S^0 - A_S\varepsilon_A] - \frac{A_S \log(x_B \times 10^{\alpha n_B(\varepsilon_B - \varepsilon_A)} + 1 - x_B)}{n_B} \qquad (15)$$

Schoenmakers Model of Reversed-Phase Chromatography

In this particular model, it is assumed that Hildebrand's concept of the solubility parameter (δ), originally formulated for liquid nonideal solutions, can also be applied to the solute and to the stationary and mobile phases of chromatographic systems.

The solubility parameter of any given substance (δ) is defined as

$$\delta = \sqrt{\frac{-E}{V}} \qquad (16)$$

where E denotes its heat of vaporization at zero pressure, $-E$ is the energy of cohesion needed for transportation of one mole of an ideal gas phase to liquid phase, and V is the molar volume of the liquid.

One of the basic retention parameters (i.e., the solute's retention coefficient k) can be expressed as a function of the solubility parameters, δ:

$$\ln k_i = \frac{v_i}{RT}(\delta_m + \delta_s - 2\delta_i)(\delta_m - \delta_s) + \ln\left(\frac{n_s}{n_m}\right) \qquad (17)$$

where v_i is the molar volume of the ith solute, δ_i, δ_s, and δ_m are respectively the solubility parameters of this solute and of the stationary and mobile phases employed, and n_s and n_m are respectively the numbers of moles of the stationary and mobile phases.

Finally, the principal equation of the Schoenmakers model [10] of solute retention in reversed-phase chromatography employing a binary aqueous mobile phase takes the parabolic form

$$\ln k = A\varphi^2 + B\varphi + C \qquad (18)$$

where the equation constants A, B, and C have a clear physicochemical significance:

$$A = \frac{v_i}{RT}(\delta_a - \delta_w)^2 \qquad (19)$$

$$B = 2\frac{v_i}{RT}(\delta_a - \delta_w)(\delta_a - \delta_i) \qquad (20)$$

$$C = \frac{v_i}{RT}(\delta_w + \delta_s - 2\delta_i)(\delta_w - \delta_s) + \ln\left(\frac{n_s}{n_m}\right) \qquad (21)$$

where δ_w and δ_a denote respectively the solubility parameters of water and of the organic modifier as the constituents of a given aqueous mobile phase.

References

1. E. C. Bate-Smith and R. G. Westall, *Biochim. Biophys. Acta 4*: 427 (1950).
2. A. J. P. Martin and R. L. M. Synge, *Biochem. J. 35*: 1358 (1941).
3. L. R. Snyder, *Principles of Adsorption Chromatography*, Marcel Dekker, Inc., New York, 1968.

4. E. Soczewinski, *Anal. Chem. 41*: 179 (1969).
5. L. R. Snyder, *Anal. Chem. 46*: 1384 (1974).
6. L. R. Snyder, J. W. Dolan, and J. R. Gant, *J. Chromatogr. 165*: 3 (1979).
7. L. R. Snyder and J. L. Glajch, *J. Chromatogr. 214*: 1 (1981).
8. J. L. Glajch and L. R. Snyder, *J. Chromatogr. 214*: 21 (1981).
9. L. R. Snyder and J. J. Kirkland, *Introduction to Modern Liquid Chromatography*, 2nd ed., John Wiley & Sons, New York, 1979.
10. P. J. Schoenmakers, H. A. H. Billiet, and L. De Galan, *J. Chromatogr. 185*: 179 (1979).

Teresa Kowalska
Wojciech Prus

Thermal FFF: Basic Introduction and Overview

Thermal field-flow fractionation (FFF) is a subtechnique of the FFF family that employs a temperature gradient as the applied field [1]. For the reasons to be outlined, thermal FFF is applied primarily to industrial polymers that are soluble in organic liquids. The molecular-weight range of thermal FFF complements that of size-exclusion chromatography (SEC). Thus, oligomers and polymers with molecular weights below about 10^4 g/mol are not generally well separated by thermal FFF unless extraordinary measures are taken. On the other hand, the resolving power of thermal FFF for polymers with molecular weights above 10^5 g/mol is generally several times that of SEC [2]. For molecular weights above 10^6 g/mol, SEC is limited by shear-induced fragmentation of the chains as they travel through the packed bed under high pressure [3]. By contrast, shear forces in the FFF channel are extremely low, therefore ultrahigh-molecular-weight polymers, gels, and particles can be separated without degradation or sample pretreatment.

Thermal FFF Retention

Like other FFF subtechniques, materials are retained in thermal FFF as a result of their field-induced concentration at one wall of the channel. In thermal FFF, that field is a temperature gradient. Several terms are used to express the movement of material in response to a temperature gradient, including thermal diffusion, thermodiffusion, thermophoresis, and the Soret effect. The term thermodiffusion is used here, as it has been adopted by the scientific committee for The International Symposium on Thermodiffusion, which is devoted to the scientific study of this phenomenon.

Like other transport processes, thermodiffusion is typically quantified by a phenomenological coefficient that defines the dependence of a mass or energy movement on a potential energy gradient. Thus, the thermodiffusion coefficient (D_T) relates the velocity (U_x) induced in a material by a temperature gradient: $U_x = D_T(dT/dx)$, where T is temperature and x represents the dimension in which the gradient is applied. This definition of thermodiffusion can be substituted into the general model of FFF retention to yield the following equation, which approximates the retention volume (V_r) of an analyte component in a thermal FFF channel:

$$\frac{V_r}{V^0} \cong \frac{\Delta T}{6} \frac{D_T}{D} \tag{1}$$

Here, ΔT is the temperature drop across the channel, which is set by the user, and D is the ordinary (mass) diffusion coefficient. The parameter V^0 is the geometric volume of the channel, which is constant for a given instrument. Note that V_r is the same parameter used to define retention in SEC and that the ratio on the left side of Eq. (1) is the number of channel volumes required to flush a sample component through the thermal FFF channel. Although Eq. (1) is an approximation, it becomes accurate to within 3% when $V_r/V^0 > 10$. More important for this discussion, Eq. (1) characterizes the influence of analyte parameters D and D_T on retention in thermal FFF.

According to Eq. (1), the retention of an analyte component is governed by the two transport coefficients D and D_T. The coefficient D scales directly with hydrodynamic volume and is the same parameter that differentiates retention in SEC. Thus, thermal FFF, like SEC, separates material according to differences in their hydrodynamic volume, which is related to molecular weight. However, V_r increases with D in SEC and decreases with D in thermal FFF, so the elution orders are opposite in the two techniques; low-molecular-weight components elute ahead of higher-molecular-weight components in thermal FFF.

Although V_r scales with the log of D in SEC, it scales directly with D in thermal FFF. As a result, molecular-weight components are distributed across a wider range of retention volumes in thermal FFF compared to SEC. Although resolving power benefits from this spread in molecular weight over V_r, thermal FFF has fewer theoretical plates than SEC. As a result, the resolving power of thermal FFF exceeds that of SEC only for molecular weights above about 10^5 g/mol, where the compression of M (due to the log scale) leads to a rapid decline in the resolving power of SEC.

The dependence of retention on thermodiffusion imparts an additional dimension to the thermal FFF separation that is not present in SEC. Although our understanding of thermodiffusion in solids and liquids is incomplete, certain aspects are clear. For example, thermodiffusion is very sensitive to the chemical composition of the polymer. As a result, thermal FFF is capable of separating components that differ in composition, even though they may have the same molecular-weight or diffusion coefficient. An example [4] of the separation of polystyrene and poly(methyl methacrylate) standards by thermal FFF is illustrated in Fig. 1. This separation cannot be accomplished with SEC because the diffusion coefficients (or hydrodynamic volumes) of the two materials are virtually identical. The ability of thermal FFF to separate materials by chemical composition has spurred additional research designed to increase our understanding of thermodiffusion [5], which, in turn, has led to the application of thermal FFF to polymer blends and copolymers [6].

Although D_T varies with the polymer–solvent system, it is independent of molecular weight in a given system, at least for random coil homopolymers. The separation of differing molecular-weight components is therefore based solely on differences in D, which means that the principles of universal calibration that are relevant to SEC are also applicable to thermal FFF. Thus, a calibration curve made with one polymer–solvent system can be applied to other systems, provided the two D_T values associated with each polymer–solvent system are available. Fortunately, accurate values of D_T can be obtained from the combination of thermal FFF with any technique that measures D, such as dynamic light scattering or even SEC (the latter is a secondary method that relies on calibration curves). Values of D_T for many polymer–solvent systems are available in the thermal FFF literature.

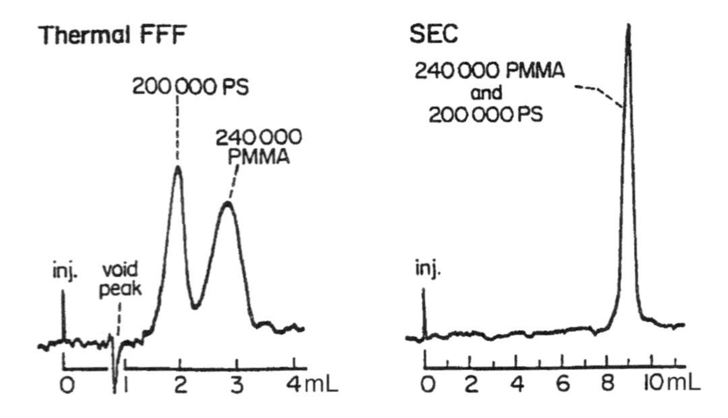

Fig. 1 Elution of similarly sized polystyrene (PS) and poly(methyl methacrylate) (PMMA) polymers by thermal FFF and SEC, illustrating the dependence of thermal FFF retention on polymer composition. (Adapted, with permission, from Ref. 2.)

Compared to SEC, a broader concept of universal calibration is possible with thermal FFF because thermal FFF channels do not contain the inherent variability associated with SEC packing materials. As a result, calibration curves are universal to all thermal FFF channels that use the same cold-wall temperature. A common cold-wall temperature is important because both D and D_T vary with temperature.

Retention in thermal FFF is directly proportional to the temperature drop (ΔT) across the channel [see Eq. (1)]. The linear relationship between V_r and ΔT holds even for moderate levels of retention ($V_r/V^0 > 3$). Having a predictable dependence of retention on ΔT means that ΔT can be efficiently tuned in order to optimize the trade-off between resolution and analysis time for each application. ΔT can also be varied over the course of a separation in order to resolve samples of extreme polydispersity in the most efficient manner. Decreasing ΔT over the course of a separation is analogous to temperature programming in gas chromatography or solvent programming in reversed-phase high-performance liquid chromatography (HPLC).

Applications of Thermal FFF

Historically, thermal FFF has been applied primarily to lipophilic polymers. The technique has not found wide applicability to hydrophilic polymers because thermodiffusion, and therefore retention, is very weak in water. Although a few hydrophilic polymers have been separated by thermal FFF, they generally must have a high molecular weight ($>10^7$ g/mol) in order to be adequately retained for characterization by thermal FFF. Alternatively, an aprotic solvent such as dimethyl sulfoxide can be used to separate hydrophilic polymers with lower molecular weights. However, flow FFF is more suited to the characterization of hydrophilic polymers.

The open FFF channel is especially suited to fragile materials, and thermal FFF has found a definite niche in its application to ultrahigh-molecular-weight polymers. Furthermore, because samples need not be filtered, thermal FFF is the technique of choice for analyzing gels, rubbers, and other materials that tend to plug SEC columns [7]. Even particles can be analyzed by thermal FFF [8], although flow and sedimentation FFF are more commonly used. Still, thermal FFF has the unique ability to separate particles by their composition [9], and as we increase our understanding of the thermodiffusion of particles, new applications are sure to emerge.

Advances in our understanding of thermodiffusion in polymer solutions have led to the application of thermal FFF to copolymers. With random copolymers, for example, the dependence of D_T on chemical composition is now predictable [6], so that compositional information can be obtained from retention measurements. With block copolymers, thermal FFF can still be used to separate components according to molecular weight, branching, and composition, but independent measurements on the separated fractions must be made in order to get quantitative information, except when "special" solvents are used. Special solvents yield a predictable dependence of D_T on composition even for block copolymers [6]. Different solvents are "special" for different copolymer systems.

As with SEC, the combination of thermal FFF with information-rich detectors is a powerful one. Detectors that yield information on the intrinsic viscosity, molecular weight, or chemical composition of eluting fractions allow detailed information on the distribution of such parameters in a complex sample. Thermal FFF can also be combined with SEC or hydrodynamic chromatography to achieve two-dimensional separations. The combination of SEC and thermal FFF is a particularly powerful one and has been used to characterize variations in chemical composition with the molecular weight of copolymers [10].

The Price of Versatility

The ability of a single thermal FFF channel to be used for the separation of lipophilic polymers, gels, rubbers, and particles makes it a very useful tool for polymer and colloid analysis. The versatility, however, comes with a price. Because all of the various applications cannot be implemented with a single field strength or carrier liquid, the user must have more than just a basic familiarity with the technique. To use thermal FFF efficiently, while taking advantage of its versatility, the user must understand the fundamentals behind the separation mechanism and assimilate a certain amount of experience.

Perhaps the most common difficulty encountered by new users arises from their tendency to use high field strengths and high sample concentrations. Unlike SEC, the elution volume is, in principle, without limit, and the application of a high field can easily lead to an elution time of hours. Thus, it is best to begin with a small temperature gradient ($\Delta T = 10-20$ K) in the development of a new application, then increase it as necessary to achieve the desired resolution.

For a detailed discussion of appropriate sample concentrations, see the entry Thermal FFF of Polymers and Particles. In general, much lower sample concentrations are used in thermal FFF compared to SEC because the sample is initially concentrated at one wall by the field. Due to the viscous nature of flow in the thin FFF channel, sample concentrations that are too high lead to poor resolution and reproducibility. On the other hand, whereas samples are initially concentrated by the field, extremely polydisperse samples are eventually diluted over a wide range of retention volumes due to the high resolving power of the technique. As a result, such samples require highly sensitive detectors. Although this requirement once posed a significant problem, the availability of highly sensitive and universal detectors, such as the evaporative mass detector, has virtually eliminated the issue of sensitivity in thermal FFF.

References

1. M. E. Schimpf, *TRIP 4*: 114 (1996).
2. J. J. Gunderson and J. C. Giddings, *Anal. Chim. Acta 189*: 1 (1986).
3. S. Lee, in *Chromatographic Characterization of Polymers: Hyphenated and Multidimensional Techniques* (T. Provder, H. Barth, and M. Urban, eds.), American Chemical Society, Washington, DC, 1995, pp. 93–107.
4. J. J. Gunderson and J. C. Giddings, in *Comprehensive Polymer Science, Vol. I, Polymer Characterizations* (C. Booth and C. Price, eds.), Pergamon Press, Oxford, 1989, pp. 279–291.
5. M. E. Schimpf, *Entropie* (in press).
6. M. E. Schimpf, C. A. Rue, G. Mercer, L. M. Wheeler, and P. F. Romeo, *J. Coat. Technol. 65*: 51 (1993).
7. S. Lee, *J. Microcol. Separ. 9*: 281 (1997).
8. P. M. Shiundu and J. C. Giddings, *Anal. Chem. 67*: 2705 (1995).
9. S. J. Jeon, A. Nyborg, and M. E. Schimpf, *Anal. Chem. 69*: 3442 (1997).
10. A. van Asten, R. J. van Dam, W. T. Kok, R. Tijssen, and H. Poppe, *J. Chromatogr. 703*: 245 (1995).

Martin E. Schimpf

Thermal FFF of Polymers and Particles

Thermal field-flow fractionation (FFF) can be applied to virtually any polymer that can be dissolved in an organic solvent (subject to low-molecular-weight limitations discussed here) [1]. Water-soluble polymers are more difficult to separate because thermodiffusion, and therefore retention, is weak in water and other protic solvents. Still, certain nonionic polymers can be separated, and with the use of mobile-phase additives, even charged materials have been retained. For example, poly(ethylene oxide) and poly(ethylene glycol) show strong retention in a range of aqueous solvents, including deionized water, whereas poly(styrene sulfonate) is moderately retained in 5 m*M* Tris-Na_2SO_4 buffer [2]. Proteins, on the other hand, have not been successfully separated by thermal FFF.

Other hydrophilic polymers that could not be retained in water have been separated in aprotic solvents. For example, a variety of polysaccharides have been separated in dimethyl sulfoxide (DMSO), including pullulans, dextrans, various starches, and Ficoll™. The latter material is a highly branched copolymer of sucrose and epichlorohydrin. Dextrans have been separated in mixtures of water and DMSO.

The major limitation of thermal FFF occurs in the separation of low-molecular-weight materials. Thus, the technique is not widely applicable to molecular weights below about 10^4 g/mol. This limit can be reduced somewhat by the use of solvent mixtures. For example, polystyrene components as small as 2500 g/mol were resolved in a mixture of tetrahydrofuran and dodecane [4]. Even lower molecular weights than 2500 g/mol have been retained, but only through the use of special channels, which were highly pressurized in order to increase the temperature gradient without boiling the solvent.

Thermal FFF has virtually no upper limit to the molecular weights that can be resolved. The channel is not packed with a stationary phase, and the flow of carrier liquid through the channel is laminar. As a result, large materials can be eluted without plugging the channel, and fragile materials are eluted without being damaged. In a comparison of thermal FFF and size-exclusion chromatography (SEC) [5], for example, it was found that SEC consistently underestimates the molecular weight of ultrahigh-molecular-weight polymers, even when extremely low flow rates are employed and a multiangle light-scattering detector is used to directly measure the molecular weight of the eluting components. In addition to the problem of shear-induced damage, SEC suffers from anomalous retention effects due to adsorptive interactions between high-molecular-weight polymers and the stationary phase. Although not completely absent, surface interactions are minimized in thermal FFF because there is no packing material. Finally, the resolution of thermal FFF for ultrahigh-molecular-weight polymers is vastly superior to that of SEC. As a result, thermal FFF enjoys a unique niche in the separation of these materials.

Another application in which thermal FFF enjoys an advantage over SEC is the analysis of high-temperature polymers. The operating temperature is limited only by the degradation temperature of the spacer used to form the channel, which for polyimides can be as high as 600 K. In the analysis of high-molecular-weight polyethylene, for example, temperatures in excess of 400 K are required for the samples to be soluble. Under these conditions, column stability and separation efficiency limit the application of SEC. By contrast, such samples can be routinely analyzed with commercially available thermal FFF channels.

Retention times in FFF are affected by the magnitude of the applied field, which can therefore be optimized for a particular molecular-weight range. For extremely broad molecular-weight distributions, the field can be programmed. Figure 1 illustrates the separation of polystyrene standards ranging in molecular weight from 9×10^3 to 5.5×10^6 g/mol in a single run. Note that the elution order is from low to high molecular weight. The field, which is expressed in Fig. 1 by the temperature difference (ΔT) between hot and cold walls, was programmed to decay exponentially from 80 to 10 K over the course of the 25-min separation.

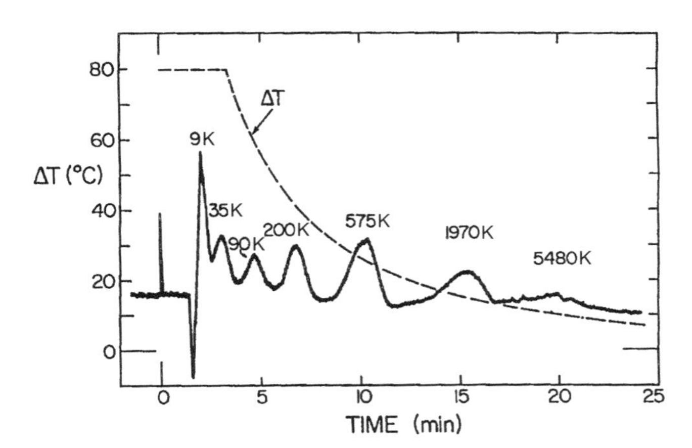

Fig. 1 Separation of seven polymers of indicated molecular weights by programmed thermal FFF. (Reproduced, with permission, from Ref. 7.)

Polymer Gels

Thermal FFF is also unique in its ability to handle gel-containing polymers. To analyze such materials by SEC, the sample solution must first be filtered to prevent damage in the form of contamination or even blockage in the column. With thermal FFF, such samples can be injected directly into the channel, and the gel content can even be characterized. The advantage of thermal FFF for analyzing samples in which the gel contains information that is critical to the analysis has been demonstrated with acrylate elastomers [5]. Thermal FFF was used to correlate the gel content of such elastomers to mechanical properties after SEC combined with viscometry and light scattering failed to elucidate any difference among the samples. In a related application, thermal FFF also proved superior in the analysis of natural rubber [8], where the ability to analyze unfiltered samples is a clear advantage.

Polymer Blends and Copolymers

The driving force for polymer retention in thermal FFF is thermodiffusion, which varies with polymer composition. As a result, polymer blends and copolymers can be separated even when the molecular weights or diffusion coefficients are identical. Furthermore, because thermodiffusion can be measured quantitatively by thermal FFF through the thermodiffusion coefficient (D_T), the compositional distribution can, in principle, be obtained directly from elution profiles, provided the dependence of D_T on composition is known. In practice, there are several complications. First, retention does not yield D_T directly, but rather the Soret coefficient, which is the ratio of D_T to the ordinary diffusion coefficient (D). Because compositional information is contained in D_T alone, an independent measure of D must be available. Second, a general model for the dependence of D_T on composition has not been established; therefore, the dependence must be determined empirically for each polymer–solvent system. Fortunately, D_T is independent of molecular weight, and for certain copolymers, the dependence of D_T on chemical composition has been established. With random copolymers, for example, D_T is a weighted average of the D_T values for the corresponding homopolymers, where the weighting factors are the mole fractions of each component in the copolymers [9]. As a result, the composition of random copolymers can be determined by combining thermal FFF with any technique that measures D.

Several techniques for measuring D have been combined with thermal FFF to obtain values of D_T and, subsequently, the composition of separated polymer and copolymer components. Besides dynamic light scattering, SEC can be used by establishing calibration curves that relate log D to retention volume. The validity of such calibration curves are based on the same arguments used for universal calibration in SEC [10]. The method was demonstrated in the characterization of both the molecular weight and composition of a styrene–isoprene copolymer [10]. Even more powerful is the combination of thermal FFF and SEC to produce two-dimensional separations, as demonstrated on a four-component blend of homopolymers and block copolymers of styrene and ethylene oxide [11]. In two-dimensional separations, the components are first separated by SEC [12]. Fractions from the SEC elution profile, which are homogeneous in D, are further separated according to chemical composition by thermal FFF. The SEC retention volume can be combined with information from mass and viscosity detectors to calculate the viscosity-average molecular weight of the components. The thermal FFF retention volume can be combined with the D value obtained from SEC to calculate D_T values and, subsequently, the composition of the separated polymer components.

With block copolymers, compositional analysis by thermal FFF is complicated by the fact that thermal diffusion is dominated by the composition of monomer units in the outer free-draining region of the polymer–solvent sphere. When block copolymers are dissolved in a selective solvent, which is a solvent that solvates certain blocks better than others, the more soluble blocks tend to segregate to the outer regions of the polymer–solvent sphere. In the extreme case of such segregation effects, the thermal diffusion of the block copolymer will mimic that of a homopolymer composed of the better-solvated component. As a result, the compositional characterization of block copolymers by thermal FFF requires the use of a nonselective solvent. In the styrene–ethylene oxide work discussed earlier [11], two solvents were combined in the proper proportion to yield a nonselective mixture and, therefore, a linear dependence of D_T on composition.

Colloids and Particles

Historically, the application of FFF to colloids and particles has been limited to flow and sedimentation FFF. However, the thermal FFF channel in not only capable of separating these materials, it is simpler in design and can be used with both aqueous and organic solvents. Furthermore, the dependence of retention on chemical composition presents unique opportunities for the separation of such materials.

The application of thermal FFF to a variety of colloids and particles has been demonstrated in both aqueous and organic carrier liquids. Figure 2 illustrates the de-

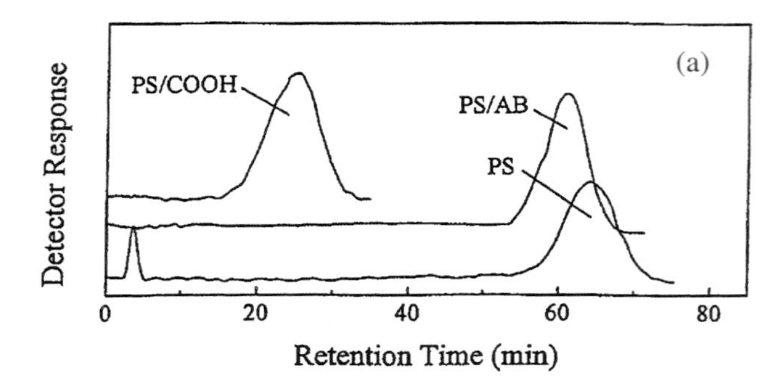

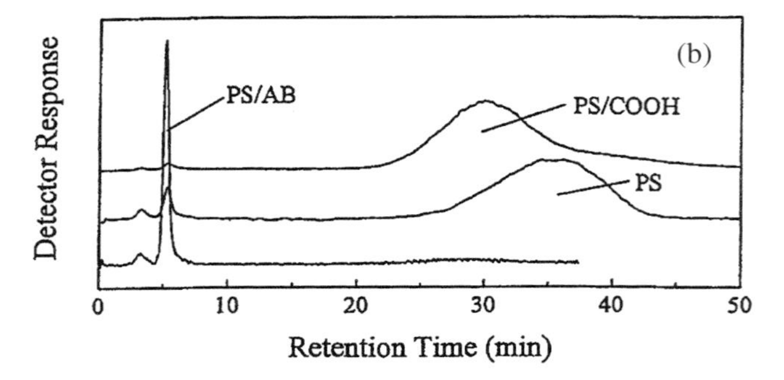

Fig. 2 A comparison of thermal FFF elution profiles for 0.2 μm particles of polystyrene (PS), carboxylated PS (PS/COOH), and aminated PS (PS/AB) in (a) an aqueous solution containing 9 mM NaN_2 and 0.05 wt% FL–70 surfactant (pH 8.5); (b) 10 mM phosphate buffer (pH 4.7). Differences in retention are due to differences in thermodiffusion, which is governed by the surface composition of the particle.

pendence of retention on the surface composition of polystyrene particles. The three particles are similar in size, but the surface of one of the samples has been carboxylated, whereas another has been aminated. The relative elution order of the three particles can be changed by modifying the carrier liquid [13].

Experimental Considerations

One of the most common mistakes made in the analysis of polymers by thermal FFF and SEC is the injection of too much sample; this is often referred to as overloading. It is important to understand that D, and therefore retention volume (V_r), varies with polymer concentration. The effect of concentration is especially large near the so-called critical concentration (c^*), where polymer solutions undergo an abrupt transition from "dilute to semidilute" behavior. In the semidilute regime, polymer coils tangle with one another and the magnitude of D drops dramatically. Therefore, it is important to keep the polymer concentration below c^* in order to avoid overloading effects, which are manifested as peak "fronting" and an increase in V_r. With excessive overloading, additional peaks occur in thermal FFF due to the formation and separation of aggregates.

Although experienced users of SEC are aware of overloading effects, they are not always aware that overloading occurs with a smaller sample load in thermal FFF. First, the channel has a relatively small volume (1–2 mL); therefore, injection volumes are smaller (3–30 μL) compared to SEC. More importantly, the sample is initially concentrated against one wall by the field. The extent of concentration varies with the sample and the magnitude of the applied field. A general rule that should be used to avoid overloading when a new method is being developed is to prepare the sample with a concentration that is one-tenth of the expected critical concentration. Of course, the critical concentration varies with molecular weight and other factors. For the experienced SEC user, a sample concentration of one-tenth that used in SEC should suffice. Sample concentrations between 0.01 and 1 mg/mL are typical. When one is uncertain of the critical concentration, the injected concentration should be varied and the elution profiles examined for indications of sample overloading. Otherwise, peak shape and reproducibility will be compromised. With recent advances in the sensitivity of detectors used for polymer analysis, sample overloading is less of a problem because more dilute samples can be analyzed, but overloading can still occur if the user is not aware of the potential.

References

1. M. E. Schimpf, *TRIP 1*: 74 (1993).
2. J. J. Kirkland and W. W. Yau, *J. Chromatogr. 353*: 95 (1986).
3. J. Lou, M. N. Myers, and J. C. Giddings, *J. Liquid Chromatogr. 17*: 3239 (1994).
4. C. A. Rue and M. E. Schimpf, *Anal. Chem. 66*: 4054 (1994).
5. S. Lee, in *Chromatography of Polymers: Characterization by SEC and FFF* (T. Provder, ed.), ACS Symposium Series Vol. 521, American Chemical Society, Washington, DC, 1993, pp. 77–88.
6. L. Pasti, S. Roccasalvo, F. Dondi, and P. Reschiglian, *J. Polym. Sci. B 33*: 1225 (1995).
7. J. C. Giddings, V. Kumar, P. S. Williams, and M. N. Myers, in *Polymer Characterization by Interdisciplinary Methods* (C. D. Craver and T. Provder, eds.), ACS Symposium Series Vol. 227, American Chemical Society, Washington, DC, 1990.
8. S. Lee and A. Molnar, *Macromolecules 28*: 6354 (1995).
9. M. E. Schimpf, L. M. Wheeler, and P. F. Romeo, in *Chromatography of Polymers: Characterization by SEC and FFF* (T. Provder, ed.), ACS Symposium Series Vol. 521, American Chemical Society, Washington, DC, 1993, pp. 63–76.

10. M. E. Schimpf, in *Chromatographic Characterization of Polymers* (T. Provder, H. G. Barth, and W. Urban, eds.), ACS Symposium Series Vol. 247, American Chemical Society, Washington, DC, 1995, pp. 183–196.
11. S. J. Jeon, and M. E. Schimpf, in *Chromatography of Polymers: Characterization by SEC, FFF, and Related Methods for Polymer Analysis* (T. Provder, ed.), American Chemical Society, Washington, DC, 2000.
12. A. C. van Asten, R. J. van Dam, W. Th. Kok, R. Tijssen, and H. Poppe, *J. Chromatogr. A 703*: 245 (1995).
13. S. J. Jeon, A. Nyborg, and M. E. Schimpf, *Anal. Chem. 67*: 3442 (1997).

Martin E. Schimpf

Thermal FFF of Polystyrene

Thermal FFF (thermal field-flow fractionation) is an elution-type separation technique applicable to the characterization of various synthetic organic polymers with molecular weights higher than about 10^4 [1]. In thermal FFF, a dilute solution of polymer sample is injected into a thin ribbon-shaped flow channel across which an external "field" (in the form of a temperature gradient) is applied. Under the influence of the temperature gradient, different components of the sample are carried down the channel at different velocities, leading to the elution of different components at different times and separation is achieved.

Molecular-Weight-Based Separation

In thermal FFF, retention time t_r is given by

$$t_r = \frac{t^0 \Delta T}{6}\left(\frac{D_T}{D}\right) \tag{1}$$

for well-retained components. Here t^0 is the channel void time (a constant for a given channel dimension), ΔT is the temperature drop across the channel, D_T is the thermal diffusion coefficient, and D is the mass diffusion coefficient. For most polymeric materials, D is related to the molecular weight M by

$$D = \frac{A}{M^b} \tag{2}$$

Combining Eqs. (1) and (2) yields

$$t_r = \frac{t^0 \Delta T D_T}{6A} M^b \tag{3}$$

It has been shown that D_T is independent of branching configuration and molecular weight for homopolymers (such as polystyrene) in a given solvent [3]. Thus, under a fixed experimental condition, the retention time of polystyrene depends only on the molecular weight, resulting in a molecular-weight-based separation (i.e., the retention time increases as the molecular weight increases).

Effect of ΔT in Polystyrene Separation

It is seen, from Eq. (3), that dt_r/dM (difference in retention time for the same molecular-weight difference) increases with ΔT, indicating the separation (and thus the resolution) between two polystyrene components increases with ΔT. However, the gain in the resolution comes at the cost of time, as the use of higher ΔT requires longer analysis times. An optimum condition for ΔT must be determined for each sample by observing the time and the profile of the elution at various ΔT values.

Finding the optimum ΔT is relatively easy for samples having narrow molecular-weight distributions. However, for samples having broad distributions, finding that the optimum ΔT may not be trivial. Field programming may be required for samples having very broad molecular-weight distributions, where ΔT starts at a high level and is gradually decreased during a run to prevent excess retention of high-molecular-weight components [4]. Figure 1 shows a separation of four polystyrene standards having nominal molecular weights ranging from 4.7×10^4 to about 1×10^6 Da.

Determination of Physicochemical Properties

The simplicity of both retention mechanism and the channel geometry of thermal FFF allows one to theoretically predict the degree of retention of a sample, once certain physicochemical properties of the sample are known. Conversely, one may determine the physicochemical properties of a sample by measuring its retention.

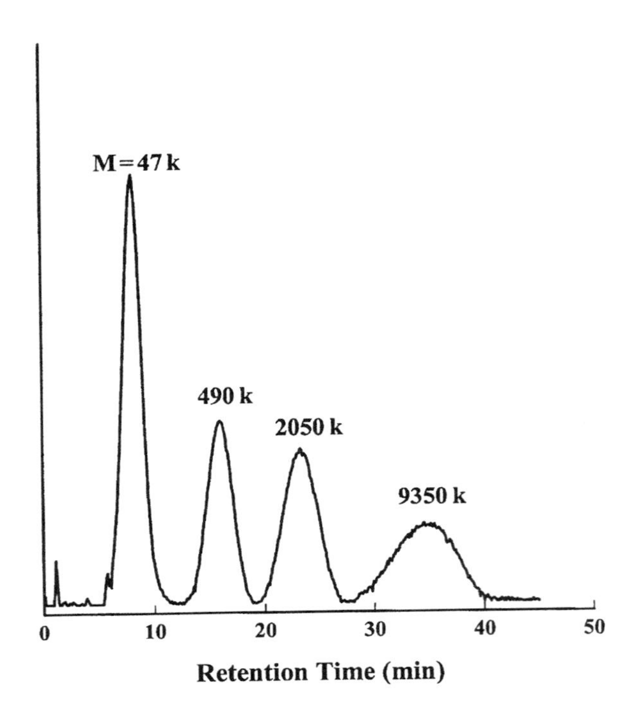

Fig. 1 Thermal FFF separation of polystyrene standards having nominal molecular weights as indicated. A power programming was used with initial $\Delta T = 80°C$, predecay time $t_l = 3$, $t_a = -6$, and the hold $\Delta T = 10°C$. The solvent/carrier was tetrahydrofuran and the flow rate was fixed at 0.2 mL/min.

Determination of Diffusion Coefficients

As seen in Eq. (1), the retention time in thermal FFF is a function of the ratio between two diffusion coefficients, D_T/D. If one of the two diffusion coefficients is known, the other can be determined directly from the retention time using Eq. (1). It is noted that the values of the diffusion coefficients may vary in various solvents.

Determination of Molecular Weight

If the values of A, b, and D_T are available, the molecular weight of a polymer can be directly determined from its retention time using Eq. (3). If D_T is not available, one may use a calibration curve [log (D/D_T) versus log M] constructed with a series of narrow polystyrene standards of known molecular weights. For the molecular-weight analysis of an unknown, the D/D_T value of the sample is first calculated from its measured retention time, and then the molecular weight is determined from the calibration curve.

Use of molecular-weight-sensitive detectors (e.g., differential viscometer or light-scattering detector) eliminates the need for calibration. With a differential viscometer used as a detector for thermal FFF, the intrinsic viscosity is first measured for each retention time and is then converted to molecular weight using the Mark–Houwink constants [5]. The use of accurate values of Mark–Houwink constants is essential in this method. Unlike the viscometer, the light-scattering detector measures the absolute molecular weight of the polymers directly. In multiangle light scattering, the scattered-light intensity is measured over a broad range of the scattering angles, allowing the determination of molecular size (radius of gyration) as well as the molecular weight [6].

Analysis of High-Molecular-Weight Polystyrene

Thermal FFF is often compared with size-exclusion chromatography (SEC), as both can be used for the same application (i.e., the molecular-weight determination of polymers). In SEC, the elution volume of the sample in a column (or in a column set) is limited to values between the interstitial volume of the column ("total exclusion limit") and the total liquid volume ("total permeation limit"). The resolving power of SEC quickly drops as the elution volume approaches these limits. By contrast, there is no theoretical limit in thermal FFF, and the applicable molecular weight is virtually unlimited to the high-molecular-weight region. Also, the openness of the channel minimizes the shear degradation and adsorption of polymers sometimes observed with SEC. Thus, thermal FFF is useful for characterizing high-molecular-weight polymers that are difficult to analyze using SEC.

It is noted, however, that for polymers of molecular weight below about 1×10^4 Da, SEC may be more useful than thermal FFF. Analysis of such low-molecular-weight samples using thermal FFF requires the application of a very high ΔT that may require the use of a pressurized system to avoid boiling of the solvent.

References

1. J. C. Giddings, *Science 260*: 1456 (1993).
2. P. M. Shiundu, E. E. Remsen, and J. C. Giddings, *J. Appl. Polym. Sci. 60*: 1695 (1996).
3. M. E. Schimpf and J. C. Giddings, *Polym. Sci. Part B: Polym. Phys. 27*: 1317 (1989).
4. P. S. Williams and J. C. Giddings, *Anal. Chem. 59*: 2038 (1987).
5. J. J. Kirkland and S. W. Rementer, *Anal. Chem. 64*: 904 (1992).
6. S. Lee and O.-S. Kwon, *Polym. Mater. Sci. Eng. 65*: 408 (1993).

Seungho Lee

Thermodynamics of GPC–SEC Separation

Partitioning

As with most other chromatographic separation methods, gel permeation chromatography–size-exclusion chromatography (GPC–SEC) is based on partitioning of analyte polymer molecules between the stationary phase (which, in the case of GPC–SEC, is contained within the pores of a porous stationary-phase support) and the mobile phase. Ideally, the mobile phase establishes a concentration equilibrium with the stationary phase at each plate in the column before being transferred to the next plate. The concentration equilibrium is dictated by an equal chemical potential of the polymer chain between the two phases [1,2]. In normal conditions of GPC–SEC, the polymer concentration c_M in the mobile phase is sufficiently low, and the solution behaves ideally dilute. Its chemical potential μ_M per molecule in the mobile phase is then given as

$$\mu_M = \mu^0 + k_B T \ln\left(\frac{c_M}{c^0}\right) \qquad (1)$$

where μ^0 and c^0 are the chemical potential and the concentration, respectively, at an appropriate reference state, k_B is the Boltzmann constant, and T is the temperature. In the stationary phase, the chemical potential μ_S has additional terms due to changes in the entropy and the enthalpy upon bringing the polymer chain into the stationary phase:

$$\mu_S = \mu_0 + k_B T \ln\left(\frac{c_S}{c^0}\right) - T\Delta S + \Delta H \qquad (2)$$

where c_S is the polymer concentration in the stationary phase, and ΔS and ΔH refer to the changes per chain. The partition coefficient K, defined as the ratio of the two concentrations, is then given as

$$K \equiv \frac{c_S}{c_M} = \exp\left(\frac{\Delta S}{k_B} - \frac{\Delta H}{k_B T}\right) \qquad (3)$$

The primary purpose of GPC–SEC is to obtain a retention curve that represents the molecular weight (MW) distribution of the analyte. Because the retention volume is a linear function of the partition coefficient, the coefficient must be a monotonically decreasing or increasing function of MW.

In GPC–SEC, porous beads of various pore sizes are used as separating media. The solution in the pore channels is the stationary phase, and the volume of liquid between the beads constitutes the mobile phase. A linear polymer chain can take different conformations in the unrestricted mobile phase. The number of conformations, W, increases in a power law of N, with N being the degree of polymerization. In the stationary phase, however, all of the monomers on a single chain have to reside within the pore space. Thus, W does not increase as rapidly with N as in the mobile phase. Because the conformational entropy of the polymer chain is calculated as $k_B \ln W$, the geometrical confinement of the pore makes $\Delta S < 0$. The way $|\Delta S|$ increases with N is determined by the geometry of the polymer chain (whether the chain is linear or branched, whether the chain is swollen or not, whether the chain is stiff or not, etc.) and the pore geometry, but $|\Delta S|$ is roughly a function of the chain dimension to the pore size, as shown below. The detailed atomic sequence in each polymer is rather of second importance. This feature is responsible for the universality of GPC–SEC.

Columns used in GPC–SEC are supposed to provide negligible ΔH regardless of MW for a given polymer. Otherwise, the dependence of ΔH on MW, different from that of ΔS, will complicate the analysis of the retention curve. Furthermore, ΔH would exhibit widely different characteristics determined by the interactions among the polymer, the pore surface, and the solvent, and thus negate the universality of GPC–SEC. With negligible ΔH, the partition coefficient is determined solely by the decrease in the conformational entropy:

$$K = \exp\left(\frac{\Delta S}{k_B}\right) \qquad (4)$$

Equivalently, K is the ratio of the probability to place the polymer chain in the pore without touching the pore walls, averaged over different conformations.

As MW increases, K decreases from 1 to 0. To have a high resolution in GPC–SEC, the dependence of K on MW needs to be sharp. Users of GPC–SEC also want the range of MW analysis in a single run to be as broad as possible, typically three to five decades. The finite range of K, however, makes the two requirements mutually exclusive, a severe restriction on the resolution. The resolution can be improved by increasing the number of theoretical plates, typically employing smaller porous beads and connecting a multiple columns in series. In normal-phase and reversed-phase high-performance liquid chromatography (HPLC), by contrast, a stationary phase that gives a large value of K, in excess of unity, offers high resolution and a broad range of analyte composition in a relatively short column.

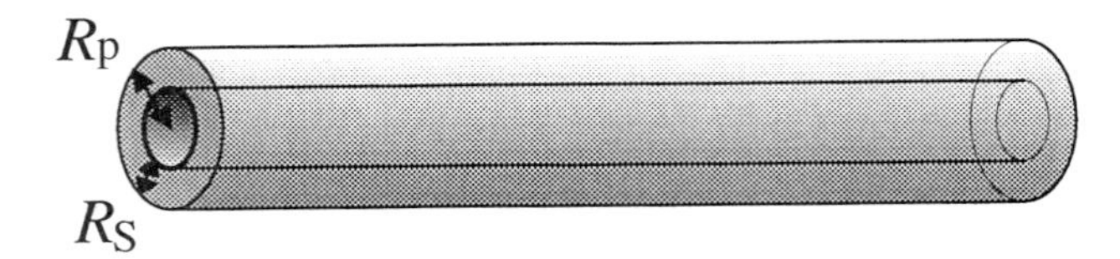

Fig. 1 The volume accessible to the center of a spherical molecule of radius R_S within a cylindrical pore of radius R_P is the interior cylinder of radius $R_S - R_P$.

Partition Coefficient

Examples for the plots of the partition coefficient as a function of the chain dimension are shown for some of the geometries of polymer molecules in a cylindrical pore of radius R_P. The simplest geometry is a sphere. The center of a spherical molecule of radius R_S is not allowed to get closer to the pore wall beyond the distance equal to the sphere radius. The ratio of the volume accessible to the sphere center to the total pore volume gives the partition coefficient. In Fig. 1, the two volumes are indicated by the interior and the exterior cylinders, respectively. Then,

$$K = \left(1 - \frac{R_S}{R_P}\right)^2 \tag{5}$$

The partition coefficient of a Gaussian chain of radius of gyration R_g was calculated by Casassa [3] and the partition coefficient of a rodlike molecule was obtained by Giddings et al. [4]. The results are compared in Fig. 2. When plotted as a function of R_g/R_P, the three polymer geometries show only a small difference, except that, at large R_g/R_P, the rodlike molecule has a larger K than the other two geometries of the molecule. The plots were converted to functions of N, in Fig. 3, for the monomer size a equal to $R_P/100$. The partition coefficient of a star-branched polymer, in which each arm takes a Gaussian conformation, was calculated by Casassa and Tagami [5].

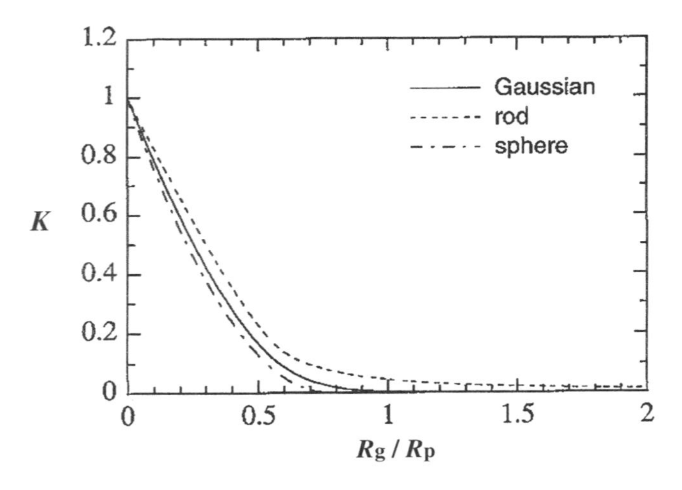

Fig. 2 Partition coefficient K plotted as a function of the radius of gyration R_g relative to the pore radius R_P for a spherical molecule, Gaussian chain, and a rodlike molecule.

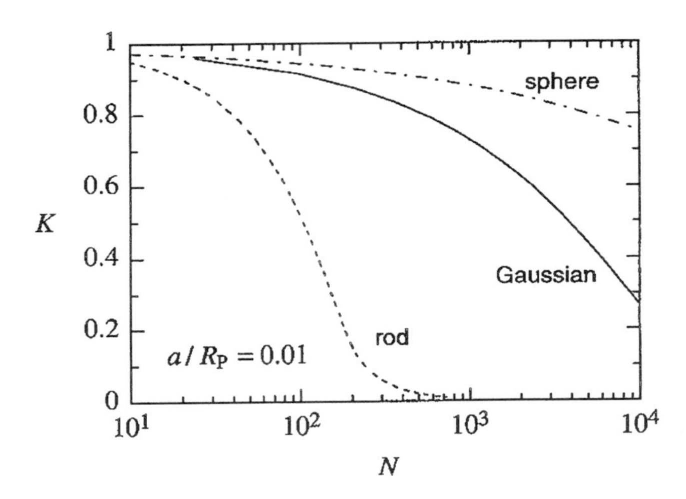

Fig. 3 Partition coefficient K plotted as a function of degree of polymerization, N, for a spherical molecule, Gaussian chain, and a rodlike molecule. The monomer size a is set to be $R_P/100$.

The partition coefficient for a real chain (excluded volume chain) has not been obtained except by the scaling theory [6]. The theory gives only a qualitative relationship between K and N for chains sufficiently longer than the pore size ($R_g \gg R_P$):

$$\frac{\Delta S}{k_B} \equiv -N\left(\frac{a}{R_P}\right)^{1/\nu} \approx -\left(\frac{R_g}{R_P}\right)^{1/\nu} \tag{6}$$

where $\nu \approx 0.588$. A computer simulation result shows that, compared at the same R_g, the real chain has a slightly larger K than the ideal chain that allows monomer overlap [7]. It is common to both models for sufficiently long chains that $K \sim \exp(-\beta M)$, where β is a numerical coefficient and M is the MW.

All of the currently used porous packing materials have a three-dimensional network structure, effectively giving rise to a pore size distribution. In these separating media, the dependence of K on N will be less sharp compared with the one in Fig. 3. It is desired by chromatographers that the retention time is a linear function of log M. Because the retention time is a linear function of K, the plot of K needs to be a linear function of log M in as broad a range of MWs as possible. A naturally occurring pore size distribution is not sufficient to cause the desired linearity.

Therefore, mixed-bed columns, packed with porous materials of different pore-size-distribution ranges, have been developed and used broadly as "linear" columns.

Effect of Intermolecular Interactions

It is often observed that, as the concentration of the polymer solution injected increases at a constant injection volume, the retention curve shifts toward a longer time [8]. This effect, called "overloading," is manifested universally when the mobile phase is a good solvent for the polymer being analyzed and is more significant for a polymer of a higher MW. Simple thermodynamics explains this effect [2]: As c_M increases, the solution starts to deviate from the ideally dilute solution, and it becomes difficult to neglect the second virial coefficient A_2. For a solution in a good solvent, a positive A_2 makes μ_M larger than the one in the ideally dilute solution of the same concentration, as given by Eq. (1). The same change applies to μ_S. Since $c_S < c_M$, the increase in μ_M is greater than the increase in μ_S, resulting in an increase in K. At higher concentrations, the analyte polymer is retained longer, thereby delaying the retention curve. The concentration at which the effect of A_2 becomes not negligible is around the overlap concentration c^* which can be defined as $c^*[\eta] = 1$, where $[\eta]$ is the intrinsic viscosity. The actual concentration of c^* is low, especially for high-MW polymer. This is why a concentration as low as 0.1 wt% sometimes exhibits a concentration-dependent chromatogram.

Another mechanism that causes a change in the retention curve with concentration is intermolecular association. In a solution in which A_2 is negative, such as a solution in a poor solvent, a polymer chain tends to associate with other chains at higher concentrations. In effect, the pore in the GPC–SEC column senses the presence of suspensions of a larger dimension, causing the retention curve to shift to a shorter time.

References

1. J. C. Giddings, *Unified Separation Science*, John Wiley & Sons, New York, 1991.
2. I. Teraoka, *Progr. Polym. Sci. 21*: 89 (1996).
3. E. F. Casassa, *J. Polym. Sci. Polym. Lett. Ed. 5*: 773 (1967).
4. J. C. Giddings, E. Kucera, C. P. Russell, and M. N. Myers, *J. Phys. Chem. 72*: 4397 (1968).
5. E. F. Casassa and Y. Tagami, *Macromolecules 2*: 14 (1969).
6. P. G. de Gennes, *Scaling Concepts in Polymer Physics*, Cornell University Press, Ithaca, NY, 1979.
7. Y. Wang and I. Teraoka, *Macromolecules 30*: 8473 (1997).
8. R. E. Boehm, D. E. Martire, D. W. Armstrong, and K. H. Bui, *Macromolecules 17*: 400 (1984).

Iwao Teraoka

Thermodynamics of Retention in Gas Chromatography

The Plate Theory shows that retention volume of a solute is directly proportional to its distribution coefficient between the two phases. Classical thermodynamics provides an expression that relates the *equilibrium constant* which, in the case of chromatographic retention, will be the distribution coefficient to the change in *standard free energy* of the solute, when transferring from one phase to the other.

The expression is

$$RT \ln K = -\Delta G_0$$

where R is the gas constant, T is the absolute temperature, and ΔG_0 is the standard free-energy change.

Now,

$$\Delta G_0 = \Delta H_0 - T\Delta S_0$$

where (ΔH_0) is the standard enthalpy change and ΔS_0 is the standard entropy change.

Thus,

$$\ln K = -\left(\frac{\Delta H_0}{RT} - \frac{\Delta S_0}{R}\right)$$

or

$$K = \exp\left[-\left(\frac{\Delta H_0}{RT} - \frac{\Delta S_0}{R}\right)\right] \tag{1}$$

Equation (1) can also be used to identify the type of retention mechanism that is taking place in a particular separation by measuring the retention volume of the solute over a range of temperatures. Rearranging Eq. (1),

$$\log K = -\frac{\Delta H_0}{RT} + \frac{\Delta S_0}{R}$$

Bearing in mind that

$$V' = KV_S$$

$$\log V' = -\frac{\Delta H_0}{RT} + \frac{\Delta S_0}{R} - \log V_S$$

It is seen that a curve relating $\log(V')$ to $1/T$ should give a straight line, the slope of which will be proportional to the *enthalpy* change during solute transfer. In a similar way, the intercept will be related to the *entropy* change and, thus, the dominant effects in any distribution system can be identified from such curves. Curves relating $\log(V')$ to $1/T$ are called van't Hoff curves, which can be used to identify the mechanism of retention and elucidate the role played by temperature in a separation (see the entries van't Hoff Curves and Chiral Separations by GC).

In the majority of distribution systems encountered in gas chromatography, the slopes of the van't Hoff curves are positive and the intercept negative. The negative value of the intercept means that the standard entropy change of the solute has resulted from the production of a less random and more orderly system during the process of distribution. More important, this entropy change *reduces* the magnitude of the distribution coefficient. This means that the greater the forces between the molecules, the greater the energy (enthalpy) contribution, the larger the distribution coefficient, and the greater the retention. In contrast, any reduction in the random nature of the molecules or an increased amount of order in the system reduces the distribution coefficient and *attenuates* the retention. Thus, in the majority of distribution systems in chromatography, the enthalpy and entropy changes oppose one another in their effect on solute retention, although one will generally dominate over the other.

Suggested Further Reading

Scott, R. P. W., *Techniques of Chromatography*, Marcel Dekker, Inc., New York, 1995.

Scott, R. P. W., *Introduction to Analytical Gas Chromatography*, Marcel Dekker, Inc., New York, 1998.

Raymond P. W. Scott

Thin-Layer Chromatographic Study of Quantitative Structure–Retention Relationships

Introduction

The study of quantitative structure–retention relationships (QSRRs) is one of the most important theoretical fields of chromatography; it has become a new investigation branch of chromatographic science.

Quantitative structure–retention relationships studies are widely investigated in high-performance liquid chromatography (HPLC), gas chromatography (GC), and thin-layer chromatography (TLC). Recently, QSRR studies in TLC have attracted more and more researchers [1]. It is known that TLC has some advantages: It is rapid, relatively simple, low cost, and easy to operation, there is a wide choice of adsorbents and solvents, and very small amounts of substance are needed. In this entry, the establishment and application of QSRR studies are reviewed.

Establishment of QSRR Equations

Nonspecific parameters, physicochemical parameters, and topological indices are the main parameters used in QSRR studies in TLC. The establishment of QSRR equations in TLC are reviewed according to these parameters.

Based on Nonspecific Parameters

Nonspecific parameters include the number of carbon atoms in a molecule, molecular volume, solvent-accessible surface, and so forth. This is relatively simple, and the most commonly used nonspecific parameter is the number of carbon atoms in the compound.

There is a relationship between the number of carbon atoms and the retention data in HPLC and GC; a similar relationship was also found in TLC. Boyce and Milborrow [2] found that there was a linear relationship between the R_m values and the number of carbon atoms in the R group of *N*-*n*-alkyltritylamines (ph_3CNHR). Janjic et al. [3] carried out a series of studies on the relationship between transition metal complexes and the number of carbon atoms in the complexes.

Because of their simplicity, nonspecific parameters can be obtained relatively conveniently, but it is not very well correlated for complex molecules.

Based on Physicochemical Parameters

Physicochemical parameters are widely used in QSRR studies. Among the many physicochemical parameters, lipophilicity is one of the most widely used parameters in TLC QSRR studies. As is well known, there is a linear relationship between the R_m [$R_m = \log(1/R_f - 1)$] values and the connection of organic modifier in mobile phase:

$$R_m = R_{m0} + bc$$

where c is the concentration of organic modifier in mobile phase and b is the slope, which is the decrease of R_m values when the concentration of organic modifier in the mobile phase increases 1%. R_{m0} is the intercept of the TLC equation which represents the extrapolated R_m value (i.e., the theoretical R_m value at 0% organic solvent). This linear relationship has been verified by many workers [2,4].

Pliska and Schmidt [5] first described the theoretical relationship between R_f and P values and presented the details of the computational procedure. At the same time, the relationship between the intercept R_{m0} and the slope b in the TLC equation was also studied by many other researchers. Biagi et al. [4] performed extensive work in this field. They found that R_{m0} values were not dependent on the nature of the organic solvent and that it does not make much difference whether the organic solvent is acetone, methanol, or acetonitrile. In addition, there is a linear relationship between the intercept (R_{m0}) and the slope (b). Biagi et al. considered that the existence of the linear relationship is a reason why the chromatographic method can be used to determine lipophilicity. The intercept of the TLC equation can be considered as a measure of the partitioning of the compounds between a polar mobile phase and a nonpolar stationary phase (i.e., as the result of the balance between the interactions with the nonpolar phase and the interactions with the polar phase); whereas the slope of the TLC equation indicates the rate at which the solubility of the compound increases in the mobile phase [4]. There are many studies dealing with this aspect.

Based on Topological Indices

Based on the plot of the suppressed hydrogens of the molecular under study, the calculation methods of molecular connectivity indices was developed. It has been used in many fields, such as the studies of correlation between structure and physicochemical and pharmacological properties, especially the relationship between structure and retention behavior, and the evaluation of the hydrophobicities of organic compounds by chromatographic methods.

The relationship between the retention behavior of benzodiazepine, sulfamides, substituted anilines, barbiturates, a group of natural phenolic derivatives, diethanolamine isomers, amino acids, sulfoether, thio-alcohol, 2,4-dinitrophenylhydrazones, and their molecular connectivity indices were studied. All of the results indicate that the R_f or R_m values are highly correlated with molecular connectivity indices.

Pyka [6–8] performed a series of experiments applying topological indices to QSRR studies in TLC. Topological indices, such as Gutman, Randic, and Wiener indices as well as other indices were used. A new optical index (I_{opt}) was proposed [6], which enables distinction between isomers of D- and L-configuration. Pyka also established a new stereoisomeric topological index (I_{STI}), which enables distinction between stereoisomers with hydroxyl groups in axial and equatorial positions. Using this index, the retention behavior of stereoisomeric menthols and thujyols have been studied.

Introducing topological indices to QSRR studies widens the possibility of correlation analysis. Molecular connectivity indices are calculated according to molecular structure; it is relatively objective. Topological indices can be calculated without synthesizing the compound. In addition, using topological indices, the molecular structure can be described by some parameters which have physicochemical significance. However, there is also a limitation in the range of application of topological indices. They can be applied as the sole parameter only when they have been correlated with other features of molecular structure, such as volume, ring size, carbon chain length, and so forth. Generally, they are combined with other parameters that characterize other properties of the compounds of interest.

Combination of Several Kinds of Parameters

As we can see, every kind of parameters have limitations; thus, the combination of several kinds of parameters is advantageous. The combination of several kinds of parameters can often completely reflect the properties of compounds. Wang et al. [9] conducted some studies in this field. They introduced several structural parameters to study the correlation between the molecular structures of *O*-ethyl-*O*-aryl-*N*-isopropyl phosphoroamidothioates, *O*-ethyl, *O*-isopropyl phosphoro(thioureido) thioates, and their retention factors in high-performance TLC (HPTLC), respectively.

Application of QSRR Studies in TLC

Prediction of Retention and Separation

From the discussion earlier, it can be seen that all the QSRR equations established in the previous part can be used to predict R_f or R_m values of compounds. This goal of QSRR is very easy to comprehend.

Determination of Lipophilicity

Many works [5] have illustrated that there is a linear relationship between R_m values and log P. R_{m0} values also can be used to measure the lipophilicities of chemical substances. In general, the chromatographic method is an excellent alternative to the traditional flask-shaking method for lipophilicity determination. It avoids the difficulties that one may encounter in the flask-shaking method. It is simple and rapid and requires only minute amounts of substances (which need not necessarily be very pure); it does not need quantitative analysis; the nature of the organic modifier does not affect the measurement of the lipophilic character, as the R_{m0} value is not affected by the organic modifier.

Evaluation of Some Physicochemical Parameters of Chemicals

The research on the evaluation physicochemical parameters using the TLC QSRR equation is very limited. Until now, only the log $E_{T(30)}$ values of some organic solvents [10] and the pK_a values of some substituted phenols [8] have been predicted using this method. In addition, topological indices are introduced to evaluate the pK_a values.

Evaluation of Biological Activities of Compounds

Although, as early as 1965, Boyce and Milborrow [2] studied the relationship between R_m values of *N*-*n*-alkyltritylamines (ph_3CNHR) and their LD_{50} values, the research on this aspect is limited. When only the R_m values is considered, the activity [log(1/C) or LD_{50}] is correlated with the square of R_m.

Topological indices were also introduced to study the biological activities of compounds. Pyka [7] studied the relationship among R_m values, topological indices, and biological activity [log(1/C)]. All of the equations he obtained show good correlation. There is a good agreement between predicted log(1/C) values and experimental log(1/C) values.

Explanation of Separation Mechanism

Theoretically speaking, a good QSRR equation can reflect the most characteristic parameters that influence the retention behavior of compounds. Thus, it can explain the retention mechanism quantitatively and qualitatively, but research on this aspect has made only poor progress until now.

Cserhati and Forgacs [11,12] did some work on this field which illustrates the high impact of steric interactions on retention. However, until now, some of his hypotheses [12] need further experimental verification to support them.

Prospects

Quantitative structure–retention relationships play a vital role in chromatographic research. The research and application fields of QSRR studies are becoming increasingly broad. Up to now, compounds used in QSRR studied are homologous series. How to establish QSRR equations that fit different kinds of compounds requires much more experimental effort.

From the viewpoint of application, QSRR equations in TLC are mainly used for retention prediction. The explanation of the separation mechanism awaits further investigation. With the application of various statistical methods, it is possible to select the primary retention-effect factors from many solute related factors which will offer explanations of separation mechanisms.

In conclusion, QSRR studies are making excellent progress in recent years, due to increased interest in this field of study. We can anticipate that QSRR studies will become more and more important in the future.

References

1. Q. S. Wang and L. Zhang, *J. Liquid Chromatogr. Related Technol. 22*: 1 (1999).
2. C. B. Boyce and B. V. Milborrow, *Nature 208*: 537 (1965).
3. T. J. Janic, G. Vuckovic, and M. B. Celap, *Chromatographia 42*: 675 (1996).
4. G. L. Biagi, A. M. Barbaro, A. Sapone, and M. Recanatini, *J. Chromatogr. A 662*: 341 (1994).
5. V. Pliska, M. Schmidt , and J.-L. Fauchere, *J. Chromatogr. 216*: 79 (1981).
6. A. Pyka, *J. Planar Chromatogr.—Mod. TLC 4*: 316 (1991).
7. A. Pyka, *J. Planar Chromatogr.—Mod. TLC 7*: 108 (1994).
8. A. Pyka, *J. Planar Chromatogr.—Mod. TLC 9*: 52 (1996).
9. Q. S. Wang, B.-W. Yan, and H.-Z. Yang, *J. Planar Chromatogr.—Mod. TLC 10*: 118 (1997).
10. A. Ahmad, Q. S. Muzaffar, A. Andrabi, and P. M. Qureshi, *J. Chromatogr. Sci. 34*: 376 (1996).
11. T. Ceserhati, E. Forgacs, *J. Liquid Chromatogr. 18*: 2783 (1995).
12. T. Ceserhati, *Anal. Chim. Acta 292*: 17 (1994).

L. Zhang
Qin-Sun Wang

Thin-Layer Chromatography–Mass Spectrometry

Introduction

There are many methods of detection of substances following separation of mixtures by thin-layer chromatography (TLC) or, in wider aspect, planar chromatography (PC). Some of the methods have been continuously used since the beginnings of TLC. These are descibed in other sections of this volume. The newest method in this field is mass spectrometry (MS) coupled with planar chromatography. The concept of using MS as a detection method for samples separated by TLC is not new; it dates almost to the beginning of organic mass spectrometry itself. It was Kaiser who suggested the possibility of using a coupled TLC–MS method in his review 30 years ago. Despite a long and successful history of application as a detector in gas chromatography (GC) and liquid–liquid chromatography (LLC), mass spectrometry has been under utilized as a detector for TLC. The need for application of this combined analytical method (i.e., mass spectrometric detection with separation by TLC) was evident by the late 1980s. The success of the method depended, above all, on the elaboration of suitable experimental means, devices, and solutions to the challenge of transfer of substances from TLC plates to the mass spectrometer.

Basic Principle of Mass Spectrometry and Its Application to TLC

In a mass spectrometer, instead of an optical spectrum in the spectrometer, we have a mass spectrum of matter radiation. The necessary condition for using a mass spectrometry detection is ionization of components of the analyzed sample. Thus, we can say that mass spectrometry is a destructive method of analysis. There are many techniques and systems for mass spectrometry, but they all have the same three elements: source of ions, ion analyzer, and ion detector (Fig. 1).

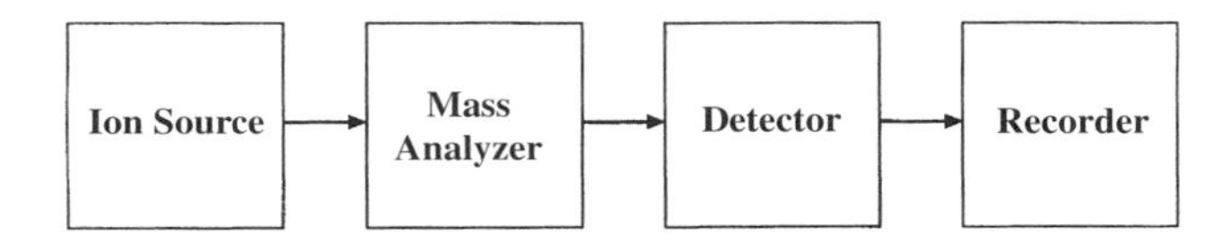

Fig. 1 Scheme of the mass spectrometry procedure.

The Source of Ions

Positive ions can appear by simple removal of one electron. This process can be demonstrated by means of the equation

$$M \rightarrow M^{+} + e$$

The M^+ ion is called a *molecular ion*. It is characterized by means of mass (m) and charge (z). In order to produce a molecular ion, there must be supplied energy at least equal to the energetic potential of the molecule. The first ionization potential for most organic molecules is between 8 and 15 eV. The studied sample is transferred to a vessel under vacuum, in which there is an anode and a cathode. The potential difference between anode and cathode is over 20 kV. With such a large potential difference and small distance between anode and cathode, there appear electrical discharges which cause the ionization of sample molecules. The positive ions are pushed away by repellent electrodes and directed toward an *acceleration gap*. Of course, it is possible that the electron would be captured by the analyzed molecule and produce a negative ion, but the probability of such a process is small. However, fragmentation of a negatively charged molecule is much easier than fragmentation of a positively charged molecule. The higher the ionization energy used, the greater will be the probability of fragmentation of a molecule.

Analysis of Ions

Magnetic fields affect charged molecules if these molecules are in motion. The direction of this action is perpendicular to the direction of the field and perpendicular to the direction of movement of the investigated ion. The strength of the force acting on an ion in a magnetic field depends on the ion charge and its velocity. The effects of the action of this force depends on mass and velocity. Finally, the ions move along arcs of circles and the deviation from their original direction of movement depends on their m/z ratio.

The most popular mass spectrometers used in chromatography are sector magnetic spectrometers. They are large, expensive, and relatively slow spectrometers. Their primary advantage is their high resolution. A constant, strong electrical field within the acceleration gap speeds the ions up to their terminal velocities. The ion beam goes into the analyzer, *in vacuum* (10^{-7}–10^{-8} Tr), and placed in the magnetic field. In order to obtain a spectrum, either the magnetic field in which the analyzer is located or the accelerating voltage between the electrodes in the acceleration gap is changed. The ion beam leads into a collector, whereas the signal is amplified by means of an electron multiplier.

Quadrupole Spectrometer

The resolution of this spectrometer is not high, but the spectrometer are of relatively low cost and fast; they collect a full spectrum of the investigated substance in about 0.3 s. In the spectrometer (also called *quadrupole filter*), there appears an electromagnetic field which causes the oscillation of ions. At given voltages, the oscillation amplitude of ions under a certain mass is larger than the distance between one of the pairs of electrodes; such ions are thrown out of the filter. Through the filter to the detector, there arrive only the ions of a determined mass.

Ion Trap

In an ion trap, in contrast to other mass spectrometers, ionization and analysis occur in the same location but at different times. A chromatographic sample is introduced directly into the space between the electrodes, where ionization takes place. An appropriate voltage across the electrodes causes appearance of an electrical field within the trap. This field keeps the ions produced during ionization inside the trap and does not allow them to leave. When ionization is completed, the voltage of the electrodes is constantly changed so that the ions leave the trap in the order of increasing m/z ratio.

Soon after the advent of fast-atom bombardment (FAB) ionization, it was realized that there is an ionization technique which is ideally suited for analyzing compounds resolved on a TLC plate. There are several early examples of the direct analysis of spots from TLC plates (without prior extraction of the adsorbed material). Detection limits of these TLC–FAB–MS measurements depend on the sample, but they are typically in the 0.1–10-μg range.

The coupled TLC–FAB–MS, without the prior recovery of analyte from the TLC plate, provides a powerful tool for the unambiguous identification of sample components. The use of tandem MS–MS represents a further refinement of this approach, enabling identification to be performed even when the separated components are incompletely resolved from each other or are subject to background interference.

General Approaches to Sample Preparation for the Coupled TLC–MS

The following are combinations to these approaches:

1. The compound is eluted from the chromatographic plate, collected, and introduced into the mass spectrometer as a discrete sample. In this method, samples collected from a TLC spot, identified with an independent method of visualization, must be sufficiently volatile to evaporate into the source of the mass spectrometer.
2. The sample and support are not separated, but both are introduced simultaneously into the source of the mass spectrometer. The spot is scraped from the support and placed on the direct insertion probe of the mass spectrometer. As the probe is heated, the more volatile sample is evaporated into the source, whereas the fairly nonvolatile chromatographic matrix remains in the probe.
3. The entire intact chromatogram is placed within the source of the mass spectrometer and analyzed.

For the last two methods, there is the assumption that extraneous material from the chromatogram does not unduly and adversely affect the quality of the mass spectrum which is recorded. The chromatogram is placed, intact, within a source housing and a spatially resolved map of organic or bio-organic compounds distributed across the surface is measured.

At the present time, the most up-to-date methods of ionization in TLC–MS are FAB and secondary ionization mass spectrometry (SIMS) in which organic molecules are sputtered from surface by impact of a stream of high-energy molecules, and laser desorption (LD), in which sputtering of organic molecules from surface of a support is due to the influence of high thermal energy produced by a laser beam onto plate surface. The ions produced this way are usually even-electron $(M^+ + H)^+$ and appear by means of chemical ionization, and the interpretation of spectrum follows the same lines.

Bush designed an excellent instrument which can be used for analyses of TLC chromatograms by FAB–MS which provides spatially discrete mass spectra directly from the surface layer.

Some Practical Problems

The use of MS as a detector for TLC concerns identification of substances in a mixture, as well their separation. A necessary condition in this approach is the purity of a sample within a spot on the chromatogram. A special problem is the large quantity of support and, sometimes, also a binding material, and mobile-phase residues. Some other factors which hinder TLC–MS might potentially be water, which is physically bound to the support, as well as, in some cases, the organic modifier of the support. A factor which complicates the detection process in TLC–MS may also be the presence of derivatives in a sample peak displayed on a chromatogram. That is why the analyzed sample should be as free of contamination with other additives as possible and the support matrix should be carefully removed before MS analysis. These indications refer, to a lesser degree, to other methods of detection which are performed after total extraction of sample molecules from sample matrix. It is a widely known fact that most of the analyses using a TLC–MS method are performed after exhaustive extraction of the sample from the matrix.

There are also some practical problems connected with the instrumental issues of MS. Most of the sources used in mass spectrometers are so small that it is possible to also have, for example, a very small inlet for the gas or liquid. In this case, such TLC–MS methods may only be used if they are based on the introduction of samples via a stream of gas. The matrix, together with sample material, may be introduced directly into the MS source. Some types of instruments even allow for the insertion of a whole chromatographic plate into the MS source.

Applications

Expansion in the field of analytical mass spectrometry has taken place in recent years; in particular, the applicability of TLC–MS in the field of chemical and biochemical science has been significantly broadened. This technique can now be used for the analysis of macromolecules such as peptides, proteins, oligosaccharides, and oligonucleotides, etc. (Refer to Fig. 2.)

The combination of TLC and MS is well established and may be used for on- and off-line analysis. The increase in popularity of this combined technology also resulted from the improvements in the separating power of TLC and the widespread availability of suitable MS techniques for difficult-to-analyze molecules. The TLC–MS is, at present, a simple and readily implemented means of

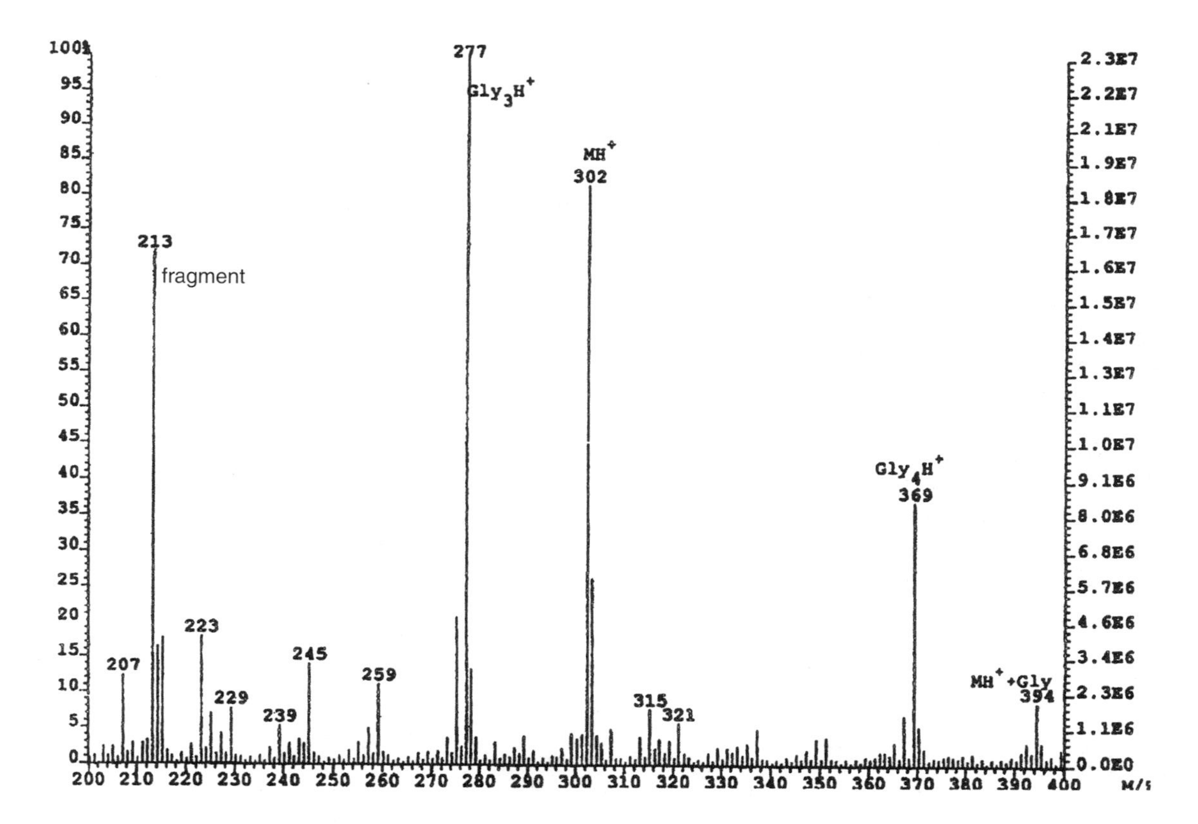

Fig. 2 TLC–FAB–MS spectrum obtained after of 1 μg deramciclane standard with an acidic mobile phase. Gly indicates glycerol clusters. (Permission of *J. Planar Chromatogr.* *10*: 90–96 (1997).)

combining a separation with mass spectrometric identification of substances. With the development of MS suited to direct TLC analysis, it seems clear that the application of TLC–MS will grow in the next decade.

Suggested Further Reading

S. M. Brown and K. L. Bush, *J. Planar Chromatogr. 4*: 198 (1991).
K. L. Bush, in *Handbook of Thin-Layer Chromatography* (J. Sherma and B. Fried, eds.), Marcel Dekker, Inc., New York, 1966, p. 183.
K. L. Bush and R. G. Cooks, *Science 218*: 247 (1982).
K. L. Bush, J. O. Mullis, and J. A. Chakel, *J. Planar Chromatogr. 5*: 9 (1992).
R. J. Day, S. E. Unger, and R. G. Cooks, *Anal. Chem. 52*: 557A (1980).
R. E. Kaiser, *Chem. Britain 5*: 54 (1969).
P. Martin, W. Morden, P. Wall, and I. Wilson, *J. Planar Chromatogr. 5*: 255 (1992).
Y. Nakagowa, *Iyo Masu Kenkuyukai Koensku 9*: 39 (1984).

Jan K. Różyło

Thin-Layer Chromatography of Natural Pigments

Various natural pigment classes, such as flavonoids, anthocyanins, carotenoids, chlorophylls and chlorophyll derivatives, porphyrins, quinones, anthraqinones, betalains, and so forth are abundant in many families of the vegetable and animal kingdoms. As consumers generally dislike the color of synthetic dyes, the concentration and composition of pigments in foods and food products exert a considerable impact on the consumer acceptance and, consequently, on the commercial value of the products. It has been proven many times that one of the main properties employed for the commercial evaluation of the quality of products is their color; that is, an adequate color is an important requirement of marketability.

Spectroscopic methods for measuring the absorbance of pigment solutions or the adsorbance of the color of product surfaces on one or more wavelengths in the visible range are excellent tools for the accurate determination of the quantity of pigments; however, they do not contain any useful information on the concentration of the individual pigment fractions.

As the stability of the various pigments against hydrolysis, oxidation, and other environmental conditions shows marked differences, the assessment of the pigment composition may help for the prediction of the shelf life of products and the elucidation of the impact of various technological steps on the individual pigment fractions resulting in more consumer-friendly processing methods. Moreover, the exact knowledge of the pigment composition may facilitate the identification of the origin of the product. The advantageous characteristics of thin-layer chromatography (TLC) (easy to use, low operating costs, no need for complicated instrumentation, manyfold possibilities of detection, etc.) make it a method of preference for the separation and, to a lesser extent, for the quantitative analysis of natural pigments.

The earlier results in the application of TLC for the analysis of natural color pigments, in general [1], and especially in plants [2], have been reviewed. Pigments are more or less strongly bonded to cellulose, protein, cell-wall components, and so forth in both plant and animal issues; therefore, the efficaceous extraction sometimes is difficult and time-consuming with the traditional extraction methods, and the recoveries sometimes are inadequate. Depending on the character of the accompanying matrix and the solubility of the pigment, a considerable number of extraction solvents or solvent mixtures were proposed and used in the TLC analysis of natural pigments.

Thus, ethanol–water (7:3 v/v) and methanol–water in various ratios for flavonoids, methanol–25% HCl (9:1 v/v), methanol–acetic acid (5%), methanol–trifluoroacetic acid (3%) for unstable anthocyanins, acetone, methanol–acetone mixtures for carotenoids, acetone and petroleum ether for chlorophylls, and so forth. The pigments are fairly stable in their natural environment, but they generally become unstable in extracts; this has to be taken into consideration in the development of new, more efficaceous extraction procedures.

The impact of the extraction conditions using various solvents on the recoveries has never been studied in detail, and the results have never been compared. The introduction of modern extraction methods, such as microwave-assisted extraction, supercritical fluid extraction, and solid-phase extraction, probably will improve the efficiency of extraction, even in the instance of unstable pigments and pigment mixtures. The majority of TLC separations were carried out on traditional silica layers. As the chemical structures and, consequently, the retention characteristics of pigments are highly different, a wide variety of eluent systems has been employed for their separation, consisting of light petroleum, ethyl formate, ethyl acetate, benzene, toluene, chloroform, methanol, *n*-butanol, formic or acetic acid, and so forth.

Besides silica, other direct phase supports such as alumina, diatomaceous earth, cellulose, MgO–diatomaceous earth, and sucrose layers were employed for the separation of various pigment mixtures. Mixed-mode supports (cyano, diol, and aminosilica layers) were also employed in both the direct and reversed-phase elution modes for the separation of natural pigments, but their performance was markedly lower that those of traditional silica layers. Reversed-phase supports (polyamide, octadecylsilica, alumina, silica, and diatomaceous earth impregnated with paraffin oil) were also applied for the separation of color pigments (carotenoids from *Capsicum annuum*, anthocyanins from red wines) and it was found that their separation capacity was commensurate with those of direct phase layers. The separation of 22 pigment fractions extracted from paprika powders on diatomaceous layers impregnated with paraffin oil, using mixtures of acetone–water, was reported and it was established that baseline separation of each fraction cannot be achieved in one run and at a single mobile-phase composition [3].

Unconventional layers have also been employed in the TLC analysis of pigments of sour cherry and blueberry. Baseline separations were carried out on corn and rice starch layers in the *n*-butanol–glacial acetic acid–water–benzene (30:20:10:0.5, v/v) mobile phase [4]. A considerable number of carotenoid standards can be separated in one run as demonstrated in Table 1, showing the R_f values of carotenoids in various eluent systems [5].

Table 1 The R_f Values Obtained for Carotenoids by Use of Different Mobile Phases

No.	Carotenoid	R_f value in mobile phase 1	2	3
1	β-Carotene	0.93	0.92	0.93
2	α-Cryptoxanthin	0.76	0.83	0.80
3	β-Cryptoxanthin	0.75	0.83	0.80
4	Zeaxanthin	0.46	0.72	0.51
5	Lutein	0.46	0.73	0.52
6	Nigroxanthin	0.52	0.78	0.63
7	α-Carotene monoepoxide	0.92	0.92	0.91
8	β-Carotene monoepoxide	0.92	0.90	0.91
9	β-Carotene diepoxide	0.91	0.90	0.91
10	Lutein epoxide	0.46	0.58	0.47
11	Antheraxanthin	0.45	0.55	0.47
12	Violaxanthin	0.41	0.46	0.44
13	Cycloviolaxanthin	0.58	0.74	0.77
14	Cucurbitaxanthin A	0.50	0.72	0.63
15	Capsanthin 3,6-epoxide	0.43	0.61	0.55
16	Capsanthin	0.38	0.60	0.44
17	Capsanthin 5,6-epoxide	0.35	0.47	0.39
18	Capsorubin	0.32	0.42	0.38
19	Capsanthol (6′*R*)	0.25	0.26	0.23
20	Capsanthol (6′*S*)	0.38	0.68	0.49
21	5,6-Diepikarpoxanthin	0.35	0.36	0.39
22	6-Epikarpoxanthin	0.26	0.38	0.24
23	5,6-Diepilatoxanthin	0.36	0.45	0.33
24	5,6-Diepicapsokarpoxanthin	0.28	0.30	0.30

Note: Mobile phase 1: petroleum ether–acetone (6:4, v/v). Mobile phase 2: petroleum ether-*tert*-butanol (8:2, v/v). Mobile phase 3: methanol–benzene–ethyl acetate (5:75:25, v/v/v).
Source: Reprinted with permission from Ref. 5.

Pigment extracts either derived from one plant or one part of a plant generally contain a considerable number of fractions with highly different retention characteristics. The separation of these extracts cannot be successfully achieved by the traditional TLC techniques. Elution methods similar to gradient elution in high-performance liquid chromatography were developed to overcome this difficulty. Stepwise gradient elution can be carried out by introducing eluent fractions with increasing eluent strength to the layer without interrupting the separation process. In programmed multiple-gradient development, the plate is developed to different distances with mobile phases of decreasing elution strength.

The plate is first developed for a given distance with the eluent having the highest elution strength and then the plates are dried. Pigments with the lowest mobility are separated. Then, a new eluent with a lower elution strength is applied for the separation of pigments with a higher mobility. The procedure can be repeatedly employed up to 25 times using eluents with decreasing elution strength. The advantages of programmed multiple development were exploited in the separation of complex plant extracts of *Radix rhei* and *Cortex frangulae* [6]. Due to the considerable number of pigment fractions in an extract, the identification of the individual pigments separated by TLC is extremely difficult. Pigments can be identified by spotting an authentic standard on the neighboring track; however, in the instance of natural pigment mixtures, authentic standards are generally not available.

Coupled spectroscopic methods such as TLC–UV (ultraviolet) and visible spectroscopy, TLC–mass spectrometry, and TLC–FTIR (Fourier transform infrared) have been developed to overcome this difficulty [7]. Their future application in the TLC analysis of natural pigments will markedly increase the information content of this simple and interesting separation technique. The automation of the various steps of TLC analysis (sample application, automated developing chambers, TLC scanners, etc.) greatly increased the reliability of the method, making it suitable for official control and legislative purposes [8].

References

1. O. M. Andersen and G. W. Francis, in *Handbook of Thin-Layer Chromatography* (J. Sherma and B. Fried, eds.), Marcel Dekker, Inc., New York, 1996, pp. 715–752.
2. J. Pothier, in *Practical Thin-Layer Chromatography*, CRC Press, Boca Raton, FL, 1996, pp. 33–49.
3. T. Cerháti, E. Forgács, and J. Holló, *J. Planar Chromatogr.—Mod. TLC 6*: 472 (1998).
4. N. Perisic-Janjic and B. Vujicic, *J. Planar Chromatogr.—Mod. TLC 10*: 447 (1998).
5. J. Deli, *J. Planar Chromatogr.—Mod. TLC 11*: 311 (1998).
6. G. Matysik, *Chromatographia 43*: 39 (1996).
7. E. Pastene, M. Montes, and M. Vega, *J. Planar Chromatogr.—Mod. TLC 10*: 362 (1997).
8. R. Brockmann, *Fleischwirtschaft 78*: 143 (1998).

Tibor Cserháti
Esther Forgács

Thin-Layer Chromatography of Synthetic Dyes

Synthetic dyes are extensively used in many up-to-date industrial processes and research, mainly in the preparation of textile, food, and leather products, as well as in cosmetics and medicine. The widespread application of synthetic dyes has resulted in serious environmental pollution: Their occurrence in ground water and wastewater and the accumulation in sediment, soil, and various biological tissues has often been observed and reported. Dyes and intermediates can cause abnormal reproductive function in males and show marked toxic effects toward bacteria. The rate of biodegradation of the majority of synthetic dyes is very low, enhancing the toxicological hazard and environmental impact.

Commercial synthetic dyes generally contain more than one color product. As the knowledge of the exact composition of dye mixtures is prerequisite for their successful application in many fields of industry and research, many efforts have been devoted for the development of various chromatographic techniques suitable for their separation and quantitative determination. Moreover, the exact determination of the composition and quantity of synthetic dye is required in the control of industrial processes, in the following of efficacy of wastewater treatment, in environmental protection studies, and in forensic science.

The chemical structures of synthetic dyes show considerable variety. They generally contain more than one aromatic group, condensed aromatic substructures or heterocyclic rings (pyrazolone, thiazole, acridine, thiazine, oxazine) which are mainly hydrophobic, and, frequently, a polar basic or cationic group which is strongly hydrophilic. Due to these structural characteristics, they readily bind both to polar adsorptive and apolar reversed-phase chromatographic supports, making their successful separation difficult. As the synthetic dyes are not volatile and volatile derivatives are not known, gas chromatographic (GC) methods cannot be employed for their analysis. Thin-layer chromatography (TLC), high-performance liquid chromatography (HPLC) and, to a lesser extent, capillary zone electrophoresis (CZE) have been equally applied for this purpose. Due to its inherent advantages (simplicity, low cost, rapidity, etc.), a large number of TLC methods have been developed and successfully applied for the analysis of synthetic dyes. Previous results obtained in the application of TLC for the separation of synthetic dyes were reviewed earlier [1]. The planar chromatography of dyes used in the leather [2] and cosmetic industries [3] was reviewed separately. The majority of TLC separations have been performed on the traditional silica layers using many solvents and solvent mixtures (chloroform, various alcohols, pyridine, *n*-hexane, toluene, ethyl acetate, etc.). The polar character of some substructures in the dyes frequently made necessary the addition of acidic (formic and acetic acid) or basic additives (ammonia, diethylamine) to the mobile phase to suppress the dissociation of these hydrophilic groups in order to increase, in this manner, the efficiency of separation and to improve spot shape. However, other than silica layers, various eluent additives have also been employed for the enhancement of the efficacy of TLC. Thus, microcrystalline cellulose thin layers were employed for the study of the retention behavior of some synthetic dyes such as methyl

Table 1 R_f Values of Azo Dyes on Microcrystalline Cellulose Thin Layers (Merck 557) with Various Eluent Additives at Room Temperature (CD = cyclodextrin)

	0.5 *M* HCl		1 *M* Na_2CO_3			1 *M* NaCl		
Azo dye	α-Cd polymer 1%	α-CD 1%	α-CD polymer 1%	α-CD 1%	Control	α-CD polymer 0.5%	β-CD polymer 0.5%	τ-CD polymer 0.5%
Methyl orange	1.00	0.77	0.89	0.72	0.10	1.00	0.80	0.79
Methyl red	0.38	0.17	0.30	0.14	0.08	0.13	0.24	0.23
Methyl yellow	0.92	0.73	0.69	0.40	0.00	0.88	0.55	0.54
Ethyl orange	1.00	0.97	0.91	0.73	0.44	1.00	0.97	1.00
Ethyl red	0.66	0.35	0.68	0.23	0.12	0.28	0.50	0.48
Alizarin yellow R	0.71	0.22	0.70	0.43	0.00	0.52	0.43	0.41
Alizarin yellow 2G	0.80	0.00	0.70	0.51	0.03	0.55	0.48	0.61

Source: Reprinted with permission from Ref. 4.

orange (4-[[4-(dimethylamino)phenyl]azo]benzene acid sodium salt), methyl red (2-[[4-(dimethylamino)phenyl]-azo]benzoic acid), methyl yellow (*N*,*N*-dimethyl-4-(phenylazo)benzeneamine), ethyl orange (4-[[4-(diethyl-amino) phenyl]azo]benzene acid sodium salt), ethyl red (2-[[4-(diethylamino)phenyl]azo]benzoic acid), alizarin yellow R (2-hydroxy-5-[(4-nitrophenyl) azo]benzoic acid) and alizarin yellow 2G (2-hydroxy-5-[(3-nitrophenyl)azo]-benzoic acid monosodium salt) [4]. The R_f values of dyes obtained in the presence of various eluent additives are compiled in Table 1. The data clearly show that eluent additives can be successfully used for the modification of the retention behavior of azo dyes, improving the separation capacity of TLC.

The separation characteristics of a considerable variety of other TLC supports were also tested using different dye mixtures (magnesia, polyamide, silylated silica, octadecyl-bonded silica, carboxymethyl cellulose, zeolite, etc.); however, these supports have not been frequently applied in practical TLC of this class of compounds. Optimization procedures such as the prisma and the simplex methods have also found application in the TLC analysis of synthetic dyes [5]. It was established that six red synthetic dyes (C.I. 15580; C.I. 15585; C.I. 15630; C.I. 15800; C.I. 15880; C.I. 15865) can be fully separated on silica high-performance TLC (HPTLC) layers in a three-solvent system calculated by the optimization models. The theoretical plate number and the consequent separation capacity of traditional TLC can be considerably enhanced by using supports of lower particle size (about 5 μm) and a narrower particle size distribution. The application of these HPTLC layers for the analysis of basic and cationic synthetic dyes has also been reviewed [6]. The advantages of overpressured (or forced flow) TLC include improved separation efficiency, lower detection limit, and lower solvent consumption, and they have also been exploited in the analysis of synthetic dyes.

The detection of synthetic dyes on any type of TLC or HPTLC support is very easy; they contain chromophore groups and can be seen without using an ultraviolet (UV) lamp or detection reagents.

The quantitative evaluation of the amount of dyes in a spot can be carried by the traditional method of scrapping the spot from the plate, dissolving the dyes in a solvent and measure the extinction of the solution. This method is time-consuming and the reproducibility is fairly poor. To date, quantitative evaluation methods such as video-densitometry and scanning densitometry have also found application in the TLC analysis of dyes. The reproducibility of these evaluation methods is higher and commensurate with those obtained by high-performance liquid chromatography; the time consumption is signifacantly lower than that of the traditional "scraping–dissolving" method. The separation and quantitative determination of the impurities of the dye Phloxim B (2′,4′,5′,7′-tetra-bromo-4,5,6,7-tetrachloro-3′,6′-dihydroxyspiro[isobenzo-furan-1(3*H*),9′-[9*H*]xanthen-3-one-sodium salt) was performed on silica HPTLC layers [7]. A lower-halogenated impurity (2′,4′,5′-tribromo derivative) was separated by the mobile-phase acetone–chloroform–*n*-butylamine (66:24:4.5 v/v). Ethyl ester impurity was separated from the main fraction of Phloxim B with chloroform–glacial acetic acid (4:1 v/v). The concentration of impurities was determined *in situ* by videodensitometry. It was established that HPTLC–videodensitometry is rapid and reliable and can be successfully used for the measurement of impurities in Phloxim B. Although the TLC analysis of synthetic dyes is well established, the number of theoretical articles dealing with the quantitative relationship between retention characterisitcs, molecular structure, and physicochemical parameters of solutes is surprisingly low. The retention behavior of seven monotetrazolium and nine ditetrazolium salts on alumina and reversed-phase alumina was recently studied and the relationship between retention and physicochemical parameters of this class of synthetic dyes was elucidated by canonical correlation analysis [8].

Calculations indicated that the retention on both TLC supports is of mixed character, including hydrophobic, electronic and steric parameters.

It can be concluded from the present state of the art of TLC analysis of synthetic dyes that the methods are mainly limited to the application of traditional TLC technique. Due to the rapidly developing instrumentation and automation of the various steps of TLC analysis and the introduction of coupled methods (TLC–mass spectrometry, TLC–Fourier transfer infrared spectrometry, etc.), their acceptance and successful application in the analysis of synthetic dyes can be expected in the near future. The use of new TLC techniques not only increases the separation capacity of the method but also considerably decreases the detection limit of solutes and may contribute to their identification.

References

1. V. K. Gupta, in *Handbook of Thin-Layer Chromatography* (J. Sherma and B. Fried, eds.), Marcel Dekker, Inc. New York, 1996, pp. 1001–1032.
2. D. Muralidharan and V. S. Sundara, *J. Soc. Leather Technol. Chem. 79*: 178 (1995).
3. L. Gagliardi, D. De Orsi, and O. Cozzoli, *Cosmet. Toiletries Ed. Ital. 16*: 34 (1995).
4. M. Lederer and H. K. H. Nguyen, *J. Chromatogr. A 723*: 405 (1996).
5. K. Morita, S. Koike, and T. Aishima, *J. Planar Chromatogr. — Mod. TLC 11*: 94 (1998).
6. A. B. Gharpure, *Text. Dyer Printer 29*: 13 (1996).
7. P. Wright, N. Richfield-Fratz, A. Rasooly, and A. Weisz, *J. Planar Chromatogr. — Mod. TLC 10*: 157 (1997).
8. T. Cserháti, A. Kósa, and S. Balogh, *J. Biochem. Biophys. Methods 36*: 131 (1998).

Tibor Cserháti
Esther Forgács

Three-Dimensional Effects in Field-Flow Fractionation: Theory

A field-flow fractionation (FFF) channel is normally ribbonlike. The ratio of its breadth b to width w is usually larger than 40. This was the reason to consider the 2D models adequate for the description of hydrodynamic and mass-transfer processes in FFF channels. The longitudinal flow was approximated by the equation for the flow between infinite parallel plates, and the influence of the side walls on mass-transfer of solute was neglected in the most of FFF models, starting with standard theory of Giddings and more complicated models based on the generalized dispersion theory [1]. The authors of Ref. 1 were probably the first to assume that the difference in the experimental peak widths and predictions of the theory may be due to the influence of the side walls.

The essence of the side walls effect follows. The flow velocity turns to zero at the side walls as well as at the main (accumulation and depletion) walls of the FFF channel. Therefore, the flow profile is nonuniform, not only along the width of the channel but also along its breadth. The size of the regions near the side walls where the flow is substantially nonuniform is of the same order as w. The nonuniformity of the flow in both directions, combined with diffusion of solute particles, leads to Taylor dispersion and peak broadening that could be different from the one predicted by the 2D models.

The first 3D model of FFF was developed in Ref. 2. The 3D diffusion–convection equation was solved with the help of generalized dispersion theory, resulting in the equations for the cross-sectional average concentration of the solute and dispersion coefficients K_1 and K_2, representing the normalized solute zone velocity and the velocity of the corresponding peak width growth, respectively. Unfortunately, only the steady-state asymptotic values of dispersion coefficients $K_1(\infty)$ and $K_2(\infty)$ were determined in Ref. 2, leading to the prediction of the solute peaks much wider than the experimental ones.

The incorrectness of the steady-state approach was noted by the authors of Ref. 3 and can be explained as follows. There are two characteristic diffusion times in an FFF channel: $t_{D1} = w^2/4D$ and $t_{D2} = b^2/4D$, where D is the diffusion coefficient of solute molecules. As the ratio $b/w > 40$, then $t_{D1}/t_{D2} > 1600$. Experimental values of retention time are usually equal to several t_{D1}, but are never as large as $1600t_{D1}$. Thus, the steady-state values of K_1 and K_2 corresponding to $t \gg t_{D2}$ are never reached during the experiment. (For the channel with $w = 200$ μm, $b = 1$ cm, and solute with $D = 10^{-6}$ cm^2/s, $t_{D1} = 100$ s and $t_{D2} = 2.5 \times 10^5$ s.)

In Ref. 3, the nonstationary solution was produced for $K_1(\tau)$ and $K_2(\tau)$, where $\tau = t/t_{D1}$, leading to much better correspondence with experimental peak widths. Similar to Ref. 2 not the exact equation for the flow profile in the rectangular channel but its approximation was used:

$$V_x(y, z) = V_{\max}\left(1 - \frac{4y^2}{w^2}\right)\left(1 - \frac{\cosh(2\sqrt{3}z/w)}{\cosh(\sqrt{3}b/w)}\right) \quad (1)$$

Only the case of uniform initial distribution of solute in the channel cross section was studied in Ref. 3.

A more general case was studied in Refs. 4 and 5, where different initial solute distributions were examined, including the distributions describing the syringe inlet with and without the stop-flow relaxation. The 3D generalized dispersion theory was used to solve the 3D diffusion–convection equation. Unlike Refs. 2 and 3, the exact equation for flow profile in the rectangular channel was used:

$$V_x(y,z) = 4V_{\max} \sum_{n=0}^{\infty} (-1)^n \left(1 - \frac{\cosh(2\varphi_n z/w)}{\cosh(\varphi_n b/w)}\right) \frac{\cos(2\varphi_n y/w)}{\varphi_n^3} \quad (2)$$

where $\varphi_n = (2n + 1)\pi/2$, $|z| \leq b/2$, and $|y| \leq w/2$, transversal coordinates (the zero of coordinates being at the axis of FFF channel).

The exact equation for the flow profile was chosen because, although approximation (1) gives very similar results for $K_1(\tau)$, the deviation for $K_2(\tau)$ could be very significant. Figure 1 presents dependences of $K_2(\tau)$ for different values of the FFF parameter λ, calculated with the help of 3D [4] and 2D [1] models. It can be seen that the difference between 2D and 3D results is growing with time τ and is larger for smaller values of λ. The growth of $K_2(\tau)$ with time leads to the nonlinear growth of the peak variance with time:

$$\sigma^2(t) = 2Dt + \frac{V_{\max}^2 w^4}{8D^2} \int_0^{4Dt/w^2} K_2(\gamma)\, d\gamma \quad (3)$$

If K_2 is constant for the typical experimental values of retention time, as predicted by the 2D theory, then $\sigma^2 \sim t$

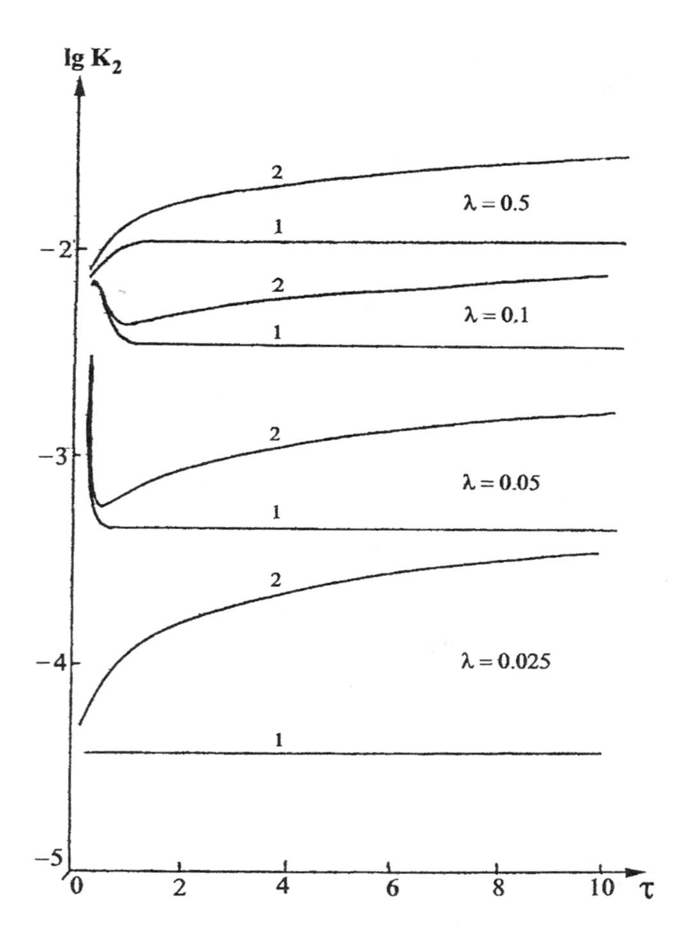

Fig. 1 Time dependences of the dispersion coefficient $K_2(\tau)$. Curve 1: 2D model; curve 2: 3D model. The solute was initially uniformly distributed in the channel cross section.

and resolution $R_s \sim \sqrt{t}$, where $R_s = \Delta X/4\sigma$ and ΔX is the distance between peak maximums for two types of solute molecule. According to the 3D model, $K_2(\tau)$ grows with time, so $\sigma^2 \sim \alpha t + \beta t^2 + \cdots$, and there exists some value of retention time for each λ, when resolution stops growing with time.

This effect is even more pronounced for the case of a syringe inlet. At the beginning of the process, the solute is distributed in the region close to $z = 0$, so it does not feel the influence of the side walls and, only after the period of time commensurate with the second diffusion time t_{D2}, the solute particles diffuse to the side walls. That is why, at the beginning of the process, the value of K_2 is close to the 2D value of K_2 and is growing toward its asymptotic 3D value slower than for the case of the solute initially distributed uniformly in the channel's cross section (presented in Fig. 1).

In Ref. 5, the time needed to separate two types of solute molecule with resolution $R_s = 1$ was studied for the cases of different initial solute distributions. It was shown, for example, that for solute molecules with $\lambda = 0.1$ and $\lambda = 0.125$, the time needed to separate them, predicted by the 3D model for uniform initial solute distribution, is 5% larger than predicted by the 2D theory, whereas, for the case of the syringe inlet, the value predicted by the 3D model is much closer to the 2D value (only 0.5% difference). The difference in 2D and 3D predictions can be more significant for closer values of λ, when the retention time needed to fractionate the solute is larger.

More complicated 3D effects were studied in Refs. 6 and 7 with the help of 3D Monte Carlo digital simulation performed with a rather powerful computer (RISK System/6000). Sedimentation FFF with different breadth-to-width channel ratios and both codirected and counterdirected rotation and flow were studied. Secondary flow forming vortexes in the y–z plane is generated in the sedimentation FFF channel, both due to its curvature, and the Coriolis force caused by the centrifuge rotation. The exact structure of the secondary flow was calculated by the numerical solution of the Navier–Stokes equations and was used in the Monte Carlo simulation of the movement of solute molecules.

It was shown that when rotation and flow are codirected, there is a considerable amount of molecules eluted prematurely before the main peak, because they are caught by the vortex of the secondary flow. This effect is practically negligible when the flow and rotation are counterdirected. So, the counterdirected mode is always preferable in sedimentation FFF. Note that this important practical finding of Ref. 7 is of pure 3D nature and cannot be done with the 2D theory.

The contribution of the channel endpieces to the peak broadening is certainly a 3D effect and should be mentioned here, although the theory used in Ref. 8 to analyze this effect was a 2D, not a 3D, one. In FFF instruments, the inlet and outlet tubes are connected with the rectangular FFF channel by triangular endpieces. So, if the solute is introduced to the channel with the flow, then "the fanning out process between the inlet and the main rectangular section of the FFF channel leads to crescent-shaped bands, a type of distortion that persists down the channel and is responsible for some incremental band broadening" [8]. The 2D (in the x–z plane) model was developed. The flow was considered to be ideal and described by the stream function, satisfying the Laplace equation. The streamlines calculated were used to simulate the solute band profile and to estimate the contribution of the endpieces effect to the theoretical plate height:

$$H_e = \frac{b^2\beta^2 \tan^2(\theta/2)}{4L} \tag{4}$$

where L is the total FFF channel length, θ is the angle of the triangular endpiece, and β is a function of θ, determined from the calculated band profiles. It was shown that $\beta^2 \tan^2(\theta/2)$ grows dramatically with θ, and is equal to 0.0226 for $\theta = 30°$ and to 0.389 for $\theta = 120°$. The role of the end effects must not be overestimated, as even for the channel with unusually large breadth $b = 6$ cm ($\theta = 75°$), the absolute value of H_e was 0.0586 cm, much smaller than the typical FFF HETP values (~1 cm). Naturally, this type of contribution to peak width is absent if the solute is introduced by the syringe into the main rectangular section of the channel.

Finally, it must be said that in most of the practical cases, retention ratio and peak width are satisfactorily predicted by the 2D models. The role of the 3D models was to estimate the contributions of the 3D effects and to determine the optimal conditions when these contributions are negligible. As it was shown, to minimize the 3D effects, one must use FFF channels with a larger breadth-to-width ratio, use the syringe inlet, and/or use the longer endpieces with the sharper angle; in the case of sedimentation FFF, one is to avoid codirected flow and rotation mode.

References

1. S. Krishnamurthy and R. S. Subramanian, Exact analysis of FFF, *Separat. Sci. Technol. 12*: 347 (1977).
2. T. Takahashi and W. N. Gill, Hydrodynamic chromatography: Three dimensional laminar dispersion in rectangular conduits with transverse flow, *Chem. Eng. Sci. 5*: 367 (1980).
3. E. K. Kim and I. J. Chung, Transient convective mass transfer in rectangular FFF channels, *Chem. Eng. Commun. 42*: 349 (1986).
4. V. P. Andreev and M. I. Khidekel, Field-flow fractionation in a rectangular channel. Three dimensional model, *Nauchn. Apparat. 4*: 123 (1989) (in Russian).
5. V. P.Andreev and M. I. Khidekel, On the influence of the channel side walls on the resolution time in FFF, *Zh. Phys. Chim. 65*: 2619 (1991).
6. M. R. Shure, Digital simulation of sedimentation FFF, *Anal. Chem. 60*: 1109 (1988).
7. M. R. Shure and S. K. Weeratunga, Coriolis-induced secondary flow in sedimentation FFF, *Anal. Chem. 63*: 2614 (1991).
8. P. S. Williams, S. B. Giddings, and J. C. Giddings, Calculation of flow properties and end effects in FFF channels by a conformal mapping procedure, *Anal. Chem. 58*: 2397 (1986).

Victor P. Andreev

TLC Immunostaining of Steroidal Alkaloid Glycosides

Introduction

The immunoassay system using monoclonal antibodies (MAbs) is indispensable to biological investigations. However, because this was rare for naturally occurring bioactive compounds having small molecular weights, we have prepared the MAbs and established assay systems using enzyme-linked immunosorbent assay (ELISA) for forskolin [1], marijuana compound [2], opium alkaloids [3], solamargine [4], ginsenoside Rbl [5], crocin [6], and glycyrrhizin [7]. Furthermore, the Western blotting method against ginseng saponins [8] and glycyrrhizin [9] have been established for the search of natural resources and for the breeding project of medicinal plants.

The natural resources of adrenocortical and sex hormones, which have been mainly supplied by diosgenin, are becoming rare in the world. The most important feature of solasodine is that it can be converted to dehydropregnenolone. Therefore, the steroidal alkaloid glycosides of solasodine type, such as solamargine, have become important as a starting material for the production of steroidal hormones. Rapid, simple, highly sensitive and reproducible assay systems are required for a large number of plants and a limited, small amount of samples, in order to select the strain of higher yielding steroidal alkaloid glycosides.

We present, here, a simple determination method for solasodine glycosides by using thin-layer chromatography (TLC)–immunostaining.

Materials and Methods

Chemicals and Immunochemicals

Bovine serum albumin (BSA) and human serum albumin (HSA) were provided by Pierce (Rockford, IL, U.S.A.). Peroxidase-labeled anti-mouse IgG was provided from Organon Teknika Cappel Pruducts (West Chester, PA, U.S.A.). Polyvinylidene difluoride (PVDF) membranes (Immobilon-N) were purchased from Millipore Corporation (Bedford, MA, U.S.A.). A glass microfiber filter sheet (GF/A) was purchased from Whatman International Ltd. (Maidstone, England). All other chemicals were standard commercial products of analytical grade.

Solamargine and solasonine were isolated from fresh fruits of *S. khasianum* as previously described [10]. Solasodine was obtained from solamargine by acid hydrolysis as previously described [10]. Solamargine (1 mg) was dissolved in MeOH containing 1*M* HCl (1 mL). The mixture was heated at 70°C for 10, 20, 30, 60, and 90 min, respectively. Individual hydrolysates were evaporated *in vacuo* and applied to TLC. Spots developed on TLC were determined by H_2SO_4 and Dragendorff reagent.

TLC

Solasodine glycosides were applied to TLC plates and developed with chloroform–methanol–ammonia solution (7:2.5:1). A developed TLC plate was dried and then sprayed with blotting solution mixture of isopropanol–methanol–water (5:20:40, by volume). It was placed on a stainless-steel plate, then covered with a PVDF membrane sheet. After covering with a glass microfiber filter sheet, the whole plate was pressed evenly for 45 s with a 130°C iron, as previously described [11], but with a modification. The PVDF membrane was separated from the plate and dried.

Immunostaining of Solasodine Glycosides on PVDF Membrane

The blotted PVDF membrane was dipped in water containing $NaIO_4$ (10 mg/mL) under stirring at room temperature for 1 h. After washing with water, 50 m*M* carbonate buffer solution (pH 9.6) containing BSA (1%) was added and stirred at room temperature for 3 h. The PVDF membrane was washed twice with phosphate buffer solution containing 0.05% of Tween 20 for 5 min, and then washed with water. The PVDF membrane was immersed in anti-solamargine MAb and stirred at room temperature for 1 h. After washing the PVDF membrane twice with TPBS and water, a 1000 times dilution of peroxidase-labeled goat anti-mouse IgG in phosphate buffer solution containing 0.2% gelatin (GPBS) was added and stirred at room temperature for 1 h. The PVDF membrane was washed twice with TPBS and water, then exposed to 1 mg/mL 4-chloro-l-naphthol–0.03% H_2O_2 in PBS solution which was freshly prepared before use for 10 min at room temperature, and the reaction was stopped by washing with water. The immunostained PVDF membrane was allowed to dry.

Results and Discussion

After solasodine glycosides were transfered to the PVDF membrane sheet from the TLC plate by heating as previously reported [11], the PVDF membrane was treated

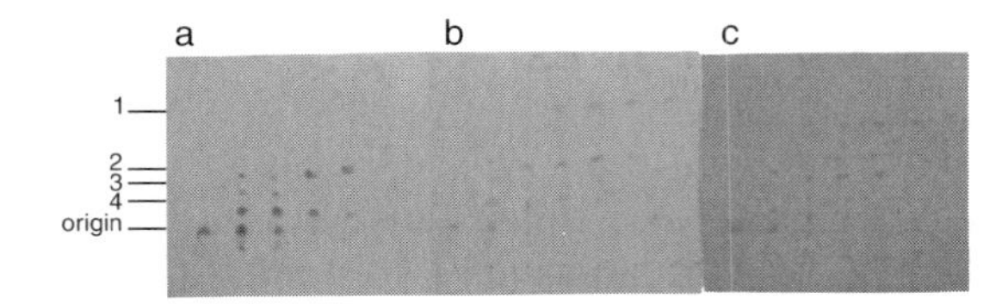

Fig. 1 Hydrolyzed products of solamargine by HCl. (a), (b), and (c) show TLC–immunostaining and the stainings with sulfuric acid and with Dragendorff reagent, respectively. Solamargine was hydrolyzed by 1*M* HCl for 10, 20, 30, 60, and 90 min, respectively. Spots 1–4 were identified with solasodine, 3-*O*-*β*-D-glucopyranosyl solasodine, L-rhamnosyl-(1→4)-*O*-3-*β*-D-glucopyranosyl solasodine, L-rhamnosyl-(1→2)-3-*β*-*O*-D-glucopyranosyl solasodine, respectively.

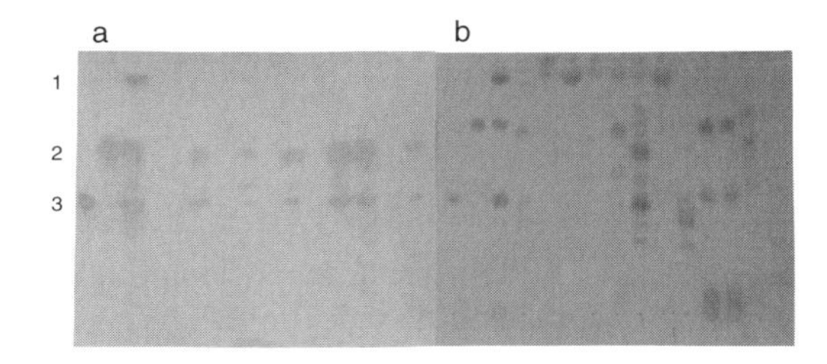

Fig. 2 TLC–immunostainings of steroidal alkaloid glycosides in the crude extracts of *Solanum* species fruits. Crude extracts were developed by a $CHCl_3$–MeOH–NH_4OH solvent system on silica gel TLC plate. After being transferred to a PVDF membrane, the membrane was treated with $NaIO_4$ and stained by MAb. Spots 1–3 were identified with khasianine, solamargine, and solasonine, respectively.

with $NaIO_4$ solution, followed by conjugation with BSA, because solasodine glycosides on PVDF membrane are washed out by buffer solution or water without the formation of conjugate with carrier protein. The PVDF membrane was immersed in anti-solamargine MAb and then peroxidase-labeled secondary MAb. When the substrate and H_2O_2 were added, clear blue spots appeared.

Figure 1 shows the immunostaining of acid hydrolysis products of solamargine hydrolyzed by 1*M* HCI for 10, 20, 30, and 60 min, respectively. Individual hydrolysates were applied to three TLC plates, then developed with a $CHCl_3$–MeOH–NH_4OH solvent system. Two plates were sprayed and colored with H_2SO_4. Figure 1 shows the immunostaining (a) and stainings by H_2SO_4 (b) and Dragendorff reagent (c). When the staining sensitivities of the three methods were compared, the immunostaining was the highest, followed by the H_2SO_4, then Dragendorff reagent. It is easily suggested that product 1 may be aglycone of solamargine, solasodine and products 2–4 might be solasodine monoglycosides and diglycosides. Therefore, products 1–4 were identified as solasodine, 3-*O*-*β*-D-glucopyranosyl-solasodine, *O*-*α*-L-rhamnosyl-(1→4)-3-*O*-*β*-D-glucopyranosyl-solasodine and *O*-*α*-L-rhamnosyl-(1→2)-3-*O*-*β*-D-glucopyranosyl-solasodine, respectively, by direct comparison with authentic samples. Compared with two stainings between immunostaining (Fig. 1a) and H_2SO_4 staining (Fig. 1b), solasodine was not detected by immunostaining despite 44% of cross-reactivity [4]; the sugar moiety was necessary in this staining process. Thus, we separated two functions, the sugar moiety of solasodine glycosides conjugates to the membrane via Schiff base and an aglycone part which is stained by MAb.

Figure 2 shows the immunostaining and H_2SO_4 staining of the crude extracts of *Solanum* species fruits which contain the higher solasodine glycosides [10]. Although the H_2SO_4 staining (Fig. 2B) detected many spots, including, probably, sugars and different types of saponins in various *Solanum* species, the immunostaining (Fig. 2A) detected only limited solasodine glycosides. Bands 1, 2, and 3 were identified to be khasianine, solamargine, and solasonine, respectively, by comparison with authentic samples. Different sensitivities between solamargine and solasonine were observed, and the sensitivity of solasonine was somewhat higher compared to that of solamargine. The detectable limit was 1.6 ng of solasonine, as previously reported.

This is the first report in which the TLC–immunostaining for solasodine glycosides is described. This assay method can be routinely used for survey of natural resources of solasodine glycosides as a simple and rapid analysis. Moreover, this methodology may be available for the assay in vitro *Solanum* plantlets; therefore, it makes it possible to study a large number of cultured plantlets, and a limited small amount of sample in vitro for the breeding of *Solanum* species containing a higher amounts of steroidal alkaloids. Furthermore, this system may be useful for the analysis of animal plasma samples of glycoside or glucronide not limited to solasodine glycosides and/or distributions in organs or tissues, because very low concentrations are expected. Although it is difficult to detect a low-molecular-weight compound by the Western blotting method, the approach described here will be particularly attractive in a wide variety of comparable situations as indicated in the distribution of solasodine glycosides in the fruit of *S. khasianum*. In the expanding studies of this result, naturally occurring pharmacologically active glycosides such as ginsenosides Rb_1, [8] and glycyrrhizin [9] have been investigated.

References

1. R. Sakata, Y. Shoyama, and H. Murakami, *Cytotechnology 16*: 101 (1994).
2. H. Tanaka, Y. Goto, and Y. Shoyama, *J. Immunoassay 17*: 321 (1996).
3. Y. Shoyama, T. Fukada, and H. Murakami, *Cytotechnology 19*: 55 (1996).
4. M. Ishiyama, Y. Shoyama, H. Murakami, and H. Shinohara, *Cytotechnology 18*: 153 (1996).
5. H. Tanaka, N. Fukuda, and Y. Shoyama, *Cytotechnology 29*: 115 (1999).
6. L. Xuan, H. Tanaka, Y. Xu, and Y. Shoyama, *Cytotechnology 29*: 65 (1999).
7. H. Tanaka and Y. Shoyama, *Biol. Pharm. Bull. 21*: 1391 (1998).
8. N. Fukuda, H. Tanaka, and Y. Shoyama, *Biol. Pharm. Bull. 22*: 219 (1999).
9. S. J. Shan, H. Tanaka, and Y. Shoyama, *Biol. Pharm. Bull. 22*: 221 (1999).
10. S. B. Mahato, N. P. Sahu, A. X. Ganguly, R. Kasai, and O. Tanaka, *Phytochernistry 19*: 2018 (1980).
11. H. Tanaka, W. Putalun, C. Tsuzaki, and Y. Shoyama, *FEBS Lett. 404*: 279 (1997).

Waraporn Putalun
Hiroyuki Tanaka
Yukihiro Shoyama

TLC Sandwich Chamber

Introduction

In column chromatography, the stationary phase exists in a closed column that is adequately packed with a suitable sorbent. The stationary phase is preequilibrated with the mobile phase before the first sample is introduced. In this case, the stationary phase cannot be in contact with vapors of the mobile phase during the chromatographic process.

Column chromatography is different from thin-layer chromatography (TLC) because, in planar chromatography, and, therefore, in TLC, development generally starts with a dry stationary phase. Therefore, the mobile-phase front moves forward in the dry stationary phase. This process takes place in the limited, confined volume of the chromatographic chamber.

The kinetics of eluent migration will depend on the degree of saturation of the chamber. The eluent front runs faster in a saturated chamber than in an unsaturated one. The problem of the reproducibility of R_f values can be seriously affected by an unsaturated chamber.

The chromatographic chamber commonly used in TLC can be divided into two categories. A normal chamber (N chamber) is a chromatographic chamber of a large volume with a distance of gas phase in front of a thin-layer greater than about 3 mm. Conversely, a sandwich chamber (S chamber) is a small chromatographic chamber having a small volume and a distance of gas space in front of the thin layer plate less than 3 mm.

The equilibration of the sorbent thin layer with the vapor phase is easily achieved in the compact S chamber. A threefold larger variation of the R_f values for experiments in unsaturated atmosphere has been observed. Also, it has been proved that, in an unsaturated chamber, the so-called "edge effect" appears, owing to the more intense evaporation at the edge of the plate. This effect may be avoided by complete saturation of an N chamber with eluent vapor or by using an S chamber.

There are two methods for development, taking into account the position of the thin-layer plate (i.e., ascending, or vertical, and horizontal).

S Chamber

The S chamber (Fig. 1a) is limited to a small volume by using a thin-layer plate as a near wall (1) to the tank. The opposite wall is a cover plate (2) of small size. This is a frame plate, which has glass strips (5 mm wide and 3 mm thick) sintered along three edges. The two plates are held together by two clamps fixed at the vertical edges of the S chamber (3). This S chamber is placed in a trough which consists of a double jacket (4) of stainless steel filled with solvent (5). The other jacket can be turned to obtain slits at varying distances, which enables plates of different widths to be introduced and ensures a tight seal for the S chamber. The S chamber has a support (6) which permits a vertical position. Only 15 mL of mobile phase are needed for

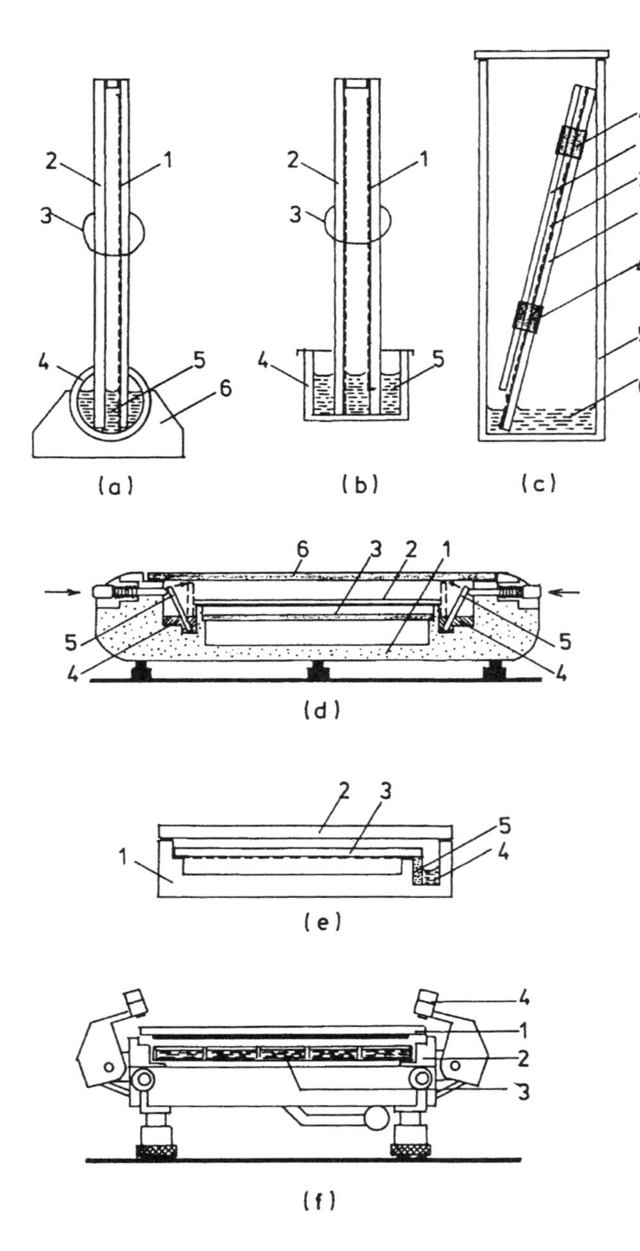

Fig. 1 Different TLC sandwich chambers: (a) diagram of the S chamber; (b) diagram of the saturation S chamber; (c) diagram of the Camag sandwich cover plate; (d) diagram of the Camag horizontal developing S chamber; (e) diagram of the Desaga H-separation chamber; (f) diagram of the Camag Vario-KS chamber.

the thin-layer plate (20 × 20 cm). Before application, a 10-mm strip of adsorbent should be removed from three sides of the TLC plate.

There is another type of S chamber which permits prior saturation of the chamber atmosphere, simultaneously with the preconditioning of the adsorbent, before starting the chromatographic process (Fig. 1b). For this purpose, the adsorbent is removed (a 1 cm width) from all the sides of the plate (1). The opposite frame plate (2) contains, on all of its surface, a thin layer of sorbent. The two plates are fixed by two clamps (3). The chamber is introduced into the trough (4) through a slit. At the beginning, the level of the eluent (5) does not touch the bottom of adsorbent on the thin-layer plate (1). When saturation of the atmosphere is achieved, a volume of eluent is added to the eluent in the trough until the adsorbent is wetted. At this moment, the separation can be started.

Method of Operation

A template is indispensable for ensuring proper spacing of the starting spots during manual sample application. The sample is applied to the layer using a glass capillary or a micropipette of appropriate size (1–10 μL). Sample may be applied as bands to the TLC plates using a band applicator such as the Camag automatic TLC sampler. The automatic TLC sampler works by mechanically moving the plate. This applies the sample automatically by a controlled nitrogen spray from the syringe, forming narrow bands or spots on the plate surface. Programming, operation, and documentation are performed under a Windows-programmed environment.

When the samples have been applied, the frame plate is placed on the top of the TLC plate and fastened by two wide clamps at the right and left edges. In this way, the S chamber is ready to be placed into the trough filled with eluent. The S chamber is kept upright by a fork-shaped holding arm fixed to the trough. Then, the trough slit is sealed by gentle rotation of the outer jacket.

The separation process can be considered finished when the eluent front touches the upper edge of the thin-layer plate.

Camag Sandwich Cover Plate

This type of S chamber has the purpose of reducing the volume of the gas phase, which is in front of the thin layer by a counterplate placed at a small distance (Fig. 1c). The cover glass (1) has two fixed glass spacers (2) located on two sides. The cover plate (1) is 2 cm shorter than the TLC plate (3) and is fixed so that it does not sit deeply into the eluent. The thin-layer plate and cover plate are held together by four stainless-steel clamps (4). The S chamber is not closed at the top edge. In this configuration, there is no need to scrape off the sorbent layer at the edges. With the Camag sandwich cover plate, the TLC plate can be developed into an N chamber (5) which contains, on the flat bottom, the eluent (6). The N chamber is used only as eluent holder. This system permits simultaneous development of two plates.

Horizontal S Chamber

The horizontal developing chamber is a sandwich-type chamber (Fig. 1d). The horizontal S chamber for TLC consists of a flat body of the chamber made from polytetrafluoroethylene (PTFE) (1) so that it is resistant to all liquids. The TLC or high-performance TLC (HPTLC) plate (2) lies on the supports with the thin layer facing downward. The glass plate (3) is used for development in the sandwich configuration (not presented with saturation configuration). A narrow trough (4) holds the eluent. Development is started by shifting the glass strips (5) to the chromatographic plate. In this way, a capillary slit is formed between glass strip and the trough wall, so that a vertical meniscus of the eluent is formed and rises instantaneously. It enters the layer evenly from its front. The glass strip (5) is tilted inward by pushing the rods (arrows). The chamber is kept covered with the glass plate (6) during both preequilibration and development.

Method of Operation

In the horizontal chamber, the HPTLC plate is developed from both opposing sides toward the middle. This permits the number of samples to be doubled, as compared with development in an N chamber. The horizontal S chamber is available for plate sizes 10 × 10 cm and 20 × 10 cm. The optimum separation distance for HPTLC silica gel, 45 mm, is still available.

The mobile phase is transported from the trough to the sorbent layer by surface tension and capillary forces and will continue to travel through the sorbent layer by capillary action. When a plate is developed from both edges simultaneously, the chamber must be kept in a perfectly horizontal position to allow the two solvent fronts the same rate of migration and to meet precisely in the middle. At this point, the capillary forces balance out and the chromatographic development ceases.

The chromatographic process takes less than 10 min per plate and the amount of solvent used is much smaller than in an N chamber. On a 20 × 20 cm HPTLC plate, it is possible to simultaneously develop, from both ends to the centers, up to 70 samples, or 35 samples in a single direction, from one end to the other.

The horizontal S chamber can be used in saturation configuration (N type) when the counter glass plate (3) is not inserted. In this way, the horizontal S chamber permits the conditioning of the adsorbent layer with the different solvent vapors by introducing several drops of the eluent or another solvent on the bottom of the chamber before placing the chromatographic plate face down.

H-Separation Chamber

The H-separation chamber (DESAGA), a mini horizontal development chamber (Fig. 1e), is made of solvent resistant PTFE (1) covered with a sheet of glass (2) 4 mm thick. The HPTLC plate (5 × 5 cm) is laid down for development with the thin layer underneath (3). The mobile phase from the trough (4) is led by a glass frit rod (5) to the thin layer of sorbent. The groove in which the frit rod sits ensures even distribution. The maximum eluent consumption is 2 mL on the plates.

Method of Operation

The sample is applied to the starting line on the layer using a capillary of appropriate size. The applied spots have to be dried completely before the plate is introduced into the separation chamber. The effective migration distance on a plate of 5 × 5 cm is only 3 cm and the migration time is 3 min.

Vario-KS Chamber

The Vario-KS chamber (Camag, Fig. 1f) is used for optimization of developing conditions for 10 × 10 cm plates by simultaneously testing of up to six different mobile phases and vapor equilibration conditions with N or S chamber conditions. Also, the Vario-KS chamber may be used for optimization of developing conditions for 20 × 20 cm TLC plates. The thin-layer plate (1) is laid down on the support (2). The eluent in the reservoir is connected to the thin layer of adsorbent by means of a filter paper strip. Under the thin-layer plate, there are many troughs (3) filled with solvents for preconditioning of the adsorbent layer. The chamber is tightly sealed by two clamps (4).

Suggested Further Reading

F. Geiss, *Fundamental of Thin-Layer Chromatography*, Hüthig, Heilbelberg, 1987.

N. Griberg (ed.), Modern Thin-Layer Chromatography, Marcel Dekker, Inc., New York, 1990.

J. G. Kirchner, *Thin-Layer Chromatography*, 2nd ed., John Wiley & Sons, New York, 1978.

J. Sherma and B. Fried (eds.), *Handbook of Thin-Layer Chromatography*, 4th ed., Marcel Dekker, Inc., New York, 1999.
E. Stahl (ed.), *Thin-Layer Chromatography*, Springer-Verlag, New York, 1969.
J. C. Touchstone, *Practice of Thin-Layer Chromatography*, 3rd ed., John Wiley & Sons, New York, 1992.
A. Zalatkis and R. E. Kaiser (eds.), *HPTLC: High Performance Thin-Layer Chromatography*, Elsevier, Amsterdam, 1977.

Simion Gocan

TLC Sorbents

Since the introduction of commercial precoated plates in the mid-1960s, continual developments with regard to the increase of selectivity and improvement of separation efficiency were pursued [i.e., ready-to-use layers suitable for high-performance thin-layer chromatography (HPTLC), polar and hydrophobic bonded phases, plates with concentrating zones].

A wide variety of TLC and HPTLC precoated plates, which give reproducible results, are commercially available today, even though it is also possible to prepare these plates in the laboratory. Home-made plates can allow access to stationary phases which are not otherwise available.

On HPTLC layers, the average particle size and the range of particle sizes are smaller than on TLC (e.g., 2–10 μm and 5–17 μm respectively for Macherey–Nagel HPTLC and TLC silica plates) and a marked fall in theoretical plate heights is observed (about one order of magnitude smaller than on standard silica layers). Consequently, smaller plate sizes (from 20 × 20 to 10 × 10 cm) and small sample volumes (from 1–3 to 0.1–0.2 μL or even less) and shorter migration distances (from 10–16 to 3–8 cm) can be employed.

A rational classification system of commonly used sorbents in thin-layer chromatography is shown in Table 1.

Hydrophilic Unmodified Sorbents

Silica, Silica Gel, or Silicic Acid

This product is by far the most frequently used sorbent in TLC and is prepared by dehydration of aqueous silicic acid generated by addition of a strong acid to a silicate solution.

Chemically, each silicon atom is surrounded by four oxygen atoms in the form of a tetrahedron. The surface of silica contains (a) siloxane groups (Si—O—Si), (b) silanol groups (Si—OH), (c) water hydrogen-bonded to the silanol groups, and (d) nonsorbed "capillary" or bulk water. The silanol groups (about 8 μmol/m^2) represent adsorption-active surface centers that are able to interact with solvent and solute molecules during the separation process.

Surface energy and surface extension together characterize the activity of silica ("activity" is the surface property of the adsorbent), and the size of the surface is reduced when covered with molecules such as water and glycol, which deactivate the surface of the sorbent. An increase in surface activity results in lower R_f values which, therefore, depend on silica porosity and humidity changes. The surface pore diameter can vary over a wide range; TLC sorbents have pores of 40, 60, 80, and 100 Å. The specific surface area of silica gel ranges from 200 to 800 m^2/g.

Activation by heating at 150–200°C removes the physically bound water. The assumption that one silica is most suitable for adsorption and another for liquid–liquid partition chromatography is questionable and, beyond that, irrelevant because pure adsorption or partition retention mechanisms generally do not occur.

Thin-layer chromatographic silica gels have specific pore volumes ranging from 0.5 to 2.0 mL/g; sorbents with the highest values are preferred for partition chromatography. Typical applications of silica gel in TLC separation of classes of organic compounds are listed in Table 2.

Alumina

The sorbents for TLC are obtained by thermal removal of water from hydrated aluminum hydroxide preparations, at low temperature (200–600°C), and the specific surface area of these aluminas ranges from 50 to 250 m^2/g. Hydroxyl groups and oxide ions are present on the surface and the latter are responsible for the basic properties of the sorbent. The use of gypsum (calcium sulfate

Table 1 Classification of the Most Used TLC Sorbents

General class	Sorbents
Polar inorganics (hydrophilic)	Silica gel or silicic acid, alumina, diatomaceous earth (Kieselguhr), magnesium silicate (Florisil)
Polar organics	Cellulose, starch, chitine, polyamide 6 or 11
Polar bonded phases	Aminopropyl-, cyanopropyl-, and diol-modified silica
Hydrophobic bonded phases	C_2-, C_8-, C_{18}-, and phenyl-modified silica; cellulose triacetate and triphenylcarbamate
Ion exchangers	
Inorganic	Zirconium phosphate, tungstate and molybdate, ammonium molybdophosphate and tungstophosphate, hydrous oxides
Organic	Polystyrene-based anion and cation exchangers, polymethacrylic acid; cellulose-based anion and cation exchangers; substance-specific complexing ligands
Impregnated layers	Silica impregnated with saturated and unsaturated hydrocarbons (squalene, paraffin oil), silicone and plant oils, complexing agents (silver ions, boric acid and borates, unsaturated and aromatic compounds), ligands (EDTA, digitonin), and transition metal salts; silanized silica gel impregnated with anionic and cationic surfactants
Gel filtration media	Cross-linked, polymeric dextran gels (Sephadex)
Chiral phases	Cellulose, cellulose triacetate, silanized silica gel impregnated with the copper(II) complex of (2*S*, 4*R*, 2′*RS*)-*N*-(2′-hydroxydodecyl)-4-hydroxyproline (CHIRALPLATE, HPTLC CHIR)

Table 2 Application in Normal- and Reversed-Phase TLC

Sorbent	Substance class
Silica gel	Aflatoxins, alkaloids, anabolic compounds, barbiturates, benzodiazepines, bile acid, carbohydrates, etheric oil components, fatty acids, flavanoids, glycosides, lipids, mycotoxins, nitroanilines, nucleotides, peptides, pesticides, steroids, sulfonamides, surfactants, sweeteners, tetracyclines, vitamins
Alumina	Aromatic hydrocarbons, herbicides, hydrazines, insecticides, metal ions, fat-soluble vitamins, lipids, lipophilic dyes, PAHs
Cellulose	Amines, amino acids, antibiotics, artificial sweeteners, carbohydrates, catechols, flavanoids, PAHs, peptides
Alkyl- and aryl-bonded phases	Alkaloids, amides, amines, amino acids, amino phenols, antibiotics, antioxidants, barbiturates, drugs, fatty acids, indole derivatives, nucleobases, oligopeptides, optical brighteners, PAHs, peptides, pharmaceuticals, phenols, phthalates, porphyrins, preservatives, steroids, surfactants, tetracyclines
Amino-modified silica gel	Nucleosides, nucleotides, pesticides, phenols, purine derivatives, steroids, vitamins
Cyano-modified silica gel	Analgesics, antibiotics, benzodiazepines, carboxylic acids, carotenoids, pesticides, phenols, steroids
Diol-modified silica gel	Nucleosides, pesticides, pharmaceuticals, phospholipids
Cellulose-based ion exchangers	DNA adducts, DNA and RNA fragments, dyes for food, inorganic ions, steroids
Polystyrene-based ion exchangers	Amines, amino acid, inorganic ions, peptides, purine and pyrimidine derivatives
Ammonium tungstophosphate	Amines, amino acids, indole derivatives, oligopeptides, polyamines, sulfonamides
Silica gel impregnated with paraffin, silicon, and plant oils	Barbiturates, carboxylic esters, fatty acid derivatives, nitrophenols, PCBs, peptides, pesticides, phenols, steroids, surfactants, triazines
Silanized silica gel impregnated with anionic and cationic surfactants	Aliphatic and aromatic amines, alkaloids, amino acids, amino sugars, carboxylic and sulfanilic acids, drugs, indole derivatives, nucleobases, nucleosides, nucleotides, peptides, dipeptides, polypeptides, phenols, phenothiazine bases, steroids, sulfonamides, water-soluble food dyes
Silver-impregnated silica gel	*cis*-Monoenoic esters, *cis*/*trans*- and *trans*/*trans*-Dienoic esters, fatty acid cholesteryl esters, positional and geometric isomers of fatty acid methyl esters, terpenoids, prostaglandins

hemihydrate) as a binder is said to neutralize the alumina surface. Aluminas with pH values of 9–10, 7–8, and 4–4.5 are designated basic, neutral, and acidic in character, respectively. Owing to the high density of hydroxyl groups (about 13 μmol/m^2), alumina tends to adsorb water and become deactivated. For this reason, it is recommended to activate aluminum oxide precoated layers, before use, by heating 10 min at 120°C.

Aluminas are of notable selectivity in adsorption chromatography of aromatic hydrocarbons; examples of separations of organic and inorganic compounds by adsorption and partition chromatography on layers of alumina are reported in Table 2.

Cellulose

Cellulose is formed by long chains of β-glucopyranose units connected one to another at the 1–4 positions. TLC sorbents are native fibrous cellulose and microcrystalline cellulose AVICEL.

The polymerization degree of native cellulose ranges from 400 to 500 glucose units and the fibers are shorter (2–20 μm) than those in paper chromatography, preventing the instantaneous spreading of solutes. The specific surface area is about 2 m^2/g.

High-purity fibrous cellulose, obtained by washing under very mild acidic conditions and, successively, with organic solvents, is also used in TLC. AVICEL is formed by dissolving the amorphous part of native cellulose by hydrolysis with hydrochloric acid.

Partition chromatographic mechanisms operate on cellulose thin layers even if adsorption effects cannot be excluded (for separation of substance classes, see Table 2).

Polyamide

Synthetic organic resins used in TLC are polyamide 6 (Nylon 6) and polyamide 11, which consist of polymeric caprolactam and undecanamide, respectively. Therefore, polyamide 6 is more hydrophilic than polyamide 11, owing to the shorter hydrophobic chain of its monomeric unit.

Polyamide precoated plates are currently used for the separation of phenols and phenolic compounds (i.e., anthocyanins, anthoxanthines, anthroquinone derivatives, and flavones) using solvents of different elution strength (DMF > formamide > acetone > methanol > water). Such eluents and solutes compete for the hydrogen bonds with the peptide groups of the polyamide.

Due to the medium polarity of polyamide 6, the sorbent can be made more or less hydrophobic than the mobile phase selecting appropriate polar and nonpolar eluents; therefore, normal- and reversed-phase chromatography and also two-dimensional technique can be developed.

Polar and Hydrophobic Bonded Phases

A variation of chromatographic selectivity in TLC has been obtained using surface-modified silica gel or cellulose.

Alkyl- and Aryl-Bonded Silica Plates

These materials are suitable for reversed-phase thin-layer chromatography owing to their lipophilic properties. Reversed phase (RP) means that the stationary phase is less polar than the mobile phase.

Although unbonded silica is still the sorbent preferred by most workers, there is an increasing demand for nonpolar plates, because they broaden the applicability of TLC.

The commonly used RP layers consist of dimethyl- (RP-2), octyl- (RP-8), octadecyl- (RP-18), and phenyl-bonded silica gel, type 60, with different mean particle sizes and particle size distributions.

The lipophilic character of the sorbent increases from RP-2 to RP-18, but it is also determined by the surface density of hydrophobic residues. Consequently, silicas are reacted to a different degree, either totally (100%) or partially (i.e., 50% of the reactive silanol groups) in order to obtain materials of various hydrophobicity and wettability.

Because aqueous–organic mixtures are commonly used as eluents, it should be noted that RP-18 plates can be developed with solvents containing a maximum water content of approximately 60% (v/v), whereas on 50% modified silica layers, water percentages as high as 80% can be employed. Wettable RP-18W plates for normal- and reversed-phase chromatography can be eluted with purely organic and aqueous–organic solvents as well with purely aqueous eluents.

Typical applications of reversed-phase chromatography are shown in Table 2. Beyond analytical applications, RP-TLC on bonded phases is also a tool for physicochemical measurements, particularly for molecular lipophilicity determination of biologically active compounds. Hydrophobicity can be measured by partition between an immiscible polar and nonpolar solvent pair, particularly in the reference system *n*-octanol–water. The partition coefficient, *P*, is frequently used to interpret quantitative structure–activity relationships (QSAR studies).

Boyce and Milborrow [1] showed that the R_M values [$R_M = \log(1/R_f - 1)$] can be used as hydrophobic parameters.

Amino-, Cyano-, and Diol-Modified Precoated Silica Layers

These sorbents possess, as functional groups, cyano, amino, and diol residues, bonded by short-chain hydrophobic spacers to the silica matrix. With respect to polarity, hydrophilic-modified silicas range between nonmodified silica and the nonpolar alkyl- or aryl-bonded phases:

$$\text{silica} > \text{diol-silica} > \text{amino-silica} > \text{cyano-silica} > \text{RP material}$$

In the case of NH_2 and CN plates, the functional groups are bonded through a trimethylene chain to the silica gel. The hydrophilic-modified layers are wetted by all solvents, including water, and are useful for the separation of polar substances which can cause problems with silica or alumina (see Table 2). The NH_2 ready-to-use plates can act as a weak basic ion exchanger.

Cellulose- and Polystyrene-Based Ion Exchangers

Several functional groups have been used to obtain cellulose anion exchangers [aminoethyl (AE), diethylaminoethyl (DEAE)], or cation exchangers [carboxymethyl (CM), phosphate (P)] for thin-layer chromatography. PEI cellulose is not a chemically modified cellulose, but a complex of cellulose with polyethyleneimine. These cellulose exchangers are particularly useful for the separation of proteins, aminoacids, enzymes, nucleobases, nucleosides, nucleotides, and nucleic acids.

Plates coated with a mixture of silica and cation- or anion-exchange resin are commercially available. These polystyrene-based ion exchangers are suited for the separation of inorganic ions and organic compounds with ionic groups (Table 2).

The large surface area of cellulose exchangers causes a large number of functional groups to be close to the surface. The distances of the active groups are longer than on exchange resins (about 50 Å and 10 Å, respectively), but cellulose ion exchangers, despite their smaller exchange capacity with respect to polystyrene-based exchangers, can be easily penetrated by large hydrophilic molecules, such as proteins, enzymes, and nucleic acids, which, therefore, interact with all the active groups.

On the contrary, the majority of ionic substituents of exchange resins do not participate in the reaction, because they are located inside the synthetic resin matrix which hydrophilic molecules cannot penetrate.

Impregnated Layers

The selectivity of sorbents can be easily improved by their impregnation with suitable organic and inorganic substances. The impregnating agent can be added to the sorbent suspension before plate preparation or, alternatively, the precoated layers may be dipped into an appropriate solution containing the impregnating agent.

Ready-to-use impregnated plates are also commercially available (i.e., caffeine- or ammonium-sulfate-impregnated silica for the separation of polynuclear aromatic hydrocarbons and surfactants, respectively).

A large number of impregnating agents have been tested, the most frequently used in TLC are silver nitrate, metal ions, cationic and anionic surfactants, silicone, and paraffin oil. Boric-acid-impregnated silica gel layers are suitable for the resolution of carbohydrates and lipids.

Argentation chromatography, in which silver is used as a π complexing metal on a silica gel support, is usually employed for the separation of organic compounds with electron-donor properties due to the presence of unsaturated groups in the molecule of the analytes (see Table 2).

Thin-layer chromatography is particularly appropriate for applying silver complexation techniques because the instability of silver causes severe limitations to column lifetime and, therefore, to HPLC methods.

The first investigation on alkyl-bonded silica layers impregnated with anionic and cationic surfactants was carried out by Lepri and co-workers [2]. The optimal concentration of the alcoholic solution of the impregnating agent was found to be 4%.

As regards the layers, ready-to-use alkyl-bonded silica plates were found to have many advantages over the previously employed home-made plates. An appropriate term proposed for this chromatographic technique is surely "dynamic ion-exchange chromatography." The method can be applied to separation of a wide variety of ionic compounds and classes of compounds (see Table 2).

Dextran Gel

Polymeric, cross-linked dextran gels, called Sephadex®, are used in size-exclusion TLC. Sephadex® gels, which are available in coarse (100–300 μm), medium (50–150 μm), fine (20–80 μm), and superfine (10–40 μm) particle size distributions, must be applied in a total swollen condition as chromatographic sorbents and eluted with the aid of continuous development techniques.

A typical application in TLC is the determination of the molecular weight of proteins [3].

Layers for Chiral Chromatography

Only cellulose, cellulose triacetate, and silanized silica gel impregnated with the copper(II) salt of derivatized L-hydroxyproline have been used as chiral stationary phases for the separation of enantiomeric solutes.

Detailed information on the topic can be drawn from the entry Enantiomer Separations by TLC.

References

1. C. B. C. Boyce and B. V. Milborrow, A simple assessment of partition data for correlating structure and biological activity using thin-layer chromatography, *Nature (London) 208*: 537 (1965).
2. L. Lepri, P. G. Desideri, and D. Heimler, Soap thin layer chromatography of some primary aliphatic amines, *J. Chromatogr. 153*: 77 (1978).
3. P. Fasella, A. Giartosio, and C. Turano, Applications of thin-layer chromatography on Sephadex to the study of proteins, in *Thin-Layer Chromatography* (G. B. Marini-Bettòlo, ed.), Elsevier, Amsterdam, 1964, pp. 205–211.

Suggested Further Reading

Geiss, F., *Fundamentals of Thin Layer Chromatography (Planar Chromatography)*, Alfred Hüthing Verlag, Heidelberg.

Grassini-Strazza, G., V. Carunchio, and A. M. Girelli, Flat-bed chromatography on impregnated layers, *J. Chromatogr. 466*: 1 (1989).

Hauck, H. E., and M. Mark, Sorbents and precoated layers in thin layer chromatography, in *Handbook of Thin Layer Chromatography* (J. Sherma and B. Fried, eds.), Marcel Dekker, Inc., New York, 1996, pp. 101–128.

Luciano Lepri
Alessandra Cincinelli

Trace Enrichment

Trace enrichment is a sample precleaning procedure which is performed prior to a sample analysis. The purpose of any sample preparation procedure is twofold. First, such a procedure must selectively collect and concentrate the components of interest. Second, the method should eliminate any other components that would either interfere with the analysis or would contaminate an analytical gas chromatography (GC) or high-performance liquid chromatography (HPLC) column to shorten its useful analytical life.

There are many sample preparation techniques listed in texts, from a simple filtration or centrifugation to many other kinds of extraction procedures, including both liquid–liquid and solid-phase extraction. When any type of sample preparation is used, it often is done manually if only a few samples are involved. If a large number of samples are to be analyzed, the entire procedure should lend itself to automation. Regardless of the number of samples, most sample preparation is done off-line; that is, the samples are prepared first with one of the methods listed, then placed into an automated sample injection system for sequential analysis of all samples.

Trace enrichment is the sample preparation procedure which is performed by passing a crude sample through a special "collection" column, but with a couple of added features. It is often done on-line, one at a time, just prior to the analytical procedure. It usually involves a minicolumn very similar to an off-line device, like a solid-phase extraction tube, but packed into a stainless-steel column and attached to the injection valve. This is called a "trace enrichment" column in many published procedures. Another name that is often given to this technique is "column switching" in the title or key words of published articles. This is because this technique involves valves that must be switched from one solvent flow stream to another. The valve, in its initial position, allows the "crude" sample to be passed through the trace enrichment column, whereas the mobile phase is being passed only through the analytical column. After the trace enrichment column has collected the sample components of interest, these components are backflushed onto the analytical column by switching the valve to its alternate position. (See Fig. 1.)

The complete procedure for using a trace enrichment (TE) column is as follows:

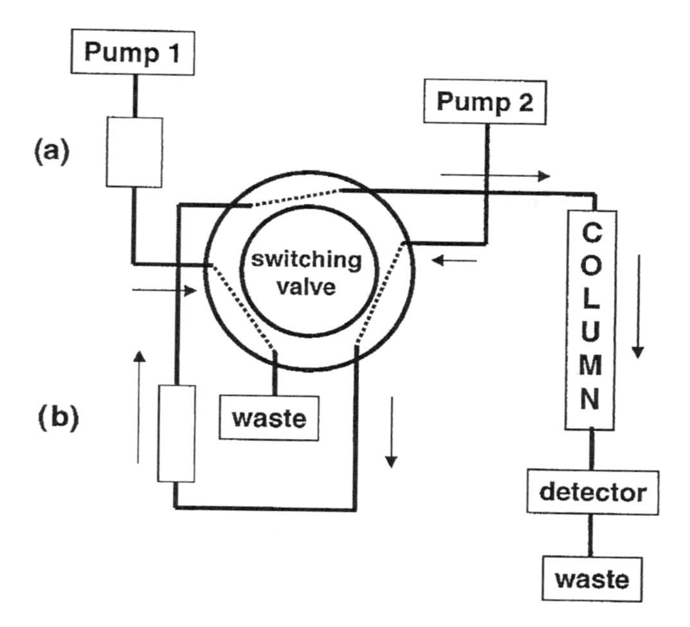

Fig. 1 Switching device.

1. The TE column is equilibrated with a minimum of two solvent combinations. The first solvent is strong, to wash the TE column free of impurities it may have accumulated from a previous sample (remember, this column is used over and over, so this step is critical so there is no sample carryover). This solvent is probably pure methanol or acetonitrile if using a reversed-phase filled TE column. The second solvent is not strong; it is used to bring the TE column to equilibrium to accept the sample components of interest. Again if a reversed-phase filled TE column is used, this could even be pure water.
2. The "crude" sample is injected or pumped onto the TE column. [Note: Even a "crude" sample requires membrane filtration and/or centrifugation to prevent shortened column life of the trace enrichment column.] The TE column will attract the components of interest, depending on the nature bonded phase in the TE column, the solvent in which the sample is dissolved, and the solvent in which the TE column has been equilibrated. The TE column, most likely, will remove more than just the components of interest and, thus, might need an additional solvent wash sequence to remove other collected impurities.
3. Finally, when sufficient sample has been passed through the TE column, the valve is switched so that the final solvent mixture is passed, with a *reversed* flow direction, through the TE column. This is usually a stronger eluting solvent combination that elutes the collected components to be chromatographed from the TE column. It is often the chromatographic mobile phase, which will also perform the actual analysis on the HPLC analytical column. Once the components are on the analytical column, the separation begins.

While the actual analytical separation is taking place, a second sample can be prepared with the same TE column. Again, a strong solvent wash, then the equilibration solvent wash precede injection of the second sample. The actual TE process on the minicolumn is usually accomplished within a few minutes, perhaps 2–3 min, and the actual separation might take 10 min. Thus, the time-limiting factor is the actual analysis, not the TE process itself. Shortening the analysis time with a shorter, more efficient column would lead to greater productivity as long as the actual analysis time is at least 1 min longer than the TE step.

The TE column ideally collects only the components of interest, permitting all unwanted impurities to pass through it. This is seldom completely true in practice, so the details of the entire trace enrichment have to be experimentally determined. These details involve determining which minicolumn packing to use and, perhaps, up to three solvent combinations. The first of these combinations is a solvent(s) which will dissolve the sample so that it carries the sample onto the TE column and simultaneously allows capture of the desired components. With experience, this first solvent combination may also remove (elute) some unwanted components.

A second solvent combination will be used to wash more of the remaining impurities out of the minicolumn. Finally, the third solvent combination is used to backflush the captured components from the TE column onto the analytical column. This latter solvent often is the mobile phase that will also perform the actual separation.

Because about 80% of the separations being performed by HPLC are done using a reversed-phase column, this is also the most used type of minicolumn used for TE columns. It could, however, be any type, because it only has to have some degree of affinity for the components of interest in the solvents chosen for the work, based on their solubility and the chemistry of the TE packing.

When beginning the development of a TE column procedure, an excellent source of solid-phase extraction (SPE) information may be found in SPE device manufacturers' literature (or on their websites). Most manufacturers offer at least 25 years of literature references on their products. Their procedures will often provide useful information about compound types and the SPE chemistries used and solvents for cleanup and elution. The same packing material may be used in the TE column, either the same size (usually 40–63 μm for SPE tubes) or it can be a 5- or 10-μm version of it. Buying the TE packing material from the same company that makes the SPE tubes will allow the solvent information to be used with the on-line work.

Other bonded-phase chemistries will give different capture capacities and efficiencies, so the solvent to capture or elute would be completely different. If in doubt about the equivalency of the chemistries, again, manufacturers are a good source of information.

One of the most difficult types of samples to cleanup are those derived from serum, plasma, urine, or other plant or animal matrices. Proteins or other high-molecular-weight biopolymers will quickly contaminate a HPLC column if they are not removed from samples. Such off-line procedures might involve multiple steps. However, another type of packing with special bonding has been used for of this type of sample cleanup. It is called the RAM or *restricted access material*. This packing material has an outer diol-bonded phase and an internal bonded reversed phase. Biopolymers cannot penetrate the smaller pores and are repelled by the diol phase and, thus, pass completely through the column. However, the components of interest, which are usually much lower-molecular-weight drug substances, are able to pass into the pores and attach themselves to the reversed phase. As mentioned earlier, it is still necessary to filter and centrifuge the samples before they are introduced onto the TE column, but these columns can often handle up to 2000, 50-μL injections before having to be replaced.

If biological samples are being purified, studies should also be conducted to see under what conditions any compounds of interest are released from the proteins being eluted from the typical RAM column mentioned earlier. This is due to the strong binding experienced with many compounds onto proteins. If 50% remain bound to the protein as it is washed away, then there is 50% less compound collected for making a measurement. One might never achieve 100% release, but an extra 10–20% might be helpful in trace work.

The size of the TE column can be as small as 2–3 mm × 10–25 mm. It simply has to have sufficient capacity to capture all, or a major percentage, of the components of interest. It should be remembered that a TE column usually captures structurally related compounds and, perhaps, some impurities; hence, it might be assumed that only 15–25% of the capacity is taken up by the compound(s) of interest. Because a relatively strong mobile phase is used for backflushing, band broadening is not generally a problem, regardless of the size of the column.

The way most of them are used with a valve–loop injector is that the sample first flows in one direction for capture, then is backflushed onto the analytical column. This process of backflushing could also push particulates resting on the inlet frit back to the inlet of the analytical column; therefore, this is another good reason for filtration/centrifugation of samples before introducing them onto the TE column. An in-line filter can also prevent particulates from migrating to places where they can cause problems.

Suggested Further Reading

Majors, R., K.-S. Boos, C.-H. Grimm, and D. Lubda, Practical guidelines for integrated sample preparation using column switching, *LC/GC Mag. 14*: 554 (1996).

Yu, Z., D. Westerlund, and K.-S. Boos, Evaluation of LC behavior of RAM precolumns in the course of direct injection of large volumes of plasma samples in column switching systems, *J. Chromatogr. 704*: 53 (1997).

Fred M. Rabel

U–V

Ultrathin-Layer Gel Electrophoresis

Introduction

Electrophoretic separation of double-stranded DNA (dsDNA) molecules, such as RFLP fragments or polymerase chain reaction (PCR) products, is usually accomplished in agarose, polyacrylamide, or composite agarose–polyacrylamide gels [1]. Agarose gels are extensively used to analyze DNA fragments from hundreds of base pairs (bp) to tens of thousands of base pairs, and polyacrylamide gels are regularly used for high-resolution DNA fragment analysis from several base pairs up to a thousand base pairs. In spite of numerous refinements in electrophoresis techniques during the past decade, agarose gel electrophoresis is still not efficient enough. The existing methodology requires multiple steps, such as gel casting, sample loading, staining, and imaging/documentation/data evaluation, making it very tedious and time-consuming because these tasks are not readily integrated and automated for high-throughput applications. Although several attempts have been made for automation, dsDNA fragment analysis in most laboratories is still done in a very conventional way: using submarine format agarose gel electrophoresis separation with ethidium bromide staining [2]. Some of the large, automated DNA sequencing systems are recently reported to be used for genotyping and STR profiling using cross-linked polyacrylamide gels and fluorescently labeled primers, and the configuration of these devices does not accommodate the use of agarose gels as separation medium.

Attempts to discover higher-resolution and faster electrophoresis separation techniques and media started in the 1960s, when miniaturized methods [3] were described as microelectrophoresis, but imaging technologies were not yet quite as adequate to be able to capture separations on such a minute scale. Later, polyacrylamide gels contained in glass capillaries were employed in order to separate both DNA and RNA molecules, but this progress was ultimately hampered by the inability to control extra-Joule heat. Recently emerging capillary-electrophoresis-based methods are much less susceptible to the effects of Joule heat, due to the ability to dissipate heat via the large surface-to-volume ratios typical of narrow-bore capillary columns. Obtaining similarly high surface-to-volume ratios in planar-format electrophoretic systems require very thin layers, preferably reaching thickness in the capillary dimension (i.e., no greater than 0.25 mm).

The most recent advances in the electrophoretic separation of nucleic acids come from exploration of new separation matrices. Linear, non-cross-linked polymers, such as polyacrylamide, derivatized celluloses, and polyethylene oxides, have all been demonstrated to be effective in the size separation of biopolymers [4]. The advantages of these non-cross-linked polymers were proven almost entirely in high-performance capillary electrophoresis applications [5]; however, it has been shown that very high-concentration non-cross-linked polymers can also be used in planar slab–gel format [6] for the separation of dsDNA molecules. Chemically modified agarose gels or composite agarose–non-cross-linked polymer gels capable of resolving differences of merely a few bases in DNA fragments of several hundreds of bases in length have also been developed [7]. The use of highly concentrated non-cross-linked polymers for DNA fragment analysis applications may be advantageous in several respects. First, it has been shown that non-cross-linked polymers may be supplied in a dessicated, dry form, providing a long shelf life [8]. Second, planar-form non-cross-linked polymer gels can be rehydrated to any of the range of final gel concentrations, buffer compositions, and/or ionic strengths. In addition to that, lower viscosity non-cross-linked polymers can be easily replaced in the separation platform; therefore, cassettes supporting repetitive work with non-cross-linked polymers can be readily used with these matrices.

Visualization by covalent labeling of nucleic acids with fluorescent tags, such as fluorescein, tetramethylrodamine, Texas Red, and so forth, has been practiced for years [9]. This approach can be used for high-sensitivity DNA fragment analysis provided that the analyte is labeled by the appropriate fluorophore prior to the separation step. The approach reported in this work uses affinity binding (e.g., intercalation) dyes for *in migratio* labeling of the dsDNA fragments during the course of their electrophoretic separation. This method is beneficial in two aspects. First, unlabeled DNA fragments can be readily analyzed, because they become labeled during the separation process. Second, the complexation phenomenon usually increases the separation selectivity, resulting in a higher resolution [10].

Large-scale, high-resolution DNA fragment analysis (e.g., mapping) requires an affordable, fully automated high-throughput agarose gel-electrophoresis-based separation device that enables rapid, high-performance separations in a wide molecular-weight range. Here, we describe a system that greatly enhances the productivity of DNA fragment analysis by automating the current manual procedure and also reducing the separation time and human intervention from sample loading to data analysis.

Separation Platform

Figure 1 shows the block diagram of the automated ultrathin-layer agarose gel electrophoresis system consisting of a high-voltage power supply (1); platinum electrodes (2 and 3); ultrathin layer separation platform with built-in buffer reservoirs (4), and a fiber-optic-bundle-based detection system (5). A lens set (7) connected to the illumination/detection system via the fiber-optic bundle, scans across the gel by means of a translation stage (6). A 532-nm frequency doubled Nd:YAG laser excitation source (8) and an avalanche photodiode detector (9) are connected to the central excitation fiber and the surrounding collecting fibers of the fiber-optic bundle, respectively. Interface electronics (10) is used to digitize the analog output of the detector and to connect the system to a personal computer. The separation platform also includes a positional heat sink holding the gel-filled cassette in horizontal position. This heat sink also eliminates local heat-spot-generated separation irregularities by means of ho-

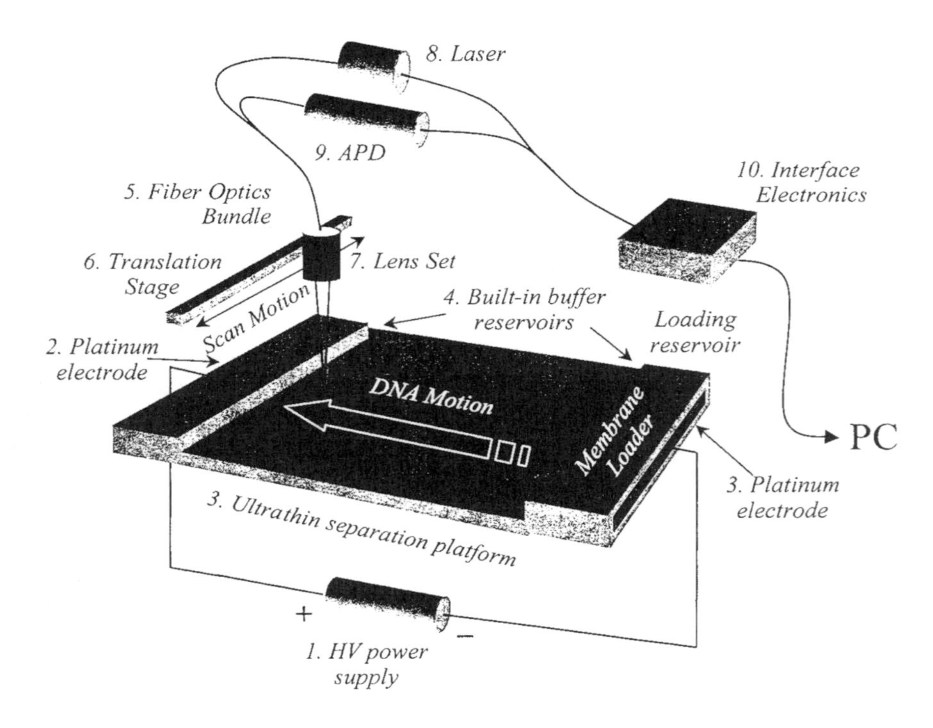

Fig. 1 Block diagram of the automated ultrathin-layer agarose gel electrophoresis system.

mogenous dissipation of any extra heat over the gel surface during the separation. Three heating cartridges and a high-speed fan control the temperature (above ambient) of the separation platform at a preprogrammed level. The preheated (40–45°C) separation cassettes (18 cm × 7.5 cm × 190 μm) are filled with melted 2% agarose. After a few minutes of cooling/solidification, the gel-filled cartridge is ready to be used. The effective separation length of the agarose-gel-filled ultrathin-layer cartridge is 6 cm. The applied separation voltage is 750 V, generating 5–7 mA of current.

Reusable Gel Cassettes and Membrane-Mediated Sample Loading

The reusable separation cassette has a flat bottom and top plate joined and secured in parallel, spaced 190 μm apart, with buffer reservoirs permanently fixed to both ends. Two spacers run along the inner-face edges of the plates to assure consistent distance between the glass plates (see the cassette in Fig. 1). For filling the cassette, a pump (e.g., large syringe) is used with a sealing applicator nozzle that matches to the top of the buffer reservoirs. To introduce the melted agarose gel into the cassette, the gel nozzle is placed at the top of one of the reservoirs, and the gel is pumped into the platform. The cassette should be preheated to approximately 40–45°C prior to the introduction of the melted agarose gel, in order to prevent premature solidification of the separation matrix. An appropriate amount of DNA staining dye [e.g., ethidium bromide (10–50 n*M*)] should be added to the melted agarose solution just before filling into the cassette.

Membrane-mediated sample loading provides a rugged, easy loading mechanism for ultrathin-layer electrophoresis gels, conveniently applicable for both vertical and horizontal formats [11]. The samples are spotted manually or automatically (robots) onto the surface of the loading membrane tabs, outside of the separation platform. The sample spotted membrane is then placed into the injection (cathode) side of the separation cassette, in intimate contact with the gel edge. By the application of the electric field, the sample components migrate into the gel. There is no need to form individual injection wells in the separation gel and loading is accomplished easily on the bench top. This novel sample injection method could be readily applied to most high-throughput gel electrophoresis based DNA analysis applications.

In Migratio Fluorescent Labeling

The use of fluorophore labeling of DNA fragments in gel electrophoresis expands the detection sensitivity and separation potential. Complex formation with fluorophores enables *in migratio* labeling [i.e., the migrating DNA fragments (negatively charged) complex with the countermigrating fluorophore intercalator dye (positively charged) which is dissolved in the separation matrix]. Using this method, high-sensitivity detection of the migrating DNA fragments as well as the high resolution of closely migrating components can be simultaneously obtained in a broad molecular-weight range. The appropriate concentration and type of the fluorophore complexing dye should be optimized for the wavelength of the excitation laser used in the illumination/detection system. For a 532-nm laser, ethidium bromide provides a good match (EX_{max} = 512 nm), but other high sensitivity dyes can also be used [12].

Agarose-Based Replaceable Separation Matrix

In the ultrathin-layer slab–gel format described in this entry, dsDNA fragment analysis was accomplished using 2% agarose gel, filled into the separation cassette at 60°C and used for electrophoresis separation at room temperature. The advantage of employing liquefied agarose above its gelling temperature is that the gel can be easily replaced in the separation platform by simply pumping fresh melted gel into the cassette (i.e., it can be easily filled, rinsed, and refilled). The inner surface of the separation platform was coated with linear polyacrylamide in order to avoid the formation of an electric double layer and concomitant electro-osmotic-flow generation [13].

Mutation Screening by Ultrathin-Layer Agarose Electrophoresis

Screening of 10 patients for the dopamine D_4 receptor (D4DR) gene polymorphism was used as a model system to demonstrate the capability of this device [14]. D4DR is the major receptor type in the limbic system. Its gene contains a 48-base-pair sequence repeat in exon 3 with a variable (twofold to eightfold) repeat number in individuals. This polymorphism is believed to be associated with personality characteristics, such as novelty seeking and extravertation/introvertation. The fragment containing the polymorphic site was amplified by PCR using two primers that anneal upstream and downstream of this region, resulting in products different in length (359 and 475 bp) which were then separated by ultrathin-layer agarose gel electrophoresis (Fig. 2).

Conclusion

In this entry, we reported the use of an automated, high-performance DNA fragment analyzer for high-through-

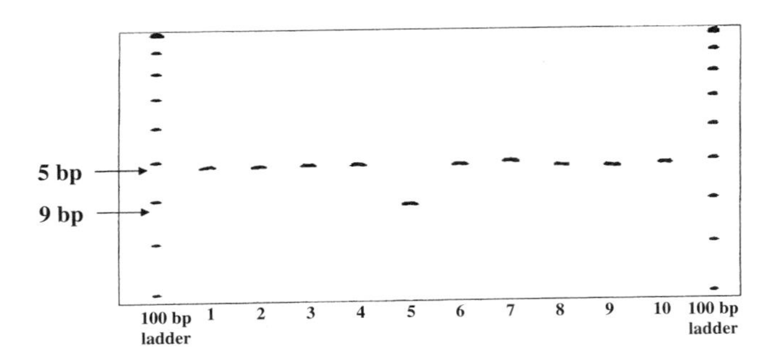

Fig. 2 Real-time electrophoretic image of the screening of 10 patients for the dopamine D_4 receptor (D4DR) gene polymorphism. Lanes 1–4 and 6–10 exhibit the 475-bp PCR product representing the common fourfold repeat alelle in homozygous form, whereas lane 5 shows a 359-bp fragment representing a patient homozygous for the twofold repeat polymorphism. [PCR was done in 35 cycles (95°C/min, 60°C/min, 72°C/min) using primers P1: 5′-GCG ACT ACG TGG TCT ACT CG-3′ and P2: 5′-AGG ACC CTC ATG GCC TTG-3′].

put applications, employing replaceable agarose gels in conjunction with a reusable ultrathin-layer separation cassette format in a horizontal configuration. Sample loading onto the ultrathin separation platform was accomplished by membrane-mediated loading technology, which also enabled robotic spotting of multiple samples [15]. The sample loading membranes also allow bar coding for identification and cataloging purposes and can be stored for several days between spotting and analysis. The analyte DNA fragments were stained during the separation process by *in migratio* complexation with appropriate fluorophore intercalator dyes and detected by a scanning fiber-optic-bundle-based integrated laser-induced fluorescence –avalanche photodiode imaging system.

Cost-effectiveness for high-throughput applications was assured by simple replacement of the separation matrix by pumping fresh melted gel composition into the cassette for each consequent analysis. After a few minutes of solidification, the gel-filled cassette was ready to be used again. When compared to regular submarine agarose gel electrophoresis, the separation performance and detection sensitivity of this new method utilizing the ultrathin-layer electrophoresis platform were found to be far superior, allowing significantly faster and higher-resolution analysis (more than 10-fold). With full system integration and automation, a high-throughput automated dsDNA analysis system can be developed by simply coupling DNA sample preparation, PCR amplification, restriction digestion, sample loading, electrophoresis separation, data analysis, and map construction.

Acknowledgments

The authors gratefully acknowledge the technical help of Doug Evans, Loi Nguyen, Swarna Ramanjulu, Nick Wilder, and Greg Theriault. This work was supported by the U.S.–Hungarian Joint Fund (J.F.No.654/96) and by the Hungarian funds FKFP 0658/1999, OTKA T022608.

References

1. A. Chrambach, *Practice of Quantitative Gel Electrophoresis*, VCH, Deerfield Beach, FL, 1985.
2. F. Sambrook, E. F. Fritch, and T. Maniatis, *Molecular Cloning*, 2nd ed., Cold Spring Harbor Laboratory Press, Plainview, NY, 1987.
3. T. Andrews, *Electrophoresis*, 2nd ed., Clarendon Press, Oxford, 1986.
4. P. Shieh, N. Cooke and A. Guttman, in *High Performance Capillary Electrophoresis* (M. G. Khaledi, ed.), John Wiley and Sons, New York, 1998, Chap. 5.
5. A. Guttman, *in Handbook of Capillary Electrophoresis* (J. P. Landers, ed.), CRC Press, Boca Raton, FL, 1994.
6. H. J. Bode, *Electrophoresis '79* Walter de Gruyter & Co., New York, 1980, p. 39.
7. D. Soto and S. Sukumar, *PCR Methods Applic 2*: 96 (1992).
8. H. Schwartz and A. Guttman, *Separation of DNA by Capillary Electrophoresis*, Beckman Instruments, Fullerton, CA, 1995.
9. L. M. Smith, J. Z. Sanders, R. J. Kaiser, P. Hughes, C. Dodd, C. R. Connell, C. Heiner, S. B. H. Kent, and L. E. Hood, *Nature 321*: 674 (1986).
10. A. Guttman and N. Cooke, *Anal. Chem. 63*: 2038 (1991).
11. S. Cassel and A. Guttman, *Electrophoresis 19*: 1341 (1998).
12. R. P. H. Haugland, *Handbook of Fluorescent Probes and Research Chemicals* (M. T. Z. Spence, ed.), Molecular Probes, Inc., Eugene, OR, 1996, Chap. 8.
13. S. Hjerten, *J. Chromatogr. 347*: 191 (1985).
14. M. Szoke, M. Sasvari-Szekely, Cs. Barta, and A.Guttman, *Electrophoresis 20*: 497, 1999.
15. J. E. Stanchfield and D. W. Batey, in *Genome Mapping and Sequencing Symposium*, 1998, p. 214.

András Guttman
Csaba Barta
Árpád Gerstner
Huba Kalász
Mária Sasvári-Székely

Unified Chromatography

Introduction

Gas chromatography (GC) was the first instrumental chromatographic mode to be demonstrated, back in the beginning of the 1950s [1]. Shortly after, Golay [2] proposed, in the same decade, the miniaturization of the technique by introducing the so-called capillary gas chromatography (CGC). It took almost two decades (1960s and 1970s) to spread out worldwide, particularly due to the problems in handling fragile glass columns, however in the 1980s and 1990s CGC became the preferred gas chromatographic mode using wall-coated open-tubular (WCOT) columns.

Supercritical fluid chromatography (SFC) was first demonstrated in the 1960s [3], but became popular only after the 1980s. Two different approaches were developed: one using packed columns of large inner diameter (i.d.) similar to those used in liquid chromatography, and the other one using nonpacked or open-tubular capillary columns similar to those used in CGC. Although open-tubular columns offer the same advantages over packed columns already perceived in CGC, packed columns are still, by far, used more than open-tubular columns in SFC.

Liquid chromatography (LC) was the first chromatographic mode to be developed almost one century ago [4]. Its instrumental version [usually referred as high-performance liquid chromatography (HPLC)] was developed during the end of the 1960s. Although its miniaturization has been discussed since then, its development has not followed the same successful route as CGC. The first step in the miniaturization of LC was the development of 1.0-mm-i.d. columns [5], nowadays usually termed "microbore columns." The next attempt to follow this direction was the introduction of capillary LC with columns of inner diameter smaller than 0.5 mm [6]. This approach is still far from being developed to a satisfactory extension in order to become a widely accepted routine technique as its GC counterpart (CGC). Although being developed slower than it should be, the miniaturization of the chromatographic techniques is mandatory for the development of unified chromatography.

The Concept of Unified Chromatography

Although the chromatographic techniques namely GC, LC, and supercritical fluid chromatography (SFC) have been developed in a completely independent way, there are no theoretical boundaries between these techniques, as pointed out by Giddings in 1965 [7]. Based on this concept Ishii later proposed and demonstrated in practice the idea of unified chromatography [8]. According to him [8], it is possible to demonstrate different-mode separations (viz. LC, SFC, and GC separation) using a single chromatographic system. According to the same author, the separation mode could be selected, changing the column temperature and pressure [8]. Shortly after the early experiments on unified chromatography by Ishii and co-workers, Yang proposed that the practice of GC, SFC, and microcolumn HPLC become more similar as the column diameter becomes smaller [9]. He also proposed that in this case, the injector; column, and detector may all be the same. In the same year, Bartle and co-workers pointed out that unified microcolumn chromatography allows high-resolution separations that make use of the same chromatographic components (injector, column, and detector) for capillary GC, microcolumn SFC, and HPLC [10].

A few reviews on unified chromatography have been published since its initial development [11–13]. To those readers interested in these techniques, it is highly recommended to consult the pioneer work described in these reviews because there is also, in the literature, some confusion about unified chromatography and multidimensional chromatography, as will be discussed next.

Unified Chromatography or Multidimensional Chromatography?

As discussed in the previous section, the concept of unified chromatography as agreed upon by the pioneers of this technique includes the use of the same chromatographic components (injector, column, and detector) performing GC, SFC, and LC separations. Because it is very common from the beginning, unified chromatography allows several combinations of chromatographic techniques, such as GC combined with LC, GC combined with SFC, SFC combined with GC, and so on. In this case, the separation is performed in such a way that the sample is introduced into the column and a fraction of it is eluted using one of the chromatographic modes (for instance, GC). After that, the conditions (in the single-substance mobile-phase mode) or the chemical nature (in the multisubstance mobile-phase mode) of the mobile phase is changed in order to perform another chromatographic mode to elute the analytes still retained in the column (for instance, using SFC).

In this example, the first step will allow the elution of the more volatile compounds in the GC mode while the less volatile will be eluted in the SFC mode. If these two modes were not enough to elute all analytes, it would have been possible to later perform a LC separation using the same instrument with a fraction of the sample still in the column. This is the original concept of unified chromatography. This concept has been confused with multidimensional chromatography that uses different columns to perform the separation. In this case, several publications deal with techniques such as GC–GC, LC–LC, SFC–SFC, GC–SFC, LC–GC, and so on. In this approach, either the total or a fraction of the analytes leaving a chromatographic column is transferred to a second column in order to obtain another dimension of analysis (thus the name "multidimensional chromatography"). In unified chromatography, there is no column transfer (switch) and the second dimension of the separation is provided by the change in the mobile-phase characteristics.

Terminology of Unified Chromatography

Several terms have been used to identify this technique. Some of them include the following:

Unified Approach to Chromatography
Unified Capillary Chromatography
Unified Microcolumn Chromatography
Unified GC and SFC
Unified Chromatography for GC, SFC, and LC
Unified Chromatography

The last term is more general and appropriate to the actual stage of the technique.

Instrumentation

The instrumentation used in unified chromatography is based on the ones used for GC, LC, and SFC with minor modifications.

Single-Substance Mobile-Phase Approach

Using a single substance as the mobile phase makes the instrumentation much easier than a multisubstance approach. In this case, the instrument could be a simple modification of HPLC equipment to also deal with a supercritical fluid, as used originally by Ishii. In this case, simply adjusting the mobile-phase pressure and temperature in order to be either in the SFC or LC mode can make the selection of the chromatographic mode (LC or SFC). This is the easiest way to perform unified chromatography but also the more limited one because it is difficult to find a single substance that presents good behavior as a chromatographic mobile phase in the GC mode as well as in the HPLC mode.

Multisubstance Mobile-Phase Approach

In this approach, the mobile-phase chemical nature is changed during the analysis when going from one chromatographic mode to the other. The elution of the sample components can start in the GC mode by introducing the mobile phase (usually a gas) into the column and providing the elution of the volatile compounds. After finishing this step the mobile-phase supply is interrupted and a second eluent (for instance, a supercritical fluid or a liquid) is introduced into the column in order to perform the second separation step. If a third mode is intended, the second eluent is interrupted and the third one is introduced into the column. As a consequence of this approach, a more complex and expensive instrument will have to be built, but also results will be more interesting because each one of the modes will have the mobile phase optimized independently. Whereas for simple samples, the first approach can be used, the multisubstance mobile phase approach might be necessary for more demanding and complex ones.

Selected Applications

Although unified chromatography still has to find its own applications niche, it has been already used for the analysis of a wide variety of samples from aromatic hydrocarbons, styrene, esters, phthalates, crude oil, amines, household wax, pesticides in vegetable oils and many others [11,14–16]. Its major application in the near future will certainly be centered in the analysis of complex samples such as environmental samples, biological fluids, forensic chemistry, and so forth. In this case, there is a need for more than one separation mode because the sample might contain volatile, semivolatile, and nonvolatile compounds of interest, thus requiring different chromatographic conditions.

Projected Enlargement of the Unified Chromatography Concept

When unified chromatography was first proposed, the chromatographic techniques were divided into GC, SFC, and LC. Nowadays, there has been a general acceptance that GC and LC are extreme mobile-phase conditions and that several intermediate conditions are possible. This includes solvated gas chromatography, enhanced liquid chromatography, and subcritical fluid chromatography, among other possibilities. Considering this fact, we can expect that in the near future, all these modes will be performed in the same instrument designed for unified chromatography. This will not only enlarge the theoretical concept of this technique but will also broaden its application range.

Limitations

The major limitation of unified chromatography is that there is no commercial instrument available up to this moment available to those interested in the techniques. The few existing instruments are basically located in universities and are built in-house to attend to the purpose of the investigator. Another considerable limitation of the techniques is the difficulty in making columns that can show good performance in more than two chromatographic modes. WCOT columns made for SFC can usually operate well in the capillary GC mode; packed columns for SFC can operate in the LC modes. However, a column, which operates in an optimized way in the GC and LC mode, is still to be developed. Although some attempts have been made in research laboratories, they are not commercially available. Another difficulty is related to the identification mode. Nowadays, mass spectrometry is the most powerful detector for all chromatographic modes. However, the interfaces for GC–MS and LC–MS (the extreme modes used in unified chromatography) are completely different. More research has to be developed in this field in order to fulfill this gap in the identification of the peaks eluted in unified chromatography.

Conclusion

Unified chromatography is an infant technique that combines all chromatographic modes in just one instrument. Although it is a fascinating task from the theoretical point of view, in practice it still presents several limitations. A severe one is the absence of commercial instruments, which hinders the development of further applications. It is expected that in the near future, the technique will be fully developed and the frontiers between the chromatographic world will be removed.

Acknowledgments

Professor Lanças wishes to acknowledge FAPESP (Fundação de Apoio à Pesquisa do Estado de São Paulo) and CNPq (Conselho Nacional de Desenvolvimento Científico e Tecnológico) for the financial support to his laboratory.

References

1. A. T. James and A. J. P. Martin, *Biochem. J. 50*: 679 (1952).
2. M. E. Golay, "Brief Report on Gas Chromatography Theory," in *Gas Chromatography* (1960 Eidinburg Symposium), R. P. Scott (ed.), Butterworths, London, 1960, p. 139.
3. E. Klesper, A. H. Corwin, and D. A. Turner, *J. Org. Chem. 27*: 700 (1962).
4. M. Tsett, *Ber. Deut. Botan. Soc. 24*: 316 (1906).
5. C. G. Horváth, B. A. Preiss, and S. R. Lipsky, *Anal. Chem. 39*: 1422 (1967).
6. D. Ishii, K. Asai, K. Hibi, T. Jonokuchi, and M. Nagaya, *J. Chromatogr. 144*: 157 (1977).
7. J. C. Giddings, "Dynamics of Chromatography," Marcel Dekker, Inc., New York, 1965.
8. D. Ishii, T. Niwa, K. Ohta, and T. Takeuchi, *J. High Resolut. Chromatogr. 11*: 800 (1988).
9. F. J. Yang, "Microbore Column Chromatography. A Unified Approach to Chromatography," Marcel Dekker, Inc., New York, 1989.
10. K. D. Bartle, I. L. Davies, M. W. Raynor, A. A. Clifford, and J. P. Kithinji, *J. Microcol. Separ. 1*: 63 (1989).
11. D. Tong, K. D. Bartle, and A. A. Clifford, *J. Chromatogr. 703*: 17 (1995).
12. T. L. Chester, *Anal. Chem. 69*: 165A (1997).
13. F. M. Lanças, "GC × SFC × LC: Towards Unified Chromatography." Proc. VIth Latin American Congress on Chromatography, COLACRO VI, Caracas, Venezuela, Intevep (ed.), 25–P1 (1996).
14. Y. Liu, F. Y. Yang, *Anal. Chem. 63*: 926 (1991).
15. I. L. Davies and F. J. Yang, *Anal. Chem. 63*: 1242 (1991).
16. D. Tong, K. Bartle, A. A. Clifford, and R. Robinson, *Analyst 120*: 2461 (1995).

Fernando M. Lanças

van't Hoff Curves

van't Hoff curves are strictly graphs of log[distribution coefficient (K)] against the reciprocal of the absolute temperature, but, in practice, log[capacity factor (k')] or log[corrected retention volume (V')] are used as alternatives (see the entry Plate Theory). van't Hoff curves can be very informative and explain both the nature of the separation and the importance of temperature in achieving an adequate resolution. An example of two van't Hoff curves relating log(V') against $1/T$ for two different types of distribution system are shown in Fig. 1. The two curves are highly exaggerated for the sake of emphasis and clarity.

It is seen that distribution system A has a large enthalpy value $(\Delta H_0/RT)_A$ and a low-entropy contribution $[(\Delta S_0/R) - V_s]_A$. The large value of $(\Delta H_0/RT)_A$ means that the distribution is *predominantly controlled by molecular forces*. The solute is preferentially distributed in the stationary phase as a result of the interactive forces between the solute molecules and those of the stationary phase being much greater than the interactive forces between the solute molecules and those of the mobile phase. Because the change in enthalpy is the major contribution to the change in free energy, *the distribution, in thermodynamic terms, is said to be "energy driven."*

In contrast, it is seen that for distribution system B, there is only a small enthalpy change $(\Delta H_0/RT)_B$, but in this case, a high-entropy contribution $[(\Delta S_0/R) - V_s]_B$. This means that the distribution is *not* predominantly controlled by molecular forces. The entropy change reflects the degree of randomness that a solute molecule experiences in a particular phase. The more random and "more free" the solute molecule is to move in a particular phase, the greater its entropy. A large entropy change means that the solute molecules are more restricted or less random in the stationary phase in system B. This loss of freedom is responsible for the greater distribution of the solute in the stationary phase and, thus, greater solute retention. Because the change in entropy in system B is the major contribution to the change in free energy, *the distribution, in thermodynamic terms, is said to be "entropically driven."*

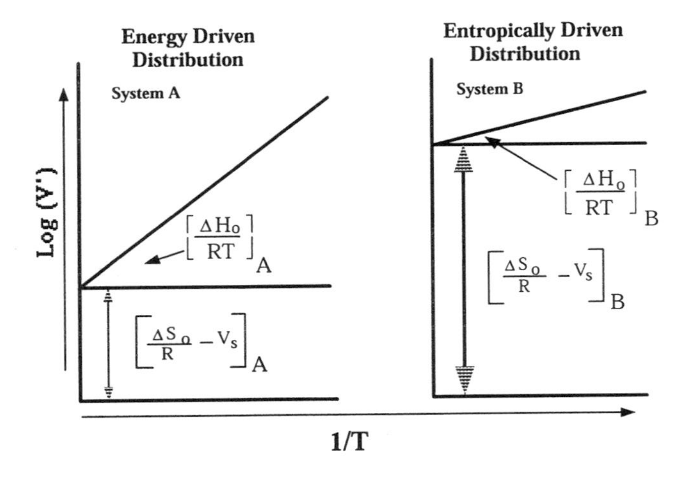

Fig. 1 van't Hoff curves for two distribution systems.

Separations dominated by size exclusion are examples of entropically driven systems. Chromatographic separation need not be exclusively "energetically driven" or "entropically driven"; in fact, very few are. In most cases, retention has both "energetic" and "entropic" components, which, by careful adjustment, can be made to achieve very difficult and subtle separations.

It is very rare, if at all, in gas chromatography (GC) that two solutes have identical standard free energies. Even when considering enantiomeric pairs, they will have differing standard free enthalpies and standard free entropies and, consequently, differing standard free energies. Accepting these facts, the van't Hoff curve provides critical information for the chromatographer. If two solutes (e.g., a pair of enantiomers) have different standard free enthalpies, then the linear van't Hoff curves must have different slopes and, therefore, they must intersect. As a consequence, there must be a temperature where the two curves intersect and there will be a specific temperature at which the two solutes must coelute. The coelution temperature may, or may not, be in the practical operating range of the chromatographic system, but it is clear that temperature is an essential operating variable. Temperature does not merely control the elution rate and allow the analysis to be achieved more rapidly, but also critically determines the column selectivity for closely eluting peaks. In fact, this is also true for liquid chromatography.

There are some examples in the literature that purport to show nonlinear van't Hoff curves. Taking an example from liquid chromatography (LC), a graph relating log(V'_r) against $1/T$ for solutes eluted by a mixed solvent from a reverse bonded phase will often be nonlinear. However, it must be emphasized that graphs relating log(V'_r) against $1/T$ can only be termed van't Hoff curves, and be treated as such if they apply to an *established distribution system*, where the nature of *both* phases does not change with *temperature*. If a mixed-solvent system is employed as the mobile phase, the *solvents themselves* are distributed between the two phases *as well* as the solute. Depending on the actual concentration of solvent, as the temperature changes, so may the relative amount of solvent adsorbed on the stationary-phase surface, and so the nature of the distribution system will also change. Consequently, the curves relating $\log(V'_r)$ against $1/T$ may not be linear and, as the distribution system is varying, will not constitute van't Hoff curves. This effect has been known for many years and an early example is afforded in some early work carried out by Scott and Lawrence [1], who investigated the effect of water vapor as a moderator on the surface of alumina in some gas–solid separations of some n-alkanes. Examples of the results obtained by those authors are shown in Fig. 2. The alumina column was moderated by a constant concentration of water vapor in the carrier gas. As the temperature of the distribution system was increased, less of the water moderator was adsorbed on the surface. As a consequence, the alumina became more active and the distribution system changed. Thus, initially, as the temperature was raised, the retention volume of each solute increased. When all the water was desorbed and the alumina surface assumed a constant interactive character and the distribution system also remained the same, the retention volume began to fall again in the expected manner and this part of the curve would then be a van't Hoff curve.

Reference

1. R. P. W. Scott and J. G. Lawrence, *J. Chromatogr. Sci. 7*: 65 (1969).

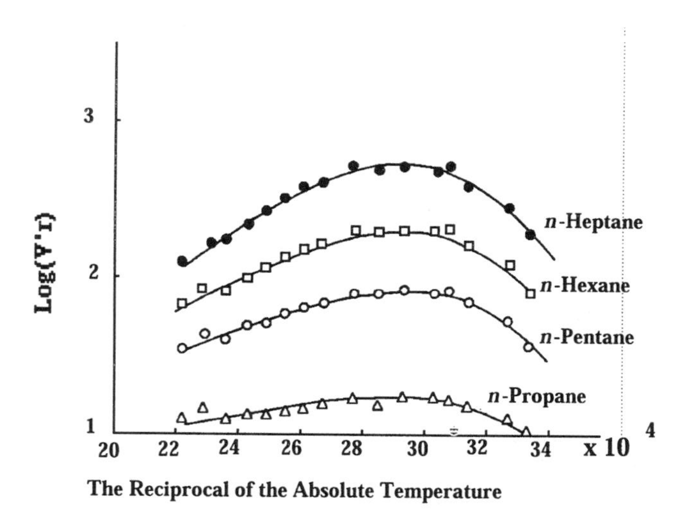

Fig. 2 Graphs of $\log(V'_r)$ against $1/T$ for some n-alkanes separated on water-vapor-moderated alumina.

Suggested Further Reading

Scott, R. P. W., *Techniques of Chromatography*, Marcel Dekker, Inc., New York, 1995.
Scott, R. P. W., *Introduction to Analytical Gas Chromatography*, Marcel Dekker, Inc., New York, 1998.

Raymond P. W. Scott

Vinyl Pyrrolidone Homopolymer and Copolymers: SEC Analysis

Introduction

Polyvinylpyrrolidone (PVP) is a synthetic polymer derived from the Reppe chemistry and is widely used in the pharmaceutical, personal care, cosmetic, agriculture, beverage, and many industrial applications. PVP is a polar and amorphous polymer which is soluble in water and some organic solvents, such as alcohols, chlorinated hydrocarbons, dimethyl formamide, dimethyl acetamide, and N-methyl pyrrolidone.

Nonionic, anionic, and cationic VP copolymers are all available commercially to enhance the hydrophilic, hydrophobic, and ionic properties of PVP for specific applications. Important comonomers include vinyl acetate (VA), acrylic acid (AA), vinyl alcohol, dimethylaminoethylmethacrylate (DMAEMA), styrene, maleic anhydride, acrylamide, methyl methacrylate, lauryl methacrylate (LM), α-olefins, methacrylamidopropyltrimethyl ammonium chloride (MAPTAC), vinyl caprolactam (VCL), and dimethylaminopropylmethacrylamide (DMAPMA).

Molecular Weight and Molecular Distribution of PVP by Size-Exclusion Chromatography

Commercial PVP are available in five grades, K-15, K-30, K-60, K-90, and K-120 with the respective range of M_w 7000–12,000, 40,000–65,000, 350,000–450,000, 900,000–1,500,000 and 2,000,000–3,000,000. Interactions between PVP and columns, such as adsorption, partition, and electrostatic interactions, have to be eliminated by prudent choice of column and mobile phase in order to obtain true separation by size and 100% recovery.

Belenkii et al. reported in 1975 [1] the size-exclusion chromatography (SEC) of PVP using Pharmacia Sephadex G-75 and G-100 columns and 0.3% sodium chloride solution as the mobile phase. Hashimoto et al. reported in 1978 [2] the SEC of PVP K-30 and K-90 using TSK-PW 3000 and two 5000 columns and $0.08M$ Tris-HCl buffer (pH = 7.94) as the mobile phase.

By using an E. Merck LiChrospher SI300 column, modified with an amide group ($—NH—CO—CH_3$) chemically bonded to the surface, Englehardt and Mathesn reported in 1979 [3] the SEC of PVP. Herman et al. synthesized monomeric diol onto E. Merck Lichrospher SI-500 and reported in 1981 [4] the SEC of PVP. Mori reported in 1983 [5] the SEC of PVP K-15 to K-90 using two Shodex AD-80M/S columns with DMF and $0.01M$ LiBr as the eluent at 60°C. Domard and Rinaudo grafted quaternized ammonium groups onto silica gels with pore diameters of 150, 300, 600, 1250, and 2000 Å and reported in 1984 [6] the SEC of PVP.

Malawer et al. [7] reported in 1984 the SEC of PVP K-15, K-30, K-60, and K-90 using diol-derivatized silica gel column sets and aqueous mobile phase modified with various polar organic solvents. Senak et al. reported in 1987 [8] the determination of the absolute molecular weight and molecular-weight distribution of PVP by SEC–LALLS and SEC with universal calibration. The column set used consists of TSK-PW 6000, 5000, 3000, and 2000 columns and a mobile phase of water–methanol (50:50, v/v) with $0.1M$ $LiNO_3$. One hundred percent recovery was reported. The results showed good agreement in M_w from SEC–LALLS and from SEC with universal calibration for PVP K-30, K-60, and K-90.

Size-exclusion chromatography of PVP K-15, K-30, K-60, K-90, and K-120 using linear aqueous columns such as the Showa Denko Shodex OH pack, Toyo Soda TSK-PW, and Waters Ultrahydrogel has been reported in 1995 by Wu et al. [9]. Using single linear columns also greatly reduces analysis time and solvent consumption, making SEC a practical method for quality assurance. A comparison of four commercial linear aqueous columns and four sets of commercial PEO standards for SEC of PVP K-15, K-30, K-60, K-90, and K-120 was reported by Wu et al. in 1999 [10]. In recent years, SEC–MALLS has also been applied in this laboratory to PVP. Figure 1 shows an overlay of SEC–MALLS chromatograms of all commercial grades of PVP using a linear aqueous column. The excellent overlap of absolute M_w versus retention volume plots for all commercial grades of PVP in Fig. 1 demonstrates that there is no significant difference in branching among all commercial grades of PVP, despite the different methods used to make them.

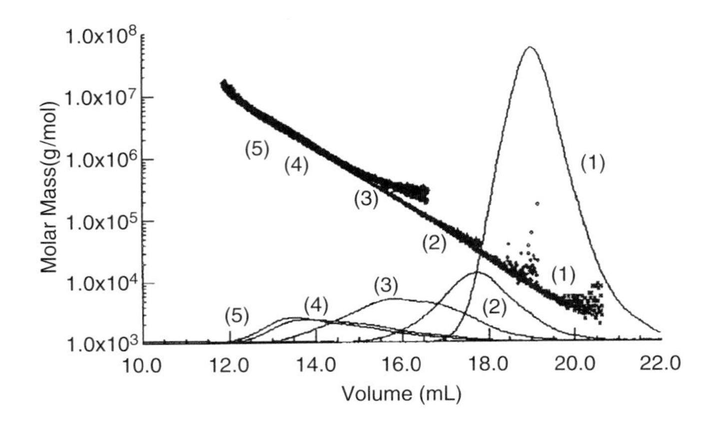

Fig. 1 Overlay of SEC chromatograms and M_w versus retention volume plots by MALLS for PVP grades: (1) K-15; (2) K-30; (3) K-60; (4) K-90; (5) K-120.

Finally, the following SEC columns, which have been introduced in recent years, have been demonstrated by the vendors to be successful with PVP: Jordi DVB Glucose BR linear column [in dimethyl sulfoxide (DMSO)], PL aquagel-OH column (in $0.1M$–$0.3M$ salt–buffer with 20% methanol); PSS Suprema column (in $0.1M$ Tris pH 7 buffer), and PSS SDV column (in DMAC, 0.1% LiBr) [11–13].

Molecular Weights and Molecular-Weight Distributions of VP-Based Copolymers by SEC

Wu et al. reported in 1991 [14] the SEC of a nonionic copolymer, such as copolymers of vinyl pyrrolidone and vinyl acetate (VP–VA) with different mole ratios and molecular weights, and a terpolymer of vinyl pyrrolidone, dimethylaminoethyl methacrylate and vinyl caprolactam (VP–DMAEMA–VC) in both aqueous and nonaqueous systems. For the aqueous system the column set used consisted of four Waters Ultrahydrogel columns of pore sizes 120, 500, 1000, and 2000 Å, and the mobile phase was water–methanol (1:1, v/v) with $0.1M$ $LiNO_3$. For the non-

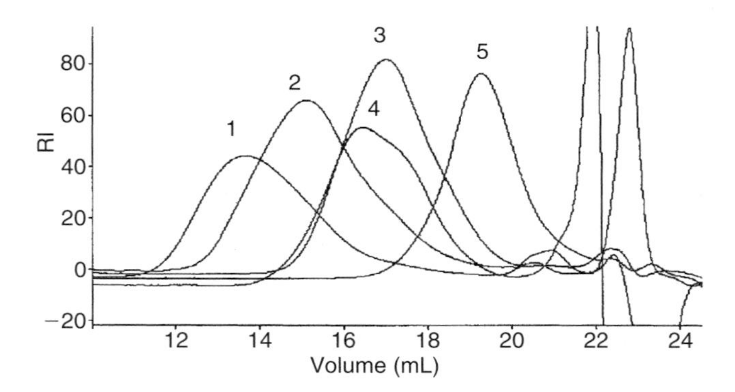

Fig. 2 Overlay of SEC chromatograms of VP-based copolymers in water–methanol (50:50) with 0.1*M* lithium nitrate: 1: VP–DMAPMA; 2: VP–DMAPMA–AA–LM; 3: VP–VCL–DMAPMA; 4: VP–AA–LM; 5: VP–VA.

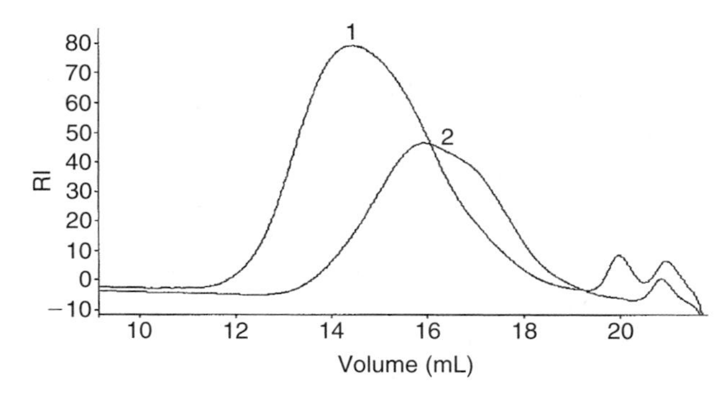

Fig. 3 Overlay of SEC chromatograms of VP-based copolymers in a pH 7 buffer: 1: high-MW VP–DMAEMA cationic copolymer; 2: medium MW VP–DMAEMA cationic copolymer.

aqueous system, the column sets were Shodex KD-80M plus Ultrahydrogel 120 Å, Shodex KD-80M plus PL gel 100 Å, and PL gel 10^4 Å plus 500 Å and the mobile phase was DMF with 0.1*M* $LiNO_3$. However, it is the aqueous system that showed the best separation at the low-molecular-weight end. Figure 2 shows the overlay of SEC chromatograms of VP-based copolymers in water–methanol (50:50) with 0.1*M* lithium nitrate and a linear Shodex OH-PAK column.

Wu and Senak reported in 1990 [15] absolute molecular-weights and molecular-weight distributions of a quaternized copolymer of vinyl pyrrolidone and dimethylaminoethylmethacrylate (PVP–DMAEMA) by SEC–LALLS and SEC with universal calibration using Waters Ultrahydrogel 120, 500, 1000, and 2000 Å columns and a 0.1*M* Tris, pH 7 buffer with 0.5*M* $LiNO_3$ as the mobile phase. Due to the cationic charges on the molecules, a much higher salt content (0.5*M* $LiNO_3$) is needed in the SEC mobile phase to improve the separation and recovery of the polymer. The water–methanol (1:1, v/v) mobile phase with 0.4*M* lithium nitrate has also been used in this laboratory for SEC of this cationic copolymers with the Shodex OH pak or Ultrahydrogel linear columns with good results. Figure 3 shows the overlay of SEC chromatograms of this cationic copolymers in a pH 7 buffer and a linear Shodex OH-PAK column.

Wu et al. reported in 1991 [14] the SEC of anionic copolymers, vinyl pyrrolidone and acrylic acid (VP–AA) with different mole ratios and molecular weights, using a 0.1*M* pH 9 Tris-buffer with 0.2*M* $LiNO_3$ as the mobile phase and the Ultrahydrogel 120, 500, 1000, and 2000 Å column set. Figure 4 shows overlay of SEC chromatograms of anionic copolymers in a pH 9 buffer and a linear Shodex OH-PAK column.

Copolymers and grafted copolymers of vinyl pyrrolidone and α-olefins, with α-olefins' contents higher than 60%, are not soluble in water or the water–methanol mixture. They should be analyzed in tetrahydrofuran using cross-linked polystyrene columns. However, copolymers and grafted copolymers of vinylpyrrolidone and α-olefins, with only 10% α-olefins, can be analyzed in the water–methanol mixture (50:50) with 0.1*M* lithium nitrate.

Summary

With a proper choice of mobile phase (aqueous or nonaqueous), many commercially columns are available for SEC of PVP and VP-based copolymers. Mobile-phase modifiers (such as methanol, salt, and buffer) are normally required to eliminate interactions with columns. A single

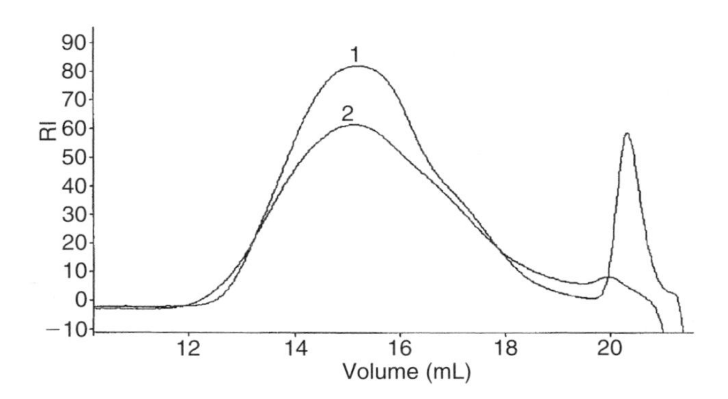

Fig. 4 Overlay of SEC chromatograms of VP-based copolymers in a pH 9 buffer: 1: VP–AA anionic copolymer; 2: VP–AA–LM anionic copolymer.

linear or mixed-bed column has been found to provide good separation of PVP and VP-based copolymers with a molecular weight range of from a few thousands to several millions. In general, the aqueous SEC system has better long-term stability and provides better separation than the nonaqueous SEC system, especially at the low-molecular-weight end. Hydroxylated methyl-methacrylate-type columns and water–methanol mobile phase (50:50, with salt) is a good first choice for SEC of PVP and many VP-based copolymers. New columns have been introduced to the marketplace in recent years; hopefully, they will offer better separation, solvent compatibility, stability, and minimal interactions.

References

1. B. G. Belenkii, L. Z. Vilenchik, V. V. Nesterov, V. J. Kolegov, and S. Y. A. Frenkel, *J. Liquid Chromatogr. 109*: 223 (1975).
2. T. Hashimoto, H. Sasaki, M. Aiura, and Y. Kato, *J. Polym. Sci. Polym. Phys. Ed. 16*: 1789 (1978).
3. H. Engelhardt and D. Mathes, *J. Chromatogr. 185*: 305 (1979).
4. D. P. Herman and L. R. Field, *J. Chromatogr. Sci. 19*: 470 (1981).
5. S. Mori, *Anal. Chem. 55*: 2414 (1983).
6. A. Domard and M. Rinaudo, *Polym. Commun. 25*: 55 (1984).
7. E. G. Malawer, J. K. DeVasto, and S. P. Frankoski, *J. Liquid Chromatogr. 7*(3): 441 (1984).
8. L. Senak, C. S. Wu, and E. G. Malawer, *J. Liquid Chromatogr. 10*(6): 1127 (1987).
9. C. Wu, J. F. Curry, E. G. Malawer, and L. Senak, in *Handbook of Size Exclusion Chromatography* (C. Wu, ed.), Marcel Dekker, Inc., New York, 1995, pp. 311–330.
10. C. Wu, L. Senak, D. Osborne, and T. Cheng, in *Column Handbook for Size Exclusion Chromatography* (C. Wu, ed.), Academic Press, New York, 1999, pp. 499–529.
11. H. Jordi, in *Column Handbook for Size Exclusion Chromatography* (C. Wu, ed.), Academic Press, New York, 1999, pp. 367–425.
12. E. Meehan, in *Column Handbook for Size Exclusion Chromatography* (C. Wu, ed.), Academic Press, New York, 1999, Chap. 12.
13. P. Kilz, in *Column Handbook for Size Exclusion Chromatography* (C. Wu, ed.), Academic Press, New York, 1999, Chap. 9.
14. C. S. Wu, J. Curry, and L. Senak, *J. Liquid Chromatogr. 14*(18): 3331 (1991).
15. C. S. Wu and L. Senak, *J. Liquid Chromatogr. 13*(5): 851 (1990).

Chi-san Wu
Larry Senak
James Curry
Edward Malawer

Viscometric Detection in GPC–SEC

Introduction

At the beginning of gel permeation chromatography (GPC), in the early 1960s, there was only one detector at the outlet of GPC columns, generally a differential refractometer, to continuously measure the concentration of eluents. This detection provided a chromatographic peak corresponding, roughly, to the mass distribution. In addition, it was possible to build a calibration curve corresponding to the response of the column set $\log(M) = F(V_e)$ (M being the molecular weight and V_e being the elution volume), which was roughly linear, by multiple injections of narrow standards with known molecular weights and very narrow molecular-weight distributions. At each increment of elution volume, the slice concentration C_i was calculated using the polymer peak intensity and the slice molecular weight M_i, using the calibration curve. Then, it was possible to calculate, by integration

$$\frac{\Sigma\, C_i}{\Sigma\, (C_i/M_i)} = M_n$$

$$\frac{\Sigma\, C_i M_i}{\Sigma\, C_i} = M_w$$

$$\left(\frac{\Sigma\, C_i M_i^a}{\Sigma\, C_i}\right)^{1/a} = M_v$$

$$\frac{\Sigma\, C_i M_i^2}{\Sigma\, C_i M_i} = M_z$$

[a being the viscosity-law (Mark–Houwink) exponent], the different average molecular weights (in number, M_n; in viscosity, M_v; in weight, M_w; in z, M_z), in a single experiment.

The use of the GPC technique spread very quickly but, very soon, it became obvious that this method was imperfect. Every polymer has its own calibration curve which differs from one polymer to another. Moreover, for copolymers or branched polymers, the molecular weight calibration curve does not work correctly.

In 1966, the "universal calibration" method was established at the University of Strasbourg, France [1]. It used, as a parameter, not molecular weight M, but the hydrodynamic volume represented by the product $[\eta]M$, $[\eta]$ being the intrinsic viscosity. This universal calibration curve $\log([\eta]M) = F(V_e)$ was supposed to be applied to every polymer and copolymer, linear or branched, as a unique calibration curve; it worked very well.

From that point, the necessity of continuously measuring viscosity, in addition to polymer concentration, became obvious. Several attempts were made to adapt existing viscometers as GPC detectors, but the problem of internal volume was critical. Ouano [2] published the first design of a single-capillary viscometer which was based on pressure measurement. Several similar designs [3–6] were published and a commercially available instrument, the Waters Model 150CV (Waters Associates, Milford, MA, U.S.A.), based on a design described in Ref. 4, became commercially available.

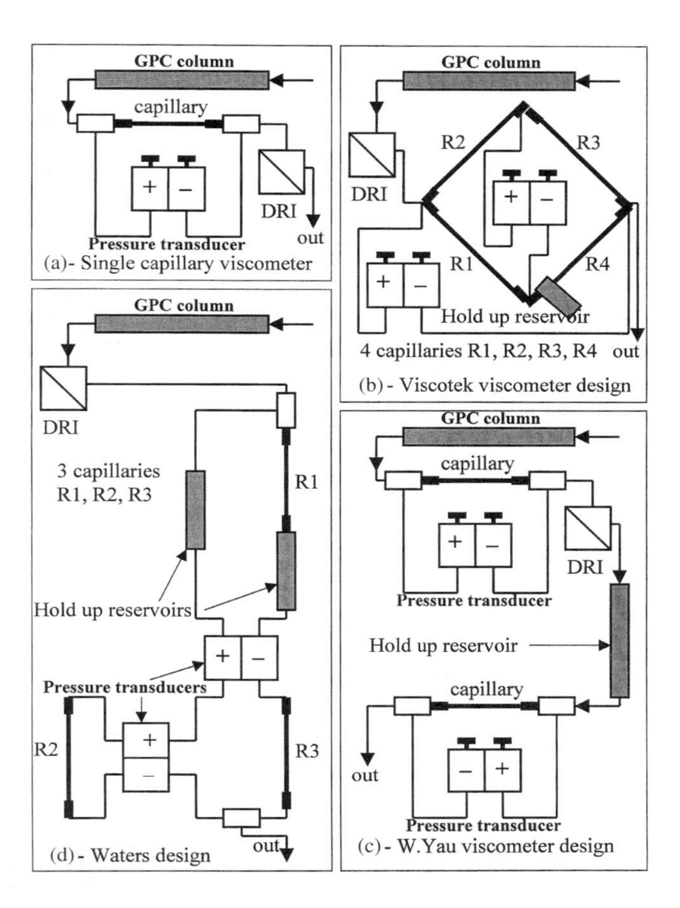

Fig. 1 Various designs of viscometric detection.

Single-Capillary Viscometer

The single-capillary viscometer (SCV) is represented in Fig. 1a. Its design is a direct extrapolation of classical viscometry measurement. It is composed of a small capillary, through which the solvent flows at a constant flow rate, and a differential pressure transducer (DPT), which measures the pressure drop across the capillary. SCV obeys Poiseuille's law and the pressure drop ΔP across the capillary depends on the geometry of the capillary, on flow rate Q, and on viscosity of the fluid η according to

$$\Delta P = \left(\frac{8}{\pi}\right)\left(\frac{l}{r^4}\right)\eta Q$$

where l and r are the length and radius of the capillary, respectively.

Classical viscometers maintain ΔP constant and calculate η by measuring the variations of flow rate Q. By contrast, the SCV uses the advantage of GPC, which already has a constant flow rate Q and allows the calculation of η by measuring the pressure drop variations ΔP across the capillary. At constant flow rate Q, the pressure drop is proportional to viscosity η, and at constant viscosity η, the pressure drop is proportional to flow rate Q. Consequently, in order to use the SCV as an accurate viscometer, the flow rate must be maintained absolutely constant during the GPC experiment. Conversely, SCV allows perfect control of flow rate and can also be used as a very accurate flow meter when only pure solvent comes out of columns. At the outlet of the GPC columns, the polymer concentration is very small and the increase of solvent viscosity η is very small as well. When measuring ΔP proportional to ηQ, Q should be maintained extremely constant; this requires pulse dampeners in the flow system to smooth small flow variations due the pumping system design.

For that reason, some other, more complicated, viscometers, which are insensitive to small flow variations, were designed using multiple capillaries and pressure transducers.

Multiple-Capillary Viscometers

The general purpose of multiple-capillary viscometers is to simultaneously measure the polymer solution in one part of the detector and pure solvent in a second part of the detector. In order to have these two parts at the same time, it is necessary to include a holdup reservoir to delay the polymer passing through the second part of the detector. One needs to wait until the delayed polymer comes out of the detector before reinjecting another sample. The basic principle is that one part of the detector measures polymer plus flow and the other part measures only flow. By obtaining the difference, the result is a signal which is proportional to polymer viscosity and insensitive to flow rate variations.

The first multiple-capillary viscometer was designed by the Viscotek Company (Houston, TX, U.S.A.) [7]. It is represented in Fig. 1b. It is composed of four identical capillaries, assembled as a bridge. Here, the difference is

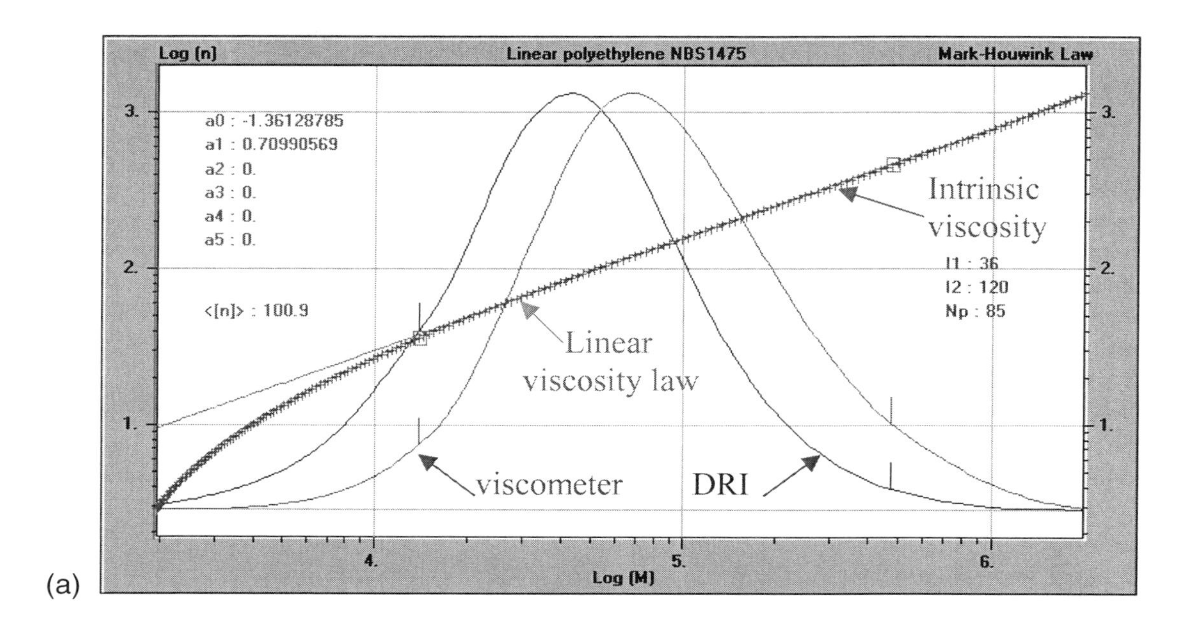

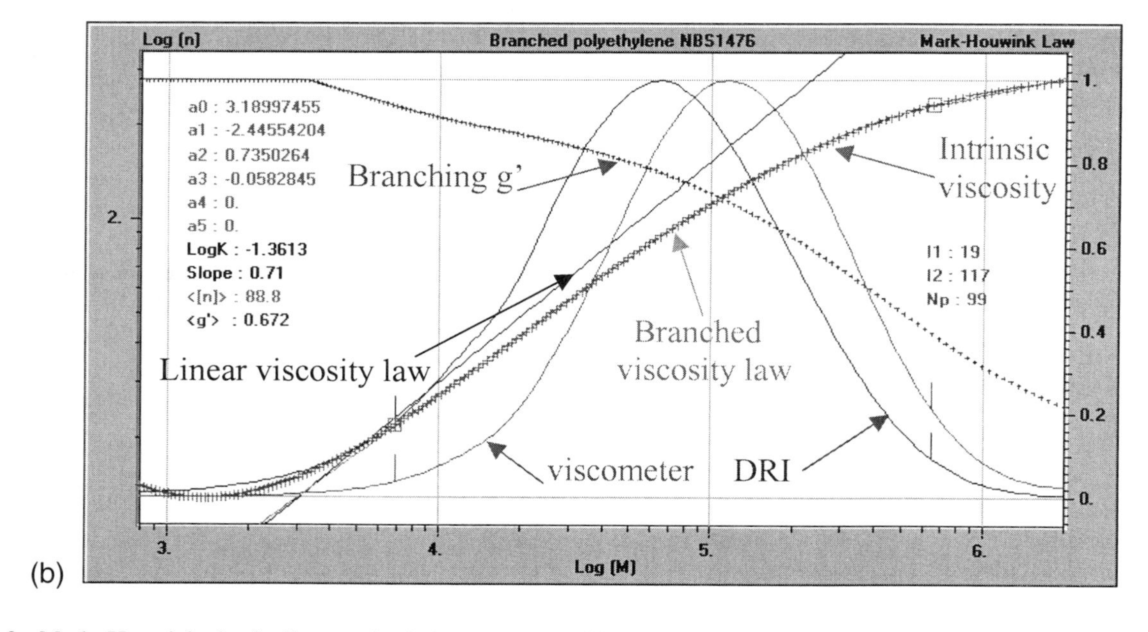

Fig. 2 Mark–Houwink plot for linear polyethylene NBS1475. (b) Mark–Houwink plot for branched polyethylene NBS1476.

generated by the bridge itself as a differential viscometer. The central DPT provides a differential viscosity signal DP and the second, DPT, provides inlet pressure IP. The intrinsic viscosity $[\eta]$ can be calculated by the formula

$$[\eta] = \frac{1}{C}\left(\frac{4\text{DP}}{\text{IP-2DP}}\right)$$

Another design was described by the Dupont Company [8]. It is represented in Fig. 1c. It is composed of two single-capillary viscometers. The first one (measure) is located as a normal SCV (Fig. 1a) and measures polymer plus flow. The second one (reference), identical to the first one, is located after a holdup reservoir which is connected after the refractometer and measures only the flow with pure solvent. It is just necessary to obtain the difference between the two signals to obtain the polymer signal with no flow contribution.

The last differential viscometer design is the Waters Corporation detector [9], which is in the Alliance GPCV2000 high-temperature instrument. It is composed of three capillaries, two differential pressure transducers, and two holdup reservoirs; it is represented in Fig. 1d. The pressure transducers are connected flow-through; this eliminates the need for frequent purges. This detector provides, at the same time, "relative viscosity information" and "relative flow information." This design does not require a perfect matching of the capillaries.

The Use of Viscometric Detection

The first purpose of viscometric detection in GPC is to use a "universal calibration curve." At every slice, the concentration C_i is obtained through refractometric data, intrinsic viscosity $[\eta]_i$ is obtained through refractometric and viscometric data, and hydrodynamic volume $[\eta]M_i$ is obtained from the universal calibration curve. It is just necessary to divide $[\eta]M_i$ by $[\eta]_i$ to get M_i and obtain the couple C_iM_i that is necessary to calculate the various average molecular weights, as shown at the beginning of this entry, independently of the chemical nature of the sample and the standards used for calibration.

The secondary use of viscometric detection is for the determination of long-chain branching. At the same molecular weight, a long-chain branched polymer has a more compact molecular structure than the linear one; consequently, it also has a smaller intrinsic viscosity. Therefore, it is eluted at a higher elution volume with a linear polymer of smaller molecular weight. That means that at a given elution volume, there is a mixture of polymers with different molecular weights and different degrees of branching eluting from the GPC column.

As intrinsic viscosity is affected by long-chain branching, viscometric detection provides branching information, just by comparing polymer viscosity laws. Figure 2a represents the viscosity-law plot of a linear polymer, polyethylene (NBS1475), with a linear viscosity law and an exponent of 0.71. Conversely, Fig. 2b represents the viscosity-law plot of a branched polymer, polyethylene (NBS1476), with a curved viscosity law, under the linear polymer law (linear viscosity law). At every molecular weight, the branching distribution g'_i is calculated by

$$g'_i = \frac{[\eta]\text{br}_i}{[\eta]\text{lin}_i}$$

where $[\eta]\text{br}_i$ and $[\eta]\text{lin}_i$ are the slice intrinsic viscosities of the branched and the linear polymer, respectively.

Conclusion

Viscometric detection in GPC–SEC is extremely useful. It allows continuous viscosity measurements, which, in turn, permits the use of a unique calibration curve for all polymers (universal calibration). In addition, the viscosity law is determined for linear polymers and long-chain-branching information can be obtained for branched polymers.

References

1. H. Benoit, Z. Grubisic, P. Rempp, D. Dekker, and G. Zilliox, *J. Chim. Phys. 63*: 1507 (1966).
2. A. C. Ouano, *J. Polym. Sci. A1 10*: 2169 (1972).
3. J. Lesec and C. Quivoron, *Analysis 4*: 399 (1976); L. Letot, J. Lesec and C. Quivoron, *J. Liquid Chromatogr. 3*: 407 (1982).
4. D. Lecacheux, J. Lesec, and C. Quivoron, *J. Appl. Polym. Sci. 27*: 4867 (1982); D. Lecacheux, J. Lesec, and R. Prechner, French Patent 82402324.6 (1982); U.S. patent 4478071 (1984).
5. F. B. Malihi, C. Kuo, M. E. Kohler, T. Provder, and A. F. Kah, *ACS Symp. Ser. 245*: 281 (1984); C. Kuo, T. Provder, M. E. Kohler, and A. F. Kah, *ACS Symp. Ser. 352*: 130 (1987).
6. J. Lesec, D. Lecacheux, and G. Marot, *J. Liquid Chromatogr. 11*: 2571 (1988); J. Lesec and G. Marot, *J. Liquid Chromatogr. 11*: 3305 (1988).
7. M. A. Haney, *J. Appl. Polym. Sci. 30*: 3037 (1985).
8. S. D. Abbot and W. W. Yau (to Dupont), U.S. Patents 4478990 and 4627271 (1986).
9. J. L. de Corral (to Waters Corp.), U.S. Patent 5637790 (1997).

James Lesec

Void Volume in Liquid Chromatography

The column dead volume can be defined as the space in the column which is not working for the chromatographic separations (i.e., not occupied by the stationary phase and its support). Its value is generally determined by an elution time or elution volume of a nonretained solute in the chromatographic system. If the column dead volume is measured by a retention time of a nonretained solute, one can refer to this as a *column dead time* t_0.

The accurate determination of the column void time, t_0, is of fundamental importance in chromatography [1]. This is explained by the fact that a reliable estimation of this quantity is essential for the correct calculation of the retention factors (some refer to this as the *capacity factor*), k, which serves as the fundamental parameter for the comparison of retention data and for the interpretation of the physicochemical phenomena taking place within a chromatographic column. However, the determination of this parameter is very sensitive to the estimated value of the column void time, as can be seen from the equation

$$k = \frac{t_r - t_0}{t_0} \tag{1}$$

where t_r is the solute retention time and t_0 is the column dead time (Fig. 1). A precise knowledge of t_0 is also essential for the proper optimization of the chromatographic system [2]. In contrast to gas chromatography (GC), where the problem of column dead-time determination has been satisfactory solved [3], determination of the true column void time in liquid chromatography (LC) presents both theoretical and practical difficulties and still remains an open question.

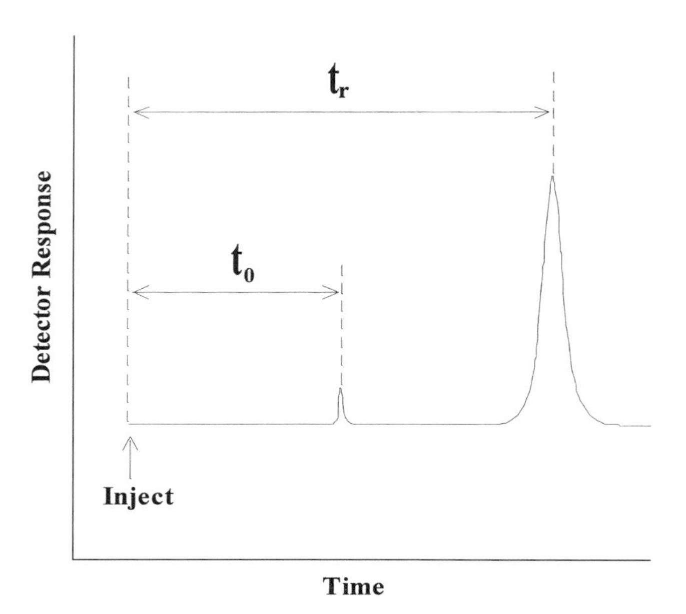

Fig. 1 Void volume determination.

Controversial opinions exist among scientists regarding the meaning of "column dead time" in LC [4]. In its broad sense, the term "column dead time" refers to the elution volume of an unretained and unexcluded solute, but it is not easy to establish which solute (if any) can be treated as both unretained and unexcluded. In LC practice, quite a large number of solutes [5] (e.g., mobile-phase components, isotopically labeled mobile-phase components, ionic and nonionic species) had been used as void time markers. The most popular method used in reversed-phase LC is to use sodium nitrate or sodium nitrite as the probe, but the use of such compounds as the marker was termed clearly dangerous by Berendsen et al. [6]. Use of small ionic species such as sodium nitrate as a void time marker was found to have the drawback that there exists a possibility of exclusion of the ionic species and, when used in small concentrations, they serve as the marker for ionic exclusion volume instead of column dead volume (Donan exclusion). Even to use higher concentrations of ionic species is now rarely utilized due to obvious shortcomings by the Donan exclusion effect. Other existing experimental methods for the determination of LC column dead time also represent a serious contradiction of opinions among scientists and, at this moment, there are no definitive methods to determine column t_0.

One promising approach to experimentally determine t_0 is the linearization method [6], although there are opinions against this method as "undesirable," pointing out some shortcomings. The linearization approach uses the homolog series of compounds, which have various alkyl chains; their retention times are plotted against their chain lengths. Then, the resultant linear curve is extrapolated to zero chain length for the t_0 value. Some people proposed various solutes in homolog series for this purpose, but there is no confirmation concerning the accuracy of the measured t_0 value and, therefore, the opinions against on this method have also appeared in many publications. On the other hand, static methods [7] have been found to represent the upper limit of column porosity, and not the column dead volume proper.

Although the above discussions have appeared numerous times in many publications, workers always need to determine the retention factor, k, in LC separations. In order to determine t_0 experimentally, one has to state what kind of method was used to determine t_0 in his experiments. This is the only best way at this moment to recommend to all chromatographers concerning column void time, although, in the near future, the "best, reliable, theoretically well-interpreted" methods will be adopted.

References

1. A. M. Krstulovic, H. Colin, and G. Guiochon, *Anal. Chem. 54*: 2438 (1982).
2. A. Alhedai, D. E. Martire, R. P. W. Scott: *Analyst 114*: 869 (1989).
3. L. S. Ettre, *Chromatographia 13*: 73 (1980).
4. J. H. Knox and R. Kaliszan, *J. Chromatogr. 349*: 211 (1985).
5. A. Malik and K. Jinno, *Chromatographia 30*: 135 (1990).
6. G. E. Berendsen, P. J. Schoenmakers, L. de Galan, G. Vigh, Z. Varga-Puchnoy, and J. Inczedy, *J. Liquid Chromatogr. 3*: 1669 (1980).
7. E. H. Slaats, J. C. Kraak, W. J. T. Brugman, and H. Poppe, *J. Chromatogr. 149*: 255 (1978).

Kiyokatsu Jinno

W–Z

Wheat Protein by Field-Flow Fractionation

The storage proteins of wheat, collectively referred to as gluten, cover a wide molecular-size range and include some of the largest known naturally occurring biological proteins [1]. The monomeric storage proteins (gliadins), as well as the salt-soluble albumins and globulins, normally have molecular weights of less than 100,000. The polymeric storage proteins (glutenin) have molecular weights that extend into the millions. The size distribution of the gluten proteins and, in particular, the polymeric wheat proteins are closely related to the dough processing characteristics and quality of wheat-based products [2]. Size-fractionation techniques such as size-exclusion–high-performance liquid chromatography (SE–HPLC), gel filtration, and sodium dodecyl sulfate–polyacrylamide gel electrophoresis (SDS–PAGE) are commonly used to characterize these proteins. These methods normally provide fractionation ranges up to molecular weights of $(0.5–1) \times 10^6$ or slightly higher. However, the larger polymeric glutenin proteins appear in the void volume with SE–HPLC and gel filtration and are not resolved [3]. Flow field-flow fractionation (FFF) provides a new tool with which to examine the polymeric wheat proteins, especially those constituting the high-molecular-weight glutenin fraction, because its resolution is not impeded by an exclusion limit [3].

Flow FFF is one of a family of techniques which allows the fractionation of macromolecules based on size-related parameters. A detailed explanation of the mechanism and theory appears elsewhere in this encyclopedia. Fractionation is achieved by the application of a field (cross-flow) perpendicular to the channel flow [4]. Components showing greater response to the force exerted by the field are displaced further from the center of channel flow and toward the channel wall. Because the laminar flow through the channel produces a parabolic flow profile, the elution order of components is determined by their relative displacement. The size-fractionation range, resolution, and run time can be adjusted by manipulating both the channel flow and the strength of the field. In symmetrical flow FFF, the eluent cross-flow is introduced through a porous frit on one side of the channel (depletion) wall and exits through a semipermeable membrane and frit which form the other (accumulation) wall. In conventional-mode flow FFF, the separation of macromolecules up to approximately 1 μm is based on their diffusion coefficient. Diffusion coefficients and the related Stokes diameters of separated species can be directly calculated based on elution time and channel characteristics [4].

The first two reports of the use of flow FFF for separating wheat proteins appeared in the literature in 1996. Researchers from two independent laboratories reported the characterization of wheat proteins from different varieties, extracted by different methods, and analyzed on two different types of flow FFF apparatus. Wahlund and co-workers [5] used asymmetrical flow FFF to examine proteins extracted from two bread wheat varieties of different protein content using sequential extraction with increasing concentrations of dilute hydrochloric acid (HCl). FFF fractograms showed an increase in the molecular size of the protein extracts concomitant with an increase in HCl concentration. Fractions extracted with lower concentrations of HCl showed peaks indicative of the presence of components corresponding in size to gliadins (about 8 nm) [5]. Fractions extracted with high concentrations of HCl showed the presence of very large components with hydrodynamic diameters in the range 15–35 nm. The calculation of molecular weights in these fractions based on hydrodynamic diameter transformations (lower limit defined by flexible random coil and upper limit by spherical protein) resulted in values consistent with those reported in the literature for glutenins (440,000 to 11 million).

Stevenson and Preston [3] used symmetrical stop-flow flow FFF to examine proteins extracted from Katepwa, a high-quality bread wheat variety, using a modification of the traditional Osborne extraction procedure. Fractions isolated by this sequential extraction procedure included wheat proteins extractable in 0.5*M* sodium chloride (albumins and globulins), 70% ethanol (gliadins), and 0.05*M* acetic acid (HAc extractable glutenins). Proteins remaining in the residue were disbursed in either 0.5*M* HAc or 0.001*M* HCl with 30-s sonication (sonicated acid extractable proteins). The albumin and globulin and the gliadin fractions showed major peaks of less than 10 nm, indicative of monomeric proteins. Analysis of the HAc extractable glutenin fraction showed a large peak at about 10 nm, representing smaller polymeric proteins and a smaller peak at 17 nm representing larger polymeric proteins. The bulk of the very large polymeric glutenin proteins were detected in the fractions extracted with sonication (broad distribution range from 11 to 36 nm with a peak at 18 nm). In both studies, the reduction of the polymeric glutenin fractions with dithiothreitol [3] or 2-mercaptoethanol [5] produced shifts in peak positions to less than 10 nm, indicating the release of their interchain disulfide-bonded constituent subunits.

A more widespread use of conventional FFF for the study of proteins has been impeded by the difficulty in obtaining reproducible fractograms which can be accurately integrated for quantification and by the lack of automation. During operation, pressure fluctuation due to the open nature of the system can result in baseline instability. This is further exacerbated by the requirement for stop-flow relaxation, where channel flow is briefly stopped/interrupted (directed via a two-way valve around the channel) when the protein enters the channel to allow equilibrium against the cross-flow. To reduce the impact of pressure fluctuations, pressure balance is maintained by manual adjustment of back-pressure valves on the channel and cross-flow outlets. The requirement for very small sample size (normally $<2\ \mu g$) to prevent overloading and the dilution effect incurred with the relatively large channel flow necessitates the use of very sensitive detector settings resulting in low signal-to-noise ratios which negatively influence baseline resolution and reproducibility.

A number of innovations in FFF design have been introduced recently which have improved resolution and sensitivity and reduced the major obstacles to automation. The introduction of cross-flow recirculation has resulted in better control of system pressures, thereby reducing baseline fluctuation and improving reproducibility [6]. The incorporation of a frit inlet (FI) into the flow FFF channel has permitted the use of hydrodynamic relaxation as a replacement for stop-flow relaxation, thus eliminating pressure fluctuations associated with the latter [7]. FI also reduces sample adhesion to the membrane on the accumulation wall, thereby reducing the likelihood of baseline drift and artifacts [7]. For the FI technique, a small piece of permeable frit material equipped with a separate frit inlet is embedded in the depletion wall of the channel just past the sample inlet. As the sample enters the channel, buffer entering through the frit inlet displaces the sample toward the accumulation wall, allowing hydrodynamic relaxation to occur [7]. Optimum results with this technique are obtained when the flow through the frit inlet is set at about 7–10 times the flow rate through the sample inlet. The incorporation of a frit outlet (FO) located at the depletion wall of the channel just before the channel outlet has facilitated the removal/recirculation of a high proportion of eluent buffer. The remaining eluent buffer containing the sample components is eluted in the channel outlet flow at much higher (up to 10 times) [6,7] effective concentrations, resulting in much higher detector signal-to-noise ratios, thereby eliminating the need for extremely sensitive detection methods.

Development and application of an automated frit inlet/frit outlet (FI/FO) flow FFF channel with recirculating frit and cross-flows to wheat protein fractionation, characterization, and quantification was described for the first time by Stevenson and co-workers [8] in 1999. Analysis of wheat protein fractions produced by the same fractionation method and from the same flour as used for previous stop-flow studies [3] showed improved separation with more well-resolved peaks, particularly with the two polymeric glutenin protein fractions (HAc extractable and sonicated HAc extractable). As can be seen in Figs. 1a and 1b, the major peaks for albumins and globulins and for gliadins indicate molecular sizes less than 10 nm. The HAc extractable proteins (Fig. 1c) show major peaks at 8 and 11 nm plus significant tailing in the 20–35-nm area, indicating the presence of larger polymeric glutenin. From these fractograms, it is evident that the bulk of the large polymeric material is extracted with sonication (Fig. 1d) and that the molecular size distribution in the sonicated fraction covers a very broad range (12–50+ nm). Using the automated FI/FO system, the average coefficient of

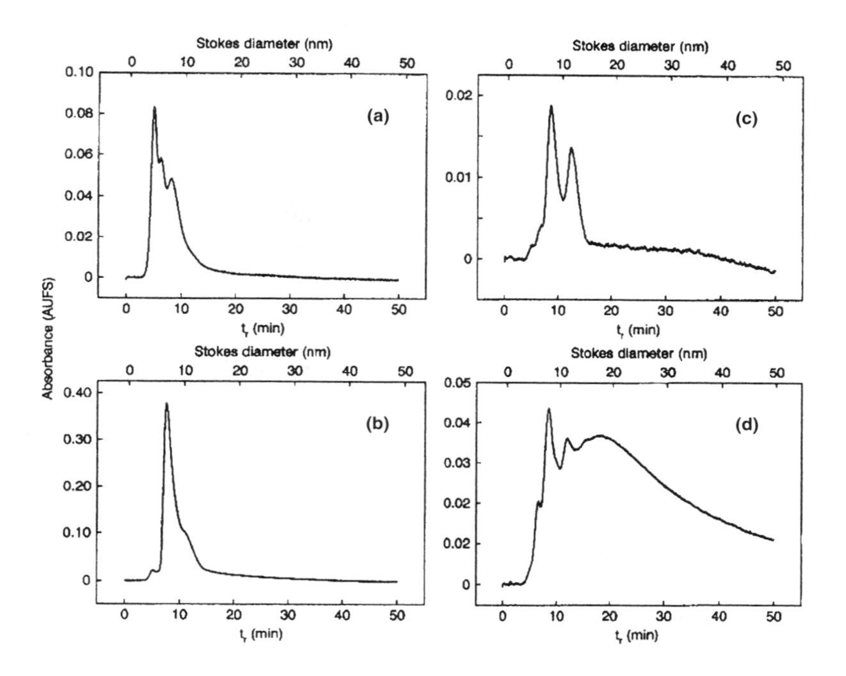

Fig. 1 Elution profiles for (a) albumin and globulin, (b) gliadin, (c) glutenin, and (d) sonicated acetic acid extractable fractions run on a FI/FO automated FFF channel. Normal operating conditions are $V_s = 0.2$, $V_f = 1.4$, and $V_c = 5$ mL/min.

variability for peak area and/or size ranges was less than 2% for six standard molecular-size marker proteins and less than 5% for wheat protein, thus permitting accurate quantification of results.

Field-flow fractionation has shown to be useful for systematically monitoring physicochemical changes in wheat proteins caused by addition of substances such as oxidants or reductants [8]. Recent studies in our laboratory (unpublished data) using automated FI/FO FFF have shown strong relationships between the size distribution of very large polymeric glutenin proteins determined by flow FFF and quality characteristics related to dough strength properties and have permitted the size fractionation of subfractions collected from the void volume peak in SE–HPLC.

References

1. C. W. Wrigley, *Nature 381*: 738 (1996).
2. F. MacRitchie, *Adv. Food Nutr. Res. 36*: 1 (1992).
3. S. G. Stevenson and K. R. Preston, *J. Cereal Sci. 23*: 121 (1996).
4. J. C. Giddings, *Science 260*: 1456 (1993).
5. K.-G. Wahlund, M. Gustavsson, F. MacRitchie, T. Nylander, and L. Wannerberger, *J. Cereal Sci. 23*: 113 (1996).
6. P. Li, M. Hansen, and J. C. Giddings, *J. Microcol. Separ. 10*: 7 (1998).
7. J. C. Giddings, *Anal. Chem. 62*: 2306 (1990).
8. S. G. Stevenson, T. Ueno, and K. R. Preston, *Anal. Chem. 71*: 8 (1999).

S. G. Stevenson
K. R. Preston

Zeta-Potential

Introduction

When an insulator is in contact with an aqueous solution of buffer components, the solid surface has a nonzero electric potential that decreases exponentially as a function of distance from that surface, to become zero at infinity. Taking silica (bulk composition silicium oxide, SiO_2), from which capillary material in capillary electrophoresis (CE) is manufactured, as an example, one can readily understand this when taking into account that on the silica surface, there are silicium hydroxide groups that are subject to an acid–base equilibrium according to

$$R-SiOH + H_2O \Leftrightarrow R-SiO^- + H_3O^+$$

indicating that the negative surface charge of the silica wall is pH dependent and becomes more negative as the buffer pH increases. At pH 2 or lower, the zeta-potential is virtually zero, as the above equilibrium is largely forced to the left due to an abundance of hydrogen ions. As the magnitude of the zeta-potential directly determines the electro-osmotic mobility, it is of importance to understand the effect of different buffer constituents on the zeta-potential and, hence, on the electro-osmotic flow (EOF). The electro-osmotic mobility μ_{EOF} is defined as

$$\mu_{EOF} = -\frac{\zeta \varepsilon}{\eta}$$

in which ζ is the zeta-potential, ε is the dielectric constant of the liquid, and η is the viscosity of the liquid at the plane of shear. The electro-osmotic velocity v_{EOF} is now

$$v_{EOF} = \mu_{EOF} E = -\frac{\zeta \varepsilon E}{\eta}$$

in which E is the electric field strength. Taking a more detailed look at the zeta-potential, one must realize that this potential is not really the value on the negatively charged silica surface, but rather the value on the plane of shear (i.e., at the interface between the stagnant layer of solvent molecules directly at the silica surface and the volume element that is subject to flow, caused by, for example, a longitudinal pressure drop). The combination of these is called the electric double layer.

There are several ways to dynamically change the zeta-potential and, thus, electro-osmosis. Adding a surfactant at a low concentration [generally smaller than the critical micelle concentration (CMC)] will alter the wall charge and zeta-potential due to selective adsorption on the capillary surface: sodium dodecyl sulfate (SDS) will make the zeta-potential more negative (and EOF more pronounced) and may even mask the pH dependence mentioned earlier. Cetyltrimethyl ammonium bromide (CTAB) can neutralize the wall charge, making the zeta-potential close to zero and, thus, eliminating EOF at a concentration of $5 \times 10^{-5} M$. A higher CTAB concentration will make the zeta-potential positive and reverse the EOF. Other additives merely decrease the zeta-potential exponentially to zero or to a small nagative value; they include polymers such as poly(vinyl alcohol), hydroxypropylmethylcellulose, and others. Changing the liquid viscosity near the electric double layer is a usual side effect of these additives, another cause of EOF suppression. Polymers like those mentioned and many others have also been used successfully for chemical modification of the capillary surface, so that addition to the buffer (so-called dynamic coating) is not necessary.

Effect on Migration Time

In uncoated fused-silica capillaries, surface charge and zeta-potential (and, consequently, EOF) can have a sig-

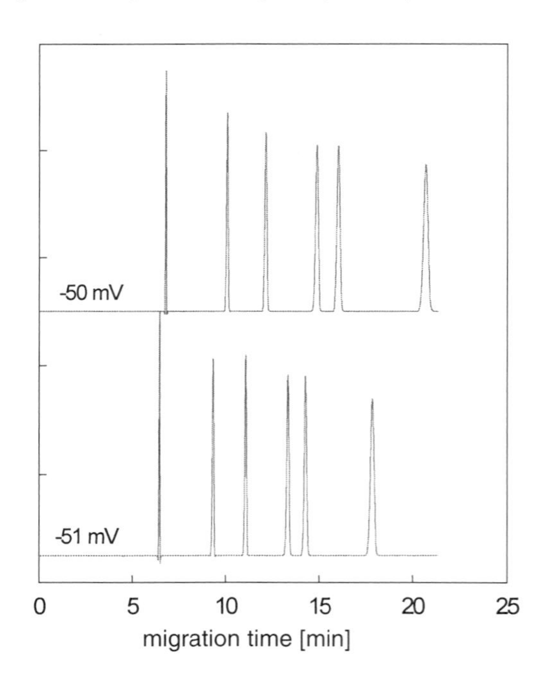

Fig. 1 Effect of capillary zeta-potential on the analysis of a mixture and anions in an uncoated fused-silica capillary in a 10 mM Tris-acetate buffer of pH 8. Normal zeta-potentials are around 50 mV at this pH.

nificant effect on migration time of anions, especially at higher pH values. As anions and EOF move in opposite directions, longer migration times are more sensitive to zeta-potential changes. This can be seen from the equation for migration time t_m:

$$t_m = \frac{L_d L_t}{(\mu_{\text{eff}} + \mu_{\text{EOF}})V}$$

where, because of opposite signs for μ_{eff} and μ_{EOF}, the denominator can be quite small. Consequently, migration time can be sensitive to slight EOF changes.

In the example in Fig. 1, a mixture of anions is analyzed in a Tris-acetate buffer of pH 8. A 2% change in zeta-potential, from −50 to −51 mV, results in a dramatic shift in migration time of peaks with longer migration times.

Suggested Further Reading

Beckers, J. L., *Isotachophoresis, some fundamental aspects*, Thesis, Eindhoven University of Technology, 1973.

Boček, P., M. Deml, P. Gebauer, and V. Dolník, *Analytical Isotachophoresis*, VCH, Weinheim, 1988.

Everaerts, F. M., J. L. Beckers, and Th. P. E. M. Verheggen, *Isotachophoresis: Theory, Instrumentation and Applications*, Elsevier, Amsterdam, 1976.

Hjertén, S. *Chromatogr. Rev. 9*: 122 (1967).

Jorgenson, J. W. and K. D. Lucaks, *Science 222*: 266 (1983).

Kenndler, E., *J. Capillary Electrophoresis 3*(4): 191 (1996).

Li, S. F. Y., *Capillary Electrophoresis – Principles, Practice and Applications*, Elsevier, Amsterdam, 1992.

VanOrman, B. B., G. G. Liversidge, G. L. McIntire, T. M. Olefirowicz, and A. G. Ewing, *J. Microcol. Separ. 2*: 176 (1990).

Jetse C. Reijenga

Zone Dispersion in Field-Flow Fractionation

Introduction

All separation processes are inherently accompanied by zone broadening, which is due to the dynamic spreading processes dispersing the concentration distribution achieved by the separation [1]. As long as the relative contributions of these dispersive processes decrease, the efficiency of the separation increases. A conventional empirical parameter describing, quantitatively, the efficiency of any separation system is the number of theoretical plates per separation unit, N, or the height equivalent to a theoretical plate (the theoretical plate height) H defined by

$$N = \left(\frac{V_R}{\sigma_V}\right)^2, \qquad H = \frac{L}{N} \tag{1}$$

where V_R is the retention volume, σ_V is the standard deviation of the elution curve (fractogram) of a uniform retained sample, expressed at the same retention volume units, and L is the length of the separation unit; in the case of the field-flow fractionation (FFF), it is the length of the fractionation channel. Figure 1a shows a model fractogram and the method of graphical determination of the efficiency of the involved separation system. A more accurate procedure for the determination of the efficiency of the separation system from the experimental fractogram consists in the numerical calculation of the statistical moments of the fractogram (viz. of the first statistical moment, which corresponds to the maximum of the fractogram and, thus, to the retention volume):

$$V_R = \frac{\sum V_i h_i}{\sum h_i} \tag{2}$$

where h_i are the heights of the fractogram at the corresponding retention volumes V_i. The square root of the second central moment corresponds to the standard deviation of the fractogram and, thus, to its width:

$$\sigma_V = \left(\frac{\sum (V_i - V_R)^2 h_i}{\sum h_i}\right)^{1/2} \tag{3}$$

It must be stressed that the terms band or zone broadening, spreading, and dispersion used here are equivalent from the viewpoint of their physical meanings.

Processes Contributing to Zone Broadening

Several dispersive processes contribute to zone broadening: longitudinal diffusion, nonequilibrium and relaxation processes, spreading due to the external parts of the whole separation system, such as the injector, detector, connecting capillaries, and so forth. It has theoretically been found [2] that the resulting efficiency of the FFF, characterized

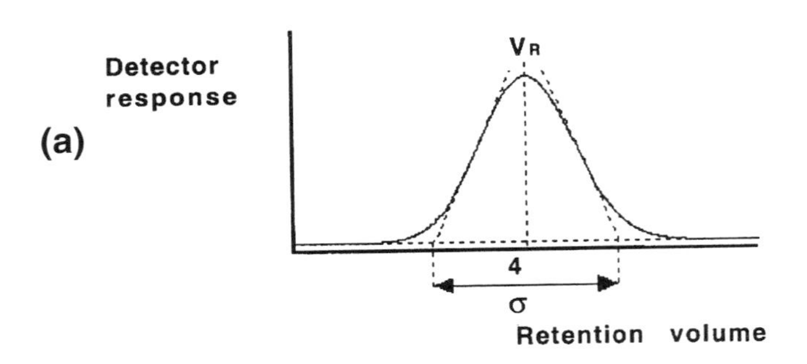

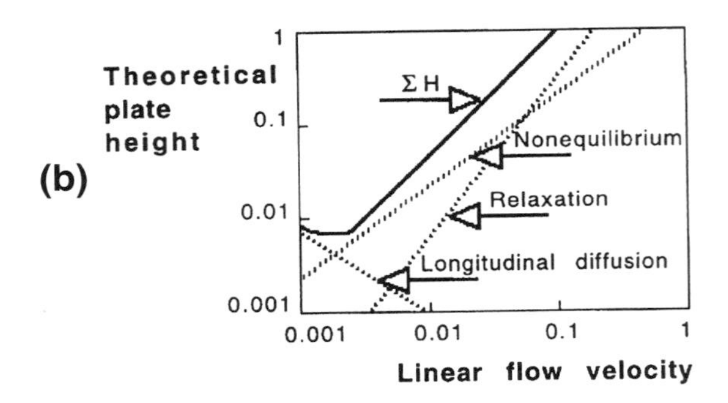

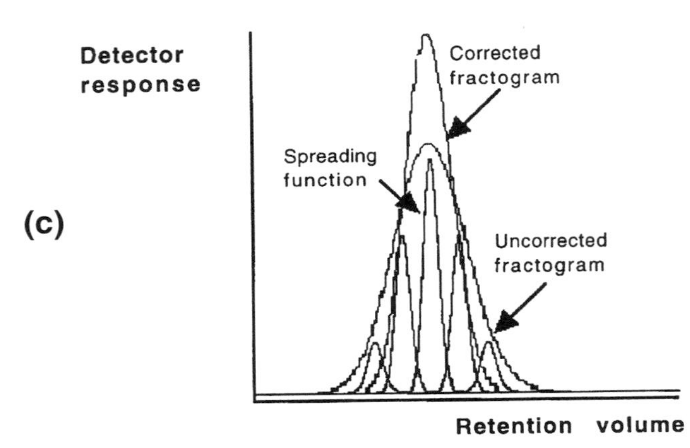

Fig. 1 Zone broadening and its correction. (a) Schematic demonstration of the evaluation of the efficiency of the separation system of FFF; (b) dependence of the theoretical plate height on the linear flow velocity; (c) schematic demonstration of the application of the correction for the zone broadening on a model fractogram.

by the height equivalent to a theoretical plate, can very accurately be described by

$$H = \frac{2D}{R\langle v(x)\rangle} + \frac{\chi w^2 \langle v(x)\rangle}{D} + \sum H_{ext} \tag{4}$$

where D is the diffusion coefficient of the retained species, R is the retention ratio, $\langle v(x)\rangle$ is the average linear velocity of the carrier liquid, w is the thickness of the fractionation channel, χ is the dimensionless parameter defined later, and ΣH_{ext} is the sum of all contributions to the zone broadening from the external parts of the separation channel (injector, detector cell, and connecting capillaries).

The retention ratio R is defined as

$$R = \frac{V_0}{V_R} \tag{5}$$

where V_0 and V_R are the retention volumes of the unretained and retained species, respectively. Graphical representation of the particular spreading processes, as a function of the average linear velocity of the carrier liquid and of their sum, which results in a curve exhibiting a minimum, are shown in Fig. 1b. At very low linear velocities of the carrier liquid, the longitudinal diffusion, represented by the first term on the right-hand side of the Eq. (4), plays a dominant role of all zone-spreading processes. The relaxation and nonequilibrium processes, represented by the second term on the right-hand side of the Eq. (4), become more important when increasing the velocity. The result is that H passes through the above-mentioned minimum. As the diffusion coefficients of the macromolecules and particles are relatively low, the contribution of the longitudinal diffusion is almost negligible and the optimal efficiency (the minimum on the resulting curve) appears at a very low flow velocity. The zone spreading due to the external elements of the fractionation channel can be minimized by reducing their volume.

The dimensionless parameter χ is a complex function of the retention parameter λ which, itself, can be approximated by the relationship

$$\lim_{\lambda \to 0} R = 6\lambda \tag{6}$$

Another approximate relationship holds between χ and λ:

$$\lim_{\lambda \to 0} \chi = 24\lambda^3 \tag{7}$$

Equation (7) is a very important relationship because, with respect to Eq. (4), it indicates that in the most practical range of the linear flow velocities, above the optimal flow, the efficiency of the separation in FFF increases very rapidly with the retention ratio. This is rather an exceptional case among separation methods and the importance of this behavior has to be regarded with respect to the fact that the FFF methods and techniques are especially convenient for the fractionation of large and polydisperse species, such as macromolecules and particles. As the retention usually increases with the molar mass or particle size in polarization FFF, the efficiency is higher in the high molar mass or large particle size domain. This is one of the reasons why the FFF methods are particularly competitive in this field of application.

Nevertheless, the difficulty to separate the large polydisperse species enough, independently of the method or technique used, impose the necessity to introduce, and apply the powerful numerical methods, which are able to evaluate the amplitude of the zone-broadening contribution to the apparent molar mass distribution (MMD) or particle size distribution (PSD), calculated by a simple data treatment of the experimental fractograms, and to correct, casually, the MMD and PSD data for zone broadening. Whenever the zone spreading attains a level not acceptable from the point of view of the analytical results obtained by a simple data treatment of the raw fractograms, a correction for the zone broadening must be applied.

Relaxation

Relaxation represents an important contribution to the zone broadening and merits mention in particular, especially because it can be reduced by an appropriate experimental procedure [3]. The concentration distribution of the fractionated sample across the channel established immediately after the injection is far from the steady state, which is formed progressively during the elution. This leads to additional zone broadening. However, if the flow is stopped after the injection for a time necessary to achieve a steady state, the shift of the retention volume and the zone spreading can substantially be reduced. So-called secondary relaxation broadening can occur when the intensity of the field varies rapidly during the elution (e.g., by programming).

Correction for the Zone Broadening

The fractogram of a polydisperse sample is a superposition of the separation and of the zone broadening. This is shown in Fig. 1c, where the spread zones of the discrete species are overlapped and the fractogram is, in fact, a convolution of all individual zones. Whenever the zone spreading is important, the accurate MMD or PSD can be calculated from the experimental fractograms only by using a correction procedure. An efficient correction method applicable in FFF [4] was derived from well-known correction procedures used in size-exclusion chromatography [5].

A raw fractogram $h(V)$ is a convolution of the fractogram corrected for the zone broadening $g(Y)$ and the spreading function $G(V, Y)$ which is a detector response to a uniform species having the elution volume Y:

$$h(V) = \int_0^\infty g(Y)G(V - Y)\, dY \tag{8}$$

The spreading function can be approximated by

$$G(V, Y) = \left(\frac{1}{2\pi\sigma_S^2}\right)^{1/2} \exp\left(-\frac{(V - Y)^2}{2\sigma_S^2}\right) \tag{9}$$

where the standard deviation σ_S should be independent of the elution volume. This independence is valid in FFF only within a not very wide range of the elution volumes. Equation (9) can correctly be applied to the fractograms of the samples with a narrow MMD or PSD only. On the other hand, whenever a nonuniform spreading function with the σ_S dependent on the retention has to be used, the correction to be applied is

$$h(V) = \int_0^\infty g(Y)\left[\frac{1}{[2\pi(\sigma_S(Y))^2]}\right]^{1/2} \exp\left(-\frac{(V - Y)^2}{[2(\sigma_S Y)]^2}\right) dY \tag{10}$$

which is convenient for samples exhibiting wide MMD or PSD. Equation (10) can be numerically solved. The graphical representation of a model result of the application of the described correction procedure is shown in Fig. 1c.

Practical utility of the described correction procedure of the experimental fractograms was demonstrated first by applying it to the real fractograms of a polymer latex obtained from sedimentation FFF [4] and was subsequently confirmed [6–8].

References

1. J. C. Giddings, *Dynamics of Chromatography*, Marcel Dekker, Inc., New York, 1965.
2. M. E. Hovingh, G. H. Thompson, and J. C. Giddings, *Anal. Chem. 42*:195 (1970).
3. J. Janča, *Field-Flow Fractionation: Analysis of Macromolecules and Particles*, Marcel Dekker, Inc., New York, 1988.
4. V. Jáhnová, F. Matulík, and J. Janča, *Anal. Chem. 59*: 1039 (1987).
5. A. E. Hamielec, in *Steric Exclusion Chromatography of Polymers* (J. Janča, ed.), Marcel Dekker, Inc., New York, 1984.
6. M. R. Schure, B. N. Barman, and J. C. Giddings, *Anal. Chem. 61*: 2735 (1989).
7. Y. Mori, K. Kimura, and M. Tanigaki, *Spec. Publ. Roy. Soc. Chem. 102*: 290 (1992).
8. J.-C. Vauthier and P. S. Williams, *J. Chromatogr. 805*: 149 (1998).

Josef Janča

Author Index

Subject Index